7 Years 143 Shifts

2025-2019 All Papers

JEE Main

Chapterwise Topicwise Solved Papers

CHEMISTRY

Only PYQ Book with
- Difficulty Level Sorting
- Important Concept Marker

7 Years 143 Shifts
2025-2019 All Papers

JEE Main

Chapterwise Topicwise Solved Papers

CHEMISTRY

Authors
Pooja Yadav
Shalini Singh

arihant
SUCCESS AND NOTHING LESS

arihant
SUCCESS AND NOTHING LESS

ARIHANT PRAKASHAN (Series), MEERUT

Administrative & Production Offices

Regd. Office

'Ramchhaya' 4577/15, Agarwal Road, Darya Ganj, New Delhi -110002
Tele: 011- 47630600, 43518550

Head Office

Kalindi, TP Nagar, Meerut (UP) - 250002
Tel: 0121-7156203, 7156204

Sales & Support Offices

Agra, Ahmedabad, Bengaluru, Bareilly, Chennai, Delhi, Guwahati, Hyderabad, Jaipur, Jhansi, Kolkata, Lucknow, Nagpur & Pune.

PRICE ₹485.00

PO No : TXT-XX-XXXXXXX-X-XX

Published by Arihant Publications (India) Ltd.

For further information about the books published by Arihant, log on to www.arihantbooks.com or e-mail at info@arihantbooks.com

Follow us on

Preface

JEE Main is a gateway examination for candidates expecting to seek admission in Bachelor in Engineering (BE), Bachelor of Technology (B.Tech) and Bachelor of Architecture (B.Arch) at Indian Institutes of Information Technology (IIITs), National Institutes of Technology (NITs), Delhi Technological University and other Centrally Funded Technical Institutes (CFTIs).

JEE Main is also an examination which is like screening examination for JEE Advanced (The gateway examination to India's most reputed Technical Institutes, Indian Institutes of Technology— IITs).

To make the students well-versed with the pattern as well as the level of the questions asked in the exam, this book contains **Chapterwise Topicwise Solutions of Questions asked in Last 7 Years' Examinations of JEE Main.**

The Unique and exclusive features of this book are:

- Questions are further sorted according to their difficulty level in each topic.
- Questions on frequently asked concepts are marked with concept name to get edge over others.

Solutions to all the questions have been kept very detailed and accurate with ingredient of special features like Key idea, Concept Enhancer & Alternate/Time saver techniques.

We hope this book would be highly beneficial for the students. We would be grateful if any discrepancy or mistake in the questions or answers is brought to our notice so that these could be rectified in subsequent editions.

Publisher

Contents

Class 11

Class 12

SYLLABUS

Chemistry

Physical Chemistry

1 UNIT — Some Basic Concepts in Chemistry

Matter and its nature, Dalton's atomic theory: Concept of atom, molecule, element, and compound:: Laws of chemical combination; Atomic and molecular masses, mole concept, molar mass, percentage composition, empirical and molecular formulae: Chemical equations and stoichiometry.

2 UNIT — Atomic Structure

Nature of electromagnetic radiation, photoelectric effect; Spectrum of the hydrogen atom. Bohr model of a hydrogen atom - its postulates, derivation of the relations for the energy of the electron and radii of the different orbits, limitations of Bohr's model; Dual nature of matter, de Broglie's relationship. Heisenberg uncertainty principle. Elementary ideas of quantum mechanics, quantum mechanics, the quantum mechanical model of the atom, and its important features. Concept of atomic orbitals as one-electron wave functions: Variation of y and y^2 with r for 1s and 2s orbitals; various quantum numbers (principal, angular momentum, and magnetic quantum numbers) and their significance; shapes of s, p, and d - orbitals, electron spin and spin quantum number: Rules for filling electrons in orbitals – Aufbau principle. Pauli's exclusion principle and Hund's rule, electronic configuration of elements, and extra stability of half-filled and completely filled orbitals.

3 UNIT — Chemical Bonding and Molecular Structure

Kossel-Lewis approach to chemical bond formation, the concept of ionic and covalent bonds.

Ionic Bonding: Formation of ionic bonds, factors affecting the formation of ionic bonds; calculation of lattice enthalpy.

Covalent Bonding: Concept of electronegativity. Fajan's rule, dipole moment: Valence Shell Electron Pair Repulsion (VSEPR) theory and shapes of simple molecules.

Quantum mechanical approach to covalent bonding: Valence bond theory - its important features, the concept of hybridization involving s, p, and d orbitals; Resonance.

Molecular Orbital Theory - Its important features. LCAOs, types of molecular orbitals (bonding, antibonding), sigma and pi-bonds, molecular orbital electronic configurations of homonuclear diatomic molecules, the concept of bond order, bond length, and bond energy. Elementary idea of metallic bonding. Hydrogen bonding and its applications.

4 UNIT — Chemical Thermodynamics

Fundamentals of thermodynamics: System and surroundings, extensive and intensive properties, state functions, entropy, types of processes.

The first law of thermodynamics - Concept of work, heat internal energy and enthalpy, heat capacity, molar heat capacity; Hess's law of constant heat summation; Enthalpies of bond dissociation, combustion, formation, atomization, sublimation, phase transition, hydration, ionization, and solution.

The second law of thermodynamics - Spontaneity of processes; ΔS of the universe and ΔG of the system as criteria for spontaneity. DG° (Standard Gibbs energy change) and equilibrium constant.

5 UNIT — Solutions

Different methods for expressing the concentration of solution - molality, molarity, mole fraction, percentage (by volume and mass both), the vapour pressure of solutions and Raoult's Law - Ideal and non-ideal solutions, vapour pressure - composition, plots for ideal and non-ideal solutions; Colligative properties of dilute solutions - a relative lowering of vapour pressure, depression of freezing point, the elevation of boiling point and osmotic pressure; Determination of molecular mass using colligative properties; Abnormal value of molar mass, Van't Hoff factor and its significance.

6
UNIT

Equilibrium

Meaning of equilibrium is the concept of dynamic equilibrium.

Equilibria involving physical processes: Solid-liquid, liquid-gas - gas and solid-gas equilibria, Henry's law. General characteristics of equilibrium involving physical processes.

Equilibrium involving chemical processes: Law of chemical equilibrium, equilibrium constants (K_p and K_c) and their significance, the significance of ΔG and $\Delta G°$ in chemical equilibrium, factors affecting equilibrium concentration, pressure, temperature, the effect of catalyst; Le Chatelier's principle.

Ionic equilibrium: Weak and strong electrolytes, ionization of electrolytes, various concepts of acids and bases (Arrhenius. Bronsted - Lowry and Lewis) and their ionization, acid-base equilibria (including multistage ionization) and ionization constants, ionization of water. pH scale, common ion effect, hydrolysis of salts and pH of their solutions, the solubility of sparingly soluble salts and solubility products, and buffer solutions.

7
UNIT

Redox Reactions and Electrochemistry

Electronic concepts of oxidation and reduction, redox reactions, oxidation number, rules for assigning oxidation number and balancing of redox reactions.

Electrolytic and metallic conduction, conductance in electrolytic solutions, molar conductivities and their variation with concentration: Kohlrausch's law and its applications.

Electrochemical cells - Electrolytic and Galvanic cells, different types of electrodes, electrode potentials including standard electrode potential, half-cell and cell reactions, emf of a Galvanic cell and its measurement: Nernst equation and its applications; Relationship between cell potential and Gibbs' energy change: Dry cell and lead accumulator; Fuel cells.

8
UNIT

Chemical Kinetics

Rate of a chemical reaction, factors affecting the rate of reactions: concentration, temperature, pressure, and catalyst; elementary and complex reactions, order and molecularity of reactions, rate law, rate constant and its units, differential and integral forms of zero and first-order reactions, their characteristics and half-lives, the effect of temperature on the rate of reactions, Arrhenius theory, activation energy and its calculation, collision theory of bimolecular gaseous reactions (no derivation).

Inorganic Chemistry

9
UNIT

Classification of Elements and Periodicity in Properties

Modem periodic law and present form of the periodic table, *s*, *p*, *d* and *f* block elements, periodic trends in properties of elements, atomic and ionic radii, ionisation enthalpy, electron gain enthalpy, valence, oxidation states, and chemical reactivity.

10
UNIT

p-block Elements

Group-13 to Group 18 Elements

General Introduction: Electronic configuration and general trends in physical and chemical properties of elements across the periods and down the groups; unique behaviour of the first element in each group.

11
UNIT

d- and *f*- Block Elements

Transition Elements

General introduction, electronic configuration, occurrence and characteristics, general trends in properties of the first-row transition elements - physical properties, ionization enthalpy, oxidation states, atomic radii, colour, catalytic behaviour, magnetic properties, complex formation, interstitial compounds, alloy formation; Preparation, properties, and uses of $K_2Cr_2O_7$, and $KMnO_4$.

Inner Transition Elements

Lanthanoids - Electronic configuration, oxidation states, and lanthanoid contraction.

Actinoids - Electronic configuration and oxidation states.

12
UNIT

Co-ordination Compounds

Introduction to coordination compounds. Werner's theory; ligands, coordination number, denticity. chelation; IUPAC nomenclature of mononuclear co-ordination compounds, isomerism; Bonding-Valence bond approach and basic ideas of Crystal field theory, colour and magnetic properties; Importance of co-ordination compounds (in qualitative analysis, extraction of metals, and in biological systems).

Organic Chemistry

13
UNIT

Purification and Characterisation of Organic Compounds

Purification - Crystallization, sublimation, distillation, differential extraction, and chromatography - principles and their applications.

Quanlitative analysis Detection of nitrogen, sulphur, phosphorus, and halogens.

Quantitative analysis (basic principles only) - Estimation of carbon, hydrogen, nitrogen, halogens, sulphur, and phosphorus. Calculations of empirical formulae and molecular formulae: Numerical problems in organic quantitative analysis.

14
UNIT

Some Basic Principles of Organic Chemistry

Tetravalency of carbon: Shapes of simple molecules - hybridization (s and p): Classification of organic compounds based on functional groups: and those containing halogens, oxygen, nitrogen, and sulphur; Homologous series: Isomerism - structural and stereoisomerism.

Nomenclature (Trivial and IUPAC)

Covalent bond fission - Homolytic and heterolytic: free radicals, carbocations and carbanions; stability of carbocations and free radicals, electrophiles and nucleophiles.

Electronic displacement in a covalent bond - Inductive effect, electromeric effect, resonance and hyperconjugation.

Common types of organic reactions - Substitution, addition, elimination, and rearrangement

15
UNIT

Hydrocarbons

Classification, isomerism, IUPAC nomenclature, general methods of preparation, properties, and reactions.

Alkanes - Conformations: Sawhorse and Newman projections (of ethane): Mechanism of halogenation of alkanes.

Alkenes - Geometrical isomerism: Mechanism of electrophilic addition: addition of hydrogen, halogens, water, hydrogen halides (Markownikoffs and peroxide effect): Ozonolysis and polymerization.

Alkynes - Acidic character: Addition of hydrogen, halogens, water, and hydrogen halides: Polymerisation.

Aromatic hydrocarbons - Nomenclature, benzene - structure and aromaticity: Mechanism of electrophilic substitution: halogenation, nitration.

Friedel-Craft's alkylation and acylation, directive influence of the functional group in monosubstituted benzene.

16
UNIT

Organic Compounds Containing Halogens

General methods of preparation, properties, and reactions: Nature of C—*X* bond; Mechanisms of substitution reactions. Uses: Environmental effects of chloroform, iodoform, freons and DDT.

17
UNIT

Organic Compounds Containing Oxygen

General methods of preparation, properties, reactions, and uses.

ALCOHOLS, PHENOLS, AND ETHERS

Alcohols: Identification of primary, secondary, and tertiary alcohols: mechanism of dehydration.

Phenols: Acidic nature, electrophilic substitution reactions: halogenation. nitration and sulphonation. Reimer - Tiemann reaction.

Ethers: Structure.

Aldehyde and Ketones: Nature of carbonyl group; Nucleophilic addition to $>C=O$ group, relative reactivities of aldehydes and ketones; Important reactions such as - Nucleophilic addition reactions (addition of HCN. NH_3, and its derivatives), Grignard reagent; oxidation: reduction (Wolf Kishner and Clemmensen); the acidity of a-hydrogen. aldol condensation, Cannizzaro reaction. Haloform reaction, Chemical tests to distinguish between aldehydes and ketones.

Carboxylic Acids

Acidic strength and factors affecting it.

18
UNIT

Organic Compounds Containing Nitrogen

General methods of preparation. Properties, reactions, and uses.

Amines: Nomenclature, classification structure, basic character, and identification of primary, secondary, and tertiary amines and their basic character.

Diazonium Salts: Importance in synthetic organic chemistry.

19 UNIT

Biomolecules

General introduction and importance of biomolecules.

CARBOHYDRATES - Classification; aldoses and ketoses: monosaccharides (glucose and fructose) and constituent monosaccharides of oligosaccharides (sucrose, lactose, and maltose).

PROTEINS - Elementary Idea of a-amino acids, peptide bond, polypeptides. Proteins: primary, secondary, tertiary, and quaternary structure (qualitative idea only), denaturation of proteins, enzymes.

VITAMINS – Classification and functions.

NUCLEIC ACIDS – Chemical constitution of DNA and RNA.

Biological functions of nucleic acids.

Hormones (General introduction)

20 UNIT

Principles Related To Practical Chemistry

Detection of extra elements (Nitrogen, Sulphur, halogens) in organic compounds; Detection of the following functional groups; hydroxyl (alcoholic and phenolic), carbonyl (aldehyde and ketones) carboxyl, and amino groups in organic compounds.

- The chemistry involved in the preparation of the following:
  Inorganic compounds; Mohr's salt, potash alum.
  Organic compounds: Acetanilide, *p*-nitro acetanilide, aniline yellow, iodoform.
- The chemistry involved in the titrimetric exercises – Acids, bases, and the use of indicators, oxalic-acid vs $KMnO_4$, Mohr's salt vs $KMnO_4$.
- Chemical principles involved in the qualitative salt analysis:

Cations — Pb^{2+}, Cu^{2+}, Al^{3+}, Fe^{3+}, Zn^{2+}, Ni^{2+}, Ca^{2+}, Ba^{2+}, Mg^{2+}, NH^{4+}.
Anions – $CO_3{}^{2-}$, S^{2-}, $SO_4{}^{2-}$, NO_2, NO_3, Cl^-, Br^-, I^- (Insoluble salts excluded).

Chemical principles involved in the following experiments:

1. Enthalpy of solution of $CuSO_4$.
2. Enthalpy of neutralisation of strong acid and strong base.
3. Preparation of lyophilic and lyophobic sols.
4. Kinetic study of the reaction of iodide ions with hydrogen peroxide at room temperature.

01 Some Basic Concepts in Chemistry

In JEE Main there is usually 1-2 questions per paper are visible from this chapter. In past 3 years 32 conceptual and 55 numerical questions were asked of moderate difficulty level from the chapter.

For detailed analysis scan QR code

Topic 1 / Fundamental Concepts and Laws of Chemical Combinations

Level 1

01 Which of the following have same number of significant figures? **[2023, 8 April Shift-II]**

A. 0.00253 B. 1.0003
C. 15.0 D. 163

Choose the correct answer from the options given below

(a) A, C and D only (b) A, B and C only
(c) C and D only (d) B and C only

02 Using the rules for significant figures, the correct answer for the expression $\frac{0.02858 \times 0.112}{0.5702}$ will be **[2022, 29 June Shift-II]**

(a) 0.005613 (b) 0.00561 (c) 0.0056 (d) 0.006

Level 2

03 The incorrect postulates of the Dalton's atomic theory are **[2024, 5 April Shift-I]**

(A) Atoms of different elements differ in mass.
(B) Matter consists of divisible atoms.
(C) Compounds are formed when atoms of different element combine in a fixed ratio.
(D) All the atoms of given element have different properties including mass.
(E) Chemical reactions involve reorganisation of atoms.

Choose the correct answer from the options given below

(a) (A), (B), (C) only (b) (C), (D), (E) only
(c) (B), (D) only (d) (B), (D), (E) only

04 Choose the incorrect statement about Dalton's atomic theory. **[2024, 4 April Shift-II]**

(a) All the atoms of a given element have identical properties including identical mass.
(b) Matter consists of indivisible atoms.
(c) Chemical reactions involve reorganisation of atoms
(d) Compounds are formed when atoms of different elements combine in any ratio.

Level 3

05 Choose the correct statements. **[2025, 29 Jan Shift-I]**

A. Weight of a substance is the amount of matter present in it.
B. Mass is the force exerted by gravity on an object.
C. Volume is the amount of space occupied by a substance.
D. Temperatures below 0°C are possible in Celsius scale, but in Kelvin scale negative temperature is not possible.
E. Precision refers to the closeness of various measurements for the same quantity.

Choose the correct answer from the options given below

(a) C, D and E only
(b) B, C and D only
(c) A, B and C only
(d) A, D and E only

Numerical Value Type Questions

06 The number of significant figures in 50000.020×10^{-3} is .... **[2021, 26 Feb Shift-I]**

07 The number of significant figures in 0.00340 is .... **[2021, 25 July Shift-II]**

Topic 2 / Atomic & Molecular Masses, Percentage Composition, Mole Concept, Empirical & Molecular Formula

Level 1

01 Mass of magnesium required to produce 220 mL of hydrogen gas at STP on reaction with excess of dil. HCl is (Given, molar mass of Mg is 24 g mol^{-1}) **[2025, 3 April Shift-II]**

(a) 236 mg (b) 2.444 g
(c) 0.24 mg (d) 235.7 g

02 2.8×10^{-3} mol of CO_2 is left after removing 10^{21} molecules from its '*x*' mg sample. The mass of CO_2 taken initially is (Given, $N_A = 6.02 \times 10^{23}$ mol^{-1}) **[2025, 23 Jan Shift-I]**

(a) 150.4 mg (b) 196.2 mg
(c) 48.2 mg (d) 98.3 mg

03 A solution is prepared by adding 2 g of "X" to 1 mole of water. Mass per cent of "X" in the solution is **[2022, 11 April Shift-II]**

(a) 20% (b) 10%
(c) 5% (d) 2%

04 The average molar mass of chlorine is 35.5 g mol^{-1}. The ratio of ^{35}Cl to ^{37}Cl in naturally occuring chlorine is close to **[2020, 6 Sep Shift-II]**

(a) 4 : 1 (b) 3 : 1
(c) 2 : 1 (d) 1 : 1

Level 2

05 On complete combustion 1.0 g of an organic compound (X) gave 1.46 g of CO_2 and 0.567 g of H_2O. The empirical formula mass of compound (X) is .... g.

(Given, molar mass in g mol^{-1} C : 12, H : 1, O : 16) **[2025, 2 April Shift-I]**

(a) 15 (b) 45
(c) 60 (d) 30

06 Among 10^{-9} g (each) of the following elements, which one will have the highest number of atoms? (Element: Pb, Po, Pr and Pt) **[2025, 3 April Shift-I]**

(a) Pr (b) Po
(c) Pt (d) Pb

07 The elemental composition of a compound is 54.2% C, 9.2% H and 36.6% O. **[2025, 24 Jan Shift-II]**

If the molar mass of the compound is 132 g mol^{-1}, the molecular formula of the compound is

[Given, the relative atomic mass of C : H : O = 2 : 1 : 16]

(a) $C_6H_{12}O_6$ (b) $C_4H_8O_2$
(c) $C_4H_9O_3$ (d) $C_6H_{12}O_3$

08 A metal chloride contains 55.0% of chlorine by weight 100 mL vapours of the metal chloride at STP weigh 0.57 g. The molecular formula of the metal chloride is (Given, atomic mass of chlorine is 35.5 u) **[2023, 12 April Shift-I]**

(a) MCl_4 (b) MCl_2
(c) MCl (d) MCl_3

09 The number of molecules and moles in 2.8375 litres of O_2 at STP are respectively **[2023, 10 April Shift-I]**

(a) 7.527×10^{23} and 0.125 mol
(b) 1.505×10^{23} and 0.250 mol
(c) 7.527×10^{22} and 0.250 mol
(d) 7.527×10^{22} and 0.125 mol

10 120 g of an organic compound that contains only carbon and hydrogen, gives 330 g of CO_2 and 270 g of water on complete combustion. The percentage of carbon and hydrogen, respectively are **[2022, 24 June Shift-II]**

(a) 25 and 75 (b) 40 and 60
(c) 60 and 40 (d) 75 and 25

11 5 moles of AB_2 weigh 125×10^{-3} kg and 10 moles of A_2B_2 weigh 300×10^{-3} kg. The molar mass of $A(M_A)$ and molar mass of $B(M_B)$ in kg mol^{-1} are **[2019, 12 April Shift-I]**

(a) $M_A = 10 \times 10^{-3}$ and $M_B = 5 \times 10^{-3}$
(b) $M_A = 50 \times 10^{-3}$ and $M_B = 25 \times 10^{-3}$
(c) $M_A = 25 \times 10^{-3}$ and $M_B = 50 \times 10^{-3}$
(d) $M_A = 5 \times 10^{-3}$ and $M_B = 10 \times 10^{-3}$

12 A 10 mg effervescent tablet containing sodium bicarbonate and oxalic acid releases 0.25 mL of CO_2 at $T = 298.15$ K and $p = 1$ bar. If molar volume of CO_2 is 25.0 L under such condition, what is the percentage of sodium bicarbonate in each tablet? [Molar mass of $NaHCO_3 = 84$ g mol^{-1}] **[2019, 11 Jan Shift-I]**

(a) 8.4 (b) 0.84
(c) 16.8 (d) 33.6

Level 3

13 0.1 M solution of KI reacts with excess of H_2SO_4 and KIO_3 solutions. According to equation

$$5I^- + IO_3^- + 6H^+ \longrightarrow 3I_2 + 3H_2O$$

Identify the correct statements. **[2025, 29 Jan Shift-II]**

(A) 200 mL of KI solution reacts with 0.004 mol of KIO_3.
(B) 200 mL of KI solution reacts with 0.006 mol of H_2SO_4.
(C) 0.5 L of KI solution produced 0.005 mol of I_2.
(D) Equivalent weight of KIO_3 is equal to $\left(\frac{\text{Molecular weight}}{5}\right)$ **[2025, 29 Jan Shift-II]**

Choose the correct answer from the options given below

(a) (A) and (D) only (b) (C) and (D) only
(c) (B) and (C) only (d) (A) and (B) only

14 A sample of $CaCO_3$ and $MgCO_3$ weighed 2.21 g is ignited to constant weight of 1.152 g. The composition of mixture is
(Given, molar mass in g mol^{-1} $CaCO_3$: 100, $MgCO_3$: 84) **[2024, 31 Jan Shift-II]**

(a) 1.187 g $CaCO_3$ + 1.187 g $MgCO_3$
(b) 1.023 g $CaCO_3$ + 1.187 g $MgCO_3$
(c) 1.187 g $CaCO_3$ + 1.023 g $MgCO_3$
(d) 1.023 g $CaCO_3$ + 1.023 g $MgCO_3$

15 Match List I with List II.

	List I		List II
A.	16 g of $CH_4(g)$	I.	Weighs 28 g
B.	1 g of $H_2(g)$	II.	60.2×10^{23} electrons
C.	1 mole of $N_2(g)$	III.	Weighs 32 g
D.	0.5 mol of $SO_2(g)$	IV.	Occupies 11.4 L volume at STP

Choose the correct answer from the options given below **[2023, 10 April Shift-II]**

(a) A-I, B-III, C-II, D-IV (b) A-II, B-IV, C-I, D-III
(c) A-II. B-III, C-IV, D-I (d) A-II, B-IV, C-III, D-I

16 Compound A contains 8.7% hydrogen, 74% carbon and 17.3% nitrogen. The molecular formula of the compound is (Given, atomic masses of C, H and N are 12, 1 and 14 amu respectively).
The molar mass of the compound A is 162 g mol^{-1}. **[2022, 28 June Shift-II]**

(a) $C_4H_6N_2$ (b) C_2H_3N
(c) C_5H_7N (d) $C_{10}H_{14}N_2$

17 Complete combustion of 1.80 g of an oxygen containing compound ($C_xH_yO_z$) gave 2.64 g of CO_2 and 1.08 g of H_2O. The percentage of oxygen in the organic compound is **[2021, 25 Feb Shift-I]**

(a) 50.33 (b) 53.33
(c) 63.53 (d) 51.63

Numerical Value Type Questions

18 X g of nitrobenzene on nitration gave 4.2 g of m-dinitrobenzene. X = ... g. (nearest integer)
[Given, molar mass (in g mol^{-1}) C : 12, H : 1, O : 16, N : 14] **[2025, 3 April Shift-II]**

19 Fortification of food with iron is done using $FeSO_4 \cdot 7H_2O$. The mass in grams of the $FeSO_4 \cdot 7H_2O$ required to achieve 12 ppm of iron in 150 kg of wheat is .... (Nearest integer)

(Given, Molar mass of Fe, S and and O respectively are 56, 32 and 16 g mol^{-1}.] **[2025, 4 April Shift-I]**

20 0.1 mol of the following given antiviral compound (P) will weigh .... $\times 10^{-1}$ g (Nearest integer)

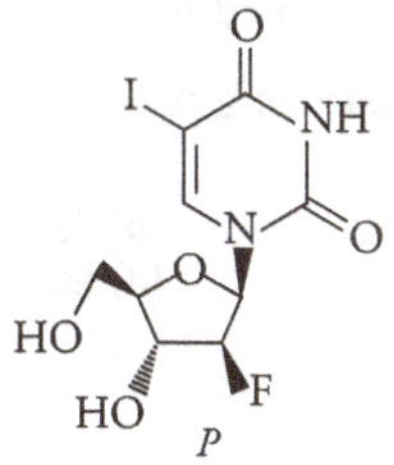

P **[2025, 2 April Shift-I]**

(Given, molar mass in g mol^{-1} H : 1, C : 12, N : 14, O : 16, F : 19, I : 127)

21 0.01 mole of an organic compound (X) containing 10% hydrogen, on complete combustion produced 0.9 g H_2O. Molar mass of (X) is ....g mol^{-1}. **[2025, 23 Jan Shift-II]**

22 Some CO_2 gas was kept in a sealed container at a pressure of 1 atm and at 273 K. This entire amount of CO_2 gas was later passed through an aqueous solution of $Ca(OH)_2$. The excess unreacted $Ca(OH)_2$ was later neutralised with 0.1 M of 40 mL HCl. If the volume of the sealed container of CO_2 was x, then x is .... cm^3 (nearest integer).

[Given, the entire amount of $CO_2(g)$ reacted with exactly half the initial amount of $Ca(OH)_2$ present in the aqueous solution.]

$A \rightarrow B$ **[2025, 22 Jan Shift-I]**

23 Quantitative analysis of an organic compound (X) shows following % composition. **[2025, 28 Jan Shift-I]**

C : 14.5% Cl : 64.46% H : 1.8%

(Empirical formula mass of the compound (X) is ...... $\times 10^{-1}$

(Given, molar mass in g mol^{-1} of C : 12, H : 1, O : 16, Cl : 35.5)

24 In the Claisen-Schmidt reaction to prepare, dibenzalacetone from 5.3g of benzaldehyde, a total of 3.51 g of product was obtained. The percentage yield in this reaction was.........%. **[2025, 29 Jan Shift-II]**

25 A solution is prepared by adding 1 mole ethyl alcohol in 9 mole water. The mass percent of solute in the solution is .... (integer answer) (Given : Molar mass in g mol^{-1}, ethyl alcohol : 46, water : 18) **[2024, 8 April Shift-II]**

Imp. *Reaction of Aniline & Percentage Yield*

26 9.3 g of pure aniline is treated with bromine water at room temperature to give a white precipitate of the product 'P'. The mass of product 'P' obtained is 26.4 g. The percentage yield is .... % **[2024, 5 April Shift-I]**

27 Molar mass of the salt from NaBr, $NaNO_3$, KI and CaF_2 which does not evolve coloured vapours on heating with concentrated H_2SO_4 is........ g mol^{-1}.

(Molar mass in g mol^{-1} : Na : 23, N : 14, K : 39, O : 16, Br : 80, I : 127, F : 19, Ca : 40) **[2024, 31 Jan Shift-I]**

28 Number of moles of methane required to produce 22g $CO_2(g)$ after combustion is $x \times 10^{-2}$ moles. The value of x is .... **[2024, 31 Jan Shift-I]**

29 0.05 cm thick coating of silver is deposited on a plate of 0.05 m^2 area. The number of silver atoms deposited on plate are .... $\times 10^{23}$.
(Atomic mass of Ag = 108, $d = 7.9$ g cm^{-3}) **[2024, 30 Jan Shift-I]**

30 1 mole of PbS is oxidised by "X" moles of O_3 to get "Y" moles of O_2. $X + Y =$ .... **[2024, 27 Jan Shift-II]**

31 A solution of sugar is obtained by mixing 200 g of its 25% solution and 500 g of its 40% solution (both by mass). The mass percentage of the resulting sugar solution is .... (Nearest integer) **[2023, 11 April Shift-I]**

32 Molar mass of the hydrocarbon (X) which on ozonolysis consumes one mole of O_3 per mole of (X) and gives one mole each of ethanal and propanone is .... g mol^{-1}. (Molar mass of C : 12 g mol^{-1}, H : 1 g mol^{-1}) **[2023, 8 April Shift-I]**

33 A protein 'A' contains 0.30% of glycine (molecular weight 75). The minimum molar mass of the protein 'A' is .... $\times 10^3$ g mol^{-1}. [Nearest integer] **[2022, 25 June Shift-II]**

34 The mass percentage of nitrogen in histamine is .... **[2022, 9 Jan Shift-I]**

35 Complete combustion of 750 g of an organic compound provides 420 g of CO_2 and 210 g of H_2O. The percentage composition of carbon and hydrogen in organic compound is 15.3 and ...... respectively (Round off to the nearest integer). **[2021, 16 March Shift-I]**

36 .... grams of 3-hydroxypropanal (MW = 74) must be dehydrated to produce 7.8 g of acrolein (MW = 56) (C_3H_4O) if the percentage yield is 64. (Rounded off to the Nearest Integer). **[2021, 18 March Shift-I]**

37 A 20.0 mL solution containing 0.2 g impure H_2O_2 reacts completely with 0.316 g of $KMnO_4$ in acid solution. the purity of H_2O_2 (in %) is .... (mol. weight of $H_2O_2 = 34$; mol weight of $KMnO_4 = 158$) **[2020, 04 Sep Shift-I]**

Topic 3 / Chemical Equations, Stoichiometry and Concentration Terms

• Level 1

01 Consider the reaction,

$$4HNO_3(l) + 3KCl(s) \longrightarrow Cl_2(g) + NOCl(g) + 2H_2O(g) + 3KNO_3(s)$$

The amount of HNO_3 required to produce 110.0 g of KNO_3 is
(Given, atomic masses of H, O, N and K are 1, 16, 14 and 39 respectively.) **[2022, 29 July Shift-II]**

(a) 32.2 g (b) 69.4 g
(c) 91.5 g (d) 162.5 g

02 SO_2Cl_2 on reaction with excess of water results into acidic mixture $SO_2Cl_2 + 2H_2O \longrightarrow H_2SO_4 + 2HCl$
16 moles of NaOH is required for the complete neutralisation of the resultant acidic mixture. The number of moles of SO_2Cl_2 used is **[2022, 25 July Shift-I]**

(a) 16 (b) 8
(c) 4 (d) 2

03 The hardness of a water sample (in terms of equivalents of $CaCO_3$) containing 10^{-3} M $CaSO_4$ is (Molar mass of $CaSO_4 = 136$ g mol^{-1}) **[2019, 12 Jan Shift-I]**

(a) 100 ppm (b) 10 ppm
(c) 50 ppm (d) 90 ppm

• Level 2

04 On combustion 0.210 g of an organic compound containing C, H and O gave 0.127 g H_2O and 0.307 g CO_2. The percentages of hydrogen and oxygen in the given organic compound respectively are **[2025, 8 April Shift-II]**

(a) 53.41, 39.6 (b) 6.72, 53.41
(c) 7.55, 43.85 (d) 6.72, 39.87

05 When a concentrated solution of sulphanilic acid and 1-naphthylamine is treated with nitrous acid (273 K) and acidified with acetic acid, the mass (g) of 0.1 mole of product formed is

(Given, molar mass in g mol^{-1} H : 1, C : 12, N : 14, O : 16, S : 32) **[2025, 2 April Shift-II]**

(a) 330 (b) 33
(c) 343 (d) 66

06 $CaCO_3(s) + 2HCl(aq) \longrightarrow CaCl_2(aq) + CO_2(g) + H_2O(l)$

Consider the above reaction, what mass of $CaCl_2$ will be formed if 250 mL of 0.76 M HCl reacts with 1000 g of $CaCO_3$?

(Given, molar mass of Ca, C, O, H and Cl are 40, 12, 16, 1 and 35.5 g mol-1, respectively)

[2025, 2 April Shift-I]

(a) 10.545 g (b) 5.272 g
(c) 3.908 g (d) 2.636 g

07 Density of 3 M NaCl solution is 1.25 g/mL. The molality of the solution is **[2025, 22 Jan Shift-II]**

(a) 2.79 m (b) 1.79 m
(c) 3 m (d) 2 m

08 Given below are two statements: **[2025, 22 Jan Shift-I]**

Statement I One mole of propyne reacts with excess of sodium to liberate half a mole of H_2 gas.

Statement II Four grams of propyne reacts with $NaNH_2$ to liberate NH_3 gas which occupies 224 mL at STP.

In the light of the above statements, choose the most appropriate answer from the options given below:

(a) Both Statement I and Statement II are correct.
(b) Both Statement I and Statement II are incorrect.
(c) Statement I is correct but Statement II is incorrect.
(d) Statement I is incorrect but Statement II is correct.

09 Combustion of glucose ($C_6H_{12}O_6$) produces CO_2 and water. The amount of oxygen (in g) required for the complete combustion of 900 g of glucose is [Molar mass of glucose in g $mol^{-1} = 180$]

[2024, 8 April Shift-I]

(a) 32 (b) 480
(c) 800 (d) 960

10 Molality (m) of 3 M aqueous solution of NaCl is (Given : Density of solution = 1.25 g mL^{-1}, Molar mass in g mol^{-1}: Na-23, Cl-35.5) **[2024, 6 April Shift-II]**

(a) 2.90 m (b) 2.79 m
(c) 1.90 m (d) 3.85 m

11 The density of 'x' M solution ('x' molar) of NaOH is 1.12 g mL^{-1}, while in molality, the concentration of the solution is 3 m(3 molal). Then x is

(Given : Molar mass of NaOH is 40 g/mol)

[2024, 6 April Shift-I]

(a) 3.5 (b) 3.0 (c) 3.8 (d) 2.8

12 The number of moles of methane required to produce 11 g $CO_2(g)$ after complete combustion is (Given, molar mass of methane in g mol^{-1} : 16)

[2024, 5 April Shift-II]

(a) 0.75 (b) 0.5 (c) 0.35 (d) 0.25

13 Given below are two statements: one is labelled as Assertion (A) and the other is labelled as Reason (R).

Assertion (A) 3.1500 g of hydrated oxalic acid dissolved in water to make 250.0 mL solution will result in 0.1 M oxalic acid solution.

Reason (R) Molar mass of hydrated oxalic acid is 126 g mol^{-1}.

In the light of the above statements, choose the correct answer from the options given below.

(a) A is true but R is false **[2023, 10 April Shift-II]**
(b) Both A and R are true and R is the correct explanation of A
(c) A is false but R is true
(d) Both A and R are true but R is not the correct explanation of A

14 If a rocket runs on a fuel ($C_{15}H_{30}$) and liquid oxygen, the weight of oxygen required and CO_2 released for every litre of fuel respectively are (Given, density of the fuel is 0.756 g/mL) **[2022, 24 June Shift-I]**

(a) 1188 g and 1296 g
(b) 2376 g and 2592 g
(c) 2592g and 2376 g
(d) 3429 g and 3142 g

15 Haemoglobin contains 0.34% of iron by mass. The number of Fe-atoms in 3.3 g of haemoglobin is (Given, atomic mass of Fe is 56 u, $N_A = 6.022 \times 10^{23}$ mol^{-1}) **[2022, 26 July Shift-II]**

(a) 1.21×10^5 (b) 12.0×10^{16}
(c) 1.21×10^{20} (d) 3.4×10^{22}

16 Production of iron in blast furnace follows the following equation,

$$Fe_3O_4(s) + 4CO(g) \longrightarrow 3Fe(l) + 4CO_2(g)$$

when 4.640 kg of Fe_3O_4 and 2.520 kg of CO are allowed to react then the amount of iron (in g) produced is

[Given, molar atomic mass (g mol^{-1}) Fe = 56 molar atomic mass (g mol^{-1}) O = 16 and molar atomic mass (g mol^{-1}) C = 12.] **[2022, 29 June Shift-I]**

(a) 1400 (b) 2200
(c) 3360 (d) 4200

17 The strengths of 5.6 volume hydrogen peroxide (of density 1 g/mL) in terms of mass percentage and molarity (M), respectively, are

(Take molar mass of hydrogen peroxide as 34 g/mol)

[2020, 3 Sep Shift-II]

(a) 0.85 and 0.25 (b) 0.85 and 0.5
(c) 1.7 and 0.5 (d) 1.7 and 0.25

• Level 3

18 What amount of bromine will be required to convert 2 g of phenol into 2, 4, 6-tribromophenol?
(Given, molar mass in g mol^{-1} of C, H, O, Br are 12, 1, 16, 80 respectively.) **[2025, 23 Jan Shift-I]**
(a) 6.0 g (b) 10.22 g
(c) 4.0 g (d) 20.44 g

19 1 L, 0.02 M solution of $[Co(NH_3)_5SO_4]Br$ is mixed with 1 L, 0.02 M solution of $[Co(NH_3)_5Br]SO_4$.
The resulting solution is divided into two equal parts (*X*) and treated with excess of $AgNO_3$ solution and $BaCl_2$ solution respectively as shown below.
1 L solution $(X) + AgNO_3$ solution (excess) $\longrightarrow Y$
1 L solution $(X) + BaCl_2$ solution (excess) $\longrightarrow Z$
The number of moles of *Y* and *Z* respectively are **[2023, 30 Jan Shift-II]**
(a) 0.02, 0.01 (b) 0.01, 0.02
(c) 0.02, 0.02 (d) 0.01, 0.01

20 250 g solution of D-glucose in water contains 10.8% of carbon by weight. The molality of the solution is nearest to (Given: Atomic weights are, H, 1u; C, 12u; O, 16u) **[2022, 27 July Shift-I]**
(a) 1.03 (b) 2.06 (c) 3.09 (d) 5.40

Imp. *Limiting Reagent*

21 $\underset{(20\text{ g})}{N_2(g)} + \underset{(5\text{ g})}{3H_2(g)} \longrightarrow 2NH_3(g)$
Consider the above reaction, the limiting reagent of the reaction and number of moles of NH_3 formed respectively are **[2022, 29 July shift-I]**
(a) H_2, 1.42 mol (b) H_2, 0.71 mol
(c) N_2, 1.42 mol (d) N_2, 0.71 mol

22 The exact volume of 1 M NaOH solution required to neutralise 50 mL of 1 M H_3PO_3 solution and 100 mL of 2 M H_3PO_2 solution respectively, are **[2021, 16 March Shift-II]**
(a) 100 mL and 100 mL (b) 100 mL and 200 mL
(c) 100 mL and 50 mL (d) 50 mL and 50 mL

23 A solution of two components containing n_1 moles of the 1st component and n_2 moles of the 2nd component is prepared. M_1 and M_2 are the molecular weights of component 1 and 2 respectively. If *d* is the density of the solution in g mL^{-1}, C_2 is the molarity and x_2 is the mole fraction of the 2nd component, then C_2 can be expressed as **[2020, 06 Sep Shift-I]**
(a) $C_2 = \dfrac{1000\, x_2}{M_1 + x_2(M_2 - M_1)}$
(b) $C_2 = \dfrac{d\, x_2}{M_2 + x_2(M_2 - M_1)}$
(c) $C_2 = \dfrac{1000 d\, x_2}{M_1 + x_2(M_2 - M_1)}$
(d) $C_2 = \dfrac{d\, x_1}{M_2 + x_2(M_2 - M_1)}$

Numerical Value Type Questions

24 Butane reacts with oxygen to produce carbon dioxide and water following the equation given below.
$$C_4H_{10}(g) + \frac{13}{2}O_2(g) \rightarrow 4CO_2(g) + 5H_2O(l)$$
If 174.0 kg of butane is mixed with 320.0 kg of O_2 the volume of water formed in liters is .... (Nearest integer)
[Given, (a) Molar mass of C, H, O are 12, 1, 16 g mol^{-1} respectively, (b) Density of water = 1 g mL^{-1}] **[2025, 7 April Shift-II]**

25
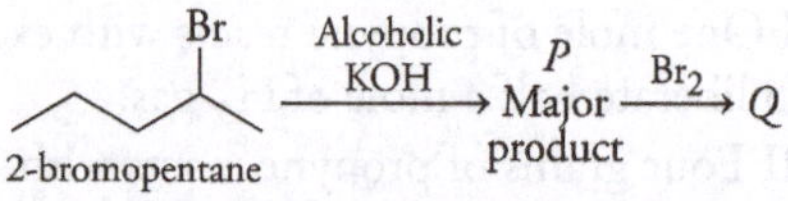

2-bromopentane
Consider the above sequence of reactions. 151 g of 2-bromopentane is made to react.
Yield of major product *P* is 80% whereas *Q* is 100%.
Mass of product *Q* obtained is .... g.
(Given, molar mass in g mol^{-1} H : 1, C : 12, O : 16, Br : 80) **[2025, 2 April Shift-II]**

26 An organic compound weighing 500 mg produced 220 mg of CO_2, on complete combustion. The percentage composition of carbon in the compound is ....%. (Nearest integer)
(Given, molar mass in g mol^{-1} of C : 12, O : 16) **[2025, 7 April Shift-I]**

27 0.5 g of an organic compound on combustion gave 1.46 g of CO_2 and 0.9 g of H_2O. The percentage of carbon in the compound is ...... (Nearest integer)
[Given, molar mass (in g mol^{-1}) C : 12, H : 1, O :16] **[2025, 3 April Shift-I]**

28 The amount of calcium oxide produced on heating 150 kg limestone (75% pure) is .... kg. (Nearest integer)
[Given, molar mass (in g mol^{-1}) of Ca-40, O-16, C-12] **[2025, 4 April Shift-II]**

29 20 mL of 2 M NaOH solution is added to 400 mL of 0.5 M NaOH solution. The final concentration of the solution is .... $\times 10^{-2}$ M. (Nearest integer) **[2025, 22 Jan Shift-II]**

30 *X*g of benzoic acid on reaction with *aq* $NaHCO_3$ released CO_2 that occupied 11.2L volume at STP. *X* is ……g. **[2025, 24 Jan Shift-I]**

31 When 81.0 g of aluminium is allowed to react with 128.0 g of oxygen gas, the mass of aluminium oxide produced in grams is......(Nearest integer)

[Given; Molar mass of Al is 27.0 g mol^{-1};
Molar mass of O is 16.0 g mol^{-1}] **[2025, 23 Jan Shift-II]**

32 Consider the following reaction occurring in the blast furnace **[2025, 24 Jan Shift-I]**

$$Fe_3O_4(s) + 4CO(g) \rightarrow 3Fe(l) + 4CO_2(g)$$

'x' kg of iron is produced when 2.32×10^3 kg Fe_3O_4 and 2.8×10^2 kg CO are brought together in the furnace. The value of 'x' is ....(Nearest integer)

[Given,
molar mass of $Fe_3O_4 = 232$ g mol^{-1}
molar mass of CO = 28 g mol^{-1}
molar mass of Fe = 56 g mol^{-1}]

33

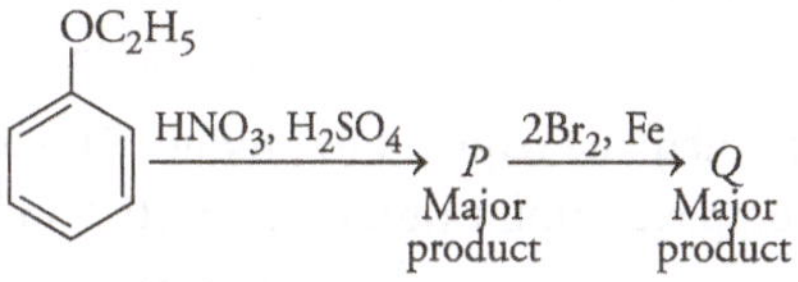

[2024,6 April Shift-II]

The ratio of number of oxygen atoms to bromine atoms in the product Q is .... $\times 10^{-1}$.

Imp. *Diazotisation Reaction & Stoichiometry*

34 9.3 g of pure aniline upon diazotisation followed by coupling with phenol gives an orange dye. The mass of orange dye produced (assume 100% yield conversion) is .... g. (Nearest integer) **[2024, 6 April Shift-I]**

35 In the Claisen-Schmidt reaction to prepare 351 g of dibenzalacetone using 87 g of acetone, the amount of benzaldehyde required is .... g. (Nearest integer) **[2024, 5 April Shift-II]**

36 X g of ethanamine was subjected to reaction with $NaNO_2$/HCl followed by hydrolysis to liberate N_2 and HCl. The HCl generated was completely neutralised by 0.2 moles of NaOH. X is .... g. **[2024, 5 April Shift-II]**

37 Only 2 mL of $KMnO_4$ solution of unknown molarity is required to reach the end point of a titration of 20 mL of oxalic acid (2 M) in acidic medium. The molarity of $KMnO_4$ solution should be...... M. **[2024, 4 April Shift-I]**

38 X g of ethylamine is subjected to reaction with $NaNO_2$ / HCl followed by water; evolved dinitrogen gas which occupied 2.24 L volume at STP. X is...... $\times 10^{-1}$ g. **[2024, 4 April Shift-I]**

39 Consider the following reaction,

$$3PbCl_2 + 2(NH_4)_3PO_4 \longrightarrow Pb_3(PO_4)_2 + 6NH_4Cl$$

If 72 m mol of $PbCl_2$ is mixed with 50 m mol of $(NH_4)_3PO_4$, then the amount of $Pb_3(PO_4)_2$ formed is .... mmol (Nearest integer). **[2024, 1 Feb Shift-I]**

40 Number of moles of H^+ ions required by 1 mole of MnO_4^- to oxidise oxalate ion to CO_2 is .... **[2024, 31 Jan Shift-II]**

41 If 50 mL of 0.5 M oxalic acid is required to neutralise 25 mL of NaOH solution, the amount of NaOH in 50 mL of given NaOH solution is .... g. **[2024, 29 Jan Shift-II]**

42 Volume of 3 M NaOH (formula weight 40 g mol^{-1}) which can be prepared from 84 g of NaOH is ....$\times 10^{-1}$ dm^3. **[2024, 27 Jan Shift-II]**

43 Mass of methane required to produce 22 g of CO_2 after complete combustion is .... g.
(Given, molar mass in g mol^{-1} C = 12.0, H = 1.0, O = 16.0) **[2024, 27 Jan Shift-I]**

44 The volume of hydrogen liberated at STP by treating 2.4 g of magnesium with excess of hydrochloric acid is .... $\times 10^{-2}$ L.

Given, molar volume of gas is 22.4L at STP.

Molar mass of magnesium is 24 g mol^{-1}. **[2023, 11 April Shift-II]**

45 If 5 moles of $BaCl_2$, is mixed with 2 moles of Na_3PO_4, the maximum number of moles of $Ba_3(PO_4)_2$ formed is .... (Nearest integer) **[2023, 6 April Shift-I]**

46 The molality of a 10% (V/V) solution of dibromine solution in CCl_4 (carbon tetrachloride) is 'x', x = ... $\times 10^{-2}$ m. (Nearest integer) **[2023, 1 Feb Shift-II]**

[Given, molar mass of Br_2 = 160 g mol^{-1}
atomic mass of C = 12 g mol^{-1}
atomic mass of Cl = 35.5 g mol^{-1}
density of dibromine = 3.2 g cm^{-3}
density of CCl_4 = 1.6 g cm^{-3}]

47 The density of 3 M solution of NaCl is 1.0 g mL^{-1}. Molality of the solution is .... $\times 10^{-2}$ m. (Nearest integer).

[Given, molar mass of Na and Cl is 23 and 35.5 g mol^{-1} respectively.] **[2023, 1 Feb Shift-I]**

48 Zinc reacts with hydrochloric acid to give hydrogen and zinc chloride. The volume of hydrogen gas produced at STP from the reaction of 11.5 g of zinc with excess HCl is .... L. (Nearest integer)

(Given, molar mass of Zn is 65.4 g mol^{-1} and molar volume of H_2 at STP = 22.7 L) **[2023, 31 Jan Shift-I]**

49 Assume carbon burns according to following equation.

$$2C(s) + O_2(g) \longrightarrow 2CO(g)$$

When 12 g carbon is burnt in 48 g of oxygen, the volume of carbon monoxide produced is .... $\times 10^{-1}$ L at STP. [Nearest integer]

[Given, assume CO as ideal gas, mass of C is 12 g mol^{-1}, mass of O is 16 g mol^{-1} and molar volume of an ideal gas at STP is 22.7 L mol^{-1}] **[2023, 31 Jan Shift-II]**

50 5 g of NaOH was dissolved in deionised water to prepare a 450 mL stock solution. What volume (in mL) of this solution would be required to prepare 500 mL of 0.1 M solution? ....

[Given, molar mass of Na, O and H is 23, 16 and 1 g mol^{-1} respectively.] **[2023, 24 Jan Shift-I]**

51 A 1.84 mg sample of polyhydric alcoholic compound '*x*' of molar mass 92.0 g/mol gave 1.344 mL of H_2 gas at STP. The number of alcoholic hydrogens present in compound '*x*' is .... **[2022, 29 July Shift-II]**

52 2 L of 0.2 M H_2SO_4 is reacted with 2 L of 0.1 M NaOH solution, the molarity of the resulting product Na_2SO_4 in the solution is .... millimolar. (Nearest integer) **[2022, 28 July Shift-II]**

53 In the given reaction,

$$X + Y + 3Z \rightleftharpoons XYZ_3$$

if one mole of each of *X* and *Y* with 0.05 mole of *Z* gives compound XYZ_3. (Given : Atomic masses of *X*, *Y* and *Z* are 10, 20 and 30 amu, respectively.) the yield of XYZ_3 is ...... g. (Nearest integer) **[2022, 28 July Shift-I]**

54 Chlorophyll extracted from the crushed green leaves was dissolved in water to make 2 L solution of Mg of concentration 48 ppm. The number of atoms of Mg in this solution is $x \times 10^{20}$ atoms. The value of *x* is .... (Nearest integer)

(Given, atomic mass of Mg is 24 g mol^{-1}; $N_A = 6.02 \times 10^{23}$ mol^{-1}) **[2022, 26 July Shift-I]**

55 When 800 mL of 0.5 M nitric acid is heated in a beaker, its volume is reduced to half and 11.5 g of nitric acid is evaporated. The molarity of the remaining nitric acid solution is $x \times 10^{-2}$ M. (Nearest integer)

(Molar mass of nitric acid is 63 g mol^{-1}) **[2022, 26 July Shift-I]**

56 The normality of H_2SO_4 in the solution obtained on mixing 100 mL of 0.1 M H_2SO_4 with 50 mL of 0.1 M NaOH is .... $\times 10^{-1}$ N. (nearest integer) **[2022, 27 July Shift-I]**

57 56.0 L of nitrogen gas is mixed with excess of hydrogen gas and it is found that 20 L of ammonia gas is produced. The volume of unused nitrogen gas is found to be .... L. **[2022, 25 July Shift-II]**

58 Number of grams of bromine that will completely react with 5.0 g of pent-1-ene is .... $\times 10^{-2}$ g. (Atomic mass of Br = 80 g/mol.) (Nearest integer) **[2022, 25 June Shift-I]**

59 The moles of methane required to produce 81 g of water after complete combustion is ...... $\times 10^{-2}$ mol. [Nearest integer] **[2022, 26 June Shift-II]**

60 0.01 M $KMnO_4$ solution was added to 20.0 mL of 0.05 M Mohr's salt solution through a burette. The initial reading of 50 mL burette is zero. The volume of $KMnO_4$ solution left in the burette after the end point is .... mL. (Nearest integer) **[2022, 28 June Shift-II]**

61 CNG is an important transportation fuel. When 100 g CNG is mixed with 208g oxygen in vehicles, it leads to the formation of CO_2 and H_2O and produced large quantity of heat during this combustion, then the amount of carbon dioxide, produced in grams is .... [Nearest integer] [Assume CNG to be methane] **[2022, 26 June Shift-II]**

62 The neutralisation occurs when 10 mL of 0.1 M acid '*A*' is allowed to react with 30 mL of 0.05 M base $M(OH)_2$. The basicity of the acid '*A*' is .... [*M* is a metal] **[2022, 25 June Shift-II]**

63 100 mL of Na_3PO_4 solution contains 3.45 g of sodium. The molarity of the solution is .... $\times 10^{-2}$ mol L^{-1}. (Nearest integer)

[Atomic masses - Na : 23.0 u, O : 16.0 u, P : 31.0 u] **[2021, 26 Aug Shift-II]**

64 10.0 mL of 0.05 M KMnO solution was consumed in a titration with 10.0 mL of given oxalic acid dihydrate solution. The strength of given oxalic acid solution is ...... $\times 10^{-2}$ g/L.

(Round off to the nearest integer) **[2021, 27 July Shift-II]**

65 When 10 mL of an aqueous solution of Fe^{2+} ions was titrated in the presence of dil. H_2SO_4 using diphenylamine indicator, 15 mL of 0.02 M solution of $K_2Cr_2O_7$ was required to get the end point. The molarity of the solution containing Fe^{2+} ions is $x \times 10^{-2}$ M. The value of x is .... (Nearest integer) **[2021, 25 July Shift-I]**

66 If the concentration of glucose $(C_6H_{12}O_6)$ in blood is 0.72 gL^{-1}, the molarity of glucose in blood is ...... $\times 10^{-3}$ M. (Nearest integer)

[Given : Atomic mass of $C = 12, H = 1, O = 16$ u] **[2021, 22 July Shift-II]**

67 4g equimolar mixture of NaOH and Na_2CO_3 contains x g of NaOH and y g of Na_2CO_3. The value of x is ...... g. (Nearest integer) **[2021, 20 July Shift-II]**

68 250 mL of 0.5 M NaOH was added to 500 mL of 1 M HCl. The number of unreacted HCl molecules in the solution after complete reaction is ...... $\times 10^{21}$. (Nearest integer) $(N_A = 6.022 \times 10^{23})$

[2021, 20 July Shift-I]

69 The density of NaOH solution is 1.2 g cm^{-3}. The molality of this solution is .... m.
(Round off to the Nearest Integer)

[Use : Atomic masses : Na : 23.0 u, O : 16.0 u, H : 1.0 u density of H_2O : 1.0 g cm^{-3}] **[2021, 27 July Shift-I]**

70 15 mL of aqueous solution of Fe^{2+} in acidic medium completely reacted with 20 mL of 0.03 M aqueous $Cr_2O_7^{2-}$. The molarity of the Fe^{2+} solution is ...... $\times 10^{-2}$ M (Round off to the nearest integer). **[2021, 17 March Shift-I]**

71 The mole fraction of a solute in a 100 molal aqueous solution ...... $\times 10^{-2}$

(Round off to the nearest integer).

[Given, atomic masses H : 1.0 u, O : 16.0 u] **[2021, 17 March Shift-I]**

72 10.0 mL of Na_2CO_3 solution is titrated against 0.2 M HCl solution. The following titre values were obtained in 5 readings. 4.8 mL, 4.9 mL, 5.0 mL, 5.0 mL and 5.0 mL based on these readings and convention of titrimetric estimation of concentration of Na_2CO_3 solution is .... mM (Round off to the nearest integer). **[2021, 18 March Shift-II]**

73 The number of chlorine atoms in 20 mL of chlorine gas at STP is .... $\times 10^{21}$. (Round off to the Nearest integer) **[2021, 17 March Shift-II]**

[Assume chlorine is an ideal gas at STP]
$R = 0.083$ L bar $mol^{-1}K^{-1}$, $N_A = 6.023 \times 10^{23}$

74 When 35 mL of 0.15 M lead nitrate solution is mixed with 20 mL of 0.12 M chromic sulphate solution .... $\times 10^{-5}$ moles of lead sulphate precipitate out. **[2021, 16 March Shift-II]**
(Round off to the Nearest Integer).

75 A reaction of 0.1 mole of Benzylamine with bromomethane gave 23 g of Benzyl trimethyl ammonium bromide. The number of moles of bromomethane consumed in this reaction are $n \times 10^{-1}$, when $n =$ ....

(Rounded off to the Nearest Integer).

[Given : Atomic masses are C : 12.0 u, H : 1.0 u, N : 14.0 u, Br : 80.0 u] **[2021, 18 March Shift-I]**

76 The $NaNO_3$ weighed out to make 50 mL of an aqueous solution containing 70.0 mg Na^+ per mL is .... g. (Rounded off to the nearest integer)

[Given : Atomic weight in g mol^{-1}, Na : 23 ; N : 14 ; O : 16]. **[2021, 26 Feb Shift-II]**

77 The formula of a gaseous hydrocarbon, which requires 6 times of its own volume of O_2 for complete oxidation and produces 4 times its own volume of CO_2 is C_xH_y. The value of y is .... **[2021, 24 Feb Shift II]**

78 1.86 g of aniline completely reacts to form acetanilide. 10% of the product is lost during purification. Amount of acetanilide obtained after purification (in g) is ....$\times 10^{-2}$. **[2021, 24 Feb Shift-II]**

79 4.5 g of compound A (MW = 90) was used to make 250 mL of its aqueous solution. the molarity of the solution is $x \times 10^{-1}$. The value of x is ....
(Rounded off to the nearest integer) **[2021, 24 Feb Shift-I]**

80 0.4 g mixture of NaOH, Na_2CO_3 and some inert impurities was first titrated with $\frac{N}{10}$ HCl using phenolphthalein as an indicator, 17.5 mL of HCl was required at the end point. After this, methyl orange was added and titrated. 1.5 mL of same HCl was required for the next end point. The weight percentage of Na_2CO_3 in the mixture is .... (Rounded-off to the nearest integer) **[2021, 25 Feb Shift-I]**

81 6.023×10^{22} molecules are present in 10 g of a substance 'x'. The molarity of a solution containing 5 g of substance 'x' in 2 L solution is ...... $\times 10^{-3}$. **[2020, 3 Sep Shift-II]**

82 A 100 mL solution was made by adding 1.43 g of $Na_2CO_3 \cdot xH_2O$. The normality of the solution is 0.1 N. The value of x is .... **[2020, 4 Sep Shift-II]**

83 The ratio of the mass percentages of 'C' and H' and 'C and O' of a saturated acyclic organic compound 'X' are 4 : 1 and 3 : 4 respectively. Then, the moles of oxygen gas required for complete combustion of two moles of organic compound 'X' is .... **[2020, 2 Sep Shift-II]**

84 The volume (in mL) of 0.1 N NaOH required to neutralise 10 mL of 0.1 N phosphonic acid is .... **[2020, 3 Sep Shift-II]**

85 The mass of ammonia in grams produced when 2.8 kg of dinitrogen quantitatively reacts with 1 kg of dihydrogen is .... **[2020, 4 Sep Shift-I]**

86 The molarity of HNO_3 in a sample which has density 1.4 g/mL and mass percentage of 63% is .... (Molecular weight of HNO_3 = 63) **[2020, 9 Jan Shift-I]**

87 The volume (in mL) of 0.125 M $AgNO_3$ required to quantitatively precipitate chloride ions in 0.3 g of $[Co(NH_3)_6]Cl_3$ is ....

$$M_{[Co(NH_3)_6]Cl_3} = 267.46 \text{ g/mol}$$
$$M_{AgNO_3} = 169.87 \text{ g/mol}$$
[2020, 8 Jan Shift-I]

88 $NaClO_3$ is used, even in spacecrafts, to produce O_2. The daily consumption of pure O_2 by a person is 492 L at 1 atm 300 K. How much amount of $NaClO_3$, in grams, is required to produce O_2 for the daily consumption of a person at 1 atm, 300 K .... ?

$$NaClO_3(s) + Fe(s) \longrightarrow O_2(g) + NaCl(s) + FeO(s)$$
$$R = 0.082 \text{ L atm mol}^{-1} \text{ K}^{-1}$$
[2020, 8 Jan Shift-II]

Key Index

Topic 1

1 (a)	2 (b)	3 (c)	4 (d)	5 (a)	6 (8)	7 (3)

Topic 2

1 (a)	2 (b)	3 (b)	4 (b)	5 (d)	6 (a)	7 (d)	8 (b)	9 (d)	10 (d)
11 (d)	12 (a)	13 (a)	14 (c)	15 (b)	16 (d)	17 (b)	18 (3)	19 (9)	20 (372)
21 (100)	22 (45)	23 (1655)	24 (60)	25 (22)	26 (80)	27 (78)	28 (50)	29 (11)	30 (8)
31 (36)	32 (70)	33 (25)	34 (37.84)	35 (3)	36 (16)	37 (85)			

Topic 3

1 (c)	2 (c)	3 (a)	4 (b)	5 (b)	6 (a)	7 (a)	8 (c)	9 (d)	10 (b)
11 (b)	12 (d)	13 (b)	14 (c)	15 (c)	16 (c)	17 (c)	18 (b)	19 (d)	20 (b)
21 (c)	22 (b)	23 (c)	24 (138)	25 (184)	26 (12)	27 (80)	28 (63)	29 (57)	30 (61)
31 (153)	32 (420)	33 (15)	34 (20)	35 (318)	36 (9)	37 (8)	38 (45)	39 (24)	40 (8)
41 (4)	42 (7)	43 (8)	44 (224)	45 (1)	46 (139)	47 (364)	48 (4)	49 (227)	50 (180)
51 (3)	52 (25)	53 (2)	54 (24)	55 (54)	56 (1)	57 (46)	58 (1143)	59 (225)	60 (30)
61 (143)	62 (3)	63 (50)	64 (1575)	65 (18)	66 (4)	67 (1)	68 (226)	69 (5)	70 (24)
71 (64)	72 (50)	73 (1)	74 (525)	75 (3)	76 (13)	77 (8)	78 (243)	79 (2)	80 (4)
81 (25)	82 (10)	83 (5)	84 (10)	85 (3400)	86 (14)	87 (27)	88 (2130)		

Detailed Explanations

Topic 1

01. *(a)* The number of significant figures of the given numbers are as follows:

(A) 0.00253 3

(B) 1.0003 5

(C) 15.0 3

(D) 163 3

Thus, (A), (C) and (D) have same number of significant figures.

02. *(b)* Answer should be reported with 3 significant figures.

$$\frac{0.02858 \times 0.112}{0.5702} = 0.00\underbrace{561}_{\text{3 significant figures}}$$

Concept Enhancer

In multiplication and division, the result must be reported with number of significant figures equal to those present in the expression with least significant number involved in the calculation.

03. *(c)* Statement given in options B and D are incorrect, while in option A, C and E are correct. The correct form of incorrect statements are

Matter consists of indivisible atoms.

All the atoms of a given element have same properties including mass.

04. *(d)* Statement given in option (d) is incorrect. Its correct form is as follows according to Dalton's atomic theory, compounds are formed when atoms of different elements combine in fixed ratio.

05. *(a)* Among the given statements, statement given in C, D and E are correct while A and B are incorrect.

Their correct forms are

Weight of substance is the force applied by gravity on body.

Mass of a substance is the amount of matter present in it.

06. *(8)* Zero's between 5 and 2 are all significant. Trailing zeroes after decimal are also significant.

$$\underline{50000.020} \times 10^{-3}$$

(Number of significant figures = 8)

07. *(3)* Total number of significant figure = 3.

Here,

$$\underbrace{0.00}_{\text{Leading zero}}\ \underbrace{34}_{\text{Non-zero digits}}\ \underbrace{0}_{\text{Trailing zero}}$$

Non-zero digits and trailing zero are considered as significant figures.

Topic 2

01. *(a)* The reaction involved is as follows

$$Mg + 2HCl \longrightarrow MgCl_2 + H_2$$

Volume of H_2 evolved = 220 mL

$$\text{Mole of } H_2 = \frac{220 \times 10^{-3}}{22.4} \times 24$$

$$= 235.7 \times 10^{-3} \text{ g} \cong 236 \text{ mg}$$

02. *(b)* Let the mass of CO_2 taken initially $= x$

No. of moles of CO_2 left

$$= 2.8 \times 10^{-3} \text{ mol}$$

No. of molecules removed

$$= 10^{21} \text{ molecules}$$

$$\text{Now, no. of initial moles} = \frac{x \times 10^{-3}}{44}$$

$$\text{No. of moles removed} = \frac{10^{21}}{6.02 \times 10^{23}}$$

No. of moles left

= initial moles − moles removed

$$2.8 \times 10^{-3} = \frac{x \times 10^{-3}}{44} - \frac{10^{21}}{6.02 \times 10^{23}}$$

$$x = 196.2 \text{ mg}$$

03. *(b)* Given,

Mass of solute $(X) = 2$g

Mass of solvent $(H_2O) = 1$ mole = 18 g

So, total mass of the solution

$$= 18 + 2 = 20 \text{ g}$$

Mass % of X

$$= \frac{\text{Mass of solute}}{\text{Total mass of the solution}} \times 100$$

$$= \frac{2}{20} \times 100 = 10\%$$

04. *(b)* Suppose the mole ratio of ^{35}Cl be x and that of ^{37}Cl be y.

$$\text{Average molar mass} = \frac{n_1 m_1 + n_2 m_2}{n_1 + n_2}$$

$$35.5 = \frac{35x + 37y}{x + y}$$

$$\Rightarrow \quad 1.5y = 0.5x \quad \Rightarrow \frac{x}{y} = \frac{3}{1}$$

05. *(d)* Number of moles of carbon

$$(C) = n_{CO_2} = \frac{1.46}{44} = 0.033$$

Moles of 'C' $= W_C = 0.033 \times 12$

Moles of hydrogen

$$(H) = W_H = 2 \times n_{H_2O} = \frac{2 \times 0.567}{18} = 0.063$$

Mass of hydrogen = 0.0063

Mass of oxygen $= 1 - (W_C + W_H)$

$$= 1 - (0.033 \times 12 + 0.063 \times 1) = 0.541 \text{g}$$

$$\text{Mole of 'O'} = \frac{0.541}{16} = 0.033$$

Empirical formula $= CH_2O$

Empirical formula mass = 30g

06. *(a)* Number of atoms can be calculated by using the formula,

Number of atoms

$$= \frac{\text{Mass in g}}{\text{Molar mass (g/mol)}} \times N_A$$

So, for the same given mass, element with least molar mass will have highest number of atoms.

Now, molar mass of Pr = 141 g/mol, Po = 209 g/mol, Pt = 195 g/mol and Pb = 207 g/mol

Thus, Pr has highest number of atoms.

07. *(d)* C : H : O

$$\frac{54.2}{12} : \frac{9.2}{1} : \frac{36.6}{16}$$

$$4.516 : 9.2 : 2.287$$

$$\frac{4.516}{2.287} : \frac{9.2}{2.287} : \frac{2.287}{2.287}$$

$$1.97 : 4.02 : 1 \approx 2 : 4 : 1$$

$C_2H_4O \Rightarrow$ empirical formula

Empirical formula mass = 24 + 4 + 16

= 44 g/mol

and molar mass = 132 g/mol

Hence, molecular formula is

$$(C_2H_4O)_3 = C_6H_{12}O_3$$

08. *(b)* Weight of metal chloride at STP is

$$= \frac{0.57}{100} \times 22400 = 127.68$$

Weight or mass of Cl = 127.68 × 0.55

in metal chloride = 70.224

Now, number of moles of Cl

$= \dfrac{\text{Mass of Cl}}{\text{Molecular mass of Cl}} = \dfrac{70.224}{35.5} \approx 2$

Thus, the molecular formula is MCl_2.

09. *(d)* No. of moles $(n) = \dfrac{\text{Given volume}}{\text{Volume at STP}}$

$\therefore$ No. of moles of $O_2 = \dfrac{2.8375}{22.7} = 0.125$

Now, number of molecules = Moles × Avagadro's number

$= 0.125 N_A$

$= (0.125) \times (6.022 \times 10^{23})$

$= 7.527 \times 10^{22}$

10. *(d)* The general equation for complete combustion of organic compound is

$$C_xH_y + \left(x + \frac{y}{4}\right)O_2 \longrightarrow xCO_2 + \frac{y}{2}H_2O$$

Moles of $CO_2 = \dfrac{330}{44} = 7.5$ moles

Moles of water $= \dfrac{270}{18} = 15$ moles

Moles of CO_2 = moles of carbon

Mass of carbon = number of moles × atomic mass

$= 7.5 \times 12 = 90$ g

% of carbon in organic compound

$= \dfrac{90}{120} \times 100 = 75\%$

% of hydrogen in organic compound

$= \left(\dfrac{30}{120}\right) \times 100 = 25\%.$

11. *(d)*

idea To find the mass of *A* and *B* in the given question concept of mole is used.

Number of moles $(n) = \dfrac{\text{Given mass } (w)}{\text{Molecular mass } (M)}$

Compound	Mass of *A* (g)	Mass of *B* (g)
AB_2	M_A	$2M_B$
A_2B_2	$2M_A$	$2M_B$

We know that,

Number of moles

$(n) = \dfrac{\text{Given mass } (w)}{\text{molecular mass } (M)}$ ...(*A*)

Using equation (*A*), it can be concluded that

$5(M_A + 2M_B) = 125 \times 10^{-3}$ kg ...(i)

$10(2M_A + 2M_B) = 300 \times 10^{-3}$ kg ...(ii)

Dividing Eq. (i) and (ii),

$\dfrac{1}{2} \dfrac{(M_A + 2M_B)}{(2M_A + 2M_B)} = \left(\dfrac{125}{300}\right)$

On solving the equation, we obtain

$M_A = 5 \times 10^{-3}$ and $M_B = 10 \times 10^{-3}$

12. *(a)* $\underset{2\text{ mol}}{2NaHCO_3} + \underset{1\text{ mol}}{H_2C_2O_4} \longrightarrow \underset{2\text{ mol}}{2CO_2} + Na_2C_2O_4 + 2H_2O$

⇒ In the reaction, number of mole of CO_2 produced.

$n = \dfrac{pV}{RT}$

$= \dfrac{1\text{ bar} \times 0.25 \times 10^{-3}\text{ L}}{0.082\text{ L atm K}^{-1}\text{mol}^{-1} \times 298.15\text{ K}}$

$= 1.02 \times 10^{-5}$ mol

Number of mole of $NaHCO_3$

$= \dfrac{\text{Weight of } NaHCO_3}{\text{Molecular mass of } NaHCO_3}$

$\therefore w_{NaHCO_3} = 1.02 \times 10^{-5} \times 84 \times 10^3$ mg

$= 0.856$ mg

$\Rightarrow \%NaHCO_3 = \dfrac{0.856}{10} \times 100$

$= 8.56\% \approx 8.4$

13. *(a)* Correct statements regarding given equation is given in statements A and D. i.e., 200 mL of KI solution reacts with 0.004 mol of KIO_3. Equivalent weight of KIO_3 is equal to $\left(\dfrac{\text{molecular weight}}{5}\right)$.

14. *(c)* The reaction involved are

$CaCO_3(s) \xrightarrow{\Delta} CaO(s) + CO_2(g)$

$MgCO_3(s) \xrightarrow{\Delta} MgO(s) + CO_2(g)$

Let the weight of $CaCO_3$ be *x* g.

$\therefore$ Weight of $MgCO_3 = (2.21 - x)$ g

Moles of $CaCO_3$ decomposed = Moles of CaO formed

$\dfrac{x}{100}$ = Moles of CaO formed

$\therefore$ Weight of CaO formed $= \dfrac{x}{100} \times 56$

Moles of $MgCO_3$ decomposed = Moles of MgO formed

$\dfrac{2.21 - x}{84}$ = Mole of MgO formed

$\therefore$ Weight of MgO formed

$= \dfrac{2.21 - x}{84} \times 40$

$\dfrac{2.21 - x}{84} \times 40 + \dfrac{x}{100} \times 56 = 1.152$

$\therefore \quad x = 1.1886 \text{ g} \approx 1.187$ g

= Weight of $CaCO_3$

Weight of $MgCO_3 = 1.0214$ g

≈ 1.023 g

Hence, option (c) is correct.

15. *(b)* The correct match is

A-II, B-IV, C-I, D-III.

(A) 16 g of $CH_4(g) = 1$ mole CH_4 contains $10 \times 6.02 \times 10^{23}$ electrons $= 60.2 \times 10^{23}$

(B) 1 g $H_2(g) = 0.5$ mole H_2 gas occupy 1 mole contains = 22.7 L

$\therefore$ 0.5 mole contains 11.35 (approx) litre volume at STP.

(C) 1 mole of $N_2(g) = 28$ g

(D) Mole of $SO_2(g) = 64$ g

0.5 mole of $SO_2(g) = 32$ g

16. *(d)* Empirical formula can be found out as follows

Elements	% composition	Atomic mass	Relative number of moles	Simple ratio of mole
Carbon (C)	74	12	$\dfrac{74}{12} = 6.17$	$\dfrac{6.17}{1.24} \approx 5$
Hydrogen (H)	8.7	1	$\dfrac{8.7}{1} = 8.7$	$\dfrac{8.7}{1.24} \approx 7$
Nitrogen (N)	17.3	14	$\dfrac{17.3}{14} = 1.24$	$\dfrac{1.24}{1.24} = 1$

So, empirical formula is C_5H_7N.

Empirical mass $= 5 \times 12 + 7 \times 1 + 14 \times 1$

$= 60 + 7 + 14 = 81$

Molecular mass = 162 (given)

Thus, $n = \dfrac{\text{Molecular mass}}{\text{Empirical mass}}$

$n = \dfrac{162}{81} = 2$

$\therefore$ Molecular formula = imperical formula

$= (C_5H_7N)_2 = C_{10}H_{14}N_2$

17. *(b)* $\underset{1.8\text{ g}}{C_xH_yO_z} + O_2 \longrightarrow \underset{2.64\text{ g}}{CO_2} + \underset{1.08\text{ g}}{H_2O}$

$n_C = \underset{\text{(Moles)}}{n_{CO_2}} = \dfrac{2.64 \text{ (Given mass)}}{44 \text{ (Molecular mass)}} = 0.06$

$n_H = 2 \times n_{H_2O} = \dfrac{1.08}{18} \times 2 = 0.12$

Weight of oxygen in $C_xH_yO_z$

$= 1.80 - 12 \times 0.06 - 0.06 \times 2$

$= 1.80 - 0.72 - 0.12 = 0.96$ g

% of oxygen by weight

$= \dfrac{0.96}{1.80} \times 100 = 53.33\%$

18. *(3)* The complete reaction is as follows.

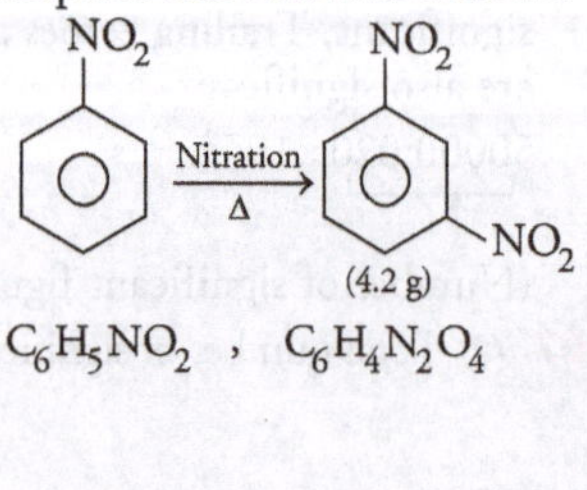

$C_6H_5NO_2$, $C_6H_4N_2O_4$

Molecular weight 123, 168

$\therefore \frac{4.2}{168} = 0.025$ mol

$\therefore$ Required gram of nitrobenzene

$= 123 \times 0.025$

$= 3.075$

$\therefore$ Nearest integer is 3.

19. *(9)* Let the mass of iron be x g.

$$\frac{x}{150 \times 10^3} \times 10^6 = 12$$

$\Rightarrow \quad x = 1.8$ g

Now, let mass of $FeSO_4 \cdot 7H_2O = y$ g

So, moles of Fe $= \frac{1.8}{56} = \frac{y}{56 + 96 + (7 \times 18)}$

$\Rightarrow \quad y = 8.935\text{ g} \approx 9$

20. *(372)* Considering the given compound

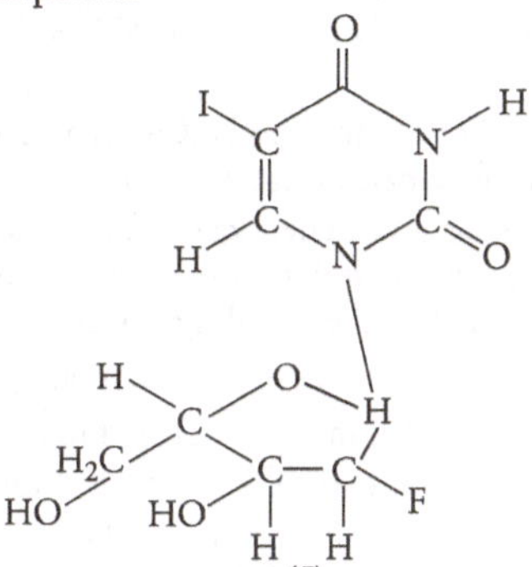

(P)

Molar mass = 372 g

1 mole = 372 g

0.1 mole has $= 372 \times 10^{-1}$ g

21. *(100)* Given, mass of H_2O on combustion $= 0.9$ g

Number of moles of organic compound $(X) = 0.01$ mol

Molar mass of $X = ?$

$$\underset{(X)}{\text{Organic compound}} \xrightarrow{\text{Combustion}} \underset{0.9\text{ g}}{H_2O}$$

$\therefore$ Mole of $H_2O = \frac{0.9}{18} = 0.05$ mole

$\therefore$ Mole of H in $H_2O = 0.05 \times 2 = 0.1$ mol

= mole of H in 0.01 mole organic compound

$\therefore$ Weight of H in 0.01 mole compound $= 0.1 \times 1 = 0.1$ g

$\therefore$ Weight of H in one mole compound

$= \frac{0.1}{0.01} = 10$ g

$\because$ Weight % of H

$$= \frac{\text{Weight of H in one mole compound}}{\text{Molar mass of compound}} \times 100$$

$$10 = \frac{10}{M} \times 100$$

$\therefore \quad M = 100$ g/mol

So, the molar mass of x is 100 g/mol.

22. *(45)* Let the number of moles of $CO_2 = y$.

Total initial moles of $Ca(OH)_2 = 2y$

Excess $Ca(OH)_2 = y$

Gram equivalent of $Ca(OH)_2$

= Gram equivalent of HCl

$$y \times 2 = \frac{0.1 \times 40}{1000} \times 1$$

$$y = 2 \times 10^{-3}$$

Volume of CO_2

$= x = 2 \times 10^{-3} \times 22400$

$= 44.8\text{ cm}^3$

$\approx 45\text{ cm}^3$.

So, the value of x is 45.

23. *(1655)*

> **idea** To solve this problem, first of all find the molar ratio of all the elements, then find the minimum integral ratio and put this value as a subscript with respective element to obtain empirical formula.

	C :	Cl :	H :	O
% mass	14.5	64.46	1.8	19.24

Molar ratio

$$= \frac{14.5}{12}, \frac{64.46}{35.5}, \frac{1.8}{1}, \frac{19.24}{16}$$

$= 1.2 \quad 1.8 \quad 1.8 \quad 1.2$

Minimum integral ratio 2 3 3 2

$\therefore$ Empirical formulae $= C_2H_3Cl_3O_2$

So, the empirical mass is $= 165.5$

Mass $= 1655 \times 10^{-1}$

24. *(60)* The complete reaction is as follows

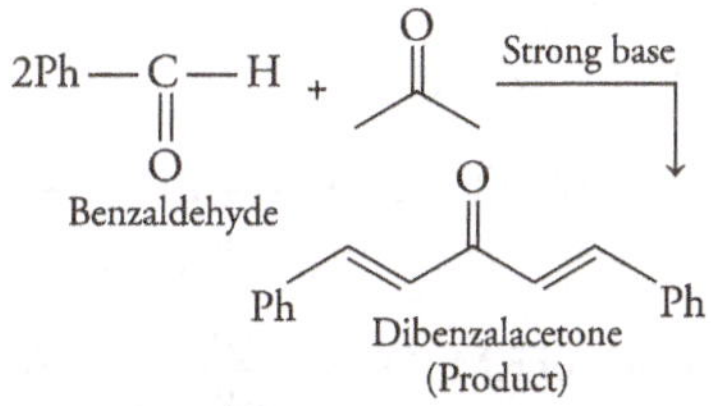

Now, 5.3 g of benzaldehyde

= 3.51 g of dibenzalacetone

$\therefore \frac{5.3}{106}$ mol of benzaldehyde

$= \frac{3.51}{234}$ mol of dibenzalacetone

or, $\frac{1}{20}$ mol (Actual) of benzaldehyde

= 0.015 mol of dibenzalacetone

Theoretical $= \frac{1}{40}$ mol

[$\because$ 2 mole sof benzaldehyde give 1 mol]

$$\%\text{ yield} = \frac{0.015}{1/40} \times 100 = 60\%$$

25. *(22)* Mass percent of alcohol

$$= \frac{\text{Mass of ethyl alcohol}}{\text{Total mass of solution}} \times 100$$

$$= \frac{1 \times 46}{1 \times 46 + (9 \times 18)} \times 100$$

$$= \frac{4600}{208} = 22.11$$

Nearest integer = 22

26. *(80)* The reaction that take place is given by

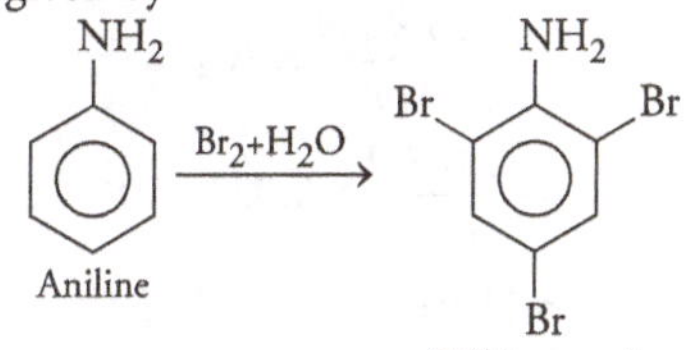

93 g of aniline produces 330 g of 2, 4, 6-tribromoaniline

Hence, 9.3 g of aniline will give 33 g of 2, 4, 6-tribromoaniline

$$\text{Percentage yield} = \frac{26.4 \times 100}{33} = 80\%$$

27. *(78)*

- $\underset{\text{(Conc.)}}{2NaBr + 2H_2SO_4} \xrightarrow{\Delta} \underset{\text{(Reddish brown vapour)}}{Br_2} + SO_2 + 2H_2O + Na_2SO_4$
- $\underset{\text{(Conc.)}}{4NaNO_3 + 2H_2SO_4} \xrightarrow{\Delta} \underset{\text{(Reddish brown vapour)}}{4NO_2} + O_2 + 2Na_2SO_4 + 2H_2O$
- $\underset{\text{(Conc.)}}{2KI + H_2SO_4} \xrightarrow{\Delta} \underset{\text{(Violet vapour)}}{I_2} + SO_2 + 2H_2O + K_2SO_4$
- $\underset{\text{(Conc.)}}{CaF_2 + H_2SO_4} \xrightarrow{\Delta} CaSO_4 + \underset{\text{(Colourless)}}{2HF}$

So, CaF_2 does not evolved coloured vapours. Hence its molar mass is 78 g/mol.

28. *(50)* $CH_4(g) + 2O_2(g) \longrightarrow CO_2(g) + 2H_2O(l)$

$$n = \frac{22}{44} = \frac{1}{2}\text{ mol of } CO_2$$

as 1 mol CH_4 form 1 mol CO_2

So, $n_{CH_4} = \frac{1}{2}$ mol $= 0.5$ mol

$n_{CH_4} = 50 \times 10^{-2}$ mol

So, the value of x is 50.

29. *(11)* Given, thickness $(d) = 0.05$ cm

Area of plate $(A) = 0.05$ m^2

Volume of silver coating

$V = A \times d = 0.05 \times 0.05 \times 10000$
$= 25$ cm^3

Mass of silver deposited $= 25 \times 7.9$ g

Mole of Ag atoms $= \frac{25 \times 7.9}{108}$

Number of silver atoms

$= \frac{25 \times 7.9}{108} \times N_A$

$= \frac{25 \times 7.9}{108} \times 6.023 \times 10^{23}$

$= 11.01 \times 10^{23} \approx 11 \times 10^{23}$

Hence, the number of Ag atoms are 11×10^{23}.

30. *(8)* The reaction involved is as follows,

$PbS + 4O_3 \longrightarrow PbSO_4 + 4O_2$

Here $X = 4$ and $Y = 4$

$\therefore X + Y = 4 + 4 = 8$

Hence, $X + Y$ is 8.

31. *(36)* Total mass of sugar in the given mixture is

$= 0.25 \times 200 + 0.40 \times 500$
$= 250$ g

Now, total mass of solution

$= 200 + 500 = 700$ g

So, mass % of sugar $= \frac{250}{700} \times 100$

$= 35.7\% \approx 36\%$.

Concept Enhancer

Empirical formula of a chemical compound is the simplest whole number ratio of atoms present in a compound.

32. *(70)* Compound (X) on ozonolysis consume one mole of O_3 and will give one mole of ethanal and one mole of propanone. Thus, (X) can be 2-methylbut-2-ene

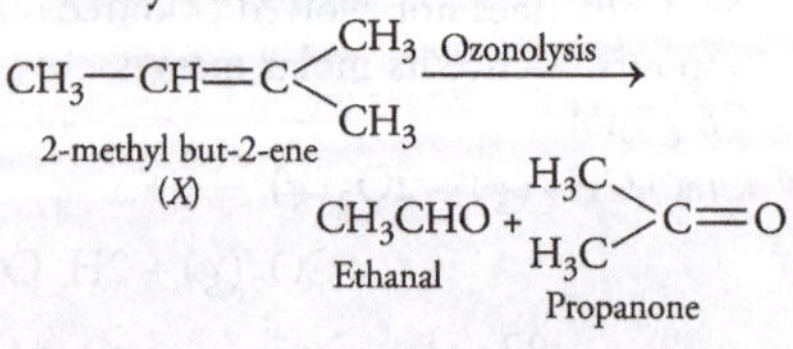

Molar mass of $X = 12 \times 5 + 1 \times 10$
$= 60 + 10 = 70$ g/mol

33. *(25)* Given, mass % of glycine $= 0.30\%$

Molecular weight of glycine $= 75$ g mol^{-1}

% of glycine

$= \left[\frac{75}{\text{Molar mass of protein}}\right] \times 100$

$0.3 = \frac{75}{M} \times 100$

$\Rightarrow M = \frac{75 \times 1000}{3} = 25 \times 10^3$ g mol^{-1}

34. *(37.84)* The structure of histamine

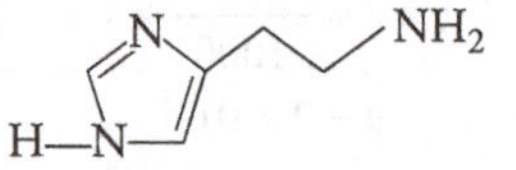

Molecular formula $= C_5H_9N_3$.

Molecular mass $= 111$

Number of nitrogen atoms per molecule $= 3$

Mass due to N $= 3 \times 14 = 42$

% of N $= \frac{42}{111} \times 100 = 37.84\%$

35. *(3)* Mass of water formed $= 210$ g

18 g of H_2O contains $= 2$ g of hydrogen

210 g of H_2O contains

$= \left(\frac{2}{18} \times 210 = \frac{70}{3}\right)$ g of hydrogen.

Given, mass of organic compound $= 750$ g

Percentage composition of hydrogen

$= \frac{70}{3 \times 750} \times 100 = 3.11\%$

Nearest integer $= 3$

36. *(16)* When 3-hydroxy propanal is heated,

$$(HO)CH_2CH_2CHO \xrightarrow[64\%]{\Delta} C_3H_4O + H_2O$$

Suppose x g $\qquad$ 7.8 g

$\frac{x}{74} \times 0.64$ mol $\qquad \frac{7.8}{56}$ mol

$\therefore \frac{x}{74} \times 0.64$ mol $= \frac{7.8}{56}$ moles

$\Rightarrow x = 16.10 \approx 16$

37. *(85)*
$$5H_2O_2 + 2\overset{(+7)}{Mn}O_4^- + 3H^+ \longrightarrow 2Mn^{2+} + 5O_2 + 8H_2O$$

Moles of $KMnO_4 = \frac{0.316}{158} = 2 \times 10^{-3}$

(Molar mass of $KMnO_4 = 158$)

Equivalents of H_2O_2 = Equivalents of $KMnO_4$

Equivalents of $KMnO_4$

$= 2 \times 10^{-3} \times 5 = 0.01$

Moles of $H_2O_2 = \frac{\text{Equivalents of } H_2O_2}{n\text{- factor of } H_2O_2}$

$= \frac{0.01}{2} = 0.005$

Mass of pure $H_2O_2 = 0.005 \times 34$
$= 0.170$ g

% purity $= \frac{0.17}{0.2} \times 100 = 85\%$

Topic 3

01. *(c)* The given reaction is

$$4HNO_3(l) + 3KCl(s) \longrightarrow Cl_2(g) + NOCl(g) + 2H_2O(g) + 3KNO_3(s)$$

Number of moles of $KNO_3 = \frac{110}{101} = 1.089$

As 3 moles of KNO_3 is produced from 4 moles or HNO_3

$\therefore$ 1.089 moles of KNO_3 is produced from $= \frac{4}{3} \times 1.089 = 1.452$ mol

Mass of HNO_3 required is

$1.452 \times 63 = 91.5$ g

02. *(c)*

$$SO_2Cl_2 + 2H_2O \longrightarrow H_2SO_4 + 2HCl$$
(given)

Number of moles of NaOH required for complete neutralisation $= 16$

Now, as given chemical equation is balanced, so stoichiometric coefficients are equal to their moles respectively.

Number of moles of $H^+ = 2$ moles [from H_2SO_4] + 2 moles [HCl] = 4 moles

Now, 1 mole of SO_2Cl_2 gives 4 moles of H^+ and 16 moles of NaOH are used in neutralisation of these 4 moles.

So, moles of SO_2Cl_2 used

$= \frac{\text{Moles of NaOH used in neutralisation}}{\text{Moles of acid } [H^+] \text{ formed from 1 mole of } SO_2Cl_2}$

$= \frac{16}{4} = 4$

03. *(a)* Hardness of water sample can be calculated in terms of ppm concentration of $CaCO_3$.

Given, molarity $= 10^{-3}$ M

i.e. 1000 mL of solution contains 10^{-3} mole of $CaCO_3$.

$\therefore$ Hardness of water = ppm of $CaCO_3$

$= \frac{10^{-3} \times 100}{1000} \times 10^6 = 100$ ppm

04. *(b)* In the combustion of organic compound, all carbon in CO_2 and all hydrogen in H_2O comes from organic compound.

The reaction is as follows,

$$\underset{0.210\ g}{C_XH_YO_Z} + O_2 \longrightarrow \underset{0.307\ g}{CO_2} + \underset{0.127\ g}{H_2O}$$

Weight of 'C' in CO_2 is

$= \frac{12}{44} \times 0.307 = 0.0837$ g

Weight of 'H' in H_2O is

$= \frac{2}{18} \times 0.127 = 0.0141$ g

% of H in compound

$= \frac{0.0141}{0.21} \times 100 = 6.72\%$

Weight of 'O' in compound

$= 0.210 - (0.0837 + 0.0141) = 0.1122$

% of 'O' in compound

$= \frac{0.1122}{0.21} \times 100 = 53.41\%$

05. *(b)* The complete reaction is as follows,

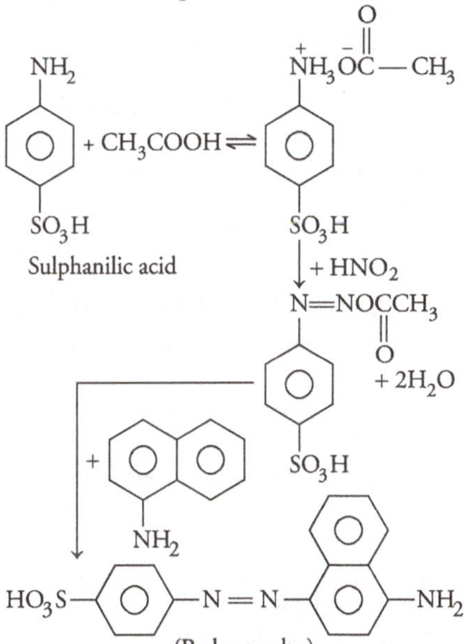

(Red - azo dye)

0.1 mole of red-azo dye (Molar mass = 327 g/mol) will have 32.7 g mass.

Nearly 33 g

06. *(a)* The reaction involved is as follows,

$CaCO_3 + 2HCl \longrightarrow CaCl_2 + CO_2 + H_2O$

Moles of $CaCO_3 = \frac{1000}{100} = 10$

Mole of HCl $= 0.76 \times \frac{250}{1000} = 0.19$

Moles of $CaCl_2$ formed $= \frac{0.19}{2}$

Mass of $CaCl_2 = \frac{0.19}{2} \times 111$

$= 10.545$ g

07. *(a)* Given, Molarity = 3M

Density = 1.25 g/mL

Molality = ?

Mass of solution = Density × Volume

Mass of solution $= 1.25 \times 1000 = 1250$ g

Now, number of moles of NaCl

= Molarity × Volume

$= 3 \times 1 = 3$ moles

Molar mass of NaCl = 58.5 g/mol

Mass of NaCl $= 3 \times 58.5 = 175.5$ g

Now mass of solvent $= 1250 - 175.5$

$= 1074.5$ g

So, molality $= \frac{\text{Number of moles} \times 1000}{\text{Mass of solvent (in g)}}$

$= \frac{3 \times 1000}{1074.5} = 2.79$ m

> **Alternate Solution**
>
> Molality $= \frac{M \times 1000}{1000d - M \times Mw}$
>
> $= \frac{3000}{1250 - 175.5} = 2.79$ m

08. *(c)* $H_3C—C \equiv CH + NaNH_2$
(4g)

$\longrightarrow H_3CC \equiv C^-Na^+ + NH_3$

Number of moles of propyne

$= \frac{4}{40} = 0.1$ mol

Number of moles of NH_3 produced $= 0.1$

Volume occupied by 0.1 mol of NH_3 at STP = 2240 mL

Thus, Statement I is correct but Statement II is incorrect.

09. *(d)* The complete reaction is as follows,

$C_6H_{12}O_6 + 6O_2 \longrightarrow 6CO_2 + 6H_2O$

Mole of glucose $= \frac{900 \text{ g}}{180 \text{ g/mol}} = 5$ mol

1 mol glucose reacts with 6 mol of O_2, so 5 mol glucose will react with 30 mol O_2.

Mole of O_2 required = 30 mol

Mass of O_2 required $= 32 \times 30$ g $= 960$ g

10. *(b)* Given,

density $= 1.25$ g mL^{-1}, $M = 3$ M

molar mass in g mol^{-1} :

Na = 23, Cl = 35.5

Molality = ?

Molality (m)

$= \frac{1000 \times M}{1000 \times d - (M \times W_{\text{Molecular weight of solute}})}$

$m = \frac{3 \times 1000}{1.25 \times 1000 - [3 \times 58.5]}$

$m = 2.79$ m

11. *(b)* Given, molality = 3

Density of x M of NaOH

$= 1.12$ g mL^{-1}

Using the formula for molality

Molality $= \frac{1000 \times M}{1000 \times d - (M \times W_{\text{solute}})}$

(Molecular weight = W)

$3 = \frac{1000 \times x}{1000 \times 1.12 - (x \times 40)}$ ...(i)

[40 = Molecular weight of NaOH]

By solving Eq. (i),

$x = 3$

12. *(d)* The reaction given in question is,

$C_nH_{2n+2} + \frac{3n+1}{2}O_2 \longrightarrow nCO_2 + (n+1)H_2O$

$CH_4 + 2O_2 \longrightarrow CO_2 + 2H_2O$
4 g — 11 g
0.25 mol — 0.25 mol

0.25 mol CH_4 gives 0.25 mol 11 g CO_2.

Hence, option (d) is correct.

> **Concept Enhancer**
>
> Generally the combustion reactions are exothermic and ΔH is negative but combustion of N_2 is endothermic because a large amount of energy is required to break $N{\equiv}N$ bond.

13. *(b)* Both A and R are true and R is the correct explanation of A.

Molarity

$= \frac{\text{Number of moles hydrated oxalic acid}}{\text{Volume of the solution}}$

$0.1 = \frac{\frac{3.1500}{\text{molar mass}} \times 1000}{250}$

$0.1 = \frac{3.1500 \times 1000}{250 \times \text{mm}}$

$\text{mm} = \frac{3.1500 \times 1000}{250 \times 0.1} = 126$ g/mol

14. *(c)* The chemical reaction involved is as follows

$C_{15}H_{30} + \frac{45}{2}O_2 \longrightarrow 15CO_2 + 15H_2O$

Mass of the fuel ($C_{15}H_{30}$)

$= 0.756 \times 1000 = 756$ g

Number of moles of fuel

$= \frac{756}{210} = 3.6$ mol

Now, weight of oxygen required

$= 3.6 \times \frac{45}{2} \times 32 = 2592$ g

and weight of CO_2 released

$= 3.6 \times 15 \times 44 = 2376$ g

15. *(c)* Since, 100 g of haemoglobin has $= 0.34$ g of Fe

Therefore, 3.3 g haemoglobin

$= \frac{0.34}{100} \times 3.3$ g of Fe

$= 0.01122$ g of Fe

Since, 56 g (1 mole) of Fe has

$= 6.022 \times 10^{23}$ atoms of Fe

Therefore, 0.01122 g of Fe has

$= \frac{6.022 \times 10^{23}}{56} \times 0.01122$ atoms of Fe

$= 1.21 \times 10^{20}$ atoms of Fe

16. *(c)* Molar mass of $Fe_3O_4 = 232$ g/mol

So, number of moles of Fe_3O_4 reacting

$= \frac{4640}{232} = 20$ moles

Similarly,

Molar mass of CO $= (12 + 16)$ g/mol $= 28$ g/mol

$\therefore$ Number of moles of CO reacting

$= \frac{2520}{28} = 90$ moles

As number of moles of Fe_3O_4 reacting is less than number of moles of CO, it is the limiting reagent and product formed will depend on number of moles of Fe_3O_4.

$\therefore$Number of moles of iron formed

$= 3 \times$ moles of Fe_3O_4

$= 3 \times 20 = 60$ moles

Weight of iron produced

$= 60 \times 56$ (in g) $= 3360$ g

17. *(c)* Volume of hydrogen $= 5.6$

Volume strength $= 11.2 \times$ molarity

$\Rightarrow$ Molarity $= \frac{\text{Volume strength}}{11.2}$

$= \frac{5.6}{11.2} = 0.5$

Suppose, that the solution taken is $= 1$ L

Mass of solution $= 1000$ mL $\times 1$ g/mL

$= 1000$ g

Mass of solute = Moles × Molar mass

$= 0.5 \times 34$

[$\because$ molar mass of $H_2O_2 = 34$]

$= 17$ g

$\Rightarrow$ Mass % $= \frac{17}{1000} \times 100 = 1.7\%$

18. *(b)* The reaction involved in conversion of phenol to 2, 4, 6-tribromophenol is as follows

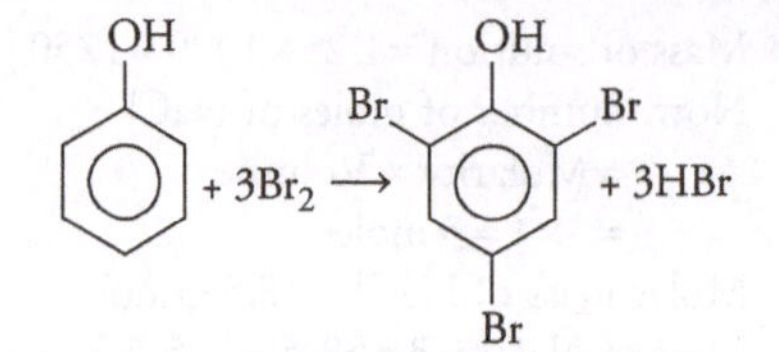

Molar mass of phenol $= 94$ g/mol

Molar mass of $Br_2 = 80$ g/mol

$\therefore$Mass of $3Br_2 = 80 \times 3 \times 2 = 480$ g

So, 94 g phenol required 480 g bromine

$\therefore$2 g phenol required

$= \frac{480}{94} \times 2$ g of bromine.

$= 10.22$ g of bromine

19. *(d)* Both complexes have equal volume and half concentration due to mixing.

$\underset{0.01\ M}{[Co(NH_3)_5SO_4]Br} + \underset{0.01\ M}{[Co(NH_3)_5Br]SO_4}$

(I) $\underset{\left(\frac{0.02}{2}\right) mol}{[Co(NH_3)_5SO_4]Br} + \underset{(Excess)}{AgNO_3} \longrightarrow \underset{\substack{0.01\ mol \\ (Y)}}{AgBr\downarrow}$

(II) $\underset{\left(\frac{0.02}{2}\right) mol}{[Co(NH_3)_5Br]SO_4} + \underset{Excess}{BaCl_2} \rightarrow \underset{\substack{0.01\ mol \\ (Z)}}{BaSO_4\downarrow}$

Thus, option (d) is correct.

20. *(b)* Percentage of carbon in glucose

$10.8 = \frac{\text{Mass of C}}{250} \times 100$

$\Rightarrow$ Mass of C $= \frac{10.8 \times 250}{100} = 27$ g

Molar mass of glucose $= 180$ g

72 g of carbon is present in 180 g of glucose.

$\therefore$ 27 g of carbon present in

$= \frac{180}{72} \times 27 = 67.5$ g of glucose

Mass of solvent $= 250\text{ g} - 67.5\text{ g}$

$= 182.5$ g

Molality $= \frac{\text{Moles of solute}}{\text{Mass of solvent}} \times 1000$

$= \frac{67.5 \times 1000}{180 \times 182.5} = 2.054$ mol/kg

21. *(c)* $\underset{(20\ g)}{N_2(g)} + \underset{(5\ g)}{3H_2(g)} \rightleftharpoons 2NH_3(g)$

Number of moles of $N_2 = \frac{20}{28}$ mol

Number of moles of $H_2 = \frac{5}{2}$ mol

Limiting reagent can be calculated using a factor

$= \frac{\text{Number of moles}}{\text{Stoichiometric coefficient}}$

For $NH_3 = \frac{20}{28}$

For $H_2 = \frac{5}{2 \times 3} = \frac{5}{6}$

Clearly, $\frac{20}{28} < \frac{5}{6}$ so, N_2 is a limiting reagent and thus decides the formation of products.

1 mole of N_2 forms 2 mole of NH_3.

So, $\frac{20}{28}$ mole of N_2 forms

$= 2 \times \frac{20}{28} = \frac{20}{14} = \frac{10}{7} = 1.42$ mol.

22. *(b)* When H_3PO_3 reacts with sodium hydroxide

$H_3PO_3 + 2NaOH \longrightarrow Na_2HPO_3 + 2H_2O$

Number of moles of $H_3PO_3 = 0.05$ mol

Number of moles of NaOH required

$= 0.05 \times 2 = 0.1$ mol

Volume of NaOH $= 0.1$ L $= 100$ mL

For H_3PO_2,

$NaOH + H_3PO_2 \longrightarrow NaH_2PO_2 + H_2O$

Number of moles of $H_3PO_2 = 0.2$ mol

Number of moles of NaOH required

$= 0.2 \times 1 = 0.2$ mol

Volume of NaOH $= 0.2$ L

$= 200$ mL

23. *(c)*

	1st component	2nd component
mole	n_1	n_2
M. weight	M_1	M_2
mass	n_1M_1	n_2M_2

Mass of solution $= n_1M_1 + n_2M_2$

Mole fraction of the 2nd component

$(x_2) = \frac{n_2}{n_1 + n_2}$

$\Rightarrow \quad n_1 = \frac{n_2(1 - x_2)}{x_2}$

Mass of solution $= n_1M_1 + n_2M_2$

$= \frac{n_2M_1(1 - x_2)}{x_2} + n_2M_2$

$= \frac{n_2}{x_2}[M_2x_2 - x_2M_1 + M_1]$

Volume of solution

$$= \frac{n_2 [M_2 x_2 - x_2 M_1 + M_1]}{1000 d x_2}$$

(where d is the density)

Thus, C_2 is represented as

$$\Rightarrow \quad C_2 = \frac{1000 d x_2}{M_1 + x_2(M_2 - M_1)}$$

24. *(138)* The reaction is as follows,

$$\underset{3\times10^3}{C_4H_{10}} + \underset{10\times10^3}{\frac{13}{2}O_2} \longrightarrow 4CO_2 + 5H_2O$$

Mole of H_2O formed

$$= n_{H_2O} = 5 \times \frac{2}{13} \times 10 \times 10^3$$

Then, $w_{H_2O} = \frac{10^5}{13} \times 18$

$$= 1.3846 \times 10^5 \text{ g}$$

Volume of H_2O will be $= 138.46$ L

25. *(184)* The complete reaction occurs as follows.

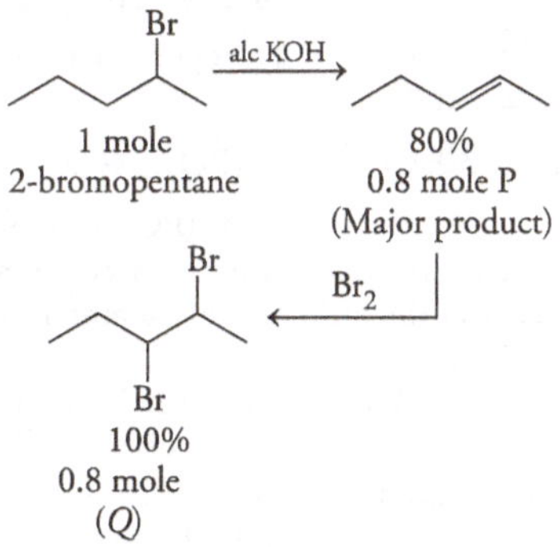

Molecular mass of $Q = 230$ g mol^{-1}

Mass of $Q = 0.8 \times 230$

$= 184$ g

26. *(12)* The reaction involved in complete combustion is as follows.

$$\text{Organic compound} \xrightarrow[\Delta]{CuO} CO_2 + H_2O$$

$$n_{CO_2} = \frac{220 \times 10^{-3}}{44} = 5 \times 10^{-3} \text{ moles}$$

$$M_C = 5 \times 10^{-3} \times 12$$

$$\% M_C = \frac{5 \times 10^{-3} \times 12}{500 \times 10^{-3}} \times 100 = 12\%$$

27. *(80)* Organic compound $\rightarrow CO_2 + H_2O$

0.50 g　　1.46 g　0.9 g

Mole of C in organic compound

$$= n_{CO_2} \times 1$$

By using principle of POAC,

So, mass of C in compound

$$= \frac{1.46}{44} \times 12$$

Now, the % of C in compound

$$= \frac{1.46}{44} \times \frac{12}{0.5} \times 100$$

$$= 79.63 \approx 80$$

28. *(63)* The reaction involved is

$$CaCO_3 \longrightarrow CaO + CO_2$$

Mass of $CaCO_3 = \frac{150 \times 75}{100} = 112.5$ kg

$= 112500$ g

$n_{CaCO_3} = 1125$

So, moles of CaO $= 1125$

So, mass of CaO $= \frac{1125 \times 56}{1000} = 63$ kg

29. *(57)* $M_F = \frac{M_1V_1 + M_2V_2}{V_1 + V_2}$

$$= \frac{2 \times 20 + 0.5 \times 400}{420}$$

$$= 0.571 \text{ M}$$

$$= 57.1 \times 10^{-2} \text{ M} \approx 57$$

30. *(61)* The complete reaction

$$\underset{X\text{ g}}{C_6H_5COOH} + NaHCO_3 \longrightarrow$$

$$C_6H_5COO^-Na^+ + H_2O + CO_2$$

Mole of $CO_2 = \frac{11.2}{22.4} = 0.5$

Mass of $C_6H_5COOH = X = 0.5 \times 122$

$= 61$ g

31. *(153)* Given,

Mass of Al $= 81$ g

Mass of oxygen $= 128$ g

Mass of $Al_2O_3 = ?$

The reaction involved is

$$4Al + 3O_2 \longrightarrow 2Al_2O_3$$

$$\frac{81}{27} = 3 \text{ mol} \qquad \frac{128}{32} = 4 \text{ mol}$$

Limiting reagent,

$\therefore$ Mole of Al_2O_3 formed $= \frac{1}{2} \times 3$ mole

$\therefore$ Weight of Al_2O_3 formed $= \frac{3}{2} \times 102$

$= 153$ g

32. *(420)*

> **idea** To solve this question first of calculate the number of moles of Fe_3O_4 and CO. Then with the help of this find out limiting reagent. After that use stoichiometry to calculate the mass of Fe.

Moles of $Fe_3O_4 = \frac{2.32 \times 10^3 \times 10^3}{232}$

$= 10000$ mol

Moles of CO $= \frac{2.8 \times 10^2 \times 10^3}{28}$

$= 10000$ mol

$$\underset{10^4\text{mol}}{Fe_3O_4} + \underset{10^4\text{mol}}{4CO} \longrightarrow 3Fe + 4CO_2$$

CO is limiting reagent

Mole of Fe $= \frac{3}{4} \times 10^4$

Mass of Fe $= \frac{3}{4} \times \frac{10^4 \times 56}{1000}$

$= 420$ kg $= x$ kg

33. *(15)* The complete reaction is as follow,

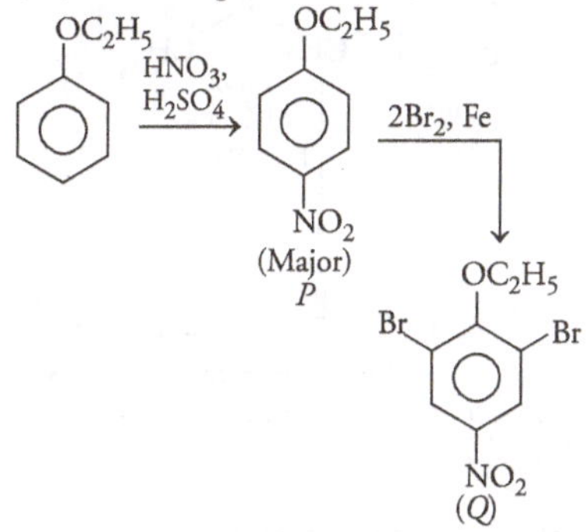

Number of O-atoms in product $Q = 3$

Number of Br atoms in product $Q = 2$

Required ratio $= \frac{3}{2} = 1.5 = 15 \times 10^{-1}$

Hence, required value is 15.

34. *(20)*

> **idea** To solve this question first of all write the complete balanced o To solve this question first of all write the complete balanced equation and then compare the number of moles of aniline with number of moles of orange dye to calculate the mass of orange dye.

The complete reaction of aniline to produce to orange dye is as follows

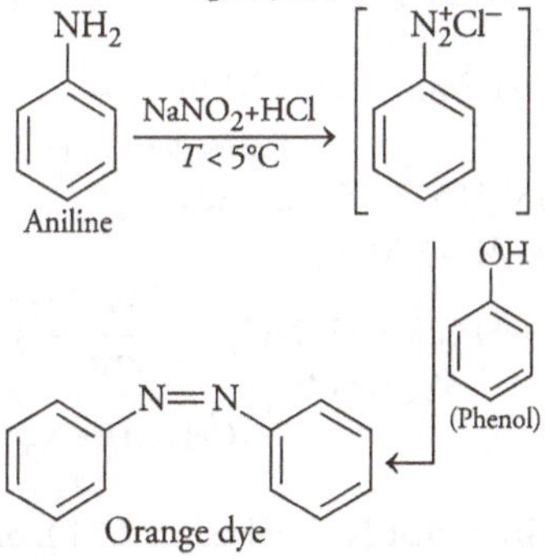

Above reaction indicates that 1 mole of aniline gives 1 mole of orange dye.

So, $(\text{Mole})_{\text{Aniline}} = (\text{Mole})_{\text{Orange dye}}$

$$\left(\frac{\text{Mass}}{\text{Molar mass}}\right)_{\text{Aniline}} = \left(\frac{\text{Mass}}{\text{Molar mass}}\right)_{\text{Orange dye}}$$

$$\frac{93\text{ g}}{93\text{ g mol}^{-1}} = \frac{\text{Mass of orange dye}}{199\text{ g mol}^{-1}}$$

Mass of orange dye = 19.9 g ≈ 20 g

35. *(318)* The complete Claisen-Schmidt reaction is,

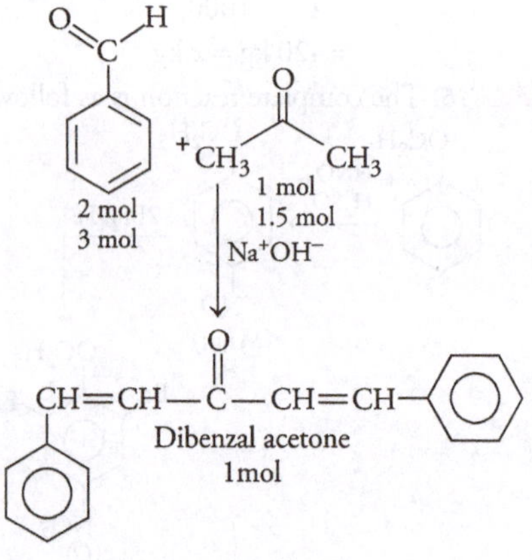

$\Rightarrow$ 351 g = 1.5 mol

Molecular weight of benzaldehyde = 106 g/mol

$106 \times 3 = 318$ g required to give 1.5 mole (or 351 g) product.

36. *(9)* The reaction is as follow,

$$\underset{\substack{\text{Ethanamine}\\0.2\text{ mol}}}{CH_3-CH_2-NH_2} \xrightarrow{NaNO_2 + HCl} CH_3-CH_2-N_2Cl \xrightarrow[-N_2]{HOH} CH_3-CH_2OH + \underset{0.2\text{ mol}}{HCl(g)}$$

Molecular weight of ethanamine = 45 g/mol

Mass = molecular weight × mole

$= 45 \times 0.2 = 9$ g

Hence, 9 g of ethanamine was subjected to reaction.

37. *(8)* Equivalent of $KMnO_4$ = Equivalent of $H_2C_2O_4$

So, $M \times 2 \times 5 = 2 \times 20 \times 2$

$\Rightarrow M = 8$ M

38. *(45)* $\underset{45\text{ g (1 mol)}}{CH_3CH_2NH_2} \xrightarrow[H_2O]{NaNO_2 + HCl} CH_3CH_2OH + \underset{28\text{ g (1 mol)}}{N_2}$

Given that N_2 evolved in 2.24 L or 0.1 mole.

So, for 0.1 mole the ethylamine $(CH_3CH_2NH_2)$ will be 4.5 g or 45×10^{-1} g.

39. *(24)* $3PbCl_2 + 2(NH_4)_3PO_4 \longrightarrow Pb_3(PO_4)_2 + 6NH_4Cl$

Limiting reagent is $PbCl_2$.

mmol of $Pb_3 \cdot (PO_4)_2$ formed

$$= \frac{\text{mmol of } PbCl_2 \text{ react}}{3}$$

$$= \frac{72\text{ mmol}^{-1}}{3}$$

$= 24$ mmol

Hence, the amount of $Pb_3(PO_4)_2$ formed is 24 mmol.

40. *(8)* The complete reaction to oxidise oxalate ion to CO_2 is,

$$2MnO_4^- + 5C_2O_4^{2-} + 16H^+ \longrightarrow 2Mn^{2+} + 10CO_2 + 8H_2O$$

$\therefore$Number of moles of H^+ ions required by 1 mole of MnO_4^- to oxidise oxalate ion to CO_2 is 8.

41. *(4)* $C_2H_2O_4 + 2NaOH \longrightarrow Na_2C_2O_4 + 2H_2O$

Equivalents of oxalic acid = Equivalents of NaOH

Volume × Molarity × *n*-factor = Volume × Molarity × *n*-factor

$50 \times 0.5 \times 2 = 25 \times M_{NaOH} \times 1$

$M_{NaOH} = 2$

Now, mass of NaOH is 50 mL solution, Molarity $= \dfrac{\text{Mass} \times 1000}{\text{Molar mass} \times V_{(\text{mL})}}$

$$\text{Mass} = \frac{50}{1000} \times 2 \times 40 = 4\text{ g}$$

42. *(7)* Given,

Weight of NaOH = 84 g

Molar mass of NaOH = 40

$$M = \frac{n_{NaOH}}{V_{sol}\text{ (in L)}}$$

$$\Rightarrow \quad 3 = \frac{\left(\frac{84}{40}\right)}{V}$$

$$V = \frac{84}{40 \times 3} = 0.7\text{ L}$$

or $V = 7 \times 10^{-1}$ L

$= 7 \times 10^{-1}\text{ dm}^3$

Hence, missing value is 7.

43. *(8)* The combustion reaction of methane is,

$$CH_4 + 2O_2 \longrightarrow CO_2 + 2H_2O$$

$$\text{Moles of } CO_2 = \frac{22}{44} = 0.5\text{ mol}$$

1 mole of CO_2 is produced from 1 mole of CH_4.

So, 0.5 mole of CO_2 will be produced from 0.5 mole of CH_4.

Mass of CH_4 formed = Moles × Molar mass

$= 0.5 \times 16 = 8$ g

44. *(224)*

$$\underset{1\text{ mol}}{Mg} + \underset{2\text{ mol}}{2HCl} \longrightarrow \underset{1\text{ mol}}{MgCl_2} + \underset{1\text{ mol}}{H_2\uparrow}$$

$$\text{No. of moles} = \frac{\text{Mass of Mg}}{\text{Molar mass of Mg}} = \frac{2.4}{24}$$

$= 0.1$ mol

So, 1 mole of Mg will give 1 mole of H_2

1 mole of gas at STP = 22.4 L

$\therefore$ 0.1 mole of gas = 0.1× 22.4

$= 2.24$ L or 224×10^{-2} L

Thus, volume of hydrogen liberated is 224×10^{-2} L.

45. *(1)* $3BaCl_2 + 2Na_3PO_4 \longrightarrow Ba_3(PO_4)_2 + 6NaCl$

The coefficient of the balanced equation tell you the ratio of the reactant and the product.

3 mol of $BaCl_2$ react with 2 mol of Na_3PO_4 to yield 1 mol of $Ba_3(PO_4)_2$

So, required ratio of $BaCl_2 : Na_3PO_4$

$= 3 : 2$

5 mol $BaCl_2$ require $2/3 \times 5 = 3{\cdot}3$ Na_3PO_4

We have only 2 mol Na_3PO_4. So, it is limiting reagent and $BaCl_2$ is in excess.

2 mol of Na_3PO_4 will yeild 1 mol of $Ba_3(PO_4)_2$.

46. *(139)* 10 mL solute in 90 mL solvent.

Mass of solute $= 10 \times 3.2 = 32$ g

Mass of solvent $= 90 \times 1.6$ g

$$m = \frac{32 \times 1000}{160 \times 90 \times 1.6} = 1.388$$

$= 138.8 \times 10^{-2} = 139 \times 10^{-2}$ m

47. *(364)*

idea In this type of question, whenever the density and molarity is given, then use this formula of molality

i.e., $m = \dfrac{1000 \times M}{1000d - M \times \text{molecular weight}}$

$$m = \frac{1000 \times M}{1000d - M \times \text{molecular weight}}$$
$$= \frac{1000 \times 3}{1000 \times 1 - (3 \times 58.5)} = 3.6385 \text{ m}$$
$$\approx 3.64 \text{ m} = 364 \times 10^{-2} \text{ m}$$

48. *(4)* $Zn + 2HCl \longrightarrow ZnCl_2 + H_2 \uparrow$

Amount of Zn = 11.5 g

Number of moles of Zn $= \frac{11.5}{65.4} = 0.176$

Number of moles of H_2 = Number of moles of Zn

Volume of H_2 liberated $= 0.176 \times 22.7$ L $= 3.99 \text{ L} \approx 4 \text{ L}$

49. *(227)* $\underset{\substack{12\text{ g} \\ 1\text{ mol}}}{2C(s)} + \underset{\substack{48\text{ g} \\ 1.5\text{ mol}}}{O_2(g)} \longrightarrow 2CO(g)$

Limiting reagent is carbon. One mole of carbon produces one mole of CO.

Thus, volume at STP is 227×10^{-1} L.

50. *(180)* Molarity $= \frac{5 \times 1000}{40 \times 450}$

Let x mL NaOH is required from $\frac{5 \times 1000}{40 \times 450}$ M NaOH solution to prepare 500 mL 0.1 M solution.

$$\therefore \left(\frac{5 \times 1000}{40 \times 450}\right) x = 0.1 \text{ M} \times 500 \text{ mL}$$
$$(\because M_1V_1 = M_2V_2)$$
$$x = \frac{0.1 \times 500 \text{ mL} \times 40 \times 450}{5 \times 1000}$$
$$= 180 \text{ mL}$$

51. *(3)* $R(OH)_y \longrightarrow H_2$

$\therefore$ Number of moles of 'OH' present in x

$= \frac{1}{2}$ (Number of moles of H_2 formed)

$$\Rightarrow y\left(\frac{1.84 \times 10^{-3} \text{ g}}{92 \text{g mol}^{-1}}\right) = \frac{1}{2} \times \frac{1.344 \text{ mL}}{22.4 \times 10^3 \text{ mL}} \times 2$$

[Volume of 1 mole of gas at STP = 22.42]

$$\Rightarrow y = \frac{1.344 \times 2 \times 92 \times 1000}{2 \times 22.4 \times 1000 \times 1.84}$$
$$\Rightarrow y = 3$$

So, number of alcoholic hydrogens present in compound 'x' is 3.

52. *(25)*

	H_2SO_4 +	$2NaOH \longrightarrow$	Na_2SO_4 +	$2H_2O$
Initial moles	0.4	0.2	0	0
Final moles	0.3	0	0.1	0.2

Limiting reagent $\rightarrow$ NaOH

$n_{Na_2SO_4} = 0.1$ mol

Total volume,

$V = V_{H_2SO_4} + V_{NaOH} = 2 + 2 = 4$ L

Molarity, $M_{Na_2SO_4} = \frac{0.1}{4} = 0.025$ M

$= 25 \times 10^{-3}$ M

$= 25$ millimolar

53. *(2)* 3 moles of Z are required to give 1 mole of XYZ_3.

So, 0.05 moles of Z are required to give

$= \frac{1}{3} \times 0.05$ mole of XYZ_3

$= 0.016$ mole of XYZ_3

Molecular weight of

$XYZ_3 = 10 + 20 + (3 \times 30) = 120$ amu

$\therefore$ Number of moles of

$$XYZ_3 = \frac{\text{Weight of } XYZ_3}{\text{Molecular weight of } XYZ_3}$$
$$0.016 = \frac{\text{Weight of } XYZ_3}{120}$$

$\therefore$ Weight of $XYZ_3 = 0.016 \times 120$

$= 1.92 \text{ g} \approx 2 \text{ g}$

54. *(24)* Given, volume of solution = 2 L

Concentration = 48 ppm

48 mg $\longrightarrow$ in 1 L

$\therefore$ 2 L contains 48 × 2 = 96 mg of Mg

24 g of Mg contains 6.02×10^{23} atoms

$\therefore$ 96 mg of Mg will contain

$$= \frac{96 \times 6.02 \times 10^{23} \times 10^{-3}}{24}$$
$$= 24 \times 10^{20}$$
$$\therefore \quad x = 24$$

55. *(54)* Moles of nitric acid

= Volume (in L) × Molarity

$= 800 \times 0.5 \times 10^{-3}$

$= 400 \times 10^{-3} = 0.4$ moles

Weight of HNO_3 = Number of moles × Molar mass

$= 0.4 \times 63 = 25.2$ g

Amount of HNO_3 remaining in the solution

$= 25.2 - 11.5 = 13.7$ g

Molarity $= \frac{13.7}{63} \times \frac{1000}{400} = 54 \times 10^{-2}$ M

56. *(1)* Reaction involved

$H_2SO_4 + 2NaOH \longrightarrow Na_2SO_4 + H_2O$

Number of equivalents of H_2SO_4

$= 100 \times 0.1 \times 2 = 20$

Number of equivalent of NaOH

$= 50 \times 0.1 = 5$

Normality

$$= \frac{\text{Number of gram equivalent}}{\text{Volume of the solution}}$$

Volume of the solution

$= 100 + 50 = 150$ mL

$$\therefore \quad N = \frac{15}{150} = 0.1 = 1 \times 10^{-1} \text{ N}$$

57. *(46)* $N_2 + 3H_2 \longrightarrow 2NH_3$

Since the reactants and product are at same condition of temperature and pressure.

So, by Avagadro's law, $V \propto n$

56.0 L of N_2 gas = 56 moles of NH_3 gas

20.0 L of NH_3 gas = 20.0 moles of NH_3 gas

To produce 20 moles of NH_3, 10 mole of N_2 required.

So, to produce 20 L of NH_3, 10 L of N_2 required.

So, unused volume of N_2 will be

$= 56 - 10 = 46$ L

46 L of N_2 will remain unused.

58. *(1143)* When pent-1-ene reacts with bromine, we get

$$\underset{70\text{ g}}{CH_3CH_2CH_2CH{=}CH_2} + \underset{160\text{g}}{Br_2} \longrightarrow CH_3CH_2CH_2\underset{\substack{| \\ Br}}{CH}\underset{\substack{| \\ Br}}{CH_2}$$

70 g of pent-1-ene reacts with 160 g Br_2

5 g of pent-1-ene will react with

$$\frac{160}{70} \times 5 \approx 11.43 \text{ g or } 1143 \times 10^{-2} \text{ g}$$

59. *(225)* $CH_4 + 2O_2 \longrightarrow CO_2 + 2H_2O$

Mole-mole analysis

$$\frac{n_{CH_4}}{1} = \frac{n_{H_2O}}{2}$$

Mass of water to be produced = 81 g

Mass of $CH_4 = \frac{1}{2} \times \frac{81}{18} \times 16 = 36$ g

So, moles of $CH_4 = \frac{36 \text{ g}}{16 \text{ g}} = 2.25$ mol

or 225×10^{-2} mol

Alternate Solution

$CH_4 + 2O_2 \longrightarrow CO_2 + 2H_2O$

Number of moles of $H_2O = \frac{81}{18} = 4.5$ moles

From balanced equation

2 moles of H_2O are produced 1 mole of CH_4.

4.5 moles of H_2O are produced by 1 mole of CH_4

$= \frac{1}{2} \times 4.5 = 2.25$

$\approx 225 \times 10^{-2}$ mol

60. *(30)* M_{eq} of oxidising agent $= M_{eq}$ of reducing agent

$(M \times V \times n_f)_{KMnO_4} = (M \times V \times n_f)_{\text{Mohr's salt}}$

$0.01 \times 20 \times 5 = 0.05 \times V \times 1$

Volume needed = 20 mL

Initial volume of $KMnO_4$ in burette is 50 mL

Thus, vol. of $KMnO_4$ left in the burette, after end point is 30 mL.

61. *(143)* $CH_4 + 2O_2 \longrightarrow CO_2 + 2H_2O$

Moles $\frac{100}{16} = 6.25$ $\quad \frac{208}{32} = 6.5$

Moles $= \frac{6.25}{1} \quad = \frac{6.5}{2}$

Stoichiometric coefficient $= 6.25 \quad = 3.25$

So, O_2 is limiting reagent.

Using mole-mole analysis

$$\frac{n_{O_2}}{2} = \frac{n_{CO_2}}{1} \Rightarrow \frac{6.5}{2} = n_{CO_2}$$

$$n_{CO_2} = 3.25$$

$\therefore$ Mass of $CO_2 = 3.25 \times 44 = 143$ g.

62. *(3)* Let basicity of acid is x.

$$\underset{\substack{0.1\text{ M}\\10\text{ mol}}}{A} + \underset{\substack{0.05\text{ M}\\30\text{ mL}}}{M(OH)_2} \longrightarrow \text{Salt} + H_2O$$

On complete neutralisation,

Milliequivalent of acid = Milliequivalent of base

$$x(0.1 \times 10) = 2(0.05 \times 30)$$
$$x = 2 \times 1.5 = 3$$

So, the basicity of acid, A is 3.

63. *(50)* Moles of sodium in 100 mL

$$= \frac{3.45}{23} = 0.15$$

Moles of sodium in 1000 mL (1 L) = 1.5

1 mole of Na_3PO_4 which contains 3 moles of Na

1.5 moles of Na will be present in $\frac{1}{2}$ moles of Na_3PO_4

(*M*) Molarity

$$= \frac{\text{Number of moles of solute}}{\text{Volume of solution in L}}$$

$$= \frac{1/2}{1} = 0.5 = 50 \times 10^{-2} \text{ mol/L}$$

64. *(1575)* For titration of $KMnO_4$ with oxalic acid,

$$n_{eq}(KMnO_4) = n_{eq}(H_2C_2O_4 \cdot 2H_2O)$$

or $\frac{10 \times 0.05}{1000} \times 5 = \frac{10 \times M}{1000} \times 2$

$$\Rightarrow M = 0.125 \text{ M}$$

Molar mass of $H_2C_2O_4 \cdot 2H_2O$

$= 2 \times 1 + 2 \times 12 + 4 \times 16 + 2(18)$

$= 126$ g/mol

Strength of oxalic acid $= (126 \times 0.125)$ g/L

$= 15.75 \text{ g/L} = 1575 \times 10^{-2}$ g/L

65. *(18)* For the given reaction,

$$Fe^{2+} + Cr_2O_7^{2-} \longrightarrow Fe^{3+} + Cr^{3+}$$

Volume of aqueous solution of Fe^{2+} ion $(V_1) = 10$ mL.

Volume of $K_2Cr_2O_7$ solution $(V_2) = 15$ mL

$$\underset{(Fe^{2+})}{n_1 M_1 V_1} = \underset{(K_2Cr_2O_7)}{n_2 M_2 V_2}$$

$$M_1 \times 10 \times 1 = 6 \times 0.02 \times 15$$

$$\Rightarrow \quad M_1 = \frac{6 \times 0.02 \times 15}{10} = 0.18$$

$$M_1 \approx 18 \times 10^{-2}$$

So, value of x is 18.

66. *(4)* Molar mass of glucose $(C_6H_{12}O_6)$

$= 12 \times 6 + 12 \times 1 + 16 \times 6 = 180 \text{ g mol}^{-1}$

$\therefore$ Molarity (*M*)

$$= \frac{\text{Concentration of glucose}}{\text{Molar mass of glucose}} = \frac{W_B}{M_B}$$

$$M = \frac{0.72}{180} = 0.004 = 4 \times 10^{-3} \text{ M.}$$

67. *(1)* Total mass = 4 g (given)

Mass of NaOH = x g

Mass of $Na_2CO_3 = y$ g

Total mass $= x + y = 4$

$$\frac{\text{Mass of NaOH}}{\text{Molar mass of NaOH}} = \frac{\text{Mass of } Na_2CO_3}{\text{Molar mass of } Na_2CO_3}$$

$$\frac{x}{40} = \frac{y}{106} \quad \text{(Equimolar)}$$

$$y = \left(\frac{106}{40}\right)x$$

$$\Rightarrow \quad x + \frac{106}{40}x = 4$$

$$x + 2.65x = 4$$

$$\Rightarrow \quad 3.65x = 4$$

$$\Rightarrow \quad x = 1.096 \text{ g} \Rightarrow x \approx 1$$

68. *(226)* We know that, number of millimoles $= V_{mL} \times$ molarity

So, millimoles of

NaOH = 250 × 0.5 = 125

Millimoles of HCl = 500 × 1 = 500

Now, reaction is

	NaOH	+ HCl	⟶ NaCl	+ H_2O
At $t = 0$,	125	500	0	0
At $t = t$,	0	375	125	125

125 millimoles of NaOH reacts with 125 millimoles of HCl.

So, millimoles of HCl left = 375

Moles of HCl $= 375 \times 10^{-3}$

Number of HCl molecules

= Avogadro's constant $(N_A) \times$ moles of HCl

$$= 6.022 \times 10^{23} \times 375 \times 10^{-3}$$
$$= 225.8 \times 10^{21} = 226 \times 10^{21}$$

Therefore, answer is 226.

69. *(5)* Take 1 L solution

mass of solution = Vol. × density

$= 1.2 \times 1000 \text{ g} = 1200$ g

Mass of 1 L solvent (water) = 1000 g

$(\because \text{density} = 1 \text{ g cm}^{-3})$

$\Rightarrow$ Mass of NaOH = (1200 − 1000) g = 200 g

$$\Rightarrow \text{Moles of NaOH} = \frac{200 \text{ g}}{40 \text{ g/mol}} = 5 \text{ mol}$$

$$\text{Molality} = \frac{\text{Number of moles of solute}}{\text{Mass of solvent (in kg)}}$$

$$\Rightarrow \quad \text{Molality} = \frac{5 \text{ mol}}{1 \text{ kg}} = 5 \text{ m}$$

70. *(24)* $\underset{+6}{Cr_2O_7^{2-}} + \underset{+2}{Fe^{2+}} \longrightarrow \underset{+3}{Cr^{3+}} + \underset{+3}{Fe^{3+}}$

Dichromate $(Cr_2O_7^{2-})$ oxidised Fe^{2+} to Fe^{3+} and itself get reduced to Cr^{3+}.

Valency factor of $Cr_2O_7^{2-} = 6$ as Cr gets reduced from +6 to +3.

Valency factor of $Fe^{2+} = 1$ as it gets oxidised from +2 to +3.

From law of equivalence,

$$\underset{(Cr_2O_7^{2-})}{M_1 V_1 n_1} = \underset{(Fe^{2+})}{M_2 V_2 n_2}$$

$$(M_1 \times V_1) \times 6 = (M_2 \times V_2) \times 1$$

$$(0.03 \times 20) \times 6 = (M_2 \times 15)1$$

$$\Rightarrow M_2 = 0.24 \text{ molar} = 24 \times 10^{-2} \text{ molar}$$

71. *(64)* Given, molality = 100

$\Rightarrow$ 100 moles of solute in 1 kg of solvent

Mole fraction of solute (χ_{Solute})

$$= \frac{\text{Moles of solute}}{\text{Total moles}} = \frac{n_{\text{Solute}}}{n_{\text{Solute}} + n_{\text{Solvent}}}$$

$$n_{\text{Solvent}} = \frac{1000}{18} = 55.5$$

$$\chi_{\text{Solute}} = \frac{100}{100 + 55.5}$$

$$= \frac{100}{155.5} = 0.643$$

$$\chi_{\text{Solute}} = 64.3 \times 10^{-2} = 64 \times 10^{-2}$$

72. *(50)* 10.0 mL of Na_2CO_3 solution is titrated against 0.2 M HCl.

$$\underset{10\text{mL}}{Na_2CO_3} + \underset{0.2\text{M HCl}}{2HCl} \longrightarrow 2NaCl + H_2O + CO_2$$

$M_{Na_2CO_3} = ?$ 5 mL

M_{eq} of $Na_2CO_3 = M_{eq}$ of HCl

$M_{Na_2CO_3} \times 10 \times 2 = 0.2 \times 5 \times 1$

$M_{Na_2CO_3} = 5 \times 10^{-2} \text{M} = 50 \times 10^{-3}$ M

$= 50$ mM

73. *(1)* Volume of Cl_2 at STP $= 20$ mL

Number of moles of chlorine gas

$$= \frac{20 \text{ mL}}{22400 \text{ mL/mol}}$$

Number of molecules of Cl_2 gas

$$= \frac{20}{22400} \times 6.023 \times 10^{23}$$

Thus, number of chlorine atoms

$$= 2 \times \frac{20}{22400} \times 6.023 \times 10^{23}$$

$$= 1.08 \times 10^{21} \approx 1 \times 10^{21}$$

74. *(525)* When lead nitrate solution is reacts with chromic sulphate solution then following reaction takes place.

$$3Pb(NO_3)_2 + Cr_2(SO_4)_3 \rightarrow 3PbSO_4 + 2Cr(NO_3)_3$$

m moles of $Pb(NO_3)_2 = 35 \times 0.15 = 5.25$

m moles of $Cr_2(SO_4)_3 = 20 \times 0.12 = 2.4$

Thus, $Pb(NO_3)_2$ is limiting reagent,

m moles of $PbSO_4$ formed $= 5.25$

$= 525 \times 10^{-5}$ moles

75. *(3)* When benzylamine reacts with bromomethane then

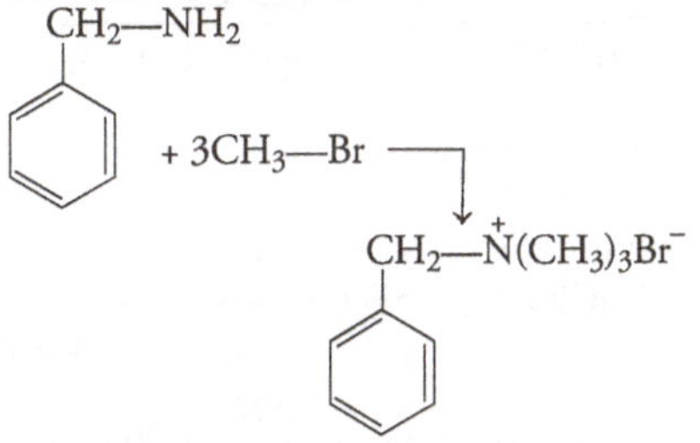

Number of moles of benzyl trimethyl ammonium bromide formed

$$= \frac{23}{230} = 0.1$$

$\therefore$ Number of moles of bromomethane consumed in the reaction

$$= 3 \times 0.1 = 3 \times 10^{-1}$$

76. *(13)* 70 mg Na^+ is present in 1 mL of $NaNO_3$ solution.

$\therefore$ 50 mL of $NaNO_3$ will contain

$$= 70 \times 50 \text{ mg} = \frac{70 \times 50}{1000} = 3.5 \text{ gm}$$

Moles of Na^+ in solution = Moles of $NaNO_3$ in solution

$$[\because NaNO_3 \longrightarrow Na^+ + NO_3^-]$$

$$= \frac{3.5}{23} \text{ mol}$$

$[\because$ Molar mass of $Na^+ = 23\text{g mol}^{-1}]$

Mass of $NaNO_3$ = mole × molar mass

$[\because$ molar mass of $NaNO_3 = 85 \text{ gmol}^{-1}]$

$$= \frac{3.5}{23} \times 85 = 12.934 \text{ g} \approx 13 \text{ g}$$

77. *(8)* Combustion reaction :

$$C_xH_y(g) + \left(x + \frac{y}{4}\right)O_2(g) \longrightarrow xCO_2(g) + \frac{y}{z}H_2O(l)$$

Suppose, volume of C_xH_y is V and volume of O_2 is 6 times greater than $C_xH_y = 6$ V

then volume of $x CO_2$ produced during complete oxidation.

$\Rightarrow \quad Vx = 4 \text{ V} \Rightarrow x = 4$

Since, $V_{O_2} = 6$ V

$$V\left(x + \frac{y}{4}\right) = 6 \text{ V}$$

$$\left(x + \frac{y}{4}\right) = 6 \quad \ldots \text{(i)}$$

Put value of $x = 4$ in Eq. (i)

We get, $4 + \frac{y}{4} = 6 \Rightarrow y = 8$

78. *(243)* Reaction

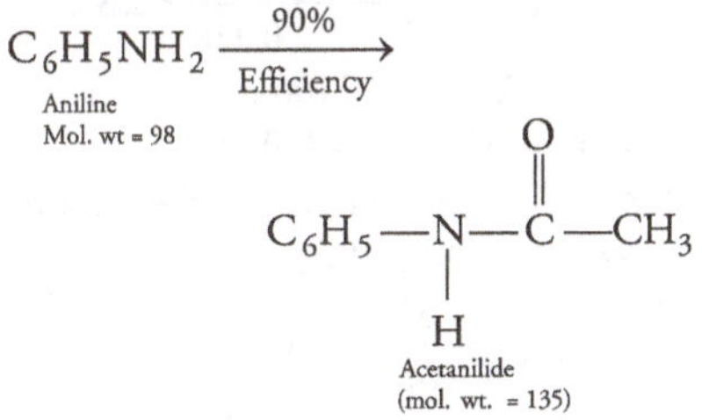

Given, weight of aniline $= 1.86$ g

Here, 1 mole of aniline gives 1 mole of acetanilide

$\therefore$ mole of aniline = mole of acetanilide

$$\Rightarrow \frac{1.86}{93} = \frac{W_{\text{Acetanilide}}}{135}$$

$$W_{\text{Acetanilide}} = \frac{1.86 \times 135}{93} \text{ g} = 2.70 \text{ g}$$

But efficiency of reaction is 90% only.

Hence, mass of acetanilide produced

$$= 2.70 \times \frac{90}{100} \text{ g}$$

$$= 2.43 \text{ g}$$

$$= 243 \times 10^{-2} \text{ g}$$

$x = 243$

79. *(2)* Molarity

$$= \frac{4.5}{90} \times \frac{1000}{250} = 0.2 = 2 \times 10^{-1}$$

So, $x = 2$

80. *(4)* For first end point :

$$\underset{(a)}{NaOH} + \underset{(b)}{Na_2CO_3} + \underset{(c)}{HCl} \longrightarrow \underset{(b)}{NaHCO_3} + NaCl + H_2O$$

Both NaOH and Na_2CO_3 are of basic character. Hence, they will not react with each other. However both of them will react with HCl.

According to law of equivalence

Equivalents of NaOH (a) + Equivalents of Na_2CO_3 (b) = Equivalents of HCl (c)

and Equivalents of Na_2CO_3 (b) = Equivalents of $NaHCO_3$ (b)

For second end point

$$\underset{(b)}{NaHCO_3} + \underset{(d)}{HCl} \longrightarrow H_2CO_3 + NaCl$$

According to law of equivalence

Equivalents of $NaHCO_3$ (b) = Equivalents of HCl (d)

$$\Rightarrow b = \frac{1}{10} \times 1.5 = 0.15$$

From first end point, equivalents of $Na_2CO_3 = 0.15$

$\Rightarrow$ Mass of Na_2CO_3 = Number of eq. × Eq. mass

$$= 0.15 \times \frac{106}{1} \times 10^{-3} \text{ g}$$

$$= 0.016 \text{ g}$$

Weight percentage of

$$Na_2CO_3 = \frac{0.016}{0.4} \times 100 = 4\%.$$

81. *(25)* Given, 6.023×10^{22} molecules are present in 10 g substance.

Number of moles

$$= \frac{\text{Number of molecules}}{6.023 \times 10^{23}}$$

$$= \frac{\text{Mass (given)}}{\text{Molar mass}}$$

$$\text{Molar mass} = \frac{10 \times 6.023 \times 10^{23}}{6.023 \times 10^{22}}$$

$$= 100 \text{ g/mol}$$

$$\text{Molarity} = \frac{\text{Number of moles of solute}}{\text{Volume of solution in litres}}$$

$$= \frac{(5/100)}{2} = \frac{5}{100 \times 2}$$

$$= 0.025 = 25 \times 10^{-3}$$

82. *(10)* Let the water of crystallisation be x.

Molar mass of $Na_2CO_3 \cdot xH_2O$.

$$= (106 + 18x)$$

Equivalent weight of $Na_2CO_3 \cdot xH_2O$

$$= \frac{\text{Molar mass}}{n\text{ factor}} = \frac{M}{2} = (53 + 9x)$$

[Here, M = molar mass and n factor = 2]

$$\text{Gram equivalent} = \frac{\text{Weight}}{\text{Equivalent weight}}$$

[Given, weight of $Na_2CO_3 \cdot xH_2O$ = 1.43 g]

Hence, gram equivalent of

$$Na_2CO_3 \cdot xH_2O = \frac{1.43}{53 + 9x}$$

$$\text{Normality} = \frac{G_{meq}}{V_{litre}};\ 0.1 = \frac{\left(\frac{1.43}{53 + 9x}\right)}{0.1}$$

{As, volume = 100 mL = 0.1 L

$$\text{So,}\quad 10^{-2} = \frac{1.43}{53 + 9x} \Rightarrow x = 10.00$$

83. *(5)* For the molecular formula,

$C_nH_{2n+2}O_2$ (M = Molar mass)

$$C\% = \frac{12n}{M} \times 100,\ H\% = \frac{2n + 2}{M} \times 100,$$

$$O\% = \frac{32}{M} \times 100$$

To be saturated and alicyclic, $n = 2$, i.e.,

$$C\% = \frac{24}{M} \times 100,\ H\% = \frac{6}{M} \times 100 \text{ and}$$

$$O\% = \frac{32}{M} \times 100$$

$$\Rightarrow \quad C\% : H\% = 24 : 6 = 4 : 1$$

$$C\% : O\% = 24 : 32 = 3 : 4$$

So, X is $C_2H_6O_2$ and its combustion equation is

$$C_2H_6O_2 + \left(2 + \frac{6}{4} - \frac{2}{2}\right)O_2 \longrightarrow 2CO_2 + \frac{6}{2}H_2O$$

$$\Rightarrow \underset{\substack{1\text{ mol}\\2\text{ mol}}}{C_2H_6O_2} + \underset{5/2\text{ mol}}{\frac{5}{2}O_2} \rightarrow 2CO_2 + 3H_2O$$

$\frac{5}{2} \times 2 = 5$ mol of O_2 will be required.

84. *(10)*

$$\underset{\text{Base}}{NaOH} + \underset{\text{Phosphonic acid}}{H_3PO_3} \longrightarrow \underset{\text{Salt}}{NaH_2PO_3} + \underset{\text{Water}}{H_2O}$$

$$\text{Given}\begin{cases} 0.1\text{ N} & 0.1\text{ N} \\ V = ? & 10\text{ mL}\end{cases}$$

On dilution

$$(N_1V_1)_{NaOH} = (N_2V_2)_{H_3PO_3}$$

$$0.1V_1 = 0.1 \times 10\text{ mL}$$

$$\Rightarrow \quad V_1 = \frac{0.1 \times 10\text{ mL}}{0.1}$$

$$V_1 = 10\text{ mL}$$

Volume of NaOH = 10 mL

85. *(3400)*

$$\underset{1\text{ mol}}{N_2} + \underset{3\text{ mol}}{3H_2} \longrightarrow \underset{2\text{ mol}}{2NH_3}$$

Molecular weight of N_2 (kg) = 0.028 kg/mol

Molecular weight of H_2 = 2 g

Here, $3H_2$ is = 3 × 2 = 6 g = 0.006 kg/mol

Molecular weight of $2NH_3$ (kg) = 0.034 kg/mol

0.028 kg N_2 require hydrogen = 0.006 kg

$$1\text{ kg } N_2 \text{ require} = \frac{0.006}{0.028}$$

2.8 kg N_2 (limiting reagent) require

$$= \frac{0.006}{0.028} \times 2.8 = 0.6\text{ kg } H_2 \text{ (Excess)}$$

Then, 0.6 kg H_2 is present in excess and N_2 will be limiting reagent.

$$0.028\text{ kg } N_2 \xrightarrow{\text{Excess}} 0.034\text{ kg of } NH_3$$

$$1\text{ kg } N_2 \longrightarrow \frac{0.034}{0.028}$$

$$2.8\text{ kg } N_2 \longrightarrow \frac{0.034}{0.028} \times 2.8$$

$$= 3.4\text{ kg or } 3400\text{ g}$$

Amount of NH_3 = 3400 g

86. *(14)* For 100 g sample of the solution, mass of HNO_3 = 63 g,

$$\therefore\ n_{HNO_3} = \frac{63}{63} = 1\text{ mol}$$

$$\text{Volume of solution} = \frac{\text{Mass}}{\text{Density}}$$

$$= \frac{100\text{ g}}{1.4\text{ g/mL}}$$

$$= \frac{100}{1.4}\text{ mL} = \frac{1}{14}\text{L}$$

$$\text{Molarity} = \frac{n}{V(\text{L})} = \frac{1}{(1/14)} = 14\text{ M}$$

87. *(27)* Given, Mass of $[Co(NH_3)_6]Cl_3$ = 0.3 g

Molar mass = 267.46 g mol^{-1}

$$\therefore \quad n = \frac{0.3}{267.46}\text{ mol}$$

From the formula, 1 mol complex will give 3 mol Cl^-.

$$\therefore \quad n_{Cl^-} = \frac{0.3}{267.46} \times 3 = \frac{0.9}{267.46}$$

The precipitation reaction takes place as follows :

$$\underset{(0.3g)}{[Co(NH_3)_6]Cl_3} + \underset{(V\text{mL}, 0.125g)}{3AgNO_3} \longrightarrow 2AgCl$$

Let volume of $AgNO_3$ (in mL) be V.

Number of moles of Cl^- = number of moles of N

$$\therefore \quad \frac{0.3 \times 3}{267.46} = 0.125 \times V \times 10^{-3}$$

$$= \frac{0.9 \times 1000}{0.125 \times 267.46}$$

$$= 26.92\text{ mL.}$$

Depending upon the accurancy of calculation this value may vary from 26.60 to 27.00.

88. *(2130)*

> **idea** To find the No. of moles of O_2 in the given question, use the formula $n = \frac{pV}{RT}$.
>
> Then, use the value of n to find out the mass of $NaClO_3$.

Given, volume of O_2 = 492 L

$$\therefore \quad n = \frac{pV}{RT}$$

$$= \frac{1\text{ atm} \times 492\text{ L}}{0.082\text{ L atm K}^{-1}\text{ mol}^{-1} \times 300\text{ K}}$$

$$= \frac{492}{8.2 \times 3}\text{ mol}$$

From the reaction,

$$NaClO_3(s) + 5Fe(s) \longrightarrow O_2(g) + NaCl(s) + 5FeO(s)$$

$$n_{NaClO_3} = n_{O_2} = \frac{492}{8.2 \times 3} = 20\text{ mol}$$

$$m_{NaClO_3} = n \times \text{molar mass of } NaClO_3$$

$$= 20 \times 106.5 = 2130\text{ g}$$

The value in different situations may range from 2120 to 2140.

02 Atomic Structure

In JEE Main there is usually 0-1 question per paper are visible from this chapter. In past 3 years, 26 conceptual and 19 numerical questions were asked of moderate difficulty level from this chapter.

For detailed analysis scan QR code

Topic 1 / Preliminary Models and Dual Nature of Light

• Level 1

01 The minimum energy that must be possessed by photons in order to produce the photoelectric effect with platinum metal is
[Given, the threshold frequency of platinum is $1.3 \times 10^{15} s^{-1}$ and $h = 6.6 \times 10^{-34}$ Js] **[2022, 25 June Shift-II]**

(a) 3.21×10^{-14} J (b) 6.24×10^{-16} J
(c) 8.58×10^{-19} J (d) 9.76×10^{-20} J

02 Hydrogen has three isotopes (A), (B) and (C). If the number of neutron(s) in (A), (B) and (C) respectively, are (x), (y) and (z), the sum of (x), (y) and (z) is **[2020, 8 Jan Shift-II]**

(a) 4 (b) 3 (c) 2 (d) 1

• Level 2

03 Match List-I with List-II. **[2024, 6 April Shift-II]**

List-I (Alkali metal)		List-II (Emission wavelength in nm)
A. Li	I.	589.2
B. Na	II.	455.5
C. Rb	III.	670.8
D. CS	IV.	780.0

Choose the correct answer from the options given below

	A	B	C	D
(a)	IV	II	I	III
(b)	II	IV	III	I
(c)	III	I	IV	II
(d)	I	IV	III	II

04 Given below are two statements, one is labelled as Assertion (A) and the other is labelled as Reason (R).

Assertion (A) In the photoelectric effect, the electrons are ejected from the metal surface as soon as the beam of light of frequency greater than threshold frequency strikes the surface.

Reason (R) When the photon of any energy strikes an electron in the atom, transfer of energy from the photon to the electron takes place.

In the light of the above statements, choose the most appropriate answer from the options given below. **[2023, 11 April Shift-I]**

(a) Both A and R are correct but R is not the correct explanation of A
(b) Both A and R are correct and R is the correct explanation of A
(c) A is correct but R is not correct
(d) A is not correct but R is correct

05 The energy of one mole of photons of radiation of wavelength 300 nm is
(Given, $h = 6.63 \times 10^{-34}$ Js, $N_A = 6.02 \times 10^{23}$ mol^{-1}, $c = 3 \times 10^8$ ms^{-1}) **[2022, 24 June Shift-II]**

(a) 235 kJ mol^{-1}
(b) 325 kJ mol^{-1}
(c) 399 kJ mol^{-1}
(d) 435 kJ mol^{-1}

Imp. *Photoelectric Effect*

06 What is the work function of the metal, if the light of wavelength 4000 Å generates photoelectron of velocity 6×10^5 ms^{-1} from it?
(Mass of electron $= 9 \times 10^{-31}$ kg
Velocity of light $= 3 \times 10^8$ ms^{-1}
Planck's constant $= 6.626 \times 10^{-34}$ Js
Charge of electron $= 1.6 \times 10^{-19}$ JeV^{-1}) **[2019, 12 Jan Shift-I]**

(a) 4.0 eV (b) 2.1 eV (c) 0.9 eV (d) 3.1 eV

07 Which of the graphs shown below does not represent the relationship between incident light and the electron ejected from metal surface? **[2019, 10 Jan Shift-I]**

(a)
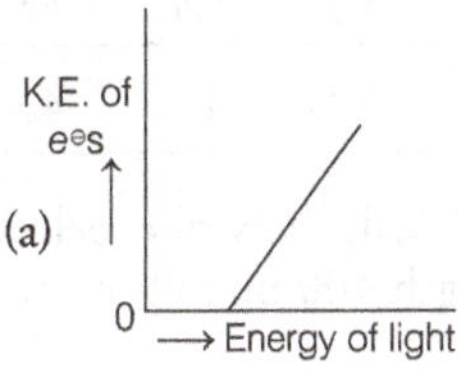

(b)
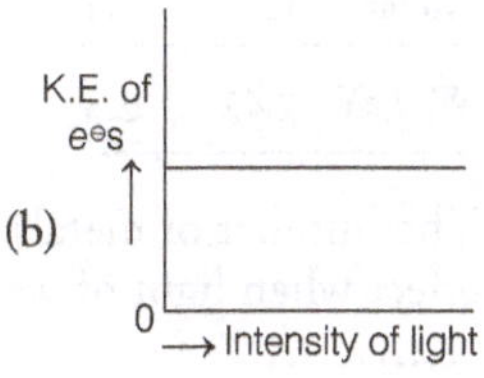

(c) 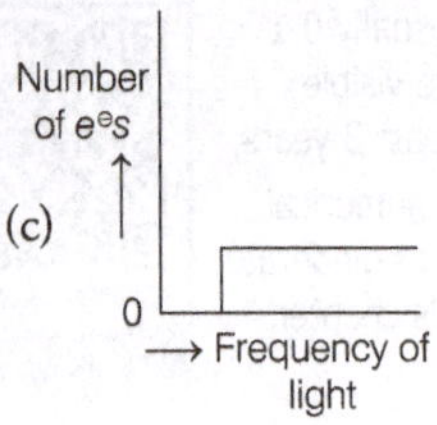

(d) 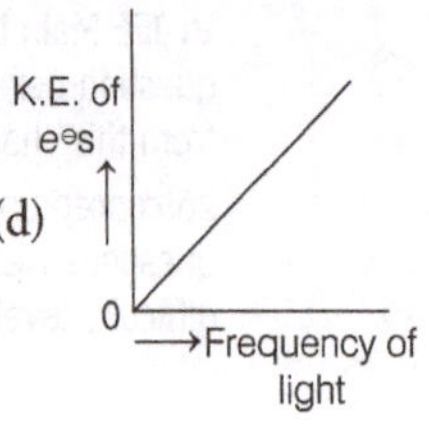

• Level 3

08 If p is the momentum of the fastest electron ejected from a metal surface after the irradiation of light having wavelength λ, then for 1.5 p momentum of the photoelectron, the wavelength of the light should be
(Assume kinetic energy of ejected photoelectron to be very high in comparison to work function) **[2019, 8 April Shift-II]**

(a) $\frac{4}{9}\lambda$ (b) $\frac{3}{4}\lambda$ (c) $\frac{2}{3}\lambda$ (d) $\frac{1}{2}\lambda$

Numerical Value Type Questions

09 A hypothetical electromagnetic wave is shown below.

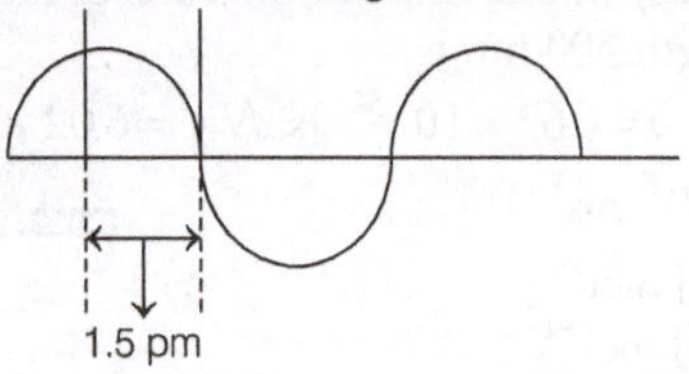

The frequency of the wave is $x \times 10^{19}$ Hz.
x = ... (nearest integer) **[2024, 8 April Shift-I]**

Imp. *IE & Electromagnetic Radiation*

10 The ionisation energy of sodium in kJ mol^{-1}, if electromagnetic radiation of wavelength 242 nm is just sufficient to ionise sodium atom is .... **[2024, 31 Jan Shift-I]**

11 The energy of one mole of photons of radiation of frequency 2×10^{12} Hz in J mol^{-1} is .... (Nearest integer) [Given, $h = 6.626 \times 10^{-34}$ Js, $N_A = 6.022 \times 10^{23}$ mol^{-1}] **[2023, 30 Jan Shift-I]**

12 Values of work function (W_0) for a few metals are given below **[2023, 12 April Shift-I]**

Metal	Li	Na	K	Mg	Cu	Ag
W_0 / eV	2.42	2.3	2.25	3.7	4.8	4.3

The number of metals which will show photoelectric effect when light of wavelength 400nm falls on it is ....

Given, $h = 6.6 \times 10^{-34}$ Js, $c = 3 \times 10^8$ ms^{-1}
$e = 1.6 \times 10^{-19}$ C

13 The number of incorrect statement/s about the black body from the following is .... **[2023, 10 April Shift-I]**
(A) Emit or absorb energy in the form of electromagnetic radiation.
(B) Frequency distribution of the emitted radiation depends on temperature.
(C) At a given temperature, intensity *vs* frequency curve passes through a maximum value.
(D) The maximum of the intensity *vs* frequency curve is at a higher frequency at higher temperature compared to that at lower temperature.

14 Consider an imaginary ion $^{48}_{22}X^{3-}$. The nucleus contains 'a'% more neutrons than the number of electrons in the ion. The value of 'a' is .... [nearest integer] **[2022, 26 July Shift-II]**

15 If the work function of a metal is 6.63×10^{-19} J, the maximum wavelength of the photon required to remove a photoelectron from the metal is ...... nm. [Nearest integer] [Given, $h = 6.63 \times 10^{-34}$ Js, and $c = 3 \times 10^8$ m s^{-1}] **[2022, 28 June Shift-I]**

16 The longest wavelength of light that can be used for the ionisation of lithium atom (Li) in its ground state is $x \times 10^{-8}$ m. The value of x is ...... [Nearest integer] (Given, energy of the electron in the first shell of the hydrogen atom is -2.2×10^{-18} J; $h = 6.63 \times 10^{-34}$ Js and $c = 3 \times 10^8$ ms^{-1}) **[2022, 25 June Shift-I]**

17 A 50 watt bulb emits monochromatic red light of wavelength of 795 nm. The number of photons emitted per second by the bulb is $x \times 10^{20}$. The value of x is .... [Given, $h = 6.63 \times 10^{-34}$ Js and $c = 3.0 \times 10^8$ ms^{-1}] **[2021, 1 Sep Shift-II]**

18 A metal surface is exposed to 500 nm radiation. The threshold frequency of the metal for photoelectric current is 4.3×10^{14} Hz. The velocity of ejected electron is .... $\times 10^5$ ms^{-1}. (Nearest integer) [Use $h = 6.63 \times 10^{-34}$ Js, $m_e = 9.0 \times 10^{-31}$ kg] **[2021, 26 Aug Shift-II]**

19 The work function of sodium metal is 4.41×10^{-19} J. If photons of wavelength 300 nm are incident on the metal, the kinetic energy of the ejected electrons will be ($h = 6.63 \times 10^{-34}$ J s; $c = 3 \times 10^8$ m/s) .... $\times 10^{-21}$ J. **[2020, 2 Sep Shift-II]**

Topic 2 / Bohr's Model and Hydrogen Spectrum

Level 1

01 If a_0 is denoted as the Bohr radius of hydrogen atom, then what is the de-Broglie wavelength (λ) of the electron present in the second orbit of hydrogen atom? [n : any integer] **[2025, 29 Jan Shift-I]**

(a) $\frac{2a_0}{n\pi}$ (b) $\frac{8\pi a_0}{n}$ (c) $\frac{4\pi a_0}{n}$ (d) $\frac{4n}{\pi a_0}$

02 Radius of the first excited state of helium ion is given as: $a_0 \rightarrow$ radius of first stationary state of hydrogen atom. **[2025, 22 Jan Shift-I]**

(a) $r = \frac{a_0}{2}$ (b) $r = 2a_0$ (c) $r = \frac{a_0}{4}$ (d) $r = 4a_0$

03 Match List-I with List-II. **[2024, 29 Jan Shift-II]**

	List I (Spectral series for hydrogen)		List II (Spectral region \ higher energy state)
A.	Lyman	I.	Infrared region
B.	Balmer	II.	UV region
C.	Paschen	III.	Infrared region
D.	Pfund	IV.	Visible region

Choose the correct answer from the options given below

(a) A-I, B-III, C-II, D-IV (b) A-II, B-IV, C-III, D-I
(c) A-I, B-II, C-III, D-IV (d) A-II, B-III, C-I, D-IV

04 Given below are two statements.

Statement I According to Bohr's model of hydrogen atom, the angular momentum of an electron in a given stationary state is quantised.

Statement II The concept of electron in Bohr's orbit, violates the Heisenberg uncertainty principle.

In the light of the above statements, choose the most appropriate answer from the options given below. **[2023, 15 April Shift-I]**

(a) Statement I is incorrect but Statement II is correct.
(b) Statement I is correct but Statement II is incorrect.
(c) Both Statement I and Statement II are incorrect.
(d) Both Statement I and Statement II are correct.

05 Which transition in the hydrogen spectrum would have the same wavelength as the Balmer type transition from $n = 4$ to $n = 2$ of He^+ spectrum **[2023, 31 Jan Shift-I]**

(a) $n = 1$ to $n = 3$ (b) $n = 1$ to $n = 2$
(c) $n = 3$ to $n = 4$ (d) $n = 2$ to $n = 1$

Imp. *Bohr's Model of H-Atom & Heisenberg Uncertainty Principle*

06 Given below are two statements.

Statement I Rutherford's gold foil experiment cannot explain the line spectrum of hydrogen atom.

Statement II Bohr's model of hydrogen atom contradicts Heisenberg's uncertainty principle.

In the light of the above statements, choose the most appropriate answer from the options given below : **[2021, 27 July Shift-I]**

(a) Statement I is false but statement II is true.
(b) Statement I is true but statement II is false.
(c) Both statement I and statement II are false.
(d) Both statement I and statement II are true.

07 The region in the electromagnetic spectrum where the Balmer series lines appear is **[2020, 4 Sep Shift-I]**

(a) infrared (b) ultraviolet
(c) microwave (d) visible

08 The radius of the second Bohr orbit in terms of the Bohr radius, a_0 in Li^{2+} is **[2020, 8 Jan Shift-II]**

(a) $\frac{2a_0}{3}$ (b) $\frac{4a_0}{3}$ (c) $\frac{4a_0}{9}$ (d) $\frac{2a_0}{9}$

09 The ratio of the shortest wavelength of two spectral series of hydrogen spectrum is found to be about 9. The spectral series are **[2019, 10 April Shift-II]**

(a) Lyman and Paschen (b) Brackett and Pfund
(c) Paschen and Pfund (d) Balmer and Brackett

Level 2

10 Which of the following postulate of Bohr's model of hydrogen atom is not in agreement with quantum mechanical model of an atom? **[2025, 3 April Shift-I]**

(a) An atom in a stationary state does not emit electromagnetic radiation as long as it stays in the same state.
(b) The electron in a H atom's stationary state moves in a circle around the nucleus.
(c) When an electron makes a transition from a higher energy stationary state to a lower energy stationary state, then it emits a photon of light.
(d) An atom can take only certain distinct energies E_1, E_2, E_3, etc. These allowed states of constant energy are called the stationary states of atom.

11 Heat treatment of muscular pain involves radiation of wavelength of about 900 nm. Which spectral line of H atom is suitable for this?

(Given, Rydberg constant $R_H = 10^5$ cm^{-1}, $h = 6.6 \times 10^{-34}$ J s, $c = 3 \times 10^8$ m/s) **[2025, 23 Jan Shift-I]**

(a) Balmer series, $\infty \rightarrow 2$ (b) Paschen series, $5 \rightarrow 3$
(c) Lyman series, $\infty \rightarrow 1$ (d) Paschen series, $\infty \rightarrow 3$

12 The energy of an electron in the first Bohr orbit of hydrogen atom is -2.18×10^{-18} J. Its energy in the third Bohr orbit is .... **[2023, 13 April Shift-I]**

(a) three times of this value (b) $\frac{1}{9}$th of this value

(c) one third of this value (d) $\frac{1}{27}$ of this value

13 If the radius of the 3rd Bohr's orbit of hydrogen atom is r_3 and the radius of 4th Bohr's orbit is r_4. Then **[2022, 26 June Shift-I]**

(a) $r_4 = \frac{9}{16} r_3$ (b) $r_4 = \frac{16}{9} r_3$ (c) $r_4 = \frac{3}{4} r_3$ (d) $r_4 = \frac{4}{3} r_3$

14 Given below are two statements.

Statement I According to Bohr's model of an atom, qualitatively the magnitude of velocity of electron increases with decrease in positive charges on the nucleus as there is no strong hold on the electron by the nucleus.

Statement II According to Bohr's model of an atom, qualitatively the magnitude of velocity of electron increases with decrease in principal quantum number.

In the light of the above statements, choose the most appropriate answer from the options given below. **[2021, 26 Aug Shift-I]**

(a) Both statement I and statement II are false.
(b) Both statement I and statement II are true.
(c) Statement I is false but statement II is true.
(d) Statement I is true but statement II is false.

Imp. *Limitations of Bohr's Model of an Atom*

15 Given below are two statements.

Statement I Bohr's theory accounts for the stability and line spectrum of Li^+ ion.

Statement II Bohr's theory was unable to explain the splitting of spectral lines in the presence of a magnetic field.

In the light of the above statements, choose the most appropriate answer from the options given below : **[2021, 18 March Shift-II]**

(a) Both statements I and II are true.
(b) Statement I is false but statement II is true.
(c) Both statements I and II are false.
(d) Statement I is true but statement II is false.

16 The difference between the radii of 3rd and 4th orbits of Li^{2+} is ΔR_1. The difference between the radii of 3rd and 4th orbits of He^+ is ΔR_2. Ratio $\Delta R_1 : \Delta R_2$ is **[2020, 5 Sep Shift-I]**

(a) 3 : 2 (b) 8 : 3
(c) 2 : 3 (d) 3 : 8

17 For the Balmer series in the spectrum of H atom, $\bar{\nu} = R_H \left\{\frac{1}{n_1^2} - \frac{1}{n_2^2}\right\}$, the correct statements among (I) to (IV) are :

I. As wavelength decreases, the lines in the series converge
II. The integer n_1 is equal to 2
III. The lines of longest wavelength corresponds to $n_2 = 3$
IV. The ionisation energy of hydrogen can be calculated from wave number of these lines **[2020, 8 Jan Shift-I]**

(a) I, II, IV (b) II, III, IV
(c) I, III, IV (d) I, II, III

18 For emission line of atomic hydrogen from $n_i = 8$ to $n_f = n$, the plot of wave number $(\bar{\nu})$ against $\left(\frac{1}{n^2}\right)$ will be (The Rydberg constant, R_H is in wave number unit) **[2019, 9 Jan Shift-I]**

(a) non linear (b) linear with slope $-R_H$
(c) linear with slope R_H (d) linear with intercept $-R_H$

19 The ground state energy of hydrogen atom is −13.6 eV. The energy of second excited state of He^+ ion in eV is **[2019, 10 Jan Shift-II]**

(a) −54.4 (b) − 3.4
(c) − 6.04 (d) − 27.2

• Level 3

20 Which one of the following about an electron occupying the Is orbital in a hydrogen atom is incorrect? (Bohr's radius is represented by a_0) **[2025, 4 April Shift-I]**

(a) The total energy of the electron is maximum when it is at a distance a_0 from the nucleus.
(b) The probability density of finding the electron is maximum at the nucleus.
(c) The electron can be found at a distance $2a_0$ from the nucleus.
(d) The 1*s* orbital is spherically symmetrical.

21

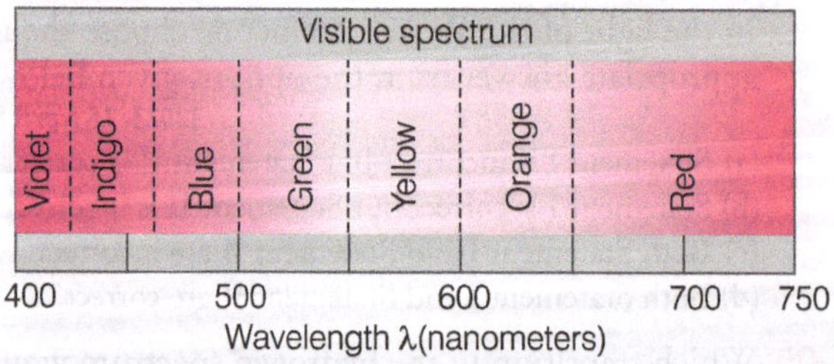

Which of the following statements are correct, if the threshold frequency of caesium is 5.16×10^{14} Hz?

A. When Cs is placed inside a vacuum chamber with an ammeter connected to it and yellow light is focused on Cs, the ammeter shows the presence of current.
B. When the brightness of the yellow light is dimmed, the value of the current in the ammeter is reduced.
C. When a red light is used instead of the yellow light, the current produced is higher with respect to the yellow light.
D. When a blue light is used, the ammeter shows the formation of current.
E. When a white light is used, the ammeter shows formation of current.

Choose the correct answer from the options given below: **[2025, 7 April Shift-I]**

(a) B, C and D Only (b) A, D and E Only
(c) A, C, D and E Only (d) A, B, D and E Only

22 According to Bohr's model of hydrogen atom, which of the following statement is incorrect? **[2025, 2 April Shift-I]**

(a) Radius of 8th orbit is four times larger than that of 4th orbit
(b) Radius of 6th orbit is three times larger than that of 4th orbit
(c) Radius of 4th orbit is four times larger than that of 2nd orbit
(d) Radius of 3rd orbit is nine times larger than that of 1st orbit

23 For hydrogen like species, which of the following graphs provides the most appropriate representation of E vs Z plot for a constant n?
[E = energy of the stationary state,
Z = atomic number, n = principal quantum number] **[2025, 29 Jan Shift-II]**

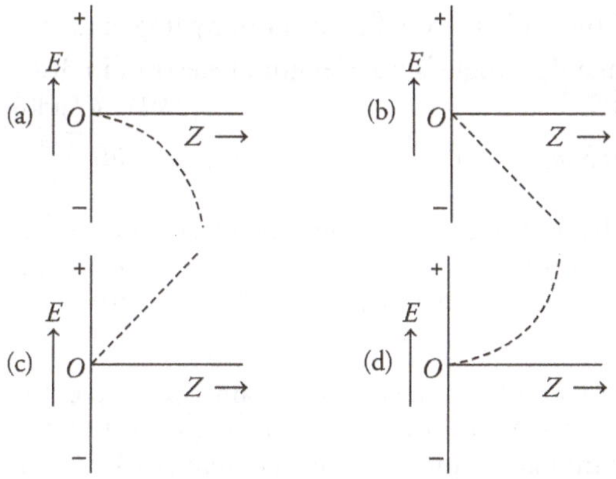

24 The shortest wavelength of hydrogen atom in Lyman series is λ. The longest wavelength in Balmer series of He^+ is **[2023, 29 Jan Shift-I]**

(a) $\frac{5}{9\lambda}$ (b) $\frac{9\lambda}{5}$
(c) $\frac{5\lambda}{9}$ (d) $\frac{36\lambda}{5}$

25 According to Bohr's atomic theory,

I. kinetic energy of electron is $\propto \frac{Z^2}{n^2}$
II. the product of velocity (v) of electron and principal quantum number (n), 'vn' $\propto Z^2$.
III. frequency of revolution of electron in an orbit is $\propto \frac{Z^3}{n^3}$
IV. coulombic force of attraction on the electron is $\propto \frac{Z^3}{n^4}$

Choose the most appropriate answer from the options given below. **[2021, 24 Feb Shift-II]**

(a) Only III (b) Only I
(c) I, III and IV (d) I and IV

26 For any given series of spectral lines of atomic hydrogen, let $\Delta\bar{v} = \bar{v}_{max} - \bar{v}_{min}$ be the difference in maximum and minimum frequencies in cm^{-1}. The ratio $\Delta\bar{v}_{Lyman} / \Delta\bar{v}_{Balmer}$ is **[2019, 9 April Shift-I]**

(a) 27 : 5 (b) 5 : 4 (c) 9 : 4 (d) 4 : 1

Numerical Value Type Questions

27 The energy of an electron in first Bohr orbit of H-atom is – 13.6 eV. The magnitude of energy value of electron in the first excited state of Be^{3+} is .... eV (Nearest integer value) **[2025, 8 April Shift-II]**

28 For hydrogen atom, energy of an electron in first excited state is −3.4 eV, KE of the same electron of hydrogen atom is x eV. Value of x is ....$\times 10^{-1}$ eV. (Nearest integer) **[2024, 6 April Shift-II]**

29 The value of Rydberg constant (R_H) is 2.18×10^{-18} J. The velocity of electron having mass 9.1×10^{-31} kg in Bohr's first orbit of hydrogen atom = ... $\times 10^5$ ms^{-1}. (Nearest integer). **[2024, 5 April Shift-I]**

30 Number of spectral lines obtained in He^+ spectra, when an electron makes transition from fifth excited state to first excited state will be .... **[2024, 30 Jan Shift-I]**

31 The number of following statement/s which is/are incorrect is **[2023, 8 April Shift-I]**

(A) line emission spectra are used to study the electronic structure.
(B) the emission spectra of atoms in the gas phase show a continuous spread of wavelength from red to violet.
(C) an absorption spectrum is like the photographic negative of an emission spectrum.
(D) the element helium was discovered in the sun by spectroscopic method.

32 The electron in the nth orbit of Li^{2+} is excited to $(n+1)$ orbit using the radiation of energy 1.47×10^{-17} J (as shown in the diagram). The value of n is .... [Given: $R_H = 2.18 \times 10^{-18}$ J]

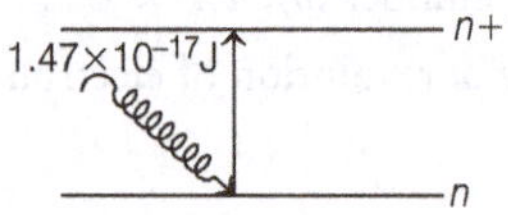

[2023, 10 April Shift-II]

33 If wavelength of the first line of the Paschen series of hydrogen atom is 720 nm, then the wavelength of the second line of this series is .... nm. (Nearest integer) **[2023, 24 Jan Shift-I]**

34 Assume that the radius of the first Bohr orbit of hydrogen atom is 0.6 Å. The radius of the third Bohr orbit of He^+ is .... picometer. (Nearest integer) **[2023, 29 Jan Shift-II]**

35 When the excited electron of a H-atom from $n=5$ drops to the ground state, the maximum number of emission lines observed are .... **[2022, 25 July Shift-II]**

36 The kinetic energy of an electron in the second Bohr orbit of a hydrogen atom is equal to $\frac{h^2}{xma_0^2}$. The value of $10x$ is ....
(a_0 is radius of Bohr's orbit) (Nearest integer)
[Given, $\pi = 3.14$] **[2021, 27 Aug Shift-I]**

Topic 3 / Dual Nature of Matter and Uncertainty Principle

• Level 1

01 Given below are two statements. **[2025, 29 Jan Shift-II]**

Statement I It is impossible to specify simultaneously with arbitrary precision, both the linear momentum and the position of a particle.

Statement II If the uncertainty in the measurement of position and uncertainty in measurement of momentum are equal for an electron, then the uncertainty in the measurement of velocity is

$$\geq \sqrt{\frac{h}{\pi}} \times \frac{1}{2m}$$

In the light of the above statements, choose the correct answer from the options given below

(a) Statement I is false but Statement II is true.
(b) Both Statement I and Statement II are false.
(c) Both Statement I and Statement II are true.
(d) Statement I is true but Statement II is false.

Imp. *Graphical Relation between p & λ of Matter*

02 According to the wave-particle duality of matter by de-Broglie, which of the following graph plot presents **most appropriate** relationship between wavelength of electron (λ) and momentum of electron (p)? **[2024, 1 Feb Shift-I]**

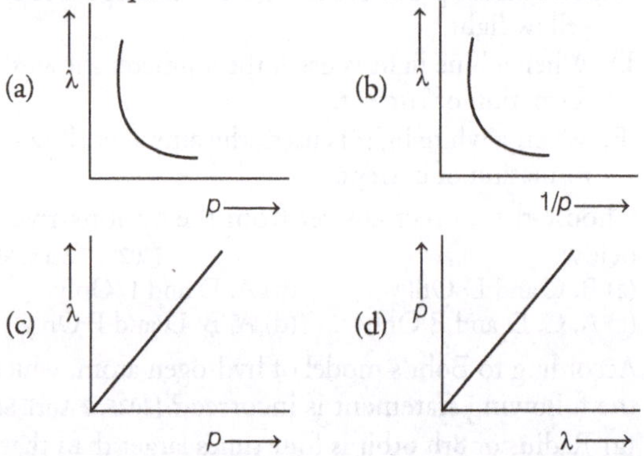

• Level 2

03 The de-Broglie wavelength (λ) associated with a photoelectron varies with the frequency (ν) of the incident radiation as, [ν_0 is threshold frequency] **[2019, 11 Jan Shift-II]**

(a) $\lambda \propto \frac{1}{(\nu - \nu_0)^{1/4}}$
(b) $\lambda \propto \frac{1}{(\nu - \nu_0)^{3/2}}$
(c) $\lambda \propto \frac{1}{(\nu - \nu_0)}$
(d) $\lambda \propto \frac{1}{(\nu - \nu_0)^{1/2}}$

• Level 3

04 If the radius of the first orbit of hydrogen atom is a_0, then de-Broglie's wavelength of electron in 3rd orbit is **[2023, 16 April Shift-II]**

(a) $3\pi a_0$ (b) $\frac{\pi a_0}{3}$ (c) $6\pi a_0$ (d) $\frac{\pi a_0}{6}$

05 The de-Broglie wavelength of an electron in the 4ᵗʰ Bohr orbit is **[2020, 9 Jan Shift-I]**

(a) $6\pi a_0$ (b) $2\pi a_0$ (c) $8\pi a_0$ (d) $4\pi a_0$

Numerical Value Type Questions

06 Based on Heisenberg's uncertainty principle, the uncertainty in the velocity of the electron to be found within an atomic nucleus of diameter 10^{-15} m is .... $\times 10^9$ ms^{-1}. (Nearest integer) **[2024, 9 April Shift-II]**

07 Frequency of the de-Broglie wave of electron in Bohr's first orbit of hydrogen atom is .... 10^{13} Hz (Nearest integer).
[Given : R_H (Rydberg constant) $= 2.18 \times 10^{-18}$ J, h(Plank's constant) $= 6.6 \times 10^{-34}$ Js] **[2024, 6 April Shift-I]**

08 The de-Broglie's wavelength of an electron in the 4th orbit is...... πa_0. (a_0 = Bohr's radius) **[2024, 4 April Shift-I]**

09 The wavelength of an electron of kinetic energy 4.50×10^{-29} J is .... $\times 10^{-5}$ m. (Nearest integer) **[2023, 6 April Shift-I]**

10 Electrons in a cathode ray tube have been emitted with a velocity of 1000 m s^{-1}. The number of following statements which is/are true about the emitted radiation is ....
[Given, $h = 6 \times 10^{-34}$ J s, $m_e = 9 \times 10^{-31}$ kg.] **[2023, 1 Feb Shift-I]**

(A) The de-Broglie wavelength of the electron emitted is 666.67 nm.
(B) The characteristic of electrons emitted depend upon the material of the electrodes of the cathode ray tube.
(C) The cathode rays start from cathode and move towards anode.
(D) The nature of the emitted electrons depends on the nature of the gas present in cathode ray tube.

11 The minimum uncertainty in the speed of an electron in an one dimensional region of length $2a_0$ (where, a_0 = Bohr radius 52.9 pm) is ...... kms^{-1}.
(Given, mass of electron $= 9.1 \times 10^{-31}$ kg. Planck's constant, $h = 6.63 \times 10^{-34}$ Js) **[2022, 29 July Shift-I]**

12 If the wavelength for an electron emitted from H-atom is 3.3×10^{-10} m, then energy absorbed by the electron in its ground state compared to minimum energy required for its escape from the atom, is .... times. (Nearest integer)
[Given, $h = 6.626 \times 10^{-34}$ Js]
[Mass of electron $= 9.1 \times 10^{-31}$ kg] **[2022, 28 July Shift-II]**

13 The wavelength of an electron and a neutron will become equal when the velocity of the electron is x times the velocity of neutron. The value of x is .... (Nearest integer) (Mass of electron is 9.1×10^{-31} kg and mass of neutron is 1.6×10^{-27} kg) **[2022, 26 July Shift-I]**

14 An accelerated electron has a speed of $5 \times 10^6 ms^{-1}$ with an uncertainty of 0.02%. The uncertainty in finding its location while in motion is $x \times 10^{-9}$m. The value of x is .... (Nearest integer)
[Use mass of electron $= 9.1 \times 10^{-31}$ kg, $h = 6.63 \times 10^{-34}$ Js, $\pi = 3.14$] **[2021, 25 July Shift-II]**

15 The wavelength of electrons accelerated from rest through a potential difference of 40 kV is $x \times 10^{-12}$ m. The value of x is ...... (Nearest integer)
Given : Mass of electron $= 9.1 \times 10^{-31}$ kg
Charge on an electron $= 1.6 \times 10^{-19}$ C
Planck's constant $= 6.63 \times 10^{-34}$ Js **[2021, 20 July Shift-II]**

16 When light of wavelength 248 nm falls on a metal of threshold energy 3.0 eV, the de- Broglie wavelength of emitted electrons is ...... Å.
[Round off to the nearest integer]
[Use : $\sqrt{3} = 1.73$, $h = 6.63 \times 10^{-34}$ Js, $m_e = 9.1 \times 10^{-31}$ kg, $c = 3.0 \times 10^8 ms^{-1}$, 1 eV $= 1.6 \times 10^{-19}$ J] **[2021, 16 March Shift-I]**

17 A ball weighing 10 g is moving with a velocity of 90 ms^{-1}. If the uncertainity in its velocity is 5%, then the uncertainty in its position is .... $\times 10^{-33}$ m (Rounded off to the nearest integer). [Given, $h = 6.63 \times 10^{-34}$ J-s] **[2021, 26 Feb Shift-II]**

Topic 4 / Quantum Mechanical Model

• Level 1

01 The correct set of four quantum numbers for the valence electron of rubidium atom ($Z = 37$) is **[2024, 29 Jan Shift-I, JEE Main 2014]**

(a) 5, 1, 1, $+\frac{1}{2}$ (b) 5, 1, 0, $+\frac{1}{2}$
(c) 5, 0, 0, $+\frac{1}{2}$ (d) 5, 0, 1, $+\frac{1}{2}$

02 The number of radial nodes for 3p-orbital is **[2024, 1 Feb Shift-II]**

(a) 1 (b) 2 (c) 4 (d) 3

03 The wave function (ψ) of 2s is given by

$$\psi_{2s} = \frac{1}{2\sqrt{2\pi}}\left(\frac{1}{a_0}\right)^{1/2}\left(2 - \frac{r}{a_0}\right)e^{-\frac{r}{2a_0}}$$

At $r = r_0$, radial node is formed. Thus, r_0 in terms of a_0 **[2023, 30 Jan Shift-II]**

(a) $r_0 = a_0$ (b) $r_0 = 2a_0$ (c) $r_0 = \frac{a_0}{2}$ (d) $r_0 = 4a_0$

04 Arrange the following orbitals in decreasing order of energy. **[2023, 31 Jan Shift-II]**

A. $n = 3, l = 0, m = 0$ B. $n = 4, l = 0, m = 0$
C. $n = 3, l = 1, m = 0$ D. $n = 3, l = 2, m = 1$

The correct option for the order is
(a) D > B > A > C (b) A > C > B > D
(c) B > D > C > A (d) D > B > C > A

05 The number of radial and angular nodes in 4d orbital are respectively **[2022, 26 June Shift-II]**

(a) 1 and 2 (b) 3 and 2
(c) 1 and 0 (d) 2 and 1

06 Given below are the quantum numbers for 4 electrons.

A. $n=3, l=2, m_1=1, m_s=+1/2$
B. $n=4, l=1, m_1=0, m_s=+1/2$
C. $n=4, l=2, m_1=-2, m_s=-1/2$
D. $n=3, l=1, m_1=-1, m_s=+1/2$

The correct order of increasing energy is **[2022, 29 July Shift-II]**

(a) D < B < A < C (b) D < A < B < C
(c) B < D < A < C (d) B < D < C < A

07 Identify the incorrect statement from the following. **[2022, 28 July Shift-I]**

(a) A circular path around the nucleus in which an electron moves is proposed as Bohr's orbit.
(b) An orbital is the one electron wave function (ψ) in an atom.
(c) The existence of Bohr's orbits is supported by hydrogen spectrum.
(d) Atomic orbital is characterised by the quantum numbers n and l only.

08 Which of the following sets of quantum numbers is not allowed? **[2022, 25 July Shift-I]**

(a) $n=3, l=2, m_l=0, s=+\frac{1}{2}$
(b) $n=3, l=2, m_l=-2, s=+\frac{1}{2}$
(c) $n=3, l=3, m_l=-3, s=-\frac{1}{2}$
(d) $n=3, l=0, m_l=0, s=-\frac{1}{2}$

09 Given below are two statements. One is labelled as Assertion (A) and the other is labelled as Reason (R).

Assertion (A) Energy of 2s orbital of hydrogen atom is greater than that of 2s orbital of lithium.

Reason (R) Energies of the orbitals in the same subshell decrease with increase in the atomic number.

In the light of the above statements, choose the correct answer from the options given below: **[2022, 27 July, Shift-I]**

(a) Both A and R are true and R is the correct explanation of A.
(b) Both A and R are true but R is not the correct explanation of A.
(c) A is true but R is false.
(d) A is false but R is true.

10 The electrons are more likely to be found **[2019, 12 April Shift-I]**

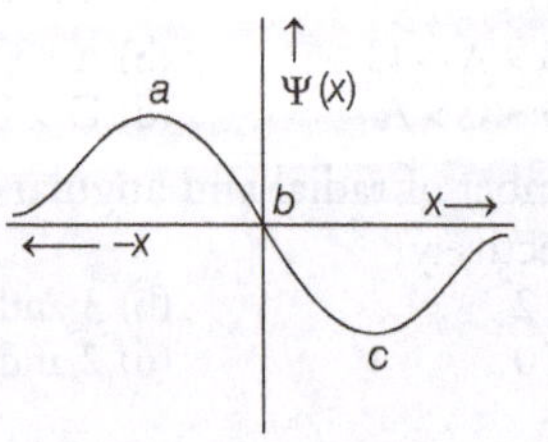

(a) in the region a and c (b) in the region a and b
(c) only in the region a (d) only in the region c

Level 2

11 For hydrogen atom, the orbital/s with lowest energy is/are **[2025, 24 Jan Shift-II]**

(A) 4s (B) $3p_x$ (C) $3d_{x^2-y^2}$ (D) $3d_{z^2}$
(E) $4p_z$

Choose the correct answer from the options given below

(a) (B), (C) and (D) only (b) (A) and (E) only
(c) (B) only (d) (A) only

12 For electrons in '2s' and '2p' orbitals, the orbital angular momentum values, respectively are **[2025, 3 April Shift-II]**

(a) $\sqrt{2}\frac{h}{2\pi}$ and 0 (b) 0 and $\sqrt{6}\frac{h}{2\pi}$
(c) $\frac{h}{2\pi}$ and $\sqrt{2}\frac{h}{2\pi}$ (d) 0 and $\sqrt{2}\frac{h}{2\pi}$

13 Consider the ground state of chromium atom ($Z=24$). How many electrons are with azimuthal quantum number $l=1$ and $l=2$ respectively? **[2025, 4 April Shift-II]**

(a) 12 and 5 (b) 16 and 5 (c) 16 and 4 (d) 12 and 4

14 Given below are two statements. **[2025, 22 Jan Shift-II]**

Statement I A spectral line will be observed for a $2p_x \rightarrow 2p_y$ transition.

Statement II $2p_x$ and $2p_y$ are degenerate orbitals.

In the light of the above statements, choose the correct answer from the options given below:

(a) Both Statement I and Statement II are true.
(b) Statement I is false but Statement II is true.
(c) Both Statement I and Statement II are false.
(d) Statement I is true but Statement II is false.

15 Which of the following is/are not correct with respect to energy of atomic orbitals of hydrogen atom?

(A) 1s < 2p < 3d < 4s
(B) 1s < 2s = 2p < 3s = 3p
(C) 1s < 2s < 2p < 3s < 3p
(D) 1s < 2s < 4s < 3d

Choose the correct answer from the options given below **[2025, 28 Jan Shift-II]**

(a) (A) and (B) only (b) (C) and (D) only
(c) (B) and (D) only (d) (A) and (C) only

16 Given below are two statements. **[2025, 23 Jan Shift-II]**

Statement I For a given shell, the total number of allowed orbitals is given by n^2.

Statement II For any subshell, the spatial orientation of the orbitals is given by $-l$ to $+l$ values including zero.

In the light of the above statements, choose the correct answer from the options given below
(a) Statement I is false but Statement II is true.
(b) Statement I is true but Statement II is false.
(c) Both Statement I and Statement II are true.
(d) Both Statement I and Statement II are false.

17 Given below are two statements. **[2024, 30 Jan Shift-I]**
Statement I The orbitals having same energy are called as degenerate orbitals.
Statement II In hydrogen atom, $3p$ and $3d$ orbitals are not degenerate orbitals.
In the light of the above statements, choose the **most appropriate** answer from the options given below
(a) Statement I is false but Statement II is true.
(b) Both Statement I and Statement II are false.
(c) Statement I is true but Statement II is false.
(d) Both Statement I and Statement II are true.

18 Compare the energies of following sets of quantum numbers for multielectron system.
(A) $n=4, l=1$ (B) $n=4, l=2$
(C) $n=3, l=1$ (D) $n=3, l=2$
(E) $n=4, l=0$
Choose the correct answer from the options given below: **[2024, 9 April Shift-I]**
(a) (E) < (C) < (D) < (A) < (B)
(b) (C) < (E) < (D) < (A) < (B)
(c) (B) > (A) > (C) > (E) > (D)
(d) (E) > (C) > (A) > (D) > (B)

19 The four quantum numbers for the electron in the outermost orbital of potassium (atomic no. 19) are **[2024, 31 Jan Shift-II]**
(a) $n=4, l=2, m=-1, s=+\frac{1}{2}$
(b) $n=2, l=0, m=0, s=+\frac{1}{2}$
(c) $n=4, l=0, m=0, s=+\frac{1}{2}$
(d) $n=3, l=0, m=1, s=+\frac{1}{2}$

20 Consider the following pairs of electrons : **[2022, 24 June Shift-I]**
(A) (a) $n=3, l=1, m_1=-1, m_s=+1/2$
(b) $n=3, l=2, m_1=1, m_s=+1/2$
(B) (a) $n=3, l=2, m_1=-2, m_s=-1/2$
(b) $n=3, l=2, m_1=-1, m_s=-1/2$
(C) (a) $n=4, l=2, m_1=2, m_s=+1/2$
(b) $n=3, l=2, m_1=2, m_s=+1/2$
The pairs of electron present in degenerate orbitals is/are
(a) Only A (b) Only B
(c) Only C (d) Both (B) and (C)

21 Which of the following is the correct plot for the probability density $\psi^2(r)$ as a function of distance 'r' of the electron from the nucleus for 2s orbital? **[2022, 29 June Shift-II]**

(a)
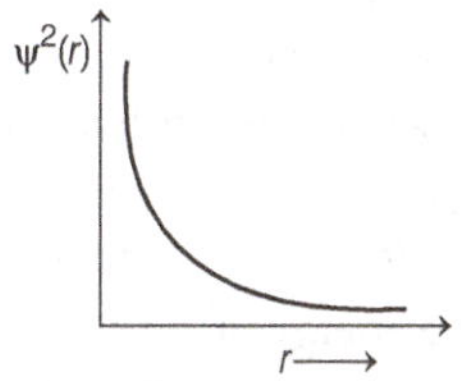

(b)
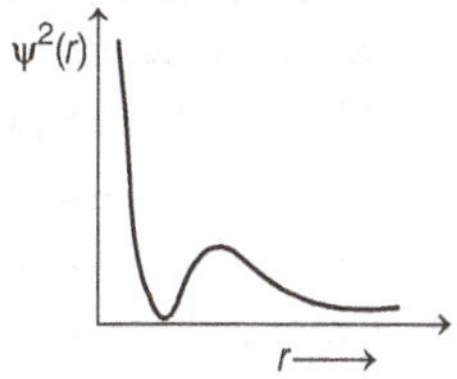

(c)
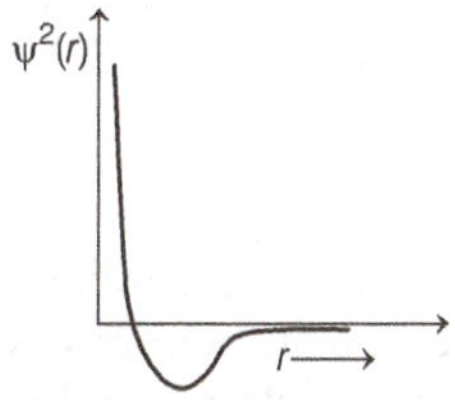

(d)
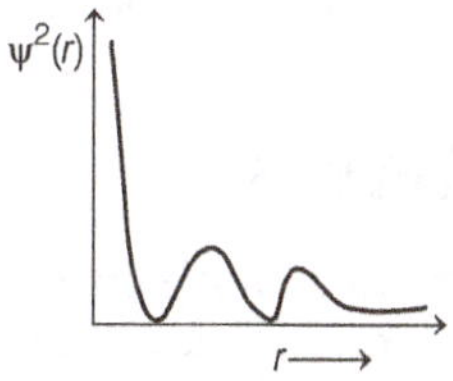

22 Which of the following statements are correct?
(A) The electronic configuration of Cr is $[Ar]3d^5 4s^1$.
(B) The magnetic quantum number may have a negative value.
(C) In the ground state of an atom, the orbitals are filled in order of their increasing energies.
(D) The total number of nodes are given by $n-2$.
Choose the most appropriate answer from the options given below **[2022, 29 June Shift-I]**
(a) (A), (C) and (D) only (b) (A) and (B) only
(c) (A) and (C) only (d) (A), (B) and (C) only

Imp. *Quantum Numbers*

23 Consider the following statements.
(A) The principal quantum number 'n' is a positive integer with values of 'n' = 1, 2, 3, ...
(B) The azimuthal quantum number 'l' for a given 'n' (principal quantum number) can have values as 'l' = 0, 1, 2,... n.
(C) Magnetic orbital quantum number 'm_l' for a particular 'l' (azimuthal quantum number) has $(2l+1)$ values.
(D) $\pm 1/2$ are the two possible orientations of electron spin.
(E) For $l=5$, there will be a total of 9 orbital.
Which of the above statements are correct? **[2022, 28 June Shift-II]**
(a) (A), (B) and (C) (b) (A), (C), (D) and (E)
(c) (A), (C) and (D) (d) (A), (B), (C) and (D)

24 The orbital having two radial as well as two angular nodes is **[2021, 26 Feb Shift-I]**
(a) $3p$ (b) $4f$ (c) $4d$ (d) $5d$

25 A certain orbital has no angular nodes and two radial nodes. The orbital is **[2021, 18 March Shift-I]**
(a) $2s$ (b) $3s$ (c) $3p$ (d) $2p$

26 The correct statement about probability density (except at infinite distance from nucleus) is **[2020, 5 Sep Shift-II]**
(a) it can be zero for 1*s*-orbital
(b) it can be negative for 2*p*-orbital
(c) it can be zero for 3*p*-orbital
(d) it can never be zero for 2*s*-orbital

27 The number of subshells associated with $n = 4$ and $m = -2$ quantum numbers is **[2020, 2 Sep Shift-II]**
(a) 8 (b) 2 (c) 16 (d) 4

28 Among the following, the energy of 2*s*-orbital is lowest in **[2019, 12 April Shift-II]**
(a) K (b) H (c) Li (d) Na

• Level 3

29 In a multielectron atom, which of the following orbitals described by three quantum numbers will have same energy in absence of electric and magnetic fields? **[2025, 28 Jan Shift-I]**
A. $n=1, l=0, m_l=0$ B. $n=2, l=0, m_l=0$
C. $n=2, l=1, m_l=1$ D. $n=3, l=2, m_l=1$
E. $n=3, l=2, m_l=0$
Choose the correct answer from the options given below.
(a) A and B only (b) D and E only
(c) C and D only (d) B and C only

30 Correct statements for an element with atomic number 9 are **[2025, 8 April Shift-II]**
A. There can be 5 electrons for which $m_s = +\frac{1}{2}$ and 4 electrons for which $m_s = -\frac{1}{2}$.
B. There is only one electron in p_z orbital.
C. The last electron goes to orbital with $n=2$ and $l=1$.
D. The sum of angular nodes of all the atomic orbitals is 1.
Choose the correct answer from the options given below.
(a) A and C only
(b) A and B only
(c) C and D only
(d) A, C and D only

31 The extra stability of half-filled subshell is due to
(A) symmetrical distribution of electrons.
(B) smaller coulombic repulsion energy.
(C) the presence of electrons with the same spin in non-degenerate orbitals.
(D) larger exchange energy.
(E) relatively smaller shielding of electrons by one another.
Indentify the correct statements: **[2025, 7 April Shift-II]**
(a) (A), (B), (D) and (E) only
(b) (B), (C) and (D) only
(c) (A), (B) and (D) only
(d) (B), (D) and (E) only

32 Which of the following statements are true? **[2025, 2 April Shift-II]**
(A) The subsidiary quantum number *l* describes the shape of the orbital occupied by the electron.
(B)
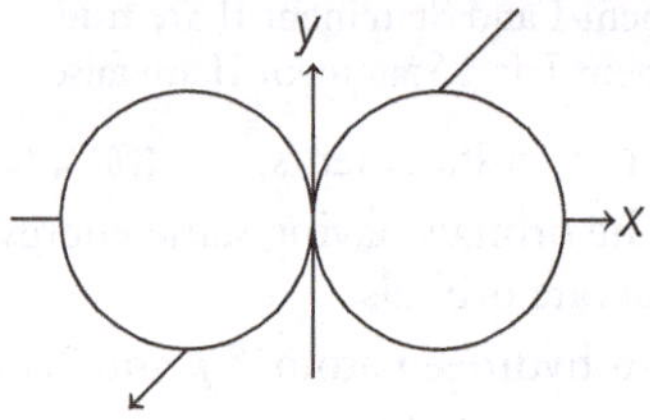

is the boundary surface diagram of the $2p_x$ orbital.
(C) The positive and negative signs in the wave function of the $2p_x$ orbital refer to charge.
(D) The wave function of $2p_x$ orbital is zero everywhere in the *xy* plane.
Choose the correct answer from the options given below :
(a) (C) and (D) only
(b) (B) and (D) only
(c) (A), (B) and (C) only
(d) (A) and (B) only

Imp. *Radial Distribution Function of Hydrogen*

33 The plots of radial distribution functions for various orbitals of hydrogen atom against '*r*' are given below. **[2021, 25 Feb Shift-I]**
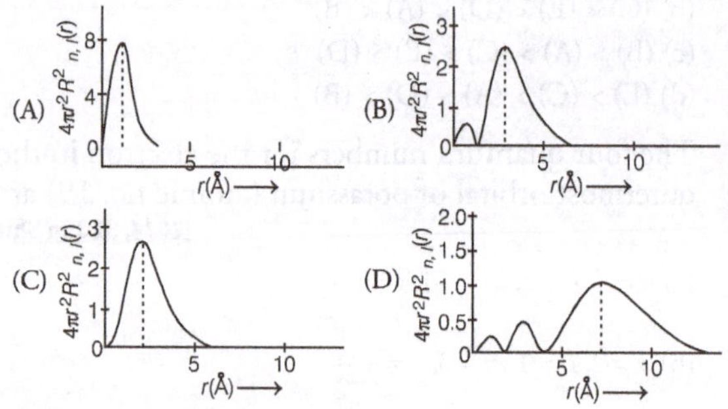

The correct plot for 3*s*-orbital is
(a) (A) (b) (B) (c) (C) (d) (D)

34 The figure that is not a direct manifestation of the quantum nature of atoms is **[2020, 2 Sep Shift-I]**

(a)
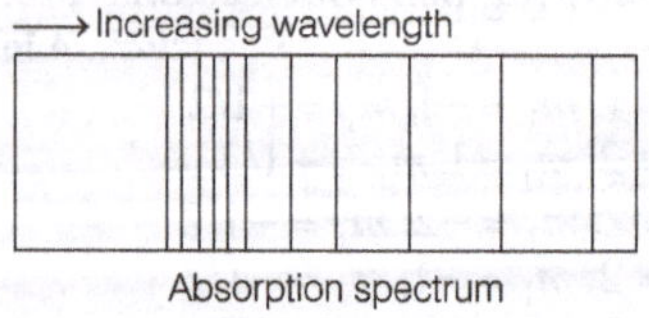

(b)
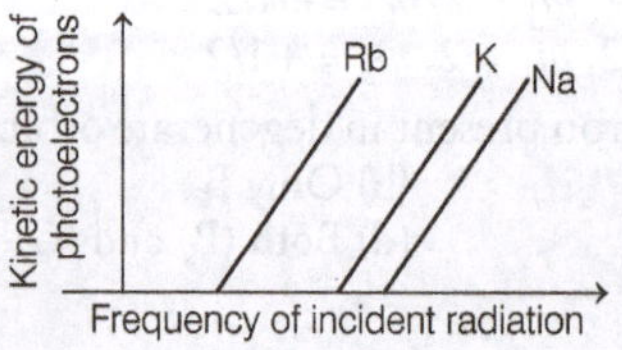

(c) 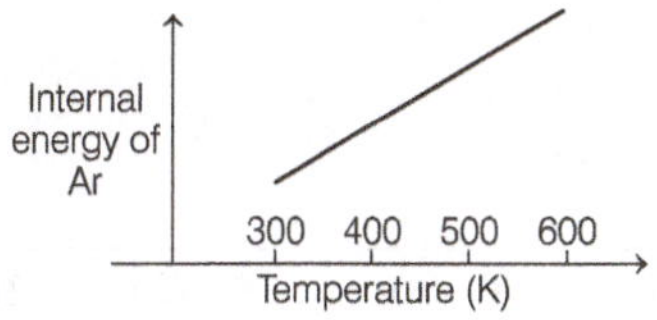

(d) 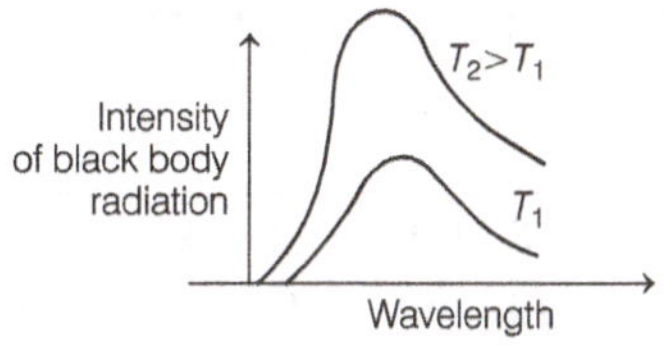

35 Which of the following combination of statements is true regarding the interpretation of the atomic orbitals?

I. An electron in an orbital of high angular momentum stays away from the nucleus than an electron in the orbital of lower angular momentum.

II. For a given value of the principal quantum number, the size of the orbit is inversely proportional to the azimuthal quantum number.

III. According to wave mechanics, the ground state angular momentum is equal to $\frac{h}{2\pi}$.

IV. The plot of ψ *vs* r for various azimuthal quantum numbers, shows peak shifting towards higher r value.

[2019, 9 Jan Shift-II]

(a) I, III
(b) II, III
(c) I, II
(d) I, IV

Numerical Value Type Questions

36 The maximum number of orbitals which can be identified with $n = 4$ and $m_l = 0$ is ... . **[2024, 4 April Shift-II]**

37 In an atom, total number of electrons having quantum numbers $n = 4$, $|m_l| = 1$ and $m_s = -\frac{1}{2}$ is .... **[2024, 5 April Shift-I]**

38 The number of given orbitals which have electron density along the axis is .... p_x, p_y, p_z, d_{xy}, d_{yz}, d_{z^2}, $d_{x^2-y^2}$ **[2023, 25 Jan Shift-II]**

39 The orbital angular momentum of an electron in 3*s* orbital is $\frac{xh}{2\pi}$. The value of x is .... (nearest integer) **[2023, 13 April Shift-II]**

40 The number of correct statements from the following is **[2023, 11 April Shift-II]**

A. For 1*s* orbital, the probability density is maximum at the nucleus.

B. For 2*s* orbital, the probability density first increases to maximum and then decreases sharply to zero.

C. Boundary surface diagrams of the orbitals encloses a region of 100% probability of finding the electron.

D. *p* and *d*-orbitals have 1 and 2 angular nodes respectively.

E. Probability density of *p*-orbital is zero at the nucleus.

41 The number of atomic orbitals from the following having 5 radial nodes is ....

7*s*, 7*p*, 6*s*, 8*p*, 8*d* **[2023, 8 April Shift-II]**

42 Consider the following set of quantum numbers.

	n	l	m_l		n	l	m_l
A.	3	3	−3	B.	3	2	−2
C.	2	1	+1	D.	2	2	+2

The number of correct sets of quantum numbers is .... **[2022, 27 June Shift-II]**

43 The number of orbitals with $n = 5$, $m_l = +2$ is .... (Round off to the nearest integer). **[2021, 16 March Shift-II]**

44 The value of magnetic quantum number of the outermost electron of Zn^+ ion is **[2021, 31 Aug Shift-II]**

45 Ge ($Z = 32$) in its ground state electronic configuration has completely filled orbitals with $m_l = 0$. The value of x is **[2021, 31 Aug Shift-I]**

46 The azimuthal quantum number for the valence electrons of Ga^+ ion is ....

(Atomic number of Ga = 31) **[2021, 20 July, Shift-I]**

47 A certain orbital has $n = 4$ and $m_l = -3$. The number of radial nodes in this orbital is ...... (Round off to the nearest integer). **[2021, 17 March Shift-I]**

Key Index

Topic 1

1 (c)	2 (b)	3 (c)	4 (c)	5 (c)	6 (b)	7 (d)	8 (a)	9 (5)	10 (494)
11 (798)	12 (3)	13 (0)	14 (4)	15 (300)	16 (4)	17 (2)	18 (5)	19 (222)	

Topic 2

1 (b)	2 (b)	3 (b)	4 (d)	5 (d)	6 (d)	7 (d)	8 (b)	9 (a)	10 (b)
11 (d)	12 (b)	13 (b)	14 (c)	15 (b)	16 (c)	17 (d)	18 (c)	19 (c)	20 (a)
21 (d)	22 (b)	23 (a)	24 (b)	25 (d)	26 (c)	27 (54)	28 (34)	29 (22)	30 (10)
31 (1)	32 (1)	33 (492)	34 (270)	35 (10)	36 (3155)				

Topic 3

1 (c)	2 (a)	3 (d)	4 (c)	5 (c)	6 (58)	7 (661)	8 (8)	9 (7)	10 (2)
11 (548)	12 (2)	13 (1758)	14 (58)	15 (6)	16 (9)	17 (1)			

Topic 4

1 (c)	2 (a)	3 (b)	4 (d)	5 (a)	6 (b)	7 (d)	8 (c)	9 (a)	10 (a)
11 (a)	12 (d)	13 (a)	14 (b)	15 (b)	16 (c)	17 (c)	18 (b)	19 (c)	20 (b)
21 (b)	22 (d)	23 (c)	24 (d)	25 (b)	26 (c)	27 (b)	28 (a)	29 (b)	30 (a)
31 (a)	32 (d)	33 (d)	34 (c)	35 (d)	36 (4)	37 (6)	38 (5)	39 (0)	40 (3)
41 (3)	42 (2)	43 (3)	44 (0)	45 (7)	46 (0)	47 (0)			

Detailed Explanations

Topic 1

01. *(c)* Minimum kinetic energy required to produce photoelectric effect is

$$E = E_0 = h\nu_0$$
$$= 6.6\times 10^{-34} \times 1.3\times 10^{15}$$
$$= 8.58\times 10^{-19}\ \text{J}$$

02. *(b)* Number of neutrons in protium ($^1_1$H) is zero (x).

Number of neutrons in deuterium ($^2_1$H or $^2_1$D) is 1 (y)

Number of neutrons in tritium ($^3_1$H or $^3_1$T) is 2 (z)

So, the sum of x, y and z is

$$x + y + z = 3$$

03. *(c)* The correct match is

A-III, B-I, C-IV, D-II

04. *(c)* A is correct but R is not correct.

Emission of electron from metal surface take place only, when the photon with energy greater than work function (w_0) strikes the metal surface. The ejected electrons leave with a specific amount of kinetic energy at a frequency $\nu > \nu_0$.

05. *(c)* Energy of a photon is given by

$$E = \frac{hc}{\lambda} = \frac{6.63\times 10^{-34}\ \text{Js}\times 3\times 10^{8}\ \text{ms}^{-1}}{300\times 10^{-9}\ \text{m}}$$

$$\Rightarrow\quad E = 6.63\times 10^{-19}\ \text{J}$$

Energy of 1 mole of photon

$$= 6.63\times 10^{-19}\times 10^{-3}\times 6.02\times 10^{23}\ \text{kJ mol}^{-1}$$
$$= 6.63\times 6.02\times 10$$
$$= 39.9\times 10 = 399\ \text{kJ mol}^{-1}$$

06. *(b)* The kinetic energy of the ejected electron is given by

$$\text{KE} = h\nu - h\nu_0$$

where, $h\nu_0 (W_0)$ = Work function

$$\frac{1}{2}m_e v^2 = \frac{hc}{\lambda} - w_0 \qquad \ldots\text{(i)}$$

On substituting all the given values in Eq. (i), we get

$$\frac{1}{2}\times 9\times 10^{-31}\ \text{kg}\times (6\times 10^{5}\ \text{ms}^{-1})^2$$
$$= \frac{6.626\times 10^{-34}\ \text{Js}\times 3\times 10^{8}\ \text{ms}^{-1}}{4000\times 10^{-10}\ \text{m}} - w_0$$
$$\therefore w_0 = -1.62\times 10^{-19}\ \text{kgm}^2\text{s}^{-2} + 4.96\times 10^{-19}\ \text{J}$$
$$= 3.34\times 10^{-19}\ \text{J}\quad [1\ \text{kg m}^2\text{s}^{-2} = 1\ \text{J}]$$
$$= 2.1\ \text{eV}$$

07. *(d)* For photoelectric effect, the graph of K.E. of electrons versus frequency of light is given as :

$$\text{KE} = h\nu - h\nu_0$$

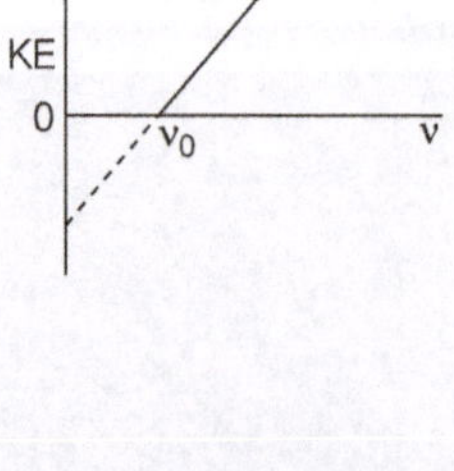

Slope = + h, intercept = – $h\nu_0$

So, option (d) is not correct. While the remaining options represents the relationship between incident light and the electron ejected from metal surface.

08. *(a)* We know that,

$$K.E. = \frac{p^2}{2m} = \frac{hc}{\lambda_{incident}}$$

Now, as per the given condition,

$$\frac{\lambda_2}{\lambda_1} = \left(\frac{p_1}{p_2}\right)^2$$

$$\Rightarrow \quad \frac{\lambda_2}{\lambda} = \left(\frac{p}{1.5 \times p}\right)^2 = \left(\frac{2}{3}\right)^2 = \frac{4}{9}$$

$$\Rightarrow \quad \lambda_2 = \frac{4}{9}\lambda \qquad \left[\because \lambda_1 = \lambda,\ p_1 = p\right]$$

09. *(5)* Wavelength of EM wave = $d \times 4$

$$\lambda = 1.5 \times 4 = 6 \text{ pm}$$

$$\nu = \frac{c}{\lambda} = \frac{3 \times 10^8}{6 \times 10^{-12}}$$

$$= 0.5 \times 10^{20} \text{ Hz} = 5 \times 10^{19} \text{ Hz}$$

Hence, the frequency of the wave is 5×10^{19} Hz.

10. *(494)*

> **idea** To obtain ionisation energy of sodium in kJ/mol. First of all use the formula $E = \frac{1240 \text{ (eV)}}{\lambda \text{ (nm)}}$.
> Then multiply the obtained value with 1.6×10^{19} to obtain energy in J/atom. After that multiply the calculated value in 2nd step with Avogadro number to get the final answer in kJ/mol.

Given, $\lambda = 242$ nm

using $E = \frac{1240 \text{ eV}}{\lambda \text{ (nm)}}$

$$E = \frac{1240}{242} \text{ eV} = 5.12 \text{ eV}$$

$$E = 5.12 \times 1.6 \times 10^{-19} \text{ J/atom}$$

$$E = 8.192 \times 10^{-19} \text{ J/atom}$$

or $\quad E = 8.192 \times 10^{-19} \times 6.022 \times 10^{23}$ atom/mol

$$= 493.32 \text{ kJ/mol}$$

$$\approx 494 \text{ kJ/mol}$$

11. *(798)* Given, $\nu = 2 \times 10^{12}$ Hz

$$E = ?$$

For one photon, $E = h\nu$

For one mole photon, $E = nh\nu$

where, h = Planck's constant

$$n = 1 \text{ mole}$$

$$\therefore \quad E = 6.022 \times 10^{23} \times 6.626 \times 10^{-34} \times 2 \times 10^{12}$$

$$= 798.03 \text{ J} \approx 798 \text{ J}$$

12. *(3)* Photoelectric effect will take place, only when the work function of given metals is lower than the energy of incident light with wavelength 400 nm.

Energy of incident light = $h\nu = \frac{hc}{\lambda}$

$$E(\text{eV}) = \frac{1240}{\lambda} = \frac{1240}{400} = 3.1 \text{ eV}$$

Thus, Li, Na, K will show photoelectric effect.

13. *(0)*

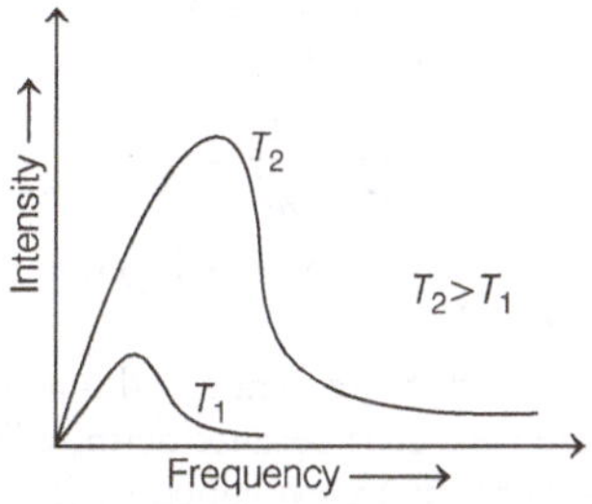

From the above graph we can see that, all statements (A), (B), (C) and (D) are also correct.

A black body can emit and absorb all the wavelength in electromagnetic spectrum. Hence, statement (A) is correct.

(B) The frequency distribution of the emitted radiation by a blackbody indeed depends on its temperature. This phenomenon is described by Planck's law. hence statement (B) is correct.

(C) The intensity *vs* frequency curve for a black body known as the planck's curve, pass through a maximum value for a given temperature. Statement (C) is correct.

(D) The frequency at which maximum intensity occurs increases with the temperature of the black body. Statement (D) is correct.

14. *(4)* The symbol ${}^{48}_{22}X^{3-}$ shows,

Atomic number = 22; mass number = 48

Total electrons = 22 + 3 = 25

Total neutrons = Mass number – Atomic number

$$= 48 - 22 = 26$$

Neutron percentage which is more than number of electrons (a)

$$= \frac{\text{Number of neutrons} - \text{Number of electrons}}{\text{Number of electrons}} \times 100$$

$$= \frac{26 - 25}{25} \times 100 = 4$$

15. *(300)* Using equation, $E = w_0 + \text{KE}$

[E = energy given, w_0 = work function, KE = kinetic energy]

As electron is just ejected, so

$$\text{KE} = 0; \ E = w_0 = \frac{hc}{\lambda}$$

$$= 6.63 \times 10^{-19}$$

$$\Rightarrow \quad \lambda = \frac{6.63 \times 10^{-34} \times 3 \times 10^8}{6.63 \times 10^{-19}}$$

$$\therefore \quad \lambda = 3 \times 10^{-7} \text{ m}$$

$$= 300 \times 10^{-9} \text{ m}$$

Hence, $\lambda = 300$ nm

16. *(4)* The potential energy of electron in nth orbit is given by,

$$E = -22 \times 10^{-19} \times Z^2 / n^2 \text{ J}$$

For Li, $Z = 3$ and $n = 2$

$$E = -22 \times 10^{-19} \times \frac{3 \times 3}{2 \times 2} \text{eV}$$

$$= 49.5 \times 10^{-19} \text{ J}$$

$$E = \frac{hc}{\lambda}$$

$$\Rightarrow \quad \lambda = \frac{6.63 \times 10^{-34} \times 3.0 \times 10^8}{49.5 \times 10^{-19}}$$

$$= 0.40 \times 10^{-7} \text{ m} = 4 \times 10^{-8} \text{ m}$$

17. *(2)* Energy of photon is given as

$$E = \frac{nhc}{\lambda} \qquad \ldots\text{(i)}$$

where n = number of photons.

On substituting the values, we get

$$50 \text{ J} = \frac{n \times 6.63 \times 10^{-34} \text{ Js} \times 3 \times 10^8 \text{ m/s}}{795 \times 10^{-9} \text{ m}}$$

$$\Rightarrow \quad n = \frac{50 \times 795 \times 10^{-9}}{6.63 \times 10^{-34} \times 3 \times 10^8}$$

$$= 1998.49 \times 10^{17} = 1.998 \times 10^{20}$$

$$\Rightarrow \quad \approx 2 \times 10^{20} \Rightarrow x = 2$$

18. *(5)* According to photoelectric effect

$$\frac{hc}{\lambda} = h\nu_0 + \frac{1}{2}mv^2$$

$$\Rightarrow \quad \frac{6.63 \times 10^{-34} \times 3 \times 10^8}{500 \times 10^{-9}}$$

$$=6.63\times10^{-34}\times4.3\times10^{14}+\frac{1}{2}mv^2$$

$$\frac{6.63\times30\times10^{-19}}{50}=6.63\times4.3\times10^{-20}+\frac{1}{2}mv^2$$

$$11.271\times10^{-20}\text{ J}=\frac{1}{2}\times9\times10^{-31}\times v^2$$

$$v=5\times10^5\text{ m/s}$$

$$\therefore\quad v=5$$

19. *(222)* For photoelectrons, its kinetic energy will be

$$\text{KE}=E-E_0=\frac{hc}{\lambda}-E_0$$

$$=\frac{6.63\times10^{-34}\times3\times10^8}{300\times10^{-9}}-4.41\times10^{-19}\text{ J}$$

$$=222\times10^{-21}\text{ J}$$

Topic 2

01. *(b)* Given, Bohr radius $=a_0$

de-Broglie wavelength $=\lambda$

Radius of nth orbit; $r_n=n^2a_0=(2)^2a_0$
$=4a_0$ $[\because n=2]$

Using the formula,

$$2\pi r_n=n\lambda\Rightarrow 2\pi(4a_0)=n\lambda$$

$$\Rightarrow\quad \lambda=\frac{8\pi a_0}{n}$$

02. *(b)* For first excited state, $n=2$

So, substituting the value of n in equation $r=\frac{a_0n^2}{Z}$ we get

$$r=\frac{(2)^2a_0}{2}=2a_0\quad(\because Z=2)$$

So, $r=2a_0$

03. *(b)* The correct match is A-II, B-IV, C-III, D-I.

A. Lyman series lies in UV region.
B. Balmer series lies in visible region.
C. Paschen and Pfund series lies in infrared region.

Time Saver Tech

Use elimination technique as time saver. Pfund with infrared region (D-I) is given only in option (b). Hence, no need to cross check other option.

04. *(d)* Both Statement I and Statement II are correct.

According to Bohr's model, angular momentum is quantised for stationary orbits. In a given stationary state, it can be expressed as

$$mvr=n\frac{h}{2\pi},\ n=1,2,3,\ldots$$

It does not follow Heisenberg's uncertainty principle.

05. *(d)* Correct option is (d) i.e., $n=2$ to $n=1$.

For He^+ ion

$$\frac{1}{\lambda(\text{H})}=R(1)^2\left[\frac{1}{n_1^2}-\frac{1}{n_2^2}\right]$$

$$\frac{1}{\lambda(\text{He}^+)}=R(2)^2\left[\frac{1}{2^2}-\frac{1}{4^2}\right]$$

Given, Balmer series $\lambda(\text{H})=\lambda(\text{He}^+)$.

$$\therefore\quad R(1)^2\left[\frac{1}{n_1^2}-\frac{1}{n_2^2}\right]=R(4)\left[\frac{1}{2^2}-\frac{1}{4^2}\right]$$

$$\frac{1}{n_1^2}-\frac{1}{n_2^2}=\frac{1}{1^2}-\frac{1}{2^2}$$

So, on comparing we will get, $n_1=2$ and $n_2=1$.

06. *(d)* Both Statement I and Statement II are true. It is because definite line in H-spectrum are explained by Bohr and not by the Rutherford.

Bohr's model of an atom is contradicted by Heisenberg uncertainty principle.

07. *(d)* Balmer series lies in the visible region of the electromagnetic spectrum.

08. *(b)* According to Bohr's model, radius of orbit is given by

$$r_n=\frac{a_0\times n^2}{Z},\text{ where }a_0=\text{Bohr's radius}$$

(Radius of 1st orbit of H-atom)

For, Li^{2+}, $Z=3$ and $n=2$

$$\therefore\quad r=\frac{2^2\times a_0}{3}=\frac{4a_0}{3}$$

09. *(a)* According to Rydberg's equation,

$$\frac{1}{\lambda}\propto\left(\frac{1}{n_1^2}-\frac{1}{n_2^2}\right)$$

For shortest wavelength, i.e. highest energy spectral line, n_2 will be (∞).

For the given spectral series, ratio of the shortest wavelength of two spectral series can be calculated as follows

$$\frac{\lambda_{\text{Paschen}}}{\lambda_{\text{Lyman}}}=\frac{\frac{1}{1^2}-\frac{1}{\infty^2}}{\frac{1}{3^2}-\frac{1}{\infty^2}}=\frac{\frac{1}{1}-0}{\frac{1}{9}-0}=9$$

Hence, the ratio of wavelength of Paschen and Lyman series of hydrogen spectrum is found to be about 9.

10. *(b)* Postulates given in option (b) is not in agreement with quantum mechanical model of an atom. In Bohr's model the electron in a H-atoms stationary state moves in a spherical orbit while electrons are not confined to specific orbits around the nucleus in a quantum mechanical model.

11. *(d)* Given, $\lambda=900$ nm

$$=9\times10^{-5}\text{ cm}$$

$$R_H=10^5\text{ cm}^{-1}$$

$$h=6.6\times10^{-34}\text{ Js}$$

$$c=3\times10^8\text{ m/s}$$

Using equation,

$$\frac{1}{\lambda}=R_HZ^2\times\left[\frac{1}{n_1^2}-\frac{1}{n_2^2}\right]$$

On substituting the given value of we get

$$\frac{1}{9\times10^{-5}}=10^5(1)^2\left[\frac{1}{n_1^2}-\frac{1}{n_2^2}\right]$$

$(\because Z=1$ for hydrogen atom)

$$\Rightarrow\frac{1}{n_1^2}-\frac{1}{n_2^2}=\frac{1}{9}$$

So, the given condition is possible only when $n_1=3$, $n_2=\infty$.

12. *(b)* Energy of hydrogen atom can be given as :

$$E_n=-2.18\times10^{-18}\left[\frac{Z^2}{n^2}\right]\text{J/atom}$$

$$E_1=-2.18\times10^{-18}\text{ J/atom}\quad\ldots\text{(i)}$$

For 3rd orbit : $(n=3)(Z=1)$

$$E_3=-2.18\times10^{-18}\left[\frac{(1)^2}{(3)^2}\right]$$

$$E_3=-2.18\times10^{-18}\left[\frac{1}{9}\right]\quad\ldots\text{(ii)}$$

Hence, the energy in the 3rd Bohr orbit is (1/9th) of the energy of electron in the first Bohr orbit.

13. *(b)* Radius of nth Bohr's orbit of hydrogen atom is given as

$$r=0.529\times\frac{n^2}{Z}\text{Å}\quad(Z=1\text{ for hydrogen})$$

So, for 3rd Bohr's orbit

$$r_3=0.529\times\frac{(3)^2}{1}$$

$$r_3=0.529\times9\quad\ldots\text{(i)}$$

For 4th Bohr's orbit $(n=4)$

$$r_4 = 0.529 \times \frac{(4)^2}{1}$$

$$r_4 = 0.529 \times 16 \quad \ldots\text{(ii)}$$

From Eqs. (i) and (ii),

$$\frac{r_4}{r_3} = \frac{0.529 \times 16}{0.529 \times 9}$$

$$\frac{r_4}{r_3} = \frac{16}{9}$$

$$\Rightarrow \quad r_4 = \frac{16}{9} r_3$$

14. *(c)* According to Bohr's atom, velocity of electron is given by,

$$v \propto \frac{Z}{n} \quad \ldots\text{(i)}$$

From Eq. (i), velocity of electron is directly proportional to atomic number of atom, corresponding to positive charge. So, as Z increases, velocity also increases. Hence, statement I is false, while statement II is true.

15. *(b)* Statement I is false, because Bohr's theory accounts for the stability and spectrum of single electronic species (e.g. He^+, Li^{2+} etc.) but Li^+ has two electrons.

Bohr's theory fails to explain splitting of spectral lines in presence of magnetic field i.e. Zeeman effect.

∴ Statement II is true.

16. *(c)* Bohr's radius is given by,

$$r = 0.529 \frac{n^2}{Z} \text{ Å}$$

For Li^{2+} $(r_{Li^{2+}})_{n=4} - (r_{Li^{2+}})_{n=3}$

$$\Rightarrow \quad \frac{0.529}{3}(16-9) = \Delta R_1 \quad \ldots\text{(i)}$$

For He^+ $(r_{He^+})_{n=4} - (r_{He^+})_{n=3}$

$$\Rightarrow \quad \frac{0.529}{2}(16-9) = \Delta R_2 \quad \ldots\text{(ii)}$$

$$\frac{\Delta R_1}{\Delta R_2} = \frac{\frac{0.529}{3}(16-9)}{\frac{0.529}{2}(16-9)} = \frac{2}{3}$$

Hence, the ratio of $\Delta R_1 : \Delta R_2$ is 2 : 3.

17. *(d)* Statements I, II and III are correct. While the statement IV is incorrect.

It's corrected form is as follows:

Ionisation energy of hydrogen can be calculated from Lyman series lines $(n = 1)$ only by using the following formula,

$$\bar{\nu} = R_H \left[\frac{1}{(1)^2} - \frac{1}{\infty^2}\right]$$

18. *(c)* According to Rydberg's formula,

$$\text{wave number } (\bar{\nu}) = R_H Z^2 \left[\frac{1}{n_f^2} - \frac{1}{n_i^2}\right]$$

Given, $n_i = 8$, $n_f = n$

[∵ it is the case of emission]

$$\bar{\nu} = R_H \times (1)^2 \left[\frac{1}{n^2} - \frac{1}{8^2}\right]$$

$$\bar{\nu} = R_H \left[\frac{1}{n^2} - \frac{1}{64}\right] = \frac{R_H}{n^2} - \frac{R_H}{64}$$

On comparing with equation of straight line, $y = mx + c$, we get

Slope $= R_H$, intercept $= \frac{-R_H}{64}$.

Thus, plot of wave number $(\bar{\nu})$ against $\frac{1}{n^2}$ will be linear with slope $(+R_H)$.

19. *(c)* The ground state energy of H-atom is -13.6 eV.

For second excited state, $n = 2 + 1 = 3$

$$\therefore E_3(He^+) = -13.6 \times \frac{Z^2}{n^2} \text{ eV}$$

[∵ for He^+, $Z = 2$]

$$= -13.6 \times \frac{2^2}{3^2} \text{ eV} = -6.04 \text{ eV}$$

20. *(a)* Incorrect statement is given in option (a). Its correct form is energy of electron is maximum at infinite distance from nucleus.

21. *(d)* Among the given statement A, B, D and E are correct, while C is incorrect. It's correct form is

Red light does not produce photoelectric effect.

22. *(b)* According to Bohr's postulate

$$r \propto n^2$$

(a) $\frac{r_8}{r_4} = \left(\frac{8}{4}\right)^2 = 4$

(b) $\left(\frac{r_4}{r_2}\right) = \left(\frac{4}{2}\right)^2 = 4$

(c) $\frac{r_6}{r_4} = \left(\frac{6}{4}\right)^2 = \frac{9}{4}$

(d) $\frac{r_3}{r_1} = \frac{(3)^2}{1} = \frac{9}{1}$

So, statement given in option (b) is incorrect.

23. *(a)* The relation between energy (E) and atomic number (Z) is

$$E = -\frac{13.6 Z^2}{n^2} \text{ eV}$$

$$E \propto -Z^2$$

So, the correct graph is

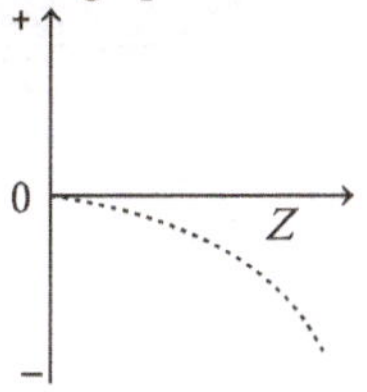

24. *(b)* As we know that, $\lambda = \frac{hc}{\Delta E}$

For, λ = minimum, i.e. shortest, ΔE = maximum

For Lyman series, $n = 1$ and for ΔE_{max} transition must be form $n = \infty$ to $n = 1$.

So, $\frac{1}{\lambda} = R_H Z^2 \left(\frac{1}{n_1^2} - \frac{1}{n_2^2}\right)$

$(\because n_1 = 1 \text{ and } n_2 = \infty)$

$$\therefore \quad \frac{1}{\lambda_1} = R_H\left(\frac{1}{1^2} - \frac{1}{\infty^2}\right) \times (1)^2$$

$$\Rightarrow \quad \lambda_1 = \frac{1}{R_H}$$

For longest wavelength,

ΔE = minimum for Balmer series, $n = 3$ to $n = 2$ will have minimum ΔE.

For He^+, $Z = 2$

So, $\frac{1}{\lambda_2} = R_H \times 2^2 \left(\frac{1}{n_1^2} - \frac{1}{n_2^2}\right)$

$$\frac{1}{\lambda_2} = R_H \times 4\left(\frac{1}{4} - \frac{1}{9}\right)$$

$$\frac{1}{\lambda_2} = R_H \times \frac{5}{9}$$

$$\Rightarrow \quad \lambda_2 = \lambda \times \frac{9}{5}$$

25. *(d)* According to Bohr's theory,

I. KE $\propto \frac{Z^2}{n^2}$ or $13.6 \times \frac{Z^2}{n^2}\frac{\text{(eV)}}{\text{(atom)}}$,

(∴ Correct)

II. Speed of electron $\propto \frac{Z}{n}$

(Here, Z = atomic number, n = number of shells)

$\therefore \quad v \times n \propto Z$ (∴ Incorrect)

III. Frequency of revolution of electron

$$= \frac{v}{2\pi r}$$

$$\text{Frequency} \propto \frac{Z^2}{n^3} \quad \left(\because v \propto \frac{Z}{n}, r \propto \frac{n^2}{Z}\right)$$

(∴ Incorrect)

IV. Coulombic force of attraction on the electron is given by

$$F = \frac{k q_1 q_2}{r^2} = \frac{kZe^2}{r^2}$$

$$F = \frac{Z}{\left(\frac{n^2}{Z}\right)^2}$$

$$F \propto \frac{Z^3}{n^4} \quad (\therefore \text{ Correct})$$

Hence, only I, and IV are correct.

26. *(c)* For any given series of spectral lines of atomic hydrogen.

Let $\Delta\bar{v} = \bar{v}_{max} - \bar{v}_{min}$ be the difference in maximum and minimum frequencies in cm^{-1}.

For Lyman series, $\Delta\bar{v} = \bar{v}_{max} - \bar{v}_{min}$

General formula:

$$\bar{v} = 109677\left[\frac{1}{n_i^2} - \frac{1}{n_f^2}\right]$$

For Lyman $n_1 = 1, n_2 = 2, 3, \ldots$

$$\bar{v}_{max} = 109,677\left(\frac{1}{1^2} - \frac{1}{\infty}\right)$$

$$= 109,677\left(\frac{1}{1} - 0\right) = 109,677$$

$$\bar{v}_{min} = 109,677\left(\frac{1}{(1)^2} - \frac{1}{(2)^2}\right)$$

$$\Delta\bar{v}_{Lyman} = \bar{v}_{max} - \bar{v}_{min}$$

$$= 109,677 - \left[\frac{109,677 \times 3}{4}\right]$$

$$= \frac{109,677}{4}$$

For Balmer series,

$$\bar{v}_{max} = 109,677\left(\frac{1}{(2)^2} - \frac{1}{\infty}\right) \Rightarrow \frac{109677}{4}$$

$$\bar{v}_{min} = 109,677\left(\frac{1}{(2)^2} - \frac{1}{(3)^2}\right)$$

$$= \frac{109677 \times 5}{36}$$

$$\Delta\bar{v} = \bar{v}_{max} - \bar{v}_{min}$$

$$\Delta\bar{v}_{Balmer} = \frac{109,677}{4} - \left[\frac{109,677}{36} \times 5\right]$$

$$= 109,677\left(\frac{1}{9}\right)$$

$$\frac{\Delta\bar{v}_{Lyman}}{\Delta\bar{v}_{Balmer}} = \frac{109,677/4}{109,677/9}$$

$$\Rightarrow \frac{\Delta\bar{v}_{Lyman}}{\Delta\bar{v}_{Balmer}} = \frac{9}{4}$$

$\therefore$ The ratio of $\frac{\Delta\bar{v}_{Lyman}}{\Delta\bar{v}_{Balmer}}$ is 9 : 4.

27. *(54)* Bohr's energy of *n*th orbit,

$$E_T = \frac{-13.6}{n^2} Z^2 \text{ eV}$$

For energy of H atom, energy of 1st Bohr orbit

$$E_1 = -13.6 \text{ eV} \qquad [Z = 1, n = 1]$$

For Be^{3+} ion, energy of 1st excited state $[Z = 4, n = 2]$

$$\frac{E_H}{E_{Be^{3+}}} = \frac{Z_1^2}{n_1^2} \times \frac{n_2^2}{Z_2^2}$$

$$\Rightarrow \frac{E_H}{E_{Be^{3+}}} = \frac{1}{1} \times \frac{4}{16}$$

$$E_{Be^{3+}} = -13.6 \times 4 = -54.4 \text{ eV}$$

$$\approx 54 \text{ eV}$$

28. *(34)* First excited state is $= -3.4$ eV

$$E_2 = -3.4 \text{ eV}$$

$$KE = -E_2$$

$$KE = 3.4 \text{ eV}$$

$$KE = 34 \times 10^{-1} \text{ eV}$$

compare it with given equation, value of *x* will be 34.

29. *(22)* Velocity in Bohr's atomic model is given by

$$v = 2.18 \times 10^6 \times \frac{Z}{n}$$

$$= 2.18 \times 10^6 \times \frac{1}{1}$$

$$\approx 22 \times 10^5 \text{ m/s}$$

Hence, missing value is 22.

30. *(10)* 5th excited state $\Rightarrow n_2 = 6$

1st excited state $\Rightarrow n_1 = 2$

$$\Delta n = n_2 - n_1 = 6 - 2 = 4$$

Maximum number of spectral lines

$$= \frac{\Delta n(\Delta n + 1)}{2}$$

where $\Delta n = n_2 - n_1 = \frac{4(4+1)}{2} = 10$

Hence, total 10 spectral lines are obtained.

31. *(1)* The emission spectra of atoms in gaseous phase do not show a continuous spread of wavelength from red to voilet, rather they emit light at specific wavelength with dark spaces between them. Hence, only statement i.e. (b) is incorrect.

32. *(1)* Given,

Radiation of energy $= 1.47 \times 10^{-17}$ J

$$R_H = 2.18 \times 10^{-18} \text{ J}$$

$$\Delta E = R_H Z^2\left(\frac{1}{n_1^2} - \frac{1}{n_2^2}\right)$$

$$1.47 \times 10^{-17} = 2.18 \times 10^{-18} \times 9\left(\frac{1}{n^2} - \frac{1}{(n+1)^2}\right)$$

$$\Rightarrow \frac{1.47}{1.962} = \frac{3}{4} = \frac{1}{n^2} - \frac{1}{(n+1)^2}$$

$$\Rightarrow n = 1$$

33. *(492)* We know that,

$$\frac{1}{\lambda} = R_H Z^2\left[\frac{1}{n_1^2} - \frac{1}{n_2^2}\right]$$

For first line of Paschen series, $n_1 = 3$, $n_2 = 4$, on substituting the values, we will get

$$\text{So, } \frac{1}{(\lambda_1)_P} = R_H Z^2\left[\frac{1}{(3)^2} - \frac{1}{(4)^2}\right]$$

$$\frac{1}{(\lambda_1)_P} = R_H Z^2\left[\frac{1}{9} - \frac{1}{16}\right] \quad \ldots(i)$$

Similarly, for second line of Paschen series

$n_1 = 3$, $n_2 = 5$

$$\frac{1}{(\lambda_2)_P} = R_H Z^2\left[\frac{1}{(3)^2} - \frac{1}{(5)^2}\right]$$

$$\frac{1}{(\lambda_2)_P} = R_H Z^2\left[\frac{1}{9} - \frac{1}{25}\right] \quad \ldots(ii)$$

From Eqs. (i) and (ii), we get

$$\frac{(\lambda_2)_P}{(\lambda_1)_P} = \frac{\frac{7}{16 \times 9}}{\frac{16}{25 \times 9}} = \frac{25 \times 7}{16 \times 16}$$

$$\Rightarrow (\lambda_2)_P = \frac{25 \times 7}{16 \times 16} \times 720$$

$$(\lambda_2)_P = 492 \text{ nm}$$

34. *(270)*

> **idea** This question includes the concept of calculating the radius of hydrogen or hydrogen like species i.e., He^+. For this, we will use the formula $r_n = r_0\left(\frac{n^2}{Z}\right)$ Å

According to Bohr model, radius of hydrogen like species is,

$$\left[r_n = r_0\left(\frac{n^2}{Z}\right) \text{Å}\right]$$

where, n = number of orbit

Z = atomic number

$\Rightarrow$ For 3rd Bohr orbit ($n = 3$)

For He^+ ($Z = 2$)

$r_0 = 0.6$ Å (Given)

$$\Rightarrow \quad r_3 = 0.6\left[\frac{(3)^2}{2}\right] \text{Å} = 2.7 \text{ Å}$$

$$= 2.7 \times 10^{-10} \text{ m}$$

$$= 2.7 \times 10^{-10} \times 10^{12} \text{ pm}$$

$$= 2.7 \times 10^{2} \text{ pm} = 270 \text{ pm}$$

$[\because 1 \text{ Å} = 10^{-10}\text{m}, 1 \text{ pm} = 10^{-12} \text{ m}]$

35. *(10)* The maximum number of emission lines when the excited electron of the H-atom from $n = 5$ drops to the ground state is given by

$$\Rightarrow \frac{n(n-1)}{2} = \frac{5(5-1)}{2} = \frac{5 \times 4}{2} = \frac{20}{2}$$

$= 10$ lines.

36. *(3155)* Bohr's radius of nth orbit is given by $r_n = a_0 n^2$.

$$mvr = \frac{nh}{2\pi}$$

For second Bohr orbit $n = 2$, $r_2 = (2)^2$

$r_0 = 4a_0$

$$\therefore \text{Kinetic energy} = \frac{n^2h^2}{8\pi^2 mr^2}$$

$$= \frac{4h^2}{8\pi^2 m(4a_0)^2} = \left(\frac{4}{8\pi^2 \times 16}\right)\frac{h^2}{ma_0^2}$$

$$= \frac{h^2}{xma_0^2} \quad \text{(Given)}$$

$$x = \frac{16 \times 8\pi^2}{4}$$

$$\Rightarrow \quad x = 315.507$$

$$10x = 3155.07 \approx 3155$$

Topic 3

01. *(c)* Both Statement I and Statement II are true.

Concept Enhancer

Heisenberg's uncertainty principle is applicable only when the location and momentum are along the same axis. i.e if Δx is uncertainty in position along the x-axis then Δp must also be uncertainty in momentum along the x-axis.

02. *(a)* According to de-Broglie hypothesis,

$$\lambda \propto \frac{1}{p} \text{ (momentum)}$$

$$\lambda = \frac{h}{p}$$

$\lambda p = h$ (constant)

So, the plot between λ and p is rectangular hyperbola.

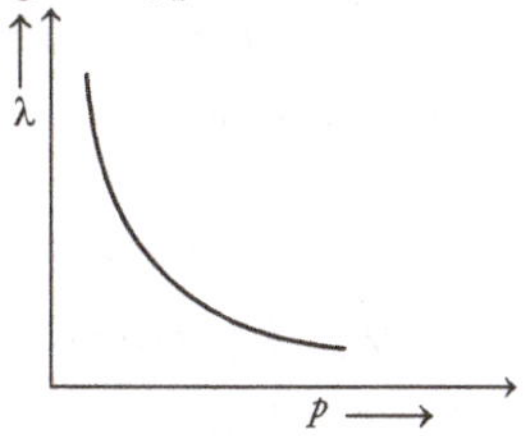

Hence, option (a) is correct.

03. *(d)* de-Broglie wavelength (λ) for electron is given by

$$\lambda = \frac{h}{\sqrt{2\,m\text{K.E}}} \quad \text{...(i)}$$

Also, according to photoelectric effect

$$\text{KE} = h\nu - h\nu_0$$

On substituting the value of KE in Eq (i), we get

$$\lambda = \frac{h}{\sqrt{2m \times (h\nu - h\nu_0)}}$$

$$\Rightarrow \quad \lambda \propto \frac{1}{(\nu - \nu_0)^{1/2}}$$

04. *(c)* Using the formula of de-Broglie wavelength, $\lambda = \frac{h}{mv}$

also, angular momentum of an electron by Bohr is : $mvr = \frac{nh}{2\pi}$

$$\text{or} \quad \frac{2\pi r}{n} = \frac{h}{mv}$$

$$\therefore \text{ de-Broglie wavelength } \lambda = \frac{2\pi r}{n}$$

de-Broglie wavelength of electron in 3rd orbit, $\lambda_3 = \frac{2\pi r_3}{n}$ ...(i)

Also radius of 3rd orbit,

$$r_3 = \text{radius of 1st orbit} \times n^2$$

$$r_3 = a_0 \times n^2$$

$$\frac{r_3}{n} = a_0 \times n \quad \text{...(ii)}$$

From Eqs. (i) and (ii) $\lambda_3 = 2\pi a_0 \times n$

for 3rd orbit, $n = 3$

$$\therefore \quad \lambda_3 = 6\pi a_0$$

05. *(c)* The de-Broglie wavelength of an electron,

$$\lambda = \frac{h}{p}, \text{ or } \frac{h}{mv}$$

But, from the Bohr's postulate of quantisation of angular momentum of electrons, $l = mvr = \frac{nh}{2\pi}$,

$$\therefore \quad mv = \frac{nh}{2\pi r}$$

Upon substitution,

$$\lambda = \frac{h}{\left(\frac{nh}{2\pi r}\right)} = \frac{2\pi r}{n},$$

where, r is the radius of orbits in Bohr model $r = \frac{n^2 a_0}{Z}$

$$\therefore \lambda = \frac{2\pi \times n^2 \times a_0}{n \times Z} = \frac{2\pi n a_0}{Z} = 8\pi a_0$$

$$\begin{Bmatrix} Z = 1 \text{ for hydrogen} \\ n = 4 \text{ (given)} \end{Bmatrix}$$

06. *(58)* Using Heisenberg's uncertainty principle

$$\Delta x \cdot m\Delta v = \frac{h}{4\pi}$$

$$\Delta v = \frac{6.626 \times 10^{-34}}{4 \times 3.14 \times (10^{-15}) \times 9.1 \times 10^{-31}}$$

$$\Delta v = 0.0579 \times 10^{12} \text{ m/s}$$

or $\Delta v = 58 \times 10^9$ m/s

07. *(661)* According to de-Broglie hypothesis

$$\lambda = \frac{h}{mv}$$

$$\Rightarrow \quad \lambda = \frac{hv}{mv^2}$$

$$\Rightarrow \quad \frac{mv^2}{h} = \frac{\nu}{\lambda} \quad \left(\frac{v}{\lambda} = \text{Frequency } (\nu)\right)$$

$$\nu = \frac{mv^2}{h} = \frac{2R_H}{h}$$

$$= \frac{2 \times 2.18 \times 10^{-18}}{6.6 \times 10^{-34}}$$

$$= 660.6 \times 10^{13} \text{ Hz}$$

Nearest integer = 661

08. *(8)* The de-Broglie's wavelength of an electron is given by

$$2\pi r_n = n\lambda_d$$

$$\text{or } 2\pi a_0 \frac{n^2}{Z} = n\lambda_d \quad \left(\because r_n = \frac{a_0 n^2}{Z}\right)$$

Here $n = 4$, $Z = 1$

Substitute the value of n and Z in above equation, we get

$$2\pi a_0 \frac{(4)^2}{1} = 4\lambda_d \Rightarrow \lambda_d = 8\pi a_0$$

09. *(7)* Given,

Kinetic energy $= 4.5 \times 10^{-29}$ J

Mass of electron $= 9 \times 10^{-31}$ kg

$h = 6.6 \times 10^{-34}$ Js

Wavelength of an electron = ?

We know that,

$$\text{wavelength } (\lambda) = \frac{h}{\sqrt{2 \text{KE } m_e}}$$

$$= \frac{6.6 \times 10^{-34} \text{ Js}}{\sqrt{2 \times 4.5 \times 10^{-29} \times 9 \times 10^{-31}}}$$

$$= 7.3 \times 10^{-5} \text{ m}$$

So, wavelength is $\lambda \approx 7 \times 10^{-5}$ m

10. *(2)* Statements A and C are correct while statements B and D are incorrect.

The characteristic of electrons emitted do not depend upon the material of the electrode of the cathode ray tube.

Nature of emitted electrons does not depend on nature of gas present in cathode ray tube.

11. *(548)* Uncertainty in position,

$$\Delta x = 2a_0 = 2 \times 52.9 \text{ pm}$$
$$= 2 \times 52.9 \times 10^{-12} \text{ m}$$

Using Heisenberg uncertainity equation,

$$\Delta x \cdot \Delta v = \frac{h}{4\pi m} = \frac{h}{4\pi m \Delta x}$$

$$= \frac{6.63 \times 10^{-34}}{4 \times 3.14 \times 9.1 \times 10^{-31} \times 2 \times 52.9 \times 10^{-12}}$$

$$= 548 \times 10^3 \text{ m/s}$$

$$= 548 \text{ km s}^{-1}$$

12. *(2)* The de-Broglie wavelength of electron is given by

$$\lambda = \frac{h}{\sqrt{2m\text{KE}}} \Rightarrow \text{KE} = \frac{h^2}{2m\lambda^2}$$

$$= \frac{(6.626 \times 10^{-34})^2}{2 \times 9.1 \times 10^{-31} \times (3.3 \times 10^{-10})^2}$$

$$= 2.215 \times 10^{-18}$$

$$E_{abs} = E_{req} + k$$

$$\frac{E_{abs}}{E_{req}} = 1 + \frac{k}{E_{req}}$$

$$= 1 + \frac{2.215 \times 10^{-18}}{13.6 \times 1.602 \times 10^{-19}}$$

$$= 2.01 \simeq 2$$

13. *(1758)* We know that wavelength,

$$\lambda = \frac{h}{mv}$$

$\therefore$ Wavelength of electron,

$$\lambda_e = \frac{h}{m_e \times v_e}$$

Wavelength of neutron,

$$\lambda_N = \frac{h}{m_N \times v_N}$$

$$\lambda_e = \lambda_N \text{ (given)}$$

when $\quad v_e = xv_N \quad$ ...(i)

$$\frac{h}{m_e \times v_e} = \frac{h}{m_N \times v_N} \quad \text{...(ii)}$$

By substituting Eq. (i) in Eq. (ii), we get

$$\frac{h}{m_e \times x \times v_N} = \frac{h}{m_N \times v_N} \Rightarrow x = \frac{m_N}{m_e}$$

m_N (mass of neutron) $= 1.6 \times 10^{-27}$ kg

m_e (mass of electron) $= 9.1 \times 10^{-31}$ kg

$$\Rightarrow \quad x = \frac{1.6 \times 10^{-27}}{9.1 \times 10^{-31}} = 1758$$

14. *(58)* Uncertainity in speed of electron

$$= \Delta v = \frac{0.02}{100} \times 5 \times 10^6 = 10^3 \text{ m/s}$$

Uncertainity in position $= x \times 10^{-9}$ m

According to Heisenberg uncertainty principle,

$$\Delta x \cdot \Delta p = \frac{h}{4\pi m}$$

$$x \times 10^{-9} \times 10^3 = \frac{6.63 \times 10^{-34}}{4 \times 3.14 \times 9.1 \times 10^{-31}}$$

$$x \times 10^{-6} = 0.058 \times 10^{-3}$$

$$\Rightarrow \quad x = \frac{0.058 \times 10^{-3}}{10^{-6}} = 58$$

15. *(6)* de-Broglie wavelength of electron

$$\lambda_e = \frac{h}{\sqrt{2m(\text{KE})}}$$

where λ_e = wavelength of electron

KE = kinetic energy of electron

also, kinetic energy $= qV$

q = charge of electron,

m = mass of electron,

V = potential.

$$\lambda_e = \frac{6.63 \times 10^{-34}}{\sqrt{2 \times 1.6 \times 10^{-19} \times 9.1 \times 10^{-31} \times 40 \times 10^3}}$$

$$\lambda_e = 6.14 \times 10^{-12} \text{ m}$$

$\therefore \quad x = 6$ (nearest integer)

16. *(9)* Given, $(\lambda) = 248$ nm $= 2480$ Å

Threshold energy = 3.0 eV

de-Broglie wavelength is given by

$$\lambda = \frac{h}{mv}$$

$$\Rightarrow \quad \lambda = \frac{h}{\sqrt{2m(\text{KE})}}$$

$$\text{or} \quad \lambda = \frac{h}{\sqrt{2m(qV)}}$$

Putting mass of electron,

$$m_e = 9.1 \times 10^{-31} \text{ kg}$$

Charge of electron, $q = 1.6 \times 10^{-19}$ C

$$\lambda = \sqrt{\frac{150}{V}} \text{ Å} \quad \text{...(i)}$$

We know,

$$E_{\text{applied}} = \phi + \text{KE}_{\text{max}} \quad \text{...(ii)}$$

$$E_{\text{applied}} = \frac{12400}{\lambda(\text{Å})} = \frac{12400}{2480} = 5 \text{ eV}$$

Putting value in Eq. (ii),

$$5 = 3 + \text{KE}_{\text{max}}$$

$$\Rightarrow \quad \text{KE}_{\text{max}} = 2 \text{ eV}$$

Voltage = 2

$$\lambda_e = \sqrt{\frac{150}{2}} = 8.6 \text{ Å}$$

Nearest integer = 9

17. *(1)* According to Heisenberg's uncertainty equation,

$$\Delta x \times m\Delta v \geq \frac{h}{4\pi}$$

$(\because m = 10 \text{ g} = 10 \times 10^{-3} \text{ kg}, v = 90 \text{ ms}^{-1})$

Uncertainty in position,

$$\Delta x = \frac{h}{4\pi} \times \frac{1}{m \times \Delta v \text{ (with uncertainty)}}$$

$$= \frac{6.63 \times 10^{-34}}{4 \times 3.14} \times \frac{1}{(10 \times 10^{-3}) \times \left(90 \times \frac{5}{100}\right)} \text{ m}$$

$$= 1.17 \times 10^{-33} \text{ m} \simeq 1 \times 10^{-33} \text{ m}$$

Topic 4

01. *(c)* $Z = 37$ (Rubidium)

$$\text{Rb} = [\text{Kr}]5s^1$$

$n = 5$ (Principal quantum number)

$l = 0$ (Azimuthal quantum number)

$m = 0$ (Magnetic quantum number)

$s = +\frac{1}{2}$, or $-\frac{1}{2}$ (spin-quantum number)

Hence, option (c) is correct.

02. *(a)* Number of radial nodes is given by

$$n - l - 1$$

For $3p$ orbital, the value of $n = 3$, $l = 1$

So, number of radial nodes

$$= 3 - 1 - 1 = 1$$

03. *(b)* We know that, at node

$$\psi_{2s} = 0$$

So, $\left(2 - \frac{r}{a_0}\right) = 0$

$\Rightarrow$ If, $r = r_0$

$\Rightarrow 2 - \frac{r_0}{a_0} = 0$

$$2 = \frac{r_0}{a_0}$$

or $\quad r_0 = 2a_0$

Thus, option (b) is correct.

04. *(d)* The orbital with greater value of $n + l$ value will have higher energy.

	$n=3$, $l=2$	> $n=4$, $l=0$	> $n=3$, $l=1$	> $n=3$, $l=0$
$n+l$	5	4	4	3
	D	> B	> C	> A

Concept Enhancer

According to $(n + l)$ rule, the orbital with more value of $(n + l)$ has more energy.

If $n + l$ value is same for two orbital, then the orbital which has greater value of n will have more energy.

05. *(a)* For $4d$ orbital the value of $n = 4$, $l = 2$

So, angular nodes $= l = 2$

Now, we know that, radial nodes

$$= (n - l - 1) = 4 - 2 - 1 = 1$$

Hence, radial node = 1, Angular node = 2

06. *(b)* The energy order of a subshell is determined by $(n + l)$ rule. The subshell with lower value of $(n + l)$ has lower energy. If the value of $n + l$ is same then lower value of n will have lower energy.

Now,

A : $n = 3$, $l = 2 \Rightarrow (n + l) = 5$

B : $n = 4$, $l = 1 \Rightarrow (n + l) = 5$

C : $n = 4$, $l = 2 \Rightarrow (n + l) = 6$

D : $n = 3$, $l = 1 \Rightarrow (n + l) = 4$

$\therefore$ The order of increasing energy is

$$D < A < B < C.$$

07. *(d)* Option (d) is incorrect. An atomic orbital is characterised by three quantum number n (principal quantum number), l (azimuthal quantum number) and m (magnetic quantum number).

08. *(c)* Given, principal quantum number $= 3$

Now, for azimuthal quantum number (l) has possible values

$$l = 0 \text{ to } n - 1$$
$$= 0, 1, 2$$

l cannot be equal to the value of n.

So, option (c) $n = 3$; $l = 3$; $m_l = -3$;

$s = -\frac{1}{2}$ is not allowed.

09. *(a)* Energies of the orbitals in the same subshell decrease with increase in the atomic number (Z), e.g., energy of $2s$ orbital of hydrogen atom is greater than that of $2s$ orbital of lithium and that of lithium is greater than that of sodium and so on, i.e.,

So, $\quad E_{2s}(\text{H}) > E_{2s}(\text{Li}) > E_{2s}(\text{Na}) > E_{2s}(\text{K}).$

10. *(a)* The electrons are more likely to be found in the region a and c. At b, wave function becomes zero and is called radial nodal surface or simply node.

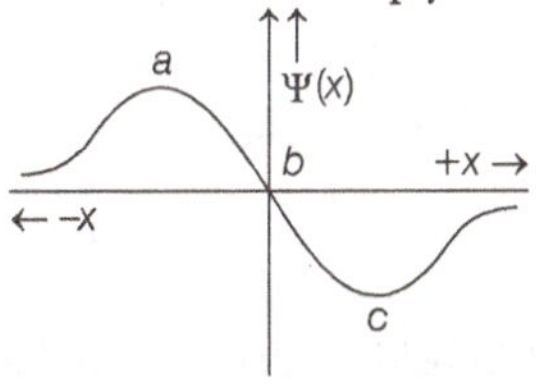

The graph between wavefunction (ψ) and distance (r) from the nucleus helps in determining the shape of orbital.

11. *(a)* For hydrogen atom, the orbital(s) with lowest energy are given in B, C and D.

In hydrogen atom the orbitals in a shell are degenerate, means energy depends only on n.

$$E_{3p_x} = E_{3d_{x^2-y^2}} = E_{3d_{z^2}}$$

12. *(d)* Orbital angular momentum is given by

$$= \sqrt{l(l+1)}\,\frac{h}{2\pi}$$

For 2s orbital; $l = 0$

$\therefore$Orbital angular momentum = 0

For $2p$ orbital ; $l = 1$

So, orbital angular momentum

$$= \sqrt{1(1+1)}\,\frac{h}{2\pi} = \sqrt{2}\,\frac{h}{2\pi}$$

13. *(a)* Electronic configuration of Cr is

$$1s^2\ 2s^2\ 2p^6\ 3s^2\ 3p^6\ 3d^5\ 4s^1$$

So, electrons with $l = 1$ are 12 (i.e., from $2p^6$ and $3p^6$)

and electrons with $l = 2$ are 5 (i.e., from $3d^5$)

So, electrons with $l = 1$ and 2 respectively are 12 and 5.

14. *(b)* Statement I is false but Statement II is true.

The correct form of statement I is :

No spectral line will be observed for a $2p_x \longrightarrow 2p_y$ transition because $2p_x$ and $2p_y$ orbitals are degenerate orbitals.

15. *(b)* Energy of atomic orbitals of hydrogen atom is given incorrect in C and D.

For single electron species energy only depends on 'n' (principal quantum number)

So, energy of $2s = 2p$ and energy of $3d < 4s$

16. *(c)* Both Statement I and Statement II are true.

17. *(c)* Statement I is true but Statement II is false.

The correct form of Statement II is :

In hydrogen atom, $3p$ and $3d$ orbitals are degenerate orbitals as in single electron species, energy depends only on principal quantum number (n).

18. *(b)* Energy level can be determined by comparing the value of $(n + l)$ value i.e. Bohr-Bury rule.

(A) $n = 4$, $l = 1 \Rightarrow (n + l) = 5$

(B) $n = 4$, $l = 2 \Rightarrow (n + l) = 6$

(C) $n = 3$, $l = 1 \Rightarrow (n + l) = 4$

(D) $n = 3$, $l = 2 \Rightarrow (n + l) = 5$

(E) $n = 4$, $l = 0 \Rightarrow (n + l) = 4$

For the same value of $n + l$, the orbital having higher values of n, will have more energy

So, the correct order is

$$C < E < D < A < B$$

19. *(c)* Electronic configuration of potassium is,

$$_{19}\text{K} = 1s^2 2s^2 2p^6 3s^2 3p^6 4s^1$$

Outermost orbital of K is $4s$ orbital.

For $4s$,

n = Principal quantum number orbital = 4

l = Azimuthal quantum number = 0

m_l = Magnetic quantum number = 0

s = Spin quantum number $= \pm\frac{1}{2}$

Hence, option (c) is correct.

20. *(b)* Electrons pair of the same shells and the same subshell have same energy and these electrons pairs are found in degenerated orbitals. So, only pairs of electron of (B) are present in degenerate orbitals as $n+l$ values in both are cases are same

(a) $n=3, l=2,$

$n+l=3+2=5$

(b) $n=3, l=2$

$n+l=3+2=5$

21. *(b)*

idea In this question we will use the concept of nodes.

The point at which the probability of finding an electron is zero is known as node or nodal surface. Number of nodes in any orbital = $n-l-1$

In case of 2*s*-orbital, number of nodes

$=2-0-1=1$

The correct plot for $\psi^2(r)$ as function of distance '*r*'.

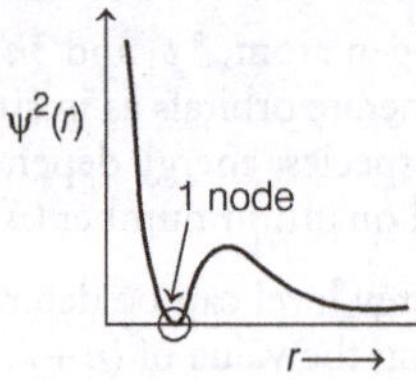

22. *(d)* Statement A, B and C are correct while statement D is incorrect. The correct explanation is given as

Total number of nodes present in atomic orbital is given by the formula $n-1$, where n is principal quantum number.

So, the formula "$n-2$" is incorrect.

23. *(c)* Statements A, C and D are correct while statements B and E are incorrect. The correct form of statements B and E are as follows

Azimuthal quantum number '*l*' can have values from 0 to $(n-1)$ for a given '*n*'. It describes the sub-shell and orbital angular momentum.

For any value of '*l*', total number of orbitals are given by $(2l+1)$ formula.

So, for $l=5$,

Total number of orbitals

$=2l+1=2\times5+1=11$

24. *(d)* Number of radial nodes

$=(n-l-1)$

[n = principal quantum number, l = azimuthal quantum number]

Number of angular nodes $=l$

Orbitals	$(n-l-1)$	l
(a) $3p(n=3, l=1)$ $\Rightarrow$	1	1
(b) $4f(n=4, l=3)$ $\Rightarrow$	0	3
(c) $4d(n=4, l=2)$ $\Rightarrow$	1	2
(d) $5d(n=5, l=2)$ $\Rightarrow$	2	2

So, 5*d*-orbital has two radial as well as two angular nodes (option-d).

25. *(b)* As, we know,

Radial nodes $=n-l-1$

For 3*p*-orbital, $n=3, l=1$

Angular node $(l)=1$

It has one angular and one radial node.

For 3*s*-orbital, $n=3, l=0$

No. of angular node $(l)=0$

No. of radial node $\Rightarrow n-l-1$

$=3-0-1=2=2$

Hence, for 3*s*-orbital has no angular node and 2 radial nodes.

Concept Enhancer

Radial nodes are spherical regions where the radial wavefunction is equal to zero whereas angular nodes are flat regions where angular wavefunction is zero.

26. *(c)*

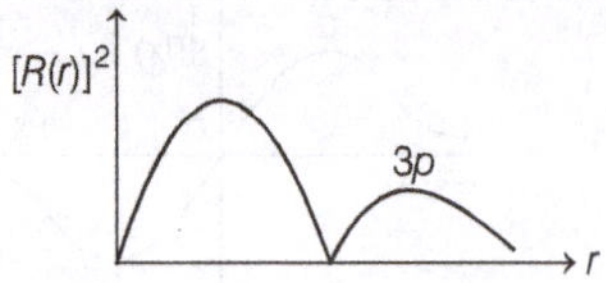

ψ^2 (Probability density) can be zero for 3*p*-orbital other than infinite distance.

Since, $n=3, l=1$ for 3*p*-orbital, the number of radial nodes $=3-1-1=1$

27. *(b)* When $n=4$

The permissible values of l and subshells are

$l=0 \rightarrow s$-subshell

$=1 \rightarrow p$-subshell

$=2 \rightarrow d$-subshell

$=3 \rightarrow f$-subshell

But, here values of l are restricted by the value of magnetic quantum number, $m=-2$, where, m depends on l as

$m=-l$ .... 0 .... $+l$

So, $m=-2$ value is possible for two subshells, viz, $l=3$ (f-subshell) and $l=2$ (d-subshell) only.

$l=3 \Rightarrow m=-3, -2-1, 0, +1, +2, +3$

$l=2 \Rightarrow m=-2, -1, 0, +1, +2$

28. *(a)* The energy of 2*s*-orbital is lowest in K(potassium). An orbital has smaller energy as the value of atomic number increases.

Correspondingly, the energy of the electron in such an orbital becomes less negative, meaning that the electron is less strongly bound and has less energy. The order energy of 2*s* orbital is H > Li > Na > K.

The graph of principal quantum number with atomic number is

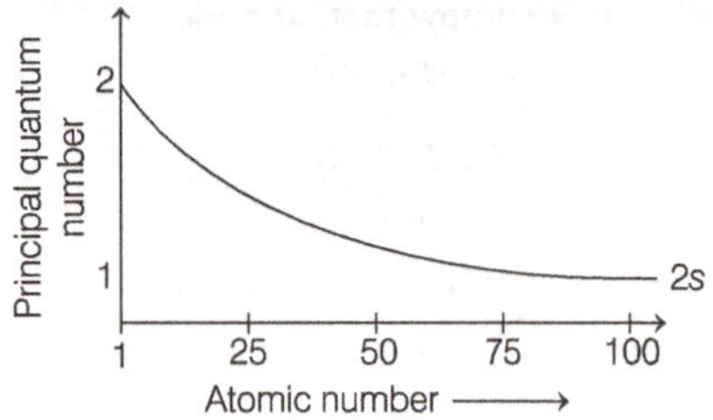

29. *(b)*

Set of quantum number	Orbitals
A : $n=1, l=0, m_l=0$	1*s*
B : $n=2, l=0, m_l=0$	2*s*
C : $n=2, l=1, m_l=1$	2*p*
D : $n=3, l=2, m_l=1$	3*d*
E : $n=3, l=2, m_l=0$	3*d*

In absence of electric and magnetic fields, all orbitals of 3*d* are degenerate.

30. *(a)* Among the given statements, A and C are correct but B and D are incorrect. Their correct forms are

- Unpaired electron can be in anyone of p_x, p_y or p_z orbital.
- Angular node for *s*-orbital = 0, while of each *p*-orbital is one.

∴ Sum of all angular nodes = 3

31. *(a)* Among the given statements, A, B, D, E are reason for extra stability of half-filled subshell.

The stability is due to,

- 'symmetrical distribution of electrons.
- large exchange of energy
- smaller coulombic repulsion
- smaller shielding of electrons by one another.

32. *(d)* Among the given statements, A and B are true, while C and D are incorrect. The correct form of C and D are as follows,

- The positive and negative sign in the more function of $2p_x$ orbital refer to the sign (phase) of the wave function, not the charge.
- The wave function of $2p_x$ orbital will be zero in *YZ* plane (Nodal plane).

33. *(d)* The correct plot for 3*s*-orbital is

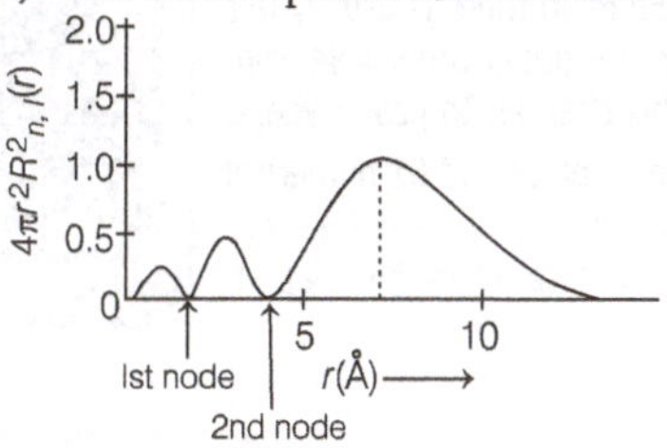

For 3*s*,

value of $l = 0$

value of $n = 3$

Number of radial nodes $= n - l - 1$

$= 3 - 2 = 2$

In graph *D*, two nodes are present, so this is the correct plot for 3*s*-orbital.

34. *(c)*

> **idea** This question includes the phenomenon associated with quantum mechanism. Quantum nature of atoms are associated with following phenomena or processes
> (i) Absorption spectrum
> (ii) Emission spectrum
> (iii) Black body radiation
> (iv) Photoelectric effect

Internal energy (U) of particles like atoms depends upon the thermodynamic variables (p, V, T) of the system. Thus, $U = f(p, T)$

or $U = f_1(T, V)$ or $U = f_2(p, V)$

So, quantum nature of atom is not associated with its internal energy.

Thus, graph (c) is not a direct manifestation of the quantum nature of atoms.

35. *(d)* Statements I and IV are correct while the statements II and III are incorrect.

Statement II

As size of an orbit $\propto$ Azimuthal quantum number (l) ($\because n =$ constant)

Statement III

As at ground state, $n = 1, l = 0$

$\Rightarrow$ Orbital angular momentum (wave mechanics)

$$= \sqrt{l(l+1)}\frac{h}{2\pi} = 0 \quad [\because l = 0]$$

36. *(4)* The possible combination of $n = 4$ and $m_l = 0$ is

$n = 4, l = 0, m_l = 0$

$n = 4, l = 1, m_l = 0$

$n = 4, l = 2, m_l = 0$

$n = 4, l = 3, m_l = 0$

So, total 4 orbitals can be identified with $n = 4$ and $m_l = 0$.

37. *(6)* For principal quantum number n, $l = 0$ to $n - 1$ and for each l, m_l will be $-l$ to $+l$.

$n = 4$

l	m_l
0	0
1	−1, 0, +1
2	−2, −1, 0, +1, +2
3	−3, −2, −1, 0, +1, +2, +3

So, number of orbital associated with $n = 4, |m_l| = 1$ are 6.

Hence, total number of electrons is 6.

38. *(5)* p_x, p_y, p_z, d_{z^2} and $d_{x^2-y^2}$ orbitals have electron density along the axes.

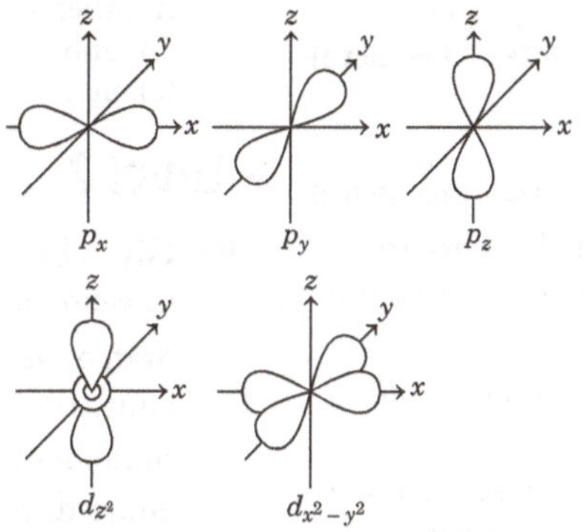

d_{xy} and d_{yz} have electron density between the axis.

39. *(0)* Orbital angular momentum (L)

$$L = \sqrt{l(l+1)}\,(h/2\pi)$$

For 3*s*-orbital

$[l = s = 0]$

So, $h = \sqrt{0(0+1)} \times h/2\pi$ or $h = 0$

Thus, the value of x is 0.

40. *(3)* Statements given in A, D and E are correct while, B and C are incorrect.

For 2*s* orbital, the probability density first decreases and then increases.

Boundary surface diagram of orbitals encloses a region of high probability (Not 100%) of finding an electron.

41. *(3)* Number of radial nodes in an orbital is given as $n - l - 1$

7*s*	:	$n = 7, l = 0$;	$7 - 0 - 1 = 6$
7*p*	:	$n = 7, l = 1$;	$7 - 1 - 1 = 5$
6*s*	:	$n = 6, l = 0$;	$6 - 0 - 1 = 5$
8*p*	:	$n = 8, l = 1$;	$8 - 1 - 1 = 6$
8*d*	:	$n = 8, l = 2$;	$8 - 2 - 1 = 5$

42. *(2)* (A) and (D) options are incorrect ones as these two set of quantum numbers are not possible since they both have principle quantum numbers equal to azimuthal quantum number which is not possible.

(B) and (C) are correct set of quantum number which represent electrons in 3*d* and 2*p*-orbitals respectively.

43. *(3)* Given, $n = 5, m_l = +2$

For $n = 5$, possible value of $l = 0, 1, 2, 3, 4$

For $l = 0, m_l = 0$

$l = 1, m_l = -1, 0, 1$

$l = 2, m_l = -2, -1, 0, 1, 2$

$l = 3, m_l = -3, -2, -1, 0, 1, 2, 3$

$l = 4$,

$m_l = -4, -3, -2, -1, 0, 1, 2, 3, 4$

Possible value of m_l for a given value of l

$= 0, \pm 1, \pm 2, \pm 3 \ldots. \pm l$

So, number of orbitals having $n = 5$ and $m_l = \pm 2$ are 3.

44. *(0)* The configuration of

$Zn^+ = 1s^2 2s^2 2p^6 3s^2 3p^6 3d^{10} 4s^1$

The outermost electron is in 4*s*-orbital. For 4*s*-orbital, principal quantum number is 4 and azimuthal quantum number (l) = 0

Magnetic quantum number, $m = -l$ to $+l$

For $l = 0, m = 0$

45. *(7)* Ge ($Z = 32$) electronic configuration is

$1s^2\ 2s^2\ 2p^6\ 3s^2 3p^6 3d^{10} 4s^2 4p^2$

Thus, it has 7 fully filled orbital with $m_l = 0$

46. *(0)* The electronic configuration of Ga^+ is $1s^2 2s^2 2p^6 3s^2 3p^6 3d^{10} 4s^2$

So, the valence electron is in *s*-orbital. Hence, Azimuthal quantum No. (l) for valence shell electron is zero.

47. *(0)* Given, $n = 4, m_l = -3$, so $l = 3$

(Possible value of $l = 0, 1, 2, 3$)

As $m = -3$ is possible only for $l = 3$

For $l = 3, m_l = -3, -2, -1, 0, 1, 2, 3$

Hence, this is 4*f*-orbital.

Number of radial nodes $= n - l - 1$

$= 4 - 3 - 1 = 0$

Hence, there is no radial node in 4*f* orbital.

03 Classification of Elements and Periodicity in Properties

In JEE Main there is usually 0-1 question per paper are visible from this chapter. In past 3 years, 38 conceptual and 01 numerical questions were asked of easy to moderate level from this chapter.

For detailed analysis scan QR code

Topic 1 / Modern Periodic Table

• Level 1

01 The element that does not belong to the same period of the remaining elements (modern periodic table) is **[2025, 23 Jan Shift-I]**

(a) platinum (b) iridium
(c) palladium (d) osmium

02 Given below are two statements: **[2024, 27 Jan Shift-I]**

Statement I The $4f$ and $5f$-series of elements are placed separately in the periodic table to preserve the principle of classification.

Statement II s-block elements can be found in pure form in nature.

In the light of the above statements, choose the **most appropriate** answer from the options given below

(a) Statement I is true but Statement II is false.
(b) Both Statement I and Statement II are false.
(c) Both Statement I and Statement II are true.
(d) Statement I is false but Statement II is true.

Imp. *Nomenclature of Element with Atomic Number > 100*

03 The IUPAC nomenclature of an element with electronic configuration $[Rn]5f^{14}6d^{1}7s^{2}$ is **[2022, 25 July Shift-I]**

(a) unnilbium (b) unnilunium
(c) unnilquadium (d) unniltrium

04 Which of the following elements is considered as a metalloid? **[2022, 26 June Shift-II]**

(a) Sc (b) Pb (c) Bi (d) Te

05 Which one of the following statements for D.I. Mendeleev, is incorrect? **[2021, 22 July Shift-II]**

(a) Element with atomic number 101 is named after him.
(b) He authored the textbook-Principles of Chemistry.
(c) He invented accurate barometer.
(d) At the time, he proposed periodic table of elements, structure of atom was known.

06 The atomic number of the element unnilennium is **[2020, 3 Sep Shift-I]**

(a) 109 (b) 102 (c) 108 (d) 119

07 The group number, number of valence electrons and valency of an element with atomic number 15, respectively, are **[2019, 12 April Shift-I]**

(a) 16, 5 and 2 (b) 15, 5 and 3
(c) 16, 6 and 3 (d) 15, 6 and 2

08 The IUPAC symbol for the element with atomic number 119 would be **[2019, 8 April Shift-II]**

(a) unh (b) uue
(c) uun (d) une

• Level 2

09 Given below are two statement abouts X-ray spectra of elements : **[2025, 23 Jan Shift-II]**

Statement I A plot of $\sqrt{v}$ (v = frequency of X-rays emitted) *vs* atomic mass is a straight line.

Statement II A plot of v(v = frequency of X-rays emitted) *vs* atomic number is a straight line.

In the light of the above statements, choose the correct answer from the options given below

(a) Statement I is true but Statement II is false.
(b) Both Statement I and Statement II are false.
(c) Statement I is false but Statement II is true.
(d) Both Statement I and Statement II are true.

10 Given below are two statements. **[2025, 28 Jan Shift-II]**

Statement I According to the law of octaves, the elements were arranged in the increasing order of their atomic number.

Statement II Meyer observed a periodically repeated pattern upon plotting physical properties of certain elements against their respective atomic numbers.

In the light of the above statements, choose the correct answer from the options given below

(a) Statement I is true but Statement II is false.
(b) Both Statement I and Statement II are true.
(c) Statement I is false but Statement II is true.
(d) Both Statement I and Statement II are false.

11 Number of elements from the following that cannot form compounds with valencies which match with their respective group valencies is.......

B, C, N, S, O, F, P, Al, Si **[2024, 4 April Shift-I]**

(a) 6 (b) 5 (c) 7 (d) 3

Imp. *Graph of X-ray Spectra*

12 Henry Moseley studied characteristic X-ray spectra of elements. The graph which represents his observation correctly is
Given v = frequency of X-ray emitted
Z = atomic number **[2023, 8 April Shift-II]**

(a) 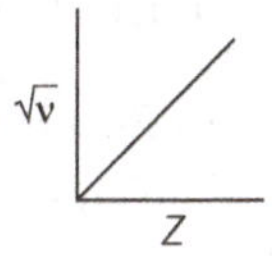

(b) 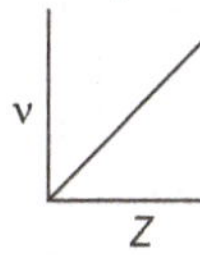

(c) 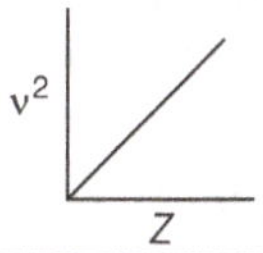

(d) 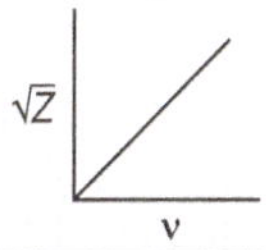

13 The characteristics of elements X, Y and Z with atomic numbers respectively 33, 53 and 83 are **[2021, 16 March Shift-II]**
(a) X is a metalloid, Y is a non-metal and Z is a metal
(b) X, Y and Z are metals
(c) X and Y are metalloids and Z is a metal
(d) X and Z are non-metals and Y is a metalloid

• Level 3

14 Match the List-I with List-II

	List-I (Family)		List-II (Symbol of Element)
A.	Pnictogen (group 15)	I.	Ts
B.	Chalcogen	II.	Og
C.	Halogen	III.	Lv
D.	Noble gas	IV.	Mc

Choose the correct answer from the options given below **[2025, 3 April Shift-II]**
(a) A-II, B-III, C-IV, D-I (b) A-IV, B-III, C-I, D-II
(c) A-III, B-I, C-IV, D-II (d) A-IV, B-I, C-II, D-III

15 Which of the following statements are not true about the periodic table? **[2025, 24 Jan Shift-I]**
A. The properties of elements are function of atomic weights.
B. The properties of elements are function of atomic numbers.
C. Elements having similar outer electronic configurations are arranged in same period.
D. An element's location reflects the quantum numbers of the last filled orbital.
E. The number of elements in a period is same as the number of atomic orbitals available in energy level that is being filled.

Choose the correct answer from the options given below
(a) B, C and Eonly (b) A, C and E only
(c) A and E only (d) D and E only

16 Element "E" belongs to the period 4 and group 16 of the periodic table. The valence shell electronic configuration of the element, which is just above "E" in the group is **[2022, 28 June Shift-I]**
(a) $3s^2, 3p^4$ (b) $3d^{10}, 4s^2, 4p^4$
(c) $4d^{10}, 5s^2, 5p^4$ (d) $2s^2, 2p^4$

17 Outermost electronic configuration of a group-13 element E is $4s^2 4p^1$. The electronic configuration of an element of p-block period-five placed diagonally to element, E is **[2021, 20 July Shift-II]**
(a) $[Kr]3d^{10}4s^2 4p^2$ (b) $[Ar]3d^{10}4s^2 4p^2$
(c) $[Xe]5d^{10}6s^2 6p^2$ (d) $[Kr]4d^{10}5s^2 5p^2$

Numerical Value Type Questions

18 If IUPAC name of an element is "ununnunnium" then the element belongs to nth group of periodic table. The value of n is .... **[2024, 30 Jan Shift-I]**

19 Total number of ions from the following with noble gas configuration is ....
$Sr^{2+}(Z=38)$, $Cs^+(Z=55)$, $La^{2+}(Z=57)$,
$Pb^{2+}(Z=82)$, $Yb^{2+}(Z=70)$ and $Fe^{2+}(Z=26)$
[2024, 27 Jan Shift-II]

20 The atomic number of unnilunium is .... **[2020, 6 Sep Shift-II]**

Topic 2 / Periodic Properties and Their Trends 1 (Atomic Radius, IE, EA, EN, Metallic Character)

• Level 1

01 The successive 5 ionisation energies of an element are 800, 2427, 3658, 25024 and 32824 kJ/mol, respectively. By using the above values predict the group in which the above element is present **[2025, 24 Jan Shift-II]**
(a) Group 2 (b) Group 14 (c) Group 4 (d) Group 13

02 Which of the following electronegativity order is incorrect ? **[2025, 22 Jan Shift-I]**
(a) Al < Mg < B < N (b) S < Cl < O < F
(c) Mg < Be < B < N (d) Al < Si < C < N

03 First ionisation enthalpy values of first four group 15 elements are given below. Choose the correct value for the element that is a main component of apatite family. **[2025, 29 Jan Shift-II]**
(a) 1402 kJ mol^{-1} (b) 834 kJ mol^{-1}
(c) 1012 kJ mol^{-1} (d) 947 kJ mol^{-1}

04 The incorrect decreasing order of atomic radii is **[2025, 28 Jan Shift-I]**
(a) Be > Mg > Al > Si (b) Si > P > Cl > F
(c) Mg > Al > C > O (d) Al > B > N > F

05 An element E has the ionisation enthalpy value of 374 kJ mol^{-1}, E reacts with elements A, B, C and D with electron gain enthalpy values of $-328, -349, -325$ and -295 kJ mol^{-1}, respectively. The correct order of the products EA, EB, EC and ED in terms of ionic character is **[2025, 29 Jan Shift-I]**
(a) $EB > EA > EC > ED$ (b) $EA > EB > EC > ED$
(c) $ED > EC > EB > EA$ (d) $ED > EC > EA > EB$

06 The element having the highest first ionisation enthalpy is **[2024, 29 Jan Shift-II]**
(a) Al (b) N (c) Si (d) C

Imp. *Order of Electronegativity*

07 The correct order of electronegativity for given elements is **[2023, 8 April Shift-I]**
(a) $P > Br > C > At$ (b) $C > P > At > Br$
(c) $Br > P > At > C$ (d) $Br > C > At > P$

08 The correct order of metallic character is **[2023, 10 April Shift-II]**
(a) $K > Be > Ca$ (b) $K > Ca > Be$
(c) $Be > Ca > K$ (d) $Ca > K > Be$

09 The first ionisation enthalpies of Be, B, N and O follow the order **[2022, 25 July Shift-II]**
(a) $O < N < B < Be$ (b) $Be < B < N < O$
(c) $B < Be < N < O$ (d) $B < Be < O < N$

10 The correct order of electron gain enthalpies of Cl, F, Te and Po is **[2022, 25 June Shift-II]**
(a) $F < Cl < Te < Po$ (b) $Po < Te < F < Cl$
(c) $Te < Po < Cl < F$ (d) $Cl < F < Te < Po$

11 Match List-I with List-II.

List-I (Electronic configuration of elements)	List-II ($\Delta_i H$ in kJ mol^{-1})
A. $1s^2 2s^2$	(i) 801
B. $1s^2 2s^2 2p^4$	(ii) 899
C. $1s^2 2s^2 2p^3$	(iii) 1314
D. $1s^2 2s^2 2p^1$	(iv) 1402

Choose the most appropriate answer from the options given below. **[2021, 26 Feb Shift-I]**

	A	B	C	D
(a)	(ii)	(iii)	(iv)	(i)
(b)	(i)	(iv)	(iii)	(ii)
(c)	(i)	(iii)	(iv)	(ii)
(d)	(iv)	(i)	(ii)	(iii)

12 The electron gain enthalpy (in kJ/mol) of fluorine, chlorine, bromine and iodine, respectively, are **[2020, 7 Jan Shift-I]**
(a) $-333, -325, -349$ and -296
(b) $-296, -325, -333$ and -349
(c) $-333, -349, -325$ and -296
(d) $-349, -333, -325$ and -296

13 The element having greatest difference between its first and second ionisation energy, is **[2019, 9 April Shift-I]**
(a) Ca (b) Sc (c) Ba (d) K

14 The size of the iso-electronic species Cl^-, Ar and Ca^{2+} is affected by **[2019, 8 April Shift-I]**
(a) Azimuthal quantum number of valence shell
(b) electron-electron interaction in the outer orbitals
(c) principal quantum number of valence shell
(d) nuclear charge

15 The correct order of atomic radius is **[2019, 12 Jan Shift-II]**
(a) $Ce > Eu > Ho > N$ (b) $Ho > N > Eu > Ce$
(c) $Eu > Ce > Ho > N$ (d) $N > Ce > Eu > Ho$

16 The correct option with respect to the Pauling electronegativity values of the elements is **[2019, 11 Jan Shift-II]**
(a) $P > S$ (b) $Si < Al$ (c) $Te > Se$ (d) $Ga < Se$

17 When the first electron gain enthalpy ($\Delta_{eg} H$) of oxygen is -141 kJ/mol, its second electron gain enthalpy is **[2019, 9 Jan Shift-II]**
(a) negative; but less negative than the first
(b) a positive value
(c) a more negative value than the first
(d) almost the same as that of the first

Level 2

18 Electronic configuration of four elements A, B, C and D are given below **[2025, 2 April Shift-II]**
(A) $1s^2 2s^2 2p^3$ (B) $1s^2 2s^2 2p^4$
(C) $1s^2 2s^2 2p^5$ (D) $1s^2 2s^2 2p^2$
Which of the following is the correct order of increasing electronegativity (Pauling's scale)?
(a) $A < C < B < D$ (b) $A < B < C < D$
(c) $A < D < B < C$ (d) $D < A < B < C$

19 The atomic number of the element from the following with lowest 1st ionisation enthalpy is: **[2025, 8 April Shift-II]**
(a) 35 (b) 87 (c) 32 (d) 19

20 Choose the incorrect trend in the atomic radii (r) of the elements. **[2025, 7 April Shift-II]**
(a) $r_{Br} < r_K$ (b) $r_{Rb} < r_{Cs}$ (c) $r_{At} < r_{Cs}$ (d) $r_{Mg} < r_{Al}$

21 Given below are two statements: **[2025, 2 April Shift-I]**
Statement I The metallic radius of Al is less than that of Ga.
Statement II The ionic radius of Al^{3+} is less than that of Ga^{3+}.
In the light of the above statements, choose the most appropriate answer from the options given below
(a) Statement I is incorrect but Statement II is correct.
(b) Statement I is correct but Statement II is incorrect.
(c) Both Statement I and Statement II are incorrect.
(d) Both Statement I and Statement II are correct.

22 The electron affinity value are negative for
A. $Be \rightarrow Be^-$ B. $N \rightarrow N^-$
C. $O \rightarrow O^{2-}$ D. $Na \rightarrow Na^-$
E. $Al \rightarrow Al^-$ **[2024, 6 April Shift-I]**

Choose the most appropriate answer from the options given below.
(a) A, B, D and E only (b) A and D only
(c) D and E only (d) A, B and C only

23 Given below are two statements: **[2024, 5 April Shift-II]**

Statement I The metallic radius of Na is 1.86 Å and the ionic radius of Na^+ is lesser than 1.86 Å.

Statement II Ions are always smaller in size than the corresponding elements.

In the light of the above statements, choose the correct answer from the options given below :
(a) Statement I is true but Statement II is false.
(b) Statement I is false but Statement II is true.
(c) Both Statement I and Statement II are true.
(d) Both Statement I and Statement II are false.

24 Given below are two statements. **[2024, 4 April Shift-II]**

Statement I The correct order of first ionisation enthalpy values of Li, Na, F and Cl is

$$Na < Li < Cl < F.$$

Statement II The correct order of negative electron gain enthalpy values of Li, Na, F and Cl is

$$Na < Li < F < Cl$$

In the light of the above statements, choose the correct answer from the options given below:
(a) Statement I is false but Statement II is true.
(b) Both Statement I and Statement II are true.
(c) Both Statement I and Statement II are false.
(d) Statement I is true but Statement II is false.

25 The correct order of first ionisation enthalpy values of the following elements is: **[2024, 4 April Shift-I]**

A. O B. N C. Be D. F
E. B

Choose the correct answer from the options given below
(a) A < B < D < C < E (b) B < D < C < E < A
(c) C < E < A < B < D (d) E < C < A < B < D

26 Given below are two statements. **[2024, 1 Feb Shift-II]**

Statement I Both metals and non-metals exist in *p* and *d*-block elements.

Statement II Non-metals have higher ionisation enthalpy and higher electronegativity than the metals.

In the light of the above statements, choose the **most appropriate** answer from the options given below
(a) Statement I is true but Statement II is false.
(b) Both Statement I and Statement II are false.
(c) Statement I is false but Statement II is true.
(d) Both Statement I and Statement II are true.

27 In case of isoelectronic species the size of F^-, Ne and Na^+ is affected by **[2024, 1 Feb Shift-I]**
(a) electron-electron interaction in the outer orbitals
(b) none of the factors because their size is the same
(c) nuclear charge (z)
(d) principal quantum number (n)

28 Consider the following elements. **[2024, 31 Jan Shift-II]**

Group ↓ $A'B'$ ⟶ Period
$C'D'$

Which of the following is are true about A', B', C' and D'?
A. Order of atomic radii : $B' < A' < D' < C'$
B. Order of metallic character : $B' < A' < D' < C'$
C. Size of the element : $D' < C' < B' < A'$
D. Order of ionic radii : $B'^+ < A'^+ < D'^+ < C'^+$

Choose the correct answer from the options given below
(a) B, C and D only (b) A only
(c) A, B and D only (d) A and B only

29 Given below are two statements : one is labelled as **Assertion (A)** and the other is labelled as **Reason (R)**.

Assertion (A) The first ionisation enthalpy decreases across a period.

Reason (R) The increasing nuclear charge outweighs the shielding across the period. **[2024, 29 Jan Shift-I]**

In the light of the above statements, choose the **most appropriate** from the options given below
(a) Both A and R are true and R is the correct explanation of A.
(b) Both A and R are true but R is not the correct explanation of A.
(c) A is false but R is true.
(d) A is true but R is false.

30 For elements B, C, N, Li, Be, O and F, the correct order of first ionisation enthalpy is **[2023, 11 April Shift-I]**
(a) B > Li > Be > C > N > O > F
(b) Li < B < Be < C < O < N < F
(c) Li < Be < B < C < O < N < F
(d) Li < Be < B < C < N < O < F

31 For electron gain enthalpies of the elements denoted as $\Delta_{eg} H$, the incorrect option is **[2023 1 Feb Shift-II]**
(a) $\Delta_{eg} H(Cl) < \Delta_{eg} H(F)$ (b) $\Delta_{eg} H(Te) < \Delta_{eg} H(Po)$
(c) $\Delta_{eg} H(Se) < \Delta_{eg} H(S)$ (d) $\Delta_{eg} H(I) < \Delta_{eg} H(At)$

32 The correct increasing order of the ionic radius is **[2023, 31 Jan Shift-I]**
(a) $Ca^{2+} < K^+ < Cl^- < S^{2-}$ (b) $S^{2-} < Cl^- < Ca^{2+} < K^+$
(c) $K^+ < S^{2-} < Ca^{2+} < Cl^-$ (d) $Cl^- < Ca^{2+} < K^+ < S^{2-}$

33 Inert gases have positive electron gain enthalpy. Its correct order is **[2023, 25 Jan Shift-I]**
(a) Xe < Kr < Ne < He (b) He < Ne < Kr < Xe
(c) He < Kr < Xe < Ne (d) He < Xe < Kr < Ne

34 The first ionisation enthalpy of Na, Mg and Si, respectively, are 496, 737 and 786 kJ mol^{-1}. The first ionisation enthalpy (kJ mol^{-1}) of Al is **[2022, 29 July Shift-I]**
(a) 487 (b) 768 (c) 577 (d) 856

35 In which of the following pairs, electron gain enthalpies of constituent elements are nearly the same or identical?

(A) Rb and Cs (B) Na and K
(C) Ar and Kr (D) I and At

Choose the correct answer from the options given below. **[2022, 28 July Shift-I]**

(a) (A) and (B) only (b) (B) and (C) only
(c) (A) and (C) only (d) (C) and (D) only

36 Given below are two statements.

Statement I In Cl_2 molecule the covalent radius is double of the atomic radius of chlorine.

Statement II Radius of anionic species is always greater than their parent atomic radius.

Choose the most appropriate answer from options given below **[2022, 26 July Shift-I]**

(a) Both Statement I and Statement II are correct.
(b) Both Statement I and Statement II are incorrect.
(c) Statement I is correct but Statement II is incorrect.
(d) Statement I is incorrect but Statement II is correct.

37 Outermost electronic configurations of four elements A, B, C, D are given below. **[2022, 27 July Shift-II]**

(A) $3s^2$ (B) $3s^2 3p^1$
(C) $3s^2 3p^3$ (D) $3s^2 3p^4$

The correct order of first ionisation enthalpy for them is

(a) (A) < (B) < (C) < (D)
(b) (B) < (A) < (D) < (C)
(c) (B) < (D) < (A) < (C)
(d) (B) < (A) < (C) < (D)

38 Given below are two statements one is labelled as Assertion (A) and the other is labelled as Reason (R).

Assertion (A) Metallic character decreases and non-metallic character increases on moving from left to right in a period.

Reason (R) It is due to increase in ionisation enthalpy and decrease in electron gain enthalpy, when one moves from left to right in a period.

In the light of the above statements, choose the most appropriate answer from the options given below.

(a) (A) is false but (R) is true. **[2021, 31 Aug Shift-I]**
(b) (A) is true but (R) is false
(c) Both (A) and (R) are correct and (R) is the correct explanation of (A).
(d) Both (A) and (R) are correct but (R) is not the correct explanation of (A).

39 The correct order of ionic radii for the ions, P^{3-}, S^{2-}, Ca^{2+}, K^+, Cl^- is **[2021, 27 Aug Shift-II]**

(a) $P^{3-} > S^{2-} > Cl^- > K^+ > Ca^{2+}$
(b) $Cl^- > S^{2-} > P^{3-} > Ca^{2+} > K^+$
(c) $P^{3-} > S^{2-} > Cl^- > Ca^{2+} > K^+$
(d) $K^+ > Ca^{2+} > P^{3-} > S^{2-} > Cl^-$

40 The ionic radii of F^- and O^{2-} respectively are 1.33 Å and 1.4 Å while the covalent radius of N is 0.74Å. The correct statement for the ionic radius of N^{3-} from the following is **[2021, 25 July Shift-II]**

(a) it is smaller than F^- and N
(b) it is bigger than O^{2-} and F^-
(c) it is bigger than F^- and N, but smaller than of O^{2-}
(d) it is smaller than O^{2-} and F^-, but bigger than of N

41 The ionic radii of K^+, Na^+, Al^{3+} and Mg^{2+} are in the order **[2021, 25 July Shift-I]**

(a) $Na^+ < K^+ < Mg^{2+} < Al^{3+}$ (b) $Al^{3+} < Mg^{2+} < Na^+ < K^+$
(c) $Al^{3+} < Mg^{2+} < K^+ < Na^+$ (d) $K^+ < Al^{3+} < Mg^{2+} < Na^+$

42 The first ionisation energy of magnesium is smaller as compared to that of elements X and Y, but higher than that of Z. The elements X, Y and Z, respectively, are **[2021, 18 March Shift-II]**

(a) chlorine, lithium and sodium
(b) argon, lithium and sodium
(c) argon, chlorine and sodium
(d) neon, sodium and chlorine

43 The ionic radius of Na^+ ion is 1.02 Å. The ionic radius (in Å) of Mg^{2+} and Al^{3+}, respectively are **[2021, 18 March Shift-I]**

(a) 1.05 and 0.99 (b) 0.68 and 0.72
(c) 0.85 and 0.99 (d) 0.72 and 0.54

44 Consider the elements Mg, Al, S, P and Si, the correct increasing order of their first ionisation enthalpy is **[2021, 24 Feb Shift-I]**

(a) Al < Mg < S < Si < P (b) Mg < Al < Si < S < P
(c) Al < Mg < Si < S < P (d) Mg < Al < Si < P < S

45 The ionic radii of O^{2-}, F^-, Na^+ and Mg^{2+} are in the order **[2020, 4 Sep Shift-I]**

(a) $O^{2-} > F^- > Mg^{2+} > Na^+$
(b) $Mg^{2+} > Na^+ > F^- > O^{2-}$
(c) $O^{2-} > F^- > Na^+ > Mg^{2+}$
(d) $F^- > O^{2-} > Na^+ > Mg^{2+}$

46 Among the statements (I-IV), **[2020, 3 Sept Shift-II]**

(I) Be has smaller atomic radius compared to Mg.
(II) Be has higher ionisation enthalpy than Al.
(III) Charge/radius ratio of Be is greater than that of Al.
(IV) Both Be and Al form mainly covalent compounds.

The correct ones are

(a) (I), (II) and (IV) (b) (I), (III) and (IV)
(c) (II), (III) and (IV) (d) (I), (II) and (III)

47 In comparison to boron, beryllium has **[2019, 12 April Shift-II]**

(a) lesser nuclear charge and lesser first ionisation enthalpy
(b) greater nuclear charge and lesser first ionisation enthalpy
(c) greater nuclear charge and greater first ionisation enthalpy
(d) lesser nuclear charge and greater first ionisation enthalpy

48 The isoelectronic set of ions is **[2019, 10 April Shift-I]**
(a) F^-, Li^+, Na^+ and Mg^{2+}
(b) N^{3-}, Li^+, Mg^{2+} and O^{2-}
(c) Li^+, Na^+, O^{2-} and F^-
(d) N^{3-}, O^{2-}, F^- and Na^+

49 The correct order of the atomic radii of C, Cs, Al and S is **[2019, 11 Jan Shift-I]**
(a) C < S < Al < Cs (b) C < S < Cs < Al
(c) S < C < Cs < Al (d) S < C < Al < Cs

50 In general, the properties that decrease and increase down a group in the periodic table, respectively are
(a) electronegativity and atomic radius **[2019, 9 Jan Shift-I]**
(b) electronegativity and electron gain enthalpy
(c) electron gain enthalpy and electronegativity
(d) atomic radius and electronegativity

Level 3

51 The correct orders among the following are
Atomic radius : B < Al < Ga < In < Tl
Electronegativity : Al < Ga < In < Tl < B
Density : Tl < In < Ga < Al < B
1st ionisation energy : In < Al < Ga < Tl < B
Choose the correct answer from the options given below. **[2025, 3 April Shift-II]**
(a) C and D only (b) A and B only
(c) B and D only (d) A and C only

52 Given below are two statements. **[2025, 29 Jan Shift-I]**
Statement I The radii of isoelectronic species increases in the order.

$$Mg^{2+} < Na^+ < F^- < O^{2-}$$

Statement II The magnitude of electron gain enthalpy of halogen decreases in the order.

$$Cl > F > Br > I$$

In the light of the above statements, choose the most appropriate answer from the options given below.
(a) Statement I is incorrect but Statement II is correct.
(b) Both Statement I and Statement II are correct.
(c) Both Statement I and Statement II are incorrect.
(d) Statement I is correct but Statement II is incorrect.

53 Consider the following elements In, Tl, Al, Pb, Sn and Ge. The most stable oxidation states of elements with highest and lowest first ionisation enthalpies, respectively, are **[2025, 28 Jan Shift-I]**
(a) +4 and +3 (b) +4 and +1
(c) +1 and +4 (d) +2 and +3

Imp. *Types of Oxides & Atomic Radius*

54 The type of oxide formed by the element among Li, Na, Be, Mg, B and Al that has the least atomic radius is **[2025, 29 Jan Shift-II]**
(a) A_2O (b) A_2O_3
(c) AO_2 (d) AO

55 Match List-I with List-II. **[2025, 22 Jan Shift-I]**

List-I	List-II
(A) $Al^{3+} < Mg^{2+} < Na^+ < F^-$	(I) Ionisation enthalpy
(B) B < C < O < N	(II) Metallic character
(C) B < Al < Mg < K	(III) Electronegativity
(D) Si < P < S < Cl	(IV) Ionic radii

Choose the correct answer from the options given below.
(a) (A)-(IV), (B)-(I), (C)-(III), (D)-(II)
(b) (A)-(II), (B)-(III), (C)-(IV), (D)-(I)
(c) (A)-(III), (B)-(IV), (C)-(II), (D)-(I)
(d) (A)-(IV), (B)-(I), (C)-(II), (D)-(III)

56 Given below are two statements. **[2025, 22 Jan Shift-II]**
Statement I An element in the extreme left of the periodic table forms acidic oxides.
Statement II Acid is formed during the reaction between water and oxide of a reactive element present in the extreme right of the periodic table.
In the light of the above statements, choose the correct answer from the options given below:
(a) Both Statement I and Statement II are true.
(b) Both Statement I and Statement II are false.
(c) Statement I is false but Statement II is true.
(d) Statement I is true but Statement II is false.

57 Match List-I with List-II.

List-I (Elements)		List-II (Properties in their respective groups)	
A.	Cl, S	I.	Elements with highest electronegativity
B.	Ge, As	II.	Elements with largest atomic size
C.	Fr, Ra	III.	Elements which show properties of both metals and non-metal
D.	F, O	IV.	Elements with highest negative electron gain enthalpy

Choose the correct answer from the options given below : **[2024, 8 April Shift-I]**
(a) A-III, B-II, C-I, D-IV (b) A-II, B-III, C-IV, D-I
(c) A-IV, B-III, C-II, D-I (d) A-II, B-I, C-IV, D-III

58 The statement(s) that are correct about the species O^{2-}, F^-, Na^+ and Mg^{2+}. **[2024, 9 April Shift-I]**
(A) All are isoelectronic
(B) All have the same nuclear charge
(C) O^{2-} has the largest ionic radii
(D) Mg^{2+} has the smallest ionic radii
Choose the most appropriate answer from the options given below:
(a) (C) and (D) only (b) (A), (B), (C) and (D)
(c) (B), (C) and (D) only (d) (A), (C) and (D) only

59 The correct sequence of electron gain enthalpy of the elements listed below is **[2024, 31 Jan Shift-I]**
A. Ar B. Br C. F D. S

Choose the most appropriate from the options given below.
(a) A > D > C > B (b) D > C > B > A
(c) C > B > D > A (d) A > D > B > C

60 Identify the elements X and Y using the ionisation energy values given below.

	Ionisation energy (Ist)	kJ/mol (IInd)
X	495	4563
Y	731	1450

[2021, 16 March Shift-II]
(a) $X = \text{Na}, Y = \text{Mg}$ (b) $X = \text{Mg}, Y = \text{F}$
(c) $X = \text{Mg}, Y = \text{Na}$ (d) $X = \text{F}, Y = \text{Mg}$

61 Within each pair of elements F and Cl, S and Se, and Li and Na, respectively, the elements that release more energy upon an electron gain are **[2020, 7 Jan Shift-II]**
(a) F, Se and Na (b) F, S and Li
(c) Cl, S and Li (d) Cl, Se and Na

62 The increasing order of the atomic radii of the following elements is
(A) C (B) O (C) F
(D) Cl (E) Br
(a) (A) < (B) < (C) < (D) < (E) **[2020, 8 Jan Shift-II]**
(b) (C) < (B) < (A) < (D) < (E)
(c) (D) < (C) < (B) < (A) < (E)
(d) (B) < (C) < (D) < (A) < (E)

Numerical Value Type Question

63 The total number of isoelectronic species from the given set is ....
O^{2-}, F^-, Al, Mg^{2+}, Na^+, O^+, Mg, Al^{3+}, F
[2023, 15 April Shift-I]

Topic 3 / Periodic Properties and Their Trends 2 (Lattice and Hydration Energy, Oxidation State and Chemical Properties)

• Level 1

01 Given below are two statements **[2024, 9 April Shift-I]**
Statement I The oxidation state of an element in a particular compound is the charge acquired by its atom on the basic of electron gain enthalpy consideration from other atoms in the molecule.
Statement II $p\pi$-$p\pi$ bond formation is more prevalent in second period elements over other periods.
In the light of the above statements, choose the most appropriate answer from the options given below.
(a) Statement I is correct but Statement II is incorrect.
(b) Both Statement I and Statement II are correct.
(c) Both Statement I and Statement II are incorrect.
(d) Statement I is incorrect and Statement II is correct.

02 The total number of acidic oxides from the following List-Is NO, N_2O, B_2O_3, N_2O_5, CO, SO_3, P_4O_{10}
[2022, 25 July Shift-II]
(a) 3 (b) 4 (c) 5 (d) 6

Imp. *Nature of Oxides*

03 Match List-I with List-II.

List-I (Oxide)	List-II (Nature)
A. Cl_2O_7	(I) Amphoteric
B. Na_2O	(II) Basic
C. Al_2O_3	(III) Neutral
D. N_2O	(IV) Acidic

Choose the correct answer from the option given below **[2022, 28 June Shift-II]**
(a) (A)–(IV), (B)–(III), (C)–(I), (D)–(II)
(b) (A)–(IV), (B)–(II), (C)–(I), (D)–(III)
(c) (A)–(II), (B)–(IV), (C)–(III), (D)–(I)
(d) (A)–(I), (B)–(II), (C)–(III), (D)–(IV)

04 Among the following basic oxide is **[2022, 28 June Shift-II]**
(a) SO_3 (b) SiO_2 (c) CaO (d) Al_2O_3

05 Given below are the oxides :
Na_2O, As_2O_3, N_2O, NO and Cl_2O_7
Number of amphoteric oxides is **[2022, 24 June Shift-I]**
(a) 0 (b) 1 (c) 2 (d) 3

06 Match list-I with list-II.

	List-I		List-II
(A)	NaOH	(i)	Acidic
(B)	$Be(OH)_2$	(ii)	Basic
(C)	$Ca(OH)_2$	(iii)	Amphoteric
(D)	$B(OH)_3$		
(E)	$Al(OH)_3$		

Choose the most appropriate answer from the options given below. **[2021, 27 July Shift-I]**
(a) (A)-(ii), (B)-(ii), (C)-(iii), (D)-(i), (E)-(iii)
(b) (A)-(ii), (B)-(iii), (C)-(ii), (D)-(i), (E)-(iii)
(c) (A)-(ii), (B)-(ii), (C)-(iii), (D)-(ii), (E)-(iii)
(d) (A)-(ii), (B)-(i), (C)-(ii), (D)-(iii), (E)-(iii)

07 The set of elements that differ in mutual relationship from those of the other sets is **[2021, 17 March Shift-II]**
(a) Li-Mg (b) B-Si (c) Be-Al (d) Li-Na

08 Which pair of oxides is acidic in nature?
[2021, 26 Feb Shift-II]
(a) B_2O_3, CaO (b) B_2O_3, SiO_2
(c) N_2O, BaO (d) CaO, SiO_2

• Level 2

09 The nature of oxide (TeO_2) and hydride (TeH_2) formed by Te, respectively are **[2025, 2 April Shift-II]**
(a) oxidising and basic (b) reducing and basic
(c) reducing and acidic (d) oxidising and acidic

10 The correct order of hydration enthalpies is

(A) K^+ (B) Rb^+

(C) Mg^{2+} (D) Cs^+

(E) Ca^{2+}

Choose the correct answer from the options given below. **[2023, 29 Jan Shift-I]**

(a) C > E > A > B > D (b) C > A > E > B > D

(c) E > C > A > B > D (d) C > E > A > D > B

11 Metals generally melt at very high temperature. Amongst the following, the metal with the highest melting point will be **[2022, 24 June Shift-II]**

(a) Hg (b) Ag

(c) Ga (d) Cs

12 The set in which compounds have different nature is **[2021, 20 July Shift-I]**

(a) $Be(OH)_2$ and $Al(OH)_3$ (b) $B(OH)_3$ and $Al(OH)_3$

(c) $B(OH)_3$ and H_3PO_3 (d) NaOH and $Ca(OH)_2$

13 Three elements X, Y and Z are in the 3rd period of the periodic table. The oxides of X, Y and Z, respectively, are basic, amphoteric and acidic. The correct order of the atomic number of X, Y and Z is **[2020, 2 Sep Shift-II]**

(a) $Z < Y < X$ (b) $X < Y < Z$

(c) $X < Z < Y$ (d) $Y < X < Z$

Level 3

14 Which of the following statements are correct?

A. The process of adding an electron to a neutral gaseous atom is always exothermic.

B. The process of removing an electron from an isolated gaseous atom is always endothermic.

C. The 1st ionisation energy of boron is less than that of beryllium.

D. The electronegativity of C is 2.5 in CH_4 and CCl_4.

E. Li is the most electropositive among elements of group I.

Choose the correct answer from the options given below: **[2025, 3 April Shift-I]**

(a) A, C and D Only (b) B and D Only

(c) B and C Only (d) B, C and E Only

Key Index

Topic 1

1	(c)	2	(a)	3	(d)	4	(d)	5	(d)	6	(a)	7	(b)	8	(b)	9	(b)	10	(d)
11	(d)	12	(a)	13	(a)	14	(b)	15	(b)	16	(a)	17	(d)	18	(11)	19	(2)	20	(101)

Topic 2

1	(d)	2	(a)	3	(c)	4	(a)	5	(a)	6	(b)	7	(d)	8	(b)	9	(d)	10	(b)
11	(a)	12	(c)	13	(d)	14	(d)	15	(c)	16	(d)	17	(b)	18	(d)	19	(b)	20	(d)
21	(a)	22	(d)	23	(a)	24	(b)	25	(d)	26	(c)	27	(c)	28	(c)	29	(c)	30	(b)
31	(c)	32	(a)	33	(d)	34	(c)	35	(c)	36	(d)	37	(b)	38	(b)	39	(a)	40	(b)
41	(b)	42	(c)	43	(d)	44	(c)	45	(c)	46	(a)	47	(d)	48	(d)	49	(a)	50	(a)
51	(c)	52	(b)	53	(a)	54	(b)	55	(d)	56	(c)	57	(c)	58	(d)	59	(d)	60	(a)
61	(c)	62	(b)	63	(5)														

Topic 3

1	(d)	2	(b)	3	(b)	4	(c)	5	(b)	6	(b)	7	(d)	8	(b)	9	(d)	10	(a)
11	(b)	12	(b)	13	(b)	14	(c)												

For Detailed Solutions scan QR Code

04 Chemical Bonding and Molecular Structure

In JEE Main there is usually 0-1 question per paper are visible from this chapter. In past 3 years, 53 conceptual and 34 numerical questions were asked of moderate difficulty level from this chapter.

For detailed analysis scan QR code

Topic 1 / Ionic and Covalent Bonding Fajan's Rule, Bond Parameters

• Level 1

01 Arrange the following compounds in increasing order of their dipole moment

HBr, H_2S, NF_3 and $CHCl_3$ **[2025, 22 Jan Shift-II]**

(a) $NF_3 < HBr < H_2S < CHCl_3$
(b) $CHCl_3 < NF_3 < HBr < H_2S$
(c) $H_2S < HBr < NF_3 < CHCl_3$
(d) $HBr < H_2S < NF_3 < CHCl_3$

02 Which of the following is least ionic? **[2024, 31 Jan Shift-II]**

(a) KCl (b) $BaCl_2$
(c) AgCl (d) $CoCl_2$

03 Choose the polar molecule from the following **[2024, 27 Jan Shift-I]**

(a) $CH_2{=}CH_2$ (b) CCl_4
(c) CO_2 (d) $CHCl_3$

04 Resonance in carbonate ion (CO_3^{2-}) is

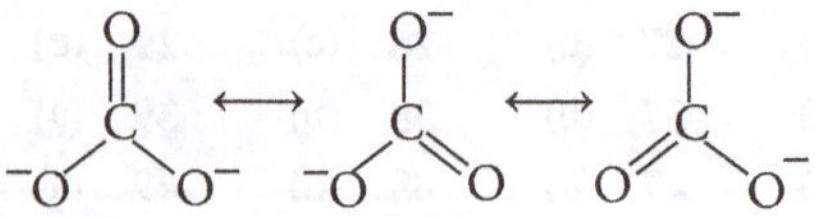

Which of the following is true? **[2023, 1 Feb Shift-I]**

(a) Each structure exists for equal amount of time.
(b) It is possible to identify each structure individually by some physical or chemical method.
(c) All these structures are in dynamic equilibrium with each other.
(d) CO_3^{2-} has a single structure i.e., resonance hybrid of the above three structures.

05 Given below are two statements.

Statement I Dipole moment is a vector quantity and by convention it is depicted by a small arrow with tail on the negative centre and head pointing towards the positive centre.

Statement II The crossed arrow of the dipole moment symbolises the direction of the shift of charges in the molecules.

In the light of the above statements, choose the most appropriate answer from the options given below. **[2023, 25 Jan Shift-II]**

(a) Both Statement I and Statement II are incorrect.
(b) Statement I is correct but Statement II is incorrect.
(c) Statement I is incorrect but Statement II is correct.
(d) Both Statement I and Statement II are correct.

06 Which of the following pair of molecules contain odd electron molecule and an expanded octet molecule? **[2022, 29 July Shift-I]**

(a) BCl_3 and SF_6 (b) NO and H_2SO_4
(c) SF_6 and H_2SO_4 (d) BCl_3 and NO

07 The potential energy curve for the H_2 molecule as a function of internuclear distance is **[2020, 5 Sep Shift-I]**

(a) 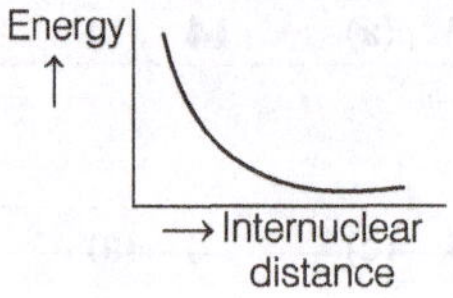

(b) 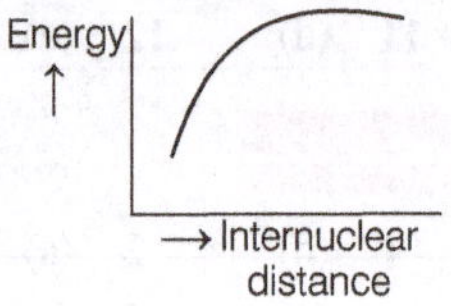

(c) 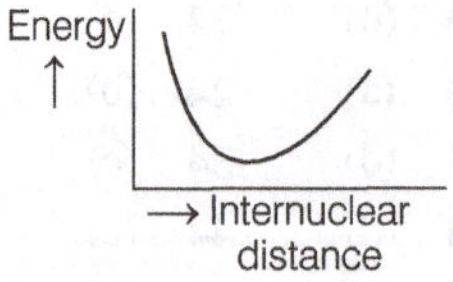

(d) 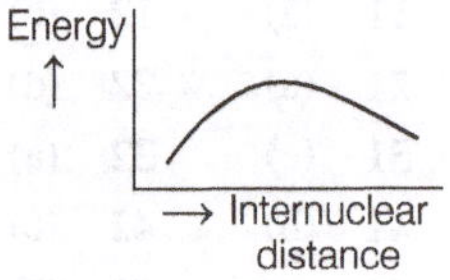

08 Arrange the following bonds according to their average bond energies in descending order

C—Cl, C—Br, C—F, C—I **[2020, 8 Jan Shift-II]**

(a) C—F > C—Cl > C—Br > C—I
(b) C—Br > C—I > C—Cl > C—F
(c) C—I > C—Br > C—Cl > C—F
(d) C—Cl > C—Br > C—I > C—F

09 The dipole moments of CCl_4, $CHCl_3$ and CH_4 are in the order **[2020, 7 Jan Shift-I]**

(a) $CCl_4 < CH_4 < CHCl_3$
(b) $CH_4 < CCl_4 < CHCl_3$
(c) $CH_4 = CCl_4 < CHCl_3$
(d) $CHCl_3 < CH_4 = CCl_4$

• Level 2

10 Given below are two statements : one is labelled as Assertion (A) and the other is labelled as Reason (R). **[2024, 5 April Shift-II]**

Assertion (A) NH_3 and NF_3 molecule have pyramidal shape with a lone pair of electrons on nitrogen atom. The resultant dipole moment of NH_3 is greater than that of NF_3.

Reason (R) In NH_3, the orbital dipole due to lone pair is in the same direction as the resultant dipole moment of the N—H bond. F is the most electronegative element.

In the light of the above statements, choose the correct answer from the options given below :

(a) (A) is true but (R) is false
(b) (A) is false but (R) is true
(c) Both (A) and (R) are true and (R) is the correct explanation of (A)
(d) Both (A) and (R) are true but (R) is not the correct explanation of (A)

11 Given below are two statements : one is labelled as Assertion (A) and the other is labelled as Reason (R).

Assertion (A) *cis*-form of alkene is found to be more polar than the *trans*-form.

Reason (R) Dipole moment of *trans*-isomer of 2-butene is zero. **[2024, 5 April Shift-I]**

In the light of the above statements, choose the correct answer from the options given below.

(a) Both A and R are true but R is not the correct explanation of A.
(b) A is false but R is true.
(c) Both A and R are true and R is the correct explanation of A.
(d) A is true but R is false.

12 Which one of the following molecules has maximum dipole moment? **[2024, 4 April Shift-I]**

(a) NF_3 (b) NH_3 (c) CH_4 (d) PF_5

Imp. *Order of Ionic Character*

13 Arrange the bonds in order of increasing ionic character in the molecules, LiF, K_2O, N_2, SO_2 and ClF_3: **[2024, 1st Feb Shift-I]**

(a) $N_2 < SO_2 < ClF_3 < K_2O < LiF$
(b) $LiF < K_2O < ClF_3 < SO_2 < N_2$
(c) $ClF_3 < N_2 < SO_2 < K_2O < LiF$
(d) $N_2 < ClF_3 < SO_2 < K_2O < LiF$

14 Given below are two statements. **[2024, 30 Jan Shift-II]**

Statement I Since fluorine is more electronegative than nitrogen, the net dipole moment of NF_3 is greater than NH_3.

Statement II In NH_3, the orbital dipole due to lone pair and the dipole moment of NH bonds are in opposite direction, but in NF_3 the orbital dipole due to lone pair and dipole moments of N—F bonds are in same direction.

In the light of the above statements, choose the **most appropriate** from the options given below :

(a) Statement I is false but Statement II is true.
(b) Statement I is true but Statement II is false.
(c) Both Statement I and Statement II are true.
(d) Both Statement I and Statement II are false.

• Level 3

15 Match the List-I with List-II. **[2025, 23 Jan Shift-I]**

List-I (Classification of molecules based on octet rule)	List-II (Example)
A. Molecules obeying octet rule	I. NO, NO_2
B. Molecules with incomplete octet	II. BCl_3, $AlCl_3$
C. Molecules with incomplete octet with odd electron	III. H_2SO_4, PCl_5
D. Molecules with expanded octet	IV. CCl_4, CO_2

Choose the correct answer from the options given below:

(a) A-II, B-IV, C-III, D-I
(b) A-III, B-II, C-I, D-IV
(c) A-IV, B-II, C-I, D-III
(d) A-IV, B-I, C-III, D-II

16 Order of covalent bond;

A. KF > KI; LiF > KF B. KF < KI; LiF > KF
C. $SnCl_4 > SnCl_2$; CuCl > NaCl
D. LiF > KF; CuCl < NaCl
E. KF < KI; CuCl > NaCl

Choose the correct answer from the options given below. **[2023, 24 Jan Shift-I]**

(a) Only B, C (b) Only A, B
(c) Only C, E (d) Only B, C, E

17 Number of electron deficient molecules among the following

PH_3, B_2H_6, CCl_4 NH_3, LiH and BCl_3 is **[2022, 25 June Shift-I]**

(a) 0 (b) 1
(c) 2 (d) 3

Numerical Value Type Questions

18 Number of molecules from the following which are exceptions to octet rule is ....

CO_2, NO_2, H_2SO_4, BF_3, CH_4, SiF_4, ClO_2, PCl_5, BeF_2, C_2H_6, $CHCl_3$, CBr_4 **[2024, 8 April Shift-II]**

19 Number of compounds from the following with zero dipole moment is ....
$HF, H_2, H_2S, CO_2, NH_3, BF_3, CH_4, CHCl_3, SiF_4, H_2O, BeF_2$ **[2024, 5 April Shift-II]**

20 In the Lewis dot structure for NO_2^-, total number of valence electrons around nitrogen is .... **[2024, 5 April Shift-I]**

21 Number of compounds/ species from the following with non-zero dipole moment is...... .
$BeCl_2, BCl_3, NF_3, XeF_4, CCl_4, H_2O, H_2S, HBr, CO_2, H_2, HCl$ **[2024, 4 April Shift-II]**

22 The total number of lone pairs of electrons on oxygen atoms of ozone is .... **[2023, 24, Jan Shift-I]**

23 The total number of molecules with zero dipole moment among CH_4, BF_3, H_2O, HF, NH_3, CO_2 and SO_2 is .... **[2022, 29 Jan Shift-II]**

24 The number of non-polar molecules from the following is ....
HF, H_2O, SO_2, H_2, CO_2, CH_4, NH_3, HCl, $CHCl_3$, BF_3 **[2022, 27 Jan Shift-II]**

25 Amongst BeF_2, BF_3, H_2O, NH_3, CCl_4 and HCl, the number of molecules with non-zero net dipole moment is .... **[2022, 25 June Shift-II]**

Topic 2 / Hybridisation, VSEPR, VBT

• Level 1

01 Match List-I with List-II

List-I (Hybridisation)		List-II (Orientation in space)	
A.	sp^3	I.	Trigonal bipyramidal
B.	dsp^2	II.	Octahedral
C.	sp^3d	III.	Tetrahedral
D.	sp^3d^2	IV.	Square planar

Choose the correct answer from the options given below. **[2024, 6 April Shift-I]**
(a) A-III, B-I, C-IV, D-II (b) A-IV, B-III, C-I, D-II
(c) A-III, B-IV, C-I, D-II (d) A-II, B-I, C-IV, D-III

02 In the structure of SF_4, the lone pair of electrons on S is in **[2022, 28 June Shift-II]**
(a) equatorial position and there are two lone pair-bond pair repulsions at 90°.
(b) equatorial position and there are three lone pair-bond pair repulsions at 90°.
(c) axial position and there are three lone pair –bond pair repulsion at 90°.
(d) axial position and there are two lone pair–bond pair repulsion at 90°.

03 Number of lone pair(s) of electrons on central atom and the shape of BrF_3 molecule respectively, are **[2022, 29 June Shift-II]**
(a) 0, triangular planar (b) 1, pyramidal
(c) 2, bent T-shape (d) 1, bent T-shape

04 Amongst the following, the linear species is **[2021, 17 March Shift-II]**
(a) NO_2 (b) Cl_2O
(c) O_3 (d) N_3^-

05 A central atom in a molecule has two lone pairs of electrons and forms three single bonds. The shape of this molecule is **[2021, 17 March Shift-I]**
(a) *see-saw* (b) planar triangular
(c) T-shaped (d) trigonal pyramidal

06 Given below are two statements : one is labelled as Assertion (A) and the other is labelled as Reason (R).
Assertion (A) The H—O—H bond angle in water molecule is 104.5°.
Reason (R) The lone pair - lone pair repulsion of electrons is higher than the bond pair - bond pair repulsion. **[2021, 16 March Shift-I]**
(a) A is false but R is true.
(b) Both A and R are true, but R is not the correct correct explanation of A.
(c) A is true but R is false.
(d) Both A and R are true, and R is the correct explanation of A.

07 The correct shape and I—I—I bond angles respectively in I_3^-, ion are **[2021, 24 Feb Shift-II]**
(a) distorted trigonal planar, 135° and 90°
(b) T-shaped, 180° and 90°
(c) Trigonal planar, 120° (d) Linear, 180°

08 In $\overset{1}{C}H_2 = \overset{2}{C} = \overset{3}{C}H - \overset{4}{C}H_3$ molecule, the hybridisation of carbon 1, 2, 3 and 4 respectively are **[2021, 26 Feb Shift-II]**
(a) sp^3, sp, sp^3, sp^3 (b) sp^2, sp^2, sp^2, sp^3
(c) sp^2, sp, sp^2, sp^3 (d) sp^2, sp^3, sp^2, sp^3

09 If AB_4 molecule is a polar molecule, a possible geometry of AB_4 is **[2020, 2 Sep Shift-I]**
(a) square pyramidal (b) square planar
(c) rectangular planar (d) tetrahedral

10 The molecular geometry of SF_6 is octahedral. What is the geometry of SF_4 (including lone pair(s) of electrons, if any)? **[2020, 2 Sep Shift-II]**
(a) Tetrahedral (b) Trigonal bipyramidal
(c) Pyramidal (d) Square planar

Level 2

11 In SO_2, NO_2^- and N_3^- the hybridisations at the central atom are respectively: **[2025, 7 April Shift-II]**

(a) sp, sp^2 and sp (b) sp^2, sp^2 and sp
(c) sp^2, sp and sp (d) sp^2, sp^2 and sp^2

12 Given below are two statements **[2025, 4 April Shift-II]**

Statement I For $\overset{\bullet\bullet}{\underset{\bullet\bullet}{Cl}}F_3$, all three possible structures may be drawn as follows.

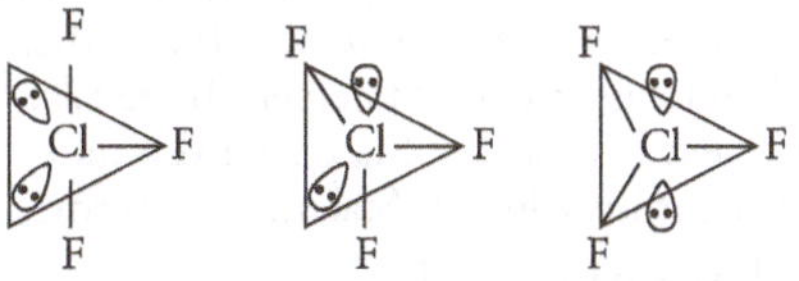

I II III

Statement II Structure III is most stable, as the orbitals having the lone pairs are axial, where the lp – bp repulsion is minimum.

In the light of the above statements, choose the most appropriate answer from the options given below:
(a) Statement I is correct but Statement II is incorrect
(b) Statement I is incorrect but Statement II is correct
(c) Both Statement I and Statement II are correct
(d) Both Statement I and Statement II are incorrect

Imp. *Geometry of Molecule/Ions*

13 The number of molecules/ions that show linear geometry among the following is ....
SO_2, $BeCl_2$, CO_2, N_3^-, NO_2, F_2O, XeF_2, NO_2^+, I_3^-, O_3 **[2025, 22 Jan Shift-I]**

14 Total number of molecules/species from following which will be paramagnetic is....
O_2, O_2^+, O_2^-, NO, NO_2, CO,
$K_2[NiCl_4]$, $[Co(NH_3)_6]Cl_3$, $K_2[Ni(CN)_4]$ **[2025, 28 Jan Shift-II]**

15 The correct increasing order for bond angles among BF_3, PF_3 and ClF_3 is **[2024, 9 April Shift-II]**
(a) $PF_3 < BF_3 < ClF_3$
(b) $ClF_3 < PF_3 < BF_3$
(c) $BF_3 = PF_3 < ClF_3$
(d) $BF_3 < PF_3 < ClF_3$

16 In which one of the following pairs the central atoms exhibit sp^2 hybridisation? **[2024, 9 April Shift-I]**
(a) H_2O and NO_2 (b) NH_2^- and H_2O
(c) BF_3 and NO_2^- (d) NH_2^- and BF_3

17 The molecule/ion with square pyramidal shape is **[2024, 30 Jan Shift-II]**
(a) BrF_5 (b) PCl_5
(c) $[Ni(CN)_4]^{2-}$ (d) PF_5

18 Identify from the following species in which d^2sp^3 hybridisation is shown by central atom. **[2024, 27 Jan Shift-II]**
(a) $[Pt(Cl)_4]^{2-}$ (b) $[Co(NH_3)_6]^{3+}$
(c) BrF_5 (d) SF_6

19 Given below are two statements.
Statement I SO_2 and H_2O both posses V-shaped structure.
Statement II The bond angle of SO_2 is less than that of H_2O.
In the light of the above statements, choose the most appropriate answer from the options given below. **[2023, 13 April Shift-II]**
(a) Both Statement I and Statement II are incorrect.
(b) Statement I is correct but Statement II is incorrect.
(c) Both Statement I and Statement II are correct.
(d) Statement I is incorrect but Statement II is correct.

20 Number of lone pairs of electrons in the central atom of SCl_2, O_3, ClF_3 and SF_6, respectively, are **[2022, 29 July Shift-I]**
(a) 0, 1, 2 and 2 (b) 2, 1, 2 and 0
(c) 1, 2, 2 and 0 (d) 2, 2, 2 and 0

21 Given below are two statements : One is labelled as Assertion (A) and the other is labelled as Reason (R).
Assertion (A) Zero orbital overlap is an out of phase overlap.
Reason (R) It results due to different orientation/direction of approach of orbitals.
In the light of the above statements, choose the correct answer from the options given below: **[2022, 28 July Shift-II]**
(a) Both A and R are true and R is the correct explanation of A.
(b) Both A and R are true and R is not the correct explanation of A.
(c) A is true but R is false.
(d) A is false but R is true.

22 Consider the species CH_4, NH_4^+ and BH_4^-. Choose the correct option with respect to these species. **[2022, 29 June Shift-II]**
(a) They are isoelectronic and only two have tetrahedral structures.
(b) They are isoelectronic and all have tetrahedral structures.
(c) Only two are isoelectronic and all have tetrahedral structures.
(d) Only two are isoelectronic and only two have tetrahedral structures.

23 The molecule in which hybrid MOs involve only one d-orbital of the central atom is **[2020, 04 Sep Shift-II]**
(a) $[Ni(CN)_4]^{2-}$ (b) BrF_5 (c) XeF_4 (d) $[CrF_6]^{3-}$

24 The compound that has the largest H—M—H bond angle (M = N, O, S, C) is **[2020, 05 Sep Shift-II]**
(a) H_2O (b) NH_3 (c) H_2S (d) CH_4

• Level 3

25 Among SO_2, NF_3, NH_3, XeF_2, ClF_3 and SF_4, the hybridisation of the molecule with non-zero dipole moment highest number of lone-pairs of electrons on the central atom is **[2025, 2 April Shift-I]**

(a) sp^3 (b) dsp^2 (c) sp^3d^2 (d) sp^3d

26 Match the List-I with List-II. **[2025, 3 April Shift-I]**

	List-I (Molecules/ion)		List-II (Hybridisation of central atom)
A.	PF_5	I.	dsp^2
B.	SF_6	II.	sp^3d
C.	$Ni(CO)_4$	III.	sp^3d^2
D.	$[PtCl_4]^{2-}$	IV.	sp^3

Choose the correct answer from the options given below:

(a) A-IV, B-I, C-II, D-III (b) A-I, B-II, C-III, D-IV
(c) A-III, B-I, C-IV, D-II (d) A-II, B-III, C-IV, D-I

27 The molecules having square pyramidal geometry are **[2025, 28 Jan Shift-I]**

(a) SbF_5 and PCl_5 (b) BrF_5 and PCl_5
(c) SbF_5 and $XeOF_4$ (d) BrF_5 and $XeOF_4$

28 Which of the following statements is true with respect to H_2O, NH_3 and CH_4? **[2025, 24 Jan Shift-I]**

A. The central atoms of all the molecules are sp^3 hybridised.
B. The H—O—H, H—N—H and H—C—H angles in the above molecules are 104.5°, 107.5° and 109.5° respectively.
C. The increasing order of dipole moment is $CH_4 < NH_3 < H_2O$.
D. Both H_2O and NH_3 are Lewis acids and CH_4 is a Lewis base.
E. A solution of NH_3 in H_2O is basic. In this solution NH_3 and H_2O act as Lowry-Bronsted acid and base respectively.

Choose the correct answer from the options given below

(a) A, D and E only (b) C, D and E only
(c) A, B, C and E only (d) A, B and C only

29 Match List-I with List-II.

	List-I (Molecule)	List-II (Shape)
A.	NH_3	I. Square pyramidal
B.	BrF_5	II. Tetrahedral
C.	PCl_5	III. Trigonal pyramidal
D.	CH_4	IV. Trigonal bipyramidal

Choose the correct answer from the options given below: **[2024, 8 April Shift-I]**

(a) A-IV, B-III, C-I, D-II (b) A-II, B-IV, C-I, D-III
(c) A-III, B-I, C-IV, D-II (d) A-III, B-IV, C-I, D-II

30 Given below are two statements. **[2024, 6 April Shift-II]**

Statement I PF_5 and BrF_5 both exhibit sp^3d hybridisation.

Statement II Both SF_6 and $[Co(NH_3)_6]^{3+}$ exhibit sp^3d^2 hybridisation.

In the light of the above statements, choose the correct answer from the options given below.

(a) Both Statement I and Statement II are true.
(b) Both Statement I and Statement II are false.
(c) Statement I is true but Statement II is false.
(d) Statement I is false but Statement II is true.

31 Match List-I with List-II.

	List-I (Compound/species)		List-II (Shape/Geometry)
A.	SF_4	I.	Tetrahedral
B.	BrF_3	II.	Pyramidal
C.	BrO_3^-	III.	See-saw
D.	NH_4^+	IV.	Bent T-Shape

Choose the correct answer from the options given below. **[2024, 6 April Shift-I]**

(a) A-III, B-IV, C-II, D-I (b) A-II, B-IV, C-III, D-I
(c) A-III, B-II, C-IV, D-I (d) A-II, B-III, C-I, D-IV

Imp. *Magnetic Property & Shape of Molecule*

32 Match List-I with List-II.

	List-I (Molecule/species)	List-II (Property/shape)
A.	SO_2Cl_2	I. Paramagnetic
B.	NO	II. Diamagnetic
C.	NO_2^-	III. Tetrahedral
D.	I_3^-	IV. Linear

Choose the correct answer from the options given below. **[2024, 6 April Shift-I]**

(a) A-II, B-III, C-I, D-IV (b) A-IV. B-I. C-III, D-II
(c) A-III, B-I, C-II, D-IV (d) A-III, B-IV, C-II, D-I

33 Match List-I with List-II.

	List-I		List-II
A.	ICl	I.	T-shape
B.	ICl_3	II.	Square pyramidal
C.	ClF_5	III.	Pentagonal bipyramidal
D.	IF_7	IV.	Linear

Choose the correct answer from the options given below **[2024, 5 April Shift-II]**

(a) A-IV, B-I, C-II, D-III (b) A-I, B-IV, C-III, D-II
(c) A-IV, B-III, C-II, D-I (d) A-I, B-III, C-II, D-IV

34 The number of species from the following that have pyramidal geometry around the central atom is......
$S_2O_3^{2-}, SO_4^{2-}, SO_3^{2-}, S_2O_7^{2-}$ **[2024, 4 April Shift-II]**

(a) 3 (b) 4 (c) 1 (d) 2

35 Number of molecules/ions from the following in which the central atom is involved in sp^3 hybridisation is...... . **[2024, 4 April Shift-I]**
$NO_3^-, BCl_3, ClO_2^-, ClO_3^-$

(a) 3 (b) 2 (c) 1 (d) 4

36 Match List-I with List-II.

	List-I (Species)		List-II (Geometry/Shape)
A.	H_3O^+	I.	Tetrahedral
B.	Acetylide anion	II.	Linear
C.	NH_4^+	III.	Pyramidal
D.	ClO_2^-	IV.	Bent

Choose the correct answer from the options given below. **[2023, 11 April Shift-I]**

(a) A-III, B-IV, C-II, D-I (b) A-III, B-I, C-II, D-IV
(c) A-III, B-IV, C-I, D-II (d) A-III, B-II, C-I, D-IV

37 Match List-I with List-II.

	List-I (Species)		List-II (Shape)
A.	XeF_4	I.	*See-saw*
B.	SF_4	II.	Square planar
C.	NH_4^+	III.	Bent T-shaped
D.	BrF_3	IV.	Tetrahedral

Choose the correct answer from the options given below **[2023, 31 Jan Shift-I]**

(a) A-II, B-I, C-IV, D-III (b) A-IV, B-I, C-II, D-III
(c) A -IV, B-III, C-II, D-I (d) A-II, B-I, C-III, D-IV

38 Match List-I with List-II.

	List-I (Compound)		List-II (Shape)
(A)	BrF_5	(I)	bent
(B)	$[CrF_6]^{3-}$	(II)	square pyramidal
(C)	O_3	(III)	trigonal bipyramidal
(D)	PCl_5	(IV)	octahedral

Choose the correct answer from the options given below : **[2022, 26 July Shift-I]**

	A	B	C	D
(a)	I	II	III	IV
(b)	IV	III	II	I
(c)	II	IV	I	III
(d)	III	IV	II	I

39 Based upon VSEPR theory, match the shape (geometry) of the molecules in List-I with the molecules in List-II and select the most appropriate option. **[2022, 27 June Shift-I]**

	List-I (Shape)		List-II (Molecules)
A.	T-shaped	I.	XeF_4
B.	Trigonal planar	II.	SF_4
C.	Square planar	III.	ClF_3
D.	*See-saw*	IV.	BF_3

	A	B	C	D
(a)	I	II	III	IV
(b)	III	IV	I	II
(c)	III	IV	II	I
(d)	IV	III	I	II

40 Match List-I with List-II.

	List-I (Compounds)		List-II (Hybrid orbitals)
(A)	$[PtCl_4]^{2-}$	(I)	sp^3d
(B)	BrF_5	(II)	d^2sp^3
(C)	PCl_5	(III)	dsp^2
(D)	$[Co(NH_3)_6]^{3+}$	(IV)	sp^3d^2

Choose the most appropriate answer from the options given below. **[2022, 24 June Shift-I]**

(a) (A) → (II), (B) → (IV), (C) → (I), (D) → (III)
(b) (A) → (III), (B) → (IV), (C) → (I), (D) → (II)
(c) (A) → (III), (B) → (I), (C) → (IV), (D) → (II)
(d) (A) → (II), (B) → (I), (C) → (IV), (D) → (III)

Imp. *Hybridisation of Molecule*

41 Match List-I and List-II.

	List-I (Species)		List-II (Hybrid orbitals)
(A)	SF_4	(I)	sp^3d^2
(B)	IF_5	(II)	d^2sp^3
(C)	NO_2^+	(III)	sp^3d
(D)	NH_4^+	(IV)	sp^3
		(V)	sp

[2021, 22 July Shift-II]

	A	B	C	D
(a)	(I)	(II)	(V)	(III)
(b)	(II)	(I)	(IV)	(V)
(c)	(III)	(I)	(V)	(IV)
(d)	(IV)	(III)	(II)	(V)

42 The hybridisations of the atomic orbitals of nitrogen in NO_2^-, NO_2^+ and NH_4^+ respectively are **[2021, 20 July Shift-II]**

(a) sp^3, sp^2 and sp (b) sp, sp^2 and sp^3
(c) sp^3, sp and sp^2 (d) sp^2, sp and sp^3

43 Which of the following are isostructural pairs ?
A. SO_4^{2-} and CrO_4^{2-} B. $SiCl_4$ and $TiCl_4$
C. NH_3 and NO_3^- D. BCl_3 and $BrCl_3$
[2021, 24 Feb Shift-I]
(a) Only A and B (b) Only A and C
(c) Only B and C (d) Only C and D

44 Which among the following species has unequal bond lengths ? **[2021, 25 Feb Shift-II]**
(a) XeF_4 (b) SiF_4
(c) SF_4 (d) BF_4^-

Numerical Value Type Questions

45 Total number of species from the following with central atom utilising sp^2 hybrid orbitals for bonding is............
NH_3, SO_2, SiO_2, $BeCl_2$, C_2H_2, C_2H_4, BCl_3, HCHO, C_6H_6, BF_3, $C_2H_4Cl_2$. **[2024, 6 April Shift-II]**

46 The number of molecules/ion/s having trigonal bipyramidal shapes is ....
PF_5, BrF_5, PCl_5, $[PtCl_4]^{2-}$, BF_3, $Fe(CO)_5$
[2024, 1 Feb Shift-I]

47 The number of species from the following in which the central atom uses sp^3 hybrid orbitals in its bonding is ....
NH_3, SO_2, SiO_2, $BeCl_2$, CO_2, H_2O, CH_4, BF_3
[2024, 31 Jan Shift-I]

48 The maximum number of lone pairs of electrons on the central atom from the following species is ....
ClO_3^-, XeF_4, SF_4 and I_3^- **[2023, 11 April Shift-II]**

49 The number of molecules from the following which contain only two lone pair of electrons is ....
H_2O, N_2, CO, XeF_4, NH_3, NO, CO_2, F_2
[2023, 10 April Shift-II]

50 The number of species from the following carrying a single lone pair on central atom xenon is ....
XeF_5^+, XeO_3, XeO_2F_2, XeF_5^-, XeO_3F_2, $XeOF_4$, XeF_4 **[2023, 8 April Shift-II]**

51 The number of bent-shaped molecule/s from the following is .....
N_3^-, NO_2^-, I_3^-, SO_2, O_3 **[2023, 10 April Shift-I]**

52 The sum of lone pairs present on the central atom of the interhalogen IF_5 and IF_7 is ....
[2023, 10 April Shift-I]

53 The number of species having a square planar shape from the following is ....
XeF_4, SF_4, SiF_4, BF_4^-, BrF_4^-, $[Cu(NH_3)_4]^{2+}$, $[FeCl_4]^{2-}$, $[PtCl_4]^{2-}$
[2023, 6 April Shift-II]

54 The number of species from the following which have square pyramidal structure is .... PF_5, BrF_4^-, IF_5 BrF_5, $XeOF_4$, ICl_4^-
[2023, 6 April Shift-I]

55 Amongst the following, the number of species having the linear shape is ....
XeF_2, I_3^+, C_3O_2, I_3^-, CO_2 , SO_2, $BeCl_2$ and $BCl_2^{\ominus}$
[2023, 31 Jan Shift-II]

56 Consider, PF_5, BrF_5, PCl_3, SF_6, $[ICl_4]^-$, ClF_3 and IF_5. Amongst the above molecule(s)/ ion(s), the number of molecule(s)/ ion(s) having sp^3d^2 hybridisation is .... **[2022, 29 July Shift-II]**

57 The number of molecule(s) or ion(s) from the following having non-planar structure is ....
NO_3^-, H_2O_2, BF_3, PCl_3, XeF_4, SF_4, XeO_3, PH_4^+, SO_3, $[Al(OH)_4]^-$ **[2022, 27 July Shift-II]**

58 The hybridisation of P exhibited in PF_5 is sp^xd^y. The value of y is .... **[2022, 28 June Shift-I]**

59 Amongst SF_4, XeF_4, CF_4 and H_2O, the number of species with two lone pair of electrons is ....
[2022, 26 June Shift-II]

60 The number of species having non–pyramidal shape among the following is **[2021, 27 Aug Shift-II]**
(i) SO_3 (ii) NO_3^- (iii) PCl_3 (iv) CO_3^{2-}

61 AB_3 is an interhalogen T-shaped molecule. The number of lone pairs of electrons on A is ....
[2021, 26 Aug Shift -I]

62 The number of lone pairs of electron on the central I atom in I_3^- is .... **[2021, 20 July Shift-I]**

63 The number of species below that have two lone pairs of electrons in their central atom is .... (Round off to the nearest integer)
SF_4, BF_4^-, ClF_3, AsF_3, PCl_5, BrF_5, XeF_4, SF_6
[2021, 18 March Shift-II]

64 Complexes (ML_5) of metals Ni and Fe have ideal square pyramidal and trigonal bipyramidal geometrics, respectively. The sum of the 90°, 120° and 180° L-M-L angles in the two complexes is .... **[2020, 8 Jan Shift-II]**

Topic 3 / MOT, Metallic and Hydrogen Bonding

• Level 1

01 When ψ_A and ψ_B are the wave functions of atomic orbitals, then σ^* is represented by **[2024, 9 April Shift-II]**
(a) $\psi_A - \psi_B$ (b) $\psi_A + \psi_B$
(c) $\psi_A - 2\psi_B$ (d) $\psi_A + 2\psi_B$

02 Match List-I with List-II.

	List-I		List-II
(A)	$\psi_{MO} = \psi_A - \psi_B$	(I)	Dipole moment
(B)	$\mu = Q \times r$	(II)	Bonding molecular orbital
(C)	$\frac{N_b - N_a}{2}$	(III)	Anti-bonding molecular orbital
(D)	$\psi_{MO} = \psi_A + \psi_B$	(IV)	Bond order

Choose the correct answer from options given below: **[2022, 27 July Shift-II]**

(a) (A)-(II), (B)-(I), (C)-(IV), (D)-(III)
(b) (A)-(III), (B)-(IV), (C)-(I), (D)-(II)
(c) (A)-(III), (B)-(I), (C)-(IV), (D)-(II)
(d) (A)-(III), (B)-(IV), (C)-(II), (D)-(I)

03 The bond order and magnetic behaviour of O_2^- ion are, respectively **[2021, 26 Aug Shift-II]**

(a) 1.5 and paramagnetic (b) 1.5 and diamagnetic
(c) 2 and diamagnetic (d) 1 and paramagnetic

04 Given below are two statements: One is labelled as Assertion (A) and the other is labelled as Reason (R).

Assertion (A) Dipole-dipole interactions are the only non-covalent interactions, resulting in hydrogen bond formation.

Reason (R) Fluorine is the most electronegative element and hydrogen bonds in HF are symmetrical.

In the light of the above statements, choose the most appropriate answer from the options given below.

(a) A is false but R is true. **[2021, 26 Feb Shift-I]**
(b) Both A and R are true and R is the correct explanation of A.
(c) A is true R is false.
(d) Both A and R are true but R is not the correct explanation of A.

05 According to molecular orbital theory, the species among the following that does not exist is **[2021, 25 Feb Shift-I]**

(a) O_2^{2-} (b) He_2^-
(c) Be_2 (d) He_2^+

06 The bond order and the magnetic characteristics of CN^- are **[2020, 7 Jan Shift-II]**

(a) 3, diamagnetic (b) $2\frac{1}{2}$, paramagnetic
(c) 3, paramagnetic (d) $2\frac{1}{2}$, diamagnetic

07 The correct order of increasing intermolecular hydrogen bond strength is **[2020, 27 June Shift-II]**

(a) $HCN < H_2O < NH_3$ (b) $HCN < CH_4 < NH_3$
(c) $CH_4 < HCN < NH_3$ (d) $CH_4 < NH_3 < HCN$

08 Among the following species, the diamagnetic molecule is **[2019, 9 April Shift-II]**

(a) CO (b) B_2 (c) NO (d) O_2

09 According to molecular orbital theory, which of the following is true with respect to Li_2^+ and Li_2^-?

(a) Both are unstable **2019, 9 Jan Shift-I]**
(b) Li_2^+ is unstable and Li_2^- is stable
(c) Both are stable
(d) Li_2^+ is stable and Li_2^- is unstable

Level 2

10 Which of the following linear combination of atomic orbitals will lead to formation of molecular orbitals in homonuclear diatomic molecules [internuclear axis in z- direction] ? **[2025, 24 Jan Shift-I]**

(A) $2p_z$ and $2p_x$ (B) $2s$ and $2p_x$
(C) $3d_{xy}$ and $3d_{x^2-y^2}$ (D) $2s$ and $2p_z$
(E) $2p_z$ and $3d_{x^2-y^2}$

Choose the correct answer from the options given below

(a) C and D only (b) A and B only
(c) D only (d) E only

11 The correct statement/s about hydrogen bonding is/are

A. Hydrogen bonding exists when H is covalently bonded to the highly electronegative atom.
B. Intermolecular H-bonding is present in *o*-nitro phenol.
C. Intramolecular H-bonding is present in HF.
D. The magnitude of H-bonding depends on the physical state of the compound.
E. H-bonding has powerful effect on the structure and properties of compounds.

Choose the correct answer from the option given below : **[2024, 4 April Shift-II]**

(a) A, D, E only (b) A, B, C only
(c) A, B, D only (d) A only

12 If the magnetic moment of a dioxygen species is 1.73 B.M, it may be. **[2020, 9 Jan Shift-II]**

(a) O_2, O_2^- or O_2^+ (b) O_2^- or O_2^+
(c) O_2 or O_2^- (d) O_2 or O_2^+

13 Among the following, the molecule expected to be stabilised by anion formation is C_2, O_2, NO, F_2. **[2019, 9 April Shift-I]**

(a) C_2 (b) F_2 (c) NO (d) O_2

14 During the change of O_2 to O_2^-, the incoming electron goes to the orbital. **[2019, 10 April Shift-I]**

(a) $\pi 2p_x$ (b) $\pi^* 2p_x$
(c) $\pi 2p_y$ (d) $\sigma^* 2p_z$

15 The bond order and magnetic property of acetylide ion are same as that of **[2023, 12 April Shift-I]**

(a) O_2^+ (b) N_2^+ (c) O_2^- (d) NO^+

• Level 3

16 Which of the following molecule(s) show/s paramagnetic behaviour? **[2025, 4 April Shift-I]**

A. O_2 B. N_2 C. F_2 D. S_2
E. Cl_2

Choose the correct answer from the options given below:

(a) A and E only (b) B only
(c) A and D only (d) A and C only

17 What is the number of unpaired electrons in the highest occupied molecular orbital of the following species N_2; N_2^+; O_2; O_2^+? **[2023, 24 Jan Shift-II]**

(a) 2, 1, 2, 1 (b) 0, 1, 0, 1
(c) 2, 1, 0, 1 (d) 0, 1, 2, 1

18 According to MO theory, the bond orders for O_2^{2-}, CO and NO^+ respectively, are **[2023, 29 Jan Shift-II]**

(a) 2, 3 and 3 (b) 1, 2 and 3
(c) 1, 3 and 3 (d) 1, 3 and 2

19 Bonding in which of the following diatomic molecule(s) become(s) stronger, on the basis of MO theory, by removal of an electron?

(A) NO (B) N_2 (C) O_2 (D) C_2
(E) B_2

Choose the most appropriate answer from the options given below. **[2022, 25 June Shift-I]**

(a) (A), (B), (C) only (b) (B), (C), (E) only
(c) (A), (C) only (d) (D) only

20 Consider the ions/molecules: $O_2^+, O_2, O_2^-, O_2^{2-}$

For increasing bond order the correct option is **[2022, 26 June Shift-I]**

(a) $O_2^{2-} < O_2^- < O_2 < O_2^+$ (b) $O_2^- < O_2^{2-} < O_2 < O_2^+$
(c) $O_2^- < O_2^{2-} < O_2^+ < O_2$ (d) $O_2^- < O_2^+ < O_2^{2-} < O_2$

21 The compound/s which will show significant intermolecular H–bonding is/are

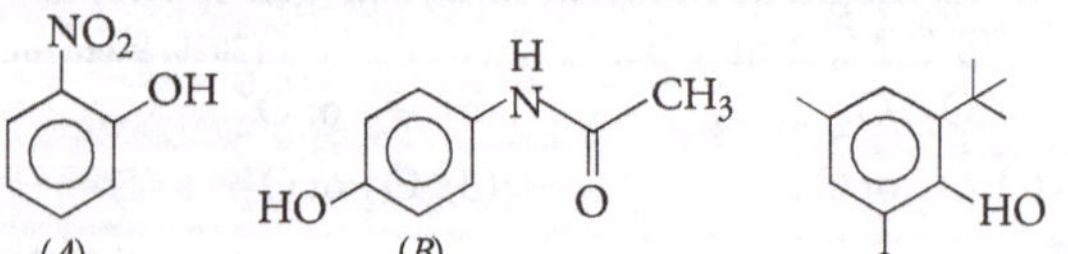

(A) (B) (C)

[2021, 27 Aug Shift-II]

(a) Only (B) (b) Only (C)
(c) (A) and (B) (d) (A), (B) and (C)

22 Match List-I with List-II.

List-I (Molecule)			List-II (Bond order)
A.	Ne_2	(i)	1
B.	N_2	(ii)	2
C.	F_2	(iii)	0
D.	O_2	(iv)	3

Choose the correct answer from the options given below. **[2021, 26 Feb Shift-II]**

	A	B	C	D		A	B	C	D
(a)	(iii)	(iv)	(i)	(ii)	(b)	(i)	(ii)	(iii)	(iv)
(c)	(ii)	(i)	(iv)	(iii)	(d)	(iv)	(iii)	(ii)	(i)

23 Of the species, NO, NO^+, NO^{2+} and NO^-, the one with minimum bond strength is **[2020, 5 Sep Shift-I]**

(a) NO^+ (b) NO
(c) NO^{2+} (d) NO^-

24 Among the following molecules/ions,

$$C_2^{2-}, N_2^{2-}, O_2^{2-}, O_2$$

Which one is diamagnetic and has the shortest bond length? **[2019, 8 April Shift-II]**

(a) C_2^{2-} (b) O_2 (c) O_2^{2-} (d) N_2^{2-}

25 Two pi and half sigma bonds are present in **[2019, 10 Jan Shift-I]**

(a) O_2^+ (b) N_2 (c) N_2^+ (d) O_2

Imp. *Magnetic Nature & Bond Order*

26 In which of the following processes, the bond order has increased and paramagnetic character has changed to diamagnetic? **[2019, 9 Jan Shift-II]**

(a) $O_2 \rightarrow O_2^+$ (b) $O_2 \rightarrow O_2^{2-}$
(c) $N_2 \rightarrow N_2^+$ (d) $NO \rightarrow NO^+$

Numerical Value Type Questions

27 Total number of electrons present in (π^*) molecular orbitals of O_2, O_2^+ and O_2^- is .... **[2024, 9 April Shift-II]**

28 The total number of species from the following in which one unpaired electron is present, is ....

$N_2, O_2, C_2^-, O_2^-, O_2^{2-}, H_2^+, CN^-, He_2^+$ **[2024, 9 April Shift-I]**

29 Number of molecules having bond order 2 from the following molecules is ....

$C_2, O_2, Be_2, Li_2, Ne_2, N_2, He_2$ **[2024, 8 April Shift-II]**

30 Number of molecules from the following, which can exhibit hydrogen bonding is .... (Nearest integer)

CH_3OH, H_2O, C_2H_6, C_6H_6, 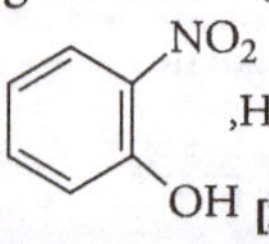 ,HF, NH_3 **[2024, 6 April Shift-I]**

31 Number of molecules/species from the following having one unpaired electron is...... .

O_2, O_2^{-1}, NO, CN^{-1}, O_2^{2-} **[2024, 4 April Shift-I]**

32 The total number of molecular orbitals formed from 2s and 2p atomic orbitals of a diatomic molecule is .... **[2024, 30 Jan Shift-I]**

33 The total number of anti-bonding molecular orbitals, formed from $2s$ and $2p$ atomic orbitals in a diatomic molecule is .... **[2024, 29 Jan Shift-II]**

34 The number of species from the following which are paramagnetic and with bond order equal to one is ....

$H_2, He_2^+, O_2^+, N_2^{2-}, O_2^{2-}, F_2, Ne_2^+, B_2$ **[2024, 29 Jan Shift-I]**

35 Sum of bond order of CO and NO^+ is ...... **[2024, 27 Jan Shift-I]**

36 Among the following species

$N_2, N_2^+, N_2^-, N_2^{2-}, O_2, O_2^+, O_2^-, O_2^{2-}$

the number of species showing diamagnetism is .... **[2022, 25 July Shift-I]**

37 The number of paramagnetic species among the following is ....

$B_2, Li_2, C_2, C_2^-, O_2^{2-}, O_2^+$ and He_2^+ **[2022, 28 July Shift-I]**

38 According to MO (molecular orbital) theory, number of species/ions from the following having identical bond order is ...... $CN^-, NO^+, O_2, O_2^+, O_2^{2+}$ **[2022, 27 July Shift-I]**

39 The spin-only magnetic moment value of B_2^+ species is .... $\times 10^{-2}$ BM. (Nearest integer) [Given, $\sqrt{3} = 1.73$] **[2021, 1 Sep Shift-II]**

40 According to molecular orbital theory, the number of unpaired electron(s) in O_2^{2-} is .... **[2021, 31 Aug Shift-II]**

41 The total number of electrons in all bonding molecular orbitals of O_2^{2-} is ....

(Round off to the nearest integer) **[2021, 27 July Shift-II]**

42 The difference between bond orders of CO and $NO^{\oplus}$ is $x/2$. where, $x =$ ....

(Round off to the nearest integer). **[2021, 27 July Shift-I]**

43 AX is a covalent diatomic molecule where A and X are second row elements of periodic table. Based on molecular orbital theory, the bond order of AX is 2.5. The total number of electron in AX is .... (Round off to the Nearest Integer). **[2021, 18 March Shift-I]**

Key Index

Topic 1

1	(a)	2	(c)	3	(d)	4	(d)	5	(b)	6	(b)	7	(c)	8	(a)	9	(c)	10	(c)
11	(c)	12	(b)	13	(a)	14	(d)	15	(c)	16	(d)	17	(c)	18	(6)	19	(6)	20	(8)
21	(5)	22	(6)	23	(3)	24	(4)	25	(3)										

Topic 2

1	(c)	2	(a)	3	(c)	4	(d)	5	(c)	6	(d)	7	(d)	8	(c)	9	(a)	10	(b)
11	(b)	12	(a)	13	(6)	14	(6)	15	(b)	16	(c)	17	(a)	18	(b)	19	(b)	20	(b)
21	(a)	22	(b)	23	(a)	24	(d)	25	(d)	26	(d)	27	(d)	28	(d)	29	(c)	30	(b)
31	(a)	32	(c)	33	(a)	34	(c)	35	(b)	36	(d)	37	(a)	38	(c)	39	(b)	40	(b)
41	(c)	42	(d)	43	(a)	44	(c)	45	(6)	46	(c)	47	(4)	48	(3)	49	(3)	50	(4)
51	(3)	52	(1)	53	(4)	54	(3)	55	(5)	56	(4)	57	(6)	58	(1)	59	(1)	60	(3)
61	(2)	62	(3)	63	(2)	64	(20)												

Topic 3

1	(a)	2	(c)	3	(a)	4	(a)	5	(c)	6	(a)	7	(c)	8	(a)	9	(d)	10	(c)
11	(a)	12	(b)	13	(a)	14	(b)	15	(d)	16	(c)	17	(d)	18	(c)	19	(c)	20	(a)
21	(a)	22	(a)	23	(d)	24	(a)	25	(c)	26	(d)	27	(6)	28	(4)	29	(2)	30	(5)
31	(2)	32	(8)	33	(4)	34	(1)	35	(6)	36	(2)	37	(4)	38	(3)	39	(173)	40	(0)
41	(10)	42	(0)	43	(15)														

Detailed Explanations

Topic 1

01. *(a)* The correct order of increasing dipole moment of given compounds is

NF_3	< HBr	< H_2S	< $CHCl_3$
(0.24 D)	(0.79 D)	(0.95 D)	(1.04 D)

02. *(c)* According to Fajan's rule, Ag and Cl bond will be covalent in nature not ionic.

The ionic character of the given compounds is in the order,

$AgCl < CoCl_2 < BaCl_2 < KCl$

Hence, AgCl is least ionic.

03. *(d)* Polar molecules are those molecules in which there exists an electronegativity difference between the bonded atom. They have non-zero dipole moment.

Hence, $CHCl_3$ is a polar molecule.

Non-polar molecules are those molecules in which electrons are shared equally. They have zero dipole moment.

$CH_2{=}CH_2, CCl_4, CO_2$ are non-polar molecules.

Hence, option (d) is correct.

04. *(d)* CO_3^{2-} has a single structure i.e., resonance hybrid of given three canonical structures.

The given structures are resonating structures of carbonate ion.

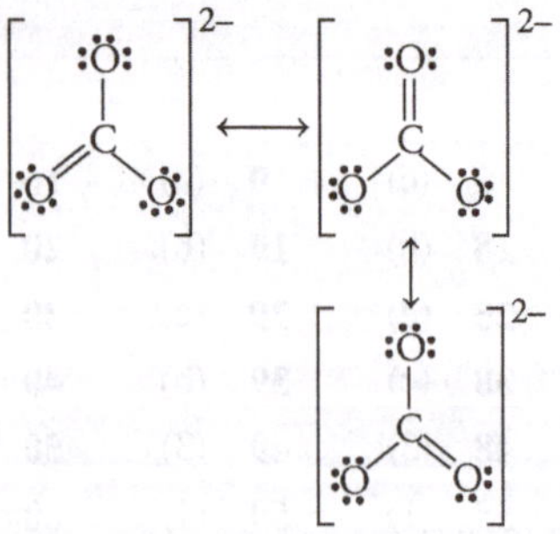

05. *(b)* Statement I is correct but Statement II is incorrect.

Dipole moment is a vector quantity i.e., it has magnitude as well as definite directions.

In chemistry, the dipole moment is represented by cross on the positive center and arrowhead on the negative center. This arrow symbolises the shift of electron density in the molecules.

06. *(b)* NO contains 15 electrons, thus it is an odd electron molecule.

Expanded octet molecule contains more than eight electrons in their valence shell. SF_6, PCl_5 and H_2SO_4 are examples of expanded octet molecules.

BCl_3 is an electron deficient molecule, i.e. it contains less than 8 electrons in its valence shell.

07. *(c)* When the distance of hydrogen nucleus is decreasing at a particular point, energy is minimum. At this point potential energy will be least. It is the point when bond formation taken place.

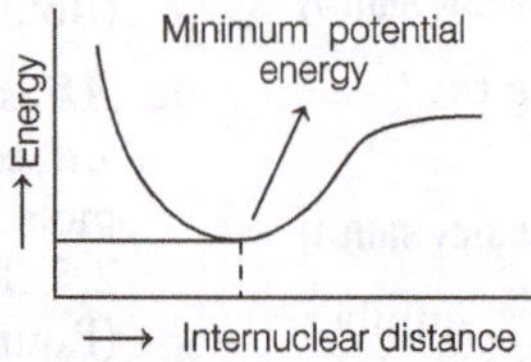

Hence, option (c) is correct.

08. *(a)* The correct descending order of bonds according to their average bond energies is as follows :

C—F > C—Cl > C—Br > C—I

Bond length is inversely proportional to bond energy. So, bond length decreases as bond energy increases.

09. *(c)* The correct order of dipole moment is $CH_4 = CCl_4 < CHCl_3$

10. *(c)* Both A and R are true and R is the correct explanation of A.

The bond structure of NF_3 and NH_3 with dipole moment direction are;

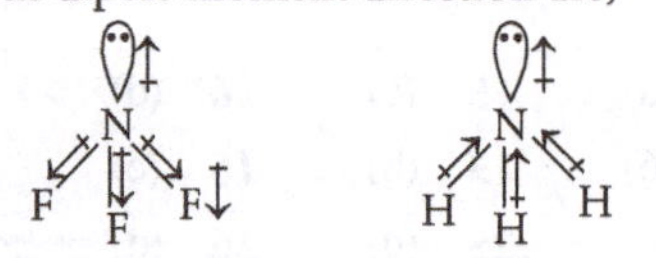

Resultant dipole moment	Resultant dipole moment
$=0.80\times10^{-30}$ Cm	$=4.9\times10^{-30}$ Cm

Hence, option (c) is correct.

11. *(c)* Both A and R are true and R is correct explanation of A.

Dipole moment is a vector quantity and for compound net dipole moment is the vector sum of all dipoles. Hence, dipole moment of *cis*-form is greater than *trans*-form.

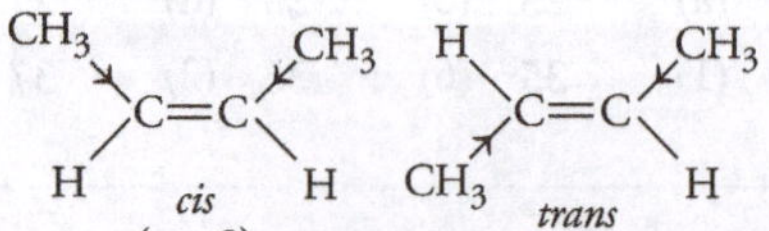

cis ($\mu > 0$) *trans* ($\mu = 0$)

12. *(b)* Among the given molecules CH_4 and PF_5 are non-polar. Hence they have net zero dipole moment while among NH_3 and NF_3, the dipole moment of NH_3 is greater than NF_3. It is because in NH_3 vector addition of bond moment and lone pair takes place while in NF_3 vector subtraction of bond moment and lone pair takes place.

13. *(a)* The ionic character in a molecule is dependent upon electronegativity difference between the constituting atoms.

Greater the electronegativity difference, greater will be the ionic character of molecules.

Therefore, the correct sequence of increasing order of ionic character is

$N_2 < SO_2 < ClF_3 < K_2O < LiF.$

Hence, option (a) is correct.

14. *(d)* Both Statement I and II are false.

Their correct form are as follows

Statement I The dipole moment of NH_3 is higher than NF_3. The reason is that flourine is more electronegative than nitrogen and thus pulls the electron towards itself. As number of NF bonds is 3, the three F-atoms pull electrons and hence resulting in decrease of dipole moment.

Statement II In NH_3, the orbital dipole due to lone pair and dipole moment of NH bonds are in same direction, but in NF_3 it will be in opposite direction.

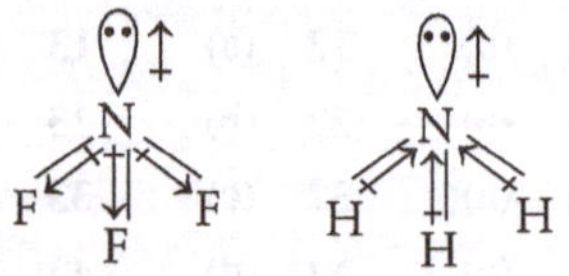

Hence, option (d) is correct.

15. *(c)* The correct match is A-IV, B-II, C-I, D-III.

16. *(d)* Greater the difference of electronegativity between the cation and anion, greater will be the ionic character and lesser will be the covalent character.

Hence, correct order is

KI > KF, LiF > KF; CuCl > NaCl

Smaller the size of the cation higher will be its polarising capacity. Since, Sn^{4+} is smaller in size hence it will have more covalent character. So, $SnCl_4 > SnCl_2$.

Alternate Solution

According to Fajan's rule, covalent character is higher for large size anion and small size cation. Thus,
(A) KF > KI is incorrect, LiF > KF is correct.
(B) KF < KI is correct, LiF > KF is correct.
(C) $SnCl_4 > SnCl_2$ is correct, CuCl > NaCl is correct.
(D) LiF > KF is correct, CuCl < NaCl is incorrect.
(E) KF < KI is correct, CuCl > NaCl is correct.
So, B, C and E are correct.

17. *(c)* B_2H_6 and BCl_3 are electron deficient species with incomplete octets.

18. *(6)* $NO_2, H_2SO_4, BF_3, ClO_2, PCl_5, BeF_2$. These are exception of octet rule. Hence, among the given molecules total 6 molecule are exception of octet rule.

19. *(6)*

idea The molecules which are symmetrical in nature have zero dipole moment.

Among the given,
$H_2, CO_2, BF_3, CH_4, SiF_4, BeF_2$ are symmetrical molecules.
Hence, total 6 molecules will show zero dipole moment.

20. *(8)* The bond structure of NO_2^- is,

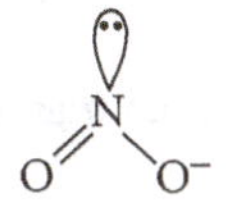

Number of valence e^- around N-atom = 8.

21. *(5)* Polar molecules will have non-zero dipole moments while non-polar molecules will have zero dipole moments. Thus, among the given species/compounds NF_3, H_2O, H_2S, HBr and HCl will have non-zero dipole moment

22. *(6)*

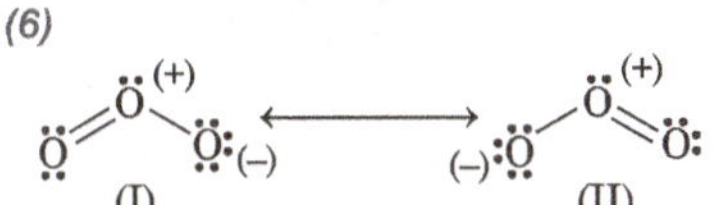

(Ozone) Resonating structures of ozone (O_3)

Ozone has 6 lone pair of electrons.

23. *(3)* Molecule with net zero dipole moment are CO_2, CH_4 and BF_3 i.e., 3 Remaining all molecules have non-zero dipole moment.

24. *(4)* The number of non-polar molecules are 4. These are CO_2, H_2, CH_4 and BF_3.
Remaining all compound are polar molecules.
$HF, H_2O, SO_2, NH_3, CHCl_3, HCl$.

25. *(3)* Out of given molecules; H_2O, NH_3 and HCl are polar molecules and BeF_2, BF_3 and CCl_4 are non-polar molecules.

Topic 2

01. *(c)* The correct match is A-III, B-IV, C-I, D-II.

Hybridisation	Orientation
sp^3	→ Tetrahedral
dsp^2	→ Square planar
sp^3d	→ Trigonal bipyramidal
sp^3d^2	→ Octahedral

02. *(a)* The hybridisation of SF_4 is sp^3d and geometry is trigonal bipyramidal. At axial position, there are 3 *lp-bp* at 90°. But at equatorial position, there are only two lone pair-bond pair repulsions at 90°.

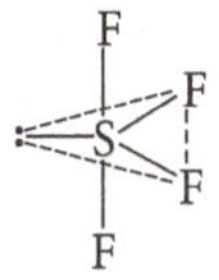

So, shape of SF_4 molecule is *see-saw*.

03. *(c)*

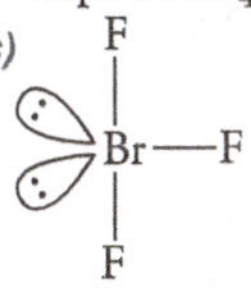

BrF_3 has 2 lone pairs of electron and it has bent T-shape.

04. *(d)* N_3^- is linear specie.

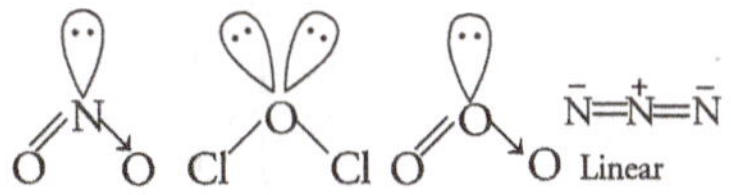

Bent shape Bent shape Bent shape

Hybridisation of N_3^- is *sp* and structure is linear.

05. *(c)* 2 lone pair + 3 bond pair
Steric number = 5
Hybridisation = sp^3d.
So, the shape is T- shape.

Concept Enhancer

Geometry of the compound having sp^3d hybridisation have tetragonal bipyramidal (TBP) geometry according to VSEPR theory. In TBP geometry lone pair occupy equatorial position.

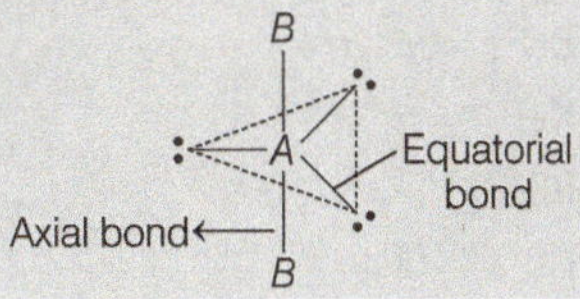

06. *(d)* Bond angle of sp^3-hybridised compound should be 109.5°.

But in case of water, it is only 104.5° due to repulsion between the lone pairs. Also, lone pair-lone pair repulsion is higher than bond pair-bond pair repulsion. (VSEPR Theory)

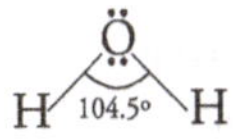

lp-lp Repulsion > bp-bp Repulsion

So, both A and R are true and R is the correct explanation of A.

07. *(d)* Hybridisation of central atom I in I_3^- is sp^3d with 3 lone pair and 2 bond pair.

Shape Linear

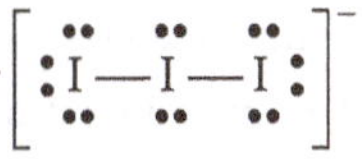

Bond angle 180° (for linear molecule)

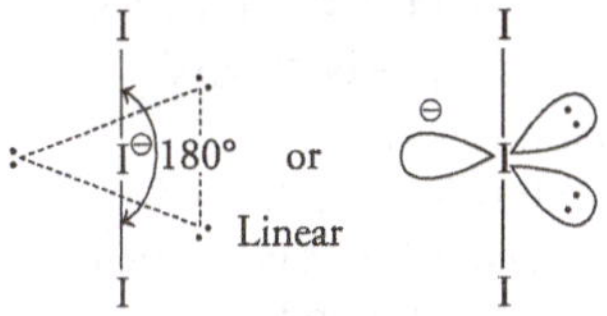

08. *(c)* $\overset{1}{C}H_2=\overset{2}{C}=\overset{3}{C}H-\overset{4}{C}H_3$

It is buta -1, 2 - diene,

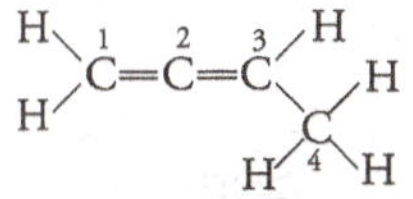

Here, $C_1 \rightarrow sp^2, C_2 \rightarrow sp, C_3 \rightarrow sp^2$ and $C_4 \rightarrow sp^3$-hybridised.
So, option (c) is the correct answer.

09. *(a)* If AB_4 molecule is a polar molecule, a possible geometry of AB_4 is square pyramidal.

10. *(b)* The geometry of SF_6 is octahedral and the geometry of SF_4 is trigonal bipyramidal.

11. *(b)*

SO_2

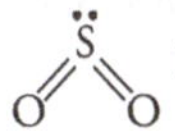

Hybridisation = sp^2

NO_2^-

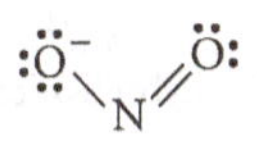

Hybridisation = sp^2

N_3^- $\overset{\ominus}{N}=\overset{\oplus}{N}=\overset{\ominus}{N}$ Hybridisation = *sp*

12. *(a)* Statement I is correct but Statement II is incorrect. It is because in sp^3d hybridisation lone pair cannot occupy axial position.

13. *(6)* Among the given species the geometry of given molecules are linear i.e.,

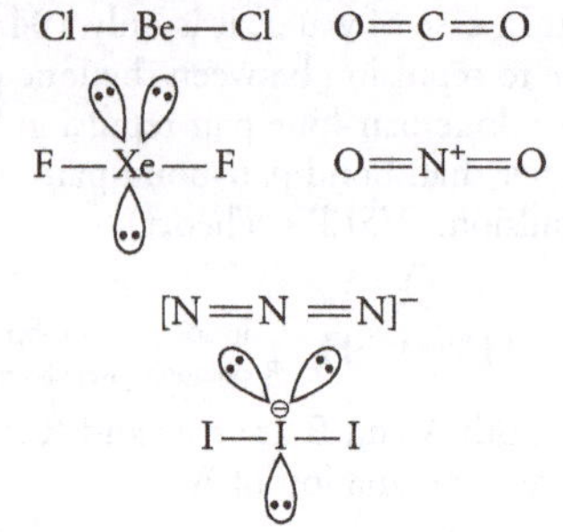

14. *(6)* According to molecular orbital theory.

$O_2 \rightarrow 2$ unpaired electron

$O_2^+ \rightarrow 1$ unpaired electron

$O_2^- \rightarrow 1$ unpaired electron

$NO \rightarrow$ odd electron species.

$NO_2 \rightarrow$ odd electron species.

$K_2[NiCl_4] \rightarrow Ni^{2+} = 3d^8$

(Weak field ligand)

Coordination number = 4

$\therefore$ Tetrahedral, paramagnetic with 2 unpaired electrons.

So, total 6 paramagnetic species are present.

15. *(b)* The structure of given compounds are as follows.

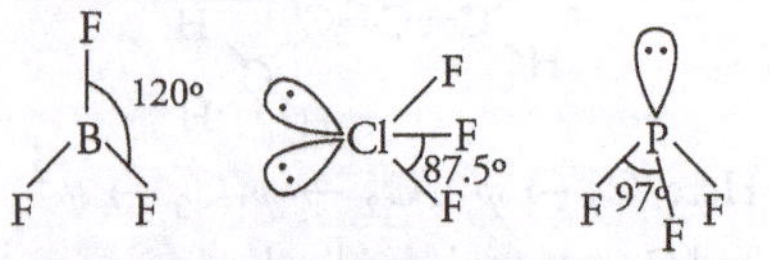

So, the correct order of bond angle is $ClF_3 < PF_3 < BF_3$.

16. *(c)* The hybridisation of central atom of given species in different options are

$BF_3, NO_2^-, NO_2 \Rightarrow sp^2$-hybridisation

H_2O and $NH_2^- \Rightarrow sp^3$-hybridisation

So, the pair of central atoms that exhibit sp^2-hybridisation is BF_3 and NO_2^-.

17. *(a)* Among the given options, BrF_5 has square pyramidal shape.

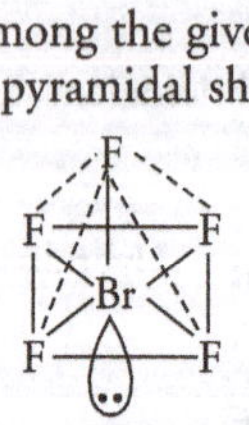

sp^3d^2 hybridisation

18. *(b)* Among the given options, only $[Co(NH_3)_6]^{3+}$ shows d^2sp^3 hybridisation.

Hybridisation of rest of the members are as follows

$BrF_5 = sp^3d^2$

$SF_6 = sp^3d^2$

$[PtCl_4]^{2-} = dsp^2$

19. *(b)* Shape of SO_2 and H_2O are given as

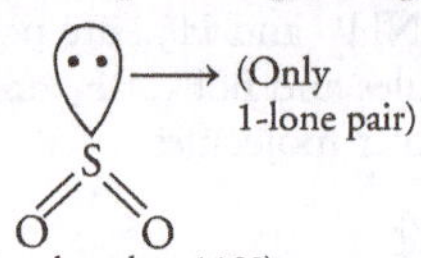

(Bond angle = 119°)

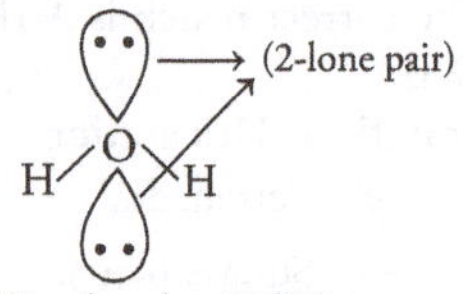

(Bond angle = 1045°)

Hence, both has V-shape, but bond angle in SO_2 is greater than H_2O, due to repulsion between lone pairs at oxygen. So, statement I is correct but statement II is incorrect.

20. *(b)* Structure of SCl_2,

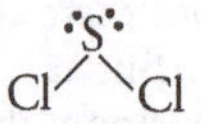

(Number of lone pairs of electron on central atom, S = 2)

Structure of O_3,

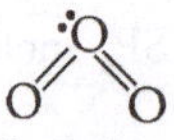

Number of lone pairs on O = 1

Structure of ClF_3,

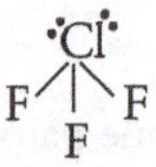

Number of lone pairs on Cl = 2

Structure of SF_6,

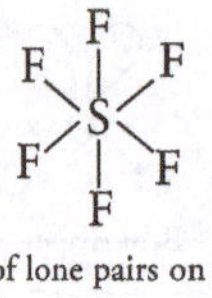

Number of lone pairs on S = 0

21. *(a)* Zero orbital overlap is out of phase overlap and it is because of the different orientation / direction of approach of orbitals.

22. *(b)* Number of electron in given species

$CH_4 = 6 + 4 = 10$

$NH_4^+ = 7 + 3 = 10$

$BH_4^- = 5 + 4 + 1 = 10$

They all have same number of electron, i.e. are isoelectronic (species with same number of electron) and also have same structure, i.e. tetrahedral.

23. *(a)* (a) $[Ni(CN)_4]^{2-} = dsp^2$

(b) $BrF_5 = sp^3d^2$

(c) $XeF_4 = sp^3d^2$

(d) $[CrF_6]^{3-} = d^2sp^2$

24. *(d)*

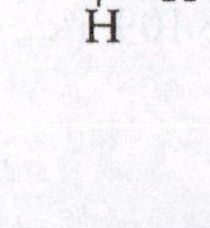

[CH_4 = hybridisation is sp^3, lone pair of electrons is 0, so that bond angle is maximum = 109°28′]

25. *(d)*

- SO_2 has sp^2 hybridisation and has non zero dipole moment. It has 1 lone pair.
- NF_3 has sp^3 hybridisation and has non-zero dipole moment. It has 1 lone pair.
- NH_3 (sp^3) Non-zero dipole moment - 1 lone pair
- XeF_2 (sp^3d) Non-zero dipole moment - 3 lone pair
- ClF_3 (sp^3d) Non-zero dipole moment - 2 lone pair
- SF_4 (sp^3d) Non-zero dipole moment - 1 lone pair.

26. *(d)* The correct match is A-II, B-III, C-IV, D-I.

27. *(d)* Among the given options, BrF_5 and $XeOF_4$ has square pyramidal geometry.

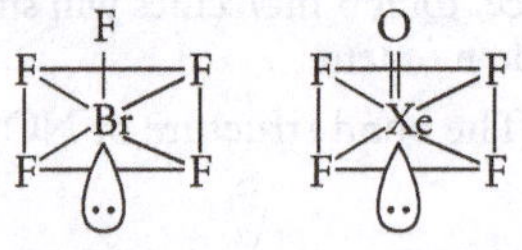

SbF_5 and PCl_5 has trigonal bipyramidal geometry.

28. *(d)* Statements given in *A*, *B* and *C* are correct while *D* and *E* are incorrect.

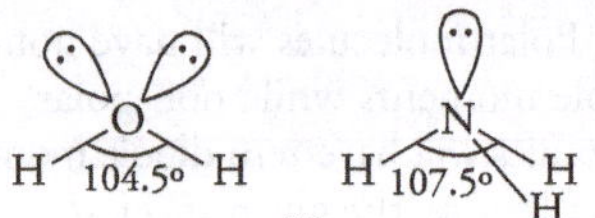

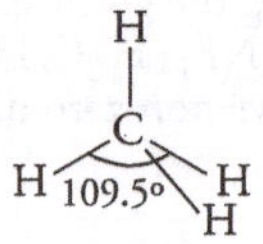

- $O \rightarrow sp^3$ $N \rightarrow sp^3$ $C \rightarrow sp^3$
- Dipole moment order : $CH_4 < NH_3 < H_2O$
- H_2O and NH_3 are Lewis base.
- NH_3 act as Lowry Bronsted base. In solution of NH_3 in H_2O.

Concept Enhancer

- According to Bronsted-Lowry acids and bases concept, the acid-base pair that differs only by one proton is called conjugate acid-base pair.
- If Bronsted acid is strong acid then its conjugate base is a weak base and *vice-versa*.

29. *(c)* The correct match is
A-III, B-I, C-IV, D-II

NH_3 = Trigonal pyramidal

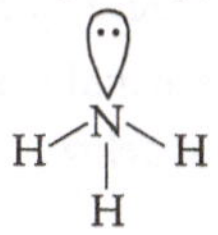

BrF_5 = Square pyramidal

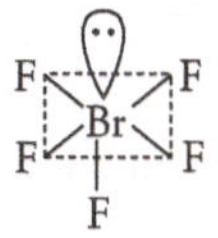

PCl_5 = Trigonal bipyramidal

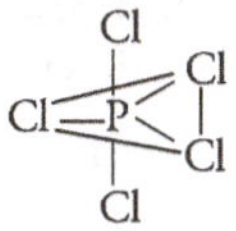

CH_4 = Tetrahedral

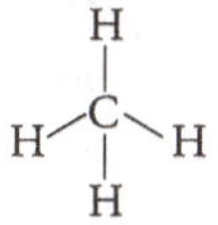

30. *(b)* Both statement I and statement II are false.
The correct form of these statements are, PF_5 has sp^3d-hybridisation while BrF_5 has sp^3d^2 hybridisation.
SF_6 has sp^3d^2-hybridisation while $[Co(NH_3)_6]^{3+}$ has d^2sp^3-hybridisation.

31. *(a)*

Molecules	Hybridisation	Shape/geometry
A. SF_4	sp^3d	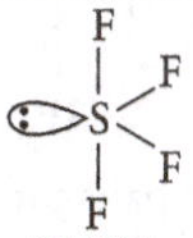 *see-saw*
B. BrF_3	sp^3d	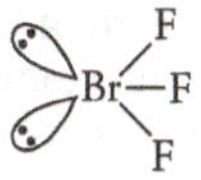 Bent T-shape
C. BrO_3^-	sp^3	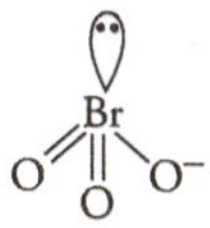 Pyramidal
D. NH_4^+	sp^3	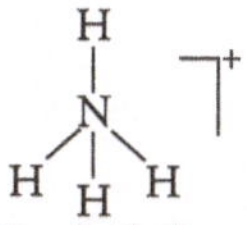 Tetrahedral

Hence, correct match is A-III, B-IV, C-II, D-I.

32. *(c)* The correct match is A-III, B-I, C-II, D-IV.
A. SO_2Cl_2 has sp^3-hybridisation and has tetrahedral geometry.
B. NO is paramagnetic in nature.
C. NO_2^- is diamagnetic in nature.
D. I_3^- has sp^3d-hybridisation and has linear geometry.

33. *(a)* The correct match is
A-IV, B-I, C-II, D-III.
ICl = I—Cl = Linear

ICl_3 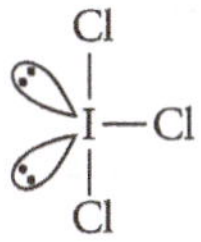 = T-shape

ClF_5 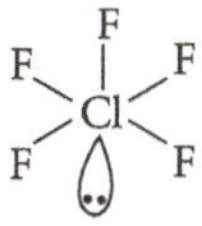 = Square pyramidal

IF_7 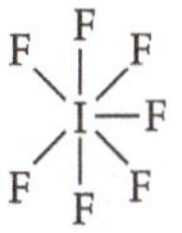 = Pentagonal bipyramidal

Hence, option (a) is correct.

34. *(c)* The geometry of the given species are as follows

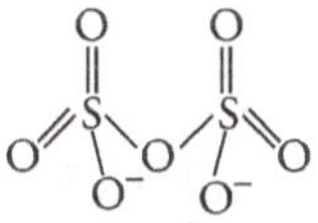
$S_2O_7^{2-}$
Tetrahedral

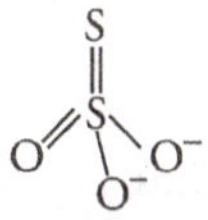
$S_2O_3^{2-}$
Tetrahedral

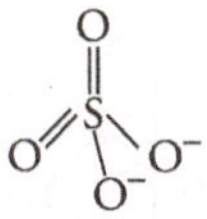
SO_4^{2-}
Tetrahedral

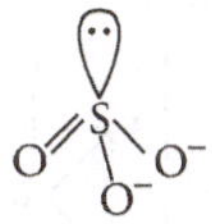
SO_3^{2-}
Pyramidal

Thus, among the given species only SO_3^{2-} has pyramidal geometry.

Time Saver Tech

The species that does not contain any lone pair of electron will have regular geometry i.e., tetrahedral while the species that contains lone pair of electron will not have regular geometry i.e., pyramidal (SO_3^{2-}).

35. *(b)* The hybridisation of central atom in the given molecules/ions are as follows

$NO_3^- \rightarrow sp^2$ 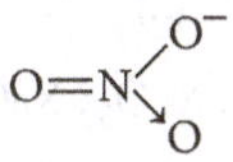

$BCl_3 \rightarrow sp^2$ 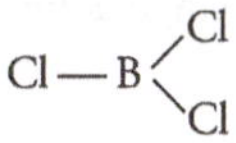

$ClO_2^- \rightarrow sp^3$ 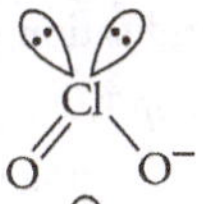

$ClO_3^- \rightarrow sp^3$ 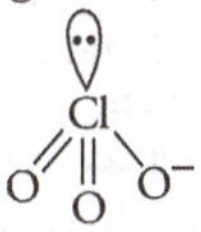

Thus, two molecules/ions i.e., ClO_2^- and ClO_3^- has sp^3 hybridisation.

36. *(d)* The correct match is
A-III, B-II, C-I, D-IV.

Species	Hybridisation	Shape	Structure
A. H_3O^+	sp^3	Pyramidal	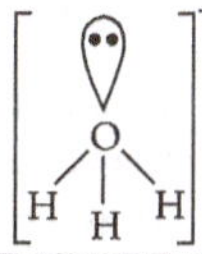
B. Acetylide anion	sp	Linear	$^{\ominus}C\equiv C^{\ominus}$
C. NH_4^+	sp^3	Tetrahedral	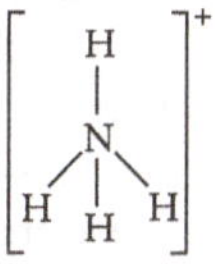
D. ClO_2^-	sp^3	Bent	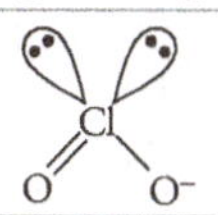

37. *(a)* The correct match is
A-II, B-I, C-IV, D-III.

38. *(c)* ∴ The correct match is A-(II), B-(IV), C-(I), D-(III).

39. *(b)* The shapes of the covalently bonded compounds are assigned on the basis of VSEPR theory.
So correct match is A-(III), B-(IV), C-(I), D-(II)

40. *(b)* The correct order is (A) → (III), (B) → (IV), (C) → (I), (D) → (II)

41. *(c)* The correct matches are :
(A)-(III), (B)-(I), (C)-(V) and (D)-(IV).

42. *(d)* In NO_2^-
In N is sp^2-hybridised.

NO_2^- 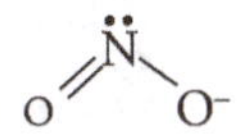

In NO_2^+, N sp-hybridised.

NO_2^+ $O=\overset{\oplus}{N}=O$

In NH_4^+, N is sp^3-hybridised.

NH_4^+ 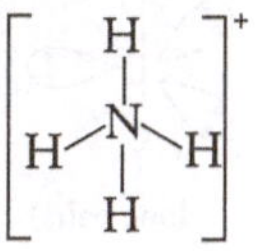

Therefore, hybridisation of N in NO_2^-, NO_2^+, NH_4^+ are sp^2, sp, sp^3 respectively.

43. *(a)* If number of sigma bond (σ), co-ordinate bond and lone pair are same for given pairs then, they are isostructural. Hence, SO_4^{2-}, CrO_4^{2-}, $SiCl_4$ and $TiCl_4$ have tetrahedral shape and are isostructural.

44. *(c)* $SF_4 \Rightarrow$

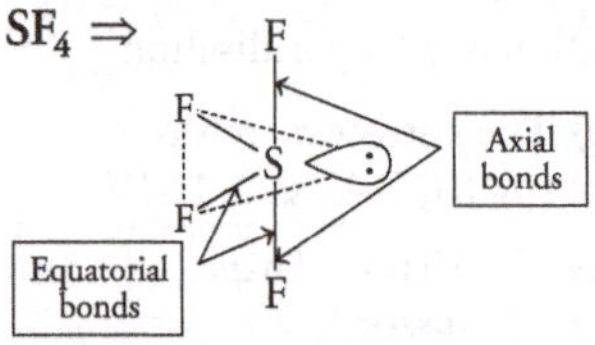

S (sp^3d-hybridised)
Geometry Trigonal bipyramidal
Shape *see-saw*
Axial S—F bonds are longer than equatorial bonds.
Hence, among given species only SF_4 has unequal bond lengths.

45. *(6)* Among the given species, total 6 species have sp^2 hybrid orbital for bonding with central atoms.
These are
SO_2, C_2H_4, BCl_3, HCHO, C_6H_6, BF_3

46. *(3)*

Molecules/Ions	Shape
PF_5, PCl_5, $Fe(CO)_5$	Trigonal bipyramidal
BrF_5	Square pyramidal
$[PtCl_4]^{2-}$	Square planar
BF_3	Trigonal planar

Hence, total 3 molecules have trigonal bipyramidal shape.

47. *(4)* The number of species that uses sp^3 hybrid orbitals in bonding are 4 i.e., NH_3, SiO_2, H_2O, CH_4.
While other species uses sp^2 and sp-hybrid orbitals which are as follows.
sp^2 — SO_2, BF_3
sp — $BeCl_2$, CO_2

48. *(3)*

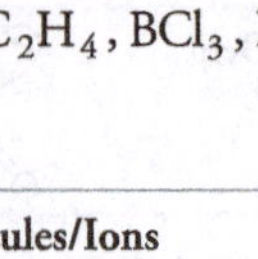
(2 lone pairs)

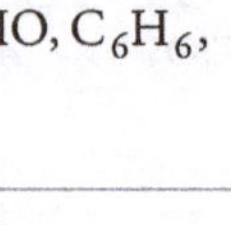
(1 lone pair)

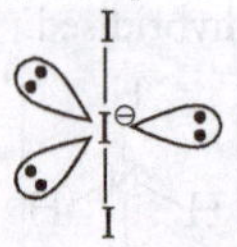
(3 lone pairs)

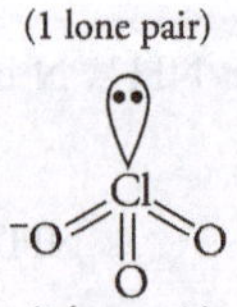
(1 lone pair)

Among the given species I_3^- contains maximum number of unpaired electrons i.e., 3.

49. *(3)*

Molecules	Shape	Lone pairs
H_2O	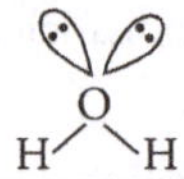	2
N_2		2
CO	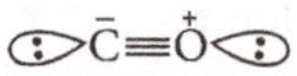	2
XeF_4	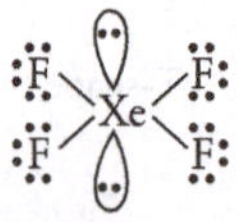	14
NH_3	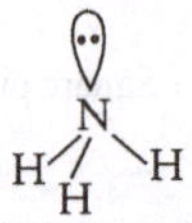	1
NO	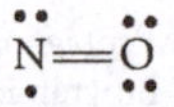	3
CO_2	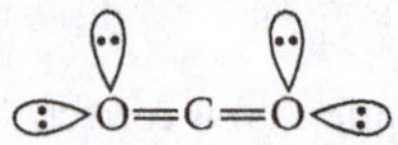	4
F_2	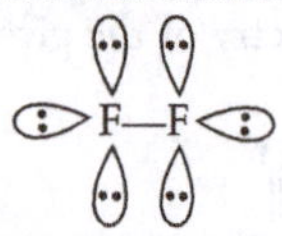	6

Thus, H_2O, N_2 and CO i.e., 3 molecules have only 2 lone pair of electrons.

50. *(4)*

XeF_5^+ 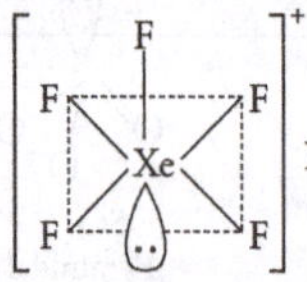

XeF_5^- 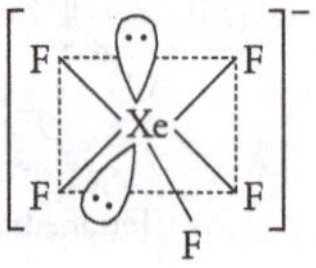

XeO_3 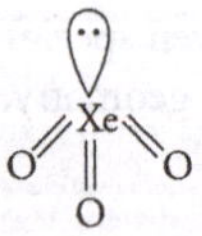

XeO_3F_2 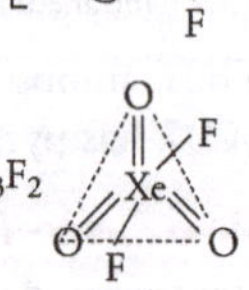

XeO_2F_2 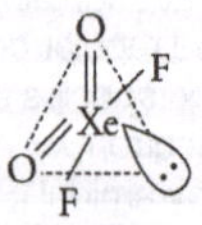

$XeOF_4$ 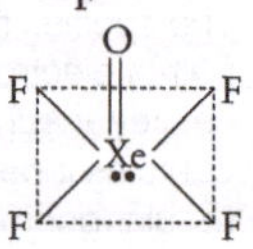

XeF_4 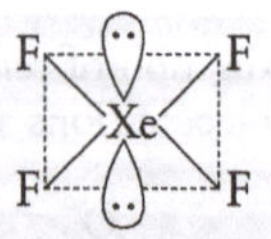

Thus, XeF_5^+, XeO_3, XeO_2F_2 and $XeOF_4$ have single lone pair on central atom.

51. *(3)* N_3^- and I_3^- are linear whereas NO_2^-, O_3 and SO_2 are bent
I_3^- is linear because its central atom is sp^3d hybridised and contains 3 lone pairs as shown below.

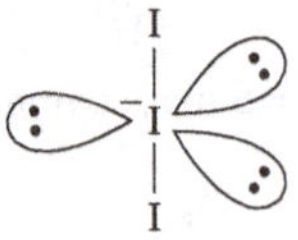

Also, N_3^- structure is linear and has sp hybridisation.

$\overset{-}{N}=\overset{+}{N}=\overset{-}{N}$
(Linear)

SO_2 is bent in shape.

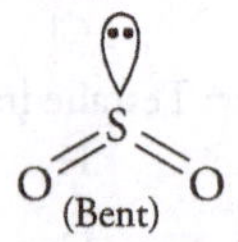
(Bent)

O_3 is bent in shape.

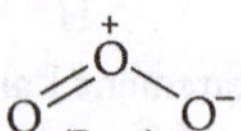
(Bent)

NO_2^-, that is nitrite ion, the N-atom have sp^2-hybridisation, thus it adopts bent geometry.

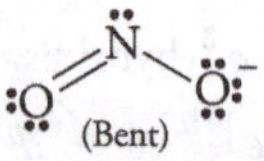
(Bent)

52. *(1)*

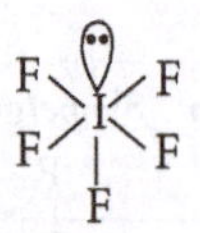
IF_5

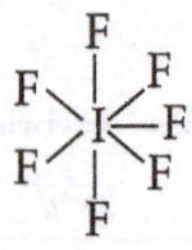
IF_7

Lone pairs present on the central atom of interhalogen, IF_5 is one and IF_7 is zero. So, sum of lone pairs present on the central atom of the interhalogen IF_5 and IF_7 is (1+0) equal to one.

53. *(4)*

Species	Hybridisation	Structure
XeF_4	sp^3d^2	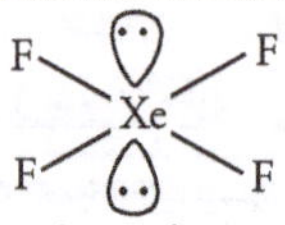 Square planar
SF_4	sp^3d	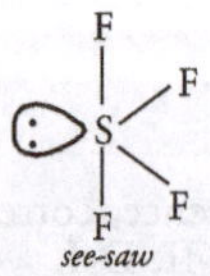 *see-saw*

SiF_4	sp^3	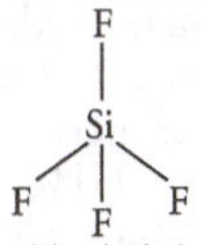 Tetrahedral
BF_4^-	sp^3	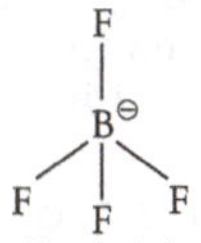 Tetrahedral
BrF_4^-	sp^3d^2	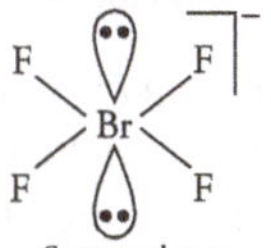 Square planar
$[Cu(NH_3)_4]^{2+}$	dsp^2	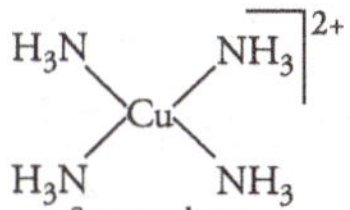 Square planar
$[FeCl_4]^{2-}$	sp^3	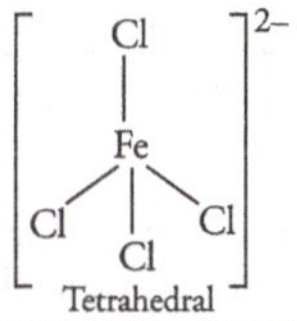 Tetrahedral
$[PtCl_4]^{2-}$	dsp^2	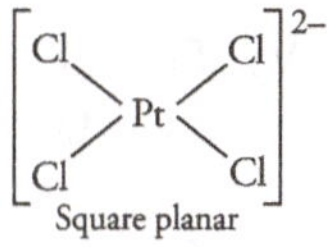 Square planar

So, total 4 species have square planar shape.

54. *(3)*

PF_5	sp^3d	Triangular pyramidal
BrF_4^-	sp^3d^2	Square planar
IF_5	sp^3d^2	Square pyramidal
BrF_5	sp^3d^2	Square pyramidal
$XeOF_4$	sp^3d^2	Square pyramidal
ICl_4^-	sp^3d^2	Square planar

In above molecules, 3 molecules (IF_5, BrF_5, $XeOF_4$) have square pyramidal structure.

55. *(5)* Five molecules have linear structure.

XeF_2 = 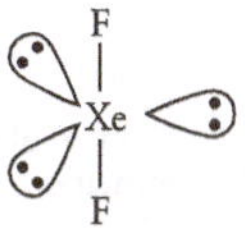 (Linear)

I_3^- = 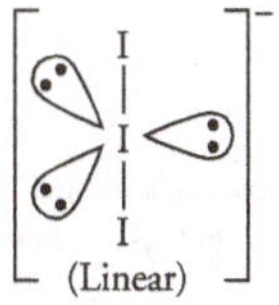 (Linear)

$BCl_2^{\ominus}$ = 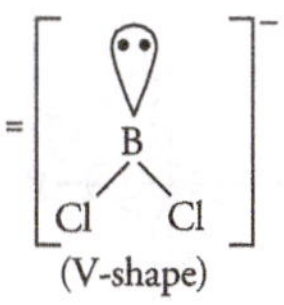 (V-shape)

C_3O_2 = [O=C=C=C=O] Linear

I_3^+ = 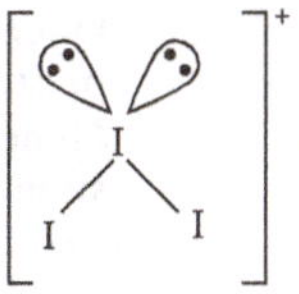 (V-shape)

CO_2 = O=C=O (Linear)

SO_2 = 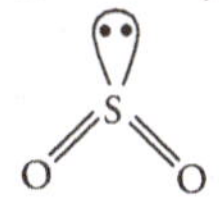 (V-shape)

$BeCl_2$ = Cl—Be—Cl (Linear)

56. *(4)* Hybridisation $= \frac{1}{2}$ (Valence electron in central atom + Number of monovalent atom attached to central atom + Negative charge – Positive charge)

$BrF_5 : \frac{1}{2}[7+5-0+0] = 6 (sp^3d^2)$

$SF_6 : \frac{1}{2}[6+6-0+0] = 6 (sp^3d^2)$

$ICl_4^- : \frac{1}{2}[7+4-0+1] = 6 (sp^3d^2)$

$IF_5 : \frac{1}{2}[7+5-0+0] = 6 (sp^3d^2)$

So, number of molecules ions having sp^3d^2 hybridisation is 4.

57. *(6)* General planar structures are linear, bent, T-shaped, trigonal planar and square planar.

$NO_3^- \rightarrow$ Trigonal planar
$H_2O_2 \rightarrow$ Open book like (3D) (non-planar)
$BF_3 \rightarrow$ Trigonal planar
$PCl_3 \rightarrow$ Pyramidal
$XeF_4 \rightarrow$ Square planar
$SF_4 \rightarrow$ *See-saw*
$XeO_3 \rightarrow$ Trigonal pyramidal
$PH_4^+ \rightarrow$ Tetrahedral
$SO_3 \rightarrow$ Trigonal planar
$[Al(OH)_4]^- \rightarrow$ Tetrahedral

Hence, only 6 have non-planar geometry.

58. *(1)* PF_5 molecule

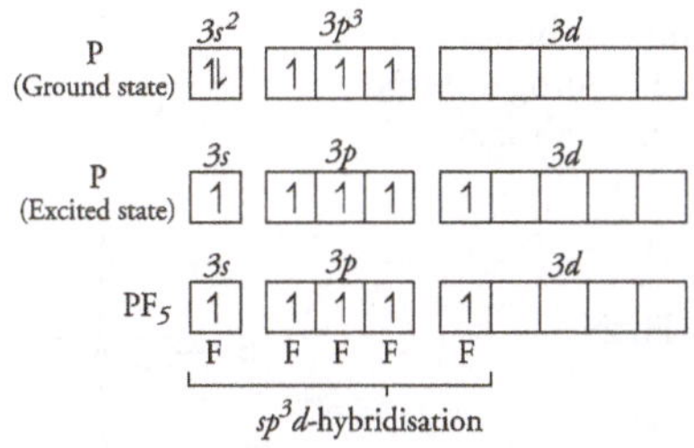

So, hybridisation of P in PF_5 is sp^3d and the value of y in sp^xd^y is 1.

59. *(1)* In H_2O,

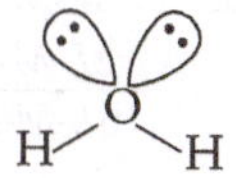

Total lone pairs = 2

60. *(3)* The structure of the given compounds are as follows :

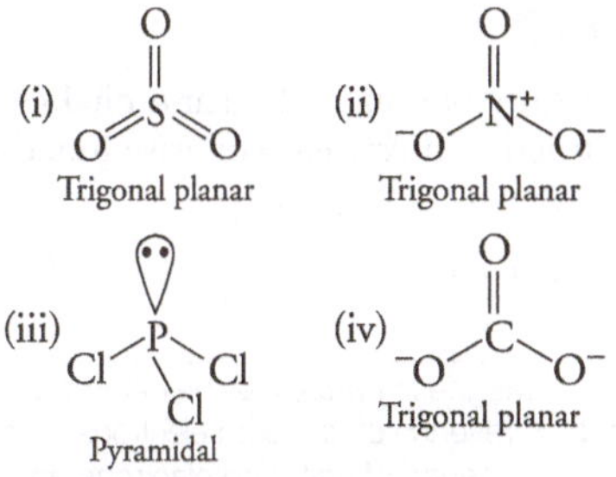

3 molecules/ions i.e. (i), (ii) and (iv) have non-pyramidal shape.

61. *(2)* AB_3 is a T-shaped molecule.

Hence, it can be ClF_3 with 2 lone pairs

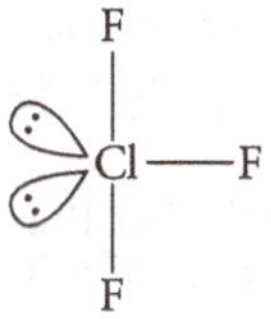

(T-shaped)

62. *(3)* Shape of I_3^- is :

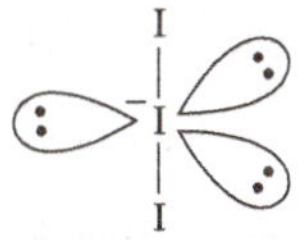

The number of lone pairs of electron on the central atom in I_3^- is 3.

63. *(2)* Number of lone pairs on central atom in given compounds are as follows

ClF_3 = 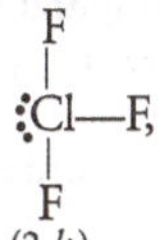 (2 *lp*)

XeF_4 = 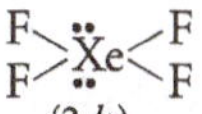 (2 *lp*)

Hence, two (lone pair) lp on central atom is ClF_3 and XeF_4.

64. *(20)*

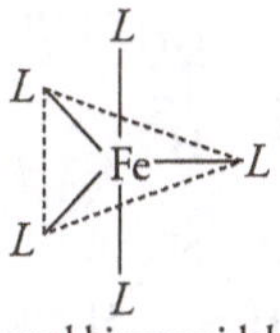

Trigonal bipyramidal

$1 L\text{-}M\text{-}L = 180°$
$6 L\text{-}M\text{-}L = 90°$
$3 L\text{-}M\text{-}L = 120°$
$10 L\text{-}M\text{-}L$ angles

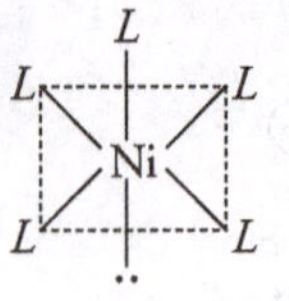

$8L\text{-}M\text{-}L = 90°$
$2L\text{-}M\text{-}L = 180°$
$10L\text{-}M\text{-}L$ angles

Square bipyramidal

$\therefore$ Sum of L-M-L angles = 20

Topic 3

01. *(a)* Antibonding molecular orbital (ABMO) are formed by destructive interference of wave function.

(ABMO) $\sigma^* = \psi_A - \psi_B$

02. *(c)* The correct match is A $\rightarrow$ III, B $\rightarrow$ I, C $\rightarrow$ IV, D $\rightarrow$ II.

03. *(a)*

> **idea** Magnetic nature can be detected by molecular orbital theory. Presence of unpaired electrons means paramagnetic and absence of unpaired electrons means diamagnetic in nature.

O_2^- (Total electrons = 17) = $\sigma 1s^2, \sigma^* 1s^2\ \sigma 2s^2\ \sigma^* 2s^2\ \sigma 2p_z^2,$

$\pi 2p_x^2 = \pi 2p_y^2, \pi^* 2p_x^2 = \pi^* 2p_y^1, \sigma^* 2p_z^0$

Number of unpaired electron = 1

So, it is paramagnetic.

Bond order $= \dfrac{N_b - N_a}{2} = \dfrac{10-7}{2} = 1.5$

04. *(a)* A is false but R is true.

Corrected form of A is, dipole-dipole interactions are purely covalent interactions of dipolar covalent bonds. Ends of dipoles of a polar covalent bonds possess partial charges (δ^+ and δ^-) which are less than unit electronic charge, $e = 1.6 \times 10^{-19}$ C.

05. *(c)* $Be_2 = \sigma 1s^2, \sigma^* 1s^2, \sigma 2s^2, \sigma^* 2s^2$

$$\text{Bond order} = \frac{4-4}{2} = 0$$

If bond order of chemical species is zero then that chemical species does not exist.

Therefore, Be_2 does not exist.

06. *(a)* CN^- has 14 electrons (6 from C, 7 from N and 1 for negative charge).

Its MO configuration must be similar to

$N_2 : \sigma 1s^2\ \sigma^* 1s^2\ \sigma 2s^2\ \sigma^* 2s^2\ \pi 2p_x^2\ \pi 2p_y^2\ \sigma 2p_z^2$

$$\therefore \quad \text{Bond order} = \frac{10-4}{2} = 3$$

As there is no unpaired electron in CN^-, thus it is diamagnetic.

07. *(c)* CH_4 doesn't show intermolecular hydrogen bonding while out of HCN and NH_3, extent of hydrogen bonding is more in NH_3 as ammonia could form 'three' hydrogen bond per molecule while HCN could form only one hydrogen bond per molecule.

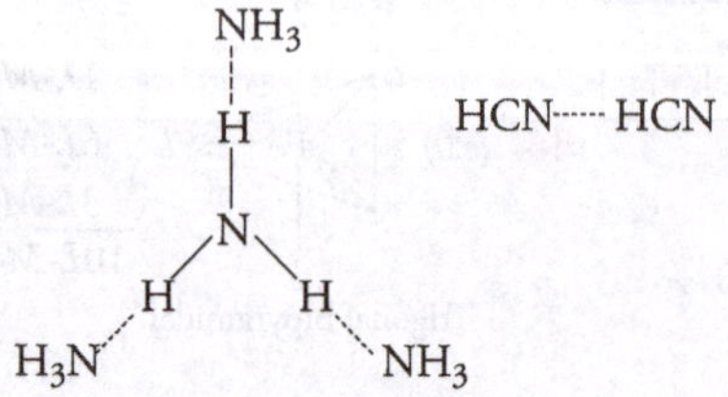

H_2O have even more stronger hydrogen bonding due to the presence of highly electronegative oxygen atom.

Hence, the correct order of increasing intermolecular H-bond strength is : $CH_4 < HCN < NH_3$.

08. *(a)* Among the given options, CO is a diamagnetic molecule. It can be proved by molecular orbital (MO) theory. The electronic configuration CO is given below.

CO (Number of electrons = 14)

Electronic configuration

$= \sigma 1s^2, \sigma^* 1s^2, \sigma 2s^2, \sigma^* 2s^2, \sigma 2p_z^2, \pi 2p_x^2 \approx \pi 2p_y^2$

Since, there is no unpaired electron in the CO molecule, so it is diamagnetic.

NO contains 1, while B_2 and O_2 contains 2 unpaired electrons each hence, all 3 of them are paramagnetic.

09. *(d)* Considering molecular orbital theory (MOT) :

The electronic configuration of

$Li_2^+ (Z = 5) = \sigma 1s^2, \sigma^* 1s^2, \sigma 2s^1$

$$\text{Bond order (BO)} = \frac{N_b - N_a}{2} = \frac{3-2}{2} = \frac{1}{2}$$

The electronic configuration of

$Li_2^- (Z = 7) = \sigma 1s^2, \sigma^* 1s^2, \sigma 2s^2, \sigma^* 2s^1$

$$\text{Bond order (BO)} = \frac{N_b - N_a}{2} = \frac{4-3}{2} = \frac{1}{2}$$

For the species having the same value of BO, the specie having lesser number of antibonding electrons $[N_a]$ will be more stable.

Here, N_a of Li_2^+ (2) < N_a of Li_2^- (3).

So, their order of stability will be $Li_2^+ > Li_2^-$.

10. *(c)*

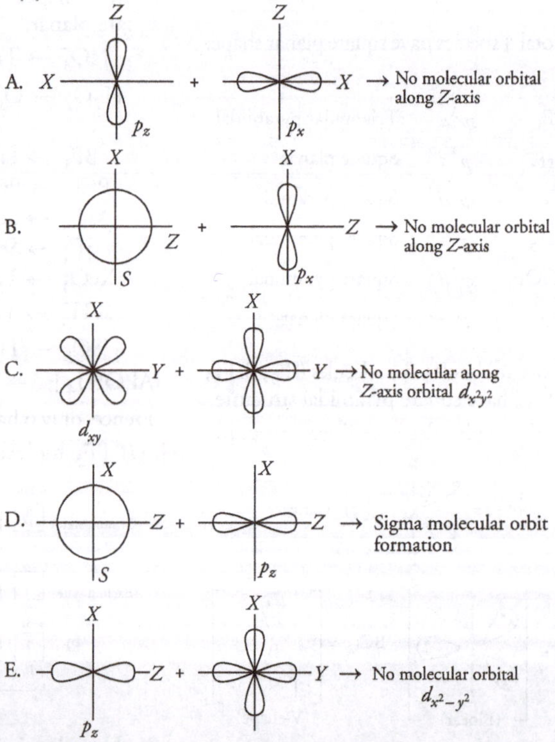

So, only $2s$ and $2p_z$ will lead to formation of molecular orbital in a homonuclear diatomic molecule.

11. *(a)* Statements given in A, D and E are correct while statements given in B and C are incorrect. Their correct form are as follows

- Intramolecular hydrogen bonding is present in *o*-nitrophenol.
- Intermolecular hydrogen bonding is present in HF.

12. *(b)*

Species and number of electrons	Molecular orbital electronic configuration	Number of unpaired electrons (n) and magnetic moment
1. O_2-16 electrons	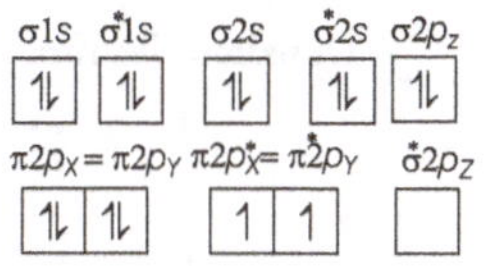	$n = 2$, $\mu = \sqrt{n(n+2)}$ $\mu = \sqrt{2 \times 4}$ BM $= 2.82$ BM
2. O_2^+-15 electrons	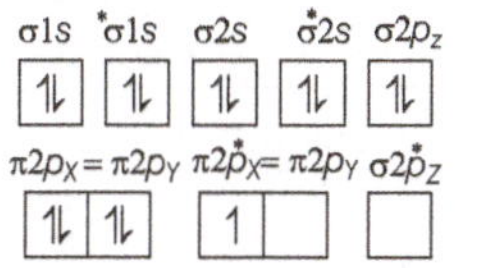	$n = 1$, $\mu = \sqrt{1 \times 3}$ $= 1.73$ BM
3. O_2^--17 electrons	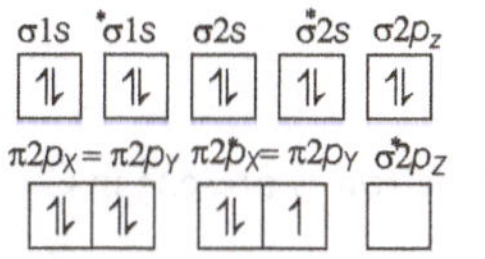	$n = 1, \mu = 1.73$ BM

13. *(a)* $C_2^- : (\sigma 1s)^2 (\sigma^* 1s)^2 (\sigma 2s)^2 (\sigma^* 2s)^2 (\pi 2p_x^2 = \pi 2p_y^2)(\sigma 2p_z^1)$

or $KK(\sigma 2s)^2 (\sigma^* 2s)^2 (\pi 2p_x^2 = \pi 2p_y^2) \sigma 2p_z^1$

$$\text{Bond order} = \frac{1}{2}(N_b - N_a) = \frac{1}{2}(9-4) = 2.5$$

$$C_2 : KK(\sigma 2s)^2 (\sigma^* 2s)^2 (\pi 2p_x^2 = \pi 2p_y^2)$$

$$\text{Bond order} = \frac{1}{2}(N_b - N_a) = \frac{1}{2}(8-4) = 2$$

The BO value of other molecules is

O_2	NO	F_2
2	2.5	1

These values will decrease with addition of electron as incoming electron will occupy AMO. As increase in BO value indicates increase in stability. Hence, C_2 will be stabilised by anion formation.

14. *(b)* The change of O_2 to O_2^- can be as follows:

$O_2 (16\, e^-)$ [Dioxygen] → Energy $\xrightarrow{+e^-}$ $O_2^- (17\, e^-)$ [Super-oxide] → Energy

$[8e^-]\ \sigma 2p_z^2$ $[8e^-]\ \sigma 2p_z^2$

$\pi 2p_x = \pi 2p_y^2 \overset{*}{\pi} 2p_x^1 = \overset{*}{\pi} 2p_y^1$ $\pi 2p_x^2 = \pi 2p_y^2\ \overset{*}{\pi} 2p_x^2 = \overset{*}{\pi} 2p_y^1$

Half-filled anti-bonding π^*-MO$_s$

So, in the formation of O_2^- from O_2, the 17th electron goes to the $\overset{*}{\pi} 2p_x$ or $\overset{*}{\pi} 2p_y$ molecular orbital (anti-bonding) which is half-filled in O_2.

15. *(d)* The bond order of acetylide ion C_2^{2-} ($^-C{\equiv}C^-$) is 3 and this ion is diamagnetic in nature.

- Bond order of $O_2^- = \frac{10-7}{2} = 1.5$ and paramagnetic
- Bond order of $O_2^+ = \frac{10-5}{2} = 2.5$ and paramagnetic
- Bond order of $N_2^+ = \frac{9-4}{2} = 2.5$ and paramagnetic
- Bond order of $NO^+ = \frac{10-4}{2} = 3$ and diamagnetic

Thus, bond order and magnetic property of acetylide ion are same as that of NO^+.

16. *(c)* Paramagnetic behaviour are shown by the species containing unpaired electrons.

Number of unpaired electrons of each species are

(A)	O_2	2
(B)	N_2	0
(C)	F_2	0
(D)	S_2	2
(E)	Cl_2	0

So, A and D are paramagnetic.

17. *(d)* $N_2 = \sigma 1s^2\ \sigma^* 1s^2\ \sigma 2s^2\ \sigma^* 2s^2 \pi 2p_x^2 = \pi 2p_y^2, \sigma 2p_z^2$

Number of unpaired electron present in HOMO = 0

$N_2^+ = \sigma 1s^2, \sigma^* 1s^2, \sigma 2s^2, \sigma^* 2s^2,$
$\pi 2p_x^2 = \pi 2p_y^2, \sigma 2p_z^1$

Number of unpaired electron present in HOMO = 1

$O_2 = \sigma 1s^2, \sigma^* 1s^2, \sigma 2s^2, \sigma^* 2s^2,$
$\sigma 2p_z^2, \pi 2p_x^2 \approx \pi 2p_y^2, \pi^* 2p_x^1 \approx \pi^* 2p_y^1$

Number of unpaired electron present in HOMO = 2

$O_2^+ = \sigma 1s^2, \sigma^* 1s^2, \sigma 2s^2, \sigma^* 2s^2,$
$\sigma 2p_z^2, \pi 2p_x^2 = \pi 2p_y^2, \pi^* 2p_x^1 = \pi^* 2p_y^0$

Number of unpaired electron present in HOMO = 1

Thus, the number of unpaired electrons are 0, 1, 2, 1.

18. *(c)* According to MO theory, the bond orders for O_2^{2-}, CO and NO^+ are 1, 3 and 3.

19. *(c)* NO : $\sigma 1s^2 \sigma^* 1s^2 \sigma 2s^2 \sigma^* 2s^2 \sigma 2p_z^2 \pi 2p_x^2 = \pi 2p_y^2 \pi^* 2p_x^1$

$$\text{Bond order} = \frac{10-5}{2} = 2.5$$

NO^+ : $\sigma 1s^2 \sigma^* 1s^2 \sigma 2s^2 \sigma^* 2s^2 \sigma 2p_z^2 \pi 2p_x^2 = \pi 2p_y^2$

$$\text{Bond order} = \frac{10-4}{2} = 3$$

O_2: $\sigma 1s^2 \sigma^* 1s^2 \sigma 2s^2 \sigma^* 2s^2 \sigma 2p_z^2 \pi 2p_x^2 = \pi 2p_y^2 \pi^* 2p_x^1 = \pi^* 2p_y^1$

$$\text{Bond order} = \frac{10-6}{2} = 2$$

O_2^+: $\sigma 1s^2 \sigma^* 1s^2 \sigma 2s^2 \sigma^* 2s^2 \sigma 2p_z^2 \pi 2p_x^2 = \pi 2p_y^2 \pi^* 2p_x^1$

$$\text{Bond order} = \frac{10-5}{2} = 2.5$$

Increase in bond order indicates increased strength of bond. Hence, correct answer is option (c).

20. *(a)*

Species	O_2^+	O_2	O_2^-	O_2^{2-}
No. of e^-	15	16	17	18
Bond order	2.5	2	1.5	1

Increasing bond order is $O_2^{2-} < O_2^- < O_2 < O_2^+$.

21. *(a)* Extensive H-bonding is observed in 2-nitrophenol as two H-atoms bonded to strongly electronegative atoms (nitrogen, oxygen) are present. H-bonding occurs when H-bonded to strongly electronegative atoms exists in the vicinity of other electronegative atoms possessing lone pairs of electron.

22. *(a)* A-(iii), B-(iv), C-(i), D-(ii)

23. *(d)* Energy order of MOs B.O. $= \dfrac{N_b - N_a}{2}$

Bond length $\propto \dfrac{1}{\text{BO (Bond order)}} \propto \dfrac{1}{\text{Bond strength}}$

So, bond $\propto$ BO strength the BO of given 4 species are :

Species	NO	NO^+	NO^{2+}	NO^-
Bond order	2.5	3	2.5	2

So,the order of bond strength : $NO^+ > NO = NO^{2+} > NO^-$

So, NO^- has minimum bond strength.

24. *(a)*

Species	MO energy order	Bond order (BO)	n, no. of unpaired e^-	Magnetic character
$C_2^{2-}(14e^-)$	$[8\bar{e}]\pi 2p_x^2 = \pi 2p_y^2 \sigma 2p_z^2$	$\dfrac{6-0}{2}=3$	0	Diamagnetic
$O_2(16e^-)$	$[8\bar{e}]\sigma 2p_z^2 \pi 2p_x^2 = \pi 2p_y^2 \pi^* 2p_x^1 = \pi^* 2p_y^1$	$\dfrac{6-2}{2}=2$	2	Paramagnetic
$O_2^{2-}(18e^-)$	$[8\bar{e}]\sigma 2p_z^2 \pi 2p_x^2 = \pi 2p_y^2 \pi 2p_x^2 = \pi 2p_y^2$	$\dfrac{6-4}{2}=1$	0	Diamagnetic
$N_2^{2-}(16e^-)$	$[8\bar{e}]\pi 2p_x^2 = \pi 2p_y^2 \sigma 2p_z^2 \overset{*}{\pi} 2p_x^1 = \overset{*}{\pi} 2p_y^1$	$\dfrac{6-2}{2}=2$	2	Paramagnetic

Bond length $\propto \dfrac{1}{\text{BO (Bond order)}}$

So order of bond length $\underset{(BO=3)}{C_2^{2-}} < \underset{(BO=2)}{O_2} = N_2^{2-} < \underset{(BO=1)}{O_2^{2-}}$

The diamagnetic species with shortest bond length is C_2^{2-} (option-a).

25. *(c)* N_2^+ $(13e^-\text{'s}) = \sigma 1s^2 \sigma^* 1s^2\ \sigma 2s^2\ \sigma^* 2s^2$

$\pi 2p_x^2 = \pi 2p_y^2\ \sigma 2p_z^1$

Thus, in case of N_2^+, two π-bonds and half σ-bond are present in the bonding MOs.

26. *(d)* $NO \longrightarrow NO^+$

Bond order of NO = 2.5 (Paramagnetic)

Bond order of NO^+ = 3 (Diamagnetic)

27. *(6)* The electronic configuration of $O_2(16e^-)$ is $\sigma 1s^2\ \sigma^* 1s^2\ \sigma 2s^2\ \sigma^* 2s^2\ \sigma 2p_z^2$

$\pi 2p_x^2 = \pi 2p_y^2,\ \pi^* 2p_x^1 = \pi^* 2p_y^1$

So, number of e^- in π^* of $O_2 = 2$

number of e^- in π^* of $O_2^+ = 1$

number of e^- in π^* of $O_2^- = 3$

So, total number of e^- in $\pi^* = 2 + 1 + 3 = 6$

28. *(4)*

Species	Unpaired electrons
N_2	0
O_2	2
C_2^-	1
O_2^-	1
O_2^{2-}	0
H_2^+	1
CN^-	0
He_2^+	1

So, one unpaired electron is present in C_2^-, O_2^-, H_2^+ and He_2^+.

29. *(2)* Formula for Bond order (B.O) $= \dfrac{N_B - N_A}{2}$

Where, N_B = Number of electron in bonding orbital.

N_A = Number of electron in antibonding orbital.

For C_2; B.O. $= \dfrac{8-4}{2} = 2$

For O_2; B.O. $= \dfrac{10-6}{2} = 2$

For Be_2; B.O. $= \dfrac{4-4}{2} = 0$

For Li_2; B.O. $= \dfrac{4-2}{2} = 1$

For Ne_2; B.O. $= \dfrac{10-10}{2} = 0$

For N_2; B.O. $= \dfrac{10-4}{2} = 6$

He_2; B.O. $= \dfrac{2-2}{2} = 0$

Hence, total 2 molecules have bond order of 2.

30. *(5)*

> **idea** This problem is based on the hydrogen bondig. It occurs when H atom is bonded with oxygen, nitrogen and fluorine.

Among the given compounds, following compounds exhibit hydrogen bonding,

CH_3OH, H_2O, 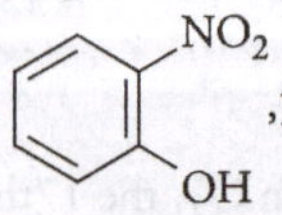 ,HF, NH_3

Hence, total 5 compounds are there.

31. *(2)* According to molecular orbital theory, number of unpaired electrons in the given molecules/species are as follows

O_2 = Number of unpaired electrons is 2
O_2^{-1} = Number of unpaired electron is 1
NO = Number of unpaired electron is 1
CN^{-1} = Number of unpaired electron is zero
O_2^{2-} = Number of unpaired electron is zero

32. *(8)* Molecular orbital formed by $2s$ atomic orbital = 2 ($\sigma 2s$ and $\sigma^* 2s$)
Molecular orbital formed by $2p$ atomic orbital
$= 6 (\sigma 2p_z$ and $\sigma^* 2p_z, \pi 2p_x\ \pi 2p_y$ and $\pi^* 2p_x, \pi^* 2p_y)$
So, total molecular orbitals = 2 + 6 = 8

33. *(4)* Antibonding molecular orbital from $2s = 1$
Antibonding molecular orbital from $2p = 3$
Therefore, total number of antibonding molecular orbitals are $= 1 + 3 = 4$

34. *(1)*

Ion/molecule	Magnetic behaviour	Bond order
H_2	Diamagnetic	1
He_2^+	Paramagnetic	0.5
N_2^{2-}	Paramagnetic	2
O_2^{2-}	Diamagnetic	1
O_2^+	Paramagnetic	2.5
F_2	Diamagnetic	1
Ne_2^+	Paramagnetic	0.5
B_2	Paramagnetic	1

Only B_2 has paramagnetic magnetic behaviour and bond order of 1.

35. *(6)* NO^+ and CO both are isoelectronic species and each of them having bond order 3.
$\therefore$ Sum of bond order of NO^+ and CO = Bond order (NO^+) + Bond order (CO) = 3 + 3 = 6

36. *(2)* $N_2 : (\sigma 1s)^2(\sigma^* 1s)^2(\sigma 2s^2)(\sigma^* 2s^2)$
$(\pi 2p_x^2 = \pi 2p_y^2)(\sigma 2p_z)^2$; Diamagnetic
O_2^{2-}:
$(\sigma 1s)^2(\sigma^* 1s)^2(\sigma 2s)^2(\sigma^* 2s)^2(\sigma 2p_z)^2$
$(\pi 2p_x^2 = \pi 2p_y^2)(\pi^* 2p_x^2 = \pi^* 2p_y^2)$;
Diamagnetic
The number of species showing diamagnetism are 2.

37. *(4)* Molecular orbital configuration of
$B_2 \Rightarrow$
$\sigma 1s^2 \sigma^* 1s^2 \sigma 2s^2 \sigma^* 2s^2 \pi 2p_x^1 = \pi 2p_y^1$
(Paramagnetic)
$C_2^- \Rightarrow$
$\sigma 1s^2 \sigma^* 1s^2 \sigma 2s^2 \sigma^* 2s^2 \pi 2p_x^2 = \pi 2p_y^2 \sigma 2p_z^1$
(Paramagnetic)
$O_2^+ \Rightarrow \sigma 1s^2 \sigma^* 1s^2 \sigma 2s^2 \sigma^* 2s^2 \sigma 2p_z^2 \pi 2p_x^2 = \pi 2p_y^2 \pi^* 2p_x^2 = \pi^* 2p_y^1$
(Paramagnetic)
$He_2^+ \Rightarrow \sigma 1s^2 \sigma^* 1s^1$ (Paramagnetic)
The number of paramagnetic species are 4.

38. *(3)* Bond order $= \dfrac{N_B - N_A}{2}$
where, N_B = total e^- in bonding molecular orbitals
N_A = total e^- in antibonding molecular orbitals.

Species/Ion	No. of e^-	Bond order
CN^-	14	$(10-4/2) = 3$
NO^+	14	$(10-4/2) = 3$
O_2	16	$(10-6/2) = 2$
O_2^+	15	$(10-5/2) = 2.5$
O_2^{2+}	14	$(10-4/2) = 3$

$\therefore$ 3 species have identical bond order.

39. *(173)* Electronic configuration of B_2^+ is as follows
$B_2^+ = \sigma 1s^2, \sigma^* 1s^2, \sigma 2s^2, \sigma^* 2s^2, \pi 2p_x^1$
$\therefore B_2^+$ has one unpaired electron,
$\mu = \sqrt{1(1+2)} = \sqrt{3} = 1.73$ BM
$= 173 \times 10^{-2}$ BM
$\therefore \quad x \times 10^{-2} = 173 \times 10^{-2}$
$\therefore \quad x = 173$
Hence, answer is 173.

40. *(0)* In O_2^{2-}, 18 electrons are present.
Configuration is given by
$\sigma 1s^2, \sigma^* 1s^2, \sigma 2s^2, \sigma^* 2s^2, \sigma 2p_z^2$,
$\pi 2p_x^2 = \pi 2p_y^2, \pi^* 2p_x^2 = \pi^* 2p_y^2$
So, number of unpaired electrons in $O_2^{2-} = 0$.

41. *(10)* Configuration of O_2^{2-} is
$(O_2^{2-}) = \sigma 1s^2, \sigma^* 1s^2, \sigma 2s^2, \sigma^* 2s^2$,
$\sigma 2p_z^2, \pi 2p_x^2 = \pi 2p_y^2\ \pi^* 2p_x^2 = \pi^* 2p_y^2$
Number of electrons in all bonding molecular orbitals = 10
Number of electrons in all anti-bonding molecular orbitals = 8

42. *(0)* Electronic configuration of CO (14 electrons).
$(\sigma 1s)^2\ (\sigma^* 1s)^2\ (\sigma 2s)^2\ (\sigma^* 2s)^2$
$(\pi 2p_x)^2 = (\pi 2p_y)^2, \sigma 2p_z^2$
Bond order
$= \dfrac{[\text{No. of bonding electron} - \text{No. of antibonding electron}]}{2}$
$= \dfrac{10-4}{2} = 3$
Electronic configuration of NO^+ (14 electrons)
$(\sigma 1s)^2\ (\sigma^* 1s)^2\ (\sigma 2s)^2\ (\sigma^* 2s)^2$
$(\pi 2p_x)^2 = (\pi 2p_y)^2\ (\sigma 2p_z)^2$
$BO = \dfrac{10-4}{2} = 3$
Bond order of both CO and $NO^{\oplus} = 3$
Difference $= 0 = \dfrac{x}{2}$; $x = 0$

43. *(15)* NO is the covalent diatomic molecule which has bond order of 2.5.
Hence, total electrons in NO = 7 + 8 = 15

05 Thermodynamics

In JEE Main there is usually 0-2 question per paper are visible from this chapter. In past 3 years, 29 conceptual and 46 numerical questions were asked of moderate difficulty level from this chapter.

For detailed analysis scan QR code

Topic 1 / Fundamentals of Thermodynamics

• Level 1

01 Among the following, the set of parameters that represents path function, is **[2019, 09 April Shift-I]**

(A) $q+W$ (B) q
(C) w (D) $H-TS$

(a) (B) and (C) (b) (A) and (D)
(c) (B), (C) and (D) (d) (A), (B) and (C)

• Level 2

02 Ice and water are placed in a closed container at a pressure of 1 atm and temperature 273.15 K. If pressure of the system is increased 2 times, keeping temperature constant, then identify correct observation from following. **[2025, 28 Jan Shift-I]**

(a) The amount of ice decreases.
(b) The solid phase (ice) disappears completely.
(c) Volume of system increases.
(d) Liquid phase disappears completely.

03 What happens when methane undergoes combustion in systems A and B respectively? **[2023, 13 April Shift-II]**

	System A (Adiabatic system)	System B (Diathermic container)
(a)	Temperature remains same	Temperature rises
(b)	Temperature falls	Temperature remains same
(c)	Temperature falls	Temperature rises
(d)	Temperature rises	Temperature remains same

04 The combination of plots which does not represent isothermal expansion of an ideal gas is **[2019, 12 Jan Shift-II]**

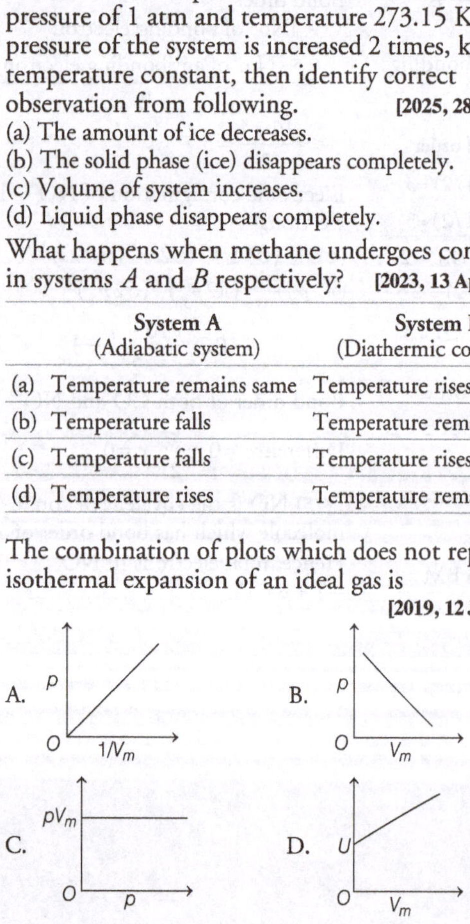

Codes
(a) A and C (b) A and D
(c) B and D (d) B and C

• Level 3

05 Which of the following properties will change when system containing solution 1 will become solution 2?

10 mol of solute x + 10 L of water	$\longrightarrow$	1 L of solution 1 + 1 mol of solute x + 1 L of water
(Solution 1)		(Solution 2)

[2025, 3 April Shift-I]

(a) Density (b) Concentration
(c) Gibbs free energy (d) Molar heat capacity

06 $C(s)+\frac{1}{2}O_2(g) \longrightarrow CO_2(g)+400\text{ kJ}$

$C(s)+\frac{1}{2}O_2(s) \longrightarrow CO(g)+100\text{ kJ}$

When coal of purity 60% is allowed to burn in presence of insufficient oxygen, 60% of carbon is converted into 'CO' and the remaining is converted into 'CO_2'. The heat generated when 0.6 kg of coal is burnt is .... **[2022, 29 july Shift-II]**

(a) 1600 kJ (b) 3200 kJ (c) 4400 kJ (d) 6600 kJ

07 Consider the reversible isothermal expansion of an ideal gas in a closed system at two different temp. T_1 and T_2 $(T_1 < T_2)$. The correct graphical depiction of the dependence of work done (W) on the final volume (V) is **[2019, 9 Jan Shift-I]**

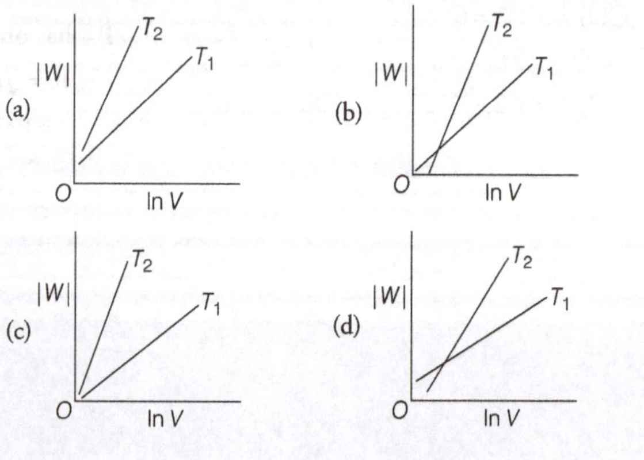

Numerical Value Type Questions

Imp. *Work done in Reversible Process*

08 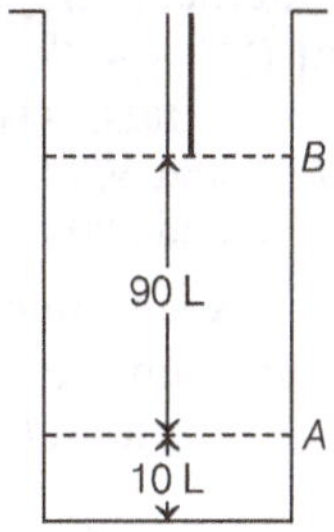

Consider the figure provided.
1 mol of an ideal gas is kept in a cylinder, fitted with a piston, at the position A, at 18°C. If the piston is moved to position B, keeping the temperature unchanged, then 'x' L atm work is done in this reversible process.
x = ... L atm (nearest integer)
[Given : Absolute temperature = °C + 273.15, $R = 0.08206$ L atm mol^{-1} K^{-1}]. **[2024, 8 April Shift-I]**

09 If 5 moles of an ideal gas expands from 10 L to a volume of 100 L at 300 K under isothermal and reversible condition then work, w, is $-x$ J. The value of x is ....
(Given, $R = 8.314$ JK^{-1} mol^{-1}) **[2024, 31 Jan Shift-II]**

10 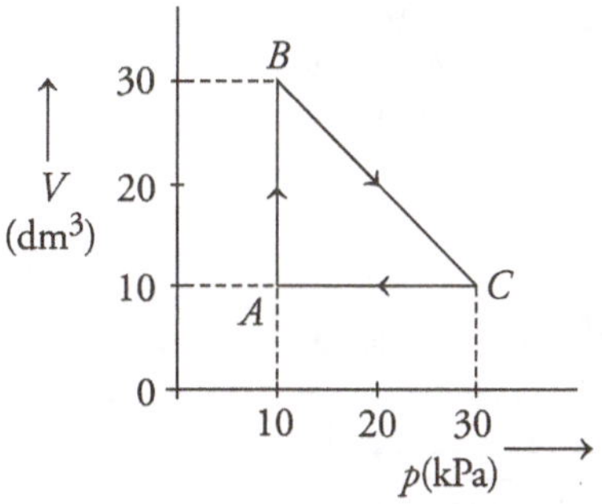

An ideal gas undergoes a cyclic transformation starting from the point A and coming back to the same point by tracing the path $A \to B \to C \to A$ as shown in the diagram above. The total work done in the process is .... J. **[2024, 30 Jan Shift-I]**

11 If three moles of an ideal gas at 300 K expand isothermally from 30 dm^3 to 45 dm^3 against a constant opposing pressure of 80 kPa, then the amount of heat transferred is .... J. **[2024, 27 Jan Shift-I]**

12 The total number of intensive properties from the following is ....
Volume, molar heat capacity, molarity, E°_{cell}, Gibbs free energy change, molar mass, mole. **[2023, 11 April Shift-II]**

13 One mole of an ideal monoatomic gas is subjected to changes as shown in the graph. The magnitude of the work done (by the system or on the system) is .... J. (Nearest integer)
Given : $\log 2 = 0.3$; $\ln 10 = 2.3$ **[2023, 24 Jan Shift-II]**

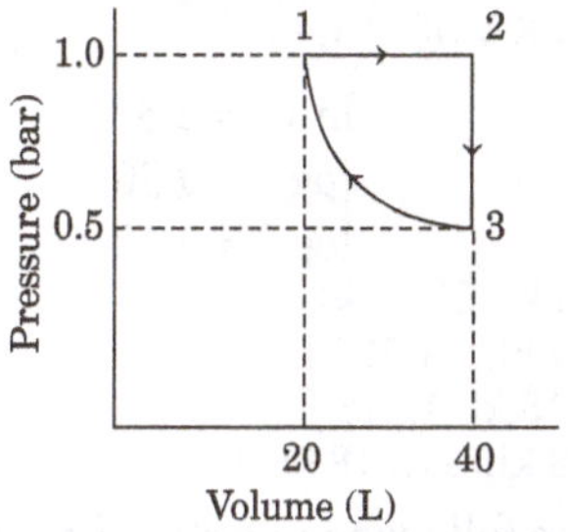

14 When 5 moles of He gas expand isothermally and reversibly at 300 K from 10 litre to 20 litre, the magnitude of the maximum work obtained is ...... J. [Nearest integer] (Given, $R = 8.3$ J K^{-1} mol^{-1} and $\log 2 = 0.3010$) **[2022, 27 June Shift-II]**

15 At 25°C, 50 g of iron reacts with HCl to form $FeCl_2$. The evolved hydrogen gas expands against a constant pressure of 1 bar. The work done by the gas during this expansion is ...... J. (Round off to the nearest integer) [Given, $R = 8.314$ J mol^{-1} K^{-1}. Assume, hydrogen is an ideal gas]
[Atomic mass of Fe is 55.85 u] **[2021, 16 March Shift-II]**

16 The magnitude of work done by a gas that undergoes a reversible expansion along the path ABC shown in the figure is .... **[2020, 8 Jan Shift-I]**

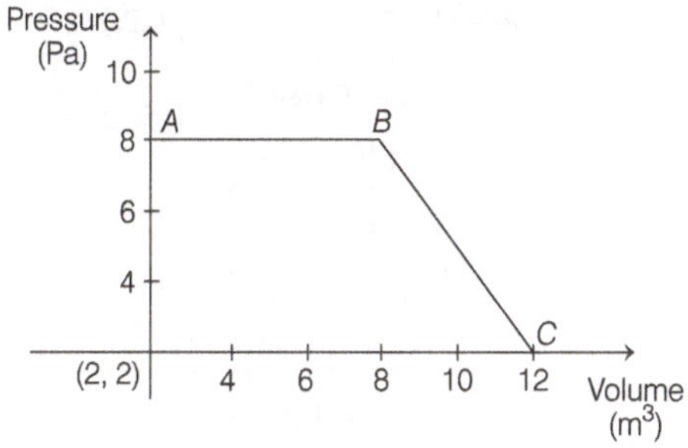

Topic 2 / First Law of Thermodynamics

• Level 1

01 A liquid when kept inside a thermally insulated closed vessel at 25°C was mechanically stirred from outside. What will be the correct option for the following thermodynamic parameters? **[2025, 22 Jan Shift-I]**
(a) $\Delta U = 0, q < 0, W > 0$
(b) $\Delta U = 0, q = 0, W = 0$
(c) $\Delta U > 0, q = 0, W > 0$
(d) $\Delta U < 0, q = 0, W > 0$

• Level 2

02 One mole of an ideal gas expands isothermally and reversibly from 10 dm^3 to 20 dm^3 at 300 K. $\Delta U, q$ and work done in the process respectively are

Given: $R = 8.3\ JK^{-1}\ mol^{-1}$ **[2025, 4 April Shift-I]**

$\ln 10 = 2.3$

$\log 2 = 0.30$

$\log 3 = 0.48$

(a) 0, 21.84 kJ, 21.84 kJ
(b) 0, 21.84 kJ, – 1.726 J
(c) 0, – 17.18 kJ, 1.718 J
(d) 0, 1.718 kJ, – 1.718 kJ

03 Arrange the following in order of magnitude of work done by the system/on the system at constant temperature. **[2025, 2 April Shift-II]**

(A) $|W_{\text{reversible}}|$ for expansion in infinite stages.
(B) $|W_{\text{irreversible}}|$ for expansion in single stage.
(C) $|W_{\text{reversible}}|$ for compression in infinite stages.
(D) $|W_{\text{irreversible}}|$ for compression in single stage.

Choose the correct answer from the options given below.

(a) C = A > D > B (b) D > C = A > B
(c) A > C > B > D (d) A > B > C > D

04

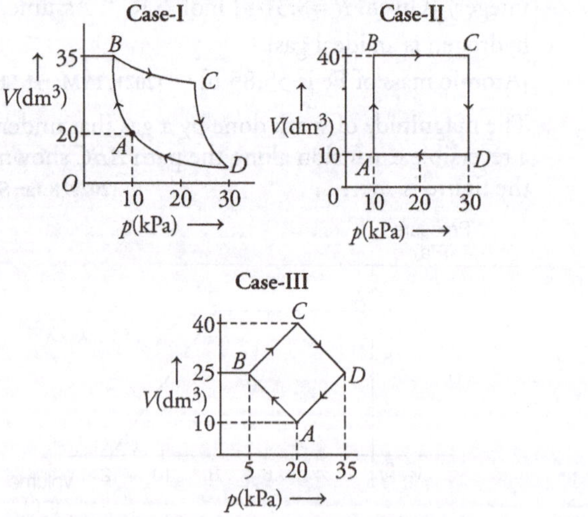

An ideal gas undergoes a cyclic transformation starting from the point A and coming back to the same point by tracing the path $A \rightarrow B \rightarrow C \rightarrow D \rightarrow A$ as shown in the three cases above. **[2025, 28 Jan Shift-II]**

Choose the correct option regarding ΔU :

(a) ΔU (Case-I) = ΔU (Case-II) = ΔU (Case-III)
(b) ΔU (Case-I) > ΔU (Case-II) > ΔU (Case-III)
(c) ΔU (Case-III) > ΔU (Case-II) > ΔU (Case-I)
(d) ΔU (Case-I) > ΔU (Case-III) > ΔU (Case-II)

05 500 J of energy is transferred as heat to 0.5 mol of Argon gas at 298 K and 1.00 atm. The final temperature and the change in internal energy respectively are (Given : $R = 8.3 J\ K^{-1}\ mol^{-1}$ **[2025, 29 Jan Shift-I]**

(a) 368 K and 500 J (b) 378 K and 300 J
(c) 378 K and 500 J (d) 348 K and 300 J

06 Choose the correct option for free expansion of an ideal gas under adiabatic condition from the following: **[2024, 1 Feb Shift-I]**

(a) $q = 0, \Delta T = 0, w = 0$
(b) $q \neq 0, \Delta T = 0, w = 0$
(c) $q = 0, \Delta T < 0, w \neq 0$
(d) $q = 0, \Delta T \neq 0, w = 0$

07 5 moles of an ideal gas at 100 K are allowed to undergo reversible compression till its temperature becomes 200 K. If $C_V = 28\ JK^{-1}\ mol^{-1}$, calculate ΔU and ΔpV for this process. ($R = 8.0\ JK^{-1}\ mol^{-1}$)

(a) $\Delta U = 2.8$ kJ; $\Delta(pV) = 0.8$ kJ **[2019, 8 April Shift-II]**
(b) $\Delta U = 14$ J; $\Delta(pV) = 0.8$ J
(c) $\Delta U = 14$ kJ; $\Delta(pV) = 4$ kJ
(d) $\Delta U = 14$ kJ; $\Delta(pV) = 18$ kJ

08 Which one of the following equations does not correctly represent the first law of thermodynamics for the given processes involving an ideal gas? (Assume non- expansion work is zero) **[2019, 8 April Shift-I]**

(a) Cyclic process : $q = -W$
(b) Adiabatic process : $\Delta U = -W$
(c) Isochoric process : $\Delta U = q$
(d) Isothermal process : $q = -W$

• Level 3

09

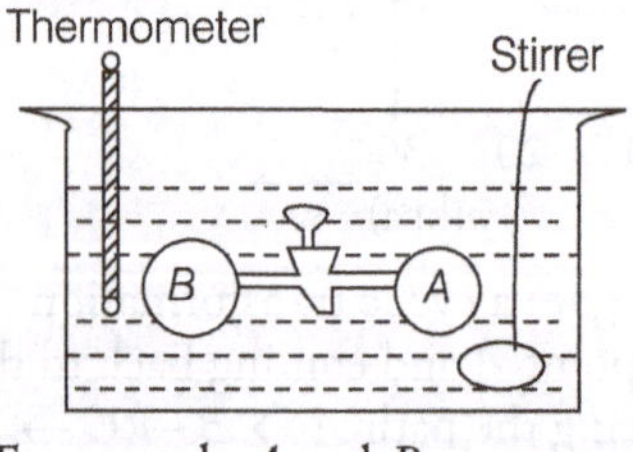

Two vessels A and B are connected *via* stopcock. The vessel A is filled with a gas at a certain pressure. The entire assembly is immersed in water and is allowed to come to thermal equilibrium with water. After opening the stopcock the gas from vessel A expands into vessel B and no change in temperature is observed in the thermometer. Which of the following statement is true? **[2025, 2 April Shift-I]**

(a) $dW \neq 0$ (b) $dU \neq 0$
(c) $dq \neq 0$
(d) The pressure in the vessel B before opening the stopcock is zero

Numerical Value Type Questions

10 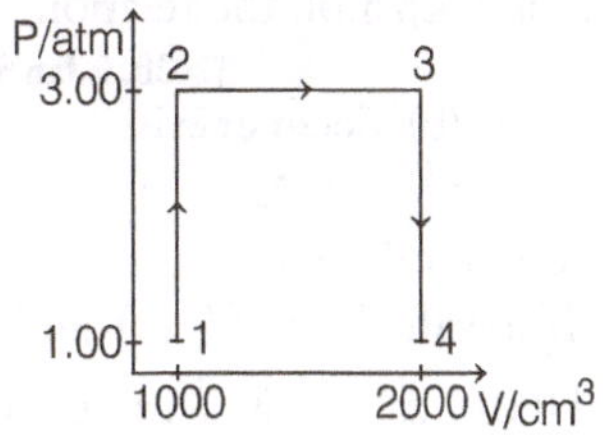

A perfect gas (0.1 mol) having $\overline{C}_v = 1.50R$ (independent of temperature) undergoes the above transformation from point 1 to point 4. If each step is reversible, the total work done (W) while going from point 1 to point 4 is (–) .... J (Nearest integer) [Given : $R = 0.082$ L atm K^{-1} mol^{-1}] **[2025, 3 April Shift-II]**

11 Three moles of an ideal gas are compressed isothermally from 60 L to 20 L using constant pressure of 5 atm. Heat exchange Q for the compression is – ...... L atm. **[2024, 4 April Shift-II]**

12 1 mole of ideal gas is allowed to expand reversibly and adiabatically from a temperature of 27°C. The work done is 3 kJ mol^{-1}. The final temperature of the gas is .... K. (Nearest integer). [Given : $C_V = 20$ J mol^{-1} K^{-1}.] **[2023, 30 Jan Shift-II]**

13 When a 60 W electric heater is immersed in a gas for 100s in a constant volume container with adiabatic walls, the temperature of the gas rises by 5°C. The heat capacity of the given gas is .... JK^{-1}. (Nearest integer) **[2023, 8 April Shift-I]**

14 When 2 L of ideal gas expands isothermally into vacuum to a total volume of 6 L, the change in internal energy is .... J. (Nearest integer) **[2023, 30 Jan Shift-I]**

15 4.0 L of an ideal gas is allowed to expand isothermally into vacuum until the total volume is 20 L. The amount of heat absorbed in this expansion is ...... L atm. **[2022, 28 June Shift-I]**

16 A system does 200 J of work and at the same time absorbs 150 J of heat. The magnitude of the change in internal energy is .... J. (Nearest integer) **[2021, 25 July Shift-II]**

Imp. *Heat Transfer*

17 Five moles of an ideal gas at 293 K is expanded isothermally from an initial pressure of 2.1 MPa to 1.3 MPa against at constant external pressure 4.3 MPa. The heat transferred in this process is .... kJ mol^{-1}. (Rounded off to the nearest integer) [$R = 8.314$ J $mol^{-1}K^{-1}$] **[2021, 25 Feb Shift-II]**

Topic 3 / Reactions Related to Enthalpies and Hess's Law

• Level 1

01 $S(g) + \frac{3}{2}O_2(g) \longrightarrow SO_3(g) + 2x$ kcal

$SO_2(g) + \frac{1}{2}O_2(g) \longrightarrow SO_3(g) + y$ kcal

The heat of formation of $SO_2(g)$ is given by **[2025, 24 Jan Shift-II]**

(a) $y - 2x$ kcal (b) $x + y$ kcal
(c) $2x + y$ kcal (d) $\frac{2x}{y}$ kcal

02 If C (diamond) $\longrightarrow$ C(graphite) $+X$ kJ mol^{-1}

C (diamond) $+ O_2(g) \longrightarrow CO_2(g) + Y$ kJ mol^{-1}

C (graphite) $+ O_2(g) \longrightarrow CO_2(g) + Z$ kJ mol^{-1}

at constant temperature. **[2025, 29 Jan Shift-II]**
Then
(a) $X = -Y + Z$ (b) $-X = Y + Z$
(c) $X = Y + Z$ (d) $X = Y - Z$

03 The difference between ΔH and ΔU ($\Delta H - \Delta U$), when the combustion of one mole of heptane (l) is carried out at a temperature T, is equal to **[2019, 10 April Shift-II]**

(a) $-4\,RT$ (b) $3\,RT$ (c) $4\,RT$ (d) $-3\,RT$

• Level 2

04 Given below are two statements: **[2025, 3 April Shift-II]**

Statement I When a system containing ice in equilibrium with water (liquid) is heated, heat is absorbed by the system and there is no change in the temperature of the system until whole ice gets melted.

Statement II At melting point of ice, there is absorption of heat in order to overcome intermolecular forces of attraction within the molecules of water in ice and kinetic energy of molecules is not increased at melting point.

In the light of the above statements, choose the correct answer from the options given below.
(a) Statement I is false but Statement II is true.
(b) Both Statement I and Statement II are false.
(c) Both Statement I and Statement II are true.
(d) Statement I is true but Statement II is false.

05 The hydration energies K^+ and Cl^- are $-x$ and $-y$ kJ/mol respectively. If lattice energy of KCl is $-z$ kJ/mol, then the heat of solution of KCl is **[2025, 7 April Shift-II]**

(a) $x + y + z$ (b) $z - (x + y)$
(c) $-z - (x + y)$ (d) $+x - y - z$

06 Let us consider a reversible reaction at temperature, T. In this reaction, both ΔH and ΔS were observed to have positive values. If the equilibrium temperature is T_e, then the reaction becomes spontaneous at: **[2025, 4 April Shift-I]**

(a) $T > T_e$ (b) $T = T_e$
(c) $T_e = 5T$ (d) $T_e > T$

Imp. *Enthalpy of Neutralisation*

07 Given below are two statements : one is labelled as **Assertion (A)** and other is labelled as **Reason (R)**. **[2024, 5 April Shift-I]**

Assertion (A) Enthalpy of neutralisation of strong monobasic acid with strong monoacidic base is always -57 kJ mol^{-1}.

Reason (R) Enthalpy of neutralisation is the amount of heat liberated when one mole of H^+ ions furnished by acid combine with one mole of OH^- ions furnished by base to form one mole of water.

In the light of the above statements, choose the **most appropriate** answer from the options given below.

(a) A is false, but R is true.
(b) Both A and R are true and R is the correct explanation of A.
(c) A is true, but R is false.
(d) Both A and R are true but R is not the correct explanation of A.

08 At 298.2 K the relationship between enthalpy of bond dissociation (in kJ mol^{-1}) for hydrogen (E_H) and its isotope, deuterium (E_D), is best described by **[2021, 25 July Shift-I]**

(a) $E_H = \frac{1}{2} E_D$ (b) $E_H = E_D$
(c) $E_H \doteq E_D - 7.5$ (d) $E_H = 2E_D$

09 Lattice enthalpy and enthalpy of solution of NaCl are 788 kJ mol^{-1} and 4 kJ mol^{-1}, respectively. The hydration enthalpy of NaCl is **[2020, 5 Sep Shift-II]**

(a) -780 kJ mol^{-1} (b) 780 kJ mol^{-1}
(c) -784 kJ mol^{-1} (d) 784 kJ mol^{-1}

10 For one mole of an ideal gas, which of these statements must be true? **[2020, 04 Sept Shift-I]**

(1) U and H each depends only on temperature
(2) Compressibility factor Z is not equal to 1
(3) $C_{P,m} - C_{V,m} = R$
(4) $dU = C_V dT$ for any processes

(a) (1) and (3)
(b) (2), (3) and (4)
(c) (3) and (4)
(d) (1), (3) and (4)

11 If enthalpy of atomisation for $Br_2(l)$ is x kJ/mol and bond enthalpy for Br_2 is y kJ/mol, the relation between them **[2020, 9 Jan Shift-II]**

(a) is $x < y$ (b) does not exist
(c) is $x > y$ (d) is $x = y$

12 Enthalpy of sublimation of iodine is 24 cal g^{-1} at 200°C. If specific heat of $I_2(s)$ and I_2(vap.) are 0.055 and 0.031 cal g^{-1} K^{-1} respectively, then enthalpy of sublimation of iodine at 250°C in cal g^{-1} is **[2019, 12 April Shift-I]**

(a) 2.85 (b) 5.7
(c) 22.8 (d) 11.4

• Level 3

13 The correct statement amongst the following is **[2025, 7 April Shift-II]**

(a) The term 'standard state' implies that the temperature is 0°C.
(b) $\Delta_f H^\theta_{298}$ is zero for O(g).
(c) The standard state of a pure gas is the pure gas at a pressure of 1 bar and temperature 273 K.
(d) $\Delta_f H^\theta_{500}$ is zero for $O_2(g)$.

14 Total enthalpy change for freezing of 1 mol of water at 10°C to ice at -10°C is ....

(Given, $\Delta_{fus} H = x$ kJ/mol
$C_p[H_2O(l)] = y$ J mol^{-1} K^{-1}
$C_p[H_2O(s)] = z$ J mol^{-1} K^{-1} **[2025, 7 April Shift-I]**

(a) $x - 10y - 10z$ (b) $10(100x + y + z)$
(c) $-10(100x + y + z)$ (d) $-x - 10y - 10z$

15 Consider the given data:

A. $HCl(g) + 10H_2O(l) \rightarrow HCl.10\,H_2O$ $\Delta H = -69.01$ kJ mol^{-1}

B. $HCl(g) + 40H_2O(l) \rightarrow HCl \cdot 40\,H_2O$ $\Delta H = -72.79$ kJ mol^{-1}

Choose the correct statement: **[2025, 4 April Shift-I]**

(a) The heat of dilution for the HCl ($HCl \cdot 10\,H_2O$ to $HCl \cdot 40\,H_2O$) is 3.78 kJ mol^{-1}.
(b) The heat of solution depends on the amount of solvent.
(c) Dissolution of gas in water is an endothermic process.
(d) The heat of formation of HCl solution is represented by both (A) and (B).

16 Reactant A converts to product D through the given mechanism (with the net evolution of heat) :

$A \rightarrow B$ slow; $\Delta H = +ve$
$B \rightarrow C$ fast; $\Delta H = -ve$
$C \rightarrow D$ fast; $\Delta H = -ve$

Which of the following represents the above reaction mechanism? **[2025, 2 April Shift-II]**

(a)
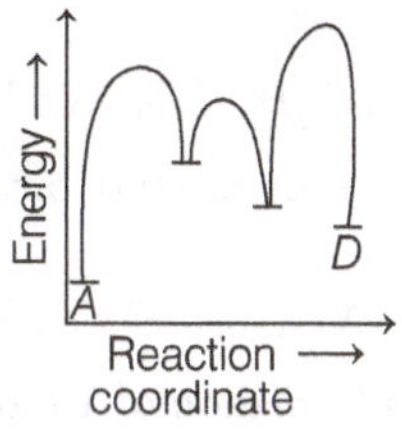

(b)
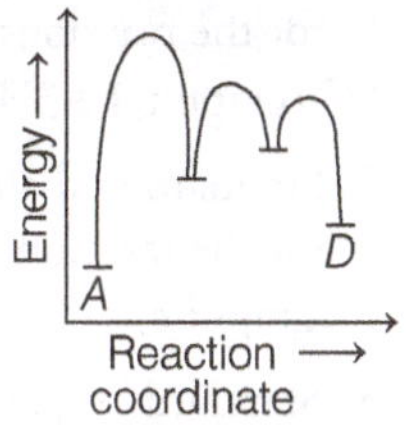

(c)
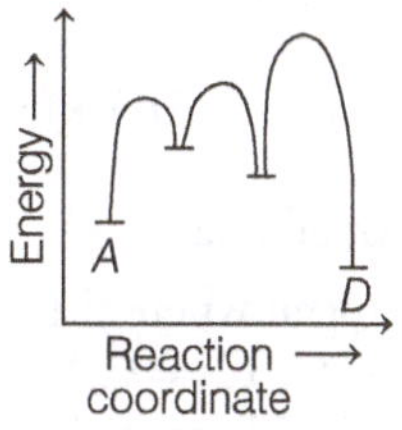

(d)
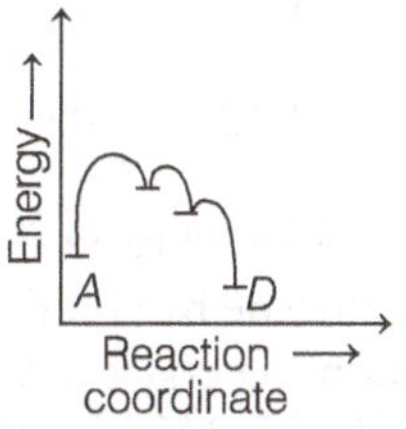

17 Given,

(A) $2CO(g) + O_2(g) \longrightarrow 2CO_2(g)$ $\quad \Delta H_1^\circ = -x$ kJ mol^{-1}

(B) C (graphite) $+ O_2(g) \longrightarrow CO_2(g)$ $\quad \Delta H_2^\circ = -y$ kJ mol^{-1}

The ΔH° for the reaction

C (graphite) $+ \frac{1}{2}O_2(g) \longrightarrow CO(g)$ is **[2023, 10 April Shift-I]**

(a) $\frac{x+2y}{2}$ (b) $\frac{x-2y}{2}$ (c) $\frac{2x-y}{2}$ (d) $2y-x$

18 At 25°C and 1 atm pressure, the enthalpies of combustion are as given below

Substance	H_2	C (graphite)	$C_2H_6(g)$
$\Delta_c H^\ominus$ (kJ mol^{-1})	−286.0	−394.0	−1560.0

The enthalpy of formation of ethane is **[2022, 24 June Shift-II]**

(a) +54.0 kJ mol^{-1} (b) −68.0 kJ mol^{-1}

(c) −86.0 kJ mol^{-1} (d) +97.0 kJ mol^{-1}

Numerical Value Type Questions

19 Resonance in X_2Y can be represented as

$$\overset{\ominus}{X} = \overset{\oplus}{X} = Y \longleftrightarrow :X \equiv \overset{\oplus}{X} - \overset{\ominus}{Y}$$

The enthalpy of formation of

$X_2Y \left(X \equiv X(g) + \frac{1}{2}Y = Y(g) \rightarrow X_2Y(g) \right)$ is 80 kJ mol^{-1}. The magnitude of resonance energy of X_2Y is .... kJ mol^{-1} (Nearest integer value)

Given: Bond energies of $X \equiv X$, $X = X$, $Y = Y$ and $X = Y$ are 940, 410, 500 and 602 kJ mol^{-1} respectively. Valence $X:3, Y:2$ **[2025, 8 April Shift-II]**

20 Given:

$\Delta H^\ominus_{sub}$ [C (graphite)] = 710 kJ mol^{-1}

$\Delta_{C-H}H^\ominus$ = 414 kJ mol^{-1}

$\Delta_{H-H}H^\ominus$ = 436 kJ mol^{-1}

$\Delta_{C=C}H^\ominus$ = 611 kJ mol^{-1}

The $\Delta H_f^\ominus$ for $CH_2 = CH_2$ is .... kJ mol^{-1}

(Nearest integer value) **[2025, 3 April Shift-I]**

21 A sample of *n*-octane (1.14 g) was completely burnt in excess of oxygen in a bomb calorimeter. whose heat capacity is 5 kJ K^{-1}. As a result of combustion reaction, the temperature of the calorimeter is increased by 5 K. The magnitude of the heat of combustion of octane at constant volume is .... kJ mol^{-1} (Nearest integer). **[2025, 3 April Shift-II]**

22 Consider the following cases of standard enthalpy of reaction (ΔH_r° in kJ mol^{-1}) **[2025, 22 Jan Shift-II]**

$C_2H_6(g) + \frac{7}{2}O_2(g) \longrightarrow 2CO_2(g) + 3H_2O(l)$; $\Delta H_1^\circ = -1550$

C (graphite) $+ O_2(g) \longrightarrow CO_2(g)$; $\Delta H_2^\circ = -393.5$

$H_2(g) + \frac{1}{2}O_2(g) \longrightarrow H_2O(l)$; $\Delta H_3^\circ = -286$

The magnitude of $\Delta H_f^\circ(C_2H_6(g))$ is .... kJ mol^{-1} (Nearest integer).

23 The formation enthalpies, $\Delta H_f^\ominus$ for H(*g*) and O(*g*) are 220.0 and 250.0 kJ mol^{-1}, respectively, at 298.15 K, and $\Delta H_f^\ominus$ for $H_2O(g)$ is −242.0 kJ mol^{-1} at the same temperature. The average bond enthalpy of the O—H bond in water at 298.15 K is .... kJ mol^{-1} (Nearest integer). **[2025, 28 Jan Shift-I]**

24 Consider the following data :

Heat of formation of $CO_2(g)$ = −393.5 kJmol^{-1}

Heat of formation of $H_2O(l)$ = −286.0 kJmol^{-1}

Heat of combustion of benzene = −3267.0 kJmol^{-1}

The heat of formation of benzene is ....kJmol^{-1}

(Nearest integer) **[2025, 28 Jan Shift-II]**

Imp. *Calculation of Bond Dissociation Enthalpy*

25 The bond dissociation enthalpy of X_2, $\Delta H^{\circ}_{\text{bond}}$ calculated from the given data is......... kJ mol^{-1}. (Nearest integer) **[2025, 23 Jan Shift-II]**

$$M^+X^-(s) \longrightarrow M^+(g) + X^-(g); \quad \Delta H^{\circ}_{\text{lattice}} = 800 \text{ kJ mol}^{-1}$$

$$M(s) \longrightarrow M(g); \Delta H^{\circ}_{\text{sub}} = 100 \text{ kJ mol}^{-1}$$

$$M(g) \longrightarrow M^+(g) + e^-; \Delta H^{\circ}_i = 500 \text{ kJ mol}^{-1}$$

$$X(g) + e^-(g) \longrightarrow X^-(g); \Delta H^{\circ}_{eg} = -300 \text{ kJ mol}^{-1}$$

$$M(s) + \frac{1}{2}X_2(g) \longrightarrow M^+X^-(s); \Delta H^{\circ}_f = -400 \text{ kJ mol}^{-1}$$

[Given, M^+X^- is a pure ionic compound and X forms a diatomic molecule X_2 in gaseous state].

26 When equal volume of 1 M HCl and 1 M H_2SO_4 are separately neutralised by excess volume of 1 M NaOH solution x and y kJ of heat is liberated respectively. The value of y/x is .... **[2024, 9 April Shift-I]**

27 The heat of solution of anhydrous $CuSO_4$ and $CuSO_4 \cdot 5H_2O$ are -70 kJ mol^{-1} and $+12$ kJ mol^{-1} respectively.
The heat of hydration of $CuSO_4$ to $CuSO_4 \cdot 5H_2O$ is $-x$ kJ. The value of x is .... (Nearest integer). **[2024, 9 April Shift-I]**

28 $\Delta_{\text{vap}}H^{\ominus}$ for water is $+40.79$ kJ mol^{-1} at 1 bar and 100 °C. Change in internal energy for this vapourisation under same condition is .... kJ mol^{-1}. (Integer answer) (Given $R = 8.3$ JK^{-1} mol^{-1}) **[2024, 8 April Shift-II]**

29 An ideal gas, $C_V = \frac{5}{2}R$, is expanded adiabatically against a constant pressure of 1 atm until it doubles in volume. If the initial temperature and pressure is 298 K and 5 atm, respectively then the final temperature is .... K (Nearest integer).
[C_V is the molar heat capacity at constant volume] **[2024, 6 April Shift-I]**

30 The heat of combustion of solid benzoic acid at constant volume is -32130 kJ at 27°C. The heat of combustion at constant pressure is $(-321.30 - xR)$ kJ, the value of x is .... **[2024, 5 April Shift-II]**

31 Combustion of 1 mole of benzene is expressed at

$$C_6H_6(l) + \frac{15}{2}O_2(g) \longrightarrow 6CO_2(g) + 3H_2O(l)$$

The standard enthalpy of combustion of 2 mole of benzene is $-$'x' kJ, x = ... **[2024, 5 April Shift-I]**

Given :

1. Standard enthalpy of formation of 1 mol of $C_6H_6(l)$, for the reactions
$6C \text{ (graphite)} + 3H_2(g) \longrightarrow C_6H_6(l)$ is 48.5 kJ mol^{-1}
2. Standard enthalpy of formation of 1 mol of $CO_2(g)$ for the reaction
$C \text{ (graphite)} + O_2(g) \longrightarrow CO_2(g)$ is -393.5 kJ mol^{-1}
3. Standard and enthalpy of formation of 1 mol of $H_2O(l)$, for the reaction
$H_2(g) + \frac{1}{2}O_2(g) \longrightarrow H_2O(l)$ is -286 kJ mol^{-1}.

32 The enthalpy of formation of ethane (C_2H_6) from ethylene by addition of hydrogen where the bond-energies of C—H, C—C, C=C, H—H are 414 kJ, 347 kJ, 615 kJ and 435 kJ respectively is − ...... kJ. **[2024, 4 April Shift-I]**

33 Standard enthalpy of vaporisation for CCl_4 is 30.5 kJ mol^{-1}. Heat required for vaporisation of 284 g of CCl_4 at constant temperature is .... kJ. (Given, molar mass in g mol^{-1}; C = 12, Cl = 35.5) **[2024, 29 Jan Shift-II]**

34 Two reactions are given below. **[2024, 30 Jan Shift-II]**

$$2Fe(s) + \frac{3}{2}O_2(g) \longrightarrow Fe_2O_3(s), \Delta H^{\circ} = -822 \text{ kJ/mol}$$

$$C(s) + \frac{1}{2}O_2(g) \longrightarrow CO(g), \Delta H^{\circ} = -110 \text{ kJ/mol}$$

Then enthalpy change for the following reaction,
$3C(s) + Fe_2O_3(s) \longrightarrow 2Fe(s) + 3CO(g)$ is ... kJ/mol.

35 The number of endothermic processes from the following is .... **[2023, 10 April Shift-II]**

A. $I_2(g) \longrightarrow 2I(g)$
B. $HCl(g) \longrightarrow H(g) + Cl(g)$
C. $H_2O(l) \longrightarrow H_2O(g)$
D. $C(s) + O_2(g) \longrightarrow CO_2(g)$
E. dissolution of ammonium chloride in water

36 Consider the following data.
Heat of combustion of $H_2(g) = -241.8$ kJ mol^{-1}
Heat of combustion of C(s) $= -393.5$ kJ mol^{-1}
Heat of combustion of $C_2H_5OH(1) = -1234.7$ kJ mol^{-1}
The heat of formation of $C_2H_5OH(l)$ is (−) .... kJ mol^{-1}. (Nearest integer) **[2023, 6 April Shift-II]**

37 For complete combustion of ethene,
$C_2H_4(g) + 3O_2(g) \longrightarrow 2CO_2(g) + 2H_2O(l)$
the amount or heat produced as measured in bomb calorimeter is 1406 kJ mol^{-1} at 300K. The minimum

value of $T\Delta S$ needed to reach equilibrium is (–) .... kJ. (Nearest integer)
Given: $R = 8.3\ \text{J K}^{-1}\ \text{mol}^{-1}$ **[2023, 8 April Shift-II]**

38 $A_2 + B_2 \longrightarrow 2AB, \Delta H_f^0 = -200\ \text{kJ mol}^{-1}$
AB, A_2 and B_2 are diatomic molecules. If the bond enthalpies of A_2, B_2 and AB are in the ratio 1 : 0.5 : 1, then the bond enthalpy of A_2 is .... kJ mol^{-1}
(Nearest integer) **[2023, 13 April Shift-I]**

Imp. *Bomb Calorimetry*

39 0.3 g of ethane undergoes combustion at 27°C in a bomb calorimeter. The temperature of calorimeter system (including the water) is found to rise by 0.5°C. The heat evolved during combustion of ethane at constant pressure is .... kJ mol^{-1}. (Nearest integer)
[Given, the heat capacity of the calorimeter system is 20 kJ K^{-1}, $R = 8.3\ \text{J K}^{-1}\ \text{mol}^{-1}$.
Assume ideal gas behaviour. Atomic mass of C and H are 12 and 1 g mol^{-1} respectively] **[2023, 01 Feb Shift-II]**

40 28.0 L of CO_2 is produced on complete combustion of 16.8 L gaseous mixture of ethene and methane at 25°C and 1 atm. Heat evolved during the combustion processes is – .... kJ. **[2023, 25 Jan Shift-II]**
[Given, $\Delta H_C(CH_4) = -900\ \text{kJ mol}^{-1}$
$\Delta H_C\ (C_2H_4) = -1400\ \text{kJ mol}^{-1}$]

41 When 600 mL of 0.2 M HNO_3 is mixed with 400 mL of 0.1 M NaOH solution in a flask, the rise in temperature of the flask is .... $\times 10^{-2}$°C.
[2022, 29 July Shift-I]
(Enthalpy of neutralisation = 57 kJ mol^{-1} and Specific heat of water = 4.2 $\text{JK}^{-1}\text{g}^{-1}$)
(Neglect heat capacity of flask)

42 A gas (molar mass = 280 g mol^{-1}) was burnt in excess O_2 in a constant volume calorimeter and during combustion the temperature of calorimeter increased from 298.0 K to 298.45 K. If the heat capacity of calorimeter is 2.5 kJ K^{-1} and enthalpy of combustion of gas is 9 kJ mol^{-1} then amount of gas burnt is .... g. (nearest integer) **[2022, 27 July Shift-II]**

43 The molar heat capacity for an ideal gas at constant pressure is 20.785 J K^{-1} mol^{-1}. The change in internal energy is 5000 J upon heating it from 300 K to 500 K. The number of moles of the gas at constant volume is .... (Nearest integer)
(Given, $R = 8.314\ \text{JK}^{-1}\ \text{mol}^{-1}$) **[2022, 27 July Shift-I]**

44 For the reaction,
$$H_2F_2(g) \longrightarrow H_2(g) + F_2(g)$$
$$\Delta U = -59.6\ \text{kJ mol}^{-1} \text{ at } 27°\text{C}.$$
The enthalpy change for the above reaction is (–) .... kJ mol^{-1} [nearest integer]
Given : $R = 8.314\ \text{JK}^{-1}\text{mol}^{-1}$ **[2022, 26 July Shift-II]**

45 2.4 g coal is burnt in a bomb calorimeter in excess of oxygen at 298 K and 1 atm pressure. The temperature of the calorimeter rises from 298 K to 300 K. The enthalpy change during the combustion of coal is $-x$ kJ mol^{-1}. The value of x is ....
(Nearest integer)
(Given, heat capacity of bomb calorimeter is 20.0 kJ K^{-1}. Assume coal to be pure carbon).
[2022, 26 July Shift-I]

46 While performing a thermodynamics experiment, a student made the following observations.
$HCl + NaOH \longrightarrow NaCl + H_2O$; $\Delta H = -57.3\ \text{kJ mol}^{-1}$
$CH_3COOH + NaOH \longrightarrow CH_3COONa + H_2O$;
$\Delta H = -55.3\ \text{kJ mol}^{-1}$
The enthalpy of ionisation of CH_3COOH as calculated by the student is .... kJ mol^{-1}.
[nearest integer] **[2022, 25 July Shift-II]**

47 The enthalpy of combustion of propane, graphite and dihydrogen at 298 K are – 2220.0 kJ mol^{-1}, – 393.5 kJ mol^{-1} and –285.8 kJ mol^{-1} respectively. The magnitude of enthalpy of formation of propane (C_3H_8) is .... kJ mol^{-1}. (Nearest integer)
[2022, 25 July Shift-I]

48 For combustion of one mole of magnesium in an open container at 300 K and 1 bar pressure, $\Delta_C H^{\ominus} = -601.70\ \text{kJ mol}^{-1}$, the magnitude of change in internal energy for the reaction is ... kJ.
(Nearest integer) (Given, $R = 8.3\ \text{JK}^{-1}\ \text{mol}^{-1}$)
[2022, 28 June Shift-II]

49 2.2 g of nitrous oxide (N_2O) gas is cooled at a constant pressure of 1 atm from 310 K to 270 K causing the compression of the gas from 217.1 mL to 167.75 mL. The change in internal energy of the process, ΔU is '$-x$' J. The value of 'x' is [Nearest integer]
(Given, atomic mass of N = 14 g mol^{-1} and of O = 16 g mol^{-1}. Molar heat capacity of N_2O is 100 $\text{JK}^{-1}\ \text{mol}^{-1}$) **[2022, 29 June Shift-II]**

50 17.0 g of NH_3 completely vapourises at $-33.42°C$ and 1 bar pressure and the enthalpy change in the process is 23.4 kJ mol^{-1}. The enthalpy change for the vaporisation of 85 g of NH_3 under the same conditions is ...... kJ. **[2022, 29 June Shift-I]**

51 A fish swimming in water body when taken out from the water body is covered with a film of water of weight 36 g. When it is subjected to cooking at 100°C, then the internal energy for vaporisation in kJ mol^{-1} is .... [Nearest integer]

[Assume steam to be an ideal gas. Given, $\Delta_{vap}H^{\ominus}$ for water at 373 K and 1 bar is 41.1 kJ mol^{-1}; $R = 8.31\ JK^{-1}\ mol^{-1}$] **[2022, 26 June Shift-II]**

52 For complete combustion of methanol

$$CH_3OH(l) + \frac{3}{2}O_2(g) \longrightarrow CO_2(g) + 2H_2O(l)$$

the amount of heat produced as measured by bomb calorimeter is 726 kJ mol^{-1} at 27°C. The enthalpy of combustion for the reaction is $-x$ kJ mol^{-1}, where x is .... (Nearest integer) (Given : $R = 8.3\ JK^{-1}mol^{-1}$) **[2022, 26 June Shift-I]**

53 For water $\Delta_{vap}H = 41$ kJ mol^{-1} at 373 K and 1 bar pressure. Assuming that water vapour is an ideal gas that occupies a much larger volume than liquid water, the internal energy change during evaporation of water is .... kJ mol^{-1}.

[Use $R = 8.3\ J\ mol^{-1}\ K^{-1}$] **[2021, 26 Aug Shift-II]**

54 200 mL of 0.2 M HCl is mixed with 300 mL of 0.1 M NaOH. The molar heat of neutralisation of this reaction is -57.1 kJ. The increase in temperature in °C of the system on mixing is $x \times 10^{-2}$. The value of x is .... (Nearest integer)

[Given, specific heat of water $= 4.18\ J\ g^{-1}\ K^{-1}$

Density of water $= 1.00\ g\ cm^{-3}$]

(Assume no volume change on mixing) **[2021, 27 Aug Shift-I]**

55 When 400 mL of 0.2 M H_2SO_4 solution is mixed with 600 mL of 0.1 M NaOH solution, the increase in temperature of the final solution is .... $\times 10^{-2}$ K. (Round off to the nearest integer).

[Use : $H^+(aq) + OH^+(aq) \longrightarrow H_2O$; $\Delta_y H = -57.1\ kJ\ mol^{-1}$]

Specific heat of $H_2O = 0.18\ J\ K^{-1}g^{-1}$, density of $H_2O = 1.0\ g\ cm^{-3}$.

Assume no change in volume of solution on mixing. **[2021, 27 July Shift-II]**

56 For water at 100°C and 1 bar,

$\Delta_{vap}H - \Delta_{vap}U = ... \times 10^2 J\ mol^{-1}$.

(Round off to the nearest integer)

[Use : $R = 8.31\ J\ mol^{-1}\ K^{-1}$]

[Assume volume of $H_2O(l)$ is much smaller than volume of $H_2O(g)$. Assume $H_2O(g)$ treated as an ideal gas] **[2021, 27 July Shift-I]**

57 At 298 K, the enthalpy of fusion of a solid (X) is 2.8 kJ mol^{-1} and the enthalpy of vaporisation of the liquid (X) is 98.2 kJ mol^{-1}. The enthalpy of sublimation of the substance (X) in kJ mol^{-1} is ...... (Nearest integer) **[2021, 25 July Shift-I]**

58 If the standard molar enthalpy change for combustion of graphite powder is -2.48×10^2 kJ mol^{-1}, the amount of heat generated on combustion of 1 g of graphite powder is ...... kJ. (Nearest integer) **[2021, 22 July Shift-II]**

59 For the reaction, $C_2H_6 \longrightarrow C_2H_4 + H_2$ the reaction enthalpy $\Delta_r H =$ .... kJ mol^{-1} (Round off to the nearest integer).

[Given : Bond enthalpies in kJ mol^{-1} : C—C = 347, C=C = 611; C—H = 414; H—H = 436] **[2021, 18 March Shift-I]**

60 The standard enthalpies of formation of Al_2O_3 and CaO are -1675 kJ mol^{-1} and -625 kJ mol^{-1} respectively. For the reaction, $3CaO + 2Al \longrightarrow 3Ca + Al_2O_3$ the standard reaction enthalpy $\Delta_r H°$ .... kJ.

(Round off to the nearest integer). **[2021, 17 March Shift-I]**

61 The average S—F bond energy in kJ mol^{-1} of SF_6 is .... (Rounded off to the nearest integer)

[Given, the values of standard enthalpy of formation of $SF_6(g)$, $S(g)$ and $F(g)$ are -1100, 275 and 80 kJ mol^{-1} respectively.] **[2021, 26 Feb Shift-II]**

62 The reaction of cyanamide, $NH_2CN(s)$ with oxygen was run in a bomb calorimeter and ΔU was found to be -742.24 kJ mol^{-1}. The magnitude of ΔH_{298} for the reaction,

$$NH_2CN(s) + \frac{3}{2}O_2(g) \longrightarrow N_2(g) + CO_2(g) + H_2O(l)$$

is .... kJ. (Rounded off to the nearest integer)

[Assume ideal gases and $R = 8.314\ J\ mol^{-1}K^{-1}$] **[2021, 25 Feb Shift-I]**

63 The internal energy change (in J) when 90 g of water undergoes complete evaporation at 100°C is ...., (Given : ΔH_{vap} for water at 373 K = 41kJ/mol, $R = 8.314\ JK^{-1}\ mol^{-1}$) **[2020, 5 Sep Shift-I]**

Topic 4 / Second Law of Thermodynamics, Entropy, Free Energy Change & Spontaneity

• Level 1

01 Which of the following relations are correct?

A. $\Delta U = q + p\Delta V$ B. $\Delta G = \Delta H - T\Delta S$

C. $\Delta S = \frac{q_{rev}}{T}$ D. $\Delta H = \Delta U - \Delta nRT$

[2023, 29 Jan Shift-II]

(a) A and C (b) B only (c) B and C (d) A and D

02 Which of the following relation is not correct? [2022, 28 July Shift-I]

(a) $\Delta H = \Delta U - p\Delta V$ (b) $\Delta U = q + W$

(c) $\Delta S_{sys} + \Delta S_{surr} \geq 0$ (d) $\Delta G = \Delta H - T\Delta S$

03 A process will be spontaneous at all temperature if [2019, 10 April Shift-I]

(a) $\Delta H > 0$ and $\Delta S < 0$ (b) $\Delta H < 0$ and $\Delta S > 0$

(c) $\Delta H < 0$ and $\Delta S < 0$ (d) $\Delta H > 0$ and $\Delta S > 0$

• Level 2

04 Ice at -5°C is heated to become vapour with temperature of 110°C at atmospheric pressure. The entropy change associated with this process can be obtained from [2025, 23 Jan Shift-I]

(a) $\int_{268K}^{383K} C_p\, dT + \frac{\Delta H_{melting}}{273} + \frac{\Delta H_{boiling}}{373}$

(b) $\int_{268K}^{383K} C_p dT + \frac{q_{rev}}{T}$

(c) $\int_{268K}^{273K} C_{p,m}\, dT + \frac{\Delta H_m, \text{fusion}}{T_f} + \frac{\Delta H_m, \text{vaporsation}}{T_b} + \int_{273K}^{373K} C_{p,m} dT + \int_{373K}^{383K} C_{p,m}\, dT$

(d) $\int_{268K}^{273K} \frac{C_{p,m}}{T}\, dT + \frac{\Delta H_m, \text{fusion}}{T_f} + \frac{\Delta H_m, \text{vaporisation}}{T_b} + \int_{273K}^{373K} \frac{C_{p,m} dT}{T} + \int_{373K}^{383K} \frac{C_{p,m}\, dT}{T}$

05 The effect of temperature on spontaneity of reactions are represented as

	ΔH	ΔS	Temperature	Spontaneity
(A)	+	−	any T	Non-spontaneous
(B)	+	+	low T	Spontaneous
(C)	−	−	low T	Non-spontaneous
(D)	−	+	any T	Spontaneous

The incorrect combinations are [2025, 23 Jan Shift-I]

(a) (B) and (D) only (b) (A) and (C) only

(c) (B) and (C) only (d) (A) and (D) only

06 Let us consider an endothermic reaction which is non-spontaneous at the freezing point of water. However, the reaction is spontaneous at boiling point of water. Choose the correct option. [2025, 24 Jan Shift-I]

(a) Both ΔH and ΔS are (−ve)

(b) Both ΔH and ΔS are (+ve)

(c) ΔH is (+ve) but ΔS is (−ve)

(d) ΔH is (−ve) but ΔS is (+ve)

07 Which of the following is not correct? [2024, 29 Jan Shift-I]

(a) ΔG is positive for a spontaneous reaction

(b) ΔG is positive for a non-spontaneous reaction

(c) ΔG is zero for a reversible reaction

(d) ΔG is negative for a spontaneous reaction

08 Match List-I with List-II.

List-I		List-II
A. Spontaneous process	I.	$\Delta H < 0$
B. Process with $\Delta p = 0$, $\Delta T = 0$	II.	$\Delta G_{T,p} < 0$
C. $\Delta H_{reaction}$	III.	Isothermal and isobaric process
D. Exothermic process	IV.	[Bond energies of molecules in reactants] − [Bond energies of product molecules]

Choose the correct answer from the options given below. [2022, 27 June Shift-I]

	A	B	C	D
(a)	III	II	IV	I
(b)	II	III	IV	I
(c)	II	III	I	IV
(d)	II	I	III	IV

09 The incorrect expression among the following is [2021, 31 Aug Shift-II]

(a) $\frac{\Delta G_{System}}{\Delta S_{Total}} = -T$ (at constant p)

(b) $\ln K = \frac{\Delta H° - T\Delta S°}{RT}$

(c) $K = e^{-\frac{\Delta G°}{RT}}$

(d) For isothermal process, $W_{reversible} = -nRT \ln \frac{V_f}{V_i}$

10 The true statement amongst the following is [2020, 9 Jan Shift-II]

(a) S is not a function of temperature but ΔS is a function of temperature.

(b) Both ΔS and S are functions of temperature.

(c) Both S and ΔS are not functions of temperature.

(d) S is a function of temperature but ΔS is not a function of temperature.

11 The process with negative entropy change is
(a) synthesis of ammonia from N_2 and H_2
(b) dissociation of $CaSO_4(s)$ to $CaO(s)$ and $SO_3(g)$
(c) dissolution of iodine in water
(d) sublimation of dry ice **[2019, 10 Jan Shift-II]**

12 The incorrect match in the following is **[2019, 12 April Shift-II]**
(a) $\Delta G° < 0, K > 1$ (b) $\Delta G° = 0, K = 1$
(c) $\Delta G° > 0, K < 1$ (d) $\Delta G° < 0, K < 1$

13 The reaction, $MgO(s) + C(s) \longrightarrow Mg(s) + CO(g)$, for which $\Delta_r H° = +491.1$ kJ mol^{-1} and $\Delta_r S° = 198.0$ JK^{-1}mol^{-1}, is not feasible at 298 K. Temperature above which reaction will be feasible is **[2019, 11 Jan Shift-II]**
(a) 2040.5 K (b) 1890.0 K
(c) 2380.5 K (d) 2480.3 K

14 For the chemical reaction, $X \rightleftharpoons Y$, the standard reaction Gibbs energy depends on temperature T (in K) as $\Delta_r G°$ (in kJ mol^{-1}) $= 120 - \frac{3}{8}T$

The major component of the reaction mixture at T is
(a) Y if $T = 280$ K **[2019, 11 Jan Shift-I]**
(b) X if $T = 350$ K
(c) X if $T = 315$ K
(d) Y if $T = 300$ K

15 A process has $\Delta H = 200$ J mol^{-1} and $\Delta S = 40$ JK^{-1} mol^{-1}. Out of the values given below, choose the minimum temperature above which the process will be spontaneous **[2019, 10 Jan Shift-II]**
(a) 20 K (b) 4 K
(c) 5 K (d) 12 K

• Level 3

16 Which of the following graphs correctly represents the variation of thermodynamic properties of Haber's process? **[2025, 2 April Shift-II]**

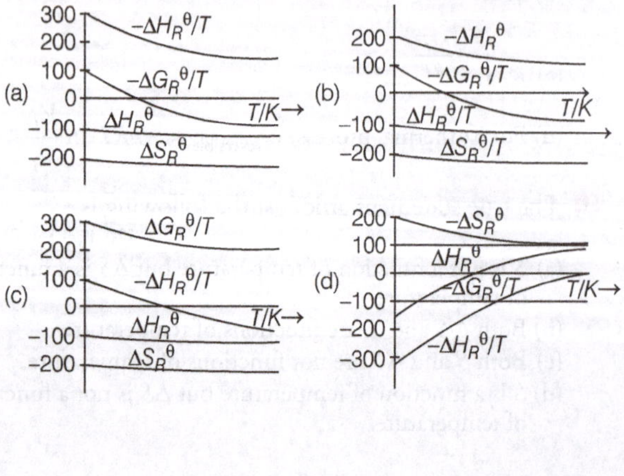

Imp. *Partial Derivatives of Thermodynamic Quantity*

17 Match List-I with List-II.

	List-I (Partial derivatives)		List-II (Thermodynamic quantity)
(A)	$\left(\frac{\partial G}{\partial T}\right)_p$	(I)	C_p
(B)	$\left(\frac{\partial H}{\partial T}\right)_p$	(II)	$-S$
(C)	$\left(\frac{\partial G}{\partial p}\right)_T$	(III)	C_V
(D)	$\left(\frac{\partial U}{\partial T}\right)_V$	(IV)	V

Choose the correct answer from the options given below **[2025, 22 Jan Shift-II]**
(a) (A)-(I), (B)-(II), (C)-(IV), (D)-(III)
(b) (A)-(II), (B)-(III), (C)-(I), (D)-(IV)
(c) (A)-(II), (B)-(I), (C)-(IV), (D)-(III)
(d) (A)-(II), (B)-(I), (C)-(III), (D)-(IV)

18 Consider the graph of Gibbs free energy G *vs* extent of reaction. The number of statements from the following which are true with respect to points (*a*), (*b*) and (*c*) is .... **[2023, 6 April Shift-I]**

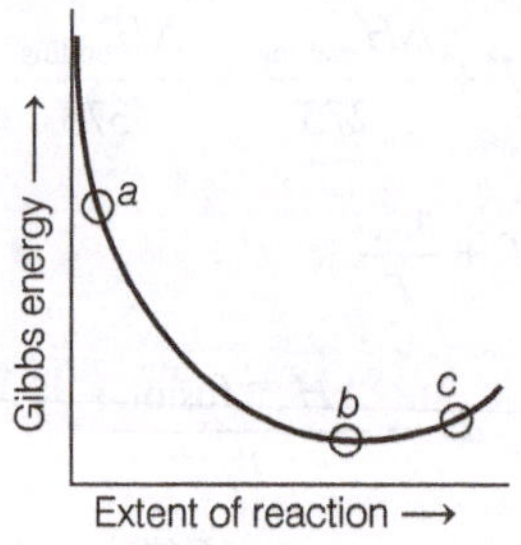

A. reaction is spontaneous at (*a*) and (*b*)
B. reaction is at equilibrium at point (*b*) and non-spontaneous at point (*c*)
C. Reaction is spontaneous at (*a*) and non-spontaneous at (*c*)
D. Reaction is non-spontaneous at (*a*) and (*b*)

19 During which of the following processes, does entropy decrease ?
A. Freezing of water to ice at 0°C.
B. Freezing of water to ice at −10°C.
C. $N_2(g) + 3H_2(g) \longrightarrow 2NH_3(g)$
D. Adsorption of CO(g) and lead surface.
E. Dissolution of NaCl in water. **[2021, 17 March Shift-II]**
(a) A, B, C and D (b) B and C
(c) A and E (d) A, C and E

Numerical Value Type Questions

20 Standard entropies of X_2, Y_2 and XY_5 are 70, 50 and 110 $JK^{-1}mol^{-1}$ respectively. The temperature in Kelvin at which the reaction **[2025, 24 Jan Shift-I]**

$\frac{1}{2}X_2 + \frac{5}{2}Y_2 \rightleftharpoons XY_5; \Delta H^\ominus = -35\ kJ\,mol^{-1}$

will be at equilibrium is...........(Nearest integer)

21 The standard enthalpy and standard entropy of decomposition of N_2O_4 to NO_2 are 55.0 kJ mol^{-1} and 175.0 J/K/mol respectively. The standard free energy change for this reaction at 25°C in J mol^{-1} is ...... (Nearest integer) **[2025, 23 Jan Shift-I]**

22 When $\Delta H_{vap} = 30$ kJ/mol and $\Delta S_{vap} = 75\ J\ mol^{-1}\ K^{-1}$, then the temperature of vapour, at one atmosphere is ... K. **[2024, 9 April Shift-II]**

23 For the reaction at 298 K, $2A + B \rightarrow C$. $\Delta H = 400\ kJ\ mol^{-1}$ and $\Delta S = 0.2\ kJ\ mol^{-1}\ K^{-1}$. The reaction will become spontaneous above .... K. **[2024, 6 April Shift-II]**

24 For a certain reaction at 300 K, $K = 10$, then ΔG° for the same reaction is .... $\times 10^{-1}$ kJ mol^{-1} (Given, $R = 8.314\ JK^{-1}\ mol^{-1}$) **[2024, 1 Feb Shift-II]**

25 For a certain thermochemical reaction, $M \rightarrow N$ at $T = 400$ K, $\Delta H^\ominus = 77.2\ kJ\ mol^{-1}$, $\Delta S = 122\ JK^{-1}$, log equilibrium constant (log K) is $-$......$\times 10^{-1}$. **[2024, 27 Jan Shift-II]**

26 The value of log K for the reaction $A \rightleftharpoons B$ at 298 K is .... (Nearest integer)
Given: $\Delta H^\circ = -54.07\ kJ\ mol^{-1}$
$\Delta S^\circ = 10\ JK^{-1}\ mol^{-1}$
(Take $2.303 \times 8.314 \times 298 = 5705$) **[2023, 6 April Shift-I]**

27 One mole of an ideal gas at 350 K is in a 2.0 L vessel of thermally conducting walls, which are in contact with the surroundings. It undergoes isothermal reversible expansion from 2.0 L to 3.0 L against a constant pressure of 4 atm. The change in entropy of the surroundings (ΔS) is ... JK^{-1}. (Nearest integer)
Given, $R = 8.314\ JK^{-1}\ mol^{-1}$ **[2023, 12 April Shift-I]**

28 Consider the following reaction approaching equilibrium at 27°C and 1 atm pressure.

$$A + B \underset{k_b = 10^2}{\overset{k_f = 10^3}{\rightleftharpoons}} C + D$$

The standard Gibb's energy change ($\Delta_r G^\circ$) at 27°C is (–) .... kJ mol^{-1}. (Nearest integer).
(Given: $R = 8.3\ J\ K^{-1}\ mol^{-1}$ and ln 10 = 2.3) **[2023, 29 Jan Shift-I]**

29 The standard free energy change (ΔG°) for 50% dissociation of N_2O_4 into NO_2 at 27°C and 1 atm pressure is $-x$ J mol^{-1}. The value of x is .... (Nearest integer) [Given : $R = 8.31\ J\ K^{-1}mol^{-1}$, log 1.33 = 0.1239 ln 10 = 2.3] **[2022, 25 June Shift-I]**

30 The standard entropy change for the reaction, $4Fe(s) + 3O_2(g) \longrightarrow 2Fe_2O_3(s)$ is $-550\ JK^{-1}$ at 298 K.
[Given, the standard enthalpy change for the reaction is -165 kJ mol^{-1}]. The temperature in K at which the reaction attains equilibrium is ... [Nearest integer] **[2022, 25 June Shift-I]**

31 For the reaction,
$2NO_2(g) \rightleftharpoons N_2O_4(g)$, when
$\Delta S = -176.0\ JK^{-1}$ and $\Delta H = -57.8\ kJ\ mol^{-1}$, the magnitude of ΔG at 298 K for the reaction is .... kJ mol^{-1}. (Nearest integer) **[2021, 1 Sep Shift-II]**

32 Data given for the following reaction is as follows.

$FeO(s) + C_{(graphite)} \longrightarrow Fe(s) + CO(g)$

Substance	ΔH° (kJ mol^{-1})	ΔS° (J mol^{-1} K^{-1})
FeO(s)	– 266.3	57.49
$C_{(graphite)}$	0	5.74
Fe(s)	0	27.28
CO(g)	– 110.5	197.6

The minimum temperature in K at which the reaction becomes spontaneous is .... (Integer answer) **[2021, 27 Aug Shift-II]**

33 For a given chemical reaction, $A \longrightarrow B$ at 300 K the free energy change is -49.4 kJ mol^{-1} and the enthalpy of reaction is 51.4 kJ mol^{-1}. The entropy change of the reaction is ...... $JK^{-1}mol^{-1}$. **[2021, 20 July Shift-II]**

34 Assuming ideal behaviour, the magnitude of log K for the following reaction at 25°C is $-x \times 10^{-1}$. The value of x is .... (Integer answer)

$$3HC{\equiv}CH(g) \rightleftharpoons C_6H_6(l)$$

[Given : $\Delta_f G^\circ(HC{\equiv}CH) = -2.04 \times 10^5\ J\ mol^{-1}$; $\Delta_f G^\circ(C_6H_6) = -1.24 \times 10^5\ J\ mol^{-1}$; $R = 8.314\ J\ K^{-1}mol^{-1}$] **[2021, 24 Feb Shift-II]**

35 For the reaction $A_{(g)} \rightarrow B_{(g)}$, the value of the equilibrium constant at 300 K and 1 atm is equal to 100.0. The value of $\Delta_r G$ for the reaction at 300 K and 1 atm in J mol^{-1} is $-xR$ where x is .... (Rounded off to the nearest integer)
[$R = 8.31$ J mol^{-1}K^{-1} and $\ln 10 = 2.3$]
[2021, 01 Feb Shift-I]

36 For a dimerisation reaction

$$2A(g) \longrightarrow A_2(g),$$

at 298 K, $\Delta U^{\ominus} = -20$ kJ mol^{-1}, $\Delta S^{\ominus} = -30$ JK^{-1} mol^{-1}, then the $\Delta G^{\ominus}$ will be - .... J.
[2020, 5 Sep Shift-II]

37 For the reaction; $A(l) \longrightarrow 2B(g)$
$\Delta U = 2.1$ kcal, $\Delta S = 20$ cal K^{-1} at 300 K.
Hence, ΔG in kcal is - ....
[2020, 8 Jan Shift-I]

Key Index

Topic 1

1 (a)	2 (b)	3 (d)	4 (c)	5 (c)	6 (d)	7 (c)	8 (55)	9 (28721)	10 (200)
11 (1200)	12 (4)	13 (620)	14 (8630)	15 (2218)	16 (48)				

Topic 2

1 (c)	2 (d)	3 (b)	4 (a)	5 (d)	6 (a)	7 (c)	8 (b)	9 (d)	10 (304)
11 (200)	12 (150)	13 (1200)	14 (0)	15 (0)	16 (50)	17 (15)			

Topic 3

1 (a)	2 (d)	3 (a)	4 (c)	5 (b)	6 (a)	7 (b)	8 (c)	9 (c)	10 (d)
11 (c)	12 (c)	13 (d)	14 (c)	15 (b)	16 (d)	17 (b)	18 (c)	19 (98)	20 (25)
21 (2500)	22 (95)	23 (466)	24 (48)	25 (200)	26 (2)	27 (82)	28 (38)	29 (274)	30 (150)
31 (6535)	32 (125)	33 (56)	34 (492)	35 (4)	36 (278)	37 (1411)	38 (800)	39 (1006)	40 (925)
41 (54)	42 (35)	43 (2)	44 (57)	45 (200)	46 (2)	47 (104)	48 (600)	49 (195)	50 (117)
51 (38)	52 (738)	53 (38)	54 (82)	55 (82)	56 (31)	57 (101)	58 (21)	59 (128)	60 (230)
61 (309)	62 (741)	63 (189494)							

Topic 4

1 (c)	2 (a)	3 (b)	4 (d)	5 (c)	6 (b)	7 (a)	8 (b)	9 (b)	10 (b)
11 (a)	12 (d)	13 (d)	14 (c)	15 (c)	16 (a)	17 (c)	18 (2)	19 (a)	20 (700)
21 (2850)	22 (400)	23 (2000)	24 (57)	25 (37)	26 (10)	27 (3)	28 (6)	29 (710)	30 (300)
31 (5)	32 (964)	33 (336)	34 (855)	35 (1380)	36 (13537.57)	37 (2.70)			

Detailed Explanations

Topic 1

01. *(a)* From change in internal energy
$[\Delta U = q + W]$ = state functions
Heat (q) = path function
Work done (W) = path function
Gibbs energy, $G = H - TS$
= state function.

02. *(b)* The correct answer is option (b).
The solid phase (ice) disappears completely.

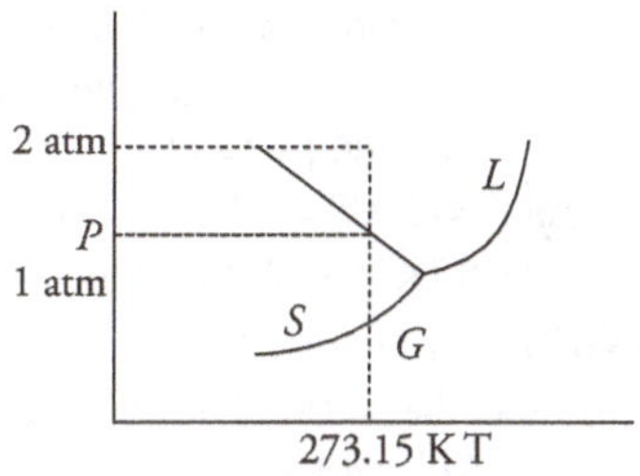

Phase diagram of H_2O.
If pressure is made two times, then mixture of ice and water will completely convert into water (liquid) form.

03. *(d)* Since in adiabatic system, no heat exchanges between system and surrounding, hence when methane undergo combustion, due to exothermic reaction, the temperature rises in system (A).

$$CH_4 + 2O_2 \longrightarrow CO_2 + 2H_2O + \text{heat}$$

In diathermic container, heat can change between system and surrounding, hence heat releases in to the surrounding and temperature of system (B) remains same.

04. *(c)* In isothermal expansion,

$$pV_m = K \text{ (constant)}$$

This relation is plotted in graph 'C'

Likewise, $p = \dfrac{K}{V_m}$

This relation is plotted in graph "A".
Thus, graph B and D are incorrect.

05. *(c)* From the given systems of solutions, both solution have same composition, which is 1 mole of x in 1 L, so all the intensive property will remain same, but as total amount is greater in solution 1 as compared to solution 2. So, all the extensive properties will be different. Hence, Gibbs free energy will be different.

06. *(d)* 0.6 kg with 60% purity of coal means

$$0.6 \text{ kg} \times \frac{60}{100} = 0.36 \text{ kg of coal}$$

Out of 0.36 kg of coal, 60% is converting to CO means :

$$0.36 \text{ kg} \times \frac{60}{100} = 0.216 \text{ kg} = 216 \text{ g}$$

Rest will be converting to CO_2, means :
$0.36 - 0.216 = 0.144 \text{ kg} = 144 \text{ g}$
From given equation:

$$C + \frac{1}{2}O_2 \longrightarrow CO + 100 \text{ kJ}$$

means 12g of carbon produces 100 kJ of energy on incomplete combustion
So, 216 g of carbon will produces

$$= \frac{100}{12} \times 216 = 1800 \text{ kJ}$$

From equation:
$C + O_2 \longrightarrow CO_2 + 400 \text{ kJ}$
12 g of carbon produces 400 kJ energy on complete combustion.
So, 144 g of carbon will produce

$$\frac{400}{12} \times 144 = 4800 \text{ kJ}$$

Total energy produces

$$= 1800 \text{ kJ} + 4800 \text{ kJ} = 6600 \text{ kJ}$$

07. *(c)* For isothermal reversible expansion,

$$|W| = nRT \ln \frac{V_f}{V_i} = nRT \ln \frac{V}{V_i}$$

where, V = final volume,
V_i = initial volume
or $|W| = nRT \ln V - nRT \ln V_i$
On comparing with equation of straight line, $y = mx + c$, we get

$$\text{slope} = m = +nRT$$
$$\text{intercept} = -nRT \ln V_i$$

Thus, plot of $|W|$ with $\ln V$ will give straight line in which slope of $2(T_2)$ is greater than slope of $1(T_1)$ which is given in all options.
Now, if we extend the line of T_1 and T_2 given in option (c) it seems to be touching the origin. If they touch the origin then y-intercept becomes zero which is not possible. Thus, it is not exactly correct answer but among the given options it is the most appropriate one.

08. *(55)* Given, $V_1 = 100 \text{ L}$
and $V_2 = 10 \text{ L}$

Work done is given,

$$W = -nRT \ln \frac{V_2}{V_1}$$
$$= -1 \times 0.08206 \times 291.15 \times 2.303 \log \frac{10}{100}$$
$$= 55 \text{ L atm}$$

Hence, final answer is 55 L atm.

09. *(28721)* It is isothermal reversible expansion, so work done is negative.
Work done is given by

$$W = -2.303nRT \log\left(\frac{V_2}{V_1}\right)$$
$$= -2.303 \times 5 \times 8.314 \times 300 \log\left(\frac{100}{10}\right)$$
$$= -28720.713 \text{ J}$$
$$\approx -28721 \text{ J}$$

Hence, the value of x is 28721.

10. *(200)* Work done is given by area enclosed in the p *vs* V cyclic graph.
Sign is +ve for clockwise cyclic process for V *vs* p graph.

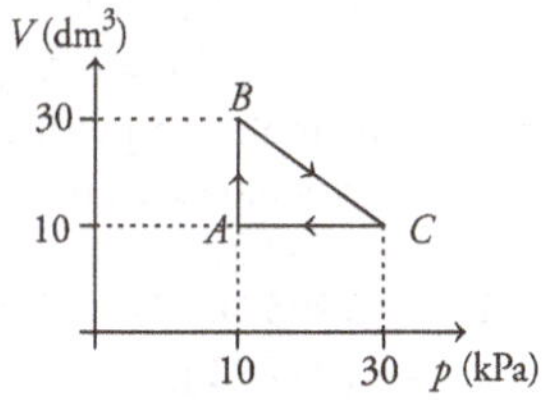

Work done = Area of triangle

$$= \frac{1}{2} \times \text{base} \times \text{height}$$
$$= \frac{1}{2} \times 20 \times 20$$
$$= 200 \text{J}$$

Hence, work done is 200J.

11. *(1200)* For isothermal expansion at constant pressure, heat is given by

$$Q = -p_{ext}(V_2 - V_1)$$
$$= -80 \times 1000 \times (45 - 30) \times 10^{-3}$$
$$= -15 \times 80 = -1200 \text{ J}$$

Negative sign indicates heat is lost.
Hence, amount of heat transfer is 1200 J.

12. *(4)* **Extensive properties** Mole, volume, Gibb's free energy change
Intensive properties Molar mass, molar heat capacity, molarity, $E°_{cell}$
Thus, total number of intensive properties are 4.

13. *(620)* $1 \rightarrow 2 \Rightarrow$ Isobaric processes

$2 \rightarrow 3 \Rightarrow$ Isochoric processes

$3 \rightarrow 1 \Rightarrow$ Isothermal processes

Work done

$$(W) = W_{1\rightarrow 2} + W_{2\rightarrow 3} + W_{3\rightarrow 1}$$

$$= \left(-p(V_2 - V_1) + 0 + \left[-p_1V_1 \ln\left(\frac{V_2}{V_1}\right)\right]\right)$$

$$= \left[-1\times(40-20) + 0 + \left[-1\times 20 \ln\left(\frac{20}{40}\right)\right]\right]$$

$$= -6.2 \text{ bar L}$$

$$|W| = 6.2 \text{ bar L} = 620 \text{ J}$$

14. *(8630)* Work done $= -nRT \ln \frac{V_2}{V_1}$

$$= -5 \times 8.3 \times 300 \ln \frac{20}{10}$$

$$= -12450 \times 2.303 \log 2$$

$$= -8630.38 \text{ J}$$

The magnitude of maximum work obtained = 8630 J (nearest integer).

15. *(2218)*

$$Fe(s) + HCl(aq) \rightarrow FeCl_2(aq) + H_2(g)$$

50g

(Moles of Fe = Moles of H_2)

moles $= \frac{50}{55.85}$ 0.8952 mol

= 0.8952 mol

Work done can be given by

$$W = -p_{ext}\Delta V = -\Delta n_g RT$$

Δ_{n_g} = (gaseous moles on product side) − (gaseous moles on reactant side)

$= 0.8952 - 0$

$$W = -0.8952 \times 8.314 \times 298$$

$$= -2217.92 \text{ J}$$

Closest integer = 2218

16. *(48)* Work done is equal to the area under pV-curve.

Work done $(W) = W_{AB} + W_{BC}$

W_{AB} = Area of square

$= 6 \times 6 = 36 \text{ Pa. m}^3$

W_{BC} = Area of triangle

$= \frac{1}{2} \times 4 \times 6 = 12 \text{ Pa. m}^3$

$\therefore$ Work done $(W) = (36+12) \text{ Pa. m}^3$

$= 48.00 \text{ Pa. m}^3 = 48.00 \text{ J}$

Topic 2

01. *(c)* For thermally insulated system

$$q = 0$$

Also from 1st law of thermodynamics

$$\Delta U = q + W$$

So, $\Delta U = W$

$\Rightarrow W > 0, \Delta U > 0$

So, the correct thermodynamic parameter is

$$\Delta U > 0, q = 0, W > 0$$

02. *(d)* Given, $T = 300 \text{K}$

$$V_1 = 10 \text{ dm}^3$$

$$V_2 = 20 \text{ dm}^3$$

$\Delta U = 3, q = ?, W = ?$

We know that, for isothermal reversible process

$$-q = W = -nRT \ln \frac{V_2}{V_1}$$

$$W = -8.3 \times 300 \times \ln\left(\frac{20}{10}\right)$$

$$W \approx -1.718 \text{ kJ}$$

$$\Rightarrow q \approx 1.718 \text{ kJ}$$

$$\Delta U = 0 \quad (\because \Delta T = 0)$$

03. *(b)* For isothermal process.

$|W_{\text{reversible}}|_{\text{expansion}}$

$= |W_{\text{reversible}}|_{\text{compression}}$

$= -nRT \ln \frac{V_f}{V_i}$

$|W_{\text{irreversible}}|_{\text{expansion}}$

$< |W_{\text{irreversible}}|_{\text{compression}}$

$d > c = a > b$

$|W_{\text{irreversible}}|_{\text{expansion}} = -p_{\text{ext}} (V_f - V_i)$

04. *(a)* As internal energy 'U' is a state function, its cyclic integral must be zero in a cyclic process.

$\therefore \Delta U$(case I) $= \Delta U$ (case II)

$= \Delta U$ (case III)

05. *(d)* Given, $\Delta q = 500$ J, $n = 0.5$ mol

$T = 298$ K $p = 1$ atm,

$R = 8.3 \text{ JK}^{-1} \text{mol}^{-1}$, $\Delta U = ?$ $T_f = ?$

Heat is given by the formula,

$$q_p = n \times C_p \times \Delta T$$

$$500 = 0.5 \times \frac{5}{2} \times 8.3(T_f - 298)$$

$$\left[\because C_p = \frac{5R}{2}\right]$$

On solving, we get,

$$T_f = 346.2 \text{ K} \simeq 348 \text{ K}$$

Also, we know that

$$\frac{\Delta H}{\Delta U} = \frac{C_p}{C_V}$$

$$= \frac{5/2}{3/2} = \frac{5}{3}$$

$$\Delta U = \frac{3}{5} \times 500 = 300 \text{ J}$$

So, the final temperature and change in internal energy, respectively are 348 K and 300 J.

06. *(a)* For free expansion, $p_{\text{external}} = 0$

$\therefore$ Work done, $w = -p_{\text{ext}}\Delta V = 0$

$\because$ Adiabatic process, $q = 0$

Also from 1st law of thermodynamics.

$$\Delta U = q + w$$

$$\Delta U = 0$$

Internal energy is a function of temperature.

If internal energy is constant

$$\Delta T = 0$$

So, for free expansion, $q = 0, \Delta T = 0, w = 0$

Hence, option (a) is correct.

07. *(c)* $\Delta U = nC_V\Delta T = nC_V (T_2 - T_1)$

$= 5 \text{ mol} \times 28 \text{ JK}^{-1} \text{ mol}^{-1} \times (200 - 100) \text{ K}$

$= 14{,}000 \text{ J} = 14 \text{ kJ}$

$\Delta pV = nR\Delta T = nR(T_2 - T_1)$

$= 5 \text{ mol} \times 8 \text{ JK}^{-1} \text{ mol}^{-1} \times (200 - 100) \text{K}$

$= 4000 \text{ J} = 4 \text{ kJ}$

08. *(b)* From the 1st law of thermodynamics,

$$\Delta U = q + w$$

(a) **Cyclic process** For cyclic process, $\Delta U = 0$

$\therefore q = -w$

Thus, option (a) is correct.

(b) **Adiabatic process** For adiabatic process,

$q = 0 \Rightarrow \Delta U = W$

Thus, option (b) is incorrect.

(c) **Isochoric process** For isochoric process,

$\Delta V = 0$. Thus, $W = 0$

$(\because W = p\Delta V)$.

$\therefore \Delta U = q$

Thus, option (c) is correct.

(d) **Isothermal process** For isothermal process,

$\Delta U = 0$

$\therefore q = -W$

Thus, option (b) is correct.

09. *(d)* From the condition mention in question, the gas follows free expansion.

$\therefore$In free expansion of gas, $p_{\text{ext}} = 0$

where $W = 0$, $q = 0$ and $\Delta U = 0$

10. *(304)* Work done from 1 to 2

$W_{1\rightarrow 2} = 0$

(As $W = p\Delta V$ and $\Delta V = 0$)

$W_{2\rightarrow 3} = -p\Delta V$

$= -3[2-1] = -3$ atm l

$W_{3\rightarrow 4} = 0$

Total work done $= -3$ atm l

$= -3 \times 101.3$ Joule

$= -304$ Joule

11. *(200)* For isothermal process, $\Delta U = 0$

So, $Q = -W$

$= -[-p_{ext}(V_2 - V_1)]$

$Q = 5[20 - 60]$

$= -200$ atm L

12. *(150)* Given,

Number of moles $= 1$ mole

Temperature $(T_i) = 27°C = 300$ K

Since, the processes is adiabatic, hence, $q = 0$

$\Rightarrow \quad W = \Delta U$

On substituting values, we will get

$W = nC_V \Delta T$

$-3 \times 10^3 = 1 \times 20[T_2 - 300]$

$T_2 - 300 = -150$ or $T_2 = 150$ K

13. *(1200)* Power of heater $= 60$ W

$= 60$ J/sec

Total energy emitted $= 60 \times 100$

$= 6000$ J

heat capacity × temperature difference = total energy emitted

heat capacity × 5°C = 6000

heat capacity $= \dfrac{6000}{5} = 1200$ J/K

14. *(0)* For an isothermal processes, the temperature remains constant. So, the change in internal energy is zero.

$pV =$ constant

15. *(0)* Change in volume,

$\Delta V = 20 - 4 = 16$ L

Work done, $W = -p_{ext}\Delta V$

but in vacuum, $p_{ext} = 0$

$\therefore \quad W = -0 \times 16 = 0$

For an isothermal expansion, $\Delta U = 0$

Using first law of thermodynamics,

$\Delta U = q + W \Rightarrow 0 = q + 0 \Rightarrow q = 0$

16. *(50)* Work done by the system,

$W = -200$ J

Heat absorbed, $q = +150$ J

According to first law of thermodynamics

Internal energy, $\Delta U = q + W$

$\Delta U = 150 + (-200)$

{ΔU = Change in internal energy}

$= -50$ J

Magnitude $= |-50\text{ J}| = |\Delta U| = 50$ J

17. *(15)* The gas performs isothermal irreversible work (W).

where, $\Delta U = 0$ (change in internal energy)

From, 1st law of thermodynamics,

$\Rightarrow \quad \Delta U = q + W$

$\Rightarrow \quad 0 = q + W \Rightarrow \quad q = -W$

Now, $\quad W = -P_{ext}(V_2 - V_1)$

$= -P_{ext}\left(\dfrac{nRT}{p_2} - \dfrac{nRT}{p_1}\right)$

$= -P_{ext} \times nRT\left(\dfrac{1}{p_2} - \dfrac{1}{p_1}\right)$

Given, $P_{ext} = 4.3$ MPa, $p_1 = 2.1$ MPa,

$p_2 = 1.3$ MPa, $n = 5$ mol,

$T = 293$ K

and $\quad R = 8.314$ J mol^{-1} K^{-1}

$= -4.3 \times 5 \times 8.314 \times 293\left(\dfrac{1}{1.3} - \dfrac{1}{2.1}\right)$

$= -15347.70$ J mol^{-1}

$= -15.347$ kJ mol^{-1}

$\simeq -15$ kJ mol^{-1}

$\Rightarrow q = 15$ kJ mol^{-1}

Topic 3

01. *(a)* The heat of formation of $SO_2(g)$.

$$SO_2(g) + \frac{1}{2}O_2(g) \longrightarrow SO_3(g), \quad \Delta H = -y$$

$\Delta_r H = (\Delta_f H)_{SO_3} - (\Delta_f H)_{SO_2}$

$-y = -2x - (\Delta H_f)_{SO_2}$

$(\Delta_f H)_{SO_2} = y - 2x$ kcal

Hence, option (a) is correct.

02. *(d)* C (diamond) $\rightarrow$ C(graphite) $+ X$ kJ/mol

C(diamond) $+ O_2(g) \longrightarrow CO_2(g) + Y$ kJ/mol

C(graphite) $+ O_2(g) \longrightarrow CO_2(g) + Z$ kJ/mol

then $X = Y - Z$ is the correct relation.

03. *(a)* The general combustion reaction of a hydrocarbon is as follows :

$$C_xH_y + \left(x + \frac{y}{4}\right)O_2 \longrightarrow xCO_2 + \frac{y}{2}H_2O.$$

For heptane, $x = 7$, $y = 16$

$\Rightarrow C_7H_{16}(l) + 11O_2(g) \longrightarrow 7CO_2(g) + 8H_2O(l)$

$\therefore \quad \Delta n_g = 7 - 11 = -4$

Now, from the principle of thermochemistry,

$\Delta H = \Delta U + \Delta n_g RT$

$\Rightarrow \Delta H - \Delta U = \Delta n_g RT$

$= -4RT$

04. *(c)* Both Statement I and II are true. At melting point when ice melts, supplied heat is utilised to overcome intermolecular attraction within the molecules so temperature remains constant.

05. *(b)* The reaction is as follows

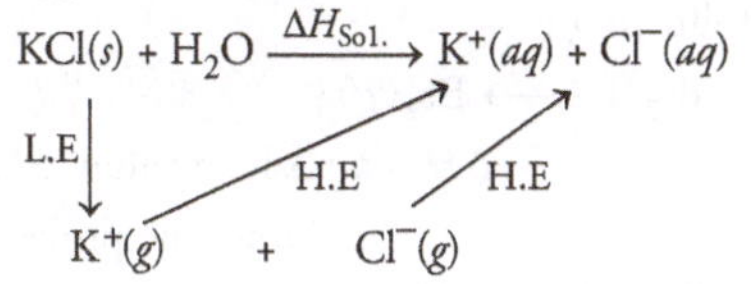

$\Delta H_{Sol.} = \text{L.E} + (\text{H.E.})_{K^+(g)} + (\text{H.E})_{Cl^-(g)}$

$= z - x - y = z - (x + y)$

06. *(a)* According to 2nd law of thermodynamics a reaction is spontaneous, if $\Delta G < 0$

or $\quad \Delta H - T\Delta S < 0$

$\Rightarrow \quad T > \dfrac{\Delta H}{\Delta S} = T_e \quad \Rightarrow T > T_e$

07. *(b)* The enthalpy of neutralisation of strong acid and strong base is always -57 kJ/mol. The reason for that is, strong monoacid gives one mole of H^+ and strong mono base gives one mole of OH^- which form one mole of water.

08. *(c)* Enthalpy of bond dissociation (kJ/mol) at 298.2 K

For hydrogen $= 435.88$ kJ/mol

For deuterium $= 443.35$ kJ/ mol

Hence, difference in energy of deuterium and hydrogen is approx. 7.50 kJ/mol

Therefore, equation is $[E_H \simeq E_D - 7.5]$

09. *(c)* The enthalpy of solution of an ionic solid is numerically equal to the sum of its hydration and lattice energies.

i.e., $\quad \Delta H^\circ_{sol} = \Delta H^\circ_{hydration} + \Delta H^\circ_{lattice}$

$\therefore$

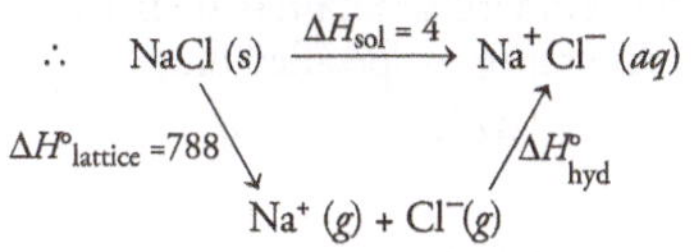

$\Delta_{sol} H^\circ = \Delta_{lattice} H^\circ + \Delta_{hyd} H^\circ$

$4 = 788 + \Delta_{hyd} H^\circ$

$\Delta H_{hyd} H^\circ = -784$ kJ/mol

10. *(d)* (1) From enthalpy, $H = U + pV$, and for ideal gas, $pV = nRT$

So, $\quad [H = U + nRT]$

and (H) depends only on temperature.

(2) For an ideal gas, compressibility factor $(Z = 1)$

(3) $C_p - C_V = R$

(4) For all processes, $[\Delta U = C_V dT]$

11. *(c)* Atomisation of $Br_2(l)$

$Br_2(l) \longrightarrow 2Br(g);$

$\Delta H = x$ kJ mol^{-1}

which can be considered as sum of following :

$Br_2(l) \longrightarrow Br_2(g)$;

$\Delta H =$ a positive value ...(i)

(endothermic process)

$Br_2(g) \longrightarrow 2Br(g)$; ΔH

(bond enthalpy of Br_2) $= y$ ...(ii)

$\therefore \; x = y +$ (Positive value), and $x > y$.

12. *(c)*

> **idea** When q is the amount of heat involved in a system then at constant pressure
>
> $q = q_p$ and $C_p\Delta T = \Delta H$

Given reaction :

$I_2(s) \longrightarrow I_2(g)$

Specific heat of $I_2(s) = 0.055$ cal g^{-1} K^{-1}.

Specific heat of

I_2(vap) $= 0.031$ cal g^{-1} K^{-1}.

Enthalpy (H_1) of sublimation of iodine $= 24$ cal g^{-1}

If q is the amount of heat involved in a system then at constant pressure $q = q_p$ and

$$\Delta H = C_p \Delta T$$

$$H_2 - H_1 = C_p(T_2 - T_1)$$

$$H_2 = H_1 + \Delta C_p(T_2 - T_1)$$

$$H_2 = 24 + (0.031 - 0.055)(250 - 200)$$

$$H_2 = 24 + (-0.024)(50)$$

$$= 24 - 1.2 = 22.8 \text{ cal/g}$$

13. *(d)* Among the given statements, option (d) is correct.

For standard state, pressure is 1 bar and temperature is specified only

$= (\Delta H_f^\circ)_{O_2(g)} = 0$

14. *(c)*

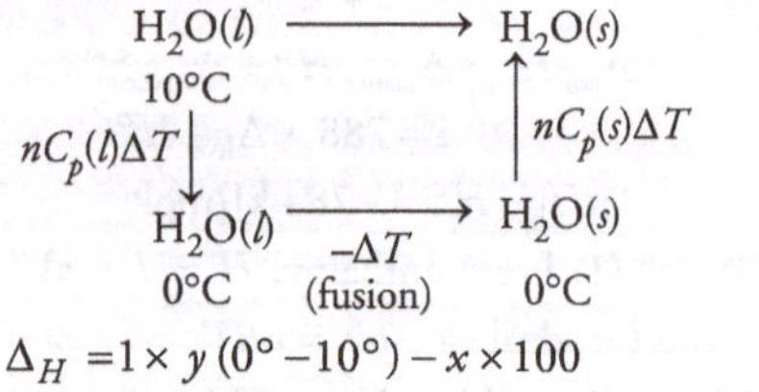

$\Delta_H = 1 \times y\,(0° - 10°) - x \times 100$

$+ 1 \times z\,(10° - 0°)$

$\Delta_H = -10\,[100x + y + z]$ Joule

15. *(b)* From the data given, ΔH is negative which implies that dissolution of HCl gas is exothermic.

$HCl(g) + 10H_2O(l) \longrightarrow HCl \cdot 10H_2O$ ...(i)

$\Delta H_1 = -69.01$ kJ/mol

$HCl(g) + 40H_2O(l) \longrightarrow HCl \cdot 40H_2O$ ...(ii)

$\Delta H_2 = -72.79$ kJ/mol

From above two equations, it is clear that heat of solution depends upon amount of solvent.

Now, subtract Eq. (ii) from (i), we get

$HCl \cdot 10H_2O + 30H_2O(l) \longrightarrow HCl \cdot 40H_2O$

so, heat of dilution $= -72.79 - (-69.01)$

$= -3.78$ kJ/mol

So, option (a) is incorrect. For heat of formation reactant should be in elemental form hence option (d) is incorrect.

16. *(d)* Consider graph (d)

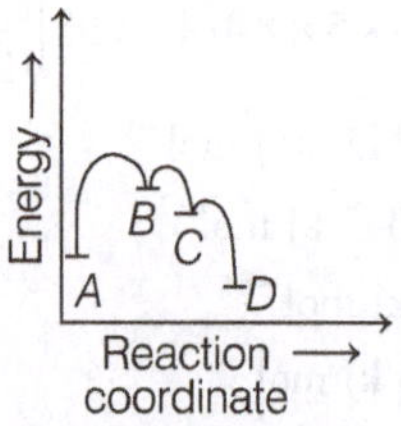

$A \rightarrow B$ (slow)

$\Delta H = +$ve, $E_{a_1} =$ high

$B \rightarrow C$ (fast)

$\Delta H = -$ ve, $E_{a_2} =$ low

$C \rightarrow D$ (fast)

$\Delta H = -$ ve, $E_{a_3} \rightarrow$ low

This graph correctly matches with given mechanism.

17. *(b)* Given equation,

(A) $2CO(g) + O_2(g) \longrightarrow 2CO_2(g)$, $\Delta H_1^\circ = -x$ kJ mol^{-1}

(B) C(graphite) $+ O_2(g) \longrightarrow CO_2(g)$, $\Delta H_2^\circ = -y$ kJ mol^{-1}

We have to find $\Delta H°$ for the reaction,

C(graphite) $+ \frac{1}{2}O_2(g) \longrightarrow CO(g)$ ...(i)

C(graphite) $+ O_2(g) \longrightarrow CO_2(g)$ ...(ii)

$\Delta H_1 = -y$ kJ/mol^{-1}

$CO_2(g) \longrightarrow CO(g) + \frac{1}{2}O_2(g)$ ...(iii)

$\Delta H_2 = \frac{x}{2}$ kJ/mole

Eq. (i) = Eq. (ii) + Eq. (iii)

$\therefore \quad \Delta H = -y + \frac{x}{2}$

$\Delta H = \frac{x}{2} - y \;\Rightarrow\; \Delta H = \frac{x - 2y}{2}$

18. *(c)* The reaction between carbon and hydrogen can be written as

$2C + 3H_2 \longrightarrow C_2H_6$ ...(R)

Combustion equation of ethane

$C_2H_6\,(g) + \frac{7}{2}O_2(g) \longrightarrow 2CO_2 + 3H_2O$ ...(i)

$\Delta H^\circ_{comb} = -1560$ kJ/mol (given)

$C(s) + O_2(g) \longrightarrow CO_2(g)$ ...(ii)

$\Delta H^\circ_{comb} = -394$ kJ/mol

$H_2(g) + \frac{1}{2}O_2 \longrightarrow H_2O(g)$ ...(iii)

$\Delta H^\circ_{comb} = -286$ kJ/mol

2 × Eq. (ii) + 3 × Eq. (iii) – Eq (i) = Eq. (R)

$= (-394 \times 2) + (-286 \times 3) - (-1560)$

$\Delta H_f^\circ = -86.0$ kJ/mol

19. *(98)* $\Delta H_{\text{Resonance energy}} = \Delta H_{f(\text{experimental})} - \Delta H_{f(\text{theoretical})}$

$\Delta H_{f(\text{exp})}$ for $X_2Y(g) = 80$ kJ / mol

For $\Delta H_{f\,(\text{theoretical})}$

$X_2(g) + \frac{1}{2}Y_2(g) \longrightarrow X_2Y(g)$; $\Delta Hf = ?$

$\Delta H_{f(\text{theoretical})} = \left[\text{BE}_{x\equiv x} + \frac{1}{2}\text{BE}_{y=y}\right] - [\text{BE}_{x=x} + \text{BE}_{x=y}]$

$= \left(940 + \frac{1}{2} \times 500\right) - (410 + 602)$

$= 178$ kJ/ mol

$\Delta H_{\text{Resonance energy}} = 80 - 178$

$= -98$ kJ / mol

$|\Delta H_{\text{Resonance energy}}| = 98$

20. *(25)* Given that,

ΔH°_{sub} [(graphite)] = 710 kJ/mol

$\Delta_{C-H}H° = 414$ kJ/mol

$\Delta_{H-H}H° = 436$ kJ/mol

$\Delta_{C=C}H° = 611$ kJ/mol

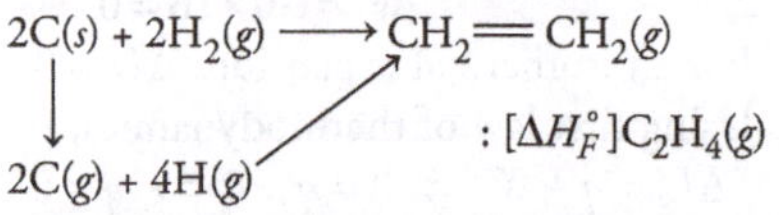

$[\Delta H_F^\circ]_{C_2H_4}(g) = 2 \times [\Delta H^\circ_{sub}]_{C(s)} + 2\Delta H^\circ_{H-H} - 1 \times \Delta H^\circ_{C=C} - 4\Delta H^\circ_{C-H}$

$[\Delta H_F^\circ]_{C_2H_4(g)} = (2 \times 710) + (2 \times 436) - 611 - 4 \times 414$

$[\Delta H_F^\circ]_{C_2H_4(g)} = 25$ kJ/mol

21. *(2500)*

Mole of octane $= \frac{1.14}{114} = 0.01$ mol

Heat evolved $= C \times \Delta T = 5 \times 5 = 25$ kJ

$\therefore$ Magnitude of heat of combustion

$= \frac{25}{0.01} = 2500$ kJ / mol

22. *(95)*

> **idea** This question is based upon the concept of Hess's law to calculate the required ΔH value for the given reaction. First obtain the required equation by multiplying Eq. (ii) by 2, (iii) by 3 then from this value subtract Eq. (i). After that calculate the value by using
> $2\Delta H_2 + 3\Delta H_3 - \Delta H_1$

Given,

$$C_2H_6(g) + \frac{7}{2}O_2(g) \longrightarrow 2CO_2(g) + 3H_2O(l);$$
$$\Delta H_1^\circ = -1550 \text{ kJ/mol} \quad ...(i)$$
$$C(\text{Graphite}) + O_2(g) \longrightarrow CO_2(g) \quad ...(ii)$$
$$\Delta H_2^\circ = -393.5 \text{ kJ/mol}$$
$$H_2(g) + \frac{1}{2}O_2(g) \longrightarrow H_2O(l) \quad ...(iii)$$
$$\Delta H_3^\circ = -286 \text{ kJ/mol}$$
$$2C(\text{graphite}) + 3H_2(g) \longrightarrow C_2H_6(g);$$
$$\Delta H_f(C_2H_6(g)) = ?$$
$$\Delta H_f = 2\Delta H_2^\circ + 3\Delta H_3^\circ - \Delta H_1^\circ$$
$$= 95 \text{ kJ/mole}$$

23. *(466)* The reaction involved is as follows

$$\frac{1}{2}H_2(g) \longrightarrow H(g);$$
$$\Delta_f H(H(g)) = 220 \text{ kJ/ mol} \quad ...(i)$$
$$\frac{1}{2}O_2(g) \longrightarrow O(g);$$
$$\Delta_f H(O_g) = 250 \text{ kJ/mol} \quad ...(ii)$$

[Eq (i) × 2 + Eq (ii)] −2 [BE(O—H)]

$$H_2(g) + \frac{1}{2}O_2(g) \longrightarrow H_2O(g)$$
$$\Delta H_f(H_2O(l)) = -242 = 440 + 250 - 2(\text{BE(O—H)})$$

BE(O—H) = 466 kJ/ mol

24. *(48)* Given, $\Delta H_f[CO_2(g)] = -393.5$ kJ/mole

$$\Delta H_f[H_2O(l)] = -286 \text{ kJ/mole}$$
$$\Delta H_C[C_6H_6] = -3267.0 \text{ kJ/mole}$$
$$\Delta H_f C_6H_6 = ?$$
$$C_6H_6 + \frac{15}{2}O_2(g) \rightarrow 6CO_2(g) + 3H_2O(l)$$
$$\Delta H_R = \Delta H_C = \Sigma\Delta H_f(P) - \Sigma\Delta H_f(R)$$
$$-3267 = 6 \times (-393.5) + 3 \times (-286) - \Delta H_f(C_6H_6)$$
$$\therefore \Delta H_f(C_6H_6) = 48 \text{ kJ/mole}$$

25. *(200)* Bond dissociation enthalpy of $X_2 = ?$

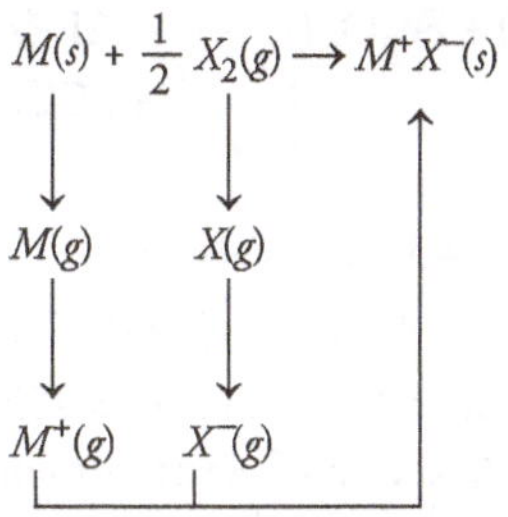

So, the value of $\Delta H_f(MX)$ can be calculated by using

$$\Delta H_f(MX) = \Delta H_{\text{sub}}(M) + \text{IE}(M) + \frac{1}{2}[\text{BE}(X-X)] + \text{EG}(X) + \text{LE}(MX)$$

Substitute the values from given data

$$-400 = 100 + 500 + \frac{1}{2}(\text{BE}) + (-300) + (-800)$$
$$\therefore \quad \text{BE} = 200 \text{ kJ/mole}$$

26. *(2)* According to question

$$H^+ + OH^- \longrightarrow H_2O \Rightarrow x$$
(for HCl)
$$2H^+ + 2OH^- \longrightarrow 2H_2O$$
$$\Rightarrow 2x = y \text{ (for } H_2SO_4) \Rightarrow \frac{y}{x} = 2$$

27. *(82)* $CuSO_4(s)\ aq \longrightarrow CuSO_4(aq);$

$$\Delta H = -70 \text{ kJ mol}^{-1} \quad ...(i)$$
$$CuSO_4.5H_2O(s) + aq \longrightarrow CuSO_4(aq) + 5H_2O;$$
$$\Delta H = +12 \text{ kJ mol}^{-1} \quad ...(ii)$$

On reversing Eq. (ii), we get

$$CuSO_4(aq) + 5H_2O \longrightarrow CuSO_4 \cdot 5H_2O(s) + aq \quad ...(iii)$$
$$\Delta H = -12 \text{ kJ mol}^{-1}$$

On adding Eq. (i)and (iii), we get

$$CuSO_4(s) + 5H_2O \longrightarrow CuSO_4 \cdot 5H_2O(s)$$
$$\therefore \quad \Delta H = -70 - 12 = -82 \text{ kJ}$$

Thus, the value of x is 82.

28. *(38)* Given,

$$\Delta_{\text{vap}}H^\ominus = +40.79 \text{ kJ/mol}$$
$$T = 100°C = 373 \text{ K}$$

Using enthalpy equation,

$$\Delta_{\text{vap}}H^\ominus = \Delta U^\circ + \Delta n_g RT$$
$$\Delta U^\circ = 40.79 - \frac{8.3 \times 373}{1000}$$
$$= 37.67 \text{ kJ/mol} \approx 38 \text{ kJ/mol}$$

29. *(274)* In adiabatic condition, $q = 0$

Then, equation for change in internal energy will be given by

$$\Delta U = q + W \quad (q = 0)$$
$$nC_V\Delta T = -p_{\text{ext}}(V_2 - V_1) \quad ...(i)$$

as $(W = p\Delta V)$

$$V_2 = 2V_1 \quad \text{(given)}$$
$$\frac{nRT_2}{p_2} = \frac{2nRT_1}{p_1} \quad ...(ii)$$
$$p_1 = 5 \text{ atm}, \ T_1 = 298 \text{ K}$$

p_2 from Eq. (ii) will be

$$p_2 = \frac{5T_2}{2 \times 298}$$

Now from Eq. (i)

(Put $p_{\text{ext}} = 1$ atm)

$$n\frac{5}{2}R(T_2 - T_1) = -1\left[\frac{nRT_2}{p_2} - \frac{nRT_1}{p_1}\right]$$
$$\frac{n5}{2}R(T_2 - T_1) = -nR\left[\frac{T_2}{p_2} - \frac{T_1}{p_1}\right]$$

Put, $T_1 = 298$ and $p_2 = \dfrac{5T_2}{2 \times 298}$

$$\frac{5}{2}(T_2 - 298) = -1\left[\frac{T_2}{\frac{5T_2}{2 \times 298}} - \frac{298}{5}\right] \quad ...(iii)$$

On solving Eq. (iii)

$$T_2 = 274.16 \text{ K}$$

Hence, nearest integer value of temperature is 274 K.

30. *(150)*

> **idea** To solve this question, first of all write the balanced chemical equation, then by using this equation, calculate the value of Δn_g after that use the equation $\Delta H = \Delta U + \Delta n_g RT$ and substitute the given values. Then compare the equation obtained with the given equation in question to obtain the value of x.

The complete reaction is as follow,

$$C_6H_5COOH(s) + \frac{15}{2}O_2(g) \longrightarrow 7CO_2(g) + 3H_2O(l)$$
$$\Delta H = \Delta U + \Delta n_g RT$$
$$= -321.30 - \frac{1}{2}R \times 300$$
$$= (-321.30 - 150R) \text{ kJ}$$

Matching with equation given in question the value of x is 150.

31. *(6535)* Given,

$$6C(\text{graphite}) + 3H_2(g) \longrightarrow C_6H_6(l);$$
$$\Delta H = 48.5 \text{ kJ/mol} \quad ...(i)$$
$$C(\text{graphite}) + O_2(g) \longrightarrow CO_2(g);$$
$$\Delta H = -393.5 \text{ kJ/mol} \quad ...(ii)$$

$H_2(g)+\frac{1}{2}O_2(g)\longrightarrow H_2O(l);$

$\Delta H=-286$ kJ/mol ...(iii)

Eq. (i) × 1 + (ii) × 6 + (iii) × 3

$=-48.5-6\times 393.5-3\times 286$

$=-3267.5$ kJ for 1mol

$=-6535$ kJ for 2 mol

32. *(125)* Given, Bond enthalpies of

C—H = 414 kJ; C—C = 347 kJ

H—H = 435 kJ; C=C = 615 kJ

Now, the formation of ethane takes place as follows

$C_2H_4(g)+H_2(g)\rightarrow C_2H_6(g)$

$\Delta H=$ B.E. (C=C) + 4B.E. (C—H) + B.E. (H—H) − B.E. (C—C) − 6B.E. (C—H)

$\Delta H=$ B.E. (C=C) + B.E. (H—H) − B.E. (C—C) − 2B.E. (C—H)

$\Rightarrow \Delta H=615+435-347-2(414)$

$\Delta H=-125$ kJ

33. *(56)* Given,

ΔH°_{vapour} CCl_4 = 30.5 kJ/mol

Mass of CCl_4 = 284 g

Molar mass of CCl_4 = 154 g/mol

Moles of $CCl_4=\frac{284}{154}=1.844$ mol

ΔH°_{vapour} for 1 mole = 30.5 kJ/mole

ΔH°_{vapour} for 1.844 mole

$=30.5\times 1.844=56.242$ kJ

Hence, final answer is 56 kJ (nearest integer).

34. *(492)*

idea This question is based upon the concept of Hess's law to calculate the required ΔH value for the given reaction. First obtain the required equation by multiplying Eq. (ii)with 3 then subtract Eq. (i) from the equation. After that calculate the value through the formula.

$\Delta H_3=3\Delta H_2-\Delta H_1$

Two given reactions are

$2Fe(s)+\frac{3}{2}O_2(g)\longrightarrow Fe_2O_3(s);$

$\Delta H^\circ=-822$ kJ/mol ...(i)

$C(s)+\frac{1}{2}O_2(g)\longrightarrow CO(g);$

$\Delta H^\circ=-110$ kJ/mol ...(ii)

Multiply Eq. (ii) by 3 and then subtract Eq. (i) from this equation to get the required equation.

$3C(s)+Fe_2O_3(s)\longrightarrow 2Fe(s)+3CO(g)$ ...(iii)

Then,

$\Delta H_3=3\times\Delta H_2-\Delta H_1$

$=3(-110)-(-822)=492$ kJ/mol

Hence, enthalpy change is 492 kJ/mol.

35. *(4)* Addition of electron to neutral atom only is exothermic process.

All others corresponding to removal of electron from a neutral atom, unipositive ion and addition of electron to uninegative ion are endothermic processes.

A → Endothermic (atomisation)

B → Endothermic (atomisation)

C → Endothermic (vapourisation)

D → Exothermic (combustion)

E → Endothermic (dissolution)

36. *(278)* $H_2+\frac{1}{2}O_2\longrightarrow H_2O$

$\Delta H=-241.8$ kJ/mol ...(i)

$C+O_2\longrightarrow CO_2$

$\Delta H=-393.5$ kJ/mol ...(ii)

$C_2H_5OH+3O_2\longrightarrow 3H_2O+2CO_2$

$\Delta H=-1234.7$ kJ/mol ...(iii)

Formation of C_2H_5OH will be

$2C(s)+3H_2(g)+\frac{1}{2}O_2(g)\longrightarrow C_2H_5OH(l)$

Heat of formation of $C_2H_5OH(l)$

= 2 × Eq. (ii) + 3 × Eq. (i) − Eq. (iii)

$=2\times(-393.5)+3\times(-241.8)-(-1234.7)$

$=-787-725.4+1234.7$

$=-277.7\approx-278$ kJ/mol

37. *(1411)* Consider the following reaction

$C_2H_4(g)+3O_2(g)\longrightarrow 2CO_2(g)+2H_2O(l)$

$\Delta U=-1406$ kJ mol^{-1}

$\Delta H=\Delta U+\Delta n_g RT$

$=-1406+\frac{(-2)\times 8.3\times 300}{1000}$

$=-1406-4.98\approx-1411$ kJ mol^{-1}

$[\because \Delta G=\Delta H-T\Delta S$ Also, $\Delta G=0]$

The minimum value of $T\Delta S$ at equilibrium, $T\Delta S=-1411$ kJ

38. *(800)* $A_2+B_2\longrightarrow 2AB$

$\Delta H_f^\circ=-200$ kJ mol^{-1}

$\Delta H_f^\circ(AB)=-200$ kJ mol^{-1}

$\therefore \Delta H_r^\circ$ for reaction $A_2+B_2\longrightarrow 2AB$ is − 400 kJ mol^{-1}

Given that, bond enthalpy of A_2, B_2 and AB is 1 : 0.5 : 1

Assume that bond enthalpy of A_2 is x kJ mol^{-1}

∴ Bond enthalpy of $B_2=0.5x$ kJ/mol

∴ Bond enthalpy of $AB=x$ kJ/mol

$A_2+B_2\longrightarrow 2AB$
$\quad\quad\quad\quad\quad\quad -2x$

$\Delta^\circ H_r=-400$ kJ mol^{-1}

$-400=x+0.5x-2x$

$-400=-0.5x$

$\Rightarrow x=800$ kJ mol^{-1}

39. *(1006)* Heat released by combustion of 1 mole of

$C_2H_6(\Delta U)=\frac{-20\times 0.5}{0.3}\times 30$

$=-1000$ kJ

$C_2H_6(g)+\frac{7}{2}O_2(g)\longrightarrow 2CO_2(g)+3H_2O(l)$

$\Delta n_g=2-\left(1+\frac{7}{2}\right)=-\frac{5}{2}$

$\Delta H=\Delta U+\Delta n_g RT$

$=-1000-\frac{5}{2}\times 8.3\times 300\times 10^{-3}$ kJ

$=-1000-6.225=-1006$ kJ

Thus, heat released = 1006 kJ/mol

40. *(925)*

idea This problem is based on the concept of stoichiometry and mole concept. The step involved to solve this question are, firstly write the balanced chemical equation, from that calculate the value of volume. After that calculate the number of moles. Using the value of number of moles, calculate the heat evolved.

Let volume of C_2H_4 is x litre,

	$C_2H_4 + 3O_2 \longrightarrow$	$2CO_2 + 2H_2O$
Initial	x	—
Final	—	$2x$

	$CH_4 + 2O_2 \longrightarrow$	$CO_2 + 2H_2O$
Initial	$(16.8-x)$	—
Final	—	$(16.8-x)$

Total volume of $CO_2=2x+16.8-x$

$\Rightarrow 28=16.8+x$ or $x=11.2$ L

Volume of $C_2H_4=11.2$ L

Volume of $CH_4=(16.8-11.2)$ L $=5.6$ L

No. of moles of $CH_4=\frac{5.6}{22.4}=0.25$ mol

No. of moles of $C_2H_4=\frac{11.2}{22.4}=0.5$ mol

So, the heat evolved will be

$\Rightarrow 0.25(-900)+0.5(-1400)$

$=-925$ kJ

41. *(54)* 600 mL of 0.2 M HNO_3 means,
$600 \times 0.2 = 120$ m mol
400 mL of 0.1 M NaOH solution means, $400 \times 0.1 = 40$ m mol
The reaction between, HNO_3 and NaOH can be written as,

$$HNO_3 + NaOH \longrightarrow NaNO_3 + H_2O$$

Start : 120 40
End : 80 0 [40 m mol]

$\Delta_r H = 40 \text{ m mol} \times 57 \times 10^3$ J/mol
or, 40×10^{-3} mol $\times 57 \times 10^3$ J/mol
or, 2280 J

$$mS\Delta T = 2280$$

$$\text{or } 1000 \text{ mL} \times \frac{1 \text{ gm}}{\text{mL}} \times 4.2 \times \Delta T = 2280$$

or, $1000 \text{ g} \times 4.2 \times \Delta T = 2280$
or, $\Delta T = 54.28 \times 10^{-2}$K
or 54.28×10^{-2}°C

42. *(35)* Given, molar mass of gas
$= 280 \text{ g mol}^{-1}$
Increase in temperature of calorimeter
$= 298.45 \text{ K} - 298.0 \text{ K}$
$\Delta T = 0.45$ K
Enthalpy of combustion
$\Delta H_{Comb} = 9 \text{ kJ mol}^{-1}$
$q = C\,\Delta T = 2.5 \times 0.45$
$= 1.125$ kJ
Considering $q = \Delta H$ (enthalpy change)
So, total enthalpy change $= q = 1.125$ kJ
Enthalpy change for 1 mole of gas,
$\Delta H = 9 \text{ kJ mol}^{-1}$

$$\text{Thus,} \quad n_{gas} = \frac{q}{\Delta H} = \frac{1.125 \text{ kJ}}{9 \text{ kJ mol}^{-1}}$$

and mass of gas taken
$= n_{gas} \times$ molar mass

$$= \frac{1.125}{9} \times 280 \approx 35 \text{ g}$$

43. *(2)* $C_V = C_p - R = 20.785 - 8.314$
$= 12.471 \text{ JK}^{-1} \text{ mol}^{-1}$
$\Delta U = nC_V\Delta T$

$$\Rightarrow \quad n = \frac{5000}{12.471 \times 200} = \frac{25}{12.471} \approx 2$$

44. *(57)* For reaction,

$$H_2F_2(g) \longrightarrow H_2(g) + F_2(g)$$

$$\Delta n_{(g)} = n_{(g)p} - n_{(g)r}$$

where,
$n_{(g)_p}$ = number of moles of product, and
$n_{(g)_r}$ = number of moles of reactants.
$\Delta n_{(g)} = 2 - 1 = 1$
$\Delta H = \Delta U + n_g RT$

$$= -59.6 + \frac{1 \times 8.314 \times 300}{1000}$$

$$= -59.6 + 2.494 = 57.10 \text{ kJ mol}^{-1}$$

45. *(200)*

$$Q = \frac{C\Delta T}{n} \quad \ldots\text{(i)}$$

Assuming coal to be pure carbon,

$$n_{coal} = \frac{2.4}{\text{Molar mass of carbon}} = \frac{2.4}{12}$$

By substituting the values in Eq. (i), we get,

$$Q = \frac{-20(300 - 298)}{\frac{2.4}{12}} = -200 \text{ kJ/mol}$$

$\therefore \quad x = 200$

46. *(2)* $HCl + NaOH \longrightarrow NaCl + H_2O$
$\Delta H = -57.3$ kJ / mol
Since, HCl, NaOH and NaCl are strong acid, strong base and salt respectively. So, are completely ionised.

$$H^+ + Cl^- + Na^+ + OH^- \longrightarrow Na^+ + Cl^- + H_2O$$

$\Delta H = -57.3$ kJ / mol

$$\Rightarrow \quad H^+ + OH^- \longrightarrow H_2O$$

$\Delta H = -57.3$ kJ/mol

$$\Rightarrow \quad H_2O \longrightarrow H^+ + OH^- \quad \ldots\text{(i)}$$

$\Delta H = +57.3$ kJ / mol

$$\underset{\text{Weak acid}}{CH_3COOH} + \underset{\text{Strong base}}{NaOH} \longrightarrow CH_3CO\overset{-}{O}\overset{+}{Na} + H_2O$$

$\Delta H = -55.3$ kJ/mol
We can also write above equation as

$$CH_3COOH + OH^- \longrightarrow CH_3COO^- + H_2O \ldots\text{(ii)}$$

$\Delta H = -55.3$ kJ/mol
On adding Eq. (i) and Eq. (ii), we get,

$$CH_3COOH + OH^- + H_2O \longrightarrow CH_3COO^- + H_2O + H^+ + OH^-$$

$$CH_3COOH \rightleftharpoons CH_3COO^{\ominus} + H^{\oplus}$$

$\Delta H = +57.3 - 55.3 = 2$ kJ/mol
So, ionisation of acid requires 2 kJ/mol of energy.

47. *(104)* Reaction for formation of propane is
3C (graphite) + $4H_2(g) \longrightarrow C_3H_8(g)$
Reaction for combustion of propane, graphite and dihydrogen are respectively.

$$C_3H_8(g) + 5O_2(g) \longrightarrow 3CO_2(g) + 4H_2O(l), \ldots\text{(i)}$$

$\Delta H_{(i)} = -2220 \text{ kJ mol}^{-1}$

$$C(\text{graphite}) + O_2(g) \longrightarrow CO_2(g) \ldots\text{(ii)}$$

$\Delta H_{(ii)} = -393.5 \text{ kJ mol}^{-1}$

$$H_2(g) + \frac{1}{2}O_2(g) \longrightarrow H_2O(l) \ldots\text{(iii)}$$

$\Delta H_{(iii)} = -285.8 \text{ kJ mol}^{-1}$

$$\Delta H_{reaction} = 3\Delta H_{(ii)} + 3\Delta H_{(iii)} - \Delta H_{(i)}$$

$= 3(-393.5) + 4(-285.8) - (-2200)$
$= -103.7 \text{ kJ mol}^{-1} \approx 104$ kJ/mol

48. *(600)* $Mg(s) + \frac{1}{2}O_2(g) \longrightarrow MgO(s)$

$$\text{Given,} \quad \Delta n_g = 0 - \frac{1}{2} = -\frac{1}{2}$$

$\Delta_C H° = -601.70 \text{ kJ mol}^{-1}$
$= -601700 \text{ JK}^{-1} \text{ mol}^{-1}$
We know that,
so, $\quad \Delta U = \Delta H - \Delta n_g RT$

$$= -601700 - \left(-\frac{1}{2}\right)8.3 \times 300$$

$= -601700 + 1245$
$= -600455 \text{ J mol}^{-1}$
Thus, $|\Delta U| = 600.455 \text{ kJ mol}^{-1}$
The nearest integer is 600.

49. *(195)* $q = nC\Delta T$

$$= \frac{2.2}{44} \times 100 \times (270 - 310) = -200 \text{ J}$$

$w = -p\Delta V$
$= -1 \times (101.3) \times (167.75 - 217.1) \times 10^{-3}$
$= 4.999$ J
$\Delta E = q + w = -200 + 4.999 \cong -195$ J
$\therefore \quad x = 195$

50. *(117)* Given, mass of NH_3 = 17.0 g
$\therefore$ Number of moles of NH_3

$$= \frac{\text{Given mass of } NH_3}{\text{Molar mass of } NH_3} = \frac{17}{17} = 1$$

number of moles equivalent to 85 g

$$= \frac{85}{17} = 5$$

$\therefore$ Enthalpy change for vaporisation of 5 moles of NH_3
$= 5 \times 23.4$ kJ/mol $= 117$ kJ

51. *(38)* $H_2O(l) \longrightarrow H_2O(g)$
Given, weight of water = 36 g

$$\text{Moles of } H_2O = \frac{36 \text{ g}}{18 \text{ g}} = 2 \text{ moles}$$

Internal energy, $\Delta U = \Delta H - \Delta n_g RT$
$(\Delta n_g = 1 - 0 = 1)$

$$= 41.1 - \frac{1 \times 8.31 \times 373}{1000} \text{ kJ / mol}$$

$= 38$ kJ/mol

52. *(738)* The relation between change in enthalpy and change in internal energy can be given by :

$$[\Delta H = \Delta U + \Delta n_g RT]$$

From given equation :

$$\left[CH_3OH(l) + \frac{3}{2}O_2(g) \longrightarrow CO_2(g) + 2H_2O(l)\right]$$

$$\Delta n_g = 1 - \frac{3}{2} = -\frac{1}{2}$$

So, $\Delta H = -726 - \frac{1}{2} \times \frac{8.3 \times 300}{100}$

or, $\Delta H = -738.45$ kJ/mol

53. *(38)* $H_2O(l) \longrightarrow H_2O(g)$

$\because \quad \Delta H = \Delta U + \Delta n_g RT$

$\therefore \quad 41\text{ kJ} = \Delta U + RT$

$\Delta U = 41 - 8.3 \times 373 \times 10^{-3}$

$= 41 - 3.095$

$= 37.90 \text{ kJ mol}^{-1} = 38 \text{ kJ mol}^{-1}$

54. *(82)* Millimoles of HCl $= 200 \times 0.2 = 40$

Millimoles of NaOH $= 300 \times 0.1 = 30$

Heat released $(q) = n \times$ molar heat

$= \frac{30}{1000} \times 57.1 \times 1000 = 1713$ J

Mass of solution $= 500 \times 1 = 500$ g

We know that,

$$\Delta T = \frac{q}{mc} = \frac{1713 \text{ J}}{500 \text{ g} \times 4.18 \text{ J/g-K}}$$

$= 0.8196 \text{ K} = 81.96 \times 10^{-2}$ K

$x \approx 82$

55. *(82)* Molarity $= \frac{n \times 1000}{V(\text{mL})}$

Moles $= \frac{\text{molarity} \times \text{volume (mL)}}{1000}$

1 mole of H_2SO_4 produces 2 moles of H^+

$n_{H^+} = \frac{400 \times 0.2}{1000} \times 2 = 0.16$ moles

and 1 mole of NaOH produces 1 mole of OH^-.

$n_{OH^-} = \frac{600 \times 0.1}{1000} = 0.06$

n_{OH^-} is low.

$$H_2SO_4 + 2NaOH \longrightarrow Na_2SO_4 + 2H_2O$$

Mol : 1 2 1 2

Now, 1 mole H_2SO_4 requires 2 moles NaOH, then 0.06 mole of NaOH will require = 0.03 moles of H_2SO_4

$\therefore$ 0.06 mole of OH^- reacts with 0.06 mole of H^+.

$m = 1 \times 1000$

Now, heat liberated from reaction = heat gained by solution.

$n \times \Delta H = m \times C_p \times \Delta T$

$0.06 \times 57.1 \times 10^3 = (1000 \times 1) \times 4.18 \times \Delta T$

$\Rightarrow \quad \Delta T = 0.8196 \text{ K} = 82 \times 10^{-2}$ K

56. *(31)* $H_2O(l) \rightleftharpoons H_2O(g)$

Using formula, $\Delta H = \Delta U + \Delta n_g RT$

For 1 mole water,

$\Delta n_g = 1$ mol

$\therefore \Delta n_g RT = 1 \text{ mol} \times 8.31 \text{ J/mol K} \times 373$ K

$= 3099.63 \text{ J} \approx 31 \times 10^2 \text{J mol}^{-1}$

57. *(101)* Given,

Enthalpy of fusion,

$\Delta H_{fus} = 2.8 \text{ kJ mol}^{-1}$

Enthalpy of vaporisation,

$\Delta H_{vap} = 98.2 \text{ kJ mol}^{-1}$

Using Hess's law,

$\Delta H_{sub} = \Delta H_{fus} + \Delta H_{vap}$

$= 2.8 + 98.2 = 101$ kJ/mol

58. *(21)* Heat generated by 1g of graphite powder

$= \frac{2.48 \times 10^2}{12} = 20.66 = 21$ kJ.

Hence, amount of heat generated is 21 kJ.

59. *(128)* This reaction involves the breaking of 2C—H bonds and formation, of 1 C$=$C bond and 1 H—H bond.

$\Delta_r H$ = [Sum of bond ethalpies of reactants] − [Sum of bond enthalpies of products]

$= [1\,\Delta_{C—C}H + 6\Delta_{C—H}H] - [1\,\Delta_{C=C}H + 4\Delta_{C—H}\,H + 1\Delta_{H—H}\,H]$

$= (347 + 6 \times 414) - [611 + 4 \times 414 + 436]$

$= 2831 - 2703 = 128$ kJ/mol.

60. *(230)*

$\Delta_r H^\circ = \Sigma \Delta_f H^\circ_{\text{Product}} - \Sigma \Delta_f H^\circ_{\text{Reactant}}$

$= \{\Delta_f H^\circ_{Al_2O_3} + 3\Delta_f H^\circ_{Ca}\} - \{3\Delta_f H^\circ_{CaO} + 2\Delta_f H^\circ_{Al}\}$

$= \Delta H^\circ_f(Al_2O_3) - 3 \times \Delta H^\circ_f(CaO)$

We know, $\Delta_f H^\circ$ for elemental state = 0

$\Rightarrow \Delta_f H^\circ_{Ca} = 0, \Delta_f H^\circ_{Al} = 0$

Putting the value,

$\Delta_r H^\circ = \{-1675 + 0\} - \{3(-625) + 0\}$

$= -1675 + 1905 = 230$ kJ

61. *(309)* $\Delta_f H^\circ[S, g] + 6 \times \Delta_f H^\circ[F, g] = \Delta_f H^\circ[SF_6, g] + 6 \times E_{S-F}$

[$\because E_{S-F}$ = Average S—F bond energy in SF_6]

$275 + 6 \times 80 = -1100 + 6 \times E_{S-F}$

$\Rightarrow \quad E_{S-F} = \frac{275 + 6 \times 80 + 1100}{6}$

$= 309.16 \text{ kJ mol}^{-1}$

$= 309 \text{ kJ mol}^{-1}$

62. *(741)* From given equation:

$$NH_2CN(s) + \frac{3}{2}O_2(g) \longrightarrow N_2(g) + CO_2(g) + H_2O(l)$$

The relation between change in enthalpy and change in internal energy is given by $\Delta H = \Delta U + \Delta n_g RT$

$$\left[\Delta n_g = 2 - \frac{3}{2} = 0.5\right]$$

So, $\Delta H = -742.24 + \frac{0.5 \times 8.314 \times 298}{1000}$

$= -741$ kJ

63. *(189494)* Given, $\Delta H = 41$ kJ/mol

$= 41 \times 1000 \times 5$ J

[For 90 g or 5 mol $H_2O(l)$]

	$H_2O(l) \longrightarrow$	$H_2O(g)$
$t = 0$	$\frac{90}{18}$ mol	—
$t = \infty$	0	5 mol

$= \Delta H - \Delta n_g$ RT

$= 41 \times 1000 \times 5 - 5 \times 8.314 \times 373$

$= 189494.39 \text{ J} \approx 189494.00$ J

Topic 4

01. *(c)* The most appropriate relations are given in B and C only.

(B) Gibbs free energy is equal to the enthalpy of the system minus the product of the temperature and entropy.

$\Delta G = \Delta H - T\Delta S$ at constant temperature.

(C) Entropy is the measure of disorderness or randomness of the particular system. It can be expressed as

$$\left[\Delta S = \frac{q_{rev}}{T}\right]$$

02. *(a)* Relation given in option (a) is not correct.

According to first law of thermodynamics, the change in heat energy of system is equal to the sum of change in internal energy of system and work done by the system i.e.,

$\Delta H = \Delta U + p\Delta V$.

03. *(b)* A process will be spontaneous when its free energy (Gibb's energy) change will be negative, i.e. $\Delta G < 0$.
$\Delta G = \Delta H - T\Delta S$ for positive/negative signs of ΔH and ΔS at any/higher/lower temperature as :

ΔH	ΔS	Comment on temperature (T)	ΔG	Comment on the process
<0	>0	at any temp.	<0	spontaneous
>0	<0	at any temp.	>0	non-spontaneous
<0	<0	at lower temp.	<0	spontaneous
>0	>0	at higher temp.	<0	spontaneous

04. *(d)*

idea To solve this problem, first of all write all the phase changes involved in the reaction. Then calculate individual entropy change for each phase. After that take sum of all the individual entropy change.

The phase change involved as per the question is given by

Ice → Ice ⇌ Water → Water ⇌ Water vapour → Water vapour

268 K (A) 273 K (B) 273 K (C) 273 K (D) 373 K (E) 383 K

So, the overall entropy change is given by

$$\Delta S_{\text{overall}} = \Delta S_A + \Delta S_B + \Delta S_C + S_D + S_E$$

$$\Delta S_A = \int_{268\text{ K}}^{273\text{ K}} C_{p,n} \cdot dT$$

$$\Delta S_B = \frac{\Delta H_{m\,\text{fusion}}}{T_f}, \quad \Delta S_C = \int_{273}^{373} \frac{C_{p,m} dT}{T}$$

$$\Delta S_D = \frac{\Delta H_{m\,\text{vaporisation}}}{T}$$

$$\Delta S_E = \int_{373}^{383} \frac{C_{p,m}\, dT}{T}$$

So, option (d) is the correct answer.

05. *(c)* The incorrect combination is given in B and C i.e option (c).
The correct match is given as

	ΔH	ΔS	Temperature	Spontanity
(B)	+	+	Low T	Non-spontaneous
(C)	−	−	Low T	Spontaneous

06. *(b)* Reaction is spontaneous at relatively high temperature and non spontaneous at low temperature.

$$\Delta G = \Delta H - T\Delta S$$

$\Delta G < 0$ is only possible when ΔH and ΔS both are positive.
Hence, option (b) is correct.

07. *(a)* Among the given options (a) is incorrect.
Its correct form is : (ΔG) is positive for non-spontaneous reaction.

08. *(b)* The correct match is
A-II, B-III, C-IV, D-I.

09. *(b)* All the expression of thermodynamics are correct except

$$\ln K = \frac{\Delta H^\circ - T\Delta S^\circ}{RT}$$

As we know,

$$\Delta G = \Delta H^\circ - T\Delta S^\circ \quad \ldots\text{(i)}$$

Also $\Delta G = -RT \ln K \quad \ldots$(ii)

∴ From (i) and (ii),

$$-RT \ln K = \Delta H^\circ - T\Delta S^\circ$$

$$\therefore \quad \ln K = \frac{-\Delta H^\circ}{RT} + \frac{\Delta S^\circ}{R}$$

The corect expression is

$$\ln K = \frac{-\Delta H^\circ}{RT} + \frac{\Delta S^\circ}{R} = -\frac{\Delta H^\circ + T\Delta S^\circ}{RT}$$

10. *(b)* Statement (b) is a true statement whereas all other statements are incorrect.
By definition,

$$dS = \frac{dQ}{T} \quad \text{and} \quad \Delta S = \frac{\int dQ}{T}$$

∴ ΔS is a function of temperature.
S is also a function of temperature.

11. *(a)* $N_2(g) + 3H_2(g) \longrightarrow 2NH_3(g)$,
$\Delta n_g = 2 - (1+3) = -2$
So, ΔS is also negative (entropy decreases)
In rest of the processes entropy change is positive.

12. *(d)* The incorrect match is $\Delta G^\circ < 0$, $K < 1$.
For an ideal gas $\Delta G^\circ = -RT \ln K$.

$$\therefore \ln K = -\frac{\Delta G^\circ}{RT} \quad \text{and} \quad K = e^{-\Delta G^\circ/RT}$$

The above equation is helpful in predicting the spontaneity of the reaction.

13. *(d)* According to Gibbs-Helmholtz equation,

$$\Delta_r G^\circ = \Delta_r H^\circ - T\Delta_r S^\circ$$

For a reaction to be feasible (spontaneous)

$$\Delta_r G^\circ < 0$$

$$\Delta_r H^\circ - T\Delta_r S^\circ < 0$$

Given, $\Delta_r H^\circ = +491.1 \text{ kJ mol}^{-1}$,
$\Delta_r S^\circ = 198 \text{ JK}^{-1} \text{ mol}^{-1}$

$$\therefore \quad 491.1 \times 10^3 - T \times 198 < 0$$

$$T > \frac{491.1 \times 10^3}{198} = 2480.3 \text{ K}$$

∴ Above 2480.3 K reaction will become spontaneous.

14. *(c)* For a given value of T,
(i) If $\Delta_r G^\circ$ becomes < 0, the forward direction will be spontaneous and then the major and minor components will be Y and X respectively.
(ii) If $\Delta_r G^\circ$ becomes > 0, the forward direction will be non-spontaneous and then the major and minor components will be X and Y respectively.

$$\Delta_r G^\circ = 120 - \frac{3}{8} \times 315 = 1.875$$

i.e. $\Delta_r G^\circ > 0$, major component $= X$

15. *(c)* $\Delta G = \Delta H - T\Delta S$
The process will be spontaneous, when
$\Delta G = -$ve, i.e. $|T\Delta S| > |\Delta H|$
Given : $\Delta H = 200 \text{ J mol}^{-1}$
and $\Delta S = 40 \text{ JK}^{-1} \text{ mol}^{-1}$

$$\Rightarrow \quad T > \frac{|\Delta H|}{|\Delta S|} = \frac{200}{40} = 5 \text{ K}$$

So, the minimum temperature for spontaneity of the process is 5 K.

16. *(a)* From Haber's process,
$N_2(g) + 3H_2(g) \longrightarrow 2NH_3$
$\Delta H^\circ = -$ ve, $\Delta S^\circ = -$ ve (As gaseous moles decreases)
(i) As temperature increases $-\frac{\Delta H^\circ}{T} R$, decreases.
(ii) $\Delta G^\circ = -RT \ln K_{eq}$

$$R \ln K_{eq} = \frac{-\Delta G^\circ}{T}$$

On increasing temperature in exothermic reaction K_{eq} decreases.
ΔH° and ΔS° are almost constant with temperature.
So, graph (a) correctly represents the variation of thermodynamic properties of Haber's process.

17. *(c)* The correct match is A-II, B-I, C-IV, D-III.
- We know that,
$dG = Vdp - SdT$
At constant pressure $Vdp = 0$
So, $dG = -SdT$
or $\left(\frac{\partial G}{\partial T}\right)_p = -S$
- $dH = (dq)_p = C_p dT$
$\left(\frac{\partial H}{\partial T}\right)_p = C_p$
- $dG = Vdp - SdT$

At constant temperature, $dG = Vdp$

$$\left(\frac{\partial G}{\partial p}\right)_T = V$$

- $dU = C_V dT = (q)_V$

$$\left(\frac{\partial U}{\partial T}\right)_V = C_V$$

18. *(2)* Graph between Gibbs free energy *vs* extent of reaction.

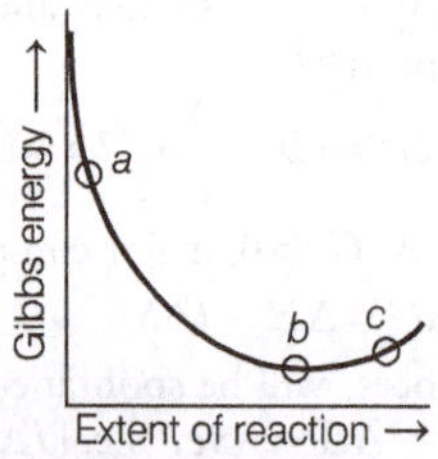

At *a* point we observe $\Delta_r G < 0$
So, reaction is spontaneous.
At *b* point $\Delta G = 0$ (equilibrium)
At *c* point Gibbs free energy $\Delta_r G > 0$
So, reaction is non-spontaneous
two statements B and C are correct (2).

19. *(a)* Entropy will decrease in A, B, C and D processes.

(A) Water $\xrightarrow{0°C}$ ice; $\Delta S = -$ ve

(B) Water $\xrightarrow{-10°C}$ ice; $\Delta S = -$ ve

(C) $N_2(g) + 3H_2(g) \longrightarrow 2NH_3(g)$; $\Delta S = -$ ve

Number of moles are decreasing

$$n = 2 - (3 + 1) \Rightarrow n = -2$$

So, entropy decreases.

(D) Adsorption; $\Delta S = -$ ve

Adsorption will lead to a decrease in the randomness of gaseous particles.

So, entropy decreases.

20. *(700)* The reaction is

$$\frac{1}{2}X_2 + \frac{5}{2}Y_2 \rightleftharpoons XY_5$$

$$\Delta H^{-} = -35 \text{ kJ/mol} = \Delta_r H°$$

$$\Delta_r S° = 110 - \left[\left(\frac{1}{2}\times 70\right) + \left(\frac{5}{2}\times 50\right)\right]$$

$$= 110 - 160 = -50 \text{ JK}^{-1}\text{mol}^{-1}$$

$\Delta G° = 0$ at equilibrium

At equilibrium

$$T = \frac{|\Delta_r H°|}{|\Delta_r S°|} = \frac{35\times 1000}{50}$$

$$T = 700 \text{ K}$$

21. *(2850)* Given, $\Delta H° = 55$ kJ/mol

$\Delta S° = 175$ J/K/mol

$T = 25$ °C $= 298$ K, $\Delta G° = ?$

Using equation,

$$\Delta G° = \Delta H° - T\Delta S°$$

On substituting the value in above equation, we get

$$\Delta G° = 55000 - 298 \times 175$$
$$= 55000 - 52150 = 2850 \text{ J/mol}$$

22. *(400)* Given,

$\Delta H_{vap} = 30$ kJ/mol

$\Delta S_{vap} = 75 \text{ J mol}^{-1} \text{ K}^{-1}$

$T = ?$

At equilibrium,

$$\Delta G = 0$$
$$\Delta H_{vap} = T\Delta S_{vap}$$

So, $30 \times 1000 = T \times 75$

$$T = 400 \text{ K}$$

23. *(2000)* From Gibb's free energy equation,

$$\Delta G = \Delta H - T\Delta S$$

For reaction to be spontaneous
$\Delta G = -$ve
For limiting case, $\Delta G = 0$

$$\Delta H = T\Delta S$$

$$T = \frac{\Delta H}{\Delta S} = \frac{400 \text{ kJ/mol}}{0.2 \text{ kJ/mol-K}} = 2000 \text{ K}$$

Hence, reaction will become spontaneous above 2000 K.

24. *(57)* $\Delta G° = -RT \ln K_{eq}$

$$= -2.303RT \log K_{eq}$$
$$= -2.303 \times 8.314 \times 300 \times \log 10$$
$$= -5744.14 \text{ J/mol}$$
$$= -5.744 \text{ kJ/mol}$$
$$= -57.44 \times 10^{-1} \text{ kJ/mol}$$

Thus, the value is 57.

25. *(37)* Given, $T = 400$ K

$$\Delta H^{\ominus} = 77.2 \text{ kJ/mol}$$
$$\Delta S = 122 \text{ JK}^{-1}$$

Free energy is given by

$$\Delta G° = \Delta H° - T\Delta S°$$
$$= 77.2 \times 10^3 - 400 \times 122$$
$$= 28400 \text{ J}$$
$$\Delta G° = -2.303RT \log K$$

$28400 = -2.303 \times 8.314 \times 400 \times \log K$

$\log K = -3.708$

or $\log K = -37.08 \times 10^{-1}$

Hence, the missing value is 37.

26. *(10)* Reaction, $A \rightleftharpoons B$ at 298 K

Given, $\Delta H° = -54.07$ kJ/mol

$\Delta S° = 10$ J/K/mol

We know that

$$\Delta G° = \Delta H° - T\Delta S°$$

or $-2.303\, RT \log K = -54070$
-298×10

$$\log K = \frac{57050}{5705}$$

So, $\log K = 10$

27. *(3)* Given $n = 1$, $R = 8.314$

$$v_2 = 3 \text{ L}, v_1 = 2 \text{ L}$$

$$\Delta S_{system} = nR \ln\left(\frac{v_2}{v_1}\right)$$

$$\Delta S_{system} = 1 \times 8.314 \ln\left(\frac{3}{2}\right)$$

$$\Delta S_{system} = 3.37 \text{ JK}^{-1}$$

$$\Delta S_{surrounding} = 3.37 \approx 3 \text{ JK}^{-1}$$

28. *(6)*

> **idea** To calculate the value of Gibbs free energy, first of all calculate K_{eq} by using the formula $K_{eq} = \frac{K_f}{K_b}$. After that substitute the value obtained, in the equation $\Delta G° = -RT\ln K_{eq}$

$$A + B \underset{K_b = 10^2}{\overset{K_f = 10^3}{\rightleftharpoons}} C + D$$

Since, equilibrium constant can be calculated as

$$K_{eq} = \frac{K_f}{K_b} \text{ or, } K_{eq} = \frac{10^3}{10^2} = 10$$

Again, the relation between $\Delta G°$ and K_{eq} can be calculated from the formula

$$\Delta G° = -RT \ln K$$

So, $\Delta G° = -8.3 \times 300 \times \ln(10)$

or $\Delta G° = -5.7$ kJ/mol

$$\approx -6 \text{ kJ/mol}$$

29. *(710)* From the given equilibrium equation:

	N_2O_4	$\rightleftharpoons$	$2NO_2$
initially ($t = 0$)	[1 mol]		
after (time $= t$)	$[1 - 0.5]$ mol $= 0.5$ mol		0.5×2 mol $= 1$ mol

Total pressure = 1 atm

$$\therefore\ K_p = \frac{[NO_2]^2}{[N_2O_4]} = \frac{\left[\frac{1}{1.5}\times 1\right]^2}{\left[\frac{0.5}{1.5}\times 1\right]}$$

$$= \frac{100}{75} = [1.33]$$

$(\Delta G°)$ for the equilibrium reaction can be given by :

$$\Delta G° = -2.303\, RT \log K$$
$$= -2.303 \times 8.31 \times 300 \times \log(1.33)$$
$$= -2.303 \times 8.31 \times 300 \times 0.1239$$
$$\Delta G° = -710.43 \text{ J/mol}$$
$$\therefore\ x = 710$$

30. *(300)* The relation between enthalpy, entropy and Gibb's free energy is given by equation

$\Delta G = \Delta H - T\Delta S$

At equilibrium, $\Delta G = 0$

$0 = -165 \times 10^3 - T(-550)$

$\Rightarrow \quad T = \frac{165 \times 10^3}{550}$

$= 300\text{ K}$

31. *(5)* Given, $\Delta H = -57.8\text{ kJ mol}^{-1}$

$\Delta S = -176\text{ JK}^{-1}\text{ mol}^{-1}$

$T = 298\text{ K}$

Using Gibb's free energy relation

$\Delta G = \Delta H - T\Delta S$

$\Delta G = 57.8\text{ kJ / mol} - [298\text{ K} \times (-176\text{ JK}^{-1}\text{mol}^{-1})]$

$= -57.8\text{ kJ / mol} - \left(298 \times \frac{-176}{1000}\text{ kJ}\right)$

$[\because 1\text{ kJ} = 1000\text{ J}]$

$= -5.352\text{ kJ / mol}$

$\Delta G = 5.352$

Hence, answer is 5.

32. *(964)*

$\Delta H° = [\Delta H^\circ_{f\,(Fe)} + \Delta H^\circ_{f\,(CO)}] - [\Delta H^\circ_{f\,(FeO)} - \Delta H^\circ_{f\,(C)}]$

$= [0 + (-110.5)] - [-266.3 - 0]$

$= 155.8\text{ kJ}^{-1}\text{ mol}^{-1}$

$\Delta S° = [\Delta S^\circ_{(Fe)} + \Delta S^\circ_{(CO)}] - [\Delta S^\circ_{(FeO)} - \Delta S^\circ_{(C)}]$

$= [27.28 + 197.6] - [57.49 + 5.74]$

$= 161.6\text{ J K}^{-1}\text{ mol}^{-1}$

The reaction becomes spontaneous when $\Delta G°$ is atleast zero or negative.

$0 = \Delta H° - T\Delta S°$

$\Rightarrow T\Delta S° = \Delta H°$

$\Rightarrow \quad T = \frac{\Delta H°}{\Delta S°}$

$= \frac{155.8\text{ kJ mol}^{-1}}{161.6\text{ J K}^{-1}\text{ mol}^{-1}}$

$= \frac{155800\text{ J mol}^{-1}}{161.6\text{ J K}^{-1}\text{ mol}^{-1}} = 964\text{ K}$

The temperature at which reaction becomes spontaneous is 964 K.

33. *(336)* For the reaction, $A \xrightarrow{300\text{ K}} B$

Given, $\Delta G = -49.4$ kJ/mol

Change in enthalpy

$\Delta H_{reaction} = 51.4$ kJ/mol

Change in entropy, $\Delta S_{reaction} = ?$

According to Gibbs free energy,

$\Delta G = \Delta H - T\Delta S$

$\Rightarrow \Delta S = \frac{\Delta H - \Delta G}{T}$

$= \frac{51.4 - (-49.4)}{300} \times 1000 \frac{\text{J}}{\text{mol K}}$

$\Delta S_{reaction} = 336$ J/mol K

34. *(855)* $\Delta G° = -2.303\, RT \log K \quad \ldots\text{(i)}$

Also,

$\Delta G° = \Sigma(\Delta G°_f)_p - \Sigma(\Delta G°_f)_R \quad \ldots\text{(ii)}$

Equate Eq. (i) and (ii)

$\therefore \; -2.303 \times 8.314 \times 298 \times \log K = -4.88 \times 10^5$

or $\quad \log K = \frac{4.88 \times 10^5}{2.303 \times 8.314 \times 298}$

$\log K \approx -85.5 \approx -855 \times 10^{-1}$

or, $\quad x = 855$

35. *(1380)* For equilibrium reaction,

$\Delta G° = -2.30\, RT \log K$

$= -2.30 \times R \times 300 \times \log 100$

$\Delta G° = -1380\, R$

36. *(13537.57)* $\Delta H = \Delta U + \Delta n_g RT$

$\Delta H = -20 \times 1000 - 1 \times 8.314\text{ J/mol.K} \times 298\text{ K}$

$= -22477.572\text{ J}$

$\Delta G = \Delta H - T\Delta S$

$\Delta G = -22477.572 + 298 \times 30$

$\Delta G = -13537.57\text{ J}$

37. *(2.70)* Given : $\Delta U = 2.1$ kcal

$\Delta S = 20\text{ cal K}^{-1}$

$= \frac{20}{1000}\text{ kcal K}^{-1}$

$A(l) \rightleftharpoons 2B(g)$

$\therefore \quad \Delta n_g = 2\text{ mol}$

$R = 2\text{ cal K}^{-1}\text{ mol}^{-1}$

$= \frac{2}{1000}\text{ kcal K}^{-1}\text{ mol}^{-1}$

As we know,

$\Delta G = \Delta H - T\Delta S$

and $\Delta H = \Delta U + \Delta n_g RT$

$\therefore \quad \Delta G = \Delta U + \Delta n_g RT - T\Delta S$

$\therefore \quad \Delta G = 2.1 + \frac{2 \times 2 \times 300}{1000} - \frac{300 \times 20}{1000}$

$= 2.1 + 1.2 - 6$

$= -2.70$

06 Chemical Equilibrium

In JEE Main there is usually 0-1 question per paper are visible from this chapter. In past 3 years, 8 conceptual and 21 numerical questions were asked of moderate difficulty level from this chapter.

For detailed analysis scan QR code

Topic 1 / Chemical Equilibrium, Law of Mass Action and Equilibrium Constant

• Level 1

01 For the reaction, **[2025, 24 Jan Shift-II]**

$$H_2(g) + I_2(g) \rightleftharpoons 2HI(g)$$

Attainment of equilibrium is predicted correctly by

(a)
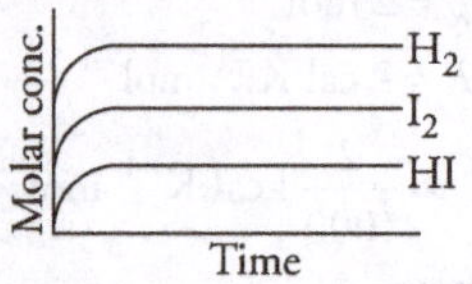

(b)
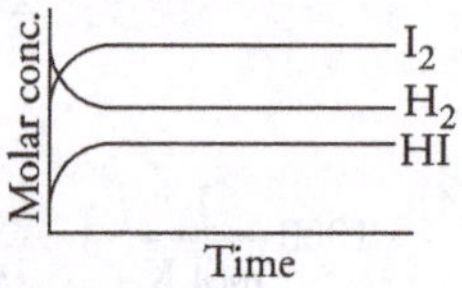

(c)
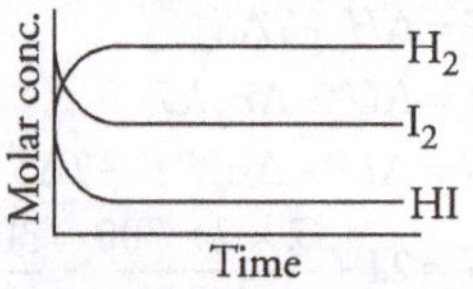

(d)
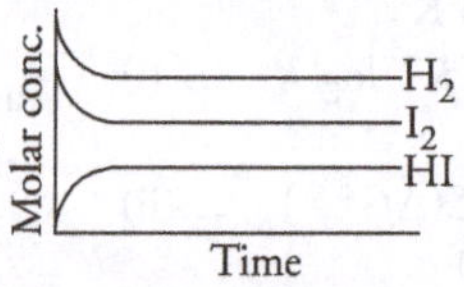

02 A vessel at 1000 K contains CO_2 with a pressure of 0.5 atm. Some of CO_2 is converted into CO on addition of graphite. If total pressure at equilibrium is 0.8 atm, then K_p is **[2025, 22 Jan Shift-I]**

(a) 0.18 atm (b) 0.3 atm
(c) 3 atm (d) 1.8 atm

03 For the given reaction, choose the correct expression of K_C from the following: **[2024, 31 Jan Shift-I]**

$$Fe^{3+}(aq) + SCN^-(aq) \rightleftharpoons (FeSCN)^{2+}(aq)$$

(a) $K_C = \dfrac{[FeSCN^{2+}]}{[Fe^{3+}][SCN^-]}$ (b) $K_C = \dfrac{[FeSCN^{2+}]^2}{[Fe^{3+}][SCN^-]}$

(c) $K_C = \dfrac{[FeSCN^{2+}]}{[Fe^{3+}]^2[SCN^-]^2}$ (d) $K_C = \dfrac{[Fe^{3+}][SCN^-]}{[FeSCN^{2+}]}$

04 For the equilibrium $A \rightleftharpoons B$, the variation of the rate of the forward (*a*) and reverse (*b*) reaction with time is given by **[2020, 4 Sep Shift-I]**

(a)
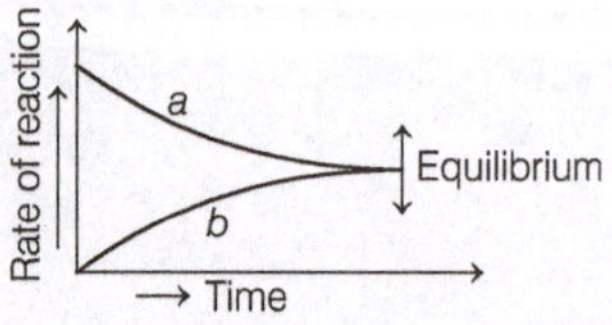

(b)
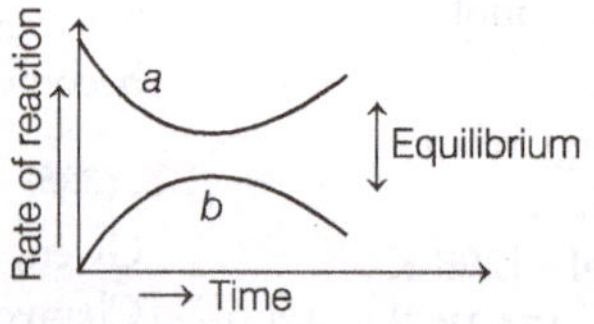

(c)
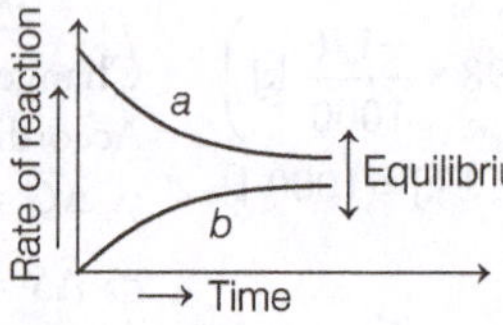

(d)
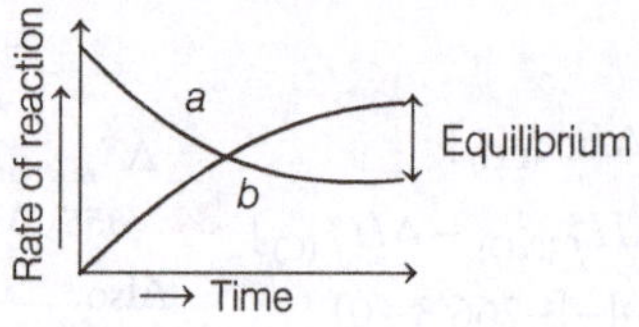

05 In the figure shown below, reactant *A* (represented by square) is in equilibrium with product *B* (represented by circle). The equilibrium constant is

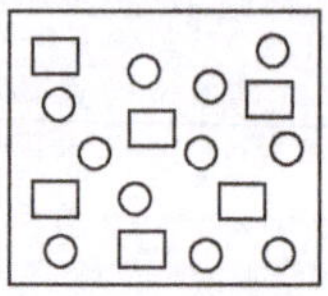

[2020, 9 Jan Shift-II]

(a) 4 (b) 2
(c) 8 (d) 1

• Level 2

06 At temperature *T*, compound $AB_2(g)$ dissociates as $AB_2(g) \rightleftharpoons AB(g) + \frac{1}{2}B_2(g)$ having degree of dissociation *x* (small compared to unity). The correct expression for *x* in terms of K_p and *p* is **[2025, 29 Jan Shift-I]**

(a) $\sqrt{K_p}$ (b) $\sqrt[4]{\dfrac{2K_p}{p}}$

(c) $\sqrt[3]{\dfrac{2K_p^2}{p}}$ (d) $\sqrt[3]{\dfrac{2K_p}{p}}$

07 Consider the reaction, **[2025, 23 Jan Shift-II]**

$$X_2Y(g) \rightleftharpoons X_2(g) + \frac{1}{2}Y_2(g)$$

The equation representing correct relationship between the degree of dissociation (x) of $X_2Y(g)$ with its equilibrium constant Kp is ....
Assume x to be very small.

(a) $x = \sqrt[3]{\frac{2Kp^2}{p}}$ (b) $x = \sqrt[3]{\frac{Kp}{p}}$ (c) $x = \sqrt[3]{\frac{2Kp}{p}}$ (d) $x = \sqrt[3]{\frac{Kp}{2p}}$

08 For the given hypothetical reactions, the equilibrium constants are as follows: **[2024, 8 April Shift-I]**

$X \rightleftharpoons Y; K_1 = 1.0$
$Y \rightleftharpoons Z; K_2 = 2.0$
$Z \rightleftharpoons W; K_3 = 4.0$

The equilibrium constant for the reaction $X \rightleftharpoons W$ is
(a) 12.0 (b) 7.0 (c) 6.0 (d) 8.0

09 The ratio $\frac{K_p}{K_c}$ for the reaction, **[2024, 6 April Shift-II]**

$$CO(g) + \frac{1}{2}O_2(g) \rightleftharpoons CO_2(g) \text{ is}$$

(a) 1 (b) $\frac{1}{\sqrt{RT}}$ (c) $(RT)^{1/2}$ (d) RT

10 The equilibrium constant for the reaction

$$SO_3(g) \rightleftharpoons SO_2(g) + \frac{1}{2}O_2(g)$$

is $K_c = 4.9 \times 10^{-2}$. The value of K_c for the reaction given below is **[2024, 4 April Shift-II]**
$2SO_2(g) + O_2(g) \rightleftharpoons 2SO_3(g)$ is.
(a) 49 (b) 41.6 (c) 4.9 (d) 416

11 4.0 moles of argon and 5.0 moles of PCl_5 are introduced into an evacuated flask of 100 litre capacity at 610 K. The system is allowed to equilibrate. At equilibrium, the total pressure of mixture was found to be 6.0 atm. The K_p for the reaction is [Given, $R = 0.082$ L atm K^{-1} mol^{-1}] **[2022, 29 June Shift-II]**

(a) 2.25 (b) 6.24
(c) 12.13 (d) 15.24

12 The value of K_C is 64 at 800 K for the reaction,

$$N_2(g) + 3H_2(g) \rightleftharpoons 2NH_3(g)$$

The value of K_C for the following reaction is

$$NH_3(g) \rightleftharpoons \frac{1}{2}N_2(g) + \frac{3}{2}H_2(g)$$ **[2020, 6 Sep Shift-II]**

(a) 1/64 (b) 8 (c) 1/4 (d) 1/8

13 The variation of equilibrium constant with temperature is given below **[2020, 06 Sep Shift-I]**

Temperature	Equilibrium constant
$T_1 = 25°C$	$K_1 = 10$
$T_2 = 100°C$	$K_2 = 100$

The values of $\Delta H°$, $\Delta G°$ at T_1 and $\Delta G°$ at T_2 (in kJ mol^{-1}) respectively, are close to
[use $R = 8.314$ J K^{-1} mol^{-1}]
(a) 28.4, –7.14 and –5.71
(b) 0.64, –7.14 and –5.71
(c) 28.4, –5.71 and –14.29
(d) 0.64, –5.71 and –14.29

14 For the reaction, $Fe_2N(s) + \frac{3}{2}H_2(g) \rightleftharpoons Fe(s) + NH_3(g)$ **[2020, 6 Sep Shift-I]**

(a) $K_C = K_p(RT)$ (b) $K_C = K_p(RT)^{-1/2}$
(c) $K_C = K_p(RT)^{1/2}$ (d) $K_C = K_p(RT)^{3/2}$

15 If the equilibrium constant for $A \rightleftharpoons B + C$ is $K_{eq}^{(1)}$ and that of $B + C \rightleftharpoons P$ is $K_{eq}^{(2)}$, the equilibrium constant for $A \rightleftharpoons P$ is **[2020, 4 Sep Shift-II]**

(a) $K_{eq}^{(1)} / K_{eq}^{(2)}$ (b) $K_{eq}^{(1)} + K_{eq}^{(2)}$
(c) $K_{eq}^{(1)} K_{eq}^{(2)}$ (d) $K_{eq}^{(2)} - K_{eq}^{(1)}$

16 An open beaker of water in equilibrium with water vapour is in a sealed container. When few grams of glucose are added to the beaker of water, the rate at which water molecules **[2020, 2 Sep Shift-I]**
(a) leaves the solution decreases
(b) leaves the solution increases
(c) leaves the vapour increases
(d) leaves the vapour decreases

17 In which one of the following equilibria, $K_p \neq K_c$?
(a) $2C(s) + O_2(g) \rightleftharpoons 2CO(g)$ **[2019, 12 April Shift-II]**
(b) $2HI(g) \rightleftharpoons H_2(g) + I_2(g)$
(c) $NO_2(g) + SO_2(g) \rightleftharpoons NO(g) + SO_3(g)$
(d) $2NO(g) \rightleftharpoons N_2(g) + O_2(g)$

18 Consider the following reversible chemical reactions,

$$A_2(g) + B_2(g) \overset{K_1}{\rightleftharpoons} 2AB(g) \quad \ldots(i)$$

$$6AB(g) \overset{K_2}{\rightleftharpoons} 3A_2(g) + 3B_2(g) \quad \ldots(ii)$$

The relation between K_1 and K_2 is **[2019, 9 Jan Shift-II]**
(a) $K_2 = K_1^3$ (b) $K_1K_2 = 3$
(c) $K_2 = K_1^{-3}$ (d) $K_1 K_2 = \frac{1}{3}$

19 For the following reactions, equilibrium constants are given :

$$S(s) + O_2(g) \rightleftharpoons SO_2(g); \ K_1 = 10^{52}$$

$$2S(s) + 3O_2(g) \rightleftharpoons 2SO_3(g); K_2 = 10^{129}$$

The equilibrium constant for the reaction,
$2SO(g) + O_2(g) \rightleftharpoons 2SO_3(g)$ is **[2019, 8 April Shift-II]**
(a) 10^{25} (b) 10^{77} (c) 10^{154} (d) 10^{181}

• Level 3

20 $A(g) \rightleftharpoons B(g) + \frac{C}{2}(g)$, the correct relationship between K_p, α and equilibrium pressure p is **[2024, 31 Jan Shift-II]**

(a) $K_p = \frac{\alpha^{3/2} p^{1/2}}{(2+\alpha)^{1/2}(1-\alpha)}$ (b) $K_p = \frac{\alpha^{1/2} p^{1/2}}{(2+\alpha)^{1/2}}$

(c) $K_p = \frac{\alpha^{1/2} p^{1/2}}{(2+\alpha)^{3/2}}$ (d) $K_p = \frac{\alpha^{1/2} p^{3/2}}{(2+\alpha)^{3/2}}$

21 Two solids dissociate as follows

$A(s) \rightleftharpoons B(g) + C(g); K_{p_1} = x \text{ atm}^2$

$D(s) \rightleftharpoons C(g) + E(g); K_{p_2} = y \text{ atm}^2$

The total pressure when both the solids dissociate simultaneously is **[2019, 12 Jan Shift-I]**

(a) $\sqrt{x+y}$ atm (b) $x^2 + y^2$ atm

(c) $(x+y)$ atm (d) $2(\sqrt{x+y})$ atm

22 Consider the reaction,

$N_2(g) + 3H_2(g) \rightleftharpoons 2NH_3(g)$

The equilibrium constant of the above reaction is K_p. If pure ammonia is left to dissociate, the partial pressure of ammonia at equilibrium is given by (Assume that $p_{NH_3} << p_{total}$ at equilibrium) **[2019, 11 Jan Shift-I]**

(a) $\frac{3^{3/2} K_p^{1/2} p^2}{4}$ (b) $\frac{3^{3/2} K_p^{1/2} p^2}{16}$

(c) $\frac{K_p^{1/2} p^2}{16}$ (d) $\frac{K_p^{1/2} p^2}{4}$

23 5.1 g NH_4SH is introduced in 3.0 L evacuated flask at 327°C. 30% of the solid NH_4SH decomposed to NH_3 and H_2S as gases. The K_p of the reaction at 327°C is ($R = 0.082$ atm mol^{-1}K^{-1}, molar mass of S = 32 g mol^{-1}, molar mass of N = 14 g mol^{-1}) **[2019, 10 Jan Shift-I]**

(a) 0.242×10^{-4} atm^2 (b) 0.242 atm^2

(c) 4.9×10^{-3} atm^2 (d) 1×10^{-4} atm^2

Numerical Value Type Questions

24 Consider the following equilibrium.

$CO(g) + 2H_2(g) \rightleftharpoons CH_3OH(g)$

0.1 mol of CO along with a catalyst is present in a 2 dm^3 flask maintained at 500 K. Hydrogen is introduced into the flask until the pressure is 5 bar and 0.04 mol of CH_3OH is formed. The K_p° is .... $\times 10^{-3}$ (Nearest integer).

Given: $R = 0.08$ dm^3 bar K^{-1} mol^{-1}

Assume only methanol is formed as the product and the system follows ideal gas behaviour. **[2025, 2 April Shift-I]**

25 37.8 g N_2O_5 was taken in a 1L reaction vessel and allowed to undergo the following reaction at 500K

$2N_2O_5(g) \rightleftharpoons 2N_2O_4(g) + O_2(g)$

The total pressure at equilibrium was found to be 18.65 bar.

Then, K_p = ...$\times 10^{-2}$ [nearest integer]

Assume N_2O_5 to behave ideally under these conditions. (Given, $R = 0.082$ bar L mol^{-1}K^{-1}) **[2025, 24 Jan Shift-I]**

26 For a certain reaction at 300 K, $K = 10$, then ΔG° for the same reaction is .... $\times 10^{-1}$ kJ mol^{-1} (Given, $R = 8.314$ JK^{-1} mol^{-1}) **[2024, 1st Feb Shift-II]**

27 Consider the following reaction at 298 K.

$\frac{3}{2}O_2(g) \rightleftharpoons O_3(g)$, $K_p = 2.47 \times 10^{-29}$, $\Delta_r G^\circ$ for the reaction is .... kJ (Given, $R = 8.314$ JK^{-1} mol^{-1}) **[2024, 31 Jan Shift-I]**

28 The following concentrations were observed at 500 K for the formation of NH_3 from N_2 and H_2. At equilibrium : $[N_2] = 2 \times 10^{-2}$ M, $[H_2] = 3 \times 10^{-2}$ M and $[NH_3] = 1.5 \times 10^{-2}$ M. Equilibrium constant for the reaction is .... **[2024, 29 Jan Shift-II]**

29 For the reaction, $N_2O_4(g) \rightleftharpoons 2NO_2(g)$. $K_p = 0.492$ atm at 300 K. K_C for the reaction at same temperature is .... $\times 10^{-2}$.

(Given, $R = 0.082$ L atm mol^{-1}K^{-1}) **[2024, 29 Jan Shift-I]**

30 For a certain thermochemical reaction, $M \rightarrow N$ at $T = 400$ K, $\Delta H^\ominus = 77.2$ kJ mol^{-1}, $\Delta S = 122$ JK^{-1}, log equilibrium constant (log K) is – ...... $\times 10^{-1}$. **[2024, 27 Jan Shift-II]**

31 A mixture of 1 mole of H_2O and 1 mole of CO is taken in a 10 litre container and heated to 725 K. At equilibrium 40% of water by mass reacts with carbon monoxide according to the equation

$CO(g) + H_2O(g) \rightleftharpoons CO_2(g) + H_2(g)$.

The equilibrium constant $K_C \times 10^2$ for the reaction is .... (Nearest integer) **[2023, 11 April Shift-I]**

32 4.5 moles each of hydrogen and iodine is heated in a sealed ten litre vessel. At equilibrium, 3 moles of HI were found. The equilibrium constant for $H_2(g) + I_2(g) \rightleftharpoons 2HI(g)$ is .... **[2023, 11 April Shift-II]**

33 $A(g) \rightleftharpoons 2B(g) + C(g)$

For the given reaction, if the initial pressure is 450 mm Hg and the pressure at time t is 720 mm Hg at a constant temperature T and constant volume V. The fraction of $A(g)$ decomposed under these conditions is $x \times 10^{-1}$. The value of x is .... (Nearest integer) **[2023, 10 April Shift-II]**

34 The equilibrium composition for the reaction, $PCl_3 + Cl_2 \rightleftharpoons PCl_5$ at 298 K is given below $[PCl_3]_{eq} = 0.2 \text{ mol L}^{-1}, [Cl_2]_{eq} = 0.1 \text{ mol L}^{-1}$, $[PCl_5]_{eq} = 0.40 \text{ mol L}^{-1}$. If 0.2 mol of Cl_2 is added at the same temperature, the equilibrium concentrations of PCl_5 is ....$\times 10^{-2}$ molL^{-1}. Given: K_c for the reaction at 298 K is 20. **[2023, 6 April Shift-II]**

35 (i) $X(g) \rightleftharpoons Y(g) + Z(g), K_{p_1} = 3$

(ii) $A(g) \rightleftharpoons 2B(g), K_{p_2} = 1$

If the degree of dissociation and initial concentration of both the reactants $X(g)$ and $A(g)$ are equal, then the ratio of the total pressure at equilibrium $\left(\frac{p_1}{p_2}\right)$ is equal to $x : 1$. The value of x is .... (Nearest integer) **[2023, 1 Feb Shift-I]**

36 At 298 K

$$N_2(g) + 3H_2(g) \rightleftharpoons 2NH_3(g); K_1 = 4 \times 10^5$$

$$N_2(g) + O_2(g) \rightleftharpoons 2NO(g); K_2 = 1.6 \times 10^{12}$$

$$H_2(g) + \frac{1}{2}O_2(g) \rightleftharpoons H_2O(g); K_3 = 1.0 \times 10^{-13}$$

Based on above equilibria, the equilibrium constant of the reaction,

$$2NH_3(g) + \frac{5}{2}O_2(g) \rightleftharpoons 2NO(g) + 3H_2O(g) \text{ is}$$

.... $\times 10^{-33}$. (Nearest integer) **[2023, 29 Jan Shift-II]**

37 At 600 K, 2 moles of NO are mixed with 1 mole of O_2.

$$2NO(g) + O_2(g) \rightleftharpoons 2NO_2(g)$$

The reaction occurring as above comes to equilibrium under a total pressure of 1 atm. Analysis of the system shows that 0.6 mole of oxygen are present at equilibrium. The equilibrium constant for the reaction is .... (Nearest integer) **[2022, 28 July Shift-II]**

38 At 298 K, the equilibrium constant is 2×10^{15} for the reaction

$$Cu(s) + 2Ag^+(aq) \rightleftharpoons Cu^{2+}(aq) + 2Ag(s)$$

The equilibrium constant for the reaction

$$\frac{1}{2}Cu^{2+}(aq) + Ag(s) \rightleftharpoons \frac{1}{2}Cu(s) + Ag^+(aq)$$

is $x \times 10^{-8}$. The value of x is .... (nearest integer) **[2022, 26 July Shift-I]**

39 40° of HI undergoes decomposition to H_2 and I_2 at 300 K. $\Delta G°$ for this decomposition reaction at one atmosphere pressure is .... J mol^{-1}. [Nearest integer]

(Use $R = 8.31 \text{ J K}^{-1}\text{mol}^{-1}$; $\log 2 = 0.3010$. In $10 = 2.3$, $\log 3 = 0.477$) **[2022, 26 June Shift-II]**

Imp. *Calculation of Equilibrium Constant K_p*

40 PCl_5 dissociates as

$$PCl_5(g) \rightleftharpoons PCl_3(g) + Cl_2(g)$$

5 moles of PCl_5 are placed in a 200 L vessel which contains 2 moles of N_2 and is maintained at 600 K. The equilibrium pressure is 2.46 atm. The equilibrium constant K_p for the dissociation of PCl_5 is ..... $\times 10^{-3}$. (Nearest integer)

(Given, $R = 0.082 \text{ L atm K}^{-1} \text{ mol}^{-1}$; assume ideal gas behaviour) **[2022, 24 June Shift-II]**

41 A box contains 0.90 g of liquid water in equilibrium with water vapour at 27°C. The equilibrium vapour pressure of water at 27°C is 32.0 Torr. When the volume of the box is increased, some of the liquid water evaporates to maintain the equilibrium pressure. If all the liquid water evaporates to maintain the equilibrium pressure, then the volume of the box must be .... litre. [Nearest integer]

(Given : $R = 0.082 \text{ L atm K}^{-1} \text{ mol}^{-1}$)

(Ignore the volume of the liquid water and assume water vapours behave as an ideal gas). **[2022, 29 June Shift-II]**

42 $2NOCl(g) \rightleftharpoons 2NO(g) + Cl_2(g)$

In an experiment, 2.0 moles of NOCl was placed in a one-litre flask and the concentration of NO after equilibrium established, was found to be 0.4 mol/L. The equilibrium constant at 30°C is ..... $\times 10^{-4}$. **[2022, 27 June Shift-I]**

43 The reaction rate for the reaction

$$[PtCl_4]^{2-} + H_2O \rightleftharpoons [Pt(H_2O)Cl_3]^- + Cl^-$$

was measured as a function of concentrations of different species. It was observed that

$$\frac{-d[[PtCl_4]^{2-}]}{dt} = 4.8 \times 10^{-5}[[PtCl_4]^{2-}] - 2.4 \times 10^{-3}[[Pt(H_2O)Cl_3]^-][Cl^-].$$

where square brackets are used to denote molar concentrations. The equilibrium constant, K = ... **[2021, 26 Aug Shift-II]**

44 The equilibrium constant for the reaction

$$A(s) \rightleftharpoons M(s) + \frac{1}{2}O_2(g)$$

is $K_p = 4$. At equilibrium, the partial pressure of O_2 is .... atm. (Round off to the nearest integer) **[2021, 27 July Shift-II]**

45 Value of K_p for the equilibrium reaction

$N_2O_4(g) \rightleftharpoons 2NO_2(g)$ at 288 K is 47.9.

The K_C for this reaction at same temperature is ...... (Nearest integer)

($R = 0.083$ L bar $K^{-1}mol^{-1}$) **[2021, 22 July Shift-II]**

46 $2SO_2(g) + O_2(g) \rightleftharpoons 2SO_3(g)$

In an equilibrium mixture, the partial pressures are $p_{SO_3} = 43$kPa; $p_{O_2} = 530$ Pa and $p_{SO_2} = 45$kPa. The equilibrium constant K_p = ... $\times 10^{-2}$. (Nearest integer) **[2021, 20 July Shift-I]**

47 The gas phase reaction, $2A(g) \rightleftharpoons A_2(g)$ at 400 K has $\Delta G° = +25.2$ kJ mol^{-1}.

The equilibrium constant K_C for this reaction is .... $\times 10^{-2}$. (Round off to the nearest integer).

[Use : $R = 8.3$ J mol^{-1} K^{-1}, In 10 = 2.3]

$\log_{10} 2 = 0.30$, 1 atm = 1 bar] [antilog (−0.3) = 0.501] **[2021, 18 March Shift-II]**

48 A homogeneous ideal gaseous reaction, $AB_2(g) \rightleftharpoons A(g) + 2B(g)$ is carried out in a 25 L flask at 27°C. The initial amount of AB_2 was 1 mole and the equilibrium pressure was 1.9 atm. The value of K_p is $x \times 10^{-2}$. The value of x is .... **[2021, 26 Feb Shift-I]**

49 At 1990 K and 1 atm pressure, there are equal number of Cl_2 molecules and Cl atoms in the reaction mixture. The value of k_p for the reaction $Cl_2(g) \rightleftharpoons 2Cl(g)$ under the above conditions is $x \times 10^{-1}$ The value of x is ....

(Rounded off to the nearest integer) **[2021, 24 Feb Shift-I]**

50 For the reaction, $A(g) \longrightarrow B(g)$, the value of the equilibrium constant at 300 K and 1 atm is equal to 100.0. The value of $\Delta_r G$ for the reaction at 300 K and 1 atm in J mol^{-1} is $-xR$, where x is ....

(Rounded off to the nearest integer)

($R = 8.31$ J mol^{-1} K^{-1} and ln10 = 2.3) **[2021, 24 Feb Shift-I]**

Topic 2 / Le-Chatelier's Principle and Factor Affecting Chemical Equilibria

• Level 1

01 Consider the equilibrium, **[2025, 29 Jan Shift-II]**

$$CO(g) + 3H_2(g) \rightleftharpoons CH_4(g) + H_2O(g)$$

If the pressure applied over the system increases by two fold at constant temperature then

(A) Concentration of reactants and products increases.

(B) Equilibrium will shift in forward direction.

(C) Equilibrium constant increases since concentration of products increases.

(D) Equilibrium constant remains unchanged as concentration of reactants and products remain same.

Choose the correct answer from the options given below

(a) (A), (B) and (C) only (b) (A) and (B) only
(c) (A),(B) and (D) only (d) (B) and (C) only

02 The following reaction occurs in the blast furnace where iron ore is reduced to iron metal

$$Fe_2O_3(s) + 3CO(g) \rightleftharpoons Fe(l) + 3CO_2(g)$$

Using the Le-chatelier's principle, predict which one of the following will not disturb the equilibrium. **[2024, 5th April Shift-I]**

(a) Removal of CO (b) Removal of CO_2
(c) Addition of CO_2 (d) Addition of Fe_2O_3

• Level 2

03 Consider the following chemical equilibrium of the gas phase reaction at a constant temperature :

$$A(g) \rightleftharpoons B(g) + C(g)$$

If p being the total pressure, K_p is the pressure equilibrium constant and α is the degree of dissociation, then which of the following is true at equilibrium? **[2025, 2 April Shift-II]**

(a) If p value is extremely high compared to K_p, α = 1.

(b) When p increases α decreases.

(c) If K_p value is extremely high compared to p, α becomes much less than unity.

(d) When p increases α increases.

04 In the following system,

$PCl_5(g) \rightleftharpoons PCl_3(g) + Cl_2(g)$ at equilibrium, upon addition of xenon gas at constant T and p, the concentration of **[2025, 3 April Shift-I]**

(a) PCl_5 will increase (b) PCl_3 will increase
(c) Cl_2 will decrease
(d) PCl_5, PCl_3 and Cl_2 remain constant

05 For a reaction, **[2025, 24 Jan Shift-I]**

$N_2O_5(g) \longrightarrow 2NO_2(g) + \frac{1}{2}O_2(g)$ in a constant volume container, no products were present initially. The final pressure of the system when 50% of reaction gets completed is

(a) 7/2 times of initial pressure
(b) 7/4 times of initial pressure
(c) 5/2 times of initial pressure
(d) 5 times of initial pressure

06 Consider the following reaction : **[2020, 5 Sep Shift-I]**

$N_2O_4(g) \rightleftharpoons 2NO_2(g); \Delta H° = +58$ kJ

For each of the following cases (A, B), the direction in which the equilibrium shifts is

(A) temperature is decreased.
(B) pressure is increased by adding N_2 at constant T.

(a) (A) towards product, (B) towards reactant
(b) (A) towards reactant, (B) no change
(c) (A) towards reactant, (B) towards product
(d) (A) towards product, (B) no change

07 For the reaction,

$$2SO_2(g) + O_2(g) \rightleftharpoons 2SO_3(g),$$

$\Delta H = -57.2$ kJ mol^{-1} and $K_c = 1.7 \times 10^{16}$. Which of the following statement is incorrect? **[2019, 10 April Shift-II]**

(a) The equilibrium constant decreases as the temperature increases
(b) The addition of inert gas at constant volume will not affect the equilibrium constant
(c) The equilibrium will shift in forward direction as the pressure increases
(d) The equilibrium constant is large suggestive of reaction going to completion and so no catalyst is required

Numerical Value Type Questions

08 The number of correct statement/s involving equilibria in physical processes from the following is **[2023, 10 April Shift-I]**

(A) Equilibrium is possible only in a closed system at a given temperature.
(B) Both the opposing processes occur at the same rate.
(C) When equilibrium is attained at a given temperature, the value of all its parameters became equal.
(D) For dissolution of solids in liquids, the solubility is constant at a given temperature.

09 Consider the following equation:

$2SO_2(g) + O_2(g) \rightleftharpoons 2SO_3(g), \Delta H = -190$ kJ

The number of factors which will increase the yield of SO_3 at equilibrium from the following is ....

A. increasing temperature B. increasing pressure
C. adding more SO_2 D. adding more O_2
E. addition of catalyst **[2023, 30 Jan Shift-II]**

Key Index

Topic 1

1 (d)	**2** (d)	**3** (a)	**4** (a)	**5** (b)	**6** (c)	**7** (a)	**8** (d)	**9** (b)	**10** (d)
11 (a)	**12** (d)	**13** (c)	**14** (c)	**15** (c)	**16** (c)	**17** (a)	**18** (c)	**19** (a)	**20** (a)
21 (d)	**22** (b)	**23** (b)	**24** (74)	**25** (962)	**26** (57)	**27** (163)	**28** (417)	**29** (2)	**30** (37)
31 (44)	**32** (1)	**33** (3)	**34** (48)	**35** (12)	**36** (4)	**37** (2)	**38** (2)	**39** (2735)	**40** (1107)
41 (29)	**42** (125)	**43** (0.02)	**44** (16)	**45** (2)	**46** (172)	**47** (2)	**48** (73)	**49** (5)	**50** (1380)

Topic 2

1 (b)	**2** (d)	**3** (b)	**4** (b)	**5** (b)	**6** (b)	**7** (d)	**8** (3)	**9** (3)

Detailed Explanations

Topic 1

01. *(d)* For the reaction

$$H_2 + I_2 \rightleftharpoons 2HI$$

Concentration of H_2 and I_2 decrease and concentration of HI increases till equilibrium condition and after equilibrium concentrations of all the reactants and the product remain constant.

02. *(d)* The reaction involved in conversion of CO_2 into CO is as follows

$$\underset{\substack{0.5 \\ 0.5-x}}{CO_2(g)} + C(s) \rightleftharpoons \underset{2x}{2CO(g)}$$

$$p_{total} = 0.5 + x = 0.8$$
$$x = 0.3$$

So, $\quad k_p = \dfrac{(0.6)^2}{0.2} = 1.8 \text{ atm}$

So, the value of k_p is 1.8 atm.

03. *(a)* K_C is given by the equation

$$K_C = \frac{\text{Concentration of product ion}}{\text{Concentration of reactant ion}}$$

So, for the reaction

$$Fe^{3+}(aq) + SCN^-(aq) \rightleftharpoons (FeSCN)^{2+}(aq)$$

The value of K_C will be

$$= \frac{[FeSCN^{2+}]}{[Fe^{3+}][SCN^-]}$$

04. *(a)* Equilibrium is the condition when a forward chemical reaction and its reverse reaction proceed at equal rates. i.e., at equilibrium, $r_a = r_b$

For the equilibrium, $A \rightleftharpoons B$

The variation of the rate of the forward and reverse reaction can be illustrated by following graph.

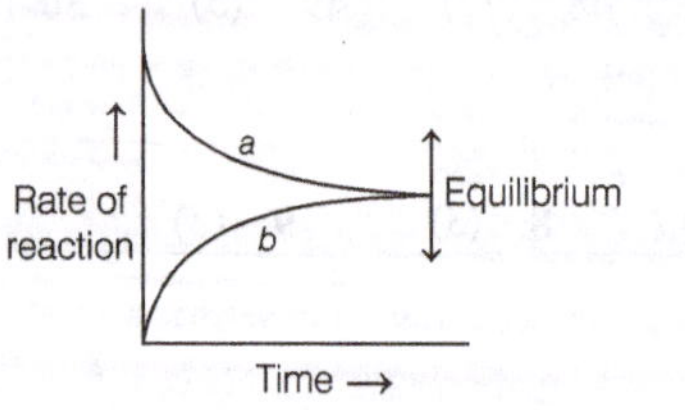

05. *(b)* Assuming the reaction to be

$$\underset{\text{(Square)}}{A} \rightleftharpoons \underset{\text{(Circle)}}{B},$$

Equilibrium constant, $K = \dfrac{[B]}{[A]}$

Number of circles in the figure $= 11$

Number of squares in the figure $= 6$

$$\therefore \quad K = \frac{11}{6} = 1.8 \approx 2$$

06. *(c)* The given reaction is as follows

$$AB_2(g) \rightleftharpoons AB(g) + \frac{1}{2}B_2(g)$$

At $t = 0$	1	0	0
At $t = t_{eq}$	$1-x$	x	$x/2$

$$\therefore K_p = \frac{\left(\dfrac{x}{1+x/2}\right)p\left(\dfrac{x/2}{1+x/2}\right)^{1/2} p^{1/2}}{\left(\dfrac{1-x}{1+x/2}\right)p}$$

$$\because \quad x << 1 \Rightarrow \frac{x}{2} + 1 \text{ and } 1 - x \approx 1$$

$$\therefore \quad K_p = \frac{(xp)\left(\dfrac{xp}{2}\right)^{1/2}}{p}$$

So, the value of x is terms of K_p and p

is $x = \sqrt[3]{\dfrac{2K_p^2}{p}}$

07. *(a)* $X_2Y(g) \rightleftharpoons X_2(g) + \dfrac{1}{2}Y_2(g)$

$$1-x \text{ mol} \qquad x\text{ mol} \qquad x/2\text{mol}$$

$$\therefore \quad p_{X_2Y} = \frac{1-x}{1+x/2} \times p$$

$$p_{X_2} = \frac{x}{1+x/2} \times p$$

$$p_{Y_2} = \frac{x/2}{1+x/2} \times p$$

Therefore, for the given reaction,

$$K_p = \frac{\left(\dfrac{x}{1+x/2}p\right)\left(\dfrac{x}{2(1+x/2)}p\right)^{1/2}}{\left(\dfrac{1-x}{1+x/2}\right)p}$$

$$\therefore \quad K_p = \left(\frac{x}{1-x}\right)\left(\frac{x}{2(1+x/2)}\right)^{1/2} \times p^{1/2}$$

Also it is given that, the value of x is very small.

So, $\quad K_p = \dfrac{x^{3/2}}{(2)^{1/2}} \times p^{1/2}$

or $\quad x^{3/2} = \dfrac{K_p \times 2^{1/2}}{p^{1/2}}$

or $\quad x = \left(\dfrac{K_p^2 \times 2}{p}\right)^{1/3}$

08. *(d)* $X \rightleftharpoons Y; K_1 = 1.0$

$$Y \rightleftharpoons Z; K_2 = 2.0$$
$$Z \rightleftharpoons W; K_3 = 4.0$$

For the equilibrium constant value of

$$X \rightleftharpoons W; K = K_1 \cdot K_2 \cdot K_3$$
$$K = 1 \times 2 \times 4$$
$$K = 8$$

Hence, option (d) is correct.

09. *(b)* Given,

$$CO(g) + \frac{1}{2}O_2(g) \rightleftharpoons CO_2(g)$$

$$\Delta n_g = 1 - \left(1 + \frac{1}{2}\right) = -\frac{1}{2}$$

$$\frac{K_p}{K_c} = (RT)^{\Delta n_g} = (RT)^{-1/2} = \frac{1}{\sqrt{RT}}$$

10. *(d)* The reaction $2SO_2 + O_2 \rightleftharpoons 2SO_3$ will be formed from the given reaction by multiplying it by coefficient 2 and then reversing it.

Thus, $K'_C = \dfrac{1}{K_C^2}$

$$= \left(\frac{1}{4.9 \times 10^{-2}}\right)^2 = 416$$

11. *(a)* $p_{total} = \dfrac{nRT}{V} = \dfrac{9 \times 0.082 \times 610}{100}$

$$= 4.5 \text{ atm}$$

$$p_{PCl_5} = \frac{5 \times 4.5}{9} = 2.5 \text{ atm}[p_A = x_A p_T]$$

$$p_{Ar} = \frac{4 \times 4.5}{9} = 2 \text{ atm}$$

Now, $\underset{2.5-p}{PCl_5} \rightleftharpoons \underset{p}{PCl_3} + \underset{p}{Cl_2}$

$$p_{total} = 2.5 - p + p + p + p_{Ar} = 6$$
$$p = 1.5$$

Now, $K_p = \dfrac{[PCl_3][Cl_2]}{[PCl_5]}$

$$= \frac{1.5 \times 1.5}{(2.5 - 1.5)} = 2.25$$

12. *(d)* Equilibrium constant for equation,

$$N_2(g) + 3H_2(g) \rightleftharpoons 2NH_3(g)$$

$$K_C = \frac{[NH_3]^2}{[N_2][H_2]^3} = 64$$

and for

$$2NH_3(g) \rightleftharpoons N_2(g) + 3H_2(g),$$

$$K'_C = \frac{1}{K_C}$$

and for equation,

$$NH_3(g) \rightleftharpoons \frac{1}{2}N_2(g) + \frac{3}{2}H_2(g)$$

$$K'_C = \frac{1}{K_C^{1/2}} = \frac{1}{K_C^{1/2}} = \frac{1}{(64)^{1/2}} = \frac{1}{8}$$

13. *(c)* $\log\left(\frac{K_2}{K_1}\right) = \frac{\Delta H^\circ}{2.303R}\left[\frac{1}{T_1} - \frac{1}{T_2}\right]$

$$\log\left(\frac{100}{10}\right) = \frac{\Delta H^\circ}{2.303 \times \left(\frac{8.314}{1000}\right)}\left[\frac{1}{298} - \frac{1}{373}\right]$$

$\Delta H^\circ = 28.4 \text{ kJ mol}^{-1}$

$\Delta G^\circ = -RT \ln K,$

ΔG° at $T_1 = (25^\circ C)$

$= -\frac{8.314}{1000} \times 298 \times 2.303 \times \log 10$

$= -5.71 \text{ kJ/mol}$

ΔG° at $T_2 (100^\circ C) = -\frac{8.314}{1000} \times 373 \times 2.303 \times \log(100)$

$= -14.29 \text{ kJ/mol}$

Hence, $\Delta H^\circ = 28.4$ kJ/mol

At T_1

$\Delta G^\circ = -5.71$ kJ/mol

At T_2

$\Delta G^\circ = -14.29$ kJ/mol

14. *(c)* $K_p = K_C(RT)^{\Delta n_g}$

Δn_g = number of gaseous product molecules − number of gaseous reactant molecules

$$Fe_2N(s) + \frac{3}{2}H_2(g) \rightleftharpoons Fe(s) + NH_3(g)$$

$$\Delta n_g = 1 - \frac{3}{2} = -\frac{1}{2}$$

Now, $K_p = K_C(RT)^{-1/2}$

$\therefore \quad K_C = \frac{K_p}{(RT)^{-1/2}} \Rightarrow K_p(RT)^{1/2}$

Hence, the correct option is (c).

15. *(c)* $A \rightleftharpoons B + C;$

$K_{eq}^{(1)} = \frac{[B][C]}{[A]}$ ... (i)

$B + C \rightleftharpoons P;\ K_{eq}^{(2)} = \frac{[P]}{[B][C]}$ ...(ii)

Equilibrium constant for $A \rightleftharpoons P$

$A \rightleftharpoons P;\ K_{eq} = \frac{[P]}{[A]}$ ...(iii)

To obtain the above equilibrium reaction add the other two equilibrium reaction. In order to obtain K_{eq}

Multiplying equation (i) and (ii)

$$K_{(eq)}^{(1)} \times K_{(eq)}^{(2)} = \frac{[P]}{[A]} = K_{eq}$$

16. *(c)* At a certain temperature, water maintains an equilibrium state with water vapour in a sealed container. When glucose is added into water, a completely miscible solution is formed in which glucose molecules occupy the inter-molecular space of solvent (H_2O) molecules.

Thus, the addition of glucose increases the total number of molecules in the solution, which increases the collision rate of water molecules with glucose molecules. This leads to an increase in the rate of evaporation of water molecules from the surface of the beaker and hence, the rate at which water molecules leave the vapour increases.

17. *(a)*

> **idea** The relationship between K_p and K_c is
>
> $K_p = K_c(RT)^{\Delta n_g}$
>
> where $\Delta n_g = \Delta n_{Products} - \Delta n_{Reactants}$
>
> If $\Delta n_g = 0$ then $K_p = K_c$
>
> If Δn_g = +ve then $K_p > K_c$
>
> If Δn_g = −ve then $K_p < K_c$

Consider the given equilibria reaction

$$2C(s) + O_2(g) \rightleftharpoons 2CO(g)$$

$$\Delta n_g = n_{product} - n_{reactant} = 2 - (1) = 1$$

$$\Delta n_g \neq 0 \quad \Rightarrow \text{So, } K_p \neq K_c$$

18. *(c)* (i) $A_2(g) + B_2(g) \overset{K_1}{\rightleftharpoons} 2AB(g)$

$$K_1 = \frac{[AB]^2}{[A_2][B_2]}$$

(ii) $6AB(g) \rightleftharpoons 3A_2(g) + 3B_2(g);$

$$K_2 = \frac{[A_2]^3[B_2]^3}{[AB]^6}$$

$$= \frac{1}{\left(\frac{[AB]^2}{[A_2][B_2]}\right)^3} = \frac{1}{K_1^3},$$

$\Rightarrow \quad K_2 = K_1^{-3}$

19. *(a)* $S + O_2 \rightleftharpoons SO_2, \quad K_1$

$\therefore \quad SO_2 \rightleftharpoons S + O_2$

$$K_1' = \frac{1}{K_1}$$

or, $\quad 2SO_2 \rightleftharpoons 2S + 2O_2,$

$K_1'' = (K_1')^2 = \frac{1}{K_1^2}$ ... (i)

$\Rightarrow \quad 2S + 3O_2 \rightleftharpoons 2SO_3, K_2$ ... (ii)

Now, [(i) + (ii)] gives

$$2SO_2 + O_2 \rightleftharpoons 2SO_3, K_3$$

The value of equilibrium constant,

$$K_3 = K_2 \times K_1'' = K_2 \times \frac{1}{K_1^2}$$

$$= 10^{129} \times \frac{1}{(10^{52})^2} = 10^{129-104} = 10^{25}$$

20. *(a)* $A(g) \rightleftharpoons B(g) + \frac{C}{2}(g)$

$t = t_{eq} \quad (1-\alpha) \qquad \alpha \qquad \frac{\alpha}{2}$

$$p_B = \frac{\alpha}{\left(1 + \frac{\alpha}{2}\right)} \cdot p,\ p_A = \frac{(1-\alpha)}{\left(1+\frac{\alpha}{2}\right)} \cdot p,$$

$$p_C = \frac{\frac{\alpha}{2}}{\left(1 + \frac{\alpha}{2}\right)} \cdot p$$

$K_p = \frac{p_B \cdot p_C^{1/2}}{p_A}$ ...(i)

Substitute the value of p_A, p_B and p_C in Eq. (i)

$$\therefore \quad K_p = \frac{(\alpha)^{3/2}(p)^{1/2}}{(1-\alpha)(2+\alpha)^{1/2}}$$

21. *(d)* The equilibrium reaction for the dissociation of two solids is given as:

$$A(s) \rightleftharpoons B(g) + C(g)$$

At equilibrium $\qquad p_1 \qquad p_1 + p_2$

$K_{p_1} = x = p_B \cdot p_C$

$= p_1(p_1 + p_2)$ ...(i)

Similarly,

$$D(s) \rightleftharpoons C(g) + E(g)$$

At equilibrium $\qquad p_1 + p_2 \qquad p_2$

$K_{p_2} = y = p_C \cdot p_E = (p_1 + p_2)p_2$ ...(ii)

On adding Eq. (i) and (ii), we get.

$K_{p_1} + K_{p_2} = x + y$

$= p_1(p_1 + p_2) + p_2(p_1 + p_2)$

$= (p_1 + p_2)^2$

or $\sqrt{x + y} = p_1 + p_2$ ...(iii)

Now, total pressure is given as

$p_T = p_B + p_C + p_E$

$= p_1 + (p_1 + p_2) + p_2$

$= 2(p_1 + p_2)$ ...(iv)

On substituting the value of $p_1 + p_2$ from Eq. (iii) to Eq. (iv), we get

$$p_T = 2\sqrt{x + y}$$

22. *(b)* $N_2(g) + 3H_2(g) \rightleftharpoons 2NH_3(g)$

At eq. $p_{N_2} = P,\ p_{H_2} = 3P,\ p_{NH_3} = 2P$

$\Rightarrow p_{(total)} = p_{N_2} + p_{H_2} + p_{NH_3}$

$$\approx p_{N_2} + p_{H_2}$$

$[\because p_{(total)} >> p_{NH_3}]$

$$p = P + 3P = 4P$$

Now, $K_p = \dfrac{p^2_{NH_3}}{p_{N_2} \times p^3_{H_2}} = \dfrac{p^2_{NH_3}}{p \times (3P)^3}$

$$= \frac{p^2_{NH_3}}{27 \times p^4} = \frac{p^2_{NH_3}}{27 \times \left(\frac{P}{4}\right)^4}$$

$[\because p = 4P]$

$$K_p = \frac{p^2_{NH_3} \times 4^4}{3^2 \times 3 \times p^4}$$

$$\Rightarrow \quad p^2_{NH_3} = \frac{3^2 \times 3 \times p^4 \times K_p}{4^4}$$

$$\Rightarrow \quad p_{NH_3} = \frac{3 \times 3^{1/2} \times p^2 \times K_p^{1/2}}{4^2}$$

$$= \frac{3^{3/2} \times p^2 \times K_p^{1/2}}{16}$$

23. *(b)* Molar mass of $NH_4SH = 18 + 33 = 51 \text{ g mol}^{-1}$

Number of moles of NH_4SH introduced in the vessel

$$= \frac{\text{Weight}}{\text{Molar mass}} = \frac{5.1}{51} = 0.1 \text{mol}$$

	$NH_4SH(s) \rightleftharpoons$	$NH_3(g) +$	$H_2S(g)$
Number of moles at $t = 0$	0.1	0	0
At $t = t_{eq}$	0.1(1 − 0.03)	30% of 0.1 = 0.03	30% of 0.1 = 0.03
Active mass (mol L^{-1})		$\frac{0.03}{3} = 0.01$	$\frac{0.03}{3} = 0.01$

$$K_C = \frac{[NH_3][H_2S]}{[NH_4HS(s)]} = \frac{0.01 \times 0.01}{1}$$

$$= 10^{-4} \text{ (mol L}^{-1})^2$$

$$\Rightarrow \quad K_p = K_C(RT)^{\Delta n_g}$$

[where, $\Delta n_g = \Sigma n_{product} - \Sigma n_{reactant}$] $= 2 - 0 = 2$

$$\therefore \quad K_p = K_C(RT)^2$$

$$= 10^{-4} \times [0.082 \times (273 + 327)]^2 \text{ atm}^2$$

$$= 0.242 \text{ atm}^2$$

24. *(74)* Consider the given reaction:

	$CO(g) +$	$2H_2(g) \rightleftharpoons$	$CH_3OH(g)$
$t = 0$	0.1 mol	a mol	−
t_{eq}	$0.1 - x$ -0.06	$a - 2x$ $0 - 0.08$ $= 0.23 - 0.08$ $= 0.15$ mole	$x = 0.04$

$V = 2\text{L}, T = 500\text{K}, p_{Total} = 5 \text{ bar}$

$n_{Total} = 0.25 = 1/4 \text{ mol}$

Using, $p_{Total} = n_{Total} \times \dfrac{RT}{V}$ ...(i)

Substituting,

$$5 = (0.06 + a - 0.08 + 0.04) \times \frac{0.08 \times 500}{2}$$

$$a = 0.23 \text{ mol}$$

$$K_p = \frac{\chi_{CH_3OH}}{\chi_{CO} \times \chi^2_{H_2}} \times \frac{1}{(p_T)^2}$$

$$= \frac{0.04}{0.06 \times (0.15)^2} \times \left[\frac{1/4}{5}\right]^2$$

$$K_p = 0.074 = 74 \times 10^{-3}$$

Hence, $n = 74$

25. *(962)* Initial pressure of N_2O_5

$$\frac{\frac{37.8}{108} \times 0.082 \times 500}{1} = 14.35 \text{ bar}$$

	$2N_2O_5 \rightleftharpoons$	$2N_2O_4 +$	O_2
at $t = 0$	14.35		
t = eq	14.35−2p	2p	p

p_{total} at eq $= 14.35 + p = 18.65$ bar

$$p = 4.3$$

$$p_{N_2O_5} = 5.75 \text{ bar}$$

$$p_{N_2O_4} = 8.6 \text{ bar}$$

$$p_{O_2} = 4.3 \text{ bar}$$

$$K_p = \frac{(8.6)^2 \times (4.3)}{(5.75)^2}$$

$$= 9.619 = X \times 10^{-2}$$

$$X = 961.9 \simeq 962$$

26. *(57)* $\Delta G° = -RT \ln K_{eq}$

$$= -2.303RT \log K_{eq}$$

$$= -2.303 \times 8.314 \times 300 \times \log 10$$

$$= -5744.14 \text{ J/mol}$$

$$= -5.744 \text{ kJ/mol}$$

$$= -57.44 \times 10^{-1} \text{ kJ/mol}$$

Thus, the value is 57.

27. *(163)* For the given reaction,

$$\frac{3}{2}O_2(g) \rightleftharpoons O_3(g)$$

$$K_p = 2.47 \times 10^{-29}$$

Using $\Delta_r G° = -RT \ln K_p$

$$= -8.314 \times 10^{-3} \times 298 \times \ln(2.47 \times 10^{-29})$$

$$\Delta_r G° = -8.314 \times 10^{-3} \times 298 \times (-65.87)$$

$$\Delta_r G° = 163.19 \text{ kJ}$$

So, the value of x is 163.

28. *(417)* The formation of NH_3 is,

$$N_2 + 3H_2 \rightleftharpoons 2NH_3$$

Equilibrium constant is given by

$$K_C = \frac{[\text{Product}]}{[\text{Reactant}]}$$

$$= \frac{[NH_3]^2}{[N_2][H_2]^3}$$

$$= \frac{(1.5 \times 10^{-2})^2}{(2 \times 10^{-2})(3 \times 10^{-2})^3}$$

$$= 417$$

Hence, equilibrium constant is 417.

29. *(2)*

> **idea** To solve this question, first calculate the value of Δn_g. Then, put this value along with the given values in the formula $K_p = K_C(RT)^{\Delta n_g}$ to calculate the value of K_C

Given reaction,

$$N_2O_4(g) \rightleftharpoons 2NO_2(g)$$

$$K_p = 0.492 \text{ atm at } 300 \text{ K}$$

The relation between K_C and K_p,

$$K_p = K_C(RT)^{\Delta n_g} \Rightarrow \Delta n_g = 1$$

$$\Rightarrow K_C = \frac{K_p}{RT} = \frac{0.492}{0.082 \times 300} = 2 \times 10^{-2}$$

Hence, the missing value in the question is 2.

30. *(37)* Given, $T = 400$ K

$$\Delta H^\ominus = 77.2 \text{ kJ/mol}$$

$$\Delta S = 122 \text{ JK}^{-1}$$

Free energy is given by

$$\Delta G° = \Delta H° - T\Delta S°$$

$$= 77.2 \times 10^3 - 400 \times 122$$

$$= 28400 \text{ J}$$

$$\Delta G° = -2.303RT \log K$$

$$28400 = -2.303 \times 8.314 \times 400 \times \log K$$

$$\log K = -3.708$$

or $\log K = -37.08 \times 10^{-1}$

Hence, the missing value is 37.

31. *(44)*

	$CO(g) +$	$H_2O(g) \rightleftharpoons$	$CO_2(g) +$	$H_2(g)$
at $t = 0$	1 mol	1 mol	0	0
at eq.	$1 - x$	$1 - x$	x	x

Given that 40% by mass, water reacts with CO at equilibrium.

$\Rightarrow \quad x = 0.4$ and $1 - x = 0.6$

So, $K_C = \dfrac{[CO_2][H_2]}{[CO][H_2O]}$

$$= \frac{0.4 \times 0.4}{0.6 \times 0.6}$$

$$= 0.44$$

or $K_C \times 10^2 = 44$

32. *(1)*

	$H_2(g) +$	$I_2(g) \rightleftharpoons$	$2HI(g)$
at $t = 0$	4.5	4.5	—
at t = eq.	$4.5 - x$	$4.5 - x$	$2x$

Given, $2x = 3 \Rightarrow x = \frac{3}{2} = 1.5$

$\therefore \ [H_2] = 4.5 - 1.5 = 3,$

$[I_2] = 4.5 - 1.5 = 3$

$$K_C = \frac{[HI]^2}{[H_2]\,[I_2]} = \frac{(3)^2}{3 \times 3} = 1$$

33. *(3)* Given,

Initial pressure = 450 mm Hg

Pressure at time t = 720 mm Hg

Constant temperature T and constant volume V.

Fraction of $A(g)$ decomposed

$= x \times 10^{-1}$

$$A(g) \rightleftharpoons 2B(g) + C(g)$$

	$A(g)$	$2B(g)$	$C(g)$
at time $t = 0$,	$p = 450$		
at time t	$P = 450 - x$	$2x$	x

$p_T = p_A + p_B + p_C$

$720 = 450 - x + 2x + x$

$720 = 450 - 2x$ or, $2x = 270$

$x = \frac{270}{2} \Rightarrow x = 135$

Fraction of A decomposed $= \frac{135}{450} = 0.3$

$= 3 \times 10^{-1}$

So, $x = 3$

34. *(48)* $PCl_3 + Cl_2 \rightleftharpoons PCl_5$

PCl_3	Cl_2	PCl_5
0.2 mol L^{-1}	0.1 mol L^{-1}	0.40 mol L^{-1}

$$K_C = \frac{[PCl_5]}{[PCl_3]\,[Cl_2]} = \frac{0.40}{0.2 \times 0.1} = 20$$

If 0.2 mol of Cl_2 is added then

$$Q_C = \frac{0.40}{0.2 \times (0.1 + 0.2)} = 6.66$$

$\therefore \ Q_C < K_C$, (Forward reaction)

$\therefore \quad PCl_3 + Cl_2 \rightleftharpoons PCl_5$

At equ. $(0.2 - x)$ $(0.3 - x)$ $0.40 + x$

$$K_C = \frac{0.4 + x}{(0.2 - x)(0.3 - x)}$$

$$\Rightarrow \quad 20 = \frac{0.4 + x}{0.06 - 0.5x + x^2}$$

$20x^2 - 11x + 0.8 = 0$

$x = 0.086$

$\therefore$ Concentration of $PCl_5 = 0.4 + 0.086$

$= 0.486$

or $48.6 \times 10^{-2} \approx 48 \times 10^{-2}$

35. *(12)* For equilibrium equation.

(i) $X(g) \rightleftharpoons Y(g) + Z(g),\ K_{p_1} = 3$

	$X(g)$	$Y(g)$	$Z(g)$
at $t = 0$	1	0	0
at t_{eq}	$1 - x$	x	x

Total number of moles

$= 1 - x + x + x = 1 + x$

Partial pressure	$\frac{(1-x)p_1}{1+x}$	$\frac{xp_1}{1+x}$	$\frac{xp_1}{1+x}$

$$\Rightarrow Kp_1 = \frac{[p_y][p_z]}{[p_x]}$$

$$= \frac{\left(\frac{xp_1}{1+x}\right)\left(\frac{xp_1}{1+x}\right)}{\left[\frac{(1-x)p_1}{1+x}\right]} = \frac{x^2}{1-x^2} \times p_1$$

(ii) $A(g) \rightleftharpoons 2B(g),\ K_{p_2} = 1$

	$A(g)$	$2B(g)$
at $t = 0$	1	0
at t_{eq}	$1 - x$	$2x$
Partial pressure	$\frac{(1-x)p_2}{1+x}$	$\frac{(2x)p_2}{1+x}$

$$\Rightarrow \quad Kp_2 = \frac{(p_B)^2}{(p_A)} = \frac{\frac{(2x)^2 p_2^2}{(1+x)^2}}{\left(\frac{1-x}{1+x}\right)p_2}$$

$$= \frac{4x^2}{1-x^2} \times p_2 \Rightarrow \frac{Kp_1}{Kp_2} = \frac{3}{1}$$

$$= \frac{x^2}{(1-x^2)} \times p_1 \times \frac{(1-x^2)}{4x^2} \times \frac{1}{p_2} = \frac{p_1}{4p_2}$$

$$\Rightarrow \quad \frac{p_1}{p_2} = \frac{12}{1} = \frac{x}{1} \qquad \text{(Given)}$$

$x = 12$

36. *(4)* $N_2(g) + 3H_2(g) \rightleftharpoons 2NH_3(g);$

$K_1 = 4 \times 10^5$ ...(i)

$N_2(g) + O_2(g) \rightleftharpoons 2NO(g);$

$K_2 = 1.6 \times 10^{12}$ ...(ii)

$H_2(g) + \frac{1}{2}O_2(g) \rightleftharpoons H_2O(g);$

$K_3 = 1 \times 10^{-13}$ ...(iii)

On solving above equation as

Eq. (ii) + 3 × Eq. (iii) − Eq. (i)

We have,

$$\left[2NH_3(g) + \frac{5}{2}O_2(g) \rightleftharpoons 2NO(g) + 3H_2O(g)\right]$$

... required

Since in Eq. (iii), 3 is multiplied, so

$3H_2 + 3/2\,O_2 \rightleftharpoons 3H_2O;$

$$K'_3 = \frac{[H_2O]^3}{[H_2]^3\,[O_2]^{3/2}} = (K_3)^3$$

Eq. (ii) + 3 × Eq. (iii) − Eq. (i)

$$K = \frac{K_2 \times (K_3)^3}{K_1}$$

or $$K = \frac{1.6 \times 10^{12} \times (10^{-13})^3}{4 \times 10^5}$$

or $K = 4 \times 10^{-33}$

37. *(2)* $2NO(g) + O_2(g) \rightleftharpoons 2NO_2(g)$

	$2NO(g)$	$O_2(g)$	$2NO_2(g)$
At $t = 0$	2mol	1 mol	0
At $t = t_{eq}$	1.2 mol	0.6 mol	0.8 mol

Total moles, $n_T = 1.2 + 0.6 + 0.8$

$= 2.6$ mol

Equilibrium constant $= K_p$

$p_T = 1$ atm

$$K_p = \frac{(p_{NO_2})^2}{(p_{NO})^2 \times p_{O_2}} \qquad (p_A = p_T x_A)$$

$$= \frac{\left(\frac{0.8}{2.6}\right)^2}{\left(\frac{1.2}{2.6}\right)^2 \times \left(\frac{0.6}{2.6}\right)} = 1.925 \approx 2$$

38. *(2)* Given,

$Cu(s) + 2Ag^+(aq) \rightleftharpoons Cu^{2+}(aq) + 2Ag(s)$

Equilibrium constant, $K = 2 \times 10^{15}$

$T = 298$ K

For the reaction,

$\frac{1}{2}Cu^{2+}(aq) + Ag(s) \rightleftharpoons \frac{1}{2}Cu(s) + Ag^+(aq)$

K' can be calculated as

$$K' = \frac{1}{(K)^{1/2}} = \frac{1}{(2 \times 10^{15})^{1/2}}$$

$= 2.2 \times 10^{-8} \quad (K' = x \times 10^{-8})$

$\therefore \quad x \approx 2$

39. *(2735)* $HI \rightleftharpoons \frac{1}{2}H_2 + \frac{1}{2}I_2$

	HI	$\frac{1}{2}H_2$	$\frac{1}{2}I_2$
t_i	1		
t_{eq}	$1 - 0.4$	$0.4/2$	$0.4/2$

$$K_p = \frac{[H_2]^{1/2}[I_2]^{1/2}}{[HI]}$$

$$K_p = \frac{(0.2)^{1/2}\,(0.2)^{1/2}}{1 - 0.4} = \frac{0.2}{0.6} = \frac{1}{3}$$

$\Delta G = \Delta G° + RT \ln K = 0$

[$\because$ at equilibrium $\Delta G = 0$]

$\Delta G° = -RT \ln K$

$= -8.31 \times 300 \times 2.3 \log\left(\frac{1}{3}\right)$

$= 2735$ J/mol

40. *(1107)*

> **idea** We can calculate the equilibrium constant for the process at temperature T, by using the relationship between $\Delta G°$ and K
>
> $\Delta G° = RT \ln K_{eq}$
>
> where $R = 8.314$ J mol^{-1} K^{-1}
>
> T = Temperature on Kelvin scale.

$$PCl_5 \rightleftharpoons PCl_3 + Cl_2$$

Initial mole	5	0	0
At equil.	5(1 − x)	5x	5x

Total moles $= 5(1-x) + 5x + 5x$
$= 5(1+x) + 2$ (Nitrogen moles)

$pV = nRT$

$$n = \frac{2.46 \times 200}{0.082 \times 600} = 10$$

$$5(1+x) + 2 = 10$$

$$x = \frac{3}{5} = 0.6$$

Partial pressure of

$$p_{PCl_5} = \frac{5(1-x)}{7+5x} = 2.46$$

Put value of $x = 0.6$

$$p_{PCl_5} = \frac{5 \times 0.4}{10} \times 2.46 \Rightarrow \frac{2.46}{5}$$

Partial pressure of p_{PCl_3} and p_{Cl_2}

$$p_{PCl_3} = p_{Cl_2} = \frac{5x}{5x+7} \times 2.46$$

Put $x = 0.6$

$$p_{PCl_3} = p_{Cl_2} = \frac{3}{10} \times 2.46$$

$$K_p = \frac{[p_{Cl_3}][p_{Cl_2}]}{p_{PCl_5}}$$

$$= \frac{\frac{3}{10} \times 2.46 \times \frac{3}{10} \times 2.46}{\frac{2}{10} \times 2.46}$$

$$= \frac{9}{10} \times \frac{2.46}{2} = 1.107 = 1107 \times 10^{-3}$$

41. *(29)* Firstly calculate moles of water

$$\text{Moles of water} = \frac{\text{Given mass}}{\text{Molecular mass}} = \frac{0.90}{18}$$

By ideal gas equation, $pV = nRT$

$$V = \frac{nRT}{P} = \frac{0.90 \times 0.082 \times 300 \times 760}{18 \times 32}$$

$$= 29.21 \approx 29$$

42. *(125)* According to the given reaction,

$$2NOCl(g) \rightleftharpoons 2NO(g) + Cl_2(g)$$

Initial moles	2	0	0
At equili moles	2 − 0.4 = 1.6	0.4	0.2

Now, equilibrium constant,

$$K_C = \frac{[NO]^2[Cl_2]}{[NOCl]^2} = \frac{(0.4)^2 \times (0.2)}{(1.6)^2}$$

$$= 0.0125 = 125 \times 10^{-4}$$

43. *(0.02)* $[PtCl_4]^{2-} + H_2O \underset{K_b}{\overset{K_f}{\rightleftharpoons}} [Pt(H_2O)Cl_3]^- + Cl^-$

Rate of a reaction = Rate of forward reaction − rate of backward reaction

$$\frac{-d[[PtCl_4]^{2-}]}{dt} = K_f [[PtCl_4]^{2-}] - K_b [[Pt(H_2O)Cl_3]^-][Cl^-]$$

Comparing this equation with the given one :

$$\Rightarrow K_{eq} = \frac{K_f}{K_b} = \frac{4.8 \times 10^{-5}}{2.4 \times 10^{-3}} = 0.02$$

44. *(16)* According to the equation,

$$A(s) \rightleftharpoons M(s) + \frac{1}{2}O_2(g)$$

Equilibrium constant, $K_p = p_{O_2}^{1/2} = 4$

where, p = partial pressure

At equilibrium, the partial pressure of O_2 is $p_{O_2} = (4)^2 = 16$ atm.

45. *(2)* Given reaction,

$$N_2O_4(g) \rightleftharpoons 2NO_2(g)$$

So, Δn_g = [moles of gaseous products− moles of gaseous reactants]

$= 2 - 1 = 1$

$$K_p = K_C(RT)^{\Delta n_g}$$

where, K_p = equilibrium constant in gaseous state = 47.9

K_C = equilibrium constant

R = gas constant

T = temperature

$$\Rightarrow K_C = \frac{47.9}{0.083 \times 288} = 2.0038$$

∴ Hence, K_C at same temperature (288 K) is 2.

46. *(172)* Given reaction,

$$2SO_2(g) + O_2(g) \rightleftharpoons 2SO_3(g)$$

$$\text{Now, } K_p = \frac{[p_{SO_3(g)}]^2}{[p_{SO_2(g)}]^2 \times [p_{O_2}]}$$

On putting given values, we get,

$$\Rightarrow K_p = \frac{(43)^2}{(45)^2 \times 0.53}$$

$$= \frac{1849}{2025 \times 0.53} = \frac{1849}{1073.25}$$

$$= 1.7228$$

$$= 172.28 \times 10^{-2} = 172$$

Hence, the equilibrium constant, $K_p = 172$.

47. *(2)* $2A(g) \rightleftharpoons A_2(g)$

Given, $\Delta G° = 25.2$ kJ mol^{-1}

$= 25200$ J mol^{-1}

$T = 400$ K

According to standard free Gibb's equation,

$$\Delta G° = -RT \ln K_p$$

$$25200 = -2.3 \times 8.3 \times 400 \log(K_p)$$

$$\log K_p = \frac{-25200}{2.3 \times 8.3 \times 400} = -3.3$$

$$K_p = 10^{-3.3} = 10^{-3} \times 0.501$$

$$K_p = 5.01 \times 10^{-4} \text{ bar}^{-1}$$

$$K_p = 5.01 \times 10^{-9} \text{ Pa}^{-1}$$

We know that,

$$K_p = K_c(RT)^{\Delta n_g}$$

$$K_p = K_c(RT)^{-1}$$

$$[\because \Delta n_g = 1 - 2 = -1]$$

$$K_p = \frac{K_C}{8.3 \times 400}$$

$$K_c = 5.01 \times 10^{-5} \times 8.3 \times 400$$

$$\Rightarrow K_c = 166 \times 10^{-5} \text{ m}^3/\text{mol}$$

$$= 1.66 \times 10^{-2} \approx 2 \times 10^{-2} \text{ L/mol}$$

According to the answer key of NTA the correct answer is 166 but that is not possible if we consider the units properly. So the correct answer is 2 (nearest integer).

48. *(73)* $AB_2(g) \rightleftharpoons A(g) + 2B(g)$

$t = 0$	1	0	0
$t = t_{eq}$	$(1-x)$	x	$2x$

Total number of moles at equilibrium
$= 1 - x + x + 2x = (1 + 2x)$

Partial pressure (atm)	$\frac{1-x}{1+2x}p$	$\frac{x}{1+2x}p$	$\frac{2x}{1+2x}p$

[p_T = Total pressure at equilibrium = 1.9 atm]

Now, at equilibrium $pV = (1+2x)RT$

$$\Rightarrow 1 + 2x = \frac{pV}{RT} = \frac{1.9 \times 25}{0.082 \times 300}$$

$$= 1.93$$

[$V = 25$ L, $R = 0.082$ L atm mol^{-1}K^{-1} $T = 300$ K]

$$\Rightarrow x = \frac{1.93 - 1}{2} = 0.465$$

$$\Rightarrow K_p = \frac{p_A \times p_B^2}{p_{AB_2}}$$

$$\Rightarrow \frac{\left(\frac{x}{1+2x}p\right) \times \left(\frac{2x}{1+2x}p\right)^2}{\left(\frac{1-x}{1+2x}p\right)}$$

$$= \frac{4x^3 \times p^2}{(1+2x)^2 \times (1-x)}$$

$$= \frac{4 \times (0.465)^3 \times (1.9)^2}{(1+2\times 0.465)^2 \times (1-0.465)}$$

$$= 0.7285 \text{ atm}$$

$$= 72.85 \times 10^{-2} \text{ atm}$$

$$\simeq 73 \times 10^{-2} = x \times 10^{-2}$$

$\therefore \quad x = 73$

49. *(5)* $Cl_2(g) \rightleftharpoons 2Cl(g)$

Let, moles of both of Cl_2 and Cl molecule be x.

(Number of moles of Cl_2 = Number of moles of Cl)

Partial pressure of Cl is,

$$p_{Cl} = \frac{x}{2x} \times 1 = \frac{1}{2}$$

Partial pressure of Cl_2 is

$$p_{Cl_2} = \frac{x}{2x} \times 1 = \frac{1}{2}$$

Now, $K_p = \dfrac{(p_{Cl})^2}{p_{Cl_2}}$

$$\Rightarrow \quad K_p = \frac{(1/2)^2}{1/2} = \frac{1}{2} = 0.5$$

or 5×10^{-1}

Hence, $x \times 10^{-1} \Rightarrow x = 5$

50. *(1380)* For a reaction, $A(g) \longrightarrow B(g)$

Formula used

$$\Delta G° = -RT \ln K_p \quad ...(i)$$

Putting given values in Eq (i), we get

$$\Delta G° = -R\,(300) \text{ In } 100$$

$$\Delta G° = -R\,(300)\,(2) \ln (10)$$

$\because \quad \ln (10) = 2.3$

$$\Delta G° = -R(300)\,(2)\,(2.3)$$

$$\Delta G° = -1380\, R$$

Hence, $\Delta G° = -x\,R$

$$x = 1380$$

Topic 2

01. *(b)* Among the given statements, A and B are correct, while C and D are incorrect.

It is because equilibrium constant remains unchanged as the temperature is constant.

02. *(d)* According to Le-Chatelier's principle, if a solid is added, then there will be no effect on the equilibrium. Fe_2O_3 is a solid compound.
Hence, option (d) is correct.

03. *(b)* The reaction at equilibrium is

	$A(g)$	$\rightleftharpoons$	$B(g)$ +	$C(g)$
$t = 0$	a		0	0
$t = t$	$a(1-\alpha)$		$a\alpha$	$a\alpha$

'a' mole of $A(g)$ taken initially and at time $t = 0$.

Now, mole fraction of $A(g)$, $B(g)$ and $C(g)$.

$$X_A = \frac{a - a\alpha}{a + a\alpha} = \frac{1-\alpha}{1+\alpha}$$

$$X_B = \frac{a\alpha}{a + a\alpha} = \frac{\alpha}{1+\alpha}$$

$$X_C = \frac{a\alpha}{a + a\alpha} = \frac{\alpha}{1+\alpha}$$

Now, if p is total pressure, then partial pressure of each A, B, C gases are

$$p_A = \left(\frac{1-\alpha}{1+\alpha}\right)p,$$

$$p_B = \left(\frac{\alpha}{1+\alpha}\right)p,$$

$$p_C = \left(\frac{\alpha}{1+\alpha}\right)p$$

$$K_p = \frac{p_B \cdot p_C}{p_A} \Rightarrow \frac{\alpha^2 p}{1-\alpha^2}$$

As K_p is only function of temperature.
So, as pressure increases, α decreases.

04. *(b)* On addition of xenon at constant p and T, reaction will move in forward direction.
So, $[PCl_5]$ will decrease and $[PCl_3]$ and $[Cl_2]$ will increase.

05. *(b)* The reaction is

$$N_2O_5(g) \longrightarrow 2NO_2(g) + \frac{1}{2}O_2(g)$$

	$N_2O_5(g)$	$2NO_2(g)$	$\frac{1}{2}O_2(g)$
at $t = 0$	p_0		
$t = t$	$p_0 - x$	$2x$	$x/2$

$$p_{total} = p_0 - \frac{p_0}{2} + p_0 + \frac{p_0}{4} \quad \left[\because x = \frac{p_0}{2}\right]$$

$$= \frac{7}{4}p_0$$

06. *(b)* $N_2O_4(g) \rightleftharpoons 2NO_2(g)$

$\Delta H = +58$ kJ

In case of an endothermic reaction ($\Delta H = +$ ve), the equilibrium constant increases with rise in temperature and hence, the reaction moves in forward direction.On adding N_2, pressure is increased at constant T, and volume would also be constant, so no change is observed.

07. *(d)* Statement given in option (a), (b) and (c) are correct while statement (d) is incorrect.

The larger value of K_c suggest that the reaction should go almost to completion. However, the oxidation of SO_2 to SO_3 is very low. Hence, the rate of reaction is increased by adding catalyst.

08. *(3)* The number of correct statement involving equilibria in physical processes from the given statements are three i.e., A, B and D.

(A) Equilibrium is possible only in a closed system at a given temperature. All measurable properties of the system remain constant.

(B) The opposing processes occur at the same rate and there is dynamic but stable condition.

(C) At equilibrium the value of parameters become constant of a given temperature and not equal.

(D) For a given solid solute and a liquid solvent solubility depends upon temperature only.

09. *(3)* $2SO_2(g) + O_2(g) \rightleftharpoons 2SO_3(g)$

$\Delta H = -190$ kJ

The reaction is exothermic in nature. On increasing pressure, equilibrium will shift in forward direction. Apart from that addition of more SO_2 and O_2 will also increase the yield of SO_3 at equilibrium while other factors decrease the yield of SO_3.

07 Ionic Equilibrium

In JEE Main there is usually 0-1 question per paper are visible from this chapter. In past 3 years, 10 conceptual and 14 numerical questions were asked of moderate difficulty level from this chapter.

For detailed analysis scan QR code

Topic 1 / Ostwald's Dilution Law & Concept of Acids and Bases

• Level 2

01 Given below are two statements one is labelled as Assertion (A) and the other is labelled as Reason (R)

Assertion (A) The amphoteric nature of water is explained by using Lewis acid/base concept.

Reason (R) Water acts as an acid with NH_3 and as a base with H_2S.

In the light of the above statements choose the correct answer from the options given below: **[2022, 25 June Shift-II]**

(a) Both A and R are true and R is the correct explanation of A.

(b) Both A and R are true but R is not the correct explanation of A.

(c) A is true but R is false.

(d) A is false but R is true.

• Level 3

02 K_{a_1}, K_{a_2} and K_{a_3} are the respective ionisation constants for the following reactions A, B, and C.

A. $H_2C_2O_4 \rightleftharpoons H^+ + HC_2O_4^-$

B. $HC_2O_4^- \rightleftharpoons H^+ + C_2O_4^{2-}$

C. $H_2C_2O_4 \rightleftharpoons 2H^+ + C_2O_4^{2-}$

The relationship between K_{a_1}, K_{a_2} and K_{a_3} is given as **[2022, 25 July Shift-II]**

(a) $K_{a_3} = K_{a_1} + K_{a_2}$

(b) $K_{a_3} = K_{a_1} - K_{a_2}$

(c) $K_{a_3} = K_{a_1} / K_{a_2}$

(d) $K_{a_3} = K_{a_1} \times K_{a_2}$

Numerical Value Type Questions

03 If 1 mM solution of ethylamine produces pH = 9, then the ionisation constant (K_b) of ethylamine is 10^{-x}. The value of x is .... (Nearest integer).

[The degree of ionisation of ethylamine can be neglected with respect to unity.] **[2025, 23 Jan Shift-I]**

04 The equilibrium constant for decomposition of $H_2O(g)$

$$H_2O(g) \rightleftharpoons H_2(g) + \frac{1}{2}O_2(g) \; (\Delta G^\circ = 92.34 \text{ kJ mol}^{-1})$$

is 8.0×10^{-3} at 2300 K and total pressure at equilibrium is 1 bar. Under this condition, the degree of dissociation (α) of water is .... $\times 10^{-2}$ (Nearest integer value).

[Assume α is negligible with respect to 1] **[2025, 8 April Shift-II]**

05 The pH of a 0.01 M weak acid HX $(K_a = 4 \times 10^{-10})$ is found to be 5. Now the acid solution is diluted with excess of water so that the pH of the solution changes to 6. The new concentration of the diluted weak acid is given as $x \times 10^{-4}$ M. The value of x is .... (Nearest integer) **[2025, 4 April Shift-I]**

06 Assuming that $Ba(OH)_2$ is completely ionised in aqueous solution under the given conditions the concentration of H_3O^+ ions in 0.005 M aqueous solution of $Ba(OH)_2$ at 298 K is ...... $\times 10^{-12}$ mol L^{-1}. (Nearest integer) **[2021, 25 July Shift-II]**

07 0.01 moles of a weak acid HA $(K_a = 2.0 \times 10^{-6})$ is dissolved in 1.0 L of 0.1M HCl solution. The degree of dissociation of HA is ...... $\times 10^{-5}$ (Round off to the nearest integer).

[Neglect volume change on adding HA. Assume degree of dissociation << 1] **[2021, 17 March Shift-I]**

Topic 2 / Solubility Product & Common Ion Effect

• Level 1

01 The molar solubility(S) of zirconium phosphate with molecular formula $(Zr^{4+})_3\ (PO_4^{3-})_4$ is given by relation **[2025, 22 Jan Shift-II]**

(a) $\left(\frac{K_{sp}}{6912}\right)^{1/7}$ (b) $\left(\frac{K_{sp}}{9612}\right)^{1/3}$

(c) $\left(\frac{K_{sp}}{8435}\right)^{1/7}$ (d) $\left(\frac{K_{sp}}{5348}\right)^{1/6}$

02 For a sparingly soluble salt AB_2, the equilibrium concentrations of A^{2+} ions and B^- ions are 1.2×10^{-4} M and 0.24×10^{-3} M, respectively. The solubility product of AB_2 is **[2024, 9 April Shift-II]**

(a) 0.069×10^{-12} (b) 6.91×10^{-12}

(c) 0.276×10^{-12} (d) 27.65×10^{-12}

03 The solubility of AgCl will be maximum in which of the following? **[2022, 29 June Shift-I]**

(a) 0.01 M KCl
(b) 0.01 M HCl
(c) 0.01 M $AgNO_3$
(d) Deionised water

04 The K_{sp} for bismuth sulphide (Bi_2S_3) is 1.08×10^{-73}. The solubility of Bi_2S_3 in mol L^{-1} at 298 K is **[2022, 25 June Shift-II]**

(a) 1.0×10^{-15} (b) 2.7×10^{-12}

(c) 3.2×10^{-10} (d) 4.2×10^{-8}

05 The solubility of $Ca(OH)_2$ in water is [Given: The solubility product of $Ca(OH)_2$ in water = 5.5×10^{-6}] **[2021, 25 Feb Shift-II]**

(a) 1.11×10^{-2} (b) 1.11×10^{-6}

(c) 1.77×10^{-2} (d) 1.77×10^{-6}

• Level 2

Imp. *Order of Solubility Product*

06 Arrange the following in increasing order of solubility product $Ca(OH)_2$, AgBr, PbS, HgS **[2025, 28 Jan Shift-II]**

(a) $Ca(OH)_2 <$ AgBr $<$ HgS $<$ PbS
(b) HgS $<$ PbS $<$ AgBr $< Ca(OH)_2$
(c) PbS $<$ HgS $< Ca(OH)_2 <$ AgBr
(d) HgS $<$ AgBr $<$ PbS $< Ca(OH)_2$

07 Which of the following happens when NH_4OH is added gradually to the solution containing [1 M] A^{2+} and [1 M] B^{3+} ions? **[2025, 23 Jan Shift-I]**

(Given, $K_{sp}[A(OH)_2] = 9 \times 10^{-10}$ and $K_{sp}[B(OH)_3] = 27 \times 10^{-18}$ at 298 K.)

(a) $A(OH)_2$ will precipitate before $B(OH)_3$
(b) $B(OH)_3$ will precipitate before $A(OH)_2$
(c) $A(OH)_2$ and $B(OH)_3$ will precipitate together
(d) Both $A(OH)_2$ and $B(OH)_3$ do not show precipitation with NH_4OH

08 K_{sp} for $Cr(OH)_3$ is 1.6×10^{-30}. What is the molar solubility of this salt in water? **[2025, 24 Jan Shift-I]**

(a) $\sqrt[4]{\frac{1.6 \times 10^{-30}}{27}}$ (b) $\frac{1.8 \times 10^{-30}}{27}$

(c) $\sqrt[5]{1.8 \times 10^{-30}}$ (d) $\sqrt[2]{1.6 \times 10^{-30}}$

09 Given below are two statements One is labelled as Assertion (A) and the other labelled as Reason (R).

Assertion (A) During the boiling of water having temporary hardness, $Mg(HCO_3)_2$ is converted to $MgCO_3$.

Reason (R) The solubility product of $Mg(OH)_2$ is greater than that of $MgCO_3$.

In the light of the above statements, choose the most appropriate answer from the options given below

(a) Both (A) and (R) are true but (R) is not the correct explanation of (A).
(b) (A) is true but (R) is false.
(c) Both A and R are true and R is the correct explanation of (A).
(d) (A) is false but (R) is true. **[2021, 18 March Shift-I]**

10 For the following Assertion and Reason, the correct option is

Assertion (A) When Cu (II) and sulphide ions are mixed, they react together extremely quickly to give a solid.

Reason (R) The equilibrium constant of $Cu^{2+}(aq) + S^{2-}(aq) \rightleftharpoons CuS(s)$ is high because the solubility product is low. **[2020, 2 Sep Shift-I]**

(a) Both (A) and (R) are false.
(b) Both (A) and (R) are true and (R) is the explanation for (A).
(c) Both (A) and (R) are true but (R) is not the explanation for (A).
(d) (A) is false and (R) is true.

11 The stoichiometry and solubility product of a salt with the solubility curve given below is, respectively : **[2020, 8 Jan Shift-I]**

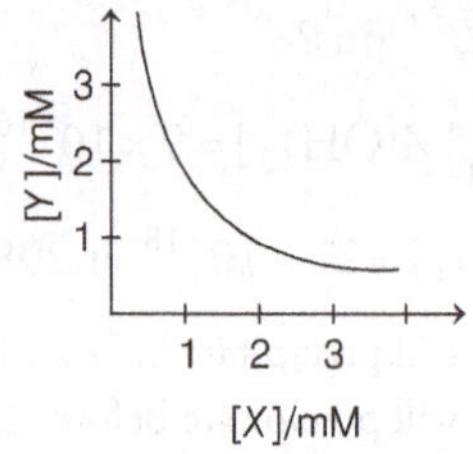

(a) $X_2Y, 2\times10^{-9}\ M^3$ (b) $XY, 2\times10^{-6}\ M^3$
(c) $XY_2, 1\times10^{-9}\ M^3$ (d) $XY_2, 4\times10^{-9}\ M^3$

12 What is the molar solubility of $Al(OH)_3$ in 0.2 M NaOH solution? Given that, solubility product of $Al(OH)_3 = 2.4\times10^{-24}$ **[2019, 12 April Shift-I]**

(a) 3×10^{-19} (b) 12×10^{-21}
(c) 3×10^{-22} (d) 12×10^{-23}

13 If solubility product of $Zr_3(PO_4)_4$ is denoted by K_{sp} and its molar solubility is denoted by S, then which of the following relation between S and K_{sp} is correct? **[2019, 8 April Shift-I]**

(a) $S=\left(\frac{K_{sp}}{144}\right)^{1/6}$ (b) $S=\left(\frac{K_{sp}}{6912}\right)^{1/7}$
(c) $S=\left(\frac{K_{sp}}{929}\right)^{1/9}$ (d) $S=\left(\frac{K_{sp}}{216}\right)^{1/7}$

• Level 3

14 A solution is 0.1 M in Cl^- and 0.001 M in CrO_4^{2-}. Solid $AgNO_3$ is gradually added to it.
Assuming that the addition does not change in volume and $K_{sp}(AgCl) = 1.7\times10^{-10}\ M^2$ and $K_{sp}(Ag_2CrO_4) = 1.9\times10^{-12} M^3$
Select correct statement from the following **[2021, 20 July Shift-II]**

(a) AgCl precipitates first because its K_{sp} is high.
(b) Ag_2CrO_4 precipitates first as its K_{sp} is low.
(c) Ag_2CrO_4 precipitates first because the amount of Ag^+ needed is low.
(d) AgCl will precipitate first as the amount of Ag^+ needed to precipitate is low.

15 If K_{sp} of Ag_2CO_3 is 8×10^{-12}, the molar solubility of Ag_2CO_3 in 0.1 M $AgNO_3$ is **[2019, 12 Jan Shift-II]**

(a) 8×10^{-12} M (b) 8×10^{-13} M
(c) 8×10^{-10} M (d) 8×10^{-11} M

Numerical Value Type Questions

16 25.0 mL of 0.050 M $Ba(NO_3)_2$ is mixed with 25.0 mL of 0.020 M NaF. K_{sp} of BaF_2 is 0.5×10^{-6} at 298 K. The ratio of $[Ba^{2+}][F^-]^2$ and K_{sp} is .... (Nearest integer) **[2023, 13 April Shift-I]**

17 The solubility product of $BaSO_4$ is 1×10^{-10} at 298K. The solubility of $BaSO_4$ in 0.1 M $K_2SO_4(aq)$ solution is ...... $\times10^{-9}$ gL^{-1} (nearest integer). Given, Molar mass of $BaSO_4$ is 233 g mol^{-1} **[2023, 8 April Shift-II]**

18 At 298 K, the solubility of silver chloride in water is 1.434×10^{-3} g L^{-1}. The value of $-\log K_{sp}$ for silver chloride is ....
(Given, mass of Ag is 107.9 g mol^{-1} and mass of Cl is 35.5 g mol^{-1}) **[2023, 31 Jan Shift-II]**

19 If the solubility product of PbS is 8×10^{-28}, then the solubility of PbS in pure water at 298 K is $x\times10^{-16}$ mol L^{-1}. The value of x is ..... (Nearest integer) [Given, $\sqrt{2}=1.41$] **[2022, 29 July Shift-I]**

20 At 310 K, the solubility of CaF_2 in water is 2.34×10^{-3} g/100 mL. The solubility product of CaF_2 is ..... $\times10^{-8}$ $(mol/L)^3$.
(Given, molar mass of CaF_2 = 78 g mol^{-1}) **[2022, 27 July Shift-I]**

21 The molar solubility of $Zn(OH)_2$ in 0.1 M NaOH solution is $x\times10^{-18}$ M. The value of x is .... (Nearest integer) (Given; The solubility product of $Zn(OH)_2$ is 2×10^{-20}). **[2021, 1 Sep Shift-II]**

22 A_3B_2 is a sparingly soluble salt of molar mass M (g mol^{-1}) and solubility x g L^{-1}. The solubility product satisfies $K_{sp}=a\left(\frac{x}{M}\right)^5$. The value of a is .... (Integer answer) **[2021, 31 Aug Shift-I]**

23 The solubility of $CdSO_4$ in water is 8.0×10^{-4} mol L^{-1}. Its solubility in 0.01 M H_2SO_4 solution is .... $\times10^{-6}$ mol L^{-1}. (Round off to the nearest integer)
(Assume that, solubility is much less than 0.01 M) **[2021, 18 March Shift-II]**

24 Two salts A_2X and MX have the same value of solubility product of 4.0×10^{-12}. The ratio of their molar solubilities i.e $\frac{S(A_2X)}{S(MX)} = ...$

(Round off to the nearest integer) **[2021, 16 March Shift-I]**

Topic 3 / pH, Buffer & Indicators Salt Hydrolysis

• Level 1

01 pH of water is 7 at 25°C. If water is heated to 80°C, it's pH will **[2025, 23 Jan Shift-II]**

(a) increase
(b) decrease
(c) H^+ concentration increases, OH^- concentration decreases
(d) remains the same

02 The pH of a 0.02 M NH_4Cl solution will be [given $K_b(NH_4OH) = 10^{-5}$ and $\log 2 = 0.301$] **[2019, 10 April Shift-II]**

(a) 2.65 (b) 4.35
(c) 4.65 (d) 5.35

• Level 2

03 An aqueous solution of HCl with pH 1.0 is diluted by adding equal volume of water (ignoring dissociation of water). The pH of HCl solution would (Given $\log 2 = 0.30$) **[2025, 7 April Shift-I]**

(a) remain same (b) reduce to 0.5
(c) increase to 1.3 (d) increase to 2

04 If equal volumes of AB_2 and XY (both are salts) aqueous solutions are mixed, which of the following combination will give a precipitate of AY_2 at 300 K? (Given K_{sp}, (at 300 K) for $AY_2 = 5.2 \times 10^{-7}$) **[2025, 2 April Shift-I]**

(a) 2.0×10^{-2} M AB_2, 2.0×10^{-2} M XY
(b) 2.0×10^{-4} M AB_2, 0.8×10^{-3} M XY
(c) 1.5×10^{-4} M AB_2, 1.5×10^{-3} M XY
(d) 3.6×10^{-3} M AB_2, 5.0×10^{-4} M XY

05 Which of the following mixing of 1 M base and 1 M acid leads to the largest increase in temperature? **[2025, 24 Jan Shift-II]**

(a) 30 mL HCl and 30 mL NaOH
(b) 30 mL CH_3COOH and 30 mL NaOH
(c) 50 mL HCl and 20 mL NaOH
(d) 45 mL CH_3COOH and 25 mL NaOH

06 Given below are two statements: **[2024, 8 April Shift-II]**

Statement I A Buffer solution is the mixture of a salt and an acid or a base mixed in any particular quantities.

Statement II Blood is naturally occurring buffer solution whose pH is maintained by $H_2CO_3/HCO_3^{\ominus}$ concentrations.

In the light of the above statements, choose the correct from the options given below:
(a) Statement I is false but Statement II is true.
(b) Both Statement I and Statement II are true.
(c) Both Statement I and Statement II are false.
(d) Statement I is true but Statement II is false.

07 Given below are two statements: **[2024, 27 Jan Shift-I]**

Statement I Aqueous solution of ammonium carbonate is basic.

Statement II Acidic/basic nature of salt solution of a salt of weak acid and weak base depends on K_a and K_b value of acid and the base forming it.

In the light of the above statements, choose the **most appropriate** answer from the options given below
(a) Statement I is correct but Statement II is incorrect.
(b) Statement I is incorrect but Statement II is correct.
(c) Both Statement I and Statement II are correct.
(d) Both Statement I and Statement II are incorrect.

08 200 mL of 0.01 M HCl is mixed with 400 mL of 0.01 M H_2SO_4. The pH of the mixture is .... **[2022, 29 July Shift-II]**

(a) 1.14 (b) 1.78
(c) 2.34 (d) 3.02

09 A compound 'X' is a weak acid and it exhibits colour change at pH close to the equivalence point during neutralisation of NaOH with CH_3COOH. Compound 'X' exists in ionised form in basic medium. The compound 'X' is **[2022, 29 July Shift-II]**

(a) methyl orange
(b) methyl red
(c) phenolphthalein
(d) erichrome Black T

10 The solubility of AgCN in a buffer solution of pH = 3 is x. The value of x is

[Assume : No cyano complex is formed; $K_{sp}(AgCN) = 2.2 \times 10^{-16}$ and $K_a(HCN) = 6.2 \times 10^{-10}$] **[2021, 25 Feb Shift-I]**

(a) 1.9×10^{-5} (b) 1.6×10^{-6}
(c) 2.2×10^{-16} (d) 0.625×10^{-6}

Imp. *pH Graph of Acid-Base Mixture*

11 100 mL of 0.1M HCl is taken in a beaker and to it 100 mL of 0.1 M NaOH is added in steps of 2 mL and the pH continuously measured. Which of the following graphs correctly depicts the change in pH? **[2020, 3 Sep Shift-II]**

(a) 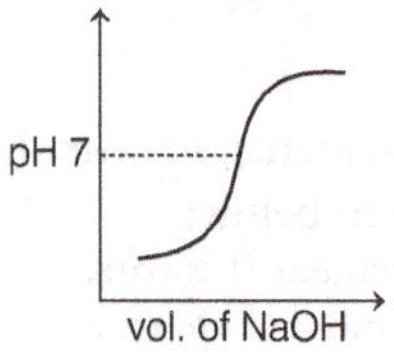

(b) 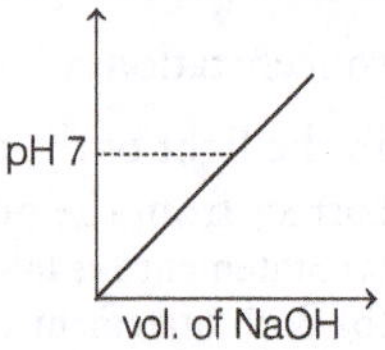

(c) 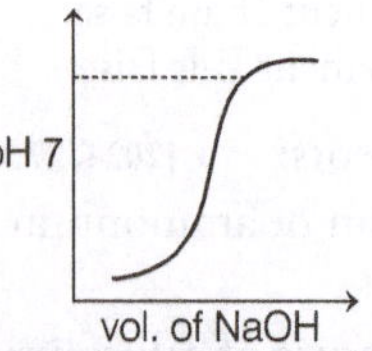

(d) 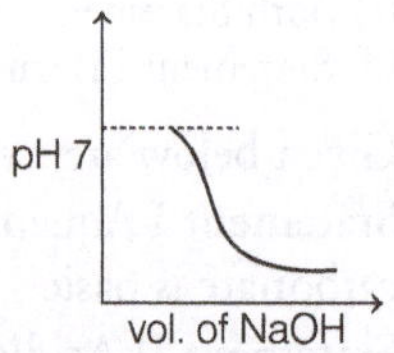

12 Arrange the following solutions in the decreasing order of pOH.
(A) 0.01 M HCl
(B) 0.01 M NaOH
(C) 0.01 M CH_3COONa
(D) 0.01 M NaCl **[2020, 6 Sep Shift-I]**

(a) (A) > (C) > (D) > (B) (b) (A) > (D) > (C) > (B)
(c) (B) > (C) > (D) > (A) (d) (B) > (D) > (C) > (A)

Imp. *pH & Dissociation of Water*

13 For the following Assertion and Reason, the correct option is
Assertion (A) The pH of water increases with increase in temperature.
Reason (R) The dissociation of water into H^+ and OH^- is an exothermic reaction. **[2020, 8 Jan Shift-II]**

(a) Assertion is not true, but Reason is true.
(b) Both assertion and Reason are true and the Reason is the correct explanation for the Assertion.
(c) Both Assertion and Reason are false.
(d) Both Assertion and Reason are true, but the Reason is not the correct explanation for the Assertion.

14 The molar solubility of $Cd(OH)_2$ is 1.84×10^{-5} M in water. The expected solubility of $Cd(OH)_2$ in a buffer solution of pH = 12 is **[2019, 12 April Shift-II]**

(a) 1.84×10^{-9} M (b) $\frac{2.49}{1.84} \times 10^{-9}$ M
(c) 6.23×10^{-11} M (d) 2.49×10^{-10} M

15 The pH of a 0.02 M NH_4Cl solution will be [Given $K_b(NH_4OH) = 10^{-5}$ and $\log 2 = 0.301$] **[2019, 10 April Shift-II]**

(a) 4.65 (b) 2.65
(c) 5.35 (d) 4.35

• Level 3

16 A weak acid H*A* has degree of dissociation *x*. Which option gives the correct expression of (pH- pK_a)? **[2025, 28 Jan Shift-I]**

(a) 0 (b) $\log(1 + 2x)$
(c) $\log\left(\frac{1-x}{x}\right)$ (d) $\log\left(\frac{x}{1-x}\right)$

17 Which of the following statement(s) is/are correct?
A. The pH of 1×10^{-8} M HCl solution is 8.
B. The conjugate base of $H_2PO_4^-$ is HPO_4^{2-}
C. K_w increases with increase in temperature.
D. When a solution of a weak monoprotic acid is titrated against a strong base at half neutralisation point, pH $= \frac{1}{2}$ pK_a. **[2023, 15 April Shift-I]**

Choose the correct answer from the options given below.
(a) B, C, D (b) B, C
(c) A, B, C (d) A, D

18 Class XII students were asked to prepare one litre of buffer solution of pH 8.26 by their chemistry teacher. The amount of ammonium chloride to be dissolved by the student in 0.2 M ammonia solution to make one litre of the buffer is (Given, pK_b (NH_3) = 4.74, Molar mass of NH_3 = 17 gmol^{-1}, Molar mass of NH_4Cl = 53.5 g mol^{-1}) **[2022, 26 July Shift-II]**

(a) 53.5 g (b) 72.3 g
(c) 107.0 g (d) 126.0 g

19 The strength of an aqueous NaOH solution is most accurately determined by titrating (Note consider that an appropriate indicator is used) **[2020, 8 Jan Shift-I]**

(a) *Aq.* NaOH in a volumetric flask and concentrated H_2SO_4 in a conical flask
(b) *Aq.* NaOH in a pipette and aqueous oxalic acid in a burette
(c) *Aq.* NaOH in a burette and aqueous oxalic acid in a conical flask
(d) *Aq.* NaOH in a burette and concentrated H_2SO_4 in a conical flask

20 Consider the following statements.

I. The pH of a mixture containing 400 mL of 0.1 M H_2SO_4 and 400 mL of 0.1 M NaOH will be approximately 1.3.

II. Ionic product of water is temperature dependent.

III. A monobasic acid with $K_a = 10^{-5}$ has a pH = 5. The degree of dissociation of this acid is 50%.

IV. The Le-Chatelier's principle is not applicable to common-ion effect.

The correct statements are **[2019, 10 April Shift-I]**

(a) I, II and IV
(b) II and III
(c) I and II
(d) I, II and III

21 20 mL of 0.1 M H_2SO_4 solution is added to 30 mL of 0.2 M NH_4OH solution. The pH of the resultant mixture is [pK_b of NH_4OH = 4.7] **[2019, 9 Jan Shift-I]**

(a) 9.3 (b) 5.0
(c) 9.0 (d) 5.2

Numerical Value Type Questions

22 x mg of $Mg(OH)_2$ (molar mass = 58) is required to be dissolved in 1.0 L of water to produce a pH of 10.0 at 298 K. The value of x is .... mg. (Nearest integer) [Given $Mg(OH)_2$ is assumed to dissociate completely in H_2O] **[2025, 4 April Shift-II]**

23 One litre buffer solution was prepared by adding 0.10 mol each of NH_3 and NH_4Cl in deionised water. The change in pH on addition of 0.05 mol of HCl to the above solution is .... $\times 10^{-2}$. (Nearest integer)

Given: pK_b of NH_3 = 4.745 and $\log_{10} 3 = 0.477$ **[2025, 7 April Shift-II]**

24 K_a for CH_3COOH is 1.8×10^{-5} and K_b for NH_4OH is 1.8×10^{-5}. The pH of ammonium acetate solution will be .... **[2024, 1st Feb Shift-I]**

25 The pH of an aqueous solution containing 1 M benzoic acid (pK_a = 4.20) and 1 M sodium benzoate is 4.5. The volume of benzoic acid solution in 300 mL of this buffer solution is .... mL. (Given, log 2 = 0.3) **[2024, 30 Jan Shift-II]**

26 The pH at which $Mg(OH)_2$ [$K_{sp} = 1 \times 10^{-11}$] begins to precipitate from a solution containing 0.10 M Mg^{2+} ions is **[2024, 30 Jan Shift-I]**

27 20 mL of 0.1 M NaOH is added to 50 mL of 0.1 M acetic acid solution. The pH of the resulting solution is .... $\times 10^{-2}$. (Nearest integer)

Given, pK_a (CH_3COOH) = 4.76
log 2 = 0.30
log 3 = 0.48 **[2023, 13 April Shift-II]**

28 The density of a monobasic strong acid (molar mass 24.2 g/mol) is 1.21 kg/L. The volume of its solution required for the complete neutralisation of 25 mL of 0.24 M NaOH is .... $\times 10^{-2}$ mL.

(Nearest integer) **[2023, 25 Jan Shift-I]**

29 If the pK_a of lactic acid is 5, then the pH of 0.005 M calcium lactate solution at 25°C is .... $\times 10^{-1}$. (Nearest integer)

Lactic acid 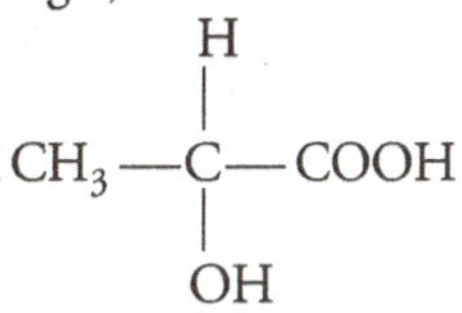

[2023, 24 Jan Shift-II]

30 K_a for butyric acid (C_3H_7COOH) is 2×10^{-5}. The pH of 0.2 M solution of butyric acid is .... $\times 10^{-1}$. (Nearest integer) (Given, log 2 = 0.30) **[2022, 28 July Shift-I]**

31 pH value of 0.001 M NaOH solution is .... **[2022, 27 June Shift-II]**

32 50 mL of 0.1 M CH_3COOH is being titrated against 0.1 M NaOH. When 25 mL of NaOH has been added, the pH of the solution will be ...... $\times 10^{-2}$. (Nearest integer)
[Given, pK_a (CH_3COOH) = 4.76]
log 2 = 0.30, log 3 = 0.48, log 5 = 0.69,
log 7 = 0.84, log 11 = 1.04 **[2022, 26 June Shift-I]**

33 The pH of a solution obtained by mixing 50 mL of 1 M HCl and 30 mL of 1 M NaOH is $x \times 10^{-4}$. The value of x is .... (Nearest integer)
[log 2.5 = 0.3979] **[2021, 31 Aug Shift-II]**

34 The OH^- concentration in a mixture of 5.0 mL of 0.0504 M NH_4Cl and 2 mL of 0.0210 M NH_3 solution is $x \times 10^{-6}$ M. The value of x is .... (Nearest integer)
[Given $K_w = 1 \times 10^{-14}$ and $K_b = 1.8 \times 10^{-5}$] **[2021, 26 Aug Shift-I]**

35 In order to prepare a buffer solution of pH 5.74, sodium acetate is added to acetic acid. If the concentration of acetic acid in the buffer is 1.0 M, the concentration of sodium acetate in the buffer is .... M. (Round off to the nearest integer). [Given, pK_a (acetic acid $= 4.74$] **[2021, 18 March Shift-I]**

36 Sulphurous acid (H_2SO_3) has $K_{a_1} = 1.7 \times 10^{-2}$ and $K_{a_2} = 6.4 \times 10^{-8}$. The pH of 0.588 M is ...... (Round off to the nearest integer) **[2021, 16 March Shift-II]**

37 The pH of ammonium phosphate solution, if pK_a of phosphoric acid and pK_b of ammonium hydroxide are 5.23 and 4.75 respectively, is .... **[2021, 26 Feb Shift-II]**

38 3 g of acetic acid is added to 250 mL of 0.1 M HCl and the solution made up to 500 mL. To 20 mL of this solution $\frac{1}{2}$ mL of 5 M NaOH is added. The pH of the solution is ....
[Given, pK_a of acetic acid $= 4.75$, molar mass of acetic acid $= 60$ g/mol, log 3 $= 0.4771$]
Neglect any changes in volume. **[2020, 7 Jan Shift-II]**

39 Two solutions, A and B, each of 100 L was made by dissolving 4 g of NaOH and 9.8 g of H_2SO_4 in water, respectively. The pH of the resultant solution obtained from mixing 40 L of solution A and 10 L of solution B is .... **[2020, 7 Jan Shift-I]**

Key Index

Topic 1

1	(d)	**2**	(d)	**3**	(7)	**4**	(5)	**5**	(25)	**6**	(1)	**7**	(2)

Topic 2

1	(a)	**2**	(b)	**3**	(d)	**4**	(a)	**5**	(a)	**6**	(b)	**7**	(b)	**8**	(a)	**9**	(d)	**10**	(c)
11	(d)	**12**	(c)	**13**	(b)	**14**	(d)	**15**	(c)	**16**	(5)	**17**	(233)	**18**	(10)	**19**	(282)	**20**	(0)
21	(2)	**22**	(108)	**23**	(64)	**24**	(50)												

Topic 3

1	(b)	**2**	(d)	**3**	(c)	**4**	(a)	**5**	(a)	**6**	(a)	**7**	(c)	**8**	(b)	**9**	(c)	**10**	(a)
11	(a)	**12**	(b)	**13**	(c)	**14**	(d)	**15**	(c)	**16**	(d)	**17**	(b)	**18**	(c)	**19**	(c)	**20**	(d)
21	(c)	**22**	(3)	**23**	(48)	**24**	(7)	**25**	(100)	**26**	(9)	**27**	(458)	**28**	(12)	**29**	(85)	**30**	(27)
31	(11)	**32**	(476)	**33**	(6021)	**34**	(3)	**35**	(10)	**36**	(1)	**37**	(7)	**38**	(5.23)	**39**	(10.60)

Detailed Explanations

Topic 1

01. *(d)* Assertion is false but Reason is true. The amphoteric nature of water can be explained by Bronsted-Lowry concept.

$$\underset{\text{Acid}}{H_2S}+\underset{\text{Base}}{H_2O} \rightleftharpoons H_3O^+ + HS^-$$

$$\underset{\text{Acid}}{H_2O}+\underset{\text{Base}}{NH_3} \rightleftharpoons NH_4^+ + \bar{O}H$$

02. *(d)* $H_2C_2O_4 \overset{K_{a_3}}{\rightleftharpoons} 2H^+ + C_2O_4^{2-}$

$$H_2C_2O_4 \overset{K_{a_1}}{\rightleftharpoons} H^+ + HC_2O_4^-$$

$$HC_2O_4^- \overset{K_{a_2}}{\rightleftharpoons} H^+ + C_2O_4^{2-}$$

$$K_{a_3} = \frac{[H^+]^2[C_2O_4^{2-}]}{[H_2C_2O_4]} \quad \ldots(i)$$

$$K_{a_1} = \frac{[H^+][HC_2O_4^-]}{[H_2C_2O_4]} \quad \ldots(ii)$$

$$K_{a_2} = \frac{[H^+][C_2O_4^{2-}]}{[HC_2O_4^-]} \quad \ldots(iii)$$

In order to obtain Eq. (i) we have to add Eq. (ii) and Eq. (iii)

$$\therefore K_{a_1} \times K_{a_2} = \frac{[H^+][HC_2O_4^-]}{[H_2C_2O_4]} \times \frac{[H^+][C_2O_4^{2-}]}{[HC_2O_4^-]}$$

$$= \frac{[H^+]^2[C_2O_4^{2-}]}{[H_2C_2O_4]} = K_{a_3}$$

$K_{a_1} \times K_{a_2} = K_{a_3}$

So, option (d) is correct.

03. *(7)* Given, pH = 9

$$C_{\text{initial}} = 1\text{ mM},\ K_b = 10^{-x},\ x = ?$$

Since pH $= 9 \Rightarrow$ pOH $= 5$

So, $[\bar{O}H] = 10^{-5}$ M

Chemical reaction involved in ionisation of ethyl amine is as follows

$$C_2H_5NH_2(aq) + H_2O \rightleftharpoons C_2H_2NH_3^+ + \bar{O}H$$

$C = 10^{-3}$ M	—	—
$C(1-\alpha)$	$C\alpha$	$C\alpha$
$\Rightarrow C = 10^{-3}$	$= 10^{-5}$	
$= 10^{-5}$		

According to question, $1-\alpha \approx 1$

Now we know that,

$$K_b = \frac{[C_2H_5NH_3^+][\bar{O}H]}{[C_2H_5NH_2]}$$

$$K_b = \frac{10^{-5} \times 10^{-5}}{10^{-3}} = 10^{-7}$$

So, the value of x is 7.

04. *(5)* Decomposition reaction in equilibrium is,

$$H_2O(g) \rightleftharpoons H_2(g) + \frac{1}{2}O_2(g)$$

	H_2O	H_2	O_2
$t = 0$	1 mole		
$t = t_{eq}$	$1-\alpha$	α	$\alpha/2$

$$n_T = 1 + \frac{\alpha}{2} = 1 \quad (\alpha <<< 1)$$

$$K_p = \frac{p_{H_2} \cdot p_{O_2}^{1/2}}{p_{H_2O}} = \frac{(\alpha \cdot p)\left(\frac{\alpha}{2}\cdot p\right)^{1/2}}{(1-\alpha)p}$$

as $p = 1$

$$8 \times 10^{-3} = \frac{\alpha^{3/2}}{\sqrt{2}}$$

$$\alpha^3 = 128 \times 10^{-6}$$

$$\alpha = 5.03 \times 10^{-2}$$

$$\approx 5 \times 10^{-2}$$

05. *(25)* pH of the new solution = 6

$$\Rightarrow [H^+] = 10^{-6} M$$

$$HX \rightleftharpoons H^+ + X^-$$

At equil. $C(1-\alpha) \quad C\alpha \quad C\alpha$

$$K_a = \frac{C\alpha^2}{1-\alpha}$$

or $\quad 8K_a = C\alpha^2$

(for extremely dilute solution)

$$[H^+] = C\alpha = \sqrt{K_a C}$$

$$10^{-6} = \sqrt{4 \times 10^{-10} \times C}$$

$$\Rightarrow \quad C = \frac{10^{-12}}{4 \times 10^{-10}} = 0.25 \times 10^{-2}$$

$$= 25 \times 10^{-4}$$

06. *(1)* $Ba(OH)_2$ is completely ionised in aqueous solution.

$$Ba(OH)_2 \longrightarrow Ba^{2+} + 2OH^-$$

$$[Ba(OH)_2] = 0.005 M$$

$$[OH^-] = 2 \times 0.005 \Rightarrow 0.01 \Rightarrow = 10^{-2}$$

At 298 K, in aqueous solution ;

$$[H_3O^+][OH^-] = 10^{-14}$$

$$[H_3O^+] = \frac{10^{-14}}{10^{-2}} = 10^{-12}$$

$$[H_3O^+] = 1 \times 10^{-12} \text{ M}$$

07. *(2)* Given, $[HA] = 0.01$

$$[HCl] = 0.1 \text{ M}$$

When strong acid (HCl) is completely dissociated,

$$\underset{0.1}{HCl} \longrightarrow \underset{0.1}{H^+(aq)} + \underset{0.1}{Cl^-(aq)}$$

For weak acid, dissociation is very less,

	HA	$\rightleftharpoons H^+(aq)$	$+ A^-(aq)$
$t = 0$	0.01	0.1	0
$t = t_{eq}$	$0.01 - 0.01\alpha$	$0.1 + 0.01\alpha$	0.01α

$$K_a = \frac{[H^+][A^-]}{[HA]} = 2 \times 10^{-6}$$

$$\Rightarrow \frac{(0.1 + 0.01\alpha)(0.01\alpha)}{(0.01 - 0.01\alpha)} = 2 \times 10^{-6}$$

As $0.01\alpha << 0.1$, $[H^+] = 0.1$

and $0.01\alpha << 0.01$, $[HA] = 0.01$

$$\therefore \quad 2 \times 10^{-6} = \frac{(0.1)(0.01\alpha)}{0.01}$$

$$\Rightarrow \quad \alpha = 2 \times 10^{-5} \Rightarrow x = 2$$

Topic 2

01. *(a)* The reaction involving dissociation of zirconium phosphate is as follows

$$Zr_3(PO_4)_4(s) \rightleftharpoons \underset{3S}{3Zr^{4+}(aq)} + \underset{4S}{4PO_4^{3-}(aq)}$$

So, K_{sp} will be given by

$$K_{sp} = (3S)^3\,(4S)^4 = 6912\,S^7$$

$$S = \left(\frac{K_{sp}}{6912}\right)^{1/7}$$

02. *(b)* For $AB_2 \rightleftharpoons A^{2+} + 2B^-$

$$K_{sp} = [A^{2+}][B^-]^2$$

$$= (1.2 \times 10^{-4})(0.24 \times 10^{-3})^2$$

$$= 0.0691 \times 10^{-10}$$

$$K_{sp} = 6.91 \times 10^{-12}$$

03. *(d)* The solubility of AgCl will be maximum in deionised water as rest of all the solutions (0.01 M KCl, 0.01 M HCl and 0.01 M $AgNO_3$) have a common ion with the solute being added (AgCl). Thus, AgCl will be most soluble in deionised water.

04. *(a)* Given, K_{sp} for

$Bi_2S_3 = 1.08 \times 10^{-73}$

$$Bi_2S_3(s) \rightleftharpoons \underset{2S}{2Bi^{3+}(aq)} + \underset{3S}{3S^{2-}(aq)}$$

$$K_{sp} = [Bi^{3+}]^2 [S^{2-}]^3$$

So, $K_{sp} = [2S]^2 [3S]^3$

$$K_{sp} = 4S^2 \times 27S^3 = 108S^5$$

$$\Rightarrow \quad S = \left(\frac{K_{sp}}{108}\right)^{1/5}$$

$$\therefore \quad S = \left(\frac{1.08 \times 10^{-73}}{108}\right)^{1/5}$$

$$= (10^{-75})^{1/5}$$

Hence, solubility,

$$S = 1.0 \times 10^{-15} \text{ mol L}^{-1}$$

05. *(a)* Let, solubility of $Ca(OH)_2$ in pure water = S mol/L

$$Ca(OH)_2 \rightleftharpoons \underset{S \text{ mol/L}}{Ca^{2+}} + \underset{2 \times S \text{ (mol/L)}}{2OH^-}$$

$$K_{sp} = [Ca^{2+}][OH^-]^2$$

$$= S \times (2S)^2$$

$$= 4S^3 \text{ (mol/L)}$$

$$S = \left(\frac{K_{sp}}{4}\right)^{1/3} = \left(\frac{5.5 \times 10^{-6}}{4}\right)^{1/3}$$

$$= 1.11 \times 10^{-2} \text{ mol/L}$$

06. *(b)* Based on K_{sp} values and salt analysis cation identification, we can say that order of K_{sp} value is

$HgS < PbS < AgBr < Ca(OH)_2$

K_{sp} values for given salts are

$HgS \rightarrow 4 \times 10^{-53}$,

$PbS \rightarrow 8 \times 10^{-28}$

$AgBr \rightarrow 5 \times 10^{-13}$,

$Ca(OH)_2 \rightarrow 5.5 \times 10^{-6}$

07. *(b)* Given, $K_{sp}[A(OH)_2] = 9 \times 10^{-10}$

$K_{sp}[B(OH)_3] = 27 \times 10^{-18}$

$[A^{2+}] = 1\text{ M}, [B^{3+}] = 1\text{ M}$

For precipitation $Q_{ip} > K_{sp}$

So, $[A^{2+}][OH^-]^2 > 9 \times 10^{-10}$

$\Rightarrow [OH^-] > 3 \times 10^{-5}$ M

For $[B(OH)_3]$

$[B^{3+}][OH^-]^3 > 27 \times 10^{-18}$

$[OH^-] > 3 \times 10^{-6}$ M

So, $B(OH)_3$ will precipitate before $A(OH)_2$.

08. *(a)* The complete reaction is

$$Cr(OH)_3(s) \rightleftharpoons Cr^{+3}(aq) + 3OH^-(aq)$$

At equilibrium $\quad S$ M $\quad 3S$ M

$$K_{sp} = (S)(3S)^3 = 27S^4$$

$$27S^4 = 1.6 \times 10^{-30}$$

$$S = \left(\frac{1.6}{27} \times 10^{-30}\right)^{1/4} \text{ M}$$

09. *(d)* During boiling, soluble $Mg(HCO_3)_2$ is converted into insoluble $Mg(OH)_2$ and $Ca(HCO_3)_2$ is converted into insoluble $CaCO_3$. This is because of high solubility product of $Mg(OH)_2$ as compared to $MgCO_3$ hence, $Mg(OH)_2$ is precipitated. These precipitate can be removed by filtration. Thus, filtrate will be obtained in soft water.

$$Mg(HCO_3)_2 \longrightarrow Mg(OH)_2 \downarrow + 2CO_2$$

K_{sp} of $Mg(OH)_2 > K_{sp}$ of $MgCO_3$ and hence, $Mg(OH)_2$ precipitated first.

10. *(c)* Both (A) and (R) are true, but (R) is not the correct explanation of (A). Because of high stability of CuS, the value of K_{eq} is high, i.e. the value of K_{sp} of CuS will be low.

11. *(d)* Units given for all options is M^3, or $(\text{mol L}^{-1})^3$, which implies that, the equilibrium has three ions. Therefore,

$$XY_2 \rightleftharpoons X^{2+}(aq) + 2Y^-(aq),$$

$$K_{sp} = [X^{2+}][Y^-]^2$$

or $\quad X_2Y \rightleftharpoons 2X^+(aq) + Y^{2-}(aq),$

$$K_{sp} = [X^+]^2[Y^{2-}]$$

Concentration values from graph is

$$[X^{2+}] = 1\text{ mM} = 1 \times 10^{-3}\text{ M}$$

and $\quad [Y^-] = 2 \times 10^{-3}$ M

(i) K_{sp} in case of XY_2,

$$K_{sp} = (1 \times 10^{-3})(2 \times 10^{-3})^2$$

$$= 4 \times 10^{-9}\text{ M}^3$$

(ii) K_{sp} in case of X_2Y,

$$K_{sp} = (1 \times 10^{-3})^2 (2 \times 10^{-3})$$

$$= 2 \times 10^{-9}\text{ M}^3$$

The only matching value of K_{sp} from options that satisfies the concentration values from graph is XY_2.

12. *(c)*

> **idea** In this question we will use the concepts of solubility. The concentration of a substance in a saturated solution is defined as its solubility (S).
>
> For $\quad A_xB_y \rightleftharpoons xA^{y+} + yB^{x-}$
>
> $K_{sp} = [A^{y+}]^x[B^{x-}]^y$

$$NaOH \longrightarrow \underset{0.2}{Na^+} + \underset{0.2}{OH^-}$$

	$Al(OH)_3$	$\rightleftharpoons$	Al^{3+}	+	$3OH^-$
Initially	1		0		0
At equilibrium	$1-S$		S		$3S + 0.2$

K_{sp} of $Al(OH)_3 = 2.4 \times 10^{-24}$ (Given)

$$K_{sp} = [Al^{3+}][OH^-]^3$$

$$2.4 \times 10^{-24} = [S][3S + 0.2]^3$$

$$[\because 0.2 >> S]$$

$$2.4 \times 10^{-24} = [S][0.008]$$

$$[S] = 3 \times 10^{-22}$$

13. *(b)* For $A_xB_y \rightleftharpoons xA^{y+} + yB^{x-}$;

$$K_{sp} = [A^{y+}]^x[B^{x-}]^y$$

For, $Zr_3(PO_4)_4$,

$$Zr_3(PO_4)_4(s) \rightleftharpoons \underset{3S}{3Zr^{4+}(aq)} + \underset{4S}{4PO_4^{3-}(aq)}$$

$$K_{sp} = [Zr^{4+}]^3 [PO_4^{3-}]^4$$

$$K_{sp} = (3S)^3(4S)^4 = 6912S^7$$

or $S = \left(\frac{K_{sp}}{6912}\right)^{1/7}$

Thus, the relation between molar solubility(S) and solubility product (K_{sp}) will be $S = \left(\frac{K_{sp}}{6912}\right)^{1/7}$

14. *(d)* $K_{sp}(AgCl) = 1.7 \times 10^{-10}\text{ M}^2$

$K_{sp}(Ag_2CrO_4) = 1.9 \times 10^{-12}\text{ M}^3$

(i) $[Ag^+]$ required to ppt. AgCl(s)

$$K_{sp} = [Ag^+][Cl^-]$$

$$\Rightarrow 1.7 \times 10^{-10} = [Ag^+][0.1]$$

$$[Ag^+] = 1.7 \times 10^{-9}\text{ M}$$

(ii) $[Ag^+]$ required to ppt. $Ag_2CrO_4(s)$

$$K_{sp} = [Ag^+]^2[CrO_4^{2-}]$$

$$\Rightarrow 1.9 \times 10^{-12} = [Ag^+]^2[0.001]$$

$$[Ag^+] = \sqrt{1.9 \times 10^{-9}}$$

$$= \sqrt{19} \times 10^{-5}$$

$$= 4.3 \times 10^{-5}\text{ M}$$

$\because [Ag^+]$ required to precipitate AgCl is low, so AgCl will precipitate first.

15. *(c)* Let the solubility of Ag_2CO_3 is S. Now, 0.1 M of $AgNO_3$ is added to this solution after which let the solubility of Ag_2CO_3 becomes S'.

$\therefore \quad [Ag^+] = S + 0.1$ and $[CO_3^{2-}] = S'$

$K_{sp} = (S+0.1)^2 (S') \quad ...(i)$

Given, $K_{sp} = 8 \times 10^{-12}$

$\because K_{sp}$ is very small, we neglect S' against S in Eq. (i)

$\therefore \quad K_{sp} = (0.1)^2 S'$

or $\quad 8 \times 10^{-12} = 0.01 S'$

or $\quad S' = 8 \times 10^{-12} \times 10^2$

$= 8 \times 10^{-10}$ M

Thus, molar solubility of Ag_2CO_3 in 0.1 M $AgNO_3$ is 8×10^{-10} M.

16. *(5)* If equal volume are mixed, concentration becomes half

$Ba(NO_3)_2 \rightleftharpoons Ba^{2+} + 2NO_3^{\ominus}$

Before mixing, $[Ba^{2+}] = 0.05$ M

After mixing, $Ba^{2+} = \dfrac{0.05}{2}$

$= 0.025$ M $\quad ...(i)$

$BaF_2 \rightleftharpoons Ba^{2+} + 2F^{\ominus}$

Before mixing, $[F^{\ominus}] = 0.02$ M

After mixing, $[F^-] = \dfrac{0.02}{2} = 0.01$ M $...(ii)$

From Eqs. (i) and (ii),

$Q = [Ba^{2+}][F^-]^2$

$= [0.025][0.01]^2 = [2.5 \times 10^{-6}]$

Since $[Q > K_{sp}]$

So, precipitation takes place and remaining solution is in equilibrium.

$\therefore \quad \dfrac{[Ba^{2+}][F^-]^2}{K_{sp}} = \dfrac{2.5 \times 10^{-6}}{0.5 \times 10^{-6}} = 5$

17. *(233)*

$\underset{}{K_2SO_4} \longrightarrow \underset{0.1}{K^+(aq)} + \underset{0.1}{SO_4^{2-}(aq)}$

$BaSO_4(s) \rightleftharpoons \underset{S}{Ba^{2+}(aq)} + \underset{S+0.1}{SO_4^{2-}(aq)}$

$K_{sp} = [Ba^{2+}][SO_4^{2-}]$

$1 \times 10^{-10} = S(S+0.1) \simeq 0.1\,S$

$S = 10^{-9}$ M

$= 233 \times 10^{-9}$ gL^{-1}

18. *(10)*

> **idea** To solve this question, the concept of solubility can be used. For this, write the equation involved in the reaction. Then use the formula of K_{sp} in terms of solubility S. Then solve the required value i.e., $-\log K_{sp}$.

$AgCl(s) \rightleftharpoons \underset{S}{Ag^+(aq)} + \underset{S}{Cl^-(aq)}$

$K_{sp} = S^2 = \left(\dfrac{1.434}{143.4} \times 10^{-3}\right)^2 = 10^{-10}$

$-\log K_{sp} = -(-10) \log 10 = 10$

19. *(282)* Given, $K_{sp}(PbS) = 8 \times 10^{-28}$

$PbS(s) \rightleftharpoons \underset{S}{Pb^{2+}(aq)} + \underset{S}{S^{2-}(aq)}$

$K_{sp} = S^2 \Rightarrow 8 \times 10^{-28} = S^2$

$\Rightarrow \quad S = \sqrt{8 \times 10^{-28}} = 2.82 \times 10^{-14}$

So, solubility, $S = 282 \times 10^{-16}$ mol L^{-1}.

$x = 282$

20. *(0)* $CaF_2 \rightleftharpoons \underset{(S)}{Ca^{2+}} + \underset{(2S)}{2F^-}$

Solubility product constant,

$K_{sp} = S \times (2S)^2 = 4S^3$

Solubility $= \dfrac{2.34 \times 10^{-3} \times 10}{78}$ mol/L

$= 3 \times 10^{-4}$ mol/L.

$K_{sp} = 4 \times (3 \times 10^{-4})^3$

$= 108 \times 10^{-12}$

$= 0.0108 \times 10^{-8}$ $(mol/L)^3$

$\simeq 0 \times 10^{-8}$ $(mol/L)^3$

21. *(2)*

> **idea** In this question, we will use the concept of common ion effect. Solubility of a sparingly soluble salt is decreased by the addition of a common ion.

$Zn(OH)_2 \rightleftharpoons \underset{S}{Zn^{2+}} + \underset{2S}{2OH^-}$

Due to common-ion effect (presence of NaOH) the concentration of OH^- will be $(2S + 0.1) \approx 0.1$ $(\because 0.1 >> 2S)$

$\therefore$ Solubility product,

$K_{sp} = (0.1)^2 \times S$

$2 \times 10^{-20} = 0.01 \times S$

$\Rightarrow \quad S = \dfrac{2 \times 10^{-20}}{0.01} = 2 \times 10^{-18}$

$\therefore \quad x = 2$

Hence, answer is 2.

22. *(108)* $A_3B_2 \rightleftharpoons \underset{3S}{3A^{2+}(aq)} + \underset{2S}{2B^{3-}(aq)}$

$K_{sp} = [A^{2+}]^3 [B^{3-}]^2$

$K_{sp} = (3S)^3 (2S)^2 = 108\,S^5$

Also, $\quad S = \dfrac{x}{M} \Rightarrow K_{sp} = 108\left(\dfrac{x}{M}\right)^5$

Given that, $K_{sp} = a\left(\dfrac{x}{M}\right)^5$

$\therefore \quad a = 108$

23. *(64)* Given, solubility in water

$(S) = 8.0 \times 10^{-4}$ mol L^{-1}

$CdSO_4(aq) \rightleftharpoons \underset{S}{Cd^{2+}(aq)} + \underset{S}{SO_4^{2-}(aq)}$

In pure water,

$K_{sp} = S^2 = (8 \times 10^{-4})^2 = 64 \times 10^{-8}$

In 0.01 M H_2SO_4

$H_2SO_4(aq) \longrightarrow \underset{0.02}{2H^+(aq)} + \underset{0.01}{SO_4^{2-}(aq)}$

$CdSO_4(s) \rightleftharpoons \underset{x}{Cd^{2+}(aq)} + \underset{(x+0.01)}{SO_4^{2-}(aq)}$

$K_{sp} = S(S+0.01) = 64 \times 10^{-8}$

$S + 0.01 \cong 0.01$ M

So, $\quad S(0.01) = 64 \times 10^{-8}$

$\Rightarrow \quad S = 64 \times 10^{-6}$ M

24. *(50)* Let the solubility of A_2X be 'S'.

$A_2X(s) \rightleftharpoons \underset{2S}{2A^+(aq)} + \underset{S}{X^{2-}(aq)}$;

$K_{sp_1} = 4 \times 10^{-12}$

$K_{sp_1} = [A^+]^2[X^{2-}]$

$4 \times 10^{-12} = (2S)^2(S)$

$4 \times 10^{-12} = 4S^3 \Rightarrow S = 10^{-4}$ M

Let the solubility of MX be 'S_1'.

$MX(s) \rightleftharpoons \underset{S_1}{M^+(aq)} + \underset{S_1}{X^-(aq)}$,

$K_{sp_2} = 4 \times 10^{-12}$

$K_{sp_2} = [M^+][X^-]$

$K_{sp_2} = (S_1)^2$

$4 \times 10^{-12} = (S_1)^2 \Rightarrow S_1 = 2 \times 10^{-6}$ M

$\dfrac{[A_2X]}{[MX]} = \dfrac{S}{S_1} = \dfrac{10^{-4}}{2 \times 10^{-6}} = 50$

Topic 3

01. *(b)* When there is an increases in temperature, the k_w of water increases. Thus, the degree of dissociation of water also increases. Hence, its pH will decrease.

02. *(d)* For salt of weak base and strong acid.

$pH = 7 - \dfrac{1}{2} pK_b - \dfrac{1}{2} \log C$

$pH = 7 - \dfrac{5}{2} - \dfrac{1}{2} \log(2 \times 10^{-2}) = 5.35$

03. *(c)* Given,

$HCl(aq)$ $pH = 1; [H^+] = 10^{-1}$

If equal volume of water is added, concentration will become half.

$[H^+]_{Sol} = \dfrac{10^{-1}}{2}$ pH = 13

So, the pH of the HCl solution would increase to 1.3.

04. *(a)* When equal volume are mixed molarity reduced to half.

For precipitation

$$[Q_{sp} > K_{sp}]$$

$$Q_{sp} = [A^{+2}][Y^{-2}]^2 > K_{sp}$$

(i) $Q_{sp} = (1.8\times10^{-3})\left(\frac{5}{2}\times10^{-4}\right)^2 < K_{sp}$

(ii) $Q_{sp} = (10^{-4})(0.4\times10^{-3})^2 < K_{sp}$

(iii) $Q_{sp} = (10^{-2})(10^{-2})^2 > K_{sp}$

(iv) $Q_{sp} = \left(\frac{1.5}{2}\times10^{-4}\right)\left(\frac{1.5}{2}\times10^{-3}\right)^2 < K_{sp}$

So, among the given combinations $2.0\times10^{-2}\,M\,AB_2$, 2.0×10^{-2}M

XY will give a ppt. of AY_2 at 300K.

05. *(a)* Complete neutralisation of a strong acid by a strong base leads to the largest increase in temperature. Hence, option (a) is correct.

06. *(a)* Statement I is false but statement II is true. The correct form of statement I is, Buffer solution is the mixture of weak acid or weak base with their conjugate ions. Hence, option (a) is correct.

07. *(c)* Both the Statement I and Statement II are correct.

08. *(b)* 200 mL of 0.01 M HCl is mixed with 400 mL of 0.01 M H_2SO_4.

$\therefore$ Total volume of solution = 200 mL + 400 mL = 600 mL

The moles of H^+ given by HCl solution = 0.01 M × 200 mL = 2 moles

The moles of H^+ given by H_2SO_4 solution = 2 × (0.01M × 400mL) = 8 moles

$$1\,H_2SO_4 \longrightarrow 2H^+$$

Total moles of H^+ = 2 + 8 = 10 moles

Total concentration of H^+,

$$[H^+] = \frac{\text{Total number of moles of } H^+}{\text{Total volume of solution}}$$

$$= \frac{10}{600} = \frac{1}{60}$$

$\therefore$pH of solution $= -\log[H^+]$

$$= -\log\left[\frac{1}{60}\right] = 1.78$$

09. *(c)* The compound X is phenolphthalein. Phenolphthalein have colour change range between pH 8.3 and 10.5 during neutralisation of weak acid and strong base.

10. *(a)* $AgCN \rightleftharpoons Ag^+ + CN^-$

Solubility of AgCN is x.

$pH = 3 \Rightarrow [H^+] = 10^{-3}$

$$HCN \rightleftharpoons H^+ + CN^-$$

$$K_a = \frac{[H^+][CN^-]}{[HCN]}$$

$$\Rightarrow \frac{[10^{-3}][CN^-]}{[HCN]} = 6.2\times10^{-10}$$

$$\Rightarrow \frac{[CN^-]}{[HCN]} = 6.2\times10^{-7}$$

Each CN–hydrolyses to give one HCN as:

$$CN^- + H_2O \rightleftharpoons HCN + OH^-$$

HCN is a weak acid, thus,

$$[HCN] >> [CN^-]$$

$$x = [Ag^+] = [CN^-] + [HCN]$$

$$\Rightarrow x = [Ag^+] \approx [HCN]$$

$$\Rightarrow [CN^-] = x \times 6.2\times10^{-7}$$

$$K_{sp}(AgCN) = [Ag^+][CN^-]$$

$$\Rightarrow 2.2\times10^{-16} = (x)(x\times6.2\times10^{-7})$$

$$\Rightarrow x^2 = 3.55\times10^{-10}$$

$$\Rightarrow x = 1.88\times10^{-5} \approx 1.9\times10^{-5}$$

11. *(a)* Volume of HCl = 100 mL

Molarity of HCl = 0.1 M

Volume of NaOH = 100 mL

Molarity of NaOH = 0.1 M

Initially when only HCl is present in the beaker, then the pH is less than 7, so the graph looks like.

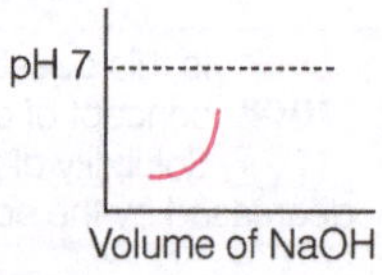

On adding 2 mL of NaOH continuously the pH value gradually increases and reach to equivalence point at pH = 7. On further addition of NaOH the pH value increases further beyond the equivalence point so, the final graph look like.

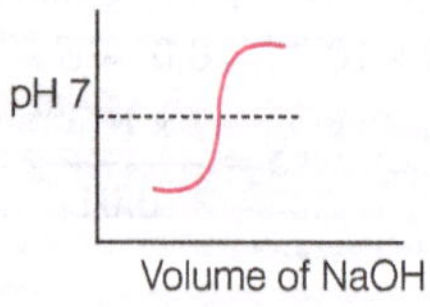

Hence, option (a) is correct.

12. *(b)* Given molar concentration are same. So strongest acid (HCl) will have Max. pOH

Strongest base (NaOH) will have Min. pOH

Neutral salt (NaCl) will have pOH equal to 7

CH_3COONa is the salt of weak acid and strong base thus having pOH value less than 7.

Thus, order of pOH value :

(A) > (D) > (C) > (B)

Hence, the correct option is (b).

13. *(c)* The dissociation of water into H^+ and OH^- is an endothermic reaction.

$\therefore$ As temperature increases, its equilibrium constant also increases. $[H^+]$ and $[OH^-]$ in water at equilibrium increases with increase in temperature.

$\therefore$ pH = – log$[H^+]$ decreases, as temperature increases.

So, both Assertion and Reason are false.

14. *(d)* $\underset{}{Cd(OH)_2} \rightleftharpoons \underset{S}{Cd^{2+}} + \underset{2S}{2OH^-}$

At equilibrium, $K_{sp} = S\,(2S)^2 = 4S^3$

$$\Rightarrow K_{sp} = 4\times(1.84\times10^{-5})^3$$

Solubility in buffer solution having pH = 12

$[OH^-] = 10^{-2}$

$$Cd(OH)_2 \rightleftharpoons \underset{S'}{Cd^{2+}} + \underset{\substack{2S+10^{-2}\\ \approx 10^{-2}}}{2OH^-}$$

$$\therefore K_{sp} = 4\times(1.84\times10^{-5})^3$$

$$= S'(10^{-2})^2$$

$$\Rightarrow S' = \frac{24.9\times10^{-15}}{10^{-4}}$$

$$= 2.49\times10^{-10}\text{ M}$$

Concept Enhancer

Buffer solution is a solution that resist change in their pH on dilution or an addition of small quantities of acids or alkalies.

15. *(c)*

idea NH_4Cl is a salt of weak base (NH_4OH) and strong acid (HCl). On hydrolysis, NH_4Cl will produce an acidic solution (pH < 7) and the expression of pH of the solution is

$$pH = 7 - \frac{1}{2}(pK_b + \log C)$$

Given, $K_b(NH_4OH) = 10^{-5}$

$\therefore pK_b = -\log K_b = -\log(10^{-5}) = 5$

C = concentration of salt solution = 0.02 M = 2×10^{-2} M

Now, $pH = 7 - \frac{1}{2}(pK_b + \log C)$

On substituting the given values in above equation, we get

$$= 7 - \frac{1}{2}[5 + \log(2\times10^{-2})]$$

$$= 7 - 1.65 = 5.35$$

16. *(d)* The dissociation of weak acid is given by

$$HA \rightleftharpoons H^+ + A^-$$

	HA	H^+	A^-
at $t = 0$	a	–	
at $t = t$	$a(1-x)$	ax	ax

$$K_a = (ax)\frac{(x)}{(1-x)}, [H^+] = ax$$

$$-\log(K_a) = -\log(ax) - \log\left(\frac{x}{1-x}\right)$$

$$pK_a = pH - \log\left(\frac{x}{1-x}\right)$$

$$\therefore pH - pK_a = \log\left(\frac{x}{1-x}\right)$$

17. *(b)* Statements (B) and (C) are correct while statements (A) and (D) are incorrect.

(A) pH of 10^{-8} M HCl is in acidic range i.e. 6.98.

(D) At half neutralisation point, half of the acid is present in the form of salt. $pH = pK_a + \log\frac{1}{1} = pK_a$.

18. *(c)* Given, [Base] $= [NH_3] = 0.2$ M

pH of buffer $= 8.26$

$pK_b\ (NH_3) = 4.74$

Molar mass of $NH_3 = 17$ g mol^{-1}

Molar mass of $NH_4Cl = 53.5$ g mol^{-1}

We know, $pOH = 14 - pH$
$= 14 - 8.26 = 5.74$

Using the expression,

$$pOH = pK_b + \log\frac{[Salt]}{[Base]}$$

or, $$5.74 = 4.74 + \log\frac{[NH_4Cl]}{[0.2]}$$

or $[NH_4Cl] = 2$ moles

Moles of NH_3 = Moles of $NH_4Cl = 2$

So, the amount of NH_4Cl required
= Number of moles × Molecular mass
$= 2 \times 53.5 = 107$ g

19. *(c)* The strength of aqueous NaOH solution is most accurately determined by titrating aqueous NaOH in burette and aqueous oxalic acid in conical flask with phenolphthalein is used as an indicator. (For weak acid-strong base titration, methyl orange cannot be used). Thus, solution turns colourless to pink at the end point.

20. *(d)* The explanation of given statements are as follows :

In statement (I), millimoles of H^+
$= 400 \times 0.1 \times 2 = 80$

Millimoles of $OH^- = 400 \times 0.1 = 40$

$\therefore$ Millimoles of H^+ left
$= 80 - 40 = 40$

$$[H^+] = \frac{40}{400+400} = \frac{1}{20} M$$

$$\Rightarrow \quad pH = -\log[H^+] = -\log\left(\frac{1}{20}\right)$$

$$= -0 + 0.301 + 1 \Rightarrow 1.30$$

Hence, the statement (I) is correct.

In statement (II), ionic product of H_2O is temperature dependent.

$K_w = [H^+][OH^-] \approx 10^{-14}$ (mol/L)^2 at 25°C

With increase in temperature, dissociation of H_2O units into H^+ and OH^- ions will also increase. As a result, the value of ionic product, $[H^+] \times [OH^-]$ will be increased.

Hence, the statement (II) is correct.

In statement (III), for a weak monobasic acid HA

$$HA \rightleftharpoons H^{\oplus} + A^{\ominus}$$
$$(1-\alpha)CM \qquad \alpha CM \qquad \alpha CM$$

$\Rightarrow$ pH of the solution is 5, i.e.

$$[H^+] = 10^{-5}\ M = \alpha C$$

$$\Rightarrow \quad K_a = \frac{\alpha C \times \alpha C}{(1-\alpha)C} = \frac{10^{-5} \times \alpha}{1-\alpha}$$

$$\Rightarrow \quad 10^{-5} = \frac{10^{-5} \times \alpha}{1-\alpha}$$

$$\Rightarrow \quad \alpha = 0.5$$

$$\Rightarrow \quad \alpha\% = 50$$

21. *(c)* Volume of $H_2SO_4\ (V_1) = 20$ mL

Concentration of $H_2SO_4\ (M_1) = 0.1$ M

Number of moles ($n_{H_2SO_4}$)
$= 0.1 \times 2 \times 20 = 4$ mmol
$[\because 1\ H_2SO_4 \rightarrow 2H^+]$

Volume of $NH_4OH\ (V_2) = 30$ mL

Concentration of $NH_4OH\ (M_2)$
$= 0.2$ M

Number of moles (n_{NH_4OH}) $= 0.2 \times 30$
$= 6$ mmol

The reaction involved is as follows:

$$H_2SO_4 + 2NH_4OH \longrightarrow (NH_4)_2SO_4 + 2H_2O$$

	H_2SO_4	$2NH_4OH$	$(NH_4)_2SO_4$	$2H_2O$
Initial	4 mmol	6 mmol	–	–
Final	0 mmol	2 mmol	4 mmol	–

So, the resulting solution is a basic buffer. According to the Henderson's equation,

$$pOH = pK_b + \log\frac{[Salt]}{[Base]}$$

$$pOH = pK_b + \log\frac{[(NH_4)_2SO_4]}{[NH_4OH]}$$

$$= 4.7 + \log\frac{4}{2} = 4.7 + 0.3 = 5$$

$$\Rightarrow \quad pH = 14 - pOH = 14 - 5 = 9$$

22. *(3)* Given, pH = 10

pOH = 4

So, $[OH^-] = 10^{-4}$

Now, number of moles of $OH^- = 10^{-4}$

number of moles of $Mg(OH)_2 = \frac{10^{-4}}{2}$
$= 5 \times 10^{-5}$

So, mass of $Mg(OH)_2$
$= 5 \times 10^{-5} \times 58 \times 10^{-3}$ mg $= 2.9 \approx 3$

23. *(48)* $pOH = pK_b + \log\frac{[NH_4^+]}{[NH_3]}$

$pOH = 4.745$

On adding 0.05 mole HCl

NH_3	$+ H^{\oplus}$	$\longrightarrow NH_4^{\oplus}$
0.1	0.05	0.1
0.05	0	0.15

$$pOH' = 4.745 + \log 3$$
$$pOH' - pOH = 0.477$$
$$14 - pH' - 14 + pH = 0.477$$
$$\Delta pH = 0.477 \approx 48 \times 10^{-2}$$

24. *(7)*

> **idea** Ammonium acetate is a salt of weak acid and weak base. So, to calculate the pH value of ammonium acetate, we have to first calculate pK_a and pK_b values of acetic acid and ammonium hydroxide. Further apply the formula
>
> $pH = 7 + \frac{1}{2}[pK_a - pK_b]$

Given,

K_a of $CH_3COOH = 1.8 \times 10^{-5}$

K_b of $NH_4OH = 1.8 \times 10^{-5}$

Using the formula,

$$pH = 7 + \frac{1}{2}[pK_a - pK_b] \quad \text{...(i)}$$

$$pK_a = -\log(K_a) = -\log(1.8 \times 10^{-5})$$
$$= +4.7447 \quad \text{...(ii)}$$

$$pK_b = -\log(K_b) = -\log(1.8 \times 10^{-5})$$
$$= 4.7447 \quad \text{...(iii)}$$

Put value of Eq. (ii) and (iii) in Eq. (i)

$$pH = 7 + \frac{1}{2}[4.7447 - 4.7447] = 7$$

25. *(100)* Given,

Solution contains = 1 M benzoic acid (V_s mL)
+ 1 M sodium benzoate (V_s mL)

Total volume of buffer solution $= 300$ mL

$pH = 4.5,\ pK_a = 4.20$

$$pH = pK_a + \log\frac{[\text{salt}]}{[\text{acid}]} \quad \ldots(i)$$

$$4.5 = 4.2 + \log\left[\frac{V_s}{V_a}\right]$$

$$\frac{V_s}{V_a} = 2 \quad \ldots(ii)$$

$$V_s + V_a = 300 \quad \ldots(iii)$$

From Eqs. (ii) and (iii)

$V_a = 100$ mL, $V_s = 200$ mL

Hence, volume of benzoic acid is 100 mL.

26. *(9)*

> **idea** When the precipitation just begins, the ionic product of the ions becomes equal to the K_{sp} of the salt. This will give the $[OH^-]$ concentration, and hence the pOH and pH value can be calculated.

When $Mg(OH)_2$ start precipitation, then $[Mg^{2+}][OH^-]^2 = K_{sp}$ of $Mg(OH)_2$

$$[0.1][OH^-]^2 = 1\times 10^{-11}$$

As $K_{sp} = 1\times 10^{-11}$ [given]

$\therefore\ [OH^-] = 10^{-5}$ M

pOH $= 5$

pH $= 14 - \text{pOH} = 14 - 5 = 9$

Hence, pH is 9.

27. *(458)* 20 mL of 0.1 M NaOH = 2 m mole of NaOH (base)

50 mL of 0.1 M CH_3COOH = 5 m mole of CH_3COOH (acid)

Now, when acid reacts with base, gives salt and water. This reaction is called neutralisation reaction.

$$\underbrace{\underset{5\text{ m mol}}{CH_3COOH} + \underset{2\text{ m mol}}{NaOH}}_{\substack{3\text{ m mol of}\\(CH_3COOH\text{ will left})}} \longrightarrow \underset{\substack{(2\text{ m mol}\\\text{of salt}\\\text{will formed})}}{CH_3COONa} + H_2O$$

Now, with weak acid and the salt of weak acid, a buffer solution will be formed, pH can be calcualted from Henderson's equation

$$pH = pK_a + \log\left[\frac{A^\ominus \text{ ion}}{HA \text{ acid}}\right]$$

$$pH = 1.76 + \log\left[\frac{2\text{ m mol}}{3\text{ m mol}}\right]$$

$$pH = 4.76 + \{\log[2] - \log(3)\}$$

$$pH = 4.76 + [0.30 - 0.48]$$

$$pH = 4.58 \text{ or } (458\times 10^{-2})$$

28. *(12)*

> **idea** To find out the volume of solution required, first of all calculate the molarity. It will be equal to the normality of solution due to its monobasic nature. After that use the formula $N_1V_1 = N_2V_2$

$$\text{Molarity} = \frac{\text{Density (g/L)}}{\text{Molar mass (g/mol)}}$$

$$\text{Molarity of acid} = \frac{1.21\times 10^3}{24.2} = \frac{1000}{20}$$

$$= 50\text{ M} = 50\text{ N}$$

$\because$ For both of HA and NaOH, n-factor $= 1$

i.e., molarity (M) = Normality (N).

$\because$ Density $= 1.21$ kg/L $= 1.21\times 10^3$ g/L

From neutralisation reaction,

$$HA + NaOH \longrightarrow NaA + H_2O$$

$$N_1V_1 = N_2V_2$$

$$\Rightarrow \quad 50\times V = 0.24\times 25$$

$$V = 0.12\text{ mL} = 12\times 10^{-2}\text{ mL}$$

29. *(85)* Concentration of lactate ion $= (2\times 0.005)$ M.

As we know that, calcium lactate is a salt of weak acid and strong base so salt hydrolysis will take place.

$$pH = 7 + \frac{1}{2}(pK_a + \log c)$$

$$= 7 + \frac{1}{2}[5 + \log(2\times 0.005)]$$

$$= 7 + \frac{3}{2} = \frac{17}{2} = 8.5 = 85\times 10^{-1}$$

Therefore, the pH of 0.005 M calcium lactate solution at 25°C is 85×10^{-1}.

30. *(27)*

$$C_3H_7COOH \rightleftharpoons C_3H_7COO^\ominus + H^+$$

	C_3H_7COOH	$C_3H_7COO^\ominus$	H^+
Initial conc.	C	0	0
Conc. after dissociation	$C - C\alpha$	$C\alpha$	$C\alpha$

Hence, concentration of $C_3H_7COOH\ (C) = 0.2$ M

So, dissociation constant for weak acid is given as

$$K_a = \frac{[C_3H_7COO^-][H^+]}{[C_3H_7COOH]}$$

$$= \frac{C\alpha\times C\alpha}{C - C\alpha} = \frac{C\alpha^2}{1-\alpha}$$

Since, butyric acid is weak, so, $1 >> \alpha$

So, $\quad K_a = C\alpha^2\ [K_a = 2\times 10^{-5}]$

$$\sqrt{\frac{2\times 10^{-5}}{0.2}} = \alpha \Rightarrow 10^{-2} = \alpha$$

Concentration of H^+ ions $= C\alpha$

$$= 0.2\times 10^{-2} = 2\times 10^{-3}$$

$$pH = -\log[H^+]$$

$$= -\log(2\times 10^{-3})$$

$$= -0.3010 + 3$$

$$pH = 2.7 \text{ or } pH = 27\times 10^{-1}$$

31. *(11)* 0.001 M NaOH will contain 0.001 M OH^- ions.

$\therefore$ Concentration of OH^- ions,

$$[OH^-] = 10^{-3}\text{ M}$$

So, $\quad pOH = -\log[OH^-]$

$$= -\log(10^{-3}) = 3$$

pH, $14 - \text{pOH} = 14 - 3 = 11$

Hence, pH value of 0.001 M NaOH solution = 11.

32. *(476)* Moles of CH_3COOH

$= $ Molarity of CH_3COOH $\times$ Volume of CH_3COOH(in L)

$$= 0.1\text{ M}\times\frac{50}{1000}\text{ L} = 0.005\text{ moles}$$

Similarly, moles of NaOH $= 0.0025$ moles

$$NaOH + CH_3COOH \longrightarrow CH_3COONa + H_2O$$

	NaOH	CH_3COOH	CH_3COONa	H_2O
Initial conc.	0.0025 mol	0.005 mol	0	0
Final conc.	0	0.0025	0.0025 mol	0.0025 mol

So, buffer is formed, according to Henderson-Hasselbalch equation

$$pH = pK_a + \log\left(\frac{[\text{Salt}]}{[\text{Acid}]}\right)$$

$$= pK_a + \log\left(\frac{\frac{0.0025}{75}}{\frac{0.0025}{75}}\right)$$

$$= pK_a = 4.76 = 476\times 10^{-2}$$

33. *(6021)* Milliequivalents of HCl

$(N_aV_a) = 50\times 1 = 50$

Milliequivalents of NaOH

$(N_bV_b) = 30\times 1 = 30$

Since, $\quad N_aV_a > N_bV_b$

and they neutralise each other

$$[H^+] = \frac{N_aV_a - N_bV_b}{V_a + V_b}$$

$$= \frac{50 - 30}{80} = 0.25 = 2.5\times 10^{-1}$$

$$pH = -\log[H^+]$$

$$= -\log(2.5\times 10^{-1})$$

$$= 1 - 0.3979 = 0.6021$$

$$pH = 6021\times 10^{-4}$$

$\therefore \quad x = 6021$

34. *(3)* $NH_3 + H_2O \rightleftharpoons NH_4^+ + OH^-$

$[NH_3] = 2 \times 0.0210 = 0.042$ m mol

$NH_4Cl \longrightarrow NH_4^+ + Cl^-$

$[NH_4Cl] = 5 \times 0.0504$

$= 0.252$ m mol

By Henderson-Hasselbalch equation,

$$pOH = pK_b + \log \frac{[Salt]}{[Base]}$$

$$= -\log(1.8 \times 10^{-5}) + \log \frac{0.252}{0.042}$$

$$= 4.74 + 0.77 = 5.51$$

$$[OH^-] = 10^{-5.51} = 3.09 \times 10^{-6}$$

$$\approx 3 \times 10^{-6}$$

The value of x is 3×10^{-6}

35. *(10)* Concentration of acetic acid in buffer = 1.0 M

Acetic acid and its conjugate base, sodium acetate makes acidic buffer.

$CH_3COOH + CH_3COONa \longrightarrow$ (Acidic buffer)

Using formula,

$$pH = pK_a + \log \frac{[CH_3COONa]}{[CH_3COOH]}$$

$$5.74 = 4.74 + \log \frac{[CH_3COONa]}{[CH_3COOH]}$$

$$\Rightarrow \frac{[CH_3COONa]}{[CH_3COOH]} = 10$$

$[\because [CH_3COOH] = 1]$

$[CH_3COONa] = [10 \times 1] = 10$

Thus, the concentration of sodium acetate in buffer is 10 M.

36. *(1)* For H_2SO_3, $K_{a_1} >> K_{a_2}$

So, we mainly consider 1st ionisation

$$H_2SO_3 \rightleftharpoons H^+ + HSO_3^-;$$

	H_2SO_3	H^+	HSO_3^-
$t = 0$	0.558	0	0
$t = t_{eq}$	$0.558 - 0.558\alpha$	0.558α	0.558α

$K_{a_1} = 1.7 \times 10^{-2}$

$$K_{a_1} = \frac{[H^+][HSO_3^-]}{[H_2SO_3]}$$

$$= \frac{(0.558\,\alpha)(0.558\,\alpha)}{0.558(1-\alpha)} = \frac{0.558\,\alpha^2}{1-\alpha}$$

$\alpha << 1$ for weak acid $(1-\alpha) \approx 1$

$$\alpha = \sqrt{\frac{K_{a_1}}{0.558}} = \sqrt{\frac{1.7 \times 10^{-2}}{0.558}}$$

$$\alpha = 1.7 \times 10^{-1} = 0.17$$

$$[H]^+ = 0.558\,\alpha = 9.5 \times 10^{-2}$$

$$pH = -\log[H^+] = -\log(9.5 \times 10^{-2})$$

$[\because \log 9.9 \approx 1]$

$$= 2 - \log 9.5 = 2 - 1$$

$$pH = 1$$

37. *(7)* Phosphoric acid is a weak tribasic acid (H_3PO_4) and NH_4OH is a weak monoacidic base. So, hydrolysis of ammonium phosphate $(NH_4)_3PO_4$ can be shown as,

$(NH_4)_3PO_4 + 3H_2O \rightleftharpoons H_3PO_4 + 3NH_4OH$

[3 : 1 type salt]

$$\text{So, } [H^+] = K_a \times \left(\frac{K_w}{K_a \times K_b}\right)^{1/2}$$

$$pH = pK_a + \frac{1}{2}[pK_w - pK_a - pK_b]$$

$$= 5.23 + \frac{1}{2}(14 - 5.23 - 4.75)$$

$$= 7.24 \approx 7 \quad [\because pK_w = 14\, H_2O]$$

Also, $pK_a = 5.23\,(H_3PO_4)$

$pK_b = 4.75\,(NH_4OH)]$

38. *(5.23)* 3g CH_3COOH $(M = 60)$

$$= \frac{3}{60} = \frac{1}{20} \text{ mol}$$

250 mL of 0.1 M HCl

$$= M \times V = \frac{250}{1000} \times 0.1$$

$$= 25 \times 10^{-3} \text{ mol}$$

After mixing solution is diluted to 500 mL.

Then from that solution, 20 mL is taken.

Number of moles of both CH_3COOH and HCl will be now only $\frac{20}{500} = \frac{1}{25}$ of original solution.

$$n_{CH_3COOH} = \frac{1}{20} \times \frac{1}{25}$$

$$= \frac{1}{500} = 2 \times 10^{-3} \text{ mol}$$

$$n_{HCl} = 25 \times 10^{-3} \times \frac{1}{25}$$

$$= 10^{-3} \text{ mol}$$

$$n_{NaOH} \text{ added} = M \times V = \frac{1}{2} \times 5 \times 10^{-3}$$

$$= 2.5 \times 10^{-3}$$

After mixing, all HCl and NaOH will be neutralised. Remaining number of moles.

$$n_{NaCl} = 10^{-3}$$

$$n_{CH_3COONa} = 1.5 \times 10^{-3}$$

$$n_{CH_3COOH} \text{ (unreacted)} = 0.5 \times 10^{-3}$$

$\therefore$ Now the solution is a buffer solution.

$$= pK_a + \log \frac{(n_{CH_3COONa})}{(n_{CH_3COOH})}$$

$$= 4.75 + \log 3 = 4.75 + 0.4771$$

$$= 5.2271 \approx 5.23$$

39. *(10.60)* (*A*) 4 g NaOH in 100 L

$$\Rightarrow \quad M = \frac{4/40}{100} = 10^{-3} \text{ mol L}^{-1}.$$

(*B*) 9.8 g H_2SO_4 in 100 L

$$\Rightarrow \quad M = \frac{9.8/98}{100} = 10^{-3} \text{ mol L}^{-1}$$

40 L of solution *A* contains

$$= 40 \times 10^{-3} \text{ mol NaOH}$$

$(\because n = M \times V)$

$$= 40 \times 10^{-3} \text{ mol } OH^-$$

10 L of solution *B* contains

$$= 10 \times 10^{-3} \text{ mol } H_2SO_4$$

$$= 20 \times 10^{-3} \text{ mol } H^+$$

When they are mixed, total volume = 50 L

and all H^+ is neutralised by OH^-.

Number of moles (n_{OH^-}) of excess OH^-.

$$= (40 - 20) \times 10^{-3} = 20 \times 10^{-3}$$

$$[OH^-] = \frac{n_{OH^-}}{V_{solution}} \text{ (in litre)}$$

$$\text{Or} \quad [OH^-] = \frac{n_{OH^-}}{\text{Total volume}}$$

$$= \frac{20 \times 10^{-3}}{50} = 4 \times 10^{-4} \text{ M}$$

As we know,

$$pOH = -\log[OH^-]$$

$$\therefore \quad pOH = -\log 4 \times 10^{-4} = 4 - \log 4$$

$$\text{or} \quad pH = 14 - pOH$$

$$= 14 - (4 - \log 4)$$

$$= 10 + \log 4 = 10.602$$

08 Redox Reactions

In JEE Main there is usually 0-1 question per paper are visible from this chapter. In past 3 years, 18 conceptual and 2 numerical questions were asked of easy to moderate level.

For detailed analysis scan QR code

Topic 1 / Oxidation-Reduction & Oxidation Number

• Level 1

01 The standard reduction potential values of some of the *p*-block ions are given below. Predict the one with the strongest oxidising capacity. **[2025, 29 Jan Shift-I]**

(a) $E^{\ominus}_{Al^{3+}/Al} = -1.66$ V (b) $E^{\ominus}_{Pb^{4+}/Pb^{2+}} = +1.67$ V

(c) $E^{\ominus}_{Sn^{4+}/Sn^{2+}} = +1.15$ V (d) $E^{\ominus}_{Tl^{3+}/Tl} = +1.26$ V

02 Based on the data given below **[2025, 24 Jan Shift-II]**

$E^{\circ}_{Cr_2O_7^{2-}/Cr^{3+}} = 1.33$ V; $E^{\circ}_{Cl_2/Cl^{(-)}} = 1.36$ V

$E^{\circ}_{MnO_4^-/Mn^{2+}} = 1.51$ V; $E^{\circ}_{Cr^{3+}/Cr} = -0.74$ V

the strongest reducing agent is

(a) Cl^- (b) Cr (c) MnO_4^- (d) Mn^{2+}

03 The reaction at cathode in the cells commonly used in clocks involves **[2024, 5 April Shift-I]**

(a) oxidation of Mn from + 2 to + 7
(b) oxidation of Mn from + 3 to + 4
(c) reduction of Mn from + 7 to + 2
(d) reduction of Mn from + 4 to + 3

04 In alkaline medium, MnO_4^- oxidises I^- to **[2024, 29 Jan Shift-I]**

(a) I_2 (b) IO_4^- (c) IO_3^- (d) IO^-

05 Which of the following options are correct for the reaction?

$$2[Au(CN)_2]^-(aq) + Zn(s) \longrightarrow 2Au(s) + [Zn(CN)_4]^{2-}(aq)$$

A. Redox reaction B. Displacement reaction
C. Decomposition reaction D. Combination reaction

Choose the correct answer from the options given below **[2023, 6 April Shift-I]**

(a) C and D only (b) A and B only
(c) A and D only (d) A only

• Level 2

06 The species which does not undergo disproportionation reaction is **[2025, 22 Jan Shift-II]**

(a) ClO_3^- (b) ClO_2^- (c) ClO^- (d) ClO_4^-

07 Match the List-I with List-II. **[2025, 28 Jan Shift-I]**

List-I (Redox reaction)	List-II (Type of redox reaction)
A. $CH_4(g) + 2O_2(g) \xrightarrow{\Delta} CO_2(g) + 2H_2O(l)$	I. Disproportionation reaction
B. $2NaH(s) \xrightarrow{\Delta} 2Na(s) + H_2(g)$	II. Combination reaction
C. $V_2O_5(s) + 5Ca(s) \xrightarrow{\Delta} 2V(s) + 5CaO(s)$	III. Decomposition reaction
D. $2H_2O_2(aq) \xrightarrow{\Delta} 2H_2O(l) + O_2(g)$	IV. Displacement reaction

Choose the correct answer from the options given below

(a) A-II, B-III, C-IV, D-I
(b) A-III, B-IV, C-I, D-II
(c) A-IV, B-I, C-II, D-III
(d) A-II, B-III, C-I, D-IV

08 Match List-I with List-II.

List-I (Reaction)	List-II (Type of redox reaction)
A. $N_2(g) + O_2(g) \rightarrow 2NO(g)$	I. Decomposition
B. $2Pb(NO_3)_2(s) \rightarrow 2PbO(s) + 4NO_2(g) + O_2(g)$	II. Displacement
C. $2Na(s) + 2H_2O(l) \rightarrow 2NaOH(aq) + H_2(g)$	III. Disproportionation
D. $2NO_2(g) + 2\bar{O}H(aq) \rightarrow NO_2^-(aq) + NO_3^-(aq) + H_2O(l)$	IV. Combination

Choose the correct answer from the options given below. **[2024, 6 April Shift-II]**

	A	B	C	D		A	B	C	D
(a)	III	II	I	IV	(b)	IV	I	II	III
(c)	I	II	III	IV	(d)	II	III	IV	I

09 The number of ions from the following that have the ability to liberate hydrogen from a dilute acid is Ti^{2+}, Cr^{2+} and V^{2+}. **[2024, 5 April Shift-II]**
(a) 1 (b) 0 (c) 2 (d) 3

Imp. *Disproportionation Reaction*

10 Given below are two statements. **[2024, 31 Jan Shift-II]**
Statement I S_8 solid undergoes disproportionation reaction under alkaline conditions to form S^{2-} and $S_2O_3^{2-}$.
Statement II ClO_4^- can undergo disproportionation reaction under acidic condition.
In the light of the above statements, choose the most appropriate answer from the options given below
(a) Both statement I and Statement II are incorrect.
(b) Statement I is correct but Statement II is incorrect.
(c) Both Statement I and Statement II are correct.
(d) Statement I is incorrect but Statement II is correct.

11 Which of the following cannot function as an oxidising agent? **[2024, 27 Jan Shift-II]**
(a) SO_4^{2-} (b) MnO_4^- (c) N^{3-} (d) BrO_3^-

12 Cerium (IV) has a noble gas configuration. Which of the following is correct statement about it? **[2022, 25 June Shift-I]**
(a) It will not prefer to undergo redox reactions.
(b) It will prefer to gain electron and act as an oxidising agent.
(c) It will prefer to give away an electron and behave as reducing agent.
(d) It acts as both, oxidising and reducing agent.

13 Identify the process in which change in the oxidation state is five **[2021, 25 July Shift-II]**
(a) $Cr_2O_7^{2-} \longrightarrow 2Cr^{3+}$ (b) $MnO_4^- \longrightarrow Mn^{2+}$
(c) $CrO_4^{2-} \longrightarrow Cr^{3+}$ (d) $C_2O_4^{2-} \longrightarrow 2CO_2$

14 The species given below that does not show disproportionation reaction is **[2021, 20 July Shift-I]**
(a) BrO_4^- (b) BrO^- (c) BrO_2^- (d) BrO_3^-

15 The compound that cannot act both as oxidising and reducing agent is **[2020, 9 Jan Shift-II]**
(a) H_2SO_3 (b) H_3PO_4 (c) HNO_2 (d) H_2O_2

• Level 3

16 Thiosulphate reacts differently with iodine and bromine in the reactions given below :

$$2S_2O_3^{2-} + I_2 \longrightarrow S_4O_6^{2-} + 2I^-$$

$$S_2O_3^{2-} + 5Br_2 + 5H_2O \longrightarrow 2SO_4^{2-} + 4Br^- + 10H^+$$

Which of the following statements justifies the above dual behaviour of thiosulphate? **[2024, 8 April Shift-I]**
(a) Thiosulphate undergoes oxidation by bromine and reduction by iodine in these reactions
(b) Bromine is a weaker oxidant than iodine
(c) Bromine is a stronger oxidant than iodine
(d) Bromine undergoes oxidation and iodine undergoes reduction in these reaction.

17 Which of the following reactions are disproportionation reactions?
A. $Cu^+ \longrightarrow Cu^{2+} + Cu$
B. $3MnO_4^{2-} + 4H^+ \longrightarrow 2MnO_4^- + MnO_2 + 2H_2O$
C. $2KMnO_4 \longrightarrow K_2MnO_4 + MnO_2 + O_2$
D. $2MnO_4^- + 3Mn^{2+} + 2H_2O \longrightarrow 5MnO_2 + 4H^+$
Choose the correct answer from the options given below **[2024, 1 Feb Shift-I]**
(a) A, D (b) A, B, C (c) B, C, D (d) A, B

18 Which of the given reactions is not an example of disproportionation reaction ? **[2022, 26 July Shift-I]**
(a) $2H_2O_2 \longrightarrow 2H_2O + O_2$
(b) $2NO_2 + H_2O \longrightarrow HNO_3 + HNO_2$
(c) $MnO_4^- + 4H^+ + 3e^- \longrightarrow MnO_2 + 2H_2O$
(d) $3MnO_4^{2-} + 4H^+ \longrightarrow 2MnO_4^- + MnO_2 + 2H_2O$

19 Which one of the following is an example of disproportionation reaction? **[2022, 26 June Shift-II]**
(a) $3MnO_4^{2-} + 4H^+ \longrightarrow 2MnO_4^- + MnO_2 + 2H_2O$
(b) $MnO_4^- + 4H^+ + 4e^- \longrightarrow MnO_2 + 2H_2O$
(c) $10I^- + 2MnO_4^- + 16H^+ \longrightarrow 2Mn^{2+} + 8H_2O + 5I_2$
(d) $8MnO_4^- + 3S_2O_3^{2-} + H_2O \longrightarrow 8MnO_2 + 6SO_4^{2-} + 2OH^-$

20 An example of a disproportionation reaction is **[2019, 12 April Shift-I]**
(a) $2MnO_4^- + 10I^- + 16H^+ \longrightarrow 2Mn^{2+} + 5I_2 + 8H_2O$
(b) $2NaBr + Cl_2 \longrightarrow 2NaCl + Br_2$
(c) $2KMnO_4 \longrightarrow K_2MnO_4 + MnO_2 + O_2$
(d) $2CuBr \longrightarrow CuBr_2 + Cu$

21 The redox reaction among the following is **[2020, 7 Jan Shift-II]**
(a) reaction of H_2SO_4 with NaOH
(b) reaction of $[Co(H_2O)_6]Cl_3$ with $AgNO_3$
(c) combination of dinitrogen with dioxygen at 2000 K
(d) formation of ozone from atmospheric oxygen in the presence of sunlight

Numerical Value Type Questions

22 The standard reduction potentials at 298 K for the following half cells are given below: **[2024, 9 April, Shift-I]**

$Cr_2O_7^{2-} + 14H^+ + 6e^- \longrightarrow 2Cr^{3+} + 7H_2O$, $E° = 1.33$ V

$Fe^{3+}(aq) + 3e^- \longrightarrow Fe$ $E° = -0.04$V

$Ni^{2+}(aq) + 2e^- \longrightarrow Ni$ $E° = -0.25$ V

$Ag^+(aq) + e^- \longrightarrow Ag$ $E° = 0.80$ V

$Au^{3+}(aq) + 3e^- \longrightarrow Au$ $E° = 1.40$ V

Consider the given electrochemical reactions,

The number of metal(s) which will be oxidised by $Cr_2O_7^{2-}$, in aqueous solution is ....

23 From the given list, the number of compounds with +4 oxidation state of sulphur is ....

$SO_3, H_2SO_3, SOCl_2, SF_4, BaSO_4, H_2S_2O_7$

[2024, 27 April, Shift-I]

24 In alkaline medium, the reduction of permanganate anion involves a gain of .... electrons.

[2023, 10 April, Shift-II]

25 In the titration of $KMnO_4$ and oxalic acid in acidic medium, the change in oxidation number of carbon at the end point is .... **[2022, 27 July Shift-I]**

26 The oxidation states of transition metal atoms in $K_2Cr_2O_7$, $KMnO_4$ and K_2FeO_4, respectively, are x, y and z. The sum of x, y and z is ....

[2020, 2 Sep Shift-II]

Topic 2 / Balancing of Chemical Equations

• Level 1

01 $2IO_3^- + xI^- + 12H^+ \longrightarrow 6I_2 + 6H_2O$

What is the value of x? **[2023, 8 April Shift-I]**

(a) 6 (b) 2 (c) 10 (d) 12

• Level 2

02 In acidic medium, $K_2Cr_2O_7$ shows oxidising action as represented in the half reaction:

$$Cr_2O_7^{2-} + XH^+ + Ye^{\ominus} \longrightarrow 2A + ZH_2O$$

X, Y, Z and A are respectively are **[2024, 1 Feb Shift-I]**

(a) 8, 4, 6 and Cr_2O_3 (b) 8, 6, 4 and Cr_2O_3

(c) 14, 7, 6 and Cr^{3+} (d) 14, 6, 7 and Cr^{3+}

• Level 3

03 Chlorine undergoes disproportionation in alkaline medium as shown below

$$a\,Cl_2(g) + b\,OH^-(aq) \longrightarrow c\,ClO^-(aq) + d\,Cl^-(aq) + e\,H_2O(l)$$

The values of a, b, c and d in a balanced redox reaction are respectively **[2024, 29 Jan Shift-I]**

(a) 1, 2, 1 and 1 (b) 2, 4, 1 and 3

(c) 3, 4, 4 and 2 (d) 2, 2, 1 and 3

Numerical Value Type Questions

04 $2MnO_4^- + bI^- + cH_2O \longrightarrow xI_2 + yMnO_2 + z\overline{O}H$

If the above equation is balanced with integer coefficients, the value of z is ....

[2024, 30 Jan Shift-I]

05 See the following chemical reaction

$$Cr_2O_7^{2-} + XH^+ + 6Fe^{2+} \longrightarrow YCr^{3+} + 6Fe^{3+} + ZH_2O$$

The sum of X, Y and Z is .... **[2023, 13 April Shift-II]**

06 $2MnO_4^- + bC_2O_4^{2-} + cH^+ \longrightarrow x\,Mn^{2+} + y\,CO_2 + zH_2O$

If the above equation is balanced with integer coefficients, the value of c is .... **[2021, 16 March Shift-I]**

07 Consider the following equations :

$$2Fe^{2+} + H_2O_2 \longrightarrow xA + yB$$

(in basic medium)

$$2MnO_4^- + 6H^+ + 5H_2O_2 \longrightarrow x'C + y'D + z'E$$

(in acidic medium)

The sum of the stoichiometric coefficients x, y, x', y' and z' for products A, B, C, D and E, respectively, is ....

[2020, 4 Sep Shift-II]

Key Index

Topic 1

1 (b)	2 (b)	3 (d)	4 (c)	5 (b)	6 (d)	7 (d)	8 (b)	9 (d)	10 (b)
11 (c)	12 (b)	13 (b)	14 (a)	15 (b)	16 (c)	17 (d)	18 (c)	19 (a)	20 (d)
21 (c)	22 (3)	23 (3)	24 (3)	25 (1)	26 (19)				

Topic 2

1 (c)	2 (d)	3 (a)	4 (8)	5 (23)	6 (16)	7 (19)

For Detailed Solutions scan QR Code

09 p-Block Elements-1

In JEE Main there is usually 0-1 question per paper are visible from this chapter. In past 3 years, 21 conceptual questions were asked from this chapter of easy to moderate level.

For detailed analysis scan QR code

Topic 1 / Group 13 Elements

Level 1

Imp. *Ionisation Enthalpy Order*

01 The correct order of the first ionisation enthalpy is **[2024, 4 April Shift-II]**

(a) B > Al > Ga (b) Ga > Al > B
(c) Al > Ga > Tl (d) Tl > Ga > Al

02 The covalency and oxidation state respectively of boron in $[BF_4]^-$, are **[2023, 13 April Shift-II]**

(a) 4 and 3 (b) 3 and 5
(c) 3 and 4 (d) 4 and 4

03 Given below are two statements.

Statement I Boron is extremely hard indicating its high lattice energy.

Statement II Boron has highest melting and boiling point compared to its other group members.

In the light of the above statements, choose the most appropriate answer from the options given below. **[2023, 12 April Shift-I]**

(a) Both Statement I and Statement II are correct.
(b) Statement I is incorrect but Statement II is correct.
(c) Statement I is correct but Statement II is incorrect.
(d) Both Statement I and Statement II are incorrect.

04 Which one of the following elements will remain as liquid inside pure boiling water? **[2023, 6 April Shift-I]**

(a) Ga (b) Li
(c) Cs (d) Br

05 The Lewis acid character of boron trihalides follows the order **[2023, 31 Jan Shift-II]**

(a) $BBr_3 > BI_3 > BCl_3 > BF_3$
(b) $BF_3 > BCl_3 > BBr_3 > BI_3$
(c) $BI_3 > BBr_3 > BCl_3 > BF_3$
(d) $BCl_3 > BF_3 > BBr_3 > BI_3$

06 Given below are two statements.

Statement I The decrease in first ionisation enthalpy from B to Al is much larger than that from Al to Ga.

Statement II The d-orbitals in Ga are completely filled.

In the light of the above statements, choose the most appropriate answer from the options given below. **[2023, 29 Jan Shift-II]**

(a) Both the statements I and II are incorrect.
(b) Statement I is correct but statement II is incorrect.
(c) Both the statements I and II are correct.
(d) Statement I is incorrect but statement II is correct.

07 Given below are two statements, one is labelled as Assertion A and the other is labelled as Reason R.

Assertion (A) Boron is unable to form BF_6^{3-}.

Reason (R) Size of B is very small.

In the light of the above statements, choose the correct answer from the options given below. **[2022, 27 July Shift-II]**

(a) Both A and R are true and R is the correct explanation of A.
(b) Both A and R are true but R is not the correct explanation of A.
(c) A is true but R is false.
(d) A is false but R is true.

Imp. *VSEPR & Chemical Reaction of Boron*

08 The geometry around boron in the product 'B' formed from the following reaction is **[2022, 25 July Shift-I]**

$$BF_3 + NaH \xrightarrow{450\text{ K}} A + NaF$$
$$A + NMe_3 \longrightarrow B$$

(a) trigonal planar (b) tetrahedral
(c) pyramidal (d) square planar

09 The metal that has very low melting point and its periodic position is closer to a metalloid is **[2022, 26 July Shift-II]**

(a) Al (b) Ga (c) Se (d) In

10 Choose the correct stability order of group 13 elements in their + 1 oxidation state. **[2022, 26 June Shift-I]**

(a) Al < Ga < In < Tl (b) Tl < In < Ga < Al
(c) Al < Ga < Tl < In (d) Al < Tl < Ga < In

11 In which one of the following molecules strongest back donation of an electron pair from halide to boron is expected? **[2021, 27 Aug Shift-I]**

(a) BCl_3 (b) BF_3 (c) BBr_3 (d) BI_3

12 The number of 2-centre-2-electron and 3-centre-2-electron bonds in B_2H_6, respectively, are **[2019, 10 Jan Shift-II]**

(a) 4 and 2 (b) 2 and 4
(c) 2 and 2 (d) 2 and 1

• Level 2

13 Given below are the pairs of group 13 elements showing their relation in terms of atomic radius.
(B < Al), (Al < Ga), (Ga < In) and (In < Tl)
Identify the elements present in the incorrect pair and in that pair find out the element (X) that has higher ionic radius (M^{3+}) than the other one. The atomic number of the element (X) is **[2025, 4 April Shift-I]**

(a) 49 (b) 31
(c) 13 (d) 81

14 The elements of Group 13 with highest and lowest first ionisation enthalpies are respectively: **[2025, 4 April Shift-II]**

(a) Tl and B (b) B and Ga
(c) B and In (d) B and Tl

15 Given below are two statements. One is labelled as **Assertion (A)** and other is labelled as **Reason (R)**.
Assertion (A) The stability order of + 1 oxidation state of Ga, In and Tl is Ga < In < Tl.
Reason (R) The inert pair effect stabilises the lower oxidation state down the group. **[2024, 8 April Shift-I]**
In the light of the above statements, choose the correct answer from the options given below.

(a) A is false, but R is true.
(b) Both A and R are true but R is not the correct explanation of A.
(c) Both A and R are true and R is the correct explanation of A.
(d) A is true, but R is false.

16 Given below are two statements : **[2024, 6 April Shift-I]**
Statement I Gallium is used in the manufacturing of thermometers.
Statement II A thermometer containing gallium is useful for measuring the freezing point (256 K) of brine solution.
In the light of the above statements. choose the correct answer from the options given below

(a) Statement I is true but Statement II is false.
(b) Statement I is false but Statement II is true.
(c) Both Statement I and Statement II are true.
(d) Both Statement I and Statement II are false.

17 Given below are two statements. **[2024, 5 April Shift-I]**
Statement I In group 13, the stability of + 1 oxidation state increases down the group.
Statement II The atomic size of gallium is greater than that of aluminium.
In the light of the above statements, choose the most appropriate answer from the options given below:

(a) Both Statement I and Statement II are correct.
(b) Statement I is incorrect but Statement II is correct.
(c) Statement I is correct but Statement II is incorrect.
(d) Both Statement I and Statement II are incorrect.

18 Given below are two statements. **[2024, 31 Jan Shift-II]**
Statement I Group 13 trivalent halides get easily hydrolysed by water due to their covalent nature.
Statement II $AlCl_3$ upon hydrolysis in acidified aqueous solution forms octahedral $[Al(H_2O)_6]^{3+}$ ion.
In the light of the above statements, choose the correct answer from the options given below

(a) Both Statement I and Statement II are false.
(b) Both Statement I and Statement II are true.
(c) Statement I is false but Statement II is true.
(d) Statement I is true but Statement II is false.

19 Given below are two statements : one is labelled as **Assertion (A)** and the other is labelled as **Reason (R)**.
Assertion (A) Melting point of boron (245.3 K) is unusually high in group 13 elements. **[2024, 1 Feb Shift-I]**
Reason (R) Solid boron has very strong crystalline lattice.
In the light of the above statements, choose the **most appropriate** answer from the options given below

(a) Both (A) and (R) are correct and (R) is the correct explanation of (A).
(b) (A) is true but (R) is false.
(c) Both (A) and (R) are correct and (R) is not the correct explanation of (A).
(d) (A) is false but (R) is true.

20 The incorrect statement from the following for borazine is **[2023, 13 April Shift-I]**

(a) it is a cyclic compound.
(b) it contains banana bonds.
(c) it has electronic delocalisation.
(d) it can react with water.

21 For compound having the formula $GaAlCl_4$, the correct option from the following is **[2023, 11 April Shift-I]**

(a) Ga is more electronegative than Al and is present as a cationic part of the salt $GaAlCl_4$.
(b) Ga is coordinated with Cl in $GaAlCl_4$.
(c) Cl forms bond with both Al and Ga in $GaAlCl_4$.
(d) Oxidation state of Ga in the salt $GaAlCl_4$ is + 3.

22 Given below are two statements.

Statement I Upon heating a borax bead dipped in cupric sulphate in a luminous flame, the colour of the bead becomes green.

Statement II The green colour observed is due to the formation of copper (I) metaborate.

In the light of the above statements, choose the most appropriate answer from the options given below. **[2023, 31 Jan Shift-II]**

(a) Statement I is true but Statement II is false.
(b) Statement I is false but Statement II is true.
(c) Both Statement I and Statement II are true.
(d) Both Statement I and Statement II are false.

23 Given below are two statements.

Statement I The chlorides of Be and Al have Cl-bridged structure. Both are soluble in organic solvents and act as Lewis bases.

Statement II Hydroxides of Be and Al dissolve in excess alkali to give beryllate and aluminate ions.

In the light of the above statements, choose the correct answer from the options given below. **[2022, 27 July Shift-I]**

(a) Both Statement I and Statement II are true.
(b) Both Statement I and Statement II are false.
(c) Statement I is true but Statement II is false.
(d) Statement I is false but Statement II is true.

Imp. *Preparation of Borazine & Stoichiometry*

24 Borazine, also known as inorganic benzene, can be prepared by the reaction of 3-equivalents of "X" with 6-equivalents of "Y". "X" and "Y", respectively are **[2022, 26 July Shift-I]**

(a) $B(OH)_3$ and NH_3 (b) B_2H_6 and NH_3
(c) B_2H_6 and NH_3 (d) NH_3 and B_2O_3

25 Aqueous solution of which of the following boron compounds will be strongly basic in nature? **[2022, 29 June Shift-II]**

(a) $NaBH_4$ (b) $LiBH_4$
(c) B_2H_6 (d) $Na_2B_4O_7$

26 Given below are the statements about diborane.

(A) Diborane is prepared by the oxidation of $NaBH_4$ and I_2.
(B) Each boron atom is in sp^2-hybridised state.
(C) Diborane has one bridged 3 centre -2-electron bond.
(D) Diborane is a planar molecule.

The option with correct statement(s) is **[2021, 22 July Shift-II]**

(a) (C) and (D) only (b) (A) only
(c) (C) only (d) (A) and (B) only

27 Given below are two statements: one is labelled as Assertion (A) and the other is labelled as Reason (R). **[2021, 26 Feb Shift-II]**

Asertion (A) In TlI_3, isomorphous to CsI_3, the metal is present in +1 oxidation state.

Reason (R) Tl metal has fourteen f electrons in its electronic configuration.

In the light of the above statements, choose the most appropriate answer from the options given below:

(a) Both A and R are correct but R is not the correct explanation of A.
(b) Both A and R are correct and R is the correct explanation of A.
(c) A is correct but R is not correct.
(d) A is not correct but R is correct.

Imp. *Structure of Diborane*

28 The correct statement about B_2H_6 is **[2021, 25 Feb Shift-I]**

(a) all B—H—B angles are of 120°
(b) the two B—H—B bonds are not of same length
(c) terminal B—H bonds have less p-character when compared to bridging bonds
(d) Its fragment, BH_3, behaves as a Lewis base

29 The reaction of $H_3N_3B_3Cl_3$ (A) with $LiBH_4$ in tetrahydrofuran gives inorganic benzene (B). Further, the reaction of (A) with (C) leads to $H_3N_3B_3(Me)_3$. Compounds (B) and (C) respectively, are **[2020, 9 Jan Shift-II]**

(a) borazine and MeBr (b) diborane and MeMgBr
(c) boron nitride and MeBr (d) borazine and MeMgBr

30 The correct statements among I to III regarding group 13 element oxides are : **[2019, 9 April Shift-II]**

I. Boron trioxide is acidic.
II. Oxides of aluminium and gallium are amphoteric.
III. Oxides of indium and thallium are basic.

(a) I, II and III (b) I and III only
(c) I and II only (d) II and III only

31 Diborane (B_2H_6) reacts independently with O_2 and H_2O to produce, respectively **[2019, 8 April Shift-I]**

(a) B_2O_3 and H_3BO_3 (b) B_2O_3 and $[BH_4]^-$
(c) H_3BO_3 and B_2O_3 (d) HBO_2 and H_3BO_3

Level 3

32 The correct statements from the following are:

(A) Tl^{3+} is a powerful oxidising agent
(B) Al^{3+} does not get reduced easily
(C) Both Al^{3+} and Tl^{3+} are very stable in solution
(D) Tl^+ is more stable than Tl^{3+}
(E) Al^{3+} and Tl^+ are highly stable

Choose the correct answer from the options given below : **[2025, 7 April Shift-II]**

(a) (A), (B), (C), (D) and (E)
(b) (A), (C) and (D) only
(c) (A), (B), (D) and (E) only
(d) (B), (D) and (E) only

Imp. *Trends in Properties of Group 13*

33 Match List-I with List-II. **[2024, 9 April Shift-II]**

	List-I		List-II
A.	Melting point [K]	I.	Tl > In > Ga > Al > B
B.	Ionic radius [M^{3+}/pm]	II.	B > Tl > Al ≈ Ga > In
C.	$\Delta_i H_1$ [kJ mol^{-1}]	III.	Tl > In > Al > Ga > B
D.	Atomic radius [pm]	IV.	B > Al > Tl > In > Ga

Choose the correct answer from the options given below.
(a) A-II, B-III, C-IV, D-I
(b) A-III, B-IV, C-I, D-II
(c) A-I, B-II, C-III, D-IV
(d) A-IV, B-I, C-II, D-III

34 The correct statements from the following are
(A) the decreasing order of atomic radii of group 13 elements is Tl > In > Ga > Al > B. **[2024, 5April Shift-II]**
(B) down the group 13 electronegativity decreases from top to bottom.
(C) Al dissolves in dil. HCl and liberates H_2 but conc. HNO_3 renders Al passive by forming a protective oxide layer on the surface.
(D) all elements of group 13 exhibits highly stable +1 oxidation state.
(E) hybridisation of Al in $[Al(H_2O_6)]^{3+}$ ion is sp^3d^2.

Choose the correct answer from the options given below.
(a) (C) and (E) only (b) (A), (C) and (E) only
(c) (A), (B), (C) and (E) only (d) (A) and (C) only

35 Identify the correct statement for B_2H_6 from those given below :
(A) In B_2H_6, all B — H bonds are equivalent.
(B) In B_2H_6 there are four 3-centre-2-electron bonds.
(C) B_2H_6 is a Lewis acid.
(D) B_2H_6 can be synthesised from both BF_3 and $NaBH_4$.
(E) B_2H_6 is a planar molecule.

Choose the most appropriate answer from the options given below : **[2022, 24 June Shift-I]**
(a) (A) and (E) only (b) (B), (C) and (E) only
(c) (C) and (D) only (d) (C) and (E) only

Topic 2 / Group 14 Elements

• Level 1

01 Which of the following material is not a semiconductor? **[2024, 6 April Shift-I]**
(a) Silicon (b) Graphite
(c) Germanium (d) Copper oxide

02 Which one of the following compounds of group-14 elements is not known? **[2021, 25 July Shift-I]**
(a) $[GeCl_6]^{2-}$ (b) $[Sn(OH)_6]^{2-}$
(c) $[SiCl_6]^{2-}$ (d) $[SiF_6]^{2-}$

03 The amorphous form of silica is **[2019, 9 April Shift-II]**
(a) tridymite (b) cristobalite (c) kieselguhr (d) quartz

04 C_{60} an allotrope of carbon contains **[2019, 9 April Shift-I]**
(a) 16 hexagons and 16 pentagons
(b) 20 hexagons and 12 pentagons
(c) 12 hexagons and 20 pentagons
(d) 18 hexagons and 14 pentagons

05 The correct order of catenation is **[2019, 10 April Shift-I]**
(a) C > Sn > Si ≈ Ge
(b) Si > Sn > C > Ge
(c) C > Si > Ge ≈ Sn
(d) Ge > Sn > Si > C

• Level 2

06 Given below are two statements: **[2025, 4 April Shift-II]**
Statement (I): The first ionisation enthalpy of group 14 elements is higher than the corresponding elements of group 13.
Statement (II): Melting points and boiling points of group 13 elements are in general much higher than those of corresponding elements of group 14.
In the light of the above statements, choose the most appropriate answer from the options given below:
(a) Both Statement I and Statement II are correct
(b) Statement I is incorrect but Statement II is correct
(c) Statement I is correct but Statement II is incorrect
(d) Both Statement I and Statement II are incorrect

07 The group 14 elements A and B have the first ionisation enthalpy values of 708 and 715 kJ mol^{-1} respectively. The above values are lowest among their group members. The nature of their ions A^{2+} and B^{4+} respectively is **[2025, 7 April Shift-I]**
(a) both oxidising (b) oxidising and reducing
(c) both reducing (d) reducing and oxidising

Imp. *Oxidising Property of Ions*

08 The number of ions from the following that are expected to behave as oxidising agent is
Sn^{4+}, Sn^{2+}, Pb^{2+}, Tl^{3+}, Pb^{4+}, Tl^{+} **[2024, 6 April Shift-II]**
(a) 1 (b) 3
(c) 4 (d) 2

09 Given below are two statements. **[2024, 1 Feb Shift-II]**
Statement I SiO_2 and GeO_2 are acidic while SnO and PbO are amphoteric in nature.
Statement II Allotropic forms of carbon are due to property of catenation and $p\pi$ - $d\pi$ bond formation.

In the light of the above statements, choose the **most appropriate** answer from the options given below
(a) Both Statement I and Statement II are false.
(b) Statement I is true but Statement II is false.
(c) Statement I is false but Statement II is true.
(d) Both Statement I and Statement II are true.

10 Consider the oxides of group 14 elements. SiO_2, GeO_2, SnO_2, PbO_2, CO and GeO. The amphoteric oxides are **[2024, 31 Jan Shift-I]**
(a) SnO_2, CO (b) SnO_2, PbO_2
(c) SiO_2, GeO_2 (d) GeO, GeO_2

11 Given below are two statements.
Statement I The electronegativity of group 14 elements from Si to Pb, gradually decreases.
Statement II Group 14 contains non-metallic, metallic, as well as metalloid elements. **[2024, 29 Jan Shift-I]**
In the light of the above statements, choose the **most appropriate** from the options given below.
(a) Statement I is false but Statement II is true.
(b) Both Statement I and Statement II are false.
(c) Statement I is true but Statement II is false.
(d) Both Statement I and Statement II are true.

12 Given below are two statements, one is labelled as Assertion (A) and the other is labelled as Reason (R).
Assertion (A) Carbon forms two important oxides —CO and CO_2, CO is neutral whereas CO_2 is acidic in nature. **[2023, 25 Jan Shift-II]**
Reason (R) CO_2 can combine with water in a limited way to form carbonic acid, while CO is sparingly soluble in water.
In the light of the above statements, choose the most appropriate answer from the options given below.
(a) Both A and R are correct but R is not the correct explanation of A.
(b) Both A and R are correct and R is the correct explanation of A.
(c) A is correct but R is not correct.
(d) A is not correct but R is correct.

13 Given below are two statements. **[2022, 29 July Shift-II]**
Statement I Stannane is an example of a molecular hydride.
Statement II Stannane is a planar molecule.
In the light of the above statement, choose the most appropriate answer from the options given below
(a) Both Statement I and Statement II are true.
(b) Both Statement I and Statement II are false.
(c) Statement I is true but Statement II is false.
(d) Statement I is false but Statement II is true.

14 The incorrect statement regarding the structure of C_{60} is **[2021, 16 March Shift-II]**
(a) the six-membered rings are fused to both six and five-membered rings
(b) each carbon atom forms three sigma bonds
(c) the five-membered rings are fused only to six-membered rings
(d) it contains 12 six-membered rings and 24 five-membered rings

Imp. *Oxidising Agent in Lead Storage Battery*

15 Compound *A* used as a strong oxidising agent is amphoteric in nature. It is the part of lead storage batteries. Compound *A* is **[2021, 26 Feb Shift-I]**
(a) Pb_3O_4 (b) $PbSO_4$
(c) PbO (d) PbO_2

16 Water does not produce CO on reacting with **[2021, 25 Feb Shift-II]**
(a) CH_4 (b) C
(c) CO_2 (d) C_3H_8

17 '*X*' melts at low temperature and is a bad conductor of electricity in both liquid and solid state. *X* is **[2020, 9 Jan Shift-II]**
(a) carbon tetrachloride (b) silicon carbide
(c) mercury (d) zinc sulphide

18 The basic structural unit of feldspar, zeolites, mica and asbestos is **[2019, 12 April Shift-I]**
(a) $(SiO_3)^{2-}$ (b) SiO_2
(c) $(SiO_4)^{4-}$ (d) $-\!\left(\overset{R}{\underset{R}{\overset{|}{\underset{|}{Si}}}}-O\right)_{n}-$ ($R =$ Me)

19 The correct statement among the following is **[2019, 12 April Shift-I]**
(a) $(SiH_3)_3N$ is planar and less basic than $(CH_3)_3N$.
(b) $(SiH_3)_3N$ is pyramidal and more basic than $(CH_3)_3N$.
(c) $(SiH_3)_3N$ is pyramidal and less basic than $(CH_3)_3N$.
(d) $(SiH_3)_3N$ is planar and more basic than $(CH_3)_3N$.

20 The number of pentagons in C_{60} and trigons (triangles) in white phosphorus, respectively, are **[2019, 10 April Shift-II]**
(a) 20 and 3 (b) 12 and 4
(c) 20 and 4 (d) 12 and 3

21 The element that shows greater ability to form $p\pi$-$p\pi$ multiple bonds, is **[2019, 12 Jan Shift-II]**
(a) Ge (b) Si (c) Sn (d) C

• Level 3

22 Evaluate the following statements related to group 14 elements for their correctness. **[2024, 6 April Shift-II]**
A. Covalent radius decreases down the group from C to Pb in a regular manner.
B. Eletronegativity decreases from C to Pb down the group gradually.
C. Maximum covalance of C is 4 whereas other elements can expand their covalence due to presence of *d*-orbitals.

D. Heavier elements do not from $p\pi$-$p\pi$ bonds.

E. Carbon can exhibit negative oxidation states.

Choose the correct answer from the options given below.

(a) C and D only (b) C, D and E only
(c) A and B only (d) A, B and C only

23 Match List-I with List-II.

List-I (Metal)	List-II (Application)
(A) Cs	(I) High temperature thermometer
(B) Ga	(II) Water repellent sprays
(C) B	(III) Photoelectric cells
(D) Si	(IV) Bullet proof vest

Choose the most appropriate answer from the options given below **[2022, 29 June Shift-I]**

(a) (A)-(III), (B)-(I), (C)-(IV), (D)-(II)
(b) (A)-(IV), (B)-(III), (C)-(II), (D)-(I)
(c) (A)-(II), (B)-(III), (C)-(IV), (D)-(I)
(d) (A)-(I), (B)-(IV), (C)-(II), (D)-(III)

24 Correct statements among (I) to (IV) regarding silicones are **[2019, 9 Jan Shift-I]**

I. They are polymers with hydrophobic character.
II. They are biocompatible.
III. In general, they have high thermal stability and low dielectric strength.
IV. Usually, they are resistant to oxidation and used as greases.

(a) I and II only (b) I, II, III only
(c) I, II, III and IV (d) I, II and IV only

Key Index

Topic 1

1	(d)	**2**	(a)	**3**	(a)	**4**	(a)	**5**	(c)	**6**	(c)	**7**	(b)	**8**	(b)	**9**	(b)	**10**	(a)
11	(b)	**12**	(a)	**13**	(b)	**14**	(c)	**15**	(c)	**16**	(a)	**17**	(c)	**18**	(b)	**19**	(a)	**20**	(b)
21	(a)	**22**	(d)	**23**	(d)	**24**	(b)	**25**	(d)	**26**	(b)	**27**	(a)	**28**	(c)	**29**	(d)	**30**	(a)
31	(a)	**32**	(c)	**33**	(d)	**34**	(a)	**35**	(c)										

Topic 2

1	(b)	**2**	(c)	**3**	(c)	**4**	(b)	**5**	(c)	**6**	(c)	**7**	(d)	**8**	(d)	**9**	(b)	**10**	(b)
11	(a)	**12**	(b)	**13**	(c)	**14**	(d)	**15**	(d)	**16**	(c)	**17**	(a)	**18**	(c)	**19**	(a)	**20**	(b)
21	(d)	**22**	(b)	**23**	(a)	**24**	(d)												

For Detailed Solutions scan QR Code

10 Purification and Characterisation of Organic Compounds

In JEE Main there is usually 0-1 question per paper are visible from this chapter. In past 3 years, 32 conceptual and 15 numerical questions were asked of moderate difficulty level from this

For detailed analysis scan QR code

Topic 1 / Methods of Purification

• Level 1

01 The purification method based on the following physical transformation is **[2025, 28 Jan Shift-II]**

$$\underset{(X)}{\text{Solid}} \xrightarrow{\text{Heat}} \underset{(Y)}{\text{Vapour}} \xrightarrow{\text{Cool}} \underset{(Z)}{\text{Solid}}$$

(a) Extraction (b) Sublimation
(c) Distillation (d) Crystallisation

02 Which of the following statement is correct for paper chromatography? **[2023, 15 April Shift-I]**

(a) Water present in the pores of the paper forms the stationary phase.
(b) Paper sheet forms the stationary phase.
(c) Water present in the mobile phase gets absorbed by the paper which then forms the stationary phase.
(d) Paper and water present in its pores together form the stationary phase.

03 Given below are two statements, one is labelled as Assertion A and the other is labelled as Reason R.

Assertion (A) Thin layer chromatography is an adsorption chromatography.

Reason (R) A thin layer of silica gel is spread over a glass plate of suitable size in thin layer chromatography which acts as an adsorbent.

In the light of the above statements, choose the correct answer from the options given below **[2022, 28 July Shift-II]**

(a) Both A and R are true and R is the correct explanation of A.
(b) Both A and R are true but R is not the correct explanation of A.
(c) A is true but R is false.
(d) A is false but R is true.

04 Which technique among the following, is most appropriate in separation of a mixture of 100 mg of *p*-nitrophenol and picric acid? **[2022, 26 July Shift-I]**

(a) Steam distillation
(b) 2-5 ft long column of silica gel
(c) Sublimation
(d) Preparative TLC (Thin Layer Chromatography)

Imp. *Thin Layer Chromatography*

05 Which one of the following techniques is not used to spot components of a mixture separated on thin layer chromatographic plate? **[2022, 28 June Shift-I]**

(a) I_2 (Solid)
(b) UV light
(c) Visualisation agent as a component of mobile phase
(d) Spraying of an appropriate reagent

06 Which of the following is structure of a separating funnel? **[2022, 27 June Shift-I]**

(a) 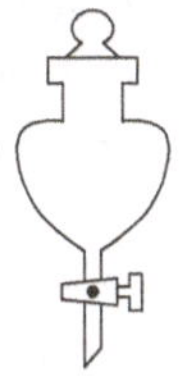

(b) 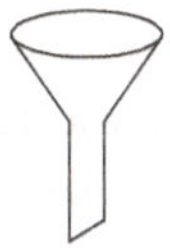

(c) 

(d) 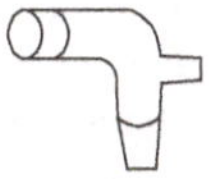

07 Which purification technique is used for high boiling organic liquid compound (decomposes near its boiling point)? **[2021, 22 July Shift-II]**

(a) Simple distillation
(b) Steam distillation
(c) Fractional distillation
(d) Reduced pressure distillation

08 The metal that can be purified economically by fractional distillation method is **[2021, 20 July Shift-I]**

(a) Fe (b) Zn
(c) Cu (d) Ni

09 In chromatography technique, the purification of compound is independent of **[2021, 16 March Shift-I]**

(a) mobility or flow of solvent system
(b) solubility of the compound
(c) length of the column or TLC plate
(d) physical state of the pure compound

10 In chromatography, which of the following statements is incorrect for R_f? **[2019, 10 April Shift-II]**
(a) R_f value depends on the type of chromatography
(b) Higher R_f value means higher adsorption
(c) R_f value is dependent on the mobile phase
(d) The value of R_f can not be more than one

• Level 2

11 The steam volatile compounds among the following are **[2025, 29 Jan Shift-I]**

A. 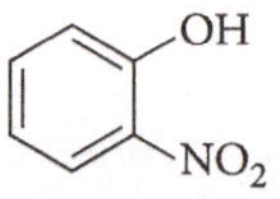

B. 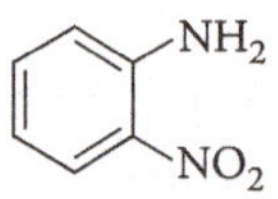

C. 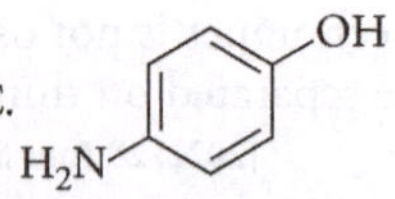

D. 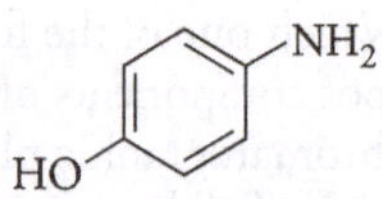

Choose the correct answer from the options given below
(a) (A) and (B) only (b) (A), (B) and (C) only
(c) (A) and (C) only (d) (B) and (D) only

12 Given below are two statements. **[2025, 29 Jan Shift-I]**
Statement I In partition chromatography, stationary phase is thin film of liquid present in the inert support.
Statement II In paper chromatography, the material of paper acts as a stationary phase.
In the light of the above statements, choose the correct answer from the options given below:
(a) Statement I is true but Statement II is false.
(b) Statement I is false but Statement II is true.
(c) Both Statement I and Statement II are false.
(d) Both Statement I and Statement II are true.

13 Methods used for purification of organic compounds are based on **[2024, 9 April Shift-I]**
(a) neither on nature of compound nor on the impurity present.
(b) nature of compound and presence of impurity.
(c) nature of compound only.
(d) presence of impurity only.

14 The adsorbent used in adsorption chromatography is/are- **[2024, 4 April Shift-II]**
A. silica gel B. alumina
C. quick lime D. magnesia
Choose the most appropriate answer from the options given below:
(a) A only (b) B only
(c) C and D only (d) A and B only

15 The fragrance of flowers is due to the presence of some steam volatile organic compounds called essential oils. These are generally insoluble in water at room temperature but are miscible with water vapour in vapour phase. A suitable method for the extraction of these oils from the flowers is **[2024, 31 Jan Shift-II]**
(a) distillation (b) crystallisation
(c) steam distillation
(d) distillation under reduced pressure

16 'Adsorption' principle is used for which of the following purification method? **[2024, 31 Jan Shift-I]**
(a) Sublimation (b) Chromatography
(c) Extraction (d) Distillation

17 Which among the following purification methods is based on the principle of "Solubility" in two different solvents? **[2024, 30 Jan Shift-II]**
(a) Column chromatography
(b) Differential extraction
(c) Distillation
(d) Sublimation

18 Chromatographic technique\s based on the principle of differential adsorption is\are **[2024, 29 Jan Shift-II]**
A. column chromatography
B. thin layer chromatography
C. paper chromatography
Choose the **most appropriate** answer from the options given below
(a) A only (b) B only (c) C only (d) A and B only

19 The technique used for purification of steam volatile water immiscibe substances is **[2024, 27 Jan Shift-II]**
(a) fractional distillation
(b) fractional distillation under reduced pressure
(c) distillation
(d) steam distillation

20 Thin layer chromatography of a mixture shows the following observation:

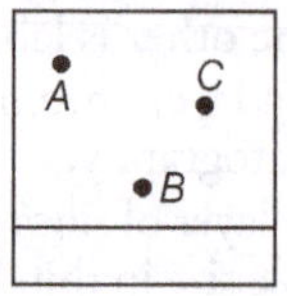

The correct order of elution in the silica gel column chromatography is **[2023, 11 April Shift-I]**
(a) *B, C, A* (b) *A, C, B* (c) *C, A, B* (d) *B, A, C*

21 From the figure of column chromatography given below, identify incorrect statements.

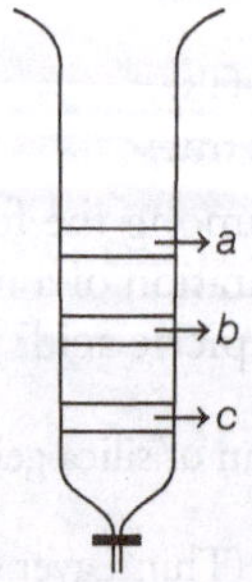

A. Compound '*c*' is more polar than '*a*' and '*b*'.
B. Compound '*a*' is least polar.
C. Compound '*b*' comes out of the column before '*c*' and after '*a*'.
D. Compound '*a*' spends more time in the column.
Choose the correct answer from the options given below: **[2023, 6 April Shift-II]**
(a) B, C and D only (b) A, B and D only
(c) A, B and C only (d) B and D only

22 Match List-I with List-II.

List-I (Mixture of compounds)		List-II (Separation technique)
A. H_2O/CH_2Cl_2	I.	Crystallisation
B. 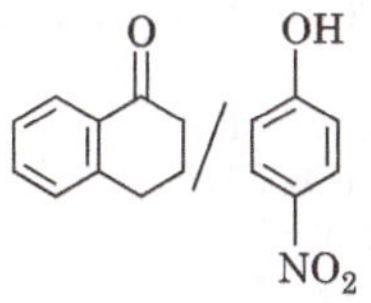	II.	Differential solvent extraction
C. Kerosene / naphthalene	III.	Column chromatography
D. $C_6H_{12}O_6/NaCl$	IV.	Fractional distillation

Correct match is **[2023, 31 Jan Shift-I]**
(a) A-I, B-III, C-II, D-IV (b) A-II, B-IV, C-I, D-III
(c) A -III, B-IV, C-II, D-I (d) A-II, B-III, C-IV, D-I

23 Match List-I with List-II. **[2023, 30 Jan Shift-II]**

List-I (Mixture)		List-II (Separation technique)
A. $CHCl_3 + C_6H_5NH_2$	I.	Steam distillation
B. $C_6H_{14} + C_5H_{12}$	II.	Differential extraction
C. $C_6H_5NH_2 + H_2O$	III.	Distillation
D. Organic compound in H_2O	IV.	Fractional distillation

(a) A-II, B-I, C-III, D-IV (b) A-IV, B-I, C-III, D-II
(c) A-III, B-I, C-IV, D-II (d) A-III, B-IV, C-I, D-II

24 Match List I with List II.

List I (Mixture)	List II (Purification process)
(A) Chloroform and aniline	(I) Steam distillation
(B) Benzoic acid and naphthalene	(II) Sublimation
(C) Water and aniline	(III) Distillation
(D) Naphthalene and sodium chloride	(IV) Crystallisation

Choose the correct answer from the options given below **[2022, 27 July Shift-II]**
(a) (A) → (IV), (B) → (III), (C) → (I), (D) → (II)
(b) (A) → (III), (B) → (I), (C) → (IV), (D) → (II)
(c) (A) → (III), (B) → (IV), (C) → (II), (D) → (I)
(d) (A) → (III), (B) → (IV), (C) → (I), (D) → (II)

25 TLC analysis of a mixture having 3 components (A, B, C) using silica gel as the stationary phase gave following R_f values; for $A = 0.72, B = 0.48, C = 0.20$. Regarding the above observations, which one of the following statements is not corect for column chromatography of the mixture? **[2022, 30 June Shift-I]**
(a) A is the most non-polar component.
(b) C is the most soluble component in mobile phase.
(c) A will be eluted first.
(d) C is the most strongly absorbed component.

26 Given below are two statements.
One is labelled as Assertion A and the other is labelled as Reason R.
Assertion (A) A mixture contains benzoic acid and naphthalene. The pure benzoic acid can be separated out by the use of benzene.
Reason (R) Benzoic acid is soluble in hot water.
In the light of the above statements, choose the most appropriate answer from the options given below. **[2022, 25 June Shift-II]**
(a) Both A and R are true and R is the correct explanation of A.
(b) Both A and R are true but R is not the correct explanation of A.
(c) A is true but R is false.
(d) A is false but R is true.

27 Phenol on reaction with dilute nitric acid, gives two products. Which method will be most effective for large scale separation? **[2022, 25 June Shift-I]**
(a) Chromatographic separation (b) Fractional crystallisation
(c) Steam distillation (d) Sublimation

Imp. *Method of Simple Distillation*

28 Given below are two statements.
One is labelled as Assertion (A) and the other is labelled as Reason (R).
Assertion (A) A simple distillation can be used to separate a mixture of propanol and propanone.
Reason (R) Two liquids with a difference of more than 20°C in their boiling points can be separated by simple distillations.
In the light of the above statements, choose the most appropriate answer from the options given below.
(a) (A) is false but (R) is true. **[2021, 31 Aug Shift-I]**
(b) Both (A) and (R) are correct but (R) is not the correct explanation of (A).
(c) (A) is true but (R) is false.
(d) Both (A) and (R) are correct and (R) is the correct explanation of (A).

29 Given below are two statements.

Statement I A mixture of chloroform and aniline can be separated by simple distillation.

Statement II When separating aniline from a mixture of aniline and water by steam distillation aniline boils below its boiling point.

In the light of the above statements, choose the most appropriate answer from the options given below. **[2021, 26 Feb Shift-I]**

(a) Statement I is false but statement II is true
(b) Both statement I and statement II are false
(c) Statement I is true but statement II is false
(d) Both statement I and statement II are true

30 Given below are two statements:

Statement I *o*-nitrophenol is steam volatile due to intramolecular hydrogen bonding.

Statement II *o*-nitrophenol has high melting due to hydrogen bonding.

In the light of the above statements, choose the most appropriate answer from the options given below. **[2021, 26 Feb Shift-I]**

(a) Statement I is true but statement II is false.
(b) Statement I is false but statement II is true.
(c) Both statement I and statement II are false.
(d) Both statement I and statement II are true.

Imp. *Technique of Fractional Distillation*

31 A flask contains a mixture of isohexane and 3-methylpentane. One of the liquids boils at 63°C while the other boils at 60°C. What is the best way to separate the two liquids and which one will be distilled out first? **[2020, 8 Jan Shift-I]**

(a) Fractional distillation 3-methylpentane
(b) Fractional distillation, isohexane
(c) Simple distillation, 3-methylpentane
(d) Simple distillation, isohexane

32 The correct match between items I and II is **[2019, 11 Jan Shift-I]**

Item I (Mixture)		Item II (Separation method)	
A.	H_2O : Sugar	P.	Sublimation
B.	H_2O : Aniline	Q.	Recrystallisation
C.	H_2O : Toluene	R.	Steam distillation
		S.	Differential extraction

(a) (A) → (Q); (B) → (R); (C) → (S)
(b) (A) → (Q); (B) → (R); (C) → (P)
(c) (A) → (S); (B) → (R); (C) → (P)
(d) (A) → (R); (B) → (P); (C) → (S)

• Level 3

33 Match List-I with List-II. **[2025, 4 April Shift-II]**

List-I (Separation of)		List-II (Separation Technique)	
(A)	Aniline from aniline-water mixture	(I)	Simple distillation
(B)	Glycerol from spent-lye in soap industry	(II)	Fractional distillation
(C)	Different fractions of crude oil in petroleum industry	(III)	Distillation at reduced pressure
(D)	Chloroform-Aniline mixture	(IV)	Steam distillation

Choose the correct answer from the options given below:

(a) (A)-(IV), (B)-(III), (C)-(II), (D)-(I)
(b) (A)-(III), (B)-(IV), (C)-(I), (D)-(II)
(c) (A)-(II), (B)-(I), (C)-(IV), (D)-(III)
(d) (A)-(I), (B)-(II), (C)-(III), (D)-(IV)

34 Match List I with list II **[2025, 2 April Shift-II]**

List I (Purification technique)		List II (Mixture of organic compounds)	
A.	Distillation (simple)	I.	Diesel + Petrol
B.	Fractional distillation	II.	Aniline + Water
C.	Distillation under reduced pressure	III.	Chloroform + Aniline
D.	Steam distillation	IV.	Glycerol + Spent-lye

Choose the correct answer from the options given below.

(a) (A)-(III), (B)-(IV), (C)-(II), (D)-(I)
(b) (A)-(III), (B)-(I), (C)-(IV), (D)-(II)
(c) (A)-(II), (B)-(IV), (C)-(I), (D)-(III)
(d) (A)-(II), (B)-(III), (C)-(IV), (D)-(I)

35 Which of the following statements are correct? **[2024, 6 April Shift-I]**

A. Glycerol is purified by vacuum distillation because it decomposes at its normal boiling point.
B. Aniline can be purified by steam distillation as aniline is miscible in water.
C. Ethanol can be separated from ethanol water mixture by azeotropic distillation because it forms azeotrope.
D. An organic compound is pure, if mixed M.P. is remained same.

Choose the most appropriate answer from the options given below.

(a) A, C, D only
(b) B, C, D only
(c) A, B, C only
(d) A, B, D only

36 Match List-I with List-II. **[2024, 31 Jan Shift-I]**

List-I (Technique)	List-II (Application)
A. Distillation	I. Separation of glycerol from spent-lye
B. Fractional distillation	II. Aniline-water mixture
C. Steam distillation	III. Separation of crude oil fractions
D. Distillation under reduced pressure	IV. Chloroform-aniline

Choose the correct answer from the options given below:

	A	B	C	D
(a)	II	III	I	IV
(b)	I	II	IV	III
(c)	IV	I	II	III
(d)	IV	III	II	I

37 The correct statement among the following for a 'chromatography' purification method is **[2024, 6 April Shift-II]**

(a) non-polar compounds are retained at top and polar compounds come down in column chromatography.
(b) organic compounds run faster than solvent in the thin layer chromatographic plate.
(c) R_f of a polar compound is smaller than that of a non-polar compound.
(d) R_f is an integral value.

Numerical Value Type Questions

38

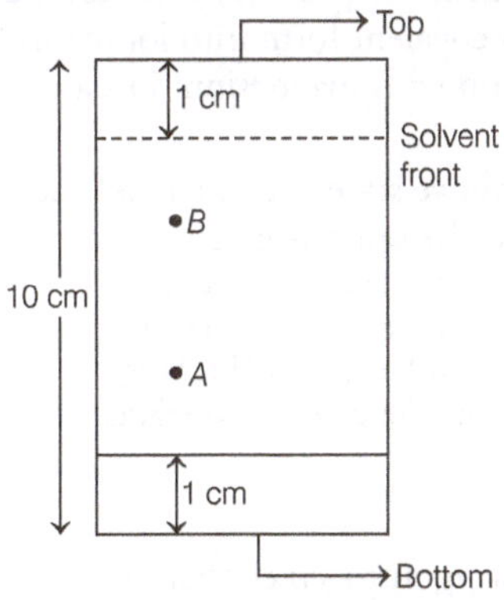

In the given TLC, the distance of spot A and B are 5 cm and 7 cm, from the bottom of TLC plate, respectively.

R_f value of B is $x \times 10^{-1}$ times more than A. The value of x is .... **[2024, 9 April Shift-II]**

39 Using the given figure, the ratio of R_f values of sample A and sample C is $x \times 10^{-2}$. Value of x is .... **[2024, 5 April Shift-II]**

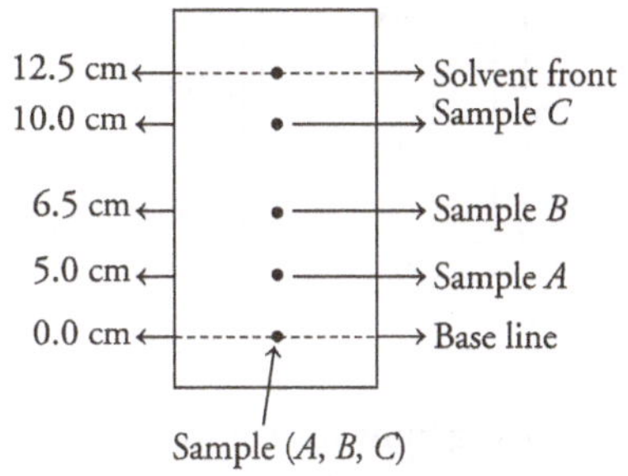

Paper chromatography of samples

40 On a thin layer chromatographic plate, an organic compound moved by 3.5 cm, while the solvent moved by 5 cm. The retardation factor of the organic compound is .... $\times 10^{-1}$. **[2024, 30 Jan Shift-I]**

41 Three organic compounds A, B and C were allowed to run in thin layer chromatography using hexane and gave the following result (see figure). The R_f value of the most polar compound is .... $\times 10^{-2}$.

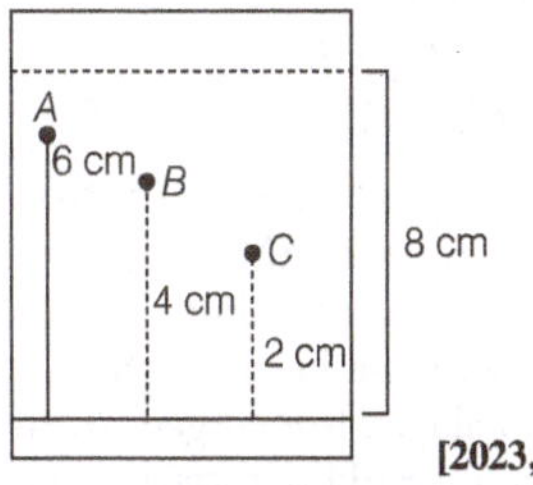

[2023, 12 April Shift-I]

42 Following chromatogram was developed by adsorption of compound 'A' on a 6 cm TLC glass plate. Retardation factor of the compound 'A' is .... $\times 10^{-1}$. **[2023, 29 Jan Shift-I]**

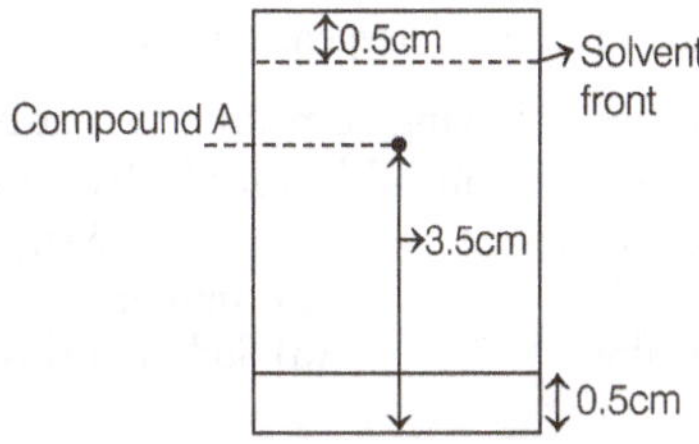

43 The separation of two coloured substances was done by paper chromatography. The distances travelled by solvent front, substance A and substance B from the base line are 3.25 cm, 2.08 cm and 1.05 cm, respectively. The ratio of R_f values of A to B is......... . **[2022, 25 July Shift-II]**

44 Using the provided information in the following paper chromatogram **[2021, 25 Feb Shift-I]**

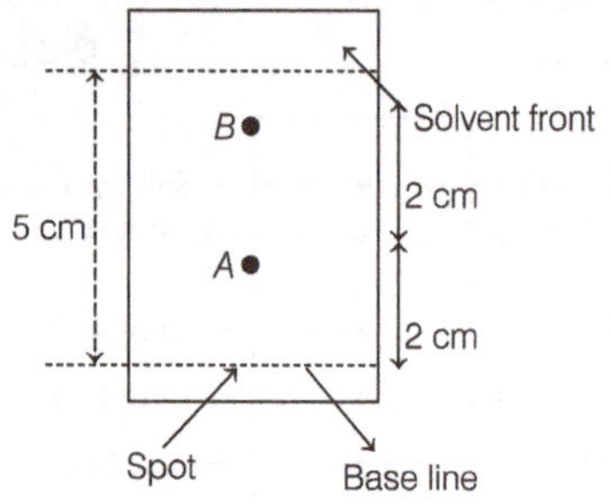

Paper chromatography for compounds A and B, then calculated R_f value of A .... $\times 10^{-1}$.

Topic 2 / Qualitative Analysis

• Level 1

01 Fe^{3+} cation gives a prussian blue precipitate on addition of potassium ferrocyanide solution due to the formation of **[2022, 27 July Shift-II]**

(a) $[Fe(H_2O)_6]_2[Fe(CN)_6]$
(b) $Fe_2[Fe(CN)_6]$
(c) $Fe_3[Fe(OH)_2(CN)_4]_2$
(d) $Fe_4[Fe(CN)_6]_3$

02 During halogen test, sodium fusion extract is boiled with concentrated HNO_3 to **[2022, 25 June Shift-II]**

(a) remove unreacted sodium
(b) decompose cyanide or sulphide of sodium
(c) extract halogen from organic compound
(d) maintain the pH of extract

03 Which one of the following set of elements can be detected using sodium fusion extract? **[2021, 26 July Shift-I]**

(a) Phosphorus, oxygen, nitrogen, halogens
(b) Nitrogen, phosphorus, carbon, sulphur
(c) Sulphur, nitrogen, phosphorus, halogens
(d) Halogens, nitrogen, oxygen, sulphur

04 Which of the following compounds is added to the sodium extract before addition of silver nitrate for testing of halogen? **[2021, 25 Feb Shift-II]**

(a) Nitric acid (b) Ammonia
(c) Hydrochloric acid (d) Sodium hydroxide

• Level 2

05 Formation of $Na_4[Fe(CN)_5NOS]$, a purple coloured complex formed by addition of sodium nitroprusside in sodium carbonate extract of salt indicates the presence of **[2025, 2 April Shift-II]**

(a) sulphate ion (b) sodium ion
(c) sulphide ion (d) sulphite ion

Imp. *Duma's Method*

06 Given below are two statements I and II. **[2025, 24 Jan Shift-I]**

Statement I Dumas method is used for estimation of "Nitrogen" in an organic compound.

Statement II Dumas method involves the formation of ammonium sulphate by heating the organic compound with conc. H_2SO_4.

In the light of the above statements, choose the correct answer from the options given below.

(a) Statement I is true but Statement II is false.
(b) Both Statement I and Statement II are false.
(c) Both Statement I and Statement II are true.
(d) Statement I is false but Statement II is true.

07 Given below are two statements. **[2025, 23 Jan Shift-I]**

Statement I In Lassaigne's test, the covalent organic molecules are transformed into ionic compounds.

Statement II The sodium fusion extract of an organic compound having N and S gives prussian blue colour with $FeSO_4$ and $Na_4[Fe(CN)_6]$.

In the light of the above statements, choose the correct answer from the options given below

(a) Statement I is false but Statement II is true.
(b) Both Statement I and Statement II is true.
(c) Both Statement I and Statement II are false.
(d) Statement I is true but Statement II is false.

08 Given below are two statements. **[2025, 22 Jan Shift-II]**

Statement I Nitrogen, sulphur, halogen and phosphorus present in an organic compound are detected by Lassaigne's test.

Statement II The elements present in the compound are converted from covalent form into ionic form by fusing the compound with magnesium in Lassaigne's test.

In the light of the above statements, choose the correct answer from the options given below:

(a) Statement I is false but Statement II is true.
(b) Both Statement I and Statement II are true.
(c) Statement I is true but Statement II is false.
(d) Both Statement I and Statement II are false.

Imp. *Lassaigne's Tests*

09 Which of the following nitrogen containing compound does not give Lassaigne's test? **[2024, 4th April Shift-I]**

(a) Hydrazine (b) Glycine
(c) Phenyl hydrazine (d)Urea

10 Compound that will give positive Lassaigne's test for both nitrogen and halogen is **[2023, 29 Jan Shift-I]**

(a) $NH_2OH \cdot HCl$ (b) NH_4Cl
(c) $N_2H_4 \cdot HCl$ (d) $CH_3NH_2 \cdot HCl$

11 Reagent, 1-naphthylamine and sulphanilic acid in acetic acid is used for the detection of **[2021, 18 March Shift-I]**
(a) N_2O (b) NO_3^- (c) NO (d) NO_2^-

Topic 3 / Quantitative Analysis

• Level 1

01 In Duma's method for estimation of nitrogen 0.4 g of an organic compound gave 60 mL of nitrogen collected at 300 K temperature and 715 mm Hg pressure. The percentage composition of nitrogen in the compound is **[2025, 3 April Shift-II]**
(Given: Aqueous tension at 300 K = 15 mm Hg)
(a) 20.95% (b) 17.46%
(c) 15.71% (d) 7.85%

02 In Duma's method for estimation of nitrogen, 0.5 g of an organic compound gave 60 mL of nitrogen collected at 300 K temperature and 715 mm Hg pressure. The percentage composition of nitrogen in the compound (Aqueous tension at 300 K = 15 mm Hg) is .... %. **[2025, 2 April Shift-II]**
(a) 20.87 (b) 18.67 (c) 12.57 (d) 1.257

03 In Duma's method for the estimation of N_2, the sample is heated with copper oxide and the gas evolved is passed over **[2023, 31 Jan Shift-II]**
(a) copper gauze (b) Pd
(c) Ni (d) copper oxide

04 In Carius method, halogen containing organic compound is heated with fuming nitric acid in the presence of **[2021, 20 July Shift-II**
(a) HNO_3 (b) $AgNO_3$
(c) $CuSO_4$ (d) $BaSO_4$

05 Nitrogen can be estimated by Kjeldahl's method for which of the following compound ? **[2021, 17 March Shift-II]**

(a) 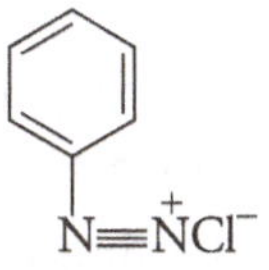

(b) 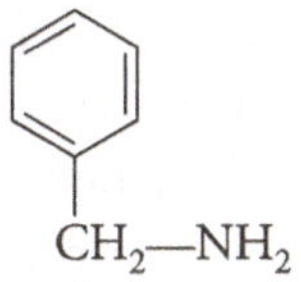

(c) 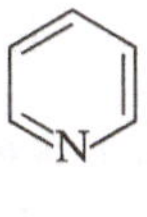

(d) 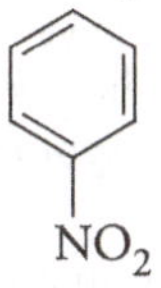

• Level 2

06 In Kjeldahl's method for estimation of nitrogen, $CuSO_4$ acts as **[2024, 1 Feb Shift-I]**
(a) hydrolysis agent (b) reducing agent
(c) catalytic agent (d) oxidising agent

07 In Carius method of estimation of halogen, 0.45 g of an organic compound gave 0.36 g of AgBr. Find out the percentage of bromine in the compound.
(Molar masses gBr = 188 g mol^{-1}; Br = 80 g mol^{-1}) **[2022, 27 July Shift-I]**
(a) 34.04% (b) 40.04% (c) 36.03% (d) 38.04%

08 120 g of an organic compound that contains only carbon and hydrogen gives 330 g of CO_2 and 270 g of water on complete combustion. The percentage of carbon and hydrogen, respectively are **[2022, 24 June Shift-II]**
(a) 25 and 75 (b) 40 and 60
(c) 60 and 40 (d) 75 and 25

09 Which of the following is a false statement? **[2021, 26 Feb Shift-I]**
(a) Kjeldahl's method is used for the estimation of nitrogen in an organic compound.
(b) Phosphoric acid produced on oxidation of phosphorus present in an organic compound is precipitated as $Mg_2P_2O_7$ by adding magnesia mixture.
(c) Carius tube is used in the estimation of sulphur in an organic compound.
(d) Carius method is used for the estimation of nitrogen in an organic compound.

10 An organic compound is estimated through Dumas method and was found to evolved 6 moles of CO_2, 4 moles of H_2O and 1 mole of nitrogen gas. The formula of the compound is **[2019, 11 Jan Shift-I]**
(a) C_6H_8N (b) $C_{12}H_8N$ (c) $C_{12}H_8N_2$ (d) $C_6H_8N_2$

• Level 3

11 Given below are two statements : **[2024, 8 April Shift-II]**
Statement I Kjeldahl method is applicable to estimate nitrogen in pyridine.
Statement II The nitrogen present in pyridine can easily be converted into ammonium sulphate in Kjeldahl method.
In the light of the above statements, choose the correct answer from the options given below:
(a) Statement I is false but Statement II is true.
(b) Both Statement I and Statement II are false.
(c) Both Statement I and Statement II are true.
(d) Statement I is true but Statement II is false.

Imp. *Carius Method*

12 In Carius tube, an organic compound '*X*' is treated with sodium peroxide to form a mineral acid '*Y*'. The solution of $BaCl_2$ is added to '*Y*' to form a precipitate '*Z*'. '*Z*' is used for the quantitative estimation of an extra element. '*X*' could be **[2023, 10 April Shift-II]**
(a) cytosine (b) methionine
(c) a nucleotide (d) chloroxylenol

13 In Carius method of estimation of halogen 0.172 g of an organic compound showed presence of 0.08 g of bromine. Which of these is the correct structure of the compound? **[2020, 2 Sep Shift-I]**

(a) $H_3C—Br$ (b) $H_3C—CH_2Br$

(c) 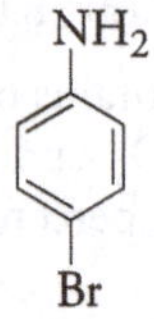 (d) 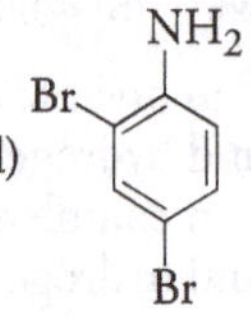

Numerical Value Type Questions

14 In Duma's method 292 mg of an organic compound released 50 mL of nitrogen gas (N_2) at 300 K temperature and 715 mm Hg pressure. The percentage composition of 'N' in the organic compound is .... % (Nearest integer)

(Aqueous tension at 300 K = 15 mm Hg) **[2025, 7 April Shift-II]**

15 During estimation of nitrogen by Duma's method of compound X (0.42 g) **[2025, 3 April Shift-I]**

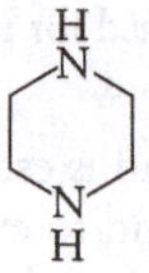

X

.... mL of N_2 gas will be liberated at STP. (Nearest integer)

(Given molar mass in g mol^{-1}: C : 12, H : 1, N : 14)

16 In Duma's method for estimation of nitrogen 1 g of an organic compound gave 150 mL of nitrogen collected at 300 K temperature and 900 mm Hg pressure. The percentage composition of nitrogen in the compound is .... % (Nearest integer)

(Aqueous tension at 300 K = 15 mm Hg) **[2025, 4 April Shift-I]**

17 In Carius method of estimation of halogen, 0.25 g of an organic compound gave 0.15 g of silver bromide (AgBr). The percentage of bromine in the organic compound is .... $\times 10^{-1}$% (Nearest integer)

(Given: Molar mass of Ag is 108 and Br is 80 g mol^{-1}) **[2025, 24 Jan Shift-II]**

18 During "S" estimation, 160 mg of an organic compound gives 466 mg of barium sulphate. The percentage of sulphur in the given compound is ....%.

(Given, molar mass in g mol^{-1} of Ba : 137, S : 32, O : 16) **[2025, 23 Jan Shift-I]**

19 In Carius method for estimation of halogens, 180 mg of an organic compound produced 143.5 mg of AgCl. The percentage composition of chlorine in the compound is ...... %.

(Given, molar mass in g mol^{-1} of Ag : 108, Cl : 35.5) **[2025, 22 Jan Shift-I]**

20 In the sulphur estimation, 0.20 g of a pure organic compound gave 0.40 g of barium sulphate. The percentage of sulphur in the compound is .... $\times 10^{-1}$%.

(Molar mass : O = 16, S = 32, Ba = 137 in g mol^{-1}) **[2025, 29 Jan Shift-II]**

21 Following Kjeldahl's method, 1g of organic compound released ammonia, that neutralised 10 mL of 2M H_2SO_4. The percentage of nitrogen in the compound is .... %. **[2024, 1st Feb Shift-II]**

22 0.400 g of an organic compound (X) gave 0.376 g of AgBr in Carius method for estimation of bromine. % of bromine in the compound (X) is ....

(Given : Molar mass AgBr = 188 g mol^{-1} Br = 80 g mol^{-1}) **[2023, 10 April Shift-II]**

23 In sulphur estimation, 0.471 g of an organic compound gave 1.4439 g of barium sulphate. The percentage of sulphur in the compound is .... (Nearest integer)

(Given : atomic mass Ba = 137 u, S = 32 u, O = 16 u) **[2023, 25 Jan Shift-I]**

24 The complete combustion of 0.492 g of an organic compound containing 'C', 'H' and 'O' gives 0.793 g of CO_2 and 0.442 g of H_2O. The percentage of oxygen composition in the organic compound is .... (Nearest integer) **[2022, 28 June Shift-II]**

25 0.25 g of an organic compound containing chlorine gave 0.40 g of silver chloride in Carius estimation. The percentage of chlorine present in the compound is .... [Nearest integer]

(Given, molar mass of Ag is 108 g mol^{-1} and that of Cl is 35.5 g mol^{-1}) **[2022, 27 June Shift-II]**

26 On complete combustion 0.30 g of an organic compound gave 0.20 g of carbon dioxide and 0.10 g of water. The percentage of carbon in the given organic compound is .... (Nearest integer) **[2022, 26 June Shift-I]**

27 0.2 g of an organic compound was subjected to estimation of nitrogen by Duma's method in which volume of N_2 evolved (at STP) was found to be 22.400 mL. The percentage of nitrogen in the compound is .... [Nearest integer]

(Given molar mass of N_2 is 28 mol^{-1}. Molar volume of N_2 at STP 22.4 L) **[2022, 24 June Shift-II]**

28 A 0.166 g of sample of an organic compound was digested with conc. H_2SO_4 and then distilled with NaOH. The ammonia gas evolved was passed through 50.0 mL of 0.5 N H_2SO_4. The used acid required 30.0 mL of 0.25 N NaOH for complete neutralisation. The mass percentage of nitrogen in the organic compound is .... **[2022, 24 June Shift-I]**

29 The transformation occurring in Duma's method is given below

$$C_2H_7N + \left(2x + \frac{y}{2}\right) CuO \longrightarrow xCO_2 + \frac{y}{2} H_2O + \frac{z}{2} N_2 + \left(2x + \frac{y}{2}\right) Cu$$

The value of y is .... (Integer answer) **[2021, 31 Aug Shift-II]**

30 In Carius method for estimation of halogens, 0.2 g of an organic compound gave 0.188 g of AgBr. The percentage of bromine in the compound is .... (Nearest integer)
[Atomic mass; Ag = 108, Br = 80] **[2021, 27 Aug Shift-I]**

31 In the sulphur estimation, 0.471 g of an organic compound gave 1.44 g of barium sulphate. The percentage of sulphur in the compound is .... %. (Nearest integer)
(Atomic mass of Ba = 137 u) **[2021, 26 Aug Shift-II]**

32 In an estimation of bromine by Carius method, 1.6 g of an organic compound gave 1.88 g of AgBr. The mass percentage of bromine in the compound is .... (Atomic mass, Ag = 108, Br = 80 g mol^{-1}) **[2020, 6 Sep Shift-I]**

Key Index

Topic 1

1 (b)	**2** (a)	**3** (a)	**4** (d)	**5** (c)	**6** (a)	**7** (d)	**8** (b)	**9** (d)	**10** (b)
11 (a)	**12** (a)	**13** (b)	**14** (d)	**15** (c)	**16** (b)	**17** (b)	**18** (d)	**19** (d)	**20** (b)
21 (c)	**22** (d)	**23** (d)	**24** (d)	**25** (b)	**26** (d)	**27** (c)	**28** (d)	**29** (d)	**30** (a)
31 (b)	**32** (a)	**33** (a)	**34** (b)	**35** (a)	**36** (d)	**37** (c)	**38** (15)	**39** (50)	**40** (7)
41 (25)	**42** (6)	**43** (2)	**44** (4)						

Topic 2

1 (d)	**2** (b)	**3** (c)	**4** (a)	**5** (c)	**6** (a)	**7** (d)	**8** (c)	**9** (a)	**10** (d)
11 (d)									

Topic 3

1 (c)	**2** (c)	**3** (a)	**4** (b)	**5** (b)	**6** (c)	**7** (a)	**8** (d)	**9** (d)	**10** (d)
11 (b)	**12** (b)	**13** (c)	**14** (18)	**15** (111)	**16** (20)	**17** (255)	**18** (40)	**19** (20)	**20** (275)
21 (56)	**22** (40)	**23** (42)	**24** (46)	**25** (40)	**26** (18)	**27** (14)	**28** (63)	**29** (7)	**30** (40)
31 (42)	**32** (50)								

For Detailed Solutions scan QR Code

11 Some Basic Principles of Organic Chemistry

In JEE Main there is usually 1-2 question per paper are visible from this chapter. In past 3 years, 44 conceptual and 16 numerical questions were asked of moderate difficulty level from this

For detailed analysis scan QR code

Topic 1 / Nomenclature, Hybridisation & Classification

• Level 1

01 The IUPAC name of the following compound is

$$HC \equiv C-CH_2-CH(OH)-CH_2-CH=CH_2$$

[2025, 4 April Shift-II]

(a) 4-hydroxyhept-6-en-1-yne
(b) 4-hydroxyhept-1-en-6-yne
(c) Hept-6-en-1-yn-4-ol
(d) Hept-1-en-6-yn-4-ol

02 What is the correct IUPAC name of

? **[2025, 8 April Shift-II]**

(a) 4-ethylcyclopent-2-en-1-ol
(b) 4-ethyl-1-hydroxycyclopent-2-ene
(c) 1-ethyl-3-hydroxycyclopent-2-ene
(d) 1-ethylcyclopent-2-en-3-ol

03 Which of the following is the correct IUPAC name of given organic compound (*X*)? **[2025, 7 April Shift-I]**

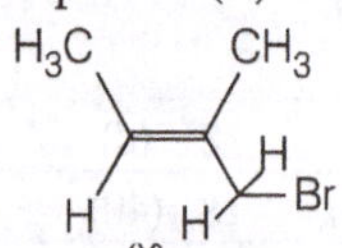

(*X*)

(a) 1-bromo-2-methylbut-2-ene
(b) 3-bromo-3-methylprop-2-ene
(c) 2-bromo-2-methylbut-2-ene
(d) 4-bromo-3-methylbut-2-ene

04 What is the correct IUPAC name of

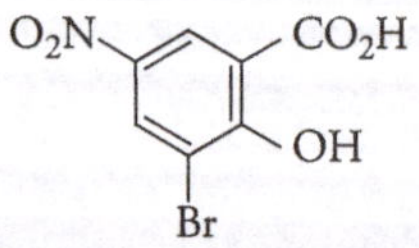

[2025, 3 April Shift-II]

(a) 3-bromo-4-hydroxy-1-nitrobenzoic acid
(b) 3-bromo-2-hydroxy-5-nitrobenzoic acid
(c) 5-nitro-3-bromo-2-hydroxybenzoic acid
(d) 2-hydroxy-3-bromo-5-nitrobenzoic acid

05 In 3,3-dimethylhex-1-en-4-yne, there are ....sp^3, ...sp^2 and ....sp hybridised carbon atoms respectively. **[2025, 2 April Shift-II]**

(a) 4, 2, 2 (b) 3, 3, 2
(c) 2, 4, 2 (d) 2, 2, 4

06 The IUPAC name of the following compound is

$$CH_3-CH(COOH)-CH_2-CH_2-CH(COOCH_3)-CH_3$$

[2025, 22 Jan Shift-I]

(a) Methyl-6-carboxy-2, 5-dimethylhexanoate
(b) Methyl-5-carboxy-2-methylhexanoate
(c) 2-carboxy-5-methoxycarbonylhexane
(d) 6-methoxycarbonyl-2, 5-dimethylhexanoic acid

07 In the given structure, number of sp and sp^2 hybridised carbon atoms present respectively are

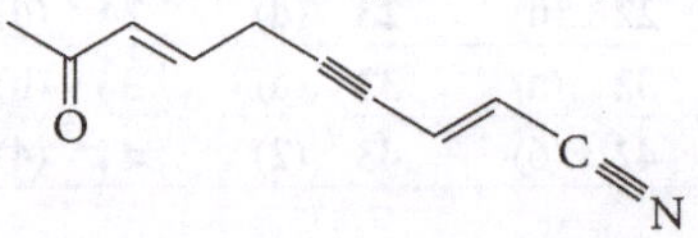

[2025, 24 Jan Shift-II]

(a) 3 and 5 (b) 3 and 6
(c) 4 and 6 (d) 4 and 5

08 IUPAC name of following hydrocarbon (*X*) is

$$CH_3-CH(CH_3)-CH_2CH_2-CH(CH_3)-CH(CH_3)-CH_2CH_3$$

(*X*)

[2024, 9 April Shift-II]

(a) 2-ethyl-2, 6-diethylheptane
(b) 3, 4, 7-trimethyloctane
(c) 2-ethyl-3, 6-dimethylheptane
(d) 2, 5, 6-trimethyloctane

09 The correct nomenclature for the following compound is

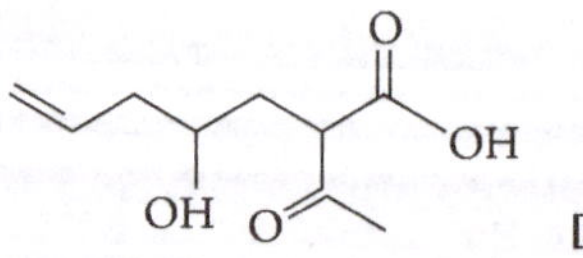

[2024, 5 April Shift-II]

(a) 2-carboxy -4-hydroxyhept-7-enal
(b) 2-formyl-4-hydroxyhept-6-enoic acid
(c) 2-formyl-4-hydroxyhept-7-enoic acid
(d) 2-carboxy-4-hydroxyhept-6-enal

10 Identify structure of 2, 3-dibromo-1-phenylpentane. **[2024, 31 Jan Shift-II]**

(a) 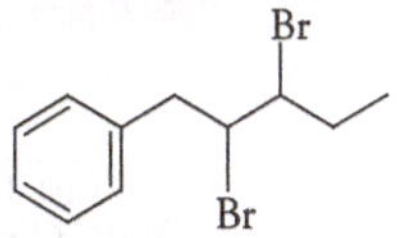
(b) 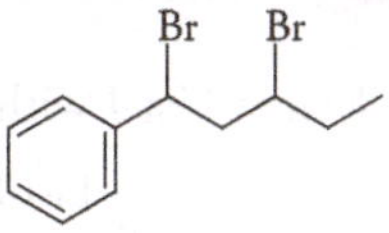
(c) 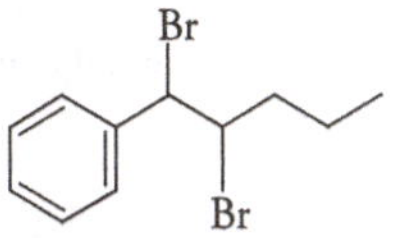
(d) 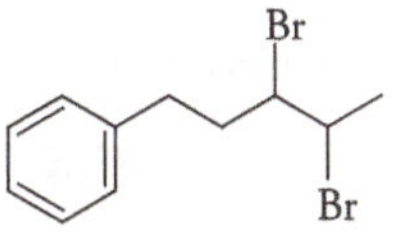

11 IUPAC name of following compound is

$CH_3—\underset{\underset{NH_2}{|}}{CH}—CH_2—CN$ **[2024, 30 Jan Shift-II]**

(a) 2-aminobutanenitrile (b) 2-aminopentanenitrile
(c) 3-aminopropanenitrile (d) 3-aminobutanenitrile

12 Bond line formula of $HOCH(CN)_2$ is **[2024, 27 Jan Shift-II]**

(a) 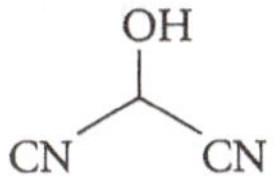
(b) 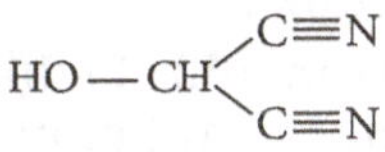
(c) 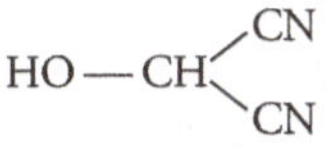
(d) 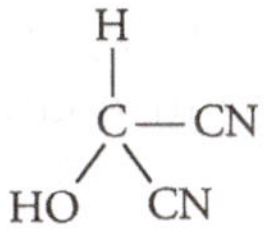

13 Cyclohexene 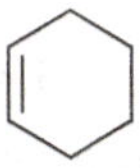 is .... type of an organic compound. **[2024, 27 Jan Shift-I]**

(a) benzenoid aromatic (b) alicyclic
(c) benzenoid non-aromatic (d) acyclic

14 IUPAC name of following compound (*P*) is

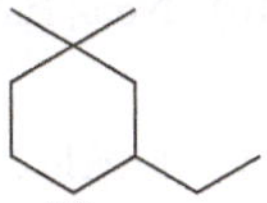
(*P*) **[2024, 27 Jan Shift-I]**

(a) 1, 1-dimethyl-3-ethylcyclohexane
(b) 3-ethyl-1, 1-dimethylcyclohexane
(c) 1-ethyl-3, 3-dimethylcyclohexane
(d) 1-ethyl-5, 5-dimethylcyclohexane

Imp. *Priority Order of Functional Group*

15 The correct decreasing order of priority of functional groups in naming an organic compound as per IUPAC system of nomenclature is **[2022, 26 July Shift-II]**

(a) $—COOH > —CONH_2 > —COCl > —CHO$
(b) $—SO_3H > —COCl > —CONH_2 > —CN$
(c) $—COOR > —COCl > —NH_2 > \rangle C{=}O$
(d) $—COOH > —COOR > —CONH_2 > —COCl$

16 The correct IUPAC name of the following compound is **[2022, 28 June Shift-II]**

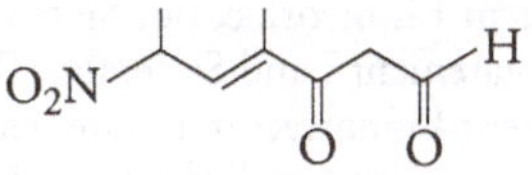

(a) 4-methyl-2-nitro-5-oxohept-3 -enal
(b) 4-methyl-5-oxo-2-nitrohept-3 -enal
(c) 4-methyl-6-nitro-3-oxohept-4 -enal
(d) 6-formyl-4-methyl-2-nitrohex- 3-enal

17 In the following molecules,

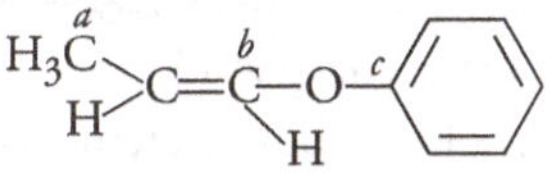

hybridisation of carbon *a*, *b* and *c* respectively are **[2021, 18 March Shift-II]**

(a) sp^3, sp, sp (b) sp^3, sp^2, sp
(c) sp^3, sp^2, sp^2 (d) sp^3, sp, sp^2

18 In $\overset{1}{C}H_2{=}\overset{2}{C}{=}\overset{3}{C}H—\overset{4}{C}H_3$ molecule, the hybridisation of carbon 1, 2, 3 and 4 respectively, are **[2021, 26 Feb Shift-II]**

(a) sp^3, sp, sp^3, sp^3 (b) sp^2, sp^3, sp^2, sp^3
(c) sp^2, sp^2, sp^2, sp^3 (d) sp^2, sp, sp^2, sp^3

19 The IUPAC name of the following compound is

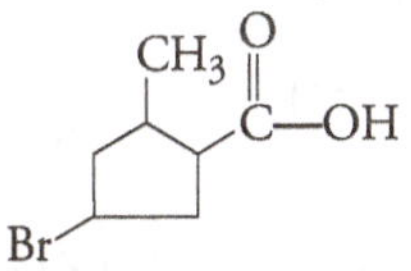
[2020, 4 Sep Shift-I]

(a) 3-bromo-5-methylcyclopentane carboxylic acid
(b) 3-bromo-5-methylcyclopentanoic acid
(c) 5-bromo-3-methylcyclopentanoic acid
(d) 4-bromo-2-methylcyclopentane carboxylic acid

• Level 2

20 Given below are two statements : **[2024, 8 April Shift-I]**

Statement I

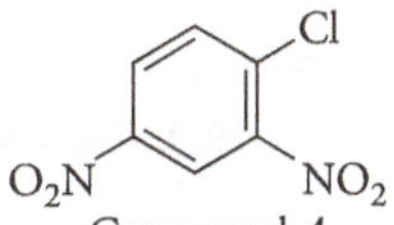
Compound *A*

IUPAC name of compound *A* is 4-chloro-1, 3-dinitrobenzene

Statement II

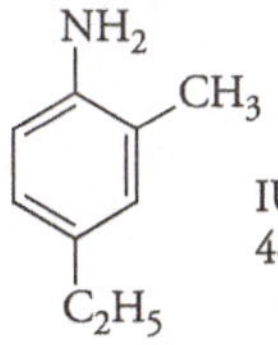
Compound *B*

IUPAC name of compound *B* is 4-ethyl-2-methylaniline

In the light of the above statements. Choose the most appropriate answer from the options given below :

(a) Statement I is incorrect but Statement II is correct.
(b) Both Statement I and Statement II are incorrect.
(c) Statement I is correct but Statement II is incorrect.
(d) Both Statement I and Statement II are correct.

21 Given below are two statements. **[2024, 31 Jan Shift-I]**

Statement I IUPAC name of $HO—CH_2—(CH_2)_3—CH_2—COCH_3$ is 7-hydroxyheptan-2-one.

Statement II 2-oxoheptan-7-ol is the correct IUPAC name for above compound.

In the light of the above statements, choose the **most appropriate** answer from the options given below.

(a) Both Statement I and Statement II are incorrect.
(b) Statement I is correct but Statement II is incorrect.
(c) Statement I is incorrect but Statement II is correct.
(d) Both Statement I and Statement II are correct.

22 Correct structure of γ-methylcyclohexane carbaldehyde is **[2022, 29 July Shift-II]**

(a) 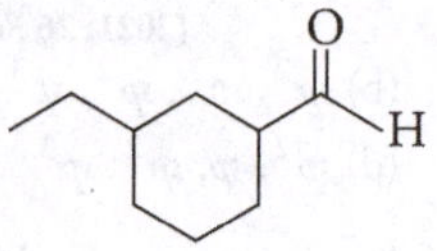 (b) 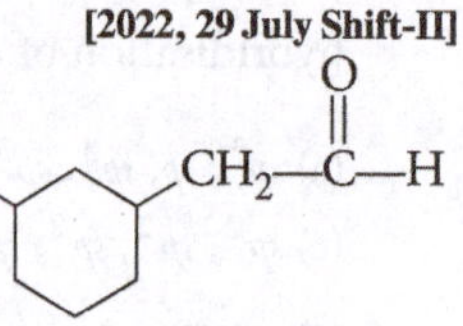

(c) 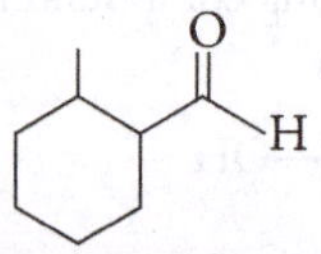 (d) 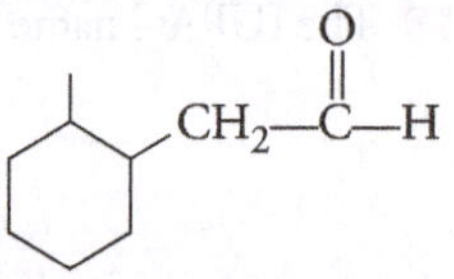

23 The IUPAC name of ethylidene chloride is **[2022, 25 June Shift-I]**

(a) 1-chloroethene (b) 1-chloroethyne
(c) 1, 2-dichloroethane (d) 1, 1-dichloroethane

24 Mesityl oxide is a common name of **[2021, 17 March Shift-I]**

(a) 2,4-dimethylpentan-3-one
(b) 3-methylcyclohexane carbaldehyde
(c) 2-methylcyclohexanone
(d) 4-methylpent-3-en-2-one

25 The correct IUPAC name of the following compound is **[2019, 9 April Shift-I]**

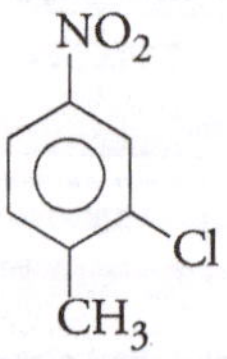

(a) 2-methyl-5-nitro-1 -chlorobenzene
(b) 3-chloro-1-methyl-1-nitrobenzene
(c) 2-chloro-1-methyl-4-nitrobenzene
(d) 5-chloro-4-methyl-1-1-nitrobenzene

• Level 3

Imp. *IUPAC Nomenclature*

26 Match List-I with List-II.

List-I (Structure)	List-II (IUPAC name)
(A) $CH_3CH_2CHCH_2CHC_2H_5$ with C_2H_5 and CH_3 branches	I. 4-methylpent-1-ene
(B) $(CH_3)_2C\,(C_3H_7)_2$	II. 3-ethyl-5-methylheptane
(C) 	III. 4,4-dimethylheptane
(D) 	IV. 2-methyl-1,3-pentadiene

Choose the correct answer from the options given below : **[2025, 29 Jan Shift-I]**

(a) (A)-(III), (B)-(II), (C)-(I), (D)-(IV)
(b) (A)-(II), (B)-(III), (C)-(I), (D)-(IV)
(c) (A)-(II), (B)-(III), (C)-(IV), (D)-(I)
(d) (A)-(III), (B)-(II), (C)-(IV), (D)-(I)

27 Match List-I with List-II. **[2022, 28 July Shift-I]**

List-I	List-II
(A) 	(I) Spiro compound
(B) 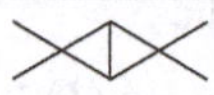	(II) Aromatic compound
(C) 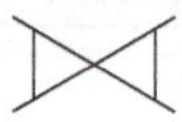	(III) Non-planar heterocyclic compound
(D) 	(IV) Bicyclo compound

Choose the correct answer from the options given below:

(a) (A)-(II), (B)-(I), (C)-(IV), (D)-(III)
(b) (A)-(IV), (B)-(III), (C)-(I), (D)-(II)
(c) (A)-(III), (B)-(IV), (C)-(I), (D)-(II)
(d) (A)-(IV), (B)-(III), (C)-(II), (D)-(I)

28 Choose the correct name for compound given below **[2021, 31 Aug Shift-I]**

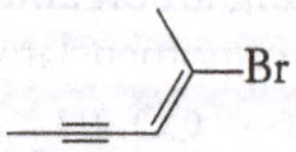

(a) (4*E*)-5-bromohex-4-en-2-yne
(b) (2*E*)-2-bromohex-4-yn-2-ene
(c) (2*E*)-2-bromohex-2-en-4-yne
(d) (4*E*)-5-bromohex-2-en-4-yne

29 The IUPAC name of the following compounds is

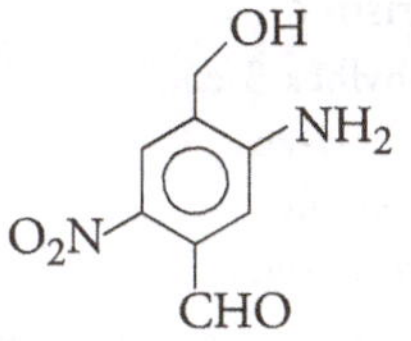

[2020, 6 Sep Shift-II]

(a) 2-nitro-4-hydroxymethyl-5-amino benzaldehyde
(b) 3-amino-4-hydroxymethyl-1-5- nitrobenzaldehyde
(c) 5-amino-4-hydroxymethyl-2-nitrobenzaldehyde
(d) 4-amino-2-formyl -5- hydroxymethyl -nitrobenzene

Numerical Values Type Questions

30 The sum of sigma (σ) and pi (π) bonds in hex-1,3-dien-5-yne is .... **[2025, 29 Jan Shift-I]**

31 Two moles of benzaldehyde and one mole of acetone under alkaline conditions using aqueous NaOH after heating gives X as the major product. The number of π-bonds in the product X is .... **[2024, 8 April Shift-II]**

32 Major product B of the following reaction has .... π-bond.

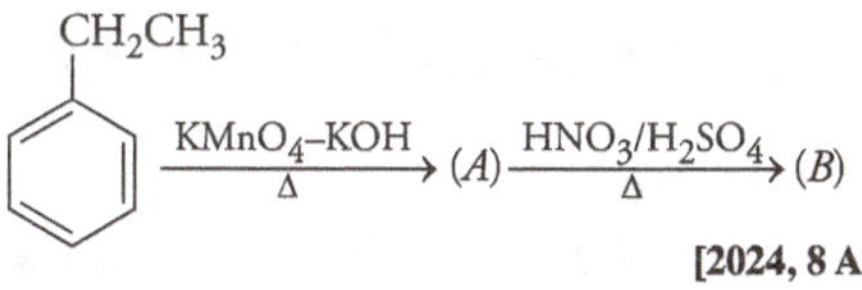

[2024, 8 April Shift-I]

Imp. *Chemical Reaction of Alkene*

33 The major products from the following reaction sequence are product A and product B. **[2024, 6 April Shift-I]**

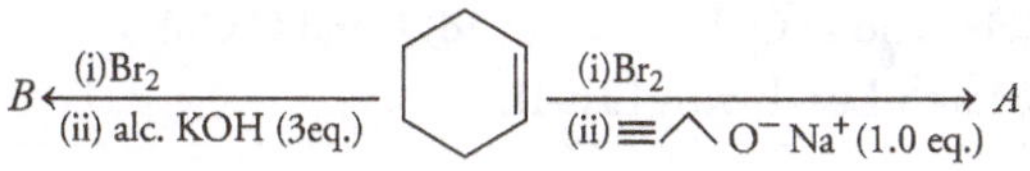

The total sum of π-electrons in product A and product B .... are (Nearest integer)

34 The total number of 'sigma' and 'Pi' bonds in 2-oxohex-4-ynoic acid is...... . **[2024, 4 April Shift-II]**

35 The number of sp^3 hybridised carbons in an acyclic neutral compound with molecular formula C_4H_5N is .... **[2022, 25 July Shift-I]**

36 The number of sigma bonds in

$$H_3C—\underset{\displaystyle H}{\underset{|}{C}}=CH—C\equiv C—H \text{ is ......}$$

[2021, 25 July Shift-I]

Topic 3 / Isomerism of Organic Compounds

Level 1

01 How many different stereoisomers are possible for the given molecule? **[2025, 22 Jan Shift-I]**

$$CH_3—\underset{\displaystyle OH}{\underset{|}{CH}}—CH=CH—CH_3$$

(a) 3 (b) 4 (c) 2 (d) 1

02 Given below are two statements.

Statement I The compound A **[2022, 29 July Shift-II]**

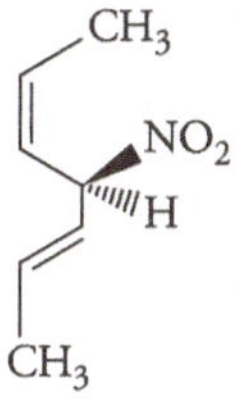

is optically active.

Statement II

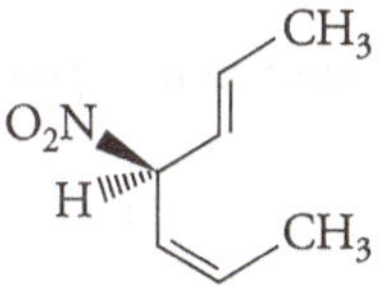

is mirror image of above compound A.

In the light of the above statements, choose the most appropriate answer from the options given below.

(a) Both statement I and statement II are correct.
(b) Both statement I and statement II are incorrect.
(c) Statement I is correct but statement II is incorrect.
(d) Statement I is incorrect but statement II is correct.

03 The number of stereoisomers possible for 1,2-dimethylcyclopropane is **[2021, 26 Aug Shift-II]**

(a) one (b) four (c) two (d) three

04 Which one of the following pairs of isomers is an example of metamerism ? **[2021, 20 July Shift-II]**

(a) $CH_3CH_2CH_2CH_2CH_3$ and $H_3C—\overset{\displaystyle CH}{\overset{|}{\underset{\displaystyle CH_3}{\underset{|}{C}}}}—CH_3$

(b) 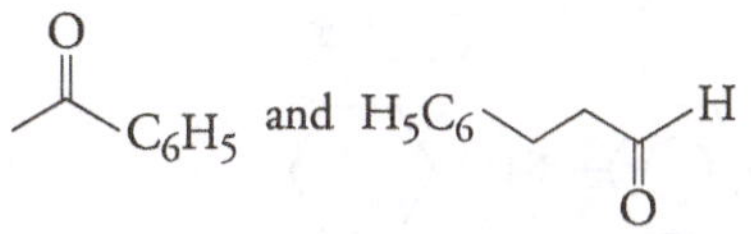

(c) 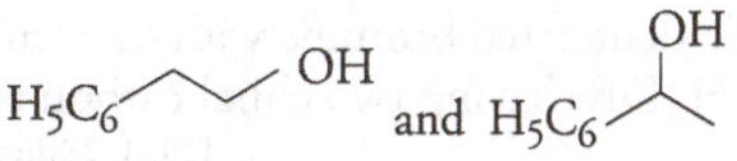

(d) 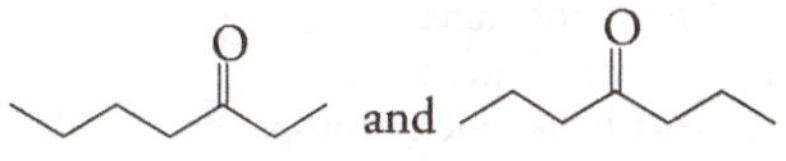

05 Compound with molecular formula C_3H_6O can show **[2021, 18 March Shift-I]**

(a) positional isomerism
(b) both positional isomerism and metamerism
(c) metamerism
(d) functional group isomerism

06 Among the following compounds, geometrical isomerism is exhibited by **[2020, 5 Sep Shift-II]**

(a) 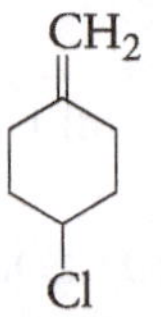
(b) 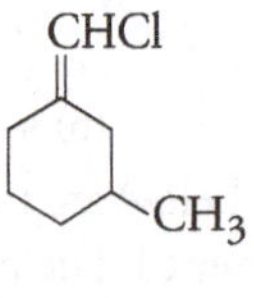
(c) 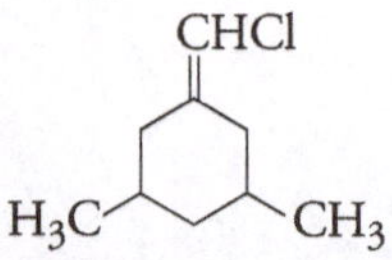
(d) 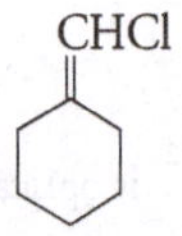

• Level 2

07 Total number of stereoisomers possible for the given structure **[2024, 9 April Shift-II]**

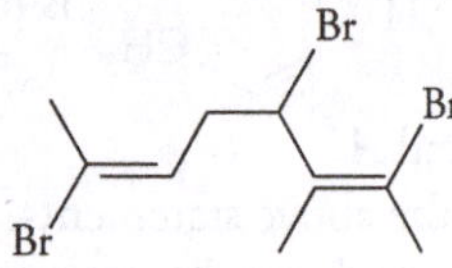

(a) 4 (b) 3 (c) 2 (d) 8

08 Which of the following is metamer of the given compound (*X*)? **[2024, 6 April Shift-I]**

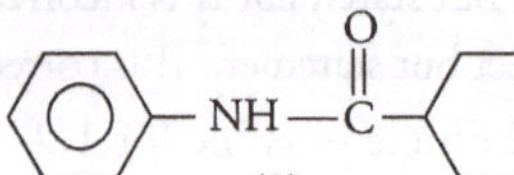
(*X*)

(a) 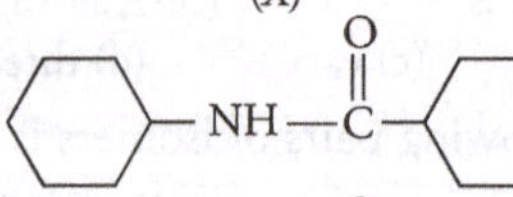
(b) 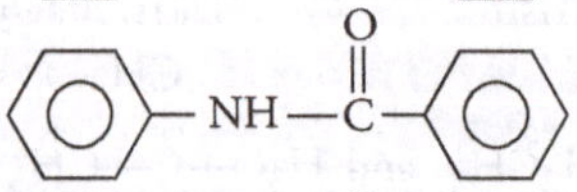
(c) 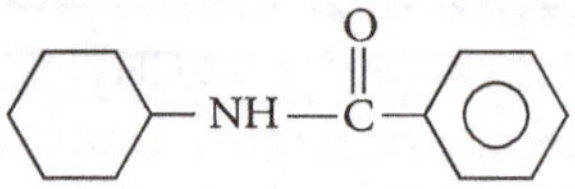
(d) 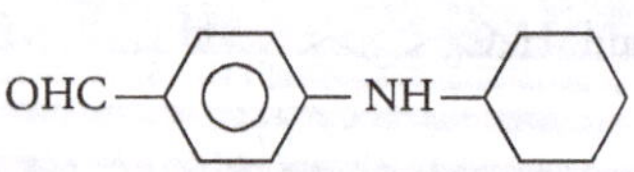

09 The isomeric deuterated bromide with molecular formula C_4H_8DBr having two chiral carbon atoms is **[2023, 25 Jan Shift-II]**

(a) 2-bromo-3-deuterobutane
(b) 2-bromo-1-deuterobutane
(c) 2-bromo-1-deutero-2-methylpropane
(d) 2-bromo-2-deuterobutane

10 Which of the following molecules does not show stereoisomerism ? **[2021, 22 July Shift-II]**

(a) 3,4-dimethylhex-3-ene
(b) 3-methylhex-1-ene
(c) 3-ethylhex-3-ene
(d) 4-methylhex-1-ene

11 Which of the following compounds produces an optically inactive compound on hydrogenation? **[2020, 3 Sep Shift-I]**

(a) 
(b) 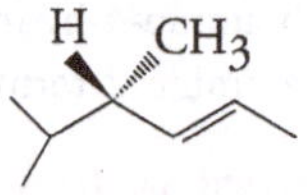
(c) 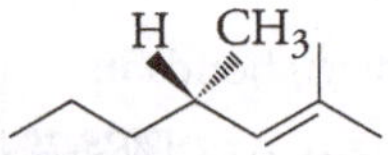
(d) 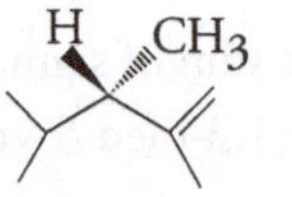

• Level 3

12 Identify the correct statements from the following.

A. 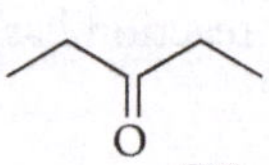 and 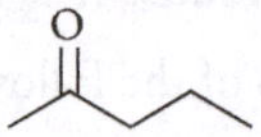 are metamers

B. 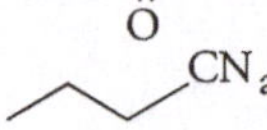 and  are functional isomers

C. 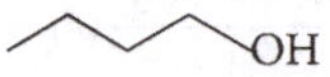 and 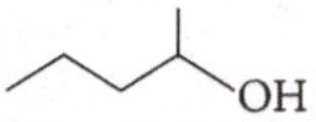 are position isomers

D. 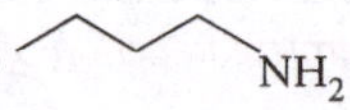 and 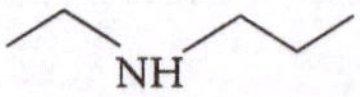 are homologous

Choose the correct answer from the options given below: **[2025, 3 April Shift-I]**

(a) A, B and C Only (b) A and B Only
(c) C and D Only (d) B and C Only

13 Match List-I with List-II.

List-I (Pair of Compounds)		List-II (Isomerism)	
A.	*n*-propanol and iso-propanol	I.	Metamerism
B.	Methoxypropane and ethoxyethane	II.	Chain isomerism
C.	Propanone and propanal	III.	Position isomerism
D.	Neopentane and iso-pentane.	IV.	Functional isomerism

Choose the correct answer from the options given below : **[2024, 5 April Shift-II]**

(a) A-II, B-I, C-IV, D-III
(b) A-III, B-I, C-IV, D-II
(c) A-III, B-I, C-II, D-IV
(d) A-I, B-III, C-IV, D-II

14 Match List-I with List-II.

List-I (Isomeric pairs)		List-II (Type of isomers)
(A) Propanamine and N-methylethanamine	I.	Metamers
(B) Hexan-2-one and hexan-3-one	II.	Positional isomers
(C) Ethanamide and hydroxyethanimine	III.	Functional isomers
(D) *o*-nitrophenol and *p*-nitrophenol	IV.	Tautomers

Choose the correct answer from the options given below. **[2023, 25 Jan Shift-II]**

(a) A-II, B-III, C-I, D-IV (b) A-III, B-I, C-IV, D-II
(c) A-III, B-IV, C-I, D-II (d) A-IV, B-III, C-I, D-II

Numerical Value Type Questions

15 The total number of structural isomers possible for the substituted benzene derivatives with the molecular formula C_9H_{12} is .... **[2025, 3 April Shift-II]**

Imp. *Isomerism*

16 The possible number of stereoisomers for 5-phenylpent-4-en-2-ol is .... **[2025, 24 Jan Shift-II]**

17 The total number of chiral compounds from the following is .... **[2023, 01 Feb Shift-I]**

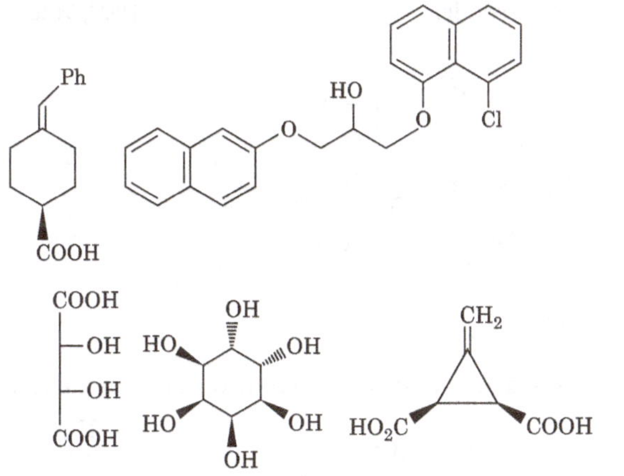

18 The number of stereoisomers formed in a reaction of (±) Ph(C=O)C(OH)(CN) Ph with HCN is .... **[2022, 29 July Shift-II]**

Imp. *Optical Activity of Enantiomeric Mixture*

19 Optical activity of an enantiomeric mixture is +12.6° and the specific rotation of (+) isomer is +30°. The optical purity is ...... %. **[2022, 27 July Shift-I]**

20 Observe structures of the following compounds. **[2022, 29 June Shift-I]**

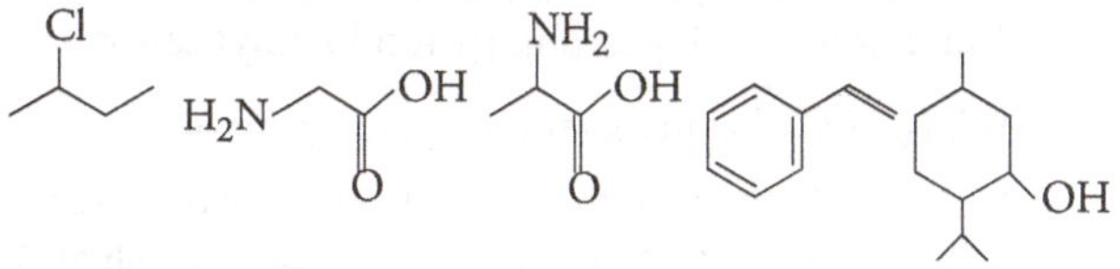

The total number of structures/ compounds which possess asymmetric carbon atoms is ....

21 Total number of possible stereoisomers of dimethyl cyclopentane is .... **[2022, 27 June Shift-I]**

22 The number of acyclic structural isomers (including geometrical isomers) for pentene are .... **[2021, 22 July Shift-II]**

23 The number of chiral carbons present in the molecule given below is .... **[2020, 2 Sep Shift-I]**

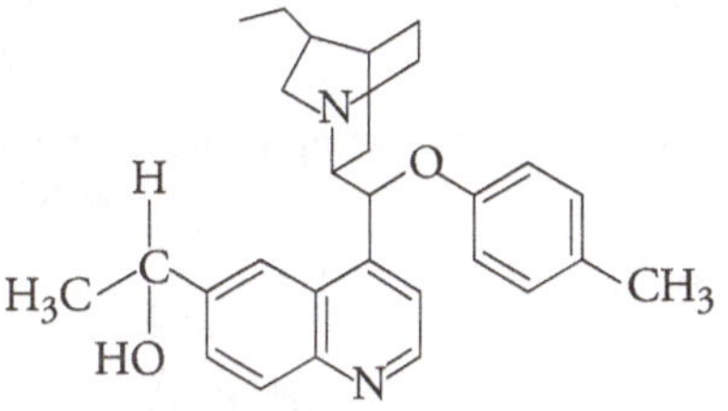

Topic 3 / Bond Fission, Electronic Displacement in Organic Molecules, Attacking Reagent & Reaction Intermediates

Level 1

01 The shape of carbocation is **[2024, 8 April Shift-II]**

(a) trigonal planar (b) diagonal
(c) tetrahedral (d) diagonal pyramidal

02 In the given compound, the number of 2° carbon atom/s is .... **[2024, 8 April Shift-I]**

$$CH_3-\underset{\displaystyle H}{\underset{|}{C}}(CH_3)-\underset{\displaystyle H}{\underset{|}{C}}H-\underset{\displaystyle H}{\underset{|}{C}}(CH_3)-CH_3$$

(a) Four (b) One
(c) Two (d) Three

03 The difference in energy between the actual structure and the lowest energy resonance structure for the given compound is **[2024, 29 Jan Shift-I]**

(a) electromeric energy (b) ionisation energy
(c) hyperconjugation energy (d) resonance energy

04 The molecular formula of second homologue in the homologous series of mono carboxylic acids is .... **[2024, 27 Jan Shift-II]**

(a) CH_2O (b) $C_3H_6O_2$
(c) $C_2H_2O_2$ (d) $C_2H_4O_2$

05 Given below are two statements.

Statement I Hyperconjugation is a permanent effect.

Statement II Hyperconjugation in ethyl cation ($CH_3—\overset{+}{C}H_2$) involves the overlapping of $C_{sp^2}—H_{1s}$ bond with empty $2p$ orbital of other carbon.

Choose the correct option **[2021, 27 July Shift-II]**

(a) Both statement I and statement II are false.
(b) Statement I is incorrect but statement II is true.
(c) Statement I is true but statement II is false.
(d) Both Statement I and statement II are true.

06 Which one among the following resonating structures is not correct ? **[2021, 25 July Shift-I]**

(a) 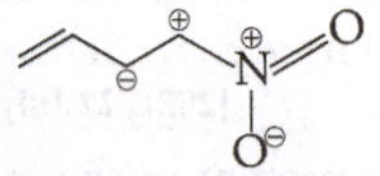
(b) 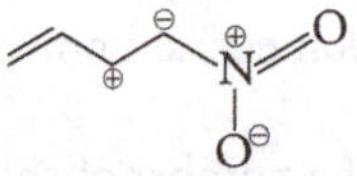
(c) 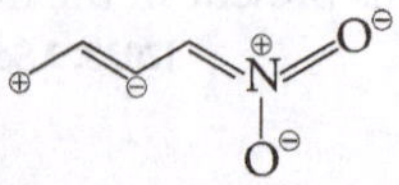
(d) 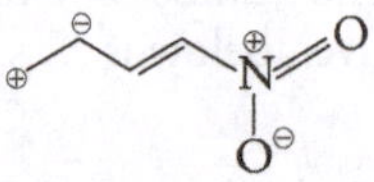

07 Which among the following is the strongest acid? **[2021, 25 July Shift-II]**

(a) $CH_3CH_2CH_2CH_3$
(b) 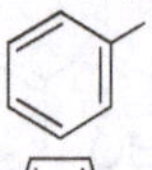
(c) 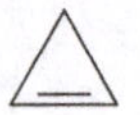
(d) 

08 Which of the following has the shortest C—Cl bond? **[2020, 8 Jan Shift-II]**

(a) $ClCH=CH—CH_3$ (b) $ClCH=CH—NO_2$
(c) $ClCH=CH—OCH_3$ (d) $ClCH=CH_2$

09 Which amongst the following is the strongest acid? **[2019, 9 Jan Shift-I]**

(a) $CHBr_3$ (b) CHI_3 (c) $CHCl_3$ (d) $CH(CN)_3$

10 The correct order for acid strength of compounds $CH\equiv CH$, $CH_3—C\equiv CH$ and $CH_2=CH_2$ is as follows **[2019, 12 Jan Shift-I]**

(a) $CH_3—C\equiv CH > CH_2=CH_2 > HC\equiv CH$
(b) $CH_3—C\equiv CH > CH\equiv CH > CH_2=CH_2$
(c) $HC\equiv CH > CH_3—C\equiv CH > CH_2=CH_2$
(d) $CH\equiv CH > CH_2=CH_2 > CH_3—C\equiv CH$

• Level 2

11. Match the List-I with List-II. **[2025, 8 April Shift-II]**

List-I		List-II	
A.	Carbocation	I.	Species that can supply a pair of electrons.
B.	C-Free radical	II.	Species that can receive a pair of electrons.
C.	Nucleophile	III.	sp^2 hybridised carbon with empty p-orbital.
D.	Electrophile	IV.	sp^2/sp^3 hybridised carbon with one unpaired electron.

Choose the correct answer from the options given below:

(a) A-III, B-IV, C-II, D-I (b) A-III, B-IV, C-1, D-II
(c) A-II, B-III, C-I, D-IV (d) A-IV, B-II, C-III, D-1

12. Given below are two statements: **[2025, 3 April Shift-II]**

Statement I Hyperconjugation is not a permanent effect.

Statement II In general, greater the number of alkyl groups attached to a positively charged C-atom, greater is the hyperconjugation interaction and stabilisation of the cation.

In the light of the above statements, choose the correct answer from the options given below.

(a) Statement I is false but Statement II is true.
(b) Both Statement I and Statement II are false.
(c) Statement I is true but Statement II is false.
(d) Both Statement I and Statement II are true.

13. Consider the following compound (*X*)

$$\overset{I}{H—C}\equiv \overset{II}{C—CH_2}—\overset{III}{\underset{\underset{CH_3}{|}}{C}H}—\overset{IV}{CH_3}$$

(*X*)

The most stable and least stable carbon radicals, respectively, produced by homolytic cleavage of corresponding C—H bond are **[2025, 2 April Shift-I]**

(a) I, IV (b) III, II
(c) II, I (d) II, IV

14 Which one of the carbocations from the following is most stable? **[2025, 24 Jan Shift-I]**

(a) 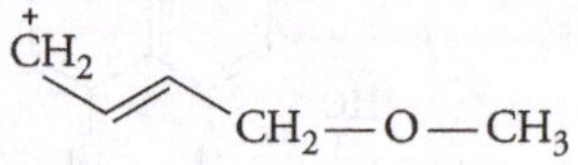
(b) 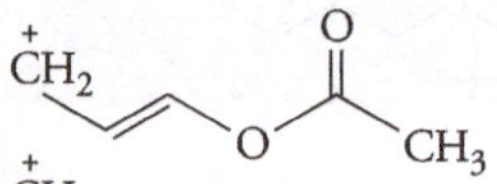
(c) 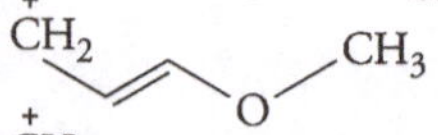
(d) 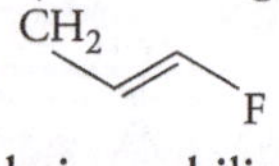

15 Relative stability of the contributing structures is

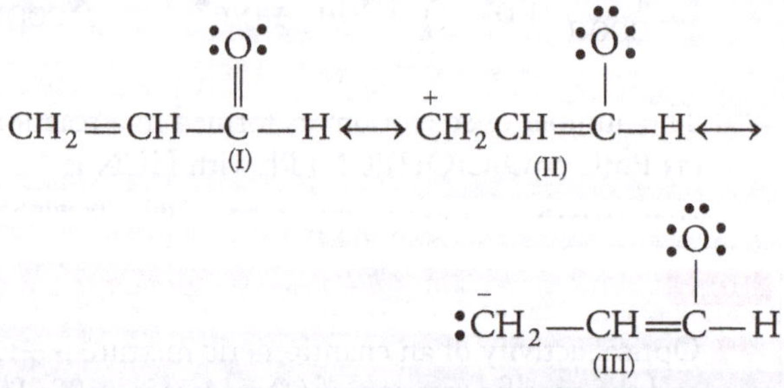

[2024, 9 April Shift-I]

(a) (II) > (I) > (III) (b) (III) > (II) > (I)
(c) (I) > (II) > (III) (d) (I) > (III) > (II)

16 Correct order of stability of carbanion is

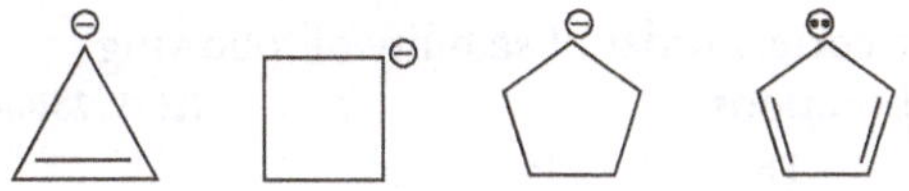

I II III IV

[2024, 4 April Shift-II]

(a) III > II > IV > I (b) I > II > III > IV
(c) IV > III > II > I (d) IV > I > III > II

17 Which among the following is incorrect statement? [2024, 4 April Shift-I]

(a) Hydrogen ion $[H^+]$ shows negative electromeric effect.
(b) Electromeric effect dominates over inductive effect.
(c) The electromeric effect is, temporary effect.
(d) The organic compound shows electromeric effect in the presence of the reagent only.

18 The functional group that shows negative resonance effect is [2024, 1 Feb Shift-II]

(a) —OR (b) —COOH (c) $—NH_2$ (d) —OH

19 A species having carbon with sextet of electrons and can act as electrophile is called [2024, 31 Jan Shift-I]

(a) pentavalent carbon (b) carbon free radical
(c) carbanion (d) carbocation

Imp. *Order of Hydride Affinity*

20 The decreasing order of hydride affinity for following carbocations is

A. $CH_2{=}CH—\overset{+}{C}(CH_3)—CH_3$ B. $C_6H_5—\overset{+}{C}(C_6H_5)—C_6H_5$

C. $CH_3—\overset{+}{C}(CH_3)—CH_3$ D. 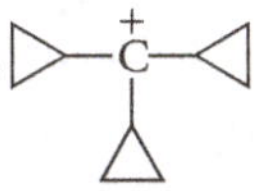

Choose the correct answer from the options given below [2023, 10 April Shift-II]

(a) C, A, B, D (b) C, A, D, B
(c) A, C, D, B (d) A, C, B, D

21 The most stable carbocation for the following is [2023, 30 Jan Shift-II]

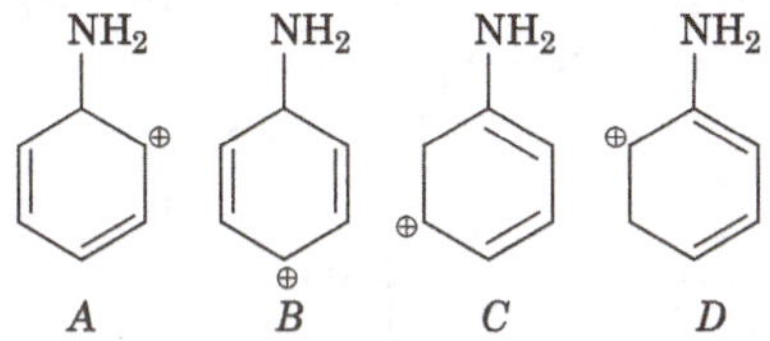

A B C D

(a) *D* (b) *C* (c) *B* (d) *A*

22 The correct order of nucleophilicity is [2022, 26 June Shift-II]

(a) $F^- > OH^-$ (b) $H_2\ddot{O} > OH^-$
(c) $R\ddot{O}H > RO^-$ (d) $NH_2^- > NH_3$

23 Arrange the following carbocations in decreasing order of stability. [2022, 24 June Shift-II]

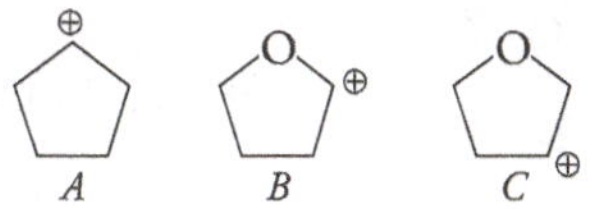

A B C

(a) $A > C > B$ (b) $A > B > C$
(c) $C > B > A$ (d) $C > A > B$

24 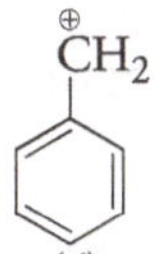 $CH_2{=}\overset{\oplus}{C}H$ $CH_3—\overset{\oplus}{C}H_2$ $H_2C{\equiv}\overset{\oplus}{C}$

(A) (B) (C) (D)

The correct order of stability of given carbocation is [2021, 27 July Shift-I]

(a) $A > C > B > D$ (b) $D > B > C > A$
(c) $D > B > A > C$ (d) $C > A > D > B$

25 Which of the following compounds does not exhibit resonance ? [2021, 22 July Shift-II]

(a) $CH_3CH_2OCH{=}CH_2$
(b) 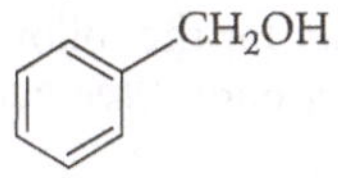
(c) $CH_3CH_2CH_2CONH_2$
(d) $CH_3CH_2CH{=}CHCH_2NH_2$

26 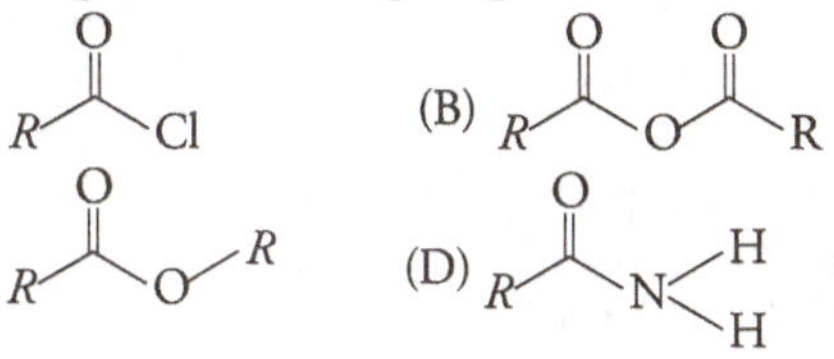

The correct order of their reactivity towards hydrolysis at room temperature is [2021, 20 July Shift-II]

(a) (A) > (B) > (C) > (D)
(b) (D) > (A) > (B) > (C)
(c) (D) > (B) > (A) > (C)
(d) (A) > (C) > (B) > (D)

27 Choose the correct statement regarding the formation of carbocations *A* and *B*. [2021, 17 March Shift-II]

$CH_3CH_2CH{=}CH_2 + HBr$

$\rightarrow CH_3—CH_2—CH_2—\overset{+}{C}H_2 + Br^-$ (*A*)

$\rightarrow CH_3—CH_2—\overset{+}{C}H—CH_3 + Br^-$ (*B*)

(a) Carbocation *B* is more stable and formed relatively at faster rate.
(b) Carbocation *A* is more stable and formed relatively at slow rate.
(c) Carbocation *B* is more stable and formed relatively at slow rate.
(d) Carbocation *A* is more stable and formed relatively at faster rate.

28 Which one of the following compounds possesses the most acidic hydrogen? **[2020, 3 Sep Shift-I]**

(a) 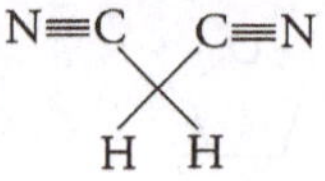   (b) $H_3C—C≡C—H$

(c) 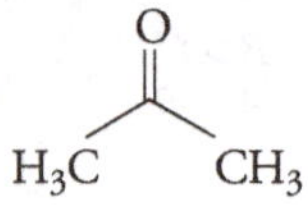   (d) 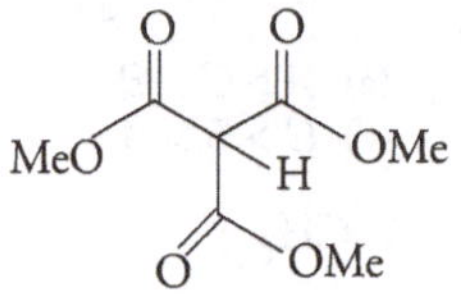

29 Arrange the following labelled hydrogens in decreasing order of acidity **[2020, 2 Sep Shift-II]**

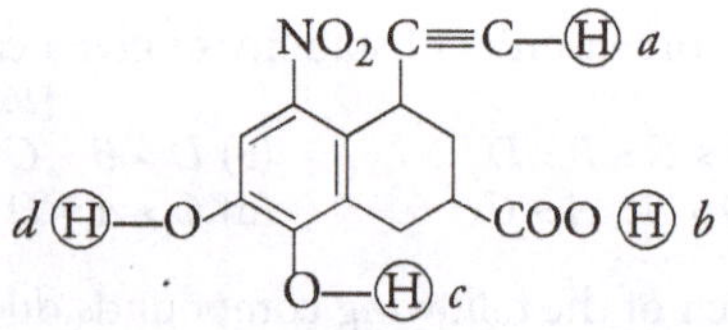

(a) $b > a > c > d$ (b) $c > b > d > a$
(c) $b > c > d > a$ (d) $c > b > a > d$

30 The increasing order of basicity for the following intermediates is (from weak to strong) **[2020, 8 Jan Shift-II]**

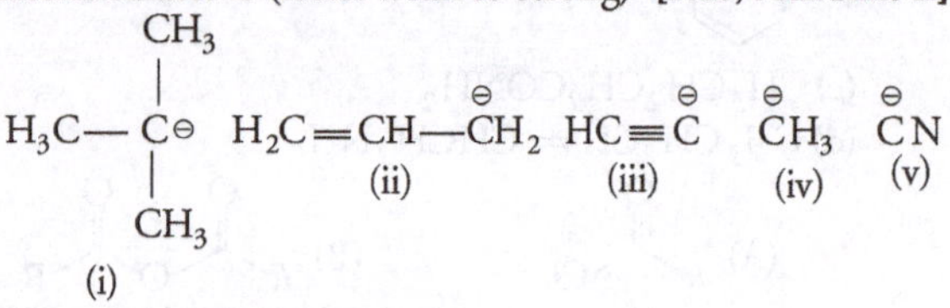

(a) (v) < (iii) < (ii) < (iv) < (i)
(b) (iii) < (i) < (ii) < (iv) < (v)
(c) (v) < (i) < (iv) < (ii) < (iii)
(d) (iii) < (iv) < (ii) < (i) < (v)

• Level 3

31. In which pairs, the first ion is more stable than the second? **[2025, 4 April Shift-II]**

A. 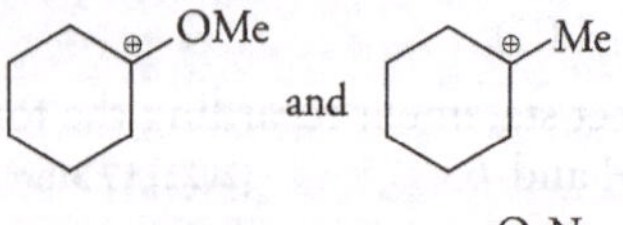

B. 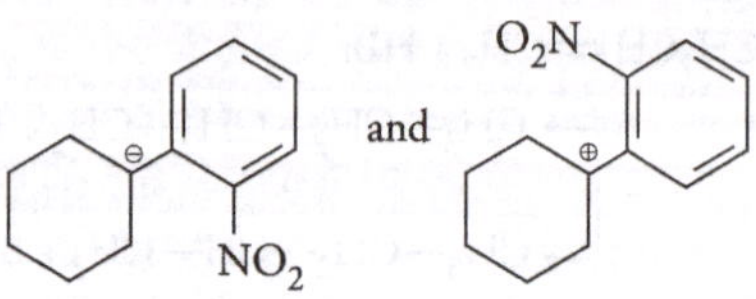

C. 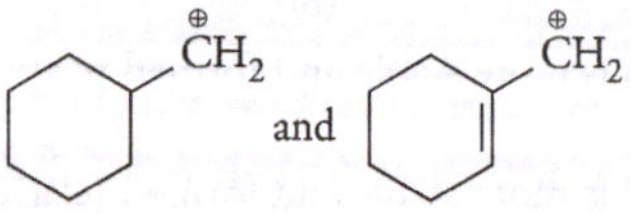

D. 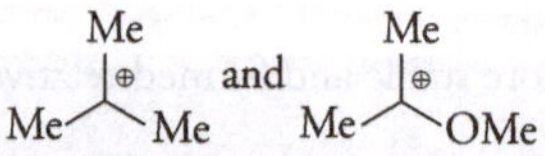

(a) (B) and (C) only (b) (A) and (C) only
(c) (B) and (D) only (d) (A) and (B) only

Imp. *Stability of Carbocation*

32 The correct order of stability of following carbocations is **[2025, 28 Jan Shift-I]**

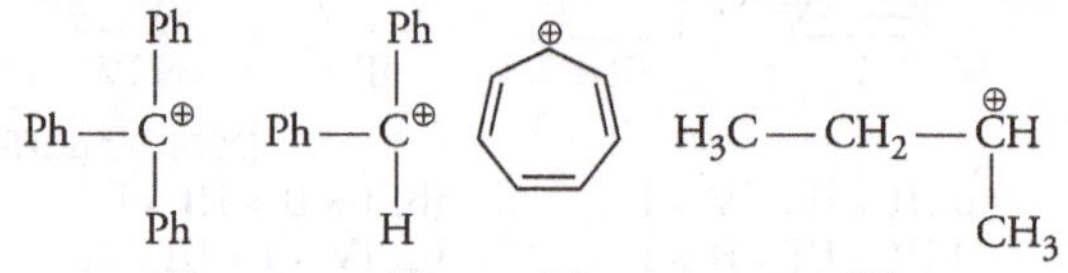

(A) (B) (C) (D)

(a) B > C > A > D (b) C > B > A > D
(c) C > A > B > D (d) A > B > C > D

33 The most stable carbocation from the following is **[2025, 22 Jan Shift-II]**

(a) 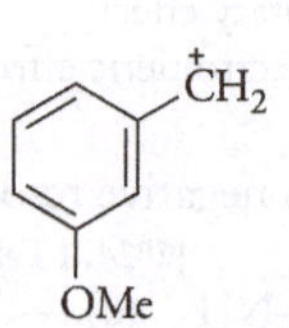 (b) 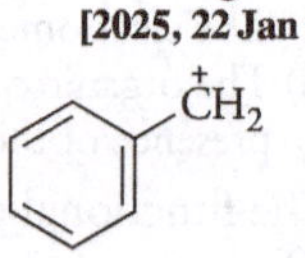

(c) 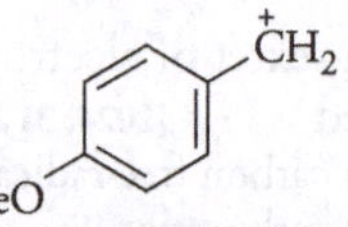 (d) 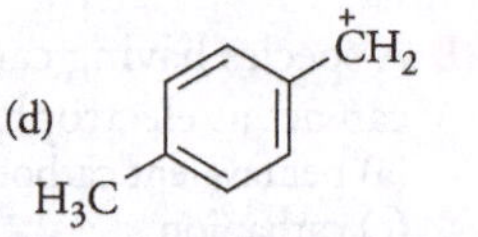

34 Total number of nucleophiles from the following is :
NH_3, PhSH, $(H_3C)_2S$, $H_2C=CH_2$,
$\overset{\ominus}{O}H^{\ominus}$, $H_3O^{\oplus}$, $(CH_3)_2CO$, 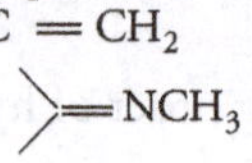 **[2025, 29 Jan Shift-I]**

(a) 5 (b) 6
(c) 7 (d) 4

35 Match List-I with List-II.

List-I (Mechanism steps)	List-II (Effect)
(A) 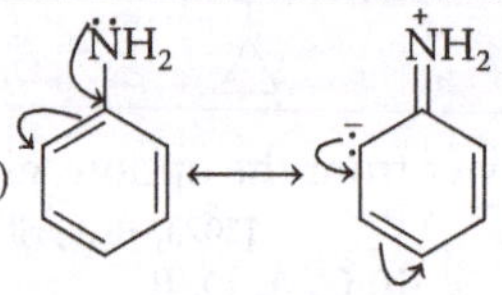	I. $-E$ effect
(B) 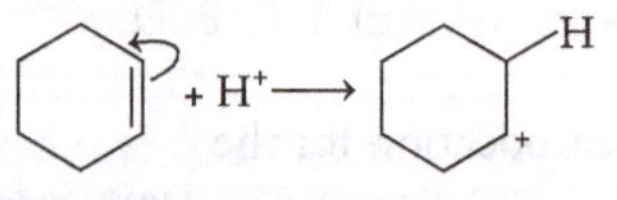	II. $-R$ effect
(C) 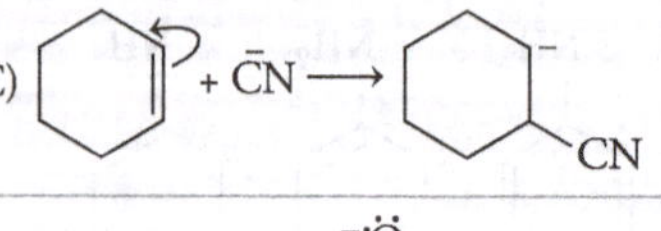	III. $+E$ effect
(D) 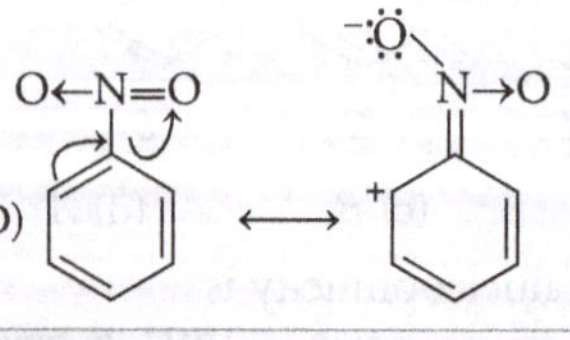	IV. $+R$ effect

Choose the correct answer from the options given below **[2024, 4 April Shift-I]**

(a) A-I, B-II, C-IV, D-III (b) A-II, B-IV, C-III, D-I
(c) A-III, B-I, C-II, D-IV (d) A-IV, B-III, C-I, D-II

36 Ionic reactions with organic compounds proceed through
(A) homolytic bond cleavage (B) heterolytic bond cleavage
(C) free radical formation (D) primary free radical
(E) secondary free radical
Choose the **correct answer** from the options given below **[2024, 1 Feb Shift-I]**
(a) (C) only (b) (A) only
(c) (B) only (d) (D) and (E) only

37 Which of the following compound will most easily be attacked by an electrophile? **[2024, 1 Feb Shift-I]**

(a) 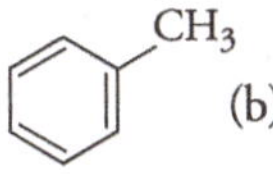 (b) 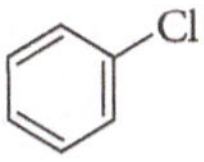 (c) 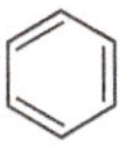 (d) 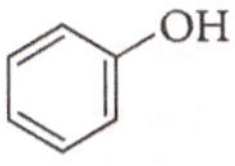

38 The correct stability order of carbocations is **[2024, 30 Jan Shift-II]**

(a) $(CH_3)_3C^+ > CH_3-\overset{+}{C}H_2 > (CH_3)_2\overset{+}{C}H > \overset{+}{C}H_3$

(b) $\overset{+}{C}H_3 > (CH_3)_2\overset{+}{C}H > CH_3-\overset{+}{C}H_2 > (CH_3)_3\overset{+}{C}$

(c) $\overset{+}{C}H_3 > CH_3-\overset{+}{C}H_2 > CH_3-\underset{CH_3}{\underset{|}{\overset{+}{C}}}H > (CH_3)_3\overset{+}{C}$

(d) $(CH_3)_3\overset{+}{C} > (CH_3)_2\overset{+}{C}H > CH_3-\overset{+}{C}H_2 > \overset{+}{C}H_3$

39 Increasing order of stability of the resonance structures is **[2023, 24 Jan Shift-I]**

(A) 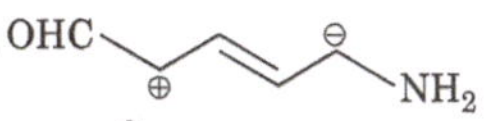 (B) 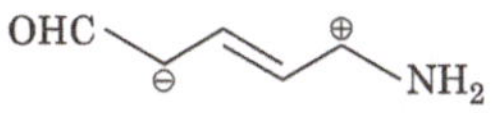

(C) 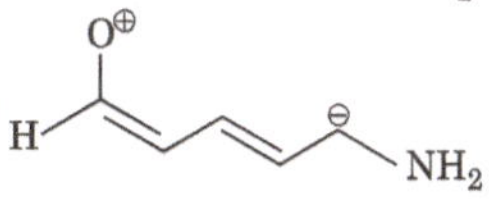 (D) 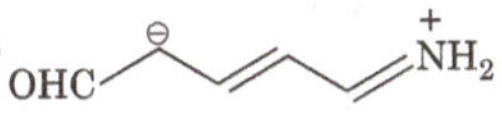

Choose the correct answer from the options given below.
(a) D, C, B, A (b) D, C, A, B
(c) C, D, B, A (d) C, D, A, B

40 Which of the following carbocations is most stable? **[2022, 29 June Shift-II]**

(a) 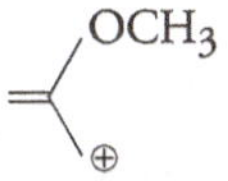 (b) 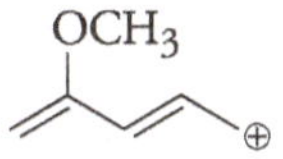

(c) 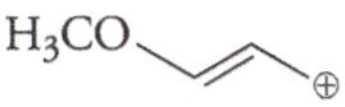 (d) 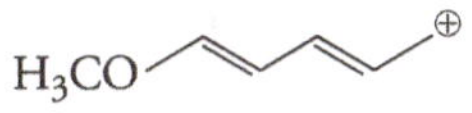

41 Which of the following is most stable? **[2022, 27 June Shift-II]**

(a)  (b)  (c)  (d) 

42 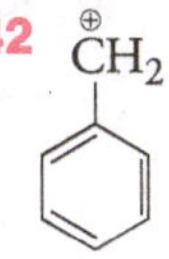 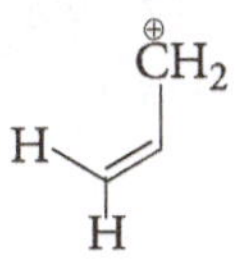 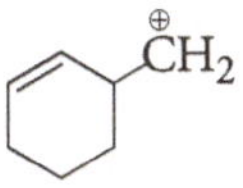 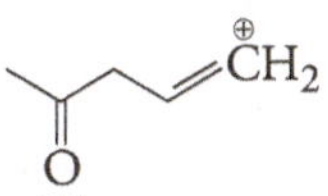

(*A*) (*B*) (*C*) (*D*)

Among the given species the resonance stabilised carbocations are **[2021, 20 July Shift-I]**
(a) (*C*) and (*D*) only (b) (*A*), (*B*) and (*D*) only
(c) (*A*) and (*B*) only (d) (*A*), (*B*) and (*C*) only

Numerical Value Type Questions

43 How many compounds among the following compounds show inductive, mesomeric as well as hyperconjugation effects? **[2024, 9 April Shift-I]**

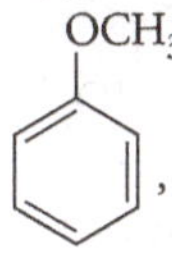, 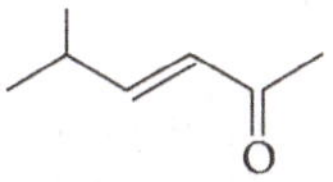, , 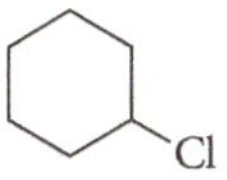

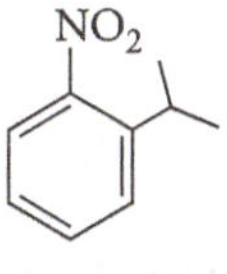, 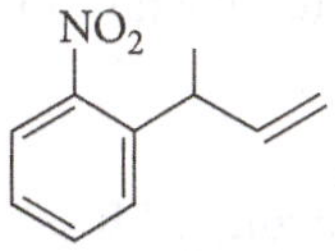, 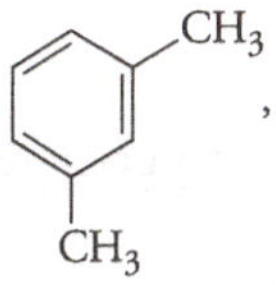, 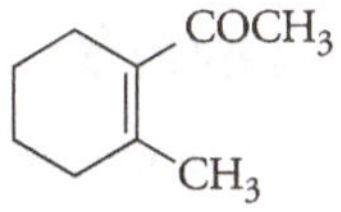

44 Number of carbocations from the following that are not stablised by hyperconjugation is **[2024, 6 April Shift-II]**

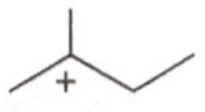, , $\overset{+}{C}H_3$
(*tert*-butyl) (*tert*-butyl)

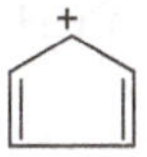, $\overset{+}{C}H_2-OCH_3$, 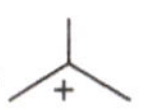, 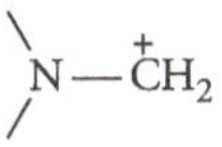

45 Number of compounds among the following which contain sulphur as heteroatom is ....
Furan, thiophene, pyridine, pyrrole, cysteine, tyrosine **[2024, 29 Jan Shift-I]**

46 Among the following, total number of *meta*-directing functional groups is .... (Integer based)
$-OCH_3$, $-NO_2$, $-CN$, $-CH_3$,
$-NHCOCH_3$, $-COR$, $-OH$, $-COOH$, $-Cl$ **[2024, 27 Jan Shift-I]**

Topic 4 / Types of Organic Reaction

• Level 1

01 The compound which will have the lowest rate towards nucleophilic aromatic substitution on treatment with OH^- is **[2023, 25 Jan Shift-I]**

(a) 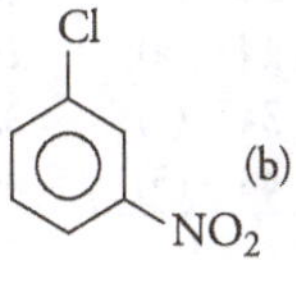 (b) 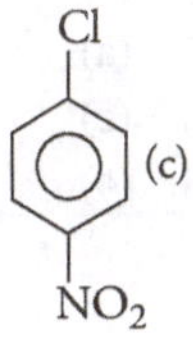 (c) 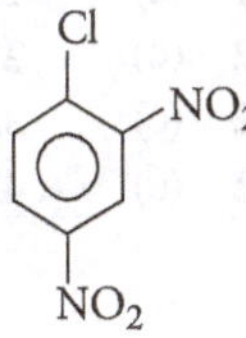 (d) 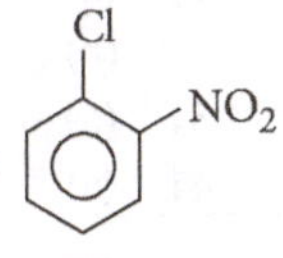

• Level 2

02 The correct order of the following compounds showing increasing tendency towards nucleophilic substitution reaction is **[2021, 24 Feb Shift-II]**

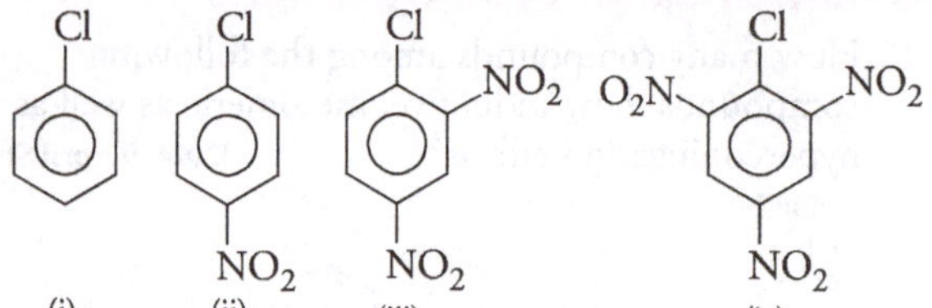

(i) (ii) (iii) (iv)

(a) (iv) < (iii) < (ii) < (i)
(b) (iv) < (i) < (ii) < (iii)
(c) (iv) < (i) < (iii) < (ii)
(d) (i) < (ii) < (iii) < (iv)

• Level 3

03 In the following given reaction, '*A*' is **[2023, 24 Jan Shift-I]**

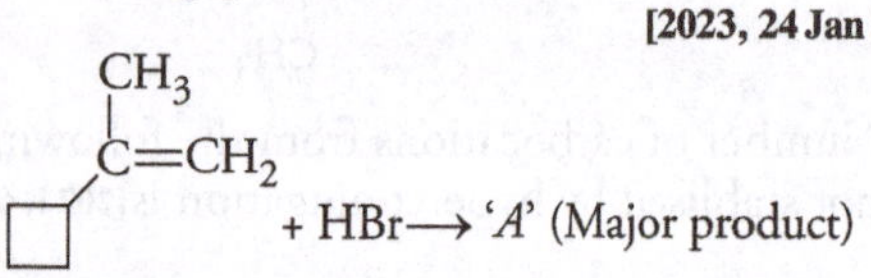

(a) 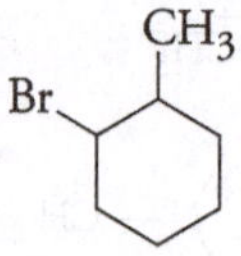

(b) 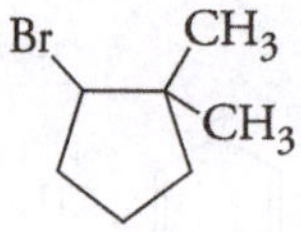

(c) 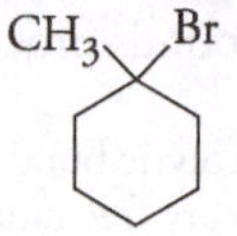

(d) 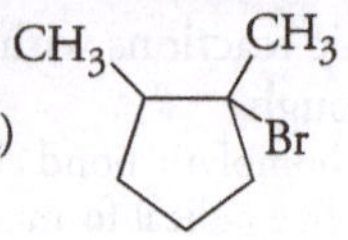

04 Arrange the following in increasing order of reactivity towards nitration.

A. *p*-xylene B. Bromobenzene
C. Mesitylene D. Nitrobenzene
E. Benzene **[2022, 28 July Shift-II]**

(a) C < D < E < A < B
(b) D < B < E < A < C
(c) D < C < E < A < B
(d) C < D < E < B < A

Numerical Value Type Question

Imp. *Number of Hyperconjugative Structure*

05 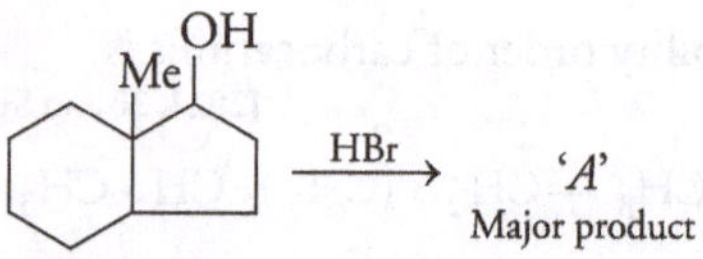

The number of hyperconjugation structure involved to stablise carbocation formed in the above reaction is .... **[2023, 11 April Shift-I]**

Key Index

Topic 1

1 (d)	2 (a)	3 (a)	4 (b)	5 (a)	6 (d)	7 (a)	8 (d)	9 (b)	10 (a)
11 (d)	12 (a)	13 (b)	14 (b)	15 (b)	16 (c)	17 (c)	18 (d)	19 (d)	20 (a)
21 (b)	22 (a)	23 (d)	24 (d)	25 (c)	26 (c)	27 (c)	28 (c)	29 (c)	30 (15)
31 (9)	32 (5)	33 (3)	34 (18)	35 (1)	36 (10)				

Topic 2

1 (b)	2 (c)	3 (d)	4 (d)	5 (d)	6 (b)	7 (d)	8 (c)	9 (a)	10 (c)
11 (d)	12 (b)	13 (b)	14 (b)	15 (8)	16 (4)	17 (2)	18 (3)	19 (42)	20 (3)
21 (5)	22 (6)	23 (5)							

Topic 3

1 (a)	2 (b)	3 (d)	4 (d)	5 (c)	6 (a)	7 (d)	8 (b)	9 (d)	10 (c)
11 (b)	12 (a)	13 (c)	14 (c)	15 (c)	16 (c)	17 (a)	18 (b)	19 (d)	20 (a)
21 (b)	22 (d)	23 (a)	24 (a)	25 (d)	26 (a)	27 (a)	28 (d)	29 (c)	30 (a)
31 (d)	32 (c)	33 (c)	34 (a)	35 (d)	36 (c)	37 (d)	38 (d)	39 (d)	40 (d)
41 (a)	42 (c)	43 (4)	44 (5)	45 (2)	46 (4)				

Topic 4

1 (a)	2 (d)	3 (d)	4 (b)	5 (7)

Detailed Explanations

Topic 1

01. *(d)* $\underset{7}{HC}\equiv\underset{6}{C}-\underset{5}{CH_2}-\underset{4}{\overset{OH}{\overset{|}{CH}}}-\underset{3}{CH_2}-\underset{2}{CH}=\underset{1}{CH_2}$

Hept-1-en-6-yn-4-ol

02. *(a)*

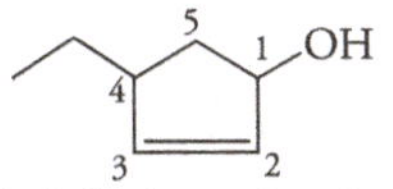

4-ethylcyclopent-2-en-1-ol

03. *(a)*

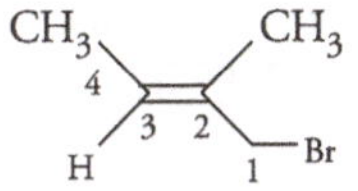

1-bromo-2 methyl but-2-ene

04. *(b)*

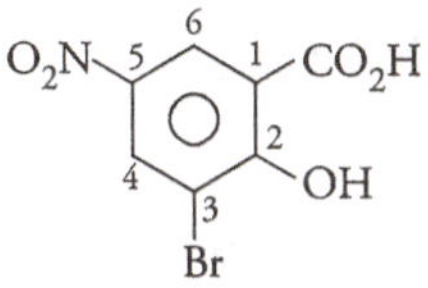

3-bromo-2 hydroxy-5-nitrobenzoic acid

05. *(a)* The structure of 3, 3-dimethylhex-1-ene-4-yne is

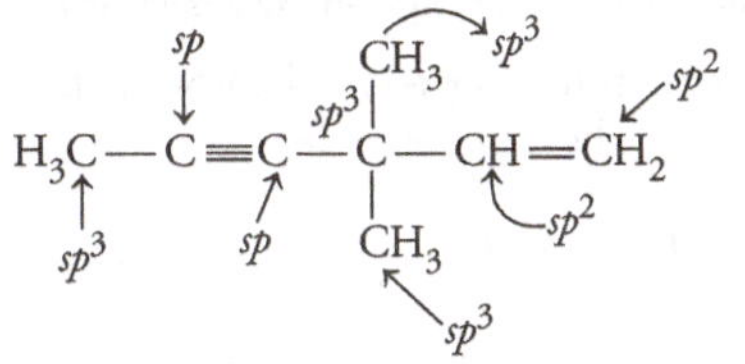

There are 4-sp^3, 2-sp^2, 2-sp hybridised carbon atoms respectively.

06. *(d)* $H_3C-\underset{2}{\overset{\overset{1}{COOH}}{\overset{|}{CH}}}-\underset{3}{CH_2}-\underset{4}{CH_2}-\underset{5}{\overset{{}_6COOCH_3}{\overset{|}{CH}}}-CH_3$

So, the IUPAC name of the given compound is 6-methoxycarbonyl -2, 5- dimethylhexanoic acid.

07. *(a)* The hybridisation of each carbon in the given compound is

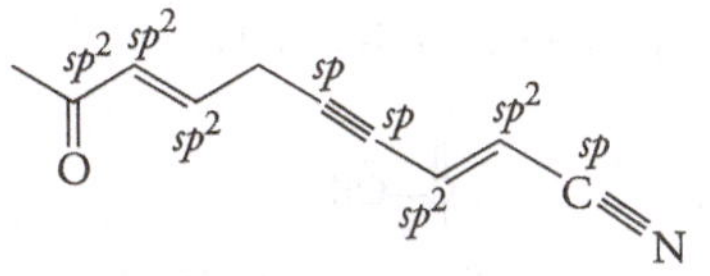

Hence, number of sp and sp^2 hybridised carbon atom are 3 and 5.

08. *(d)* The IUPAC name of given hydrocarbon (X) is 2, 5, 6-trimethyloctane.

09. *(b)* 2-formyl-4-hydroxyhept-6-enoic acid. Hence, option (b) is correct.

10. *(a)* The structure of 2, 3-dibromo-1-phenylpentane is

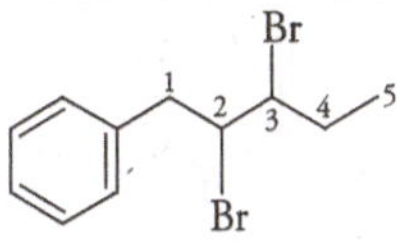

11. *(d)* $\overset{4}{CH_3}-\underset{NH_2}{\underset{|}{\overset{3}{CH}}}-\overset{2}{CH_2}-\overset{1}{CN}$

IUPAC name of the given compound is 3-aminobutanenitrile.

12. *(a)* The bond line formula for $HOCH(CN)_2$ is

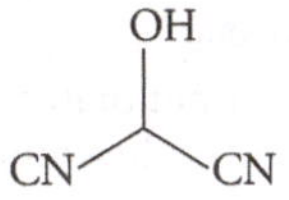

13. *(b)*

Cyclohexene is an alicyclic compound.
An alicyclic compound are the organic compounds that are both aliphatic and cyclic. These are either saturated or unsaturated but do not have aromatic character.

14. *(b)*

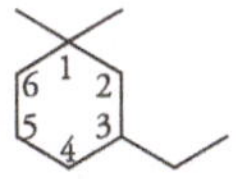

3-ethyl-1,1-dimethylcyclohexane.

15. *(b)* For naming the polyfunctional organic compounds, IUPAC has given an order of decreasing priority for the functional groups as follows

$-COOH > -SO_3H > -COOR > -COCl > -CONH_2 > -CN > -CHO > \rangle C=O > -OH > -NH_2 > \rangle C=C\langle > -C\equiv C-$

16. *(c)* Priority order for functional groups present in the given compound is

aldehyde > ketone > alkene > nitro.

The carbon atom of aldehyde group get lowest possible number in longest carbon chain of seven carbon atoms.

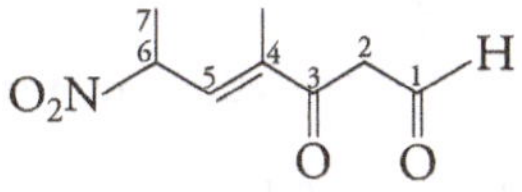

So, correct IUPAC name is 4-methyl-6-nitro-3-oxohept-4-enal.
Thus, option (c) is the correct answer.

17. *(c)* Hybridisation of carbon a, b and c in the given molecule respectively are sp^3, sp^2 and sp^2.

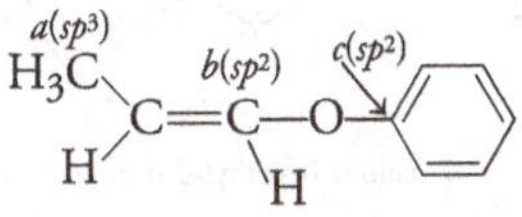

18. *(d)* The hybridisation of carbon 1, 2, 3 and 4 respectively are

$$\overset{1}{C}H_2=\overset{2}{C}=\overset{3}{C}H-\overset{4}{C}H_3$$
$$\downarrow sp^2 \quad \downarrow sp \quad \downarrow sp^2 \quad \downarrow sp^3$$

19. *(d)* The IUPAC name of the given compound,

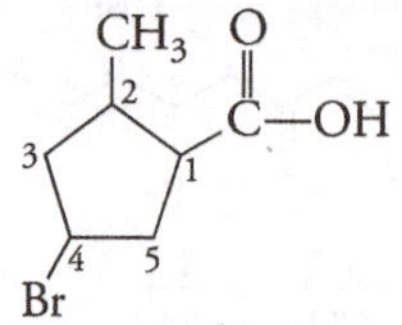

is 4-bromo-2-methylcyclopentane carboxylic acid.

20. *(a)* Statement I is incorrect but Statement II is correct.
The correct form of statement I is,
IUPAC name of compound (*A*)
1-chloro-2,4-dinitrobenzene.

21. *(b)* Statement I is correct but Statement II is incorrect. It is because the IUPAC name of the compound is

$$\overset{7}{HOCH_2}\overset{6}{CH_2}\overset{5}{CH_2}\overset{4}{CH_2}\overset{3}{CH_2}\underset{2}{\overset{O}{\overset{\|}{C}}}\overset{1}{CH_3}$$

7-hydroxyheptan-2-one

22. *(a)*

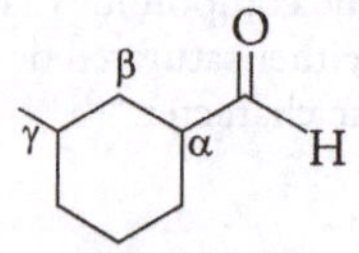

γ-methylcyclohexane carbaldehyde

Here, cyclohexane is considered as parent chain, aldehyde is functional group but since it is not included in principal chain, it is written as carbaldehyde.
Third position to the aldehyde group is called γ-position.

23. *(d)* The structure of ethylidene chloride is

$$\overset{2}{H_3C}-\overset{1}{HC}\begin{matrix} Cl \\ Cl \end{matrix}$$

The IUPAC name is 1,1-dichloroethane.

24. *(d)* Mesityl oxide is a common name of 4-methyl pent-3-en-2-one.

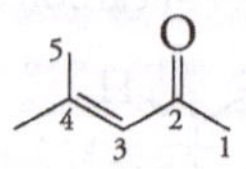

(4-methylpent-3-en-2-one)

25. *(c)* The IUPAC name of the given compound is 2-chloro-1-methyl-4-nitrobenzene .

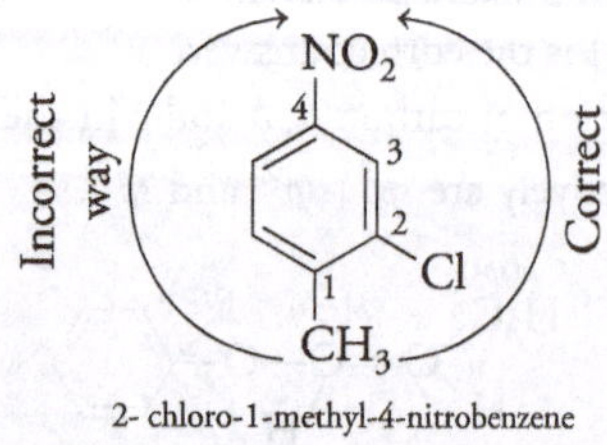

2- chloro-1-methyl-4-nitrobenzene

26. *(c)* The correct match is A-II, B-III, C-IV, D-I.

27. *(c)* The correct match is A-(III), B-(IV), C-(I), D-(II).

28. *(c)*

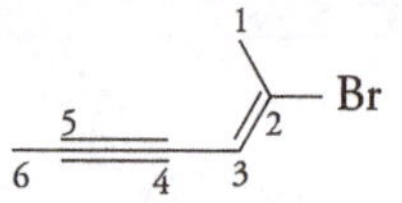

The IUPAC name of above compound is 2-bromohex-2-en-4-yne.
Now, to assign *E*/*Z* first assign priority to groups along the double bond.

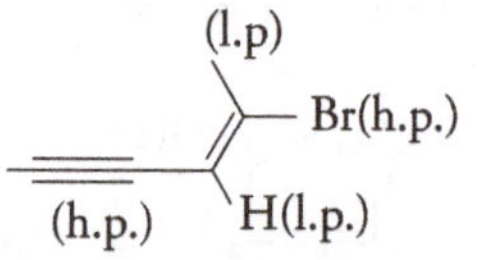

[h.p.=high priority
l.p.=low priority]

Since, h.p. groups are in opposite direction. So, *E* configuration will be assigned to double bond at second position.
Hence, full IUPAC name of given compound is (2*E*)-2-bromohex-2-en-4-yne.
Hence, correct option is (c).

29. *(c)*

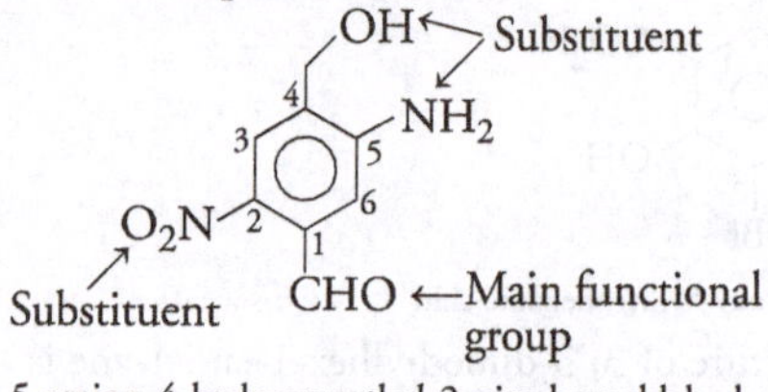

5-amino-4-hydroxymethyl-2-nitrobenzaldehyde

30. *(15)* The structure of hex-1, 3-dien-5-yne is

$$H-\overset{H}{\overset{|}{C}}=\underset{H}{\underset{|}{C}}-\underset{H}{\underset{|}{C}}=\underset{H}{\underset{|}{C}}-C\equiv C-H$$

So, σ bond $=11$, π bond $=4$
Thus, total $\sigma+\pi=11+4=15$

31. *(9)* The reaction mentioned in question is as follows,

$$2Ph-CHO + CH_3-\overset{O}{\overset{\|}{C}}-CH_3 \xrightarrow[\Delta]{aq.\ NaOH} Ph-CH=CH-\overset{O}{\overset{\|}{C}}-CH=CH-Ph$$

Product *X*

Total number of π bonds in $X=9$.

32. *(5)* The complete reaction is as follow,

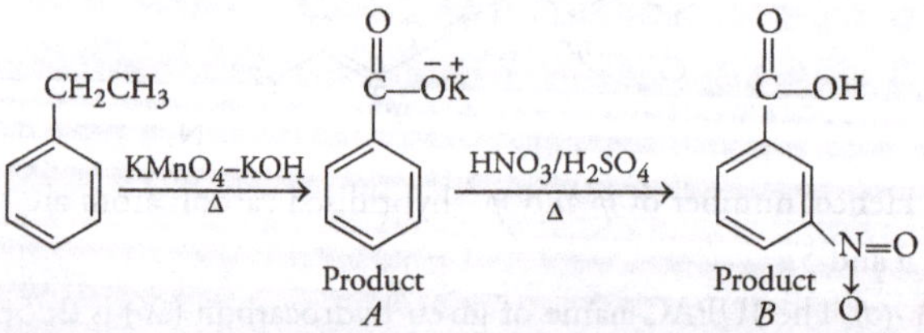

Total number of π bond in product $B=5$.

33. *(3)* The complete reaction is as follows,

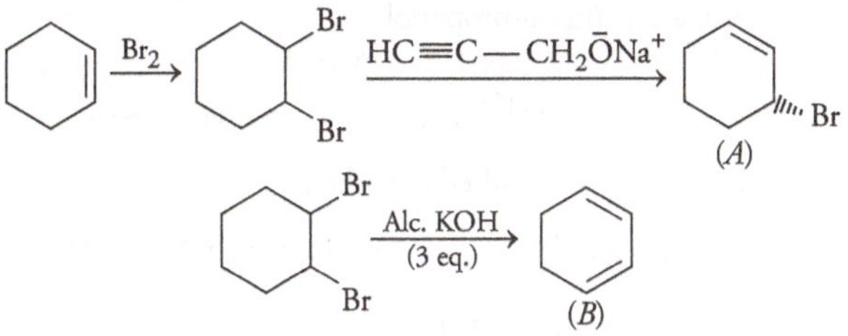

So, compound A contains 1 π bond while compound B contains 2π bond. Thus total π bond in A and B are three.

34. *(18)* The structure of 2-oxohex-4-ynoic acid is

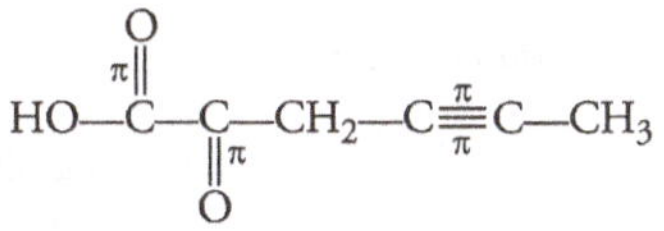

Thus, the given compound has 4π bonds and 14 sigma bonds. Hence, sum of sigma and pi bonds is 18.

35. *(1)* Molecular formula $= C_4H_5N$

$$\text{Degree of unsaturation} = nC - \frac{nH}{2} + \frac{nN}{2} - \frac{nX}{2} + 1$$

$$= 4 - \frac{5}{2} + \frac{1}{2} + 1 = 3$$

Compound has one nitrogen, so it must contain CN and it is acyclic. Thus, the possible structure could be.

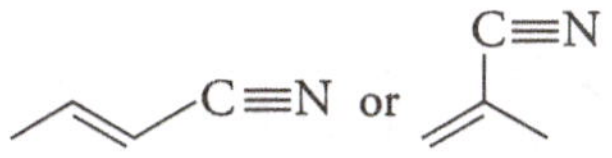

Number of sp^3 hybridised carbon is 1 in both the cases.

36. *(10)* According to given structure,

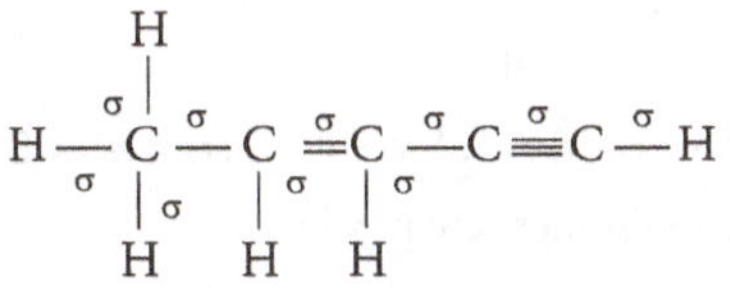

Total number of σ bonds $= 10$

Topic 2

01. *(b)* For the given molecule total 4 stereoisomers are possible i.e., *R cis*, *R trans*, *S cis* and *S trans*.

02. *(c)* Compound A does not contain axes of symmetry and hence, it is optically active.

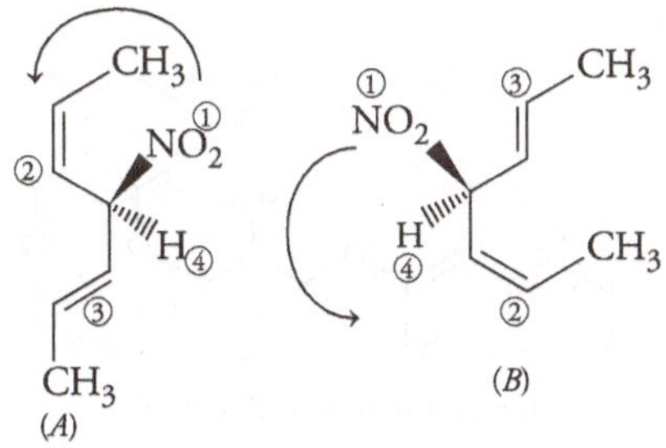

Both compounds (A) and (B) have same S-configurations, so they are not mirror images.

So, statement I is correct but statement II is incorrect.

Concept Enhancer

Optically active compounds are those compounds which have chiral carbon or absence of axes of symmetry or whose mirror images are non super imposable on each other.

03. *(d)* 1,2-dimethylcyclopropane is

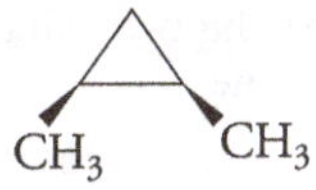

Hence, stereoisomers are as follows

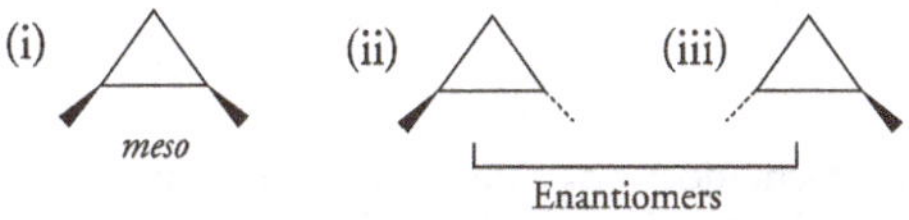

meso-form in optically inactive, whereas enantiomeric pairs are optically active. Therefore, total number of stereoisomers are three (3).

04. *(d)* Metamers are compounds which have different alkyl groups present along both side of polyvalent functional group.

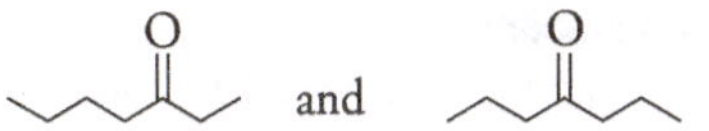

3-heptanone 4-heptanone

(Metamers)

05. *(d)* Functional isomers are those which have same molecular formula but different functional groups. C_3H_6O molecule can exists as aldehyde as well as ketone as follows

$$\underset{\text{(Aldehyde)}}{CH_3-CH_2-CH=O} \quad \text{and} \quad \underset{\text{(Ketone)}}{CH_3-\overset{\displaystyle O}{\overset{\|}{C}}-CH_3}$$

So, they are functional isomers of each other.

06. *(b)* Geometrical isomers are molecule that are locked into their spatial positions with respect to one another due to double bond or a ring structure. The presence of locking or restricted rotation is due to double bond or a ring structure. Two different groups must be attached to any two carbon atoms of restricted rotation.

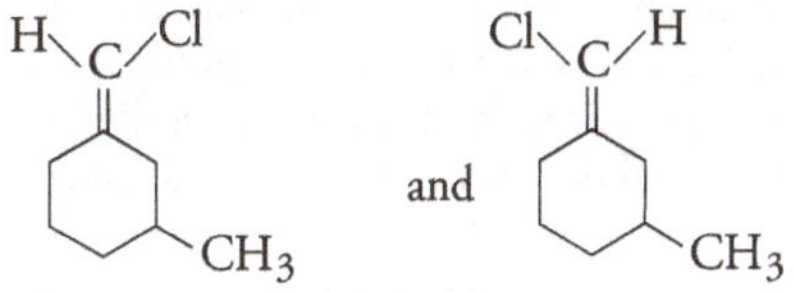

show geometrical (possible *cis* and *trans* isomer)

07. *(d)* For the structure.

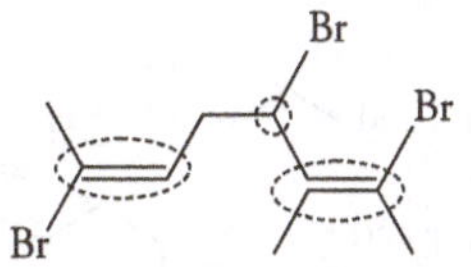

Total 3 stereocentres are there.

So, total stereoisomers possible are $2^3 = 8$.

08. *(c)* Metamer are those isomers having same molecular formula, same functional group, but different alkyl/aryl groups on either side of functional group.

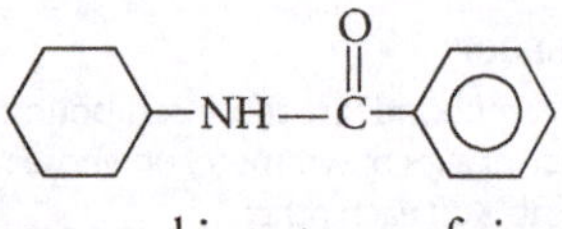

Above drawn compound is metamer of given compound (X).

09. *(a)* The isomeric deuterated bromide with molecular formula C_4H_8DBr having two chiral atoms is 2-bromo-3-deuterobutane

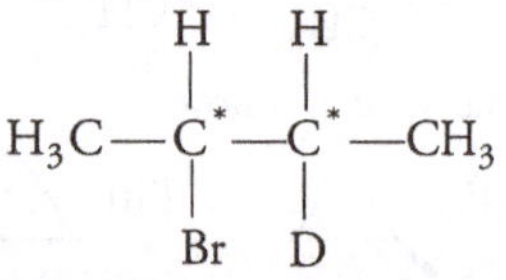

10. *(c)* 3-ethylhex-3-ene does not show stereoisomerism as it optically inactive due to absence of chiral carbon centre.

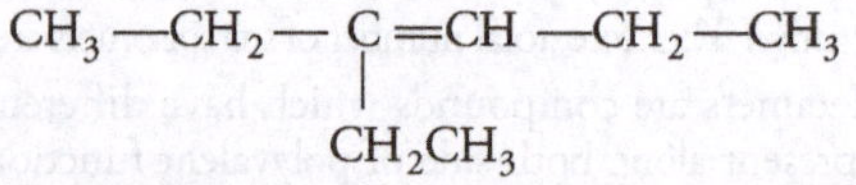

Concept Enhancer

Compounds having the same molecular formula (constitution) and the same connectivity of atoms in their molecules but different-3-dimensional orientation of their atoms in space are called stereoisomers.

11. *(d)* Reaction of all options are as follows

(a) 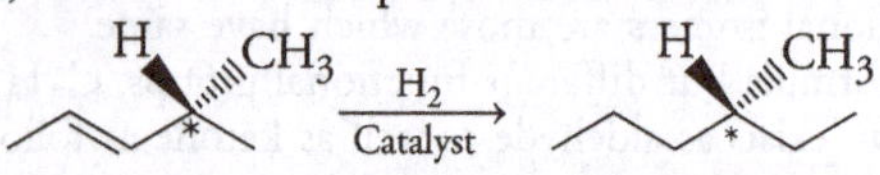

(b) 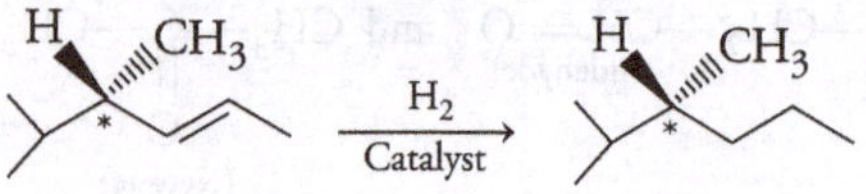

(c) 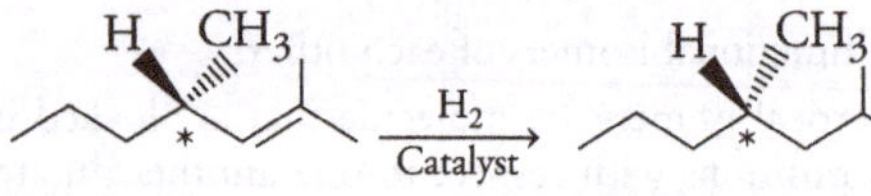

(d) 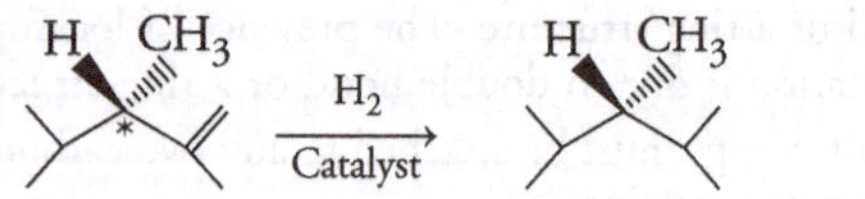

No change in chirality or optical activity due to hydrogenation of (a), (b) and (c) but in (d) the substrate (alkene) on hydrogenation becomes optically inactive as the product obtained does not have any chiral atom.

12. *(b)*

(A) 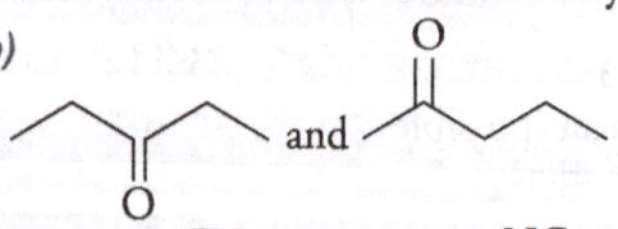 are metamers

(B) 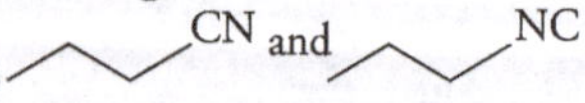 are functional isomers

(C) 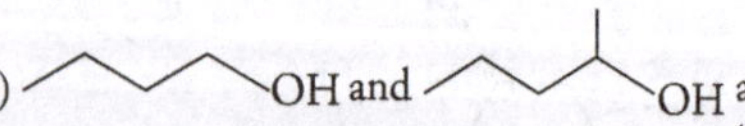 are homologous to each other

(D) 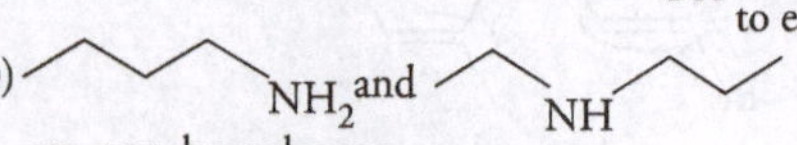

are non-homologous

So, A and B, i.e. option (b) is correct.

13. *(b)* The correct match is A-III, B-I, C-IV, D-II.

- *n*-propanol and *iso*-propanol

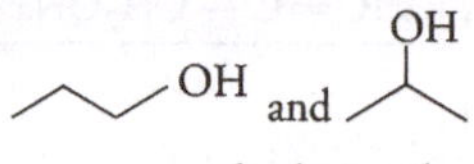 = Position isomers

- Methoxypropane and ethoxyethane

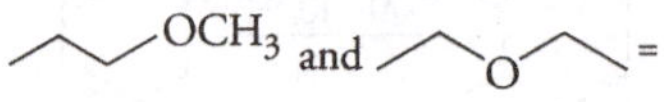 = Metamers

- Propanone and propanal

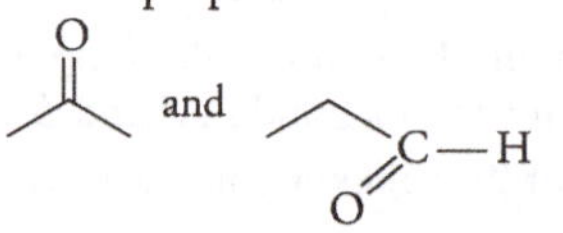 = Functiornal isomers

- Neopentane and *iso*-pentane

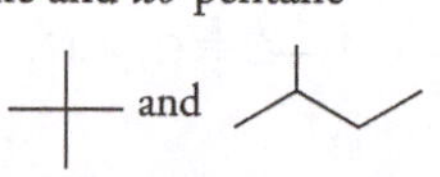 = Chain isomers

14. *(b)* The correct match is A-III, B-I, C-IV, D-II.

15. *(8)* There are total eight structural isomers possible for C_9H_{12}.

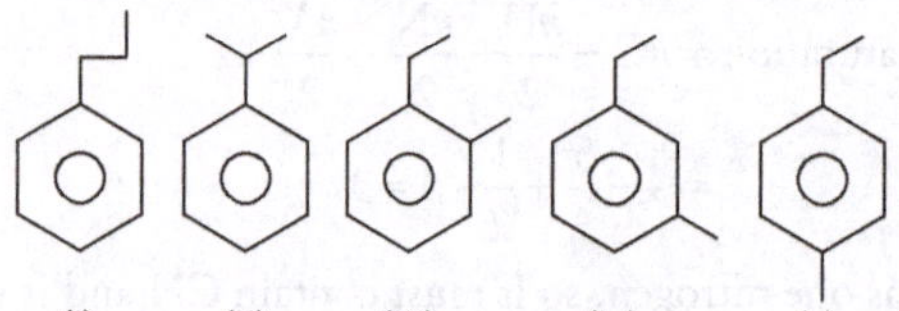

(i) (ii) (iii) (iv) (v)

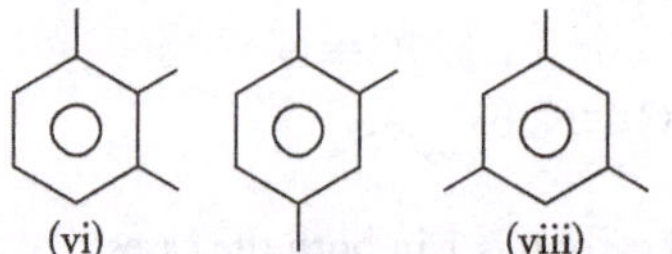

(vi) (vii) (viii)

16. *(4)*

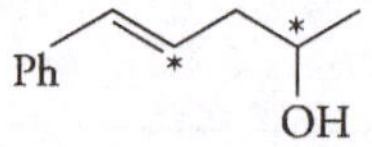

n (stereogenic unit) = 2, Stereoisomers $= n^2 = 2^2 = 4$

Thus, 4 stereoisomers are possible.

17. *(2)* There are two chiral compounds [(i) and (ii)].

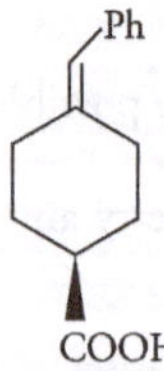

[No, plane of symmetry, axes of symmetry and central of symmetry]

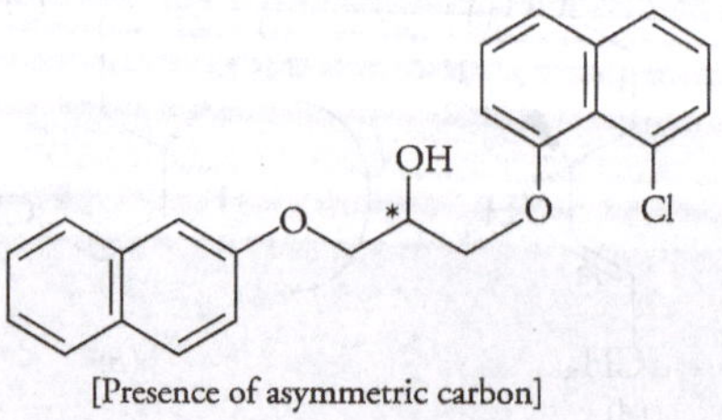

[Presence of asymmetric carbon]

Concept Enhancer

The objects which are non-superimposable on their mirror image are said be chiral while the objects that are superimposable on each other are called as achiral. The other criteria for the chirality is the presence of asymmetric carbon atom.

18. *(3)*

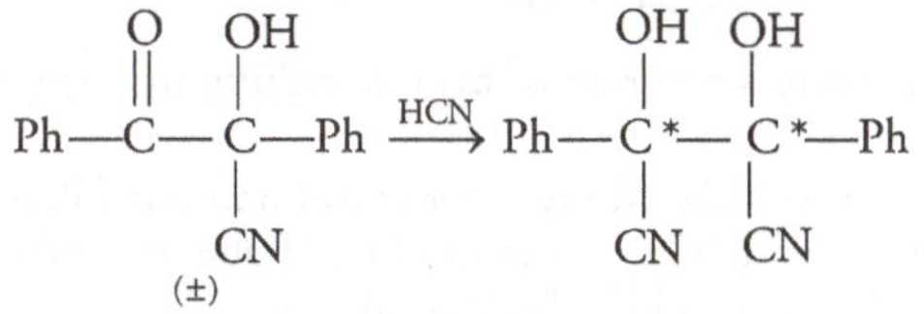

The compound formed has two chiral centres.

So, it forms three stereoisomers (2 enantiomers + 1 *meso* isomer)

19. *(42)* $\%\ \text{Optical purity} = \dfrac{\text{Observed rotation of mixture}}{\text{Rotation of pure enantiomer}} \times 100$

$$= \frac{+12.6}{+30} \times 100 = 42\%$$

20. *(3)* An asymmetric carbon atom is a carbon atom which is attached to four different atoms or group of atoms. There are three compounds which possess asymmetric carbon atoms.

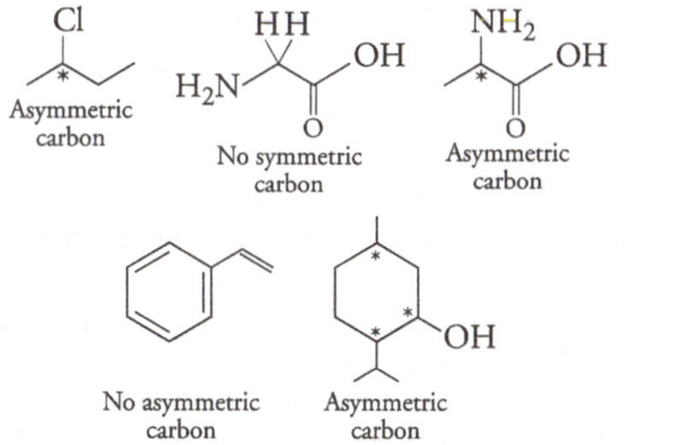

21. *(5)* Total number of possible stereoisomers of dimethyl cyclopentane is 5 which are as follows

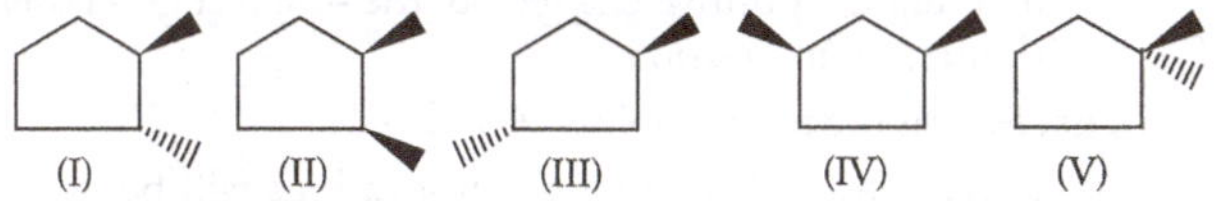

22. *(6)* (i) Pent-1-ene

(ii) Pent-2-ene it shows geometrical isomerism, i.e. *cis* and *trans*

(iii) 2-methyl but-1-ene

(iv) 2-methyl but-2-ene

(v) 3-methyl but-1-ene

Total number of acyclic structural isomers (including geometrical isomers) for pentene is 6.

23. *(5)*

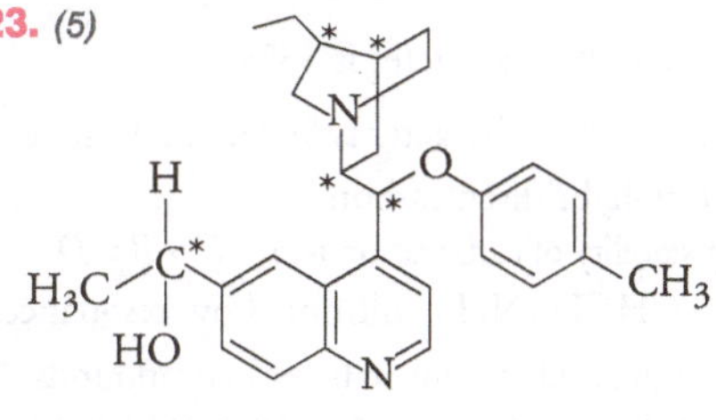

The molecule contains five chiral carbon (C*) atoms.

Topic 3

01. *(a)* Carbocation is sp^2 hybridised, hence its shape is trigonal planar.

02. *(b)*

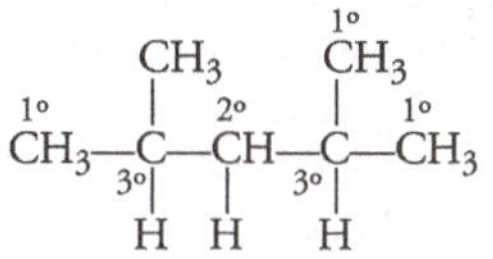

Only one 2° carbon is present in this compound.

Hence, option (b) is correct.

03. *(d)* The difference in energy between the actual structure and the lowest energy resonance structure for the given compound is called as resonance energy. Hence, option (d) is correct.

04. *(d)* The molecular formula of second homologue in the homologous series of mono-carboxylic acid is CH_3COOH i.e., $C_2H_4O_2$.

05. *(c)* Hyperconjugation is a permanent effect.

Ethyl cation ($CH_3-\overset{+}{C}H_2$) involves overlapping of $C_{sp^3}-H_{1s}$ bond with empty $2p$-orbital of other carbon.

∴ Statement I is true and statement II is false.

06. *(a)* Stability of a resonating structure depends on

(i) Charge separation between opposite charges should be minimum.

(ii) Charge separation between same charges should be maximum.

Thus, in option (a) two cationic charges are very close to each other. So, these charge repel each other. Therefore, structure (a) is unstable.

07. *(d)* The acid whose conjugate base is most stable will be strongest acid.

Conjugate base of molecule (d) is most stable as an aromatic compound is formed after donating a proton (conjugative base).

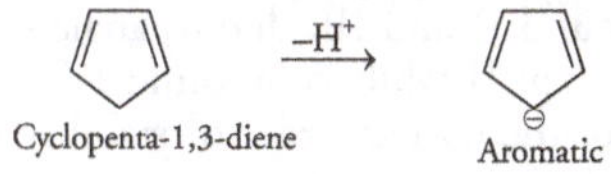

08. *(b)* Due to $-R$ (or $-M$) effect of 'NO_2', there is stronger double bond character for the 'C—Cl' bond in $Cl-CH=CH-NO_2$. (option-b)

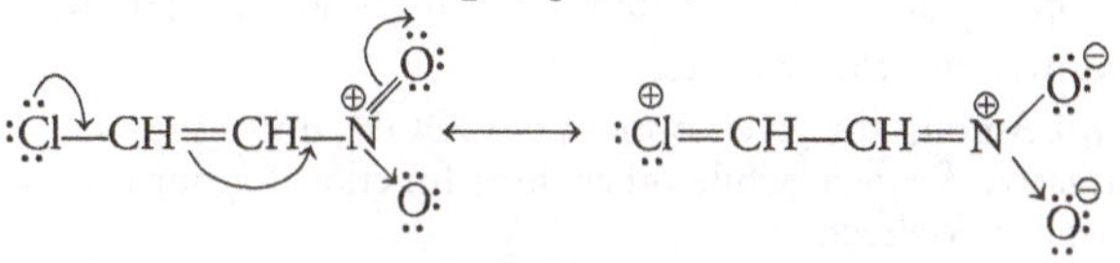

∴ The bond length is shortest in (b).

09. *(d)* We know, a stronger acid produces its stable or weaker conjugate base. Here, $CH(CN)_3$ produces the most stable conjugate base $(NC)_3C^-$. Stronger $-R$ and $-I$ effects of the CN^- group, make the carbanion (conjugate base) very stable.

10. *(c)* Ethene ($H_2C = CH_2$) is sp^2-hybridised and ethyne ($HC \equiv CH$) is sp-hybridised. In ethyne, the sp-hybridised carbon atom possesses maximum s-character and hence, maximum electronegativity. Due to which, it attracts the shared electron pair of C—H bond to a greater extent and makes the removal of proton easier. Hence, alkyne is much more acidic than alkene. Further, among all the acidic alkynes $CH \equiv CH$ is most acidic.

Among $CH_2 = CH_2$ and $CH_3 - C \equiv CH$, $CH_2 = CH_2$ is more acidic as in $CH_3 - C \equiv CH$ the ERG i.e., CH_3 — is attached. Thus, the order for acidic strength of the compounds is $CH \equiv CH > CH_2 = CH_2 > CH_3 - C \equiv CH$

11. *(b)* The correct match is A-III, B-IV, C-I, D-II.

12. *(a)* Statement I is false but Statement II is true.

The correct form of Statement I is

Hyperconjugation is permanent effect because external reagent is not required.

13. *(c)* The given compound is

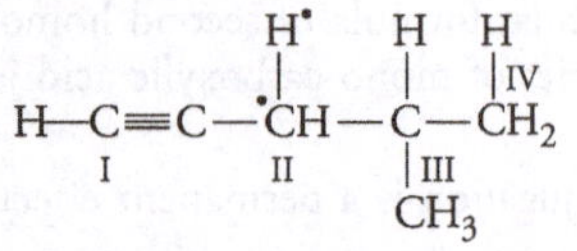

II carbon is most stable carbon radical due to resonance stabilising factor.

I carbon is least stable carbon radical due to no stabilising factor.

14. *(c)* Carbocation intermediate is stabilised by $+I$, $+M$ and hyperconjugation effect. Since in

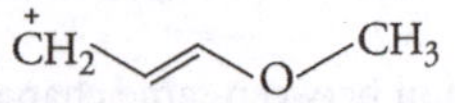

Carbocation is in conjugation with stronger $+M$ with group $-OCH_3$, hence it is most stable.

15. *(c)* Generally, neutral structures are more stable than charged ones. Therefore, I is more stable than II and III. While positive charge on less electronegative atom is more stable i.e. C^+ is more stable than O^+.

Thus, the correct order is I > II > III.

16. *(c)* The compound I is antiaromatic hence it is least stable. Among compound II and III, due to angle strain theory, III is more stable than II while compound IV is aromatic hence most stable. So, the correct order of stability of given carbanion i.e.

IV > III > II > I

17. *(a)* Among the given statements, incorrect statement is given in option (a). Its correct form is hydrogen ion [H^+] shows positive electromeric effect.

18. *(b)* Among the given options —COOH group shows negative *R*-effect, while other three functional groups shows positive *R*-effect.

e.g.

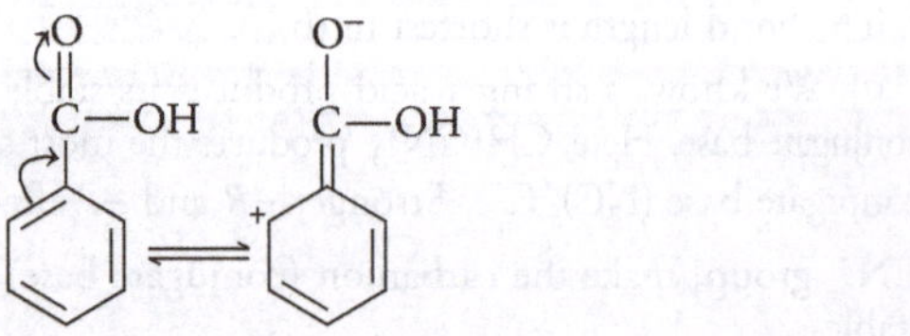

19. *(d)* Carbocations are the species that have sextet of electrons and can act as an electrophile i.e.,

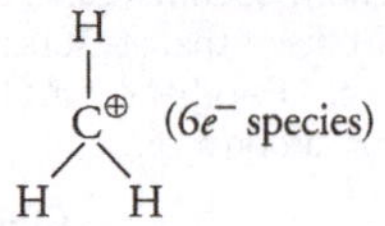

20. *(a)* The decreasing order of hydride affinity for the given carbocations is C > A > B > D.

The correct hydride affinity order of carbocations will be decided by the stability of carbocation. Higher the stability of carbocation, lower will be hydride affinity.

21. *(b)*

> **idea** Use the concept of stability of carbocation to solve this problem. This stability can be judged on the basis of resonance and hyperconjugation.

Carbocation (*C*) is most stable because positive charge delocalises with 2 π bonds and lone pair of N-atom in NH_2 group and due to hyperconjugation.

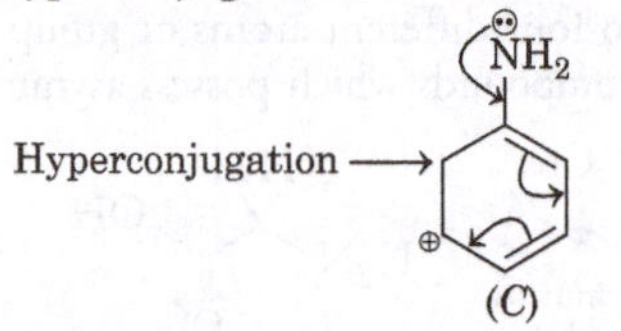

Thus, option (b) is correct.

22. *(d)* NH_2^- is more nucleophilic than NH_3 due to the presence of the presence of more electron density of nitrogen atom.

23. *(a)*

(*A*) No electron withdrawing group is attached to it so it is the most stable structure among them.

(*B*) Oxygen is electron withdrawing group and directly attached to the carbon containing the positive charge which makes it least stable.

(*C*) In this structure oxygen is indirectly attached to carbon containing the positive charge. So, the $-I$ effect gets lesser with increase in position.

So, decreasing order of stability $A > C > B$.

24. *(a)* Correct order of stability of carbocations will be

(A) 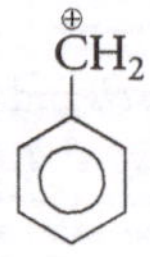 Benzyl carbocation get stabilise by resonance.

(C) $CH_3 - \overset{\oplus}{C}H_2$ gets stabilised due to hyperconjugation.

Electronegativity of carbon increases with increase in % *s*-character.

For sp^3 -hybridised carbon, % *s*-character is 33%.

For sp^3 -hybridised carbon, % *s*-character is 50%. So, positive charge is less stable on *sp*-hybridised carbon.

So, the correct order of stability of carbocation is $A > C > B > D$.

25. *(d)* $CH_3CH_2CH = CHCH_2NH_2$ will not show resonance as it does not have conjugated system, whereas compounds (or their parts) in option a, b and c have conjugated system. So, they exhibit resonance.

26. *(a)* Rate of hydrolysis is directly proportional to positive charge present on carbon of $>C{=}O$ group.

Therefore,

Order of rate of hydrolysis → Acid chloride > Acid anhydride > Ester > Amide.

27. *(a)* Carbocation *B* (2°) is more stable due to more hyperconjugation and formed relatively at faster rate compared to *A* (1°).

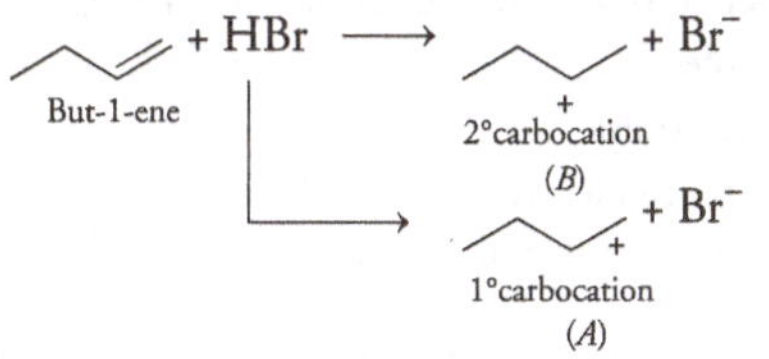

28. *(d)* Acidity of H-atom ∝ Attachment with more and stronger electron withdrawing groups.

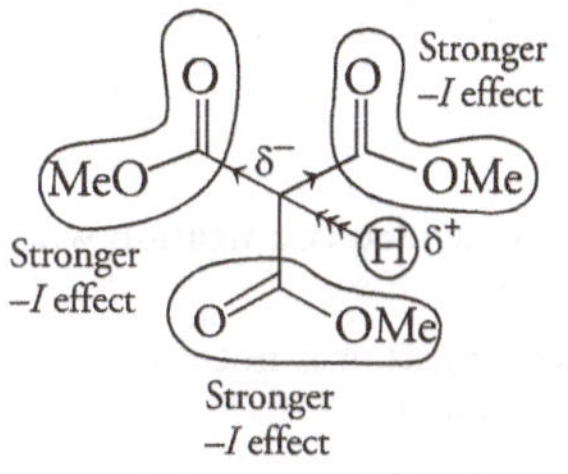

⟶ It is most acidic, because the carbon atom bearing this H-atom has stronger –*I*-effects by three ester groups $\left(-\overset{\overset{O}{\|}}{C}-OMe.\right)$

29. *(c)* The H-atom (*a*) attacked with *sp*-carbon ($s\% = 50$) is also acidic but lesser acidic than H-atoms (*b*, *c* and *d*) who are attached with more electronegative O-atoms.

∴Electronegativity : $O(sp^3) > C(sp)$

Among *c* and *d* the acidity of *c* is more as $—NO_2$ group shows –*M* effect and it is directly impacting it at *p*-position

Hence, the order of acidity *b* > *c* > *d* > *a*.

30. *(a)* Basicity order for the following in intermediates is

$$\underset{(v)}{CN^-} < \underset{(iii)}{CH{\equiv}C^{\ominus}} < \underset{(ii)}{H_2C{=}CH{-}\overset{\ominus}{C}H_2} < \underset{(iv)}{\overset{\ominus}{C}H_3} < \underset{(i)}{CH_3-\overset{CH_3}{\underset{CH_3}{C^{\ominus}}}}$$

which is also the opposite of the order of stability of carbanions.

$$:\overset{\ominus}{C}{\equiv}N: > \underset{(sp\text{-carbon})}{CH{\equiv}\overset{\ominus}{C}} > \underset{\text{(resonance) stabilised}}{CH_2{=}CH{-}\overset{\ominus}{C}H_2} > \overset{\ominus}{C}H_3 > \underset{\text{(less stable due to }+I\text{ effect of }CH_3\text{ groups)}}{\overset{\ominus}{C}(CH_3)_3}$$

31. *(d)* The order of stability for all the given options are as follows

(A) 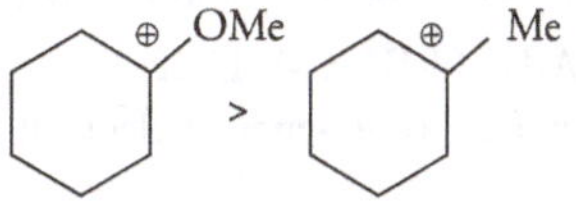

Stable by backbonding

(B) 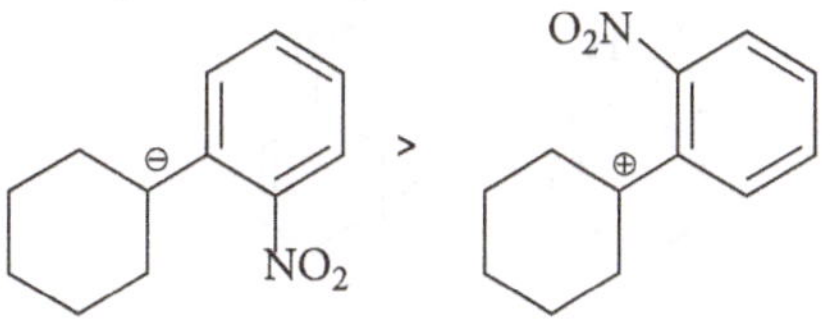

Stable due to –*M*, –*I* effect of NO_2 group

(C) 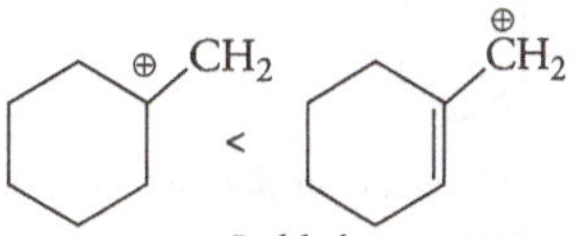

Stable by resonance

(D) 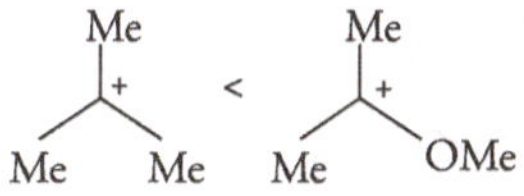

Stable by backbonding

32. *(c)*

$A = Ph-\overset{\overset{Ph}{|}}{\underset{\underset{Ph}{|}}{C^{\oplus}}}$, $B = Ph-\overset{\overset{Ph}{|}}{\underset{\underset{H}{|}}{C^{\oplus}}}$

C =  and $D = H_3C{-}CH_2{-}\overset{\oplus}{\underset{\underset{CH_3}{|}}{C}H}$

Now, C is aromatic due to positive charge and hence it is most stable.

A has more resonance structure.

B has less resonance structure

D has only hyperconjugation.

∴Stability is in order,

Aromaticity > Resonance > Hyper conjugation

Hence, the correct order of stability is C > A > B > D.

33. *(c)* The order of stability of given carbocations are as follows.

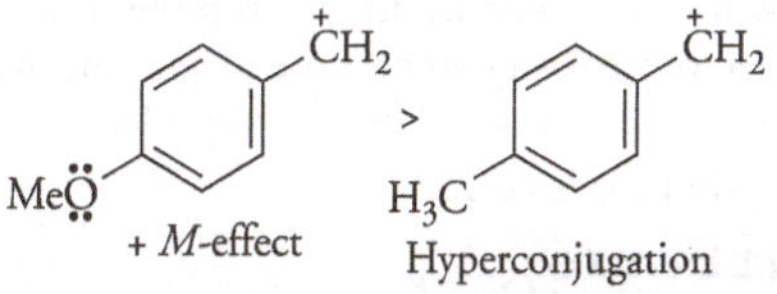

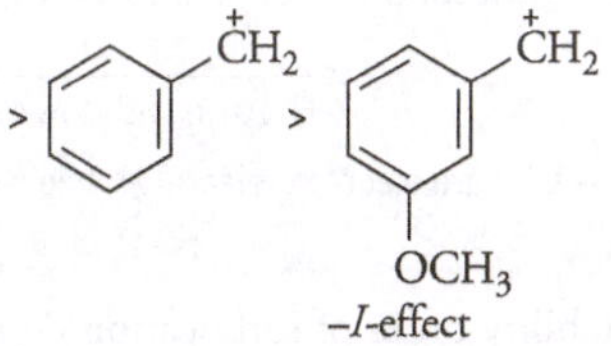

So, carbocation given in option (c) is most stable.

34. *(a)* Among the given options, there are five nucleophiles i.e, NH_3, PhSH, $(H_3C)_2S$, $\overset{\ominus}{O}H$ and $CH_2=CH_2$

35. *(d)* The correct match is A-IV, B-III, C-I, D-II.

(A) Here, $+R$ effect is shown due to presence of electron donating group. $\ddot{N}H_2$.

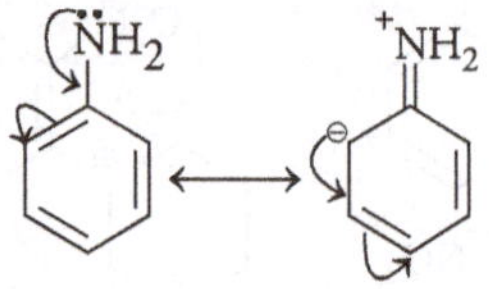

(B) 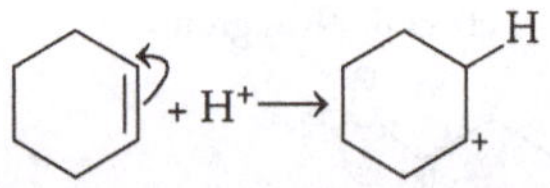

Here, due to presence of H^+, electron deficient species, $+E$ effect is observed in the given mechanism.

(C) 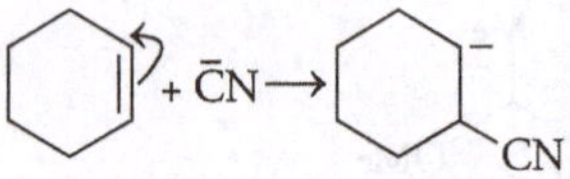

Here, $-E$ effect is observed as the shared pair of electron is transferred away from the attacking reagent.

(D) 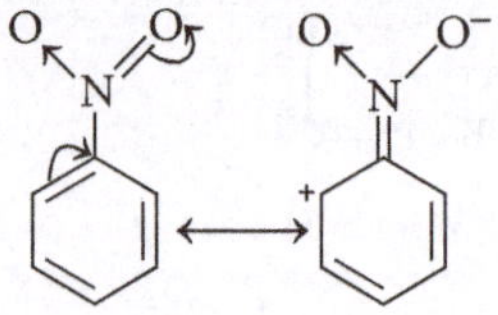

Here $-R$ effect is shown due to presence of electron withdrawing group NO_2.

Time Saver Tech

Use elimination technique as time saver. As all the given option are different for *A*, *B*, *C*, *D*. So, you can solve for any one reaction and can choose your answer. The presence of electron donating group will show $+R$ effect so, A-IV is given in option (d) only.

36. *(c)* Ionic reaction with organic compound proceed through heterolytic bond cleavage.
Hence, option (c) is correct.

37. *(d)* Greater the electron density in the benzene ring more easily it will be attacked by an electrophile. Phenol has highest electron density among the given options, therefore it will be most easily attacked by electrophile.
Hence, option (d) is correct.

Concept Enhancer

Electrophilic substitution $\propto$ Electron releasing group

$$\propto \frac{1}{\text{Electron withdrawing group}}$$

—OH and —CH_3 are electron releasing while —Cl is e^- withdrawing in nature.

38. *(d)* The stability order of carbocation depends on the number of hyperconjugable hydrogens.

$(CH_3)_3C^+$ has highest hyperconjugable hydrogen then $(CH_3)_2\overset{+}{C}H$, than $CH_3-\overset{+}{C}H_2$ and $\overset{+}{C}H_3$ has lowest among the given option.
Therefore, the correct order is

$$\underset{(9\alpha\text{-hydrogen})}{(CH_3)_3\overset{+}{C}} > \underset{(6\alpha\text{-hydrogen})}{(CH_3)_2\overset{+}{C}H} > \underset{3\alpha\text{-hydrogen}}{CH_3-\overset{+}{C}H_2} > \underset{(\text{zero }\alpha-\text{H})}{\overset{+}{C}H_3}$$

Hence, option (d) is correct.

39. *(d)* Presence of (+) ve charge on electronegative atom (O) is unfavourable.
$\therefore$(C) is least stable.
'O' being more electronegative than N. Presence of '+ ve' charge on 'O' is more unfavourable than 'N' as in the case of (D).
Between A and B in (B) '−ve' charge is able to delocalise on both C=O and C=C. Hence, stability increases of (B) than (A). The increasing order of stability of the resonance structure is $C < D < A < B$

40. *(d)* 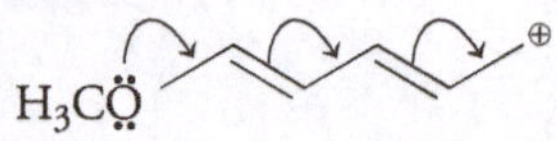

is most stable carbocation due to extended conjugation which also involves lone pair of oxygen.
In rest of the carbocations, the conjugation is not that extended as it is in (d). So are less stable.

41. *(a)* 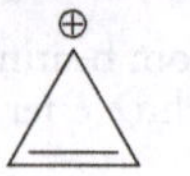 is most stable, since it is aromatic compound, follows Huckel rule $(4n+2)$

$$4n+2=2$$
$$4n=0$$
$$n=0$$

42. *(c)* Resonance is shown by compound (*A*) and (*B*) only, hence they form stable carbocations.
Compound (*C*) and compound (*D*) do not show resonance because double bond and positive charge are not in conjugation.

43. *(4)* Among the given compounds, total four compounds shows inductive, mesomeric as well as hyperconjugation effects i.e.

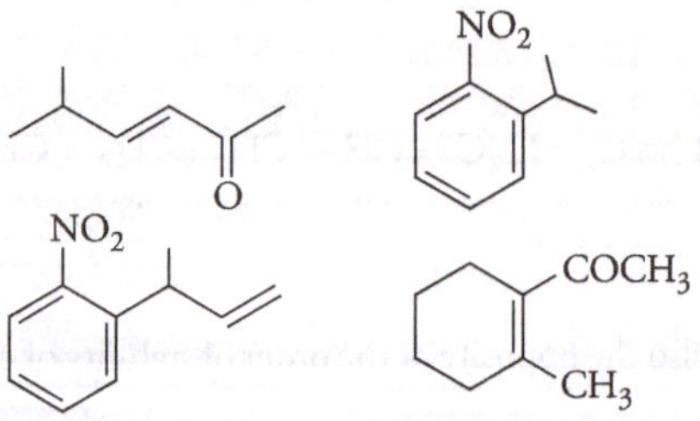

44. *(5)*

idea If carbocation has no α-hydrogen, it will not be stabilised by hyperconjugation.

The carbocation 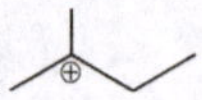 has 8α-hydrogen

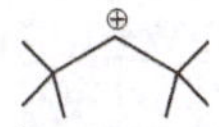 has no α-hydrogen

$\overset{+}{C}H_3$ has no α-hydrogen.

 has no α-hydrogen

$\overset{+}{C}H_2$ —OCH_3 has no α-hydrogen.

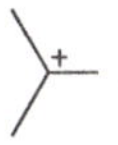 has 9 α-hydrogen.

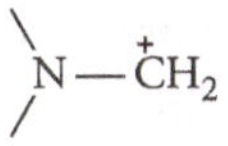 has no α-hydrogen.

Hence, total 5 carbocations are not stabilised by hyperconjugation.

45. *(2)* The number of compounds which contain sulphur as heteroatom in question is 2. These are thiophene and cysteine.

Thiophene ⟶ 

Cysteine ⟶ 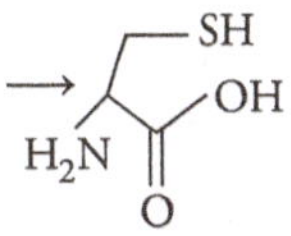

46. *(4)* Among the given functional groups, —NO_2, —C≡N, —COR, —COOH are *meta*-directing. Hence, total 4 functional groups are *meta*-directing.

Topic 4

01. *(a)* Nitrochlorobenzene with strong E.W.G. (i.e. —NO_2) undergoes nucleophilic substitution at faster rate with OH^- as nucleophile.

3-chloronitrobenzene has weaker E.W.G. nature of NO_2 (*–I* only) as compared to other compounds. So, it gives lowest rate towards nucleophilic aromatic substitution reaction (ArS_N2).

Thus, option (a) is correct.

02. *(d)* Rate of nucleophilic substitution in aromatic halide are as follows

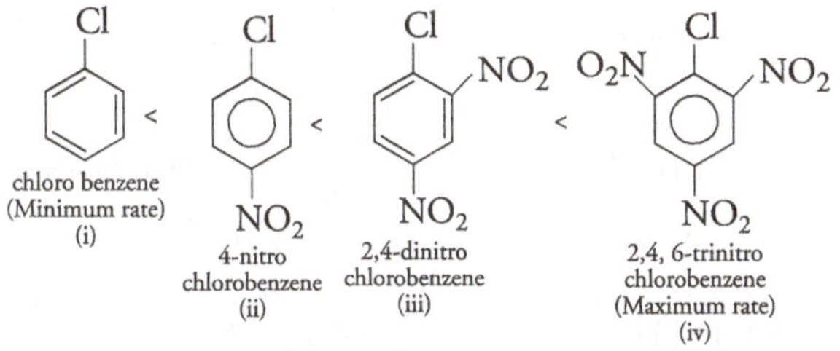

Greater the number of nitro (—NO_2) group in chlorobenzene, greater the rate of nucleophilic substitution reaction in aromatic halides.

When a nitro group is present at *ortho* and *para* position, it can withdraw electrons from the benzene ring. This facilitates, the attack of nucleophiles on haloarenes.

03. *(d)*

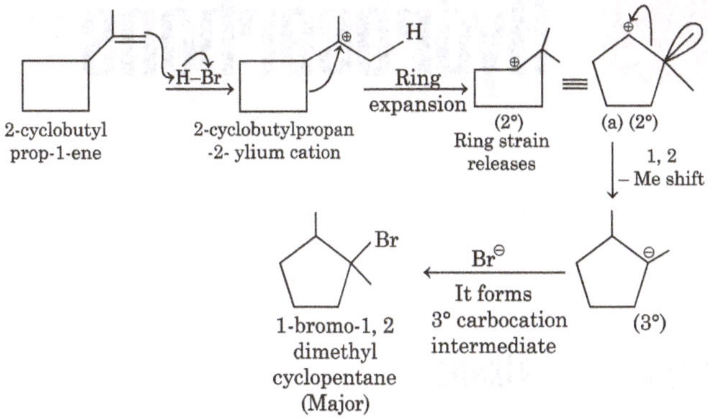

Concept Enhancer

2-cyclobutylprop-1-ene reacts with HBr to give a 3° carbocation, which undergoes ring expansion to release ring strain then again it rearranges to 3° carbocation *via* 1,2-methyl shift attack of Br^- to give major product.

04. *(b)* Nitration of aromatic compounds proceeds *via* electrophilic substitution reaction. It means, higher the electron density on benzene, more will be the reactivity.

The order of electron density on the benzene ring is

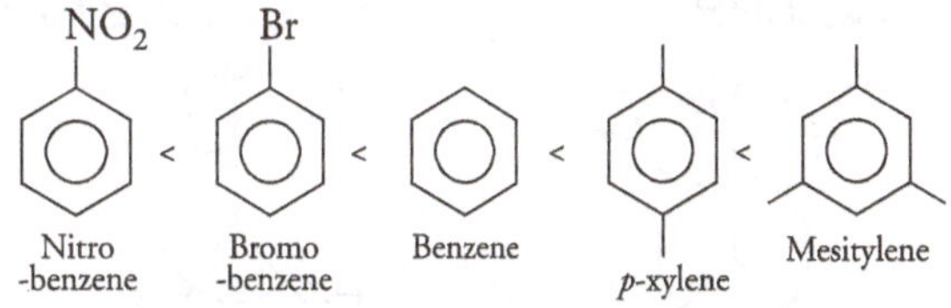

Hence, the increasing order of reactivity towards nitration is

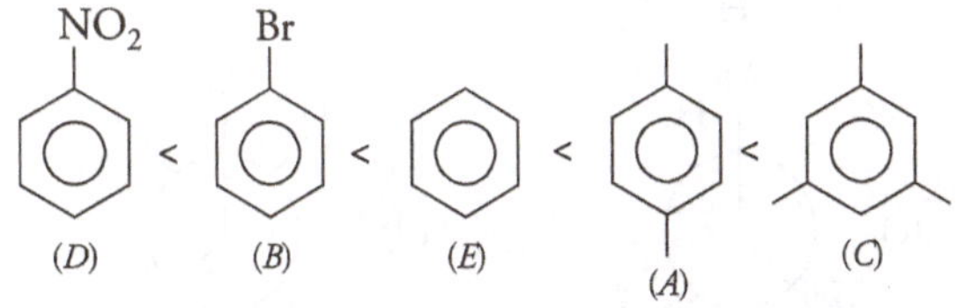

05. *(7)* To stabilise the given carbocation 7 hyperconjugation (including final product) structures are involved.

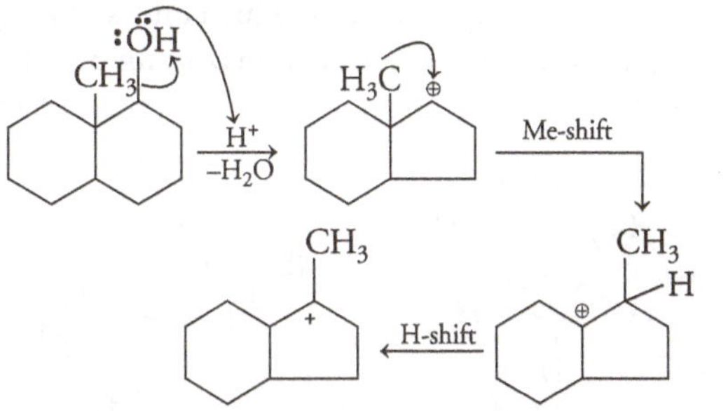

12 Hydrocarbons

In JEE Main there is usually 0-2 question per paper are visible from this chapter. In past 3 years, 35 conceptual and 12 numerical questions were asked of moderate difficulty level from this chapter.

For detailed analysis scan QR code

Topic 1 / Alkanes

• Level 1

01 The alkane from below having two secondary hydrogens is **[2025, 22 Jan Shift-II]**

(a) 4-ethyl-3,4-dimethyloctane
(b) 2,2,3,3-tetramethylpentane
(c) 2,2,4,4-tetramethylhexane
(d) 2,2,4,5-tetramethylheptane

02 Which of the following conformations will be the most stable? **[2023, 25 Jan Shift-I]**

(a) 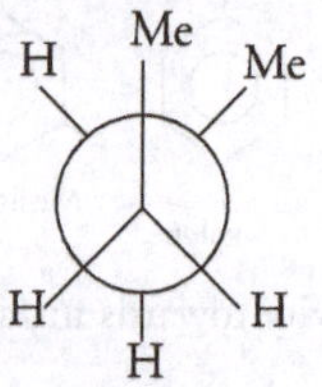
(b) 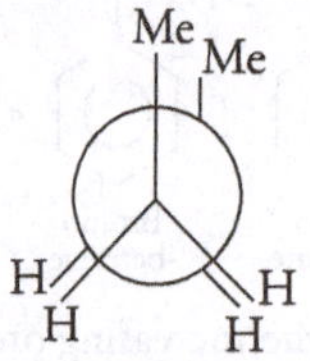
(c) 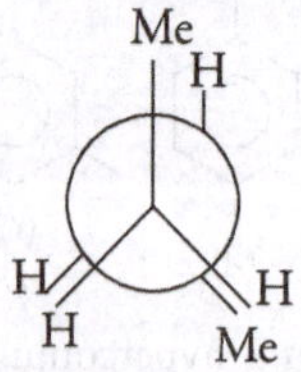
(d) 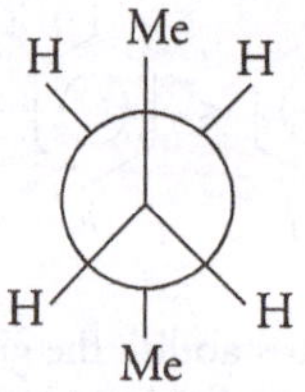

03 In the following structures, which one is having staggered conformation with maximum dihedral angle? **[2022, 25 June Shift-I]**

(a) 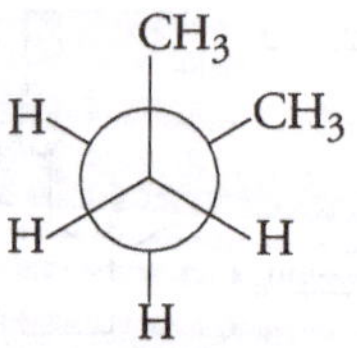
(b) 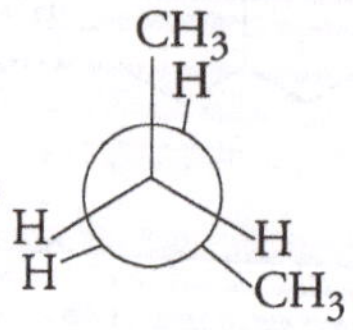
(c) 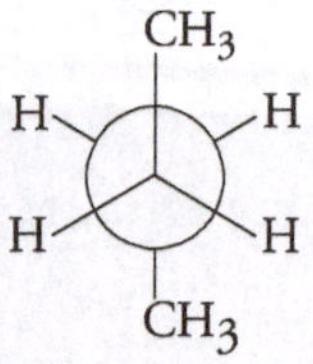
(d) 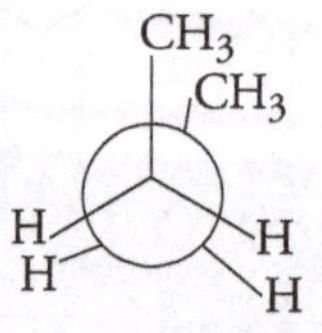

04 Excess of isobutane on reaction with Br_2 in presence of light at 125°C gives which one of the following as the major product? **[2021, 26 Aug Shift-I]**

(a) 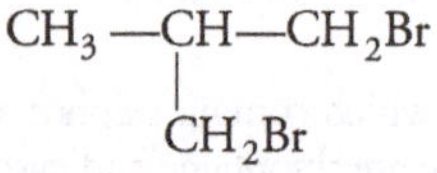
(b) 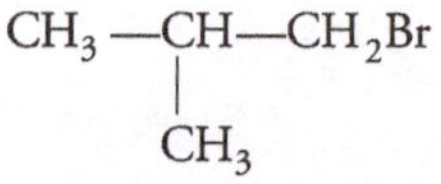
(c) 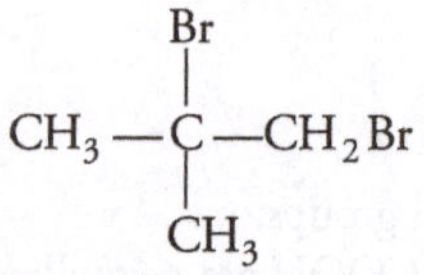
(d) 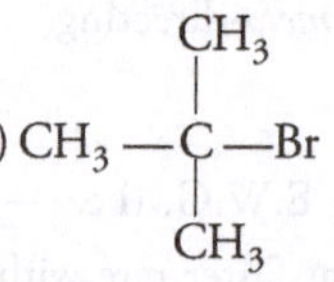

05 Staggered and eclipsed conformers of ethane are **[2021, 27 July Shift-I]**

(a) polymers
(b) mirror images
(c) rotamers
(d) enantiomers

06 For the given reaction

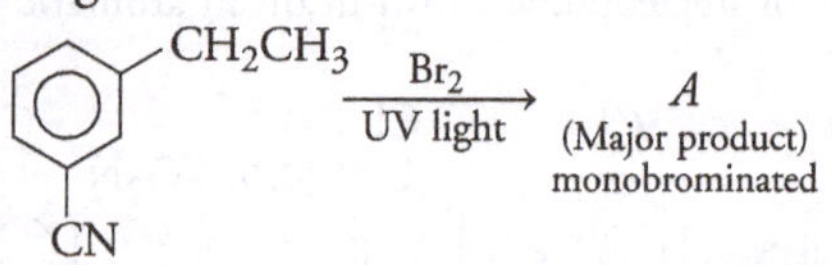

What is '*A*'? **[2021, 26 Feb Shift-I]**

(a) 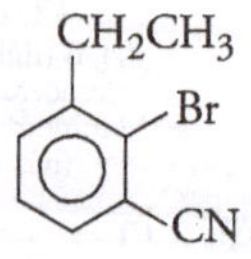
(b) 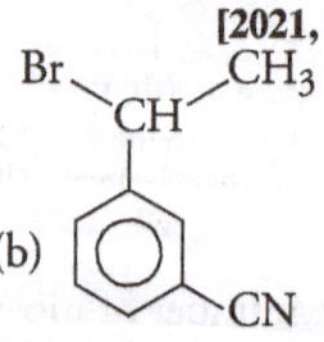
(c) 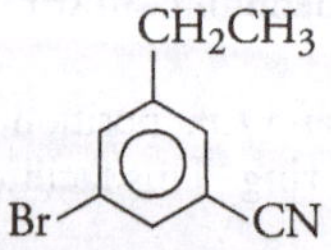
(d) 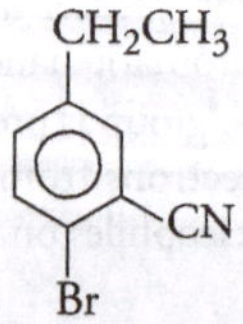

Imp. *Order of Conformational Isomers*

07 Arrange the following conformational isomers of *n*-butane in order of their increasing potential energy **[2021, 31 Aug Shift-II]**

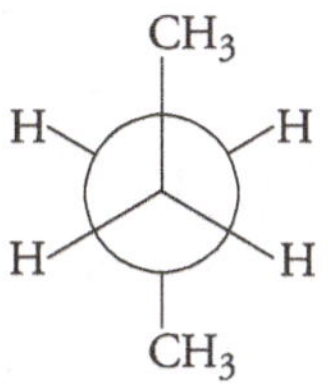

I

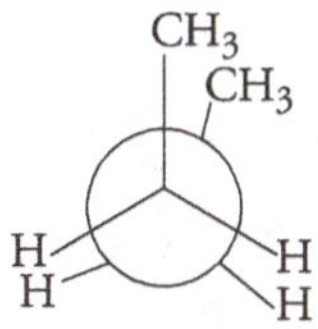

II

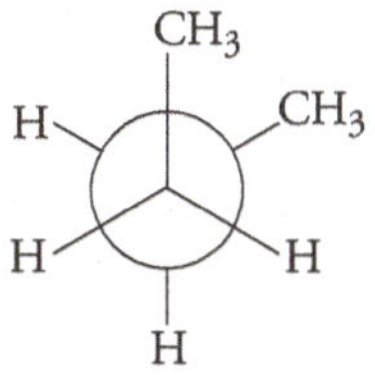

III

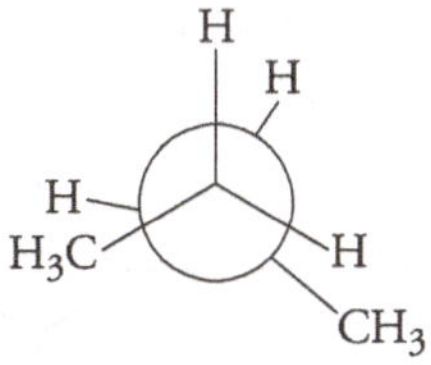

IV

(a) I < III < IV < II
(b) I < IV < III < II
(c) II < IV < III < I
(d) II < III < IV < I

08 Which of the following reagent is used for the following reaction? **[2021, 24 Feb Shift-I]**

$$CH_3CH_2CH_3 \xrightarrow{?} CH_3CH_2CHO$$

(a) Copper at high temperature and pressure
(b) Molybdenum oxide
(c) Potassium permanganate
(d) Manganese acetate

09 In the following skew conformation of ethane, H′—C—C—H″ dihedral angle is

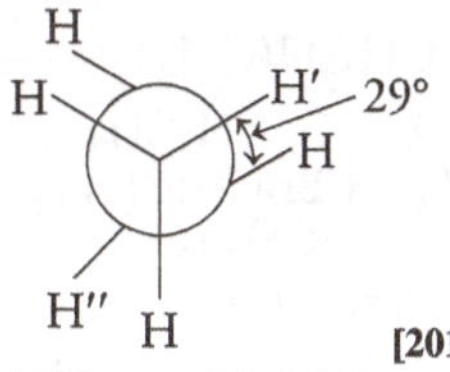

[2019, 12 April Shift-II]

(a) 58° (b) 149° (c) 151° (d) 120°

• Level 2

10 Given below are two statements. **[2025, 2 April Shift-II]**

Statement I Neopentane forms only one monosubstituted derivative.

Statement II Melting point of neopentane is higher than *n*-pentane.

In the light of the above statements, choose the most appropriate answer from the options given below.

(a) Statement I is correct but Statement II is incorrect.
(b) Both Statement I and Statement II are incorrect.
(c) Both Statement I and Statement II are correct.
(d) Statement I is incorrect but Statement II is correct.

11 Given below are two statements.

Statement-I 2-methylbutane on oxidation with $KMnO_4$ gives 2-methylbutan-2-ol.

Statement-II *n*-alkanes can be easily oxidised to corresponding alcohols with $KMnO_4$.

Choose the correct option. **[2021, 17 March Shift-II]**

(a) Both Statements I and II are correct.
(b) Both Statements I and II are incorrect.
(c) Statement I is correct but Statement II is incorrect.
(d) Statement I is incorrect but Statement II is correct.

• Level 3

12 The incorrect statement regarding conformations of ethane is **[2024, 27 Jan Shift-II]**

(a) the conformations of ethane are inter-convertible to one-another.
(b) the dihedral angle in staggered conformation is 60°.
(c) eclipsed conformation is the most stable conformation.
(d) ethane has infinite number of conformations.

Numerical Value Type Questions

13 The number of different chain isomers for C_7H_{16} is .... **[2024, 4 April Shift-I]**

14 Total number of isomeric compounds (including stereoisomers) formed by monochlorination of 2-methylbutane is .... **[2024, 31 Jan, 1st Feb Shift-I]**

15 Number of alkanes obtained on electrolysis of a mixture of CH_3COONa and C_2H_5COONa is .... **[2024, 31 Jan Shift-I]**

16 The total number of monobromo derivatives formed by the alkanes with molecular formula C_5H_{12} is (excluding stereo isomers) **[2022, 25 July Shift-II]**

17 Total number of possible stereoisomers of dimethyl cyclopentane is .... **[2022, 27 june Shift-I]**

18 The dihedral angle in staggered form of Newman projection of 1, 1, 1-trichloro ethane is .... degree. (Round off to the nearest integer) **[2021, 27 July Shift-II]**

Topic 2 / Alkenes

• Level 1

01 The major product of the following reaction is

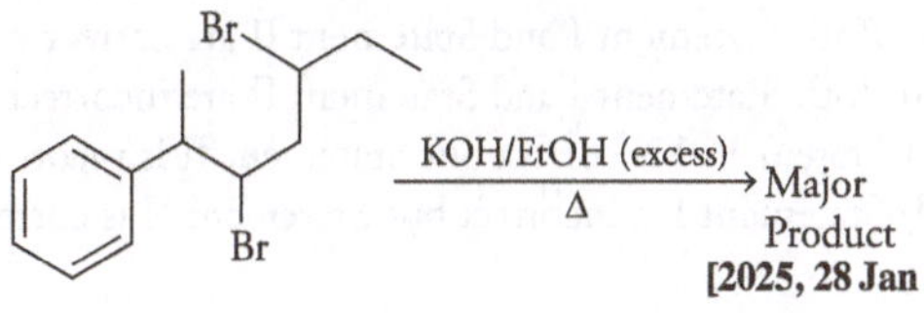

[2025, 28 Jan Shift-II]

(a) 6-phenylhepta -2, 4-diene
(b) 6-phenylhepta -3, 5-diene
(c) 2-phenylhepta -2, 4-diene
(d) 2-phenylhepta -2, 5-diene

02 Number of σ and π-bonds present in ethylene molecule is respectively **[2024, 5 April Shift-I]**
(a) 5 and 1 (b) 4 and 1 (c) 3 and 1 (d) 5 and 2

03 Identify the correct statement for the below given transformation.

$$CH_3CH_2CH_2-\underset{\underset{\overset{+}{N}(CH_3)_3}{|}}{CH}-CH_3 \xrightarrow[C_2H_5OH]{C_2H_5ONa} \underset{(Major)}{A} + \underset{(Minor)}{B}$$

[2022, 28 July Shift-I]

(a) A- $CH_3CH_2CH{=}CH{-}CH_3$, B- $CH_3CH_2CH_2CH{=}CH_2$ Saytzeff products
(b) A- $CH_3CH_2CH{=}CH{-}CH_3$, B- $CH_3CH_2CH_2CH{=}CH_2$ Hofmann products
(c) A- $CH_3CH_2CH_2CH{=}CH_2$, B- $CH_3CH_2CH{=}CH{-}CH_2$ Hofmann poducts
(d) A- $CH_3CH_2CH_2CH{=}CH_2$, B- $CH_3CH_2CH{=}CH{-}CH_3$ Saytzeff products

04 The major product (P) in the reaction

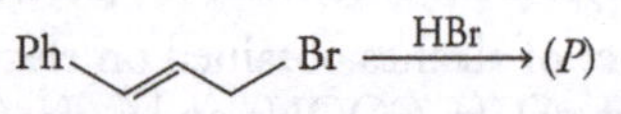

[Ph is $-C_6H_5$] is **[2022, 28 June Shift-I]**

(a) 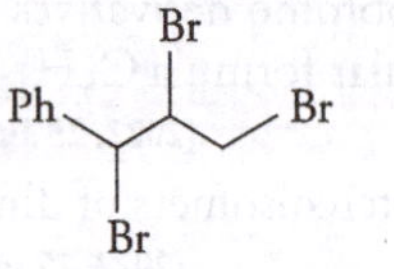 (b) 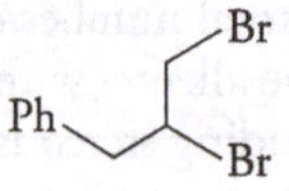

(c) 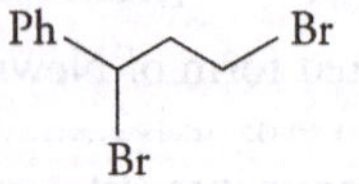 (d) 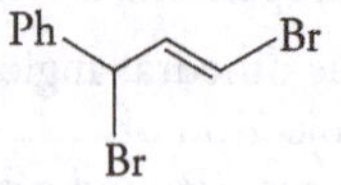

05 Given below are two statements.
Statement I The presence of weaker π-bonds make alkenes less stable than alkanes.
Statement II The strength of the double bond is greater than that of carbon-carbon single bond.
In the light of the above statements, choose the correct answer from the options given below. **[2022, 24 June Shift-II]**
(a) Both Statement I and Statement II are correct.
(b) Both Statement I and Statement II are incorrect.
(c) Statement I is correct but Statement II is incorrect.
(d) Statement I is incorrect but Statement II is correct.

06 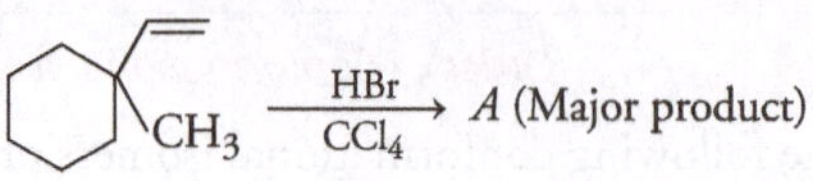

Product "A" in the above chemical reaction is **[2021, 17 March Shift-I]**

(a) 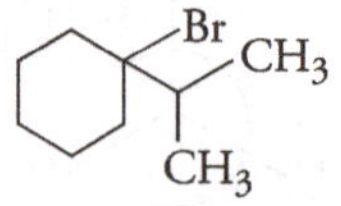 (b) 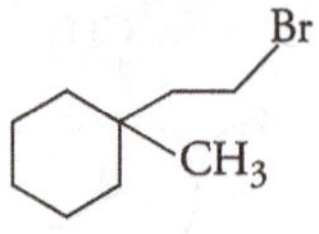

(c) 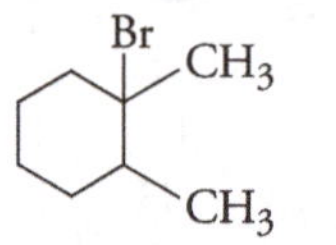 (b) 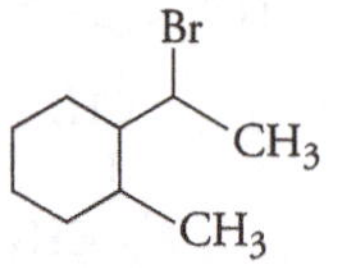

07 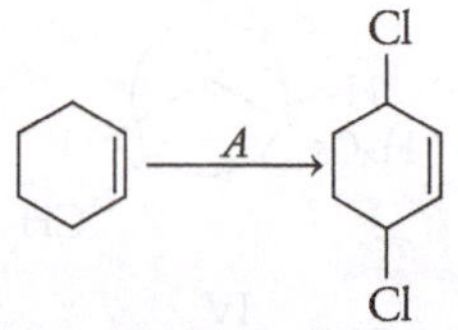

Identify the reagent(s) 'A' and condition(s) for the reaction. **[2021, 16 March Shift-II]**
(a) $A = HCl, ZnCl_2$
(b) $A = HCl$; anhydrous $AlCl_3$
(c) $A = Cl_2$; dark, anhydrous $AlCl_3$
(d) $A = Cl_2$; UV light

08 Which of the following is Lindlar catalyst?
(a) Cold dilute solution of $KMnO_4$ **[2021, 16 March Shift-I]**
(b) Partially deactivated palladised charcoal
(c) Zinc chloride and HCl
(d) Sodium and liquid NH_3

09 The major product formed in the following reaction is **[2020, 05 Sept Shift-II]**

$$CH_3CH{=}CHCH(CH_3)_2 \xrightarrow{HBr}$$

(a) $CH_3CH(Br)CH_2CH(CH_3)_2$
(b) $CH_3CH_2CH(Br)CH(CH_3)_2$
(c) $Br(CH_2)_3CH(CH_3)_2$
(d) $CH_3CH_2CH_2C(Br)(CH_3)_2$

10 The major product in the following reaction is

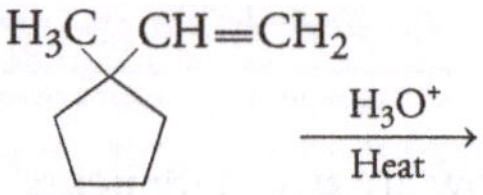

[2020, 2 Sep Shift-I]

(a) 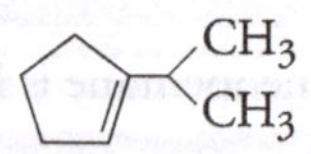 (b) 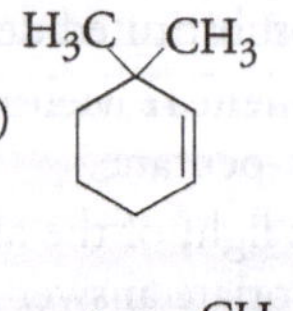

(c) 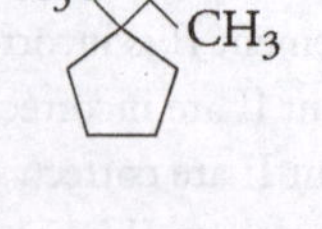 (d) 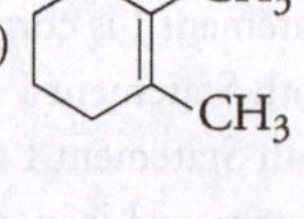

11 The correct order of heat of combustion for following alkadienes is **[2020, 9 Jan Shift-I]**

(*A*) 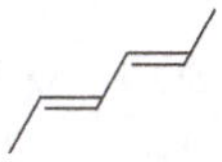

(*B*) 

(*C*) 

(a) (*B*) < (*C*) < (*A*)
(b) (*A*) < (*B*) < (*C*)
(c) (*A*) < (*C*) < (*B*)
(d) (*C*) < (*B*) < (*A*)

12 But-2-ene on reaction with alkaline $KMnO_4$ at elevated temperature followed by acidification will give **[2019, 12 April Shift-I]**

(a) $CH_3-CH(OH)-CH(OH)-CH_3$
(b) one molecule of CH_3CHO and one molecule of CH_3COOH
(c) 2 molecules of CH_3COOH
(d) 2 molecules of CH_3CHO

Level 2

13 Which compound would give 3-methyl-6-oxoheptanal upon ozonolysis? **[2025, 3 April Shift-I]**

(a) 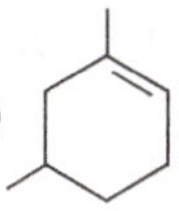

(b) 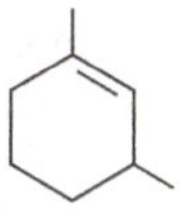

(c) 

(d) 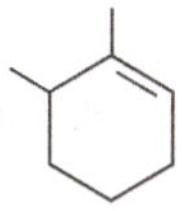

14 The number of optically active products obtained from the complete ozonolysis of the given compound is:

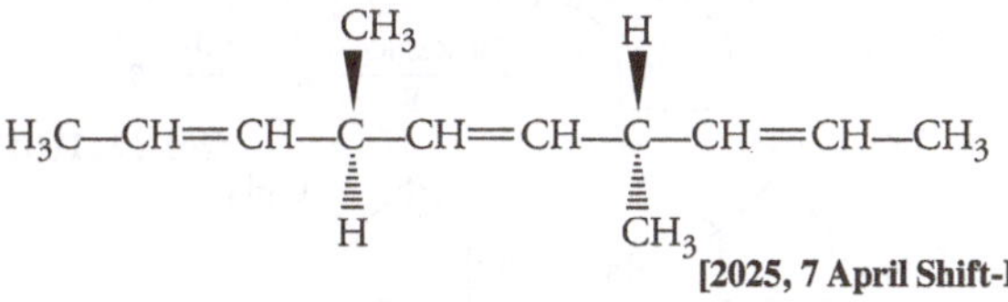

[2025, 7 April Shift-II]

(a) 2 (b) 4 (c) 0 (d) 1

15 Given below are two statements: **[2025, 7 April Shift-I]**

Statement I Ozonolysis followed by treatment with Zn, H_2O of *cis*-2-butene gives ethanal.

Statement II The product obtained by ozonolysis followed by treatment with Zn, H_2O of 3, 6-dimethyloct-4-ene has no chiral carbon atom.

In the light of the above statements, choose the correct answer from the options given below.

(a) Statement I is false but Statement II is true.
(b) Statement I is true but Statement II is false.
(c) Both Statement I and Statement II are true.
(d) Both Statement I and Statement II are False.

Imp. *Ozonolysis & Aldol Condensation*

16 Aman has been asked to synthesise the molecule

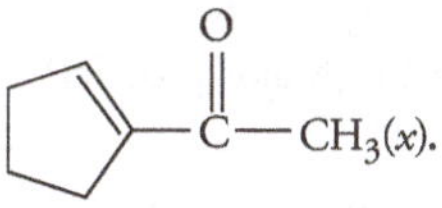

(*x*). He thought of preparing the molecule using an aldol condensation reaction. He found a few cyclic alkenes in his laboratory. He thought of performing ozonolysis reaction on alkene to produce a dicarbonyl compound followed by aldol reaction to prepare "*x*". Predict the suitable alkene that can lead to the formation of "*x*". **[2025, 24 Jan Shift-I]**

(a) 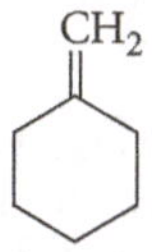

(b) 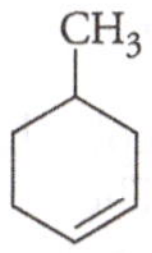

(c) 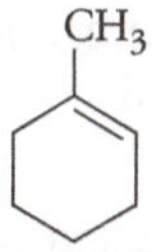

(d) 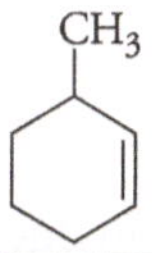

17 Following are the four molecules "*P*", "*Q*", "*R*" and "*S*". **[2025, 24 Jan shift-I]**

Which one among the four molecules will react with H – Br(*aq*) at the fastest rate?

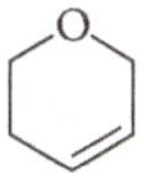
P

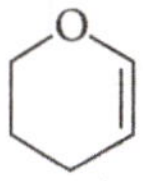
Q

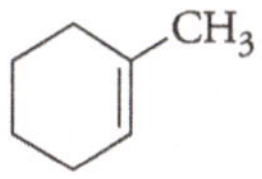
R

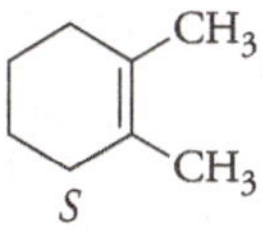
S

(a) *R* (b) *S* (c) *P* (d) *Q*

18 The incorrect statement regarding ethyne is **[2024, 9 April Shift-II]**

(a) ethyne is linear
(b) both carbons are *sp*-hybridised
(c) the C—C bonds in ethyne is shorter than that in ethene
(d) the carbon-carbon bonds in ethyne is weaker than that in ethene

19 Consider the given chemical reaction :

 $\xrightarrow[\text{Heat}]{KMnO_4 - H_2SO_4}$ Product *A* **[2024, 5 April Shift-II]**

Product *A* is

(a) adipic acid
(b) oxalic acid
(c) acetic acid
(d) picric acid

20 A hydrocarbon '*X*' with formula C_6H_8 uses two moles of H_2 on catalytic hydrogenation of its one mole. On ozonolysis, '*X*' yields two moles of methane dicarbaldehyde. The hydrocarbon '*X*' is **[2023, 31 Jan Shift-II]**

(a) 1-methylcyclopenta-l, 4-diene
(b) cyclohexa - 1, 3 - diene
(c) hexa-1, 3, 5-triene
(d) cyclohexa-1, 4-diene

21 In the following given reaction, '*A*' is **[2023, 24 Jan Shift-I]**

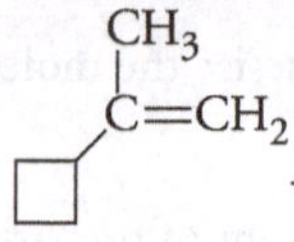

+ HBr ⟶ '*A*' (Major product)

(a) 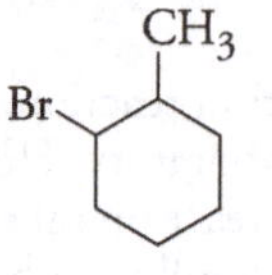
(b) 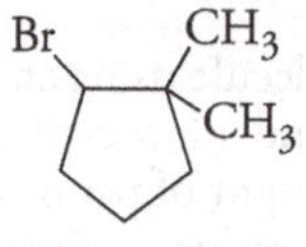
(c) 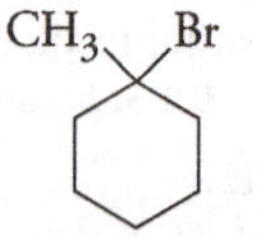
(d) 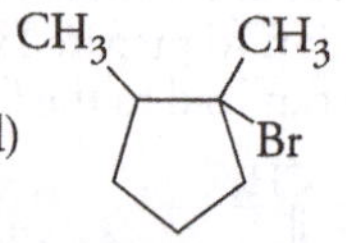

22 When ethanol is heated with conc. H_2SO_4, a gas is produced. The compound formed, when this gas is treated with cold dilute aqueous solution of Baeyer's reagent, is **[2022, 29 July Shift-II]**

(a) formaldehyde (b) formic acid
(c) glycol (d) ethanoic acid

23 The major product (*P*) of the given reaction is (where, Me is —CH_3) **[2022, 28 June Shift-II]**

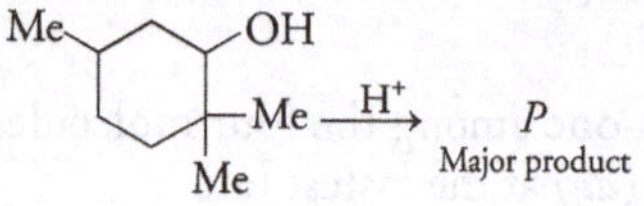

(a) 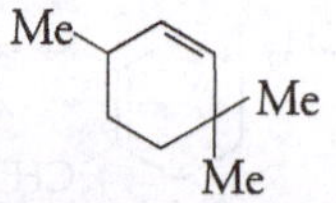
(b) 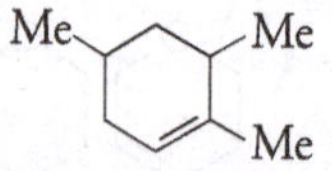
(c) 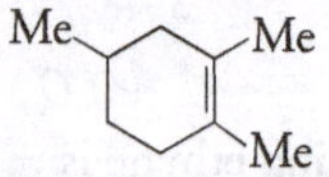
(d) 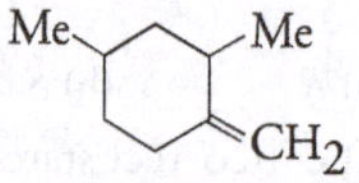

24 Amongst the following, the major product of the given chemical reaction is **[2022, 25 June Shift-II]**

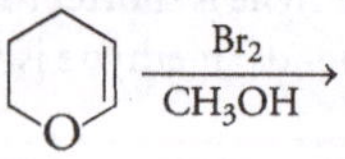

Major product

(a) 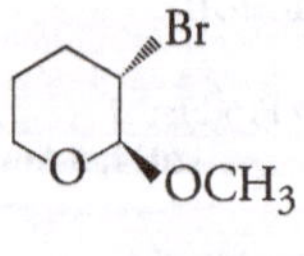
(b) 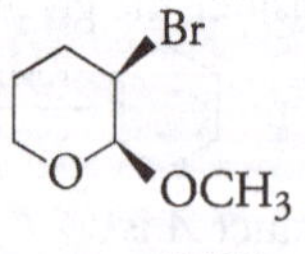
(c) 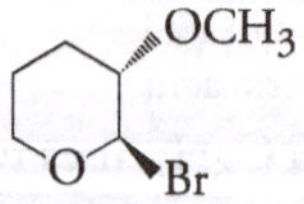
(d) 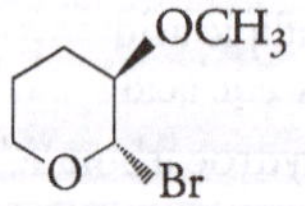

Imp. *Electrophilic Addition of Bromine*

25 The stereoisomers that are formed by electrophilic addition of bromine to *trans*-but-2-ene is/are **[2021, 01 Sept Shift-II]**

(a) 2 identical mesomers
(b) 2 enantiomers
(c) 2 enantiomers and 2 mesomers
(d) 1 racemic and 2 enantiomers

26 The major product formed in the following reaction is **[2021, 31 Aug Shift-I]**

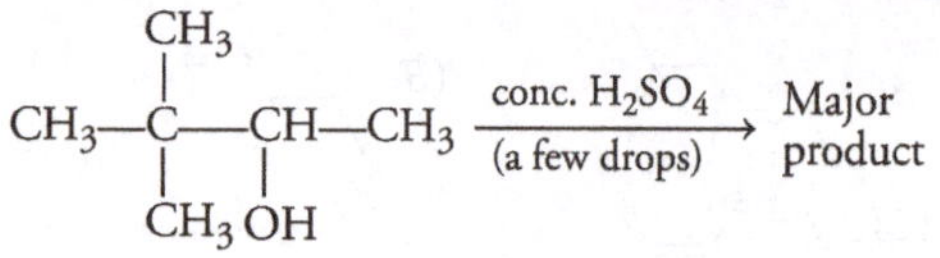

(b) 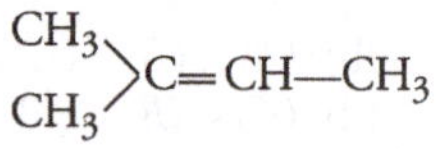
(b) 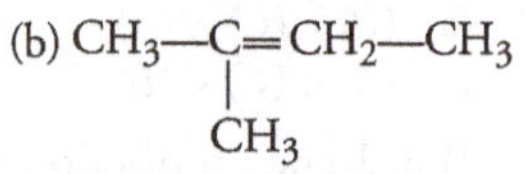
(c) 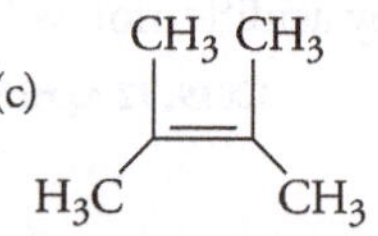
(d) 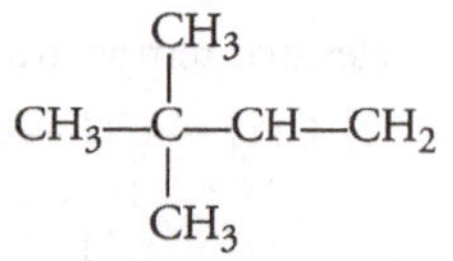

27 Which of the following molecules does not show stereo isomerism? **[2021, 22 July Shift-II]**

(a) 3,4-dimethylhex-3-ene (b) 3-methylhex-1-ene
(c) 3-ethylhex-3-ene (d) 4-methylhex-1-ene

28 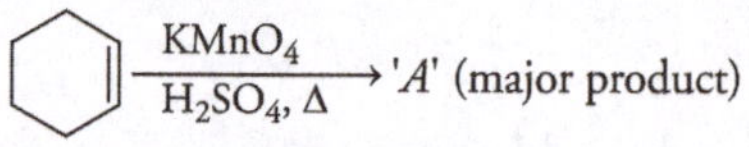

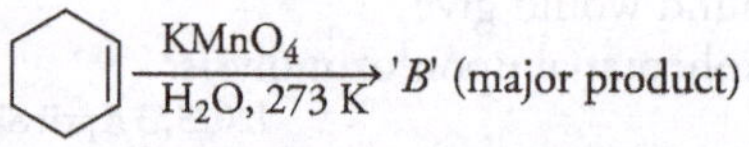

For above chemical reactions, identify the correct statement from the following. **[2021, 20 July Shift-I]**

(a) Compound '*A*' is diol and compound '*B*' is dicarboxylic acid.
(b) Both compound '*A*' and compound '*B*' are dicarboxylic acids.
(c) Compound '*A*' is dicarboxylic acid and compound '*B*' is diol.
(d) Both compound '*A*' and compound '*B*' are diols.

29 Identify products *A* and *B*.

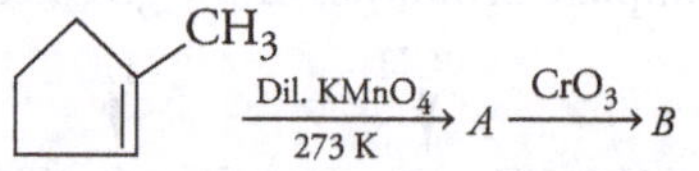

[2021, 24 Feb Shift-I]

(a) *A* : 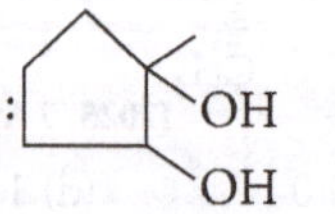 *B* : 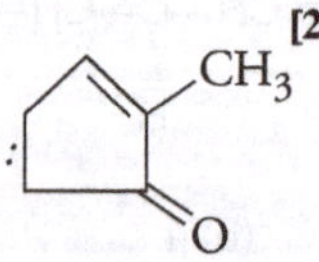
(b) *A* : 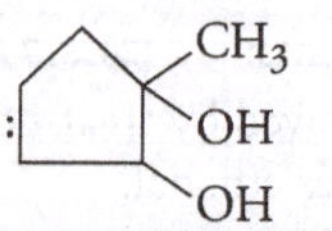 *B* : 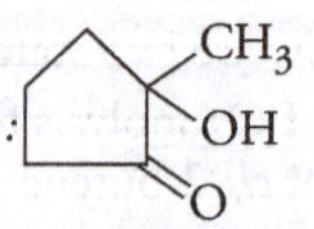
(c) *A* : 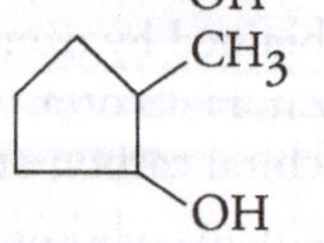 *B* : 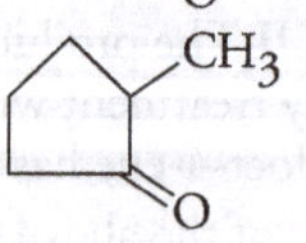
(d) *A* : $OHC—CH_2CH_2CH_2—\overset{O}{\overset{\|}{C}}—CH_3$
B : $HOOC—CH_2CH_2CH_2—\overset{O}{\overset{\|}{C}}—CH_3$

30 Which of the following compounds shows geometrical isomerism? **[2020, 06 Sept Shift-I]**
(a) 2-methylpent-2-ene (b) 4-methylpent-2-ene
(c) 4-methylpent-1-ene (d) 2-methylpent-1-ene

31 When *neo*-pentyl alcohol is heated with an acid, it slowly converted into an 85 : 15 mixture of alkenes *A* and *B*, respectively. What are these alkenes? **[2020, 4 Sep Shift-I]**

(a) 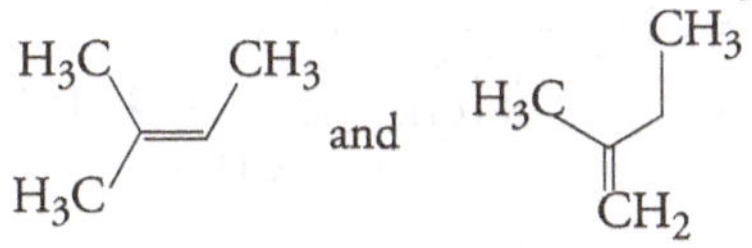

(b) 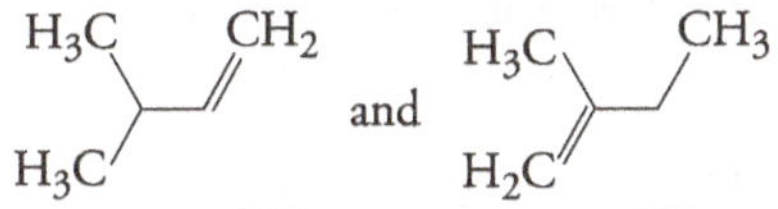

(c) 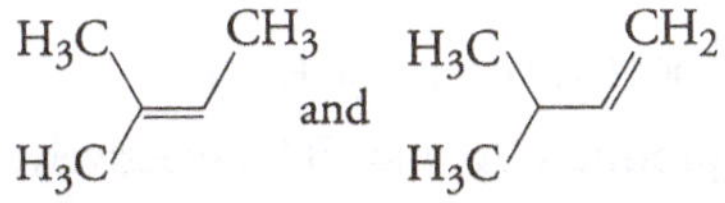

(d) 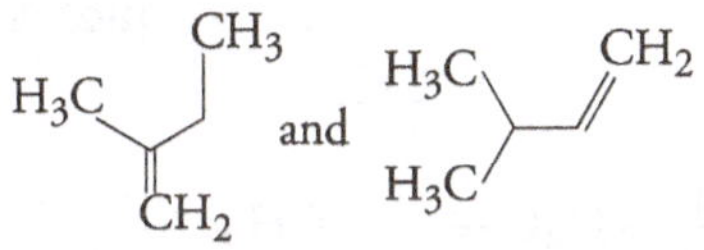

32 Which of the following compounds produces an optically inactive compound on hydrogenation? **[2020, 03 Sept Shift-I]**

(a) 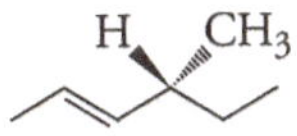 (b) 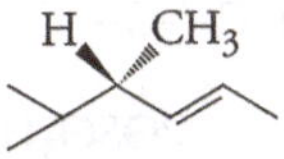

(c) 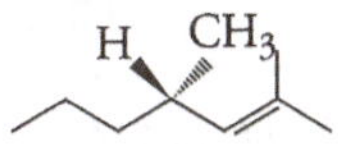 (d) 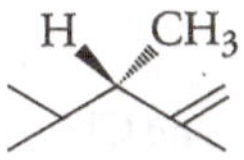

33 An unsaturated hydrocabon *X* absorbs two hydrogen molecules on catalytic hydrogenation and also gives following reaction

$$X \xrightarrow[Zn/H_2O]{O_3} A \xrightarrow{[Ag(NH_3)_2]^+} B\text{(3-oxo-hexanedicarboxylic acid)}$$

X will be **[2020, 08 Jan Shift-II]**

(a) 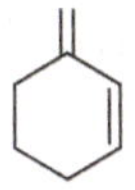 (b) 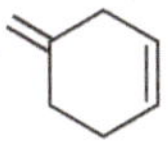

(c) 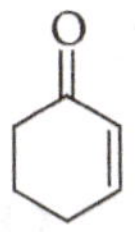 (d) 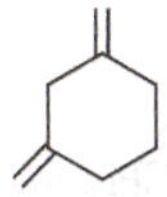

34 The major product of the following addition reaction is

$$H_3C{-}CH{=}CH_2 \xrightarrow{Cl_2/H_2O}$$ **[2019, 12 April Shift-I]**

(a) $CH_3{-}\underset{\large Cl}{\underset{|}{C}H}{-}\underset{\large OH}{\underset{|}{C}H_2}$ (b) $CH_3{-}\underset{\large OH}{\underset{|}{C}H}{-}\underset{\large Cl}{\underset{|}{C}H_2}$

(c) 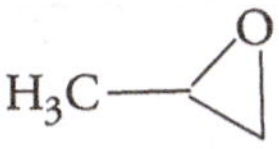 (d) 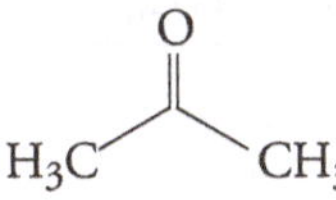

• Level 3

35 Consider the following molecule (*X*).
The structure of *X* is **[2025, 4 April Shift-II]**

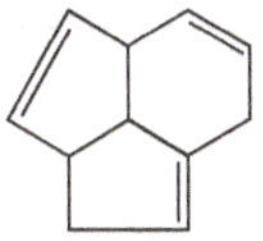

(*X*)

The major product formed when the given molecule (*X*) is treated with HBr (1 eq) is

(a) 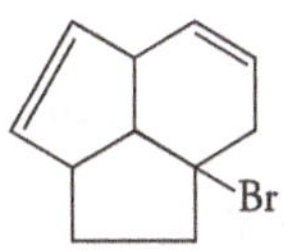 (b) 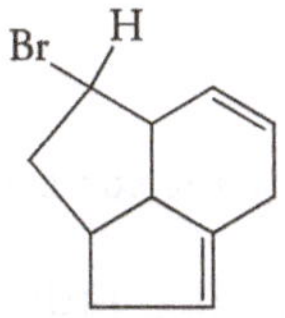

(c) 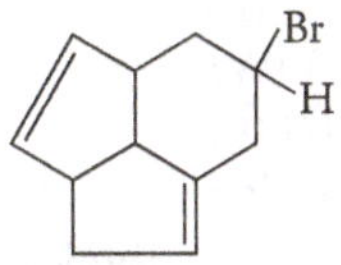 (d) 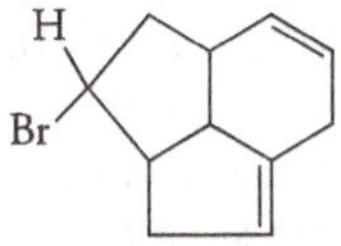

36 The reactions which cannot be applied to prepare an alkene by elimination, are **[2025, 7 April Shift-I]**

(A) 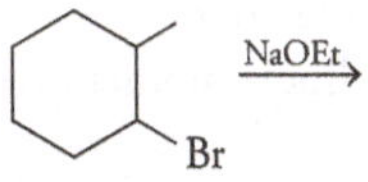

(B) $CH_3{-}CH_2{-}\underset{\large Br}{\underset{|}{C}H}{-}CH_3 \xrightarrow{KOH(aq)}$

(C) $CH_3{-}\underset{\large CH_3}{\overset{\large CH_3}{\underset{|}{\overset{|}{C}}}}{-}Br \xrightarrow{NaOMe}$

(D) 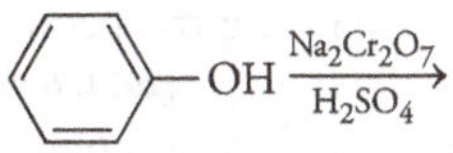

(E) $CH_3{-}\underset{\large CH_3}{\overset{\large CH_3}{\underset{|}{\overset{|}{C}}}}{-}OH \xrightarrow[573\,K]{Cu}$

Choose the correct answer from the options given below.
(a) A, C and D only (b) B and D only
(c) B, C and D only (d) B and E only

37 Match List-I with List-II. **[2025, 23 Jan Shift-II]**

List-I (Isomers of $C_{10}H_{14}$)		List-II (Ozonolysis product)
(a) 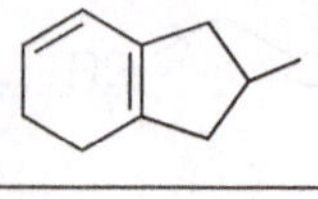	I.	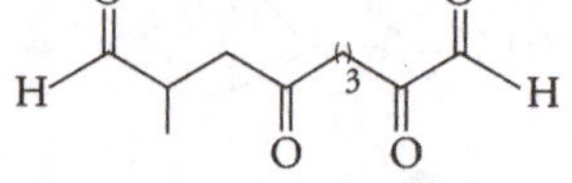
(b) 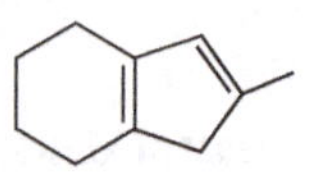	II.	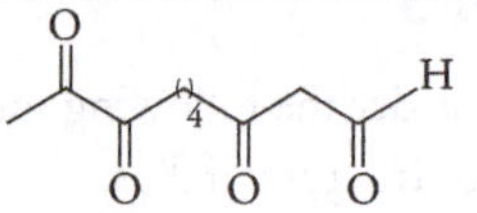
(c) 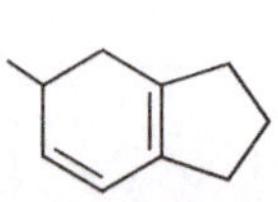	III.	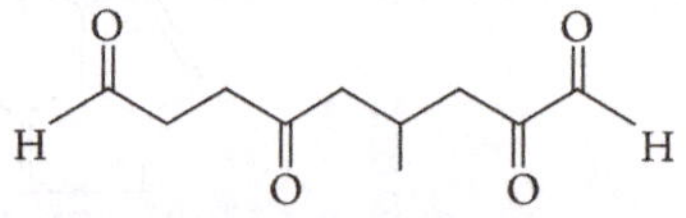
(d) 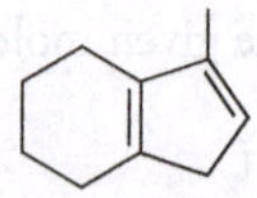	IV.	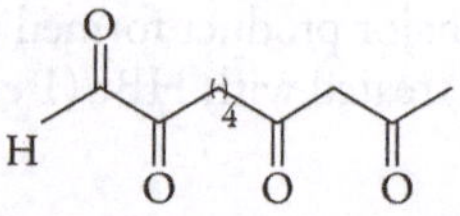

Choose the **correct** answer from the options given below

(a) (A)-(II), (B)-(III), (C)-(I), (D)-(IV)
(b) (A)-(I), (B)-(IV), (C)-(III), (D)-(II)
(c) (A)-(III), (B)-(II), (C)-(I), (D)-(IV)
(d) (A)-(III), (B)-(IV), (C)-(I), (D)-(II)

38 The incorrect statements regarding geometrical isomerism are **[2025, 22 Jan Shift-I]**

(A) Propene shows geometrical isomerism.
(B) *trans* isomer has identical atoms/groups on the opposite sides of the double bond.
(C) *cis*-but-2-ene has higher dipole moment than *trans*-but-2-ene.
(D) 2-methylbut-2-ene shows two geometrical isomers.
(E) *trans*-isomer has lower melting point than *cis* isomer.

Choose the correct answer from the options given below.

(a) (A), (D) and (E) Only (b) (B) and (C) Only
(c) (A) and (E) Only (d) (C), (D) and (E) Only

39 The incorrect statement regarding the geometrical isomers of 2-butene is **[2024, 6 April Shift-II]**

(a) *trans*-2-butene is more stable than *cis*-2-butene.
(b) *cis*-2-butene and *trans*-2 butene are not interconvertible at room temperature.
(c) *cis*-2-butene has less dipole moment than *trans*-2-butene.
(d) *cis*-2-butene and *trans*-2-butene are stereoisomers.

40 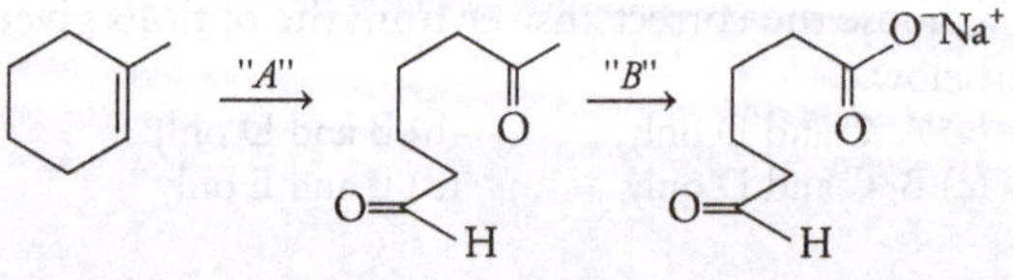

In the above chemical reaction sequence "*A*" and "*B*" respectively are **[2024, 4 April Shift-II]**

(a) H_2O, H^+ and $KMnO_4$
(b) O_3, Zn/H_2O and $KMnO_4$
(c) O_3, Zn/H_2O and $NaOH(alc)/I_2$
(d) H_2O, H^+ and $NaOH(alc)/I_2$

41 The product formed in the following multi-step reaction is

$$CH_3CH{=}CH_2 \xrightarrow[\text{(iii) PCC, (iv) } CH_3MgBr]{\text{(i) } B_2H_6, \text{ (ii) } H_2O_2, NaOH}$$

[2023, 15 April Shift-I]

(a) $CH_3CH_2CH_2CH_2OH$ (b) $CH_3-C(OH)(CH_3)-CH_3$
(c) $CH_3CH_2COCH_3$ (d) $CH_3CH_2CH(OH)CH_3$

42 The major products '*A*' and '*B*' respectively, are **[2023, 30 Jan Shift-I]**

$$\text{'}A\text{'} \xleftarrow[H_2SO_4]{\text{Cold}} H_3C-C(CH_3){=}CH_2 \xrightarrow[80°C]{H_2SO_4} \text{'}B\text{'}$$

(a) $H_3C-C(CH_3)(OSO_3H)-CH_3$ and $CH_3-CH(CH_3)-CH_2CH_2-HC(CH_3)-CH_3$
(b) $CH_3-C(CH_3){=}CH-C(CH_3)_2-CH_3$ and $CH_3-C(CH_3)(OSO_3H)-CH_3$
(c) $CH_3-CH(CH_3)-CH_2CH_2-CH(CH_3)-CH_3$ and $H_3C-C(CH_3)(OSO_3H)-CH_3$
(d) $H_3C-C(CH_3)(OSO_3H)-CH_3$ and $CH_3-C(CH_3){=}CH-C(CH_3)_2-CH_3$

43 The one giving maximum number of isomeric alkenes on dehydrohalogenation reaction is (excluding rearrangement) **[2023, 29 Jan Shift-II]**

(a) l-bromo-2-methylbutane
(b) 2-bromopentane
(c) 2-bromo-3 ,3-dimethylpentane
(d) 2-bromopropane

44 A compound *A* on reaction with *X* and *Y* produces the same major product but different by-product *a* and *b*. Oxidation of *a*, gives a substance produced by ants,

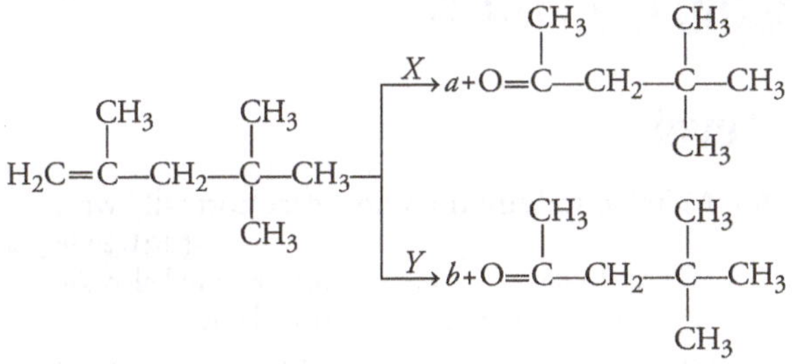

X and *Y* respectively are **[2022, 25 July Shift-I]**

(a) $KMnO_4/H^+$ and dil. $KMnO_4$, 273 K
(b) $KMnO_4$ (dilute), 273 K and $KMnO_4/H^+$
(c) $KMnO_4/H^+$ and O_3, H_2O/Zn
(d) O_3, H_2O/Zn and $KMnO_4/H^+$

45 Two isomers '*A*' and '*B*' with molecular formula C_4H_8 give different products on oxidation with $KMnO_4$ in acidic medium. Isomer '*A*' on reaction with $KMnO_4/H^+$ results in effervescence of a gas and gives ketone. The compound '*A*' is **[2022, 29 June Shift-I]**

(a) but-1-ene (b) *cis*-but-2-ene
(c) *trans*-but-2-ene (d) 2-methylpropene

46 *A* and *B* respectively are

$$A \xrightarrow[(2)\ Zn\text{-}H_2O]{(1)\ O_3} \text{ethane-1,2-dicarbaldehyde} + \text{glyoxal/oxaldehyde}$$

$$B \xrightarrow[(2)\ Zn\text{-}H_2O]{(1)\ O_3} \text{5-oxohexanal}$$

[2022, 27 June Shift-I]

(a) 1-methylcyclohex-1,3-diene and cyclopentene
(b) cyclohex-1,3-diene and cyclopentene
(c) 1-methylcyclohex-1,4-diene and 1-methylcyclopent-1-ene
(d) cyclohex-1,3-diene and 1-methylcyclopent-1-ene

47 The major product in the following reaction **[2022, 26 June Shift-II]**

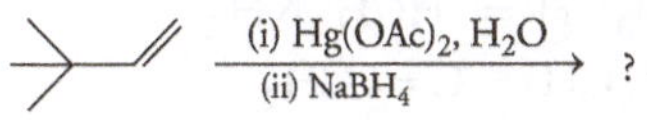

(a) 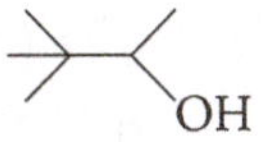 (b) 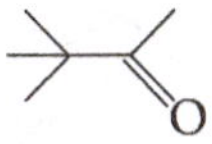

(c) 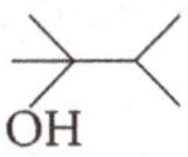 (d) 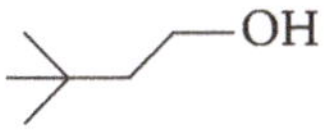

48

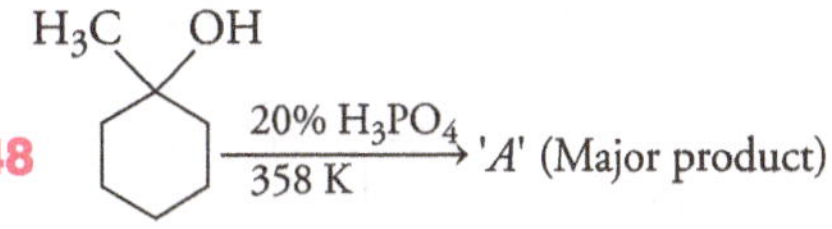

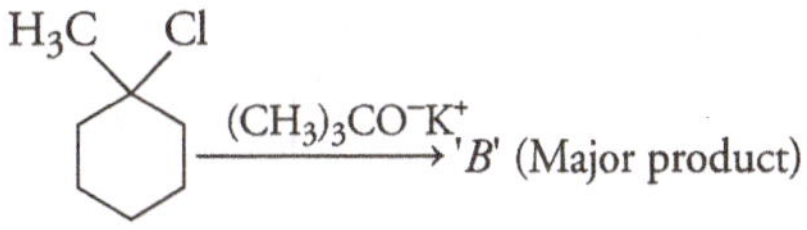

The products "*A*" and "*B*" formed in above reactions are **[2021, 16 March Shift-I]**

(a) 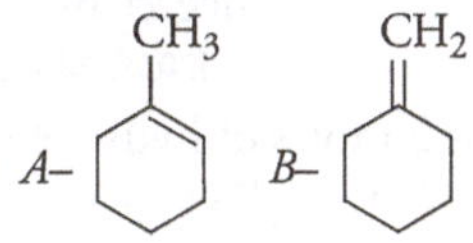 (b) 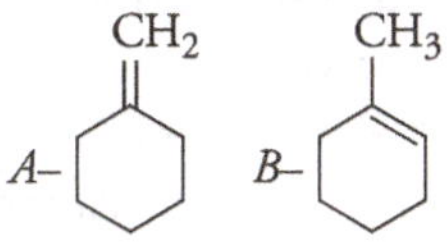

(c) 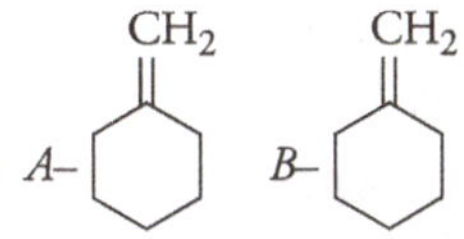 (d) 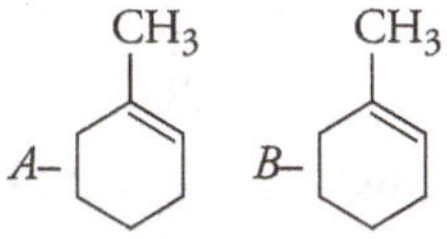

49 The major product [*B*] in the following sequence of reactions is

$$CH_3-C(CH(CH_3)_2)=CH-CH_2CH_3 \xrightarrow[(ii)\ H_2O_2,\ \overset{\ominus}{O}H]{(i)\ B_2H_6} [A] \xrightarrow[\Delta]{\text{Dil. } H_2SO_4} [B]$$

[2020, 8 Jan Shift-II]

(a) $CH_3-C(CH(CH_3)_2)-CH=CH-CH_3$
(b) $CH_3-C(CH(CH_3)_2)=CH-CH_2CH_3$
(c) $CH_3-C(=C(CH_3)_2)-CH_2CH_2CH_3$
(d) $CH_2=C(CH(CH_3)_2)-CH_2CH_2CH_3$

50 The major products *A* and *B* in the following reaction are **[2020, 08 Jan Shift-I]**

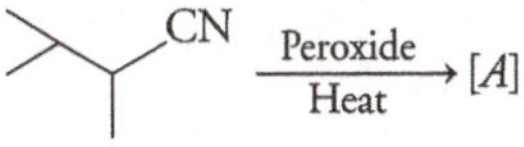

[A] +  B

(a) A = 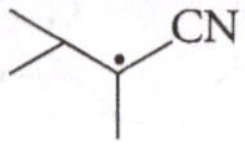 and B = 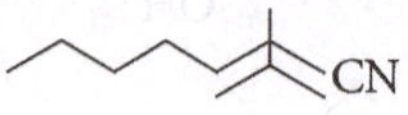

(b) A = 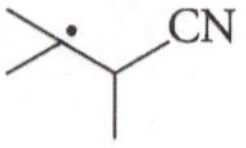 and B = 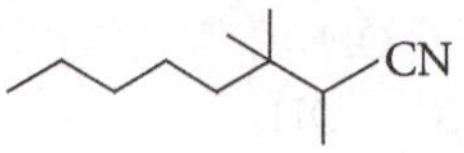

(c) A = 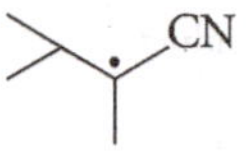 and B = 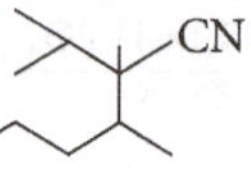

(d) A = 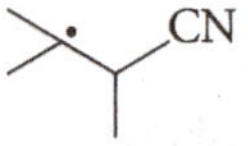 and B = 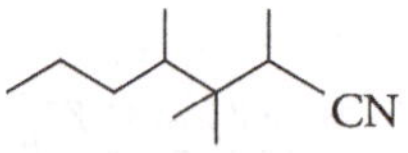

Numerical Value Type Questions

51 The hydrocarbon (X) with molar mass 80 g mol^{-1} and 90% carbon has .... degree of unsaturation. **[2025, 24 Jan Shift-II]**

52 A compound 'X' absorbs 2 moles of hydrogen and 'X' upon oxidation with $KMnO_4|H^+$ gives

$CH_3-\overset{\overset{\large O}{\|}}{C}-CH_3$, $CH_3-\overset{\overset{\large O}{\|}}{C}-OH$

and $CH_3-\overset{\overset{\large O}{\|}}{C}-CH_2CH_2-\overset{\overset{\large O}{\|}}{C}-OH$

The total number of σ-bonds present in the compound 'X' is .... **[2025, 23 Jan Shift-II]**

53 3-methylhex-2-ene on reaction with HBr in presence of peroxide forms and addition product (A). The number of possible stereoisomers for 'A' is .... **[2024, 27 Jan Shift-I]**

54 A hydrocarbon 'X' is found to have molar mass of 80. A 10.0 mg of compound 'X' on hydrogenation consumed 8.40 mL of H_2 gas (measured at STP). Ozonolysis compound 'X' yields only formaldehyde and dialdehyde. The total number of fragments/molecules produced from the ozonolysis of compound 'X' is .... **[2022, 30 June Shift-I]**

55 The major product of the following reaction contains .... bromine atom(s).

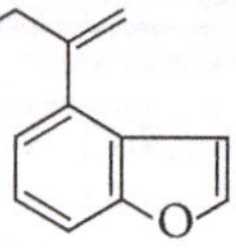 $\xrightarrow[h\nu]{Br_2}$ Major product

[2022, 28 june Shift-II]

56 The major product 'A' of the following given reaction has .... sp^2 hybridised carbon atoms.

2,7-dimethyl-2,6-octadiene $\xrightarrow{H^+}$ A (Major product)

[2022, 24 June Shift-I]

57 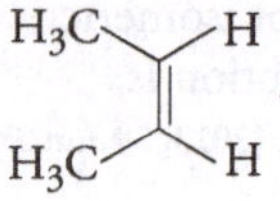 + Br_2 $\xrightarrow{CCl_4}$ Product (P)

Consider the above chemical reaction. The total number of stereoisomers possible for product P is ...... **[2021, 25 July Shift-II]**

58 The number of acyclic structural isomers (including, geometrical isomers) for pentene are .... **[2021, 22 July Shift-II]**

Topic 3 / Alkynes

• Level 1

01 Metallic sodium does not react normally with **[2021, 20 July Shift-II]**

(a) but-2-yne
(b) *tert*-butyl alcohol
(c) gaseous ammonia
(d) ethyne

02 The major product obtained from the following reaction is **[2020, 6 Sep Shift-I]**

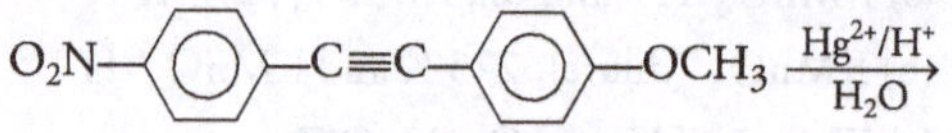

(a) 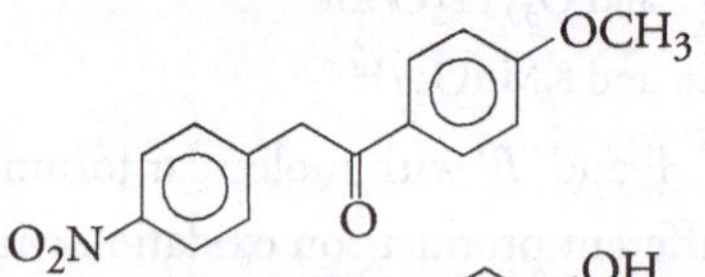

(b) 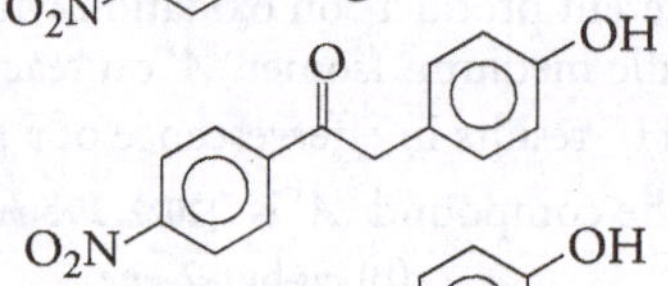

(c) 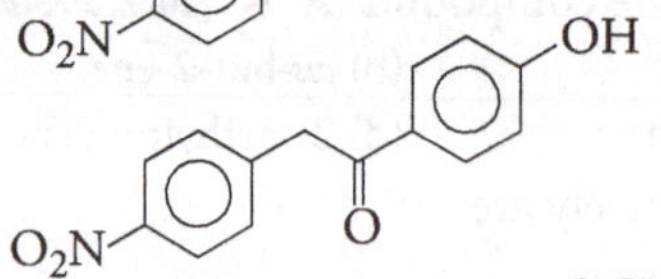

(d) 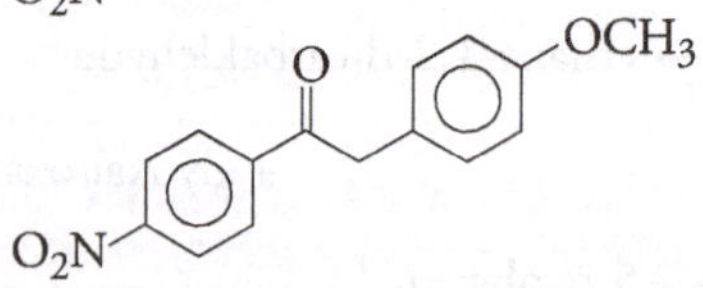

03 The major product of the following reaction is

$CH_3CH_2\underset{\underset{\large Br}{|}}{C}H-\underset{\underset{\large Br}{|}}{C}H_2 \xrightarrow[\text{(ii) } NaNH_2 \text{ in liq. } NH_3]{\text{(i) alc. KOH}}$ **[2019, 12 Jan Shift-II]**

(a) $CH_3CH_2\underset{\underset{\large NH_2}{|}}{C}H-\underset{\underset{\large NH_2}{|}}{C}H_2$

(b) $CH_3CH{=}CHCH_2NH_2$

(c) $CH_3CH{=}C{=}CH_2$

(d) $CH_3CH_2C\equiv CH$

• Level 2

04 Choose the correct set of reagents for the following conversion. **[2023, 31 Jan Shift-I]**

trans (Ph—CH= CH—CH$_3$) ⟶ *cis* (Ph—CH= CH—CH$_3$)

(a) Br_2, aq. KOH, $NaNH_2$, H_2 Lindlar catalyst
(b) Br_2, alc. KOH, $NaNH_2$, H_2 Lindlar catalyst
(c) Br_2, aq. KOH, $NaNH_2$, Na(liq. NH_3)
(d) Br_2, alc. KOH, $NaNH_2$, Na(liq. NH_3)

05 What will be the major product of following sequence of reactions? **[2022, 27 June Shift-II]**

$$n\text{-Bu}—\equiv \xrightarrow[\text{(ii) Lindlar catalysis, } H_2]{\text{(i) } n\text{-BuLi, } n\text{-}C_5H_{11}Cl}$$

(a) 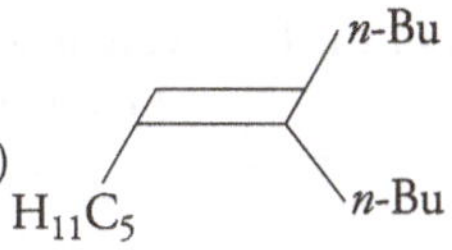

(b) 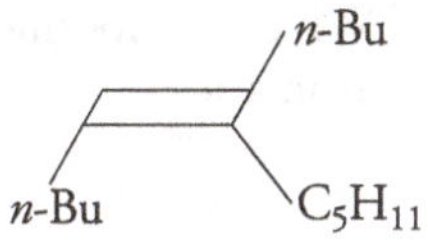

(c) 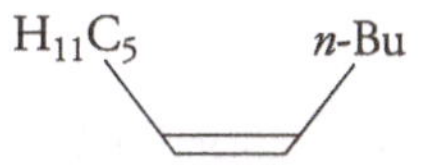

(d) 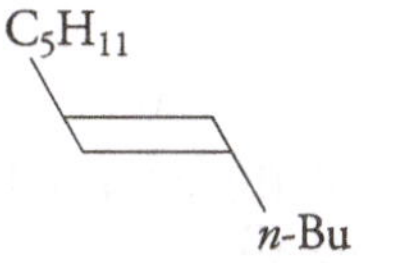

06 An unsaturated hydrocarbon *X* on ozonolysis gives *A*. Compound *A* when warmed with ammoniacal silver nitrate forms a bright silver mirror along the sides of the test tube. The unsaturated hydrocarbon *X* is **[2021, 16 March Shift-II]**

(a) HC≡C—CH$_2$—CH$_3$
(b) CH$_3$—C(CH$_3$)=◁ (cyclopropylidene)
(c) CH$_3$—C≡C—CH$_3$
(d) CH$_3$—C(CH$_3$)=C(CH$_3$)—CH$_3$

07 The major product (*Y*) in the following reactions is **[2020, 09 Jan Shift-I]**

$$CH_3—CH(CH_3)—C\equiv CH \xrightarrow[H_2O]{HgSO_4,\ H_2SO_4} (X) \xrightarrow[\text{(ii) Conc. } H_2SO_4/\Delta]{\text{(i) } C_2H_5MgBr,\ H_2O} (Y)$$

(a) CH$_3$—C(CH$_3$)=C(CH$_2$CH$_3$)—CH$_3$
(b) CH$_3$CH(CH$_3$)—C(CH$_3$)=CHCH$_3$
(c) H$_3$C—C(=CH$_2$)—CH(C$_2$H$_5$)—CH$_3$
(d) CH$_3$—CH(CH$_3$)—C(CH$_2$CH$_3$)=CH$_2$

08 The major product of the following reaction is

$$CH_3C\equiv CH \xrightarrow[\text{(ii) DI}]{\text{(i) DCl (1 equiv.)}}$$ **[2019, 9 April Shift-I]**

(a) CH$_3$CD(Cl)CHD(I)
(b) CH$_3$CD$_2$CH(Cl) (I)
(c) CH$_3$CD (I)CHD (Cl)
(d) CH$_3$C (I) (Cl)CHD$_2$

• Level 3

09 In the given reactions identify *A* and *B*.

$$H_2 + A \xrightarrow{Pd/C} (CH_3)(H)C=C(C_2H_5)(H)$$

$$CH_3—C\equiv C—CH_3 + H_2 \xrightarrow{\text{Na/Liquid } NH_3} B$$

[2024, 01 Feb Shift-II]

(a) *A* : 2-pentyne *B* : *trans*-2-butene
(b) *A* : *n*-pentane *B* : *cis*-2-butene
(c) *A* : *n*-pentane *B* : *trans*-2-butene
(d) *A* : 2-pentyne *B* : *cis*-2-butene

10 Compound *A* formed in the following reaction reacts with *B* gives the product *C*. Find out *A* and *B*.

$$CH_3—C\equiv CH + Na \longrightarrow A \xrightarrow{B} CH_3C—C\equiv C—CH_2—CH_2 + NaBr$$

(with CH$_3$ attached below the CH$_2$) (*C*)

[2024, 30 Jan Shift-I]

(a) *A* = CH$_3$—CH$_2$CH$_3$, *B* = CH$_3$—C≡CH
(b) *A* = CH$_3$—CH=CH$_2$, *B* = CH$_3$CH$_2$—CH$_2$Br
(c) *A* = CH$_3$—C≡$\overset{-}{C}$ $\overset{+}{Na}$, *B* = CH$_3$CH$_2$—CH$_2$Br
(d) *A* = CH$_3$—C≡$\overset{-}{C}$ $\overset{+}{Na}$, *B* = CH$_3$—CH$_2$—CH$_3$

Imp. *Chemical Reactions of Ethyne*

11 The major product [*R*] in the following sequence of reactions is **[2020, 04 Sept Shift-II]**

$$HC\equiv CH \xrightarrow[\text{(ii) } (H_3C)((CH_3)_2CH)CH—Br]{\text{(i) } LiNH_2\text{/ether}} [P] \xrightarrow[\text{(ii) } NaBH_4]{\text{(i) } HgSO_4/H_2SO_4} [Q] \xrightarrow[\Delta]{\text{Conc. } H_2SO_4} [R]$$

(a) (H$_3$C)((CH$_3$)$_2$CH)C=C—CH$_2$—CH$_3$
(b) (H$_3$C)((CH$_3$)$_2$CH)CH—CH=CH$_2$
(c) (H$_3$C)(H$_3$CCH$_2$)C=C(CH$_3$)$_2$
(d) (H$_3$C)((CH$_3$)$_2$CH)C=CH$_2$—CH$_3$

12 Consider the following reactions,

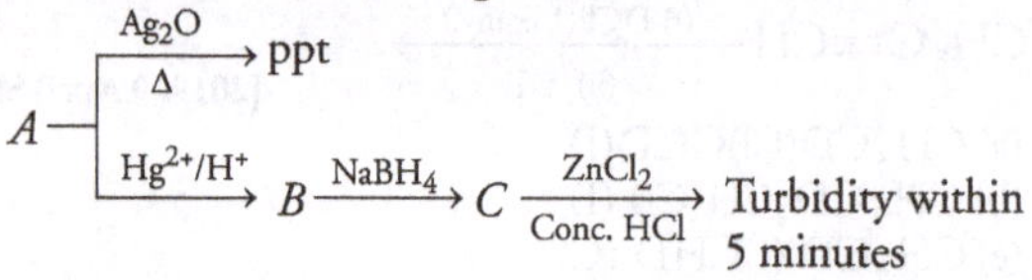

A is **[2019, 12 April Shift-II]**

(a) $CH \equiv CH$
(b) $CH_3C \equiv CCH_3$
(c) $CH_3 — C \equiv CH$
(d) $CH_2 = CH_2$

Numerical Value Type Questions

13 The compound with molecular formula C_6H_6, which gives only one monobromo derivative and takes up four moles of hydrogen per mole for complete hydrogenation has .... π-electrons. **[2025, 22 Jan shift-II]**

14 The major product of the following reaction is P.

$$CH_3C \equiv C—CH_3 \xrightarrow[\text{(ii) dil. } KMnO_4\text{, 273 K}]{\text{(i) Na/liq. } NH_3} \text{'}P\text{'}$$

Number of oxygen atoms present in product 'P' is .... (Nearest integer) **[2024, 6 April Shift-I]**

Topic 4 / Aromatic Hydrocarbons

• Level 1

01 Which of the following is not an example of benzenoid compound? **[2022, 26 July shift-II]**

(a) 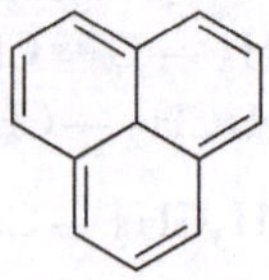

(b) 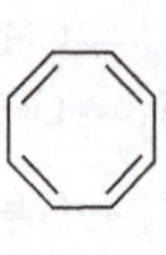

(c) 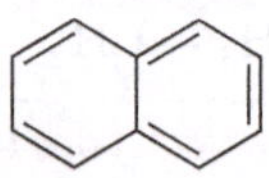

(d) 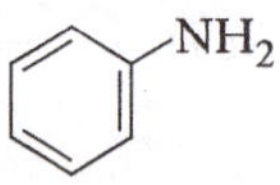

02  $+ CH_3CH_2CH_2Cl \xrightarrow{\text{Anhydrous, } AlCl_3} \underset{\text{Major product}}{A}$

The stable carbocation formed in the above reaction is **[2022, 29 June Shift-II]**

(a) $CH_3CH_2\overset{\oplus}{C}H_2$
(b) $CH_3\overset{\oplus}{C}H_2$
(c) $CH_3—\overset{\oplus}{C}H—CH_3$
(d) 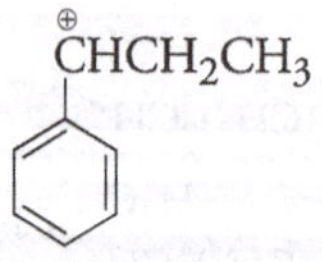

03 Benzene on nitration gives nitrobenzene in presence of HNO_3 and H_2SO_4 mixutre, where **[2021, 20 July Shift-II]**

(a) both H_2SO_4 and HNO_3 act as a bases
(b) HNO_3 acts as an acid and H_2SO_4 acts as a base
(c) both H_2SO_4 and HNO_3 act as an acids
(d) HNO_3 acts as a base and H_2SO_4 acts as an acid

04 Which of the following is an aromatic compound? **[2021, 17 March Shift-I]**

(a) 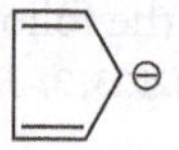
(b) 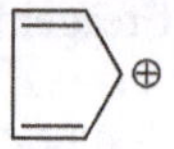
(c) 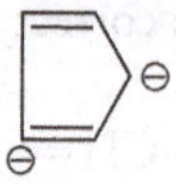
(d) 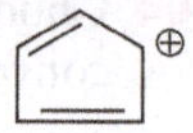

05 The increasing order of the reactivity of the following compounds towards electrophilic aromatic substitution reaction is **[2019, 10 April Shift-I]**

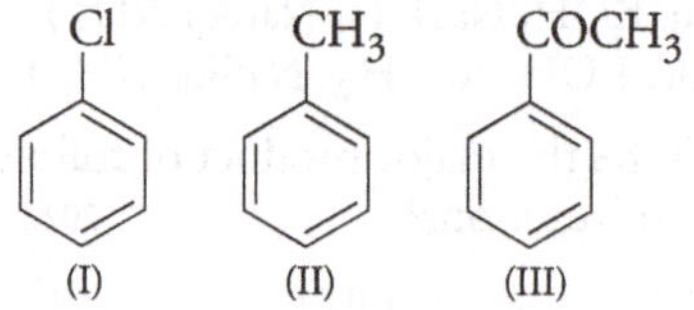

(a) III < I < II
(b) II < I < III
(c) III < II < I
(d) I < III < II

06 Which compound (s) out of the following is/are not aromatic? **[2019, 11 Jan Shift-I]**

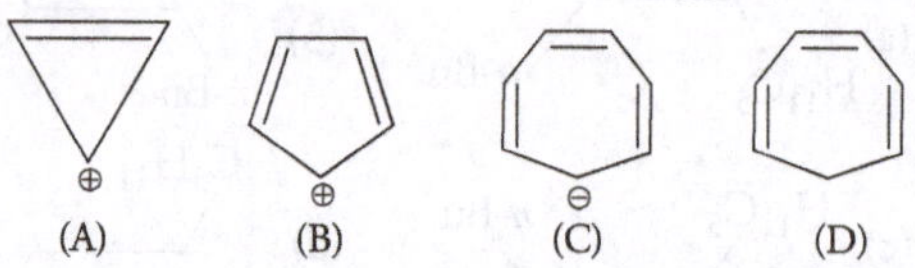

(a) (B), (C) and (D)
(b) (C) and (D)
(c) (B)
(d) (A) and (C)

07 Which of the following compounds is not aromatic? **[2019, 09 Jan Shift-II]**

(a) 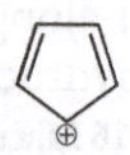
(b) 
(c) 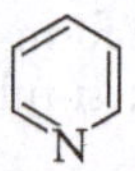

(d) 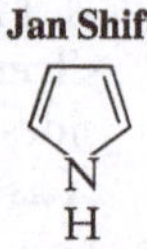

• Level 2

08 The correct stability order of the following species/molecules is **[2025, 23 Jan Shift-I]**

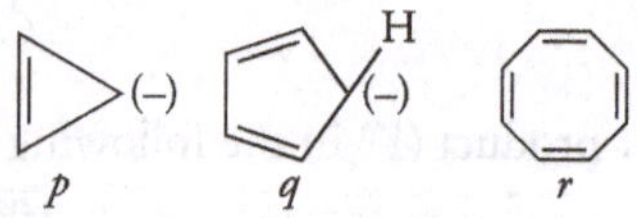

(a) $q > p > r$
(b) $p > q > r$
(c) $q > r > p$
(d) $r > q > p$

09 Which of the following are aromatic?

A. 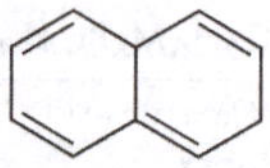

B. 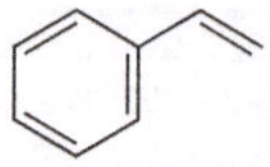

C. 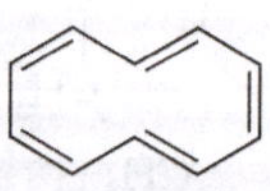

D. 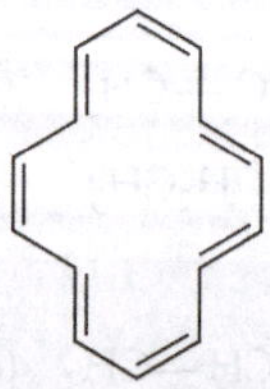

[2024, 8 April Shift-I]

(a) C and D only
(b) B and D only
(c) A and C only
(d) A and B only

10 Which of the following molecule/species is most stable? **[2024, 30 Jan Shift-I]**

(a) 
(b) 
(c) 
(d) 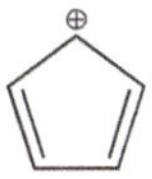

Imp. *Aromaticity of Tropolone*

11 Given below are two statements.

Statement I Tropolone is an aromatic compound and has 8 π electrons.

Statement II π electrons of $\rangle C{=}O$ group in tropolone are involved in aromaticity.

In the light of the above statements, choose the correct answer from the options given below. **[2023, 13 April Shift-II]**

(a) Both Statement I and Statement II are true.
(b) Statement I is true but Statement II is false.
(c) Statement I is false but Statement II is true.
(d) Both Statement I and Statement II are false.

12 'X' is : 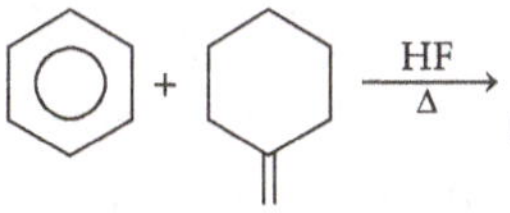 X Major product

[2023, 01 Feb Shift-II]

(a) 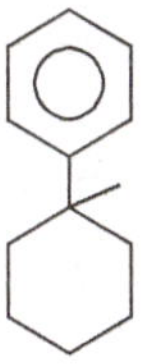
(b) 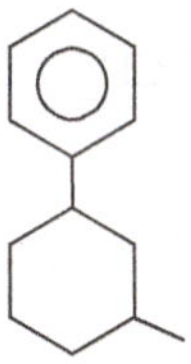
(c) 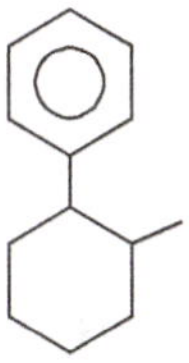
(d) 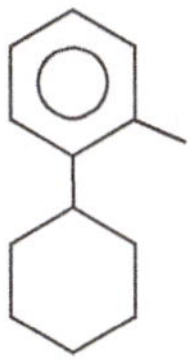

13 Given below are two statements, one is labelled as Assertion (A) and, the other is labelled as Reason (R).

Assertion (A) [6] Annulene, [8] Annulene and *cis*-[10] Annulene, are respectively aromatic, non-aromatic and aromatic. **[2022, 27 July shift-I]**

[6] Annulene 

[8] Annulene 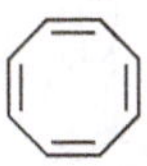

cis-[10] Annulene 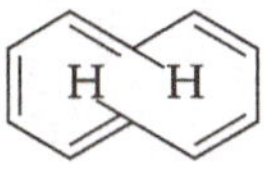

Reason (R) Planarity is one of the requirements of aromatic systems.

In the light of the above statements, choose the most appropriate answer from the options given below.

(a) Both A and R are correct and R is the correct explanation of A.
(b) Both A and R are correct but R is not the correct explanation of A.
(c) A is correct but R is not correct.
(d) A is not correct but R is correct.

14 Which of the following compounds is non-aromatic? **[2022, 26 July shift-I]**

(a) 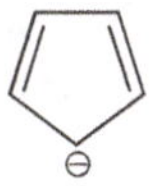
(b) 
(c) 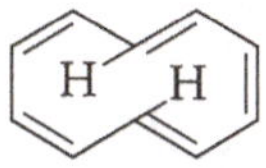
(d) 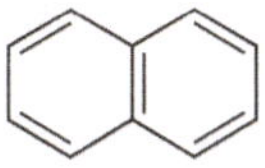

15 The major product of the following reaction is **[2022, 26 July shift-I]**

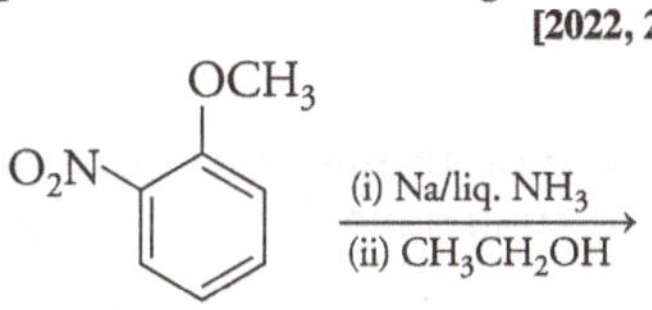

(a) 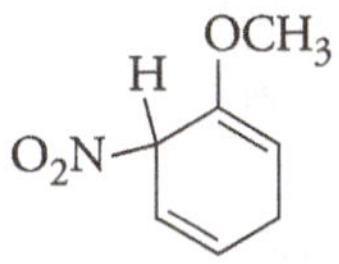
(b) 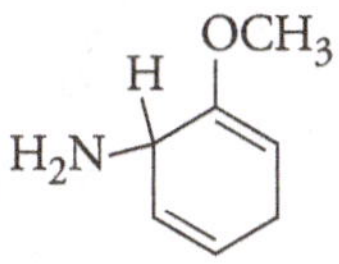
(c) 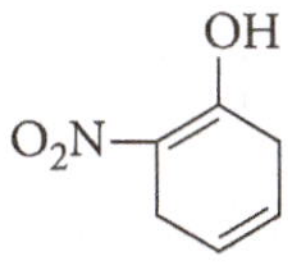
(d) 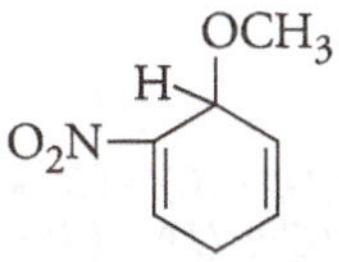

16 Which one of the following compound(s) is/are aromatic in nature ? **[2021, 1 Sep Shift-II]**

(a) 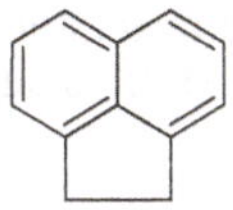
(b) 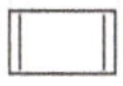
(c) 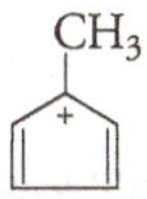
(d) 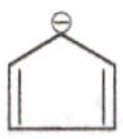

17 For the following

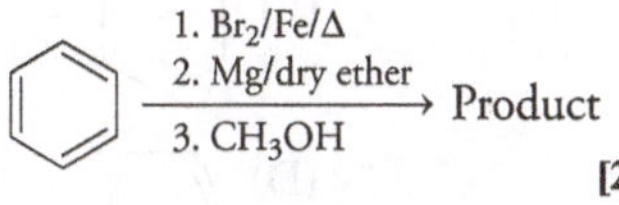

[2021, 31 Aug Shift-II]

(a) 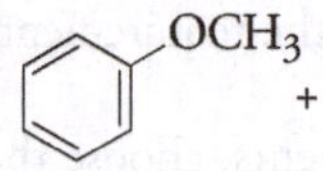 + HMgBr

(b) 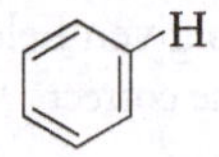 + Mg(OCH$_3$)Br

(c) 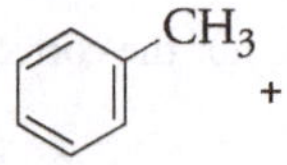 + Mg(OH)Br

(d) 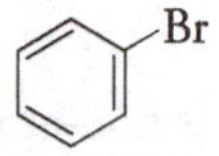 + Mg(OH)(OCH$_3$)

18 Which one of the following compounds is not aromatic ? **[2021, 26 Aug Shift-II]**

(a) 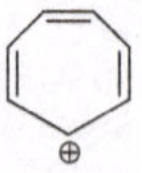

(b) 

(c) 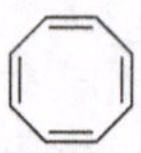

(d) 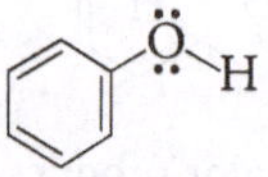

Imp. *Decreasing Order of Density*

19 The correct decreasing order of densities of the following compounds is **[2021, 25 July Shift-II]**

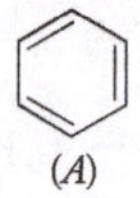
(A)

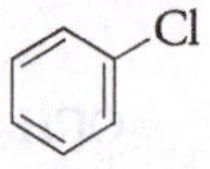
(B)

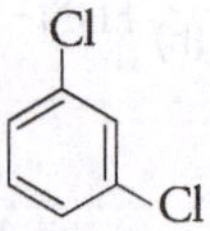
(C)

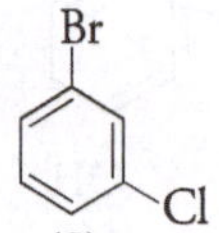
(D)

(a) (D) > (C) > (B) > (A)
(b) (C) > (D) > (A) > (B)
(c) (C) > (B) > (A) > (D)
(d) (A) > (B) > (C) > (D)

20 Which among the following is the strongest acid? **[2021, 25 July Shift-II]**

(a) 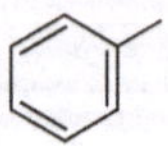

(b) 

(c) 

(d) $CH_3CH_2CH_2CH_3$

21 Among the following, the aromatic compounds are **[2021, 16 March Shift-I]**

(A) 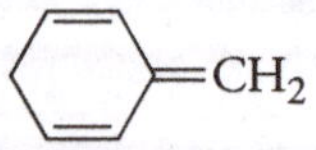

(B) 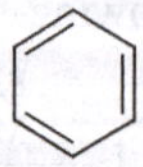

(C) 

(D) 

Choose the correct answer from the following options.

(a) Only (A) and (B)
(b) Only (B) and (C)
(c) (B), (C) and (D)
(d) (A), (B) and (C)

22 The major product of the following reaction is **[2021, 25 Feb Shift-II]**

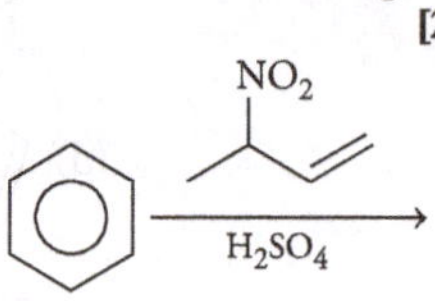

(a) 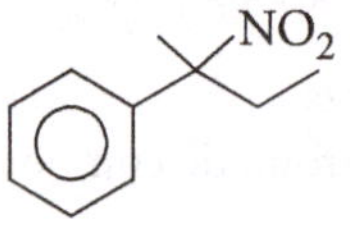

(b) 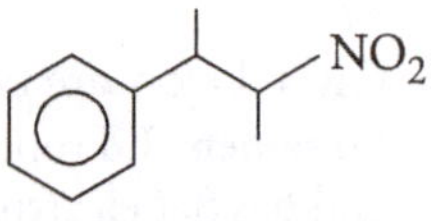

(c) 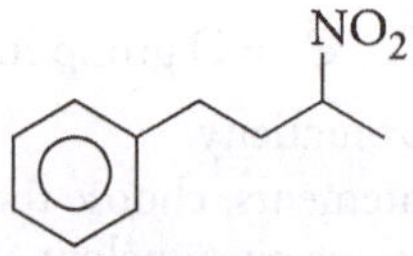

(d) 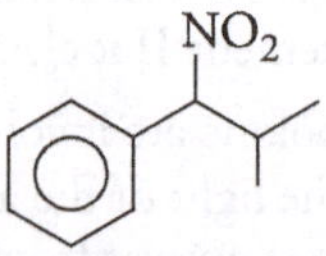

23 Which one of the following compounds is non-aromatic ? **[2021, 24 Feb Shift-II]**

(a) 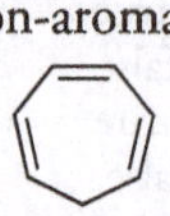

(b) 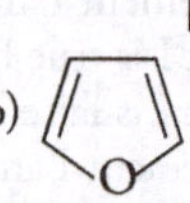

(c) 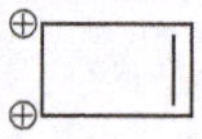

(d) 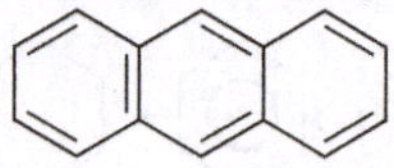

24 The major product obtained in the given reaction is

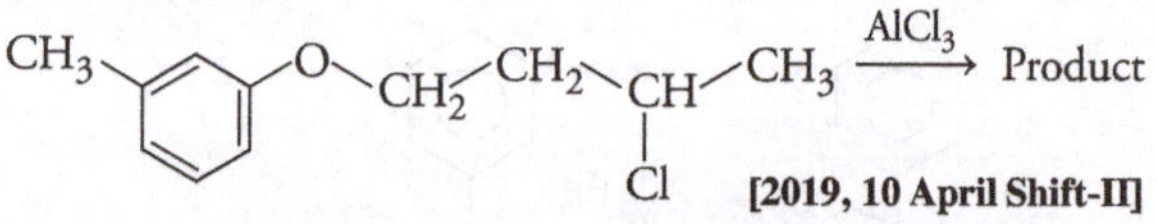

[2019, 10 April Shift-II]

(a) 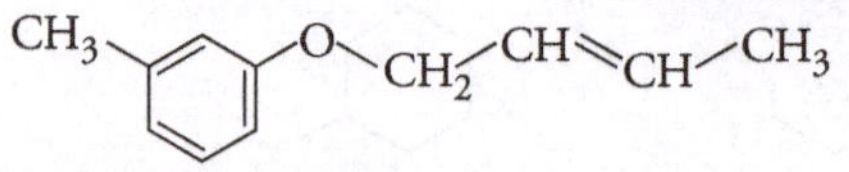

(b) 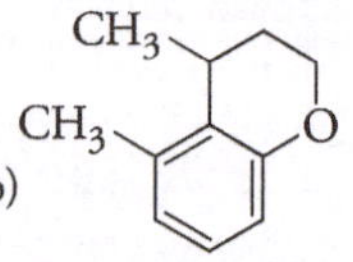

(c) 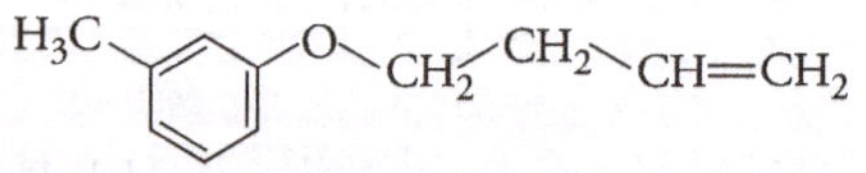

(d) 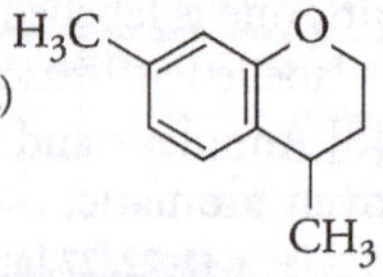

25 The increasing order of reactivity of the following compounds towards aromatic electrophilic substitution reaction is **[2019, 9 April Shift-I]**

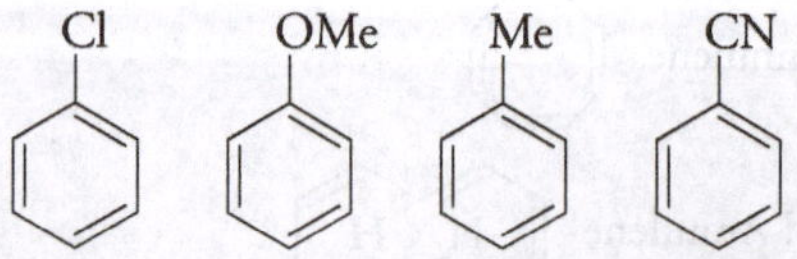

A *B* *C* *D*

(a) $A < B < C < D$
(b) $B < C < A < D$
(c) $D < A < C < B$
(d) $D < B < A < C$

26 Among the following four aromatic compounds, which one will have the lowest melting point? **[2019, 12 Jan Shift-II]**

(a) 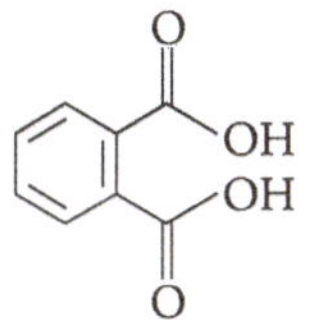 (b) 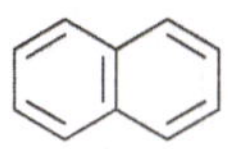

(c) 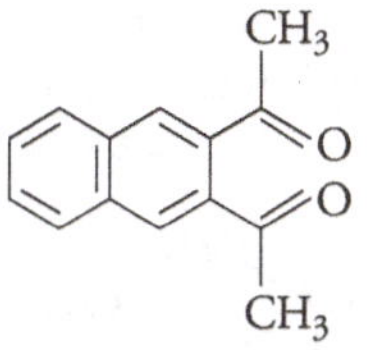 (d) 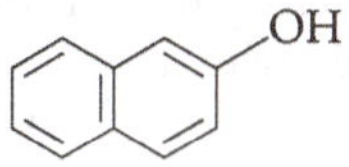

27 The major product of the following reaction is

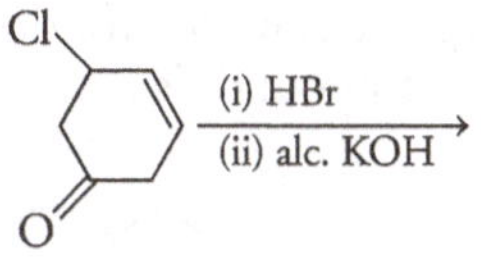

[2019, 11 Jan Shift-I]

(a) 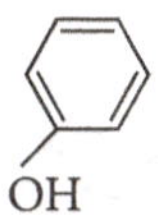 (b) 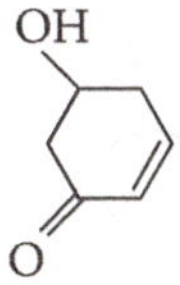

(c) 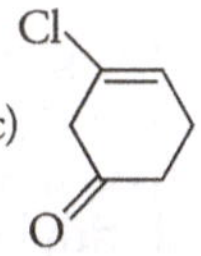 (d) 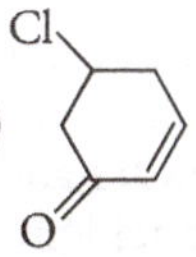

28 The major product of the following reaction is

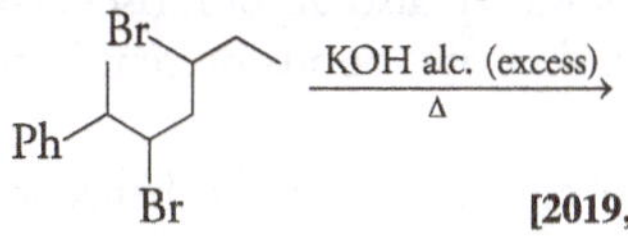

[2019, 10 Jan Shift-I]

(a) 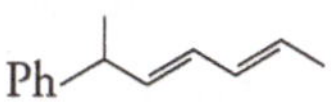 (b) 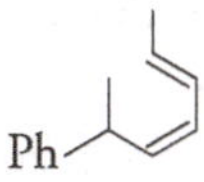

(c) 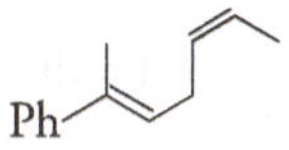 (d) 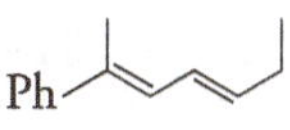

29 The major product of the following reaction is

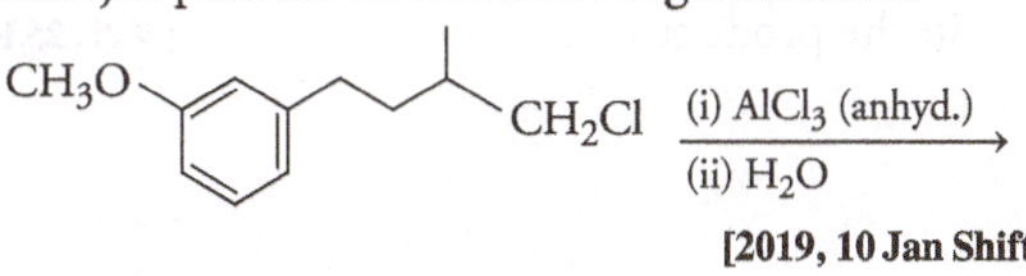

[2019, 10 Jan Shift-I]

(a) 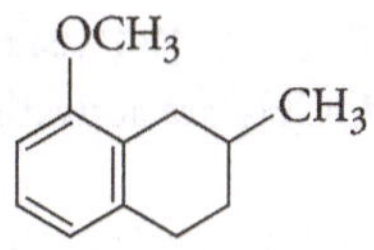 (b) 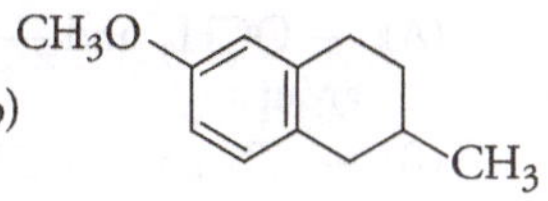

(c) 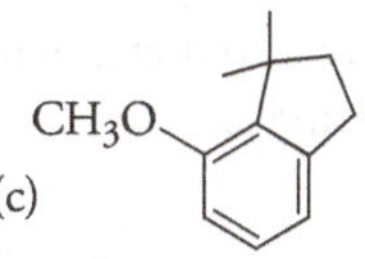 (d) 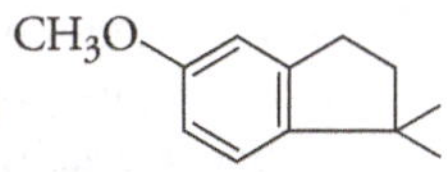

30 The major product of the following reaction is

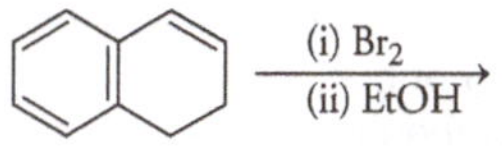

[2019, 09 Jan Shift-I]

(a) 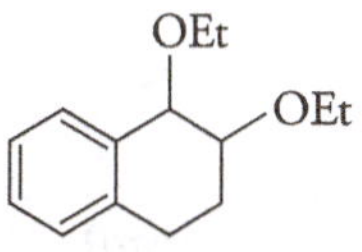 (b) 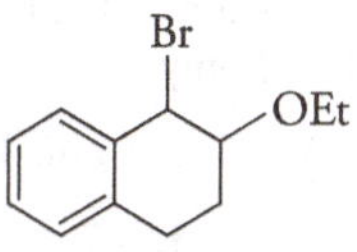

(c) 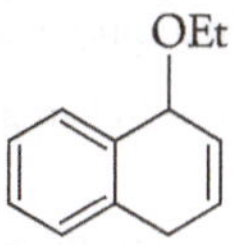 (d) 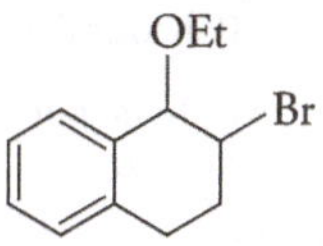

• Level 3

31 Designate whether each of the following compounds is aromatic or not aromatic. **[2025, 2 April Shift-I]**

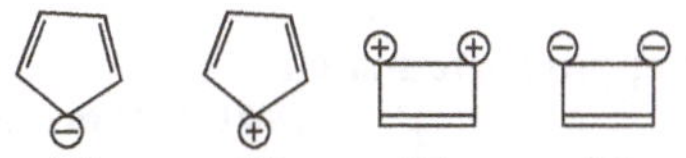

(A) (B) (C) (D)

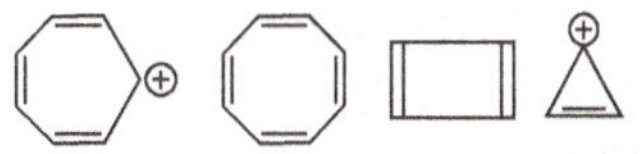

(E) (F) (G) (H)

(a) A, C, D, E, H aromatic and B, F, G not aromatic
(b) E, G aromatic and A, B, C, D, F, H not aromatic
(c) B, E, F, G aromatic and A, C, D, H not aromatic
(d) A, B, C, D aromatic and E, F, G, H not aromatic

32 Given below are two statements. **[2025, 29 Jan Shift-II]**

Statement I On nitration of *m*-xylene with HNO_3, H_2SO_4 followed by oxidation, 4-nitrobenzene-1,3-dicarboxylic acid is obtained as the major product.

Statement II —CH_3 group is *o*/*p*-directing while —NO_2 group is *m*-directing group.

In the light of the above statements, choose the correct answer from the options given below:
(a) Both Statement I and Statement II are false.
(b) Statement I is false but Statement II is true.
(c) Statement I is true but Statement II is false.
(d) Both Statement I and Statement II are true.

33 Identify correct statement/s **[2025, 24 Jan Shift-II]**

(A) — OCH_3 and — $NHCOCH_3$ are activating groups.
(B) — CN and — OH are *meta* directing groups.
(C) — CN and — SO_3H are *meta* directing groups.
(D) Activating groups act as *ortho* and *para* directing groups.
(E) Halides are activating groups.

Choose the correct answer from the options given below

(a) (A), (B) and (E) only
(b) (A), (C) and (D) only
(c) (A) and (C) only
(d) (A) only

34 Match List-I and List-II.

List-I (Chemicals)	List-II (Use/Preparation/ Constituent)
(A) Alcoholic potassium hydroxide	(i) Electrodes in batteries
(B) $Pd/BaSO_4$	(ii) Obtained by addition reaction
(C) BHC (Benzene hexachloride)	(iii) Used for β-elimination reaction
(D) Polyacetylene	(iv) Lindlar's catalyst

Choose the most appropriate match **[2021, 18 March Shift-I]**

(a) (A)-(ii), (B)-(iv), (C)-(i), (D)-(iii)
(b) (A)-(iii), (B)-(iv), (C)-(ii), (D)-(i)
(c) (A)-(iii), (B)-(i), (C)-(iv), (D)-(ii)
(d) (A)-(ii), (B)-(i), (C)-(iv), (D)-(iii)

35 The correct sequence of reagents used in the preparation of 4-bromo-2-nitroethyl benzene from benzene is **[2021, 25 Feb Shift-II]**

(a) $CH_3COCl/AlCl_3$, $Br_2/AlBr_3$, HNO_3/H_2SO_4, Zn/HCl
(b) $CH_3COCl/AlCl_3$, Zn-Hg/HCl, $Br_2/AlBr_3$, HNO_3/H_2SO_4
(c) $Br_2/AlBr_3$, $CH_3COCl/AlCl_3$, HNO_3/H_2SO_4, Zn/HCl
(d) HNO_3/H_2SO_4, $Br_2/AlCl_3$, $CH_3COCl/AlCl_3$, Zn-Hg/HCl

Numerical Value Type Questions

36 Total number of aromatic compounds among the following compounds is .... **[2024, 8 April Shift-II]**

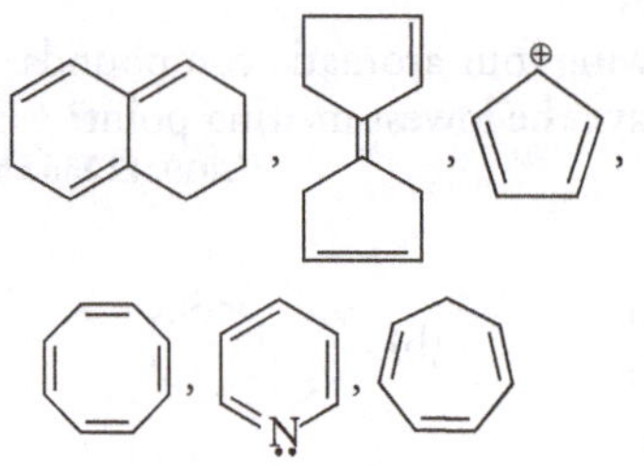

37 Total number of deactivating groups in aromatic electrophilic substitution reaction among the following is .... **[2024, 1st Feb Shift-I]**

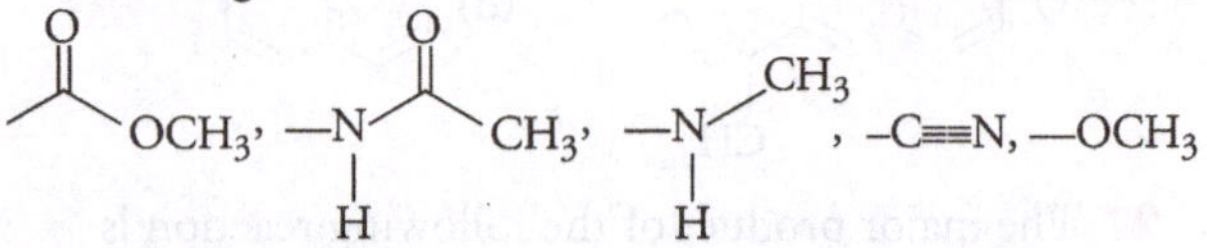

38 Among the given organic compounds, the total number of aromatic compounds is .... **[2024, 27 Jan Shift-I]**

A. 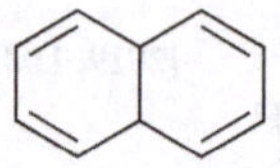
B. 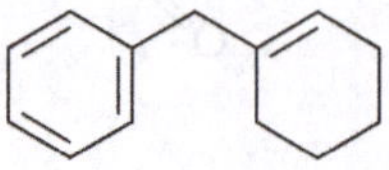
C. 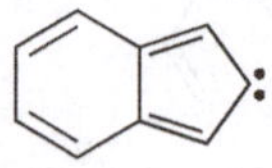
D. 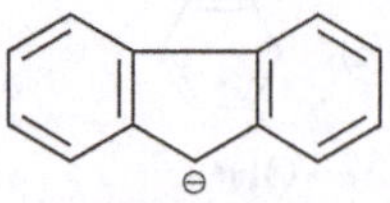

39 A trisubstituted compound '*A*', $C_{10}H_{12}O_2$ gives neutral $FeCl_3$ test positive. Treatment of compound '*A*' with NaOH and CH_3Br gives $C_{11}H_{14}O_2$, with hydroiodic acid gives methyl iodide and with hot conc. NaOH gives a compound *B*, $C_{10}H_{12}O_2$. Compound '*A*' also decolourises alkaline $KMnO_4$. The number of π bonds present in the compound '*A*' is .... **[2023, 30 Jan Shift-I]**

40 In the presence of sunlight, benzene reacts with Cl_2 to give product, *X*. The number of hydrogens in *X* is .... **[2022, 26 July Shift-I]**

41 Consider the following chemical reaction.

$$CH{\equiv}CH \xrightarrow[\text{(2) CO, HCl, AlCl}_3]{\text{(1) Red hot Fe tube, 873 K}} \text{Product}$$

The number of sp^2 hybridised carbon atom(s) present in the product is .... **[2021, 25 Feb Shift-I]**

Detailed Explanations

Topic 1

1 (b)	**2** (d)	**3** (c)	**4** (d)	**5** (c)	**6** (b)	**7** (a)	**8** (b)	**9** (b)	**10** (c)
11 (c)	**12** (c)	**13** (9)	**14** (6)	**15** (3)	**16** (8)	**17** (6)	**18** (60)		

Topic 2

1 (c)	**2** (a)	**3** (c)	**4** (c)	**5** (a)	**6** (a)	**7** (d)	**8** (b)	**9** (d)	**10** (d)
11 (b)	**12** (c)	**13** (c)	**14** (c)	**15** (b)	**16** (c)	**17** (d)	**18** (d)	**19** (a)	**20** (d)
21 (d)	**22** (c)	**23** (c)	**24** (a)	**25** (a)	**26** (c)	**27** (c)	**28** (c)	**29** (d)	**30** (b)
31 (a)	**32** (d)	**33** (b)	**34** (b)	**35** (a)	**36** (b)	**37** (d)	**38** (a)	**39** (c)	**40** (c)
41 (d)	**42** (d)	**43** (b)	**44** (d)	**45** (d)	**46** (d)	**47** (a)	**48** (a)	**49** (c)	**50** (a)
51 (3)	**52** (27)	**53** (4)	**54** (4)	**55** (1)	**56** (2)	**57** (2)	**58** (6)		

Topic 3

1 (a)	**2** (a)	**3** (d)	**4** (b)	**5** (c)	**6** (a)	**7** (a)	**8** (d)	**9** (a)	**10** (c)
11 (c)	**12** (c)	**13** (8)	**14** (2)						

Topic 4

1 (b)	**2** (c)	**3** (d)	**4** (a)	**5** (a)	**6** (a)	**7** (a)	**8** (c)	**9** (b)	**10** (c)
11 (d)	**12** (a)	**13** (d)	**14** (c)	**15** (a)	**16** (a,d)	**17** (b)	**18** (c)	**19** (a)	**20** (b)
21 (b)	**22** (b)	**23** (a)	**24** (d)	**25** (c)	**26** (b)	**27** (a)	**28** (d)	**29** (d)	**30** (d)
31 (a)	**32** (d)	**33** (b)	**34** (b)	**35** (b)	**36** (1)	**37** (2)	**38** (3)	**39** (4)	**40** (6)
41 (7)									

Detailed Explanations

Topic 1

01. *(b)* The structure of given alkanes are as follows.

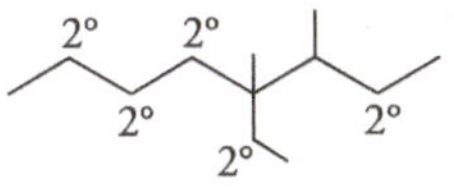

⇒ 10 secondary hydrogens are present.

4-ethyl-3, 4-dimethyloctane

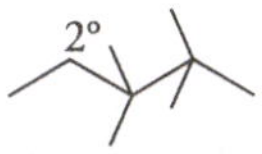

⇒ 2 secondary hydrogens are present.

2, 2, 3, 3- tetramethylpentane

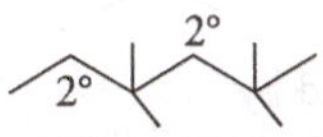

⇒ 4 secondary hydrogens are present.

2, 2, 4, 4- tetramethylhexane

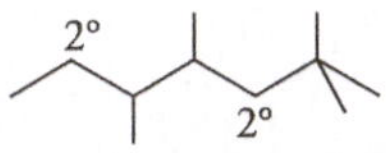

⇒ 4 secondary hydrogens are present.

2, 2, 4, 5- tetramethylheptane

So, 2, 2, 3, 3-tetramethylpentane has 2 secondary hydrogen.

02. *(d)* Following are the conformations of *n*-butane with their stability

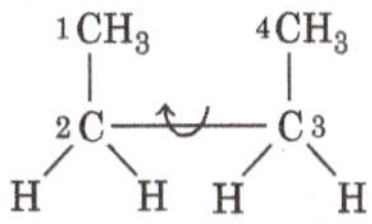

n-butane

(a)

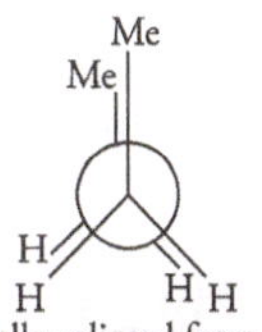

Fully eclipsed form

Fully eclipsed form has maximum steric repulsion between alkyl group (CH_3). So, it is least stable.

(b)

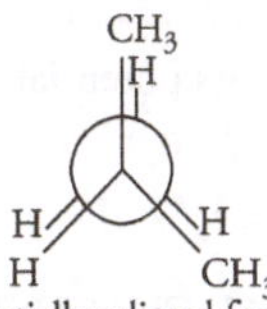

Partially eclipsed form

Partially eclipsed form has lesser steric repulsion than the fully eclipsed form. So, it is more stable than the fully eclipsed form.

(c)

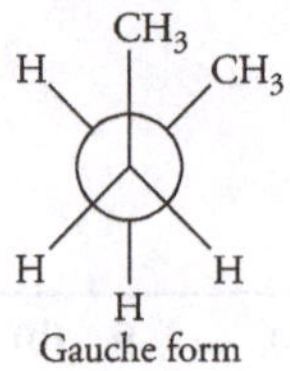

Gauche form

Gauche form is more stable than the partially eclipsed form.

(d)

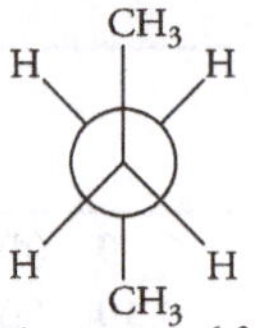

Anti or staggered form

Antistaggered or staggered form is most stable due to lowest steric and torsional strains.

03. *(c)* Among the given conformation(a) is gauche and (c) is staggered configurations (d) is eclipsed configurations. Whereas (b) is partially eclipsed form. In the staggered or anti-conformation (c), the dihedral angle is the largest (180°) between the two methyl groups. Thus, structure (c) has staggered conformation with maximum dihedral angle.

04. *(d)*

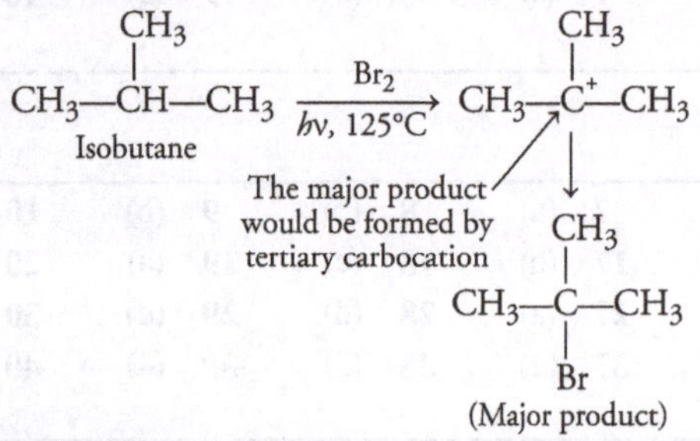

05. *(c)* Staggered and eclipsed conformers of ethane are also called as rotamers.

Concept Enhancer

Staggered conformation is that conformation in which the substituents attached to the carbon atoms are far apart. While Eclipsed conformation is that in which the substituents are closely attached as possible.

06. *(b)*

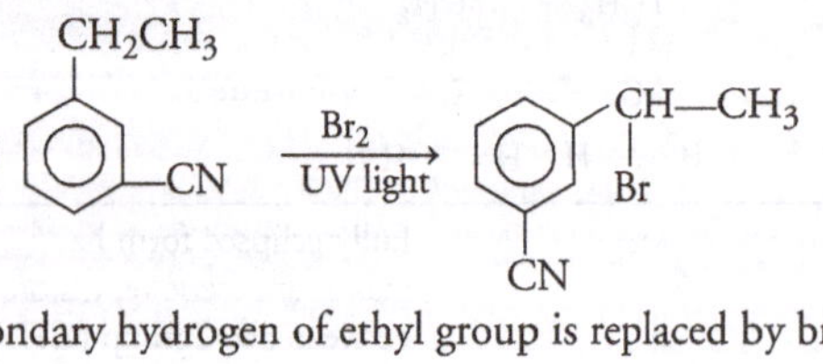

Secondary hydrogen of ethyl group is replaced by bromine atom.

07. *(a)*

idea Stability of any compound is inversely proportional to its energy. To achieve stability energy should be minimised.

Higher the stability, lower will be its potential energy.

Order of stability : I > III > IV > II

Order of potential energy I < III < IV < II.

08. *(b)* $CH_3CH_2CH_3 \xrightarrow{Mo_2O_3} CH_3CH_2CHO$

Hence, correct option is (b).

09. *(b)* A dihedral angle is the angle between two C—H bonds projected on a plane orthogonal to the C—C bond. In the given skew conformation, having Newman's projection the dihedral angle is

$$H'—C—C—H'' = (H'—C—C—H_a) + (H_aC—C—H'')$$
$$= 29° + 120° = 149°$$

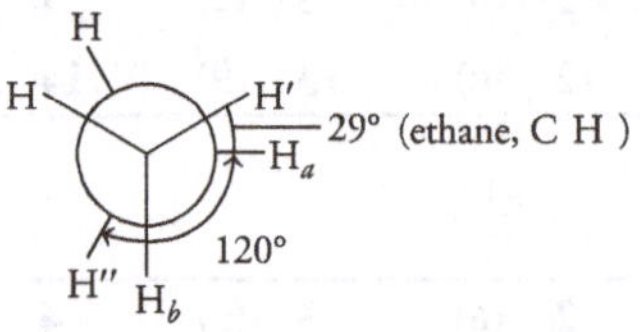

10. *(c)* Both Statement I and Statement II are correct.

$$CH_3—\underset{CH_3}{\overset{CH_3}{C}}—CH_3 + X_2 \xrightarrow{h\nu} CH_3—\underset{CH_3}{\overset{CH_3}{C}}—CH_2—X$$

(X = Cl, Br)

Melting point

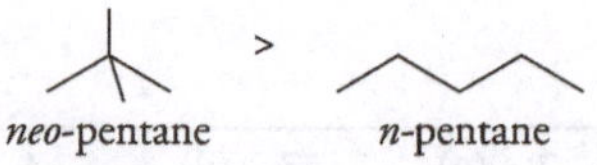

11. *(c)* Statement I is correct, whereas Statement II is incorrect. Corrected statement is as follows:

Tertiary alkanes can be oxidised to alcohol by treating with $KMnO_4$.

n-alkane are very less reactive and have almost no reaction with $KMnO_4$.

$$n\text{-alkanes} \xrightarrow{KMnO_4} \text{No reaction}$$

12. *(c)* The statement given in option (a), (b) and (d) are correct, while statement (c) is incorrect. The correct form of statement (c) is :

Eclipsed conformation is the least stable conformation of ethane.

Concept Enhancer

It is interesting to note that in all the conformations, the bond angles and the bond lengths remain the same.

13. *(9)* Total nine different chain isomers for C_7H_{16} is possible. These are as follows.

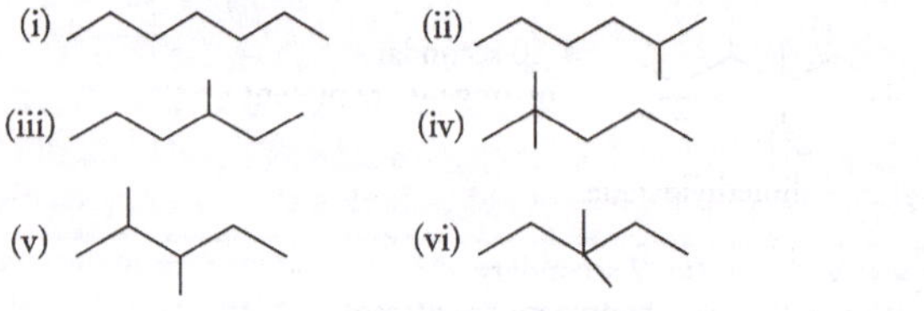

14. *(6)* Total 6 isomeric compounds are formed by monochlorination of 2-methylbutane. These are as follows

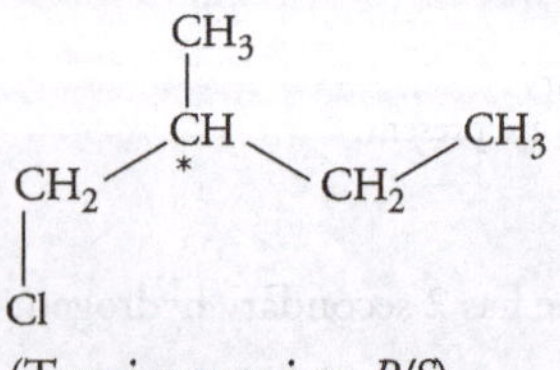

(Two isomers i.e., *R/S*)

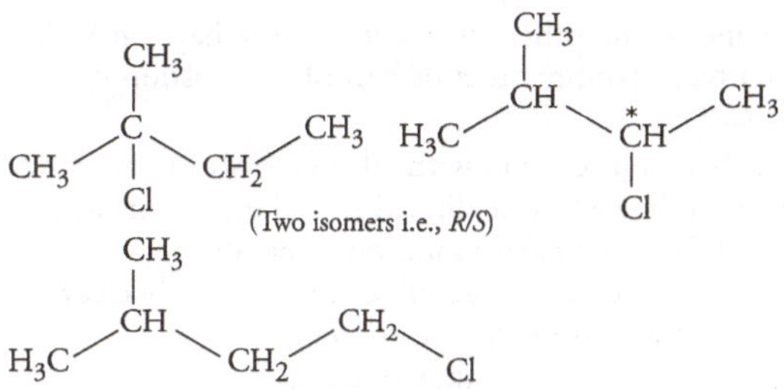

15. *(3)* Alkanes obtained on electrolysis of a mixture of CH_3COONa and C_2H_5COONa are CH_3-CH_3, $CH_3-CH_2-CH_2-CH_3$ and $CH_3-CH_2-CH_3$. Thus, total three alkanes are obtained.

16. *(8)* The possible structures of alkanes with formula C_5H_{12} are

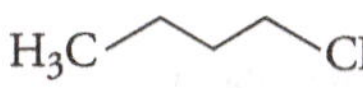
= 3 mono substituted products

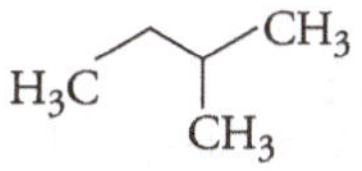
= 4 mono substituted product

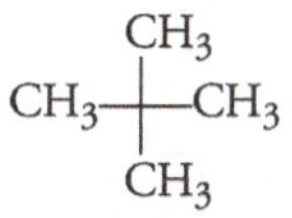
= 1 mono substituted product

Total number of monobromo derivatives formed by alkane with formula C_5H_{12} are $= 3 + 4 + 1 = 8$

17. *(6)*

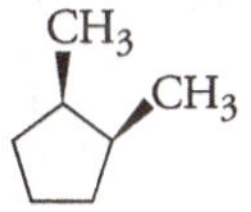
It has 2 stereoisomers.

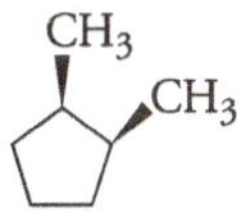
It has only one form due to plane of symmetry.

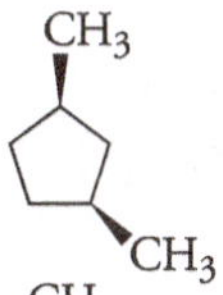
It has only one stereo form due to plane of symmetry.

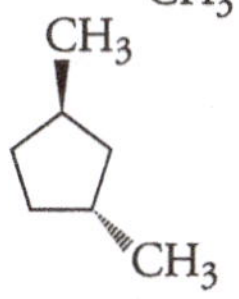
It has 2 stereoisomers.

Total stereoisomers of dimethyl cyclopentane $= 2 + 1 + 1 + 2 = 6$

Concept Enhancer

A structure of a compound has plane of symmetry if it can be divided into two halves by a plane and have each half be a reflection of the other across the plane.

18. *(60)* Newman staggered form of 1,1,1-trichloro ethane (CCl_3-CH_3) is as follows.

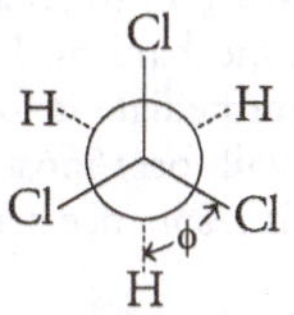

It's dihedral angles (ϕ) is 60°.

Topic 2

01. *(c)* The complete reactions is as follow

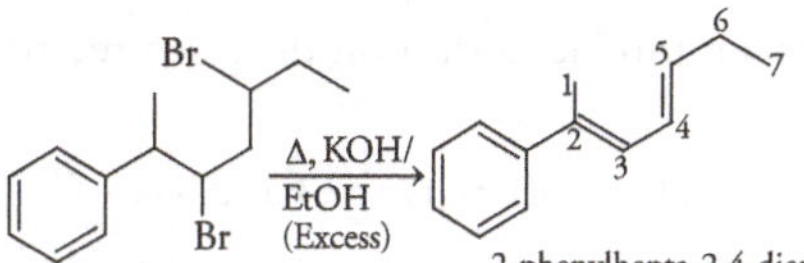

2-phenylhepta-2,4-diene
(Major product)

02. *(a)* The bond structure of ethylene is

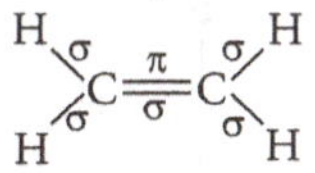

It has 5σ bonds and 1π bond.

03. *(c)* This is β-elimination reaction of quaternary ammonium salts. It is also known as Hofmann elimination as the major alkene formed will be a Hofmann product i.e., less substituted alkene.

$$\underset{\substack{|\\ \overset{\oplus}{N}(CH_3)_3}}{CH_3CH_2\overset{\beta}{C}H_2\overset{\alpha}{C}H\overset{\beta}{C}H_3} \xrightarrow[C_2H_5OH]{C_2H_5ONa} \underset{\substack{(A)\\ \text{Hoffmann product}\\ \text{(Major)}}}{CH_3CH_2CH_2CH{=}CH_2}$$

$$+\underset{\text{(Minor) }(B)}{CH_3CH_2CH{=}CH{-}CH_3}$$

04. *(c)* The major product is formed according to Markownikoff's rule.

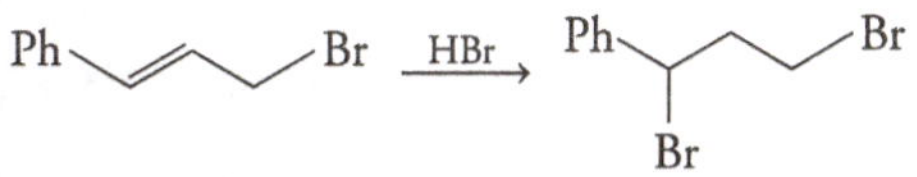

(P)

05. *(a)* Due to presence of π-bonds, alkenes are less stable than alkanes.

Stability order of hydrocarbons is $C{\equiv}C < C{=}C < C-C$

The strength of hydrocarbons is $C{\equiv}C > C{=}C > C-C$

This is because bond length is minimum in alkyne and maximum in alkane.

So, both Statement I and Statement II are correct.

06. *(a)*

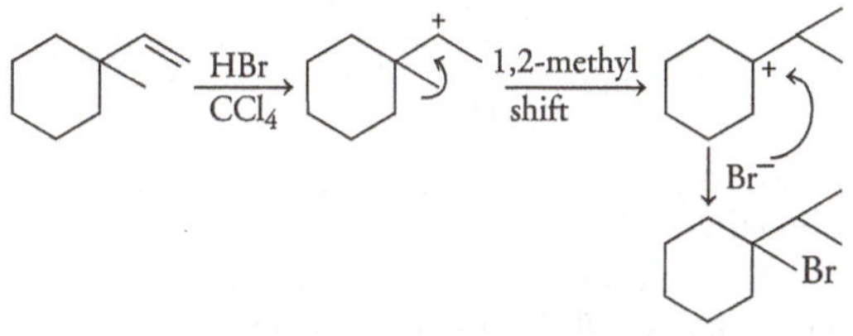

(A)
(Major product)

It shows Markownikoff's addition.

07. *(d)*

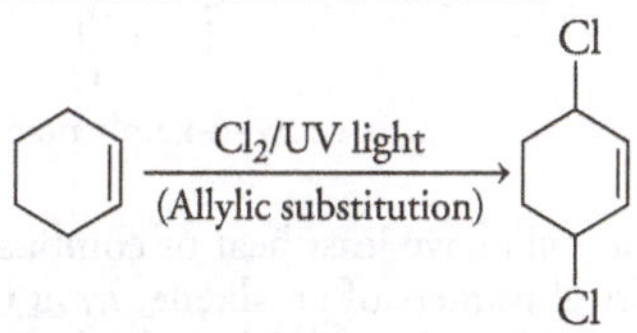

(Cl group attach to the allylic position.)

08. *(b)* Lindlar's catalyst consist of palladium catalyst poisoned with traces of lead and quinoline.
i.e., $Pd/CaCO_3 + (CH_3COO)_2Pb + \text{Quinoline}$.

09. *(d)* The major product formed in the given reaction is an alkane.

$$CH_3CH_2{=}CH{-}CH(CH_3)_2 \xrightarrow{H^+} CH_3CH_2{-}\overset{+}{C}HCH(CH_3)_2$$

$$\downarrow \text{1, 2 hydride shift}$$

$$CH_3{-}CH_2{-}CH_2{-}\overset{+}{C}{-}(CH_3)_2$$

$$CH_3{-}CH_2{-}CH_2{-}\underset{CH_3}{\overset{Br}{C}}{-}CH_3$$

Concept Enhancer

According to Markownikoff rule the negative part of the addition gets attached to that carbon atom which possess lesser number of hydrogen atom.

10. *(d)* Complete mechanism is as follows

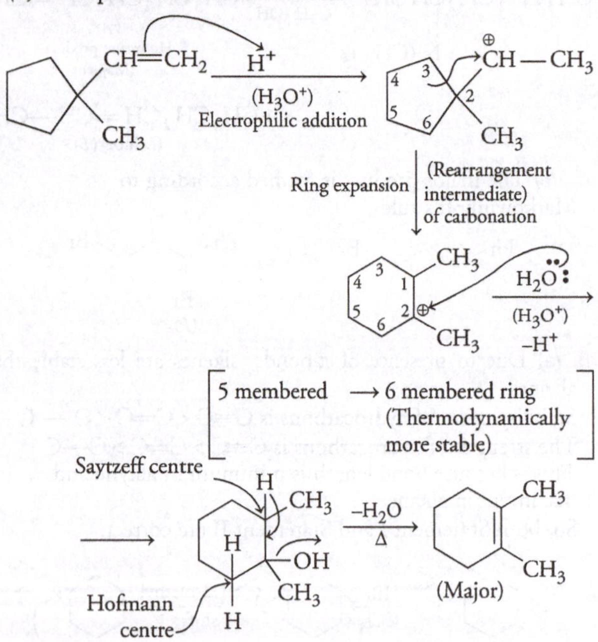

The major product obtained in the given reaction is 1, 2-dimethyl cyclohexene.

11. *(b)* All given three alkadienes are isomers. Upon combustion, all of them gives same product.

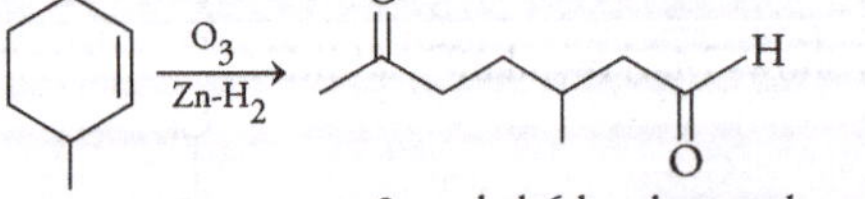

3-methyl-6-ketoheptanal

The most stable isomer will have least heat of combustion. Between two geometrical isomers of an alkene, *trans* is more stable than *cis*. In the given options, (A) have both double bonds that are *trans*, (B) have one *cis* and other one *trans* and (C) have both *cis*. So, the correct stability order of heat of combustion is $A < B < C$.

12. *(c)* But-2-ene on reaction with alkaline $KMnO_4$ at elevated temperature followed by acidification will give acetic acid (CH_3COOH). Hot alkaline solution of potassium permanganate followed by acidification oxidatively cleaved alkenes. The reaction proceed as follows :

$$\underset{\text{But-2-ene}}{CH_3{-}CH{=}CH{-}CH_3} \xrightarrow[H_3O^+]{\text{Alk. }KMnO_4\text{, heat}} \underset{\text{Acetic acid}}{2CH_3COOH}$$

13. *(c)*

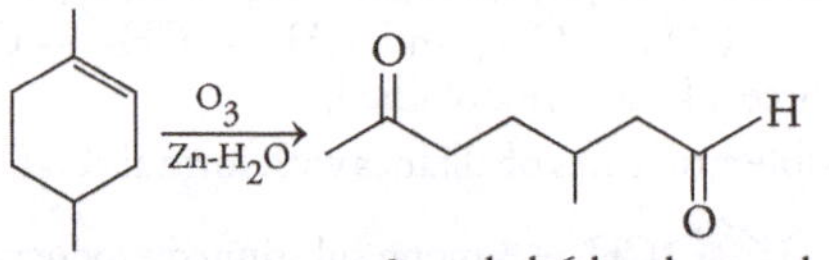

3-methyl-6-ketoheptanal

14. *(c)*

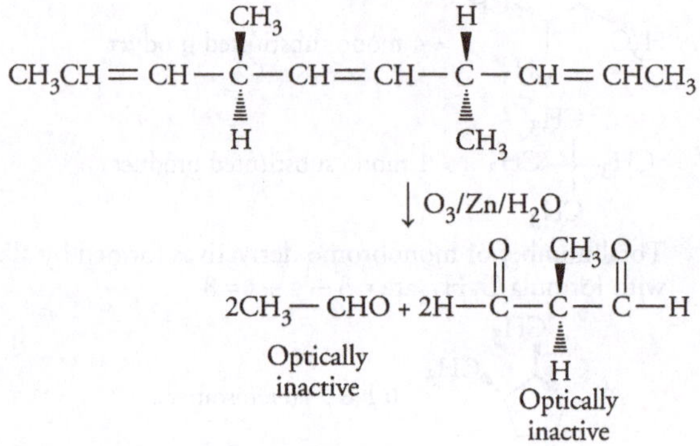

15. *(b)* Statement I is true but Statement II is false:
The reaction involved to obtain product by ozonolysis followed by treatment with Zn, H_2O.

$$H_3C{-}CH_2{-}\underset{CH_3}{CH}{-}HC{=}CH{-}\underset{CH_3}{C}{-}CH_2CH_3$$

$$\downarrow \text{(i) } O_3/Zn\text{, (ii) } H_2O$$

$$2[CH_3{-}CH_2{-}\overset{*}{\underset{CH_3}{CH}}{-}\underset{O}{\overset{\|}{C}}{-}]$$

It has chiral centre.

16. *(c)* The complete reaction is as follows

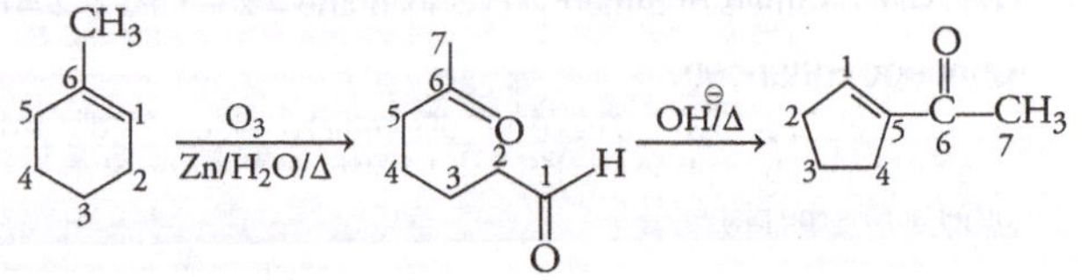

So, the suitable alkene is given in option (c).

17. *(d)* Addition of H—Br (*aq*) to an alkene follows electrophilic addition mechanism. In the rate determining step a carbocation intermediate is formed. Among *P*, *Q*, *R*, and *S* compound, *Q* will form most stable carbocation intermediate since it is resonance stabilised.

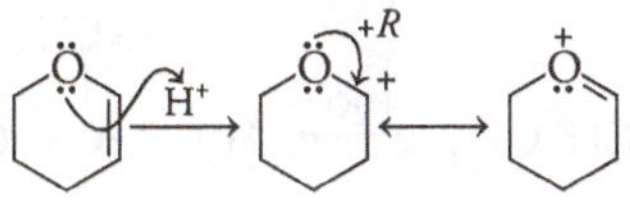

18. *(d)*

> **idea** Use Bond strength $\propto \frac{1}{\text{Bond length}}$

Statement given in option *(d)* is incorrect. Its correct form is, the carbon-carbon bonds in ethyne is stronger than that in ethene.

19. *(a)* The complete reaction is as follows,

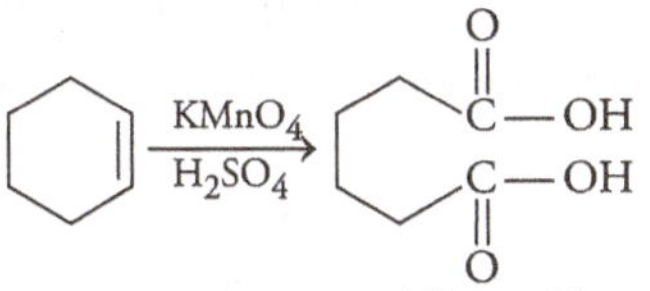

Adipic acid
(*A*)

Hence, option (a) is correct.

20. *(d)*

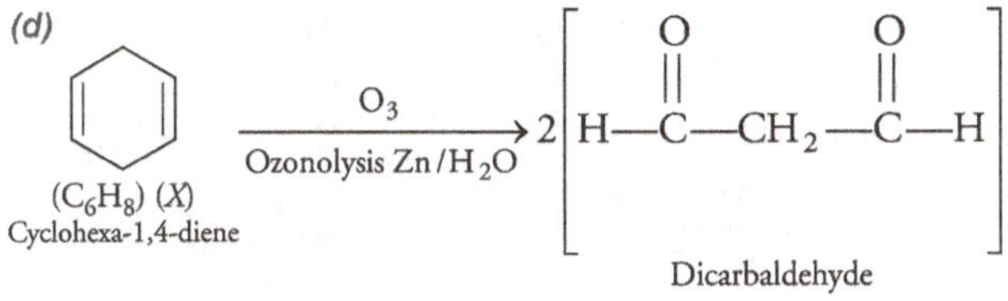

21. *(d)*

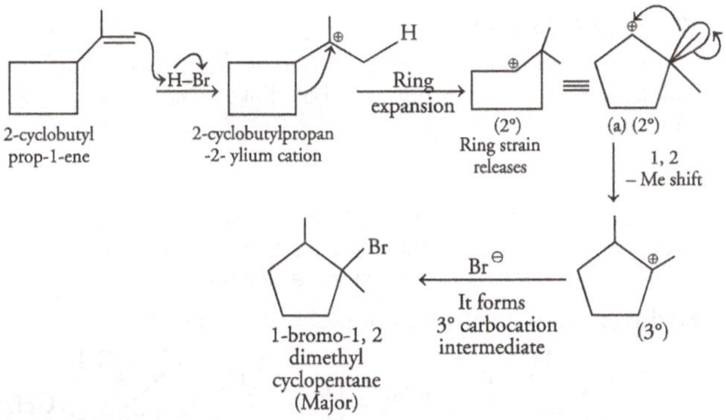

> **Concept Enhancer**
> 2-cyclobutylprop-1-ene reacts with HBr to give a 3° carbocation, which undergoes ring expansion to release ring strain then again it rearranges to 3° carbocation *via* 1,2-methyl shift attack of Br^- to give major product.

22. *(c)*

$$\underset{\text{Ethanol}}{CH_3CH_2OH} \xrightarrow{\text{Conc. } H_2SO_4} \underset{\text{Ethene}}{CH_2=CH_2} \xrightarrow[\text{(Baeyer's reagent)}]{\text{Cold/dil. } KMnO_4} \underset{\text{Ethan-1,2-diol}}{\underset{OH\quad OH}{CH_2—CH_2}}$$

In presence of conc. H_2SO_4 and Baeyer's reagent ethanol gets converted into diol.

23. *(c)* The alcoholic (—OH) group attacks / takes the proton (H^+) and water leaves to form carbocation.

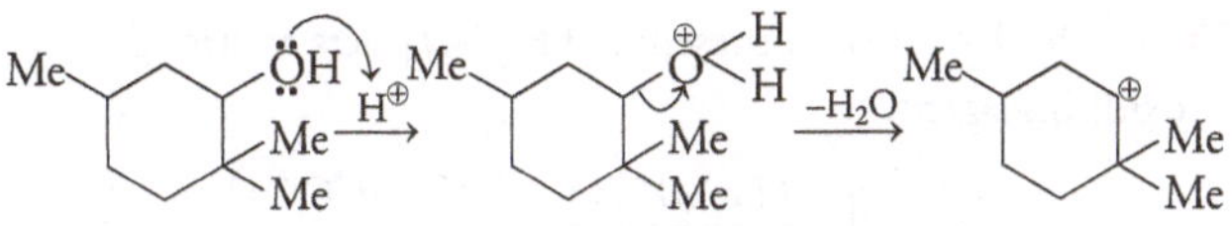

Now, rearrangement of carbocation takes place by 1,2-methyl shift to form more stable carbocation (3°-carbocation) followed by removal of H^+ ion to given more substituted alkene (major product).

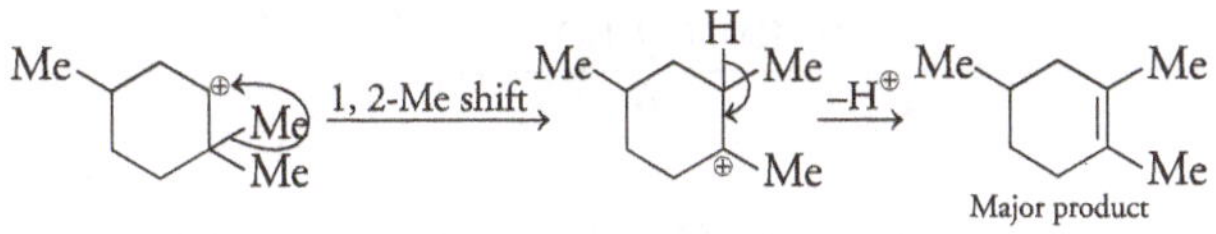

24. *(a)* This reaction proceeds in two steps :

Step 1. *syn*-addition of bromine or formation of bromonium ion.

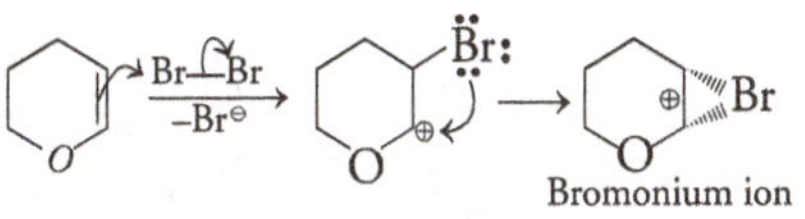

Step 2. Attack of methanol on carbon next to oxygen as it is the most electrophilic centre.

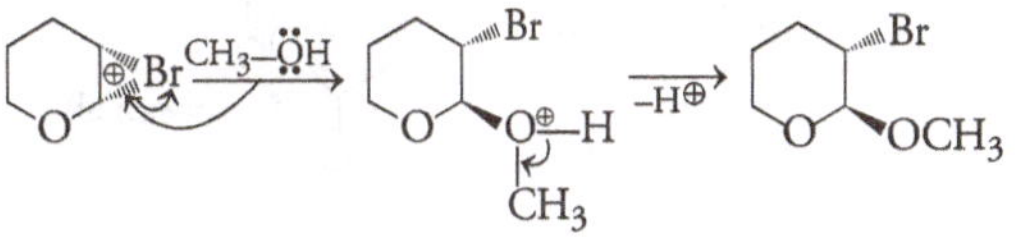

25. *(a)*

> **idea** Addition of bromine to an alkene proceeds as anti addition.

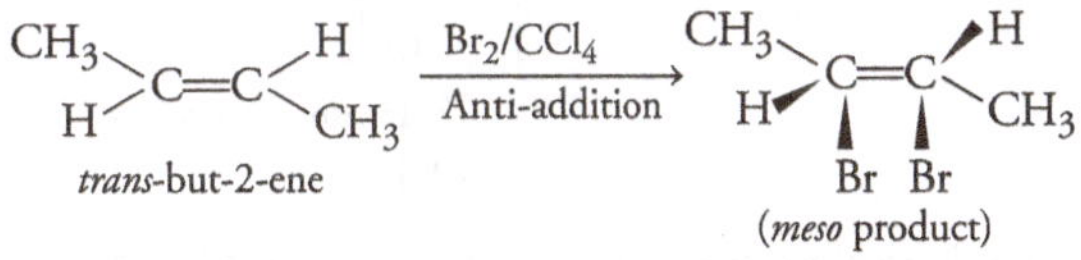

26. *(c)* The major product formed by 3° carbocation intermediate.

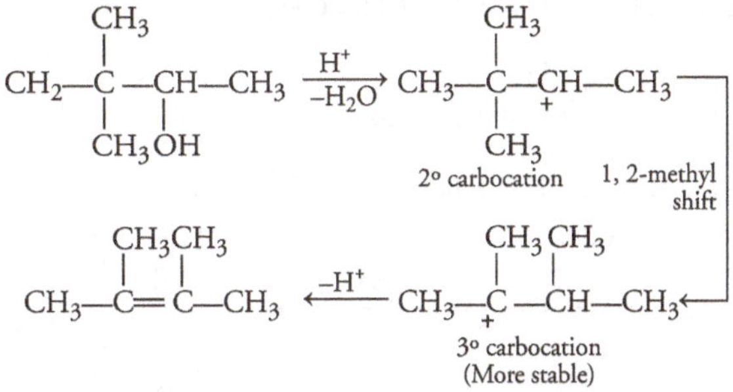

27. *(c)* Stereoisomers will be exhibited by those alkenes which have two different subtituent groups attached to each carbon of double bond.

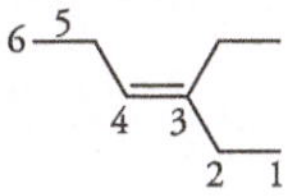

Thus, 3 ethylhex-3-ene do not show stereoisomerism.

28. *(c)* Hot $KMnO_4$ in presence of H_2SO_4 acts as strong oxidising agent.

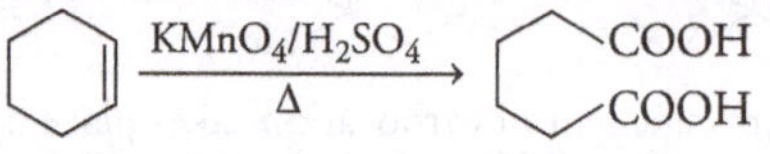

$KMnO_4$ in water at room temperature is known as Baeyer's reagent and acts as a weak oxidising agent.

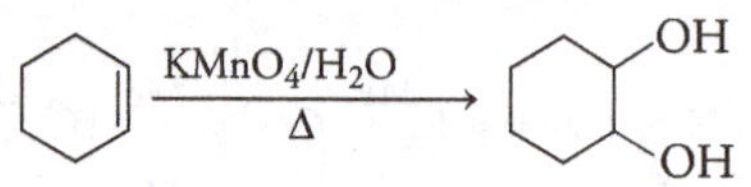

29. *(d)* In first step, $KMnO_4$ (dil.) help to break alkene in ketone and aldehyde and in second step, CrO_3 is used for selective over oxidation of aldehyde only. CrO_3 will form aldehyde to acid in product (*B*).

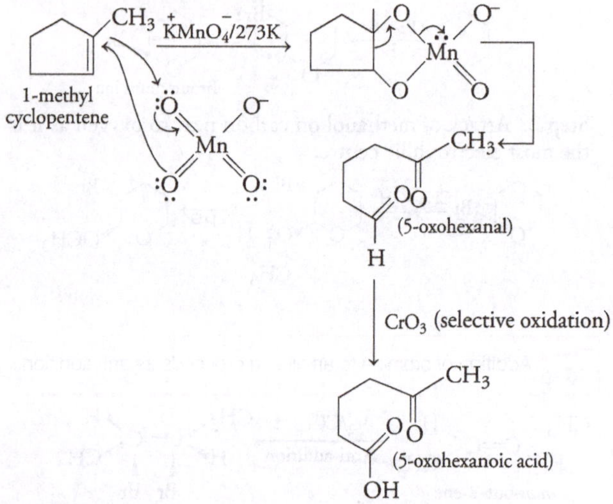

30. *(b)* Geometrical isomerisms is shown by those alkenes which have different type of groups attached to the carbon atoms around the double bond.

e.g., $CH_3-\underset{\underset{CH_3}{|}}{CH}-CH=CH-CH_3$

4-methylpent-2-ene

31. *(a)* Reaction is as follows

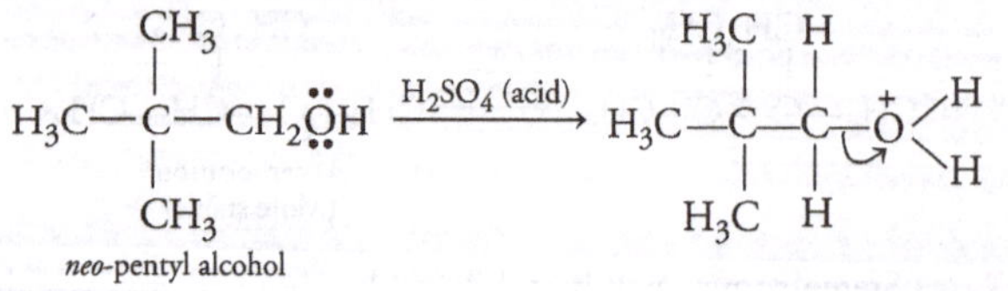

Here O being eletronegative contains positive charge thus it is unstable. Then H_2O molecule will remove from here.

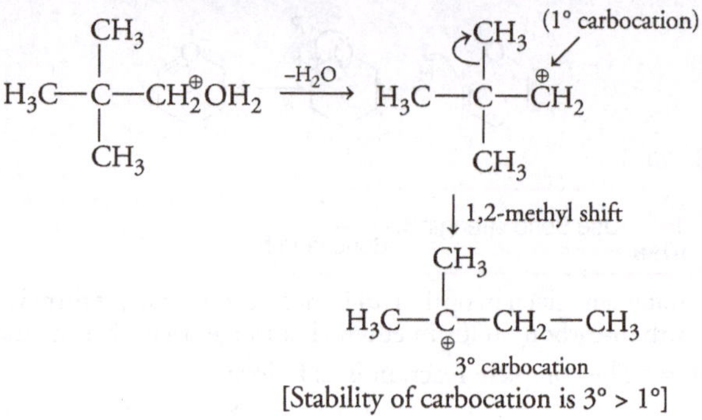

On heating 3° carbocation, 2 possible alkenes are possible. These are as follows

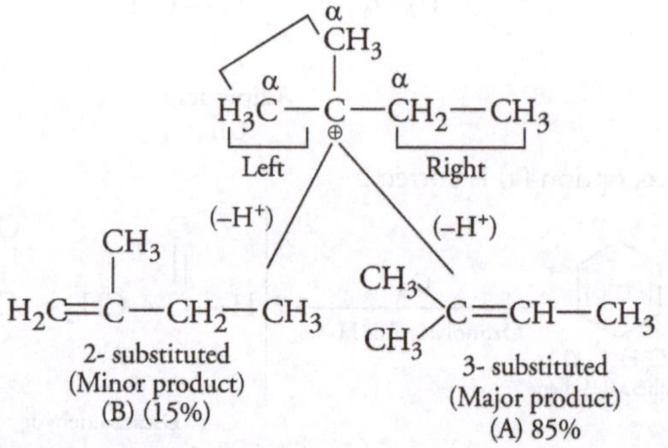

So, correct option is (a).

Concept Enhancer

The major and minor product (alkene) formed is governed by Saytzeff rule according to which a more substituted alkene will form a major product.

32. *(d)*

idea In this question we will use the concept of optical activity. Compound which contain a chiral carbon or doesn't contain axes of symmetry is an optically active compound.

Hydrogenation involve addition of hydrogen atom.

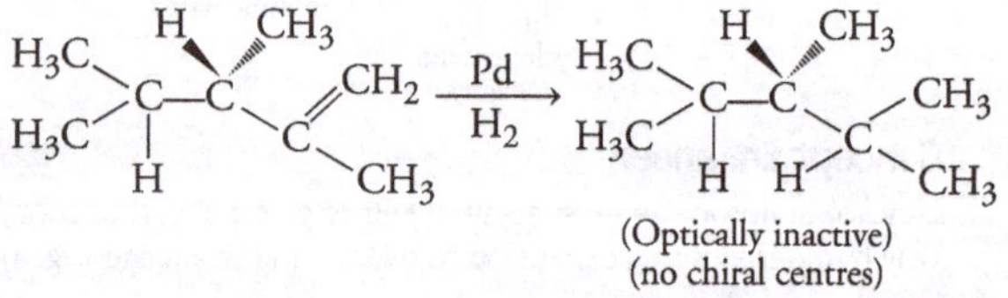

33. *(b)*

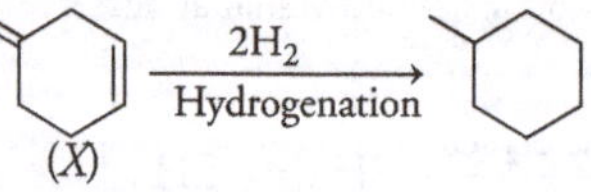

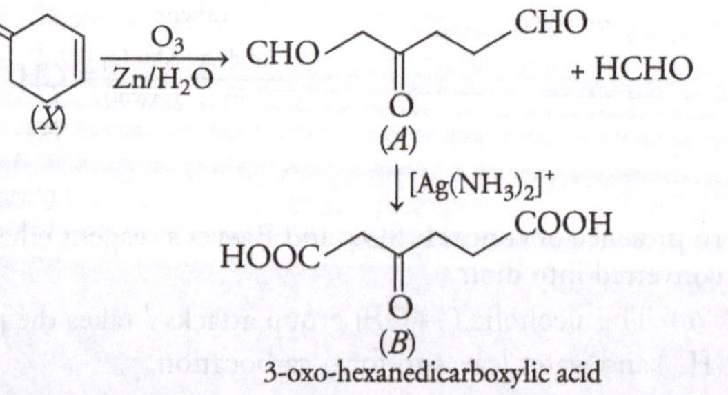

The correct option is (b).

34. *(b)* The major product of the given addition reaction is

$$H_3C-\underset{OH}{\underset{|}{CH}}-\underset{Cl}{\underset{|}{CH_2}}.$$

The reaction proceeds through following mechanism

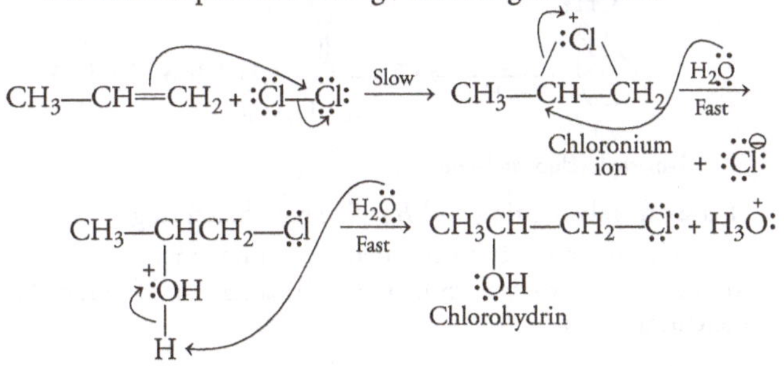

35. *(a)* Complete reaction sequence involved when molecule (*X*) reacts with 1 eq of HBr is as follows.

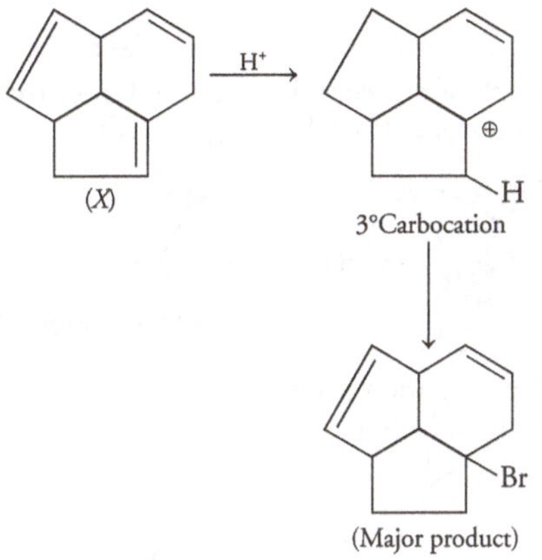

36. *(b)* Reaction given in B and D will not form alkene by elimination. The product formed in each reactions are as follows.

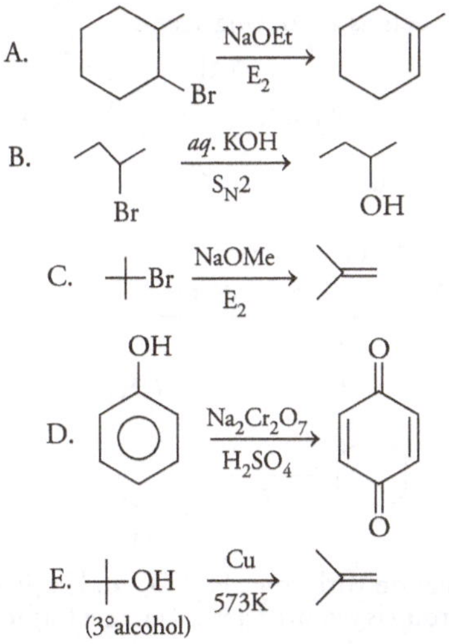

37. *(d)* The correct match is A-III, B-IV, C-I, D-II.

38. *(a)* Statements given in A, D and E are incorrect while B and C are correct. The correct form of incorrect statements are

(A) Propene does not show geometrical isomerism.

(D) 2-methylbut-2-ene does not show geometrical isomerism.

(E) *trans* isomer has higher melting point than *cis* isomer.

39. *(c)* Statement given in option (c) is incorrect. Its correct form is,

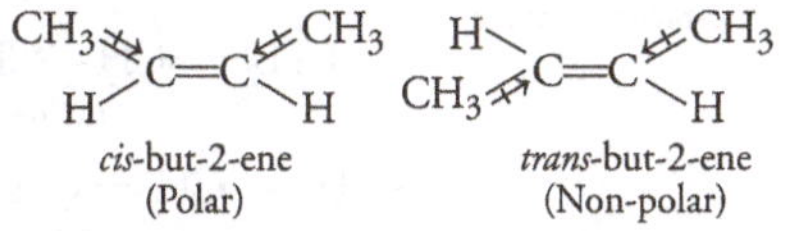

cis but-2-ene has higher dipole moment that *trans* but-2-ene.

40. *(c)*

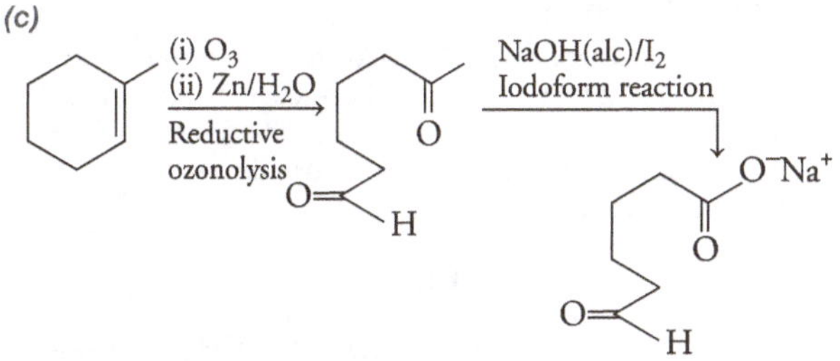

Thus, *A* and *B* in the above chemical sequence is O_3, Zn/H_2O and NaOH (alc.)/I_2.

41. *(d)* Complete reaction is as follows

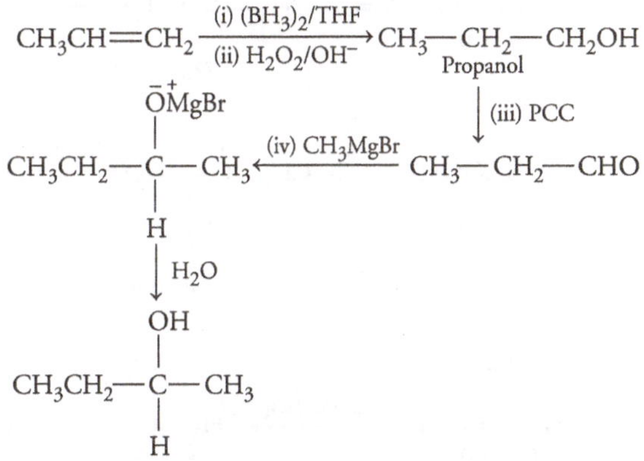

42. *(d)*

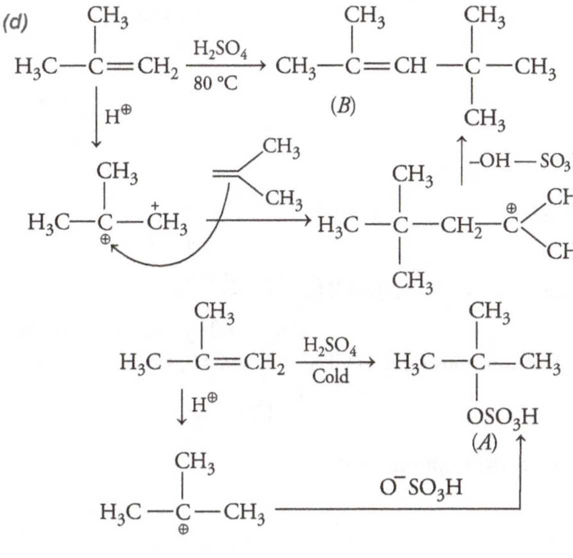

43. *(b)* The one giving maximum number of isomeric alkane on dehydrohalogenation will be

$$CH_3-\underset{\underset{Br}{|}}{CH}-CH_2-CH_2-CH_3 \xrightarrow{\text{Alc. KOH }-HBr}$$

2-bromopentane

$$CH_2{=}CH-CH_2CH_2CH_3 \text{ (Minor)}$$
$$+\,CH_3-CH{=}CH-CH_2CH_3 \text{ (Major)}$$

Concept Enhancer

Dehydrohalogenation is an elimination reaction, that removes hydrogen halide from a substrate. The reaction is usually associated with the synthesis of alkenes. Generally, halides having no C—H bond on an adjacent carbon are not suitable to undergo dehydrohalogenation.

44. *(d)*

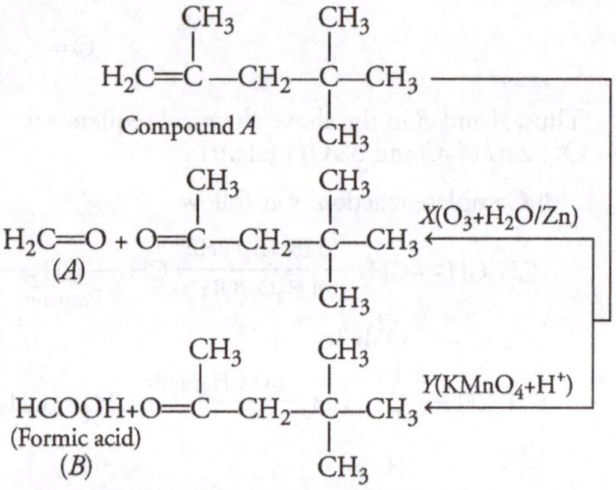

Oxidation of '*a*', i.e., HCHO produces formic acid which is produced by ants.

45. *(d)* $CH_3-\overset{\overset{CH_3}{|}}{C}{=}CH_2 \xrightarrow{\text{Acidic } KMnO_4} CH_3-\overset{\overset{CH_3}{|}}{C}{=}O + HCOOH$

(C_4H_8) Acetone (ketone group)

$\downarrow \Delta$

$H_2O + CO_2\uparrow$ Effervescence

$$CH_3-\overset{H}{\overset{|}{C}}{=}\overset{H}{\overset{|}{C}}-CH_3 \xrightarrow{\text{Acidic } KMnO_4} 2CH_3CHO \text{ Ethanal (Aldehyde)}$$

$$CH_3-CH_2-CH{=}CH_2 \xrightarrow{\text{Acidic } KMnO_4} CH_3-CH_2-CHO + HCOOH$$

The compound (*A*) is $CH_3-\underset{\underset{CH_3}{|}}{C}{=}CH_2$ (2-methylpropene)

So, correct option is (d).

46. *(d)*

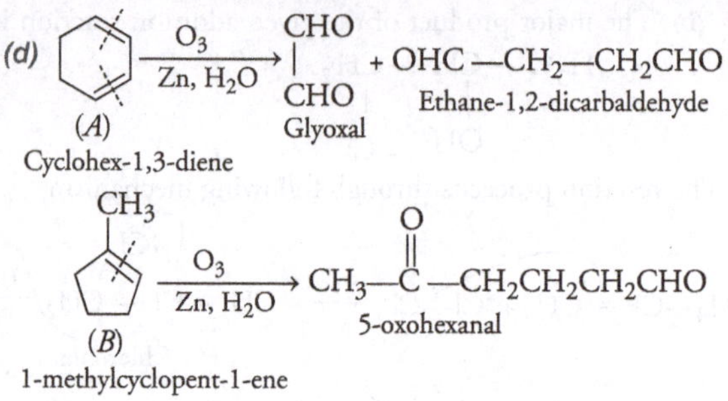

A is six membered ring and *B* is five membered ring.

47. *(a)* This is oxymercuration- demercuration reaction.

Step 1 Alkene reacts with mercury (II) acetate and forms cyclic mercurinium ion.

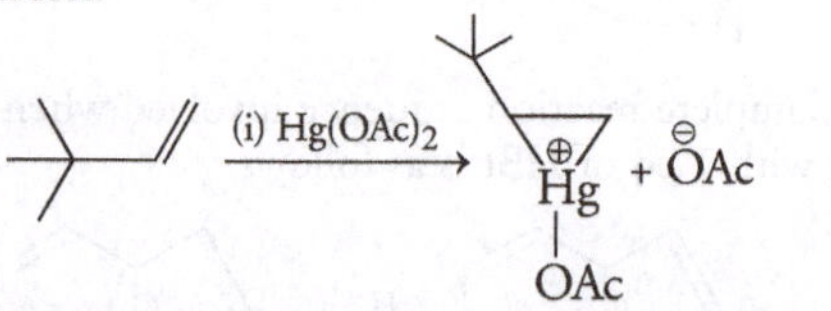

Step 2 Attack of nucleophile occurs and this mechanism follows Markownikoff's rule of regioselectivity.

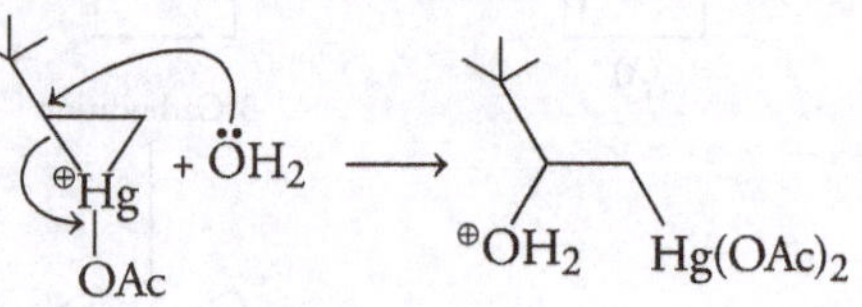

Step 3 Deprotonation takes place in presence of base acetate ion which gives alcohol.

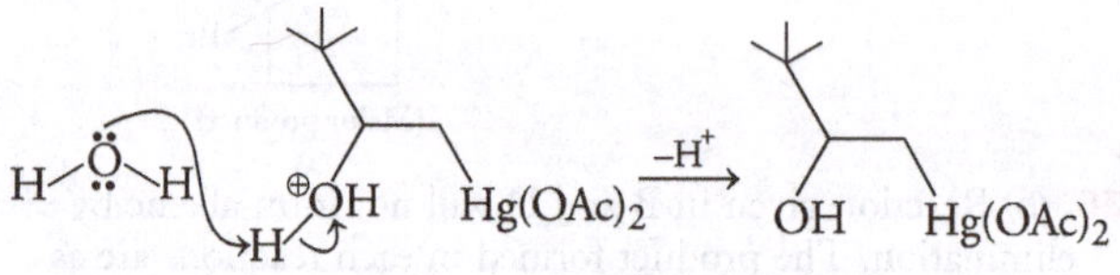

Step 4 Addition of $NaBH_4$ reduces the acetyl mercury by a hydrogen atom. $NaBH_4$ acts as a reducing agent.

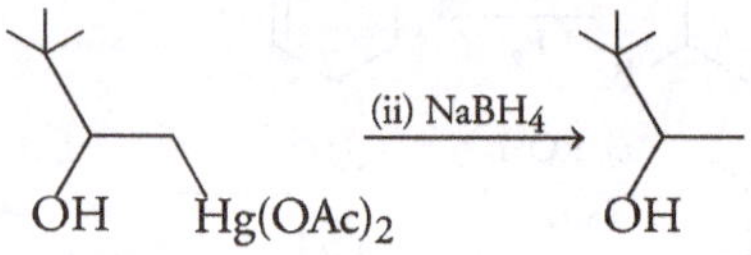

48. *(a)*

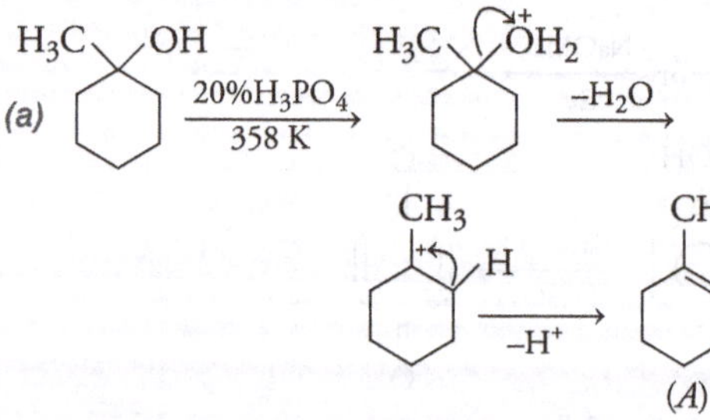

Alcohols gives E1 elimination with conc. H_2SO_4 and H_3PO_4 on heating. Thus, OH group is removed and alkene get formed.

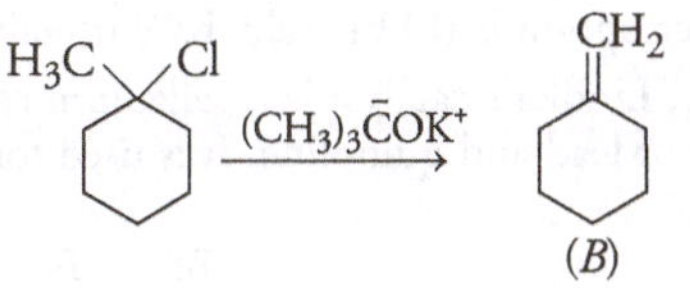

t-BuO$^-$ is a bulky base and will form Hoffmann alkene via elimination.

49. *(c)*

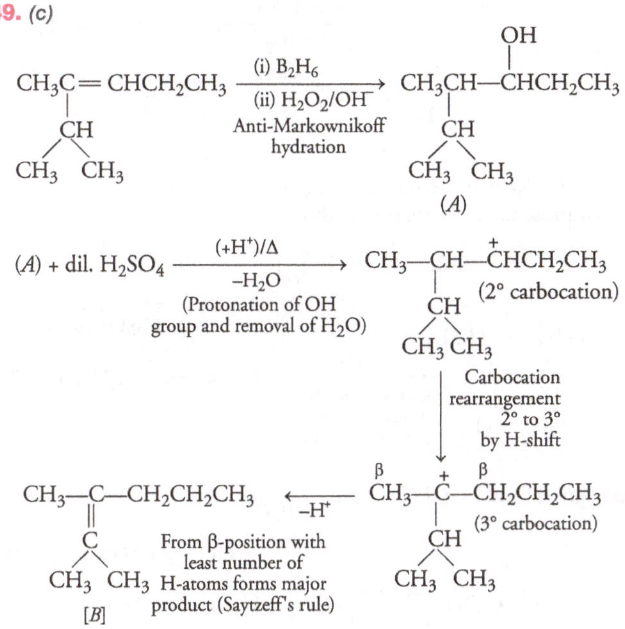

50. *(a)*

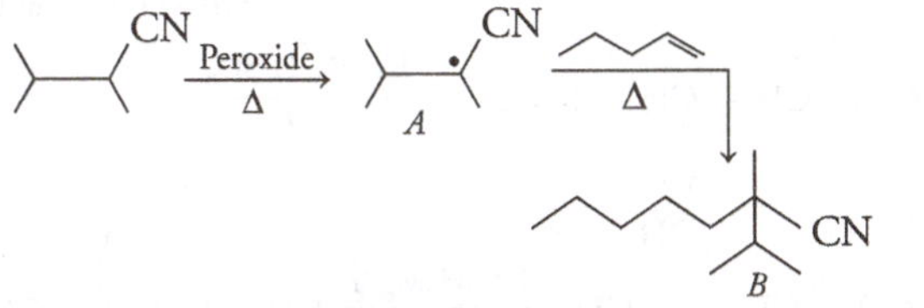

Reaction proceed *via* free radical formation.

51. *(3)* Mass of carbon $= \frac{80 \times 90}{100} = 72$ g

Number of C-atoms $= \frac{72}{12} = 6$

Mass of hydrogen $= \frac{80 \times 10}{100} = 8$ g

Number of H-atoms $= \frac{8}{1} = 8$

So, molecular formula C_6H_8.

Hence, degree of unsaturation $= 6 + 1 - \frac{8}{2} = 7 - 4 = 3$

52. *(27)*

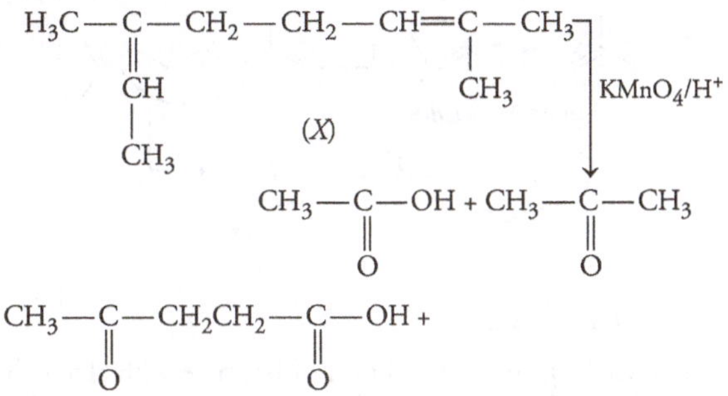

Thus, in the compound (X) total 27 σ-bonds are present.

53. *(4)* The complete reaction is as follow

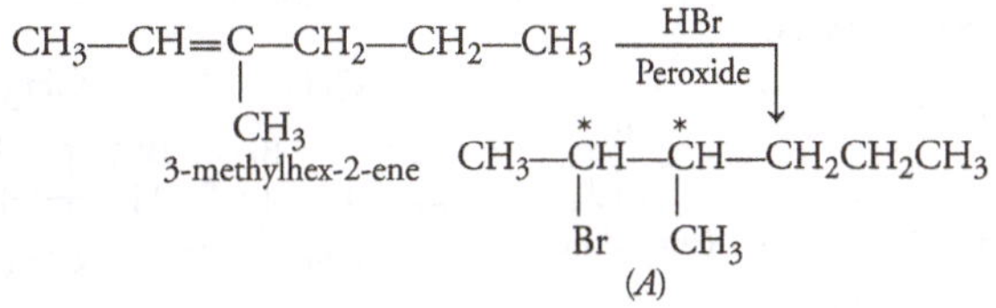

Number of stereocentre (n) = 2

Number of possible stereoisomers $= 2^n = 2^2 = 4$

54. *(4)* Moles of $X = \frac{10 \text{ mg}}{80} = 0.125$ mol

Moles consumed of $H_2 = \frac{8.4}{22.4} = 0.375$ m mol

$$\frac{n_{H_2}}{n_X} = \frac{0.375}{0.125} = 3$$

So, the compound X have 3 double bond.

The compound (X) is

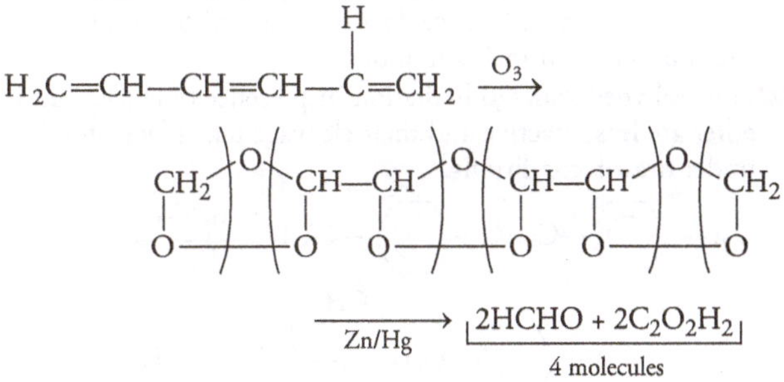

55. *(1)* The reaction is allylic substitution and the major product is

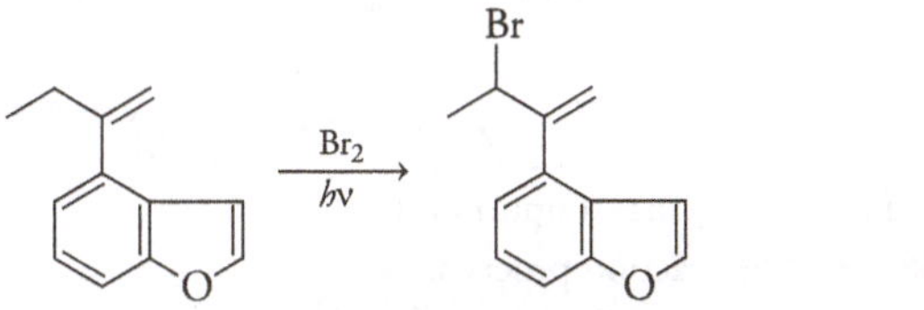

Number of Br atoms = 1

56. *(2)* The major product formed has two sp^2 hybridised carbon.

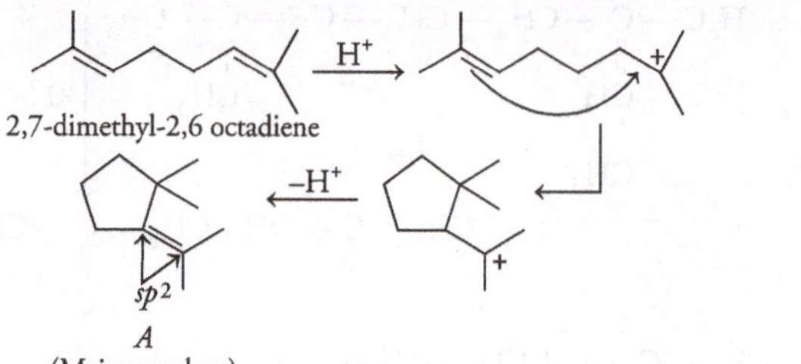

57. *(2)* Br_2/CCl_4 gives anti-addition of bromine with alkene but product formed in the reaction depends upon the stereo of alkene (*E* or *Z* alkene).
Here, *cis* alkene /Z-alkene is given, so we get racemic mixture of two isomers.

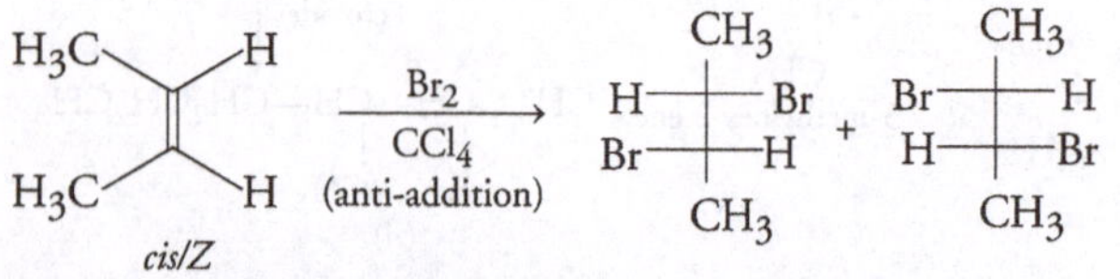

The total number of products possible = 2 [Enantiomers]

58. *(6)* There are 6 acyclic structural isomer for pentane

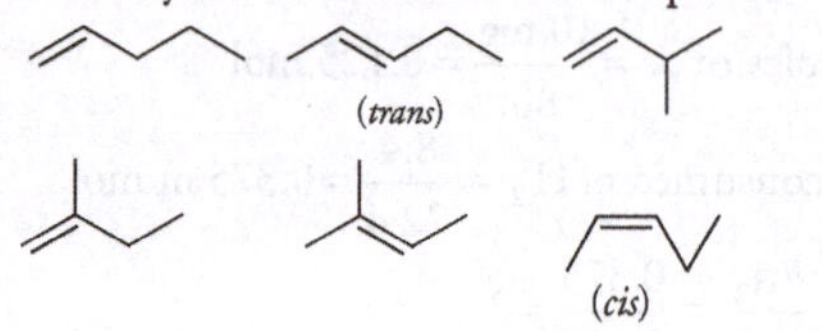

Topic 3

01. *(a)* But-2-yne ($CH_3-C\equiv C-CH_3$)does not have any acidic hydrogen and hence does not react with metallic sodium under normal conditions.

02. *(a)* Alkynes undergo hydration in presence of $HgSO_4$ and ether are least reactive and their cleavage take place only under drastic conditions.

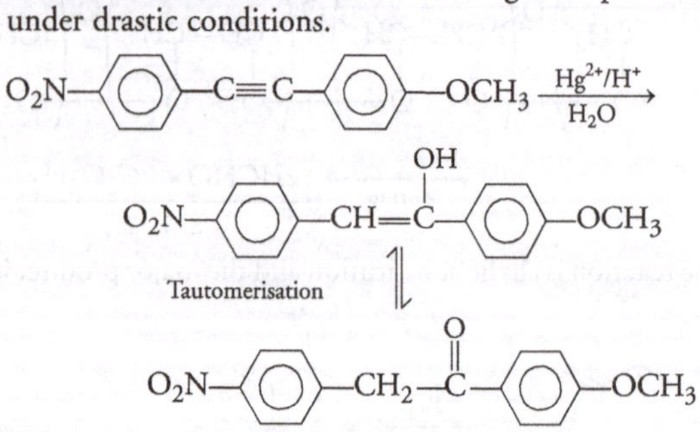

Hence, the correct option is (a).

03. *(d)* The reaction proceeds as

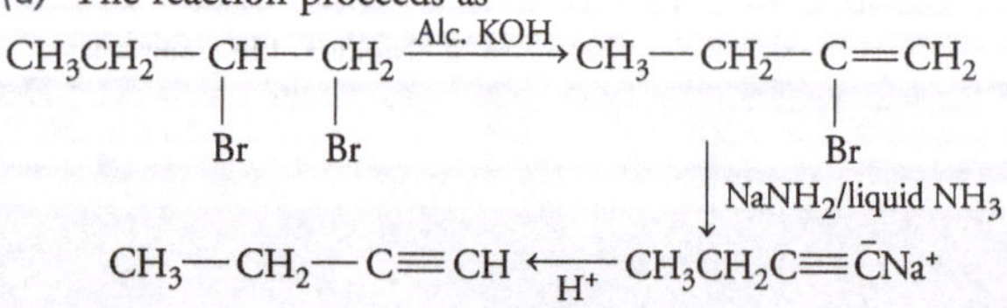

Thus, option (d) is the correct answer.

04. *(b)* Correct option is (b) Br_2, alc. KOH, $NaNH_2$, Lindlar catalyst H_2. Lindlar's catalyst is a palladium catalyst poisoned with traces of lead and quinoline. It is used for formation of *cis*-product.

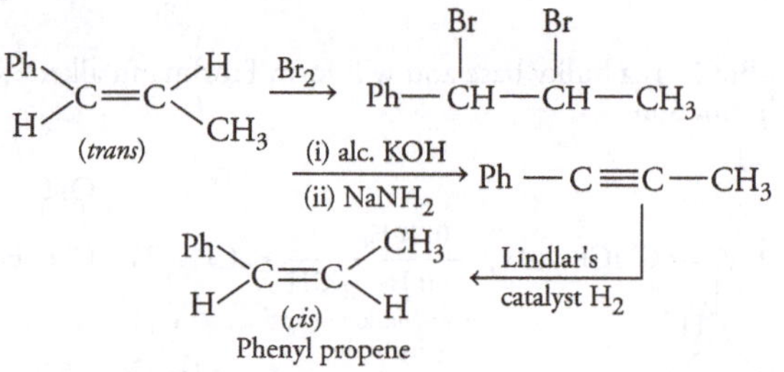

05. *(c)* Alkyne undergoes hydrogenation reaction to give alkene in presence of Lindlar's catalyst.

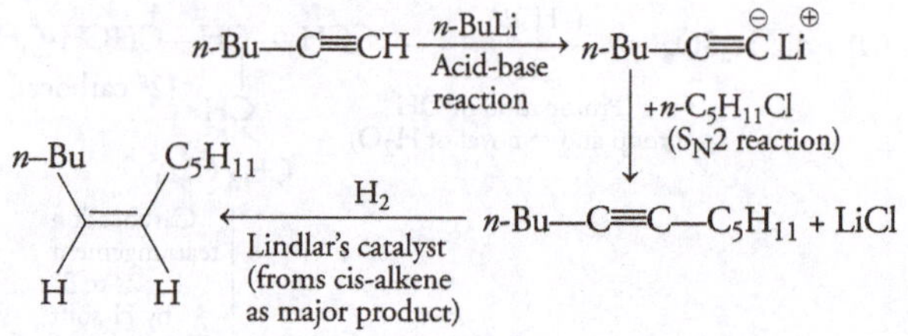

06. *(a)* $\underset{(X)}{HC\equiv C-CH_2-CH_3} \xrightarrow[H_2O]{O_3} CH_3CH_2COOH + \underset{(A)}{HCOOH}$

$$HCOOH \xrightarrow[{[Ag(NH_3)_2]^+}]{\text{Tollen's test}} CO_2 + H_2O + 2Ag\downarrow$$

(Bright silver mirror)

07. *(a)*

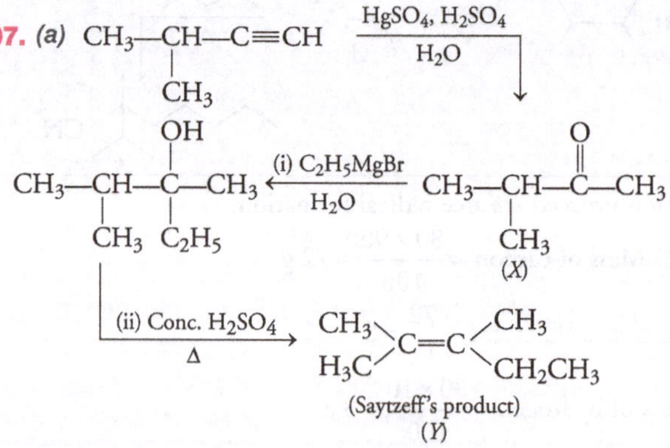

08. *(d)* The major product obtained in the given reaction is $CH_3C\,(I)\,(Cl)\,CHD_2$.
Reaction proceed as follows

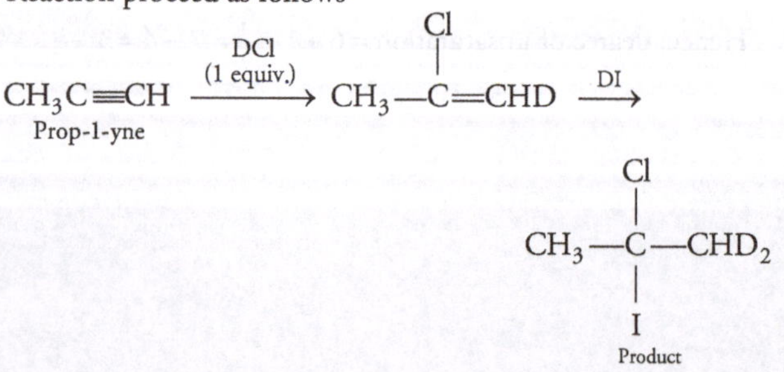

09. *(a)* A will be 2-pentyne i.e.,

$H_3C-C\equiv C-CH_2-CH_3$

as *cis*-2-pentene is formed on reacting with Pd/C.

The complete reaction is given as,

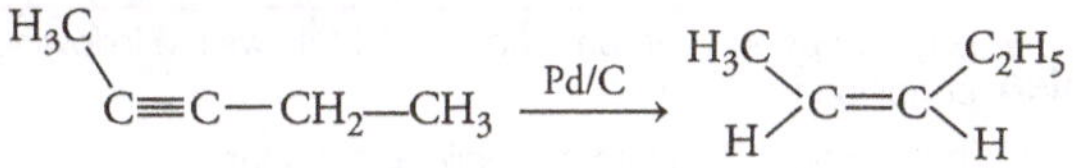

Also, B is *trans*-2-butene as Na/Liq. NH_3 is used as reducing agent.

The complete reaction is

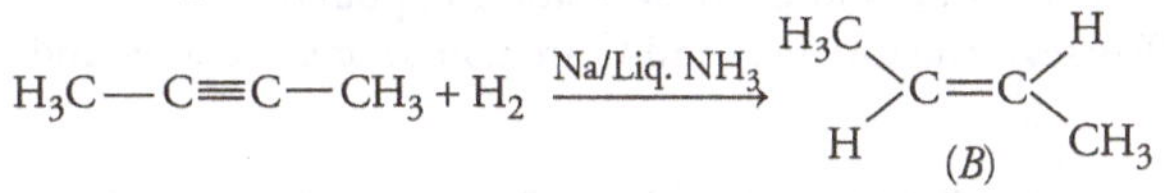

(*B*)
trans-2-butene

10. *(c)* The complete reaction is as follows,

$$CH_3-C\equiv CH \xrightarrow{Na} \underset{(A)}{CH_3-C\equiv C^-Na^+} \xrightarrow[(B)]{CH_3CH_2CH_2Br} CH_3-C\equiv C-CH_2CH_2CH_3 + NaBr$$

Therefore, $A = CH_3-C\equiv C^-Na^+$

$B = CH_3CH_2CH_2Br$

Hence, option (c) is correct.

11. *(c)*

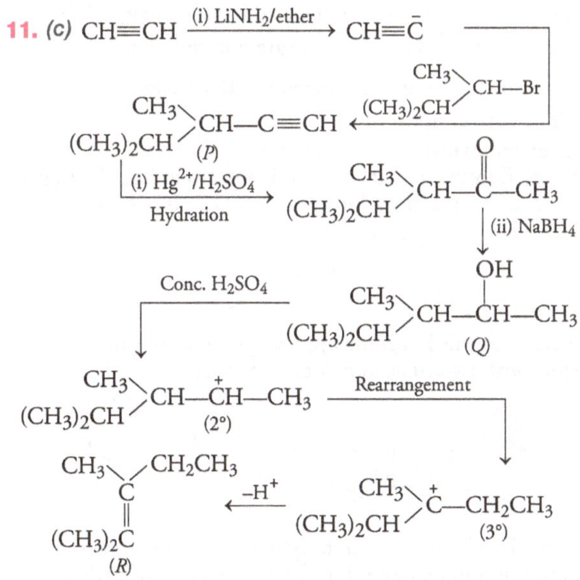

12. *(c)* According to the given conditions, the compound should be alkyne with triple bond present at the terminal. Secondly, appearance of turbidity in 5 minutes indicates the formation of a 2° alcohol as compound *C*. The chemical reactions involved are as follows

Step 1 $$\underset{\text{Prop-1-yne}}{CH_3-C\equiv CH} \xrightarrow{Ag_2O} \underset{\text{(White precipitates)}}{CH_3-C\equiv C-Ag}$$

Step 2

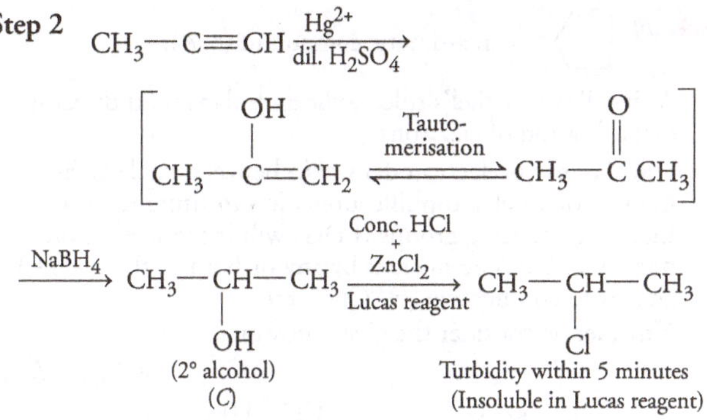

13. *(8)* The reaction involved is as follows.

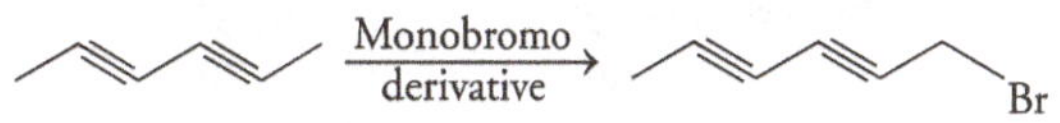

Number of $\pi e^- = 8$

14. *(2)* The complete reaction is as follows,

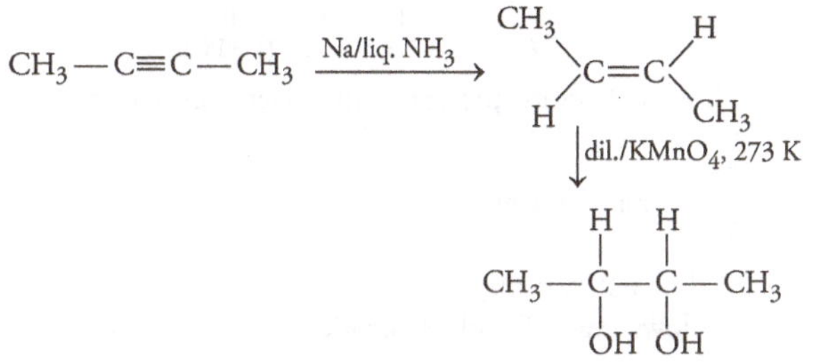

Final product (*P*)

Hence, the number of oxygen atom present in product *P* is 2.

Topic 4

01. *(b)* The benzenoid compounds have at least one benzene ring in the molecule whereas non-benzenoid compounds have no benzene rings in their structure. Compound (b) is non-benzenoid as it does not have benzene ring.

02. *(c)*

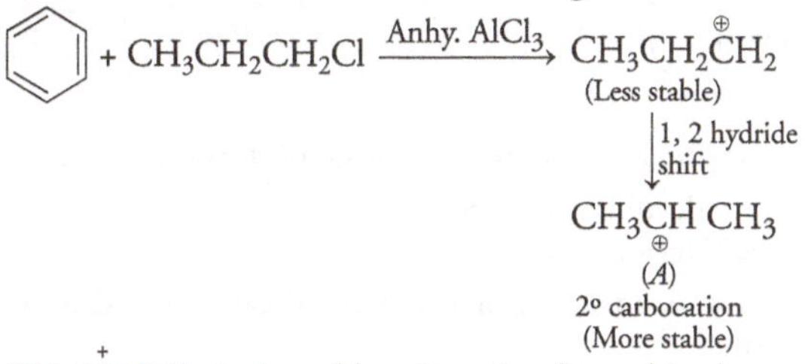

Hence, $CH_3\overset{+}{C}HCH_3$ is the stable carbocation formed in the given reaction.

03. *(d)* Benzene on nitration gives nitrobenzene in presence of HNO_3 and H_2SO_4.

$$\underset{\text{(Acid)}}{H_2SO_4} + \underset{\text{(Base)}}{HNO_3} \rightleftharpoons HSO_4^{\ominus} + H_2\overset{\oplus}{N}O_3$$

$$H_2\overset{\oplus}{N}O_3 \rightleftharpoons H_2O + \overset{\oplus}{N}O_2$$

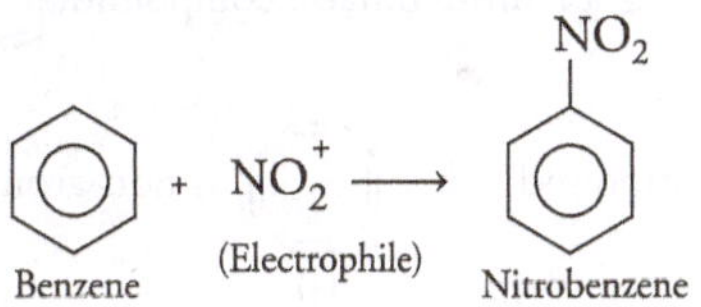

04. *(a)* 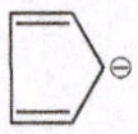 is an aromatic compound ($6\pi e^-$).

As it follows huckel's rule, cyclic and planar and threat is delocalisation of electrons.

05. *(a)* Higher is electron density in benzene nucleus faster will the rate of an electrophilic aromatic substitute reaction. Electron-donating groups (EDG) will increase electron density in benzene nucleus by any or both of their $+R/+M$ and hyperconjugative (HPC) effects.
Now, let us consider the given substrate.

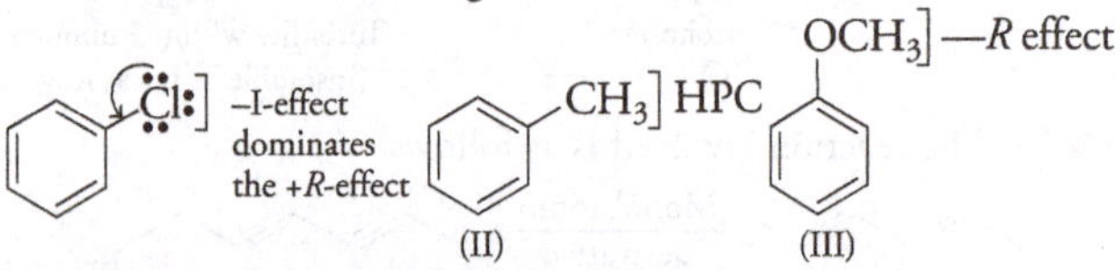

As $-R$ effect is more powerful than $-I$ effect.
Thus, the order of their reactivity of given compounds towards an electrophilic aromatic substitution will be

$$\underset{(-R)}{\text{III}} < \underset{(-I>+R)}{\text{I}} < \underset{(+\text{HPC})}{\text{II}}$$

06. *(a)* The π electrons present in the given compounds are as follows

(A) 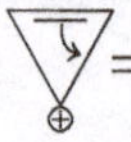 $\Rightarrow 2\pi\text{-}e^-$ system

$[(4n+2)\pi, n=0]$

$\Rightarrow$ Follows Huckel's rule Aromatic

(B)  $\Rightarrow 4\pi\text{-}e^-$ system

$[4n\pi, n=1] \Rightarrow$ Anti-aromatic

(C) 

$\Rightarrow 8\pi\text{-}e^-$ system $[4n\pi, n=2]$ $\Rightarrow$ Anti-aromatic

(D) 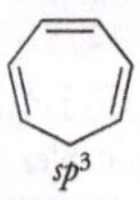

$\Rightarrow$The sp^3-carbon does not support conjugation and also devoids planarity.
$\Rightarrow$ Non-aromatic

07. *(a)* Aromaticity of a compound can be decided by Huckel's rule.

Hence,  is anti-aromatic. It does not follow Huckel's rule as it has conjugated 4π-electron ($4n\,\pi, n=1$) system. Rest of the species are aromatic as each of them belongs to 6π-electrons [$(4n+2)\,\pi, n=1$] system and thus follow Huckel's rule.

08. *(c)* 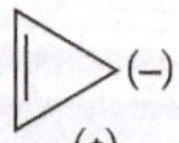 (*p*) is an anti-aromatic compound, 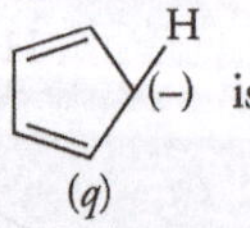 (*q*) is an aromatic compound, while 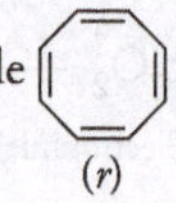 (*r*) is non-aromatic compound. As we know more is the aromatic character higher is the stability. Thus, the correct order of stability is $q>r>p$.

09. *(b)*

> **idea** Any compound can be aromatic if it follows the following conditions
> - The compound should be cyclic and planar.
> - The compound should follow Huckle rule i.e., have $(4n+2)\pi$ electrons.

Structure in B and D are aromatic compound.
Whereas structure in A and C are non-aromatic compound.

10. *(c)*  is most stable due to aromatic character. It has $2\pi e^-$ and follow $(4n+2)\ \pi e^-$ (Huckel's rule).

11. *(d)* 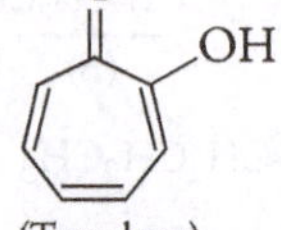

(Tropolone)

- Non-benzoid aromatic compound.
- Follows $(4n+2)\pi$ rule.
- Only $6\pi\, e^-$ of the ring are involved in aromaticity while π-electrons of $>C=O$ group are not involved.

Thus, both statement I and statement II are false.

12. *(a)* The correct option is (a).
Mechanism involve in the reaction of cycloalkene with acid hydrogen fluoride so that a proton is abstracted from hydrogen fluoride to form carbocation of cyclohexene is as follows

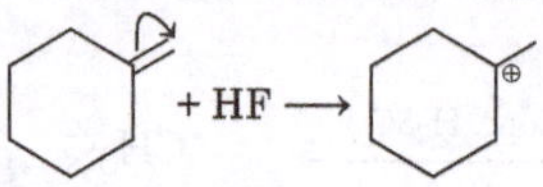

The benzene attacks on the cyclohexyl cation to form the product with two rings joined to each other.

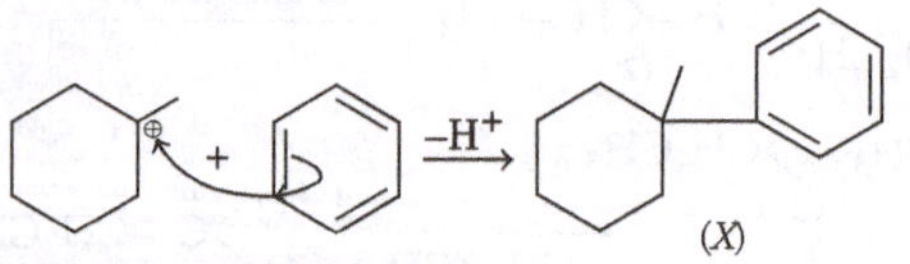

The positive charge on cyclohexyl ring is stabilised by the delocalisation in benzene ring. The abstraction of hydrogen from benzene ring gives final product as shown above.

13. *(d)* 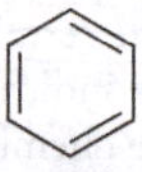 [6] annulene (aromatic)

 (Non aromatic)

[8] annulene

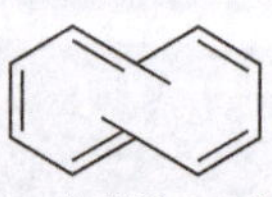 (Non aromatic)

cis-[10] annulene

In cis [10] annulene → the H-atoms at the 1 and 6 position interfere with each other and force the molecule out of planarity. Thus, it is ion aromatic planarity is one of the requirements of aromatic system.

14. *(c)*

> **idea** In this question, we use concept of aromaticity. There are three requirements for a compound to be aromatic which are as follows
> (i) It should be cyclic and planar.
> (ii) Follows's Huckel's rule i.e., $(4n + 2)\ \pi$ electrons.

Among the given options only (c) is non-aromatic while other are aromatic. This is because it doesn't follows Huckel's rule and is also non-planar.

Non - planar
10 πe^-
(Non - aromatic)

15. *(a)* The given reaction is Birch reduction.

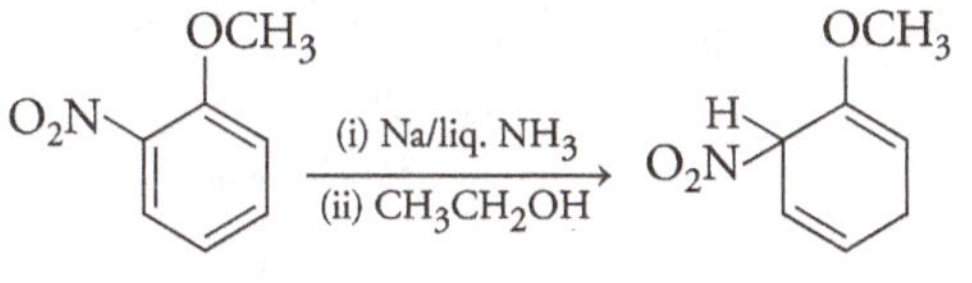

16. *(a,d)* Aromatic compounds are conjugated planar ring system having delocalised π-electron clouds and follows $(4n + 2)\pi$-electron rule, i.e. Huckel's rule. $\{n = 0, 1, 2, \ldots\}$

(a) 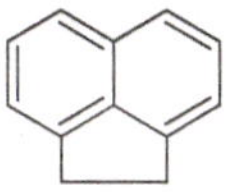

⟶ $10\pi e^-$ in cyclic conjugation place.
It is aromatic.

(d)  $6\pi e^-$ in cyclic conjugation.
It is aromatic.

17. *(b)* Benzene on reaction with Br_2/Fe gives bromo benzene as follows

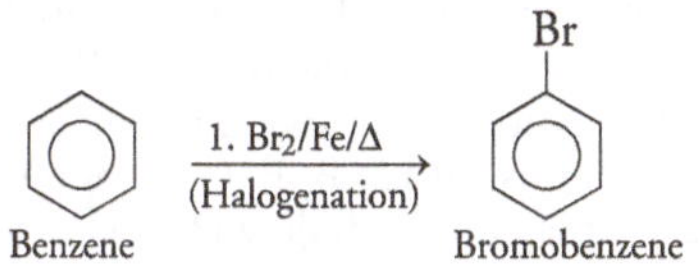

Bromobenzene on reaction with Mg / dry ether followed by CH_3OH gives benzene as follows

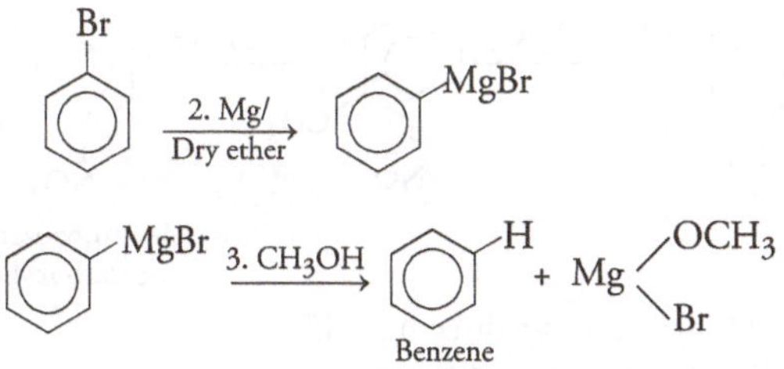

18. *(c)* The compound which follows Huckel's rule, has planar structure and has delocalisation of πe^-s is aromatic in nature.

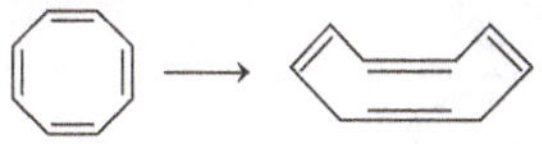

It exist in a non-planar tube shape and does not follows Huckle rule, so it is non-aromatic.

19. *(a)* Density is directly proportional to mass of the halogen. On moving down the group, mass increases.
Order of decreasing density in benzene ring by halogen is I > Br > Cl > F.
Thus, density order of the given compounds is

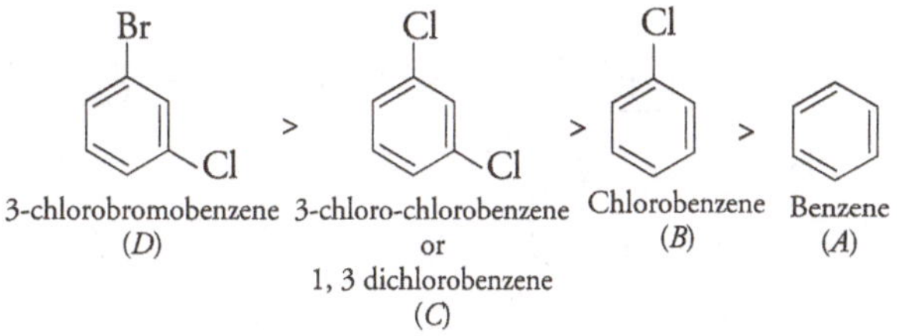

20. *(b)* Acidic strength of an acid ∝ stability of conjugate base.

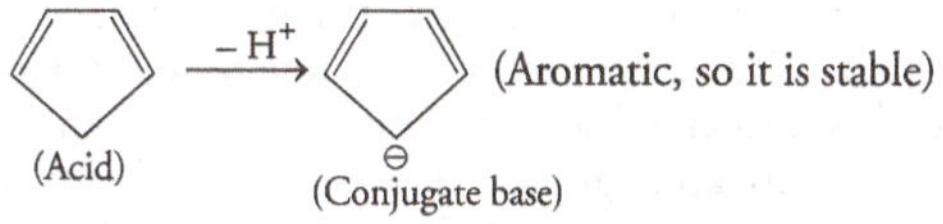

The negative charge on ring makes stable conjugate base and thus, compound (b) is the strongest acid.

21. *(b)*

(B) 

(Planar) $6\pi e^-$
(Huckel's rule followed, i.e. aromatic)

(C) 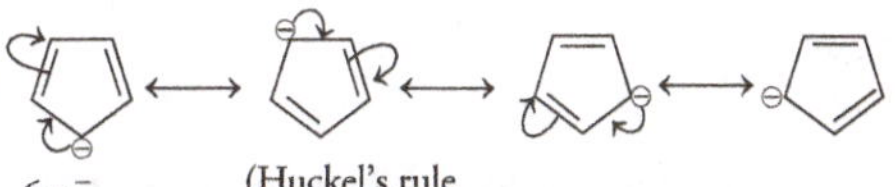

$6\pi e^-$
planar
(Huckel's rule followed, i.e. aromatic)

22. *(b)* The reaction take place as follows

Step I 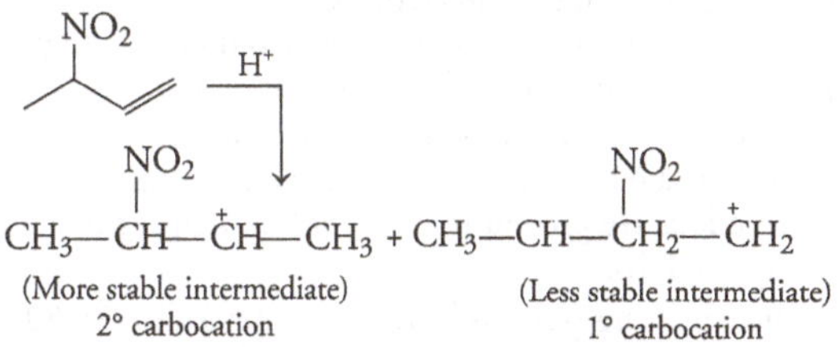

Step II 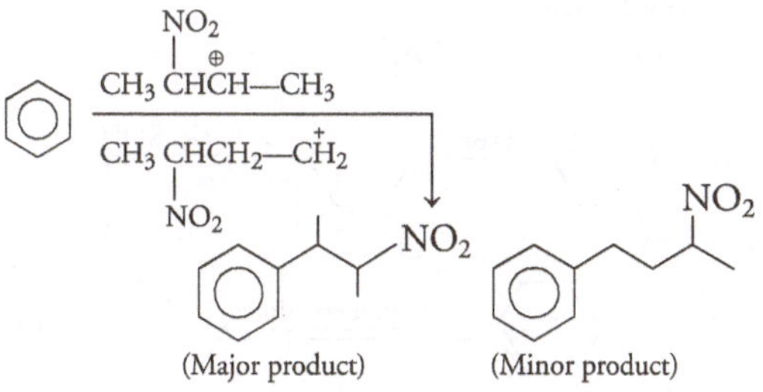

23. *(a)* Compound which contain one or more sp^3 carbon in its structure will be non-aromatic compound.

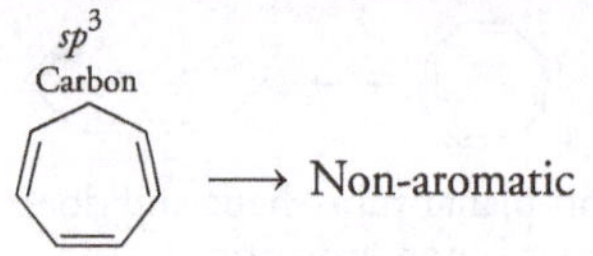

Only (a) is non-aromatic and rest all are aromatic.

24. *(d)*

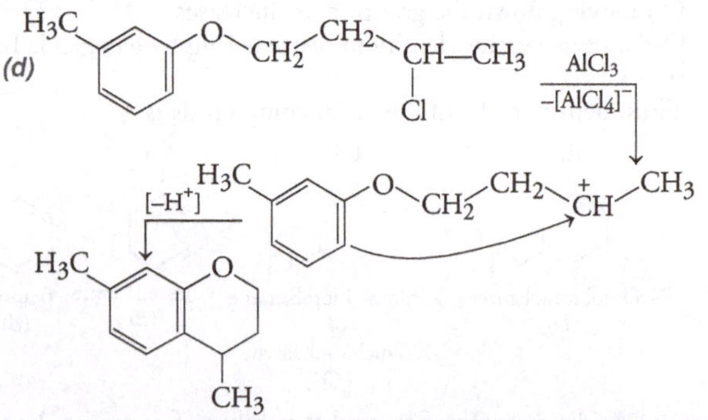

25. *(c)* Electron withdrawing groups (EWGs) reduces the electron density in the benzene ring due to its strong $-I$ effect and makes further substitution difficult. Hence, they are called as deactivating groups (like —CN, —Cl etc). While electron releasing groups (ERGs) increases the electron density in the benzene due to resonance. Therefore, activates the benzene ring for the attack by an electrophile. Hence, they are called as activating groups (like —Me, —OCH_3, etc.)
So, the correct order is

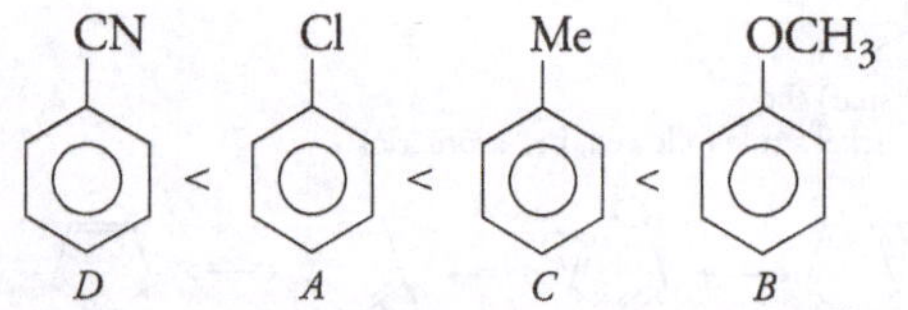

26. *(b)* Melting point of a compound depends on size and force of attraction between the molecules. Compound (c) has largest size and also possess weak intermolecular association due to dipole-dipole interactions.
Compound (a) is a dicarboxylic acid and possess high melting point due to intermolecular hydrogen bonding.
As a result, it exist as associated molecule. Compound (d) is an alcohol and possess intermolecular H-bonding. No interactions are present in hydrocarbon (naphthalene) compound (b).
Hence, melting point is lowest for naphthalene (≈80°C), i.e. compound (b).

27. *(a)*

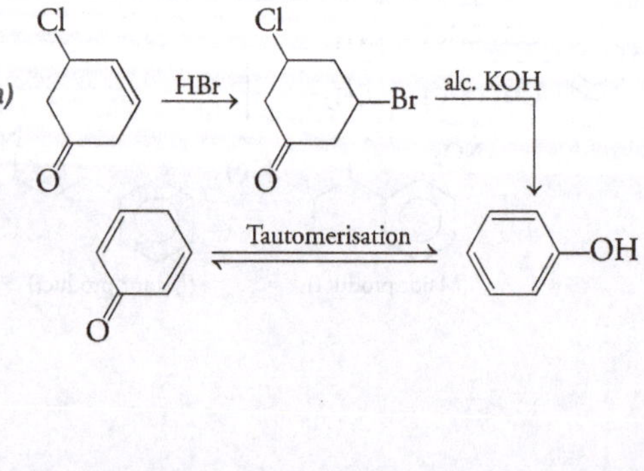

28. *(d)* The reaction follows α, β-elimination mechanism to give a more substituted stable alkene as a major product. As the substrate is a α, γ -dibromo (1, 3-) compound it gives a conjugated diene.

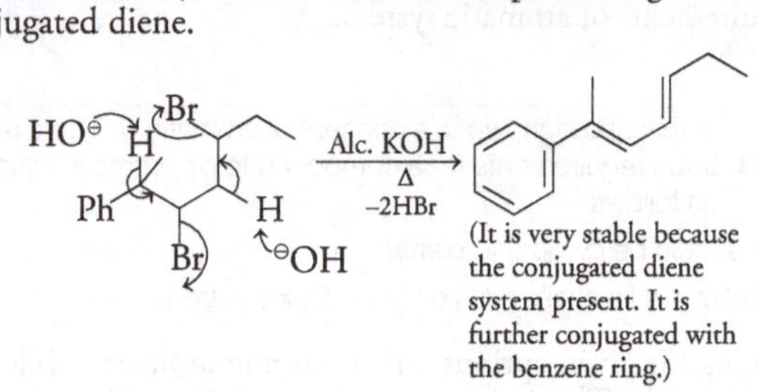

29. *(d)*

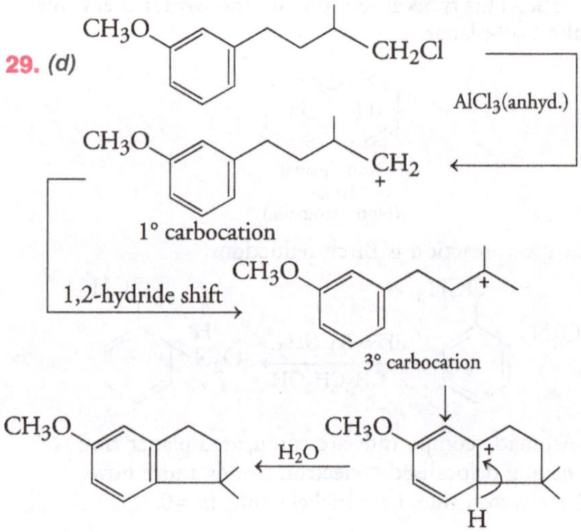

30. *(d)*

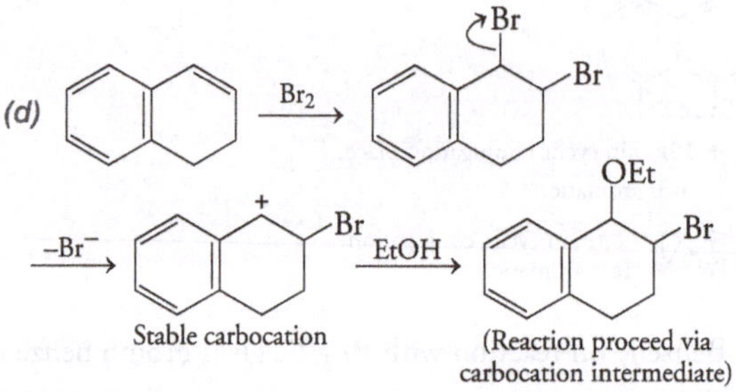

31. *(a)* According to Huckel's rule, planar, cyclic fully conjugated molecules is aromatic if it contains $(4n+2)\pi$ electrons. A, C, D, E and H follows Huckel's rule hence they are aromatic compounds. While B, F, G are not aromatic compounds. As they do not follow Huckel's rule.

32. *(d)* Both Statement I and II are true. Reaction for statement I is

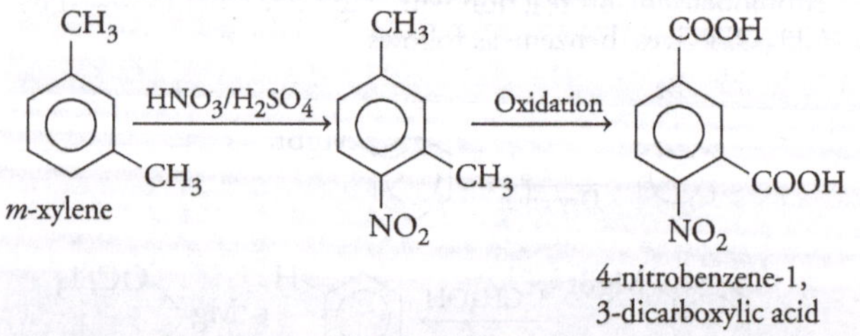

— CH_3 group is *o*/*p* directing while
—NO_2 group is *meta* directing.

33. *(b)* Among the given statements, A, C and D are correct. While B and E are incorrect.
Their correct forms are
- —CN is *meta* directing but —OH is *ortho*/*para* directing.
- Halides are deactivating group.

34. *(b)* The correct match is
A → (iii), B → (iv), C → (ii), D → (i)

35. *(b)* Benzene on reaction with $CH_3COCl / AlCl_3$ gives acetophenone which on reduction with Zn-Hg/HCl gives ethyl benzene. Bromination of ethylbenzene give 4-bromo-ethylbenzene which upon nitration gives 4-bromo-2-nitroethylbenzene.

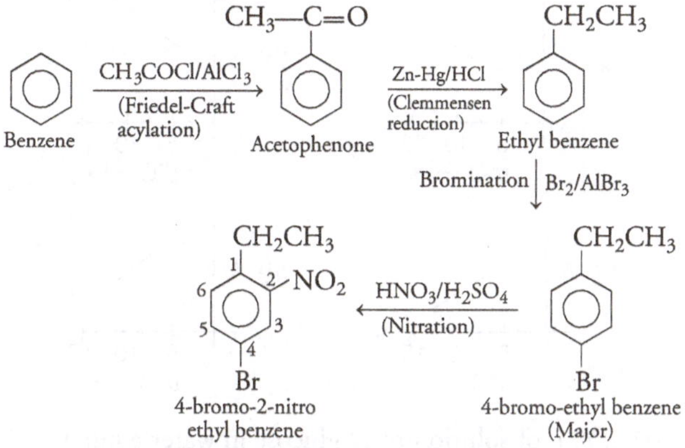

[–*I*-effect of Br directs NO_2^- at its *meta*-position]
So, option (b) is the correct sequence of reagents used.

36. *(1)* There is only one aromatic compound among the given compounds.

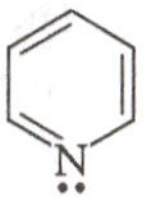

37. *(2)* Total number of deactivating groups in aromatic electrophilic substitution reaction is 2.

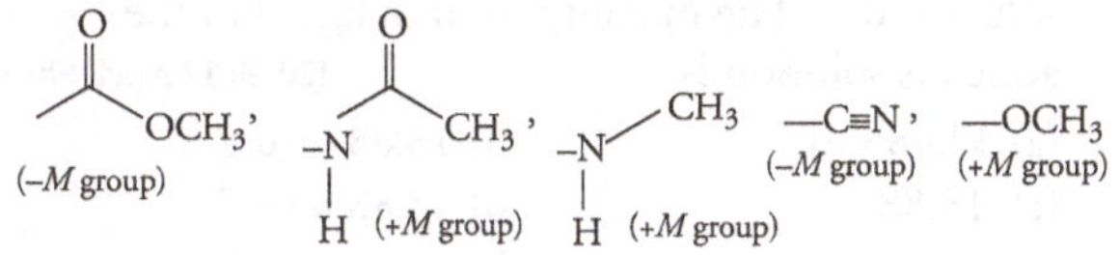

38. *(3)* Aromatic compounds are chemical compounds that consist of conjugated planar ring systems accompanied by delocalised π-electron clouds in place of individual alternating double and single bonds.
These compounds should follows huckel rule i.e., $(4n+2)\,\pi e^-$ rule.
Thus among the given compounds. B, C, D are aromatic compounds.
Hence, total number of aromatic compound is 3.

39. *(4)* It is given that the given compound (*A*) gives neutral $FeCl_3$ test that means it is a phenol secondly, it decolourises the alkaline $KMnO_4$ that means an unsaturated carbon is present (may be double bond or triple bond). By using these information there are two following possibility 1.

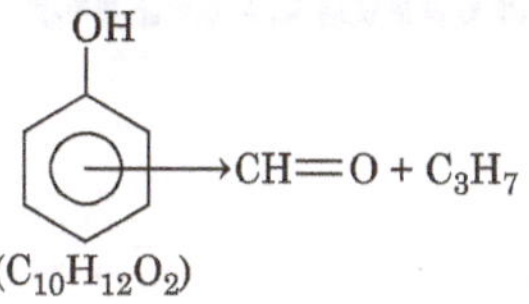

$(C_{10}H_{12}O_2)$
(*A*)

Possibility II (More possible case as C—C double or triple bond decolourises the alkaline $KMnO_4$)

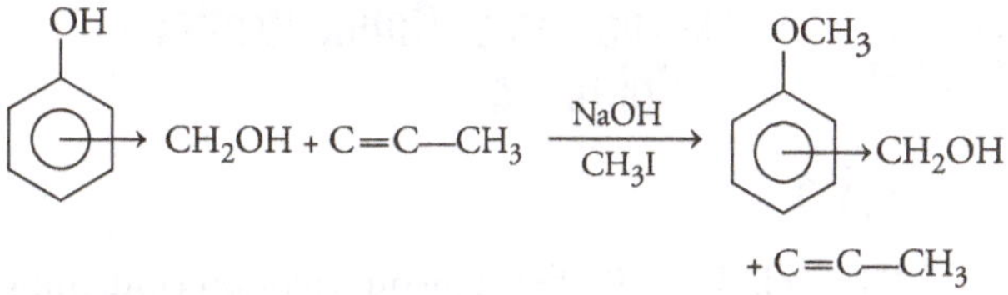

However, from both the possibility the answer will be same i.e., 4π bond.

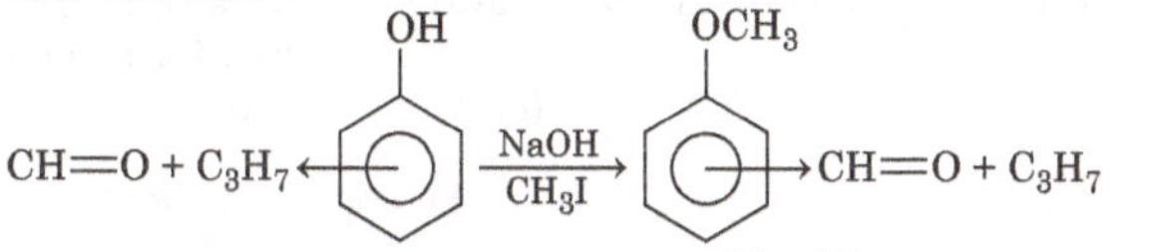

Degree of unsaturation (DU) of $A = \frac{22-12}{2} = 5$
1 DU is due to benzene ring. 4π-bonds will be there (3π-bonds in ring and 1π-bond outside ring) as it decolourises alkaline $KMnO_4$.

40. *(6)*

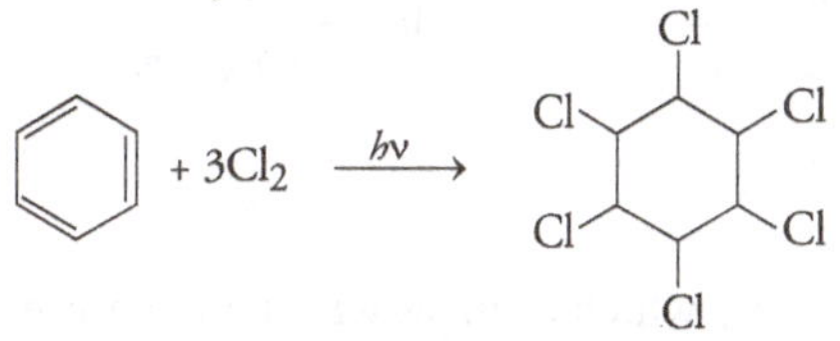

(B.H.C.)
(*X*)

Product '*X*' contains 6 hydrogen atom at all the six position of hexane ring.

41. *(7)* The reaction take place as follows

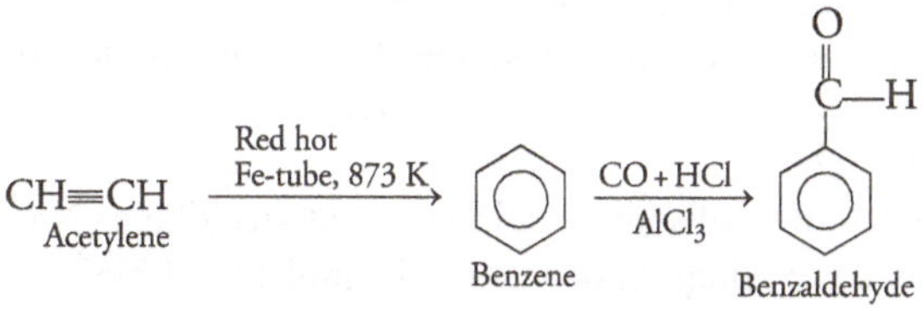

Every carbon atom found in benzaldehyde (6 carbon in benzene ring +1 in aldehyde group), forms three bonds with neighbouring atoms-(one double bond + two single bond). Therefore, in benzaldehyde total number of sp^2 'C' are 7.

13 Solutions

In JEE Main there is usually 0-1 question per paper are visible from this chapter. In past 3 years, 32 conceptual and 39 numerical questions were asked of moderate difficulty level from this chapter.

For detailed analysis scan QR code

Topic 1 / Expression of Concentrations of Solutions

• Level 1

01 The molarity (M) of an aqueous solution containing 5.85 g of NaCl in 500 mL water is
(Given, Molar mass Na: 23 and Cl: 35.5 gmol^{-1}) **[2024, 4 April Shift-I]**

(a) 4 (b) 0.2 (c) 2 (d) 20

Imp. *Relation of Mole Fraction*

02 If a substance 'A' dissolves in solution of a mixture of 'B' and 'C' with their respective number of moles as n_A, n_B and n_C. Mole fraction of C in the solution is **[2024, 30 Jan Shift-I]**

(a) $\dfrac{n_C}{n_A \times n_B \times n_C}$ (b) $\dfrac{n_C}{n_A - n_B - n_C}$

(c) $\dfrac{n_C}{n_A + n_B + n_C}$ (d) $\dfrac{n_B}{n_A + n_B}$

03 The quantity which changes with temperature is **[2024, 27 Jan Shift-II]**

(a) molality (b) mole fraction
(c) mass percentage (d) molarity

04 What would be the molality of 20% (mass/mass) aqueous solution of KI?
(Molar mass of KI = 166 g mol^{-1}) **[2019, 9 April Shift-II]**

(a) 1.48 (b) 1.51
(c) 1.35 (d) 1.08

05 A solution of sodium sulphate contains 92 g of Na^+ ions per kilogram of water. The molality of Na^+ ions in that solution in mol kg^{-1} is **[2019, 9 Jan Shift-I]**

(a) 16 (b) 4 (c) 132 (d) 8

• Level 2

06 'x' g of NaCl is added to water in a beaker with a lid. The temperature of the system is raised from 1°C to 25°C. Which out of the following plots, is best suited for the change in the molarity (M) of the solution with respect to temperature?
[Consider the solubility of NaCl remains unchanged over the temperature range] **[2025, 2 April Shift-II]**

(a)
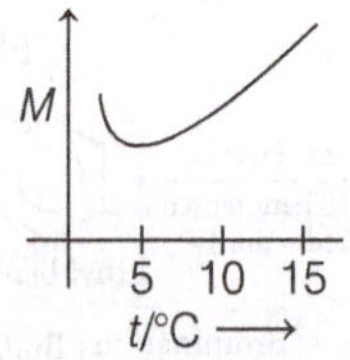

(b)
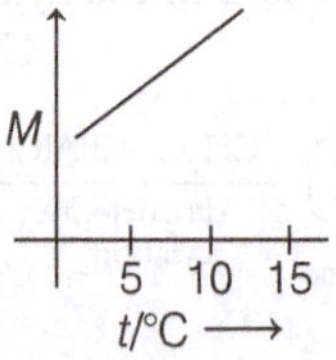

(c)
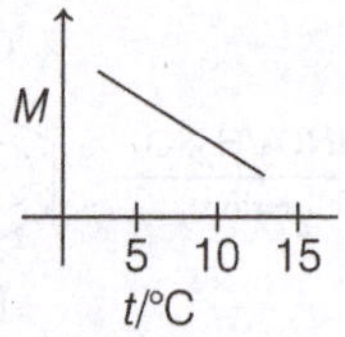

(d)
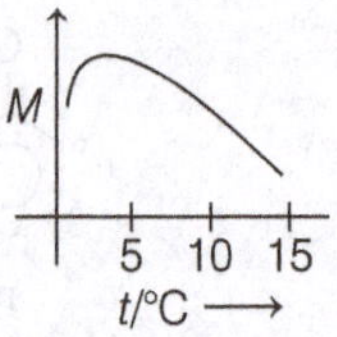

07 250 g of solution of D-glucose in water contains 10.8% of carbon by weight. The molality of the solution is nearest to (Given : Atomic weights are H : 1u, C : 12 u, O : 16 u) **[2022, 27 July Shift-I]**

(a) 1.03 (b) 2.06
(c) 3.09 (d) 5.40

08 The mole fraction of a solvent in aqueous solution of a solute is 0.8. The molality (in mol kg^{-1}) of the aqueous solution is **[2019, 12 April Shift-I]**

(a) 13.88×10^{-2} (b) 13.88×10^{-1}
(c) 13.88 (d) 13.88×10^{-3}

• Level 3

09 Concentrated nitric acid is labelled as 75% by mass. The volume in mL of the solution which contains 30g of nitric acid is......... . **[2025, 28 Jan, Shift-II]**
(Given, density of nitric acid solution is 1.25 g/mL.)

(a) 32 (b) 55
(c) 45 (d) 40

10 A commercially sold conc. HCl is 35% HCl by mass. If the density of this commercial acid is 1.46 g/mL, the molarity of this solution is
(Atomic mass : Cl = 35.5 amu, H = 1 amu) **[2022, 26 June, Shift-I]**

(a) 10.2 M (b) 12.5 M
(c) 14.0 M (d) 18.2 M

11 8 g of NaOH is dissolved in 18 g of H_2O. Mole fraction of NaOH in solution and molality (in mol kg^{-1}) of the solution respectively are **[2019, 12 Jan Shift-II]**

(a) 0.2, 11.11 (b) 0.167, 22.20
(c) 0.2, 22.20 (d) 0.167, 11.11

Numerical Value Type Questions

12 Sea water, which can be considered as a 6 molar (6 M) solution of NaCl, has a density of 2 g mL^{-1}. The concentration of dissolved oxygen (O_2) in sea water is 5.8 ppm. Then the concentration of dissolved oxygen (O_2) in sea water is $x \times 10^{-4}$ m.

$x =$ .... (Nearest integer)

Given: Molar mass of NaCl is 58.5 g mol^{-1}

Molar mass of O_2 is 32 g mol^{-1} **[2025, 4 April shift-II]**

13 20 mL of sodium iodide solution gave 4.74 g silver iodide when treated with excess of silver nitrate solution. The molarity of the sodium iodide solution is .... M. (Nearest Integer value)

Given Na = 23, I = 127, Ag = 108, N = 14, O = 16 g mol^{-1}) **[2025, 8 April Shift-II]**

14 The molarity of a 70% (mass/mass) aqueous solution of a monobasic acid (X) is ...... $\times 10^{-1}$ M (Nearest integer) **[2025, 28 Jan, Shift-I]**

[Given, density of aqueous solution of (X) is 1.25 g mL^{-1}

Molar mass of the acid is 70 g mol^{-1}]

15 Molarity (M) of an aqueous solution containing x g of anhyd. $CuSO_4$ in 500 mL solution at 32°C is 2×10^{-1} M. Its molality will be .... $\times 10^{-3}$ m. (Nearest integer).

[Given density of the solution = 1.25 g/mL.] **[2024, 9 April Shift-I]**

16 Molality of an aqueous solution of urea is 4.44 m. Mole fraction of urea in solution is $x \times 10^{-3}$. Value of x is .... (Integer answer) **[2024, 8 April Shift-II]**

17 Molality of 0.8 M H_2SO_4 solution (density 1.06 g cm^{-3}) is .... $\times 10^{-3}$ m. **[2024, 29 Jan Shift-II]**

18 A solution of H_2SO_4 is 31.4% H_2SO_4 by mass and has a density of 1.25 g/mL. The molarity of the H_2SO_4 solution is .... M (nearest integer).

[Given, molar mass of H_2SO_4 = 98 g mol^{-1}] **[2024, 29 Jan Shift-I]**

19 Some amount of dichloromethane (CH_2Cl_2) is added to 671.141 mL of chloroform ($CHCl_3$) to prepare 2.6×10^{-3} M solution of CH_2Cl_2 (DCM). The concentration of DCM is .... ppm (by mass).

Given, atomic mass C = 12, H = 1, Cl = 35.5, density of $CHCl_3$ = 1.49 g cm^{-3} **[2023, 30 Jan Shift-I]**

20 The number of units, which are used to express concentration of solutions from the following is .... mass per cent, mole, mole-fraction, molarity, ppm, molality **[2023, 24 Jan Shift-II]**

21 When 800 mL of 0.5 M nitric acid is heated in a beaker, its volume is reduced to half and 11.5 g of nitric acid is evaporated. The molarity of the remaining nitric acid solution is $x \times 10^{-2}$ M. (Nearest integer) **[2022, 26 July Shift-I]**

22 If 80 g of copper sulphate $CuSO_4 \cdot 5H_2O$. is dissolved in deionised water to make 5 L of solution. The concentration of the copper sulphate solution is $x \times 10^{-3}$ mol L^{-1}. The value of x is ....

[Atomic masses Cu = 63.54 u, S = 32 u, O = 16 u, H = 1 u] **[2021, 1 Sep Shift-II]**

23 An aqueous KCl solution of density 1.20 g mL^{-1} has a molality of 3.30 mol kg^{-1}. The molarity of the solution in mol L^{-1} is ....

[Molar mass of KCl = 74.5] **[2021, 26 Aug Shift-I]**

24 The density of NaOH solution is 1.2 g cm^{-3}. The molality of this solution is .... m. (Round off to the nearest integer)

[Use : Atomic masses : Na = 23.0 u, O = 16.0 u, H = 1.0 u. Density of H_2O = 1.0 g cm^{-3}] **[2021, 27 July Shift-I]**

25 The mole fraction of a solute in a 100 molal aqueous solution is .... $\times 10^{-2}$.

(Round off to the Nearest Integer)

[Given : Atomic masses : H : 1.0 u, O : 16.0 u] **[2021, 17 March Shift-I]**

26 A 6.50 molal solution of KOH (aq) has a density of 1.89 g cm^{-3}. The molarity of the solution is ...... mol dm^{-3} (Round off to the nearest integer).

[Atomic masses: K : 39.0 u, O : 16.0u, H : 1.0 u] **[2021, 16 March Shift-I]**

27 The hardness of a water sample containing 10^{-3} M $MgSO_4$ expressed as $CaCO_3$ equivalents (in ppm) is...... (molar mass of $MgSO_4$ is 120.37 g/mol) **[2020, 9 Jan Shift-II]**

28 10.30 mg of O_2 is dissolved into a litre of sea water of density 1.03 g/mL. The concentration of O_2 in ppm is **[2020, 9 Jan Shift-II]**

Topic 2 / Vapour Pressure, Henry's Law & Raoult's Law

• Level 1

01 A solution is made by mixing one mole of volatile liquid A with 3 moles of volatile liquid B. The vapour pressure of pure A is 200 mm Hg and that of the solution is 500 mm Hg. The vapour pressure of pure B and the least volatile component of the solution, respectively, are: **[2025, 2 April Shift-I]**

(a) 600 mm Hg, A (b) 1400 mm Hg, B
(c) 1400 mm Hg, A (d) 600 mm Hg, B

• Level 2

02 When a non-volatile solute is added to the solvent, the vapour pressure of the solvent decreases by 10 mm of Hg. The mole fraction of the solute in the solution is 0.2. What would be the mole fraction of the solvent if decrease in vapour pressure is 20 mm of Hg? **[2025, 23 Jan Shift-II]**

(a) 0.4 (b) 0.2
(c) 0.8 (d) 0.6

03 Liquid A and B form an ideal solution. The vapour pressures of pure liquids A and B are 350 and 750 mm Hg respectively at the same temperature. If x_A and x_B are the mole fraction of A and B in solution while y_A and y_B are the mole fraction of A and B in vapour phase then, **[2024, 7 April Shift-II]**

(a) $\frac{x_A}{x_B} = \frac{y_A}{y_B}$ (b) $(x_A - y_A) < (x_B - y_B)$

(c) $\frac{x_A}{x_B} > \frac{y_A}{y_B}$ (d) $\frac{x_A}{x_B} < \frac{y_A}{y_B}$

04 A graph of vapour pressure and temperature for three different liquids X, Y, and Z is shown below : **[2020, 8 Jan Shift-I]**

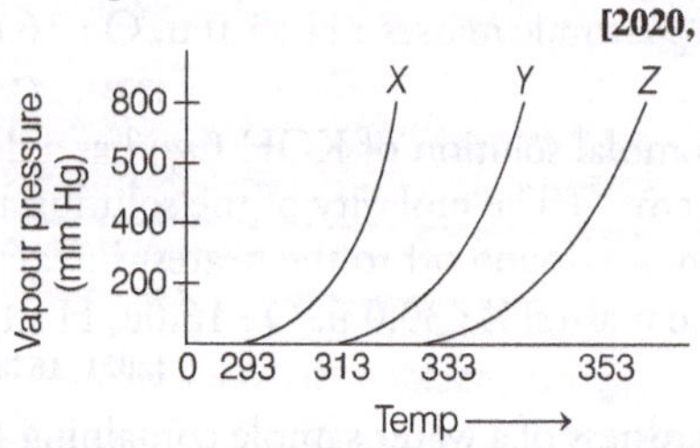

The following inference are made :

A. X has higher intermolecular interactions compared to Y.

B. X has lower intermolecular interactions compared to Y.

C. Z has lower intermolecular interactions compared to Y.

The correct inference(s) is/are :

(a) (C) (b) (B)
(c) (A) (d) (A) and (C)

05 At 35°C, the vapour pressure of CS_2 is 512 mm Hg and that of acetone is 344 mm Hg. A solution of CS_2 in acetone has a total vapour pressure of 600 mm Hg. The false statement amongst the following is **[2020, 7 Jan Shift-I]**

(a) Raoult's law is not obeyed by this system
(b) CS_2 and acetone are less attracted to each other than to themselves
(c) a mixture of 100 mL CS_2 and 100 mL acetone has a volume < 200 mL
(d) heat must be absorbed in order to produce the solution at 35°C

06 Liquid M and liquid N form an ideal solution. The vapour pressures of pure liquids M and N are 450 and 700 mmHg, respectively, at the same temperature. Then correct statement is **[2019, 9 April Shift-I]**

x_M = mole fraction of M in solution;
x_N = mole fraction of N in solution;
y_M = mole fraction of M in vapour phase;
y_N = mole fraction of N in vapour phase

(a) $\frac{x_M}{x_N} > \frac{y_M}{y_N}$ (b) $\frac{x_M}{x_N} = \frac{y_M}{y_N}$

(c) $\frac{x_M}{x_N} < \frac{y_M}{y_N}$ (d) $(x_M - y_M) < (x_N - y_N)$

07 For the solution of the gases w, x, y and z in water at 298 K, the Henry's law constants (K_H) are 0.5, 2, 35 and 40 K bar, respectively. The correct plot for the given data is **[2019, 8 April Shift-II]**

(a)
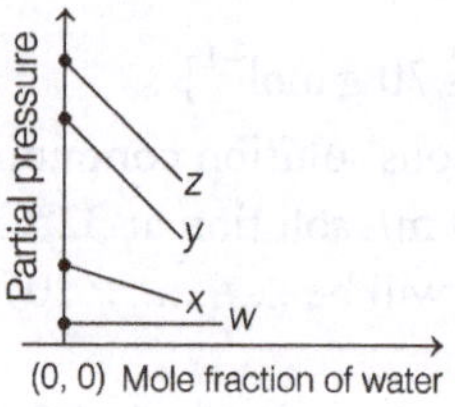

(b)
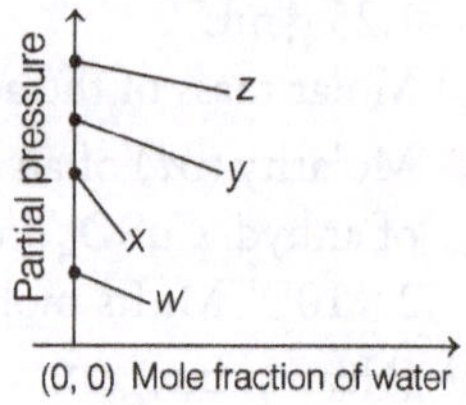

(c)
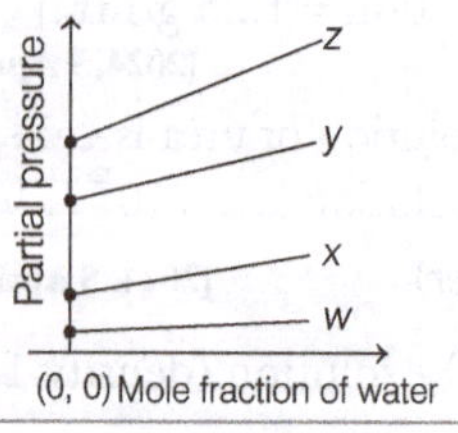

(d)
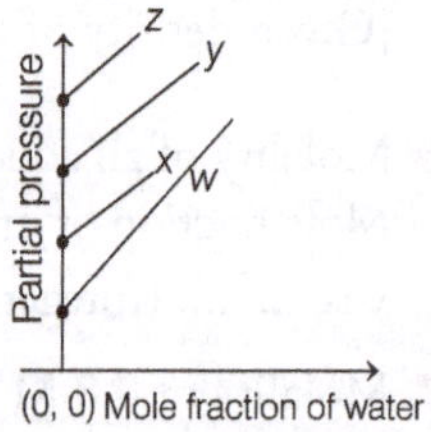

Imp. *Henry's Law Statement*

08 Which one of the following statements regarding Henry's law is not correct?

(a) Different gases have different K_H (Henry's law constant) values at the same temperature
(b) Higher the value of K_H at a given pressure, higher is the solubility of the gas in the liquids
(c) The value of K_H increases with increase of temperature and K_H is function of the nature of the gas
(d) The partial pressure of the gas in vapour phase is proportional to the mole fraction of the gas in the solution **[2019, 9 Jan Shift-I]**

• Level 3

09 Which of the following binary mixture does not show the behaviour of minimum boiling azeotropes? **[2025, 8 April Shift-II]**

(a) $CS_2 + CH_3COCH_3$ (b) $CH_3OH + CHCl_3$
(c) $H_2O + CH_3COC_2H_5$ (d) $C_6H_5OH + C_6H_5NH_2$

10 Match List - I with List - II. **[2025, 7 April Shift-II]**

	List-I		List-I
(A)	Solution of chloroform and acetone	(I)	Minimum boiling azeotrope
(B)	Solution of ethanol and water	(II)	Dimerizes
(C)	Solution of benzene and toluene	(III)	Maximum boiling azeotrope
(D)	Solution of acetic acid in benzene	(IV)	$\Delta V_{mix} = 0$

Choose the correct answer from the options given below:

(a) (A)-(III), (B)-(I), (C)-(IV), (D)-(II)
(b) (A)-(II), (B)-(IV), (C)-(I), (D)-(III)
(c) (A)-(II), (B)-(I), (C)-(IV), (D)-(III)
(d) (A)-(III), (B)-(IV), (C)-(I), (D)-(II)

11 At the sea level, the dry air mass percentage composition is given as nitrogen gas: 70.0, oxygen gas: 27.0 and argon gas: 3.0. If total pressure is 1.15 atm, then calculate the ratio of following respectively:

(i) partial pressure of nitrogen gas to partial pressure of oxygen gas.
(ii) partial pressure of oxygen gas to partial pressure of argon gas.

(Given: Molar mass of N, O and Ar are 14, 16 and 40 g mol^{-1} respectively) **[2025, 7 April Shift-I]**

(a) 5.46, 17.8
(b) 2.96, 11.2
(c) 2.59, 11.85
(d) 4.26. 19.3

12 Which of the following graph correctly represents the plots of K_H at 1 bar for gases in water *versus* temperature? **[2025, 2 April Shift-I]**

(a)
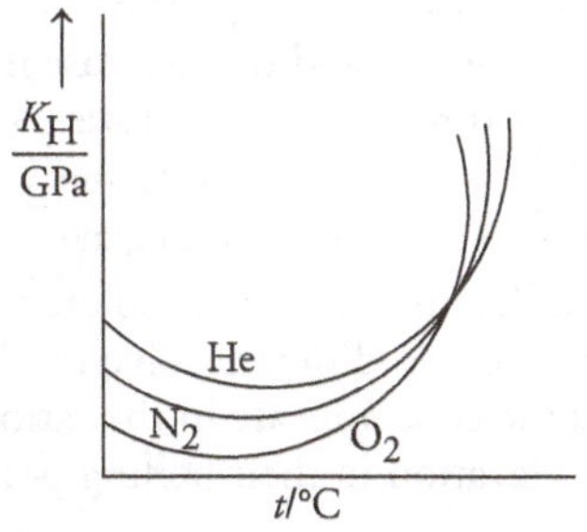

(b)
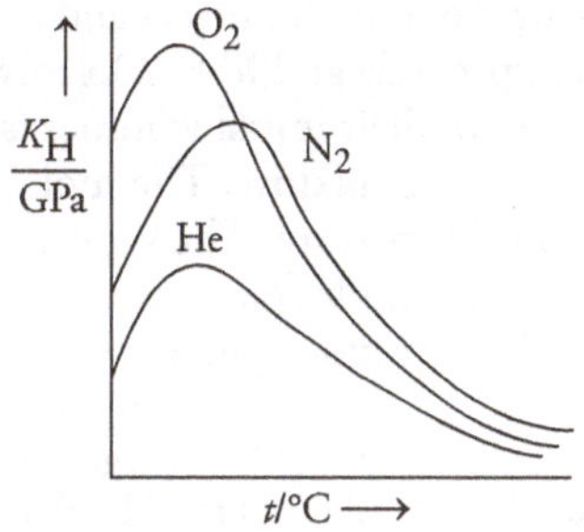

(c)
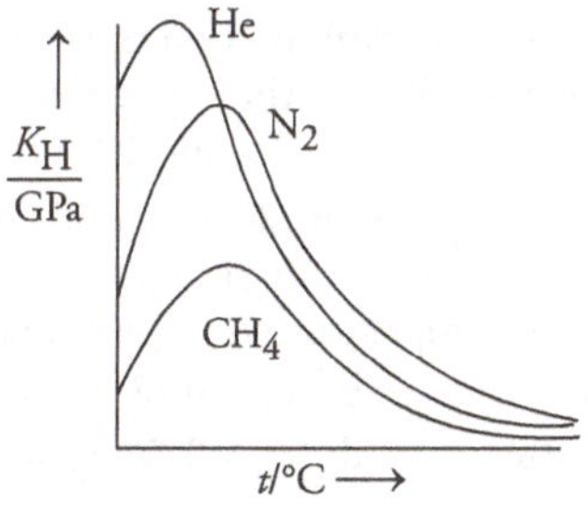

(d)
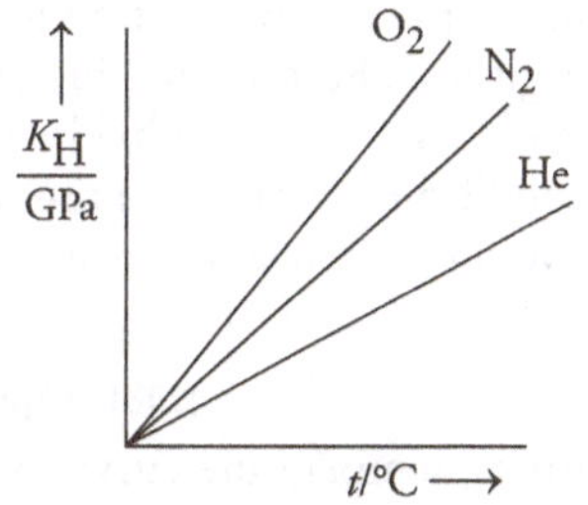

13 Consider a binary solution of two volatile liquid components 1 and 2. x_1 and y_1 are the mole fractions of component 1 in liquid and vapour phase, respectively. The slope and intercept of the linear plot of $1/x_1$ *vs* $1/y_1$ are given respectively as **[2025, 23 Jan Shift-II]**

(a) $\frac{p_2^\circ}{p_1^\circ}, \frac{p_2^\circ - p_1^\circ}{p_2^\circ}$ (b) $\frac{p_1^\circ}{p_2^\circ}, \frac{p_1^\circ - p_2^\circ}{p_2^\circ}$

(c) $\frac{p_2^\circ}{p_1^\circ}, \frac{p_1^\circ - p_2^\circ}{p_2^\circ}$ (d) $\frac{p_1^\circ}{p_2^\circ}, \frac{p_2^\circ - p_1^\circ}{p_2^\circ}$

14 Identify the mixture that shows positive deviations from Raoult's law. **[2024, 31 Jan Shift-I]**

(a) $(CH_3)_2CO + CS_2$
(b) $CHCl_3 + C_6H_6$
(c) $(CH_3)_2CO + C_6H_5NH_2$
(d) $CHCl_3 + (CH_3)_2CO$

15 A solution of two miscible liquids showing negative deviation from Raoult's law will have **[2024, 27 Jan Shift-I]**

(a) increased vapour pressure, increased boiling point
(b) decreased vapour pressure, increased boiling point
(c) increased vapour pressure, decreased boiling point
(d) decreased vapour pressure, decreased boiling point

16 The vapour pressures of pure liquids A and B are 400 and 600 mmHg, respectively at 298 K. On mixing the two liquids, the sum of their initial volumes is equal to the volume of the final mixture. The mole fraction of liquid B is 0.5 in the mixture. The vapour pressure of the final solution, the mole fractions of components A and B in vapour phase, respectively are **[2019, 8 April Shift-I]**

(a) 450 mmHg, 0.4, 0.6 (b) 500 mmHg, 0.5, 0.5
(c) 450 mmHg, 0.5,0.5 (d) 500 mmHg, 0.4,0.6

17 Liquids A and B form an ideal solution in the entire composition range. At 350 K, the vapour pressures of pure A and pure B are 7×10^3 Pa and 12×10^3 Pa, respectively. The composition of the vapour in equilibrium with a solution containing 40 mole percent of A at this temperature is **[2019, 10 Jan Shift-I]**

(a) $x_A = 0.76$; $x_B = 0.24$ (b) $x_A = 0.28$; $x_B = 0.72$
(c) $x_A = 0.4$; $x_B = 0.6$ (d) $x_A = 0.37$; $x_B = 0.63$

Numerical Value Type Questions

18 The vapour pressure of pure benzene and methyl benzene at 27°C is given as 80 torr and 24 torr, respectively. The mole fraction of methyl benzene in vapour phase, in equilibrium with an equimolar mixture of those two liquids (ideal solution) at the same temperature is .... $\times 10^{-2}$. (Nearest integer) **[2024, 9 April Shift-II]**

19 The vapour pressure *vs*, temperature curve for a solution solvent system is shown below.

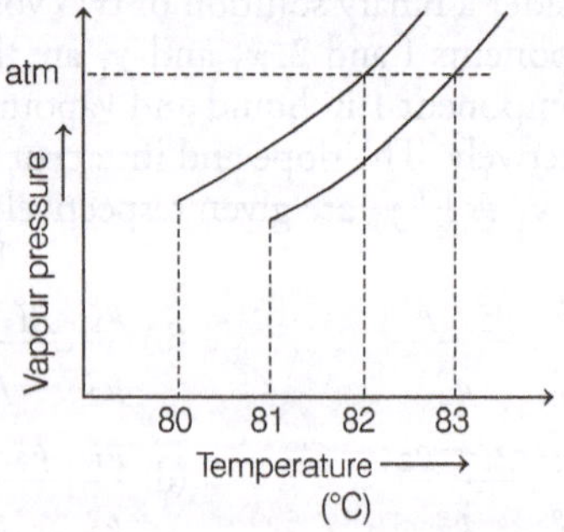

The boiling point of the solvent is .... °C. **[2023, 8 April Shift-I]**

20 The total pressure observed by mixing two liquids A and B is 350 mm Hg when their mole fractions are 0.7 and 0.3 respectively.

The total pressure becomes 410 mm Hg if the mole fractions are changed to 0.2 and 0.8 respectively for A and B. The vapour pressure of pure A is .... mm Hg. (Nearest integer)

Consider the liquids and solutions behave ideally. **[2023, 24 Jan Shift-II]**

21 'x' g of molecular oxygen (O_2) is mixed with 200 g of neon (Ne). The total pressure of the non-reactive mixture of O_2 and Ne in the cylinder is 25 bar. The partial pressure of Ne is 20 bar at the same temperature and volume. The value of 'x' is ....

[Given : Molar mass of $O_2 = 32 \text{ g mol}^{-1}$.
Molar mass of Ne = 20 g mol^{-1}] **[2022, 29 July Shift-II]**

22 If O_2 gas is bubbled through water at 303 K, the number of millimoles of O_2 gas that dissolve in 1 litre of water is .... (Nearest integer).

(Given, Henry's law constant for O_2 at 303 K is 46.82 k bar and partial pressure of O_2 = 0.920 bar)

(Assume, solubility of O_2 in water is too small, nearly negligible.) **[2022, 29 July Shift-I]**

23 CO_2 gas is bubbled through water during a soft drink manufacturing process at 298 K. If CO_2 exerts a partial pressure of 0.835 bar then x m mol of CO_2 would dissolve in 0.9 L of water. The value of x is .... (Nearest integer)

(Henry's law constant for CO_2 at 298 K is 1.67×10^3 bar) **[2021, 25 July Shift-I]**

24 At 363 K the vapour pressure of A is 21 kPa and that of B is 18 kPa. One mole of A and 2 moles of B are mixed. Assuming that this solution is ideal, the vapour pressure of the mixture is ...... kPa. [Round off to the Nearest Integer]. **[2021, 16 March Shift-II]**

25 At 300 K, the vapour pressure of a solution containing 1 mole of n-hexane and 3 moles of n-heptane is 550 mm of Hg. At the same temperature, if one more mole of n-heptane is added to this solution, the vapour pressure of the solution increases by 10 mm of Hg. What is the vapour pressure in mm Hg of n-heptane in its pure state .... ? **[2020, 4 Sep Shift-I]**

Topic 3 / Colligative Properties

• Level 1

01 What is the freezing point depression constant of a solvent, 50 g of which contain 1 g non-volatile solute (molar mass 256 g mol^{-1}) and the decrease in freezing point is 0.40 K? **[2025, 28 Jan Shift-I]**

(a) $5.12 \text{ K kg mol}^{-1}$ (b) $3.72 \text{ K kg mol}^{-1}$
(c) $4.43 \text{ K kg mol}^{-1}$ (d) $1.86 \text{ K kg mol}^{-1}$

02 1 g of a non-volatile, non-electrolyte solute is dissolved in 100 g of two different solvents A and B, whose ebullioscopic constants are in the ratio of 1 : 5. The ratio of the elevation in their boiling points, $\frac{\Delta T_b(A)}{\Delta T_b(B)}$, is **[2019, 10 April Shift-II]**

(a) 5 : 1 (b) 10 : 1 (c) 1 : 5 (d) 1 : 0.2

• Level 2

03 2 moles each of ethylene glycol and glucose are dissolved in 500 g of water. The boiling point of the resulting solution is **[2025, 3 April Shift-I]**

(Given, ebullioscopic constant of water = 0.52 K kg mol^{-1})

(a) 375.3 K (b) 377.3 K (c) 277.3 K (d) 379.2 K

04 $HA(aq) \rightleftharpoons H^+(aq) + A^-(aq)$ **[2025, 8 April Shift-II]**

The freezing point depression of a 0.1 in aqueous solution of a monobasic weak acid HA is 0.20 °C. The dissociation constant for the acid is

Given, $K_f(H_2O) = 1.8$ K kg mol^{-1}, molality ≡ molarity

(a) 1.90×10^{-3} (b) 1.89×10^{-1} (c) 1.1×10^{-2} (d) 1.38×10^{-3}

Imp. *Graph of VP vs T*

05 Consider the given plots of vapour pressure (VP) *vs* temperature (T / K). Which amongst the following options is correct graphical representation showing ΔT_f depression in the freezing point of a solvent in a solution? **[2025, 24 Jan Shift-I]**

(a)
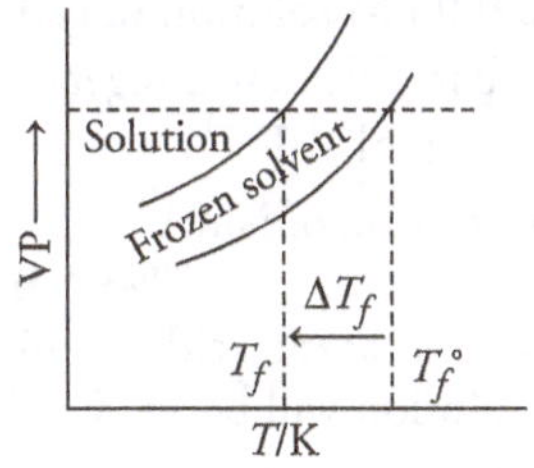

(b)
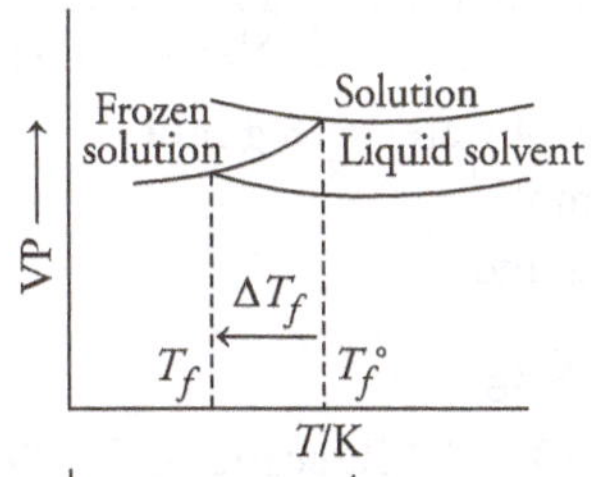

(c)
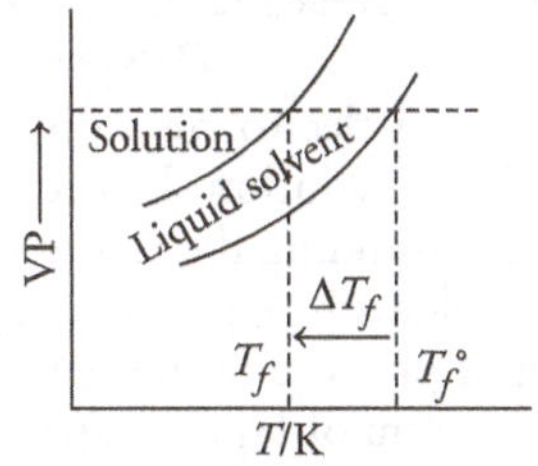

(d)
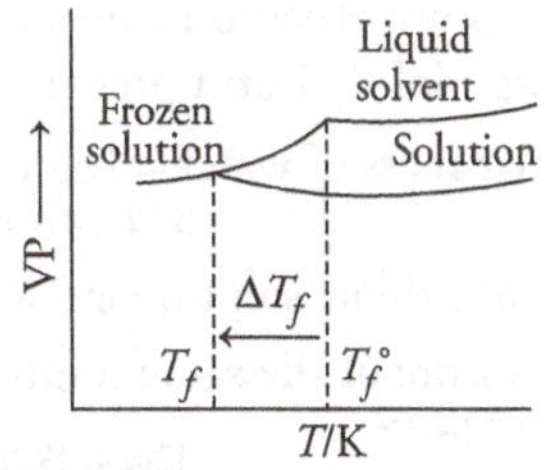

06 Given below are two statements. **[2025, 29 Jan Shift-II]**

Statement I NaCl is added to the ice at 0°C, present in the ice cream box to prevent the melting of ice cream.

Statement II On addition of NaCl to ice at 0°C, there is a depression in freezing point.

In the light of the above statements, choose the correct answer from the options given below:

(a) Both Statement I and Statement II are false.
(b) Statement I is false but Statement II is true.
(c) Statement I is true but Statement II is false.
(d) Both Statement I and Statement II are true.

07 What happens to freezing point of benzene when small quantity of naphthalene is added to benzene? **[2024, 30 Jan Shift-I]**

(a) First decreases and then increases
(b) Remains unchanged
(c) Increases
(d) Decreases

08 What weight of glucose must be dissolved in 100 g of water to lower the vapour pressure by 0.20 mm Hg ? (Assume, dilute solution is being formed)

Given, vapour pressure of pure water is 54.2 mm Hg at room temperature. Molar mass of glucose is 180 g mol^{-1} **[2023, 11 April Shift-II]**

(a) 3.69 g (b) 3.59 g
(c) 2.59 g (d) 4.69 g

09 In the depression of freezing point experiment.

A. Vapour pressure of the solution is less than that of pure solvent
B. Vapour pressure of the solution is more than that of pure solvent
C. Only solute molecules solidify at the freezing point
D. Only solvent molecules solidify at the freezing point

Choose the most appropriate answer from the options given below. **[2023, 24 Jan Shift-I]**

(a) B and C only (b) A and D only
(c) A and C only (d) A only

10 Two solutions A and B are prepared by dissolving 1 g of non-volatile solutes X and Y, respectively in 1 kg of water. The ratio of depression in freezing points for A and B is found to be 1 : 4. The ratio of molar masses of X and Y is **[2022, 25 July Shift-II]**

(a) 1 : 4 (b) 1 : 0.25
(c) 1 : 0.20 (d) 1 : 5

11 A solution is prepared by dissolving 0.6 g of urea (molar mass = 60 g mol^{-1}) and 1.8 g of glucose (molar mass = 180 g mol^{-1}) in 100 mL of water at 27°C. The osmotic pressure of the solution is (R = 0.08206 L atm K^{-1} mol^{-1}) **[2019, 12 April Shift-II]**

(a) 8.2 atm (b) 2.46 atm
(c) 4.92 atm (d) 1.64 atm

12 At room temperature, a dilute solution of urea is prepared by dissolving 0.60 g of urea in 360 g of water. If the vapour pressure of pure water at this temperature is 35 mm Hg, lowering of vapour pressure will be (Molar mass of urea $= 60 \text{ g mol}^{-1}$) **[2019, 10 April Shift-I]**

(a) 0.027 mmHg (b) 0.031 mmHg
(c) 0.017 mmHg (d) 0.028 mmHg

13 Elevation in the boiling point for 1 molal solution of glucose is 2 K. The depression in the freezing point for 2 molal solution of glucose in the same solvent is 2 K. The relation between K_b and K_f is **[2019, 10 Jan Shift-II]**

(a) $K_b = 1.5\, K_f$ (b) $K_b = 0.5\, K_f$
(c) $K_b = K_f$ (d) $K_b = 2K_f$

• Level 3

14 Given below are two statements: **[2025, 4 April Shift-II]**

Statement I Molal depression constant K_f is given by $\frac{M_1 RT_f}{\Delta S_{\text{fus}}}$, where symbols have their usual meaning.

Statement II K_f for benzene is less than the K_f for water.

In the light of the above statements, choose the most appropriate answer from the options given below:
(a) Both Statement I and Statement II are correct
(b) Statement I is incorrect but Statement II is correct
(c) Statement I is correct but Statement II is incorrect
(d) Both Statement I and Statement II are incorrect

15 $CrCl_3 \cdot xNH_3$ can exist as a complex. 0.1 molal aqueous solution of this complex shows a depression in freezing point of 0.558°C. Assuming 100% ionisation of this complex and coordination number of Cr is 6, the complex will be
(Given, $K_f = 1.86 \text{ K kg mol}^{-1}$) **[2025, 23 Jan Shift-I]**

(a) $[Cr(NH_3)_3Cl_3]$ (b) $[Cr(NH_3)_4Cl_2]Cl$
(c) $[Cr(NH_3)_6]Cl_3$ (d) $[Cr(NH_3)_5Cl]Cl_2$

16 0.05 M $CuSO_4$ when treated with 0.01 M $K_2Cr_2O_7$ gives colour solution of $Cu_2Cr_2O_7$. The two solutions are separated as shown below.
[SPM : Semi-permeable membrane] **[2024, 9 April Shift-I]**

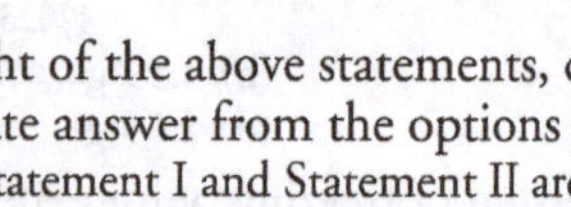

Due to osmosis
(a) molarity of $K_2Cr_2O_7$ solution is lowered.
(b) green colour formation observed on side *X*.
(c) molarity of $CuSO_4$ solution is lowered.
(d) green colour formation observed on side *Y*.

17 Boiling point of a 2% aqueous solution of a non-volatile solute *A* is equal to the boiling point of 8% aqueous solution of a non-volatile solute *B*. The relation between molecular weights of *A* and *B* is **[2022, 27 July Shift-I]**

(a) $M_A = 4M_B$ (b) $M_B = 4M_A$
(c) $M_A = 8M_B$ (d) $M_B = 8M_A$

Numerical Value Type Questions

18 When $x \times 10^{-2}$ mL methanol (molar mass $= 32 \text{ g mol}^{-1}$: density $= 0.792 \text{ g/cm}^2$) is added to 100 mL water (density $= 1 \text{ g/cm}^3$), the following diagram is obtained.

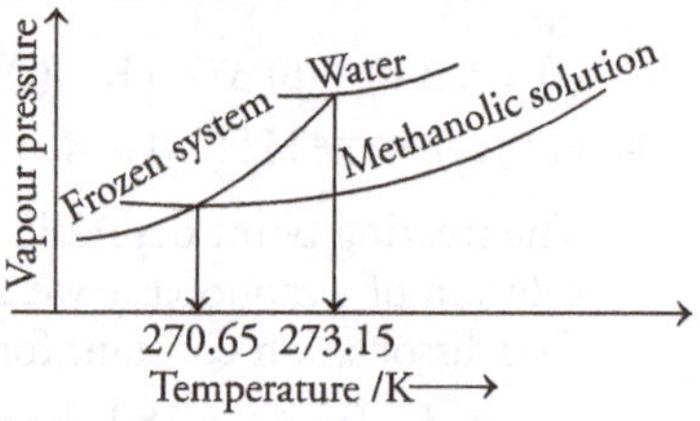

[2024, 6 April Shift-II]

$x = \ldots\ldots\ldots$ (nearest integer)
[Given : Molal freezing point depression constant of water at 273.15 K is $1.86 \text{ K kg mol}^{-1}$]

19 An artificial cell is made by encapsulating 0.2 M glucose solution within a semipermeable membrane. The osmotic pressure developed when the artificial cell is placed within a 0.05 M solution of NaCl at 300 K is ...... $\times 10^{-1}$ bar. (Nearest integer).
[Given : $R = 0.083 \text{ L bar mol}^{-1} \text{ K}^{-1}$]
Assume complete dissociation of NaCl. **[2024, 5 April Shift-I]**

20 2.7 kg of each of water and acetic acid are mixed. The freezing point of the solution will be $-x$°C. Consider the acetic acid does not dimerise in water, nor dissociates in water, $x = \ldots\ldots\ldots$ (Nearest integer)
[given: Molar mass of water $= 18 \text{ g mol}^{-1}$, acetic acid $= 60 \text{ g mol}^{-1}$
$K_{f_{H_2O}}$: $1.86 \text{ K kg mol}^{-1}$
$K_{f\ \text{acetic acid}}$: $3.90 \text{ K kg mol}^{-1}$
Freezing point: H_2O = 273 K, acetic acid = 290 K] **[2024, 4 April Shift-II]**

21 2.5 g of a non-volatile, non-electrolyte is dissolved in 100g of water at 25°C. The solution showed a boiling point elevation by 2°C. Assuming the solute concentration is negligible with respect to the solvent concentration, the vapor pressure of the resulting aqueous solution is...... mm of Hg. (Nearest integer)
[Given: Molal boiling point elevation constant of water $(K_b) = 0.52 \text{ K kg mol}^{-1}$, 1 atm pressure =760 mm of Hg, molar mass of water = 18g mol^{-1}] **[2024, 4 April Shift-I]**

22 The osmotic pressure of a dilute solution is 7×10^5 Pa at 273 K. Osmotic pressure of the same solution at 283 K is .... $\times 10^4 \text{ Nm}^{-2}$. **[2024, 29 Jan Shift-I]**

23 The vapour pressure of 30% (w/v) aqueous solution of glucose is .... mm Hg at 25°C.
[Given, the density of 30% (w/v), aqueous solution of glucose is 1.2 g cm^{-3} and vapour pressure of pure water is 24 mm Hg.] (Molar mass of glucose is 180 g mol^{-1}) **[2023, 15 April Shift-I]**

24 If the boiling points of two solvents X and Y (having same molecular weights) are in the ratio 2:1 and their enthalpy of vaporisations are in the ratio 1:2, then the boiling point elevation constant of X is m times the boiling point elevation constant of Y. The value of m is .... (nearest integer). **[2023, 8 April Shift-II]**

25 Mass of urea (NH_2CONH_2) required to be dissolved in 1000 g of water in order to reduce the vapour pressure of water by 25% is .... g. (Nearest integer)
Given, molar mass of N, C, O and H are 14, 12, 16 and 1g mol^{-1} respectively. **[2023, 6 April Shift-I]**

26 At 27°C, a solution containing 2.5 g of solute in 250.0 mL of solution exerts an osmotic pressure of 400 Pa. The molar mass of the solute is .... g mol^{-1}. (Nearest integer) **[2023, 31 Jan Shift-I]**
(Given : $R = 0.083$ L bar K^{-1} mol^{-1})

27 A solution containing 2 g of a non-volatile solute in 20 g of water boils at 373.52 K. The molecular mass of the solute is .... g mol^{-1}. (Nearest integer)
[Given : water boils at 373 K, K_b for water $= 0.52$ K kg mol^{-1}] **[2023, 30 Jan Shift-I]**

28 The osmotic pressure of solutions of PVC in cyclohexanone at 300K are plotted on the graph. The molar mass of PVC is .... g mol^{-1}. (Nearest integer) **[2023, 25 Jan Shift-I]**

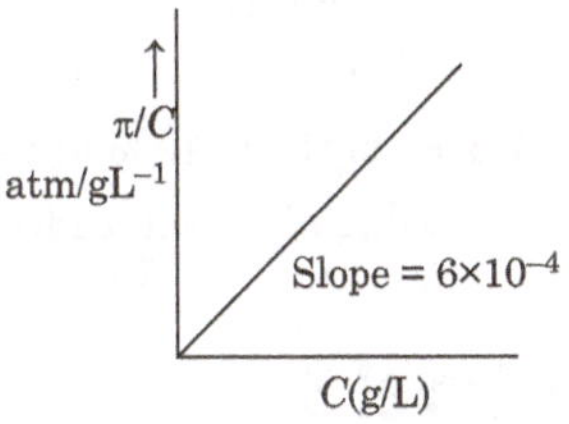

(Given : $R = 0.083$ L atm K^{-1} mol^{-1}).

29 150 g of acetic acid was contaminated with 10.2 g ascorbic acid ($C_6H_8O_6$) to lower down its freezing point by ($x \times 10^{-1}$)°C. The value of x is .... (Nearest integer) [Given $K_f = 3.9$ K kg mol^{-1}; molar mass of ascorbic acid = 176 g mol^{-1}] **[2022, 28 July Shift-I]**

30 When a certain amount of solid A is dissolved in 100 g of water at 25°C to make a dilute solution, the vapour pressure of the solution is reduced to one-half of that of pure water. The vapour pressure of pure water is 23.76 mmHg. The number of moles of solute A added is .... (nearest integer) **[2022, 27 July Shift-II]**

31 The elevation in boiling point for 1 molal solution of non-volatile solute A is 3 K. The depression in freezing point for 2 molal solution of A in the same solvent is 6 K. The ratio of K_b and K_f i.e., K_b / K_f is 1 : x. The value of x is......... [nearest integer] **[2022, 26 July Shift-II]**

32 A solution containing 2.5×10^{-3} kg of a solute dissolved in 75×10^{-3} kg of water boils at 373.535 K. The molar mass of the solute is .... g mol^{-1}. [Nearest integer] (Given, $K_b(H_2O) = 0.52$ K kg mol^{-1} and boiling point of water $= 373.15$ K). **[2022, 27 June Shift-II]**

33 2 g of a non-volatile non-electrolyte solute is dissolved in 200 g of two different solvents A and B whose ebullioscopic constants are in the ratio of 1:8. The elevation in boiling points of A and B are in the ratio $\frac{x}{y}$ ($x : y$). The value of y is .... (Nearest integer) **[2022, 27 June Shift-I]**

34 The osmotic pressure exerted by a solution prepared by dissolving 2.0 g of protein of molar mass 60 kg mol^{-1} in 200 mL of water at 27°C is ...... Pa. [Integer value] (Use, $R = 0.083$ L bar mol^{-1} K^{-1}) **[2022, 26 June Shift-II]**

35 The osmotic pressure of blood is 7.47 bar at 300 K. To inject glucose to a patient intravenously, it has to be isotonic with blood. The concentration of glucose solution in gL^{-1} is ....
(Molar mass of glucose = 180 g mol^{-1} $R = 0.083$ L bar K^{-1} mol^{-1}) (Nearest integer) **[2022, 24 June, Shift-I]**

36 40 g of glucose (Molar mass = 180) is mixed with 200 mL of water. The freezing point of solution is .... K. (Nearest integer)
[Given, $K_f = 1.86$ K kg mol^{-1}, density of water = 1.00 g cm^{-3}, freezing point of water = 273.15 K] **[2021, 27 Aug Shift-II]**

37 1 kg of 0.75 molal aqueous solution of sucrose can be cooled up to −4°C before freezing. The amount of ice (in g) that will be separated out is .... (Nearest integer)
[Given, $K_f(H_2O) = 1.86$ K kg mol^{-1}] **[2021, 27 Aug Shift-I]**

38 1.46 g of a biopolymer dissolved in a 100 mL water at 300 K exerted an osmotic pressure of 2.42×10^{-3} bar. The molar mass of the biopolymer is ... $\times 10^4$ g mol^{-1}. (Round off to the nearest integer)
[Use : $R = 0.083$ L bar mol^{-1} K^{-1}] **[2021, 27 July Shift-I]**

39 When 3.00 g of a substance '*X*' is dissolved in 100 g of CCl_4, it raises the boiling point by 0.60 K. The molar mass of the substance '*X*' is .... g mol^{-1}. (Nearest integer)
[Given, K_b for CCl_4 is 5.0 K kg mol^{-1}]
[2021, 25 July Shift-II]

40 A 1 molal $K_4Fe(CN)_6$ solution has a degree of dissociation of 0.4. Its boiling point is equal to that of another solution which contains 18.1 weight per cent of a non- electrolytic solute *A*. The molar mass of *A* is .... u. (Round off to the nearest integer).
[Density of water = 1.0 g cm^{-3}] **[2021, 17 March Shift-II]**

41 A cylinder containing an ideal gas (0.1 mol of 1.0 dm^3) is in thermal equilibrium with a large volume of 0.5 molal aqueous solution of ethylene glycol at its freezing point. If the stoppers S_1 and S_2 (as shown in the figure) are suddenly withdrawn, the volume of the gas in litres after equilibrium is achieved will be ....
(Given, K_f (water) = 2.0 K kg mol^{-1}, R = 0.08 dm^3 atm $K^{-1} mol^{-1}$) **[2020, 9 Jan Shift-II]**

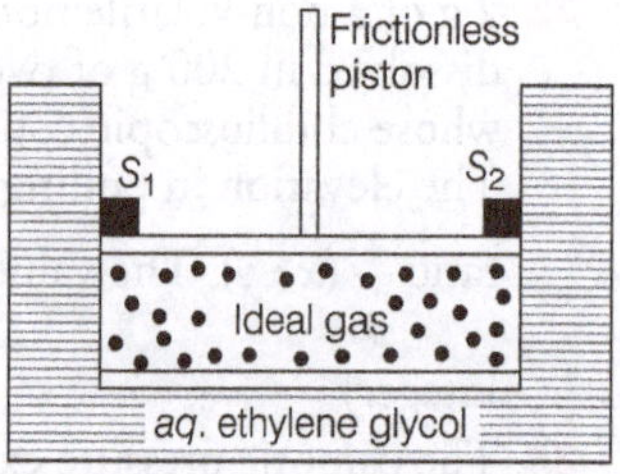

Topic 4 / Abnormal Molecular Masses and van't Hoff Factor

• Level 1

01 Which one of the following 0.10 M aqueous solutions will exhibit the largest freezing point depression ? **[2021, 31 Aug Shift-I]**
(a) Hydrazine (b) Glucose (c) Glycine (d) $KHSO_4$

• Level 2

02 The solution from the following with highest depression in freezing point/lowest freezing point is **[2024, 30 Jan Shift-II]**
(a) 180 g of acetic acid dissolved in water
(b) 180 g of glucose dissolved in water
(c) 180 g of acetic acid dissolved in benzene
(d) 180 g of benzoic acid dissolved in benzene

03 We have three aqueous solutions of NaCl labelled as 'A', 'B' and 'C' with concentration 0.1 M, 0.01 M and 0.001 M, respectively. The value of van't Hoff factor (*i*) for these solutions will be in the order **[2024, 1 Feb Shift-I]**
(a) $i_A = i_B = i_C$ (b) $i_A > i_B > i_C$
(c) $i_A < i_C < i_B$ (d) $i_A < i_B < i_C$

Imp. *Colligative Properties and Azeotropes*

04 Match List-I with List-II

List-I		List-II	
A.	van't Hoff factor, *i*	I.	Cryoscopic constant
B.	k_f	II.	Isotonic solutions
C.	Solutions with same osmotic pressure	III.	$\frac{\text{Normal molar mass}}{\text{Abnormal molar mass}}$
D.	Azeotropes	IV.	Solutions with same composition of vapour above it

Choose the correct answer from the options given below : **[2023, 29 Jan Shift-II]**
(a) A-III, B-I, C-IV, D-II (b) A-I, B-III, C-II, D-IV
(c) A-III, B-I, C-II, D-IV (d) A-III, B-II, C-I, D-IV

05 Which one of the following 0.06 M aqueous solutions has lowest freezing point? **[2021, 22 July Shift-II]**
(a) $Al_2(SO_4)_3$ (b) $C_6H_{12}O_6$ (c) KI (d) K_2SO_4

06 Molal depression constant for a solvent is 4.0 K kg mol^{-1}. The depression in the freezing point of the solvent for 0.03 mol kg^{-1} solution of K_2SO_4 is (Assume complete dissociation of the electrolyte) **[2019, 9 April Shift-II]**
(a) 0.18 K (b) 0.36 K (c) 0.12 K (d) 0.24 K

07 The osmotic pressure of a dilute solution of an ionic compound *XY* in water is four times that of a solution of 0.01 M $BaCl_2$ in water. Assuming complete dissociation of the given ionic compounds in water, the concentration of *XY* (in mol L^{-1}) in solution is **[2019, 9 April Shift-I]**
(a) 4×10^{-2} (b) 16×10^{-4} (c) 4×10^{-4} (d) 6×10^{-2}

08 K_2HgI_4 is 40% ionised in aqueous solution. The value of its van't Hoff factor (*i*) is **[2019, 11 Jan Shift-II]**
(a) 1.6 (b) 1.8 (c) 2.2 (d) 2.0

• Level 3

09 *XY* is the membrane/partition between two chambers 1 and 2 containing sugar solutions of concentration c_1 and $c_2 (c_1 > c_2)$ mol L^{-1}. For the reverse osmosis to take place identify the correct condition.
(Here p_1 and p_2 are pressures applied on chamber 1 and 2). **[2025, 4 April Shift-I]**

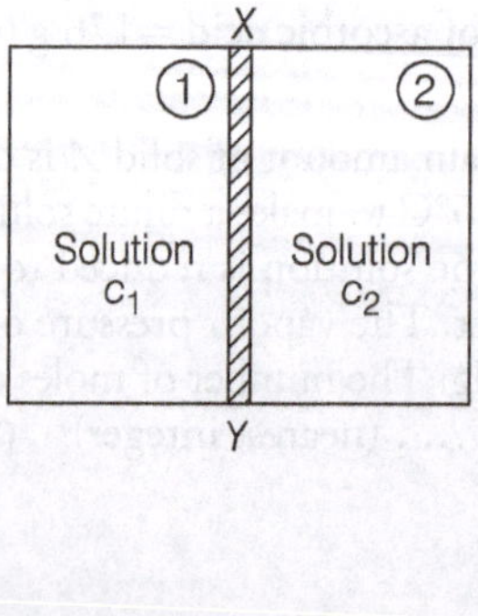

A. Membrane/Partition: Cellophane, $p_1 > \pi$
B. Membrane/Partition: Porous, $p_2 > \pi$
C. Membrane/Partition: Parchment paper, $p_1 > \pi$
D. Membrane/Partition: Cellophane, $p_2 > \pi$

Choose the correct answer from the option given below:

(a) A and C Only (b) B and D Only
(c) C Only (d) A and D Only

Imp. *Order of Boiling Point*

10 Arrange the following solutions in order of their increasing boiling points. **[2025, 22 Jan Shift-I]**

(i) 10^{-4} M NaCl (ii) 10^{-4} M Urea
(iii) 10^{-3} M NaCl (iv) 10^{-2} M NaCl

(a) (ii) < (i) < (iii) < (iv) (b) (ii) < (i) < (iii) < (iv)
(c) (iv) < (iii) < (i) < (ii) (d) (i) < (ii) < (iii) < (iv)

11 1.24 g of AX_2 (molar mass 124 g mol^{-1}) is dissolved in 1 kg of water to form a solution with boiling point of 100.0156°C, while 25.4 g of AY_2 (molar mass 250 g mol^{-1}) in 2 kg of water constitutes a solution with a boiling point of 100.0260°C. **[2025, 29 Jan Shift-I]**

[$K_b(H_2O) = 0.52$ K kg mol^{-1}]

Which of the following is correct?

(a) AX_2 and AY_2 (both) are fully ionised.
(b) AX_2 is fully ionised while AY_2 is completely unionised.
(c) AX_2 is completely unionised while AY_2 is fully ionised.
(d) AX_2 and AY_2 (both) are completely unionised.

12 Evaluate the following statements for their correctness.

A. The elevation in boiling point temperature of water will be same for 0.1 M NaCl and 0.1 M urea.
B. Azeotropic mixtures boil without change in their composition.
C. Osmosis always takes place from hypertonic to hypotonic solution.
D. The density of 32% H_2SO_4 solution having molarity 4.09 M is approximately 1.26 g mL^{-1}.
E. A negatively charged sol is obtained when KI solution is added to silver nitrate solution.

Choose the correct answer from the options given below. **[2023, 31 Jan Shift-II]**

(a) B and D (b) A and C
(c) A, B and D (d) B, D and E

13 The depression in freezing point observed for a formic acid solution of concentration 0.5 mol L^{-1} is 0.0405°C. Density of formic acid is 1.05 g mol L^{-1}. The van't Hoff factor of the formic acid solution is nearly (Given, for water $K_f = 1.86$ K kg mol^{-1}) **[2022, 25 July Shift-I]**

(a) 0.8 (b) 1.1 (c) 1.9 (d) 2.4

14 Solute A associates in water. When 0.7 g of solute A is dissolved in 42.0 g of water, it depresses the freezing point by 0.2°C. The percentage association of solute A in water, is

[Given, molar mass of $A = 93$ g mol^{-1}. Molal depression constant of water is 1.86 K kg mol^{-1}] **[2022, 25 June Shift-II]**

(a) 50% (b) 60%
(c) 70% (d) 80%

Numerical Value Type Questions

15 The percentage dissociation of a salt (MX_3) solution at given temperature (van't Hoff factor $i = 2$) is .... % (Nearest integer) **[2025, 7 April Shift-I]**

16 When 1 g each of compounds AB and AB_2 are dissolved in 15 g of water separately, they increased the boiling point of water by 2.7 K and 1.5 K respectively. The atomic mass of A (in amu) is .... $\times 10^{-1}$ (Nearest integer)

(Given : Molal boiling point elevation constant is 0.5 K kg mol^{-1}) **[2025, 2 April Shift-II]**

17 The observed and normal molar masses of compound MX_2 are 65.6 and 164 respectively. The percent degree of ionisation of MX_2 is ....%. (Nearest integer) **[2025, 24 Jan Shift-II]**

18 If A_2B is 30% ionised in an aqueous solution, then the value of van't Hoff factor (i) is .... $\times 10^{-1}$ **[2025, 29 Jan Shift-I]**

19 A solution containing 10 g of an electrolyte AB_2 in 100 g of water boils at 100.52°C. The degree of ionisation of the electrolyte (α) is .... $\times 10^{-1}$. (Nearest integer)

[Given : Molar mass of $AB_2 = 200$ g mol^{-1}, K_b (Molal boiling point elevation const. of water) = 0.52 K kg mol^{-1}, boiling point of water = 100°C, AB_2 ionises as $AB_2 \longrightarrow A^{2+} + 2B^-$] **[2024, 8 April Shift-II]**

20 Mass of ethylene glycol (antifreeze) to be the added to 18.6 kg of water to protect the freezing point at –24°C is .... kg (Molar mass in g mol^{-1} for ethylene glycol is 62, K_f of water = 1.86 K kg mol^{-1}) **[2024, 1 Feb Shift-II]**

21 80 mole per cent of $MgCl_2$ is dissociated in aqueous solution. The vapour pressure of 1.0 molal aqueous solution of $MgCl_2$ at 38°C is .... mm Hg. (Nearest integer)

(Given, vapour pressure of water at 38°C is 50 mm Hg) **[2023, 12 April Shift-I]**

22 Sea water contains 29.25% NaCl and 19% $MgCl_2$ by weight of solution. The normal boiling point of the sea water is .... °C (Nearest integer)

Assume 100% ionisation for both NaCl and $MgCl_2$. (Given, $K_b(H_2O) = 0.52$ K kg mol^{-1}, Molar mass of NaCl and $MgCl_2$ is 58.5 and 95 g mol^{-1} respectively.) **[2023, 10 April Shift-II]**

23 If the degree of dissociation of aqueous solution of weak monobasic acid is determined to be 0.3, then the observed freezing point will be ....% higher than the expected/theoretical freezing point. (Nearest integer) **[2023, 10 April Shift-I]**

Imp. *Isotonic Solutions*

24 Consider the following pairs of solution which will be isotonic at the same temperature. The number of pairs of solutions is/are .... **[2023, 6 April Shift-II]**
A. 1 M aq. NaCl and 2 M aq. urea
B. 1 M aq. $CaCl_2$ and 1.5 M aq. KCl
C 1.5 M aq. $AlCl_3$ and 2 M aq. Na_2SO_4
D. 2.5 M aq. KCl and 1 M aq. $Al_2(SO_4)_3$

25 20% of acetic acid is dissociated when its 5 g is added to 500 mL of water. The depression in freezing point of such water is .... $\times 10^{-3}$°C. Atomic mass of C, H and O are 12, 1 and 16 amu, respectively. [Given, molal depression constant and density of water are 1.86 K kg mol^{-1} and 1g cm^{-3} respectively. **[2023, 1 Feb Shift-II]**

26 25 mL of an aqueous solution of KCl was found to require 20 mL of 1 M $AgNO_3$ solution when titrated using K_2CrO_4 as an indicator. What is the depression in freezing point of KCl solution of the given concentration? .... (Nearest integer) (Given, $K_f = 2.0$ K kg mol^{-1})

Assume
1. 100% ionisation and
2. density of the aqueous solution as 1 g mL^{-1} **[2023, 1 Feb Shift-I]**

27 Lead storage battery contains 38% by weight solution of H_2SO_4. The van't Hoff factor is 2.67 at this concentration. The temperature in Kelvin at which the solution in the battery will freeze is .... (Nearest integer). **[2023, 30 Jan Shift-II]**
[Given : $K_f = 1.8$ K kg mol^{-1}]

28 The number of pairs of the solutions having the same value of the osmotic pressure from the following is .... (Assume 100% ionisation)
A. 0.500 M $C_2H_5OH(aq)$ and 0.25 M KBr(aq)
B. 0.100 M $K_4[Fe(CN)_6](aq)$ and 0.100 M $FeSO_4(NH_4)_2SO_4(aq)$
C. 0.05 M $K_4[Fe(CN)_6](aq)$ and 0.25 M NaCl(aq)
D. 0.15 M NaCl(aq) and 0.1 M $BaCl_2(aq)$
E. 0.02 M KCl · $MgCl_2 \cdot 6H_2O(aq)$ and 0.05 M KCl(aq) **[2023, 25 Jan Shift-II]**

29 1.2 mL of acetic acid is dissolved in water to make 2.0 L of solution. The depression in freezing point observed for this strength of acid is 0.0198°C. The percentage of dissociation of the acid is .... (Nearest integer)
[Given, density of acetic acid is 1.02 g mL^{-1}, molar mass of acetic acid is 60 g mol^{-1} $K_f(H_2O) = 1.85$ K kg mol^{-1}] **[2022, 29 June Shift-I]**

30 A 0.5 per cent solution of potassium chloride was found to freeze at – 0.24°C. The percentage dissociation of potassium chloride is .... (Nearest integer)
(Molal depression constant for water is 1.80 K kg mol^{-1} and molar mass of KCl is 74.6 g mol^{-1}) **[2022, 26 June Shift-I]**

31 83 g of ethylene glycol dissolved in 625 g of water. The freezing point of the solution is ....K. (Nearest integer)
[Use, molal freezing point depression constant of water = 1.86 K kg mol^{-1}, Freezing point of water = 273 K and Atomic masses : C = 12.0 u, O = 16.0 u, H = 1.0 u] **[2021, 26 Aug Shift-II]**

32 Of the following four aqueous solutions, total number of those solutions whose freezing point is lower than that of 0.10 M C_2H_5OH is .... (Integer answer)
(i) 0.10 M $Ba_3(PO_4)_2$ (ii) 0.10 M Na_2SO_4
(iii) 0.10 M KCl (iv) 0.10 M Li_3PO_4 **[2021, 26 Aug Shift-I]**

33 In a solvent 50% of an acid *HA* dimerises and the rest dissociates. The van't Hoff factor of the acid is .... $\times 10^{-2}$.
(Round off to the nearest integer) **[2021, 27 July Shift-II]**

34 A solute *A* dimerises in water. The boiling point of a 2 molar solution of *A* is 100.52°C. The percentage association of *A* is .... (Round off to the nearest integer)
[Use : K_b for water = 0.52 K kg mol^{-1}, boiling point of water = 100°C] **[2021, 18 March Shift-II]**

35 2 molal solution of a weak acid *HA* has a freezing point of 3.885°C. The degree of dissociation of this acid is .... $\times 10^{-3}$. (Round off to the nearest integer).
[Given : Molal depression constant of water = 1.85 K kg mol^{-1}, freezing point of pure water = 0°C] **[2021, 18 March Shift-I]**

36 When 12.2 g of benzoic acid is dissolved in 100 g of water, the freezing point of solution was found to be $-0.93°C$ ($K_f(H_2O) = 1.86$ K kg mol^{-1}). The number (n) of benzoic acid molecules associated (assuming 100% association) is ....
[2021, 26 Feb Shift-II]

37 If a compound AB dissociates to the extent of 75% in an aqueous solution, the molality of the solution which shows a 2.5 K rise in the boiling point of the solution is .... molal. (Rounded off to the nearest integer) [$K_b = 0.52$ K kg mol^{-1}] **[2021, 25 Feb Shift-II]**

38 1 molal aqueous solution of an electrolyte A_2B_3 is 60% ionised. The boiling point of the solution at 1 atm is .... K (Rounded off to the nearest integer). [Given, K_b for $(H_2O) = 0.52$ K kg mol^{-1}]
[2021, 25 Feb Shift-I]

39 C_6H_6 freezes at 5.5°C. The temperature at which a solution 10 g of C_4H_{10} in 200 g of C_6H_6 freeze is ....°C. (The molal freezing point depression constant of C_6H_6 is 5.12°C/m.) **[2021, 24 Feb Shift-II]**

40 The elevation of boiling point of 0.10 m aqueous $CrCl_3 \cdot x NH_3$ solution is two times that of 0.05 m aqueous $CaCl_2$ solution. The value of x is .... [Assume 100% ionisation of the complex and $CaCl_2$, coordination number of Cr as 6, and that all NH_3 molecules are present inside the coordination sphere]
[2020, 6 Sep Shift-I]

41 How much amount of NaCl should be added to 600 g of water ($\rho = 1.00$ g/mL) to decrease the freezing point of water to $-0.2°C$? ...... (The freezing point depression constant for water = 2 K kg mol^{-1})
[2020, 9 Jan Shift-I]

Key Index

Topic 1

1 (b)	**2** (c)	**3** (d)	**4** (b)	**5** (b)	**6** (d)	**7** (b)	**8** (c)	**9** (a)	**10** (c)
11 (d)	**12** (2)	**13** (1)	**14** (125)	**15** (164)	**16** (74)	**17** (815)	**18** (4)	**19** (148)	**20** (5)
21 (54)	**22** (64)	**23** (3.17)	**24** (5)	**25** (64)	**26** (9)	**27** (100)	**28** (10)		

Topic 2

1 (a)	**2** (d)	**3** (c)	**4** (b)	**5** (c)	**6** (a)	**7** (a)	**8** (b)	**9** (d)	**10** (a)
11 (b)	**12** (c)	**13** (d)	**14** (a)	**15** (b)	**16** (d)	**17** (b)	**18** (23)	**19** (82)	**20** (314)
21 (80)	**22** (1)	**23** (25)	**24** (19)	**25** (600)					

Topic 3

1 (a)	**2** (c)	**3** (b)	**4** (d)	**5** (d)	**6** (d)	**7** (d)	**8** (a)	**9** (b)	**10** (b)
11 (c)	**12** (c)	**13** (d)	**14** (c)	**15** (d)	**16** (c)	**17** (b)	**18** (543)	**19** (25)	**20** (31)
21 (707)	**22** (73)	**23** (23)	**24** (8)	**25** (833)	**26** (62250)	**27** (100)	**28** (41500)	**29** (15)	**30** (6)
31 (1)	**32** (45)	**33** (8)	**34** (415)	**35** (54)	**36** (271)	**37** (518)	**38** (15)	**39** (250)	**40** (85)
41 (2.18)									

Topic 4

1 (d)	**2** (a)	**3** (d)	**4** (c)	**5** (a)	**6** (b)	**7** (d)	**8** (b)	**9** (a)	**10** (b)
11 (b)	**12** (a)	**13** (c)	**14** (d)	**15** (33)	**16** (25)	**17** (75)	**18** (16)	**19** (5)	**20** (15)
21 (48)	**22** (116)	**23** (30)	**24** (4)	**25** (372)	**26** (3)	**27** (243)	**28** (4)	**29** (5)	**30** (98)
31 (269)	**32** (4)	**33** (125)	**34** (100)	**35** (50)	**36** (2)	**37** (3)	**38** (375)	**39** (1)	**40** (5)
41 (1.76)									

Detailed Explanations

Topic 1

01. *(b)* Given,
Mass of NaCl = 5.85 g
Volume of water = 500 mL = 0.5 L
Using, Molarity

$$= \frac{\text{No. of moles of NaCl}}{\text{Volume of solution (in L)}}$$

$$M = \frac{\frac{5.85}{58.5}}{0.5}$$

$$\Rightarrow M = 0.2\text{ M}$$

02. *(c)* Mole fraction

$$= \frac{\text{Number of moles of component}}{\text{Total number of moles of all the component}}$$

From above definition, mole fraction of *C* will be

$$= \frac{n_C}{n_A + n_B + n_C}$$

where n_A, n_B, n_C are number of moles of substance *A*, *B* and *C*.
Hence, option (c) is correct.

03. *(d)* Concentration terms such as mass percentage, mole fraction and molality are independent of temperature, However, molarity is a function of temperature. Since it depend on the total volume of the solution.

$$\text{Molarity} = \frac{\text{Mole of solute}}{\text{Volume of solution}}$$

> **Concept Enhancer**
> On increasing the temperature, volume of solution decreases and molarity increases.

04. *(b)* This question involves the concept of molality.

> **idea** Molality is defined as number of moles of solute per kg of solvent.
>
> $$m = \frac{w_2}{Mw_2} \times \frac{1000}{w_1}$$
>
> w_2 = mass of solute;
> Mw_2 = molecular mass of solute
> w_1 = mass of solvent

20% aqueous solution of KI means that 20 g of KI is present in 80 g solvent.

$$m = \frac{20}{166} \times \frac{1000}{80}$$

$$= 1.506 \approx 1.51\text{ mol/kg}$$

05. *(b)* Molality $(m) = \dfrac{w_{Na^+} \times 1000}{M_{Na^+} \times w_{H_2O}}$

$$= \frac{92 \times 1000}{23 \times 1000} = 4\text{ mol kg}^{-1}$$

06. *(d)* The best suited graph is given in option (d). The reason is explained through the given flow chart.

$$\underset{1°C}{\text{Water}} \xrightarrow[\text{decrease}]{\text{Volume}} \underset{4°C}{\text{Water}} \xrightarrow[\text{Thermal expansion}]{\text{Volume increases}} \underset{25°C}{\text{Water}}$$

$$\text{Molarity} = \frac{\text{No. of moles of solute}}{\text{Volume of solution in litre}}$$

Hence,

$$1°C \xrightarrow[\text{Increases}]{\text{Molarity}} 4°C \xrightarrow[\text{decreases}]{\text{Molarity}} 25°C$$

07. *(b)* The molecular mass of glucose is 180 g/mol
Mass of carbon in 250 gram solution is

$$= \frac{250 \times 10.8}{100} = 27\text{ g}$$

We know that, 72 g of C is present in 180 g glucose
so, 27 g of carbon is present in $\dfrac{180}{72} \times 27$ g glucose

$\Rightarrow$ Weight of glucose

$$= \frac{180}{72} \times 27 = 67.5\text{ g}$$

So, mass of solvent

$$= 250 - 67.5 = 182.5\text{ g}$$

$$\Rightarrow \text{Molality} = \frac{67.5 \times 1000}{180 \times 182.5}$$

$$= 2.055 \approx 2.06\text{ m}$$

08. *(c)* $X_{solvent} = 0.8$ (Given)
It means that $n_{solvent}\,(n_1) = 0.8$
and $n_{solute}\,(n_2) = 0.2$
Using formula, $m = n_2 \times \dfrac{1000}{n_1 \times M_1}$

$$= 0.2 \times \frac{1000}{0.8 \times 18} = 13.88\text{ mol kg}^{-1}$$

09. *(a)* % *w*/*w* of HNO_3 = 75%,
Means that 100 g, of solution contains 75 g of HNO_3

and $\left(\dfrac{g}{mL}\right)_{\text{Solution}} = 1.25 = \dfrac{100}{V}$ g

$\because$ 75 g of HNO_3 present in $\dfrac{100}{1.25}$ mL solution.

$\therefore$ 30 g of HNO_3 present in

$$\frac{100}{1.25 \times 75} \times 30 = 32\text{ mL solution}$$

10. *(c)* Given, density of commercial acid = 1.46 g/mL
So, mass of commercial acid

$$= \text{Density} \times \text{Volume}$$

$$= 1.46 \times 1000 = 1460\text{ g}$$

$\therefore$ Mass of given 35% HCl

$$= \frac{35}{100} \times 1460 = 511\text{ g}$$

Molarity of solution

$$= \frac{\text{Number of moles of acid}}{\text{Volume of acid (in litre)}}$$

$$= \frac{\text{Weight of acid}}{\text{M.M of acid} \times \text{Volume of acid (in litre)}}$$

$$= \frac{511}{36.5 \times 1\text{ L}} = 14\text{ M}$$

11. *(d)* $\chi_{Solute} = \dfrac{\dfrac{w_{Solute}}{Mw_{Solute}}}{\dfrac{w_{Solute}}{Mw_{Solute}} + \dfrac{w_{Solvent}}{Mw_{Solvent}}}$

Given, $w_{Solute} = w_{NaOH} = 8$ g
$Mw_{Solute} = Mw_{NaOH} = 40\text{ g mol}^{-1}$
$w_{Solvent} = w_{H_2O} = 18$ g
$Mw_{Solvent} = 18\text{ g mol}^{-1}$

$$\therefore \quad \chi_{Solute} = \chi_{NaOH} = \frac{\frac{8}{40}}{\frac{8}{40} + \frac{18}{18}}$$

$$= \frac{0.2}{0.2 + 1} = \frac{0.2}{1.2} = 0.167$$

$$m = \frac{\frac{w_{Solute}}{Mw_{Solute}}}{w_{Solvent}\text{ (in g)}} \times 1000$$

$$= \frac{\frac{8}{40}}{18} \times 1000 = \frac{0.2}{18} \times 1000$$

$$= 11.11\text{ mol kg}^{-1}$$

Thus, mole fraction of NaOH in solution and molality of the solution respectively are 0.167 and 11.11 mol kg^{-1}

12. *(2)* Given that, sea water is 6M solution of NaCl

So, 1000 ml of sea water contains 6mol of NaCl

Mass of solution = Volume × density

$$=1000\times 2=2000\text{ g}$$

$$\text{ppm}=\frac{\text{Mass of } O_2}{2000}\times 10^6$$

$$\text{Mass of } O_2=5.8\times 2\times 10^{-3}$$

$$=1.16\times 10^{-2}\text{ g}$$

$$\text{Molality for } O_2=\frac{1.16\times 10^{-2}/32\times 1000}{(2000-6\times 58.5)}$$

$$=\frac{1.16\times 10}{32\times 1649}=2.19\times 10^{-4}$$

$$\approx 2$$

13. *(1)* The reaction involved is as follows

$$\underset{\text{(M) 20mL}}{NaI(aq)}+\underset{\text{Excess}}{AgNO_3(aq)}\longrightarrow \underset{4.74\text{ g}}{AgI(s)}+NaNO_3(aq)$$

Moles of I^- in NaI = moles of I^- in AgI

$$=\frac{4.74}{235}$$

$$\text{Moles of NaI}=\frac{4.74}{235}$$

$$\text{Molarity [NaI]}=\frac{4.74}{235\times 0.02}=1.008\approx 1$$

14. *(125)* Given, $\frac{\text{Mass}}{\text{Mass}}\%=70\%$

Molarity = 2

Let 100 g solution contain 70 g solute.

Volume of 100 g solution will be $\frac{100}{1.25}$ mL

$$\text{Molarity}=\frac{70/70}{100/1.25}\times 1000$$

$$=12.5\text{ or } 125\times 10^{-1}$$

15. *(164)* $M_{soln}=V_{soln}\times d_{soln}$

$$=500\times 1.25=625\text{ g}$$

Mass of solute $(x)=0.2\times 0.5\times 159.5$

$$=15.95\text{ g}$$

$$n_{solute}=\frac{15.95\text{ g}}{159.5\text{ g/mol}}=0.1$$

Mass of solvent = Mass of solution – Mass of solute

$$=625-15.95=609.05\text{ g}$$

Now using formula of molality

$$\text{Molality}=\frac{\text{No. of moles of solute}\times 1000}{\text{Mass of solvent (in kg)}}$$

$$m=\frac{0.1\times 1000}{609.05}$$

$$m=0.164\text{ m}$$

$$m=164\times 10^{-3}\text{ m}$$

Answer is different from NTA i.e., 81.

16. *(74)* $\frac{x_B}{x_A}=\frac{m.M_A}{1000}$

Where x_B, x_A = Mole fraction,

m = molality

M_A = Molar mass of solvent

$$\frac{x_B}{1-x_B}=\frac{4.44\times 18}{1000}\quad [x_A+x_B=1]$$

$$\frac{x_B}{1-x_B}=0.08$$

$$x_B=0.074=74\times 10^{-3}$$

Compare it with given value, $x=74$.

17. *(815)* Molality is given by formula

$$m=\frac{M\times 1000}{d_{sol}\times 1000-M\times \text{Molar mass solute}}\quad \ldots\text{(i)}$$

$M=0.8$ M [given]

$d_{sol}=1.06\text{ g/cm}^3$

Molar mass of H_2SO_4 = 98 g/mol

Put all the value in Eq. (i),

$$=\frac{0.8\times 1000}{1.06\times 1000-0.8\times 98}$$

$$\text{Molality}=815\times 10^{-3}\text{ m}$$

18. *(4)* Given, let mass of solution = 100 g

Mass of H_2SO_4 in solution = 31.4%

density = 1.25 g/mL

Molar mass of H_2SO_4 = 98 g/mol

$$\text{Molarity, } M=\frac{n_{solute}}{V}\times 1000$$

$$=\frac{(31.4/98)}{(100/1.25)}\times 1000=4.005\approx 4\text{ M}$$

Hence, nearest integer is 4.

19. *(148)* Molarity

$$=\frac{\text{No. of moles of solute}}{\text{Volume of solution } (L)}$$

$$2.6\times 10^{-3}=\frac{x/85}{0.67141}$$

$$x=0.148\text{ g}$$

Conc. of dichloromethane in ppm

$$=\frac{0.148}{1.49\times 671.141}\times 10^6$$

$$=148\text{ ppm}$$

20. *(5)* Mass per cent, mole fraction, molarity, ppm and molality are the units which are used to express concentration of solution.

So, total number of units = 5.

21. *(54)* Moles of nitric acid

$$=800\times 10^{-3}\times 0.5=0.4$$

Moles of niric acid evaporated

$$=\frac{11.5}{63}=0.182$$

Moles left = 0.4 - 0.182 = 0.218

$$\text{Molarity}=\frac{0.218\times 10^3}{400}$$

$$=0.545\text{ M}=54.5\times 10^{-2}\text{ M}$$

22. *(64)* Given, mass of $CuSO_4\cdot 5H_2O$ = 80 g

The concentration of copper sulphate solution is $x\times 10^{-3}$ mol/L.

Molar mass of

$$CuSO_4\cdot 5H_2O=63.54+32+16\times 4+5\times 18$$

$$=249.54\text{ g / mol}$$

Number of moles of solute

$$=\frac{\text{Weight of solute}}{\text{Molecular mass of solute}}$$

$$=\frac{80\text{ g}}{249.54\text{ g / mol}}=0.32\text{ mol}$$

Volume of solution = 5 L

$$\text{Molarity}=\frac{\text{Number of moles of solute}}{\text{Volume of solution (in L)}}$$

$$\therefore \text{Molarity}=\frac{0.32}{5}$$

$$=64.11\times 10^{-3}\text{ mol / L}$$

$$\therefore \quad x=64.11\text{ or } x\approx 64$$

Hence, answer is 64.

23. *(3.17)* Using the formula of molality

$$m=\frac{M\times 1000}{(1000\,d)-(M\times M_{solute})}$$

$$3.3=\frac{M\times 1000}{1000\times 1.2-74.5\times M}$$

$$3960-245.8\,M=1000\,M$$

$$1245.85\text{ M}=3960$$

$$\Rightarrow \quad M=3.17\text{ M}$$

24. *(5)* Given density = 1.2 g/cm^3

So, mass of solvent

$$=1\text{ cm}^3\times 1\text{ g/cm}^3=1\text{ g}$$

Mass of solute = 1.2 – 1 = 0.2 g

$$m=\frac{0.2}{40}\times 1000=5$$

25. *(64)* 100 molal aqueous solution means 100 mole solute in 1000 g water.

$$\therefore \text{mole fraction of solute}=\frac{100}{100+\frac{1000}{18}}$$

$$=\frac{1800}{2800}=64.28\times 10^{-2}$$

26. *(9)* Molality

$$(m) = \frac{M \times 1000}{\{(1000\,d) - M \times M_{solute}\}}$$

M_{solute} = molar mass of solute

Putting the values,

$$6.50 = \frac{M \times 1000}{(1000 \times 1.89) - (M \times 56)}$$

On solving, $M = \frac{12285}{1364}$

$M = 9 \text{ mol dm}^{-3}$

27. *(100)* Hardness in ppm

$$= \frac{\text{Hardness causing substances} \times 10^6}{\text{Mass of water sample}}$$

Here, $[Mg^{2+}] = 10^{-3}$ M

$n_{Mg^{2+}} = 10^{-3}$ per 1 L water $\cong$ 1000 g

Mass of 10^{-3} moles of $CaCO_3$

$= 10^{-3} \times 100 \text{ g} = 0.1 \text{ g}$

$\therefore \text{ppm} = \frac{0.1 \times 10^6}{1000} = 100 \text{ ppm}$

28. *(10)* Mass of O_2 = 10.30 mg (Given)

Mass of sea water solution

= volume × density

$= 1 \text{ L} \times 1.03 \text{ g/mL}$

$= 1000 \text{ mL} \times 1.03 \text{ g mL}^{-1}$

$= 1030$ g

$$\text{ppm} = \frac{\text{mass of } O_2}{\text{mass of solution}} \times 10^6$$

$$= \frac{10.30 \times 10^{-3} \text{ g} \times 10^6}{1030 \text{ g}}$$

= 10 ppm

Topic 2

01. *(a)* Using Dalton's Law of partial pressure

$p_S = p_A^o X_A + p_B^o X_B$

$500 = 200 \times \frac{1}{4} + p_B^o \times \frac{3}{4}$

$p_B^o = 600$ mmHg

As $p_A^o < p_B^o \Rightarrow A$ is least volatile.

02. *(d)* Given, $\Delta p_1 = 10$ mm Hg

Initial mole fraction of solute $\chi_{B_1} = 0.2$

$\Delta p_2 = 20$ mm Hg

Final mole fraction of solute, $\chi_{B_2} = ?$

Mole fraction of solvent when vapour pressure is 20 mm Hg

$\chi_{A_2} = ?$

$$\frac{\Delta p_1}{p^\circ} = \chi_{B_1} \Rightarrow \frac{\Delta p_2}{p^\circ} = \chi_{B_2}$$

$$\frac{\Delta p_1}{\Delta p_2} = \frac{\chi_{B_1}}{\chi_{B_2}}$$

$$\Rightarrow \quad \frac{10}{20} = \frac{0.2}{\chi_{B_2}}$$

$\chi_{B_2} = 0.2 \times 2 = 0.4$

$\chi_{A_2} = 1 - \chi_{B_2} = 1 - 0.4 = 0.6$

03. *(c)* Given, $\overset{\circ}{p}_A < \overset{\circ}{p}_B$

$$\frac{\overset{\circ}{p}_A}{\overset{\circ}{p}_B} < 1$$

$$\frac{y_A}{y_B} = \frac{\overset{\circ}{p}_A\, x_A}{\overset{\circ}{p}_B\, x_B}$$

$$\frac{\frac{y_A}{y_B}}{\frac{x_A}{x_B}} < 1 \Rightarrow \frac{y_A}{y_B} < \frac{x_A}{x_B}$$

04. *(b)* From given graph it can be seen that, X has higher vapour pressure as compared to Y at lower temperature (or to attain same vapour pressure, Y need higher temperature than X). It means that intermolecular interactions must be weaker in X as compared to Y.

05. *(c)* According to Raoult's law, vapour pressure of the solution p_{ideal} (ideal vapour pressure) $= \overset{\circ}{p}_A + (\overset{\circ}{p}_B - \overset{\circ}{p}_A) x_B$

(A = solvent : acetone, B = solute : CS_2)

Given; $\overset{\circ}{p}_A = 344$ mm Hg.

$\overset{\circ}{p}_B = 512$ mm Hg.

$\therefore \quad p_{ideal} = 344 + (512 - 344) x_B$

or $\quad p_{ideal} = 344 + 168\, x_B$

So, $\quad 0 < x_B < 1$.

Also, total vapour pressure (p) = 600 mmHg

$\therefore$ For any value of x_B, $p > p_{ideal}$.

Option (c) is incorrect. It is because, mixture of pure CS_2 and CH_3COCH_3 shows positive deviation from Raoult's law. For such solution $\Delta V_{min} > 0$ and $\Delta H_{mix} > 0$.

06. *(a)*

> **idea** For a solution of volatile liquids the partial vapour pressure of each component of the solution is directly proportional to its mole fraction present in solution. This is known as Raoult's law.

Liquid M and N form an ideal solution. Vapour pressures of pure liquids M and N are 450 and 700 mm Hg, respectively.

$\therefore \quad p^\circ{}_N > p^\circ{}_M$

So, by using Raoult's law

$y_N > x_N$ ...(i)

and $x_M > y_M$ ...(ii)

Multiplying (i) and (ii) we get

$y_N\, x_M > y_M\, x_N$

$$\therefore \quad \frac{x_M}{x_N} > \frac{y_M}{y_N}$$

Thus, correct relation is (a).

07. *(a)* According to Henry's law (at constant temperature)

$p_{gas} = K_H \times \chi_{gas\,(solute)}$

$= K_H \times [1 - \chi_{H_2O\,(solvent)}]$

$p_{gas} = K_H - K_H\, \chi_{H_2O}$

Comparing the above equation with the equation of straight line $y = mx + c$

Slope $= -K_H$, intercept $= K_H$

So, higher the value of K_H, more (–ve) will be the slope.

Therefore, the intercept on Y-axis is increases from w to z.

08. *(b)* Henry's law constant (K_H)

$$K_H = \frac{\text{Pressure}}{\text{Solubility of a gas in a liquid}} = \frac{p}{S}$$

Thus, higher the value of K_H at a given pressure, the lower is the solubility of the gas in the liquids.

09. *(d)* Binary mixture of C_6H_5OH and $C_6H_5NH_2$ shows negative deviation from Raoult's law.

So, vapour pressure of solution is less than vapour pressure of pure C_6H_5OH and $C_6H_5NH_2$.

Thus, boiling point of solution is greater than boiling point of pure C_6H_5OH and $C_6H_5NH_2$.

So, it shows maximum boiling azeotrope.

10. *(a)* The correct match is A-III, B-I, C-IV, D-II

- Solution of chloroform and acetone shows –ve deviation, so maximum boiling azeotrope.
- Solution of ethanol and water shows +ve deviation, so minimum boiling azeotrope.
- Solution of benzene and toluene from ideal solution. $\Delta V_{mix} = 0$.
- Acetic acid in benzene form dimer.

11. *(b)* Using Dalton's Law of partial pressure,

$$\frac{p_{N_2}}{p_{O_2}} = \frac{x_{N_2} \cdot p_T}{x_{O_2} \cdot p_T} = \frac{n_{N_2}}{n_{O_2}} = \frac{\frac{70}{28}}{\frac{27}{32}} = 2.96$$

$$\frac{p_{O_2}}{p_{Ar}} = \frac{n_{O_2}}{n_{Ar}} = \frac{\frac{27}{32}}{\frac{3}{40}} = 11.25 \approx 11.2$$

12. *(c)* When we increase the temperature of solutions, first solubility decreases than increases.
Hence K_H first increase than decrease.
Also at moderate temperature K_H value is $He > N_2 > CH_4$.

13. *(d)* For the given binary solution of two volatile liquid components 1 and 2.

$$p_1 = p_T y_1 = p_1^\circ x_1$$

$$\frac{p_T}{x_1} = \frac{p_1^\circ}{y_1}$$

So, $$\frac{p_2^\circ + x_1(p_1^\circ - p_2^\circ)}{x_1} = \frac{p_1^\circ}{y_1}$$

$\therefore$ $$\frac{p_2^\circ}{x_1} + (p_1^\circ - p_2^\circ) = \frac{p_1^\circ}{y_1}$$

$\therefore$ $$\frac{1}{x_1} = \left(\frac{p_1^\circ}{p_2^\circ}\right)\left(\frac{1}{y_1}\right) + \left(\frac{p_2^\circ - p_1^\circ}{p_2^\circ}\right)$$

$\therefore$ Slope $= \left(\frac{p_1^\circ}{p_2^\circ}\right)$

So, Intercept $= \left(\frac{p_2^\circ - p_1^\circ}{p_2^\circ}\right)$

Thus, option (d) is the correct answer.

14. *(a)* Among the given options $(CH_3)_2CO + CS_2$ i.e., shows positive deviation from Raoult's law. Since acetone-CS_2 attractions are weaker than acetone-acetone attractions and $CS_2 - CS_2$ attractions.

Concept Enhancer
The liquid solutions which shows positive deviation from the Raoult's law form minimum boiling azeotropes.

15. *(b)* The $A - B$ interactions in negative deviation solutions are stronger than the $A - A$ and $B - B$ interactions in the two liquids.
Due to greater interactions, A and B types of molecules have a lower tendency to escape from solutions than pure liquids.
As a result based on Raoult's law, each component of solution has a partial pressure lower than expected.
If vapour pressure decreases so boiling point increases.

16. *(d)* According to Dalton's law of partial pressure, $p_{total} = p_A + p_B$

$$= p_A^\circ \chi_A + p_B^\circ \chi_B \quad \ldots(i)$$

Given, $p_A^\circ = 400$ mm Hg,

$p_B^\circ = 600$ mm Hg

$\chi_B = 0.5$, $\chi_A + \chi_B = 1$

$\therefore$ $\chi_A = 0.5$

On substituting the given values in Eq. (i), We get,

$$p_{total} = 400 \times 0.5 + 600 \times 0.5 = 500 \text{ mm Hg}$$

Mole fraction of A in vapour phase,

$$= \frac{0.5 \times 400}{500} = 0.4$$

Mole fraction of B in vapour phase,

$$Y_A + Y_B = 1$$
$$Y_B = 1 - 0.4 = 0.6$$

17. *(b)* For ideal solution,

$$p = x'_A p^\circ_A + x'_B p^\circ_B \quad \ldots(i)$$

$\because$ $x'_A = 0.4$, $x'_B = 0.6$

$p^\circ_A = 7 \times 10^3$ Pa,

$p^\circ_B = 12 \times 10^3$ Pa

On substituting the given values in Eq. (i), we get

$$p = 0.4 \times 7 \times 10^3 + 0.6 \times 12 \times 10^3 = 10 \times 10^3 \text{ Pa} = 1 \times 10^4 \text{ Pa}$$

In vapour phase,

$$x_A = \frac{0.4 \times 7 \times 10^3}{1 \times 10^4} = 0.28$$

$\therefore$ $x_B = 1 - 0.28 = 0.72$ $[\because x_A + x_B = 1]$

18. *(23)* $$y_A = \frac{p_{\text{methyl benzene}}}{p_{total}}$$

$$y_A = \frac{24}{80 + 24} = 0.2307$$

$$y_A = 23.07 \times 10^{-2}$$

$$y_A \approx 23 \times 10^{-2}$$

19. *(82)*

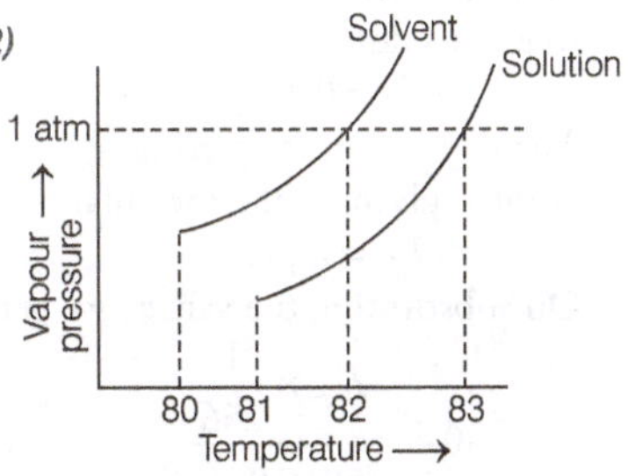

From the graph at 1 atm pressure, boiling point of solvent = 82°C

20. *(314)* Given;
Mole fraction of $A_1 = 0.7$
Mole fraction of $B_1 = 0.3$
Total pressure, $p_1 = 350$ mm Hg
Mole fraction of $A_2 = 0.2$
Mole fraction of $B_2 = 0.8$
Total pressure, $p_2 = 410$ mm Hg
Let vapour pressure of pure A be p_A° and B be p_B°.
When, mole fraction of

$A_1 = \chi_{A_1} = 0.7$

and mole fraction of B_1,

$= \chi_{B_1} = 0.3$

and total pressure $p_1 = 350$.

$$\Rightarrow (p_A^\circ \times 0.7) + (p_B^\circ \times 0.3) = 350 \quad \ldots(i)$$

When, mole fraction of A_2,

$= \chi_{A_2} = 0.2$

mole fraction of B_2, $= \chi_{B_2} = 0.8$

and, total pressure $p_2 = 410$

$$(p_A^\circ \times 0.2) + (p_B^\circ \times 0.8) = 410 \quad \ldots(ii)$$

Now, solving Eq. (i) and (ii), we get

$$0.7p_A^\circ + 0.3p_B^\circ = 350 \quad \ldots(i)$$
$$0.2p_A^\circ + 0.8p_B^\circ = 410 \quad \ldots(ii)$$

Multiply Eq. (i) by 0.2 and Eq. (ii) by 0.7, we get

$$0.14p_A^\circ + 0.06p_B^\circ = 70 \quad \ldots(iii)$$
$$0.14p_A^\circ + 0.56p_B^\circ = 287 \quad \ldots(iv)$$
$$- \quad - \quad -$$
$$0 - 0.5p_B^\circ = -217$$

$$\Rightarrow \quad -0.5p_B^\circ = -217$$

$$\Rightarrow \quad p_B^\circ = \frac{217}{0.5} = 434 \text{ mm Hg}$$

Now put the value of p_B° in Eq. (i),

$$0.7p_A^\circ + 0.3(434) = 350$$
$$0.7p_A^\circ = 350 - 130.2$$
$$0.7p_A^\circ = 219.8$$
$$p_A^\circ = \frac{219.8}{0.7} = 314$$

Therefore, the vapour pressure of pure 'A' is 314 mm Hg.

21. *(80)* Total pressure of cylinder,

$p_{total} = 25$bar

Partial pressure of Ne, $p_{Ne} = 20$ bar

$\therefore$ Partial pressure of O_2

$$p_{O_2} = p_{total} - p_{Ne} = 25 - 20 = 5 \text{ bar}$$

Moles of $O_2 = \frac{x}{32}$ mol

Mass of Ne = 200 g

Moles of Ne $= \frac{200}{20}$ mol

Now, putting value of p_{O_2}

$$\Rightarrow \quad 5 = \frac{\frac{x}{32}}{\frac{x}{32} + \frac{200}{20}} \times 25$$

$$\Rightarrow \quad \frac{1}{5} = \frac{\frac{x}{32}}{\frac{x}{32} + 10} \Rightarrow 5x = x + 320$$

$$\Rightarrow \quad 4x = 320 \Rightarrow x = 80 \text{ g}$$

So, the value of 'x' is 80.

22. *(1)* Given, Henry's constant,

$$K_H = 46.82 \text{ k bar} = 46.82 \times 10^3 \text{ bar}$$

Partial pressure of O_2, $p_{O_2} = 0.92$ bar

Using the Henry's equation,

$p_{O_2} = K_H \cdot \chi_{O_2}$ [χ = mole fraction]

So, $\chi_{O_2} = \frac{p_{O_2}}{K_H}$

$$\chi_{O_2} = \frac{0.92 \times 10^{-3}}{46.82} = 0.02 \times 10^{-3}$$

Mole fraction of water,

$\chi_{H_2O} = 1 - \chi_{O_2} = 1 - 0.02 \times 10^{-3} = 0.99$

1 L water = 1000 gm ($\rho_{H_2O} - 1 \text{ gmL}^{-1}$)

Moles of water $= \frac{1000}{18} = 55.55$ mol

So, moles of O_2,

$$n_{O_2} = \frac{55.55 \times 0.02 \times 10^{-3}}{0.99}$$

$$= 1.13 \times 10^{-3} \text{ mol}$$
$$= 1.13 \text{ m mol} \approx 1 \text{ m mol}$$

23. *(25)* According to Henry's law,

$$p_{CO_2} = K_H \chi_{CO_2}$$

p_{CO_2}, K_H and χ_{CO_2} are pressure of CO_2 gas, Henry's constant and mole fraction of CO_2 respectively.

$$\Rightarrow \quad \chi_{CO_2} = \frac{p_{CO_2}}{K_H} \quad \ldots \text{(i)}$$

Given, $K_H = 0.835 \times 2 \times 10^3$ bar

$= 1.67 \times 10^3$ bar

$p_{CO_2} = 0.835$ bar

Put all values in equation (i), we get

$$\chi_{CO_2} = \frac{0.835}{0.835 \times 10^3 \times 2} = 0.5 \times 10^{-3}$$

Volume of water = 0.9 L = 900 mL

Density of water = 1 g/mL

Mass of water = density × volume

= 900 g

So, number of mole of water

$$= \frac{900}{18} = 50 \text{ moles}$$

So, $\frac{n_{CO_2}}{n_{CO_2} + n_{H_2O}} = 0.5 \times 10^{-3}$

$n_{CO_2} << n_{H_2O} = 50$ moles

So, $n_{CO_2} + n_{H_2O} = n_{H_2O}$

$n_{CO_2} = 0.5 \times 10^{-3} \times 50$

$= 25 \times 10^{-3}$ mol

Millimoles of $CO_2 = 0.025 \times 1000 = 25$

24. *(19)* Given, $p_A^\circ = 21$ kPa, $p_B^\circ = 18$ kPa

$n_A = 1$ mole, $n_B = 2$ mole

$$x_A = \frac{1}{3} \quad x_B = \frac{2}{3}$$

$$P_T = 21 \times \frac{1}{3} \text{ kPa} + 18 \times \frac{2}{3} \text{ kPa}$$
$$= 7 + 12 = 19 \text{ kPa}$$

25. *(600)* $p_{Total} = x_{hexane} p^\circ_{hexane} + x_{heptane} p^\circ_{heptane}$

Given that, $p_{total} = 550$ mm

$$x_{hexane} = \frac{1}{1+3} = \frac{1}{4}, \; x_{heptane} = \frac{3}{3+1} = \frac{3}{4}$$

Substituting the values in above formula

$$550 = \frac{1}{4} \times p^\circ_{hexane} + \frac{3}{4} \times p^\circ_{heptane} \quad \ldots \text{(i)}$$

After mixing 1 more mole of n-heptane

$$560 = \frac{1}{5} \times p^\circ_{hexane} + \frac{4}{5} \times p^\circ_{heptane} \quad \ldots \text{(ii)}$$

From Eq. (i) and (ii)

$p^\circ_{heptane} = 600$ mm Hg.

Topic 3

01. *(a)* Given molar mass of non-volatile, Solute = 256 g/mol

Mass of solvent = 50 g

Mass of solute = 1 g

$\Delta T_f = 0.40$ g

The depression in freezing point is given by the formula,

$$\Delta T_f = K_F m$$

On substituting the values, we get

$$0.4 = \frac{K_F \times \frac{1}{256}}{50 \times 10^{-3}}$$

$$\Rightarrow \quad K_F = 5.12 \text{ K kg/mol}$$

02. *(c)* The expression of elevation of boiling point,

$$\Delta T_b = K_b \times m$$

So, the expression becomes,

$$\Delta T_b \propto K_b$$

$$\Rightarrow \quad \frac{\Delta T_b(A)}{\Delta T_b(B)} = \frac{K_b(A)}{K_b(B)} = \frac{1}{5}$$

$$\left[\text{Given } \frac{K_b(A)}{K_b(B)} = \frac{1}{5}\right]$$

03. *(b)* Given $K_b = 0.52$ K kg mol^{-1}

Mass of water = 500 g = 0.5 kg

Number of moles of ethylene glycol = 2

Number of moles of water = 2

Using $\Delta T_b = i_1 m_1 k_b + i_2 m_2 k_b$

$$= 1 \times \frac{2}{0.5} \times 0.52 + \frac{1 \times 2}{0.5} \times 0.52$$
$$= 4.16$$

$(T_b)_{Solution} = 373.16 + 4.16 = 377.3$ K

04. *(d)* Depression in freezing point is given by

$$\Delta T_f = iK_f m$$
$$0.2 = i \times 1.8 \times 0.1$$
$$i = \frac{20}{18} = \frac{10}{9}$$

For, $HA(aq) \rightleftharpoons H^+(aq) + A^-(aq)$

$t = 0$	1		
$t = t_{eq}$	$1 - \alpha$	α	α

$i = 1 + \alpha$

$$\frac{10}{9} = 1 + \alpha \Rightarrow \alpha = \frac{1}{9}$$

$$K_{eq} = \frac{[H^+][A^-]}{[HA]} = \frac{C\alpha^2}{1-\alpha} = \frac{0.1\left(\frac{1}{9}\right)^2}{1 - \frac{1}{9}}$$

$$K_{eq} = 1.38 \times 10^{-3}$$

05. *(d)* When a non-volatile solute is added to solvent, freezing point of solution decreases.

$T_f < T_f^\circ$ i.e.,

Freezing point of solution < freezing point of pure solvent.

Also vapour pressure of solution decreases on adding a non-volatile solute in a solvent.

Hence graph (d) is correct.

06. *(d)* Both Statement I and Statement II are true.

07. *(d)* When small amount of naphthalene is added to benzene, depression in freezing point takes place and freezing point of benzene decreases.

08. *(a)*

> **idea** This question involves the concept of Raoult's law and mole concept. Use Raoult's law to calculate the number of moles of glucose. Then use this value to find out the weight by multiplying it with molar mass of glucose.

According to Raoult's law, for dilute solution

$$\frac{p^\circ - p}{p^\circ} = \frac{n}{N}$$

[where n = number of moles of glucose, N = number of moles of H_2O]

On substituting the values

$$\frac{0.2}{54.2} = \frac{n \times 18}{100} \Rightarrow n = \frac{100}{271 \times 18}$$

Now weight of glucose = number of moles × molar mass

$$w = \frac{100}{271 \times 18} \times 180 \Rightarrow w = 3.69 \text{ g}$$

09. *(b)* Upon addition of non-volatile solute, less number of solvent particle can evaporate. Hence, vapour pressure of solution molecule decreases than pure solvent and hence, vapour pressure of pure solvent is more than that of solution. So, in order to get equal vapour pressure at freezing point only solvent molecule solidifies.

10. *(b)* Using the equation, $\Delta T_f = K_f m$

$$\therefore \frac{\Delta T_{f_A}}{\Delta T_{f_B}} = \frac{K_f m_A}{K_f m_B} = \frac{m_A}{m_B} = \frac{M_Y}{M_X} = \frac{1}{4}$$

So, $M_X : M_Y = 4 : 1 = \dfrac{4 \times 0.25}{1 \times 0.25} = \dfrac{1}{0.25}$

$M_X : M_Y = 1 : 0.25$

So, option (b) is correct.

11. *(c)*

> **idea** Osmotic pressure is proportional to the molarity (C) of the solution at a given temperature (T). So, use the formula $\pi = CRT$ to calculate the value of π.

For the relation,

$$\pi = CRT = \frac{n}{V} RT$$

Given, mass of urea = 0.6 g

Molar mass of urea = 60 g mol^{-1}

Mass of glucose = 1.8 g

Molar mass of glucose = 180 g mol^{-1}

$$\pi = \frac{\left(\frac{0.6}{60} + \frac{1.8}{180}\right)}{100} \times 1000 \times 0.0821 \times 300$$

$$= (0.01 + 0.01) \times 10 \times 0.0821 \times 300$$

$$\pi = 4.92 \text{ atm}$$

12. *(c)* χ_B = mole fraction of solute (urea)

$$= \frac{n_B}{n_A + n_B} = \frac{0.60/60}{\frac{360}{18} + \frac{0.60}{60}} = \frac{0.01}{20 + 0.01}$$

$$= \frac{0.01}{20.01} = 0.0005$$

Now, according to Raoult's law,

$$\Delta p = \chi_B \times p^\circ$$

On substituting the above given values, we get

$$\Delta p = 0.0005 \times 35 = 0.0175 \text{ mm Hg}$$

13. *(d)* Elevation in boiling point

$$(\Delta T_b) = K_b \times m$$

Depression in freezing point

$$(\Delta T_f) = K_f \times m$$

where, m = molality

$$\Delta T_b^{1m} = \Delta T_f^{2m} = 2K$$

So, $K_b \times 1 = K_f \times 2$

$\Rightarrow K_b = 2K_f$

14. *(c)* Statement I is correct but Statement II is incorrect.

The correct form of Statement II is k_f for benzene is more than k_f for water.

15. *(d)* Given, $K_f = 1.86$ K kg mol^{-1}

$\Delta T_f = 0.558$ ℃

Molality = 0.1 molal

Using formula,

$$\Delta T_f = iK_f m$$
$$0.558 = i \times 1.86 \times 0.1$$

On solving, we get

$$i = 3$$

So, among the given options, compound given in option (d) have the value of $i = 3$. So, the complex is $[Cr(NH_3)_5Cl]Cl_2$.

16. *(c)* Due to osmosis, net movement of water molecules takes place from low osmotic pressure to high osmotic pressure. Since the value of concentration of 0.05 M $CuSO_4$ solution is higher hence, it has higher osmotic pressure. So, movement of water molecule will take towards $CuSO_4$ solution. Due to which molarity of $CuSO_4$ is lowered.

17. *(b)* $\Delta T_b = K_b \times m$

$$(\Delta T_b)_A = (\Delta T_b)_B \text{ (given)}$$

$\therefore \quad m_A = m_B$

Molality

$$= \frac{\text{Mass of solute} \times 1000}{\text{M.M. of solute} \times \text{Weight of solvent}}$$

$$m_A = \frac{2 \times 1000}{M_A \times 98};$$

$$m_B = \frac{8 \times 1000}{M_B \times 92}$$

$$\frac{2 \times 1000}{M_A \times 98} = \frac{8 \times 1000}{M_B \times 92}$$

$\therefore \quad M_B = 4M_A$

18. *(543)* Given,

$\Delta T_f = 2.5°C$ (from diagram)

$K_f = 1.86$ K kg mol^{-1}

$$\Delta T_f = K_f \times m$$

$$2.5 = 1 \times 1.86 \times \frac{n_B}{0.1}$$

$n_B = 0.1344$ mole

Mass of methanol = 0.1344 × 32 = 4.3 g

$$d = \frac{m}{V} \Rightarrow V = \frac{m}{d}$$

$$V = \frac{4.3}{0.792} = 5.43 \text{ mL}$$

$$= 543 \times 10^{-2} \text{ mL}$$

Thus, the value of x is 543.

19. *(25)* $NaCl \rightarrow Na^+ + Cl^-$

0.05 M   0.05 M   0.05 M

Total C_1 = 0.05 + 0.05 = 0.1 M (NaCl)

C_2 = 0.2 M (glucose)

Osmotic pressure is given by,

$$\pi = (C_2 - C_1)RT$$
$$= (0.2 - 0.1) \times 0.083 \times 300 = 2.49 \text{ bar}$$
$$= 24.9 \times 10^{-1} \text{ bar}$$
$$\approx 25 \times 10^{-1} \text{ bar}$$

20. *(31)*

As the number of moles of water is greater than the moles of acetic acid. Thus, water acts as a solvent.

Now, calculating the molality of acetic

acid $= \dfrac{2700}{60} \times \dfrac{1}{2.7}$ mol/kg

= 16.667 mol/kg

Now, using $\Delta T_f = K_f m$

$= 1.86 \times 16.667 = 31$ K

21. *(707)*

> **idea** To solve this question, use the formula $\Delta T_b = k_b m$ to calculate the value of molality (m). After that use this value of molality in the equation $\dfrac{\Delta p}{p^\circ} = \dfrac{m}{100} \times M_{solvent}$, as the solution is diluted, to calculate Δp. Then using this Δp calculate the required value of p_s.

Given that,

$\Delta T_b = 2°C$

$K_b = 0.52$ K kg mol^{-1}

$p^\circ = 760$ mm Hg

$M_{solvent} = 18$ g mol^{-1}

Now using $\Delta T_b = K_b m$

$$2 = 0.52 \times m \Rightarrow m = \frac{2}{0.52}$$

Now, according to question, since the solution is diluted much. Hence,

$$\frac{\Delta p}{p^\circ} = \frac{n_{\text{solute}}}{n_{\text{solvent}}} \Rightarrow \frac{\Delta p}{p^\circ} = \frac{m}{1000} \times M_{\text{solvent}}$$

$$\Rightarrow \Delta p = 760 \times \frac{\left(\frac{2}{0.52}\right)}{1000} \times 18$$

$\Delta p = 52.615$

So, $p_s = 760 - 52.615$

$p_s = 707.385$ mm Hg

Hence, nearest integer is 707.

22. *(73)* Given, $T_1 = 273$ K

Osmotic pressure at $T_1 = 7 \times 10^5$ Pa

$T_2 = 283$ K

Osmotic pressure, $\pi = CRT$

$\pi_1 = CRT_1$ and $\pi_2 = CRT_2$

$$\frac{\pi_1}{\pi_2} = \frac{T_1}{T_2} \Rightarrow \pi_2 = \frac{\pi_1 T_2}{T_1}$$

$$\Rightarrow \pi_2 = \frac{7 \times 10^5 \times 283}{273}$$

$\pi_2 = 72.56 \times 10^4$ N/m^2

$\approx 73 \times 10^4$ N/m^2

Hence, nearest integer value is 73.

23. *(23)* Given,

weight of solute $= 30$ g

volume of solution $= 100$ mL

weight of solution $= 1.2 \times 100 = 120$ g

weight of solvent $= 120 - 30 = 90$ g

$$m = \frac{30 \times 1000}{180 \times 90} = 1.85$$

$$\frac{24 - p_S}{p_S} = \frac{1.85 \times 18}{1000}$$

$\Rightarrow \quad 24 - p_S = 0.0333 p_S$

$p_S (1.033) = 24$

$\Rightarrow \quad p_S = 23.22$

24. *(8)* The boiling point elevation constant (K_b) of a volatile liquid is given by the following expression

$$K_b = \frac{RT_b^2 M}{\Delta Hv}$$

where, $T_b =$ boiling point of the liquid

$M =$ molar mass of the liquid

$\Delta Hv =$ enthalpy of vaporisation

Given, $M_x = M_y$

$(Tb)_x = 2(Tb)_y$

$2(\Delta Hv)_x = (\Delta Hv)_y$

$$\therefore \frac{(K_b)_x}{(K_b)_y} = \left(\frac{2}{1}\right)^2 \times \frac{2}{1} = \frac{8}{1} = m$$

$\therefore \quad m = 8$

25. *(833)* Given, mass of water $= 1000$ g

Molar mass of mass of urea $= 60$ g mol^{-1}

Mass of urea $= ?$

We know that, vapour pressure decreases by 25%.

$$\text{So, } \frac{p^\circ - p}{p^\circ} = \frac{\text{Moles of urea}}{\text{Moles of water}}$$

$$\frac{100 - 75}{100} = \frac{\frac{x}{60}}{\frac{1000}{18}}$$

$x = 833.3 \text{ g} \approx 833$

26. *(62250)* Osmotic pressure $(\pi) = CRT$

Given : solute mass $= 2.5$ g

Volume of solution $= 250$ ml

osmotic pressure $= 400$ pa

$$\text{So, } \frac{400\text{Pa}}{10^5} = \frac{2.5\text{g}/M}{250/1000} \times 0.83 \frac{\text{L-bar}}{\text{K mol}} \times 300\text{K}$$

$\therefore \quad M = 62250$ gmol^{-1}

27. *(100)* Given, mass of solute $= 2\text{g} = 2 \times 10^{-3}$ kg

Mass of solvent $= 20\text{g} = 20 \times 10^{-3}$ kg

Temperature $= 373.52$ K

Elevation in boiling point,

$\Delta T_b = 373.52 \text{ K} - 373 \text{ K} = 0.52$ K

Use the formula,

$\Delta T_b = K_b \times m$

Ebullioscopic constant of water

$(K_b) = 0.52$ K kg mol^{-1}

Mass of solute $= 2\text{g} = 0.002$ kg

$$\text{So, } 0.52 = 0.52 \times \frac{2}{\text{molar mass}} \times \frac{1}{20 \times 10^{-3}}$$

$$\text{Molar mass} = \frac{0.52 \times 2 \times 1}{0.52 \times 20 \times 10^{-3}}$$

$= 100$ g/mol

28. *(41500)*

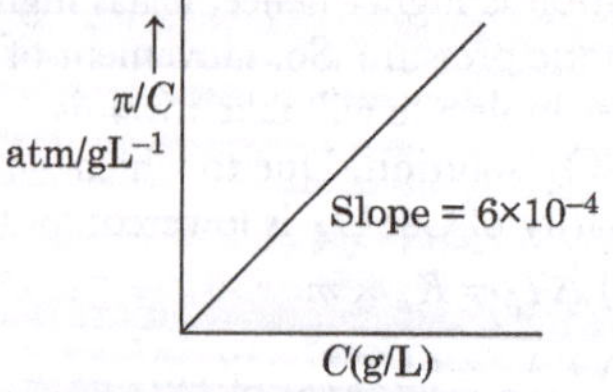

$R = 0.0831$ L atm/K/mol

Osmotic pressure, $\pi =$ Molar concentration $\times RT$

$$\pi = \frac{C}{M} RT$$

$$\Rightarrow \quad \frac{\pi}{C} = \frac{RT}{M}$$

Assuming π/C *vs* C graph

$$\frac{RT}{\text{Slope}} = \frac{RT}{M} = \frac{0.083 \times 300}{6 \times 10^{-4}}$$

$M = 41500$ g/mol

29. *(15)* We know that,

$\Delta T_f = K_f \times m$

$$= 3.9 \times \frac{10.2 \times 1000}{176 \times 150} = 1.5°\text{C}$$

or 15×10^{-1} °C

$\therefore x = 15$

30. *(6)* Vapour pressure of pure water,

$p^\circ_{H_2O} = 23.76$ mmHg

Vapour pressure of solution,

$$p_s = \frac{p^\circ_{H_2O}}{2} = 11.88 \text{ mmHg}$$

According to the expression of relative lowering of vapour pressure.

$$\frac{p^\circ_{H_2O} - p_s}{p^\circ_{H_2O}} = \chi_A$$

$$\frac{23.76 - 11.88}{23.76} = \frac{n_A}{n_A + n_{H_2O}} \quad \left[n_{H_2O} = \frac{100}{18} = 5.55\right]$$

$$\frac{1}{2} = \frac{n_A}{n_A + 5.55} \text{ or } n_A \approx 6$$

31. *(1)* Given,

$\Delta T_b = K_b \times m = 3$

or, $\quad \Delta T_b = K_b \times 1 = 3$

$\Delta T_f = K_f \times m = 6$

or, $\quad \Delta T_f = K_f \times 2 = 6$

$$\frac{\Delta T_b}{\Delta T_f} = \frac{K_b \times 1}{K_f \times 2}$$

$$\frac{3}{6} = \frac{K_b}{K_f \times 2}$$

$$\Rightarrow \quad \frac{6}{6} = \frac{K_b}{K_f}$$

According to data given,

$$\frac{K_b}{K_f} = \frac{1}{1} = \frac{1}{x} \Rightarrow x = 1$$

32. *(45)* Mass of solute $= 2.5 \times 10^{-3}$ kg

Mass of solvent $= 75 \times 10^{-3}$ kg

Boiling point of solution $= 373.535$ K

Boiling point of water $= 373.15$ K

$\therefore$Elevation in boiling point ΔT_b

$=$ Boiling point of solution $-$ Boiling point of water

$= 373.535 - 373.15$

$= 0.385$ K

Also, $\Delta T_b = K_b \times$ molality

$$0.385\ K = 0.52 \times \left(\frac{2.5\times 10^{-3}}{M\times 75\times 10^{-3}}\right)$$

$M = 0.045$ kg/mol or $M = 45$ g/mol

Hence, molar mass of solute is 45 g/mol.

33. *(8)* We know that $\Delta T_b \propto K_b$, as

$$\Delta T_b = K_b m$$

$$\Rightarrow \frac{(\Delta T_b)_A}{(\Delta T_b)_B} = \frac{(K_b)_A}{(K_b)_B} \Rightarrow \frac{x}{y} = \frac{1}{8}$$

$\therefore \quad y = 8$

34. *(415)* Osmotic pressure, $\pi = CRT$

Given, mass of protein = 2 g

Molar mass of protein = 60 kg mol^{-1}

= 60,000 g mol^{-1}

Volume of water = 200 mL = 0.2 L

Concentration,

$$C = \frac{\text{Given mass}}{\text{Molar mass}} \times \frac{1}{\text{Volume solution}}$$

$$\pi = \frac{2\times 0.083\times 300}{60000\ g/mol \times 0.2\ L}$$

$\pi = 0.00415$ bar (1 bar = 10^5 Pascal)

So, $\pi = 0.00415 \times 10^5$ Pa

$\Rightarrow \pi = 415$ Pa

35. *(54)* Molar mass of glucose

= 180 g mol^{-1}.

We know, $\pi = CRT$

$$7.47 = C \times 0.083 \times 300$$

$$\Rightarrow C = \frac{7.47}{0.083\times 300} = 0.3\ M$$

Concentration of glucose solution = 0.3 M

Now, concentration in gL^{-1}

$= 0.3 \times 180 = 54\ gL^{-1}$

36. *(271)* 40 g of glucose mixed with 200 mL of water

180 g of glucose = 1 moles of glucose

40 g of glucose = 0.22 mol of glucose

1 mL of water = 1 g of water

$[d = 1\ g/cm^2, 1\ mL = 1\ cm^3]$

200 mL of water = 200 g of water

$$\Delta T_f = K_f m$$

where,

ΔT_f = depression in freezing point,

K_f = molal elevation constant

= 1.86 K kg mol^{-1} and

m = molality of solution

$$\Delta T_f = \frac{1.86\times 0.22}{200} \times 1000$$

$(\because 1000\ g = 1\ kg)$

$\Delta T_f = 2\ K$

$\therefore (T_f - T_f') = 2$

$\left[\begin{array}{l} T_f = \text{freezing point of water (273 K)} \\ T_f' = \text{freezing point of solution} \end{array}\right]$

$273\ K - T_f' = 2\ K$

$T_f' = (273-2) K = 271\ K$

37. *(518)* Let mass of water (initially present) = x g

Mass of sucrose = $(1000 - x)$ g

$$\text{Moles of sucrose} = \frac{1000-x}{342}$$

$$0.75 = \frac{\left(\frac{1000-x}{342}\right)}{\left(\frac{x}{1000}\right)}$$

$$\frac{x}{1000} = \frac{1000-x}{342\times 0.75}$$

$$256.5x = 10^6 - 1000x$$

$\Rightarrow \quad x = 795.86$ g

$$\text{Moles of sucrose} = \frac{1000-795.86}{342} = 0.5969$$

New mass of $H_2O = a$ kg

Depression in freezing point

$$\Delta T_f = K_f \times m$$

$$4 = \frac{0.5969}{a} \times 1.86 \Rightarrow a = 0.2775\ kg$$

Ice separated = 795.86 − 277.5

= 518.3 g ≈ 518 g.

38. *(15)* Using formula; $\pi = CRT$

$$\text{Molarity} = \frac{1.46\times 1000}{M\times 100(\text{mL})} = \frac{1460}{100\times M}$$

$$\therefore \quad \pi = \frac{1460}{100\times M} \times 0.083\times 300$$

Given, $\pi = 2.42\times 10^{-3}$ bar

Therefore, $M = 15.02\times 10^4$ g/mol

39. *(250)* Weight of substance 'X' = 3.00 g

Elevation in boiling point, $\Delta T_b = 0.60$

K_b (boiling constant) of CCl_4

= 5 K kg mol^{-1}

$$m = \frac{\frac{3}{M}}{\frac{100}{1000}} \quad \ldots(i)$$

Now using the formula,

$\Delta T_b = K_b \times$ molality

Put value of (i) we get,

$$0.60 = 5\times \frac{\frac{3}{M}}{\frac{100}{1000}}$$

$$\Rightarrow 0.60 = 5\times \frac{3\times 1000}{M\times 100}$$

$$0.60\,M = 5\times 30 \Rightarrow M = \frac{150}{0.60} = 250$$

$\therefore$ Molar mass of substance 'X' is 250 g mol^{-1}.

40. *(85)*

$$K_4Fe(CN)_6 \rightleftharpoons 4K^+ + [Fe(CN)_6]^{4-}$$

	$K_4Fe(CN)_6$	$4K^+$	$[Fe(CN)_6]^{4-}$
Initial conc.	1 m	0	0
Final conc.	(1 − 0.4) m = 0.6 m	4 × 0.4 = 1.6 m	0.4 m

Effective molality

= 0.6 + 1.6 + 0.4 = 2.6 m

Molality of non-electrolyte solution = molality of $K_4[Fe(CN)_6]$ = 2.6 m

Now, 18.1 weight per cent solution means 18.1 g solute is present in 100 g solution and hence, (100 − 18.1 =) 81.9 g water.

Molality

$$\text{Now, } 2.6 = \frac{18.1/M}{81.9/1000}$$

Molar mass of solute, $M = 85$

41. *(2.18)* $K_f = 2.0$ (Given) $m = 0.5$ m (Given)

$\Delta T_f = K_f \times m = 2\times 0.5 = 1$

$\therefore T_f$ = Actual freezing point of water $- \Delta T_f$.

= 273 − 1 = 272 K.

Pressure of gas after the stoppers are withdrawn = 1 atm (atmospheric pressure)

$$\therefore \quad V = \frac{nRT}{p} = \frac{0.1\times 0.08\times 272}{1\ \text{atm}}$$

= 2.176 L ≈ 2.18 L

Topic 4

01. *(d)* The van't Hoff factor (i) is highest for $KHSO_4$ as it dissociate into its ions. Hydrazine, glucose and glycine will not dissociate to produce ions.

Hence, correct option is (d).

02. *(a)* Highest depression in freezing point (colligative properties) is inversely proportional to experimental molar mass of non-volatile solute.

- On dissociation, experimental molar mass decreases.
- On association, experimental molar mass increases.

03. *(d)* As dilution increases concentration of solution decreases and degree of dissociation increases, resulting in increase of van't Hoff factor.

So, the correct order will be

$i_A < i_B < i_C$.

Hence, option (d) is correct.

04. *(c)* The correct match is

A-III, B-I, C-II, D-IV

05. *(a)* van't Hoff factor of $Al_2(SO_4)_3$ is 5 (maximum), so this will have lowest freezing point.

Depression in freezing point,

$$\Delta T_f = iK_f m$$

For freezing point to be lowest, ΔT_f should be highest and the i values are as follows

Solute	i
$C_6H_{12}O_6$	1
$Al_2(SO_4)_3$	$5 \rightarrow i$ highest, ΔT_f maximum, T_f minimum
KI	2
K_2SO_4	3

Hence, for $Al_2(SO_4)_3$ freezing point is lowest.

06. *(b)* $K_f = 4.0$ K kg mol^{-1} (Given)

$m = 0.03$ mol kg^{-1} (Given)

$\Delta T_f = ?$

For K_2SO_4, $i = 3$

It can be verified by the following equation :

$$K_2SO_4 \rightleftharpoons 2K^+ + SO_4^{2-}$$

Using formula

$$\Delta T_f = iK_f \times m$$
$$= 3 \times 4 \times 0.03 = 0.36 \text{ K}$$

07. *(d)*

> **idea** Osmotic pressure is proportional to the molarity (C) of the solution at a given temperature, $\pi = CRT$.

Concentration of

$BaCl_2 = 0.01$ M (Given)

$\pi_{XY} = 4\pi_{BaCl_2}$ (Given)

$i \times CRT = 4 \times i \times CRT$ ...(i)

For the calculation of i,

$XY \longrightarrow X^+ + Y^-$ (Here, $i = 2$)

$BaCl_2 \longrightarrow Ba^{2+} + 2Cl^-$ (Here, $i = 3$)

Putting the values of i in (i)

$$2 \times [XY] = 4 \times 3 \times [BaCl_2]$$
$$2 \times [XY] = 12 \times 0.01$$
$$[XY] = \frac{12 \times 0.01}{2}$$

So, the concentration of XY

$= 0.06$ mol L^{-1} $= 6 \times 10^{-2}$ mol L^{-1}

08. *(b)* The ionisation of K_2HgI_4 in aqueous solution is as follows:

$$K_2[HgI_4] \rightleftharpoons 2K^+ + [HgI_4]^{2-}$$

van't Hoff factor (i) for ionisation reaction is given as, $i = 1 + \alpha(n-1)$

where, n = number of ions,

α = degree of ionisation or dissociation

From above equation, it is clear that $n = 3$

$$i = 1 + 0.4(3-1) = 1.8$$

[Given, % α = 40% or $\alpha = 0.4$]

09. *(a)* Given that, $C_1 > C_2$

So, normal osmosis will take place from 2 to 1 while reverse osmosis will take place from 1 to 2.

Pressure, $p_1 > \pi$

So, statement A and C are correct.

10. *(b)* We know that,

$$\Delta T_b = i \times m \times k_b$$
$$\Rightarrow \quad \Delta T_b \propto iM$$

(Where, M = concentration)

For NaCl, the value of i is 2 while for urea the value of i is 1.

(i) For 10^{-4} M NaCl, the value of $\Delta T_b \propto 10^{-4} \times 2$

(ii) For 10^{-4} M urea, the value of $\Delta T_b \propto 10^{-4} \times 1$

(iii) For 10^{-3} M NaCl, the value of $\Delta T_b \propto 10^{-3} \times 2$

(iv) For 10^{-2} M NaCl, the value of $\Delta T_b \propto 10^{-2} \times 2$

So, the correct increasing order of boiling points is

(ii) < (i) < (iii) < (iv)

11. *(b)* The correct statement is AX_2 is fully ionised while AY_2 is completly unionised.

For AX_2, $\Delta T_b = K_b \times m \times i$

Given, $T_b = 100.0156°C = 373.0156$ K

$$\therefore \Delta T_b = T_b - T_b^\circ$$
$$= (373.0156 - 373) \text{ K}$$
$$= 0.0156 \text{ K}$$
$$K_b = 0.52 \text{ K kg mol}^{-1},$$

$$\text{Molality } (m) = \frac{1.24 \text{ g}/124 \text{ g mol}^{-1}}{1 \text{ kg}}$$

$$0.0156 = 0.52 \times \frac{0.01}{1} \times i_{AX_2}$$

$$\Rightarrow \quad i_{AX_3} = 3$$

$\Rightarrow$ Completely ionised

For AY_2

$\therefore \quad \Delta T_b = K_b \times m \times i$

Given, $T_b = 100.0260°C = 373.0260$ K

$$\therefore \Delta T_b = T_b - T_b^\circ$$
$$= (373.0260 - 373) \text{ K} = 0.0260 \text{ K}$$

$$\text{Molality } (m) = \frac{25.4 \text{ g}/250 \text{ g mol}^{-1}}{2 \text{ kg}}$$
$$= 0.0508 \text{ mol kg}^{-1}$$

$$0.026 = 0.52 \times 0.0508 \times i_{AY_2}$$
$$= i_{AY_2} \approx 1$$

$\Rightarrow$ Completely unionised.

12. *(a)* Statement B and D are correct.

The corrected form of Statements A, C and E are respectively

(A) $\Delta T_b \propto i \times C$.

$NaCl \rightleftharpoons Na^+ + Cl^-$ $(i = 2)$

$NH_2CONH_2 \longrightarrow$ Non-ionic solution

(C) Osmosis always takes place from hypotonic to hypertonic solution.

(E) When KI solutions is added to $AgNO_3$ solution, positively charged solution results due to adsorption of Ag^+ ions from dispersion medium.

AgI/Ag^+
Positively charged

13. *(c)* Given, concentration of formic acid = 0.5 mol L^{-1}

Density of formic acid = 1.05 g mL^{-1}.

$$K_f = 1.86 \text{ K kg mol}^{-1}$$
$$\Delta T_f = 0.0405°C$$

Now, $\quad d = \frac{M}{V} \Rightarrow M = d \cdot V$

$= 1.05 \times 0.5 = 0.525$ g

According to formula,

$\Delta T_f = iK_f m$ (m = molality)

Let us take, mass of solution = 1 kg = 1000 g

Mass of solvent = 1000 − 0.525 = 999.475 g (formic acid)

So $\quad m = \frac{0.525 \times 1000}{46 \times 999.475} = 0.0114$

$$\Delta T_f = iK_f\,(0.0114)$$
$$\Rightarrow \quad i = \frac{0.0405}{1.86 \times 0.0114} = 1.92$$

14. *(d)* Given, Mass of A = 0.7 g

Molar mass of A = 93 g mol^{-1}

$$\text{Moles of } A = \frac{0.7}{93} \text{ mol}$$

Mass of solvent = 42 g

$$\text{Molality} = \frac{0.7 \times 1000}{93 \times 42}$$

Depression in freezing point,

$$\Delta T_f = 0.2°C$$

Molal depression constant,

$K_f = 1.86 \text{ K kg mol}^{-1}$

Using the equation,

$\Delta T_f = iK_f m$

$0.2 = i \times 1.86 \times \frac{0.7 \times 1000}{93 \times 42}$

$i = \frac{0.2 \times 93 \times 42}{1.86 \times 0.7 \times 1000} = 0.6$

For association,

$\beta = \frac{i-1}{\left(\frac{1}{n}-1\right)}$

So, $\beta = \frac{0.6-1}{\frac{1}{2}-1} = \frac{-0.4}{-0.5}$

$\beta = 0.8 \Rightarrow \%\beta = 0.8 \times 100 = 80\%$

15. *(33)* The dissociation reaction is

$MX_3 \longrightarrow M^{3+} + 3X^-$

$i = 1 + (n-1)\alpha$

$i = 1 + (4-1)\alpha = 2$

$\alpha = \frac{1}{3} = 33.33 \text{ or } 33\%$

16. *(25)* For AB

Change in boiling point, $\Delta T_b = 2.7$ K

Using formula, $\Delta T_b = iK_b m$

$2.7 = 1 \times 0.5 \times m \Rightarrow m = \frac{27}{5}$

Let molar mass of $AB = X$

So, $\frac{1/X}{15} \times 1000 = \frac{27}{5}$

$X = 12.34$

For AB_2

$\Delta T_b = 1.5$ K

$1.5 = 1 \times 0.5 \times m$

$m = 3$

Let molar mass of $AB_2 = Y$

$\frac{\frac{1}{Y}}{15} \times 1000 = 3 \Rightarrow Y = 22.22$

Now, $A + b = 12.34$ ...(i)

$A + 2b = 22.27$ ...(ii)

Where, a and b be atomic mass of A and B

$b = 9.88$

$a = 24.6 \times 10^{-1}$

$\simeq 25 \times 10^{-1}$

17. *(75)* The dissociation takes place as follows

$MX_2 \longrightarrow M^{2+} + 2X^-$

$i = \frac{\text{Normal molar mass}}{\text{Observed molar mass}}$

$i = \frac{164}{65.6}$

$1 + (3-1)\alpha = \frac{164}{65.6}$

$2\alpha = \frac{98.4}{65.6}$

$\alpha = 0.75$

Per cent degree of ionisation = 75%

18. *(16)* The reaction involved is as follows

$A_2B \longrightarrow 2A^+ + B^{2-}, y = 3$

$\alpha = 0.3$

$i = 1 + (y-1)\alpha$

On substituting the value of y and α in above equation, we get

$i = 1 + (3-1)(0.3) = 1.6 \text{ or } 16 \times 10^{-1}$

19. *(5)* $\Delta T_b = 0.52°\text{C}$

$\Delta T_b = i \times K_b \times m$

$0.52 = i \times 0.52 \times \frac{10}{200 \times 0.1}$

$i = 2$ ...(i)

$\alpha = \frac{i-1}{n-1}$

$AB_2 \longrightarrow A^{2+} + 2B^-$

$n = 3$

$\alpha = \frac{2-1}{3-1} \Rightarrow \alpha = \frac{1}{2} = 0.5$

$\alpha = 5 \times 10^{-1}$

Hence, degree of ionisation is 5×10^{-1}.

20. *(15)*

idea This problem is based on the concept of depression in freezing point. To solve this, one should know about temperature difference (ΔT_f) and van't Hoff factor (i) as $\Delta T_f = iK_f m$

In this case, the value of i is 1 as ethylene glycol is a non-electrolyte.

Using, $\Delta T_f = iK_f \times m$

$i = 1$ (for ethylene glycol)

$\Delta T_f = 0 - (-24)$

$= 24°\text{C or } 24 \text{ K}$

On substituting the given values in above formula we get

$24 = \frac{1.86 \times x \times 1000}{62} \times \frac{1}{18.6}$

$x = \frac{24 \times 62 \times 18.6}{1860}$

$= 14.88 \approx 15 \text{ kg}$

21. *(48)* Molality = m, means that 1 mol of $MgCl_2$ is present in 1 kg of solvent (H_2O)

$p_S = ?$

$MgCl_2 \longrightarrow Mg^{2+} + 2Cl^-$

$n = 3$

$i = 1 - \alpha + n\alpha$

$i = 1 - 0.8 + 3(0.8) = 2.6$

$\frac{p_A^\circ - p_S}{p_S} = i \times \frac{n_B \text{ (moles of solute)}}{n_A \text{ (moles of solvent)}}$

$\frac{50 - p_S}{p_S} = 2.6 \times \frac{18}{1000}$

$[\because p°_A = 50 \text{ (given)}]$

$\Rightarrow p_S = 47.76 \approx 48 \text{ mm Hg}$

22. *(116)* Elevation in boiling point

$= [\Delta T_b = i \times K_b \times \text{molality}]$

Since two solute are present which dissociates completely, hence for NaCl ($i = 2$),

$MgCl_2$, ($i = 3$)

$\Rightarrow$ Mass of NaCl = 29.25% (by weight)

= 29.25 g

$\Rightarrow$ Mass of $MgCl_2$ = 19 g

So, total mass of solute = 29.25 + 19

= 48.25 g ...(i)

So, mass of solvent = 100 − 48.25

= 51.75 g ...(ii)

$\Delta T_b = K_b \left[\frac{i \times n_1 + i \times n_2}{\text{Mass of solvent (kg)}}\right]$

$= 0.52 \left[\frac{2 \times \frac{29.25}{58.5} + 3 \times \frac{19}{95}}{\frac{51.75}{1000}}\right]$

$\Rightarrow \frac{0.52(1 + 0.6)}{0.05175}$

$\Rightarrow [16.07]$

So, boiling point of solution

$= 100 + 16.07$

$= 116°\text{C}$

23. *(30)* Given degree of dissociation of aqueous solution of weak monobasic acid = 0.3

$HA \rightleftharpoons H^+ + A^-$

$(1-a) \quad (a) \quad (a)$

$i = 1 + a \Rightarrow a = 0.3$ (given)

So, $i = 1 + 0.3 = 1.3$

$(\Delta T_f)_{obs} = 1.3 \times K_f \times m$

$(\Delta T_f)_{cal} = 1 \times K_f \times m$

$\Rightarrow \frac{(\Delta T_f)_{obs} - (\Delta T_f)_{cal}}{(\Delta T_f)_{cal}} \times 100$

$= \frac{0.3}{1} \times 100 = 30\%$

24. *(4)* For isotonic solutions, osmotic pressure should be same or equal,

(A) 1 M aq. NaCl and 2 M aq. urea

0.1 M aq NaCl, $i = 2$

Osmotic pressure $\pi = iCRT$

$= 2 \times 1 \times RT$

$= 2RT$

2 M aq. urea, $i = 1$

$\pi = iCRT$

$= 1 \times 2\,RT = 2RT$

It is pair of isotonic solution.

(B) 1 M aq. $CaCl_2$ and 1.5 M aq KCl

1 M aq. $CaCl_2$, $i = 3$

$\pi = iCRT = 3RT$

1.5 M aq. KCl, $i = 2$

$\pi = iCRT = 2 \times 1.5\,RT = 3RT$

$\therefore$ It is pair of isotonic solution.

(C) 1.5 M aq. $AlCl_3$ and 2 M aq. Na_2SO_4

1.5 M aq. $AlCl_3$, $i = 4$

$\pi = 6RT$

2 M aq. Na_2SO_4, $i = 3$

$\pi = 6RT$

$\therefore$ It is a pair of isotonic solution.

(D) 2.5 M aq. KCl and 1 M aq. $Al_2(SO_4)_3$

2.5 M aq. KCl, $i = 2$

$\pi = 5RT$

1 M aq. $Al_2(SO_4)_3$, $i = 5$

$\pi = 5RT$

$\therefore$ It is pair of isotonic solution.

25. *(372)* $i = 1 + (n-1)\alpha$

$[i = 1 + 0.2(2-1) = 1.2]$

$\Delta T_f = iK_b m$

$= 1.2 \times 1.86 \times \dfrac{5 \times 1000}{60 \times 500}$

$= 0.372 = 372 \times 10^{-3}$

Thus, the depression in freezing point of water is 372×10^{-3} °C.

26. *(3)* Reaction equation of titration is

$$\underset{\substack{V = 25\text{ mL} \\ 1\text{ M}}}{KCl} + \underset{\substack{20\text{ mL} \\ 1\text{ M}}}{AgNO_3} \longrightarrow AgCl + KNO_3$$

At equivalent point

millimole of KCl = millimole of $AgNO_3$ = 20 millimole

(i) Volume of KCl solution = 25 mL

(ii) Mass of KCl solution = 25 (g) (assuming, $d = 1$ g/mL)

(iii) Mass of KCl (solute) $= (20 \times 10^{-3} \times 74.5)$

[$\because$ molar mass of KCl = 39 + 35.5 = 74.5 g/mol]

(iv) Mass of solvent = mass of solution − mass of solute

$= 25 - [20 \times 10^{-3} \times 74.5]$

$= 23.51$ g

$= 23.51 \times 10^{-3}$ kg

$m_{KCl} = \dfrac{20 \times 10^{-3}}{23.51 \times 10^{-3}} = 0.85$ m

Depression of freezing point of the KCl solution,

$\Delta T_f = i \times K_f \times m$

[$\because i$ for KCl = 2 assuming it gets 100% dissociated in water]

$= 2 \times 2 \times 0.85 = 3.4 \approx 3$

27. *(243)*

> **idea** Consider the percentage concentration in grams and use the formula $\Delta T = i_{max} \times K_f \times$ molality.

Given, $W_{H_2SO_4} = 38$ g, (van't Hoff factor, $i = 2.67$)

$K_f = 1.8$ K Kg/mol

Let us consider 100 g solution, then the mass of solvent will be

$= 100 - 38 = 62$ g

From equation $\Delta T = i_{max} \times K_f \times$ molality

$= 2.67 \times 1.8 \times \dfrac{38 \times 1000}{98 \times 62} = 30$°C

Freezing point of liquid in battery is − 30°C.

Thus, the temperature in Kelvin at which the solution in the battery will freeze is 243 K.

28. *(4)* 'A', 'B', 'D' and 'E' have same value of osmotic pressure.

Osmotic pressure can be calculated with the help of following formula :

$\pi = iCRT$, also $\pi \propto iC$

A. $(iC)_{C_2H_5OH} = 0.5$

$\therefore (iC)_{KBr} = 2 \times 0.25 = 0.5$

Osmotic pressure will be same.

B. $(iC)_{K_4[Fe(CN)_6]} = 0.1 \times 5 = 0.5$

$(iC)_{FeSO_4(NH_4)_2SO_4} = 0.1 \times 5 = 0.5$

Osmotic pressure will be same.

C. $(iC)_{K_4[Fe(CN)_6]} = 5 \times 0.05 = 0.25$

$(iC)_{NaCl} = 0.25 \times 2 = 0.5$

Osmotic pressure will not be same.

D. $(iC)_{NaCl} = 0.15 \times 2 = 0.3$

$(iC)_{BaCl_2} = 0.1 \times 3 = 0.3$

Osmotic pressure will be same.

E. $(iC)_{KCl \cdot MgCl_2 \cdot 6H_2O} = 0.02 \times 5 = 0.1$

$(iC)_{KCl} = 0.05 \times 2 = 0.1$

Osmotic pressure will be same.

29. *(5)* Molar mass of acetic acid $= 60$ g mol^{-1}

Number of moles of acetic acid

$= \dfrac{V \times \rho}{\text{Molar mass}}$

$= \dfrac{(1.2)\text{mL} \times (1.02)\text{g/mL}}{(60)\text{g/mol}}$

$= 0.0204$ mol

Molality

$(m) = \dfrac{\text{Moles of solute}}{\text{Mass of solvent }(g)} \times 1000$

$= \dfrac{0.0204}{2000} \times 1000$

(As 2 L solution in water is prepared so mass of solvent becomes 2000 g).

$= 0.0102$

$\Delta T_f = i \times K_f \times m$

ΔT_f = Depression in freezing point $= 0.0198$°C

$i = \dfrac{\Delta T_f}{K_f \times m} = \dfrac{0.0198}{1.85 \times 0.0102}$

$= 1.049$

$$CH_3COOH \rightleftharpoons CH_3COO^- + H^+$$

	CH_3COOH	CH_3COO^-	H^+
Initial	n	0	0
Final	$n(1-x)$	nx	nx

[n = number of moles, x = degree of dissociation]

Total moles $= n - nx + 2nx$

$= n - nx$

$i = \dfrac{n(1-x) + nx + nx}{n}$

$= \dfrac{n[1 - x + x + x]}{n}$

$\Rightarrow \quad i = 1 + x$

$\therefore \quad 1.049 = 1 + x$

x (degree of dissociation) $= 1.049 - 1$

$= 0.049$

≈ 0.05

% degree of dissociation $= 0.05 \times 100$

$= 5\%$

30. *(98)* We are given with 0.5% of KCl solution.

$\therefore$ Molality of KCl solution,

$m = \dfrac{0.5}{74.6} \times \dfrac{100}{99.5}$

$0.24 = i \times 1.80 \times \dfrac{0.5}{74.6} \times \dfrac{1000}{99.5}$

$\Rightarrow \quad i = \dfrac{0.24 \times 74.6 \times 99.5}{0.5 \times 1.80 \times 1000} = 1.979$

$$KCl \rightleftharpoons K^+ + Cl^-$$

	KCl	K^+	Cl^-
Initial	1	0	0
At equilibrium	$1-x$	x	x

{x is the amount of KCl dissociated}

$i = x + x + (1-x)$

$1.979 = 1 + x$

$x = 0.979 \Rightarrow \%x = 97.9\% \approx 98\%$

$\therefore$ Percentage of dissociation of KCl is 98%

31. *(269)* $\Delta T_f = iK_f \times m$

For ethylene glycol, $i = 1$

$$\text{Molality} = \frac{\text{Number of moles of solute}}{\text{Weight of solvent (kg)}}$$

$$\Delta T_f = 1 \times 1.86 \times \frac{83 \times 1000}{62 \times 625} \simeq 3.98$$

Freezing point of solution

$= \text{Freezing point of water} - \Delta T_f$

$= 273 - 3.98 \approx 269$ K.

32. *(4)* As 0.10 M C_2H_5OH is non-dissociative in nature and all salts [0.10 M $Ba_3(PO_4)_2$, 0.10 M, Na_2SO_4, 0.10 M KCl, 0.10 M Li_3PO_4] are electrolyte and dissociative in nature.

$$\Delta T_f = iK_f m$$

Where i is van't Hoff factor.

$i = 1$ for C_2H_5OH (no dissociation)

$i > 1$ for all other electrolytes where dissociation takes place.

So, for all these electrolytes effective molarity is more than 0.10. So, all the elecrolytes given will have low freezing point.

33. *(125)* Dimerisation $2HA \rightleftharpoons H_2A_2$

	$2HA$	H_2A_2
Initial	0.5	0
Final	0	0.25

Dissociation

	HA	H^+	A^-
Initial	0.5	0	0
Final	0	0.5	0.5

$$i = \frac{0.25 + 0.5 + 0.5}{1}$$

$= 1.25 = 125 \times 10^{-2}$

34. *(100)* Given : $T_b = 100.52°C$

$T_b^\circ = 100\,°C \Rightarrow K_b = 0.52$ K kg mol^{-1}

Elevation in boiling point,

$\Delta T_b = T_b - T_b^\circ$

$\Delta T_b = 100.52 - 100$

$\Delta T_b = 0.52°C$

For dimerisation (van't Hoff factor)

$$i = \left(1 - \frac{\alpha}{2}\right)$$

$\because \quad \Delta T_b = i \times K_b \times m$

$$0.52 = \left(1 - \frac{\alpha}{2}\right) \times 0.52 \times 2$$

$\alpha = 1$

So, percentage association $= 100\%$

35. *(50)* Given,

$K_f = 1.85$ K kg mol^{-1}

$T_f^\circ = 0°C$

$T_f = 3.885°C$

$\Delta T_f = T_f - T_f^\circ$

$= (3.885 - 0)°C$

$= 3.885°C$

$\Delta T_f = 3.885°C$

Depression in freezing point,

$\Delta T_f = i \times K_f \times m$

$3.885 = i \times 1.85 \times 2$

but, $\quad i = (1 + \alpha)$

$3.885 = (1 + \alpha) \times 1.85 \times 2$

$$\Rightarrow \quad \alpha = \frac{0.185}{3.7}$$

$\Rightarrow \quad \alpha = 0.05 \Rightarrow \alpha = 50 \times 10^{-3}$

36. *(2)*

$$\underset{\text{Benzoic acid}}{n\,\text{PhCOOH}} \xrightarrow{\text{Association}} (\text{PhCOOH})_n$$

Assuming 100% association ($\alpha = 1$),

$$\Rightarrow \quad i = 1 - \alpha\left(1 - \frac{1}{n}\right) = \frac{1}{n} \qquad [\because \alpha = 1]$$

Now, $\Delta T_f = K_f \times m \times i$

$$0 - (0.93) = 1.86 \times \frac{w_B \times 1000}{w_A \times M_B} \times \frac{1}{n}$$

[$\because w_B$ = mass of PhCOOH = 12.2 g

w_A = mass of H_2O = 100 g

M_B = molar mass of PhCOOH

= 122g mol^{-1}]

$$0.93 = 1.86 \times \frac{12.2 \times 1000}{100 \times 122} \times \frac{1}{n}$$

$$\Rightarrow \quad n = \frac{1.86 \times 12.2 \times 1000}{0.93 \times 100 \times 122} = 2$$

$\therefore$ Number of benzoic acid molecules associated, $n = 2$

37. *(3)* As AB is a binary electrolyte,

$\therefore \quad AB \rightleftharpoons A^+ + B^-, n = 2$

$$i = 1 + \alpha(n - 1) = 1 + \frac{75}{100}(2 - 1) = 1.75$$

Given, $\Delta T_b = 2.5$ K

$K_b = 0.52$ K kg mol^{-1}

$\therefore \quad \Delta T_b = K_b \times m \times i$

$$\Rightarrow \quad m = \frac{\Delta T_b}{K_b \times i} = \frac{2.5}{0.52 \times 1.75}$$

$= 2.74 \simeq 3$ mol / kg

38. *(375)* Given,

$K_b(H_2O) = 0.52$ K kg mol^{-1}

$$A_2B_3 \longrightarrow 2A^{+3} + 3B^{-2}$$

No. of ions, $n = 5$, concentration, $m = 1$ molal (Given)

$$\alpha = \frac{60}{100} = 0.6$$

$\Delta T_b = i \cdot K_f \cdot m$

$= [(1 + (n - 1)\alpha] \times K_f \times m$

$\Delta T_b = [(1 + (5 - 1)0.6] \times 0.52 \times 1$

$= (1 + 2.4) \times 0.52$

$\Delta T_b = 1.768$

$T_b = 1.768 + 373.15$

$= 375.268 = 375$ K

39. *(1)* Pure solvent : $C_6H_6(l)$

Given, $T_f^\circ = 5.5\,°C$

$K_f = 5.12°$ C/m, $m_{\text{solvent}} = 200$ g

$m_{\text{solute}} = 10$ g

Molar mass of solute

$C_4H_{10} = 12 \times 4 + 10 = 58$

Solute (C_4H_{10}) is non-dissociative;

$\therefore \quad i = 1$

$\therefore \quad \Delta T_f = iK_f\, m$

$$\Rightarrow \quad (T_f^\circ - T_f^1) = 1 \times 5.12 \times \frac{(10/58)}{(200/1000)}$$

$$5.5 - T_f^1 = \frac{5.12 \times 5 \times 10}{58}$$

$\Rightarrow \quad T_f^1 = 1.086°C$ or $T_f^1 \approx 1°C$

40. *(5)* $\Delta T_b = i \times m \times K_b$

For $CaCl_2$, $i = 3$, $m_{(CaCl_2)} = 0.05$ m

$\therefore \quad \Delta T_b = 3 \times 0.05 \times K_b = 0.15K_b$

Molality of $CrCl_3 \cdot x\,NH_3 = 0.1$ m

$\therefore \quad \Delta T_b' = i \times m \times K_b = 0.1\,iK_b$

Given that : $0.1\,iK_b = 2 \times 0.15 \times K_b$

$\therefore i = 3$ and coordination number is 6.

Thus, complex will be $[Cr(NH_3)_5Cl]Cl_2$

$\therefore \quad x = 5$

Hence, the correct answer is 5.

41. *(1.76)*

$$\Delta T_f = \frac{i \times K_f \times w_{\text{NaCl}} \times 1000}{\underset{(=58.5)}{M_{\text{NaCl}}} \times w_{H_2O}}$$

$$(m = \text{molality}) = \frac{w_B \times 1000}{M_B \times w_A}$$

$\Delta T_f = 0.2$ because freezing point $= -0.2°C$

K_f of water 2 K kg mol^{-1} (Given)

i = van't Hoff factor, for NaCl = 2

$$\therefore w_{\text{NaCl}} = \frac{\Delta T_f \times M_{\text{NaCl}} \times w_{H_2O}}{i \times K_f \times 1000}$$

$$= \frac{0.2 \times 58.5 \times 600}{2 \times 2 \times 1000}$$

$= 1.755$ g ≈ 1.76 g

14 Electrochemistry

In JEE Main there is usually 0-1 question per paper are visible from this chapter. In past 3 years, 34 conceptual and 32 numerical questions were asked of moderate difficulty level from this chapter.

For detailed analysis scan QR code

Topic 3 / Conductance and Electrolysis

• Level 1

01 A solution of aluminium chloride is electrolysed for 30 minutes using a current of 2A. The amount of the aluminium deposited at the cathode is ....
[Given, molar mass of aluminium and chlorine are 27 g mol^{-1} and 35.5 g mol^{-1} respectively. Faraday constant = 96500 C mol^{-1}] **[2025, 22 Jan Shift-I]**

(a) 1.007 g (b) 0.336 g (c) 0.441 g (d) 1.660 g

02 Which of the following electrolyte can be used to obtain $H_2S_2O_8$ by the process of electrolysis? **[2025, 22 Jan Shift-I]**

(a) Acidified dilute solution of sodium sulphate.
(b) Dilute solution of sulphuric acid.
(c) Concentrated solution of sulphuric acid.
(d) Dilute solution of sodium sulphate.

03 The variation of molar conductivity with concentration of an electrolyte (X) in aqueous solution is shown in the given figure.

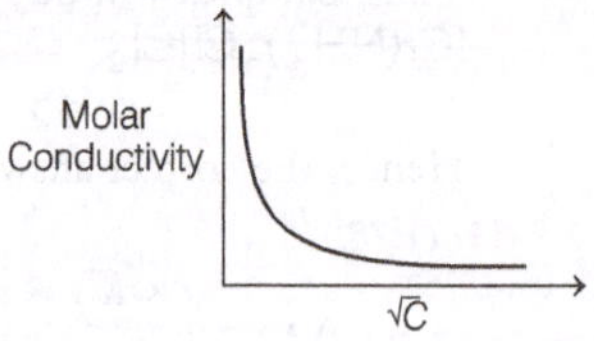

The electrolyte X is **[2020, 5 Sep Shift-II]**

(a) HCl (b) NaCl (c) KNO_3 (d) CH_3COOH

• Level 2

04 Correct order of limiting molar conductivity for cations in water at 298 K is **[2025, 3 April Shift-I]**

(a) $Mg^{2+} > H^+ > Ca^{2+} > K^+ > Na^+$
(b) $H^+ > Ca^{2+} > Mg^{2+} > K^+ > Na^+$
(c) $H^+ > Na^+ > K^+ > Ca^{2+} > Mg^{2+}$
(d) $H^+ > Na^+ > Ca^{2+} > Mg^{2+} > K^+$

05 The molar conductivity of a weak electrolyte when plotted against the square root of its concentration, which of the following is expected to be observed? **[2025, 29 Jan Shift-I]**

(a) Molar conductivity decreases sharply with increase in concentration.
(b) A small decrease in molar conductivity is observed at infinite dilution.
(c) A small increase in molar conductivity is observed at infinite dilution.
(d) Molar conductivity increases sharply with increase in concentration.

Imp. *Graphical Representation of Strong & Weak Electrolyte*

06 The molar conductivity for electrolytes A and B are plotted against $C^{1/2}$ as shown below. Electrolytes A and B respectively are **[2024, 9 April Shift-I]**

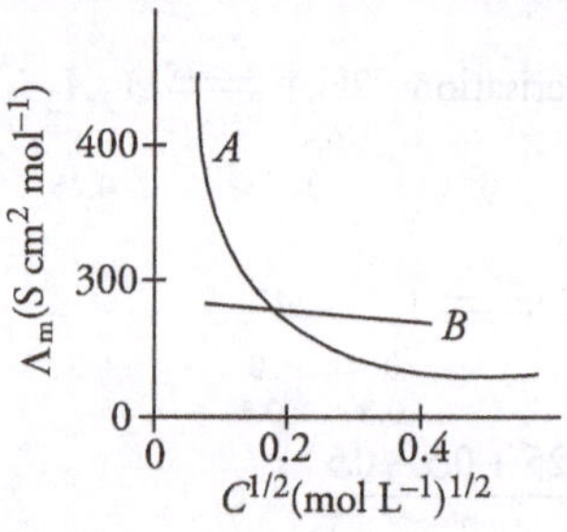

	A	B
(a)	Weak electrolyte	Strong electrolyte
(b)	Weak electrolyte	Weak electrolyte
(c)	Strong electrolyte	Weak electrolyte
(d)	Strong electrolyte	Strong electrolyte

07 A conductivity cell with two electrodes (dark side) are half-filled with infinitely dilute aqueous solution of a weak electrolyte. If volume is doubled by adding more water at constant temperature, the molar conductivity of the cell will **[2024, 6 April Shift-I]**

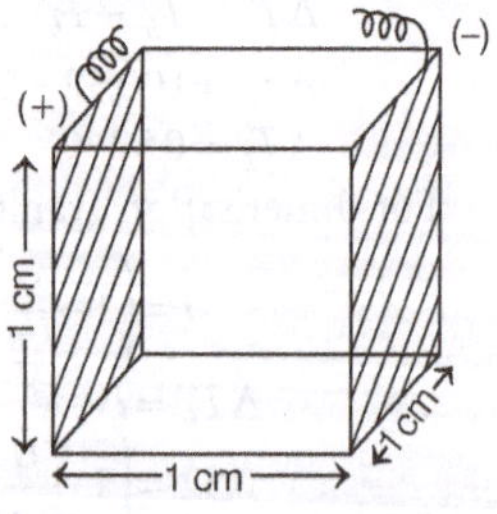

(a) depend upon type of electrolyte
(b) decrease sharply
(c) increase sharply
(d) remain same or can not be measured accurately

08 The quantity of silver deposited when one coulomb charge is passed through $AgNO_3$ solution **[2024, 5 April Shift-II]**
(a) 1 electrochemical equivalent of silver
(b) 1 chemical equivalent of silver
(c) 1 g of silver
(d) 0.1 g atom of silver

09 Molar ionic conductivities of divalent cation and anion are $57 \text{ S cm}^2 \text{ mol}^{-1}$ and $73 \text{ S cm}^2 \text{ mol}^{-1}$ respectively. The molar conductivity of solution of an electrolyte with the above cation and anion will be **[2024, 5 April Shift-I]**
(a) $130 \text{ S cm}^2 \text{ mol}^{-1}$ (b) $187 \text{ S cm}^2 \text{ mol}^{-1}$
(c) $260 \text{ S cm}^2 \text{ mol}^{-1}$ (d) $65 \text{ S cm}^2 \text{ mol}^{-1}$

Imp. *Unit of Molar Conductivity*

10 For a strong electrolyte, a plot of molar conductivity against (concentration)$^{1/2}$ is a straight line, with a negative slope, the correct unit for the slope is **[2024, 4 April Shift-II]**
(a) $\text{S cm}^2 \text{mol}^{\frac{-3}{2}} \text{L}^{\frac{-1}{2}}$ (b) $\text{S cm}^2 \text{mol}^{-1} \text{L}^{\frac{1}{2}}$
(c) $\text{S cm}^2 \text{mol}^{\frac{-3}{2}} \text{L}^{\frac{1}{2}}$ (d) $\text{S cm}^2 \text{mol}^{\frac{-3}{2}} \text{L}$

11 One of the commonly used electrode is calomel electrode. Under which of the following categories, calomel electrode comes? **[2024, 4 April Shift-I]**
(a) Gas- ion electrodes
(b) Oxidation-reduction electrodes
(c) Metal- insoluble salt-anion electrodes
(d) Metal ion-metal electrodes

12 Identify the factor from the following that does not affect electrolytic conductance of a solution. **[2024, 31 Jan Shift-I]**
(a) The nature of the electrode used.
(b) Concentration of the electrolyte.
(c) The nature of the electrolyte added.
(d) The nature of solvent used.

13 Which one of the following statements is correct for electrolysis of brine solution? **[2023, 31 Jan Shift-I]**
(a) Cl_2 is formed at cathode.
(b) OH^- is formed at cathode.
(c) O_2 is formed at cathode.
(d) H_2 is formed at anode.

14 Choose the correct representation of conductometric titration of benzoic acid *vs* sodium hydroxide. **[2023, 24 Jan Shift-II]**

(a)
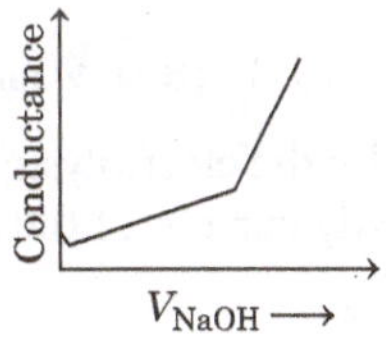

(b)
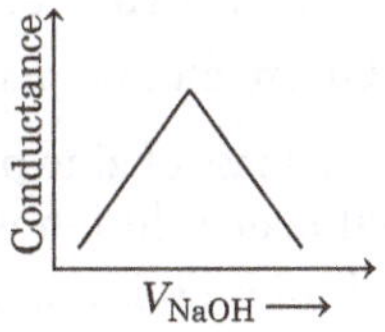

(c)
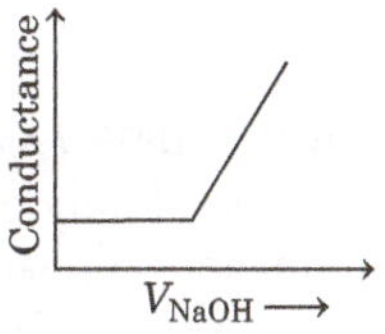

(d)
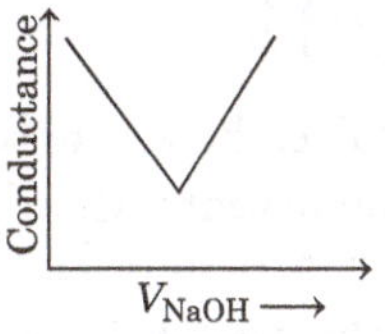

15 Given below are two statements.
Statement I For KI, molar conductivity increases steeply with dilution.
Statement II For carbonic acid, molar conductivity increases slowly with dilution.
In the light of the above statements, choose the correct answer from the options given below. **[2022, 27 July Shift-II]**
(a) Both Statement I and Statement II are true.
(b) Both Statement I and Statement II are false.
(c) Statement I is true but Statement II is false.
(d) Statement I is false but Statement II is true.

16 The molar conductivity of a conductivity cell filled with 10 moles of 20 mL NaCl solution is Λ_{M_1} and that of 20 moles another identical cell having 80 mL NaCl solution is Λ_{M_2}. The conductivities exhibited by these two cells are same. The relationship between Λ_{M_2} and Λ_{M_1} is **[2022, 25 July Shift-II]**
(a) $\Lambda_{M_2} = 2\Lambda_{M_1}$ (b) $\Lambda_{M_2} = \Lambda_{M_1}/2$
(c) $\Lambda_{M_2} = \Lambda_{M_1}$ (d) $\Lambda_{M_2} = 4\Lambda_{M_1}$

17 Given below are two statements.
Statement I The limiting molar conductivity of KCl (strong electrolyte) is higher compared to that of CH_3COOH (weak electrolyte).
Statement II Molar conductivity decreases with decrease in concentration of electrolyte.
In the light of the above statements, choose the most appropriate answer from the options given below **[2021, 26 Aug Shift-I]**
(a) Statement I is true but Statement II is false.
(b) Statement I is false but Statement II is true.
(c) Both Statement I and Statement II are true.
(d) Both Statement I and Statement II are false.

18 A solution of $Ni(NO_3)_2$ is electrolysed between platinum electrodes using 0.1 Faraday electricity. How many mole of Ni will be deposited at the cathode? **[2019, 9 April Shift-II]**
(a) 0.20 (b) 0.10
(c) 0.15 (d) 0.05

19 The anodic half-cell of lead-acid battery is recharged using electricity of 0.05 Faraday. The amount of $PbSO_4$ electrolysed in g during the process is (Molar mass of $PbSO_4 = 303 \text{ g mol}^{-1}$) **[2019, 9 Jan Shift-I]**
(a) 11.4 (b) 7.6
(c) 15.2 (d) 22.8

• Level 3

20 Given below are two statements: **[2025, 7 April Shift-I]**

Statement I Mohr's salt is composed of only three types of ions-ferrous, ammonium and sulphate.

Statement II If the molar conductance at infinite dilution for ferrous, ammonium and sulphate ions are x_1, x_2 and x_3 S cm^2 mol^{-1}, respectively then the molar conductance for Mohr's salt solution at infinite dilution would be given by $x_1 + x_2 + 2x_3$

In the light of the given statements, choose the correct answer from the options given below:

(a) Both Statements I and Statement II are true.
(b) Both Statements I and Statement II are false.
(c) Statement I is false but Statement II is true.
(d) Statement I is true but Statement II are false.

21 Which out of the following is a correct equation to show change in molar conductivity with respect to concentration for a weak electrolyte, if the symbols carry their usual meaning : **[2024, 9 April Shift-II]**

(a) $\Lambda_m - \Lambda_m^\circ - AC^{1/2} = 0$
(b) $\Lambda_m^2 C - K_a \Lambda_m^{\circ 2} + K_a \Lambda_m \Lambda_m^\circ = 0$
(c) $\Lambda_m - \Lambda_m^\circ + AC^{1/2} = 0$
(d) $\Lambda_m^2 C + K_a \Lambda_m^{\circ 2} - K_a \Lambda_m \Lambda_m^\circ = 0$

22 Match List-I with List-II.

List-I (Parameter)	List-II (Unit)
A. Cell constant	1. S cm^2 mol^{-1}
B. Molar conductivity	2. Dimensionless
C. Conductivity	3. m^{-1}
D. Degree of dissociation of electrolyte	4. $\Omega^{-1} m^{-1}$

Choose the most appropriate answer from the options given below **[2021, 31 Aug Shift-II]**

	A	B	C	D		A	B	C	D
(a)	3	1	4	2	(b)	3	1	2	4
(c)	1	4	3	2	(d)	2	1	3	4

23 Λ°_m for NaCl, HCl and Na*A* are 126.4, 425.9 and 100.5 S cm^2 mol^{-1}, respectively. If the conductivity of 0.001 M H*A* is 5×10^{-5} S cm^{-1}, degree of dissociation of H*A* is **[2019, 12 Jan Shift-II]**

(a) 0.25
(b) 0.50
(c) 0.75
(d) 0.125

Numerical Value Type Questions

24 The molar conductance of an infinitely dilute solution of ammonium chloride was found to be 185 S cm^2 mol^{-1} and the ionic conductance of hydroxyl and chloride ions are 170 and 70 S cm^2 mol^{-1}, respectively. If molar conductance of 0.02 M solution of ammonium hydroxide is 85.5 S $cm^2 mol^{-1}$, its degree of dissociation is given by $x \times 10^{-1}$.

The value of x is ....(Nearest integer) **[2025, 4 April Shift-II]**

25 Given below is the plot of the molar conductivity *vs* $\sqrt{\text{concentration}}$ for KCl in aqueous solution. **[2025, 28 Jan Shift-I]**

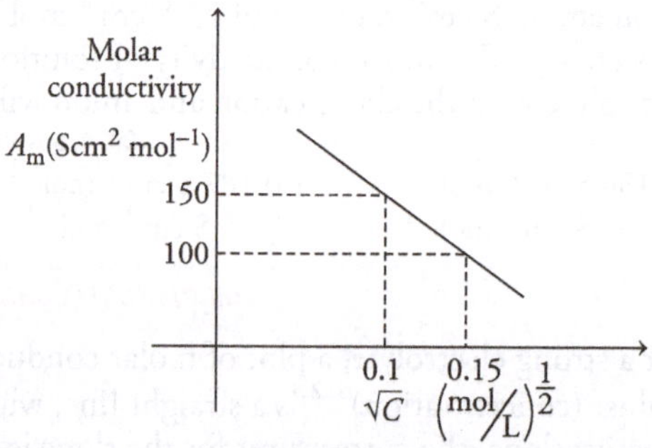

If, for the higher concentration of KCl solution, the resistance of the conductivity cell is 100 Ω, then the resistance of the same cell with the dilute solution is 'x' Ω

The value of x is .... (Nearest integer)

26 Electrolysis of 600 mL aqueous solution of NaCl for 5 min changes the pH of the solution to 12. The current in Amperes used for the given electrolysis is......(Nearest integer). **[2025, 28 Jan Shift-II]**

27 The amount of electricity in coulomb required for the oxidation of 1 mol of H_2O to O_2 is .... $\times 10^5$ C. **[2024, 1 Feb, Shift-I]**

28 The value of conductivity of some materials at 298.15 K in Sm^{-1} are 2.1×10^3, 1.0×10^{-16}, 1.2×10, 3.91, 1.5×10^{-2}, 1×10^{-7}, 1.0×10^3. The number of conductors among the materials is ... . **[2024, 31 Jan, Shift-II]**

29 One Faraday of electricity liberates $x \times 10^{-1}$ gram atom of copper from copper sulphate, x is .... **[2024, 31 Jan, Shift-I]**

30 A constant current was passed through a solution of $AuCl_4^-$ ion between gold electrodes. After a period of 10.0 minutes, the increase in mass of cathode was 1.314 g. The total charge passed through the solution is .... $\times 10^{-2}$ F.

(Given, atomic mass of Au = 197) **[2024, 29 Jan, Shift-II]**

31 The mass of zinc produced by the electrolysis of zinc sulphate solution with a steady current of 0.015 A for 15 minutes is .... $\times 10^{-4}$ g.

(Atomic mass of zinc = 65.4 amu) **[2024, 29 Jan, Shift-I]**

32 The number of correct statement from the following is .... **[2023, 15 April, Shift-I]**

A. Conductivity always decreases with decrease in concentration for both strong and weak electrolytes.
B. The number of ions per unit volume that carry in a solution increases on dilution.
C. Molar conductivity increases with decrease in concentration.
D. The variation in molar conductivity is different for strong and weak electrolytes.
E. For weak electrolytes, the change in molar conductivity with dilution is due to decrease in degree of dissociation.

33 A metal surface of 100 cm^2 area has to be coated with nickel layer of thickness 0.001 mm. A current of 2A was passed through a solution of $Ni(NO_3)_2$ for 'x' seconds to coat the desired layer. The value of x is .... (Nearest integer) (ρ_{Ni} density of nickel) is 10 g mL^{-1}. Molar mass of nickel is 60 g mol^{-1} $F = 96500 \text{ C mol}^{-1}$) **[2023, 13 April Shift-I]**

34 The specific conductance of 0.0025 M acetic acid is $5 \times 10^{-5} \text{ S cm}^{-1}$ at a certain temperature. The dissociation constant of acetic acid is .... $\times 10^{-7}$. (Nearest integer)
Consider limiting molar conductivity of CH_3COOH as $400 \text{ S cm}^2 \text{ mol}^{-1}$. **[2023, 10 April Shift-II]**

35 1×10^{-5} M $AgNO_3$ is added to 1 L of saturated solution of AgBr. The conductivity of this solution at 298 K is .... $\times 10^{-8} \text{ S m}^{-1}$.
[Given, $K_{sp}(AgBr) = 4.9 \times 10^{-13}$ at 298 K

$\lambda^\circ_{Ag^+} = 6 \times 10^{-3} \text{ S m}^2 \text{ mol}^{-1}$

$\lambda^\circ_{Br^-} = 8 \times 10^{-3} \text{ S m}^2 \text{ mol}^{-1}$

$\lambda^\circ_{NO_3^-} = 7 \times 10^{-3} \text{ S m}^2 \text{ mol}^{-1}$ **[2023, 1 Feb Shift-II]**

36 The resistivity of a 0.8 M solution of an electrolyte is $5 \times 10^{-3} \Omega$ cm. Its molar conductivity is .... $\times 10^4$ $\Omega^{-1} \text{cm}^2 \text{ mol}^{-1}$. (Nearest integer) **[2023, 31 Jan Shift-II]**

37 Resistance of a conductivity cell (cell constant 129 m^{-1}) filled with 74.5 ppm solution of KCl is 100Ω (labelled as solution 1). When the same cell is filled with KCl solution of 149 ppm, the resistance is 50Ω (labelled as solution 2). The ratio of molar conductivity of solution 1 and solution 2 is i.e.

$\frac{\Lambda_1}{\Lambda_2} = x \times 10^{-3}$. The value of x is ....
(Nearest integer) **[2022, 29 July Shift-I]**

38 The quantity of electricity in Faraday needed to reduce 1 mole of $Cr_2O_7^{2-}$ to Cr^{3+} is .... **[2022, 28 June Shift-I]**

39 The limiting molar conductivities of NaI, $NaNO_3$ and $AgNO_3$ are 12.7, 12.0 and 13.3 $\text{mS m}^2 \text{ mol}^{-1}$, respectively (all at 25°C). The limiting molar conductivity of AgI at this temperature is ...... $\text{mS m}^2 \text{ mol}^{-1}$. **[2022, 27 June Shift-I]**

40 A solution of $Fe_2(SO_4)_3$ is electrolysed for 'x' min with a current of 1.5 A to deposit 0.3482 g of Fe. The value of x is .... [Nearest integer]
Given, $1 \text{ F} = 96500 \text{ C mol}^{-1}$,
atomic mass of Fe = 56 g mol^{-1} **[2022, 25 June Shift-II]**

41 The molar conductivities at infinite dilution of barium chloride, sulphuric acid and hydrochloric acid are 280, 860 and 426 $\text{Scm}^2 \text{ mol}^{-1}$ respectively. The molar conductivity at infinite dilution of barium sulphate is ... $\text{S cm}^2 \text{ mol}^{-1}$ (Round off to the nearest Integer). **[2021, 18 March Shift-II]**

42 A KCl solution of conductivity 0.14 S m^{-1} shows a resistance of 4.19Ω in a conductivity cell. If the same cell is filled with an HCl solution, the resistance drops to 1.03Ω. The conductivity of the HCl solution is .... $\times 10^{-2} \text{ S m}^{-1}$ (Round off to the nearest integer) **[2021, 17 March Shift-II]**

43 A 5.0 m mol dm^{-3} aqueous solution of KCl has a conductance of 0.55 mS when measured in a cell constant 1.3 cm^{-1}. The molar conductivity of this solution is ...... $\text{mSm}^2 \text{ mol}^{-1}$.
(Round off to the nearest integer) **[2021, 16 March Shift-II]**

44 Consider the following reaction,
$MnO_4^- + 8H^+ + 5e^- \rightarrow Mn^{2+} + 4H_2O, \; E^\circ = 1.51V.$
The quantity of electricity required in Faraday to reduce five moles of MnO_4^- is .... ***[2021, 26 Feb Shift-I]***

45 Potassium chlorate is prepared by the electrolysis of KCl in basic solution
$6OH^- + Cl^- \longrightarrow ClO_3^- + 3H_2O + 6e^-$. If only 60% of the current is utilised in the reaction, the time (rounded to the nearest hour) required to produce 10 g of $KClO_3$ using a current of 2 A is ....
(Given : $F = 96,500 \text{ C mol}^{-1}$; molar mass of $KClO_3 = 122 \text{ g mol}^{-1}$) **[2020, 6 Sep Shift-I]**

Topic 2 / Electrochemical Series, Cell & Their EMF

• Level 1

01 For a Mg $|Mg^{2+}(aq)||Ag^{+}(aq)|$ Ag the correct Nernst equation is **[2025, 29 Jan Shift-I]**

(a) $E_{cell} = E^{\circ}_{cell} - \frac{RT}{2F} \ln \frac{[Mg^{2+}]}{[Ag^{+}]}$

(b) $E_{cell} = E^{\circ}_{cell} + \frac{RT}{2F} \ln \frac{[Ag^{+}]^2}{[Mg^{2+}]}$

(c) $E_{cell} = E^{\circ}_{cell} - \frac{RT}{2F} \ln \frac{[Ag^{+}]}{[Mg^{2+}]}$

(d) $E_{cell} = E^{\circ}_{cell} - \frac{RT}{2F} \ln \frac{[Ag^{+}]^2}{[Mg^{2+}]}$

02 The reaction

$$\frac{1}{2}H_2(g) + AgCl(s) \rightleftharpoons H^{+}(aq) + Cl^{-}(aq) + Ag(s)$$

occurs in which of the given galvanic cell. **[2024, 8 April Shift-II]**

(a) Pt $|H_2(g)|$ HCl(sol$''$) $|$AgCl$(s)|$ Ag

(b) Pt $|H_2(g)|$ KCl(sol$''$)$|$ AgCl$(s)|$ Ag

(c) Ag $|$ AgCl$(s)|$ KCl(sol$''$)$|$ $AgNO_3$ $|$ Ag

(d) Pt $|H_2(g)|$ HCl(sol$''$) $|$ $AgNO_3$ (sol)$|$ Ag

03 The standard electrode potential of M^{+}/M in aqueous solution does not depend on **[2023, 6 April Shift-I]**

(a) ionisation of a solid metal atom

(b) ionisation of a gaseous metal atom

(c) hydration of a gaseous metal ion

(d) sublimation of a solid metal

Imp. *Calculation of Gibbs Energy*

04 The standard Gibbs energy for the given cell reaction in kJ mol^{-1} at 298 K is

$$Zn(s) + Cu^{2+}(aq) \longrightarrow Zn^{2+}(aq) + Cu(s),\ E^{\circ} = 2\text{ V at } 298\text{ K}$$

(Faraday's constant, F = 96000 C mol^{-1}) **[2019, 9 April Shift-I]**

(a) 384 (b) 192 (c) −384 (d) −192

• Level 2

05 $FeO_4^{2-} \xrightarrow{+2.0\text{ V}} Fe^{3+} \xrightarrow{0.8\text{ V}} Fe^{2+} \xrightarrow{-0.5\text{ V}} Fe^{0}$

In the above diagram, the standard electrode potentials are given in volts (over the arrow).

The value of $E^{\circ}_{FeO_4^{2-}/Fe^{2+}}$ is **[2025, 23 Jan Shift-I]**

(a) 2.1 V (b) 1.2 V

(c) 1.7 V (d) 1.4 V

06 How can an electrochemical cell be converted into an electrolytic cell? **[2024, 6 April Shift-II]**

(a) Reversing the flow of ions in salt bridge.

(b) Exchanging the electrodes at anode and cathode.

(c) Applying an external opposite potential greater than E°_{cell}

(d) Applying an external opposite potential lower than E°_{cell}

07 For the electrochemical cell

$$M|M^{2+}||X|X^{2-}$$

If, $E^{\circ}_{(M^{2+}/M)} = 0.46$ V

and $E^{\circ}_{(X/X^{2-})} = 0.34$ V.

Which of the following is correct? **[2024, 5 April Shift-II]**

(a) $E_{cell} = 0.80$ V

(b) $E_{cell} = -0.80$ V

(c) $M + X \rightarrow M^{2+} + X^{2-}$ is a spontaneous reaction

(d) $M^{2+} + X^{2-} \longrightarrow M + X$ is a spontaneous reaction

08 What pressure (bar) of H_2 would be required to make emf of hydrogen electrode zero in pure water at 25°C? **[2024, 4 April Shift-I]**

(a) 0.5 (b) 10^{-14}

(c) 1 (d) 10^{-7}

09 The $\left(\frac{\partial E}{\partial T}\right)_p$ of different types of half-cells are as follows

A	B	C	D
1×10^{-4}	2×10^{-4}	0.1×10^{-4}	0.2×10^{-4}

(where, E is the electromotive force.)

Which of the above half-cells would be preferred to be used as reference electrode? **[2022, 26 June Shift-I]**

(a) A (b) B (c) C (d) D

10

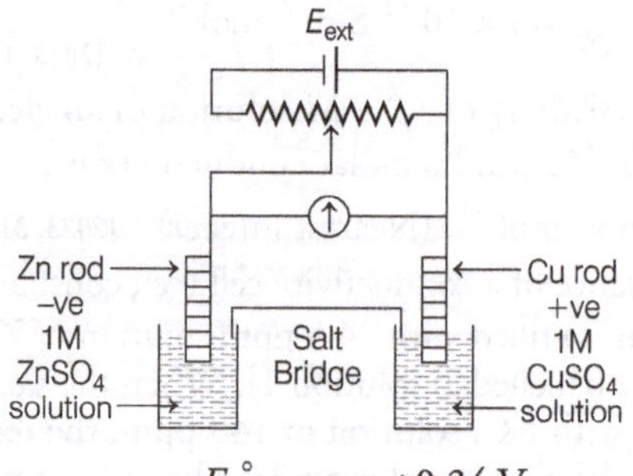

$E^{\circ}_{Cu^{2+}|Cu} = +0.34$ V

$E^{\circ}_{Zn^{2+}|Zn} = -0.76$ V

Identify the incorrect statement from the options below for the above cell. **[2020, 4 Sep Shift-I]**

(a) If $E_{ext} > 1.1$ V, Zn dissolves at Zn electrode and Cu deposits at Cu electrode.

(b) If $E_{ext} = 1.1$ V, no flow of electrons or current occurs.

(c) If $E_{ext} > 1.1$ V, electrons flows from Cu to Zn.
(d) If $E_{ext} < 1.1$ V, Zn dissolves at anode and Cu deposits at cathode.

11 Given that the standard potentials ($E°$) of Cu^{2+}/Cu and Cu^{+}/Cu are 0.34 V and 0.522 V respectively, the $E°$ of Cu^{2+}/Cu^{+} is **[2020, 7 Jan Shift-I]**
(a) -0.158 V (b) $+0.158$ V
(c) -0.182 V (d) 0.182 V

12 If the standard electrode potential for a cell is 2V at 300 K, the equilibrium constant (K) for the reaction, $Zn(s) + Cu^{2+}(aq) \rightleftharpoons Zn^{2+}(aq) + Cu(s)$ at 300 K is approximately ($R = 8$ $JK^{-1}mol^{-1}$, F = 96000 C mol^{-1}) **[2019, 9 Jan Shift-II]**
(a) e^{-160} (b) e^{160}
(c) e^{-80} (d) e^{320}

• Level 3

13 Given below are two statements. **[2025, 7 April Shift-II]**
1 M aqueous solutions of each of $Cu(NO_3)_2$, $AgNO_3$, $Hg_2(NO_3)_2$, $Mg(NO_3)_2$ are electrolysed using inert electrodes. Given: $E^{\theta}_{Ag^+/Ag} = 0.80$ V, $E^{\theta}_{Hg_2^{2+}/Hg} = 0.79$ V, $E^{\theta}_{Cu^{2+}/Cu} = 0.24$ V and $E^{\theta}_{Mg^{2+}/Mg} = -2.37$ V.

Statement I With increasing voltage, the sequence of deposition of metals on the cathode will be Ag, Hg and Cu.

Statement II Magnesium will not be deposited at cathode instead oxygen gas will be evolved at the cathode.

In the light of the above statements, choose the most appropriate answer from the options given below:
(a) Statement I is incorrect but Statement II is correct.
(b) Both Statement I and Statement II are incorrect.
(c) Both Statement I and Statement II are correct.
(d) Statement I is correct but Statement II is incorrect.

14 The standard cell potential ($E^{\ominus}_{cell}$) of a fuel cell based on the oxidation of methanol in air that has been used to power television relay station is measured as 1.21 V. The standard half cell reduction potential for O_2 ($E^{\circ}_{O_2/H_2O}$) is 1.229 V. **[2025, 3 April Shift-II]**
Choose the correct statement.
(a) The standard half cell reduction potential for the reduction of CO_2 ($E^{\circ}_{CO_2/CH_3OH}$) is 19 mV.
(b) Oxygen is formed at the anode.
(c) Reactants are fed at one go to each electrode.
(d) Reduction of methanol takes place at the cathode.

15 For the given cell
$$Fe^{2+}(aq) + Ag^{+}(aq) \longrightarrow Fe^{3+}(aq) + Ag(s)$$
The standard cell potential of the above reaction is (Given :
$Ag^{+} + e^{-} \rightarrow Ag$ $E° = x$V
$Fe^{2+} + 2e^{-} \rightarrow Fe$ $E° = y$V
$Fe^{3+} + 3e^{-} \rightarrow Fe$ $E° = z$V) **[2025, 24 Jan Shift-I]**
(a) $y - 2x$ (b) $x + 2y - 3z$
(c) $x + y - z$ (d) $x + 2y$

16 Standard electrode potentials for a few half cells are mentioned below.
$$E^{\circ}_{Cu^{2+}/Cu} = 0.34 \text{ V}, E^{\circ}_{Zn^{2+}/Zn} = -0.76 \text{ V}$$
$$E^{\circ}_{Ag^{+}/Ag} = 0.80 \text{ V}, E^{\circ}_{Mg^{2+}/Mg} = -2.37 \text{ V}$$
Which one of the following cells given the most negative value of $\Delta G°$? **[2025, 23 Jan Shift-II]**
(a) $Ag|Ag^{+}(1M)||Mg^{2+}(1M)|Mg$
(b) $Cu|Cu^{2+}(1M)||Ag^{+}(1M)|Ag$
(c) $Zn|Zn^{2+}(1M)||Ag^{+}(1M)|Ag$
(d) $Zn|Zn^{2+}(1M)||Mg^{2+}(1M)|Mg$

17 The emf of cell $Tl|\underset{(0.001\text{ M})}{Tl^{+}}||\underset{(0.01\text{ M})}{Cu^{2+}}|Cu$ is 0.83 V at 298 K. It could be increased by **[2024, 8 April Shift-II]**
(a) decreasing concentration of both Tl^{+} and Cu^{2+} ions
(b) increasing concentration of Tl^{+} ions
(c) increasing concentration of both Tl^{+} and Cu^{2+} ions
(d) increasing concentration of Cu^{2+} ions

18 The correct order of reduction potentials of the following pairs is
A. Cl_2/Cl^{-} B. I_2/I^{-} C. Ag^{+}/Ag
D. Na^{+}/Na E. Li^{+}/Li
Choose the correct answer from the options given below. **[2022, 25 June Shift-II]**
(a) A > C > B > D > E (b) A > B > C > D > E
(c) A > C > B > E > D (d) A > B > C > E > D

Numerical Value Type Questions

19 Consider the following half cell reaction
$$Cr_2O_7^{2-}(aq) + 6e^{-} + 14H^{+}(aq) \longrightarrow 2Cr^{3+}(aq) + 7H_2O(l)$$
The reaction was conducted with the ratio of $\frac{[Cr^{3+}]^2}{[Cr_2O_7^{2-}]} = 10^{-6}$. The pH value at which the EMF of the half cell will become zero is .... (nearest integer value)
[Given : standard half cell reduction potential $E^{\circ}_{Cr_2O_7^{2-}, H^{+}/Cr^{3+}} = 1.33$ V, $\frac{2.303RT}{F} = 0.059$ V.] **[2025. 8 April Shift-II]**

20 1 Faraday electricity was passed through Cu^{2+} (1.5 M, 1 L)/ Cu and 0.1 Faraday was passed through Ag^+(0.2 M, 1 L)/Ag electrolytic cells. After this the two cells were connected as shown below to make an electrochemical cell. The emf of the cell thus formed at 298 K is .... mV (Nearest integer)

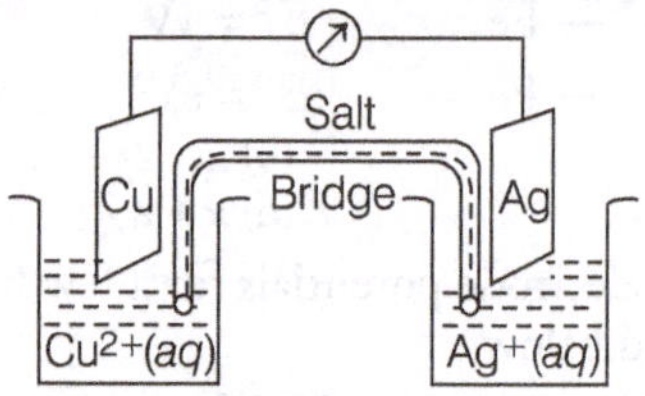

Given, $E^\circ_{Cu^{2+}/Cu} = 0.34\text{ V} \Rightarrow E^\circ_{Ag^+/Ag} = 0.8\text{ V}$

$\frac{2.303RT}{F} = 0.06\text{ V}$ **[2025, 7 April Shift-I]**

21 0.2% (*w*/*V*) solution of NaOH is measured to have resistivity 870.0 mΩ m. The molar conductivity of the solution will be .... $\times 10^2$ mS dm^2 mol^{-1}. (Nearest integer) **[2025, 2 April Shift-II]**

22 Consider the following electrochemical cell at standard condition. **[2025, 2 April Shift-I]**

$Au(s)|QH_2, Q\,|NH_4X\ (0.01\text{ M})\,||\,Ag^+\ (1\text{ M})\,|\,Ag(s)\ E_{cell} = +0.4\text{ V}$

The couple QH_2/Q represents quinhydrone electrode, the half cell reaction is given below:

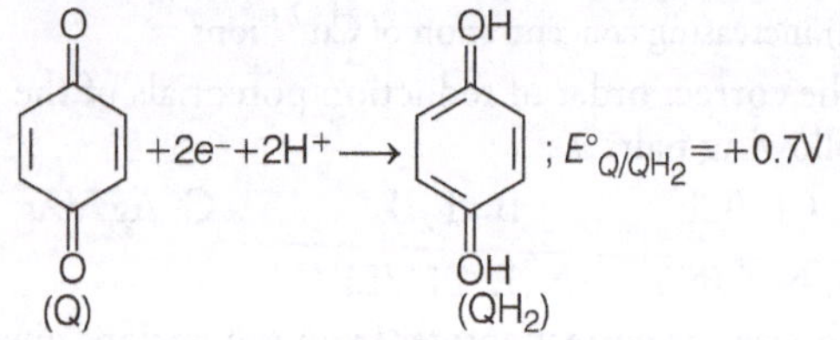

[Given: $E^\circ_{Ag^+/Ag} = +0.8$ V and $\frac{2.303RT}{F} = 0.06$ V]

The pK_b value of the ammonium halide salt (NH_4X) used here is .... (Nearest integer)

23 Consider the following redox reaction,

$MnO_4^- + H^+ + H_2C_2O_4 \longrightarrow Mn^{2+} + H_2O + CO_2$

The standard reduction potentials are given as below (E°_{red}).

$$E^\circ_{MnO_4^-/Mn^{2+}} = +1.51\text{ V}$$

$$E^\circ_{CO_2/H_2C_2O_4} = -0.49\text{ V}$$

If the equilibrium constant of the above reaction is given as $K_{eq} = 10^x$, then the value of x = ... (Nearest integer) **[2024, 1 Feb Shift-II]**

24 The potential for the given half-cell at 298 K is (–) .... $\times 10^{-2}$ V.

$2H^+(aq) + 2e^- \longrightarrow H_2(g)$

$[H^+] = 1M,\ p_{H_2} = 2$ atm

(Given, $2.303\,RT/F = 0.06$ V, $\log 2 = 0.3$) **[2024, 1 Feb Shift-I]**

25 The hydrogen electrode is dipped in a solution of pH = 3 at 25°C. The potential of the electrode will be .... $\times 10^{-2}$ V. $\left(\frac{2.303RT}{F} = 0.059\text{ V}\right)$ **[2024, 27 Jan Shift-II]**

26 At 298 K, the standard reduction potential for Cu^{2+}/Cu electrode is 0.34 V.

Given : $K_{sp}Cu(OH)_2 = 1 \times 10^{-20}$

Take $\frac{2.303\,RT}{F} = 0.059$ V

The reduction potential at pH = 14 for the above couple is (–)$x \times 10^{-2}$ V.

The value of x is .... **[2023, 13 April Shift-II]**

27 The number of correct statements from the following is ......

A. E_{cell} is an intensive parameter.

B. A negative $E^\ominus$ means that the redox couple is a stronger reducing agent than the H^+/H_2 couple.

C. The amount of electricity required for oxidation or reduction depends on the stoichiometry of the electrode reaction.

D. The amount of chemical reaction which occurs at any electrode during electrolysis by a current is proportional to the quantity of electricity passed through the electrolyte. **[2023, 11 April Shift-II]**

28 In an electrochemical reaction of lead, at standard temperature, if $E^\circ_{(Pb^{2+}/Pb)} = m$ volt and $E^\circ_{(Pb^{4+}/Pb)} = n$ volt, then the value of $E^\circ_{(Pb^{2+}/Pb^{4+})}$ is given by $m - xn$. Then value of x is .... (Nearest integer). **[2023, 11 April Shift-I]**

29 $FeO_4^{2-} \xrightarrow{+2.2\text{ V}} Fe^{3+} \xrightarrow{+0.70\text{ V}} Fe^{2+} \xrightarrow{-0.45\text{ V}} Fe^0$

$E^\circ_{FeO_4^{2-}/Fe^{2+}}$ is $x \times 10^{-3}$ V. The value of x is .... **[2023, 10 April Shift-I]**

30 The number of incorrect statements from the following is ....

A. the electrical work that a reaction can perform at constant pressure and temperature is equal to the reaction Gibbs energy.

B. E°_{cell} is dependent on the pressure.

C. $\frac{dE^\circ_{cell}}{dT} = \frac{\Delta_r S^\circ}{nF}$

D. a cell is operating reversibly if the cell potential is exactly balanced by an opposing source of potential difference. **[2023, 8 April Shift-II]**

31 The standard reduction potentials at 298 K for the following half cells are given below

$NO_3^- + 4H^+ + 3e^- \longrightarrow NO(g) + 2H_2O$ $E^\theta = 0.97$ V

$V^{2+}(aq) + 2e^- \longrightarrow V$ $E^\theta = -1.19$ V

$Fe^{3+}(aq) + 3e^- \longrightarrow Fe$ $E^\theta = -0.04$ V

$Ag^+(aq) + e^- \longrightarrow Ag(s)$ $E^\theta = 0.80$ V

$Au^{3+}(aq) + 3e^- \longrightarrow Au(s)$ $E^\theta = 1.40$ V

The number of metal(s) which will be oxidised by NO_3^- in aqueous solution is .... **[2023, 6 April Shift-II]**

32 At what pH, given half-cell $MnO_4^-(0.1\ M) \mid Mn^{2+}(0.001\ M)$ will have electrode potential of 1.282 V? .... (Nearest integer)

Given, $E^\circ_{MnO_4/Mn^{2+}} = 1.54$ V, $\frac{2.303RT}{F} = 0.059$ V **[2023, 1 Feb Shift-I]**

33 The logarithm of equilibrium constant for the reaction $Pd^{2+} + 4Cl^- \rightleftharpoons PdCl_4^{2-}$ is .... (Nearest integer)

Given : $\frac{2.303RT}{F} = 0.06$ V

$Pd^{2+}(aq) + 2e^- \rightleftharpoons Pd(s)$, $E^\ominus = 0.83$ V

$PdCl_4^{2-}(aq) + 2e^- \rightleftharpoons Pd(s) + 4Cl^-(aq)$,

$E^\ominus = 0.65$ V **[2023, 31 Jan Shift-I]**

34 The electrode potential of the following half cell at 298 K $X \mid X^{2+}$ (0.001 M) $\|$ Y^{2+} (0.01 M) $\mid Y$ is .... $\times 10^{-2}$ V. (Nearest integer).

[Given : $E^\circ_{X^{2+}/X} = -2.36$ V $\Rightarrow E^\circ_{Y^{2+}/Y} = +0.36$ V

$\frac{2.303RT}{F} = 0.06$ V] **[2023, 30 Jan Shift-II]**

35 Consider the cell

$Pt(s) \mid H_2$ (g, 1 atm) $\mid H^+$ (aq, 1 M) $\|$ $Fe^{3+}(aq)$, $Fe^{2+}(aq) \| Pt(s)$

When the potential of the cell is 0.712 V at 298 K, the ratio $[Fe^{2+}]/[Fe^{3+}]$ is .... (Nearest integer)

[Given : $Fe^{3+} + e^- \rightleftharpoons Fe^{2+}$,

$E^\circ_{Fe^{3+}, Fe^{2+}/Pt} = 0.771$ V

$\frac{2.303RT}{F} = 0.06$ V **[2023, 30 Jan Shift-I]**

36 The equilibrium constant for the reaction

$Zn(s) + Sn^{2+}(aq) \rightleftharpoons Zn^{2+}(aq) + Sn(s)$ is 1×10^{20} at 298 K. The magnitude of standard electrode potential of Sn / Sn^{2+} if $E^\circ_{Zn^{2+}/Zn} = -0.76$ V is .... $\times 10^{-2}$ V. (Nearest integer).

Given, $\frac{2.303RT}{F} = -0.059$ V **[2023, 29 Jan Shift-II]**

37 $Pt(s) \mid H_2(g)$ (1 bar) $\mid H^+(aq)$ (1 M) $\|$ $M^{3+}(aq) \mid M^+(aq) \mid Pt(s)$

The E_{cell} for the given cell is 0.1115 V at 298 K when $\frac{[M^+(aq)]}{[M^{3+}(aq)]} = 10^a$

The value of a is ....

[Given, $E^\circ_{M^{3+}/M^+} = 0.2$ V

$\frac{2.303RT}{F} = 0.059$ V] **[2023, 25 Jan Shift-II]**

38 At 298 K, a 1 L solution containing 10 m mol of $Cr_2O_7^{2-}$ and 100 mmol of Cr^{3+} shows a pH of 3.0.

Given : $Cr_2O_7^{2-} \longrightarrow Cr^{3+}$; $E^\circ = 1.330$ V and $\frac{2.303RT}{F} = 0.059$ V

The potential for the half-cell reaction is $x \times 10^{-3}$ V. The value of x is .... **[2023, 24 Jan Shift-I]**

39 For a cell,

$Cu(s) \mid Cu^{2+}$ (0.001 M) $\|$ Ag^+ (0.01 M $\mid Ag(s)$

the cell potential is found to be 0.43 V at 298 K. The magnitude of standard electrode potential for Cu^{2+}/Cu is .... $\times 10^{-2}$ V.

[Given : $E^\circ_{Ag^+/Ag} = 0.80$ V and $\frac{2.303\ RT}{F} = 0.06$ V] **[2022, 29 July Shift-II]**

40 The cell potential for $Zn \mid Zn^{2+}(aq) \| Sn^{x+} \mid Sn$ is 0.801 V at 298 K. The reaction quotient for the above reaction is 10^{-2}. The number of electrons involved in the given electrochemical cell reaction is ....

(Given, $E^\circ_{Zn^{2+}/Zn} = -0.763$ V, $E^\circ_{Sn^{x+}/Sn} = +0.008$ V and $\frac{2.303\ RT}{F} = 0.06$ V) ***[2022, 25 July Shift-I]***

41 The cell potential for the given cell at 298 K $Pt|H_2(g, 1\text{ bar})|H^+(aq)||Cu^{2+}(aq)|Cu(s)$ is 0.31 V. The pH of the acidic solution is found to be 3, whereas the concentration of Cu^{2+} is 10^{-x} M. The value of x is ....

(Given, $E^{\ominus}_{Cu^{2+}/Cu} = 0.34$ V and $\dfrac{2.303\,RT}{F} = 0.06$ V)

[2022, 29 June Shift-II]

42 For the given reactions, $Sn^{2+} + 2e^- \longrightarrow Sn$

$Sn^{4+} + 4e^- \longrightarrow Sn$

The electrode potentials are $E^{\circ}_{Sn^{2+}/Sn} = -0.140$ V and $E^{\circ}_{Sn^{4+}/Sn} = 0.010$ V. The magnitude of standard electrode potential for Sn^{4+}/Sn^{2+} i.e., $E^{\circ}_{Sn^{4+}/Sn^{2+}}$ is .... $\times 10^{-2}$ V. (Nearest integer) **[2022, 28 June shift-II]**

43 For the reaction taking place in the cell:

$Pt(s)\,|\,H_2(g)\,|H^+(aq)||Ag^+(aq)|Ag(s)$

$E^{\circ}_{cell} = +0.5332$ V.

The value of $\Delta_f G^{\ominus}$ is – .... kJ mol^{-1}.

[Nearest integer] **[2022, 27 June Shift-II]**

44 $Cu(s) + Sn^{2+}(0.001M) \longrightarrow Cu^{2+}(0.01\text{ M}) + Sn(s)$

The Gibb's free energy change for the above reaction at 298 K is $x \times 10^{-1}$ kJ mol^{-1}.

The value of x is .... [Nearest integer]

[Given, $E^{\ominus}_{Cu^{2+}/Cu} = 0.34$ V; $E^{\ominus}_{Sn^{2+}/Sn} = -0.14$ V; $F = 96500$ C mol^{-1}]. **[2022, 26 June Shift-II]**

45 In a cell, the following reactions takes place **[2022, 25 June Shift-I]**

$Fe^{2+} \longrightarrow Fe^{3} + e^-;\ E^{\circ}_{Fe^{3+}/Fe^{2+}} = 0.77$ V

$2I^- \longrightarrow I_2 + 2e^-;\ E^{\circ}_{I_2/I^-} = 0.54$ V

The standard electrode potential for the spontaneous reaction in the cell is $x \times 10^{-2}$ V 298 K. The value of x is ... [Nearest integer]

46 Consider the following cell reaction,

$$Cd(s) + Hg_2SO_4(s) + \frac{9}{5}H_2O(l) \rightleftharpoons CdSO_4 \cdot \frac{9}{5}H_2O(s) + 2Hg(l)$$

The value of E°_{cell} is 4.315 V at 25°C. If $\Delta H^{\circ} = -825.2$ kJ mol^{-1}, the standard entropy change ΔS° in J K^{-1} is .... (Nearest integer)

[Given, Faraday constant = 96487 C mol^{-1}]

[2021, 31 Aug Shift-I]

47 For the galvanic cell,

$Zn(s) + Cu^{2+}(0.02\text{ M}) \longrightarrow Zn^{2+}(0.04M) + Cu(s)$,

$E_{Cell} = \ldots\ldots\ldots \times 10^{-2}$ V.

(Nearest integer)

[Use $E^{\circ}_{Cu/Cu^{2+}} = -0.34$ V, $E^{\circ}_{Zn/Zn^{2+}} = +0.76$ V, $\dfrac{2.303\,RT}{F} = 0.059$ V]

[2021, 26 Aug Shift-II]

48 These are physical properties of an element.

A. Sublimation enthalpy
B. Ionisation enthalpy
C. Hydration enthalpy
D. Electron gain enthalpy

The total number of above properties that affect the reduction potential is .... (Integer answer)

[2021, 26 Aug Shift-I]

49 For the cell

$Cu(s)\,|Cu^{2+}(aq)\,(0.1)\text{ M})\,||\,Ag^+(aq)\,(0.01M)\,|Ag(s)$

the cell potential, $E_1 = 0.3095$ V

For the cell,

$Cu(s)\,|Cu^{2+}(aq)\,(0.01\text{ M})\,||\,Ag^+(aq)\,(0.001M)\,|\,Ag(s)$

the cell potential = ...$\times 10^{-2}$ V.

(Round off the nearest integer).

$\left[\text{Use}: \dfrac{2.303\,RT}{F} = 0.059\right]$

[2021, 27 July Shift-II]

50 Consider the cell at 25° C

$Zn|Zn^{2+}(aq),\,(1M)\,||Fe^{3+}(aq),\,Fe^{2+}(aq)\,|\,Pt(s)$

The fraction of total iron present as Fe^{3+} ion at the cell potential of 1.500 V is $x \times 10^{-2}$. The value of x is .... (Nearest integer)

(Given, $E^{\circ}_{Fe^{3+}/Fe^{2+}} = 0.77$ V, $E^{\circ}_{Zn^{2+}/Zn} = -0.76$ V)

[2021, 25 July Shift-I]

51 Assume a cell with the following reaction

$Cu(s) + 2Ag^+(1 \times 10^{-3}M) \longrightarrow Cu^{2+}\,(0.250M) + 2Ag(s),\ E^{\ominus}_{cell} = 2.97$ V

E_{cell} for the above reaction is .... V. (Nearest integer)

[Given : log 2.5 = 0.3979, $T = 298$ K]

[2021, 22 July Shift -II]

52 For the reaction,

$2Fe^{3+}(aq) + 2I^-(aq) \longrightarrow 2Fe^{2+}(aq) + I_2(s)$

The magnitude of the standard molar Gibbs free energy change, $\Delta_r G^{\circ}_m = -$ .... kJ (Round off to the nearest integer).

$$\left[\begin{array}{ll} E^\circ_{Fe^{2+}/Fe(s)} = -0.440\text{ V}; & E^\circ_{Fe^{3+}/Fe(s)} = -0.036\text{ V} \\ E^\circ_{I_2/2I^-} = 0.539\text{ V}; & F = 96500\text{ C} \end{array}\right]$$

[2021, 18 March Shift-I]

53 Emf of the following cell at 298 K in V is $x \times 10^{-2}$,
$Zn \mid Zn^{2+}$ (0.1 M) $\|$ Ag^+ (0.01 M) $\mid Ag$
The value of x is ....
(Rounded off to the nearest integer).
[Given, $E^\circ_{Zn^{2+}/Zn} = -0.76$ V, $E^\circ_{Ag^+/Ag} = +0.80$ V, $\frac{2.303RT}{F} = 0.059$] **[2021, 26 Feb Shift-II]**

54 The magnitude of the change in oxidising power of the MnO_4^- / Mn^{2+} couple is $x \times 10^{-4}$ V, if the H^+ concentration is decreased from 1 M to 10^{-4} M at 25°C. (Assume concentration of MnO_4^- and Mn^{2+} to be same on change in H^+ concentration). The value of x is .... (Rounded off to the nearest integer).
$\left[\text{Given, } \frac{2.303\,RT}{F} = 0.059\right]$ **[2021, 24 Feb Shift-II]**

55 An oxidation-reduction reaction in which 3 electrons are transferred has a ΔG° of 17.37 kJ mol^{-1} at 25° C. The value of E°_{cell} (in V) is – ...... $\times 10^{-2}$.
(1 F = 96,500 C mol^{-1}) **[2020, 5 Sep Shift-I]**

56 The Gibbs energy change (inJ) for the given reaction at $[Cu^{2+}] = [Sn^{2+}] = 1$M and 298 K is
$Cu(s) + Sn^{2+}(aq) \longrightarrow Cu^{2+}(aq) + Sn(s)$;
($E^\circ_{Sn^{2+}/Sn} = -0.16$ V, $E^\circ_{Cu^{2+}/Cu} = 0.34$ V,
Take, $F = 96500$ C mol^{-1}) **[2020, 2 Sep Shift-I]**

57 For the disproportionation reaction
$2Cu^+(aq) \rightleftharpoons Cu(s) + Cu^{2+}(aq)$ at 298 K, In K (where K is the equilibrium constant) is ...... $\times 10^{-1}$.
Given : ($E^\circ_{Cu^{2+}/Cu^+} = 0.16$ V,
$E^\circ_{Cu^+/Cu} = 0.52$ V and $\frac{RT}{F} = 0.025$) **[2020, 2 Sep Shift-II]**

Topic 3 / Batteries, Fuel Cell & Corrosion

• Level 2

01 On charging the lead storage battery, the oxidation state of lead changes from x_1 to y_1 at the anode and from x_2 to y_2 at the cathode. The values of x_1, y_1, x_2, y_2 are respectively: **[2025, 4 April Shift-I]**
(a) + 4, + 2, 0, + 2 (b) 0, + 2, + 4, + 2
(c) + 2, 0, 0, + 4 (d) + 2, 0, + 2, + 4

02 Given below are two statements:
Statement I Corrosion is an electrochemical phenomenon in which pure metal acts as an anode and impure metal as a cathode.
Statement II The rate of corrosion is more in alkaline medium than in acidic medium.
In the light of the above statements, choose the correct answer from the options given below: **[2025, 22 Jan Shift-II]**
(a) Statement I is true but Statement II is false.
(b) Both Statement I and Statement II are true.
(c) Statement I is false but Statement II is true.
(d) Both Statement I and Statement II are false.

03 Compound A used as a strong oxidising agent is amphoteric in nature. It is the part of lead storage batteries. Compound A is **[2021, 26 Feb Shift-I]**
(a) PbO_2 (b) PbO (c) $PbSO_4$ (d) Pb_3O_4

• Level 3

Imp. *Application of Cell and Batteries*

04 Match List-I with List-II.

List-I (Applications)	List-II (Batteries/Cell)
A. Transistors	I. Anode – Zn/Hg; Cathode –HgO+C
B. Hearing aids	II. Hydrogen fuel cell
C. Invertors	III. Anode-Zn;Cathode-Carbon
D. Apollo space ship	IV. Anode-Pb;Cathode-Pb|PbO_2

Choose the correct answer from the options given below **[2025, 29 Jan Shift-II]**
(a) (A)-(III), (B)-(II), (C)-(IV), (D)-(I)
(b) (A)-(II), (B)-(III), (C)-(IV), (D)-(I)
(c) (A)-(III), (B)-(I), (C)-(IV), (D)-(II)
(d) (A)-(IV), (B)-(III), (C)-(II), (D)-(I)

05 Match List-I with List-II.

List-I (Cell)	List-II (Use/Property/Reaction)
A. Leclanche cell	I. Converts energy of combustion into electrical energy
B. Ni – Cd cell	II. Does not involve any ion in solution and is used in hearing aids
C. Fuel cell	III. Rechargeable
D. Mercury cell	IV. Reaction at anode $Zn \longrightarrow Zn^{2+} + 2e^-$

Choose the correct answer from the options given below **[2024, 9 April Shift-II]**
(a) A-III, B-I, C-IV, D-II
(b) A-I, B-II, C-III, D-IV
(c) A-II, B-III, C-IV, D-I
(d) A-IV, B-III, C-I, D-II

06 Fuel cell, using hydrogen and oxygen as fuels,

A. has been used in spaceship
B. has as efficiency of 40% to produce electricity
C. uses aluminum as catalysts
D. is eco-friendly
E. is actually a type of Galvanic cell only.

Choose the correct answer from the options given below: **[2024, 4 April Shift-II]**

(a) A, B, C only (b) A, B, D only
(c) A, B, D, E only (d) A, D, E only

07 For lead storage battery pick the correct statements.

A. During charging of battery, $PbSO_4$ on anode is converted into PbO_2.
B. During charging of battery, $PbSO_4$ on cathode is converted into PbO_2.
C. Lead storage battery consists of grid of lead packed with PbO_2 as anode.
D. Lead storage battery ~ 38% solution of sulphuric acid as an electrolyte. **[2023, 12 April Shift-I]**

Choose the correct answer from the options given below.

(a) B, D only (b) B, C only
(c) B, C, D only (d) A, B, D only

08 Match List-I with List-II.

	List-I		List-II
(A)	$Cd(s) + 2Ni(OH)_3(s) \longrightarrow CdO(s) + 2Ni(OH)_2(s) + H_2O(l)$	(I)	Primary battery
(B)	$Zn(Hg) + HgO(s) \longrightarrow ZnO(s) + Hg(l)$	(II)	Discharging of secondary battery
(C)	$2PbSO_4(s) + 2H_2O(l) \longrightarrow Pb(s) + PbO_2(s) + 2H_2SO_4(aq)$	(III)	Fuel cell
(D)	$2H_2(g) + O_2(g) \longrightarrow 2H_2O(l)$	(IV)	Charging of secondary battery

Choose the correct answer from the options given below. **[2022, 28 July Shift-I]**

(a) (A) → (I), (B) → (II), (C) → (III), (D) → (IV)
(b) (A) → (IV), (B) → (I), (C) → (II), (D) → (III)
(c) (A) → (II), (B) → (I), (C) → (IV), (D) → (III)
(d) (A) → (II), (B) → (I), (C) → (III), (D) → (IV)

Key Index

Topic 1

1 (b)	2 (c)	3 (d)	4 (b)	5 (a)	6 (a)	7 (d)	8 (a)	9 (a)	10 (c)
11 (c)	12 (a)	13 (b)	14 (a)	15 (b)	16 (a)	17 (d)	18 (d)	19 (b)	20 (d)
21 (b)	22 (a)	23 (d)	24 (3)	25 (150)	26 (2)	27 (2)	28 (4)	29 (5)	30 (2)
31 (46)	32 (3)	33 (161)	34 (66)	35 (14)	36 (25)	37 (1000)	38 (6)	39 (14)	40 (20)
41 (288)	42 (57)	43 (14)	44 (25)	45 (11)					

Topic 2

1 (b)	2 (a)	3 (a)	4 (c)	5 (c)	6 (c)	7 (d)	8 (b)	9 (c)	10 (a)
11 (b)	12 (b)	13 (d)	14 (a)	15 (b)	16 (c)	17 (d)	18 (a)	19 (10)	20 (400)
21 (23)	22 (6)	23 (338)	24 (1)	25 (18)	26 (25)	27 (4)	28 (2)	29 (1825)	30 (1)
31 (3)	32 (3)	33 (6)	34 (275)	35 (10)	36 (17)	37 (3)	38 (917)	39 (34)	40 (4)
41 (7)	42 (16)	43 (51)	44 (983)	45 (23)	46 (25)	47 (109)	48 (3)	49 (28)	50 (24)
51 (3)	52 (45)	53 (147)	54 (3776)	55 (6)	56 (96500)	57 (144)			

Topic 3

1 (d)	2 (a)	3 (a)	4 (a)	5 (d)	6 (d)	7 (a)	8 (c)

Detailed Explanations

Topic 1

01. *(b)* From Faraday's law,
gram equivalent of Al deposited

$$= \frac{It}{96500}$$

$$\frac{w}{27} \times 3 = \frac{2 \times 30 \times 60}{96500}$$

so $\quad w = 0.336$ g

Thus, the amount of aluminium deposited at cathode is 0.336 g.

02. *(c)* Concentrated solution of sulphuric acid is used as an electrolyte to obtain $H_2S_2O_8$ by the process of electrolysis.

03. *(d)* The given graph of molar conductivity *vs* $\sqrt{C}$ is for weak electrolyte. Hence among the given options CH_3COOH is the correct answer, as it is a weak electrolyte (X).

04. *(b)* The value of limiting molar conductivities for given cations in water are

H^+	349.6 S cm^2 mol^{-1}
Na^+	50.1 S cm^2 mol^{-1}
K^+	73.5 S $cm^2 mol^{-1}$
Ca^{2+}	119 S $cm^2 mol^{-1}$
Mg^{2+}	106 S $cm^2 mol^{-1}$

05. *(a)* The relation between molar conductivity and concentration is as follows

$$\text{Molar conductivity} \propto \frac{1}{\sqrt{C}}$$

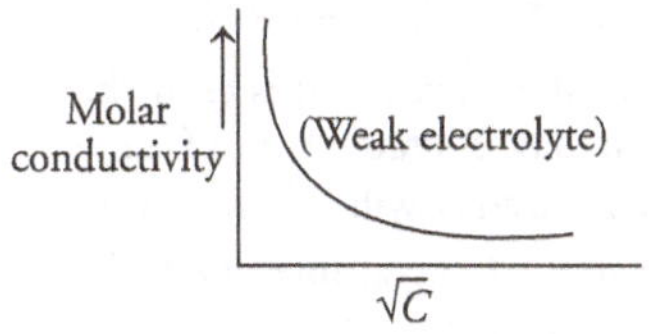

Thus, for weak electrolyte molar conductivity decreases sharply as there is an increases in concentration. Thus, option (a) is the correct answer.

06. *(a)* The electrolytes A and B are weak electrolyte and strong electrolyte respectively.

07. *(d)* As the given solution is already infinitely dilute hence no change in molar conductivity is observed upon addition of water.

08. *(a)* Amount of substance deposited is given by Faraday's law,

$$W = Zit$$

$$W = ZQ \qquad [Q = it]$$

$$Q = \frac{W}{Z}$$

$W = QZ$ when $Q = 1$ then $W = Z$ where, Z is electrochemical equivalent

Hence, option (a) is correct.

09. *(a)* Given,
Ionic conductivities,

$$\lambda_{C^{2+}} = 57 \text{ Scm}^2/\text{mol}$$

$$\lambda_{A^{2-}} = 73 \text{ Scm}^2/\text{mol}$$

Molar conductivity of solution is given by

$$\Lambda_{\text{solution}} = \lambda_{C^{2+}} + \lambda_{A^{2-}}$$

$$= 57 + 73 = 130 \text{ Scm}^2/\text{mol}$$

Hence, option (a) is correct.

10. *(c)* Units of molar conductivity $= \text{Scm}^2\text{mol}^{-1}$.

Unit of $(\text{concentration})^{1/2} = \text{mol}^{1/2}\text{L}^{-1/2}$

Units for slope

$$= \frac{\text{Unit of molar conductivity}}{\text{Unit of (concentration)}^{1/2}}$$

$$= \text{Scm}^2\text{mol}^{-3/2}\text{L}^{1/2}$$

11. *(c)* Calomel electrodes comes in the category of metal-insoluble salt-anion electrodes.

12. *(a)* Among the given options only nature of electrode does not affect electrolytic conductance of a solution.

13. *(b)* Reaction involved in electrolysis of brine solution is as follows

$$NaCl(aq) \longrightarrow Na^+(aq) + Cl^-(aq)$$

At anode $2Cl^-(aq) \longrightarrow Cl_2(g) + 2e^-$

$$2H_2O(l) \longrightarrow O_2(g) + 4H^+ + 4e^-$$

At cathode

$$2H_2O(l) + 2e^- \longrightarrow H_2(g)\uparrow + 2OH^-(aq)$$

$$2Na^+ + 2OH^- \longrightarrow 2NaOH$$

Thus, OH^- is formed at cathode.

14. *(a)* The correct representation of conductometric titration of benzoic acid *vs* sodium hydroxide is (a).

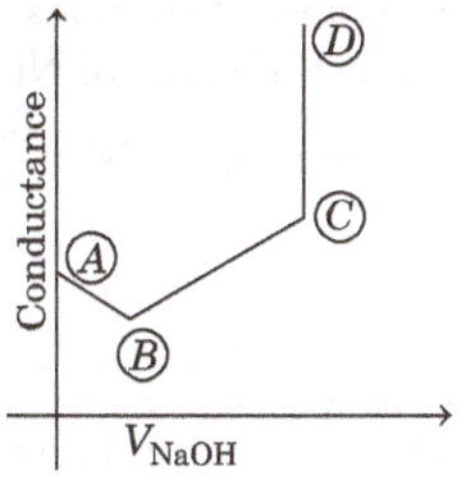

In A to B, free H^+ ions are replaced by Na^+ which decreases conductance.
In B to C, un-dissociated benzoic acid reacts with NaOH and forms salt which increases number of ions and thus conductance increases.
In C to D, NaOH added further increases Na^+ and OH^- ions which further increases the conductance.

15. *(b)* Molar conductivity of strong electrolyte (i.e., KI) increases slowly on dilution, while molar conductivity of weak electrolyte (i.e., H_2CO_3) increases steeply with dilution.

16. *(a)* Molar conductivity,

$$\Lambda_M = \frac{\kappa \times 1000}{M}$$

$$\Lambda_{M_1} = \frac{\kappa_1 \times 1000}{\frac{10}{20} \times 1000} = 2\kappa_1$$

$$\Lambda_{M_2} = \frac{\kappa_2 \times 1000}{\frac{20}{80} \times 1000} = 4\kappa_2$$

Given, $\quad \kappa_1 = \kappa_2 = \kappa$

$\therefore \quad \Lambda_{M_1} = 2\kappa; \; \Lambda_{M_2} = 4\kappa$

$$\frac{\Lambda_{M_2}}{\Lambda_{M_1}} = \frac{4\kappa}{2\kappa} = \frac{4}{2}$$

$\Rightarrow \quad \Lambda_{M_2} = 2\Lambda_{M_1}$

17. *(d)* Limiting molar conductivity of CH_3COOH is more than KCl. Hence, statement I is false.
With decrease in concentration of electrolyte, molar conductivity increases. Hence, statement II is also false.

18. *(d)* A solution of $Ni(NO_3)_2$ is electrolysed between platinum electrodes using 0.1 Faraday electricity. It means that 0.1 equivalent of Ni^{2+} will be discharged.

Electrolysis of $Ni(NO_3)_2$ gives

$$Ni^{2+} + 2e^- \longrightarrow Ni$$

(Atomic mass of Ni = 58.7)

Number of equivalents = Number of moles × number of electrons transferred

0.1 = Number of moles of Ni × 2

$\therefore$ Number of moles of Ni $= \frac{0.1}{2} = 0.05$

19. *(b)*

idea This question is based upon Faraday's first law which states that "Mass of any substance deposited or liberated at any electrode is directly proportional to the quantity of electricity passed."

During charging :

$$Pb + SO_4^{2-} \longrightarrow PbSO_4 + 2e^-$$

$$\Rightarrow \quad Q = nF$$

$$Q = 2F$$

2F gives 303 g/mol

Amount of $PbSO_4$ electrolysed by 0.05 F is

$$\therefore \quad = \frac{303}{2} \times 0.05 \text{ g of } PbSO_4$$

$$= 7.575 \approx 7.6 \text{ g of } PbSO_4$$

20. *(d)* Statement I is true but Statement II is false.

Mohr's salt molecular formula is

$$FeSO_4 \cdot (NH_4)_2SO_4 \cdot 6H_2O$$

Using Kohlrausch law, molar conductance is

λ_m^∞ (Mohr's salt) $= x_1 + 2x_2 + 2x_3$.

21. *(b)* For a weak electrolyte

$$HA(aq) \rightleftharpoons H^+(aq) + A^-(aq)$$

$$K_a = \frac{\alpha^2 C}{1-\alpha}$$

$$\alpha^2 C = K_a - K_a\alpha$$

$$\alpha^2 C + K_a\alpha - K_a = 0$$

$$\left(\frac{\Lambda_m}{\Lambda_m^\circ}\right)^2 C + K_a\left(\frac{\Lambda_m}{\Lambda_m^\circ}\right) - K_a = 0$$

$$\Lambda_m^2 C + K_a\Lambda_m\Lambda_m^\circ - K_a\Lambda_m^{\circ 2} = 0$$

22. *(a)* The correct match is

A → 3, B → 1, C → 4, D → 2.

23. *(d)* According to Kohlrausch's law,

$$\Lambda_m^\circ(HA) = [\Lambda_m^\circ(H^+) + \Lambda_m^\circ(Cl^-)] + [\Lambda_m^\circ(Na^+) + \Lambda_m^\circ(A^-)] - [\Lambda_m^\circ(Na^+) + \Lambda_m^\circ(Cl^-)]$$

$$= 425.9 + 100.5 - 126.4$$

$$= 400 \text{ S cm}^2 \text{ mol}^{-1}$$

Also, molar conductivity at given concentration is given as,

$$\Lambda_m = \frac{1000 \times \kappa}{M}$$

$$\therefore \quad \Lambda_m = \frac{1000 \times 5 \times 10^{-5} \text{ S cm}^{-1}}{10^{-3} \text{ M}}$$

$$= 50 \text{ S cm}^2 \text{ mol}^{-1}$$

Therefore, degree of dissociation (α), of HA is,

$$\alpha = \frac{\Lambda_m}{\Lambda_m^\circ} = \frac{50 \text{ S cm}^2\text{mol}^{-1}}{400 \text{ S cm}^2\text{mol}^{-1}}$$

$$= 0.125$$

24. *(3)* Given,

$$\lambda_m^\circ \text{ of } NH_4Cl = 185$$

$$(\lambda_m^\circ)_{NH_4^+} + (\lambda_m^\circ)_{Cl^-} = 185$$

$$(\lambda_m^\circ)_{NH_4^+} = 185 - 70 = 115 \text{ Scm}^2\text{mol}^{-1}$$

$$(\lambda_m^\circ)_{NH_4OH} = (\lambda_m^\circ)_{NH_4^+} + (\lambda_m^\circ)_{OH^-}$$

$$= 115 + 170$$

$$(\lambda_m^\circ)_{NH_4OH} = 285 \text{ Scm}^2\text{mol}^{-1}$$

$$\text{Degree of dissociation} = \frac{(\lambda_m)_{NH_4OH}}{(\lambda_m^\circ)_{NH_4OH}}$$

$$= \frac{85.5}{285} = 0.3$$

$$= 3 \times 10^{-1}$$

25. *(150)* Given, $R = 100\,\Omega$

Resistance of same cell with dilute solution $x = ?$

The resistance is given by the formule

$$R = \rho\frac{l}{A}$$

$$\kappa = G \times G^* \text{ (where, } G = \frac{1}{R})$$

$$\text{and } G^* \text{ (Cell constant)} = \frac{l}{A}$$

$$\frac{\kappa_C}{\kappa_d} = \frac{R_d}{R_C};$$

$$\lambda_m = \frac{\kappa \times 1000}{C}$$

$$\frac{\kappa_C}{\kappa_d} = \frac{(\lambda_m.C)_c}{(\lambda_m.C)_d} = \frac{R_d}{R_C}$$

$C \rightarrow$ concentrated solution,

$d \rightarrow$ dilute solution.

$$\frac{100 \times (0.15)^2}{150 \times (0.1)^2} = \frac{R_d}{100}$$

$$\Rightarrow \quad R_d = 150\Omega$$

So, the value of x is 150.

26. *(2)*

idea To solve this problem, first of all find the value of $[OH^-]$ by using the pH value. After that use Faraday's law to calculate the value of current in ampere.

Electrolysis of NaCl is

$$NaCl + H_2O(aq) \longrightarrow NaOH(aq) + \frac{1}{2}Cl_2(g) + \frac{1}{2}H_2(g)$$

Since during electrolysis pH changes to 12.

So, $[OH^-] = 10^{-2}$ and $[H^+] = 10^{-12}$

So, according to Faraday law,

Gram amount of substance deposited

= Amount of electricity passed

$$10^{-2} \times \frac{600}{1000} \times 96500 = I \times t$$

$$\frac{10^{-2} \times 600}{1000} \times 96500 = I \times 5 \times 60$$

$$I = 1.93 \text{ A} \simeq 2 \text{ ampere.}$$

27. *(2)* $H_2O \longrightarrow 2H^+ + \frac{1}{2}O_2 + 2e^-$

So, we can say, oxidation of 1 mole of water required 96500 × 2C

$$= 96500 \times 2$$

$$= 193000$$

$$= 1.93 \times 10^5 \text{ C}$$

$$\approx 2 \times 10^5 \text{ C}$$

28. *(4)* Conductivity (Sm^{-1})

2.1×10^3, 1.2×10, 3.91, 1×10^3 } Conductor at 298.15 K

1×10^{-16} Insulator at 298.15 K

29. *(5)* $Cu^{2+} + 2e^- \longrightarrow Cu$

From the reaction, it is clear that 2 Faraday liberates 1 mol Cu.

So, 1 Faraday will librate 0.5 mol Cu.

0.5 mol = 0.5 g atom $= 5 \times 10^{-1}$

Thus, the value of $x = 5$

30. *(2)* Given, increase in mass at cathode,

$W = 1.314$ g

$$Au^{3+} + 3e^- \longrightarrow Au$$

$$\frac{W}{E} = \frac{\text{Charge}}{1}$$

$$\frac{1.314}{\frac{197}{3}} = \frac{Q}{1\text{ F}} \quad \text{[At. mass of Au = 197]}$$

$$Q = 0.02 \text{ F} = 2 \times 10^{-2} \text{ F.}$$

31. *(46)*

> **idea** This problem is based on the faraday's first law of electrolysis. To solve this question, first write the electrolysis reaction of zinc sulphate. Using this equation calculate the value of *n*. Then, put this value along with given values in the formula $W = Zit$, to calculate the value of *w*.

The electrolysis reaction of zinc sulphate is, $Zn^{2+} + 2e^- \longrightarrow Zn$

$$SO_4^{2-} + H_2O \longrightarrow H_2SO_4$$

$$n = 2$$

$$\underset{\text{(mass)}}{W} = Z \times i \times t$$

$$= \frac{\text{Atomic mass} \times i \times t}{nF}$$

Mass of zinc deposit is

$$= \frac{65.4}{2 \times 96500} \times 0.015 \times 15 \times 60$$

[Given, $i = 0.015$, $t = 15 \times 60$, Atomic mass of $Zn = 65.4$ amu]

$$W = 45.75 \times 10^{-4} \text{ g} \approx 46 \times 10^{-4} \text{ g}$$

Hence, the nearest integer is 46.

32. *(3)* Only three statements A, C and D are correct whereas all other statements B, E are incorrect.

The correct form of B and E respectively are :

(B) The number of ions per unit volume in a solution decreases on dilution, because the total number of ions in solution remains the same, but the volume increases.

(E) For weak electrolyte, the change in molar conductivity with dilution is due to an increase in the degree of dissociation.

33. *(161)* Volume of Ni deposited $= v \times d$

$$= (10^{-4})(100) \text{ cm}^3 = 10^{-2} \text{ cm}^3$$

$\therefore$ Weight of Ni deposited

$$= (10^{-2}) \times (10)$$

$$= 0.1 \text{ gram}$$

No. of moles of (Ni) deposited

$$= \frac{\text{Given mass}}{\text{Molar mass}} = \frac{0.1}{60} = \left[\frac{1}{6} \times 10^{-2}\right]$$

$$Ni^{2+} + 2e^{\ominus} \longrightarrow Ni$$

$\Rightarrow$ Charge required $= \frac{1}{6} \times 10^{-2} \times 2F$

$$= i \times t$$

$$t = \frac{1}{6} \times 10^{-2} \times \frac{2 \times 96500}{2}$$

$$t = 161 \text{ s}$$

34. *(66)* Given,

$C = 0.0025$ M acetic acid

(κ) specific conductance of 0.0025 M acetic acid $= 5 \times 10^{-5}$ S cm mol^{-1}

Limiting molar conductivity of CH_3COOH, $\lambda_m^{\infty} = 400$ S cm mol^{-1}

Dissociation constant, $\Lambda_m = \frac{\kappa}{C} \times 1000$

$$= \frac{5 \times 10^{-5} \times 10^{-3}}{0.0025}$$

$$= \frac{5 \times 10^{-2}}{2.5 \times 10^{-3}}$$

$$= 20 \text{ S cm}^2 \text{ mol}^{-1}$$

$$\alpha = \lambda_m / \lambda_m^{\infty} = \frac{20}{400} = \frac{1}{20}$$

$$K_a = \frac{C\alpha^2}{1-\alpha} = \frac{0.0025 \times \frac{1}{20} \times \frac{1}{20}}{\frac{19}{20}}$$

$$= \frac{0.0025}{19 \times 20} = 6.6 \times 10^{-6} = 66 \times 10^{-7}$$

35. *(14)* $\underset{10^{-5}}{AgNO_3} \longrightarrow \underset{10^{-5}}{Ag^+} + \underset{10^{-5}}{NO_3^-}$

$$AgBr \rightleftharpoons \underset{S}{Ag^+} + \underset{S}{Br^-}$$

$$K_{sp} = [Ag^+][Br^-] = 4.9 \times 10^{-13}$$

$\Rightarrow$ $S^2 = 4.9 \times 10^{-13}$

$\Rightarrow$ $S = 7 \times 10^{-7}$ M

$\therefore$ $[Ag^+] = 7 \times 10^{-7} + 100 \times 10^{-7}$

$$= 107 \times 10^{-7} \text{ M}$$

$\therefore$ $[Br^-] = \frac{K_{sp}}{[Ag^+]}$

$$= \frac{4.9 \times 10^{-13}}{107 \times 10^{-7}} = 4.58 \times 10^{-8}$$

Conductivity

$$= [Ag^+]\lambda^{\circ}_{Ag^+} + [Br^-]\lambda^{\circ}_{Br^-} + [NC_3^-]\lambda^{\circ}_{NO_3^-}$$

$$= (107 \times 10^{-7} \times 6 \times 10^{-3}) + (4.58 \times 10^{-8} \times 8 \times 10^{-3}) + (10^{-5} \times 7 \times 10^{-3})$$

$$= (642 \times 10^{-10}) + (36.64 \times 10^{-11}) + (7 \times 10^{-8})$$

$$= (642 + 3.664 + 700) \times 10^{-10}$$

$$= 1346 \times 10^{-10} \text{ Sm}^{-1}$$

or 13.46×10^{-8} Sm^{-1} $\approx 14 \times 10^{-8}$ Sm^{-1}

36. *(25)*

> **idea** This question can be solved by using the concept of molar conductivity for this, use the formula $\Lambda_m = \frac{\kappa \times 1000}{M}$, where κ can be expressed in terms of resistivity as $\frac{1}{\rho}$

$$\Lambda_m = \frac{\kappa \times 1000}{M} = \frac{1}{\rho} \times \frac{1000}{M}$$

$$= \frac{1}{5 \times 10^{-3}} \times \frac{1000}{0.8}$$

$$= 25 \times 10^4 \ \Omega^{-1} \text{ cm}^2 \text{ mol}^{-1}$$

37. *(1000)* In case of solution 1, cell constant

$$\frac{l}{A} = 129 \text{ m}^{-1}$$

Resistance, $R = 100\,\Omega$

Specific conductivity,

$$\kappa = \frac{1}{R} \cdot \frac{l}{A} = \frac{1}{100} \times 129$$

$$= 1.29\Omega^{-1} \text{m}^{-1}$$

Concentration, $C = 74.5$ ppm

Molarity, $M_1 = \frac{74.5}{\text{molar mass} \times 1000}$

$$= \frac{74.5}{74.5 \times 1000} = 10^{-3} \text{ M}$$

Molar conductivity $\Lambda_1 = \frac{\kappa}{M}$

$$\Lambda_1 = \frac{1.29}{10^{-3}}$$

$$= 1290\Omega^{-1} \text{ m}^{-1} \text{ M}^{-1}$$

or $\frac{129 \times 10^3}{100}$

In case of solution 2,

Resistance, $R = 50\Omega$

$$\kappa = \frac{1}{50} \times 129 = \frac{129}{50}\Omega^{-1}\text{m}^{-1}$$

$$M = \frac{149}{74.5 \times 1000}$$

$$= 2 \times 10^{-3} \text{ M}$$

$$\Lambda_2 = \frac{\kappa}{M} = \frac{129 \times 10^3}{50 \times 2}\Omega^{-1}\text{m}^{-1}\text{M}^{-1}$$

$$\frac{\Lambda_1}{\Lambda_2} = \frac{129 \times 10^3 \times 50 \times 2}{100 \times 129 \times 10^3} = 1$$

$$\frac{\Lambda_1}{\Lambda_2} = 1$$

$$= 1000 \times 10^{-3}.$$

Thus, the value of *x* is 1000.

38. *(6)* Reaction involved is as follows

$$Cr_2O_7^{2-} + 14H^{\oplus} + 6e^{\ominus} \longrightarrow 2Cr^{3+} + 7H_2O$$

So, 1 mole $Cr_2O_7^{2-}$ requires 6 moles electrons

1 mole $= nF$

1 mole electron = 1 Faraday

6 moles electrons = 6 Faraday

39. *(14)* Given,

$$\Lambda^{\infty}_{AgI} = \Lambda^{\infty}_{AgNO_3} + \Lambda^{\infty}_{NaI} - \Lambda^{\infty}_{NaNO_3}$$
$$= 13.3 + 12.7 - 12.0$$
$$= 14 \text{ mS m}^2 \text{ mol}^{-1}.$$

40. *(20)* Given, time, $t = x$ min

Using Faraday's first law of electrolysis,

$$w = ZIt = \frac{EIt}{96500}$$
$$\frac{0.3482 \times 96500 \times 3}{56 \times 1.5} = t$$
$$\Rightarrow t = 1200 \text{ s}.$$

So, time, $x = 20$ min.

41. *(288)* Molar conductivity of $BaCl_2 = 280 \text{ Scm}^2\text{mol}^{-1}$

Molar conductivity of $H_2SO_4 = 860 \text{ Scm}^2\text{mol}^{-1}$

Molar conductivity of HCl $= 426 \text{ S cm}^2\text{mol}^{-1}$

Molar conductivity of $BaSO_4$ = ?

From Kohlrausch's law,

$$\Lambda^{\infty}_m(BaSO_4) = \lambda^{\infty}_m(Ba^{2+}) + \lambda^{\infty}_m(SO_4^{2-})$$
$$\Lambda^{\infty}_m(BaSO_4) = \Lambda^{\infty}_m(BaCl_2) + \Lambda^{\infty}_m(H_2SO_4) - 2\Lambda^{\infty}_m(HCl)$$

On putting given values we get,

$$= 280 + 860 - 2(426)$$
$$= 288 \text{ S cm}^2 \text{ mol}^{-1}$$

Concept Enhancer

According to Kohlrausch law of independent migration of ions, the limiting molar conductivity of an electrolyte can be represented as the sum of the individual contribution of the anion and cation of the electrolyte.

42. *(57)* We know that,

$$\kappa = 1/R \cdot G^*$$

$\because$ The conductivity cell is same, G^* is constant and hence $\kappa \cdot R$ = constant

$$\therefore \quad \kappa_{KCl} \cdot R_{KCl} = \kappa_{HCl} \cdot R_{HCl}$$

Therefore, $0.14 \times 4.19 = \kappa \times 1.03$

$$\text{or, } \kappa \text{ of HCl solution} = \frac{0.14 \times 4.19}{1.03}$$
$$= 0.5695 \text{ Sm}^{-1}$$
$$\approx 57 \times 10^{-2} \text{ Sm}^{-1}$$

43. *(14)*

idea To find out molar conductivity, first of all we will calculate the conductivity by using the formula $\kappa = G\, l/A$ then we will use this value in the formula of molar conductivity

$$\Lambda_m = \frac{\kappa\,(\text{S cm}^{-1}) \times 1000\,(\text{cm}^3/\text{L})}{\text{Molarity (mol/L)}}$$

Using the formula

$$\kappa = G\frac{l}{A} = 0.55 \times 10^{-3} \times 1.3 \text{ S cm}^{-1}$$

Now using,

$$\Lambda_m = \frac{\text{Conductivity (S cm}^{-1}) \times 1000\,(\text{cm}^3/\text{L})}{\text{Molarity (mol/L)}}$$
$$= \frac{0.55 \times 10^{-3} \times 1.3 \times 10^3}{5 \times 10^{-3}}$$
$$= 143 \text{ S cm}^2 \text{ mol}^{-1}$$

$\therefore$ 1 S cm^2 mol^{-1} = 10^{-4} S m^2 mol^{-1}

$$\Rightarrow 143 \text{ S cm}^2 \text{ mol}^{-1}$$
$$= 143 \times 10^{-4} \text{ S m}^2 \text{ mol}^{-1}$$
$$= 143 \times 10^{-4} \times 10^3 \text{ m Sm}^2 \text{ mol}^{-1}$$
$$\Rightarrow 14.3 \text{ m Sm}^2 \text{ mol}^{-1}$$
$$\approx 14 \text{ m Sm}^2 \text{ mol}^{-1}$$

44. *(25)* In the reduction half-cell,

$$MnO_4^- + 8H^+ + 5e^- \longrightarrow Mn^{2+} + 4H_2O$$

As one mole of MnO_4^- required 5F of charge,

$\therefore$ 5 moles of MnO_4^- will require charge, $Q = 5 \times 5$ Faraday = 25 Faraday

45. *(11)* For 1 mole of ClO_3^-, 6 F charge required but current efficiency is only 60%. So, to obtain 1 mol ClO_3^-, charge required is 10 F.

$$\text{Now, moles of } KClO_3 = \frac{10 \text{ g}}{122}$$
$$\therefore \text{ Charge required} = \frac{10}{122} \times 10 \text{ F}$$

$\therefore$ Current = 2 A

Apply, $Q = i \cdot t$

$$\therefore \quad 100 \times \frac{96500}{122} = 2t$$
$$\Rightarrow t = 39549.18 \text{ sec} = 10.98 \text{ h} \approx 11 \text{ h}$$

Hence, the correct answer is (11).

Topic 2

01. *(b)*

idea To solve this question, first of all write the balanced chemical equation, then using this find the value of change in number of e$^-$. After that use Nernst-equation to obtain the correct option.

According to Nernst equation,

$$E_{cell} = E^{\circ}_{cell} - \frac{RT}{nF} \ln Q$$

The cell reaction given in question is

$$Mg(s) + 2Ag^+(aq) \rightleftharpoons 2Ag(s) + Mg^{2+}(aq)$$
$$Q = \frac{[Mg^{2+}]}{[Ag^+]^2}$$

Also, as there is change in 2 number of electrons. So, $n = 2$

$$\therefore \quad E = E^{\circ}_{Cell} - \frac{RT}{2F} \ln\left[\frac{[Mg^{2+}]}{[Ag^+]^2}\right]$$
$$\text{or} \quad E = E^{\circ}_{Cell} + \frac{RT}{2F} \ln\left[\frac{[Ag^+]^2}{[Mg^{2+}]}\right]$$

02. *(a)* For the reaction,

$$\frac{1}{2}H_2(g) + AgCl(s) \rightleftharpoons H^+(aq) + Cl^-(aq) + Ag(s)$$

At anode : $\frac{1}{2}H_2(g) \rightleftharpoons H^+(aq) + e^-$

At cathode :

$$AgCl(s) + e^- \longrightarrow Ag(s) + Cl^-(aq)$$

$\therefore$ Cell representation.

$$Pt|H_2(g)|HCl(sol^n) \,|\, AgCl(s)|Ag$$

03. *(a)* The standard electrode potential of M^+/M in aqueous solution does not depend on ionisation of a solid metal atom. So, option (a) is correct.

04. *(c)* To find the value of ΔG° (kJ mol), we use the formula

$$\Delta G^{\circ} = -nFE^{\circ}_{cell}$$
$$\Delta G^{\circ} = -2 \times 96000 \times 2$$
$$= -384000 \text{ J/mol}$$

In terms of kJ/mol,

$$\Delta G^{\circ} = \frac{-384000}{1000}$$
$$= -384 \text{ kJ/mol}$$

05. *(c)*

$$FeO_4^{2-} \xrightarrow[n_1 = 3]{E_1^{\circ} = +2.0 \text{ V}} Fe^{3+} \xrightarrow[n_2 = 1]{E_2^{\circ} = 0.8 \text{ V}} Fe^{2+} \xrightarrow{E_3^{\circ} = -0.5 \text{ V}} Fe^0$$

($FeO_4^{2-} \to Fe^{2+}$: $E_4^{\circ} = ?$, $n_4 = 4$)

$$-n_4 FE_4^{\circ} = -n_1 FE_1^{\circ} - n_2 FE_2^{\circ}$$

On substituting the values, we get

$$+4E_4^{\circ} = 3 \times 2 + (1 \times 0.8)$$
$$E_4^{\circ} = \frac{6.8}{4} = 1.7 \text{ V}$$

So, the value of $E^{\circ}_{FeO_4^{2-}/Fe^{2+}} = 1.7$ V

06. *(c)* An electrochemical cell can be converted into an electrolytic cell by applying an external potential greater than E°_{cell}.

The entire cell reaction will be reversed. The oxidised species will be reduced and the reduced species will be oxidised.

07. *(d)* $M | M^{+2} || X | X^{2-}$

The cell potential is given by

$E^\circ_{cell} = E^\circ_{M/M^{+2}} + E^\circ_{X/X^{-2}}$

$= -0.46 + 0.34 = -0.12$ V

As E°_{cell} is negative, so anode becomes cathode and cathode becomes anode.

Spontaneous reaction will be

$M^{+2} + X^{2-} \rightarrow M + X$

Hence, option (d) is correct.

08. *(b)*

idea To solve this question, first of all write the balanced chemical equation to find the value of *n*, then use the equation $E = E^\circ - \frac{0.059}{n} \log \frac{p_{H_2}}{[H^+]^2}$ to calculate the pressure of H_2.

The required reaction is

$2e^- + 2H^+(aq) \rightarrow H_2(g)$

Using, $E = E^\circ - \frac{0.059}{n} \log \frac{p_{H_2}}{[H^+]^2}$

$0 = 0 - \frac{0.059}{2} \log \frac{p_{H_2}}{(10^{-7})^2}$

$\log \frac{p_{H_2}}{(10^{-7})^2} = 0 \Rightarrow \frac{p_{H_2}}{(10^{-7})^2} = 1$

$p_{H_2} = 10^{-14}$ bar

09. *(c)* A cell with less value of $\left(\frac{\partial E}{\partial T}\right)_p$ is preferred to be used as reference electrode. Because it can be used for wider range of temperature without much deviation from standard value.

Thus, half cell *C* would be preferred to be used as reference electrode.

10. *(a)* $E^\circ_{cell} = E^\circ_{Cu^{2+}/Cu} - E^\circ_{Zn^{2+}/Zn}$

$= 0.34 - (-0.76) = 1.10$ volt

If $E_{ext} < E^\circ_{cell}$, then e^-s flow from Zn to Cu and Zn dissolves at anode and Cu deposites at cathode.

Then, statement (a) is incorrect whereas all other statements are correct.

11. *(b)* The half cell reactions are :

(1) $Cu^{2+} + 2e^- \rightleftharpoons Cu$; $E^\circ = 0.34$ V

(2) $Cu^+ + e^- \rightleftharpoons Cu$; $E^\circ = 0.522$ V

(3) $Cu^{2+} + e^- \rightleftharpoons Cu^+$; $E^\circ = x$

As we know, $\Delta G^\circ = -nFE$

∴ To obtain the E° of required equation, subtract ΔG° of equation (2) from ΔG° of (1) as

ΔG° of (3) = ΔG° of (1) $-\Delta G^\circ$ of (2)

or $-1 \times F \times x =$

$(-2 \times F \times 0.34) - (-1 \times F \times 0.522)$

∴ $x = 2 \times 0.34 - 0.522$

$= 0.68 - 0.522 = 0.158$ V

12. *(b)* We know that,

$E^\circ = \frac{RT}{nF} \ln K \Rightarrow K = e^{nFE^\circ/RT}$

Given, $n = 2$, $F = 96000$ C mol^{-1}

$E^\circ = 2$ V, $R = 8$ JK^{-1} mol^{-1},

$T = 300$ K

On substituting the values,

∴ $K = e^{\frac{2 \times 96000 \times 2}{8 \times 300}} = e^{160}$

13. *(d)* Statement I is correct but Statement II is incorrect.

The correct form of Statement II is,

At cathode, instead of Mg, $H_2O(l)$ will reduce and evolve H_2 gas.

14. *(a)* Among the given statements only (a) is correct

Cell potential is given by

$E^\circ_{cell} = E^\circ_{cathode} - E^\circ_{anode}$

$1.21 = 1.229 - E^\circ_{anode}$

$= 19$ mV

∴Fuel cell involves oxidation of methanol which will occur at anode and reduction of O_2 will occur at cathode.

15. *(b)* For given cell;

$Fe^{2+}(aq) + Ag^+(aq) \longrightarrow Fe^{3+}(aq) + Ag(s)$

$Fe \xrightarrow{\Delta G_1^\circ} Fe^{2+} \xrightarrow{\Delta G_2^\circ} Fe^{3+}$

$\Delta G_3^\circ = \Delta G_1^\circ + \Delta G_2^\circ$

$-3F(-z) = -2F(-y) + \Delta G_2^\circ$

$\Delta G_2^\circ = 3Fz - 2Fy$

Also $\Delta G_2^\circ = -nFE^\circ_{Fe^{2+}/Fe^{3+}}$

$3F_z - 2F_y = -IF(E^0_{Fe^{2+}/Fe^{+3}})$

$E^\circ_{Fe^{2+}/Fe^{3+}} = 2y - 3z$

E°_{cell} for reaction will be

$E^\circ_{Ag^+/Ag} + E^\circ_{Fe^{2+}/Fe^{3+}} = x + 2y - 3z$

16. *(c)*

idea To solve this problem first of all calculate the value of E° of corresponding option. After that use this value of E° into the equation, $\Delta G^\circ = -nFE^\circ$ to obtain the value of ΔG°.

Using the formula,

$\Delta G^\circ = -nFE^\circ$

For option (a),

$E^\circ = -2.37 - 0.8 = -3.17$ V

$\Delta G^\circ = -2 \times F \times (-3.17) = +6.34\ FV$

For option (b),

$E^\circ = 0.8 - 0.34 = 0.46$ V

$\Delta G^\circ = -2 \times F \times 0.46$

$= -0.92\ FV$

For option (c),

$E^\circ = 0.8 + 0.76 = 1.56$ V

Now, $\Delta G^\circ = -2 \times F \times 1.56 = -3.12\ FV$

For option (d),

$E^\circ = -2.37 + 0.76 = -1.61$ V

Now, $\Delta G^\circ = -2 \times F \times (-1.61)$

$= +3.22\ FV$

Thus, the most negative value of ΔG° is obtained for option (c).

17. *(d)* Anode reaction

$[Tl(s) \longrightarrow Tl^+(aq) + e^-] \times 2$

Cathode reaction

$Cu^{+2}(aq) + 2e^- \longrightarrow Cu(s)$

Overall redox reaction

$2Tl(s) + Cu^{+2}(aq) \longrightarrow 2Tl^+(aq) + Cu(s)$

$E_{cell} = E^\circ_{cell} - \frac{0.0591}{2} \log \frac{[Tl^+]^2}{[Cu^{+2}]}$

E_{cell} increases by increasing concentration of $[Cu^{+2}]$.

18. *(a)* The values of reduction potentials are as follows

A. Cl_2 / Cl^-; $E^\circ_{Cl_2/Cl^-} = +1.36$ V

B. I_2 / I^-; $E^\circ_{I_2/I^-} = +0.54$ V

C. Ag^+ / Ag; $E^\circ_{Ag^+/Ag} = +0.80$ V

D. Na^+ / Na; $E^\circ_{Na^+/Na} = -2.71$ V

E. Li^+ / Li; $E^\circ_{Li^+/Li} = -3.05$ V

Hence, the correct order of reduction potentials is

A > C > B > D > E.

Concept Enhancer

Reduction potential represents the tendency of a substance to loose or gain electrons from an electrode. Standard reduction potential $= E^\circ_{Cathode} - E^\circ_{Anode}$.

19. *(10)* The balanced cell reaction is as follows,

$$Cr_2O_7^{2-}(aq)+14H^+(aq)+6e^- \longrightarrow 2Cr^{3+}(aq)+7H_2O(l)$$

Using nernst equation

$$E_R=E_R^o-\frac{0.059}{6}\log\frac{[Cr^{3+}]^2}{[Cr_2O_7^{2-}][H^+]^{14}}$$

$$0=1.33-\frac{0.059}{6}\log\frac{10^{-6}}{[H^+]^{14}}$$

$$141.254=14\,pH \Rightarrow pH=10.08$$

20. *(400)* Cell reactions are as follows,

$$Cu^{2+} + 2e^- \longrightarrow Cu$$

(1 Faraday = charge on 1 mole electrons)

$t=0$	1.5	1 mole	
$t=t$	1	–	0.5 mole

$[Cu^{2+}]=1$ M after electrolysis.

$$Ag^+ + e^- \longrightarrow Ag$$

$t=0$	0.2	0.1 mole	
$t=t$	0.1	–	–

$[Ag^+]=0.1$ M after electrolysis.

Cell reaction:

$$Cu(s)+2Ag^+(aq)\longrightarrow Cu^{2+}(aq)+2Ag(s)$$

$$E=E^\circ-\frac{0.06}{n}\log\frac{[Cu^{2+}]}{[Ag^+]^2}$$

$$E=(0.8-0.34)-\frac{0.06}{2}\log\frac{(1)}{(0.1)^2}$$

$$=0.4\text{ V}=\text{or }400\text{ mV}$$

21. *(23)* Concentration of NaOH

$=0.2\%\ (w/V)$

$\therefore$ 0.2 g of NaOH in 100 mL of solution

Molarity of NaOH solution

$$=\frac{\text{Mole of solute}}{V_{mL}}\times1000$$

$$\frac{\frac{0.2}{40}}{100}\times1000=\frac{2}{40}\text{ M}$$

Given, resistivity of solution

$=870$ m ohm m $=8.7$ ohm dm

Now conductivity,

$$\kappa=\frac{1}{\rho}=\frac{1}{8.7}\text{ ohm}^{-1}\text{ dm}^{-1}$$

Molar conductivity,

$$\lambda_m=\frac{\kappa}{M}=\frac{\frac{1}{8.7}}{\frac{2}{40}}=2.29\text{ S dm}^2\text{ mol}^{-1}$$

or 23×10^2 m S dm^2 mol^{-1}

22. *(6)* Consider the given electrochemical cell reaction,

$$QH_2+2Ag^+\longrightarrow 2Ag+Q+2H^+$$

$$E=E^\circ\frac{-0.06}{2}\log[H^+]^2$$

$$E=E^\circ-0.06\times\log[H^+]$$

$$pH=-\log[H^+]=\frac{E-E^\circ}{0.06}$$

$$=\frac{0.4-0.1}{0.06}=\frac{0.3}{0.06}=5$$

$$pH+NH_4X=7-\frac{1}{2}pK_b-\frac{1}{2}\log C$$

$$5=7-\frac{1}{2}\times pK_b-\frac{1}{2}\log(10^{-2})$$

$$pK_b=6$$

23. *(338)* $[Mn^{+7}+5e^-\rightarrow Mn^{+2}]\times2$

$$[C_2O_4^{-2}\rightarrow 2CO_2+2e^-]\times5$$

$$2Mn^{+7}+5C_2O_4^{-2}\rightarrow 2Mn^{+2}+10CO_2$$

$$E^\circ_{cell}=0.49+1.51=2.00\text{ V}$$

At equilibrium, $E_{cell}=0$

So, $E^\circ_{cell}=\frac{0.0591}{n}\log K_{eq}$

$$\log K_{eq}=\frac{2\times10}{0.0591}$$

$$\log K_{eq}=338.40$$

$$\log K_{eq}\approx338$$

$$K_{eq}=10^{338}$$

Thus, $x=338$

24. *(1)* $2H^+(aq)+2e^-\longrightarrow H_2(g)$

$$E_{cell}=E^\circ-\frac{RT}{nF}\log\frac{[\text{Product}]}{[\text{Reactant}]}$$

$$E_{cell}=E^\circ_{H^+/H_2}-\frac{0.06}{2}\log\frac{pH_2}{[H^+]^2}$$

$$=0.00-\frac{0.06}{2}\log\frac{2}{[1]^2}$$

$$=-0.03\times0.3$$

$$=-0.9\times10^{-2}\text{ V}$$

Nearest integer value, -1×10^{-2} V

25. *(18)* The reaction in cell is,

$$2H^+(aq)+2e^-\longrightarrow H_2(g)$$

$$E_{cell}=E^\circ-\frac{0.059}{2}\log\left[\frac{p_{H_2}}{[H^+]^2}\right]$$

$$=0-0.059\times3=-0.177\text{ V}$$

or $=-17.7\times10^{-2}$ V

$$\approx-18\times10^{-2}\text{ V}$$

Hence, the missing term will be 18.

26. *(25)* For the reaction,

$$Cu^{2+}+2e^-\longrightarrow Cu$$

pH = 14

means, $p_{OH}=0$ or $[OH^-]=1$ M

Also, $K_{sp}=[Cu^{2+}][OH^-]^2$

Hence, $1\times10^{-20}=[Cu^{2+}]\,[1]^2$

or, $[Cu^{2+}]=1\times10^{-20}$ ...(1)

The reduction potential of copper can be written as

$$E_{cell}=E^\circ_{cell}-\frac{0.0591}{2}\times\log[1/Cu^{2+}]$$

$$E_{cell}=E^\circ_{cell}+\frac{0.059}{2}\times\log[Cu^{2+}]$$

$$E_{cell}=0.34+0.0295\times\log[10^{-20}]$$

$$E_{cell}+0.34+[0.0295\times(-20)]$$

$$=0.34+[-0.59]=-0.25$$

or -25×10^{-2} V

Thus, the value of $x=25$

27. *(4)* All the given statements i.e., *A*, *B*, *C* and *D* are correct.

28. *(2)* $Pb^{2+}+2e^-\longrightarrow Pb;$

$$\Delta G^\circ_1=-2FE^\circ_1$$

$$Pb^{4+}+4e^-\longrightarrow Pb;$$

$$\Delta G^\circ_2=-4FE^\circ_2$$

$$Pb^{2+}\longrightarrow Pb^{4+}+2e^-$$

$$\Delta G^\circ_3=-2FE^\circ_3$$

$$[\because \Delta G^\circ=-nFE^\circ]$$

$$\Rightarrow\ \Delta G^\circ_3=\Delta G^\circ_1-\Delta G^\circ_2$$

$$-2FE^\circ_3=2F(2n-m)$$

$$\begin{bmatrix}\because E^\circ_{Pb^{4+}/Pb}=n\\ E^\circ_{Pb^{2+}/Pb}=m\end{bmatrix}$$

$$\Rightarrow\ E^\circ_3=m-2n=m-xn$$

$$\Rightarrow\ x=2$$

29. *(1825)* Given,

For the reaction,

$$FeO_4^{2-}+4e^-\longrightarrow Fe^{2+}$$

$$E^\circ_{FeO_4^{2-}/Fe^{2+}}=x\times10^{-3}\text{ V}$$

$$FeO_4^{2-}\xrightarrow[n=3]{2.2V}Fe^{3+}\xrightarrow[n=1]{0.7\text{ V}}Fe^{2+}\xrightarrow{-0.45V}Fe$$

$$-nFE^\circ_{FeO_4^{2-}/Fe^{2+}}=-nFE^\circ_{FeO_4^{2-}/Fe^{3+}}+(-nFE^\circ_{Fe^{3+}/Fe^{2+}})$$

$$\Rightarrow\ 4\times x=3\times2.2+1+0.7$$

$$\Rightarrow\ x=\frac{7.3}{4}=1.825\text{ V}$$

$$=1825\times10^{-3}\text{ V}$$

30. *(1)* Only statement B is incorrect.

E°_{cell} of an electrochemical cell depends on equilibrium constant of the cell reaction, nature of the reactant as well as temperature and does not depend on the pressure of gas involved in the cell.

31. *(3)* For the redox reaction to be feasible, E°_{cell} must be positive. Therefore, E°_{cell} will be positive in case of 3 metals i.e., V, Fe and Au.

(i) $E^\circ_{cell} = E^\circ_{reduction} - E^\circ_{oxidation}$

$$E^\circ_{cell} = E^\circ_{NO_3^-/NO} - E^\circ_{V/V^{2+}}$$
$$= 0.97 - (-1.19) = 2.16 V$$

(ii) $E^\circ_{cell} = E^\circ_{NO_3^-/NO} - E^\circ_{Fe/Fe^{3+}}$

$$= 0.97 - (-0.04) = 1.01 V$$

(iii) $E^\circ_{cell} = E^\circ_{NO_3^-/NO} - E^\circ_{Ag/Ag^+}$

$$= 0.97 - (0.80) = 0.17\ V$$

Alternate Solution

Metals having lower standard reduction potential than that of given redox reaction i.e. 0.97 V will be oxidised by NO_3^-. So, among the given options metals with SRP value −1.19 V, −0.04 V and 0.80 V will be oxidised by NO_3^-.

32. *(3)* Reaction equation is

$$MnO_4^- + 8H^+ + 5e^- \rightleftharpoons Mn^{2+} + 4H_2O$$

$$\Rightarrow E = E^\circ - \frac{0.059}{5} \log \frac{[Mn^{2+}]}{[MnO_4^-][H^+]^8}$$

$$\Rightarrow 1.282 = 1.54 - \frac{0.059}{5} \log \frac{10^{-3}}{10^{-1} \times [H^+]^8}$$

$$\Rightarrow 1.282 = 1.54 - 0.0118 \log \frac{10^{-2}}{[H^+]^8}$$

$$\Rightarrow -0.258 = -0.0118 \log \frac{10^{-2}}{[H^+]^8}$$

$$\Rightarrow \frac{0.258}{0.0118} = \log \frac{10^{-2}}{[H^+]^8}$$

$$\Rightarrow 21.86 = -2 - 8 \log [H^+]$$

$$\Rightarrow -\log[H^+] = 2.9825 \simeq 3$$

$$\Rightarrow pH = 3$$

33. *(6)*

idea To find out the value of $\log k$, first of all we will calculate the value of n and E° from the given equation. Then we will substitute these values in the equation $\Delta G^\circ = -RT \ln K$.

$$Pd^{+2}(aq) + 2e^- \rightleftharpoons Pd(s),$$
$$E^\circ_{Cathode,\ (reduction)} = 0.83\ V$$

$$Pd(s) + 4Cl^-(aq) \rightleftharpoons PdCl_4^{2-}(aq) + 2e^-,$$
$$E^\circ_{Anode,\ (oxidation)} = 0.65$$

Net reaction is

$$Pd^{2+}(aq) + 4Cl^-(aq) \rightleftharpoons PdCl_4^{2-}(aq)$$

$$E^\circ_{cell} = E^\circ_{Cathode,\ (reduction)} - E^\circ_{Anode,\ (oxidation)}$$

$$E^\circ_{cell} = 0.83 - 0.65 = 0.18 \quad \ldots(i)$$

Also, $n = 2 \quad \ldots(ii)$

$$\Delta G^\circ = -RT \ln K_{eq}$$
$$-nFE^\circ_{cell} = -RT \times 2.303 (\log_{10} K_{eq})$$

$$\frac{E^\circ_{cell}}{0.06} \times n = \log K_{eq} \quad \ldots(iii)$$

Using Eq. (i), (ii) and (iii), we get

$$\frac{0.18 \times 2}{0.06} = \log K_{eq} \Rightarrow \log K = 6$$

34. *(275)* $E^\circ_{cell} = E^\circ_{Y^{2+}/Y} - E^\circ_{X^{2+}/X}$

$$= 0.36 - (-2.36) = 2.72$$

From the Nernst equation,

$$E_{cell} = E^\circ_{cell} - \frac{0.06}{2} \log \frac{[X^{2+}]}{[Y^{2+}]}$$

$$= 2.72 - 0.03 \log \left(\frac{0.001}{0.01}\right)$$

$$= 2.72 - 0.03 \log (10^{-1})$$
$$= 2.72 - 0.03 \times (-1)$$
$$= 2.72 + 0.03 = 2.75$$
$$= 275 \times 10^{-2}\ V$$

35. *(10)* Reactions involved in the cell are

At anode : $H_2 \longrightarrow 2H^+ + 2e^-$

At cathode :

$$(Fe^{3+} + e^- \longrightarrow Fe^{2+}) \times 2$$

Overall equation

$$H_2 + 2Fe^{3+} \longrightarrow 2H^+ + 2Fe^{2+}$$

$$E_{cell} = E^\circ_{cell} - \frac{0.059}{2} \log \left(\frac{Fe^{2+}}{Fe^{3+}}\right)^2$$

$$0.712 = 0.771 - 0.059 \log \left(\frac{Fe^{2+}}{Fe^{3+}}\right)^2$$

$$0.712 - 0.771 = -0.059 \log \left(\frac{Fe^{2+}}{Fe^{3+}}\right)^2$$

$$\log \left(\frac{Fe^{2+}}{Fe^{3+}}\right)^2 = \frac{0.059}{0.059} = 1$$

$$\Rightarrow \frac{Fe^{2+}}{Fe^{3+}} = 10$$

36. *(17)* $Zn(s) + Sn^{2+}(aq) \rightleftharpoons Zn^{2+}(aq) + Sn(s)$

$$\Rightarrow \Delta G^\circ = -2.303\ RT \log_{10} K_{eq}$$

$$-nE^\circ_{cell} = -\frac{2.303\ RT}{F} \log_{10} K_{eq}$$

$$\Rightarrow E^\circ_{Zn/Zn^{2+}} + E^\circ_{Sn^{2+}/Sn} = \frac{0.059}{2} \log_{10} K_{eq} \quad [\because n = 2]$$

$$0.76 + E^\circ_{Sn^{2+}/Sn} = \frac{0.059}{2} \log_{10} 10^{20}$$

$$[\because E^\circ_{Zn^{2+}/Zn} = -0.76\ V]$$

$$0.76 + E^\circ_{Sn^{2+}/Sn} = \frac{0.059 \times 20}{2}$$

$$\Rightarrow E^\circ_{Sn^{2+}/Sn} = 0.59 - 0.76$$

$$E^\circ_{Sn^{2+}/Sn} = -0.17$$

$$\Rightarrow E^\circ_{Sn/Sn^{2+}} = -17 \times 10^{-2}\ V$$

37. *(3)*

idea To calculate the value of a, Nernst equation is used. First determine the value of E°_{cell} from given $E^\circ_{X^{2+}/X}$ and $E^\circ_{Y^{2+}/Y}$ value. Then use this value in Nernst Equation.

Given, $E^\circ_{M^{3+}/M^+} = 0.2\ V$

$$\frac{2.303\ RT}{F} = 0.059\ V$$

$$\frac{[M^+(aq)]}{[M^{3+}(aq)]} = 10^a$$

$E_{cell} = 0.1115$ V at 298 K

Overall reaction of the cell is given by

$$H_2(g) + M^{3+}(aq) \longrightarrow M^+(aq) + 2H^+(aq)$$

According to Nernst Equation

$$E_{cell} = (E^\circ_{Cathode} - E^\circ_{Anode}) - \frac{0.059}{2} \log \frac{[M^+]}{[M^{+3}]}$$

$$0.1115 = (E^\circ_{M^{3+}/M^+}) - \frac{0.059}{2} \log \frac{[M^+]}{[M^{+3}]}$$

$$0.1115 = 0.2 - \frac{0.059}{2} \log 10^a$$

$$\left(\because \frac{[M^+]}{[M^{+3}]} = 10^a\right)$$

$$\Rightarrow \frac{0.059}{2} \log 10^a = 0.2 - 0.115$$

$$\Rightarrow \frac{0.059}{2} \log 10^a = 0.0885$$

$$\Rightarrow \log 10^a = \frac{0.0885 \times 2}{0.059}$$

$$\Rightarrow \log 10^a = 3$$

$$\Rightarrow a = 3$$

So, the value of 'a' is 3.

38. *(917)*

> **idea** To calculate the potential for the half-cell reaction. First of all, write the balanced chemical equation. Then calculate the value of n from it. After that use Nernst equation to calculate potential for half-cell reaction.

$[Cr_2O_7]^{-2} = 10$ m mol; $[Cr^{+3}] = 100$ m mol $= 0.1$ mol $= 10^{-1}$ mol

$$\overset{+6}{Cr_2}O_7^{-2} \longrightarrow Cr^{+3};\ E° = 1.33\ V$$

$$\overset{+6}{Cr_2}O_7^{-2} + 14H^+ + 6e^- \xrightarrow{\text{Reduction}} 2Cr^{+3} + 7H_2O$$

$\therefore n = 6$

Using Nernst Equation

$$E_{cell} = 1.33 - \frac{0.059}{n}\log\frac{[Cr^{3+}]^2}{[Cr_2O_7^{2-}][H^+]^{14}}$$

$$= 1.33 - \frac{0.059}{6}\log\left[\frac{(0.1)^2}{(10^{-2})(10^{-3})^{14}}\right]$$

$$E = 1.33 - \frac{0.059}{6} \times 42 = 0.917$$

$$E = 917 \times 10^{-3} \Rightarrow x = 917$$

39. *(34)* $Cu(s)\,|Cu^{2+}(0.001)\ M\,||\,Ag^+(0.01\ M)|Ag(s);\ E_{cell} = 0.43\ V$

At anode, $Cu \longrightarrow Cu^{2+} + 2e^-$

At cathode, $2Ag^+ + 2e^- \longrightarrow 2Ag$

Net, cell reaction is

$$Cu + 2Ag^+ \longrightarrow Cu^{2+} + 2Ag$$

According to Nernst equation,

$$E_{cell} = E°_{cell} - \frac{2.303\,RT}{nF}\log\frac{[Cu^+]}{[Ag^+]^2}$$

$$E_{cell} = E°_{cell} - \frac{0.06}{n}\log\frac{[Cu^{2+}]}{[Ag^+]^2}$$

[n = number of e^- involved]

$$\Rightarrow\quad 0.43 = E°_{cell} - \frac{0.06}{2}\log\frac{(0.001)}{(0.01)^2}$$

$$\Rightarrow\quad 0.43 = E°_{cell} - 0.03\log\frac{10^{-3}}{10^{-4}}$$

$$\Rightarrow\quad 0.43 = E°_{cell} - 0.03\log 10$$

$$\Rightarrow\quad E°_{cell} = 0.43 + 0.03 = 0.46\ V$$

Now, $E°_{cell} = E°_{cathode} - E°_{anode}$

$$\therefore\quad E°_{cell} = E°_{Ag^+/Ag} - E°_{Cu^{2+}/Cu}$$

$$\Rightarrow\quad 0.46 = 0.80 - E°_{Cu^{2+}/Cu}$$

$$\Rightarrow E°_{Cu^{2+}/Cu} = 0.34\ V$$

$$= 34 \times 10^{-2}\ V$$

40. *(4)* Reactions of half-cell are given as

At anode $Zn \longrightarrow Zn^{2+} + 2e^-$;

At cathode : $Sn^{x+} + xe^- \longrightarrow Sn$

$$Zn + Sn^{x+} + xe^- \rightarrow Sn + Zn^{2+} + 2e^-;$$

$$E°_{cell} = E°_{Sn^{x+}/Sn} - E_{Zn/Zn^{2+}}$$

$$= 0.008 - (-0.763)$$

$$E°_{cell} = 0.771\ V$$

Applying Nernst equation,

$$E_{cell} = E°_{cell} - \frac{2.303\,RT}{nF}\log Q$$

$$0.801 = 0.771 - \frac{0.06}{n}\log 10^{-2}$$

$$0.801 - 0.771 = \frac{-0.06}{n}(-2\log 10)$$

$$0.03 = \frac{0.12}{n}$$

$$\therefore\quad n = 4$$

41. *(7)* The two half-reactions are

$$H_2 \longrightarrow 2H^+ + 2e^-\quad E° = 0$$

$$Cu^{2+} + 2e^- \longrightarrow Cu\quad E° = 0.34$$

$$H_2 + Cu^{2+} \longrightarrow 2H^+ + Cu(s)\ E° = 0.34$$

Using Nernst Equation

$$E_{cell} = E°_{cell} - \frac{0.0591}{2}\log\frac{[H^+]^2}{[Cu^{2+}]\,p_{H_2}}$$

$$0.31 = 0.34 - 0.03\log\frac{[10^{-3}]^2}{Cu^{2+} \times 1}$$

$$0.03\log\frac{10^{-6}}{Cu^{2+}} = 0.03$$

$$\log\frac{10^{-6}}{Cu^{2+}} = 1$$

Taking antilog,

$$\frac{10^{-6}}{Cu^{2+}} = 10$$

$\Rightarrow\quad Cu^{2+} = 10^{-7}$ ... (i)

Given $[Cu^{2+}] = 10^{-x}M$ ...(ii)

$\therefore\quad x = 7$ [from Eq. (i) and (ii)]

42. *(16)* Given,

$Sn^{2+}(aq) + 2e^- \longrightarrow Sn(s)$;

$E°_{Sn^{2+}/Sn} = -0.140$ V

$$\Delta G_1° = -nFE°_{Sn^{2+}/Sn}$$

$$= -2 \times (-0.14)\ F = 0.28\ F\ \ldots(i)$$

$Sn^{4+}(aq) + 4e^- \longrightarrow Sn(s)$;

$E°_{Sn^{4+}/Sn} = 0.010$ V

$$\Delta G_2° = -nFE°_{Sn^{4+}/Sn}$$

$$= -4 \times 0.01\ F = -0.04\ F \quad \ldots(ii)$$

$Sn^{4+}(aq) + 2e^- \longrightarrow Sn^{2+}(aq);\ \Delta G_3°$ ...(iii)

To get Eqs. (iii), subtract Eq. (i) from (ii)

So, we get

$$Sn^{4+} + 4e^- \longrightarrow Sn;\quad \Delta G_2° = -0.04\ F$$

$$Sn \rightarrow Sn^{2+} + 2e^-;\ -\Delta G_1° = -0.28F$$

$$Sn^{4+} + 2e^- \longrightarrow Sn^{2+};$$

$$\Delta G_3° = \Delta G_2° - \Delta G_1°$$

$$-nFE°_{Sn^{4+}/Sn^{2+}} = -0.04F - 0.28\ F$$

$$-2FE°_{Sn^{4+}/Sn^{2+}} = -0.32\ F$$

$$E°_{Sn^{4+}/Sn^{2+}} = \frac{0.32\ F}{2\ F} = 0.16\ V$$

$$E°_{Sn^{4+}/Sn^{2+}} = 16 \times 10^{-2}\ V$$

> ***Alternate Solution***
>
> This question can be solved using Latimer diagram,
>
> $$Sn^{4+} \xrightarrow{E°_{Sn^{4+}/Sn^{2+}}} Sn^{2+} \xrightarrow{-0.140\ V} Sn$$
> ($Sn^{4+} \to Sn$: $0.010\ V$)
>
> $$nE°_{Sn^{4+}/Sn} = nE°_{Sn^{4+}/Sn^{2+}} - nE°_{Sn^{2+}/Sn}$$
>
> So, $\quad 4 \times 0.01 = 2E°_{Sn^{4+}/Sn^{2+}} - 2 \times -0.14$
>
> $$2E°_{Sn^{4+}/Sn^{2+}} = 0.04 + 0.28 = 0.32$$
>
> $$E°_{Sn^{4+}/Sn^{2+}} = \frac{0.32}{2} = 16 \times 10^{-2}\ V$$

43. *(51)* The reaction taking place in cell is written as

$$\frac{1}{2}H_2 + Ag^+ \longrightarrow H^+ + Ag$$

Standard Gibb's free energy is given as

$$\Delta G^- = -nFE°_{cell}$$

$$= -1 \times 96500\ J \times 0.5332\ V$$

$$= -51.35\ kJ \approx -51\ kJ$$

44. *(983)* The given reaction is as follows

$$Cu(s) + Sn^{2+}(0.001\ M) \longrightarrow Cu^{2+}(0.01\ M) + Sn(s)$$

$$E°_{cell} = E°_{cathode} - E°_{anode}$$

$$= (-0.14) - (0.34) = -0.48\ V$$

$$E_{cell} = E°_{cell} - \frac{0.059}{2}\log\frac{[Cu^{2+}]}{[Sn^{2+}]}$$

$$= -0.48 - 0.0295\log\frac{0.01}{0.001}$$

$$= -0.5095\ V$$

$$\Delta G = -nFE_{cell}$$

$$= -2 \times 96500 \times (-0.5095)$$

$$= 98333.5\ J/mol$$

$$= 98.3335\ kJ/mol$$

or 983.335×10^{-1} kJ/mol

$$\Delta G \approx 983 \times 10^{-1} \text{ kJ mol}^{-1}$$

45. *(23)* From the following oxidation reactions, the spontaneous cell reaction will be in which Fe^{3+} gets reduced and I^- gets oxidised as the reduction potential of Fe^{3+} is higher than I_2.

Reduction (cathode) half-cell reaction

$$Fe^{3+} + e^- \longrightarrow Fe^{2+}$$

$$E^\circ_{Fe^{3+}/Fe^{2+}} = 0.77 \text{ V}$$

Oxidation (anode) half-cell reaction

$$2I^- \longrightarrow I_2 + 2e^-; \; E^\circ_{I_2/I^-} = 0.54 \text{ V}$$

The cell reaction

$$Fe^{3+} + 2I^- \longrightarrow Fe^{2+} + I_2$$

$$E^\circ_{cell} = E^\circ_{cathode} - E^\circ_{anode}$$
$$= 0.77 \text{ V} - 0.54 \text{ V}$$
$$= 0.23 \text{ V} = 23 \times 10^{-2} \text{ V}$$

46. *(25)* $\Delta G^\circ = \Delta H^\circ - T\Delta S^\circ$

Also, $\Delta G^\circ = -nFE^\circ$

$$\therefore \quad -nFE^\circ = \Delta H^\circ - T\Delta S^\circ$$

$$\Rightarrow \quad \Delta S^\circ = \frac{nFE^\circ + \Delta H^\circ}{T}$$

$$= \frac{(2 \times 96487 \text{ C mol}^{-1} \times 4.315 \text{ V}) + (-825.2 \text{ kJ / mol})}{298 \text{ K}}$$

$$\Delta S^\circ = 25.11 \text{ J / K mol} \simeq 25 \text{ J/K mol}$$

47. *(109)* Cell reaction,

$$\underset{0.02\text{ M}}{Cu^{2+}(aq)} + Zn(s) \rightleftharpoons Cu(s) + \underset{0.04\text{ M}}{Zn^{2+}(aq)}$$

$$E^\circ_{cell} = E^\circ_{Cu^{2+}/Cu} - E^\circ_{Zn^{2+}/Zn}$$
$$= 0.34 - (-0.76) = 1.1 \text{ V}$$

$$E_{cell} = E^\circ_{cell} - \frac{0.059}{2} \log \frac{[Zn^{2+}]}{[Cu^{2+}]}$$
$$= 1.1 - \frac{0.059}{2} \log \frac{[0.04]}{[0.02]}$$
$$= 1.1 - 0.03 \log 2$$
$$= 1.1 - 0.03 \times 0.30$$
$$= 1.09 \text{ V} = 109 \times 10^{-2} \text{ V}$$

48. *(3)* The physical properties of an element that affects the reduction potential is sublimation enthalpy, ionisation enthalpy and hydration enthalpy. So, there are total 3 properties that will affect the reduction potential.

49. *(28)* Cell reaction,

$$Cu(s) + 2Ag^+(aq) \longrightarrow Cu^{2+}(aq) + 2Ag(s)$$

$$E_{cell} = E^\circ_{cell} - \frac{0.0591}{2} \log \frac{[Cu^{2+}]}{[Ag^+]^2}$$

$$E_1 = 0.3095$$
$$= E^\circ_{cell} - \frac{0.0591}{2} \log \frac{(0.1)}{(0.01)^2}$$

$$E^\circ_{cell} = 0.3095 + \frac{0.0591}{2} \log(1000)$$
$$= 0.3095 + \frac{0.0591}{2} \log 10^3$$

$$E^\circ_{cell} = 0.3095 + \frac{0.0591}{2} \times 3 \quad \ldots \text{(i)}$$

$$E_2 = E^\circ_{cell} - \frac{0.0591}{2} \log \frac{(0.01)}{(0.001)^2}$$
$$= E^\circ_{cell} - \frac{0.0591}{2} \log 10^4$$
$$E_2 = E^\circ_{cell} - \frac{0.0591}{2} \times 4$$

From Eq. (i), we substitute E°_{cell}

$$E_2 = 0.3095 + \frac{0.0591}{2} \times 3 - \frac{0.0591}{2} \times 4$$
$$= 0.3095 - 0.0295 = 0.28 = 28 \times 10^{-2} \text{ V}$$

50. *(24)* Reaction cell

$Zn \mid Zn^{2}(aq), (1M) \| Fe^{3+}(aq) |, Fe^{2+}(aq), Pt(s)$

(**Anode**) $Zn \longrightarrow Zn^{2+} + 2e^-$

(**Cathode**) $2Fe^{3+} + 2e^- \longrightarrow 2Fe^{2+}$

Overall $Zn + 2Fe^{3+} \longrightarrow Zn^{2+} + 2Fe^{2+}$

So, $E^\circ_{cell} = E^\circ_{Fe^{3+}/Fe^{2+}} - E^\circ_{Zn^{2+}/Zn}$
$= 0.77 - (-0.76) = 1.53$ V

Using Nernst equation,

$$E_{cell} = E^\circ_{cell} - \frac{0.0591}{n} \log \frac{[P]}{[R]}$$

$$1.50 = 1.53 - \frac{0.0591}{2} \log \left(\frac{Fe^{2+}}{Fe^{3+}}\right)^2$$

$$\log \left(\frac{Fe^{2+}}{Fe^{3+}}\right)^2 = \frac{0.06}{0.06}$$

$$\Rightarrow \quad \frac{[Fe^{2+}]}{[Fe^{3+}]} = 10^{1/2} = \sqrt{10}$$

$$\frac{[Fe^{3+}]}{[Fe^{2+}]} = \frac{1}{\sqrt{10}}$$

Fractin of total iron present as Fe^{3+} ion.

$$= \frac{[Fe^{3+}]}{[Fe^{2+}] + [Fe^{3+}]} = \frac{1}{1 + \sqrt{10}} = \frac{1}{4.16}$$
$$= 0.2402 = 24 \times 10^{-2}$$

Thus, the value of $x = 24$

51. *(3)* Given $E^\circ_{cell} = 2.97$ V

$n = 2$ (for 2 electrons)

Using nernst equation,

$$E_{cell} = E^\circ_{cell} - \frac{0.0591}{n} \log \frac{[Cu^{2+}]}{[Ag^+]^2}$$
$$= 2.97 - \frac{0.0591}{2} \log \frac{0.250}{(10^{-3})^2}$$
$$= 2.81 \approx 3 \text{ (approx.)}$$

Hence, E°_{cell} of the reaction is 3V.

52. *(45)* Given reaction is

$$2Fe^{3+}(aq) + 2I^-(aq) \longrightarrow 2Fe^{2+}(aq) + I_2(s)$$

Now, $Fe^{3+} \xrightarrow{E_1^\circ} Fe^{2+} \xrightarrow{E_2^\circ} Fe$ (overall $Fe^{3+} \xrightarrow{E_3^\circ} Fe$)

$$\Delta G_1^\circ + \Delta G_2^\circ = \Delta G_3^\circ$$
$$nE_1^\circ + nE_2^\circ = nE_3^\circ$$

[n = number of electron transferred]

$$1E_1^\circ + 2E_2^\circ = 3E_3^\circ$$

$$E^\circ_{Fe^{3+}/Fe^{2+}} + 2E^\circ_{Fe^{2+}/Fe\,(s)} = 3E^\circ_{Fe^{3+}/Fe\,(s)}$$

$$E^\circ_{Fe^{3+}/Fe^{2+}} + (-0.440) \times 2 = (-0.036) \times 3$$

$$E^\circ_{Fe^{3+}/Fe^{2+}} = 0.772 \text{ V}$$

The standard electrode potential for the above given reaction,

$$E^\circ_{cell} = E^\circ_{Fe^{3+}|Fe^{2+}} - E^\circ_{I^-|I_2}$$

$$E^\circ_{cell} = E^\circ_{cathode} - E^\circ_{anode}$$
$$= 0.772 - 0.539 = 0.233 \text{ V}$$

The magnitude of the standard molar Gibb's free energy,

$$\Delta G^\circ = -nFE^\circ_{cell}$$
$$= -2 \times 96500 \times 0.233$$
$$\Delta G^\circ = -44969 \text{ J} = -44.9 \text{ kJ} \simeq -45 \text{ kJ}$$

53. *(147)*

$Zn | Zn^{2+}(0.1M) \;\|\; Ag^+(0.01M) | Ag$

Cell reaction

$$Zn(s) + \underset{0.01M}{2Ag^+} \longrightarrow \underset{0.1M}{Zn^{2+}} + 2Ag(s)$$

$$\Rightarrow \frac{[P]}{[R]} = \frac{[Zn^{2+}]\,[Ag]^2}{[Zn]\,[Ag^+]^2} \; ([Ag] = [Zn] = 1)$$
$$= \frac{0.1 \times 1^2}{1 \times (0.01)^2} = 10^3$$

$$n = 2$$

and $E^\circ_{cell} = E^\circ_{Ag^+/Ag} - E^\circ_{Zn^{2+}/Zn}$
$$= 0.8 - (-0.76) = 1.56 \text{ V}$$

$$\Rightarrow E_{cell} = E^\circ_{cell} - \frac{0.059}{n}\log\frac{[Zn^{2+}]}{[Ag^+]}$$

$$= 1.56 - \frac{0.059}{2}\log 10^3$$

$$= 1.4715 \text{ V}$$

$$= 147.15\times10^{-2} \text{ V}$$

$$= 147\times10^{-2} \text{ V}$$

$$x = 147$$

54. *(3776)* Reaction,

$$MnO_4^- + H^+ + 5e^- \longrightarrow Mn^{2+} + 4H_2O$$

$$n = 5$$

Applying Nernst equation,

$$E_{cell} = E^\circ_{cell} - \frac{0.0591}{n}\log\frac{[P]}{[R]}$$

or $E_{cell} = E^\circ_{cell}$

$$-\frac{0.0591}{5}\log\frac{[Mn^{2+}]}{[MnO_4^-]}\left[\frac{1}{H^+}\right]^8$$

Case (I) Given, $[H^+] = 1$ M

$$E_1 = E^\circ - \frac{0.0591}{5}\log\frac{[Mn^{2+}]}{[MnO_4^-]}$$

Case (II) Now, $[H^+] = 10^{-4}$ M

$$E_2 = E_1 - \frac{0.0591}{5}\log\frac{[Mn^{2+}]}{[MnO_4^-]}\times\frac{1}{(10^{-4})^8}$$

∴ Magnitude of change in oxidising power of $MnO_4^- | Mn^{2+}$

$$|E_1 - E_2| = \frac{0.0591}{5}\times 32$$

$$= 0.3776 \text{ V}$$

$$= 3776\times10^{-4}$$

$$\Rightarrow x = 3776$$

55. *(6)* We know that,

$$\Delta G^\circ = -nFE^\circ_{cell}$$

Here, $\Delta G^\circ = 17.37 \text{ kJ mol}^{-1}$

$$17.37\times1000 = -3\times96500\times E^\circ_{cell}$$

$$E^\circ_{cell} = \frac{-17.37\times1000}{3\times96500}$$

$$E^\circ_{cell} = -0.06 = -6\times10^{-2}$$

56. *(96500)*

$$E^\circ_{cell} = (E^\circ_{Sn^{2+}/Sn} - E^\circ_{Cu^{2+}/Cu}) - \frac{0.0591}{2}\log\frac{[Cu^{2+}]}{[Sn^{2+}]}$$

$$= (-0.16 - 0.34) - \frac{0.0591}{2}\log\frac{1}{1}$$

$$= -0.50 \text{ V}$$

$$\Delta G = -nFE^\circ_{cell}$$

$$= -2\times96500\times(-0.50) \text{ J}$$

$$= 96500.00 \text{ J}$$

57. *(144)*

$$2Cu^+(aq) \rightleftharpoons Cu(s) + Cu^{2+}(aq)$$

For the cell reaction,

$$E^\circ_{cell} = E^\circ_{Cu^+/Cu} - E^\circ_{Cu^{2+}/Cu^+}$$

$$= (0.52 - 0.16) \text{ V}$$

$$= 0.36 \text{ V}$$

$$\Rightarrow \Delta G^\circ = -nFE^\circ_{cell}$$

$$\Delta G = -RT\ln K$$

$$\Rightarrow -RT\ln K = -nFE^\circ_{cell}$$

$$\Rightarrow \ln K = n\times\frac{E}{RT}\times E^\circ_{cell}$$

$$= 1\times\frac{1}{0.025}\times0.36$$

$$\ln K = 14.4 = 144\times10^{-1}$$

Topic 3

01. *(d)* On charging the lead storage battery, following reactions takes place.

$$2PbSO_4(s) + 2H_2O(l) \longrightarrow Pb(s) + PbO_2(s) + 2H_2SO_4(aq)$$

At anode $PbSO_4$ is reduced to Pb while at cathode $PbSO_4$ is oxidised to PbO_2.

Thus,. $x_1 = +2, y_1 = 0$

and $x_2 = +2, y_2 = 4$

02. *(a)* Statement I is true but Statement II is false. It is because corrosion is more favourable in acidic medium than alkaline medium. So, the rate of corrosion is high in acidic medium than alkaline.

03. *(a)* PbO_2 or Pb_3O_4 can be the compound *A*. But out of these two compounds only PbO_2 is used in lead storage batteries where a grid of lead packed with PbO_2 acts as cathode and also it is amphoteric in nature. It reacts with both acids and alkali.

(i) $\overset{+4}{PbO_2} + 4H\overset{-1}{Cl} \longrightarrow \overset{+2}{Pb}Cl_2 + \overset{0}{Cl_2} + 2H_2O$

Here, PbO_2 acts as a basic oxide as well as an oxidising agent.

(ii) $PbO_2 + 2NaOH \longrightarrow Na_2PbO_2 + H_2O$

Here, PbO_2 acts as an acidic oxide.

So, the compound *A* is PbO_2 (option-a).

04. *(a)* The correct match is A-III, B-II, C-IV, D-I

05. *(d)* The correct match is A-IV, B-III, C-I, D-II

06. *(d)* Statements given in A, D and E are correct regarding fuel cells while B and C are incorrect.

It is because fuel cells produces electricity with an efficiency of about 70% and aluminium is not used a catalyst in it.

07. *(a)* Statements given in B, D only are correct.

On charging the lead storage battery, the reaction is reversed and $PbSO_4(s)$ on anode is converted into Pb instead of PbO_2.

Lead storage battery consists of lead anode and a grid of lead packed with lead oxide as cathode, a 38% solution of H_2SO_4 is used as an electrolyte.

08. *(c)* The correct match is A-II, B-I, C-IV, D-III.

15 Chemical Kinetics

In JEE Main there is usually 0-1 question per paper are visible from this chapter. In past 3 years, 27 conceptual and 40 numerical questions were asked of moderate difficulty level from this chapter.

For detailed analysis scan QR code

Topic 1 / Rate of Chemical Reaction & Rate Expression

• Level 1

01 Rate law for a reaction between A and B is given by

$$r = k[A]^n [B]^m$$

If concentration of A is doubled and concentration of B is halved from their initial value, the ratio of new rate of reaction to the initial rate of reaction $\left(\frac{r_2}{r_1}\right)$ is **[2025, 4 April Shift-I]**

(a) $2^{(n-m)}$ (b) $\frac{1}{2^{m+n}}$

(c) $(n-m)$ (d) $(m+n)$

02 Consider an elementary reaction,

$$A(g) + B(g) \longrightarrow C(g) + D(g)$$ **[2025, 28 Jan Shift-II]**

If the volume of reaction mixture is suddenly reduced to $\frac{1}{3}$ of its initial volume, the reaction rate will become 'x' times of the original reaction rate. The value of x is

(a) 1/9 (b) 1/3 (c) 9 (d) 3

03 For the reaction $2A + 3B + \frac{3}{2}C \rightarrow 3P$, which statement is correct? **[2020, 3 Sep Shift-II]**

(a) $\frac{dn_A}{dt} = \frac{2}{3}\frac{dn_B}{dt} = \frac{3}{4}\frac{dn_C}{dt}$ (b) $\frac{dn_A}{dt} = \frac{3}{2}\frac{dn_B}{dt} = \frac{3}{4}\frac{dn_C}{dt}$

(c) $\frac{dn_A}{dt} = \frac{dn_B}{dt} = \frac{dn_C}{dt}$ (d) $\frac{dn_A}{dt} = \frac{2}{3}\frac{dn_B}{dt} = \frac{4}{3}\frac{dn_C}{dt}$

04 For the reaction

$$2H_2(g) + 2NO(g) \longrightarrow N_2(g) + 2H_2O(g)$$

the observed rate expression is, rate $= K_f[NO]^2[H_2]$. The rate expression for the reverse reaction is **[2020, 7 Jan Shift-II]**

(a) $K_b[N_2][H_2O]^2/[H_2]$ (b) $K_b[N_2][H_2O]$

(c) $K_b[N_2][H_2O]^2/[NO]$ (d) $K_b[N_2][H_2O]^2$

• Level 2

05 For a reaction $A \xrightarrow{K_1} B \xrightarrow{K_2} C$

If the rate of formation of B is set to be zero then the concentration of B is given by : **[2024, 8 April Shift-II]**

(a) $\left(\frac{K_1}{K_2}\right)[A]$ (b) $(K_1 + K_2)[A]$

(c) $(K_1 - K_2)[A]$ (d) $K_1K_2[A]$

06 NO_2 required for a reaction is produced by the decomposition of N_2O_5 in CCl_4 as per the equation,

$$2N_2O_5(g) \longrightarrow 4NO_2(g) + O_2(g)$$

The initial concentration of N_2O_5 is 3.00 mol L^{-1} and it is 2.75 mol L^{-1} after 30 minutes. The rate of formation of NO_2 is **[2019, 12 April Shift-II]**

(a) 4.167×10^{-3} mol L^{-1} min^{-1}

(b) 1.667×10^{-2} mol L^{-1} min^{-1}

(c) 8.333×10^{-3} mol L^{-1} min^{-1}

(d) 2.083×10^{-3} mol L^{-1} min^{-1}

• Level 3

07 A person's wound was exposed to some bacteria and then bacterial growth started to happen at the same place. The wound was later treated with some antibacterial medicine and the rate of bacterial decay(r) was found to be proportional with the square of the existing number of bacteria at any instance. Which of the following set of graphs correctly represents the 'before' and 'after situation of the application of the medicine? **[2025, 7 April Shift-I]**

[Given, N = No. of bacteria, t = time, bacterial growth follows 1st order kineties.]

(a) 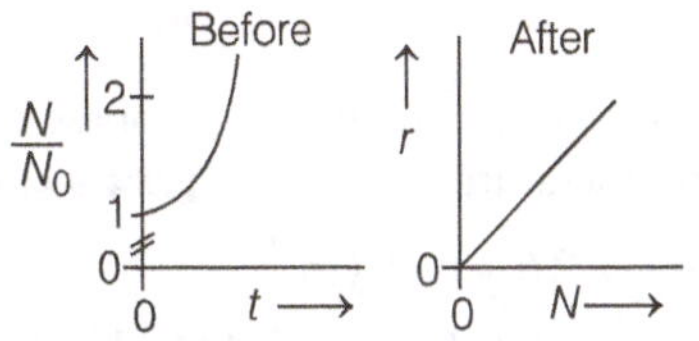

(b) 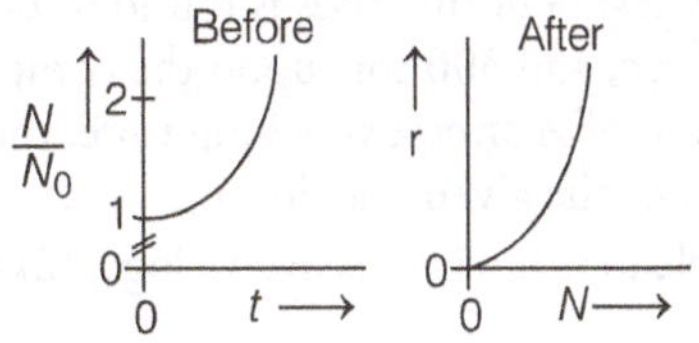

(c) 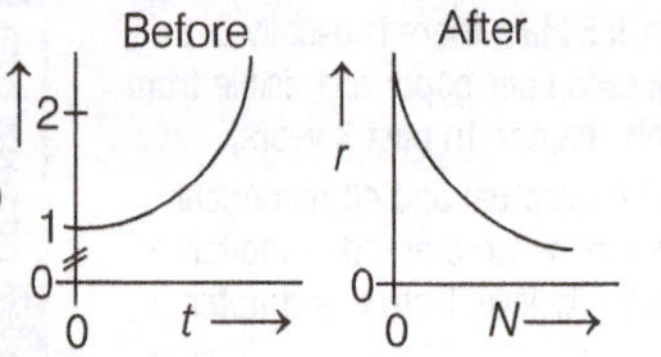

(d) 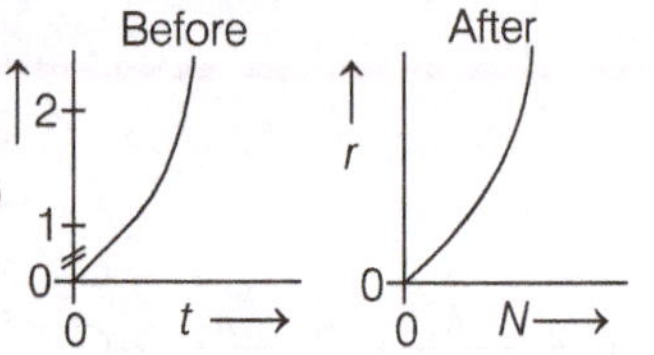

08 For an elementary chemical reaction, $A_2 \underset{k_{-1}}{\overset{k_1}{\rightleftharpoons}} 2A$, the expression for $\frac{d[A]}{dt}$ is **[2019, 10 Jan Shift-II]**

(a) $2k_1[A_2] - k_{-1}[A]^2$

(b) $k_1[A_2] - k_{-1}[A]^2$

(c) $2k_1[A_2] - 2k_{-1}[A]^2$

(d) $k_1[A_2] + k_{-1}[A]^2$

Imp. *Graphical Representation of Concentration vs Time*

09 A bacterial infection in an internal wound grows as $N'(t) = N_0 \exp(t)$, where the time t is in hours. A dose of antibiotic, taken orally, needs 1 hour to reach the wound. Once it reaches there, the bacterial population goes down as $\frac{dN}{dt} = -5N^2$. What will be the plot of $\frac{N_0}{N}$ *vs* t after 1 hour? **[2019, 10 April Shift-I]**

(a) 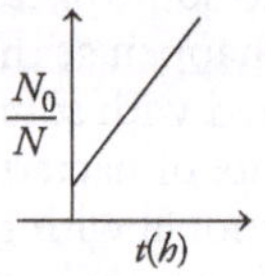

(b) 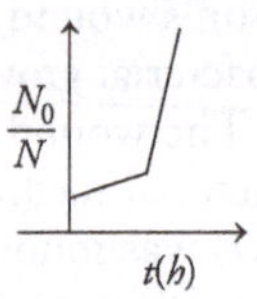

(c) 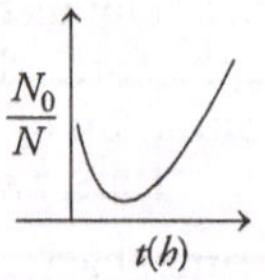

(d) 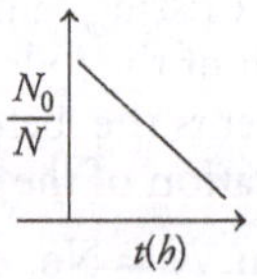

Numerical Value Type Questions

10 Consider the following first order gas phase reaction at constant temperature **[2024, 9 April Shift-II]**

$$A(g) \longrightarrow 2B(g) + C(g)$$

If the total pressure of the gases is found to be 200 torr after 23 sec, and 300 torr upon the complete decomposition of A after a very long time, then the rate constant of the given reaction is .... $\times 10^{-2} s^{-1}$ (Nearest integer) [Given : $\log_{10}(2) = 0.301$]

11 Given below are two statements: **[2024, 9 April Shift-I]**

Statement I The rate law for the reaction $A + B \longrightarrow C$ is rate $(r) = k[A]^2[B]$. When the concentration of both A and B is doubled, the reaction rate is increased "x" times.

Statement II

The figure is showing "the variation in concentration against time plot" for a "y" order reaction.

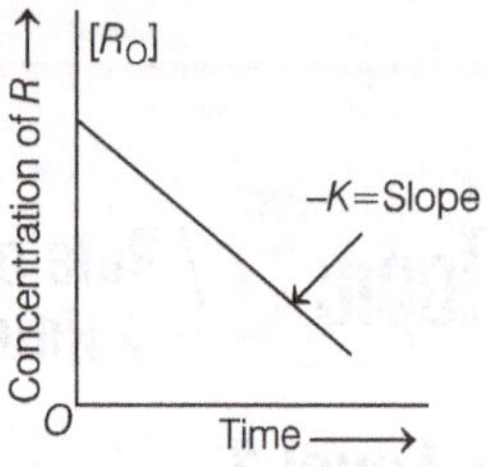

The value of $x + y$ is ....

12 NO_2 required for a reaction is produced by decomposition of N_2O_5 in CCl_4 as by equation

$$2N_2O_5(g) \longrightarrow 4NO_2(g) + O_2(g)$$

The initial concentration of N_2O_5 is 3 mol L^{-1} and it is 2.75 mol L^{-1} after 30 minutes.

The rate of formation of NO_2 is $x \times 10^{-3}$ mol L^{-1}min^{-1}, value of x is ....

(nearest integer) **[2024, 30 Jan Shift-II]**

13 The following data were obtained during the first order thermal decomposition of a gas A at constant volume. **[2024, 1 Feb Shift-II]**

$$A(g) \longrightarrow 2B(g) + C(g)$$

S. No.	Time/(s)	Total pressure/(atm)
1.	0	0.1
2.	115	0.28

The rate constant of the reaction is .... 10^{-2} s^{-1} (nearest integer)

14 For a given chemical reaction,

$$\gamma_1 A + \gamma_2 B \longrightarrow \gamma_3 C + \gamma_4 D$$

concentration of C changes from 10 m mol dm^{-3} to 20 mmol dm^{-3} in 10 seconds. Rate of appearance of D is 1.5 times the rate of disappearance of B which is twice the rate of disappearance A. The rate of appearance of D has been experimentally determined to be 9 m mol dm^{-3} s^{-1}. Therefore, the rate of reaction is ... mmol dm^{-3} s^{-1}. [Nearest integer] **[2022, 25 June Shift-I]**

15 The reaction that occurs in a breath analyser, a device used to determine the alcohol level in a person's blood stream is

$$2K_2Cr_2O_7 + 8\,H_2SO_4 + 3\,C_2H_6O \longrightarrow$$
$$2Cr_2(SO_4)_3 + 3\,C_2H_4O_2 + 2\,K_2SO_4 + 11\,H_2O$$

If the rate of appearance of $Cr_2(SO_4)_3$ is 2.67 mol min^{-1} at a particular time, the rate of disappearance of C_2H_6O at the same time is .... mol min^{-1}. (Nearest integer) **[2021, 27 Aug Shift-I]**

16 For a chemical reaction $A \longrightarrow B$, it was found that concentration of B is increased by 0.2 mol L^{-1} in 30 min. The average rate of the reaction is ...... $\times 10^{-1}$ mol L^{-1} h^{-1}. (Nearest integer) **[2021, 25 July Shift-II]**

17 The reaction $2A + B_2 \longrightarrow 2AB$ is an elementary reaction. For a certain quantity of reactants, if the volume of the reaction vessel is reduced by a factor of 3, the rate of the reaction increases by a factor of .... (Round off to the nearest integer). **[2021, 17 March Shift-II]**

Topic 2 / Order, Molecularity & Half-life Period

• Level 1

01 Which of the following graphs most appropriately represents a zero order reaction? **[2025, 23 Jan Shift-II]**

(a)
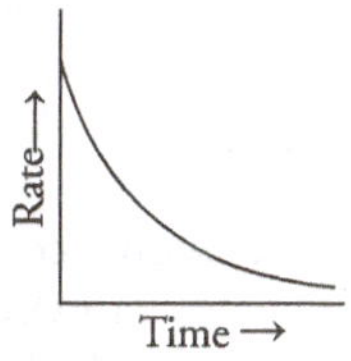

(b)
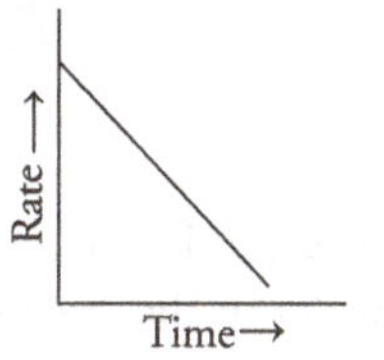

(c)
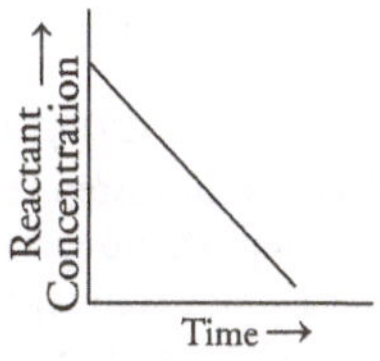

(d)
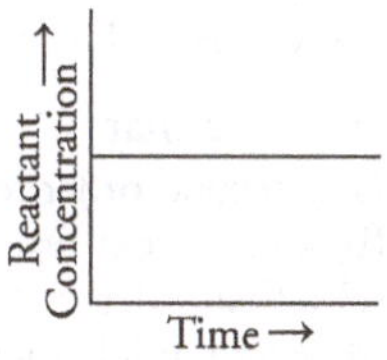

Imp. *Unit of Rate Constant*

02 For a reaction of order n, the unit of the rate constant is **[2021, 27 July Shift-I]**

(a) $mol^{1-n}\ L^{1-n}\ s$ (b) $mol^{1-n}\ L^{2n}\ s^{-1}$

(c) $mol^{1-n}\ L^{n-1}\ s^{-1}$ (d) $mol^{1-n}\ L^{1-n}\ s^{-1}$

• Level 2

03 In a first order decomposition reaction, the time taken for the decomposition of reactant to one fourth and one eighth of its initial concentration are t_1 and t_2(s), respectively. The ratio $\frac{t_1}{t_2}$ will be **[2025, 8 April Shift-II]**

(a) $\frac{3}{2}$ (b) $\frac{4}{3}$ (c) $\frac{3}{4}$ (d) $\frac{2}{3}$

04 $A(g) \rightarrow B(g) + C(g)$ is a first order reaction.

Time	t	∞
p_{system}	p_t	p_∞

The reaction was started with reactant A only. Which of the following expression is correct for rate constant k? **[2025, 7 April Shift-II]**

(a) $k = \frac{1}{t} \ln \frac{p_\infty}{p_t}$ (b) $k = \frac{1}{t} \ln \frac{p_\infty}{(p_\infty - p_t)}$

(c) $k = \frac{1}{t} \ln \frac{p_\infty}{2(p_\infty - p_t)}$ (d) $k = \frac{1}{t} \ln \frac{2(p_\infty - p_t)}{p_t}$

05 Half-life of zero order reaction $A \rightarrow$ product is 1 hour, when initial concentration of reactant is 2.0 mol L^{-1}. The time required to decrease concentration of A from 0.50 to 0.25 mol L^{-1} is **[2025, 4 April Shift-II]**

(a) 0.5 hour (b) 4 hour (c) 15 min (d) 60 min

06 In a reaction $A + B \rightarrow C$, initial concentrations of A and B are related as $[A]_0 = 8[B]_0$. The half-lifes of A and B are 10 min and 40 min, respectively. If they start to disappear at the same time, both following first order kinetics, after how much time will the concentration of both the reactants be same ? **[2025, 3 April Shift-I]**

(a) 40 min (b) 80 min (c) 20 min (d) 60 min

07 Drug X becomes ineffective after 50% decomposition. The original concentration of drug in a bottle was 16 mg/ mL which becomes 4 mg/mL in 12 months. The expiry time of the drug in months is ....

Assume that the decomposition of the drug follows first order kinetics. **[2025, 29 Jan Shift-II]**

(a) 2 (b) 6 (c) 12 (d) 3

08 The reaction, $A_2 + B_2 \longrightarrow 2AB$ follows the mechanism

$$A_2 \underset{k_{-1}}{\overset{k_1}{\rightleftharpoons}} A + A \text{ (Fast)}$$

$$A + B_2 \xrightarrow{k_2} AB + B \text{ (Slow)}$$

$$A + B \longrightarrow AB \text{ (Fast)}$$

The overall order of the reaction is **[2025, 29 Jan Shift-I]**

(a) 2 (b) 1.5 (c) 2.5 (d) 3

09 Which of the following statement is not true for radioactive decay? **[2025, 22 Jan Shift-I]**

(a) Amount of radioactive substance remained after three Half-lifes is 1/8th of original amount.

(b) Half-life is ln2 times of $\frac{1}{\text{rate constant}}$.

(c) Decay constant does not depend upon temperature.

(d) Decay constant increases with increase in temperature.

Imp. *Graph of First Order Reactions*

10 Given below are two statements. **[2025, 24 Jan Shift-II]**

Statement I

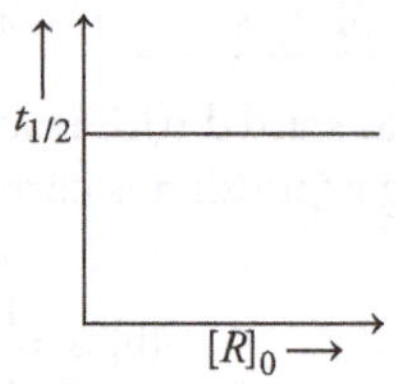

is valid for first order reaction.

Statement II

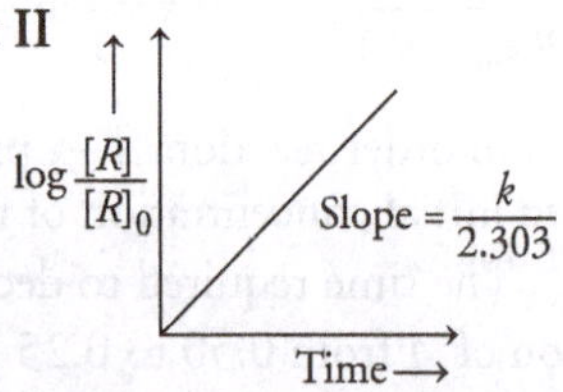

is valid for first order reaction.

In the light of the above statements, choose the correct answer from the options given below

(a) Statement I is true but Statement II is false.
(b) Both Statement I and Statement II are false.
(c) Both Statement I and Statement II are true.
(d) Statement I is false but Statement II is true.

11 For bacterial growth in a cell culture, growth law is very similar to the law of radioactive decay. Which of the following graphs is most suitable to represent bacterial colony growth? **[2025, 28 Jan Shift-II]**
Where N-number of bacteria at any time, N_0 –Initial number of bacteria.

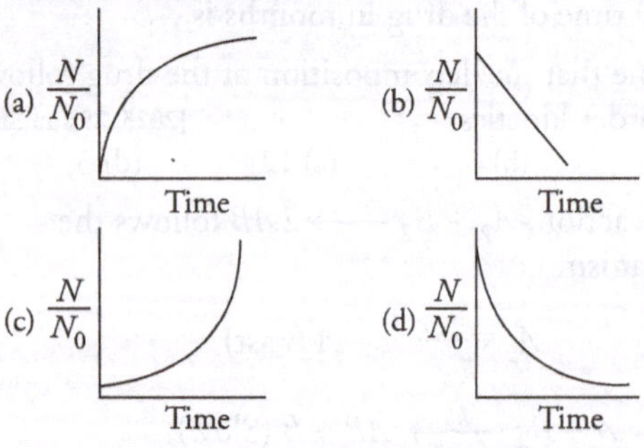

12 Integrated rate law equation for a first order gas phase reaction is given by (where p_i is initial pressure and p_t is total pressure at time t) **[2024, 31 Jan Shift-I]**

(a) $k = \frac{2.303}{t} \times \frac{p_i}{(2p_i - p_t)}$

(b) $k = \frac{2.303}{t} \times \log \frac{p_i}{(2p_i - p_t)}$

(c) $k = \frac{2.303}{t} \times \log \frac{2p_i}{(2p_i - p_t)}$

(d) $k = \frac{2.303}{t} \times \log \frac{(2p_i - p_t)}{p_i}$

13 A student has studied the decomposition of a gas AB_3 at 25°C. He obtained the following data.

p (mm Hg)	50	100	200	400
Relative $t_{1/2}$ (s)	4	2	1	0.5

The order of the reaction is **[2023, 24 Jan Shift-II]**

(a) 1 (b) 0 (zero) (c) 0.5 (d) 2

Imp. *Half-Life of 1st Order Reaction*

14 For a first order reaction, the time required for completion of 90% reaction is 'x' times the half-life of the reaction. The value of 'x' is (Given, ln10 = 2.303 and log 2 = 0.3010) **[2022, 24 June Shift-II]**

(a) 1.12 (b) 2.43
(c) 3.32 (d) 33.31

15 Consider the following reactions
$A \rightarrow$ P1; $B \rightarrow$ P2; $C \rightarrow$ P3; $D \rightarrow$ P4,
The order of the above reactions are a, b, c and d, respectively. The following graph is obtained when log[rate] vs log[conc.] are plotted :

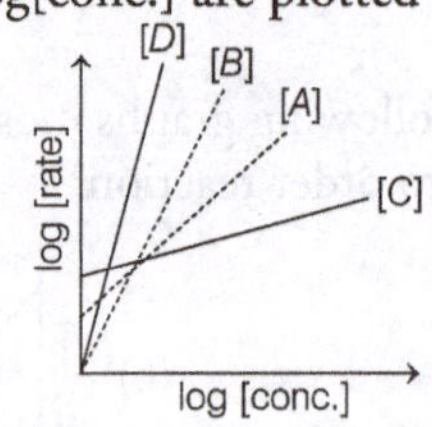

Among the following, the correct sequence for the order of the reactions is **[2020, 6 Sep Shift-I]**

(a) $D > A > B > C$ (b) $A > B > C > D$
(c) $C > A > B > D$ (d) $D > B > A > C$

16 It is true that **[2020, 3 Sep Shift-I]**

(a) a second order reaction is always a multistep reaction.
(b) a zero order reaction is a multistep reaction.
(c) a first order reaction is always a single step reaction.
(d) a zero order reaction is a single step reaction.

17 The reaction, $2X \rightarrow B$ is a zeroth order reaction. If the initial concentration of X is 0.2 M, the half-life is 6 h. When the initial concentration of X is 0.5 M, the time required to reach its final concentration of 0.2 M will be **[2019, 11 Jan Shift-II]**

(a) 7.2 h (b) 18.0 h (c) 12.0 h (d) 9.0 h

• Level 3

18 Reaction $A(g) \rightarrow 2B(g) + C(g)$ is a first order reaction. It was started with pure A **[2025, 7 April Shift-I]**

t/min	Pressure of system at time t/time Hg
10	160
∞	240

Which of the following option is incorrect?
(a) The reaction never goes to completion.
(b) Initial pressure of A is 80 mm Hg
(c) Rate constant of the reaction is 1.693 min^{-1}
(d) Partial pressure of A after 10 minute is 40 mm Hg

19

$[A]_0$ / mol L^{-1}	$t_{1/2}$ / min
0.100	200
0.025	100

For a given reaction $R \to P$, $t_{1/2}$ is related to $[A]_0$ as given in table.

Given, $\log 2 = 0.30$

Which of the following is true?
A. The order of the reaction is 1/2
B. If $[A]_0$ is 1M, then $t_{1/2}$ is $200\sqrt{10}$ min.
C. The order of the reaction changes to 1 if the concentration of reactant changes from 0.100 M to 0.500 M.
D. $t_{1/2}$ is 800 min for $[A]_0 = 1.6$ M

Choose the correct answer from the options given below. **[2025, 28 Jan Shift-I]**
(a) C and D only (b) A and C only
(c) A, B and D only (d) A and B only

20 For a chemical reaction, $A + B \longrightarrow$ Product, the order is 1 with respect to A and B.

Rate mol L^{-1} s^{-1}	$[A]$ mol L^{-1}	$[B]$ mol L^{-1}
0.10	20	0.5
0.40	x	0.5
0.80	40	y

What is the value of x and y? **[2023, 11 April Shift-II]**
(a) 80 and 4 (b) 80 and 2
(c) 40 and 4 (d) 160 and 4

21 At 30°C, the half-life for the decomposition of AB_2 is 200 s and is independent of the initial concentration of AB_2. The time required for 80% of the AB_2 to decompose is
(Given, $\log 2 = 0.30$, $\log 3 = 0.48$) **[2022, 26 July Shift-II]**
(a) 200 s (b) 323 s (c) 467 s (d) 532 s

22 For the following graphs

(A)
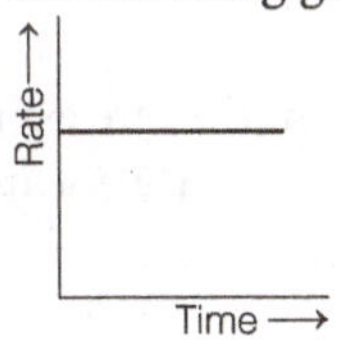

(B)
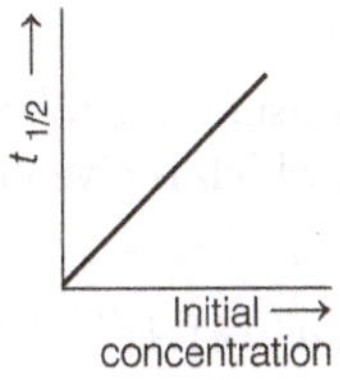

(C)
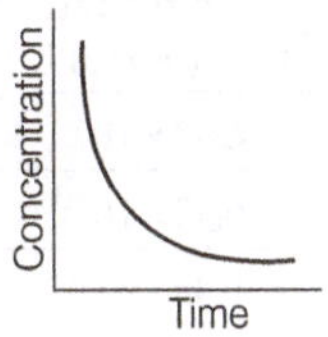

(D)
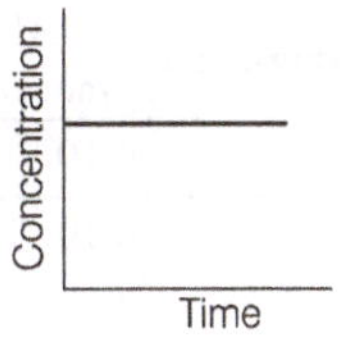

(E)
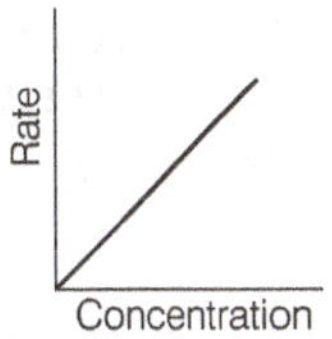

Choose from the options given below, the correct one regarding order of reaction is **[2021, 25 July Shift-I]**
(a) (B) zero order (C) and (E) first order
(b) (A) and (B) zero order (E) first order
(c) (B) and (D) zero order (E) first order
(d) (A) and (B) zero order (C) and (E) first order

23 The results given in the below table were obtained during kinetic studies of the following reaction

$$2A + B \longrightarrow C + D$$

Experiment	$[A]$/mol L^{-1}	$[B]$/mol L^{-1}	Initial rate/mol L^{-1} min^{-1}
I	0.1	0.1	6.00×10^{-3}
II	0.1	0.2	2.40×10^{-2}
III	0.2	0.1	1.20×10^{-2}
IV	X	0.2	7.20×10^{-2}
V	0.3	Y	2.88×10^{-1}

X and Y in the given table are respectively
(a) 0.4, 0.4 **[2020, 2 Sep Shift-II]**
(b) 0.4, 0.3
(c) 0.3, 0.4
(d) 0.3, 0.3

24 For the following reaction, $2A + B \longrightarrow$ products

When concentration of both (A and B) becomes double, then rate of reaction increases from 0.3 mol L^{-1} s^{-1} to 2.4 mol L^{-1} s^{-1}.

When concentration of only A is doubled, the rate of reaction increases from 0.3 mol L^{-1} s^{-1} to 0.6 mol L^{-1} s^{-1}.

Which of the following is true? **[2019, 9 Jan Shift-II]**
(a) The whole reaction is of 4th order
(b) The order of reaction w.r.t. B is one
(c) The order of reaction w.r.t. B is 2
(d) The order of reaction w.r.t. A is 2

25 The following results were obtained during kinetic studies of the reaction;

$$2A + B \longrightarrow \text{Products}$$

Experiment	$[A]$ $(mol\ L^{-1})$	$[B]$ $(mol\ L^{-1})$	Initial rate of reaction (in $mol\ L^{-1}\ min^{-1}$)
I.	0.10	0.20	6.93×10^{-3}
II.	0.10	0.25	6.93×10^{-3}
III.	0.20	0.30	1.386×10^{-2}

The time (in minutes) required to consume half of A is **[2019, 9 Jan Shift-I]**

(a) 5 (b) 10 (c) 100 (d) 1

Numerical Value Type Questions

26 For the reaction $A \rightarrow B$ the following graph was obtained. The time required (in seconds) for the concentration of A to reduce to $2.5\ g\ L^{-1}$ (if the initial concentration of A was $50\ g\ L^{-1}$) is .... (Nearest integer)

Given : $\log 2 = 0.3010$ **[2025, 2 April Shift-II]**

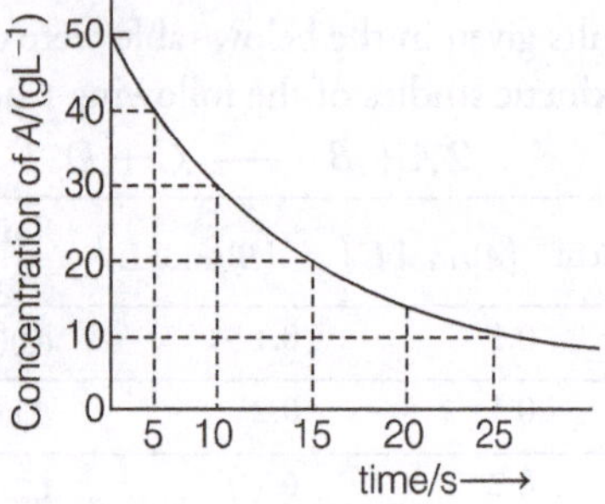

27 For the reaction $A \rightarrow$ products

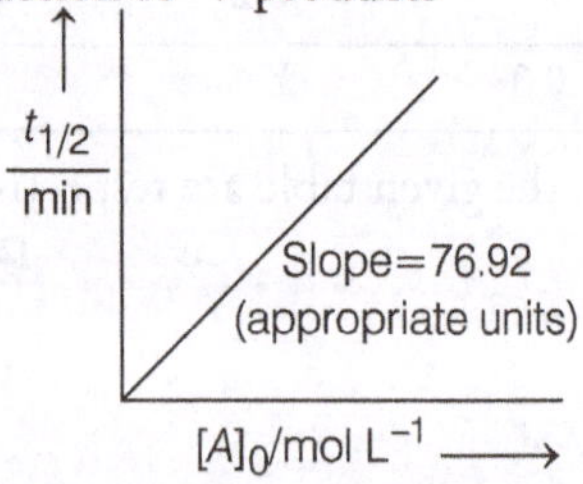

The concentration of A at 10 minutes is .... $\times 10^{-3}$ mol L^{-1} (nearest integer). The reaction was started with 2.5 mol L^{-1} of A. **[2025, 2 April Shift-I]**

28 For the thermal decomposition of $N_2O_5\ (g)$ at constant volume, the following table can be formed. For the reaction mentioned below. **[2025, 23 April Shift-I]**

$$2N_2O_5(g) \longrightarrow 2N_2O_4(g) + O_2(g)$$

S.No.	Time	Total pressure (atm)
1.	0	0.6
2.	100	'x'

$x =$ .... $\times 10^{-3}$ atm [Nearest integer]

(Given, rate constant for the reaction is $4.606 \times 10^{-2} s^{-1}$.)

29 Consider the following reaction **[2024, 8 April Shift-I]**

$$A + B \longrightarrow C$$

The time taken for A to become $1/4^{th}$ of its initial concentration is twice the time taken to become 1/2 of the same. Also, when the change of concentration of B is plotted against time, the resulting graph gives a straight line with a negative slope and a positive intercept on the concentration axis.

The overall order of the reaction is ....

30 Consider the two different first order reactions given below. **[2024, 6 April Shift-II]**

$$A + B \longrightarrow C \text{ (Reaction 1)}$$
$$P \longrightarrow Q \text{ (Reaction 2)}$$

The ratio of the half-life of reaction 1 : reaction 2 is 5 : 2. If t_1 and t_2 represent the time taken to complete $\frac{2}{3}$rd and $\frac{4}{5}$th of reaction 1 and reaction 2, respectively, then the value of the ratio $t_1 : t_2$ is .... $\times 10^{-1}$ (Nearest integer).

[Given, $\log_{10}(3) = 0.477$ and $\log_{10}(5) = 0.699$]

31 Time required for 99.9% completion of a first order reaction is .... times the time required for completion of 90% reaction. (Nearest integer) **[2024, 6 April Shift-I]**

32 Consider the following single step reaction in gas phase at constant temperature. **[2024, 5 April Shift-II]**

$$2A(g) + B(g) \longrightarrow C(g)$$

The initial rate of the reacion is recorded as r_1, when the reaction starts with 1.5 atm pressure of A and 0.7 atm pressure of B. After some time, the rate r_2, is recorded when the pressure of C becomes 0.5 atm. The ratio $r_1 : r_2$ is .... $\times 10^{-1}$ (Nearest integer)

33 During kinetic study of reaction, $2A + B \longrightarrow C + D$, the following results were obtained. **[2024, 5 April Shift-I]**

	A[M]	B[M]	Initial rate of formation of D
I.	0.1	0.1	6.0×10^{-3}
II.	0.3	0.2	7.2×10^{-2}
III.	0.3	0.4	2.88×10^{-1}
IV.	0.4	0.1	2.40×10^{-2}

Based on above data, overall order of the reaction is ....

34 Consider the following reaction, the rate expression of which is given below **[2024, 4 April Shift-II]**

$$A + B \rightarrow C$$

$$\text{Rate} = k[A]^{1/2}[B]^{1/2}$$

The reaction is initiated by taking 1 M concentration of A and B each. If the rate constant (k) is $4.6 \times 10^{-2} s^{-1}$, then the time taken for A to become 0.1 M is ...... sec. (Nearest integer)

35 The ratio of $\frac{^{14}C}{^{12}C}$ in a piece of wood is $\frac{1}{8}$ part that of atmosphere. If half-life of ^{14}C is 5730 years, the age of wood sample is .... years. **[2024, 1 Feb Shift-I]**

36 $r = k[A]$ for a reaction, 50% of A is decomposed in 120 minutes. The time taken for 90% decomposition of A is .... minutes. **[2024, 31 Jan Shift-II]**

37 The rate of first order reaction is 0.04 mol $L^{-1}s^{-1}$ at 10 minutes and 0.03 mol $L^{-1}s^{-1}$ at 20 minutes after initiation. Half-life of the reaction is .... minutes. (Given, $\log 2 = 0.3010$, $\log 3 = 0.4771$) **[2024, 30 Jan Shift-I]**

38 The half-life of radioisotope bromine-82 is 36 hours. The fraction which remains after one day is .... $\times 10^{-2}$.
(Given, antilog 0.2006 = 1.587) **[2024, 29 Jan Shift-II]**

39 Time required for completion of 99.9% of a first order reaction is .... times of half-life ($t_{1/2}$) of the reaction. **[2024, 27 Jan Shift-II]**

40 Consider the following data for the given reaction

$$2HI(g) \longrightarrow H_2(g) + I_2(g)$$

	1	2	3
HI (mol L^{-1})	0.05	0.01	0.02
Rate (mol $L^{-1} s^{-1}$)	7.5×10^{-4}	3.0×10^{-3}	1.2×10^{-2}

The order of the reaction is .... **[2024, 27 Jan Shift-I]**

41 $A(g) \longrightarrow 2B(g) + C(g)$ is a first order reaction. The initial pressure of the system was found to be 800 mm Hg which increased to 1600 mm Hg after 10 min. The total pressure of the system after 30 min will be .... mm Hg. (Nearest integer) **[2023, 13 April Shift-II]**

Imp. *Relation between $t_{87.5}$ and t_{50} of Ist Order Reaction*

42 $t_{87.5}$ is the time required for the reaction to undergo 87.5% completion and t_{50} is the time required for the reaction to undergo 50% completion. The relation between $t_{87.5}$ and t_{50} for a first order reaction is $t_{87.5} = x \times t_{50}$
The value of x is .... (Nearest integer) **[2023, 13 April Shift-I]**

43 The reaction $2NO + Br_2 \longrightarrow 2NOBr$ takes places through the mechanism given below

$$NO + Br_2 \rightleftharpoons NOBr_2 \text{(fast)}$$

$$NOBr_2 + NO \longrightarrow 2NOBr \text{ (Slow)}$$

The overall order of the reaction is .... **[2023, 12 April Shift-I]**

44 The number of incorrect statements from the following is **[2023, 10 April Shift-II]**

A. The successive half lives of zero order reactions decreases with time.
B. A substance appearing as reactant in the chemical equation may not affect the rate of reaction
C. Order and molecularity of a chemical reaction can be a fractional number
D. The rate constant units of zero and second order reaction are mol $L^{-1}s^{-1}$ and mol $L^{-1} s^{-1}$ respectively

45 A molecule undergoes two independent first order reactions whose respective half lives are 12 min and 3 min. If both the reactions are occurring then the time taken for the 50% consumption of the reactant is .... min. (Nearest integer) **[2023, 10 April Shift-I]**

46 $A \rightarrow B$

The above reaction is of zero order. Half-life of this reaction is 50 min. The time taken for the concentration of A to reduce to one-fourth of its initial value is .... min. (Nearest integer) **[2023, 1 Feb Shift-II]**

47 A and B are two substances undergoing radioactive decay in a container. The half-life of A is 15 min and that of B is 5 min. If the initial concentration of B is 4 times that of A and they both start decaying at the same time, how much time will it take for the concentration of both of them to be same? .... min. **[2023, 1 Feb Shift-I]**

48 The rate constant for a first order reaction is 20 min^{-1}. The time required for the initial concentration of the reactant to reduce to its $\frac{1}{32}$ level is .... $\times 10^{-2}$ min. (Nearest integer)
(Given, $\ln 10 = 2.303$, $\log 2 = 0.3010$) **[2023, 31 Jan Shift-II]**

49 An organic compound undergoes first order decomposition. If the time taken for the 60% decomposition is 540 s, then the time required for 90% decomposition will be is .... s. (Nearest integer) [Given : $\ln 10 = 2.3$; $\log 2 = 0.3$] **[2023, 30 Jan Shift-II]**

50 If compound A reacts with B following first order kinetics with rate constant $2.011 \times 10^{-3} s^{-1}$. The time taken by A (in seconds) to reduce from 7 g to 2 g will be .... (Nearest integer) [$\log 5 = 0.698$, $\log 7 = 0.845$, $\log 2 = 0.301$] **[2023, 30 Jan Shift-I]**

51 For conversion of compound $A \rightarrow B$, the rate constant of the reaction was found to be 4.6×10^{-5} L mol^{-1} s^{-1}. The order of the reaction is ....

[2023, 29 Jan Shift-II]

52 For certain chemical reaction $X \rightarrow Y$, the rate of formation of product is plotted against the time as shown in the figure. The number of correct statements from the following is .... **[2023, 29 Jan Shift-I]**

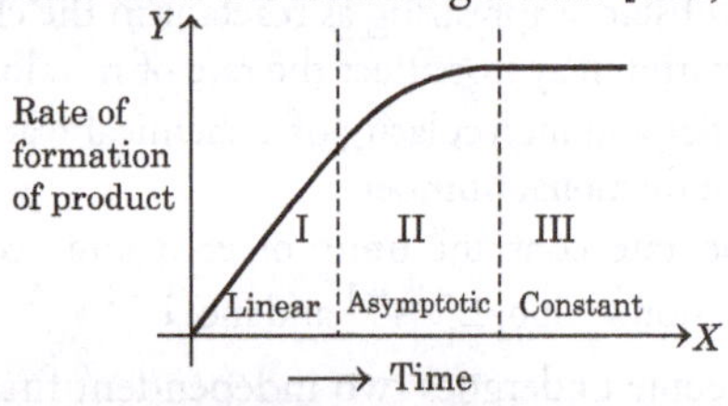

(A) over all order of this reaction is one.

(B) order of this reaction can't be determined.

(C) in region I and III, the reaction is of first and zero order respectively.

(D) in region-II, the reaction is of first order.

(E) in region-II, the order of reaction is in the range of 0.1 to 0.9.

53 A first order reaction has the rate constant, $k = 4.6 \times 10^{-3}$ s^{-1}. The number of correct statements from the following is/are ....

[Given, $\log 3 = 0.48$] **[2023, 25 Jan Shift-II]**

A. reaction completes in 1000 s.

B. the reaction has a half-life of 500 s.

C. the time required for 10% completion is 25 times the time required for 90% completion.

D. the degree of dissociation is equal to $(1 - e^{-kt})$.

E. the rate and the rate constant have the same unit.

54 For the first order reaction $A \rightarrow B$, the half-life is 30 min. The time taken for 75% completion of the reaction is .... min. (Nearest integer).

[Given : $\log 2 = 0.3010$
$\log 3 = 0.4771$
$\log 5 = 0.6989$] **[2023, 25 Jan Shift-I]**

55 Assuming 1 μg of trace radioactive element X with a half-life of 30 years is absorbed by a growing tree. The amount of X remaining in the tree after 100 years is .... $\times 10^{-1}$ μg.

[Given : $\ln 10 = 2.303$; $\log 2 = 0.30$] **[2022, 29 July Shift-II]**

56 The reaction between X and Y is first order with respect to X and zero order with respect to Y.

Experiment	$[X]$ $(mol\ L^{-1})$	$[Y]$ $(mol\ L^{-1})$	Initial rate $(mol\ L^{-1}\ min^{-1})$
I.	0.1	0.1	2×10^{-3}
II.	L	0.2	4×10^{-3}
III.	0.4	0.4	$M \times 10^{-3}$
IV.	0.1	0.2	2×10^{-3}

Examine the data of table and calculate ratio of numerical values of M and L. (Nearest integer)

[2022, 29 July Shift-I]

57 For the given first order reaction $A \rightarrow B$, the half-life of the reaction is 0.3010 min. The ratio of the initial concentration of reactant to the concentration of reactant at time 2.0 min will be equal to .... (Nearest integer) **[2022, 28 July Shift-I]**

58 $[A] \longrightarrow [B]$

Reactant Product

If formation of compound $[B]$ follows the first order of kinetics and after 70 minutes the concentration of $[A]$ was found to be half of its initial concentration. Then the rate constant of the reaction is $x \times 10^{-6}\ s^{-1}$. The value of x is .... (nearest integer)

[2022, 27 July Shift-II]

59 $2NO + 2H_2 \longrightarrow N_2 + 2H_2O$

The above reaction has been studied at 800°C. The related data are given in the table below.

Reaction serial number	Initial pressure of H_2/kPa	Initial pressure of NO/kPa	Initial rate $\left(\frac{-dp}{dt}\right)$/ (kPa/s)
1	65.6	40.0	0.135
2	65.6	20.1	0.033
3	38.6	65.6	0.214
4	19.2	65.6	0.106

The order of the reaction with respect to NO is...... .

[2022, 27 July Shift-I]

60 For a reaction, $A \longrightarrow 2B + C$ the half-lifes are 100 s and 50 s when the concentration of reactant A is 0.5 and 1.0 mol L^{-1} respectively. The order of the reaction is .... (Nearest integer)

[2022, 26 July Shift-I]

61 For the decomposition of azomethane, $CH_3N_2CH_3(g) \longrightarrow CH_3CH_3(g) + N_2(g)$, a first order reaction, the variation in partial pressure with time at 600 K is given as

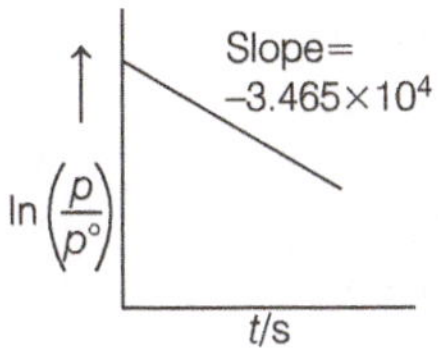

The half-life of the reaction is ...... $\times 10^{-5}$ s. **[2022, 25 July Shift-II]**

62 For a first order reaction, $A \to B$, the rate constant, $k = 5.5 \times 10^{-14}\ s^{-1}$. The time required for 67% completion of reaction is $x \times 10^{-1}$ times the half-life of reaction. The value of x is .... [Nearest integer] (Given, log 3 = 0.4771). **[2022, 28 June Shift-I]**

63 A flask is filled with equal moles of A and B. The half-lives of A and B are 100 s and 50 s respectively and are independent of the initial concentration. The time required for the concentration of A to be four times that of B is ...... s.
(Given, ln 2 = 0.693) **[2022, 26 June Shift-I]**

64 At 345 K, the half-life for the decomposition of a sample of a gaseous compound initially at 55.5 kPa was 340 s. When the pressure was 27.8 kPa, the half-life was found to be 170 s. The order of the reaction is .... [Nearest integer] **[2022, 25 June Shift-II]**

65 For a first order reaction, the ratio of the time for 75% completion of a reaction to the time for 50% completion is ...... (Integer answer) **[2021, 31 Aug Shift-I]**

66 The following data was obtained for chemical reaction given below at 975 K.

$$2NO(g) + 2H_2(g) \longrightarrow N_2(g) + 2H_2O(g)$$

	[NO] mol L^{-1}	[H_2] mol L^{-1}	[Rate] mol L^{-1}
A.	8×10^{-5}	8×10^{-5}	7×10^{-9}
B.	24×10^{-5}	8×10^{-5}	2.1×10^{-8}
C.	24×10^{-5}	32×10^{-5}	8.4×10^{-8}

The order of the reaction with respect to NO is .... [Integer answer] **[2021, 26 Aug Shift-I]**

67 For the first order reaction, $A \longrightarrow 2B$, 1 mole of reactant A gives 0.2 moles of B after 100 minutes. The half-life of the reaction is .... min. (Round off to the nearest integer). [Use : ln 2 = 0.69, ln 10 = 2.3; properties of logarithms : $\ln x^y = y \ln x$; $\ln\left(\frac{x}{y}\right) = \ln x - \ln y$]
(Round off to the nearest integer) **[2021, 27 July Shift-II]**

68 The inactivation rate of a viral preparation is proportional to the amount of virus. In the first minute after preparation, 10% of the virus is inactivated. The rate constant for viral inactivation is ...... $\times 10^{-3}\ min^{-1}$. (Nearest integer)
[Use : ln 10 = 2.303; $\log_{10} 3 = 0.477$; property of logarithm : $\log x^y = y \log x$] **[2021, 20 July Shift-I]**

69 A reaction has a half-life of 1 min. The time required for 99.9% completion of the reaction is .... min (Round off to the nearest integer).
[Use : ln 2 = 0.69, ln 10 = 2.3] **[2021, 18 March Shift-II]**

70 $2NO(g) + Cl_2(g) \rightleftharpoons 2NOCl(s)$

This reaction was studied at −10°C and the following data was obtained **[2021, 18 March Shift-I ; 2019, 8 April Shift-I]**

Run	$[NO]_0$	$[Cl_2]_0$	r_0
1	0.10	0.10	0.18
2	0.10	0.20	0.35
3	0.20	0.20	1.40

$[NO]_0$ and $[Cl_2]_0$ are the initial concentrations and r_0 is the initial reaction rate.
The overall order of the reaction is ....
(Rounded off to the nearest integer)

71 For a certain first order reaction 32% of the reactant is left after 570 s. The rate constant of this reaction is ...... $\times 10^{-3} s^{-1}$. (Round off to the nearest integer).
[Given, $\log_{10} 2 = 0.301$, ln 10 = 2.303] **[2021, 17 March Shift-I]**

72 A and B decompose *via* first order kinetics with half-lives 54.0 min and 18.0 min respectively. Starting from an equimolar non-reactive mixture of A and B, the time taken for the concentration of A to become 16 times that of B is .... min.
(Round off to the nearest integer). **[2021, 16 March Shift-II]**

73 Sucrose hydrolyses in acid solution into glucose and fructose following first order rate law with a half-life of 3.33 h at 25°C. After 9 h, the fraction of sucrose remaining is f. The value of $\log_{10}\left(\frac{1}{f}\right)$ is ... $\times 10^{-2}$
(Rounded off to the nearest integer).
[Assume, ln 10 = 2.303, ln 2 = 0.693] **[2021, 24 Feb Shift-II]**

74 Gaseous cyclobutene isomerises to butadiene in a first order process which has a 'k' value of $3.3 \times 10^{-4}\ s^{-1}$ at 153°C. The time in minutes it takes for the isomerisation to proceed 40 % to completion at this temperature is ....
(Rounded off to the nearest integer) **[2021, 24 Feb Shift-I]**

75 If 75% of a first order reaction was completed in 90 minutes, 60% of the same reaction would be completed in approximately (in minutes) .... (Take : log 2 = 0.30; log 2.5 = 0.40) **[2020, 4 Sep Shift-I]**

76 During the nuclear explosion, one of the products is ^{90}Sr with half-life of 6.93 years. If 1 μg of ^{90}Sr was absorbed in the bones of a newly born baby in place of Ca, how much time, in years, is required to reduce it by 90% if it is not lost metabolically .... **[2020, 7 Jan Shift-I]**

Topic 3 / Arrhenius Theory, Activation Energy, Collision & Related Theories

• Level 1

01 Consider the given figure and choose the correct option **[2025, 22 Jan Shift-II]**

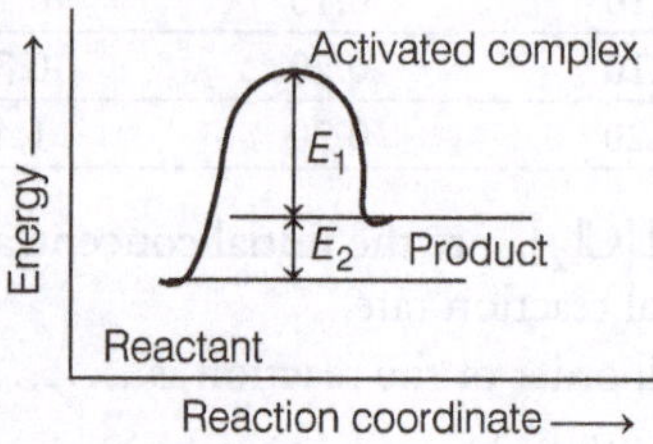

(a) Activation energy of forward reaction is $E_1 + E_2$ and product is less stable than reactant.
(b) Activation energy of both forward and backward reaction is $E_1 + E_2$ and reactant is more stable than product.
(c) Activation energy of forward reaction is $E_1 + E_2$ and product is more stable than reactant.
(d) Activation energy of backward reaction is E_1 and product is more stable than reactant.

Imp. *Energy Profile Graph of Positive Catalyst*

02 The correct reaction profile diagram for a positive catalyst reaction is **[2023, 8 April Shift-II]**

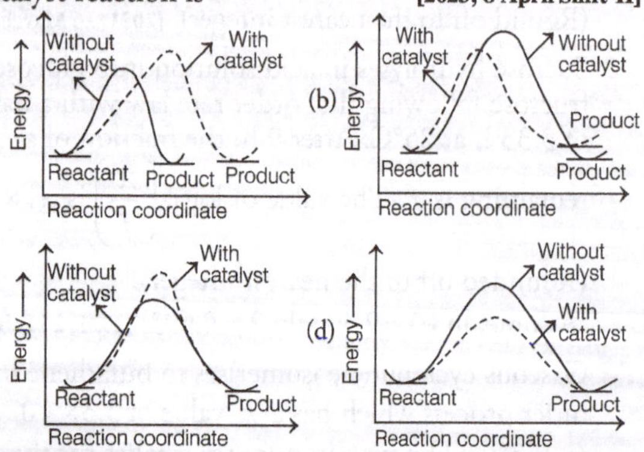

03 Which one of the following given graphs represents the variation of rate constant (k) with temperature (T) for an endothermic reaction? **[2021, 1 Sep Shift-II]**

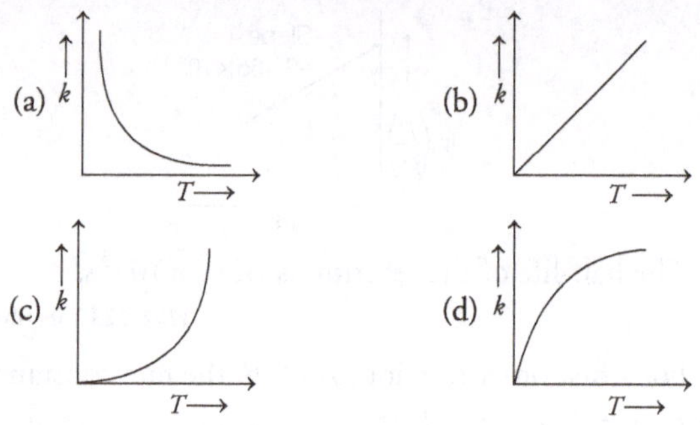

• Level 2

04 Consider a complex reaction taking place in three steps with rate constants k_1, k_2 and k_3 respectively. The overall rate constant k is given by the expression $k = \sqrt{\frac{k_1 k_3}{k_2}}$.

If the activation energies of the three steps are 60, 30 and 10 kJ mol^{-1} respectively, then the overall energy of activation in kJ mol^{-1} is .... (Nearest integer) **[2025, 24 Jan Shift-II]**

05 The molecule A changes into its isomeric from B by following a first order kinetics at a temperature of 1000 K.

If the energy barrier with respect to reactant energy for such isomeric transformation is 191.48 kJ mol^{-1} and the frequency factor is 10^{20}, the time required for 50% molecules of A to become B is .... picoseconds (nearest integer).

$[R = 8.314 \text{ J K}^{-1}\text{mol}^{-1}]$ **[2025, 22 Jan Shift-I]**

06 Consider the following reaction that goes from A to B in three steps as shown below

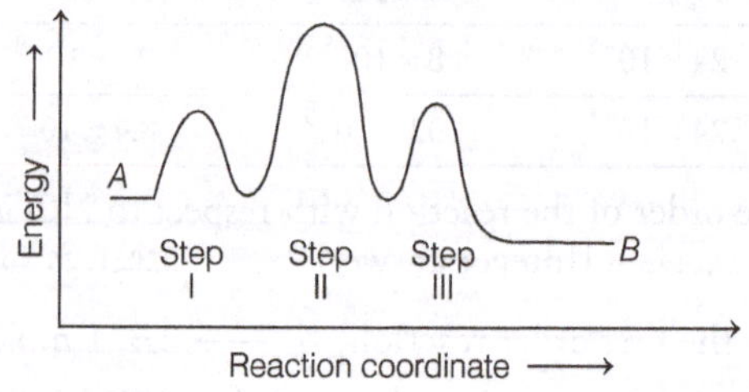

Choose the correct option **[2023, 6 April Shift-II]**

	Number of intermediates	Number of activated complexes	Rate determining step
(a)	2	3	III
(b)	3	2	II
(c)	2	3	II
(d)	2	3	I

07 The rate constant (k) of a reaction is measured at different temperature (T), and the data are plotted in the given figure. The activation energy of the reaction in kJ mol^{-1} is (R is gas constant)

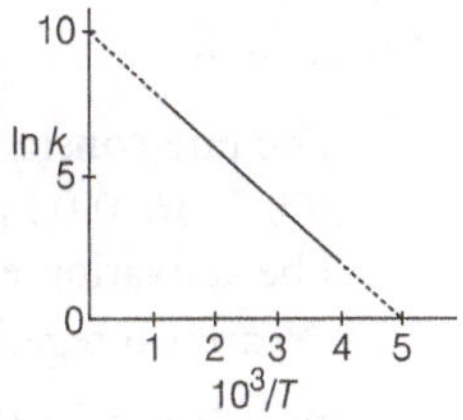

[2020, 5 Sep Shift-II]

(a) $\frac{2}{R}$ (b) $\frac{1}{R}$ (c) R (d) $2R$

08 Consider the following plots of rate constant *versus* $1/T$ for four different reactions. Which of the following orders is correct for the activation energies of these reactions?

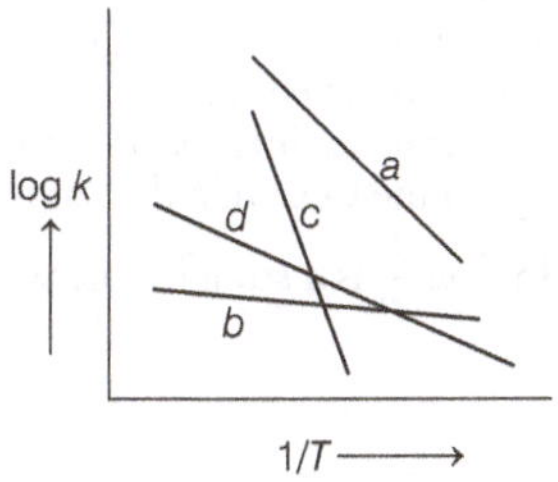

[2020, 8 Jan Shift-II]

(a) $E_b > E_a > E_d > E_c$ (b) $E_a > E_c > E_d > E_b$
(c) $E_b > E_d > E_c > E_a$ (d) $E_c > E_a > E_d > E_b$

09 The rate of a certain biochemical reaction at physiological temperature (T) occurs 10^6 times faster with enzyme than without. The change in the activation energy upon adding enzyme is [2020, 8 Jan Shift-I]

(a) $+6\,RT$ (b) $-6(2.303)\,RT$
(c) $+6(2.303)\,RT$ (d) $-6\,RT$

10 For the reaction of H_2 with I_2, the rate constant is $2.5 \times 10^{-4}\ dm^3 mol^{-1} s^{-1}$ at 327°C and $1.0\ dm^3\ mol^{-1}\ s^{-1}$ at 527°C.
The activation energy for the reaction, in kJ mol^{-1} is ($R = 8.314\ JK^{-1}\ mol^{-1}$) [2019, 10 April Shift-II]

(a) 59 (b) 72 (c) 150 (d) 164

11 For a reaction, consider the plot of ln k *versus* $1/T$ given in the figure. If the rate constant of this reaction at 400 K is $10^{-5}\ s^{-1}$, then the rate constant at 500 K is

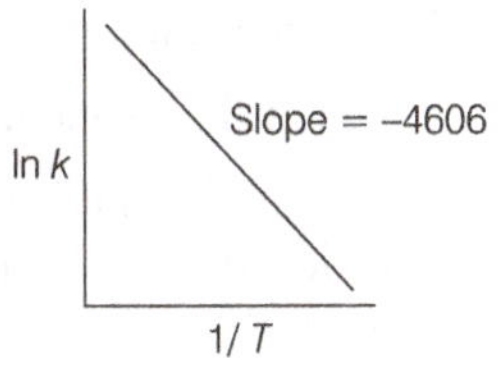

[2019, 12 Jan Shift-II]

(a) $4 \times 10^{-4}\ s^{-1}$ (b) $10^{-6}\ s^{-1}$
(c) $10^{-4}\ s^{-1}$ (d) $2 \times 10^{-4}\ s^{-1}$

• Level 3

12 Consider the following plots of log of rate constant k (log k) *vs* $\frac{1}{T}$ for three different reactions. The correct order of activation energies of these reactions is:

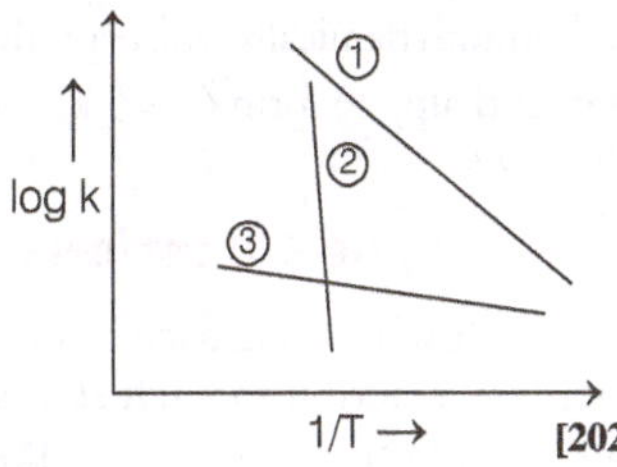

[2025. 4 April Shift-II]

(a) $Ea_3 > Ea_2 > Ea_1$ (b) $Ea_1 > Ea_3 > Ea_2$
(c) $Ea_2 > Ea_1 > Ea_3$ (d) $Ea_1 > Ea_2 > Ea_3$

13 Consider the following statements related to temperature dependence of rate constants.
Identify the correct statements. [2025, 3 April Shift-II]

A. The Arrhenius equation holds true only for an elementary homogeneous reaction.
B. The unit of A is same as that of k in Arrhenius equation.
C. At a given temperature, a low activation energy means a fast reaction.
D. A and E_a as used in Arrhenius equation depend on temperature.
E. When $E_a >> RT$, A and E_a become interdependent.

Choose the correct answer from the options given below.

(a) B and C only (b) B, D and E only
(c) A and B only (d) A, C and D only

14 For $A_2 + B_2 \rightleftharpoons 2AB$

E_a for forward and backward reaction are 180 and 200 kJ mol^{-1} respectively. [2025, 4 April Shift-I]

If catalyst lowers E_a for both reaction by 100 kJ mol^{-1}.
Which of the following statement is correct?

(a) Catalyst does not alter the Gibbs energy change of a reaction.
(b) Catalyst can cause non-spontaneous reactions to occur.
(c) The enthalpy change for the reaction is + 20 kJ mol^{-1}.
(d) The enthalpy change for the catalysed reaction is different from that of uncatalysed reaction.

15 Consider the given plot of enthalpy of the following reaction between A and B.

$$A + B \longrightarrow C + D$$

Identify the incorrect statement. [2019, 9 April Shift-II]

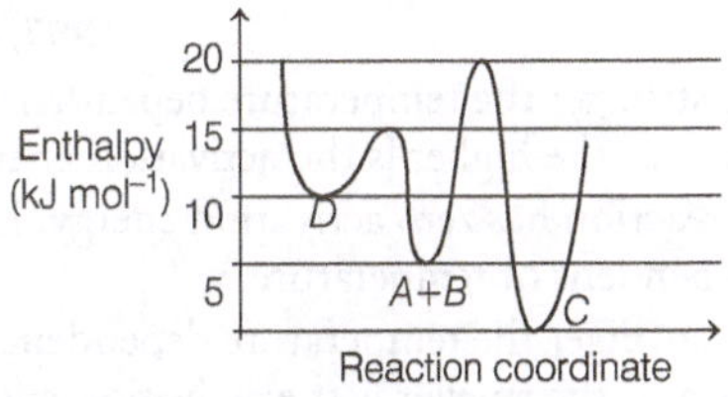

(a) D is kinetically stable product.
(b) Formation of A and B from C has highest enthalpy of activation.

(c) C is the thermodynamically stable product.
(d) Activation enthalpy to form C is 5 kJ mol^{-1} less than that to form D.

Numerical Value Type Questions

16 Consider the following transformation involving first order elementary reaction in each step at constant temperature as shown below. **[2024, 4 April Shift-I]**

$$A + B \underset{\text{Step 3}}{\overset{\text{Step 1}}{\rightleftharpoons}} C \xrightarrow{\text{Step 2}} P$$

Some details of the above reactions are listed below.

Step	Rate constant (sec^{-1})	Activation energy (kJ mol^{-1})
1	k_1	300
2	k_2	200
3	k_3	Ea_3

If the overall rate constant of the above transformation (k) is given as $k = \frac{k_1 k_2}{k_3}$ and the overall activation energy (E_a) is 400 kJ mol^{-1}, then the value of E_{a_3} is..... kJ mol^{-1}. (Nearest integer)

17 For a reaction taking place in three steps at same temperature, overall rate constant $k = \frac{k_1 k_2}{k_3}$. If E_{a_1}, E_{a_2} and E_{a_3} are 40, 50 and 60 kJ/mol respectively, the overall E_a is .... kJ/mol. **[2024, 29 Jan Shift-I]**

18 For a reversible reaction $A \rightleftharpoons B$, the $\Delta H_{\text{forward reaction}} = 20$ kJ mol^{-1}. The activation energy of the uncatalysed forward reaction is 300 kJ mol^{-1}. When the reaction is catalysed keeping the reactant concentration same, the rate of the catalysed forward reaction at 27°C is found to be same as that of the uncatalysed reaction at 327°C is found to be same as that of the uncatalysed reaction at 327°C. The activation energy of the catalysed backward reaction is .... kJ mol^{-1}. **[2023, 15 April Shift-I]**

19 The number of given statement/s which is/are correct is **[2023, 8 April Shift-I]**
(A) The stronger the temperature dependence of the rate constant, the higher is the activation energy.
(B) If a reaction has zero activation energy, its rate is independent of temperature.
(C) The stronger the temperature dependence of the rate constant, the smaller is the activation energy.
(D) If there is no correlation between the temperature and the rate constant then it means that the reaction has negative activation energy.

20 $A \rightarrow B$

The rate constants of the above reaction at 200 K and 300 K are 0.03 min^{-1} and 0.05 min^{-1} respectively. The activation energy for the reaction is .... J. (Nearest integer)[Given : ln 10 = 2.3, R = 8.3 J K^{-1} mol^{-1}, log 5 = 0.70, log 3 = 0.48, log 2 = 0.30] **[2023, 31 Jan Shift-I]**

Imp. *Characteristic of Activation Energy*

21 The number of correct statements from the following is .... **[2023, 24 Jan Shift-I]**
(A) larger the activation energy, smaller is the value of the rate constant.
(B) the higher is the activation energy, higher is the value of the temperature coefficient.
(C) at lower temperature, increase in temperature causes more change in the value of k than at higher temperature.
(D) a plot of ln k *vs* $\frac{1}{T}$ is a straight line with slope equal to $-\frac{E_a}{R}$

22 For a reaction, given below is the graph of ln k *vs* $1/T$. The activation energy for the reaction is equal to .... cal mol^{-1}. (Nearest integer) (Given, R = 2 cal K^{-1} mol^{-1})

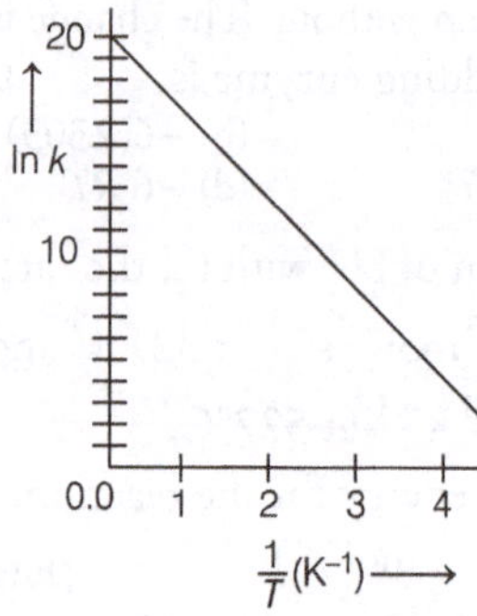

[2022, 28 July Shift-II]

23 For the reaction $P \rightarrow B$, the values of frequency factor A and activation energy E_A are 4×10^{11} s^{-1} and 8.3 kJ mol^{-1} respectively. If the reaction is of first order, the temperature at which the rate constant is 2×10^{-6} s^{-1} is .... $\times 10^{-1}$ K. (Given, ln 10 = 2.3, R = 8.3 J K^{-1} mol^{-1}, log 2 = 0.30) **[2022, 30 June Shift-I]**

24 It has been found that for a chemical reaction with rise in temperature by 9K the rate constant gets doubled. Assuming a reaction to be occurring at 300 K, the value of activation energy is found to be .... kJ mol^{-1}. [Nearest integer] (Given, ln 10 = 2.3, R = 8.3 JK^{-1}mol^{-1}, log 2 = 0.30) **[2022, 27 June Shift-II]**

25 The rate constant for a first order reaction is given by the following equation $\ln k = 33.24 - \frac{2.0 \times 10^4 \text{ K}}{T}$

The activation energy for the reaction is given by …… kJ mol^{-1}. [In nearest integer]
(Given, $R = 8.3$ J K^{-1} mol^{-1}) **[2022, 27 June Shift-I]**

26 Catalyst A reduces the activation energy for a reaction by 10 kJ mol^{-1} at 300 K. The ratio of rate constants, $\frac{K', \text{catalysed}}{K, \text{uncatalysed}}$ is e^x. The value of x is …… .
[Nearest integer]
[Assume that, the pre-exponential factor is same in both the cases. Given, $R = 8.31 \text{JK}^{-1}$ mol^{-1}] **[2022, 26 June Shift-II]**

27 The rate constants for decomposition of acetaldehyde have been measured over the temperature range 700-1000 K. The data has been analysed by plotting In k vs $\frac{10^3}{T}$ graph. The value of activation energy for the reaction is …… kJ mol $^{-1}$. [Nearest integer]

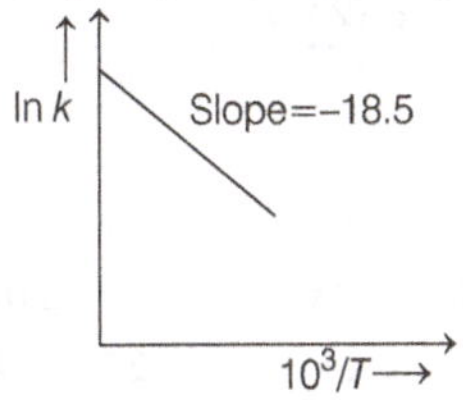

(Given, R = 8.31 J K^{-1} mol^{-1}) **[2022, 24 June Shift-I]**

28 For the reaction $A \rightarrow B$, the rate constant k (in s^{-1}) is given by $\log_{10} k = 20.35 - \frac{(2.47 \times 10^3)}{T}$.

The energy of activation in kJ mol^{-1} is ……… (Nearest integer)
[Given : $R = 8.314$ J K^{-1} mol^{-1}] **[2021, 31 Aug Shift-II]**

29 The first order rate constant for the decomposition of $CaCO_3$ at 700 K is 6.36×10^{-3} s^{-1} and activation energy is 209 kJ mol^{-1}. Its rate constant (in s^{-1}) at 600 K is $x \times 10^{-6}$. The value of x is ………. .
(Nearest integer)
[Given, $R = 8.31$ $JK^{-1}mol^{-1}$, $\log 6.36 \times 10^{-3} = -2.19$, $10^{-4.79} = 1.62 \times 10^{-5}$] **[2021, 27 Aug Shift-II]**

30 The decomposition of formic acid on gold surface follows first order kinetics. If the rate constant at 300K is $1.0 \times 10^{-3} s^{-1}$ and the activation energy, $E_a = 11.488$ kJ mol^{-1}, the rate constant at 200 K is …… $\times 10^{-5} s^{-1}$ (Round off to the nearest integer).
[Given, R = 8.314 J mol^{-1} K^{-1}] **[2021, 16 March Shift-I]**

31 If the activation energy of a reaction is 80.9 kJ mol^{-1}, the fraction of molecules at 700 K, having enough energy to react to form products is e^{-x}. The value of x is ……… . (Rounded off to the nearest integer)
[Use, $R = 8.31$ JK^{-1} mol^{-1}]. **[2021, 26 Feb Shift-II]**

32 An exothermic reaction $X \rightarrow Y$ has an activation energy 30 kJ mol^{-1}. If energy change ΔE during the reaction is –20 kJ mol^{-1}, then the activation energy for the reverse reaction in kJ is ………… . **[2021, 26 Feb Shift-I]**

33 The rate constant of a reaction increases by five times on increase in temperature from 27°C to 52°C. The value of activation energy in kJ mol^{-1} is ………
(Rounded off to the nearest integer)
[R = 8.314 J K^{-1} mol^{-1}] **[2021, 25 Feb Shift-II]**

34 For the reaction, $aA + bB \rightarrow cC + dD$, the plot of $\log k$ vs $\frac{1}{T}$ is given below

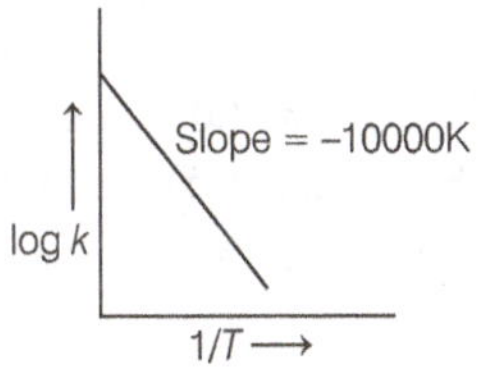

The temperature at which the rate constant of the reaction is 10^{-4} s^{-1} is ……… K
(Rounded off to the nearest integer).
[Given, The rate constant of the reaction is 10^{-5} s^{-1}at 500 K] **[2021, 25 Feb Shift-I]**

35 The rate of a reaction decreased by 3.555 times when the temperature was changed from 40° C to 30° C. The activation energy (in kJ mol^{-1}) of the reaction is ……… .
(Take; $R = 8.314$ J $mol^{-1}K^{-1}$, $\ln 3.555 = 1.268$) **[2020, 6 Sep Shift-II]**

36 The number of molecules with energy greater than the threshold energy for a reaction increases five fold by a rise of temperature from 27 °C to 42 °C. Its energy of activation in J/mol is ……… .
(Take $\ln 5 = 1.6094$; $R = 8.314$ J mol^{-1} K^{-1}) **[2020, 4 Sep Shift-II]**

37 A sample of milk splits after 60 min at 300 K and after 40 min. at 400 K when the population of *lactobacillus acidophilus* in it doubles. The activation energy (in kJ/mol) for this process is closest to
(Given, $R = 8.3$ J $mol^{-1}K^{-1}$
$\ln\left(\frac{2}{3}\right) = 0.4$, $e^{-3} = 4.0$) ***[2020, 9 Jan Shift-II]***

Key Index

Topic 1

1 (a)	2 (c)	3 (d)	4 (a)	5 (a)	6 (b)	7 (b)	8 (c)	9 (a)	10 (3)
11 (8)	12 (17)	13 (2)	14 (1)	15 (4)	16 (4)	17 (27)			

Topic 2

1 (c)	2 (c)	3 (d)	4 (c)	5 (c)	6 (a)	7 (b)	8 (b)	9 (d)	10 (a)
11 (c)	12 (b)	13 (d)	14 (c)	15 (d)	16 (b)	17 (b)	18 (c)	19 (c)	20 (b)
21 (c)	22 (b)	23 (c)	24 (c)	25 (b)	26 (43)	27 (2435)	28 (897)	29 (1)	30 (17)
31 (3)	32 (315)	33 (33)	34 (50)	35 (17190)	36 (399)	37 (24)	38 (63)	39 (10)	40 (2)
41 (2200)	42 (3)	43 (3)	44 (1)	45 (2)	46 (75)	47 (15)	48 (17)	49 (1350)	50 (623)
51 (2)	52 (2)	53 (2)	54 (60)	55 (1)	56 (40)	57 (100)	58 (165)	59 (2)	60 (2)
61 (2)	62 (16)	63 (200)	64 (0)	65 (2)	66 (1)	67 (656)	68 (106)	69 (10)	70 (3)
71 (2)	72 (108)	73 (81)	74 (26)	75 (60)	76 (23.03)				

Topic 3

1 (a)	2 (d)	3 (c)	4 (20)	5 (69)	6 (c)	7 (d)	8 (d)	9 (b)	10 (d)
11 (c)	12 (c)	13 (a)	14 (a)	15 (d)	16 (100)	17 (30)	18 (130)	19 (2)	20 (2520)
21 (3)	22 (8)	23 (225)	24 (59)	25 (166)	26 (4)	27 (154)	28 (47)	29 (16)	30 (10)
31 (14)	32 (50)	33 (52)	34 (526)	35 (100)	36 (84297.47)	37 (3.98)			

Detailed Explanations

Topic 1

01. *(a)* $r_1 = k[A]^n[B]^m$

Given that, A is doubled and B is halved in concentration.

So, $r_2 = k2^n[A]^n \dfrac{[B]^m}{2^m} \Rightarrow \dfrac{r_2}{r_1} = 2^{n-m}$

02. *(c)* The rate of reaction is given as

$$R_1 = K[A]^1[B]^1$$

$$R_1 = K\left[\frac{n_A}{V}\right]^1\left[\frac{n_B}{V}\right]^1$$

Now, for R_2 rate is given as

$$R_2 = K\left[\frac{3n_A}{V}\right]^1\left[\frac{3n_B}{V}\right]^1 = 9R_1$$

03. *(d)* Rate of reaction

$$\frac{-1}{2}\frac{d[A]}{dt} = \frac{-1}{3}\frac{d[B]}{dt} = \frac{-2}{3}\frac{d[C]}{dt} = \frac{1}{3}\frac{d[P]}{dt}$$

$$\frac{dn_A}{dt} = \frac{2}{3}\frac{dn_B}{dt} = \frac{4}{3}\frac{dn_C}{dt}$$

04. *(a)* For the reaction,

$$2H_2(g) + 2NO(g) \longrightarrow N_2(g) + 2H_2O(g)$$

Rate law for forward reaction is given as:

$$\text{rate} = K_f\,[NO]^2\,[H_2]^2$$

Rate of forward reaction is equal to rate of reverse reaction at equilibrium and also,

at equilibrium $(K) = \dfrac{K_f}{K_b}$

where, K (equilibrium constant)

$$= \frac{[N_2]\,[H_2O]^2}{[H_2]^2\,[NO]^2}$$

$$\Rightarrow \quad \frac{K_f}{K_b} = \frac{[N_2]\,[H_2O]^2}{[H_2]^2\,[NO]^2}$$

On rearranging :

$$K_f[H_2]\,[NO]^2 = K_b[N_2][H_2O]^2/[H_2]$$

At Equilibrium, $r_f = r_b$

∴ Rate of reverse reaction

$$= K_b\,[N_2]\,[H_2O]^2/[H_2]$$

05. *(a)* Given reaction,

$$A \xrightarrow{K_1} B \xrightarrow{K_2} C$$

Rate of formation of B is,

$$\frac{d[B]}{dt} = K_1[A] - K_2[B]$$

$$0 = K_1[A] - K_2[B]$$

$$\left[\frac{K_1}{K_2}\right][A] = [B]$$

06. *(b)*

idea The rate of formation of product i.e. NO_2 can be calculated by equating the rate of disappearance of reactant i.e. N_2O_5 with rate of appearance of product, NO_2.

$$2N_2O_5(g) \longrightarrow 4NO_2(g) + O_2(g)$$

$t = 0$ 3.0 M

$t = 30$ 2.75 M

From the equation, rate of reaction can be concluded that

$$\frac{1}{2} \times \frac{-\Delta[N_2O_5]}{\Delta t} = \frac{1}{4} \times \frac{\Delta[NO_2]}{\Delta t}$$

and $\frac{\Delta[NO_2]}{\Delta t} = -2\frac{\Delta(N_2O_5)}{\Delta t}$

$\Rightarrow \frac{\Delta[NO_2]}{\Delta t} = 2\times\frac{0.25}{30}$

$= 1.667\times 10^{-2}$ mol L^{-1} min^{-1}

07. *(b)* Before applying medicine,

Rate law for first order growth.

$$\frac{dA}{dt} = k[A]$$

$$\frac{A}{A_0} = \frac{N}{N_0} = e^{kt}$$

After applying medicine

Active bacteria ⟶ Inactive bacteria
(A) (I)

$$r = \frac{-dA}{dt} = k[A]^2 \text{ (Rate law)}$$

$$y = kx^2$$

Hence, a parabolic curve will form.

08. *(c)*

(i) Rate of the reaction,

$$A_2 \underset{K_{-1}}{\overset{K_1}{\rightleftharpoons}} 2A$$

$$r = r_{forward} - r_{backward}$$
$$= k_1[A_2] - k_{-1}[A]^2 \quad \ldots \text{(i)}$$

(ii) Also differential, rate of the reaction can be expressed as,

$$r = -\frac{d[A_2]}{dt} = +\frac{1}{2}\frac{d[A]}{dt}$$

So, the rate of appearance of A,

i.e. $\frac{d[A]}{dt} = 2r$

[From Eq. (i)]

$$\frac{d[-A]}{dt} = 2k_1[A_2] - 2k_{-1}[A]^2$$

09. *(a)* The expression for bacterial growth is

$$N = N_0 e^t$$

From 0 to 1 hour $N'(t) = N_0 e^t$

From 1 hour onwards, $\frac{dN}{dt} = -5N^2$

On differentiating the above equation from N' to N we get.

$$\int_{eN_0}^{N} N^{-2} dN = -5\int_1^t dt$$

[$\because$ At 1 hour, $N' = eN_0$]

$$\left[\frac{1}{N} - \frac{1}{eN_0}\right] = 5(t-1)$$

Multiply both sides by N_0, we get

$$\frac{N_0}{N} - \frac{1}{e} = 5N_0(t-1)$$

$$\frac{N_0}{N} = 5N_0 t + \left[\frac{1}{e} - 5N_0\right]$$

On comparing the above equation with equation of straight line, $y = mx + c$

We get $m = 5N_0 \Rightarrow c = \frac{1}{e} - 5N_0$

$\therefore$ Plot of $\frac{N_0}{N}$ *vs* t is shown in option (a).

10. *(3)* The given reaction is

	$A(g) \longrightarrow$	$2B(g)$	$+ C(g)$
At time $t = 0$	$a - x$	$2x$	x
On completion	0	$2a$	a

Given that, at 23 seconds

$(p)_t = a - x + 2x + x = 200$ torr ...(i)

at completion $2a + a = 300$ torr ...(ii)

$3a = 300$ torr

$[p]_0 = a = 100$ torr

From Eq. (i),

$a + 2x = 200$

$100 + 2x = 200$

$\Rightarrow x = 50$ torr

Now, $K = \frac{2.303}{t}\log\frac{a}{a-x}$

$= \frac{2.303}{23}\log\frac{100}{100-50}$

$K = 3\times 10^{-2}$ s^{-1}

11. *(8)* Given $r = k[A]^2[B]$

If concentration is doubled

$r' = k[2A]^2[2B]^1$

$r' = 8r \Rightarrow x = 8$

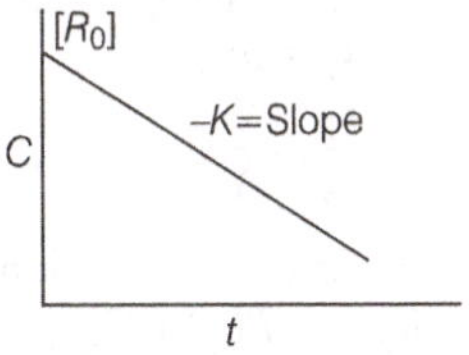

From the graph, the concentration of reactant R decrease linearly with time, which indicates a zero order reaction thus, $y = 0$

So $x + y = 8$

12. *(17)* For reaction,

$$2N_2O_5(g) \rightleftharpoons 4NO_2(g) + O_2(g)$$

Rate of reaction (ROR)

$$= -\frac{1}{2}\frac{\Delta[N_2O_5]}{\Delta t}$$

$$= \frac{1}{4}\frac{\Delta[NO_2]}{\Delta t} = \frac{\Delta[O_2]}{\Delta t}$$

$$\text{ROR} = -\frac{1}{2}\frac{\Delta[N_2O_5]}{\Delta t}$$

$$= -\frac{1}{2}\frac{(2.75-3)}{30} \text{ mol L}^{-1}\text{min}^{-1}$$

$$\text{ROR} = -\frac{1}{2}\frac{(-0.25)}{30}$$

$$= \frac{1}{240} \text{ mol L}^{-1} \text{ min}^{-1}$$

Rate of formation of NO_2

$$= \frac{\Delta[NO_2]}{\Delta t} = 4\times\text{ROR}$$

$$= \frac{4}{240}$$

$$= 16.66\times 10^{-3} \text{ mol L}^{-1} \text{ min}^{-1}$$

$$\approx 17\times 10^{-3} \text{ mol L}^{-1} \text{ min}^{-1}$$

13. *(2)*

	$A(g) \longrightarrow$	$2B(g)$	$+ C(g)$
At $t = 0$	p_0	0	0
$t = t$	$p_0 - x$	$2x$	x

Total pressure after time t,

$p_t = p_0 - x + x + 2x$

$= p_0 + 2x$

$p_t = 0.1 + 2x$

$x = \frac{0.28 - 01}{2} = 0.09$

$k = \frac{2.303}{t}\log\frac{p_0}{p_0 - x}$

$= \frac{2.303}{115}\log\frac{0.1}{0.1 - 0.09}$

$= \frac{2.303}{115}\log 10$

$= 0.020 = 2\times 10^{-2}$ s^{-1}

Thus, the nearest integer is 2.

14. *(1)* Rate of reaction,

$$= -\frac{1}{\gamma_1}\frac{dA}{dt} = -\frac{1}{\gamma_2}\frac{dB}{dt}$$

$$= \frac{1}{Y_3}\frac{dC}{dt} = \frac{1}{\gamma_4}\frac{dD}{dt}$$

$$= -\frac{1}{\gamma_1}\times 3 = -\frac{1}{\gamma_2}\times 6$$

$$= \frac{1}{\gamma_3}\times 1 = \frac{1}{\gamma_4}\times 9$$

Let us assume $\gamma_1 = 1$ then

$$\gamma_2 = 2; \gamma_3 = \frac{1}{3}; \gamma_4 = 3$$

Substituting in the reaction,

$$1A + 2B \longrightarrow \frac{1}{3}C + 3D$$

Multiplying the coefficients by 3 to remove fraction 1/3 gives values of $\gamma_1 = 3; \gamma_2 = 6; \gamma_3 = 1$ and $\gamma_4 = 9$

$$3A + 6B \longrightarrow C + 9D$$

Rate of reaction

$$= \frac{1}{\gamma_4}\frac{dD}{dt} = \frac{1}{9} \times 9 \text{ m mol dm}^{-3}\text{ s}^{-1}$$

$$= 1 \text{ m mol dm}^{-3}\text{ s}^{-1}$$

15. *(4)* For the given reaction,

$$\text{rate of reaction} = -\frac{1}{2}\frac{d[K_2Cr_2O_7]}{dt}$$

$$= \frac{-1}{8}\frac{d[H_2SO_4]}{dt} = \frac{-1}{3}\frac{d[C_2H_6O]}{dt}$$

$$= \frac{1}{2}\frac{d[Cr_2(SO_4)_3]}{dt} = \frac{1}{3}\frac{d[C_2H_4O_2]}{dt}$$

$$= \frac{1}{2}\frac{d[K_2SO_4]}{dt} = \frac{1}{11}\frac{d[H_2O]}{dt}$$

Rate of appearence of 2 moles of $Cr_2(SO_4)_3$ is given as

$$\frac{d}{dt}[Cr_2(SO_4)_3] = 2.67 \text{ mol/min}$$

Rate of disappearence of C_2H_6O can be calculated as

$$\frac{1}{3}\frac{d[C_2H_6O]}{dt} = \frac{+1}{2}\frac{d[Cr_2(SO_4)_3]}{dt}$$

$$= \frac{3}{2}\frac{d[Cr_2(SO_4)_3]}{dt}$$

$$= \frac{2.67}{2} \times 3$$

$$= 4.005 = 4 \text{ mol/min}$$

16. *(4)* Reaction, $A \longrightarrow B$

		A	B
At	$t = 0$		0
At	$t = 30$ min		0.2 M

Average rate of reaction

$$= -\frac{\Delta[A]}{\Delta t} = \frac{\Delta[B]}{\Delta t} = \frac{(0.2 - 0)}{\frac{1}{2}}$$

$$= 0.4 \text{ mol L}^{-1}\text{ h}^{-1}$$

$$= 4 \times 10^{-1} \text{ mol L}^{-1}\text{ h}^{-1}$$

17. *(27)* Given, the reaction

$2A + B_2 \longrightarrow 2AB$ is an elementary reaction.

So, rate of reaction $= k[A]^2[B_2]$

Initial rate $= k(a/V)^2(b/V)^1$

where, a and b are the number of moles of A and B respectively and V is the volume.

On reducing the volume by a factor of 3, the concentrations of A and B_2 will become 3 times

Final rate $= k(3a/V)^2(3b/V)^1$

$$3^2 \times 3k\left(\frac{a}{V}\right)^2\left(\frac{b}{V}\right)^1 = 27 \times \text{initial rate}$$

Hence, the rate becomes = 27 times of initial rate.

Topic 2

01. *(c)* For zero order reaction,

$$A \longrightarrow P, \text{ rate} = k$$

$$[A]_t = [A]_0 - kt$$

So, the most suitable graph is given in option (c).

i.e.,

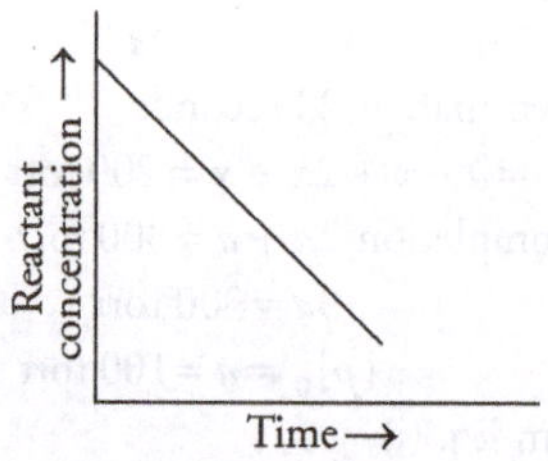

02. *(c)* $\Rightarrow \quad k = \dfrac{\text{Rate of reaction}}{[\text{Reactant}]^n}$

$$= \frac{\text{mol}^1\text{ L}^{-1}\text{s}^{-1}}{[\text{mol}^n/\text{L}^n]}$$

$$= \text{mol}^{1-n}\text{L}^{n-1}\text{s}^{-1}$$

Therefore, option (c) is correct.

03. *(d)* For 1st order reaction,

When concentration is C_t at given time t

$$C_t = \frac{C_0}{4}$$

$$t_1 = 2t_{50\%}$$

When $\quad C_t = C_0/8$

$$t_2 = 3t_{50\%}$$

$$\therefore \quad \frac{t_1}{t_2} = \frac{2}{3}$$

04. *(c)* Given first order reaction is,

	$A(g) \rightarrow$	$B(g) +$	$C(g)$
$t = 0$	$p°$	0	0
$t = t$	$p° - x$	x	x
$t = \infty$	0	$p°$	$p°$

$$p_t = p° + x \Rightarrow x = p_t - p° = p_t - \frac{p_\infty}{2}$$

$$p_\infty = 2p° \Rightarrow p° = \frac{p_\infty}{2}$$

$$k = \frac{1}{t}\ln\frac{p°}{p° - x}$$

$$\Rightarrow \quad k = \frac{1}{t}\ln\frac{p_\infty}{2(p_\infty - p_t)}$$

05. *(c)* For zero order reaction, half-life is given by

$$\text{half-life} = \frac{A_0}{2k}$$

$$60 = \frac{2}{2k} \Rightarrow k = \frac{1}{60}\text{ M/min}$$

Now, we know that $A_t = A_0 - kt$

$$\Rightarrow \quad t = \frac{A_0 - A_t}{k} = \frac{0.5 - 0.25}{1/60} = 15 \text{ min}$$

$$\Rightarrow \quad t = 15 \text{ min}$$

06. *(a)* Given, $\quad [A_0] = 8[B_0]$

$$[t_{1/2}]_A = 10 \text{ min}$$

$$[t_{1/2}]_B = 40 \text{ min}$$

A and B follows first order kinetics

$$t = ?$$

$$[A]_t = [B]_t$$

$$-k_A t = -k_B t$$

or $$\frac{[A_0]}{[B_0]} = e^{(k_A - k_B)t}$$

$$8 = e^{(k_A - k_B)t}$$

$$\ln 8 = (k_A - k_B)t$$

$$\ln 8 = \ln 2\left(\frac{1}{(t_{1/2})_A} - \frac{1}{(t_{1/2})_B}\right) \times t$$

$$2.08 = 0.693\left(\frac{1}{10} - \frac{1}{40}\right)t$$

$$\Rightarrow \quad t = 40 \text{ min}$$

07. *(b)* Given, $[A] = 16$ mg/mL

$[A_0] = 4$ mg/mL, $T = 12$ months

Using the formula of rate constant for 1st order reaction.

$$k = \frac{2.303}{t}\log\frac{A}{[A_0]} \quad \ldots\text{(i)}$$

$$= \frac{2.303}{12}\log\frac{16}{4}$$

On solving, we get

$$k = 0.1155$$

Now, since 50% decomposition takes place. So,

$$[A] = [A_0] - \frac{50}{100}[A_0]$$

$$[A] = 0.5[A_0]$$

Now, substitute the obtained values in Eq. (i) to calculate the expiry time

$$0.1155 = \frac{2.303}{t}\log\frac{[A_0]}{[0.5A_0]}$$

$$t = 6 \text{ months}$$

08. *(b)* The rate of reaction is given by

$$\text{Rate} = k_2[A_2][B_2] \quad \ldots\text{(i)}$$

Also $\left(\frac{k_1}{k_{-1}}\right) = \left(\frac{[A]^2}{[A_2]}\right)$

Or $[A] = \sqrt{\frac{k_1}{k_{-1}}}\sqrt{A_2}$

Substituting the value of [A] in Eq (i),

$$\text{rate} = k_2 \sqrt{\frac{k_1}{k_{-1}}} \cdot [A_2]^{1/2}[B_2]$$

So, the order will be given by $\frac{1}{2} + 1 = \frac{3}{2}$

$\Rightarrow$ Order = 1.5

09. *(d)* Statement given in option (d) is not true for radioactive decay. Its correct form is decay constant is independent of temperature.

10. *(a)* Statement I is true but Statement II is false.

The correct explanation for Statement II is :

For first order reaction, $t_{1/2} = \frac{\ln 2}{k}$

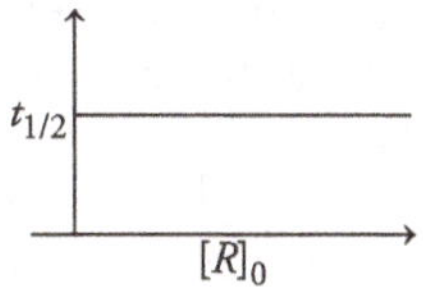

For first order reaction,

$$\log \frac{[R]_0}{[R]} = \frac{1}{2.303} kt$$

$$\log \frac{[R]_0}{[R]} = \left[\frac{k}{2.303}\right]t$$

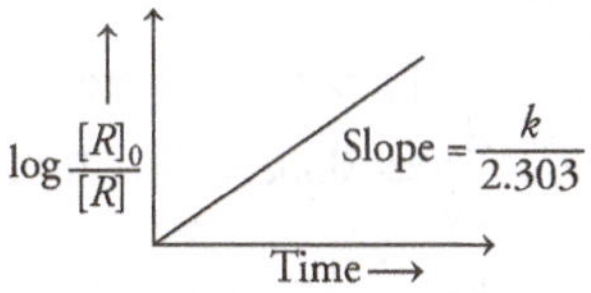

11. *(c)* Initial number of bacteria = N_0

Number of bacteria at any given time = N

Bacterial growth is given as

$$N = N_0 e^{kt}$$

Where, k = growth constant for bacterial growth.

The most suitable graph to show bacterial growth is

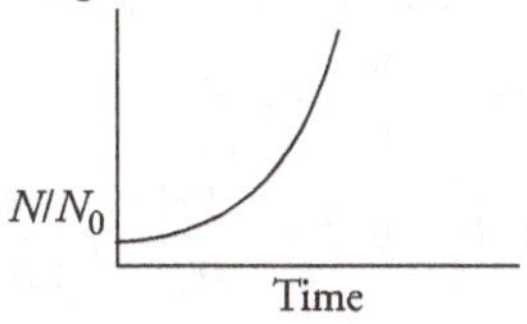

12. *(b)* $A(g) \longrightarrow B(g) + C(g)$

	$A(g)$	$B(g)$	$C(g)$
At $t = 0$	p_i	0	0
At $t = t$	$p_i - x$	x	x

$$p_t = p_i - x + x + x$$
$$p_t = p_i + x$$
$$p_i - x = p_i - p_t + p_i$$
$$= 2p_i - p_t$$

So, $k = \frac{2.303}{t} \log \frac{p_i}{(2p_i - p_t)}$

13. *(d)*

> **idea** In this question, we will use the concept of half life. It is the time in which the concentration of a reactant is reduced to one half of its initial concentration.

$$AB_3(g) \longrightarrow A(g) + 3B(g)$$

As decomposition reaction of $AB_3(g)$ is not given, we assume that p (mm Hg) is for $AB_3(g)$ only.

So, half-life $t_{1/2} \propto (p_0)^{1-n}$

$$\frac{(t_{1/2})_1}{(t_{1/2})_2} = \frac{(p_0)_1^{1-n}}{(p_0)_2^{1-n}}$$

$$\Rightarrow \left(\frac{4}{2}\right) = \left(\frac{50}{100}\right)^{1-n} \Rightarrow 2 = \left(\frac{1}{2}\right)^{1-n}$$

$$\Rightarrow n = 1 + 1 \Rightarrow n = 2$$

Therefore, the order of reaction is 2.

14. *(c)* Given, $t_{90\%} = x\, t_{50\%}$

$$t_{90\%} = \frac{2.303}{k} \log\left(\frac{100}{100-90}\right)$$

$$t_{50\%} = \frac{2.303}{k} \log(2) \quad \text{...(ii)}$$

Eq. (i) divided by Eq. (ii), we get

$$\frac{t_{90\%}}{t_{50\%}} = \frac{\frac{2.303}{k}\log 10}{\frac{2.303}{k}\log 2} = \frac{1}{0.3010}$$

So, $\frac{t_{90\%}}{t_{50\%}} = 3.32 \Rightarrow x = 3.32$

15. *(d)* We know that,

Rate = k (concentration)^n

Taking log both side,

log (Rate) = log k + n log [concentration]

Slope of graph is the order of reaction greater the slope, greater is the order of reaction.

∴Correct sequence for the order of reaction is $D > B > A > C$.

Hence, the correct option is (d).

16. *(b)* Only statement (b) is true whereas all other statements are incorrect.

17. *(b)* For zero order reaction,

$$[A_0] - [A_t] = kt \quad \text{...(i)}$$

Also, for zero order reaction, $t_{1/2} = \frac{[A_0]}{2k}$

$\therefore \quad 6 = \frac{0.2}{2k}$

or $\quad k = \frac{0.2}{2 \times 6} = \frac{1}{60}$

Now, from Eq. (i)

$$[A_0] - [A_t] = kt$$

Given, $[A_0] = 0.5$ M, $[A_t] = 0.2$ M

$\therefore \quad 0.5 - 0.2 = \frac{1}{60} \times t \quad \left[\because k = \frac{1}{60}\right]$

$$0.3 = \frac{1}{60} \times t$$

$\Rightarrow \quad t = 0.3 \times 60 = 18$ h

18. *(c)* Among the given statements, statement given in option C is incorrect. It is because

$$A(g) \longrightarrow 2B(g) + C(g)$$

	$A(g)$	$2B(g)$	$C(g)$
at $t = 0$	p_0	—	—
$t = \infty$	0	$2p_0$	p_0

$p_\infty = 3p_0 = 240 \Rightarrow p_0 = 80$ mm Hg

$$kt = \ln\left[\frac{p_\infty - p_0}{p_\infty - p_t}\right]$$

$$\Rightarrow k \times 10 = \ln\left[\frac{240 - 80}{240 - 160}\right]$$

$$k = 0.0693 \text{ min}^{-1}$$

So, the rate constant of reaction is 0.0693 min^{-1}.

19. *(c)* Among the given statements, A, B and D are correct, while statement C is incorrect.

The correct form of statement C is :

1st order kinetics have half-life ($t_{1/2}$) independent of concentration. So, upon changing the concentration $t_{1/2}$ should not change for 1st order reaction.

20. *(b)* We know that

$$r = k[A]^1[B]^1$$

From the given data

$$0.1 = k[20]^1[0.5]^1 \quad \text{...(i)}$$
$$0.40 = k[x]^1[0.5]^1 \quad \text{...(ii)}$$
$$0.80 = k[40]^1[y]^1 \quad \text{...(iii)}$$

Dividing Eqs. (i) and (ii)

$$\frac{0.1}{0.40} = \frac{K[20]^1\,[0.5]^1}{K[x]^1\,[0.5]^1}$$

$$\frac{10}{40} = \frac{[20]^1}{[x]^1}$$

$\therefore \quad x = 80$

Putting the value of x in Eq. (ii), we get

$$0.40 = K\,[80]^1\,[0.5]^1$$

Dividing the above Eq. by Eq. (iii) we get

$$\frac{0.40 = K\,[80]^1\,[0.5]^1}{0.80 = K\,[40]^1\,[y]^1}$$

$$\frac{1}{2} = \frac{2\,[0.5]^1}{[y]^1} \Rightarrow y = 2$$

21. *(c)* It is a first order reaction because the half-life is independent of initial concentration of reactants.

For first order reaction,

$$k = \frac{2.303}{t}\log\frac{[A]_0}{[A]}$$

For 50% reaction,

$$k = \frac{2.303}{t}\log\frac{100}{(100-50)}$$

or, $$k = \frac{2.303}{200}\log 2$$

$$\Rightarrow \quad k = 0.0034545$$

For 80% reaction,

$$k = \frac{2.303}{t}\log\frac{100}{100-80}$$

or, $$0.0034545 = \frac{2.303}{t}\log\frac{100}{20}$$

or, $$t = 467\text{ s}$$

22. *(b)* From graph (A) it can be interpreted that the rate of the reaction is independent of time. Thus (A) is a zero order as its

$$\text{Rate} = K\,[\text{Reactant}]^0 = K.$$

From graph (B) it can be interpreted that the half-life is directly proportional to initial concentration of reactant. Thus, (B) is also a zero order. As half life of zero order is given by $t_{\frac{1}{2}} = \frac{[R_0]}{2K}$.

From graph (E) it can be interpreted that it is first order as the rate of reaction is increasing with increase in time.

$$\text{Rate} = K\,[\text{Reactant}]^1$$

23. *(c)* Let, the rate law is : $r \propto [A]^a[B]^b$

$$\frac{\text{Experiment II}}{\text{Experiment I}} = \left(\frac{0.1}{0.1}\right)^a \times \left(\frac{0.2}{0.1}\right)^b$$

$$\Rightarrow \quad \frac{2.4\times10^{-2}}{6\times10^{-3}} = 1\times 2^b$$

$$\Rightarrow \quad 2^b = 4 \Rightarrow b = 2$$

$$\frac{r_{\text{III}}}{r_{\text{I}}} = \left(\frac{0.2}{0.1}\right)^a \times \left(\frac{0.1}{0.1}\right)^2$$

$$\Rightarrow \quad \frac{1.2\times10^{-2}}{6\times10^{-3}} = 2^a \times 1$$

$$\Rightarrow \quad 2^a = 2 \Rightarrow a = 1$$

$$\Rightarrow \quad r_1 = k\times(0.1)^1\,(0.1)^2$$

$$\Rightarrow 6\times10^{-3} = k\times10^{-3}$$

$$\Rightarrow \quad k = 6\text{ L}^2\text{ mol}^{-2}\text{ min}^{-1}$$

Experiment IV : $r_{\text{II}} = 6\times(X)^1\times(0.2)^2$

$$\Rightarrow \quad 7.2\times10^{-2} = 6\times(X)\times(4\times10^{-2})$$

$$\Rightarrow \quad X = 0.3\text{ mol L}^{-1}$$

$\Rightarrow$ Experiment V: $r_{\text{V}} = 6\times(0.3)^1\times(Y)^2$

$$\Rightarrow \quad 2.88\times10^{-1} = 6\times0.3\times(Y)^2$$

$$\Rightarrow \quad Y = 0.4\text{ mol L}^{-1}$$

24. *(c)* For the reaction,

$$2A + B \longrightarrow \text{products.}$$

Let, the rate expression is $r \propto [A]^a[B]^b$

Let the initial rate of the reaction

$$r_1 = [A]^a[B]^b \ldots\text{(i)}$$

where a and b are the respective orders with respect to reactants A and B.

Rate of reaction when conc. of both A and B gets doubled,

$$r_2 = [2A]^a[2B]^b$$

$$\frac{r_2}{r_1} = \left(\frac{2A}{A}\right)^a\left(\frac{2B}{B}\right)^b$$

$$\Rightarrow \quad \frac{2.4}{0.3} = 2^a\times2^b \Rightarrow 2^3 = 2^{a+b}$$

$$\Rightarrow \quad 3 = a + b \ \ldots\text{(i)}$$

Rate of reaction when only the concentration of A gets doubled.

$$r_3 = K\,[2A]^a[B]^b$$

$$\frac{r_3}{r_1} = \left(\frac{2A}{A}\right)^a\left(\frac{B}{B}\right)^b \Rightarrow \frac{0.6}{0.3} = 2^a\times1$$

$$\Rightarrow \quad 2^1 = 2^a \Rightarrow a = 1 \qquad \ldots\text{(ii)}$$

$\therefore$ From Eq. (i), $1 + b = 3 \Rightarrow b = 2$

$\Rightarrow$ Order of the reaction

$(n) = a + b = 1 + 2 = 3$

$\Rightarrow$ Order of the reaction wrt $A = 1$

$\Rightarrow$ Order of the reaction wrt $B = 2$

25. *(b)* Let, the rate expression is $r \propto [A]^a\,[B]^b$.

From experiment I,

$$\frac{r_2}{r_1} = \left(\frac{0.1}{0.1}\right)^a \times \left(\frac{0.25}{0.20}\right)^b$$

$$\Rightarrow \quad \frac{6.93\times10^{-3}}{6.93\times10^{-3}} = 1\times\left(\frac{5}{4}\right)^b$$

$$\Rightarrow \quad 1 = \left(\frac{5}{4}\right)^b \Rightarrow \left(\frac{5}{4}\right)^0 = \left(\frac{5}{4}\right)^b \Rightarrow b = 0$$

From experiment II,

$$\frac{r_3}{r_1} = \left(\frac{0.2}{0.1}\right)^a \times \left(\frac{0.30}{0.20}\right)^b$$

$$\Rightarrow \quad \frac{1.386\times10^{-2}}{0.693\times10^{-2}} = (2)^a\times(1.5)^0$$

$$\Rightarrow \quad 2 = 2^a\times1 \Rightarrow 2^1 = 2^a \Rightarrow a = 1$$

$\Rightarrow$ So, $\quad r \propto [A]^1[B]^0 \Rightarrow r \propto [A]$

Order of the reaction $(n) = 1$

$\Rightarrow$ Now, for the 1st experiment,

$$r_1 = k\cdot[A]$$

$$\Rightarrow \quad k = \frac{r_1}{[A]} = \frac{6.93\times10^{-3}}{0.1}$$

$$= 6.93\times10^{-2}\text{ s}^{-1}$$

$$\Rightarrow \quad t_{50} = \frac{0.693}{k} = \frac{0.693}{6.93\times10^{-2}} = 10\text{ s}$$

26. *(43)* For specific time interval 0-5 sec, 5-10 sec and 10-15 sec, order comes to be zero, but graph is not straight line. Assuming 1st order kinetics,

$$k = \frac{1}{t}\ln\frac{A_0}{A_t} \Rightarrow k = \frac{1}{t}\ln\frac{50}{2.5}$$

$$\frac{1}{10}\ln 2 = \frac{1}{t}\ln 20$$

$$\Rightarrow \quad t = \frac{1.3010\times10}{0.3010} = 43.3\text{ sec} \approx 43\text{ sec}$$

27. *(2435)* Half-life is directly proportional to $[A]_0$

$\therefore t_{1/2} \propto [A]_0 \Rightarrow$ order = zero

$$t_{1/2}\frac{A_0}{2k} \Rightarrow \text{Slope} = \frac{1}{2k} = 76.92$$

$$k = \frac{1}{2\times76.92}$$

$$\Rightarrow [A]_{10} = -kt + A_0$$

$$= -\frac{1}{2\times76.92}\times10 + 2.5 = 2.435$$

$$= 2435\times10^{-3}\text{ mol/L}$$

28. *(897)*

$$2N_2O_5(g) \longrightarrow 2N_2O_4(g) + O_2(g)$$

	$2N_2O_5(g)$	$2N_2O_4(g)$	$O_2(g)$
p_i	0.6	0	0
p_t	$0.6-p$	p	$p/2$

Using $k = \frac{2.303}{t}\log\frac{p_i}{p_t}$

$$4.606\times10^{-2} = \frac{2.303}{100}\log\frac{0.6}{0.6-p}$$

$$10^2 = \frac{0.6}{0.6-p} \Rightarrow 10^4 = \frac{0.6}{0.6-p}$$

On solving, we get $p = 0.594$

$$p_{\text{total}} = 0.6 + \frac{p}{2}$$
$$= 0.6 + 0.297$$
$$p_{\text{Total}} = 897 \times 10^{-3}$$

So, the value of approx is 897.

29. *(1)* Given reaction,

$$A + B \longrightarrow C$$
$$A \xrightarrow{t = x} \frac{A}{2} \xrightarrow{t = 2x} \frac{A}{4}$$

Order w.r.t. $A = 1$

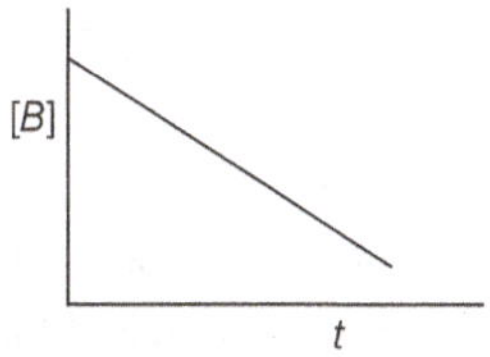

$$[B] = -kt + [B]_0$$

Order w.r.t. B = zero order

Overall order $= 1 + 0 = 1$

Hence, the overall order is 1.

30. *(17)*

$$A + B \rightarrow C \quad \ldots\text{(i)}$$
$$P \rightarrow Q \quad \ldots\text{(ii)}$$
$$t_{1/2} \propto \frac{1}{K}$$

We know that

$$\frac{(t_{1/2})_1}{(t_{1/2})_2} = \frac{K_2}{K_1} = \frac{5}{2}$$
$$\frac{t_{2/3}}{t_{4/5}} = \frac{K_2 \log 3}{K_1 \log 5}$$
$$= \left(\frac{5}{2}\right)\frac{0.477}{0.699} = 1.7$$
$$= 17 \times 10^{-1}$$

Comparing it with given equation the value is 17.

31. *(3)* Reaction rate,

$$k = \frac{1}{\text{time required}} \ln\left[\frac{\text{initial concentration}}{\text{final concentration}}\right]$$

According to question,

$$k = \frac{1}{t_{99.9\%}} \ln\left[\frac{100}{0.1}\right] = \frac{1}{t_{90\%}} \ln\left(\frac{100}{10}\right)$$
$$t_{99.9\%} = t_{90\%} \frac{\ln [10^3]}{\ln [10]}$$
$$t_{99.9\%} = 3 \times t_{90\%}$$

Hence, time required for 99.9% completion of first order reaction is three times the time required for 90%.

32. *(315)* The reaction given to us is

	$2A(g)$ +	$B(g)$ $\longrightarrow$	$C(g)$
r_1	1.5 atm	0.7 atm	
r_2	0.5 atm	0.2 atm	0.5 atm

We know that,

Reaction rate, $r = k\,[p_A]^2\,[p_B]$

$$r_1 = k\,[1.5]^2\,[0.7] \quad \ldots\text{(i)}$$
$$r_2 = k\,[0.5]^2\,[0.2] \quad \ldots\text{(ii)}$$

Divide equation (i) by (ii)

$$\frac{r_1}{r_2} = [3]^2\left[\frac{0.7}{0.2}\right] = 31.5 = 315 \times 10^{-1}$$

Hence, the value is 315.

33. *(3)* Reaction rate, $r = k\,[A]^x\,[B]^y$

for I, $r = 6 \times 10^{-3}$, $A = 0.1$, $B = 0.1$

$$6 \times 10^{-3} = k\,[0.1]^x\,[0.1]^y$$

Similarly for IV,

$$2.4 \times 10^{-2} = [0.4]^x\,[0.1]^y$$

Now divide IV by I

$$4 = 4^x \Rightarrow x = 1$$

For III,

$$2.88 \times 10^{-1} = k\,[0.3]^x\,[0.4]^y$$

For II,

$$7.2 \times 10^{-2} = k\,[0.3]^x\,[0.2]^y$$

III/II, $4 = 2^y \Rightarrow y = 2$

Overall order $= x + y = 1 + 2 = 3$

34. *(50)* Given $[A_0] = 1$

$$[A] = 0.1$$
$$K = 4.6 \times 10^{-2}\ \text{s}^{-1}$$

Since the given reaction is of first order. Hence, using

$$t = \frac{2.303}{K} \log\frac{[A_0]}{[A]}$$

On substituting the values

$$t = \frac{2.303}{4.6 \times 10^{-2}} \log\frac{1}{0.1} \Rightarrow t = 50 \text{ sec}$$

35. *(17190)*

$$\lambda t = \ln \frac{(^{14}\text{C}/^{12}\text{C})_{\text{atmosphere}}}{(^{14}\text{C}/^{12}\text{C})_{\text{wood sample}}} \quad \ldots\text{(i)}$$

According to data given in question,

$$\frac{(^{14}\text{C}/^{12}\text{C})_{\text{wood}}}{(^{14}\text{C}/^{12}\text{C})_{\text{atmosphere}}} = \frac{1}{8}$$

Then, Eq. (i) will become,

$$\lambda t = \ln 8 \Rightarrow \frac{\ln 2}{t_{1/2}} t = \ln 8$$
$$t = 3 \times t_{1/2} = 3 \times 5730$$
$$= 17190 \text{ years}$$

Hence, the age of wood sample is 17190 years.

36. *(399)* Rate of reaction $r = k[A]$

So, order of reaction = 1

$$t_{1/2} = 120 \text{ min}$$

For 90% completion of reaction

$$k = \frac{2.303}{t} \log\left(\frac{a}{a - x}\right)$$
$$\frac{0.693}{t_{1/2}} = \frac{2.303}{t} \log\left(\frac{100}{10}\right)$$
$$t = 398.7 \approx 399 = 399 \text{ min}$$

Hence, 399 min is required for 90% decomposition.

37. *(24)* Rate of first order reaction is given by

$$R = k[A]_0 e^{-kt} \quad \ldots\text{(i)}$$
$$= 0.04 \text{ mol L}^{-1}\text{s}^{-1}\ 10 \text{ minutes}$$

substitute in Eq. (i)

$$0.04 = k[A]_0 e^{-k \times 10 \times 60} \quad \ldots\text{(ii)}$$
$$R = 0.03 \text{ mol L}^{-1}\text{s}^{-1} \text{ for } 20 \text{ min}$$
$$0.03 = k[A]_0 e^{-k \times 20 \times 60} \quad \ldots\text{(iii)}$$

Divide Eq. (ii) by (iii),

$$\frac{4}{3} = e^{600(k)(2-1)}$$
$$\frac{4}{3} = e^{600k} \Rightarrow \ln\frac{4}{3} = 600k$$
$$\ln\frac{4}{3} = 600 \times \frac{\ln 2}{t_{1/2}}$$
$$\Rightarrow t_{1/2} = \frac{600 \times \log 2}{\log 4 - \log 3} \text{ s}$$
$$= \frac{10 \times 0.3010}{0.6020 - 0.477} \text{ min}$$
$$= 24.08 \text{ min} \approx 24 \text{ min}$$

Hence, half-life of reaction is 24 min.

38. *(63)* Given, half-life of Br = 36 hrs

$$36 = \frac{0.693}{k} \Rightarrow k = 0.01925 \text{ hr}^{-1}$$

Ist order kinetic equation

$$t = \frac{2.303}{k} \log\frac{a}{a - x}$$
$$\log\frac{a}{a - x} = \frac{t \times k}{2.303} \quad (t = 1 \text{ day} = 24 \text{ hrs})$$
$$\log\frac{a}{a - x} = \frac{24 \times 0.01925}{2.303}$$
$$\Rightarrow \frac{a}{a - x} = 1.587$$

If $a = 1 \Rightarrow \frac{1}{1 - x} = 1.587$

$1 - x = 0.6301 = 63 \times 10^{-2}$

(Fractional remain after 1 day)

Hence, missing term is 63.

Time Saver Tech

n = Number of half-life $= \frac{24}{36} = \frac{2}{3}$

Amount left $= \frac{N_0}{2^{2/3}} = 63 \times 10^{-2} N_0$

So, the fraction left $= 63 \times 10^{-2}$

Thus, the missing term is 63.

39. *(10)*

idea To solve this problem first put the given values in the formula, $t_{1/2} = \frac{2.303}{k} \log 2$ and $t_{99.9\%} = \frac{2.303}{k} \log\left(\frac{a}{a-x}\right)$ then compare these equation to get the required relation between $t_{1/2}$ and $t_{99.2}$.

$t_{1/2}$ is given by $= \frac{2.303}{k} \log 2$

$$t_{99.9\%} = \frac{2.303}{k} \log\left(\frac{a}{a-x}\right)$$

$$\frac{t_{99.9\%}}{t_{1/2}} = \frac{\log\left(\frac{a}{a-x}\right)}{\log 2}$$

$$\frac{t_{99.9}}{t_{1/2}} = \frac{\log\left(\frac{100}{100-99.9}\right)}{\log 2}$$

$$\frac{t_{99.9}}{t_{1/2}} = \frac{\log 10^3}{\log 2} = \frac{3}{0.3}$$

$$t_{99.9} = 10 t_{1/2}$$

40. *(2)* The reaction is

$$2HI(g) \longrightarrow H_2(g) + I_2(g)$$

The rate of reaction R is given by

$$R = k[HI]^n$$

By using data of (1) and (2) we get,

$$\frac{3 \times 10^{-3}}{7.5 \times 10^{-4}} = \left(\frac{0.01}{0.005}\right)^n$$

$$n = 2$$

Hence, the order of reaction is 2.

41. *(2200)* The value of x is 3.

$$A(g) \longrightarrow 2B(g) + C(g)$$

Initially (t = 0) 800 – –

When (t = 10 min) $(800 - p)$ $2p$ p

Total pressure $(p_t) = 800 - p + 2p + p$

$= 800 + 2p = 1600$

or $[p = 400 \text{ mm}]$ ...(i)

Since pressure reduce to half of its initial pressure. Hence, 10 min will be the half life.

So, $[t_{1/2} = 10 \text{ min}]$

So,

$$\underbrace{p \xrightarrow{10\text{ min}} p/2\, p \xrightarrow{10\text{ min}} p/4\, p \xrightarrow{10\text{ min}} p/8}_{\text{Total} = 30\text{ min}}$$

Hence, after 30 min pressure will left or $= (p - p/8) = (7p/8)$

$$A \longrightarrow 2B + C$$

$T = 0$ 800 – –

$T = 10$ min $(800 - p)$ $2p$ p

$T = 30$ min $\left(800 - \frac{7p}{8}\right)$ $(14p/8)$ $(7p/8)$

$$p_T = p/8 + 1 + 14p/8 + 7p/8$$

$$p_T = 22p/8 = \frac{22 \times 800}{8} = 2200 \text{ mm Hg}$$

42. *(3)* $t_{87.5} = \frac{1}{K} \ln\left(\frac{100}{100 - 87.5}\right)$

$$= \frac{1}{K} \ln(8) = \frac{3 \ln 2}{K} \quad \text{...(i)}$$

$$t_{50\%} = \frac{1}{K} \ln\left(\frac{100}{100-50}\right) = \frac{1}{K} \ln 2 \quad \text{...(ii)}$$

$$\therefore \frac{t_{87.5}}{t_{50}} = \frac{\frac{3 \ln 2}{K}}{\frac{\ln 2}{K}} = [3]$$

or $[t_{87.5} = 3 \times t_{50}]$

The value of x is 3.

43. *(3)* Slowest step is the rate determining step.

$$NOBr_2 + NO \longrightarrow 2NOBr$$

$$r = K\, [NOBr_2]\, [NO] \quad \text{...(i)}$$

$$NO + Br_2 \rightleftharpoons NOBr_2$$

$$K_{eq} = \frac{[NOBr_2]}{[NO]\, [Br_2]}$$

$$[NOBr_2] = K_{eq} [NO][Br_2] \quad \text{...(ii)}$$

Subtituting Eqs. (ii) in (i)

$$r = K \cdot K_{eq}\, [NO]\, [Br_2]\, [NO]$$

$$r = K^1 [NO]^2 [Br_2]$$

Thus, the overall order = 3

44. *(1)*

(A) For zero order, $t_{1/2} = \frac{[A]_0}{2K}$ as concentration decreases, half-life decreases. So, statement 'A' is correct.

(B) If order w.r.t that reactant is zero then it will not affect rate of reaction. So statement 'B' is correct.

(C) Order can be fractional but molecularity cannot be. So statement 'C' is incorrect.

(D) For zero order reaction unit is $mol\,L^{-1}s^{-1}$ and for second order reaction unit is $mol^{-1}Ls^{-1}$. So statement 'D' is correct.

45. *(2)* Given, $(t_{1/2})_A = 12$ minute, $(t_{1/2})_B = 3$ minute

$$\frac{1}{(t_{1/2})_C} = \frac{1}{(t_{1/2})_A} + \frac{1}{(t_{1/2})_B}$$

$$= \frac{1}{12} + \frac{1}{3} = \frac{1+4}{12} = \frac{5}{12}$$

$$\Rightarrow (t_{1/2})_C = \frac{12}{5} \text{ minutes} = 2.4 \text{ minutes}$$

46. *(75)* Let initial conc. of $A = 1$ mol

For zero order reaction half-life period,

$$t_{1/2} = \frac{a}{2k}$$

$$\Rightarrow \quad k = \frac{a}{2t_{1/2}} = \frac{a}{2 \times 50} = \frac{1}{2 \times 50}$$

For 75% completion

$$\left(a - \frac{a}{4}\right) = kt \Rightarrow \frac{4a - a}{4} = kt \Rightarrow \frac{3a}{4} = kt$$

$$t = \frac{3}{4}\frac{a}{k} = \frac{3}{4} \times \frac{100}{1} = 75 \text{ min}$$

47. *(15)* Given that $[B] = 4[A]$

For A decay, $[A] \xrightarrow{15\text{ min}} \frac{[A]}{2}$

For B decay, $[B]$

$$4[A] \xrightarrow{5\text{ min}} 2[A] \xrightarrow{5\text{ min}} [A] \xrightarrow{5\text{ min}} \frac{[A]}{2}$$

So, after $(3 \times 5$ min$)$ concentration of A and B will be same.

48. *(17)* From first order

$$C = \frac{C_0}{2^n} = \frac{C_0}{32} = \frac{C_0}{2^5} \Rightarrow n = 5$$

$$t = 5t_{1/2} = \frac{5 \times 0.693}{20}$$

$$= \frac{0.693}{4} = 0.17325 \text{ min}$$

$$= 17.325 \times 10^{-2} \text{ min}$$

49. *(1350)* $c_{t(60\%)} = 100 - 60 = 40$

$c_{t(90\%)} = 100 - 90 = 10$

From first order reaction,

$$t = \frac{1}{k} \ln \frac{c_0}{c_t}$$

So, in cases of 60% and 90% decomposition,

$$t_{60\%} = 540 = \frac{1}{k} \ln \frac{100}{40} \quad ...(i)$$

$$t_{90\%} = \frac{1}{k} \ln \frac{100}{10} \quad ...(ii)$$

From Eq. (i) and (ii), we get

$$\frac{540}{t_{90\%}} = \frac{\ln \frac{100}{40}}{\ln \frac{100}{10}}$$

On substituting the values, of ln gives in the question we obtained

$$\frac{540}{t_{90\%}} = \frac{2-1.6}{2-1} = 0.4$$

$$t_{90\%} = \frac{540}{0.4} = 1350 \text{ seconds}$$

50. *(623)* $A + B \longrightarrow P$

$t = 0 \quad 7\text{g}$

$t = t \quad 2\text{g}$

For Ist order

$$t = \frac{2.303}{k} \log \frac{[A]_0}{[A]_t}$$

$$= \frac{2.303}{2.011 \times 10^{-3}} \log \frac{7}{2}$$

$$= \frac{2.303 \times 10^3}{2.011}(0.845 - 0.301)$$

$$= 622.99 \approx 623 \text{ sec}$$

51. *(2)*

> **idea** To approach this type of questions, simply solve the unit of rate constant by using the general expression $[\text{mol L}^{-1}]^{1-n}$ sec^{-1} for determining the order of a reaction.

The unit of rate constant can be the expressed as

$$[\text{mol L}^{-1}]^{1-n} \text{s}^{-1}$$

On putting ($n = 2$), unit of K becomes $(\text{mol L}^{-1})^{1-2}\text{s}^{-1}$ or $(\text{mol L}^{-1})^{-1}\text{s}^{-1}$

or $\quad [\text{mol}^{-1}\text{L s}^{-1}]$

Hence, it is a second order reaction.

52. *(2)* Region (I) is the straight line for first order reaction and Region (III) is the constant line for zero order reaction, since for zero order, rate do not depend on the time but on the condition. (catalyst, $h\nu$) etc.

First order reaction

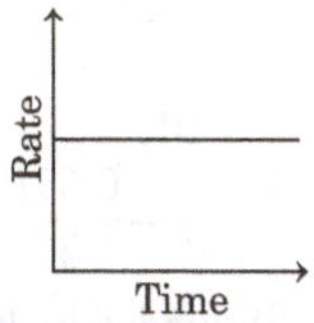

Zero order reaction

Hence, statement (C) is correct.

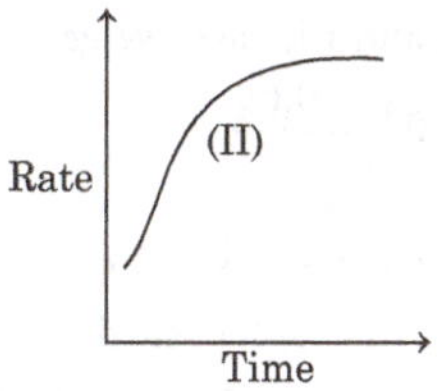

Since region (II) is asymptotic region it shows the running time, hence order cannot be determined.

Thus, statement A and C are correct.

53. *(2)* Time required for 10% completion,

$$t_{10\%} = \frac{1}{\text{Rate constant }(k)} \ln\left(\frac{a}{a-x}\right)$$

$$\Rightarrow \quad t_{10\%} = \frac{1}{k} \ln\left(\frac{100}{100-10}\right)$$

$$t_{10\%} = \frac{2.303}{k} \ln\left(\frac{100}{90}\right)$$

$$t_{10\%} = \frac{2.303}{k} \times (0.04)$$

Similarly, $t_{90\%} = \frac{1}{k} \ln\left(\frac{a}{a-x}\right)$

$$t_{90\%} = \frac{1}{k} \ln\left(\frac{100}{100-90}\right) = \frac{1}{k} \ln\left(\frac{100}{10}\right)$$

$$t_{90\%} = \frac{2.303}{k}$$

Now, $\frac{t_{90\%}}{t_{10\%}} = \frac{\frac{2.303}{k}}{\frac{2.303 \times 0.04}{k}} = \frac{1}{0.04} = 25$

So, statement C is correct.

$$e^{kt} = \frac{a}{a-x}$$

By reversing the above equation

$$\frac{a-x}{a} = e^{-kt} \text{ or } 1 - \frac{x}{a} = e^{-kt}$$

$$\Rightarrow \quad x = a(1 - e^{-kt})$$

$$\Rightarrow \quad d = \frac{x}{a} = (1 - e^{-kt})$$

Where, d is degree of dissociation.

So, statement D is also correct.

54. *(60)* Given, $t_{1/2} = 30$ min

For first order $K = \frac{0.693}{t_{1/2}}$

for 75% completion

$$t = \frac{2.303}{K} \log \frac{100}{25}$$

$$= \frac{2.303}{K} \log 2^2 = 2t_{1/2}$$

$$\Rightarrow t_{75\%} = 2t_{1/2} \text{ (for first order reaction)}$$

$$t_{75\%} = 2 \times 30 = 60 \text{ min}$$

55. *(1)* Given, $t_{1/2} = 30$ years

Initial amount of elements x, $a = 1$ μg

Now, for first order reaction

$$t_{1/2} = \frac{0.693}{k}$$

$$\Rightarrow \quad k = \frac{0.693}{t_{1/2}} = \frac{0.693}{30} = 0.0231 \text{ years}^{-1}$$

$$t = \frac{2.303}{k} \log\left(\frac{a}{a-x}\right)$$

$$\Rightarrow 100 = \frac{2.303}{0.0231} \log\left(\frac{1}{w}\right)$$

[let weight remaining be 'w']

$$\Rightarrow \quad \log \frac{1}{w} = 1 \Rightarrow \log \frac{1}{w} = \log 10$$

$$\Rightarrow \quad \frac{1}{w} = 10$$

$$\Rightarrow \quad w = 0.1 \text{ μg} = 1 \times 10^{-1} \text{ μg}.$$

56. *(40)* Rate $= k[X][Y]^0 = k[X]$

Using data of experiment I and II,

$$\frac{\text{Rate}_\text{I}}{\text{Rate}_\text{II}} = \frac{k[X_\text{I}]}{k[X_\text{II}]}$$

$$\frac{2 \times 10^{-3}}{4 \times 10^{-3}} = \frac{0.1}{L}$$

$$\Rightarrow \quad L = 0.1 \times 2 = 0.2$$

Using experiment III and IV,

$$\frac{\text{Rate}_\text{III}}{\text{Rate}_\text{IV}} = \frac{K[X_\text{III}]}{K[X_\text{IV}]}$$

$$\frac{M \times 10^{-3}}{2 \times 10^{-3}} = \frac{0.4}{0.1}$$

$$M = 4 \times 2 = 8$$

So, $\quad M : L = \frac{8}{0.2} = \frac{80}{2} = 40.$

57. *(100)* For a first order reaction,

$$\text{(half-life) } t_{1/2} = \frac{0.693}{k}$$

$$= \frac{0.693}{0.3010} \quad [\text{Given, } t_{1/2} = 0.3010]$$

$$= 2.303$$

The rate law for 1st order reaction is given as

$$k = \frac{2.303}{t} \log \frac{[A_0]}{[A_t]}$$

$$\therefore 2.303 = \frac{2.303}{2} \log \frac{[A_0]}{[A_t]}$$

$$\Rightarrow \quad 2 = \log \frac{[A_0]}{[A_t]}$$

$\Rightarrow \quad \dfrac{[A_0]}{[A_t]} = \text{antilog } (2)$

$\Rightarrow \quad \dfrac{[A_0]}{[A_t]} = 100$

58. *(165)* Since after 70 minutes, the concentration of reactant $[A]$ is halved.

So, $t_{1/2} = 70$ min

Half-life of first order reaction is given as

$$t_{1/2} = \frac{0.6932}{k}$$

[k = rate constant for 1st order reaction]

$$k = \frac{0.6932}{t_{1/2}} = \frac{0.6932}{70}$$

$$= 0.0099 \text{ min}^{-1}$$

$$= \frac{0.0099}{60}\text{s}^{-1} = 165 \times 10^{-6}\text{s}^{-1}$$

The value of x is 165.

59. *(2)* Rate of reaction $(r) = k[\text{NO}]^x[\text{H}_2]^y$

From I data,

$$0.135 = k[40]^x[65.6]^y \quad \ldots\text{(i)}$$

From II data,

$$0.033 = k[20.1]^x[65.6]^y \quad \ldots\text{(ii)}$$

On dividing Eq. (i) by Eq. (ii),

$$\frac{0.135}{0.033} = \left[\frac{40}{20.1}\right]^x$$

$$4 = (2)^x \Rightarrow x = 2$$

$\therefore$ Order of reaction with respect to NO is 2.

60. *(2)* The half-life for nth order

$$t_{1/2} \propto \frac{1}{(a_0)^{n-1}}$$

when $t_{1/2}$ is 100 s then the equation is

$$100 \propto \frac{1}{(0.5)^{n-1}} \quad \ldots\text{(i)}$$

when $t_{1/2}$ is 50 s then the equation is

$$50 \propto \frac{1}{(1)^{n-1}} \quad \ldots\text{(ii)}$$

Dividing Eq. (i) by Eq. (ii), we get,

$$\frac{100}{50} = \left(\frac{1}{0.5}\right)^{n-1} \Rightarrow 2 = (2)^{n-1}$$

$$\therefore \quad n - 1 = 1 \Rightarrow n = 1 + 1 = 2$$

61. *(2)* For first order reaction,

$$\ln A = \ln A_0 - kT$$

As slope $-k = -3.465 \times 10^4 \quad \ldots\text{(i)}$

Also, $\quad k = \dfrac{0.69}{t_{1/2}} \quad \ldots\text{(ii)}$

From (i) and (ii) case, we get

$$3.465 \times 10^4 = \frac{0.69}{t_{1/2}}$$

$$\Rightarrow \quad t_{1/2} = 2 \times 10^{-5} \text{ s}.$$

62. *(16)* For first order reaction,

$$t = \frac{2.303}{k}\log\frac{A_0}{A_0 - A_t}$$

$$t_{67\%} = \frac{2.303}{k}\log\left(\frac{100}{33}\right) \quad \ldots\text{(i)}$$

$$t_{50\%} = \frac{0.693}{k} \quad \ldots\text{(ii)}$$

Dividing Eqs. (i) by (ii),

$$\frac{t_{67\%}}{t_{50\%}} = \frac{2.303 \times k}{k \times 2.303 \times 0.3} \times \log\frac{100}{33}$$

$$= \frac{0.4814}{0.3} = 1.604$$

So, $\quad t_{67\%} = 16.04 \times 10^{-1} t_{50\%}$

$$= x \times 10^{-1} t_{50\%}$$

Hence, value of x is 16.

63. *(200)* Since, half-lives are independent of initial concentration, that means reaction is Ist order.

So, $\quad t_{1/2} = \dfrac{0.6932}{k}$

$$k_A = \frac{0.6932}{100}, \; k_B = \frac{0.6932}{50}$$

Rate law for Ist order reaction is

$C_t = C_0 e^{-kt}$

For A, $\quad A_t = A_0 e^{\left(\frac{-0.6932}{100} \times t\right)}$

For B, $\quad B_t = B_0 e^{\left(\frac{-0.6932}{50} \times t\right)}$

$$A_0 = B_0 \text{ and } A_t = 4B_t$$

$$e^{\left(\frac{-0.6932}{100} \times t\right)} = 4e^{\left(\frac{-0.6932}{50} \times t\right)}$$

Taking ln both sides,

$$\left(-\frac{0.6932}{100} \times t\right) = \ln 4 + \left(\frac{-0.6932}{50} \times t\right)$$

$$\frac{-0.6932\,t}{100} + \frac{0.6932\,t}{50} = 2 \ln 2$$

$$+\,0.6932t\left[\frac{-1}{100} + \frac{1}{50}\right] = 2 \times 0.6932$$

$$= 2 \times 100 = 200 \text{ s}$$

64. *(0)* Using the equation,

$$\frac{(t_{1/2})_1}{(t_{1/2})_2} = \left(\frac{p_1}{p_2}\right)^{1-n}$$

$$\frac{340}{170} = \left(\frac{55.5}{27.8}\right)^{1-n}$$

$$(2)^1 = (2)^{1-n}$$

So, $\quad 1 = 1 - n$

$\Rightarrow \quad n = 1 - 1 = 0$

65. *(2)* For 50% completion,

$$k = \frac{2.303}{t_{50\%}}\log\frac{100}{100-50} \quad \ldots\text{(i)}$$

For 75% completion,

$$k = \frac{2.303}{t_{75\%}}\log\frac{100}{100-75} \quad \ldots\text{(ii)}$$

From Eqs. (i) and (ii),

$$\frac{2.303}{t_{50\%}}\log\frac{100}{50} = \frac{2.303}{t_{75\%}}\log\frac{100}{25}$$

$$\frac{0.3010}{t_{50\%}} = \frac{0.6020}{t_{75\%}}$$

$$t_{75\%} = 2t_{50\%}$$

$$\therefore \quad \frac{t_{75\%}}{t_{50\%}} = \frac{2}{1}$$

66. *(1)* Rate of the reaction

$$= K[\text{NO}]^x\,[\text{H}_2]^y$$

where K is rate constant and x, y are order of reaction w.r.t. NO and H_2 respectively.

From observations A and B

$$7 \times 10^{-9} = K(8 \times 10^{-5})^x(8 \times 10^{-5})^y \ldots\text{(i)}$$

and

$$2.1 \times 10^{-8} = K(24 \times 10^{-5})^x\,(8 \times 10^{-5})^y \quad \ldots\text{(ii)}$$

Dividing Eqs. (i) by (ii)

$$\frac{7 \times 10^{-9}}{2.1 \times 10^{-8}} = \frac{K(8 \times 10^{-5})^x\,(8 \times 10^{-5})^y}{K(24 \times 10^{-5})^x\,(8 \times 10^{-5})^y}$$

$$\frac{1}{3} = \left(\frac{1}{3}\right)^x \Rightarrow x = 1$$

$\therefore$ Order of reaction w.r.t. NO is 1.

67. *(656)* $\quad A \longrightarrow 2B$

	A	$2B$
Initial	1 mol	0
Final	$1 - x$	$2x$

Given, $\quad 2x = 0.2 \Rightarrow x = 0.1$

$\therefore$ Final $\quad 0.9 \quad 0.2$

Now, according to first order kinetics

$$t = \frac{2.303}{k}\log_{10}\frac{[A_0]}{[A_t]} \quad \ldots\text{(i)}$$

$$t_{1/2} = \frac{0.691}{k} \quad \ldots\text{(ii)}$$

Eq. (i) divided by Eq. (ii).

$$\frac{t}{t_{1/2}} = \frac{2.303}{0.691}\log_{10}\frac{(1)}{(0.9)} = \frac{1}{0.3}\log_{10}\left(\frac{10}{9}\right)$$

$$\frac{t}{t_{1/2}} = \frac{0.0457}{0.3}$$

$$t_{1/2} = \frac{3\times t}{0.457} = \frac{3\times 100}{0.457} \approx 656$$

68. *(106)* Unit of rate constant is $\min^{-1}$, so it must be a first order reaction.

For first order reaction,

$$k\times t = 2.303 \log \frac{A_0}{A_t}$$

Where, $A_0 = 100$, $A_t = 90$ in 1 min

So, $k\times 1 = 2.303\times \log \frac{100}{90}$

$$= 2.303\times (\log 10 - 2\log 3)$$
$$= 2.303\times (1 - 2\times 0.477)$$
$$= 0.10593$$
$$= 105.93\times 10^{-3}\ \min^{-1}$$
$$\approx 106\times 10^{-3}\ \min^{-1}$$

Hence, the rate constant for viral inactivation is $106\times 10^{-3}\ \min^{-1}$.

69. *(10)* The rate constant for first order reaction is given by

$$k = \frac{1}{t}\ln\left(\frac{C_0}{C_t}\right)$$

As $k = \frac{\ln 2}{t_{1/2}}$

$t_{1/2} = 1\,\text{min}$ [Given]

For 99.9% completion the rate constant will be

$$k = \frac{1}{t}\ln\frac{100}{(100-99.9)}$$

$$\therefore \quad \frac{\ln 2}{1} = \frac{1}{t}\ln\left(\frac{100}{0.1}\right)$$

$$\Rightarrow \quad t = \frac{\ln 1000}{\ln 2} \Rightarrow t = \frac{3\times 2.3}{0.69} = 10$$

70. *(3)* General form of rate law is given by $r = k[\text{NO}]^x[\text{Cl}_2]^y$

Using the values of concentrations and rate constants from all 3 experiments

$(r_1)\ 0.18 = k[0.10]^x[0.10]^y$ ...(i)

$(r_2)\ 0.35 = k[0.10]^x[0.20]^y$ ...(ii)

$(r_3)\ 1.40 = k[0.20]^x[0.20]^y$ ...(iii)

Dividing Eq. (i) and (ii),

$$\frac{0.18}{0.35} = \frac{[0.10]^x}{[0.10]^x}\ \frac{[0.10]^y}{[0.20]^y}$$

$$\Rightarrow \quad \frac{0.18}{0.35} = \left(\frac{1}{2}\right)^y \Rightarrow y = 1$$

So, order w.r.t $[\text{Cl}_2] = 1$

Dividing Eq. (ii) and (iii) we get

$$\frac{0.35}{1.40} = \frac{[0.10]^x[0.20]^y}{[0.20]^x[0.20]^y}$$

$$\Rightarrow \quad \frac{0.35}{1.40} = \left[\frac{1}{2}\right]^x \Rightarrow x = 2$$

Order w.r.t. $[\text{NO}] = 2$

$$\therefore \quad r = k[\text{NO}]^2[\text{Cl}_2]^1$$

71. *(2)* For first order reaction,

$$k\times t = \ln\frac{A_0}{A_t} \quad \text{...(i)}$$

$$k\times 570 = \ln\frac{100}{32}$$

$$k\times 570 = 2.303\ (\log 10^2 - \log 2^5)$$

$$\Rightarrow \quad k = \frac{2.303}{570}(2 - 5\log 2)$$

$$= \frac{2.303}{570}\times 0.5$$

$$k = 2\times 10^{-3}\ \text{s}^{-1}$$

72. *(108)* Given, $(t_{1/2})_1 = 54$ min

$(t_{1/2})_2 = 18$ min

A	B
$t=0$, 'x' M	$t=0$, 'x' M

For Ist order reaction, $[A_t] = \frac{A_0}{(2)^n}$

n = number of half-lives

According to question,

$$[A_t] = 16\times[B_t]$$

$$\frac{x}{(2)^{n_1}} = \frac{x}{(2)^{n_2}}\times 16$$

$$(2)^{n_2} = (2)^{n_1}\times(2)^4$$

$$\Rightarrow \quad n_2 = n_1 + 4$$

$$\frac{t}{(t_{1/2})_2} = \frac{t}{(t_{1/2})_1} + 4$$

$$\Rightarrow t\left(\frac{1}{18} - \frac{1}{54}\right) = 4$$

$$\Rightarrow \quad t = \frac{4\times 18\times 54}{36} = 108\ \text{min}$$

73. *(81)* Given,

$$\underset{\text{Sucrose}}{C_{12}H_{22}O_{11}} + H_2O \xrightarrow{\text{1 st order}} \underset{\text{Glucose}}{C_6H_{12}O_6} + \underset{\text{Fructose}}{C_6H_{12}O_6}$$

$$t_{1/2} = \frac{10}{3}\ \text{h}$$

At $t = 0$, $a = [A]_0$ (initial conc.)

$t = 9$ h $a - x = [A]_t$ [conc. at time t]

On using 1st order equation,

$$k = \frac{2.303}{t}\log\frac{[A]_0}{[A]_t}$$

$$\Rightarrow \quad \frac{k\times t}{2.303} = \log\frac{[A]_0}{[A]_t}$$

$$\frac{\ln 2\times 9}{10/3\times 2.303} = \log\left(\frac{1}{f}\right)\left(\because k = \frac{\ln 2}{t_{1/2}}\right)$$

$$\Rightarrow \quad \log\left(\frac{1}{f}\right) = 0.8124$$

$$\log\left(\frac{1}{f}\right) = 81.24\times 10^{-2}$$

$$x = 81.24$$

or $x \approx 81$

> **Concept Enhancer**
> The sucrose hydrolysis is an example of pseudo first order reaction. These reactions are not Ist order reaction but under certain conditions becomes Ist order.

74. *(26)* Integrated rate law for 1st order reaction is

$$kt = \ln\frac{[A]_0}{[A]_t} \quad \text{...(i)}$$

Given, $k = 3.3\times 10^{-4}\ \text{s}^{-1}$

$[A]_0 = 100$

$\Rightarrow \quad [A]_t = 100 - 40 = 60$

Put values in Eq. (i), we get

$$3.3\times 10^{-4}\,\text{s}^{-1}\times t = \ln\frac{100}{60}$$

$$t = 1547.95\ \text{s}$$
$$= 25.79\ \text{min}$$

$$\left(1\ \text{min} = 60\ \text{s or } 1\ \text{s} = \frac{1}{60}\ \text{min}\right)$$

$$= 26\ \text{minutes}$$

75. *(60)* For first order reaction,

$$t = \frac{2.303}{k}\log\left(\frac{a_0}{a}\right)$$

For 75%,

$$t_{75\%} = \frac{2.303}{k}\log\left(\frac{100}{25}\right) \quad \text{...(i)}$$

For 60%,

$$t_{60\%} = \frac{2.303}{k}\log\left(\frac{100}{40}\right) \quad \text{...(ii)}$$

On dividing Eqs. (ii) by (i),

$$\frac{t_{60\%}}{t_{75\%}} = \frac{\frac{2.303}{k}\log\left(\frac{100}{40}\right)}{\frac{2.303}{k}\log\left(\frac{100}{25}\right)}$$

$$t_{60\%} = t_{75\%} \times \left(\frac{0.3979}{0.602}\right)$$

$$\Rightarrow \quad t_{60\%} = 90 \times \frac{0.3979}{0.602}$$

$$t_{60\%} = 59.48 \text{ minute} \approx 60 \text{ min}$$

76. *(23.03)* All the radioactive reactions follow Ist order kinetics. So, time taken for decay from N_0 to N_t is (N = number of nuclei)

$$t = \frac{1}{\lambda}\ln\frac{N_0}{N_t}$$

$$t = \frac{1}{\lambda} \times 2.303 \log\frac{N_0}{N_t} \quad \text{...(i)}$$

Also, we know λ = (decay constant)

$$= \frac{0.693}{t_{1/2}}$$

where, $t_{1/2} = 6.93$ yr (given)

Also, we know 90% nuclei are decayed

$$\therefore \quad \frac{N_0}{N_t} = \frac{100}{10} = 10$$

Put the values of λ and $\frac{N_0}{N_t}$ in Eq. (i), we get

$$\therefore t = \frac{6.93}{0.693} \times 2.303 \times \log 10 = 23.03 \text{ (yr)}$$

Topic 3

01. *(a)* For the given graph, activation energy of forward reaction is $E_1 + E_2$. Also energy of product > Energy of reactant.

So, order of stability is reactant > product.

02. *(d)* A catalyst provides an alternate pathway or reaction mechanism by reducing the activation energy between reactants and products and hence lowering the potential energy barrier. Hence, the correct reaction profile diagram is as follows.

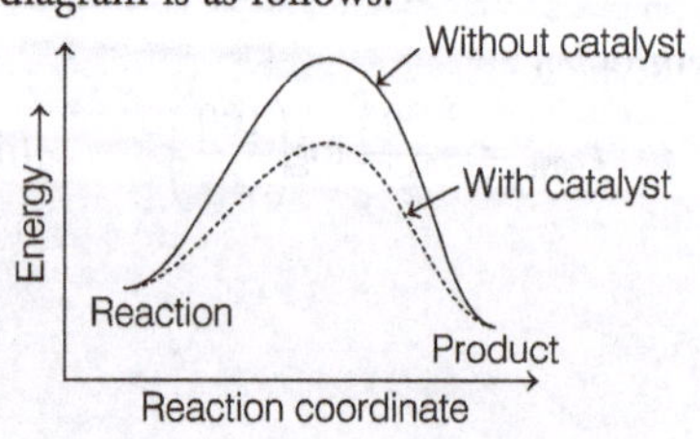

03. *(c)* According to Arrhenius equation,

$$k = Ae^{-E_a/RT}$$

As the temperature increases, rate constant k also increases which will generally speed up the process (increase the rate). So, graph will exponentially increases as in option (c).

04. *(20)* The resultant constant is given by

$$k = \sqrt{\frac{k_1 k_3}{k_2}}$$

$$Ae^{-E_a/RT} = \sqrt{\frac{A_1 e^{-E_{a_1}/RT} \times A_3 e^{-E_{a_3}/RT}}{A_2 e^{-E_{a_2}/RT}}}$$

By comparing exponential term

$$\frac{E_a}{RT} = \frac{1}{2} \times \left[\frac{E_{a_1}}{RT} + \frac{E_{a_3}}{RT} - \frac{E_{a_2}}{RT}\right]$$

$$E_a = \frac{(E_{a_1} + E_{a_3} - E_{a_2})}{2}$$

$$= \frac{(60 + 10 - 30)}{2} = 20 \text{ kJ/mol}$$

05. *(69)* Given, $A = 10^{20}$

$E_a = 191.48$ kJ/mol

$T = 1000$ K

$R = 8.314 \text{ JK}^{-1} \text{ mol}^{-1}$

$t_{1/2} = ?$

We know that, $t_{1/2} = \frac{0.693}{k}$

and k can be calculated by using,

$$= Ae^{\frac{-E_a}{RT}}$$

$$= 10^{20} \times e^{-\frac{191.48 \times 10^3}{8.314 \times 1000}}$$

$$= 10^{20} \times e^{-23.031}$$

$$= \frac{10^{20}}{10^{10}} = 10^{10} s^{-1}$$

Now, substitute this value in half-life equation

$$t_{1/2} = \frac{0.693}{10^{10}} = 6.93 \times 10^{-11} \text{ s}$$

or $\quad t_{1/2} = 693 \times 10^{-12}$ s

So, time required for 50% molecule of A to become molecule B is 69 picoseconds.

06. *(c)*

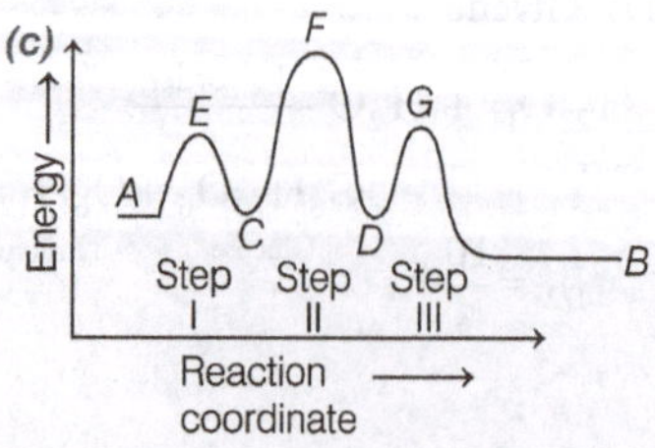

This is a 3-step reaction. Total number of intermediates are 2 (C and D) and transition states are 3 (E, F and G). The rate determining step is step II. Since it has highest activation energy and higher energy correspond to slowest step.

07. *(d)* $k = Ae^{-E_a/RT}$

Taking natural logarithm on both sides, the Arrhenius equation becomes.

$$\Rightarrow \quad \ln k = \ln A - \left(\frac{E_a}{R}\right)\frac{1}{T}$$

$$\Rightarrow \quad \ln k = \ln A - \left(\frac{E_a}{R \times 10^3}\right) \times \frac{10^3}{T}$$

$$\text{Slope of a graph} = \frac{-E_a}{R \times 10^3} = \frac{-10}{5}$$

$$E_a = 2R \times 10^3 \text{ J mol}^{-1}$$

$$\Rightarrow \quad E_a = 2R \text{ kJ}$$

08. *(d)* According to Arrhenius equation

$$k = Ae^{-E_a/RT}$$

$$\Rightarrow \quad \log k = \log A - \frac{E_{\text{act}}}{2.303RT}$$

Slope of the graph of '$\log k$' $vs \frac{1}{T}$ is $\frac{-E_{\text{act}}}{2.303R}$

Greater the value of E_{act}, more will be the negative slope.

From the graph,

most $-ve$ slope = c, followed by a, d and b.

$\therefore$ Order of activation energies must be

$$E_c > E_a > E_d > E_b$$

09. *(b)* Rate constant of reaction without catalyst is given by Arrhenius equation.

$$k = Ae^{-E_a/RT}$$

Taking log on both sides, we get

$$\log k = \log A - \frac{E_a}{RT(2.303)} \quad \text{...(i)}$$

With catalyst, rate constant is 10^6 times more

$$\therefore \quad 10^6 k = Ae^{\frac{-E_a}{RT}}$$

$$\log k + \log 10^6 = \log A - \frac{E_a^*}{RT(2.303)}$$

$$\therefore \quad \log k = \log A - \frac{E_a^*}{RT(2.303)} - 6 \quad \text{...(ii)}$$

Comparing Eqs. (i) and (ii)

$$\frac{E_a}{RT(2.303)} = \frac{E_a^*}{RT(2.303)} + 6$$

$$\therefore \quad E_a^* - E_a = -6(2.303)RT$$

10. *(d)*

> **idea** To calculate the activation energy, use Arrhenius equation for rate constant at two different temperatures (in K) i.e.
> $$\log\frac{k_2}{k_1}=\frac{E_a}{2.303\,R}\left[\frac{T_2-T_1}{T_1T_2}\right]$$

For the reaction,

$$H_2+I_2\longrightarrow 2HI$$

Given $k_1=2.5\times10^{-4}\ \text{dm}^3\,\text{mol}^{-1}\text{s}^{-1}$

$$T_1=(273+327)\,\text{K}=600\,\text{K}$$
$$k_2=1\,\text{dm}^3\,\text{mol}^{-1}\,\text{s}^{-1}$$

at $T_2=(273+527)\,\text{K}=800\,\text{K}$

Now, $\log\frac{k_2}{k_1}=\frac{E_a}{2.303R}\left(\frac{T_2-T_1}{T_1T_2}\right)$

$$\Rightarrow \log\frac{1}{2.5\times10^{-4}}=\frac{E_a}{2.303\times8.314\times10^{-3}}\left(\frac{800-600}{600\times800}\right)$$

$$\Rightarrow \log\frac{(10\times10^3)}{2.5}=\frac{E_a}{0.019}\times\frac{200}{48\times10^4}$$

$$\Rightarrow \log4+3\log10\simeq E_a\times0.022$$

$$\Rightarrow E_a=\frac{2\times\log2+3}{0.022}$$

$$=\frac{3.6}{0.022}\simeq163.6$$

$$\approx164\ \text{kJ mol}^{-1}$$

11. *(c)* Using the Arrhenius equation,

$$k=Ae^{-E_a/RT}\qquad\ldots\text{(i)}$$

Taking natural logarithm on both sides, the Arrhenius equation becomes,

$$\ln k=\ln A-\frac{E_a}{RT}$$

On comparing the above equation with the straight line equation.

$$y_{\text{axis}}=\ln K$$
$$x_{\text{axis}}=\frac{1}{T},$$
$$\text{Slope}=\frac{-E_a}{R},$$
$$I=\ln A$$

$\Rightarrow$ From given graph,

$$\frac{-Ea}{R}=-4606$$

$$\therefore\quad \ln\frac{k_2}{k_1}=\frac{E_a}{R}\left(\frac{1}{T_1}-\frac{1}{T_2}\right)\qquad\ldots\text{(ii)}$$

On substituting all the given values in Eq. (ii), we get

$$\ln\frac{k_2}{10^{-5}}=4606\left(\frac{1}{400}-\frac{1}{500}\right)$$

$$\ln\frac{k_2}{10^{-5}}=2.303$$

$$\frac{k_2}{10^{-5}}=10\ \Rightarrow\ k_2=10^{-4}\ \text{s}^{-1}$$

Therefore, rate constant for the reaction at 500 K is $10^{-4}\ \text{s}^{-1}$.

12. *(c)* We know that,

$$k=Ae^{-E_a/RT}$$

On taking log on both side we get

$$\log k=\log A-\frac{E_a}{2.303\,RT}$$

For graph between $\log k$ with $\frac{1}{T}$

Slope of curve is given by $\frac{E_a}{2.303\,R}$

So, the magnitude of slope will be

$$E_{a_2}>E_{a_1}>E_{a_3}$$

13. *(a)* Among the given statements, only B and C are correct, while A, D and E are incorrect. Their correct form are as follows.

- Arrhenius equation hold true for elementry as well as complex reactions.
- Unit of A is same as unit of k, rate of reaction is high if activation energy is low.
- A and E_a are temperature independent.

14. *(a)* For reaction, $A_2+B_2\rightleftharpoons 2AB$

Given, $E_F=180\text{kJ/mol}$

$$E_B=200\ \text{kJ/mol}$$
$$\Delta H=E_F-E_B=-20\ \text{kJ/mol}$$

In presence of catalyst,

$$E_F=180-100=80\ \text{kJ/mol}$$
$$E_B=200-100=100\ \text{kJ/mol}$$

Also, catalyst does not change ΔH and ΔG of a reaction. Thus, statement given in option (a) is correct.

15. *(d)* Only statement (d) is incorrect. Corrected statement is "Activation enthalpy to form C is 15 kJ mol^{-1} which is more than 5 kJ mol^{-1} that is required to form D."

16. *(100)* Given that, $k=\frac{k_1k_2}{k_3}$

$$\text{or } Ae^{-E_a/RT}=\frac{A_1e^{-\frac{E_{a_1}}{RT}}A_2e^{-\frac{E_{a_2}}{RT}}}{A_3\,e^{-E_{a_3}/RT}}$$

$$Ae^{-E_a/RT}=\frac{A_1A_2}{A_3}e^{-\frac{(E_{a_1}+E_{a_2}-E_{a_3})}{RT}}$$

$$E_a=E_{a_1}+E_{a_2}-E_{a_3}$$
$$400=300+200-E_{a_3}$$
$$E_{a_3}=100\ \text{kJ/mole}$$

17. *(30)* Given,

Overall rate constant, $k=\frac{k_1k_2}{k_3}$

$$E_{a_1}=40\ \text{kJ/mol},\ E_{a_2}=50\ \text{kJ/mol}$$
$$E_{a_3}=60\ \text{kJ/mol}$$
$$k=\frac{k_1\cdot k_2}{k_3}$$

$$Ae^{-\frac{E_a}{RT}}=\frac{A_1A_2}{A_3}e^{-\left(\frac{E_{a_1}+E_{a_2}-E_{a_3}}{RT}\right)}$$

$$A=\frac{A_1A_2}{A_3},\ \text{and}\ E_a=E_{a_1}+E_{a_2}-E_{a_3}$$

$$E_a=40+50-60=30\ \text{kJ/mol}$$

18. *(130)* Given, $E_a=300$ kJ/mol

$$\frac{E_a}{T}=\frac{E_a'}{T'}\quad(\because\ \text{Rate of catalysed and uncatalysed reaction is same})$$

$$\frac{300}{600}=\frac{E_{a\,\text{Forward}}}{300}$$

$$\Rightarrow\ E_{a\,\text{Forward}}=150$$

Now, $20=150-E_{a\,\text{Backward}}$

$$\Rightarrow\ E_{a\,\text{Backward}}=130$$

19. *(2)* Among the given statements A and B are correct.

We know that

$$K=Ae^{-E_a/RT}$$

$$\text{or }\log K=\log A-\frac{E_a}{2.303R}\times\frac{1}{T}$$

Thus, higher is the E_a values, stronger will be the temperature dependence of K. Also, if activation energy (E_a) of a reaction is zero, then the rate is independent of temperature.

20. *(2520)* From Arrhenius equation of kinetics

$$\ln\left(\frac{k_2}{k_1}\right)=\frac{E_a}{R}\left(\frac{1}{T_1}-\frac{1}{T_2}\right)$$

$$\log\left(\frac{0.05}{0.03}\right)=\frac{E_a}{2.3\times8.3}\left[\frac{1}{200}-\frac{1}{300}\right]$$

$$\log(1.66)=\frac{E_a}{2.3\times8.3}\left[\frac{300-200}{300\times200}\right]$$

$$\Rightarrow0.22=\frac{E_a}{2.3\times8.3}\left(\frac{1}{600}\right)$$

$$E_a=0.22\times2.3\times8.3\times600$$
$$=2519.88\ \text{J}\approx2520\ \text{J}$$

21. *(3)* Among the given statements A, B and D are correct. However, statement C is incorrect.

The correct statement for Statement (C) is that the value of k may be greater or lesser depending upon the temperature.

22. *(8)* Using Arrhenius equation,

$$k = Ae^{-E_a/RT}$$

So, $\ln k = \ln A - \frac{E_a}{RT}$

Thus, slope of the graph $\ln k$ vs $\frac{1}{T}$ is $-\frac{E_a}{R}$

$$\Rightarrow \left(\frac{0-20}{5-0}\right) = \frac{-E_a}{R}$$

$$\Rightarrow \frac{20}{5} = \frac{E_a}{R}$$

$$\Rightarrow E_a = 4R = 4 \times 2 = 8 \text{ cal mol}^{-1}$$

23. *(225)* Using Arrhenius equation,

$$K = Ae^{-E_a/RT}$$

Taking ln on both sides,

$$\ln k = \ln A - \frac{E_a}{RT}$$

$$\Rightarrow \log(2\times10^{-6}) = \log(4\times10^{13}) - \frac{8.3\times10^3}{8.3\times T\times 2.3}$$

$$\log 2 - 6 = 2\log 2 + 13 - \frac{8.3\times10^3}{8.3\times T\times 2.3}$$

$$-19.3 = \frac{-10^3}{T\times 2.3}$$

$$\Rightarrow T = 225\times10^{-1}\text{ K}$$

24. *(59)* According to Arrhenius equation,

$$k = Ae^{-E_a/RT}$$

Taking ln both sides

$$\ln k = \ln A - \frac{E_a}{RT}$$

Rate constant when reaction occur at 300 K:

$$\ln k_1 = \ln A - \frac{E_a}{R\times300} \quad \text{... (i)}$$

Rate constant when reaction occur at temperature 309 K

$$\ln k_2 = \ln A - \frac{E_a}{R\times309} \quad \text{...(ii)}$$

$\because \quad k_2 = 2k_1$ (given)

$$\ln(2k_1) = \ln A - \frac{E_a}{309R} \quad \text{...(iii)}$$

Subtract Eqs. (i) from (iii),

$$\ln(2k_1) - \ln k_1 = \frac{-E_a}{309R} + \frac{E_a}{300R}$$

$$\log(2) = \frac{E_a}{2.303\,R}\left[\frac{9}{300\times309}\right]$$

$$0.3 = \frac{E_a}{2.303\times8.3}\times\frac{9}{300\times309}$$

$$\Rightarrow E_a = 59065.04\text{ J} = 59.06\text{ kJ}$$

25. *(166)* Given,

$$\ln k = 33.24 - \frac{2.0\times10^4}{T}$$

Also, $\ln k = \ln A - \frac{E_a}{RT}$

On comparing both above equations,

$$\frac{E_a}{R} = 2.0\times10^4$$

$$\therefore \quad E_a = 2.0\times10^4\times R$$

$$= 2.0\times10^4\times8.3\text{ J}$$

$$= 16.6\times10^4\text{ J} = 166\text{ kJ}$$

26. *(4)* For catalyst A,

$$K_{\text{catalysed}} = Ae^{-E'_a/RT}$$

(From Arrhenius equation)

For uncatalysed, $K_{\text{uncatalysed}} = Ae^{-E_a/RT}$

So,

$$\frac{K_{\text{catalysed}}}{K_{\text{uncatalysed}}} = e^{(E_a - E'_a)/RT}$$

$$= e^{10000\text{ J/mol}/(8.314\times300)}$$

$$= e^{4.009} = e^x$$

$$\therefore \quad x = 4$$

27. *(154)* According to given Arrhenius plot,

$$\ln k = \ln A - \frac{E_a}{10^3\,RT}\times10^3$$

$$\ln k = \ln A + \frac{10^3}{T}\left[\frac{-E_a}{10^3\,R}\right]$$

where, $\frac{-E_a}{10^3\,R}$ represent slope.

$$\therefore \text{ Slope} = \frac{-E_a}{10^3\,R} = -18.5$$

$$\therefore \quad E_a = 18.5\times10^3\times8.31$$

$$= 153.735\text{ kJ/mol}$$

$$\approx 154\text{ kJ/mol}$$

28. *(47)* According to Arrhenius theory,

$$\log k = \log A - \frac{E_a}{2.303RT} \quad \text{...(i)}$$

$$\log_{10} k = 20.35 - \frac{2.47\times10^3}{T} \quad \text{...(ii)}$$

On comparing Eqs. (i) and (ii),

$$\Rightarrow \frac{-2.47\times10^3}{T} = -\frac{E_a}{2.303RT}$$

$$E_a = 2.47\times2.303\times8.314$$

$$= 47.29\text{ kJ/mol} \simeq 47\text{ kJ/mol}$$

29. *(16)* According to Arrhenius theory,

$$\log\frac{k_1}{k_2} = \frac{E_a}{2.303\,R}\left[\frac{1}{T_2} - \frac{1}{T_1}\right]$$

At, $T_1 = 700$ K, $k_1 = 6.36\times10^{-3}\text{ s}^{-1}$,

$E_a = 209\text{ kJ mol}^{-1}$,

$R = 8.314\text{ J K}^{-1}\text{ mol}^{-1}$

At, $T_2 = 600$ K, $k_2 = x\times10^{-6}\text{ s}^{-1}$

$$\log\frac{6.36\times10^{-3}}{x\times10^{-6}} = \frac{209\times10^3}{2.303\times8.314}\left(\frac{1}{600} - \frac{1}{700}\right)$$

$$= \frac{209\times10^3}{2.303\times8.314}\left(\frac{100}{420000}\right)$$

$$= 16\times10^{-6}\text{ s}^{-1}$$

$$x = 16\text{ s}^{-1}$$

30. *(10)* We know, from Arrhenius equation,

$$\log\frac{k_2}{k_1} = \frac{E}{2.303\,R}\left(\frac{1}{T_1} - \frac{1}{T_2}\right)$$

Putting the all given values,

$$\log\frac{10^{-3}}{k_1} = \frac{11.488\times10^3}{2.303\times8.314}\left(\frac{1}{200} - \frac{1}{300}\right)$$

$$\log\frac{10^{-3}}{k_1} = 1$$

$$\Rightarrow \log\frac{10^{-3}}{k_1} = \log 10$$

$$\Rightarrow k_1 = 10^{-4} = 10\times10^{-5}\text{ s}^{-1}$$

31. *(14)* According to Arrhenius equation, the factor $e^{-E_a/RT}$ corresponds to the fraction of molecules that have kinetic energy greater than E_a (activation energy).

$$\Rightarrow e^{-E_a/RT} = e^{\frac{-80.9\times1000}{8.31\times700}} = e^{-13.90}$$

$$\simeq e^{-14} = e^{-x}$$

$$x = 14$$

32. *(50)* $X \longrightarrow Y$, it is an exothermic reaction whose ΔE or

$$\Delta H = -20\text{ kJ mol}^{-1}$$

$$E_a^f = 30\text{ kJ mol}^{-1} \quad \text{(Given)}$$

Activation energy of the backward or reverse reaction ($Y \rightarrow X$),

E_a^b can be calculated as,

$$\Delta H = E_a^f - E_a^b$$

$$\Rightarrow \quad E_a^b = E_a^f - \Delta H$$

$$= 30 - (-20) = 50 \text{ kJ mol}^{-1}$$

33. *(52)* $T_1 = (273 + 27) = 300$ K,

$T_2 = (273 + 52) = 325$ K

Given, temperature coefficient of the reaction,

$$\alpha_T = \frac{K_{325}}{K_{300}} = 5$$

From Arrhenius equation,

$$\log \frac{K_{T_2}}{K_{T_1}} = \frac{E_a}{2.303R} \times \left(\frac{T_2 - T_1}{T_1 T_2}\right)$$

$$\log \frac{K_{325}}{K_{300}} = \frac{E_a}{2.303 \times 8.314}\left(\frac{325 - 300}{300 \times 325}\right)$$

$$\log 5 = \frac{E_a}{2.303 \times 8.314} \times \frac{25}{300 \times 325}$$

$$E_a = 52271 \text{ J}$$

$$= 52.271 \text{ kJ mol}^{-1}$$

$$\simeq 52 \text{ kJ mol}^{-1}$$

34. *(526)* According to Arrhenius equation,

$$\log K = \log A - \frac{E_a}{2.303\, RT}$$

Given, Slope $= \frac{E_a}{2.303\, R} = 10{,}000$

$$K_1 = 10^{-5}, K_2 = 10^{-4}$$

$$T_1 = 500 \text{ K}$$

$$\log_{10}\left[\frac{K_1}{K_2}\right] = \frac{Ea}{2.303R}\left[\frac{1}{T_1} - \frac{1}{T_2}\right]$$

$$\log\left[\frac{10^{-4}}{10^{-5}}\right] = 10000\left[\frac{1}{500} - \frac{1}{T_2}\right]$$

$$\log 10 = 10000\left[\frac{1}{500} - \frac{1}{T_2}\right]$$

$$T_2 = 526.31 \simeq 526 \text{ K}$$

35. *(100)* The Arrhenius Equation is

$$\ln\left(\frac{k_2}{k_1}\right) = \frac{E_a}{R}\left[\frac{1}{T_1} - \frac{1}{T_2}\right]$$

Given, $\frac{k_2}{k_1} = 3.555$

$T_1 = 30 + 273 = 303$ K;

$T_2 = 40 + 273 = 313$ K

Substituting the given values in the Arrhenius equation,

$$\ln (3.555) = \frac{E_a}{8.314}\left[\frac{1}{303} - \frac{1}{313}\right]$$

$$1.268 = \frac{E_a}{8.314}\left[\frac{313 - 303}{303 \times 313}\right]$$

$$E_a = \frac{1.268 \times 8.314 \times 303 \times 313}{10}$$

$$= 99980.7 \text{ J mol} = 99.98 \text{ kJ/mol}$$

$$\simeq 100 \text{ kJ/ mol}$$

36. *(84297.47)* $T_1 = 273 + 27 = 300$ K

$T_2 = 273 + 42 = 315$ K

Now, using (Arrhenius equation)

$k = Ae^{-E_a/RT}$

Taking ln on both sides

$$\ln\left(\frac{k_2}{k_1}\right) = \frac{E_a}{R}\left(\frac{1}{T_1} - \frac{1}{T_2}\right)$$

Given, $\frac{k_2}{k_1} = 5$

$$\ln 5 = \frac{E_a}{R}\left(\frac{1}{300} - \frac{1}{315}\right)$$

$$1.6094 = \frac{E_a}{8.314} \times \frac{15}{300 \times 315}$$

$$E_a = \frac{1.6094 \times 8.314 \times 300 \times 315}{15}$$

$$E_a = 84297.47\,50 \text{ J/ mol}$$

37. *(3.98)* Time taken for milk to

$$\text{split} \propto \frac{1}{k}$$

where, k = rate constant of the process and, k at 2 different temperatures and activation energy are related.

$$\ln \frac{k_2}{k_1} = \frac{-E_a}{R}\left(\frac{1}{T_1} - \frac{1}{T_2}\right)$$

$$\therefore \quad \ln\left(\frac{60}{40}\right) = \frac{-E_a}{8.3}\left(\frac{1}{400} - \frac{1}{300}\right)$$

$$E_a = 0.4 \times 8.3 \times 1200$$

$$= 3984 \text{ J mol}^{-1}$$

$$= 3.984 \text{ kJ mol}^{-1}$$

Note Given, $\ln 2/3 = 0.4 \Rightarrow \ln\frac{3}{2} = -0.4$

(These are wrong values, because $\ln 2/3$ is –ve and $\ln 3/2$ is +ve, but assume them to be correct).

16 p-Block Elements-2

In JEE Main, there is usually 0-1 question per paper are visible from this chapter. In past 3 years, 40 conceptual and 5 numerical questions were asked of moderate difficulty level from this chapter.

For detailed analysis scan QR code

Topic 1 / Group 15 Elements

• Level 1

01 The maximum covalency of a non-metallic group 15 element 'E' with weakest $E—E$ bond is **[2025, 22 Jan Shift-I]**

(a) 4 (b) 3 (c) 6 (d) 5

02 Concentrated HNO_3 reacts with iodine to give **[2022, 28 July Shift-II]**

(a) HI, NO_2 and H_2O (b) HIO_2, N_2O and H_2O
(c) HIO_3, NO_2 and H_2O (d) HIO_4, N_2O and H_2O

03 The gas produced by treating an aqueous solution of ammonium chloride with sodium nitrite is **[2022, 27 June Shift-II]**

(a) NH_3 (b) N_2 (c) N_2O (d) Cl_2

04 PCl_5 is well known but NCl_5 is not, because **[2022, 24 June Shift-II]**

(a) nitrogen is less reactive than phosphorus.
(b) nitrogen doesn't have d-orbitals in its valence shell.
(c) catenation tendency is weaker in nitrogen than phosphorus.
(d) size of phosphorus is larger than nitrogen.

05 The most stable trihalide of nitrogen is **[2022, 24 June Shift-I]**

(a) NF_3 (b) NCl_3 (c) NBr_3 (d) NI_3

06 Which one of the following group-15 hydride is the strongest reducing agent? **[2021, 22 July Shift-II]**

(a) AsH_3 (b) BiH_3 (c) PH_3 (d) SbH_3

07 A group 15 element, which is a metal and forms a hydride with strongest reducing power among group 15 hydrides. The element is **[2021, 16 March Shift-I]**

(a) Sb (b) P (c) As (d) Bi

Imp. *Structure of Pyrophosphoric Acid*

08 In a molecule of pyrophosphoric acid, the number of P—OH, P=O and P—O—P bonds/moiety(ies) respectively are **[2020, 3 Sep Shift-I]**

(a) 2, 4 and 1 (b) 3, 3, and 3
(c) 4, 2, and 0 (d) 4, 2, and 1

• Level 2

09 Given below are two statements: **[2025, 4 April Shift-I]**

Statement I Nitrogen forms oxides with +1 to +5 oxidation states due to the formation of $p\pi$-$p\pi$ bond with oxygen.

Statement II Nitrogen does not form halides with +5 oxidation state due to the absence of d-orbital in it.

In the light of given statements, choose the correct answer from the options given below.

(a) Statement I is true but Statement II is false.
(b) Both Statement I and Statement II are true.
(c) Both Statement I and Statement II are False.
(d) Statement I is false but Statement II are true.

10 On reaction of lead sulphide with dilute nitric acid which of the following is not formed? **[2024, 9 April Shift-I]**

(a) Sulphur (b) Nitric oxide
(c) Lead nitrate (d) Nitrous oxide

11 Given below are two statements. **[2024, 8 April Shift-I]**

Statement I $N(CH_3)_3$ and $P(CH_3)_3$ can act as ligands to form transition metal complexes.

Statement II As N and P are from same group, the nature of bonding of $N(CH_3)_3$ and $P(CH_3)_3$ is always same with transition metals.

In the light of the above statements, choose the most appropriate answer from the options given below:

(a) Statement I is correct but Statement II is incorrect.
(b) Statement I is incorrect but Statement II is correct.
(c) Both Statement I and Statement II are incorrect.
(d) Both Statement I and Statement II are correct.

12 The strongest reducing agent among the following is **[2024, 1 Feb Shift-II]**

(a) NH_3 (b) PH_3 (c) BiH_3 (d) SbH_3

13 Given below are two statements : one is labelled as **Assertion (A)** and the other is labelled as **Reason (R)**.

Assertion (A) PH_3 has lower boiling point than NH_3.

Reason (R) In liquid state NH_3 molecules are associated through van der Waals' forces, but PH_3 molecules are associated through hydrogen bonding.

[2024, 1 Feb Shift-I]

In the light of the above statements, choose the **most appropriate** answer from the options given below
(a) A is false but R is true.
(b) Both A and R are true but R is not the correct explanation of A.
(c) Both A and R are true and R is the correct explanation of A.
(d) A is true but R is false.

14 Given below are two statements : one is labelled as Assertion (A) and the other is labelled as Reason (R).

Assertion (A) There is a considerable increase in covalent radius from N to P. However from As to Bi only a small increase in covalent radius is observed.

Reason (R) Covalent and ionic radii in a particular oxidation state increases down the group. **[2024, 30 Jan Shift-I]**

In the light of the above statements, choose the **most appropriate** answer from the options given below
(a) Both A and R are true and R is the correct explanation of A.
(b) A is true but R is false.
(c) A is false but R is true.
(d) Both A and R are true but R is not the correct explanation of A.

15 Which of the phosphorus oxoacid can create silver mirror from $AgNO_3$ solution? **[2023, 24 Jan, Shift-I]**
(a) $H_4P_2O_6$ (b) $H_4P_2O_7$
(c) $(HPO_3)_n$ (d) $H_4P_2O_5$

16 Dinitrogen and dioxygen, the main constituents of air do not react with each other in atmosphere to form oxides of nitrogen because **[2022, 28 July Shift-II]**
(a) N_2 is unreactive in the condition of atmosphere
(b) oxides of nitrogen are unstable
(c) reaction between them can occur in the presence of a catalyst
(d) the reaction is endothermic and require very high temperature

17 White phosphorus reacts with thionyl chloride to give **[2022, 28 July, Shift-II]**
(a) PCl_5, SO_2 and S_2Cl_2 (b) PCl_3, SO_2 and S_2Cl_2
(c) PCl_3, SO_2 and Cl_2 (d) PCl_5, SO_2 and Cl_2

18 $A \xrightarrow{573\text{ K}}$ Red phosphorus $\xrightarrow[\text{Under pressure}]{\text{Heat ; 803 K}} B$

Red phosphorus is obtained by heating "A" at 573 K, and can be converted to "B" by heating at 803 K under pressure. A and B, respectively, are **[2022, 30 June, Shift-I]**
(a) β-black phosphorus and white phosphorus
(b) White phosphorus and β-black phosphorus
(c) α-black phosphorus and white phosphorus
(d) White phosphorus and α-black phosphorus

19 The oxoacid of phosphorus that is easily obtained from a reaction of alkali and white phosphorus and has two P—H bonds, is **[2022, 29 June Shift-I]**
(a) phosphonic acid (b) phosphinic acid
(c) pyrophosphorus acid (d) hypophosphoric acid

20 Nitrogen gas is obtained by thermal decomposition of **[2022, 28 June, Shift-I]**
(a) $Ba(NO_3)_2$ (b) $Ba(N_3)_2$
(c) $NaNO_2$ (d) $NaNO_3$

21 Heating white phosphorus with conc. NaOH solution gives mainly **[2022, 27 June Shift-I]**
(a) Na_3P and H_2O (b) H_3PO and NaH
(c) $P(OH)_3$ and NaH_2PO_4 (d) PH_3 and NaH_2PO_2

22 The oxide which contains an odd electron at the nitrogen atom is **[2022, 26 June Shift-II]**
(a) N_2O (b) NO_2 (c) N_2O_3 (d) N_2O_5

23 The oxide without nitrogen- nitrogen bond is **[2021, 1 Sep Shift-II]**
(a) N_2O (b) N_2O_4 (c) N_2O_3 (d) N_2O_5

24 Which one of the following is formed (mainly) when red phosphorus is heated in a sealed tube at 803 K? **[2021, 27 Aug Shift-II]**
(a) White phosphorus (b) Yellow phosphorus
(c) β-black phosphorus (d) α-black phosphorus

25 Chemical nature of the nitrogen oxide compound obtained from a reaction of concentrated nitric acid and P_4O_{10} (in 4 : 1 ratio) is **[2021, 20 July Shift-I]**
(a) acidic (b) basic
(c) amphoteric (d) neutral

Imp. *Order of Oxidation States of Nitrogen*

26 The oxidation states of nitrogen in NO, NO_2, N_2O and NO_3^- are in the order of **[2021, 18 March Shift-II]**
(a) $NO_3^- > NO_2 > NO > N_2O$
(b) $NO_2 > NO_3^- > NO > N_2O$
(c) $N_2O > NO_2 > NO > NO_3^-$
(d) $NO > NO_2 > N_2O > NO_3^-$

27 The set that represents the pair of neutral oxides of nitrogen is **[2021, 17 March Shift-II]**
(a) NO and N_2O (b) N_2O and N_2O_3
(c) N_2O and NO_2 (d) NO and NO_2

28 Which of the following compound cannot act as a Lewis base? **[2021, 17 March Shift-I]**
(a) NF_3 (b) PCl_5 (c) SF_4 (d) ClF_3

29 **Statement I** Sodium hydride can be used as an oxidising agent.

Statement II The lone pair of electrons on nitrogen in pyridine makes it basic.

Choose the correct answer from the options given below. [2021, 16 March Shift-II]

(a) Both Statement I and Statement II are true
(b) Both Statement I and Statement II are false
(c) Statement I is true but Statement II is false
(d) Statement I is false but Statement II is true

30 The reaction of NO with N_2O_4 at 250 K gives [2020, 6 Sep Shift-II]

(a) N_2O (b) NO_2
(c) N_2O_3 (d) N_2O_5

31 Reaction of ammonia with excess Cl_2 gives [2020, 5 Sep Shift-II]

(a) NH_4Cl and N_2 (b) NH_4Cl and HCl
(c) NCl_3 and NH_4Cl (d) NCl_3 and HCl

32 *Aqua-regia* is used for dissolving noble metals (Au, Pt, etc.). The gas evolved in this processes is [2020, 03 Sep, Shift-I]

(a) NO (b) N_2O_5 (c) N_2 (d) N_2O_3

33 The number of P—OH bonds and the oxidation state of phosphorus atom in pyrophosphoric acid ($H_4P_2O_7$) respectively are [2020, 03 Sep, Shift-I, 2017]

(a) four and four (b) five and four
(c) five and five (d) four and five

34 An organic compound *A* is oxidised with Na_2O_2 followed by boiling with HNO_3. The resultant solution is then treated with ammonium molybdate to yield a yellow precipitate.
Based on above observation, the element present in the given compound is [2019, 12 April Shift-I]

(a) nitrogen (b) phosphorus
(c) fluorine (d) sulphur

35 Good reducing nature of H_3PO_2 is attributed to the presence of [2019, 9 Jan Shift-II]

(a) two P— H bonds (b) one P— H bond
(c) two P— OH bonds (d) one P — OH bond

• Level 3

36 Given below are two statements:

Statement I The N—N single bond is weaker and longer than that of P—P single bond.

Statement II Compounds of group 15 elements in +3 oxidation states readily undergo disproportionation reactions. [2025, 3 April Shift-II]

In the light of the above statements, choose the correct answer from the options given below

(a) Statement I is true but Statement II is false.
(b) Both Statement I and Statement II are false.
(c) Both Statement I and Statement II are true.
(d) Statement I is false but Statement II is true.

37 The incorrect statement among the following is [2025, 23 Jan Shift-I]

(a) PH_3 shows lower proton affinity than NH_3.
(b) NO_2 can dimerise easily.
(c) SO_2 can act as an oxidising agent, but not as a reducing agent.
(d) PF_5 exists but NF_5 does not.

38 Identify the incorrect statements about group 15 elements:

(A) Dinitrogen is a diatomic gas which acts like an inert gas at room temperature.
(B) The common oxidation states of these elements are −3, +3 and +5.
(C) Nitrogen has unique ability to form $p\pi$-$p\pi$ multiple bonds.
(D) The stability of +5 oxidation states increases down the group.
(E) Nitrogen shows a maximum covalency of 6.

Choose the correct answer from the options given below: [2024, 8 April Shift-II]

(a) (A), (C), (E) only (b) (D) and (E) only
(c) (B), (D), (E) only (d) (A), (B), (D) only

39 Choose the correct statements about the hydrides of group 15 elements. [2024, 30 Jan Shift-II]

A. The stability of the hydrides decreases in the order $NH_3 > PH_3 > AsH_3 > SbH_3 > BiH_3$.
B. The reducing ability of the hydride increases in the order $NH_3 < PH_3 < AsH_3 < SbH_3 < BiH_3$.
C. Among the hydride, NH_3 is strong reducing agent while BiH_3 is mild reducing agent.
D. The basicity of the hydrides increases in the order $NH_3 < PH_3 < AsH_3 < SbH_3 < BiH_3$.

Choose the **most appropriate** from the options given below :

(a) A and D only (b) B and C only
(c) C and D only (d) A and B only

40 The number of P—O—P bonds in $H_4P_2O_7$, $(HPO_3)_3$ and P_4O_{10} are respectively [2023, 15 April Shift-II]

(a) 1, 2, 4 (b) 0, 3, 4 (c) 0, 3, 6 (d) 1, 3, 6

41 Reaction of thionyl chloride with white phosphorus forms a compound [*A*], which on hydrolysis gives [*B*], a dibasic acid. [*A*] and [*B*] are respectively [2023, 25 Jan, Shift-I]

(a) PCl_3 and H_3PO_3 (b) PCl_5 and H_3PO_4
(c) $POCl_3$ and H_3PO_4 (d) P_4O_6 and H_3PO_3

42 Dinitrogen is a robust compound, but reacts at high altitude to form oxides. The oxide of nitrogen that can damage plant leaves and retard photosynthesis is [2022, 29 July, Shift-II]

(a) NO (b) NO_3^- (c) NO_2 (d) NO_2^-

43 Which oxo-acid of phosphorus has the highest number of oxygen atoms present in its chemical formula? **[2022, 27 July Shift-I]**

(a) Pyrophosphorus acid (b) Hypophosphoric acid
(c) Phosphoric acid (d) Pyrophosphoric acid

44 Among the given oxides of nitrogen; N_2O, N_2O_3, N_2O_4 and N_2O_5, the number of compounds having N—N bond is **[2022, 28 June Shift-II]**

(a) 1 (b) 2
(c) 3 (d) 4

45 Given below are two statements. **[2022, 28 June Shift-I]**

Statement I The pentavalent oxide of group-15 element, E_2O_5, is less acidic than trivalent oxide, E_2O_3 of the same element.

Statement II The acidic character of trivalent oxide of group 15 elements, E_2O_3, decreases down the group.

In light of the above statements, choose most appropriate answer from the options given below

(a) Both Statement I and Statement II are true.
(b) Both Statement I and Statement II are false.
(c) Statement I is true but Statement II is false.
(d) Statement I is false but Statement II is true.

46 Consider the following reaction,

$A + \text{alkali} \rightarrow B$ (major product)

If B is an oxoacid of phosphorus with no P—H bond, then A is **[2022, 26 June Shift-I]**

(a) white P_4 (b) red P_4
(c) P_2O_3 (d) H_3PO_3

47 The number of ionisable hydrogens present in the product obtained from a reaction of phosphorus trichloride and phosphonic acid is **[2021, 18 March Shift-I]**

(a) 3 (b) 0
(c) 2 (d) 1

48 Match List-I with List-II. **[2021, 16 March, Shift-I]**

List-I (Name of oxo acid)		List-II (Oxidation state of 'P')	
(A)	Hypophosphorus acid	(I)	+5
(B)	Orthophosphoric acid	(II)	+4
(C)	Hypophosphoric acid	(III)	+3
(D)	Orthophosphorus acid	(IV)	+2
		(V)	+1

Choose the correct answer from the options given below

(a) (A)-(IV), (B)-(V), (C)-(II), (D)-(III)
(b) (A)-(V), (B)-(IV), (C)-(II), (D)-(III)
(c) (A)-(V), (B)-(I), (C)-(II), (D)-(III)
(d) (A)-(IV), (B)-(I), (C)-(II), (D)-(III)

49 Find A, B and C in the following reactions

$$NH_3 + A + CO_2 \longrightarrow (NH_4)_2CO_3$$
$$(NH_4)_2CO_3 + H_2O + B \longrightarrow NH_4HCO_3$$
$$NH_4HCO_3 + NaCl \longrightarrow NH_4Cl + C$$

[2021, 26 Feb Shift-I]

(a) $A-O_2$, $B-CO_2$, $C-Na_2CO_3$
(b) $A-H_2O$, $B-O_2$, $C-Na_2CO_3$
(c) $A-H_2O$, $B-O_2$, $C-NaHCO_3$
(d) $A-H_2O$, $B-CO_2$, $C-NaHCO_3$

50 On heating, lead (II) nitrate gives a brown gas (A). The gas (A) on cooling changes to a colourless solid-liquid (B). (B) on heating with NO changes to a blue solid (C). The oxidation number of nitrogen in solid (C) is **[2020, 04 Sep, Shift-I]**

(a) +5 (b) +2 (c) +3 (d) +4

51 The correct order of the oxidation states of nitrogen in NO, NO_2, NO_2 and N_2O_3 is **[2019, 9 April Shift-I]**

(a) $NO_2 < NO < N_2O_3 < N_2O$
(b) $N_2O < NO < N_2O_3 < NO_2$
(c) $O_2 < N_2O_3 < NO < N_2O$
(d) $N_2O < N_2O_3 < NO < NO_2$

52 The pair that contains two P—H bonds in each of the oxoacids is **[2019, 10 Jan Shift-II]**

(a) $H_4P_2O_5$ and $H_4P_2O_6$ (b) H_3PO_3 and H_3PO_2
(c) $H_4P_2O_5$ and H_3PO_3 (d) H_3PO_2 and $H_4P_2O_5$

Numerical Value Type Questions

53 A group 15 element forms $d\pi - d\pi$ bond with transition metals. It also forms hydride, which is a strongest base among the hydrides of other group members that form $d\pi - d\pi$ bond. The atomic number of the element is....... **[2025, 28 Jan Shift-II]**

54 The ratio of σ and π bonds present in pyrophosphoric acid is .... **[2023, 8 April Shift-II]**

55 The number of non-ionisable protons present in the product B obtained from the following reaction is ....

$$C_2H_5OH + PCl_3 \longrightarrow C_2H_5Cl + A$$
$$A + PCl_3 \longrightarrow B$$

[2022, 26 July, Shift-II]

56 Consider the following reactions,

$$PCl_3 + H_2O \longrightarrow A + HCl$$
$$A + H_2O \longrightarrow B + HCl$$

The number of ionisable protons present in the product B is .... **[2022, 25 June Shift-II]**

57 The reaction of white phosphorus on boiling with alkali in inert atmosphere resulted in the formation of product A. The reaction of 1 mol of A with excess of $AgNO_3$ in aqueous medium gives ...... moles of Ag (Round off to the nearest integer). **[2021, 17 March Shift-I]**

Topic 2 / Group 16 Elements

• Level 1

01 Anomalous behaviour of oxygen is due to its
(a) large size and high electronegativity **[2024, 29 Jan Shift-II]**
(b) small size and low electronegativity
(c) small size and high electronegativity
(d) large size and low electronegativity

02 The correct order of melting point of hydrides of group 16 elements is **[2022, 26 June Shift-I]**
(a) $H_2S < H_2Se < H_2Te < H_2O$
(b) $H_2O < H_2S < H_2Se < H_2Te$
(c) $H_2S < H_2Te < H_2Se < H_2O$
(d) $H_2Se < H_2S < H_2Te < H_2O$

03 In polythionic acid, $H_2S_xO_6 (x = 3 \text{ to } 5)$ the oxidation states of sulphur is/are **[2021, 27 Aug Shift-I]**
(a) only +5 (b) only +6
(c) +3 and +5 (d) 0 and +5

04 Chalcogen group elements are **[2021, 26 Aug Shift-II]**
(a) Se, Tb and Pu (b) Se, Te and Po
(c) S, Te and Pm (d) O, Ti and Po

05 The correct order of electron gain enthalpy is **[2021, 26 Feb Shift-II]**
(a) $S > Se > Te > O$ (b) $Te > Se > S > O$
(b) $O > S > Se > Te$ (d) $S > O > Se > Te$

• Level 2

06 Given below are two statements: **[2025, 8 April Shift-II]**
Statement I H_2Se is more acidic than H_2Te.
Statement II H_2Se has higher bond enthalpy for dissociation than H_2Te.
In the light of the above statements, choose the correct answer from the options given below
(a) Both Statement I and Statement II are true.
(b) Statement I is false but Statement II is true.
(c) Both Statement I and Statement II are false.
(d) Statement I is true but Statement II is false.

07 Which among the following molecules is (a) involved in sp^3d hybridisation, (b) has different bond lengths and (c) has lone pair of electrons on the central atom? **[2025, 2 April Shift-II]**
(a) PF_5 (b) XeF_4 (c) XeF_2 (d) SF_4

08 Given below are two statements. **[2025, 24 Jan Shift-II]**
Statement I Experimentally determined oxygen- oxygen bond lengths in the O_3 are found to be same and the bond length is greater than that of a O═O (double bond) but less than that of a single (O—O) bond.
Statement II The strong lone pair-lone pair repulsion between oxygen atoms is solely responsible for the fact that the bond length in ozone is smaller than that of a double bond (O═O) but more than that of a single bond (O—O).
In the light of the above statements, choose the correct answer from the options given below
(a) Both Statement I and Statement II are false.
(b) Statement I is false but Statement II is true.
(c) Statement I is true but Statement II is false.
(d) Both Statement I and Statement II are true.

09 Given below are two statements : one is labelled as **Assertion (A)** and the other is labelled as **Reason (R)**. **[2024, 30 Jan Shift-II]**
Assertion (A) H_2Te is more acidic than H_2S.
Reason (R) Bond dissociation enthalpy of H_2Te is lower than H_2S.
In the light of the above statements, choose the **most appropriate** answer from the options given below:
(a) A is false but R is true.
(b) Both A and R are true but R is not the correct explanation of A.
(c) A is true but R is not true.
(d) Both A and R are true and R is the correct explanation of A.

10 Given below are two statements. **[2024, 27 Jan Shift-I]**
Statement I Oxygen being the first member of group 16 exhibits only −2 oxidation state.
Statement II Down the group 16 stability of +4 oxidation state decreases and +6 oxidation state increases.
In the light of the above statements, choose the **most appropriate** answer from the options given below.
(a) Statement I is correct but Statement II is incorrect.
(b) Both Statement I and Statement II are incorrect.
(c) Statement I is incorrect but Statement II is correct.
(d) Both Statement I and Statement II are correct.

11 Sulphur dioxide is one of the components of polluted air. SO_2 is also a major contributor to acid rain. The correct and complete reaction to represent acid rain caused by SO_2 is **[2022, 29 June Shift-II]**
(a) $2SO_2 + O_2 \longrightarrow 2SO_3$
(b) $SO_2 + O_3 \longrightarrow SO_3 + O_2$
(c) $SO_2 + H_2O_2 \longrightarrow H_2SO_4$
(d) $2SO_2 + O_2 + 2H_2O \longrightarrow 2H_2SO_4$

12 The depositions of X and Y on ground surfaces is referred to as wet and dry depositions, respectively. X and Y are **[2021, 31 Aug Shift-II]**
(a) X = Ammonium salts, $Y = CO_2$
(b) $X = SO_2$, Y = Ammonium salts
(c) X = Ammonium salts, $Y = SO_2$
(d) $X = CO_2$, $Y = SO_2$

13 If the boiling point of H_2O is 373 K, the boiling point of H_2S will be **[2020, 03 Sep, Shift-I]**
(a) less than 300 K
(b) equal to 373 K
(c) more than 373 K
(d) greater than 300 K but less than 373 K

• Level 3

14 The large difference between the melting and boiling points of oxygen and sulphur may be explained on the basis of **[2025, 24 Jan Shift-I]**
(a) electronegativity (b) atomic size
(c) electron gain enthalpy (d) atomicity

15 Bond dissociation energy of "E—H" bond of the "H_2E" hydrides of group 16 elements (given below), follows order. **[2023, 30 Jan, Shift-II]**
A. O B. S C. Se D. Te

Choose the correct from the options given below:
(a) $A > B > C > D$ (b) $D > C > B > A$
(c) $B > A > C > D$ (d) $A > B > D > C$

16 Which of the following oxoacids of sulphur contains "S" in two different oxidation states? **[2022, 28 June Shift-II]**
(a) $H_2S_2O_3$ (b) $H_2S_2O_6$
(c) $H_2S_2O_7$ (d) $H_2S_2O_8$

17 The number of S=O bonds present in sulphurous acid, peroxodisulphuric acid and pyrosulphuric acid, respectively are **[2021, 31 Aug Shift-II ; 2017]**
(a) 2, 3 and 4 (b) 1, 4 and 3 (c) 2, 4 and 3 (d) 1, 4 and 4

18 The product obtained from the electrolytic oxidation of acidified sulphate solutions, is **[2021, 27 July Shift-I]**
(a) HSO_4^- (b) HO_3SOOSO_3H
(c) HO_2SOSO_2H (d) HO_3SOSO_3H

Imp. *Application of Industrial Process*

19 Match List -I with List II.

List-I (Industrial process)	List-II (Application)
A. Haber's process	(i) HNO_3 synthesis
B. Ostwald's process	(ii) Aluminium extraction
C. Contact process	(iii) NH_3 synthesis
D. Hall-Heroult process	(iv) H_2SO_4 synthesis

Choose the correct answer from the options given below **[2021, 16 March Shift-I]**
(a) A-(ii), B-(iii), C-(iv), D-(i)
(b) A-(iii), B-(iv), C-(i), D-(ii)
(c) A-(iii), B-(i), C-(iv), D-(ii)
(d) A-(iv), B-(i), C-(ii), D-(iii)

20 Given below are two statements.
Statement I α and β-forms of sulphur can change reversibly between themselves with slow heating or slow cooling.
Statement II At room temperature, the stable crystalline form of sulphur is monoclinic sulphur.
In the light of the above statements, choose the correct answer from the options given below.
(a) Both statements I and II are true. **[2021, 25 Feb Shift-II]**
(b) Both statements I and II are false.
(c) Statement I is true but statement II is false.
(d) Statement I is false but statement II is true.

21 Choose the correct statements from the following.
A. All group 16 elements form oxides of general formula EO_2 and EO_3, where E = S, Se, Te and Po. Both the types of oxides are acidic in nature.
B. TeO_2 is an oxidising agent while SO_2 is reducing in nature.
C. The reducing property decreases from H_2S to H_2Te down the group.
D. The ozone molecule contains five lone pairs of electrons.

Choose the correct answer from the options given below. **[2020, 31 Jan, Shift-I]**
(a) A and B only (b) B and C only
(c) C and D only (d) A and D only

22 Reaction of an inorganic sulphite X with dilute H_2SO_4 generates compound Y. Reaction of Y with NaOH gives X. Further, the reaction of X with Y and water affords compound Y and Z. respectively are **[2020, 06 Sep, Shift-II]**
(a) SO_2 and Na_2SO_3 (b) SO_3 and $NaHSO_3$
(c) SO_2 and $NaHSO_3$ (d) S and Na_2SO_3

23 The number of bonds between sulphur and oxygen atoms in $S_2O_8^{2-}$ and the number of bonds between sulphur and sulphur atoms in rhombic sulphur, respectively, are **[2020, 8 Jan Shift-I]**
(a) 4 and 6 (b) 8 and 8
(c) 4 and 8 (d) 8 and 6

Numerical Value Type Questions

Imp. *Structure of Oxoacids*

24 Consider the following sulphur based oxoacids.
H_2SO_3, H_2SO_4, $H_2S_2O_8$ and $H_2S_2O_7$.
Amongst these oxo-acids, the number of those with peroxo (O—O) bond is .... **[2022, 29 July Shift-II]**

25 Consider the sulphides HgS, PbS, CuS, Sb_2S_3, As_2S_3 and CdS. Number of these sulphides soluble in 50% HNO_3 is **[2021, 31 Aug Shift-I]**

26 Among the following allotropic forms of sulphur, the number of allotropic forms, which will show paramagnetism is .... **[2021, 24 Feb Shift-II]**
A. α-sulphur B. β-sulphur
C. S_2-form

27 Number of oxygen atoms present in chemical formula of fuming sulphuric acid is .... **[2020, 9 April Shift-II]**

Topic 3 / Group 17 Elements

• Level 1

01 Which halogen is known to cause the reaction given below?

$$2Cu^{2+} + 4X^- \longrightarrow Cu_2X_2(s) + X_2$$

[2023, 8 April Shift-I]

(a) all halogens (b) only chlorine
(c) only iodine (d) only bromine

02 Element not present in Nessler's reagent is **[2023, 6 April Shift-II]**
(a) N (b) Hg
(c) K (d) I

03 Which element is not present in Nessler's reagent? **[2023, 1 Feb Shift-II]**
(a) Oxygen (b) Iodine
(c) Potassium (d) Mercury

04 Formula for Nessler's reagent is **[2023, 30 Jan Shift-II]**
(a) $KHgI_3$ (b) HgI_2
(c) K_2HgI_4 (d) KHg_2I_2

05 Given below are two statements. One is labelled as Assertion A and the other is labelled as Reason R.

Assertion (A) Fluorine forms one oxoacid.

Reason (R) Fluorine has smallest size amongst all halogens and is highly electronegative.

In the light of the above statements, choose the most appropriate answer from the options given below. **[2022, 27 June Shift-II]**

(a) Both A and R are correct and R is the correct explanation of A.
(b) Both A and R are correct but R is not the correct explanation of A.
(c) A is correct but R is not correct.
(d) A is not correct but R is correct.

06 The absolute value of the electron gain enthalpy of halogens satisfies **[2021, 17 March Shift-I]**
(a) I > Br > Cl > F (b) Cl > Br > F > I
(c) Cl > F > Br > I (d) F > Cl > Br > I

• Level 2

07 Among the following halogens F_2, Cl_2, Br_2 and I_2 Which can undergo disproportionation reactions? **[2024, 8 April Shift-II]**

(a) F_2, Cl_2 and Br_2 (b) F_2 and Cl_2
(c) Only I_2 (d) Cl_2, Br_2 and I_2

08 Given below are two statements: **[2024, 29 Jan Shift-II]**

Statement I Fluorine has most negative electron gain enthalpy in its group.

Statement II Oxygen has least negative electron gain enthalpy in its group.

In the light of the above statements, choose the **most appropriate** from the options given below

(a) Statement I is false but Statement II is true.
(b) Both Statement I and Statement II are false.
(c) Statement I is true but Statement II is false.
(d) Both Statement I and Statement II are true.

09 Element not showing variable oxidation state is **[2024, 27 Jan Shift-II]**
(a) bromine (b) chlorine (c) fluorine (d) iodine

Imp. *Oxidation of Halide Ion in Acidic Medium*

10 The correct group of halide ions which can be oxidised by oxygen in acidic medium is **[2023, 13 April Shift-II]**
(a) I^- only (b) Br^- only
(c) Cl^-, Br^- and I^- only (d) Br^- and I^- only

11 Given below are two statements.

Statement I $SbCl_5$ is more covalent than $SbCl_3$.

Statement II The higher oxides of halogens also tend to be more stable than the lower ones.

In the light of the above statements, choose the correct answer from the options given below. **[2023, 12 April Shift-I]**

(a) Statement I is incorrect but Statement II is correct.
(b) Both statement I and Statement II are correct.
(c) Statement I is correct but Statement II is incorrect.
(d) Both Statement I and Statement II are incorrect.

12 The interhalogen compound formed from the reaction of bromine with excess of fluorine is a **[2022, 25 July Shift-I]**
(a) hypohalite (b) halate
(c) perhalate (d) halite

13 Which one of the following correctly represents the order of stability of oxides, X_2O (X = halogen) ? **[2021, 31 Aug Shift-II]**
(a) Br > Cl > I (b) Br > I > Cl
(c) Cl > I > Br (d) I > Cl > Br

14 Which one of the following is used to remove most of plutonium from spent nuclear fuel? **[2021, 27 Aug Shift-II]**
(a) ClF_3 (b) O_2F_2 (c) I_2O_5 (d) BrO_3

15 Choose the incorrect statement. **[2021, 26 Aug Shift-I]**
(a) Cl_2 is more reactive than ClF.
(b) F_2 is more reactive than ClF.
(c) On hydrolysis ClF froms HOCl and HF.
(d) F_2 is a stronger oxidising agent than Cl_2 in aqueous solution.

16 Number of Cl=O bonds in chlorous acid, chloric acid and perchloric acid respectively are **[2021, 27 July Shift-II]**
(a) 3, 1 and 1 (b) 4, 1 and 0
(c) 1, 1 and 3 (d) 1, 2 and 3

17 HF has highest boiling point among hydrogen halides, because it has **[2019, 10 April Shift-I]**
(a) lowest ionic character
(b) strongest van der Waals' interactions
(c) strongest hydrogen bonding
(d) lowest dissociation enthalpy

18 What products are expected from the disproportionation reaction of hypochlorous acid ? **[2019, 8 April Shift I]**
(a) $HClO_3$ and Cl_2O (b) $HClO_2$ and $HClO_4$
(c) HCl and Cl_2O (d) HCl and $HClO_3$

19 Among the following reactions of hydrogen with halogens, the one that requires a catalyst is **[2019, 10 Jan Shift-II]**
(a) $H_2 + Cl_2 \longrightarrow 2HCl$ (b) $H_2 + I_2 \longrightarrow 2HI$
(c) $H_2 + F_2 \longrightarrow 2HF$ (d) $H_2 + Br_2 \longrightarrow 2HBr$

• Level 3

20 The property/properties that show irregularity in first four elements of group-17 is/are: **[2025, 2 April Shift-I]**
A. covalent radius
B. electron affinity
C. ionic radius
D. first ionisation energy

Choose the correct answer from the options given below:
(a) B only (b) A, B, C and D
(c) A and C only (c) B and D only

21 Match List-I with List-II, match the gas evolved during each reaction.

List-I	List-II
(A) $(NH_4)_2Cr_2O_7 \xrightarrow{\Delta}$	(I) H_2
(B) $KMnO_4 + HCl \longrightarrow$	(II) N_2
(C) $Al + NaOH + H_2O \longrightarrow$	(III) O_2
(D) $NaNO_3 \longrightarrow$	(IV) Cl_2

Choose the correct answer from the options given below. **[2022, 28 July Shift-I]**
(a) (A) → (II), (B) → (III), (C) → (I), (D) → (IV)
(b) (A) → (III), (B) → (I), (C) → (IV), (D) → (II)
(c) (A) → (II), (B) → (IV), (C) → (I), (D) → (III)
(d) (A) → (III), (B) → (IV), (C) → (I), (D) → (II)

22 The correct order of bond dissociation enthalpy of halogens is **[2021, 25 Feb Shift-II]**
(a) $F_2 > Cl_2 > Br_2 > I_2$ (b) $I_2 > Br_2 > Cl_2 > F_2$
(c) $Cl_2 > Br_2 > F_2 > I_2$ (d) $Cl_2 > F_2 > Br_2 > I_2$

Numerical Value Type Questions

23 Among the following oxides of *p*-block elements, number of oxides having amphoteric nature is
Cl_2O_7, CO, PbO_2, N_2O, NO, Al_2O_3, SiO_2, N_2O_5, SnO_2 **[2024, 1 Feb, Shift-I]**

24 In the following reactions, the total number of oxygen atoms in X and Y is ....

$$Na_2O + H_2O \longrightarrow 2X$$
$$Cl_2O_2 + H_2O \longrightarrow 2Y$$
[2023, 10 April Shift-I]

25 The number of interhalogens from the following having square pyramidal structure is
ClF_3, IF_7, BrF_5, BrF_3, I_2Cl_6, IF_5, ClF, ClF_5 **[2022, 28 July Shift-I]**

26 The number of halogens forming halic (V) acid is .... **[2021, 31 Aug Shift-I]**

27 Among the following, the number of halides which is/are inert to hydrolysis is .... **[2021, 25 Feb Shift-I]**
(i) BF_3 (ii) $SiCl_4$
(iii) PCl_5 (iv) SF_6

28 The number of Cl = O bonds in perchloric acid is ".........". **[2020, 6 Sep Shift-I]**

29 Chlorine reacts with hot and concentrated NaOH and produces compounds (X) and (Y). Compound (X) gives white precipitate with silver nitrate solution. The average bond order between Cl and O atoms in (Y) is .... **[2020, 7 Jan Shift-I]**

Topic 4 / Group 18 Elements

• Level 1

01 The noble gas that does not occur in the atmosphere is **[2019, 10 April Shift-II]**
(a) Ra (b) Kr
(c) He (d) Ne

Level 2

02 Given below are two statements.

Statement I Noble gases have very high boiling points.

Statement II Noble gases are monoatomic gases. They are held together by strong dispersion forces. Because of this they are liquefied at very low temperature. Hence, they have very high boiling points.

In the light of the above statements, choose the correct answer from the options given below.

(a) Statement I is true but Statement II is false.
(b) Statement I is false but Statement II is true.
(c) Both Statement I and Statement II are true.
(d) Both Statement I and Statement II are false.

03 The difference between electron gain enthalpies will be maximum between **[2023, 6 April Shift-I]**

(a) Ar and Cl
(b) Ar and F
(c) Ne and Cl
(d) Ne and F

04 Match List-I with List-II.

List-I (Molecule)		List-II (Hybridisation shape)
A. XeO_3	I.	sp^3d; linear
B. XeF_2	II.	sp^3 ; pyramidal
C. $XeOF_4$	III.	sp^3d^3; distorted octahedral
D. XeF_6	IV.	sp^3d^2; square pyramidal

Choose the correct answer from the options given below. **[2022, 25 July Shift-II]**

(a) A-II, B-I, C-IV, D-III
(b) A-II, B-IV, C-III, D-I
(c) A-IV, B-II, C-III, D-I
(d) A-IV, B-II, C-I, D-III

05 The shape/structure of $[XeF_5]^-$ and XeO_3F_2, respectively, are **[2020, 2 Sep Shift-II]**

(a) pentagonal planar and trigonal bipyramidal
(b) octahedral and square pyramidal
(c) trigonal bipyramidal and pentagonal planar
(d) trigonal bipyramidal and trigonal bipyramidal

06 The type of hybridisation and number of lone pair(s) of electrons of Xe in $XeOF_4$, respectively, are **[2019, 10 Jan Shift-I]**

(a) sp^3d^2 and 1
(b) sp^3d and 2
(c) sp^3d and 1
(d) sp^3d^2 and 2

Level 3

07 Match the List-I with List-II. **[2025, 7 April Shift-I]**

List-I (Molecule/ion)	List-II Bond pair : lone pair (on the central atom)
A. ICl_2^-	I. 4 : 2
B. H_2O	II. 4 : 1
C. SO_2	III. 2 : 3
D. XeF_4	IV. 2 : 2

Choose the correct answer from the options given below:

(a) A-III, B-IV, C-I, D-II
(b) A-III, B-IV, C-II, D-I
(c) A-IV, B-III, C-II, D-I
(d) A-II, B-I, C-IV, D-III

08 A molecule with the formula AX_4Y has all it's elements from *p*-block. Element *A* is rarest monoatomic, non-radioactive from its group and has the lowest ionisation enthalpy value among *A*, *X* and *Y*. Elements *X* and *Y* have first and second highest electronegativity values respectively among all the known elements. The shape of the molecule is: **[2025, 2 April Shift-I]**

(a) pentagonal planar
(b) trigonal bipyramidal
(c) square pyramidal
(d) octahedral

09 Match List-I with List-II.

List-I (Species)	List-II (Number of lone pair of electrons on the central atom)
A. XeF_2	1. 0
B. XeO_2F_2	2. 1
C. XeO_3F_2	3. 2
D. XeF_4	4. 3

Choose the most appropriate answer from the options given below **[2021, 27 Aug Shift-I]**

(a) A-4, B-1, C-2, D-3
(b) A-3, B-4, C-2, D-1
(a) A-3, B-2, C-4, D-1
(b) A-4, B-2, C-1, D-3

10 The reaction in which the hybridisation of the underlined atom is affected is **[2020, 4 Sep Shift-II]**

(a) $\underline{Xe}F_4 + SbF_5 \longrightarrow$
(b) $H_2\underline{S}O_4 + NaCl \xrightarrow{420\ K}$
(c) $H_3\underline{P}O_2 \xrightarrow{\text{Disproportionation}}$
(d) $\underline{N}H_3 \xrightarrow{H^+}$

Numerical Value Type Questions

11 The difference in the oxidation state of Xe between the oxidised product of Xe formed on complete hydrolysis of XeF_4 and XeF_4 is .... **[2023, 10 April Shift-II]**

12 XeF_4 reacts with SbF_5 to form $[XeFm]^n\ [SbF_y]^{z-}$

$m + n + y + z = ...$ **[2023, 8 April Shift-I]**

13 The sum of number of lone pairs of electrons present on the central atoms of XeO_3, $XeOF_4$ and XeF_6, is .... **[2022, 25 July Shift-II]**

14 A xenon compound '*A*' upon partial hydrolysis gives XeO_2F_2. The number of lone pair of electrons present in compound *A* is .... (Round off to the nearest integer). **[2021, 18 March Shift-II]**

Key Index

Topic 1

1 (a)	2 (c)	3 (b)	4 (b)	5 (a)	6 (b)	7 (d)	8 (d)	9 (b)	10 (d)
11 (a)	12 (c)	13 (d)	14 (d)	15 (d)	16 (d)	17 (b)	18 (d)	19 (b)	20 (b)
21 (d)	22 (b)	23 (d)	24 (d)	25 (a)	26 (a)	27 (a)	28 (b)	29 (d)	30 (c)
31 (d)	32 (a)	33 (d)	34 (b)	35 (a)	36 (b)	37 (c)	38 (b)	39 (d)	40 (d)
41 (a)	42 (c)	43 (d)	44 (c)	45 (d)	46 (b)	47 (c)	48 (c)	49 (d)	50 (c)
51 (b)	52 (d)	53 (15)	54 (6)	55 (2)	56 (2)	57 (4)			

Topic 2

1 (c)	2 (a)	3 (d)	4 (b)	5 (a)	6 (b)	7 (d)	8 (c)	9 (d)	10 (b)
11 (d)	12 (c)	13 (a)	14 (d)	15 (a)	16 (a)	17 (d)	18 (b)	19 (c)	20 (c)
21 (a)	22 (c)	23 (b)	24 (1)	25 (4)	26 (1)	27 (7)			

Topic 3

1 (c)	2 (a)	3 (a)	4 (c)	5 (a)	6 (c)	7 (d)	8 (a)	9 (c)	10 (a)
11 (b)	12 (b)	13 (d)	14 (b)	15 (a)	16 (d)	17 (c)	18 (d)	19 (b)	20 (a)
21 (c)	22 (c)	23 (3)	24 (5)	25 (3)	26 (3)	27 (1)	28 (3)	29 (1.67)	

Topic 4

1 (a)	2 (d)	3 (c)	4 (a)	5 (a)	6 (a)	7 (b)	8 (c)	9 (d)	10 (a)
11 (2)	12 (11)	13 (3)	14 (19)						

For Detailed Solutions
scan QR Code

17 *d* and *f*–Block Elements

In JEE Main, there is usually 1-2 question per paper is visible from this chapter. In past 3 years, 41 conceptual and 6 numerical questions were asked of moderate difficulty level from this chapter.

For detailed analysis scan QR code

Topic 1 / Properties of Transition Elements

• Level 1

01 The correct decreasing order of spin only magnetic moment values (BM) of Cu^+, Cu^{2+}, Cr^{2+} and Cr^{3+} ions is **[2025, 8 April Shift-II]**
(a) $Cr^{2+} > Cr^{3+} > Cu^{2+} > Cu^+$
(b) $Cr^{3+} > Cr^{2+} > Cu^+ > Cu^{2+}$
(c) $Cu^+ > Cu^{2+} > Cr^{3+} > Cr^{2+}$
(d) $Cu^{2+} > Cu^+ > Cr^{2+} > Cr^{3+}$

02 The metal that shows highest and maximum number of oxidation state is **[2024, 5 April Shift-I]**
(a) Ti (b) Fe
(c) Co (d) Mn

03 Which of the following 3*d*-metal ion will give the lowest enthalpy of hydration ($\Delta_{hyd} H$), when dissolved in water ? **[2022, 29 July Shift-II]**
(a) Cr^{2+} (b) Mn^{2+} (c) Fe^{2+} (d) Co^{2+}

Imp. *Electronic Configuration of 5d-block Element*

04 The electronic configuration of Pt (atomic number 78) is **[2022, 29 June Shift-I]**
(a) $[Xe]\,4f^{14}\,5d^9\,6s^1$ (b) $[Kr]4f^{14}\,5d^{10}$
(c) $[Xe]\,4f^{14}\,5d^{10}$ (d) $[Xe]\,4f^{14}\,5d^8\,6s^2$

05 In 3*d*-series, the metal having the highest M^{2+}/M standard electrode potential is **[2022, 27 June Shift-II]**
(a) Cr (b) Fe
(c) Cu (d) Zn

06 Metals generally melt at very high temperature. Amongst the following, the metal with highest melting point will be **[2022, 24 June Shift-II]**
(a) Hg (b) Ag (c) Ga (d) Cs

07 The common positive oxidation states for an element with atomic number 24, are **[2021, 17 March Shift-II]**
(a) + 2 to + 6 (b) + 1 and + 3 to + 6
(c) + 1 and + 3 (d) + 1 to + 6

08 What is the spin only magnetic moment value (BM) of a divalent metal ion with atomic number 25, in its aqueous solution? **[2021, 17 March Shift-I]**
(a) 5.26 (b) 5.0 (c) 5.92 (d) Zero

09 The major components of German silver are **[2021, 25 Feb Shift-II]**
(a) Cu, Zn and Ag (b) Cu, Zn and Ni
(c) Ge, Cu and Ag (d) Zn, Ni and Ag

10 The major components in Gun metal are **[2021, 24 Feb Shift-I]**
(a) Cu, Sn and Zn (b) Cu, Zn and Ni
(c) Cu, Ni and Fe (d) Al, Cu, Mg and Mn

11 The electrode potential of M^{2+}/M of 3*d*-series elements shows positive value for **[2021, 24 Feb Shift-I]**
(a) Fe (b) Co (c) Zn (d) Cu

12 The atomic radius of Ag is closest to **[2020, 7 Jan Shift-I]**
(a) Hg (b) Au (c) Ni (d) Cu

13 The statement that is incorrect about the interstitial compounds is **[2019, 8 April Shift-II]**
(a) they are very hard
(b) they have metallic conductivity
(c) they have high melting points
(d) they are chemically reactive

14 The element that usually does not show variable oxidation states is **[2019, 1 Jan Shift-I]**
(a) Sc (b) Cu (c) Ti (d) V

• Level 2

15 The metal ions that have the calculated spin-only magnetic moment value of 4.9. BM are
A. Cr^{2+} B. Fe^{2+} C. Fe^{3+}
D. Co^{2+} E. Mn^{3+}

Choose the correct answer from the options given below: **[2025, 3 April Shift-I]**
(a) A, B and E Only (b) A, D and E Only
(c) B and E Only (d) A, C and E Only

16 The incorrect relationship in the following pairs in relation to ionisation enthalpies is **[2025, 4 April Shift-II]**
(a) $Fe^{2+} < Fe^{3+}$ (b) $Mn^{2+} < Fe^{2+}$
(c) $Mn^+ < Cr^+$ (d) $Mn^+ < Mn^{2+}$

17 The first transition series metal 'M' has the highest enthalpy of atomisation in its series. One of its aquated ion (M^{n+}) exists in green colour. The nature of the oxide formed by the above (M^{n+}) ion is **[2025, 7 April Shift-I]**

(a) neutral (b) amphoteric
(c) acidic (d) basic

18 The number of valence electrons present in the metal among Cr, Co, Fe and Ni which has the lowest enthalpy of atomisation is **[2025, 7 April Shift-I]**

(a) 8 (b) 10 (c) 6 (d) 9

19 Pair of transition metal ions having the same number of unpaired electrons is **[2025, 4 April Shift-I]**

(a) V^{2+}, Co^{2+} (b) Ti^{3+}, Mn^{2+}
(c) Ti^{2+}, Co^{2+} (d) Fe^{3+}, Cr^{2+}

20 Given below are two statements. **[2025, 3 April Shift-II]**

Statement I CrO_3 is a stronger oxidising agent than MoO_3.

Statement II Cr(VI) is more stable than Mo(VI).

In the light of the above statements, choose the correct answer from the options given below.

(a) Statement I is false but Statement II is true.
(b) Both Statement I and Statement II are false.
(c) Both Statement I and Statement II are true.
(d) Statement I is true but Statement II is false.

21 The amphoteric oxide among V_2O_3, V_2O_4 and V_2O_5, upon reaction with alkali leads to formation of an oxide anion. The oxidation state of V in the oxide anion is **[2025, 28 Jan Shift-II]**

(a) +4 (b) +7 (c) +5 (d) +3

22 Given below are two statements. **[2024, 9 April Shift-II]**

Statement I The higher oxidation states are more stable down the group among transition elements unlike *p*-block elements.

Statement II Copper cannot liberate hydrogen from weak acids.

In the light of the above statements, choose the correct answer from the options given below.

(a) Both Statement I and Statement II are false.
(b) Statement I is true but Statement II is false.
(c) Statement I is false but Statement II is true.
(d) Both Statement I and Statement II are true.

23 The electronic configuration of Cu(II) is $3d^9$ whereas that of Cu(I) is $3d^{10}$. Which of the following is correct? **[2024, 9 April Shift-I]**

(a) Cu(II) is more stable
(b) Cu(II) is less stable
(c) Stability of Cu(I) and Cu(II) depends on nature of copper salts
(d) Cu(I) and Cu(II) are equally stable

24 A first row transition metal in its +2 oxidation state has a spin-only magnetic moment value of 3.86 BM. The atomic number of the metal is **[2024, 4 April Shift-II]**

(a) 25 (b) 26 (c) 22 (d) 23

25 The element which shows only one oxidation state other than its elemental form is **[2024, 4 April Shift-I]**

(a) Nickel (b) Cobalt (c) Titanium (d) Scandium

26 Choose the correct option having all the elements with d^{10} electronic configuration from the following. **[2024, 27 Jan Shift-II]**

(a) ^{27}Co, ^{28}Ni, ^{26}Fe, ^{24}Cr (b) ^{28}Ni, ^{24}Cr, ^{26}Fe, ^{29}Cu
(c) ^{46}Pd, ^{28}Ni, ^{26}Fe, ^{24}Cr (d) ^{29}Cu, ^{30}Zn, ^{48}Cd, ^{47}Ag

27 Identify the incorrect pair from the following:

(a) Haber process—Iron **[2024, 27 Jan Shift-II]**
(b) Polythene preparation—$TiCl_4$, $Al(CH_3)_3$
(c) Wacker process—$PtCl_2$
(d) Photography—AgBr

28 Given below are two statements.

One is labelled as Assertion (A) and the other is labelled as Reason (R).

Assertion (A) The first ionisation enthalpy of $3d$ series elements is more than that of group 2 metals.

Reason (R) In $3d$ series of elements successive filling of *d*-orbitals takes place.

In the light of the above statements, choose the correct answer from the options given below.

(a) (A) is true but (R) is false. **[2023, 31 Jan Shift-II]**
(b) (A) is false but (R) is true.
(c) Both (A) and (R) are true and (R) is the correct explanation of (A).
(d) Both (A) and (R) are true but (R) is not the correct explanation of (A).

29 Which of the following has least tendency to liberate H_2 from mineral acids? **[2022, 28 July Shift-I]**

(a) Cu (b) Mn (c) Ni (d) Zn

30 Among the following, which is the strongest oxidising agent? **[2022, 25 June Shift-I]**

(a) Mn^{3+} (b) Fe^{3+} (c) Ti^{3+} (d) Cr^{3+}

31 Identify the elements for which electronic configuration in +3 oxidation state is $[Ar]3d^5$. **[2021, 1 Sep Shift-II]**

(a) Ru (b) Mn (c) Co (d) Fe

32 Which one of the following when dissolved in water gives coloured solution in nitrogen atmosphere? **[2021, 26 Aug Shift-I]**

(a) $CuCl_2$ (b) AgCl (c) $ZnCl_2$ (d) Cu_2Cl_2

33 The spin only magnetic moments (in BM) for free Ti^{3+}, V^{2+} and Sc^{3+} ions respectively are (Atomic number Sc : 21, Ti : 22, V : 23) **[2021, 25 July Shift-II]**
(a) 1.73, 0, 3.87 (b) 3.87, 1.73, 0
(c) 0, 3.87, 1.73 (d) 1.73, 3.87, 0

34 In which of the following pairs, the outer most electronic configuration will be the same? **[2021, 25 Feb Shift-I]**
(a) Ni^{2+} and Cu^{+} (b) Fe^{2+} and Co^{+}
(c) Cr^{+} and Mn^{2+} (d) V^{2+} and Cr^{+}

35 The set that contains atomic numbers of only transition elements, is **[2020, 6 Sep Shift-I]**
(a) 37, 42, 50, 64 (b) 21, 25, 42, 72
(c) 9, 17, 34, 38 (d) 21, 32, 53, 64

36 The pair that has similar atomic radii is **[2019, 12 April Shift-II]**
(a) Mn and Re (b) Ti and Hf
(c) Sc and Ni (d) Mo and W

• Level 3

37 The correct set of ions (aqueous solution) with same colour from the following is **[2025, 23 Jan Shift-I]**
(a) V^{2+}, Cr^{3+}, Mn^{3+} (b) Zn^{2+}, V^{3+}, Fe^{3+}
(c) Sc^{3+}, Ti^{3+}, Cr^{2+} (d) Ti^{4+}, V^{4+}, Mn^{2+}

38 The correct option with order of melting points of the pairs (Mn, Fe), (Tc, Ru) and (Re, Os) is **[2025, 29 Jan Shift-I]**
(a) Mn < Fe, Tc < Ru and Os < Re
(b) Fe < Mn, Ru < Tc and Os < Re
(c) Mn < Fe, Tc < Ru and Re < Os
(d) Fe < Mn, Ru < Tc and Re < Os

39 Match List-I with List-II. **[2025, 24 Jan Shift-II]**

List-I (Transition metal ion)	List-II (Spin only magnetic moment (BM))
(A) Ti^{3+}	(I) 3.87
(B) V^{2+}	(II) 0.00
(C) Ni^{2+}	(III) 1.73
(D) Sc^{3+}	(IV) 2.84

Choose the correct answer from the options given below
(a) (A) - (IV), (B) - (II), (C)-(III), (D) - (I)
(b) (A) - (III), (B) - (I), (C) - (IV), (D) - (II)
(c) (A) - (III), (B) - (I), (C) - (II), (D) - (IV)
(d) (A) - (II), (B) - (IV), (C) - (I), (D)-(III)

40 Consider 'n' is the number of lone pair of electrons present in the equatorial position of the most stable structure of ClF_3. The ions from the following with 'n' number of unpaired electrons are
A. V^{3+} B. Ti^{3+} C. Cu^{2+} D. Ni^{2+}
E. Ti^{2+}

Choose the correct answer from the options given below **[2025, 28 Jan Shift-I]**
(a) B and C only (b) A, D and E only
(c) B and D only (d) A and C only

Imp. *Calculation of Number of Unpaired Electrons*

41 Arrange the following elements in the increasing order of number of unpaired electrons in it.
A. Sc B. Cr C. V
D. Ti E. Mn

Choose the correct answer from the options given below **[2024, 6 April Shift-II]**
(a) C < E < B < A < D (b) A < D < C < B < E
(c) A < D < C < E < B (d) B < C < D < E < A

42 Which of the following electronic configuration would be associated with the highest magnetic moment? **[2024, 27 Jan Shift-I]**
(a) $[Ar]\,3d^7$ (b) $[Ar]\,3d^3$ (c) $[Ar]\,3d^6$ (d) $[Ar]\,3d^8$

43 In chromyl chloride, the number of d-electrons present on chromium is same as in (Given atomic number of Ti : 22, V : 23, Cr : 24, Mn : 25, Fe : 26) **[2023, 8 April Shift-I]**
(a) V (IV) (b) Ti (I) (c) Mn (VII) (d) Fe (II)

44 In following pairs, the one in which both transition metal ions are colourless is **[2022, 29 July Shift-I]**
(a) Sc^{3+}, Zn^{2+} (b) Ti^{4+}, Cu^{2+}
(c) V^{2+}, Ti^{3+} (d) Zn^{2+}, Mn^{2+}

45 The metal ion (in gaseous state) with lowest spin only magnetic moment value is **[2022, 25 June Shift-II]**
(a) V^{2+} (b) Ni^{2+} (c) Cr^{2+} (d) Fe^{2+}

Imp. *Colour and Magnetic Properties of d-block Species*

46 The set having ions which are coloured and paramagnetic both is **[2021, 22 July Shift-II]**
(a) Cu^{+}, Zn^{2+}, Mn^{4+} (b) Ni^{2+}, Mn^{7+}, Hg^{2+}
(c) Cu^{2+}, Cr^{3+}, Sc^{+} (d) Sc^{3+}, V^{5+}, Ti^{4+}

47 Match List-I and List-II.

List-I		List-II	
(A)	Chlorophyll	(i)	Ruthenium
(B)	Vitamin-B_{12}	(ii)	Platinum
(C)	Anticancer drug	(iii)	Cobalt
(D)	Grubbs catalyst	(iv)	Magnesium

Choose the most appropriate answer from the options given below. **[2021, 18 March Shift-I]**
(a) (A)-(iv), (B)-(ii), (C)-(iii), (D)-(i)
(b) (A)-(iv), (B)-(iii) (C)-(i), (D)-(ii)
(c) (A)-(iii), (B)-(ii), (C)-(iv), (D)-(i)
(d) (A)-(iv), (B)-(iii), (C)-(ii), (D)-(i)

48 Consider the hydrated ions of Ti^{2+}, V^{2+}, Ti^{3+} and Sc^{3+}. The correct order of their spin-only magnetic moment is **[2019, 10 April 12 Shift-I]**

(a) $Sc^{3+} < Ti^{3+} < Ti^{2+} < V^{2+}$
(b) $Sc^{3+} < Ti^{3+} < V^{2+} < Ti^{2+}$
(c) $Ti^{3+} < Ti^{2+} < Sc^{3+} < V^{2+}$
(d) $V^{2+} < Ti^{2+} < Ti^{3+} < Sc^{3+}$

Numerical Value Type Questions

49 Among Sc, Mn, Co and Cu. identify the element with highest enthalpy of atomisation. The spin only magnetic moment value of that element in its +2 oxidation state is .... BM (in Nearest integer) **[2025, 3 April Shift-II]**

50 Niobium (Nb) and ruthenium (Ru) have 'x' and 'y' number of electrons in their respective 4*d*-orbitals. The value of $x + y$ is .... **[2025, 22 Jan Shift-II]**

51 The spin only magnetic moment (μ) value (BM) of the compound with strongest oxidising power among Mn_2O_3, TiO and VO is .... BM (Nearest integer). **[2025, 28 Jan Shift-II]**

52 A transition metal '*M*' among Sc, Ti, V, Cr, Mn and Fe has the highest second ionisation enthalpy. The spin-only magnetic moment value of M^+ ion is .... BM. (Nearest integer)

(Given, atomic number Sc : 21, Ti : 22, V : 23, Cr : 24, Mn : 25, Fe : 26) **[2024, 9 April Shift-II]**

53 The 'spin only' magnetic moment value of MO_4^{2-} is .... BM. (Where *M* is a metal having least metallic radii among Sc, Ti, V, Cr, Mn and Zn).

(Given atomic number : Sc = 21, Ti = 22, V = 23, Cr = 24, Mn = 25 and Zn = 30) **[2024, 8 April Shift-II]**

54 The spin-only magnetic moment value of the ion among Ti^{2+}, V^{2+}, Co^{3+} and Cr^{2+}, that acts as strong oxidising agent in aqueous solution is .... BM (Nearest integer).

(Given, atomic numbers : Ti : 22, V : 23, Cr : 24, Co : 27) **[2024, 5 April Shift-I]**

55 A first row transition metal with highest enthalpy of atomisation, upon reaction with oxygen at high temperature forms oxides of formula M_2O_n (where $n = 3, 4, 4$). The 'spin-only' magnetic moment value of the amphoteric oxide from the above oxides is...... BM. (Nearest integer)

(given, atomic number: Sc : 21, Ti : 22, V : 23, Cr : 24, Mn : 25, Fe : 26, Co : 27, Ni : 28, Cu : 29, Zn : 30) **[2024, 4 April Shift-II]**

56 Consider the following reaction

$$MnO_2 + KOH + O_2 \longrightarrow A + H_2O$$

Product '*A*' in neutral or acidic medium disproportionate to give products '*B*' and '*C*' along with water. The sum of spin-only magnetic moment values of *B* and *C* is...... BM. (Nearest integer). (Given atomic number of Mn is 25) **[2024, 4 April Shift-I]**

57 The spin - only magnetic moment value of M^{3+} ion (in gaseous state) from the pairs Cr^{3+} / Cr^{2+}, Mn^{3+} / Mn^{2+}, Fe^{3+} / Fe^{2+} and Co^{3+} / Co^{2+} that has negative standard electrode potential, is .... BM. [nearest integer] **[2022, 25 July Shift-II]**

58 Among Co^{3+}, Ti^{2+}, V^{2+} and Cr^{2+} ions, one if used as a reagent cannot liberate H_2 from dilute mineral acid solution, its spin-only magnetic moment in gaseous state is .... BM (Nearest integer) **[2022, 25 JulyShift-I]**

59 Number of electrons that vanadium ($Z = 23$) has in *p*-orbitals is equal to .... **[2021, 22 July Shift-II]**

60 In the ground state of atomic Fe($Z = 26$), the spin-only magnetic moment is .... $\times 10^{-1}$ BM. (Round off to the nearest integer). [Given : $\sqrt{3} = 1.73$, $\sqrt{2} = 1.41$] **[2021, 17 March Shift-II]**

Topic 2 / Important Compounds of Transition Elements

• Level 1

01 Dihydrogen reacts with CuO to give **[2022, 28 June Shift-I]**

(a) CuH_2 (b) Cu
(c) Cu_2O (d) $Cu(OH)_2$

02 The nature of oxides V_2O_3 and CrO is indexed as '*X*' and '*Y*' type respectively. The correct set of *X* and *Y* is **[2021, 27 Aug Shift-I]**

(a) *X* = acidic *Y* = acidic
(b) *X* = basic *Y* = amphoteric
(c) *X* = basic *Y* = basic
(d) *X* = amphoteric *Y* = basic

03 In the given chemical reaction, colours of the Fe^{2+} and Fe^{3+} ions, are respectively

$$5Fe^{2+} + MnO_4^- + 8H^+ \longrightarrow Mn^{2+} + 4H_2O + 5Fe^{3+}$$

[2021, 1 Sep Shift-II]

(a) yellow, orange
(b) yellow, green
(c) green, orange
(d) green, yellow

Level 2

04 Find the compound '*A*' from the following reaction sequences. **[2025, 24 Jan Shift-II]**

$$A \xrightarrow{\text{Aqua-regia}} B \xrightarrow[\text{(ii) AcOH}]{\text{(i) } KNO_2/[NH_4OH]} \text{yellow ppt.}$$

(a) ZnS (b) MnS (c) CoS (d) NiS

05 Consider the following reactions, **[2025, 23 Jan Shift-II]**

$$K_2Cr_2O_7 \xrightarrow[-H_2O]{KOH} [A] \xrightarrow[-H_2O]{H_2SO_4} [B] + K_2SO_4$$

The products [*A*] and [*B*], respectively are

(a) K_2CrO_4 and $K_2Cr_2O_7$ (b) $K_2Cr(OH)_6$ and Cr_2O_3
(c) K_2CrO_4 and CrO (d) K_2CrO_4 and Cr_2O_3

06 The orange colour of $K_2Cr_2O_7$ and purple colour of $KMnO_4$ is due to **[2024, 30 Jan Shift-II]**

(a) $d \to d$ transitions in $KMnO_4$ and charge transfer transitions in $K_2Cr_2O_7$
(b) $d \to d$ transitions in $K_2Cr_2O_7$ and charge transfer transitions in $KMnO_4$
(c) Charge transfer transition in both
(d) $d \to d$ transitions in both

07 Alkaline oxidative fusion of MnO_2 gives "*A*" which on electrolytic oxidation in alkaline solution produces *B*. *A* and *B* respectively are **[2024, 30 Jan Shift-II]**

(a) MnO_4^{2-} and MnO_4^- (b) Mn_2O_3 and MnO_4^{2-}
(c) Mn_2O_7 and MnO_4^- (d) MnO_4^{2-} and Mn_2O_7

08 Highest oxidation state of Mn is exhibited in Mn_2O_7. The correct statements about Mn_2O_7 are

A. Mn is tetrahedrally surrounded by oxygen atoms.
B. Mn is octahedrally surrounded by oxygen atoms.
C. contains Mn—O—Mn bridge.
D. contains Mn—Mn bond.

Choose the correct answer from the options given below. **[2023, 1 Feb Shift-I]**

(a) B and D only (b) B and C only
(c) A and C only (d) A and D only

09 When Cu^{2+} ion is treated with KI, a white precipitate, *X* appears in solution. The solution is titrated with sodium thiosulphate, the compound *Y* is formed. *X* and *Y* respectively are **[2023, 31 Jan Shift-I 2023]**

(a) $X = Cu_2I_2$ $Y = Na_2S_4O_5$
(b) $X = CuI_2$ $Y = Na_2S_2O_3$
(c) $X = CuI_2$ $Y = Na_2S_4O_6$
(d) $X = Cu_2I_2$ $Y = Na_2S_4O_6$

Imp. *Structure of Mn_2O_7*

10 The total number of Mn ═O bonds in Mn_2O_7 is **[2022, 27 July Shift-I]**

(a) 4 (b) 5 (c) 6 (d) 3

11 The products obtained from a reaction of hydrogen peroxide and acidified potassium permanganate are **[2022, 26 July Shift-II]**

(a) Mn^{4+}, H_2O only (b) Mn^{2+}, H_2O only
(c) Mn^{4+}, H_2O, O_2 only (d) Mn^{2+}, H_2O, O_2 only

12 In the structure of the dichromate ion, there is a **[2021, 31 Aug Shift-I]**

(a) linear symmetrical Cr—O—Cr bond
(b) non-linear unsymmetrical Cr—O—Cr bond
(c) non-linear symmetrical Cr—O—Cr bond
(d) linear unsymmetrical Cr—O—Cr bond

13 The oxide that shows magnetic property is **[2021, 18 March Shift-II]**

(a) SiO_2 (b) Mn_3O_4 (c) Na_2O (d) MgO

14 Fex_2 and Fey_3 are known when *x* and *y* are **[2021, 16 March Shift-II]**

(a) *x* = F, Cl, Br, I and *y* = F, Cl and Br
(b) *x* = F, Cl, Br and *y* = F, Cl, Br and I
(c) *x* = Cl, Br, I and *y* = F, Cl, Br and I
(d) *x* = F, Cl, Br, I and *y* = F, Cl, Br and I

Level 3

Imp. *Oxidation Reaction of $K_2Cr_2O_7$ and $KMnO_4$*

15 Which of the following oxidation reactions are carried out by both $K_2Cr_2O_7$ and $KMnO_4$ in acidic medium? **[2025, 28 Jan Shift-I]**

A. $I^- \to I_2$ B. $S^{2-} \to S$ C. $Fe^{2+} \to Fe^{3+}$
D. $I^- \to IO_3^-$ E. $S_2O_3^{2-} \to SO_4^{2-}$

Choose the correct answer from the options given below

(a) C, D and E only (b) A, B and C only
(c) B, C and D only (d) A, D and E only

16 Preparation of potassium permanganate from MnO_2, involves two step process in which the 1st step is a reaction with KOH and KNO_3 to produce **[2025, 24 Jan Shift-I]**

(a) K_3MnO_4 (b) $K_4[Mn(OH)_6]$
(c) K_2MnO_4 (d) $KMnO_4$

17 Match List -I with List -II. **[2025, 23 Jan Shift-II]**

List-I		List-II	
(A)	Bronze	(I)	Cu, Ni
(B)	Brass	(II)	Fe, Cr, Ni, C
(C)	UK silver coin	(III)	Cu, Zn
(D)	Stainless steel	(IV)	Cu, Sn

Choose the correct answer from the options given below.

(a) (A)-(IV), (B)-(III), (C)-(I), (D)-(II)
(b) (A)-(III), (B)-(I), (C)-(IV), (D)-(II)
(c) (A)-(III), (B)-(IV), (C)-(II), (D)-(I)
(d) (A)-(IV), (B)-(II), (C)-(III), (D)-(I)

Imp. *Reactions of $KMnO_4$*

18 During the reaction of permanganate with thiosulphate, the change in oxidation of manganese occurs by value of 3. Identify which of the below medium will favour the reaction. **[2023, 6 April Shift-II]**
(a) Both aqueous acidic and faintly alkaline
(b) Aqueous acidic
(c) Aqueous neutral
(d) Both aqueous acidic and neutral

19 In neutral or faintly alkaline medium, $KMnO_4$ being a powerful oxidant can oxidise, thiosulphate almost quantitatively, to sulphate. In this reaction overall change in oxidation state of manganese will be **[2022, 29 July Shift-I]**
(a) 5 (b) 1 (c) 0 (d) 3

20 Given below are two statements.
Statement I Iron (III) catalyst, acidified $K_2Cr_2O_7$ and neutral $KMnO_4$ have the ability to oxidise I^- to I_2 independently.
Statement II Manganese ion is paramagnetic in nature and involves $p\pi - p\pi$ bonding.
In the light of the above statements, choose the correct answer from the options given below. **[2022, 27 July Shift-I]**
(a) Both Statement I and Statement II are true.
(b) Both Statement I and Statement II are false.
(c) Statement I is true but Statement II is false.
(d) Statement I is false but Statement II is true.

21 The dark purple colour of $KMnO_4$ disappears in the titration with oxalic acid in acidic medium. The overall change in the oxidation number of manganese in the reaction is **[2022, 26 July Shift-I]**
(a) 5 (b) 1 (c) 7 (d) 2

22 In which one of the following sets all species show disproportionation reaction **[2021, 31 Aug Shift-II]**
(a) ClO_2^-, F_2, MnO_4^{2-} and $Cr_2O_7^{2-}$
(b) $Cr_2O_7^{2-}, MnO_4^-, ClO_2^-$ and Cl_2
(c) ClO_2^-, Cl_2 and Mn^{3+}
(d) $ClO_4^-, MnO_4^{2-}, ClO_2^-$ and F_2

23 Potassium permanganate on heating at 513 K gives a product which is **[2021, 27 Aug Shift-II]**
(a) paramagnetic and colourless
(b) diamagnetic and green
(c) diamagnetic and colourless
(d) paramagnetic and green

24 The correct order of following $3d$-metal oxides, according to their oxidation numbers is
(A) CrO_3 (B) Fe_2O_3
(C) MnO_2 (D) V_2O_5
(E) Cu_2O **[2021, 25 July Shift-I]**
(a) (D) > (A) > (B) > (C) > (E)
(b) (A) > (C) > (D) > (B) > (E)
(c) (A) > (D) > (C) > (B) > (E)
(d) (C) > (A) > (D) > (E) > (B)

25 Given below are two statements
Statement I Potassium permanganate, on heating at 573 K forms potassium manganate.
Statement II Both potassium permanganate and potassium manganate are tetrahedral and paramagnetic in nature.
In the light of the above statements, choose the most appropriate answer from the options given below **[2021, 17 March Shift-I]**
(a) Both statement I and statement II are true.
(b) Both statement I and statement II are false.
(c) Statement I is false but statement II is true.
(d) Statement I is true but statement II is false.

26 The incorrect statement among the following is **[2021, 24 Feb Shift-II]**
(a) red colour of ruby is due to the presence of Co^{3+}
(b) RuO_4 is an oxidizing agent
(c) Cr_2O_3 is an amphoteric oxide
(d) $VOSO_4$ is a reducing agent

27 The incorrect statement is : **[2020, 3 Sep Shift-II]**
(a) Manganate ion is green in colour and permanganate ion is purple in colour.
(b) Manganate and permanganate ions are paramagnetic.
(c) In manganate and permanganate ions, the π-bonding takes place by overlap of p-orbitals of oxygen and d-orbitals of manganese.
(d) Manganate and permanganate ions are tetrahedral.

28 $$\underline{A} \xrightarrow{4\,KOH,\,O_2} \underset{\text{(Green)}}{2\underline{B}} + 2H_2O$$
$$3\underline{B} \xrightarrow{4\,HCl} \underset{\text{(Purple)}}{2\underline{C}} + MnO_2 + 2H_2O$$
$$2\underline{C} \xrightarrow{H_2O,\,KI} 2\underline{A} + 2KOH + \underline{D}$$
In the above sequence of reactions, $\underline{A}$ and $\underline{D}$, respectively, are **[2019, 11 Jan Shift-II]**
(a) KI and $KMnO_4$ (b) MnO_2 and KIO_3
(c) KI and K_2MnO_4 (d) KIO_3 and MnO_2

Numerical Value Type Questions

29 Consider the following reactions. **[2025, 3 April Shift-I]**
$$A + \underset{\substack{\text{Little}\\ \text{amount}}}{NaCl} + H_2SO_4 \longrightarrow CrO_2Cl_2 + \text{side products}$$
$$CrO_2Cl_2(\text{Vapour}) + NaOH \longrightarrow B + NaCl + H_2O$$
$$B + H^+ \longrightarrow C + H_2O$$

The number of terminal 'O' present in the compound 'C' is ....

30 $KMnO_4$ acts as an oxidising agent in acidic medium. 'X' is the difference between the oxidation states of Mn in reactant and product. 'Y' is the number of 'd' electrons present in the brown red precipitate formed at the end of the acetate ion test with neutral ferric chloride. The value of $X+Y$ is **[2025, 4 April Shift-I]**

Imp. *Calculation of μ in chemical of $KMnO_4$*

31 The difference in the 'spin-only' magnetic moment values of $KMnO_4$ and the manganese product formed during titration of $KMnO_4$, against oxalic acid in acidic medium is .... BM. (Nearest integer) **[2024, 6 April Shift-I]**

32 Among CrO, Cr_2O_3 and CrO_3 the sum of spin-only magnetic moment values of basic and amphoteric oxides is .... 10^{-2} BM (Nearest integer).
(Given, atomic number of Cr is 24) **[2024, 6 April Shift-I]**

33 The fusion of chromite ore with sodium carbonate in the presence of air leads to the formation of products A and B along with the evolution of CO_2. The sum of spin-only magnetic moment values of A and B is .... B.M. (Nearest integer)
[Given, atomic number : C : 6, Na : 11, O : 8, Fe : 26, Cr : 24] **[2024, 5 April Shift-II]**

34 The total change in the oxidation state of manganese involved in the reaction of $KMnO_4$ and potassium iodide in the acidic medium is .... **[2023, 15 April Shift-I]**

35 The spin-only magnetic moment value of the compound with strongest oxidising ability among MnF_4, MnF_3 and MnF_2 is .... BM. [nearest integer] **[2022, 26 July Shift-II]**

36 In chromyl chloride, the oxidation state of chromium is (+) .... **[2023, 15 April Shift-I]**

37 The oxidation states of transition metal atoms in $K_2Cr_2O_7$, $KMnO_4$ and K_2FeO_4, respectively, are x, y and z. The sum of x, y and z is .... **[2022, 2 Sept Shift-II]**

38 The disproportionation of MnO_4^{2-} in acidic medium resulted in the formation of two manganese compounds A and B. If the oxidation state of Mn in B is smaller than that of A, then the spin-only magnetic moment (μ) value of B in BM is .... (Nearest integer) **[2022, 28 July Shift-I]**

39 The number of terminal oxygen atoms present in the product B obtained from the following reaction is .... **[2022, 29 June Shift-I]**

$$FeCr_2O_4 + Na_2CO_3 + O_2 \longrightarrow A + Fe_2O_3 + CO_2$$
$$A + H^+ \longrightarrow B + H_2O + Na^+$$

40 An acidified manganate solution undergoes disproportionation reaction. The spin-only magnetic moment value of the product having manganese in higher oxidation state is ...... BM. (Nearest integer) **[2022, 29 June Shift-I]**

41 The number of statement(s) correct from the following for copper (Atomic number 29) is/are ...... **[2022, 27 June Shift-I]**
(A) Cu(II) complexes are always paramagnetic.
(B) Cu(I) complexes are generally colourless.
(C) Cu(I) is easily oxidised.
(D) In Fehling solution, the active reagent has Cu(I).

42 The spin–only magnetic moment value of the most basic oxide of vanadium among V_2O_3, V_2O_4 and V_2O_5 is ...... BM. (Nearest integer) **[2022, 26 June Shift-I]**

43 Manganese (VI) has ability to disproportionate in acidic solution. The difference in oxidation states of two ions it forms in acidic solution is .... **[2022, 24 June Shift-II]**

44 The difference in oxidation state of chromium in chromate and dichromate salt is .... **[2022, 24 June Shift-I]**

45 In mild alkaline medium, thiosulphate ion is oxidised by MnO_4^- to "A". The oxidation state of sulphur in "A" is .... **[2021, 26 Feb Shift-II]**

46 Dichromate ion is treated with base, the oxidation number of Cr in the product formed is .... **[2021, 26 Feb Shift-I]**

47 The sum of the total number of bonds between chromium and oxygen atoms in chromate and dichromate ions is .... **[2020, 9 Jan Shift-II]**

Topic 3 / Inner Transition Elements (Lanthanoids and Actinoids)

• Level 1

01 Which of the following ions is the strongest oxidising agent? (Atomic Number of Ce = 58, Eu = 63, Tb = 65, Lu = 71) **[2025, 24 Jan Shift-I]**
(a) Lu^{3+} (b) Ce^{3+} (c) Tb^{4+} (d) Eu^{2+}

02 The electronic configuration of Einsteinium is (Given atomic number of Einsteinium = 99) **[2024, 9 April Shift-II]**

(a) [Rn] $5f^{12}\,6d^0\,7s^2$ (b) [Rn] $5f^{11}6d^0 7s^2$
(c) [Rn] $5f^{13}6d^0 7s^2$ (d) [Rn] $5f^{10}6d^0 7s^2$

03 The number of element from the following that do not belong to lanthanoids is
Eu, Cm, Er, Tb, Yb and Lu **[2024, 6 April Shift-I]**

(a) 5 (b) 4
(c) 3 (d) 1

04 The electronic configuration for neodymium is [Atomic number for neodymium 60] **[2024, 27 Jan Shift-I]**

(a) [Xe] $5f^7 7s^2$ (b) [Xe] $4f^4 6s^2$
(c) [Xe] $4f^1\,5d^1\,6s^2$ (d) [Xe] $4f^6 6s^2$

05 Given below are two statements, one is labelled as Assertion (A) and the other is labelled as Reason (R).

Assertion (A) $5f$-electrons can participate in bonding to a far greater extent than $4f$-electrons.

Reason (R) $5f$-orbitals are not as buried as $4f$-orbitals.

In the light of the above statements, choose the correct answer from the options given below. **[2023, 12 April Shift-I]**

(a) A is true but R is false
(b) Both A and R are true and R is the correct explanation of A
(c) Both A and R are true but R is not the correct explanation of A
(d) A is false but R is true

06 Nd^{2+} = ... **[2023, 31 Jan Shift-I]**

(a) $4f^3$ (b) $4f^4$ (c) $4f^4 6s^2$ (d) $4f^2 6s^2$

07 Which one amongst the following are good oxidising agents?

A. Sm^{2+} B. Ce^{2+} C. Ce^{4+} D. Tb^{4+}

Choose the most appropriate answer from the options given below. **[2023, 24 Jan Shift-II]**

(a) C and D (b) A and B (c) Only C (d) Only D

08 The most common oxidation state of lanthanoid elements is +3. Which of the following is likely to deviate easily from +3 oxidation state? **[2022, 26 June Shift-II]**

(a) Ce (58) (b) La (57)
(c) Lu (71) (d) Gd (64)

09 The Eu^{2+} ion is a strong reducing agent in spite of its ground state electronic configuration (outermost) [Atomic number of Eu = 63] **[2021, 31 Aug Shift-II]**

(a) $4f^7 6s^2$ (b) $4f^6$ (c) $4f^7$ (d) $4f^6 6s^2$

10 Misch metal is an alloy consisting mainly of

(a) lanthanoid metals **[2020, 6 Sep Shift-II]**
(b) actinoid and transition metals
(c) lanthanoid and actinoid metals
(d) actinoid metals

11 The lanthanoid that does not show +4 oxidation state is **[2020, 6 Sep Shift-I]**

(a) Dy (b) Ce (c) Eu (d) Tb

12 The elements with atomic numbers 101 and 104 belong to, respectively, **[2020, 4 Sep Shift-I]**

(a) actinoids and group 4
(b) actinoids and group 6
(c) group 6 and actinoids
(d) group 11 and group 4

13 The highest possible oxidation states of uranium and plutonium, respectively, are **[2019, 10 April Shift-II]**

(a) 7 and 6 (b) 6 and 7
(c) 6 and 4 (d) 4 and 6

14 The correct order of atomic radii is **[2019, 12 Jan Shift-II]**

(a) Ho > N > Eu > Ce (b) N > Ce > Eu > Ho
(c) Eu > Ce > Ho > N (d) Ce > Eu > Ho > N

15 The effect of lanthanoid contraction in the lanthanoid series of elements by and large means **[2019, 10 Jan Shift-I]**

(a) increase in atomic radii and decrease in ionic radii
(b) decrease in both atomic and ionic radii
(c) increase in both atomic and ionic radii
(d) decrease in atomic radii and increase in ionic radii

• Level 2

16 Lanthanoid ions with $4f^7$ configuration are

(A) Eu^{2+} (B) Gd^{3+} (C) Eu^{3+} (D) Tb^{3+} (E) Sm^{2+}

Choose the correct answer from the options given below **[2025, 22 Jan Shift-I]**

(a) (A) and (D) only
(b) (A) and (B) only
(c) (B) and (E) only
(d) (B) and (C) only

17 Given below are two statements. **[2024, 27 Jan Shift-II]**

Statement I In the lanthanoids, the formation of Ce^{4+} is favoured by its noble gas configuration.

Statement II Ce^{4+} is a strong oxidant reverting to the common +3 state.

In the light of the above statements, choose the **most appropriate** answer from the options given below.

(a) Statement I is false but Statement II is true.
(b) Statement I is true but Statement II is false.
(c) Both Statement I and Statement II are true.
(d) Both Statement I and Statement II are false.

Imp. *Electronic Configuration of Lanthanoids*

18 The 'f' orbitals are half and completely filled, respectively in lanthanide ions
(Given, atomic number : Eu 63; Sm 62; Tm 69; Tb 65; Yb 70; Dy 66] **[2022, 27 June Shift-II]**
(a) Eu^{2+} and Tm^{2+} (b) Sm^{2+} and Tm^{3+}
(c) Tb^{4+} and Yb^{2+} (d) Dy^{3+} and Yb^{3+}

19 Cerium (IV) has a noble gas configuration. Which of the following is correct statement about it? **[2022, 25 June Shift-I]**
(a) It will not prefer to undergo redox reactions.
(b) It will prefer to gain electron and act as an oxidising agent.
(c) It will prefer to give away an electron and behaves as reducing agent.
(d) It acts as both, oxidising and reducing agent.

20 Given below are two statements.
Statement I The $E°$ value of Ce^{4+}/Ce^{3+} is +1.74 V.
Statement II Ce is more stable in Ce^{4+} state than Ce^{3+} state.
In the light of the above statements, choose the most appropriate answer from the options given below. **[2021, 16 March Shift-I]**
(a) Both statement I and statement II are correct.
(b) Statement I is incorrect but statement II is correct.
(c) Both statement I and statement II are incorrect.
(d) Statement I is correct but statement II is incorrect.

21 Given below are two statements. One is labelled as Assertion (A) and the other is labelled as Reason (R).
Assertion (A) Size of Bk^{3+} ion is less than Np^{3+} ion.
Reason (R) The above is a consequence of the lanthanoid contraction. In the light of the above statements, choose the correct answer from the options given below **[2021, 16 March Shift-I]**
(a) A is false but R is true.
(b) Both A and R are true but R is not the correct explanation of (a)
(c) Both A and R are true and R is the correct explanation of (a)
(d) A is true but R is false.

22 Which one of the following lanthanoids does not form MO_2?
[M is lanthanoid metal] **[2021, 26 Feb Shift-I]**
(a) Pr (b) Dy (c) Nd (d) Yb

23 The correct electronic configuration and spin-only magnetic moment (BM) of Gd^{3+} $(Z=64)$, respectively, are **[2020, 5 Sep Shift-I]**
(a) [Xe] $4f^7$ and 8.9 (b) [Xe] $4f^7$ and 7.9
(c) [Xe] $5f^7$ and 8.9 (d) [Xe] $5f^7$ and 7.9

24 The maximum number of possible oxidation states of actinides are shown by **[2019, 9 April Shift-II]**
(a) berkelium (Bk) and californium (Cf)
(b) neptunium (Np) and plutonium (Pu)
(c) nobelium (No) and lawrencium (Lr)
(d) actinium (Ac) and thorium (Th)

Level 3

25 The pair of lanthanides in which both elements have high third - ionisation energy **[2023, 13 April Shift-I]**
(a) Lu, Yb (b) Dy, Gd
(c) Eu, Yb (d) Eu, Gd

26 Which of the following pair is not isoelectronic species?
(Atomic number : Sm 62; Er 68; Yb 70; Lu 71; Eu 63; Tb 65; Tm 69) **[2022, 28 July Shift-II]**
(a) Sm^{2+} and Er^{3+} (b) Yb^{2+} and Lu^{3+}
(c) Eu^{2+} and Tb^{4+} (d) Tb^{2+} and Tm^{4+}

27 Which one of the lanthanoids given below is the most stable in divalent form? **[2022, 28 June Shift-I]**
(a) Ce (atomic number 58)
(b) Sm (atomic number 62)
(c) Eu (atomic number 63)
(d) Yb (atomic number 70)

28 Which one of the following lanthanides exhibits +2 oxidation state with diamagnetic nature ?
(Given, Z for Nd = 60, Yb = 70, La = 57, Ce = 58) **[2021, 31 Aug Shift-I]**
(a) Nd (b) Yb
(c) La (d) Ce

29 Given below are two statements
Statement I CeO_2 can be used for oxidation of aldehydes and ketones.
Statement II Aqueous solution of $EuSO_4$ is a strong reducing agent.
In the light of the above statements, choose the correct answer from the options given below. **[2021, 25 Feb Shift-I]**
(a) Both statement I and statement II are true.
(b) Both statement I and statement II are false.
(c) Statement I is true but statement II is false.
(d) Statement I is false but statement II is true.

30 The incorrect statement(s) among (1)-(3) is (are)
(1) W(VI) is more stable than Cr(VI).
(2) In the presence of HCl, permanganate titrations provide satisfactory results.
(3) Some lanthanoid oxides can be used as phosphorus. **[2020, 4 Sep Shift-II]**
(a) (2) and (3) only
(b) (1) and (2) only
(c) (2) only
(d) (1) only

Numerical Value Type Questions

31 Number of colourless lanthanoid ions among the following is ....
Eu^{3+}, Lu^{3+}, Nd^{3+}, La^{3+}, Sm^{3+} **[2024, 9 April Shift-I]**

32 The number of f-electrons in the ground state electronic configuration of Np ($Z = 93$) is .... (Nearest integer) **[2021, 27 Aug Shift-I]**

33 The number of $4f$ electrons in the ground state electronic configuration of Gd^{2+} is .... [Atomic number of Gd is 64.] **[2021, 26 Aug Shift-I]**

34 Number of electrons present in $4f$ orbital of Ho^{3+} ion is .... (Given, atomic number of Ho = 67) **[2021, 25 July Shift-II]**

Key Index

Topic 1

1	(a)	**2**	(d)	**3**	(b)	**4**	(a)	**5**	(c)	**6**	(b)	**7**	(a)	**8**	(c)	**9**	(b)	**10**	(a)
11	(d)	**12**	(b)	**13**	(d)	**14**	(a)	**15**	(a)	**16**	(b)	**17**	(d)	**18**	(c)	**19**	(a)	**20**	(d)
21	(c)	**22**	(d)	**23**	(a)	**24**	(d)	**25**	(d)	**26**	(d)	**27**	(c)	**28**	(c)	**29**	(a)	**30**	(a)
31	(d)	**32**	(a)	**33**	(d)	**34**	(c)	**35**	(b)	**36**	(d)	**37**	(a)	**38**	(a)	**39**	(b)	**40**	(b)
41	(c)	**42**	(c)	**43**	(c)	**44**	(a)	**45**	(b)	**46**	(c)	**47**	(d)	**48**	(a)	**49**	(4)	**50**	(11)
51	(5)	**52**	(6)	**53**	(0)	**54**	(5)	**55**	(0)	**56**	(4)	**57**	(4)	**58**	(5)	**59**	(12)	**60**	(49)

Topic 2

1	(b)	**2**	(c)	**3**	(d)	**4**	(c)	**5**	(a)	**6**	(c)	**7**	(a)	**8**	(c)	**9**	(d)	**10**	(c)
11	(d)	**12**	(c)	**13**	(b)	**14**	(a)	**15**	(b)	**16**	(c)	**17**	(a)	**18**	(c)	**19**	(d)	**20**	(b)
21	(a)	**22**	(c)	**23**	(d)	**24**	(c)	**25**	(d)	**26**	(a)	**27**	(b)	**28**	(b)	**29**	(6)	**30**	(10)
31	(6)	**32**	(877)	**33**	(6)	**34**	(5)	**35**	(5)	**36**	(6)	**37**	(19)	**38**	(4)	**39**	(6)	**40**	(0)
41	(3)	**42**	(3)	**43**	(3)	**44**	(0)	**45**	(6)	**46**	(6)	**47**	(12)						

Topic 3

1	(c)	**2**	(b)	**3**	(d)	**4**	(b)	**5**	(b)	**6**	(b)	**7**	(a)	**8**	(a)	**9**	(c)	**10**	(a)
11	(c)	**12**	(a)	**13**	(b)	**14**	(c)	**15**	(b)	**16**	(b)	**17**	(c)	**18**	(c)	**19**	(b)	**20**	(d)
21	(d)	**22**	(d)	**23**	(b)	**24**	(b)	**25**	(c)	**26**	(a,d)	**27**	(c)	**28**	(b)	**29**	(a)	**30**	(c)
31	(2)	**32**	(18)	**33**	(7)	**34**	(10)												

For Detailed Solutions
scan QR Code

18 Coordination Compounds

In JEE Main there is usually 0-3 question per paper are visible from this chapter. In past 3 years, 85 conceptual and 17 numerical questions were asked of moderate difficulty level from this chapter.

For detailed analysis scan QR code

Topic 1 / Werner's Theory, Coordination Number, Nomenclature and Isomerism of Coordination Compounds

• Level 1

01 The correct IUPAC name of $[PtBr_2(PMe_3)_2]$ is **[2024, 6 April Shift-II]**

(a) *bis*(trimethylphosphine)dibrompolatinum (II)
(b) dibromo*bis*(trimethylphosphine) platinum(II)
(c) *bis*[bromo(trimethylphosphine)] platinum (II)
(d) dibromodi(trimethylphosphine) platinum (II)

02 The correct IUPAC name of K_2MnO_4 is **[2024, 29 Jan Shift-II]**

(a) potassium tetraoxopermanganate (VI)
(b) dipotassium tetraoxidomanganate (VII)
(c) potassium tetraoxidomanganate (VI)
(d) potassium tetraoxidomanganese (VI)

03 The IUPAC name of $K_3[Co(C_2O_4)_3]$ is **[2023, 6 April Shift-II]**

(a) potassium trioxalatocobaltate (III)
(b) potassium tris(oxalato)cobalt (III)
(c) potassium tris(oxalato)cobaltate (III)
(d) potassium trioxalatocobalt (III)

04 The correct structures of *trans*-$[NiBr_2(PPh_3)_2]$ and meridionial-$[Co(NH_3)_3(NO_2)_3]$, respectively are **[2021, 18 March Shift-I]**

(a) 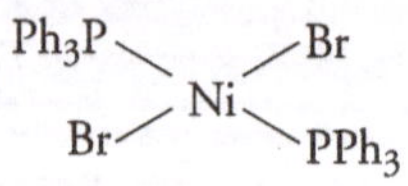

and 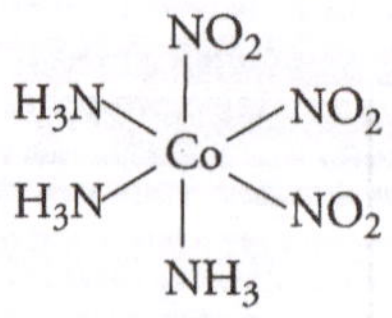

(b) 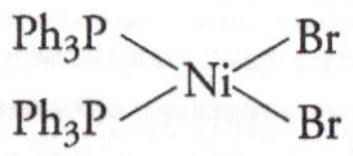

and 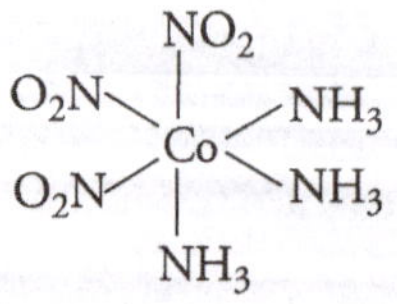

(c) 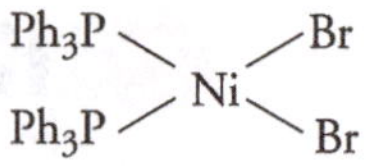

and 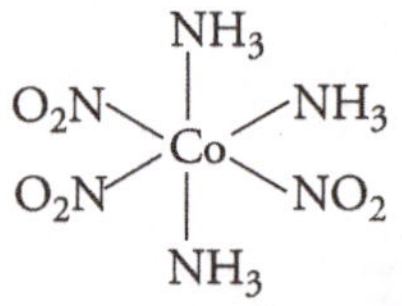

(d) 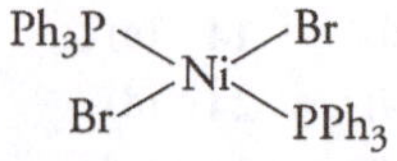

and 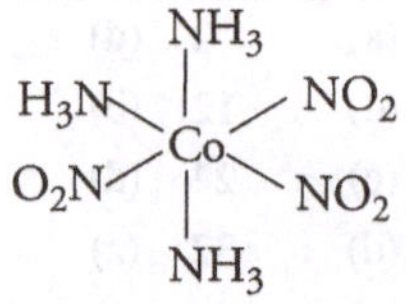

• Level 2

05 '*X*' is the number of acidic oxides among VO_2, V_2O_3, CrO_3, V_2O_5 and Mn_2O_7. The primary valency of cobalt in $[Co(H_2NCH_2CH_2NH_2)_3]_2(SO_4)_3$ is Y. The value of $X + Y$ is .... **[2025, 7 April Shift-II]**

(a) 2 (b) 4
(c) 3 (d) 5

06 Given below are two statements: **[2025, 8 April Shift-II]**

Statement I A homoleptic octahedral complex, formed using monodentate ligands, will not show stereoisomerism.

Statement II *cis* - and *trans* - platin are heteroleptic complexes of Pd.

In the light of the above statements, choose the correct answer from the options given below

(a) Statement I is false but Statement II is true.
(b) Both Statement I and Statement II are true.
(c) Statement I is true but Statement II is false.
(d) Both Statement I and Statement II are false.

07 Number of stereoisomers possible for the complexes, $[CrCl_3(py)_3]$ and $[CrCl_2(ox)_2]^{3-}$ are respectively (py = pyridine, ox = oxalate) **[2025, 4 April Shift-I]**

(a) 2 and 2 (b) 2 and 3
(c) 1 and 2 (d) 3 and 3

08 The complex that shows *facial - meridional* isomerism is **[2025, 23 Jan Shift-I]**
(a) $[Co(NH_3)_4Cl_2]^+$ (b) $[Co(NH_3)_3Cl_3]$
(c) $[Co(en)_2Cl_2]^+$ (d) $[Co(en)_3]^{3+}$

09 One mole of the octahedral complex compound $Co(NH_3)_5Cl_3$ gives 3 moles of ions on dissolution in water. One mole of the same complex reacts with excess of $AgNO_3$ solution to yield two moles of AgCl(*s*). The structure of the complex is **[2025, 24 Jan Shift-I]**
(a) $[Co(NH_3)_4Cl_2]Cl \cdot NH_3$
(b) $[Co(NH_3)_3Cl_3] \cdot 2NH_3$
(c) $[Co(NH_3)_5Cl]Cl_2$
(d) $[Co(NH_3)_4Cl] \cdot Cl_2 \cdot NH_3$

10 Given below are two statements : one is labelled as Assertion (A) and other is labelled as Reason (R). **[2024, 9 April Shift-I]**

Assertion (A) The total number of geometrical isomers shown by $[Co(en)_2Cl_2]^+$ complex ion is three.

Reason (R) $[Co(en)_2Cl_2]^+$ complex ion has an octahedral geometry.

In the light of the above statements, choose the most appropriate answer from the options given below.
(a) A is not correct but R is correct.
(b) Both A and R are correct but R is not the correct explanation of A.
(c) A is correct but R is not correct.
(d) Both A and R are correct but R is the correct explanation of A.

11 An octahedral complex with the formula $CoCl_3 .nNH_3$ upon reaction with excess of $AgNO_3$ solution gives 2 moles of AgCl. Consider the oxidation state of Co in the complex is '*x*'. The value of "$x + n$" is .... **[2024, 8 April Shift-I]**
(a) 3 (b) 6 (c) 5 (d) 8

Imp. *Geometrical Isomers*

12 The metal atom present in the complex *MABXL* (where *A, B, X* and *L* are unidentate ligands and *M* is metal) involves sp^3-hybridisation. The number of geometrical isomers exhibited by the complex is **[2024, 5 April Shift-II]**
(a) 0 (b) 4 (c) 2 (d) 3

13 Which of the following complex is homoleptic? **[2024, 1 Feb Shift-I]**
(a) $[Ni(NH_3)_2Cl_2]$ (b) $[Fe(NH_3)_4Cl_2]^+$
(c) $[Co(NH_3)_4Cl_2]^+$ (d) $[Ni(CN)_4]^{2-}$

14 The total number of stereoisomers for the complex $[Cr(ox)_2ClBr]^{3-}$ (where ox = oxalate) is **[2023, 13 April Shift-II]**
(a) 1 (b) 3 (c) 2 (d) 4

15 Which of the following complex has a possibility to exist as meridional isomer? **[2023, 11 April Shift-II]**
(a) $[Pt(NH_3)_2Cl_2]$ (b) $[Co(NH_3)_3(NO_2)_3]$
(c) $[Co(en)_2Cl_2]$ (d) $[Co(en)_3]$

16 The set which does not have ambidentate ligand(s) is **[2023, 11 April Shift-I]**
(a) NO_2^-, $C_2O_4^{2-}$, $EDTA^{4-}$
(b) $EDTA^{4-}$, NCS^-, $C_2O_4^{2-}$
(c) $C_2O_4^{2-}$, NO_2^-, NCS^-
(d) $C_2O_4^{2-}$, ethylene diammine, H_2O

17 The complex cation which has two isomers is **[2023, 1 Feb Shift-II]**
(a) $[Co(NH_3)_5Cl]^+$ (b) $[Co(H_2O)_6]^{3+}$
(c) $[Co(NH_3)_5NO_2]^{2+}$ (d) $[Co(NH_3)_5Cl]^{2+}$

18 Which of the following are the example of double salt? **[2023, 1 Feb Shift-I]**
A. $FeSO_4 \cdot (NH_4)_2SO_4 \cdot 6H_2O$
B. $CuSO_4 \cdot 4NH_3 \cdot H_2O$
C. $K_2SO_4 . Al_2(SO_4)_3 . 24H_2O$
D. $Fe(CN)_2 . 4KCN$

Choose the correct answer given below.
(a) A and B only (b) A, B and D only
(c) A and C only (d) B and D only

19 The primary and secondary valencies of cobalt respectively in $[Co(NH_3)_5Cl]Cl_2$ are **[2023, 24 Jan Shift-I]**
(a) 3 and 6 (b) 2 and 6
(c) 3 and 5 (d) 2 and 8

20 White precipitate of AgCl dissolves in aqueous ammonia solution due to formation of: **[2022, 25 June Shift-I]**
(a) $[Ag(NH_3)_4]Cl_2$ (b) $[Ag(Cl)_2(NH_3)_2]$
(c) $[Ag(NH_3)_2]Cl$ (d) $[Ag(NH_3)Cl]Cl$

21 Indicate the complex/complex ion which did not show any geometrical isomerism. **[2021, 26 Aug Shift-II]**
(a) $[CoCl_2(en)_2]$ (b) $[Co(CN)_5(NC)]^{3-}$
(c) $[Co(NH_3)_3(NO_2)_3]$ (d) $[Co(NH_3)_4Cl_2]^+$

22 Which one of the following metal complexes is most stable? **[2021, 25 July Shift-II]**
(a) $[Co(en)(NH_3)_4]Cl_2$ (b) $[Co(en)_3]Cl_2$
(c) $[Co(en)_2(NH_3)_2]Cl_2$ (d) $[Co(NH_3)_6]Cl_2$

23 Match List-I and List-II.

List-I	List-II
A. $[Co(NH_3)_6][Cr(CN)_6]$	1. Linkage isomerism
B. $[Co(NH_3)_3(NO_2)_3]$	2. Solvate isomerism
C. $[Cr(H_2O)_6]Cl_3$	3. Co-ordination isomerism
D. *cis*-$[CrCl_2(ox)_2]^{3-}$	4. Optical isomerism

Choose the correct answer from the options given below. **[2021, 17 March Shift-II]**

	A	B	C	D		A	B	C	D
(a)	3	1	2	4	(b)	4	2	3	1
(c)	2	1	3	4	(d)	1	2	3	4

24 The number of isomers possible for $[Pt(en)(NO_2)_2]$ is **[2020, 4 Sep Shift-I]**

(a) 3 (b) 2
(c) 4 (d) 1

25 The one that is not expected to show isomerism is **[2020, 2 Sep Shift-II]**

(a) $[Ni(NH_3)_4(H_2O)_2]^{2+}$ (b) $[Ni(en)_3]^{2+}$
(c) $[Ni(NH_3)_2Cl_2]$ (d) $[Pt(NH_3)_2Cl_2]$

Imp. *Optical Isomers*

26 The number of possible optical isomers for the complexes MA_2B_2 with sp^3 and dsp^2 hybridised metal atom, respectively, is

Note : *A* and *B* are unidentate neutral and unidentate monoanionic ligands, respectively. **[2020, 7 Jan Shift-II]**

(a) 0 and 0 (b) 0 and 1
(c) 2 and 2 (d) 0 and 2

27 The complex that can show optical activity is **[2020, 3 Sep Shift-I]**

(a) *trans*-$[Cr(Cl_2)(ox)_2]^{3-}$
(b) *trans*-$[Fe(NH_3)_2(CN)_4]^-$
(c) *cis*-$[Fe(NH_3)_2(CN)_4]^-$
(d) *cis*-$[CrCl_2(ox)_2]^{3-}$ (ox = oxalate)

28 The complex that can show *fac*- and *mer*-isomers is **[2020, 8 Jan Shift-I]**

(a) $[Pt(NH_3)_2Cl_2]$
(b) $[Co(NH_3)_3(NO_2)_3]$
(c) $[Co(NH_3)_4Cl_2]^+$
(d) $[CoCl_2(en)_2]$

29 The species that can have a *trans*-isomer is (en = ethane -1, 2-diamine, ox = oxalate) **[2019, 10 April Shift-I]**

(a) $[Pt(en)Cl_2]$ (b) $[Cr(en)_2(ox)]^+$
(c) $[Pt(en)_2Cl_2]^{2+}$ (d) $[Zn(en)Cl_2]$

• Level 3

30 '*X*' is the number of electrons in t_{2g} orbitals of the most stable complex ion among $[Fe(NH_3)_6]^{3+}$, $[FeCl_6]^{3-}$, $[Fe(C_2O_4)_3]^{3-}$ and $[Fe(H_2O)_6]^{3+}$. The nature of oxide of vanadium of the type V_2O_X is **[2025, 4 April Shift-II]**

(a) amphoteric (b) neutral
(c) basic (d) acidic

31 An octahedral complex having molecular composition $Co \cdot 5NH_3 \cdot Cl \cdot SO_4$, has two isomers *A* and *B*. The solution of *A* gives a white precipitate with $AgNO_3$ solution and the solution of *B* gives white precipitate with $BaCl_2$, solution. The type of isomerism exhibited by the complex is, **[2025, 7 April Shift-I]**

(a) ionisation isomerism (b) linkage isomerism
(c) geometrical isomerism (d) co-ordinate isomerism

32 Match List -I with List - II. **[2025, 7 April Shift-II]**

	List -I (Complex)		List -II (Primary valency and Secondary valency)	
(A)	$[Co(en)_2Cl_2]Cl$	(I)	3	6
(B)	$[Pt(NH_3)_2Cl(NO_2)]$	(II)	3	4
(C)	$Hg[Co(SCN)_4]$	(III)	2	6
(D)	$[Mg(EDTA)]^{2-}$	(IV)	2	4

Choose the correct answer from the options given below:

(a) (A)-(III), (B)-(I), (C)-(II), (D)-(IV)
(b) (A)-(II), (B)-(III), (C)-(IV), (D)-(I)
(c) (A)-(I), (B)-(III), (C)-(II), (D)-(IV)
(d) (A)-(I), (B)-(IV), (C)-(II), (D)-(III)

33 Identify the coordination complexes in which the central metal ion has d^4 configuration.

(A) $[FeO_4]^{2-}$ (B) $[Mn(CN)_6]^{3-}$
(C) $[Fe(CN)_6]^{3-}$
(D) $Cr_2(O-\overset{O}{\overset{\|}{C}}-Me)_4(H_2O)_2$
(E) $[NiF_6]^{2-}$

Choose the correct answer from the option given below. **[2025, 23 Jan Shift-II]**

(a) (A), (B) and (E) only (b) (C) and (E) only
(c) (B) and (D) only (d) (B), (C) and (D) only

34 Match List-I with List-II.

	List-I (Complex)		List-II (Hybridisation of central metal ion)
(A)	$[CoF_6]^{3-}$	I.	d^2sp^3
(B)	$[NiCl_4]^{2-}$	II.	sp^3
(C)	$[Co(NH_3)_6]^{3+}$	III.	sp^3d^2
(D)	$[Ni(CN)_4]^{2-}$	IV.	dsp^2

Choose the correct answer from the options given below. **[2025, 28 Jan Shift-II]**

(a) (A)-(III), (B)-(II), (C)-(I), (D)-(IV)
(b) (A)-(I), (B)-(II), (C)-(III), (D)-(IV)
(c) (A)-(I), (B)-(IV), (C)-(III), (D)-(II)
(d) (A)-(III), (B)-(IV), (C)-(I), (D)-(II)

35 Identify the homoleptic complexes with odd number of *d*-electrons in the central metal **[2025, 29 Jan Shift-II]**

(A) $[FeO_4]^{2-}$ (B) $[Fe(CN)_6]^{3-}$

(C) $[Fe(CN)_5NO]^{2-}$ (D) $[CoCl_4]^{2-}$

(E) $[Co(H_2O)_3F_3]$

Choose the correct answer from the options given below

(a) (A), (B) and (D) only (b) (C) and (E) only
(c) (B) and (D) only (d) (A), (C) and (E) only

36 The F^- ions make the enamel on teeth much harder by converting hydroxyapatite (the enamel on the surface of teeth) into much harder fluoroapatite having the formula. **[2024, 9 April Shift-I]**

(a) $[3(Ca_3(PO_4)_2 \cdot CaF_2]$
(b) $[3(Ca_3(PO_4)_2) \cdot Ca(OH)_2]$
(c) $[3(Ca_2(PO_4)_2) \cdot Ca(OH)_2]$
(d) $[3(Ca_3)(PO_4)_3) \cdot CaF_2]$

37 Correct formula of the compound which gives a white precipitate with $BaCl_2$ solution, but not with $AgNO_3$ solution is **[2022, 30 June Shift-I]**

(a) $[Co(NH_3)_5Br]SO_4$ (b) $[Co(NH_3)_5SO_4]Br$
(c) $[Pt(NH_3)_4Cl_2]Br_2$ (d) $[Pt(NH_3)_4Br_2]Cl_2$

38 The number of geometrical isomers found in the metal complexes $[PtCl_2(NH_3)_2]$, $[Ni(CO)_4]$, $[Ru(H_2O)_3Cl_3]$ and $[CoCl_2(NH_3)_4]^+$ respectively, are **[2021, 27 July Shift-I]**

(a) 1, 1, 1, 1 (b) 2, 1, 2, 2
(c) 2, 0, 2, 2 (d) 2, 1, 2, 1

Imp. *Mole Concept & Formula of Coordination Compounds*

39 Complex (*A*) has a composition of $H_{12}O_6Cl_3Cr$. If the complex on treatment with conc. H_2SO_4 loses 13.5% of its original mass, the correct molecular formula of (*A*) is[Given : atomic mass of Cr = 52 amu and Cl = 35 amu] **[2020, 3 Sep Shift-II]**

(a) $[Cr(H_2O)_4Cl_2]Cl \cdot 2H_2O$
(b) $[Cr(H_2O)_6]Cl_3$
(c) $[Cr(H_2O)_5Cl]Cl_2 \cdot H_2O$
(d) $[Cr(H_2O)_3Cl_3] \cdot 3H_2O$

40 Among (A) - (D), the complexes that can display geometrical isomerism are

(A) $[Pt(NH_3)_3Cl]^+$ (B) $[Pt(NH_3)Cl_5]^-$

(C) $[Pt(NH_3)_2Cl(NO_2)]$ (D) $[Pt(NH_3)_4ClBr]^{2+}$

[2020, 8 Jan Shift-II]

(a) (D) and (A) (b) (C) and (D)
(c) (A) and (B) (d) (B) and (C)

Numerical Value Type Questions

41 The number of optical isomers exhibited by the iron complex (*A*) obtained from the following reaction is ....

$$FeCl_3 + KOH + H_2C_2O_4 \rightarrow A$$ **[2025, 3 April Shift-I]**

42 A metal complex with a formula $MCl_4.3NH_3$ is involved in sp^3d^2 hybridisation. It upon reaction with excess of $AgNO_3$ solution gives '*x*' moles of AgCl. Consider '*x*' is equal to the number of lone pairs of electron present in central atom of BrF_5. Then the number of geometrical isomers exhibited by the complex is .... **[2025, 4 April Shift-II]**

43 Number of ambidentate ligands among the following is ....

NO_2^-, SCN^-, $C_2O_4^{2-}$, NH_3, CN^-, SO_4^{2-}, H_2O

[2024, 9 April Shift-I]

44 Number of complexes which show optical isomerism among the following is ....

cis-$[Cr(ox)_2Cl_2]^{3-}$, $[Co(en)_3]^{3+}$, *cis*-$[Pt(en)_2Cl_2]^{2+}$, *cis*-$[Co(en)_2Cl_2]^+$, *trans*-$[Pt(en)_2Cl_2]^{2+}$, *trans* -$[Cr(ox)_2Cl_2]^{3-}$ **[2024, 30 Jan Shift-II]**

45 Total number of moles of AgCl precipitated on addition of excess of $AgNO_3$ to one mole each of the following complexes $[Co(NH_3)_4Cl_2]Cl$, $[Ni(H_2O_6]Cl_2$, $[Pt(NH_3)_2Cl_2]$ and $[Pd(NH_3)_4]Cl_2$ is .... **[2023, 25 Jan Shift-II]**

46 Sum of oxidation state (magnitude) and coordination number of cobalt in $Na[Co(bpy)Cl_4]$ is ....

(Given, bpy = 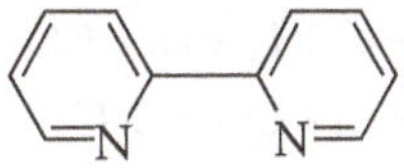) **[2022, 29 July Shift-II]**

47 For the reaction given below

$CoCl_3 \cdot xNH_3 + AgNO_3\,(aq) \longrightarrow$

If two equivalents of AgCl precipitate out, then the value of *x* will be .... **[2022, 29 June Shift-II]**

48 (i) $CoCl_3 \cdot 4NH_3$, (ii) $CoCl_3 \cdot 5NH_3$, (iii) $CoCl_3 \cdot 6NH_3$ and (iv) $CoCl(NO_3)_2 \cdot 5NH_3$

Number of complex(es) which will exist in *cis-trans* form is/are .... **[2022, 28 June Shift-II]**

49 The sum of oxidation states of two silver ions in $[Ag(NH_3)_2][Ag(CN)_2]$ complex is ....

[2021, 1 Sep Shift-II]

Imp. *Werner's Theory*

50 3 moles of metal complex with formula $Co(en)_2Cl_3$ gives 3 moles of silver chloride on treatment with excess of silver nitrate. The secondary valency of Co in the complex is ....
(Round off to the nearest integer) **[2021, 27 July Shift-II]**

51 The number of geometrical isomers possible in triamminetrinitrocobalt (III) is *X* and in trioxalatochromate (III) is *Y*. Then, the value of $X + Y$ is .... **[2021, 27 July Shift-I]**

52 The equivalents of ethylene diamine required to replace the neutral ligands from the coordination sphere of the *trans*-complex of $CoCl_3 \cdot 4NH_3$ is ..... (Round off to the nearest integer). **[2021, 16 March Shift-I]**

53 The total number of coordination sites in ethylenediaminetetraacetate $[EDTA^{4-}]$ is .... . **[2020, 5 Sep Shift-I]**

Topic 2 / Valence Bond Theory and CFT

• Level 1

01 In which of the following complexes the CFSE, Δ_0 will be equal to zero? **[2025, 22 Jan Shift-I]**
(a) $[Fe(NH_3)_6]Br_2$ (b) $K_3[Fe(SCN)_6]$
(c) $[Fe(en)_3]Cl_3$ (d) $K_4[Fe(CN)_6]$

02 Which of the following is correct order of ligand field strength? **[2023, 31 Jan Shift-I]**
(a) $NH_3 < en < CO < S^{2-} < C_2O_4^{2-}$
(b) $CO < en < NH_3 < C_2O_4^{2-} < S^{2-}$
(c) $S^{2-} < C_2O_4^{2-} < NH_3 < en < CO$
(d) $S^{2-} < NH_3 < en < CO < C_2O_4^{2-}$

03 Low oxidation state of metals in their complexes are common when ligands **[2022, 27 July Shift-II]**
(a) have good π-accepting character
(b) have good σ-donor character
(c) are having good π-donating ability
(d) are having poor σ-donating ability

04 Spin only magnetic moment of an octahedral complex of Fe^{2+} in the presence of a strong field ligand in BM is **[2021, 20 July Shift-II]**
(a) 0 (b) 4.89 (c) 2.82 (d) 3.46

05 According to the valence bond theory the hybridisation of central metal atom is dsp^2 for which one of the following compounds? **[2021, 20 July Shift-I]**
(a) $NiCl_2 \cdot 6H_2O$ (b) $K_2[Ni(CN)_4]$
(c) $[Ni(CO)_4]$ (d) $Na_2[NiCl_4]$

06 For a d^4 metal ion in an octahedral field, the correct electronic configuration is **[2020, 6 Sep Shift-II]**
(a) $t_{2g}^3 e_g^1$ when $\Delta_0 < P$ (b) $t_{2g}^3 e_g^1$ when $\Delta_0 > P$
(c) $t_{2g}^4 e_g^0$ when $\Delta_0 < P$ (d) $t_g^2 e_{2g}^2$ when $\Delta_0 < P$

07 The species that has a spin-only magnetic moment of 5.9 BM, is (T_d = tetrahedral) **[2020, 6 Sep Shift-I]**
(a) $[Ni(CN)_4]^{2-}$ (b) $[NiCl_4]^{2-}(T_d)$
(c) $[Ni(CO)_4](T_d)$ (d) $[MnBr_4]^{2-}(T_d)$

08 The crystal field stabilisation energy (CFSE) of $[CoF_3(H_2O)_3]$ $(\Delta_0 < P)$ is **[2020, 4 Sep Shift-II]**
(a) $-0.8\,\Delta_o$ (b) $-0.8\Delta_o + 2P$
(c) $-0.4\Delta_o$ (d) $-0.4\Delta_o + P$

09 The *d*-electron configuration of $[Ru(en)_3]Cl_2$ and $[Fe(H_2O)_6]Cl_2$, respectively are **[2020, 3 Sep Shift-II]**
(a) $t_{2g}^6 e_g^0$ and $t_{2g}^6 e_g^0$ (b) $t_{2g}^4 e_g^2$ and $t_{2g}^6 e_g^0$
(c) $t_{2g}^6 e_g^0$ and $t_{2g}^4 e_g^2$ (d) $t_{2g}^4 e_g^2$ and $t_{2g}^4 e_g^2$

10 The correct statements among I to III are:
(I) Valence bond theory cannot explain the colour exhibited by transition metal complexes.
(II) Valence bond theory can predict quantitatively the magnetic properties of transition metal complexes.
(III) Valence bond theory cannot distinguish ligands as weak and strong field one. **[2019, 9 April Shift-II]**
(a) (II) and (III) only (b) (I), (II) and (III)
(c) (I) and (III) only (d) (I) and (II) only

11 In Wilkinson's catalyst, the hybridisation of central metal ion and its shape are respectively: **[2019, 10 Jan Shift-I]**
(a) sp^3d, trigonal bipyramidal
(b) d^2sp^3, octahedral
(c) dsp^2, square planar
(d) sp^3, tetrahedral

• Level 2

12 The correct order of $[FeF_6]^{3-}$, $[CoF_6]^{3-}$, $[Ni(CO)_4]$ and $[Ni(CN)_4]^{2-}$ complex species based on the number of unpaired electrons present is **[2025, 4 April Shift-II]**
(a) $[FeF_6]^{3-} > [CoF_6]^{3-} > [Ni(CN)_4]^{2-} > [Ni(CO)_4]$
(b) $[CoF_6]^{3-} > [FeF_6]^{3-} > [Ni(CO)_4] > [Ni(CN)_4]^{2-}$
(c) $[Ni(CN)_4]^{2-} > [FeF_6]^{3-} > [CoF_6]^{3-} > [Ni(CO)_4]$
(d) $[FeF_6]^{3-} > [CoF_6]^{3-} > [Ni(CN)_4]^{2-} = [Ni(CO)_4]$

13 The number of species from the following that are involved in sp^3d^2 hybridisation is $[Co(NH_3)_6]^{3+}$, SF_6, $[CrF_6]^{3-}$, $[CoF_6]^{3-}$, $[Mn(CN)_6]^{3-}$ and $[MnCl_6]^{3-}$ **[2025, 8 April Shift-II]**
(a) 6 (b) 3 (c) 5 (d) 4

14 Which one of the following complexes will have $\Delta_0 = 0$ and $\mu = 5.96$ B.M? **[2025, 4 April Shift-I]**
(a) $[Co(NH_3)_6]^{3+}$ (b) $[Fe(CN)_6]^{4-}$
(c) $[FeF_6]^{4-}$ (d) $[Mn(SCN)_6]^{4-}$

15 The type of hybridisation and the magnetic property of $[MnCl_6]^{3-}$ are **[2025, 2 April Shift-II]**
(a) sp^3d^2, paramagnetic with four unpaired electrons.
(b) d^2sp^3, paramagnetic with four unpaired electrons.
(c) sp^3d^2, paramagnetic with two unpaired electrons.
(d) d^2sp^3, paramagnetic with two unpaired electrons.

16 Given below are two statements. **[2025, 2 April Shift-I]**
Statement I In octahedral complexes, when $\Delta_0 < P$ high spin complexes are formed. When $\Delta_0 > P$ low spin complexes are formed.
Statement II In tetrahedral complexes because of $\Delta_t < P$, low spin complexes are rarely formed.
In the light of the above statements, choose the most appropriate answer from the options given below:
(a) Statement I is correct but Statement II is incorrect.
(b) Both Statement I and Statement II are incorrect.
(c) Both Statement I and Statement II are correct.
(d) Statement I is incorrect but Statement II is correct.

17 The d-electronic configuration of an octahedral Co(II) complex having magnetic moment of 3.95 BM is **[2025, 23 Jan Shift-I]**
(a) $e^4t_2^3$ (b) $t_{2g}^6e_g^1$
(c) $t_{2g}^3e_g^0$ (d) $t_{2g}^5e_g^2$

18 The conditions and consequence that favours the $t_{2g}^3e_g^1$ configuration in a metal complex are **[2025, 24 Jan Shift-II]**
(a) weak field ligand, low spin complex
(b) strong field ligand, high spin complex
(c) weak field ligand, high spin complex
(d) strong field ligand, low spin complex

19 The calculated spin-only magnetic moments of $K_3[Fe(OH)_6]$ and $K_4[Fe(OH)_6]$ respectively are **[2025, 29 Jan Shift-II]**
(a) 3.87 and 4.90 BM (b) 4.90 and 5.92 BM
(c) 4.90 and 4.90 BM (d) 5.92 and 4.90 BM

20 When ethane-1,2-diamine is added progressively to an aqueous solution of nickel (II) chloride, the sequence of colour change observed will be **[2025, 24 Jan Shift-II]**
(a) Violet → Blue → Pale blue → Green
(b) Green → Pale blue → Blue → Violet
(c) Pale blue → Blue → Green → Violet
(d) Pale blue → Blue → Violet → Green

21 The coordination environment of Ca^{2+} ion in its complex with $EDTA^{4-}$ is **[2024, 9 April Shift-II]**
(a) tetrahedral
(b) trigonal prismatic
(c) square planar
(d) octahedral

22 If an iron (III) complex with the formula $[Fe(NH_3)_x(CN)_y]^-$ has no electron in its e_g orbital, then the value of $x + y$ is **[2024, 4 April Shift-II]**
(a) 5 (b) 3
(c) 4 (d) 6

23 The correct sequence of ligands in the order of decreasing field strength is: **[2024, 4 April Shift-I]**
(a) $NCS^- > EDTA^{4-} > CN^- > CO$
(b) $CO > H_2O > F^- > S^{2-}$
(c) $^-OH > F^- > NH_3 > CN^-$
(d) $S^{2-} > ^-OH > EDTA^{4-} > CO$

24 $[Co(NH_3)_6]^{3+}$ and $[CoF_6]^{3-}$ are respectively known as **[2024, 1 Feb Shift-II]**
(a) spin paired complex, spin free complex
(b) inner orbital complex, spin paired complex
(c) outer orbital complex, inner orbital complex
(d) spin free complex, spin paired complex

25 Given below are two statements. **[2024, 1 Feb Shift-I]**
Statement I A solution of $[Ni(H_2O)_6]^{2+}$ is green in colour.
Statement II A solution of $[Ni(CN)_4]^{2-}$ is colourless.
In the light of the above statements, choose the **most appropriate** answer from the options given below.
(a) Statement I is correct but Statement II is incorrect.
(b) Both Statement I and Statement II are incorrect.
(c) Both Statement I and Statement II are correct.
(d) Statement I is incorrect but Statement II is correct.

26 Select the option with correct property. **[2024, 31 Jan Shift-II]**
(a) $[Ni(CO)_4]$ and $[NiCl_4]^{2-}$ both paramagnetic.
(b) $[Ni(CO)_4]$ and $[NiCl_4]^{2-}$ both diamagnetic.
(c) $[Ni(CO)_4]$ diamagnetic, $[NiCl_4]^{2-}$ paramagnetic.
(d) $[NiCl_4]^{2-}$ diamagnetic, $[Ni(CO)_4]$ paramagnetic.

27 The coordination geometry around the manganese in decarbonyldimanganese (0) is **[2024, 30 Jan Shift-II]**
(a) square planar (b) octahedral
(c) trigonal bipyramidal (d) square pyramidal

28 The octahedral diamagnetic low spin complex among the following is **[2023, 10 April Shift-I]**
(a) $[CoF_6]^{3-}$ (b) $[Co(NH_3)_6]^{3+}$
(c) $[NiCl_4]^{2-}$ (d) $[CoCl_6]^{3-}$

29 The correct order of spin only magnetic moments for the following complex ions is **[2023, 8 April Shift-I]**

(a) $[Fe(CN)_6]^{3-} < [CoF_6]^{3-} < [MnBr_4]^{2-} < [Mn(CN)_6]^{3-}$
(b) $[Fe(CN)_6]^{3-} < [Mn(CN)_6]^{3-} < [CoF_6]^{3-} < [MnBr_4]^{2-}$
(c) $[MnBr_4]^{2-} < [CoF_6]^{3-} < [Fe(CN)_6]^{3-} < [Mn(CN)_6]^{3-}$
(d) $[CoF_6]^{3-} < [MnBr_4]^{2-} < [Fe(CN)_6]^{3-} < [Mn(CN)_6]^{3-}$

30 A solution of $FeCl_3$ when treated with $K_4[Fe(CN)_6]$ gives a prussian blue precipitate due to the formation of **[2023, 1 Feb Shift-I]**

(a) $K[Fe_2(CN)_6]$ (b) $Fe_4[Fe(CN)_6]_3$
(c) $Fe_3[Fe(CN)_6]_2$ (d) $Fe[Fe(CN)_6]$

Imp. *Colour and Geometry of the Coordination Compounds*

31 Cobalt chloride when dissolved in water forms pink coloured complex $\underline{X}$ which has octahedral geometry. This solution on treating with conc. HCl forms deep blue complex, $\underline{Y}$ which has a $\underline{Z}$ geometry. X, Y and Z, respectively, are **[2023, 31 Jan Shift-I]**

(a) $X = [Co(H_2O)_4Cl_2]^+, Y = [CoCl_4]^{2-}, Z =$ tetrahedral
(b) $X = [Co(H_2O)_6]^{2+}, Y = [CoCl_6]^{3-}, Z =$ octahedral
(c) $X = [Co(H_2O)_6]^{2+}, Y = [CoCl_4]^{4-}, Z =$ tetrahedral
(d) $X = [Co(H_2O)_6]^{3+}, Y = [CoCl_6]^{3+}, Z =$ octahedral

32 Correct order of spin only magnetic moment of the following complex ions is
(Given : Atomic number Fe = 26, Co = 27) **[2023, 29 Jan Shift-II]**

(a) $[CoF_6]^{3-} > [FeF_6]^{3-} > [Co(C_2O_4)_3]^{3-}$
(b) $[Co(C_2O_4)_3]^{3-} > [CoF_6]^{3-} > [FeF_6]^{3-}$
(c) $[FeF_6]^{3-} > [CoF_6]^{3-} > [Co(C_2O_4)_3]^{3-}$
(d) $[FeF_6]^{3-} > [Co(C_2O_4)_3]^{3-} > [CoF_6]^{3-}$

33 The hybridisation and magnetic behaviour of cobalt ion in $[Co(NH_3)_6]^{3+}$ complex, respectively is **[2023, 24 Jan Shift-II]**

(a) d^2sp^3 and diamagnetic (b) d^2sp^3 and paramagnetic
(c) sp^3d^2 and diamagnetic (d) sp^3d^2 and paramagnetic

34 If Ni^{2+} is replaced by Pt^{2+} in the complex $[NiCl_2Br_2]^{2-}$, which of the following properties are expected to get changed?

A. Geometry
B. Geometrical isomerism
C. Optical isomerism
D. Magnetic properties **[2023, 11 April Shift-II]**

(a) A, B and D (b) A and D
(c) A, B and C (d) B and C

35 Given below are two statements, one is labelled as Assertion (A) and the other is labelled as Reason (R).

Assertion (A) $[CoCl(NH_3)_5]^{2-}$ absorbs at lower wavelength of light with respect to $[Co(NH_3)_5(H_2O)]^{3+}$.

Reason (R) It is because the wavelength of the light absorbed depends on the oxidation state of the metal ion.

In the light of the above statements, choose the most appropriate answer from the options given below. **[2023, 11 April Shift-II]**

(a) Both A and R are true and R is the correct explanation of A
(b) A is true but R is false
(c) Both A and R are true but R is not the correct explanation of A
(d) A is false but R is true

Imp. *Jahn-Teller Distortion*

36 Octahedral complexes of copper (II) undergo structural distortion (Jahn-Teller). Which one of the given copper (II) complexes will show the maximum structural distortion? (en–ethylenediamine; $H_2N—CH_2—CH_2—NH_2$) **[2022, 29 July Shift-II]**

(a) $[Cu(H_2O)_6]SO_4$ (b) $[Cu(en)(H_2O)_4]SO_4$
(c) *cis*-$[Cu(en)_2Cl_2]$ (d) *trans*-$[Cu(en)_2Cl_2]$

37 The correct order of energy of absorption for the following metal complexes is

A. $[Ni(en)_3]^{2+}$ B. $[Ni(NH_3)_6]^{2+}$
C. $[Ni(H_2O)_6]^{2+}$ **[2022, 25 July Shift-II]**

(a) C < B < A (b) B < C < A
(c) C < A < B (d) A < C < B

38 Given below are two statements.

Statement I In $CuSO_4 \cdot 5H_2O$, Cu—O bonds are present.

Statement II In $CuSO_4 \cdot 5H_2O$, ligands coordinating with Cu(II) ion are O-and S-based ligands.

In the light of the above statements, choose the correct answer from the options given below **[2022, 29 June Shift-II]**

(a) Both Statement I and Statement II are correct.
(b) Both Statement I and Statement II are incorrect.
(c) Statement I is correct but Statement II is incorrect.
(d) Statement I is incorrect but Statement II is correct.

39 Given below are two statements :

Statement I $[Ni(CN)_4]^{2-}$ is square planar and diamagnetic complex with dsp^2 hybridisation for Ni but $[Ni(CO)_4]$ is tetrahedral paramagnetic and with sp^3-hybridisation for Ni.

Statement II $[NiCl_4]^{2-}$ and $[Ni(CO)_4]$ both have same d-electron configuration have same geometry and are paramagnetic.

In light of the above statements, choose the correct answer from the options given below: **[2022, 28 June Shift-I]**

(a) Both Statement I and Statement II are correct.
(b) Both Statement I and Statement II are incorrect.
(c) Statement I is correct but Statement II is incorrect.
(d) Statement I is incorrect but Statement II is correct.

40 Arrange the following coordination compounds in the increasing order of magnetic moments. (Atomic numbers : Mn = 25 ; Fe = 26)

(A) $[FeF_6]^{3-}$ (B) $[Fe(CN)_6]^{3-}$
(C) $[MnCl_6]^{3-}$ (high spin) (D) $[Mn(CN)_6]^{3-}$

Choose the correct answer from the options given below : **[2022, 27 June Shift-II]**

(a) A < B < D < C (b) B < D < C < A
(c) A < C < D < B (d) B < D < A < C

Imp. *Crystal Field Splitting*

41 Transition metal complex with highest value of crystal field splitting (Δ_o) will be **[2022, 24 June Shift-II]**

(a) $[Cr(H_2O)_6]^{3+}$ (b) $[Mo(H_2O)_6]^{3+}$
(c) $[Fe(H_2O)_6]^{3+}$ (d) $[Os(H_2O)_6]^{3+}$

42 Arrange the following cobalt complexes in the order of increasing crystal field stabilisation energy (CFSE) value.

$\underset{(A)}{[CoF_6]^{3-}}$, $\underset{(B)}{[Co(H_2O)_6]^{3+}}$, $\underset{(C)}{[Co(NH_3)_6]^{3+}}$ and $\underset{(D)}{[Co(en)_3]^{3+}}$

Choose the correct option. **[2021, 26 Aug Shift-II]**

(a) $A < B < C < D$ (b) $B < A < C < D$
(c) $B < C < D < A$ (d) $C < D < B < A$

43 Which one of the following complexes is violet in colour? **[2021, 26 Aug Shift-I]**

(a) $[Fe(CN)_6]^{4-}$ (b) $[Fe(SCN)_6]^{4-}$
(c) $Fe_4[Fe(CN_6)]_3 \cdot H_2O$ (d) $[Fe(CN)_5NOS]^{4-}$

44 The type of hybridisation and magnetic property of the complex $[MnCl_6]^{3-}$, respectively, are **[2021, 27 July Shift-I]**

(a) sp^3d^2 and diamagnetic (b) d^2sp^3 and diamagnetic
(c) d^2sp^3 and paramagnetic (d) sp^3d^2 and paramagnetic

45 Which one of the following species responds to an external magnetic field **[2021, 25 July Shift-I]**

(a) $[Fe(H_2O)_6]^{3+}$ (b) $[Ni(CN)_4]^{2-}$
(b) $[Co(CN)_6]^{3-}$ (d) $[Ni(CO)_4]$

46 The correct order of intensity of colors of the compound is **[2021, 20 July 20 Shift-I]**

(a) $[Ni(CN)_4]^{2-} > [NiCl_4]^{2-} > [Ni(H_2O)_6]^{2+}$
(b) $[Ni(H_2O)_6]^{2+} > [NiCl_4]^{2-} > [Ni(CN)_4]^{2-}$
(c) $[NiCl_4]^{2-} > [Ni(H_2O)_6]^{2+} > [Ni(CN)_4]^{2-}$
(d) $[NiCl_4]^{2-} > [Ni(CN)_4]^{2-} > [Ni(H_2O)_6]^{2+}$

47 The hybridisation and magnetic nature of $[Mn(CN)_6]^{4-}$ and $[Fe(CN)_6]^{3-}$, respectively are **[2021, 25 Feb Shift-I]**

(a) d^2sp^3 and paramagnetic (b) sp^3d^2 and diamagnetic
(c) d^2sp^3 and diamagnetic (d) sp^3d^2 and paramagnetic

48 The calculated magnetic moments (spin only value) for species $[FeCl_4]^{2-}$, $[Co(C_2O_4)_3]^{3-}$ and MnO_4^{2-} respectively are **[2021, 24 Feb Shift-II]**

(a) 5.82, 0 and 0 BM (b) 4.90, 0 and 1.73 BM
(c) 5.92, 4.90 and 0 BM (d) 4.90, 0 and 2.83 BM

49 The molecule in which hybrid MOs involve only one d-orbital of the central atom is **[2020, 4 Sep Shift-II]**

(a) XeF_4 (b) $[CrF_6]^{3-}$ (c) BrF_5 (d) $[Ni(CN)_4]^{2-}$

50 The pair in which both the species have the same magnetic moment (spin only) is **[2020, 4 Sep Shift-I]**

(a) $[Mn(H_2O)_6]^{2+}$ and $[Cr(H_2O)]^{2+}$
(b) $[Cr(H_2O)_6]^{2+}$ and $[Fe(H_2O)_6]^{2+}$
(c) $[Co(OH)_4]^{2-}$ and $[Fe(NH_3)_6]^{2+}$
(d) $[Cr(H_2O)_6]^{2+}$ and $[CoCl_4]^{2-}$

51 The electronic spectrum of $[Ti(H_2O)_6]^{3+}$ shows a single broad peak with a maximum at 20,300 cm^{-1}. The crystal field stabilisation energy (CFSE) of the complex ion, in kJ mol^{-1}, is (1 kJ mol^{-1} = 83.7 cm^{-1}) **[2020, 3 Sep Shift-I]**

(a) 145.5 (b) 242.5 (c) 83.7 (d) 97

52 Consider that a d^6 metal ion (M^{2+}) forms a complex with *aqua* ligands and the spin only magnetic moment of the complex is 4.90 BM. The geometry and the crystal field stabilisation energy of the complex is **[2020, 2 Sep Shift-I]**

(a) tetrahedral and $-1.6\Delta_t + 1P$
(b) octahedral and $-2.4\,\Delta_0 + 2P$
(c) octahedral and $-1.6\,\Delta_0$
(d) tetrahedral and $-0.6\,\Delta_t$

53 The correct order of the calculated spin only magnetic moments of complexes (A) to (D) is **[2020, 8 Jan Shift-II]**

(A) $Ni(CO)_4$ (B) $[Ni(H_2O)_6]Cl_2$
(C) $Na_2[Ni(CN)_4]$ (D) $PdCl_2(PPh_3)_2$

(a) (A) ≈ (C) < (B) ≈ (D) (b) (C) ≈ (D) < (B) < (A)
(c) (C) < (D) < (B) < (A) (d) (A) ≈ (C) ≈ (D) < (B)

Imp. *Splitting Pattern of Octachedral Complex*

54 Complete removal of both the axial ligands (along the z-axis) from an octahedral complex leads to which of the following splitting patterns? (relative orbital energies not on scale). **[2019, 12 April Shift-I]**

(a) 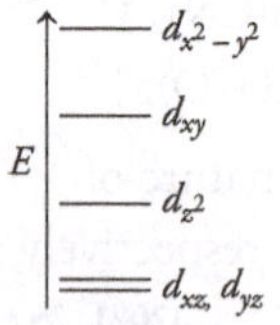

(b) 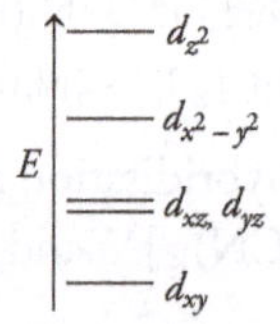

(c) 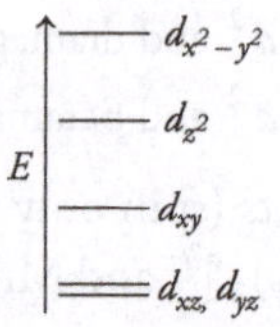

(d) 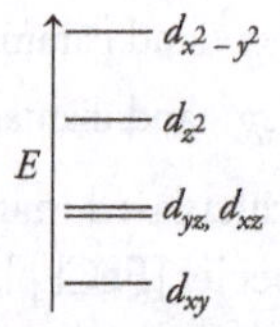

55 The crystal field stabilisation energy (CFSE) of $[Fe(H_2O)_6]Cl_2$ and $K_2[NiCl_4]$, respectively, are **[2019, 10 April Shift-II]**

(a) $-0.4\ \Delta_o$ and $-1.2\ \Delta_t$
(b) $-0.4\ \Delta_o$ and $-0.8\ \Delta_t$
(c) $-2.4\ \Delta_o$ and $-1.2\ \Delta_t$
(d) $-0.6\ \Delta_o$ and $-0.8\ \Delta_t$

56 The correct order of the spin only magnetic moment of metal ions in the following low spin complexes, $[V(CN)_6]^{4-}$, $[Fe(CN)_6]^{4-}$, $[Ru(NH_3)_6]^{3+}$, and $[Cr(NH_3)_6]^{2+}$, is **[2019, 8 April Shift-I]**

(a) $Cr^{2+} > Ru^{3+} > Fe^{2+} > V^{2+}$
(b) $V^{2+} > Cr^{2+} > Ru^{3+} > Fe^{2+}$
(c) $V^{2+} > Ru^{3+} > Cr^{2+} > Fe^{2+}$
(d) $Cr^{2+} > V^{2+} > Ru^{3+} > Fe^{2+}$

Imp. *Colour of Coordination Compounds*

57 Two complexes $[Cr(H_2O)_6]Cl_3$ (*A*) and $[Cr(NH_3)_6]Cl_3$ (*B*) are violet and yellow coloured, respectively. The incorrect statement regarding them is **[2019, 9 Jan Shift-I]**

(a) Δ_o value for (*A*) is less than that of (*B*)
(b) both absorb energies corresponding to their complementary colours
(c) Δ_o values of (*A*) and (*B*) are calculated from the energies of violet and yellow light, respectively
(d) both are paramagnetic with three unpaired electrons

• Level 3

58 The correct order of the complexes $[Co(NH_3)_5(H_2O)]^{3+}$ (*A*), $[Co(NH_3)_6]^{3+}$ (*B*), $[Co(CN)_6]^{3-}$ (*C*) and $[CoCl(NH_3)_5]^{2+}$ (*D*) in terms of wavelength of light absorbed is **[2025, 3 April Shift-I]**

(a) $D > C > B > A$ (b) $C > B > D > A$
(c) $D > A > B > C$ (d) $C > B > A > D$

59 The number of unpaired electrons responsible for the paramagnetic nature of the following complex species are respectively:
$[Fe(CN)_6]^{3-}$, $[FeF_6]^{3-}$, $[CoF_6]^{3-}$, $[Mn(CN)_6]^{3-}$ **[2025, 7 April Shift-II]**

(a) 1, 5, 4, 2 (b) 1, 5, 5, 2
(c) 1, 1, 4, 2 (d) 1, 4, 4, 2

60 Match the List-I with List-II. **[2025, 8 April Shift-II]**

List-I (Complex/ Species)	List-II (Shape and magnetic moment)
A. $[Ni(CO)_4]$	I. Tetrahedral, 2.8 BM
B. $[Ni(CN)_4]^{2-}$	II. Square planar, 0 BM
C. $[NiCl_4]^{2-}$	III. Tetrahedral, 0 BM
D. $[MnBr_4]^{2-}$	IV. Tetrahedral, 5.9 BM

Choose the correct answer from the options given below:

(a) A-III, B-IV, C-II, D-I
(b) A-I, B-II, C-III, D-IV
(c) A-IV, B-I, C-III, D-II
(d) A-III, B-II, C-I, D-IV

61 Identify the diamagnetic octahedral complex ions from below : **[2025, 3 April Shift-II]**

A. $[Mn(CN)_6]^{3-}$
B. $[Co(NH_3)_6]^{3+}$
C. $[Fe(CN)_6]^{4-}$
D. $[Co(H_2O)_3F_3]$

Choose the correct answer from the options given below.

(a) A and D only (b) B and C only
(c) A and C only (d) B and D only

62 The *d*-orbital electronic configuration of the complex among $[Co(en)_3]^{3+}$, $[CoF_6]^{3-}$, $[Mn(H_2O)_6]^{2+}$ and $[Zn(H_2O)_6]^{2+}$ that has the highest CFSE is **[2025, 2 April Shift-II]**

(a) $t_{2g}^6 e_g^0$ (b) $t_{2g}^3 e_g^2$
(c) $t_{2g}^4 e_g^2$ (d) $t_{2g}^6 e_g^4$

63 From the magnetic behaviour of $[NiCl_4]^{2-}$ (paramagnetic) and $[Ni(CO)_4]$ (diamagnetic), choose the correct geometry and oxidation state. **[2025, 22 Jan Shift-I]**

(a) $[NiCl_4]^{2-}$: Ni^{II}, square planar
$[Ni(CO)_4]$: Ni(0), square planar

(b) $[NiCl_4]^{2-}$: Ni^{II}, tetrahedral
$[Ni(CO)_4]$: Ni^{II}, square planar

(c) $[NiCl_4]^{2-}$: Ni(0), tetrahedral
$[Ni(CO)_4]$: Ni(0), square planar

(d) $[NiCl_4]^{2-}$: Ni^{II}, tetrahedral
$[Ni(CO)_4]$: Ni(0), tetrahedral

64 Identify the homoleptic complex(es) that is/are low spin. **[2025, 22 Jan Shift-II]**

(A) $[Fe(CN)_5NO]^{2-}$ (B) $[CoF_6]^{3-}$
(C) $[Fe(CN)_6]^{4-}$ (D) $[Co(NH_3)_6]^{3+}$
(E) $[Cr(H_2O)_6]^{2+}$

Choose the correct answer from the options given below:

(a) (C) and (D) only (b) (A) and (C) only
(c) (C) only (d) (B) and (E) only

Imp. *Order of CFSE of Coordination Compounds*

65 The correct order of the following complexes in terms of their crystal field stabilisation energies is **[2025, 22 Jan Shift-II]**

(a) $[Co(NH_3)_4]^{2+} < [Co(NH_3)_6]^{2+} < [Co(NH_3)_6]^{3+} < [Co(en)_3]^{3+}$

(b) $[Co(en)_3]^{3+} < [Co(NH_3)_6]^{3+} < [Co(NH_3)_6]^{2+} < [Co(NH_3)_4]^{2+}$

(c) $[Co(NH_3)_6]^{2+} < [Co(NH_3)_6]^{3+} < [Co(NH_3)_4]^{2+} < [Co(en)_3]^{3+}$

(d) $[Co(NH_3)_4]^{2+} < [Co(NH_3)_6]^{2+} < [Co(en)_3]^{3+} < [Co(NH_3)_6]^{3+}$

66 The correct increasing order of stability of the complexes based on Δ_o value is **[2025, 29 Jan Shift-I]**

I. $[Mn(CN)_6]^{3-}$ II. $[Co(CN)_6]^{4-}$
III. $[Fe(CN)_6]^{4-}$ IV. $[Fe(CN)_6]^{3-}$

(a) I < II < IV < III (b) II < III < I < IV
(c) IV < III < II < I (d) III < II < IV < I

67 Match List-I with List-II. **[2025, 29 Jan Shift-I]**

List-I (Complex)		List-II (Hybridisation and Magnetic characters)	
(A)	$[MnBr_4]^{2-}$	I.	d^2sp^3 and diamagnetic
(B)	$[FeF_6]^{3-}$	II.	sp^3d^2 and paramagnetic
(C)	$[Co(C_2O_4)_3]^{3-}$	III.	sp^3 and diamagnetic
(D)	$[Ni(CO)_4]$	IV.	sp^3 and paramagnetic

Choose the correct answer from the options given below

(a) (A)-(IV), (B)-(II), (C)-(I), (D)-(III)
(b) (A)-(III), (B)-(II), (C)-(I), (D)-(IV)
(c) (A)-(IV), (B)-(I), (C)-(II), (D)-(III)
(d) (A)-(III), (B)-(I), (C)-(II), (D)-(IV)

68 Match List-I with List-II.

List-I (Complex ion)		List-II (Spin only magnetic moment in B.M.)	
A.	$[Cr(NH_3)_6]^{3+}$	I.	4.90
B.	$[NiCl_4]^{2-}$	II.	3.87
C.	$[CoF_6]^{3-}$	III.	0.0
D.	$[Ni(CN_4)]^{2-}$	IV.	2.83

Choose the correct answer from the options given below: **[2024, 8 April Shift-II]**

(a) A-I, B-IV, C-II, D-III (b) A-II, B-III, C-I, D-IV
(c) A-II, B-IV, C-I, D-III (d) A-IV, B-III, C-I, D-II

69 Match List-I with List-II

List-I	List-II
A. $K_2[Ni(CN)_4]$	I. sp^3
B. $[Ni(CO)_4]$	II. sp^3d^2
C. $[Co(NH_3)_6]Cl_3$	III. dsp^2
D. $Na_3[CoF_6]$	IV. d^2sp^3

Choose the correct answer from the options given below: **[2024, 9 April Shift-II]**

(a) A-III, B-I, C-II, D-IV (b) A-III, B-I, C-IV, D-II
(c) A-I, B-III, C-II, D-IV (d) A-III, B-II, C-IV, D-I

70 Number of complexes with even number of electrons in t_{2g} orbitals is $[Fe(H_2O)_6]^{2+}$, $[Co(H_2O)_6]^{2+}$, $[Co(H_2O)_6]^{3+}$, $[Cu(H_2O)_6]^{2+}$, $[Cr(H_2O)_6]^{2+}$ **[2024, 8 April Shift-I]**

(a) 5 (b) 3 (c) 1 (d) 2

71 Match List-I with List-II **[2024, 6 April Shift-II]**

List-I (Tetrahedral complex)		List-II (Electronic configuration)
A.	$TiCl_4$	I. e^2, t_2^0
B.	$[FeO_4]^{2-}$	II. e^4, t_2^3
C.	$[FeCl_4]^-$	III. e^0, t_2^0
D.	$[CoCl_4]^{2-}$	IV. e^2, t_2^3

Choose the correct answer from the options given below.

(a) A-I, B-III, C-IV, D-II (b) A-III, B-IV, C-II, D-I
(c) A-III, B-I, C-IV, D-II (d) A-IV, B-III, C-I, D-II

72 Consider the following complexes. **[2024, 6 April Shift-I]**

$[CoCl(NH_3)_5]^{2+}$ (*A*), $[Co(CN)_6]^{3-}$ (*B*),

$[Co(NH_3)_5(H_2O)]^{3+}$ (*C*), $[Cu(H_2O)_4]^{2+}$ (*D*)

The correct order of *A, B, C* and *D* in terms of wave number of light absorbed is :

(a) $B < C < A < D$ (b) $D < A < C < B$
(c) $C < D < A < B$ (d) $A < C < B < D$

73 The number of complexes from the following with no electrons in the t_2 orbital is ....

$TiCl_4$, $[MnO_4]^-$, $[FeO_4]^{2-}$, $[FeCl_4]^-$, $[CoCl_4]^{2-}$ **[2024, 5 April Shift-II]**

(a) 3 (b) 1 (c) 4 (d) 2

74 Which one of the following complexes will exhibit the least paramagnetic behaviour?
[Atomic number, Cr = 24, Mn = 25, Fe = 26, Co = 27] **[2024, 5 April Shift-I]**

(a) $[Mn(H_2O)_6]^{2+}$ (b) $[Fe(H_2O)_6]^{2+}$
(c) $[Co(H_2O)_6]^{2+}$ (d) $[Cr(H_2O)_6]^{2+}$

75 Match List-I with List-II. **[2024, 31 Jan Shift-II]**

	List-I (Complex ion)		List-II (Electronic configuration)
A.	$[Cr(H_2O)_6]^{3+}$	I.	$t_{2g}^2 e_g^0$
B.	$[Fe(H_2O)_6]^{3+}$	II.	$t_{2g}^3 e_g^0$
C.	$[Ni(H_2O)_6]^{2+}$	III.	$t_{2g}^3 e_g^2$
D.	$[V(H_2O)_6]^{3+}$	IV.	$t_{2g}^6 e_g^2$

Choose the correct answer from the options given below.

(a) A-IV, B-III, C-I, D-II (b) A-III, B-II, C-IV, D-I
(c) A-IV, B-I, C-II, D-III (d) A-II, B-III, C-IV, D-I

76 The correct statements from following are: **[2024, 31 Jan Shift-I]**

A. The strength of anionic ligands can be explained by crystal field theory.
B. Valence bond theory does not give a quantitative interpretation of kinetic stability of coordination compounds.
C. The hybridisation involved in formation of $[Ni(CN)_4]^{2-}$ complex is dsp^2.
D. The number of possible isomer(s) of *cis*-$[PtCl_2(en)_2]^{2+}$ is one

Choose the correct answer from the options given below:

(a) B, D only (b) B, C only
(c) A, D only (d) A, C only

77 Consider the following complex ions

$P = [FeF_6]^{3-}$, $Q = [V(H_2O)_6]^{2+}$, $R = [Fe(H_2O)_6]^{2+}$

The correct order of the complex ions, according to their spin only magnetic moment values (in BM) is **[2024, 27 Jan Shift-I]**

(a) $R < P < Q$ (b) $R < Q < P$
(c) $Q < R < P$ (d) $Q < P < R$

78 Match List-I with List-II.

	List-I (Complex)		List-II (Colour)
A.	$Mg(NH_4)PO_4$	I.	brown
B.	$K_3[Co(NO_2)_6]$	II.	white
C.	$MnO(OH)_2$	III.	yellow
D.	$Fe_4[Fe(CN)_6]_3$	IV.	blue

Choose the correct answer from the options given below. **[2023, 11 April Shift-II]**

(a) A-III, B-IV, C-II, D-I
(b) A-II, B-III, C-IV, D-I
(c) A-II, B-IV, C-I, D-III
(d) A-II, B-III, C-I, D-IV

79 The complex with highest magnitude of crystal field splitting energy (Δ_0) is **[2023, 15 April Shift-I]**

(a) $[Ti(OH_2)_6]^{3+}$ (b) $[Cr(OH_2)_6]^{3+}$
(c) $[Fe(OH_2)_6]^{3+}$ (d) $[Mn(OH_2)_6]^{3+}$

80 Which of the following complexes will exhibit maximum attraction to an applied magnetic field? **[2023, 13 April Shift-II]**

(a) $[Co(H_2O)_6]^{2+}$ (b) $[Co(en)_3]^{3+}$
(c) $[Ni(H_2O)_6]^{2+}$ (d) $[Zn(H_2O)_6]^{2+}$

81 Match List-I with List-II.

	List-I (Complex)		List-II (CFSE (Δ_0))
A.	$[Cu(NH_3)_6]^{2+}$	I.	−0.6
B.	$[Ti(H_2O)_6]^{3+}$	II.	−2.0
C.	$[Fe(CN)_6]^{3-}$	III.	−1.2
D.	$[NiF_6]^{4-}$	IV.	−0.4

Choose the correct answer from the options given below. **[2023, 12 April Shift-I]**

(a) A-I, B-II, C-IV, D-III
(b) A-II, B-III, C-I, D-IV
(c) A-I, B-IV, C-II, D-III
(d) A-III, B-IV, C-I, D-II

82 Match List-I with List-II.

List-I (Complex)		List-II (Crystal field splitting energy (Δ_0))	
A.	$[Ti(H_2O)_6]^{2+}$	I.	-1.2
B.	$[V(H_2O)_6]^{2+}$	II.	-0.6
C.	$[Mn(H_2O)_6]^{3+}$	III.	0
D.	$[Fe(H_2O)_6]^{3+}$	IV.	-0.8

Choose the correct answer from the options given below **[2023, 10 April Shift-II]**

(a) A-II, B-IV, C-I, D-III (b) A-IV, B-I, C-III, D-II
(c) A-IV, B-I, C-II, D-III (d) A-II, B-IV, C-III, D-I

83 The correct order of the number of unpaired electrons in the given complexes is

A. $[Fe(CN)_6]^{3-}$ B. $[FeF_6]^{3-}$
C. $[CoF_6]^{3-}$ D. $[Cr(oxalate)_3]^{3-}$
E. $[Ni(CO)_4]$

Choose the correct answer from the options given below : **[2023, 10 April Shift-II]**

(a) $E < A < B < D < C$ (b) $A < E < C < B < D$
(c) $A < E < D < C < B$ (d) $E < A < D < C < B$

84 Match List-I with List-II.

List-I (Coordination complex)		List-II (Number of unpaired electrons)	
A.	$[Cr(CN)_6]^{3-}$	I.	0
B.	$[Fe(H_2O)_6]^{2+}$	II.	3
C.	$[Co(NH_3)_6]^{3+}$	III.	2
D.	$[Ni(NH_3)_6]^{2+}$	IV.	4

Choose the correct answer from the options given below: **[2023, 8 April Shift-II]**

(a) A-II, B-IV, C-I, D-III (b) A-III, B-IV, C-I, D-II
(c) A-II, B-I, C-IV, D-III (d) A-IV, B-III, C-II, D-I

85 Match List-I with List-II. **[2023, 30 Jan, Shift-II]**

List-I (Complexes)		List-II (Hybridisation)	
A.	$[Ni(CO)_4]$	I.	sp^3
B.	$[Cu(NH_3)_4]^{2+}$	II.	dsp^2
C.	$[Fe(NH_3)_6]^{2+}$	III.	sp^3d^2
D.	$[Fe(H_2O)_6]^{2+}$	IV.	d^2sp^3

(a) A-II, B-I, C-III, D-IV (b) A-II, B-I, C-IV, D-III
(c) A-I, B-II, C-IV, D-III (d) A-I, B-II, C-III, D-IV

86 Match List-I with List-II.

List-I (Coordination entity)		List-II (Wavelength of light absorbed in nm)	
A.	$[CoCl(NH_3)_5]^{2+}$	I.	310
B.	$[Co(NH_3)_6]^{3+}$	II.	475
C.	$[Co(CN)_6]^{3-}$	III.	535
D.	$[Cu(H_2O)_4]^{2+}$	IV.	600

Choose the correct answer from the options given below. **[2023, 25 Jan Shift-II]**

(a) A-II, B-III, C-IV, D-I
(b) A-III, B-II, C-I, D-IV
(c) A-IV, B-I, C-III, D-II
(d) A-III, B-I, C-II, D-IV

87 Match List-I with List-II.

List-I (Complex)		List-II (Hybridisation)	
A.	$Ni(CO)_4$	I.	sp^3
B.	$[Ni(CN)_4]^{2-}$	II.	sp^3d^2
C.	$[Co(CN)_6]^{3-}$	III.	d^2sp^3
D.	$[CoF_6]^{3-}$	IV.	dsp^2

Choose the correct answer from the options given below **[2022, 28 July Shift-II]**

(a) A-IV, B-I, C-III, D-II (b) A-I, B-IV, C-III, D-II
(c) A-I, B-IV, C-II, D-III (d) A-IV, B-I, C-II, D-III

88 The metal complex that is diamagnetic is (Atomic number Fe, 26; Cu, 29) **[2022, 26 July Shift-II]**

(a) $K_3[Cu(CN)_4]$ (b) $K_2[Cu(CN)_4]$
(c) $K_3[Fe(CN)_6]$ (d) $K_4[FeCl_6]$

89 Arrange the following metal complex/compounds in the increasing order of spin only magnetic moment. Presume all the three, high spin system.
(Atomic numbers, Ce = 58, Gd = 64 and Eu = 63.)

A. $(NH_4)_2[Ce(NO_3)_6]$
B. $Gd(NO_3)_3$ and
C. $Eu(NO_3)_3$ **[2021, 16 March Shift-II]**

(a) (B) < (A) < (C) (b) (C) < (A) < (B)
(c) (A) < (B) < (C) (d) (A) < (C) < (B)

90 In which of the following order the given complex ions are arranged correctly with respect to their decreasing spin only magnetic moment? **[2021, 25 Feb Shift-II]**

(i) $[FeF_6]^{3-}$ (ii) $[Co(NH_3)_6]^{3+}$
(iii) $[NiCl_4]^{2-}$ (iv) $[Cu(NH_3)_4]^{2+}$

(a) (i) > (iii) > (iv) > (ii) (b) (ii) > (iii) > (i) > (iv)
(c) (iii) > (iv) > (ii) > (i) (d) (ii) > (i) > (iii) > (iv)

Imp. *Absorption Spectra of Complex in Relation with Wavelength*

91 Simplified absorption spectra of three complexes [(i), (ii) and (iii)] of M^{n+} ion are provided below; their λ_{max} values are marked as A, B and C respectively. The correct match between the complexes and their λ_{max} values is

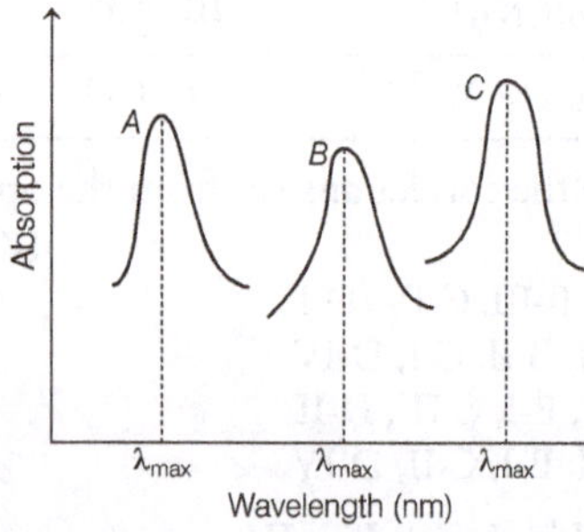

(i) $[M(NCS)_6]^{(-6+n)}$ (ii) $[MF_6]^{(-6+n)}$
(iii) $[M(NH_3)_6]^{n+}$ **[2020, 2 Sep Shift-II]**
(a) A-(iii), B-(i), C-(ii) (b) A-(ii), B-(i), C-(iii)
(c) A-(ii), B-(iii), C-(i) (d) A-(i), B-(ii), C-(iii)

92 $[Pd(F)(Cl)(Br)(I)]^{2-}$ has n number of geometrical isomers. Then, the spin-only magnetic moment and crystal field stabilisation energy [CFSE] of $[Fe(CN)_6]^{n-6}$ respectively, are **[2020, 9 Jan Shift-II]**
[**Note** Ignore the pairing energy]
(a) 1.73 BM and $-2.0\Delta_o$ (b) 2.84 BM and $-1.6\Delta_o$
(c) 0 BM and $-2.4\Delta_o$ (d) 5.92 BM and 0

93 The correct order of the spin only magnetic moments of the following complexes is **[2020, 9 Jan Shift-II]**
(I) $[Cr(H_2O)_6]Br_2$ (II) $Na_4[Fe(CN)_6]$
(III) $Na_3[Fe(C_2O_4)_3](\Delta_0 > P)$
(IV) $(Et_4N)_2[CoCl_4]$
(a) (II) ≈ (I) > (IV) > (III) (b) (I) > (IV) > (III) > (II)
(c) (III) > (I) > (IV) > (II) (d) (III) > (I) > (II) > (IV)

94 The complex ion that will lose its crystal field stabilisation energy upon oxidation of its metal to +3 state is **[2019, 12 April Shift-I]**

(Phen = 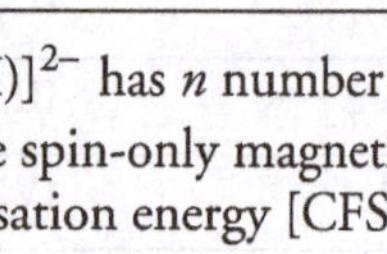

(a) $[Co(phen)_3]^{2+}$ (b) $[Ni(phen)_3]^{2+}$
(c) $[Zn(phen)_3]^{2+}$ (d) $[Fe(phen)_3]^{2+}$

95 The difference in the number of unpaired electrons of a metal ion in its high-spin and low-spin octahedral complexes is two. The metal ion is: **[2019, 10 Jan Shift-II]**
(a) Ni^{2+} (b) Fe^{2+} (c) Co^{2+} (d) Mn^{2+}

Numerical Value Type Questions

96 Consider the following low-spin complexes $K_3[Co(NO_2)_6]$, $K_4[Fe(CN)_6]$, $K_3[Fe(CN)_6]$, $Cu_2[Fe(CN)_6]$ and $Zn_2[Fe(CN)_6]$.
The sum of the spin-only magnetic moment values of complexes having yellow colour is ...... BM (answer in nearest integer). **[2025, 29 Jan Shift-II]**

97 The number of paramagnetic complex species among $[Co(NH_3)_6]^{3+}$, $[Co(C_2O_4)_3]^{3-}$, $[MnCl_6]^{3-}$, $[Mn(CN)_6]^{3-}$, $[CoF_6]^{3-}$, $[Fe(CN)_6]^{3-}$ and $[FeF_6]^{3-}$ with same number of unpaired electrons is .... **[2025, 7 April Shift-II]**

98 The number of paramagnetic complexes among $[FeF_6]^{3-}$, $[Fe(CN)_6]^{3-}$, $[Mn(CN)_6]^{3-}$, $[Co(C_2O_4)_3]^{3-}$, $[MnCl_6]^{3-}$, and $[CoF_6]^{3-}$, which involved d^2sp^3 hybridisation is .... **[2025, 7 April Shift-I]**

99 The spin-only magnetic moment value of M^{n+} ion formed among Ni, Zn, Mn and Cu that has the least enthalpy of atomisation is .... (in nearest integer)
Here n is equal to the number of diamagnetic complexes among $K_2[NiCl_4]$, $[Zn(H_2O)_6]Cl_2$, $K_3[Mn(CN)_6]$ and $[Cu(PPh_3)_3I]$. **[2025, 2 April Shift-II]**

100 A transition metal (M) among Mn, Cr, Co and Fe has the highest standard electrode potential (M^{3+}/M^{2+}). It forms a metal complex of the type $[M(CN)_6]^{4-}$. The number of electrons present in the e_g orbital of the complex is .... **[2025, 2 April Shift-I]**

101 The complex of Ni^{2+} ion and dimethyl glyoxime contains .... number of hydrogen (H) atoms. **[2025, 22 Jan Shift-II]**

102 The spin only magnetic moment for $[Ni(NH_3)_6]^{2+}$ is ....$\times 10^{-1}$ BM.
(Given, atomic number of Ni = 28) **[2024, 31 Jan Shift-I]**

103 The spin only magnetic moment value of square planar complex $[Pt(NH_3)_2Cl(NH_2CH_3)]Cl$ is .... B.M. (Nearest integer)
(Given, atomic number for Pt = 78) **[2024, 27 Jan Shift-II]**

104 The homoleptic and octahedral complex of Co^{2+} and H_2O has .... unpaired electron(s) in the t_{2g} set of orbitals. **[2023, 15 April Shift-I]**

105 In potassium ferrocyanide, there are .... pairs of electrons in the t_{2g} set of orbitals. **[2023, 10 April Shift-I]**

106 The observed magnetic moment of the complex $[Mn(NCS)_6]^{x-}$ is 6.06 BM. The numerical value of x is .... **[2023, 8 April Shift-II]**

107 If the CFSE of $[Ti(H_2O)_6]^{3+}$ is – 96.0 kJ/mol, this complex will absorb maximum at wavelength .... nm. (Nearest integer) **[2023, 31 Jan Shift-II]**

[Assume Planck's constant $(h) = 6.4 \times 10^{-34}$ Js, speed of light $(c) = 3.0 \times 10^8$ m/s and Avogadro's constant $(N_A) = 6 \times 10^{23}$/mol]

108 The number of paramagnetic species from the following is .... **[2023, 25 Jan Shift-I]**
$[Ni(CN)_4]^{2-}$, $[Ni(CO)_4]$, $[NiCl_4]^{2-}$, $[Fe(CN)_6]^{4-}$, $[Cu(NH_3)_4]^{2+}$ $[Fe(CN)_6]^{3-}$ and $[Fe(H_2O)_6]^{2+}$

109 $[Fe(CN)_6]^{3-}$ should be an inner orbital complex. Ignoring the pairing energy, the value of crystal field stabilisation energy for this complex is (–) .... Δ_o. (Nearest integer) **[2022, 29 July Shift-I]**

110 The difference between spin only magnetic moment value of $[Co(H_2O)_6]Cl_2$ and $[Cr(H_2O)_6]Cl_3$ is .... **[2022, 26 July Shift-I]**

111 Consider the following metal complexes
$[Co(NH_3)_6]^{3+}$; $[CoCl(NH_3)_5]^{2+}$
$[Co(CN)_6]^{3-}$; $[Co(NH_3)_5(H_2O)]^{3+}$
The spin-only magnetic moment value of the complex that absorbs light with shortest wavelength is .... BM. (Nearest integer) **[2022, 25 July Shift-I]**

112 Reaction of $[Co(H_2O)_6]^{2+}$ with excess ammonia and in the presence of oxygen results into a diamagnetic product. Number of electrons present in t_{2g}-orbitals of the product is .... **[2022, 26 June Shift-II]**

113 Amongst $FeCl_3 \cdot 3H_2O$, $K_3[Fe(CN)_6]$ and $[Co(NH_3)_6]Cl_3$, the spin-only magnetic moment value of the inner-orbital complex that absorbs light at shortest wavelength is .... BM. [Nearest integer] **[2022, 25 June Shift-II]**

114 If $[Cu(H_2O)_4]^{2+}$ absorbs a light of wavelength 600 nm for d-d transition, then the value of octahedral crystal field splitting energy for $[Cu(H_2O)_6]^{2+}$ will be $\ldots \times 10^{-21}$ J. [Nearest integer]
(Given, $h = 6.63 \times 10^{-34}$ Js and $c = 3.08 \times 10^8$ ms^{-1}) **[2022, 25 June Shift-I]**

115 The total number of unpaired electrons present in $[Co(NH_3)_6]Cl_2$ and $[Co(NH_3)_6]Cl_3$ is .... **[2021, 22 July Shift-II]**

116 An aqueous solution of $NiCl_2$ was heated with excess sodium cyanide in presence of strong oxidising agent to form $[Ni(CN)_6]^{2-}$. The total change in number of unpaired electrons on metal centre is .... **[2021, 20 July Shift-II]**

117 The spin only magnetic moment value for the complex $[Co(CN)_6]^{4-}$ is ...... BM.
[Atomic number of Co = 27] **[2021, 20 July Shift-I]**

118 The total number of unpaired electrons present in the complex $K_3[Cr(oxalate)_3]$ is .... **[2021, 18 March Shift-I]**

119 $[Ti(H_2O)_6]^{3+}$ absorbs light of wavelength 498 nm during a d-d transition. The octahedral splitting energy for the above complex is .... $\times 10^{-1}$ J (Round off to the nearest intger).
$[h = 6.626 \times 10^{-34}$ Js, $c = 3 \times 10^8$ ms$^{-1}]$ **[2021, 16 March Shift-II]**

120 Considering that $\Delta_0 > P$, the magnetic moment (in BM) of $[Ru(H_2O)_6]^{2+}$ would be .... **[2020, 5 Sep Shift-II]**

Topic 3 / Application of Coordination Compounds and Organometallic Compounds

• Level 1

01 The compound used in the treatment of lead poisoning is **[2019, 12 April Shift-II]**
(a) D-penicillamine (b) desferrioxime-B
(c) *cis*-platin (d) EDTA

02 The compound that inhibits the growth of tumors is **[2019, 8 April Shift-II]**
(a) *trans*-$[Pt(Cl)_2(NH_3)_2]$ (b) *cis*-$[Pd(Cl)_2(NH_3)_2]$
(c) *cis*-$[Pt(Cl)_2(NH_3)_2]$ (d) *trans*-$[Pd(Cl)_2(NH_3)_2]$

03 $Mn_2(CO)_{10}$ is an organometallic compound due to the presence of **[2019, 12 Jan Shift-I]**
(a) Mn—C bond (b) Mn—Mn bond
(c) Mn—O bond (d) C—O bond

• Level 2

04 The mismatched combinations are

A. chlorophyll - Co

B. water hardness - EDTA

C. photography -$[Ag(CN)_2]^-$

D. Wilkinson catalyst - $[(Ph_3P)_3 RhCl]$

E. chelating ligand -D-penicillamine

Choose the correct answer from the options given below **[2023, 13 April Shift-I]**

(a) A and C only (b) D and E only

(c) A, C, and E only (d) A and E only

05 The complex that dissolves in water is **[2023, 11 April Shift-I]**

(a) $[Fe_3(OH)_2(OAc)_6]Cl$ (b) $Fe_4[Fe(CN)_6]_3$

(c) $K_3[Co(NO_2)_6]$ (d) $(NH_4)_3[As(Mo_3O_{10})_4]$

• Level 3

06 Choose the correct statements from the following. **[2024, 30 Jan Shift-I]**

(A) Ethane-1, 2-diamine is a chelating ligand.

(B) Metallic aluminium is produced by electrolysis of aluminium oxide in presence of cryolite.

(C) Cyanide ion is used as ligand for leaching of silver.

(D) Phosphine act as a ligand in Wilkinson catalyst.

(E) The stability constants of Ca^{2+} and Mg^{2+} are similar with EDTA complexes.

Choose the correct answer from the options given below.

(a) (A), (B), (C) only

(b) (B), (C), (E) only

(c) (C), (D), (E) only

(d) (A), (D), (E) only

07 In which one of the following metal carbonyls, CO forms a bridge between metal atoms? **[2024, 29 Jan Shift-I]**

(a) $[Mn_2(CO)_{10}]$ (b) $[Ru_3(CO)_{12}]$

(c) $[Os_3(CO)_{12}]$ (d) $[Co_2(CO)_8]$

08 The number of bridging CO ligand(s) and Co—Co bond(s) in $Co_2(CO)_8$, respectively are **[2019, 11 Jan Shift-II]**

(a) 2 and 0 (b) 0 and 2

(c) 4 and 0 (d) 2 and 1

Numerical Value Type Questions

09 In the cobalt-carbonyl complex $[Co_2(CO)_8]$, number of Co — Co bonds is "X" and terminal CO ligands is "Y". $X + Y = \ldots\ldots$. **[2022, 24 June Shift-I]**

10 Number of complexes which will exhibit synergic bonding amongst, $[Cr(CO)_6]$, $[Mn(CO)_5]$ and $[Mn_2(CO)_{10}]$ is …… . **[2022, 28 June Shift-I]**

11 The spin only magnetic moment of the complex present in Fehling's reagent is ……….. BM. (nearest integer). **[2022, 27 July Shift-II]**

12 Number of bridging CO ligands in $[Mn_2(CO)_{10}]$ is ………… **[2021, 26 Feb Shift-I]**

Imp. *Stability Constant of the Coordination Compounds*

13 The stepwise formation of $[Cu(NH_3)_4]^{2+}$ is given below

$$Cu^{2+} + NH_3 \overset{K_1}{\rightleftharpoons} [Cu(NH_3)]^{2+}$$

$$[Cu(NH_3)]^{2+} + NH_3 \overset{K_2}{\rightleftharpoons} [Cu(NH_3)_2]^{2+}$$

$$[Cu(NH_3)_2]^{2+} + NH_3 \overset{K_3}{\rightleftharpoons} [Cu(NH_3)_3]^{2+}$$

$$[Cu(NH_3)_3]^{2+} + NH_3 \overset{K_4}{\rightleftharpoons} [Cu(NH_3)_4]^{2+}$$

The value of stability constants K_1, K_2, K_3 and K_4 are 10^4, 1.58×10^3, 5×10^2 and 10^2 respectively. The overall equilibrium constants for dissociation of $[Cu(NH_3)_4]^{2+}]$ is $x \times 10^{-12}$. The value of x is …….. . (Rounded off to the nearest integer) **[2021, 24 Feb Shift-I]**

Key Index

Topic 1

1	(b)	**2**	(c)	**3**	(a)	**4**	(d)	**5**	(d)	**6**	(c)	**7**	(b)	**8**	(b)	**9**	(c)	**10**	(a)
11	(d)	**12**	(a)	**13**	(d)	**14**	(b)	**15**	(b)	**16**	(d)	**17**	(c)	**18**	(c)	**19**	(a)	**20**	(c)
21	(b)	**22**	(b)	**23**	(a)	**24**	(a)	**25**	(c)	**26**	(a)	**27**	(d)	**28**	(b)	**29**	(c)	**30**	(a)
31	(a)	**32**	(d)	**33**	(c)	**34**	(a)	**35**	(c)	**36**	(a)	**37**	(a)	**38**	(c)	**39**	(a)	**40**	(b)
41	(2)	**42**	(2)	**43**	(3)	**44**	(4)	**45**	(5)	**46**	(9)	**47**	(5)	**48**	(1)	**49**	(2)	**50**	(6)
51	(2)	**52**	(2)	**53**	(6)														

Topic 2

1	(b)	**2**	(c)	**3**	(a)	**4**	(a)	**5**	(b)	**6**	(a)	**7**	(d)	**8**	(c)	**9**	(c)	**10**	(c)
11	(c)	**12**	(d)	**13**	(b)	**14**	(d)	**15**	(a)	**16**	(c)	**17**	(d)	**18**	(c)	**19**	(d)	**20**	(b)
21	(d)	**22**	(d)	**23**	(b)	**24**	(a)	**25**	(c)	**26**	(c)	**27**	(b)	**28**	(b)	**29**	(b)	**30**	(b)
31	(c)	**32**	(c)	**33**	(a)	**34**	(a)	**35**	(d)	**36**	(a)	**37**	(a)	**38**	(c)	**39**	(b)	**40**	(b)
41	(d)	**42**	(a)	**43**	(d)	**44**	(d)	**45**	(a)	**46**	(c)	**47**	(a)	**48**	(b)	**49**	(d)	**50**	(b)
51	(d)	**52**	(d)	**53**	(d)	**54**	(a)	**55**	(b)	**56**	(b)	**57**	(c)	**58**	(c)	**59**	(a)	**60**	(d)
61	(b)	**62**	(a)	**63**	(d)	**64**	(a)	**65**	(a)	**66**	(a)	**67**	(a)	**68**	(c)	**69**	(b)	**70**	(b)
71	(c)	**72**	(b)	**73**	(a)	**74**	(c)	**75**	(d)	**76**	(b)	**77**	(c)	**78**	(d)	**79**	(b)	**80**	(a)
81	(c)	**82**	(c)	**83**	(d)	**84**	(a)	**85**	(c)	**86**	(b)	**87**	(b)	**88**	(a)	**89**	(d)	**90**	(a)
91	(a)	**92**	(a)	**93**	(b)	**94**	(d)	**95**	(c)	**96**	(0)	**97**	(2)	**98**	(2)	**99**	(0)	**100**	(1)
101	(14)	**102**	(28)	**103**	(0)	**104**	(1)	**105**	(3)	**106**	(4)	**107**	(480)	**108**	(4)	**109**	(2)	**110**	(0)
111	(0)	**112**	(6)	**113**	(2)	**114**	(766)	**115**	(1)	**116**	(2)	**117**	(2)	**118**	(3)	**119**	(4)	**120**	(0)

Topic 3

1	(d)	**2**	(c)	**3**	(a)	**4**	(a)	**5**	(a)	**6**	(a)	**7**	(d)	**8**	(d)	**9**	(7)	**10**	(3)
11	(2)	**12**	(0)	**13**	(1)														

Detailed Explanations

Topic 1

01. *(b)* The correct IUPAC name of $[PtBr_2(PMe_3)_2]$ is dibromo*bis* (trimethylphosphine) platinum (II).

02. *(c)* Compound $= K_2MnO_4$

Oxidation state of Mn,

$$0 = 2 \times 1 + x - 4 \times 2$$
$$x = +6$$

Therefore, IUPAC name

= Potassium tetraoxidomanganate (VI)

03. *(a)* $C_2O_4^{2-}$ is an oxalato ion which is a bidentate ligand. In this complex, let the oxidation state of cobalt is

$$\begin{bmatrix} +1 \times 3 + x + (-2) \times 3 = 0 \\ x = +3 \end{bmatrix}.$$

So, the IUPAC name of $K_3[Co(C_2O_4)_3]$ is potassium trioxalatocobaltate (III).

04. *(d)* The correct structure *trans* $[Ni\,Br_2(PPh_3)_2]$ is

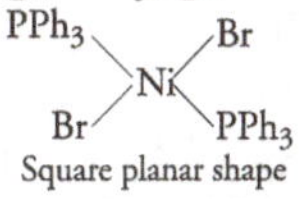

Square planar shape

and of meridional $[Co(NH_3)_3(NO_2)_3]$ is

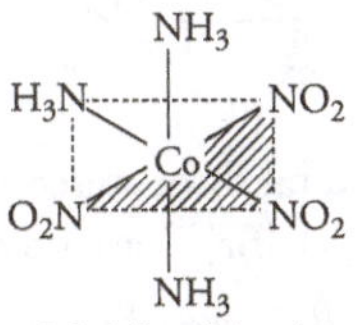

Meridional (mer)
[*trans*]

Concept Enhancer

Meridional isomer exist for those complexes having $[Ma_3b_3]$ type possessing octahedral geometry. Its other isomer is facial. The facial isomer of $[Co(NH_3)_3(NO_2)_3]$ can be drawn as

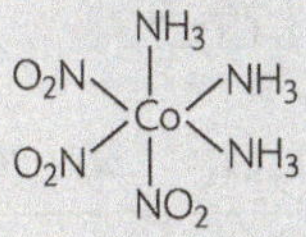

05. *(d)* CrO_3 is acidic nature.
Mn_2O_7 is acidic in nature.

$\therefore \quad X = 2$

The reaction is,

$$[Co(H_2NCH_2CH_2NH_2)_3]_2(SO_4)_3 \rightleftharpoons 2[Co(H_2NCH_2CH_2NH_2)_3]^{3+} + 3SO_4^{2-}$$

$\therefore$ Primary valency = 3

$\therefore \quad X + Y = 5$

06. *(c)* Statement I is true but Statement II is false.

Homoleptic complex of type $[Ma_6]$ (where *a* is monodentate ligand) cannot show geometrical as well as optical isomerism.

cis-platin and *trans*-platin has formula $[Pt(NH_3)_2Cl_2]$ which is a heteroleptic complex of platinum.

07. *(b)* For complex $[CrCl_3(py)_3]$ the stereoisomers are

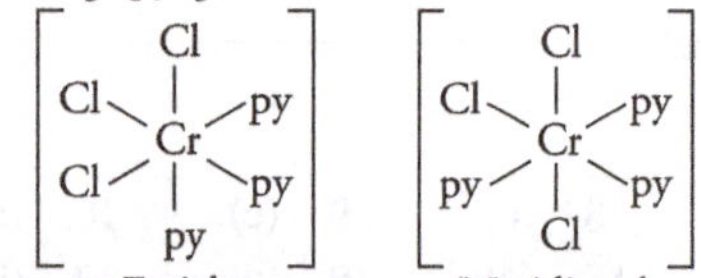

Facial Meridional

So, total 2 stereoisomers.

$[CrCl_3(ox)_2]^{3-}$

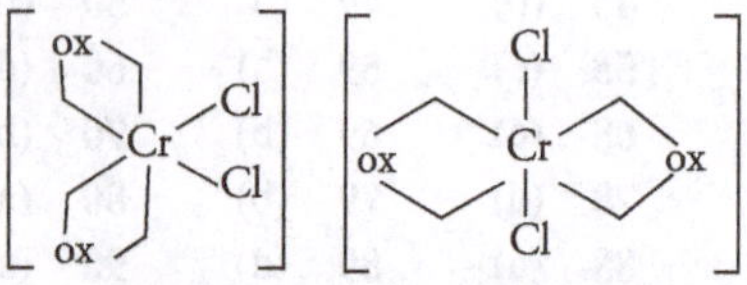

cis-isomer *trans*-isomer

So, geometricaly isomer = 2 (1 *cis* + 1 *trans*)

Optical isomer =3 (2 optically active + 1 optically inactive). So, total 3 stereoisomer are there.

08. *(b)* Complexes of the type Ma_3b_3 shows facial-meridional isomerism.

So, among the given options, compound given in option (b) i.e., $[Co(NH_3)_3Cl_3]$ is of type Ma_3b_3. Thus, it will show facial-meridional isomerism.

> **idea** Coordination compound of the type $[Ma_3b_3]$ shows facial-meridional isomerism.
> So, among the given options only option (b) is of type $[Ma_3b_3]$.

09. *(c)* The complete reaction is,

$[Co(NH_3)_5Cl]Cl_2$

$\downarrow$

$\underbrace{[Co(NH_3)_5Cl]^{2+}(aq) + 2Cl^-(aq)}_{\text{3 ions in water}}$

$[Co(NH_3)_5Cl]Cl_2(aq) + 2AgNO_3)(aq)$
$\longrightarrow [Co(NH_3)_5Cl](NO_3)_2(aq)$
$+ 2AgCl(s)$

Thus, the structure of the complex is $[Co(NH_3)_5Cl]Cl_2$.

10. *(a)* A is not correct but R is correct. The correct form of A is the total number of geometrical isomers shown by $[Co(en)_2Cl_2]^+$ complex ion is two.

> **Concept Enhancer**
> Geometrical isomerism arises in heteroleptic complexes with coordination number 4 and 6 like $[MX_2L_2]$, where X and L are unidentate, *MABXL* (where A, B, X and L are unidentate), MX_2L_4 and $[MX_2(L-L)_2]$ where *L-L* are didentate ligand.
> Note that tetrahedral complexes do not show geometrical isomerism. The complex with coordination number 4 are square planar complexes.

11. *(d)* $CoCl_2 \cdot nNH_3 \xrightarrow[\text{Excess}]{AgNO_3} 2$ mol of AgCl

It means $2Cl^-$ ions will be present in ionisation sphere. So coordination compound should be $[Co(NH_3)_5Cl]Cl_2$.

Oxidation state of cobalt is +3

$$x = +3, n = 5 \Rightarrow x + n = 8$$

12. *(a)* The metal atom present, have sp^3 hybridisation. The complex will have tetrahedral structure. Tetrahedral complex does not show geometrical isomerism.

Hence, option (a) is correct.

13. *(d)* Homoleptic complex are those complex in which all the ligand attached with central atom are same.

Among the given options, only $[Ni(CN)_4]^{2-}$ is a homoleptic complex.

Hence, option (d) is correct.

14. *(b)* $[Cr(ox_2)ClBr)]^3 = [M(AA)_2bc]$ type gives 2-Geometrical isomers - *cis* and *trans* where the *cis*- form is optically active

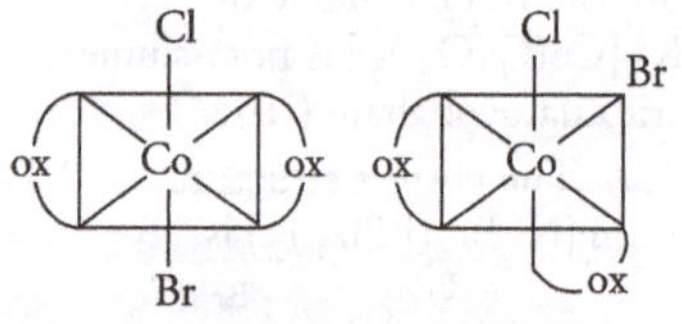

(*trans*) [(+) *cis*]

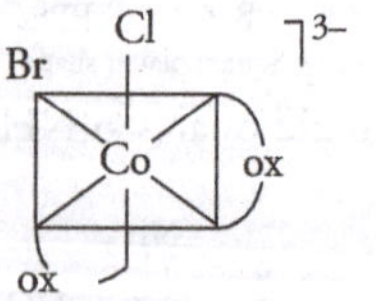

[(−) *cis*]

Thus, the total three stereoisomers for the $[Cr(ox)_2ClBr]^{3-}$ are possible.

15. *(b)* $[MA_3B_3]$ type of compound exists as facial and meriodional isomer. Thus $[Co(NH_3)_3(NO_2)_3]$ will exist as meriodional isomer.

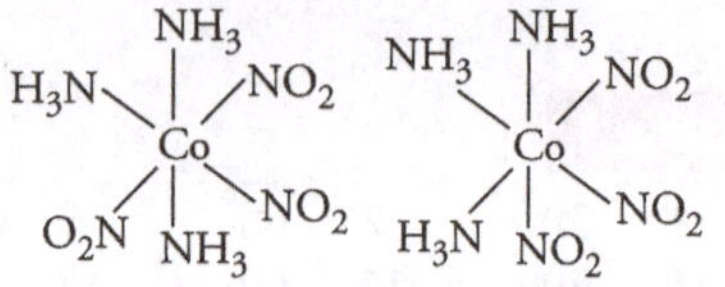

[mer-] [fac-]

16. *(d)* Ambidentate ligand has two different donor atoms and either of the two donor atom is attached to the metal during complex formation.

Among the given ligands, only NCS^- and NO_2^- are ambidentate ligand while $C_2O_4^{2-}$, ethylene diamine, H_2O and EDTA are not ambidentate ligand.

17. *(c)* The correct option is (c).

The complex $[Co(NH_3)_5NO_2]^{2+}$ shows linkage isomerism. NO_2 group is an ambidentate ligand which can bind to a metal atom either through N or O-atom and hence two different isomers are formed.

18. *(c)* The examples of double salt are given in A and C only.

$FeSO_4 \cdot (NH_4)_2SO_4 \cdot 6H_2O$ is hydrated ferrous ammonium sulphate known as Mohr's salt.

$K_2SO_4 \cdot Al_2(SO_4)_3 \cdot 24H_2O$, potash alum is a double salt of K_2SO_4 and $Al_2(SO_4)_3$.

19. *(a)* Let, oxidation state of Co is *x*.

$[Co(NH_3)_5Cl]Cl_2$

$\therefore x + 0 - 1 = +2$ (NH_3 = Neutral ligand)

$x = +3$ (Primary valency)

$\therefore$ Coordination number of Co = 6 (Secondary valency)

Hence, primary valency is 3 and secondary valency is 6.

20. *(c)* $AgCl + 2NH_3 \longrightarrow \underset{\text{soluble}}{[Ag(NH_3)_2]^+Cl^-}$

Thus, white ppt of AgCl is soluble in NH_3 due to the formation of $[Ag(NH_3)_2]Cl$.

21. *(b)* $[Co(CN)_5(NC)]^{3-}$ is of MA_5B type complex, which will not show any isomerism and the rest complexes will show geometrical isomerism as compounds of MA_4B_2, $M(AA)_2B_2$ and MA_3B_3 shows geometrical isomerism.

$\therefore$ Option (b) is correct.

22. *(b)* Stability of a complex depends on the chelating effect. Complex $[Co(en)_3]Cl_2$ is most stable complex among the given complex compounds because more number of chelate rings are present in this complex as compare to others.

23. *(a)* The correct match is (A)-3, (B)-1, (C)-2, (D)-4.

24. *(a)* Total number of possible isomers for $[Pt(en)(NO_2)_2]$ are 3.
These are $[Pt(en)(NO_2)_2]$, $[Pt(en)(ONO)(NO_2)]$ and $[Pt(en)(ONO)_2]$.

25. *(c)* In $[Ni(NH_3)_2Cl_2]$, Ni(II) is sp^3-hybridised. So, the complex has tetrahedral geometry. So, stereoisomers (geometrical and optical) as well as structural isomerism will not be shown by the complex.
On the other hand, rest of them shows geometrical isomerism.

26. *(a)* There are clearly many planes of symmetry in the tetrahedral (M is sp^3 in MA_2B_2) structure and the structure is superimposable on its mirror image.
∴ It has no optical isomers.
An MA_2B_2 complex of dsp^2-hybridised metal (M) has square planar geometry.

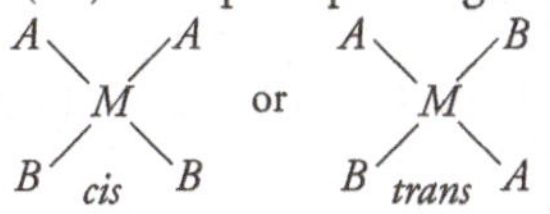

There are two geometrical isomers *cis* and *trans*.
But, since they are planar and therefore they always have a plane of symmetry.
∴ No optical isomerism also.

27. *(d)* The *cis*-$[CrCl_2(ox)_2]^{3-}$ will give non-super imposable mirror image.
So, it is optical active.
Whereas other complexes will give super imposable mirror image hence they are optically inactive.
So, option (d) is correct.

28. *(b)* Octahedral coordination entities of the type $[Ma_3b_3]$ shows geometrical isomers: *fac* and *mer* isomers. Among the given complexes, the complex with general formula $[Ma_3b_3]$ is $[Co(NH_3)_3(NO_2)_3]$. Thus, it can show *fac* and *mer* isomers.

29. *(c)* $[Pt(en)_2(Cl_2)]^{2+}$ with formula $[M(AA)_2b_2]$ will show geometrical isomerism as follows :

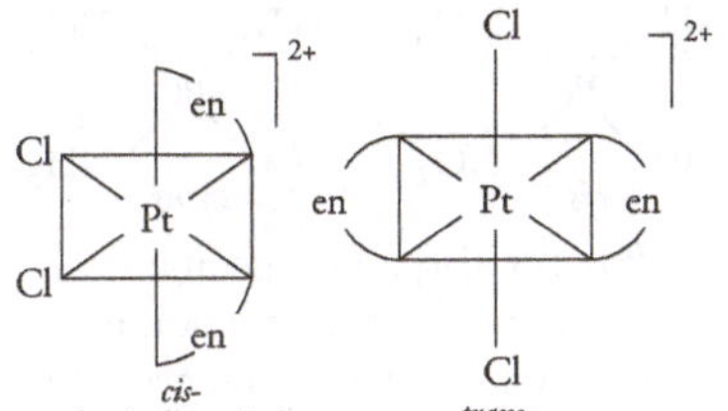

cis- (optically active) (I)   *trans*- (optically inactive)

30. *(a)* Due to Chelation effect most stable complex is $[Fe(C_2O_4)_3]^{3-}$
$[Fe(C_2O_4)_3]^{3-}$ electronic configuration is given by

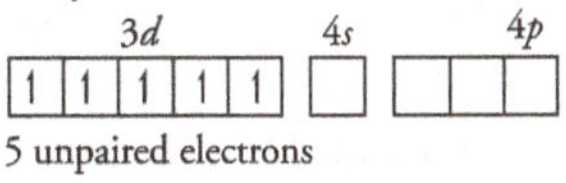

5 unpaired electrons

So, $x = 5$
thus, V_2O_5 is amphoteric.

31. *(a)*
(A) Complex is $[Co(NH_3)_5(SO_4)]Cl$
(B) Complex is $[Co(NH_3)_5Cl]SO_4$
Both (A) and (B) are ionisation isomers. So, they show ionisation isomerism.

32. *(d)* The correct match is
A-I, B-IV, C-II, D-III

33. *(c)* The electronic configuration of the central metal ion in the given complex with their respective oxidation state are as follows

$$[A]Fe^{6+} = [Ar]\,3d^2$$
$$[B]Mn^{3+} = [Ar]\,3d^4$$
$$[C]Fe^{3+} = [Ar]\,3d^5$$
$$[D]Cr^{2+} = [Ar]\,3d^4$$
$$[E]\,Ni^{4+} = [Ar]\,3d^6$$

So, among the given complexes $[Mn(CN)_6]^{3-}$ and $Cr_2(O{-}\overset{O}{\overset{\|}{C}}{-}Me)_4(H_2O)_2$ has their central metal in d^4 configuration.

34. *(a)* The correct match is
A-III, B-II, C-I, D-IV
$A \rightarrow [CoF_6]^{3-}$
Here, Co is in +3 oxidation state. Thus, outer electronic configuration is
$Co^{3+} \longrightarrow 3d^6$

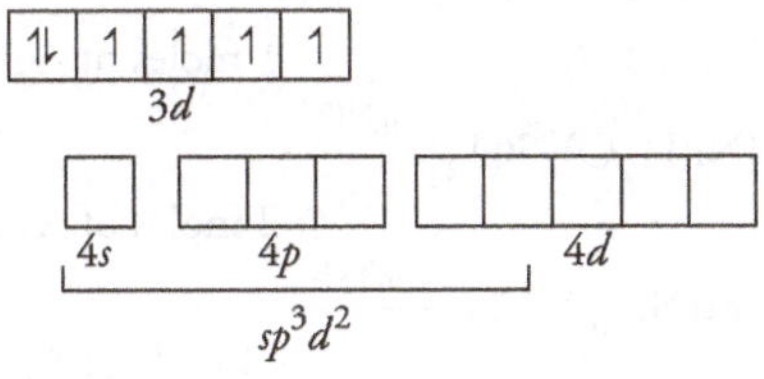

(No pairing will take place as F^- is a weak field ligand.)

$B \longrightarrow [NiCl_4]^{2-}$
Here, Ni is in +2 oxidation state. Thus, outer electronic configuration is
$Ni^{2+} \longrightarrow 3d^8$

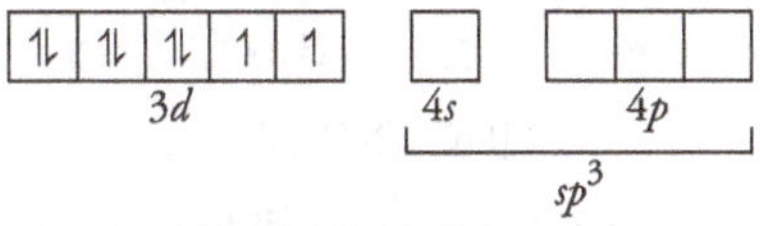

(No pairing will take place as Cl^- is a weak field ligand.)

$C \longrightarrow [Co(NH_3)_6]^{3+}$
Here, Co is in +3 oxidation state. Thus, outer electronic configuration is
$Co^{3+} \longrightarrow 3d^6$

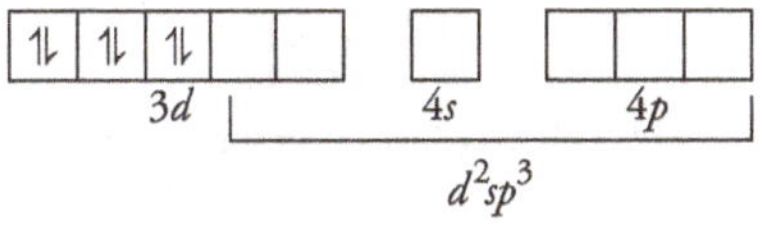

(Pairing will take place as NH_3 will act as strong field ligand as Co is in +3 oxidation state.)

$D \longrightarrow [Ni(CN)_4]^{2-}$
Here, Ni is in +2 oxidation state, thus outer electronic configuration is
$Ni^{2+} \longrightarrow 3d^8$

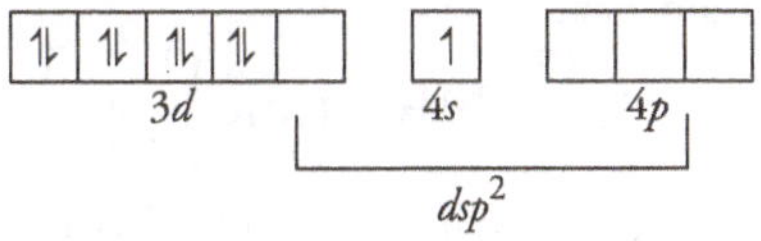

(Pairing will take place as CN is a strong field ligand.)

35. *(c)*

> **idea** Homoleptic complexes are those in which all the ligand which are connected with central atom are same. So, look for the complex in which same ligands are present in central atom.

Among the given complexes, the one which is homoleptic complex with odd number of d-electrons in the central metal are $[Fe(CN)_6]^{3-}$ with d^5 and $[CoCl_4]^{2-}$ with d^7 configuration i.e., B and D only.

36. *(a)* The formula of fluoroapatite is $[3(Ca_3(PO_4)_2 \cdot CaF_2)]$.

37. *(a)* Only the counter ions ate exchanged by other reagents.

$$[Co(NH_3)_5Br]SO_4 + BaCl_2 \longrightarrow [Co(NH_3)_5Br]Cl_2 + \underset{\text{White ppt.}}{BaSO_4\downarrow}$$

It does not have chloride ion as counter ion so it will not give white ppt with $AgNO_3$.

38. *(c)* Number of geometrical isomers of $[PtCl_2(NH_3)_2]$ = two geometrical isomers.

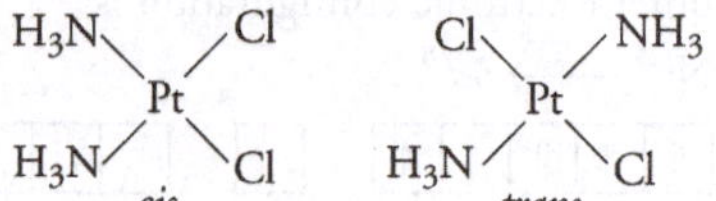

cis *trans*

$[PtCl_2(NH_3)_2]$

In $[Ni(CO)_4]$ complex all ligands are same. Therefore, it does not show geometrical isomerism.

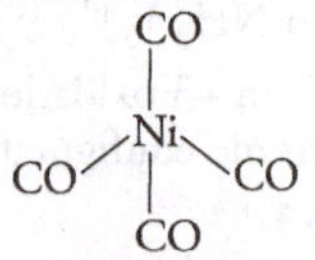

$[Ni(CO)_4]$
(no isomerism)

$[Ru(H_2O)_3Cl_3]$: two geometrical isomers.

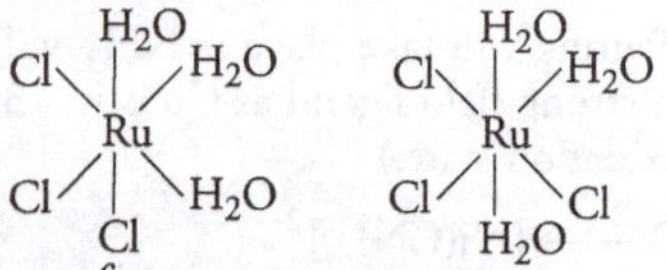

fac *mer*

$[Ru(H_2O)Cl_3]$

$[CoCl_2(NH_3)_4]^+$: two geometrical isomerism.

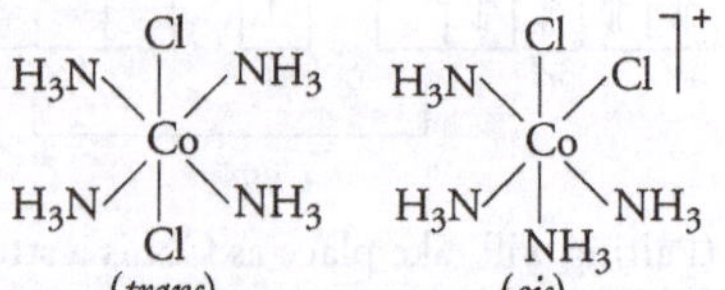

(*trans*) (*cis*)

$[CoCl_2(NH_3)_4]^+$

39. *(a)* Complex
$(A) = [Cr(H_2O)_4Cl_2]Cl \cdot 2H_2O$
Molar mass of complex = 266.5 g/mol
On reaction with conc. H_2SO_4 the mass lost by complex $= \frac{13.5}{100} \times 265$
$= 35.78 \approx 36$ g
It means that 2 moles of H_2O are present in complex.
[∵ Molar mass of H_2O = 18 g/mol]
Then, formula of complex is
$[Cr(H_2O)_4Cl_2]Cl \cdot 2H_2O$

40. *(b)* Pt^{2+} complexes with coordination number equals to 4 are square planar. In $[Pt(NH_3)_2Cl(NO_2)]$ two NH_3 ligands could arrange in *cis* or *trans* positions.
∴ There are two geometrical isomers.

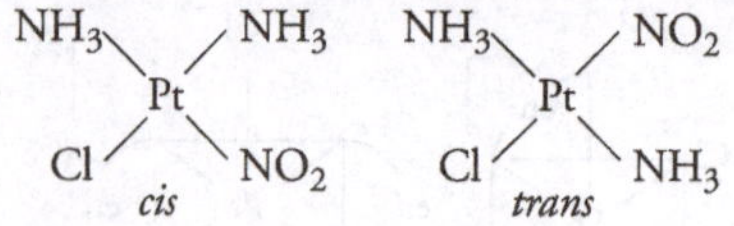

cis *trans*

$[Pt(NH_3)_4ClBr]$: Two geometrical isomers are possible coordination number = 6 and Octahedral geometry Cl and Br can be arranged *cis* or *trans*
∴ 2 geometrical isomers

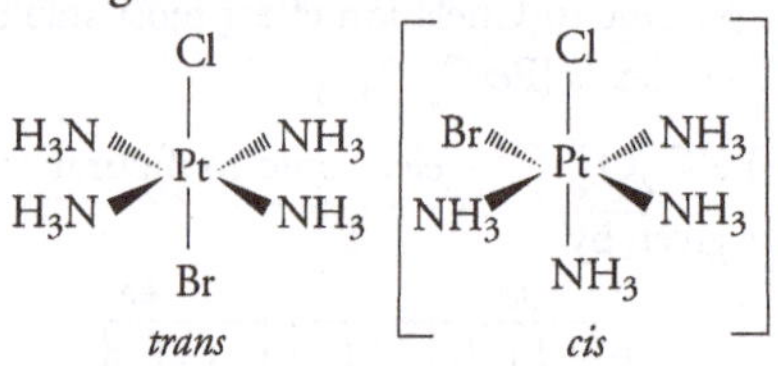

trans *cis*

Hence, (C) and (D) will show geometrical isomerism.

41. *(2)* $FeCl_3 + KOH + H_2C_2O_4 \rightarrow K_3(Fe(C_2O_4)_3]$
(*A*)

Here, $[Fe(C_2O_4)_3]^{3-}$ is $[M(AA)_3]$ type complex.
Hence, total 2 optical isomers are there.

42. *(2)* The number of lone pairs in BrF_5 are one.

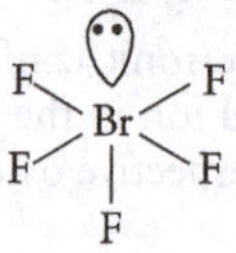

Thus, 1 mole of AgCl complex is $[M(NH_3)_3Cl_3]Cl$.
It shows 2 geometrical isomers (Ma_3b_3 type) i.e., facial and meridional.

43. *(3)* Ligands which have two different donor sites but at a time connects with only one donor site to central metal are called as ambidentate ligands. Among the given ligands ambidentate ligands are NO_2^-, SCN^- and CN^-.

44. *(4)*

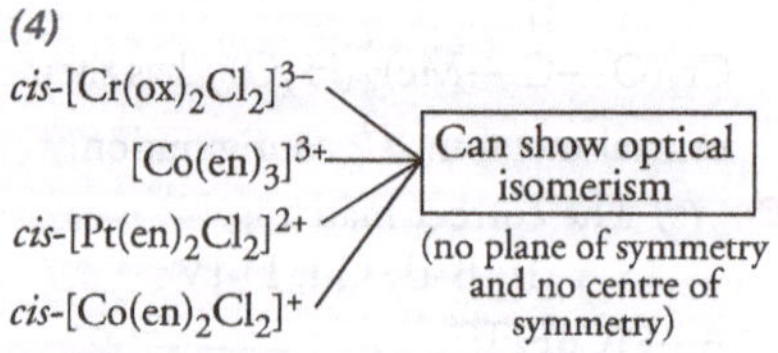

Hence, total 4 complex shows optical isomerism.

45. *(5)* $[Co(NH_3)_4Cl_2]Cl \xrightarrow{AgNO_3}$ 1 moles of AgCl

$[Ni(H_2O)_6]Cl_2 \xrightarrow{AgNO_3}$ 2 moles of AgCl

$[Pt(NH_3)_2Cl_2] \xrightarrow{AgNO_3}$ zero mole of AgCl

$[Pd(NH_3)_4]Cl_2 \xrightarrow{AgNO_3}$ 2 moles of AgCl

Therefore, total number of moles of AgCl precipitated on addition of excess of $AgNO_3$ to one mole each of $[Co(NH_3)_4Cl_2]Cl$, $[Ni(H_2O)_6]Cl_2$, $[Pt(NH_3)_2Cl_2]$ and $[Pd(NH_3)_4]Cl_2$ is = 1 mole + 2 moles + 0 moles + 2 moles = 5 moles of AgCl.

46. *(9)* Here, bpy is bidentate ligand. So, coordination number of cobalt is 6.
Let the oxidation state of Co in $[Co(bpy)Cl_4]^-$ be *x*.
∴ $x + 0 + 4(-1) = -1 \Rightarrow x = 3$
∴ Oxidation state of Co = 3
Sum of oxidation state and coordination number of cobalt is 6 + 3 i.e., 9.

47. *(5)* ∴ $Co^{3+} = [Ar]\,4s^1 3d^5$
i.e. it can accomodate total 6 ligands.
As
$CoCl_3 \cdot x\,NH_3 + Ag\,NO_3\,(aq) \longrightarrow$ 2 moles of AgCl
∴ 2 chloride ions in $CoCl_3 \cdot x\,NH_3$ are primary/ionisable valency (present outside the square bracket)
$[CoCl(NH_3)_x]Cl_2$
Now, as total number of ligands that can bond to cobalt are 6, out of which one is Cl.
∴ $x = 5$
$[Co(Cl)(NH_3)_5]Cl_2$. might be the complex.

48. *(1)* For the complex $CoCl_3 \cdot 4NH_3$ with formula $[Co(NH_3)_4Cl_2]Cl$ geometrical isomers i.e., *cis* and *trans* are possible.
Their structure are as follows

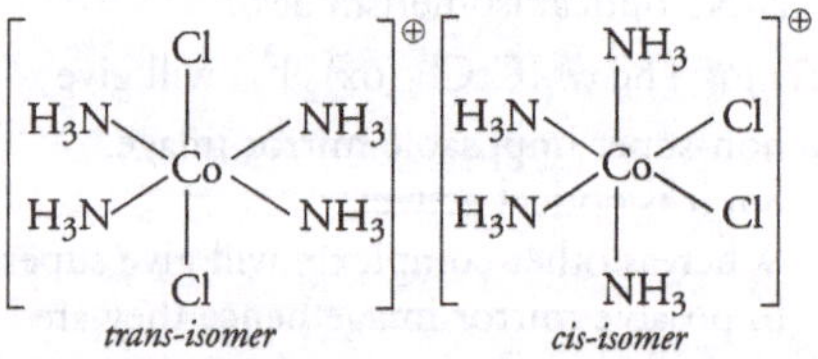

trans-isomer *cis-isomer*

49. *(2)* $\overset{+1}{[Ag(NH_3)_2]^+}\,\overset{+1}{[Ag(CN)_2]^-}$
Total sum of oxidation states of two silver ions = 1 + 1 = 2

50. *(6)* 1 mole of $[Co(en)_2Cl_3]$ produces 1 mole of AgCl.
3 moles of $[Co(en)_2Cl_3]$ produces 3 moles of AgCl.
∴ Given compound is $[Co(en)_2Cl_2]Cl$.
$[Co(en)_2Cl_2]Cl\,(aq) + \underset{(excess)}{AgNO_3\,(aq)} \rightarrow [Co(en)_2Cl_2]\,NO_3\,(aq) + \underset{(white)}{AgCl\,(s)\downarrow}$

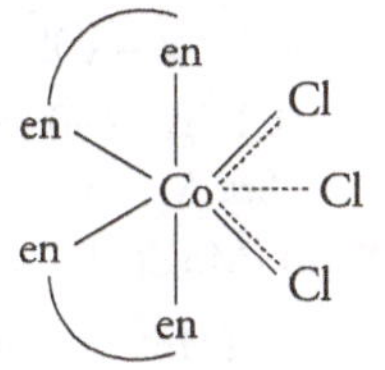

Dotted line represents primary valency = 3.
Solid line represents secondary valency (coordinated ions) of Co = 6.

Concept Enhancer

Primary valences are normally ionisable and are satisfied by negative ions.

51. *(2)* Formula of triamminetrinitrocobalt (III) is $[Co(NO_2)_3(NH_3)_3]$. Its geometrical isomers are as follows :
Thus $X = 2$

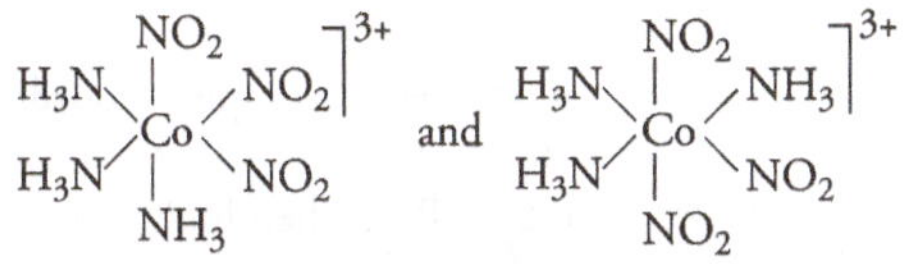

The formula of trioxalatochromate (III) ion is $[Cr(C_2O_4)_3]^{3-}$. It has no geometrical isomers. Thus, $Y = 0$

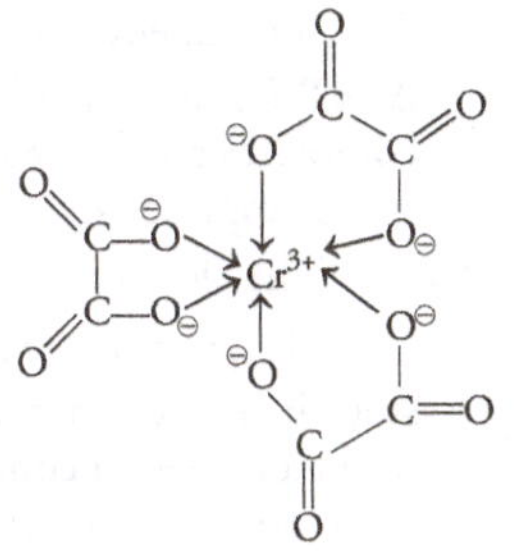

Thus, the value of $X + Y = 2 + 0 = 2$

52. *(2)* $[Co(NH_3)_4Cl_2]Cl + 2en \longrightarrow [Co(en)_2Cl_2] + 4NH_3$

NH_3 is the neutral monodentate ligand. Ethylene diamine is a neutral didentate ligand.

$H_2\ddot{N}—CH_2—CH_2—\ddot{N}H_2$ (en)

So, two ethylene diamine equivalent are required to replace four 'NH_3' ligands.

53. *(6)*

idea $EDTA^{4-}$ is ethylenediaminetetraacetate anion. It is a polyprotic acid containing four carboxylic acid groups and two amine groups with lone-pair electrons that chelate calcium and several other metal ions, it has six coordination sites.

The structure of $EDTA^{-4}$ is

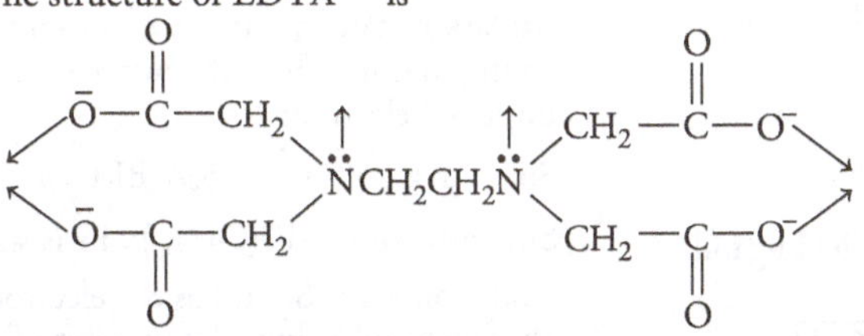

It is a hexadentate ligand. Thus it has six coordination sites.

Topic 2

01. *(b)* Electronic configuration of Fe^{3+} ion in complex $K_3[Fe(SCN)_6]$ is given by

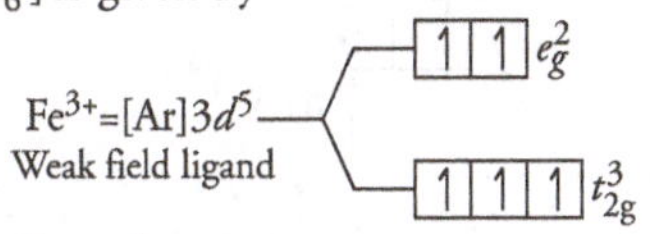

So, using the formula,

$$= (-0.4 \times t_{2g}e^- + 0.6 \times e_g e^-)\Delta_0$$
$$= (-0.4 \times 3 + 0.6 \times 2)\Delta_0 = 0\Delta_0$$

02. *(c)* In general, ligands can be ordered in the following sequence in terms of increasing field strength and in accordance with the spectrochemical series.

$I^- < Br^- < SCN^- < Cl^- < S^{2-} < F^- < OH^- < C_2O_4^{2-} < H_2O < NCS^- < EDTA^{4-} < NH_3 < en < CN^- < CO.$

According to this series the ligand is correctly arranged in option (c).

Concept Enhancer

Spectrochemical series is a series in which ligands are arranged in the order of increasing field strength. It is an experimentally determined series based on the absorption of light by complexes with different ligands.

03. *(a)* Low oxidation state of metals in their complexes is favoured when ligands have good π-accepting ability in addition to sigma bonding (Synergic effect).

04. *(a)* $Fe^{2+} = [Ar]\,3d^6$

In presence of strong field ligand, low spin complex is formed. Hence, all electrons get paired with the configuration. The electronic configuration of $Fe^{+2} = t_{2g}^6\, e_g^0$

Number of unpaired electrons = 0

$\mu = \sqrt{n(n+2)} = 0$ BM

05. *(b)* The complex $K_2[Ni(CN)_4]$ has dsp^2 hybridisation of Ni(II).

$$Ni^{2+} = [Ar]\,3d^8$$

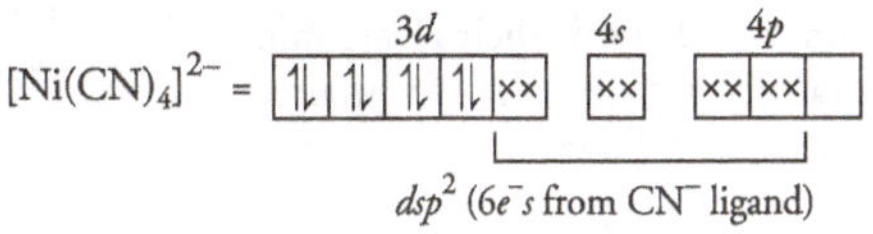

06. *(a)* In octahedral complexes

If $\Delta_0 < P$ (weak field, high spin situation).

The fourth electron enters one of the e_g orbitals giving the configuration $t_{2g}^3 e_g^1$ (metal ion : sp^3d^2).

07. *(d)* Given that $\sqrt{n(n+2)} = 5.91$ BM.

∴ $n = 5$

	Complex	Electronic configuration	Unpaired electron	Ligand
(a)	$[Ni(CN)_4]^{2-}$	$Ni^{2+} = [Ar]3d^8$	0	$(CN)^-$ Strong
(b)	$[NiCl_4]^{2-}$	$Ni^{2+} = [Ar]3d^8$	2	(Cl^-) Weak
(c)	$[Ni(CO)_4]$	$Ni = [Ar]3d^8 4s^2$	2	(CO) Strong
(d)	$[MnBr_4]^{2-}$	$Mn^{2+} = [Ar]3d^5$	5	(Br^-) Weak

Thus, 5 unpaired electrons are present in Mn.

08. *(c)* So the electronic configuration of Co (III) will be $t_{2g}^4 e_g^2$, i.e., it has four unpaired electrons.

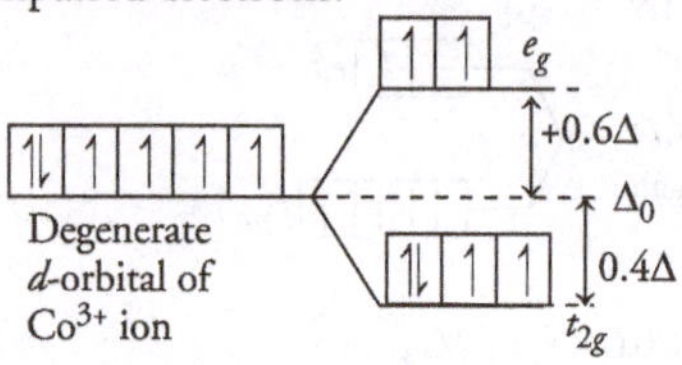

Here, CFSE can be calculated as

$CFSE = [-04.x + 0.6y]\Delta_0$

$= [-0.4 \times 4 + 0.6 \times 2]\Delta_0$

$= [-1.6 + 1.2]\Delta_0 = [-0.4\Delta_0]$

Hence, the crystal field stabilisation energy (CFSE) of $[CoF_3(H_2O)_3]$ will be $-0.4\Delta_0$.

09. *(c)* The electronic configuration is

$[Ru(en)_3]^{2+} \Rightarrow Ru^{2+} = 4d^6 = t_{2g}^6, e_g^0$

$[Fe(H_2O)_6]^{2+} \Rightarrow Fe^{2+} = 3d^6 = t_{2g}^4, e_g^2$

As, en is strong field whereas H_2O is a weak field ligand.

So, correct answer is (c).

10. *(c)* Valence bond theory cannot distinguish between weak field ligands and strong field ligands. It also can't explain the colour exhibits by transition metal complex.

11. *(c)* Wilkinson catalyst is $[RhCl(PPh_3)_3]$.

$[RhCl(PPh_3)_3] \equiv [\overset{}{Rh}\ \overset{(-1)}{Cl}(\overset{(-1)}{P}\overset{(0)}{Ph_3})_3]$

$Rh^+ \rightarrow [Kr]$

The hybridisation of central metal ion is dsp^2 and its shape is square planar.

12. *(d)* Compounds with their electronic configuration and number of unpaired electrons are as follows.

$[FeF_6]^{3-}$	$[Ar]3d^5$	5 unpaired e^-
$[CoF_6]^{3-}$	$[Ar]3d^6$	4 unpaired e^-
$[Ni(CN)_4]^{2-}$	$[Ar]3d^8$	zero unpaired e^-
$[Ni(CO)_4]$	$[Ar]3d^{10}$	zero unpaired e^-

So, the correct order is

$[FeF_6]^{3-} > [CoF_6]^{3-} > [Ni(CN)_4]^{2-} = [Ni(CO)_4]$

13. *(b)* In $[Co(NH_3)_6]^{3+}$, Co^{3+}: $[Ar]3d^6$

NH_3 is strong field ligand,

$Co^{3+} = d^2sp^3$

- $[CrF_6]^{3-}$, Cr^{3+}: $[Ar]3d^3$

Hybridisation of $Cr^{3+} = d^2sp^3$.

- $SF_6 = sp^3d^2$ of sulphur
- $[CoF_6]^{3-}$, Co^{3+} : $[Ar]3d^6$, F^- is strong field ligand.

  Hybridisation of $Co^{3+} = sp^3d^2$
- $[Mn(CN)_6]^{3-}$, Mn^{3+} : $[Ar]3d^4 CN^-$ is strong field ligand.

  Hybridisation of $Mn^{3+} = d^2sp^3$.
- $[MnCl_6]^{3-}$, $Mn^{3+} = [Ar]3d^4$, Cl^- is weak field ligand

  Hybridisation of Cl^- is sp^3d^2.

  Hence, total three sp^3d^2 hybridised molecules.

14. *(d)* $[Mn(SCN)_6]^{4-}$

(a) $Co^{3+}[Ar]3d^6 4s^0$

Also, NH_3 is a SFL thus

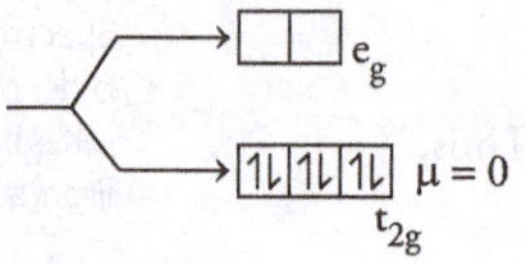

Now, $CFSE = [-0.4 \times 6 + 0.6 \times 0]\Delta_o$

$= -2.4\Delta_0$

(b) $[Fe(CN)_6]^{4-}$

$Fe^{2+} = [Ar]3d^6 4s^0$

Also, $\bar{C}N$ is a SFL

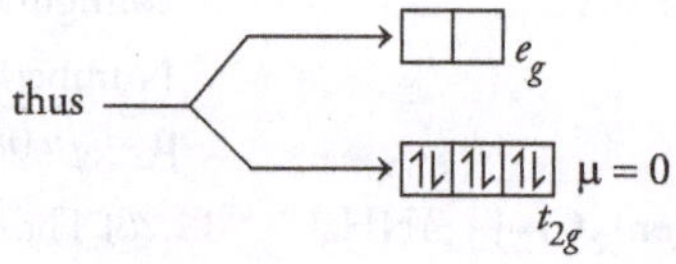

$\mu = 0$

$CFSE = -2.4\Delta_o$

(c) $[FeF_6]^{4-}$

$Fe^{2+} = [Ar]3d^6 4s^0$

Also, F^- is a weak field ligand

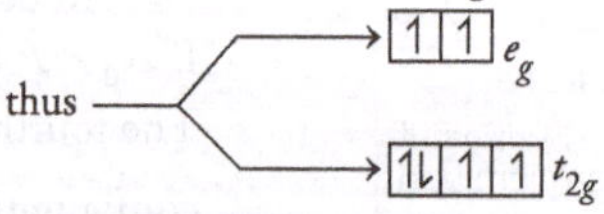

$\mu = \sqrt{24}\ BM = 4.89\ BM$

$CFSE = (-0.4 \times 4 + 0.6 \times 2)\Delta_0$

$= -1.2\Delta_o$

(d) $[Mn(SCN)_6]^{4-}$

$Mn^{2+} = [Ar]3d^5 4s^0$

Also, $\bar{S}CN$ is a weak field ligand. Then,

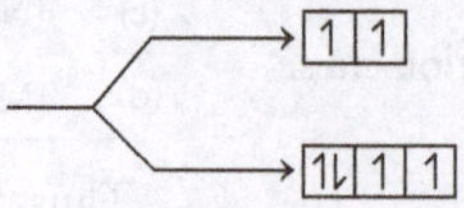

$\mu = \sqrt{35} = 5.96\ BM$

$CFSE = (-0.4 \times 3 + 0.6 \times 2)\Delta_o$

So, $\Delta_o = 0$

15. *(a)* $[MnCl_6]^{3-}$ contains Mn^{3+}

$[Mn^{3+}] = [Ar]3d^4$

Ligand $= Cl^- =$ weak field ligand

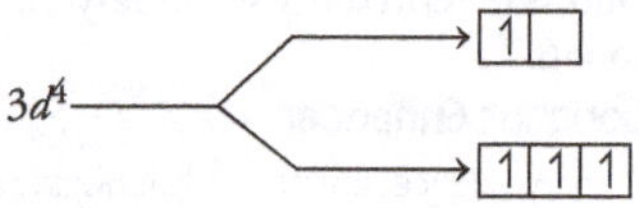

Hybridisation $= sp^3d^2$ and 4 unpaired electrons.

16. *(c)* Both Statement I and II are correct.

In octahedral complex, coordination number is 6.

If $\Delta_0 <$ P.E., then high spin complexes are formed.

If $\Delta_0 >$ P.E., then low spin complexes are formed.

But in tetrahedral complex (CN = 4) $\Delta_t <$ P.E., then mainly high spin complexes are formed and rarely low spin complexes are formed.

17. *(d)* The electronic configuration of Co^{2+} is given by $[Ar]3d^7 4s^0$. Also, since it is given that magnetic moment of the octahedral complex of Co(II) is 3.95. So, it has 3 unpaired electrons. Thus, its configuration is $t_{2g}^5 e_g^2$.

18. *(c)* For $3d^4$,

if ligand is strong field ligand : $t_{2g}^4 e_g^0$ (low spin)

If ligand is weak field ligand : $t_{2g}^3 e_g^1$ (high spin)

Concept Enhancer

Ligands for which $\Delta_o < P$ are called as weak field ligands, they form high spin complexes.

Ligands for which $\Delta_o > P$ are called as strong filed ligands, they form low spin complexes.

19. *(d)* In $K_3[Fe(OH)_6]$, Fe is in +3 oxidation state. So, it has d^5 electronic configuration. Thus, the number of unpaired electrons = 5

So, $\mu = \sqrt{5(5+2)} = 5.92\ BM$

Similarly, in $K_4[Fe(OH)_6]$, Fe is +2 oxidation state. So, it has d^4 electronic configuration. Thus, the number of unpaired electrons = 4

So, $\mu = \sqrt{4(4+2)} = 4.90\ BM$

Thus, the calculated spin-only magnetic moment of $K_3[Fe(OH)_6]$ and $K_4[Fe(OH)_6]$, respectively are 5.92 and 4.90 BM.

20. *(b)* When ethane -1, 2-diamine is added progressively to an aqueous solution of nickel (II) chloride.

- $[Ni(H_2O)_6]^{2+}(aq) + en\,(aq)$ (Green)

$$\downarrow$$

$$[Ni(H_2O)_4(en)]^{2+}(aq) + 2H_2O$$

- $[Ni(H_2O)_4 en]^{2+}(aq) + en\,(aq)$ (Pale blue)

$$\downarrow$$

$$[Ni(H_2O)_2(en)_2]^{2+}(aq) + 2H_2O$$

- $[Ni(H_2O)_2 en]^{+2}(aq) + en(aq)$ (Blue/purple)

$$\downarrow$$

$$[Ni(en)_3]^{2+}(aq) + 2H_2O$$

Violet

21. *(d)* The coordination environment of Ca^{2+} ion in its complex with $EDTA^{4-}$ is octahedral. As $EDTA^{4-}$ is hexadentate ligand.

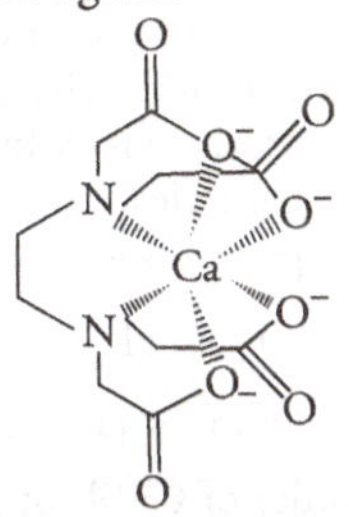

22. *(d)* For the given complex $[Fe(NH_3)_x(CN)_y]^-$ if it has no electron in its e_g orbital then, it will become $[Fe^{III}(NH_3)_2(CN)_4]^-$

So, $x = 2$, $y = 4$

Hence, $x + y = 6$.

23. *(b)* According to spectrochemical series the correct sequence of ligands in the order of decreasing field strength is $CO > H_2O > F^- > S^{2-}$.

24. *(a)* In complex $[Co(NH_3)_6]^{3+}$, Co is present in +3 oxidation state. So, configuration of Co^{3+} in presence of strong field ligand (i.e., NH_3) is $3d^6(t_{2g}^6 e_g^0)$.

Its hybridisation is d^2sp^3 and is an inner orbital complex *i.e.*, (spin paired complex).

In $[CoF_6]^{3-}$, Co is present in +3 oxidation state. So, configuration of Co^{3+} in presence of weak field ligand (i.e. F^-) is $3d^6(t_{2g}^4 e_g^2)$.

Its hybridisation is sp^3d^2 and is an outer orbital complex *i.e.*, (spin free complex).

25. *(c)* Both Statement I and Statement II are correct.

$[Ni(H_2O)_6]^{2+}$ is green colour solution because of *d-d* transition.

$[Ni(CN_4)]^{2-}$ is diamagnetic in nature and is colourless.

Hence, option (c) is correct.

26. *(c)* $[Ni(CO)_4] \rightarrow$ Diamagnetic, sp^3-hybridisation, number of unpaired electrons $= 0$

$[NiCl_4]^{2-} \rightarrow$ Paramagnetic, sp^3-hybridisation number of unpaired electrons $= 2$

27. *(b)* The coordination geometry around the manganese in decacarbonyldimanganese (0) is octahedral. The structure of $Mn_2(CO)_{10}$ is as follows.

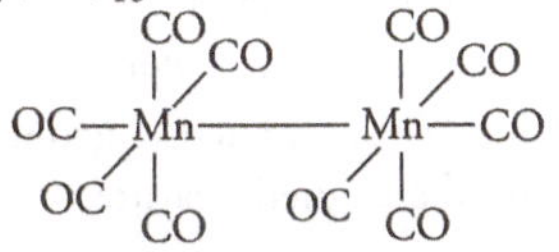

Hence, option (b) is correct.

28. *(b)* $[Co(NH_3)_6]^{3+}$ is octahedral diamagnetic low spin complex.

NH_3 is a strong field ligand which causes larger splitting of *d*-orbital and pairing of electrons is favoured. Hence, $[Co(NH_2)_6]^{3+}$ is a low spin complex as well as diamagnetic.

In $[Co(NH_3)_6]^{3+}$

Electronic configuration after complex formation is

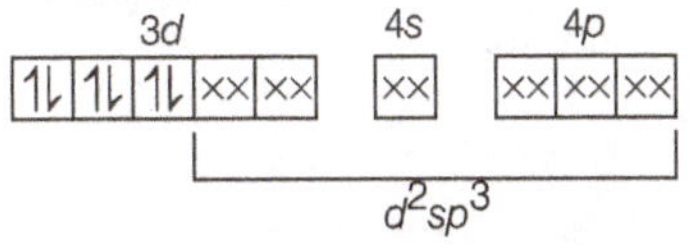

Number of unpaired electrons $= 0$

Geometry of complex ion $=$ Octahedral.

29. *(b)*

Complex	Configuration of *d*-orbital	Unpaired electron	Spin moment, $\mu = \sqrt{n(n+2)}$
$[Fe(CN)_6]^{3-}$	$Fe^{3+} \Rightarrow t_{2g}^5 e_g^0$	1	1.732
$[Mn(CN)_6]^{3-}$	$Mn^{3+} \Rightarrow t_{2g}^4 e_g^0$	2	2.828
$[CoF_6]^{3-}$	$Co^{3+} \Rightarrow t_{2g}^4 e_g^2$	4	4.899
$[MnBr_4]^{2-}$	$Mn^{2+} \Rightarrow e^2 t_2^3$	5	5.916

Hence, correct order of spin only magnetic moment is

$$[Fe(CN)_6]^{3-} < [Mn(CN)_6]^{3-} < [CoF_6]^{3-} < [MnBr_4]^{2-}.$$

30. *(b)* $4FeCl_3 + 3K_4[Fe(CN)_6] \longrightarrow 12KCl + \overset{III}{Fe_4}[\overset{II}{Fe}(CN)_6]_3$ (Prussian blue)

Prussian blue colour is due to formation of coordination compound of iron $Fe_4[Fe(CN)_6]_3$ (ferri-ferrocyanide).

31. *(c)*

$$CoCl_2 + 6H_2O \longrightarrow \underset{\text{Pink }(X)\text{ octahedral}}{[Co(H_2O)_6]Cl_2} \xrightarrow{\text{(conc.) HCl}} \underset{(Y)\text{ Blue solution, }(Z)\text{ Tetrahedral}}{[CoCl_4]^{2-}}$$

Thus, $X = [Co(H_2O)_6]^{2+}$,

$Y = [CoCl_4]^{2-}$, $Z =$ tetrahedral

> **Concept Enhancer**
>
> Crystal field theory attributes to the colour of the coordination complex due to *d-d* transition of electron. It should be noted that in the absence of ligand, crystal field splitting does not takes place. Hence, the substance is colourless. Although in some complexes the colour is due to charge transfer spectra.

32. *(c)* Spin only magnetic moment can be calculated by using the formula

$$\mu = \sqrt{n(n+2)}\text{ BM}$$

For $[FeF_6]^{3-}$

Fe^{3+} : $_{18}[Ar]3d^5$ so, 5 unpaired e^-.

$\Rightarrow \quad \mu = \sqrt{5(5+2)} = 5.91$ BM

For $[CoF_6]^{3-}$:

Co^{3+} : $_{18}[Ar]3d^6$ So, 4 unpaired e^-

$\Rightarrow \quad \mu = \sqrt{4(4+2)} = 4.89$ BM

For $[Co(C_2O_4)_3]^{3-}$

Co^{3+} : $_{18}[Ar]3d^6$ but due to pairing no unpaired e^- are present

$\Rightarrow \quad \mu = \sqrt{0(0+2)} = 0$ BM

$\therefore$Correct order of spin only magnetic moment's

$[FeF_6]^{3-} > [CoF_6]^{3-} > [Co(C_2O_4)_3]^{3-}$

33. *(a)* In $[Co(NH_3)_6]^{3+}$, the oxidation state of cobalt is + 3. Ammonia is a strong field ligand so it will pair up four unpaired electron present in 3*d*-orbitals.

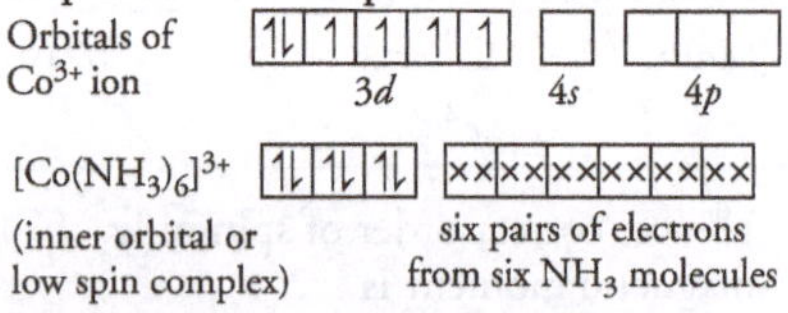

Since, it has no unpaired electrons, so, $[Co(NH_3)_6]^{3+}$ is diamagnetic and its hybridisation is d^2sp^3.

34. *(a)* In the complex $[NiCl_2Br_2]^{2-}$, the geometry is tetrahedral due to presence of weak field ligands and do not show geometrical isomerism.

- However for complex $[PtCl_2Br_2]^{2-}$ the geometry is square planar and it shows two geometrical isomers.
- Both the complex species are optically inactive.
- Also they contain different number of unpaired electron. Hence, their magnetic properties varies.

35. *(d)* A is false but R is true.

Correct statement for Assertion is that, $[CoCl(NH_3)_5]^{2+}$ absorbs at higher wavelength of light with respect to $[Co(NH_3)_5(H_2O)]^{3+}$.

Presence of weak field ligand like Cl^- as compared to H_2O, decreases the CFSE for $[CoCl(NH_3)_5]^{2+}$. Thus it absorbs at higher wavelength.

36. *(a)* Jahn-Teller distortion is shown in those octahedral complexes which have unsymmetrically filled t_{2g} or e_g orbital. In case of $[Cu(H_2O)_6]SO_4$, uneven filling of electron in e_g orbital $(t_{2g}^6\, e_g^3)$ leads to structural distortion of octahedral geometry.

37. *(a)* As strength of ligand increases, crystal field splitting energy increases. In same way energy of absorption for the metal complexes also increases.

Thus, the correct order is

$\underset{(C)}{[Ni(H_2O)_6]^{2+}} < \underset{(B)}{[Ni(NH_3)_6]^{2+}} < \underset{(A)}{[Ni(en)_3]^{2+}}$

Because the order of field strength of the ligands is $H_2O < NH_3 <$ (en).

38. *(c)*

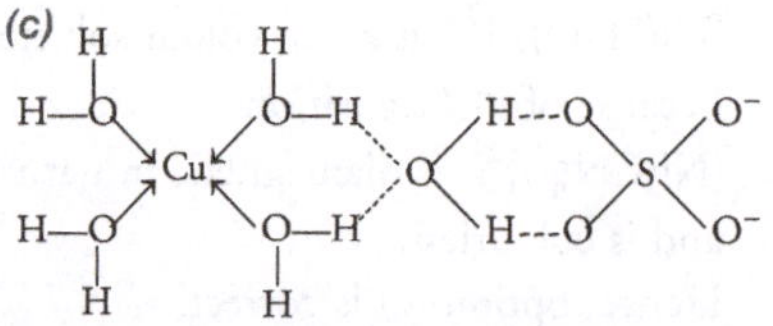

In $CuSO_4 \cdot 5H_2O$, Cu—O bonds are present. It is a correct statement.

In $CuSO_4 \cdot 5H_2O$, ligands coordinating with Cu(II) ion are O - and S-based ligands. The statement is incorrect. The ligands coordinating with Cu(II) ion is O-based and not S-based. It is a incorrect statement.

39. *(b)* $[Ni(CN)_4]^{2-}$: Ni^{2+} has d^8 configuration, Due to strong field ligand it will have sq. planar geometry (dsp^2). It has all paired electrons. Therefore is diamagnetic.

$[Ni(CO)_4]$: Ni has d^{10} configuration also it has, SFL, tetrahedral geometry (sp^3) and diamagnetic nature.

$[NiCl_4]^{2-}$: has d^8 configuration, due to weak field ligand, tetrahedral geometry (sp^3), paramagnetic (2 unpaired e^-)

40. *(b)* As the number of unpaired electrons (*n*) increases magnetic moment of complex also increases.

(A) $[FeF_6]^{3-}$ is a high-spin complex. So, pairing of electrons will not occur.

$Fe^{3+} \Rightarrow t_{2g}^3\, e_g^2 \ (n=5)$

Where *n* = number of unpairs electrons

(B) $[Fe(CN)_6]^{3-}$ is a low-spin complex. So, pairing of electrons occur is *d*-orbital.

$Fe^{3+} \Rightarrow t_{2g}^5\, e_g^0 \ (n=1)$

(C) $[MnCl_6]^{3-}$ is a high spin complex. So, pairing of electrons will not occur.

$Mn^{3+} \Rightarrow t_{2g}^3\, e_g^1 \ (n=4)$

(D) $[Mn(CN)_6]^{3-}$ is a low-spin complex. So, pairing of electrons take place.

$Mn^{2+} \Rightarrow t_{2g}^3\, e_g \ (n=2)$

So, increasing order of magnetic moment : B < D < C < A.

41. *(d)* All the transition metals have same oxidation states.

$[Cr(H_2O)_6]^{3+} \longrightarrow 3d$ series

$[Mo(H_2O)_6]^{3+} \longrightarrow 4d$ series

$[Fe(H_2O)_6]^{3+} \longrightarrow 3d$ series

$[Os(H_2O)_6]^{3+} \longrightarrow 5d$ series

For the same oxidation state, CFSE increasing order will be $3d < 4d < 5d$.

So, $[Os(H_2O)_6]^{3+}$ has maximum CFSE (Δ_o).

42. *(a)* For the same central metal ion, the CFSE increases with increase in strength of the ligand. The order of strength of ligand is given by the spectrochemical series :

$I^- < Br^- < SCN^- < Cl^- < S^{2-} < F^- < OH^- < C_2O_4^{2-} < H_2O < NCS^- < EDTA^{4-} < NH_3 <$ en $< CN^- < CO$

Thus, the order of CFSE of given complexes will be :

$\underset{(A)}{[CoF_6]^{3-}} < \underset{(B)}{[Co(H_2O)_6]^{3+}} < \underset{(C)}{[Co(NH_3)_6]^{3+}} < \underset{(D)}{[Co(en)_3]^{3+}}$.

43. *(d)* The complex $[Fe(CN)_5NOS]^{4-}$ is violet in colour. It is formed when sodium nitroprusside reacts with sulphide ion to give violet colour as follows

$$S^{2-} + [Fe(CN)_5NO]^{2-} \longrightarrow \underset{\text{(Violet colour)}}{[Fe(CN)_5NOS]^{4-}}$$

44. *(d)* $[MnCl_6]^{3-}$

Mn^{3+} in $[MnCl_6]^{3-} \Rightarrow$

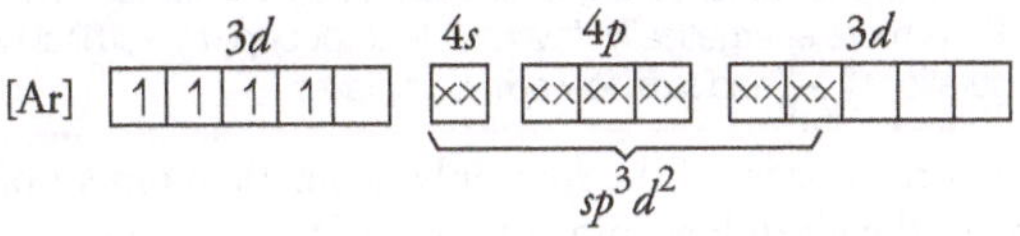

Cl^- is a weak field ligand, so Mn^{3+} forms high spin complex or outer orbital complex. Hence, it will be sp^3d^2 hybridised and is paramagentic.

45. *(a)* Only paramagnetic substance or complex response to external magnetic field.

For $[Fe(H_2O)_6]^{3+}$; Electronic configuration of

$Fe^{3+} = [Ar]3d^5 4s^0$

All electrons are unpaired as H_2O is a weak field ligand

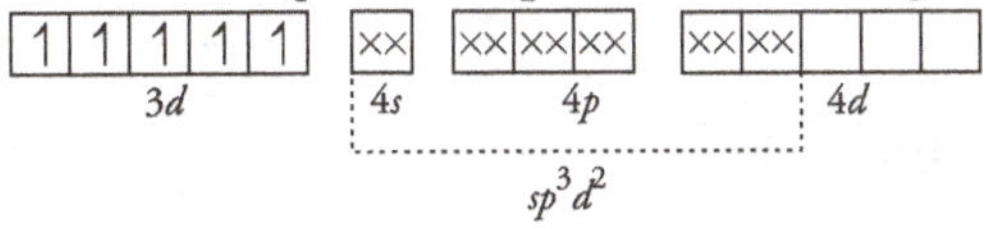

Unpaired electrons = 5

Magnetic nature = Paramagnetic (so this complex response to external magnetic field)

46. *(c)* The CFSE order of different geometric is

$\Delta_t < \Delta_0 < \Delta_{sq.}$

$[NiCl_4]^{2-} < [Ni(H_2O)_6]^{2+} < [Ni(CN)_4]^2$

CFSE $\propto$ Energy of absorption $\propto$ Intensity of color

$\Rightarrow$ Intensity of (complementary) colour of compounds will be $[NiCl_4]^{2-} > [Ni(H_2O)_6]^{2+} > [Ni(CN)_4]^2$

47. *(a)* $[Mn(CN)_6]^{4-}$ In this complex, oxidation state of Mn is + 2.

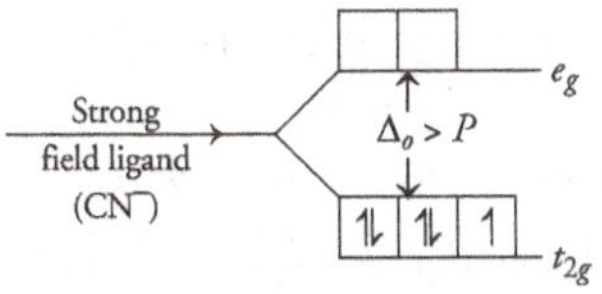

$Mn^{2+} = [Ar]3d^5$

Hybridisation- d^2sp^3

Magnetic nature-Paramagnetic

$[Fe(CN)_6]^{-3}$

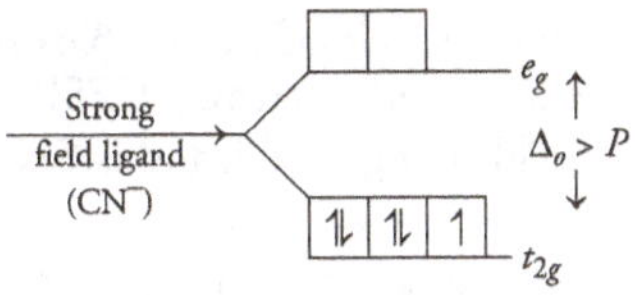

$Fe^{3+} = [Ar]3d^5$

Hybridisation- d^2sp^3

Magnetic nature-Paramagnetic

48. *(b)* So, magnetic moment

$$(\mu) = \sqrt{n(n+2)} \text{ BM}$$

$$n = 4$$

(i) $[FeCl_4]^{2-} \longrightarrow Fe^{2+} \longrightarrow [Ar]\,3d^6$

$$= \sqrt{4(4+2)} \text{ BM} = \sqrt{24} \text{ BM} = 4.90 \text{ BM}$$

(ii) $[Co(C_2O_4)_3]^{3-} \longrightarrow Co^{3+} \longrightarrow [Ar]3d^6$

All electrons are paired, $n = 0$

hence, $\mu = 0$

(iii) $MnO_4^{2-} \rightarrow Mn^{+6} \rightarrow [Ar]3d^1$

$n = 1$

$$= \sqrt{n(n+2)} \text{ BM} = \sqrt{1(1+2)} \text{ BM} = \sqrt{3} \text{ BM} = 1.73 \text{ BM}$$

49. *(d)*

	Molecule	Hybridisation
(a)	XeF_4	sp^3d^2
(b)	$[CrF_6]^{3-}$	d^2sp^3
(c)	BrF_5	sp^3d^2
(d)	$[Ni(CN)_4]^{2-}$	dsp^2

In $[Ni(CN)_4]^{2-}$ molecule hybrid MOs involve only one *d*-orbital of the central atom.

50. *(b)* The pairs in which both the species have same magnetic moment is $[Cr(H_2O)_6]^{2+}$ $(3d^4)$ and $[Fe(H_2O)_6]^{2+}$ $3d^6$ due to presence of same number of unpaired electron. i.e., $n = 4$

51. *(d)* Under octahedral crystal field, the configuration of Ti^{3+} will be, $t_{2g}^1 e_g^0$.

So, CFSE $= [(-0.4) \times 1 + (+0.6) \times 0]\,\Delta_0$

$= -0.4 \times \Delta_0 = -0.4 \times 20{,}300 \text{ cm}^{-1}$ $[\because \Delta_0 = 20{,}300 \text{ cm}^{-1}]$

$= -0.4 \times \dfrac{20{,}300}{83.7} \text{ kJ mol}^{-1}$ $[\because 1 \text{ kJ mol}^{-1} = 83.7 \text{ cm}^{-1}]$

$= -97.013 \text{ kJ mol}^{-1} \Rightarrow$ CFSE $= 97 \text{ kJ mol}^{-1}$

52. *(d)* $\underset{(d^6)}{M^{2+}} + mH_2O \longrightarrow [M(H_2O)_m]^{2+}$

Here, H_2O is a weak field ligand and it should give a high spin, paramagnetic ($\mu = 4.90$ BM) complex ion.

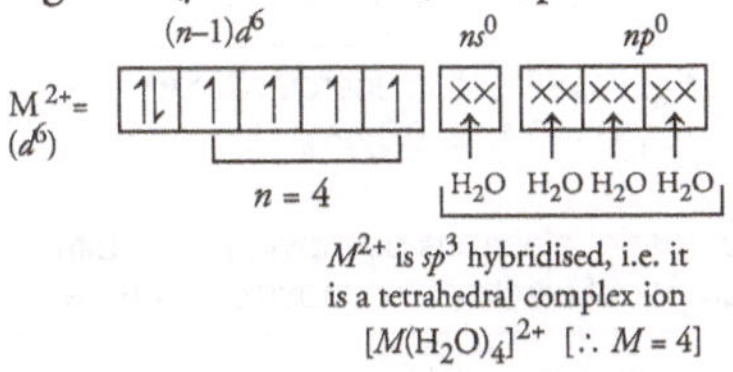

So, the complex will experience splitting of $(n-1)\,d$-orbitals in tetrahedral field. $(e^3\,t_g^3)$

CFSE $= [(-0.6) \times 3 + (0.4) \times 3]\Delta_t = -0.6\,\Delta_t$

So, the geometry and the crystal field stabilisation energy of the complex is tetrahedral and $-0.6\,\Delta_t$ respectively.

53. *(d)* (A) In $Ni(CO)_4$ Ni has $3d^{10}4s^2$ configuration (sp^3d^2) and tetrahedral (strong field ligand).

$\therefore$ Unpaired electrons = 0, $\mu = 0$ BM

(B) In $[Ni(H_2O)_6]Cl_2$ Ni^{2+} has $(3d^8)$ configuration (sp^3d^2) and has two unpaired electrons.

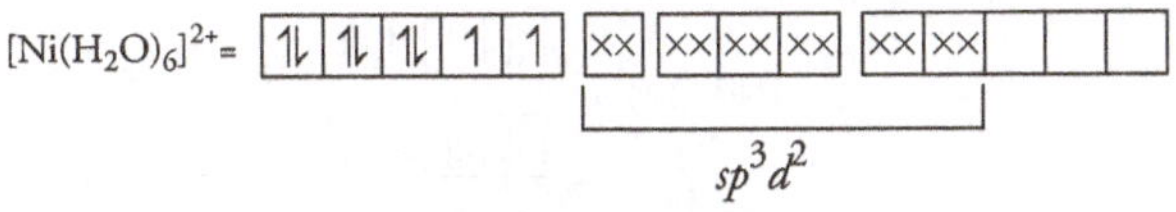

$$\mu = \sqrt{n(n+2)} = \sqrt{2 \times 4}$$

$$= \sqrt{8} \text{ B.M} = 2.82 \text{ BM}$$

(C) In $Na_2[Ni(CN)_4]$ Ni^{2+} has (d^8) configuration and square planar (strong field ligand).

$\therefore$ dsp^2 hybridisation and no unpaired electron

$\therefore$ $\mu = 0$ BM

(D) In $PdCl_2(PPh_3)_2$ Pd^{2+} has ($3d^8$) configuration and coordination number = 4

Complexes of Pd^{2+} are square planar.

$\therefore$ dsp^2 hybridised.

No unpaired electron, $\mu = 0$ BM

The correct order of the calculated spin only magnetic moments of complexes is

$$(A) \approx (C) \approx (D) < (B)$$

54. *(a)* Complete removal of both the axial ligands (along the *z*-axis) from an octahedral complex leads to the following splitting pattern.

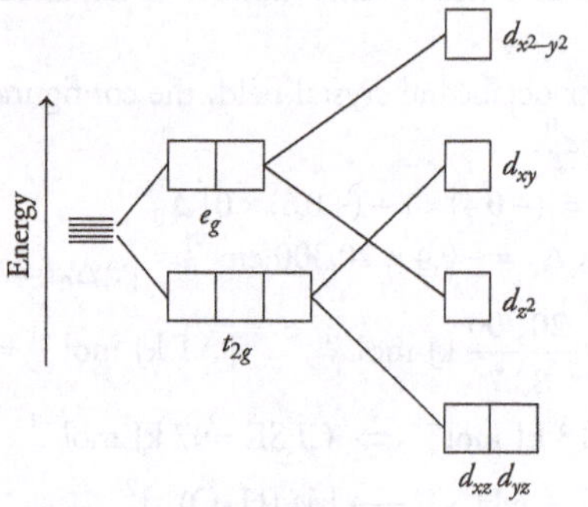

A square planar complex is formed, the attempt to form an octahedral complex being unsuccessful.

55. *(b)*

> **idea** Crystal field stabilisation energy (CFSE) for octahedral complexes = $(-0.4x + 0.6y)\Delta_o$
> where,
> x = number of electrons occupying t_{2g} orbital.
> y = number of electrons occupying e_g orbital.

In $[Fe(H_2O)_6]Cl_2$, H_2O is a weak field ligand, so it is a high spin (outer orbital) octahedral complex of Fe^{2+}.

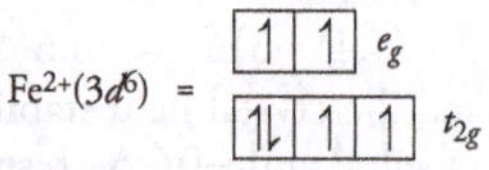

$\therefore$ CFSE $= (-0.4x + 0.6y)\Delta_o$
$= [-0.4 \times 4 + 0.6 \times 2]\Delta_o$
$= -0.4\Delta_o$

In $K_2[NiCl_4]$, Cl^- is a weak field ligand, so it is a high spin tetrahedral complex of Ni^{2+}.

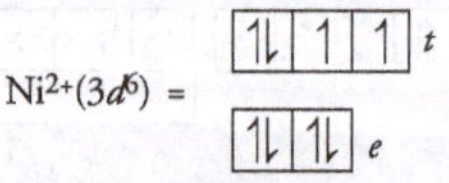

$\therefore$ CFSE $= (-0.6 \times 4 + 0.4 \times 4)\Delta_t$
$= -0.8\Delta_t$

56. *(b)*

> **idea** In presence of strong field ligands, $\Delta_o > P$. for fourth electron it is more energetically favourable to occupy t_{2g} orbital with configuration $t_{2g}^4 e_g^0$ and form low spin complexes.

The correct order of the spin only magnetic moment of metal ions in the given low-spin complexes is
$V^{2+} > Cr^{2+} > Ru^{3+} > Fe^{2+}$.

Complex	Oxidation state	Configuration	Orbital splitting	No. of unpaired electrons
$[V(CN)_6]^{4-}$	V^{2+}	$t_{2g}^3\, e_g^0$	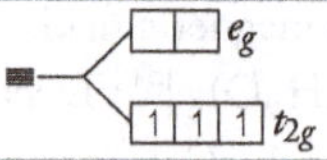	3
$[Cr(NH_3)_6]^{2+}$	Cr^{2+}	$t_{2g}^4\, e_g^0$	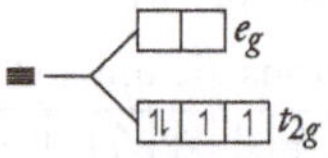	2
$[Ru(NH_3)_6]^{3+}$	Ru^{3+}	$t_{2g}^5\, e_g^0$	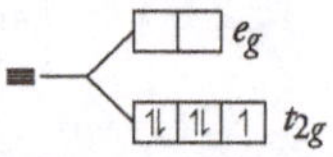	1
$[Fe(CN)_6]^{4-}$	Fe^{2+}	$t_{2g}^6\, e_g^0$	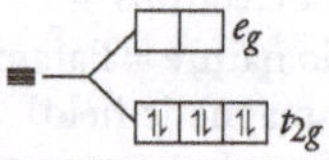	0

57. *(c)* '*A*' absorbs yellow light of less energy and emits violet light of high energy (complementary colour) because H_2O is a weak field ligand. But in case of '*B*', due to presence of strong field ligand (NH_3), it absorbs high energy violet light and emits low energy complementary yellow colour.

$\Rightarrow$ Both the complexes contain three unpaired electrons. Therefore, both are paramagnetic.

$\Rightarrow$ $A(t_{2g}^3\, e_g^0)$; CFSE $= (-0.4 \times 3)\,\Delta_0 = -1.2\Delta_0$

$B\,(t_{2g}^3\, e_g^0)$; CFSE $= [(-0.4 \times 3) + P]\,\Delta_0 = [-1.2 + P]\,\Delta_0$

$\therefore |\text{CFSE}|_A < |\text{CFSE}|_B$
(absorbs, yellow light) (absorbs violet light)

58. *(c)* As we know that $E = \frac{hc}{\lambda}$

$$E \propto \frac{1}{\lambda}$$

Here in all the species Co is in +3 oxidation state. So, as the ligand field strength increases, CFSE also increases. So, the order of field strength of ligand is $CN^- > NH_3 > H_2O > Cl^-$
(*C*) (*B*) (*A*) (*D*)

So, the order of wavelength is $D > A > B > C$

59. *(a)*

- $[Fe(CN)_6]^{3-}$ Fe^{3+} $3d^5$ $t_{2g}^{2,2,1}$ $e_g^{0,0}$ unpaired $e^- = 1$
- $[FeF_6]^{3-}$ Fe^{3+} $3d^5$ $t_{2g}^{1,1,1}$ $e_g^{1,1}$ unpaired $e^- = 5$
- $[CoF_6]^{3-}$ Co^{3+} $3d^6$ $t_{2g}^{2,1,1}$ e_g^{1} unpaired $e^- = 4$
- $[Mn(CN)_6]^{3-}$ Mn^{3+} $3d^4$ $t_{2g}^{2,1,1}$ $e_g^{0,0}$ unpaired $e^- = 2$

60. *(d)* The correct match is A-III, B-II, C-I, D-IV

- $[Ni(CO)_4]$, Ni^0 : $[Ar]3d^8 4s^2$

Valence orbitals of Ni^0 in pre-hydridisation state :

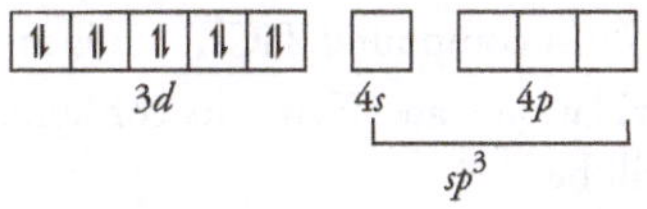

Tetrahedral, diamagnetic, $\mu_s = 0$ BM.

- $[Ni(CN)_4]^{2-}$, $Ni^{2+}: [Ar]3d^8 4s^0$

Valence orbitals of Ni^{2+} in pre-hybridisation state.

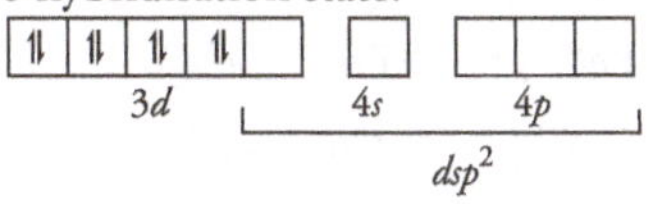

Square planar, diamagnetic, $\mu_s = 0$ B.M.

- $[NiCl_4]^{2-}$, Ni^{2+} ; $[Ar]3d^8 4s^0$

Valence Orbitals of Ni^{2+} in ground state:

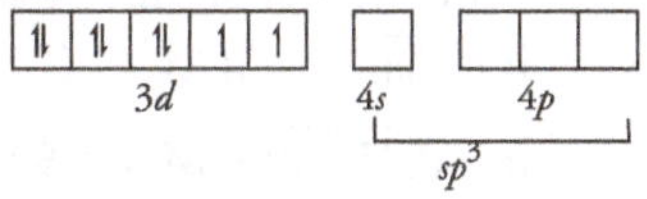

Tetrahedral, paramagnetic, $\mu_s = \sqrt{8} = 2.8$ B.M.

- $[MnBr_4]^{2-}$, $Mn^{2+}: [Ar]3d^5$

Valence orbitals of Mn^{2+} in ground state

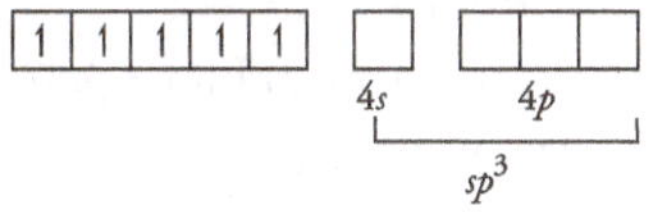

Tetrahedral, paramagnetic, $\mu_s = \sqrt{35} = 5.9$ B.M.

61. *(b)*

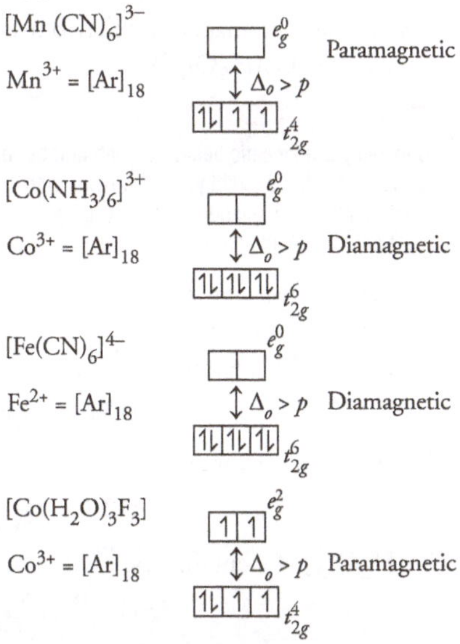

Thus, among the given complexes, diamagnetic octahedral complex ions are $[Co(NH_3)_6]^{3+}$ and $[Fe(CN)_6]^{4-}$.

62. *(a)* In $[Co(en)_3]^{3+}$, strong field ligand is present and hence highest value of CFSE.

In rest complexes, weak field ligand are present and hence CFSE is low.

$[Co(en)_3]^{3+} \Rightarrow Co^{3+}$ and en(SFL)

$Co^{3+} \Rightarrow [Ar]3d^6$

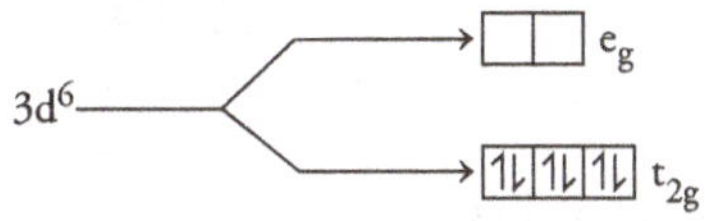

$\therefore$ CFSE is $t_{2g}^6 e_g^0$.

63. *(d)* In complex $[NiCl_4]^{2-}$, Ni is in + 2 oxidation state. So, configuration of $Ni^{2+} = [Ar]3d^8$. Also, since Cl is a weak field ligand. So, the geometry of compound is tetrahedral with sp^3 hybridisation.

Similarly in complex $[Ni(CO)_4]$, Ni is in zero oxidation state. So, configuration of $Ni^0 = [Ar]3d^8 4s^2$. Also, since CO is a strong field ligand. So, after rearrangement the configuration of Ni $= [Ar]3d^{10} 4s^0$ So, the geometry of the compound is tetrahedral with sp^3 hybridisation.

So, option (d) is the correct answer.

64. *(a)*

(A) $[Fe(CN)_5NO]^{2-}$ is a heteroleptic complex with d^2sp^3 hybridisation, low spin (as strong field ligand).

(B) $[CoF_6]^{3-}$ is a homoleptic complex with sp^3d^2 hybridisation. It is high spin (as weak field ligand).

(C) $[Fe(CN)_6]^{4-}$ is a homoleptic complex, with d^2sp^3 hybridisation low spin (as strong field ligand).

(D) $[Co(NH_3)_6]^{3+}$ is a homoleptic complex with, d^2sp^3 hybridisation low spin (as Co is in + 3 oxidation state. Thus, NH_3 will act (as strong field ligand).

(E) $[Cr(H_2O)_6]^{2+}$ is a homoleptic complex with d^2sp^3 hybridisation high spin (as H_2O is a weak field ligand).

Thus, option a is correct.

65. *(a)* As the charge on complex increases, the value of CFSE increases.

Also, as the strength of ligand increases the value of CFSE increases.

Thus, the correct order of CFSE is

$[Co(NH_3)_4]^{2+} < [Co(NH_3)_6]^{2+} < [Co(NH_3)_6]^{3+} < [Co(en)_3]^{3+}$

66. *(a)* The less the value of CFSE the less stable will be the compound, so using the formula,

$$CFSE = -0.4\,\Delta_0 \text{ (number of } e^- \text{ in } t_{2g} \text{ orbital)} + 0.6\,\Delta_0 \text{ (number of } e^- \text{ in } e_g \text{ orbital)}$$

For $[Mn(CN)_6]^{3-}$ the electronic configuration is $t_{2g}^4 e_g^0$.

On substituting the values in above formula, we get

$$CFSE = -1.6\,\Delta_0$$

Similarly for $[Co(CN)_6]^{4-}$, $[Fe(CN)_6]^{4-}$, $[Fe(CN)_6]^{3-}$ the value of CFSE are $-1.8\Delta_0$, $-2.4\Delta_0$, $-2.0\Delta_0$ respectively

Thus, the correct increasing order of stability is

$[Mn(CN)_6]^{3-} < [Co(CN)_6]^{4-} < [Fe(CN)_6]^{3-} < [Fe(CN)_6]^{4-}$

i.e., I < II < IV < III

67. *(a)* The correct match is

A-IV, B-II, C-I, D-III.

A. $[MnBr_4]^{2-}$

In the given complex Mn is in +2 oxidation state. So, electronic configuration of Mn^{2+} is

$$Mn^{2+} \rightarrow [Ar]\,3d^5$$

In presence of weak field ligand, Br

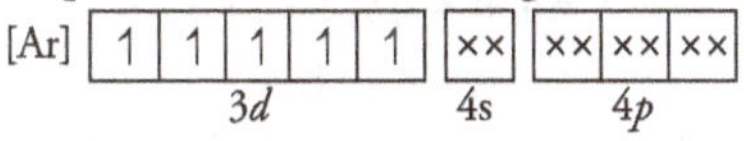

$= sp^3$ hybridisation, paramagnetic

B. $[FeF_6]^{3-}$

In the complex Fe is in +3 oxidation state. So, electronic configuration of Fe^{3+} is

$$Fe^{3+} = [Ar]\,3d^5$$

In presence of weak field ligand, F

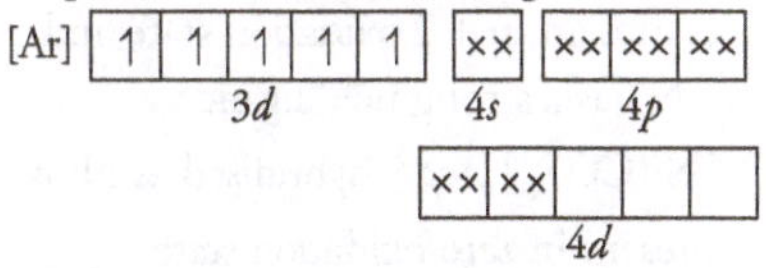

sp^3d^2 hybridisation, paramagnetic

C. $[Co(C_2O_4)_3]^{3-}$

In the complex Co is in +3 oxidation state. So, electronic configuration of Co^{3+} is

$$Co^{3+} = [Ar]3d^6$$

In presence of strong field ligand, $(C_2O_4)^{2-}$

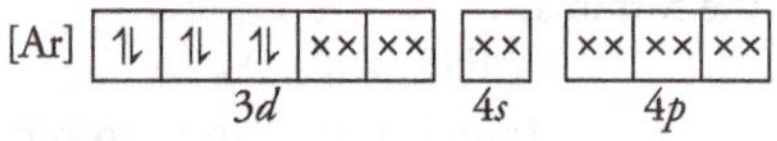

d^2sp^3 hybridisation, diamagnetic

D. $[Ni(CO)_4]$

In this complex Ni is present in zero oxidation state. So, electronic configuration of Ni^0 is

$$Ni^0 = [Ar]\,3d^8 4s^2$$

In presence of strong field ligand, CO

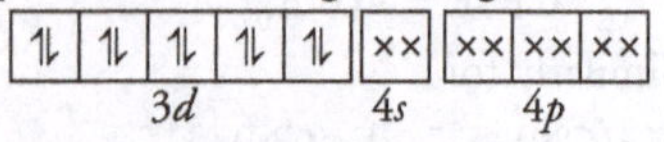

sp^3 hybridisation, diamagnetic

68. *(c)* The correct match is A-II, B-IV, C-I, D-III

A. $[Cr(NH_3)_6]^{3+} \Rightarrow Cr^{3+}; 3d^3$

unpaired electron, $n = 3$

Using the formula, spin only magnetic moment

$$= \sqrt{n(n+2)} \text{ B.M.}$$

$$\mu = 3.87 \text{ B.M.}$$

B. $[NiCl_4]^{2-} \Rightarrow Ni^{2+} = 3d^8$

unpaired electron, $n = 2$,

$$\mu = \sqrt{2(2+2)} \text{ B.M.}$$

$$\mu = 2.83 \text{ B.M.}$$

C. $[CoF_6]^{3-} \Rightarrow Co^{3+}; 3d^6$

unpaired electron,

$$n = 4, \mu = \sqrt{4(4+2)} \text{ B.M.}$$

$$\mu = 4.90 \text{ B.M.}$$

D. $[Ni(CN)_4]^{2-} \Rightarrow Ni^{2+}; 3d^8$

unpaired electron, $n = 0$,

$$\mu = \sqrt{0(0+2)} \text{ B.M.}$$

$$\mu = 0$$

69. *(b)* The correct match is A-III, B-I, C-IV, D-II

- $K_2[Ni(CN)_4]$ is dsp^2 hybridised as Ni is present in + 2 oxidation state and CN^- is a strong field ligand.
- $[Ni(CO)_4]$ is sp^3 hybridised as Ni is present in zero oxidation state.
- $[Co(NH_3)_6]Cl_3$, Co is d^2sp^3 hybridised as NH_3 acts as strong field ligand for Co^{3+}.
- $Na_3[CoF_6]$ is sp^3d^2 hybridised, F act as weak field ligand.

70. *(b)* $[Fe(H_2O)_6]^{2+}$

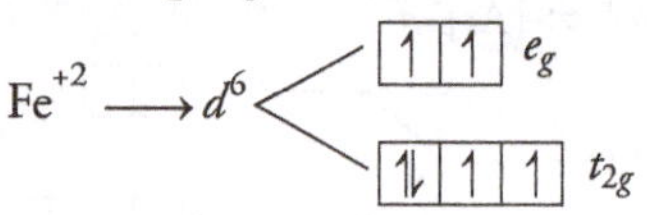

Electron in $t_{2g} = 4$ (even)

$[Co(H_2O)_6]^{2+}$

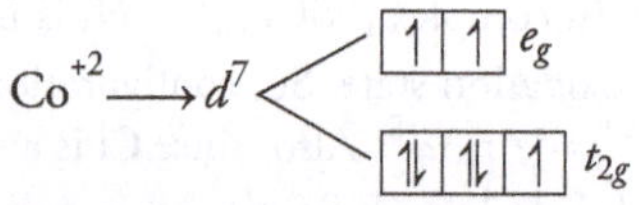

Electron in $t_{2g} = 5$ (odd)

$[Co(H_2O)_6]^{3+}$

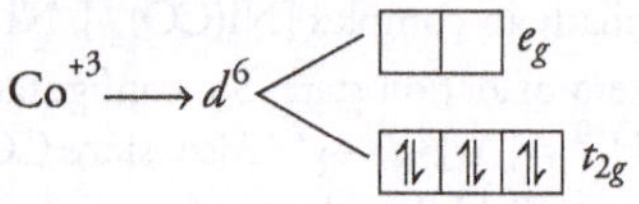

Electron in $t_{2g} = 6$ (even)

$[Cu(H_2O)_6]^{2+}$

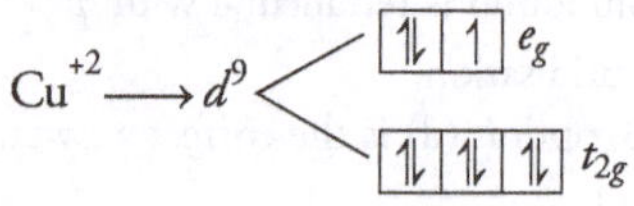

Electron in $t_{2g} = 6$ (even)

$[Cr(H_2O)_6]^{2+}$

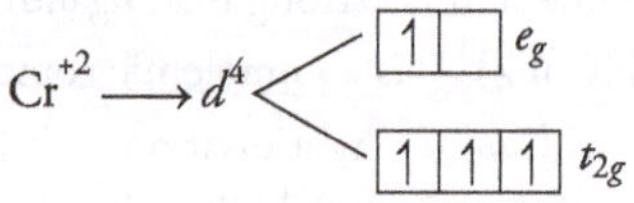

Electron in $t_{2g} = 3$ (odd)

Hence, total 3 complexes have even no. of electrons.

71. *(c)* The correct match is, A-III, B-I, C-IV, D-II

72. *(b)* The correct order of wave number of light absorbed is

$[Cu(H_2O)_4]^{2+} < [CoCl(NH_3)_5]^{2+} < [Co(NH_3)_5(H_2O)]^{3+} < [Co(CN)_6]^{3-}$

The reason is, as ligand field strength increases, light of more energy is absorbed.

Energy $\propto$ wave number $(\bar{\nu})$

73. *(a)* In compound $TiCl_4$, Ti is in +4 oxidation state. Hence its configuration will be $e^0 t_2^0$.

Similarly, for other complexes their configuration is given as

$[MnO_4]^- = Mn^{7+} = e^0 t_2^0$

$[FeO_4]^{2-} = Fe^{6+} = e^2 t_2^0$

$[FeCl_4]^{2-} = Fe^{2+} = e^3 t_2^3$

$[CoCl_4]^{2-} = Co^{2+} = e^4 t_2^3$

Hence, in $[MnO_4]^-$, $[FeO_4]^{2-}$ and $TiCl_4$ does not have any electron in t_2 orbital. Hence, option (a) is correct.

74. *(c)* $[Mn(H_2O)_6]^{2+} = 5$

(number of unpaired electrons)

$$\mu = \sqrt{n(n+2)} \text{ B.M}$$

$$\mu = \sqrt{5(5+2)} \text{ B.M} = 5.92 \text{ B.M}$$

$[Fe(H_2O)_6]^{2+} = 4$

(number of unpaired electrons)

$[Co(H_2O)_6]^{2+} = 3$

(number of unpaired electrons)

$$\mu = \sqrt{n(n+2)} \text{ B.M}$$

$$= \sqrt{3(3+2)} \text{ B.M} = 3.87 \text{ B.M}$$

$[Cr(H_2O)_6]^{2+} = 4$

(number of unpaired electrons)

$$\mu = \sqrt{n(n+2)} \text{ B.M}$$

$$= \sqrt{4(4+2)} \text{ B.M} = 4.89 \text{ B.M}$$

Hence, least paramagnetic behaviour is shown by = $[Co(H_2O)_6]^{2+}$

> **Time Saver Tech**
>
> Connect paramagnetic behaviour with number of unpaired electrons directly. The complex with least number of unpaired electrons will show least paramagnetic behaviour.

75. *(d)* The correct match is A-II, B-III, C-IV, D-I

$[Cr(H_2O)_6]^{3+}$ contains Cr^{3+} :

$$[Ar]\,3d^3 : t_{2g}^3 e_g^0$$

$[Fe(H_2O)_6]^{3+}$ contains Fe^{3+} :

$$[Ar]\,3d^5 : t_{2g}^3 e_g^2$$

$[Ni(H_2O)_6]^{2+}$ contains Ni^{2+} :

$$[Ar]\,3d^8 : t_{2g}^6 e_g^2$$

$[V(H_2O)_6]^{3+}$ contains V^{3+} :

$$[Ar]\,3d^2 : t_{2g}^2 e_g^0$$

76. *(b)* Statement given in B and C are correct while statements A and D are incorrect. Their correct form are as follows

- The strength of anionic ligands cannot be explained by crystal field theory.
- The number of possible isomers of *cis* $[PtCl_2(en)_2]^{2+}$ is more than one.

77. *(c)* Magnetic moment is given by

$\mu_{\text{spin only}} = \sqrt{n(n+2)}$ BM

where, n is number of unpaired electrons.

(*P*) $[FeF_6]^{3-}$ has Fe in +3 oxidation state.

$Fe^{3+} : 3d^5$

So, it has 5 unpaired electrons as F^- is a weak field ligand.

$\therefore \ \mu = \sqrt{5(5+2)}$

$= \sqrt{35}$ BM (*P*)

(*Q*) $[V(H_2O_6)]^{2+}$ has V in + 2 oxidation state.

$V^{2+} : 3d^3$

So, it has 3 unpaired electrons also, H_2O is a weak field ligand.

$\mu = \sqrt{3(3+2)}$

$= \sqrt{15}$ BM (*Q*)

(R) $[Fe(H_2O)_6]^{2+}$ has Fe in +2 oxidation state.

$Fe^{2+} : 3d^6$

So, it has 4 unpaired electrons,

$\mu = \sqrt{4(4+2)}$

$= \sqrt{24}$ BM (*R*)

Hence, the correct sequence of spin only magnetic moment is $Q < R < P$.

78. *(d)* The correct match is A-II, B-III, C-I, D-IV.

	Complex		Colour
(A)	$Mg(NH_4)PO_4$	II.	White
(B)	$K_3[Co(NO_2)_6]$	III.	Yellow
(C)	$MnO(OH)_2$	I.	Brown
(D)	$Fe_4[Fe(CN)_6]_3$	IV.	Blue

79. *(b)* (a) $[Ti(OH_2)_6]^{3+} \Rightarrow t_{2g}^1 e_g^0$

$\Rightarrow$ CFSE

$= [-0.4 \times 1 + 0.6 \times 0]\,\Delta_0 = -0.4\Delta_0$

(b) $[Cr(OH_2)_6]^{3+} \Rightarrow t_{2g}^3 e_g^0$

$\Rightarrow$ CFSE $= [-0.4 \times 3 + 0.6 \times 0]\,\Delta_0$

$= -1.2\,\Delta_0$

(c) $[Fe(OH_2)_6]^{3+}$

$\Rightarrow t_{2g}^3 e_g^2$

$\Rightarrow$ CFSE $= [-0.4 \times 3 + 0.6 \times 2]\,\Delta_0 = 0$

(d) $[Mn(OH_2)_6]^{3+} \Rightarrow t_{2g}^3 e_g^1$

$\Rightarrow$ CFSE $= [-0.4 \times 3 + 0.6 \times 1]\,\Delta_0$

$= -0.6\,\Delta_0$

$\Rightarrow$ So, $|\text{CFSE}|_{max}$ is for option (b).

80. *(a)* More the number of unpaired electron in the *d*-orbital more it will be affected by magnetic field.

$[Co(H_2O)_6]^{2+} \longrightarrow Co^{2+} = 3d^7$

$\Rightarrow t_{2g}^5 e_g^2$

$\Rightarrow$ unpaired electron = 3

$[Ni(H_2O)_6]^{2+} \longrightarrow Ni^{2+}$

$\Rightarrow 3d^8 \Rightarrow t_{2g}^6 e_g^2$

$\Rightarrow$ unpaired electron = 2

$[Zn(H_2O)_6]^{2+} \longrightarrow Zn^{2+} \Rightarrow 3d^{10}$

$\Rightarrow t2_g^6 e_g^4$

$\Rightarrow$ unpaired electron = 0

$[Co(en)_3]^{3+} \longrightarrow Co^{3+} \Rightarrow 3d^6$

$\Rightarrow t_{2g}^6 e_g^0$

$\Rightarrow$ unpaired electron = 0

81. *(c)* The correct match is A-I, B-IV, C-II, D-III.

Use formula CFSE

$= (-0.4\, nt_{2g} + 0.6 ne_g)\,\Delta_0$

where,

nt_{2g} = number of electrons t_{2g} orbitals

ne_g = number of electrons in ne_g orbitals

(A) $[Cu(NH_3)_6]^2$ has $9e^-$ in *d*-orbital

i.e. $t_{2g}^6 e_g^3$

CFSE $= (-0.4 \times 6 + 0.6 \times 3)\,\Delta_0$

$= -0.6\Delta_0$

(B) $[Ti(H_2O)_6]^{3-}$ has $1e^-$ in *d*-orbital

i.e. $t_{2g}^1 e_g^0$

CFSE $= (-0.4 \times 1 + 0.6 \times 0)\,\Delta_0$

$= -0.4\,\Delta_0$

(C) $[Fe(CN)_6]^{3-}$ has $5e^-$ in *d*-orbital,

i.e. $t_{2g}^5 e_g^0$

CFSE $= (-0.4 \times 5 + 0.6 \times 0) = -2\Delta_0$

(D) $[NiF_6]^{4-}$ has $8e^-$ in *d*-orbital, i.e.

$t_{2g}^6 e_g^2$

CFSE $= (-0.4 \times 6 + 0.6 \times 2)$

$= -1.2\,\Delta_0$.

82. *(c)* The correct match is A-IV, B-I, C-II, D-III.

(A) $[Ti(H_2O)_6]^{2+}$

$Ti^{2+} \Rightarrow 3d^2 4s^0$

$\Rightarrow \quad t_{2g}e^- = 2,\ e_g e^- = 0$

CFSE $= [-0.4 \times 2 + 0.6 \times 0]\,\Delta_0$

$= -0.8\,\Delta$

(B) $[V(H_2O)_6]^{2+}$

$V^{2+} \Rightarrow 3d^3 4s^0$

$\Rightarrow \quad t_{2g}e^- = 3$

$\Rightarrow \quad e_g e^- = 0$

CFSE $= [-0.4 \times 3 + 0.6 \times 0]\,\Delta_0$

$= -1.2\,\Delta_0$

(C) $[Mn(H_2O)_6]^{3+}$

$Mn^{3+} \Rightarrow 3d^4 4s^0$

$\Rightarrow \quad t_{2g}e^- = 3,\ e_g e^- = 1$

CFSE $= [-0.4 \times 3 + 0.6 \times 1]\,\Delta_0$

$= -0.6\,\Delta_0$

(D) $[Fe(H_2O)_6]^{3+}$

$Fe^{3+} \Rightarrow 3d^5 4s^0$

$\Rightarrow \quad t_{2g} = 3, e = 2$

CFSE $= [-0.4 \times 3 + 0.6 \times 2]\,\Delta_0 = 0$

83. *(d)* The correct order of the number of unpaired electrons in the given complexes is E < A < D < C < B.

(A) $[Fe(CN)_6]^{3-}$	$n = 1$
(B) $[FeF_6]^{3-}$	$n = 5$
(C) $[CoF_6]^{3-}$	$n = 4$
(D) $[Cr(oxalate)_3]^{3-}$	$n = 3$
(E) $[Ni(CO)_4]$	$n = 0$

84. *(a)*

Complex	Configuration	Number of unpaired electrons
(A) $[Cr(CN)_6]^{3-}$	$3d^3$: [1][1][1][xx][xx]; $4s^0$: [xx]; $4p$: [xx][xx][xx]	3 (II)
(B) $[Fe(H_2O)_6]^{2+}$ Weak field ligand	$3d^6$: [1⇂][1][1][1][1]; $4s^0$: [xx]; $4p$: [xx][xx][xx]; $4d$: [xx][xx][ ][ ][ ]	4 (IV)
(C) $[Co(NH_3)_6]^{3+}$ Strong field ligand	$3d^6$: [1⇂][1⇂][1⇂][xx][xx]; $4s^0$: [xx]; $4p$: [xx][xx][xx]	0 (I)
(D) $[Ni(NH_3)_6]^{3-}$ Strong field ligand	$3d^8$: [1⇂][1⇂][1⇂][1][1]; $4s^0$: [xx]; $4p$: [xx][xx][xx]; $4d$: [xx][xx][ ][ ][ ]	2 (III)

85. *(c)* The correct match is A-I, B-II, C-IV, D-III.

86. *(b)* The correct match is A-III, B-II, C-I, D-IV.

Ligand field strength :

$$(CN^- > NH_3 > H_2O > Cl^-) \propto E_{abs} \propto \frac{1}{\lambda_{abs}}$$

87. *(b)* The correct match is A-I, B-IV, C-III, D-II.

88. *(a)*

Metal complex	Configuration of central metal atom	Magnetic behaviour
$K_3[Cu(CN)_4]$	$Cu^+(3d^{10}4s^0)$	Diamagnetic (All electrons paired)
$K_2[Cu(CN)_4]$	$Cu^{2+}(3d^9 4s^0)$	Paramagnetic (1 unpaired electron)
$K_3[Fe(CN)_6]$	$Fe^{3+}(3d^5 4s^0)$	Paramagnetic (1 unpaired electron)
$K_4[FeCl_6]$	$Fe^{2+}(3d^6 4s^0)$ $(t_{2g}^4 e^2)$	Diamagnetic (4 unpaired electrons)

89. *(d)* Magnetic moment $\propto$ No. of unpaired electrons

(A) $_{58}Ce \longrightarrow [Xe]\,4f^2 5d^0 6s^2$

In this complex $Ce^{+4} \longrightarrow [Xe]\,4f^0 5d^0 6s^0$

There is no unpaired electron, so $\mu_m = 0$ BM

(B) $_{64}Gd \longrightarrow [Xe]4f^7 5d^1 6s^2$

In this complex, $_{64}Gd^{2+} \longrightarrow [Xe]4f^7 5d^1$

There are 8 unpaired electrons.

(C) $_{63}Eu \longrightarrow [Xe]4f^9 5d^0 6s^0$

$_{63}Eu^{+3} \longrightarrow [Xe]\,4f^6 5d^0 6s^0$

In this complex six unpaired electrons are present.

Hence, order of spin only magnetic moment $A < C < B$.

90. *(a)* $\mu \propto n$.

(i) $[FeF_6]^{3-}$

$\Rightarrow Fe^{3+} = (3d^5)$

(weak field ligand). $(n = 5)$

(ii) $[Co(NH_3)_6]^{3+} \Rightarrow Co^{3+} = (3d^6)$ (strong field ligand).

Thus, pairing of electron takes place. $(n = 0)$

(iii) $[NiCl_4]^{2-} \Rightarrow Ni^{2+} = (3d^8)$ (weak field ligand).

Thus, pairing of electron will not takes place. $(n = 2)$

(iv) $[Cu(NH_3)_4]^{2+} \Rightarrow Cu^{2+}(3d^9)$ (strong field ligand).

Thus, pairing of electron takes place. $(n = 1)$

So, the decreasing order of μ is

$$\underset{(n=5)}{\mu_i} > \underset{(n=2)}{\mu_{iii}} > \underset{(n=1)}{\mu_{iv}} > \underset{(n=0)}{\mu_{ii}}$$

91. *(a)* We have to compare their CFSE (Δ_0) as well as wavelength (λ) values, where

$\Delta E(\text{CFSE}) = \frac{hc}{\lambda}$ or, $\Delta_0 \propto \frac{1}{\lambda}$

Power of ligand $\propto \Delta_0 \propto \frac{1}{\lambda}$,

$F^- < NCS^- < NH_3$; $\bar{F} > NC\bar{S} > NH_3$

$\longrightarrow$ Power of Ligand; $\longleftarrow \lambda \longrightarrow \Delta_o$

So, for the given complexes, it is evidented from the plot (Absorption vs. λ)

$$\underset{\substack{(C)\\ [F^-]}}{\lambda_{max}^{(ii)}} > \underset{\substack{(B)\\ [NCS^-]}}{\lambda_{max}^{(i)}} > \underset{\substack{(A)\\ [NH_3]}}{\lambda_{max}^{(iii)}}$$

92. *(a)* Pd^{2+} complexes with coordination number = 4 are of square planar geometry, irrespective of the field strength of ligands.

For square planar complex with four different ligands, three, geometrical isomers are possible.

$[Fe(CN)_6]^{n-6} = [Fe(CN)_6]^{-3}$. So, oxidation state of iron in $[Fe(CN)_6]^{-3}$ is +3. $Fe^{3+} = d^5$.

It forms a low spin complex (CN^- is a strong field ligand).

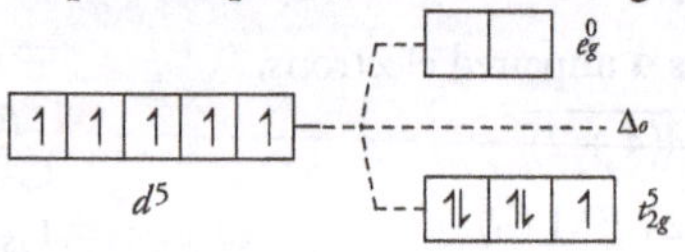

$\therefore$ Number of unpaired electrons $(n) = 1$

Spin only magnetic moment $\mu_s = \sqrt{n(n+2)}$ BM

$= \sqrt{3} = 1.73$ BM

$\text{CFSE} = (t_{2g}e^- \times -0.4\Delta_o) + (e_g e^- \times 0.6\Delta_o)$

$= 5 \times -0.4\Delta_o = -2.0\Delta_o$.

93. *(b)*

Complex	Electronic configuration	Number of unpaired electron	Magnetic moment (BM)
$[Cr(H_2O)_6]Br_2$	$t_{2g}^3 e_g^1$	4	$\sqrt{24}$
$Na_4[Fe(CN)_6]$	$t_{2g}^6 e_g^0$	0	0
$Na_3[Fe(C_2O_4)_3]$ $(\Delta_0 > P)$	$t_{2g}^5 e_g^0$	1	1.73
$[Et_4N)_2[CoCl_4]$	$e^4 t_2^3$	3	$\sqrt{15}$

Order is I > IV > III > II.

94. *(d)* The complex ion that will lose its crystal field stabilisation energy upon oxidation of its metal to +3 state is $[Fe(phen)_3]^{2+}$.

$$[Fe(phen)_3]^{2+} \xrightarrow{-e^-} [Fe(phen)_3]^{3+}$$

In $[Fe(phen)_3]^{2+}$, electronic configuration of Fe^{2+} is $3d^6 4s^0$.

Phenanthrene is a strong field symmetrical bidentate ligand. The splitting of orbital in Fe^{2+} is as follows

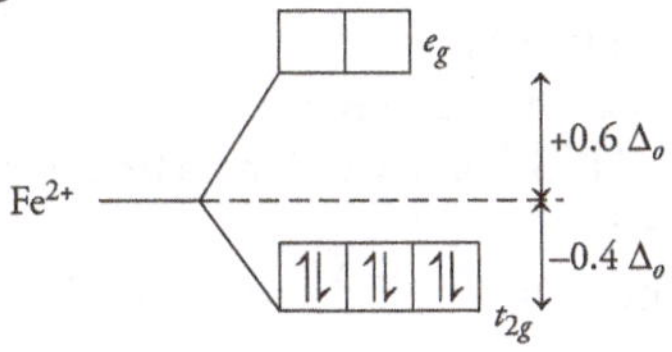

$CFSE = 6 \times -0.4\Delta_o = -2.4\,\Delta_o$.

The splitting of orbital and arrangement of electrons in Fe^{3+} is as follows

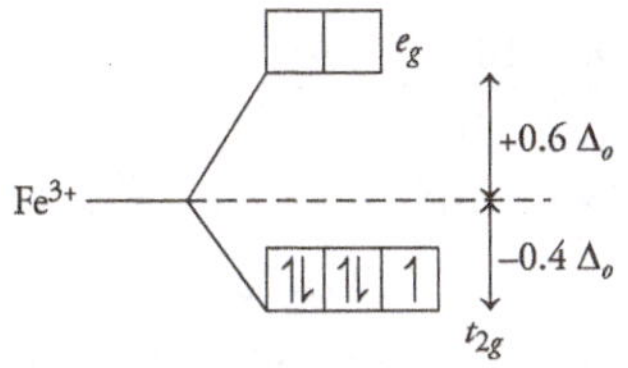

$CFSE = 5 \times -0.4\,\Delta_o = -2.0\,\Delta_o$

Fe^{2+} upon oxidation of its metal to +3 state lose its CFSE from $-2.4\,\Delta_o$ to $-2.0\Delta_o$.

95. *(c)*

Metal ion	No. of unpaired elec. High spin	No. of unpaired elec. Low spin	Difference in the unpaired electrons
$Ni^{2+}(3d^8)$	2	2	0
$Mn^{2+}(3d^5)$	5	1	4
$Fe^{2+}(3d^6)$	4	0	4
$Co^{2+}(3d^7)$	3	1	2

Thus, Co^{2+} has difference of 2.

96. *(0)* $K_3[Co(NO_2)_6] \rightarrow$ Zero unpaired electron.

$K_4[Fe(CN)_6] \rightarrow$ Zero unpaired electron.
$K_3[Fe(CN)_6] \rightarrow$ One unpaired electron.
$Cu_2[Fe(CN)_6] \rightarrow$ Zero unpaired electron.
$Zn_2[Fe(CN)_6] \rightarrow$ Zero unpaired electron.
Yellow colour compounds are
$K_3[Co(NO_2)_6]$ and $K_4[Fe(CN)_6]$

Now, in $K_3[Co(NO_2)_6]$, the outer electronic configuration of Co^{3+} is $[Ar]3d^6$

In presence of strong field ligand, NO_2

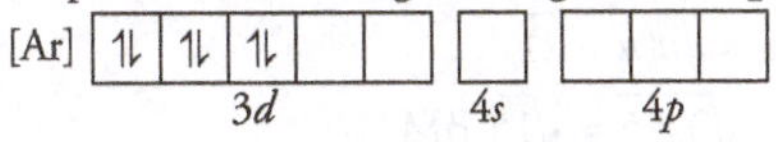

Magnetic moment $(\mu) = \sqrt{n(n+2)}$ BM
$= 0$ BM

In $K_4[Fe(CN)_6]$, Fe^{2+} is $[Ar]3d^6$

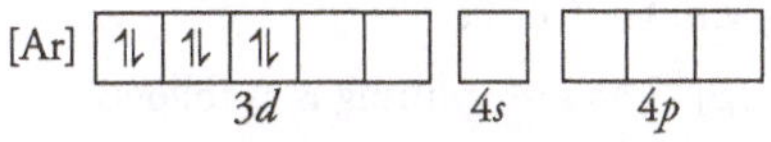

$\mu = \sqrt{n(n+2)}$ BM $= 0$ BM

So, total spin only magnetic moment value of complex with yellow colour is zero.

97. *(2)*

- $[Co(NH_3)_6]^{3+}$ Co^{3+} $3d^6$ $t_{2g}^{2,2,2}$ $e_g^{0,0}$
  Diamagnetic [as unpaired electron $= 0$]
- $[Co(C_2O_4)_3]^{3-}]$ Co^{3+} $3d^6$ $t_{2g}^{2,2,2}$ $e_g^{0,0}$
  Diamagnetic [unpaired electron $= 0$]
- $[MnCl_6]^{3-}$ Mn^{3+} $3d^4$ $t_{2g}^{1,1,1}$ $e_g^{1,0}$
  Paramagnetic [unpaired electron $= 4$]
- $[Mn(CN)_6]^{3-}$ Mn^{3+} $3d^4$ $t_{2g}^{2,1,1}$ $e_g^{0,0}$
  Paramagnetic [unpaired electron $= 2$]
- $[CoF_6]^{3-}$ Co^{3+} $3d^6$ $t_{2g}^{2,1,1}$ $e_g^{1,1}$
  Paramagnetic [unpaired electron $= 4$]
- $[Fe(CN)_6]^{3-}$ Fe^{3+} $3d^5$ $t_{2g}^{2,2,1}$ $e_g^{0,0}$
  Paramagnetic [unpaired electron $= 1$]
- $[FeF_6]^{3-}$ Fe^{3+} $3d^5$ $t_{2g}^{1,1,1}$ $e_g^{1,1}$
  Paramagnetic [unpaired electron $= 5$]

Hence, total two complex have same number of unpaired electron.

98. *(2)*

Complex	Magnetic nature	Hybridisation
$[FeF_6]^{3-}$	Paramagnetic	sp^3d^2
$[Fe(CN_6)]^{3-}$	Paramagnetic	d^2sp^3
$[Mn(CN)_6]^{3-}$	Paramagnetic	d^2sp^3
$[Co(C_2O_4)_3]^{3-}$	Diamagnetic	d^2sp^3
$[MnCl_6]^{3-}$	Paramagnetic	sp^3d^2
$[CoF_6]^{3-}$	Paramagnetic	sp^3d^2

Only $[Fe(CN)_6]^{3-}$ and $[Mn(CN)_6]^{3-}$ are paramagnetic with d^2sp^3 hybridisation.

99. *(0)* $K_2[NiCl_4] = sp^3$, paramagnetic

$[Zn(H_2O)_6]Cl_2 = sp^3d^2$, diamagnetic

$[K_3[Mn(CN)_6] = d^2sp^3$, paramagnetic

$[Cu(PPh_3)_3I] = sp^3$, diamagnetic

Hence, the value of $n = 2$.

Least value of enthalpy of atomisation among Ni, Zn, Mn and Cu is Zn.

Zn^{2+} : $[Ar]3d^{10}$

$$\mu = \sqrt{n(n+2)}\ (n = 0) \Rightarrow \mu = 0$$

100. *(1)* Co has highest standard electrode potential (M^{3+}/M^{2+}) among Mn, Cr, Co, Fe

∴Complex is $[Co(CN)_6]^{4-}$ and its splitting is as follows

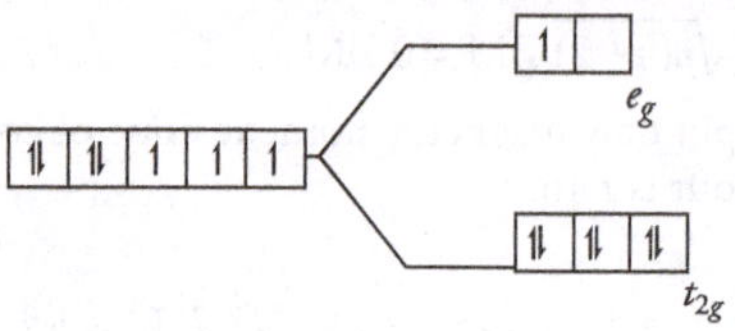

∴ Electron in e_g orbital is one.

101. *(14)* The structure of complex Ni^{2+} ion and dimethyl glyoxime is

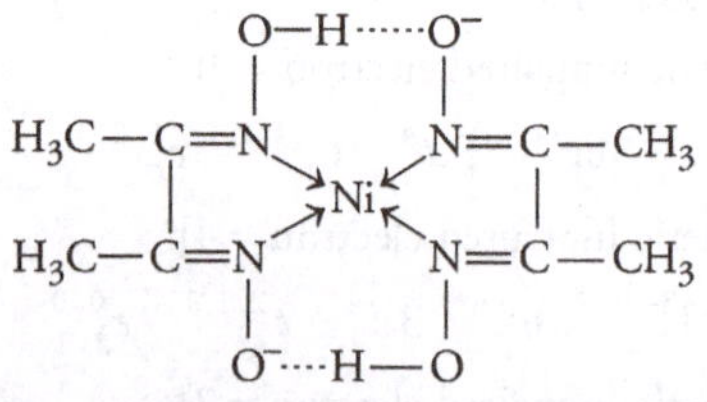

So, the number of H-atom = 14

102. *(28)* The electronic configuration of Ni^{2+} is $[Ar]\,3d^8 4s^0$.

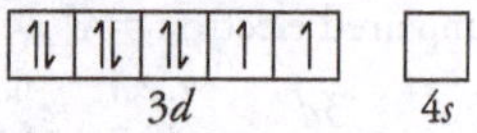

So, no. of unpaired electrons = 2

using $\mu = \sqrt{n(n+2)}$ BM

$= \sqrt{2(2+2)}$ BM $= \sqrt{8}$ BM

$= 2.8$ BM $= 28 \times 10^{-1}$ BM

103. *(0)* The electronic configuration of Pt^{2+} is $5d^8 6s^0$.

Pt in $[Pt(NH_3)_2Cl(NH_2CH_3)]Cl$ =

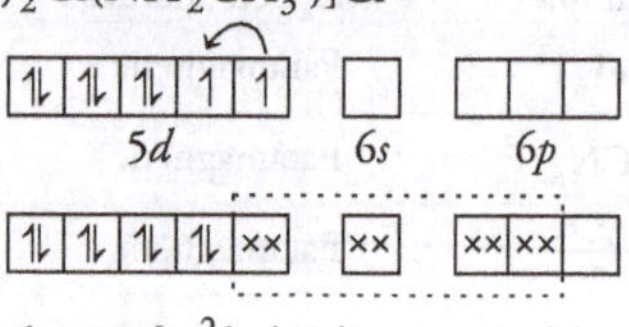

Thus, the complex is dsp^2 hybridisation and have no unpaired electrons $(n = 0)$

Magnetic moment, $\mu = \sqrt{n(n+2)} = 0$

Hence, magnetic moment is zero.

104. *(1)* $Co^{2+} = [Ar]\,3d^7 4s^0$

Only one unpaired electron in t_{2g} set of orbital is present.

105. *(3)* In potassium ferrocyanide $K_4[Fe(CN)_6]$, there are three pairs of electrons in the t_{2g} set of orbitals.

In $K_4[Fe(CN)_6]$, oxidation state of iron is +2.

In case of strong field ligand i.e., CN^-, all electron are paired up then the configuration of this complex will be t_{2g}^6.

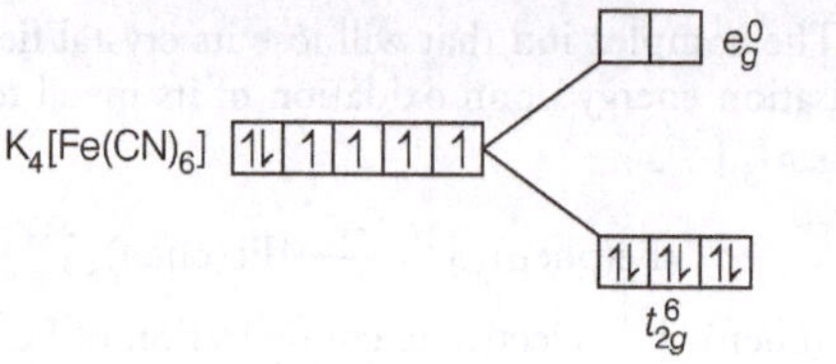

t_{2g} contains 6 electrons so it have 3 pairs.

106. *(4)* For the complex $[Mn(NCS)_6]^{x-}$

$\mu = 6.06$ BM $\simeq 5.92$ BM

Number of unpaired electrons = 5

Here, ligand is $NCS^{\ominus}$ and Mn^{2+} has electronic configuration of $3d^5$. So, $x = 4$.

107. *(480)* In $[Ti(H_2O)_6]^{3+}$, Ti^{3+} has outer electronic configuration $3d^1$

CFSE for $[Ti^{3+}] = -0.4 \times \Delta_o = -\dfrac{96 \times 10^3}{N_A}$ J

$$\Rightarrow \quad \Delta_o = \frac{96 \times 10^3}{0.4 \times 6 \times 10^{23}} = \frac{hc}{\lambda}$$

$$\therefore \quad \lambda = \frac{0.4 \times 6 \times 10^{23} \times 6.4 \times 10^{-34} \times 3 \times 10^8}{96 \times 10^3}$$

$= 0.48 \times 10^{-6}$ m $= 480 \times 10^{-9} = 480$ nm

108. *(4)*

(i) $[NiCl_4]^{2-}$; $Ni^{2+} = 3d^8$, sp^3 (Paramagnetic)

(ii) $[Fe(CN)_6]^{3-}$; $Fe^{3+} = 3d^5$, d^2sp^3 (Paramagnetic)

(iii) $[Fe(H_2O)_6]^{2+}$; $3d^6$, sp^3d^2 (Paramagnetic)

(iv) $[Cu(NH_3)_4]^{2+}$; $3d^9$, dsp^2 (Paramagnetic)

109. *(2)* In $[Fe(CN)_6]^{3-}$

Oxidation state of Fe is +3.

As complex is inner sphere, so electrons are paired.

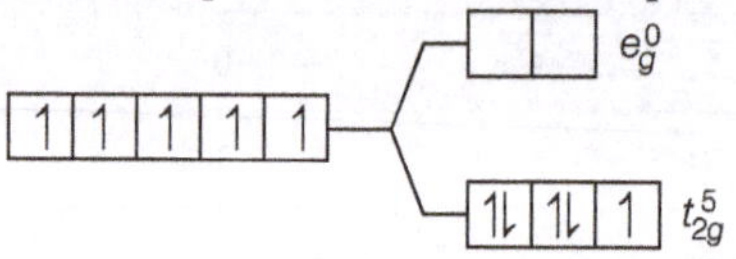

CFSE $= -0.4 \times 5\Delta_o = -2\Delta_o$

So, answer is 2.

110. *(0)* $\overset{+2}{[Co}(H_2O)_6]Cl_2$

Electronic configuration of

$Co^{2+} = [Ar]\,4s^0 3d^7$

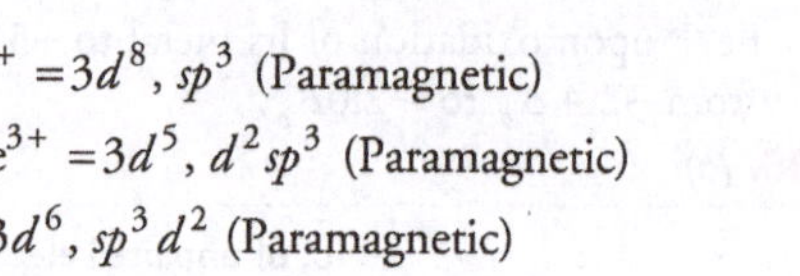

(e^- pair from H_2O)

i.e., there are 3 unpaired electrons.

Magnetic moment, $\mu = \sqrt{n(n+2)}$

∴ $\mu_1 = \sqrt{3(3+2)} = \sqrt{3 \times 5} = \sqrt{15}$ BM

In complex $\overset{+3}{[Cr(H_2O)_6]}Cl_3$

Electronic configuration of

$Cr^{3+} \rightarrow [Ar]\, 4s^0\, 3d^3$

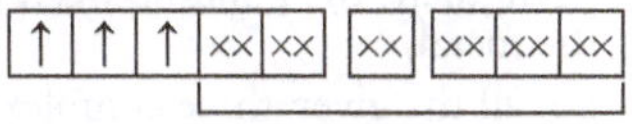

(e^- pair from H_2O)

$$n = 3 \Rightarrow \mu_2 = \sqrt{15}\text{ BM}$$

Hence, $\mu_1 - \mu_2 = 0$

111. *(0)* Oxidation state of Co in all the given complexes = + 3.

The inner orbital and outer orbital nature of complex depends on the chemical nature of ligands and also crystal field splitting depends on strength of ligands.

Ligand strength order is

$$C\overset{\ominus}{N} > NH_3 > H_2O > Cl^-$$

$$\Delta_0 \propto \frac{1}{\lambda}$$

So, shortest wavelength is corresponds to $[Co(CN)_6]^{3-}$

The electronic configuration of Co^{3+} in the given complex is $3d^6 4s^0$

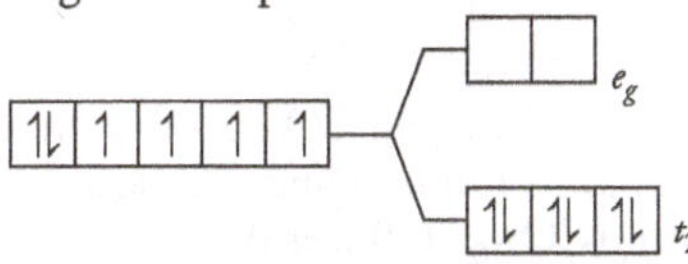

Hence, number of unpaired electrons $(n) = 0$

So, spin only magnetic moment

$$= \sqrt{n(n+2)}$$

$$\mu_{spin} = 0$$

112. *(6)*

$$[Co(H_2O)_6]^{2+} + NH_3\,(\text{excess}) \xrightarrow{O_2} \underset{\substack{\text{Diamagnetic} \\ \text{(means no unpaired electrons)}}}{[Co(NH_3)_6]^{3+}} + 6H_2O$$

$[Co(NH_3)_6]^{3+}$ is low spin complex and diamagnetic in nature. So, configuration of in this complex

$$Co^{3+} = [Ar]3d^6\, 4s^0$$

$$= [Ar]\,(t_{2g}^6 e_g^0)\, 4s^0$$

$\therefore$ Total number of electrons in t_{2g} orbital of $[Co(NH_3)_6]^{3+}$ is 6.

113. *(2)* The order of strength of ligand is

$$C\overset{\ominus}{N} > NH_3 > H_2O > Cl$$

So, splitting in $K_3[Fe(CN)_6]$ is highest and thus, it absorbs light of more energy and shortest wavelength.

$$Fe^{3+} \longrightarrow 3d^5$$

$K_3[Fe(CN)_6] =$

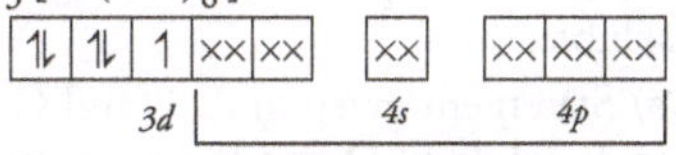

e^- pair from CN^- ligand

So, number of unpaired electrons n is 1.

$$\mu\text{ (spin only)} = \sqrt{n(n+2)}$$

$$= \sqrt{1(1+2)} = \sqrt{3}$$

$$\mu_{spin} = 1.73\text{ BM} \approx 2\text{ BM}$$

114. *(766)* $\Delta_t = \frac{4}{9}\Delta_0$ where Δ_t and Δ_0 are crystal field splitting energy for tetrahedral and octahedral complexes respectively.

$$\Delta_t \propto \frac{1}{\lambda_t};\ \Delta_0 \propto \frac{1}{\lambda_0}$$

$$\frac{1}{\lambda_t} = \frac{4}{9}\frac{1}{\lambda_0}$$

$$\lambda_0 = \frac{4}{9} \times 600\text{ nm}$$

$$= 266.6\text{ nm}$$

$$\Delta_0 = \frac{hc}{\lambda_0} = \frac{6.63 \times 10^{-34} \times 3.08 \times 10^8}{266.6}$$

$$= 765.95 \times 10^{-21}\text{ J} \approx 766 \times 10^{-21}\text{ J}$$

115. *(1)* (i) $[Co(NH_3)_6]\,Cl_2$

$\Rightarrow$ Oxidation state of cobalt Co^{2+} is +2.

$$Co^{2+} = [Ar]\, 4s^0 3d^7 \text{ i.e.}$$

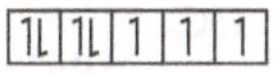

NH_3 acts as SFL (strong field ligand) due to small size of cobalt with +2 charge, so pairing will took place.

$$\Rightarrow \quad t_{2g}^6\, e_g^1$$

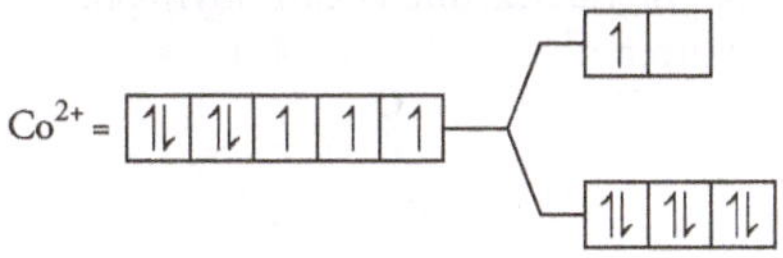

$$\therefore \quad n = 1$$

(ii) $[Co(NH_3)_6]\,Cl_3 \Rightarrow$ Oxidation state of cobalt (Co^{2+}) is +3.

So, NH_3 acts as strong field ligand.

Hence, $Co^{3+} \longrightarrow 3d^6\, 4s^0$

$$\Rightarrow \quad t_{2g}^6\, e_g^0$$

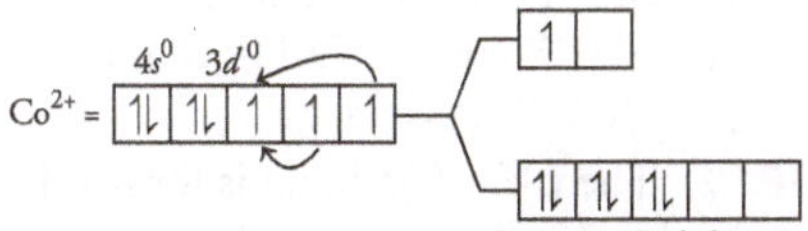

Zero unpaired electron

$$\therefore \quad n = 0$$

So, total (1 + 0) = 1 unpaired electron.

116. *(2)* In $[Ni(CN)_6]^{2-}$, Ni is in +4 oxidation state.

$$Ni^{+4} \rightarrow 3d^6 = t_{2g}^6, e_g^0$$

(Presence of strong field)

Unpaired electron = 0

Similarly, $NiCl_2 \longrightarrow Ni^{2+} = t_{2g}^6, e_g^2$

(Two unpaired electrons)

Number of unpaired electron in $NiCl_2$ and $[Ni(CN)_6]^{2-}$ are two and zero respectively.

Change in number of unpaired electrons $\Rightarrow$ 2 to 0.

Change = 2

117. *(2)* Electronic configuration of Co^{2+} : $[Ar]\, 3d^7$ and CN^- is a strong field ligand which can pair electron of central atom.

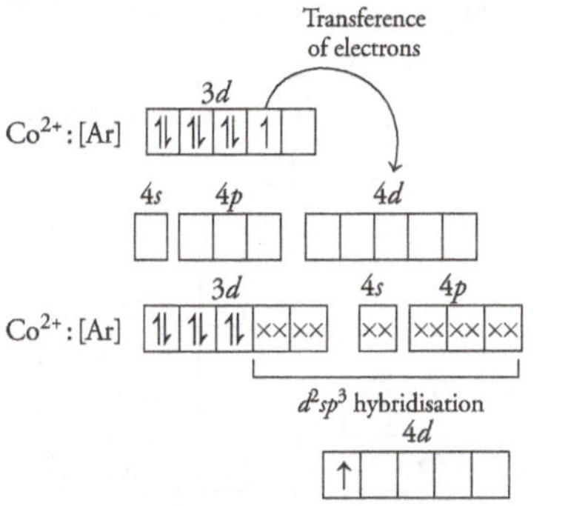

So, spin only magnetic moment

$$(\mu) = \sqrt{n(n+2)}\text{ BM}$$

where, n = number of unpaired electrons

$$= \sqrt{1(1+2)}\text{ BM} = \sqrt{3}\text{ BM}$$

$$\mu = \sqrt{3}\text{ BM}$$

$$= 1.73\text{ BM}$$

Nearest integer = 2

118. *(3)* $K_3[Cr(ox)_3]$ the oxidation state of Cr in $K_3[Cr(C_2O_4)_3]$ is + 3.

$Cr \longrightarrow 3d^5\, 4s^1$ (ground state electronic configuration)

$Cr^{3+} \longrightarrow 3d^3\, 4s^0$ (Excited state)

$K_3[Cr(ox)_3] =$

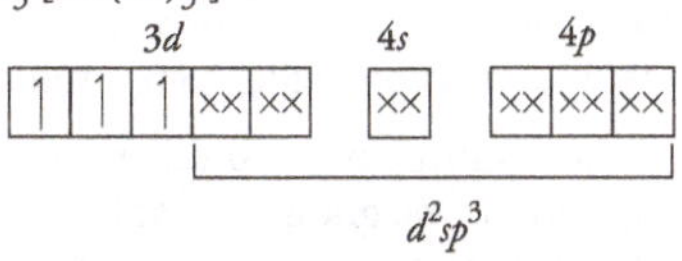

$$\Rightarrow t_{2g}^3\, e_g^0 \qquad CN = 6$$

$\therefore$ oxalate is strong field ligand

$\therefore$ 3 unpaired electron in d-orbital. The hybridisation of chromium in the complex is d^2sp^3.

119. *(4)* $[Ti(H_2O)_6]^{3+}$ is an octahedral complex.

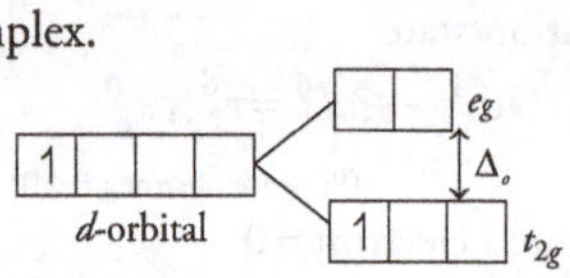

For transition of electron from t_{2g} to e_g orbital, Δ_o amount of energy is required.

Given, $\lambda = 498 \text{ nm} = 498 \times 10^{-9}$ m

$$\Delta_o = hc/\lambda$$

where, h = Planck's constant
c = speed of light
λ = wavelength of radiation

$$\text{So, } \Delta_o = \frac{6.626 \times 10^{-34} \times 3 \times 10^8}{498 \times 10^{-9}}$$

$$= 3.99 \times 10^{-19} \text{ J} = 4 \times 10^{-19} \text{ J}$$

120. *(0)* Magnetic moment (in BM) of $[Ru(H_2O)_6]^2$ would be; while considering that $\Delta_0 > P$,

$Ru_{(44)}$; $[Kr]4d^7 5s^1$ (in ground state)

$\Rightarrow Ru^{2+}$ in complex

$\Rightarrow 4d^6 \Rightarrow (t_{2g})^6 (e_g)^0$

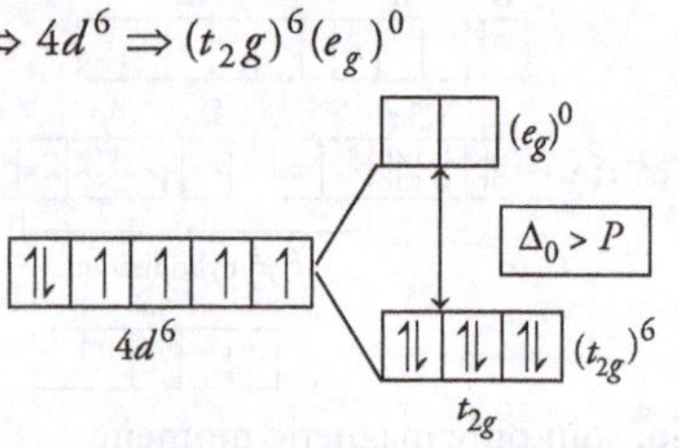

Here, number of unpaired electrons is zero and hence

$$\mu = \sqrt{n(n+2)} \text{ BM} = 0 \text{ BM}$$

Topic 3

01. *(d)* The compound used in the treatment of lead poisoning is EDTA. Medication occurs through chelation therapy.

02. *(c)* *cis*-platin is used as an anti-cancer drug which helps in treatment of cancer.

03. *(a)* Compounds having atleast one carbon metal (Mn—C) bond are known as organometallic compounds.

04. *(a)* Chlorophyll is a coordination compound of magnesium (Mg).

$[Ag(S_2O_3)_2]^3$ is used in photography.

Hence, (A) and (C) are mismatched combinations.

05. *(a)* Among the given options only $[Fe_3(OH)_2(OAc)_6]Cl$ is water soluble, while all other options are water insoluble.

06. *(a)* Statement given in A, B and C are correct, while in D and E are incorrect. The correct form of D and E are: In (D) Wilkinson's catalyst $[Rh(PPh_3)_3Cl]$, phosphine does not act as a ligand.

(E) Ca^{2+} ion form more stable complex with EDTA.

07. *(d)*

$[Co_2(CO)_8]$

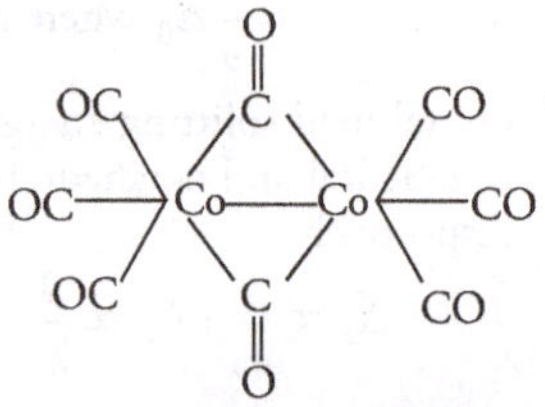

$[Mn_2(CO)_{10}]$

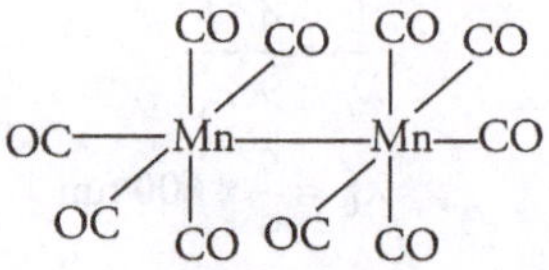

$[Os_3(CO)_{12}]$

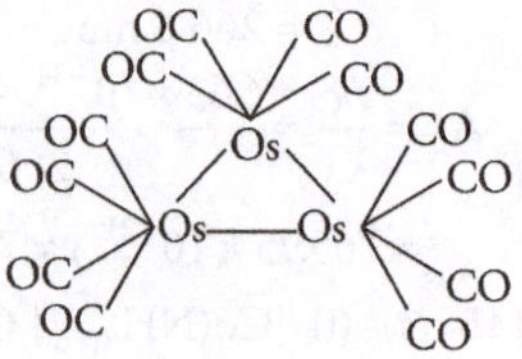

$[Ru_3(CO)_{12}]$

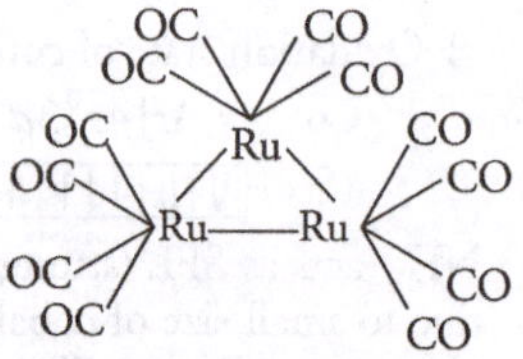

08. *(d)* The structure of $Co_2(CO)_8$ (a polynuclear metal carbonyl) can be written as

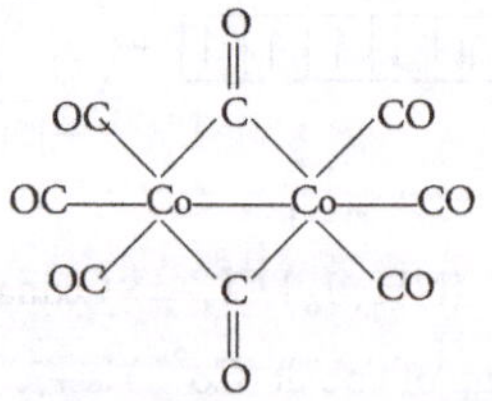

Total number of bridging CO ligands = 2 and the Co—Co bond = 1

09. *(7)* The structure of $[Co_2(CO)_8]$ is

Number of Co—Co bond $(X) = 1$

Number of terminal CO ligands $(Y) = 6$

$\therefore \quad X + Y = 1 + 6 = 7$

10. *(3)* The $(M - C)\pi$-bond is formed by the donation of a pair of electrons from a filled *d*-orbital of metal to the vacant antibonding, π^*-orbital of carbon of carbon ligand. Thus, CO acts as a σ-donor (OC⟶*M*) and a π-acceptor (*M*⟶CO), with the two interaction creating a synergic effect which strengthens the bond between CO and the metal.

So, all the given three complexes have CO ligands and all will exhibit synergic bonding.

11. *(2)* Fehling's solution contains

$Cu^{2+} \rightarrow [Ar]3d^9$ configuration

= | 1l | 1l | 1l | 1l | 1 |

$\therefore$ Number of unpaired e^- $(n) = 1$

So, spin magnetic moment (μ).

$$= \sqrt{n(n+2)} \text{ BM}$$

$$= \sqrt{1(3)} = 1.73 \text{ BM} \approx 2 \text{ BM}$$

12. *(0)* $[Mn_2(CO)_{10}]$ is a polynuclear metal carbonyl compound. It is decarboxyldimanganese (0) which is made up of two square pyramidal $Mn(CO)_5$ units joined by a Mn—Mn bond. The structure of $[Mn_2(CO)_{10}]$ is

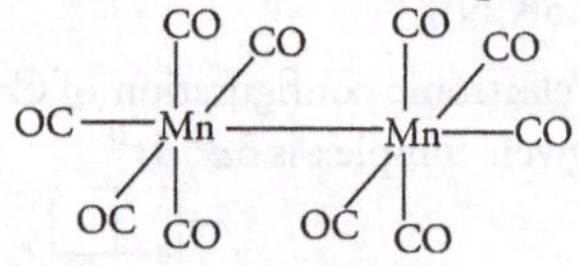

Here, no bridging CO ligand is present. So, number of bridging CO ligands in $[Mn_2(CO)_{10}]$ is 0 (zero).

13. *(1)* Given, stability constant value,

$K_1 = 10^4$; $K_2 = 1.58 \times 10^3$

$K_3 = 5 \times 10^2 \Rightarrow K_4 = 10^2$

On adding all equations,

$$Cu^{2+} + 4NH_3 \overset{K}{\rightleftharpoons} [Cu(NH_3)_4]^{2+}$$

$\therefore$ The overall reaction constant (K) or equilibrium constant for formation of $[Cu(NH_3)_4]^{2+}$ is

$$K = K_1 \times K_2 \times K_3 \times K_4$$

$$K = 10^4 \times 1.58 \times 10^3 \times 5 \times 10^2 \times 10^2$$

$$K = 7.9 \times 10^{11}$$

where, K = equilibrium constant for formation of $[Cu(NH_3)_4]^{2+}$

So, equilibrium constant K' for dissociation of $[Cu(NH_3)_4]^{2+}$ is $\frac{1}{K}$.

$$K' = \frac{1}{K} = \frac{1}{7.9 \times 10^{11}} = 1.26 \times 10^{-12}$$

Hence, $K' = x \times 10^{-12} \Rightarrow x = 1.26 \approx 1$

19 Organic Compounds Containing Halogens

In JEE Main there is usually 0-1 question per paper are visible from this chapter. In past 3 years, 56 conceptual and 9 numerical questions were asked of moderate difficulty level from this chapter.

For detailed analysis scan QR code

Topic 1 / Haloalkanes

• Level 1

01 When *sec*-butylcyclohexane reacts with bromine in the presence of sunlight, the major product is **[2025, 22 Jan Shift-II]**

(a) 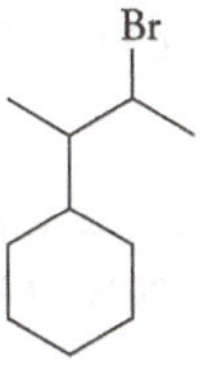

(b) 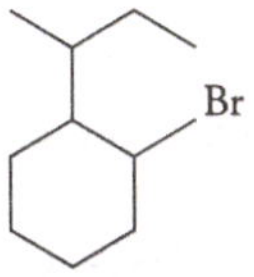

(c) 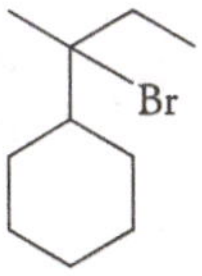

(d) 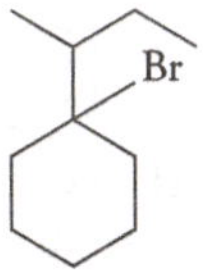

02 Total number of stereoisomers possible for the given structure **[2024, 9 April Shift-II]**

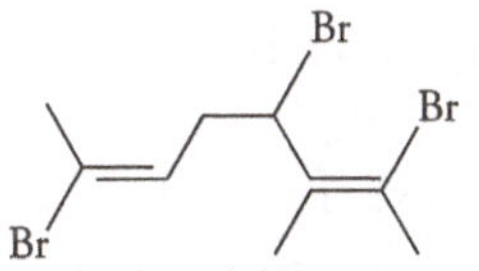

(a) 4
(b) 3
(c) 2
(d) 8

03 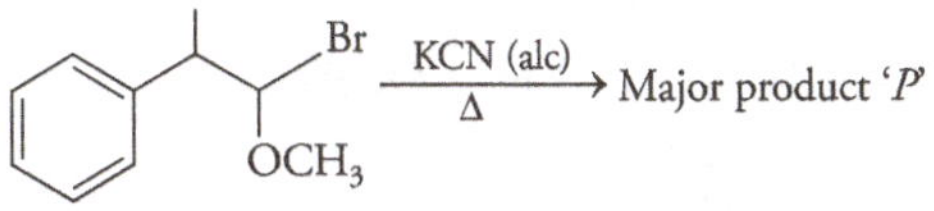

In the above reaction product '*P*' is **[2024, 9 April Shift-II]**

(a) 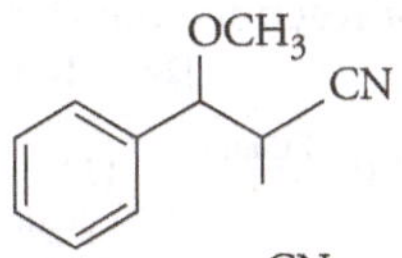

(b) 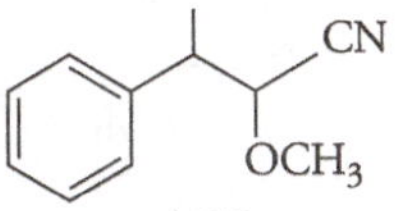

(c) 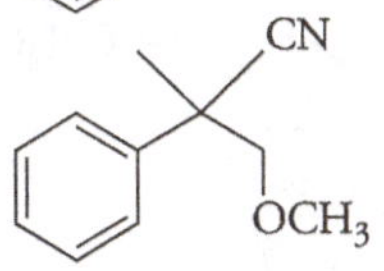

(d) 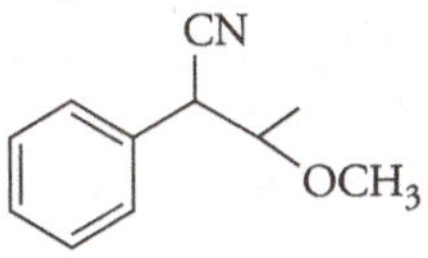

04 Identify the major product in the following reaction. **[2024, 5 April Shift-II]**

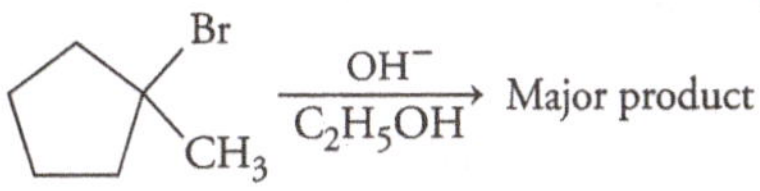

(a) 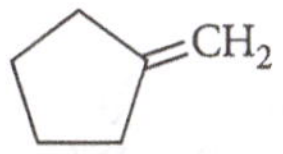

(b) 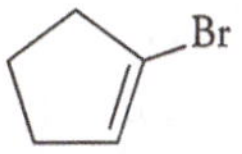

(c) 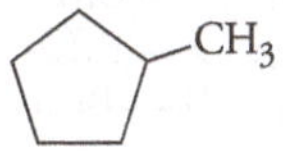

(d) 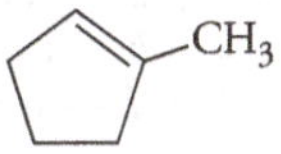

05 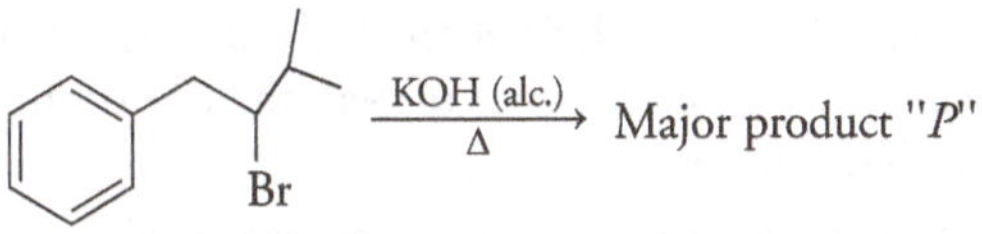

Product *P* is **[2024, 4 April Shift-II]**

(a) 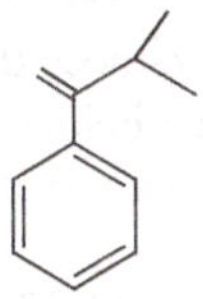

(b) 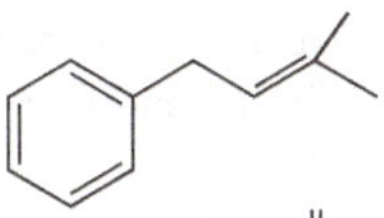

(c) 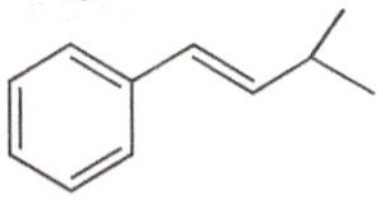

(d) 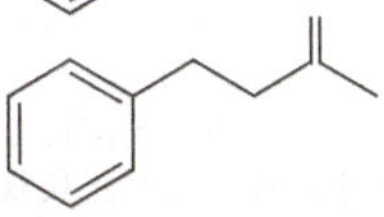

06 Identify structure of 2, 3-dibromo-1-phenylpentane. **[2024, 31 Jan Shift-II]**

(a) 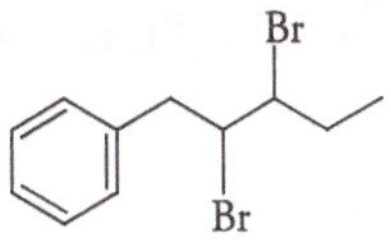

(b) 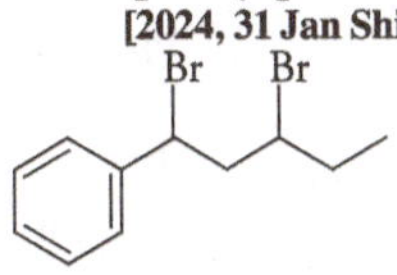

(c) 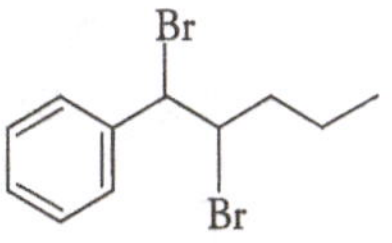

(d) 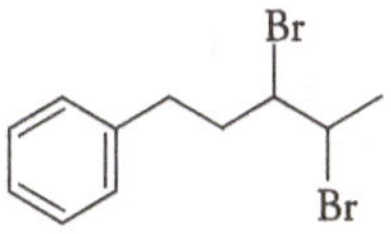

07 Example of vinylic halide is **[2024, 30 Jan Shift-I]**

(a) 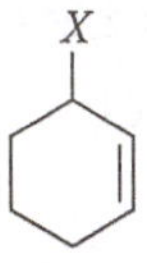

(b) 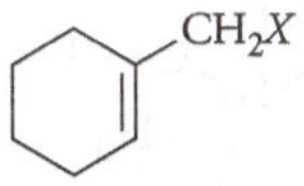

(c) 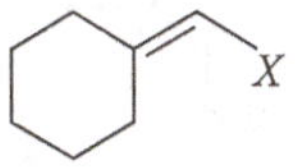

(d) 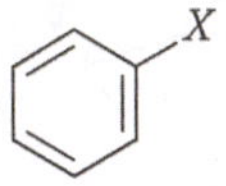

08 Alkyl halide is converted into alkyl isocyanide by reaction with **[2024, 29 Jan Shift-II]**

(a) KCN (b) NaCN (c) NH_4CN (d) AgCN

Imp. *Reactivity of Halogens towards S_N1 Reaction*

09 Choose the halogen which is most reactive towards S_N1 reaction in the given compounds (*A, B, C* and *D*) **[2023, 8 April Shift-I]**

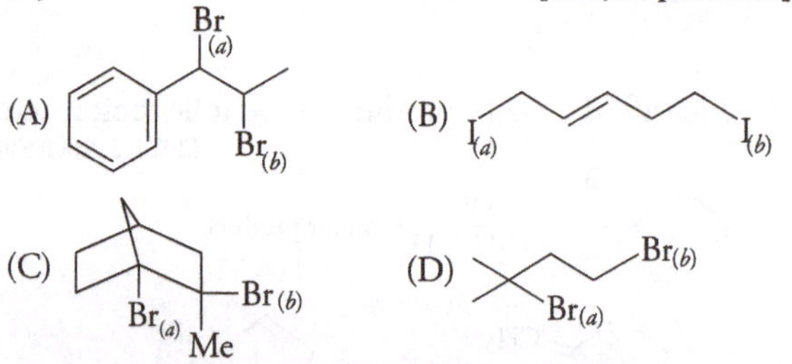

(a) *A*—Br(*a*); *B*—I(*a*); *C*—Br(*b*); *D*—Br(*a*)
(b) *A*—Br(*b*); *B*—I(*b*); *C*—Br(*b*); *D*—Br(*b*)
(c) *A*—Br(*b*); *B*—I(*a*); *C*—Br(*a*); *D*—Br(*a*)
(d) *A*—Br(*a*); *B*—I(*a*); *C*—Br(*a*); *D*—Br(*a*)

10 **Assertion (A)** Hydrolysis of an alkyl chloride is a slow reaction but in the presence of NaI, the rate of the hydrolysis increases.

Reason (R) I^- is a good nucleophile as well as a good leaving group. **[2023, 24 Jan Shift-I]**

In the light of the above statements, choose the correct answer from the options given below.

(a) Both A and R are true and R is the correct explanation of A.
(b) Both A and R are true but R is not the correct explanation of A.
(c) A is false but R is true.
(d) A is true but R is false.

11 The major product (*P*) in the reaction. **[2022, 28 June Shift-I]**

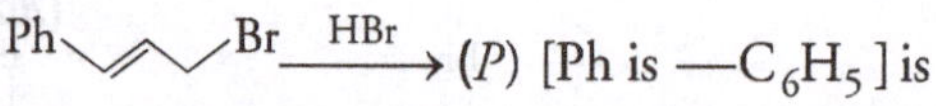

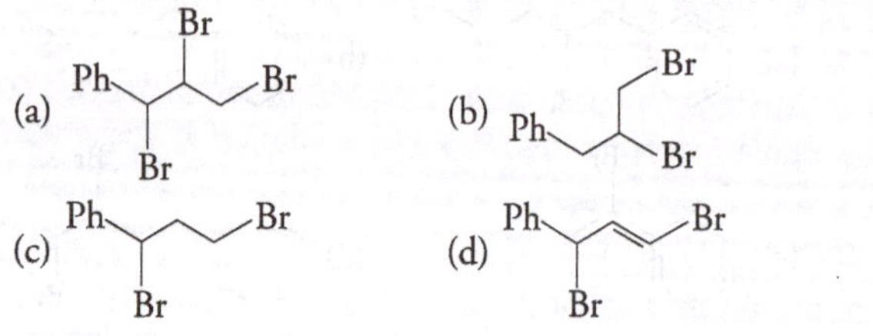

12 Which one of the following compounds is inactive towards S_N1 reaction? **[2022, 28 June Shift-I]**

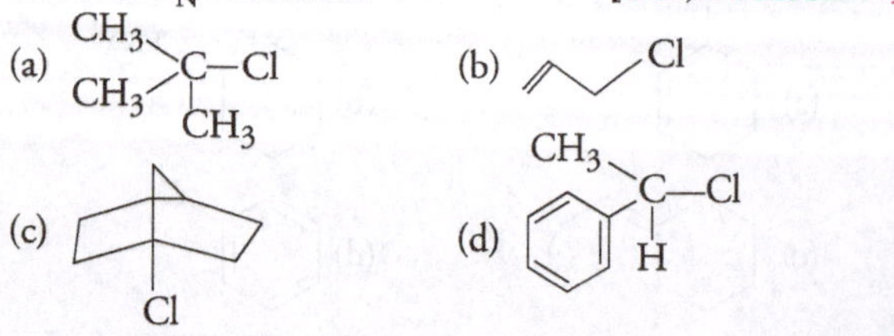

13 The major product in the reaction **[2022, 25 June Shift-I]**

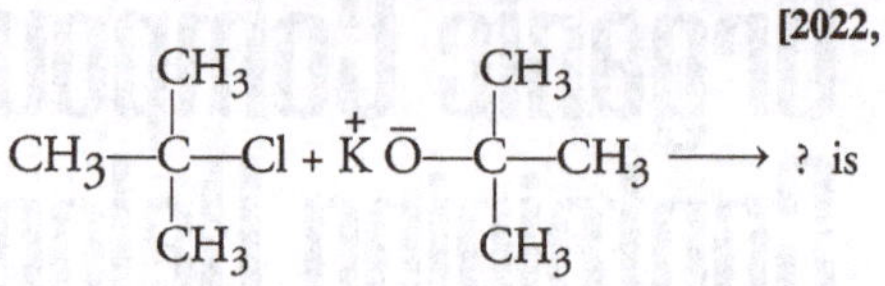

(a) *t*-butyl ethyl ether (b) 2,2 dimethyl butane
(c) 2-methyl pent-1-ene (d) 2-methylprop-1-ene

14 The major product of the following reaction is **[2021, 31 Aug Shift-II]**

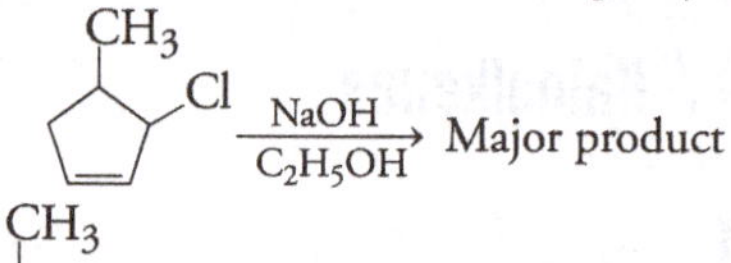

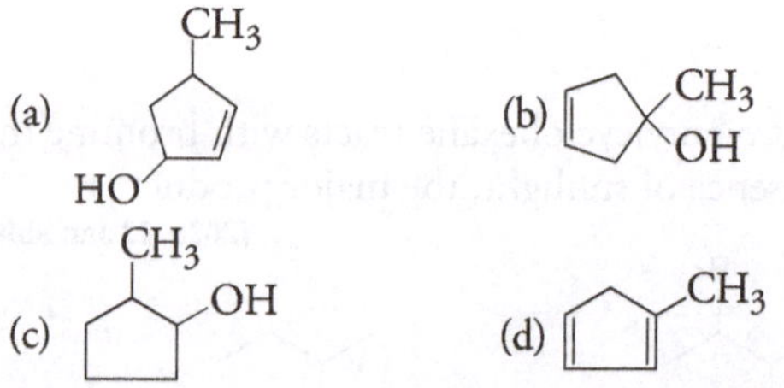

15 The major product (*A*) formed in the reaction given below is **[2021, 27 Aug Shift-II]**

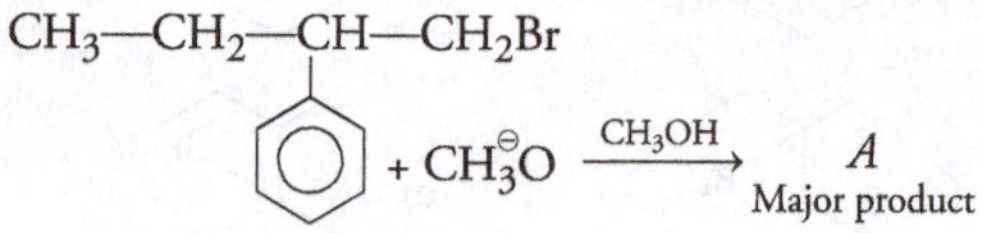

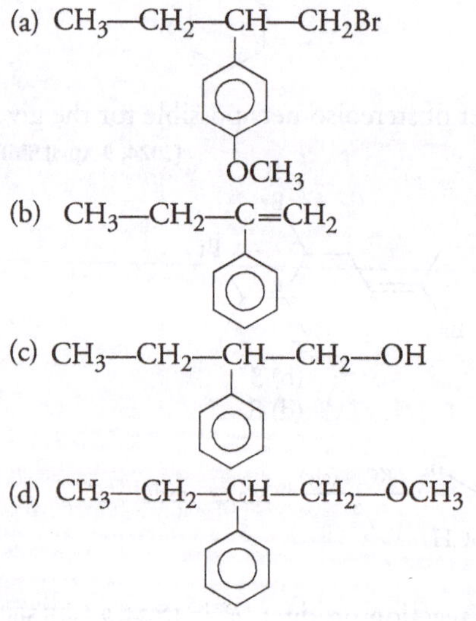

16 In the following sequence of reactions the *P* is **[2021, 27 Aug Shift-I]**

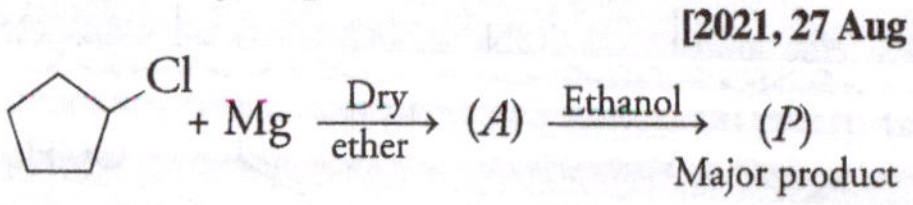

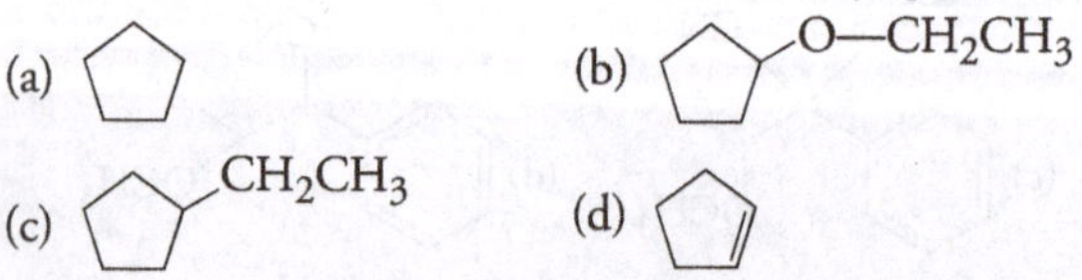

17 The major product formed in the following reaction is **[2021, 26 Aug Shift-I]**

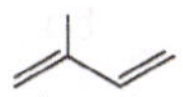
(excess) $\xrightarrow{HBr}$ Major product

(a) 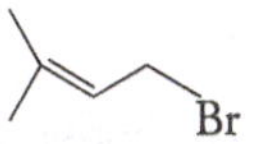
(b) 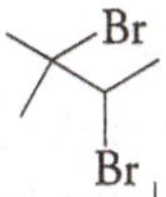
(c) 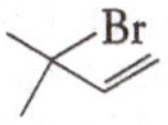
(d) 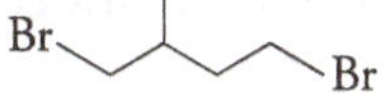

18 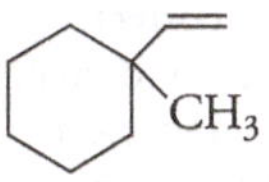 $\xrightarrow[CCl_4]{HBr}$ A (Major product)

Product A is **[2021, 17 March Shift-I]**

(a) 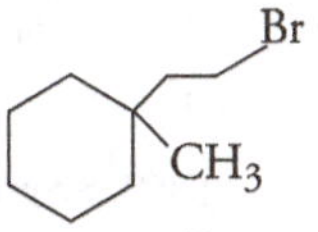
(b) 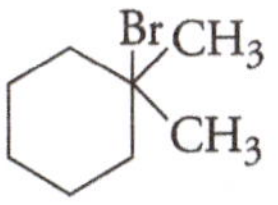
(c) 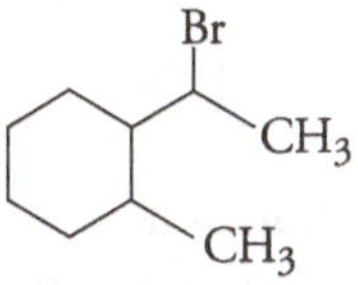
(d) 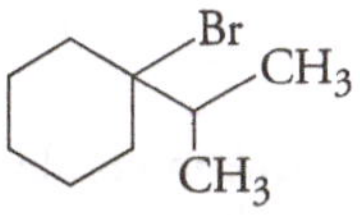

19 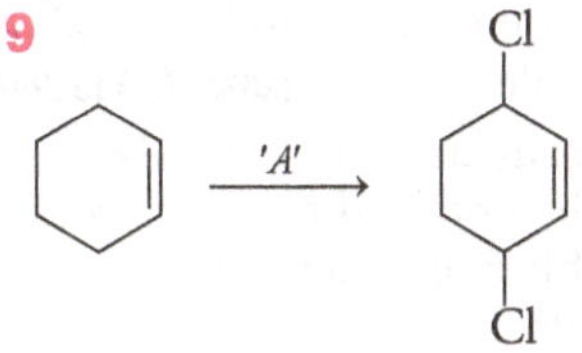

Identify the reagent(s) 'A' and condition(s) for the reaction **[2021, 16 March Shift-II]**

(a) $A = HCl$, anhydrous $AlCl_3$
(b) $A = HCl$, $ZnCl_2$
(c) $A = Cl_2$, UV light
(d) $A = Cl_2$, dark, anhydrous $AlCl_3$

20 What is the major product formed by HI on reaction with

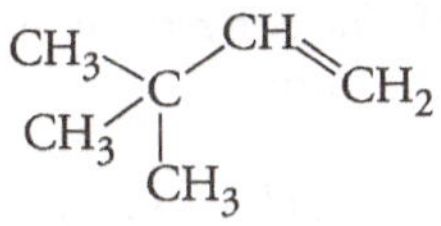

[2021, 24 Feb Shift-I]

(a) 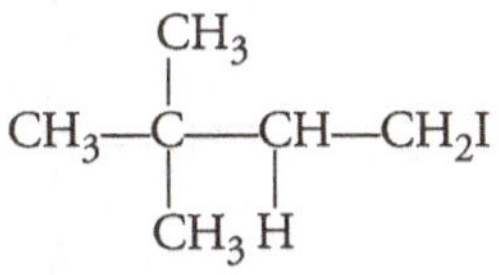
(b) 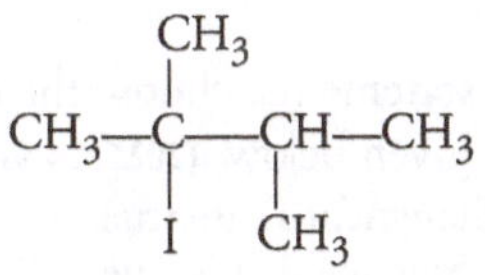
(c) 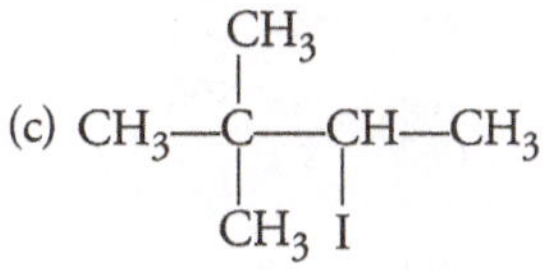
(d) 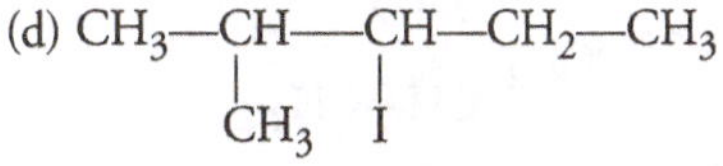

21 The major product formed in the following reaction is

$CH_3CH{=}CHCH(CH_3)_2 \xrightarrow{HBr}$ **[2020, 5 Sep Shift-II]**

(a) $CH_3CH(Br)CH_2CH(CH_3)_2$
(b) $CH_3CH_2CH(Br)CH(CH_3)_2$
(c) $Br(CH_2)_3CH(CH_3)_2$
(d) $CH_3CH_2CH_2C(Br)(CH_3)_2$

22 Among the following compounds, which one has the shortest C—Cl bond? **[2020, 4 Sep Shift-II]**

(a) $H_3C—Cl$
(b) 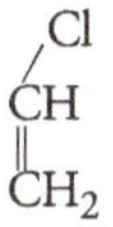
(c) 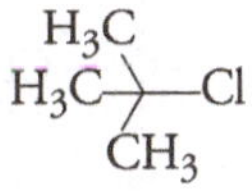
(d) 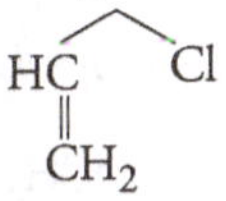

23 The major product in the following reaction is

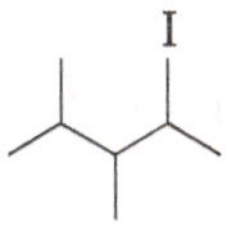 $\xrightarrow[Heat]{t\text{-}BuOH}$

[2020, 3 Sep Shift-II]

(a) 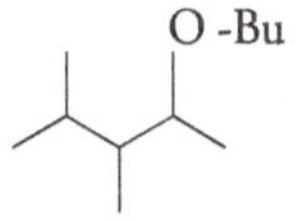
(b) 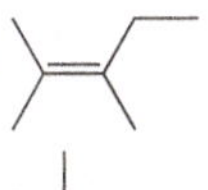
(c) 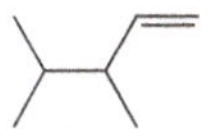
(d) 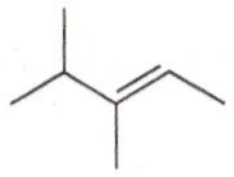

24 The major product obtained in the given reaction is

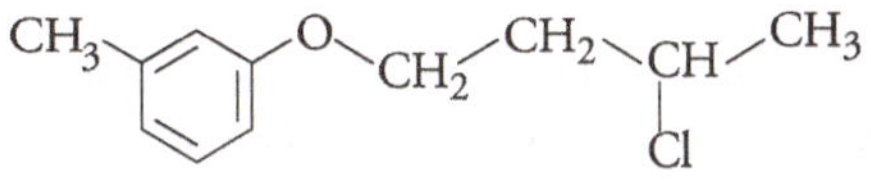 $\xrightarrow{AlCl_3}$ Product

[2019, 10 April Shift II]

(a) 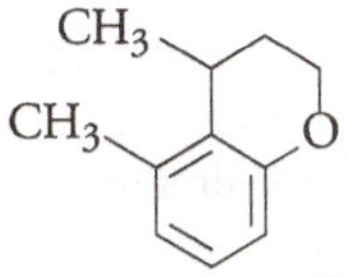
(b) 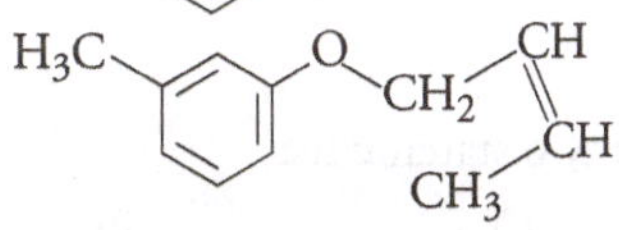

(c) 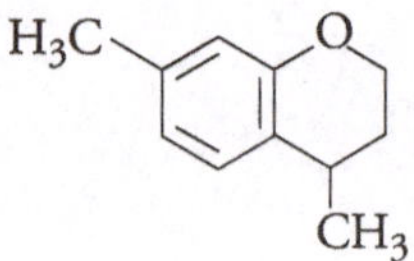

(d) 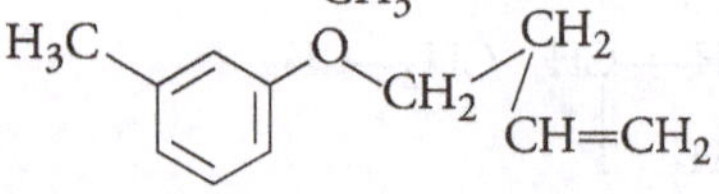

25 The major product of the following reaction is

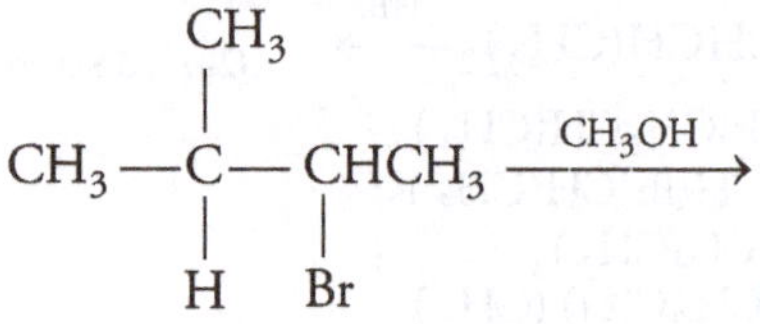

[2019, 10 April Shift-I]

(a) 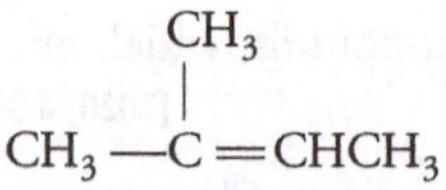 (b) 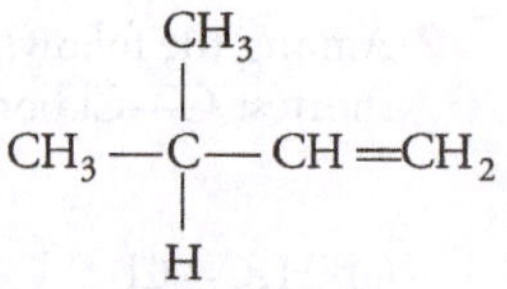

(c) 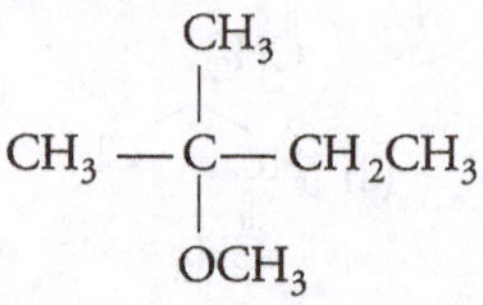 (d) 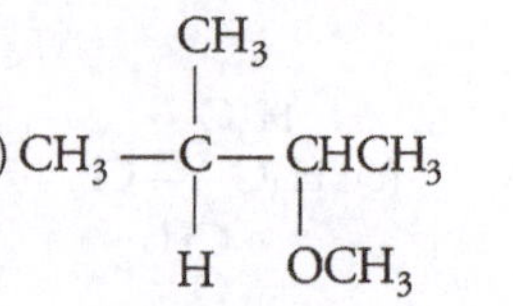

26 The major product in the following conversion is **[2019, 12 Jan Shift-II]**

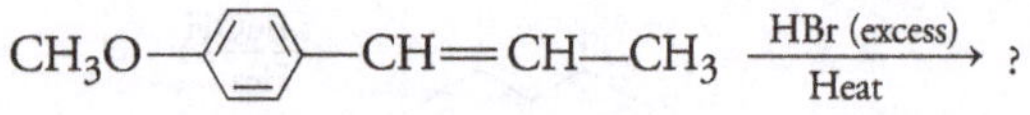

(a) 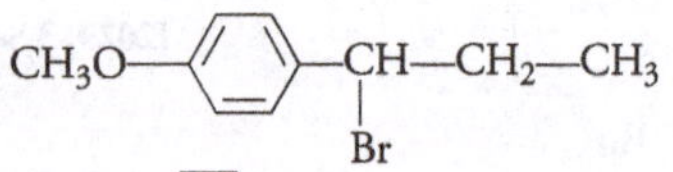

(b) 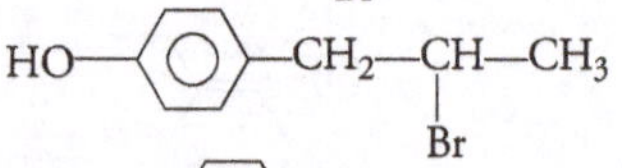

(c) 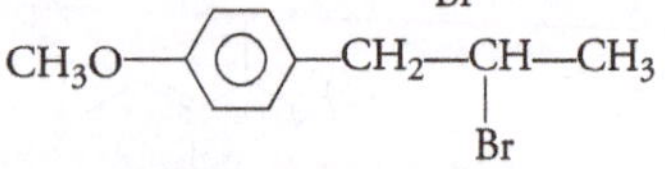

(d) 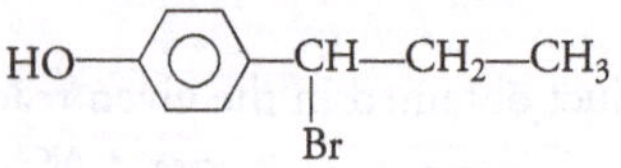

27 Which hydrogen in compound (*E*) is easily replaceable during bromination reaction in presence of light?

$$\overset{\delta}{CH_3}-\overset{\gamma}{CH_2}-\overset{\beta}{CH}=\overset{\alpha}{CH_2}$$

(*E*) **[2019, 10 Jan Shift-I]**

(a) β-hydrogen (b) δ-hydrogen

(c) γ-hydrogen (d) α-hydrogen

• Level 2

28 Given below are two statements:

Statement I 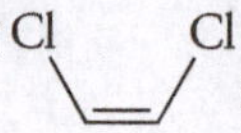 is more polar than 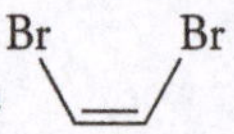

Statement (II): Boiling point of 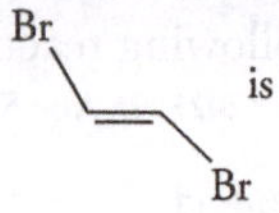 is lower than  but it is more polar than 

In the light of the above statements, choose the most appropriate answer from the options given below: **[2025, 7 April Shift-II]**

(a) Statement I is incorrect but Statement II is correct.
(b) Both Statement I and Statement II are correct.
(c) Statement I is correct but Statement II is incorrect.
(d) Both Statement I and Statement II are incorrect.

29 1,2-dibromocyclooctane $\xrightarrow[\substack{\text{(ii) } NaNH_2 \\ \text{(iii) } Hg^{2+}/H^+ \\ \text{(iv) } Zn-Hg/H^+}]{\text{(i) KOH (alc.)}}$ $\underset{\text{product}}{\underset{\text{Major}}{P}}$

'*P*' is **[2025, 8 April Shift-II]**

(a) 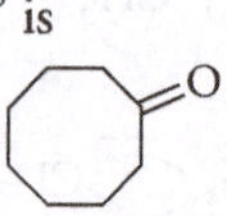 (b) 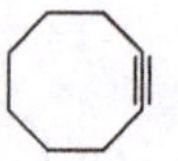 (c) 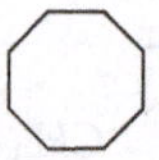 (d) 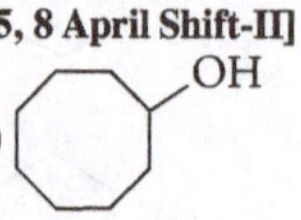

30 Propane molecule on chlorination under photochemical condition gives two di-chloro products. "*x*" and "*y*". Amongst "*x*" and "*y*", "*x*" is an optically active molecule. **[2025, 23 Jan Shift-I]**
How many tri-chloro products (consider only structural isomers) will be obtained from "*x*" when it is further treated with chlorine under the photochemical condition?

(a) 4 (b) 2 (c) 3 (d) 5

31 Given below are two statements:
Statement I The conversion proceeds well in the less polar medium.

$$CH_3-CH_2-CH_2-CH_2-Cl \xrightarrow{HO^-} CH_3-CH_2-CH_2-CH_2-OH + Cl^{\ominus}$$

Statement II The conversion proceeds well in the more polar medium.

$$CH_3-CH_2-CH_2-CH_2-Cl \xrightarrow{R_3\ddot{N}}$$

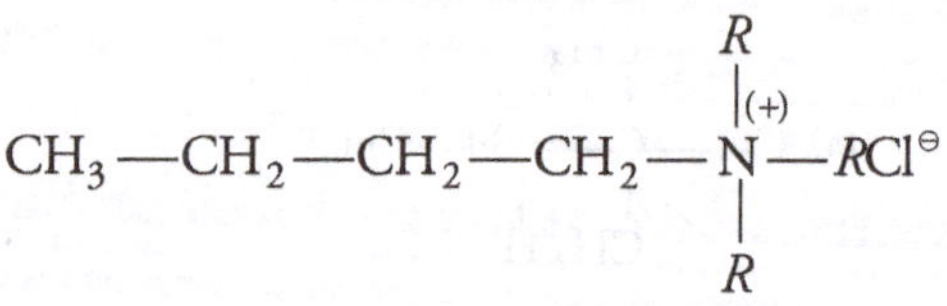

In the light of the above statements, choose the correct answer from the options given below **[2025, 24 Jan Shift-I]**

(a) Both Statement I and Statement II are true.
(b) Statement I is false but Statement II is true.
(c) Statement I is true but Statement II is false.
(d) Both Statement I and Statement II are false.

32 Given below are two statements: **[2025, 28 Jan Shift-I]**

Statement I 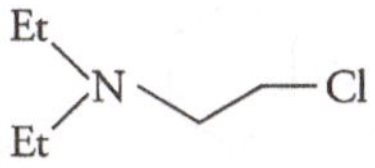 will undergo alkaline hydrolysis at a faster rate than 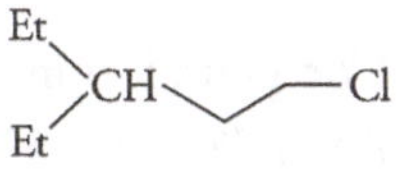

Statement II In 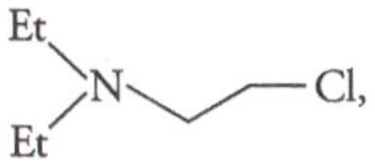 intramolecular substitution takes place first by involving lone pair of electrons on nitrogen.

In the light of the above statements, choose the most appropriate answer from the options given below

(a) Statement I is incorrect but Statement II is correct.
(b) Statement I is correct but Statement II is incorrect.
(c) Both Statement I and Statement II are incorrect.
(d) Both Statement I and Statement II are correct.

33 Given below are two statements. **[2025, 22 Jan Shift-I]**

Statement I CH_3-O-CH_2-Cl will undergo S_N1 reaction though it is a primary halide.

Statement II $$\begin{array}{c} CH_3 \\ | \\ CH_3-C-CH_2-Cl \\ | \\ CH_3 \end{array}$$ will not undergo S_N2 reaction very easily though it is a primary halide.

In the light of the above statements, choose the most appropriate answer from the options given below

(a) Both Statement I and Statement II are correct.
(b) Statement I is incorrect but Statement II is correct.
(c) Statement I is correct but Statement II is incorrect.
(d) Both Statement I and Statement II are incorrect.

34 Given below are two statements : **[2024, 8 April Shift-II]**

Statement I S_N2 reactions are 'stereospecific', indicating that they result in the formation of only one *stereo*-isomer as the product.

Statement II S_N1 reactions generally result in formation of product as racemic mixtures.

In the light of the above statements, choose the correct answer from the options given below:

(a) Statement I is true but statement II is false.
(b) Statement I is false but statement II is true.
(c) Both statement I and statement II are true.
(d) Both statement I and statement II are false.

35 Which among the following compounds will undergo fastest S_N2 reaction. **[2024, 8 April Shift-I]**

(a) 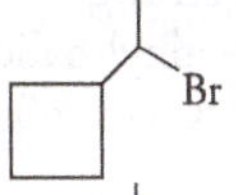 (b) 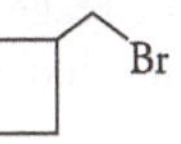

(c) 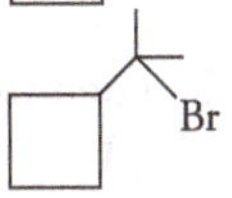 (d) 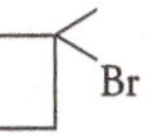

36 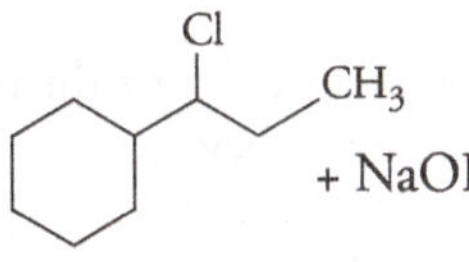 $+ NaOH \xrightarrow{H_2O}$ Major product "*A*"

Consider the above chemical reaction. Product '*A*' is **[2024, 6 April Shift-II]**

(a) 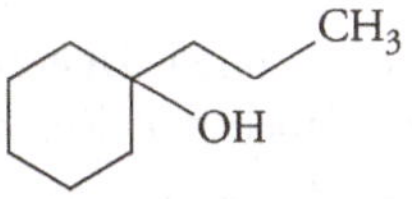 (b) 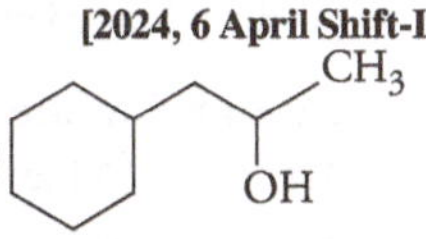

(c) 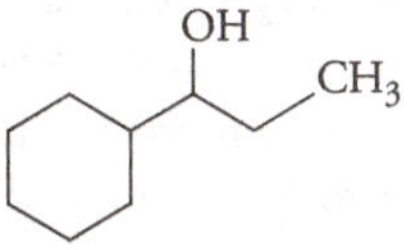 (d) 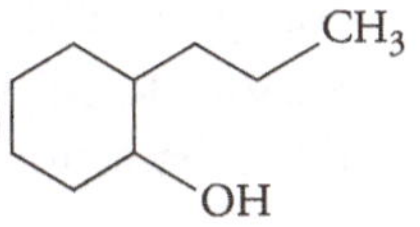

37 Identify the correct set of reagents or reaction conditions '*X*' and '*Y*' in the following set of transformation. **[2024, 4 April Shift-I]**

$$CH_3-CH_2-CH_2-Br \xrightarrow{'X'} \text{Product} \xrightarrow{'Y'} \begin{array}{c} CH_3-CH-CH_3 \\ | \\ Br \end{array}$$

(a) X =conc. alc. NaOH, 80°C, Y =HBr/Acetic acid
(b) X =dil. aq. NaOH, 20°C, Y =HBr/Acetic acid
(c) X =dil. aq. NaOH, 20°C, $Y = Br_2/CHCl_3$
(d) X =conc. alc. NaOH, 80°C, $Y = Br_2/CHCl_3$

38 Given below are two statements : one is labelled as **Assertion (A)** and other is labelled as **Reason (R)**. **[2024, 1 Feb Shift-I]**

Assertion (A) Haloalkanes react with KCN to form alkyl cyanides as a main product while with AgCN form isocyanide as the main product.

Reason (R) KCN and AgCN both are highly ionic compounds.

In the light of the above statements, choose the **most appropriate** answer from the options given below.

(a) Both (A) and (R) are correct but (R) is not the correct explanation of (A).
(b) (A) is correct but (R) is not correct.
(c) Both (A) and (R) are correct and (R) is the correct explanation of (A).
(d) (A) is not correct but (R) is correct.

39 Given below are two statements. **[2024, 30 Jan Shift-II]**

Statement I High concentration of strong nucleophilic reagent with secondary alkyl halides which do not have bulky substituents will follow S_N2 mechanism.

Statement II A secondary alkyl halide when treated with a large excess of ethanol follows S_N1 mechanism.

In the light of the above statements, choose the **most appropriate** from the options given below :

(a) Both Statement I and Statement II are true.
(b) Statement I is false but Statement II is true.
(c) Statement I is true but Statement II is false.
(d) Both Statement I and Statement II are false.

40 Given below are two statements : one is labelled as **Assertion (A)** and the other is labelled as **Reason (R)**.

Assertion (A) $CH_2{=}CH-CH_2-Cl$ is an example of allyl halide. **[2024, 30 Jan Shift-I]**

Reason (R) Allyl halides are the compounds in which the halogen atom is attached to sp^2-hybridised carbon atom.

In the light of the above statements, choose the **most appropriate** answer from the options given below:

(a) A is false but R is true.
(b) Both A and R are true and R is the correct explanation of A.
(c) Both A and R are true but R is not the correct explanation of A.
(d) A is true but R is false.

41 Identify product *A* and product *B*. **[2024, 29 Jan Shift-I]**

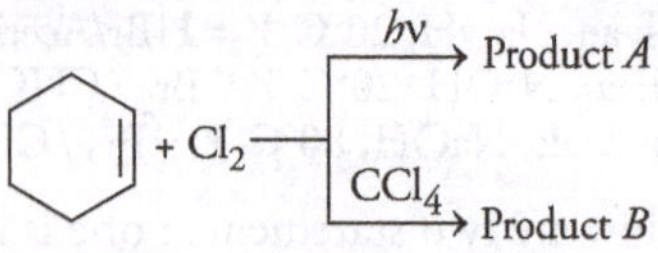

(a) *A* : 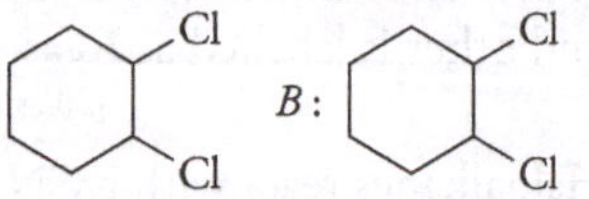 *B* :

(b) *A* : 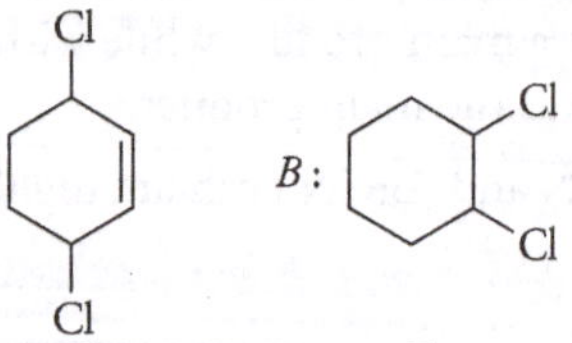 *B* :

(c) *A* : 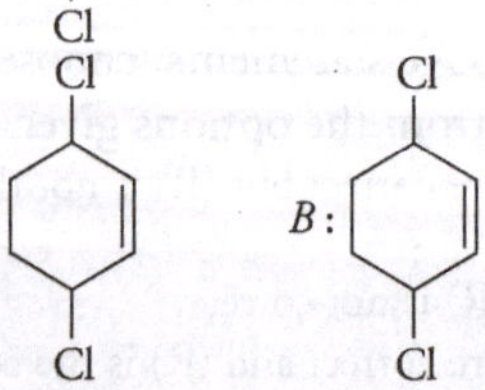 *B* :

(d) *A* : 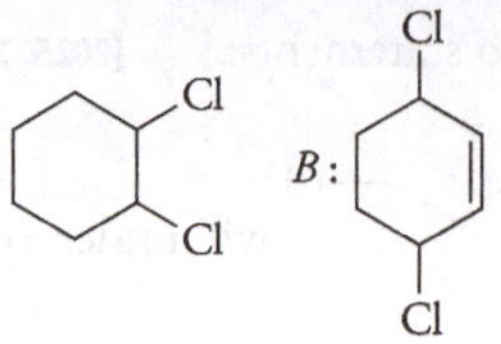 *B* :

42 Compound from the following that will not produce precipitate on reaction with $AgNO_3$ is **[2023, 11 April Shift-II]**

(a) 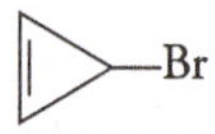
(b) 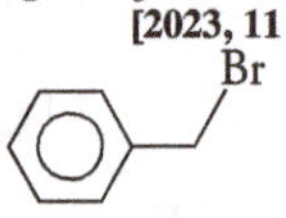
(c) 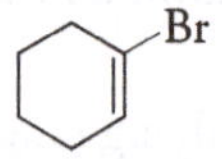
(d) 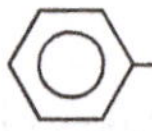$-CH{=}CHCH_2Br$

43 Major product '*P*' formed in the following reaction is

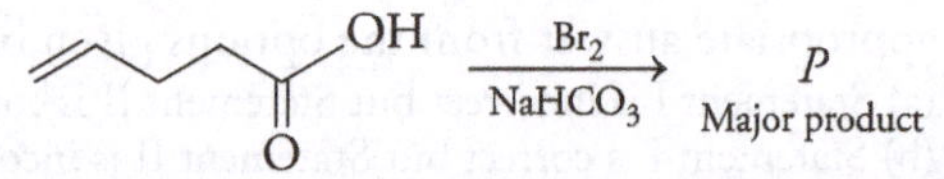

[2023, 8 April Shift-II]

(a) 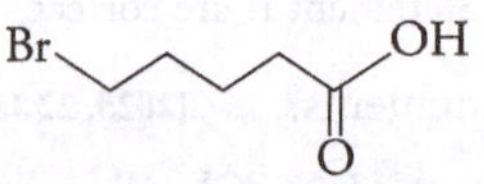
(b) 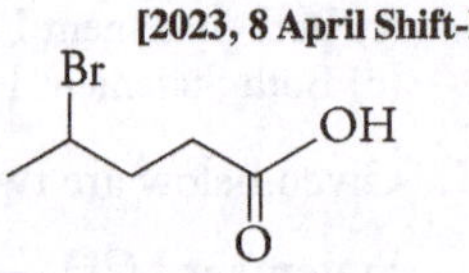
(c) 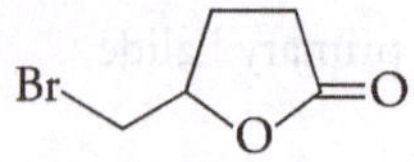
(d) 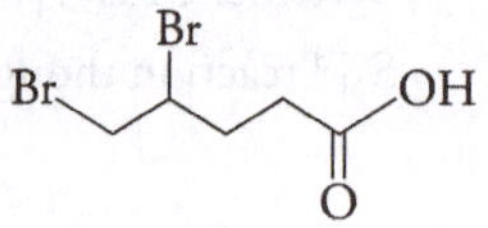

44 For the reaction,

$$RCH_2Br + I^- \xrightarrow{\text{Acetone}} \underset{\text{Major}}{RCH_2I} + Br^-$$

The correct statement is **[2023, 6 April Shift-I]**

(a) Br^- can act as competing nucleophile.
(b) The solvent used in the reaction solvates the ions formed in rate determining step.
(c) The transition state formed in the above reaction is less polar than the localised anion.
(d) The reaction can occur in acetic acid also.

45 Compound '*A*' undergoes following sequence of reactions to give compound '*B*'. The correct structure and chirality of compound '*B*' is [where, Et is $-C_2H_5$]

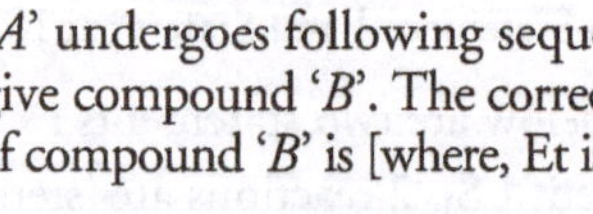

[2022, 29 July Shift-II]

(a) 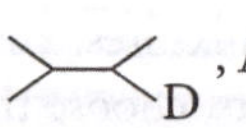, Achiral
(b) 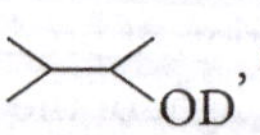, Chiral
(c) 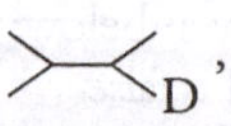, Chiral
(d) 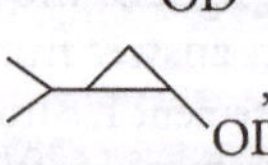, Achiral

Imp. *Reaction of Haloalkanes with AgCN and NaCN*

46 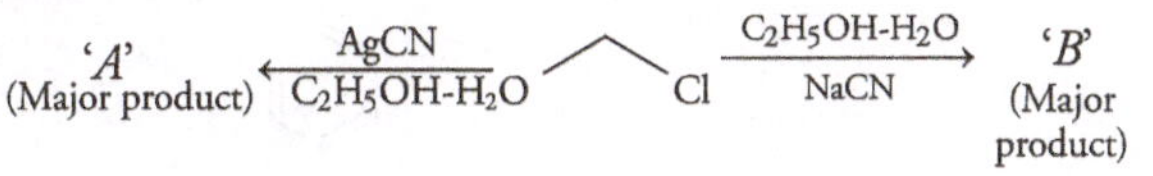

Considering the above reactions, the compound 'A' and compound 'B' respectively are **[2022, 29 July shift-I]**

(a) 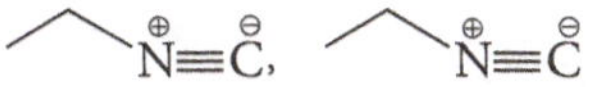

(b) 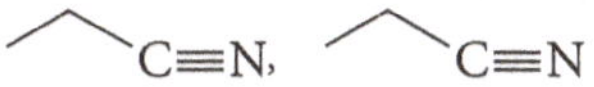

(c) 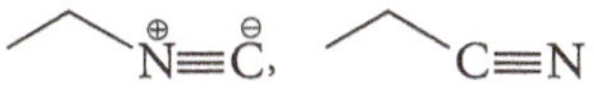

(d) 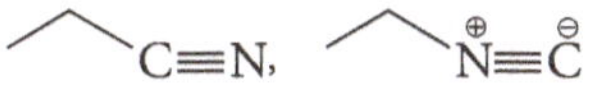

47 The major product in the given reaction is **[2022, 28 July shift-II]**

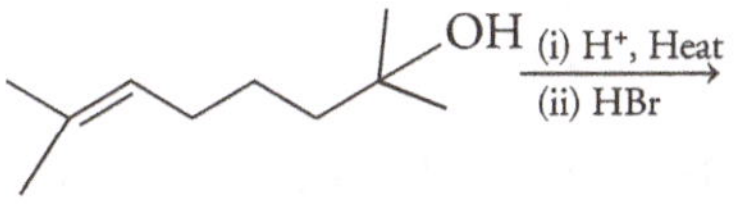

(a) 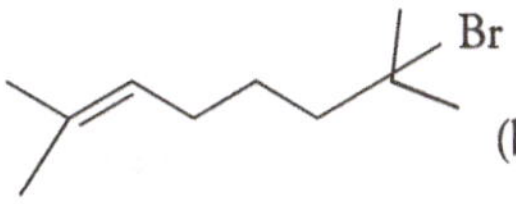

(b) 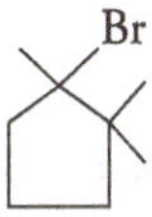

(c) 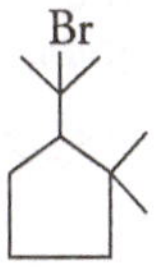

(d) 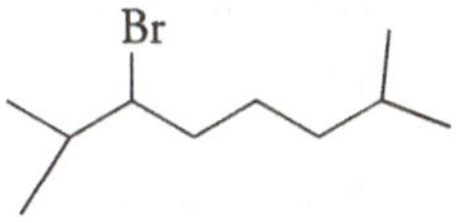

48 Major product 'B' of the following reaction sequence is **[2022, 27 July shift-II]**

$$CH_3-\underset{\substack{|\\CH_3}}{C}=CH-CH_3 \xrightarrow[CH_3OH]{Br_2} A \xrightarrow{HI} \underset{\text{(Major product)}}{B}$$

(a) $CH_3-\underset{\substack{|\\CH_3}}{\overset{\substack{HO\\|}}{C}}-\overset{\substack{Br\\|}}{CH}-CH_3$ (b) $CH_3-\underset{\substack{|\\CH_3}}{\overset{\substack{I\\|}}{C}}-\overset{\substack{Br\\|}}{CH}-CH_3$

(c) $CH_3-\underset{\substack{|\\CH_3}}{\overset{\substack{Br\\|}}{C}}-\overset{\substack{OH\\|}}{CH}-CH_3$ (d) $CH_3-\underset{\substack{|\\CH_3}}{\overset{\substack{Br\\|}}{C}}-\overset{\substack{I\\|}}{CH}-CH_3$

49 The correct order of reactivity of the given chlorides with acetate in acetic acid is **[2021, 31 Aug Shift-I]**

(a) 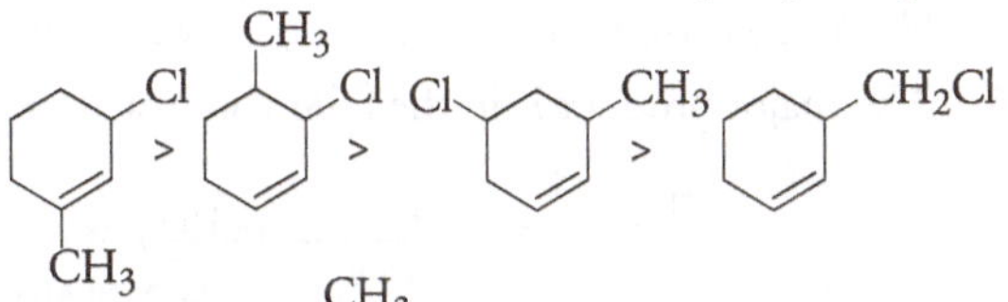

(b) 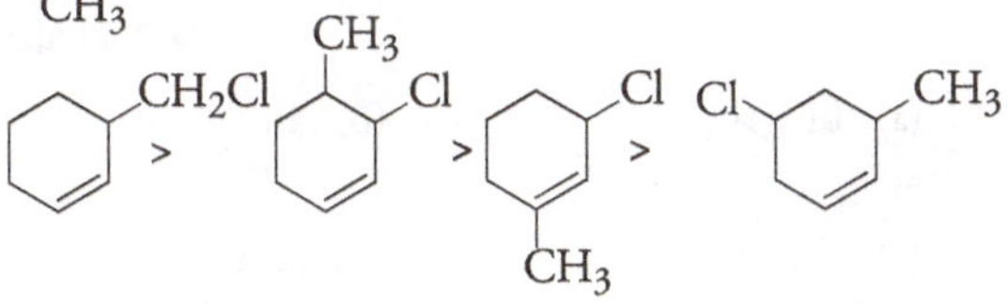

(c) 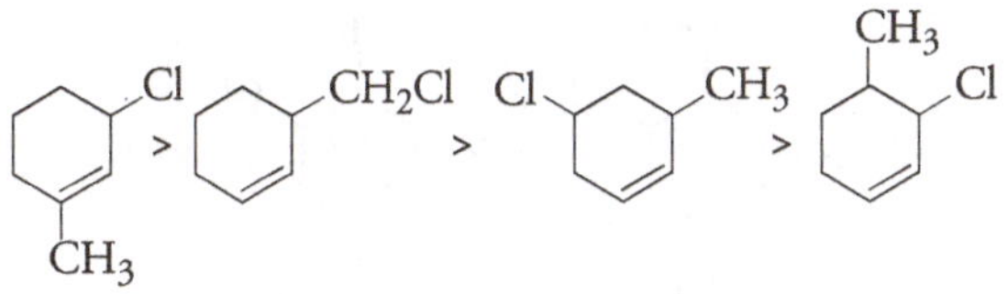

(d) 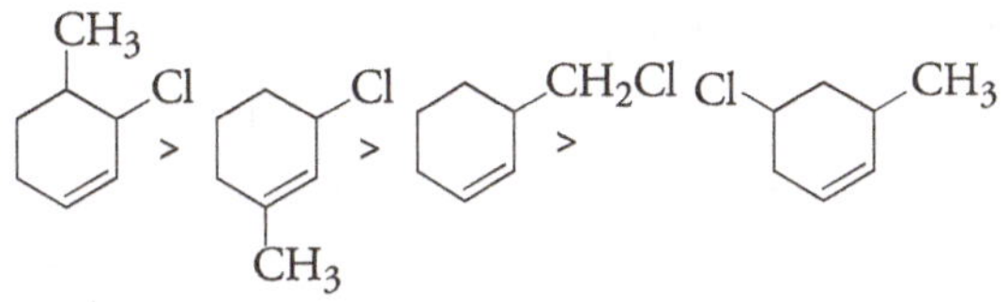

50 Excess of isobutane on reaction with Br_2 in presence of light at 125°C gives which one of the following, as the major product? **[2021, 26 Aug Shift-I]**

(a) $CH_3-\underset{\substack{|\\CH_3}}{\overset{\substack{Br\\|}}{C}}-CH_2Br$ (b) $CH_3-\underset{\substack{|\\CH_2Br}}{CH}-CH_2Br$

(c) $CH_3-\underset{\substack{|\\CH_3}}{CH}-CH_2Br$ (d) $CH_3-\underset{\substack{|\\CH_3}}{\overset{\substack{CH_3\\|}}{C}}-Br$

51 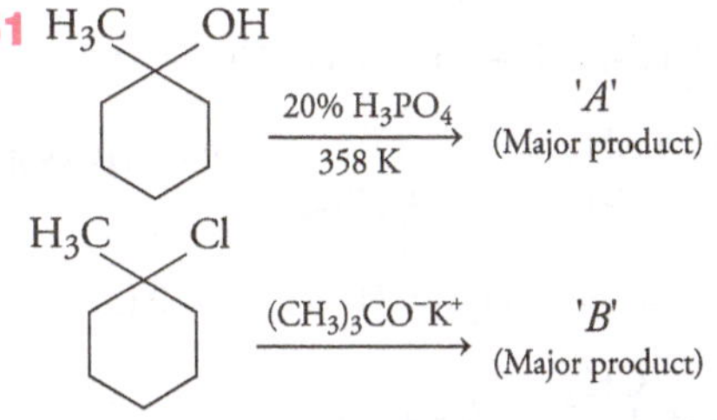

The product A and B formed in above reactions are **[2021, 16 March Shift-I]**

(a) 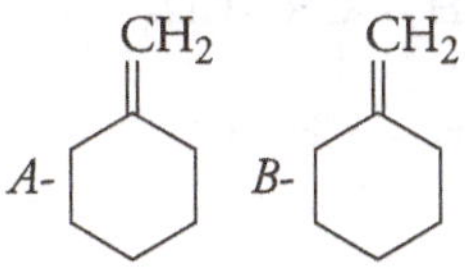

(b) 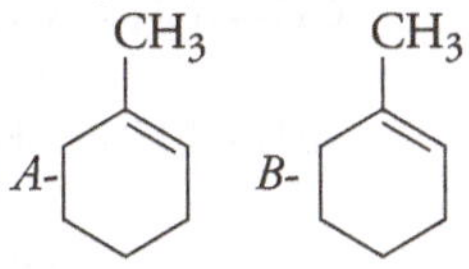

(c) 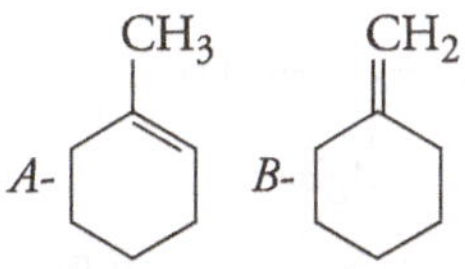

(d) 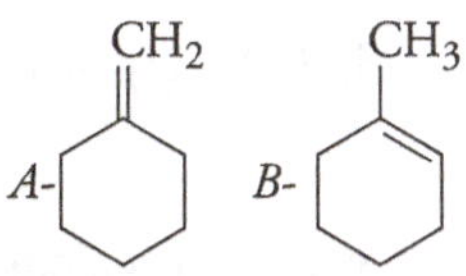

52 Identify A and B in the chemical reaction.

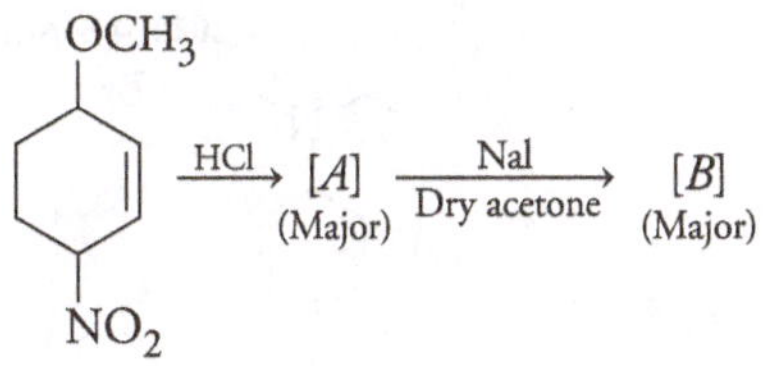

[2021, 25 Feb Shift-I]

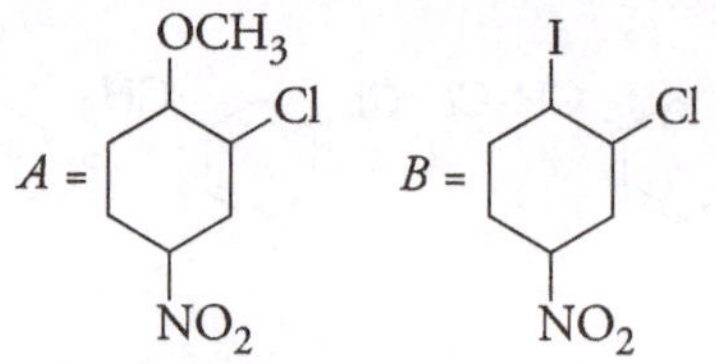

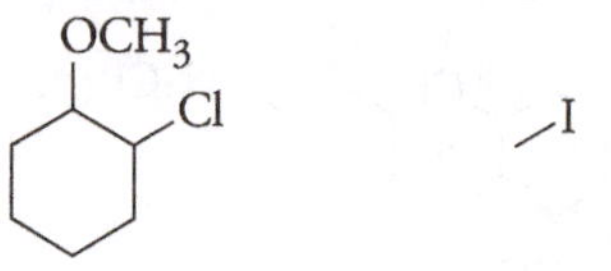

(c) 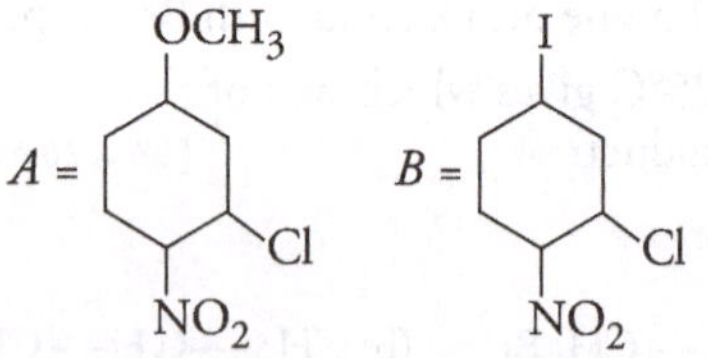

(d) 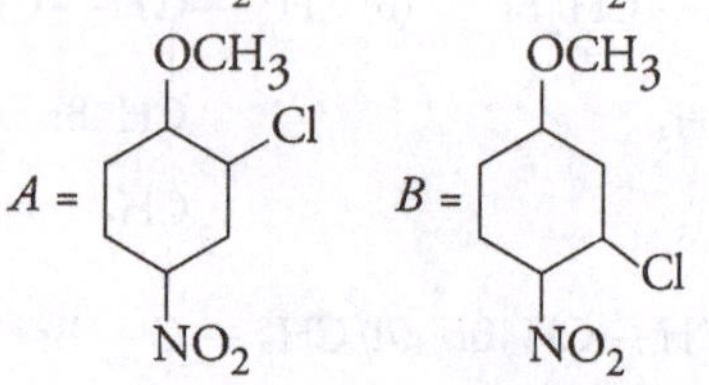

53 The product formed in the first step of the reaction of

$CH_3-CH_2-\underset{}{\overset{Br}{|}}{CH}-CH_2-\overset{Br}{\overset{|}{CH}}-CH_3$ with excess Mg/Et_2O ($Et = C_2H_5$) is **[2021, 24 Feb Shift-I]**

(a) $CH_3-CH_2-CH-CH_2-CH-CH_3$
$\qquad\quad|\qquad\qquad|$
$CH_3-CH-CH_2-CH-CH_2-CH_3$

(b) 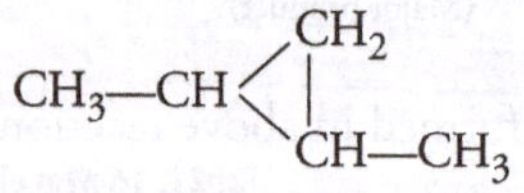

(c) $CH_3-CH_2-CH-CH_2-CH-CH_3$
$\qquad\quad|\qquad\qquad|$
$CH_3-CH_2-CH-CH_2-CH-CH_3$

(d) $CH_3CH_2-\overset{MgBr}{\overset{|}{CH}}-CH_2-\underset{MgBr}{\underset{|}{CH}}-CH_3$

54 Which of the following compounds will form the precipitate with *aq*. $AgNO_3$ solution most readily? **[2020, 4 Sep Shift-II]**

(a) 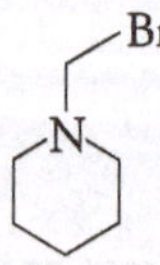

(b) 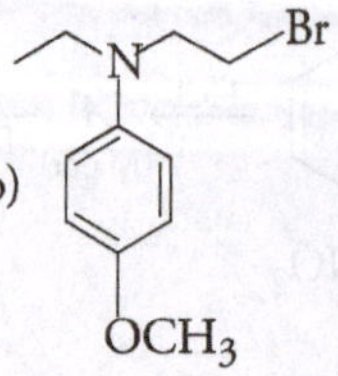

(c) 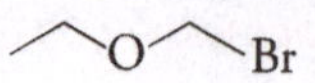

(d) 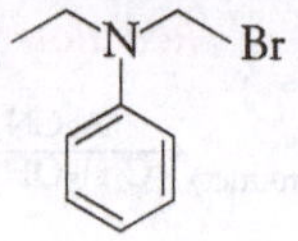

55 Which of the following reactions will not produce a racemic product? **[2020, 9 Jan Shift-II]**

(a) 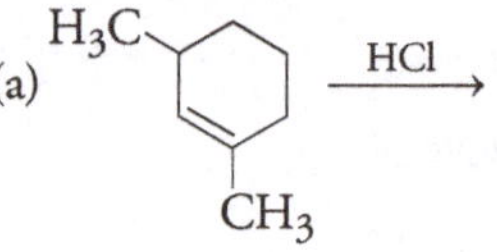

(b) $CH_3-\overset{O}{\overset{||}{C}}CH_2CH_3 \xrightarrow{HCN}$

(c) $CH_3CH_2CH=CH_2 \xrightarrow{HBr}$

(d) $CH_3-\underset{H}{\underset{|}{\overset{CH_3}{\overset{|}{C}}}}-CH=CH_2 \xrightarrow{HCl}$

56 The decreasing order of reactivity towards dehydrohalogenation (E_1) reaction of the following compounds is **[2020, 8 Jan Shift-I]**

(A) 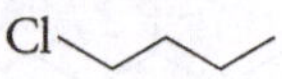

(B) 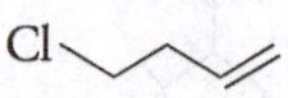

(C) 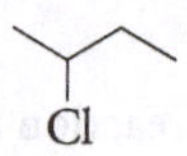

(D) 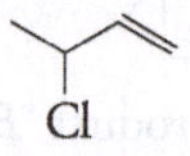

(a) B > A > D > C
(b) B > D > A > C
(c) D > B > C > A
(d) B > D > C > A

57 An Assertion and a Reason are given below. Choose the correct answer from the following options. **[2019, 12 April Shift-II]**

Assertion (A) Vinyl halides do not undergo nucleophilic substitution easily.

Reason (R) Even though the intermediate carbocation is stabilised by loosely held π-electrons, the cleavage is difficult because of strong bonding.

(a) Both (A) and (R) are wrong statements.
(b) Both (A) and (R) are correct statements and (R) is correct explanation of (A).
(c) Both (A) and (R) are correct statements but (R) is not the correct explanation of (A).
(d) (A) is a correct statement but (R) is a wrong statement.

58 The major product *Y* in the following reaction is

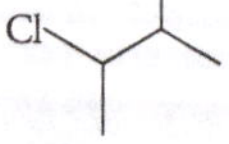 $\xrightarrow[\text{Heat}]{\text{EtONa}} X \xrightarrow{HBr} Y$

[2019, 10 April Shift-II]

(a) 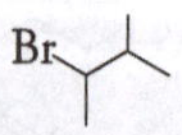

(b) 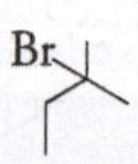

(c) 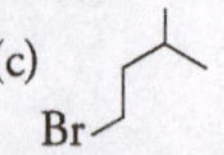

(d) 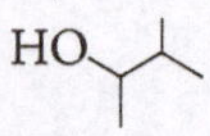

• Level 3

59 Predict the major product of the following reaction sequence: **[2025, 4 April Shift-I]**

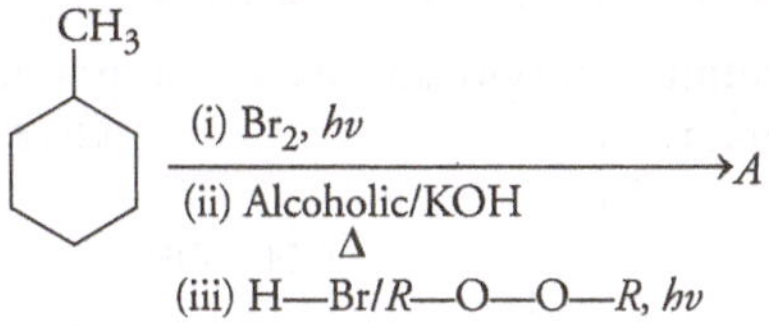

(a) 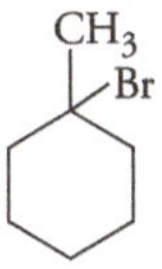
(b) 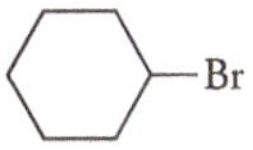
(c) 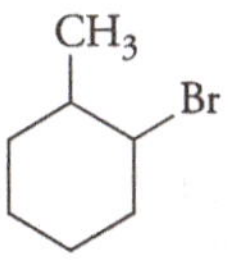
(d) 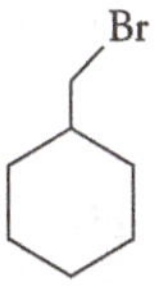

60 Match the List-I with List-II. **[2025, 23 Jan Shift-I]**

List-I (Name reaction)	List-II (Product obtainable)
A. Swarts reaction	I. Ethyl benzene
B. Sandmeyer's reaction	II. Ethyl iodide
C. Wurtz Fittig reaction	III. Cyanobenzene
D. Finkelstein reaction	IV. Ethyl fluoride

Choose the correct answer from the options given below:
(a) A-IV, B-I, C-III, D-II (b) A-IV, B-III, C-I, D-II
(c) A-II, B-III, C-I, D-IV (d) A-II, B-I, C-III, D-IV

61 The ascending order of relative rate of solvolysis of following compounds is **[2025, 23 Jan Shift-II]**

A. 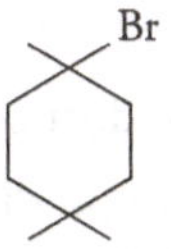
B. 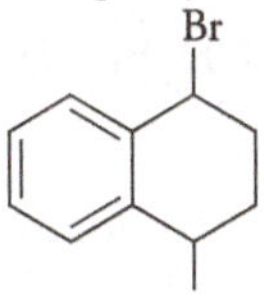
C. 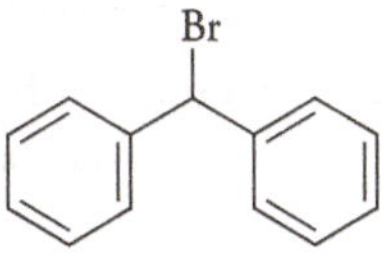
D. 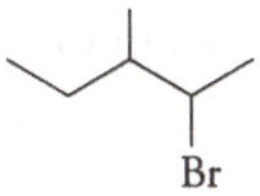

(a) (C) < (B) < (A) < (D) (b) (D) < (B) < (A) < (C)
(c) (C) < (D) < (B) < (A) (d) (D) < (A) < (B) < (C)

62 $RBr \xrightarrow[\text{(ii) } H_2O]{\text{(i) Mg, dry ether}}$ 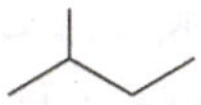 2- methylbutane **[2025, 22 Jan Shift-II]**

The maximum number of *R*Br producing 2-methylbutane by above sequence of reactions is .... (Consider the structural isomers only)
(a) 5 (b) 3 (c) 4 (d) 1

63 $C_2H_5Br \xrightarrow{\text{Alc. KOH}} A \xrightarrow[CCl_4]{Br_2} B \xrightarrow[\text{Excess}]{KCN} C \xrightarrow[\text{Excess}]{H_3O^+} D$

Acid *D* formed in above reaction is **[2024, 1 Feb Shift-II]**
(a) oxalic acid (b) malonic acid
(c) gluconic acid (d) Succinic acid

Imp. *Reaction of Haloalkanes with Alcoholic & Aq. KOH*

64 The product (*C*) in the below mentioned reaction is

$$CH_3—CH_2—CH_2—Br \xrightarrow[\Delta]{\text{KOH (alc.)}} A \xrightarrow{HBr} B \xrightarrow[\text{KOH}(aq)]{\Delta} C$$

[2024, 31 Jan Shift-I]
(a) Propan-1-ol (b) Propan-2-ol
(c) Propene (d) Propyne

65 Which of the following reaction is correct? **[2024, 29 Jan Shift-II]**

(a) 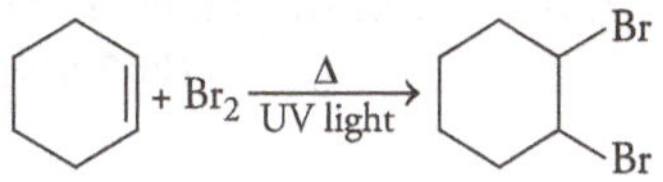

(b) $C_2H_5CONH_2 + Br_2 + NaOH \longrightarrow C_2H_5CH_2NH_2 + Na_2CO_3 + NaBr + H_2O$

(c) 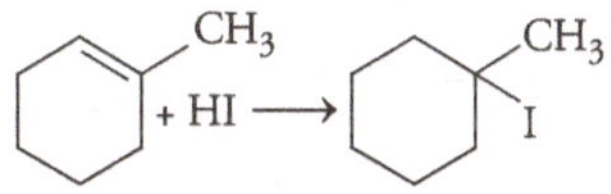

(d) $CH_3CH_2CH_2NH_2 \xrightarrow[H_2O]{HNO_2,\ 0°C} CH_3CH_2OH + N_2 + HCl$

66 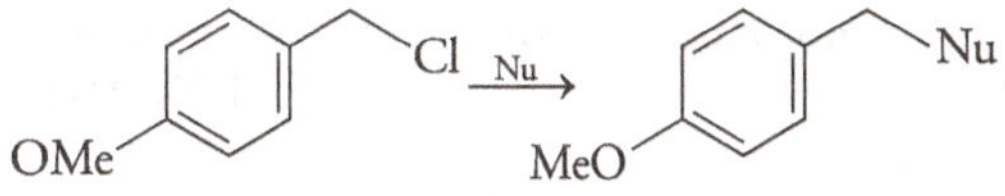

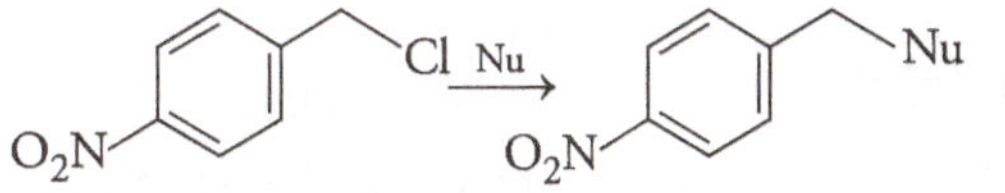

where, Nu = Nucleophile **[2023, 11 April Shift-I]**

Find out the correct statement from the options given below for the above 2 reactions.
(a) Reaction (I) and (II) both are of 1st order
(b) Reactions (I) and (II) both are of 2nd order
(c) Reactions (I) is of 2nd order and reaction (II) is of 1st order
(d) Reaction (I) is of 1st order and reaction (II) is of 2nd order

67 The pair from the following pairs having both compounds with net non-zero dipole moment is **[2023, 10 April Shift-I]**
(a) 1, 4-dichlorobenzene, 1,3-dichlorobenzene
(b) CH_2Cl_2, $CHCl_3$
(c) Benzene, anisidine
(d) *cis*-butene, *trans*-butene

68 In the given reaction sequence, the major product C is

$$C_8H_{10} \xrightarrow[H_2SO_4]{HNO_3} A \xrightarrow[\Delta]{Br_2} B \xrightarrow[KOH]{\text{Alcoholic}} C$$

[2022, 24 June Shift-I]

(a) 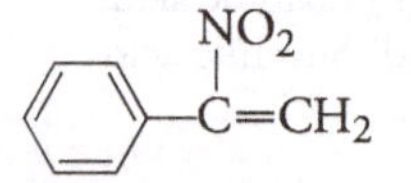

(b) 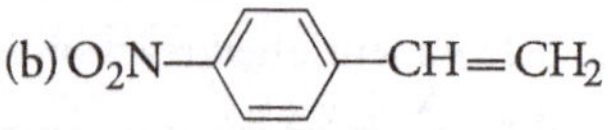

(c) 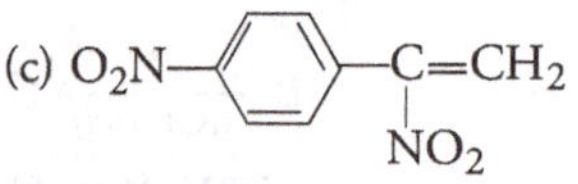

(d) 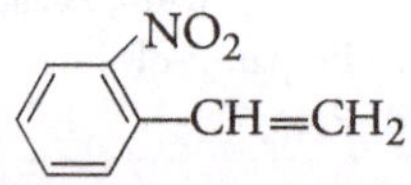

69 The increasing order of the boiling points of the major products A, B and C of the following reactions will be [2020, 6 Sep Shift-II]

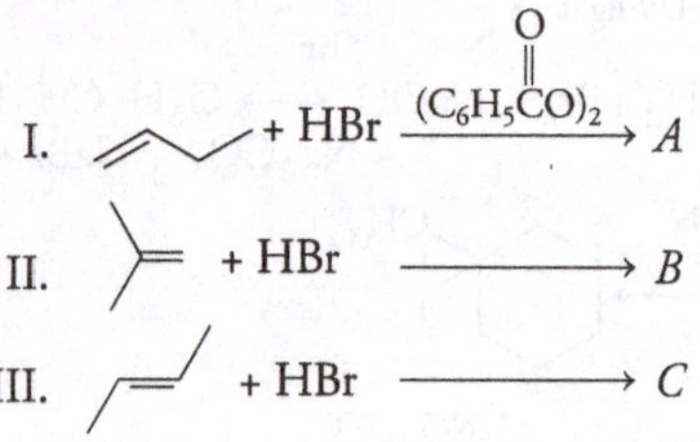

(a) II < III < I (b) III < I < II (c) I < II < III (d) I < III < II

70 The decreasing order of reactivity of the following organic molecules towards $AgNO_3$ solution is [2020, 4 Sep Shift-I]

(A)  (B) 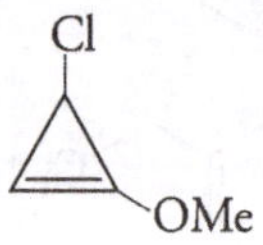

(C) $CH_3CH(Cl)CH_3$ (D) $CH_3CH(Cl)CH_2NO_2$

(a) (A) > (B) > (C) > (D) (b) (C) > (D) > (A) > (B)
(c) (B) > (A) > (C) > (D) (d) (A) > (B) > (D) > (C)

71 The mechanism of S_N1 reaction is given as

$$RX \longrightarrow \underset{\text{Ion pair}}{R^{\oplus}X^{\ominus}} \longrightarrow \underset{\text{Solvent Separated ion pair}}{R^{\oplus}||X^{\ominus}} \xrightarrow{Y^{\ominus}} RY + X^{\ominus}$$

A student writes general characteristics based on the given mechanism as :
(A) The reaction is favoured by weak nucleophiles.
(B) $R^{\oplus}$ would be easily formed if the substituents are bulky.
(C) The reaction is accompanied by recemisation.
(D) The reaction is favoured by non-polar solvents.

Which observations are correct? [2020, 3 Sep Shift-I]
(a) (A) and (B) (b) (A) and (C)
(c) (A), (B) and (C) (d) (B) and (D)

Numerical Value Type Questions

72 Total number of optically active compounds from the following is .... [2024, 8 April Shift-II]

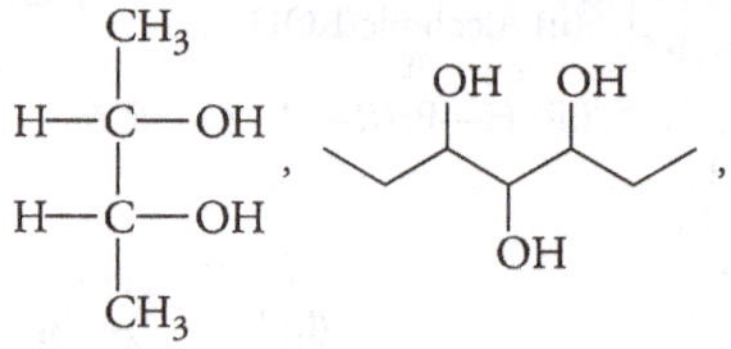

$CH_3—CH_2—CH_2—CH_2—OH$,
$CH_3—CH_2—CH(Cl)—CH_3$

$CH_3—CH_2—CH_2—CH_2—Cl$
$(CH_3)_2CH—CH_2—CH_2—Cl$

73 The number of optical isomers in following compound is ...... [2024, 8 April Shift-I]

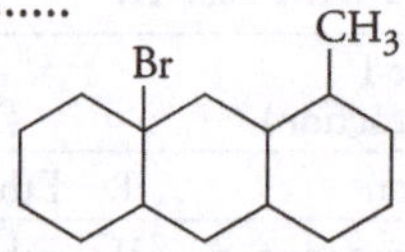

74 Total number of isomeric compounds (including stereoisomers) formed by monochlorination of 2-methylbutane is .... [2024, 1 Feb Shift-II]

75 Number of optical isomers possible for 2-chlorobutane .... [2024, 1 Feb Shift-I]

76

$$CH_3CH_2Br + NaOH \begin{cases} \xrightarrow{C_2H_5OH} \text{Product } A \\ \xrightarrow{H_2O} \text{Product } B \end{cases}$$

The total number of hydrogen atoms in product A and product B is .... [2024, 31 Jan Shift-I]

77 Total number of compounds with chiral carbon atoms from following is .... [2024, 27 Jan Shift-II]

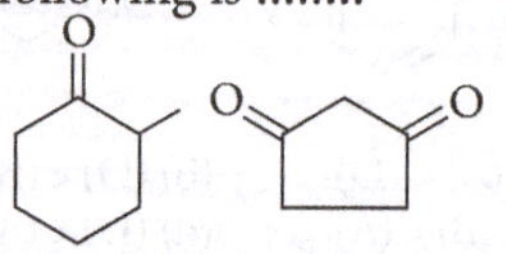

$CH_3—CH_2—CH(NO_2)—COOH$
$CH_3—CH_2—CHBr—CH_2—CH_3$
$CH_3—CH(I)—CH_2—NO_2$
$CH_3—CH_2—CH(OH)—CH_2OH$
$CH_3—CH(I)—CH(I)—C_2H_5$

Topic 2 / Haloarenes

• Level 1

01 The structure of the major product formed in the following reaction is **[2025, 24 Jan Shift-II]**

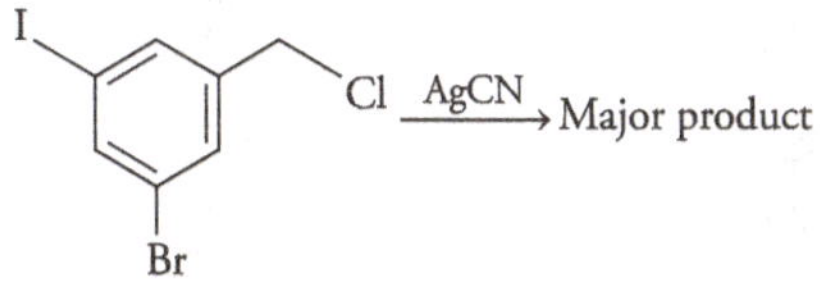

(a) 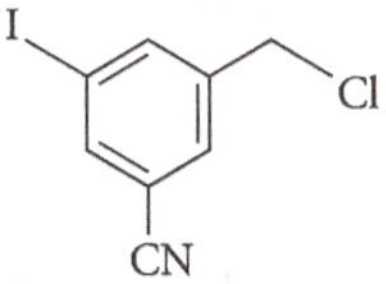 (b) 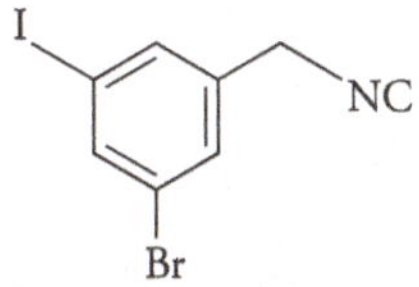

(c) 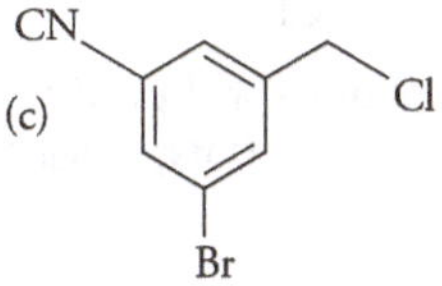 (d) 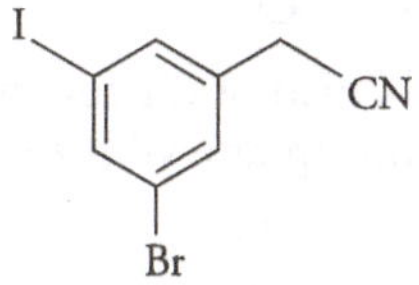

02 Which among the following halides will generate the most stable carbocation in the nucleophilic substitution reaction? **[2025, 29 Jan Shift-II]**

(a) 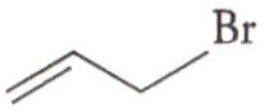 (b) 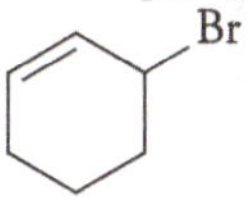

(c) 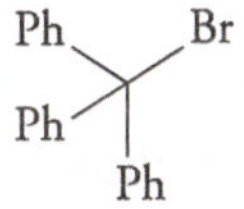 (d) 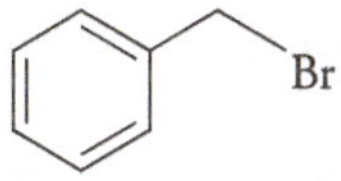

03 In the following substitution reaction,

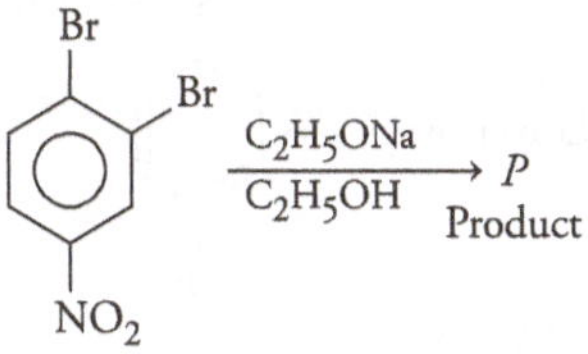

Product P formed is **[2025, 29 Jan Shift-I]**

(a) 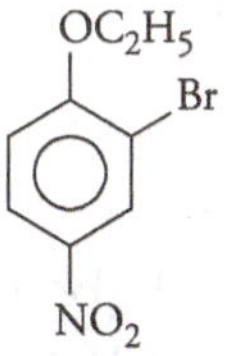 (b) 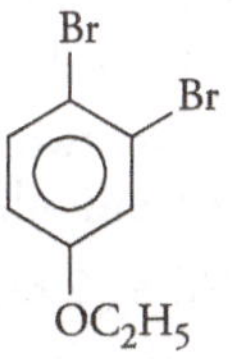

(c) 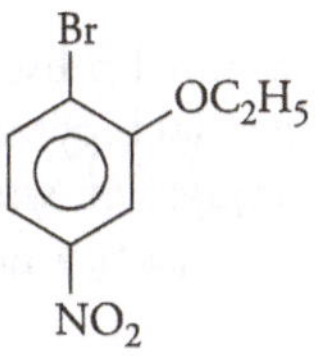 (d) 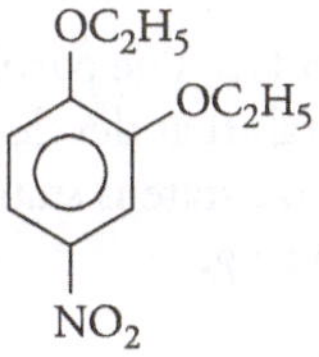

04 The major product formed in the Friedel-Craft acylation of chlorobenzene is **[2023, 15 April Shift-I]**

(a) 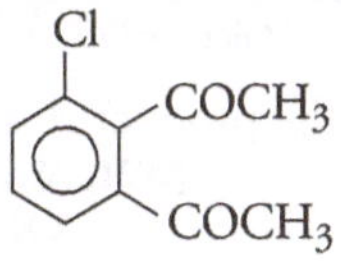 (b) 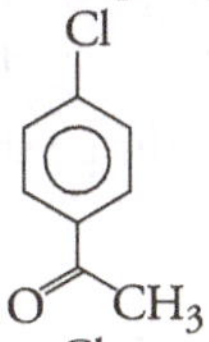

(c) 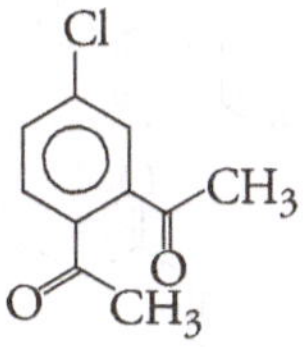 (d) 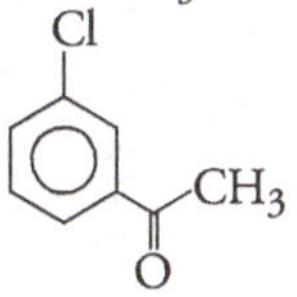

05 The compound which will have the lowest rate towards nucleophilic aromatic substitution on treatment with OH^- is **[2023, 25 Jan Shift-I]**

(a) 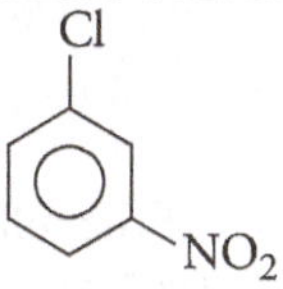 (b) 

(c) 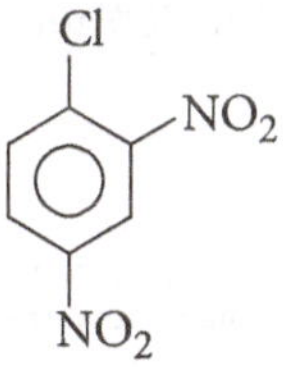 (d) 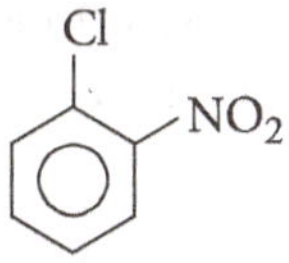

06 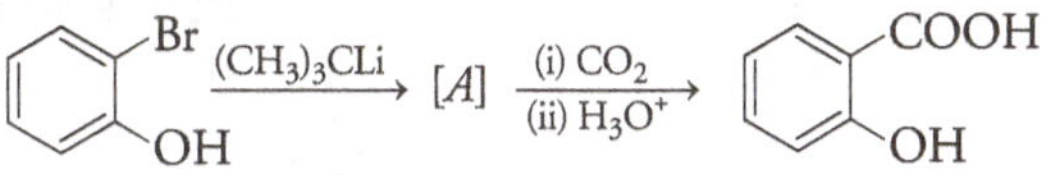

In the given conversion the compound A is **[2022, 29 June Shift-I]**

(a) 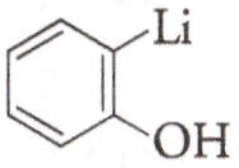 (b) 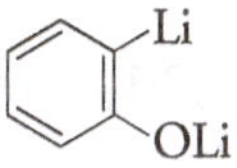

(c) 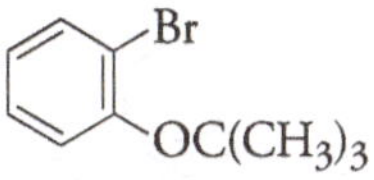 (d) 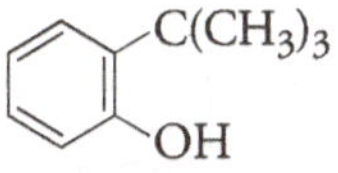

07 Halogenation of which one of the following will yield *m*-substituted product with respect to methyl group as a major product? **[2022, 26 June Shift-II]**

(a) 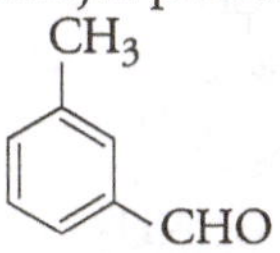 (b) 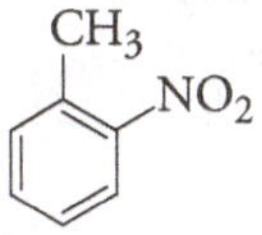

(c) 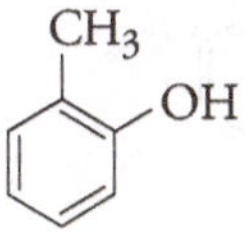 (d) 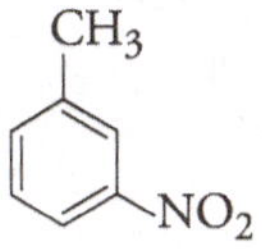

08 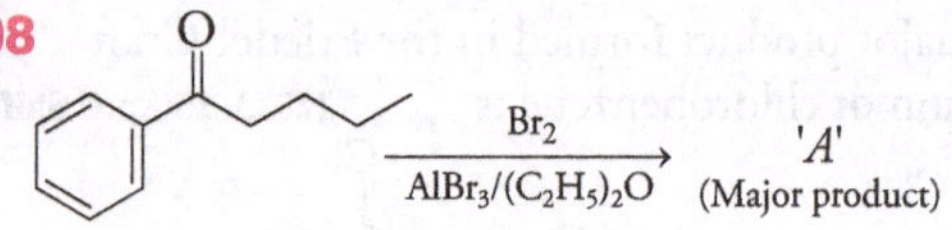

Consider the given reaction, the product A is **[2021, 26 Aug Shift-II]**

(a) 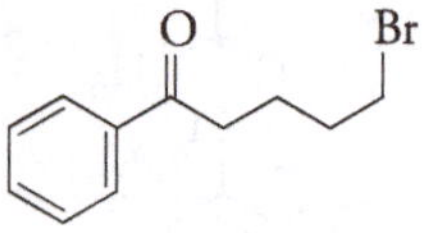 (b) 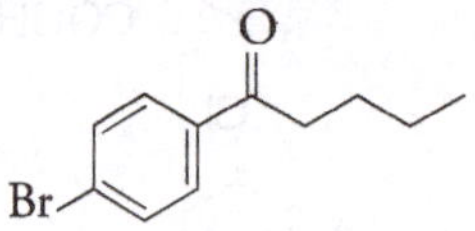

(c) 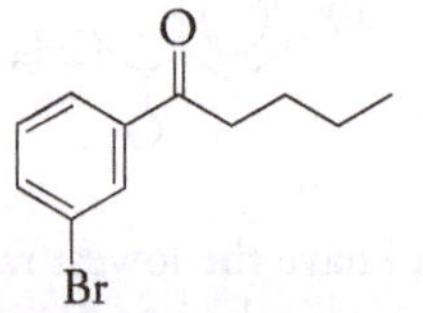 (d) 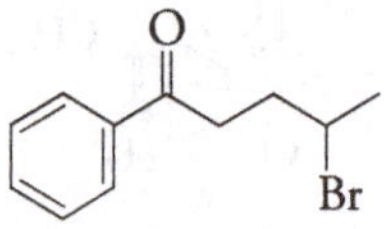

09 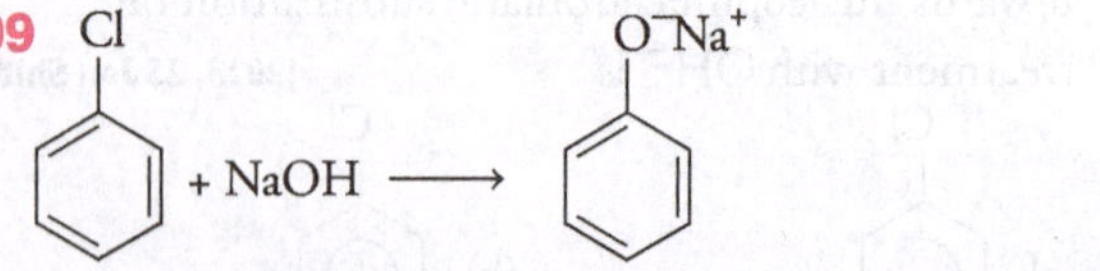

The above reaction requires which of the following reaction conditions? **[2021, 17 March Shift-I]**

(a) 573 K, Cu, 300 atm (b) 623 K, Cu, 300 atm
(c) 573 K, 300 atm (d) 623 K, 300 atm

10 For the given reaction,

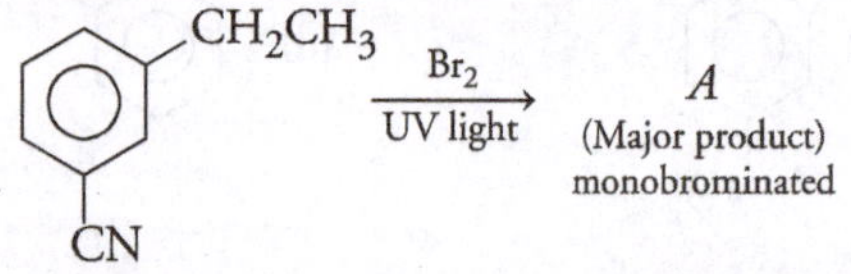

What is A? **[2021, 26 Feb Shift-I]**

(a) 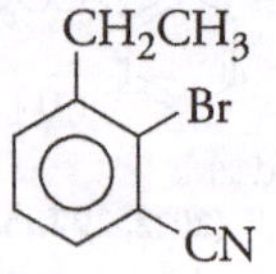 (b) 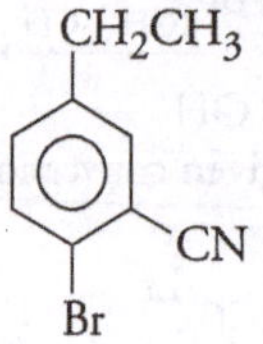

(c) 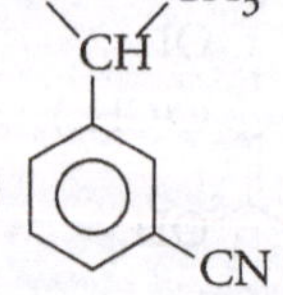 (d) 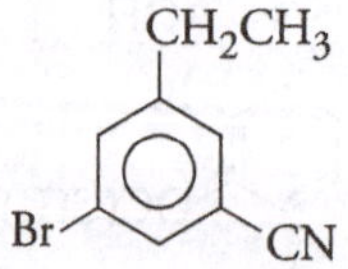

11 Which of these will produce the highest yield in Friedel-Crafts reaction? **[2020, 9 Jan Shift-I]**

(a) 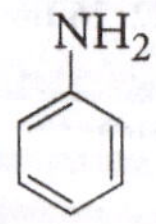 (b) 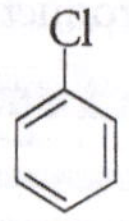

(c) 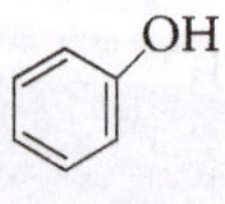 (d) 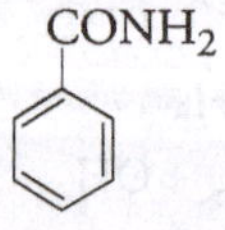

12 The major product of the following reaction is

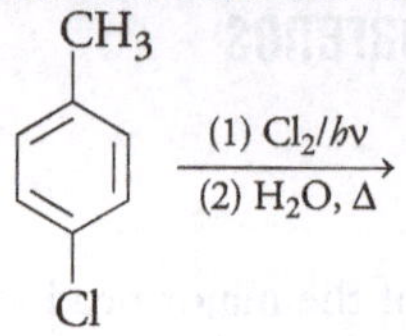

[2019, 8 April Shift-II]

(a) 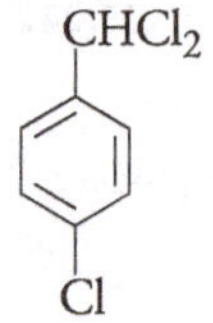 (b) 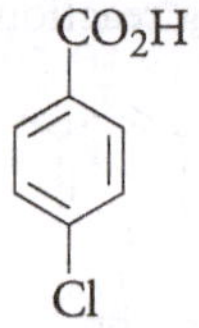

(c) 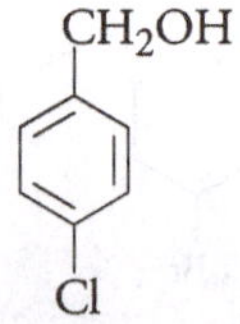 (d) 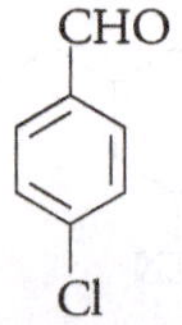

13 Which of the following compounds will produce a precipitate with $AgNO_3$? **[2019, 11 Jan Shift-II]**

(a) 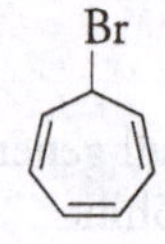 (b) 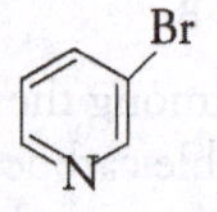

(c) 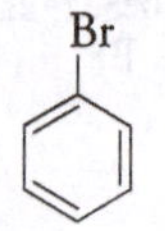 (d) 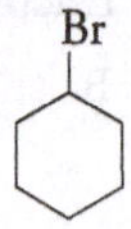

• Level 2

14. Choose the correct set of reagents for the following conversion. **[2025, 8 April Shift-II]**

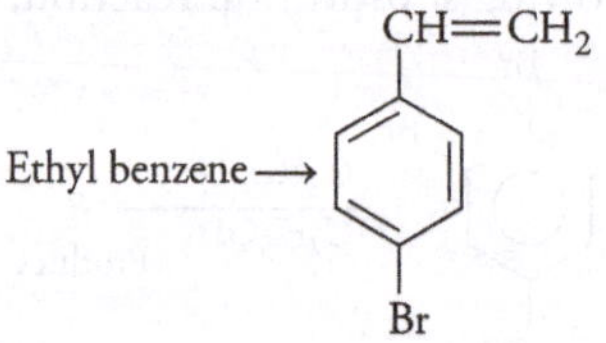

(a) Br_2 / Fe ; Cl_2, Δ; alc . KOH
(b) Cl_2 / anhy . $AlCl_3$; Br_2 / Fe; alc . KOH
(c) Br_2 / anhy . $AlCl_3$; Cl_2, Δ; *aq* . KOH
(d) Cl_2 / Fe; Br_2 / anhy . $AlCl_3$; *aq* . KOH

15 Given below are two statements : one is labelled as Assertion (A) and other is labelled as Reason (R).

Assertion (A) S_N2 reaction of $C_6H_5CH_2Br$ occurs more readily than the S_N2 reaction of CH_3CH_2Br.

Reason (R) The partially bonded unhybridised *p*-orbital that develops in the trigonal bipyramidal transition state is stabilised by conjugation with the phenyl ring. **[2024, 9 April Shift-I]**

In the light of the above statements, choose the most appropriate answer from the options given below.

(a) Both A and R are correct but R is not the correct explanation of A.
(b) A is not correct, but R is correct.
(c) Both A and R are correct but R is the correct explanation of A.
(d) A is correct but R is not correct.

16 Given below are two statements one is labelled as **Assertion (A)** and the other is labelled as **Reason (R).** **[2024, 29 Jan Shift-II]**

Assertion (A) Aryl halides cannot be prepared by replacement of hydroxyl group of phenol by halogen atom.

Reason (R) Phenols react with halogen acids violently.

In the light of the above statements, choose the **most appropriate** from the options given below

(a) Both A and R are true but R is not the correct explanation of A.
(b) Both A and R are true and R is the correct explanation of A.
(c) A is false but R is true.
(d) A is true but R is false.

17 The correct order of reactivity of following haloarenes towards nucleophilic substitution with aqueous NaOH is **[2023, 8 April Shift-II]**

(A) 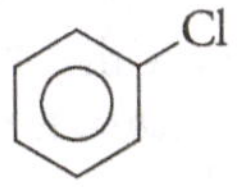

(B) 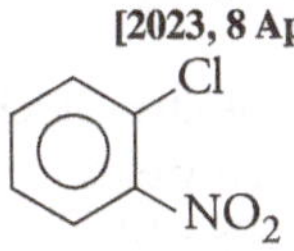

(C) 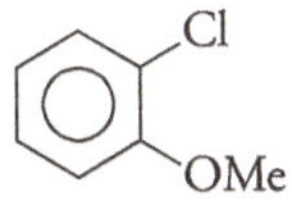

(D) 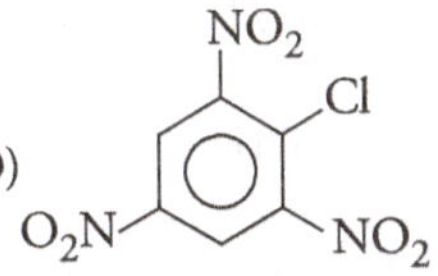

Choose the correct answer from the options given below:

(a) C > A > D > B (b) A > B > D > C
(c) D > B > A > C (d) D > C > B > A

18 The correct order of melting points of dichlorobenzenes is **[2023, 31 Jan Shift-I]**

(a) 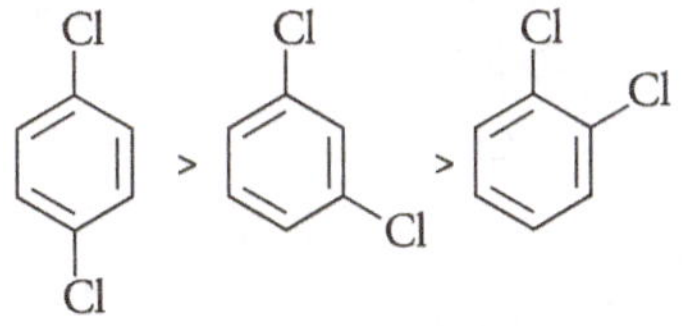

(b) 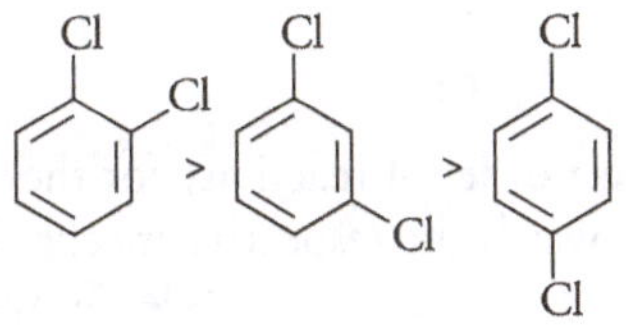

(c) 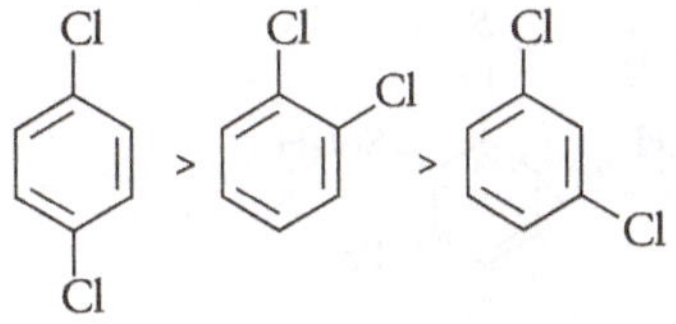

(d) 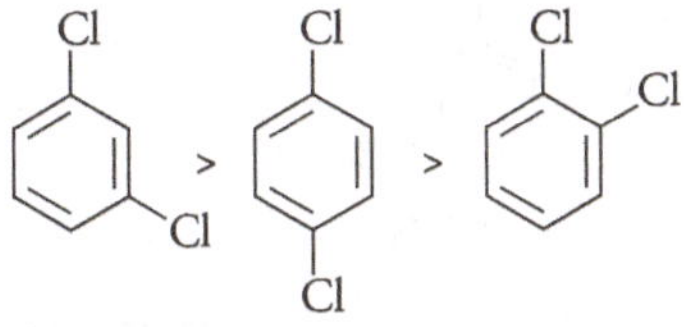

19 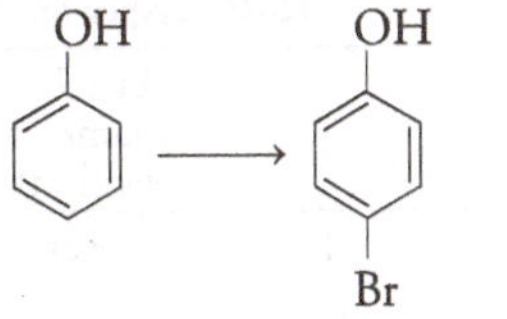

(Major product)

The given reaction can occur in the presence of

(A) bromine water (B) Br_2 in CS_2, 273 K
(C) Br_2 / $FeBr_3$ (D) Br_2 in $CHCl_3$, 273 K

Choose the correct answer from the options given below **[2021, 22 July Shift-I]**

(a) (B) and (D) only (b) (A) and (C) only
(c) (B), (C) and (D) only (d) (A), (B) and (D) only

20 The compounds *A* and *B* in the following reaction are, respectively

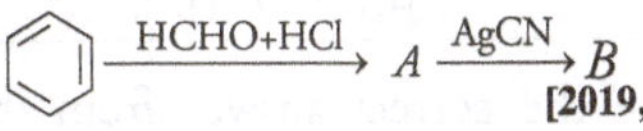

[2019, 9 Jan Shift-I]

(a) $A =$ Benzyl alcohol, $B =$ Benzyl isocyanide
(b) $A =$ Benzyl alcohol, $B =$ Benzyl cyanide
(c) $A =$ Benzyl chloride, $B =$ Benzyl isocyanide
(d) $A =$ Benzyl chloride, $B =$ Benzyl cyanide

• Level 3

21 In the following series of reactions identify the major products *A* and *B* respectively **[2025, 3 April Shift-II]**

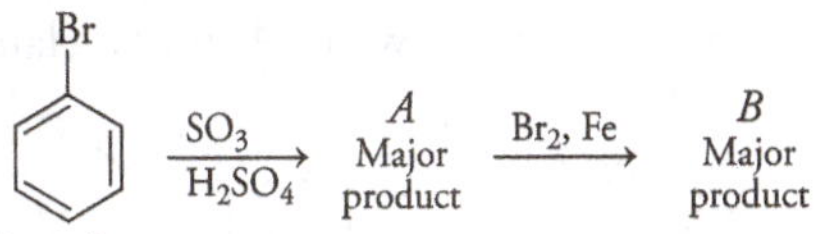

A B

(a) 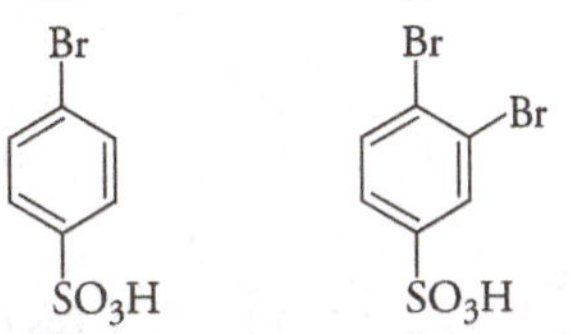

(b) 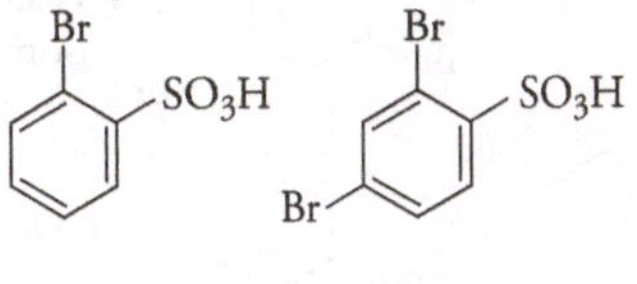

(c) 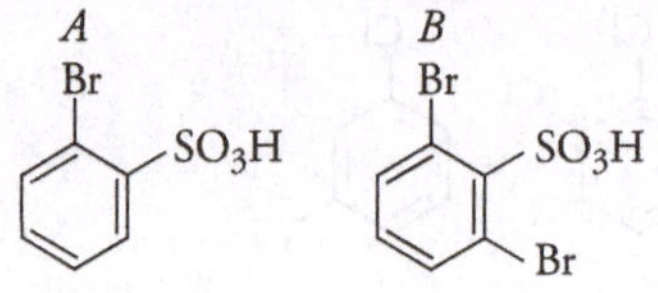

(d) 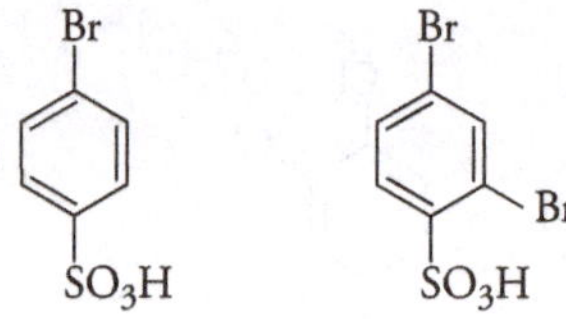

22 Match List-I with List-II. **[2025, 2 April Shift-II]**

	List-I (Reaction)		List-II (Name of reaction)
A.	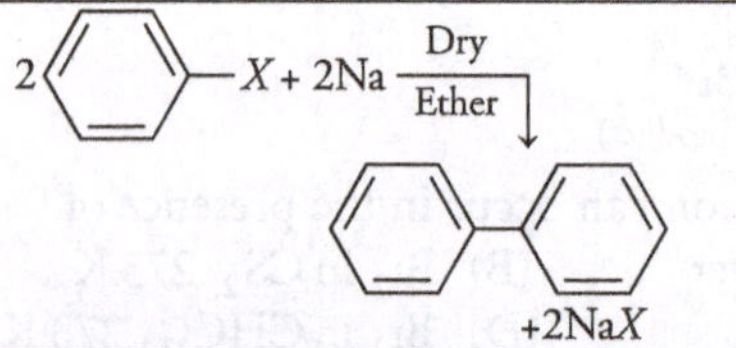	I.	Lucas reaction
B.	$ArN_2^+X^- \xrightarrow[HCl]{Cu} ArCl + N_2\uparrow + CuX$	II.	Finkelstein reaction
C.	$C_2H_5Br + NaI \xrightarrow[Acetone]{Dry} C_2H_5I + NaBr$	III.	Fittig reaction
D.	$CH_3C(OH)(CH_3)CH_3 \xrightarrow[ZnCl_2]{HCl} CH_3C(Cl)(CH_3)CH_3$	IV.	Gattermann reaction

Choose the correct answer from the options given below.

(a) (A)-(III), (B)-(II), (C)-(IV), (D)-(I)
(b) (A)-(IV), (B)-(I), (C)-(II), (D)-(III)
(c) (A)-(IV), (B)-(III), (C)-(I), (D)-(II)
(d) (A)-(III), (B)-(IV), (C)-(II), (D)-(I)

23 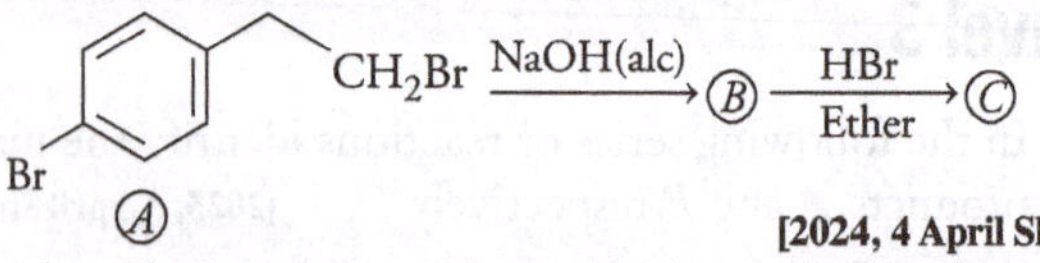

[2024, 4 April Shift-I]

Identify *B* and *C* and how are *A* and *C* related?

(a) 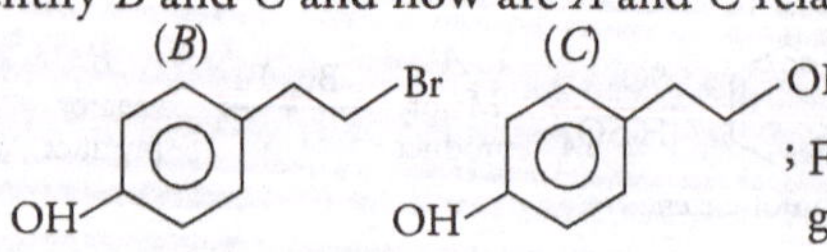 ; Functional group isomers

(b) 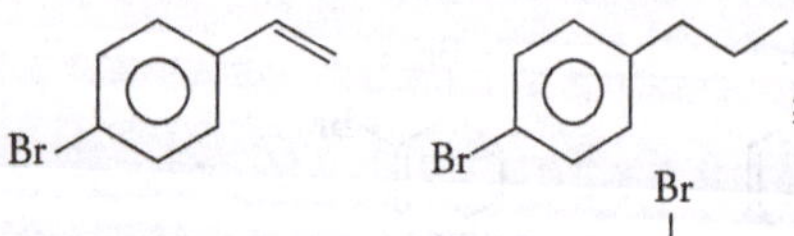 ; Chain isomers

(c) 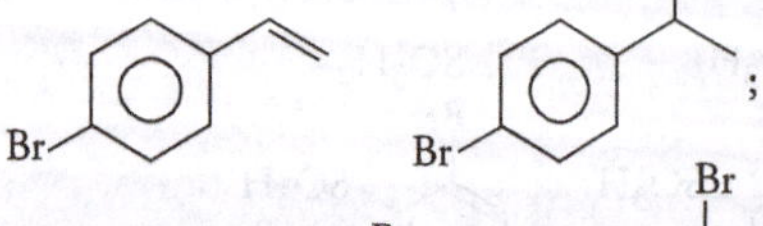 ; Position isomers

(d) 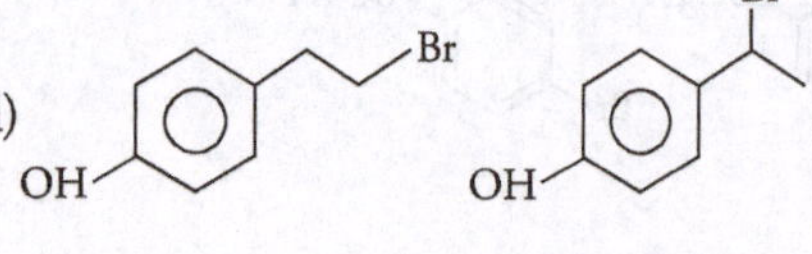 ; Derivative

24 Identify *A* and *B* in the following reaction sequence.

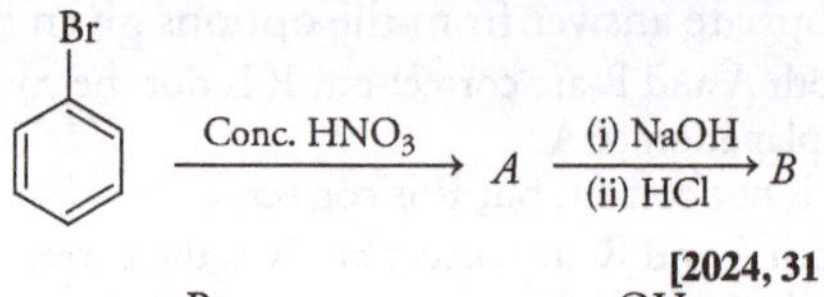

[2024, 31 Jan Shift-II]

(a) 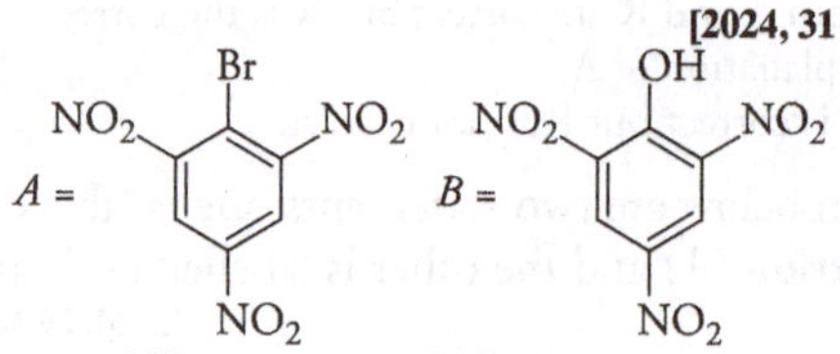

(b) 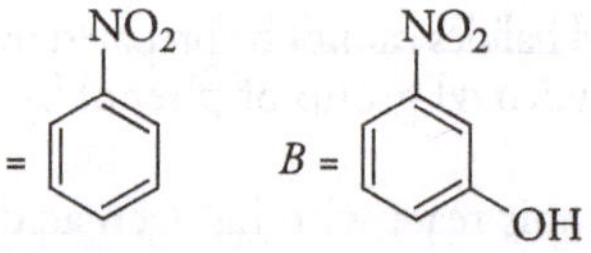

(c) 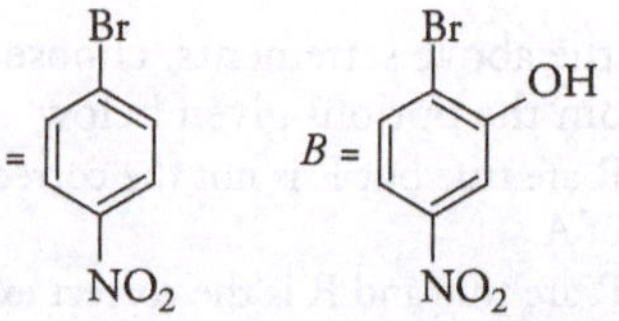

(d) 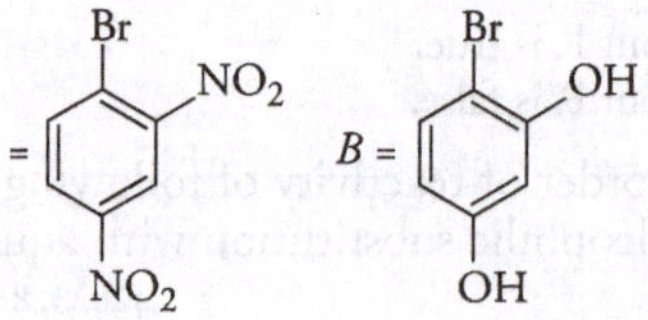

25 *m*-chlorobenzaldehyde on treatment with 50% KOH solution yields **[2024, 30 Jan Shift-II]**

(a) 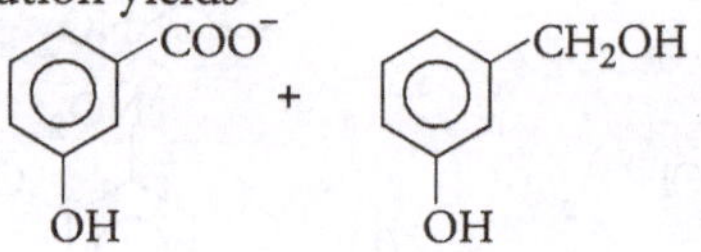

(b) 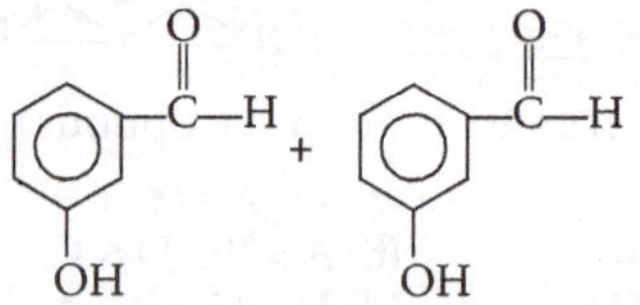

(c) 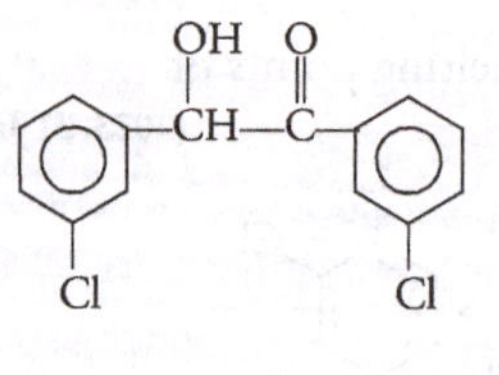

(d) 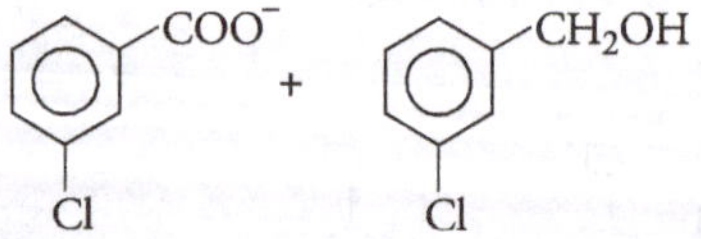

26 Identify the correct order of reactivity for the following pairs towards the respective mechanism **[2023, 10 April Shift-I]**

(A) S_N2 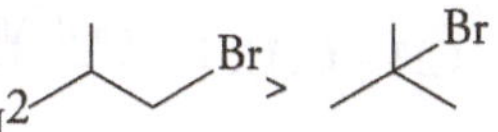

(B) S_N1 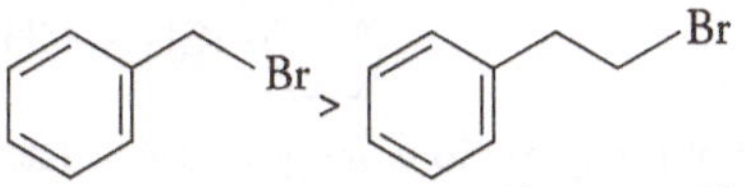

(C) Electrophilic substitution 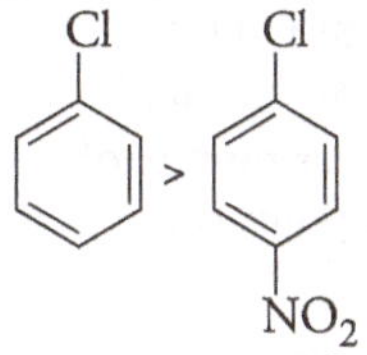

(D) Nucleophilic substitution 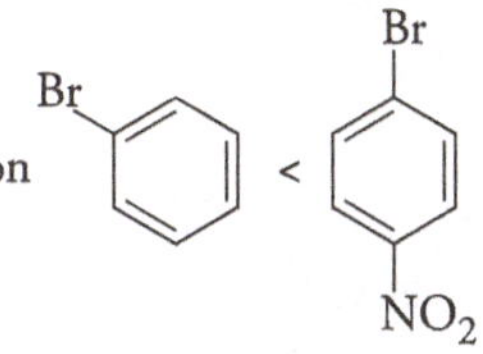

Choose the correct answer from the options given below:
(a) (A), (C) and (D) only
(b) (A), (B), (C) and (D) only
(c) (A), (B) and (D) only
(d) (B), (C) and (D) only

27 Match List-I with List-II.

	List-I	List-II
A	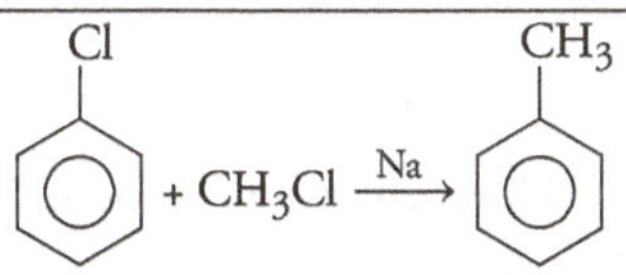	I. Fittig reaction
B	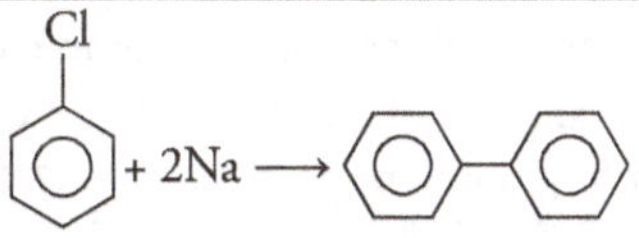	II. Wurtz Fittig reaction
C	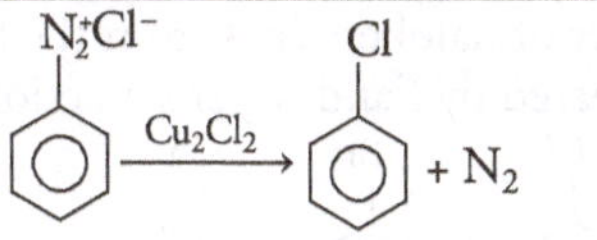	III. Finkelstein reaction
D	$C_2H_5Cl + NaI \longrightarrow C_2H_5I + NaCl$	IV. Sandmeyer reaction

Choose the correct answer from the options given below.
[2023, 30 Jan Shift-I]
(a) A-II, B-I, C-IV, D-III (b) A-III, B-II, C-IV, D-I
(c) A -II, B-I, C-III, D-IV (d) A-IV, B-II, C-III, D-I

28 Two isomers (*A*) and (*B*) with molar mass 184 g/mol and elemental composition *C*, 52.2%; H, 4.9% and Br, 42.9% gave benzoic acid and *p*-bromobenzoic acid, respectively on oxidation with $KMnO_4$. Isomer *A* is optically active and gives a pale yellow precipitate when warmed with alcoholic $AgNO_3$. Isomers *A* and *B* are, respectively **[2022, 29 June Shift-II]**

(a) $H_3C—CHBr—C_6H_5$ and 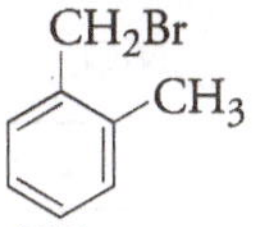

(b) 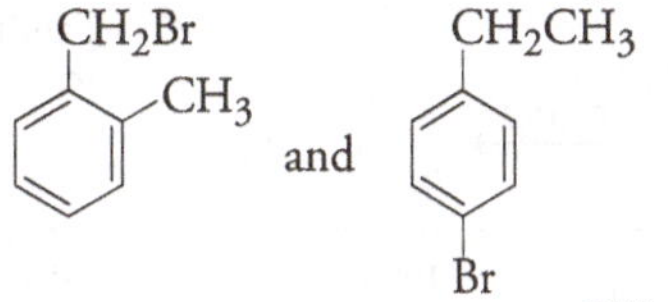

(c) $H_3C—CHBr—C_6H_5$ and 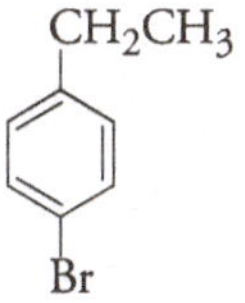

(d) 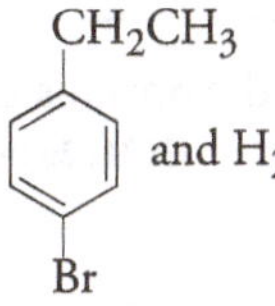 and $H_2C—CHBr—C_6H_5$

29 $A \xrightarrow[\text{(ii) } CN^- \text{ (iii) } H_2O/H^+]{\text{(i) } Cl_2, \Delta}$ 4-bromophenyl acetic acid.

In the above reaction '*A*' is **[2022, 28 June Shift-II]**

(a) 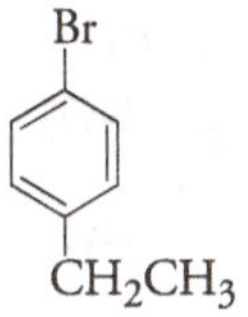 (b) 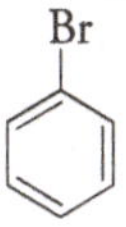

(c) 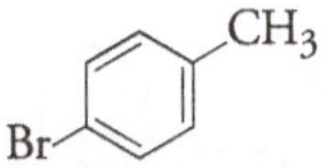 (d) 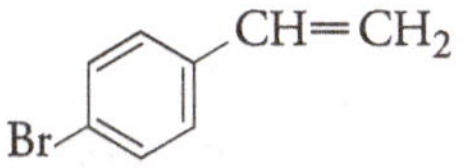

Imp. *Reagents that forms Yellow Precipitates with $AgNO_3$*

30 Among the following compounds I-IV, which one forms a yellow precipitate on reacting sequentially with
(i) NaOH (ii) dil. HNO_3
(iii) $AgNO_3$? **[2021, 26 Aug Shift-I]**

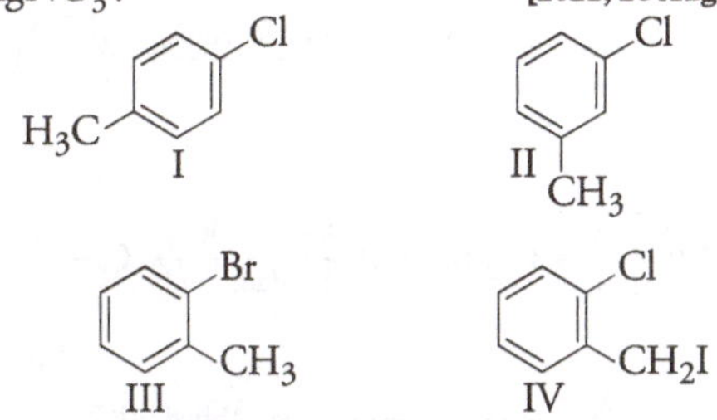

(a) II (b) IV (c) I (d) III

31 Match List-I with List-II.

List-I		List-II
A. 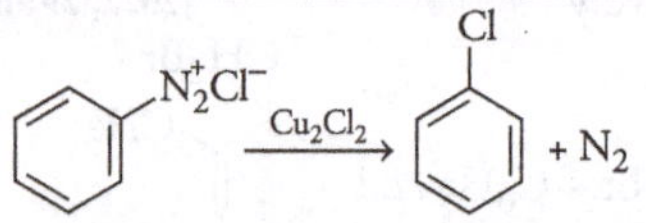	1.	Wurtz reaction
B. 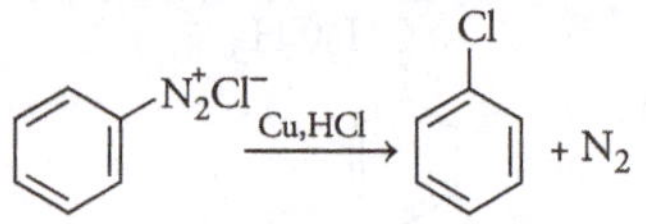	2.	Sandmeyer reaction
C. $2CH_3CH_2Cl + 2Na \xrightarrow{\text{Ether}} C_2H_5—C_2H_5 + 2NaCl$	3.	Fittig reaction
D. $2C_6H_5Cl + 2Na \xrightarrow{\text{Ether}} C_6H_5—C_6H_5 + 2NaCl$	4.	Gattermann reaction

Choose the correct answer from the options given below. **[2021, 26 Feb Shift-II]**

(a) A-3, B-1, C-4, D-2
(b) A-2, B-1, C-4, D-3
(c) A-2, B-4, C-1, D-3
(d) A-3, B-4, C-1, D-2

32 The decreasing order of reactivity of the following compounds towards nucleophilic substitution (S_N2) reaction is

(I) 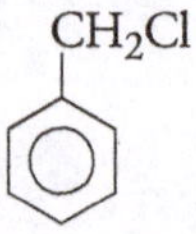

(II) 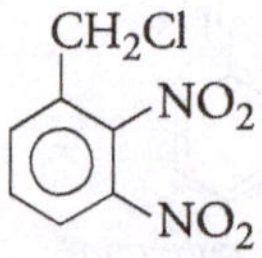

(III) 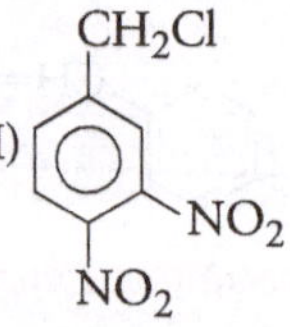

(IV) 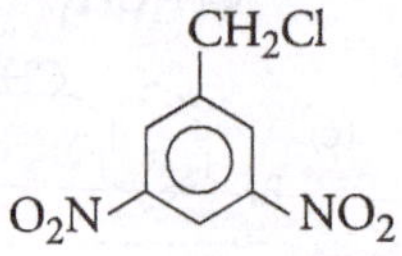

[2020, 3 Sep Shift-II]

(a) (II) > (III) > (IV) > (I)
(b) (IV) > (II) > (III) > (I)
(c) (III) > (II) > (IV) > (I)
(d) (II) > (III) > (I) > (IV)

33 Consider the following reactions :

(A) 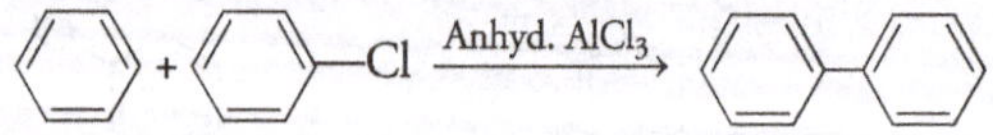

(B) 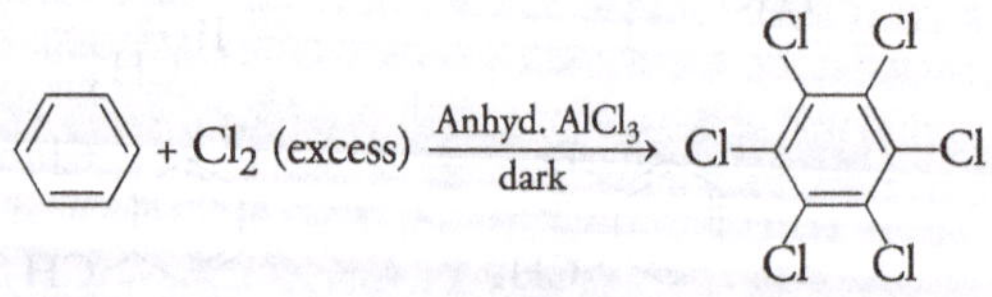

(C) 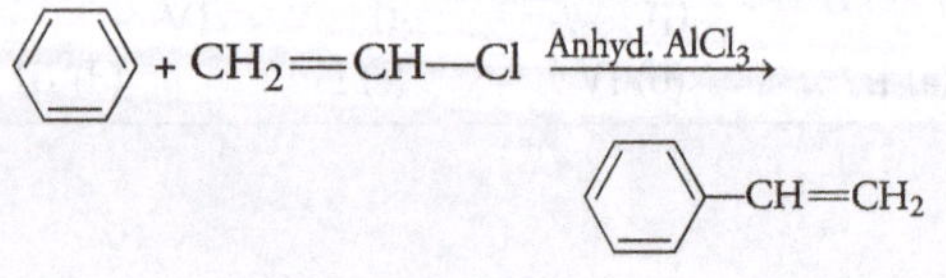

(D) 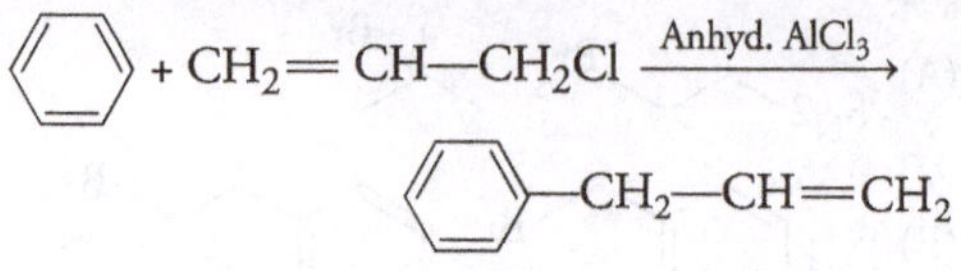

Which of these reactions are possible? **[2020, 7 Jan Shift-II]**

(a) (B) and (D)
(b) (A) and (D)
(c) (B), (C) and (D)
(d) (A) and (B)

34 Increasing rate of S_N1 reaction in the following compounds is **[2019, 10 April Shift-I]**

(A) 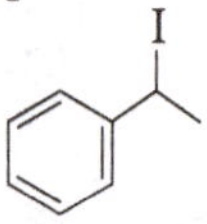

(B) 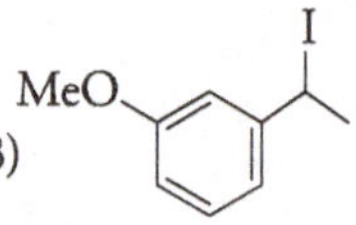

(C) 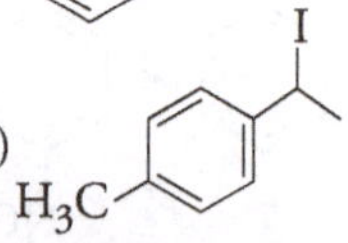

(D) 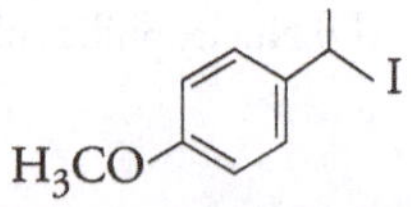

(a) (A) < (B) < (C) < (D)
(b) (B) < (A) < (C) < (D)
(c) (A) < (B) < (D) < (C)
(d) (B) < (A) < (D) < (C)

Numerical Value Type Questions

Imp. *Reaction of Chlorobenzene & Stoichiometry*

35 Consider the following sequence of reactions

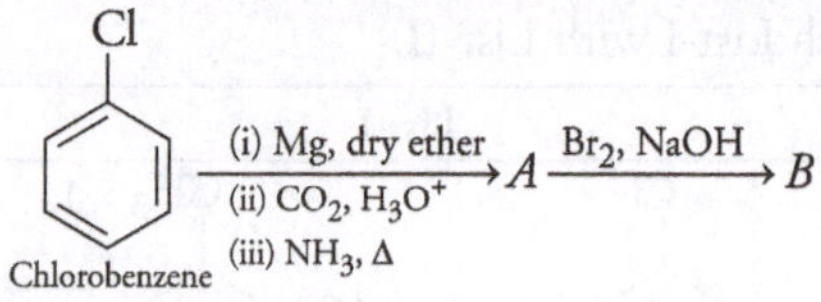

11.25 mg of chlorobenzene will produce .... $\times 10^{-1}$ mg of product *B*.

(Consider the reactions result in complete conversion.)

[Given, molar mass of C, H, O, N and Cl as 12, 1, 16, 14 and 35.5 g mol^{-1} respectively.] **[2025, 28 Jan Shift-I]**

36 The number of halobenzenes from the following that can be prepared by Sandmeyer's reaction is ....

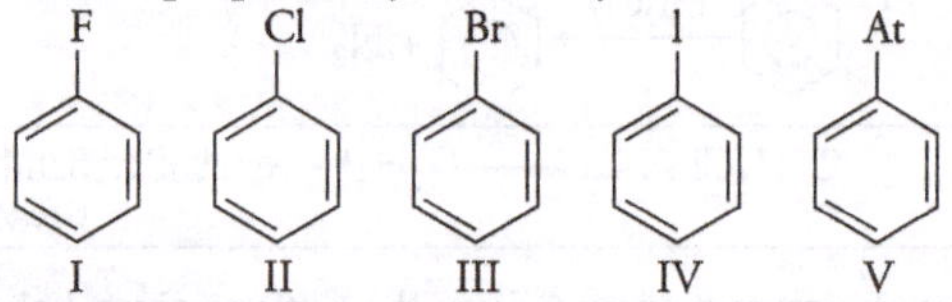

I II III IV V

[2024, 5 April Shift-I]

37 In the presence of sunlight, benzene reacts with Cl_2 to give product *X*. The number of hydrogens in *X* is .... **[2022, 26 July Shift-I]**

38 The major product of the following reaction contains .... bromine atom (s). **[2022, 28 June Shift-II]**

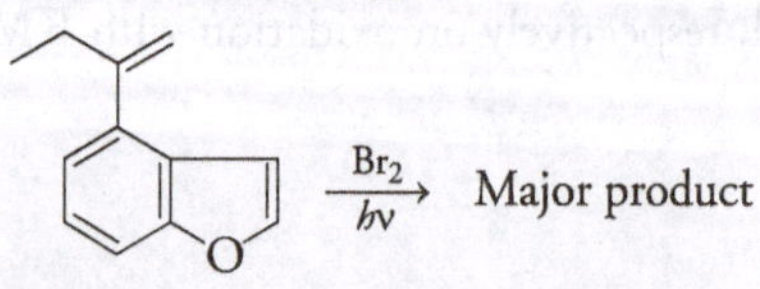

Topic 3 / Some Important Polyhalogen Compounds

• Level 1

01 In the following halogenated organic compounds the one with maximum number of chlorine atoms in its structure is **[2023, 31 Jan Shift-II]**

(a) gammaxene (b) chloropicrin
(c) freon-12 (d) chloral

• Level 2

02 In the following sequence of reaction, the major products *B* and *C* respectively are

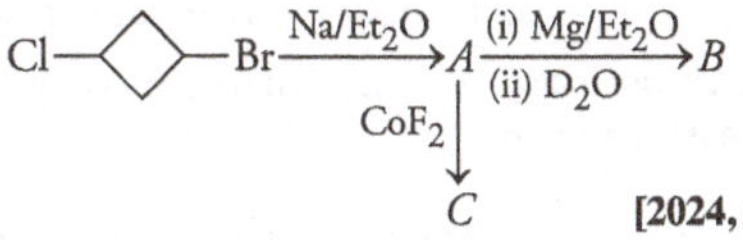

[2024, 9 April Shift-I]

(a) 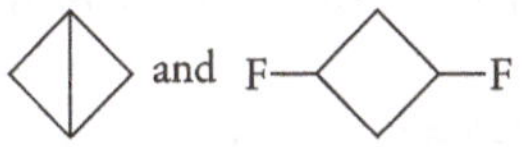

(b) 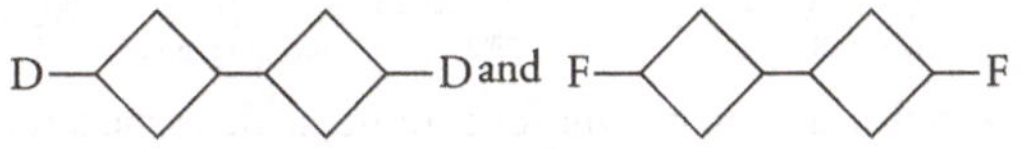

(c) 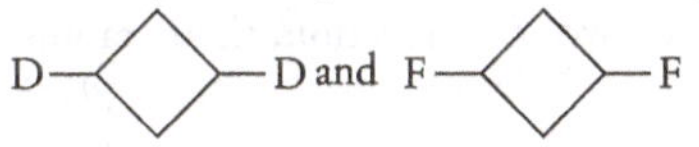

(d) 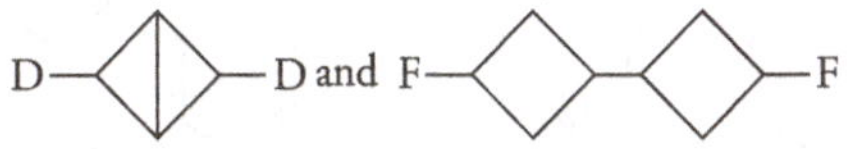

• Level 3

03 Match List-I with List-II.

	List-I (Compound)		List-II (Uses)
A.	Iodoform	I.	Fire extinguisher
B.	Carbon tetrachloride	II.	Insecticide
C.	CFC	III.	Antiseptic
D.	DDT	IV.	Refrigerants

Choose the correct answer from the options given below. **[2024, 6 April Shift-I]**

(a) A-II, B-IV, C-I, D-III (b) A-III, B-I, C-IV, D-II
(c) A-III, B-II, C-IV, D-I (d) A-I, B-II, C-III, D-IV

04 Match List-I with List-II

	List-I (Compound)		List-II (Use)
A.	Carbon tetrachloride	I.	Paint remover
B.	Methylene chloride	II.	Refrigerators and air conditioners
C.	DDT	III.	Fire extinguisher
D.	Freons	IV.	Non-biodegradable insecticide

Choose the correct answer from the options given below **[2024, 1 Feb Shift-II]**

(a) A-IV, B-III, C-II, D-I (b) A-III, B-I, C-IV, D-II
(c) A-II, B-III, C-I, D-IV (d) A-I, B-II, C-III, D-IV

Key Index

Topic 1

1	(a)	2	(d)	3	(b)	4	(d)	5	(c)	6	(a)	7	(c)	8	(d)	9	(a)	10	(a)
11	(c)	12	(c)	13	(d)	14	(d)	15	(b)	16	(a)	17	(a)	18	(d)	19	(c)	20	(b)
21	(d)	22	(b)	23	(b)	24	(c)	25	(c)	26	(d)	27	(c)	28	(c)	29	(c)	30	(c)
31	(a)	32	(d)	33	(a)	34	(c)	35	(b)	36	(a)	37	(a)	38	(b)	39	(a)	40	(d)
41	(b)	42	(c)	43	(c)	44	(c)	45	(c)	46	(c)	47	(c)	48	(b)	49	(a)	50	(d)
51	(c)	52	(b)	53	(d)	54	(a)	55	(d)	56	(c)	57	(c)	58	(b)	59	(c)	60	(b)
61	(d)	62	(c)	63	(d)	64	(b)	65	(c)	66	(d)	67	(b)	68	(b)	69	(a)	70	(c)
71	(c)	72	(1)	73	(32)	74	(6)	75	(2)	76	(10)	77	(5)						

Topic 2

1	(b)	2	(c)	3	(a)	4	(b)	5	(a)	6	(b)	7	(c)	8	(c)	9	(d)	10	(c)
11	(b)	12	(d)	13	(a)	14	(a)	15	(c)	16	(d)	17	(c)	18	(c)	19	(c)	20	(c)
21	(a)	22	(d)	23	(c)	24	(a)	25	(d)	26	(b)	27	(a)	28	(c)	29	(c)	30	(b)
31	(c)	32	(a)	33	(a)	34	(b)	35	(93)	36	(2)	37	(6)	38	(1)				

Topic 3

1	(a)	2	(b)	3	(b)	4	(b)

Detailed Explanations

Topic 1

01. *(a)* The reaction involved is as follows.

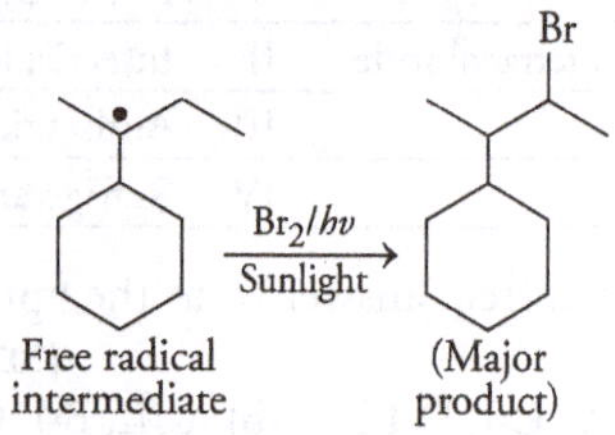

02. *(d)* For the structure.

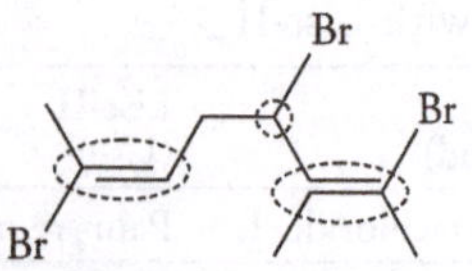

Total 3 stereocentres are there

So, total stereoisomers possible are $2^3 = 8$.

03. *(b)* The complete reaction takes place as follows

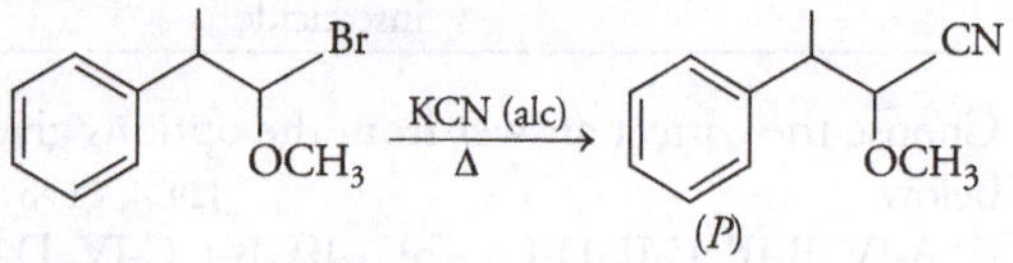

Nucleophilic substitution reaction will occur.

04. *(d)* The complete reaction is as follows

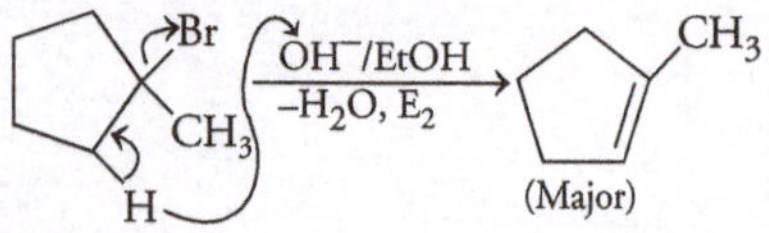

Hence, option (d) is correct.

05. *(c)* The complete reaction proceeds as follows

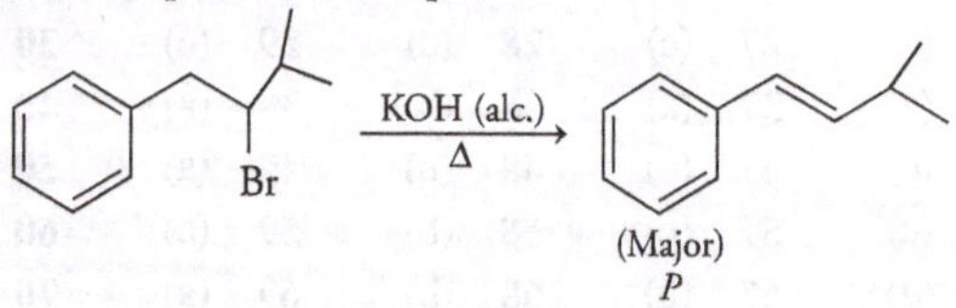

Here, more stable double bond will be formed, as double bond is in the resonance with the ring.

06. *(a)* The structure of 2, 3-dibromo-1-phenylpentane is

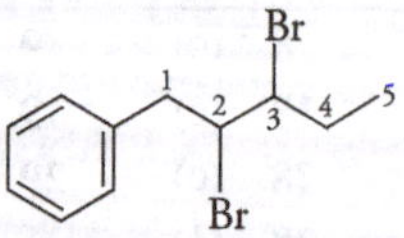

07. *(c)* Vinyl carbon is sp^2 hybridised aliphatic carbon.

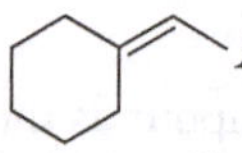 X is vinyl halide.

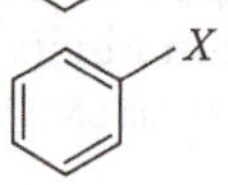 X is aryl halide.

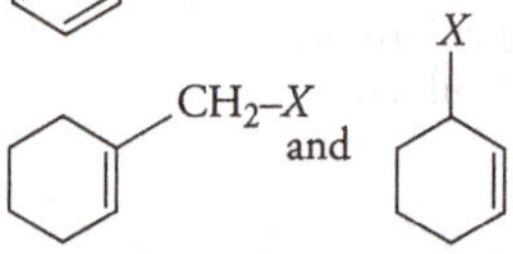

are allyl halide.

> **Concept Enhancer**
>
> Vinylic halides are the compounds in which the halogen atom is bonded to a sp^2 hybridised carbon atom of C ═ C. In such compounds, if lone pair is present, then they are always in conjugation with π-bond.

08. *(d)* Alkyl halide is converted into alkyl isocyanide by reaction with AgCN.

$$\underset{\text{Alkyl halide}}{R\text{—}X} + \text{AgCN} \xrightarrow[\text{reaction}]{\text{Substitution}} \underset{\text{Alkyl isocyanide}}{R\text{—NC}} + \text{Ag}X$$

09. *(a)* Substrate which will form most stable carbocation will be highly reactive towards S_N1 reaction, their stability orders.

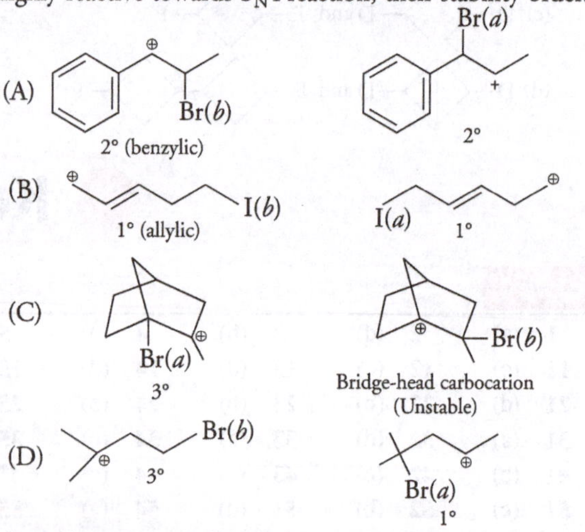

So, the carbocation resulted from in option (a)

10. *(a)* Both A and R are true and R is the correct explanation of A.

> **Concept Enhancer**
>
> When alkyl chloride undergoes hydrolysis, the reaction is slow because Cl^- is a poor leaving group, hence the rate of S_N2 reaction is slow. But in presence of NaI, $I^{\ominus}$ is a good leaving group that facilitates the hydrolysis. Hence, rate of hydrolysis increases.

11. *(c)* The given reaction proceeds in two steps:

Step 1. Addition of H^+ or attack of alkene on H^+ (electrophilic addition)

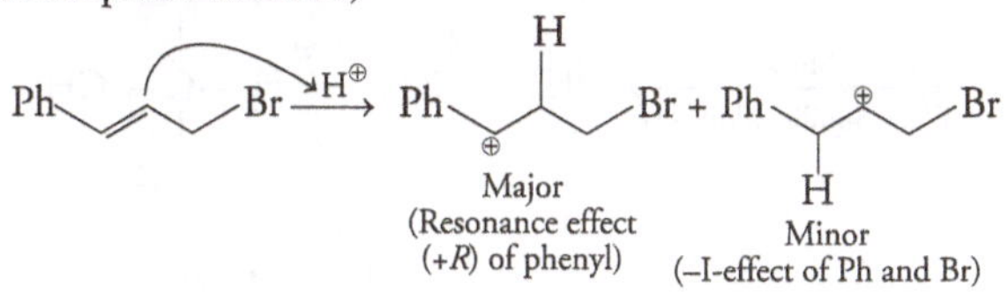

Step 2. Addition of Br^- to carbocation.

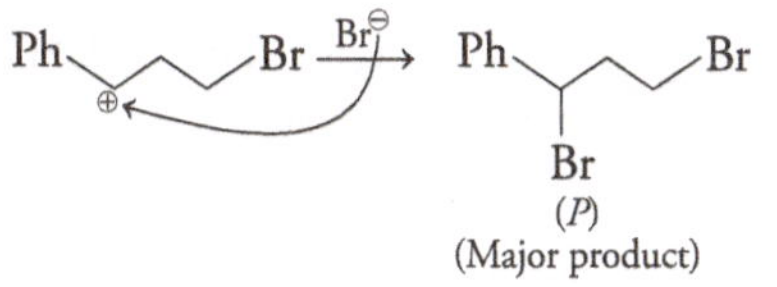

12. *(c)* Rate of S_N1 reaction depends on stability of carbocation.

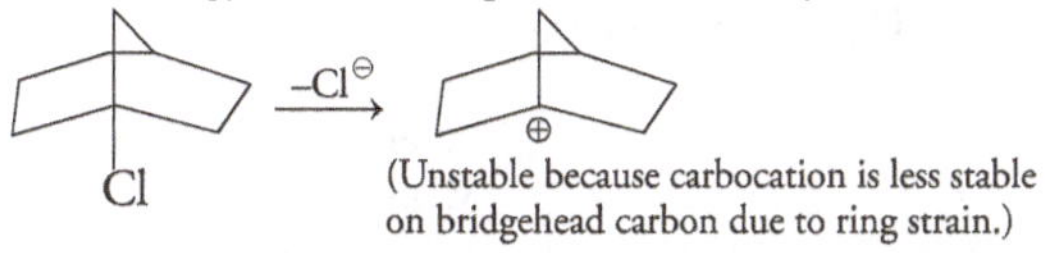

Hence, compound of option (c) is inactive towards S_N1 reaction. However, carbocation resultant from other options are stabilised by + *R*-effect and hyperconjugation.

13. *(d)* In 3° alkyl halides E2 elimination will dominate with most nucleophiles (even if they are weak bases). No S_N2 substitution take place due to sterically hindered butoxide base.

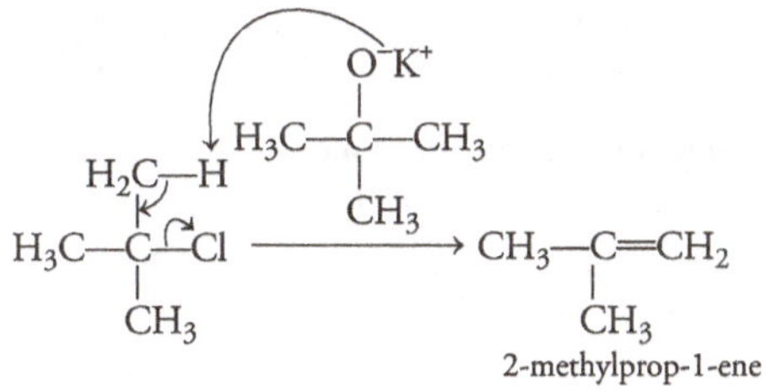

14. *(d)* In the given reaction E2 elimination reaction takes place in which two substituents are removed from a molecule to form double bond (alkene conjugate diene).

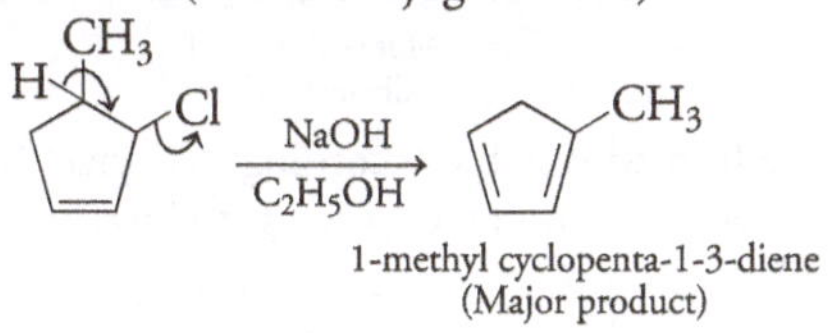

15. *(b)* On reaction of haloalkane with methoxide ion, alkene is formed.

Mechanism

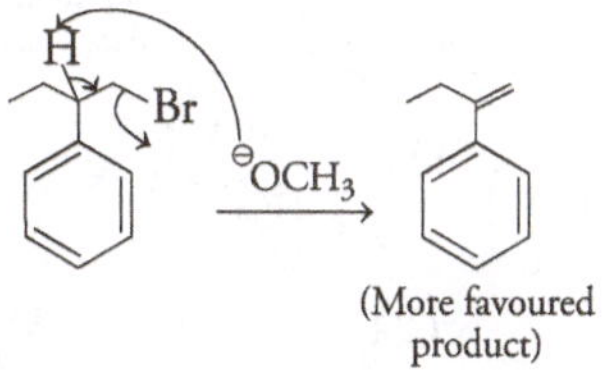

- $\overset{\ominus}{O}CH_3$ will act as a base due to its small size and high electron density and therefore, abstracts proton to form double bond which is in conjugation with aromatic ring.

16. *(a)* In the first step, Grignard reagent is obtained. In the second step, acid-base reaction occurs. Due to presence of an acidic hydrogen in alcohol, neutralisation reaction takes place that produces alkane and water.

The reaction is

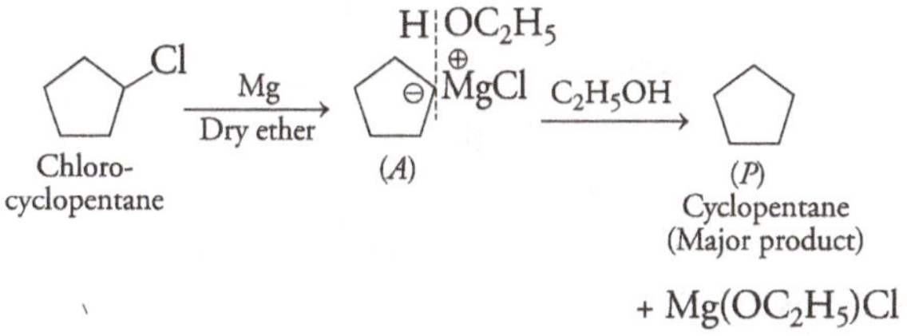

17. *(a)*

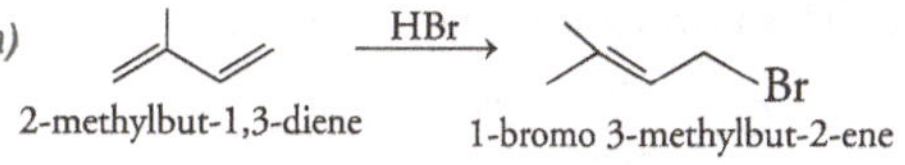

This addition is known as 1, 4-addition.

Mechanism This reaction can undergoes two pathways

Path *A*

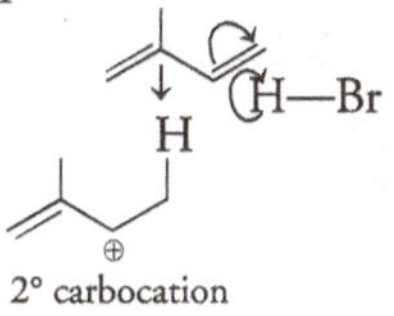

Path *B*

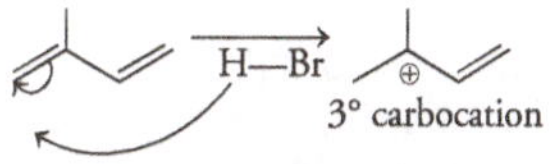

Out of the two intermediates formed in two paths, the path *B* intermediate is more stable as it has more stable carbocation. So, major product will be formed because of path *B*.

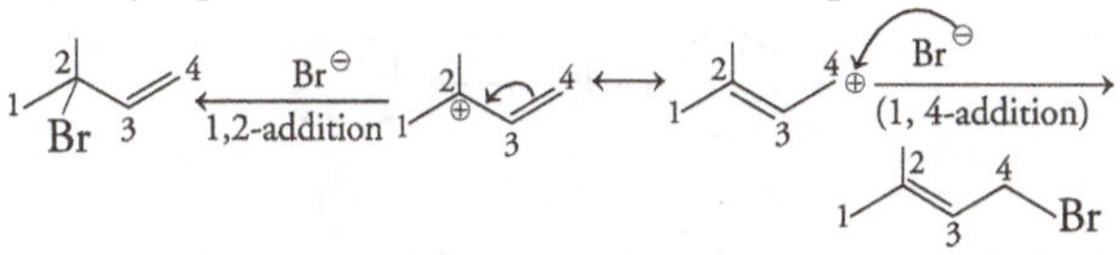

1, 2-addition product is formed at low temperature and will be less stable as double bond is less substituted. 1, 4-addition is thermodynamically stable product as double bond is more substituted.

18. *(d)* Alkene undergoes electrophilic addition with HBr to give a major product.

Step 1. Addition of electrophile to form carbocation.

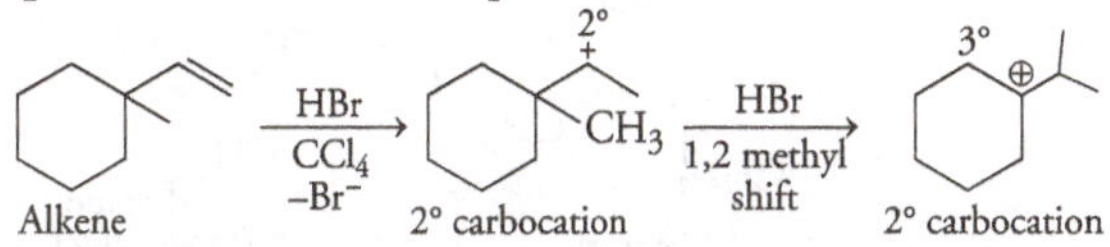

2° carbocation undergoes α-methyl shift to form more stable 3° carbocation.

Step 2. Attack of nucleophile on carbocation.

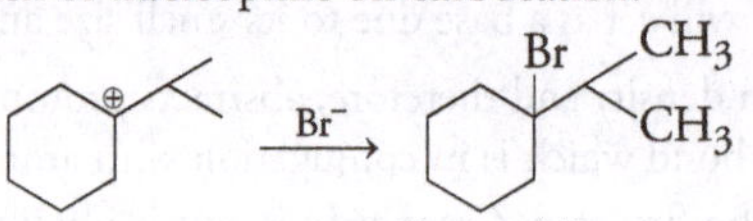

Product (*A*)

19. *(c)* The reaction can be achieved by Cl_2/UV light (*A*).
Given, reaction follows free radical substitution mechanism. (allylic)

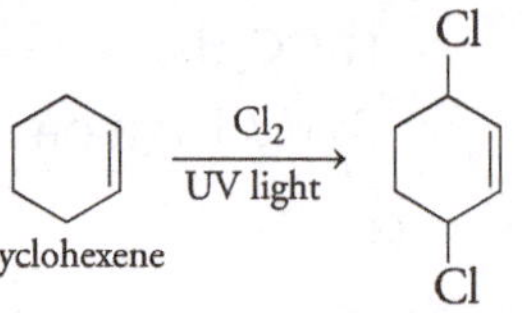

Cyclohexene 1, 4-dichloro cyclohexene

Mechanism

(I) Initiation This involves formation of radicals.

$$Cl_2 \xrightarrow[\text{light}]{\text{UV}} 2Cl^{\bullet}$$

(II) Propagation : Reactive free radicals react with stable molecules to form new free radicals.

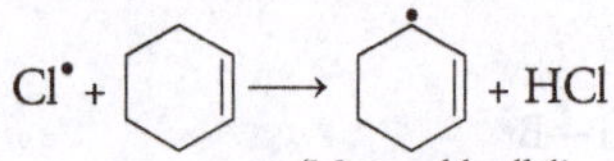

(More stable allylic free radical is formed)

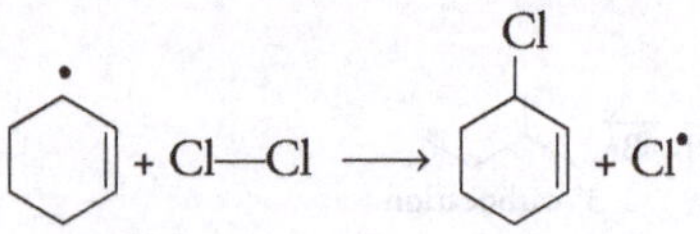

Similarly second Cl atom will also be attached at another allylic position.

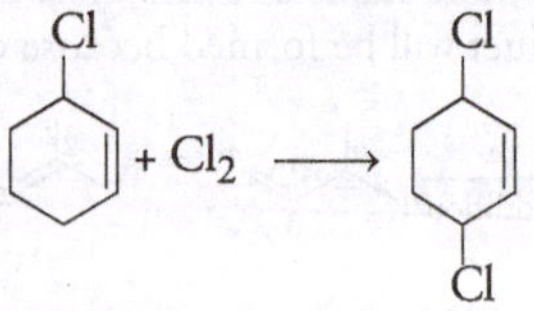

20. *(b)* The major product formed is 2-iodo-2, 3-dimethylbutane.
Mechanism Involved in the reaction is as follows

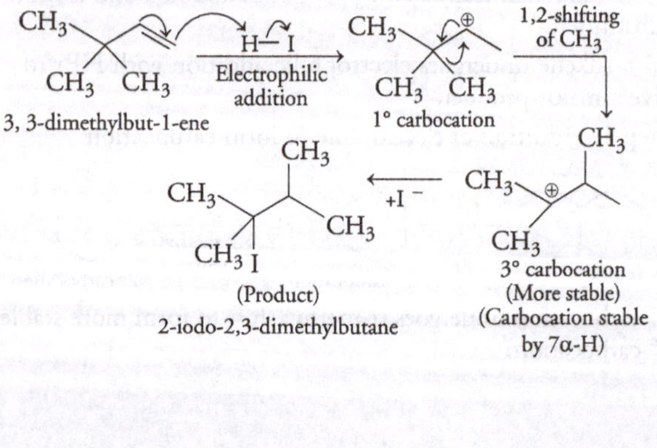

21. *(d)* $CH_3CH{=}CHCH(CH_3)_2 \xrightarrow[-Br^-]{HBr}$

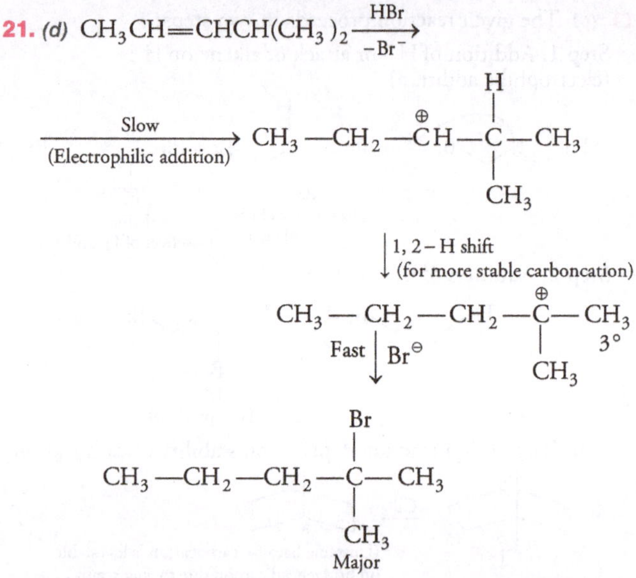

22. *(b)* In option (b) vinyl chloride, C—Cl bond is shortest due to resonance of lone pair of — Cl (chlorine atom) (+*R*).
Due to resonance C—Cl bond aquire partial double bond character.
Hence, vinyl chloride have shortest C—Cl bond length.

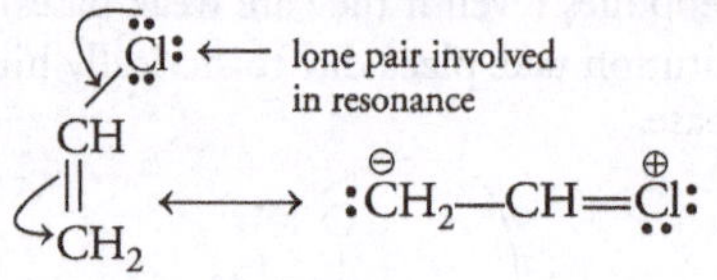

23. *(b)* Complete reaction is as follows

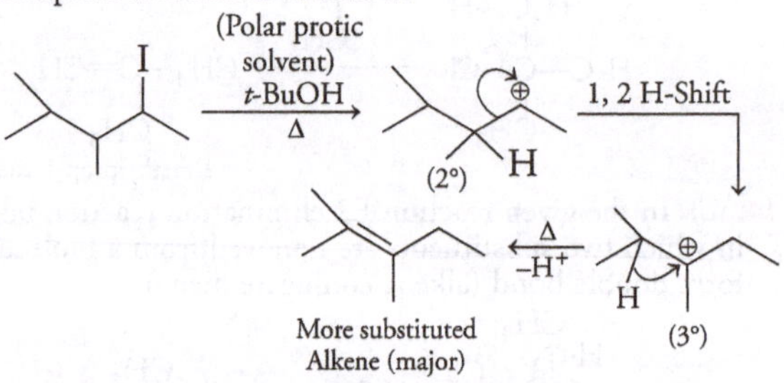

The alkene formed is stable according to Saytzeff's rule.

24. *(c)* The given reaction takes place as follows

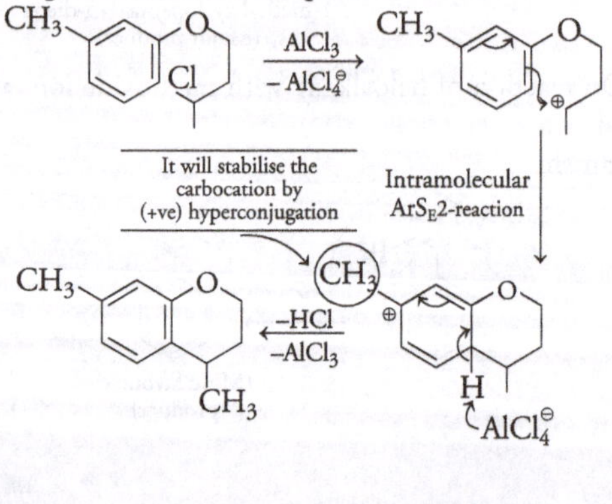

25. *(c)* In the given question, the substrate is a 2°-halide (bromide) and the medium, CH_3OH (as well as a poor nucleophile) is protic in nature, So, the reaction will follow mainly S_N1 pathways via the formation of a carbocation intermediate (I).

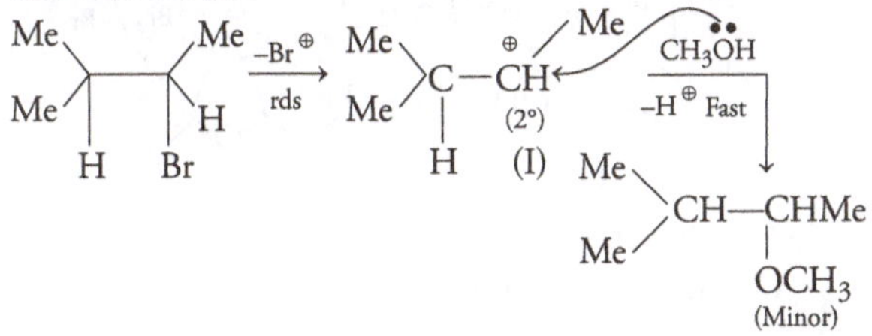

The intermediate, I can be rearranged into the more stable form I′ (3°) by α-hydride shift.

I′ will give the major product.

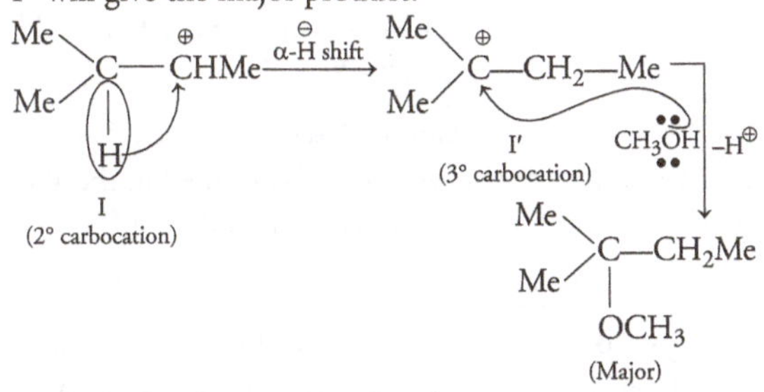

26. *(d)*

> **idea** The excess of HBr and high temperature in given reaction serves for dual purpose:
> (i) Hydrolysis of ether *via* S_N2 mechanism.
> (ii) Markownikoff addition at double bond of the branch.

The complete reaction is as follows:

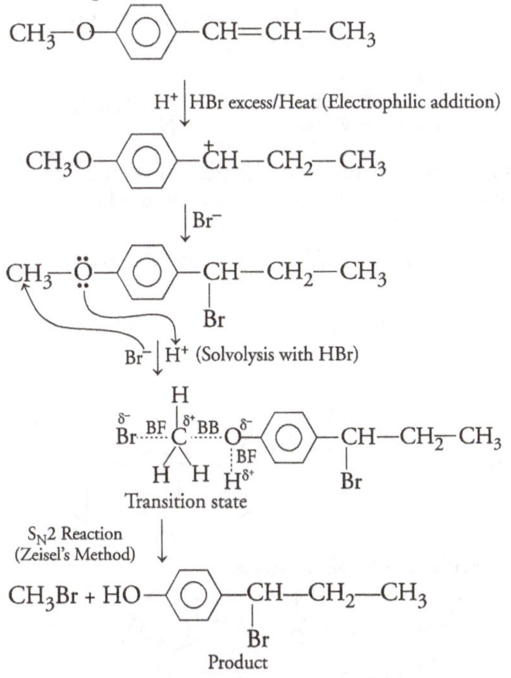

Here, BB = Bond breakage BF = Bond formation

27. *(c)* The compound (*E*) has two allyl-hydrogen atoms (γ). When *E* reacts with Br_2 / *hν*, it readily undergoes allylic free radical substitution and forms 3, 3-dibromobut-1-ene

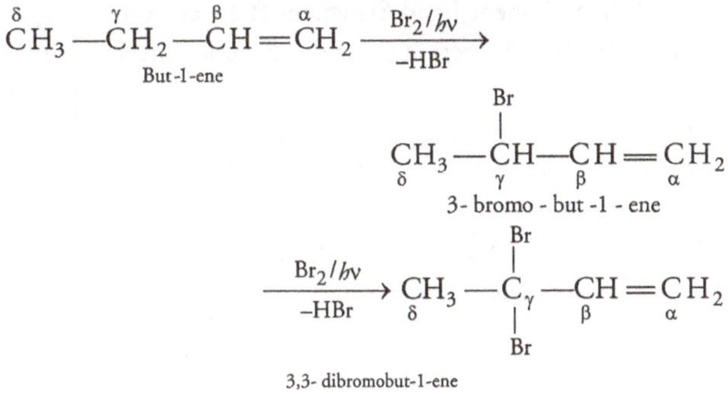

28. *(c)* Statement I is correct but Statement II is incorrect.

(i) 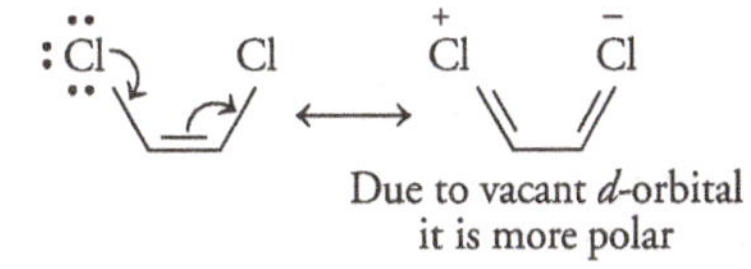

(ii) 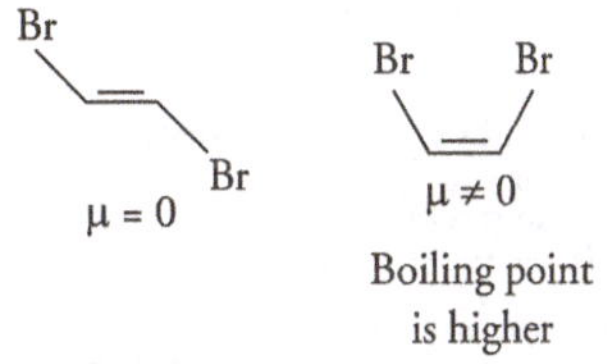

29. *(c)* The complete reaction is as follows

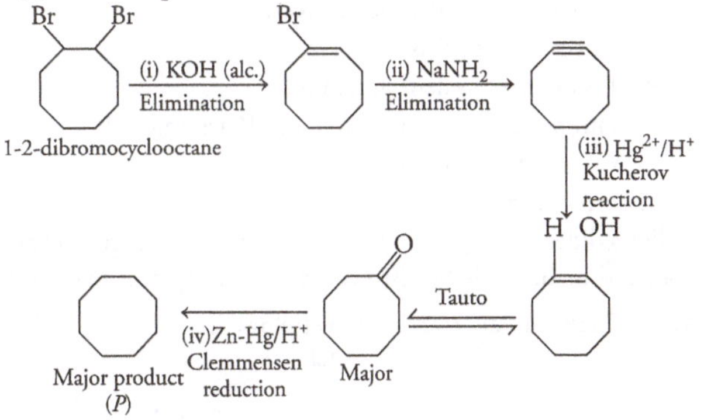

30. *(c)* According to question, the product 'x' formed is

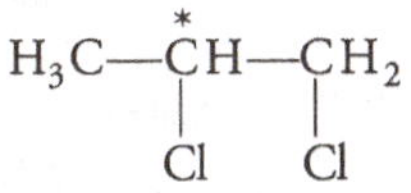

So, reaction of 'x' with chlorine under the photochemical condition is as follows

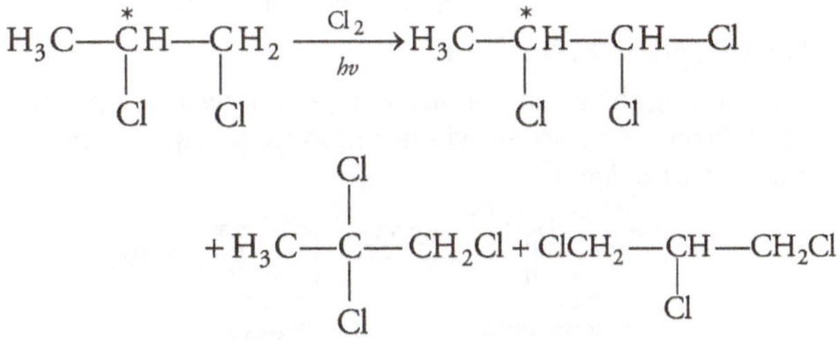

So, total '3' tri-chloro products are obtained.

31. *(a)* Both Statement I and Statement II are true.

32. *(d)* Both Statement I and Statement II are correct.

33. *(a)* Both Statement I and Statement II are correct.

34. *(c)* Both Statement I and Statement II are correct.

35. *(b)* Rate of S_N2 reaction depends on steric hindrance, less the steric hindrance more will be rate of reaction

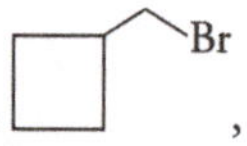

, it will undergo S_N2 reaction fastest.

Rate of S_N2 reaction is

$Me — X > 1° — X > 2° — X > 3° — X$

36. *(a)* The complete reaction is as follow,

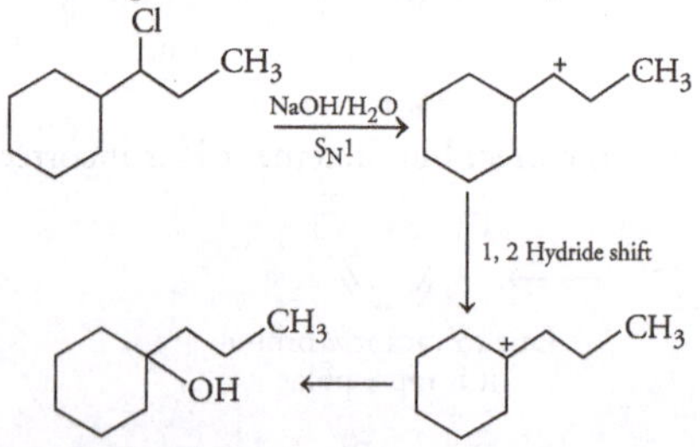

Product *A*

37. *(a)* The complete reaction takes place as follows

$$CH_3—CH_2—CH_2—Br \xrightarrow[80°C\ (X)]{\text{Conc. alc. NaOH}}$$

$$CH_3—CH{=}CH_2 \xrightarrow{\text{HBr/Acetic acid } (Y)} CH_3—CHBr—CH_3$$

38. *(b)* A is correct but R is not correct.

The correct form of R is :

AgCN is mainly covalent in nature.

39. *(a)* Both Statement I and Statement II are true.

40. *(d)* A is true but R is false.

The correct form of R is :

Allyl halides are the compounds in which the halogen atom is bonded to an sp^3 hybridised carbon atom adjacent to C$=$C.

41. *(b)* The complete reaction is as follow

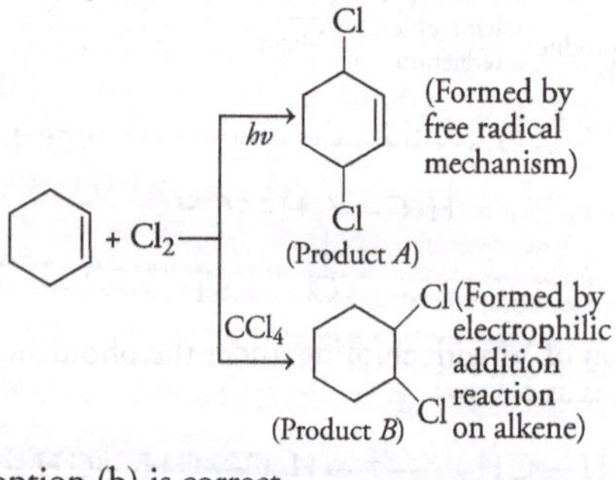

Hence, option (b) is correct.

42. *(c)* Among the given options, compound given in option (c) i.e. 1-bromocyclohexene will not produce precipitate on reaction with $AgNO_3$.

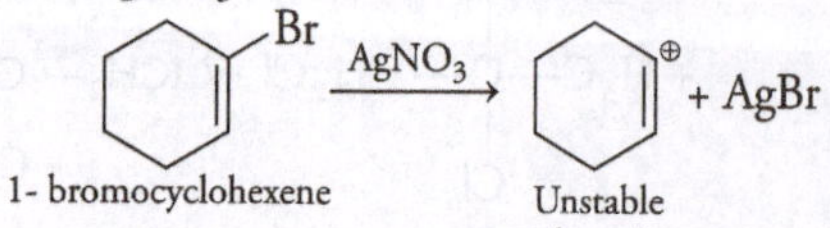

43. *(c)* The complete reaction is as follows.

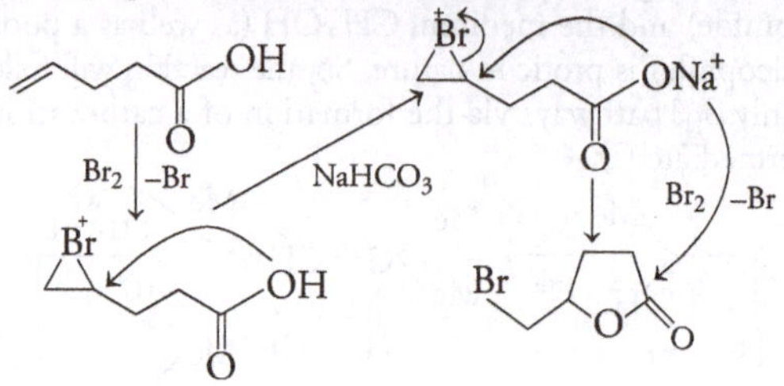

44. *(c)* For the reaction

$$RCH_2—Br + I^- \xrightarrow{\text{Liquid acetone}}$$

Finkelstein reaction (halogen transfer reaction)

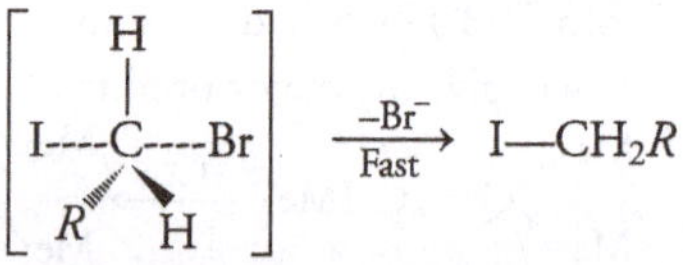

Penta-coordinated TS

During the reaction, transition state (TS) formed in the above reaction is less polar than the localised anion Br^- or I^-.

45. *(c)*

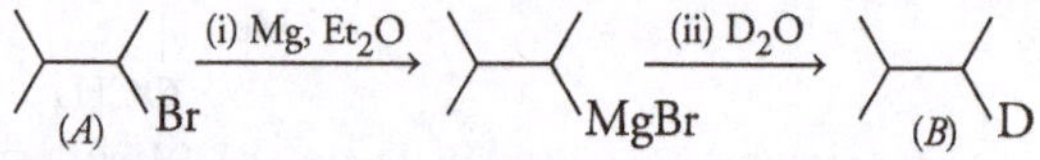

Compound (*A*) reacts with magnesium to form Grignard reagent which undergoes acid-base reaction with D_2O to form compound *B*.

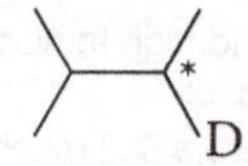

Compound *B* has one chiral centre (*) therefore, it is chiral in nature.

46. *(c)* NaCN is ionic and provides cyanide ion (CN^-) in solution. The attack (nucleophilic) takes place mainly through carbon atom and not through nitrogen atom.

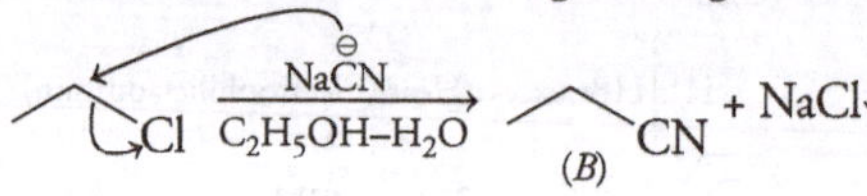

AgCN is covalent in nature and nitrogen is free to donate electron pair forming isocyanide as the main product.

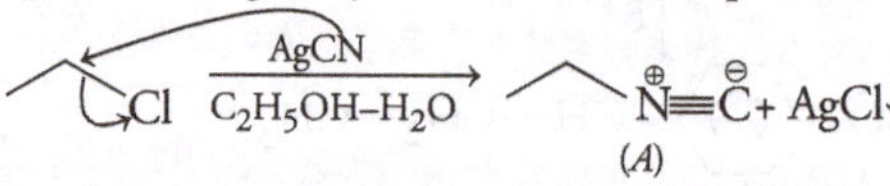

47. *(c)* The major product formed is given in option (c).

Mechanism

1. Formation of carbocation

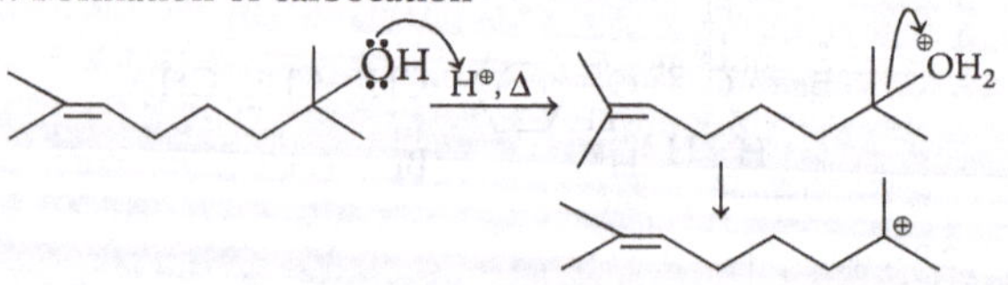

2. Intramolecular nucleophilic attack

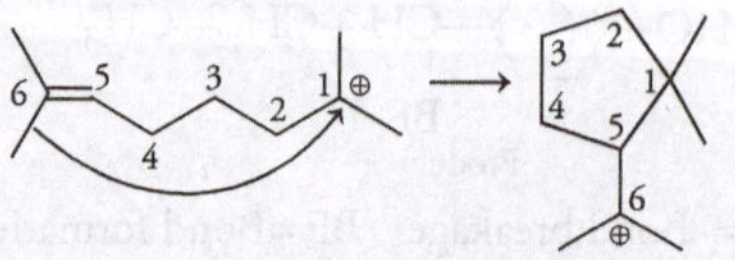

3. Nucleophilic attack of $Br^{\ominus}$

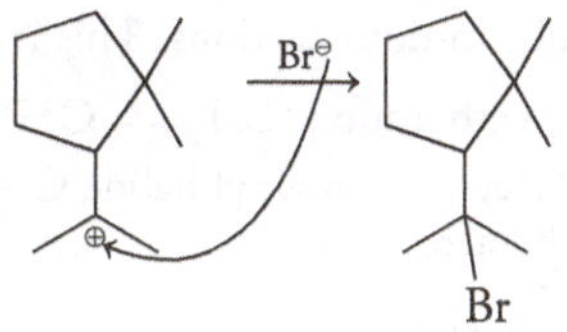

48. *(b)* **Step I** Addition of CH_3O^- and Br^- on unsaturated part of alkene.

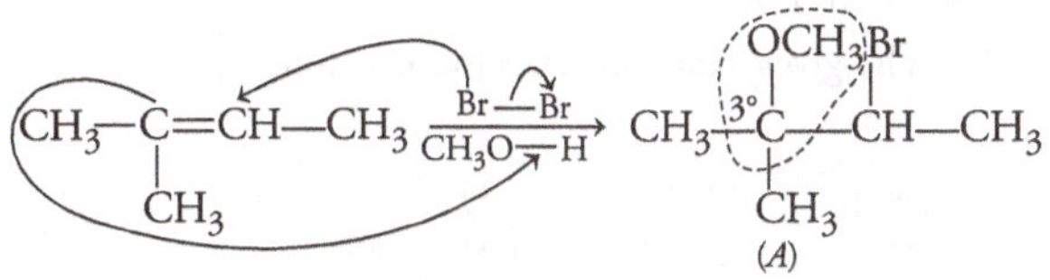

Step II S_N1 solvolysis of the ether with HI (1 eqv.) follows on

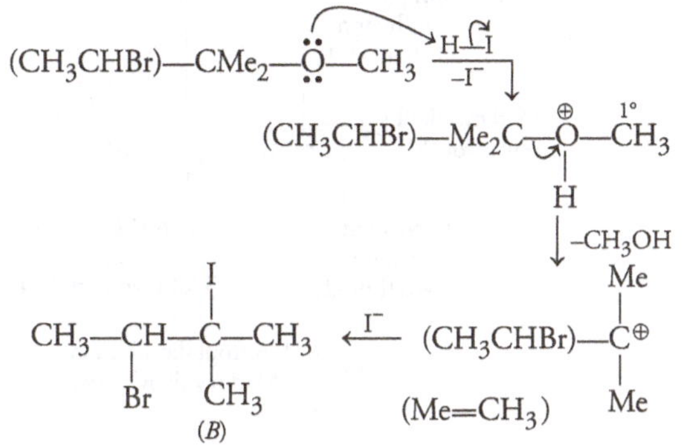

49. *(a)* The correct order of reactivity of chlorides with acetate in acetic acid is

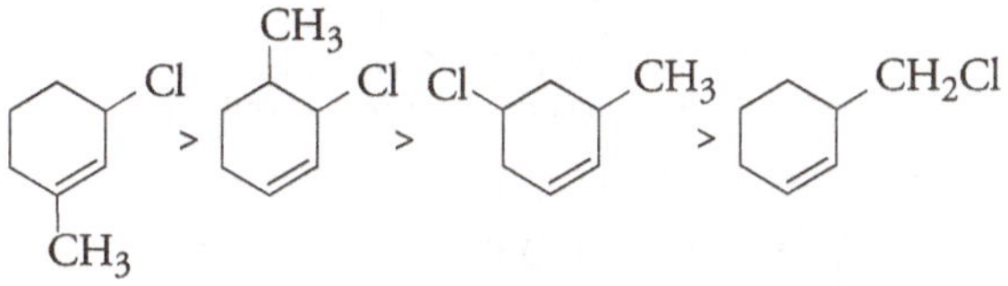

The given chlorides undergoes S_N1 reaction. So, as the stability of carbocation formed increases, rate of reaction increases. Stability of carbocation is as follows

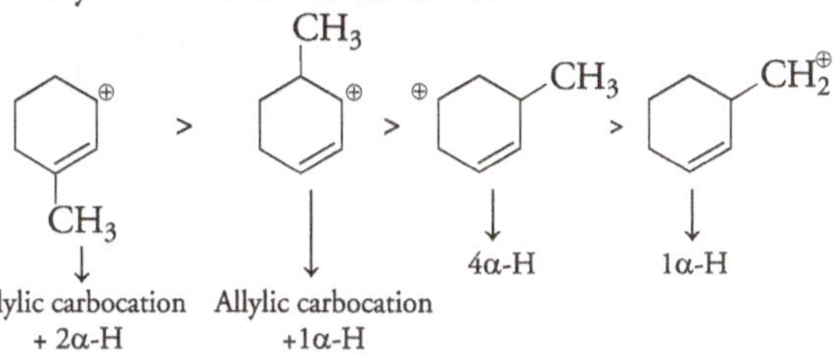

Hence, correct option is (a).

50. *(d)* Excess of isobutane reacts with Br_2 in presence of light at 125°C gives 2-bromo-2-methyl propane as major product.

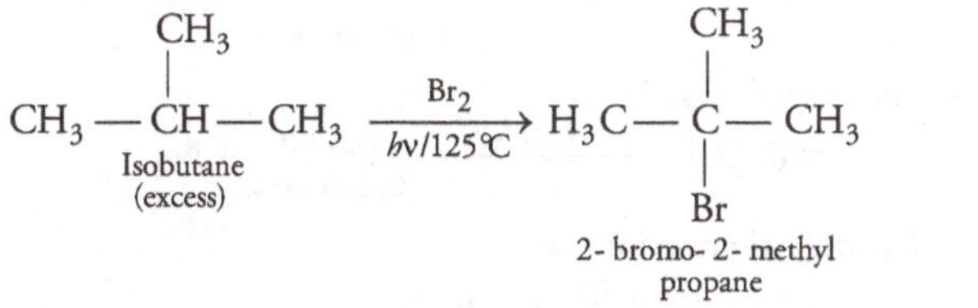

Mechanism The above reaction is halogenation of alkane *via* free radical substitution reaction.

Initiation step

$$Br_2 \xrightarrow{h\nu} Br^{\bullet} + Br^{\bullet}$$

Propagation step

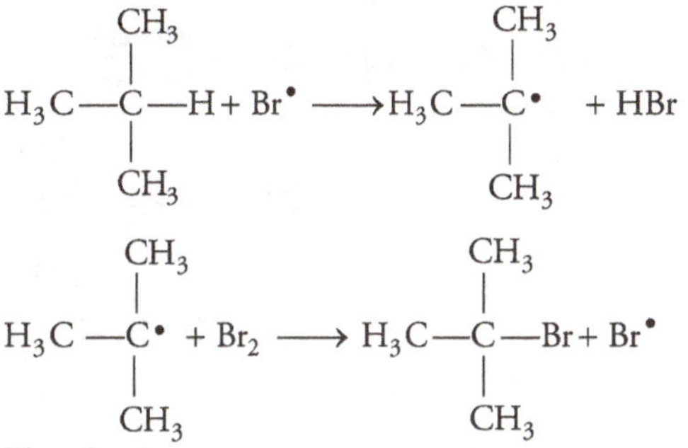

Termination step

$$Br^{\bullet} + Br^{\bullet} \longrightarrow Br_2$$

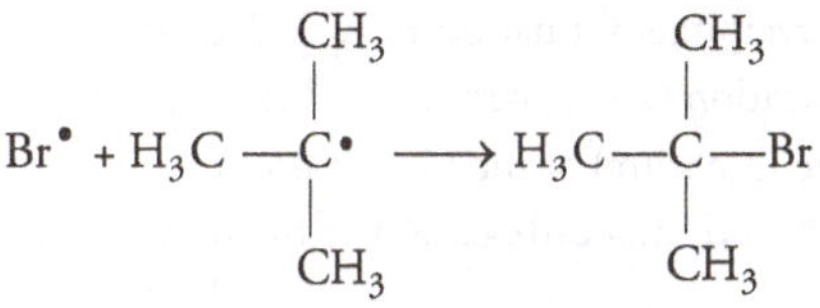

(Major product)

Since, isobutane is in excess, so dibromination of single isobutane is not favourable reaction. This make (a) and (b) incorrect options.

51. *(c)* Alcohols undergo dehydration (removal of water molecule) in the presence of protic acids (conc. H_2SO_4, H_3PO_4) at high temperature to give Saytzeff alkene (more alkylated) as major product.

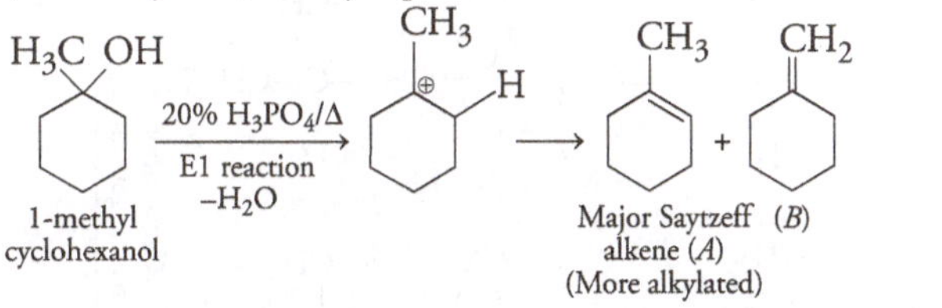

Bulky bases like *tert*-butoxide, due to steric hindrance, a abstracts a less hindered β-proton, that leads to formation of the less substituted product as, the major one, called the Hofmann product.

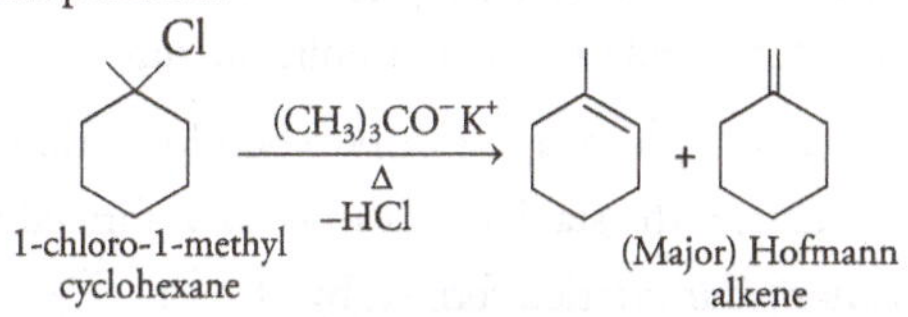

52. *(b)*

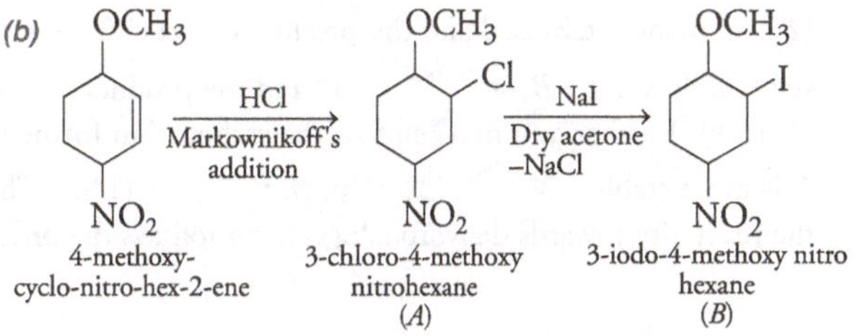

Ist reaction is electrophilic addition of HCl on double bond while 2nd reaction is halide substitution by Finkelstein reaction in which chlorine get displaced by iodine.

53. *(d)* Here, only one equiv. of Mg/ Et_2O attacks on one bromine.

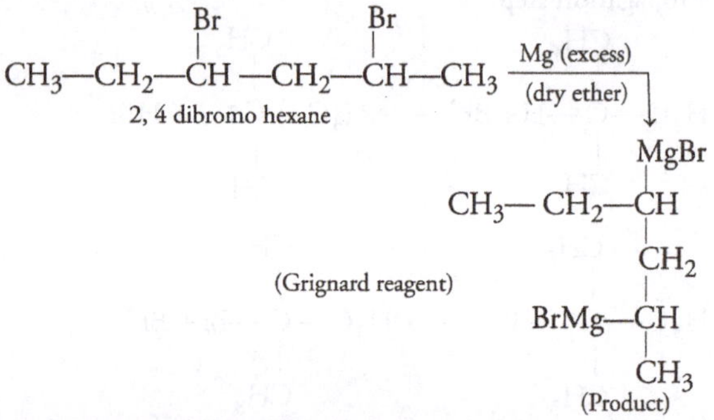

54. *(a)* All given species in the question are alkyl halide. They react with aqueous $AgNO_3$ solution with different rate. Hence, rate of precipitate formation of Ag*X* depends on stability of carbocation (R^+), obtained by heterolysis of the most polar (ionic in nature) of the C—Br bond

Among the given option(s) carbocation of the bromide (a) is most stable due to stronger + *M* effect, so 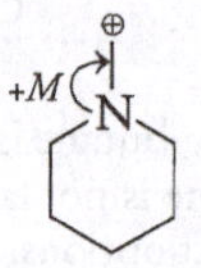 give precipitate with aqueous $AgNO_3$ solution most readily.

55. *(d)* Reaction (d) will not produce a racemic product whereas other reactions will form racemic mixture.

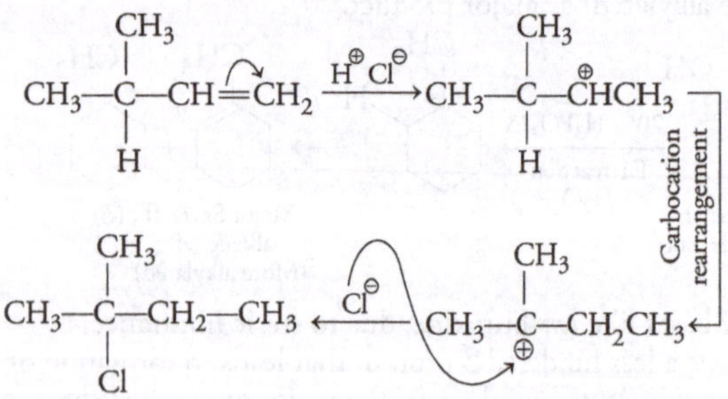

There is no chiral carbon atom present in the product (major). Thus, product would not be a racemic mixture.

56. *(c)* E_1 reaction proceeds *via* carbocation formation, therefore greater the stability of carbocation formed faster is the reaction. Carbocation formed by *D*, 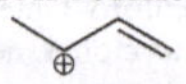 (2°, resonance stabilised and the product is diene) is most stable followed by *B*, 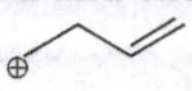 (1° and the product formed is diene). Further among C and *A*, the carbocation formed by *C* is more stable 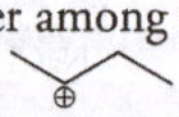 (2°) than *A*,  (1°). Thus, the reactivity towards dehydrohalogenation follows the order :

$$D > B > C > A$$

57. *(c)* Vinyl halide ($CH_2=CH-Cl$)do not undergo nucleophilic substitution reactions. This is because it forms highly unstable carbocation ($CH_2=\overset{\oplus}{C}H$). It cannot delocalise its π-electron. In vinyl halide C—Cl bond possess double bond character also.

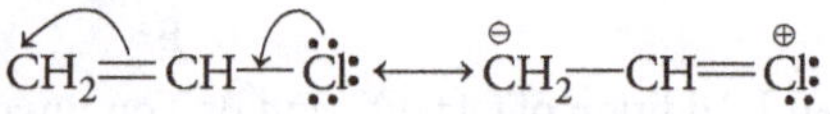

Thus, both A and R are correct but R is not the correct explanation of A.

58. *(b)* The given reaction takes place as follows

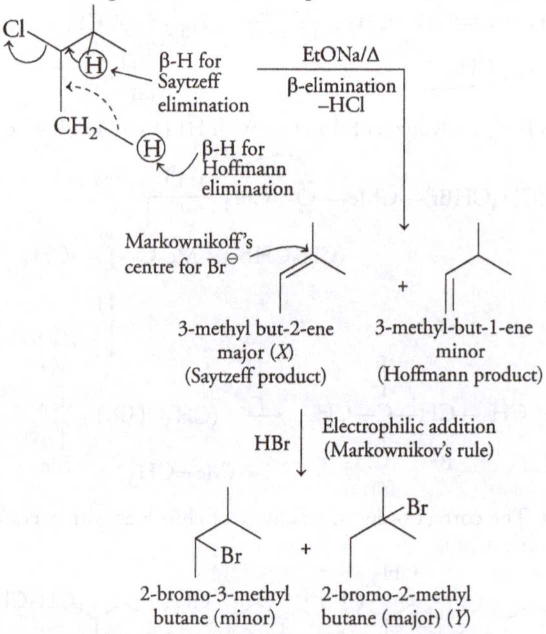

59. *(c)* Complete reaction sequence is as follows

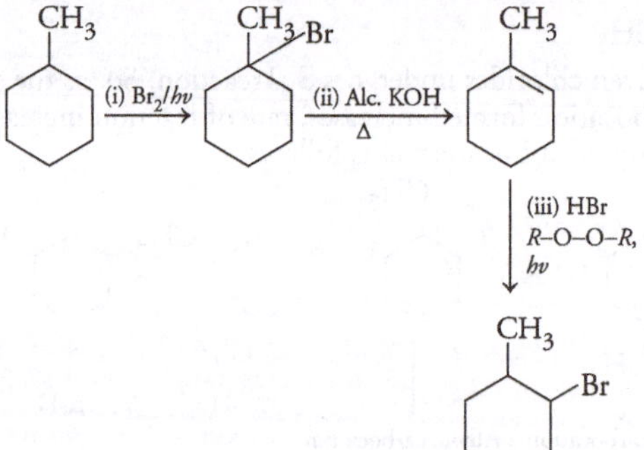

60. *(b)* The correct match is A-IV, B-III, C-I, D-II.

- Swarts reaction →

$$C_2H_5-Br + AgF \longrightarrow \underset{\text{Ethyl fluoride}}{C_2H_5-F} + AgBr$$

- Sandmeyer's reaction →

$$PhN_2^+Cl^- \xrightarrow{CuCN/KCN} \underset{\text{Cyanobenzene}}{PhCN} + N_2$$

- Wurtz-Fittig reaction →

$$Ph-Cl + EtCl \xrightarrow[\text{Ether}]{Na} \underset{\text{Ethyl benzene}}{Ph\text{-}Et}$$

- Finkelstein reaction →

$$\text{Et-Cl} \xrightarrow[\text{Acetone}]{\text{NaI}} \underset{\text{Ethyl iodide}}{\text{Et-I}} + \text{NaCl}$$

61. *(d)* Rate of solvolysis depends on stability of carbocation. As the stability of carbocation increases the rate of solvolysis also increases. The order of stability of the carbocations of the given compounds is

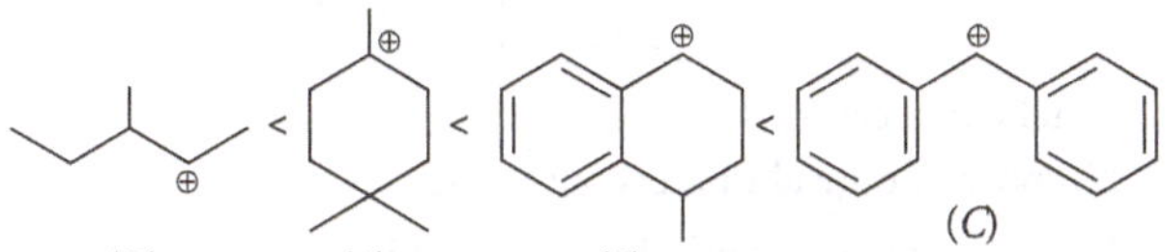

(D) (A) (B) (C)

So, the order of solvolysis is $D < A < B < C$.

62. *(c)* For the given reaction

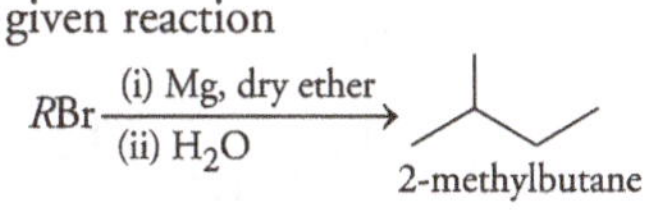

The maximum number of *R*Br are

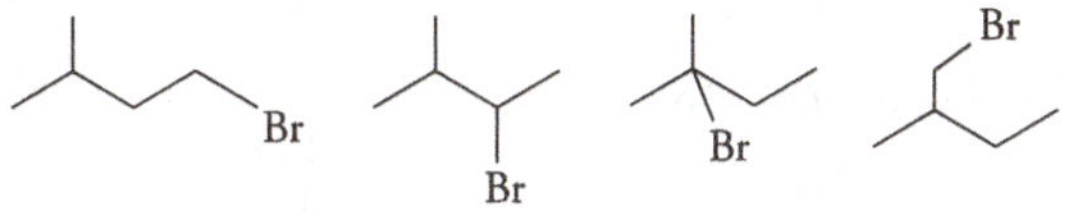

Thus, total 4 possible structures are possible.

63. *(d)* The complete reaction sequence is as follows

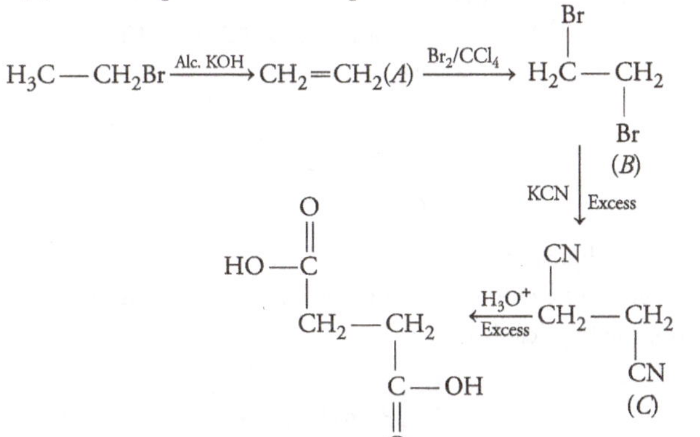

(D) Succinic acid

64. *(b)*

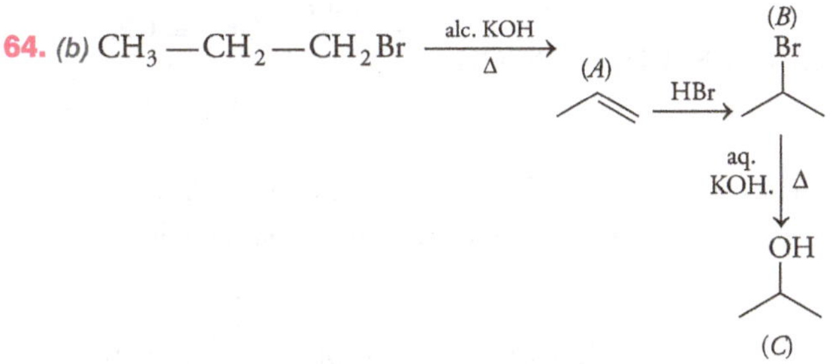

Propan 2- ol

65. *(c)* Reaction given in option (c) is correct while reactions in (a), (b) and (d) are incorrect. According to Markownikoff's rule when reagent adds to an unsymmetrical alkene, the more positive part of the reagent goes to the carbon that have more number of hydrogen atoms.

Reaction in option (c) is an example of Markownikoff addition reaction.

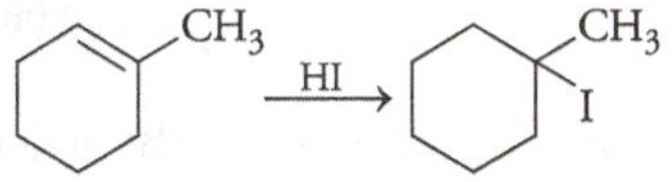

Markownikoff's addition

66. *(d)* Reaction (I) *via* S_N1 mechanism which involves the formation of an intermediate. It is a 1st order reaction.

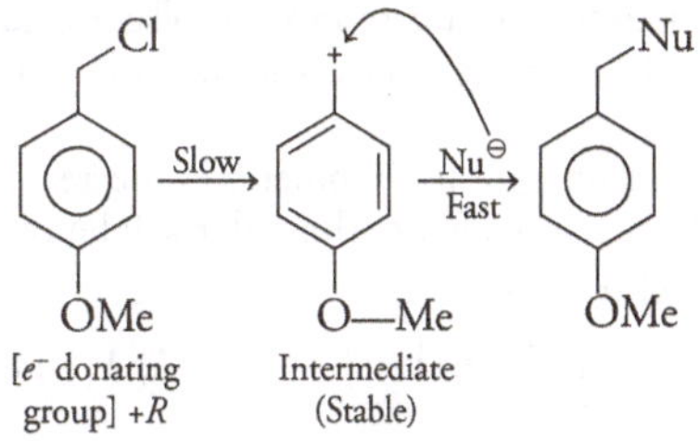

Reaction (II) proceeds *via* S_N2 mechanism. It is a 2nd order reaction.

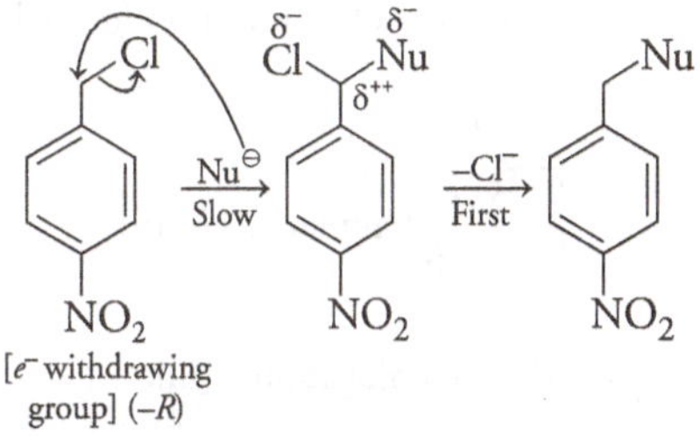

67. *(b)* In CH_2Cl_2, there are 2C—H bonds and 2C—Cl bonds.

Here the resultant of 2C—H dipole bonds are reduced by the 2 C—Cl dipole bond

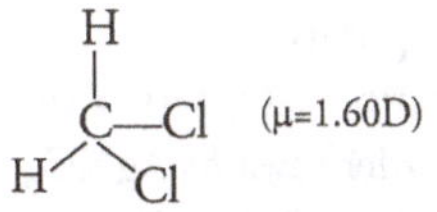

This result in non-zero dipole moment of CH_2Cl_2.

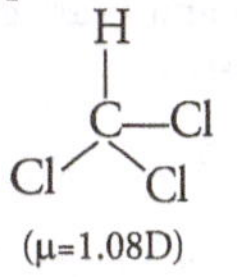

Similarly, the net dipole moment of $CHCl_3$ is non-zero.

68. *(b)*

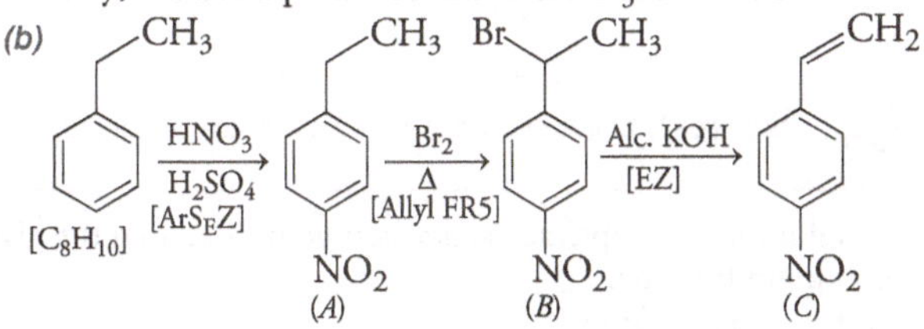

69. *(a)* Complete reaction (peroxide effect, free radical addition) of I, II and III are as follow:

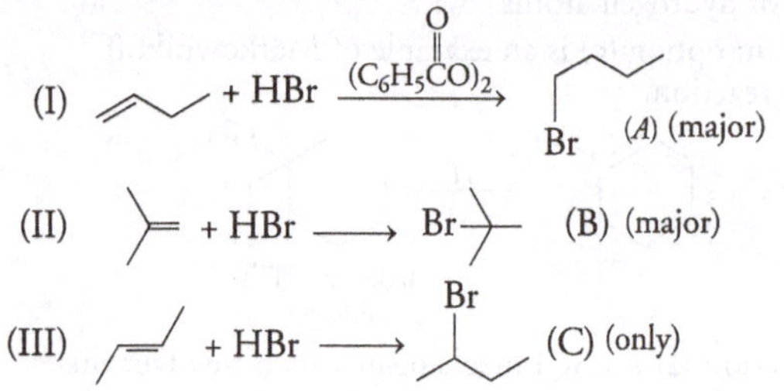

Branching decreases the surface area as in alkane because the branched *C* chains are more spherical -like, which results in lower boiling point.

Hence, the boiling point of isomeric *RX* decreases with increase in branching, hence the order will be II < III< I

70. *(c)* General reaction :

$$AgNO_3 + X^- \text{ (Halogen halide)} \longrightarrow AgX + NO_3^{2-}$$

It follows S_N1 reaction.

In S_N1 reaction,

Rate of reaction ∝ stability of carbocation formed

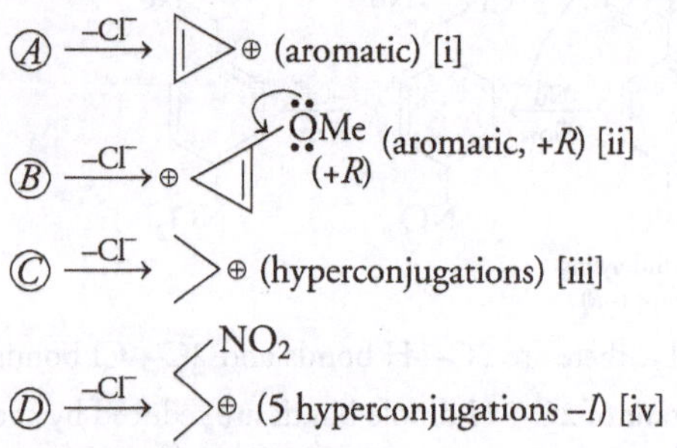

Among the given options.

Stability of carbocation is (ii) > (i) > (iii) > (iv).

So, the reactivity order towards $AgNO_3$ solution is

$$(B) > (A) > (C) > (D).$$

71. *(c)* Observations *A*, *B* and *C* are correct while observation *D* is incorrect. It is because

$$R—X \longrightarrow \overset{\oplus}{R}X^{\ominus} \longrightarrow \overset{\oplus}{R}||X^{\ominus} \xrightarrow{Y^{\ominus}} R—Y + X^{\ominus}$$

$R^{\oplus}$ (carbocation) formation takes place and $R^{\oplus}$ got stabilised by electronic factors and polar solvent molecules.

72. *(1)*

> **idea** All optically active compounds are chiral asymmetrical.

Total number of optically active compounds among the given compounds is one.

$$CH_3—CH_2—\underset{\underset{Cl}{|}}{C}H—CH_3$$

73. *(32)*

> **idea** To calculate the total number of optical isomers, first of all look for number of chiral centres, then use the formula 2^n, where *n* = no. of chiral centre.

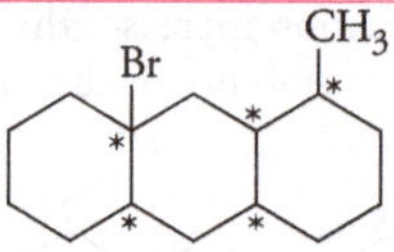

Total chiral centre = 5

Number of optical isomers = 2^5 = 32

Hence, number of optical isomers is 32.

74. *(6)* Total 6 isomeric compounds are formed by monochlorination of 2-methylbutane. These are as follows

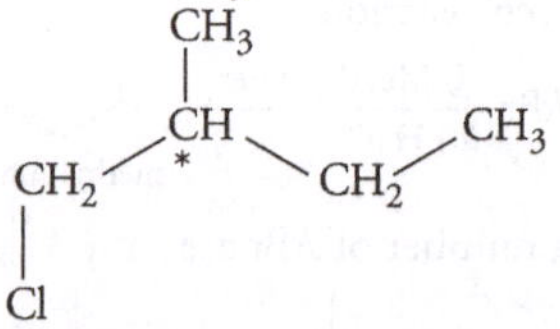

(Two isomers i.e., *R/S*)

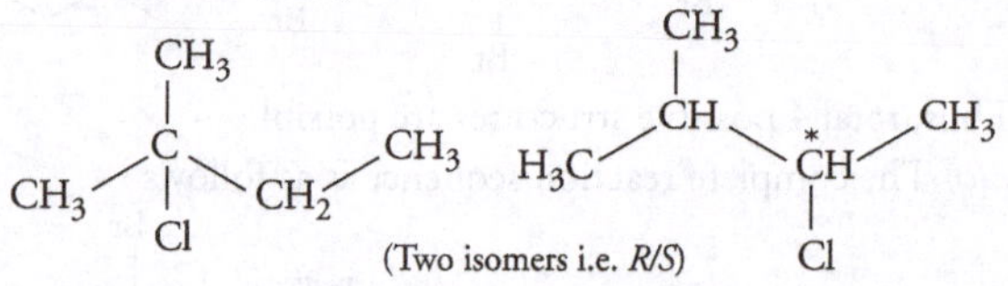

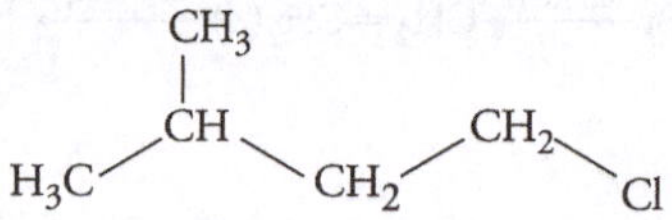

75. *(2)* 2-chlorobutane 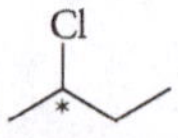

There is one chiral centre present in given compound.

So, total optical isomer = 2

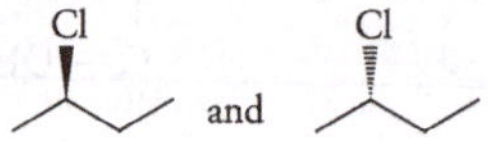

Hence, total isomer is 2.

76. *(10)*

$$CH_3CH_2Br + NaOH \begin{cases} \xrightarrow{C_2H_5OH} \underset{(A)}{CH_2{=}CH_2} \\ \xrightarrow[H_2O]{} \underset{(B)}{CH_3—CH_2OH} \end{cases}$$

Thus total number of hydrogen atom in *A* and *B* is 10.

77. *(5)* The total number of chiral carbon compounds are 5.

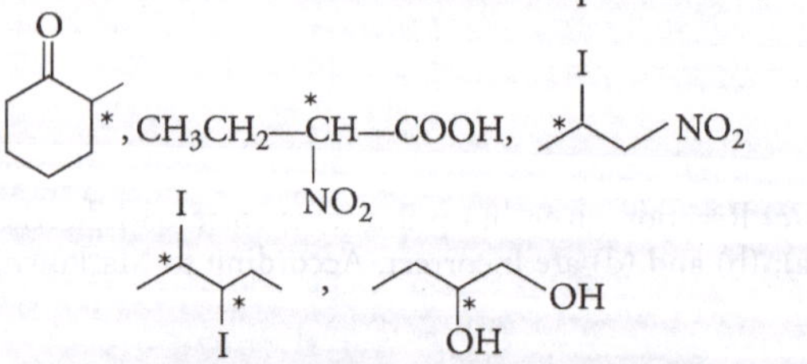

Chiral carbons are marked by (*).

Topic 2

01. *(b)* The complete reaction is as follows

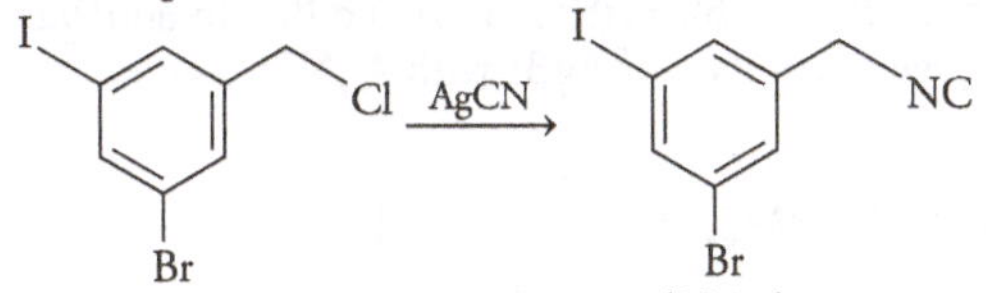

(Major)

02. *(c)* Stability order of carbocation is

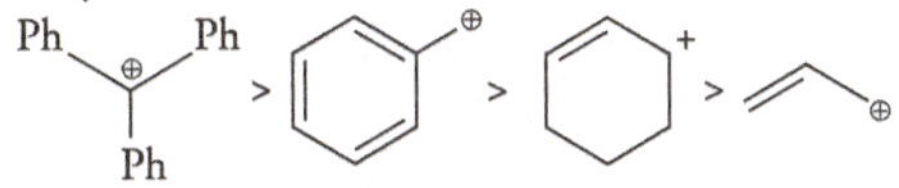

So, the most stable carbocation is given in option (c).

03. *(a)* The given reaction is an example of nucleophilic aromatic substitution reaction and gives the following product '*P*'.

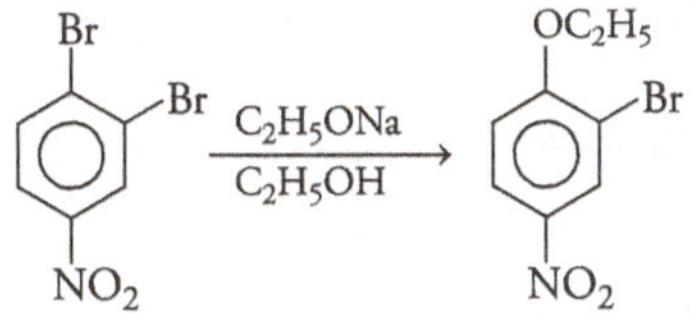

(*P*)

04. *(b)* Complete reaction is as follows.

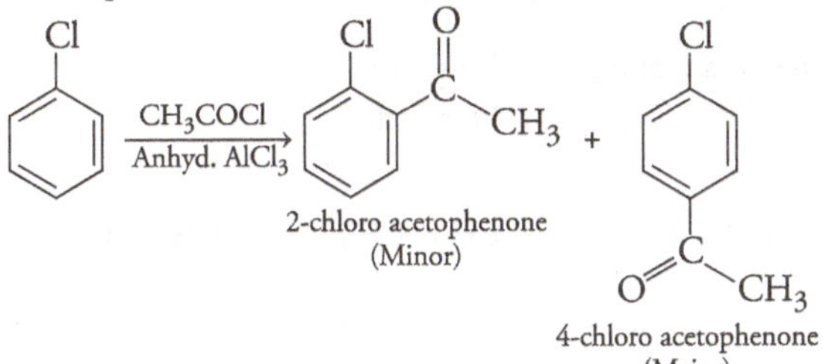

This reaction is called Friedel-Crafts reaction.

05. *(a)*

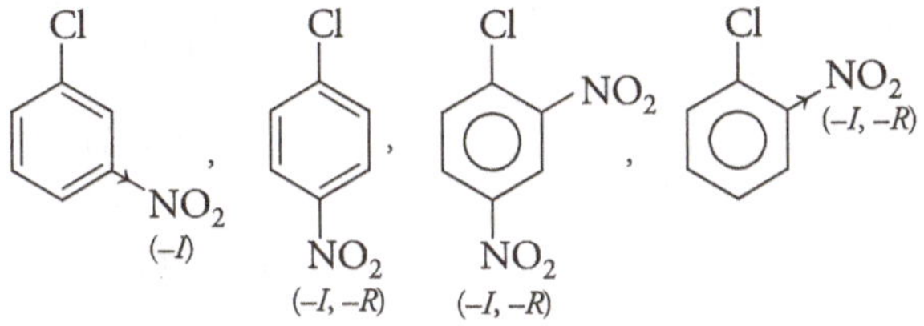

3-chloronitrobenzene has weaker electron withdrawing nature of NO_2 ($-I$ only) as compared to other compounds. So, it gives lowest rate towards nucleophilic aromatic substitution reaction (ArS_N2).

Thus, option (a) is correct.

06. *(b)*

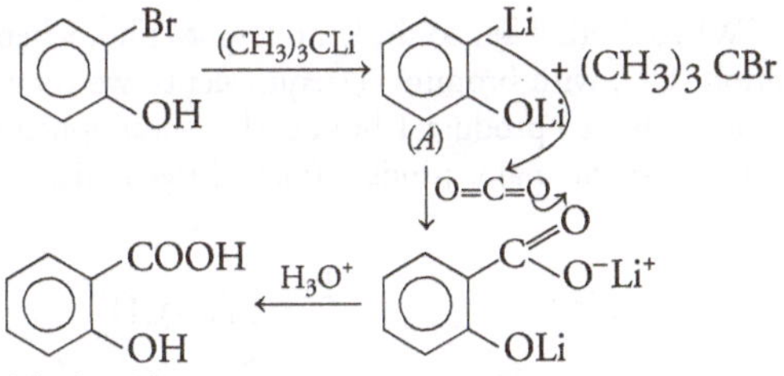

Salicylic acid

07. *(c)* Since —OH is a better electron releasing group than —CH_3.

So, electrophile will attack at *ortho* and *para*-position with respect to —OH group.

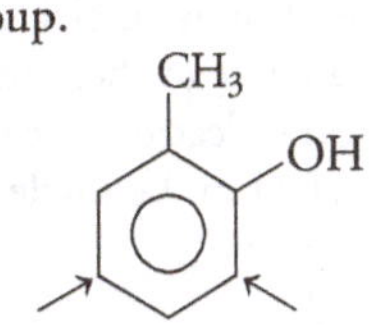

para-position with respect to —OH group is the *meta*-position with respect to —CH_3 group.

—CHO and —NO_2 are electron withdrawing groups which lead to electrophilic attack at *meta*-position.

08. *(c)* In presence of Lewis acid, electrophilic halogenation reaction takes place at *meta* position to give

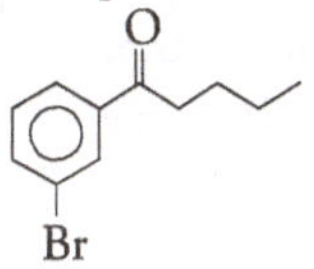

as follows

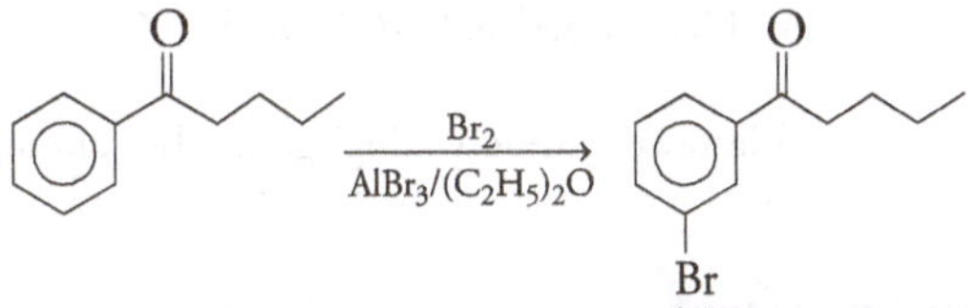

(*A*)
(Major product)

09. *(d)* When chlorobenzene is treated with NaOH at 623 K temperature and 300 atm pressure, sodium phenoxide is formed by nucleophilic substitution.

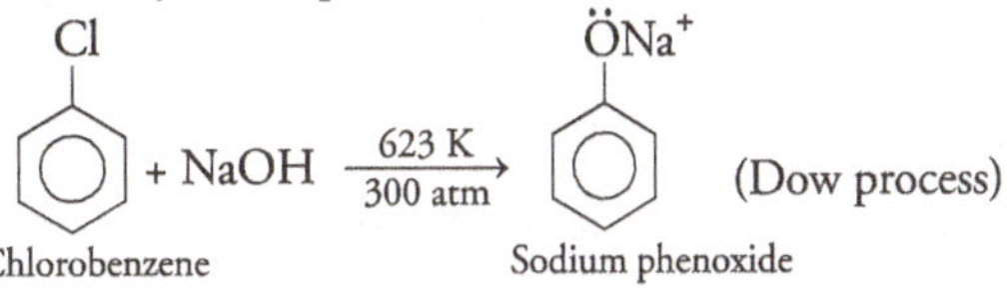

Nucleophilic substitution is not preferred in aromatic compounds. So, to do such reactions high temperature and pressure are required.

10. *(c)* When 3-ethyl benzonitrile undergoes photochemical reaction (UV) with bromine (1 equivalent), we get a monobrominated product 1 bromo -1 (3-cyanophenyl) ethane (*A*) as the major product (benzyl free radical substitution).

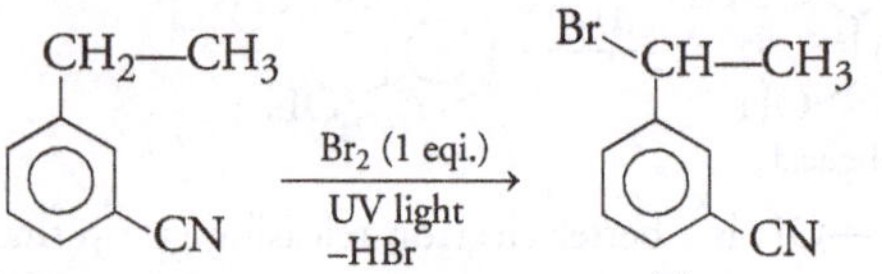

3-ethyl benzonitrile (*A*)

11. *(b)* Friedel-Crafts reaction is an electrophilic substitution reaction. In general, the rate of these reactions is highest when benzene have some electron releasing groups, (i.e. $—NH_2$, —OH) attached to the nucleus.

But in 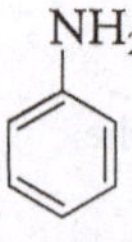 and 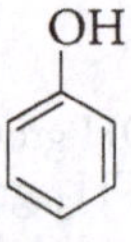, $—\ddot{N}H_2$ and $—\ddot{\underset{\cdot\cdot}{O}}H$ groups are deactivated because the Lewis acid (anhydrous $AlCl_3$) used in these reactions form an adduct with the '$—NH_2$' or '—OH' groups and becomes ineffective. '$—\overset{O}{\overset{\|}{C}}—NH_2$' is an electron withdrawing (deactivating) group.

∴ 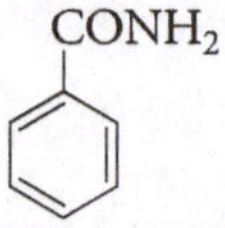 is not highly reactive. '—Cl' in 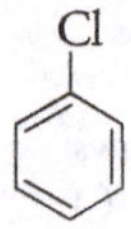 is weakly electron withdrawing (deactivating) group, but among the given options, it is most reactive towards Friedel-Crafts reaction.

> **Concept Enhancer**
> Friedel-Crafts reaction is an organic coupling reaction involving an electrophilic aromatic substitution that is used for the attachment of substitutents to aromatic rings. The two primary types of friedel-crafts reactions are the alkylation and acylation reactions.

12. *(d)*

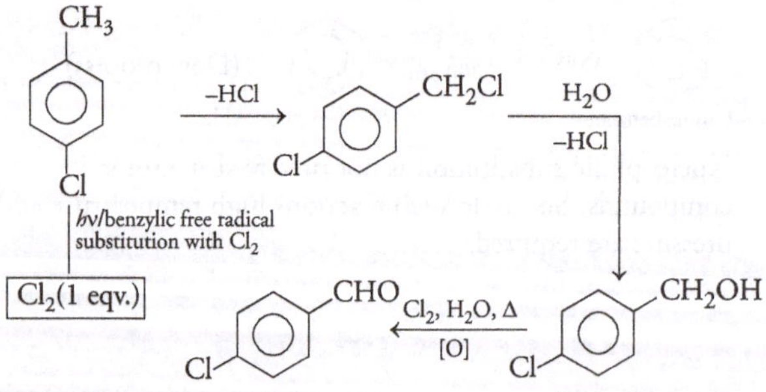

(Option d) (Option-c)

So, considering the second reaction condition, Cl^-, H_2O and Δ in the statement of the question, the correct answer is option (d).

13. *(a)* Only ionic halides (X^-) give precipitate of Ag*X* with $AgNO_3$ solution. So, an organic bromide will be able to produce $R^{\oplus}$ (stable carbocation) and Br^- in aqueous solution will give precipitate of AgBr with $AgNO_3$.

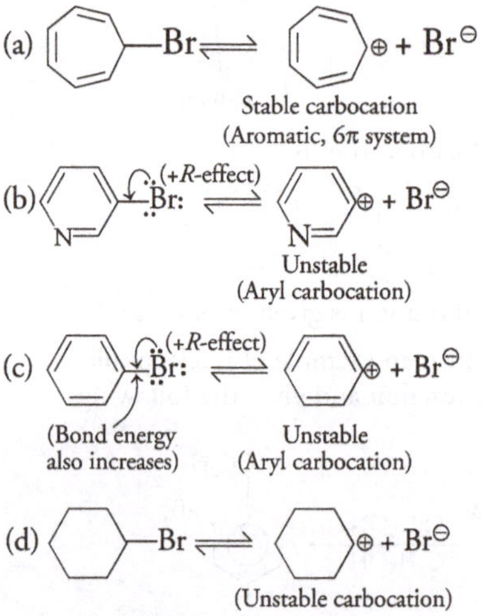

So, only 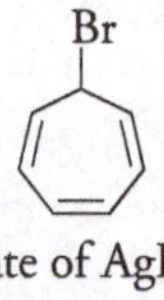 produces a precipitate of AgBr with $AgNO_3$ solution.

14. *(a)* The complete reaction is as follows,

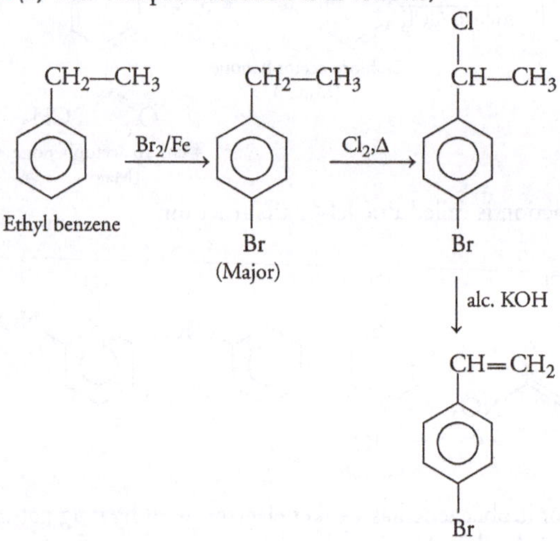

15. *(c)* Both A and R are correct and R is the correct explanation of A.

16. *(d)* A is true but R is false.

The correct form of R is : No reaction occurs between phenol and halogen acid. It is due to the presence of the double bond character in C—O bond.

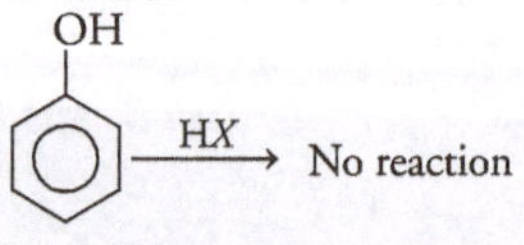

Hence, option (d) is correct.

17. *(c)* The correct reactivity of haloarenes towards nucleophilic substitutions with *aq*. NaOH is

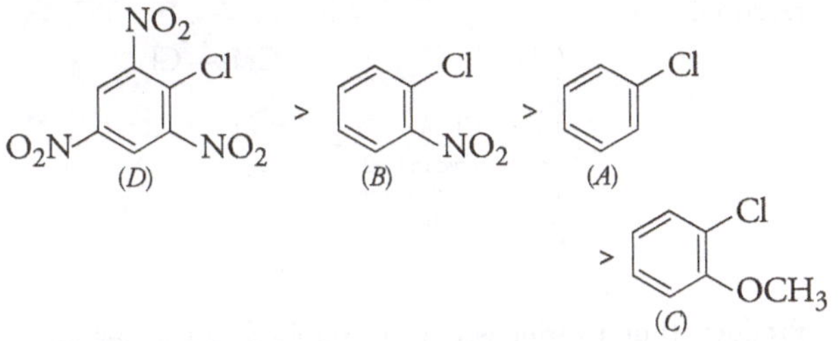

Generally, aryl halides are less reactive towards nucleophilic substitutions with aq. NaOH. Due to presence of electron releasing group i.e., ($—OCH_3$) further decreases its reactivity in (c) towards nucleophilic substitution whereas the presence of an electron withdrawing group ($-R$) i.e., ($—NO_2$) at *ortho* and *para* position increases its reactivity towards nucleophilic substitution. As the number of electron withdrawing groups increases, reactivity increases.

18. *(c)* correct option is (c)

Melting point of molecule depends on packing efficiency. If a molecule is symmetrical then close packing is possible due to better crystal fitting, which results in strong intermolecular attractive force. Thus, more energy is required to break the bond, which results in increased melting point.

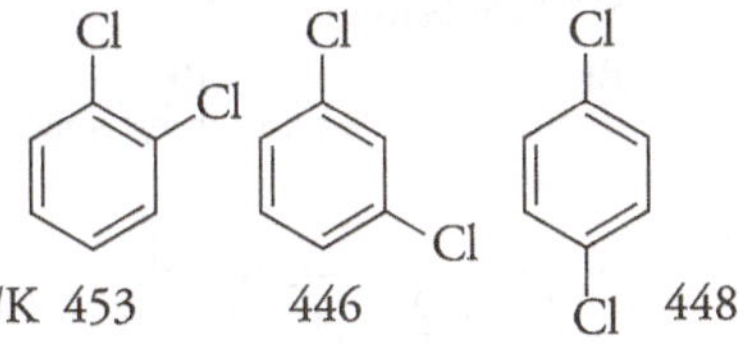

Boiling point/K	453	446	448
Meting point/K	256	249	323

19. *(c)* Only bromine water gives tribromo products, while all other gives monobromo products in which *para* bromo is major product.

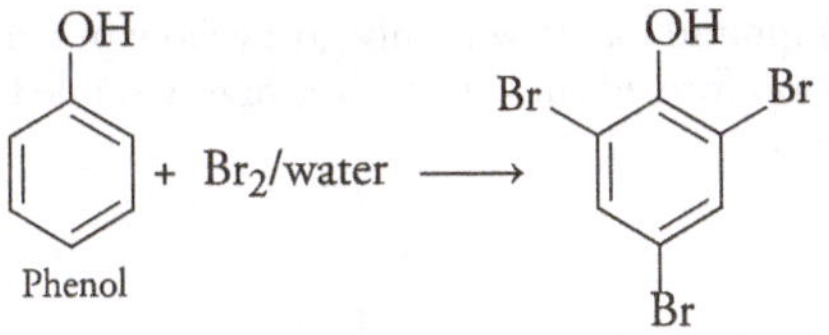

When phenol reacts with bromine in the presence of CS_2 or $FeBr_3$ or $CHCl_3$, it results in the formation of two products.

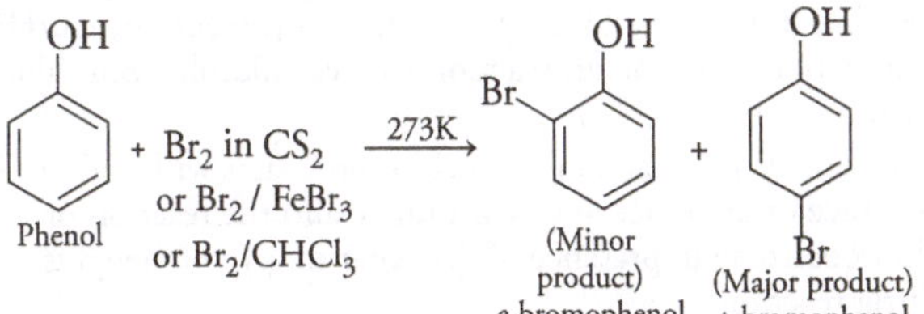

20. *(c)* The mechanism of the given reaction is as follows

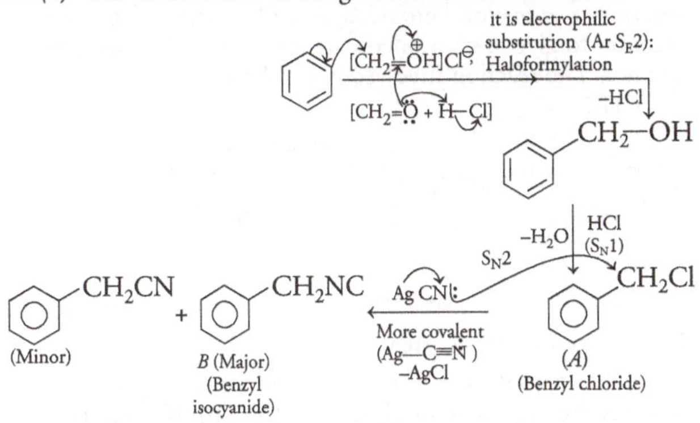

Thus, both benzyl cyanide and benzyl isocyanide are the products of reaction but benzyl isocyanide being the major product gives the correct option as (c).

21. *(a)* The complete reaction sequence is as follows,

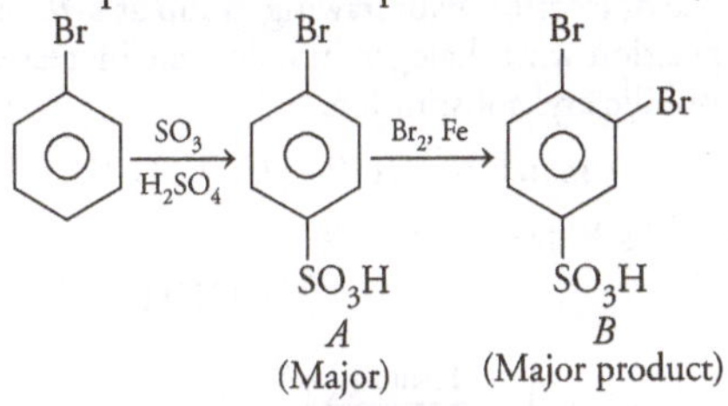

Both reactions are electrophilic substitution reaction, Ist (*A* major) is sulphonation and IInd (*B* major) is halogenation.

22. *(d)* The correct match is A-III, B-IV, C-II, D-I

23. *(c)* The complete reaction takes place as follows

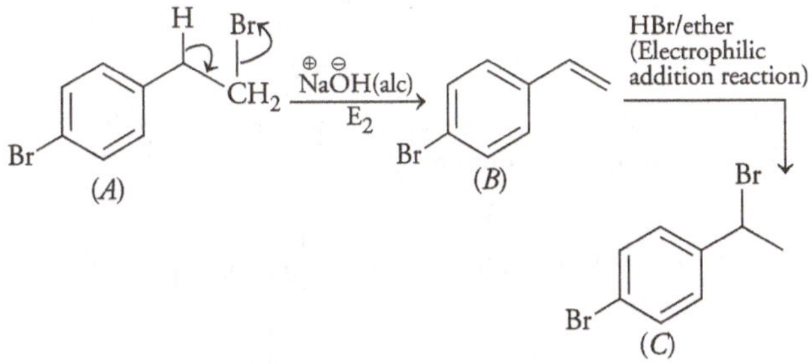

Here A and C are position isomers.

24. *(a)* The complete reaction is as follow,

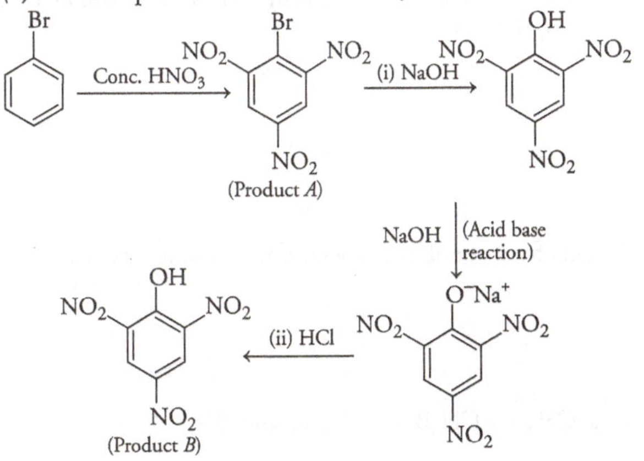

25. *(d)* When *m*-chlorobenzaldehyde is treated with 50% KOH solution, *meta*-chlorobenzaldehyde will undergo Cannizzaro reaction to give *m*-chlorobenzoate ion and *m*-chlorobenzyl alcohol. The reaction involved is as follows

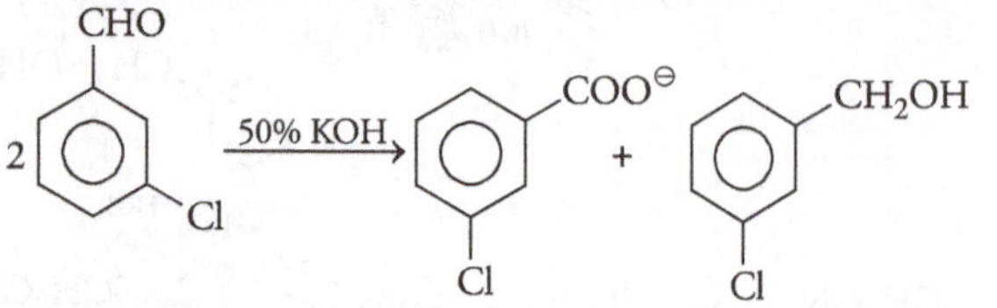

m-chlorobenzoate *m*-chlorobenzyl alcohol

Hence, option (d) is correct.

26. *(b)* The correct order of reactivity for the given pairs towards the respective mechanism is (A), (B), (C) and (D).

(A) S_N2 reaction decreases with increase in steric crowding.
(B) S_N1 reaction increases with stability of carbocation.
(C) EAS reaction decreases with decreases in electron density.
(D) Presence of electron withdrawing group at *ortho* and *para*-position w.r.t. halogen in haloarene increases nucleophilic aryl substitution.

27. *(a)* The correct match is A-II, B-I, C-IV, D-III.

28. *(c)* According to given statement,

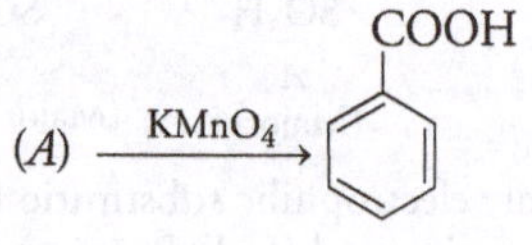

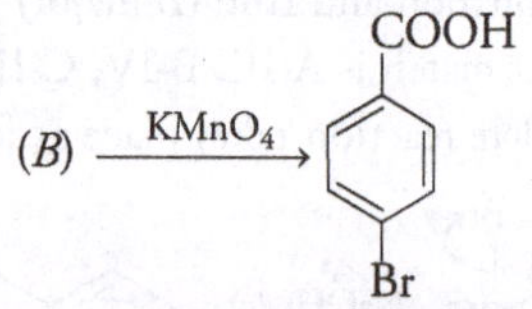

(*A*) is optically active and gives yellow ppt. with $AgNO_3$.
So, (*A*) can be either option (a) or (c).

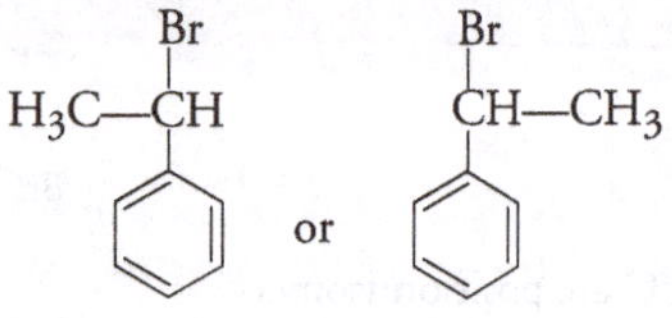

But as *B* gives *p*-bromobenzene on oxidation, it will be option (b) or option (c).

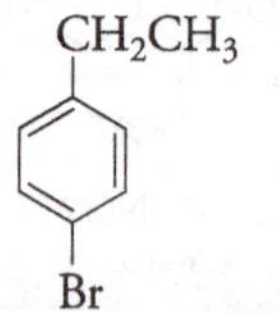

Hence, from the above conclusion, option (c) is correct, i.e.,

A is CH_3 — CH(Br)— C_6H_5 and *B* is

29. *(c)* On heating *p*-bromotoluene (*A*) with Cl_2 gives *p*-bromo benzyl chloride(I) product by free radical substitution reaction.

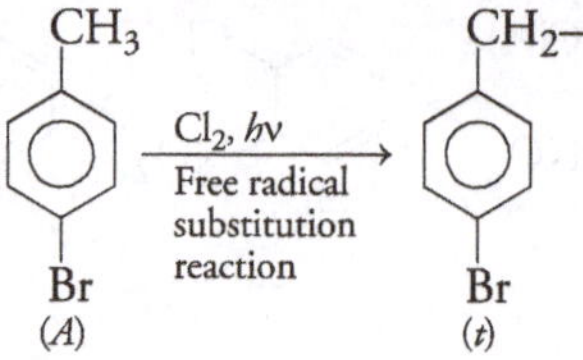

Product (I) on treating with CN^- shows S_N2 reaction at benzylic carbon. CN^- replace Cl^- to form *p*-bromo benzylic nitrile (II).

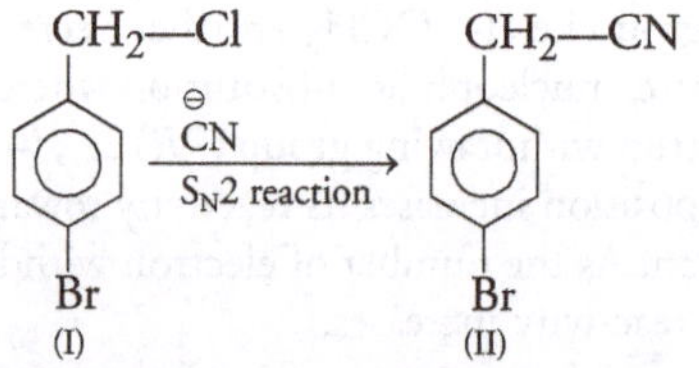

Benzyl nitrile (II) on complete hydrolysis with acidic water gives 4-bromophenyl acetic acid.

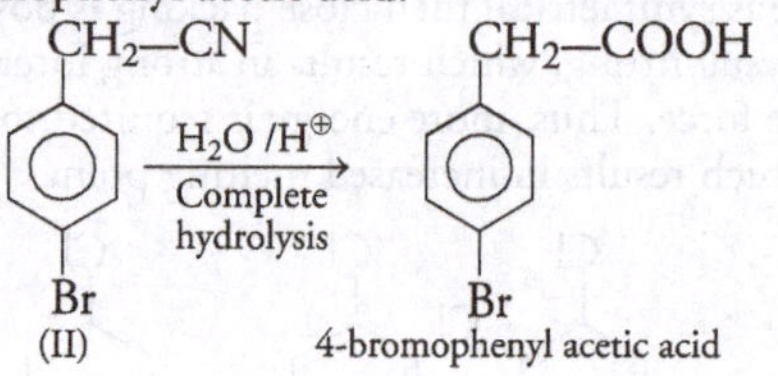

30. *(b)*

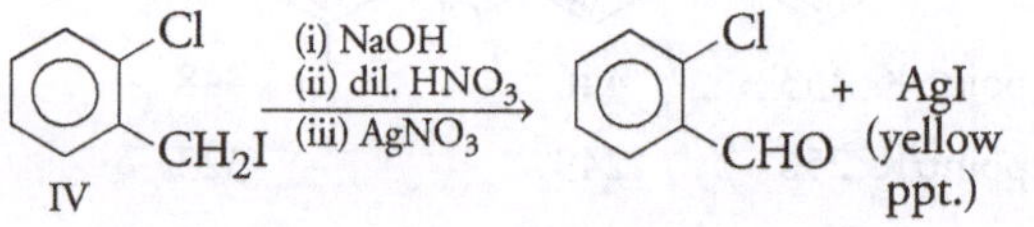

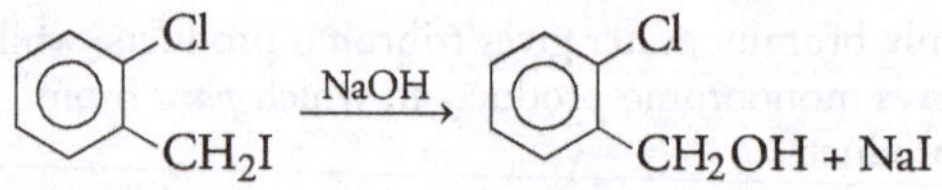

This compound halide will only give yellow ppt. as benzyl carbocation formed shown below is highly stable by conjugation.

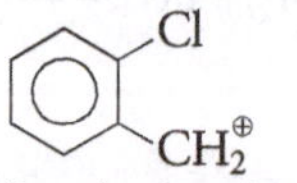

Benzyl carbocation

While other compounds halide cannot be removed because their corresponding carbocation is highly unstable.

31. *(c)* Benzene diazonium chloride undergoes Sandmeyer reaction in presence of Cu_2Cl_2 while in presence of Cu/HCl it undergo. Gattermann reaction to give chlorobenzene along with nitrogen gas (N_2).
While in Wurtz reaction ethyl chloride reacts with sodium in presence of dry ether to give *n*-butane and the reaction of chlorobenzene in presence of dry ether to give biphenyl is Fittig reaction.
So, correct match is A-(2), B-(4), C-(1) and D-(3).

32. *(a)* Rate of S_N2 reaction $\propto \dfrac{1}{\text{Stability of carbocation}}$

Stability of carbocation $\propto \dfrac{1}{-I/-M}$ (effective group)

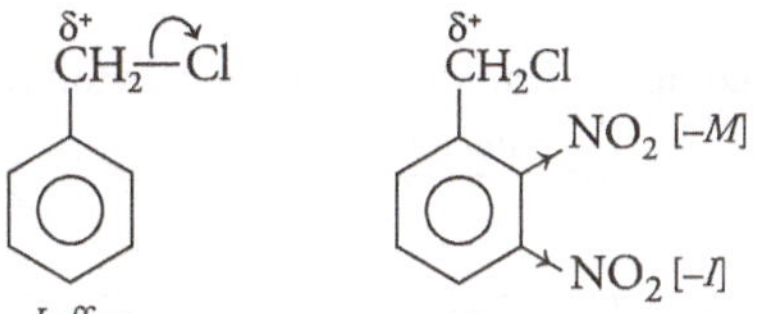

–*I*-effect (I)

(II) (Distance dominating)

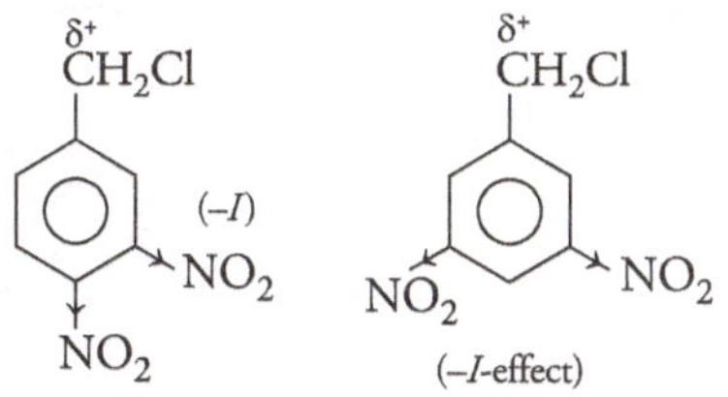

(–*M*-effect –*I*-effect) (III)

(–*I*-effect) (IV)

So, the decreasing order of reactivity of the given compounds toward S_N2 reaction is

(II) > (III) > (IV) > (I).

33. *(a)* B and D are possible while (A) and (C) are not possible. $AlCl_3$ do not act as electrophiles to give Friedel-Crafts reaction with benzene. Therefore, A and C reactions are not possible.

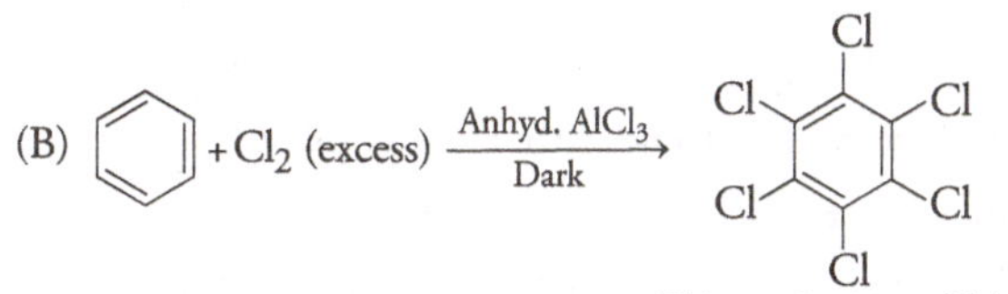

(This reaction is possible)

(Allyl halide easily form electrophile with $AlCl_3$ and can react with benzene to give Friedel-Craft's reaction)

(This is also possible)

34. *(b)* More stable the carbocation intermediate, higher will be the rate of S_N1 reaction.

The reaction involving carbocation intermediate formation for the given compounds are as follows

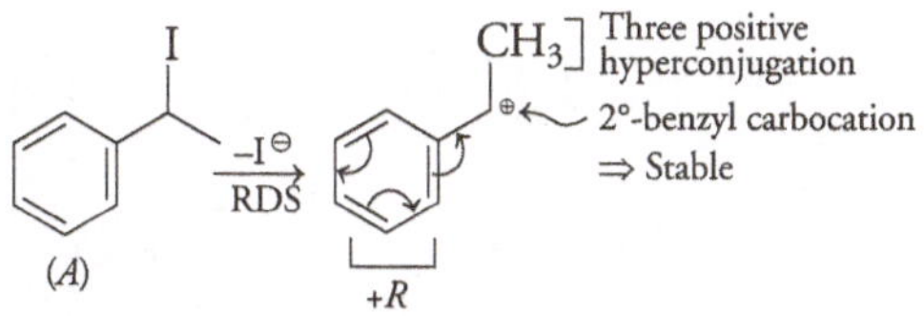

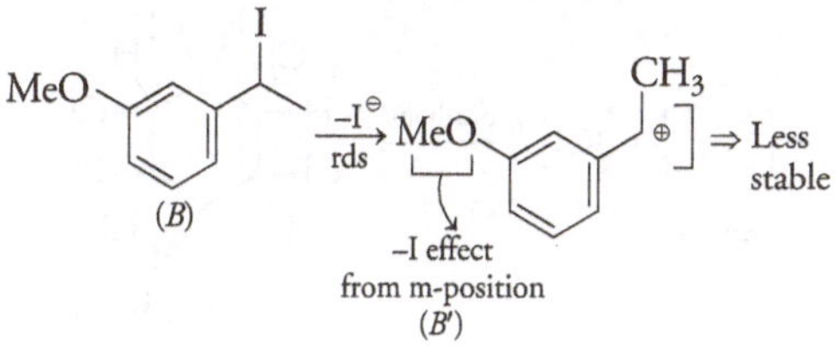

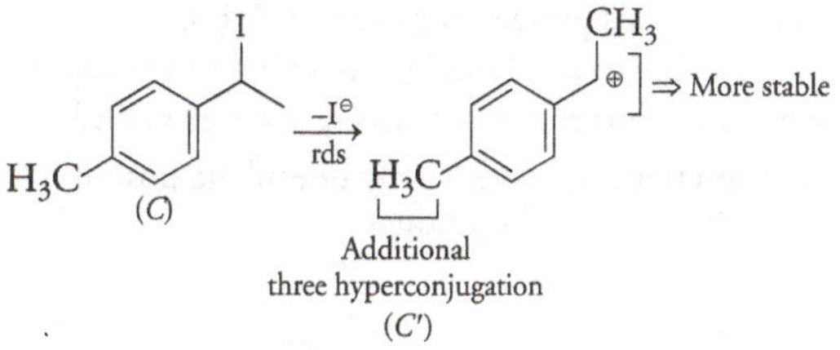

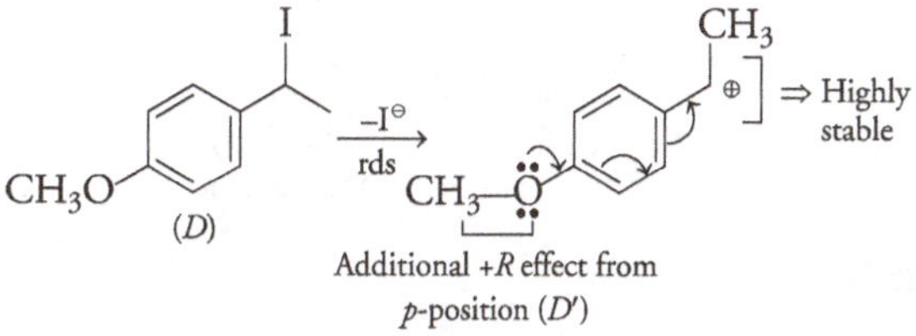

Stability of intermediates will follow the order

$B' < A' < C' < D'$
(–*I*) (–HPC=3) (+HPC=6) (+HPC=3, +*R*=1)

So, the rate of S_N1 reaction with the given compounds will be, $B < A < C < D$

35. *(93)* The complete reaction is as follows

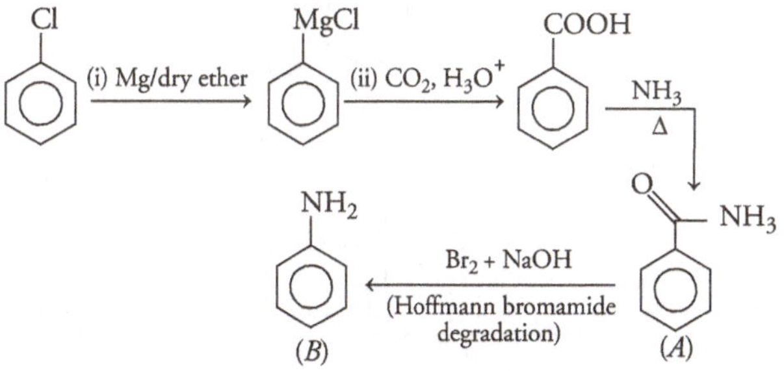

Number of moles of 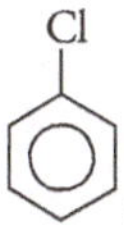 = Number of moles of 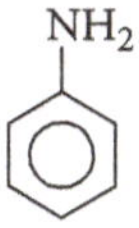

$$\frac{11.25\times10^{-3}}{112.5}=\frac{x\times10^{-1}\times10^{-3}}{93}$$

$$x = 93 \text{ mg}$$

36. *(2)* In Sandmeyer's reaction only bromobenzene and chlorobenzene are prepared.

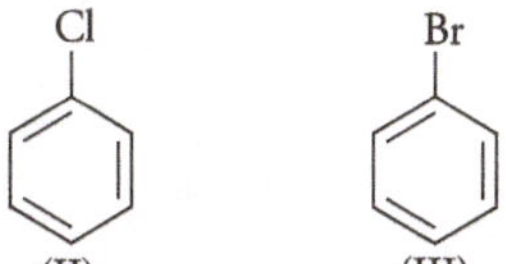

(II) (III)

Hence, the number of halobenzene is 2.

37. *(6)*

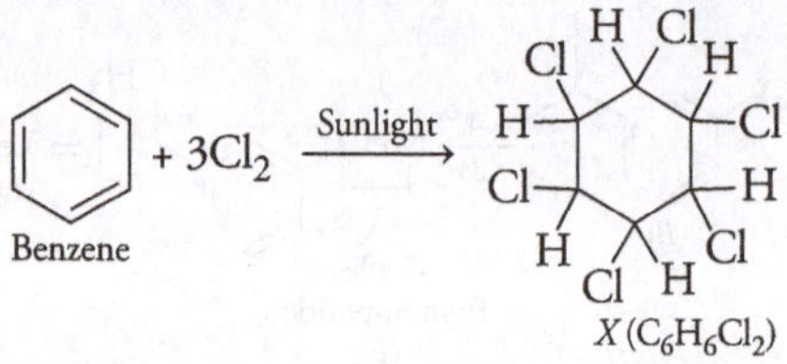

Number of hydrogens present in product (X) is 6.

38. *(1)* Bromine in presence of sunlight is added to the allylic sp^3 carbon atom. This reaction proceeds *via* free radical substitution reaction. As there is only one allylic position(*). So, only one Br is present in product.

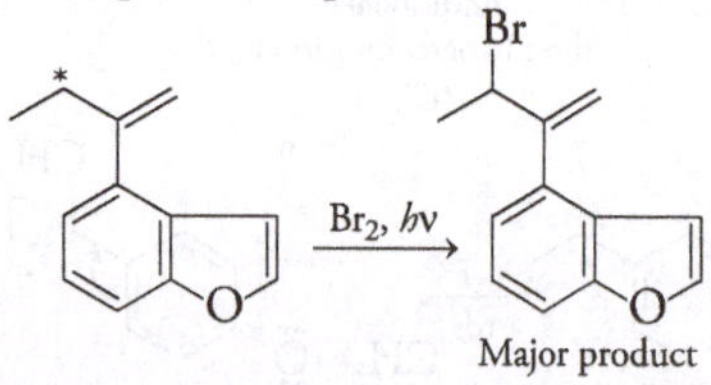

Topic 3

01. *(a)* Correct option is (a) gammaxene

Chloral	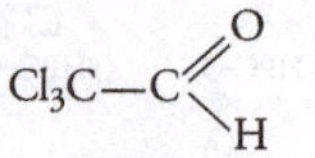	3 Cl-atoms
Gammaxene	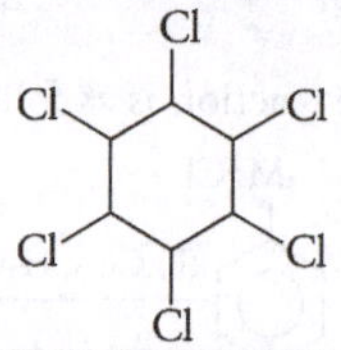	6 Cl-atoms
Chloropicrin	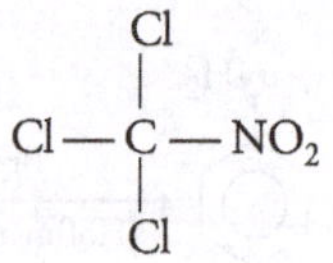	3 Cl-atoms
Freon-12	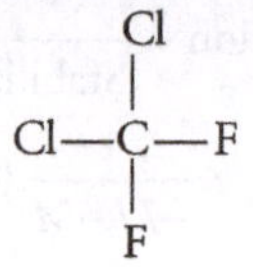	2 Cl-atoms

02. *(b)* The complete reaction sequence is as follows

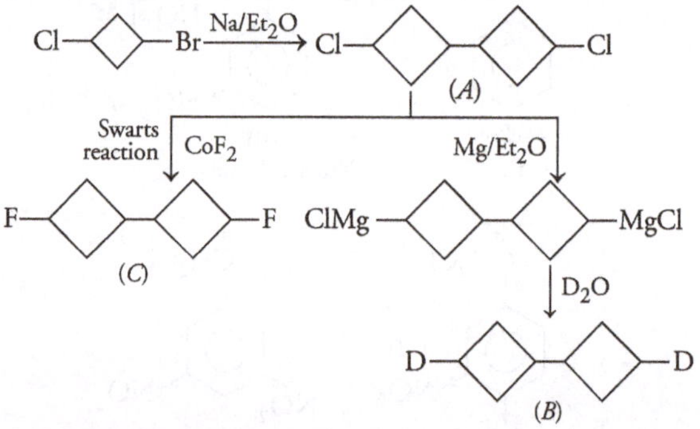

03. *(b)*

Time Saver Tech

Use elimination technique as time saver. DDT with insecticide (D-II) is given in option (b) only.

The correct match is A-III, B-I, C-IV, D-II.

- Iodoform is used as antiseptic.
- Carbon tetrachloride is used in fire extinguisher.
- CFC are used in refrigerants.
- DDT is used as insecticide in farming.

04. *(b)* The correct match is A-III, B-I, C-IV, D-II.

Compound	Use
(A) Carbon tetrachloride	(III) Fire extinguisher
(B) Methylene chloride	(I) Paint remover
(C) DDT	(IV) Non-biodegradable insecticide
(D) Freons	(II) Refrigerators and air conditioners

20 Alcohols, Phenols and Ethers

In JEE Main there is usually 0-2 questions per paper are visible from this chapter. In past 3 years, 47 conceptual and 2 numerical questions were asked of moderate difficulty level from this chapter.

For detailed analysis scan QR code

Topic 1 / Preparation, Properties and Uses of Alcohols

• Level 1

01 Incorrect method of preparation of alcohols from the following is **[2023, 10 April Shift-II]**

(a) reaction of ketone with *R*MgBr followed by hydrolysis.
(b) hydroboration-oxidation of alkene.
(c) ozonolysis of alkene.
(d) reaction of alkyl halide with aqueous NaOH.

02 The water gas on reacting with cobalt as a catalyst forms **[2023, 8 April Shift-I]**

(a) methanoic acid (b) ethanol
(c) methanal (d) methanol

03 Given below are two statements: One is labelled as Assertion A and the other is labelled as Reason R.

Assertion (A) Butan-1-ol has higher boiling point than ethoxyethane.

Reason (R) Extensive hydrogen bonding leads to stronger association of molecules. **[2023, 8 April, Shift-I]**

In the light of the above statements, choose the correct answer from the options given below

(a) A is true but R is false
(b) Both A and R are true but R is not the correct explanation of A
(c) Both A and R are true and R is the correct explanation of A
(d) A is false but R is true

04 A solution of CrO_5 in amyl alcohol has a .... colour. **[2023, 29 Jan Shift-I]**

(a) orange-red (b) yellow
(c) green (d) blue

Imp. *Oxymercuration-Demerucuration of Alkenes*

05 The major product in the following reaction

 $\xrightarrow[\text{(ii) NaBH}_4]{\text{(i) Hg(OAc)}_2\text{, H}_2\text{O}}$?

[2022, 26 June Shift-II]

(a) 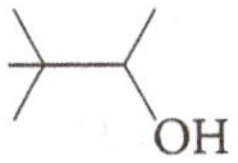
(b) 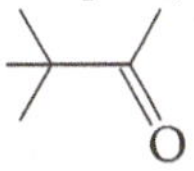
(c) 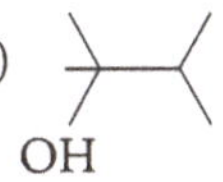
(d) 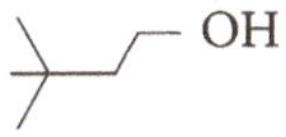

06 What is '*X*' in the given reaction?

$$\begin{matrix} CH_2OH \\ | \\ CH_2OH \end{matrix} + \text{oxalic acid} \xrightarrow{210°C} \underset{\text{Major product}}{X}$$

[2021, 25 Feb Shift-II]

(a) $\begin{matrix} CHO \\ | \\ CHO \end{matrix}$ (b) $\begin{matrix} CH_2 \\ \| \\ CH_2 \end{matrix}$ (c) $\begin{matrix} CHOH \\ \| \\ CH_2 \end{matrix}$ (d) $\begin{matrix} CH_2OH \\ | \\ CHO \end{matrix}$

07 The major product of the following reaction is

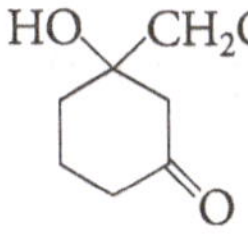 $\xrightarrow{H_2SO_4}$

[2020, 5 Sep Shift-II]

(a) 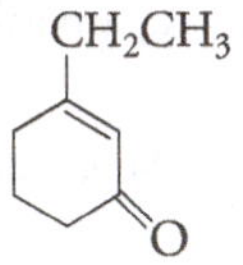
(b) 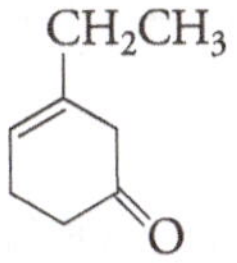
(c) 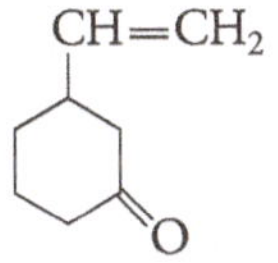
(d) 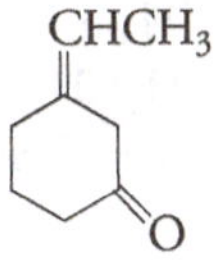

08 When neopentyl alcohol is heated with an acid, it is slowly converted into an 85 : 15 mixture of alkenes *A* and B, respectively. What are these alkenes? **[2020, 04 Sep Shift-I]**

(a) 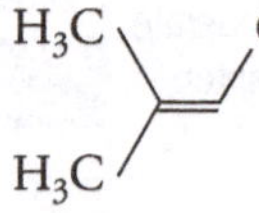 and 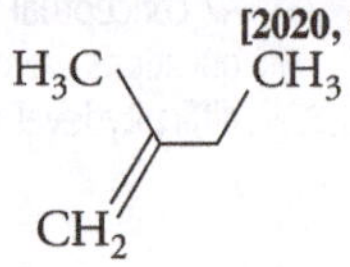

(b) 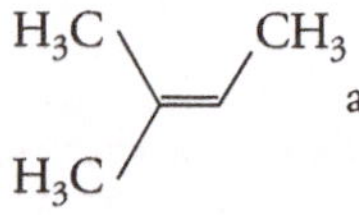 and 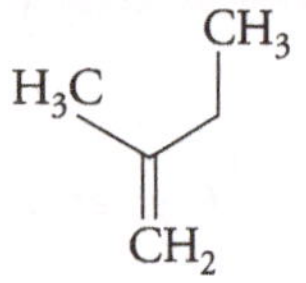

(c) 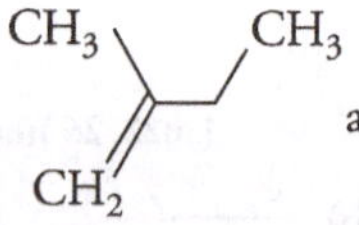 and 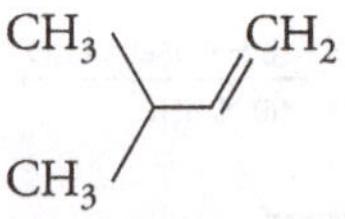

(d) 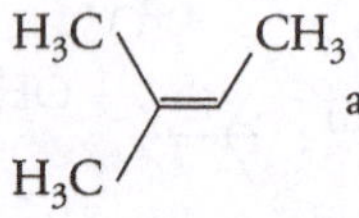 and 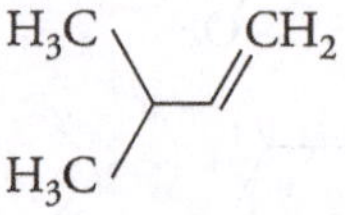

09 Consider the following reaction

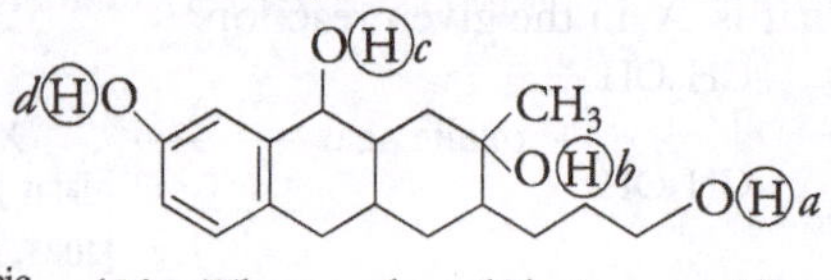

$\xrightarrow[\text{anhydride}]{\text{Chromic}}$ '*P*' The product '*P*' gives positive ceric ammonium nitrate test. This is because of the presence of which of these —OH group(s)? **[2020, 3 Sep Shift-II]**

(a) (*b*) and (*d*) (b) (*d*) only (c) (*b*) only (d) (*c*) and (*d*)

10 $CH_3CH_2—C(OH)(Ph)—CH_3$ cannot be prepared by **[2019, 12 Jan Shift-I]**

(a) $CH_3CH_2COCH_3 + PhMgX$
(b) $PhCOCH_3 + CH_3CH_2MgX$
(c) $PhCOCH_2CH_3 + CH_3MgX$
(d) $HCHO + PhCH(CH_3)CH_2MgX$

• Level 2

11 Identify the products [*A*] and [*B*], respectively in the following reaction.

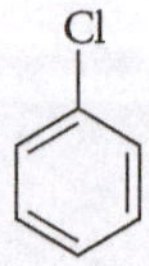 $\xrightarrow[\text{(ii) } H^+]{\text{(i) NaOH, 623 K, 300 atm}} [A] \xrightarrow[H_2SO_4]{Na_2Cr_2O_7} [B]$

[2025, 23 Jan Shift-II]

(a) (A) 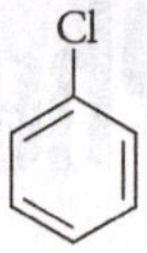, (B) 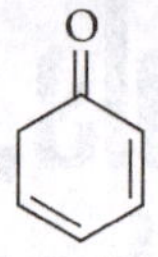 (b) (A) 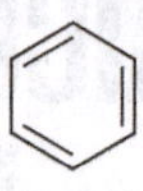, (B) 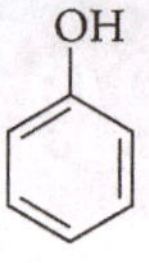

(c) (A) 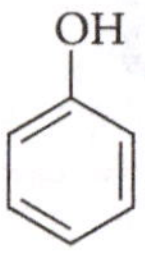, (B) 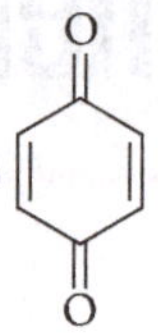 (d) (A) 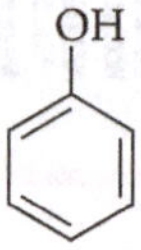, (B) 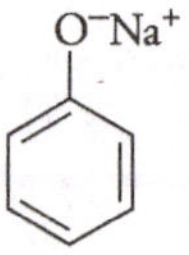

12 Identify the product *A* and product *B* in the following set of reactions. **[2024, 9 April Shift-I]**

$CH_3—CH{=}CH_2 \xrightarrow{H_2O, H^+}$ Major product *A*

$CH_3—CH{=}CH_2 \xrightarrow[H_2O, H_2O_2, \overset{\ominus}{O}H]{(BH_3)_2}$ Major product *B*

(a) *A*-$CH_3CH_2CH_3$, *B*-$CH_3CH_2CH_3$
(b) *A*-$CH_3—CH(OH)—CH_3$, *B*-$CH_3CH_2CH_2—OH$
(c) *A*-$CH_3CH_2CH_2—OH$, *B*-$CH_3—CH(OH)—CH_3$
(d) *A*-$CH_3CH_2CH_2—OH$, *B*-$CH_3CH_2CH_2—OH$

13 $CH_3CH_2—OH \xrightarrow[\text{(ii) } KMnO_4 \text{ (iii) NaOH, CaO, } \Delta]{\text{(i) Jone's reagent}} P$

Consider the above reaction sequence and identify the major product *P*. **[2024, 5 April Shift-II]**

(a) Methoxymethane
(b) Methanoic acid
(c) Methane
(d) Methanal

Imp. *Properties of Alcohols*

14 Given below are two statements : one is labelled as **Assertion (A)** and the other is labelled as **Reason (R)**.

Assertion (A) Alcohols react both as nucleophiles and electrophiles. **[2024, 31 Jan Shift-I]**

Reason (R) Alcohols react with active metals such as sodium, potassium and aluminium to yield corresponding alkoxides and liberate hydrogen.

In the light of the above statements, choose the correct answer from the options given below.

(a) A is false but R is true.
(b) A is true but R is false.
(c) Both A and R are true but R is not the correct explanation of A.
(d) Both A and R are true and R is the correct explanation of A.

15 Products A and B formed in the following set of reactions are **[2024, 30 Jan Shift-II]**

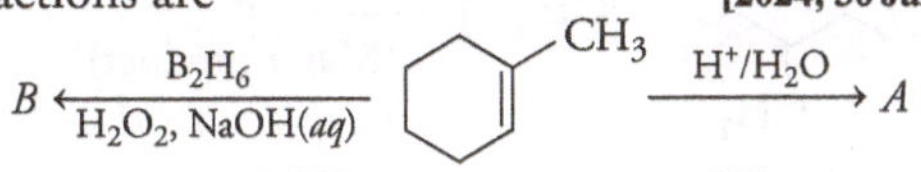

(a) 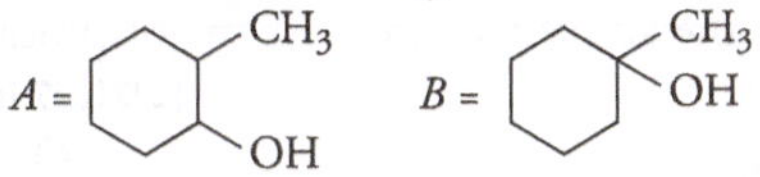

(b) 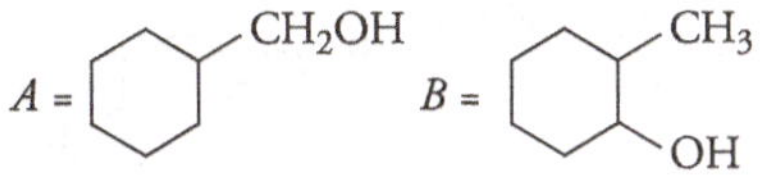

(c) 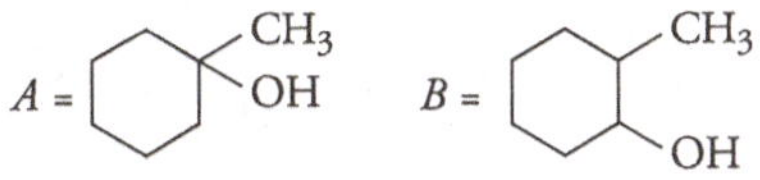

(d) 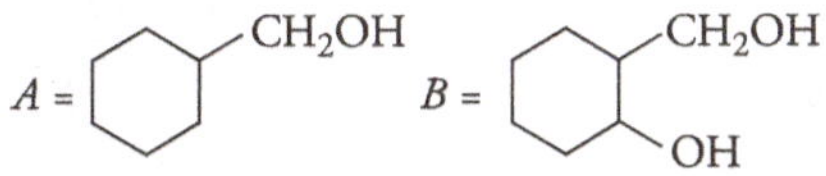

16 The major product for the following reaction is

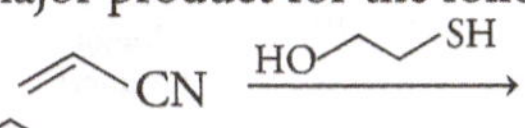

[2023, 13 April Shift-II]

(a) 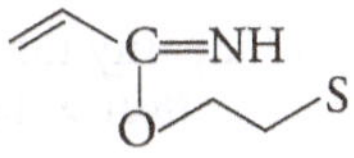
(b) 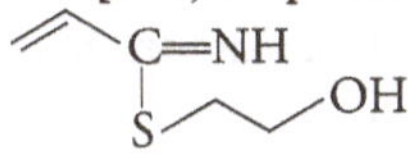

(c) 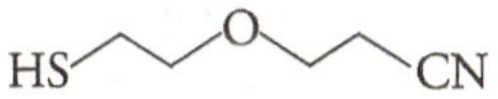
(d) 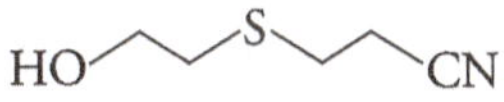

17 Given below are two statements, one is labelled as Assertion (A) and the other is labelled as Reason (R).

Assertion (A) Order of acidic nature of the following compounds is A > B > C.

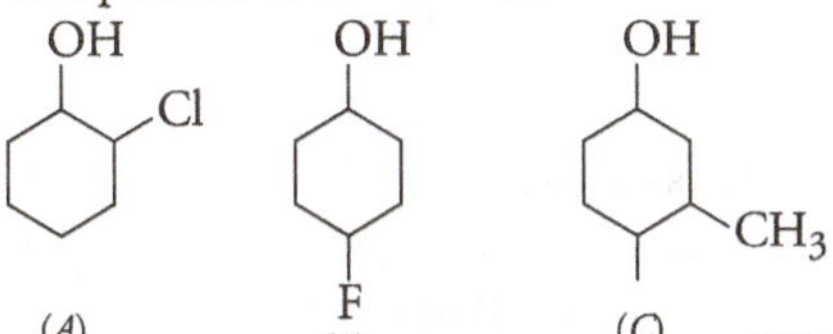

(A) (B) (C) **[2023, 13 April Shift-II]**

Reason (R) Fluoro is a stronger electron withdrawing group than chloro group.

In the light of the above statements, choose the most appropriate answer from the options given below.

(a) A is true but R is false
(b) Both A and R are correct and R is the correct explanation of A
(c) Both A and R are correct but R is not the correct explanation of A
(d) A is false but R is true

18 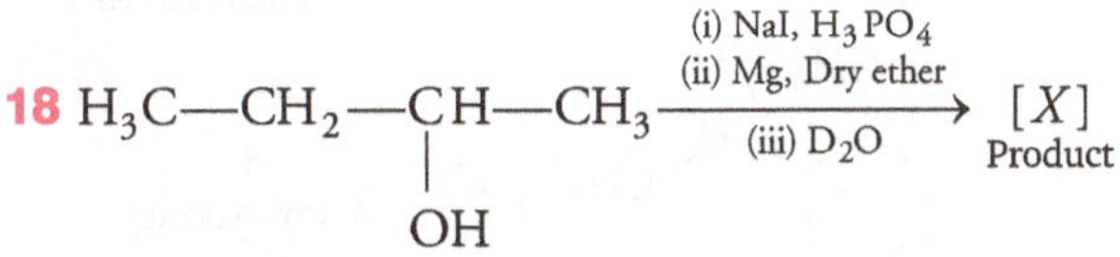

Product [X] formed in the above reaction is **[2023, 11 April Shift-II]**

(a) $H_3C—CH=CH—CH_3$
(b) $H_3C—CH_2—CH=CH_2$
(c) 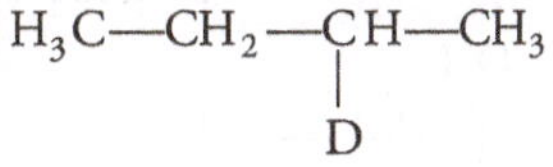
(d) 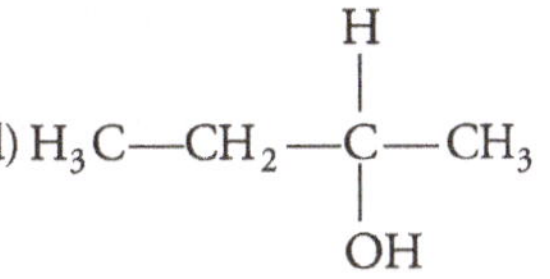

19 Decreasing order of dehydration of the following alcohols is **[2023, 1 Feb Shift-II]**

(i) 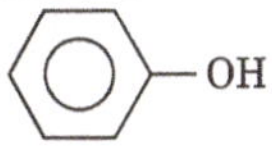
(ii) 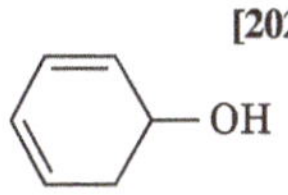
(iii) 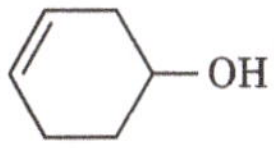
(iv) 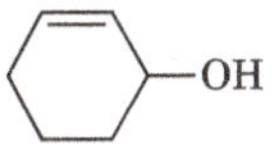

(a) (iv) > (ii) > (iii) > (i)
(b) (i) > (iv) > (ii) > (iii)
(c) (ii) > (i) > (iv) > (iii)
(d) (ii) > (iv) > (iii) > (i)

20 Find out the major product from the following reaction.

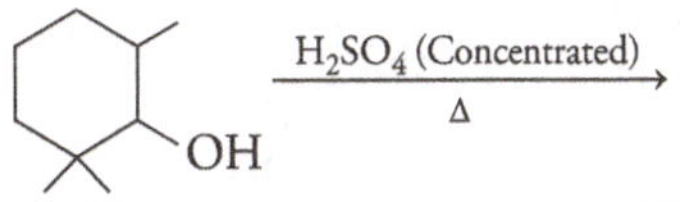

[2023, 25 Jan Shift-II]

(a) 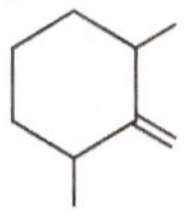
(b) 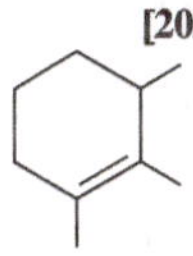
(c) 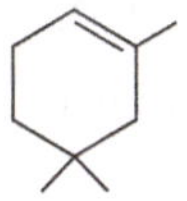
(d) 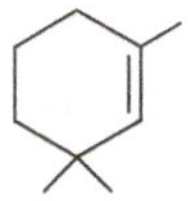

21 When ethanol is heated with conc. H_2SO_4, a gas is produced. The compound formed, when this gas is treated with cold dilute aqueous solution of Baeyer's reagent, is **[2022, 29 July Shift-II]**

(a) formaldehyde
(b) formic acid
(c) glycol
(d) ethanoic acid

22 Which reactant will give the following alcohol on reaction with one mole of phenyl magnesium bromide (PhMgBr) followed by acidic hydrolysis?

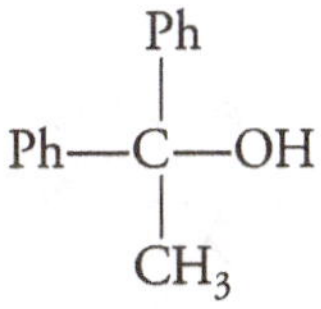

[2022, 26 July Shift-I]

(a) $CH_3—C\equiv N$
(b) $Ph—C\equiv N$
(c) $CH_3—\overset{O}{\overset{\|}{C}}—O—Ph$
(d) $Ph—\underset{O}{\underset{\|}{C}}—CH_3$

23 The major product (P) of the given reaction is (where, Me is —CH_3) **[2022, 28 June Shift-II]**

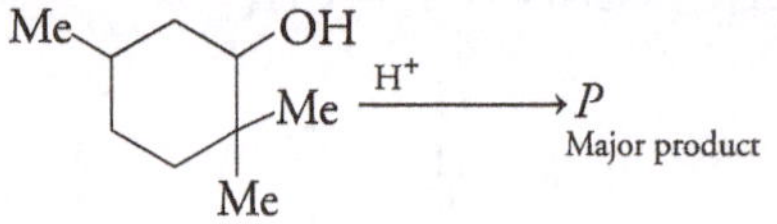

(a) 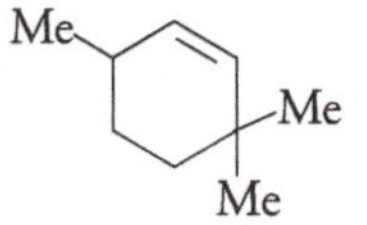 (b) 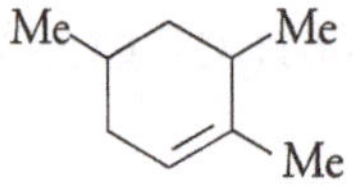

(c) 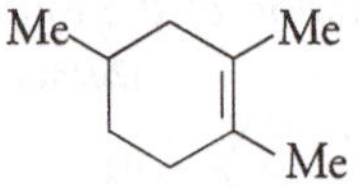 (d) 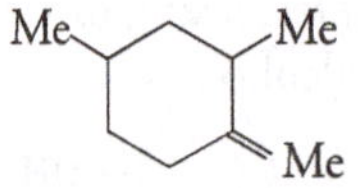

24 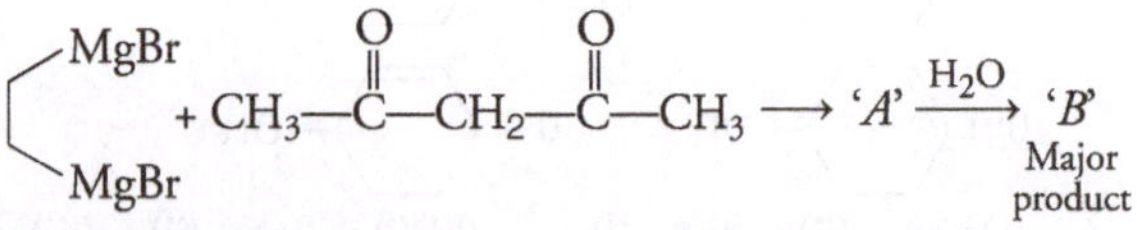

Consider the above reaction sequence and identify the product B. **[2022, 26 June Shift-I]**

(a) 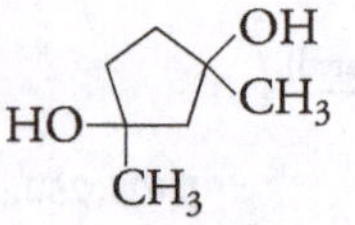 (b) 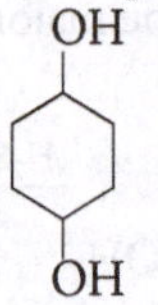

(c) 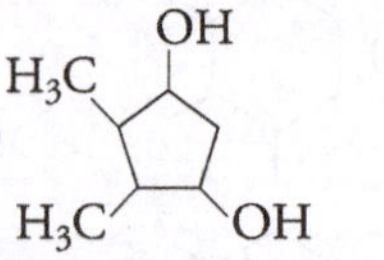 (d) 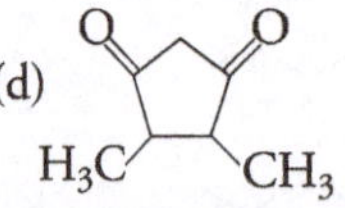

25 The major product formed in the following reaction is

$$CH_3-\underset{CH_3}{\overset{CH_3}{\underset{|}{\overset{|}{C}}}}-\underset{OH}{\underset{|}{CH}}-CH_3 \xrightarrow[\text{a few drops}]{\text{Conc.}H_2SO_4} \text{Major product}$$

[2021, 31 Aug Shift-I]

(a) $CH_3-\underset{CH_3}{\underset{|}{C}}=CH-CH_2CH_3$ (b) 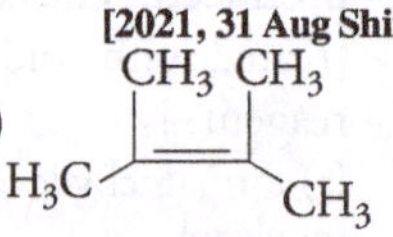

(c) $\begin{matrix}CH_3\\CH_3\end{matrix}\!\!>C=CH-CH_3$ (d) $CH_3-\underset{CH_3}{\overset{CH_3}{\underset{|}{\overset{|}{C}}}}-CH=CH_2$

26 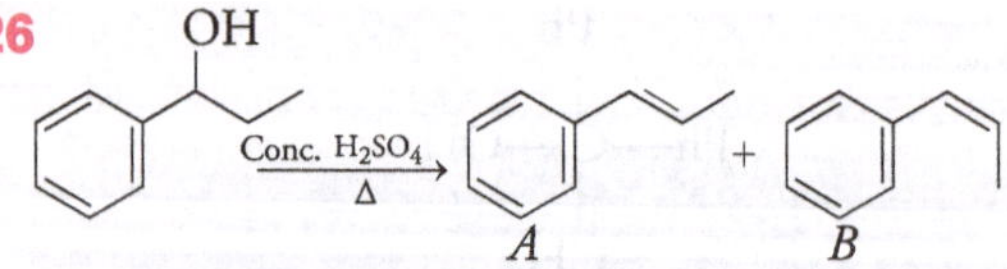

Consider the above reaction, and choose the correct statement. **[2021, 27 July Shift-II]**

(a) The reaction is not possible in acidic medium.
(b) Both compounds A and B are formed equally.
(c) Compound A will be the major product.
(d) Compound B will be the major product.

27 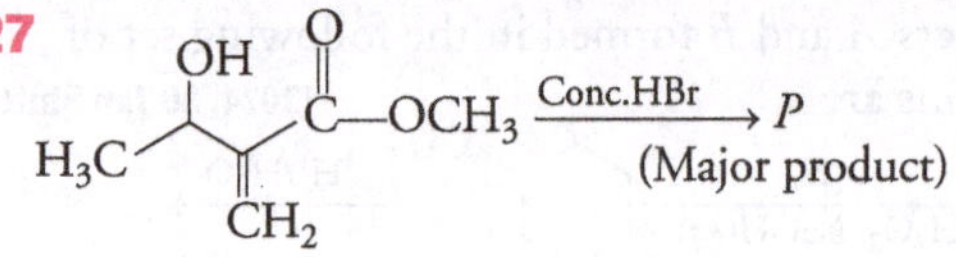

Consider the above reaction, the major product P formed is **[2021, 27 July Shift-II]**

(a) 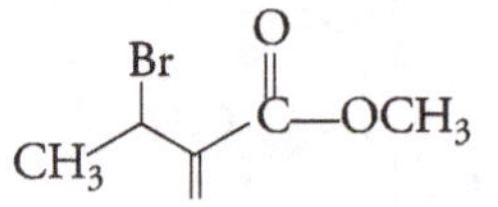 (b) 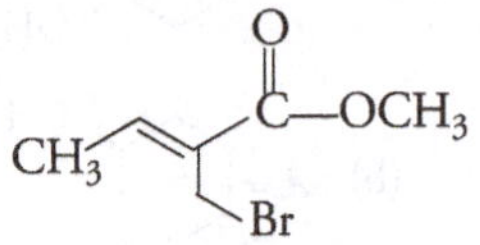

(c) 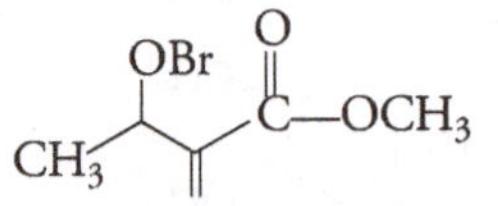 (d) 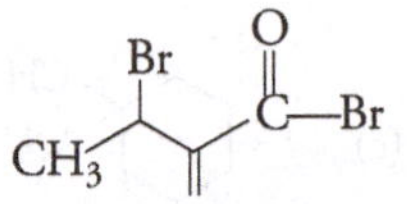

Imp. *Chemical Reactions of Alcohols*

28 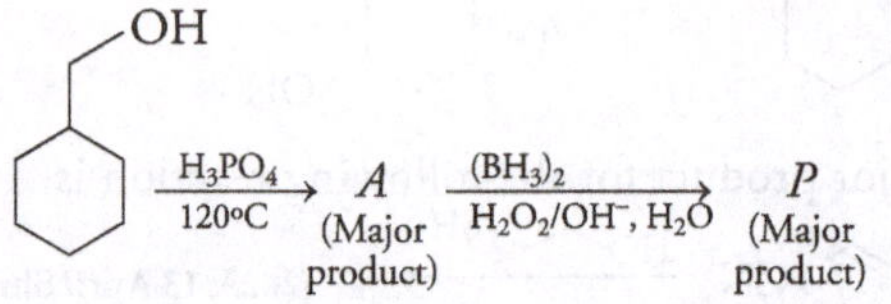

Consider the above reaction and identify the product (P). **[2021, 27 July Shift-I]**

(a) 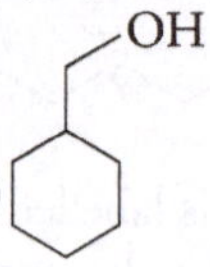 (b) 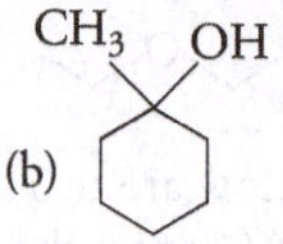

(c) 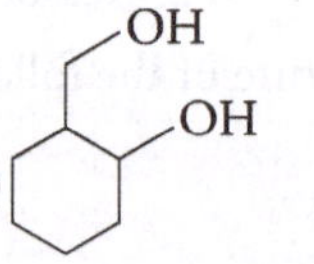 (d) 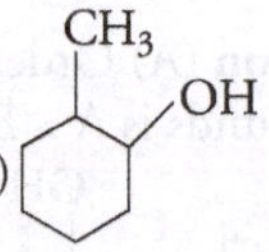

29 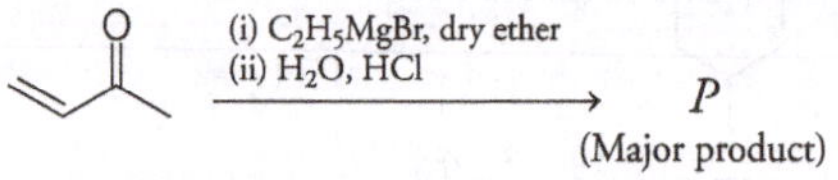

Consider the above reaction, the major product P is **[2021, 25 July Shift-I]**

(a) 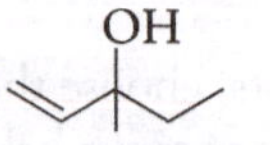 (b) 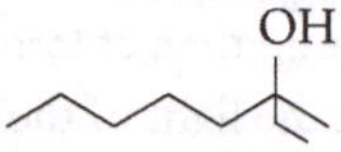

(c) 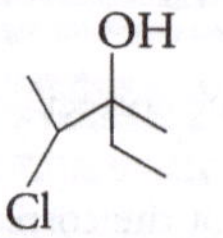 (d) 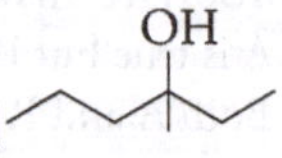

30 What is the final product (major) 'A' in the given reaction? **[2021, 24 Feb Shift-I]**

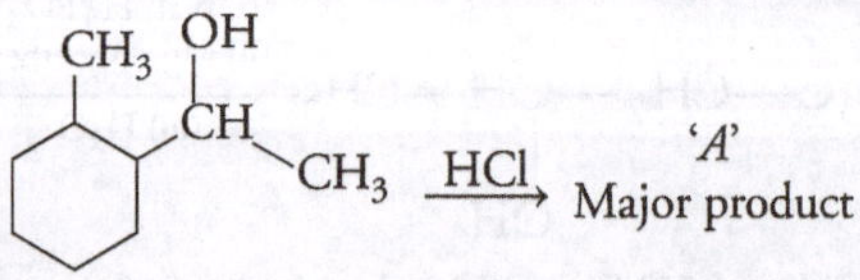

(a) 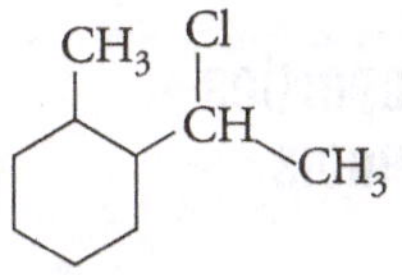 (b) 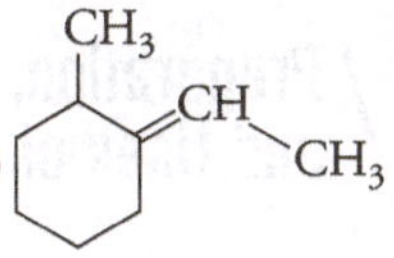

(c) 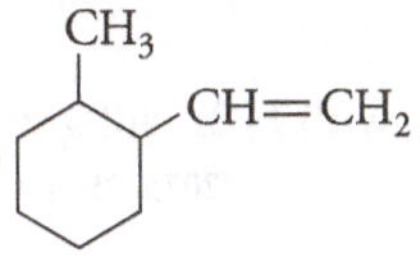 (d) 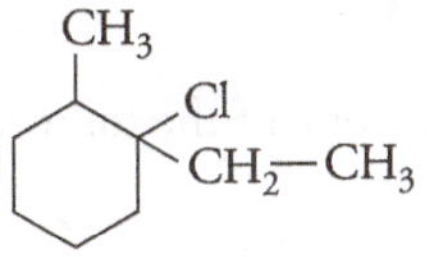

31 The major product of the following reaction is

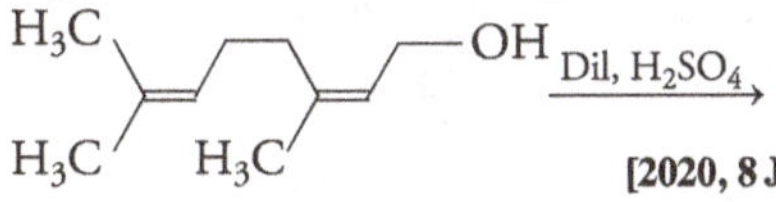

[2020, 8 Jan Shift-I]

(a) 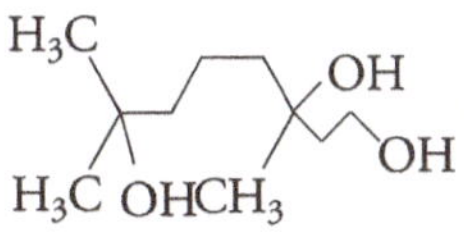 (b) 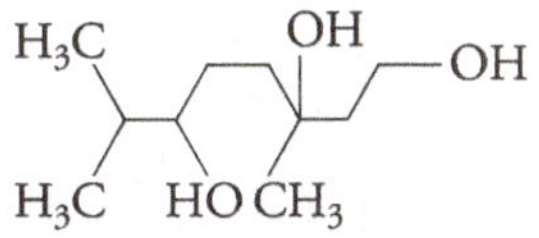

(c) 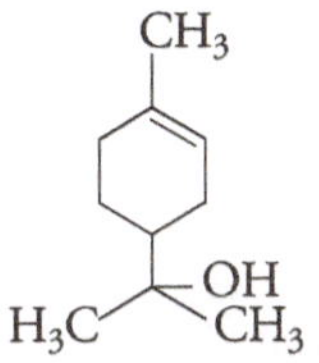 (d) 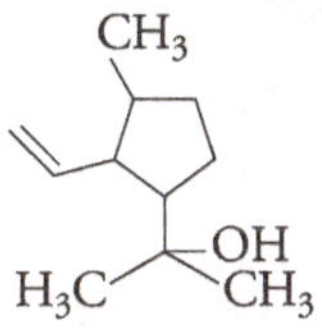

• Level 3

32. Given below are two statements: **[2025, 4 April Shift-II]**

Statement I Alcohols are formed when alkyl chlorides are treated with aqueous potassium hydroxide by elimination reaction.

Statement II In alcoholic potassium hydroxide, alkyl chlorides form alkenes by abstracting the hydrogen from the β-carbon.

In the light of the above statements, choose the most appropriate answer from the options given below:

(a) Both Statement I and Statement II are incorrect.
(b) Both Statement I and Statement II are correct.
(c) Statement I is incorrect but Statement II is correct.
(d) Statement I is correct but Statement II is incorrect.

33. A molecule ("P") on treatment with acid undergoes rearrangement and gives ("Q"). ("Q") on ozonolysis followed by reflux under alkaline condition gives ("R"). The structure of ("R") is given below.

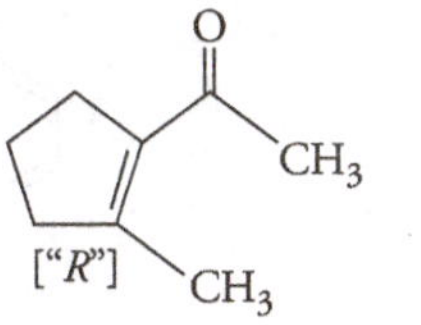

["R"]

The structure of ("P") is **[2025, 28 Jan Shift-I]**

(a) 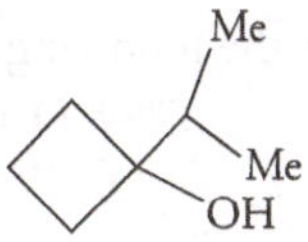 (b) 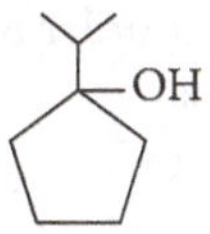

(c) 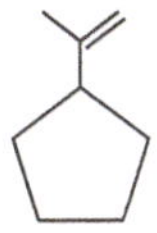 (d) 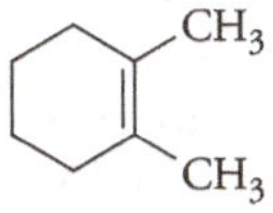

34 $CH_3—CH_2—CH_2—Br + NaOH \xrightarrow{C_2H_5OH}$ Product 'A'

Product A $\xrightarrow[H^+]{H_2O}$ Product "B"

Product A $\xrightarrow[H_2O/H_2O_2/OH^-]{\text{Diborane}}$ Product "C"

Consider the above reactions, identify product B and product C. **[2024, 4 April Shift-II]**

(a) $B = C =$ 2- Propanol
(b) $B = C =$1- Propanol
(c) $B =$ 2- Propanol $C =$1- Propanol
(d) $B =$1-Propanol $C =$ 2- Propanol

35 Most stable product of the following reaction is **[2022, 25 July Shift-I]**

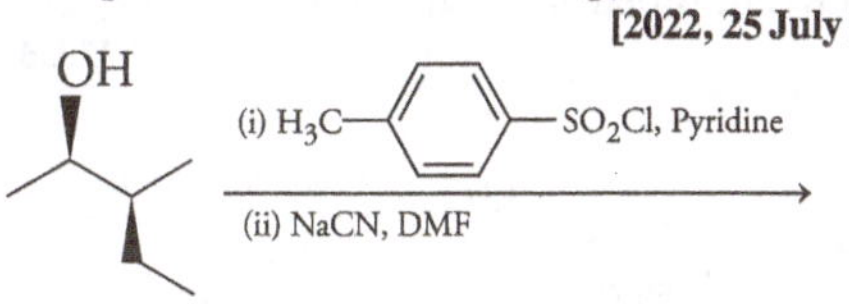

(a) 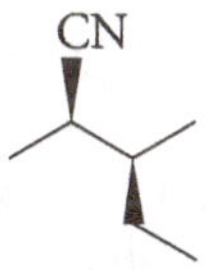 (b) 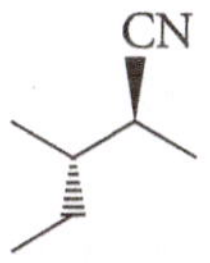

(c) 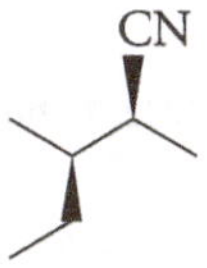 (d) 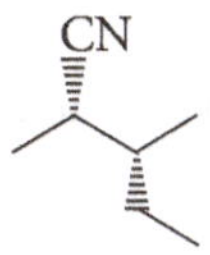

36 In the following sequence of reactions, the final product D is

$CH_3—C{\equiv}C—H + NaNH_2 \longrightarrow A$

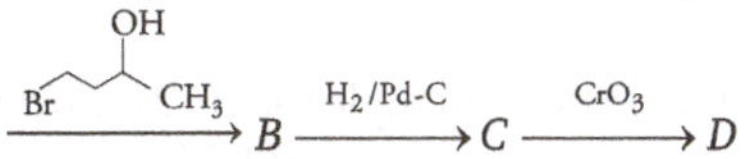

[2021, 27 Aug Shift-I]

(a) $CH_3—CH_2—CH_2—CH_2—CH_2—\overset{O}{\overset{\|}{C}}—CH_3$

(b) $H_3C—CH{=}CH—CH(OH)—CH_2—CH_2CH_3$

(c) $CH_3—CH{=}CH—CH_2—CH_2—CH_2—COOH$

(d) $H_3C—CH_2—CH_2—CH_2—CH_2—\overset{O}{\overset{\|}{C}}—H$

37 The correct order of stability for the following alkoxides is [2020, 7 Jan Shift-II]

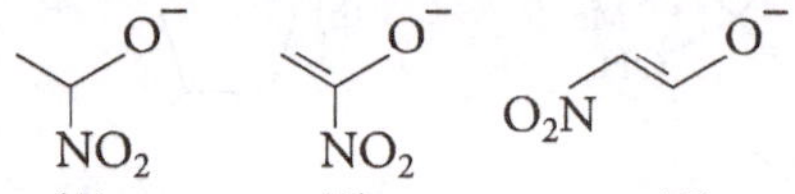

(A) (B) (C)

(a) (C) > (B) > (A)
(b) (B) > (C) > (A)
(c) (B) > (A) > (C)
(d) (C) > (A) > (B)

Numerical Value Type Questions

Imp. *Chemical Reactions of Alcohols & Stoichiometry*

38

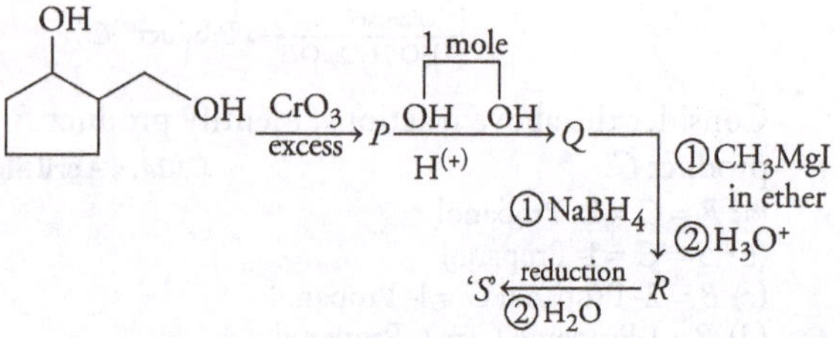

0.1 mole of compound 'S' will weigh ...... g.
(Given, molar mass in g mol^{-1} C : 12, H : 1, O : 16) [2025, 29 Jan Shift-I]

39

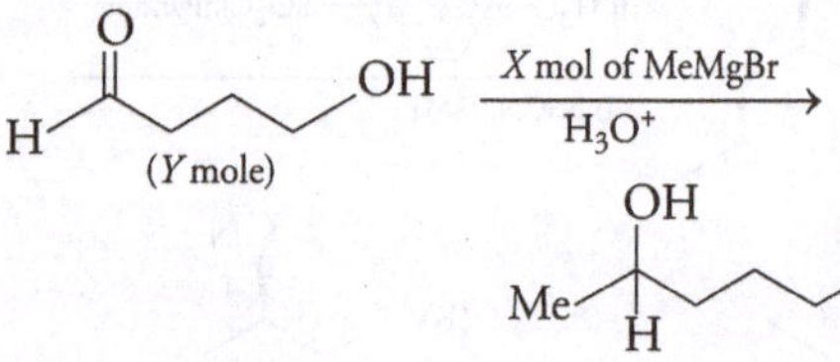

[2023, 11 April Shift-I]

The ratio X / Y on completion of the above reaction is ....

40 In the given reaction, [2022, 29 June Shift-II]

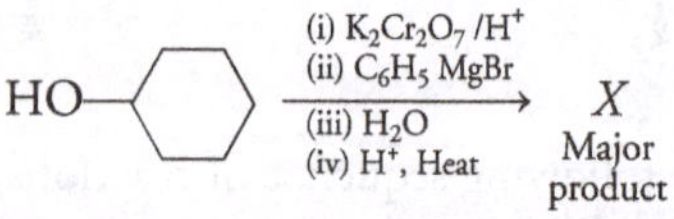

the number of sp^2 hybridised carbon(s) in compound 'X' is ....

41 To synthesise 1.0 mole of 2-methylpropan-2-ol from Ethylethanoate ________ equivalents of CH_3MgBr reagent will be required. (Integer value) [2021, 20 July Shift-I]

Topic 2 / Preparation, Properties and Uses of Phenols

• Level 1

01 Which one of the following, with HBr will give a phenol? [2025, 29 Jan Shift-II]

(a) 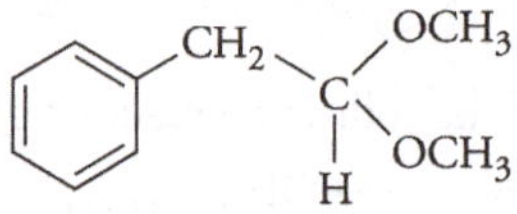

(b) 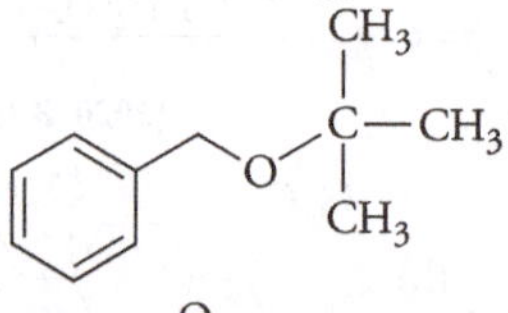

(c) 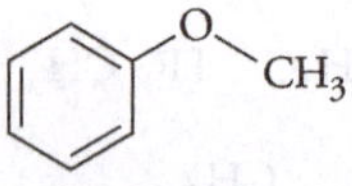

(d) 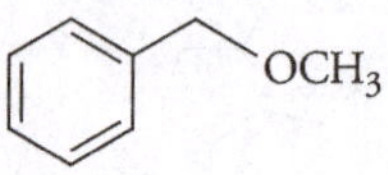

Imp. *Intermediate Involved in Reimer Tiemann Reaction*

02 In Reimer-Tiemann reaction, phenol is converted into salicylaldehyde through an intermediate. The structure of intermediate is [2024, 6 April Shift-I]

(a) 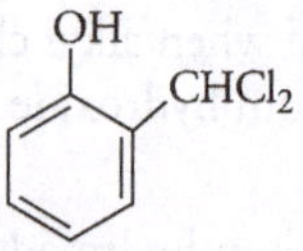

(b) 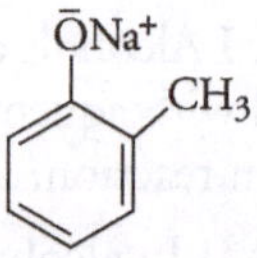

(c) 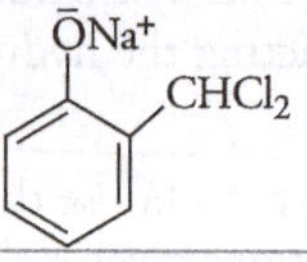

(d) 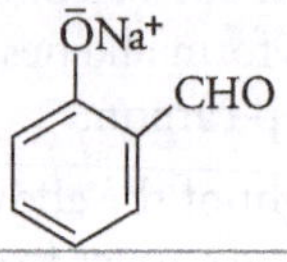

03 Salicylaldehyde is synthesised from phenol, when reacted with [2024, 30 Jan Shift-II]

(a) CCl_4, NaOH
(b) $HCCl_3$, NaOH
(c) CO_2, NaOH
(d) 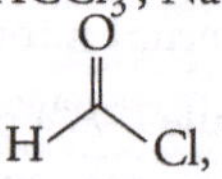 , NaOH

04 The strongest acid from the following is [2023, 6 April Shift-II]

(a) 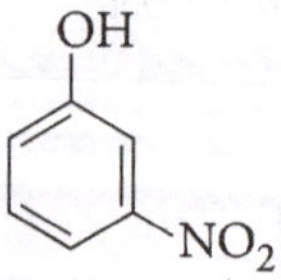

(b) 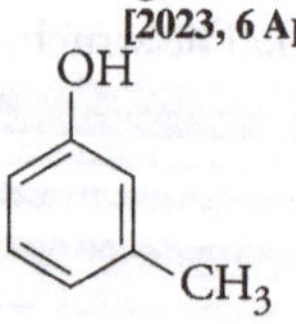

(c) 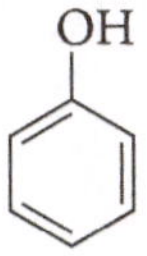

(d) 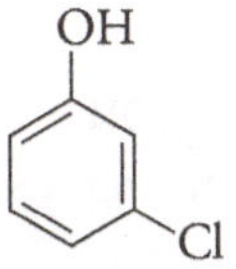

05 The difference in the reaction of phenol with bromine in chloroform and bromine in water medium is due to **[2022, 26 July Shift-I]**

(a) hyperconjugation in substrate
(b) polarity of solvent
(c) free radical formation
(d) electromeric effect of substrate

06 Given below are two statements. **[2022, 29 June Shift-I]**

Statement I Phenols are weakly acidic.

Statement II Therefore they are freely soluble in NaOH solution and are weaker acids than alcohols and water.

Choose the most appropriate option.

(a) Both Statement I and Statement II are correct.
(b) Both Statement I and Statement II are incorrect.
(c) Statement I is correct but Statement II is incorrect.
(d) Statement I is incorrect but Statement II is correct.

07 The intermediate *X*, in the reaction **[2022, 25 June Shift-I]**

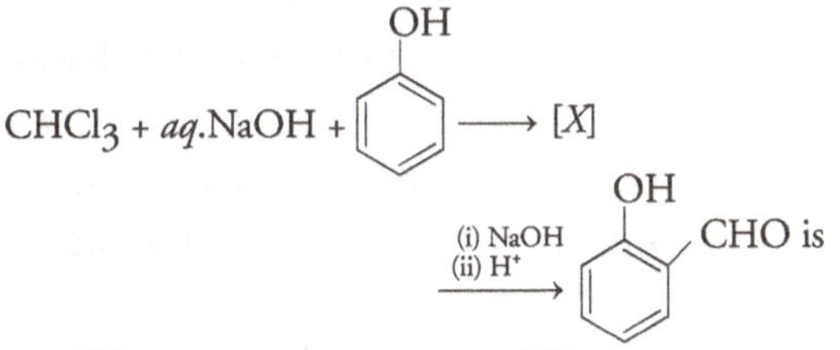

(a) 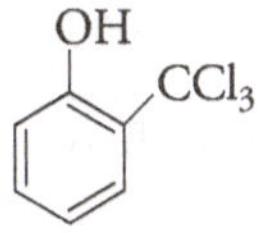

(b) 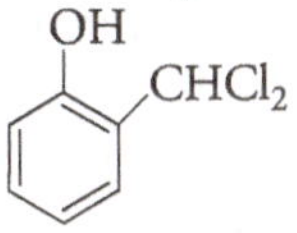

(c) 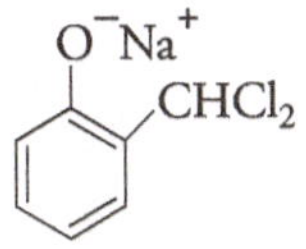

(d) 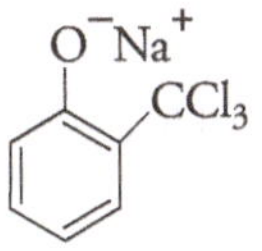

08 Which one of the following phenols does not give colour when condensed with phthalic anhydride in presence of conc. H_2SO_4 ? **[2021, 26 Aug Shift-II]**

(a) 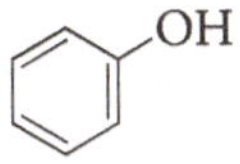

(b) 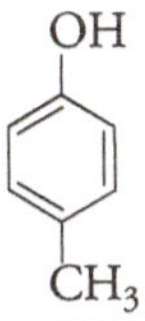

(c) 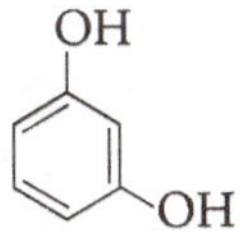

(d) 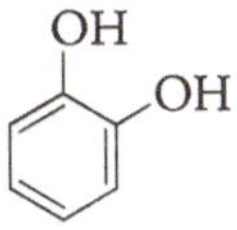

• Level 2

09 Given below are two statetments. **[2024, 6 April Shift-I]**

Statement I Picric acid is 2, 4, 6-trinitrotoluene

Statement II Phenol-2, 4-disulphonic acid is treated with conc. HNO_3, to get picric acid.

In the light of the above statements, choose the most appropriate answer from the options given below.

(a) Both statement I and Statement II are incorrect.
(b) Statement I is incorrect but Statement II is correct.
(c) Both Statement I and Statement II are correct.
(d) Statement I is correct but Statement II is incorrect.

10 Which one of the following reaction is not possible ? **[2024, 5 April Shift-II]**

(a) 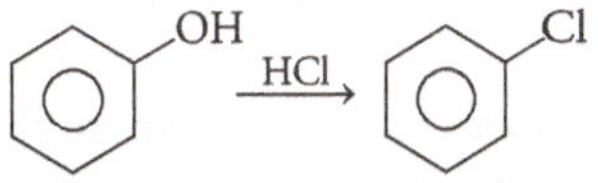

(b) 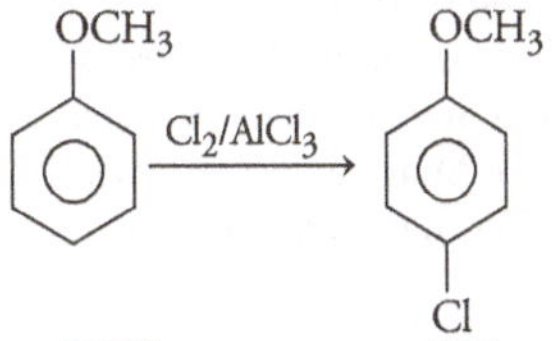

(c) 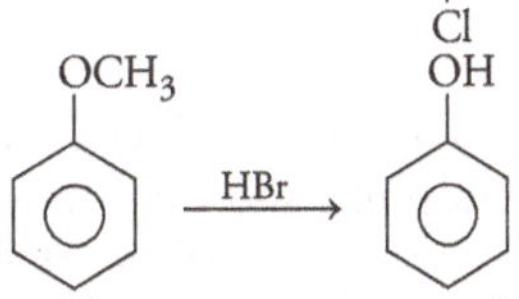

(d) 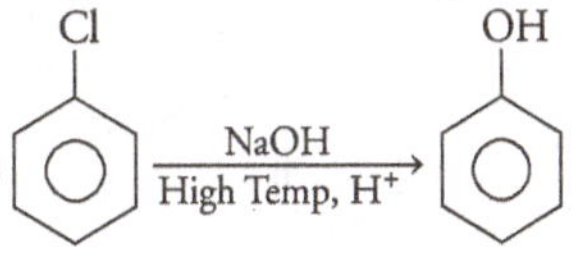

11 Given below are two statement. **[2024, 5 April Shift-I]**

Statement I Bromination of phenol in solvent with low polarity such as $CHCl_3$ or CS_2 requires Lewis acid catalyst.

Statement II The lewis acid catalyst polarises the bromine to generate Br^+.

In the light of the above statements, choose the correct appropriate answer from the options given below:

(a) Both Statement I and Statement II are false.
(b) Statement I is false but Statement II is true.
(c) Statement I is true but Statement II is false.
(d) Both Statement I and Statement II are true.

12 The major product (*P*) in the following reaction is **[2024, 29 April Shift-I]**

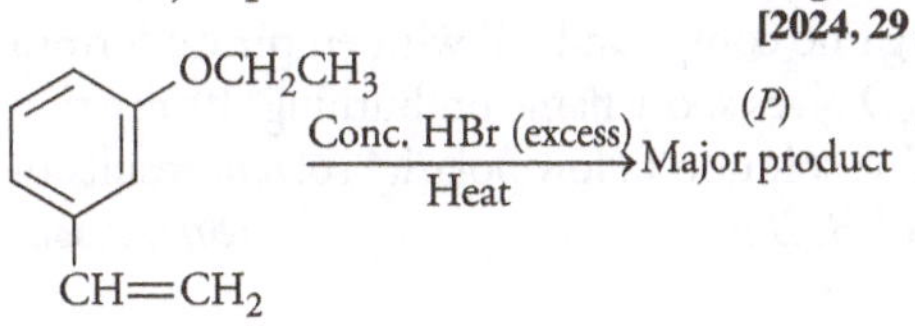

(a) 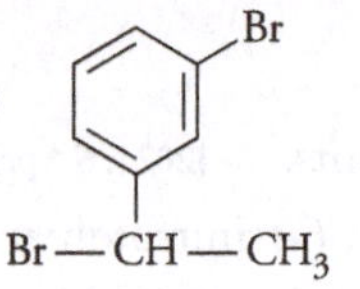
(b) 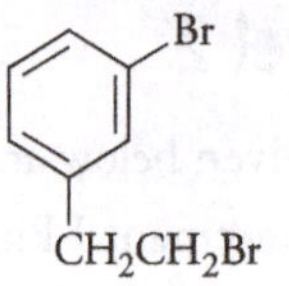
(c) 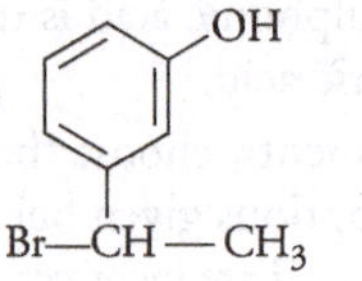
(d) 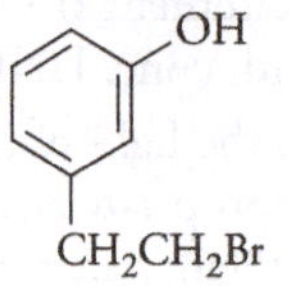

13 Given below are two statements. **[2024, 27 Jan Shift-I]**

Statement I *p*-nitrophenol is more acidic than *m*-nitrophenol and *o*-nitrophenol.

Statement II Ethanol will give immediate turbidity with Lucas reagent.

In the light of the above statements, choose the **correct** answer from the options given below

(a) Both Statement I and Statement II are true.
(b) Both Statement I and Statement II are false.
(c) Statement I is true but Statement II is false.
(d) Statement I is false but Statement II is true.

Imp. *Order of Electrophilic Substitution Reaction of Alcohols*

14 Arrange the following compounds in increasing order of rate of aromatic electrophilic substitution reaction **[2023, 11 April Shift-I]**

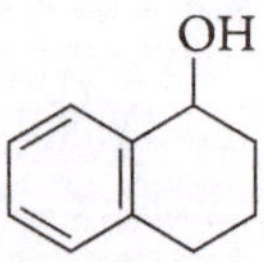
a
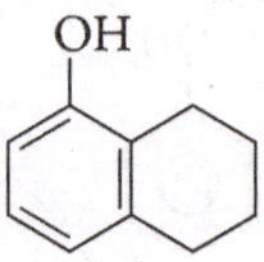
b
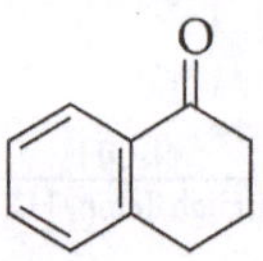
c
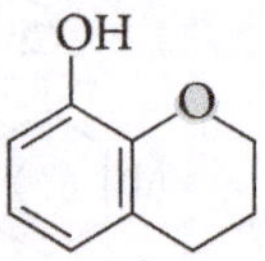
d

(a) *d, b, c, a* (b) *b, c, a, d*
(c) *c, a, b, d* (d) *d, b, a, c*

15 The correct order for acidity of the following hydroxyl compound is

A. CH_3OH *B*. $(CH_3)_3COH$

C. 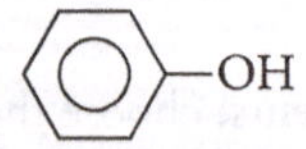 *D*. 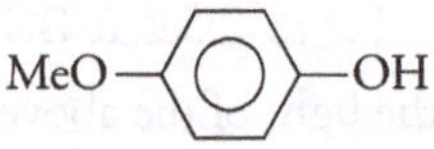

E. 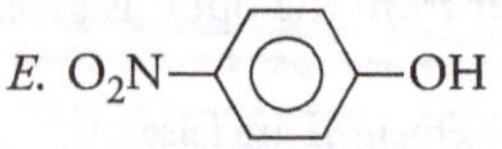

Choose the correct answer from the options given below **[2023, 10 April Shift-II]**

(a) $C > E > D > B > A$ (b) $E > D > C > B > A$
(c) $E > C > D > A > B$ (d) $D > E > C > A > B$

16 An organic compound '*A*' with empirical formula C_6H_6O gives sooty flame on burning. Its reaction with bromine solution in low polarity solvent results in high yield of *B*. *B* is **[2023, 31 Jan Shift-I]**

(a) 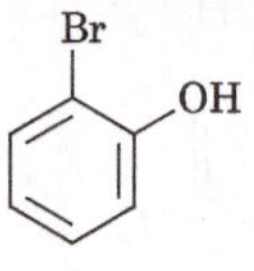
(b) 
(c) 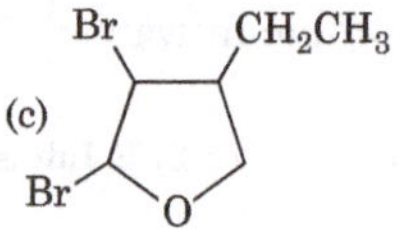
(d) 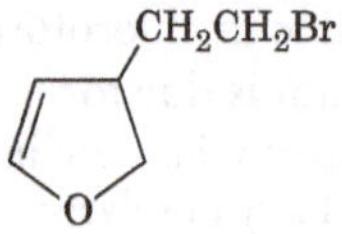

17 The correct order of pK_a values for the following compounds is **[2023, 31 Jan Shift-I]**

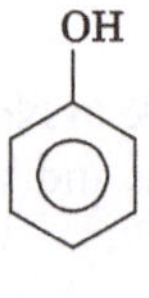
A
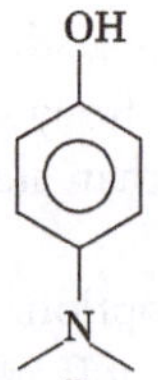
B
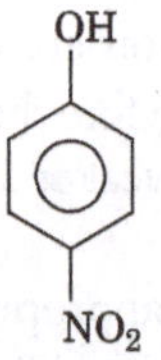
C
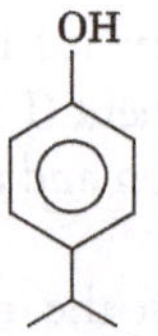
D

(a) $B > D > A > C$ (b) $B > A > D > C$
(c) $C > A > D > B$ (d) $A > B > C > D$

Imp. *Order of pK_a of Phenols*

18 The increasing order of pK_a for the following phenols is **[2023, 29 Jan Shift-I]**

(A) 2, 4 - dinitrophenol (B) 4 - nitrophenol
(C) 2, 4, 5 - trimethylphenol (D) phenol
(E) 3-chlorophenol

Choose the correct answer from the option given below.

(a) A, E, B, D, C (b) A, B, E, D, C
(c) C, D, E, B, A (d) C, E, D, B, A

19 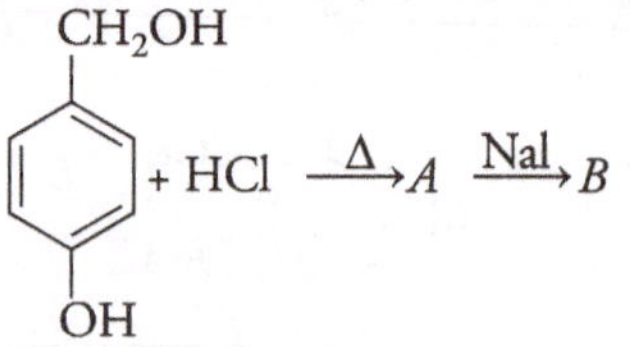

[2022, 27 July Shift-I]

(a) 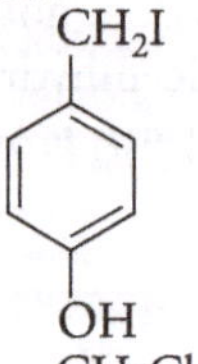
(b) 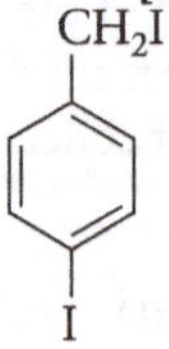
(c) 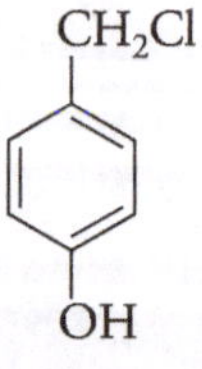
(d) 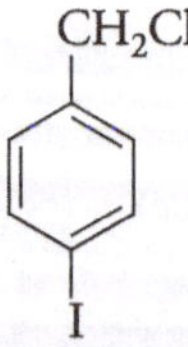

20 In the given reaction

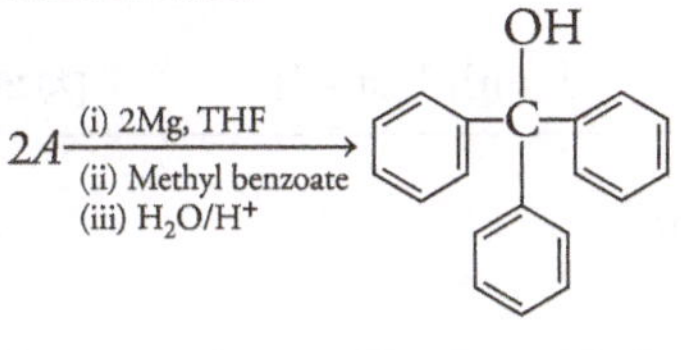

A can be **[2022, 25 June Shift-II]**

(a) benzyl bromide (b) bromobenzene
(c) cyclohexyl bromide (d) methyl bromide

21 In the following reaction, **[2022, 25 June Shift-I]**

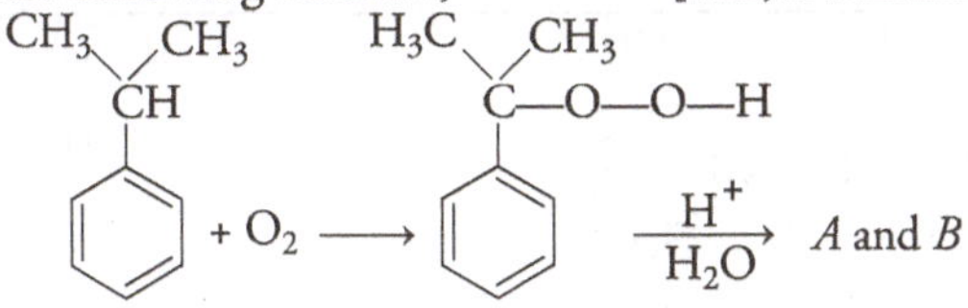

The compounds A and B respectively are

(a) 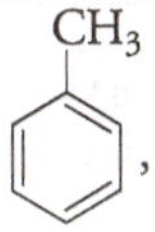, CH_3COOH (b) 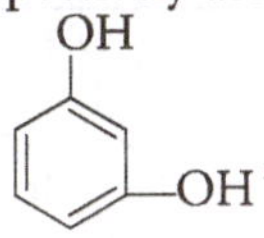, CH_3COOH

(c) 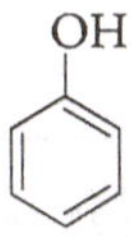, CH_3COCH_3 (d) 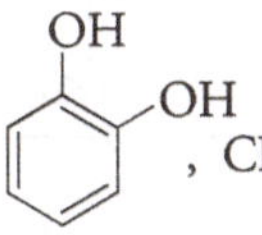, CH_3COCH_3

22 The major product of the following reaction, if it occurs by S_N2 mechanism is **[2021, 27 Aug Shift-II]**

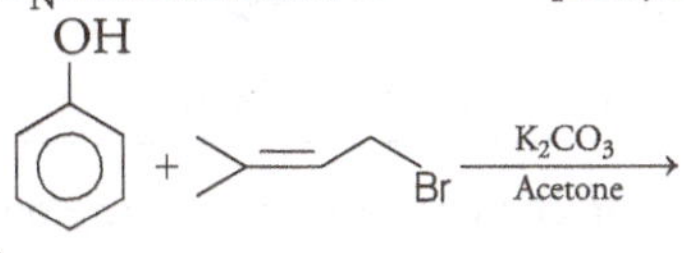

(a) 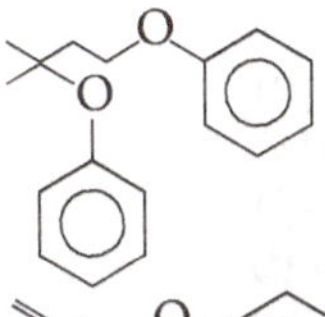 (b) 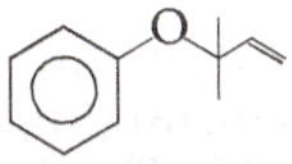

(c) 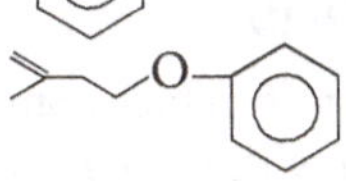 (d) 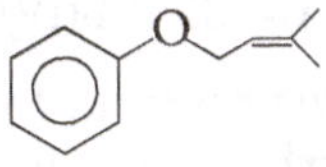

23 The correct options for the products A and B of the following reactions are **[2021, 26 Aug Shift-I]**

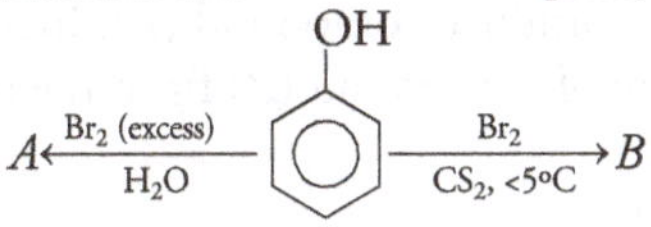

(a) A = 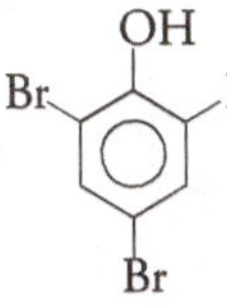, B = 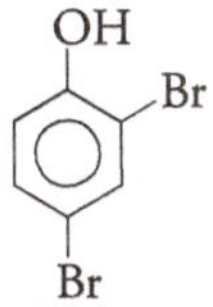

(b) A = 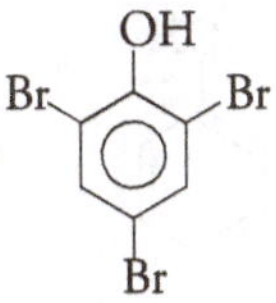, B = 

(c) A = 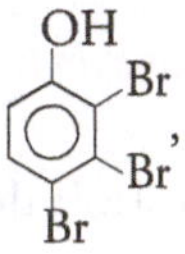, B = 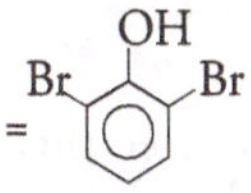

(d) A = 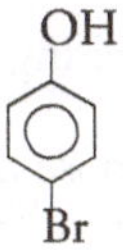, B = 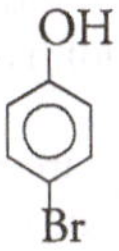

24 Identify A in the given reaction.

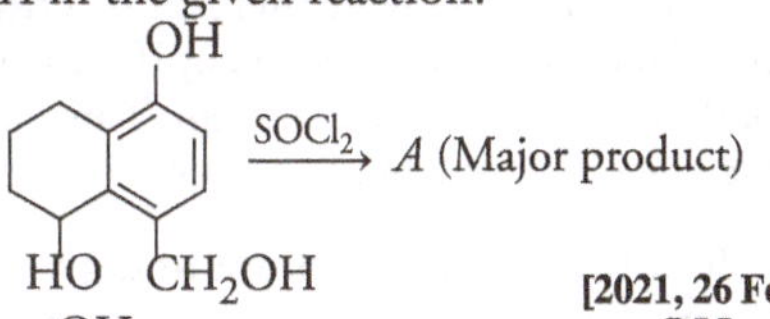

[2021, 26 Feb Shift-II]

(a) 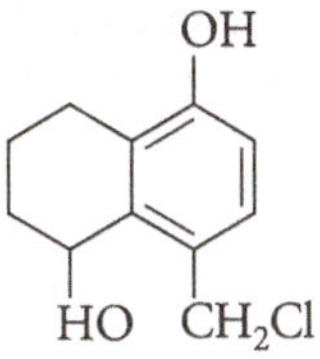 (b) 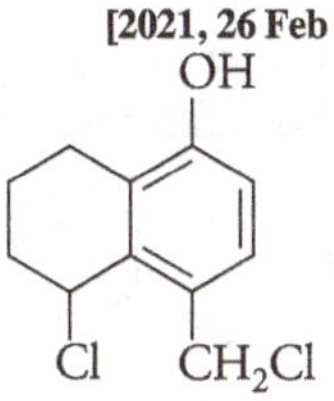

(c) 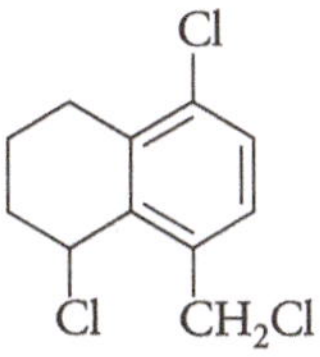 (d) 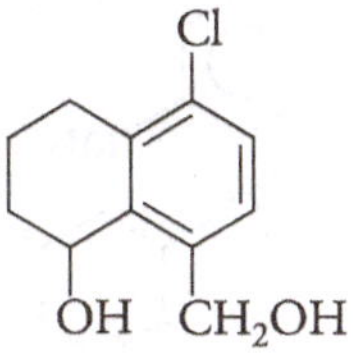

25 Identify the major products A and B respectively in the following reactions of phenol. **[2021, 26 Feb Shift-I]**

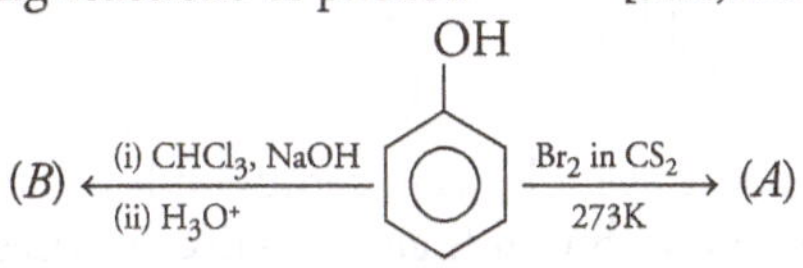

(a) 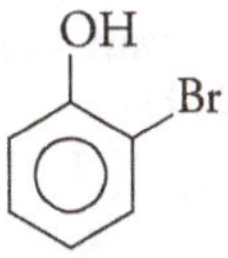 and 

(b)  and 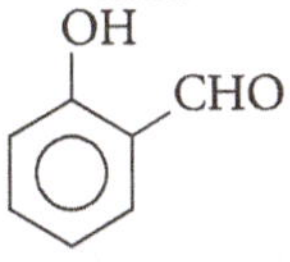

(c) 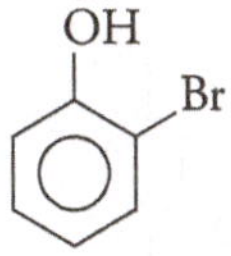 and 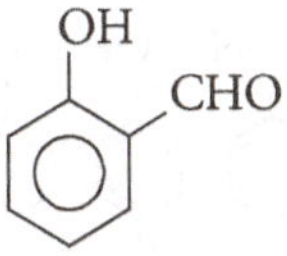

(d)  and 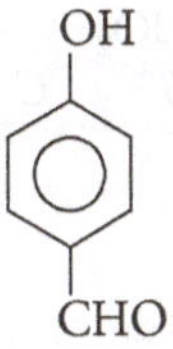

Imp. *Types of H-bonding in Nitrophenols*

26 Given below are two statements
Statement I *o*-nitrophenol is steam volatile due to intramolecular hydrogen bonding.
Statement II *o*-nitrophenol has high melting due to hydrogen bonding.
In the light of the above statements, choose the most appropriate answer from the options given below. **[2021, 26 Feb Shift-I]**
(a) Statement I is false but Statement II is true
(b) Both Statement I and Statement II are true
(c) Both Statement I and Statement II are false
(d) Statement I is true but Statement II is false

27 Which of the following compound gives pink colour on reaction with phthalic anhydride in conc. H_2SO_4 followed by treatment with NaOH? **[2021, 24 Feb Shift-I]**

(a) 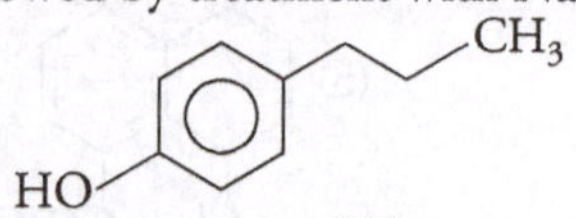

(b) 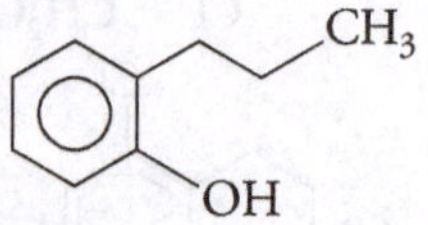

(c) 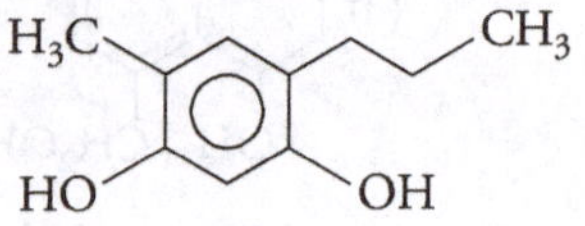

(d) 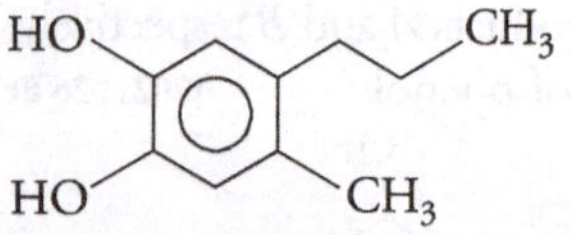

28 Which of the following compounds reacts with ethyl magnesium bromide and also decolourises bromine water solution **[2019, 11 Jan Shift-II]**

(a) 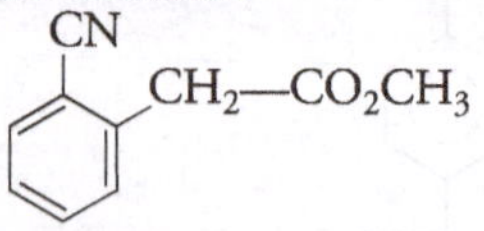
(b) 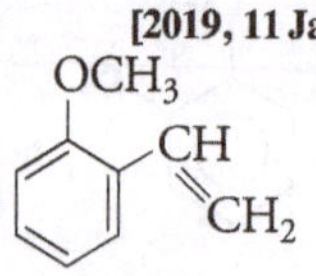

(c) 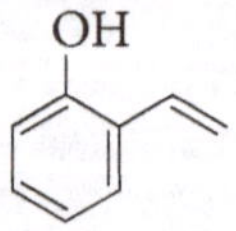
(d) 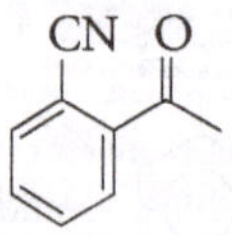

29 The increasing order of the pK_a values of the following compounds is **[2019, 10 Jan Shift-I]**

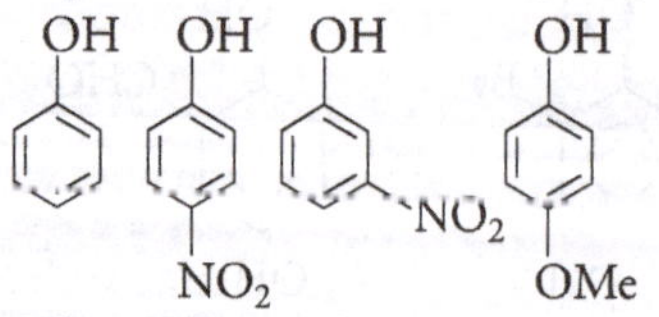

A *B* *C* *D*

(a) $D < A < C < B$
(b) $B < C < A < D$
(c) $C < B < A < D$
(d) $B < C < D < A$

• Level 3

30. Match List - I with List - II. **[2025, 7 April Shift-II]**

	List - I (Conversion)		List - II (Reagents, conditions used)
A.	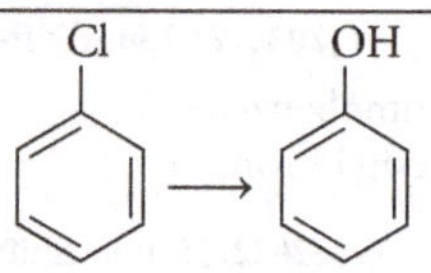	I.	Warm, H_2O
B.	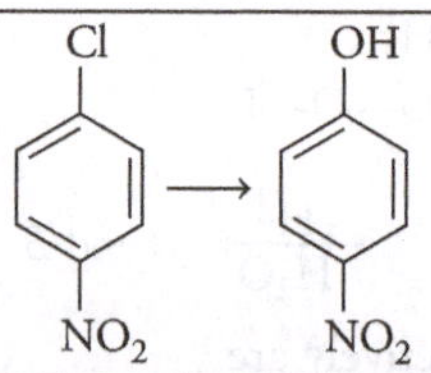	II.	(a) NaOH, 368 K; (b) H_3O^+
C.	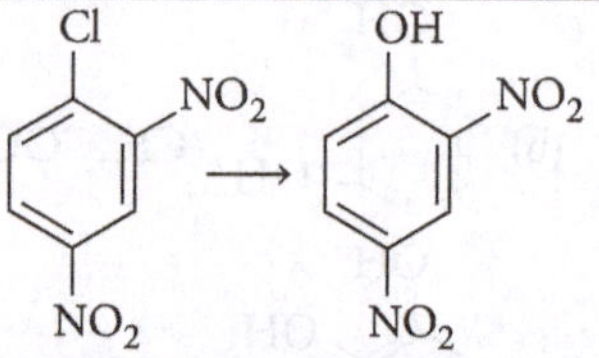	III.	(a) NaOH, 443 K; (b) H_3O^+
D.	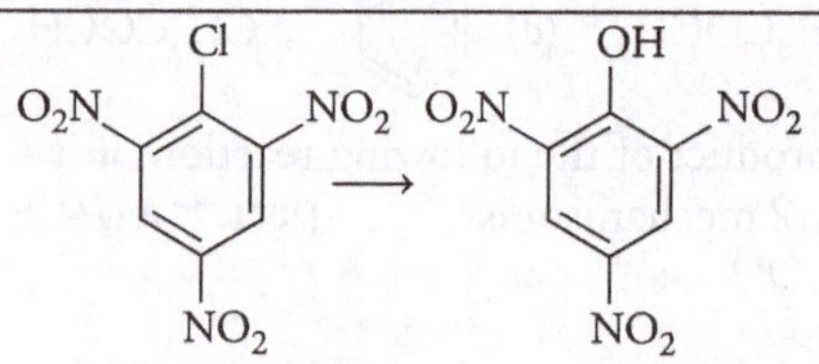	IV.	(a) NaOH, 623 K, 300 atm; (b) H_3O^+

Choose the correct answer from the options given below:
(a) (A)-(IV), (B)-(III), (C)-(I), (D)-(II)
(b) (A)-(II), (B)-(III), (C)-(I), (D)-(IV)
(c) (A)-(IV), (B)-(III), (C)-(II), (D)-(I)
(d) (A)-(III), (B)-(IV), (C)-(II), (D)-(I)

31. Benzene is treated with oleum to produce compound (*X*) which when further heated with molten sodium hydroxide followed by acidification produces compound (*Y*). The compound *Y* is treated with zinc metal to produce compound (*Z*). Identify the structure of compound (*Z*) from the following option. **[2025, 4 April Shift-I]**

(a) 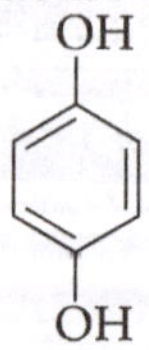
(b) 

(c) 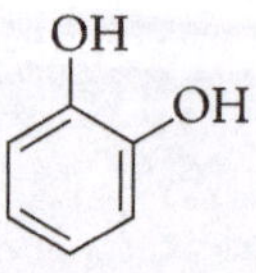
(d) 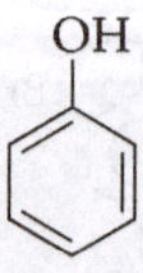

32 Match List-I with List-II. **[2024, 8 July Shift-II]**

List-I (Reactions)		List-II (Products)	
A.	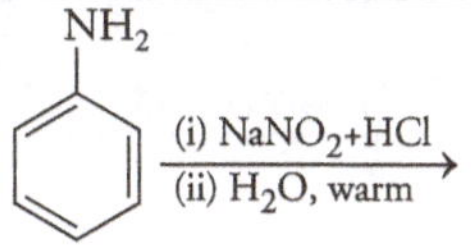	I.	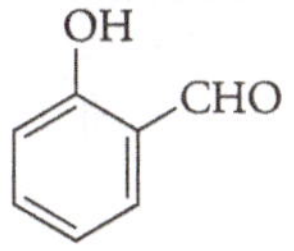
B.	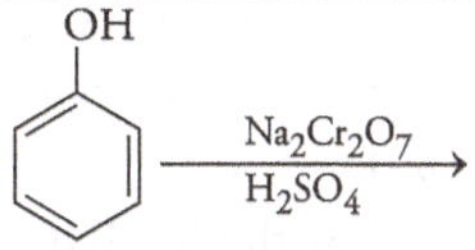	II.	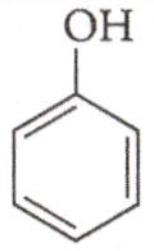
C.	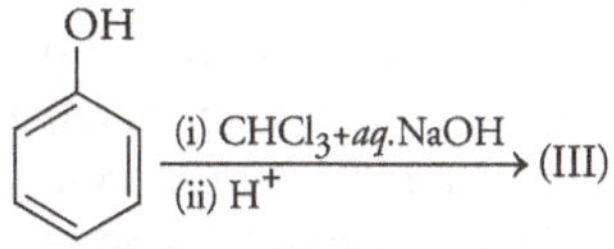	III.	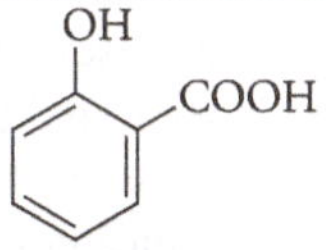
D.	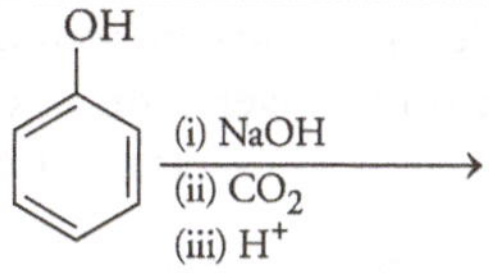	IV.	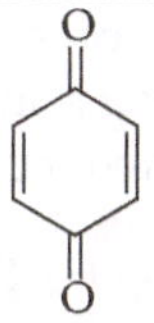

Choose the correct answer from the options given below

(a) A-I, B-IV, C-II, D-III (b) A-III, B-II, C-I, D-IV
(c) A-IV, B-II, C-III, D-I (d) A-II, B-IV, C-I, D-III

Imp. *Chemical Reactions of Phenol*

33 Match List-I with List-II. **[2024, 1 Feb Shift-II]**

List-I (Reactants)	List-II (Product)
A. Phenol, Zn / Δ	I. Salicylaldehyde
B. Phenol, $CHCl_3$, NaOH, HCl	II. Salicylic acid
C. Phenol, CO_2, NaOH, HCl	III. Benzene
D. Phenol, Conc. HNO_3	IV. Picric acid

Choose the correct answer from the options given below

(a) A-IV, B-I, C-II, D-III (b) A-III, B-IV, C-I, D-II
(c) A-III, B-I, C-II, D-IV (d) A-IV, B-II, C-I, D-III

34 Match List-I with List-II. **[2024, 29 Jan Shift-II]**

List-I (Compound)	List-II (pK_a value)
A. Ethanol	I. 10.0
B. Phenol	II. 15.9
C. *m*-nitrophenol	III. 7.1
D. *p*-nitrophenol	IV. 8.3

Choose the correct answer from the options given below

(a) A-IV, B-I, C-II, D-III (b) A-I, B-II, C-III, D-IV
(c) A-III, B-IV, C-I, D-II (d) A-II, B-I, C-IV, D-III

35 Match List-I with List-II. **[2024, 27 Jan Shift-II]**

List-I (Reaction)		List-II (Reagent(s))
A.	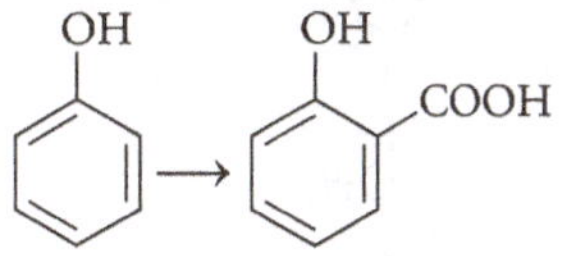	I. $Na_2Cr_2O_7$, H_2SO_4
B.	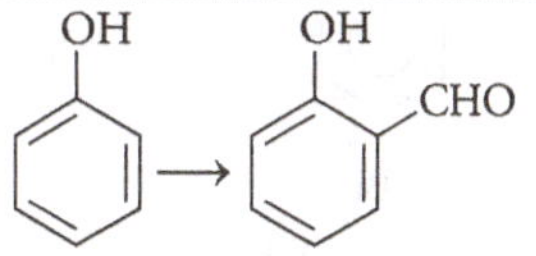	II. (i) NaOH (ii) CH_3Cl
C.	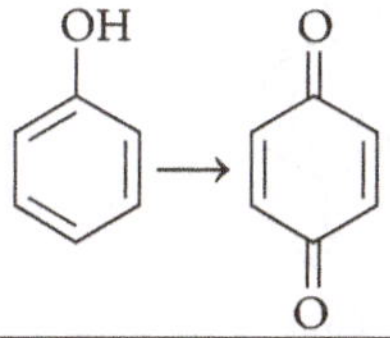	III. (i) NaOH, $CHCl_3$ (ii) NaOH (iii) HCl
D.	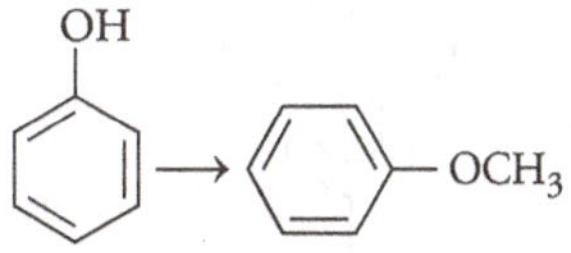	IV. (i) NaOH (ii) CO_2 (iii) HCl

Choose the correct answer from the options given below

(a) A-II, B-III, C-I, D-IV (b) A-II, B-I, C-III, D-IV
(c) A-IV, B-I, C-III, D-II (d) A-IV, B-III, C-I, D-II

36 The ascending order of acidity of —OH group in the following compounds is **[2024, 27 Jan Shift-I]**

A. Bu—OH

B. 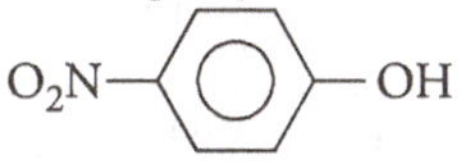

C. 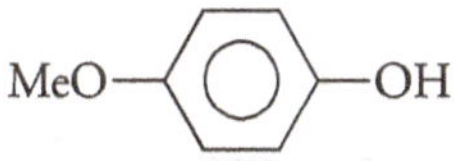

D. 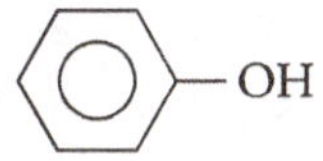

E. 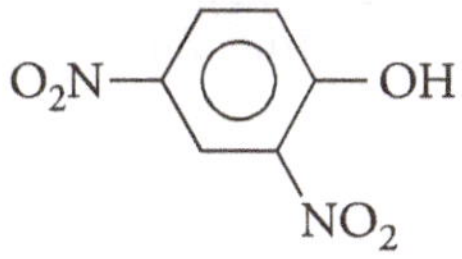

Choose the correct answer from the options given below:

(a) A < C < D < B < E
(b) A < D < C < B < E
(c) C < D < B < A < E
(d) C < A < D < B < E

37 Identify the major products *A* and *B* for the below given reaction sequence. **[2022, 28 July Shift-I]**

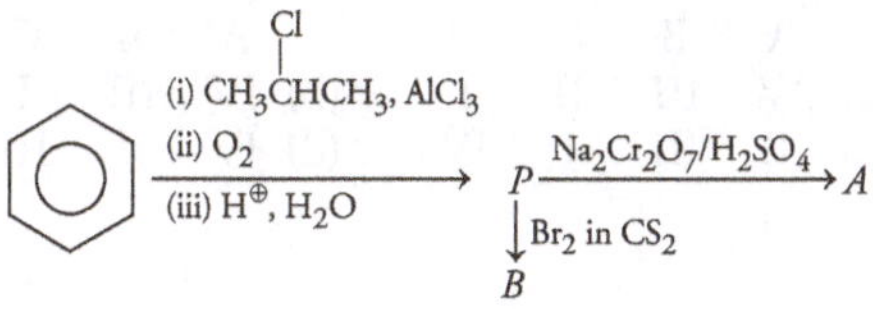

	(A)		(B)
(a)	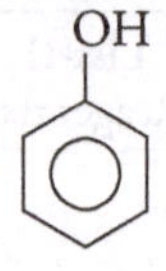	and	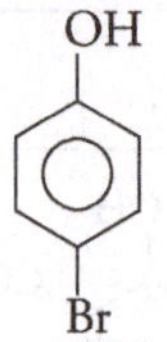
(b)	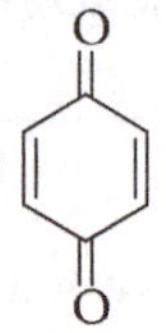	and	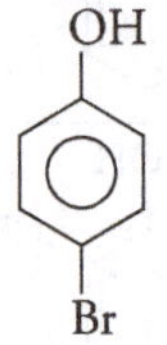
(c)	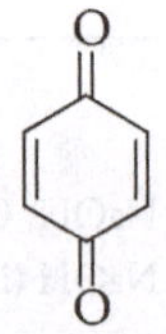	and	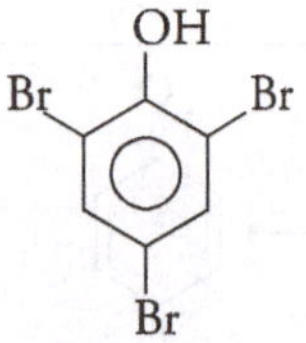
(d)	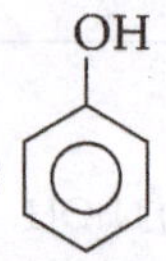	and	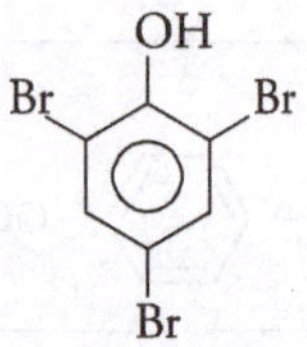

38 Match List-I with List-II. **[2022, 27 June Shift-II]**

List-I		List-II	
A.	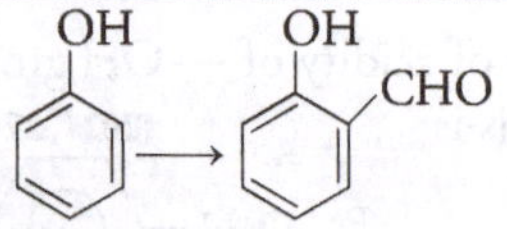	I.	Br_2 in CS_2
B.	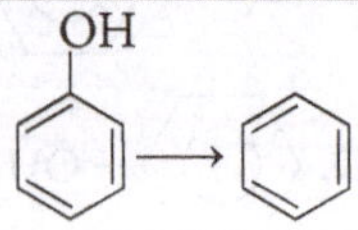	II.	$Na_2Cr_2O_7/H_2SO_4$
C.	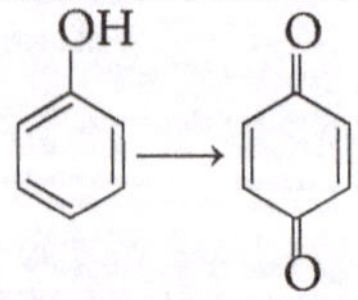	III.	Zn
D.	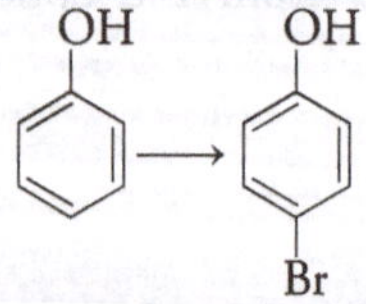	IV.	$CHCl_3$ / NaOH

Choose the correct answer from the options given below.

	A	B	C	D		A	B	C	D
(a)	IV	III	II	I	(b)	IV	III	I	II
(c)	II	III	I	IV	(d)	IV	II	III	I

Numerical Value Type Questions

39 In the following reaction, **[2022, 27 July Shift-I]**

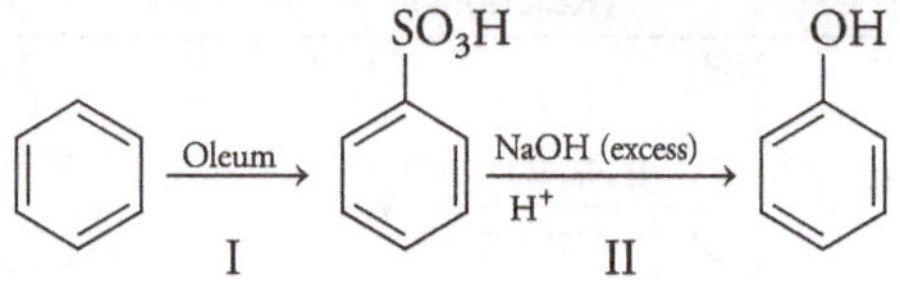

the % yield for reaction I is 60% and that of reaction II is 50%. The overall yield of the complete reaction is ....%. [Nearest integer]

40 Compound (*P*) on nitration with dil. HNO_3 yields two isomers (*A*) and (*B*). These isomers can be separated by steam distillation. Isomers (*A*) and (*B*) show the intramolecular and intermolecular hydrogen bonding respectively. Compound (*P*) on reaction with conc. HNO_3 yields a yellow compound (*C*), a strong acid. The number of oxygen atoms present in compound (*C*) is. **[2022, 26 June Shift-I]**

Topic 3 / Preparation, Properties and Uses of Ethers

Level 1

01 Suitable reaction condition for preparation of methyl phenyl ether is **[2023, 10 April Shift-I]**

(a) $Ph\overset{\ominus}{O}Na^{\oplus}$, MeOH
(b) Ph—Br, $Me\overset{\ominus}{O}Na^{\oplus}$
(c) Benzene, MeBr
(d) $Ph\overset{\ominus}{O}Na^{\oplus}$, MeBr

02 Given below are two statements, one is labelled as Assertion (A) and the other is labelled as Reason (R).

Assertion (A) Acetal/ketal is stable in basic medium.

Reason (R) The high leaving tendency of alkoxide ion gives the stability to acetal/ketal in basic medium.

In the light of the above statements, choose the correct answer from the options given below : **[2023, 25 Jan Shift-I]**

(a) A is true but R is false.
(b) A is false but R is true.
(c) Both A and R are true but R is not the correct explanation of A.
(d) Both A and R are true and R is the correct explanation of A.

03 The major product of the following reaction is

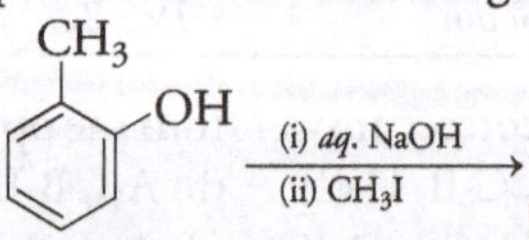

[2019, 10 Jan Shift-II]

(a) 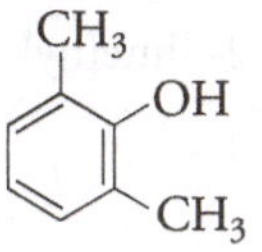

(b) 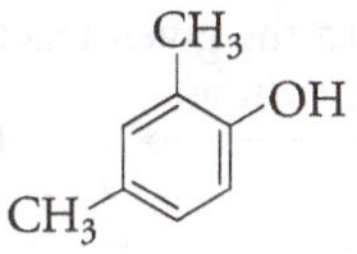

(c) 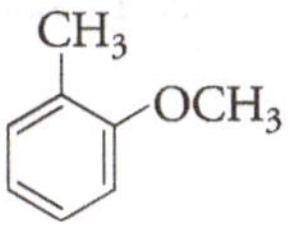

(d) 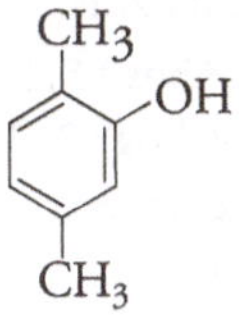

• Level 2

04 Which one of the following reactions will not lead to the desired ether formation in major proportion?
(*iso*-Bu ⇒ *iso*butyl, *sec*-Bu ⇒ *sec*-butyl, *n*Pr ⇒ *n*-propyl, ^{t}Bu ⇒ *tert*-butyl, Et ⇒ ethyl) **[2025, 8 April Shift-II]**

(a) *iso*-BuO$^-$ Na$^+$ + *sec*-BuBr ⟶ *sec*-Bu—O—*iso*-Bu

(b) 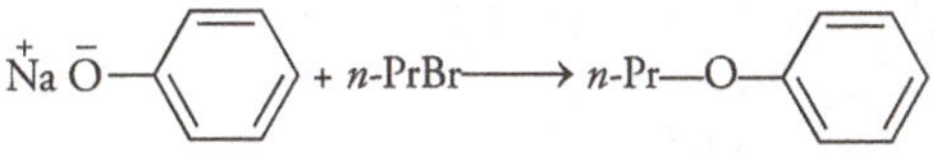

(c) 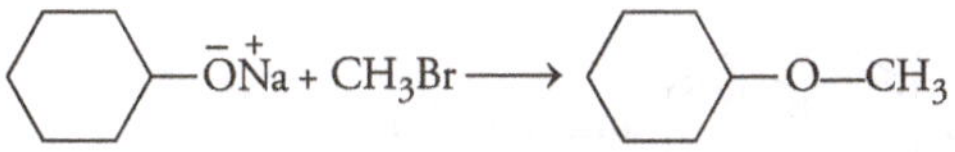

(d) ^{t}BuO$^-$ Na$^+$ + EtBr ⟶ ^{t}Bu—O—Et

05 Given below are two statements: **[2025, 7 April Shift-I]**

Statement I Dimethyl ether is completely soluble in water. However, diethyl ether is soluble in water to a very small extent.

Statement II Sodium metal can be used to dry diethyl ether and not ethyl alcohol.

In the light of given statements, choose the correct answer from the options given below

(a) Statement I is true but Statement II is false.
(b) Both Statement I and Statement II are true.
(c) Both Statement I and Statement II are false.
(d) Statement I is false but Statement II are true.

06 Given below are two statements. **[2025, 23 Jan Shift-II]**

Statement I The boiling points of alcohols and phenols increase with increase in the number of C-atoms.

Statement II The boiling points of alcohols and phenols are higher in comparison to other class of compounds such as ethers, haloalkanes.

In the light of the above statements, choose the correct answer from the options given below

(a) Statement I is false but Statement II is true.
(b) Both Statement I and Statement II are true.
(c) Statement I is true but Statement II is false.
(d) Both Statement I and Statement II are false.

07 Identify the major products *A* and *B* respectively in the following set of reactions. **[2024, 8 April Shift-I]**

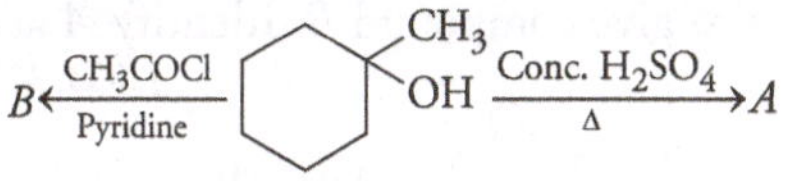

(a) *A* = 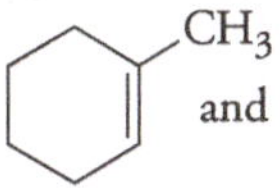 and *B* = 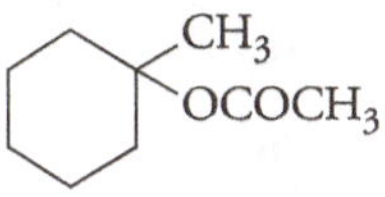

(b) *A* = 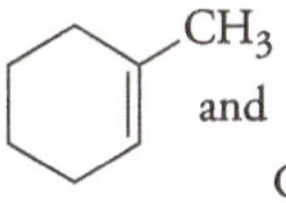 and *B* = 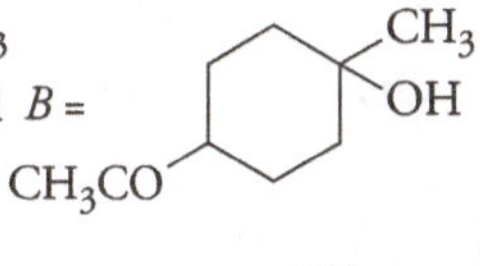

(c) *A* = 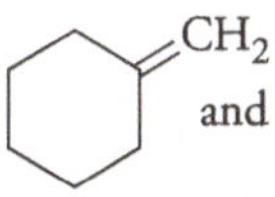 and *B* = 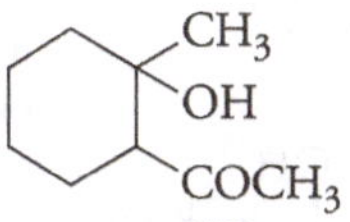

(d) *A* = 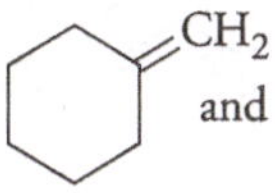 and *B* = 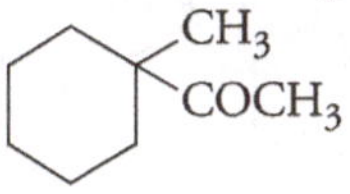

08 The major products formed

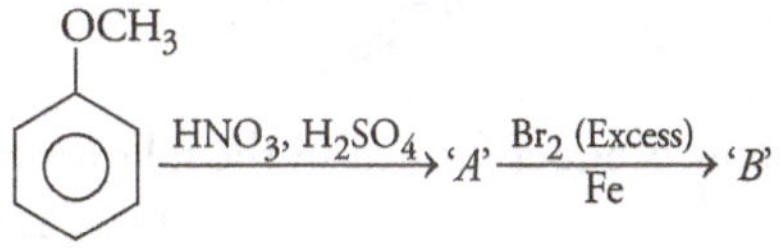

A and *B* respectively are **[2024, 6 April Shift-II]**

(a) 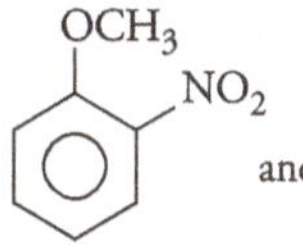 and 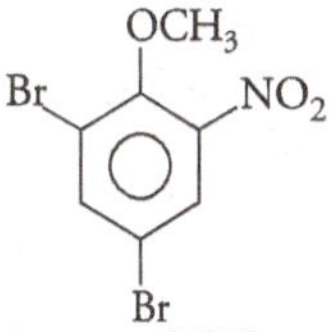

(b) 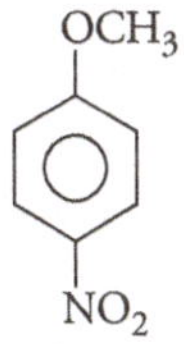 and 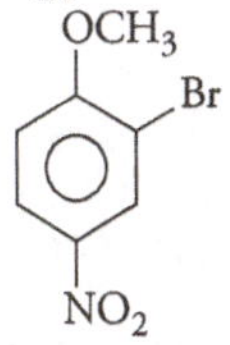

(c) 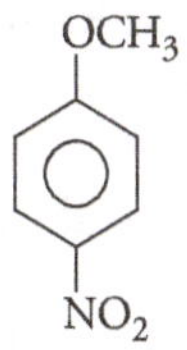 and 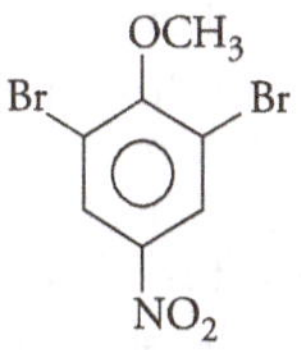

(d) 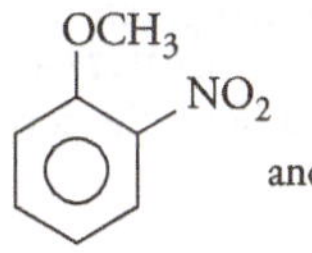 and 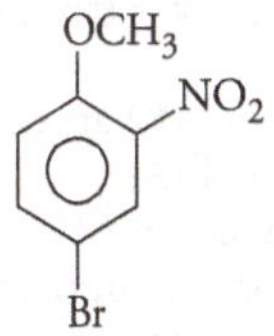

09 Compound I is heated with conc. HI to give a hydroxy compound A which is further heated with Zn dust to give compound B. Identify A and B.

[2022, 28 July Shift-II]

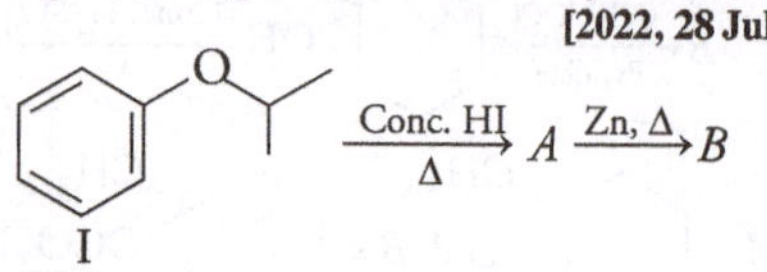

(a) A = 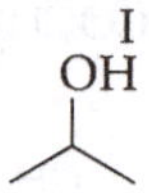 B = 

(b) A = 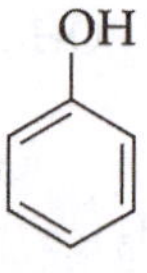 B = 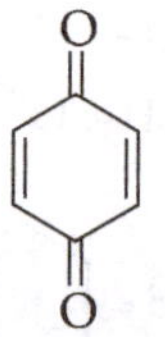

(c) A = 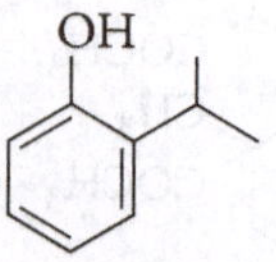 B = 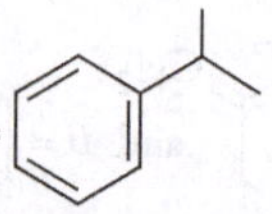

(d) A = 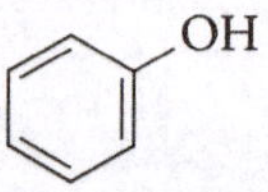 B = 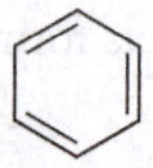

10 The major product formed in the following reaction, is

[2022, 25 June Shift-II]

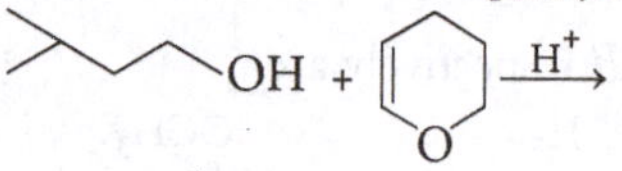

(a) 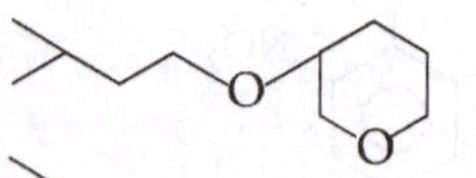

(b) 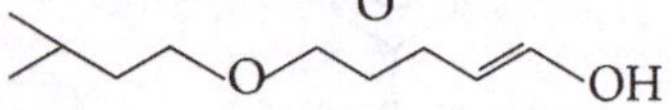

(c) 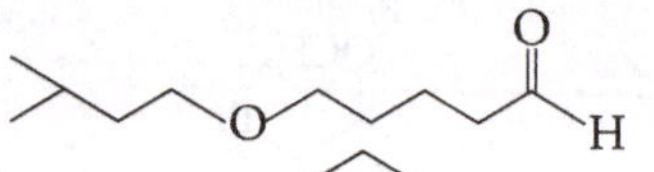

(d) 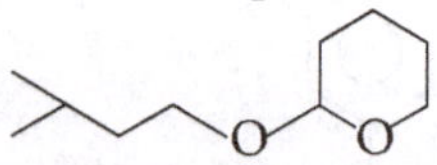

11 Given below are two statements one is labelled as Assertion (A) and the other is labelled as Reason (R).

Assertion (A) Synthesis of ethyl phenyl ether may be achieved by Williamson synthesis.

Reason (R) Reaction of bromobenzene with sodium ethoxide yields ethyl phenyl ether.

In the light of the above statements, choose the most appropriate answer from the options given below

[2021, 27 Aug Shift-I]

(a) Both (A) and (R) are correct and (R) is the correct explanation of (A)
(b) (A) is correct but (R) is incorrect
(c) (A) is incorrect but (R) is correct
(d) Both (A) and (R) are correct but (R) is not the correct explanation of (A)

12 In the given reaction, 3-bromo-2, 2-dimethyl butane $\xrightarrow{C_2H_5OH}$ (A) .
Major product

Product A is **[2021, 20 July Shift-I]**

(a) 2-ethoxy-3, 3-dimethyl butane
(b) 1-ethoxy-3, 3-dimethyl butane
(c) 2-ethoxy-2, 3-dimethyl butane
(d) 2-hydroxy-3, 3-dimethyl butane

13 Main products formed during a reaction of 1-methoxy naphthalene with hydroiodic acid are

[2021, 18 March Shift-II]

(a) 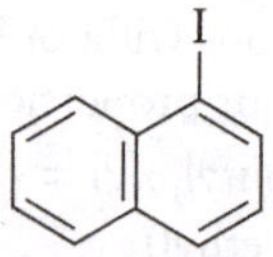 and CH_3I

(b) 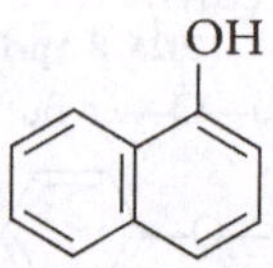 and CH_3I

(c) 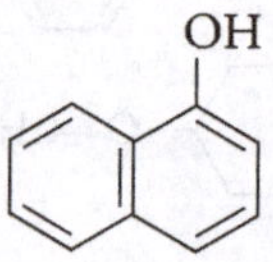 and CH_3OH

(d) 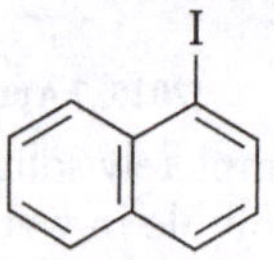 and CH_3OH

14 Arrange the following compounds in increasing order of C—OH bond length methanol, phenol, *p*-ethoxyphenol **[2020, 8 Jan Shift-I]**

(a) phenol < methanol < *p*-ethoxyphenol
(b) phenol < *p*-ethoxyphenol < methanol
(c) methanol < *p*-ethoxyphenol < phenol
(d) methanol < phenol < *p*-ethoxyphenol

15 The major product of the following reaction is

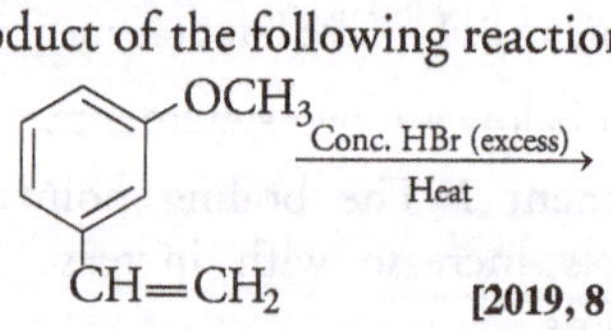

[2019, 8 April Shift-I]

(a) 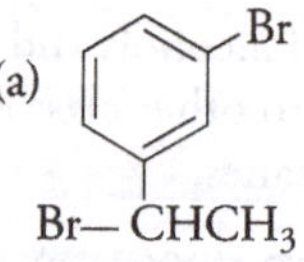

(b) 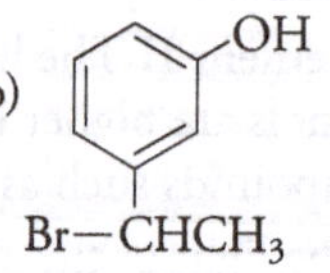

(c) 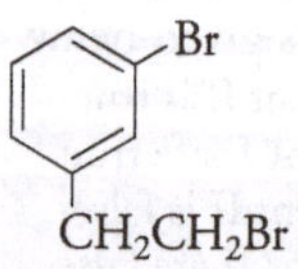

(d) 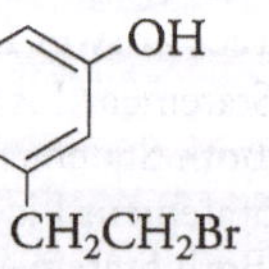

• Level 3

16 A toxic compound '*A*' when reacted with NaCN in aqueous acidic medium yields an edible cooking component and food preservative 'B'. 'B' is converted to '*C*' by diborane and can be used as an additive to petrol to reduce emission. '*C*' upon reaction with oleum at 140°C yields an inhalable anesthetic '*D*'. Identify '*A*', '*B*', '*C*' and '*D*', respectively: **[2025, 4 April Shift-II]**

(a) Acetaldehyde; 2-hydroxypropanoic acid; propanoic acid; dipropyl ether
(b) Methanol; formaldehyde; methyl chloride; chloroform
(c) Methanol; acetic acid; ethanol; diethyl ether
(d) Ethanol; acetonitrile; ethylamine; ethylene

17 2-methyl propyl bromide reacts with C_2H_5O and gives '*A*' whereas on reaction with C_2H_5OH it gives '*B*'. The mechanism followed in these reactions and the products '*A*' and '*B*' respectively are **[2023, 13 April Shift-I]**

(a) S_N1, A = *tert*-butyl ethyl ether; S_N2, B = *iso*-butyl ethyl ether
(b) S_N2, A = 2-butyl ethyl ether; S_N2, B = iso-butyl ethyl ether
(c) S_N2, A = iso-butyl ethyl ether; S_N1, B = tert-butyl ethyl ether
(d) S_N1, A = iso-butyl ethyl ether; S_N1, B = 2-butyl ethyl ether

18 The major product of the following reaction is

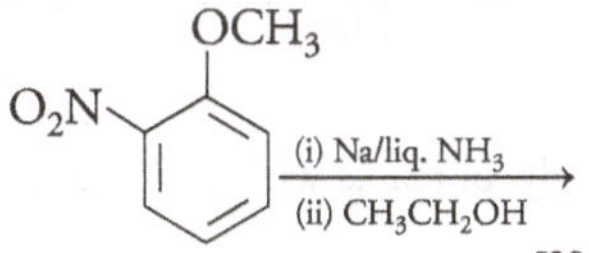

[2022, 26 July Shift-I]

(a) 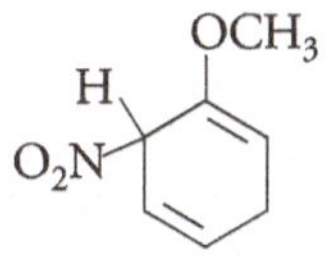

(b) 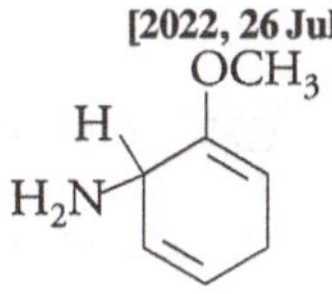

(c) 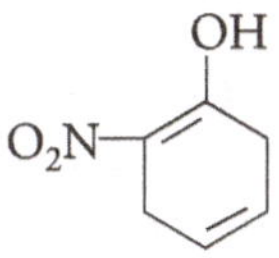

(d) 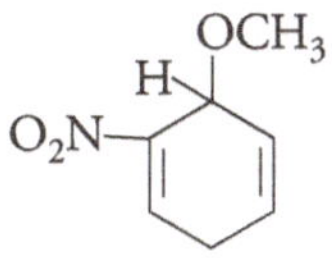

19 Among the compounds *A* and *B* with molecular formula $C_9H_{18}O_3$, *A* is having higher boiling point the *B*. The possible structures of *A* and *B* are **[2020, 8 Jan Shift-II]**

(a) 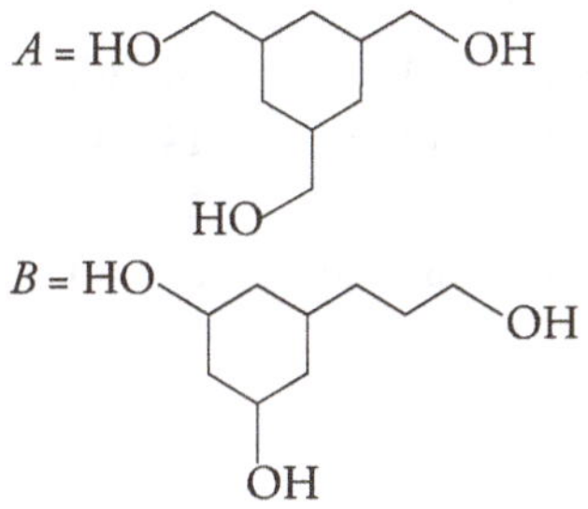

(b) 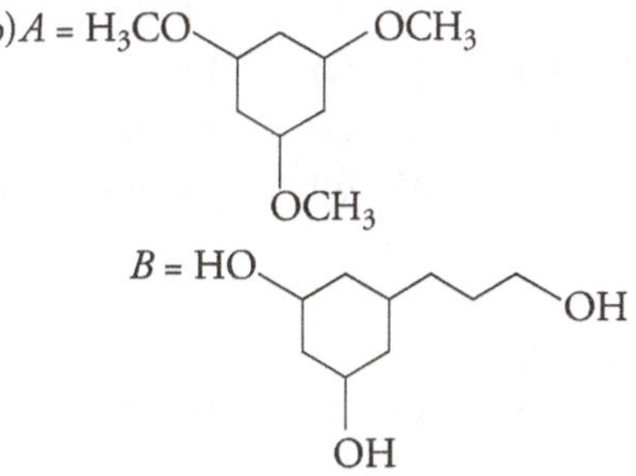

(c) 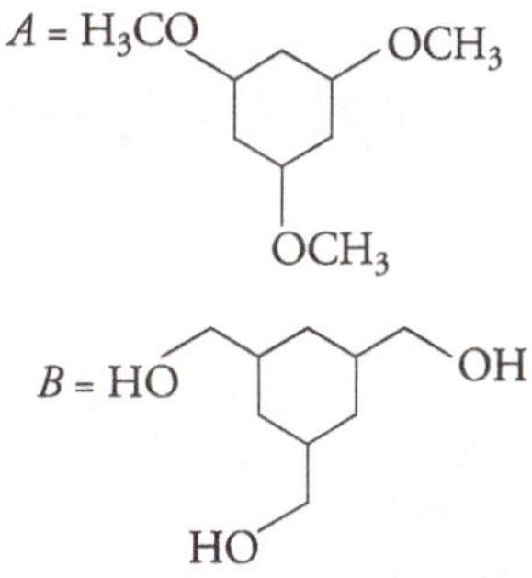

(d) 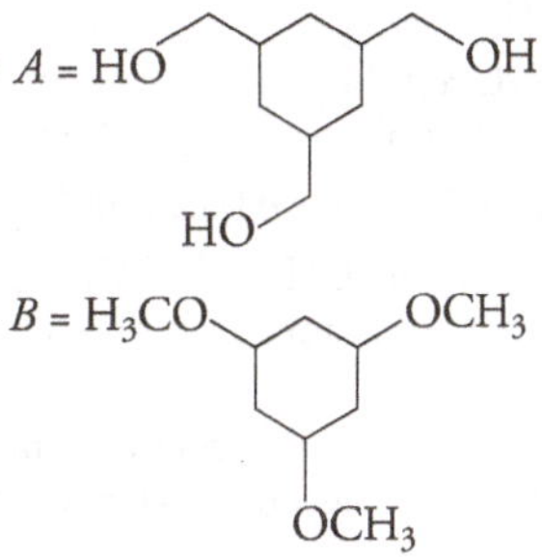

Key Index

Topic 1

1	(c)	2	(d)	3	(c)	4	(d)	5	(a)	6	(b)	7	(a)	8	(b)	9	(c)	10	(d)
11	(c)	12	(b)	13	(c)	14	(c)	15	(c)	16	(d)	17	(c)	18	(c)	19	(d)	20	(b)
21	(c)	22	(d)	23	(c)	24	(a)	25	(b)	26	(c)	27	(b)	28	(d)	29	(c)	30	(d)
31	(c)	32	(c)	33	(b)	34	(c)	35	(b)	36	(a)	37	(a)	38	(13)	39	(2)	40	(8)
41	(2)																		

Topic 2

1	(c)	2	(c)	3	(b)	4	(a)	5	(b)	6	(c)	7	(c)	8	(b)	9	(b)	10	(a)
11	(b)	12	(c)	13	(c)	14	(c)	15	(c)	16	(b)	17	(a)	18	(b)	19	(a)	20	(b)
21	(c)	22	(d)	23	(b)	24	(b)	25	(b)	26	(d)	27	(b)	28	(c)	29	(b)	30	(c)
31	(b)	32	(d)	33	(c)	34	(d)	35	(d)	36	(a)	37	(b)	38	(a)	39	(30)	40	(7)

Topic 3

1	(d)	2	(a)	3	(c)	4	(a)	5	(b)	6	(b)	7	(a)	8	(c)	9	(d)	10	(d)
11	(b)	12	(c)	13	(b)	14	(b)	15	(b)	16	(c)	17	(c)	18	(a)	19	(d)		

Detailed Explanations

Topic 1

01. *(c)* Among the given options, the incorrect method of preparation of alcohols is from ozonolysis of alkene. It is because on ozonolysis, alkenes gives aldehyde, ketones and carboxylic acids instead of alcohol.

02. *(d)* The water gas on reaction with cobalt as a catalyst will form methanol (CH_3OH).

$$CO + 2H_2 \xrightarrow{Co} \underset{\text{Methanol}}{CH_3OH}$$

03. *(c)* Butan-1-ol ($CH_3CH_2CH_2CH_2OH$) can undergo H—bonding while ethoxyethane ($CH_3CH_2OCH_2CH_3$) has no hydrogen attached to oxygen atom and it doesn't undergo H—bonding.

More is the extent of intermolecular H—bonding, more will be the association of molecules. Thus, butan-1-ol has higher boiling point.

Hence, both A and R are true and R is the correct explanation of A.

04. *(d)* A solution of CrO_5 in amyl alcohol has a blue colour.

When an acidified cold solution of potassium chromate and H_2O_2 is layered with an equal volume of amyl alcohol, a blue alcohol layer is obtained. The blue colour is due to the formation of a (higher top layer of amyl alcohol) chromium (VI) compound, chromium peroxide (CrO_5).

05. *(a)* The given reaction involves Markownikov addition. It is known as oxymercuration-demercuration reaction

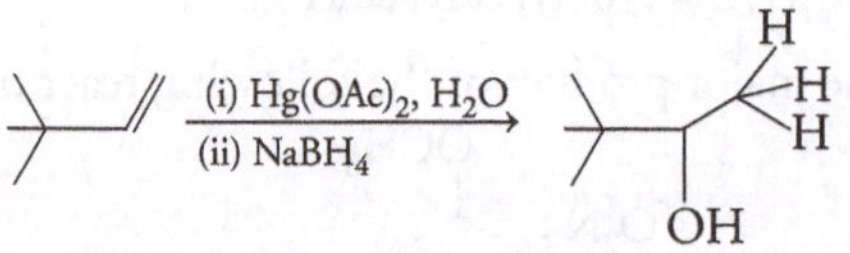

06. *(b)* Ethylene glycol reacts with oxalic acid at 210°C to form ethene.

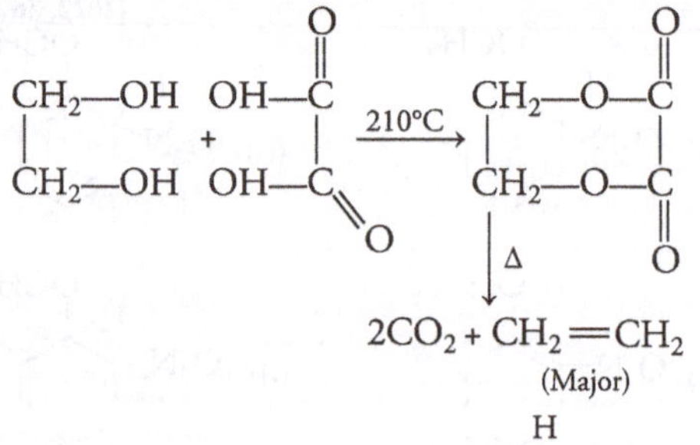

07. *(a)*

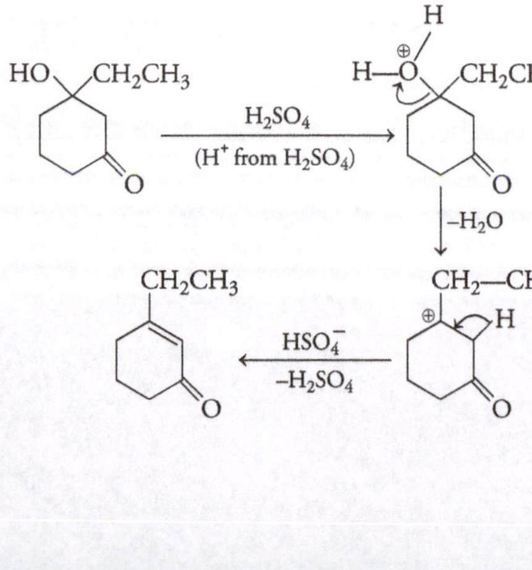

08. *(b)* As β-hydrogen is not present therefore E-2 elimination is not possible.

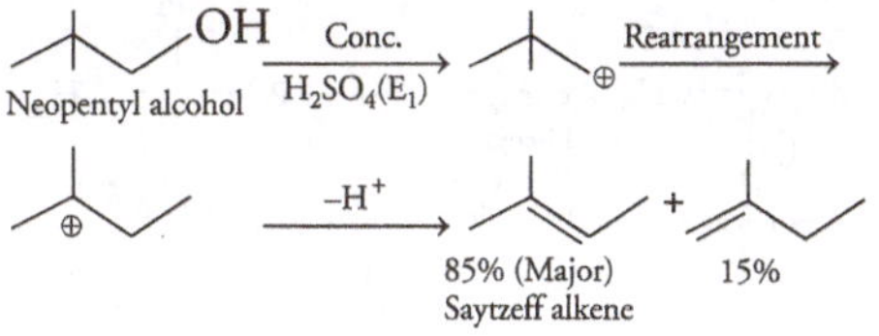

09. *(c)* Only 3° alcohol give positive ceric ammonium nitrate test.

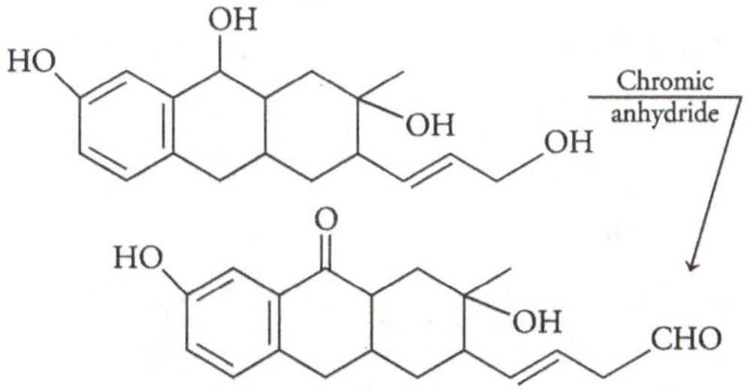

Product 'P'
(3° alcohol gives red colour with ceric ammonium nitrate)

10. *(d)* $CH_3CH_2—C(OH)(Ph)—CH_3$ cannot be prepared by HCHO and $PhCH(CH_3)CH_2MgX$. This can be easily illustrated by following reaction.

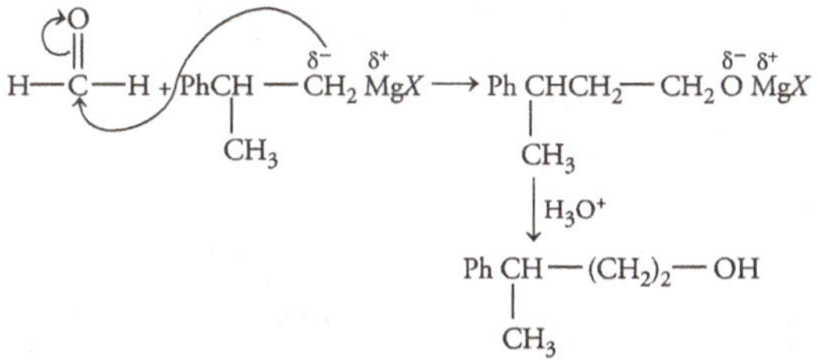

The obtained product is not the required substance. While option (a), (b) and (c) can readily prepare the required substance.

11. *(c)* The complete reaction sequence is as follows.

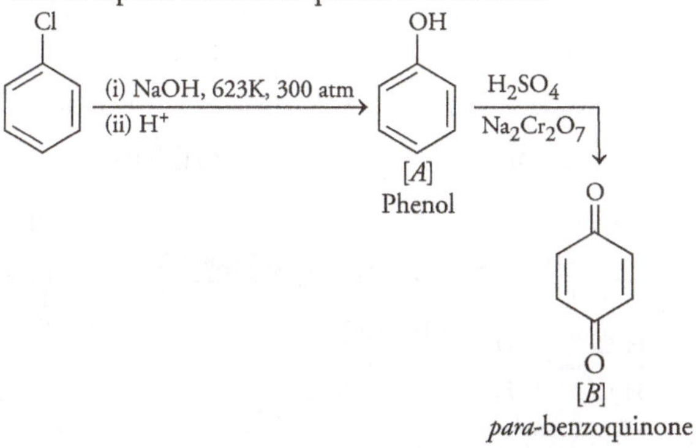

So, option (c) is the correct answer.

12. *(b)* The complete sequence of the given reaction is as follows

$$CH_3—CH{=}CH_2 \xrightarrow{H_2O,\ H^+} CH_3—CH(OH)—CH_3$$

(Markownikoff product) (*A*)

$$CH_3—CH{=}CH_2 \xrightarrow{(BH_3)_2} (CH_3CH_2CH_2)_3B \xrightarrow{H_2O,\ H_2O_2,\ OH^-} CH_3CH_2CH_2OH$$

(Hydroboration-oxidation reaction) (*B*)

13. *(c)* The complete reaction is as follow,

$$CH_3—CH_2OH \xrightarrow[\text{(ii) } KMnO_4]{\text{(i) Jones reagent } (CrO_3 + H^+)} CH_3—\overset{O}{\overset{\|}{C}}—OH \xrightarrow[\text{(iii) NaOH, CaO, }\Delta]{\text{Soda lime process}} CH_4 + Na_2CO_3$$

The major product is methane.
Hence, option (c) is correct.

14. *(c)* Both A and R are true but R is not the correct explanation of A.

Concept Enhancer

Alcohols behave as electrophiles and nucleophiles both due to the presence of oxygen in them. Their nucleophilic nature is due to the ability of oxygen to donate its lone pair. On the other hand, its electrophilic activity is due to the polarity in the C—O bond.

15. *(c)* The complete reaction is as follows,

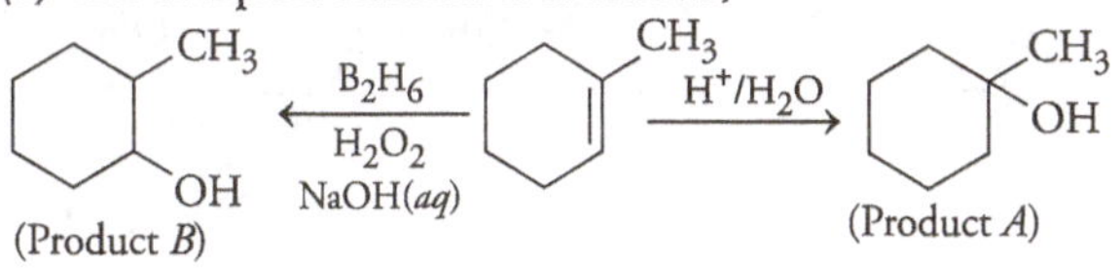

Hence, option (c) is correct.

16. *(d)*

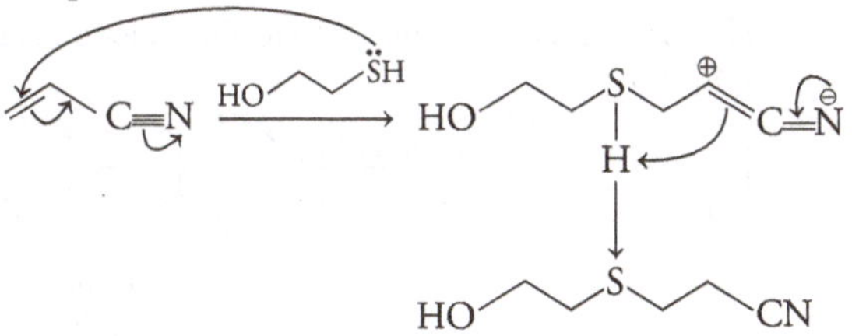

17. *(c)*

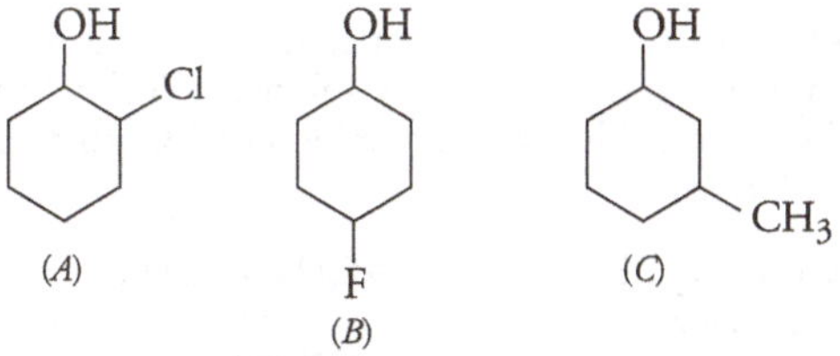

Acidic character depends upon the presence of electron withdrawing group which shows $(-I)$ effect with less distance.
Hence, (*A*) is more acidic than (*B*).
CH_3 shows $(+I)$ effect hence least acidic.
Hence both Assertion and Reason are correct but Reason is not correct explanation of Assertion.

18. *(c)* The product [*X*] formed in the given reaction is given in option (c).

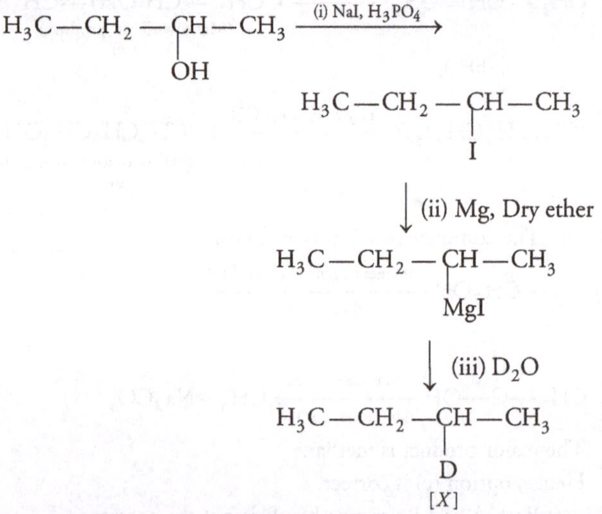

19. *(d)* The correct order is (ii) > (iv) > (iii) > (i).

Dehydration of an alcohol in the presence of an acid involves the loss of —OH functional group as H_2O after protonation of —OH group and formation of carbocation. Its reactivity is directly proportional to the stability of the carbocation. So, according to stability of a carbocation the order of dehydration will be

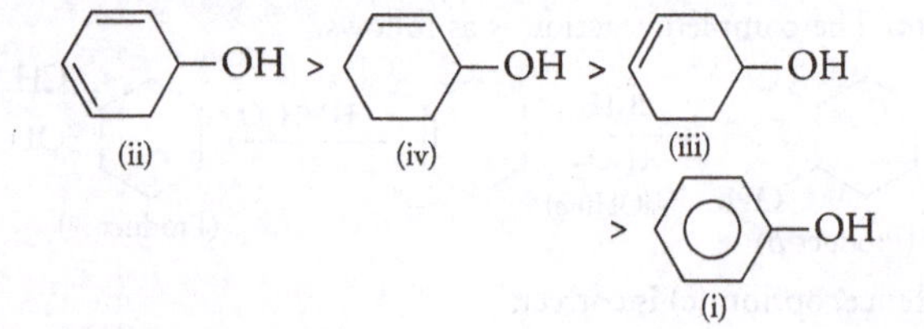

20. *(b)* The major product from the following reaction is given in option (b).

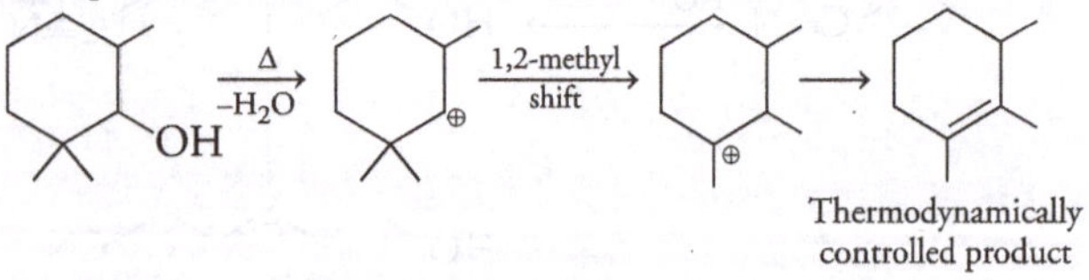

21. *(c)* When ethanol is heated with conc. H_2SO_4 dehydration reaction occurs and acetylene gas is formed which when treated with cold dilute aqueous solution of Baeyer's reagent glycol is formed.

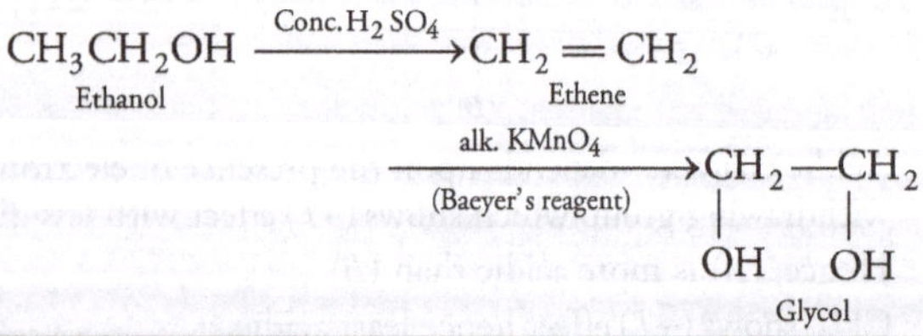

Concept Enhancer

Baeyer's reagent is a cold potassium permanganate alkaline solution ($KMnO_4$). It is a strong oxidising agent that is used in qualitative organic analysis to check for unsaturation.

22. *(d)* Diphenylethanol can be obtained from acetophenone and phenyl magnesium bromide as shown below.

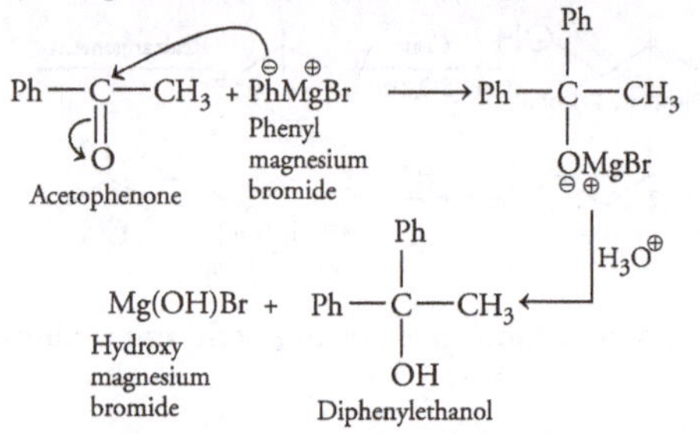

23. *(c)* The following reaction sequence are involved

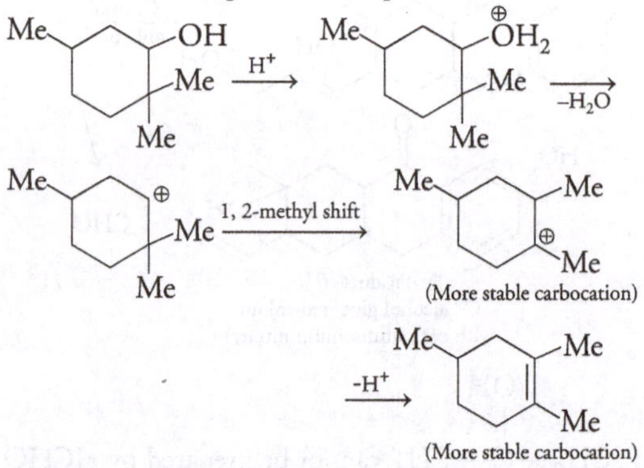

24. *(a)* According to the given options, the reaction would be nucleophilic addition reaction. Hydrolysis of the product formed gives diol.

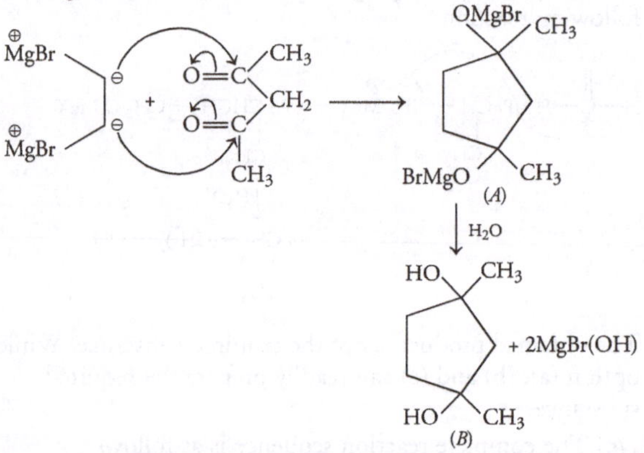

25. *(b)*

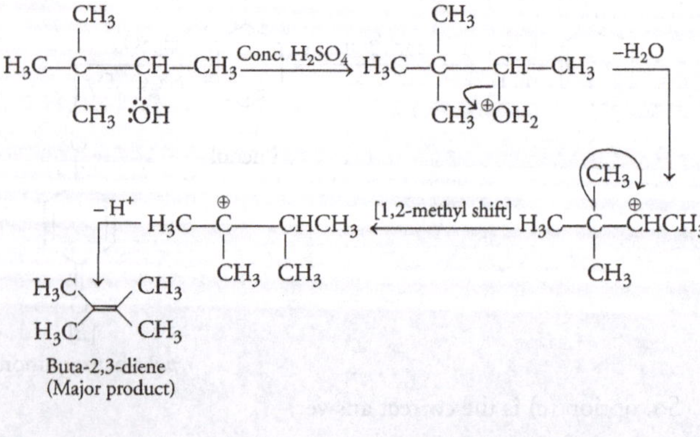

26. *(c)* Complete mechanism of given reaction is as follows **Hydrolysis process.** Acid catalysed hydration of alcohol group.

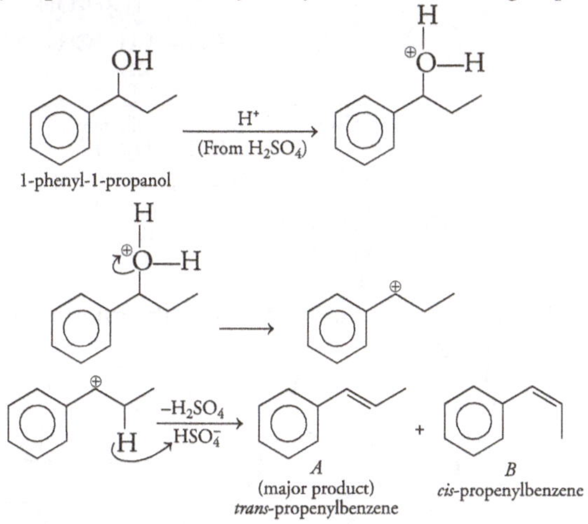

trans alkenes are more stable than *cis* alkene, so product *A* is major product formed in the reaction.

27. *(b)* Complete reaction with mechanism is as follows

(i) **Hydrolysis process** Attack of H^+ on OH^- ion.

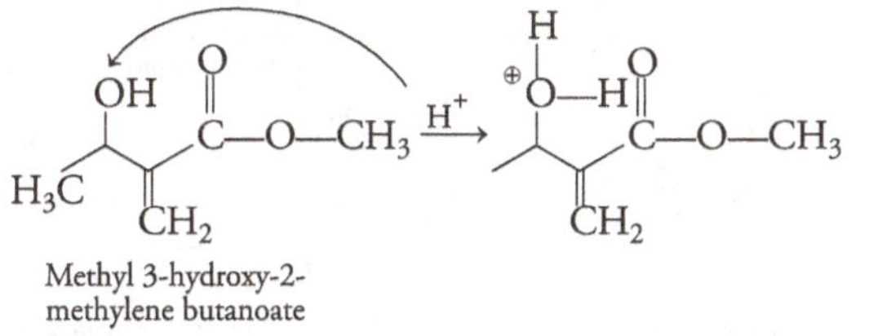

(ii) Removal of water molecule/formation of carbocation.

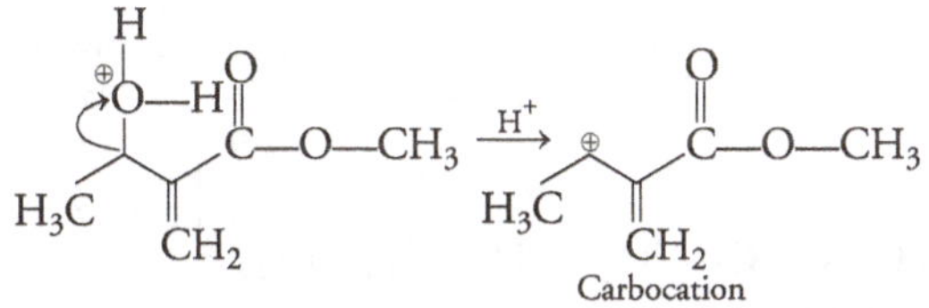

(iii) **Resonance** Movement of double bond/shifting of carbocation.

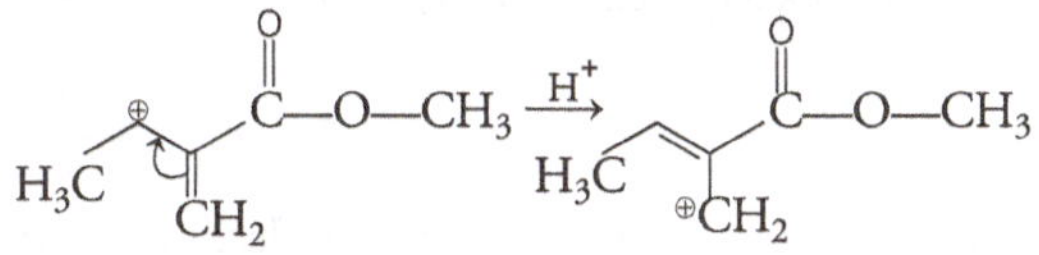

(iv) Addition of bromide ion.

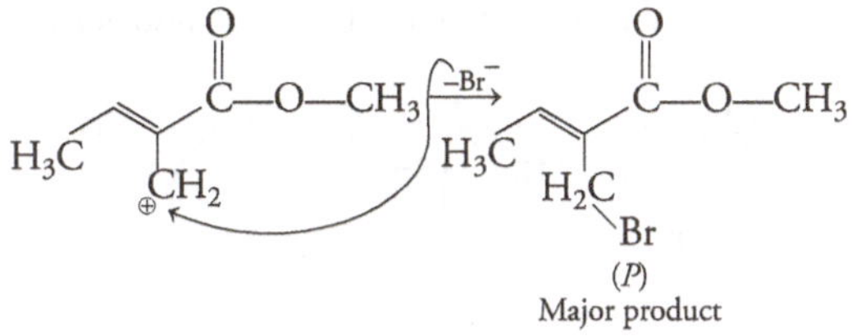

28. *(d)*

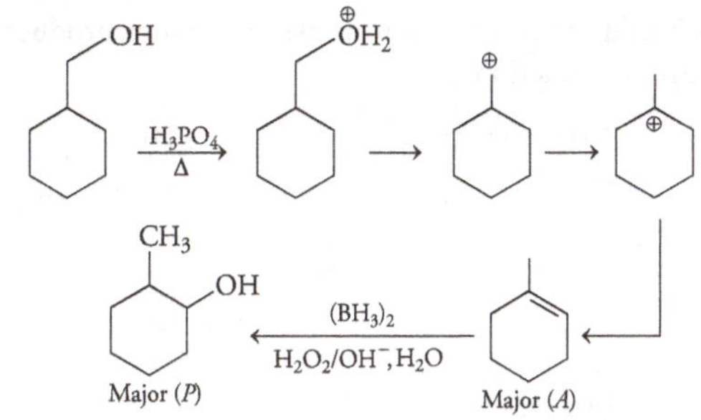

Conversion of compound *A* to compound *P* in undergoes through hydroboration-oxidation reaction that converts an alkene into an alcohol.

29. *(c)* In the given reaction, Grignard reagent is used for 1,2-addition in the carbonyl carbon and form alcohol as a major product.

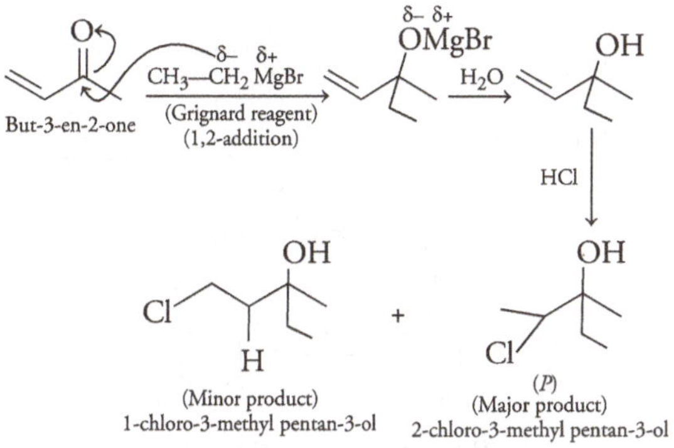

Addition of HCl at the alkene gives 2-chloro-3-methylpentan-3-ol as major product.

30. *(d)* Steps involved in this reaction are as follows

Step 1 $HCl \longrightarrow H^+ + Cl^-$

H^+ attacks on —OH (lone pair) and formed $\overset{\oplus}{O}H_2$ ion. Here, $\overset{\oplus}{O}H_2$ is the good leaving group.

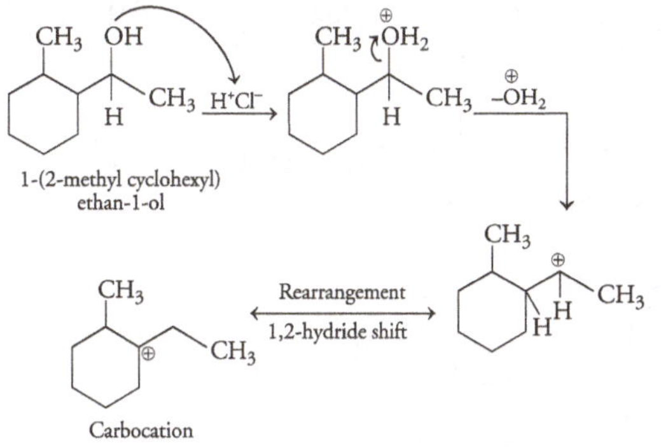

Step 2 Elimination reaction always give minor product than substitution reaction.

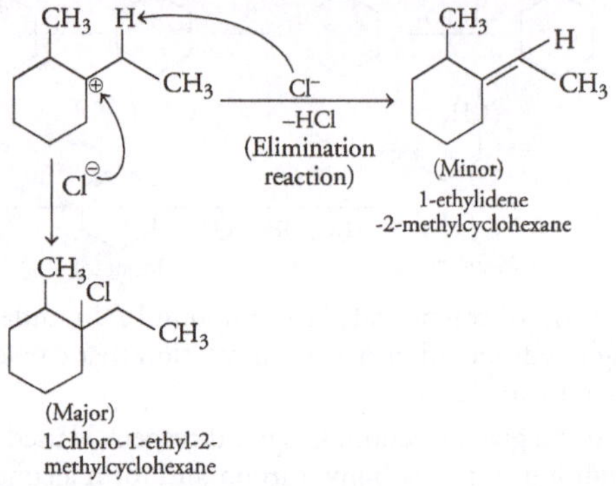

31. *(c)* The reaction proceeds as follows:

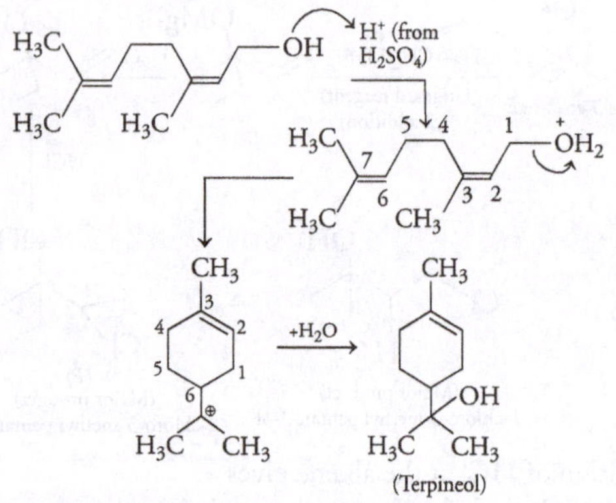

32. *(c)* Statement I is incorrect but Statement II is correct. The corrected form of statement I is

Alkene is formed when alkyl chlorides are treated with aqueous potassium hydroxide by elimination reaction.

33. *(b)* The given reaction involved in the given question is as follows

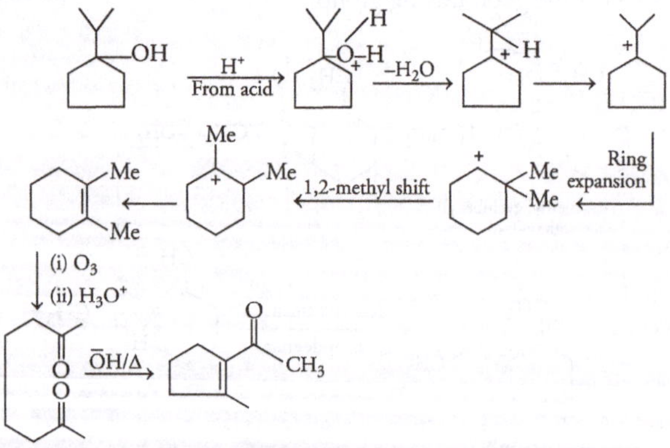

So, option (b) is the correct answer.

34. *(c)* The given reaction proceeds as follows

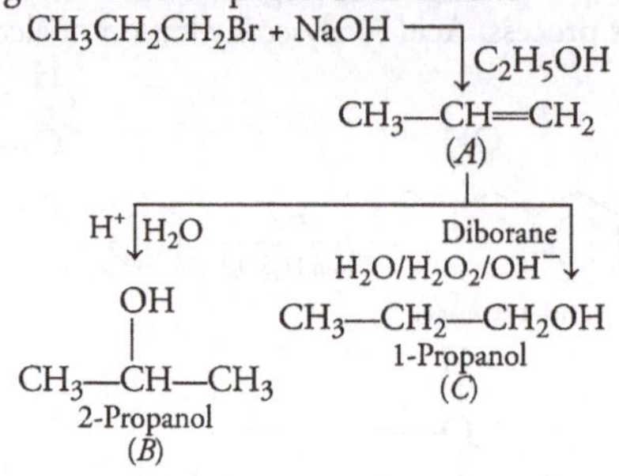

Thus, the compound B and C are 2-Propanol and 1-Propanol respectively.

35. *(b)*

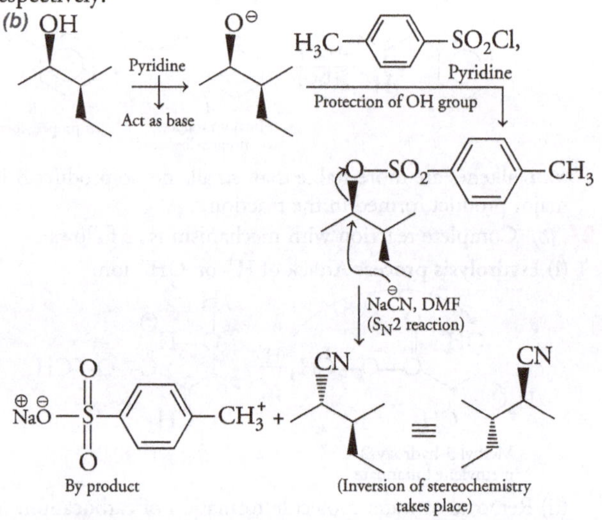

36. *(a)* In the given reaction sequence compounds involved are :

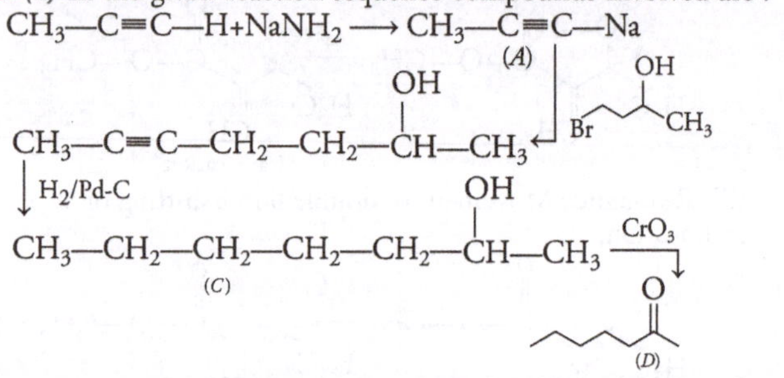

37. *(a)* (C) is most stable due to strong $-I$effect as well as $-R$ (or $-M$) effect of '—NO_2' group. The negative charge is delocalised as represented by the following resonance structure

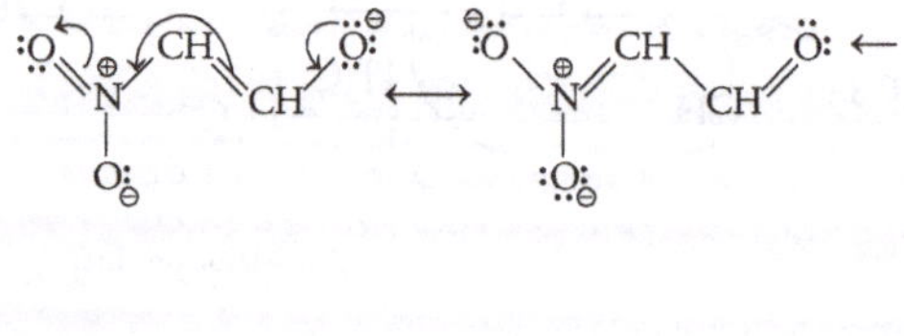

In (*A*), there is no resonance delocalisation of negative charge, and hence it is least stable. In (*B*), negative charge is delocalised by resonance as shown :

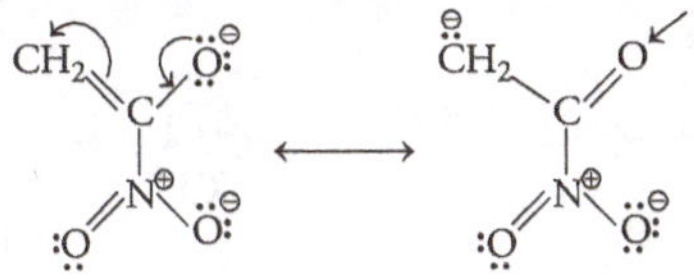

But the contribution of structure on RHS is relatively less, so stability of (*B*) is more than (*A*) but less than (*C*).

38. *(13)* The complete reaction is as follows

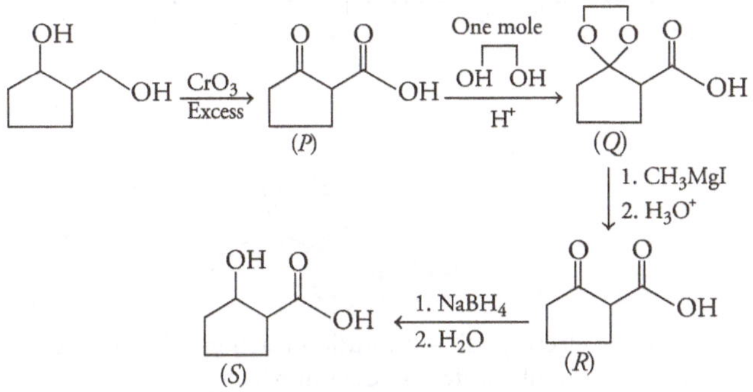

0.1 mol of compound (*S*) weight in g = 0.1 × molar mass of compound (*S*) = 0.1 × 130 = 13 g.

39. *(2)*

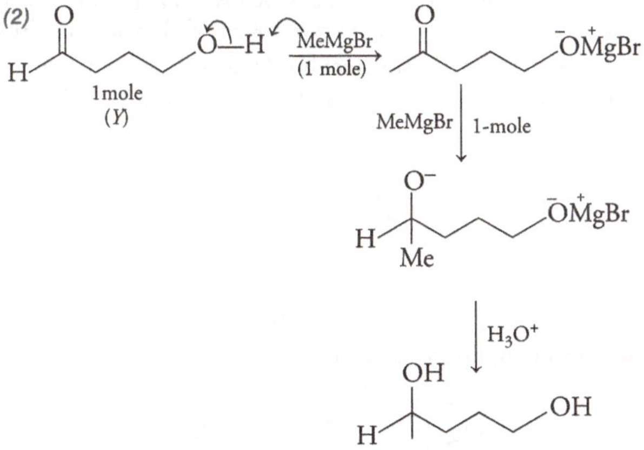

∴ 2 moles of MeMgBr are involved in this reaction and 1 mole of reactant. Thus $X = 2$ and $Y = 1$

$$\Rightarrow \quad \frac{X}{Y} = \frac{2}{1} = 2$$

40. *(8)* The complete reaction is as follows

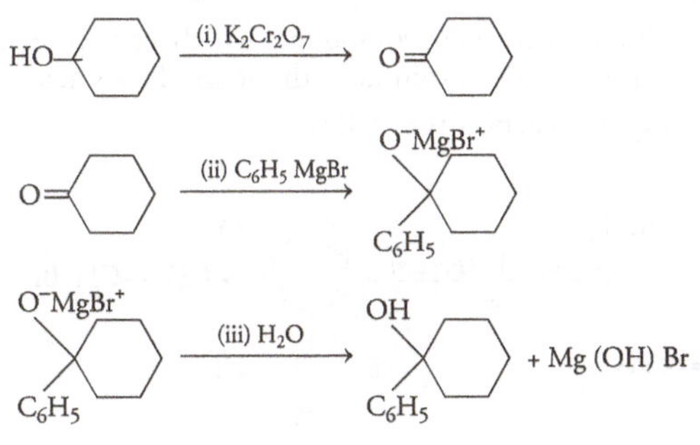

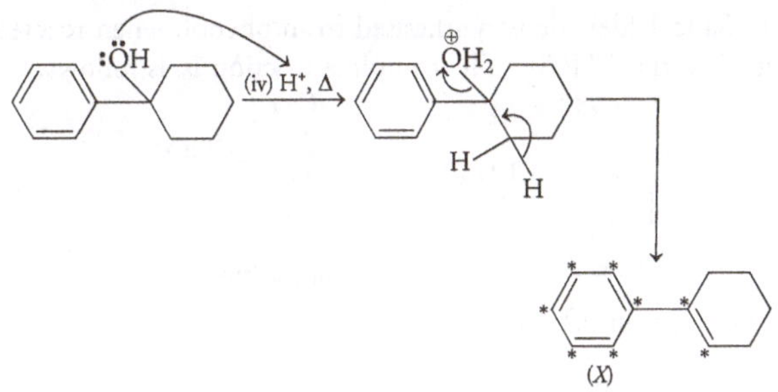

Number of sp^2 hybridised carbons in compound *X* is 8.

41. *(2)* Two moles of CH_3MgBr is required as follows.

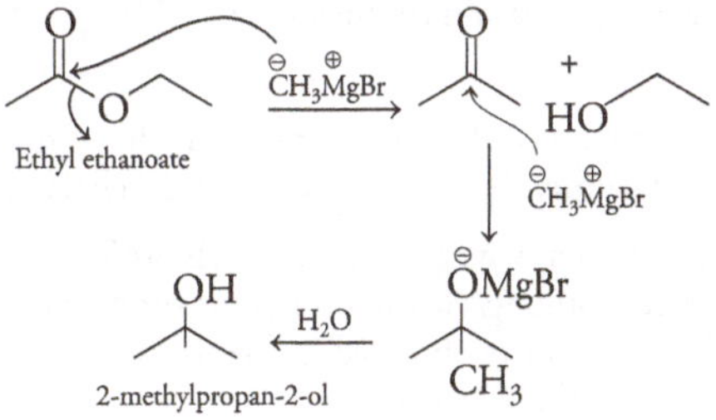

Topic 2

01. *(c)* Among the given compounds, compound given in option (c) will react with HBr to give a phenol.

The complete reaction is as follows,

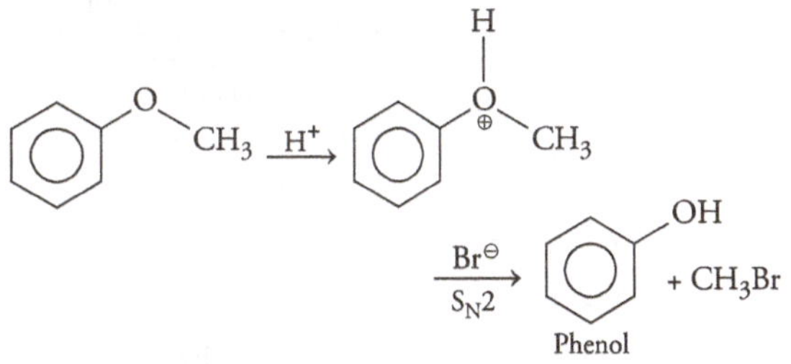

02. *(c)* The complete Reimer-Tiemann reaction is

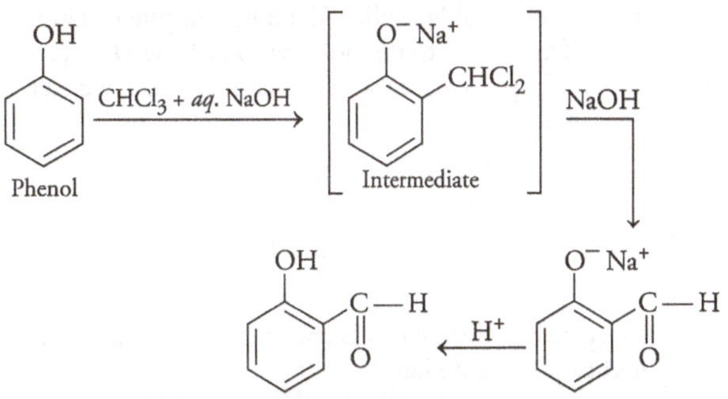

Hence, option (c) is correct.

Concept Enhancer

Riemer-Tiemann reaction is an electrophilic substitution reaction. In this reaction, dichlorocarbene acts as an electrophile this dichlorocarbene is formed by abstraction of hydrogen from chloroform that creates a trichlorocarbanion, which loses a chloride ion to form dichlorocarbene.

03. *(b)* Salicylaldehyde is synthesised from phenol, when reacted with NaOH, $CHCl_3$. The complete reaction is as follows,

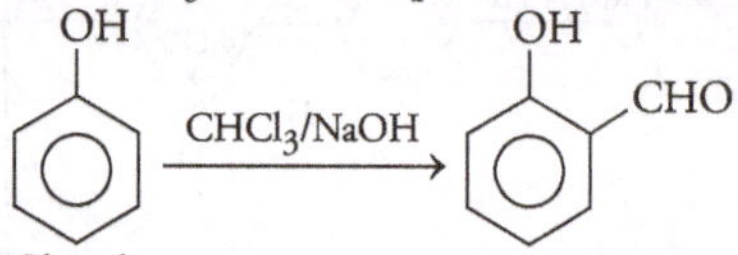

Phenol Salicylaldehyde

Hence, option (b) is correct.

04. *(a)* Among the given options *m*-nitrophenol i.e.,

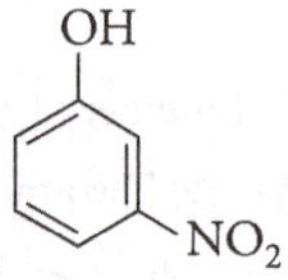

is most acidic due to presence of —NO_2 group which has more electron withdrawing nature. Hence, increases the acidic character.

05. *(b)* Phenol reacts differently with bromine in chloroform and bromine in water forming/yielding different products. The difference in the products obtained is due to the difference in polarity of the solvents.

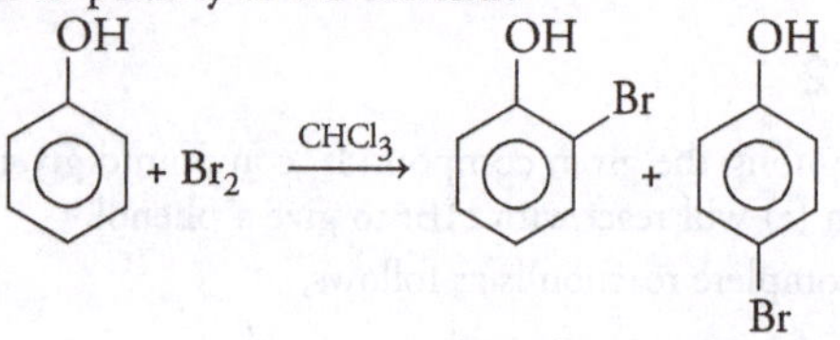

o-substituted *p*-substituted

Mixture

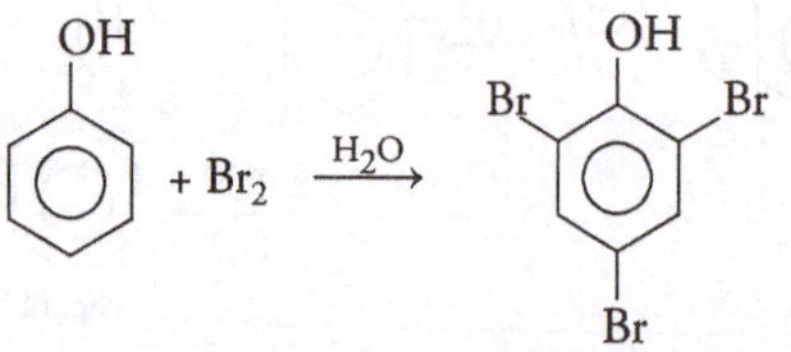

tri-substituted product

06. *(c)* (I) Phenol is weakly acidic. Alcohols and phenols are weak acids. This is due to the polarisation of the O—H bond making H partially positively charged. It is a weaker acid as compared to carboxylic acid.

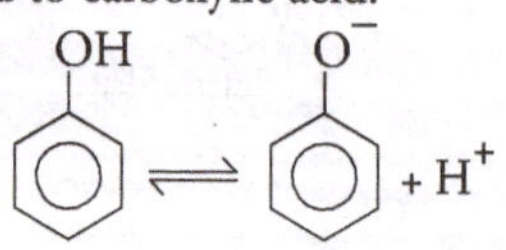

Phenoxide ion

The conjugate base formed by carboxylic acid is more stable as compared to phenoxide ion.

(II) Phenols are freely soluble in NaOH solution and are stronger acids than alcohols and water. This is due to more stability of phenoxide ion (conjugate base formed) as negative charge on the oxygen atom is delocalised on the ring (resonance) as compared to stability of alkoxide (conjugate base of alcohol) and hydroxide ion (conjugate base of water)

Hence, Statement I is correct but Statement II is incorrect.

07. *(c)* Phenol in presence of sodium hydroxide reacts with chloroform to form salicylaldehyde. The reaction is known as Reimer-Tiemann reaction.

Mechanism

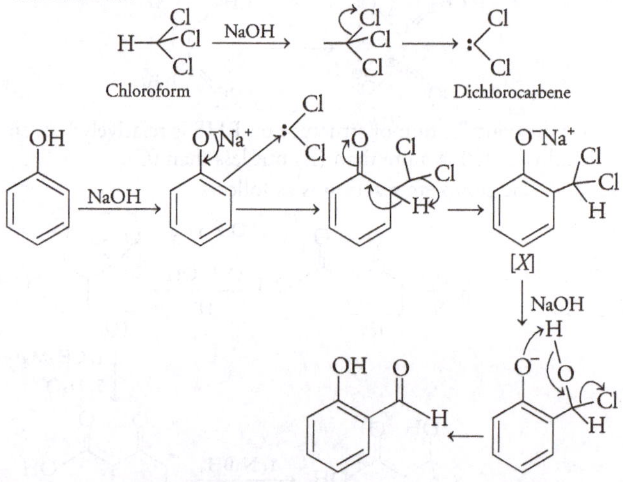

08. *(b)* In the electrophilic substitution reaction, phenol gives *ortho* and *para* substituted product in which *para* is the major product. The electrophilic substitution reaction of phthalic anhydride with phenol is taking place at *para* position of the phenol. All the given molecules have free *para* position except in *p*-cresol. Hence, it does not give colour of phthalic anhydride.

09. *(b)* Statement I is incorrect but Statement II is correct. The correct form of Statement I is, picric acid is 2, 4, 6-trinitrophenol.

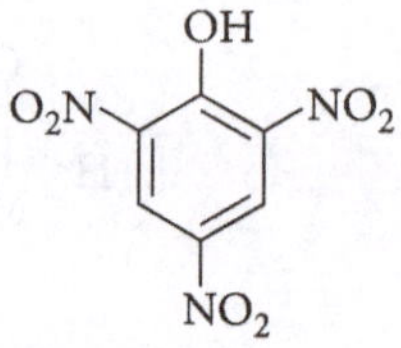

10. *(a)* Among the given reaction,

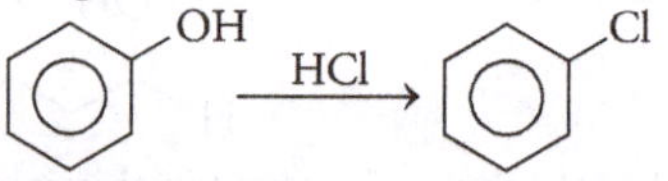

is not possible as due to resonance C—O bond has partial double bond character. So, bond strength will increase and substitution will be less likely.

11. *(b)* Statement I is false but Statement II is true. The correct form of Statement I is,

Phenol is a highly activated compound which can undergo bromination directly with bromine without any Lewis acid.

12. *(c)* The complete reaction is as follows

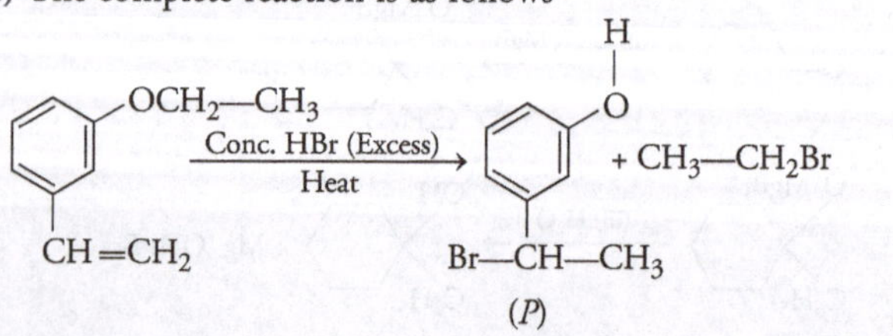

(*P*)

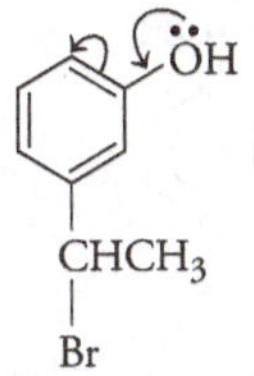

does not react further with HBr due to double bond character between C—O bond.

Hence, option (c) is correct.

13. *(c)* Statement I is true but Statement II is false.

The correct form of Statement II is :

Ethanol will not give immediate turbidity with Lucas reagent. The reason is that ethanol is a primary alcohol.

Tertiary alcohol gives immediate turbidity while secondary alcohol give turbidity within five minutes.

14. *(c)* The reactivity of benzene increases towards electrophilic substitution reaction when any substituent increases the electron density.

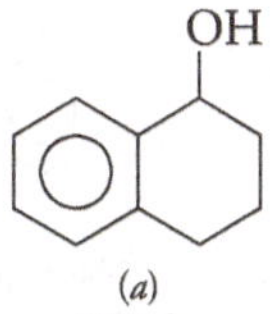

(*a*)
$-CH_2$ has +H effect

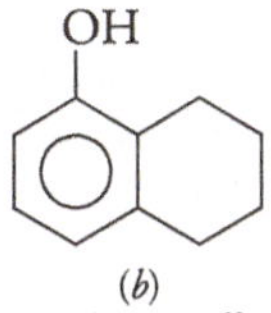

(*b*)
–OH has +*R* effect while $-CH_2-$ group has +H effect

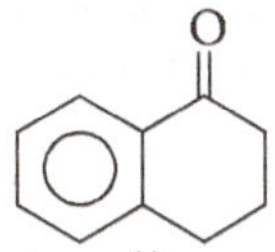

(*c*)

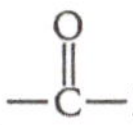

has, – *R* effect

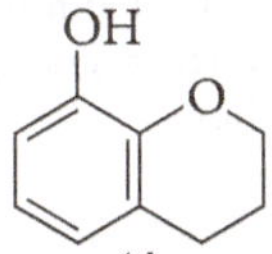

(*d*)
–OH and –O– both shows +*R* effect

Thus, the correct order is $c < a < b < d$.

15. *(c)* The correct order for acidity of the given hydroxyl compound is $E > C > D > A > B$.

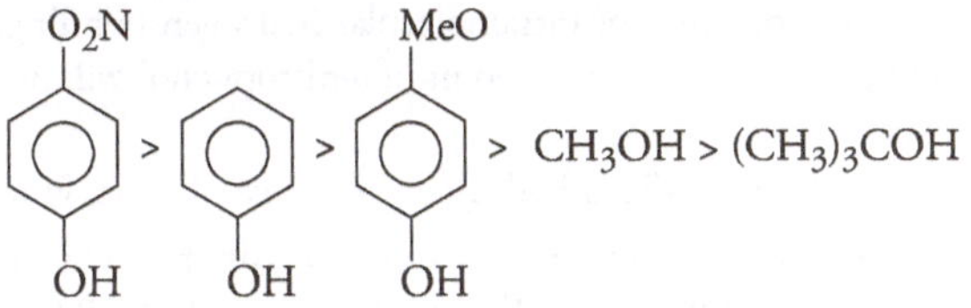

16. *(b)* Correct option is (b).

The organic compound that burns with sooty flame is phenol. Saturated hydrocarbon undergo complete combustion and give clean blue flame whereas unsaturated hydrocarbons burn with sooty flame.

Thus,

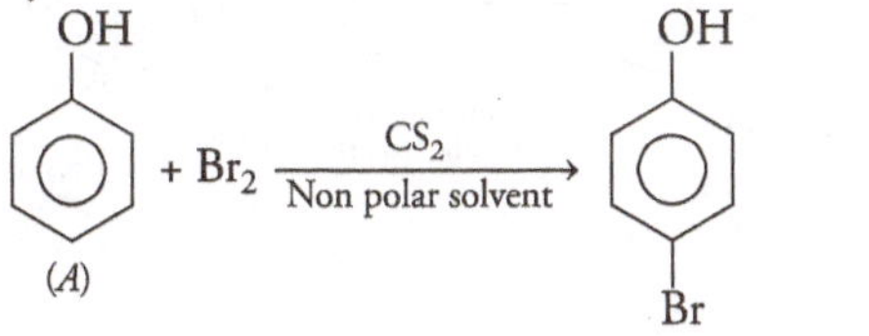

(*B*) (Major)
4-bromophenol

Concept Enhancer

Unsaturated carbon compounds do not burn completely. Hence, they give yellow coloured flame that tends to pollute, such flames are called sooty flames. While the saturated compounds, burns with blue coloured non-sooty flame.

17. *(a)*

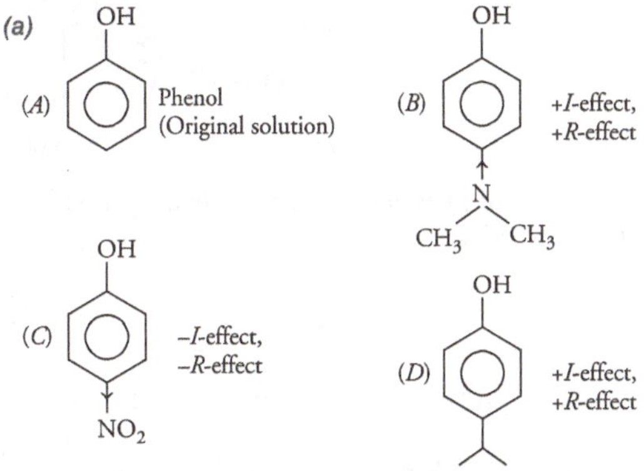

K_a value increases by electron withdrawing group i.e., NO_2. Hence, the order of K_a value will be $C > A > D > B$.

Also, we know that, $pK_a \propto \frac{1}{K_a}$

So the order of, pK_a value is $B > D > A > C$.

18. *(b)* The increasing order of pK_a for the given phenol is A, B, E, D, C. Presence of electron withdrawing group increases the acidic nature of the molecule and hence decreases the pK_a value for an acid.

$$\left[\text{Acidic nature} \propto \text{electron withdrawing group} \propto \frac{1}{pK_a}\right]$$

Hence, correct order for increasing pK_a value
[2, 4-dinitrophenol, 4-nitrophenol, 3-chlorophenol, phenol, 2, 4, 5-trimethylphenol.]

19. *(a)*

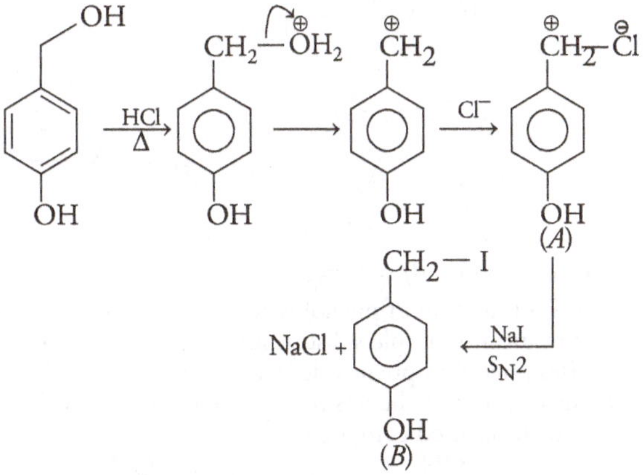

The final product formed in the given reaction is 4-iodomet hylphenol.

20. *(b)* In the given reaction, starting compound is bromobenzene.

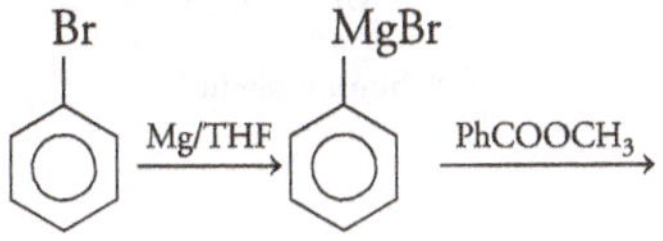

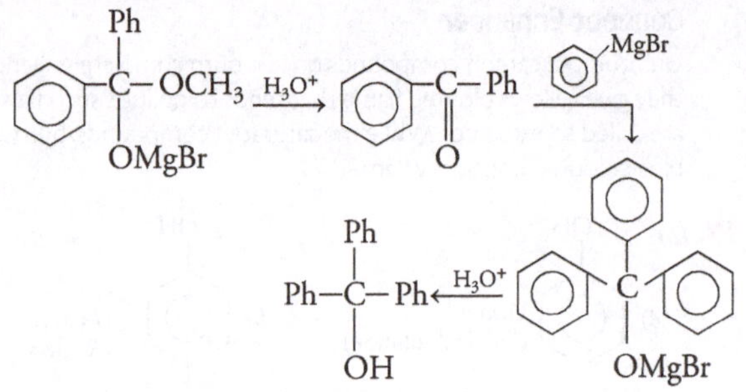

21. *(c)* Preparation of phenol from cumene hydroperoxide through air oxidation.

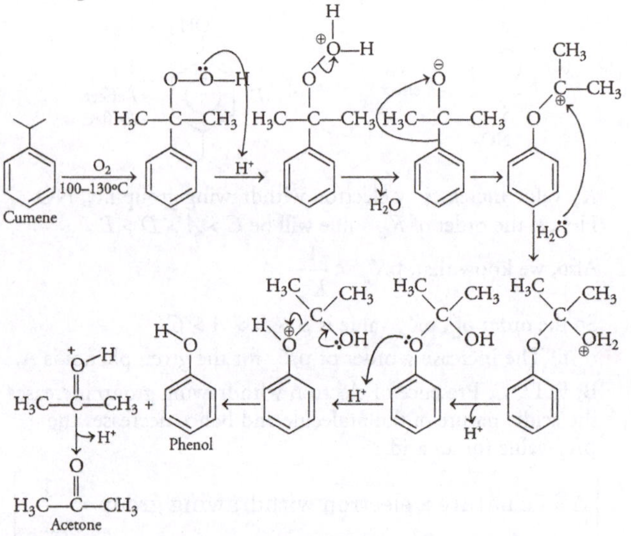

22. *(d)* Here, in the question

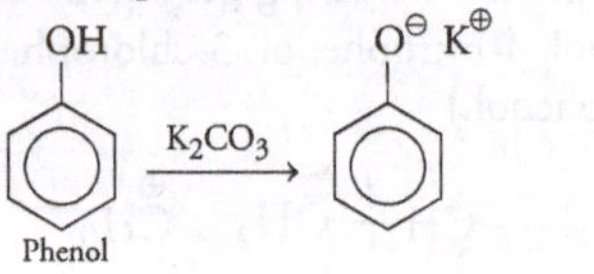

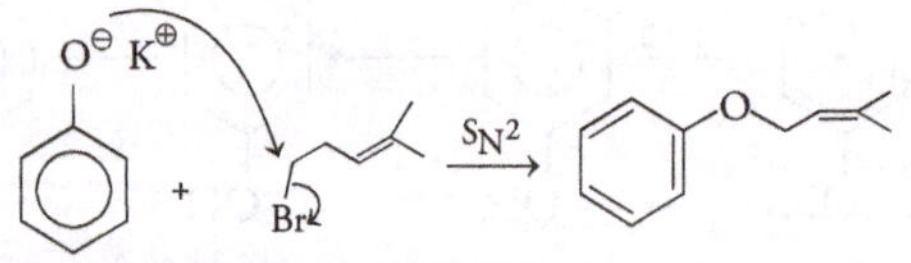

So, here option (d) is correct.

23. *(b)* Bromination of phenol with excess of bromine and water will result in multiple substitution around ring as water ionises phenol to phenoxide ion, which activities the ring for trisubstitution, while less polar solvent, i.e. CS_2 will give mono-substituted compound.

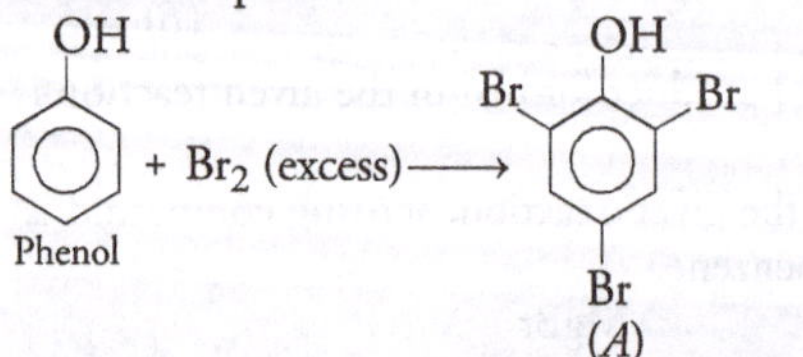

(A)
2,4,5 tribromophenol

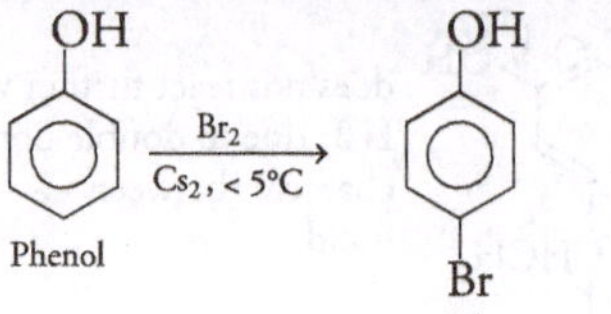

(B)
Mono-substituted product

24. *(b)*

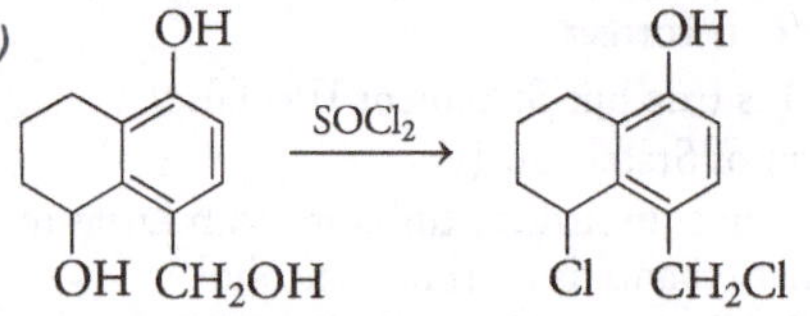

25. *(b)* Phenol on reaction with Br_2 in CS_2 / 273 K undergoes an electrophilic substitution reaction by Br^+ (electrophile) in aprotic solvent CS_2 to give 4-bromophenol as the major product.

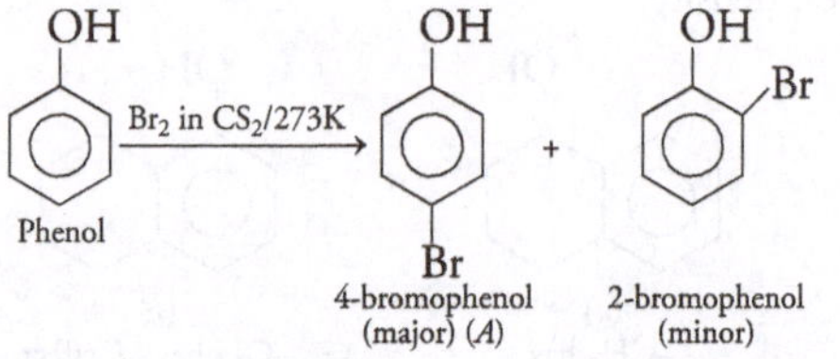

Phenol on reaction with $CHCl_3$, NaOH followed by hydrolysis gives salicylaldehyde as a major product. It is Reimer-Tiemann reaction. It is also an electrophilic substitution reaction of phenol by dichlorocarbene CCl_2 (electrophilic).

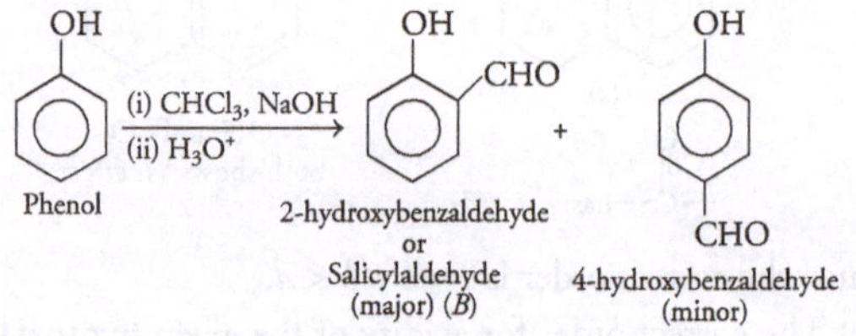

26. *(d)* Statement I is true. While statement II is false, because, due to the presence of intramolecular hydrogen bonding, boiling point and melting point of *o*-nitrophenol will be lower.

27. *(b)* Firstly phthalic anhydride in presence of conc. H_2SO_4 undergoes protonation to give an intermediate carbocation. This carbocation reacts with 2-propylphenol in presence of NaOH to give pink colour compound.

Mechanism

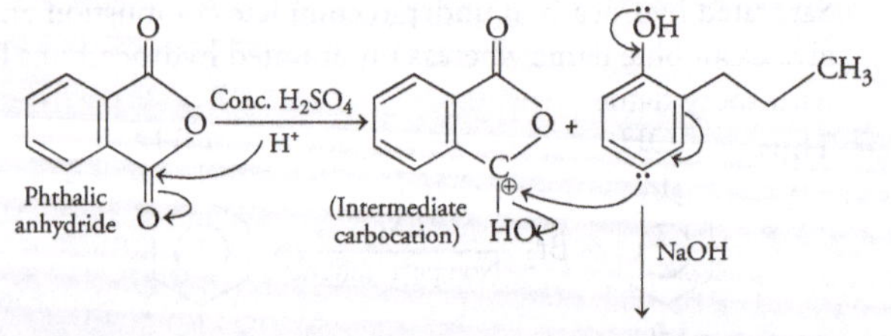

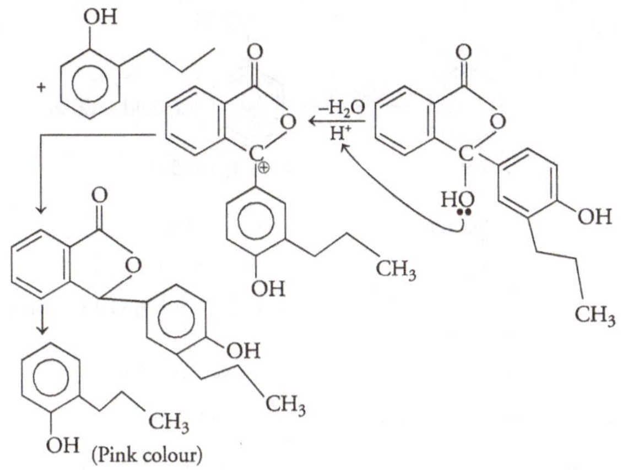

(Pink colour)

28. *(c)* Ethyl magnesium bromide is a Grignard reagent (GR), it constitutes $C_2H_5^-$ $[C_2H_5^- \overset{\oplus}{M}gBr$ in ether/aprotic medium] which can act as nucleophile as well as strong base. Bromine water (Br_2/H_2O) gets decolourised with phenol derivatives.

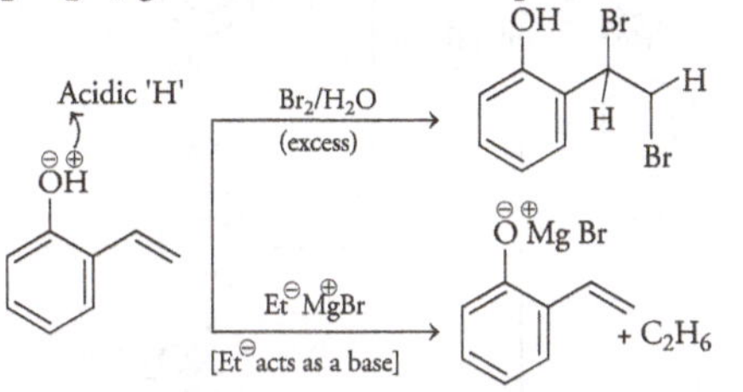

29. *(b)* Acidic strength is inversely proportional to pK_a value.
The acidity of phenols is due to greater resonance stabilisation of phenoxide ion relative to phenol.
Thus, order of acidic strength of nitrophenol is *p*-nitrophenol > *m*-nitrophenol and the correct order of the pK_a values of give option is

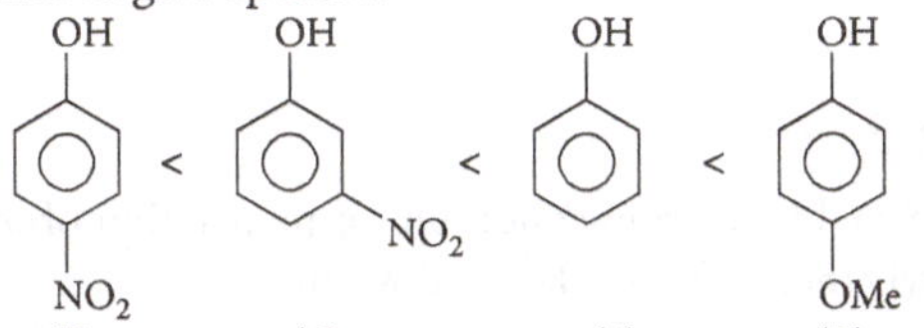

(B) (C) (A) (D)

30. *(c)* The correct match is A-IV, B-III, C-II, D-I

31. *(b)* Complete reaction sequence is as follows

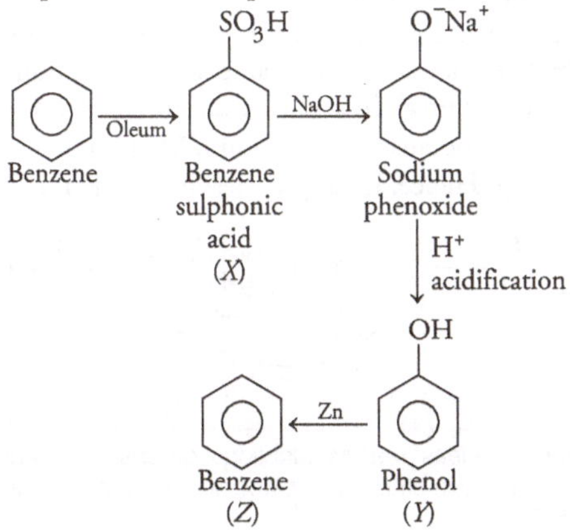

32. *(d)*

> **Time Saver Tech**
> Use elimination technique as time saver. As preparation of phenol from aniline i.e. A-II is given is option (d) only.

The correct match is A-II, B-IV, C-I, D-III.

33. *(c)* The correct match is A-III, B-I, C-II, D-IV.

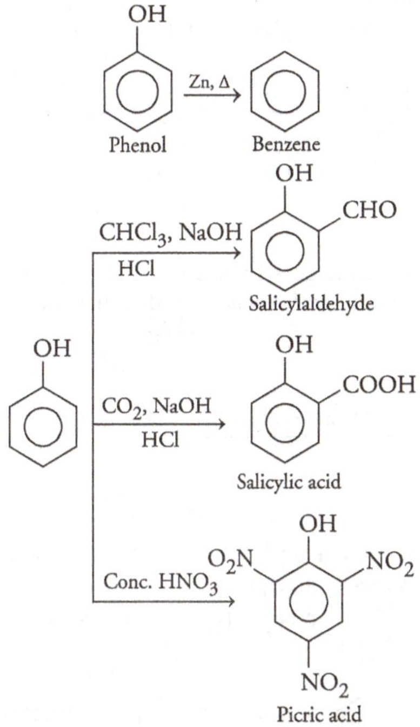

34. *(d)* The correct match is A-II, B-I, C-IV, D-III.

Ethanol $pK_a = 15.9$ Phenol $pK_a = 10.0$
*m*nirophenol $pK_a = 83$ *p*-nitrophenol $pK_a = 7.1$

> **Concept Enhancer**
> The acidic strength of an acid can be determined with the help of pK_a value it is the long inverse of dissociation constant of an acid. Hence, smaller is the value of pK_a more strong is the acid.

35. *(d)* The correct match is A-IV, B-III, C-I, D-II.

(A) 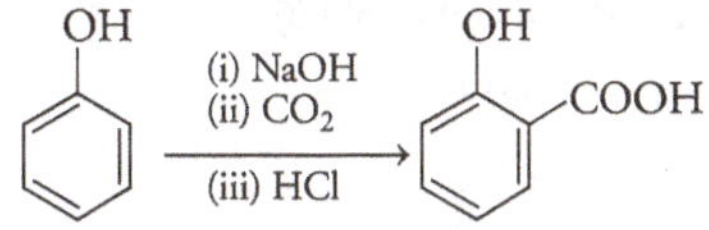

Kolbe-Schmidt reaction

(B) 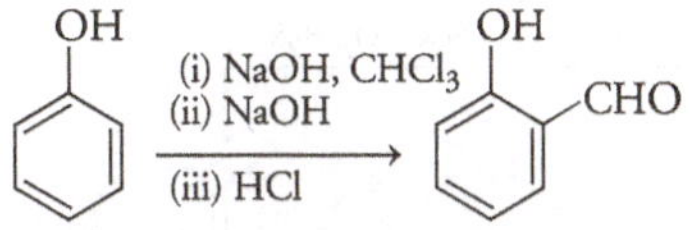

Reimer-Tiemann reaction

(C) 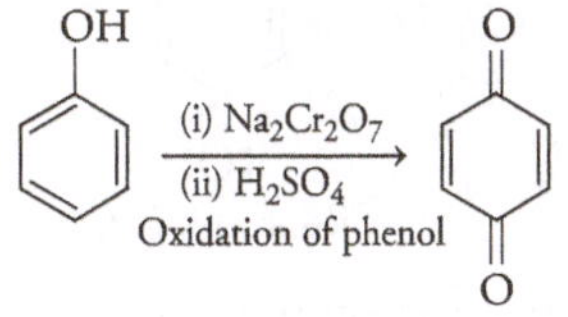

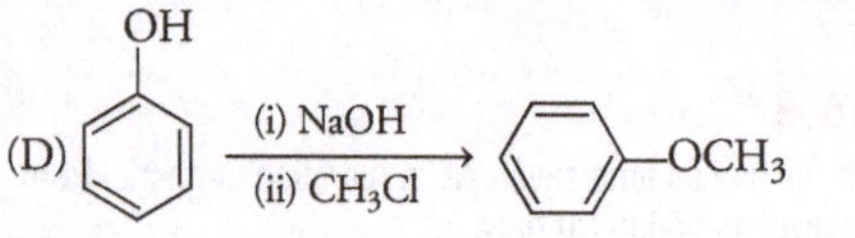

Hence, option (d) is correct.

36. *(a)* The correct sequence of ascending order of acidity of —OH group is

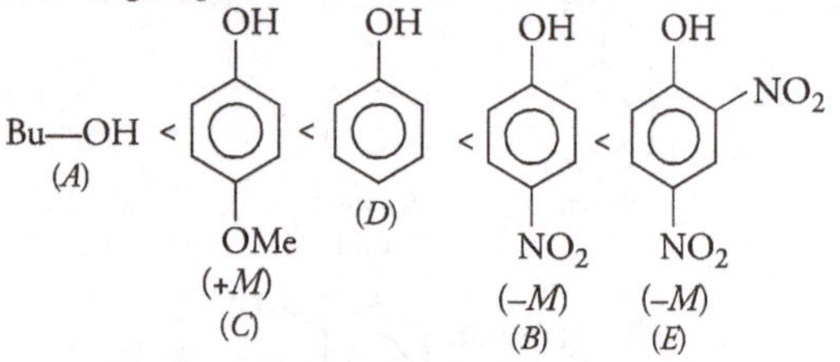

37. *(b)* Benzene undergoes Friedel-Craft alkylation reaction in presence of $AlCl_3$ and reacts with alkyl halide.

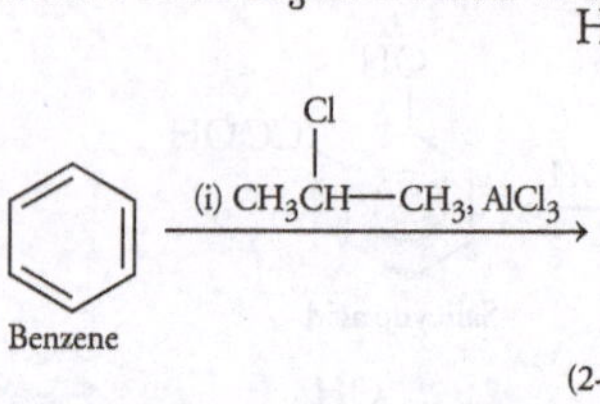

In next step, cumene undergoes oxidation followed by addition of water to give phenol.

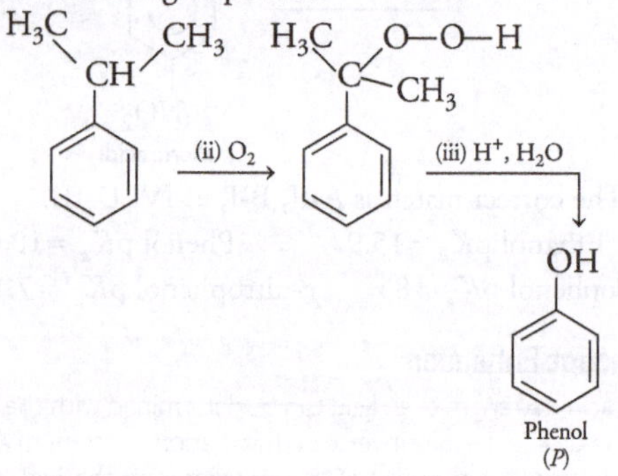

Now, phenol reacts with sodium chromate in presence of H_2SO_4 to give benzoquinone (*A*). When phenol reacts with bromine in CS_2 it gives *p*-bromophenol (*B*).

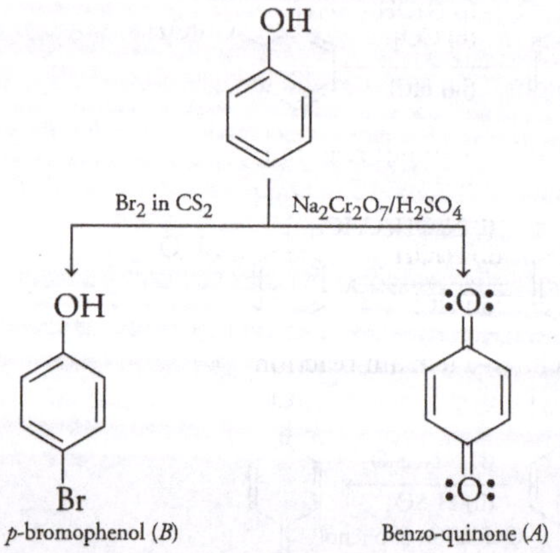

38. *(a)* Correct matching is A-IV, B-III, C-II, D-I.

39. *(30)*

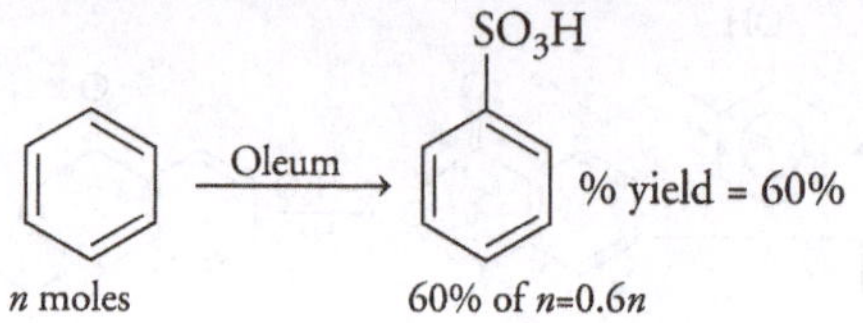

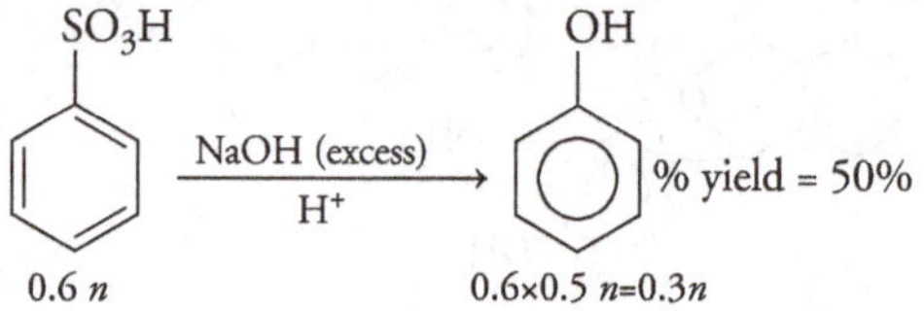

$$\text{Overall \% yield of reactions} = \frac{0.3n}{n} \times 100 = 30$$

40. *(7)* Phenol on reaction with mixed acid forms picric acid. Picric acid contains seven oxygen atoms.

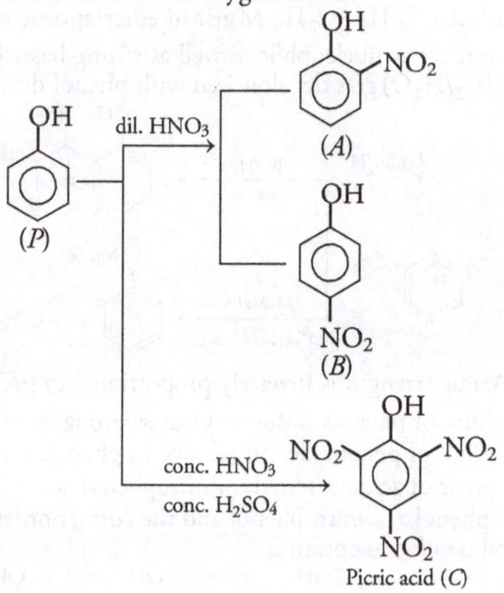

Topic 3

01. *(d)* Suitable reaction condition for preparation of methyl phenyl ether is PhO^-Na^+ and MeBr.

$$PhO^{\ominus}Na^{\oplus} + Me\text{—}Br \xrightarrow{S_N2} Ph\text{—}O\text{—}Me$$

02. *(a)* A is true but R is false.

It is because that acetals and ketals are *gem*-di ethers (Lewis bases). Hence, they should be stable in basic medium. However, it will break in acidic medium. Hence, A is true, while R is false because alkoxide ion $[RO^-]$ is not a good leaving group. Hence, it has poor leaving tendency.

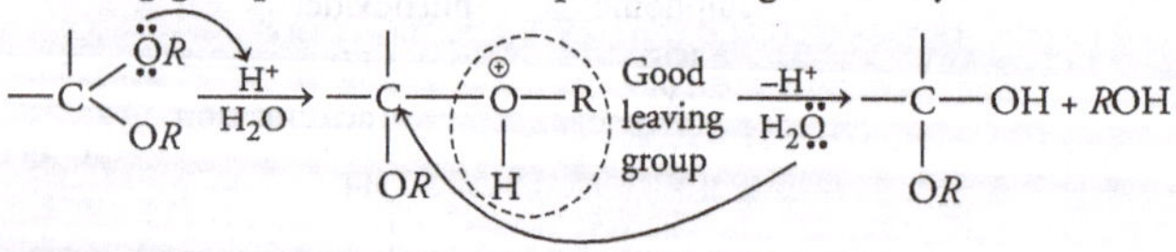

03. *(c)*

> **idea** Ethers are least reactive functional groups. The cleavage of C—O bond in ethers take place under drastic conditions with excess of HX.

Substituted phenols react with aq. NaOH to form sodium phenoxides which on reaction with CH_3I undergoes S_N2 reaction to give 2-methoxy-1-methyl benzene.

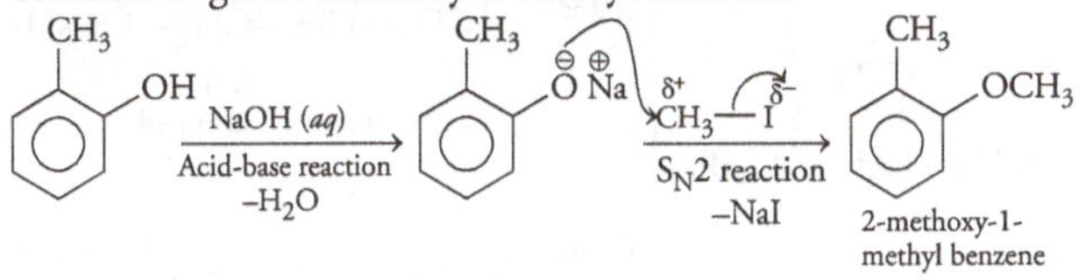

04. *(a)* Reaction mention in (a) will not lead to desired ether formation in major proportion.

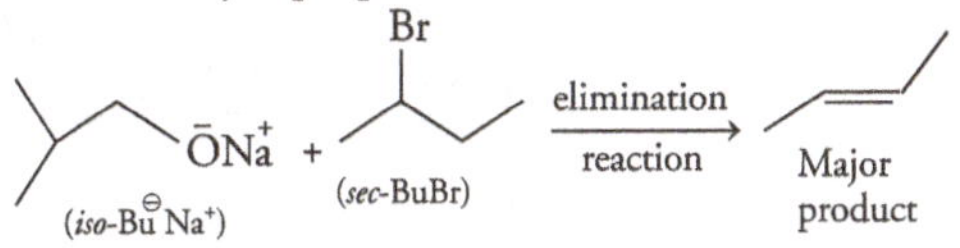

05. *(b)* Both Statement I and Statement II are true.

06. *(b)* Both Statement I and Statement II are true.

07. *(a)* The complete reaction is as follows,

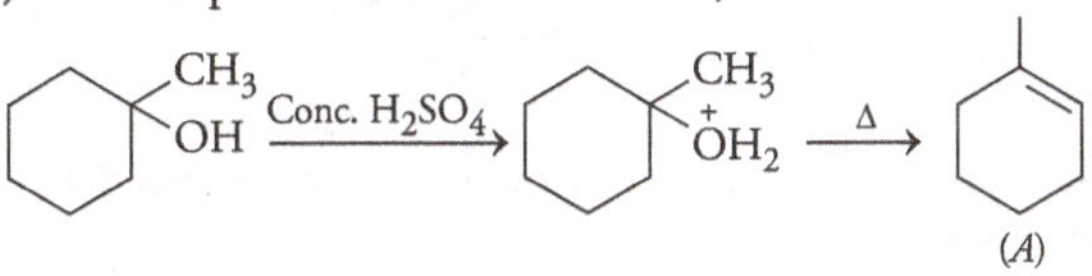

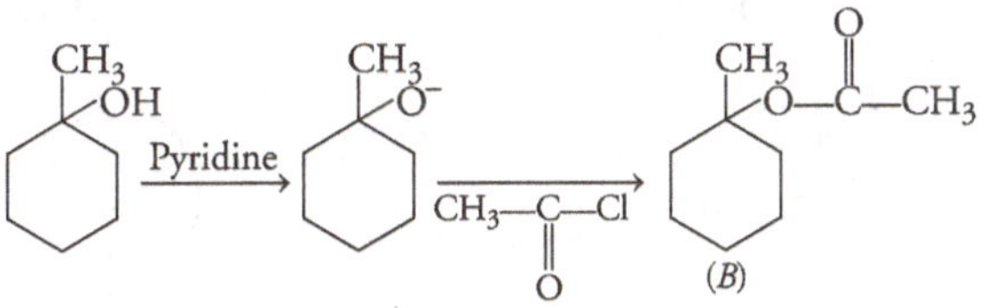

Hence, option (a) is correct.

08. *(c)* The complete reaction is as follows,

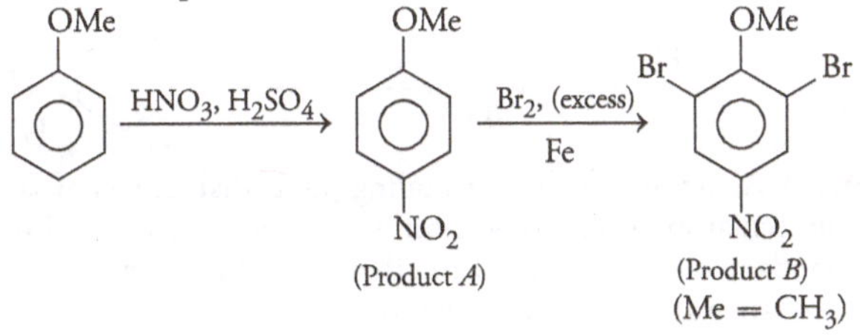

09. *(d)* **Reaction mechanism**

1. Proton capture

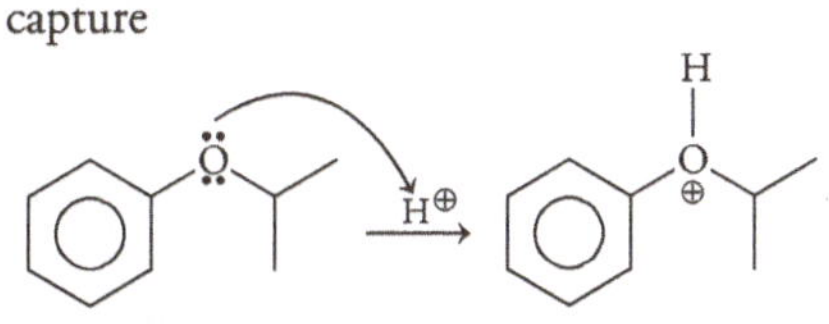

2. Nucleophilic attack of $I^{\ominus}$

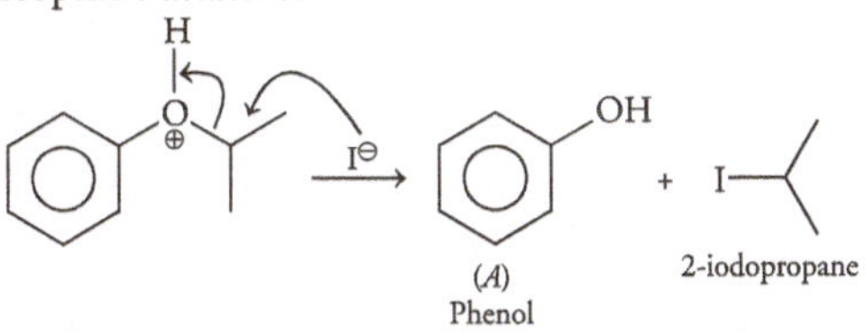

3. Phenol on heating with Zn-metal is reduced to benzene.

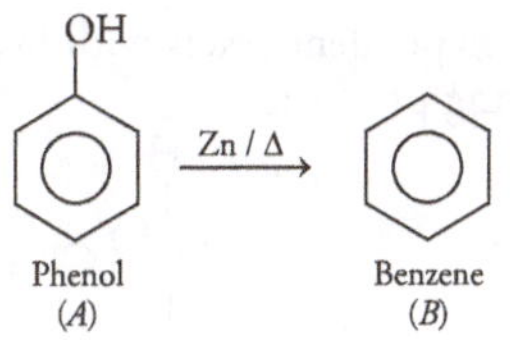

10. *(d)* The mechanism of the given reaction is as follows :

Step 1.

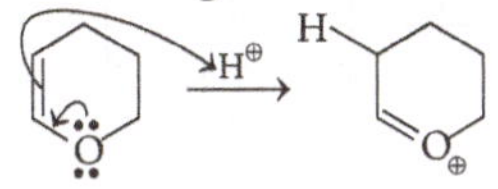

Step 2.

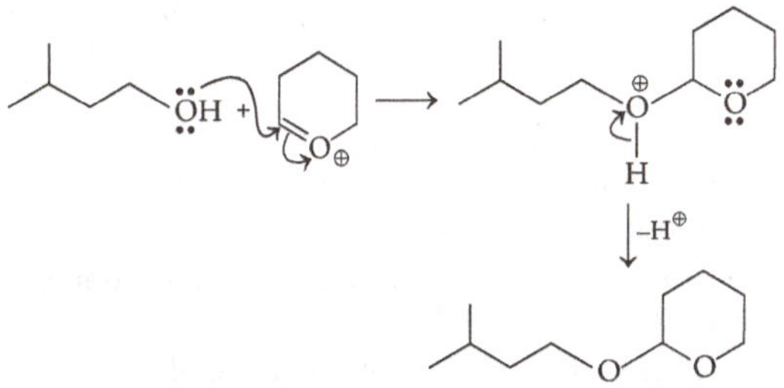

11. *(b)* Williamson's synthesis is used to prepare mixed ethers as well as simple ether. To prepare ethyl phenyl ether following reaction take place.

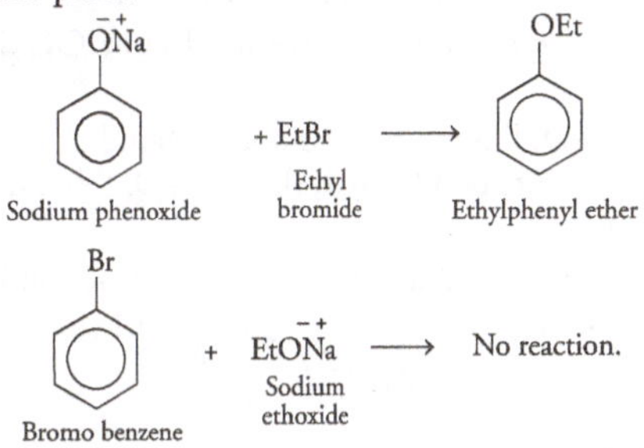

This reaction is not possible as bromine has partial double bond character with benzene ring, so it cannot be dissociated. So, Assertion is correct but Reason is not correct.

12. *(c)* The given reaction proceeds via S_N2 reaction mechanism as follows

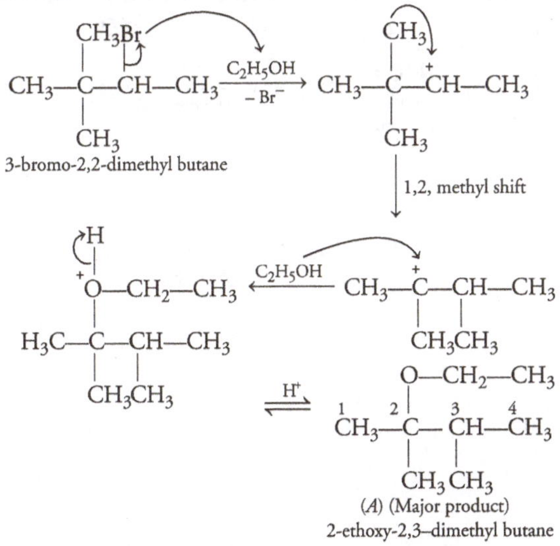

∴ Therefore, product *A* is 2-ethoxy-2, 3-dimethyl butane.

13. *(b)* 1-methoxy napthalene reacts with hydroiodic acid to form the following products.

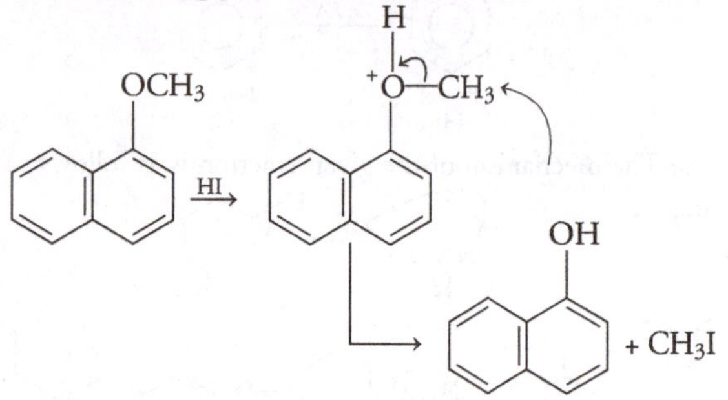

14. *(b)* The increasing order of C—OHbond length in the given compounds is : phenol < *p*-ethoxy phenol < methanol.
It is explained as follows
Methanol has no resonance and no partial double bond character in C—OH bond.
∴It has maximum C—OH bond length.
In both phenol and *p*-ethoxyphenol, resonance is involved.
∴ C—OH bond has partial double bond character and hence C—OH bond length is less than methanol.
But in *p*-ethoxyphenol, there is +*R*-effect of 'OC_2H_5' group also.
So the 'C—OH' bond has less double bond character and slightly more bond length than C—OH' bond length in phenol.

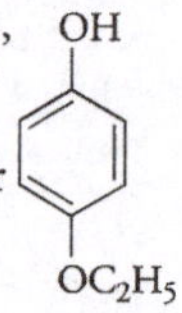

15. *(b)* The major product obtained in the reaction is as follows

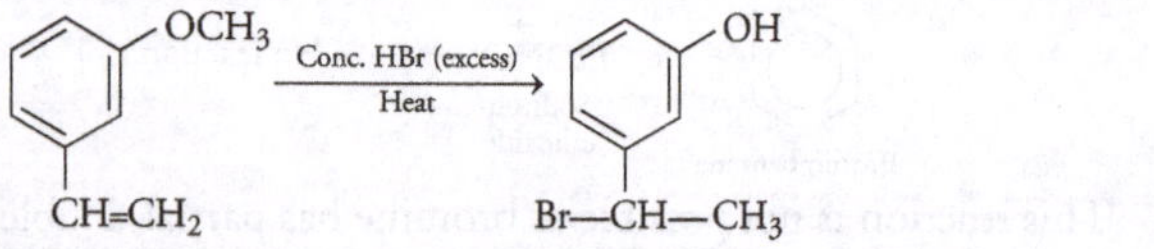

As conc. HBr is in excess. So, reaction will take place at both the substituents.

16. *(c)* Complete reaction sequence is as follows.
Methanol → Acetic acid → Ethanol → Diethylether
(*A*) (*B*) (*C*) (*D*)

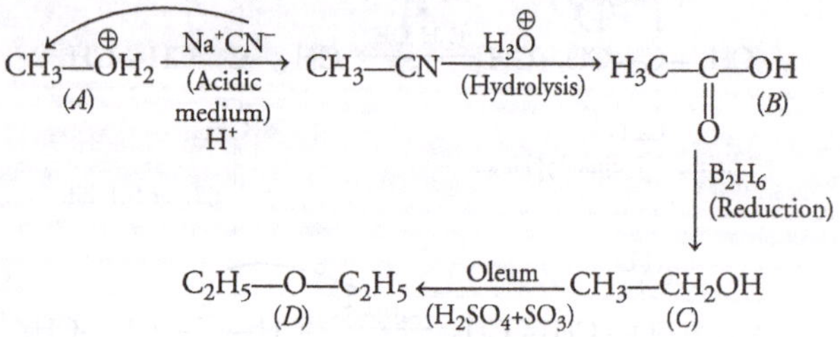

17. *(c)*

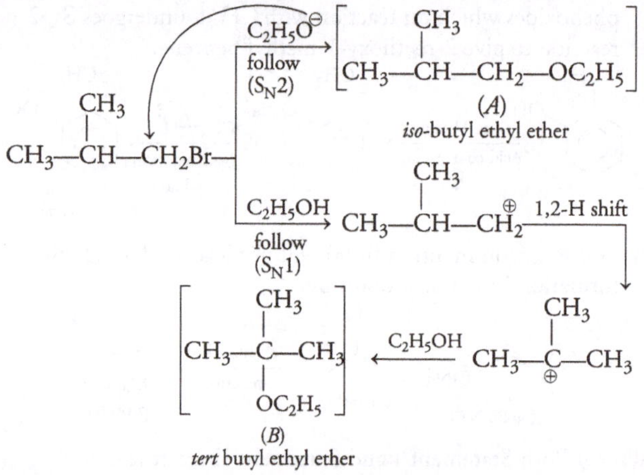

18. *(a)*

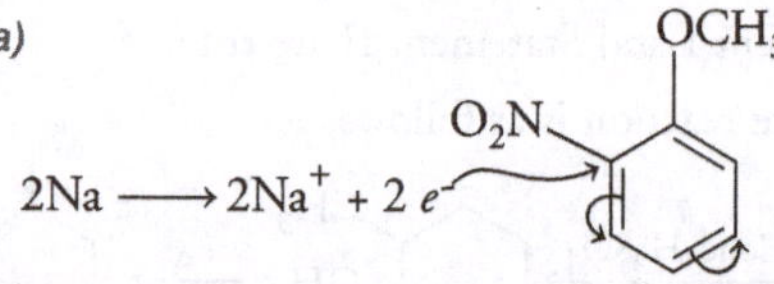

Sodium will donate electrons to the more electron deficient position.

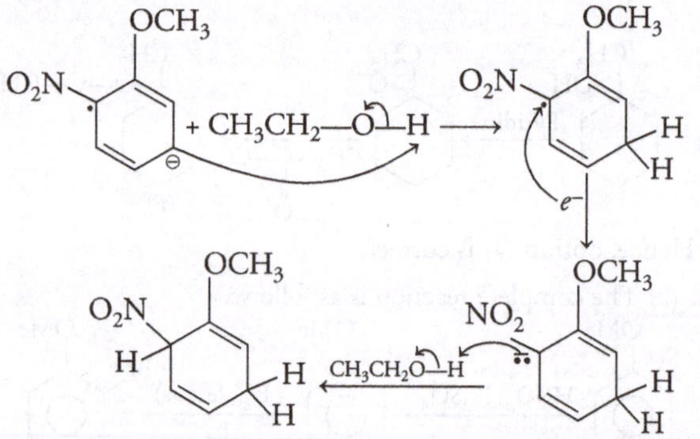

19. *(d)* Alcohols have higher boiling point than ethers of same molecular mass due to possibility of intermolecular hydrogen bonds in alcohols and not in ethers. So, *A* is alcohol derivative and *B* is ether derivative.

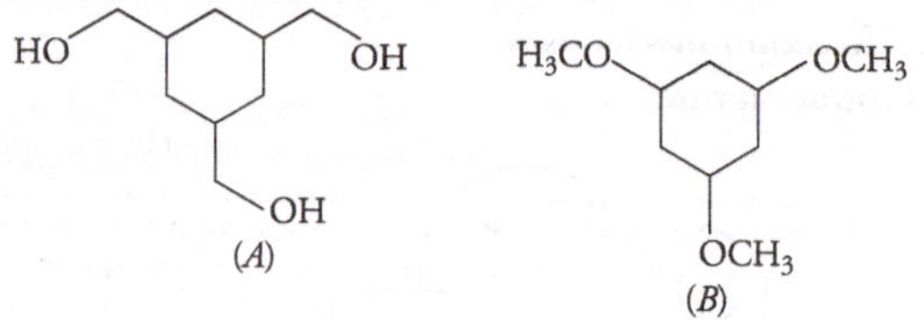

21 Aldehydes, Ketones & Carboxylic Acids

In JEE Main there is usually 0-4 question per paper are visible from this chapter. In past 3 years, 52 conceptual and 3 numerical questions were asked of hard to moderate level from this chapter.

For detailed analysis scan QR code

Topic 1 / Preparation, Properties and Uses of Aldehydes

• Level 1

01 The major product of the following reaction is

$$CH_3CH_2CH{=}O \xrightarrow[\text{Reflux}]{\text{Excess HCHO alkali}} ?$$

[2025, 23 Jan Shift-I]

(a) $CH_3—C(CH_2—OH)_3$ (i.e. $CH_3—C$ bearing three $CH_2—OH$ groups)

(b) $CH_3—CH(CH_2—OH)—CH{=}O$

(c) $CH_3—CH_2—CH_2—OH$

(d) $CH_3—C({=}CH_2)—CH{=}O$

02 The product (P) formed in the following reaction is

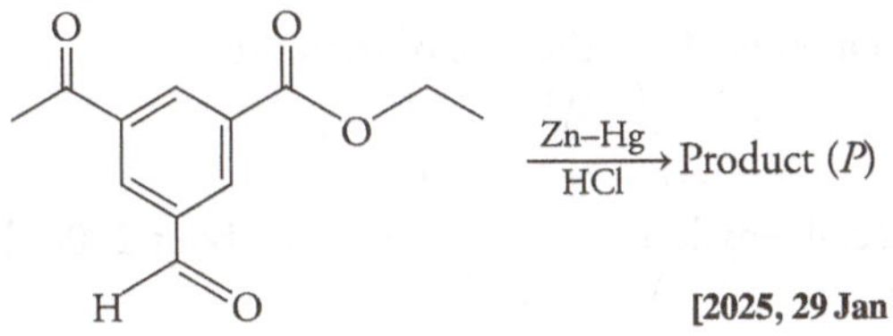

[2025, 29 Jan Shift-I]

(a)

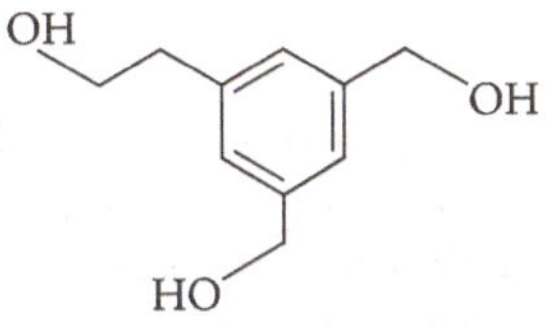

(b)

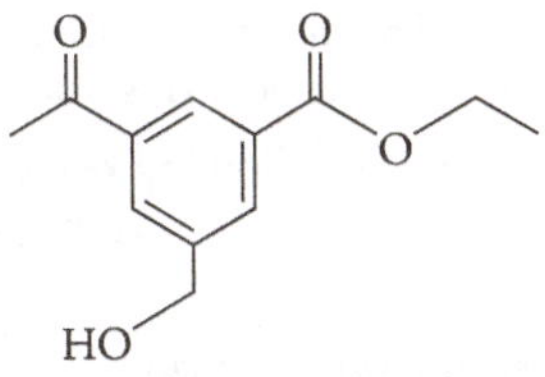

(c)

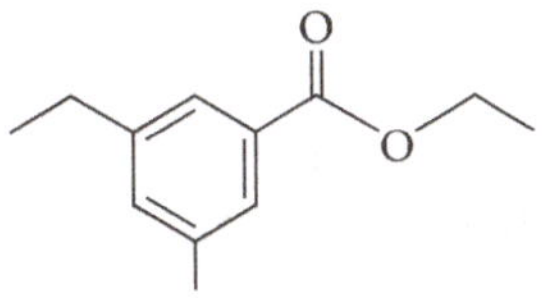

(d)

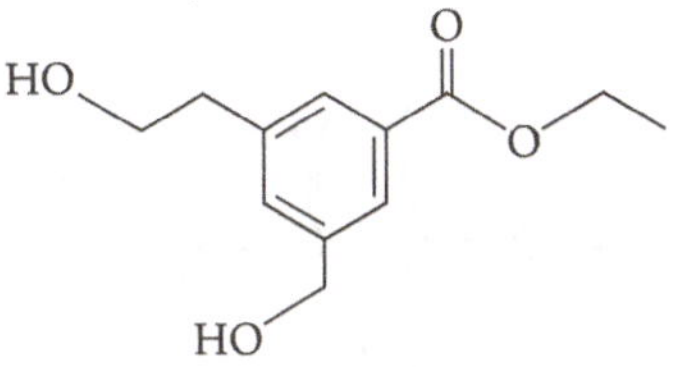

03 Identify major product "X" formed in the following reaction: **[2024, 9 April Shift-I]**

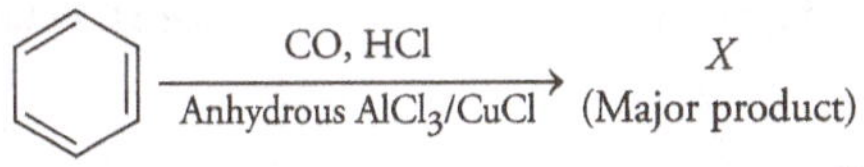

(a)

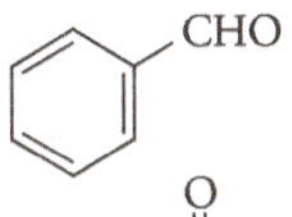

(b)

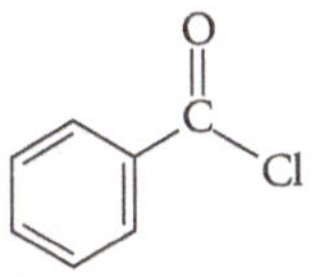

(c)

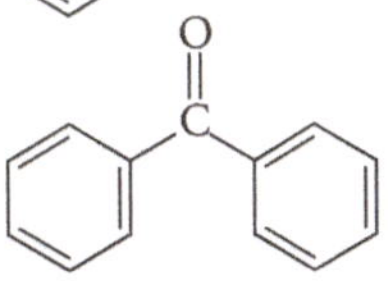

(d)

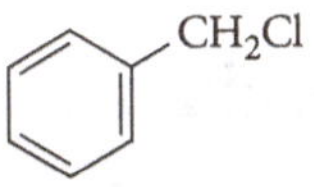

04 Which among the following aldehydes is most reactive towards nucleophilic addition reactions? **[2024, 6 April Shift-I]**

(a) $C_2H_5—\overset{O}{\overset{\|}{C}}—H$ (b) $H—\overset{O}{\overset{\|}{C}}—H$

(c) $C_3H_7—\overset{O}{\overset{\|}{C}}—H$ (d) $CH_3—\overset{O}{\overset{\|}{C}}—H$

05 Identify the name reaction. **[2024, 31 Jan Shift-II]**

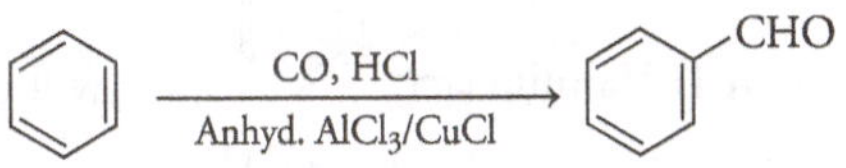

(a) Etard reaction (b) Gatterman-Koch reaction

(c) Stephen reaction (d) Rosenmund reduction

06 Structure of 4-methylpent-2-enal is **[2024, 30 Jan Shift-I]**

(a) 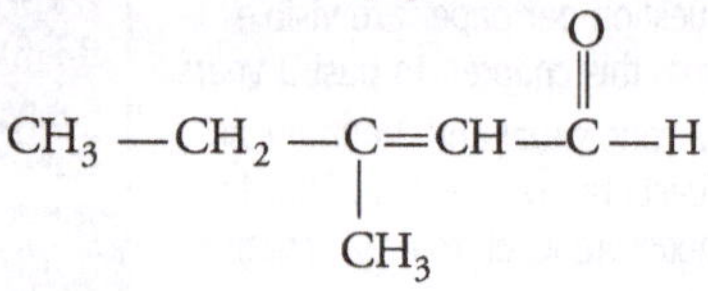

(b) 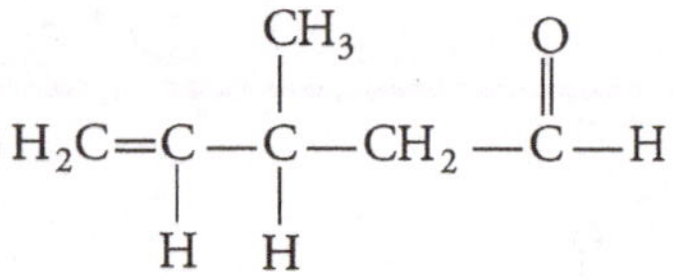

(c) 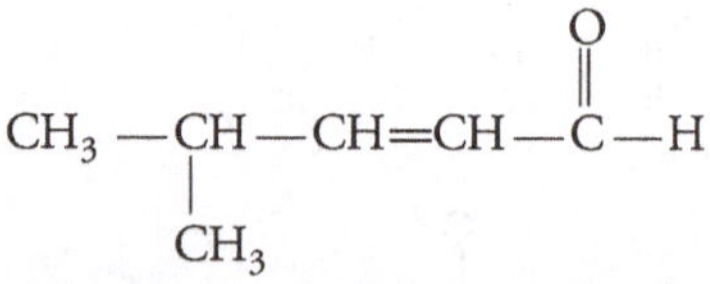

(d) 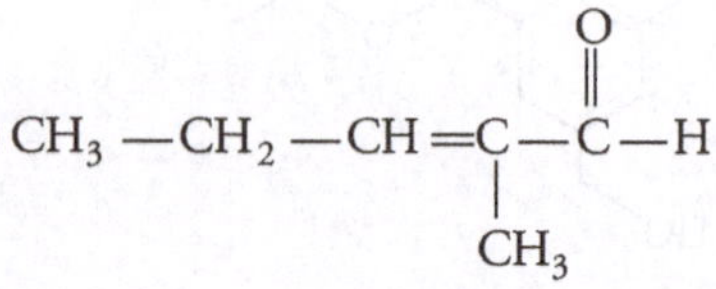

07 Correct structure of γ-methylcyclohexane carbaldehyde is **[2022, 29 July Shift-II]**

(a) 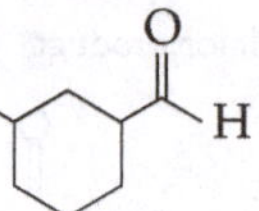
(b) 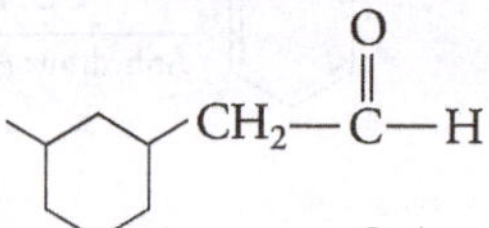

(c) 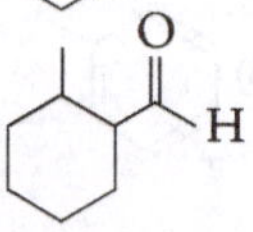
(d) 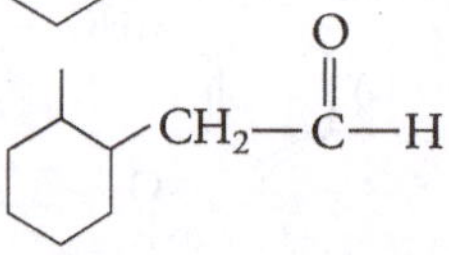

• Level 2

08 Given below are two statements. **[2025, 2 April Shift-I]**

Statement I Vanillin 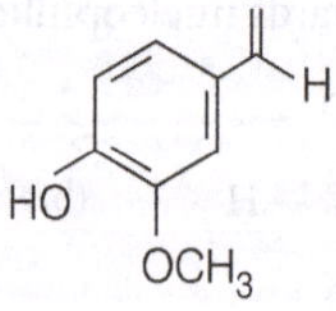 will react with NaOH and also with Tollen's reagent.

Statement II Vanillin 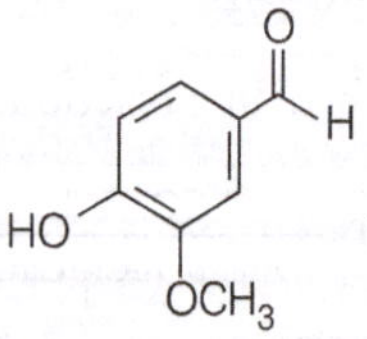 will undergo self aldol condensation very easily.

In the light of the above statements, choose the most appropriate answer from the options given below

(a) Statement I is correct but Statement II is incorrect.
(b) Statement I is incorrect but Statement II is correct.
(c) Both Statement I and Statement II are correct.
(d) Both Statement I and Statement II are incorrect.

09 Given below are two statements. **[2025, 4 April Shift-I]**

Statement I The dipole moment of $\overset{4}{C}H_3-\overset{3}{C}H=\overset{2}{C}H-\overset{1}{C}H=O$ is greater than $\overset{4}{C}H_3-\overset{3}{C}H_2-\overset{2}{C}H_2-\overset{1}{C}H=O$

Statement II C_1-C_2 bond length of $\underset{4}{C}H_3\underset{3}{C}H=\underset{2}{C}H-\underset{1}{C}H=O$ is greater than C_1-C_2 bond length of $\underset{4}{C}H_3-\underset{3}{C}H_2-\underset{2}{C}H_2-\underset{1}{C}H=O$

In the light of the above statements, choose the correct answer from the options given below.

(a) Both Statement I and Statement II are false.
(b) Both Statement I and Statement II are true.
(c) Statement I is false but Statement II is true.
(d) Statement I is true but Statement II is false.

10 Given below are two statements

Consider the following reaction, **[2025, 23 Jan Shift-II]**

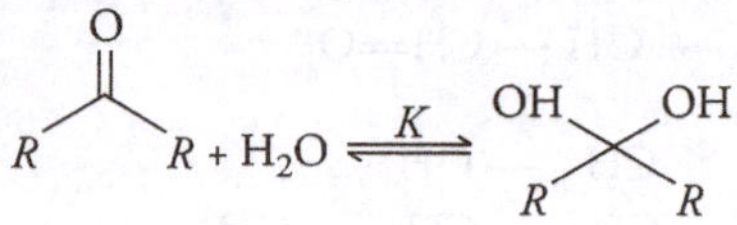

Statement I In the case of formaldehyde 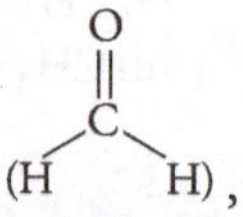, K is about 2280, due to small substituents hydration is faster.

Statement II In the case of trichloro acetaldehyde 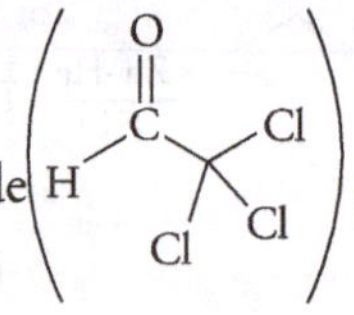, K is about 2000 due to $-I$ effect of –Cl.

In the light of the above statements, choose the correct answer from the options given below

(a) Statement I is true but Statement II is false.
(b) Both Statement I and Statement II are false.
(c) Both Statement I and Statement II are true.
(d) Statement I is false but Statement II is true.

11 The products formed in the following reaction, A and B are **[2022, 26 July Shift-I]**

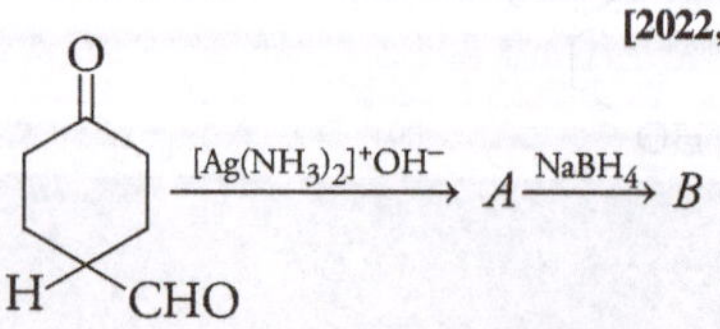

(a) $A =$ 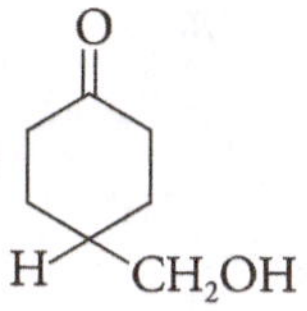 $B =$ 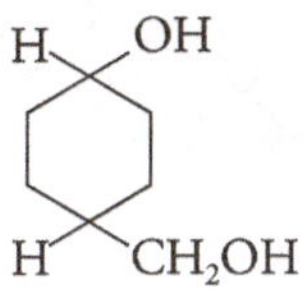

(b) $A =$ 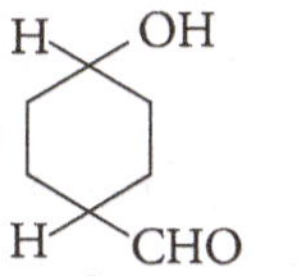 $B =$ 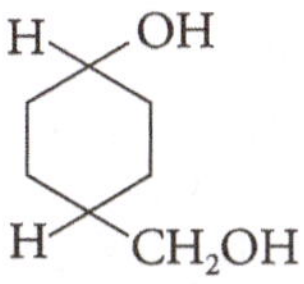

(c) $A =$ 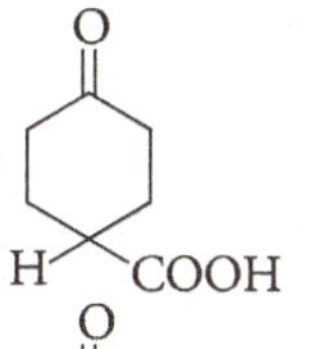 $B =$ 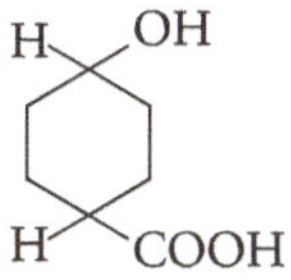

(d) $A =$ 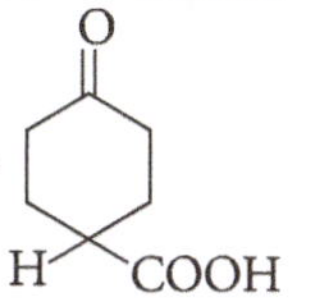 $B =$ 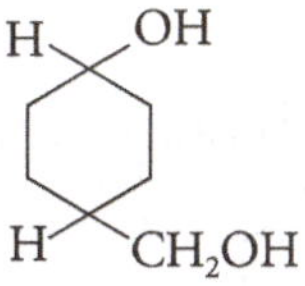

12 What is the major product of the following reaction? **[2022, 25 July Shift-II]**

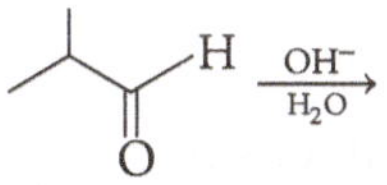

(a) 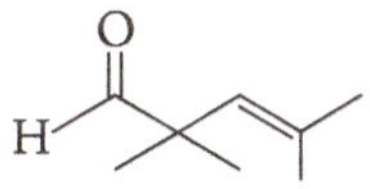

(b) 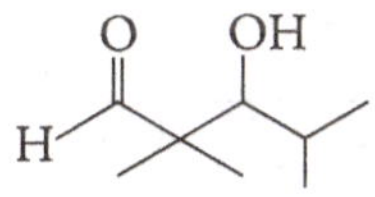

(c) 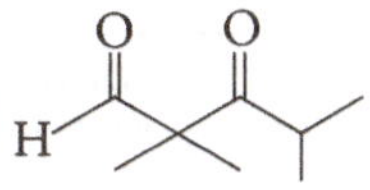

(d) 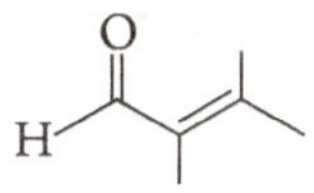

13 Oxidation of toluene to benzaldehyde can be easily carried out with which of the following reagents? **[2022, 26 June Shift-II]**

(a) CrO_3/acetic acid, H_3O^+
(b) CrO_3/acetic anhydride, H_3O^+
(c) $KMnO_4$/HCl, H_3O^+
(d) CO/HCl, anhydrous $AlCl_3$

14 Which of the following reagents/reactions will convert 'A' to 'B'? **[2022, 24 June Shift-II]**

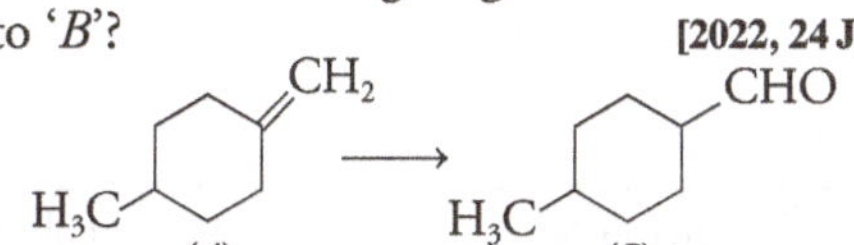

(a) PCC oxidation (b) Ozonolysis
(c) BH_3, H_2O_2/OH^- followed by PCC oxidation
(d) HBr, hydrolysis followed by oxidation by $K_2Cr_2O_7$

15 $R—CN \xrightarrow[\text{(ii) } H_2O]{\text{(i) DIBAL–H}} R—Y$

Consider the above reaction and identify Y. **[2021, 27 July Shift-II]**

(a) $—CH_2NH_2$ (b) $—CONH_2$
(c) —CHO (d) —COOH

16 An unsaturated hydrocarbon X on ozonolysis gives A. Compound A when warmed with ammoniacal silver nitrate forms a bright silver mirror along the sides of the test tube. The unsaturated hydrocarbon X is **[2021, 17 March Shift-II]**

(a) $CH_3-\underset{CH_3}{\underset{|}{C}}=\underset{CH_3}{\underset{|}{C}}-CH_3$

(b) 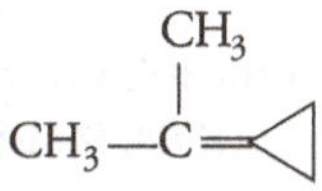

(c) $HC\equiv C—CH_2—CH_3$
(d) $CH_3—C\equiv C—CH_3$

17

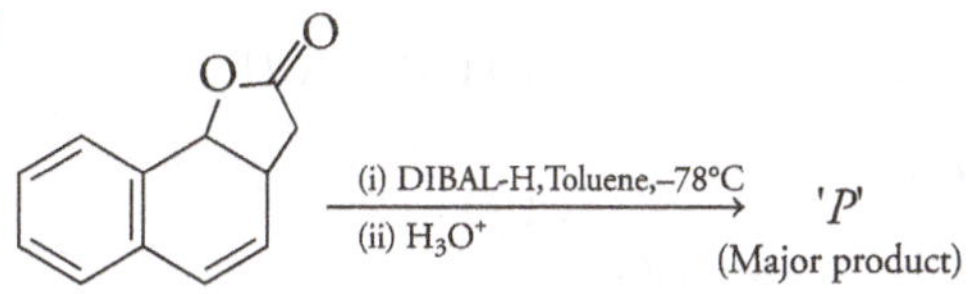

The product P in the above reaction is **[2021, 16 March Shift-I]**

(a) 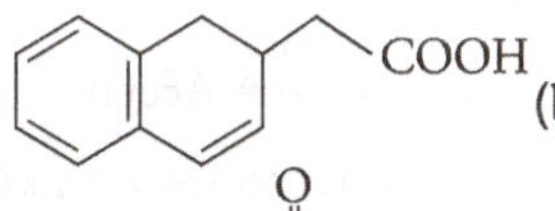

(b) 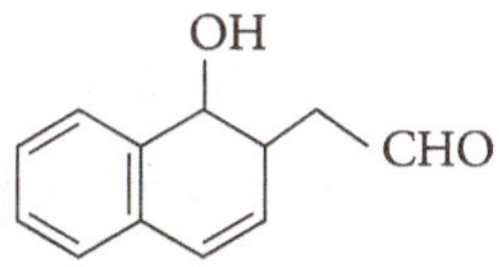

(c) 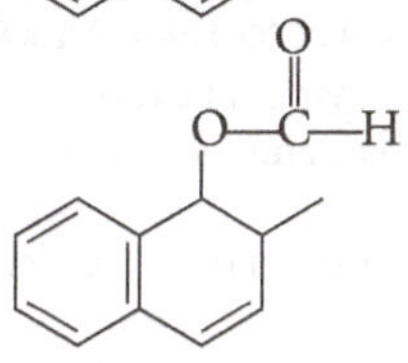

(d) 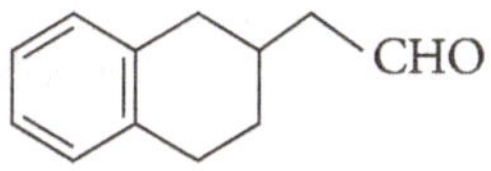

18 The major product of the following reaction is

$CH_3CH_2CH=CH_2 \xrightarrow[\text{Rh catalyst}]{H_2/CO}$ **[2021, 25 Feb Shift-II]**

(a) $CH_3CH_2CH_2CH_2CHO$
(b) $CH_3CH_2CH_2CHO$
(c) $CH_3CH_2CH=CHCHO$
(d) $CH_3CH_2\underset{CHO}{\underset{|}{C}}=CH_2$

19 Which one of the following reactions will not form acetaldehyde? **[2021, 25 Feb Shift-I]**

(a) $CH_2=CH_2+O_2 \xrightarrow[H_2O]{\text{Pd (II)/Cu (II)}}$

(b) $CH_3CH_2OH \xrightarrow{CrO_3\text{-}H_2SO_4}$

(c) $CH_3CN \xrightarrow[\text{(ii) } H_2O]{\text{(i) DIBAL -H}}$

(d) $CH_3CH_2OH \xrightarrow[573\text{ K}]{Cu}$

20 Which of the following reagent is used for the following reaction ? **[2021, 24 Feb Shift-I]**

$$CH_3CH_2CH_3 \xrightarrow{?} CH_3CH_2CHO$$

(a) Copper at high temperature and pressure
(b) Molybdenum oxide
(c) Manganese acetate
(d) Potassium permanganate

21 The major product(s) obtained in the following reaction is/are **[2019, 12 April Shift-I]**

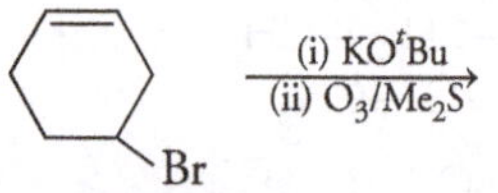

(a) 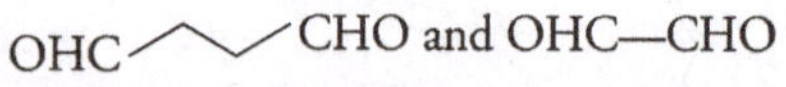

(b) 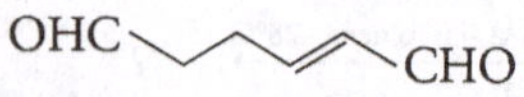

(c) 

(d) 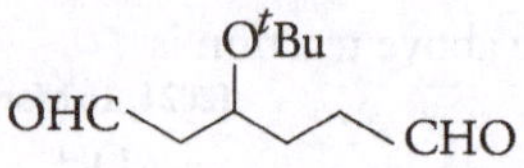

22 In the following reaction,

$$\text{Carbonyl compound} + MeOH \underset{}{\overset{HCl}{\rightleftharpoons}} \text{Acetal}$$

Rate of the reaction is the highest for **[2019, 9 April Shift-II]**
(a) Acetone as substrate and methanol in excess
(b) Propanal as substrate and methanol in stoichiometric amount
(c) Acetone as substrate and methanol in stoichiometric amount
(d) Propanal as substrate and methanol in excess

23 The aldehydes which will not form Grignard product with one equivalent Grignard reagents are

(A) 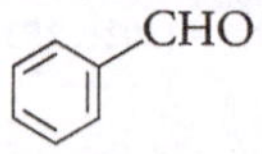

(B) 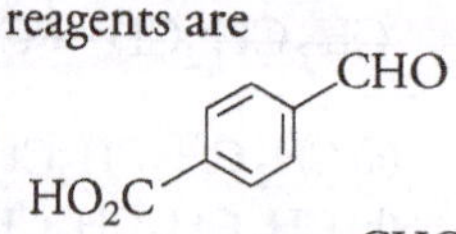

(C) 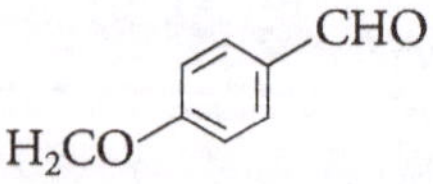

(D) 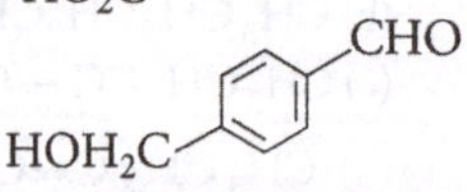

[2019, 12 Jan Shift-II]

(a) (C), (D) (b) (B), (D)
(c) (B), (C), (D) (d) (B), (C)

• Level 3

24

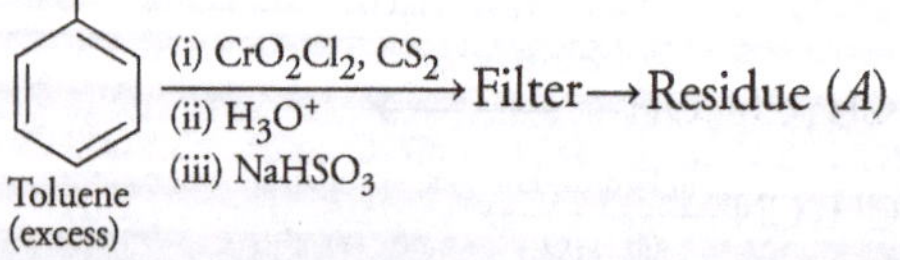

Residue (A) + HCl(dil.) $\longrightarrow$ Compound (B)

Structure of residue (A) and compound (B) formed respectively is **[2025, 22 Jan Shift-II]**

	[A]	[B]
(a)	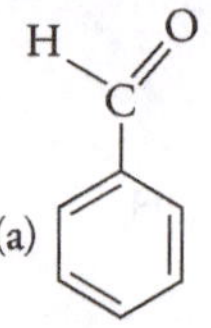	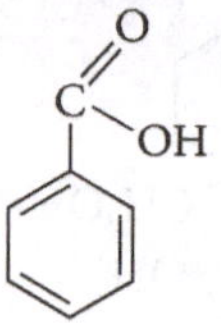
(b)	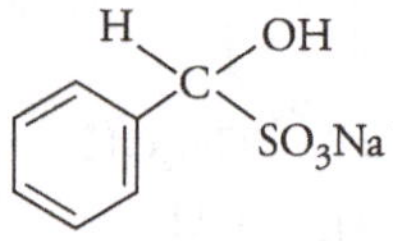	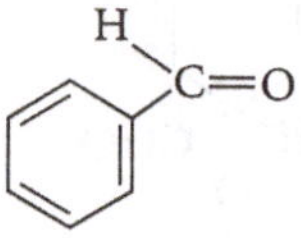
(c)	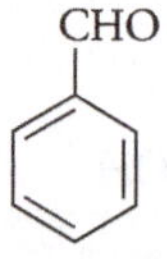	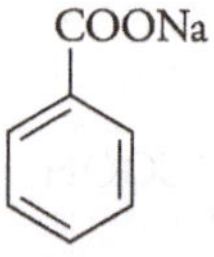
(d)	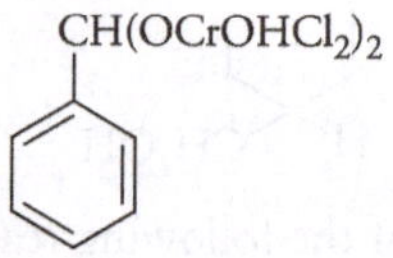	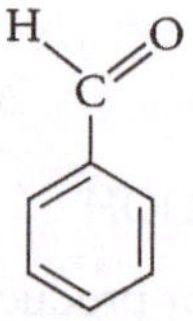

Imp. *Preparation of Benzaldehyde*

25 Match List-I with List-II. **[2025, 24 Jan Shift-II]**

	List-I		List-II
A.	$RCN \xrightarrow[\text{(ii) } H_3O^+]{\text{(i) } SnCl_2, HCl} RCHO$	I.	Etard reaction
B.	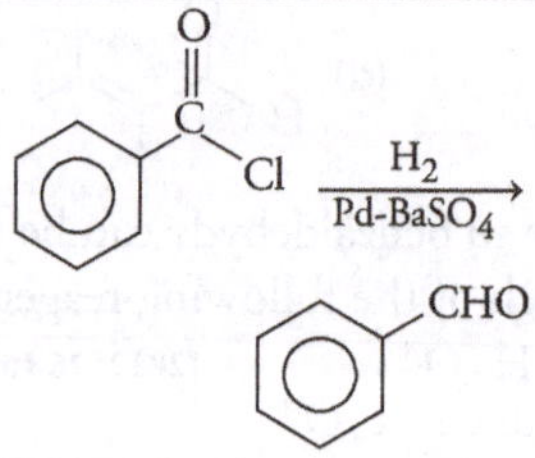	II.	Gattermann-Koch reaction
C.	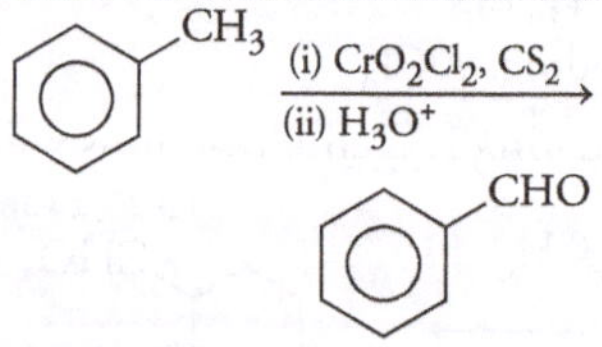	III.	Rosenmund reduction
D.	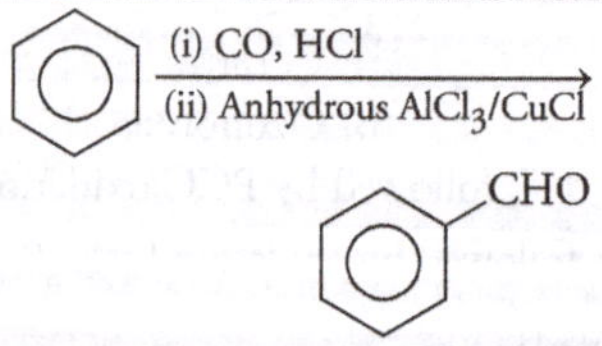	IV.	Stephen reaction

Choose the correct answer from the options given below.
(a) A-III, B-IV, C-I, D-II (b) A-III, B-IV, C-II, D-I
(c) A-I, B-III, C-II, D-IV (d) A-IV, B-III, C-I, D-II

26 Identify product [*A*], [*B*] and [*C*] in the following reaction sequence **[2025, 28 Jan Shift-II]**

$$CH_3—C\equiv CH \xrightarrow[H_2]{Pd/C} [A] \xrightarrow[(ii)\ Zn,\ H_2O]{(i)\ O_3} [B]+[C]$$

(a) [*A*] : $CH_3—CH=CH_2$, [*B*] : CH_3CHO, [C] : HCHO

(b) [*A*] : $CH_2=CH_2$, [*B*] : $H_3C—\overset{O}{\overset{\|}{C}}—CH_3$, [*C*] : HCHO

(c) [*A*] : $CH_3CH_2CH_3$, [*B*] : CH_3CHO, [*C*] : HCHO

(d) [*A*] : $CH_3—CH=CH_2$, [*B*] CH_3CHO, [*C*] : CH_3CH_2OH

27 The structures of major products *A*, *B* and *C* in the following reaction sequence are

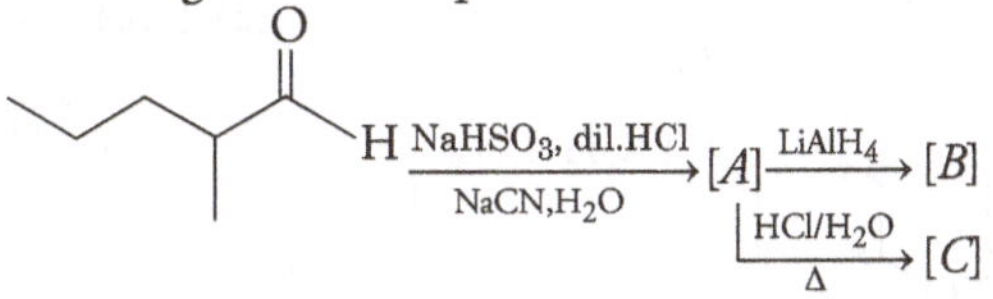

[2023, 1 Feb Shift-II]

(a) *A* = 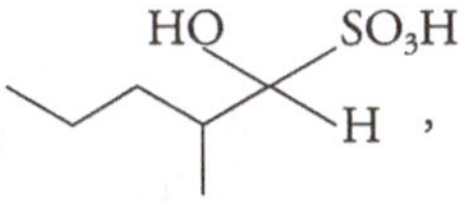 , *B* = 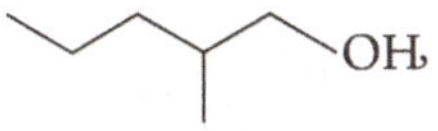

C = 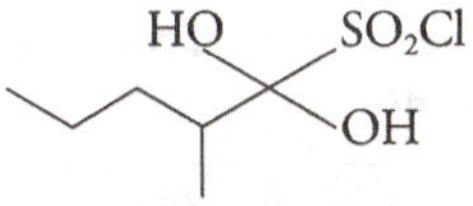

(b) *A* = 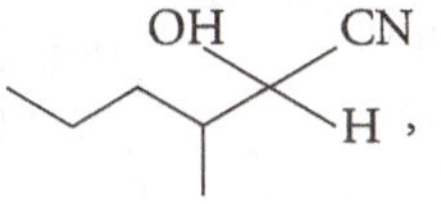 , *B* = 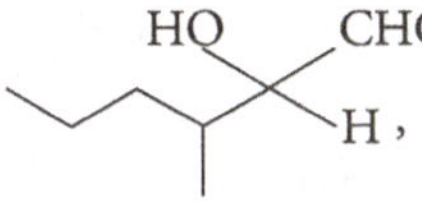

C = 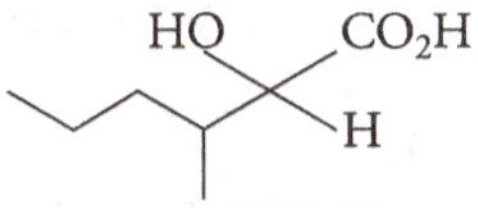

(c) *A* = 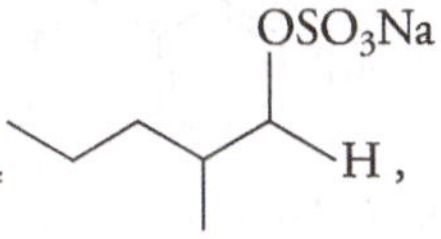 , *B* = 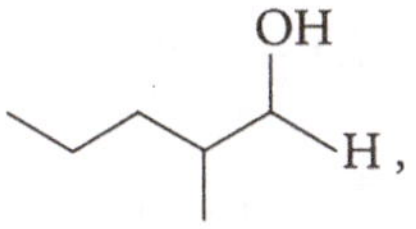

C = 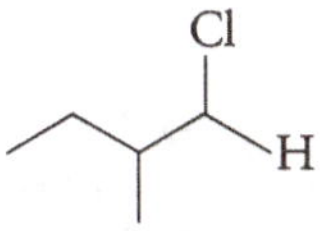

(d) *A* = 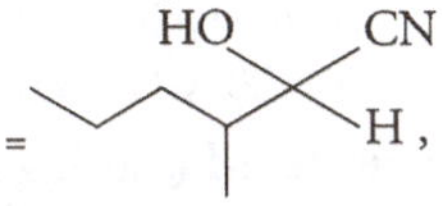 , *B* = 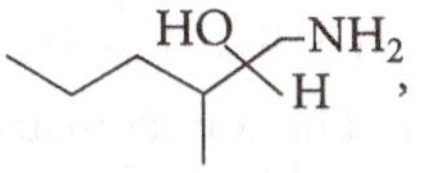

C = 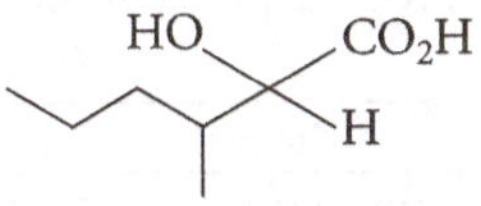

28 Compound (*X*) undergoes following sequence of reactions to give the lactone (*Y*).

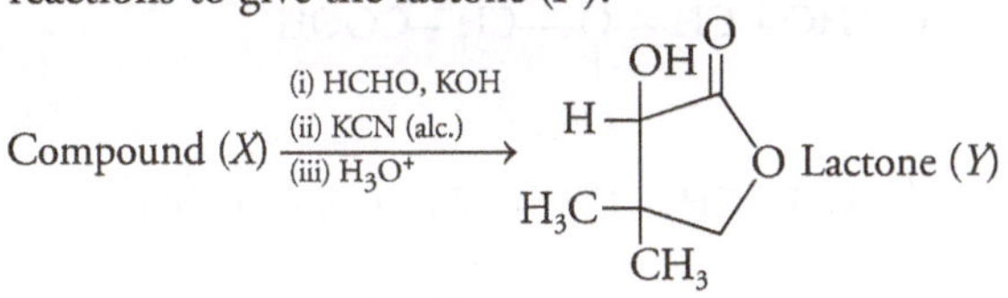

Compound (*X*) is **[2023, 24 Jan Shift-I]**

(a) $HOH_2C—C(CH_3)_2—CHO$

(b) $H_2C(OH)CH_2CH_2CH_2CHO$

(c) $H—C(CH_3)_2—CHO$

(d) $H_2C(CH_3)—CH_2—CHO$

29

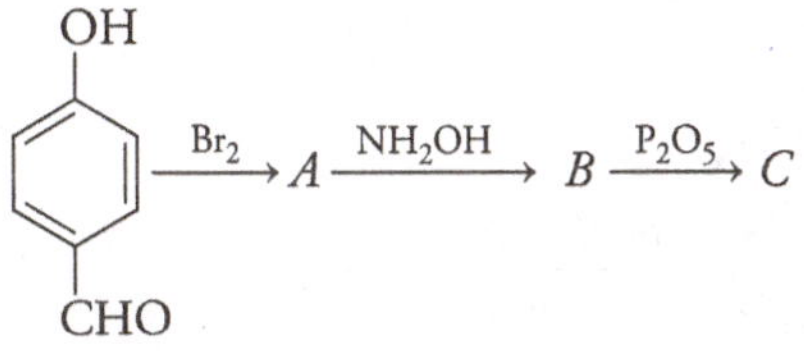

Consider the above reaction sequence, the product '*C*' is **[2022, 29 July Shift-I]**

(a) 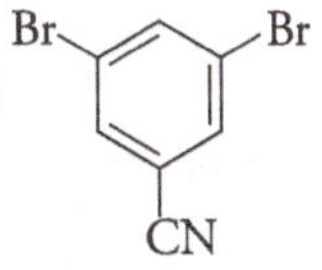

(b) 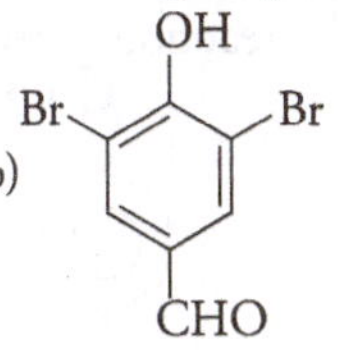

(c) 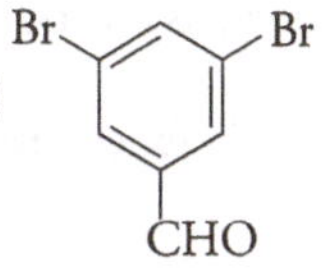

(d) 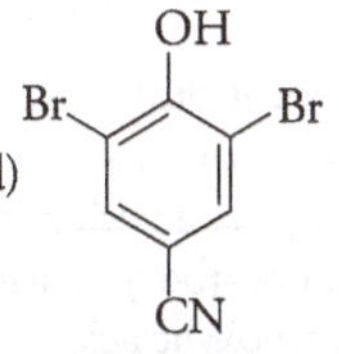

30 Isobutyraldehyde on reaction with formaldehyde and K_2CO_3 gives compound *A*. Compound *A* reacts with KCN and yields compound *B*, which on hydrolysis gives a stable compound *C*. The compound *C* is **[2022, 28 June Shift-II]**

(a) 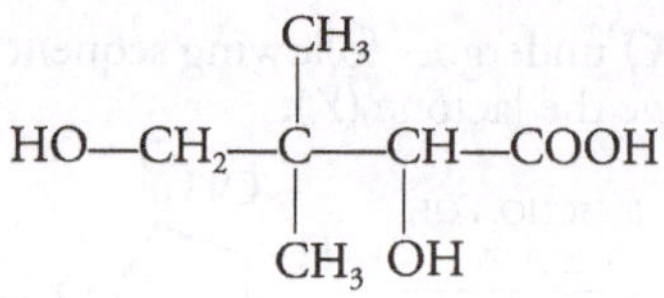

(b) 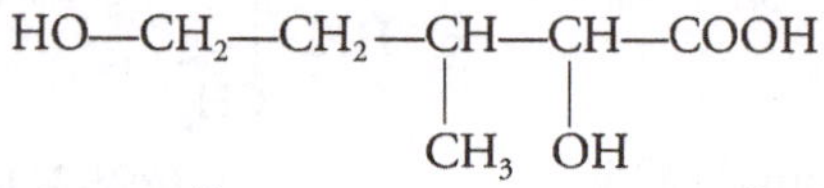

(c) 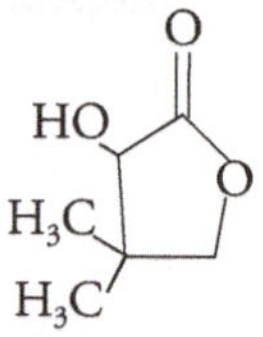 (d) 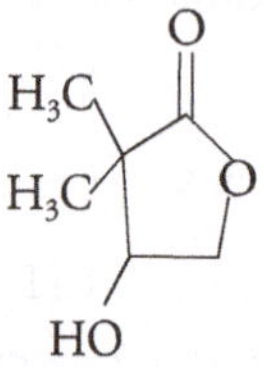

31 Which of the following reactions will yield benzaldehyde as a product? **[2022, 27 June Shift-I]**

(A) 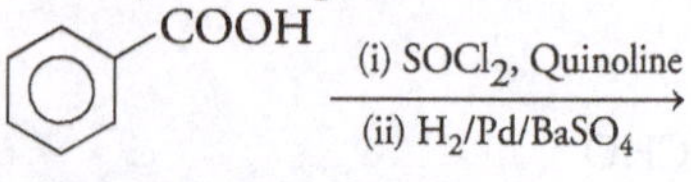

(B) 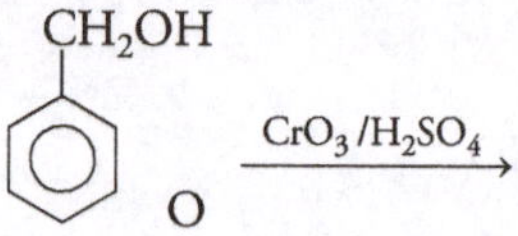

(C) 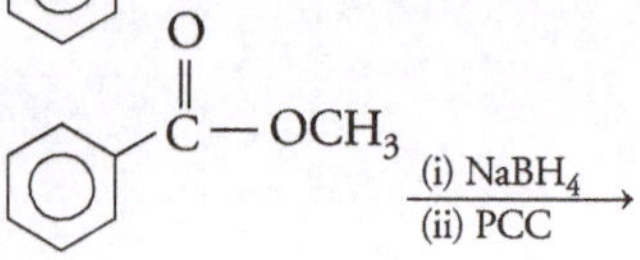

(D) 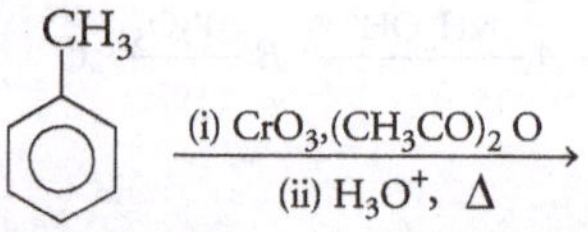

(a) (B) and (C) (b) (C) and (D)
(c) (A) and (D) (d) (A) and (C)

32 For the reaction given below **[2021, 31 Aug Shift-II]**

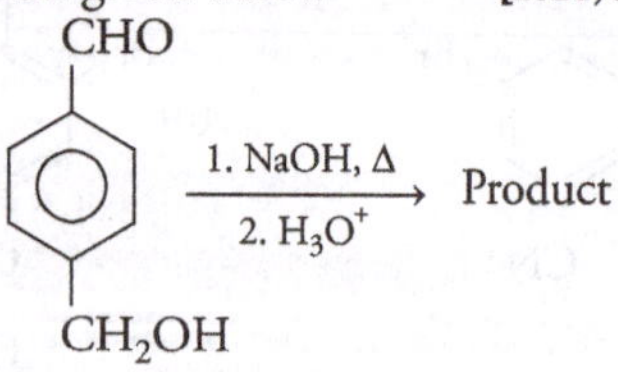

The compound which is not formed as a product in the reaction is a
(a) compound with both alcohol and acid functional groups
(b) monocarboxylic acid
(c) dicarboxylic acid
(d) diol

33 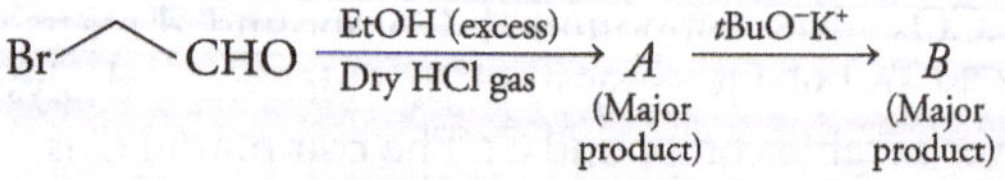

[where, Et = C_2H_5, tBu = $(CH_3)_3C$—]
Consider the above reaction sequence, product A and product B formed respectively are **[2021, 25 July Shift-II]**

(a) 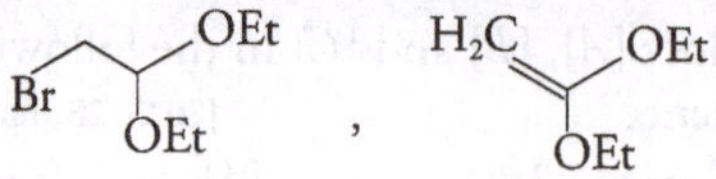

(b) 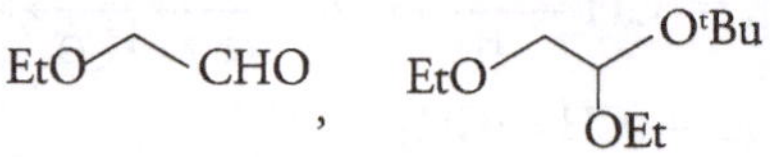

(c) 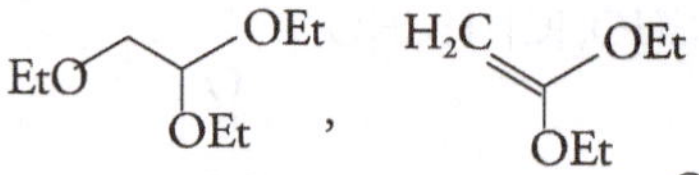

(d) 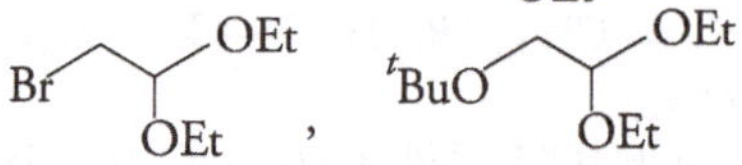

34 Identify A in the following chemical reaction.

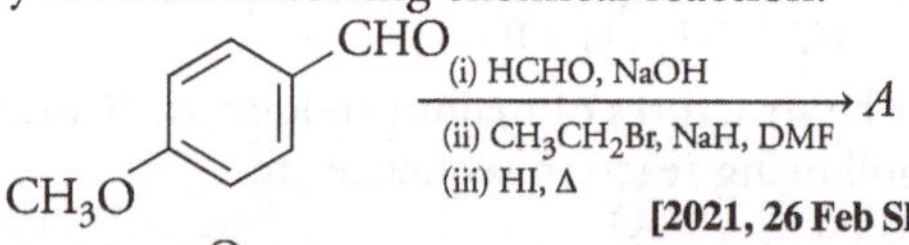

[2021, 26 Feb Shift-II]

(a) 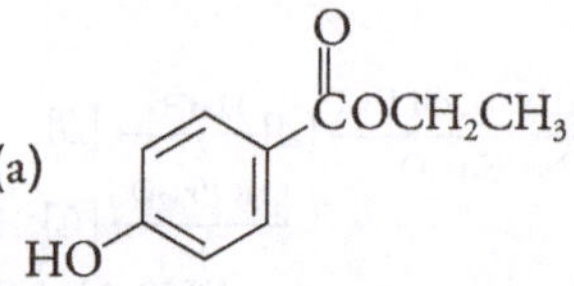 (b) 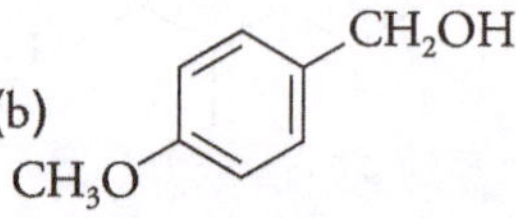

(c) 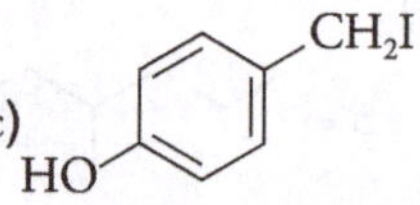 (d) 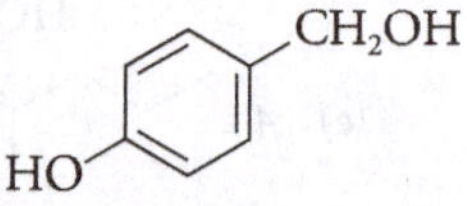

35 Identify A in the given chemical reaction,

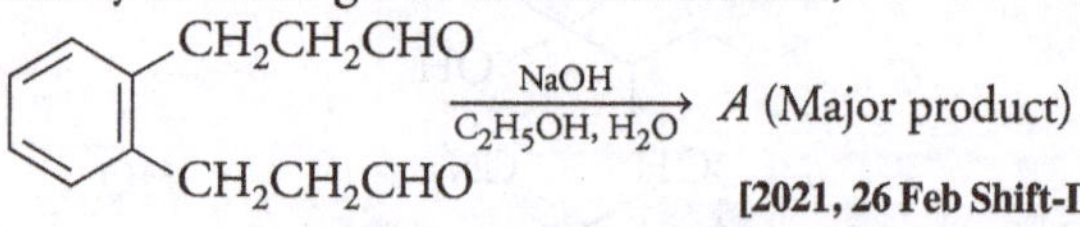

[2021, 26 Feb Shift-II]

(a) 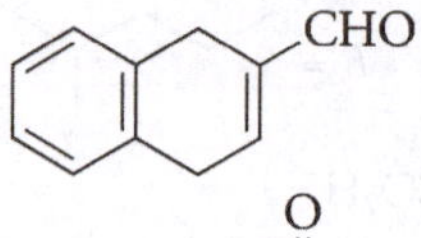 (b) 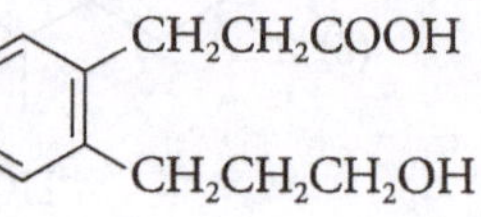

(c) 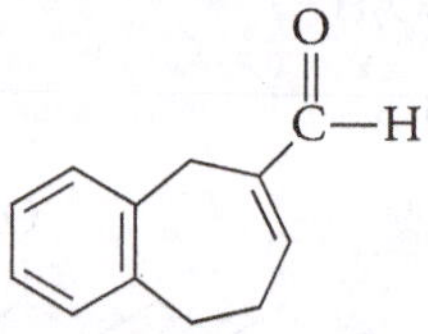 (d) 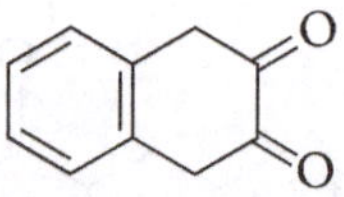

36 The major product of the following chemical reaction is

$$CH_3CH_2CN \xrightarrow[\text{(iii) Pd/BaSO}_4\text{, }H_2]{\text{(i) }H_3O^+\text{, }\Delta\text{ (ii) }SOCl_2} ?$$

[2021, 25 Feb Shift-I]

(a) $CH_3CH_2CH_2OH$ (b) $(CH_3CH_2CO)_2O$
(c) $CH_3CH_2CH_3$ (d) CH_3CH_2CHO

37 The correct match between Item-I (starting material) and Item-II (reagent) for the preparation of benzaldehyde is **[2020, 6 Sep Shift-II]**

	Item-I		Item-II
I.	Benzene	(P)	HCl and $SnCl_2$, H_3O^+
II.	Benzonitrile	(Q)	H_2, Pd-$BaSO_4$, S and quinoline
III.	Benzoyl chloride	(R)	CO, HCl and $AlCl_3$

(a) I-(Q), II-(R) and III-(P) (b) I-(P), II-(Q) and III-(R)
(c) I-(R), II-(P) and III-(Q) (d) I-(R), II-(Q) and III-(P)

38 The increasing order of the following compounds towards HCN addition is **[2020, 2 Sep Shift-I]**

(i) 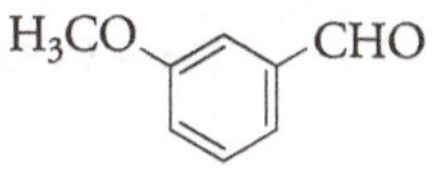
(ii) 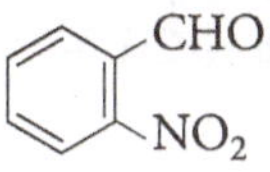
(iii) 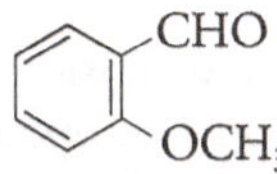
(iv) 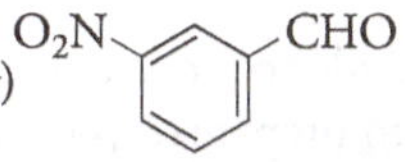

(a) (i) < (iii) < (iv) < (ii) (b) (iii) < (iv) < (i) < (ii)
(c) (iii) < (iv) < (ii) < (i) (d) (iii) < (i) < (iv) < (ii)

39 In the following reaction *A* is **[2020, 9 Jan Shift-II]**

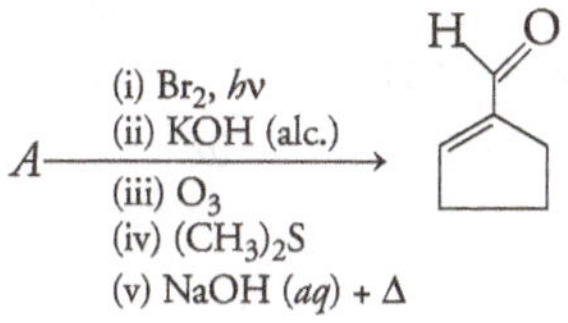

(a) 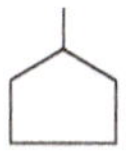 (b) 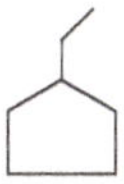
(c)  (d) 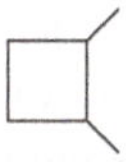

Numerical Value Type Questions

40 The product of the following reaction is *P*. **[2024, 31 Jan Shift-I]**

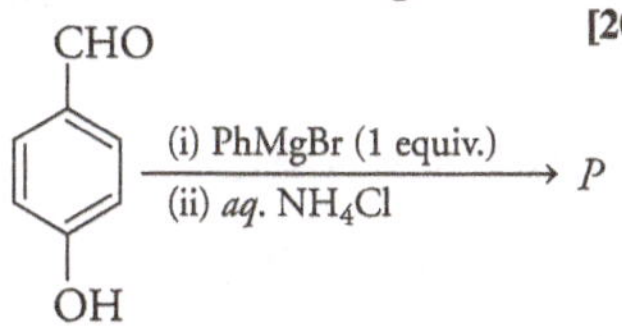

The number of hydroxyl groups present in the product *P* is ....

Topic 2 / Preparation, Properties and Uses of Ketones

• Level 1

01 Identify the product in the following reaction:

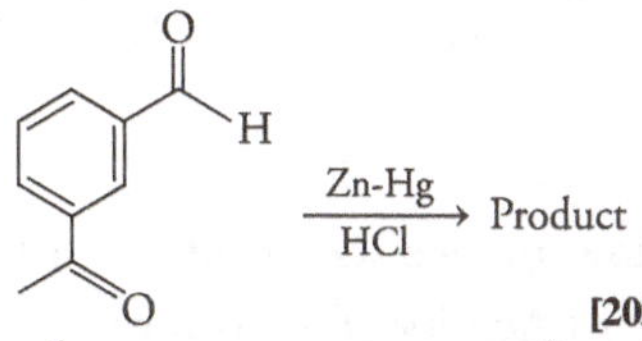

[2024, 4 April Shift-I]

(a) 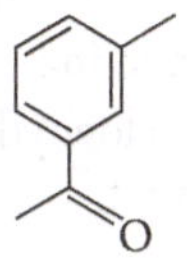 (b) 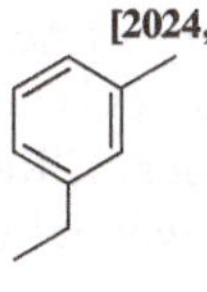

(c) 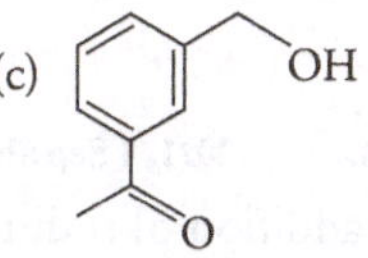 (d) 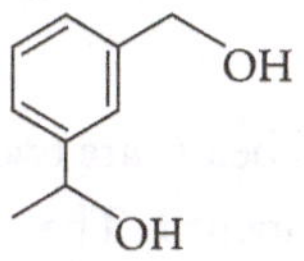

02 Given below are two statements, one is labelled as Assertion (A) and the other is labelled as Reason (R).

Assertion (A) 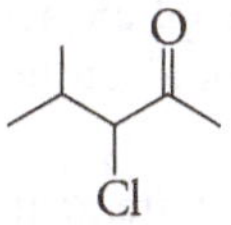 can be subjected to Wolff-Kishner reduction to give 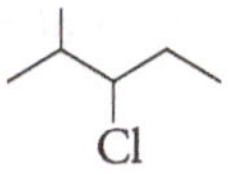.

Reason (R) Wolff-Kishner reduction is used to convert 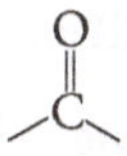 into 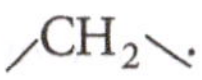.

In the light of the above statements, choose the most appropriate answer from the options given below. **[2023, 11 April Shift-II]**

(a) Both A and R are true and R is the correct explanation of A.
(b) A is true but R is false.
(c) A is false but R is true.
(d) Both A and R are true but R is not the correct explanation of A.

Imp. *Basicity of Ketones*

03 Which will undergo deprotonation most readily in basic medium? **[2023, 24 Jan Shift-II]**

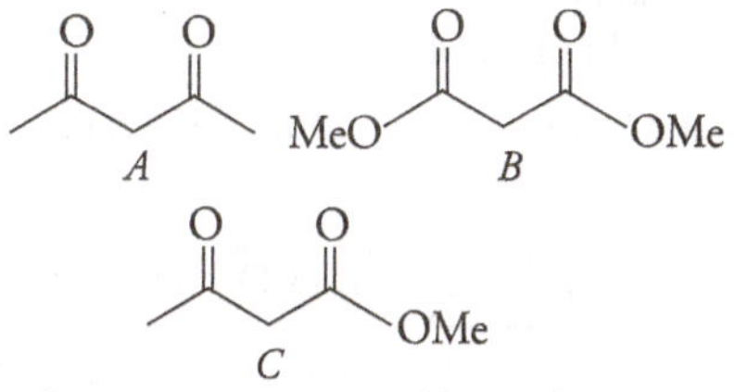

(a) *C* only (b) *B* only
(c) Both *A* and *C* (d) *A* only

04 Which of the following is an example of conjugated diketone? **[2022, 24 June Shift-I]**

(a) $CH_3—\overset{O}{\overset{\|}{C}}—CH_2—CH_2—\overset{O}{\overset{\|}{C}}—CH_2—CH_3$

(b) 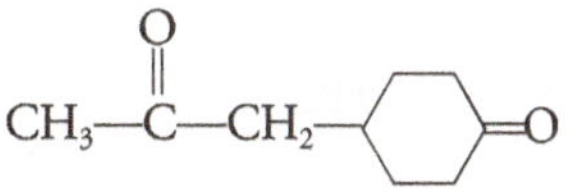

(c) 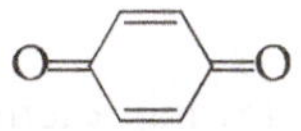

(d) 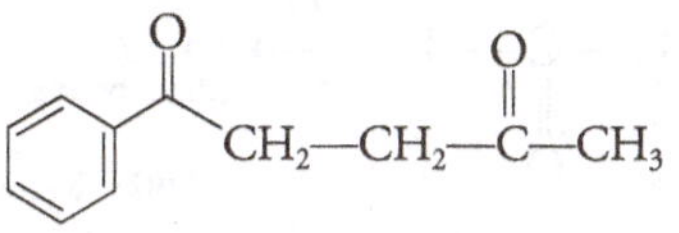

Imp. *Nucleophilic Addition Reaction*

05 Given below are two statements. **[2021, 1 Sep Shift-II]**

Statement I The nucleophilic addition of sodium hydrogen sulphite to an aldehyde or a ketone involves proton transfer to form a stable ion.

Statement II The nucleophilic addition of hydrogen cyanide to an aldehyde or a ketone yields amine as final product.

In the light of the above statements, choose the most appropriate answer from the options given below.

(a) Both statement I and statement II are true.
(b) Statement I is false but statement II is true.
(c) Statement I is true but statement II is false.
(d) Both statement I and statement II are false.

06 **Assertion (A)** Enol form of acetone $[CH_3COCH_3]$ exists in < 0.1% quantity. However, the enol form of acetyl acetone $[CH_3COCH_2OCCH_3]$ exists in approximately 15% quantity.

Reason (R) Enol form of acetyl acetone is stabilised by intramolecular hydrogen bonding, which is not possible in enol form of acetone.

Choose the correct statement. **[2021, 16 March Shift-I]**

(a) A is false but R is true.
(b) Both A and R are true and R is the correct explanation of A.
(c) Both A and R are true but R is not the correct explanation of A.
(d) A is true but R is false.

07 The major product in the following reaction is

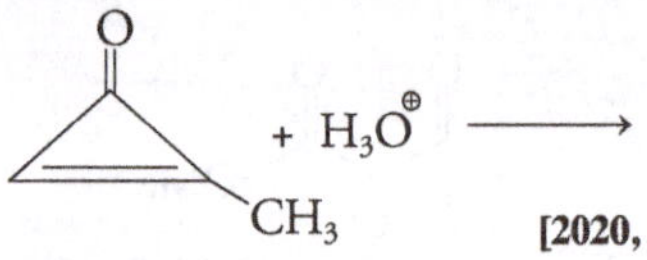

[2020, 8 Jan Shift-II]

(a) 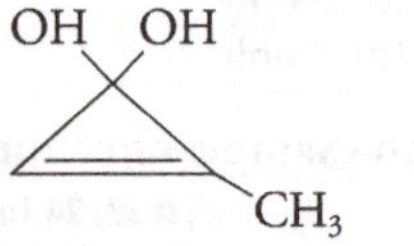

(b) 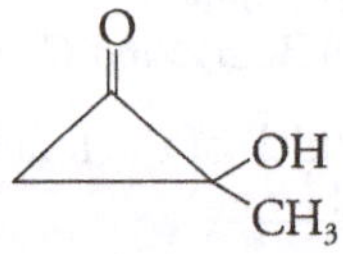

(c) 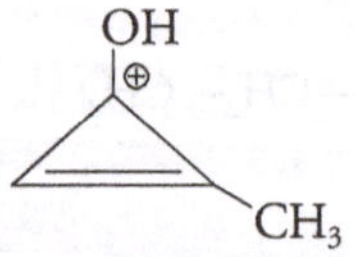

(d) 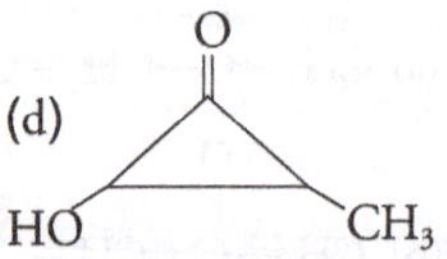

• Level 2

08 The major product (P) in the following reaction is

$$Ph—\overset{\underset{\|}{O}}{C}—\overset{\underset{\|}{O}}{C}—H \xrightarrow[\Delta]{KOH} \underset{\text{Major product}}{P}$$

[2025, 3 April Shift-II]

(a) $Ph—C(=O)—CH_2OH$ (b) $Ph—C(=O)—COO^-K^+$

(c) $Ph—CH(OH)—CH_2OH$ (d) $Ph—CH(OH)—COO^-K^+$

09 Aldol condensation is a popular and classical method to prepare α, β-unsaturated carbonyl compounds. This reaction can be both intermolecular and intramolecular. Predict which one of the following is not a product of intramolecular aldol condensation? **[2025, 4 April Shift-I]**

(a) 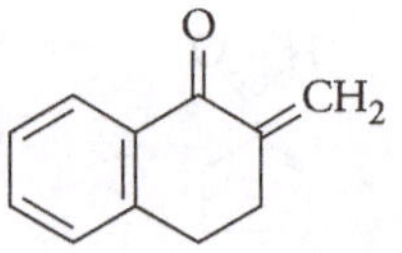

(b) 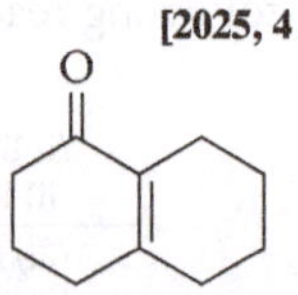

(c) 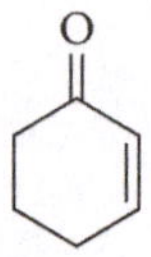

(d) 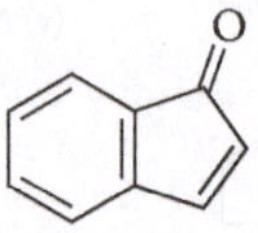

10 'P' is an optically active compound with molecular formula $C_6H_{12}O$. When 'P' is treated with 2, 4-dinitrophenylhydrazine, it gives a positive test. However, in presence of Tollens reagent, 'P' gives a negative test. Predict the structure of 'P'. **[2025, 7 April Shift-II]**

(a) $H—C(=O)—CH_2—CH(CH_2—CH_3)—CH_3$

(b) $H_3C—C(=O)—CH(CH_2—CH_3)—CH_3$

(c) $CH_3—C(=O)—CH_2—CH_2—CH_2—CH_3$

(d) $CH_3—C(=O)—CH_2—CH(CH_3)—CH_3$

11 Given below are two statements **[2024, 4 April Shift-I]**

Statements I Acidity of α-hydrogens of aldehydes and ketones is responsible for aldol reaction.

Statements II Reaction between benzaldehyde and ethanal will not give cross-aldol product.

In the light of the above statements, choose the most appropriate answer from the options given below:
(a) Both Statement I and Statement II are incorrect.
(b) Both Statement I and Statement II are correct.
(c) Statement I is incorrect but Statement II is correct.
(d) Statement I is correct but Statement II is incorrect.

12 Which of the following compound can give positive idoform test when treated with aqueous KOH solution followed by potassium hypoiodite. **[2024, 9 April Shift-II]**

(a) $CH_3CH_2-C(Cl)_2-CH_3$

(b) $CH_3CH_2-C(=O)-CH_2CH_3$

(c) $CH_3\,CH_2\,CH_2\,CHO$

(d)
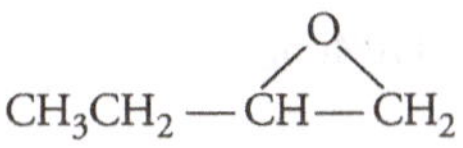

13 Major product of the following reaction is

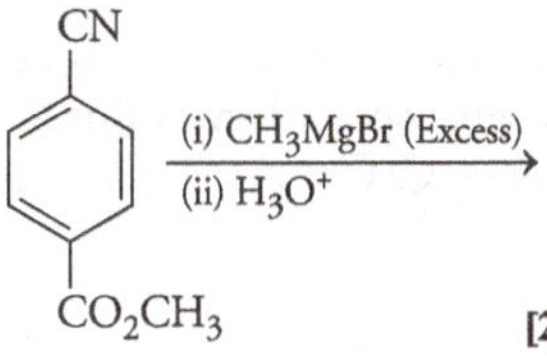

[2024, 9 April Shift-II]

(a)
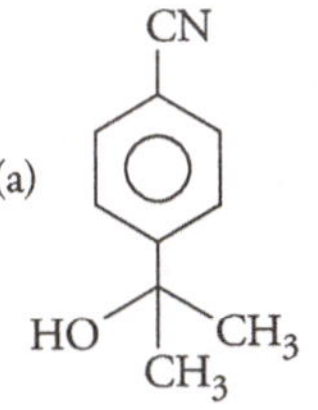

(b)
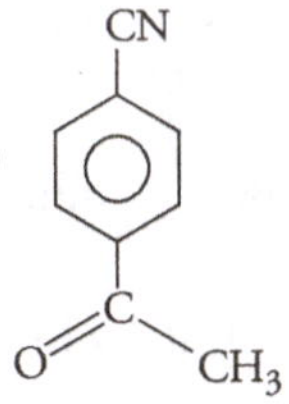

(c)
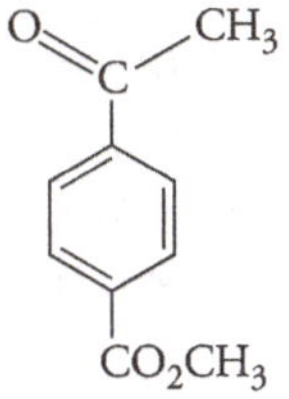

(d)
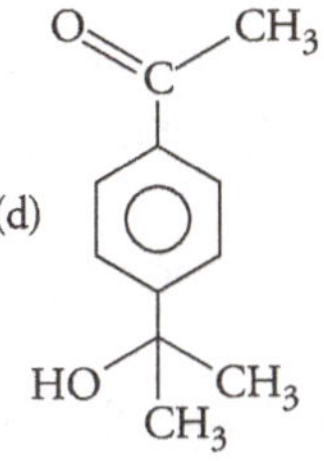

14 Identify major product '*P*' formed in the following reaction. **[2024, 31 Jan Shift-II]**

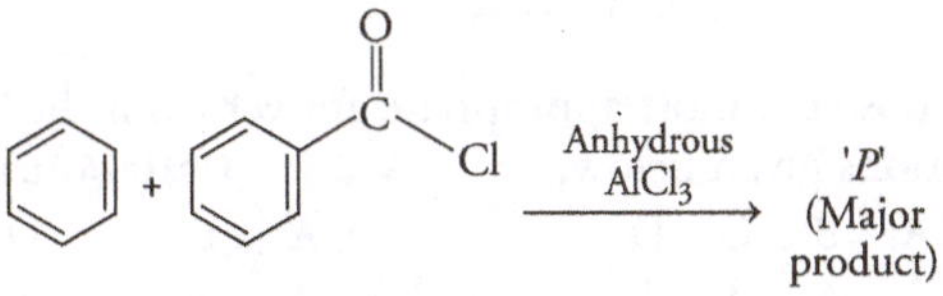

(a)
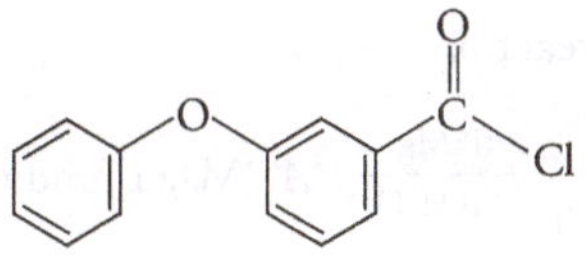

(b)
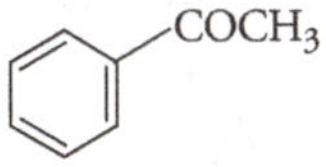

(c)
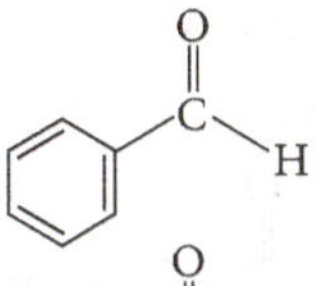

(d)
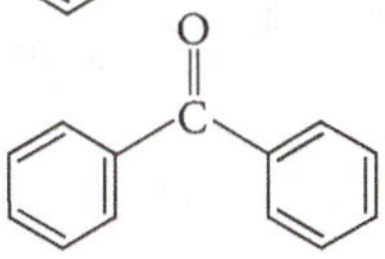

15 Which of the following has highly acidic hydrogen? **[2024, 27 Jan Shift-I]**

(a)
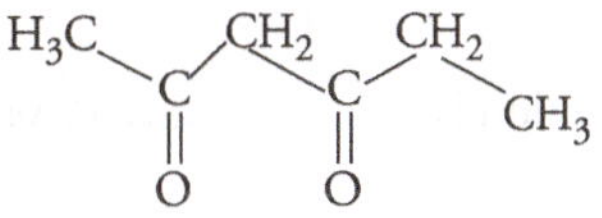

(b)
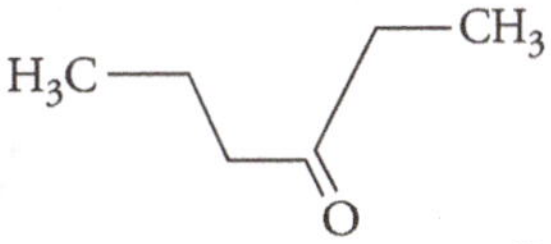

(c)
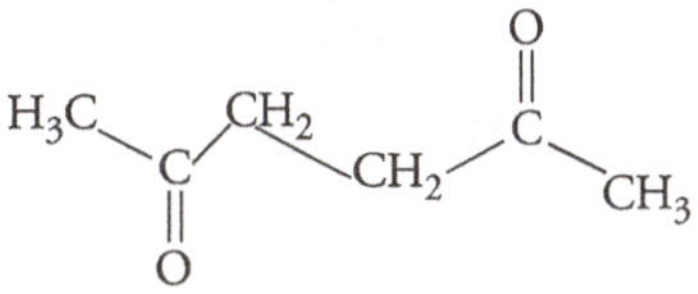

(d)
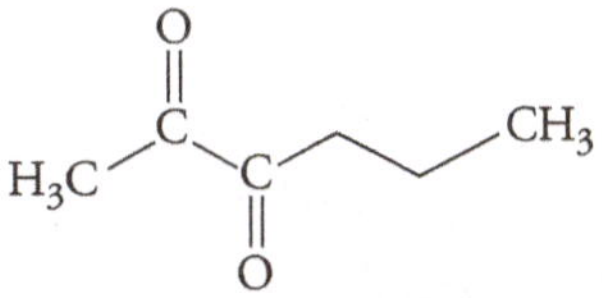

16 Correct statements for the given reaction are

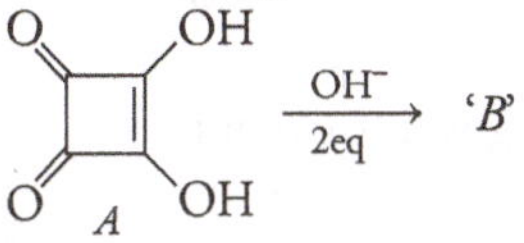

A. Compound '*B*' is aromatic.
B. The completion of above reaction is very slow.
C. '*A*' shows tautomerism.
D. The bond lengths of C—C in compound *B* are found to be same.

Choose the correct answer from the options given below. **[2023, 12 April Shift-I]**

(a) B, C and D only
(b) A, C and D only
(c) A, B and D only
(d) A, B and C only

17 In the following reaction

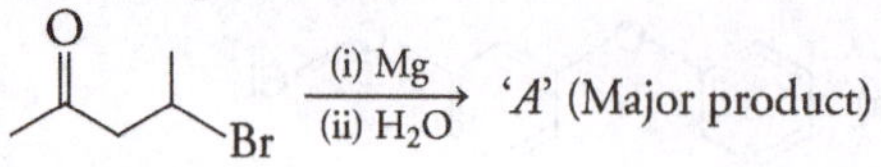

A is **[2023, 12 April Shift-I]**

(a) 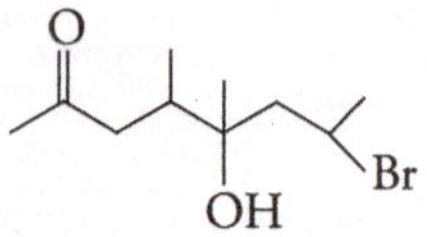
(b) 

(c) 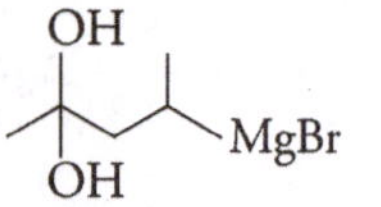
(d) 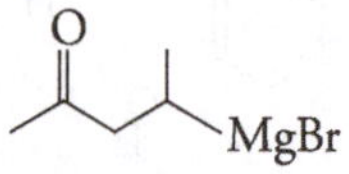

18

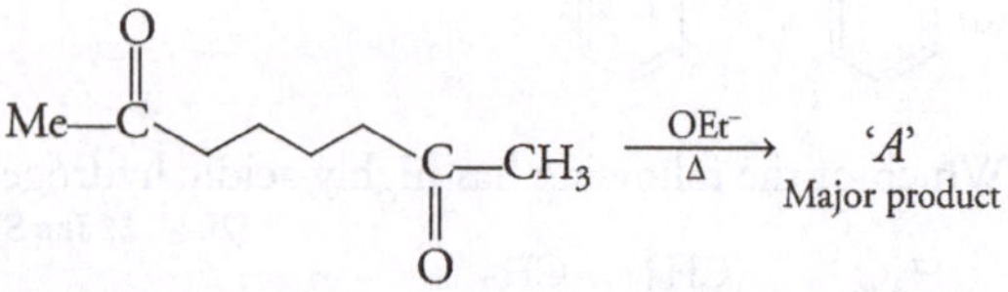

A in the above reaction is **[2023, 12 April Shift-I]**

(a) 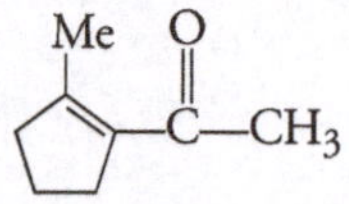

(b) 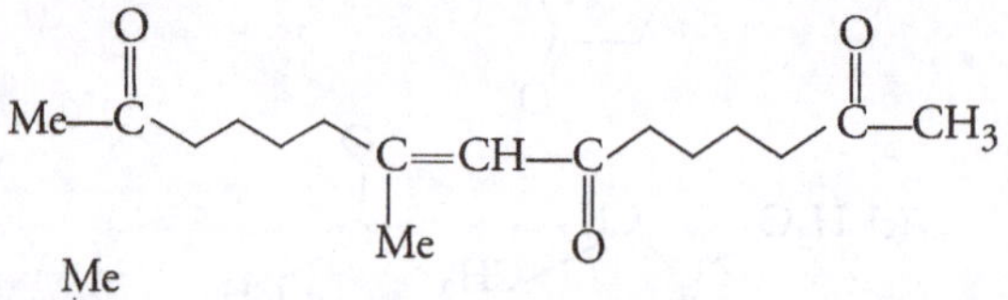

(c) 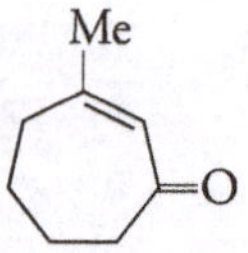

(d) 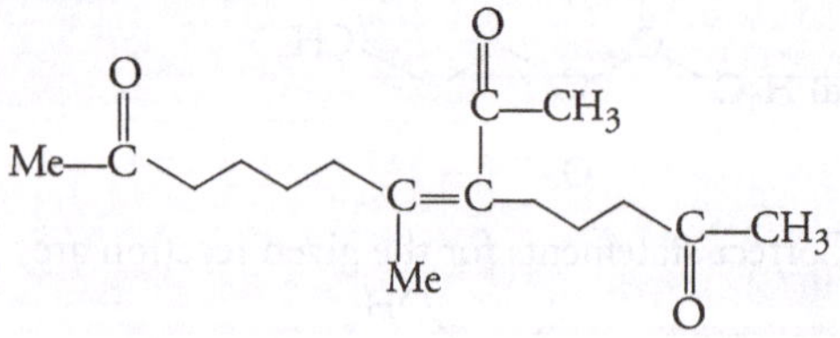

19 'A' in the given reaction is **[2023, 25 Jan Shift-II]**

(a) 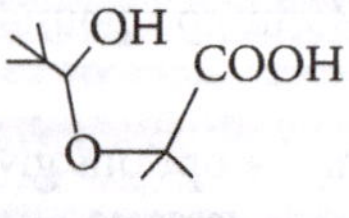
(b) 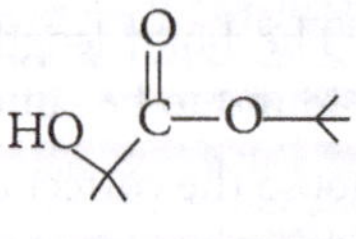

(c) 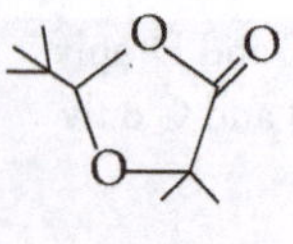
(d) 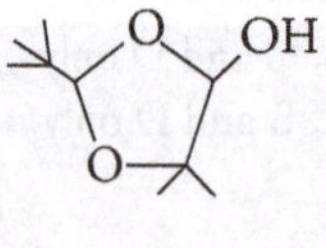

20 $(C_7H_5O_2)_2 \xrightarrow{h\nu} [X] \longrightarrow 2C_6H_5 + 2CO_2$

Consider the above reaction and identify the intermediate X. **[2022, 26 June, Shift-I]**

(a) $C_6H_5-\overset{O}{\overset{\|}{C}}{}^{+}$
(b) $C_6H_5-\overset{O}{\overset{\|}{C}}-O^{-}$

(c) $C_6H_5-\overset{O}{\overset{\|}{C}}{}^{+}-\ddot{\underset{..}{O}}$
(d) $C_6H_5-\overset{O}{\overset{\|}{C}}-\ddot{\underset{..}{O}}\cdot$

21 Which of the following conditions or reaction sequence will not give acetophenone as the major product? **[2022, 25 June Shift-II]**

(a) I. $C_6H_5-\overset{O}{\overset{\|}{C}}-H + CH_3MgBr$ II. $Na_2Cr_2O_7, H^+$

(b) I. $H_3C-\overset{O}{\overset{\|}{C}}-H + C_6H_5MgBr$ II. PCC, DCM

(c) $C_6H_5-\overset{O}{\overset{\|}{C}}-OC_2H_5 + 2CH_3MgBr$

(d) $C_6H_5-\overset{O}{\overset{\|}{C}}-Cl + CH_3MgBr + CdCl_2$

22 Which of the following ketone will not give enamine on treatment with secondary amines?
[where t-Bu is $-C(CH_3)_3$] **[2022, 25 June Shift-II]**

(a) $C_2H_5-\overset{O}{\overset{\|}{C}}-C_2H_5$
(b) $C_2H_5-\overset{O}{\overset{\|}{C}}-CH_3$

(c) $t\text{-Bu}-\overset{O}{\overset{\|}{C}}-t\text{-Bu}$
(d) 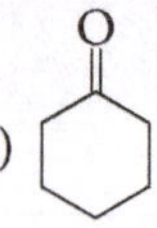

23 Match List-I with List-II.

List-I (Chemical reaction)	List-II (Reagent used)
A. $CH_3COOC_2H_5 \longrightarrow C_2H_5OH$	1. CH_3MgBr/H_3O^+ (1.equivalent)
B. $CH_3COOCH_3 \longrightarrow CH_3CHO$	2. H_2SO_4/H_2O
C. $CH_3C\equiv N \longrightarrow CH_3CHO$	3. DIBAL-H/H_2O
D. $CH_3C\equiv N \longrightarrow CH_3-\overset{O}{\overset{\|}{C}}-CH_3$	4. $SnCl_2$, HCl/H_2O

Choose the most appropriate answer from the options given below. **[2021, 26 Aug Shift-II]**

	A	B	C	D
(a)	2	4	3	1
(b)	4	2	3	1
(c)	2	3	4	1
(d)	3	2	1	4

24 The major products formed in the following reaction sequence A and B are **[2021, 26 Aug Shift-I]**

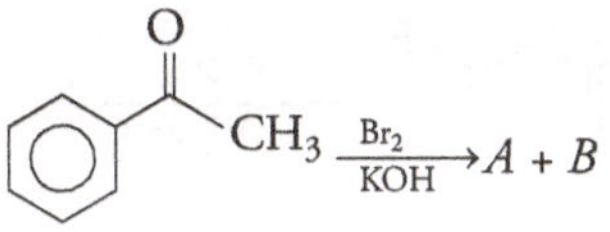

(a) $A =$ 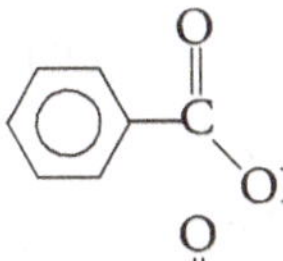 , $B = CHBr_3$

(b) $A =$ 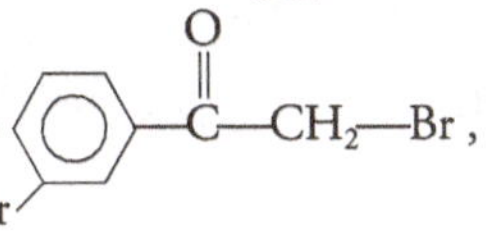 , $B =$ 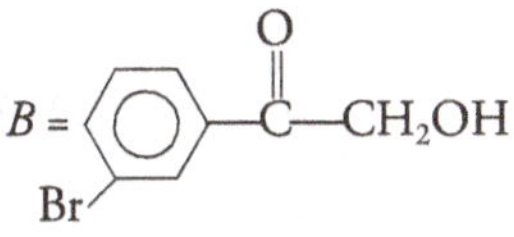

(c) $A =$ 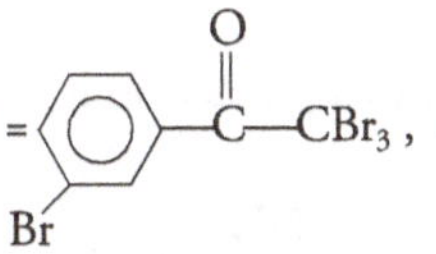 , $B =$ 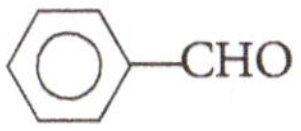

(d) $A =$ 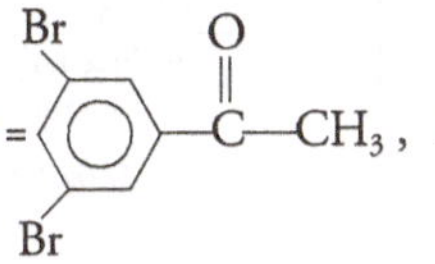 , $B =$ 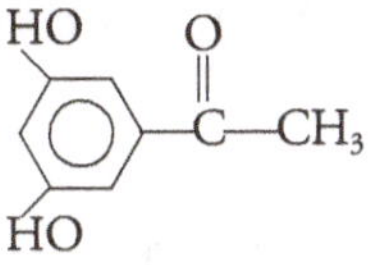

25 Which one of the following compounds will give orange precipitate when treated with 2, 4 dinitrophenyl hydrazine? **[2021, 27 July Shift-I]**

(a) 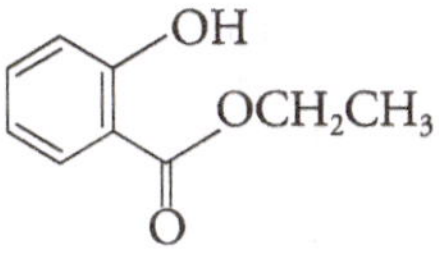 (b) 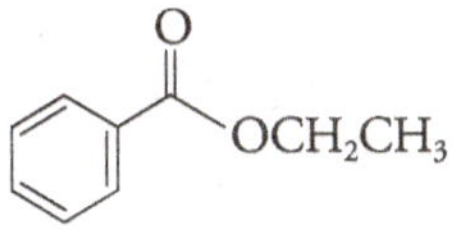

(c) 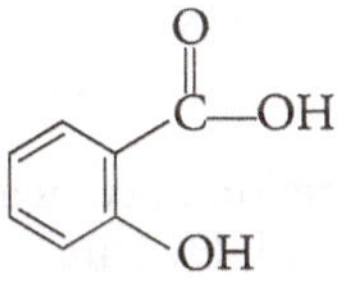 (d) 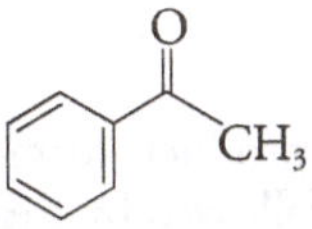

26 A reaction of benzonitrile with one equivalent CH_3MgBr followed by hydrolysis produces a yellow liquid P. The compound P will give positive **[2021 25 July Shift-II]**

(a) iodoform test (b) Schiff's test
(c) ninhydrin's test (d) Tollen's test

27

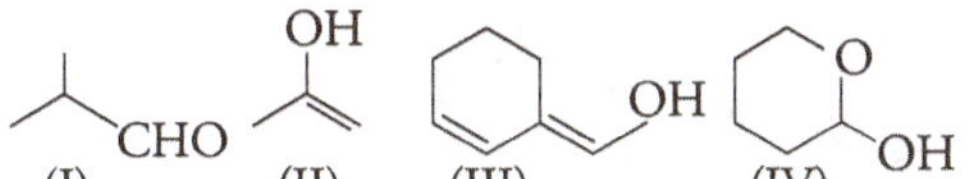

(I) (II) (III) (IV)

Which among the above compound/s does/do not form silver mirror when treated with Tollen's reagent?

(a) (I), (III) and (IV) only **[2021, 20 July Shift-I]**
(b) Only (IV)
(c) Only (II)
(d) (III) and (IV) only

28

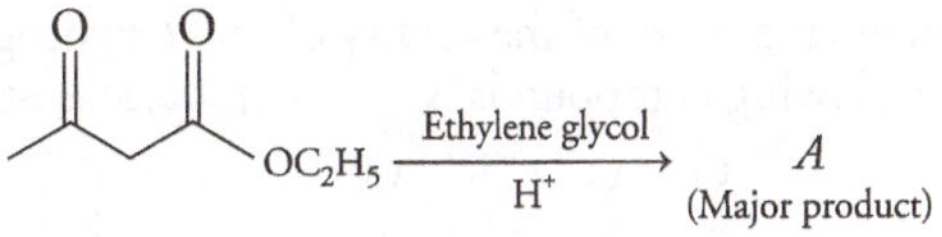

The product in the above reaction is **[2021, 17 March Shift-I]**

(a) 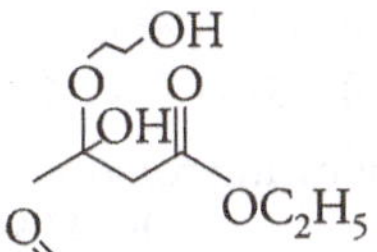 (b) 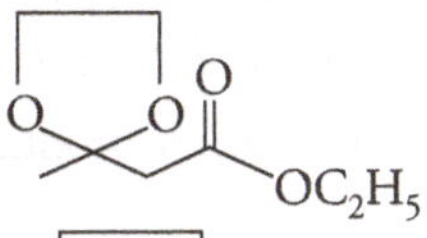

(c) 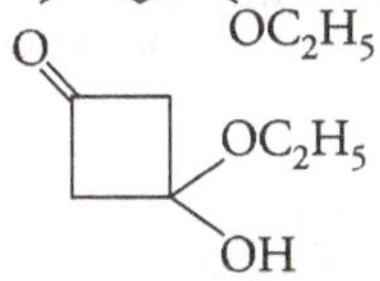 (d) 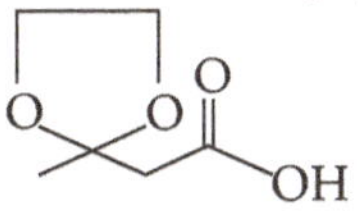

29.

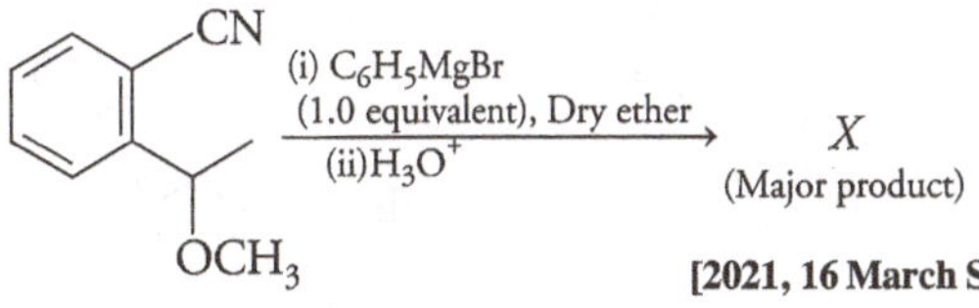

[2021, 16 March Shift-II]

(a) 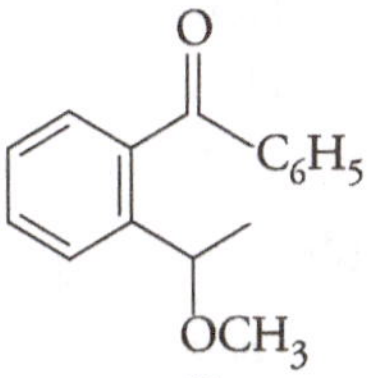 (b) 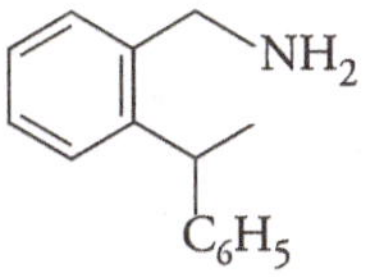

(c) 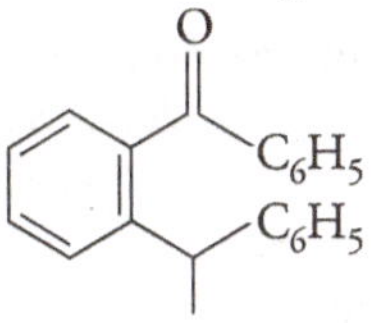 (d) 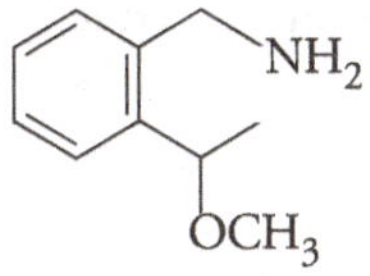

Imp. *Preparation of Ketone*

30 The major product obtained from the following reaction is **[2020, 6 Sep Shift-I]**

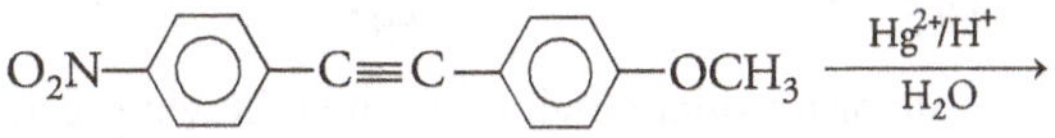

(a) 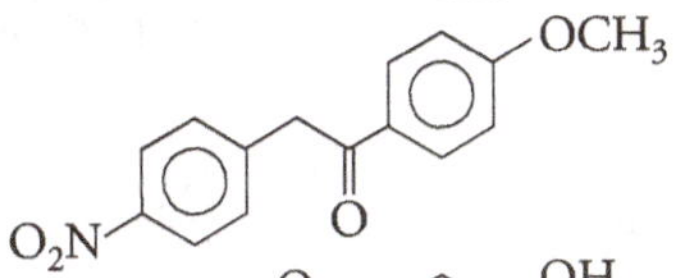

(b) 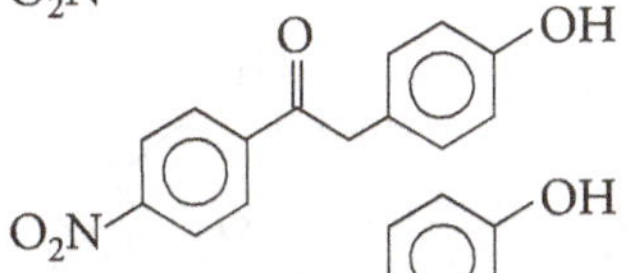

(c) 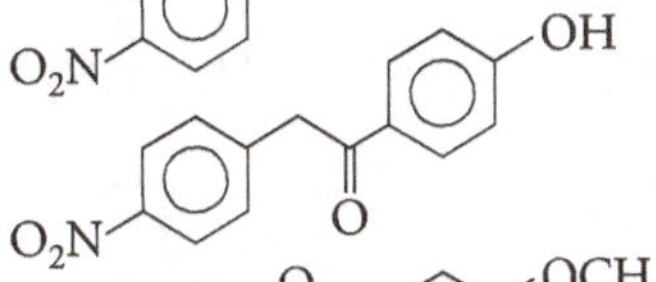

(d) 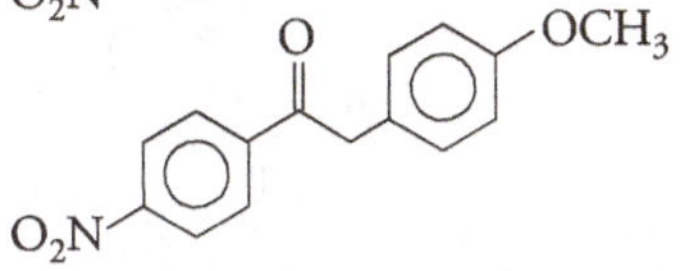

31 The increasing order of the acidity of the α-hydrogen of the following compounds is **[2020, 5 Sep Shift-I]**

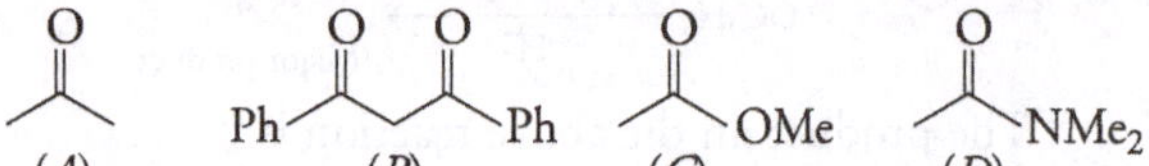

(*A*) (*B*) (*C*) (*D*)

(a) (*D*) < (*C*) < (*A*) < (*B*) (b) (*C*) < (*A*) < (*B*) < (*D*)
(c) (*A*) < (*C*) < (*D*) < (*B*) (d) (*B*) < (*C*) < (*A*) < (*D*)

32 The major product of the following reaction is **[2019, 12 April Shift-I]**

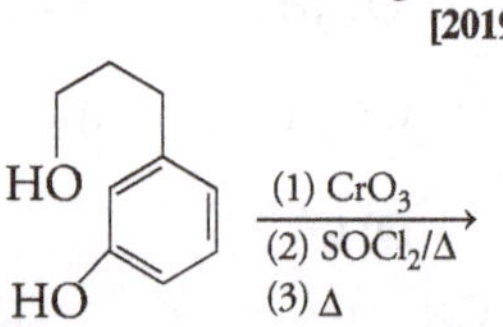

(a) 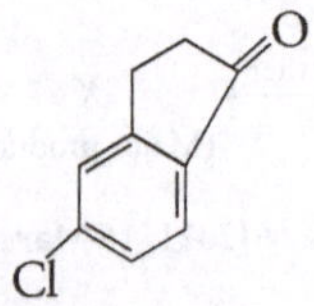 (b) 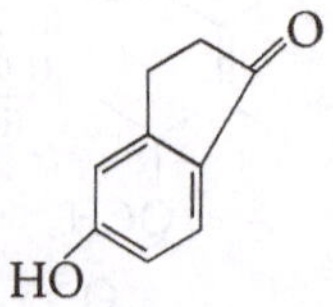

(c) 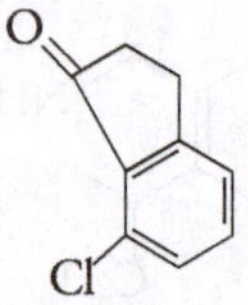 (d) 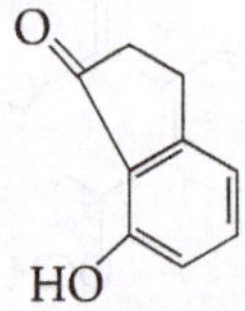

33 *p*-hydroxybenzophenone upon reaction with bromine in carbon tetrachloride gives **[2019, 9 April Shift-II]**

(a) 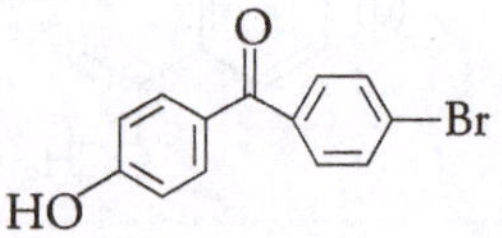 (b) 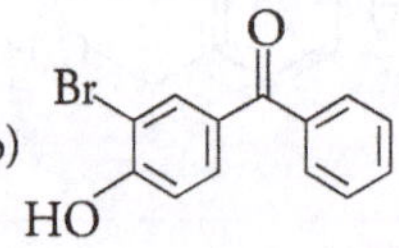

(c) 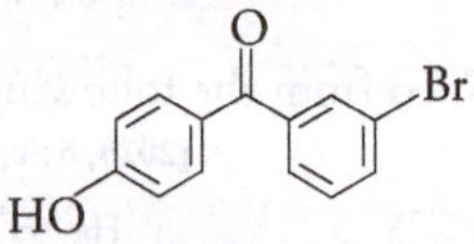 (d) 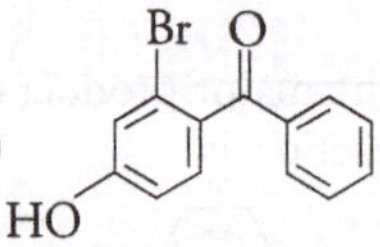

34 The major product obtained in the following reaction is **[2019, 8 April Shift-II]**

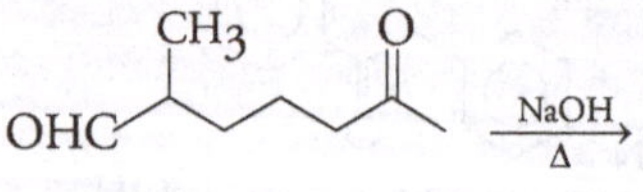

(a) 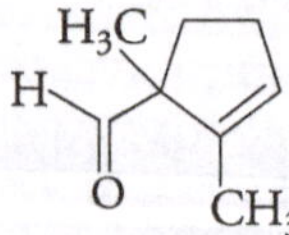 (b) 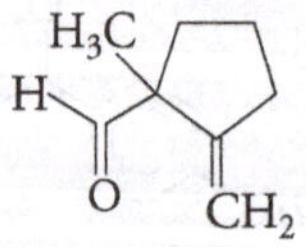

(c) 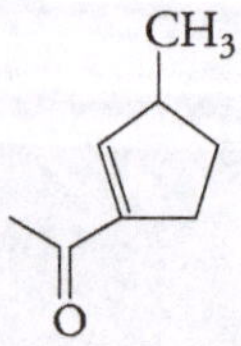 (d) 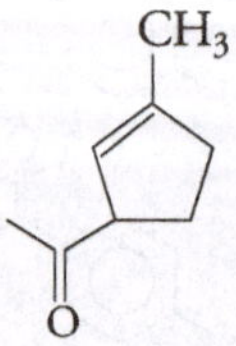

35 The major product of the following reaction is

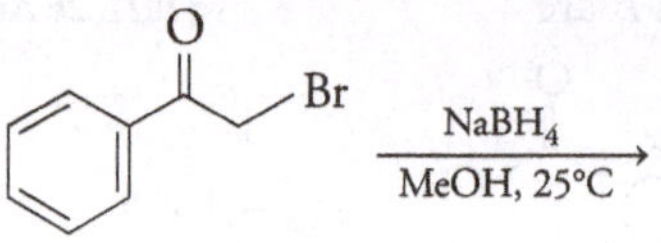

[2019, 8 April Shift-I]

(a) 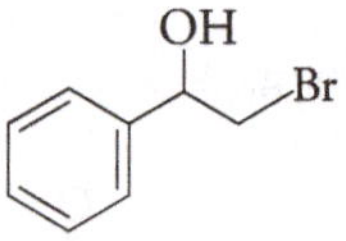 (b) 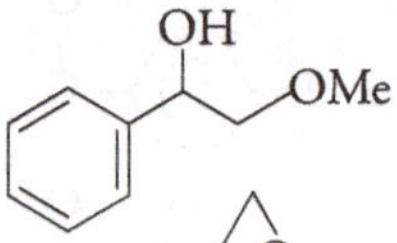

(c) 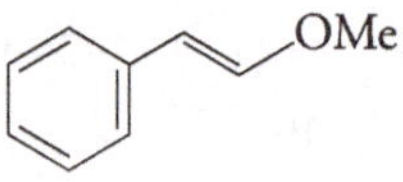 (d) 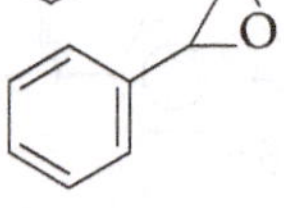

36 Major product of the following reaction is

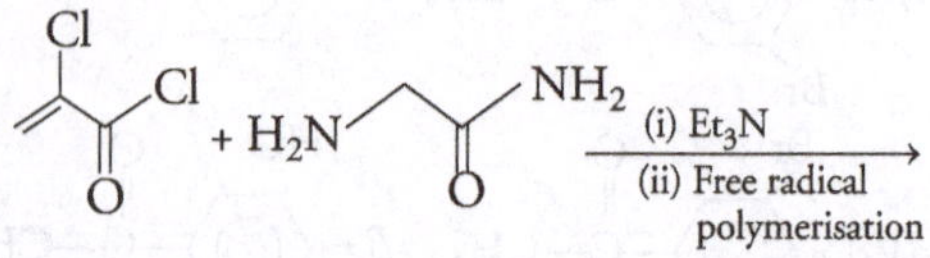

[2019, 9 Jan Shift-I]

(a) 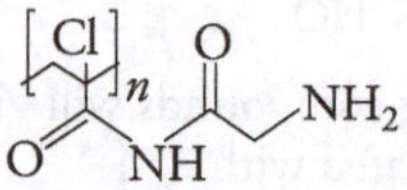 (b) 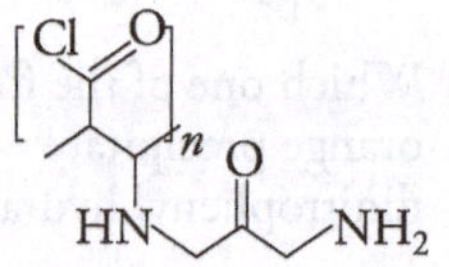

(c) 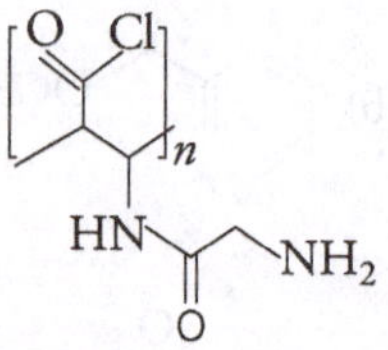 (d) 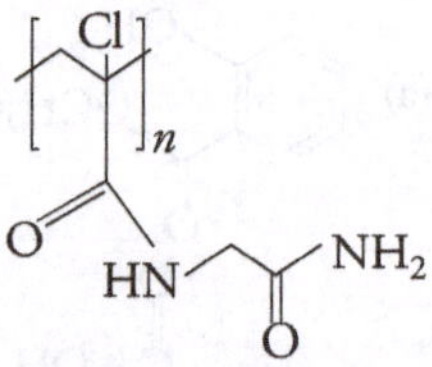

37 The products formed in the reaction of cumene with O_2 followed by treatment with dil. HCl are **[2019, 9 Jan Shift-I]**

(a) 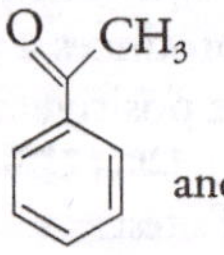 and CH_3—OH

(b) 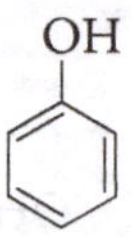 and 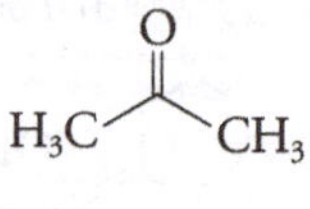

(c)  and 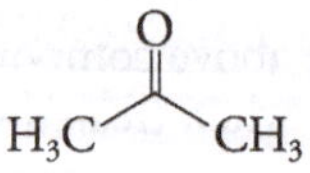

(d) 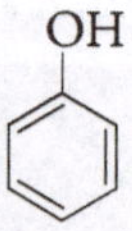 and 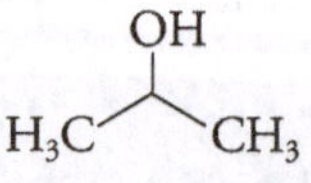

• Level 3

38 An optically active alkyl halide $C_4H_9Br[A]$ reacts with hot KOH dissolved in ethanol and forms alkene [*B*] as major product which reacts with bromine to give dibromide [*C*]. The compound [*C*] is converted into a gas [*D*] upon reacting with alcoholic $NaNH_2$. During hydration 18 gram of water is added to 1 mole of gas [*D*]. On warming with mercuric sulphate and dilute acid at 333 K to form compound [*E*]. The IUPAC name of compound [*E*] is **[2025, 2 April Shift-I]**
(a) But-2-yne
(b) Butan-2-one
(c) Butan-1-al
(d) Butan-2-ol

39 An organic compound (*X*) with molecular formula C_3H_6O is not readily oxidised. On reduction it gives C_3H_8O (*Y*) which reacts with HBr to give a bromide (*Z*) which is converted to Grignard reagent. This Grignard reagent on reaction with (*X*) followed by hydrolysis give 2, 3-dimethylbutan-2-ol. Compounds (*X*), (*Y*) and (*Z*) respectively are: **[2025, 4 April Shift-I]**
(a) CH_3COCH_3, $CH_3CH_2CH_2OH$, $CH_3CH(Br)CH_3$
(b) CH_3COCH_3, $CH_3CH(OH)CH_3$, $CH_3CH(Br)CH_3$
(c) CH_3CH_2CHO, $CH_3CH=CH_2$, $CH_3CH(Br)CH_3$
(d) CH_3CH_2CHO, $CH_3CH_2CH_2OH$, $CH_3CH_2CH_2Br$

40 When 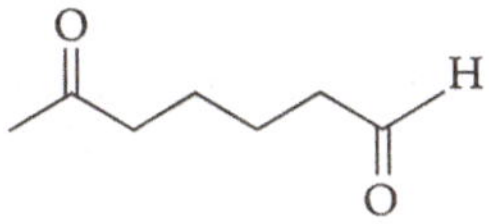 undergoes intramolecular aldol condensation, the major product formed is **[2025, 8 April Shift-II]**

(a) 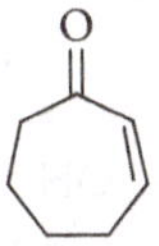
(b) 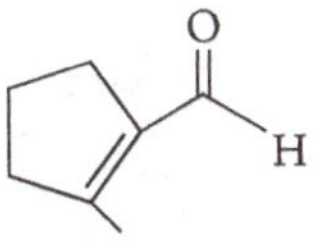
(c) 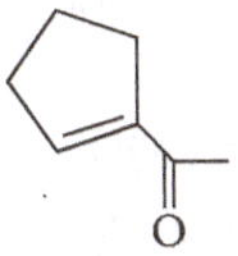
(d) 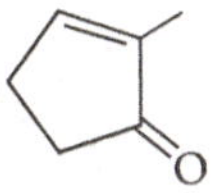

41 The correct order of basicity for the following molecules is **[2025, 4 April Shift-II]**

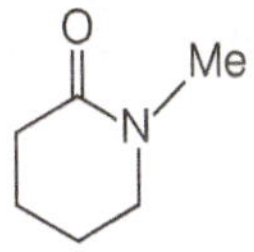
(*P*)
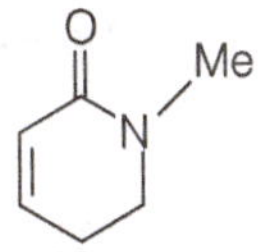
(*Q*)
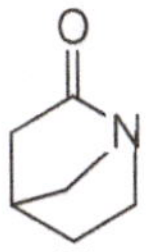
(*R*)

(a) $R > Q > P$
(b) $P > Q > R$
(c) $Q > P > R$
(d) $R > P > Q$

42 Consider the following molecules: **[2025, 2 April Shift-I]**

p. $CH_3-CH_2-\overset{O}{\overset{\|}{C}}-Cl$

q. $CH_3-CH_2-\overset{O}{\overset{\|}{C}}-O-\overset{O}{\overset{\|}{C}}-CH_3$

r. $CH_3-CH_2-\overset{O}{\overset{\|}{C}}-O-CH_2-CH_3$

s. $CH_3-CH_2-\overset{O}{\overset{\|}{C}}-NH_2$

The correct order of rate of hydrolysis is
(a) $p > q > r > s$
(b) $q > p > r > s$
(c) $r > q > p > s$
(d) $p > r > q > s$

43 The compounds which give positive Fehling's test are

(A) 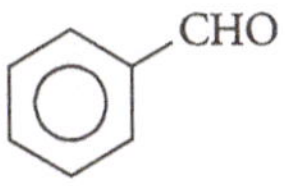
(B) 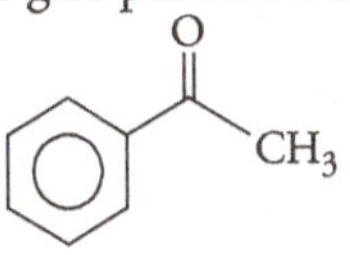

(C) $HOCH_2-CO-(CHOH)_3-CH_2-OH$

(D) $CH_3-\overset{O}{\overset{\|}{C}}-H$
(E) 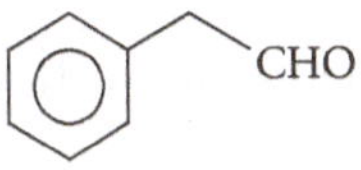

Choose the correct answer from the options given below. **[2025, 22 Jan Shift-I]**
(a) (A), (B) and (C) only
(b) (C), (D) and (E) only
(c) (A), (C) and (D) only
(d) (A), (D) and (E) only

Imp. *Order of Nucleophilic Addition Reaction in Aldehydes*

44 Which of the following arrangements with respect to their reactivity in nucleophilic addition reaction is correct? **[2025, 24 Jan Shift-I]**
(a) acetophenone < *p*-tolualdehyde < benzaldehyde < *p*-nitrobenzaldehyde
(b) *p*-nitrobenzaldehyde < benzaldehyde < *p*-tolualdehyde < acetophenone
(c) benzaldehyde < acetophenone <*p*- nitrobenzaldehyde < *p*-tolualdehyde
(d) acetophenone < benzaldehyde < *p*-tolualdehyde < *p*-nitrobenzaldehyde

45 Both acetaldehyde and acetone (individually) undergo which of the following reactions? **[2025, 28 Jan Shift-I]**
A. Iodoform reaction
B. Cannizzaro reaction
C. Aldol condensation
D. Tollen's test
E. Clemmensen reduction

Choose the correct answer from the options given below.

(a) A, B and D only (b) A, C and E only
(c) B, C and D only (d) C and E only

46 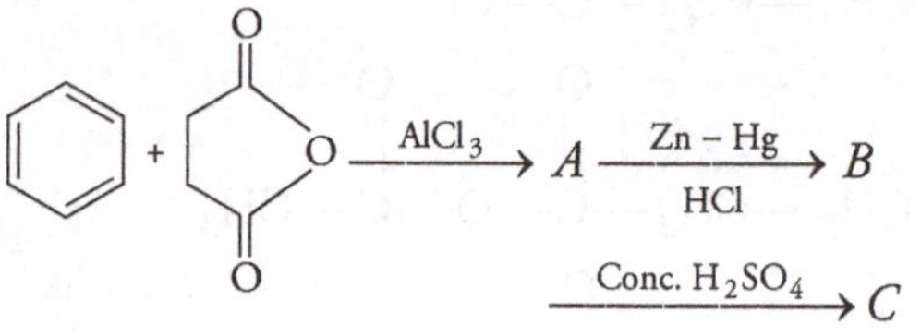

What is the structure of C? [2024, 9 April Shift-I]

(a) 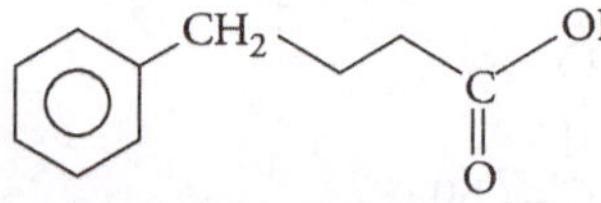

(b) 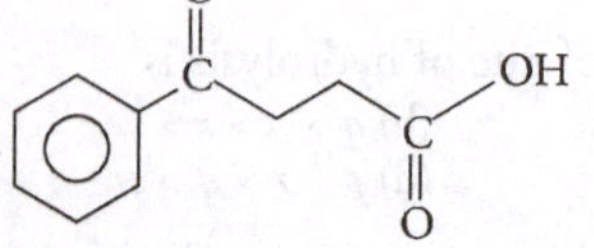

(c) 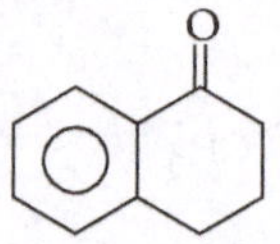

(d) 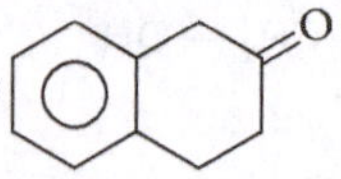

47 Match List-I with List-II. [2024, 1 Feb Shift-I]

List-I (Reactions)	List-II (Reagents)
A. $CH_3(CH_2)_5—C(=O)—OC_2H_5 \longrightarrow CH_3(CH_2)_5CHO$	I. CH_3MgBr, H_2O
B. $C_6H_5COC_6H_5 \longrightarrow C_6H_5CH_2C_6H_5$	II. Zn(Hg) and conc. HCl
C. $C_6H_5CHO \longrightarrow C_6H_5CH(OH)CH_3$	III. $NaBH_4, H^+$
D. $CH_3COCH_2COOC_2H_5 \longrightarrow CH_3C(OH)(H)CH_2COOC_2H_5$	IV. DIBAL-H, H_2O

Choose the correct answer from the options given below.

	A	B	C	D
(a)	III	IV	II	I
(b)	III	IV	I	II
(c)	IV	II	III	I
(d)	IV	II	I	III

48 Find out the major product from the following reaction.

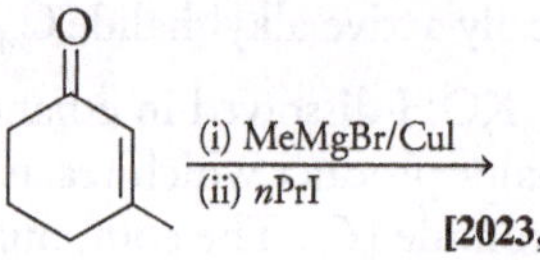

[2023, 6 April Shift-II]

(a) 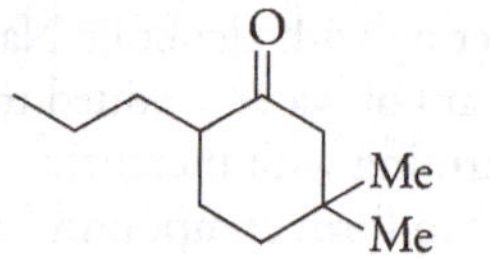

(b) 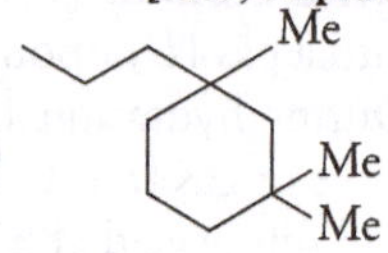

(c) 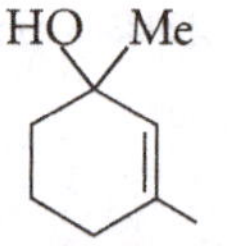

(d) 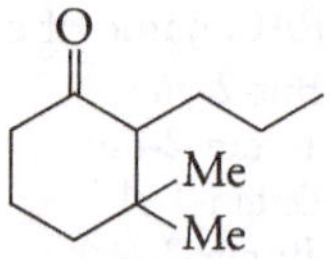

49 Cyclohexylamine when treated with nitrous acid yields (*P*). On treating (*P*) with PCC results in (*Q*). When (*Q*) is heated with dil. NaOH we get (*R*). The final product (*R*) is [2023, 31 Jan Shift-II]

(a) 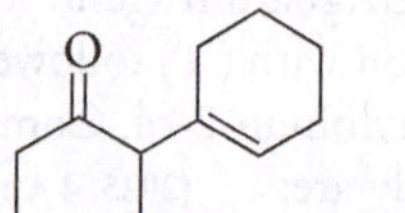

(b) 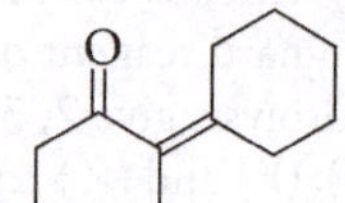

(c) 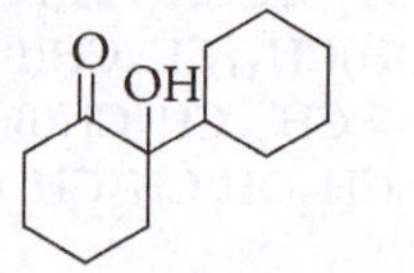

(d) 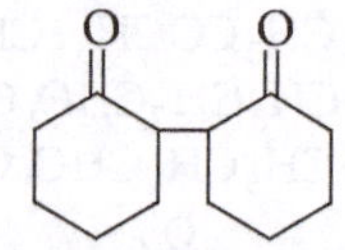

50 Find out the major products from the following reaction sequence. [2023, 29 Jan Shift-II]

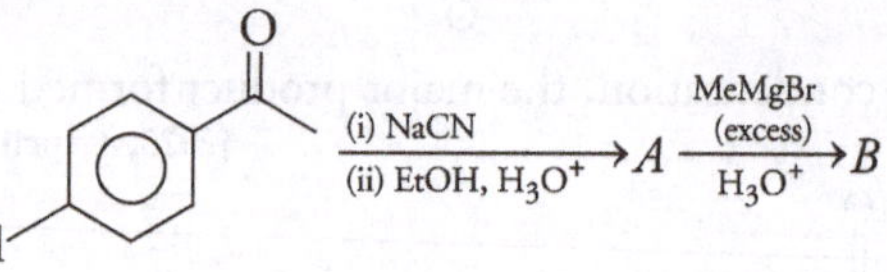

(a) *A* = 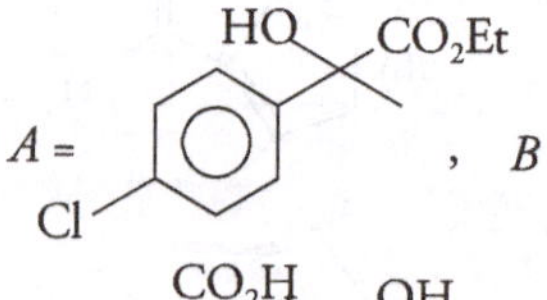 , *B* = 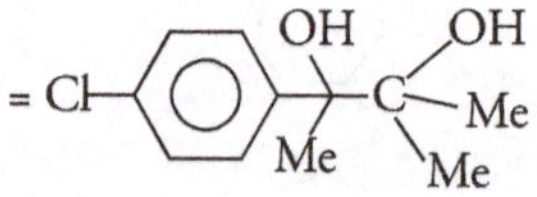

(b) *A* = 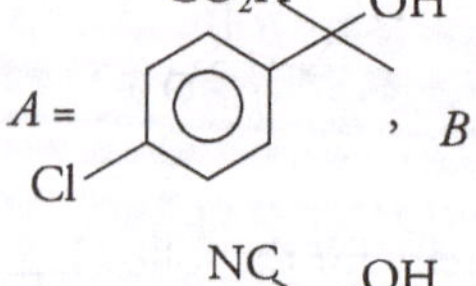 , *B* = 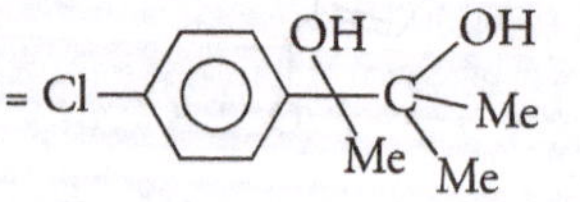

(c) *A* = 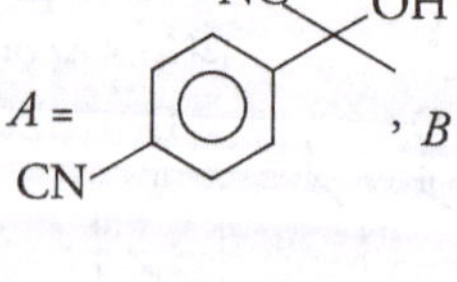 , *B* = 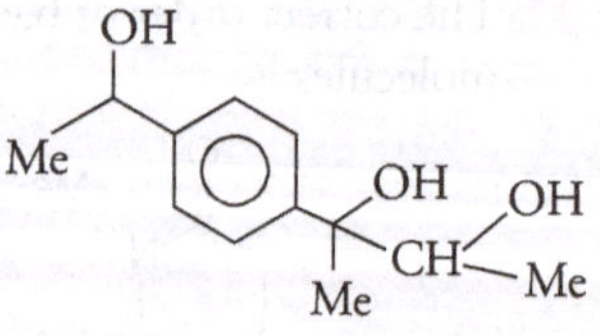

(d) *A* = 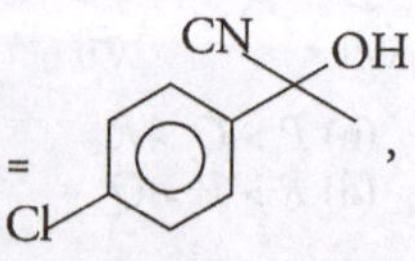 , *B* = 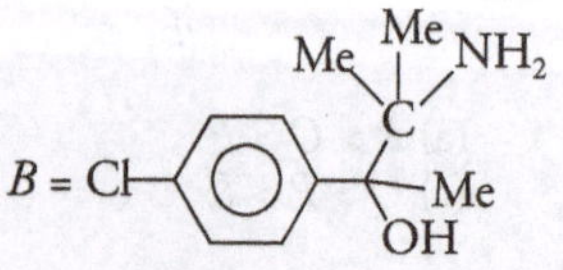

51 'R' formed in the following sequence of reaction is

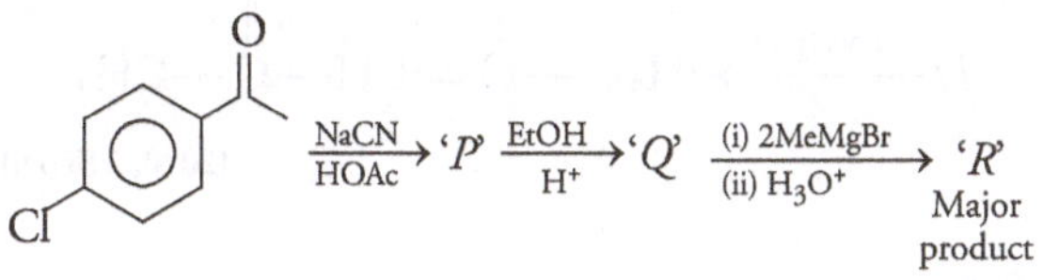

[2023, 24 Jan Shift-I]

(a) 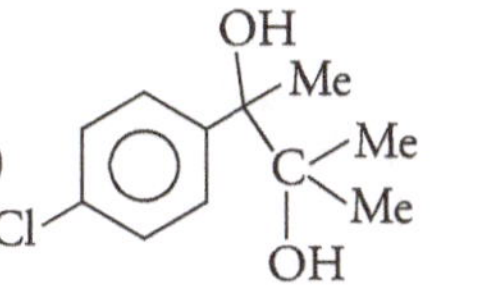
(b) 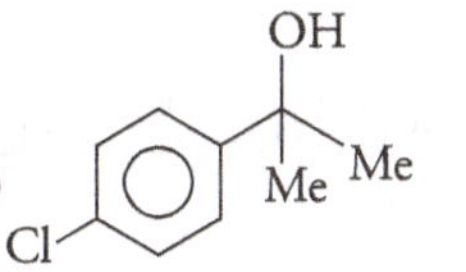
(c) 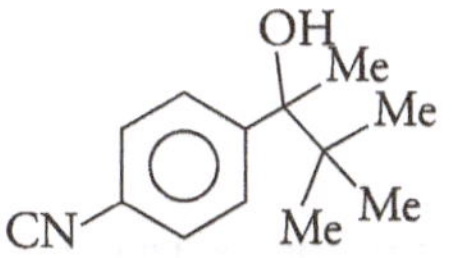
(d) 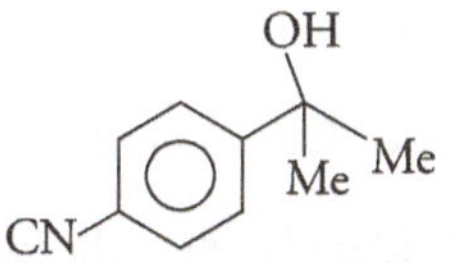

52 $CH_3-CH_2C\equiv CN \xrightarrow[\text{Ether}]{CH_3MgBr} A \xrightarrow{H_3O^+} B \xrightarrow[\text{HCl}]{\text{Zn}-\text{Hg}} C$

The correct structure of C is **[2022, 25 July Shift-II]**

(a) $CH_3CH_2CH_2CH_3$

(b) $CH_3CH_2-\overset{\overset{O}{\|}}{C}-CH_3$

(c) $CH_3CH_2\overset{\overset{OH}{|}}{CH}CH_3$

(d) $CH_3CH_2CH=CH_2$

53 Which one of the following is the major product of the given reaction? **[2021, 27 Aug Shift-II]**

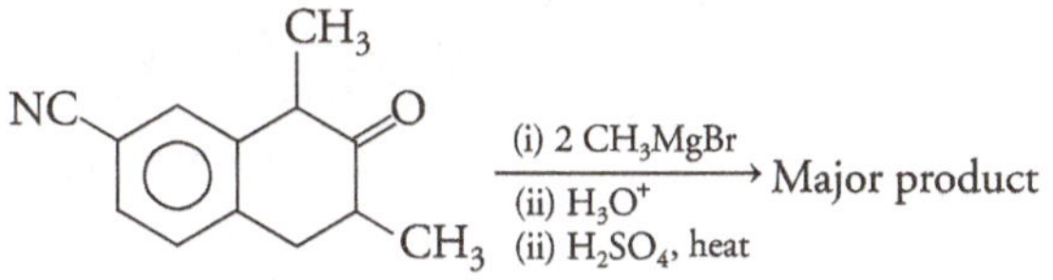

(a) 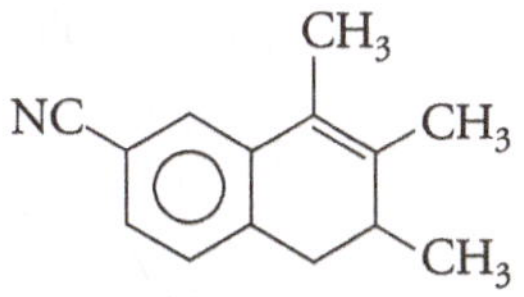
(b) 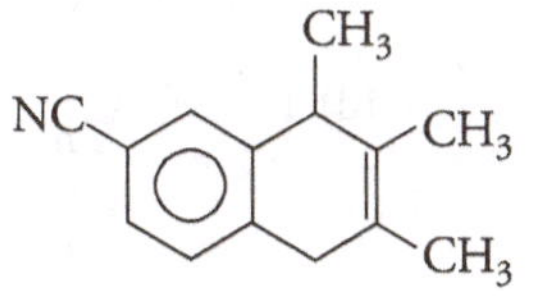
(c) 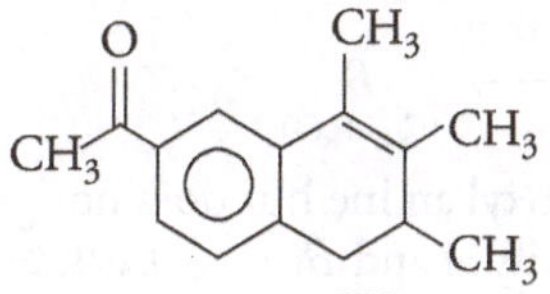
(d) 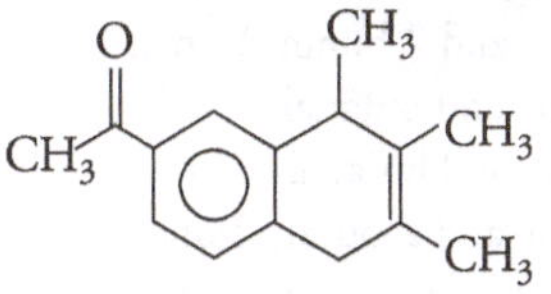

54 In the following sequence of reactions, the final product D is **[2021, 27 Aug Shift-I]**

$CH_3-C\equiv C-H + NaNH_2 \longrightarrow A$ 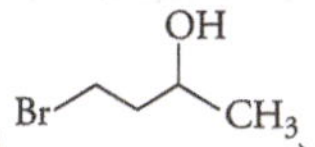 $\longrightarrow B \xrightarrow{H_2/Pd-C} C \xrightarrow{CrO_3} D$

(a) $H_3C-CH_2-CH_2-CH_2-CH_2-\overset{\overset{O}{\|}}{C}-H$

(b) $CH_3-CH=CH-CH_2-CH_2-CH_2-COOH$

(c) $H_3C-CH=CH-CH(OH)-CH_2-CH_2CH_3$

(d) $CH_3-CH_2-CH_2-CH_2-CH_2-\overset{\overset{O}{\|}}{C}-CH_3$

55 The major product P in the following reaction is

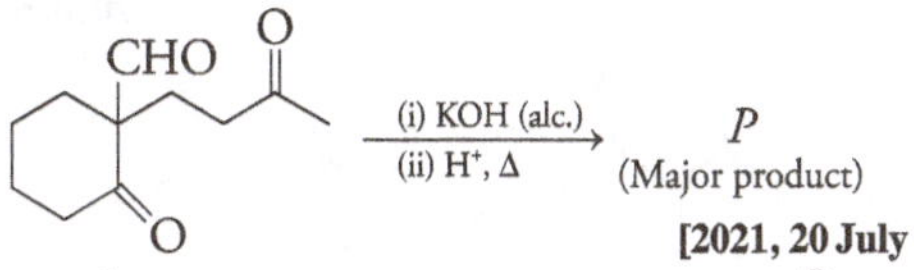

[2021, 20 July Shift-II]

(a) 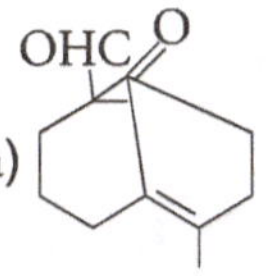
(b) 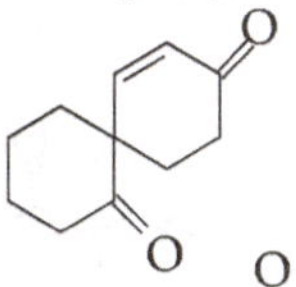
(c) 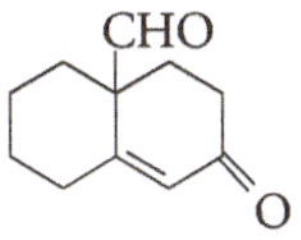
(d) 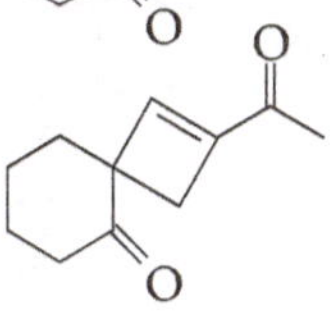

56 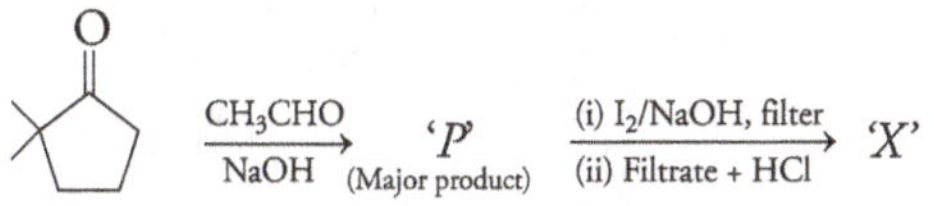

Consider the given reaction, the product X is **[2021, 22 July Shift-I]**

(a) 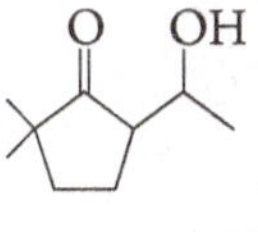
(b) 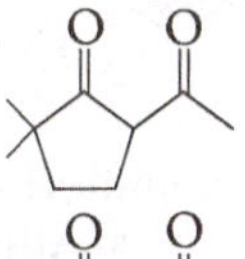
(c) 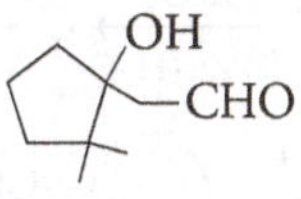
(d) 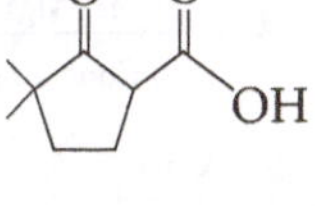

57 $\underset{(C_4H_8Cl_2)}{A} \xrightarrow[373\,K]{\text{Hydrolysis}} \underset{(C_4H_8O)}{B}$

B reacts with hydroxyl amine but does not give Tollen's test. Identify *A* and *B*. **[2021, 26 Feb Shift-I]**

(a) 1,1-dichlorobutane and 2-butan-2-one
(b) 2,2-dichlorobutane and butanal
(c) 1,1-dichlorobutane and butanal
(d) 2,2-dichlorobutane and 2-butan-2-one

58 Which one of the following carbonyl compounds cannot be prepared by addition of water on an alkyne in the presence of $HgSO_4$ and H_2SO_4? **[2021, 24 Feb Shift-II]**

(a) $CH_3-\overset{O}{\overset{\|}{C}}-H$

(b) 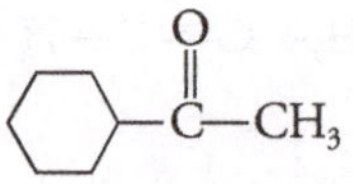

(c) $CH_3-CH_2-\overset{O}{\overset{\|}{C}}-H$

(d) $CH_3-\overset{O}{\overset{\|}{C}}-CH_2CH_3$

59 In the following reaction sequence the major products *A* and *B* are

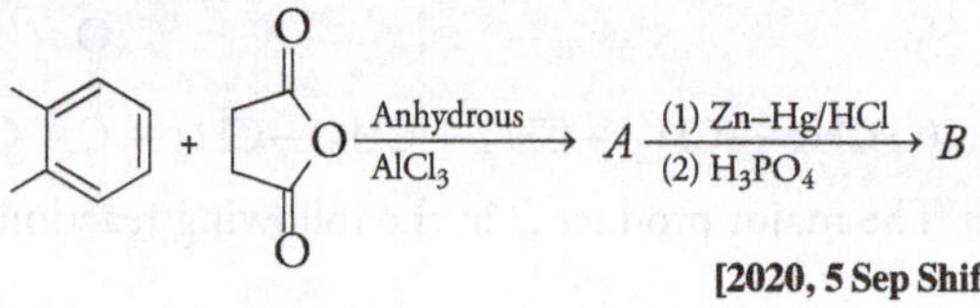

[2020, 5 Sep Shift-I]

(a) A = 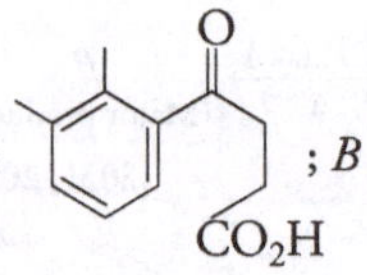 ; B = 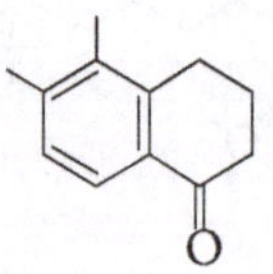

(b) A = 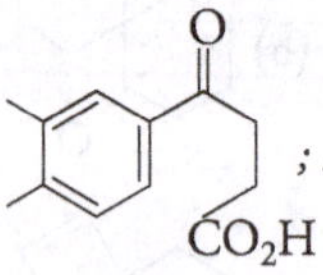 ; B = 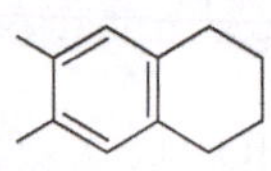

(c) A = 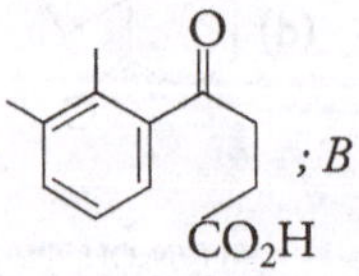 ; B = 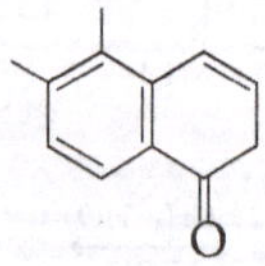

(d) A = 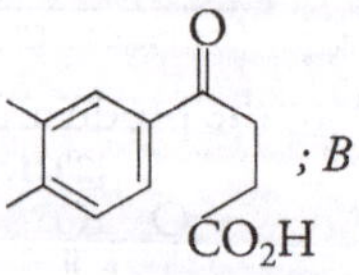 ; B = 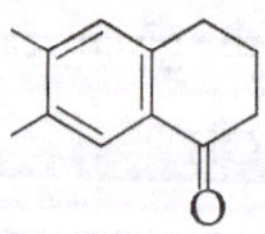

60 The compound, *A* in the following

$A \xrightarrow[\text{(ii) Conc. } H_2SO_4/\Delta]{\text{(i) } CH_3MgBr/H_2O} B \xrightarrow[\text{(ii) } Zn/H_2O]{\text{(i) } O_3} C + D$

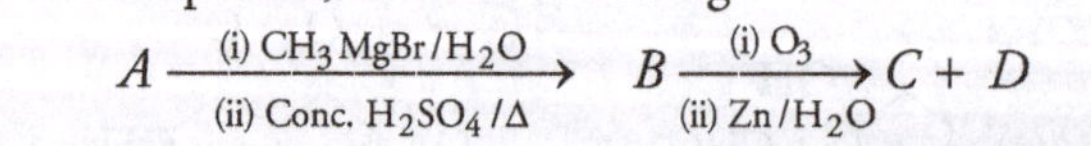

$D \xrightarrow[\Delta]{Ba(OH)_2} H_3C-\overset{CH_3}{\overset{|}{C}}=CH-\overset{O}{\overset{\|}{C}}-CH_3$

[2020, 3 Sep Shift-II]

(a) $C_6H_5-CH_2-\overset{O}{\overset{\|}{C}}-CH_3$

(b) $C_6H_5-\overset{O}{\overset{\|}{C}}-CH_3$

(c) $C_6H_5-\overset{O}{\overset{\|}{C}}-CH_2CH_3$

(d) $C_6H_5-\overset{O}{\overset{\|}{C}}-CH(CH_3)_2$

61 Identify (*A*) in the following reaction sequence :

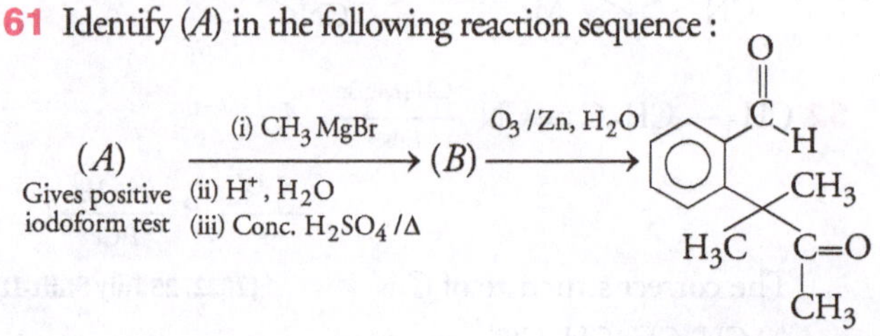

[2020, 9 Jan Shift-I]

(a) 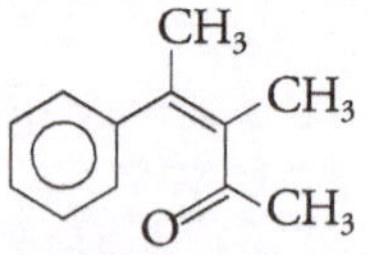

(b) 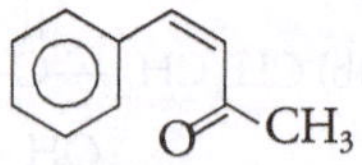

(c) 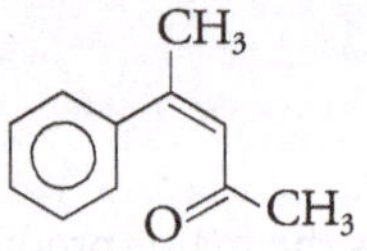

(d) 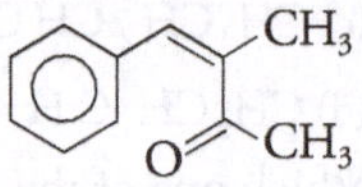

62 The major product *Y* in the following reaction is

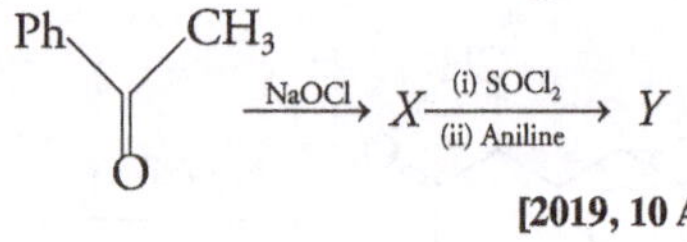

[2019, 10 April Shift-II]

(a) 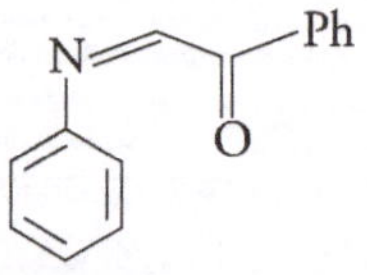

(b) 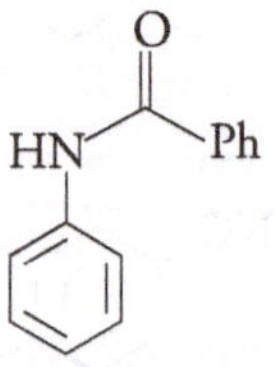

(c) 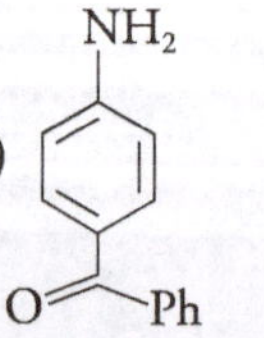

(d) 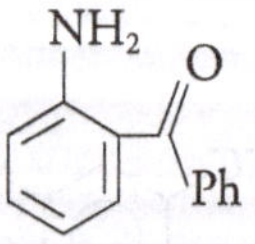

63 In the following reactions, products *A* and *B* are

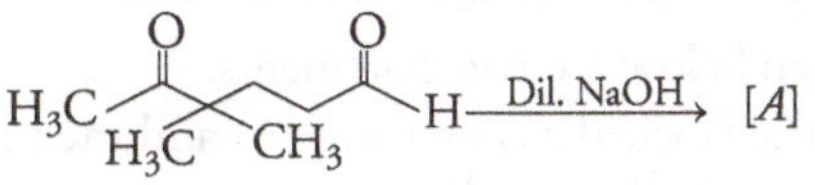

$[A] \xrightarrow[\Delta]{H_3O^+} [B]$

[2019, 12 Jan Shift-I]

(a) 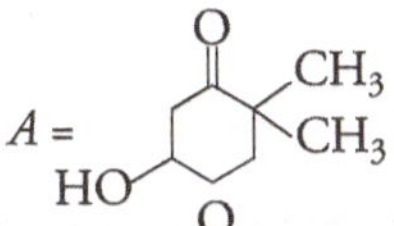 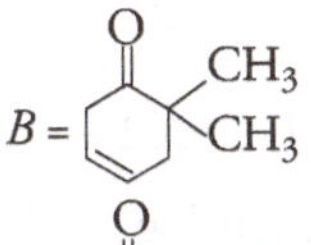

(b) 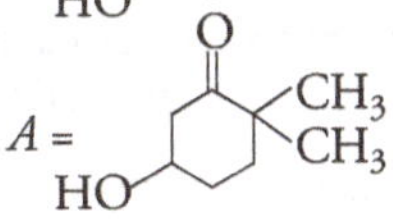 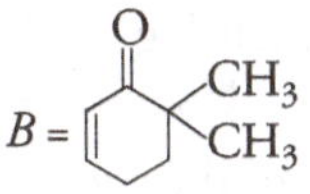

(c) 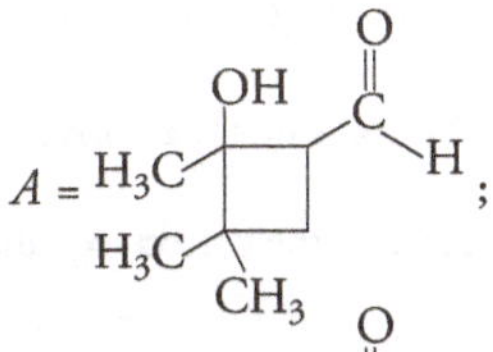 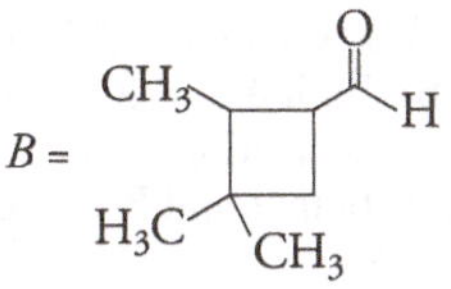

(d) 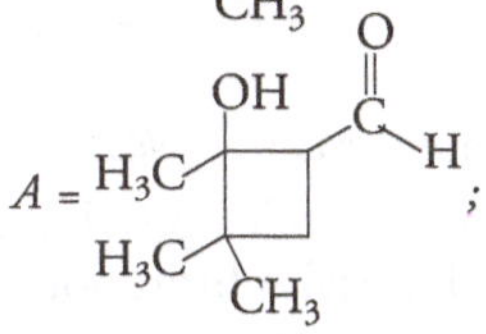 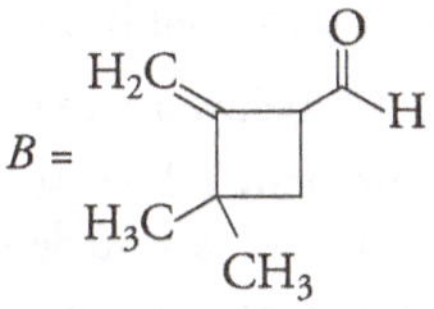

64 The major product obtained in the following reaction is

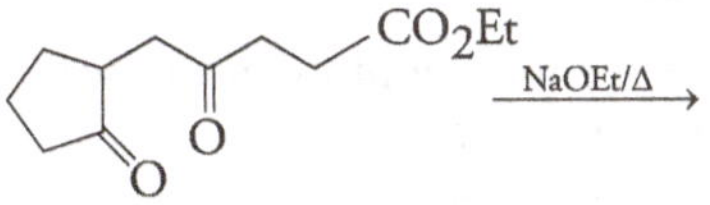

[2019, 10 Jan Shift-II]

(a) 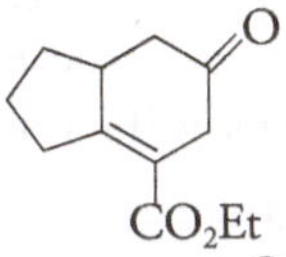

(b) 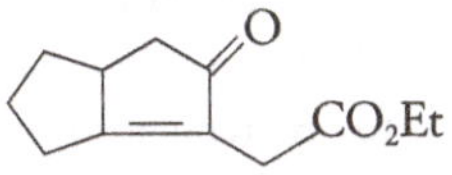

(c) 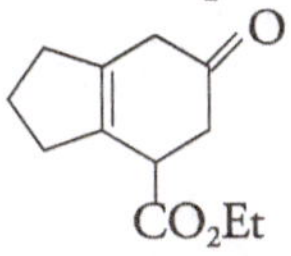

(d) 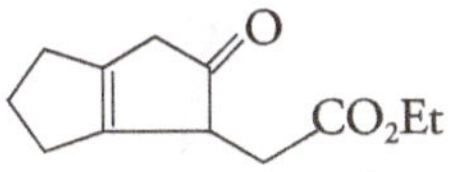

Numerical Value Type Questions

65 Identify the structure of the final product (*D*) in the following sequence of the reactions:

$$Ph-\overset{O}{\overset{||}{C}}-CH_3 \xrightarrow[\Delta]{PCl_5} A \xrightarrow{3\ eq. NaNH_2/NH_3}$$

$$B \xrightarrow{Acidify} C \xrightarrow[2.\ H_2O_2/OH^-]{1.\ B_2H_6} D$$

Total number of sp^2 hybridised carbon atoms in product *D* is .... [2025, 7 April Shift-II]

66 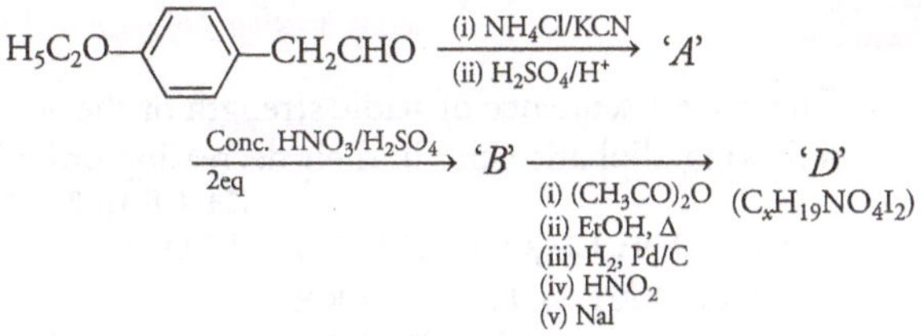

The value of *x* in compound '*D*' is ....

[2023, 12 April Shift-I]

67 The number of molecules which gives haloform test among the following molecules is ....

[2023, 31 Jan Shift-II]

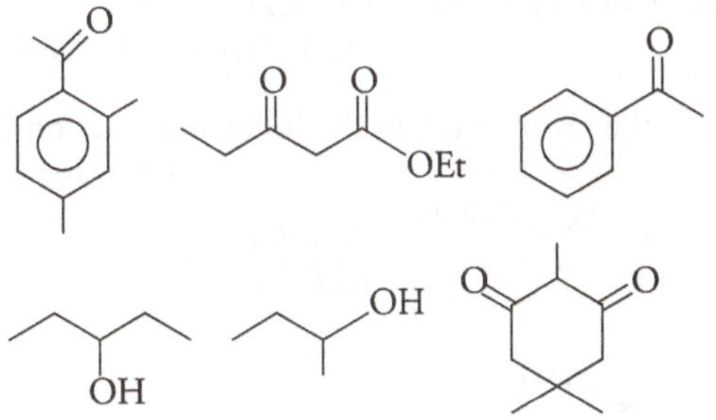

68 In the given reaction [2022, 25 July Shift-I]

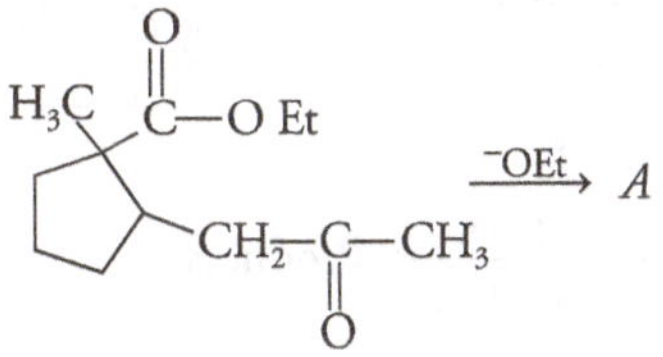

(where, Et is $-C_2H_5$)

The number of chiral carbon/s in product *A* is ......

Topic 3 / Preparation, Properties and Uses of Carboxylic Acids

• Level 1

01. Which of the following compounds is least likely to give effervescence of CO_2 in presence of *aq.* $NaHCO_3$? [2024, 7 April shift-I]

(a) $Ph-\overset{\oplus}{N}H_3\,\overset{\ominus}{Cl}$

(b) 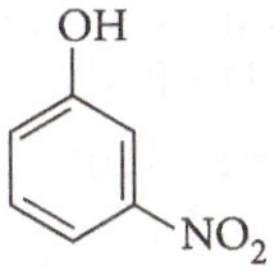

(c) 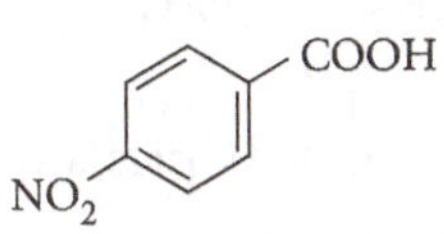

(d) 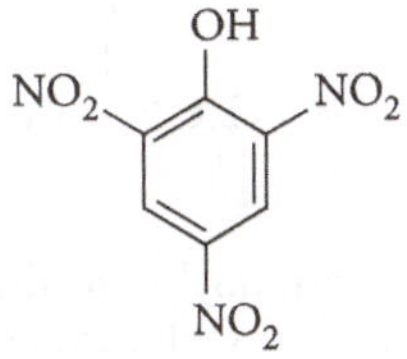

Imp. *Acidic Strength of Aliphatic Acid*

02 The correct sequence of acidic strength of the following aliphatic acids in their decreasing order is **[2024, 8 April Shift-II]**

CH_3CH_2COOH, CH_3COOH, $CH_3CH_2CH_2COOH$, $HCOOH$

(a) $CH_3COOH > CH_3CH_2COOH > CH_3CH_2CH_2COOH > HCOOH$

(b) $HCOOH > CH_3COOH > CH_3CH_2COOH > CH_3CH_2CH_2COOH$

(c) $HCOOH > CH_3CH_2CH_2COOH > CH_3CH_2COOH > CH_3COOH$

(d) $CH_3CH_2CH_2COOH > CH_3CH_2COOH > CH_3COOH > HCOOH$

03 Identify the product (*P*) in the following reaction : **[2024, 8 April Shift-I]**

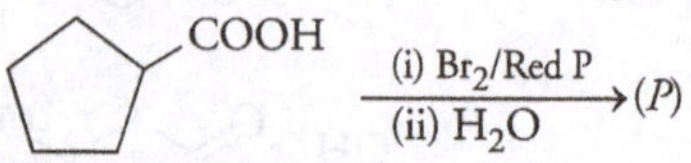

(a) 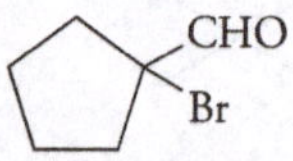

(b) 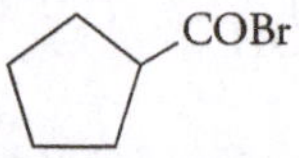

(c) 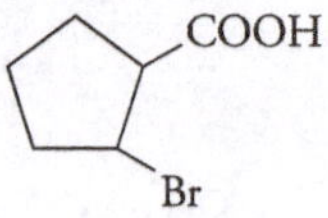

(d) 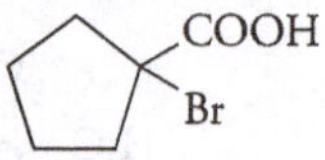

04 2-hexene $\xrightarrow[\text{(ii) } H_2O]{\text{(i) } O_3}$ Products

The two products formed in above reaction are **[2023, 12 April Shift-I]**

(a) butanoic acid and acetaldehyde

(b) butanal and acetic acid

(c) butanal and acetaldehyde

(d) butanoic acid and acetic acid

05 The descending order of acidity for the following carboxylic acid is

A. CH_3COOH B. $F_3C—COOH$

C. $ClCH_2—COOH$ D. $FCH_2—COOH$

E. $BrCH_2—COOH$

Choose the correct answer from the options given below: **[2023, 8 April Shift-II]**

(a) D > B > A > E > C (b) B > D > C > E > A

(c) E > D > B > A > C (d) B > C > D > E > A

06 The correct IUPAC nomenclature for the following compound is

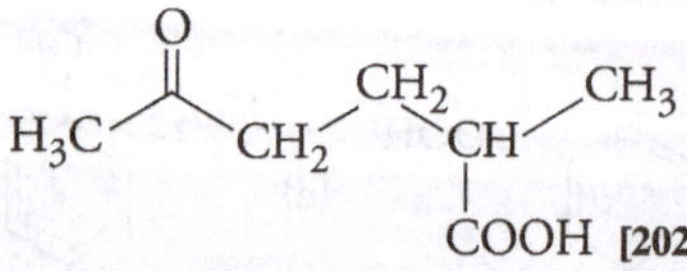

[2023, 8 April Shift-II]

(a) 5-formyl -2-methylhexanoic acid

(b) 5-methyl-2-oxohexan-6-oic acid

(c) 2-methyl-5-oxohexanoic acid

(d) 2-formyl -5-methylhexan-6-oic acid

07 Given below are two statements.

One is labelled as Assertion (A) and the other is labelled as Reason (R).

Assertion (A) 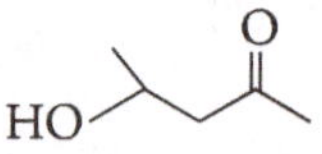

can be easily reduced using Zn-Hg / HCl to 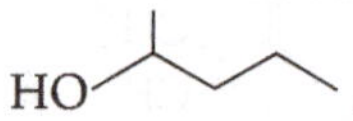

.

Reason (R) Zn-Hg/HCl is used to reduce carbonyl group to $—CH_2—$ group. **[2023, 30 Jan Shift-II]**

In the light of the above statements, choose the correct answer from the options given below.

(a) A is false but R is true

(b) Both A and R are true but R is not the correct explanation of A

(c) Both A and R are true and R is the correct explanation of A

(d) A is true but R is false

08 Given below are two statements : **[2022, 29 July Shift-I]**

Statement I The esterification of carboxylic acid with an alcohol is a nucleophilic acyl substitution.

Statement II Electron withdrawing groups in the carboxylic acid will increase the rate of esterification reaction.

Choose the most appropriate answer from the options given below :

(a) Both Statement I and Statement II are correct.

(b) Both Statement I and Statement II are incorrect.

(c) Statement I is correct but Statement II is incorrect.

(d) Statement I is incorrect but Statement II is correct.

09 Consider the given chemical reaction.

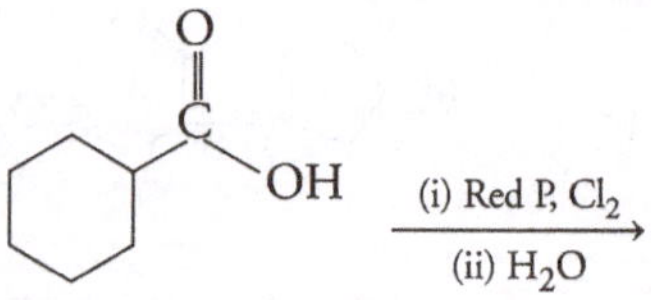

Identify the product *P*. **[2022, 30 June Shift-I]**

(a) 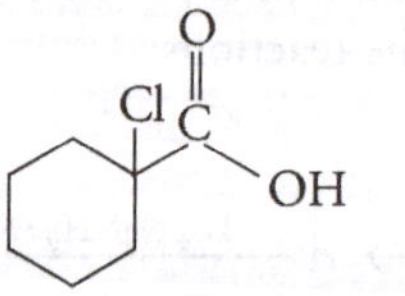

(b) 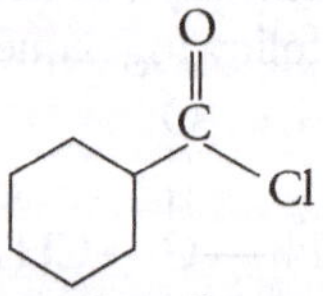

(c) 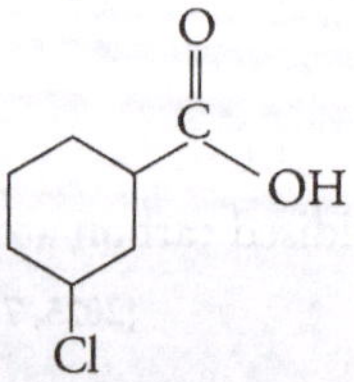

(d) 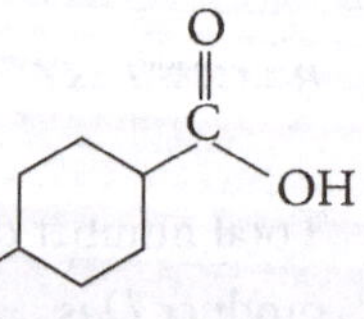

10 The reagent, from the following, which converts benzoic acid to benzaldehyde in one step is **[2022, 26 June Shift-II]**

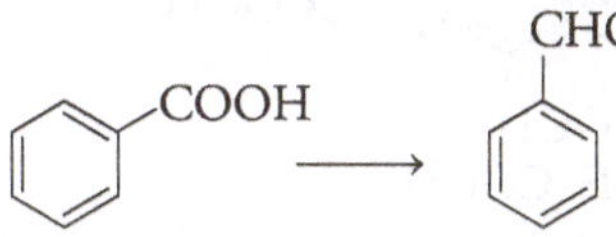

(a) $LiAlH_4$ (b) $KMnO_4$ (c) MnO (d) $NaBH_4$

11 Two statements are given below. **[2022, 24 June Shift-I]**

Statement I The melting point of monocarboxylic acid with even number of carbon atoms is higher than that of with odd number of carbon atoms acid immediately below and above it in the series.

Statement II The solubility of monocarboxylic acids in water decreases with increase in molar mass.

Choose the most appropriate option.
(a) Both Statement I and Statement II are true.
(b) Both Statement I and Statement II are false.
(c) Statement I is true but Statement II is false.
(d) Statement I is false but Statement II is true.

12 The structure of the starting compound *P* used in the reaction given below is **[2021, 27 Aug Shift-I]**

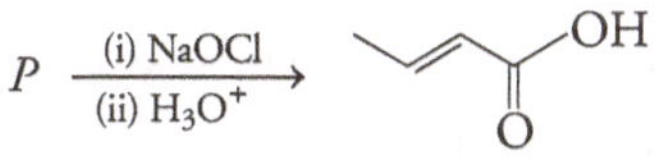

(a) 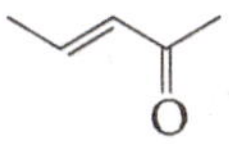
(b) 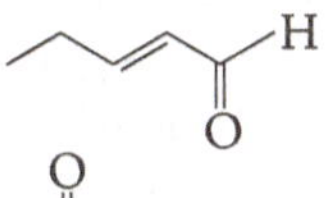
(c) 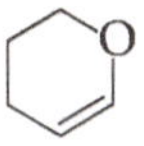
(d) 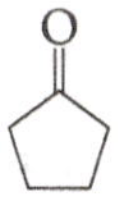

13 (A) 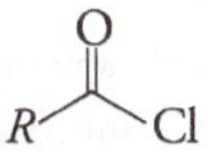
(B) 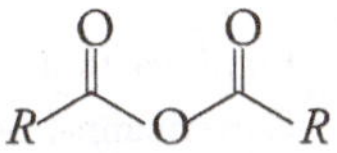
(C) 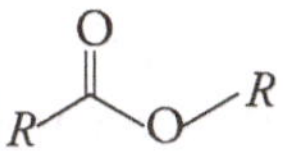
(D) 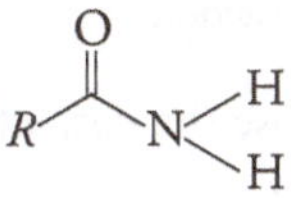

The correct order of their reactivity towards hydrolysis at room temperature is **[2021, 20 July Shift-II]**
(a) (D) > (A) > (B) > (C) (b) (A) > (B) > (C) > (D)
(c) (A) > (C) > (B) > (D) (d) (D) > (B) > (A) > (C)

14 Arrange the following labelled hydrogens in decreasing order of acidity: **[2020, 2 Sep Shift-II]**

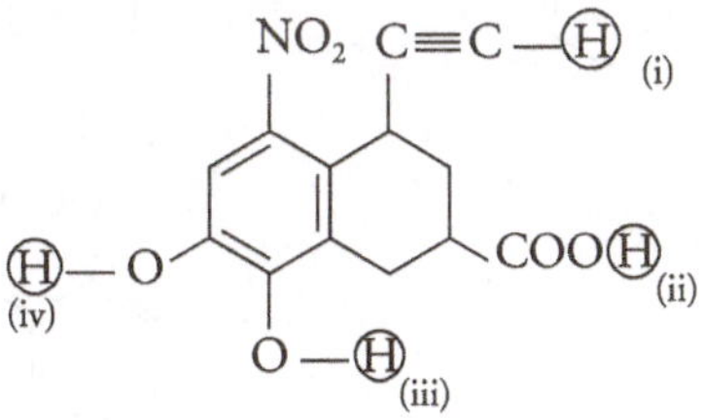

(a) (ii) > (i) > (iii) > (iv)
(b) (iii) > (ii) > (iv) > (i)
(c) (ii) > (iii) > (iv) > (i)
(d) (ii) > (iii) > (i) > (iv)

15 The IUPAC name of the following compound is

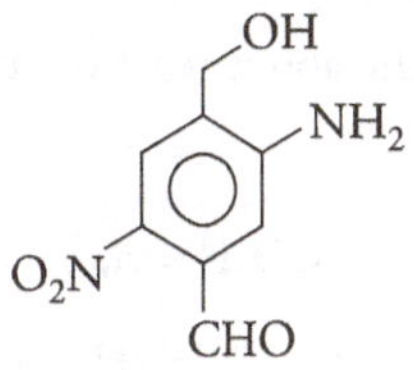

[2020, 2 Sep Shift-II]
(a) 2-nitro-4-hydroxymethyl-5-aminobenzaldehyde
(b) 3-amino-4-hydroxymethyl-5-nitrobenzaldehyde
(c) 5-amino-4-hydroxymethyl-2-nitrobenzaldehyde
(d) 2-amino-2-formyl-5-hydroxymethylnitrobenzene

16 The correct decreasing order for acid strength is **[2019, 9 Jan Shift-I]**
(a) $FCH_2COOH > NCCH_2COOH > NO_2CH_2COOH > ClCH_2COOH$
(b) $CNCH_2COOH > O_2NCH_2COOH > FCH_2COOH > ClCH_2COOH$
(c) $NO_2CH_2COOH > NCCH_2COOH > FCH_2COOH > ClCH_2COOH$
(d) $NO_2CH_2COOH > FCH_2COOH > CNCH_2COOH > ClCH_2COOH$

Level 2

17 The least acidic compound, among the following is: **[2025, 3 April shift-I]**

(A) 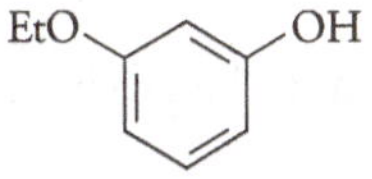
(B) 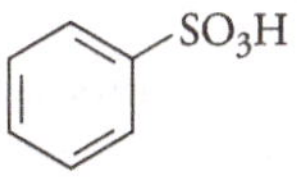
(C) 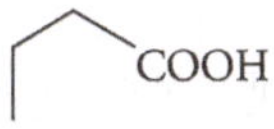
(D) $EtO_2C-C\equiv C-H$

(a) D (b) C
(c) A (d) B

18 In the reaction given below

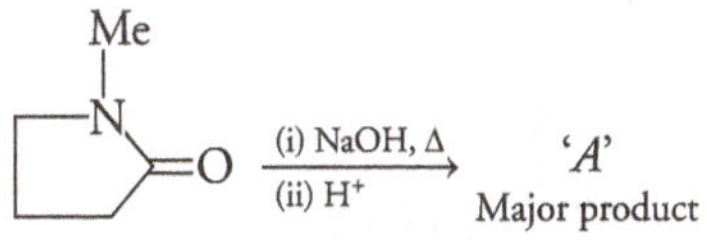

'*A*' is **[2023, 13 April Shift-I]**

(a) 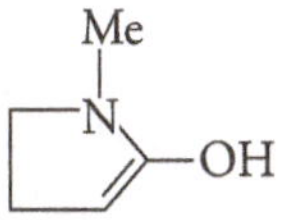
(b) 
(c) 
(d) 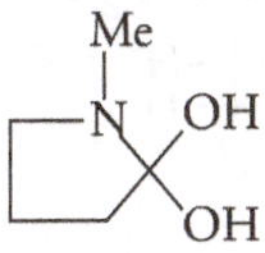

19

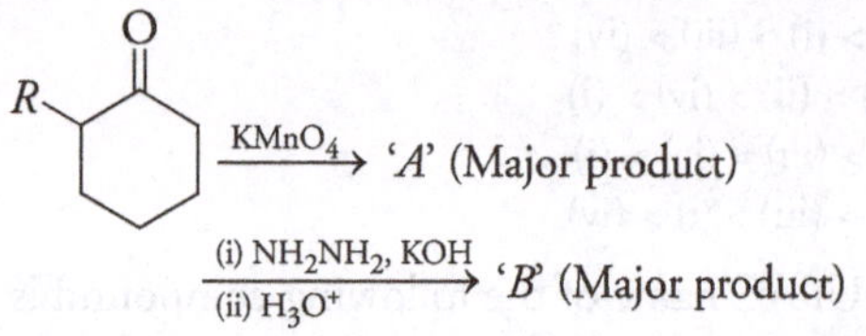

'A' and 'B' in the above reaction are **[2023, 11 April Shift-I]**

(a)
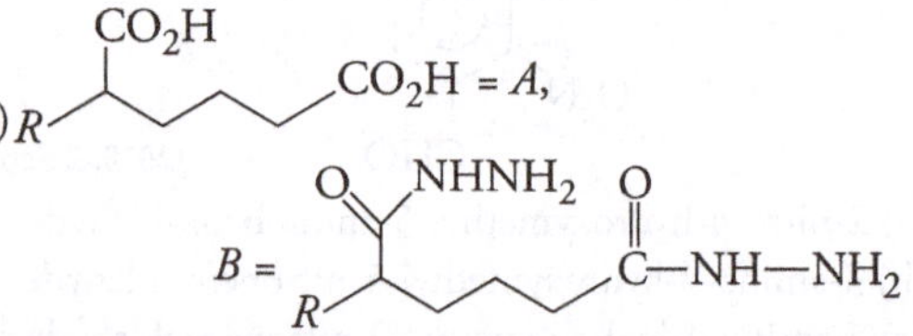

(b)
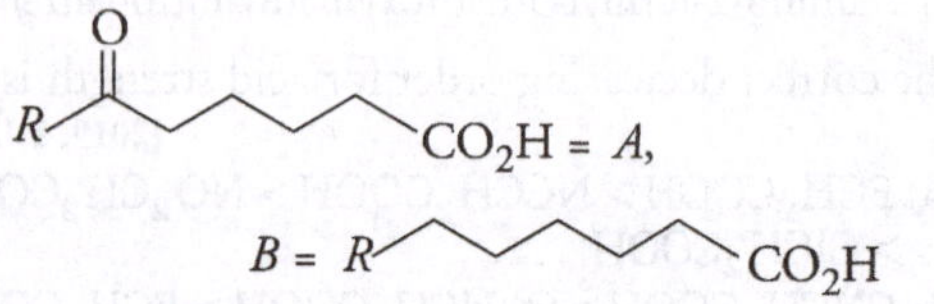

(c)
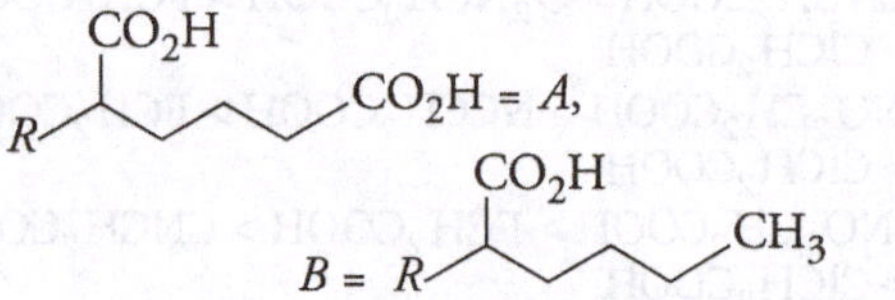

(d)
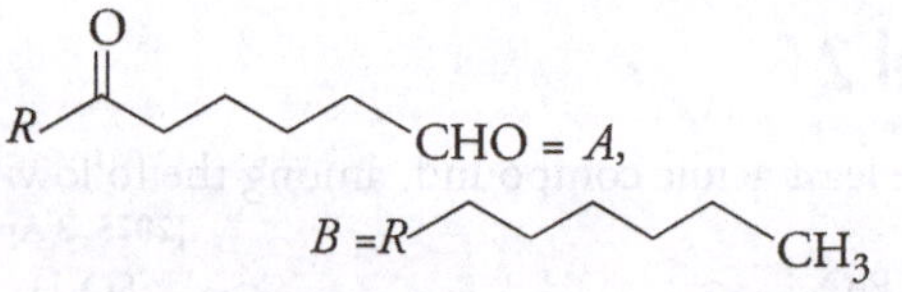

20 The major product 'P' formed in the given reaction is

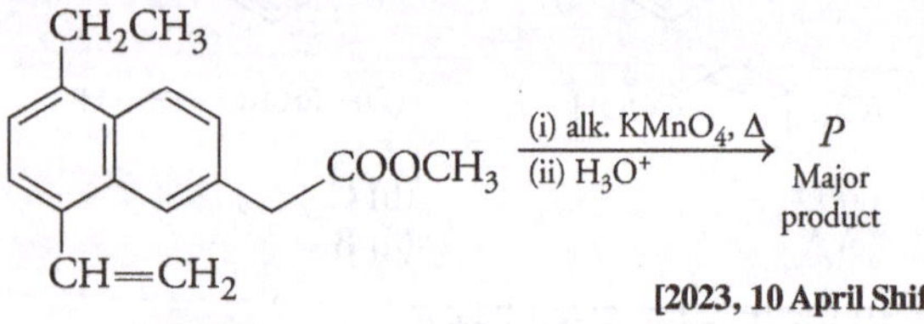

[2023, 10 April Shift-I]

(a)
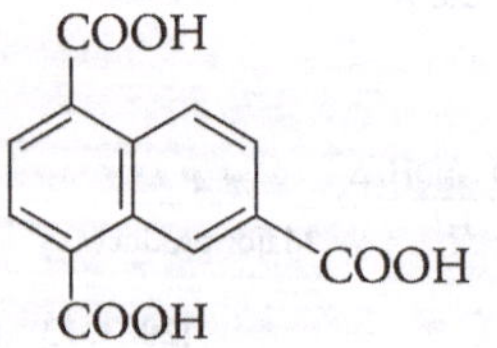

(b)
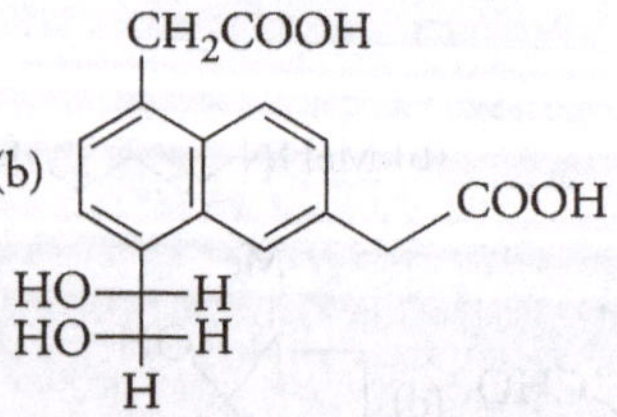

(c)
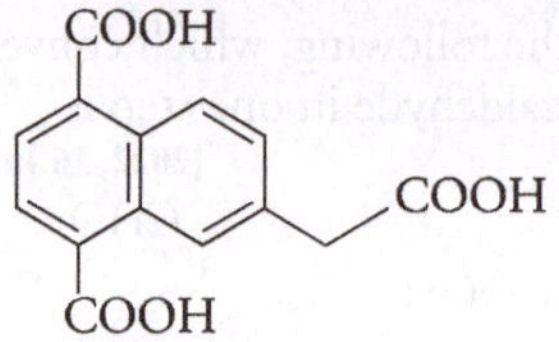

(d)
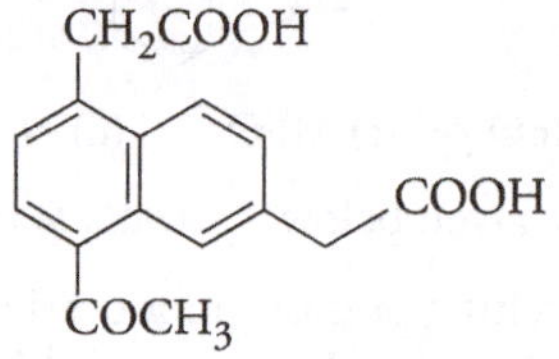

21 The major product formed in the following reaction is

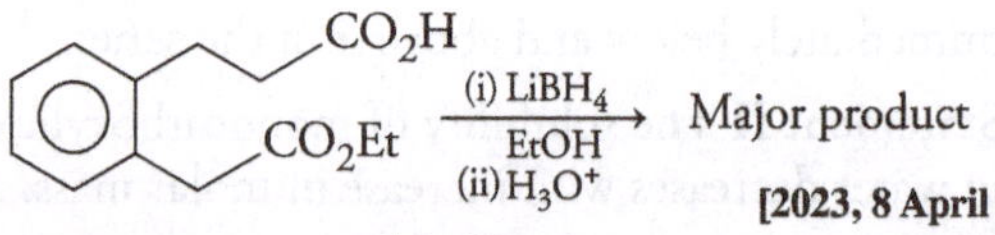

[2023, 8 April Shift-I]

(a)
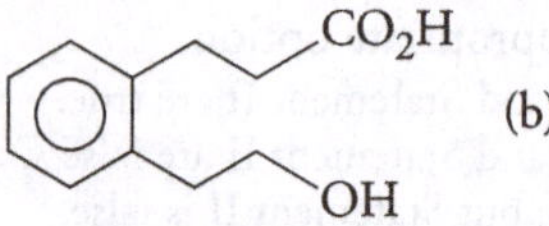

(b)
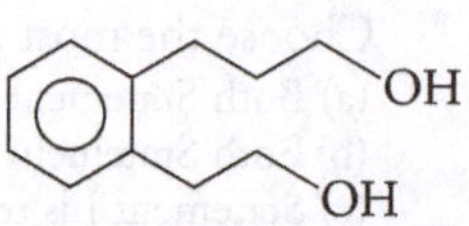

(c)
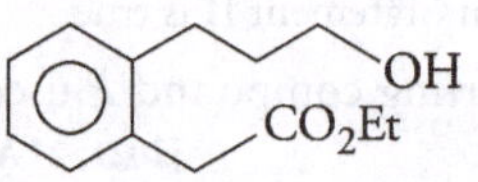

(d)
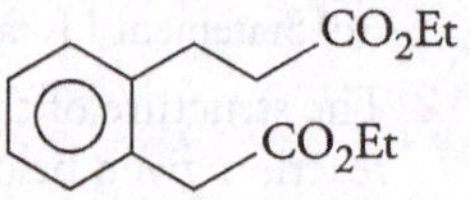

Imp. *Reagent Used in Name Reactions*

22 Match List-I with List-II.

	List-I (Name of reaction)		List-II (Reagent used)
A.	Hell-Volhard-Zelinsky reaction	I.	NaOH + I_2
B.	Iodoform reaction	II.	(i) CrO_2Cl_2, CS_2 (ii) H_2O
C.	Etard reaction	III.	Br_2/red phosphorus
D.	Gattermann-Koch reaction	IV.	CO, HCl, anhyd. $AlCl_3$

Choose the correct answer from the options given below **[2023, 6 April Shift-I]**

(a) A-I, B-II, C-III, D-IV
(b) A-III, B-I, C-II, D-IV
(c) A-III, B-II, C-I, D-IV
(d) A-III, B-I, C-IV, D-II

23 In a reaction,

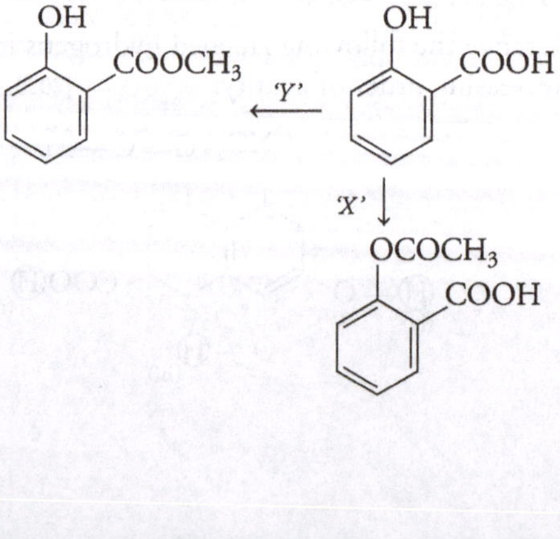

reagents '*X*' and '*Y*' respectively are **[2023, 1 Feb Shift-II]**

(a) $CH_3OH/H^+, \Delta$ and $(CH_3CO)_2O/H^+$
(b) $CH_3OH/H^+, \Delta$ and $CH_3OH/H^+, \Delta$
(c) $(CH_3CO)_2O/H^+$ and $CH_3OH/H^+, \Delta$
(d) $(CH_3CO)_2O/H^+$ and $(CH_3CO)_2O/H^+$

24 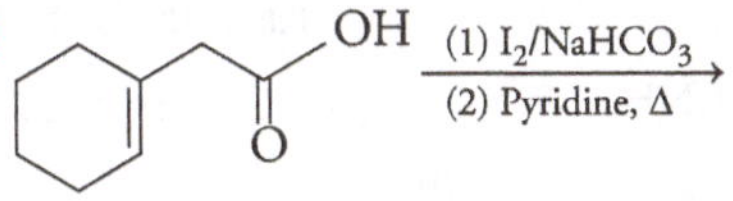

Find out the major product for the above reaction. **[2022, 28 July Shift-II]**

(a) 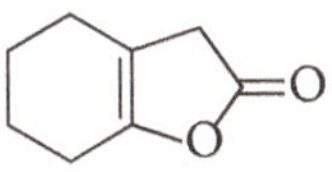 (b) 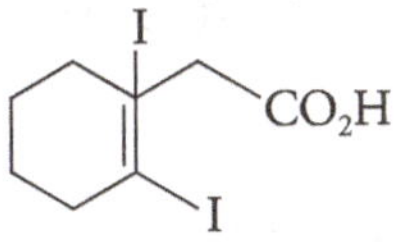

(c) 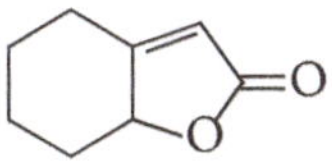 (d) 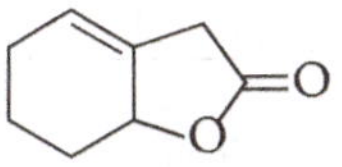

25 Among the following marked proton which compound shows lowest pK_a value? **[2022, 28 July Shift-I]**

(a)  (b) 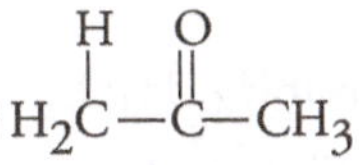

(c) 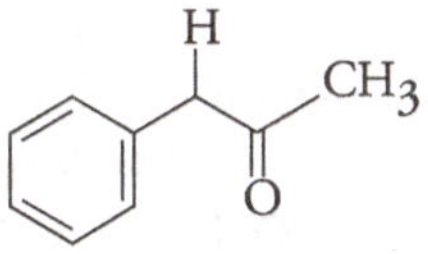 (d) 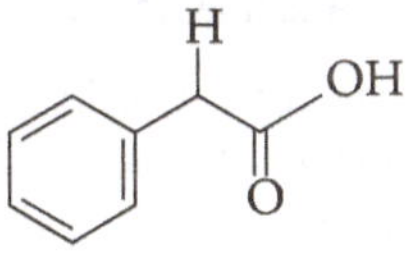

26 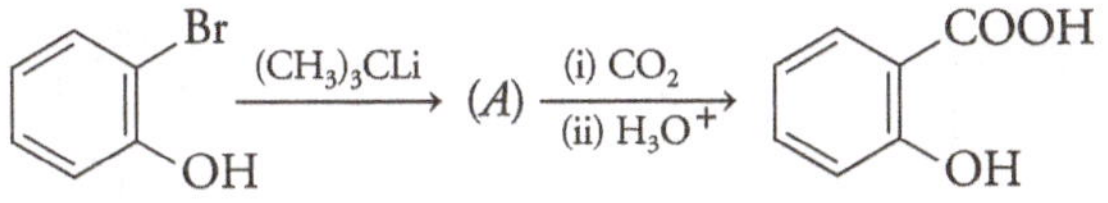

In the given conversion the compound A is: **[2022, 29 June Shift-I]**

(a) 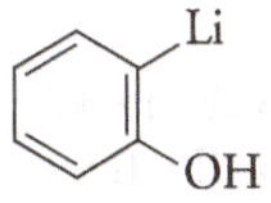 (b) 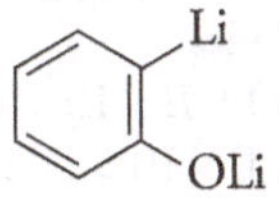

(c) 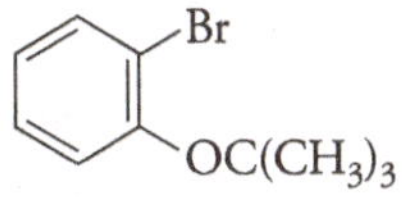 (d) 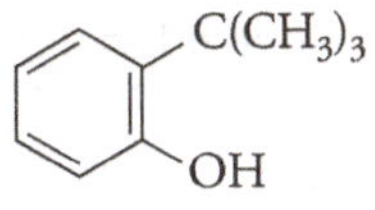

27 $(A) \xrightarrow[\text{(iii) } H_2O/H^+]{\text{(i) } Cl_2, \Delta \text{ (ii) } CN^-}$ 4-Bromophenyl acetic acid.

In the above reaction '*A*' is **[2022, 28 June Shift-II]**

(a) 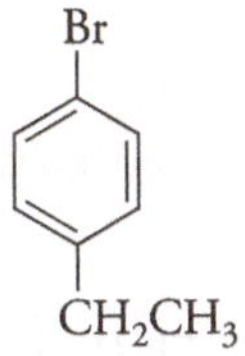 (b) 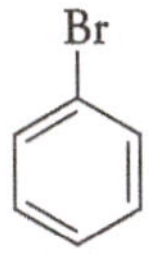

(c) 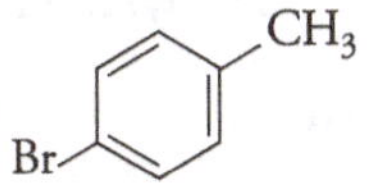 (d) 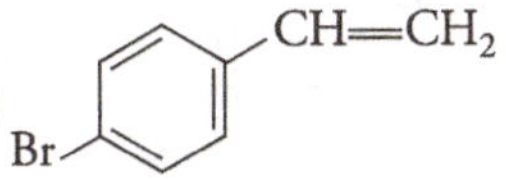

28 The final product '*A*' in the following reaction sequence **[2022, 26 June Shift-II]**

$$CH_3CH_2-\underset{\underset{O}{\|}}{C}-CH_3 \xrightarrow{HCN} ? \xrightarrow[\text{Heat}]{95\%\ H_2SO_4} A$$

(a) $CH_3-CH=\overset{\overset{CH_3}{|}}{C}-COOH$
(b) $CH_3-CH=\underset{\underset{CH_3}{|}}{C}-CN$
(c) $CH_3-CH_2-\overset{\overset{OH}{|}}{\underset{\underset{CH_3}{|}}{C}}-COOH$
(d) $CH_3-CH=\underset{\underset{CH_3}{|}}{C}-CONH_2$

29 In the following sequence of reactions a compound *A*, (molecular formula $C_6H_{12}O_2$) with a straight chain structure gives a C_4 carboxylic acid. *A* is

$$A \xrightarrow[H_3O^+]{LiAlH_4} B \xrightarrow{\text{Oxidation}} C_4 \text{ carboxylic acid}$$

[2021, 01 Sep Shift-II]

(a) $CH_3-CH_2-COO-CH_2-CH_2-CH_3$
(b) $CH_3-CH_2-\overset{\overset{OH}{|}}{C}H-CH_2-O-CH=CH_2$
(c) $CH_3-CH_2-CH_2-COO-CH_2-CH_3$
(d) $CH_3-CH_2-CH_2-O-CH=CH-CH_2-OH$

30 Which one of the following reactions will not yield propionic acid? **[2021, 27 Aug Shift-II]**

(a) $CH_3CH_2CH_2Br + Mg, CO_2$ dry ether/ H_3O^+
(b) $CH_3CH_2CH_3 + KMnO_4$ (Heat), OH^-/H_3O^+
(c) $CH_3CH_2CCl_3 + OH^-/H_3O^+$
(d) $CH_3CH_2COCH_3 + OI^-/H_3O^+$

31 Given below are two statements.

Statement I Ethyl pent-4-yn-oate on reaction with CH_3MgBr gives a 3° alcohol.

Statement II In this reaction one mole of ethyl pent-4- yn-oate utilizes two moles of CH_3MgBr.

In the light of the above statements, choose the most appropriate answer from the options given below: **[2021, 27 Aug Shift-II]**

(a) Both Statement I and Statement II are false.
(b) Statement I is false but Statement II is true.
(c) Statement I is true but Statement II is false.
(d) Both Statement I and Statement II are true.

32 The major product formed in the following reaction is

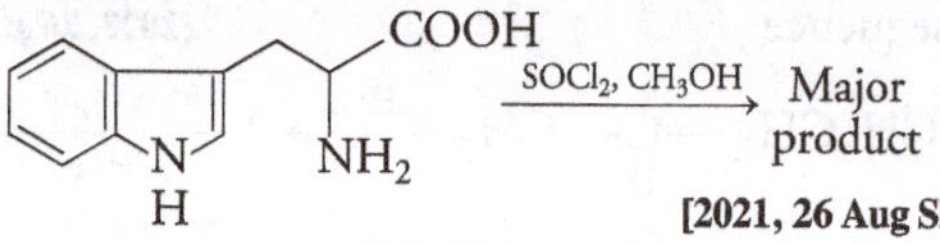

[2021, 26 Aug Shift-I]

(a) 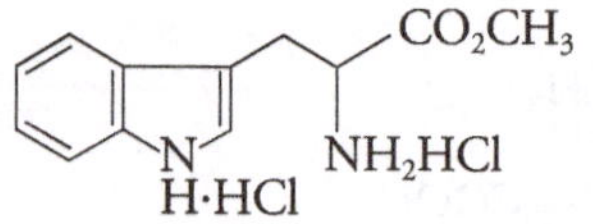

(b) 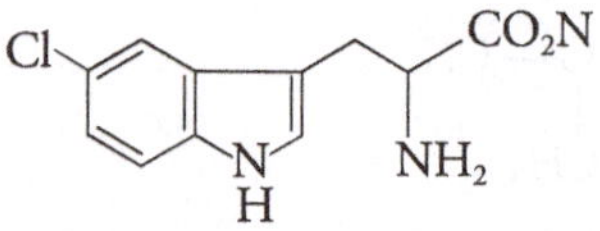

(c) 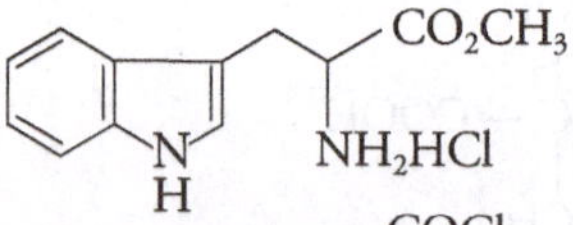

(d) 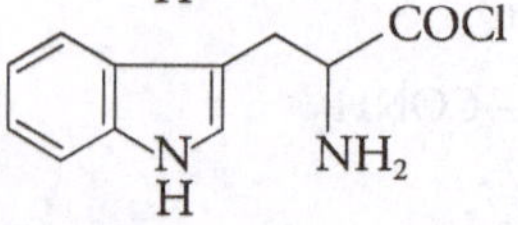

33 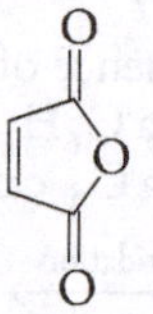

(Maleic anhydride)

Maleic anhydride can be prepared by **[2021, 25 July Shift-II]**

(a) heating *trans*-but-2-enedioic acid

(b) heating *cis*-but-2-enedioic acid

(c) treating *cis*-but-2-enedioic acid with alcohol and acid

(d) treating *trans*-but-2-enedioic acid with alcohol and acid

34 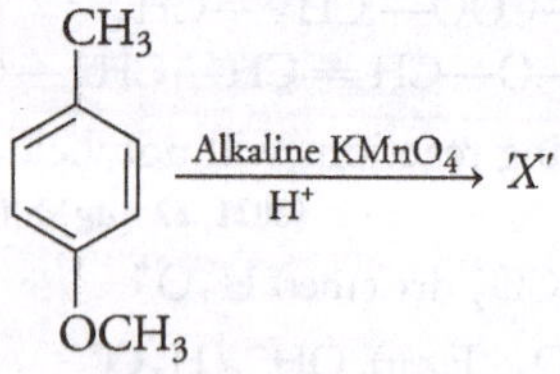

Considering the above chemical reaction, identify the product '*X*'. **[2021, 18 March Shift-I]**

(a) *X*- 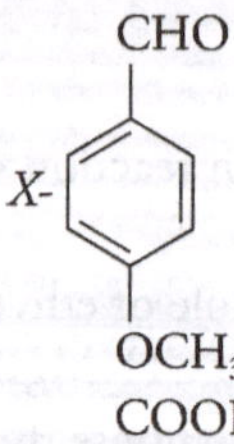

(b) *X*- 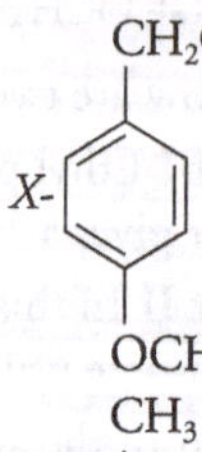

(c) *X*- 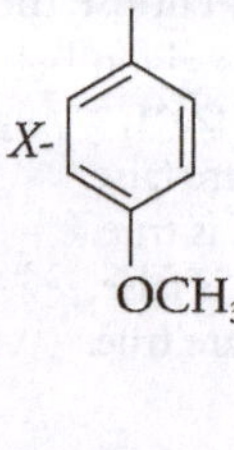

(d) *X*- 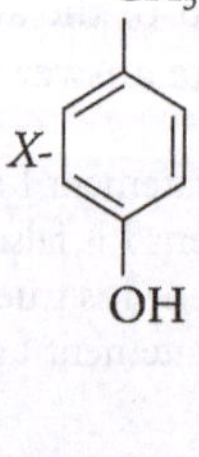

35 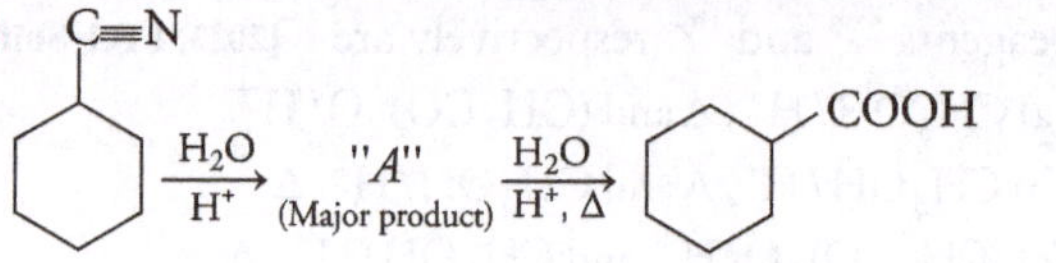

Consider the above chemical reaction and identify product "*A*". **[2021, 18 March Shift-I]**

(a) 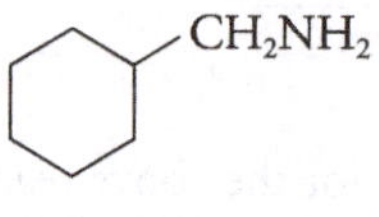

(b) 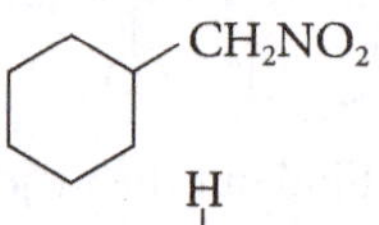

(c) 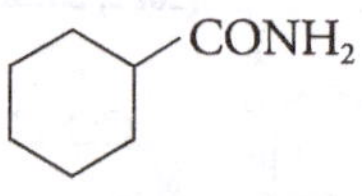

(d) 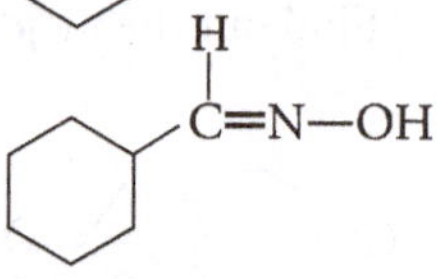

36 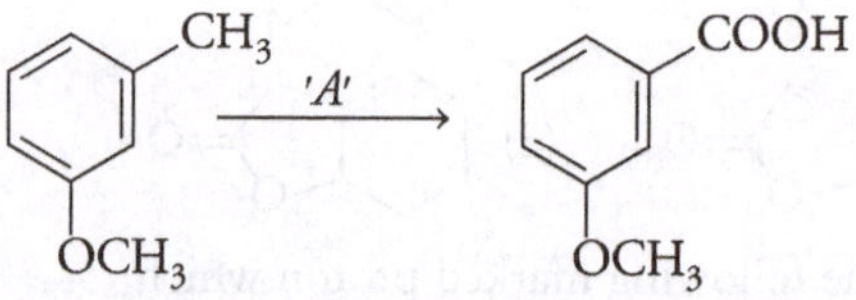

In the above reaction, the reagent '*A*' is **[2021, 16 March Shift-II]**

(a) $NaBH_4, H_3O^+$ (b) $LiAlH_4$

(c) Alkaline $KMnO_4, H^+$ (d) HCl, Zn - Hg

37 The correct order of acid character of the following compounds is **[2021, 25 Feb Shift-II]**

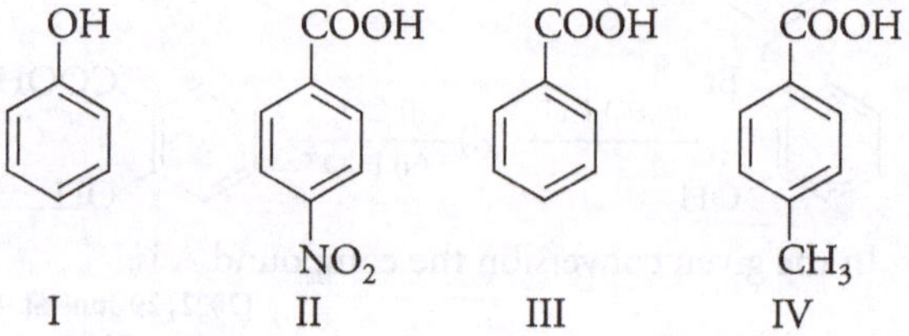

I II III IV

Codes

(a) I > II > III > IV (b) III > II > I > IV

(c) II > III > IV > I (d) IV > III > II > I

38 Match List-I with List-II.

	List-I	List-II
A.	$R{-}C({=}O){-}Cl \longrightarrow R{-}CHO$	1. Br_2 /NaOH
B.	$R{-}CH_2{-}COOH \longrightarrow R{-}CH(Cl){-}COOH$	2. H_2 /Pd - $BaSO_4$
C.	$R{-}C({=}O){-}NH_2 \longrightarrow R{-}NH_2$	3. Zn(Hg)/Conc.HCl
D.	$R{-}C({=}O){-}CH_3 \longrightarrow RCH_2CH_3$	4. Cl_2 / Red P, H_2O

Choose the correct answer from the options given below. **[2021, 24 Feb Shift-II]**

	A	B	C	D		A	B	C	D
(a)	2	1	4	3	(b)	3	4	1	2
(c)	2	4	1	3	(d)	3	1	4	2

39 Consider the following molecules and statements related to them.

(A) 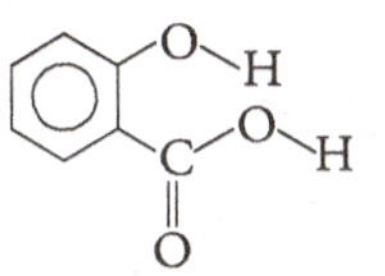
(B) 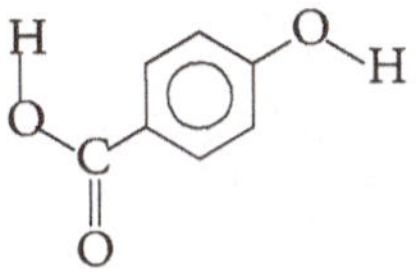

(I) (*B*) is more likely to be crystalline than (*A*).
(II) (*B*) has higher boiling point than (*A*).
(III) (*B*) dissolves more readily than (*A*) in water.

Identify the correct option from below : **[2020, 3 Sep Shift-II]**

(a) (I) and (III) are true. (b) Only (I) is true.
(c) (II) and (III) are true. (d) (I) and (II) are true.

40 An unsaturated hydrocarbon X absorbs two hydrogen molecules on catalytic hydrogenation, and also gives following reaction:

$$X \xrightarrow[Zn/H_2O]{O_3} A \xrightarrow{[Ag(NH_3)_2]^+} B$$ (3-oxo-hexanedicarboxylic acid)

X will be **[2020, 08 Jan Shift-II]**

(a) 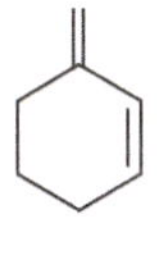
(b) 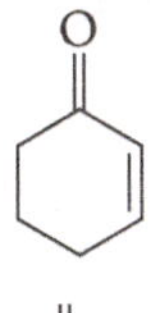
(c) 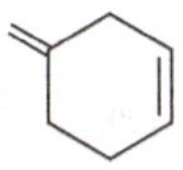
(d) 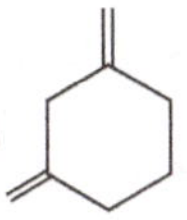

41 The most suitable reagent for the given conversion is

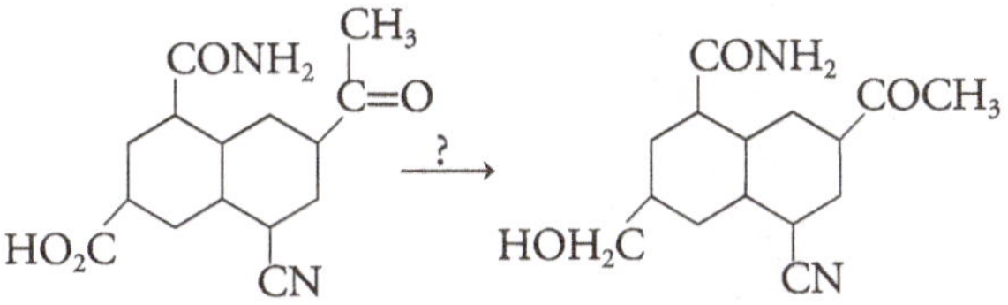

[2020, 8 Jan Shift-I]

(a) B_2H_6 (b) $LiAlH_4$ (c) $NaBH_4$ (d) H_2/Pd

42 The major product of the following reaction is

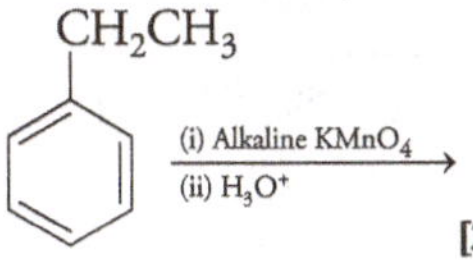

[2019, 9 April Shift-I]

(a) 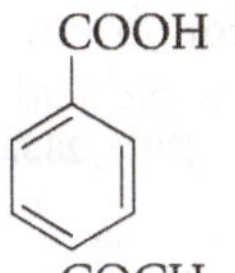
(b) 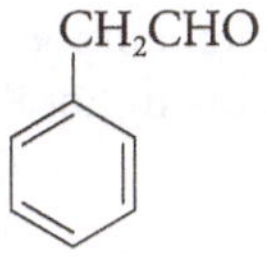
(c) 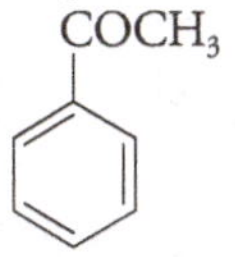
(d) 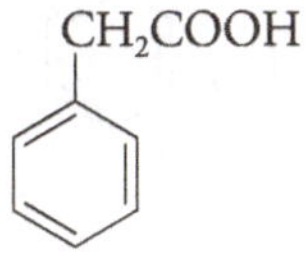

43 The decreasing order of ease of alkaline hydrolysis for the following esters is **[2019, 10 Jan Shift-I]**

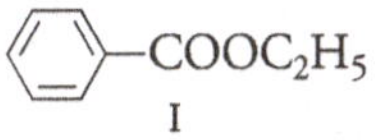
I

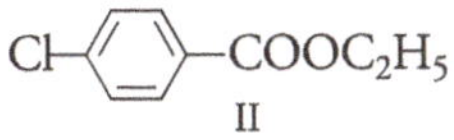
II

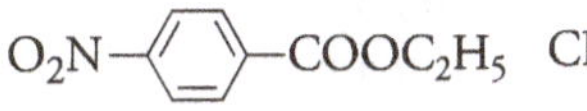
III

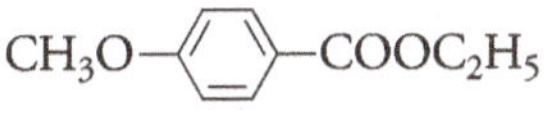
IV

(a) III > II > IV > I (b) III > II > I > IV
(c) II > III > I > IV (d) IV > II > III > I

44 Which dicarboxylic acid in presence of a dehydrating agent is least reactive to give an anhydride? **[2019, 10 Jan Shift-I]**

(a) 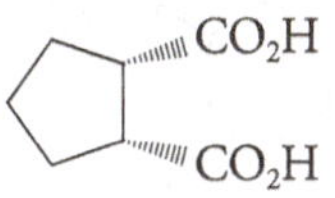
(b) 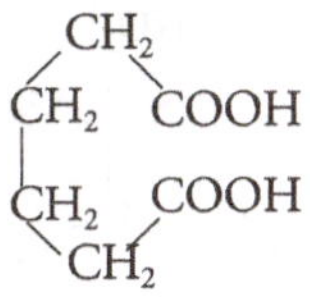
(c) 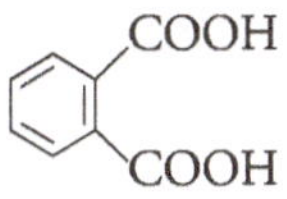
(d) 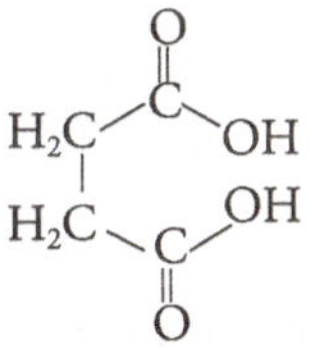

• Level 3

45 Consider the following reactions. From these reactions which reaction will give carboxylic acid as a major product? **[2025, 2 April Shift-II]**

(A) $R—C{\equiv}N \xrightarrow[\text{Mild condition}]{\text{(i) } H^+/H_2O}$

(B) $R—MgX \xrightarrow[\text{(ii) } H_3O^+]{\text{(i) } CO_2}$

(C) $R—C{\equiv}N \xrightarrow[\text{(ii) } H_3O^+]{\text{(i) } SnCl_2/HCl}$

(D) $RCH_2OH \xrightarrow{PCC}$

(E) 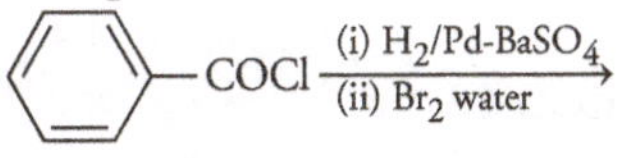

Choose the correct answer from the options given below.

(a) B and E only (b) B, C and E only
(c) A and D only (d) A, B and E only

46 The total number of compounds from below when treated with hot $KMnO_4$ giving benzoic acid is **[2025, 28 Jan Shift-II]**

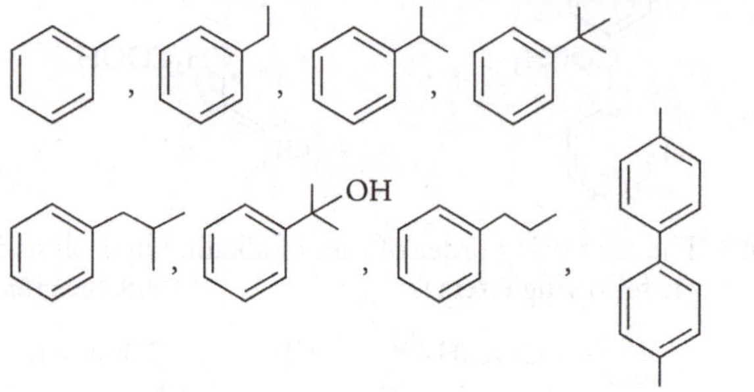

(a) 5 (b) 6
(c) 3 (d) 4

47 The compounds that produce CO_2 with aqueous $NaHCO_3$ solution are **[2025, 28 Jan Shift-I]**

(A) 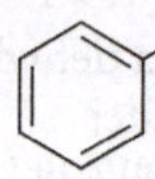 (B) 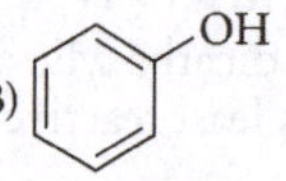

(C) 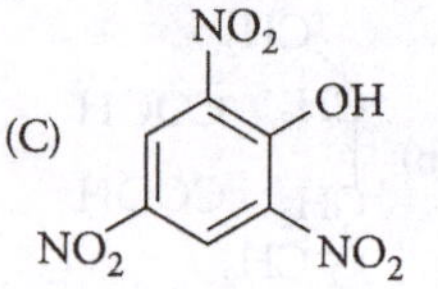 (D) 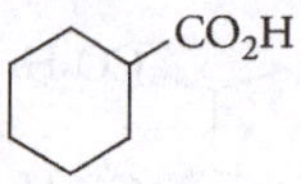

(E) 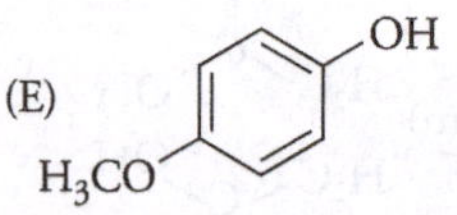

Choose the correct answer from the options given below.

(a) A and B only
(b) A, B and E only
(c) A, C and D only
(d) A and C only

48 Consider the given reaction, identify the major product *P*. **[2024, 6 April Shift-II]**

$$CH_3-COOH \xrightarrow[\text{(iv) } H_2O/\bar{O}H, \Delta]{\text{(i) } LiAlH_4 \text{ (ii) PCC (iii) } HCN/\bar{O}H} \text{'}P\text{'}$$

(a) $CH_3-CH_2-CH_2-OH$

(b) $CH_3-CH_2-\overset{\overset{O}{\|}}{C}-NH_2$

(c) $CH_3-\underset{}{\overset{\overset{OH}{|}}{C}H}-COOH$

(d) $CH_3-\overset{\overset{O}{\|}}{C}-CH_2CH_3$

49 Identify *A* and *B* in the given chemical reaction sequence : **[2024, 5 April Shift-II]**

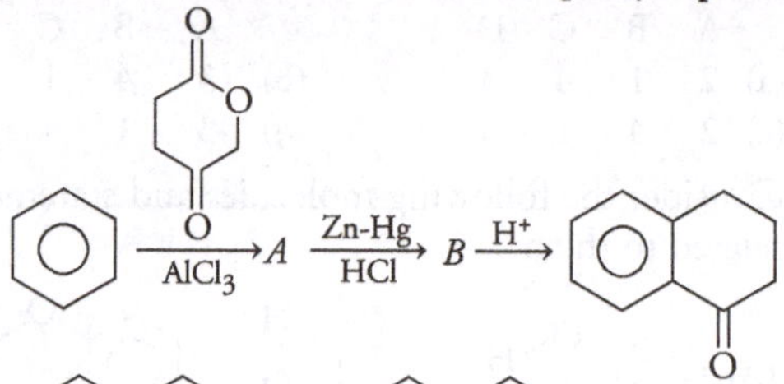

(a) *A*- 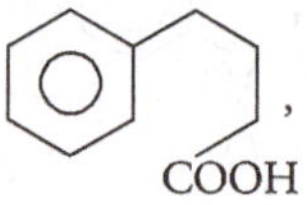, *B*- 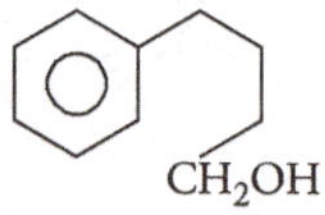

(b) *A*- 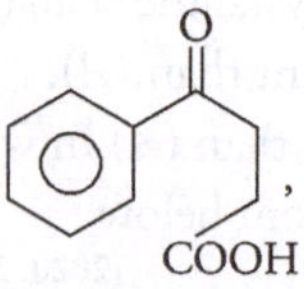, *B*- 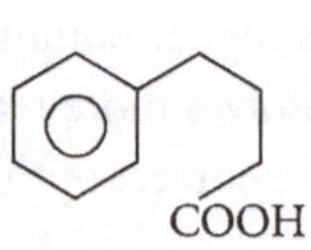

(c) *A*- 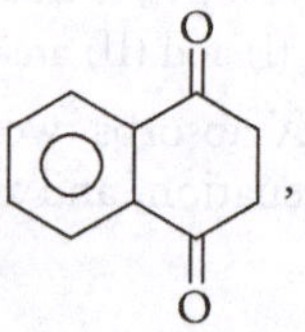, *B*- 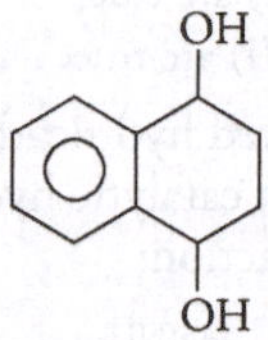

(d) *A*- 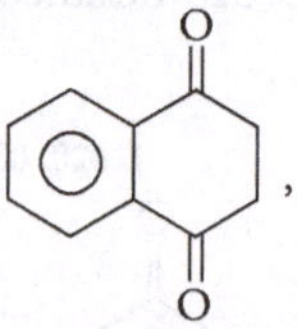, *B*- 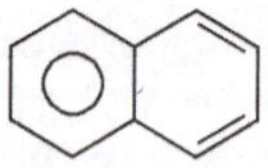

50 Identify *A* and *B* in the following sequence of reaction.

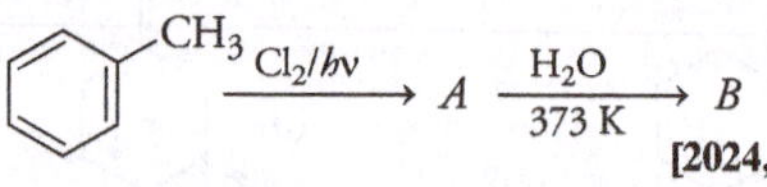

[2024, 1 Feb Shift-I]

(a) *A* = 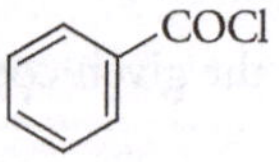 *B* = 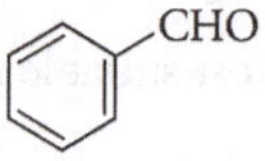

(b) *A* = 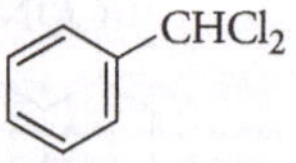 *B* = 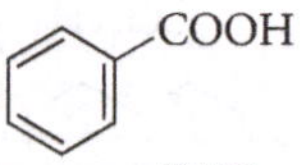

(c) *A* = 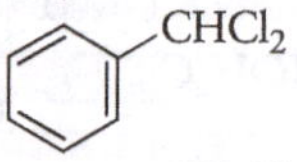 *B* = 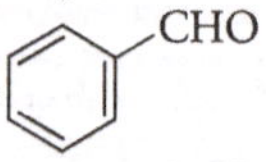

(d) *A* = 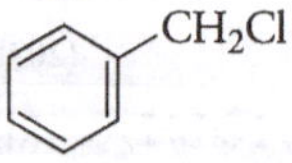 *B* = 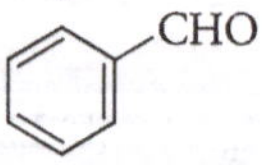

51 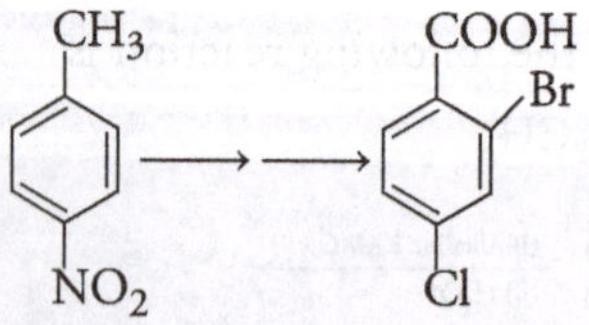

In the above conversion the correct sequence of reagents to be added is **[2023, 15 April Shift-I]**

(a) (i) Br_2/Fe, (ii) Fe/H^+, (iii) HONO, CuCl, (iv) $KMnO_4$

(b) (i) Br_2/Fe, (ii) Fe/H^+, (iii) $KMnO_4$, (iv) Cl_2

(c) (i) $KMnO_4$, (ii) Br_2/Fe, (iii) Fe/H^+, (iv) Cl_2

(d) (i) Fe/H^+, (ii) HONO, (iii) CuCl, (iv) $KMnO_4$, (v) Br_2

52 The major product '*P*' formed in the following sequence of reactions is

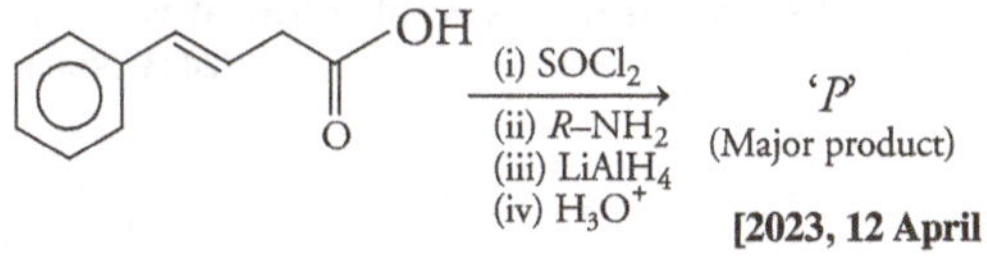

[2023, 12 April Shift-I]

(a) 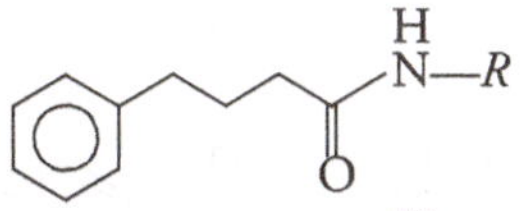

(b) 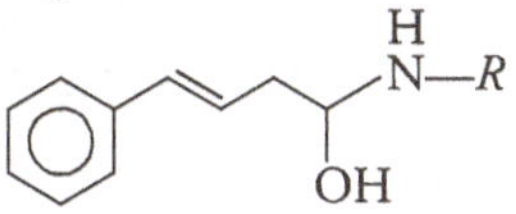

(c) 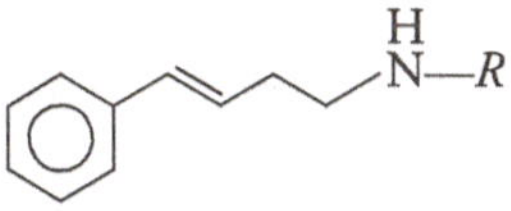

(d) 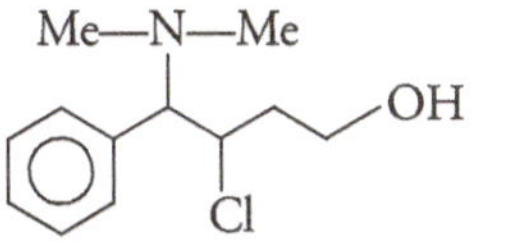

53 Two isomers (*A*) and (*B*) with molar mass 184 g/mol and elemental composition '*C*', 52.2%; H, 4.9% and Br, 42.9% gave benzoic acid and *p*-bromobenzoic acid, respectively on oxidation with $KMnO_4$. Isomer '*A*' is optically active and gives a pale yellow precipitate when warmed with alcoholic $AgNO_3$. Isomers '*A*' and '*B*' are, respectively **[2022, 29 June Shift-II]**

(a) $H_3C—CHBr—C_6H_5$ and 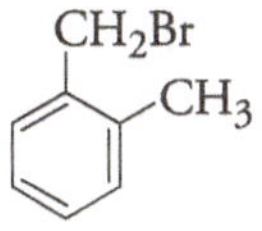

(b) 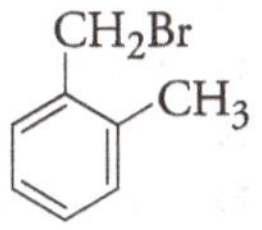 and 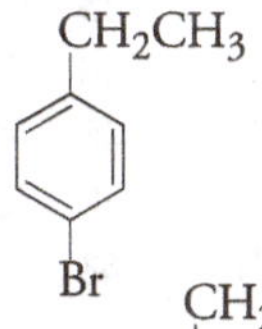

(c) $H_3C—CHBr—C_6H_5$ and 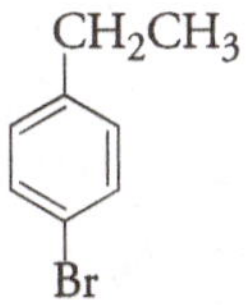

(d) 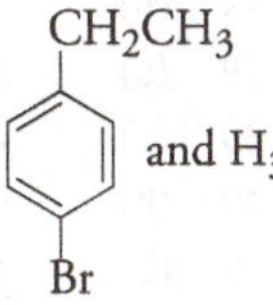 and $H_3C—CHBr—C_6H_5$

54 The structure of product *C*, formed by the following sequence of reactions is

$$CH_3COOH + SOCl_2 \longrightarrow A \xrightarrow[AlCl_3]{\text{Benzene}} B \xrightarrow[-OH]{KCN} C$$

[2021, 31 Aug Shift-I]

(a) 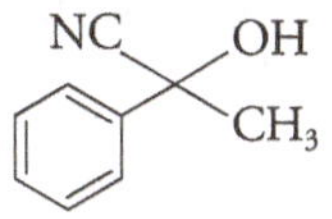

(b) 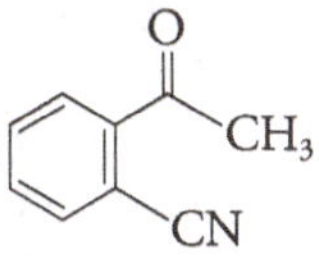

(c) 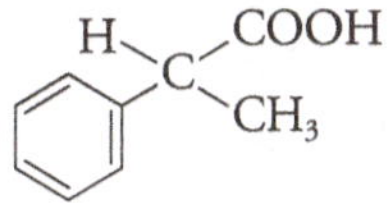

(d) 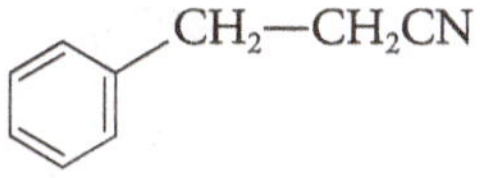

55 The correct sequential addition of reagents in the preparation of 3-nitrobenzoic acid from benzene is **[2021, 26 Aug Shift-I]**

(a) $Br_2/AlBr_3$, HNO_3/H_2SO_4, Mg/ether, CO_2, H_3O^+

(b) $Br_2/AlBr_3$, NaCN, H_3O^+, HNO_3/H_2SO_4

(c) $Br_2/AlBr_3$, HNO_3/H_2SO_4, NaCN, H_3O^+

(d) HNO_3/H_2SO_4, $Br_2/AlBr_3$, Mg/ether, CO_2, H_3O^+

56 Which one of the following compounds will liberate CO_2, when treated with $NaHCO_3$? **[2021, 25 July Shift-I]**

(a) $(CH_3)_3\overset{\oplus}{N}H\overset{\ominus}{Cl}$ (b) $(CH_3)_3\overset{\oplus\ominus}{N}OH$

(c) $CH_3—\overset{O}{\overset{\|}{C}}—NH_2$ (d) CH_3NH_2

57 Compound(s) which will liberate carbon dioxide with sodium bicarbonate solution is/are **[2021, 25 Feb Shift-I]**

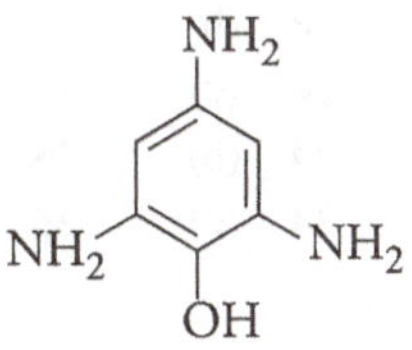
(*A*)

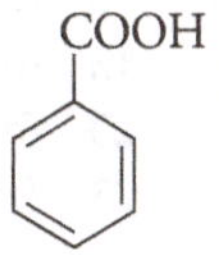
(*B*)

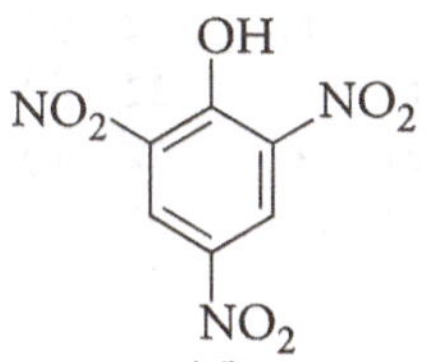
(*C*)

(a) *A* and *B* (b) *C* only (c) *B* and *C* (d) *B* only

58 The major products of the following reactions are

$$CH_3—\underset{}{\overset{CH_3}{\overset{|}{C}H}}—\underset{OSO_2CH_3}{\underset{|}{C}H}—CH_3 \xrightarrow[\text{(ii) } O_3/H_2O_2]{\text{(i) } KO^tBu/\Delta}$$

[2020, 6 Sep Shift-I]

(a) 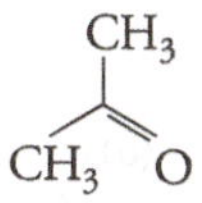 + CH_3CHO

(b) 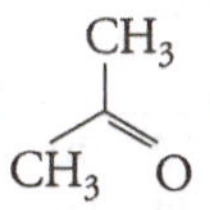 + CH_3COOH

(c) 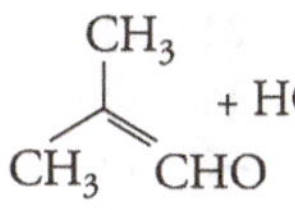 + HCHO

(d) $H_3C—\overset{CH_3}{\overset{|}{C}H}—COOH$ + HCOOH

59 In the following reaction sequence the major products *A* and *B* are : **[2020, 5 Sep Shift-I]**

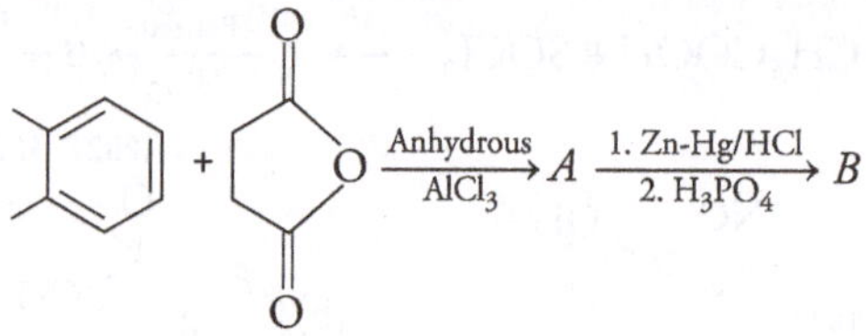

(a) A = 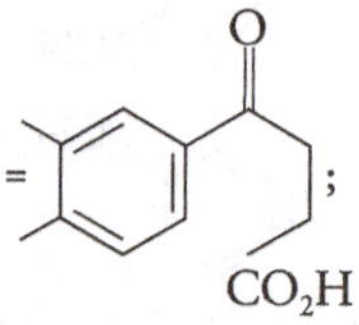 ; B = 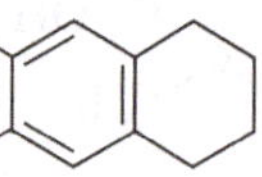

(b) A = 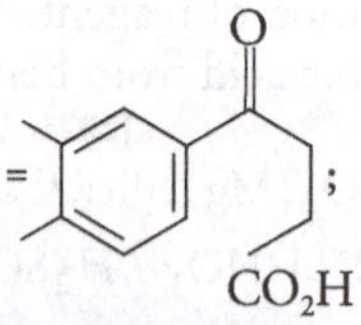 ; B = 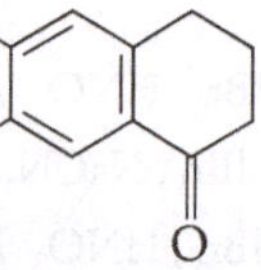

(c) A = 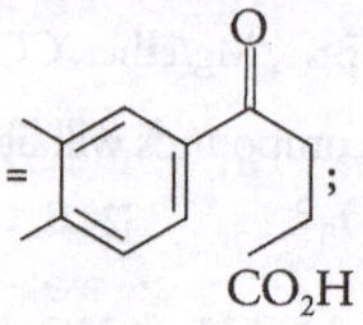 ; B = 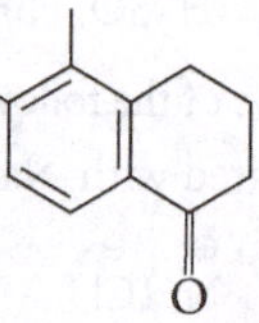

(d) A = 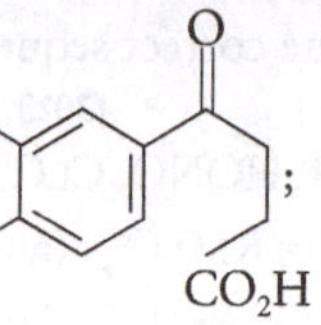 ; B = 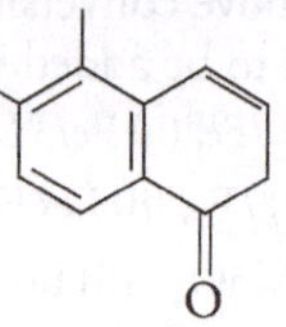

Numerical Value Type Questions

60 Number of compounds from the following which will not dissolve in cold $NaHCO_3$ and NaOH solutions but will dissolve in hot NaOH solution is .... **[2023, 30 Jan Shift-II]**

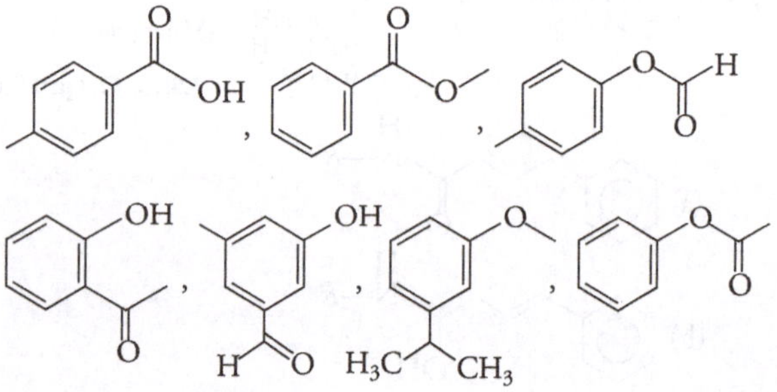

Key Index

Topic 1

1 (a)	2 (c)	3 (a)	4 (b)	5 (b)	6 (c)	7 (a)	8 (a)	9 (d)	10 (c)
11 (c)	12 (b)	13 (b)	14 (c)	15 (c)	16 (c)	17 (b)	18 (a)	19 (b)	20 (b)
21 (a)	22 (d)	23 (b)	24 (b)	25 (d)	26 (a)	27 (d)	28 (c)	29 (d)	30 (c)
31 (c)	32 (c)	33 (a)	34 (c)	35 (c)	36 (d)	37 (c)	38 (d)	39 (c)	40 (0)

Topic 2

1 (b)	2 (a)	3 (d)	4 (c)	5 (c)	6 (b)	7 (c)	8 (d)	9 (a)	10 (b)
11 (d)	12 (a)	13 (d)	14 (d)	15 (a)	16 (b)	17 (a)	18 (a)	19 (c)	20 (d)
21 (c)	22 (c)	23 (c)	24 (a)	25 (d)	26 (a)	27 (c)	28 (b)	29 (a)	30 (a)
31 (a)	32 (b)	33 (b)	34 (c)	35 (d)	36 (d)	37 (b)	38 (b)	39 (b)	40 (c)
41 (a)	42 (a)	43 (b)	44 (a)	45 (b)	46 (c)	47 (d)	48 (d)	49 (b)	50 (a)
51 (a)	52 (a)	53 (c)	54 (d)	55 (b)	56 (d)	57 (d)	58 (c)	59 (d)	60 (a)
61 (d)	62 (b)	63 (b)	64 (b)	65 (7)	66 (15)	67 (3)	68 (2)		

Topic 3

1 (b)	2 (b)	3 (d)	4 (d)	5 (b)	6 (c)	7 (c)	8 (a)	9 (a)	10 (c)
11 (a)	12 (a)	13 (b)	14 (c)	15 (c)	16 (c)	17 (a)	18 (b)	19 (b)	20 (a)
21 (a)	22 (b)	23 (c)	24 (c)	25 (c)	26 (b)	27 (c)	28 (a)	29 (c)	30 (a)
31 (c)	32 (c)	33 (b)	34 (c)	35 (c)	36 (c)	37 (c)	38 (c)	39 (c)	40 (c)
41 (a)	42 (a)	43 (b)	44 (b)	45 (a)	46 (a)	47 (c)	48 (c)	49 (b)	50 (c)
51 (a)	52 (c)	53 (c)	54 (a)	55 (d)	56 (a)	57 (c)	58 (d)	59 (b)	60 (3)

Detailed Explanations

Topic 1

01. *(a)* Complete reaction sequence is as follows

$$CH_3CH_2CH{=}O \xrightarrow[\text{Alkali}]{2HCHO} H_3C{-}\underset{CH_2OH}{\overset{CH_2OH}{C}}{-}CHO$$

$$\xrightarrow{\text{HCHO, Alkali}} HCOO^- + CH_3{-}\underset{CH_2OH}{\overset{CH_2OH}{C}}{-}CH_2{-}OH$$

(Major product)

So, the major product obtained is given in option (a).

02. *(c)* The given reaction proceeds via Clemmensen reduction in which carbonyl group is reduced to methylene group. The complete reaction involved is as follow

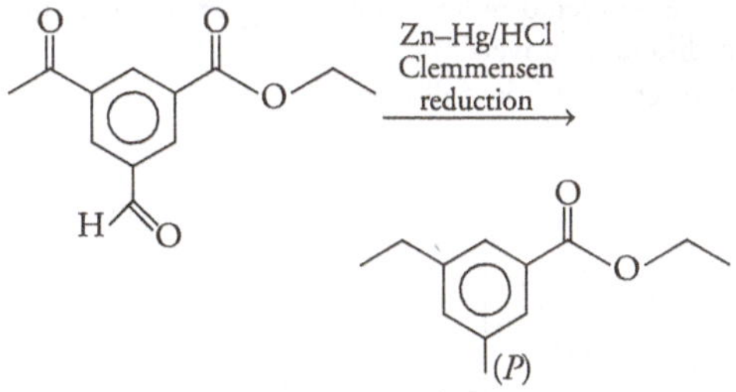

03. *(a)* The given reaction proceeds as follows

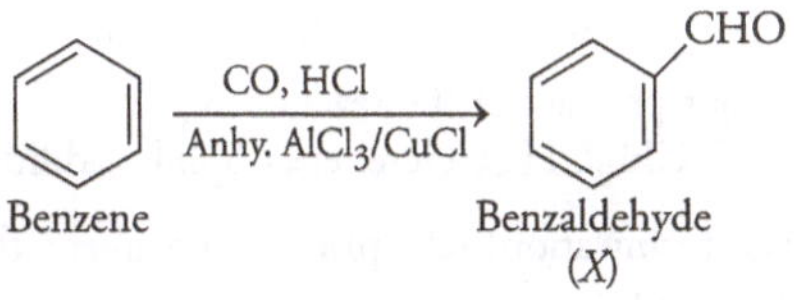

This reaction is called as Gatterman-Koch reaction.

04. *(b)* Among the given aldehydes, $H{-}\overset{O}{\overset{\|}{C}}{-}H$ is most reactive towards nucleophilic addition reaction.

The reason is that it has low steric hinderance at carbonyl carbon and high partial positive charge at carbonyl carbon.

05. *(b)* The given reaction is Gatterman-Koch reaction. In this reaction, benzene is treated with carbon monoxide in the acidic medium in presence of anhydrous aluminium chloride or cuprous chloride to give benzaldehyde.

The complete reaction is as follow,

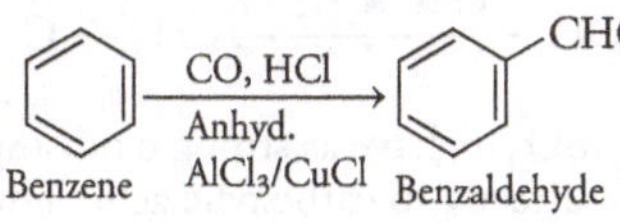

06. *(c)* Structure of 4-methylpent-2-enal is

$$\overset{5}{C}H_3{-}\underset{CH_3}{\overset{4}{C}H}{-}\overset{3}{C}H{=}\underset{2}{C}H{-}\underset{1}{\overset{O}{\overset{\|}{C}}}{-}H$$

Hence, option (c) is correct.

07. *(a)* The structure of γ-methylcyclohexane carbaldehyde is given as

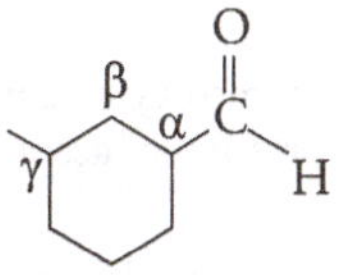

08. *(a)* Statement I is correct, whereas Statement II is incorrect.

The correct form of statement II is,

Vanillin does not give self aldol reaction due to lack of acidic H for condensation.

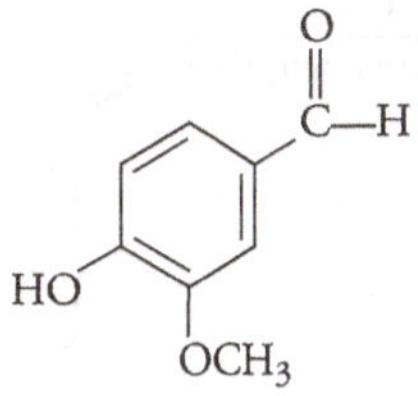

Phenolic group soluble in NaOH benzaldehyde derivative reacts with Tollen's reagent.

09. *(d)* Statement I is true but Statement II is false.

For statement II,

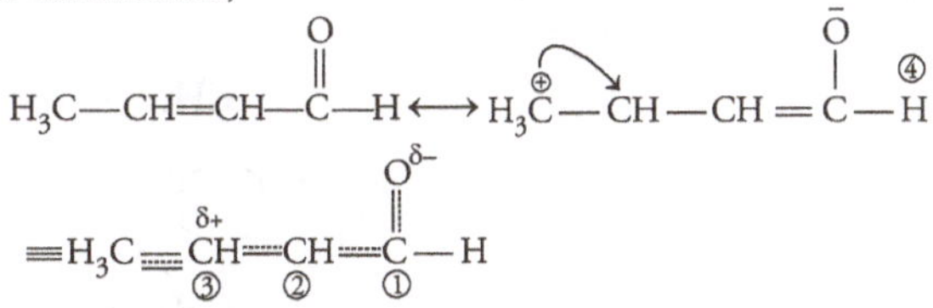

$C_1{-}C_2$ bond has partial double bond character that means lesser bond than $C_1{-}C_2$ bond of other compounds.

10. *(c)* Both Statement I and Statement II are true.

11. *(c)*

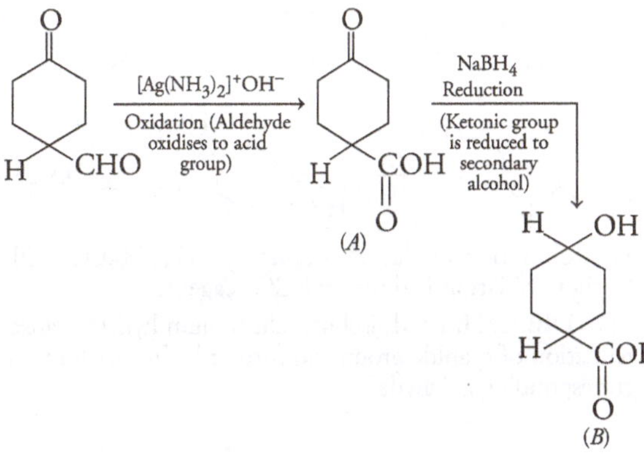

12. *(b)*

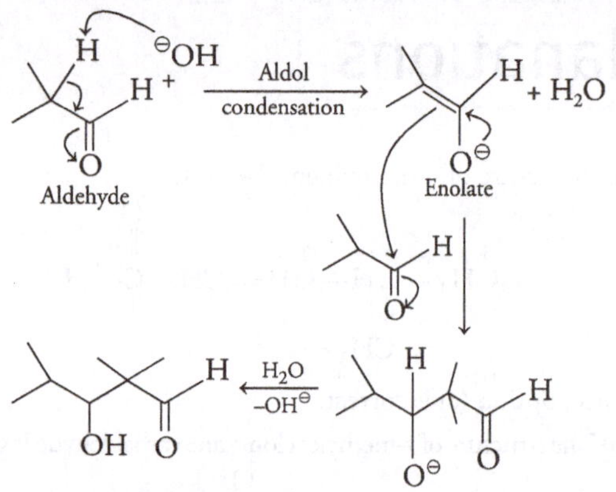

13. *(b)* Benzaldehyde is formed by oxidation of toluene with chromic oxide in the presence of acetic anhydride, followed by acidic hydrolysis.

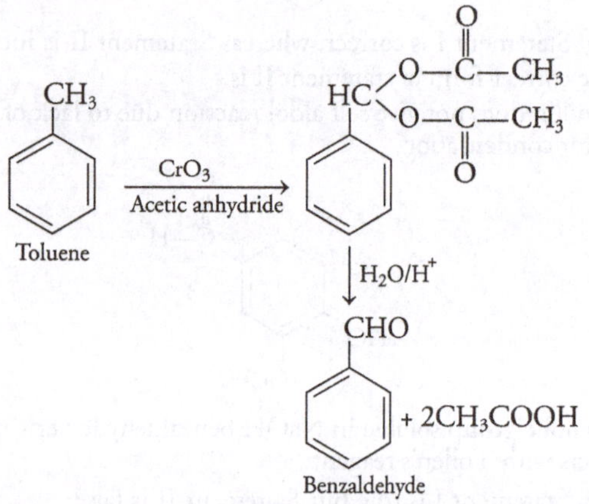

14. *(c)* Reagents used in the given reaction are BH_3, H_2O_2 / OH^- and PCC.

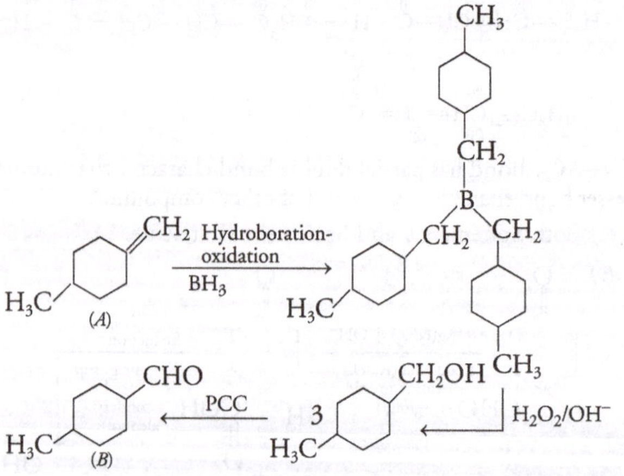

Hydroboration-oxidation is done by BH_3, H_2O_2 / OH^- and further oxidation is done by PCC reagent.

15. *(c)* DIBAL-H, i.e. di-isobutyl aluminium hydride causes partial reduction of cyanide group and further hydrolysis forms corresponding aldehyde.

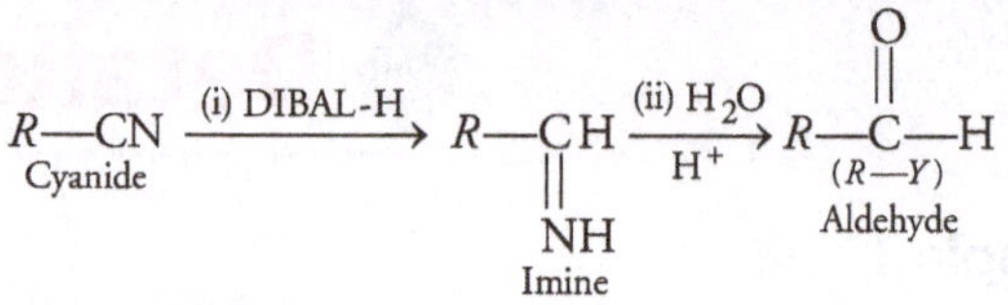

Hence, Y is —CHO group.

16. *(c)* On ozonolysis, unsaturated hydrocarbon X will form carbonyl compound, which gives positive Tollen's test. So, it should be aldehyde.

Butyne reacts with O_3 and Zn/H_2O to give 2-oxobutyraldehyde. The aldehyde will give Tollen's test to give silver mirror.

$$\underset{\text{Butyne}}{CH\equiv C-CH_2CH_3} \xrightarrow[Zn/H_2O]{O_3} \underset{\text{2-oxobutyraldehyde}}{OHC-\overset{O}{\overset{\|}{C}}-CH_2-CH_3} \xrightarrow[\text{reagent}]{\text{Tollen's}} \text{Silver mirror}$$

Note Tollen's reagent contain silver nitrate ($AgNO_3$), ammonia and sodium hydroxide (NaOH) which react with aldehydes to give Ag (silver mirror).

Only product of option (c) contains aldehyde group.

17. *(b)* DIBAL-H is added to produce aldehyde from ester by keeping the temperature low while at higher temperature it can perform different conversions.

Complete reaction is as follows:

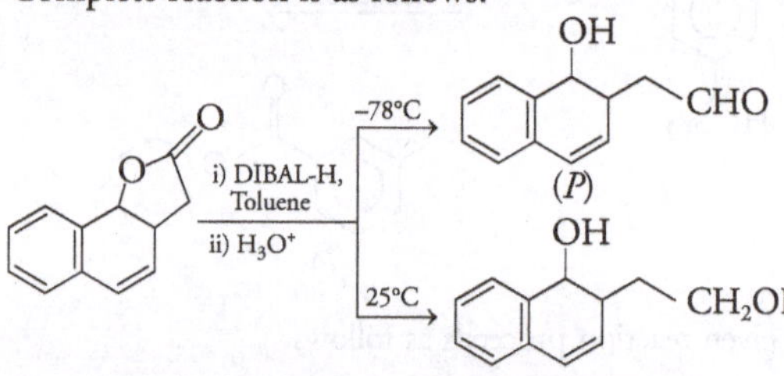

Note DIBAL-H does not reduce double bond.

18. *(a)* The major product of the reaction is $CH_3CH_2CH_2CH_2CHO$. Here, electrophilic addition of $H-\overset{+}{C}=O$ (formylation) take place to the alkene through Markownikoff addition.

(i) $H_2 + \underset{\text{Water gas}}{:C\equiv\ddot{O}:} \xrightarrow[\text{(Catalyst)}]{Rh} [H-\overset{\oplus}{C}=O]\,H^{\ominus}$

(ii) $\underset{\text{But-1-ene}}{CH_3CH_2-CH=CH_2} \xrightarrow[\text{Slow}]{H-\overset{\oplus}{C}=O}$

$CH_3CH_2-\overset{\oplus}{C}H-CH_2-CHO \xrightarrow[\text{Fast}]{H^{\ominus}}$

$\underset{\text{Pentanal}}{CH_3CH_2CH_2CH_2CHO}$ (major)

19. *(b)* $\underset{\text{Ethanol}}{CH_3CH_2OH} \xrightarrow{CrO_3-H_2SO_4} \underset{\text{Ethanoic acid}}{CH_3-\overset{O}{\overset{\|}{C}}-OH}$

Since, $CrO_3 \cdot H_2SO_4$ behave as strong oxidising agent and it converts alcohol directly to carboxylic acid. Thus, reaction (b) will not form acetaldehyde.

20. *(b)* Molybdenum oxide (Mo_2O_3) is used for oxidising alkanes to aldehyde. It is used to manufacture molybdenum metal, which serves as an additive to steel and corrosive resistant alloys.

$$\underset{\text{Propane}}{CH_3-CH_2-CH_3} \xrightarrow[\text{Oxidising agent}]{Mo_2O_3} \underset{\text{Propanal}}{CH_3-CH_2-CHO}$$

21. *(a)* The reaction involved and the major product is given below

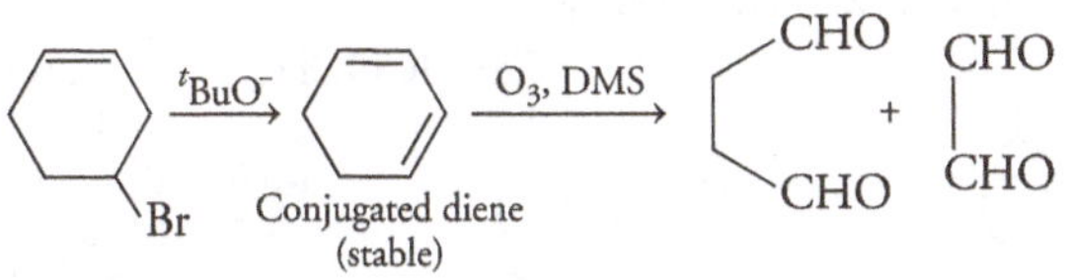

22. *(d)* For the reaction,

$$\text{Carbonyl compound} + \text{MeOH} \underset{}{\overset{HCl}{\rightleftharpoons}} \text{Acetal}$$

Rate of reaction is highest for propanal as substrate and methanol in excess. Propanal is an aldehyde and more reactive than ketones. When MeOH is taken in excess then reaction moves in the forward direction that results in the formation of acetal. Reaction involved is as follows :

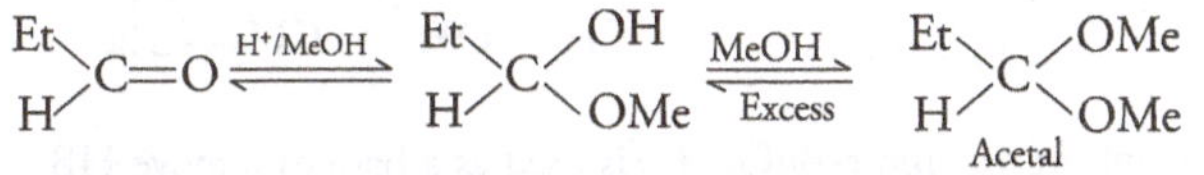

23. *(b)* The question is related to above reaction only with the condition that the consumption of *RMgX* will be more than 1 equivalent in some of the given cases.

Among the given compounds *B*, i.e.

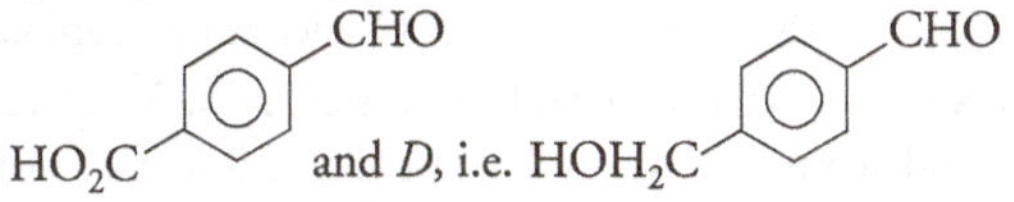

contain additional groups which can give active hydrogens.

24. *(b)* The complete reaction sequence involved is as follows.

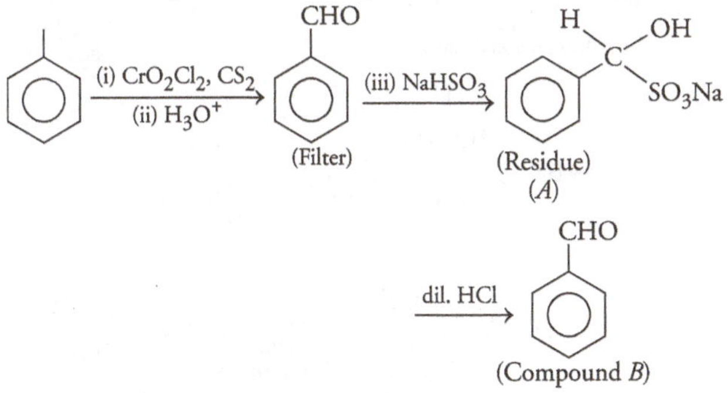

Thus option (b) is the correct answer.

25. *(d)* The correct match is A-IV, B-III, C-I, D-II

(A) $RCN \xrightarrow[\text{(ii) } H_3O^+]{\text{(i) } SnCl_2, HCl} RCHO$

(Stephen reaction)

(B)

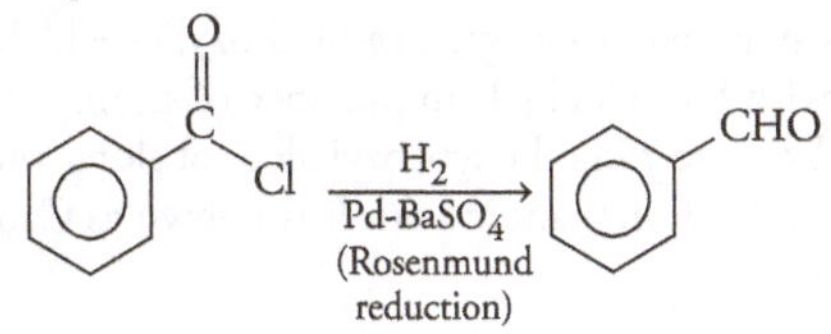

(C)

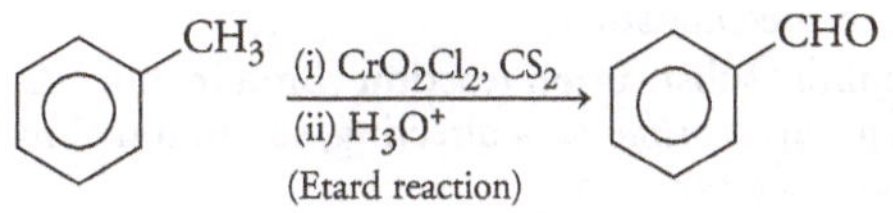

(D)

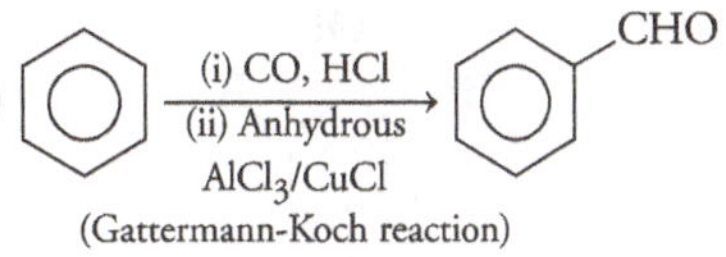

26. *(a)* The complete reaction is as follows.

$$CH_3-C\equiv CH \xrightarrow[H_2]{Pd/C} \underset{(A)}{CH_3-CH=CH_2} \xrightarrow[\text{(ii) } Zn, H_2O]{\text{(i) } O_3} \underset{(B)}{CH_3-CHO} + \underset{(C)}{HCHO}$$

27. *(d)*

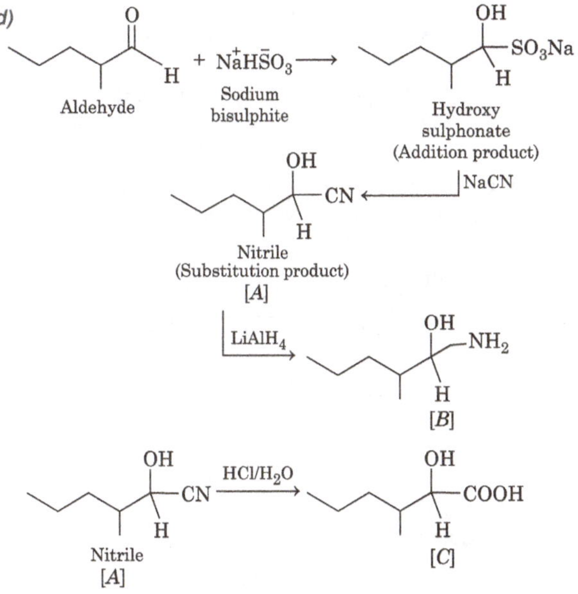

28. *(c)*

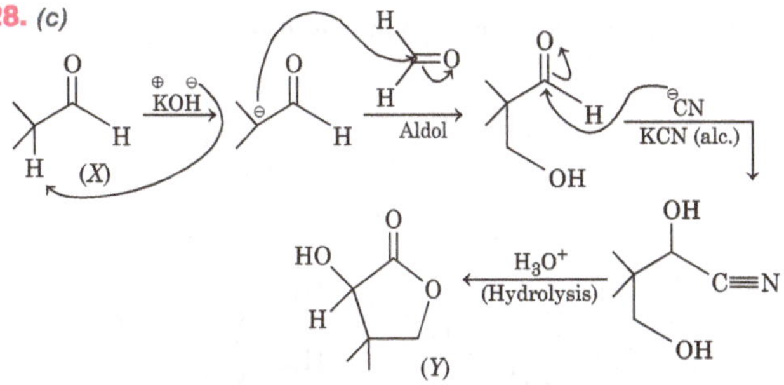

Compound [*X*] on reaction with —HCHO, KOH, gives aldol reaction. The product of aldol reaction upon reaction with cyanohydrin, gives nucleophilic addition reaction on $>C=O$ of —CHO group, upon reaction with H_3O^+; $-CN^-$ hydrolysis to $-CO_2H$ and a ring closing reaction takes place to give product [*Y*].

29. *(d)* **Reaction mechanism**

1. **Electrophilic substitution reaction** (Bromination) As —OH group is *ortho, para* directing, so bromination occurs at *ortho*-position.

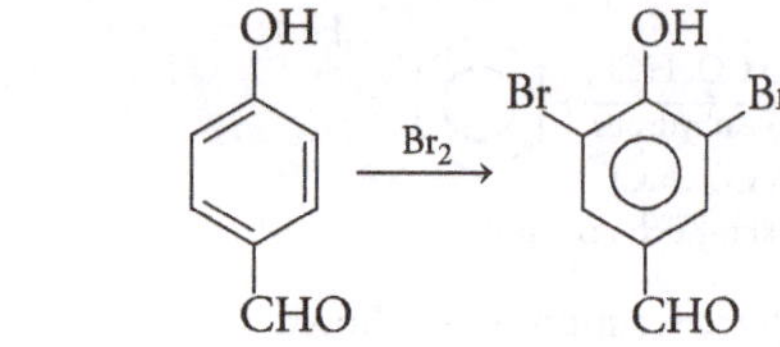

(A)

2. Hydroxylamine (NH_2OH) reacts with carbonyl group to form oximes.

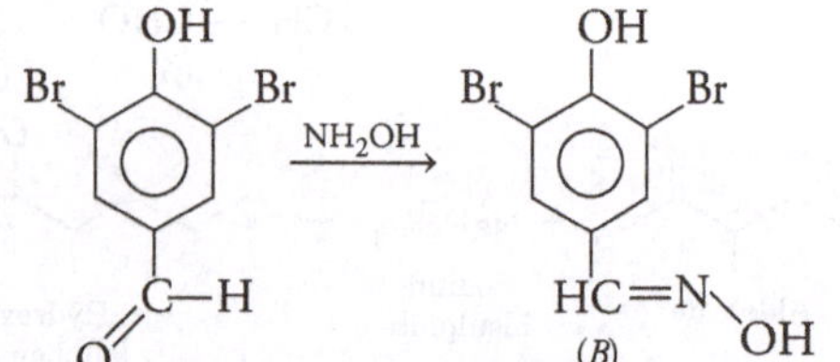

(B) (Oxime)

3. P_2O_5 is a dehydrating agent and upon heating, it removes water from —CH= NOH to form cyanide group.

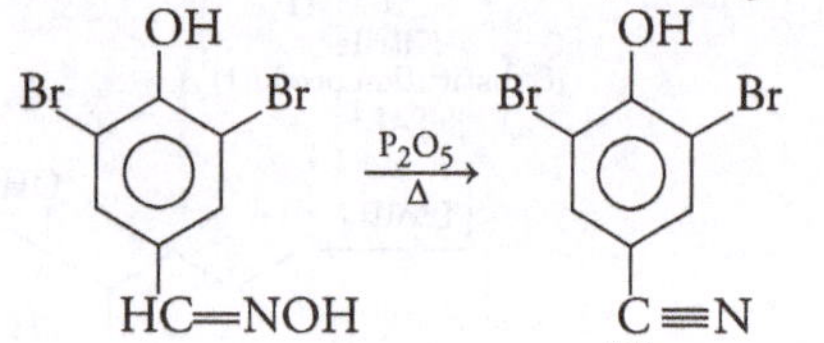

(C)

Hence, the final product *C* is

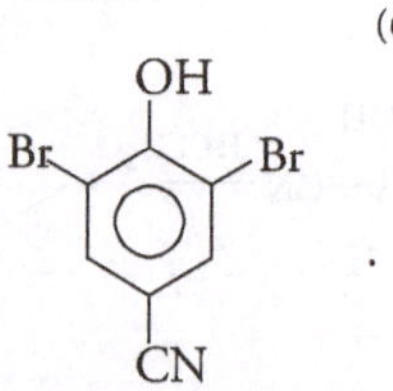

30. *(c)* The reaction sequence given is as follow :

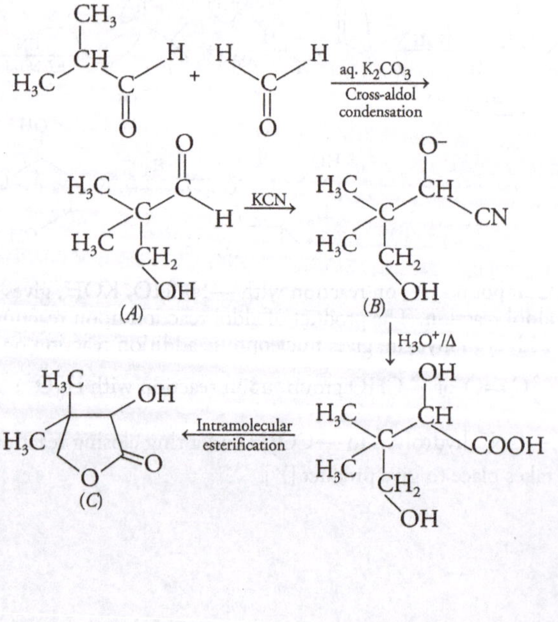

31. *(c)* Benzoic acid in presence of $SOCl_2$ and quinoline followed by reduction gives benzaldehyde.

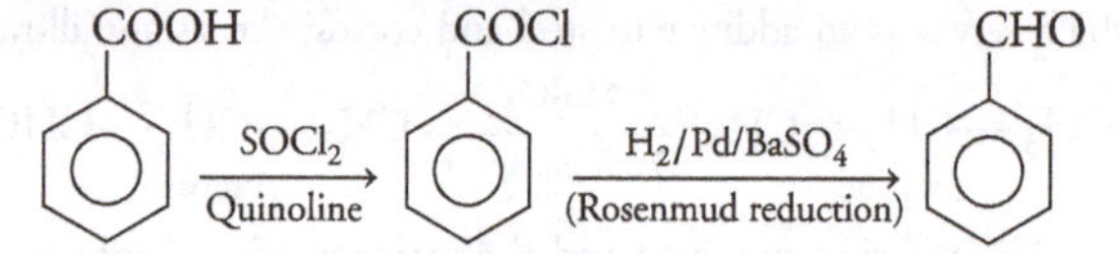

Toluene in presence of chromium oxide followed by hydrolysis gives benzaldehyde

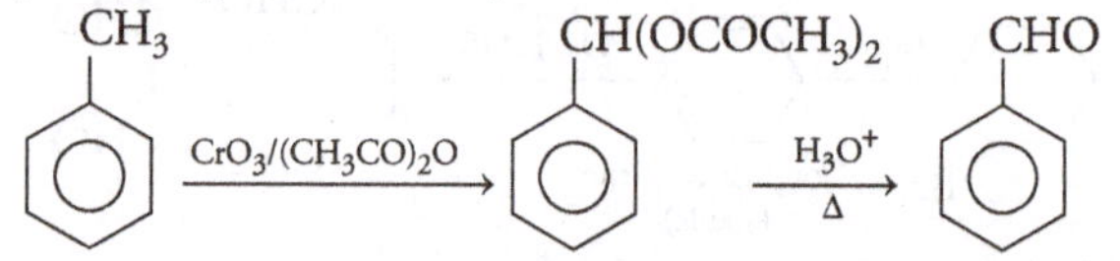

32. *(c)* The given compound does not form dicarboxylic acid on reaction with sodium hydroxide followed by hydrolysis in any case.

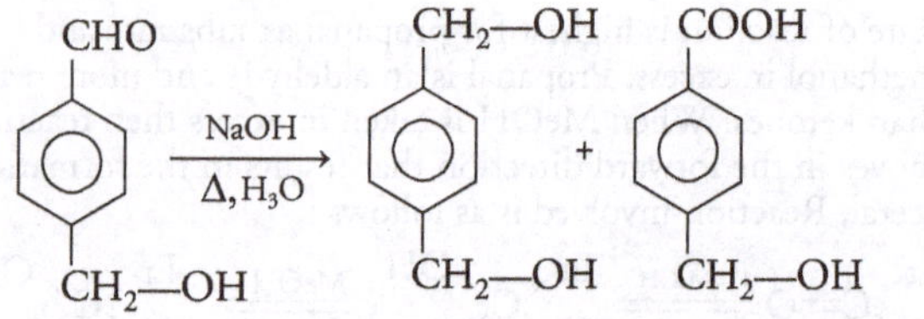

33. *(a)* In reaction t-BuO$^-$H$^+$ is used as a base to remove HBr and form alkene as a major product *(B) via* E2 elimination mechanism.

Mechanism

Firstly, 1-bromoacetaldehyde undergoes protonation which is then react with t-BuO$^-$H$^+$. (t-BuO$^-$ act as nucleophile and attack on carbonyl carbon). In next step, t-BuO$^\ominus$K$^+$ act as base and accepts proton and at last alkene is formed as major product.

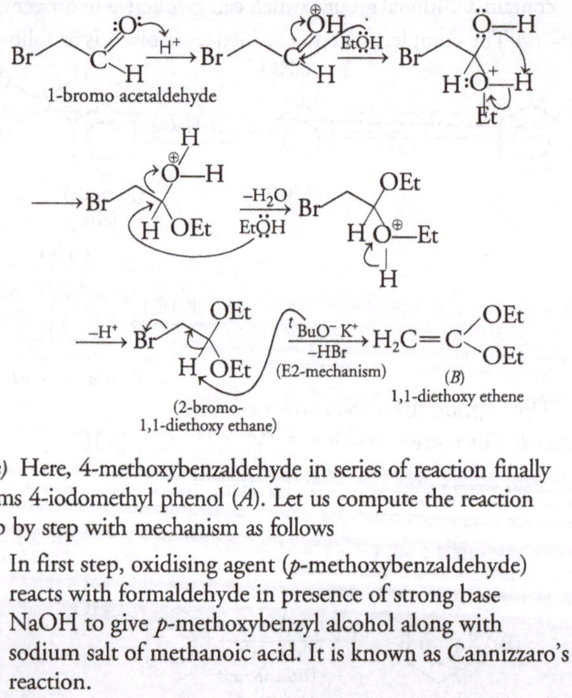

34. *(c)* Here, 4-methoxybenzaldehyde in series of reaction finally forms 4-iodomethyl phenol *(A)*. Let us compute the reaction step by step with mechanism as follows

(i) In first step, oxidising agent (*p*-methoxybenzaldehyde) reacts with formaldehyde in presence of strong base NaOH to give *p*-methoxybenzyl alcohol along with sodium salt of methanoic acid. It is known as Cannizzaro's reaction.

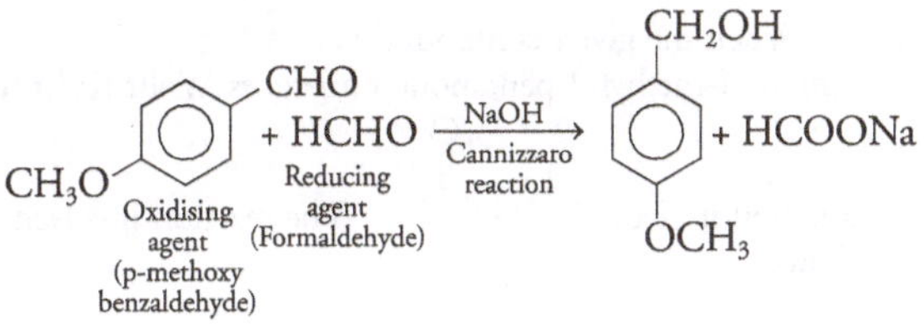

(ii) In second step, deprotonation of *p*-methoxybenzyl alcohol in presence of sodium hydride (NaH) to form alkoxide which further react with bromoethane in presence of DMF to give 1-ethoxymethyl-4-methoxybenzene.

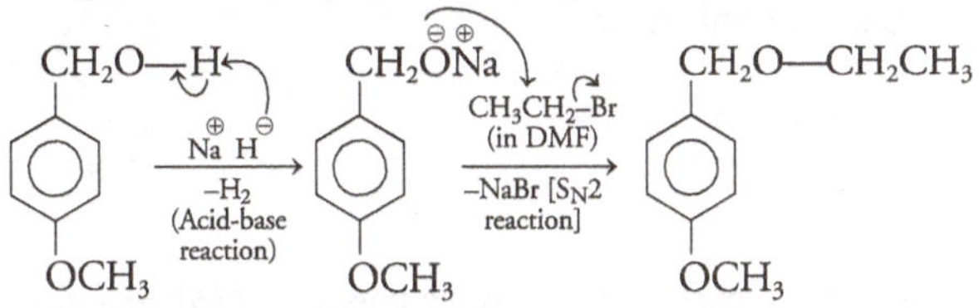

(iii) In last step, 1-ethoxymethyl-4-methoxybenzene undergoes solvolysis reaction followed by S_N2 attack of two iodide ion (I^-) to give 4-iodomethyl phenol, iodoform and iodoethane.

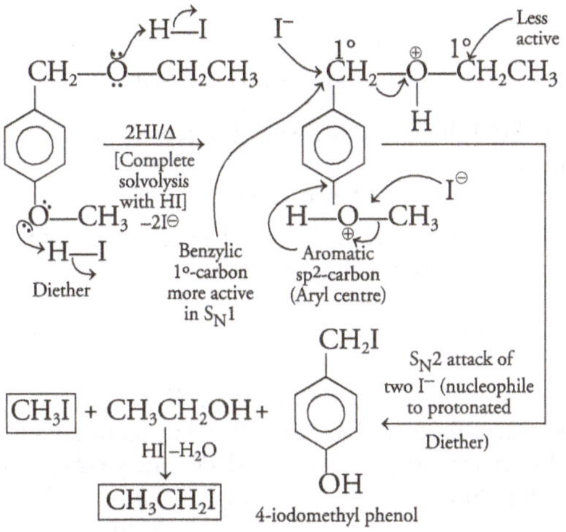

35. *(c)* It is an example of intramolecular aldol condensation. α - hydrogen atom of one — CHO group gets abstracted by NaOH form enolate ion which then attacks of another — CHO group to form aldol or β - hydroxy aldehyde. The aldol on heating with C_2H_5OH/H_2O forms the final product (α, β-unsaturated aldehyde) as the major product.

Mechanism

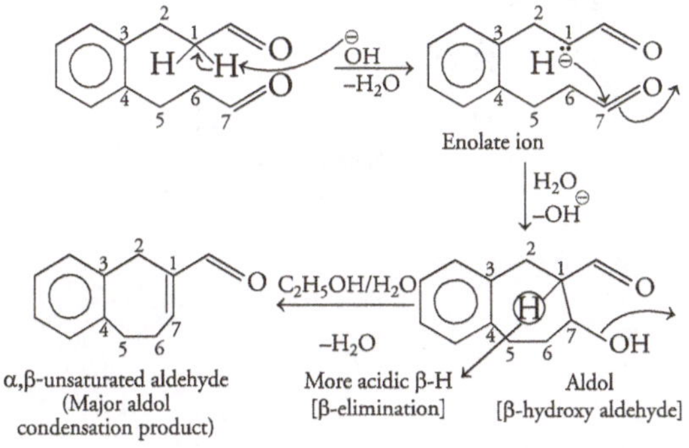

36. *(d)* The complete reaction take place as follows

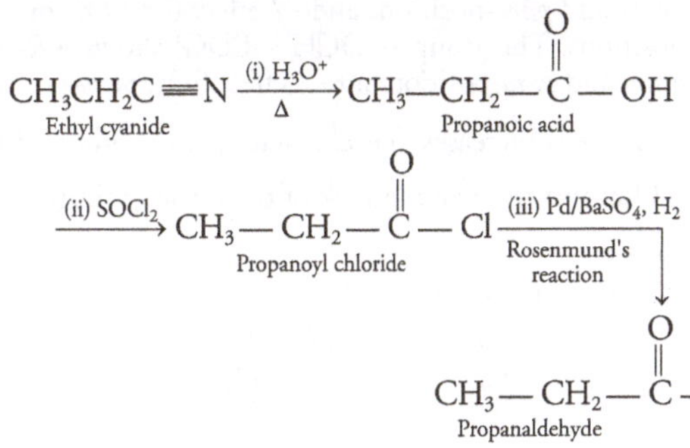

Final product of reaction is propanaldehyde.

37. *(c)*

(I) Given, reaction is Gattermann -Koch reaction. It involves the formation of an aryl aldehyde using CO and HCl as reactants. The reaction catalysed by $AlCl_3$.

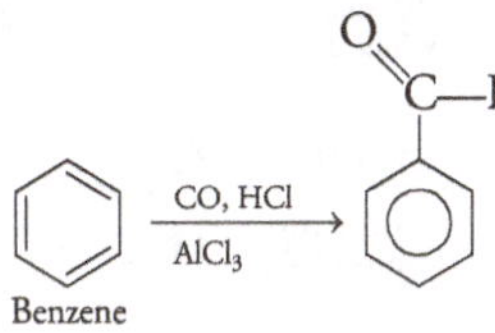

(II) Given reaction is Stephen reduction reaction. It involves the preparation of aldehydes from nitriles using tin (II) chloride, hydrochloric acid and quenching the resulting minimum salt with water.

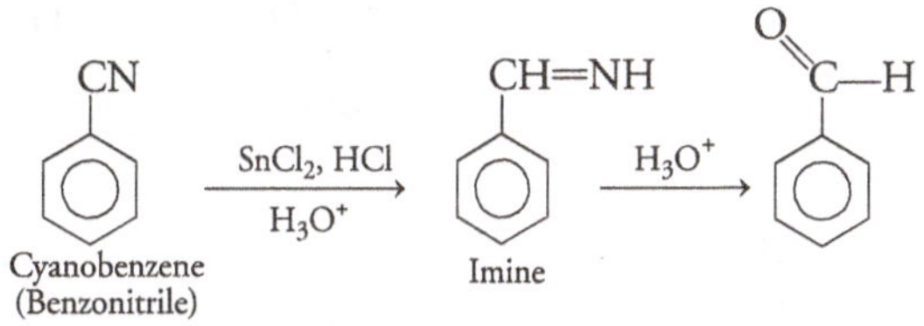

(III) Given reaction is Rosenmund reduction. It involves hydrogenation process in which an acyl chloride is selectively reduced to an aldehyde.

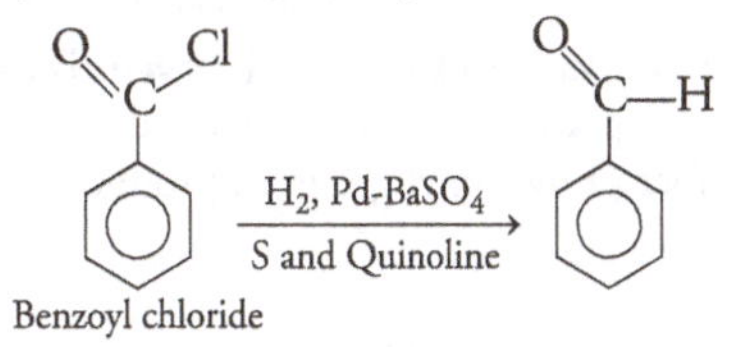

Correct match of Item-I and Item-II will be (I) - (R), (II) - (P) and (III) - (Q).

38. *(d)* Rate of nucleophilic addition ∝ Polarity of the $>\overset{\delta+}{C}=\overset{\delta-}{O}$ group ∝ Attachment of electron withdrawing group (EWG), i.e., $-R$ and $-I$ groups, (where $-R > -I$)

$$\propto \frac{1}{\text{Attachment of electron donating group (EDG) (i.e., } +R \text{ and } +I \text{ groups)}}$$

In substituted benzaldehyde, —NO_2 (EWG) shows $-R$ effect from *ortho*-and *para*-positions and $-I$ effect (only) from *meta*-positions. The group —OCH_3 (EDG) shows $+R$ effect from *ortho*-and *para*-position but $-I$ effect from *meta*-positions.

EWG ($-R > -I$) increases 'δ+' character of sp^2 carbon of the >C=O group and hence attack of the nucleophile (CN^-) will be easier.

So, the increasing order of HCN addition is :

(iii) < (i) < (iv) < (ii)
(+R) (−I) (−I) (−R)

− I effect : $NO_2 > OCH_3$

39. *(c)*

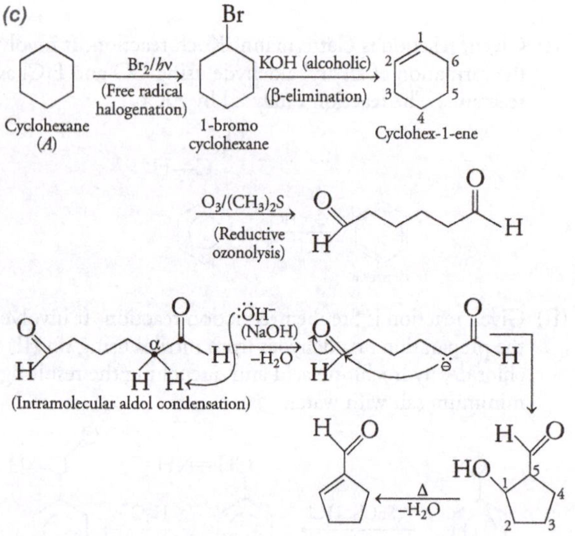

40. *(0)*

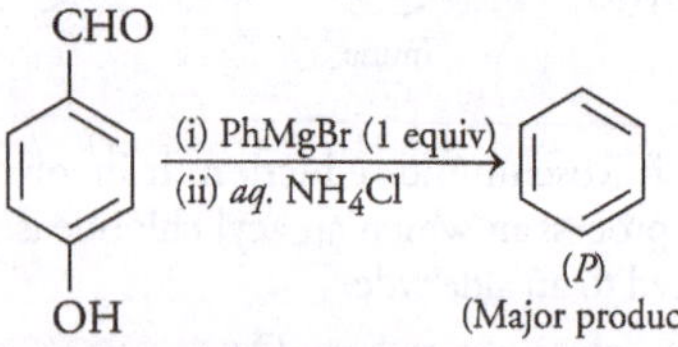

Since only 1 equivalent of Grignard reagent is used it will attack OH group and benzene is obtained as product *P*. So, the number of hydroxyl group in product *P* is zero.

Topic 2

01. *(b)* The given reaction is a Clemmensen reduction. The product formed is as follows

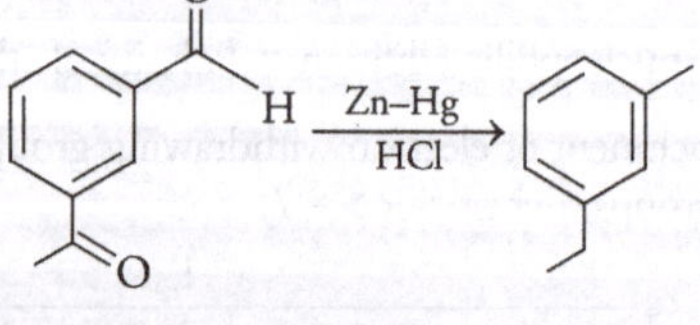

02. *(a)* When the given compound i.e. 3-chloro-4-methyl-2-pentanone undergoes Wolff Kishner reduction it gives 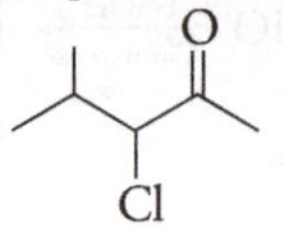 . The reaction involved is as follows.

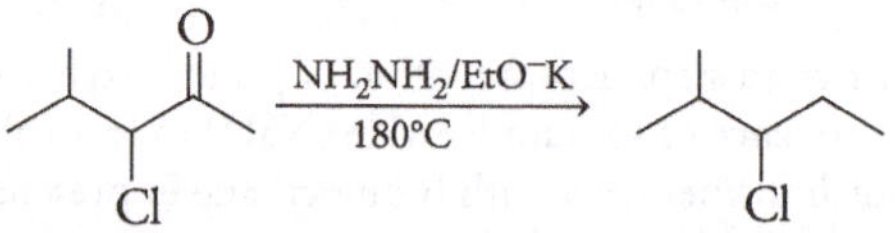

Concept Enhancer

In Wolff-Kishner reaction the carbonyl group of aldehydes and ketones is reduced to >CH_2 group on treatment with hydrazine followed by heating with sodium or potassium hydroxide in high boiling solvent like ethylene glycol.

03. *(d)* Deprotonation will be easily possible for (*A*) because it is more resonance stabilised.

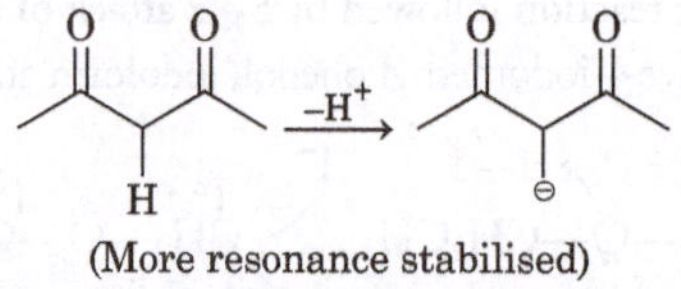

(*B*) and (*C*) shows less tendency for deprotonation due to less conjugation.

Hence, option (d) is the correct answer.

04. *(c)* Conjugated diketone has two ketone group directly attached to carbon atom containing double bond.

Hence, 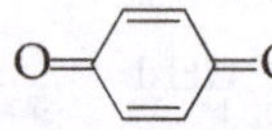 is a conjugated diketone.

05. *(c)* Statement I is true but statement II is false.

The correct form is : Nucleophilic addition of HCN (hydrogen cyanide) to an aldehyde/ketone yield cyanohydrin as final product.

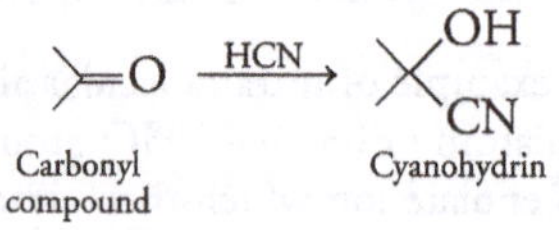

06. *(b)* Both A and R are true and R is the correct explanation of A.

07. *(c)* Option (c) is the major product formed.

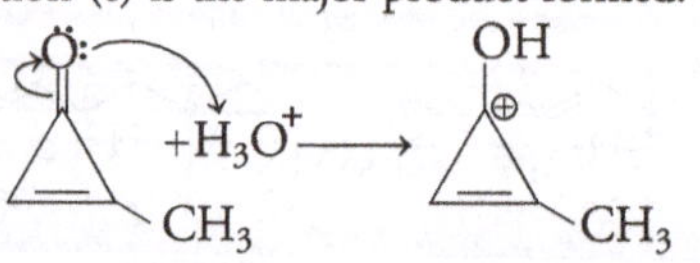

Aromatic product

08. *(d)* The complete reaction is as follows,

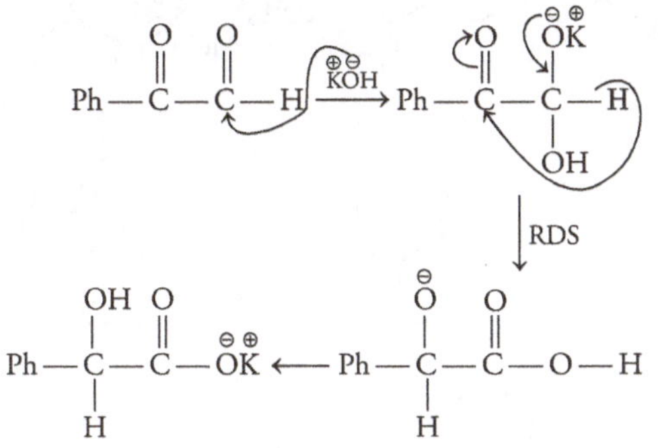

[P]

09. *(a)*

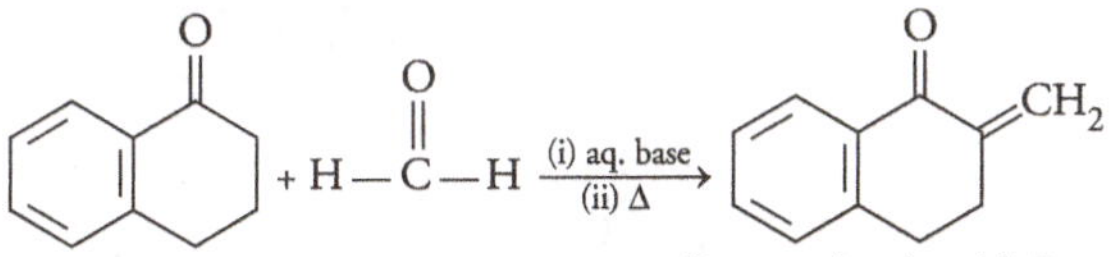

(Intermolecular aldol)

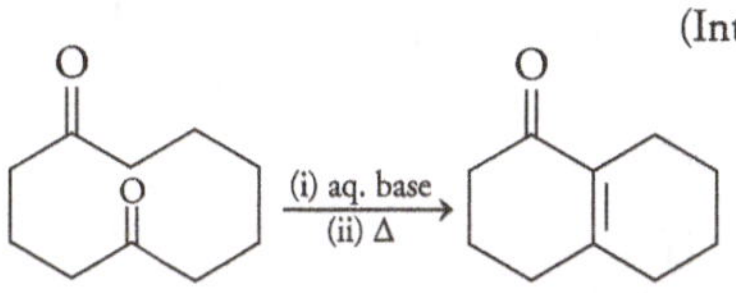

(Intramolecular aldol)

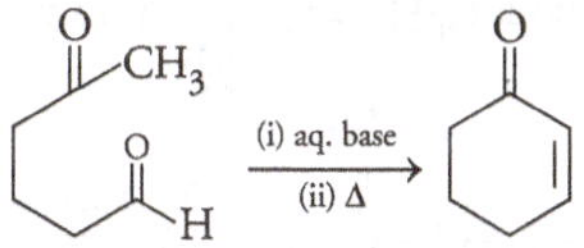

(Intermolecular aldol)

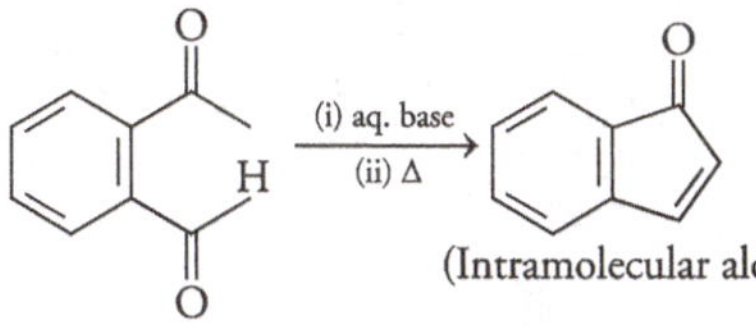

(Intramolecular aldol)

10. *(b)* The complete reaction is as follows,

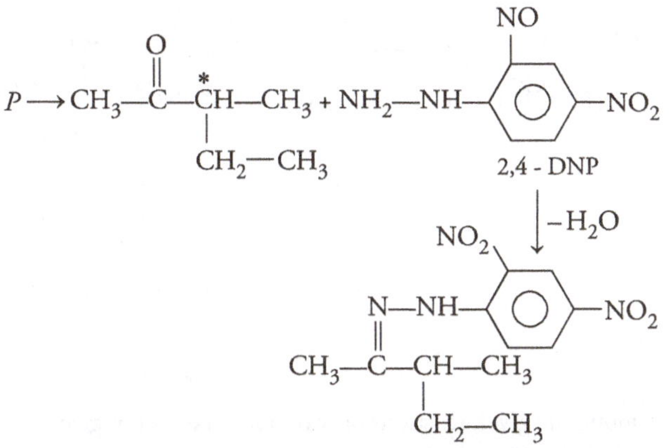

All ketones except α-hydroxy ketones chiral does not give Tollen's test.

Gives test with 2,4-DNP.

11. *(d)* Statement I is correct but Statement II is incorrect. The correct form of Statement II is reaction between benzaldehyde and ethanal will give cross-aldol product.

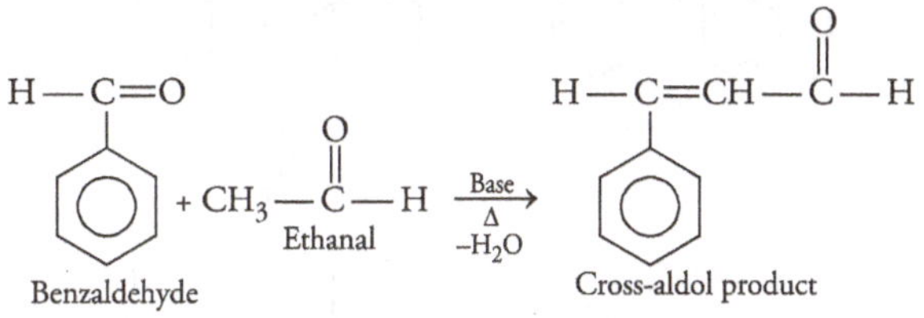

12. *(a)* The reaction sequence involved is as follows

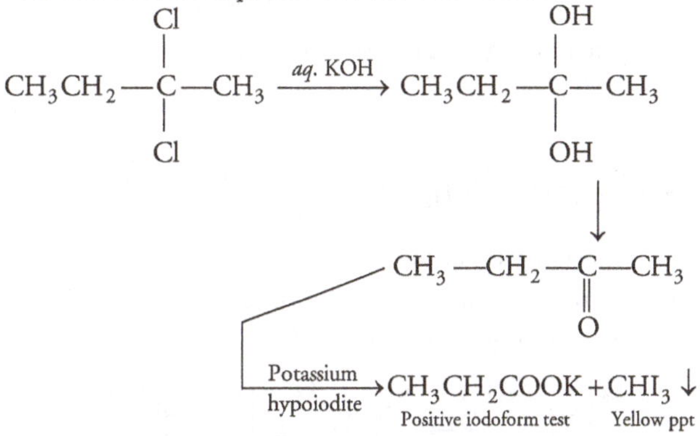

Concept Enhancer

Positive iodoform test can be given by those compounds which have $CH_3-\overset{O}{\overset{\|}{C}}-$ and $CH_3-\underset{|}{\overset{OH}{\overset{|}{C}H}}$ group or their halogen derivatives in which halogen is substituted from $-CH_3$ group.

13. *(d)* The complete reaction takes place as follows

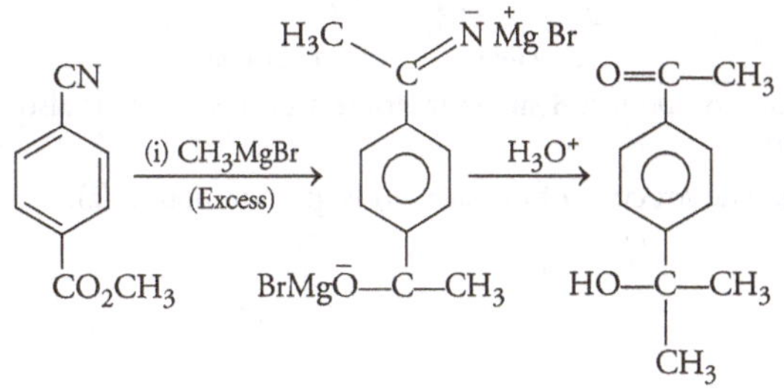

14. *(d)* The complete reaction is as follow,

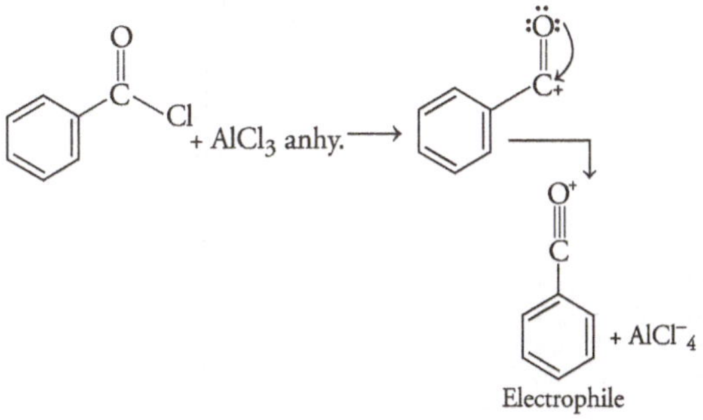

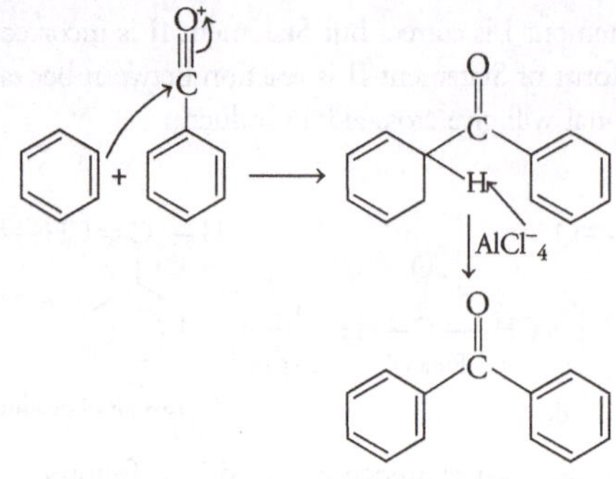

(Major product *P*)

15. ***(a)*** Among the given options, option (a) has most acidic hydrogen.

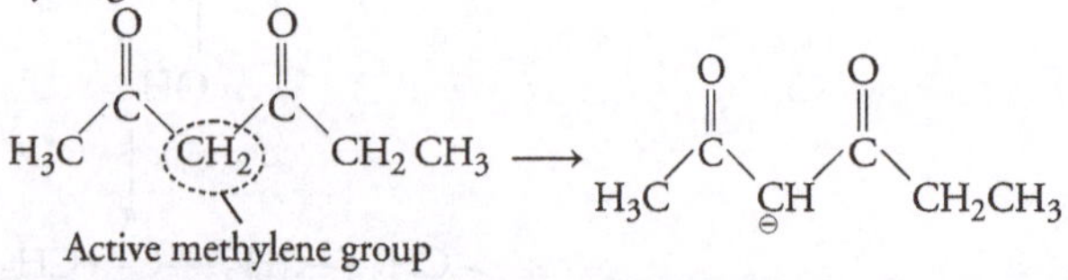

Conjugate base is more stable due to resonance.

16. ***(b)*** Statements given in A, C and D are correct.

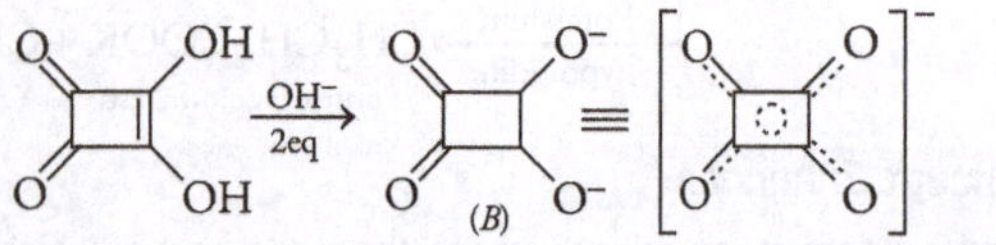

Resonance hybrides of *B*

The above structures shows that *B* is an aromatic compound. Since it follows Huckel rule. Also, the resonance hybrid of *B* shows that all C—C bond length are same. Thus, A and D are correct. Now,

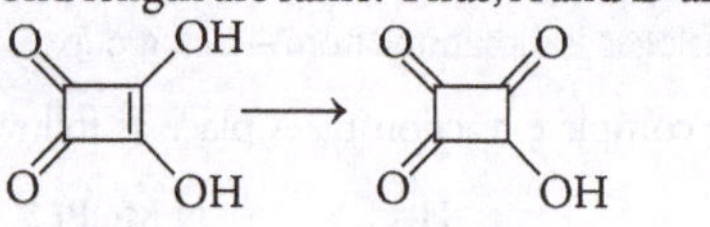

Enol form Keto form

Thus, compound *A* shows tautomerism. Hence, C is also correct.

17. ***(a)*** The structure of product (*A*) is given in option (a).

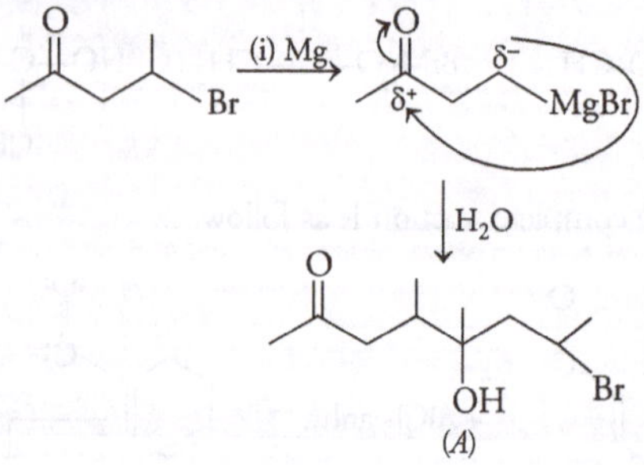

18. ***(a)***

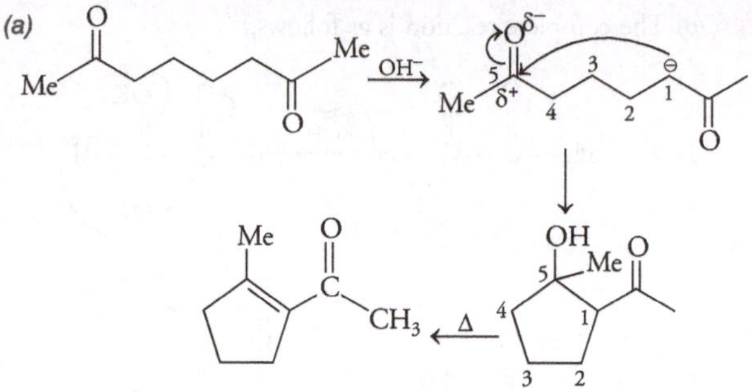

19. ***(c)***

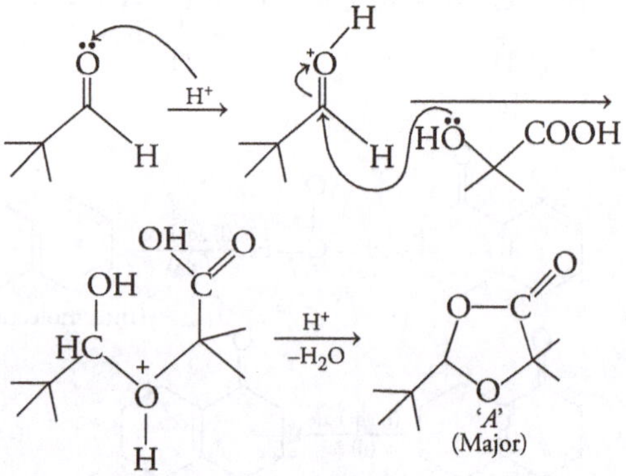

20. ***(d)*** Benzoate radical, $C_6H_5—C(=O)\ddot{O}\cdot$ is formed as intermediate.

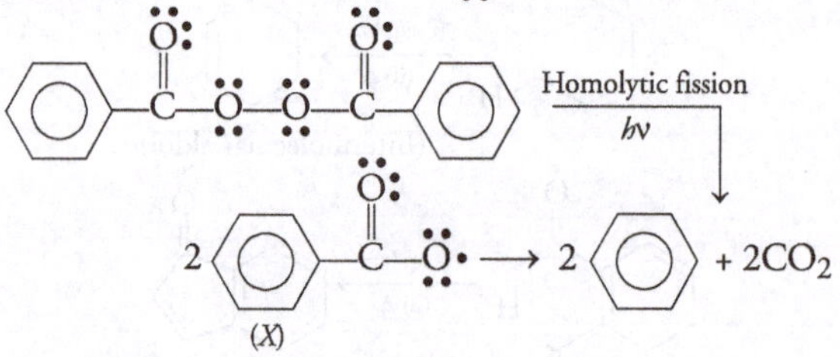

(*X*)

Hence, *X* is $C_6H_5—\overset{O}{\overset{\|}{C}}—\dot{O}$:

21. ***(c)***

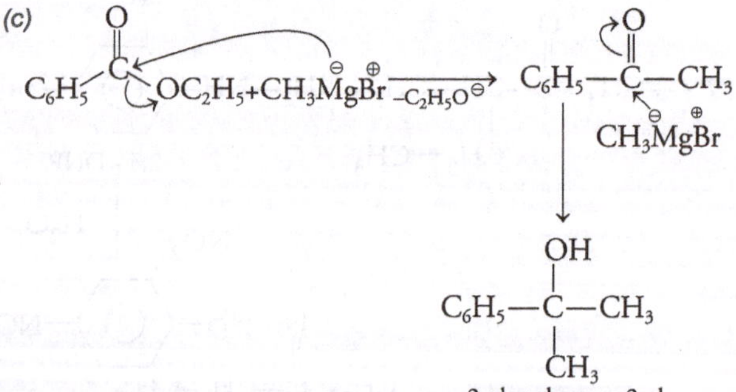

2-phenylpropan-2-ol

Among the given reactions only option (c) does not give acetophenone as the major product.

22. *(c)* Enamine formation takes place by nucleophilic addition of 1°/2° amines to the carbonyl group.

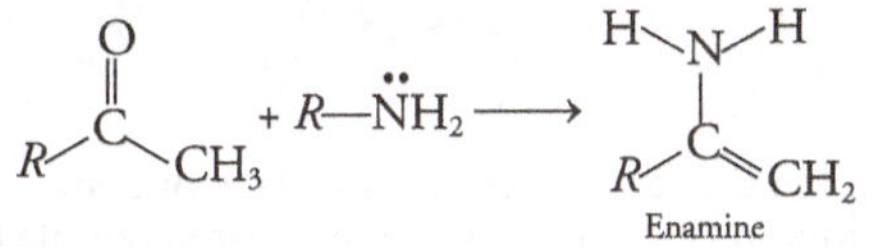

As t-Bu—C(=O)—t-Bu has no α—H and also it has more steric crowd around the carbonyl group, so nucleophilic addition by 2° amine is not possible.

23. *(c)* The correct match is (c).

$A \rightarrow 2, B \rightarrow 3, C \rightarrow 4, D \rightarrow 1.$

24. *(a)* Acetophenone is a methyl ketone which on reaction with Br_2 and KOH will give conjugate base of an acid and methyl group will turn into haloform. This reaction is called haloform reaction.

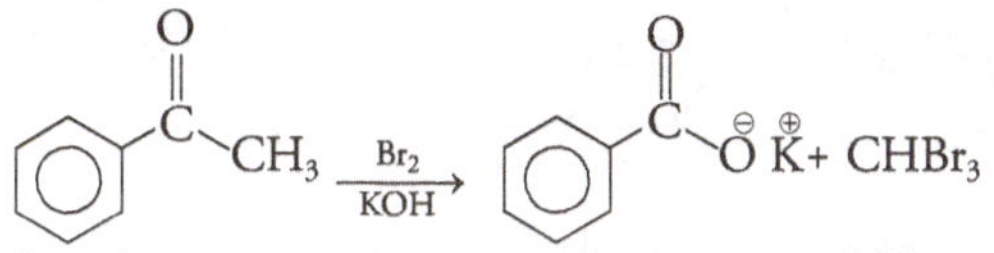

Acetophenone — (A) Conjugate base of acid — (B) Haloform (Bromoform)

25. *(d)* 2, 4-dinitrophenylhydrazine can be used for qualitative identification of ketone or aldehyde functional group.
Among the given options, only option (d) give orange precipitate when treated with 2,4-DNP.

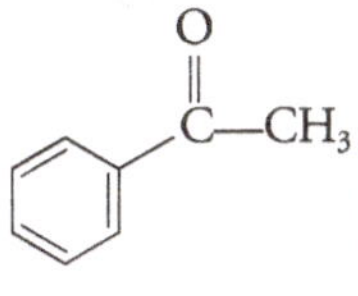

– Presence of ketonic group, so it will give orange ppt. when treated with 2, 4-DNP.

26. *(a)* Treatment of benzonitrile with Grignard reagent followed by acid hydrolysis gives acetophenone product.

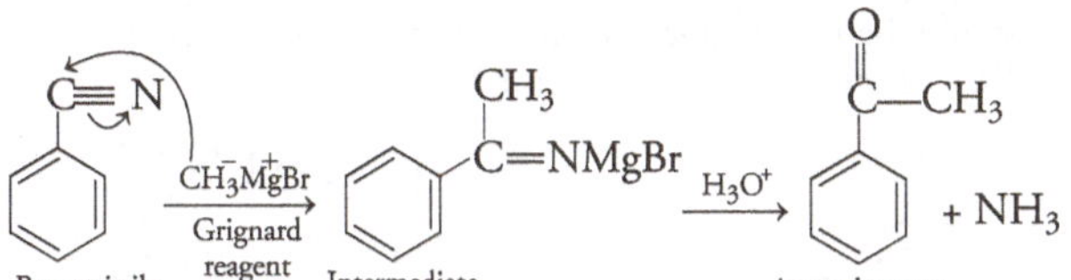

Benzonitrile — Intermediate in ionic form — Acetophenone (It gives positive iodoform test)

Iodoform test is used to check the presence of carbonyl compound with the structure R—CO—CH_3 or alcohols with the structure R—CH(OH)—CH_3 in a given unknown substance.

This conclude that acetophenone give iodoform test.

27. *(c)* Only compound (II) which has ketonic group does not form silver mirror when treated with Tollen's reagent.

Because Tollen's reagent oxidises an aldehyde into the corresponding carboxylic acid. Ketones are not oxidised by Tollen's reagent, so the treatment of a ketone with Tollen's reagent in a glass test tube does not result in a silver mirror.
Aldehydes give positive Tollen's test (silver mirror test).

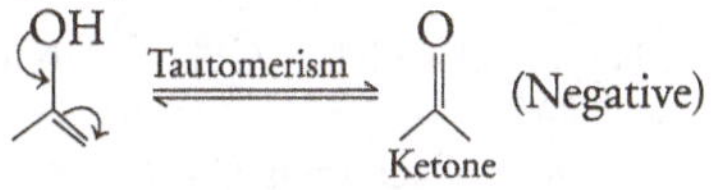

28. *(b)* Nucleophilic addition of ethylene glycol in ketone to give ketal.

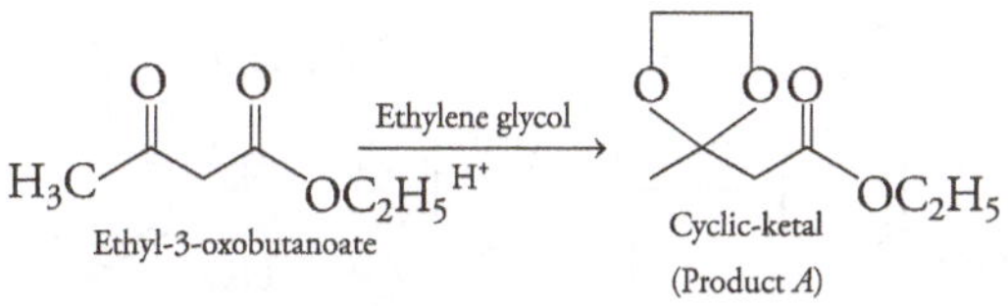

Addition of one molecule of alcohol on ketone give hemiketal and addition of two molecules give ketal.

29. *(a)* The reaction is represented as

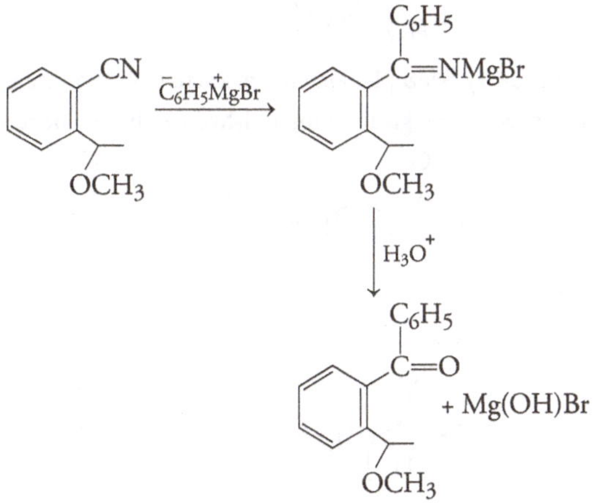

30. *(a)* Compound given in option (a) is formed. The steps involved are as follows:

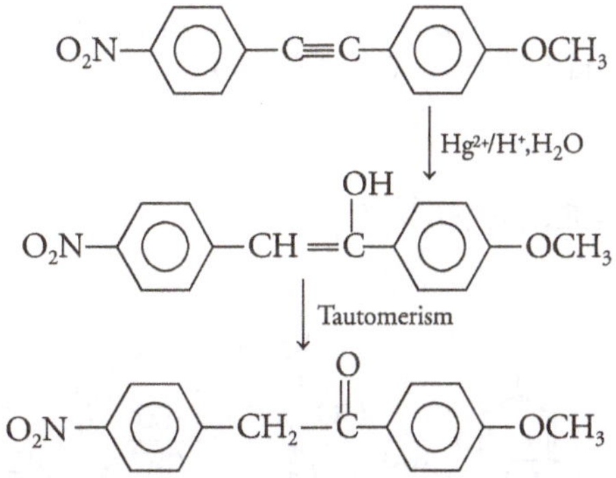

31. *(a)* Stability of conjugate determines the acidity of the given compounds

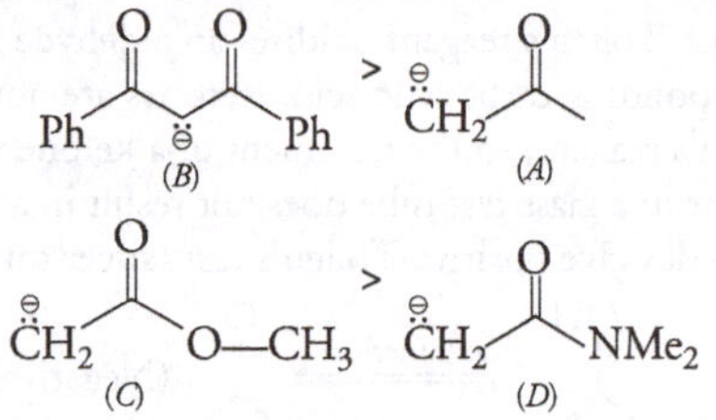

Thus increasing order of acidity is $D < C < A < B$.

32. *(b)* In the given reaction cyclisation occurs and the aliphatic alcoholic groups get converted to keto group as follows:

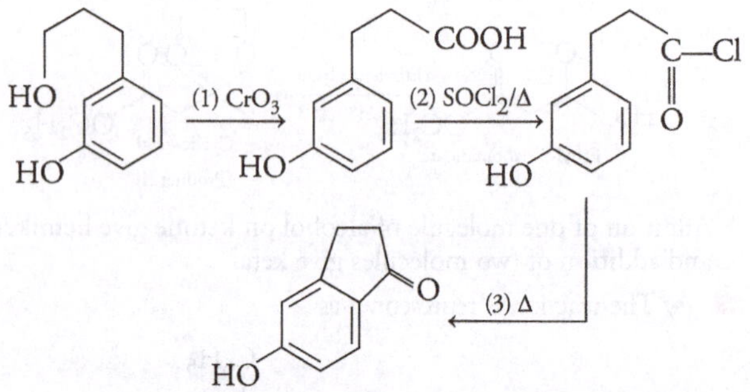

33. *(b)* *p*-hydroxy benzophenone upon reaction with bromine in carbon tetrachloride gives 3-bromo-4-hydroxy benzophenone.

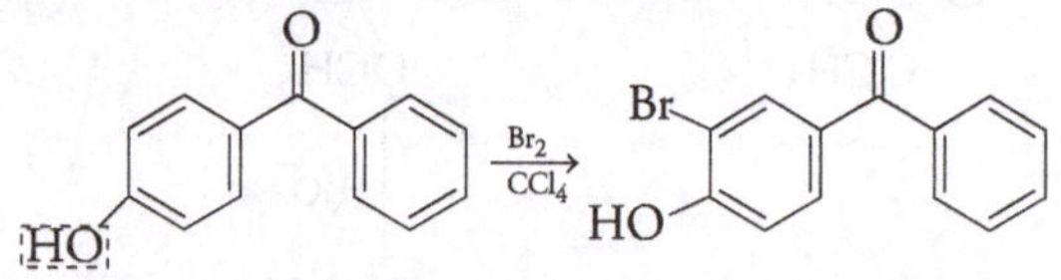

34. *(c)* It is an intermolecular aldol condensation reaction.

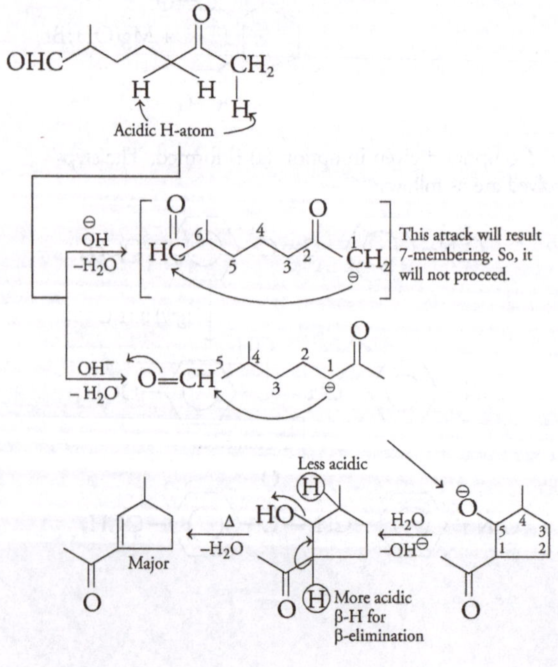

35. *(d)* Major product obtained in the given reaction is

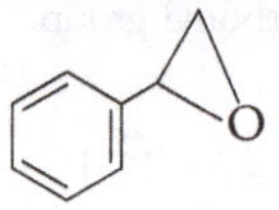

$NaBH_4$ in the reaction is used for the reduction by addition of a hydride ion and a proton. Carbon-oxygen double bonds are easily reduced by sodium borohydride. The actual reducing agent in these reductions is hydride ion (H^-). Hydride ion adds to the carbonyl carbon and the alkoxide ion that is formed is subsequently protonated by water. In other words, the carbonyl group is reduced by adding an H^- followed by an H^+. The mechanism of the given reaction is as follows :

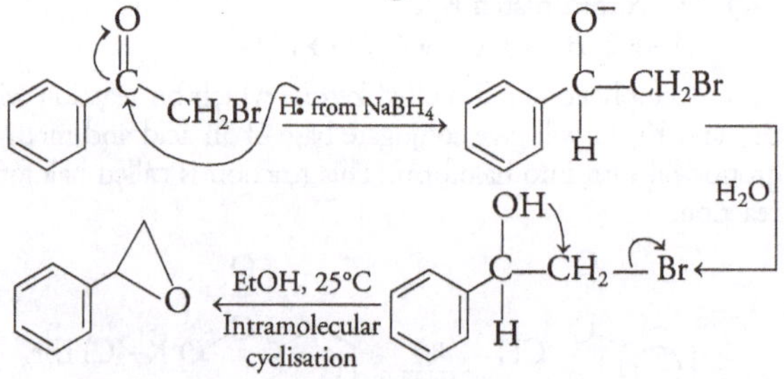

36. *(d)* The reaction can take place as follows

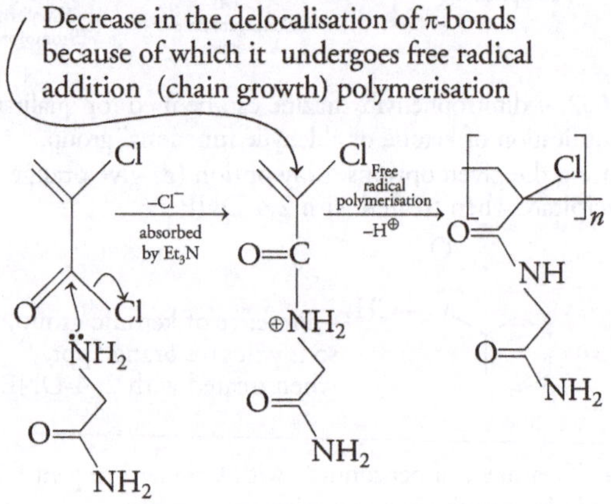

37. *(b)* The given process is cumene process (Hock process) to synthesise phenol and acetone industrially. In Hock process, Ph— group migrate and release H_2O.

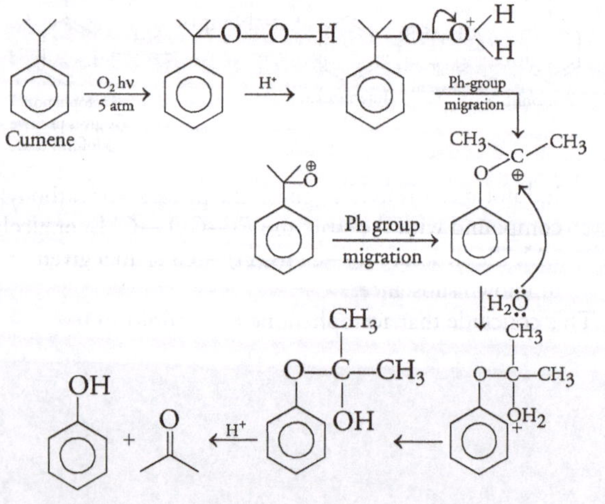

38. *(b)* The complete reaction is as follows,

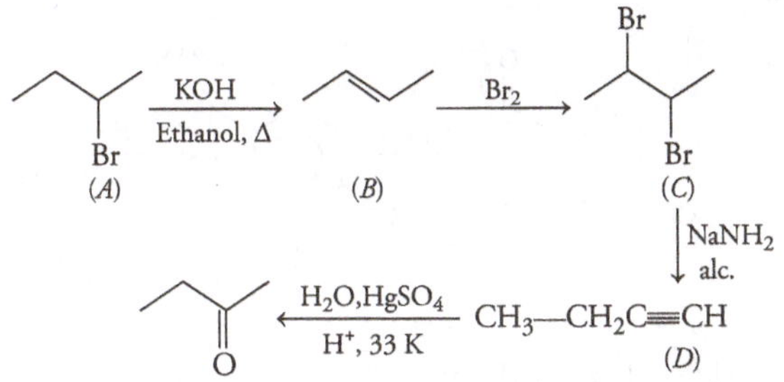

Butan-2-one
(E)

39. *(b)* Complete reaction sequence is as follows

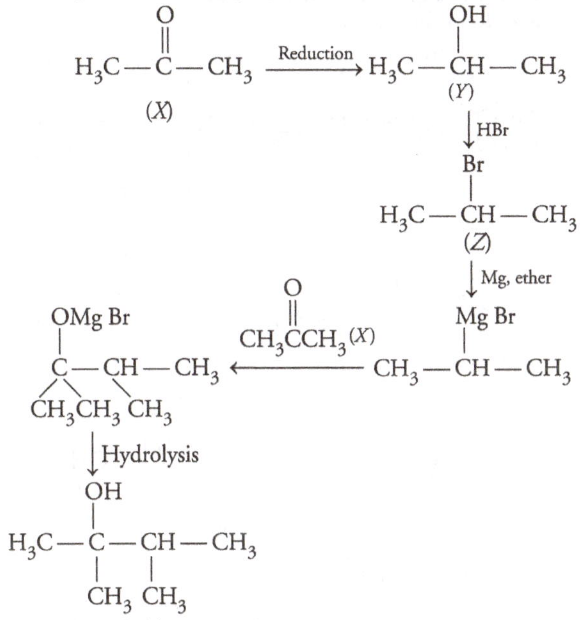

2,3 - dimethyl butan -2-ol

40. *(c)* The complete aldol condensation reaction is as follows,

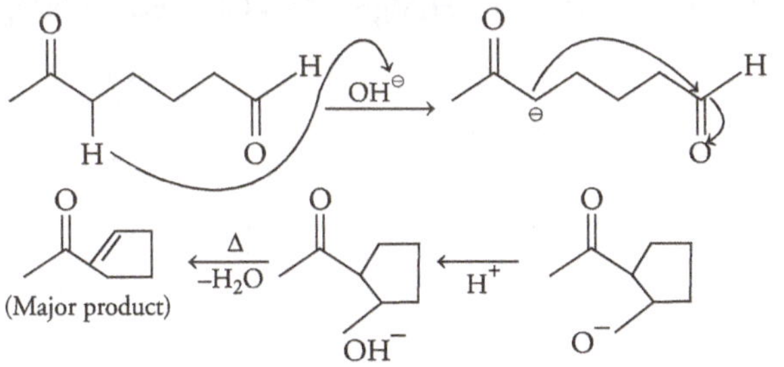

41. *(a)*

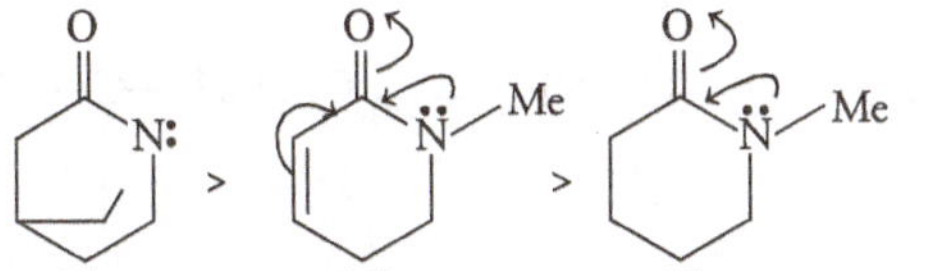

(R) It has delocalised lone pair according to Bredt's rule.
(Q) Cross conjugation
(P)

Thus, the correct order of basicity is $R > Q > P$.

42. *(a)* The correct order of hydrolysis is $p > q > r > s$

As we known that rate of hydrolysis is directly proportional to leaving group ability.

43. *(b)* Positive Fehling's test is given by aldehydes having alpha hydrogen.

Also, ketones generally do not react until they are α-hydroxy ketones.

So, among the given options, compounds given in C, D and E i.e.,

$HOCH_2—CO—(CHOH)_3—CH_2—OH$,

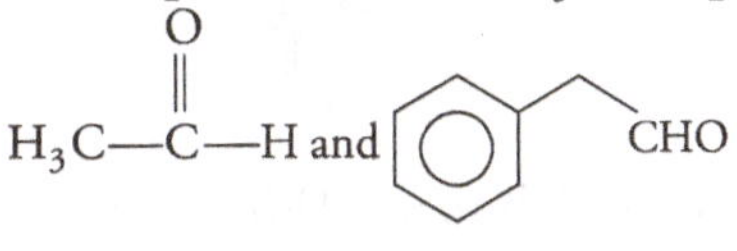

will give positive Fehling's test.

44. *(a)* The rate of nucleophilic addition reaction decreases due to steric crowding around carbonyl carbon and increases by electron withdrawing group if the steric crowding is same then the reactivity towards nucleophilic addition will be

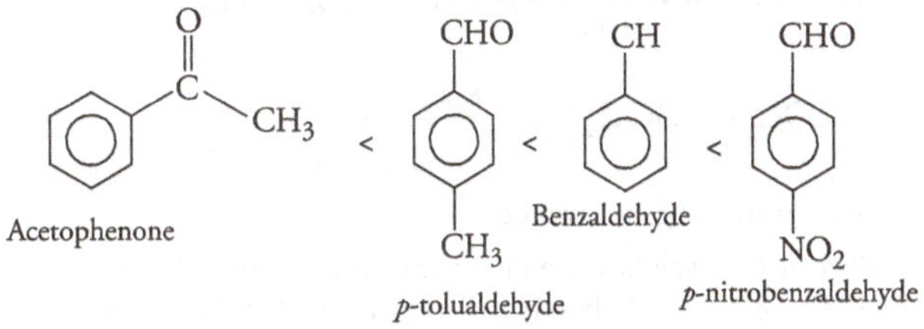

45. *(b)* Both acetaldehyde and acetone (individually) undergoes
(A) Iodoform reaction (C) Aldol condensation
(E) Clemmensen reduction

Name of reaction	Acetaldehyde	Acetone
Iodoform	(+ ve)	(+ ve)
Cannizzaro	(− ve)	(−ve)
Aldol condensation	(+ ve)	(+ ve)
Tollen's test	(+ ve)	(−ve)
Clemmensen reduction	(+ ve)	(+ ve)

Concept Enhancer

- Positive **Iodoform test** can be given by those compounds that have

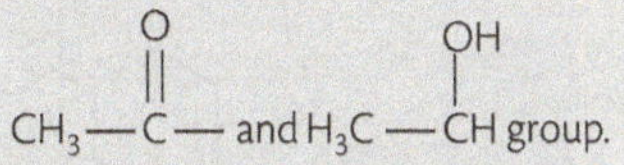

- **Cannizzaro reaction** Aldehydes that do not have an α-hydrogen undergoes self oxidation and reduction reaction of heating with concentrated alkali.
- **Aldol condensation** Aldehydes and ketones with at least one α-hydrogen undergo a reaction in the presence of dilute alkali as catalyst to form β-hydroxy aldehyde.
- **Tollen's test** It is a mild oxidising agent that converts aldehydes into corresponding carboxylate anion.

Clemmensen reduction In this reaction, carbonyl group of aldehydes and ketones is reduced to CH_2 group on treatment with zinc amalgam and concentrated hydrochloric acid

46. *(c)* The complete reaction sequence is as follows

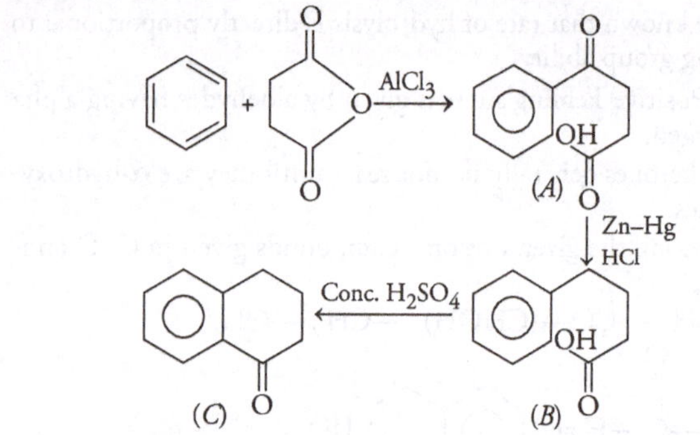

47. *(d)* The correct match is, A-IV, B-II, C-I, D-III.

The reactions are

$$CH_3(CH_2)_5COOC_2H_5 \xrightarrow[H_2O]{DIBAL\text{-}H} CH_3(CH_2)_5CHO$$

$$C_6H_5COC_6H_5 \xrightarrow{Zn(Hg)\text{ and Conc. }HCl} C_6H_5CH_2C_6H_5$$

$$C_6H_5CHO \xrightarrow[H_2O]{CH_3MgBr} C_6H_5CH(OH)CH_3$$

$$CH_3COCH_2COOC_2H_5 \xrightarrow{NaBH_4,\ H^+} CH_3CH(OH)CH_2COOC_2H_5$$

Hence, option (d) is correct.

48. *(d)* 3-methylcyclohex-2-enone reacts with Grignard's reagent in presence of copper iodide (CuI) to give '*A*'. '*A*' further reacts with *n*-propyliodide to give 3,3-dimethyl-2-propylcyclohexanone as final product.

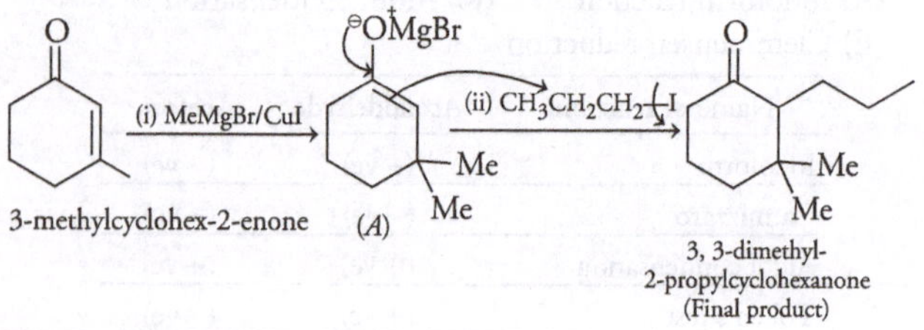

49. *(b)*

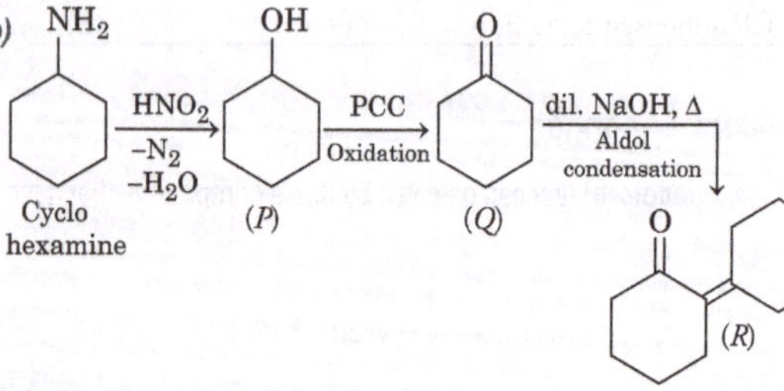

50. *(a)*

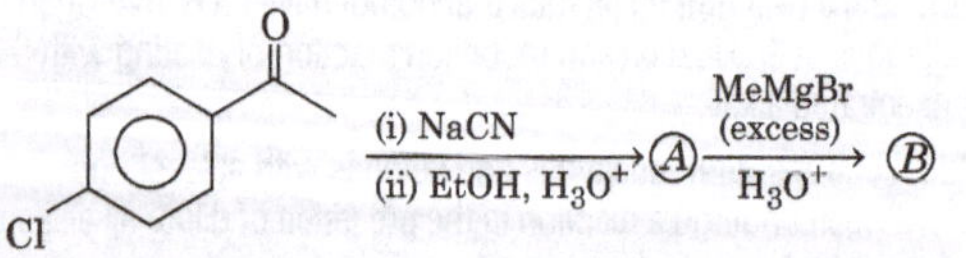

In step (i), nucleophile CN^- of NaCN will attack the electrophilic centre (δ^+) group of $>C=O$ and form cyanohydrin.

In presence of water (polar solvent) cyanide converts into acid.

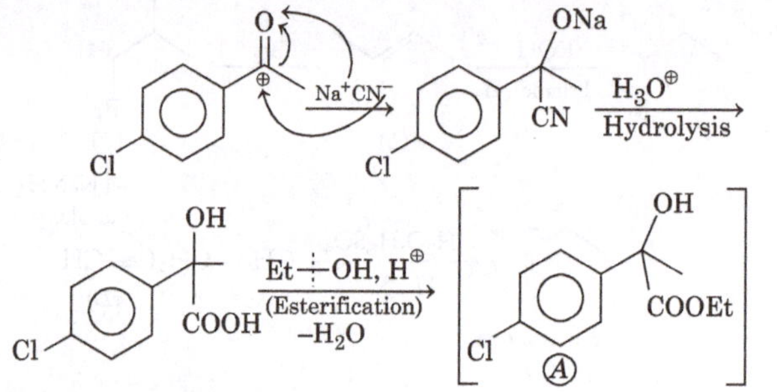

(*A*) on reacting with methyl magnesium bromide in excess shows following reaction.

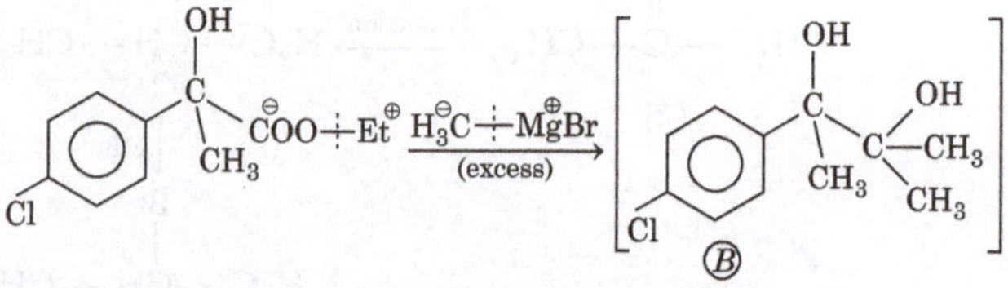

51. *(a)*

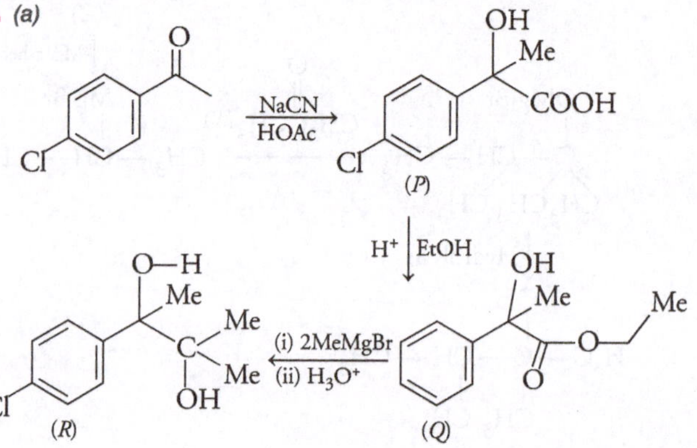

52. *(a)*

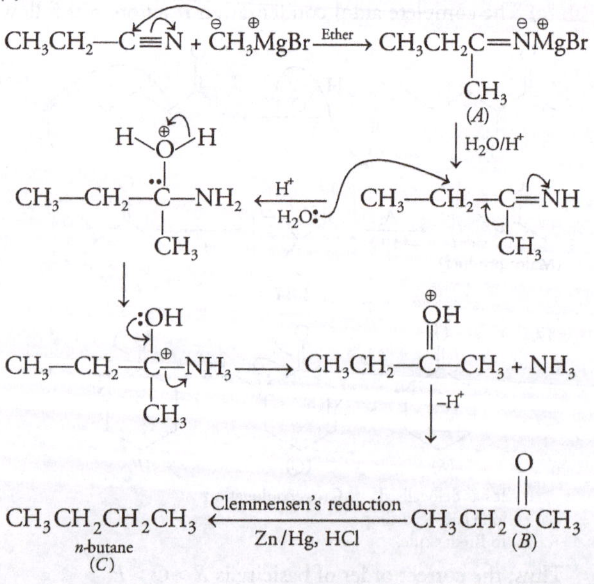

53. *(c)* The major product is given in option (c) which involves the following pathway.

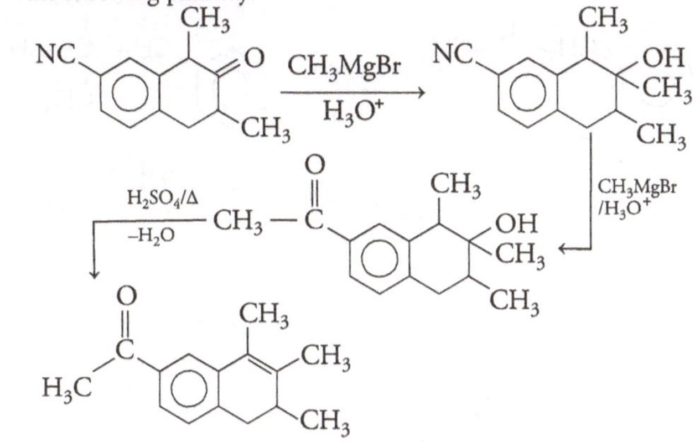

54. *(d)* The complete reaction take place as follows.

$CH_3—C{\equiv}CH + NaNH_2 \rightarrow CH_3—C{\equiv}\overset{\ominus}{C}Na^+ + NH_3$

Propyne (A) Acetylide ion

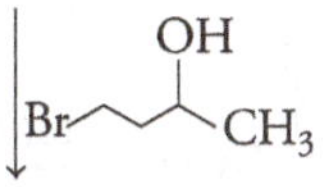

$CH_3—C{\equiv}C—CH_2—CH_2—\underset{OH}{CH}—CH_3$

(B) Hept-5-yn-2-ol

Reduction ↓ H_2 / Pd - C

$CH_3—CH_2—CH_2—CH_2—CH_2—\overset{OH}{CH}—CH_3$

(C) Heptan-2-ol (2° alcohol)

Oxidation ↓ CrO_3

$CH_3—(CH_2)_4—\underset{\|}{\overset{}{C}}—CH_3$ (C=O)

(D) Heptan-2-one

55. *(b)* In this reaction, substrate undergoes intramolecular aldol condensation in which condensation reaction of two aldehyde groups or ketone groups occur in same molecule.

Mechanism

Step I Abstraction of proton by base KOH to form enolate.

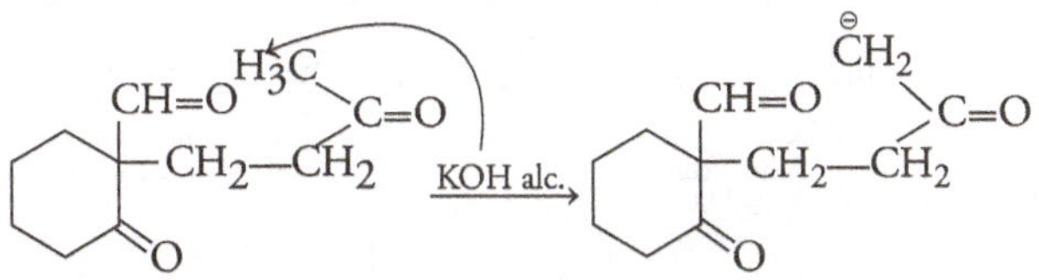

Enolate

Step II Reaction of enolate and aldehyde group on substrate to form β-hydroxy ketone.

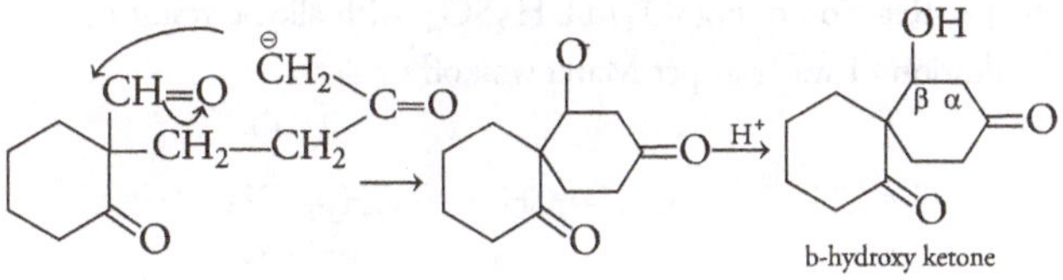

b-hydroxy ketone

Step III β-hydroxy ketone undergoes dehydrate to form a conjugated enone.

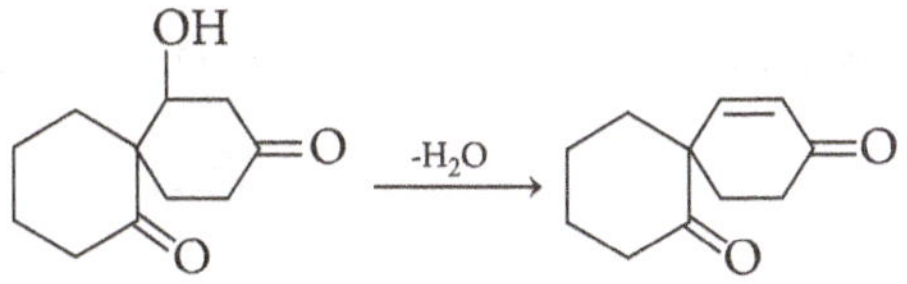

Conjugated enone

56. *(d)* In first step, aldol reaction is involved in reaction mechanism and in second step, (I_2, NaOH) is used for iodoform test, which gives CHI_3 (yellow precipitate) as by product.

Reaction Mechanism

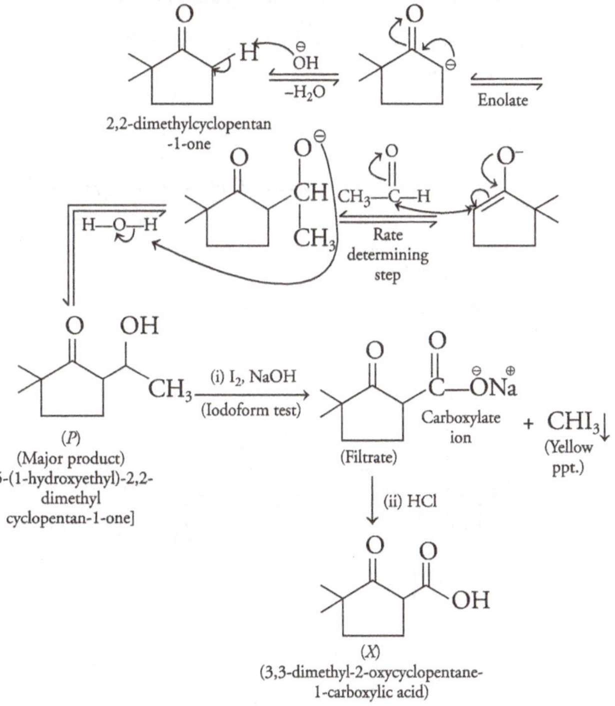

57. *(d)* Compound $B\,(C_4H_8O)$ reacts with hydroxylamine (NH_2OH). So, compound B is an aldehyde or a ketone. Again, B does not give Tollen's test which indicates that B is a ketone but not an aldehyde. So, B is

$$CH_3—\overset{O}{\overset{\|}{C}}—CH_2CH_3 \quad (C_4H_8O)$$

Butan-2-one

Compound $A\,(C_4H_8Cl_2)$ is a dihalide which undergoes hot hydrolysis (H_2O/ 373 K) to give B, a ketone. So, A is a non-terminal *geminal* or *gem* dichloride and B is 2-butan-2-one.

58. *(c)* Reaction of $HgSO_4$/dil. H_2SO_4 with alkyne result in addition of water as per Markownikoff's rule.

(a) $HC{\equiv}CH \xrightarrow[H_2SO_4]{HgSO_4} CH_2{=}CH(OH) \rightleftharpoons CH_3{-}CHO$

Acetylene → Acetaldehyde

(b)

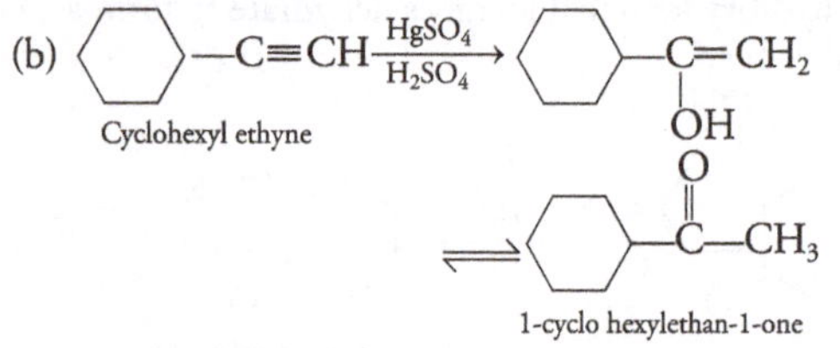

(c) $CH_3{-}C{\equiv}CH \xrightarrow[H_2SO_4]{HgSO_4} CH_3{-}C(OH){=}CH_2 \rightleftharpoons CH_3{-}CO{-}CH_3$

Prop-1-yne → Acetone

Hence, $CH_3{-}CH_2{-}CHO$ cannot be form.

(d) $CH_3C{\equiv}C{-}CH_3 \xrightarrow[H_2SO_4]{HgSO_4} CH_3{-}C(OH){=}CH{-}CH_3 \rightleftharpoons CH_3{-}CO{-}CH_2{-}CH_3$

But - 2- yne → But-2- one

59. *(d)*

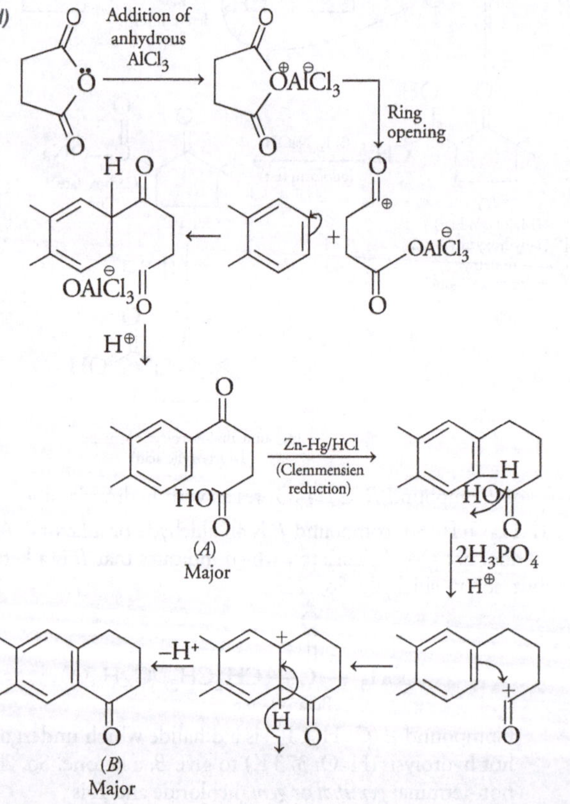

60. *(a)* Complete reaction is as follows

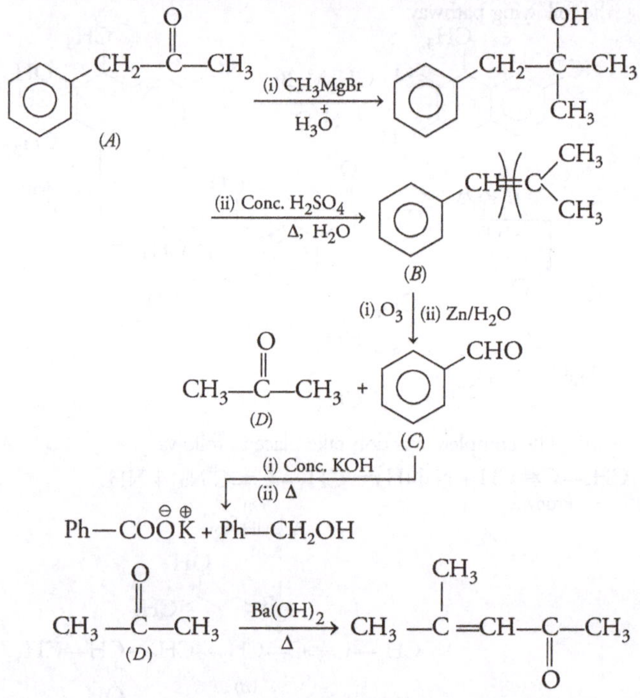

61. *(d)*

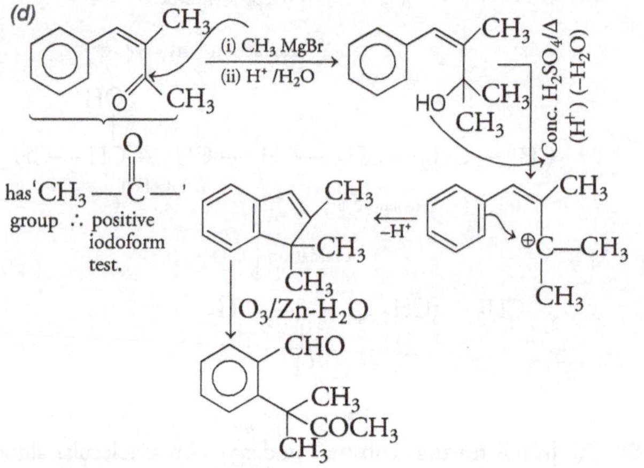

62. *(b)* NaOCl (sodium hypochlorite) is the reagent of haloform (chloroform formation) reaction.

The given reaction takes place as follows

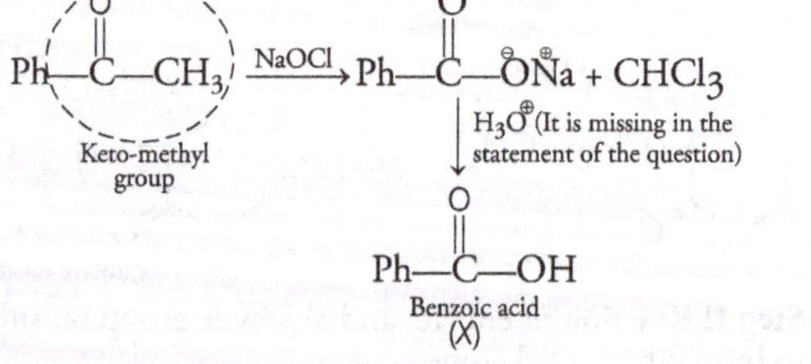

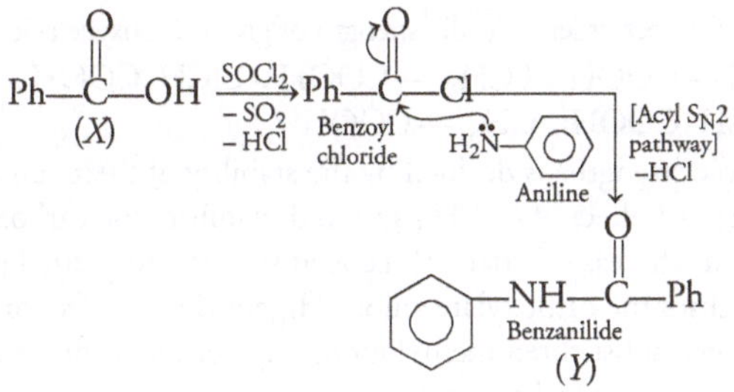

63. *(b)* The reactant in presence of dil. NaOH undergoes intramolecular aldol condensation reaction.

As a result of this, β-hydroxyketone (*A*) is obtained which on hydrolysis followed by heating produces α, β-unsaturated ketone (*B*).

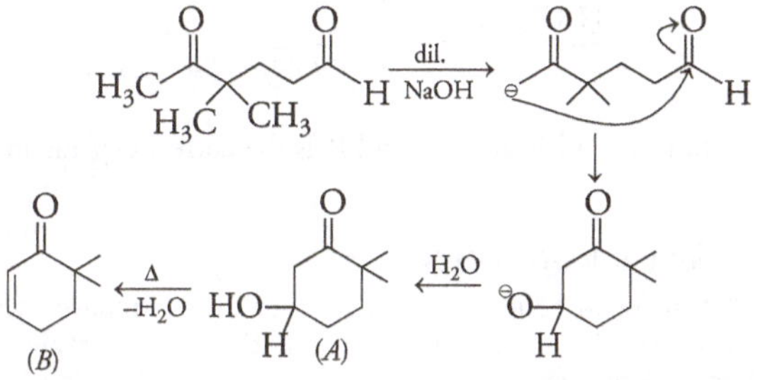

64. *(b)* In presence of strong base, acidic H can easily be removed that result in formation of anion. The resulting anion undergoes intramolecular nucleophilic addition which on hydrolysis followed by heating gives the required product.

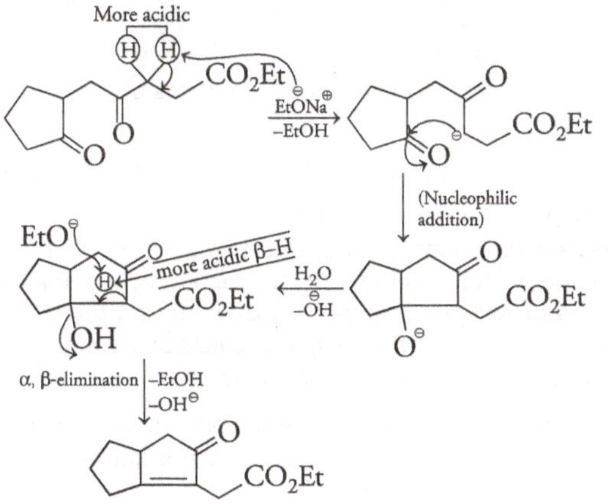

65. *(7)* The complete reaction is as follows,

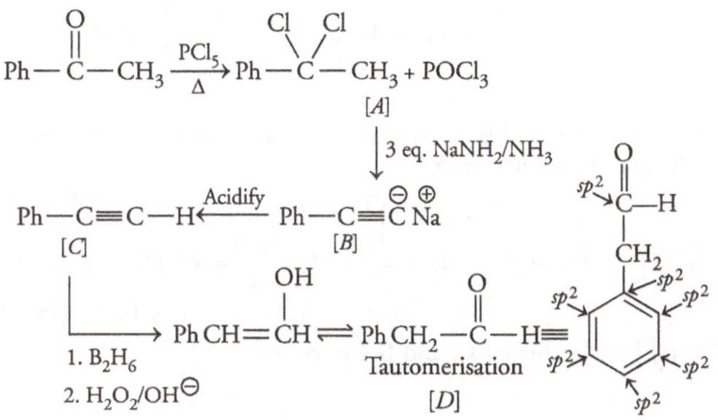

Thus, total number of sp^2-hybridised carbon atom in product *D* is 7.

66. *(15)*

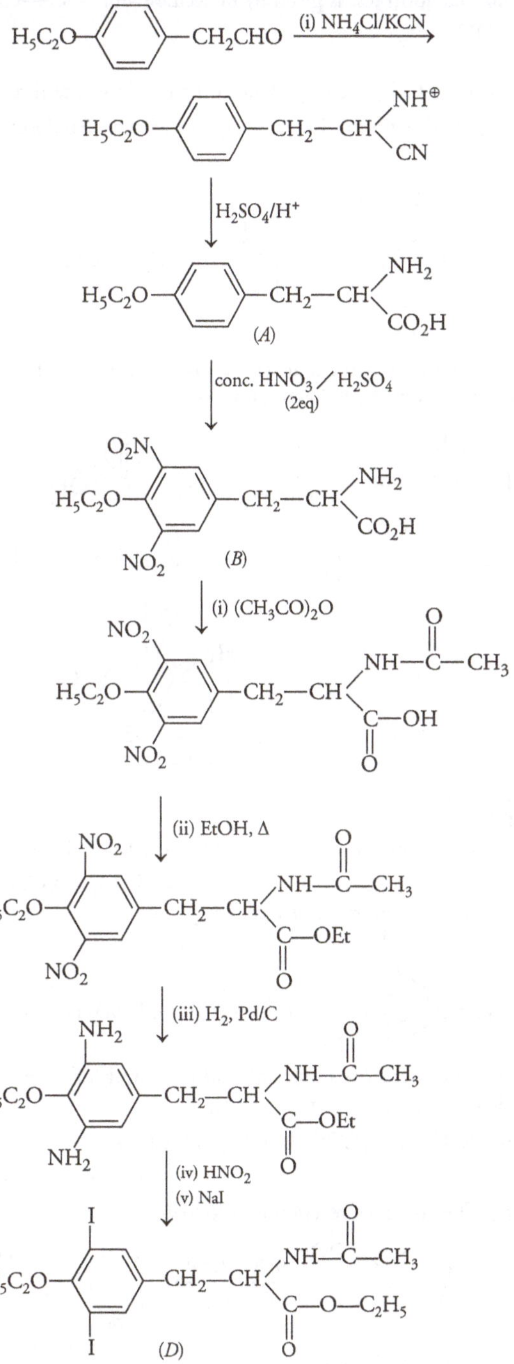

[Molecular formula $C_{15}H_{19}NO_4I_2$]

Thus, value of *X* is 15.

67. *(3)* Haloform test is given by molecules with $-\overset{O}{\overset{\|}{C}}-CH_3$ and $-\underset{}{\overset{OH}{\overset{|}{C}}}H-CH_3$ group present in them. Thus total three compounds with this functional group will give haloform test i.e.,

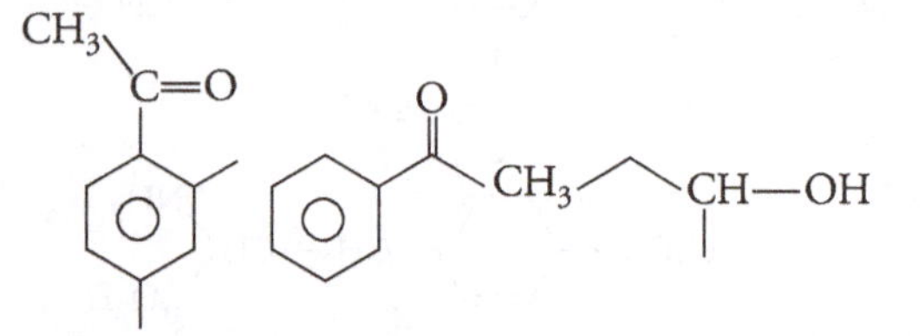

68. *(2)* There are two chiral centres present in the product obtained from the following reaction.

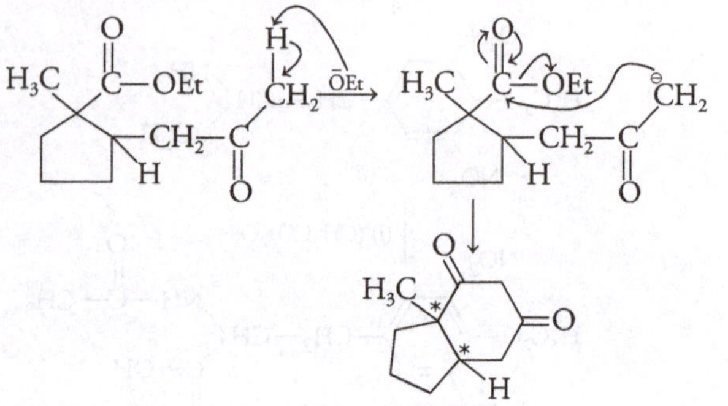

2 chiral carbons

Topic 3

01. *(b)* Compounds which are more acidic than H_2CO_3 can give effervescence of CO_2 with *aq.* $NaHCO_3$. Options (a), (c), (d) gives effervescence of CO_2 gas.

Where as 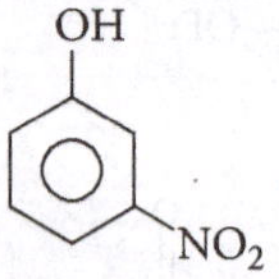 does not give effervescence.

02. *(b)* The correct sequence of acidic strength of given aliphatic acids in their decreasing order is;

$HCOOH > CH_3COOH > CH_3CH_2COOH > CH_3CH_2CH_2COOH$

03. *(d)* The complete reaction is as follows,

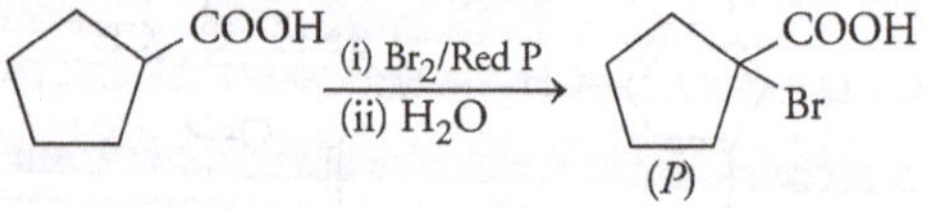

(*P*)

This is HVZ reaction.

04. *(d)* In the given reaction, product is formed by oxidative ozonolysis.

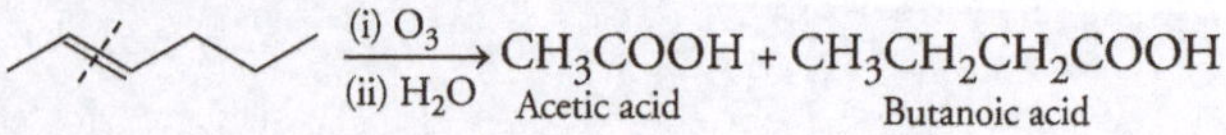

05. *(b)* Correct order of acidic strength of given carboxylic acid is:

$F_3C-COOH > FCH_2-COOH > ClCH_2COOH > BrCH_2COOH > CH_3-COOH$

Acidic strengths is decided by the stability of their conjugate bases. $+I$ effect of $-CH_3$ group destabilises the carboxylate anion whereas $-I$ effect of halogen substituted methyl group stabilises the carboxylate anion. Higher the $-I$ effect of halogen substituted methyl group, higher the acidic strength of the corresponding acid.

06. *(c)* IUPAC name of given compound is 2-methyl-5-oxohexanoic acid.

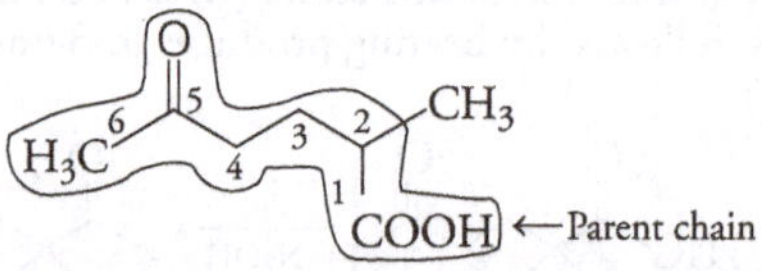

07. *(c)* Both A and R are true and R is the correct explanation of the A.

> **Concept Enhancer**
>
> Clemmensen's reduction is most commonly used to convert acyl benzene to alkyl benzene in the presence of zinc amalgam and hydrochloric acid.
>
> 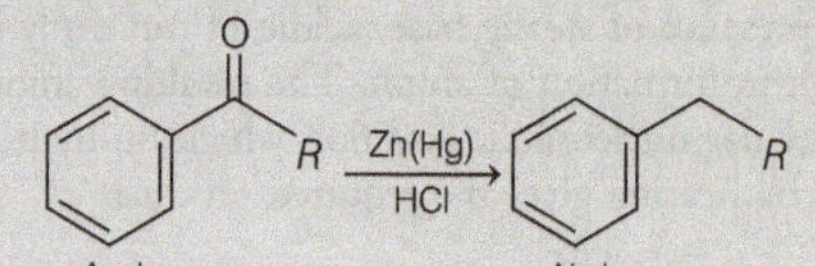
>
> Acyl benzene → Akyl benzene
>
> It involves the formation of saturated hydrocarbons from carbonyl compounds by the removal of oxygen.

08. *(a)* The esterification of carboxylic acid with an alcohol is a nucleophilic acyl substitution reaction.
Electron withdrawing groups in the carboxylic acid will increase the rate of esterification reaction by making carbonyl carbon of acyl group more electron deficient, thereby increasing the rate of nucleophilic attack by alcohol.
Hence, both the statements are correct.

09. *(a)* Benzoic acid in presence of red phosphorus undergoes halogenation at α-position. This reaction is known as Hell-Volhard Zelinsky reaction.

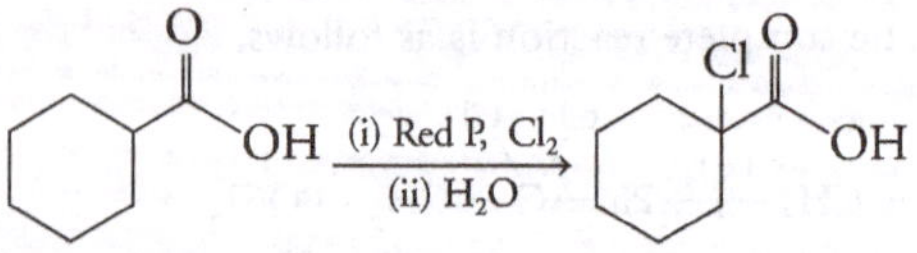

10. *(c)* Benzoic acid is converted to benzaldehyde in one step by the use of reagent MnO.

$$C_6H_5-\overset{O}{\overset{\|}{C}}-OH + HO-\overset{O}{\overset{\|}{C}}-H \xrightarrow[\Delta]{MnO} C_6H_5-\overset{O}{\overset{\|}{C}}-H + CO_2 + H_2O$$

11. *(a)* Both Statements I and II are correct.

12. *(a)* Sodium hypochlorite (NaOCl) is a strong oxidising agent that will convert ketone to carboxylic acid.
Since, product is carboxylic acid, therefore reactant (*P*) would be ketone.

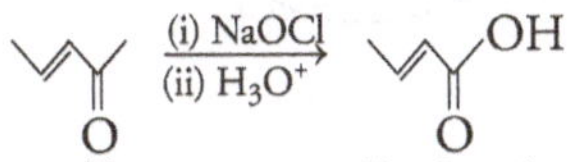

(*P*) Pent-3-en-2-one   But-2-enoic acid

13. *(b)* Rate of hydrolysis is directly proportional to positive charge present on carbon of >C=O group.
Rate of hydrolysis : Acid chloride > Acid anhydride > ester > amide.
Hence, the order is $A > B > C > D$.

14. *(c)* Greater is stability of conjugate base more will be the acidic strength.
General order of acidic strength is
R—COOH > Ph—OH > R—C≡CH
In between (iii) and (iv), (iii) is more acidic due to -M effect of $—NO_2$.
Thus, decreasing order of acidity is (ii) > (iii) > (iv) > (i).

15. *(c)* The IUPAC name of the given compound is

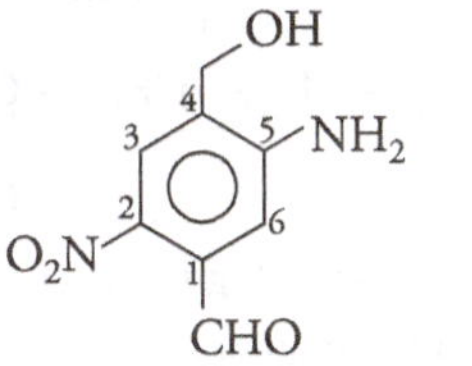

5-amino-4-hydroxymethyl-2-nitrobenzaldehyde.

16. *(c)* All the given compounds are α-monosubstituted acetic acid derivatives and the α-substitutions have been made by strong–*I* groups/atoms. More powerful the – *I* group, stronger will be the acid. –*I* power of different groups is as follows
$—NO_2$ > —CN > —F > —Cl.
Thus, the correct decreasing order for acid strength is:
$NO_2CH_2COOH > NCCH_2COOH > FCH_2COOH > ClCH_2COOH$.

17. *(a)* Among the given compounds EtO_2C—≡—H is least acidic. It is because conjugate base of terminal alkyne will be *sp*-hybridised and localised, while in other compounds conjugate base will be resonance stabilised.

18. *(b)*

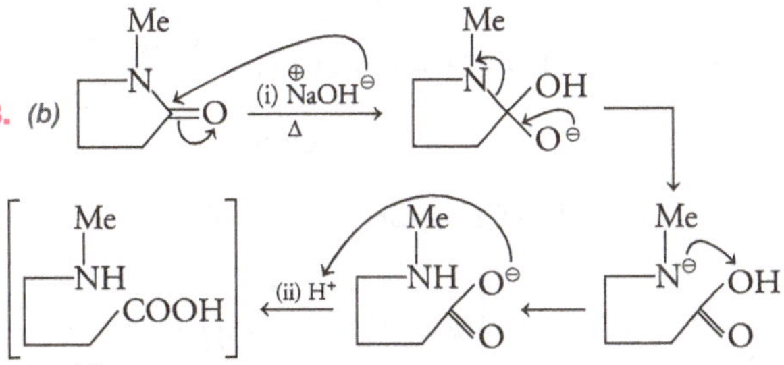

(*A*)

19. *(b)*

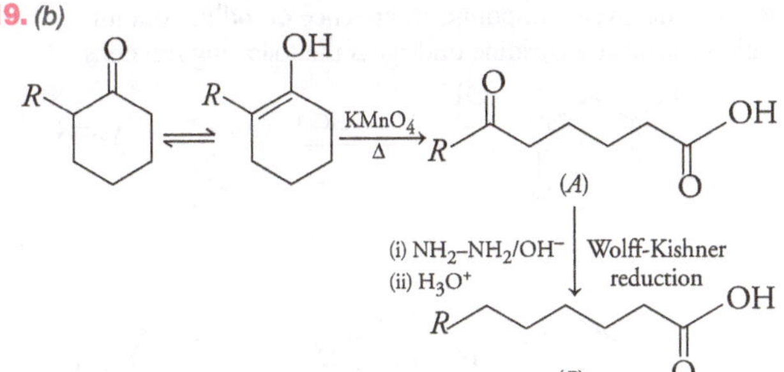

20. *(a)*

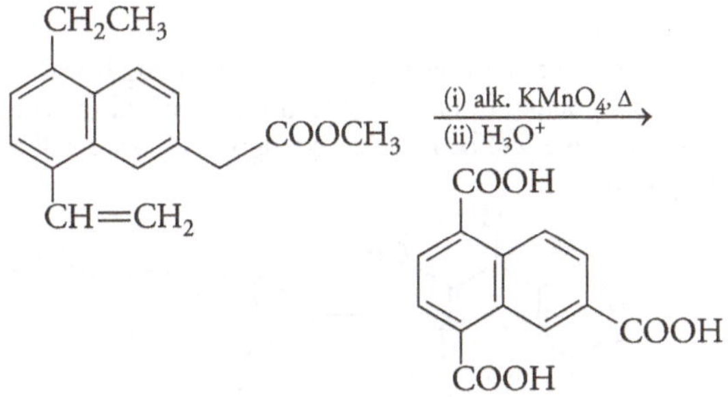

In the above reaction $KMnO_4$ oxidises benzylic carbon containing at least one α-hydrogen atom to —COOH.

21. *(a)* In first step, $LiBH_4$ can reduce ester selectively but not carboxylic acid. In second step, protonation of acid occur to give the final product.

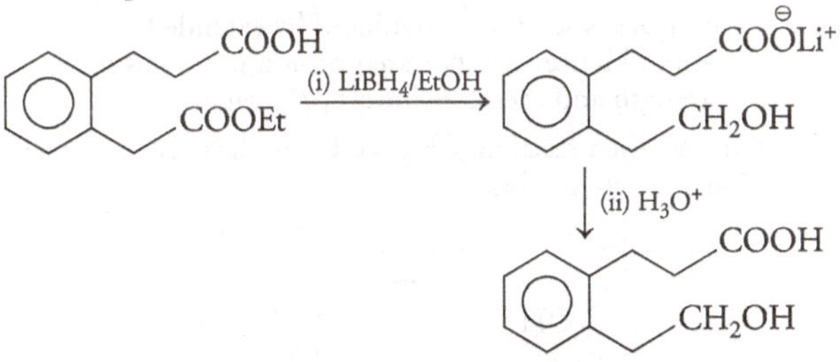

22. *(b)* The correct match is A-III, B-I, C-II, D-IV.
A. Hell-Volhard-Zelinsky reaction (III) reagent-Br_2/Red P
B. Iodoform reaction (I) NaOH + I_2
C. Etard reaction (II) CrO_2Cl_2, CS_2 and H_2O
D. Gattermann-Koch reaction (IV) CO, HCl, Anh. $AlCl_3$

23. *(c)* Correct option is (c).

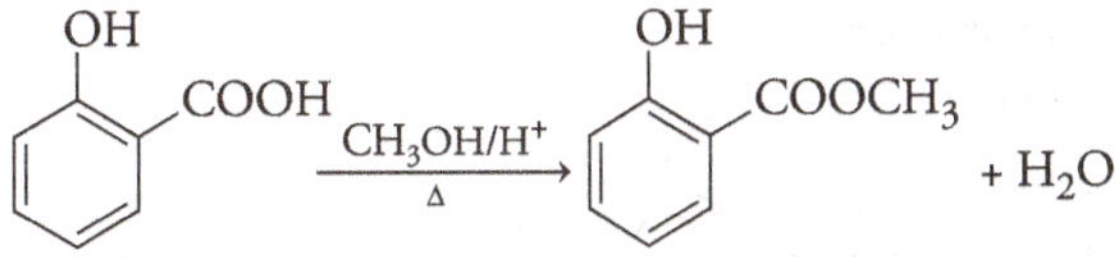

(Salicylic acid)   (*Y*)

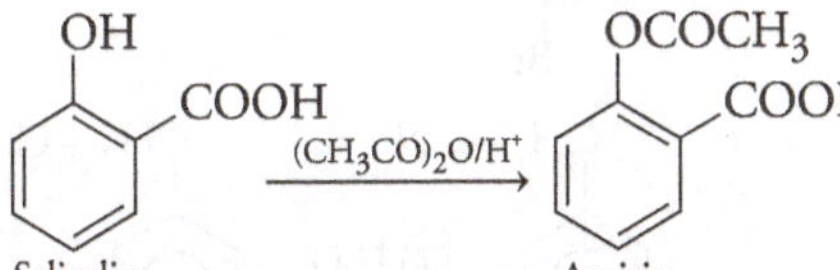

Salicylic acid   Aspirin (*X*)

Thus '*X*' = $(CH_3CO)_2O/H^+$ and '*Y*' = CH_3OH/H^+, Δ

24. *(c)* The given compound in presence of iodine sodium bicarbonate and pyridine undergoes the following reactions

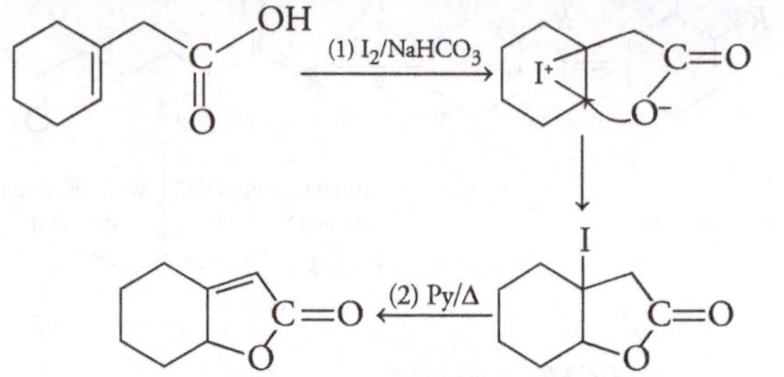

25. *(c)* The compound having higher acidic strength will have lower pK_a value. In other words, more stabilised the conjugate base, the more will be its acidic strength and lower will be the pK_a value.

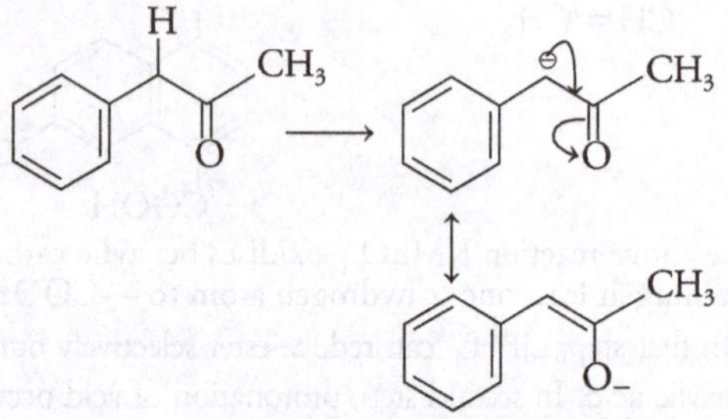

Extended conjugation

The conjugate base of (c) is stabilised by extended conjugation. Hence, the indicated proton in (c) has highest acidic strength and therefore lowest pK_a value.

26. *(b)* In the given reaction $(CH_3)_3CLi$ give halogen exchange reaction and acts as a base.

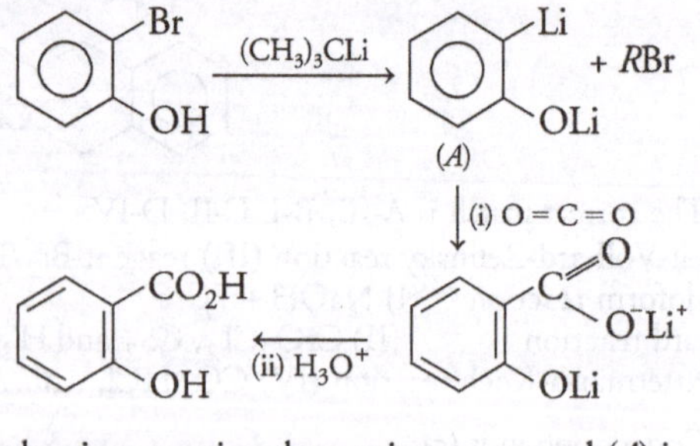

27. *(c)* In the given reaction the starting compound (*A*) is *p*-bromotoluene.

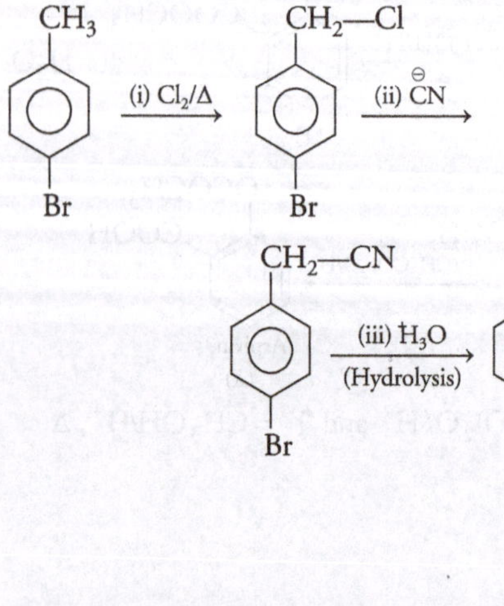

28. *(a)*

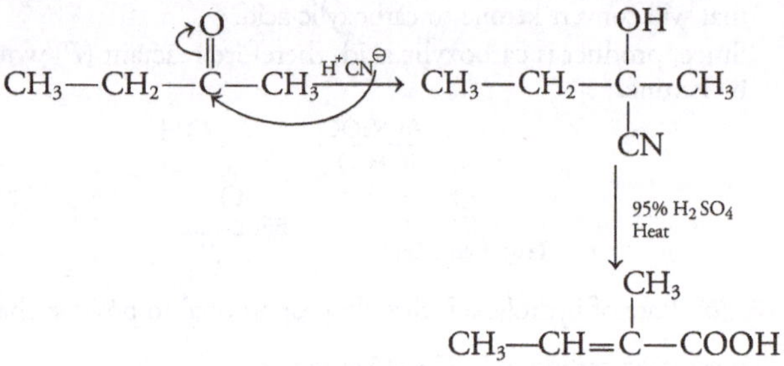

29. *(c)* The complete reaction is as follows

$$\underset{\text{Ethylbutanoate }(A)}{CH_3-CH_2-CH_2-\overset{\overset{\displaystyle O}{||}}{C}-O-CH_2-CH_3}$$

$$\downarrow LiAlH_4,\ H_3O^+$$

$$\underset{\text{Butanol }(B)}{CH_3-CH_2-CH_2-CH_2-OH} + \underset{\text{Ethanol}}{CH_3-CH_2-OH}$$

$$\downarrow \text{Oxidation}$$

$$CH_3-CH_2-CH_2-\underset{\underset{\displaystyle O}{||}}{C}-OH$$

30. *(a)* Bromopropane with Mg in presence of CO_2 and dry ether form butanoic acid as follows :

$$CH_3CH_2CH_2Br \xrightarrow[\text{dry ether}]{Mg} CH_3CH_2CH_2MgBr$$

$$\downarrow CO_2$$

$$CH_3-CH_2CH_2-\underset{\underset{\displaystyle O}{||}}{C}-OMgBr$$

$$\downarrow H_3O^+$$

$$\underset{\text{Butanoic acid}}{CH_3CH_2CH_2COOH} + Mg\begin{matrix}Br\\OH\end{matrix}$$

31. *(c)* Statement I is true but Statement II is false.
In this reaction one mole of Ethyl pent-4-yn-oate utilizes three moles of CH_3MgBr as follows :

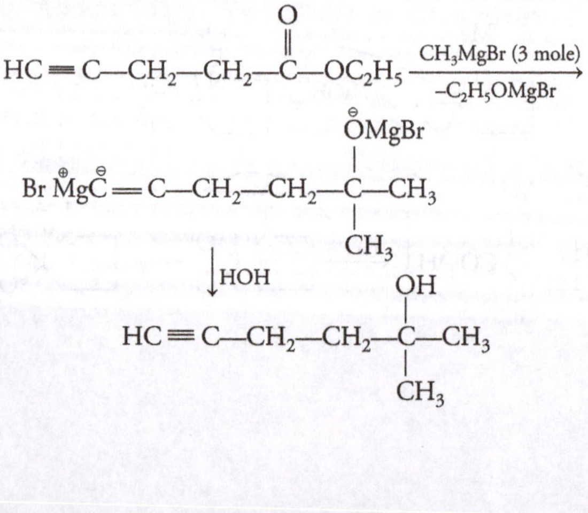

32. *(c)* Esterification of carboxylic group (—COOH) take place in presence of $SOCl_2$ and CH_3OH as follows.

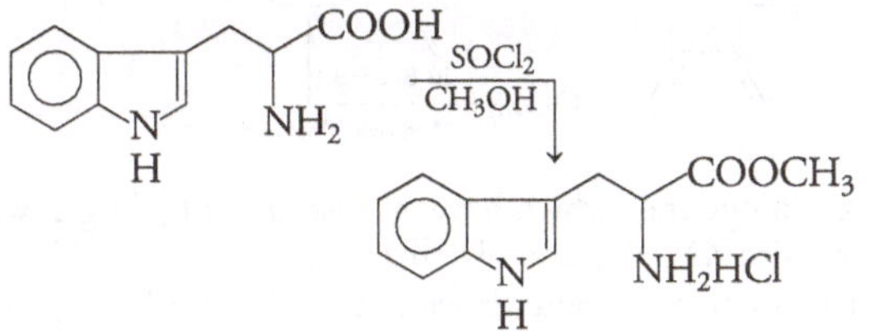

33. *(b)* Maleic anhydride can be prepared by heating *cis*-but-2-enedioic acid.

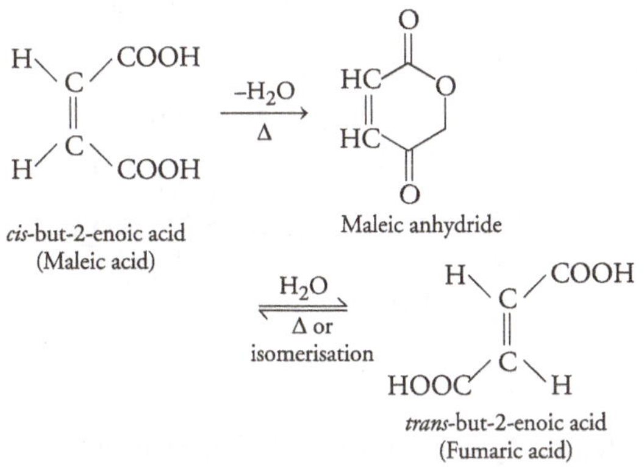

34. *(c)* 4-methoxy toluene in the presence of alkaline $KMnO_4$ to give 4-methoxy benzoic acid.
Chemical reaction is as follows :

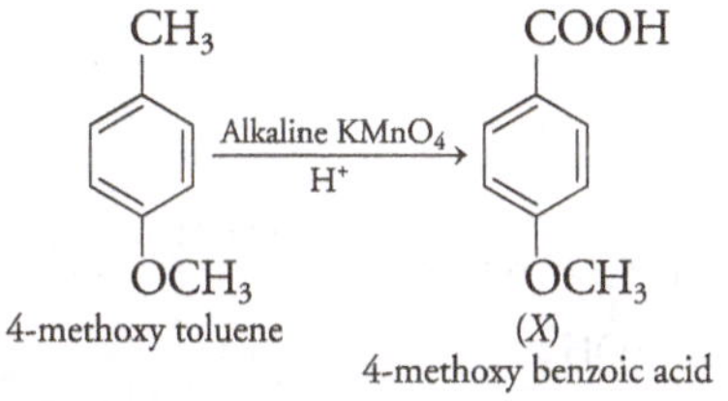

35. *(c)* 1-cyano cyclo hexane on complete hydrolysis gives 1-cyclo hexanoic acid. Intermediate 1- methanamide cyclohexane (A) is formed by partial hydrolysis.
Complete reaction is as follows

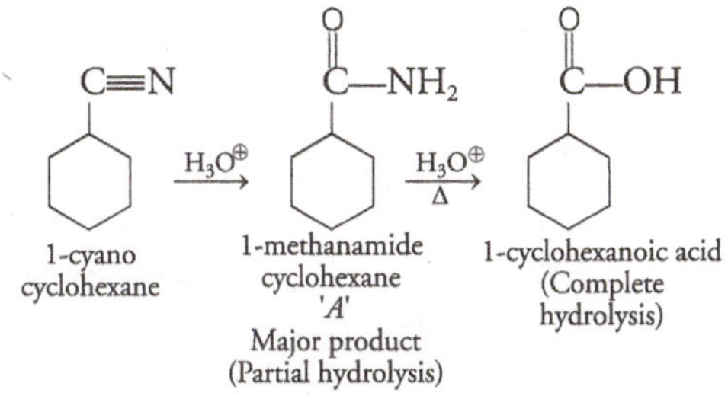

36. *(c)* In the given reaction, A is alkaline $KMnO_4, H^+$. Alkaline $KMnO_4/H^+$ is a strong oxidising agent and oxidises alkyl benzene to benzoic acid. While $NaBH_4$, $LiAlH_4$ and Zn-Hg/HCl are reducing agent.

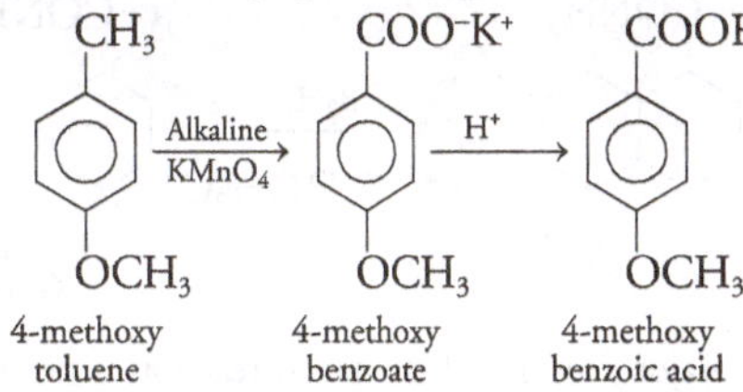

37. *(c)*

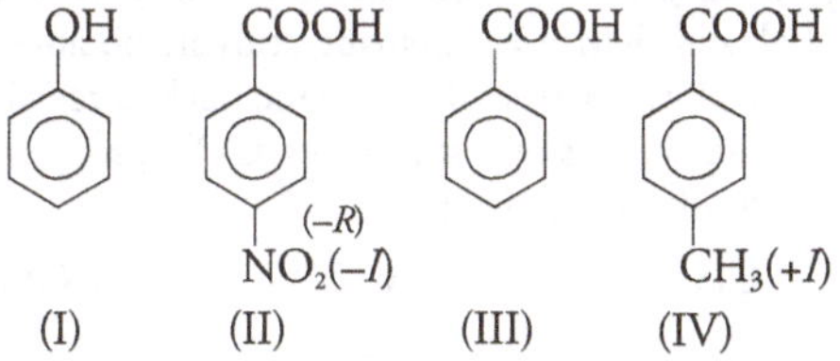

Acidity of phenol (I) is weaker than any carboxylic acid.

Electron withdrawing nature $(-R, -I)$, of —NO_2 group at *para* position increases acidic strength of (II), whereas $+I$ effect of —CH_3 group at *para* position decreases acidic strength of (IV).

So, the order of acid character is II > III > IV > I.

38. *(c)* The correct match is (A)–2, (B)–4, (C)–1, (D)–3.

39. *(c)* Given, molecules (A) and (B) undergoes intramolecular H-bonding and intermolecular H-bonding respectively. It is shown as follows :

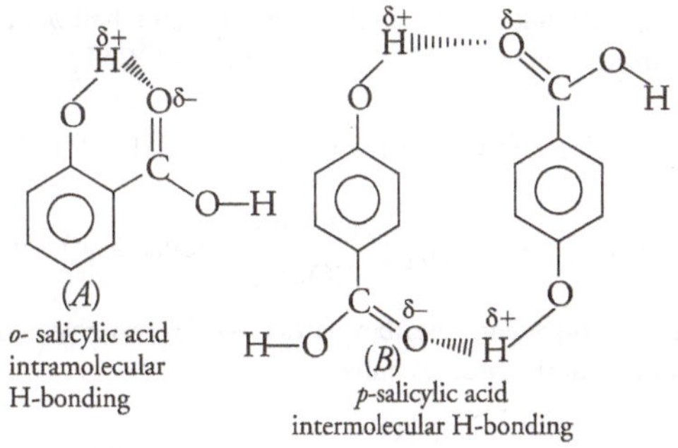

Among the given statements, II and III are true whereas I is false. Corrected statement I, (A) is more likely to crystalline than (B).

40. *(c)* Following reaction is involved in the given conditions

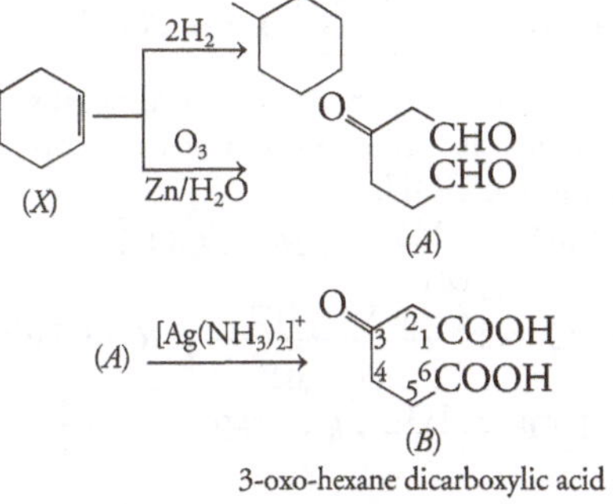

41. *(a)* The most suitable reagent for the given conversion is B_2H_6.

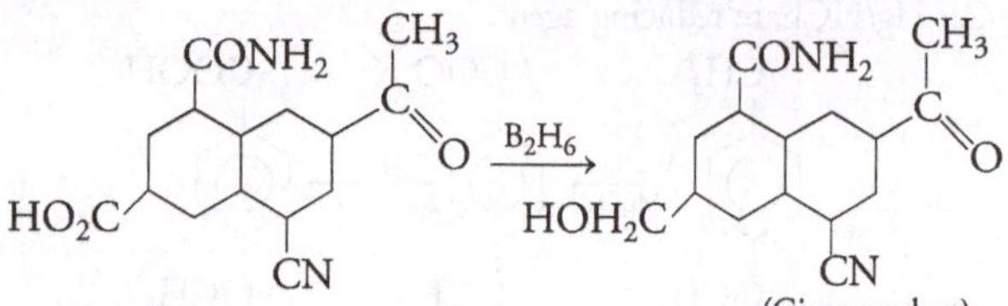

(Given product)

42. *(a)* The major product of the given reaction is benzoic acid (C_6H_5COOH). On vigorous oxidation of alkyl benzene with acidic or alkaline $KMnO_4$, aromatic acids are obtained. During oxidation of alkyl benzene, the aromatic nucleus remains intact and the entire chain is oxidised to —COOH group irrespective of the length of carbon chain.

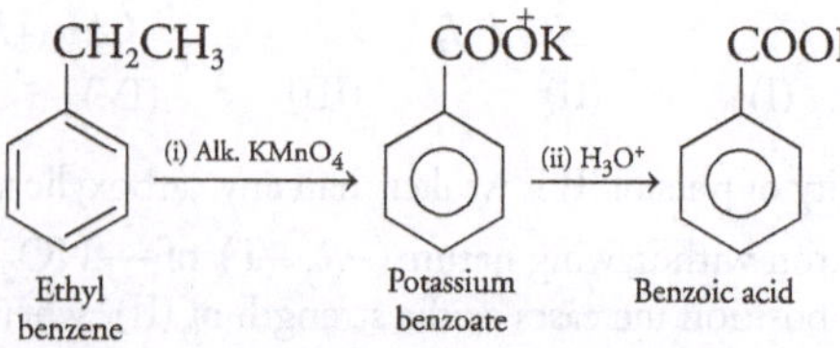

43. *(b)* Alkaline hydrolysis of an ester (carboxylic acid derivative) follows acyl S_N2 mechanism.

Rate of S_N2 mechanism depends on the polarity of group of —COO*R* group. Electron withdrawing group $(-R > -I)$ increases the rate of S_N2 reaction whereas electron donating group $(+R > +I)$ decreases the rate of S_N2 reaction.

Here, the nature of functional groups attached *para* to the benzene ring are $\underset{(-R)}{—NO_2} > \underset{(-I)}{—Cl} > \underset{(+R)}{—OCH_3}$

So, the order of hydrolysis will be, $\text{III} > \underset{(-R)}{\text{II}} > \underset{(-I)}{\text{I}} > \underset{(+R)}{\text{IV}}$

44. *(b)* Heating of $(CH_2)_4<^{COOH}_{COOH}$ (adipic acid) with a dehydrating agent, decarboxylates $(—CO_2)$ to give a ketone (cyclopentanone), not an anhydride.

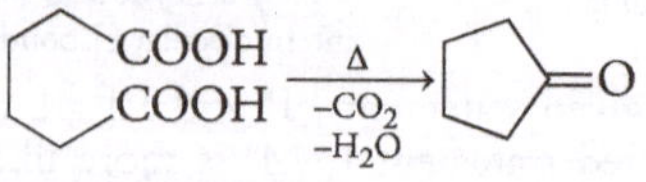

45. *(a)* Reaction B and E gives carboxylic acid as a major product. The complete reaction are as follows.

(A) $R—C \equiv N \xrightarrow[\text{Mild condition}]{\text{(i) } H^+/H_2O} R—CONH_2$

Under mild condition amide is formed because this reaction is typically slow if further more heat will supplied then it converted into COOH.

(B) $R—MgX \xrightarrow[\text{(ii) } H_3O^+]{\text{(i) } CO_2} R—COOH$

(C) $R—C \equiv N \xrightarrow[\text{(ii) } H_3O^+]{\text{(i) } SnCl_2/HCl} R—CHO$

(D) $RCH_2OH \xrightarrow{PCC} R—CHO$

(E) 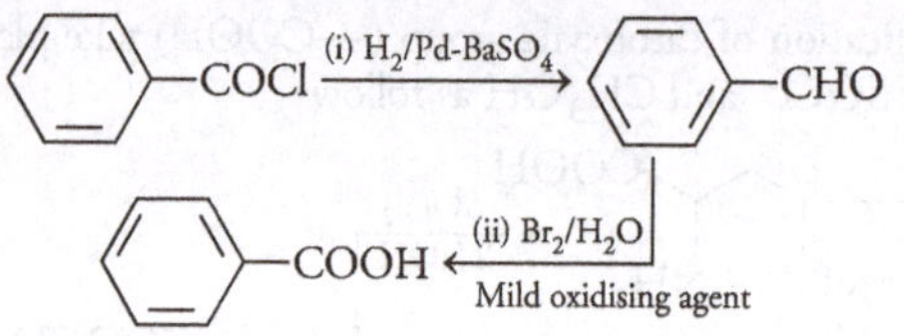

46. *(a)* Compounds which have at least one α-hydrogen will react with $KMnO_4$ and give benzoic acid.

There are five compounds with at least one α-hydrogen, these are

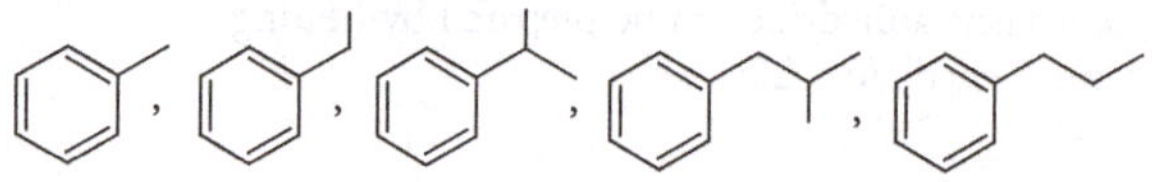

47. *(c)* Among the given options,

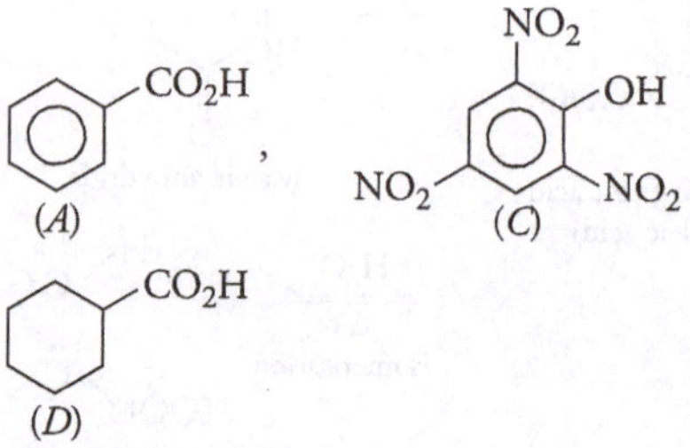

produces CO_2 with aqueous $NaHCO_3$ solution.

48. *(c)* The complete reaction is as follows:

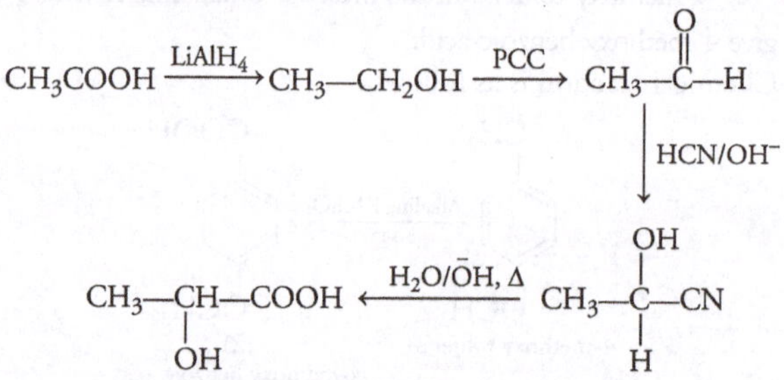

Hence, final product is $CH_3—CH(OH)—COOH$

49. *(b)* The complete reaction is as follows,

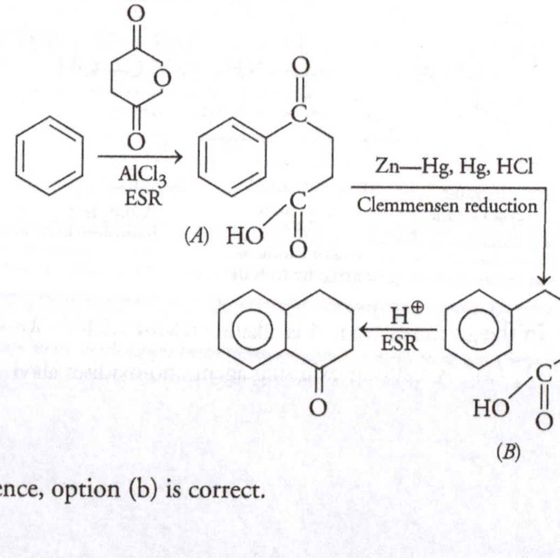

Hence, option (b) is correct.

50. *(c)* The complete reaction is as follow,

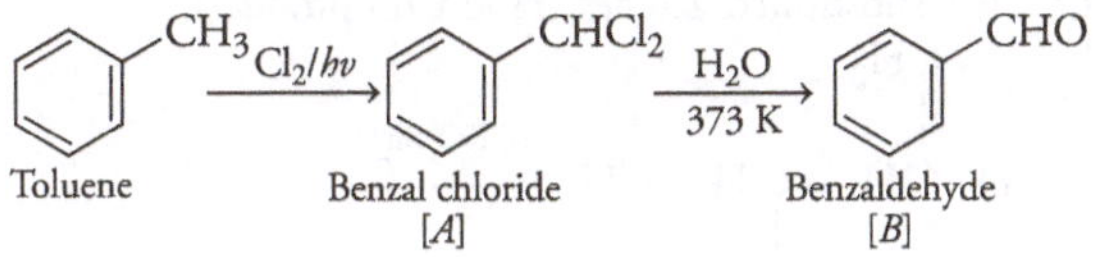

51. *(a)* Complete reaction is as follows:

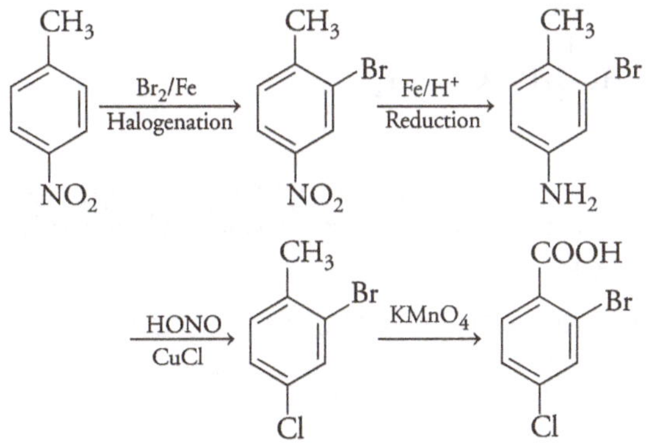

52. *(c)* The major product '*P*' is given in option (c).

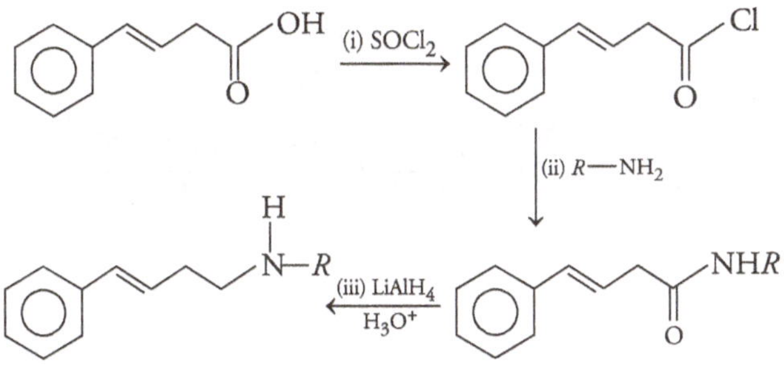

53. *(c)*

Element	%	Mass	Mole	Simplest Ratio
C	52.2	52.2	4.35	8.1 ≈ 8
H	4.9	4.9	4.9	9.12 ≈ 9
Br	42.9	42.9	0.537	1

Empirical formula $= C_8H_9Br$

Empirical formula mass = 184.9 g/mol

Molar mass = 184 g/mol

Now, molecular formula $= C_8H_9Br$

Hence, *A* is an optically active compound.

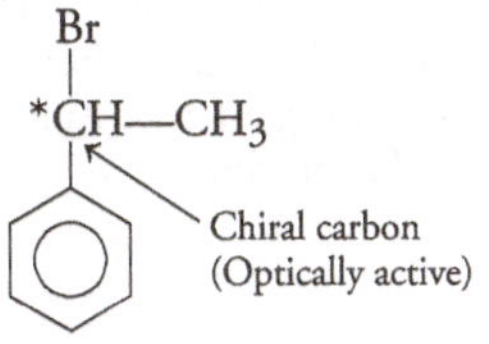

(*A*)

B produces *p*-bromobenzoic acid on oxidation with $KMnO_4$ (Potassium permagnate)

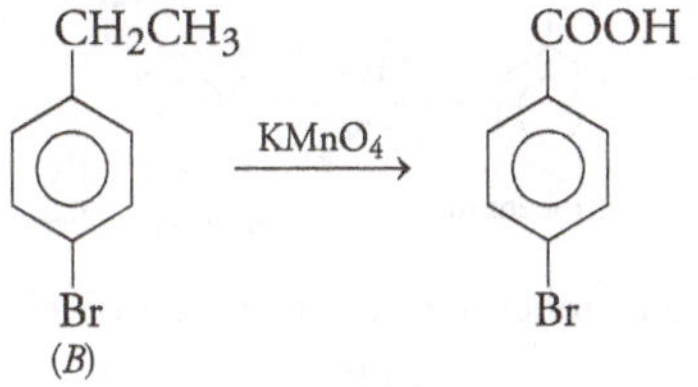

A gives benzoic acid on oxidation with $KMnO_4$ as well as it gives pale yellow ppt. on reaction with alc. $AgNO_3$ as

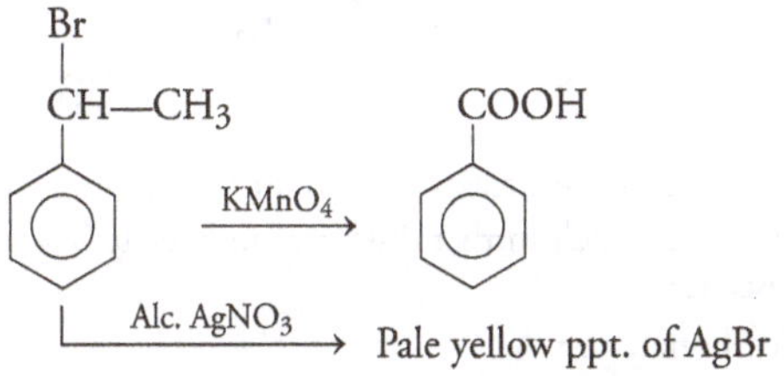

54. *(a)* Acetic acid on reaction with $SOCl_2$ gives acetyl chloride (*A*).

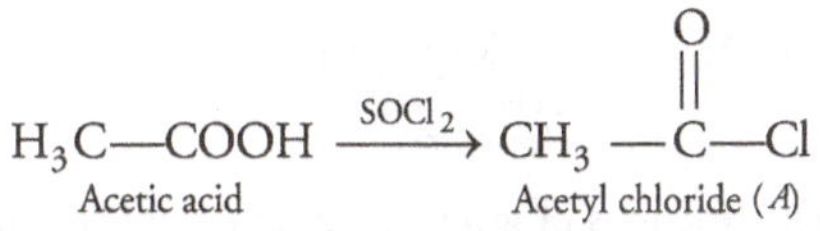

Acetyl chloride undergoes Friedel-Craft acylation in presence of anhyd. $AlCl_3$ and benzene to form acetophenone (*B*).

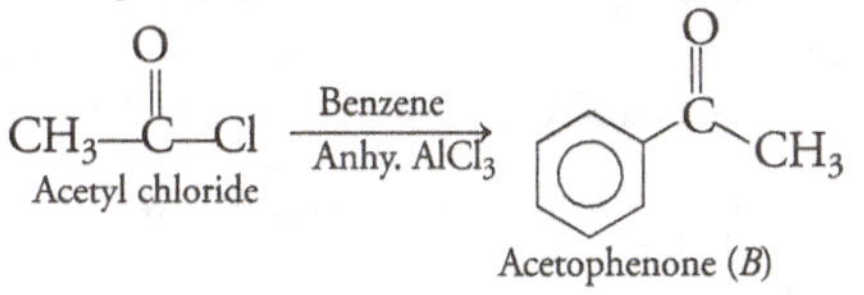

Acetophenone reacts with KCN to give (*C*) the final product, i.e. 2-hydroxy-2-phenolpropionitrile.

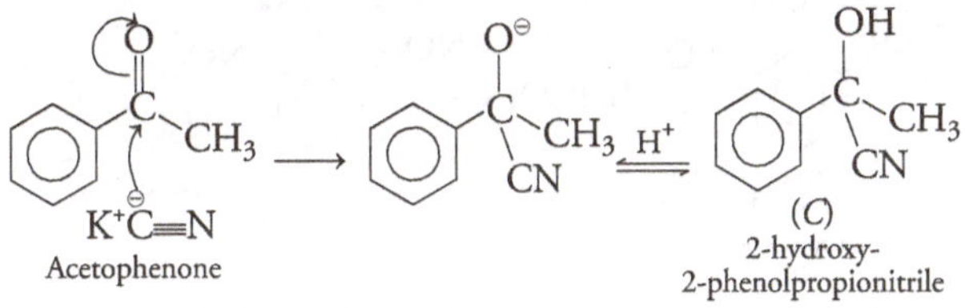

Hence, correct option is (a).

55. *(d)* The preparation of 3-nitrobenzoic acid from benzene is as follows

- Nitration of benzene

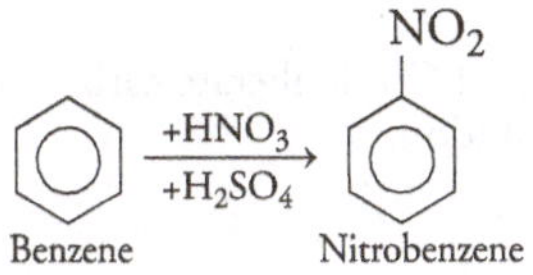

Electrophilic substitution of nitrobenzene

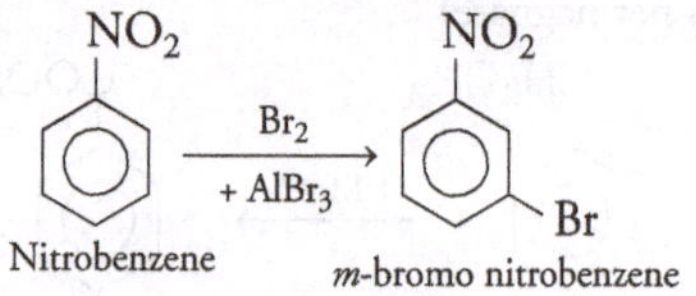

m-bromo nitrobenzene is converted to 3-nitrobenzoic acid.

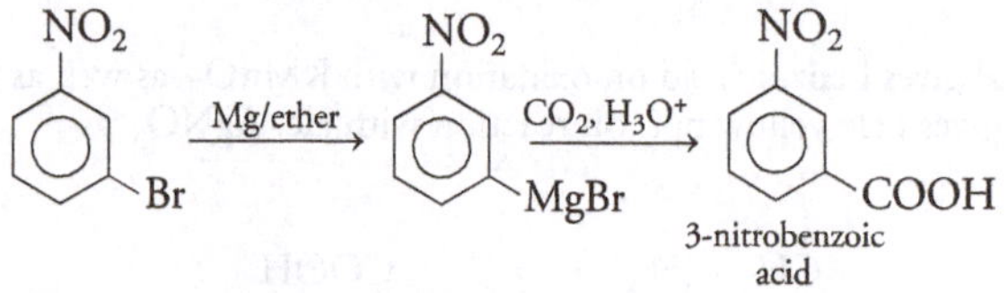

56. *(a)* Compound (*a*), reacts with sodium bicarbonate to form carbonic acid which further dissociate to liberate CO_2 effervescence.

Bicarbonate test

$$(CH_3)_3\overset{\oplus}{N}HCl^- + NaHCO_3 \longrightarrow \underset{\text{Carbonic acid}}{H_2CO_3} + \underset{\text{Trimethyl-amine}}{(CH_3)_3N} + NaCl$$

(Sodium bicarbonate)

$$H_2CO_3 \longrightarrow \underset{\substack{\text{Carbon dioxide}\\\text{(effervescence)}}}{CO_2} + H_2O$$

57. *(c)* The reactions of given compound with sodium bicarbonate solution are as follows

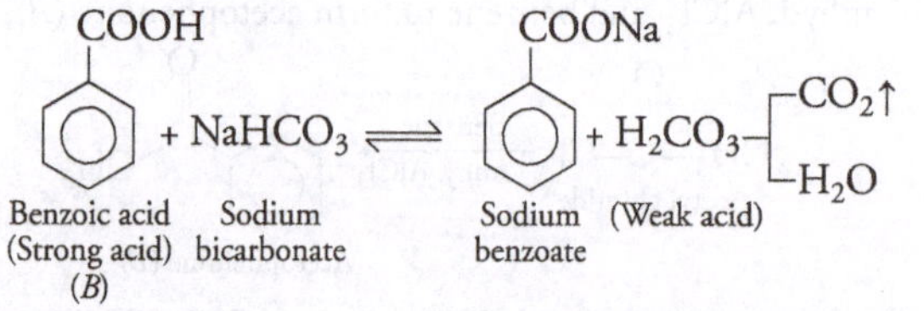

As H_2CO_3 is weak acid it dissociate to liberate CO_2 gas and H_2O.

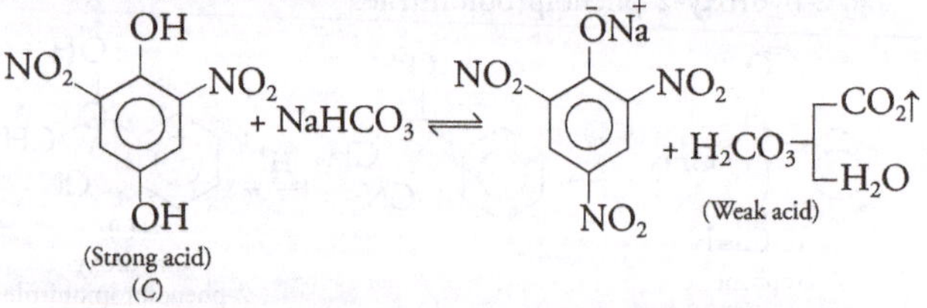

Equilibrium favours forward direction and CO_2 is liberated. In the above two reactions, H_2CO_3 is comparatively weak acid.

∴ CO_2 gas is liberated

Equilibrium favours backward direction and CO_2 is not liberated.

Thus, only *B* and *C* will liberate carbon dioxide with sodium bicarbonate solution.

58. *(d)* Tertiary butoxide is a bulky base, which favour elimination to form less substituted alkenes (Hofmann-product).

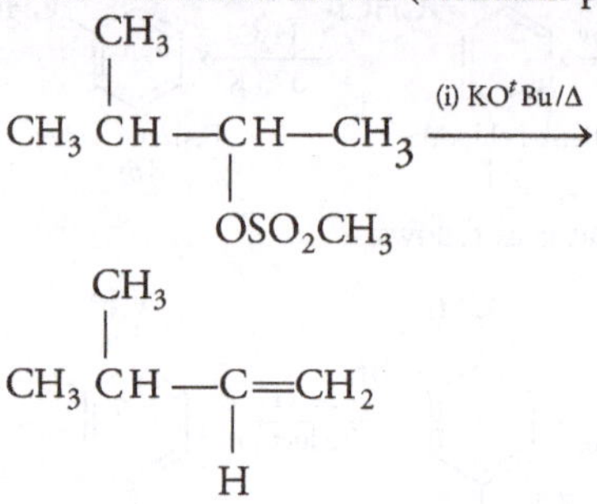

Now, alkene in presence of O_3/H_2O_2 undergoes oxidative ozonolysis to produce 2-methyl propionic acid and formic acid.

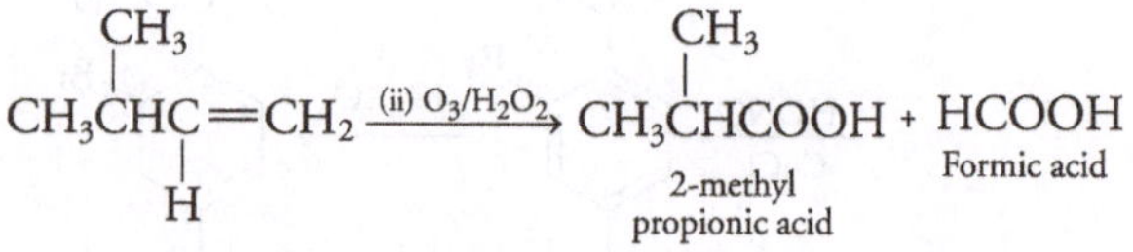

59. *(b)* The reaction sequence is given as :

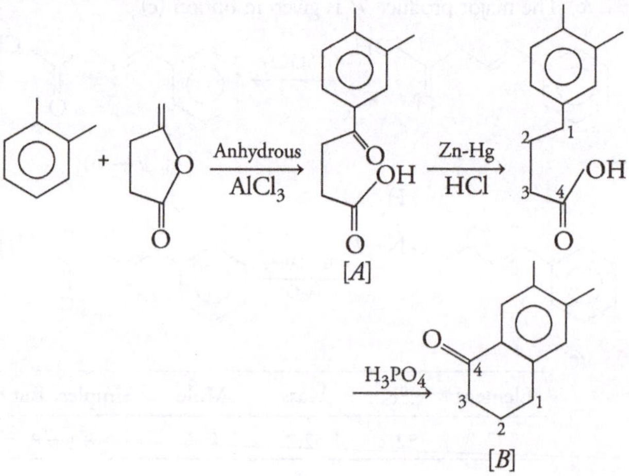

60. *(3)*

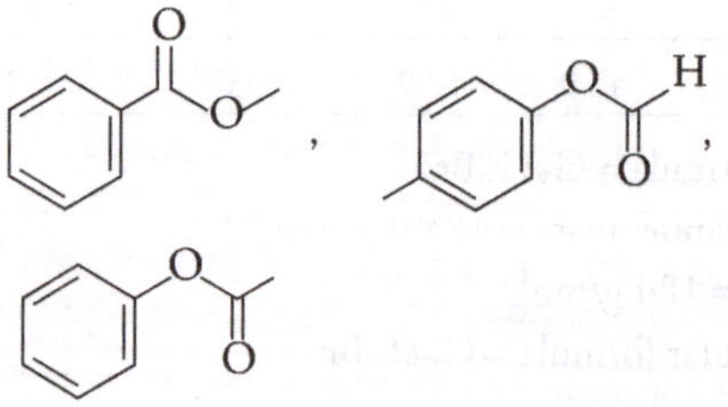

Those compounds dissolved in cold $NaHCO_3$ and NaOH solution that contain carboxylic acid and phenol group. But the compounds containing an ester functional group will only dissolve in hot NaOH solution.

22 Organic Compounds Containing Nitrogen

In JEE Main there is usually 0-2 question per paper are visible from this chapter. In past 3 years, 52 conceptual and 7 numerical questions were asked of moderate to difficulty level from this chapter.

For detailed analysis scan QR code

Topic 1 / Aliphatic Amines

• Level 1

01 Reaction of propanamide with Br_2 / KOH(*aq*) produces **[2023, 29 Jan Shift-II]**

(a) propyl amine (b) ethylamine
(c) ethylnitrile (d) propanenitrile

02 The major product '*P*' for the following sequence of reactions is

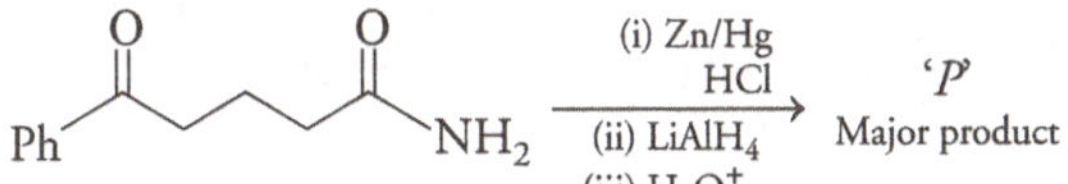

[2023, 29 Jan Shift-I]

(a) 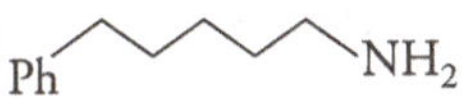 (b) 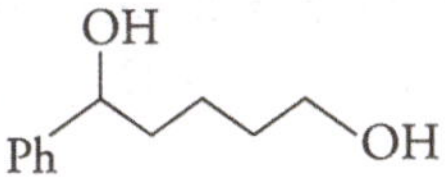

(c) 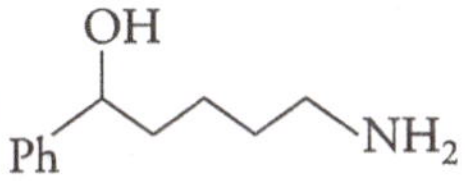 (d) 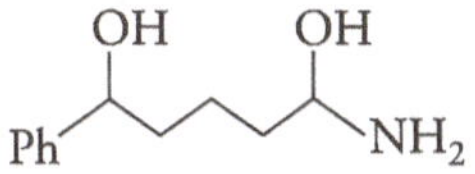

03 The correct order in aqueous medium of basic strength in case of methyl substituted amines is **[2023, 25 Jan Shift-I]**

(a) $Me_2NH > Me_3N > MeNH_2 > NH_3$
(b) $NH_3 > Me_3N > MeNH_2 > Me_2NH$
(c) $Me_2NH > MeNH_2 > Me_3N > NH_3$
(d) $Me_3N > Me_2NH > MeNH_2 > NH_3$

Imp. *Hofmann Degradation Reaction*

04 Given below are two statements. **[2022, 27 June Shift-I]**

Statement I In Hofmann degradation reaction, the migration of only an alkyl group takes place from carbonyl carbon of the amide to the nitrogen atom.

Statement II The group is migrated in Hofmann degradation reaction to electron deficient atom.

In the light of the above statements, choose the correct answer from the options given below.

(a) Both Statement I and Statement II are correct.
(b) Both Statement I and Statement II are incorrect.
(c) Statement I is correct but Statement II is incorrect.
(d) Statement I is incorrect but Statement II is correct.

05 A primary aliphatic amine on reaction with nitrous acid in cold (273 K) and there after raising temperature of reaction mixture to room temperature (298 K) gives **[2022, 28 June Shift-I]**

(a) nitrile (b) alcohol
(c) diazonium salt (d) secondary amine

06 What is the major product *P* of the following reaction ?

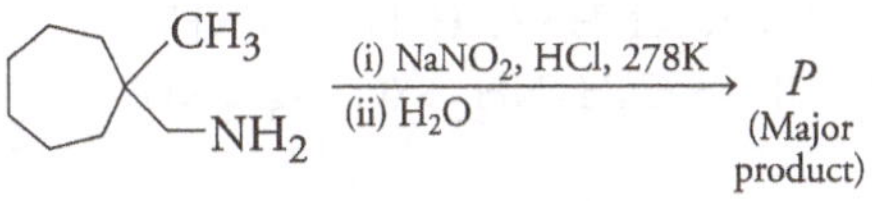

[2021, 25 July Shift-II]

(a) 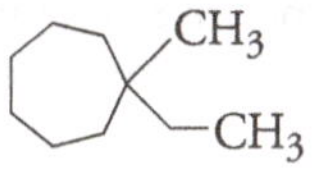 (b) 

(c) 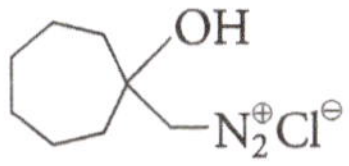 (d) 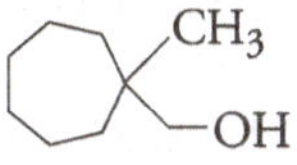

07 Compound *A* is converted to *B* on reaction with $CHCl_3$ and KOH. The compound *B* is toxic and can be decomposed by *C*. *A*, *B* and *C* respectively are **[2021, 20 July Shift-I]**

(a) primary amine, nitrile compound, conc. HCl
(b) secondary amine, isonitrile compound, conc. NaOH
(c) primary amine, isonitrile compound, conc. HCl
(d) secondary amine, nitrile compound, conc. NaOH

08 In the reaction of hypobromite with amide, the carbonyl carbon is lost as **[2021, 18 March Shift-II]**

(a) CO_3^{2-} (b) HCO_3^-
(c) CO_2 (d) CO

09 An organic compound "*A*" on treatment with benzene sulphonyl chloride gives compound *B*. *B* is soluble in dil. NaOH solution. Compound *A* is **[2021, 18 March Shift-II]**

(a) $C_6H_5—N—(CH_3)_2$
(b) $C_6H_5—NHCH_2CH_3$
(c) $C_6H_5—CH_2NHCH_3$
(d) $C_6H_5—\underset{\displaystyle CH_3}{\underset{|}{CH}}—NH_2$

10 Which of the following is least basic? **[2021, 16 March Shift-II]**

(a) $(CH_3CO)\ddot{N}HC_2H_5$

(b) $(C_2H_5)_3\ddot{N}$

(c) $(CH_3CO)_2\ddot{N}H$

(d) $(C_2H_5)_2\ddot{N}H$

11 The most appropriate reagent for conversion of C_2H_5CN into $CH_3CH_2CH_2NH_2$ is **[2020, 5 Sep Shift-I]**

(a) $Na(CN)BH_3$ (b) $LiAlH_4$

(c) $NaBH_4$ (d) CaH_2

Imp. *Order of Basic Strength*

12 In the following compounds, the decreasing order of basic strength will be **[2019, 8 April Shift-I]**

(a) $C_2H_5NH_2 > NH_3 > (C_2H_5)_2NH$

(b) $(C_2H_5)_2NH > NH_3 > C_2H_5NH_2$

(c) $(C_2H_5)_2NH > C_2H_5NH_2 > NH_3$

(d) $NH_3 > C_2H_5NH_2 > (C_2H_5)_2NH$

13 A compound 'X' on treatment with $Br_2/NaOH$,, provided C_3H_9N, which gives positive carbylamine test. Compound 'X' is **[2019, 11 Jan Shift-II]**

(a) $CH_3COCH_2NHCH_3$

(b) $CH_3CH_2CH_2CONH_2$

(c) $CH_3CON(CH_3)_2$

(d) $CH_3CH_2COCH_2NH_2$

14 The major product of following reaction is

$$R—C\equiv N \xrightarrow[\text{(ii) } H_2O]{\text{(i) } AlH(i\text{-}Bu)_2} ?$$

[2019, 9 Jan Shift-I]

(a) RCHO (b) $RCONH_2$

(c) RCOOH (d) RCH_2NH_2

• Level 2

15 Find out the major product formed from the following reaction. [Me: $—CH_3$] **[2024, 4 April Shift-II]**

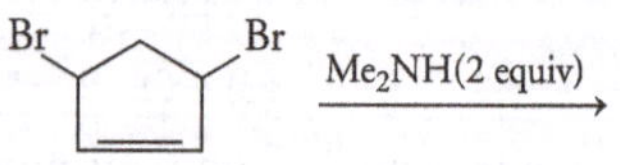

(a) 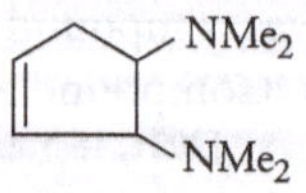 (b) 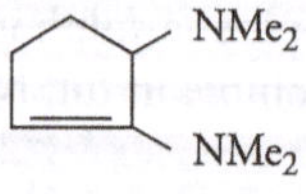

(c) 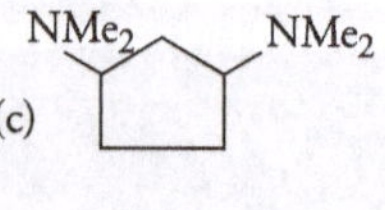 (d) 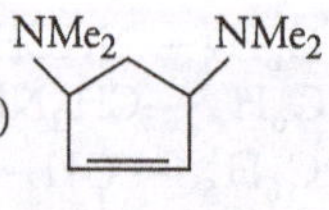

16 Identify *B* formed in the reaction. **[2024, 27 Jan Shift-II]**

$$Cl(CH_2)_4Cl \xrightarrow{\text{Excess } NH_3} A \xrightarrow{NaOH} B + H_2O + NaCl$$

(a) 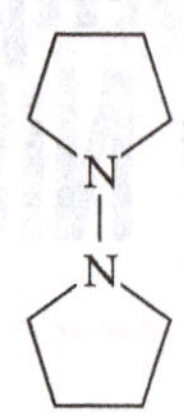 (b) 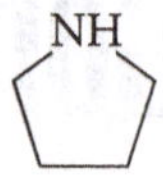

(c) $H_2N—(CH_2)_4—NH_2$ (d) $Cl\overset{+}{N}H_3(CH_2)_4\overset{+}{N}H_3Cl^-$

17 The major product formed in the following reaction is

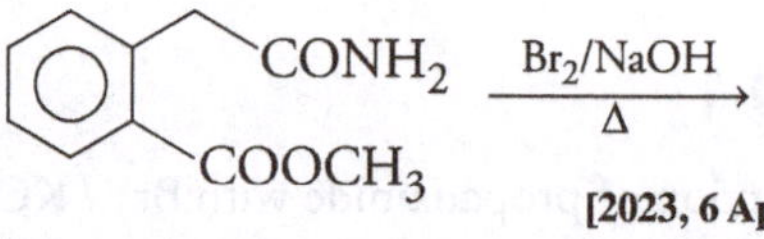

[2023, 6 April Shift-I]

(a) 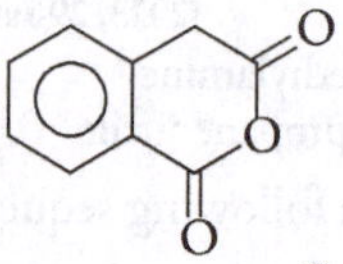 (b) 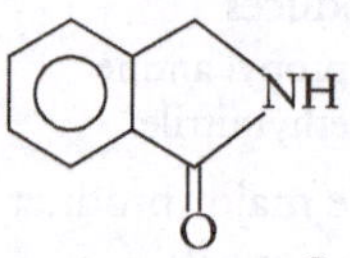

(c) 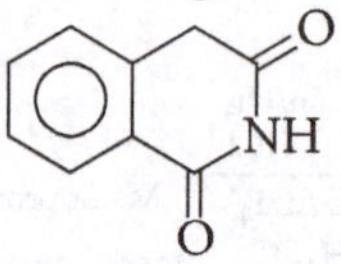 (d) 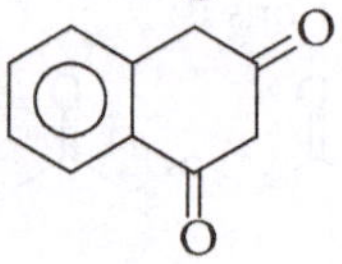

18 In the following reaction, 'A' is

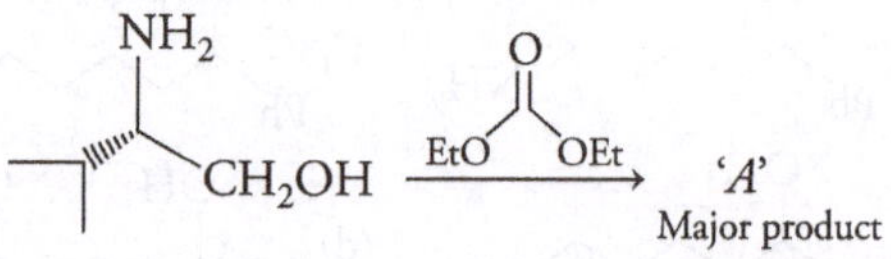

[2023, 1 Feb Shift-I]

(a) 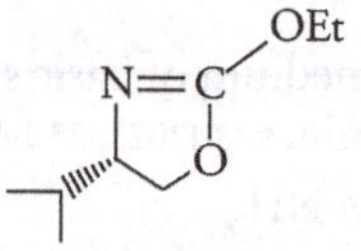 (b) 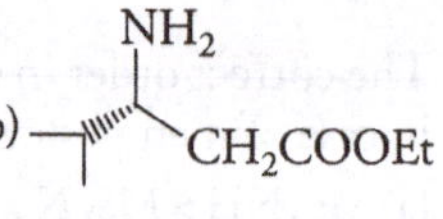

(c) 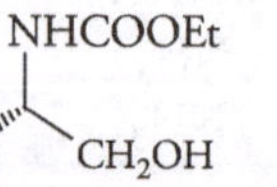 (d) 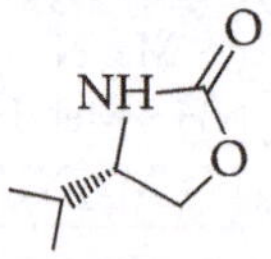

19 An organic compound $[A](C_4H_{11}N)$, shows optical activity and gives N_2 gas on treatment with HNO_2. The compound $[A]$ reacts with $PhSO_2Cl$ producing a compound which is soluble in KOH. The structure of *A* is **[2023, 31 Jan Shift-II]**

(a) 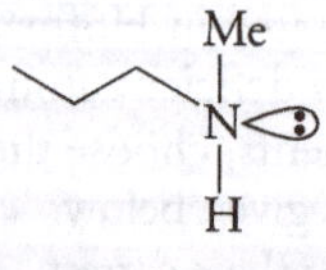 (b) 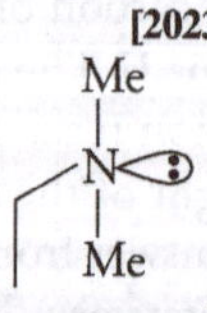

(c) 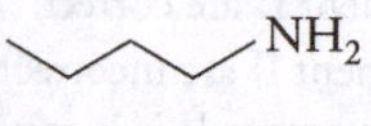 (d) 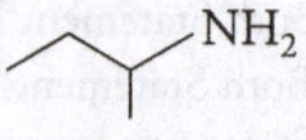

20 Match List-I with List-II.

List-I		List-II	
A.	Benzene sulphonyl chloride	I.	Test for primary amines
B.	Hofmann bromamide reaction	II.	Anti Saytzeff
C.	Carbylamine reaction	III.	Hinsberg reagent
D.	Hofmann orientation	IV.	Known reaction of isocyanates

Choose the correct answer from the options given below. **[2022, 27 July Shift-I]**

(a) A-IV, B-III, C-II, D-I
(b) A-IV, B-II, C-I, D-II
(c) A-III, B-IV, C-I, D-II
(d) A-IV, B-III, C-I, D-II

21 The reaction of $R—\underset{\underset{O}{\|}}{C}—NH_2$ with bromine and KOH gives RNH_2 as the end product. Which one of the following is the intermediate product formed in this reaction? **[2022, 25 June Shift-I]**

(a) $R—\underset{\underset{O}{\|}}{C}—NH—Br$
(b) $R—NH—Br$
(c) $R—N{=}C{=}O$
(d) $R—\underset{\underset{O}{\|}}{C}—NBr_2$

22 Which of the following is not a correct statement for primary aliphatic amines? **[2021, 27 Aug Shift-I]**

(a) The intermolecular association in primary amines is less than the intermolecular association in secondary amines.
(b) Primary amines on treating with nitrous acid solution form corresponding alcohols except methyl amine.
(c) Primary amines are less basic than the secondary amines.
(d) Primary amines can be prepared by the gabriel phthalimide synthesis.

23 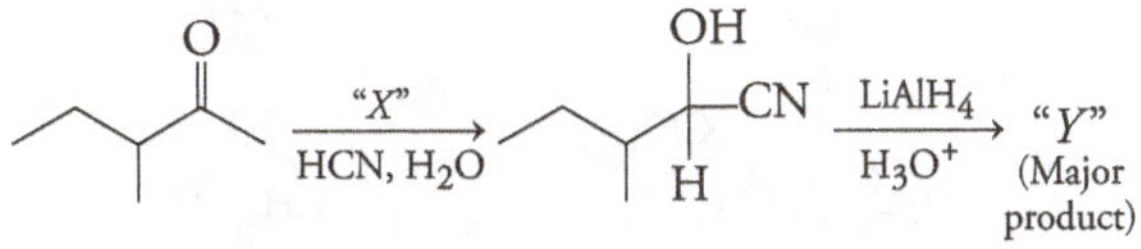

Consider the given reaction, identify 'X' and 'Y'. **[2021, 26 Aug Shift-II]**

(a) X – NaOH Y – 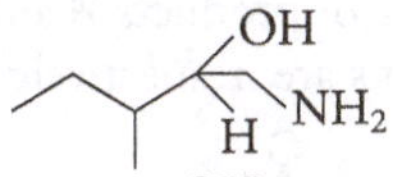

(b) X – NaOH Y – 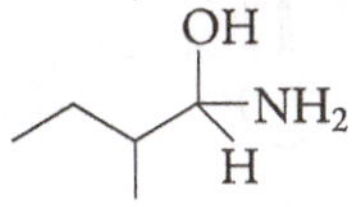

(c) X – HNO_3 Y – 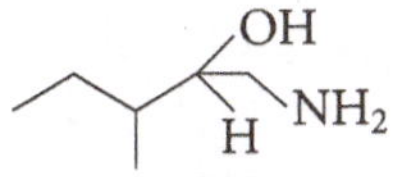

(d) X – HNO_3 Y – 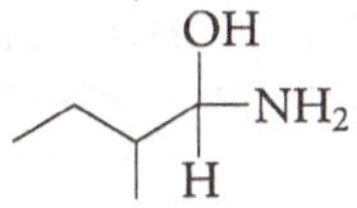

24 What is A in the following reaction?

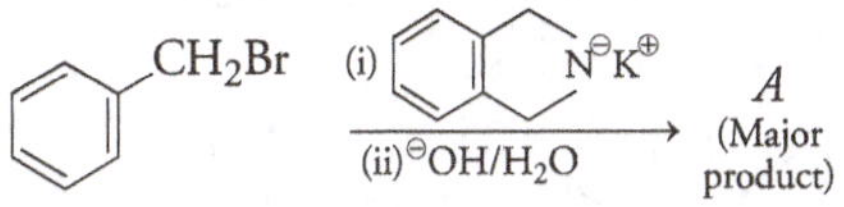

[2021, 27 July Shift-II]

(a) 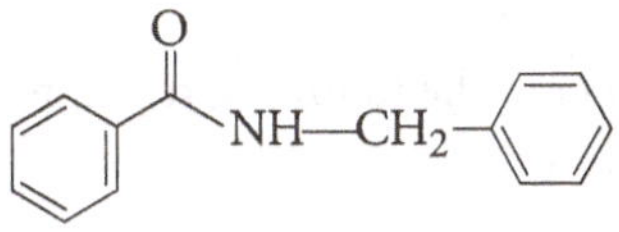

(b) 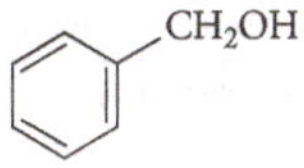

(c) 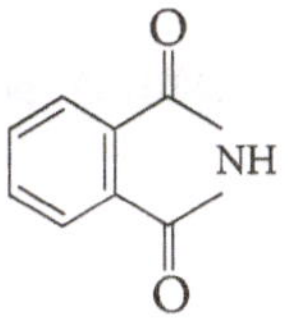

(d) 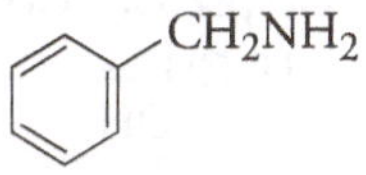

25 Ammonolysis of alkyl halides followed by the treatment with NaOH solution can be used to prepare primary, secondary and tertiary amines. The purpose of NaOH in the reaction is **[2021, 16 March Shift-II]**

(a) to remove basic impurities
(b) to activate NH_3 used in the reaction
(c) to remove acidic impurities
(d) to increase the reactivity of alkyl halide

26 The increasing order of pK_b for the following compounds will be

$NH_2—CH{=}NH$, (A) 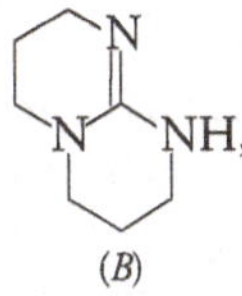

(B), $CH_3—NH—CH_3$ (C)

[2020, 7 Jan Shift-I]

(a) (B) < (C) < (A)
(b) (C) < (A) < (B)
(c) (A) < (B) < (C)
(d) (B) < (A) < (C)

27 The major products *A* and *B* for the following reactions are, respectively **[2019, 9 April Shift-II]**

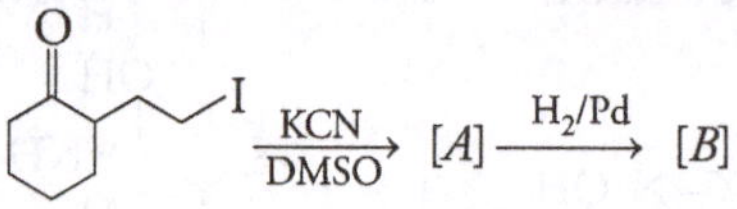

(a) 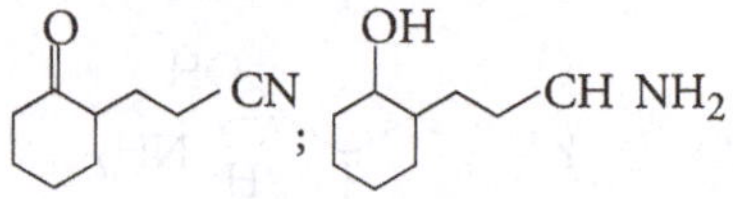

(b) 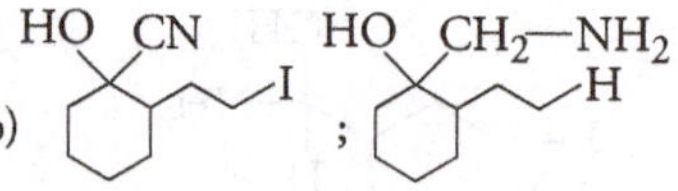

(c) 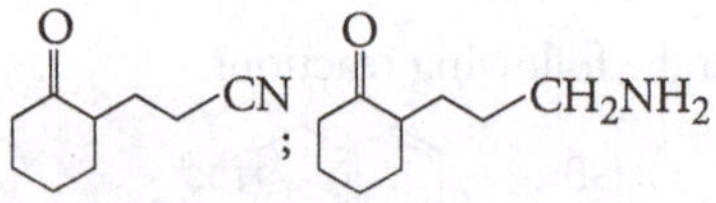

(d) 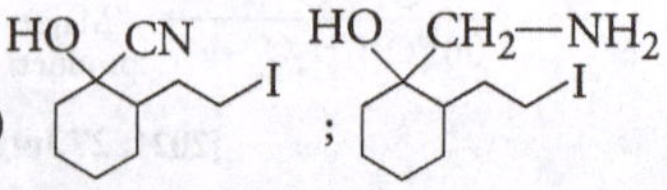

28 The major product of the following reaction is

$$CH_3CH(OH)CH_2CH_2NH_2 \xrightarrow[\text{Triethylamine}]{\text{Ethyl formate (1 equiv.)}}$$

(a) $CH_3CH(OH)CH_2CH_2NHCHO$ **[2019, 10 April Shift-I]**

(b) $CH_3CH{=}CH{-}CH_2NH_2$

(c) $CH_3{-}CH(OH){-}CH{=}CH_2$

(d) 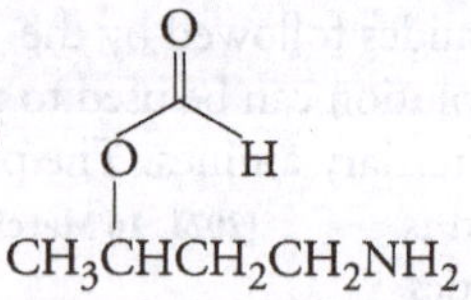

• Level 3

29 The correct order of basic nature in aqueous solution for the bases **[2025, 2 April Shift-I]**

NH_3, $H_2N{-}NH_2$, $CH_3CH_2NH_2$, $(CH_3CH_2)_2NH$ and $(CH_3CH_2)_3N$ is:

(a) $NH_3 < H_2N{-}NH_2 < (CH_3CH_2)_3N < CH_3CH_2NH_2 < (CH_3CH_2)_2NH$

(b) $H_2N{-}NH_2 < NH_3 < (CH_3CH_2)_3N < CH_3CH_2NH_2 < (CH_3CH_2)_2NH$

(c) $NH_2{-}NH_2 < NH_3 < CH_3CH_2NH_2 < (CH_3CH_2)_3N < (CH_3CH_2)_2NH$

(d) $NH_3 < H_2N{-}NH_2 < CH_3CH_2NH_2 < (CH_3CH_2)_2NH < (CH_3CH_2)_3N$

30 The product (*A*) formed in the following reaction sequence is **[2025, 24 Jan Shift-I]**

$$CH_3{-}C{\equiv}CH \xrightarrow[\text{(iii) } H_2/Ni]{\text{(i) } Hg^{2+}, H_2SO_4 \text{ (ii) HCN}} \underset{\text{Product}}{(A)}$$

(a) $CH_3{-}CH_2{-}CH(NH_2){-}CH_2{-}OH$

(b) $CH_3{-}C(NH_2)(CH_3){-}CH_2{-}OH$

(c) $CH_3{-}C(OH)(CH_3){-}CH_2{-}NH_2$

(d) $CH_3{-}CH_2{-}CH(OH){-}CH_2{-}NH_2$

31 In the reaction given below

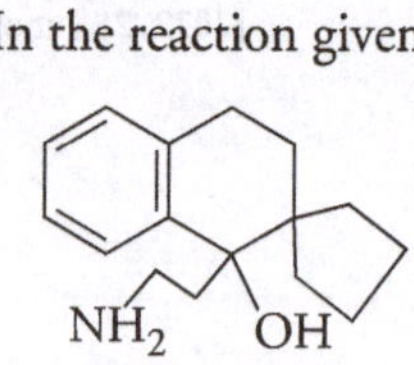

[2023, 13 April Shift-I]

(a) 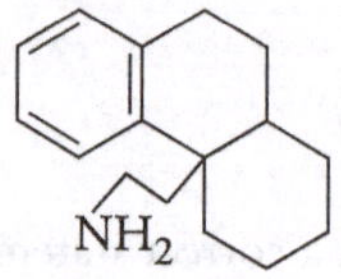

(b) 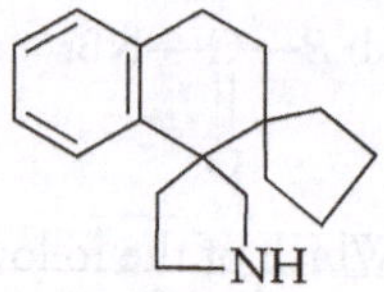

(c) 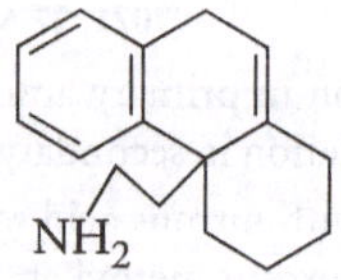

(d) 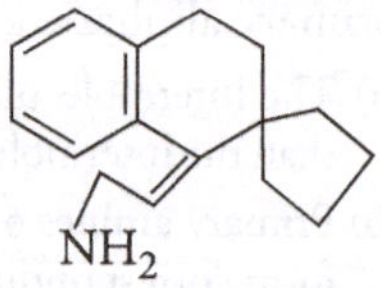

32 Which among the following is the strongest Bronsted base? **[2022, 29 July Shift-I]**

(a) 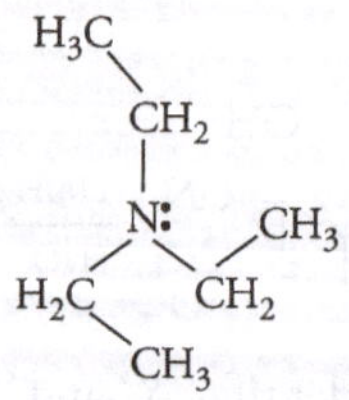

(b) 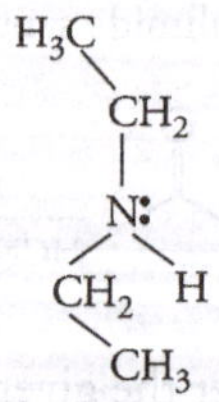

(c) 

(d) 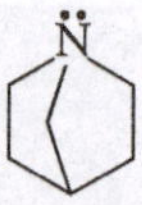

33 The conversion of propan-1-ol to *n*-butylamine involves the sequential addition of reagents. The correct sequential order of reagents is **[2022, 24 June Shift-II]**

(a) (i) $SOCl_2$ (ii) KCN (iii) $H_2/Ni, Na(Hg)/C_2H_5OH$
(b) (i) HCl (ii) H_2/Ni, $Na(Hg)/C_2H_5OH$
(c) (i) $SOCl_2$ (ii) KCN (iii) CH_3NH_2
(d) (i) HCl (ii) CH_3NH_2

34 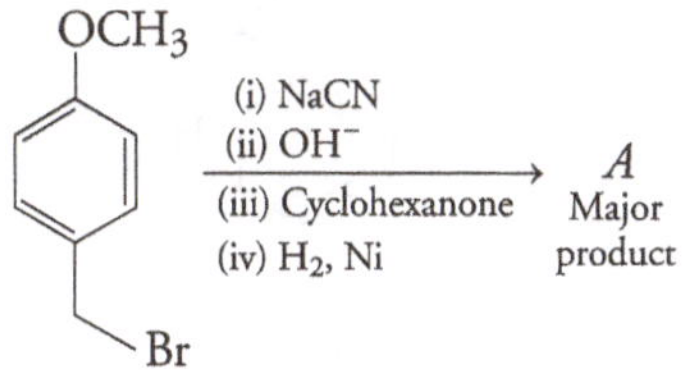

The major product of the above reaction is **[2022, 24 June Shift-I]**

(a) 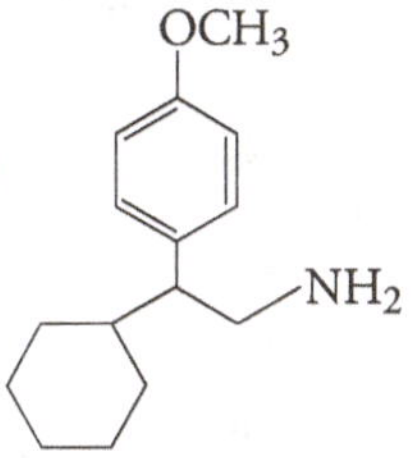

(b) 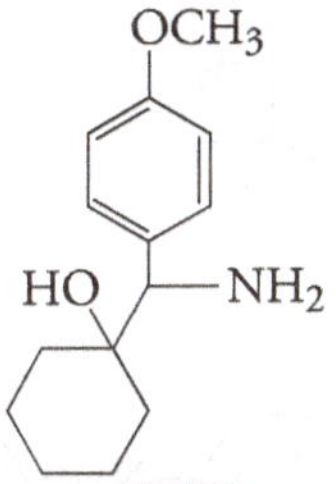

(c) 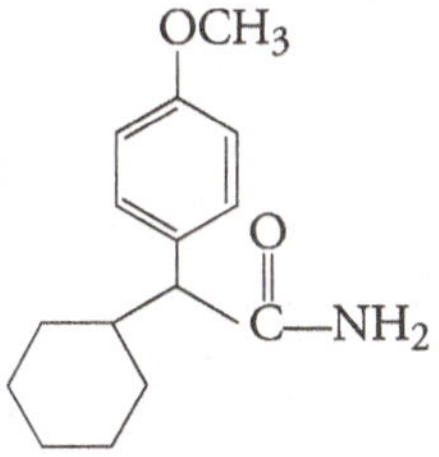

(d) 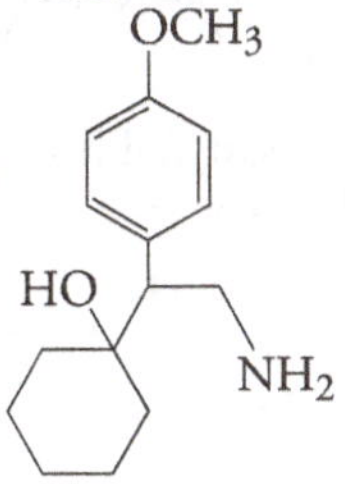

35 The major product of the following reaction is

$$CH_3-\underset{\displaystyle CH_3}{CH}-CH_2CH_2-\overset{\displaystyle O}{\overset{\|}{C}}-Cl \xrightarrow[\substack{\text{(iii) } NaNO_2, HCl \\ \text{(iv) } H_2O}]{\substack{\text{(i) Alcoholic } NH_3 \\ \text{(ii) } NaOH, Br_2}} \text{Major product}$$

[2021, 27 Aug Shift-I]

(a) $CH_3-\underset{\displaystyle CH_3}{CH}-\overset{\displaystyle Br}{CH}-CH_2OH$

(b) $CH_3-\underset{\displaystyle CH_3}{CH}-CH_2-CH_2-CH_2OH$

(c) $CH_3-\underset{\displaystyle CH_3}{CH}-CH_2-CH_2OH$

(d) $CH_3-\underset{\displaystyle CH_3}{CH}-CH_2-CH_2-Cl$

36 Which of the following reaction does not involve Hoffmann bromamide degradation? **[2021, 16 March Shift-I]**

(a) 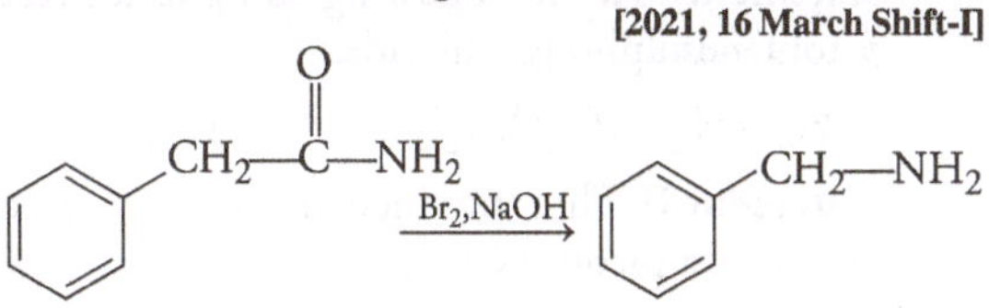

(b) 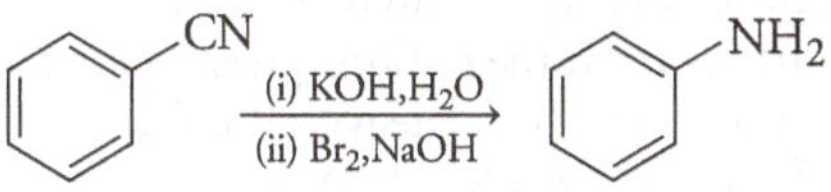

(c) 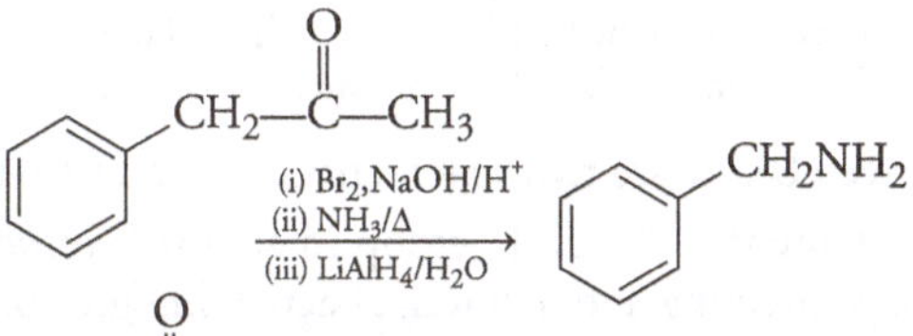

(d) 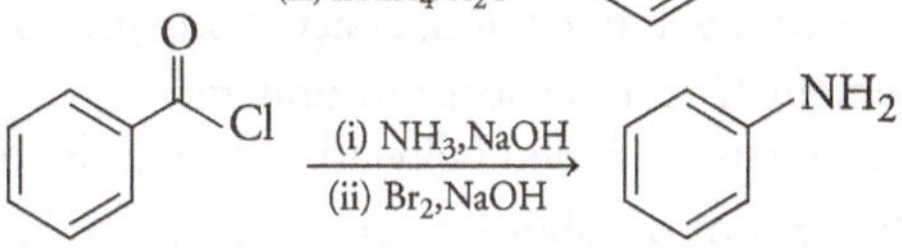

37 Which of the following is not a correct method of the preparation of benzylamine from cyanobenzene? **[2019, 10 April Shift-II]**

(a) H_2/Ni
(b) (i) HCl/H_2O (ii) $NaBH_4$
(c) (i) $LiAlH_4$ (ii) H_3O^+
(d) (i) $SnCl_2$ + HCl (gas) (ii) $NaBH_4$

Numerical Value Type Questions

38 A compound (X) with molar mass 108 g mol^{-1} undergoes acetylation to give product with molar mass 192 g mol^{-1}. The number of amino groups in the compound (X) is .... **[2024, 31 Jan Shift-II]**

39 In gaseous triethyl amine, the "—C—N—C—" bond angle is .... degree. **[2021, 27 July Shift-I]**

Topic 2 / Aromatic Amines

• Level 1

01 Correct order of basic strength of pyrrole , pyridine 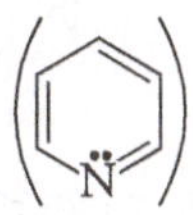 and piperidine 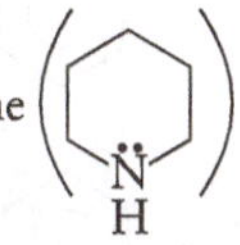 is **[2024, 9 April Shift-I]**

(a) pyridine > piperidine > pyrrole
(b) piperidine > pyridine > pyrrole
(c) pyrrole > piperidine > pyridine
(d) pyrrole > pyridine > piperidine

02 Given below are two statements : **[2024, 8 April Shift-II]**

Statement I All the following compounds react with *p*-toluenesulphonyl chloride.

$C_6H_5NH_2$ $(C_6H_5)_2NH$ $(C_6H_5)_3N$

Statement II Their products in the above reaction are soluble is aqueous NaOH.

In the light of the above statements, choose the correct answer from the options given below:

(a) Statement I is true but Statement II is false.
(b) Both Statement I and Statement II are true
(c) Both Statement I and Statement II are false.
(d) Statement I is false but Statement II is true.

03 Given below are two statements. **[2024, 1 Feb Shift-I]**

Statement I The NH_2 group in aniline is *ortho* and *para* directing and a powerful activating group.

Statement II Aniline does not undergo Friedel-Craft's reaction (alkylation and acylation).

In the light of the above statements, choose the **most appropriate** answer from the options given below.

(a) Both Statement I and Statement II are correct.
(b) Statement I is correct but Statement II is incorrect.
(c) Statement I is incorrect but Statement II is correct.
(d) Both Statement I and Statement II are incorrect.

04 Given below are two statements. **[2024, 1 Feb Shift-I]**

Statement I Aminobenzene and aniline are same organic compounds.

Statement II Aminobenzene and aniline are different organic compounds.

In the light of the above statements, choose the **most appropriate** answer from the options given below.

(a) Statement I is incorrect but Statement II is correct.
(b) Both Statement I and Statement II are correct.
(c) Both Statement I and Statement II are incorrect.
(d) Statement I is correct but Statement II is incorrect.

05 The product *A* formed in the following reaction is

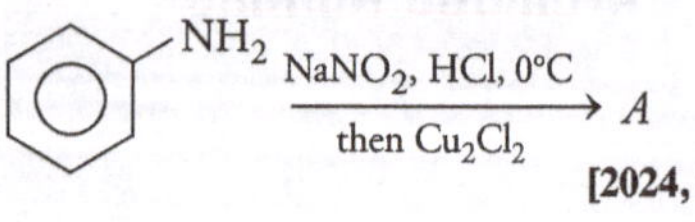

[2024, 29 Jan Shift-II]

(a) 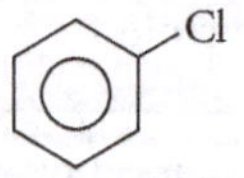
(b) 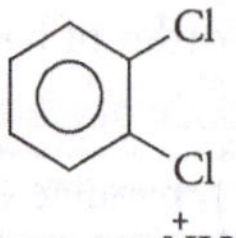
(c) 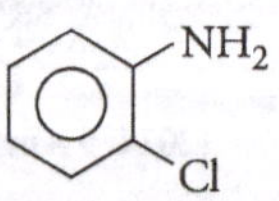
(d) 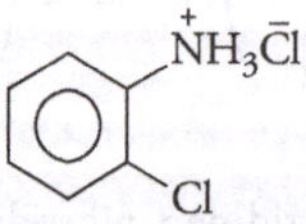

06 The arenium ion which is not involved in the bromination of aniline is .... **[2024, 29 Jan Shift-I]**

(a) 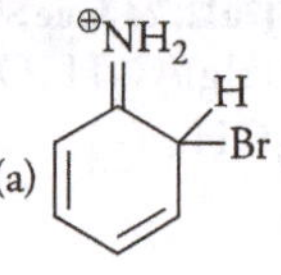
(b) 
(c) 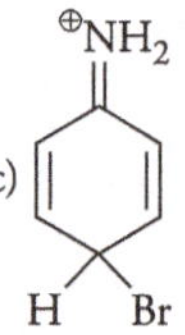
(d) 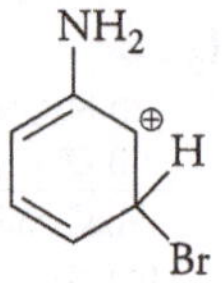

07 Which of the following is strongest bronsted base? **[2024, 27 Jan Shift-II]**

(a) 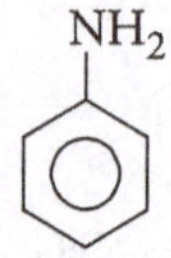
(b) 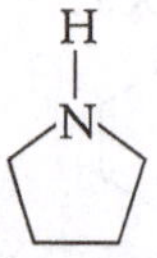
(c) 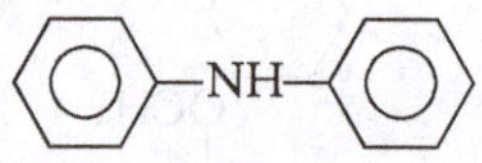
(d) 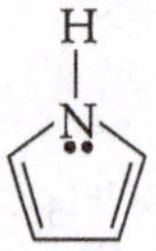

08 The most stable carbocation among the following is **[2023, 30 Jan Shift-II]**

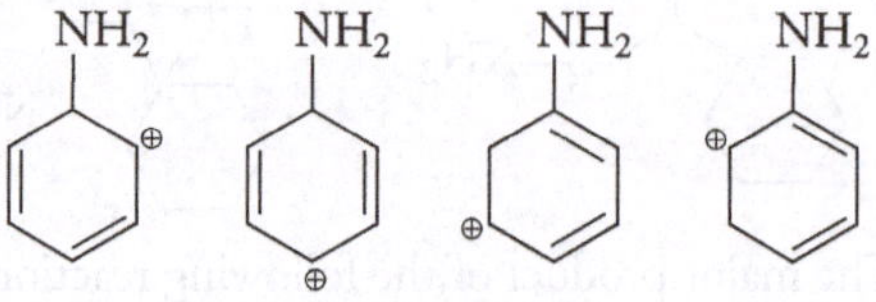

A *B* *C* *D*

(a) *D* (b) *C*
(c) *B* (d) *A*

09 Given below are two statements :

Statement I Pure aniline and other arylamines are usually colourless.

Statement II Arylamines get coloured on storage due to atmospheric reduction.

In the light of the above statements, choose the most appropriate answer form the options given below. **[2023, 24 Jan Shift-II]**

(a) Both Statement I and Statement II are incorrect.
(b) Both Statement I and Statement II are correct.
(c) Statement I is correct but Statement II is incorrect.
(d) Statement I is incorrect but Statement II is correct.

10 The Hinsberg reagent is **[2022, 29 July Shift-II]**

(a) 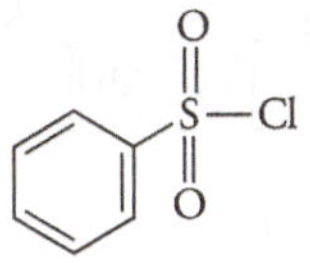 (b) 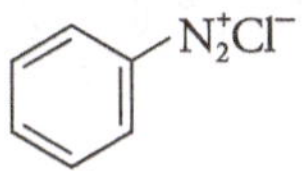

(c) 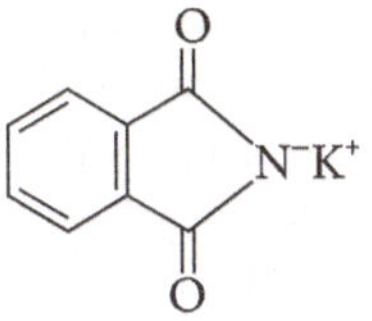 (d) 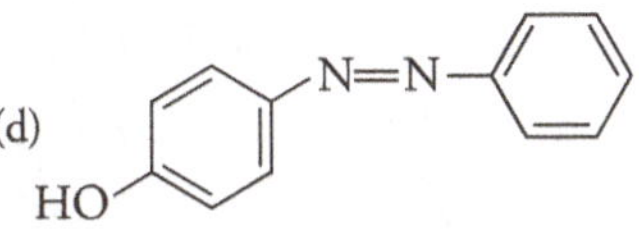

11 Given below are two statements, one is labelled as Assertion (A) and the other is labelled as Reason (R).

Assertion (A) Aniline on nitration yields *ortho*, *meta* and *para* nitro derivatives of aniline.

Reason (R) Nitrating mixture is a strong acidic mixture.

In the light of the above statements, choose the correct answer from the options given below. **[2022, 28 July Shift-II]**

(a) Both A and R are true and R is the correct explanation of A
(b) Both A and R are true but R is not the correct explanation of A
(c) A is true but R is false
(d) A is false but R is true

12 Which statement is not correct for *p*-toluene sulphonyl chloride? **[2022, 26 June Shift-II]**

(a) It is known as Hinsberg's reagent.
(b) It is used to distinguish primary and secondary amines.
(c) On treatment with secondary amine, it leads to a product, that is soluble in alkali.
(d) It doesn't react with tertiary amines.

13 In Friedel-Craft's alkylation of aniline, one gets **[2022, 29 June Shift-II]**

(a) alkylated product with *ortho* and *para* substitution.
(b) secondary amine after acidic treatment.
(c) an amide product.
(d) positively charged nitrogen at benzene ring.

14 Given below are two statements.

Statement I Aniline is less basic than acetamide.

Statement II In aniline, the lone pair of electrons on nitrogen atom is delocalised over benzene ring due to resonance and hence less available to a proton.

Choose the most appropriate option **[2021, 27 July Shift-I]**

(a) Statement I is true but Statement II is false.
(b) Statement I is false but Statement II is true.
(c) Both Statement I and Statement II are true.
(d) Both Statement I and Statement II are false.

15 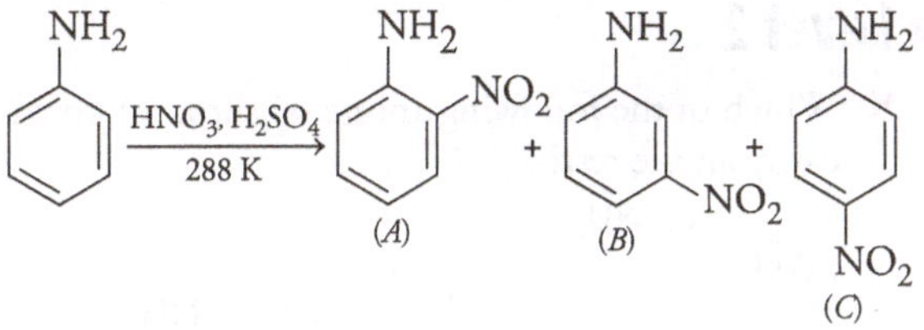

Consider the given reaction, percentage yield of **[2021, 18 March Shift-II]**

(a) (*C*) > (*A*) > (*B*) (b) (*B*) > (*C*) > (*A*)
(c) (*A*) > (*C*) > (*B*) (d) (*C*) > (*B*) > (*A*)

16 Primary, secondary and tertiary amines can be separated using **[2021, 17 March Shift-II]**

(a) *para*-toluene sulphonyl chloride
(b) chloroform and KOH
(c) benzene sulphonic acid
(d) acetyl amide

17 An amine on reaction with benzene sulphonyl chloride produces a compound insoluble in alkaline solution. This amine can be prepared by ammonolysis of ethyl chloride. The correct structure of amine is **[2021, 26 Feb Shift-I]**

(a) 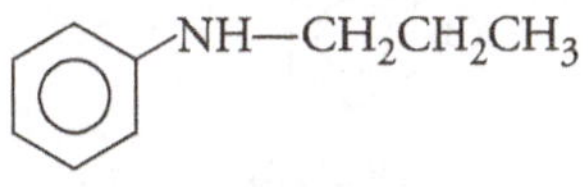

(b) $CH_3CH_2NH_2$
(c) $CH_3CH_2CH_2NHCH_3$
(d) $CH_3CH_2CH_2NH—CH_2CH_3$

18 In the following reaction the reason why *meta*-nitro product also formed is **[2021, 24 Feb Shift-I]**

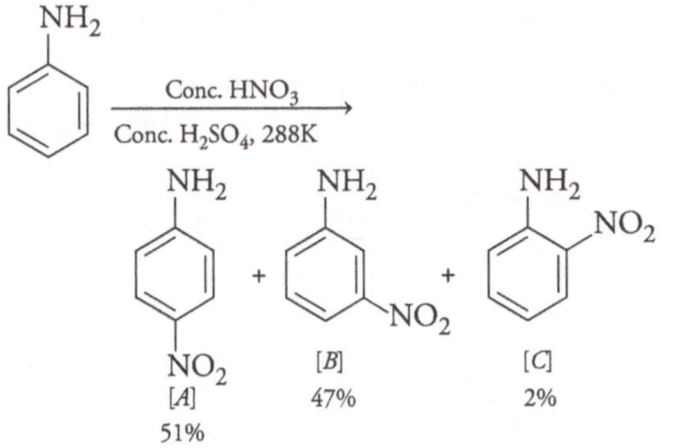

(a) $—NH_2$ group is highly *meta*-directive
(b) $—NO_2$ substitution always takes place at *meta*-position
(c) Formation of anilinium ion
(d) Low temperature

19 Arrange the following amines in the decreasing order of basicity : **[2019, 9 Jan Shift-I]**

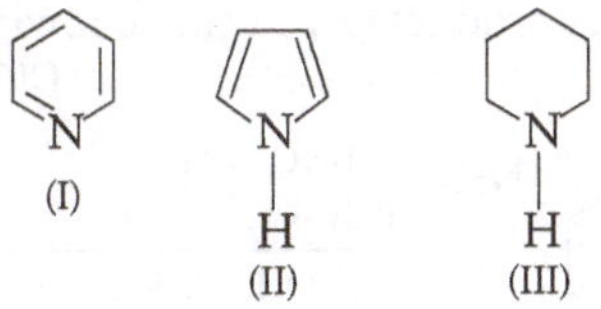

(a) I > II > III (b) III > II > I
(c) I > III > II (d) III > I > II

• Level 2

20 Which of the following amine (s) show (s) positive carbylamine test?

(A) 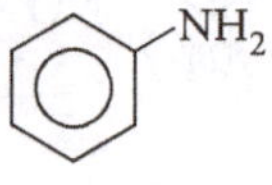

(B) $(CH_3)_2NH$

(C) CH_3NH_2

(D) $(CH_3)_3N$

(E) 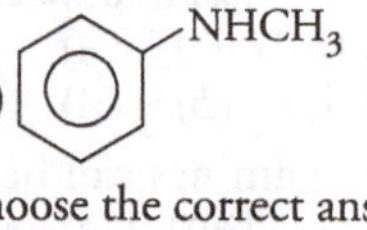

Choose the correct answer from the options given below: **[2025, 7 April Shift-I]**

(a) B, C and D only (b) C only
(c) A and E only (d) A and C only

Imp. *Test of Amines*

21 Which among the following react with Hinsberg's reagent? **[2025, 23 Jan Shift-I]**

(A) 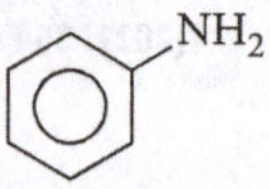

(B) 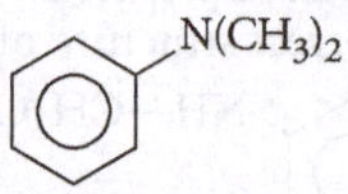

(C) $CH_3—NH_2$

(D) $N(CH_3)_3$

(E) 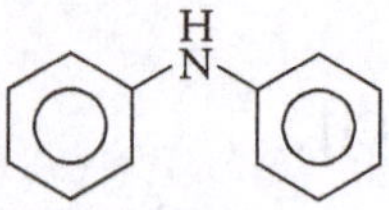

Choose the correct answer from the options given below
(a) C and D only (b) B and D only
(c) A, C and E only (d) A, B and E only

22 The product *B* formed in the following reaction sequence is **[2025, 28 Jan Shift-II]**

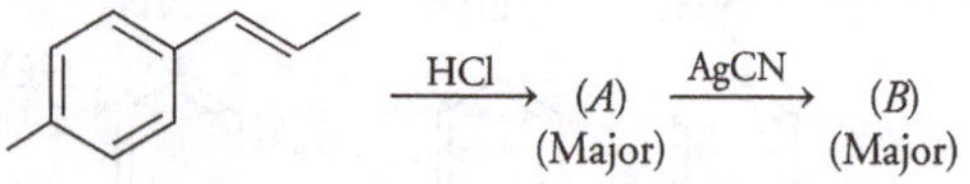

(a) 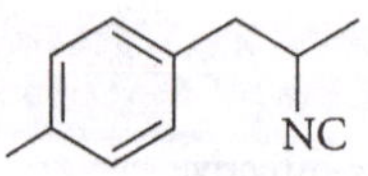

(b) 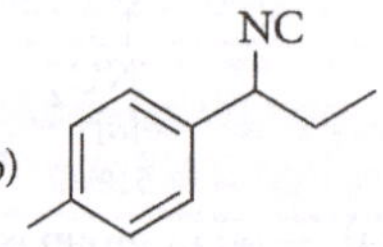

(c) 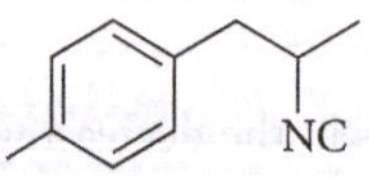

(d) 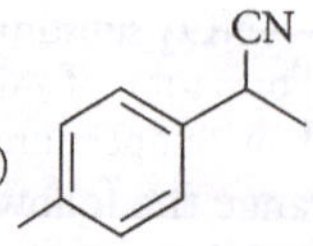

23 Identify the product (*A*) in the following reaction. **[2024, 6 April Shift-II]**

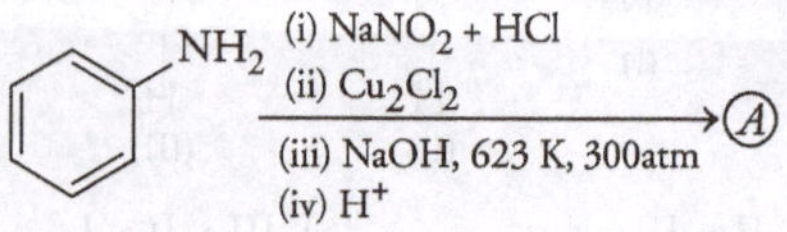

(a) 

(b) 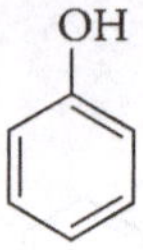

(c) 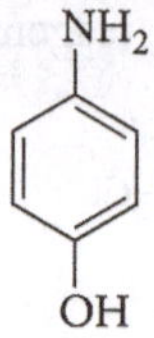

(d) 

24 Decreasing order of reactivity towards electrophilic substitution for the following compounds is

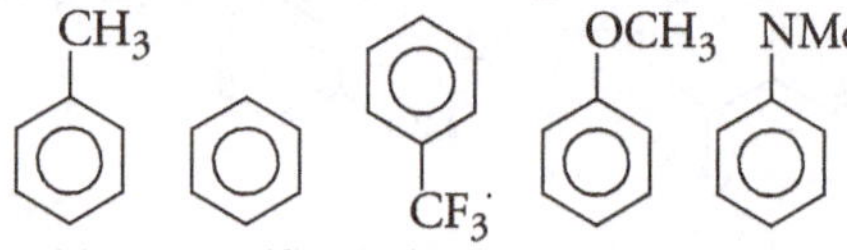

(*a*) (*b*) (*c*) (*d*) (*e*)

[2023, 15 April Shift-I]

(a) $a > d > e > b > c$ (b) $e > d > a > b > c$
(c) $c > b > a > d > e$ (d) $d > a > e > c > b$

25 The major products *A* and *B* from the following reactions are **[2023, 6 April Shift-I]**

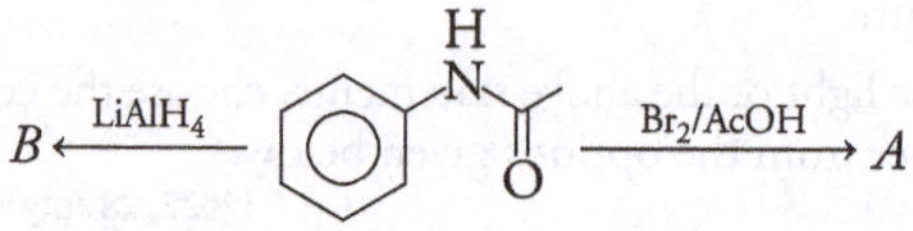

(a) $A =$ 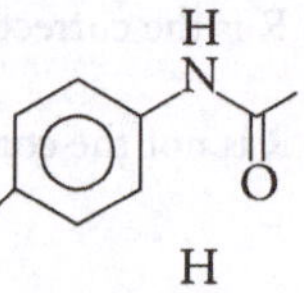 , $B =$ 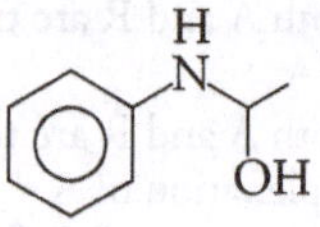

(b) $A =$ 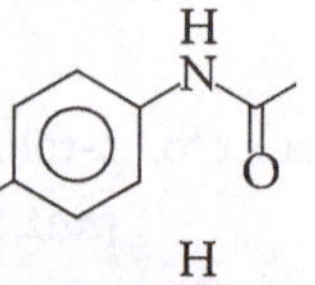 , $B =$ 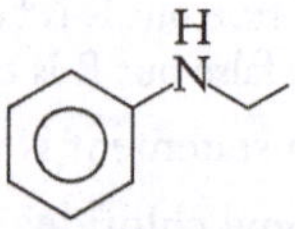

(c) $A =$ 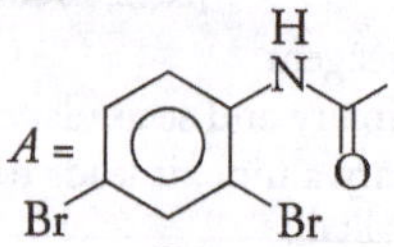 , $B =$ 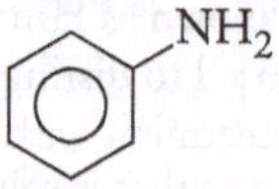

(d) $A =$ 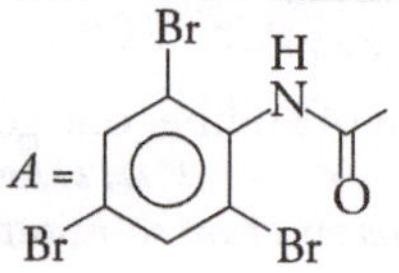 , $B =$ 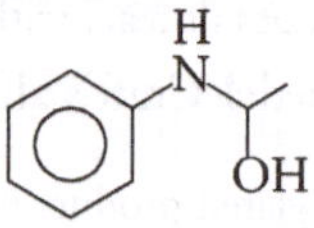

Imp. *pK_b Value of Amines*

26 Match List-I with List-II.

List-I (Amines)		List-II (pK_b)	
A.	Aniline	I.	3.25
B.	Ethanamine	II.	3.00
C.	N-ethylethanamine	III.	9.38
D.	N, N-diethylethanamine	IV.	3.29

Choose the correct answer from the options given below. **[2023, 25 Jan Shift-II]**

(a) A-III, B-II, C-I, D-IV (b) A-III, B-IV, C-II, D-I
(c) A-I, B-IV, C-II, D-III (d) A-III, B-II, C-IV, D-I

27 Given below are two statements, one is labelled as Assertion (A) and the other is labelled as Reason (R).

Assertion (A) Experimental reaction of CH_3Cl with aniline and anhydrous $AlCl_3$ does not give *o* and *p*-methylaniline. **[2022, 26 July Shift-I]**

Reason (R) The —NH_2 group of aniline becomes deactivating because of salt formation with anhydrous $AlCl_3$ and hence yields *m*-methyl aniline as the product.

In the light of the above statements, choose the most appropriate answer from the options given below

(a) Both (A) and (R) are true and (R) is the correct explanation of (A).
(b) Both (A) and (R) are true but (R) is not the correct explanation of (A).
(c) (A) is true, but (R) is false.
(d) (A) is false, but (R) is true.

28 ‘*A*’ ($C_8H_6Cl_2O$) $\xrightarrow{NH_3}$ C_8H_8ClNO $\xrightarrow[NaOH]{Br_2}$ 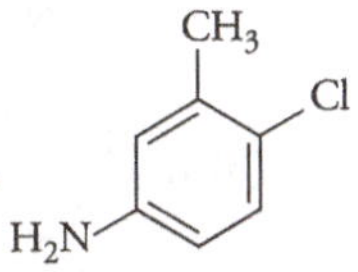

Consider the above reaction, the compound ‘*A*’ is **[2022, 29 July Shift-I]**

(a) 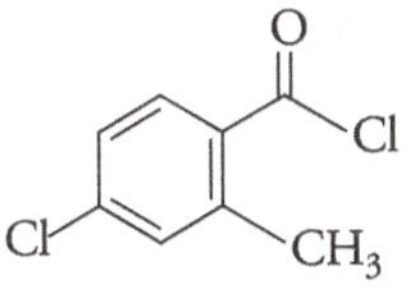
(b) 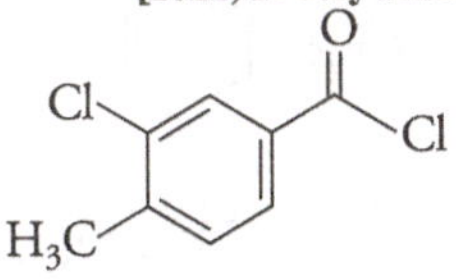
(c) 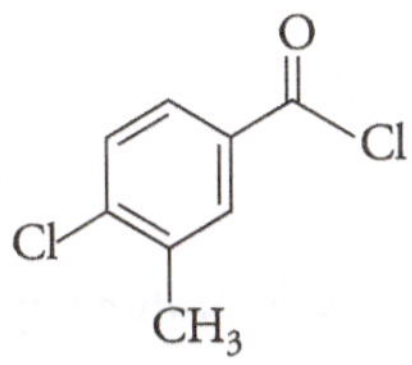
(d) 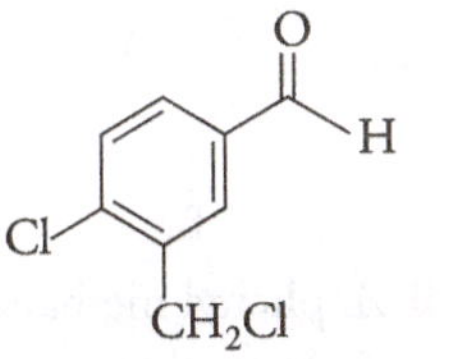

29 With respect to the following reaction, consider the given statements. **[2022, 28 June Shift-II]**

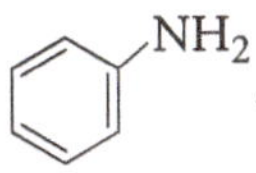 $\xrightarrow[H_2SO_4,\ 288\ K]{HNO_3}$ Products

(A) *o*-nitroaniline and *p*-nitroaniline are the predominant products.
(B) *p*-nitroaniline and *m*-nitroaniline are the predominant products.
(C) HNO_3 acts as an acid.
(D) H_2SO_4 acts as an acid.

Choose the correct option.

(a) (A) and (C) are correct statements.
(b) (A) and (D) are correct statements.
(c) (B) and (D) are correct statements.
(d) (B) and (C) are correct statements.

30 Which one of the products of the following reactions does not react with Hinsberg reagent to form sulphonamide? **[2021, 25 July Shift-I]**

(a) 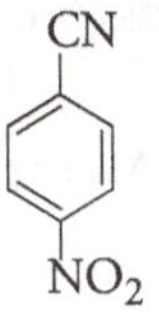 + Na/Hg $\xrightarrow{C_2H_5OH}$

(b) 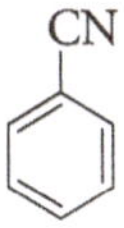 + $SnCl_2$ + HCl $\longrightarrow$

(c) 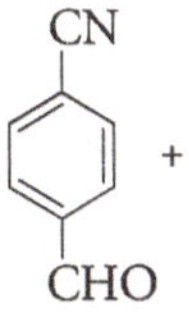 + $LiAlH_4$ $\xrightarrow{H_3O^+}$

(d) 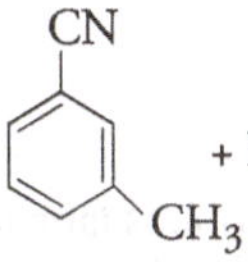 + H_2/Ni $\longrightarrow$

31 The major products *A* and *B* formed in the following reaction sequence are **[2021, 31 Aug Shift-II]**

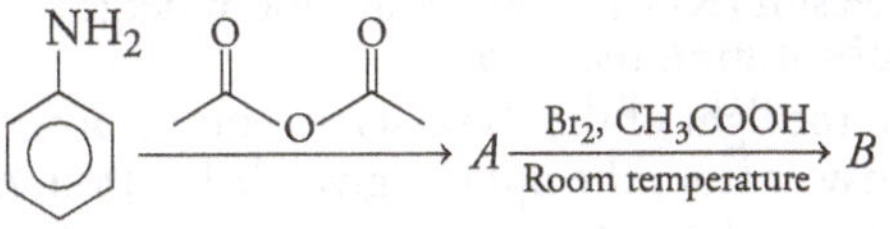

(a) *A* = 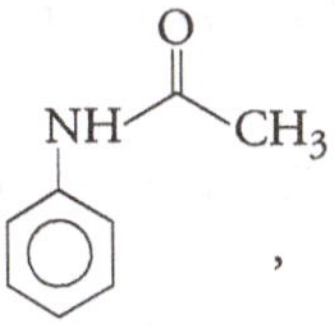 , *B* = 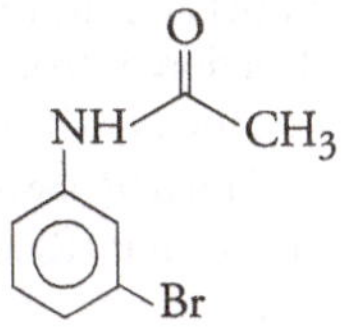

(b) *A* = 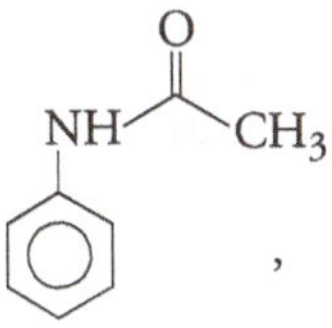 , *B* = 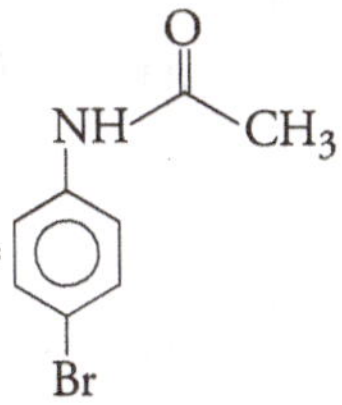

(c) *A* = 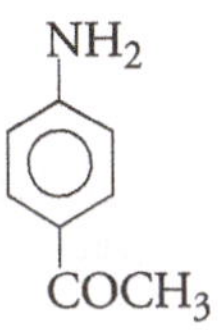 , *B* = 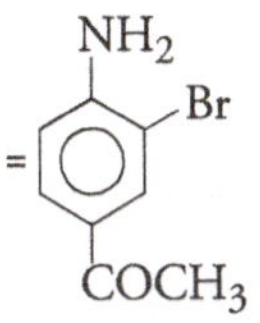

(d) *A* = 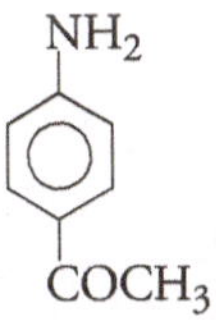 , *B* = 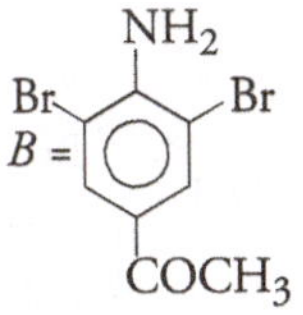

32 The major product formed in the following reaction is **[2021, 26 Aug Shift-I]**

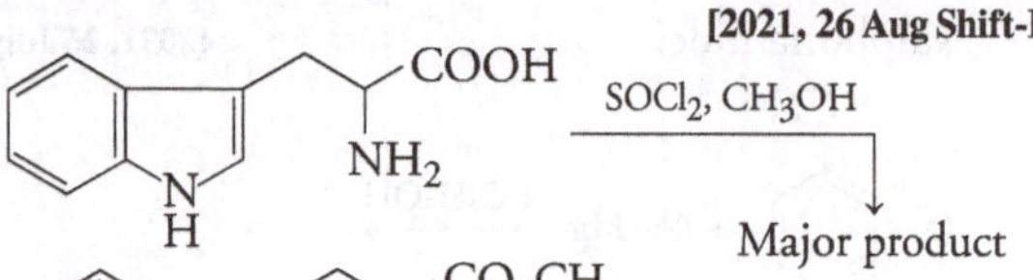

(a) 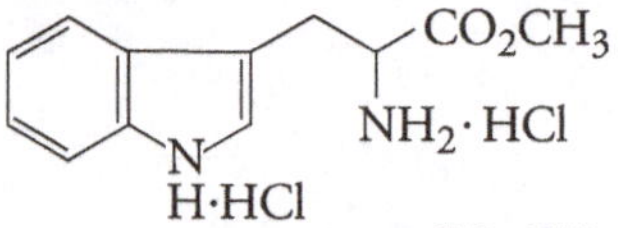

(b) 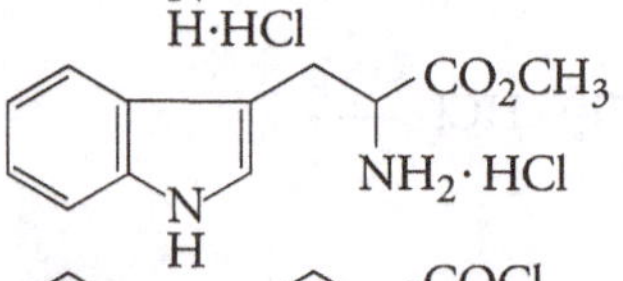

(c) 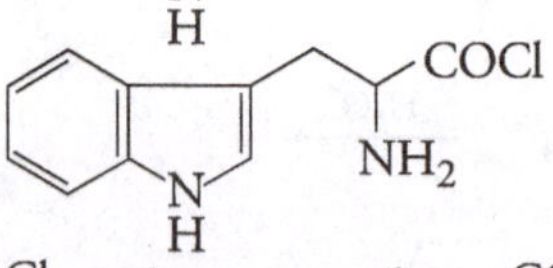

(d) 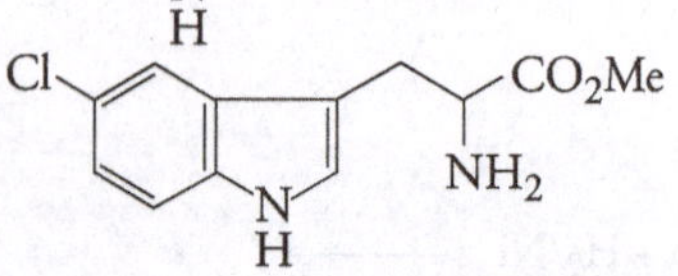

33 Given below are two statements, one is labelled as Assertion (A) and other is labelled as Reason (R).

Assertion (A) Gabriel phthalimide synthesis cannot be used to prepare aromatic primary amines.

Reason (R) Aryl halides do not undergo nucleophilic substitution reaction.

In the light of the above statements, choose the correct answer from the options given below **[2021, 22 July Shift-I]**

(a) Both (A) and (R) are true but (R) is not the correct explanation of (A).
(b) (A) is false but (R) is true.
(c) Both (A) and (R) are true and (R) is correct explanation of (A).
(d) (A) is true but (R) is false.

34

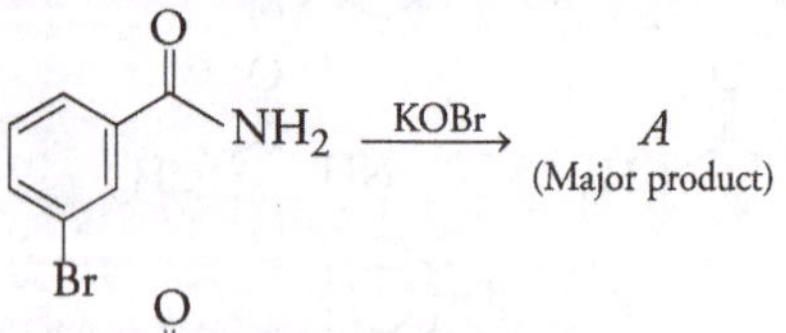

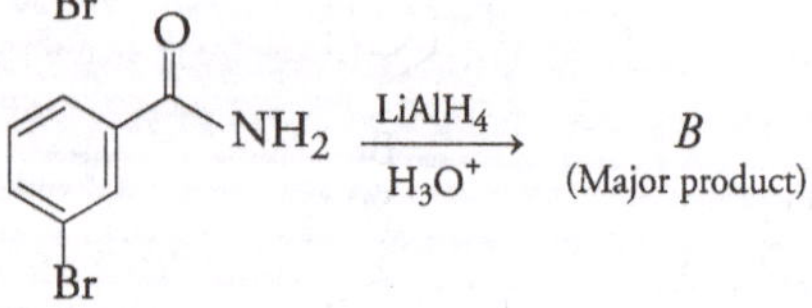

In the above reactions, products *A* and *B* respectively are **[2021, 20 July Shift-II]**

(a) 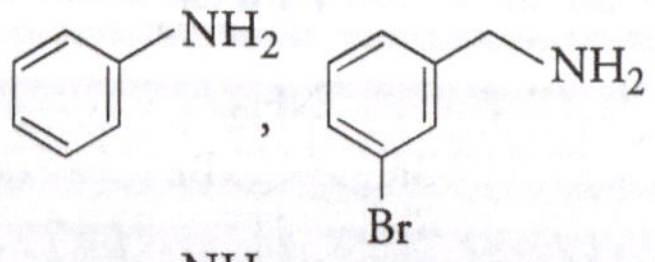

(b) 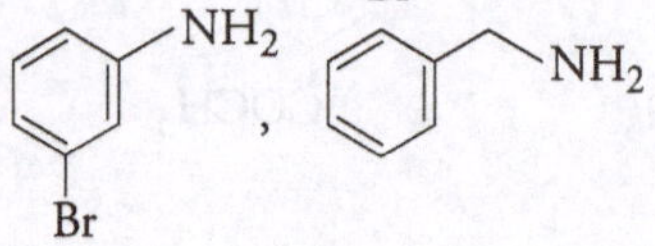

(c) 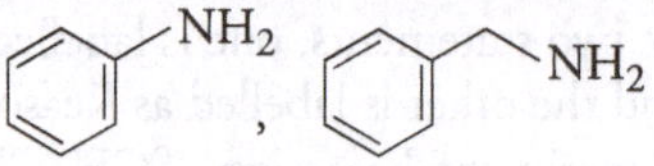

(d) 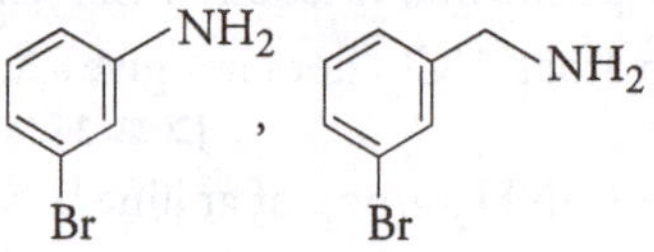

35 Hofmann bromamide degradation of benzamide gives product *A*, which upon heating with $CHCl_3$ and NaOH gives product *B*. The structures of *A* and *B* are **[2021, 17 March Shift-I]**

(a) *A* = 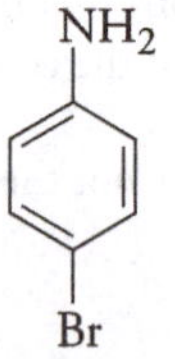 *B* = 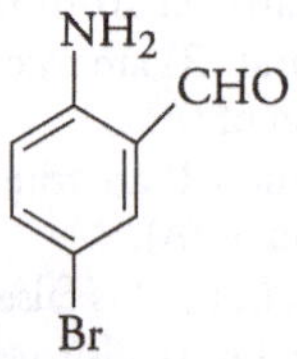

(b) *A* = 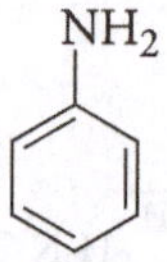 *B* = 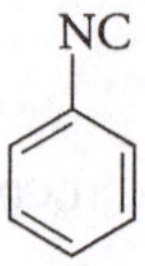

(c) *A* = 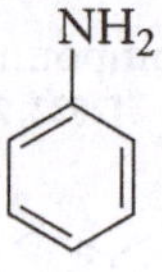 *B* = 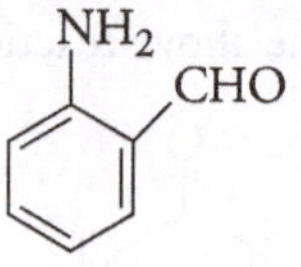

(d) *A* = 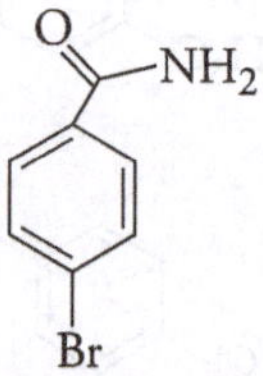 *B* = 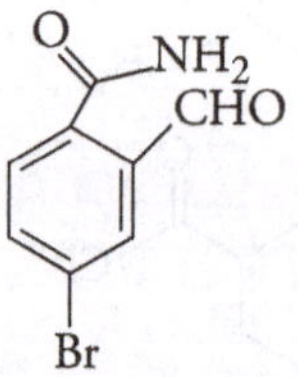

36 *A*. phenyl methanamine *B*. N,N-dimethylaniline
C. N-methyl aniline *D*. Benzenamine

Choose the correct order of basic nature of the above amines. **[2021, 26 Feb Shift-II]**

(a) $A > C > B > D$
(b) $D > C > B > A$
(c) $D > B > C > A$
(d) $A > B > C > D$

37 Carbylamine test is used to detect the presence of primary amino group in an organic compound. Which of the following compound is formed when this test is performed with aniline? **[2021, 25 Feb Shift-II]**

(a) 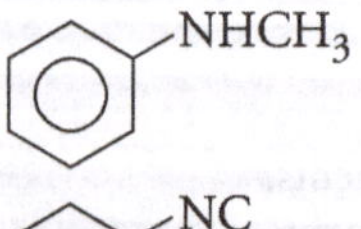

(b) 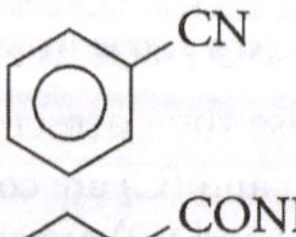

(c) 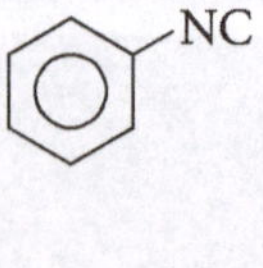

(d) 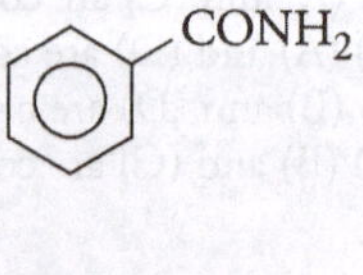

38 The increasing order of basicity of the following compounds is **[2020, 5 Sep Shift-I]**

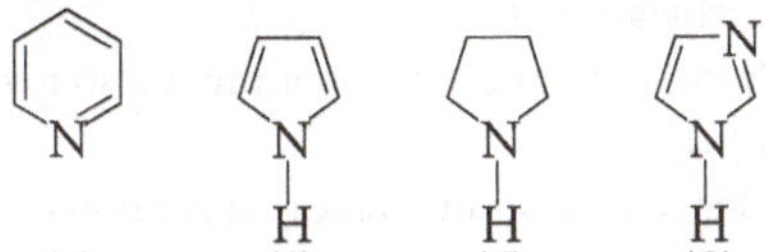

(*A*) (*B*) (*C*) (*D*)

(a) (*A*) < (*B*) < (*C*) < (*D*)
(b) (*D*) < (*A*) < (*B*) < (*C*)
(c) (*B*) < (*A*) < (*D*) < (*C*)
(d) (*B*) < (*A*) < (*C*) < (*D*)

39 The increasing order of pK_b values of the following compounds is **[2020, 6 Sep Shift-I]**

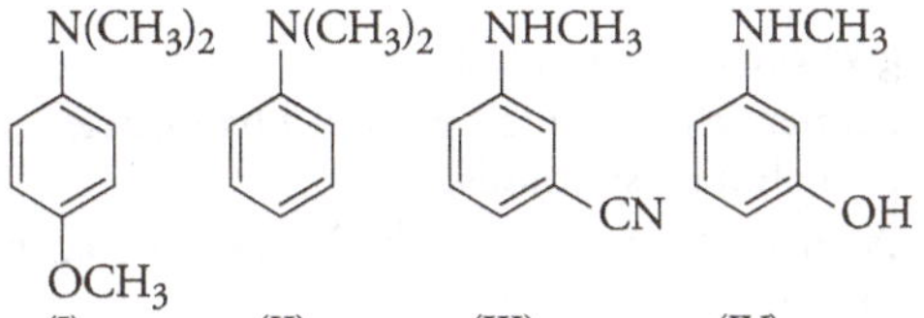

(I) (II) (III) (IV)

(a) II < IV < III < I (b) I < II < IV < III
(c) II < I < III < IV (d) I < II < III < IV

40 The decreasing order of basicity of the following amines is **[2020, 9 Jan Shift-II]**

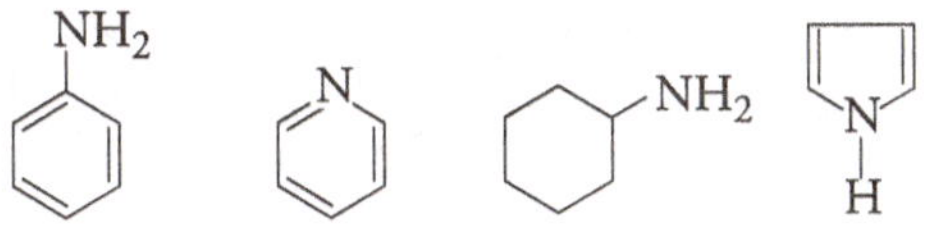

(I) (II) (III) (IV)

(a) (III) > (I) > (II) > (IV)
(b) (III) > (II) > (I) > (IV)
(c) (I) > (III) > (IV) > (II)
(d) (II) > (III) > (IV) > (I)

41 The increasing order of reactivity of the following compounds towards reaction with alkyl halides directly is **[2019, 12 Jan Shift-I]**

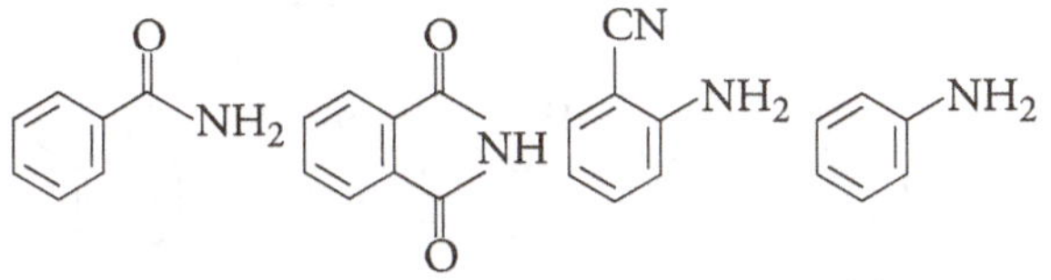

(*A*) (*B*) (*C*) (*D*)

(a) (*A*) < (*C*) < (*D*) < (*B*) (b) (*B*) < (*A*) < (*C*) < (*D*)
(c) (*B*) < (*A*) < (*D*) < (*C*) (d) (*A*) < (*B*) < (*C*) < (*D*)

42 The major product of the following reaction is

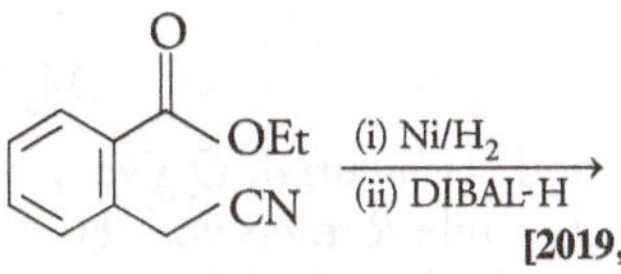

[2019, 11 Jan Shift-I]

(a) 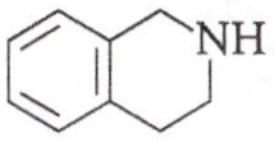 (b) 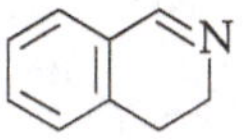

(c) 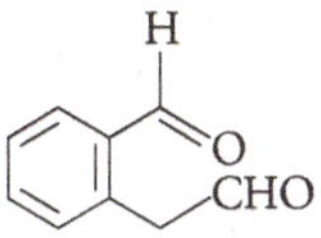 (d) 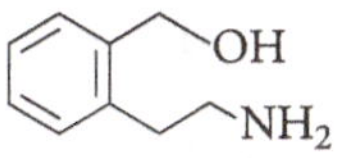

43 An aromatic compound '*A*' having molecular formula $C_7H_6O_2$ on treating with aqueous ammonia and heating forms compound '*B*'. The compound '*B*' on reaction with molecular bromine and potassium hydroxide provides compound '*C*' having molecular formula C_6H_7N. The structure of '*A*' is **[2019, 10 Jan Shift-II]**

(a) 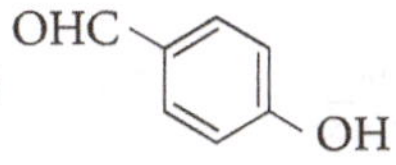 (b) 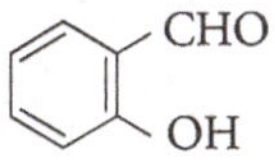

(c) 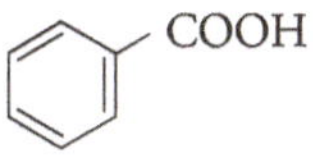 (d) 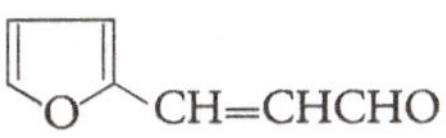

• Level 3

44 The descending order of basicity of following amines is: **[2025, 7 April Shift-II]**

(A) 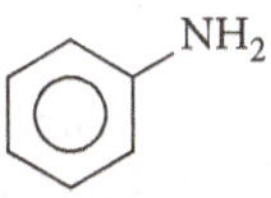 (B) 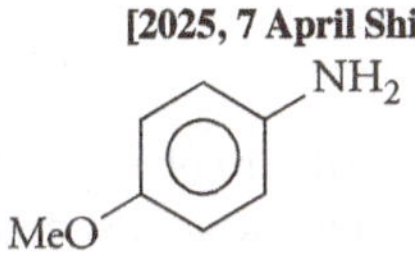

(C) 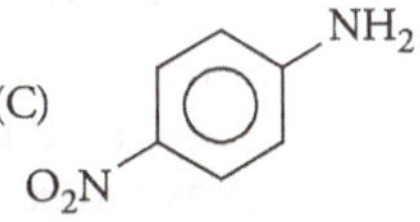 (D) CH_3NH_2

(E) $(CH_3)_2NH$

Choose the correct answer from the options given below:
(a) E > D > B > A > C
(b) E > D > A > B > C
(c) E > A > D > C > B
(d) B > E > D > A > C

45 The major product (*A*) formed in the following reaction sequence is **[2025, 4 April Shift-I]**

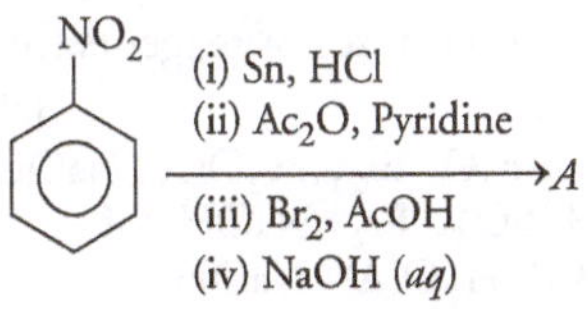

(a) 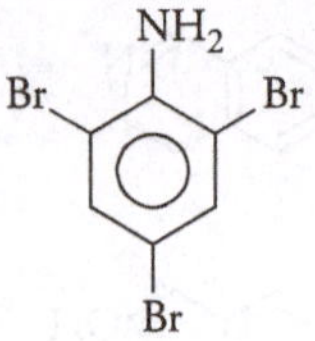 (b) 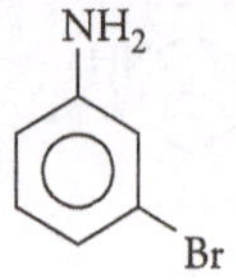

(c) 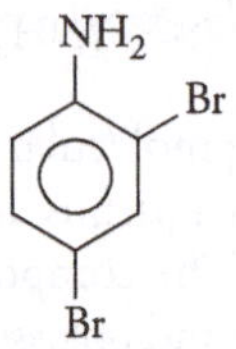 (d) 

46 The sequence from the following that would result in giving predominantly 3, 4, 5-tribromoaniline is **[2025, 3 April Shift-II]**

(a) 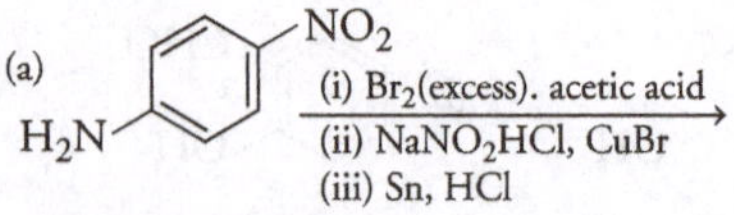

(b) 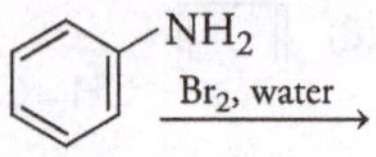

(c) 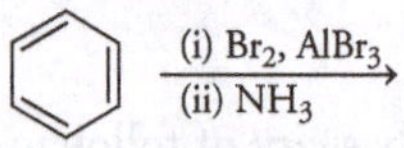

(d) 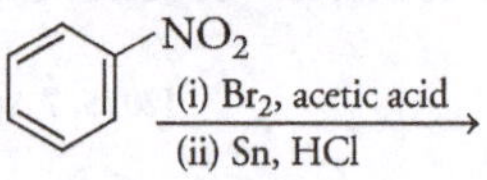

47 Mixture of 1 g each of chlorobenzene, aniline and benzoic acid is dissolved in 50 mL ethyl acetate and placed in a separating funnel. 5 M NaOH (30 mL) was added in the same funnel. The funnel was shaken vigorously and then kept aside. The ethyl acetate layer in the funnel contains: **[2025, 7 April Shift-II]**

(a) benzoic acid
(b) benzoic acid and aniline
(c) chlorobenzene and aniline
(d) benzoic acid and chlorobenzene

48 For reaction,

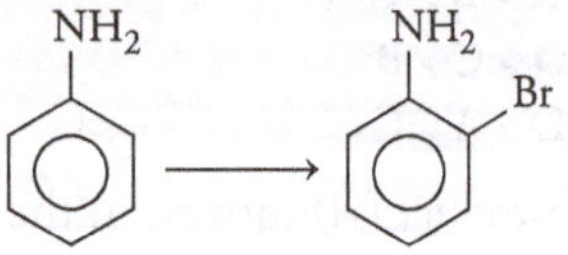

(Major)

The correct order of set of reagents for the above conversion is **[2025, 24 Jan Shift-II]**

(a) H_2SO_4, Ac_2O, Br_2, $H_2O(\Delta)$, NaOH
(b) Ac_2O, H_2SO_4, Br_2, NaOH
(c) Br_2 |$FeBr_3$, $H_2O(\Delta)$, NaOH
(d) Ac_2O, Br_2, $H_2O(\Delta)$, NaOH

49 Identify correct statements.

A. Primary amines do not give diazonium salts when treated with $NaNO_2$ in acidic condition.
B. Aliphatic and aromatic primary amines on heating with $CHCl_3$ and ethanolic KOH form carbylamines.
C. Secondary and tertiary amines also give carbylamine test.
D. Benzenesulphonyl chloride is known as Hinsberg's reagent.
E. Tertiary amines reacts with benzenesulphonyl chloride very easily.

Choose the correct answer from the options given below. **[2025, 28 Jan Shift-II]**

(a) (D) and (E) only (b) (B) and (C) only
(c) (B) and (D) only (d) (A) and (B) only

50 Isomeric amines with molecular formula $C_8H_{11}N$ give the following tests

Isomer (*P*) ⇒ can be prepared by Gabriel phthalimide synthesis.
Isomer (*Q*) ⇒ reacts with Hinsberg's reagent to give solid insoluble in NaOH.
Isomer (R) ⇒ reacts with HONO followed by β-naphthol in NaOH to give red dye.
Isomers (*P*), (*Q*) and (*R*) respectively are **[2023, 10 April Shift-I]**

P *Q* *R*

(a) 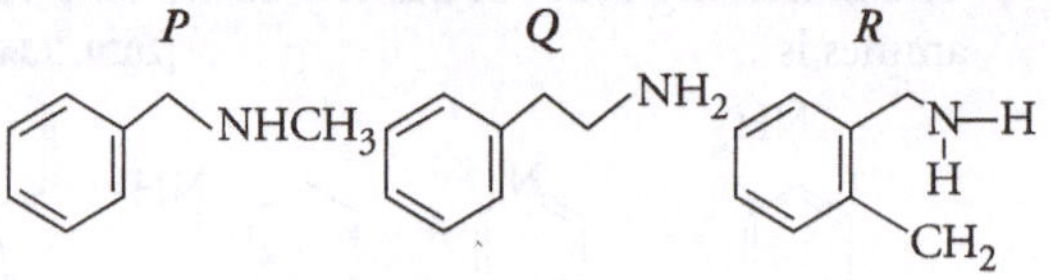

(b) 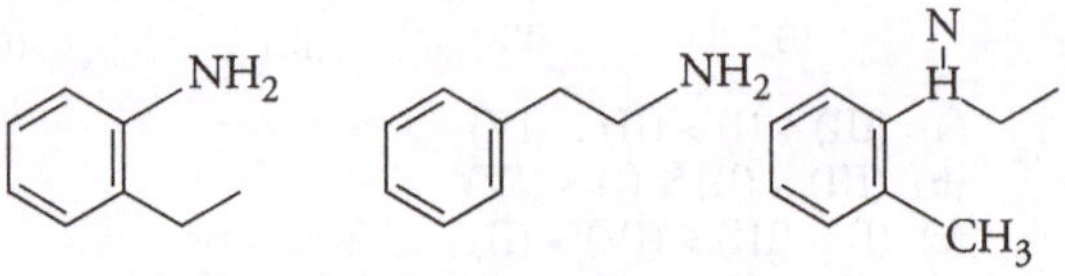

P *Q* *R*

(c) 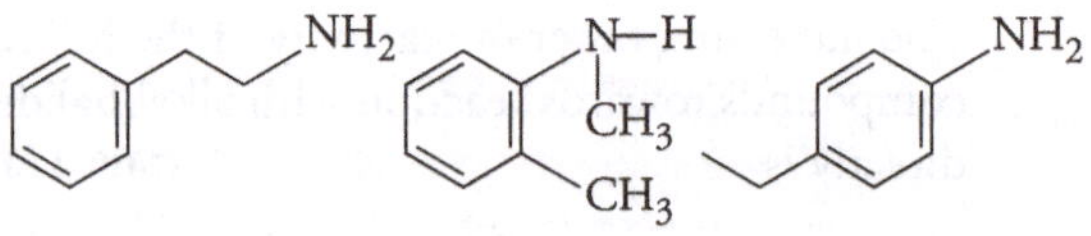

(d) 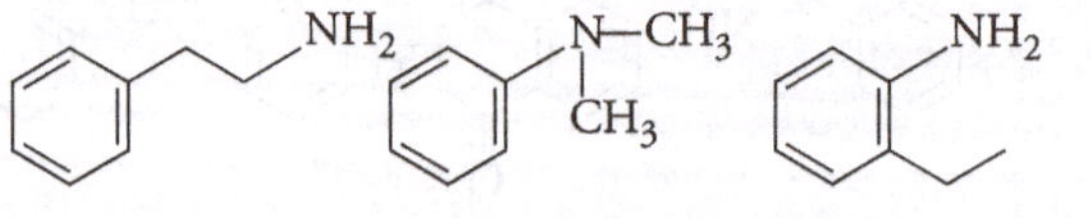

51 Compound *P* $\xrightarrow{HCl, \Delta}$ Filter → Residue *Q* ; → Filtrate $\xrightarrow{NaOH}$ Oily liquid *R*

(M.F. $C_{14}H_{13}ON$; M.F = Molecular formula)

Compound *P* is neutral, *Q* gives effervescence with $NaHCO_3$ while *R* reacts with Hinsbergs reagent to give solid soluble in NaOH. Compound *P* is **[2023 6 April Shift-I]**

(a) 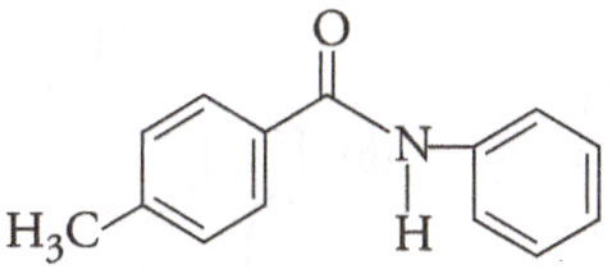

(b) 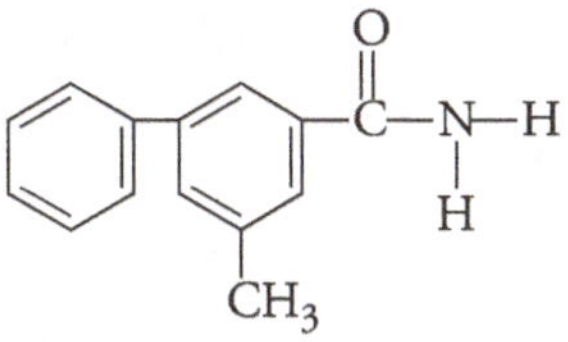

(c) 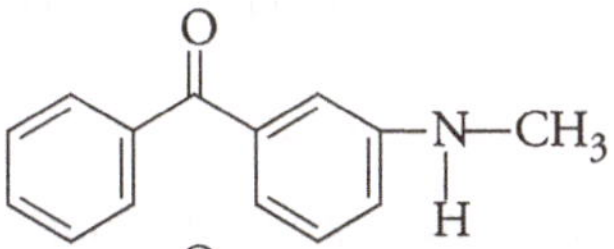

(d) 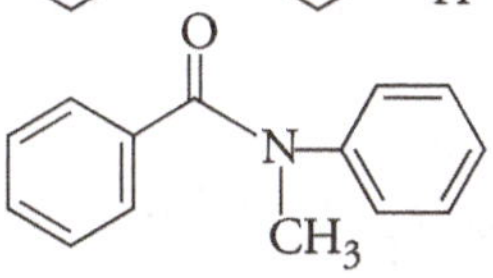

52 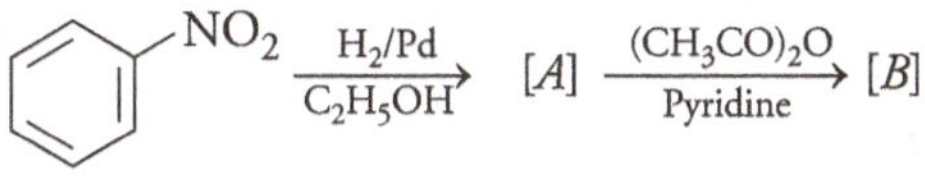

Consider the above reaction and identify the product *B*. **[2023, 31 Jan Shift-I]**

(a) 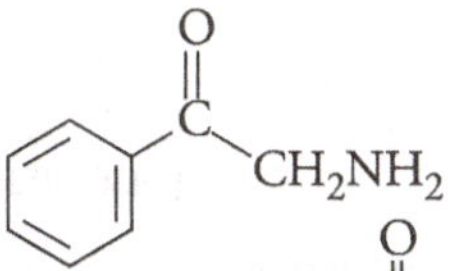

(b) 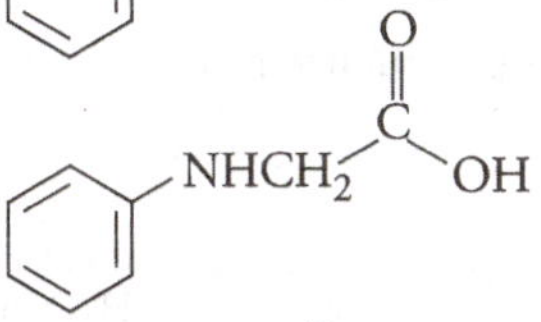

(c) 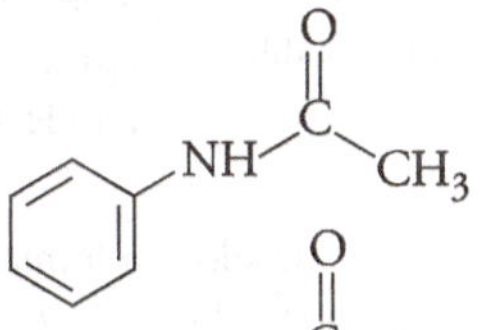

(d) 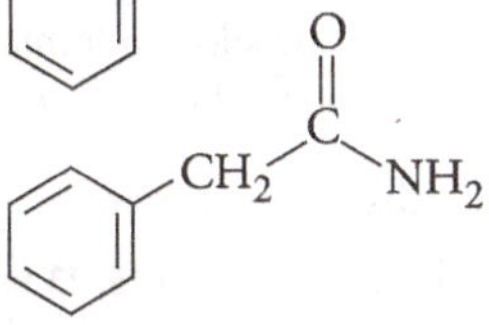

53 Match List-I with List-II.

List-I (Reaction)		List-II (Reagents)
A. Hoffmann degradation	I.	Conc. KOH, Δ
B. Clemmensen reduction	II.	$CHCl_3$, $NaOH/H_3O^+$
C. Cannizzaro reaction	III.	Br_2, NaOH
D. Reimer-Tiemann reaction	IV.	Zn-Hg/HCl

Choose the correct answer from the options given below. **[2023, 29 Jan Shift-I]**

(a) A-II, B-I, C-III, D-IV (b) A-II, B-IV, C-I, D-III
(c) A-III, B-IV, C-I, D-II (d) A-III, B-IV, C-II, D-I

54 Consider the following reaction sequence

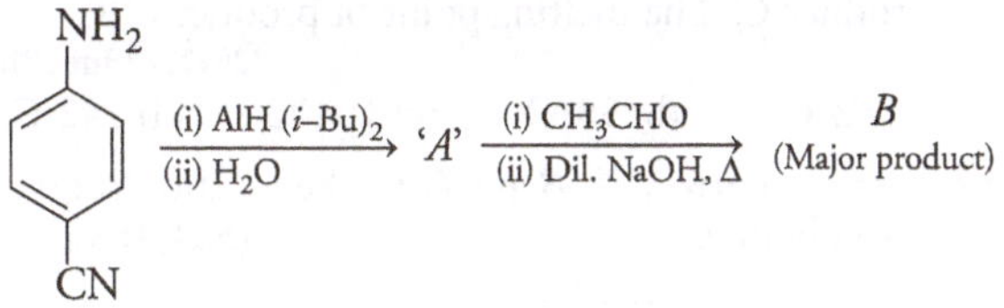

The product '*B*' is **[2022, 29 July Shift-I]**

(a) 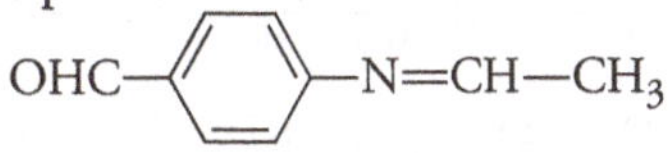

(b) 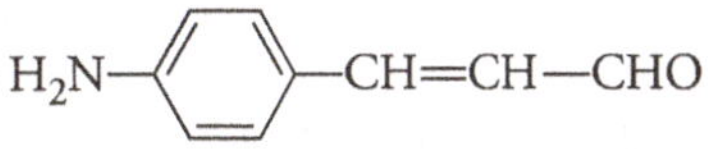

(c) 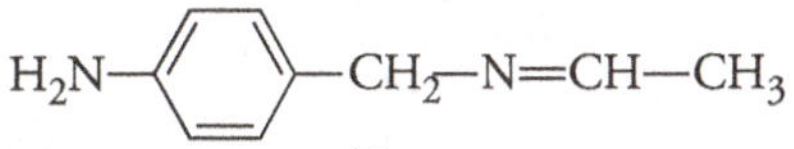

(d) 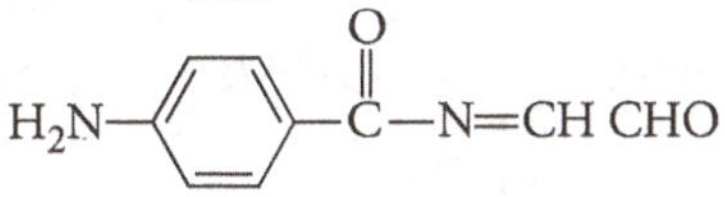

55 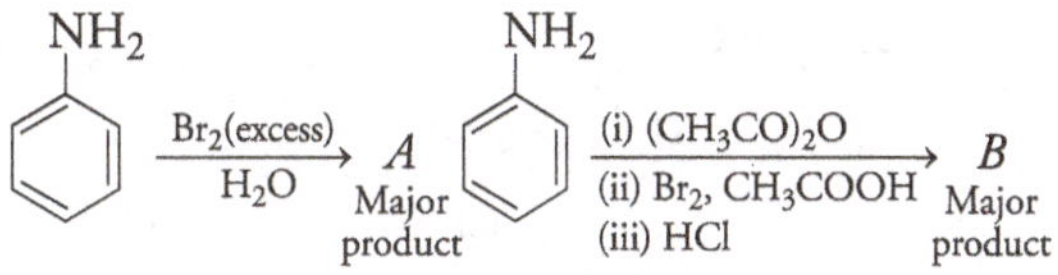

Consider the above reactions, the product *A* and product *B* respectively are **[2022, 29 June Shift-I]**

(a) 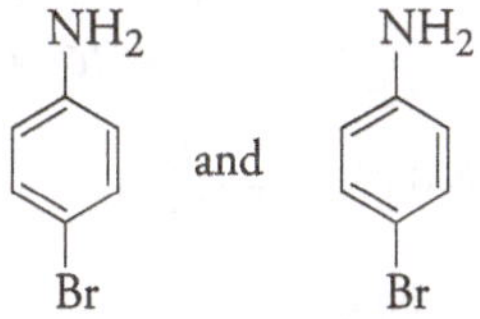

(b) 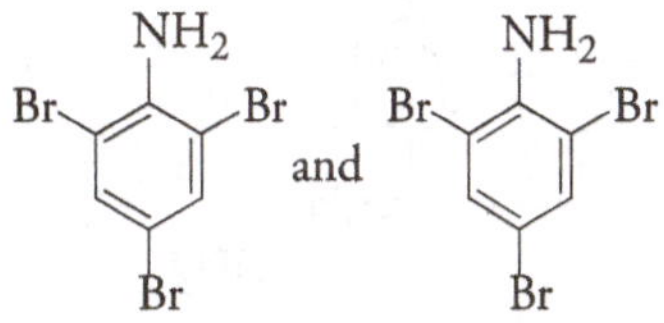

(c) 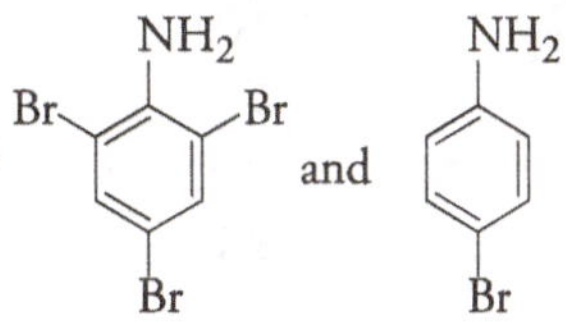

(d) 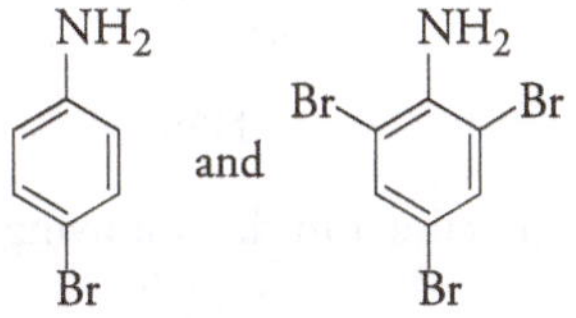

56 Decarboxylation of all six possible forms of diaminobenzoic acid $C_6H_3(NH_2)_2COOH$ yields three products *A*, *B* and *C*. Three acids give a product

A, two acids give a product '*B*' and one acid gives a product *C*. The melting point of product *C* is **[2022, 27 June Shift-II]**

(a) 63°C (b) 90°C (c) 104°C (d) 142°C

57 Identify correct *A*, *B* and *C* in the reaction sequence given below. **[2021, 31 Aug Shift-II]**

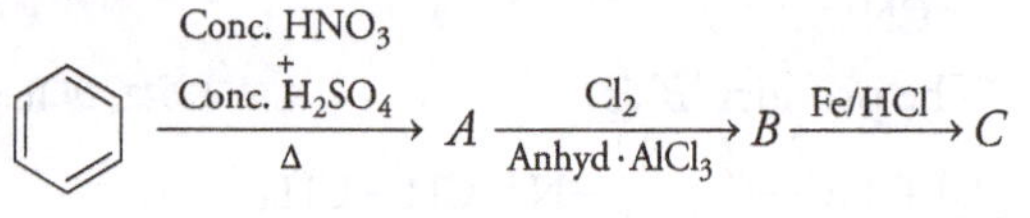

(a) 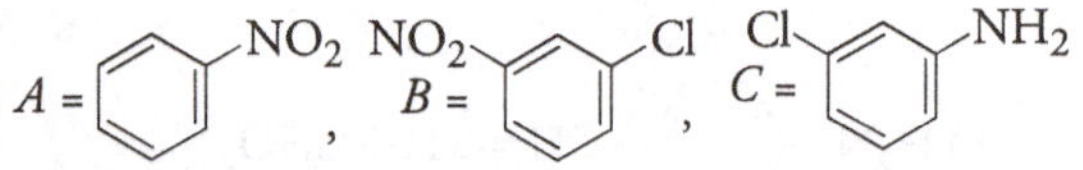

(b) 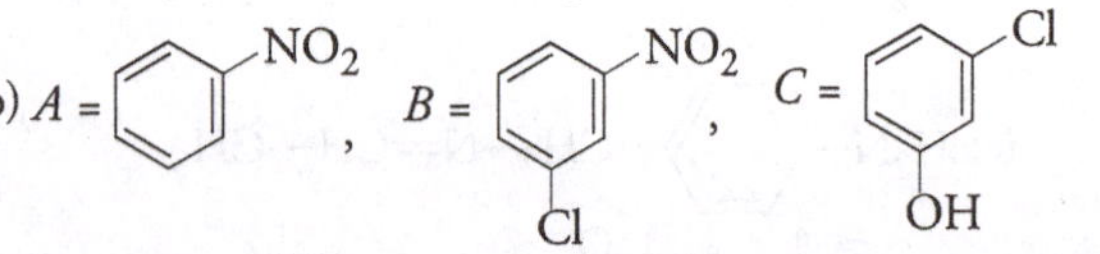

(c) 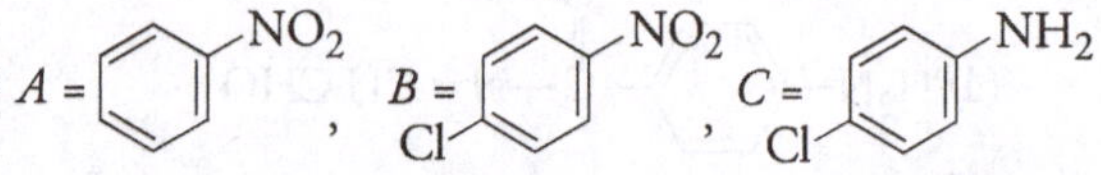

(d) 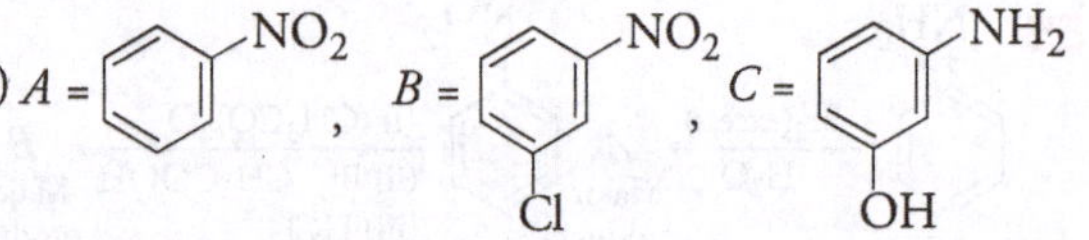

58 Which one of the following reactions does not occur? **[2021, 22 July Shift-II]**

(a) 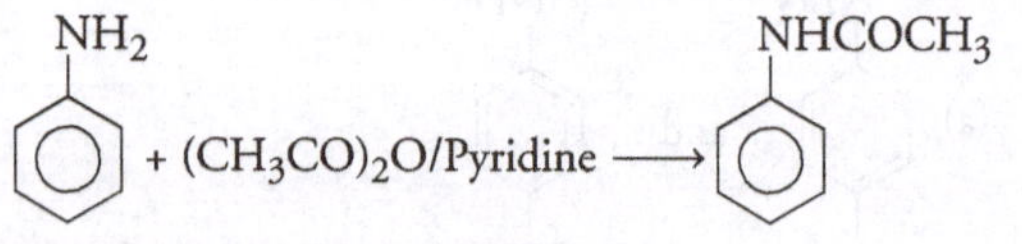

(b) 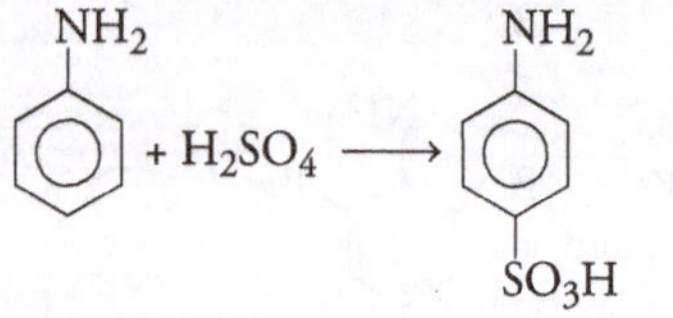

(c) 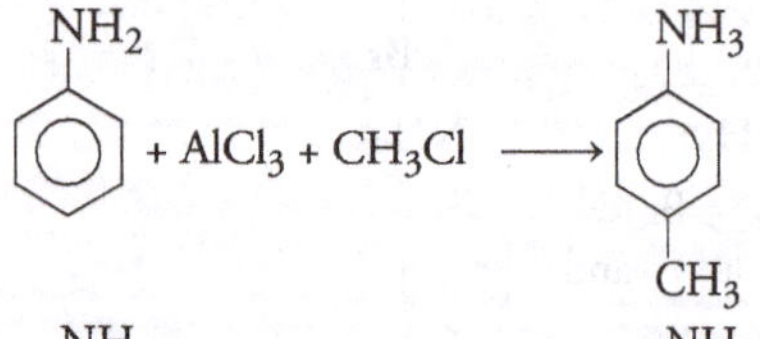

(d) 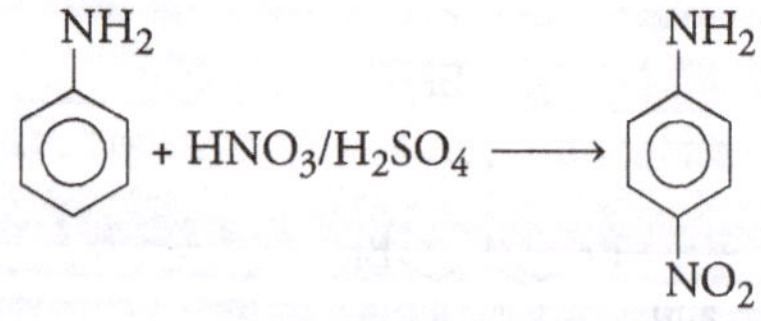

59 The major product obtained in the following reaction is

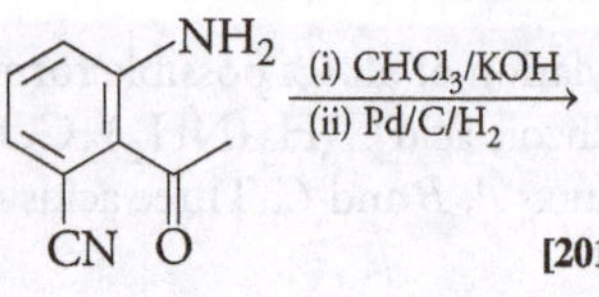

[2019, 8 April Shift-II]

(a) 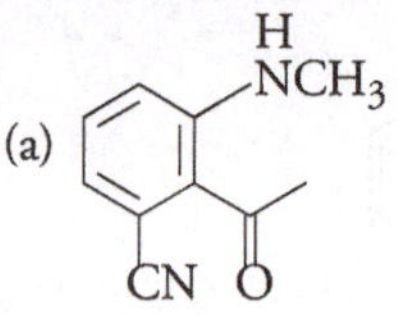

(b) 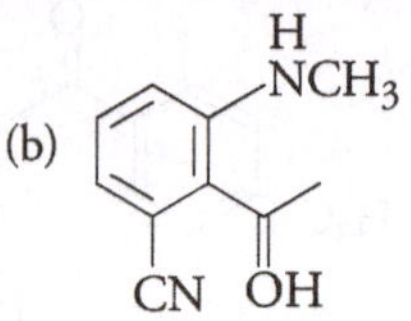

(c) 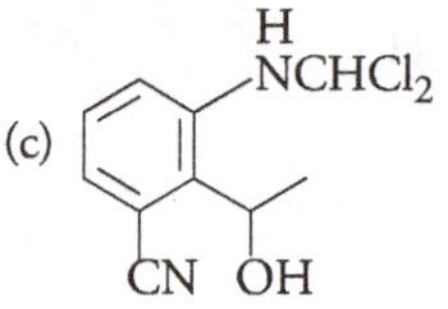

(d) 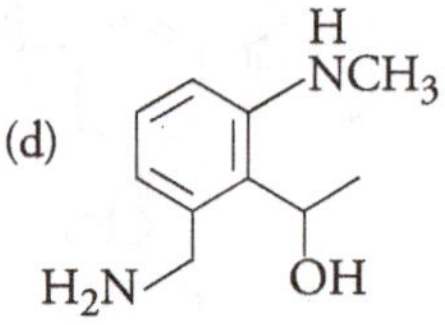

Numerical Value Type Questions

60 Given below are some nitrogen containing compounds

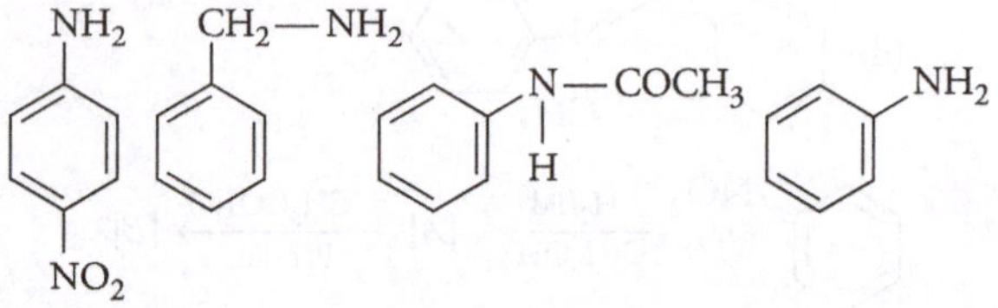

Each of them is treated with HCl separately. 1.0 g of the most basic compound will consume .... mg of HCl.

(Given, molar mass in g mol^{-1} C : 12, H : 1, O : 16, Cl : 35.5) **[2025, 29 Jan Shift-I]**

61 Consider the following sequence of reactions.

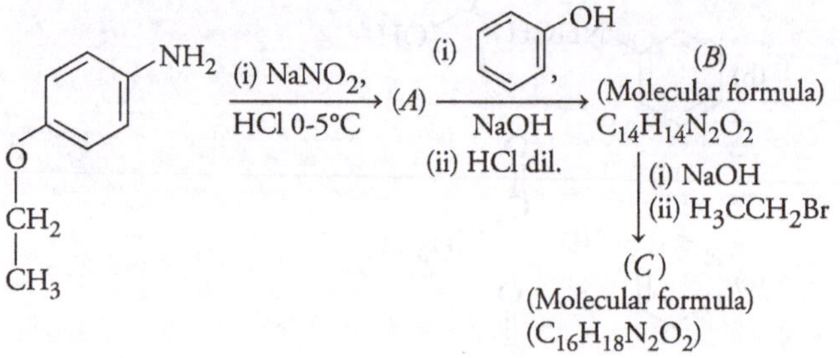

Total number of sp^3 hybridised carbon atoms in the major product *C* formed is .... **[2025, 23 Jan Shift-II]**

62 Number of amine compounds from the following giving solids which are soluble in NaOH upon reaction with Hinsberg's reagent is .... **[2024, 8 April Shift-I]**

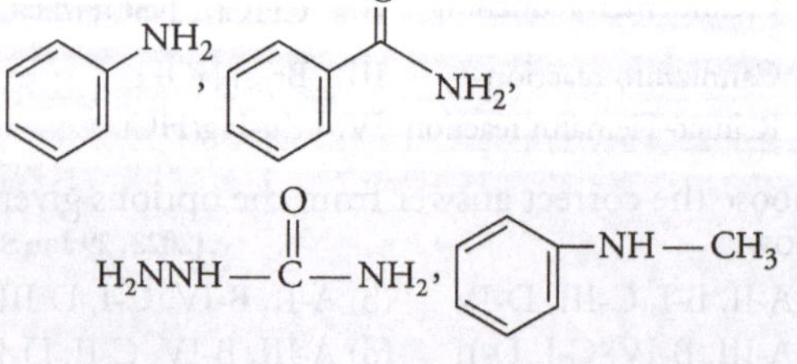

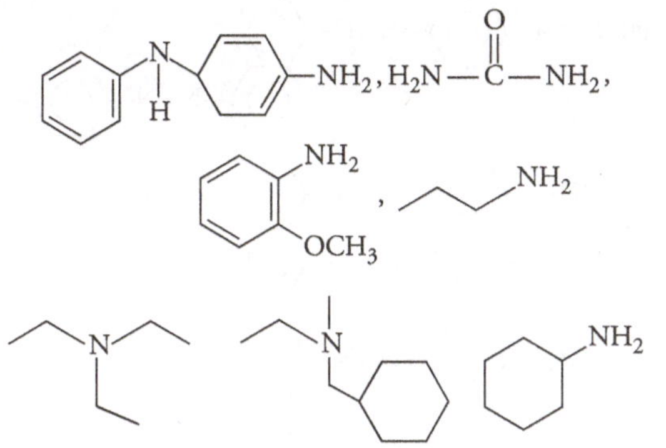

63 Number of compounds which give reaction with Hinsberg's reagent is .... **[2024, 1 Feb Shift-II]**

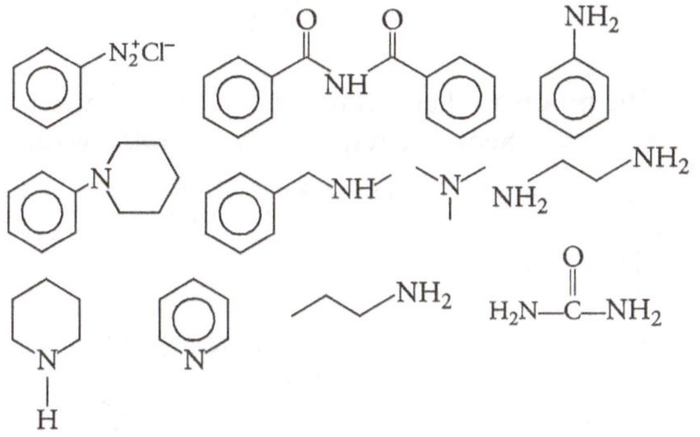

64 How many of the transformations given below would result in aromatic amines? **[2023, 31 Jan Shift-I]**

(a) 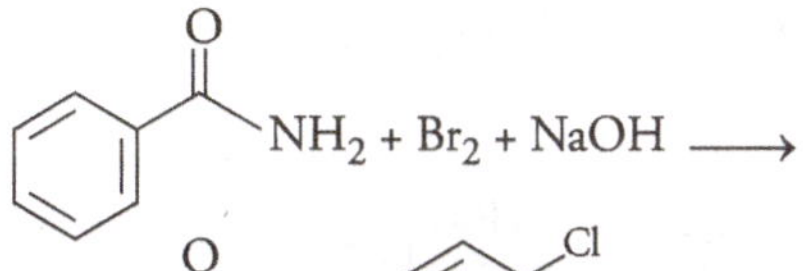

(b) 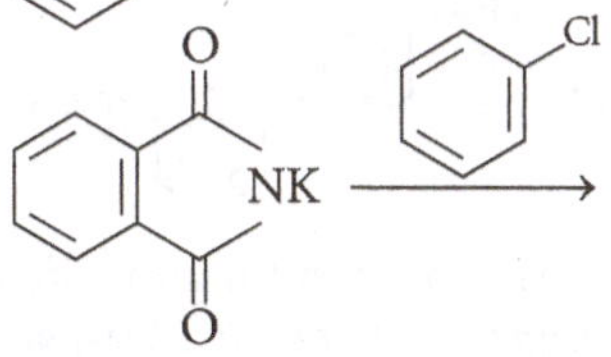

(c) 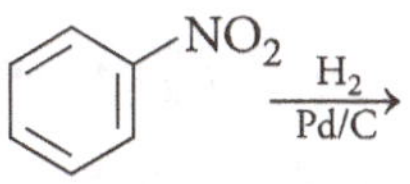

(d) 

65 The total number of reagents from those given below, that can convert nitrobenzene into aniline is .... (Integer answer) **[2021, 31 Aug Shift-I]**

I. Sn – HCl II. Sn – NH_4OH
III. Fe – HCl IV. Zn – HCl
V. H_2 – Pd
VI. H_2-Raney nickel

66 The total number of amines among the following which can be synthesised by Gabriel synthesis is .... **[2021, 24 Feb Shift-II]**

(a) $(CH_3)_2CHCH_2NH_2$

(b) $CH_3CH_2NH_2$

(c) 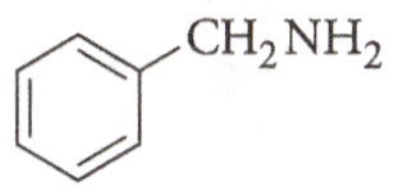

(d) 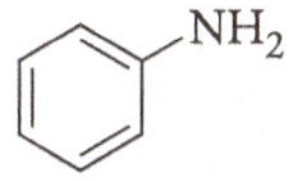

Topic 3 / Diazonium Salts and Other Nitrogen Containing Compounds

• Level 1

01 *o*-phenylenediamine $\xrightarrow{HNO_2}$ '*X*' (Major)
Product '*X*' is **[2023, 11 April Shift-I]**

(a) 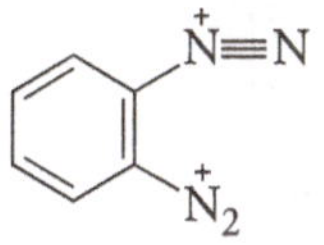

(b) 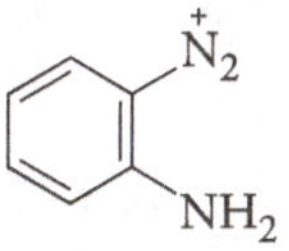

(c) 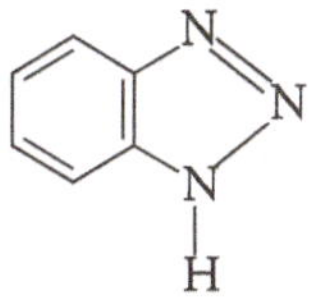

(d) 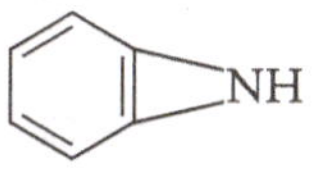

02 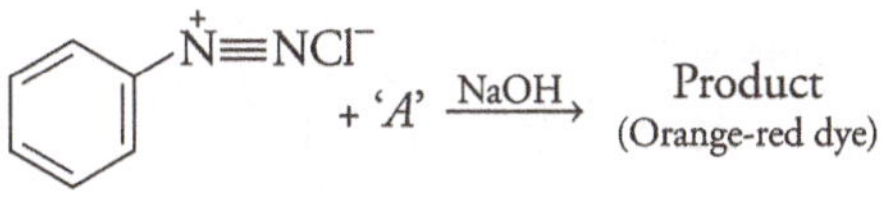

Which among the following represent reagent '*A*'? **[2022, 29 July Shift-I]**

(a) 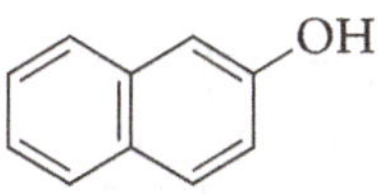

(b) 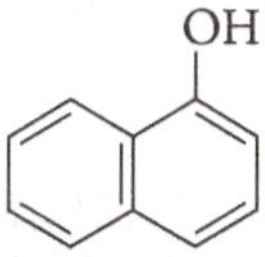

(c) 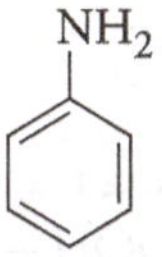

(d) 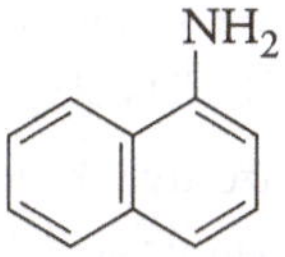

03 Hydrolysis of which compound will give carbolic acid? **[2022, 26 July Shift-II]**

(a) Cumene
(b) Benzenediazonium chloride
(c) Benzal chloride
(d) Ethyene glycol ketal

04 The correct stability order of the following diazonium salt is **[2022, 26 July Shift-I]**

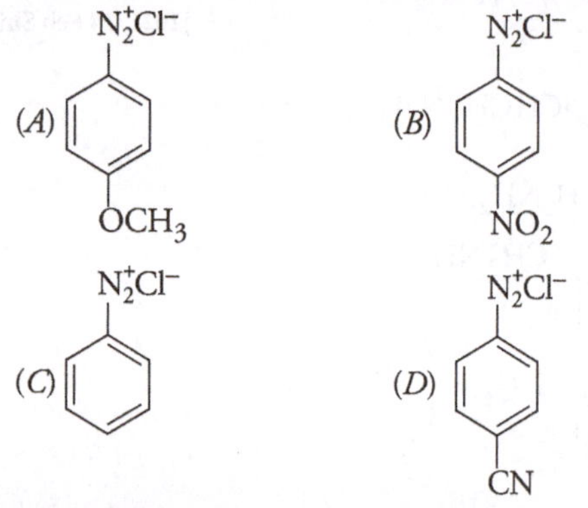

(a) (A) > (B) > (C) > (D) (b) (A) > (C) > (D) > (B)
(c) (C) > (A) > (D) > (B) (d) (C) > (D) > (B) > (A)

• Level 2

05 In the following reactions, which one is NOT correct? **[2025, 3 April Shift-I]**

(a) 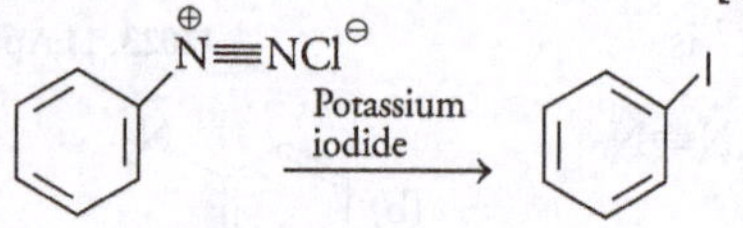

(b) 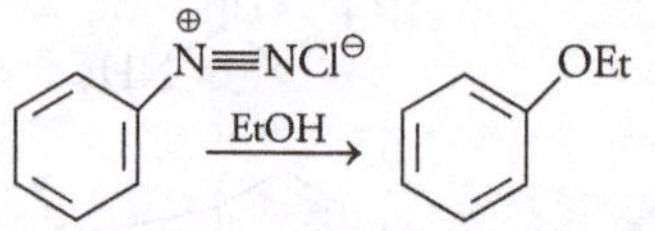

(c) 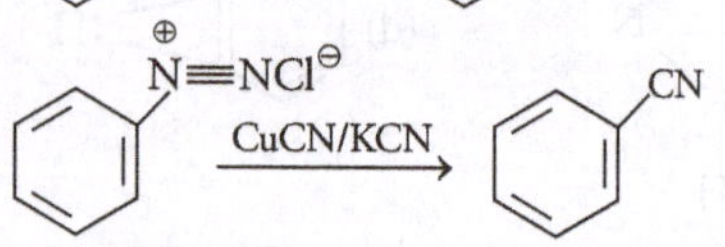

(d) 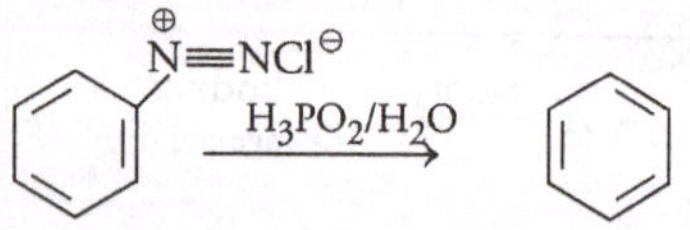

06 The products *A* and *B* in the following reactions, respectively are **[2025, 28 Jan Shift-I]**

$$A \xleftarrow{Ag-NO_2} CH_3—CH_2—CH_2—Br \xrightarrow{AgCN} B$$

(a) $CH_3CH_2CH_2NO_2$, $CH_3CH_2CH_2NC$
(b) $CH_3CH_2CH_2ONO$, $CH_3CH_2CH_2CN$
(c) $CH_3CH_2CH_2ONO$, $CH_3CH_2CH_2NC$
(d) $CH_3CH_2CH_2NO_2$, $CH_3CH_2CH_2CN$

07 The azo-dye (*Y*) formed in the following reactions is:
Sulphanilic acid + $NaNO_2$ + $CH_3COOH \longrightarrow X$

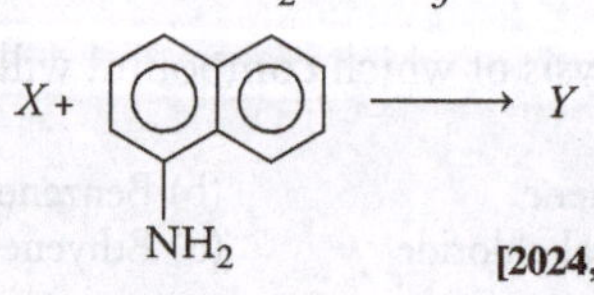

[2024, 31 Jan Shift-II]

(a) 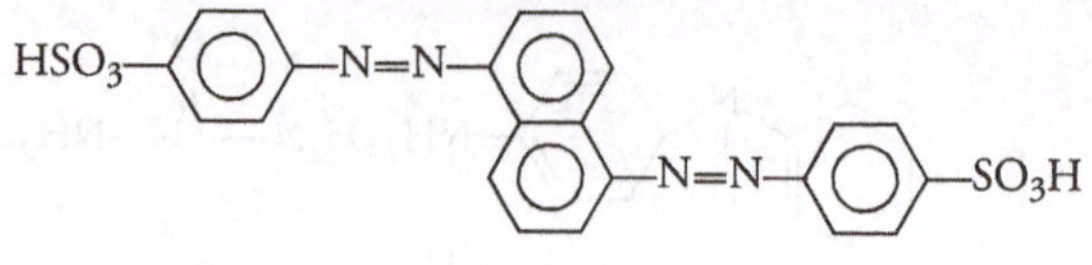

(b) 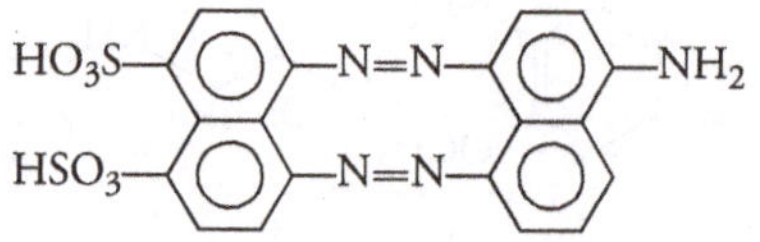

(c) 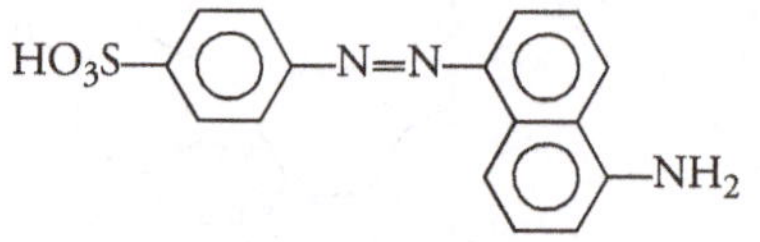

(d) 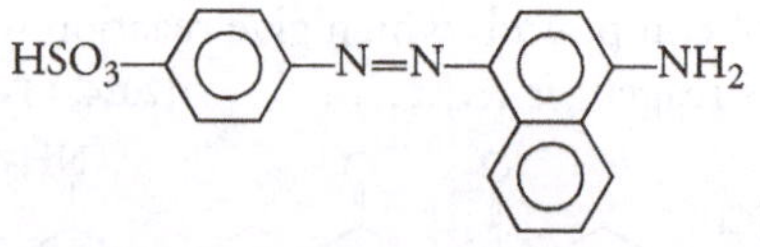

08 The products *A* and *B* formed in the following reaction scheme are respectively **[2024, 30 Jan Shift-II]**

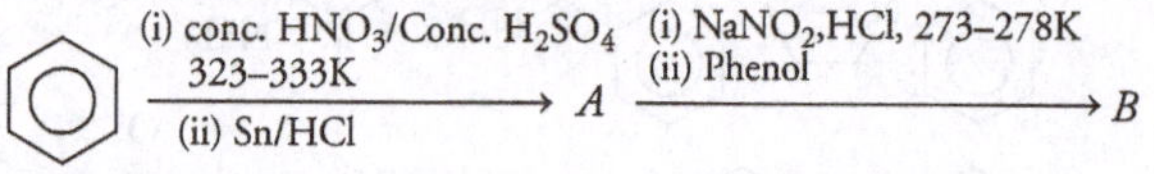

(a) 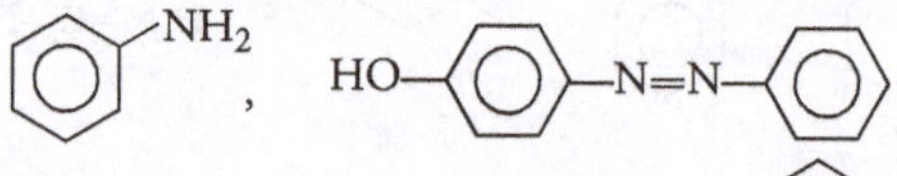

(b) 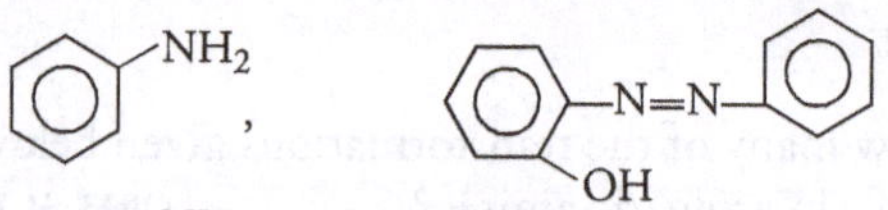

(c) 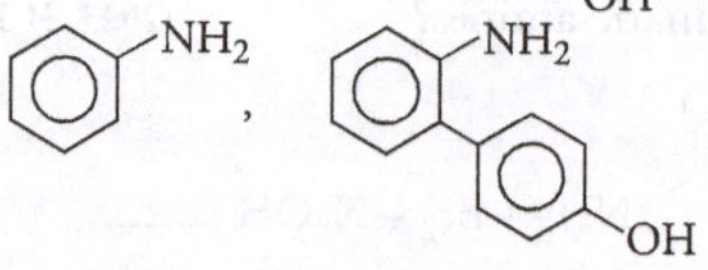

(d) 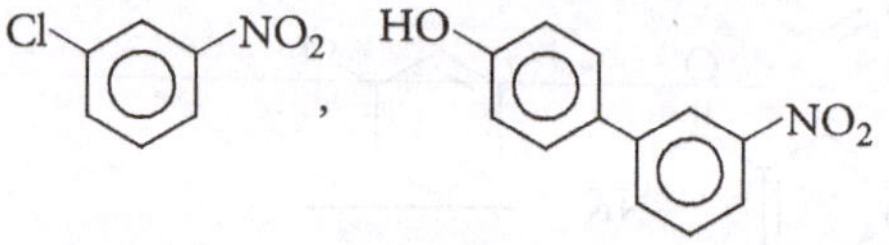

09 Following is a confirmatory test for aromatic primary amines. Identify reagent (*A*) and (*B*). **[2024, 30 Jan Shift-I]**

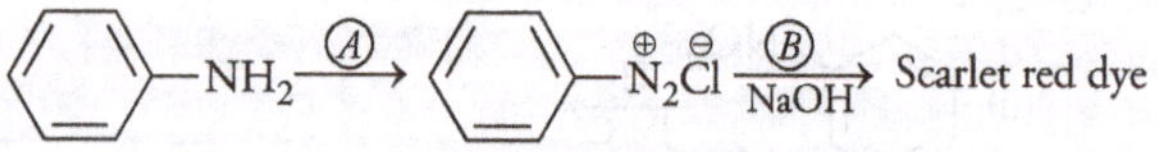

(a) A = $NaNO_2$+HCl, 0-5°C; B = 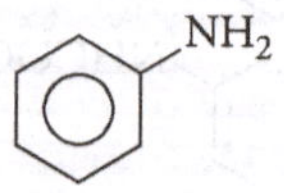

(b) A = $NaNO_2$+HCl, 0-5°C; B = 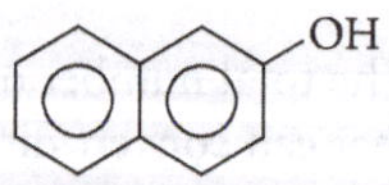

(c) A = HNO_3/H_2SO_4 B = 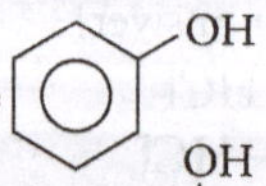

(d) A = $NaNO_2$+HCl, 0-5°C; B = 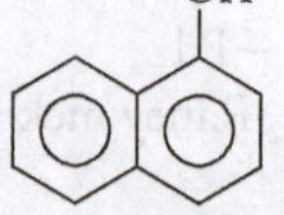

10 Consider the following sequence of reactions

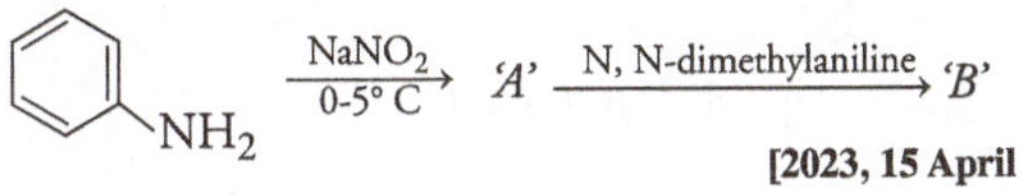

[2023, 15 April Shift-I]

The product '*B*' is

(a) 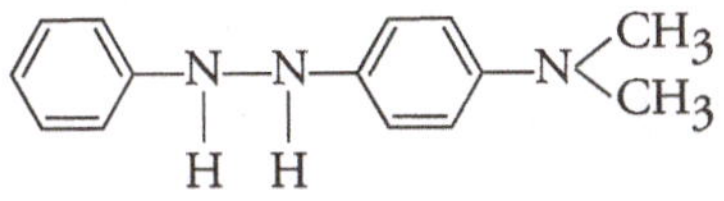

(b) 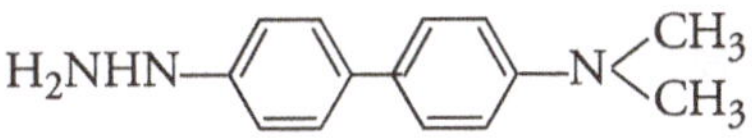

(c) 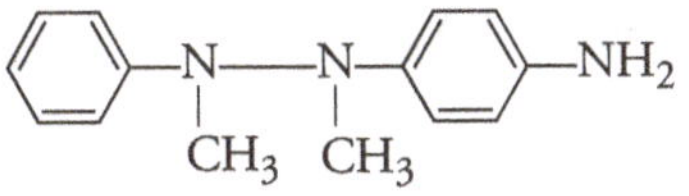

(d) 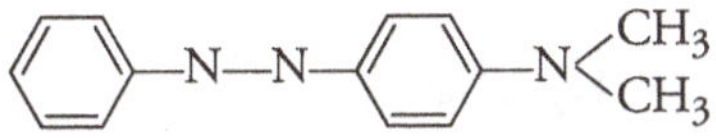

11 Which one of the following gives the most stable diazonium salt ? [2021, 01 Sep Shift-II]

(a) $CH_3—CH_2—CH_2NH_2$ (b) 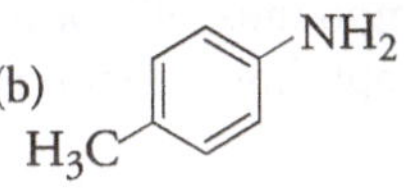

(c) 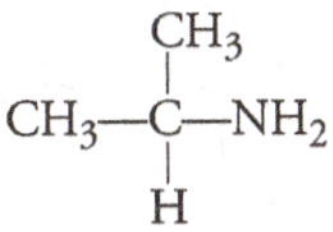 (d) 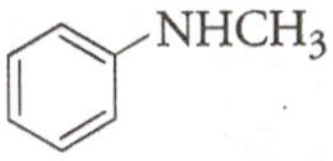

12 $C_6H_5NO_2 \xrightarrow{Sn + HCl} A \xrightarrow[H^+]{C_6H_5N_2^+Cl^-} P$ (Yellow coloured compound)

Consider the above reaction, the product *P* is [2021, 25 July Shift-II]

(a) 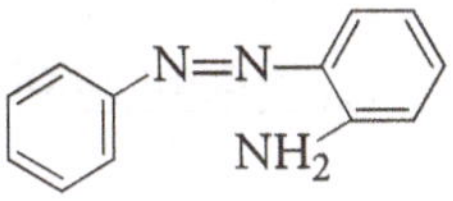

(b) 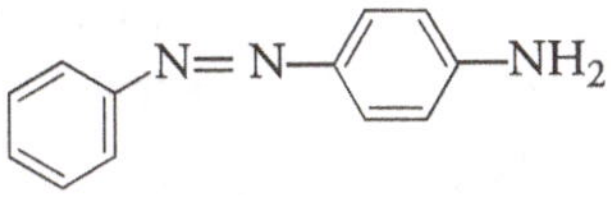

(c) 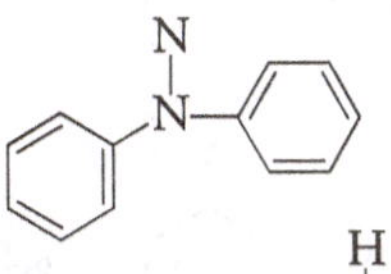

(d) 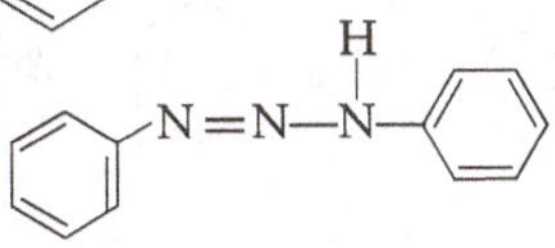

Imp. *Reaction of Diazonium Salts*

13 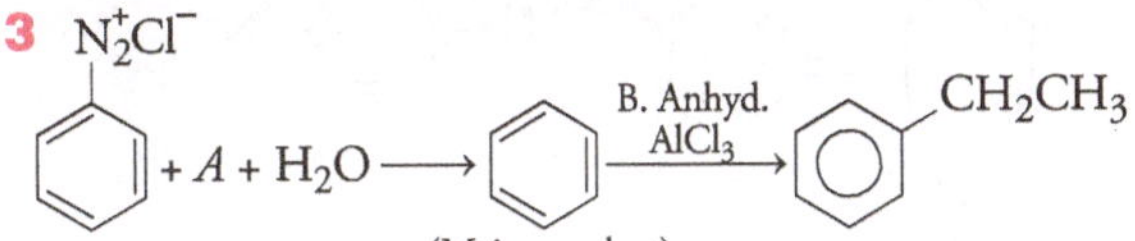

(Major product)

In the chemical reactions given above *A* and *B* respectively are [2021, 22 July Shift-II]

(a) H_3PO_2 and CH_3CH_2Cl
(b) CH_3CH_2OH and H_3PO_2
(c) H_3PO_2 and CH_3CH_2OH
(d) CH_3CH_2Cl and H_3PO_2

14 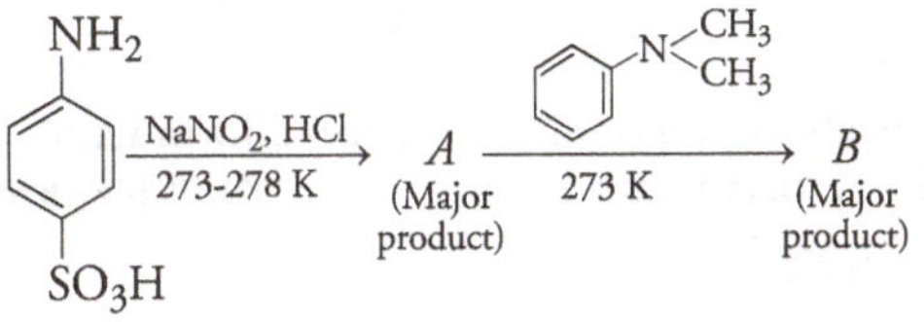

Consider the above reaction, compound *B* is [2021, 20 July Shift-II]

(a) 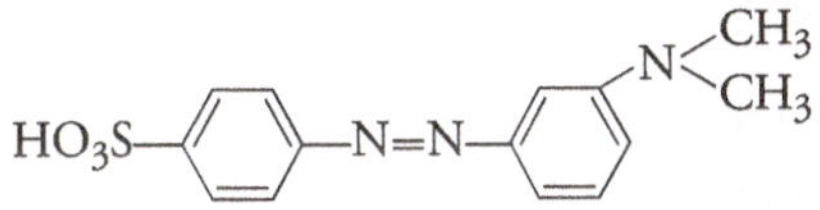

(b) 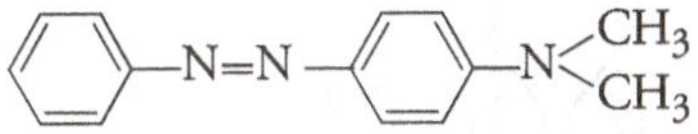

(c) 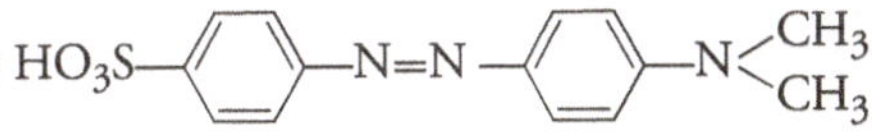

(d) 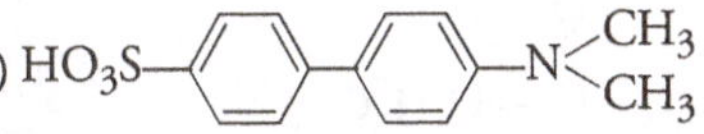

15 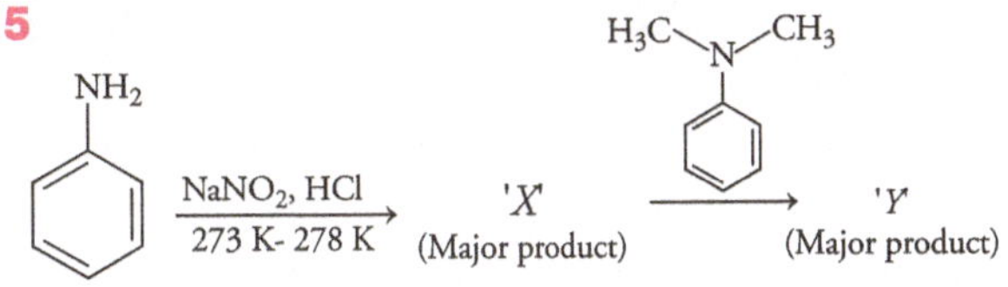

The product *X* and *Y* are [2021, 18 March Shift-I]

(a) 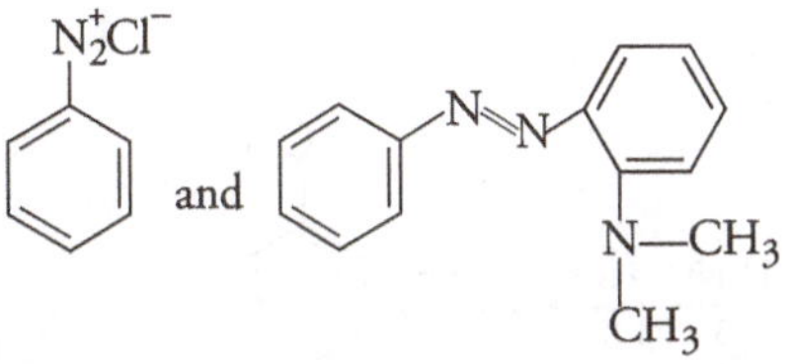

(b) 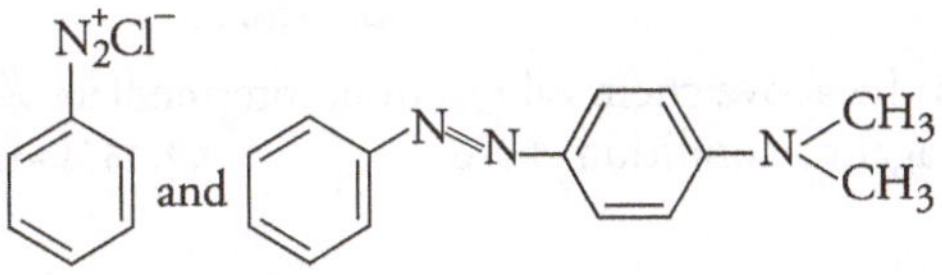

(c) 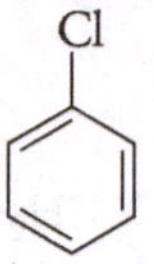 and 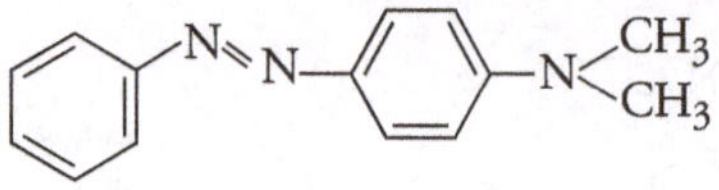

(d) 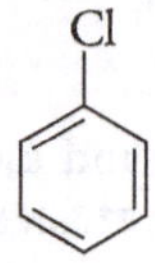 and 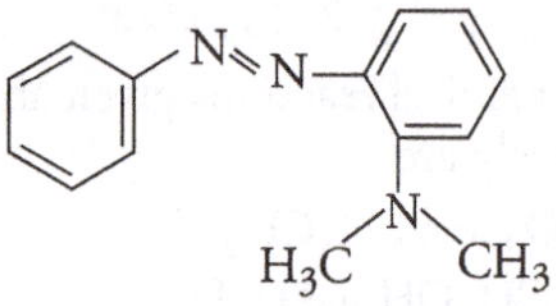

16 $C_7H_7N_2OCl + C_2H_5OH \longrightarrow$ 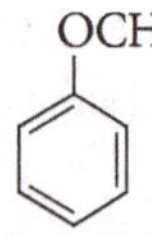 $+ N_2 + X + Y$
(*A*)

In the above reaction, the structural formula of *A*, *X* and *Y* respectively are **[2021, 17 March Shift-II]**

(a) 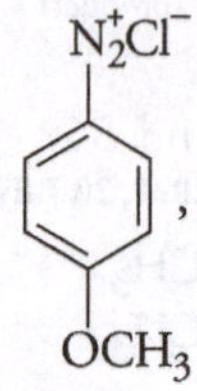 , $CH_3-\overset{O}{\overset{\|}{C}}-H$, HCl

(b) 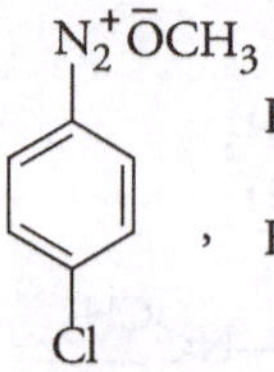 , 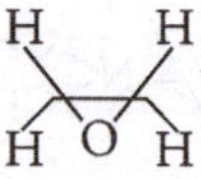 , HCl

(c) 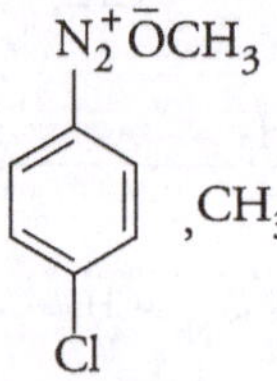 , $CH_3-\overset{O}{\overset{\|}{C}}-H$, H_2O

(d) 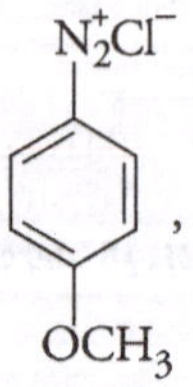 , 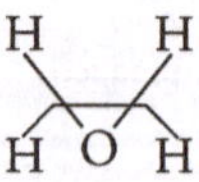 , H_2O

17 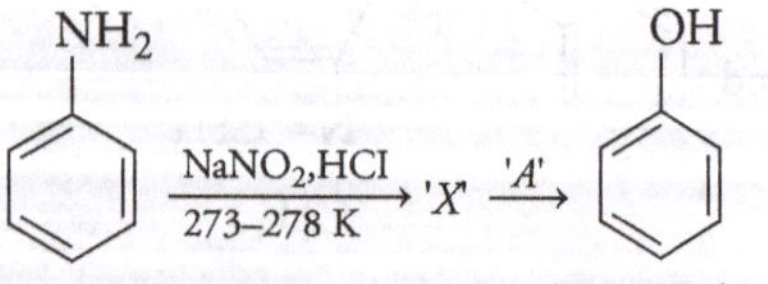

(Major product)

In the above chemical reaction, intermediate *X* and reagent / condition *A* are **[2021, 16 March Shift-I]**

(a) *X*- 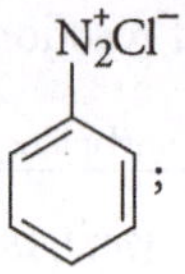 ; *A* -H_2O/NaOH

(b) *X*- 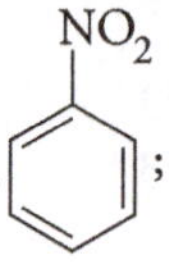 ; *A* -H_2O/Δ

(c) *X*- 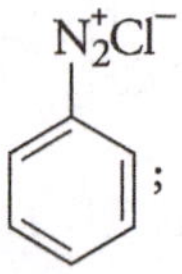 ; *A* -H_2O/Δ

(d) *X*- 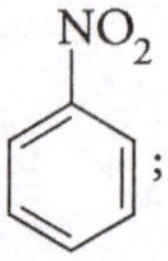 ; *A* -H_2O/NaOH

18 The diazonium salt of which of the following compounds will form a coloured dye on reaction with β-naphthol in NaOH ? **[2021, 24 Feb Shift-II]**

(a) 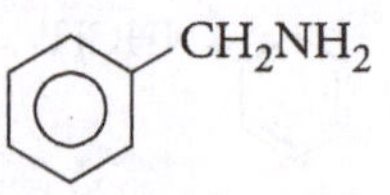
(b) 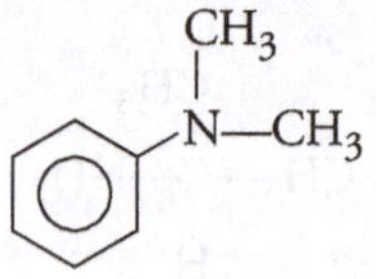
(c) 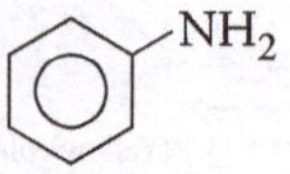
(d) 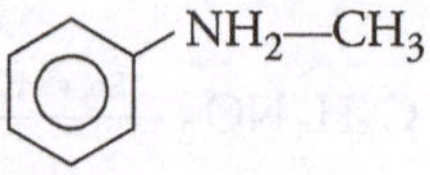

19 '*A*' and '*B*' in the following reaction are

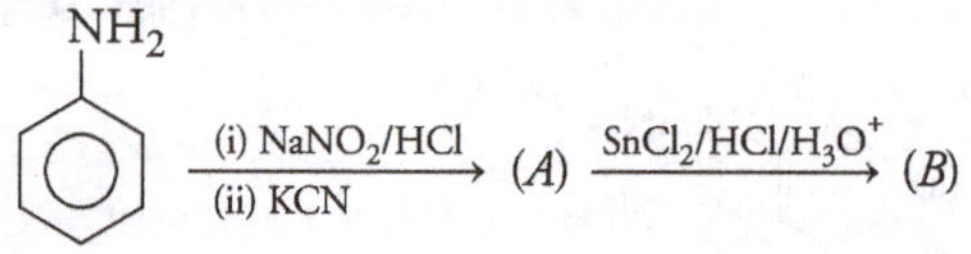

[2021, 24 Feb Shift-I]

(a) (*A*) 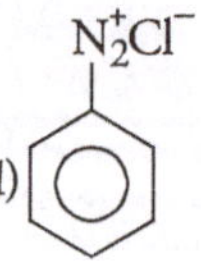 (*B*) 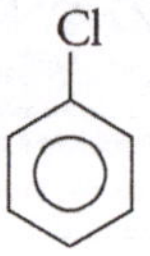
(b) (*A*) 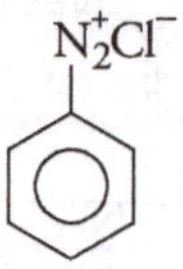 (*B*) 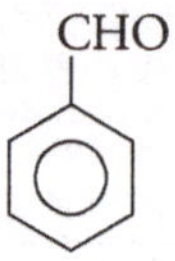
(c) (*A*) 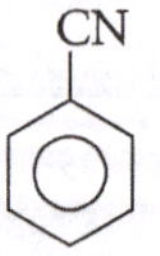 (*B*) 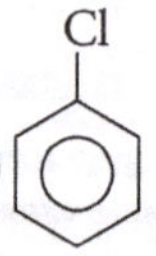
(d) (*A*) 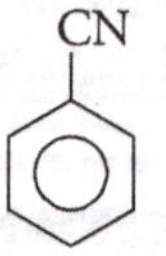 (*B*) 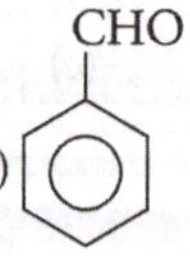

• Level 3

20 Identify [*A*] [*B*] and [*C*], respectively in the following reaction sequence: **[2025, 3 April Shift-I]**

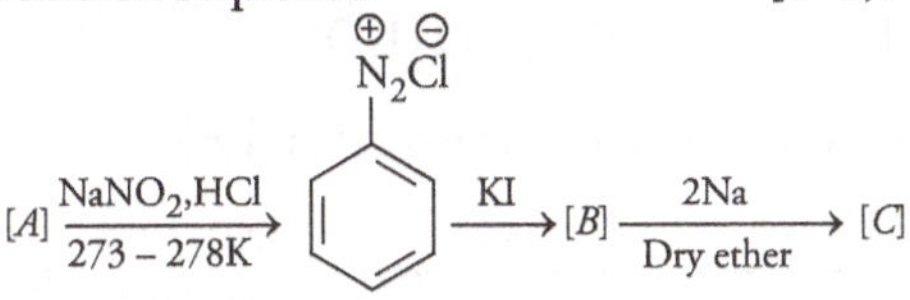

(a) [A] [*B*] [*C*]

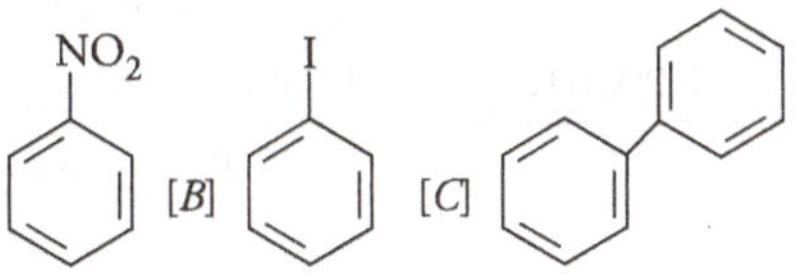

(b) [A] [*B*] [*C*]

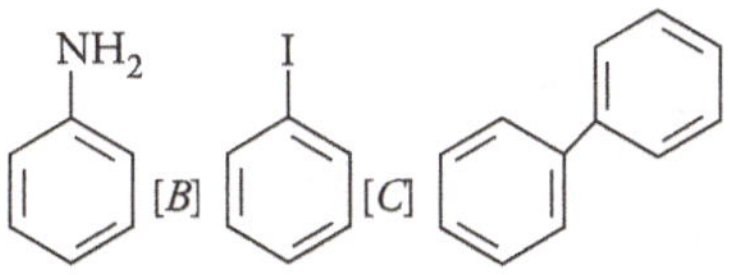

(c) [A] [*B*] [*C*]

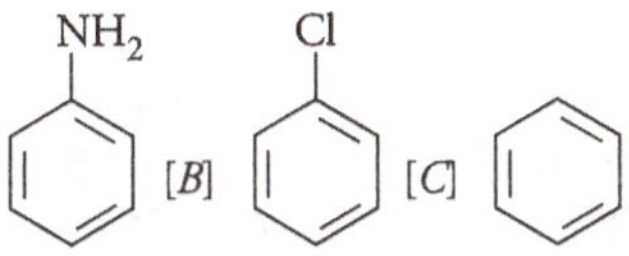

(d) [A] [*B*] [*C*]

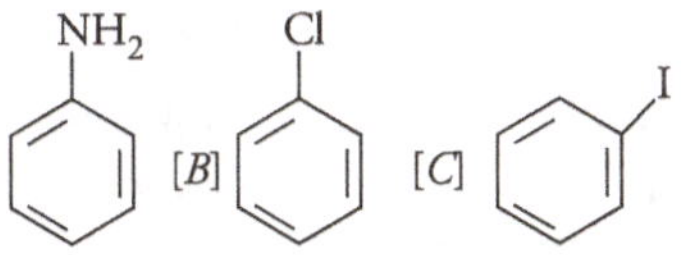

21 The products formed in the following reaction sequence are **[2025, 22 Jan Shift-I]**

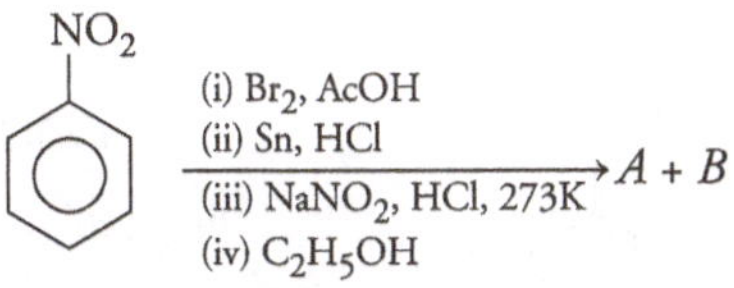

(a) 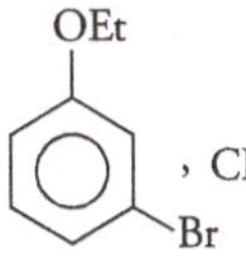 , CH_3—COOH

(b) 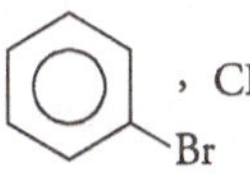 , CH_3—CHO

(c) 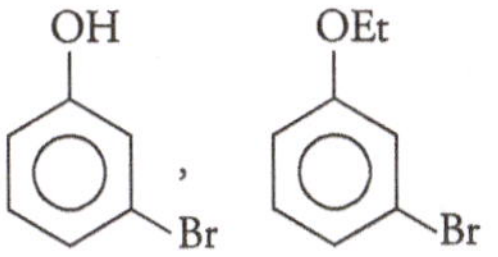

(d) 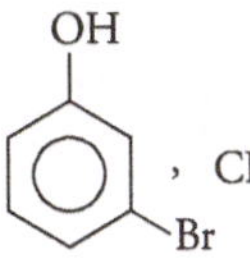 , CH_3—CHO

22 Match the compounds (List-I) with the appropriate Catalyst/Reagents (List-II) for their reduction into corresponding amines. **[2025, 22 Jan Shift-II]**

	List-I (Compounds)		List-II (Catalyst/Reagents)
(A)	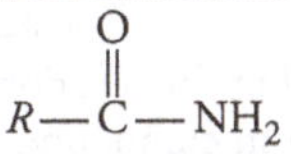	(I)	NaOH (aqueous)
(B)	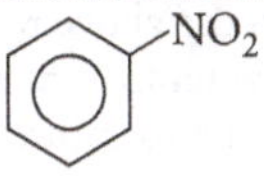	(II)	H_2 / Ni
(C)	R—C≡N	(III)	$LiAlH_4$, H_2O
(D)	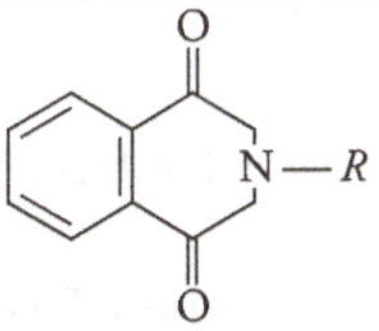	(IV)	Sn, HCl

Choose the correct answer from the options given below:

(a) (A)-(II), (B)-(I), (C)-(III), (D)-(IV)
(b) (A)-(III), (B)-(II), (C)-(IV), (D)-(I)
(c) (A)-(III), (B)-(IV), (C)-(II), (D)-(I)
(d) (A)-(II), (B)-(IV), (C)-(III), (D)-(I)

Imp. *Preparation of Azo Dye*

23 Which one of the following reaction sequences will give an azo dye? **[2025, 29 Jan Shift-II]**

(a) 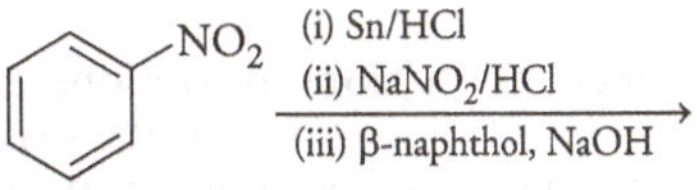

(b) 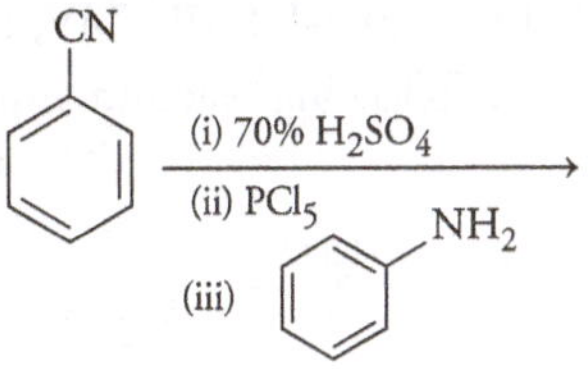

(c) 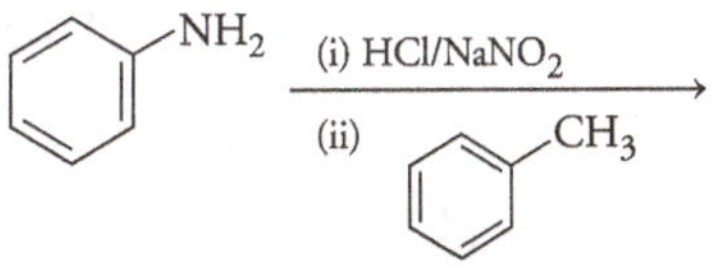

(d) 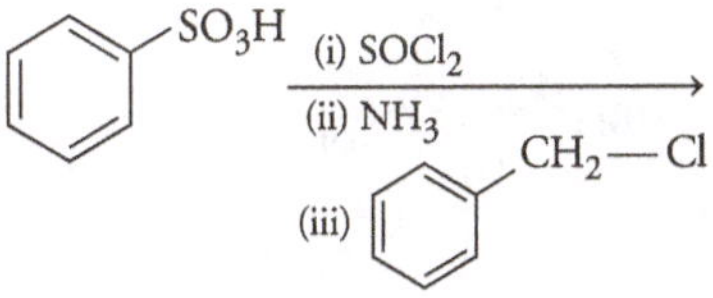

24 The incorrect statement regarding the reaction given below is

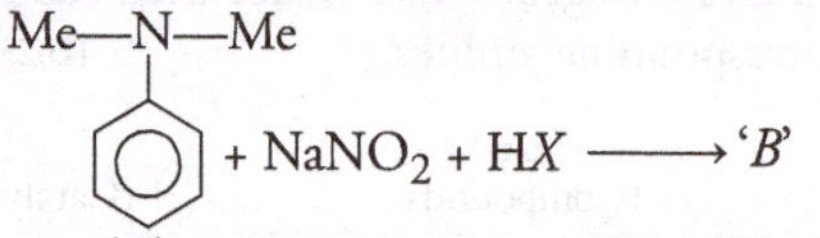

'A' **[2023, 12 April Shift-I]**

(a) The electrophile involved in the reaction is NO^+.
(b) 'B' is *N*-nitroso ammonium compound.
(c) The product 'B' formed in the above reaction is *p*-nitroso compound at low temperature.
(d) The reaction occurs at low temperature.

25 Match List-I with List-II.

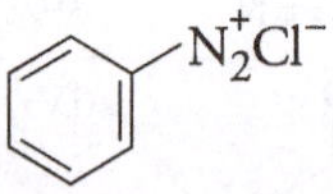

is reacted with reagents in List-I to form products in List-II

List-I (Reagent)	List-II (Product)
A. 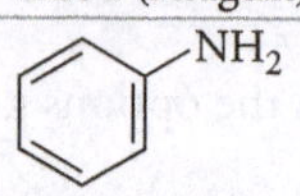	I. 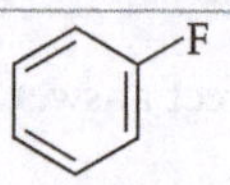
B. HBF_4, Δ	II. 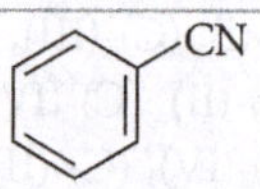
C. Cu, HCl	III. 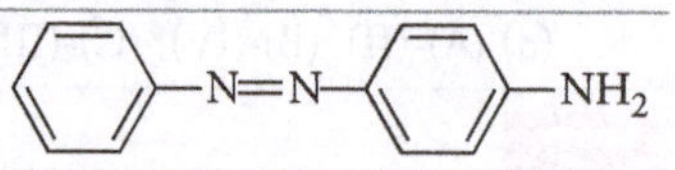
D CuCN/KCN	IV. 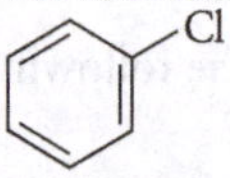

Choose the correct answer from the options given below **[2023, 8 April Shift-I]**

(a) A-IV, B-III, C-II, D-I (b) A-III, B-I, C-IV, D-II
(c) A-III, B-I, C-II, D-IV (d) A-I, B-III, C-IV, D-II

26 Compound *A* from the following reaction sequence is

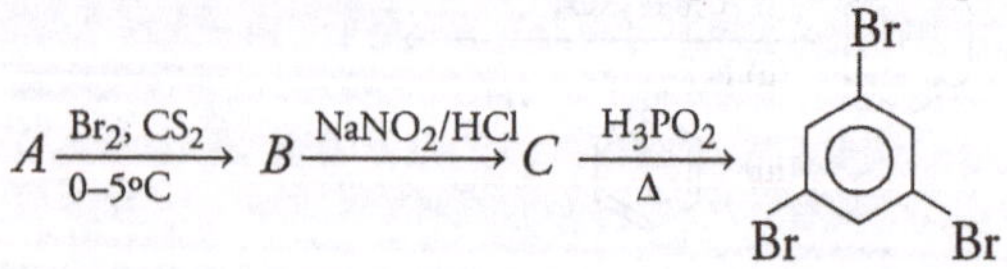

[2023, 13 April Shift-II]

(a) aniline (b) benzoic acid
(c) salicylic acid (d) phenol

27

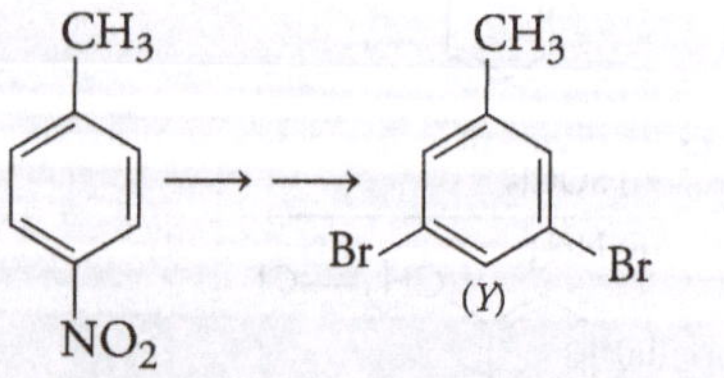

In the above conversion of compound (*X*) to product (*Y*), the sequence of reagents to be used will be **[2023, 30 Jan Shift-II]**

(a) (i) Fe, H^+ (ii) $Br_2(aq)$ (iii) HNO_2 (iv) CuBr
(b) (i) Fe, H^+ (ii) $Br_2(aq)$ (iii) HNO_2 (iv) H_3PO_2
(c) (i) Br_2, Fe (ii) Fe, H^+ (iii) $LiAlH_4$
(d) (i) $Br_2(aq)$ (ii) $LiAlH_4$ (iii) H_3O^+

28 Identify the products formed (*A* and *E*).

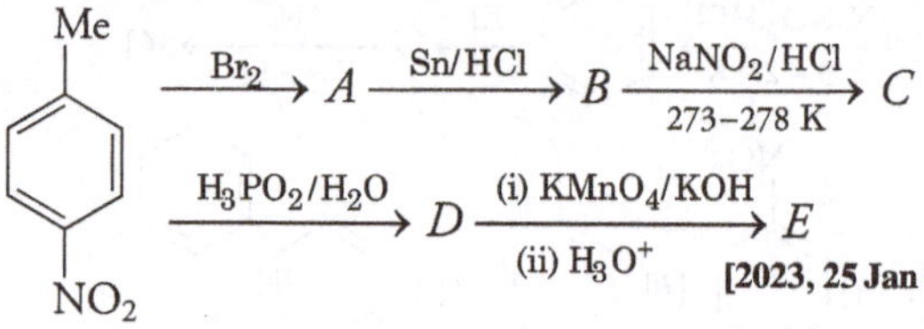

[2023, 25 Jan Shift-I]

(a) A = 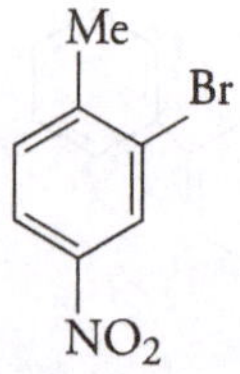 , E = 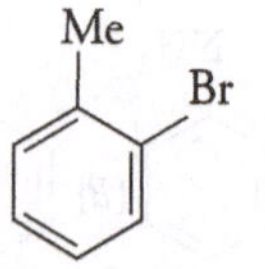

(b) A = 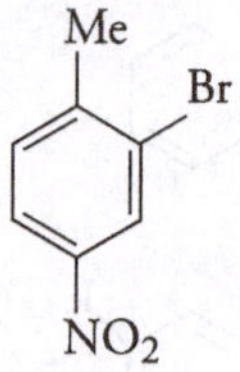 , E = 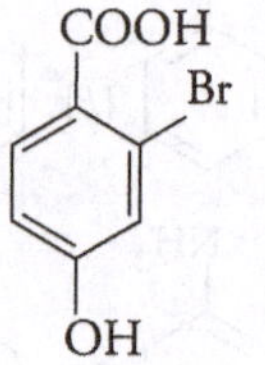

(c) A = 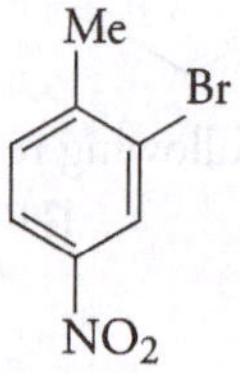 , E = 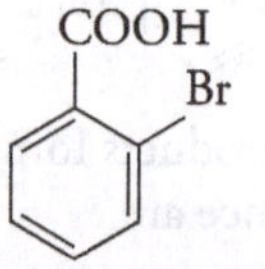

(d) A = 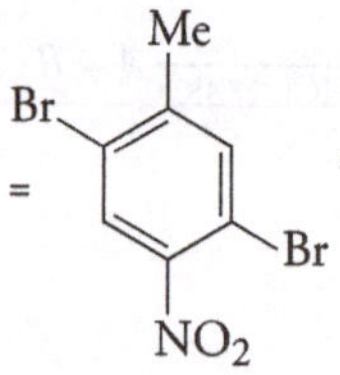 , E = 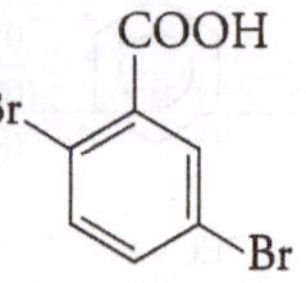

29 The final product '*C*' in the following series of reactions. **[2022, 26 June Shift-II]**

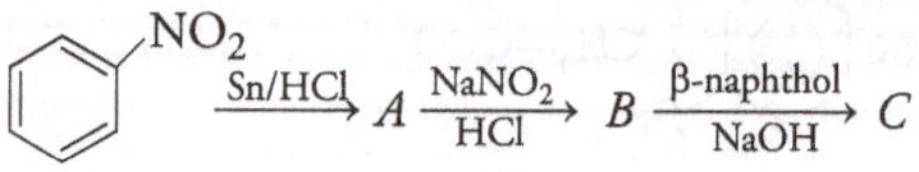

(a) 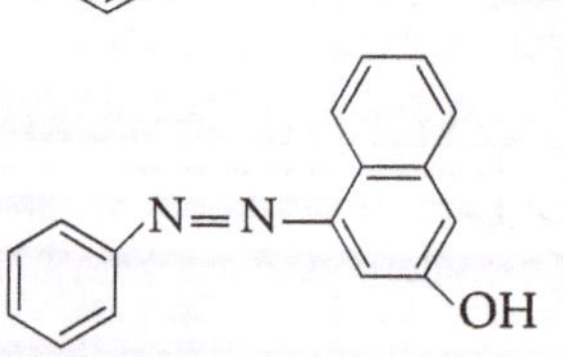

(b) 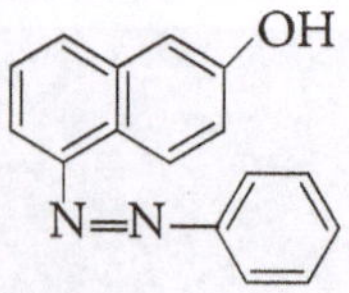

(c) 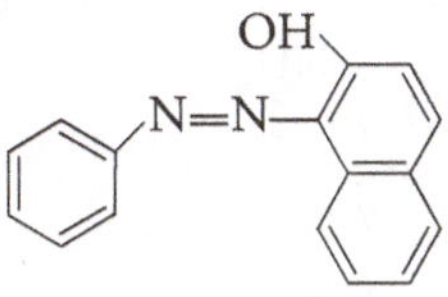

(d) 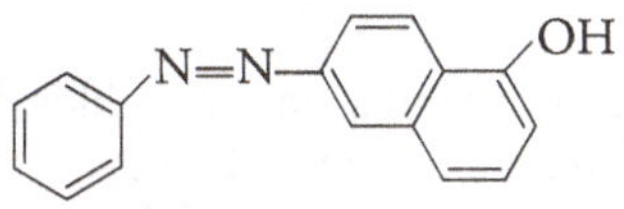

30 An organic compound '*A*' contains nitrogen and chlorine. It dissolves readily in water to give a solution that turns litmus red. Titration of compound '*A*' with standard base indicates that the molecular weight of '*A*' is 131 ± 2. When a sample of '*A*' is treated with aq. NaOH, a liquid separates which contains N but not Cl. Treatment of the obtained liquid with nitrous acid followed by phenol gives orange precipitate. The compound *A* is **[2022, 27 July Shift-II]**

(a) 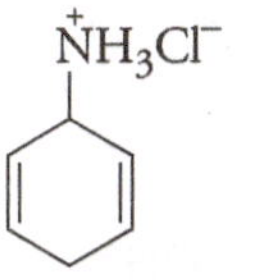

(b) 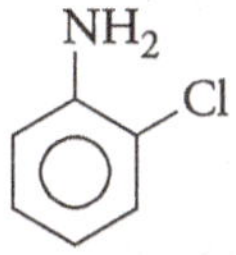

(c) 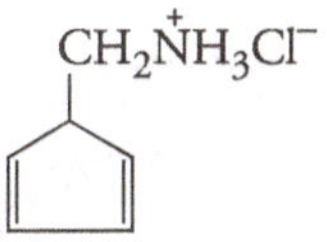

(d) 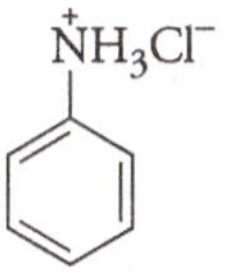

31 Identify the major product formed in the following sequence of reactions. **[2022, 28 June Shift-I]**

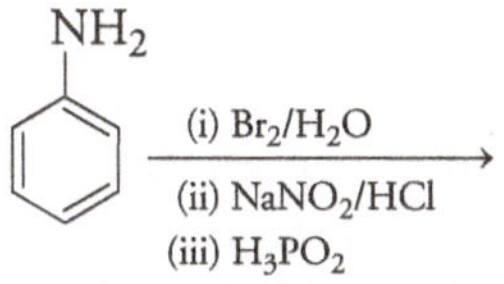

(a) 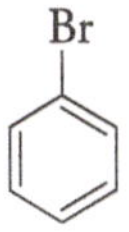

(b) 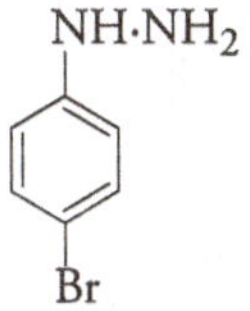

(c) 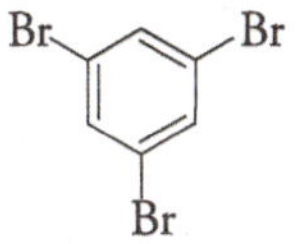

(d) 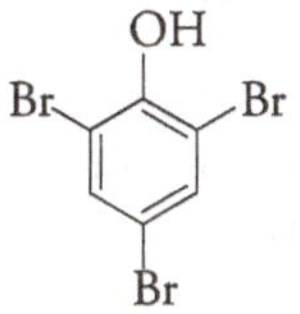

32 The major product of the following reaction is **[2021, 27 Aug Shift-I]**

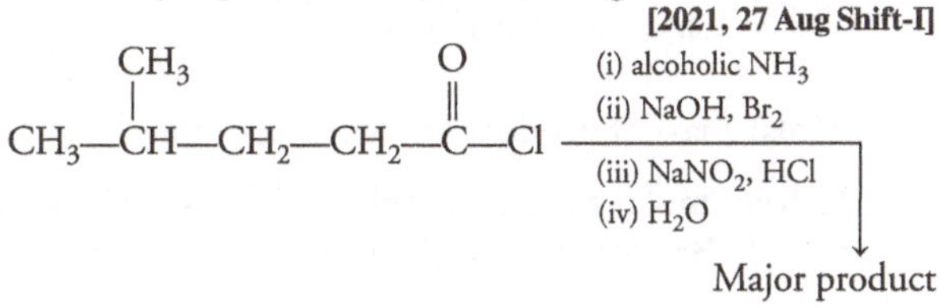

(a) $CH_3—CH(CH_3)—CH(Br)—CH_2OH$

(b) $CH_3—CH(CH_3)—CH_2—CH_2—CH_2OH$

(c) $CH_3—CH(CH_3)—CH_2—CH_2OH$

(d) $CH_3—CH(CH_3)—CH_2—CH_2—Cl$

33 Which of the following reaction(s) will not give *p*-aminoazobenzene? **[2021, 25 Feb Shift-I]**

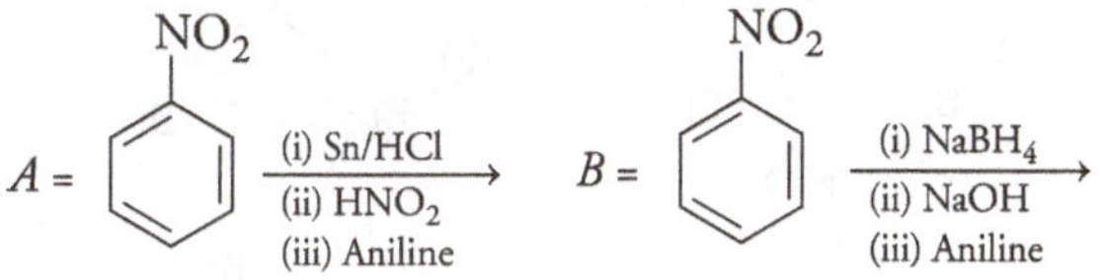

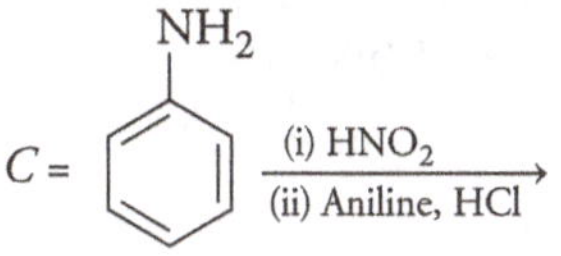

(a) *A* only (b) *C* only (c) *B* only (d) *A* and *B*

34 What is the correct sequence of reagents used for converting nitrobenzene into *m*-dibromobenzene?

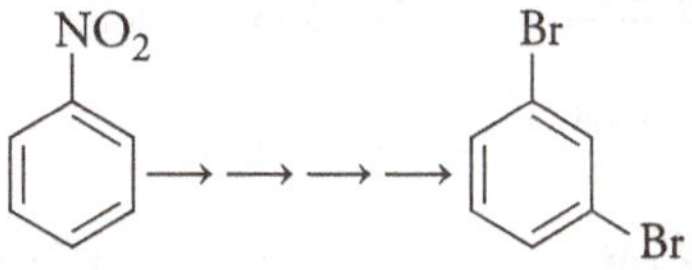

[2021, 24 Feb Shift-II]

(a) $\xrightarrow{NaNO_2}$ / $\xrightarrow{HCl}$ / $\xrightarrow{KBr}$ / $\xrightarrow{H^+}$

(b) $\xrightarrow{Br_2/Fe}$ / $\xrightarrow{Sn/HCl}$ / $\xrightarrow{NaNO_2/HCl}$ / $\xrightarrow{CuBr/HBr}$

(c) $\xrightarrow{Sn/HCl}$ / $\xrightarrow{KBr}$ / $\xrightarrow{Br_2}$ / $\xrightarrow{H^+}$

(d) $\xrightarrow{Sn/HCl}$ / $\xrightarrow{Br_2}$ / $\xrightarrow{NaNO_2}$ / $\xrightarrow{NaBr}$

35 In the following reaction sequence, [*C*] is **[2020, 04 Sept Shift-II]**

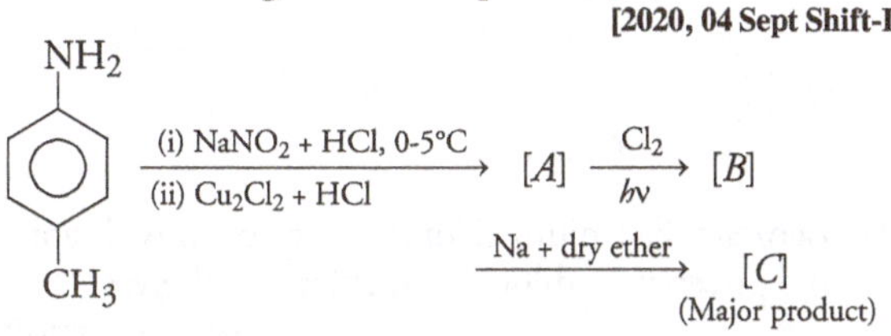

(a) 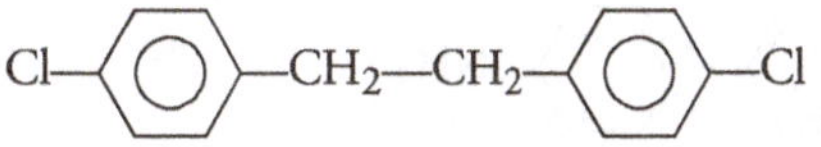

(b) 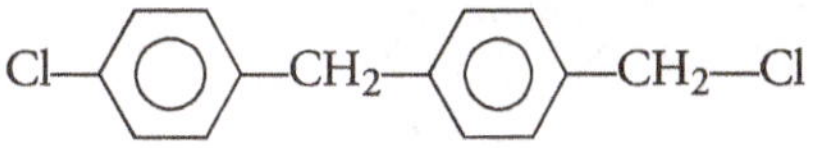

(c) 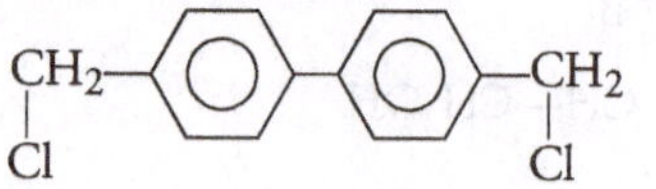

(d) 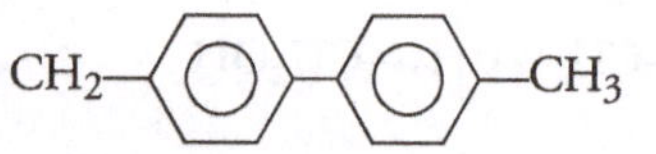

36 The major product Z obtained in the following reaction scheme is **[2020, 9 Jan Shift-I]**

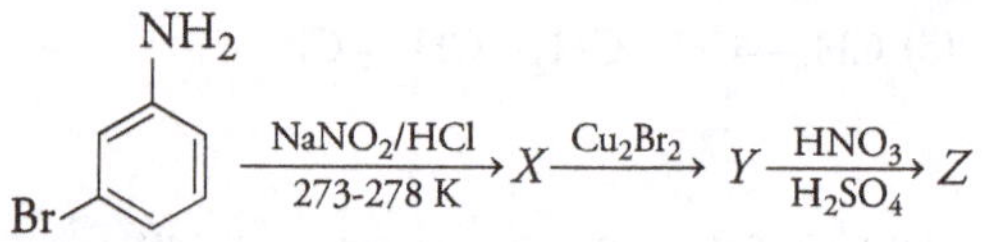

(a) 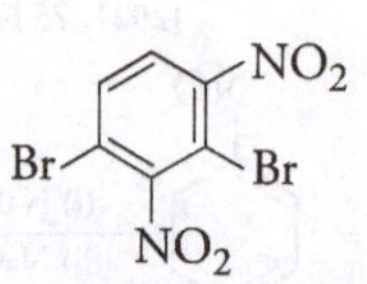

(b) 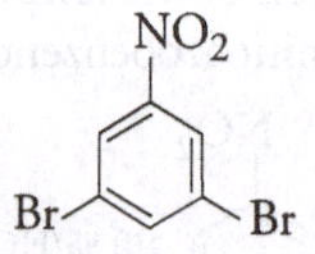

(c) 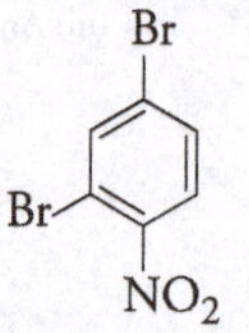

(d) 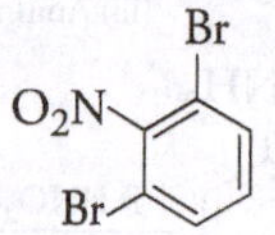

37 Consider the following reactions

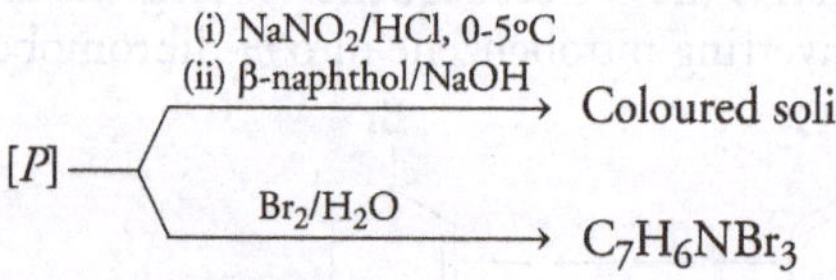

The compound [P] is **[2020, 9 Jan Shift-II]**

(a) 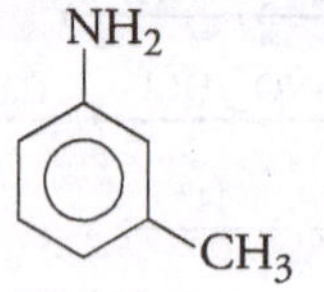

(b) 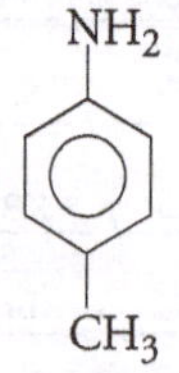

(c) 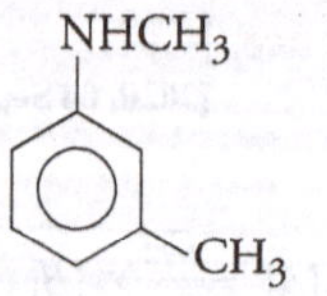

(d) 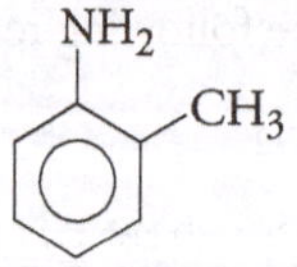

38 Benzene diazonium chloride on reaction with aniline in the presence of dilute hydrochloric acid gives **[2019, 12 April Shift-II]**

(a) 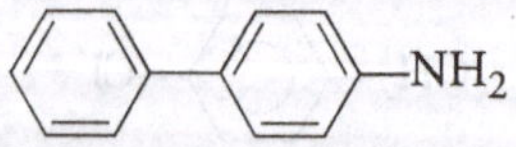

(b) 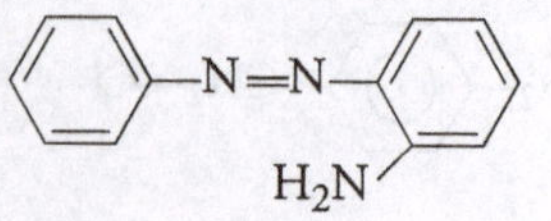

(c) 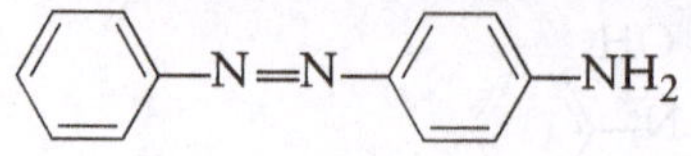

(d) 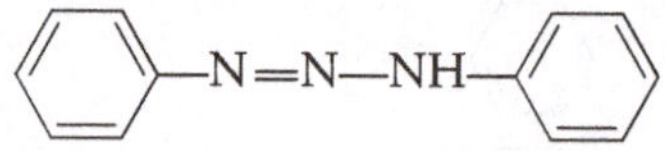

39 Coupling of benzene diazonium chloride with 1-naphthol in alkaline medium will give **[2019, 8 April Shift-I]**

(a) 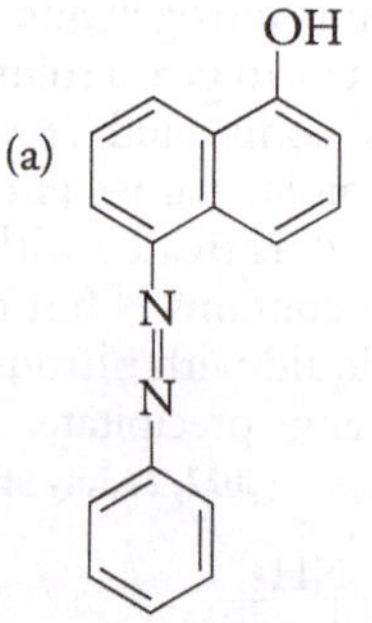

(b) 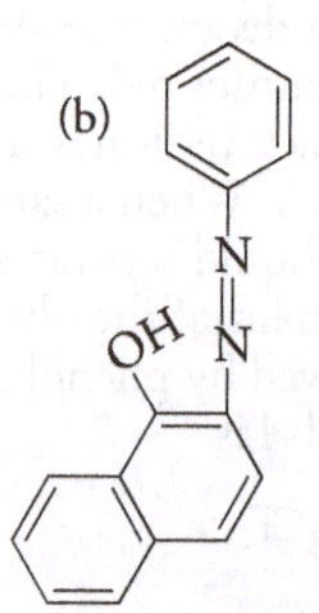

(c) 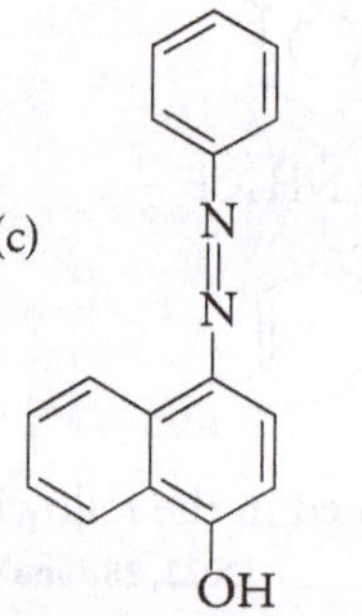

(d) 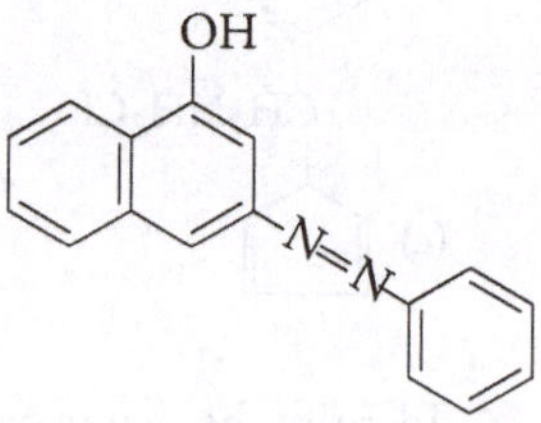

Numerical Values Type Questions

40 Consider the following sequence of reactions,

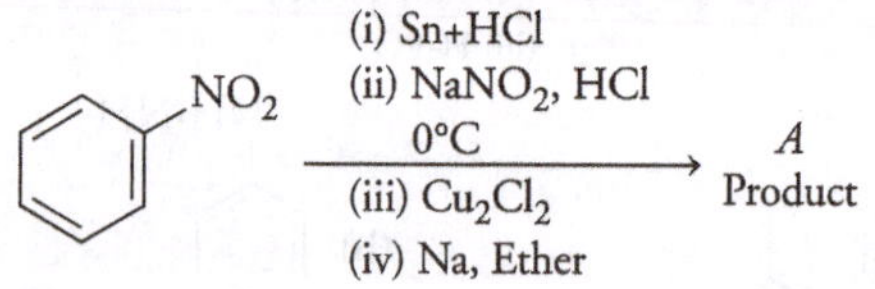

Molar mass of the product formed (A) is ...... g mol^{-1}. **[2025, 22 Jan Shift-I]**

41 Consider the following sequence of reactions to produce major product (A)

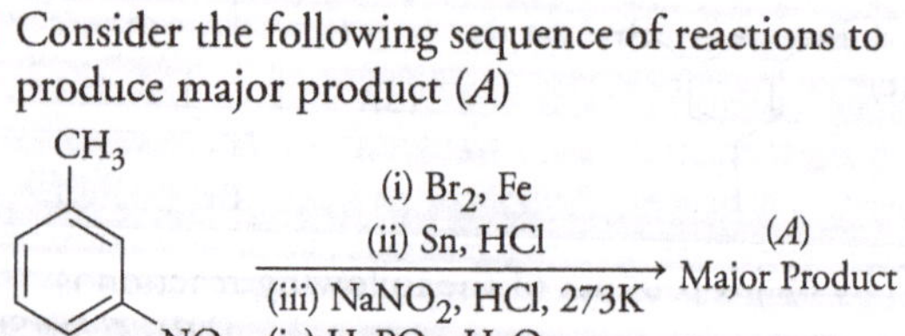

Molar mass of product (A) is .... g mol^{-1}.

(Given, molar mass in g mol^{-1} of C : 12, H : 1, O : 16, Br : 80, N : 14, P : 31)

Key Index

Topic 1

1 (b)	2 (a)	3 (c)	4 (d)	5 (b)	6 (d)	7 (c)	8 (a)	9 (d)	10 (c)
11 (b)	12 (c)	13 (b)	14 (a)	15 (a)	16 (c)	17 (b)	18 (d)	19 (d)	20 (c)
21 (c)	22 (a)	23 (a)	24 (d)	25 (c)	26 (d)	27 (c)	28 (a)	29 (c)	30 (c)
31 (c)	32 (d)	33 (a)	34 (d)	35 (c)	36 (c)	37 (b)	38 (2)	39 (108)°	

Topic 2

1 (b)	2 (c)	3 (a)	4 (d)	5 (a)	6 (d)	7 (b)	8 (b)	9 (c)	10 (a)
11 (a)	12 (c)	13 (d)	14 (b)	15 (d)	16 (a)	17 (d)	18 (c)	19 (d)	20 (d)
21 (c)	22 (b)	23 (b)	24 (b)	25 (b)	26 (b)	27 (c)	28 (c)	29 (c)	30 (b)
31 (b)	32 (b)	33 (c)	34 (d)	35 (b)	36 (d)	37 (c)	38 (c)	39 (b)	40 (b)
41 (b)	42 (b)	43 (c)	44 (a)	45 (d)	46 (a)	47 (c)	48 (a)	49 (c)	50 (c)
51 (a)	52 (c)	53 (c)	54 (b)	55 (c)	56 (d)	57 (a)	58 (c)	59 (d)	60 (341)
61 (4)	62 (5)	63 (5)	64 (3)	65 (5)	66 (3)				

Topic 3

1 (c)	2 (a)	3 (b)	4 (b)	5 (b)	6 (a)	7 (d)	8 (a)	9 (b)	10 (d)
11 (b)	12 (b)	13 (a)	14 (c)	15 (b)	16 (a)	17 (c)	18 (c)	19 (d)	20 (b)
21 (b)	22 (c)	23 (a)	24 (b)	25 (b)	26 (a)	27 (b)	28 (c)	29 (c)	30 (d)
31 (c)	32 (c)	33 (c)	34 (b)	35 (a)	36 (c)	37 (a)	38 (c)	39 (c)	40 (154)
41 (171)									

Detailed Explanations

Topic 1

01. *(b)* $\underset{\text{Propanamide}}{CH_3—CH_2—\overset{O}{\overset{||}{C}}—NH_2} \xrightarrow[\substack{-K_2CO_3 \\ -KBr \\ -H_2O}]{Br_2/KOH} \underset{\text{Ethylamine}}{CH_3CH_2NH_2}$

The above reaction is an example of Hoffmann bromamide reaction.

02. *(a)* The Clemmensen reduction is used to reduce aldehydes/ketones into alkanes using zinc amalgam and HCl. Also, $LiAlH_4$ is a good reducing agent, which reduces amides into amines.

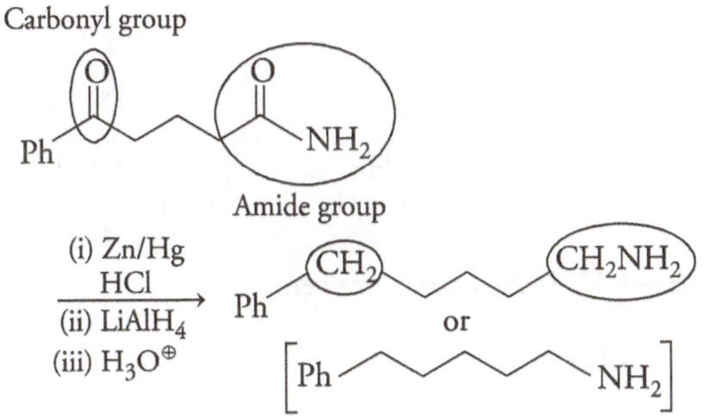

03. *(c)* In aqueous medium basic strength of methylsubstituted amines is dependent on inductive effect, solvation effect (H-bonding) and steric hinderance.

So, correct order is

$(CH_3)_2NH > CH_3—NH_2 > (CH_3)_3N > NH_3$

or $\quad Me_2NH > MeNH_2 > Me_3N > NH_3$.

04. *(d)* Hofmann degradation reaction involves the migration of an alkyl as well as aryl group from the carbonyl carbon atom of the amide to the nitrogen atom.

Therefore, statement I is incorrect but statement II is correct.

05. *(b)* Primary amines react with nitrous acid at 273 K to form diazonium salt which is very unstable. It breaks to give nitrogen gas and alcohol on raising temperature to 298 K.

$$R—NH_2 \xrightarrow{HNO_2,\ 273K}$$

$$\underset{\text{Diazonium salt}}{R—\overset{+}{N}_2Cl^-} \xrightarrow[H_2O]{298\ K} \underset{\text{Alcohol}}{R—OH}$$

06. *(d)*

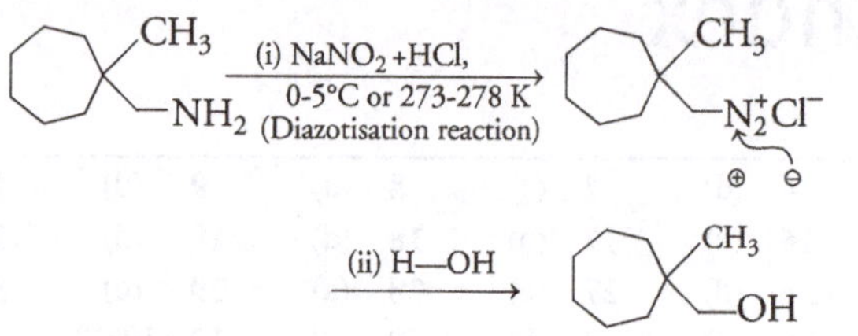

(1-methylcyclohexyl) methanol
(*P*)
(Major product)

07. *(c)* Chemical reaction is as follows

$$\underset{\substack{1^\circ \text{ amine} \\ (A)}}{RNH_2} \xrightarrow{CHCl_3} \underset{\substack{\text{Isonitrile} \\ (B)}}{R—\overset{\oplus}{N}\equiv\overset{\ominus}{C}} \xrightarrow[HCl\ (C)]{H_3O^+} RNH_2 + HCOOH$$

(*B*) ↓ Toxic

Hence, *A*-primary amine, *B*-isonitrile compound and *C*-conc. HCl.

08. *(a)* Hypobromite transforms the primary amide to carbamic acid which further loses carbonyl carbon as CO_3^{2-}, yielding the amine product.

$$R—\overset{O}{\overset{\|}{C}}—NH_2 + BrO^- \longrightarrow R—NH_2 + CO_3^{2-} + Br^-$$

09. *(d)*

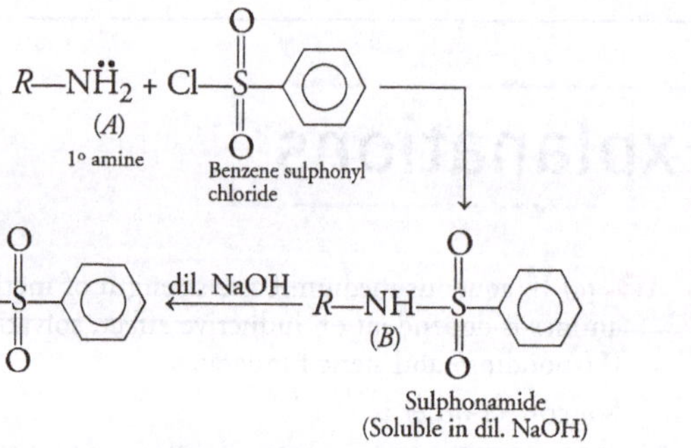

Hence, *A* is 1° amine i.e. $C_6H_5—\underset{\underset{CH_3}{|}}{CH}—NH_2$

10. *(c)* There is delocalisation of electron in both options (a) and (c). In option (c) there is delocalisation of electron with both carbonyl group and hence the lone pair is less available for donation. Hence, it is weakest base.

11. *(b)* For the conversion of C_2H_5CNinto $CH_3CH_2CH_2NH_2$, a strong reducing agent is required among the given reducing agent. Strong reducing agent is $LiAlH_4$. It is the most appropriate reagent for this reduction.

$$C_2H_5CN \xrightarrow{LiAlH_4} CH_3CH_2CH_2NH_2$$

12. *(c)* In first case, on increasing the size of alkyl group + *I* effect increases and the positive charge of ammonium cation gets dispersed more easily, Therefore, order of basicity is 2° amine > 1° amine > ammonia.

$(C_2H_5)_2NH > C_2H_5NH_2 > NH_3$

Concept Enhancer

Basic strength can be compared by the reaction of an alkanamine and ammonia with proton.

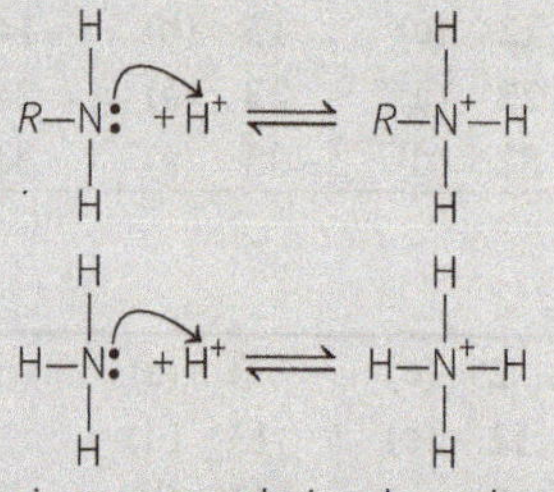

Basicity of an amine in aqueous solution depends upon the solubility of ammonium cation formed by accepting proton from water. The stability of ammonium cation in aqueous medium depends upon the following factors.

(a) +*I*-effect (b) Steric effect (c) Solvation effect

13. *(b)* $X \xrightarrow{Br_2/NaOH} C_3H_9N \xrightarrow[KOH]{CHCl_3} CH_3CH_2CH_2NC$

The first reaction is Hoffmann bromamide degradation reaction and the second reaction is carbylamine reaction. Since the compound C_3H_9N gives positive carbylamine test, this means it is 1° amine. Thus, the compound *X* is $CH_3CH_2CH_2CONH_2$.

$$\underset{(X)}{CH_3CH_2CH_2CONH_2} \xrightarrow[NaOH]{Br_2} CH_3CH_2CH_2NH_2$$

$$\xrightarrow{CHCl_3 + KOH} CH_3CH_2CH_2NC + 3KCl + 3H_2O$$

14. *(a)*

idea $AlH(i\text{-}Bu)_2$ also known DIBAL-H is di isobutyl aluminium hydride, $[(CH_3)_2CHCH_2]_2AlH$. It is a selective reducing agent. It reduces carboxylic acids, carboxylic acid derivatives and nitriles into aldehydes. It is an electrophilic reducing agent.

The mechanism of the reaction is as follows

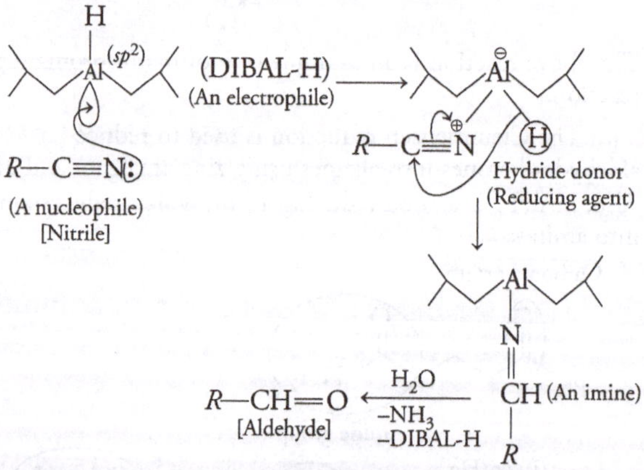

So, $R—\overset{}{\underset{\underset{O}{\|}}{C}}H$ is the correct answer.

15. *(a)* The complete reaction, takes place as follows

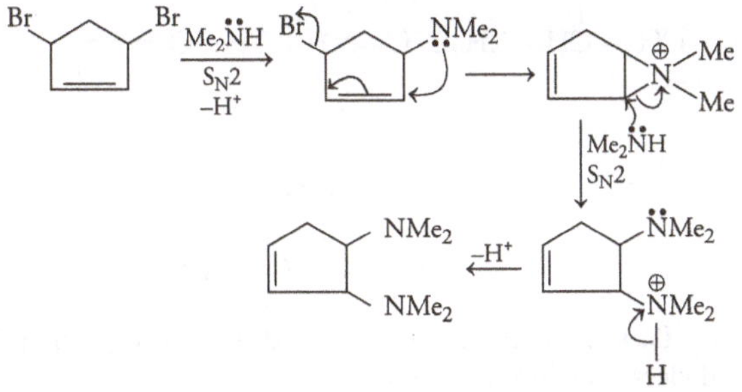

16. *(c)* The complete reaction is as follows

$$Cl(CH_2)_4Cl \xrightarrow{\text{Excess } NH_3} Cl^-NH_3^+ -(CH_2)_4 -NH_3^+Cl^- \; (A)$$

$$\xrightarrow{NaOH} H_2N-(CH_2)_4-NH_2 + 2NaCl + 2H_2O \; (B)$$

17. *(b)* The reaction will takes place as follows

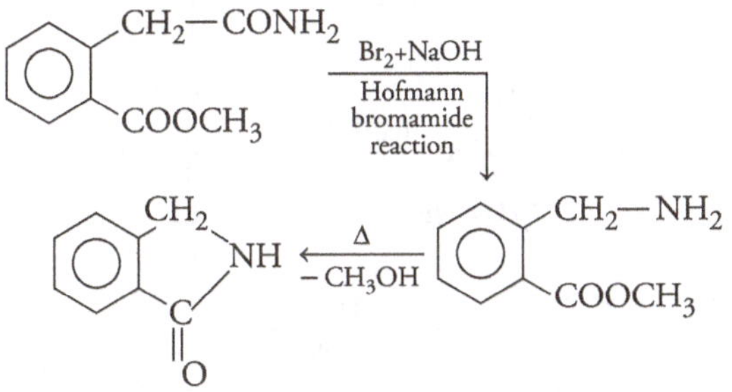

Option (b) is correct.

18. *(d)*

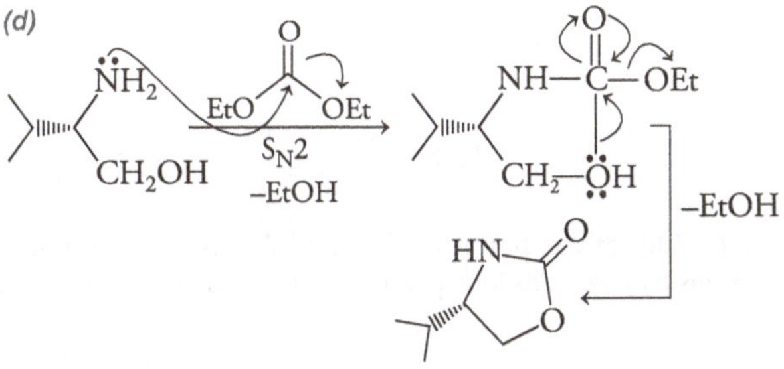

Initially, the lone pair electrons of $-NH_2$ will attack on electrophilic carbon of $O{=}C(OEt)_2$, after that lone pair electrons of oxygen of $-OH$ group, will attack intramolecularly, leading to formation of the cyclic compound.

19. *(d)*

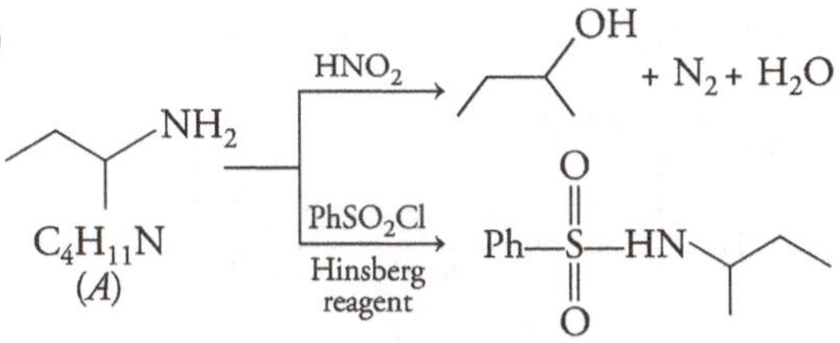

$C_4H_{11}N$ since it releases N_2 with HNO_2, it is a primary amine. After reacting with Hinsberg reagent it forms a compound which is soluble in KOH.

20. *(c)* The correct match is A-(III), B-(IV), C-(I), D-(II).

21. *(c)* Both bromoamide and isocyanate are isolated as intermediate.

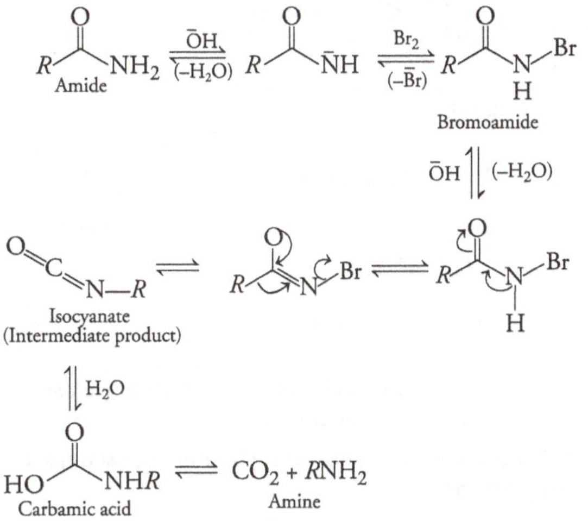

22. *(a)* The intermolecular association is more in primary amines than in secondary amines as there is only one hydrogen atom is available for formation of hydrogen bond in secondary amines, while primary amines contain two hydrogen atoms. So, statement (a) is not correct.

23. *(a)*

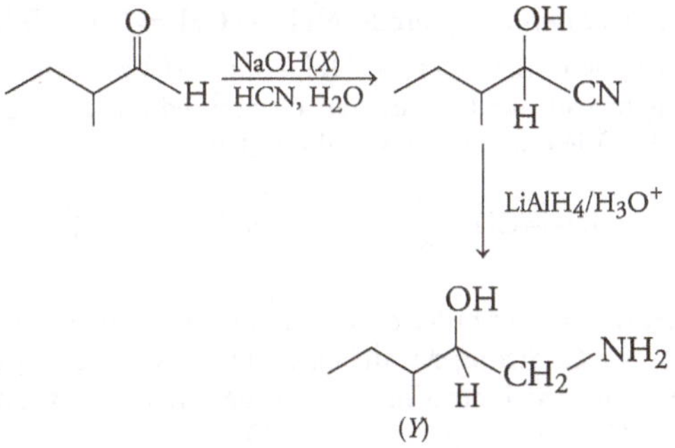

24. *(d)* Complete reaction is as follows

(i) **Removal of Br$^-$**

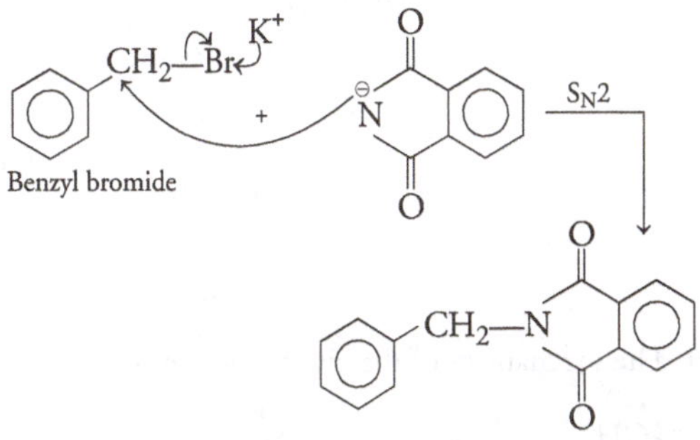

(ii) **Hydrolysis process in basic medium**

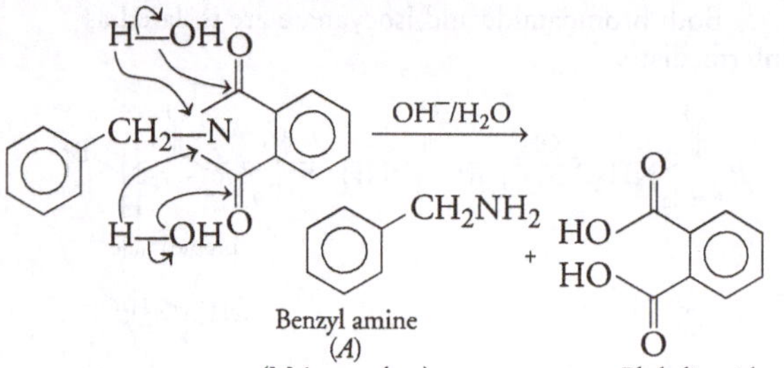

Benzyl amine (A) (Major product)

Phthalic acid

25. *(c)* Ammonolysis of alkyl halides

$$R{-}X \xrightarrow[-HX]{NH_3} R{-}NH_2 \xrightarrow[-HX]{R-X} R_2NH \xrightarrow[-HX]{R-X} R_3N \xrightarrow{RX} R_4\overset{\oplus\ominus}{NX}$$

During the reaction HX (acid) is formed. Hence, we use NaOH to remove this acidic impurities.

26. *(d)* (B) is most basic because of high resonance stabilisation of its conjugate acid.

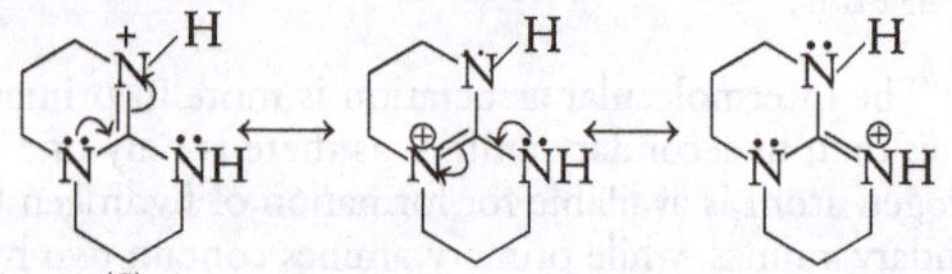

(B)

Among rest two compounds, $NH_2-CH=\ddot{N}H$ (A) has higher basicity than $CH_3-NH-CH_3$ (C) because conjugate acid of (A) is resonance stabilised but conjugate acid of (C) is not resonance stabilised at all.

$$\ddot{N}H_2-CH=\overset{\oplus}{N}H \longleftrightarrow \overset{\oplus}{N}H_2=CH-\ddot{N}H$$

(A)

Hence, the correct order of basicity of the given compounds is, $B > A > C$. Stronger is the base, smaller will be its pK_b value. Thus, the increasing order of pK_b values is $B < A < C$.

27. *(c)*

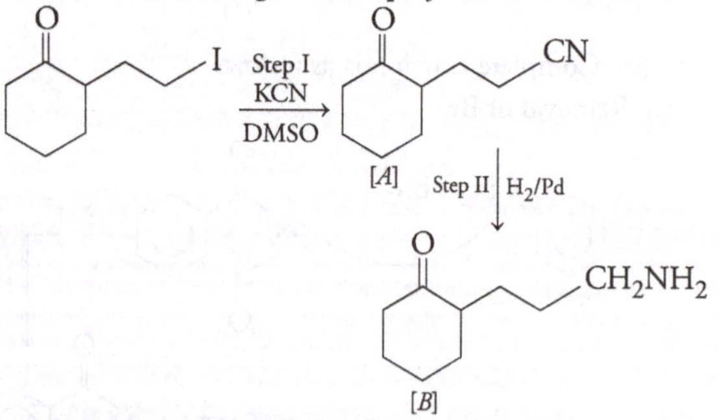

28. *(a)* The mechanism of the given reaction is as follows :

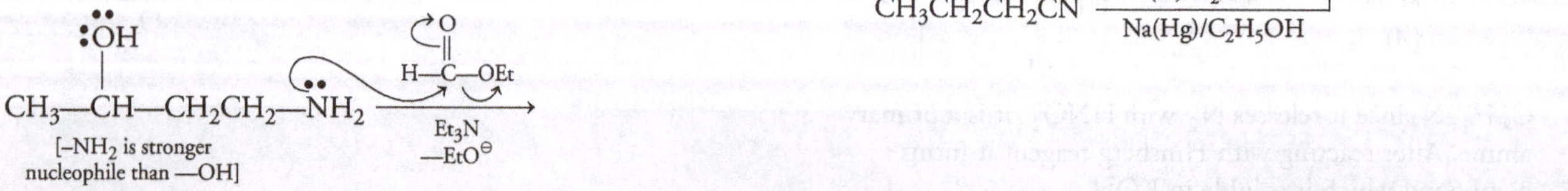

29. *(c)* Basic strength of amine depends on hydrogen bonding and electronic inductive effect.

Hence, the correct order is

$$NH_2-NH_2 < NH_3 < CH_3CH_2NH_2 < (CH_3CH_2)_3N < (CH_3CH_2)_2NH$$

30. *(c)* The complete reaction involved is as follows

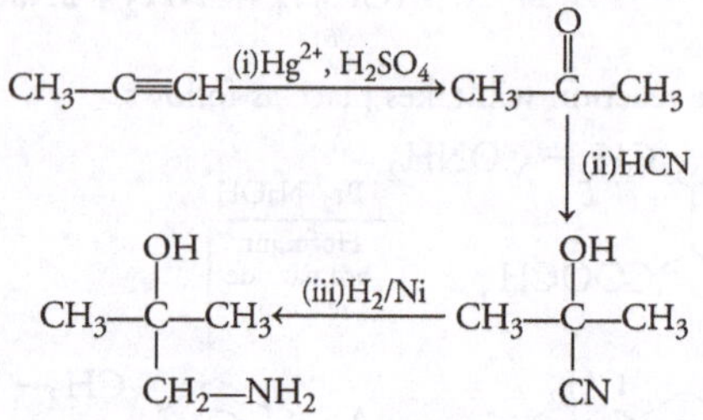

31. *(c)*

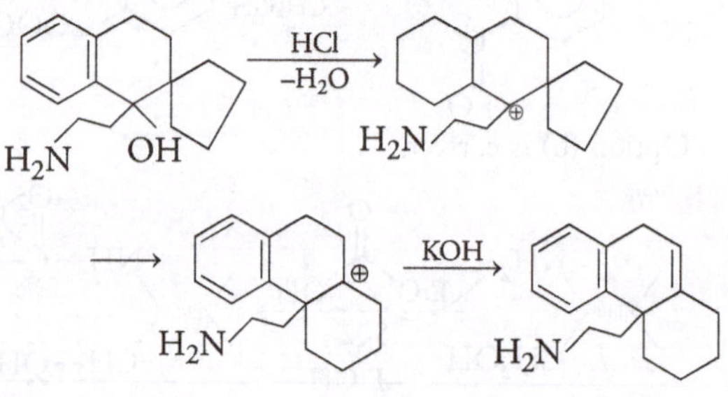

32. *(d)* Out of compounds a, b, c and d, d is the strongest Bronsted base as its lone pair of electrons can easily be donated.

33. *(a)*

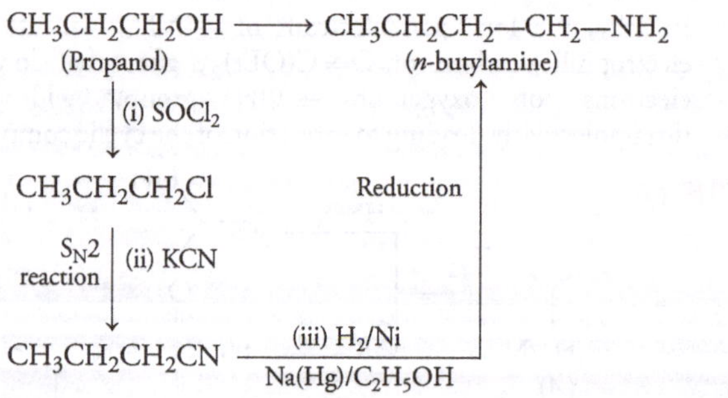

34. *(d)*

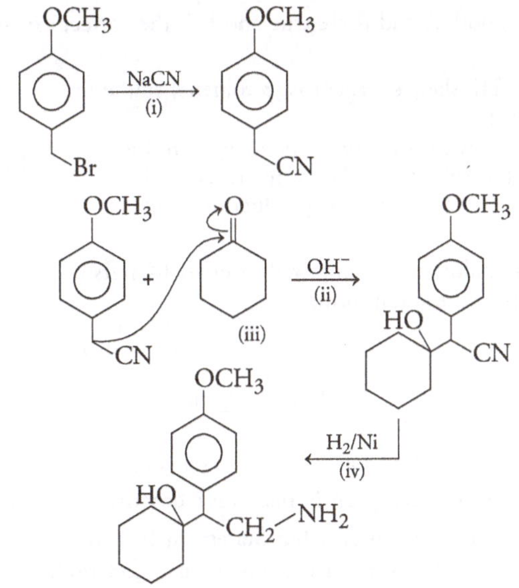

35. *(c)* The complete reaction is

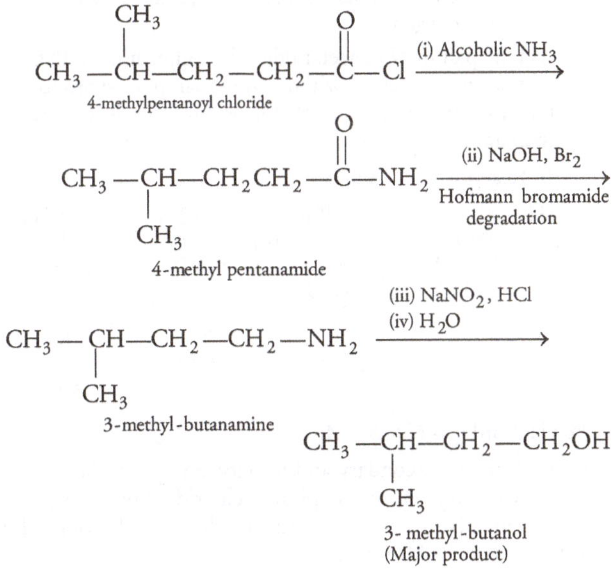

36. *(c)*

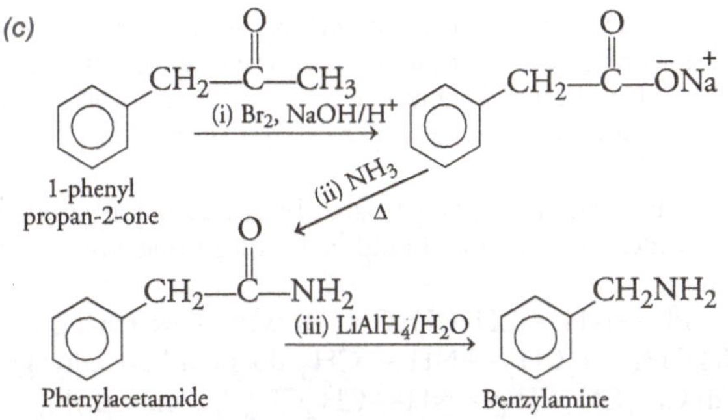

37. *(b)*

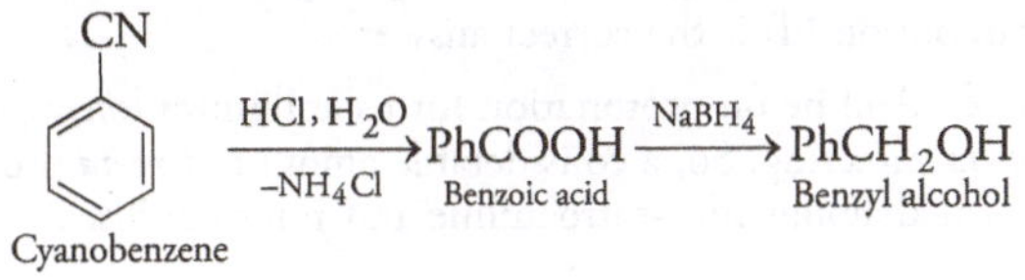

Thus, option (b) is incorrect method of the preparation of benzylamine from cyanobenzene..

38. *(2)* The acetylation reaction is as follows

$$R\text{—}NH_2 + CH_3\text{—}\overset{\overset{O}{\|}}{C}\text{—}Cl \longrightarrow R\text{—}NH\text{—}\overset{\overset{O}{\|}}{C}\text{—}CH_3$$

Compound (X) molar mass $= 108$ g/mol

Product molar mass $= 192$ g mol^{-1}

Gain in molecular weight after acetylation with one $—NH_2$ group is 42.

Total increase in molecular weight $= 84$

$\therefore$ Number of amino group in $X = \frac{84}{42} = 2$

Hence, 2 amino groups are present in compound X.

39. *(108°)*

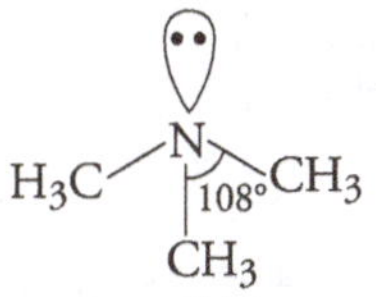

Pyramidal shape

Topic 2

01. *(b)* Among the given compounds is weakest as in this the lone pair is completely delocalised *via* cyclic resonance to show aromaticity.
While among pyridine and piperidine due to $+I$ effect of substituents piperidine is more basic. Hence, the correct order of basic strength is

Piperidine > Pyridine > Pyrrole

02. *(c)* Both Statement I and Statement II are false. The reason is that, 1° and 2° amines only reacts with *p*-toluenesulphonyl chloride and product of 1° amine is only soluble in aqueous NaOH.

03. *(a)* Both Statement I and Statement II are correct.
The NH_2 group in aniline is *ortho* and *para* directing and a powerful activating group as NH_2 has strong $+M$ effect.
Thus, aniline does not undergo Friedel-Craft's reaction because it will form complex with $AlCl_3$ that will deactivate the benzene ring.
Hence, option (a) is correct.

04. *(d)* Statement I is correct but Statement II is incorrect.
Correct form of Statement II is
Aniline is also known as amino benzene. It is a liquid organic compound formed by bond formation between benzene and ammonia. Chemical formula $= C_6H_5NH_2$
Hence, option (d) is correct.

05. *(a)* The complete reaction takes place as follows,

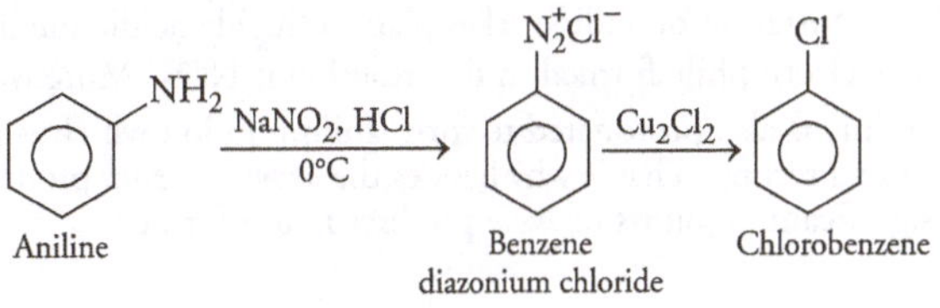

06. *(d)* As $-\ddot{N}H_2$ group is *o*/*p*-directing, hence arenium ion will be formed by attack at meta position i.e.

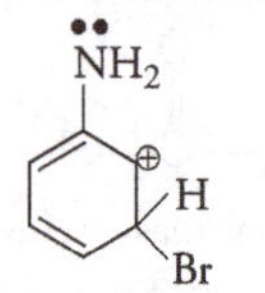

Hence, option (d) is correct.

07. *(b)* In option (a), (c) and (d) lone pairs is delocalised due to resonance.
While option (b) has highest availability of lone pair due to absence of resonance. Thus, it act as best proton acceptor or strongest bronsted base.

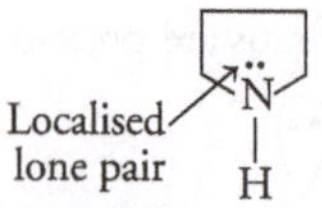

08. *(b)*

> **idea** Use the concept of stability of carbocation to solve this problem. This stability can be judged on the basis of resonance and hyperconjugation.

Carbocation (*C*) is most stable because positive charge delocalises with 2 π bonds and lone pair of N-atom in NH_2 group and due to hyperconjugation.

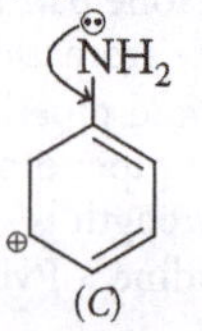

(*C*)

Thus, option (b) is correct.

09. *(c)* Statement I is correct but Statement II is incorrect.
Aniline and other aryl amines are generally colourless, but they get coloured if stored outdoors due to atmospheric oxidation. Aniline becomes dark brown due to action of air and light [oxidation].

10. *(a)* Hinsberg reagent is another name used for benzene sulphonyl chloride.

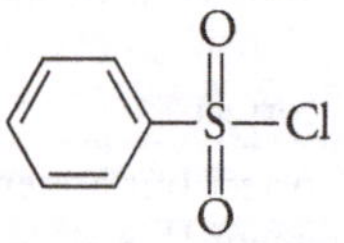

Concept Enhancer

Hinsberg reagent name is given due to its use in the hinsberg test for the detection & distinction of primary, secondary and tertiary amines in a given sample. It is an organo-sulphur compound with chemical formula $C_6H_5SO_2Cl$.

11. *(a)* Nitration of aniline takes place in highly acidic medium. The electrophile formed in this reaction is NO_2^+. Moreover, aniline is also protonated to form anilinium ion which is *meta*-directing. That is why besides the *ortho* and *para* products, significant amounts of *meta* product is also formed.
Thus, both A and R are true and R is the correct explanation of A.

12. *(c)* Hinsberg's reagent is (*p*-toluene) sulphonyl chloride.
It reacts with primary amine to give product which is soluble in alkali while reacts with secondary amine to give product insoluble in alkali.

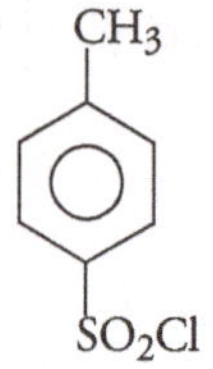

13. *(d)* Aniline when subjected to Friedel-Crafts alkylation reaction forms

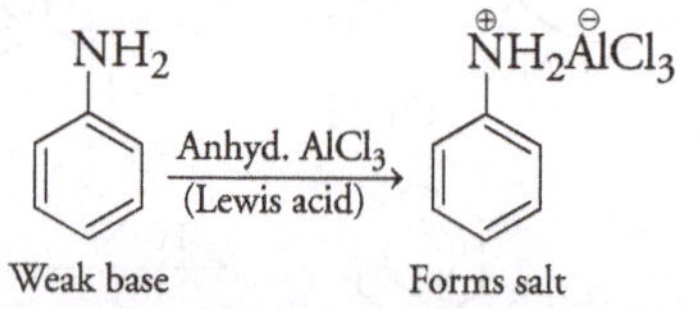

The positive charge on N makes it a deactivating group.

14. *(b)* Statement I is false but Statement II is true.
Aniline ($C_6H_5NH_2$) is more basic than acetamide (CH_3CONH_2) because in acetamide lone pair of electron present on nitrogen is delocalised to more electronegative element, i.e. oxygen.
As lone pair of N in acetamide is in conjugation with C=O that is strong electron withdrawing than phenyl group, so lone pair of N is unavailable for proton, so it is less basic than aniline.

15. *(d)* Reaction is as follows

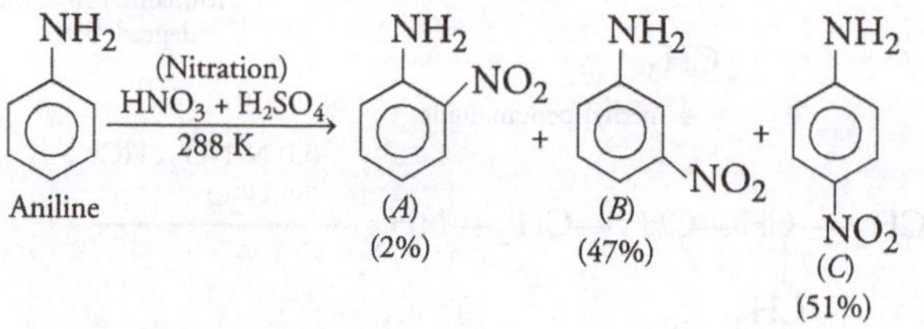

% yield order is $C > B > A$.

16. *(a)* Primary, secondary and tertiary amines can be separated using benzene sulphonyl chloride. Here it is present in its derivative form i.e., *p*-toluene sulphonyl chloride. This test is known as Hinsberg test.

17. *(d)* The amine on reaction with benzene sulphonyl chloride (Heisenberg reagent) produces a compound insoluble in alkali. It indicates the amine is a 2° amine. i.e. all options are possible except option (b) which is a 1° amine ($CH_3CH_2NH_2$).
As this 2° amine can be prepared by ammonolysis of ethyl chloride, the 2°-amine should have at least one ethyl (C_2H_5) group.
(a) Ph—NH—$CH_2CH_2CH_3$ does not have ethyl group.
(c) $CH_3CH_2CH_2$—NH—CH_3 does not have ethyl group.
(d) $CH_3CH_2CH_2$—NH—CH_2CH_3 has one ethyl group.
So, option (d) is the correct answer.

18. *(c)* Aniline on protonation forms anilinium ion, which is *meta*-directing. So, a considerable amount of *meta* product is formed alongwith *o*-nitroaniline and *p*-nitroaniline.

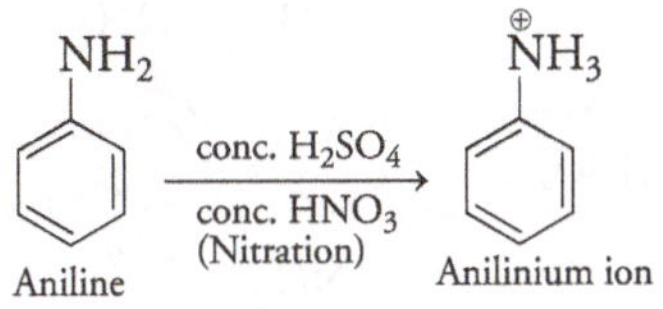

When aniline is treated with nitrating mixture *ortho*, *meta* and *para* nitroanilines are obtained.
Aniline being basic in nature, reacts with acids to form anilinium ion which is *meta* directing.

19. *(d)*

> **idea** Basicity ∝ Ease of donation of lone pair of electrons
> $\propto \frac{1}{\%\text{ of s-character of N}} \propto \frac{1}{\text{Electronegativity of N}}$

The % of *s*-character in the given amines are as follows:

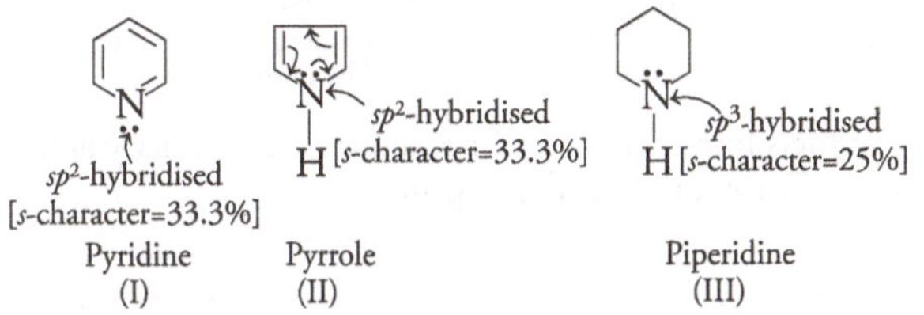

Therefore, piperidine (III) having minimum % s-character is most basic. Between, pyridine (I) and pyrrole (II) the lone pair of electrons of N in pyrrole (II) is involved in delocalisation and follows $(4n+2)\,\pi$ aromatic $(n=1)$ system. So, the N-atom of pyrrole (II) will show least basicity.
Thus, the order of basicity is as follows:

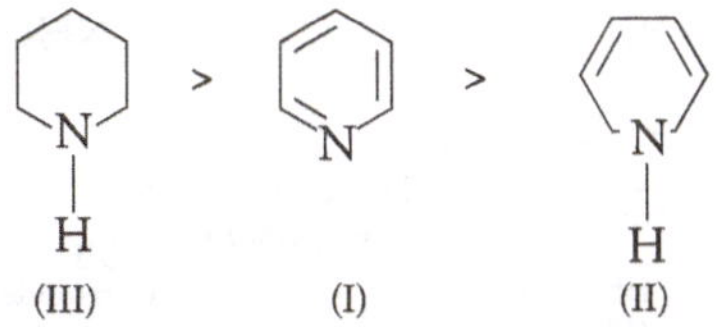

20. *(d)* Only primary amines (1°) gives carbylamine test. The reaction is as follows,

$$R—NH_2 \xrightarrow{CHCl_3 + KOH} R—N \rightleftarrows C$$

Alkyl isocyanide (Pungent)

Option (A) and (C) are primary amine and therefore will give positive carbylamine test.

21. *(c)* Hinsberg reagent reacts with primary and secondary amines to form sulphonamides. But Hinsberg reagent do not react with tertiary amines. So, among the given options compound given in A, C and E will react Hinsberg's reagent while compound given in B and D will not react.

22. *(b)* The complete reaction is as follows.

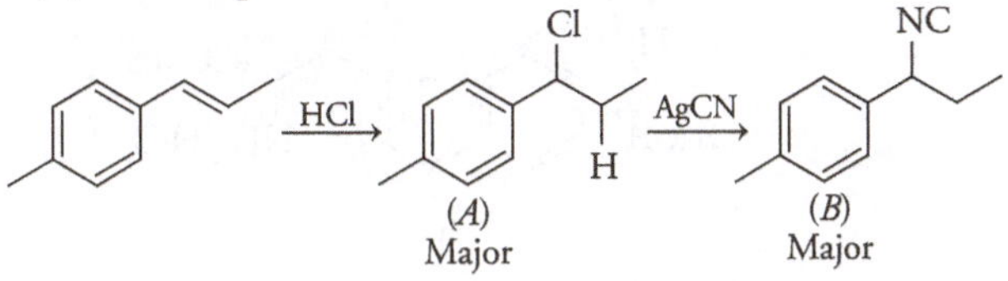

23. *(b)* The complete reaction is as follows:

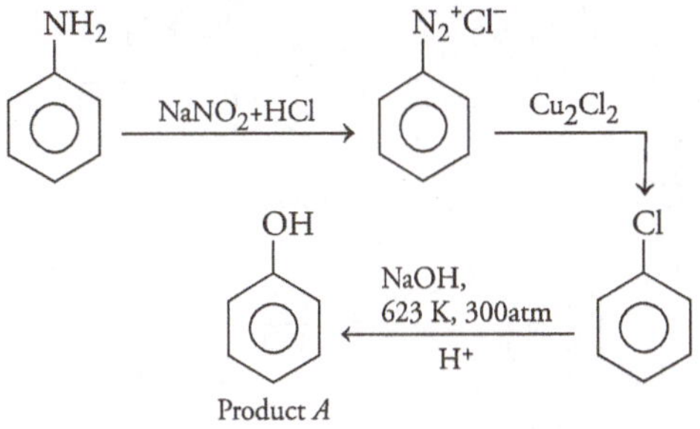

24. *(b)* Correct order of rate of electrophilic aromatic substitution is

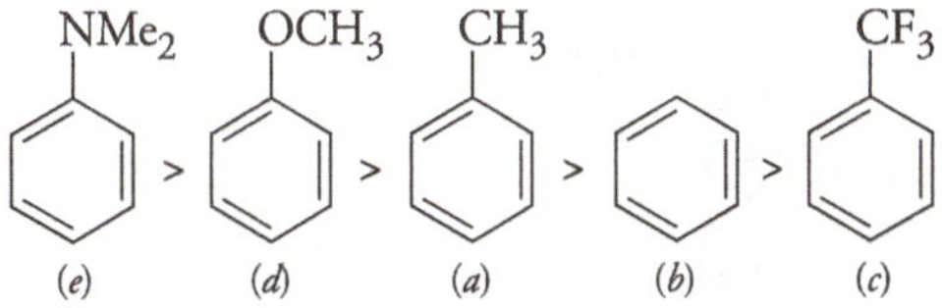

Rate of electrophilic aromatic substitution is directly proportional to the electron density on benzene ring. Electron releasing groups ($NMe_2 >$ —OCH_3) increases electron density whereas electron withdrawing groups (—$CF_3 >$ —NO_2) decreases electron density.

25. *(b)* Option (b) is correct.

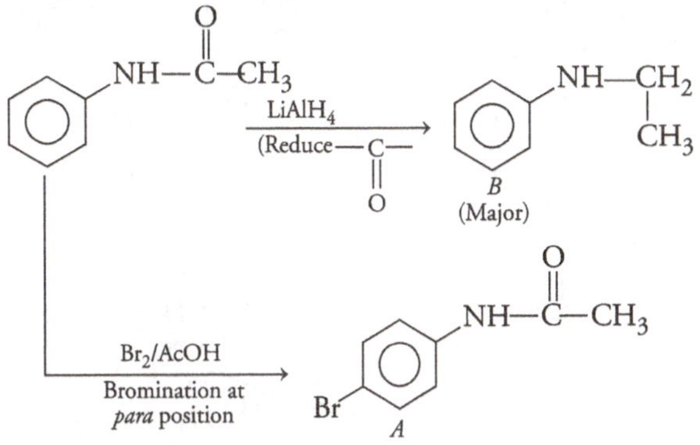

26. *(b)* The correct match is A-III, B-IV, C-II, D-I.

27. *(c)*

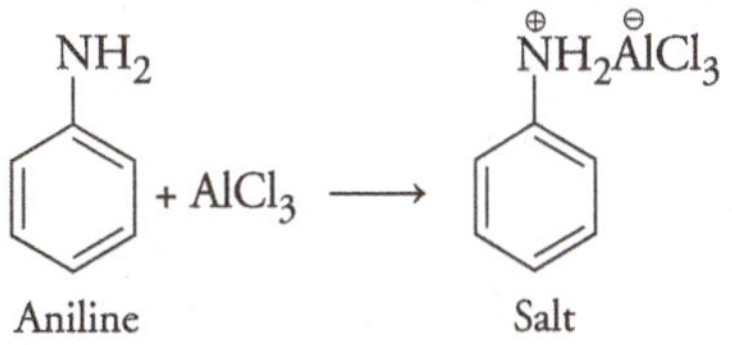

Aniline does not show Friedel Craft reaction as it is a weak base and reacts with $AlCl_3$ (electron deficient) which act as Lewis acid, forming salt instead of *o*- or *p*-substituted product.
∴ (A) is true but (R) is false.

28. *(c)*

1. **Addition-elimination reaction** $\ddot{N}H_3$ attacks on carbonyl carbon and eliminates Cl to form amide.

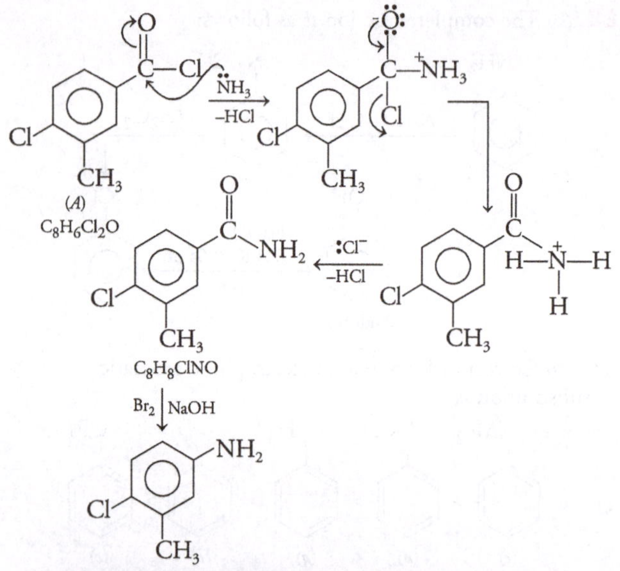

2. **Hoffmann bromamide degradation** Amide with X_2 in presence of NaOH undergoes Hoffmann bromamide degradation to form amine.

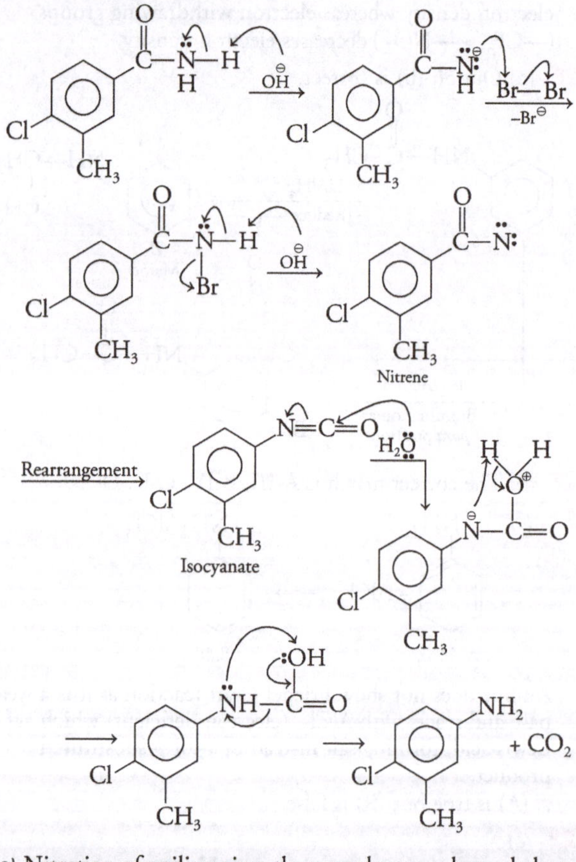

29. *(c)* Nitration of aniline gives three products as shown below in presence of HNO_3 and H_2SO_4 at 288 K.

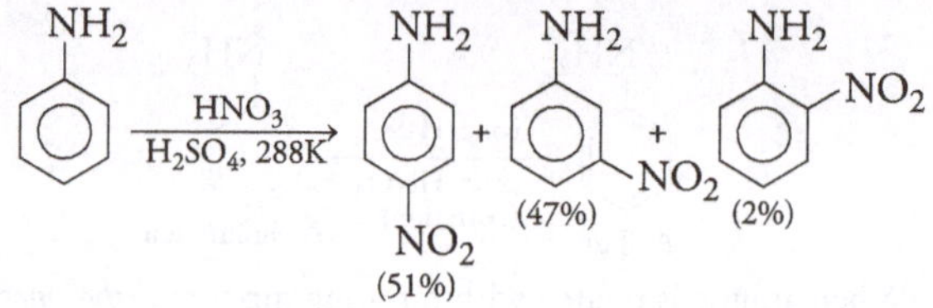

Here, H_2SO_4 acts as an acid.

$$H_2SO_4 + HNO_3 \longrightarrow NO_2^{\oplus} + HSO_4^{\ominus} + H_2O$$

Hence, statements (B) and (D) are correct.

30. *(b)* In reaction (b), imino group formed does not react with Hinsberg reagent.

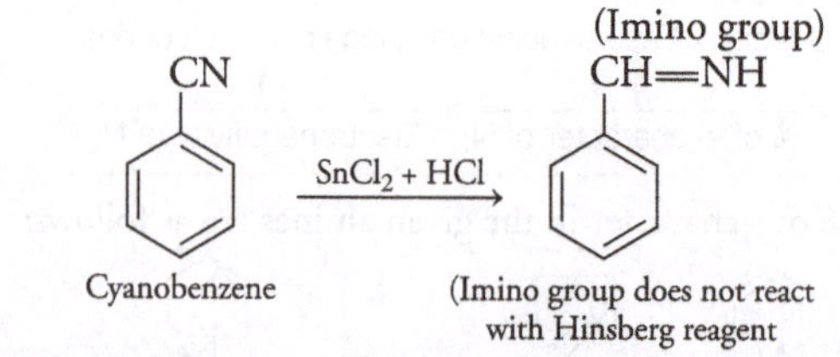

Products formed in all other reactions are a primary amine so, they react with Hinsberg reagent.

31. *(b)*

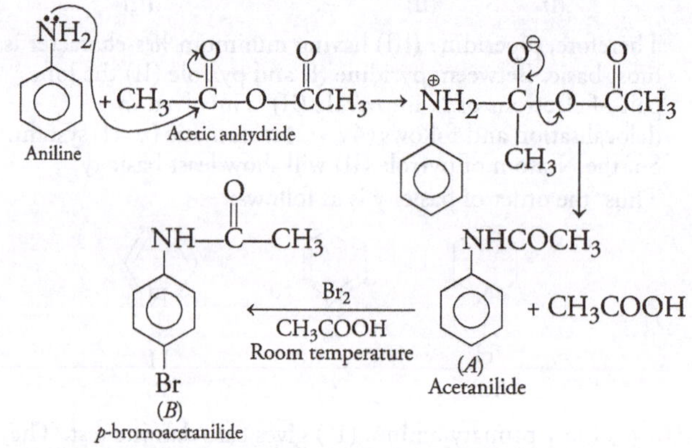

32. *(b)*

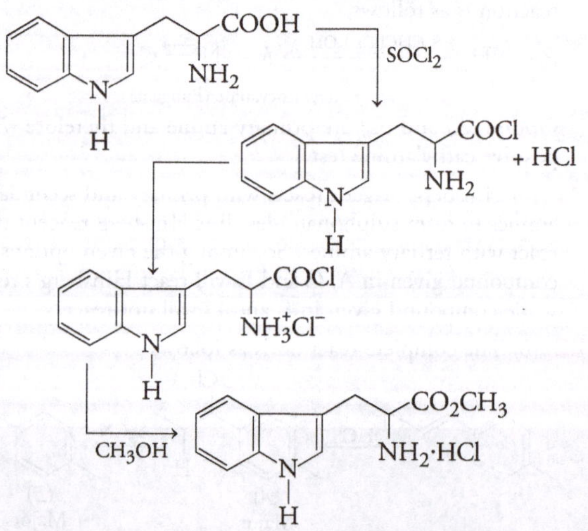

33. *(c)* Both (A) and (R) are true and (R) is correct explanation of (A).

34. *(d)* Reaction I is known as Hofmann bromamide reaction. Hofmann reagent reduces amide to amine with one carbon less.

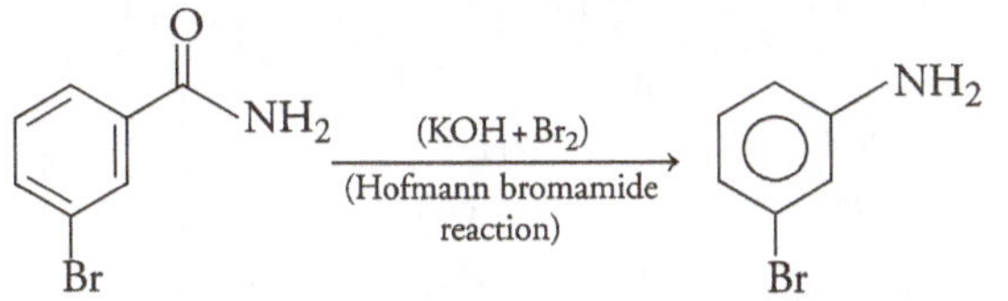

(A)

In reaction II, $LiAlH_4$ reduces amide to primary amine with same number of carbon atom.

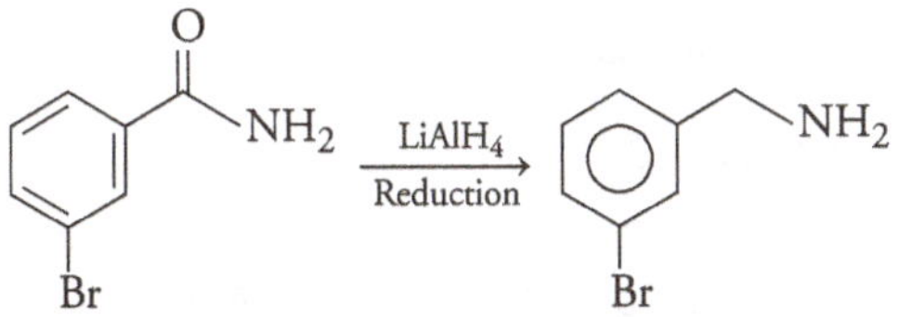

(B)

35. *(b)*

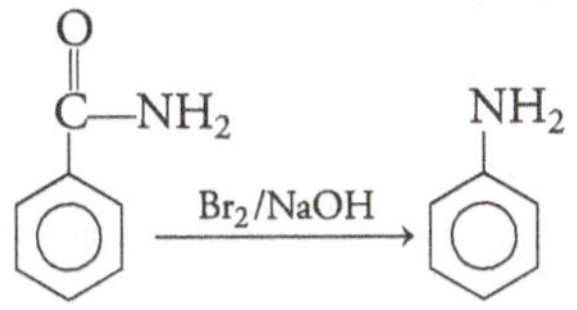

Benzamide Aniline (A)

Hofmann bromamide reaction

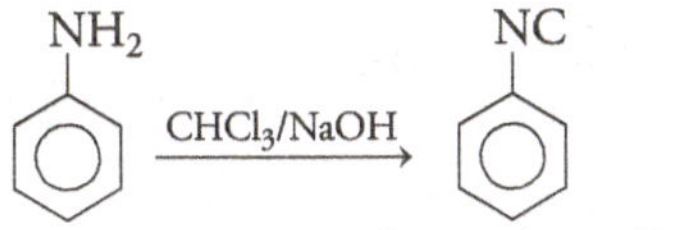

Benzene isocyanide (B)

Carbylamine reaction

Note Secondary and tertiary amines do not show this reaction. This reaction is known as carbylamine reaction or isocyanide test and is used as a test for primary amines.

36. *(d)* In phenyl methanamine, the lone pair on nitrogen of $—NH_2$ group is localised and does not undergoes resonance as attached to sp^3-hybridised C-atom.

A is purely aliphatic 1°-amine. *B* is aromatic 3°-amine with more basic nature (due $—CH_3$ groups). *C* is aromatic 2°-amine with less basic nature (due one $—CH_3$ group). *D* is purely aromatic 1°-amine. Hence basicity order is

$$A > B > C > D.$$

37. *(c)* Aniline on carbylamine reaction produces a foul smelling gas, phenyl isocyanide (C_6H_5NC).

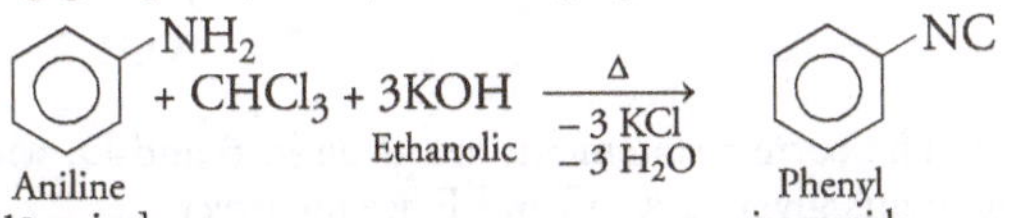

Aniline [1° amine] Phenyl isocyanide

Carbylamine test is used to detect aliphatic and aromatic primary amines.

38. *(c)* Availability of free lone pair of electron for donation, hence more basic.

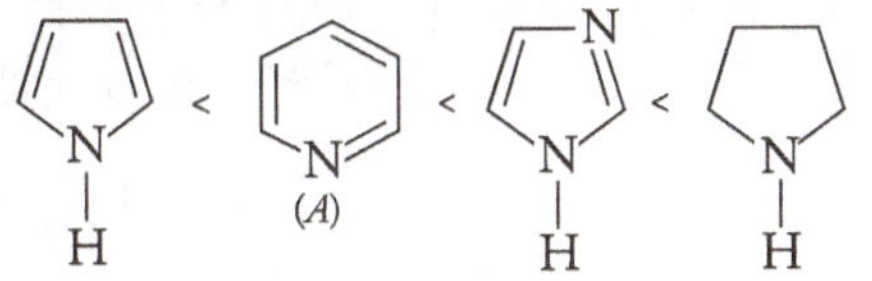

(B) Least basic (A) (D) (C) Most basic

Hence, correct increasing order of basicity is

$$(B) < (A) < (D) < (C)$$

39. *(b)*

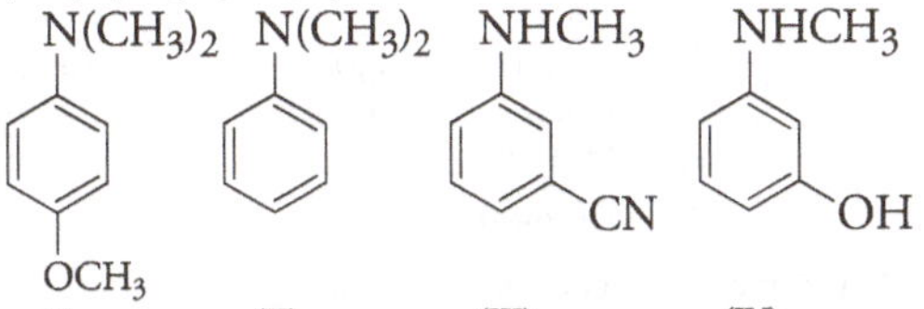

(I) (II) (III) (IV)

(+M, +R, –I) effect of —OCH_3 group; (–*I* effect of —CN group); (–*I* effect of —OH group)

pK_b is inversely proportional to basic strength and as the electron density increases in the ring, K_b will increases and pK_b value decreases. Electron density within the ring is maximum in (I) thus it has least pK_b value. So, option (a), (c) and (d) are incorrect.

Thus decreasing order for K_b is (I) > (II) > (IV) > (III)

∴ Increasing order of pK_b : (I) < (II) < (IV) < (III).

Hence, the correct option is (b).

40. *(b)* The correct order of basicity is III > II > I > IV. This is because, in compound III, nitrogen is sp^3 hybridised has less *s* character and more basic. In compound II, lone pair of electrons are not implicated in resonance. Hence easily available for donation. On the other hand, in compound I, resonance is caused by lone pair of electrons and hence not available for donation and in compound IV, lone of electrons is involved in generating conjugated system of π-electrons.

41. *(b)* The reaction of alkyl halide with amine or amides is a nucleophilic substitution reaction.

More nucleophilic nitrogen is more reactive with alkyl halide. Compound (*A*) is benzamide and lone pair of 'N' of it, is not available in this compound.

In compound (*D*), lone pair of 'N' are available but in compound (*C*), the lone pair of 'N' are not readily available due to the presence of electron withdrawing group (—CN) attach to benzene ring.

In compound (*B*), i.e. pthalimide, 'N' is not nucleophilic due to the presence of two polar groups attached to it that pulls lone pair towards them.

Hence, the correct order is : (*B*) < (*A*) < (*C*) < (*D*).

42. *(b)* Ni/H_2 can reduce $—C{\equiv}N$ into $—CH_2—NH_2$ (1°-amine) but cannot reduce an ester group ($—CO_2Et$) whereas DIBAL-H, di-isobutylaluminium hydride, $[(CH_3)_2CH]_2AlH$ reduces the ester group ($—CO_2Et$) into —CHO (an aldehyde).

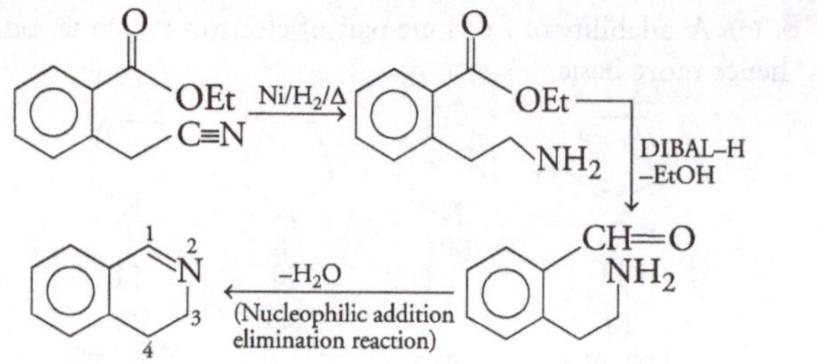

43. *(c)* Using retro-synthesis, to get the required compounds, *A*, *B* and *C*.

C_6H_7N (*C*) [1°-amine] (No. of C=6, 1 carbon less) $\xleftarrow[\text{KOH}]{Br_2}$ Hoffmann bromamide reaction (step down) *B* [An amide] (No.of C=7) $\xleftarrow{NH_3/\Delta}$ *A* [Carboxylic acid $C_7H_6O_2$]

So, the reactions can be shown as:

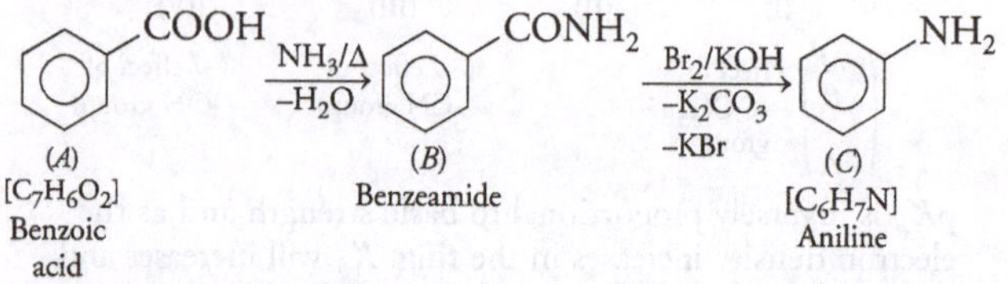

44. *(a)* The correct sequence is
E > D > B > A > C

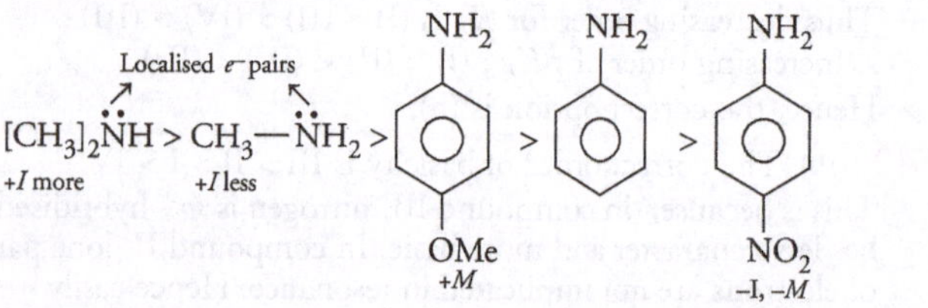

45. *(d)* Complete reaction sequence takes place as follows,

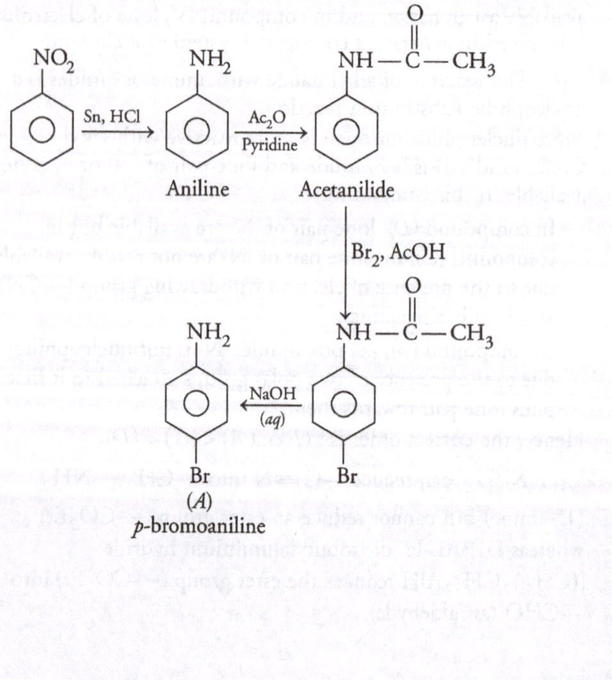

46. *(a)* Among the given options, reaction given in option (a) will give 3,4,5-tribromoaniline.

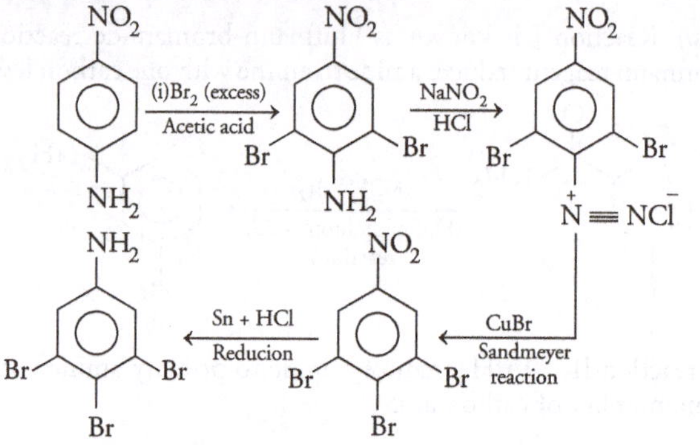

3,4,5-tribromoaniline

47. *(c)*

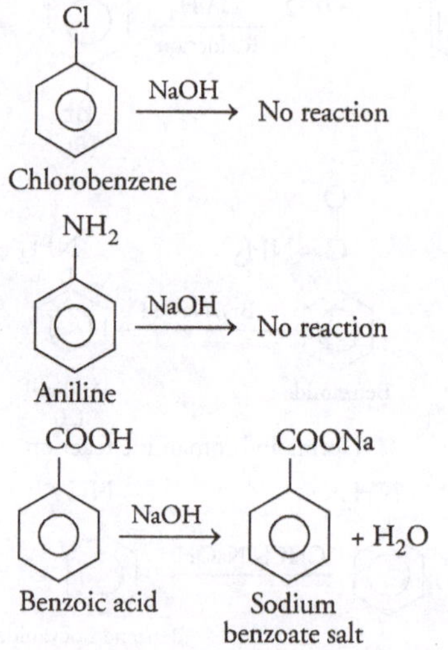

Organic layer in funnel are mixture of chlorobenzene and aniline.

48. *(a)* The complete reaction is as follows

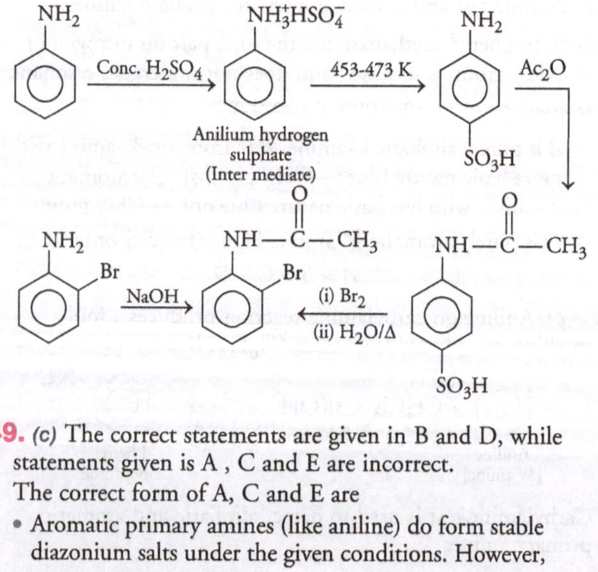

49. *(c)* The correct statements are given in B and D, while statements given is A, C and E are incorrect.

The correct form of A, C and E are

- Aromatic primary amines (like aniline) do form stable diazonium salts under the given conditions. However, aliphatic primary amines form diazonium salts that are highly unstable and decompose rapidly.

- The carbylamine reaction is specific to primary amines. Secondary and tertiary amines do not give this test.
- Tertiary amines do not react with benzenesulphonyl chloride, as they lack a hydrogen atom to participate in the reaction.

Concept Enhancer

Benzene sulphonyl chloride ($C_6H_5SO_2Cl$) is an organosulphur compound. It reacts with compounds that have reactive O—H and N—H bonds. This reagent is used as a test for amines.

50. *(c)* Isomers (*P*), (*Q*) and (*R*) respectively are

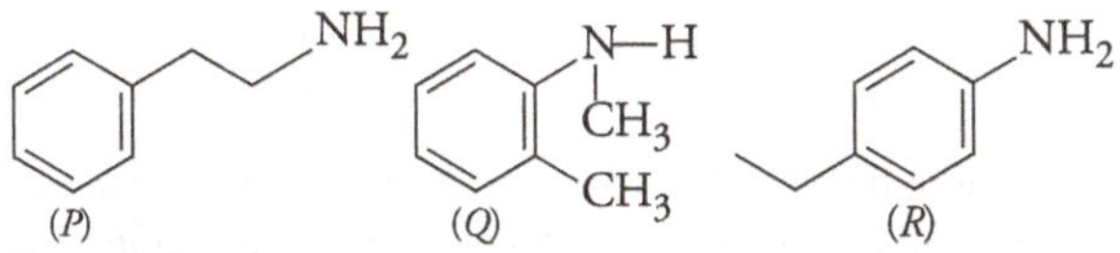

(*P*) Gabriel phthalimide synthesis is used for the preparation of aliphatic primary amines.

(*Q*) 2°-amines reacts with Hinsberg's reagent to give solid insoluble in NaOH.

(*R*) Aromatic primary amine react with nitrous acid at low temperature (273 K – 298 K) to form diazonium salts, which form red dye with β-naphthol.

51. *(a)*

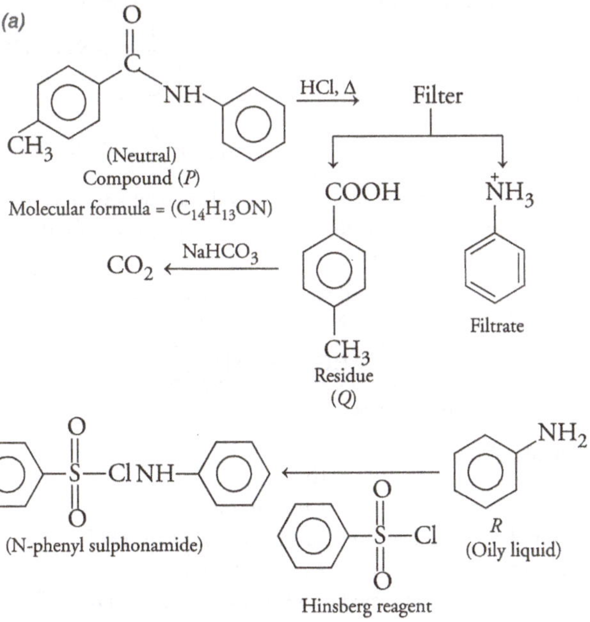

52. *(c)*

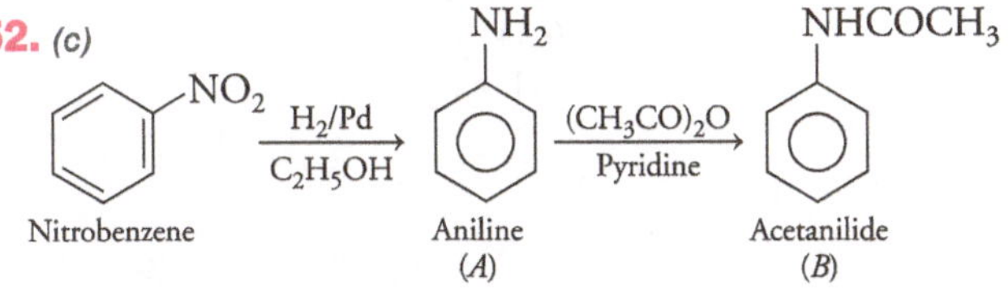

Amino group of aniline on acetylation with acetic anhydride/pyridine gives acetanilide.

53. *(c)* The correct match is A-III, B-IV, C-I, D-II.

54. *(b)* **Reaction mechanism**

1. DIBAL-H i.e. ($AlH(i\text{-}Bu)_2$) reagent is used to convert CN group to aldehyde group.

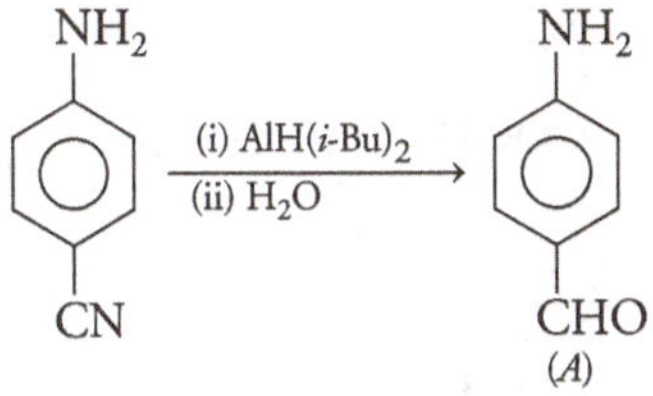

2. **Cross aldol condensation**

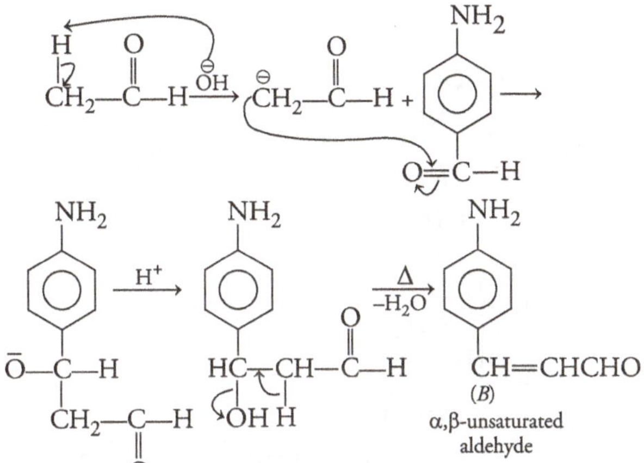

55. *(c)*

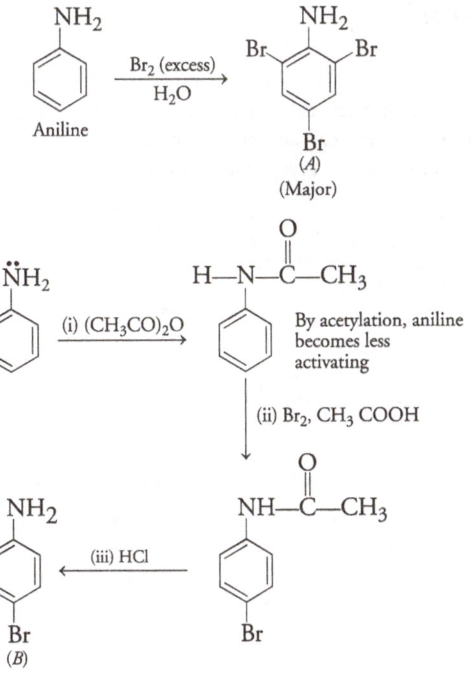

56. *(d)* 6 possible isomers of $C_6H_3(NH_2)_2COOH$ are

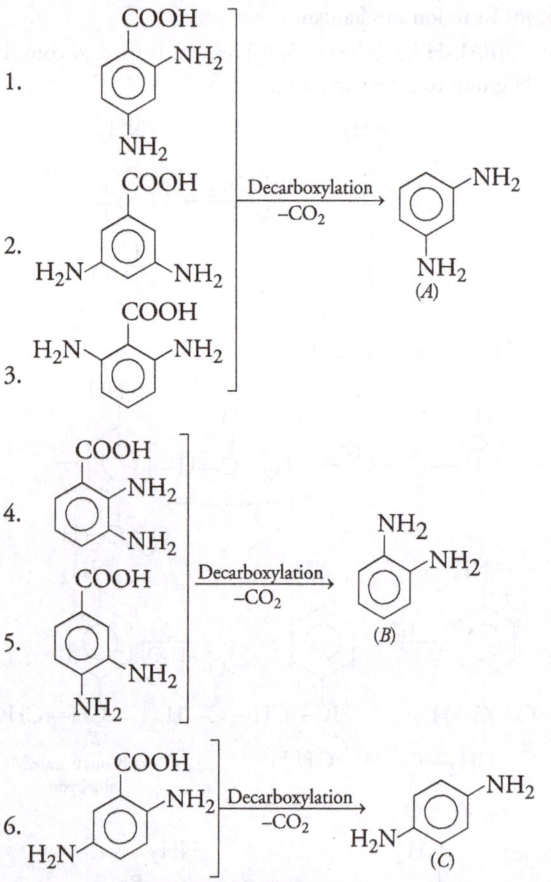

Product (*C*) have high melting point, i.e. 142°C more than products (*A*) and (*B*) because of the symmetry. It is packed more tightly in the crystal lattice. Hence, more heat is required to break the crystal lattice of product (*C*).

57. *(a)*

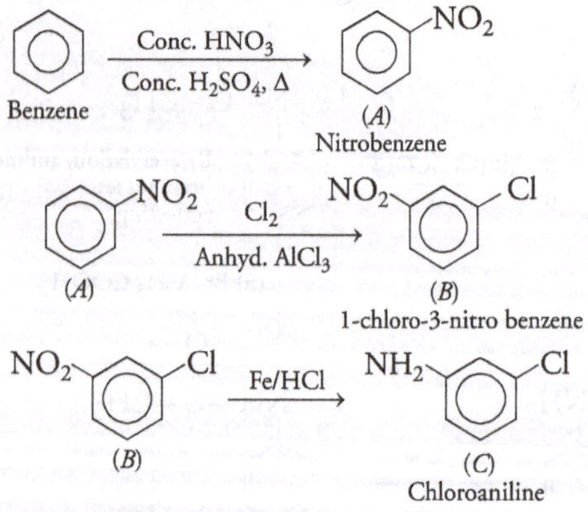

58. *(c)* Friedel-Craft reaction is not possible in highly activated ring, i.e. aniline because NH_2 group forms complex with Lewis acid and deactivates the ring.

Chemical reaction is as follows :

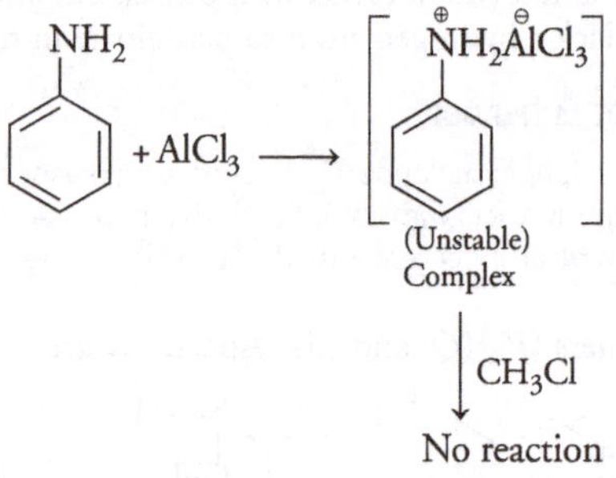

Due to lone pair of electrons on N, aniline is a strong Lewis base. So, an acid base reaction occur between aniline and $AlCl_3$ leading to anilinium salt formation (complex).

59. *(d)*

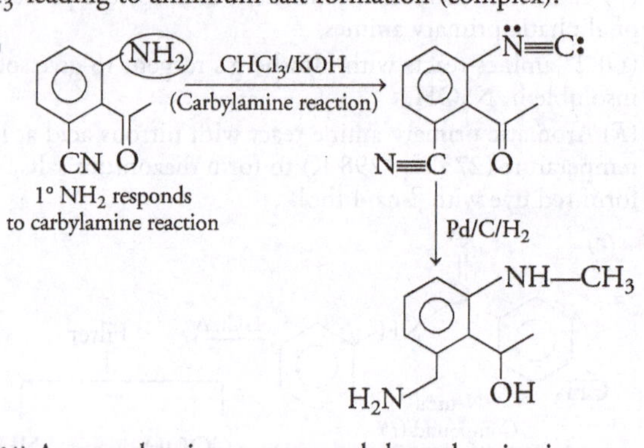

60. *(341)* Among the given compounds benzyl amine is most basic due to localised lone pair. The reaction involved as per question is as follow.

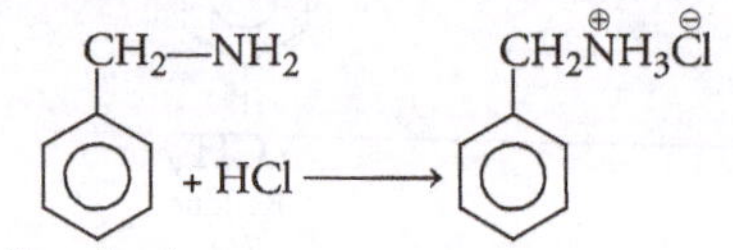

Benzyl amine

Mole of benzyl amine $= \frac{1}{107}$

$= 0.00934 \text{mol}$

1 mol of benzyl amine consumed 1 mole of HCl.

So, mole of HCl consumed = 0.00934 mol.

mass of HCl consumed = 0.00934 × molar mass of HCl

$= 0.00934 \times 36.5$

$= 0.341 \text{ g}$

$= 341 \text{ mg}$

61. *(4)* The complete reaction involved is as follows

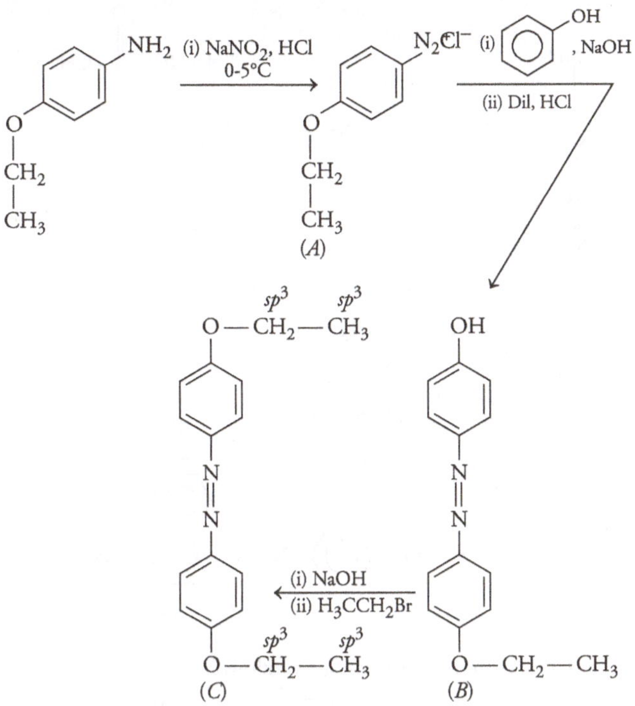

So, total four sp^3 hybridised carbon are present in product *C*.

62. *(5)*

> **idea** The compound that contains one hydrogen atom attached to nitrogen will give alkali insoluble product on reaction with Hinsberg's reagent.

—NH_2 group containing compound can give solid with Hinsberg's reagent, which is soluble in NaOH solution.

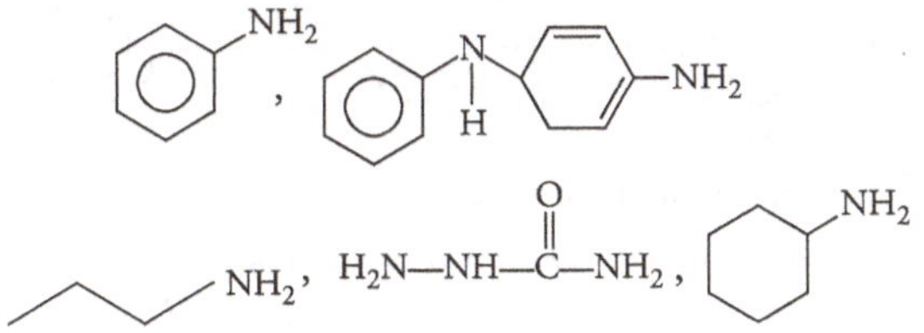

can give solid with Hinsberg reagent, which is soluble in NaOH solution.

63. *(5)* Both 1° and 2° amines give ppt. with Hinsberg's reagent while tertiary amines do not react with it.
Thus, among the given compounds only 5 gives Hinsberg's reagent test. These compounds are

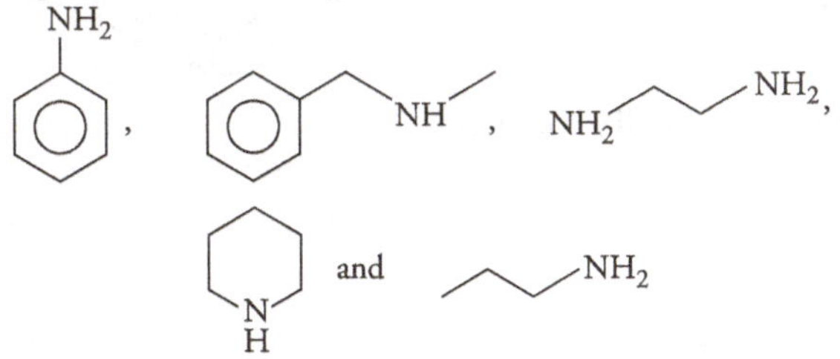

Concept Enhancer
The Hinsberg reagent is an alternative name of benzene sulphonyl chloride ($C_6H_5SO_2Cl$). This reagent undergoes a reaction with compounds containing O—H and N—H bonds that are reactive in nature like $C_2H_5NH_2$.

64. *(3)*

(a)

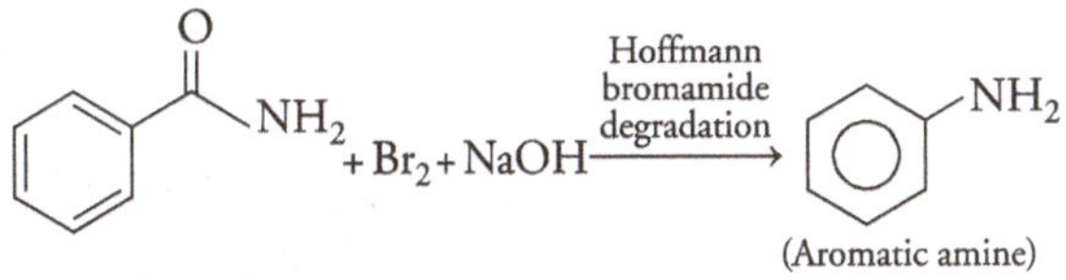

(b) In Gabriel phthalimide synthesis chlorobenzene is poor substrate for S_N2 thus no reaction will take place.

(c)

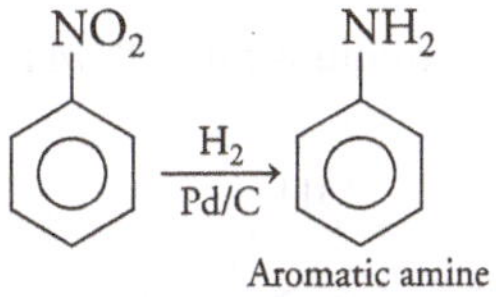

(d)

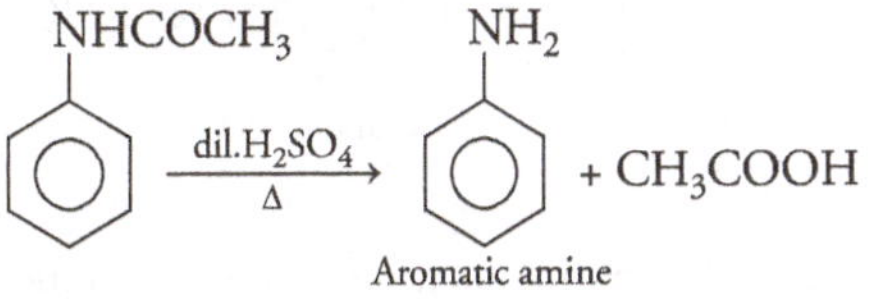

Aromatic amines will be formed in (a), (c) and (d).

65. *(5)* The nitrobenzene can be reduced to aniline *via* following reagents.
(I) Sn-HCl (V) H_2-Pd (III) Fe-HCl
(VI) H_2 (Raney Ni) (IV) Zn-HCl

66. *(3)*

> **idea** To find out the total number of amines formed by Gabriel synthesis. Simply check for the compounds containing α-carbon. It is because Gabriel phthalimide synthesis is used to prepare 1° aliphatic and alicyclic amine. Hence, amine which can be synthesised by Gabriel phthalimide synthesis method contains α-carbon.

Aniline ($C_6H_5NH_2$) does not contain α -C. Hence, cannot be prepared by Gabriel reaction.
Rest all amines all contain α-C in its respective position, hence they can easily give Gabriel phthalimide reaction.
∴ Three amines out of given four options can be prepared by Gabriel synthesis.

Topic 3

01. *(c)*

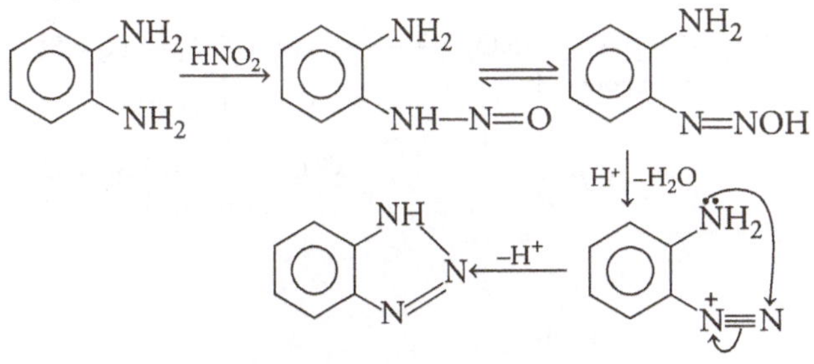

02. *(a)* Diazonium salts react with β- naphthol to form azo-compounds which are orange red dyes.

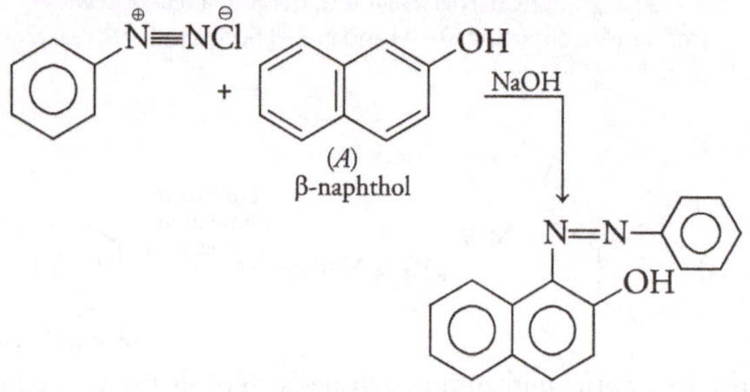

Red-orange dye

03. *(b)* Hydrolysis of benzene diazonium chloride will give carbolic acid/phenol.

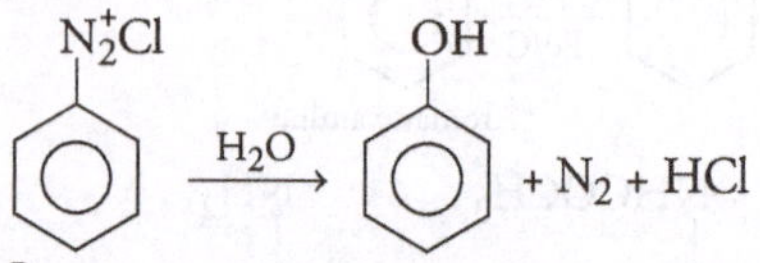

Benzene diazonium chloride Carbolic acid

04. *(b)* Presence of electron donating groups in conjugation to diazonium salt containing aryl group stabilises the positive charge on —N due to resonance. Electron withdrawing groups when linked to aryl system containing diazonium group have a destabilising effect as it intensifies the positive charge.

Thus, the correct order of stability will be

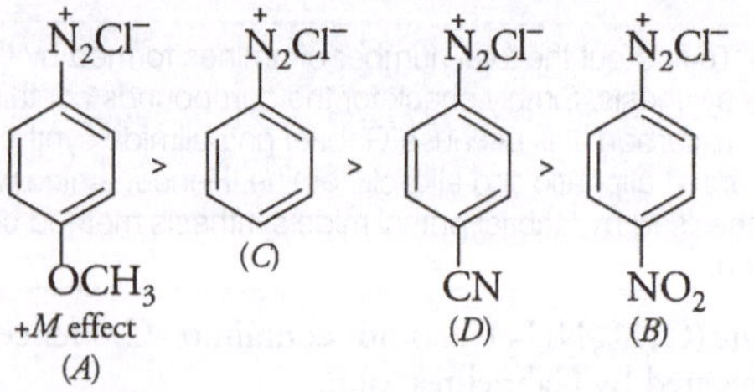

05. *(b)* Incorrect reaction is given in option (b). The correct reaction is as follows,

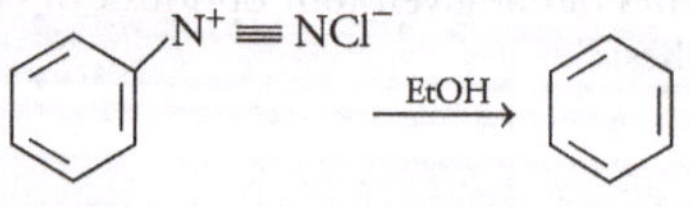

06. *(a)* The complete reaction involved is as follows.

$$\underset{\text{(Compound A)}}{CH_3-CH_2-CH_2-NO_2} \xleftarrow{Ag-NO_2} CH_3-CH_2-CH_2-Br$$

$$CH_3-CH_2-CH_2-Br \xrightarrow{Ag-CN} \underset{\text{(Compound B)}}{CH_3-CH_2-CH_2-NC}$$

07. *(d)* The complete reaction is as follow,

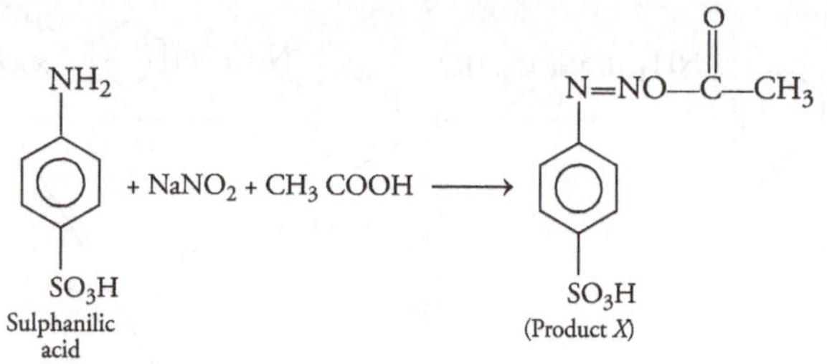

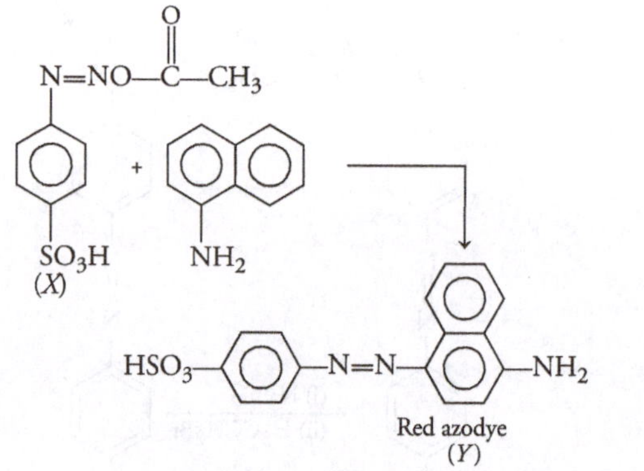

08. *(a)* The complete reaction is as follows:

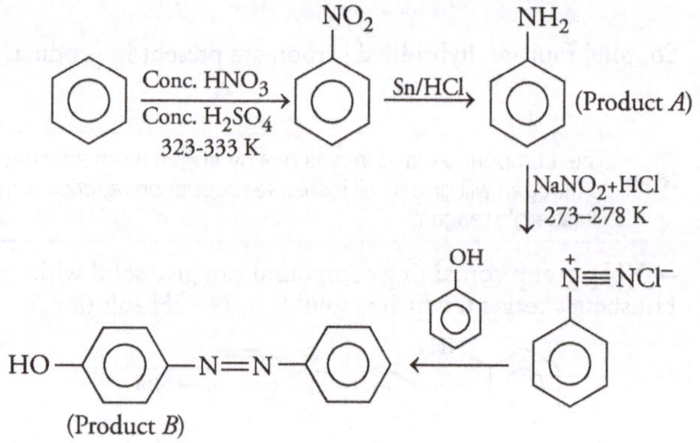

Hence, option (a) is correct.

09. *(b)* The complete reaction is as follows,

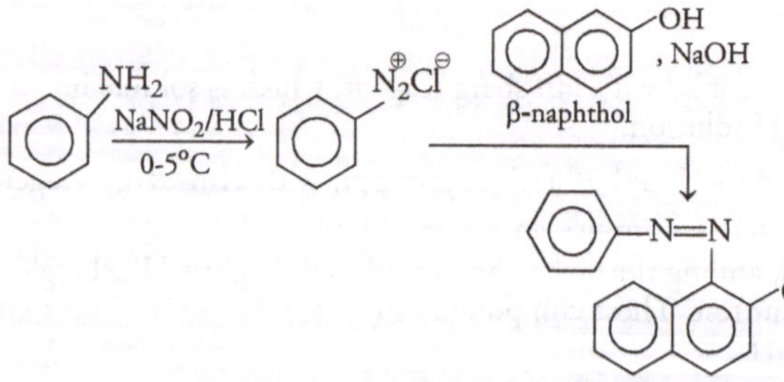

Scarlet red dye

Hence, option (b) is correct.

10. *(d)* Complete reaction is as follows.

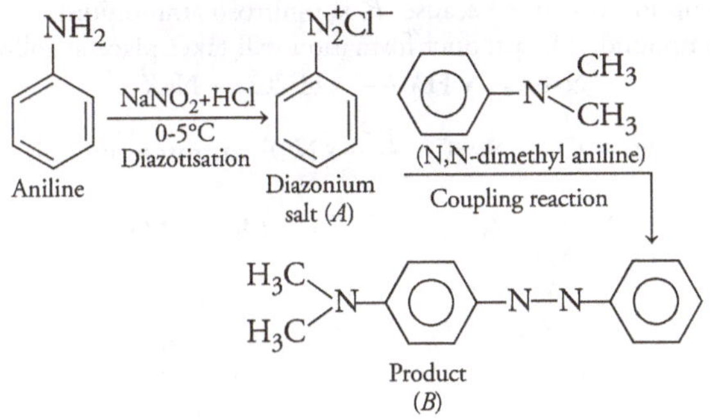

11. *(b)*

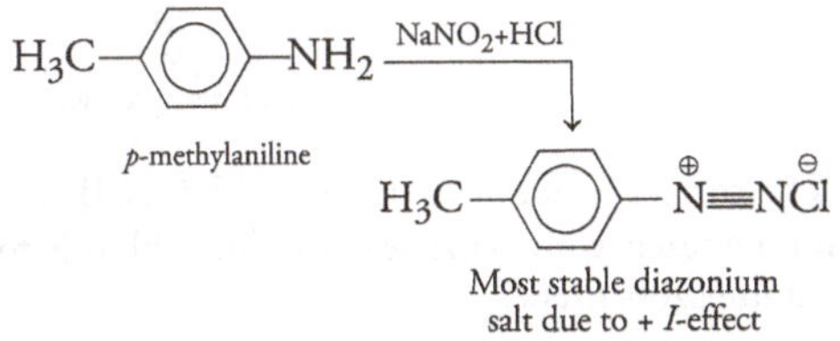

∴ In option (b) the most stable diazonium salt is formed.

12. *(b)* Sn + HCl is used for reduction of —NO_2 group into —NH_2 group. In next step, $PhN_2^+Cl^-$ is used to form yellow coloured azo dye with aniline. This is a coupling reaction.

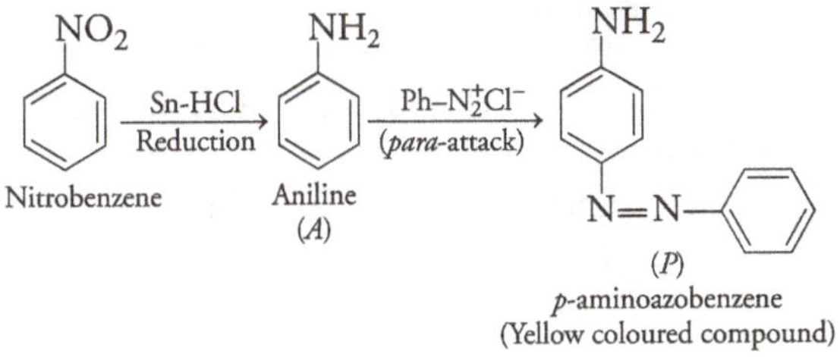

13. *(a)* H_3PO_2 is compound *A* and CH_3CH_2Cl is compound *B*.

14. *(c)*

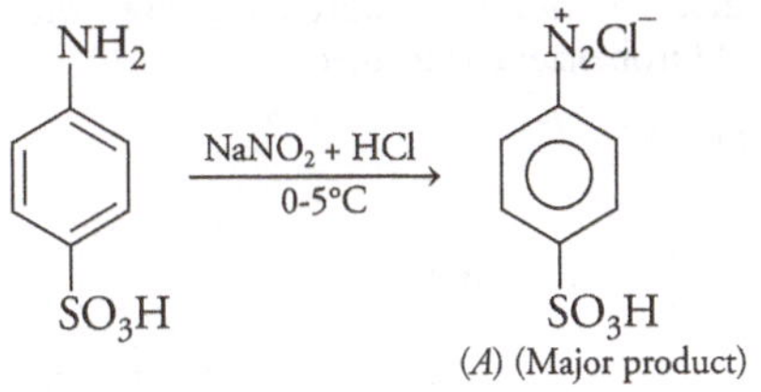

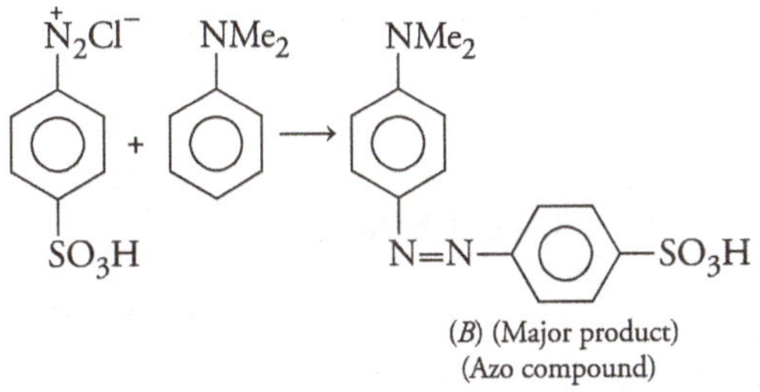

15. *(b)* Chemical reactions are as follows

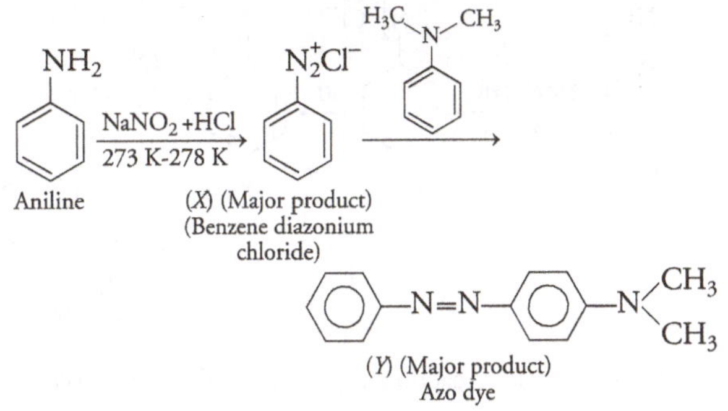

16. *(a)* The structural formula of *A*, *X* and *Y* are *p*-methoxy benzene diazonium chloride, acetaldehyde and hydrochloric acid respectively.

17. *(c)* Aniline reacts with $NaNO_2$, HCl at 273-278 K to give benzene diazonium salt ($C_6H_5\overset{+}{N}_2\overset{-}{Cl} = X$) which on hydrolysis ($H_2O/\Delta$) gives phenol.

So, $X =$ 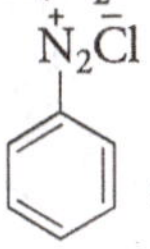 and $A = H_2O/\Delta$

18. *(c)*

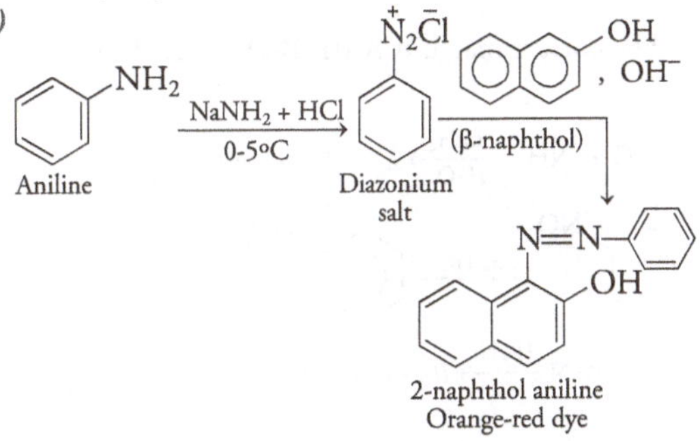

19. *(d)*

(i) 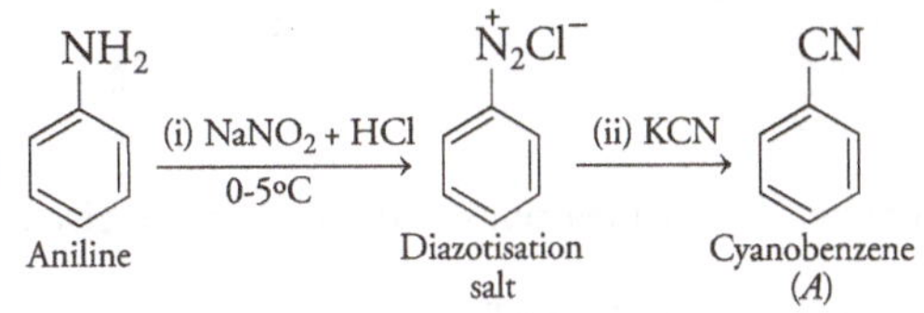

(ii) 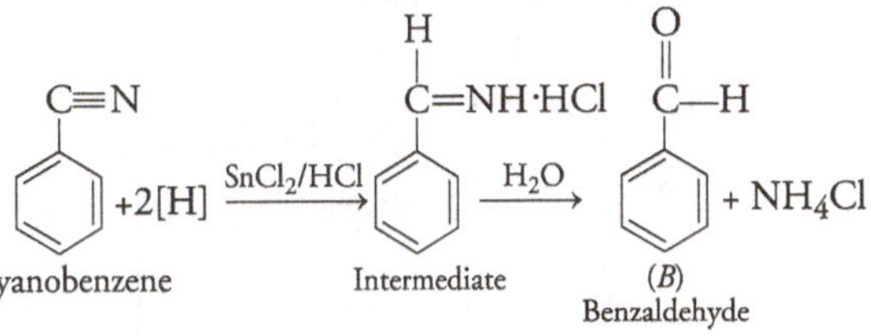

20. *(b)* Complete reaction sequence is as follows

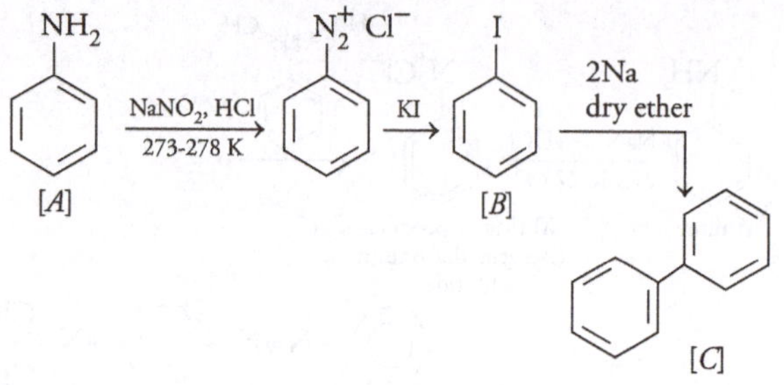

21. *(b)* The complete reaction sequence involved is as follows.

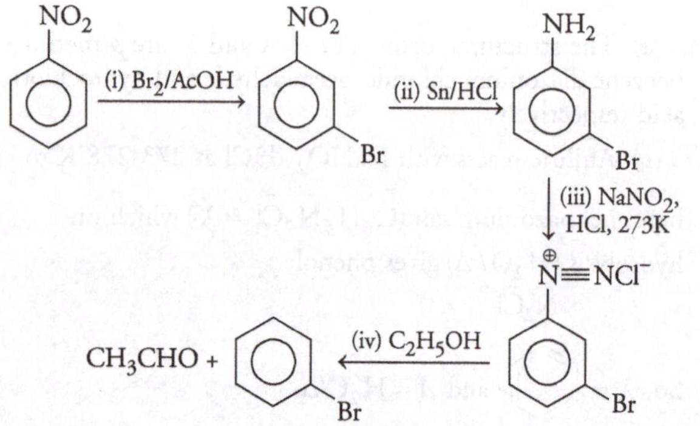

22. *(c)* The correct match is A-III, B-IV, C-II, D-I.

(A) $R{-}\overset{O}{\overset{\|}{C}}{-}NH_2 \xrightarrow[H_2O]{LiAlH_4} R{-}CH_2{-}NH_2$

(B)

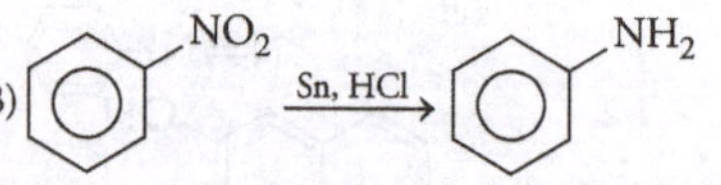

(C) $R{-}C{\equiv}N \xrightarrow{H_2/Ni} R{-}CH_2NH_2$

(D)

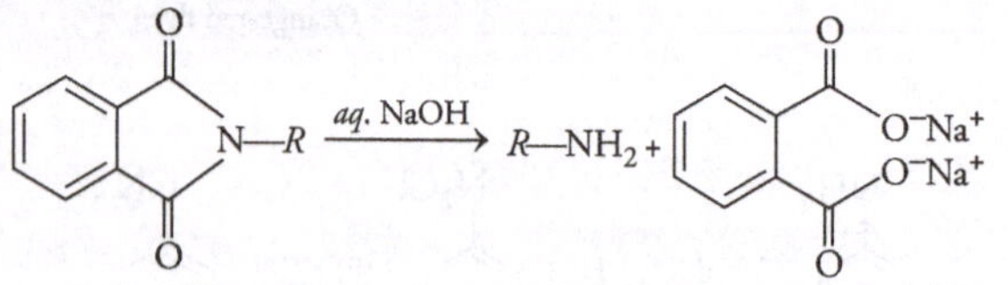

23. *(a)* Among the given reactions, nitrobenzene will give an azo dye.

The complete reaction is as follows :

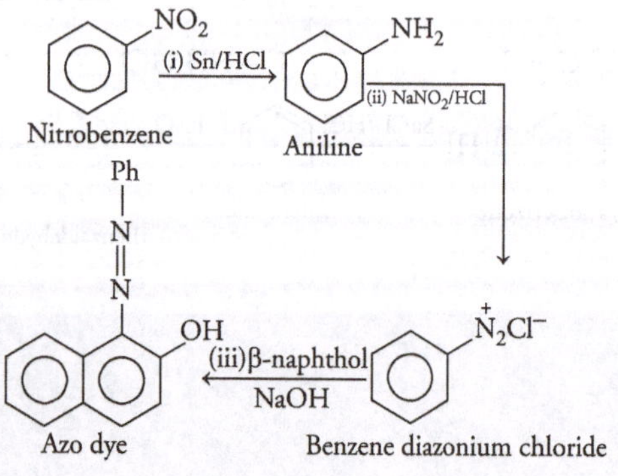

24. *(b)* Among the given options, incorrect statement is given in option (b). It is because '*B*' is *p*-nitroso ammonium compound. The product formation will takes place as follows

$$NaNO_2 + HX \longrightarrow HNO_2 + NaX$$

$$H{-}O{-}N{=}O \xrightarrow[-H_2O]{H^{\oplus}} NO^+ \text{ (nitrosonium ion)}$$

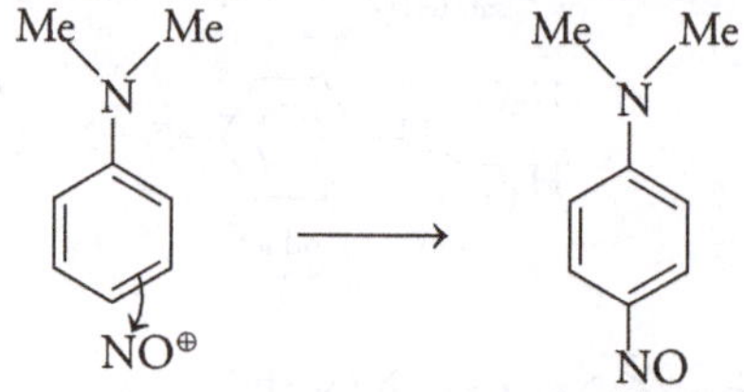

(*B*)
p-nitroso product

25. *(b)* The correct match is A-III, B-I, C-IV, D-II.

A. Aniline reacts with benzene diazonium chloride to give *p*-aminoazobenzene.

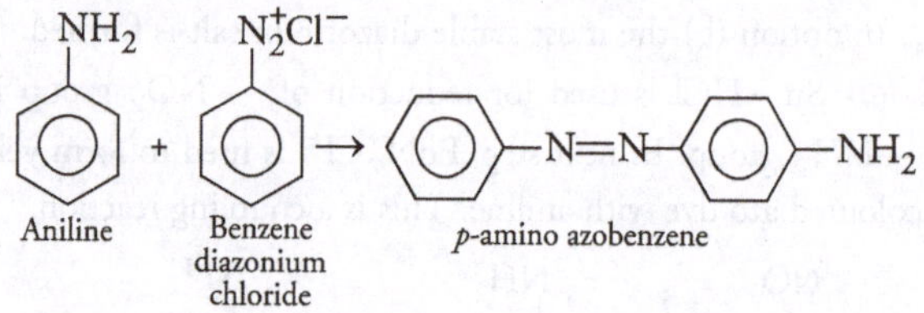

B. Benzene diazonium chloride reacts with HBF_4 to given fluorobenzene.

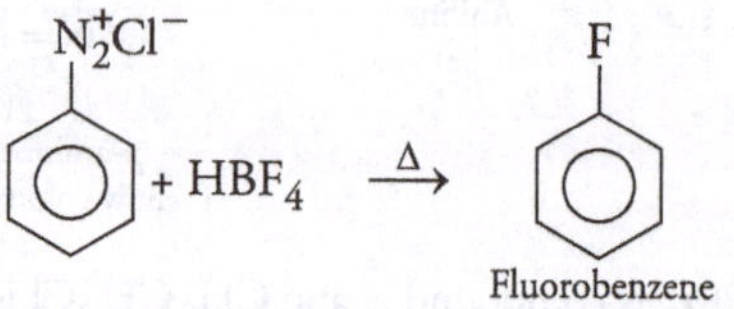

Fluorobenzene

C. When benzene diazonium chloride is treated with copper in HCl, chlorobenzene is formed.

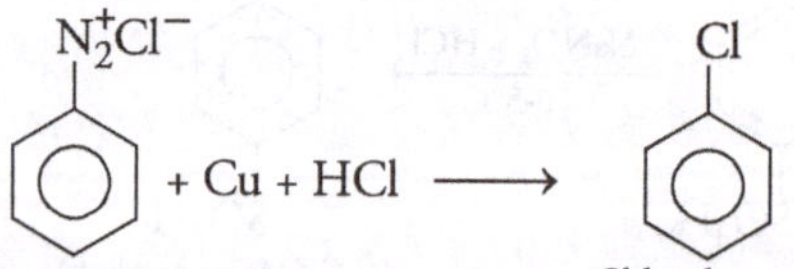

Chlorobenzene

D. Benzene diazonium chloride reacts with CuCN/KCN to give benzene nitrile.

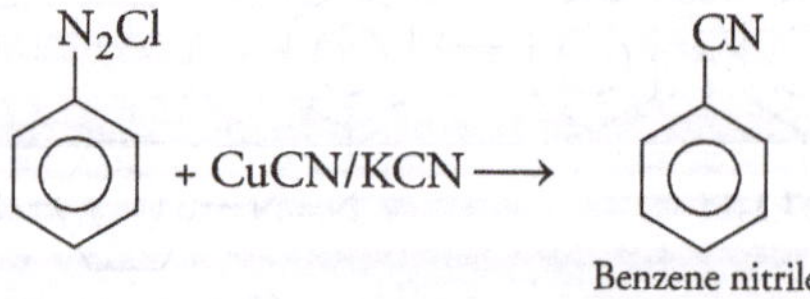

Benzene nitrile

26. *(a)*

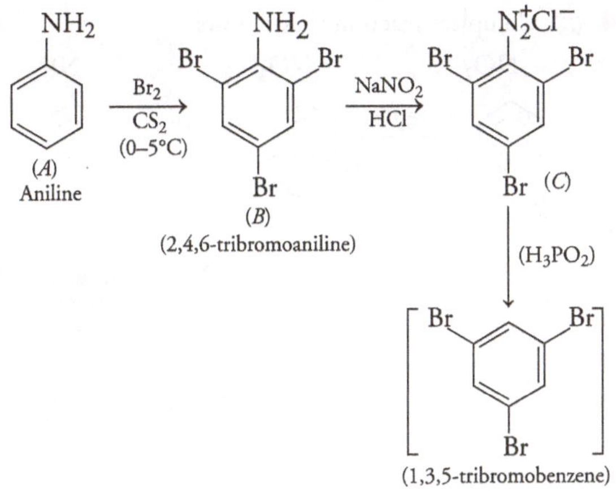

27. *(b)* *p*-nitrotoluene converts into 3,5-dibromotoluene as follows :

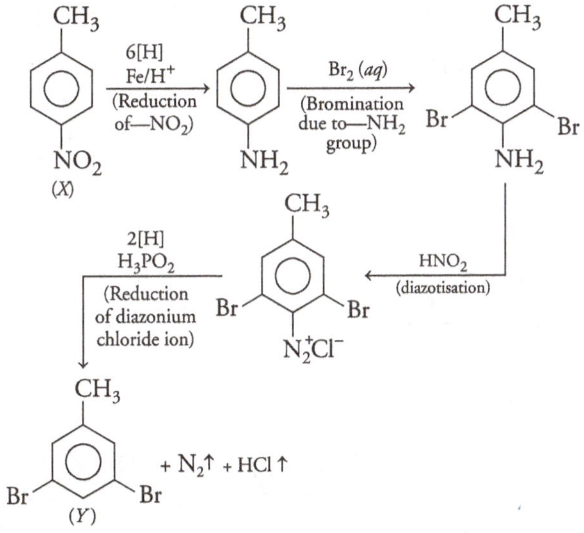

Thus, option (b) is correct.

28. *(c)* In reaction,

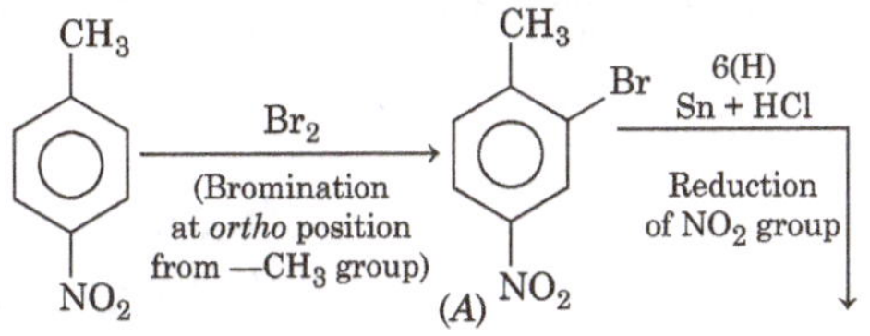

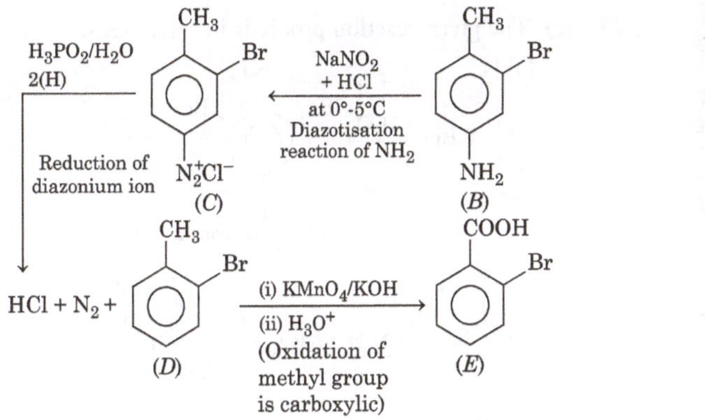

29. *(c)*

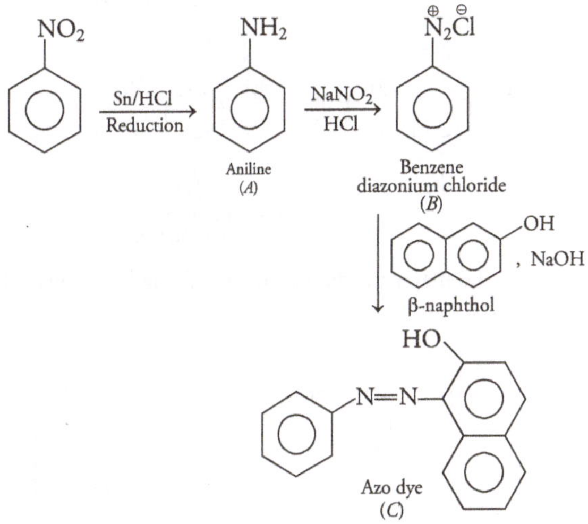

30. *(d)* Reactions involved are as follows :

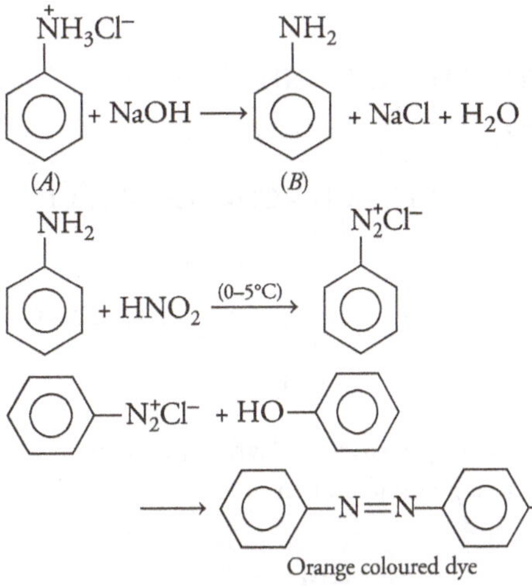

31. *(c)* The given reaction proceeds in three steps;

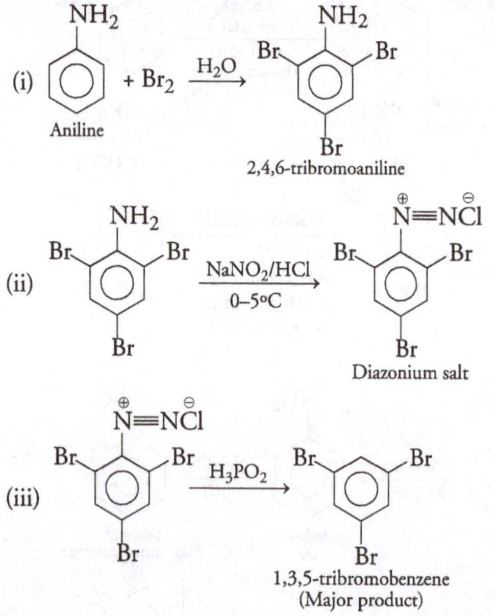

32. *(c)* The major product formed is 3-methyl propanol as follows :

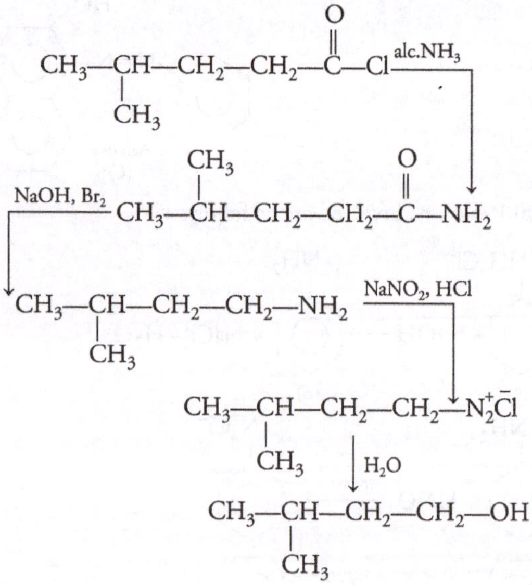

33. *(c)* Nitrobenzene in presence of $NaBH_4$, NaOH and aniline will not give *p*-amino azobenzene.

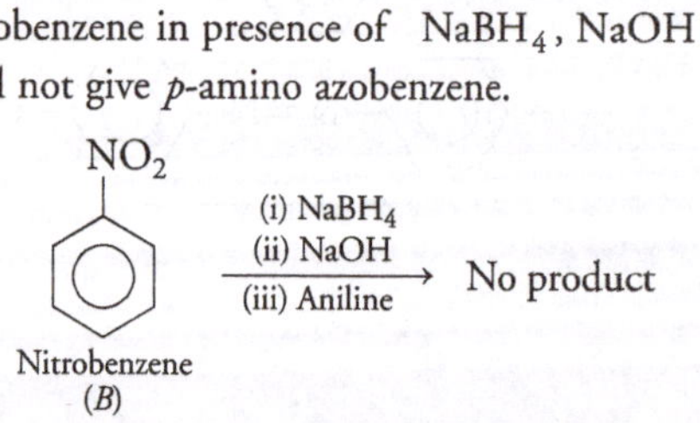

34. *(b)* Complete reaction is as follows

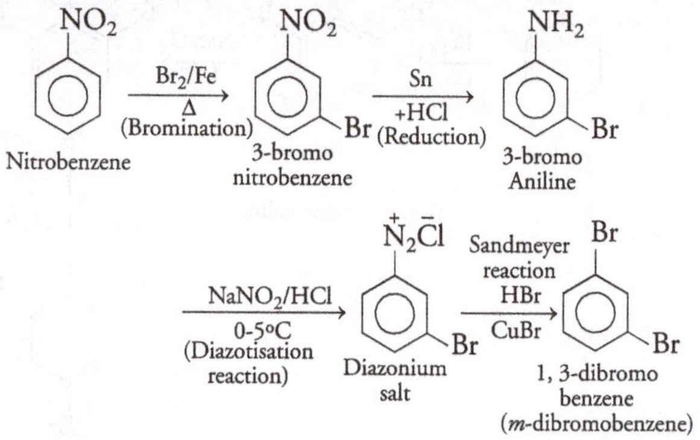

35. *(a)*

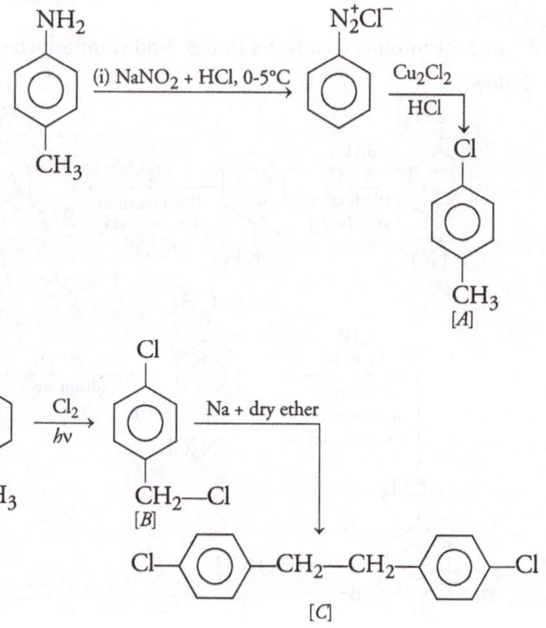

36. *(c)*

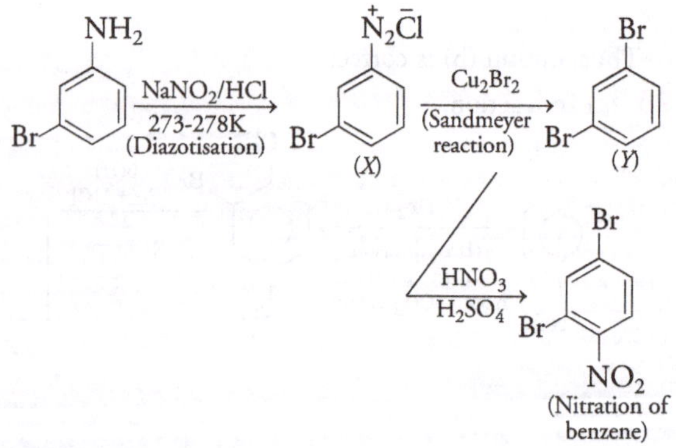

Br is *o*, *p*-directing group.

∴ NO_2 is attached at the *ortho* position of one of the Br atoms which is also *para* position with respect to the other Br atom.

37. *(a)* Compound [*P*] is 3-methylaniline.

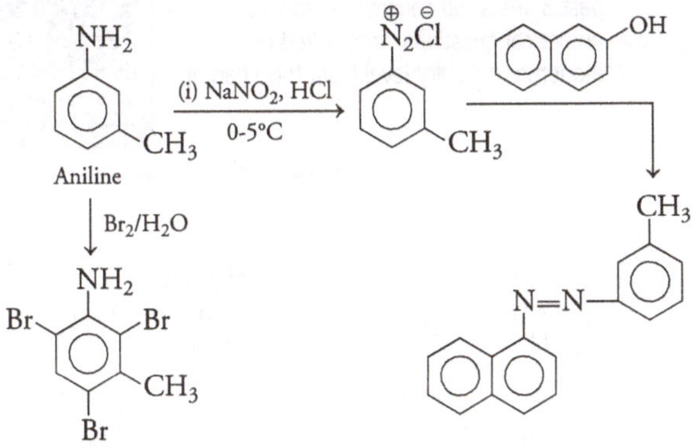

38. *(c)* Benzene diazonium chloride on reaction with aniline in the presence of dilute hydrochloric acid undergoes coupling reaction and produces *p*-amino azobenzene.

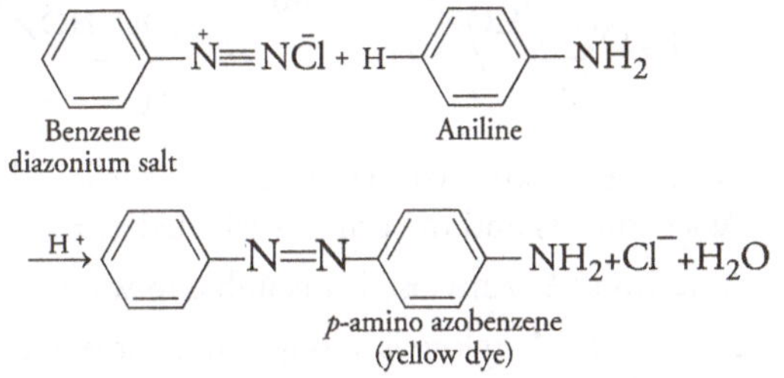

39. *(c)* Coupling of benzene diazonium chloride with 1-naphthol in alkaline medium will give the following coupling reaction.

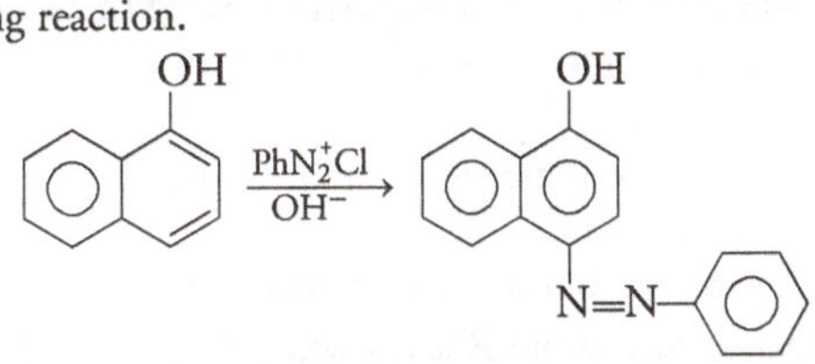

Concept Enhancer

When benzene diazonium chloride reacts with phenol in which the phenol molecule is at its *para* position is coupled with the diazonium salt to form *p*-hydroxybenzene. This type of reaction is called a coupling reaction.

Following conditions are required for coupling reaction to take place.

(a) The solution must be so alkaline that the concentration of diazonium ion is too low.

(b) It must not be so acidic that the concentration of free phenoxide ion is too low.

40. *(154)* Complete reaction sequence involved is as follows

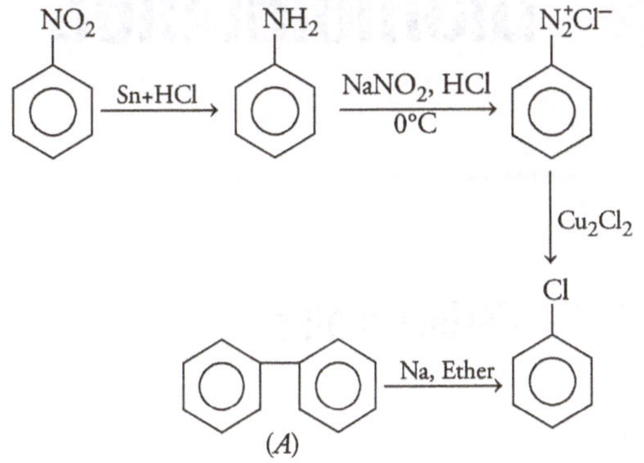

So, the molar mass of the product (A) is 154 g/mol.

41. *(171)* The complete reaction sequence involved is as follows

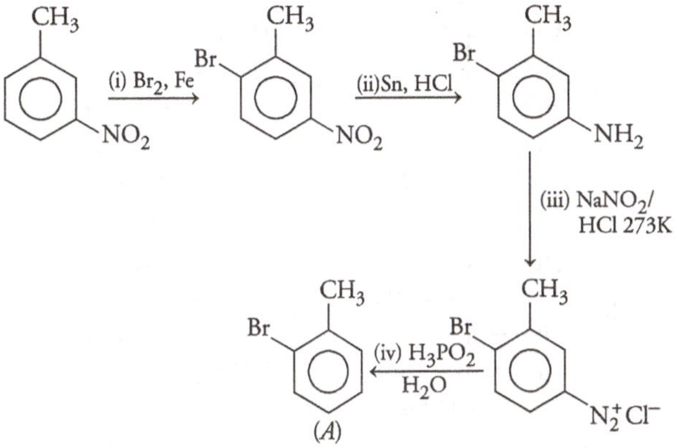

So, the compound A is C_7H_7Br

The molar mass of C_7H_7Br (A)

$$= 7 \times 12 + 7 \times 1 + 80$$

$$= 171\ g/mol$$

So, the molar mass of compound A is 171 g/mol.

23 Biomolecules

In JEE Main there is usually 0-1 question per paper are visible from this chapter. In past 3 years, 50 conceptual and 11 numerical questions were asked of moderate difficulty level from this chapter.

For detailed analysis scan QR code

Topic 1 / Carbohydrates

• Level 1

01 Which of the following is the correct structure of L-glucose? **[2024, 4 April Shift-I]**

(a) 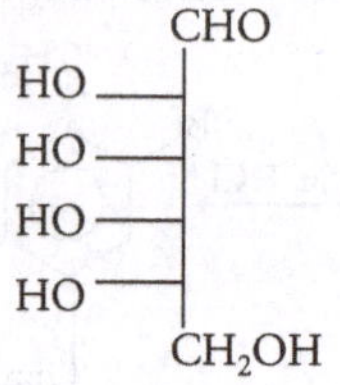

(b) 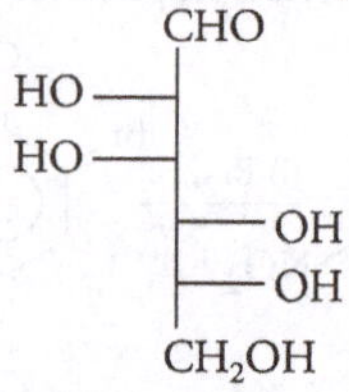

(c) 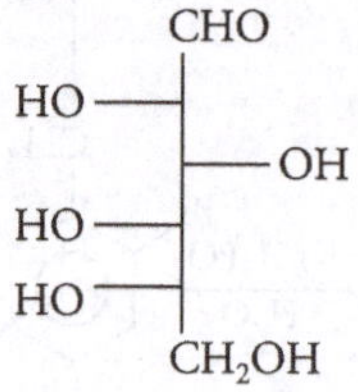

(d) 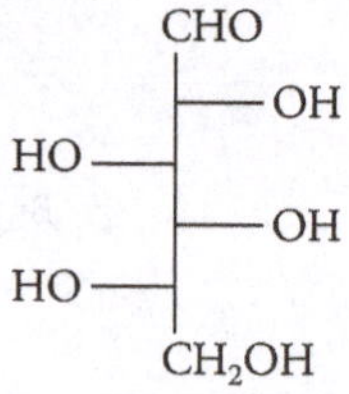

02 Which of the following is the correct structure of L-Fructose? **[2023, 03 April Shift-I]**

(a) 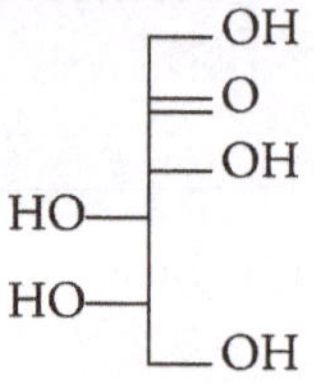

(b) 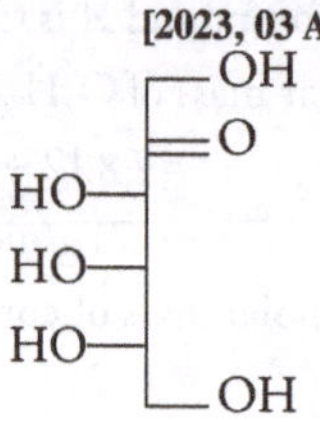

(c) 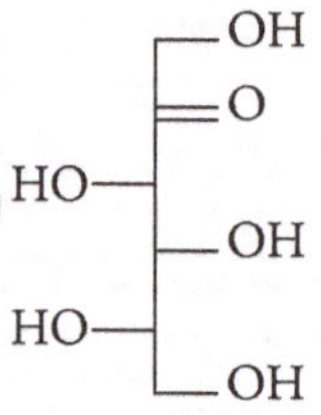

(d) 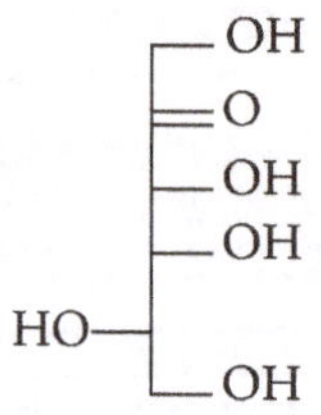

03 The correct representation in six membered pyranose form for the following sugar $[X]$ is **[2023, 01 Feb Shift-I]**

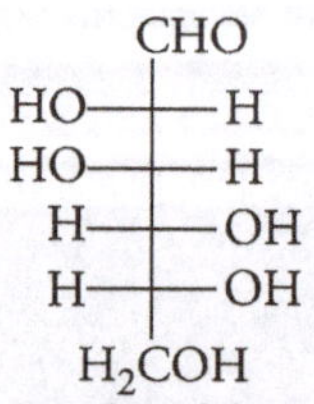

Sugar $[X]$

(a) 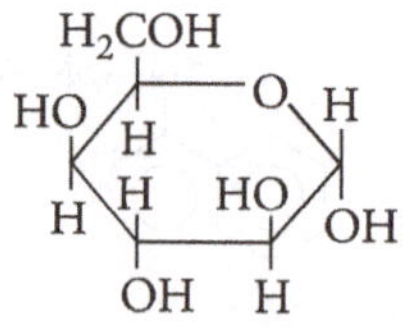

(b) 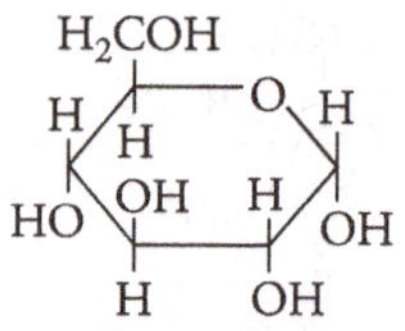

(c) 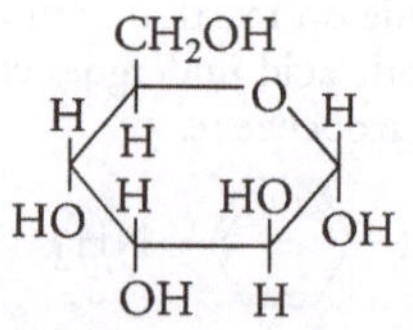

(d) 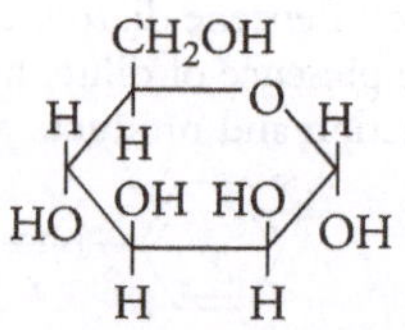

04 Given below are two statements. One is labelled as Assertion (A) and the other is labelled as Reason (R).

Assertion (A) Amylose is insoluble in water.

Reason (R) Amylose is a long linear molecule with more than 200 glucose units.

In the light of the above statements, choose the correct answer from the options given below. **[2022, 29 July Shift-II]**

(a) Both A and R are correct and R is the correct explanation of A.

(b) Both A and R are correct and R is not the correct explanation of A.

(c) A is correct but R is not correct.

(d) A is incorrect but R is correct.

05 Which one of the following is a reducing sugar? **[2022, 26 July Shift-I]**

(a) 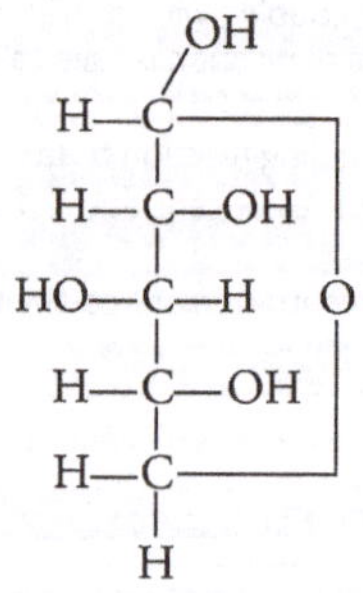

(b) 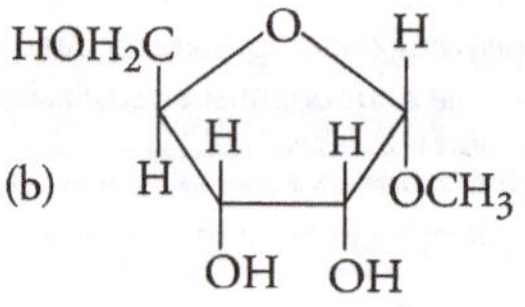

(c) 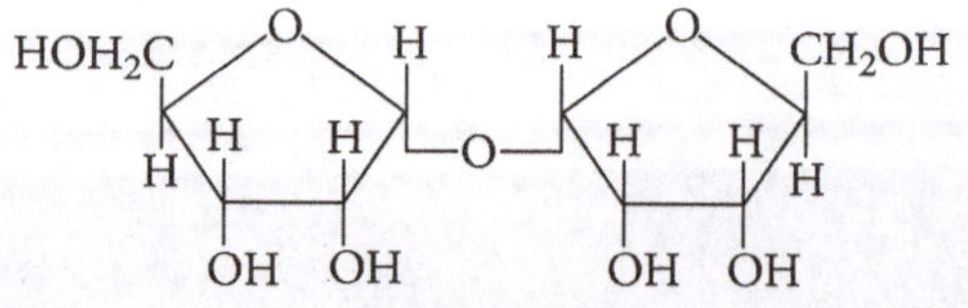

(d) 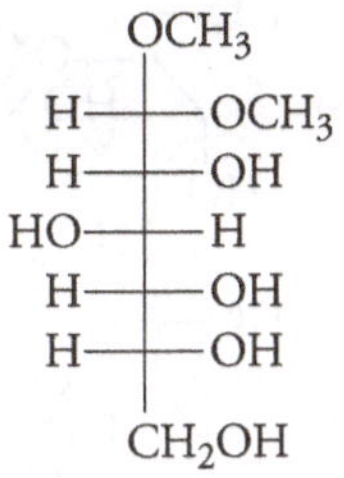

06 Glycosidic linkage between C_1 of α-glucose and C_2 of β-fructose is found in **[2022, 25 July Shift-II]**
(a) maltose (b) sucrose
(c) lactose (d) amylose

07 Hydrolysis of sucrose gives **[2021, 27 Aug Shift-II]**
(a) α-D-(–)-glucose and β-D(–)-fructose
(b) α-D-(+)-glucose and β-D-(+)-fructose
(c) α-D-(–)-glucose and βD-(+)-fructose
(d) α-D-(+)-glucose and βD-(–)-fructose

08 Which one among the following chemical tests is used to distinguish monosaccharide from disaccharide? **[2021, 27 July Shift-I]**
(a) Seliwanoff's test (b) Iodine test
(c) Barfoed test (d) Tollen's test

09 A non-reducing sugar "*A*" hydrolysis to give two reducing monosaccharides. Sugar *A* is **[2021, 18 March Shift-I]**
(a) fructose (b) galactose
(c) glucose (d) sucrose

10 Fructose is an example of **[2021, 17 March Shift-II]**
(a) pyranose (b) ketohexose
(c) aldohexose (d) heptose

11 Match List-I with List-II.

List-I		List-II	
A.	Sucrose	(i)	β-D-galactose and β-D-glucose
B.	Lactose	(ii)	α-D-glucose and β-D-fructose
C.	Maltose	(iii)	α-D-glucose and α-D-glucose

Choose the correct answer from the options given below. **[2021, 26 Feb Shift-II]**
(a) A → (i), B → (iii), C → (ii)
(b) A → (iii), B → (i), C → (iii)
(c) A → (ii), B → (i), C → (iii)
(d) A → (iii), B → (ii), C → (i)

12 Two monomers in maltose are **[2020, 8 Jan Shift-II]**
(a) α-D-glucose and α-D-glucose
(b) α-D-glucose and α-D-galactose
(c) α-D-glucose and β-D-glucose
(d) α-D-glucose and α-D-fructose

13 Glucose and galactose are having identical configuration in all the positions except position **[2019, 12 April Shift-II]**
(a) C-3 (b) C-4 (c) C-2 (d) C-5

• Level 2

Imp. *Properties of Fructose*

14 Given below are two statements: **[2025, 23 Jan Shift-I]**
Statement I Fructose does not contain an aldehydic group but still reduces Tollen's reagent.
Statement II In the presence of base, fructose undergoes rearrangement to give glucose.
In the light of the above statements, choose the correct answer from the options give below
(a) Both Statement I and Statement II are true.
(b) Statement I is false but Statement II is true.
(c) Both Statement I and Statement II are false.
(d) Statement I is true but Statement II is false.

15 Given below are two statements. **[2025, 28 Jan Shift-I]**
Statement I D-glucose penta acetate reacts with 2, 4-dinitrophenylhydrazine.
Statement II Starch, on heating with concentrated sulphuric acid at 100°C and 2-3 atmosphere pressure produces glucose.
In the light of the above statements, choose the correct answer from the options given below
(a) Statement I is true but Statement II is false.
(b) Statement I is false but Statement II is true.
(c) Both Statement I and Statement II are false.
(d) Both Statement I and Statement II are true.

16 The incorrect statement about glucose is
(a) Glucose is an aldohexose. **[2024, 9 April Shift-II]**
(b) Glucose is one of the monomer unit in sucrose.
(c) Glucose is soluble in water because of having aldehyde functional group.
(d) Glucose remains in multiple isomeric form in its aqueous solution.

17 Sugar which does not give reddish brown precipitate with Fehling's reagent, is **[2024, 30 Jan Shift-I]**
(a) maltose (b) lactose (c) glucose (d) sucrose

18 L-isomer of tetrose $X(C_4H_8O_4)$ gives positive Schiff's test and has two chiral carbons. On acetylation, '*X*' yields triacetate. '*X*' also undergoes following reactions

$$\text{'}A\text{'} \xleftarrow{HNO_3} \text{'}X\text{'} \xrightarrow{NaBH_4} \underset{\text{Chiral compound}}{\text{'}B\text{'}}$$

'*X*' is **[2023, 11 April Shift-I]**

(a) 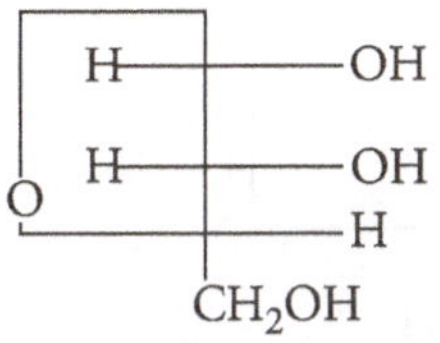
(b) 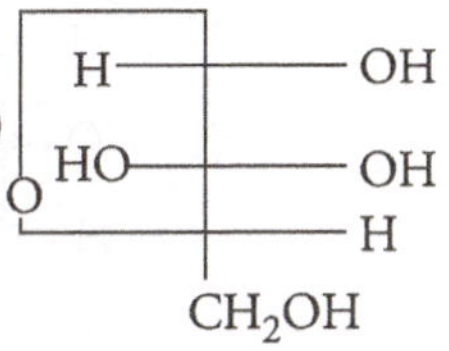
(c) 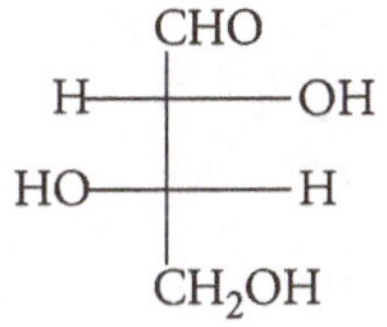
(d) 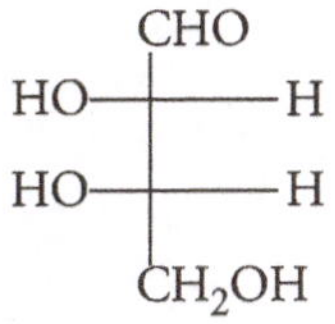

19 Match items of Row-I with those of Row-II.

Row I

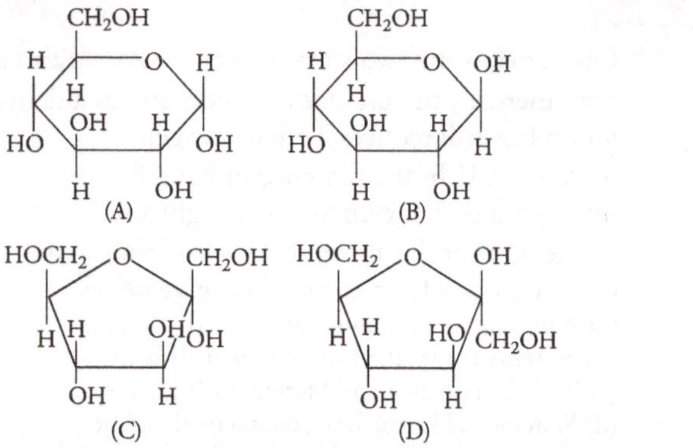

Row II

(i) α-D-(–)-fructofuranose,
(ii) β-D-(–)-fructofuranose
(iii) α-D-(–) glucopyranose,
(iv) β-D-(–)-glucopyranose

Correct match is **[2023, 25 Jan Shift-I]**

(a) A-(iv), B-(iii), C-(i), D-(ii)
(b) A-(iii), B-(iv), C-(i), D-(ii)
(c) A-(i), B-(ii), C-(iii), D-(iv)
(d) A-(iii), B-(iv), C-(ii), D-(i)

20 The formulas of A and B for the following reaction sequence **[2022, 28 July Shift-II]**

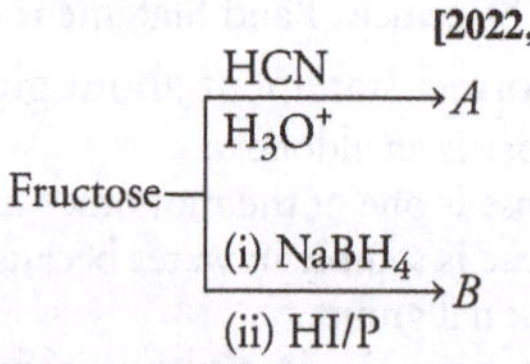

(a) $A = C_7H_{14}O_8$, $B = C_6H_{14}$
(b) $A = C_7H_{13}O_7$, $B = C_7H_{14}O$
(c) $A = C_7H_{12}O_8$, $B = C_6H_{14}$
(d) $A = C_7H_{14}O_8$, $B = C_6H_{14}O_6$

21 For the below given cyclic hemiacetal (X), the correct pyranose structure is **[2022, 28 July Shift-I]**

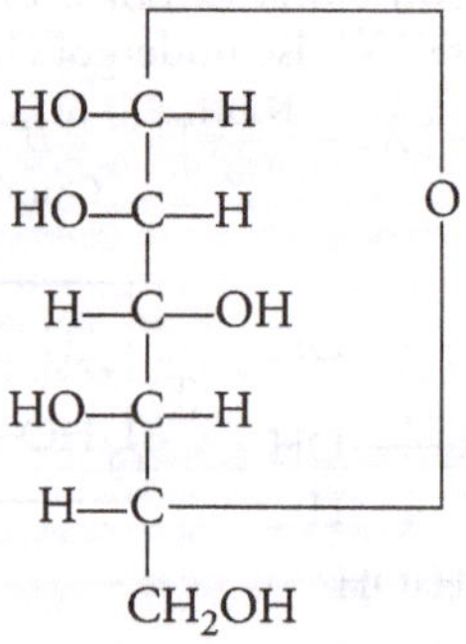

(X)

(a)
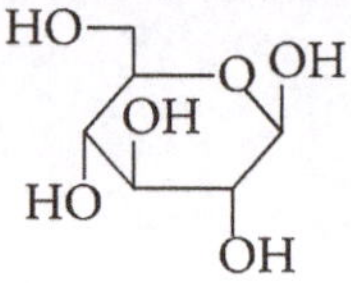

(b)
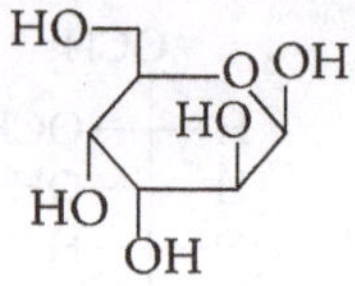

(c)
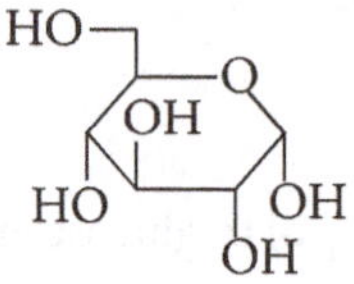

(d)
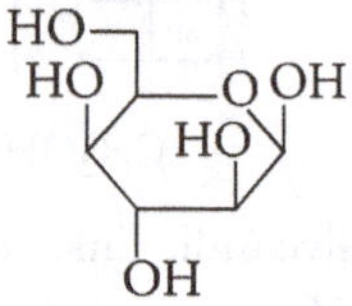

Imp. *Chemical Reactions of Glucose*

22 Match List-I with List-II.

List-I	List-II
(A) Glucose + HI	(I) Gluconic acid
(B) Glucose + Br_2 water	(II) Glucose pentacetate
(C) Glucose + acetic anhydride	(III) Saccharic acid
(D) Glucose + HNO_3	(IV) Hexane

Choose the correct answer from the options given below: **[2022, 27 July Shift-II]**

(a) (A)-(IV), (B)-(I), (C)-(II), (D)-(III)
(b) (A)-(IV), (B)-(III), (C)-(II), (D)-(I)
(c) (A)-(III), (B)-(I), (C)-(IV), (D)-(II)
(d) (A)-(I), (B)-(III), (C)-(IV), (D)-(II)

23 Given below are two statements.

Statement I Maltose has two α-D-glucose units linked at C_1 and C_4 and is a reducing sugar.

Statement II Maltose has two monosaccharides: α-D-glucose and β-D-glucose linked at C_1 and C_6 and it is a non-reducing sugar.

In the light of the above statements, choose the correct answer from the options given below. **[2022, 27 June Shift-II]**

(a) Both Statement I and Statement II are true.
(b) Both Statement I and Statement II are false.
(c) Statement I is true but Statement II are false.
(d) Statement I is false but Statement II is true.

24 Which one of the following compounds contain β-C_1—C_4 glycosidic linkage ? **[2021, 31 Aug Shift-I]**

(a) Lactose (b) Sucrose (c) Maltose (d) Amylose

25 Given below are two statements : One is labelled as Assertion (A) and other is labelled as Reason (R).

Assertion (A) Sucrose is a disaccharide and a non-reducing sugar.

Reason (R) Sucrose involves glycosidic linkage between C_1 of β-glucose and C_2 of α-fructose.

Choose the most appropriate answer from the options given below. **[2021, 26 Aug Shift-II]**

(a) Both (A) and (R) are true but (R) is not the correct explanation of (A).
(b) (A) is false but (R) is true.
(c) (A) is true but (R) is false.
(d) Both (A) and (R) are true and (R) is the correct explanation of (A).

26 Compound *A* gives D-galactose and D-glucose on hydrolysis. The compound *A* is **[2021, 27 July Shift-II]**
(a) amylose (b) sucrose (c) maltose (d) lactose

27 Identify the incorrect statement from the following. **[2021, 20 July Shift-I]**
(a) Amylose is a branched chain polymer of glucose.
(b) Starch is a polymer of α-D-glucose .
(c) β-glycosidic linkage makes cellulose polymer.
(d) Glycogen is called as animal starch.

28 Which of the glycosidic linkage between galactose and glucose is present in lactose? **[2021, 25 Feb Shift-I]**
(a) C-1 of galactose and C-4 of glucose
(b) C-1 of galactose and C-6 of glucose
(c) C-1 of glucose and C-4 of galactose
(d) C-1 of glucose and C-6 of galactose

29 What are the functional groups present in the structure of maltose? **[2020, 4 Sep Shift-I]**
(a) One ketal and one hemiketal
(b) One acetal and one ketal
(c) One acetal and one hemiacetal
(d) Two acetals

30 Which of the following statement is not true for glucose? **[2020, 8 Jan Shift-I]**
(a) The pentaacetate of glucose does not react with hydroxylamine to give oxime
(b) Glucose exists in two crystalline forms α and β
(c) Glucose gives Schiff's test for aldehyde
(d) Glucose reacts with hydroxylamine to form oxime

31 Which of the following statements is correct? **[2020, 7 Jan Shift-II]**
(a) Gluconic acid is obtained by oxidation of glucose with HNO_3
(b) Gluconic acid is a dicarboxylic acid
(c) Gluconic acid can form cyclic (acetal/hemiacetal) structure
(d) Gluconic acid is a partial oxidation product of glucose

32 Which of the given statements is incorrect about glycogen? **[2019, 12 April Shift-II]**
(a) It is straight chain polymer similar to amylose
(b) Only α-linkages are present in the molecule
(c) It is present in animal cells
(d) It is present in some yeast and fungi

33 Maltose on treatment with dilute HCl gives **[2019, 8 April Shift-I]**
(a) D-glucose and D-fructose (b) D-fructose
(c) D-galactose (d) D-glucose

Level 3

34 Given below are two statements

Statement I *D*-(+)-glucose + *D*-(+) fructose $\xrightarrow{-H_2O}$ Sucrose

Sucrose $\xrightarrow{\text{Hydrolysis}}$ *D*-(+) glucose + *D*-(+) fructose

Statement II Invert sugar is formed during sucrose hydrolysis.

In the light of the above statements, choose the correct answer from the options given below **[2025, 07 April Shift-I]**
(a) Both Statement I and Statement II are true.
(b) Both Statement I and Statement II are false.
(c) Statement I is true but Statement II is false.
(d) Statement I is false but Statement II are true.

35 Given below are two statements.

Statement I Wet cotton clothes made of cellulose based carbohydrate takes comparatively longer time to get dried than wet nylon polymer based clothes.

Statement II Intermolecular hydrogen bonding with water molecule is more in nylon-based clothes than in the case of cotton clothes.

In the light of above statements, choose the correct answer from the options given below. **[2025, 03 April Shift-II]**
(a) Statement I is false but Statement II is true.
(b) Statement I is true but Statement II is false.
(c) Both Statement I and Statement II are true.
(d) Both Statement I and Statement II are false.

36 The carbohydrate "Ribose" present in DNA, is
A. a pentose sugar
B. present in pyranose form
C. in "D" configuration
D. a reducing sugar, when free
E. in α-anomeric form

Choose the correct answer from the options given below: **[2025, 24 Jan Shift-I]**
(a) A, C and D only (b) A, B and E only
(c) B, D and E only (d) A, D and E only

37 Match List-I with List-II. **[2025, 29 Jan Shift-I]**

	List-I (Carbohydrate)		List-II (Linkage source)
(A)	Amylose	I.	$\beta-C_1-C_4$, plant
(B)	Cellulose	II.	$\alpha-C_1-C_4$, animal
(C)	Glycogen	III.	$\alpha-C_1-C_4$, $\alpha-C_1-C_6$, plant
(D)	Amylopectin	IV.	$\alpha-C_1-C_4$, plant

Choose the correct answer from the options given below :
(a) (A)-(IV), (B)-(I), (C)-(III), (D)-(II)
(b) (A)-(IV), (B)-(I), (C)-(II), (D)-(III)
(c) (A)-(III), (B)-(II), (C)-(I), (D)-(IV)
(d) (A)-(II), (B)-(III), (C)-(I), (D)-(IV)

38 Identify the number of structure/s from the following which can be correlated to D-glyceraldehyde. **[2025, 22 Jan Shift-II]**

(A) 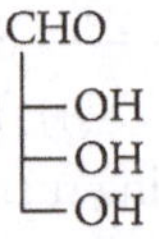

(B) 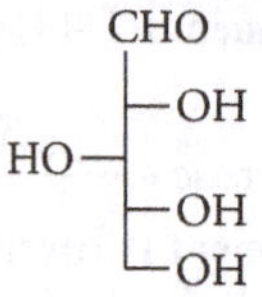

(C) 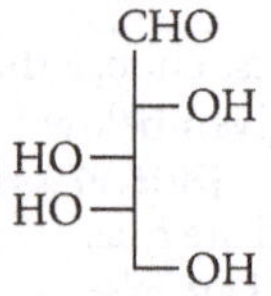

(D) 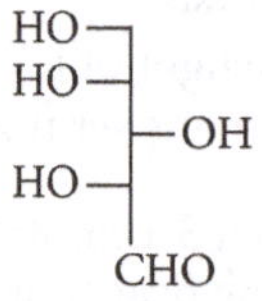

(a) four (b) one (c) two (d) three

Imp. *Acidic Hydrolysis of Carbohydrates*

39 Identify correct conversion during acidic hydrolysis from the following. **[2025, 28 Jan Shift-II]**

A. Starch gives galactose.
B. Cane sugar gives equal amount of glucose and fructose.
C. Milk sugar gives glucose and galactose.
D. Amylopectin gives glucose and fructose.
E. Amylose gives only glucose.

Choose the correct answer from the options given below

(a) (A), (B) and (C) only (b) (B), (C) and (D) only
(c) (C), (D) and (E) only (d) (B), (C) and (E) only

40 Match List-I with List-II. **[2025, 28 Jan Shift-II]**

	List-I (Saccharides)		List-II (Glycosidic-linkages found)
(A)	Sucrose	I.	$\alpha 1-4$
(B)	Maltose	II.	$\alpha 1-4$ and $\alpha 1-6$
(C)	Lactose	III.	$\alpha 1-\beta 2$
(D)	Amylopectin	IV.	$\beta 1-4$

Choose the correct answer from the options given below :

(a) (A)-(III), (B)-(I), (C)-(IV), (D)-(II)
(b) (A)-(II), (B)-(IV), (C)-(III), (D)-(I)
(c) (A)-(IV), (B)-(II), (C)-(I), (D)-(III)
(d) (A)-(I), (B)-(II), (C)-(III), (D)-(IV)

41 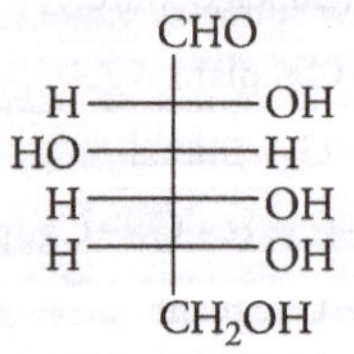

The incorrect statement regarding the given structure is **[2024, 8 April Shift-I]**

(a) will coexist in equilibrium with 2 other cyclic structure
(b) has 4 asymmetric carbon atom
(c) despite the presence of —CHO does not give Schiff's test
(d) can be oxidised to a dicarboxylic acid with Br_2 water

42 Match List I with List II. **[2024, 4 April Shift-II]**

	List I		List II
A.	α- Glucose and α- Galactose	I.	Functional isomers
B.	α- Glucose and β- Glucose	II.	Homologous
C.	α- Glucose and α- Fructose	III.	Anomers
D.	α- Glucose and α- Ribose	IV.	Epimers

Choose the correct answer from the options given below:

(a) A- III, B- IV, C- II, D- I
(b) A- III, B- IV, C- I, D- II
(c) A- IV, B- III, C- II, D- I
(d) A- IV, B-III, C- I, D- II

43 Match List-I with List-II. **[2024, 31 Jan Shift-I]**

	List-I		List-II
A.	Glucose/ $NaHCO_3$ / Δ	I.	Gluconic acid
B.	Glucose/HNO_3	II.	No reaction
C.	Glucose HI / Δ	III.	*n*-hexane
D.	Glucose/bromine water	IV.	Saccharic acid

Choose the correct answer from the options given below:

	A	B	C	D
(a)	II	IV	III	I
(b)	IV	I	III	II
(c)	III	II	I	VI
(d)	I	IV	III	II

Imp. *Optical Activity of Carbohydrate*

44 D-(+)-Glyceraldehyde $\xrightarrow[\text{(ii) } H_2O\ /\ H^+ \quad \text{(iii) } HNO_3]{\text{(i) HCN}}$

The products formed in the above reaction are **[2023, 13 April Shift-I]**

(a) one optically inactive products
(b) two optically active products
(c) two optically inactive products
(d) one optically active and one meso product

45 A sugar '*X*' dehydrates very slowly under acidic condition to give furfural which on further reaction with resorcinol gives the coloured product after sometime. Sugar '*X*' is **[2022, 27 July Shift-I]**

(a) aldopentose (b) aldotetrose
(c) oxalic acid (d) ketotetrose

46 When sugar '*X*' is boiled with dilute H_2SO_4 in alcoholic solution, two isomers '*A*' and '*B*' are formed. '*A*' on oxidation with HNO_3 yields saccharic acid whereas '*B*' is laevorotatory. The compound '*X*' is **[2022, 28 June Shift-II]**

(a) maltose (b) sucrose
(c) lactose (d) starch

Imp. *Preparation Reactions and Test of Carbohydrates*

47 L-isomer of a compound (A) ($C_4H_8O_4$) gives a positive test with $[Ag(NH_3)_2]^+$. Treatment of A with acetic anhydride yields triacetate derivative. Compound (A) produces an optically active compound (B) and an optically inactive compound (C) on treatment with bromine water and HNO_3 respectively. Compound (A) is **[2022, 27 June Shift-I]**

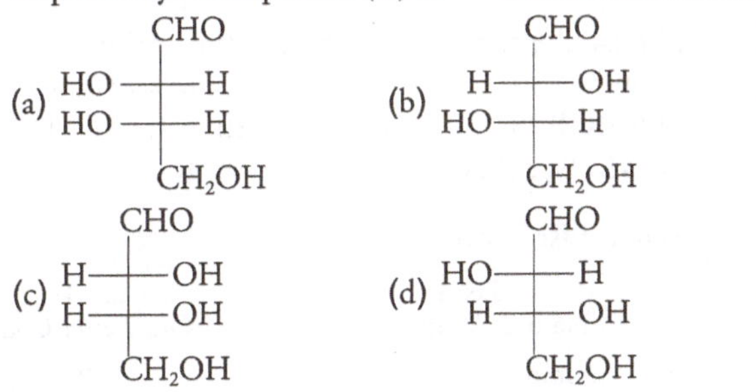

48 A polysaccharide 'X' on boiling with dil. H_2SO_4 at 393 K under 2-3 atm pressure yields 'Y'. 'Y' on treatment with bromine water gives gluconic acid. 'X' contains β-glycosidic linkages only. Compound 'X' is **[2022, 24 June Shift-I]**

(a) starch (b) cellulose
(c) amylose (d) amylopectin

49 The correct observation in the following reactions, is

$$\text{Sucrose} \xrightarrow[\text{Cleavage (Hydrolysis)}]{\text{Glycosidic bond}} A + B \xrightarrow[\text{reagent}]{\text{Seliwanoff's}}$$

[2020, 2 Sep Shift-II]

(a) Formation of blue colour (b) Gives no colour
(c) Formation of red colour (d) Formation of violet colour

Numerical Value Type Questions

50 Number of compounds from the following which will not produce orange red precipitate with Benedict solution is ………… .

Glucose, maltose, sucrose, ribose, 2-deoxyribose, amylose, lactose. **[2023, 11 April Shift-II]**

51 $C_6H_{12}O_6 \xrightarrow{\text{Zymase}} A \xrightarrow[\Delta]{\text{NaOI}} B + CHI_3$

The number of carbon atoms present in the product B is ……… . **[2022, 29 June Shift-I]**

Topic 2 / Proteins and Enzymes

Level 1

01 Type of amino acids obtained by hydrolysis of proteins is **[2024, 29 Jan Shift-I]**

(a) γ (b) α (c) β (d) δ

02 During the denaturation of proteins, which of these structures will remain intact? **[2022, 25 July Shift-I]**

(a) Primary (b) Secondary
(c) Tertiary (d) Quaternary

03 The water soluble protein is **[2021, 22 July Shift-I]**

(a) fibrin (b) albumin
(c) myosin (d) collagen

04 Which of the following is correct structure of tyrosine? **[2021, 17 March Shift-I]**

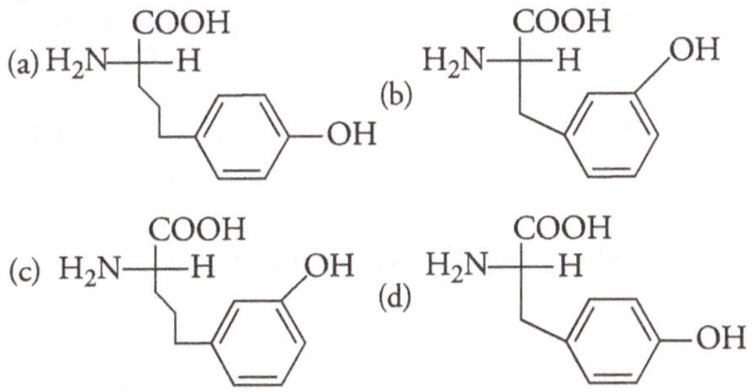

05 The secondary structure of protein is stabilised by **[2021, 16 March Shift-II]**

(a) peptide bond (b) glycosidic bond
(c) hydrogen bonding (d) van der Waals' forces

Level 2

06 A dipeptide, 'x' on complete hydrolysis gives 'y' and 'z'. 'y' on treatment with aq. HNO_2, produces lactic acid. On the other hand 'z' on heating gives the following cyclic molecule.

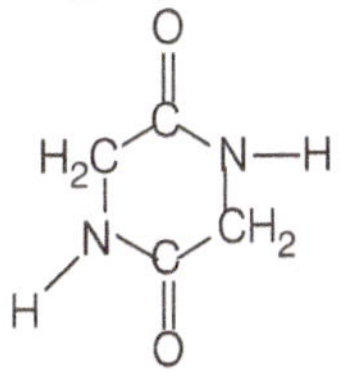

Based on the information given, the dipeptide X is **[2025, 04 April Shift-II]**

(a) alanine-alanine (b) valine-leucine
(c) valine-glycine (d) alanine-glycine

07 Given below are two statements:

Statement (I) On hydrolysis, oligo peptides give rise to fewer number of α-amino acids while proteins give rise to a large number of β-amino acids.

Statement (II) Natural proteins are denatured by acids which convert the water soluble form of fibrous proteins to their water insoluble form.

In the light of the above statements, choose the most appropriate answer from the options given below **[2025, 07 April Shift-II]**

(a) Statement I is incorrect but Statement II is correct.
(b) Both Statement I and Statement II are correct.
(c) Statement I is correct but Statement II is incorrect.
(d) Both Statement I and Statement II are incorrect.

08 A tetrapeptide, 'x' on complete hydrolysis produced glycine (Gly), alanine (Ala), valine (Val), leucine (Leu) in equimolar proportion each. The number of tetrapeptides (sequences) possible involving each of these amino acids is **[2025, 02 April Shift-II]**

(a) 8 (b) 32
(c) 24 (d) 16

09 Identify the pair of reactants that upon reaction, with elimination of HCl will give rise to the dipeptide Gly-Ala. **[2025, 04 April Shift-I]**

(a) NH_2-CH_2-COOH and $NH_2-\underset{\displaystyle CH_3}{\overset{}{C}}H-COOH$ (with CH_3 on the CH)

(b) NH_2-CH_2-COCl and $\overset{\oplus}{N}H_3-CH(CH_3)-COCl$

(c) NH_2-CH_2-COOH and $NH_2-CH(CH_3)-COCl$

(d) NH_2-CH_2-COCl and $NH_2-CH(CH_3)-COOH$

10 Coagulation of egg, on heating is because of **[2024, 5 April Shift-II]**

(a) biological property of protein remains unchanged
(b) the secondary structure of protein remains unchanged
(c) breaking of the peptide linkage in the primary structure of protein occurs
(d) denaturation of protein occurs.

11 Which of the following gives a positive test with ninhydrin? **[2024, 5 April Shift-I]**

(a) Cellulose
(b) Polyvinyl chloride
(c) Egg albumin
(d) Starch

12 Which structure of protein remains intact after coagulation of egg white on boiling? **[2024, 27 Jan Shift-II]**

(a) Secondary (b) Primary
(c) Quaternary (d) Tertiary

13 Which is not true for arginine? **[2023, 15 April Shift-I]**

(a) It is associated with more than one pK_a values.
(b) It is a crystalline solid.
(c) It has high solubility in benzene
(d) It has a fairly high melting point

14 The naturally occurring amino acid that contains only one basic functional group in its chemical structure is **[2023, 13 April Shift-II]**

(a) histidine (b) asparagine
(c) lysine (d) arginine

Imp. *Symbols of Amino Acids*

15 Match List I with List II.

List I (Natural amino acid)		List II (One letter code)
A. Glutamic acid	I.	Q
B. Glutamine	II.	W
C. Tyrosine	III.	E
D. Tryptophan	IV.	Y

Choose the correct answer from the options given below : **[2023, 8 April Shift-II]**

(a) A-III, B-IV, C-I, D-II (b) A-III, B-I, C-IV, D-II
(c) A-II, B-I, C-IV, D-III (d) A-IV, B-III. C-I.D-I

16 Match List I with List II.

List I (Natural Amino acid)		List II (One Letter Code)
A. Arginine	I.	D
B. Aspartic acid	II.	N
C. Asparagine	III.	A
D. Alanine	IV.	R

Choose the correct answer from the options given below **[2023, 6 April Shift-II]**

(a) A-III, B-I, C-II, D-IV (b) A-IV, B-I, C-III, D-II
(c) A-I, B-III, C-IV, D-II (d) A-IV, B-I, C-II, D-III

17 Match List I with List II

List I (Enzymatic reaction)		List II (Enzyme)
A. Sucrose $\rightarrow$ Glucose and fructose	I.	Zymase
B. Glucose $\rightarrow$ ethyl alcohol and CO_2	II.	Pepsin
C. Starch $\rightarrow$ Maltose	III.	Invertase
D. Proteins $\rightarrow$ Amino acids	IV.	Diastase

Choose the correct answer from the options given below **[2023, 6 April Shift-I]**

(a) A-I, B-IV, C-III, D-II (b) A-I, B-II, C-IV, D-III
(c) A-III, B-I, C-II, D-IV (d) A-III, B-I, C-IV, D-II

18 The one that does not stabilise 2° and 3° structures of proteins is **[2023, 4 April Shift-I]**

(a) H-bonding (b) —S—S—linkage
(c) van der Waals forces (d) —O—O— linkage

19 Given below are two statements, one is labelled as Assertion (A) and the other is labelled as Reason (R).

Assertion (A) α-halocarboxylic acid on reaction with dil. NH_3 gives a good yield of α-amino carboxylic acid whereas the yield of amines is very low when prepared from alkyl halides.

Reason (R) Amino acids exist in Zwitter ion form in aqueous medium.
In the light of the above statements, choose the correct answer from the options given below. **[2023, 01 Feb Shift-II]**
(a) A is incorrect but R is correct
(b) Both A and R are correct and R is the correct explanation of A
(c) A is correct but R is incorrect
(d) Both A and R are correct but R is not the correct explanation of A

20 Number of cyclic tripeptides formed with 2 amino acids *A* and *B* is **[2023, 29 Jan Shift-I]**
(a) 3 (b) 4 (c) 2 (d) 5

21 Two statements in respect of drug-enzyme interaction are given below
Statement I Action of an enzyme can be blocked only when an inhibitor blocks the active site of the enzyme.
Statement II An inhibitor can form a strong covalent bond with the enzyme.
In the light of the above statements, choose the correct answer from the options given below **[2022, 28 July Shift-II]**
(a) Both Statement I and Statement II are true
(b) Both Statement I and Statement II are false
(c) Statement I is true but Statement II is false
(d) Statement I is false but Statement II is true

22 Stability of α-helix structure of proteins depends upon **[2022, 28 June Shift-I]**
(a) dipolar interaction (b) H-bonding interaction
(c) van der Waals' forces (d) π-stacking interaction

23 Which of the following is not an example of fibrous protein ? **[2021, 31 Aug Shift-II]**
(a) Keratin (b) Albumin
(c) Collagen (d) Myosin

24 Out of the following, which type of interaction is responsible for the stabilisation of α-helix structure of proteins ? **[2021, 24 Feb Shift-I]**
(a) van der Waals' forces (b) Covalent bonding
(c) Ionic bonding (d) Hydrogen bonding

25 Which of the following is not an essential amino acid? **[2020, 5 Sep Shift-I]**
(a) Leucine (b) Valine
(c) Lysine (d) Tyrosine

26 Among the following compounds, most basic amino acid is **[2019, 12 Jan Shift-I]**
(a) serine (b) asparagine
(c) lysine (d) histidine

Imp. *pK_a Order of Amino Acids*

27 The increasing order of pK_a of the following amino acids in aqueous solution is Gly, Asp, Lys, Arg **[2019, 9 Jan Shift-II]**
(a) Asp < Gly < Arg < Lys (b) Arg < Lys < Gly < Asp
(c) Gly < Asp < Arg < Lys (d) Asp < Gly < Lys < Arg

• Level 3

28 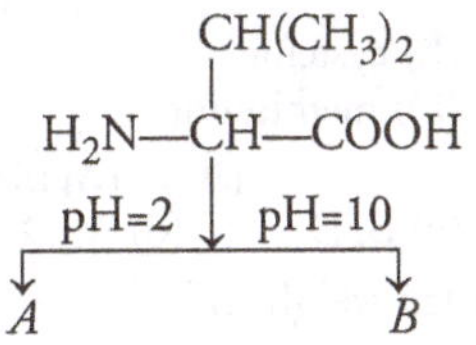

Choose the correct option for structures of *A* and *B*, respectively **[2025, 08 April Shift-II]**
(a) $H_2N-\underset{CH(CH_3)_2}{\underset{|}{C}H}-CO\bar{O}$ and $H_3\overset{+}{N}-\underset{CH(CH_3)_2}{\underset{|}{C}H}-CO\bar{O}$
(b) $H_2N-\underset{CH(CH_3)_2}{\underset{|}{C}H}-CO\bar{O}$ and $H_3\overset{+}{N}-\underset{CH(CH_3)_2}{\underset{|}{C}H}-COOH$
(c) $H_3\overset{+}{N}-\underset{CH(CH_3)_2}{\underset{|}{C}H}-COOH$ and $H_2N-\underset{CH(CH_3)_2}{\underset{|}{C}H}-CO\bar{O}$
(d) $H_3\overset{+}{N}-\underset{CH(CH_3)_2}{\underset{|}{C}H}-CO\bar{O}$ and $H_3\overset{+}{N}-\underset{CH(CH_3)_2}{\underset{|}{C}H}-COOH$

29 Identify the correct statement among the following **[2025, 02 April Shift-I]**
(a) All naturally occurring amino acids except glycine contain one chiral centre.
(b) All naturally occurring amino acids are optically active.
(c) Glutamic acid is the only amino acid that contains a —COOH group at the side chain.
(d) Amino acid, cysteine can easily undergo dimerisation due to the presence of free SH group.

30 Identify the essential amino acids from below
(A) Valine (B) Proline
(C) Lysine (D) Threonine
(E) Tyrosine
Choose the correct answer from the options given below **[2025, 29 Jan Shift-II]**
(a) (A), (C) and (E) only
(b) (A), (C) and (D) only
(c) (C), (D) and (E) only
(d) (B), (C) and (E) only

31 The incorrect statements regarding enzymes are **[2024, 6 April Shift-II]**
A. Enzymes are biocatalysts.
B. Enzymes are non-specific and can catalyse different kinds of reactions.
C. Most enzymes are globular proteins.
D. Enzyme-oxidase catalyses the hydrolysis of maltose into glucose.
Choose the correct answer from the options given below
(a) B and D (b) B and C
(c) A, B and C (d) B, C and D

32 Sulphur (S) containing amino acids from the following are :
(A) isoleucine (B) cysteine
(C) lysine (D) methionine
(E) glutamic acid **[2023, 8 April Shift-I]**
(a) A, B, C (b) B, C (c) B, D (d) A, D

33 A protein 'X' with molecular weight of 70,000 u, on hydrolysis gives amino acids. One of these amino acid is **[2023, 31 Jan Shift-I]**
(a) $NH_2-CH_2-CH(CH_3)-CH_2CH_2COOH$
(b) $CH_3-CH(CH_3)-CH_2-CH(NH_2)-COOH$
(c) $CH_3-C(CH_3)(NH_2)-CH_2-CH_2COOH$
(d) $CH_3-CH(CH_3)-CH(NH_2)-CH_2-COOH$

34 The peptide that gives positive ceric ammonium nitrate and carbylamine tests is **[2019, 9 April Shift-II]**
(a) Lys-Asp (b) Ser-Lys (c) Gln-Asp (d) Asp-Gln

35 The correct structure of product 'P' in the following reaction is **[2019, 10 Jan Shift-I]**
$$\text{Asn} - \text{Ser} + \underset{\text{(Excess)}}{(CH_3CO)_2O} \xrightarrow{NEt_3} P$$
(a)
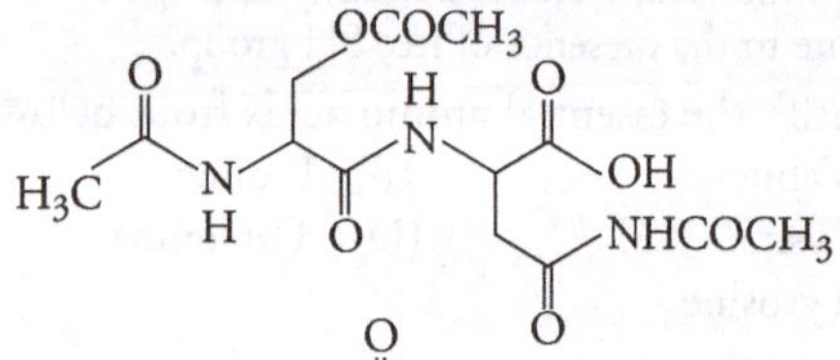
(b)
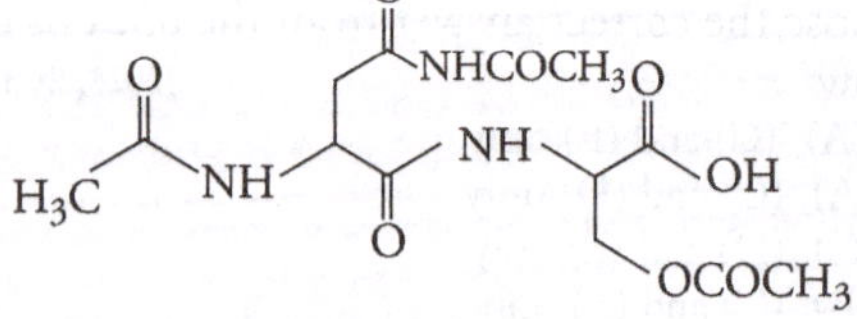
(c)
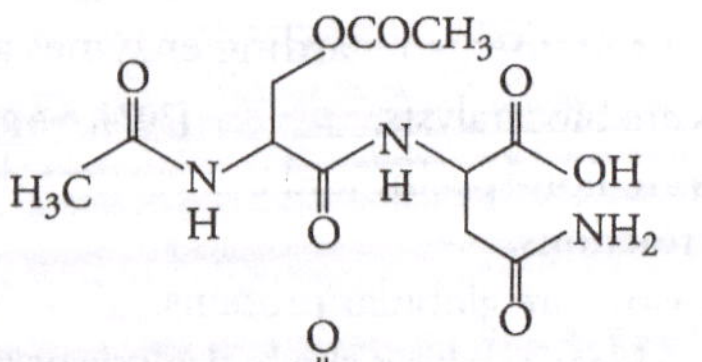
(d)
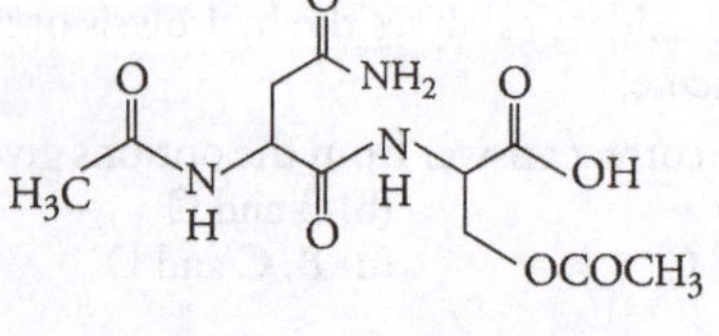

Imp. *Structure of Amino Acids*

36 The correct sequence of amino acids present in the tripeptide given below is **[2019, 9 Jan Shift-II]**
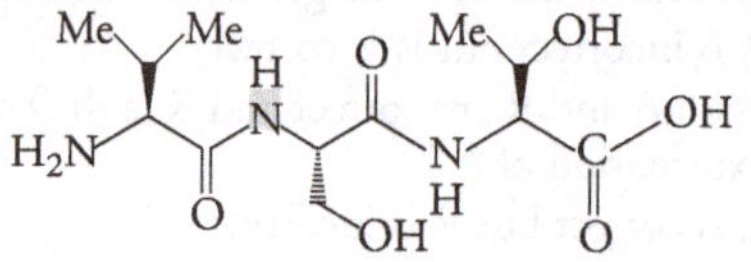
(a) Thr - Ser - Leu (b) Leu - Ser - Thr
(c) Val - Ser - Thr (d) Thr - Ser - Val

Numerical Value Type Questions

37 Total number of essential amino acid among the given list of amino acids is ....
Arginine, Phenylalanine, Aspartic Acid, Cysteine, Histidine, Valine, Proline **[2024, 9 April Shift-I]**

38 The total number of carbon atoms present in tyrosine, an amino acid is ...... **[2024, 8 April Shift-II]**

39 The number of tripeptides formed by three different amino acids using each amino acid once is .... **[2024, 1 Feb Shift-I]**

Imp. *Number of Peptide Linkages in Amino Acids*

40 A short peptide on complete hydrolysis produces 3 moles of glycine (G), two moles of leucine (L) and two moles of valine (V) per mole of peptide. The number of peptide linkages in it are .... **[2023, 30 Jan Shift-II]**

41 Total number of tripeptides possible by mixing of valine and proline is ...... **[2023, 24 Jan Shift-II]**

42 In a linear tetrapeptide (Constituted with different amino acids), (number of amino acids) - (number of peptide bonds) is .... **[2022, 29 July Shift-I]**

43 In alanylglycylleucylalanylvaline, the number of peptide linkages is .... **[2022, 24 June Shift-II]**

44 How many of the given compounds will give a positive biuret test ...... ? **[2022, 25 June Shift-II]**
Glycine, Glycylalanine, Tripeptide, Biuret

45 A peptide synthesised by the reactions of one molecule each of glycine, leucine, aspartic acid and histidine will have .... peptide linkages. **[2021, 01 Sep Shift-II]**

46 The total number of negative charge in the tetrapeptide Gly-Glu-Asp-Tyr, at pH 12.5 will be .... **[2021, 26 Aug Shift-I]**

47 The number of chiral carbon(s) present in peptide, Ile-Arg-Pro, is .... **[2020, 5 Sep Shift-I]**

48 The number of chiral centres present in threonine is .... **[2020, 04 Sept Shift-II]**

49 The number of $\rangle C=O$ groups present in a tripeptide Asp-Glu-Lys is .... **[2020, 3 Sep Shift-II]**

Topic 3 / Vitamins, Nucleic Acids and Hormones

• Level 1

01 Which of the following acids is a vitamin? **[2025, 22 Jan Shift-I]**

(a) Aspartic acid (b) Ascorbic acid
(c) Saccharic acid (d) Adipic acid

02 All structures given below are of vitamin C. Most stable of them is **[2023, 01 Feb Shift-II]**

(a) 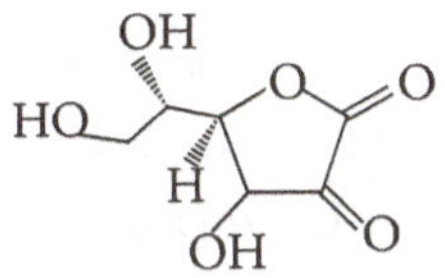

(b) 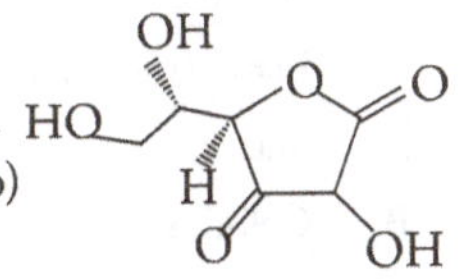

(c) 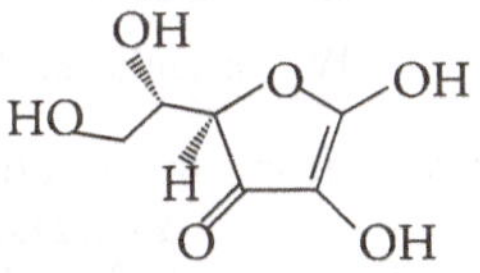

(d) 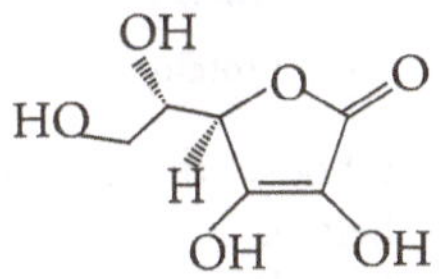

03 Which one of the following is a water soluble vitamin, that is not excreted easily? **[2022, 26 June Shift-II]**

(a) Vitamin B_2 (b) Vitamin B_1
(c) Vitamin B_6 (d) Vitamin B_{12}

04 Thiamine and pyridoxine are also known respectively as **[2021, 22 July Shift-II]**

(a) vitamin B_2 and vitamin E
(b) vitamin E and vitamin B_2
(c) vitamin B_6 and vitamin B_2
(d) vitamin B_1 and vitamin B_6

05 Deficiency of vitamin K causes **[2021, 18 March Shift-II]**

(a) increase in blood clotting time
(b) increase in fragility of RBC's
(c) cheilosis
(d) decrease in blood clotting time

06 Which of the following vitamin is helpful in delaying the blood clotting ? **[2021, 26 Feb Shift-I]**

(a) Vitamin C (b) Vitamin B
(c) Vitamin E (d) Vitamin K

• Level 2

07 DNA molecule contains 4 bases whose structures are shown below. One of the structures is not correct. identify the incorrect base structure. **[2024, 6 April Shift-I]**

(a) 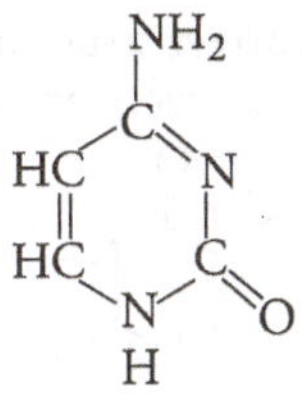

(b) 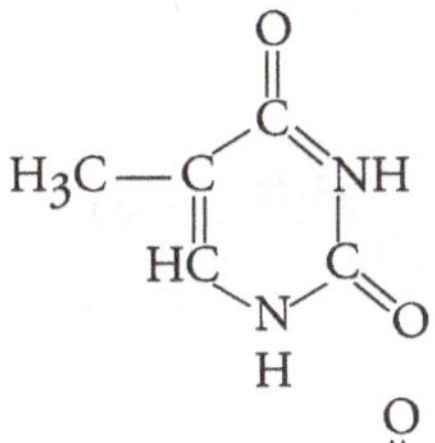

(c) 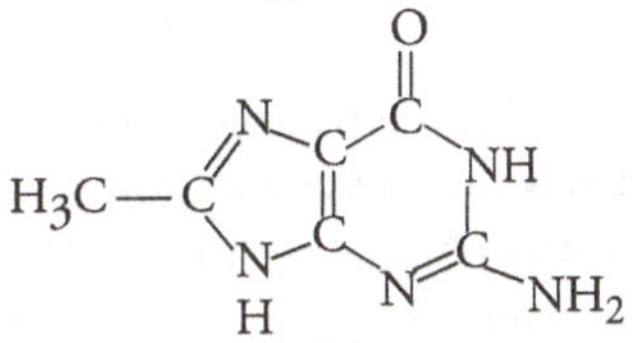

(d) 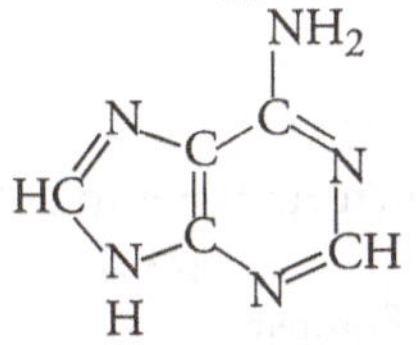

08 If one strand of a DNA has the sequence ATGCTTCA, sequence of the bases in complementary strand is **[2024, 1 Feb Shift-II]**

(a) ATGCGACT (b) CATTAGCT
(c) TACGAAGT (d) GTACTTAC

09 Two nucleotides are joined together by a linkage known as **[2024, 27 Jan Shift-I]**

(a) peptide linkage
(b) phosphodiester linkage
(c) disulphide linkage
(d) glycosidic linkage

10 Sugar moiety in DNA and RNA molecules respectively are **[2022, 29 June Shift-I]**

(a) β-D-2-deoxyribose, β-D-deoxyribose
(b) β-D-2-deoxyribose, β-D-ribose
(c) β-D-ribose, β-D-2-deoxyribose
(d) β-D-deoxyribose, β-D-2-deoxyribose

11 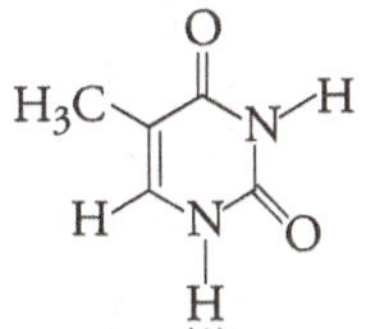

(A)

The compound '*A*' is a complementary base of .... in DNA strands. **[2021, 27 July Shift-I]**

(a) uracil (b) guanine
(c) adenine (d) cytosine

12 Which one of the following is correct structure for cytosine? **[2021, 25 July Shift-II]**

(a) 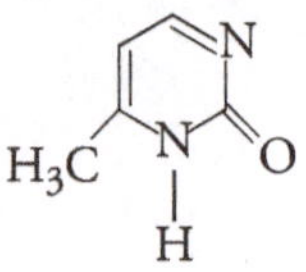

(b) 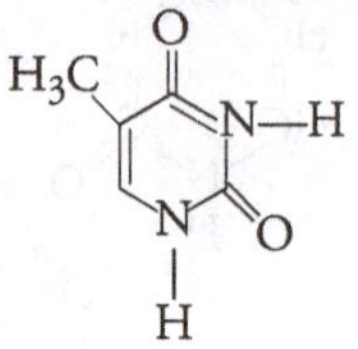

(c) 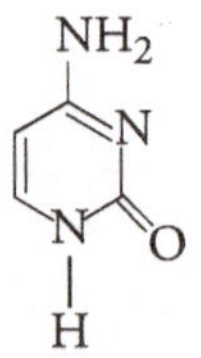

(d) 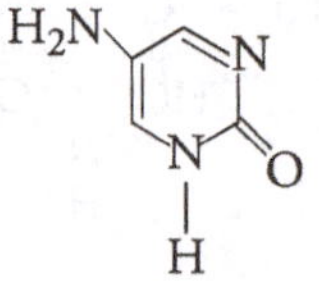

13 Which among the following pair of vitamins is stored in our body relatively for longer duration ? **[2021, 16 March Shift-I]**

(a) Thiamine and vitamin A
(b) Vitamin A and vitamin D
(c) Thiamine and ascorbic acid
(d) Ascorbic acid and vitamin D

14 Which of the following statement is not true about RNA? **[2019, 12 April Shift-I]**

(a) It controls the synthesis of protein
(b) It has always double stranded α-helix structure
(c) It usually does not replicate
(d) It is present in the nucleus of the cell

• Level 3

15 Fat soluble vitamins are

A. Vitamin B_1
B. Vitamin C
C. Vitamin E
D. Vitamin B_{12}
E. Vitamin K

Choose the correct answer from the options given below. **[2025, 03 April Shift-II]**

(a) B and C only (b) C and D only
(c) C and E only (d) A and B only

16 Match List - I with List - II. **[2025, 24 Jan Shift-II]**

	List - I		List - II
(A)	Adenine	(I)	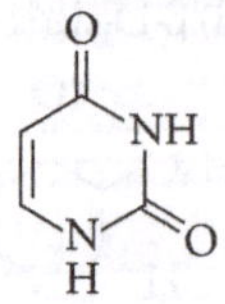
(B)	Cytosine	(II)	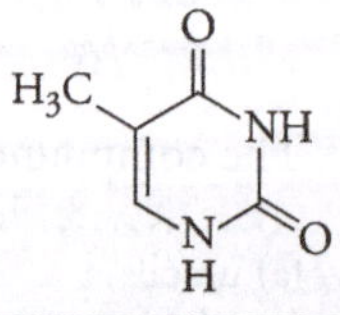
(C)	Thymine	(III)	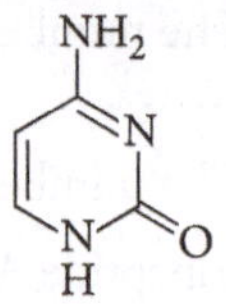
(D)	Uracil	(IV)	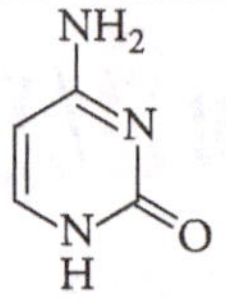

Choose the correct answer from the options given below

(a) (A) - (III), (B) - (I), (C) - (IV), (D) - (II)
(b) (A) - (IV), (B)-(III), (C) - (II), (D) - (I)
(c) (A) - (III), (B) - (IV), (C) - (II), (D) - (I)
(d) (A) - (III), (B)-(IV), (C) - (I), (D) - (II)

17 Match List-I with List-II.

	List-I (Biopolymer)		List-II (Monomer)
A.	Starch	I.	Nucleotide
B.	Cellulose	II.	α-glucose
C.	Nucleic acid	III.	β-glucose
D.	Protein	IV.	α-amino acid

Choose the correct answer from the options given below. **[2024, 29 Jan Shift-II]**

(a) A-IV, B-II, C-I, D-III (b) A-II, B-I, C-III, D-IV
(c) A-I, B-III, C-IV, D-II (d) A-II, B-III, C-I, D-IV

18 Which of the following will react with $CHCl_3$ + alc. KOH ? **[2020, 4 Sep Shift-I]**

(a) Adenine and lysine (b) Thymine and proline
(c) Adenine and thymine (d) Adenine and proline

19 Match the following : **[2020, 7 Jan Shift-I]**

(i)	Riboflavin	(A)	Beriberi
(ii)	Thiamine	(B)	Scurvy
(iii)	Pyridoxine	(C)	Cheilosis
(iv)	Ascorbic acid	(D)	Convulsions

(a) (i)-(C), (ii)-(D), (iii)-(A), (iv)-(B)
(b) (i)-(C), (ii)-(A), (iii)-(D), (iv)-(B)
(c) (i)-(D), (ii)-(B), (iii)-(A), (iv)-(C)
(d) (i)-(A), (ii)-(D), (iii)-(C), (iv)-(B)

Numerical Value Type Questions

20 Thyroxine, the hormone has given below structure **[2025, 7 April Shift-I]**

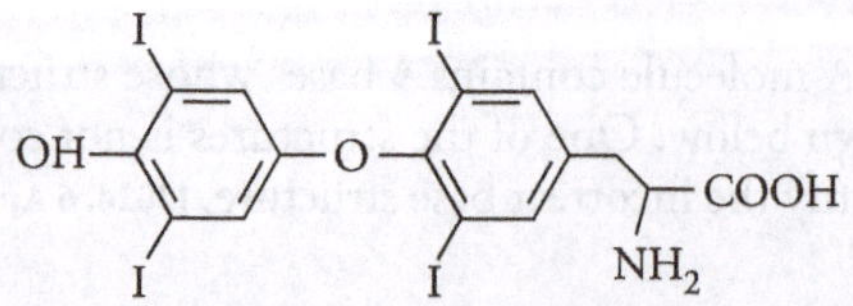

The percentage of iodine in thyroxine is ....%. (Nearest integer)
(Given molar mass in g mol^{-1} C : 12, H : 1, O : 16, N : 14, 1 : 127)

21 The total number of hydrogen bonds of a DNA-double Helix strand whose one strand has the following sequence of bases is
5′-G-G-C-A-A-A-T-C-G-G-C-T-A-3′
[2025, 4 April Shift-I]

22 From the vitamins A, B_1, B_6, B_{12}, C, D, E and K, the number of vitamins that can be stored in our body is .... **[2024, 31 Jan Shift-II]**

23 The total number of correct statements, regarding the nucleic acids is .... **[2024, 30 Jan Shift-II]**

A. RNA is regarded as the reserve of genetic information
B. DNA molecule self-duplicates during cell division
C. DNA synthesises proteins in the cell.
D. The message for the synthesis of particular proteins is present in DNA
E. Identical DNA strands are transferred to daughter cells

24 Uracil is a base present in RNA with the following structure, % of N in uracil is ....

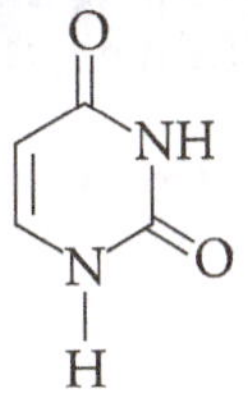

[2023, 24 Jan Shift-I]

[Given : Molar mass N = 14g mol^{-1}
O = 16 g mol^{-1}, C = 12 g mol^{-1}, H = 1 g mol^{-1}]

25 The number of oxygens present in a nucleotide formed from a base, that is present only in RNA is .... **[2022, 26 June Shift-I]**

Key Index

Topic 1

1 (c)	2 (a)	3 (d)	4 (d)	5 (a)	6 (b)	7 (d)	8 (c)	9 (d)	10 (b)
11 (c)	12 (a)	13 (b)	14 (a)	15 (b)	16 (c)	17 (d)	18 (c)	19 (b)	20 (a)
21 (d)	22 (a)	23 (c)	24 (a)	25 (c)	26 (d)	27 (a)	28 (a)	29 (c)	30 (c)
31 (d)	32 (a)	33 (d)	34 (d)	35 (b)	36 (a)	37 (b)	38 (d)	39 (d)	40 (a)
41 (d)	42 (d)	43 (a)	44 (d)	45 (a)	46 (b)	47 (a)	48 (b)	49 (c)	50 (2)
51 (1)									

Topic 2

1 (b)	2 (a)	3 (b)	4 (d)	5 (c)	6 (d)	7 (d)	8 (c)	9 (d)	10 (d)
11 (c)	12 (b)	13 (c)	14 (b)	15 (b)	16 (d)	17 (d)	18 (d)	19 (d)	20 (b)
21 (d)	22 (b)	23 (b)	24 (d)	25 (d)	26 (d)	27 (d)	28 (c)	29 (d)	30 (b)
31 (a)	32 (c)	33 (b)	34 (b)	35 (d)	36 (c)	37 (4)	38 (9)	39 (6)	40 (6)
41 (8)	42 (1)	43 (4)	44 (2)	45 (3)	46 (4)	47 (4)	48 (2)	49 (5)	

Topic 3

1 (b)	2 (d)	3 (d)	4 (d)	5 (a)	6 (d)	7 (c)	8 (c)	9 (b)	10 (b)
11 (c)	12 (c)	13 (b)	14 (b)	15 (c)	16 (c)	17 (d)	18 (a)	19 (b)	20 (65)
21 (33)	22 (5)	23 (3)	24 (25)	25 (9)					

For Detailed Solutions
scan QR Code

24 Principles Related to Practical Chemistry

In JEE Main there is usually 0-1 question per paper are visible from this chapter. In past 3 years, 39 conceptual and 7 numerical questions were asked of moderate difficulty level from this chapter.

For detailed analysis scan QR code

Topic 1 / Detection of Extra Elements & Functional Groups

Level 1

01 Lassaigne's test is used for detection of **[2024, 1 Feb Shift-II]**
(a) nitrogen and sulphur only
(b) phosphorus and halogens only
(c) nitrogen, sulphur and phosphorus only
(d) nitrogen, sulphur, phosphorus and halogens

02 Phenolic group can be identified by a positive **[2024, 27 Jan Shift-II]**
(a) Tollen's test (b) carbylamine test
(c) phthalein dye test (d) Lucas test

03 The formula of the purple colour formed in Lassaigne's test for sulphur using sodium nitroprusside is **[2022, 28 June Shift-I]**
(a) $NaFe[Fe(CN)_6]$ (b) $Na[Cr(NH_3)_2(NCS)_4]$
(c) $Na_2[Fe(CN)_5(NO)]$ (d) $Na_4[Fe(CN)_5(NOS)]$

04 Which one of the following test used for the identification of functional groups in organic compounds does not use copper reagent ? **[2021, 27 Aug Shift-II]**
(a) Barfoed's test (b) Seliwanoff's test
(c) Benedict's test (d) Biuret test for peptide bond

Imp. *Test for Detection of Alcohol & Amine*

05 Ceric ammonium nitrate and $CHCl_3$ /alc. KOH are used for the identification of functional groups present in .... and ...... respectively. **[2021, 26 Feb Shift-II]**
(a) alcohol, phenol (b) amine, alcohol
(c) alcohol, amine (d) amine, phenol

06 Which one of the following complexes is violet in colour? **[2021, 26 Aug Shift-I]**
(a) $Fe_4[Fe(CN)_6]_3 \cdot H_2O$ (b) $[Fe(CN)_5NOS]^{4-}$
(c) $[Fe(SCN)_6]^{4-}$ (d) $[Fe(CN)_6]^{4-}$

07 Acidic ferric chloride solution on treatment with excess of potassium ferrocyanide gives a prussian blue coloured colloidal species. It is **[2021, 27 Aug Shift-I]**
(a) $HFe[Fe(CN)_6]$ (b) $K_5Fe[Fe(CN)_6]_2$
(c) $Fe_4[Fe(CN)_6]_3$ (d) $KFe[Fe(CN)_6]$

Imp. *Silver Nitrate Test for Halogen Detection*

08 Which of the following compound is added to the sodium extract before addition of silver nitrate for testing of halogens? **[2021, 25 Feb Shift-II]**
(a) Hydrochloric acid (b) Sodium hydroxide
(c) Nitric acid (d) Ammonia

09 Which of the following test cannot be used for identifying amino acids? **[2019, 10 Jan Shift-II]**
(a) Biuret test (b) Barfoed test
(c) Ninhydrin test (d) Xanthoproteic test

Level 2

10 Number of molecules from below which cannot give iodoform reaction is
Ethanol, Isopropyl alcohol, Bromoacetone, 2-Butanol, 2-Butanone, Butanal, 2-Pentanone, 3-Pentanone, Pentanal and 3-Pentanol. **[2025, 3 April, Shift-I]**
(a) 2 (b) 5 (c) 4 (d) 3

11 Which of the following compounds will give silver mirror with ammoniacal silver nitrate?
A. Formic acid B. Formaldehyde
C. Benzaldehyde D. Acetone
Choose the correct answer from the options given below. **[2024, 9 April Shift-II]**
(a) A only (b) A, B and C only
(c) C and D only (d) B and C only

12 Appearance of blood red colour, on treatment of the sodium fusion extract of an organic compound with $FeSO_4$ in presence of concentrated H_2SO_4 indicates the presence of elements **[2024, 29 Jan Shift-I]**
(a) N and S (b) N (c) Br (d) S

13 Match List-I with List-II.

List-I (Element detected)	List-II (Reagent used/Product formed)
A. Nitrogen	I. $Na_2[Fe(CN)_5NO]$
B. Sulphur	II. $AgNO_3$
C. Phosphorus	III. $Fe_4[Fe(CN)_6]_3$
D. Halogen	IV. $(NH_4)_2MoO_4$

Choose the correct answer from the options given below **[2023, 6 April Shift-I]**

(a) A-II, B-IV, C-I, D-III (b) A-III, B-I, C-IV, D-II
(c) A-IV, B-II, C-I, D-III (d) A-II, B-I, C-IV, D-III

14 During halogen test, sodium fusion extract is boiled with concentrated HNO_3 to **[2022, 25 June Shift-II]**

(a) remove unreacted sodium
(b) decompose cyanide or sulphide of sodium
(c) extract halogen from organic compound
(d) maintain the pH of extract

• Level 3

15 $A \xrightarrow[\text{(ii) } H_3O^+]{\text{(i) NaOH}} B \xrightarrow[\text{(ii) } H_2SO_4,\ \Delta]{\text{(i) EtOH}} C$

'*A*' shows positive Lassaigne's test for N and its molar mass is 121.

'*B*' gives effervescence with aq. $NaHCO_3$.

'*C*' gives fruity smell.

Identify *A*, *B* and *C* from the following. **[2025, 8 April Shift-II]**

(a) 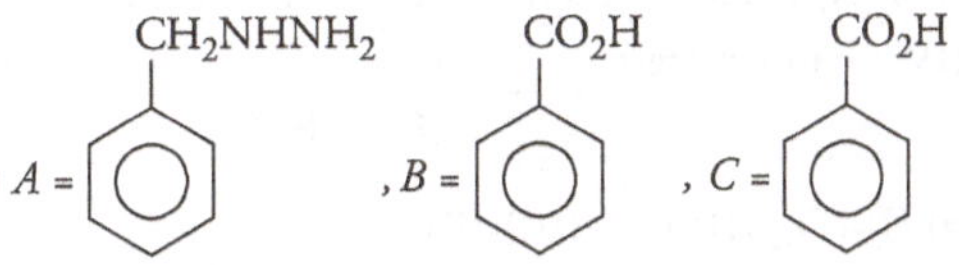

(b) 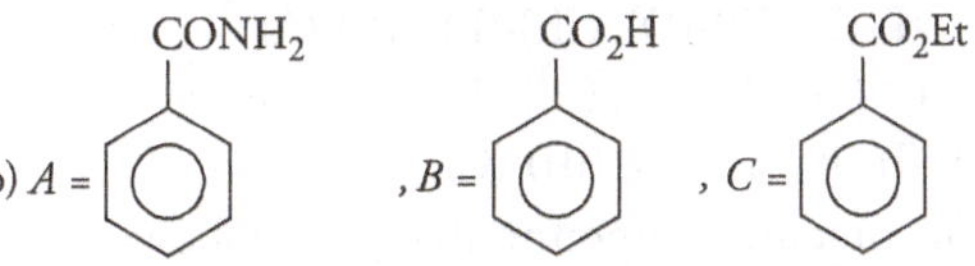

(c) 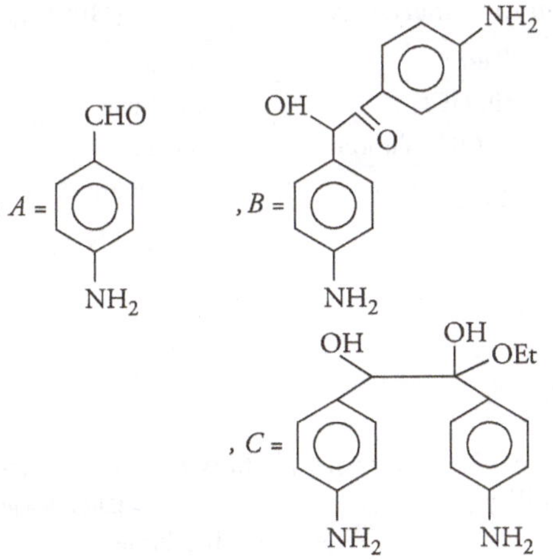

(d) 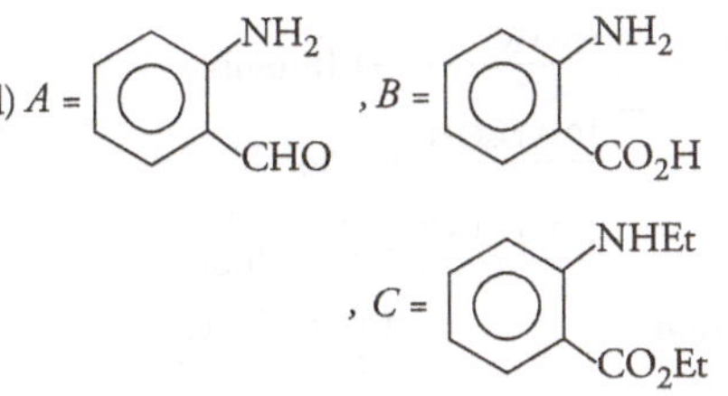

16 Which among the following compounds give yellow solid when reacted with NaOI/NaOH? **[2025, 4 April, Shift-II]**

(*A*) $CH_3-CH(OH)-C_2H_5$ (*B*) $CH_3-CH_2-CH_2-OH$

(*C*) $CH_3-C(=O)-C_2H_5$ (*D*) $CH_3-C(=O)-OH$

(*E*) $CH_3-CH_2-C(=O)-OH$

Choose the correct answer from the options given below

(a) (*A*), (*C*) and (*D*) only (b) (*B*), (*C*) and (*E*) only
(c) (*C*) and (*D*) only (d) (*A*) and (*C*) only

17 Match the List-I with List-II **[2025, 8 April Shift-II]**

	List-I (Reagent)		List-II (Functional group detected)
A.	Sodium bicarbonate solution	I.	Double bond unsaturation
B.	Neutral ferric chloride	II.	Carboxylic acid
C.	Ceric ammonium nitrate	III.	Phenolic - OH
D.	Alkaline $KMnO_4$	IV.	Alcoholic - OH

Choose the correct answer from the options given below

(a) A-II, B-III, C-IV, D-I (b) A-III, B-II. C-IV, D-I
(c) A-II, B-III, C-I, D-IV (d) A-II, B-IV, C-III, D-I

18 Match List-I with List-II. **[2024, 9 April Shift-II]**

	List-I (Test)		List-II (Observation)
A.	Br_2 water test	I.	Yellow orange or orange red precipitate formed
B.	Ceric ammonium nitrate test	II.	Reddish orange colour disappears
C.	Ferric chloride test	III.	Red colour appears
D.	2, 4-DNP test	IV.	Blue, green, violet or red colour appear

Choose the correct answer from the options given below.

(a) A-III, B-IV, C-I, D-II (b) A-I, B-II, C-III, D-IV
(c) A-II, B-III, C-IV, D-I (d) A-IV, B-I, C-II, D-III

19 Match List-I with List-II. **[2024, 8 April Shift-II]**

	List-I (Test)		List-II (Identification)
A.	Bayer's test	I.	Phenol
B.	Ceric ammonium nitrate test	II.	Aldehyde
C.	Phthalein dye test	III.	Alcoholic —OH group
D.	Schiff's test	IV.	Unsaturation

Choose the correct answer from the options given below:
(a) A-IV, B-I, C-II, D-III
(b) A-III, B-I, C-IV, D-II
(c) A-II, B-III, C-IV, D-I
(d) A-IV, B-III, C-I, D-II

20 Match List-I with List-II. **[2024, 8 April Shift-I]**

	List-I (Compound)	List-II (Colour)
A.	$Fe_4[Fe(CN)_6]_3 \cdot xH_2O$	I. Violet
B.	$[Fe(CN)_5NOS]^{4-}$	II. Blood red
C.	$[Fe(SCN)]^{2+}$	III. Prussian blue
D.	$(NH_4)_3PO_4 \cdot 12MoO_3$	IV. Yellow

Choose the correct answer from the options given below:
(a) A-II, B-III, C-IV, D-I
(b) A-IV, B-I, C-II, D-III
(c) A-I, B-II, C-III, D-IV
(d) A-III, B-I, C-II, D-IV

Imp. *Reagents Used for Functional Group Detection*

21 Match List-I with List-II

	List-I (Reagents used)	List-II (Compound with Functional group detected)
A.	Alkaline solution of copper sulphate and sodium citrate	I. 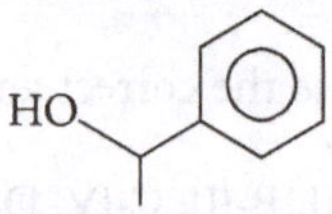
B.	Neutral $FeCl_3$, solution	II. 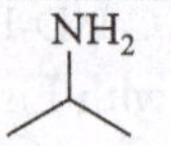
C.	Alkaline chloroform solution	III. 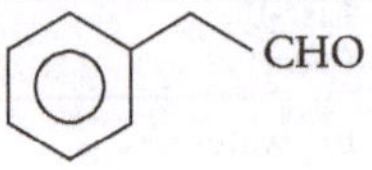
D.	Potassium iodide and sodium hypochlorite	IV. 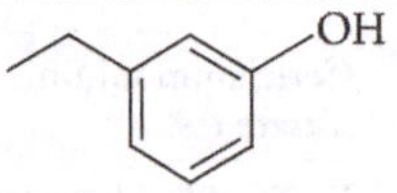

Choose the correct answer from the options given below: **[2023, 8 April Shift-I]**
(a) A-III, B-IV, C-I, D-II
(b) A-II, B-IV, C-III, D-I
(c) A-IV, B-I, C-II, D-III
(d) A-III, B-IV, C-II, D-I

22 A compound 'X' is acidic and it is soluble in NaOH solution, but insoluble in $NaHCO_3$ solution. Compound 'X' also gives violet colour with neutral $FeCl_3$ solution. The compound 'X' is **[2022, 29 July Shift-I]**

(a) 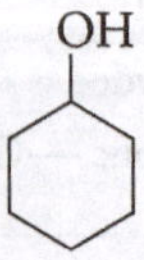 (b) 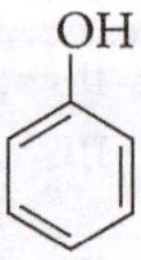 (c) 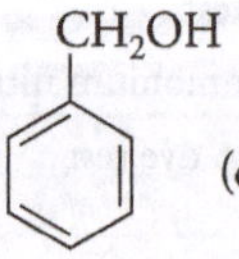 (d) 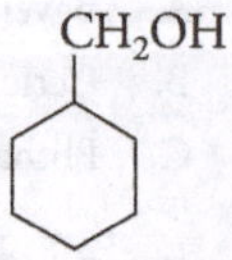

23 Given below are two statements.

Statement I : In Lassaigne's test, when both nitrogen and sulphur are present in an organic compound, sodium thiocyanate is formed.

Statement II : If both nitrogen and sulphur are present in an organic compound, then the excess of sodium used in sodium fusion will decompose the sodium thiocyanate formed to give NaCN and Na_2S.

In the light of the above statements, choose the most appropriate answer from the options given below. **[2022, 26 June Shift-I]**

(a) Both Statement I and Statement II are correct.
(b) Both Statement I and Statement II are incorrect.
(c) Statement I is correct but Statement II is incorrect.
(d) Statement I is incorrect but Statement II is correct.

24 Match the following. **[2020, 6 Sep Shift-II]**

	Column I (Test/Method)		Column II (Reagent)
(I)	Lucas test	(A)	$C_6H_5SO_2Cl$/ aq. KOH
(II)	Dumas method	(B)	$HNO_3/AgNO_3$
(III)	Kjeldahl's method	(C)	CuO/CO_2
(IV)	Hinsberg test	(D)	Conc. HCl and $ZnCl_2$
		(E)	H_2SO_4

(a) (I)-(D), (II)-(C), (III)-(B), (IV)-(A)
(b) (I)-(B), (II)-(D), (III)-(E), (IV)-(A)
(c) (I)-(D), (II)-(C), (III)-(E), (IV)-(A)
(d) (I)-(B), (II)-(A), (III)-(C), (IV)-(D)

25 The organic compound that gives following qualitative analysis is **[2019, 9 April Shift-I]**

	Test	Inference
(A)	dil. HCl	Insoluble
(B)	NaOH solution	Soluble
(C)	Br_2/water	Decolourisation

(a) 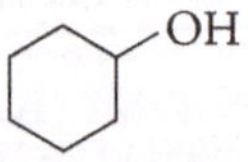 (b) 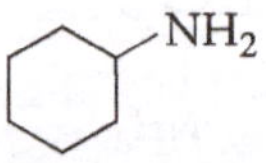

(c) 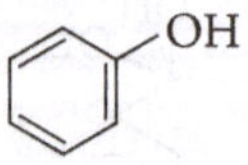 (d) 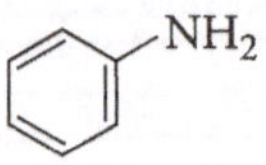

26 An organic compound X showing the following solubility profile is **[2019, 8 April Shift-I]**

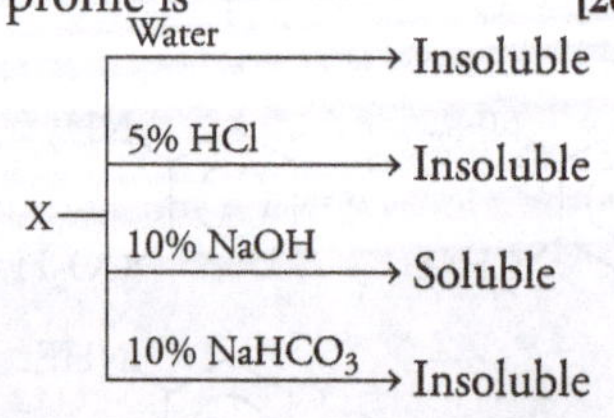

(a) o-toluidine
(b) oleic acid
(c) m-cresol
(d) benzamide

Numerical Value Type Questions

27 Isomeric hydrocarbons $\longrightarrow$ negative Baeyer's test (Molecular formula C_9H_{12})
The total number of isomers from above with four different non-aliphatic substitution sites is **[2025, 29 Jan Shift-II]**

Topic 2 / Chemistry Involved in the Preparation of Organic & Inorganic Compounds

01 While preparing crystals of Mohr's salt dil. H_2SO_4. is added to a mixture of ferrous sulphate and ammonium sulphate, before dissolving this mixture in water, dil. H_2SO_4, is added here to **[2024, 5 April Shift-II]**

(a) make the medium strongly acidic
(b) increase the rate of formation of crystals
(c) prevent the hydrolysis of ammonium sulphate
(d) prevent the hydrolysis of ferrous sulphate

02 The test performed on compound *X* and their inferences are

	Test	Inference
(i)	2,4-DNP test	Coloured precipitate
(ii)	Iodoform test	Yellow precipitate
(iii)	Azo-dye test	No dye formation

Compound '*X*' is **[2019, 9 Jan Shift-II]**

(a) 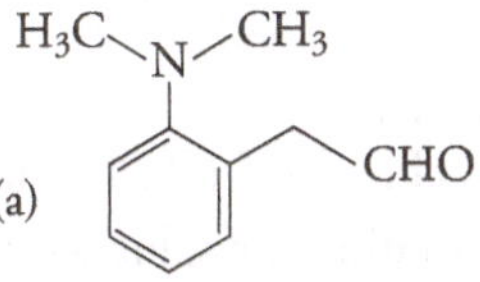

(b) 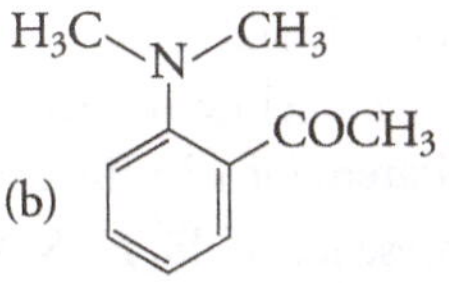

(c) 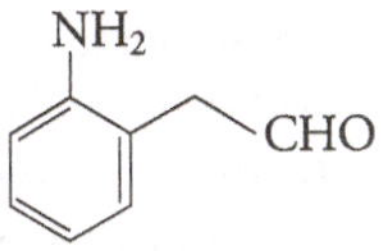

(d) 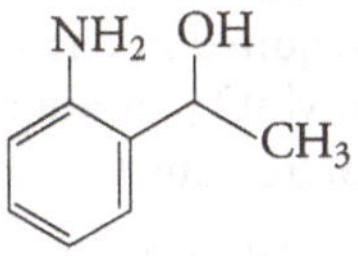

Topic 3 / Chemistry Involved in Titrimetric Experiments

Level 1

01 Given below are two statements, one is labelled as Assertion A and the other is labelled as Reason R.
Assertion (A) Permanganate titrations are not performed in presence of hydrochloric acid.
Reason (R) Chlorine is formed as a consequence of oxidation of hydrochloric acid.
In the light of the above statements, choose the correct answer from the options given below. **[2022, 28 July Shift-II]**

(a) Both A and R are true and R is the correct explanation of A
(b) Both A and R are true but R is not the correct explanation of A
(c) A is true but R is false
(d) A is false but R is true

02 In base *vs* acid titration, at the end point methyl orange is present as **[2022, 25 July Shift-II]**

(a) quinonoid form
(b) heterocyclic form
(c) phenolic form
(d) benzenoid form

Level 2

03 10 mL of 2 M NaOH solution is added to 20 mL of 1 M HCl solution kept in a beaker. Now, 10 mL of this mixture is poured into a volumetric flask of 100 mL containing 2 moles of HCl and made the volume upto the mark with distilled water. The solution in this flask is **[2025, 3 April Shift-II]**

(a) Neutral solution
(b) 10 M HCl solution
(c) 0.2 M NaCl solution
(d) 20 M HCl solution

04 A compound *X* when treated with phthalic anhydride in presence of concentrated H_2SO_4 yields '*Y*'. '*Y*' is used as an acid-base indicator. '*X*' and '*Y*' are respectively **[2023, 8 April Shift-II]**

(a) carbolic acid, phenolphthalein
(b) salicylaldehyde, phenolphthalein
(c) toludine, phenolphthalein
(d) anisole, methyl orange

05 Given below are two statements. **[2023, 8 April Shift-II]**

Statement I Methyl orange is a weak acid.

Statement II The benzenoid form of methyl orange is more intense/deeply coloured than the quinonoid form.

In the light of the above statement, choose the most appropriate answer from the options given below

(a) Both Statement I and Statement II are correct
(b) Statement I is incorrect but Statement II is correct
(c) Both Statement I and Statement II are incorrect
(d) Statement I is correct but Statement II is incorrect

Imp. *Properties of Phenolphthalein*

06 Given below are two statements.
One is labelled as Assertion A and the other is labelled as Reason R.

Assertion (A) Phenolphthalein is a pH dependent indicator, remains colourless in acidic solution and gives pink colour in basic medium.

Reason (R) Phenolphthalein is a weak acid. It doesn't dissociate in basic medium.
In the light of the above statements, choose the most appropriate answer from the options given below. **[2022, 26 July shift-II]**

(a) Both A and R are true and R is the correct explanation of A.
(b) Both A and R are true but R is not the correct explanation of A.
(c) A is true but R is false.
(d) A is false but R is true.

07 The strength of an aqueous NaOH solution is most accurately determined by titrating
(Note : consider that an appropriate indicator is used.) **[2020, 8 Jan Shift-I]**

(a) *aq*. NaOH in a pipette and aqueous oxalic acid in a burette.
(b) *aq*. NaOH in a volumetric flask and concentrated H_2SO_4 in a conical flask.
(c) *aq*. NaOH in a burette and concentrated H_2SO_4 in a conical flask.
(d) *aq*. NaOH in a burette and aqueous oxalic acid in a conical flask.

08 In an acid-base titration, 0.1 M HCl solution was added to the NaOH solution of unknown strength. Which of the following correctly shows the change of pH of the titration mixture in this experiment? **[2019, 9 April Shift-II]**

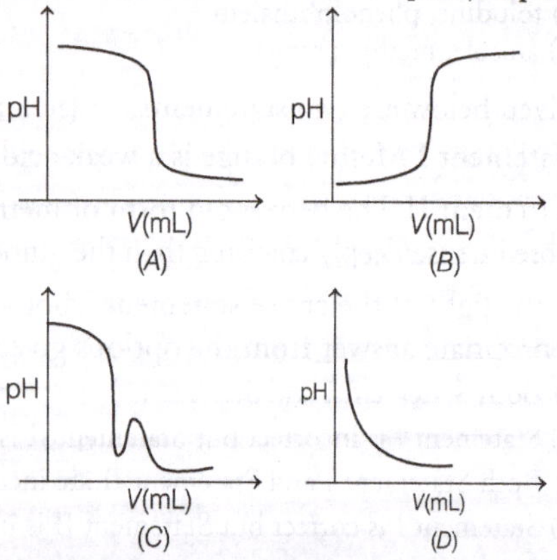

(a) (*D*) (b) (*A*)
(c) (*B*) (d) (*C*)

Level 3

09 40 mL of a mixture of CH_3COOH and HCl (aqueous solution) is titrated against 0.1 M NaOH solution conductometrically. Which of the following statement is correct? **[2025, 3 April Shift-II]**

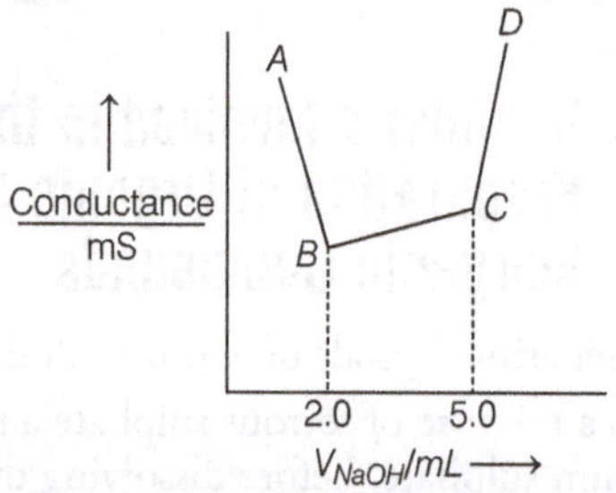

(a) The concentration of CH_3COOH in the original mixture is 0.005 M.
(b) Point '*C*' indicates the complete neutralisation of HCl.
(c) The concentration of HCl in the original mixture is 0.005 M.
(d) CH_3COOH is neutralised first followed by neutralisation of HCl.

10 Compounds that should not be used as primary standards in titrimetric analysis are

A. $Na_2Cr_2O_7$ B. Oxalic acid
C. NaOH D. $FeSO_4 \cdot 6H_2O$
E. Sodium tetraborate

Choose the most appropriate answer from the options given below. **[2025, 3 April Shift-II]**

(a) B and D only (b) D and E only
(c) C, D and E only (d) A, C and D only

11 Given below are two statements:
Statement I In the oxalic acid *vs* $KMnO_4$ (in the presence of dil. H_2SO_4) titration the solution needs to be heated initially to 60°C, but no heating is required in ferrous ammonium sulphate (FAS) *vs* $KMnO_4$ titration (in the presence of dil. H_2SO_4).
Statement II In oxalic acid *vs* $KMnO_4$ titration, the initial formation of $MnSO_4$ takes place at high temperature, which then acts as catalyst for further reaction. In the case of FAS *vs* $KMnO_4$, heating oxidises Fe^{2+} into Fe^{3+} by oxygen of air and error may be introduced in the experiment.
In the light of the above statements, choose the correct answer from the options given below **[2025, 28 Jan Shift-I]**

(a) Statement I is true but Statement II is false.
(b) Statement I is false but Statement II is true.
(c) Both Statement I and Statement II are false.
(d) Both Statement I and Statement II are true.

12 Identify the incorrect statements regarding primary standard of titrimetric analysis. **[2024, 9 April Shift-I]**

(A) It should be purely available in dry form.
(B) It should not undergo chemical change in air.

(C) It should be hygroscopic and should react with another chemical instantaneously and stoichiometrically.
(D) It should be readily soluble in water.
(E) $KMnO_4$ and NaOH can be used as primary standard.
Choose the correct answer from the options below.
(a) (A) and (B) only (b) (B) and (E) only
(c) (C) and (D) only (d) (C) and (E) only

13 A compound 'X' is a weak acid and it exhibits colour change at pH close to the equivalence point during neutralisation of NaOH with CH_3COOH. Compound 'X' exists in ionised form in basic medium. The compound 'X' is **[2022, 29 July Shift-II]**
(a) methyl orange (b) methyl red
(c) phenolphthalein (d) erichrome black T

Numerical Value Type Questions

14 Only 2 mL of $KMnO_4$ solution of unknown molarity is required to reach the end point of a titration of 20 mL of oxalic acid (2 M) in acidic medium. The molarity of $KMnO_4$ solution should be...... M. **[2024, 4 April Shift-I]**

15 20 mL of calcium hydroxide was consumed when it was reacted with 10 mL of unknown solution of H_2SO_4. Also 20 mL standard solution of 0.5 M HCl containing 2 drops of phenolphthalein was titrated with calcium hydroxide, the mixture showed pink colour when burette displayed the value of 35.5 mL whereas the burette showed 25.5 mL initially. The concentration of H_2SO_4 is .... M. (Nearest integer) **[2023, 13 April Shift-I]**

16 $KMnO_4$ is titrated with ferrous ammonium sulphate hexahydrate in presence of dilute H_2SO_4. Number of water molecules produced for 2 molecules of $KMnO_4$ is .... **[2023, 13 April Shift-I]**

17 When 10 mL of an aqueous solution of $KMnO_4$ was titrated in acidic medium, equal volume of 0.1 M of an aqueous solution of ferrous sulphate was required for complete discharge of colour. The strength of $KMnO_4$ in g/L is .... $\times 10^{-2}$. (Nearest integer)
[Atomic mass of K = 39, Mn = 55, O = 16] **[2021, 27 Aug Shift-I]**

18 10.0 mL of 0.05 M $KMnO_4$ solution was consumed in a titration with 10.0 mL of given oxalic acid dihydrate solution. The strength of given oxalic acid solution is .... $\times 10^{-2}$ g/L. (Round off to the Nearest Integer). **[2021, 27 July Shift-II]**

19 When 10 mL of an aqueous solution of Fe^{2+} ions was titrated in the presence of dil. H_2SO_4 using diphenylamine indicator, 15 mL of 0.02 M solution of $K_2Cr_2O_7$ was required to get the end point. The molarity of the solution containing Fe^{2+} ions is $x \times 10^{-2}$ M. The value of x is .... (Nearest integer) **[2021, 25 July Shift-I]**

20 A 10 mL of Na_2CO_3 solution is titrated against 0.2 M HCl solution. The following titre values were obtained in 5 readings
4.8 mL, 4.9 mL, 5.0 mL and 5.0 mL and 5.0 mL
Based on these readings, and convention of titrimetric estimation, the concentration of Na_2CO_3 solution is .... mM. (Round off to the Nearest Integer). **[2021, 18 March Shift-II]**

21 Consider titration of NaOH solution *versus* 1.25 M oxalic acid solution. At the end point following burette readings were obtained.
(i) 4.5 mL (ii) 4.5 mL
(iii) 4.4 mL (iv) 4.4 mL
(v) 4.4 mL
If the volume of oxalic acid taken was 10.0 mL, then the molarity of the NaOH solution is .... M. (Rounded off to the nearest integer) **[2021, 25 Feb Shift-II]**

Topic 4 / Qualitative Salt Analysis

01 The compound that is white in colour is **[2024, 31 Jan Shift-I]**
(a) lead sulphate
(b) ammonium arsinomolybdate
(c) ammonium sulphide
(d) lead iodide

02 A reagent which gives brilliant red precipitate with nickel ions in basic medium is **[2024, 29 Jan Shift-II]**
(a) *meta*-dinitrobenzene (b) dimethyl glyoxime
(c) sodium nitroprusside (d) neutral $FeCl_3$

03 In the wet tests for detection of various cations by precipitation, Ba^{2+} cations are detected by obtaining precipitate of **[2023, 13 April Shift-II]**
(a) Ba(*ox*) : Barium oxalate (b) $BaSO_4$
(c) $Ba(OAc)_2$ (d) $BaCO_3$

Imp. *Confirmatory Test of Lead*

04 Formation of which complex, among the following, is not a confirmatory test of Pb^{2+} ions **[2023, 6 April Shift-II]**
(a) lead chromate (b) lead nitrate
(c) lead sulphate (d) lead iodide

05 In the wet tests for identification of various cations by precipitation. Which transition element cation doesn't belong to group IV in qualitative inorganic analysis? **[2023, 30 Jan Shift-I]**
(a) Ni^{2+} (b) Co^{2+} (c) Fe^{3+} (d) Zn^{2+}

06 Fe^{3+} cation gives a prussian blue precipitate on addition of potassium ferrocyanide solution due to the formation of [2022, 27 July Shift-II]
(a) $[Fe(H_2O)_6]_2[Fe(CN)_6]$ (b) $Fe_2[Fe(CN)_6]_2$
(c) $Fe_3[Fe(OH)_2(CN)_4]_2$ (d) $Fe_4[Fe(CN)_6]_3$

07 Which statement is not true with respect to nitrate ion test? [2022, 26 June Shift-I]
(a) A dark brown ring is formed at the junction of two solutions.
(b) Ring is formed due to nitroferrous sulphate complex.
(c) The brown complex is $[Fe(H_2O)_5(NO)SO_4$.
(d) Heating the nitrate salt with conc. H_2SO_4, light brown fumes are evolved.

08 Which one of the following when dissolved in water gives coloured solution in nitrogen atmosphere? [2021, 26 Aug Shift-I]
(a) Cu_2Cl_2 (b) $ZnCl_2$ (c) $CuCl_2$ (d) AgCl

09 Cu^{2+} salt reacts with potassium iodide to give [2021, 20 July Shift-II]
(a) Cu_2I_2 (b) CuI (c) $Cu(I_3)_2$ (d) Cu_2I_3

Level 2

10 When a salt is treated with sodium hydroxide solution it gives gas *X*. On passing gas *X* through reagent *Y* a brown coloured precipitate is formed. *X* and *Y* respectively, are
(a) $X = NH_3$ and $Y = HgO$
(b) $X = NH_4Cl$ and $Y = KOH$
(c) $X = HCl$ and $Y = NH_4Cl$
(d) $X = NH_3$ and $Y = K_2HgI_4 + KOH$

Imp. *Colour of Salt (Salt Analysis)*

11 Identify the inorganic sulphides that are yellow in colour
(A) $(NH_4)_2S$ (B) PbS
(C) CuS (D) As_2S_3
(E) As_2S_5
Choose the correct answer from the options given below : [2025, 28 Jan Shift-II]
(a) (A) and (C) only (b) (D) and (E) only
(c) (A) and (B) only (d) (A), (D) and (E) only

12 The metal ion whose electronic configuration is not affected by the nature of the ligand and which gives a violet colour in non-luminous flame under hot condition in borax bead test is [2025, 28 Jan Shift-I]
(a) Ti^{3+} (b) Ni^{2+} (c) Mn^{2+} (d) Cr^{3+}

13 During the detection of acidic radical present in a salt, a student gets a pale yellow precipitate soluble with difficulty in NH_4OH solution when sodium carbonate extract was first acidified with dil. HNO_3 and then $AgNO_3$ solution was added. This indicates presence of [2024, 6 April Shift-II]
(a) Cl^- (b) CO_3^{2-} (c) I^- (d) Br^-

14 Given below are two statements. [2024, 30 Jan Shift-I]
Statement I The gas liberated on warming a salt with dil. H_2SO_4, turns a piece of paper dipped in lead acetate into black, it is a confirmatory test for sulphide ion.
Statement II In statement-I the colour of paper turns black because of formation of lead sulphite.
In the light of the above statements, choose the **most appropriate** answer from the options given below
(a) Statement I is true but Statement II is false.
(b) Both Statement I and Statement II are true.
(c) Both Statement I and Statement II are false.
(d) Statement I is false but Statement II is true.

15 A white precipitate was formed when $BaCl_2$ was added to water extract of an inorganic salt. Further, a gas *X* with characteristic odour was released when the formed white precipitate was dissolved in dilute HCl. The anion present in the inorganic salt is [2022, 29 June Shift-II]
(a) I^- (b) SO_3^{2-} (c) S^{2-} (d) NO_2^-

Imp. *Flame Test*

16 In the flame test of a mixture of salts, a green flame with blue centre was observe(d) Which one of the following cations may be present? [2022, 24 June Shift-II]
(a) Cu^{2+} (b) Sr^{2+} (c) Ba^{2+} (d) Ca^{2+}

17 The addition of dilute NaOH to Cr^{3+} salt solution will give [2021, 27 Aug Shift-II]
(a) a solution of $[Cr(OH)_4]^-$
(b) precipitate of $[Cr(OH)_6]^{3-}$
(c) precipitate of $Cr_2O_3(H_2O)_n$
(d) precipitate of $Cr(OH)_3$

18 To an aqueous solution containing ions such as Al^{3+}, Zn^{2+}, Ca^{2+}, Fe^{3+}, Ni^{2+}, Ba^{2+} and Cu^{2+} conc. HCl, was added followed by H_2S. The total number of cations precipitated during this reaction is/are [2021, 27 July Shift-II]
(a) 1 (b) 3 (b) 4 (d) 2

19 Reagent 1-napthylamine and sulphanilic acid in acetic acid is used for the detection of [2021, 18 March Shift-I]
(a) N_2O (b) NO_2^- (c) NO_3^- (d) NO

Level 3

20 Choose the correct tests with respective observations.
(A) $CuSO_4$ (acidified with acetic acid) $+ K_4[Fe(CN)_6]$ $\rightarrow$ Chocolate brown precipitate.
(B) $FeCl_3 + K_4[Fe(CN)_6] \rightarrow$ Prussian blue precipitate.

(C) $ZnCl_2 + K_4[Fe(CN)_6]$, neutralised with $NH_4OH \rightarrow$ White or bluish white precipitate.
(D) $MgCl_2 + K_4[Fe(CN)_6] \rightarrow$ Blue precipitate.
(E) $BaCl_2 + K_4[Fe(CN)_6]$, neutralised with $NaOH \rightarrow$ White precipitate.

Choose the correct answer from the options given below [2025, 2 April Shift-I]

(a) A, D and E only (b) A, B and C only
(c) C, D and E only (d) B, D and E only

21 Identify *A*, *B* and *C* in the given below reaction sequence

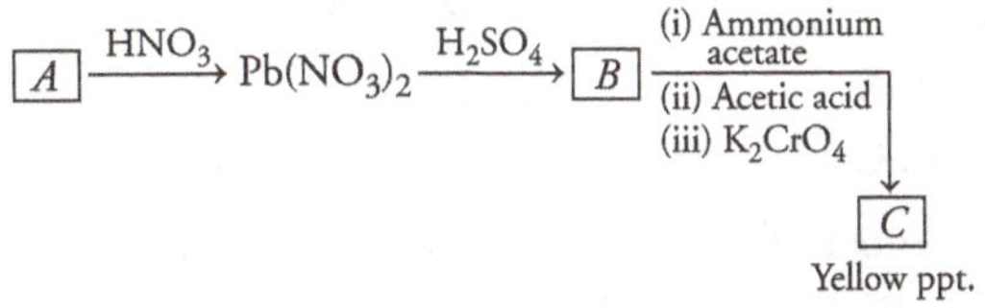

[2025, 23 Jan Shift-II]

(a) $PbCl_2, PbSO_4, PbCrO_4$ (b) $PbS, PbSO_4, PbCrO_4$
(c) $PbS, PbSO_4, Pb(CH_3COO)_2$
(d) $PbCl_2, Pb(SO_4)_2, PbCrO_4$

22 Match List-I with List-II. [2024, 8 April Shift-I]

	List-I (Name of the test)		List-II (Reaction sequence involved) [*M* is metal]
A.	Borax bead test	I.	$MCO_3 \rightarrow MO \xrightarrow[+\Delta]{Co(NO_3)_2} CoO \cdot MO$
B.	Charcoal cavity test	II.	$MCO_3 \rightarrow MCl_2 \rightarrow M^{2+}$
C.	Cobalt nitrate test	III.	$MSO_4 \xrightarrow[D]{Na_2B_4O_7} M(BO_2)_2 \rightarrow MBO_2 \rightarrow M$
D.	Flame test	IV.	$MSO_4 \xrightarrow[\Delta]{Na_2CO_3} MCO_3 \rightarrow MO \rightarrow M$

Choose the correct answer from the options given below

(a) A-III, B-II, C-IV, D-I (b) A-III, B-IV, C-I, D-II
(c) A-III, B-I, C-II, D-IV (d) A-III, B-I, C-IV, D-II

23 Match List-I with List-II.

	List-I (Precipitating reagent and conditions)		List-II (Cation)
A.	$NH_4Cl + NH_4OH$	I.	Mn^{2+}
B.	$NH_4OH + Na_2CO_3$	II.	Pb^{2+}
C.	$NH_4OH + NH_4Cl + H_2S$ gas	III.	Al^{3+}
D.	Dilute HCl	IV.	Sr^{2+}

Choose the correct answer from the options given below [2024, 6 April Shift-I]

(a) A-III, B-IV, C-I, D-II (b) A-III, B-IV, C-II, D-I
(c) A-IV, B-III, C-I, D-II (d) A-IV, B-III, C-II, D-I

24 When a solution of mixture having two inorganic salts was treated with freshly prepared ferrous sulphate in acidic medium, a dark brown ring was formed whereas on treatment with neutral $FeCl_3$, it gave deep red colour which disppeared on boiling and a brown red ppt was formed. The mixture contains [2023, 11 April Shift-I]

(a) CH_3COO^- and NO_3^- (b) $C_2O_4^{2-}$ and NO_3^-
(c) SO_3^{2-} and $C_2O_4^{2-}$ (d) SO_3^{2-} and CH_3COO^-

25 Match List-I with List-II.

	List-I (Element)		List-II (Colour imparted to the flame)
A.	K	I.	Brick red
B.	Ca	II.	Violet
C.	Sr	III.	Apple green
D.	Ba	IV.	Crimson red

Choose the correct answer from the options given below [2023, 25 Jan Shift-I]

(a) A-II, B-I, C-IV, D-III (b) A-II, B-IV, C-I, D-III
(c) A-IV, B-III, C-II, D-I (d) A-II, B-I, C-III, D-IV

26 Match List-I with List-II.

	List-I (Cations)		List-II (Group reagents)
A.	Pb^{2+}, Cu^{2+}	I.	H_2S gas in presence of dilute HCl
B.	Al^{3+}, Fe^{3+}	II.	$(NH_4)_2CO_3$ in presence of NH_4OH
C.	Co^{2+}, Ni^{2+}	III.	NH_4OH in presence of NH_4Cl
D.	Ba^{2+}, Ca^{2+}	IV.	H_2S in presence of NH_4OH

Correct match is [2023, 25 Jan Shift-I]

(a) A-III, B-I, C-IV, D-II (b) A-I, B-III, C-IV, D-II
(c) A -I, B-III, C-II, D-IV (d) A-IV, B-II, C-III, D-I

27 During the qualitative analysis of salt with cation Y^{2+}, addition of a reagent (*X*) to alkaline solution of the salt gives a bright red precipitate. The reagent (*X*) and the cation (Y^{2+}) present respectively are

(a) dimethylglyoxime and Ni^{2+} [2022, 24 June Shift-I]
(b) dimethylglyoxime and Co^{2+}
(c) Nessler's reagent and Hg^{2+}
(d) Nessler's reagent and Ni^{2+}

28 Match List-I with List-II.

List-I (Metal iron)	List-II (Group in qualitative analysis)
(A) Mn^{2+}	(i) Group-III
(B) As^{3+}	(ii) Group-IIA
(C) Cu^{2+}	(iii) Group-IV
(D) Al^{3+}	(iv) Group-IIB

Choose the most appropriate answer from the options given below **[2021, 31 Aug Shift-II]**

(a) (A)-(iii), (B)-(iv), (C)-(ii), (D)-(i)
(b) (A)-(iv), (B)-(ii), (C)-(iii), (D)-(i)
(c) (A)-(i), (B)-(iv), (C)-(ii), (D)-(iii)
(d) (A)-(i), (B)-(ii), (C)-(iii), (D)-(iv)

29 An inorganic compound 'X' on treatment with concentrated H_2SO_4 produces brown fumes and gives dark brown ring with $FeSO_4$ in presence of concentrated H_2SO_4. Also compound 'X' give precipitate 'Y', when its solution in dilute HCl is treated with H_2S gas. The precipitate 'Y' on treatment with concentrated HNO_3 followed by excess of NH_4OH further gives deep blue coloured solution, compound 'X' is **[2021, 20 July Shift-II]**

(a) $Pb(NO_3)_2$ (b) $Pb(NO_2)_2$
(c) $Cu(NO_3)_2$ (d) $Co(NO_3)_2$

Numerical Value Type Questions

30 Among the following cations, the number of cations which will give characteristic precipitate in their identification tests with $K_4[Fe(CN)_6]$ is.....

$Cu^{2+}, Fe^{3+}, Ba^{2+}, Ca^{2+}, NH_4^+, Mg^{2+}, Zn^{2+}$ **[2025, 24 Jan Shift-I]**

31 The number of white coloured salts, among the following is .... **[2024, 1 Feb Shift-I]**

(a) $SrSO_4$ (b) $Mg(NH_4)PO_4$
(b) $BaCrO_4$ (d) $Mn(OH)_2$
(e) $PbSO_4$ (f) $PbCrO_4$
(g) AgBr (g) PbI_2
(i) CaC_2O_4 (j) $[Fe(OH)_2(CH_3COO)]$

32 Number of metal ions characterised by flame test among the following is ....

$Sr^{2+}, Ba^{2+}, Ca^{2+}, Cu^{2+}, Zn^{2+}, Co^{2+}, Fe^{2+}$ **[2024, 30 Jan Shift-II]**

Key Index

Topic 1

1	(d)	2	(c)	3	(d)	4	(b)	5	(c)	6	(b)	7	(d)	8	(c)	9	(b)	10	(c)
11	(b)	12	(a)	13	(b)	14	(b)	15	(b)	16	(d)	17	(a)	18	(c)	19	(d)	20	(d)
21	(d)	22	(b)	23	(a)	24	(c)	25	(c)	26	(c)	27	(2)						

Topic 2

1	(d)	2	(b)

Topic 3

1	(a)	2	(a)	3	(d)	4	(a)	5	(d)	6	(c)	7	(d)	8	(b)	9	(c)	10	(d)
11	(d)	12	(d)	13	(c)	14	(8)	15	(1)	16	(68)	17	(316)	18	(1575)	19	(18)	20	(50)
21	(6)																		

Topic 4

1	(a)	2	(b)	3	(d)	4	(b)	5	(c)	6	(d)	7	(b)	8	(c)	9	(a)	10	(d)
11	(d)	12	(b)	13	(d)	14	(a)	15	(b)	16	(a)	17	(c)	18	(a)	19	(b)	20	(b)
21	(b)	22	(b)	23	(a)	24	(a)	25	(a)	26	(b)	27	(a)	28	(a)	29	(c)	30	(4)
31	(5)	32	(4)																

For Detailed Solutions scan QR Code